NEUROSCIENCE

Exploring the Brain

NEUROSCIENCE

Exploring the Brain

Enhanced Fourth Edition

Mark F. BEAR, PhD
Picower Professor of Neuroscience
The Picower Institute for Learning and Memory
Department of Brain and Cognitive Sciences
Massachusetts Institute of Technology
Cambridge, Massachusetts

Barry W. CONNORS, PhD
L. Herbert Ballou University Professor
Professor of Neuroscience and Chair
Department of Neuroscience
Brown University
Providence, Rhode Island

Michael A. PARADISO, PhD
Sidney A. Fox and Dorothea Doctors Fox Professor of
Opthamology and Visual Science
Department of Neuroscience
Brown University
Providence, Rhode Island

World Headquarters
Jones & Bartlett Learning
5 Wall Street
Burlington, MA 01803
978-443-5000
info@jblearning.com
www.jblearning.com

Jones & Bartlett Learning books and products are available through most bookstores and online booksellers. To contact Jones & Bartlett Learning directly, call 800-832-0034, fax 978-443-8000, or visit our website, www.jblearning.com.

Substantial discounts on bulk quantities of Jones & Bartlett Learning publications are available to corporations, professional associations, and other qualified organizations. For details and specific discount information, contact the special sales department at Jones & Bartlett Learning via the above contact information or send an email to specialsales@jblearning.com.

21881-7

Production Credits
VP, Product Management: Amanda Martin
Director of Product Management: Cathy L. Esperti
Product Specialist: Ashley Malone
Product Coordinator: Elena Sorrentino
Digital Project Specialist: Angela Dooley
Director of Marketing: Andrea DeFronzo
Marketing Manager: Dani Burford
Production Services Manager: Colleen Lamy
VP, Manufacturing and Inventory Control: Therese Connell
Product Fulfillment Manager: Wendy Kilborn
Composition: S4Carlisle Publishing Services
Project Management: S4Carlisle Publishing Services
Cover & Text Design: Kristin E. Parker
Senior Media Development Editor: Troy Liston
Rights Specialist: Rebecca Damon
Cover Image: Courtesy of Satrajit Ghosh and John Gabrieli, McGovern Institute for Brain Research and Department of Brain and Cognitive Sciences, Massachusetts Institute of Technology.
Part I Chapter Opener: Courtesy of Sebastian Seung, Princeton University, and Alex Norton, EyeWire.
Part II Chapter Opener: Courtesy of Shane Crandall, Saundra Patrick, and Barry Connors, Department of Neuroscience, Brown University.
Part III Chapter Opener: Courtesy of Arthur Toga and Paul Thompson, Keck School of Medicine, University of Southern California.
Part IV Chapter Opener: Courtesy of Miquel Bosch and Mark Bear, The Picower Institute for Learning and Memory and Department of Brain and Cognitive Sciences, Massachusetts Institute of Technology.
Printing and Binding: LSC Communications

Library of Congress Cataloging-in-Publication Data
Library of Congress Cataloging-in-Publication Data unavailable at time of printing

LCCN: 2020931308

6048

Printed in the United States of America
24 23 22 21 20 10 9 8 7 6 5 4 3 2 1

DEDICATION

Anne, David, and Daniel
Ashley, Justin, and Kendall

Brian and Jeffrey

Wendy, Bear, and Boo

PREFACE

THE ORIGINS OF *NEUROSCIENCE: EXPLORING THE BRAIN*

For over 30 years, we have taught a course called Neuroscience 1: An Introduction to the Nervous System. "Neuro 1" has been remarkably successful. At Brown University, where the course originated, approximately one out of every four undergraduates takes it. For a few students, this is the beginning of a career in neuroscience; for others, it is the only science course they take in college.

The success of introductory neuroscience reflects the fascination and curiosity everyone has for how we sense, move, feel, and think. However, the success of our course also derives from the way it is taught and what is emphasized. First, there are no prerequisites, so the elements of biology, chemistry, and physics required for understanding neuroscience are covered as the course progresses. This approach ensures that no students are left behind. Second, liberal use of commonsense metaphors, real-world examples, humor, and anecdotes remind students that science is interesting, approachable, exciting, and fun. Third, the course does not survey all of neurobiology. Instead, the focus is on mammalian brains and, whenever possible, the human brain. In this sense, the course closely resembles what is taught to most beginning medical students. Similar courses are now offered at many colleges and universities by psychology, biology, and neuroscience departments.

The first edition of *Neuroscience: Exploring the Brain* was written to provide a suitable textbook for Neuro 1, incorporating the subject matter and philosophy that made this course successful. Based on feedback from our students and colleagues at other universities, we expanded the second edition to include more topics in behavioral neuroscience and some new features to help students understand the structure of the brain. In the third edition, we shortened chapters when possible by emphasizing principles more and details less and made the book even more user-friendly by improving the layout and clarity of the illustrations. We must have gotten it right because the book now ranks as one of the most popular introductory neuroscience books in the world. It has been particularly gratifying to see our book used as a catalyst for the creation of new courses in introductory neuroscience.

NEW IN THE FOURTH EDITION

The advances in neuroscience since publication of the third edition have been nothing short of breathtaking. The elucidation of the human genome has lived up to its promise to "change everything" we know about our brains. We now have insight into how neurons differ at the molecular level, and this knowledge has been exploited to develop revolutionary technologies to trace their connections and interrogate their functions. The genetic basis for many neurological and psychiatric diseases has been revealed. The methods of genetic engineering have made it possible to create animal models to examine how genes and genetically defined circuits contribute to brain function. Skin cells derived from patients have been transformed into stem cells, and these have been transformed into neurons that reveal how cellular functions go awry in diseases and how the brain might be repaired. New imaging and computational methods now put within reach the dream of creating a "wiring diagram" for the entire brain. A goal for the fourth edition was to make these and other exciting new developments accessible to the first-time neuroscience student.

We authors are all active neuroscientists, and we want our readers to understand the allure of brain research. A unique feature of our book is the *Path of Discovery* boxes, in which famous neuroscientists tell stories about their own research. These essays serve several purposes: to give a flavor of the thrill of discovery; to show the importance of hard work and patience, as well as serendipity and intuition; to reveal the human side of science; and to entertain and amuse. We have continued this tradition in the fourth edition, with contributions from 26 esteemed scientists. Included in this illustrious group are Nobel laureates Mario Capecchi, Eric Kandel, Leon Cooper, May-Britt Moser, and Edvard Moser.

AN OVERVIEW OF THE BOOK

Neuroscience: Exploring the Brain surveys the organization and function of the human nervous system. We present material at the cutting edge of neuroscience in a way that is accessible to both science and nonscience students alike. The level of the material is comparable to an introductory college text in general biology.

The book is divided into four parts: Part I, Foundations; Part II, Sensory and Motor Systems; Part III, The Brain and Behavior; and Part IV, The Changing Brain. We begin Part I by introducing the modern field of neuroscience and tracing some of its historical antecedents. Then we take a close look at the structure and function of individual neurons, how they communicate chemically, and how these building blocks are arranged to form a nervous system. In Part II, we go inside the brain to examine the structure and function of the systems that serve the senses and command voluntary movements. In Part III, we explore the neurobiology of human behavior, including motivation, sex, emotion, sleep, language, attention, and mental illness. Finally, in Part IV, we look at how the environment modifies the brain, both during development and in adult learning and memory.

The human nervous system is examined at several different scales, ranging from the molecules that determine the functional properties of neurons to the large systems in the brain that underlie cognition and behavior. Many disorders of the human nervous system are introduced as the book progresses, usually within the context of the specific neural system under discussion. Indeed, many insights into the normal functions of neural systems have come from the study of diseases that cause specific malfunctions of these systems. In addition, we discuss the actions of drugs and toxins on the brain using this information to illustrate how different brain systems contribute to behavior and how drugs may alter brain function.

Organization of Part I: Foundations (Chapters 1–7)

The goal of Part I is to build a strong base of general knowledge in neurobiology. The chapters should be covered sequentially, although Chapters 1 and 6 can be skipped without a loss of continuity.

In Chapter 1, we use an historical approach to review some basic principles of nervous system function and then turn to the topic of how neuroscience research is conducted today. We directly confront the ethics of neuroscience research, particularly that which involves animals.

In Chapter 2, we focus mainly on the cell biology of the neuron. This is essential information for students inexperienced in biology, and we find that even those with a strong biology background find this review helpful. After touring the cell and its organelles, we go on to discuss the structural features that make neurons and their supporting cells unique, emphasizing the correlation of structure and function. We also introduce some of the feats of genetic engineering that neuroscientists now use routinely to study the functions of different types of nerve cell.

Chapters 3 and 4 are devoted to the physiology of the neuronal membrane. We cover the essential chemical, physical, and molecular properties that enable neurons to conduct electrical signals. We discuss the principles behind the revolutionary new methods of optogenetics. Throughout the chapter, we appeal to students' intuition by using a commonsense approach, with a liberal use of metaphors and real-life analogies.

Chapters 5 and 6 cover interneuronal communication, particularly chemical synaptic transmission. Chapter 5 presents the general principles of chemical synaptic transmission, and Chapter 6 discusses the neurotransmitters and their modes of action in greater detail. We also describe many of the modern methods for studying the chemistry of synaptic transmission. Later chapters do not assume an understanding of synaptic transmission at the depth of Chapter 6, however, so this chapter can be skipped at the instructor's discretion. Most coverage of psychopharmacology appears in Chapter 15, after the general organization of the brain and its sensory and motor systems have been presented. In our experience, students wish to know where, in addition to how, drugs act on the nervous system and behavior.

Chapter 7 covers the gross anatomy of the nervous system. Here we focus on the common organizational plan of the mammalian nervous system by tracing the brain's embryological development. (Cellular aspects of development are covered in Chapter 23.) We show that the specializations of the human brain are simple variations on the basic plan that applies to all mammals. We introduce the cerebral cortex and the new field of connectomics.

Chapter 7's appendix, An Illustrated Guide to Human Neuroanatomy, covers the surface and cross-sectional anatomy of the brain, the spinal cord, the autonomic nervous system, the cranial nerves, and the blood supply. A self-quiz will help students learn the terminology. We recommend that students become familiar with the anatomy in the guide before moving on to Part II. The coverage of anatomy is selective, emphasizing the relationship of structures that will be covered in later chapters. We find that students love to learn the anatomy.

Organization of Part II: Sensory and Motor Systems (Chapters 8–14)

Part II surveys the systems within the brain that control sensation and movement. In general, these chapters do not need to be covered sequentially, except for Chapters 9 and 10 on vision and Chapters 13 and 14 on the control of movement.

We chose to begin Part II with a discussion of the chemical senses—smell and taste—in Chapter 8. These are good systems for illustrating the general principles and problems in the encoding of sensory information, and the transduction mechanisms have strong parallels with other systems.

Chapters 9 and 10 cover the visual system, an essential topic for all introductory neuroscience courses. Many details of visual system organization are presented, illustrating not only the depth of current knowledge but also the principles that apply across sensory systems.

Chapter 11 explores the auditory system, and Chapter 12 introduces the somatic sensory system. Audition and somatic sensation are such important parts of everyday life; it is hard to imagine teaching introductory neuroscience without discussing them. The vestibular sense of balance is covered in a separate section of Chapter 11. This placement offers instructors the option to skip the vestibular system at their discretion.

In Chapters 13 and 14, we discuss the motor systems of the brain. Considering how much of the brain is devoted to the control of movement, this more extensive treatment is clearly justified. However, we are well aware that the complexities of the motor systems are daunting to students and instructors alike. We have tried to keep our discussion sharply focused, using numerous examples to connect with personal experience.

Organization of Part III: The Brain and Behavior (Chapters 15–22)

Part III explores how different neural systems contribute to different behaviors, focusing on the systems where the connection between the brain and behavior can be made most strongly. We cover the systems that control visceral function and homeostasis, simple motivated behaviors such as eating and drinking, sex, mood, emotion, sleep, consciousness, language, and attention. Finally, we discuss what happens when these systems fail during mental illness.

Chapters 15–19 describe a number of neural systems that orchestrate widespread responses throughout the brain and the body. In Chapter 15, we focus on three systems that are characterized by their broad influence and their interesting neurotransmitter chemistry: the secretory hypothalamus, the autonomic nervous system, and the diffuse modulatory systems of the brain. We discuss how the behavioral manifestations of various drugs may result from disruptions of these systems.

In Chapter 16, we look at the physiological factors that motivate specific behaviors, focusing mainly on recent research about the control of eating habits. We also discuss the role of dopamine in motivation and addiction, and we introduce the new field of "neuroeconomics." Chapter 17 investigates the influence of sex on the brain, and the influence of the brain on sexual behavior. Chapter 18 examines the neural systems believed to underlie emotional experience and expression, specifically emphasizing fear and anxiety, anger, and aggression.

In Chapter 19, we investigate the systems that give rise to the rhythms of the brain, ranging from the rapid electrical rhythms during sleep and wakefulness to the slow circadian rhythms controlling hormones, temperature, alertness, and metabolism. We next explore aspects of brain processing that are highly developed in the human brain. Chapter 20 investigates the neural basis of language and Chapter 21 discusses changes in brain activity associated with rest, attention, and consciousness. Part III ends with a discussion of mental illness in Chapter 22. We introduce the promise of molecular medicine to develop new treatments for serious psychiatric disorders.

Organization of Part IV: The Changing Brain (Chapters 23–25)

Part IV explores the cellular and molecular basis of brain development and learning and memory. These subjects represent two of the most exciting frontiers of modern neuroscience.

Chapter 23 examines the mechanisms used during brain development to ensure that the correct connections are made between neurons. The cellular aspects of development are discussed here rather than in Part I for several reasons. First, by this point in the book, students fully appreciate that normal brain function depends on its precise wiring. Because we use the visual system as a concrete example, the chapter must also follow a discussion of the visual pathways in Part II. Second, we survey aspects of experience-dependent development of the visual system that are regulated by behavioral state, so this chapter is placed after the early chapters of Part III. Finally, an exploration of the role of the sensory environment in brain development in Chapter 23 is followed in the next two chapters by discussions of how experience-dependent modifications of the brain form the basis for learning and memory. We see that many of the mechanisms are similar, illustrating the unity of biology.

Chapters 24 and 25 cover learning and memory. Chapter 24 focuses on the anatomy of memory, exploring how different parts of the brain contribute to the storage of different types of information. Chapter 25 takes a deeper look into the molecular and cellular mechanisms of learning and memory, focusing on changes in synaptic connections.

HELPING STUDENTS LEARN

Neuroscience: Exploring the Brain is not an exhaustive study. It is intended to be a readable textbook that communicates to students the important principles of neuroscience clearly and effectively. To help students learn neuroscience, we include a number of features designed to enhance comprehension:

- **Chapter Outlines and Introductory and Concluding Remarks.** These elements preview the organization of each chapter, set the stage, and place the material into broader perspective.
- **Of Special Interest Boxes.** These boxes are designed to illuminate the relevance of the material to the students' everyday lives.
- **Brain Food Boxes.** More advanced material that might be optional in many introductory courses is set aside for students who want to go deeper.
- **Path of Discovery Boxes.** These essays, written by leading researchers, demonstrate a broad range of discoveries and the combination of hard work and serendipity that led to them. These boxes both personalize scientific exploration and deepen the reader's understanding of the chapter material and its implications.
- **Key Terms and Glossary**. Neuroscience has a language of its own, and to comprehend it, one must learn the vocabulary. In the text of each chapter, important terms are highlighted in boldface type. To facilitate review, these terms appear in a list at the end of each chapter in the order in which they appeared in the text, along with page references. The same terms are assembled at the end of the book, with definitions, in a glossary.
- **Review Questions.** At the end of each chapter, a brief set of questions for review are specifically designed to provoke thought and help students integrate the material.
- **Further Reading.** We include a list of several recent review articles at the end of each chapter to guide study beyond the scope of the textbook.
- **Internal Reviews of Neuroanatomical Terms.** In Chapter 7, where nervous system anatomy is discussed, the narrative is interrupted periodically with brief self-quiz vocabulary reviews to enhance understanding. In Chapter 7's appendix, an extensive self-quiz is provided in the form of a workbook with labeling exercises.

- **References and Resources.** At the end of the book, we provide selected readings and online resources that will lead students into the research literature associated with each chapter. Rather than including citations in the body of the chapters, where they would compromise the readability of the text, we have organized the references and resources by chapter and listed them at the end of the book.
- **Full-Color Illustrations.** We believe in the power of illustrations—not those that "speak a thousand words" but those that each make a single point. The first edition of this book set a new standard for illustrations in a neuroscience text. The fourth edition reflects improvements in the pedagogical design of many figures from earlier editions and includes many superb new illustrations as well.

USER'S GUIDE

Succeed in your course and discover the excitement of the dynamic, rapidly changing field of neuroscience with this fourth edition of ***Neuroscience: Exploring the Brain***. This user's guide will help you discover how to best use the features of this book.

Chapter Outline
This "road map" to the content outlines what you will learn in each chapter and can serve as a valuable review tool.

CHAPTER ONE

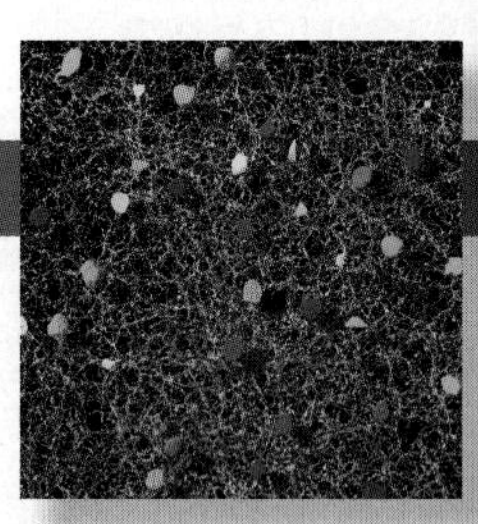

Neuroscience: Past, Present, and Future

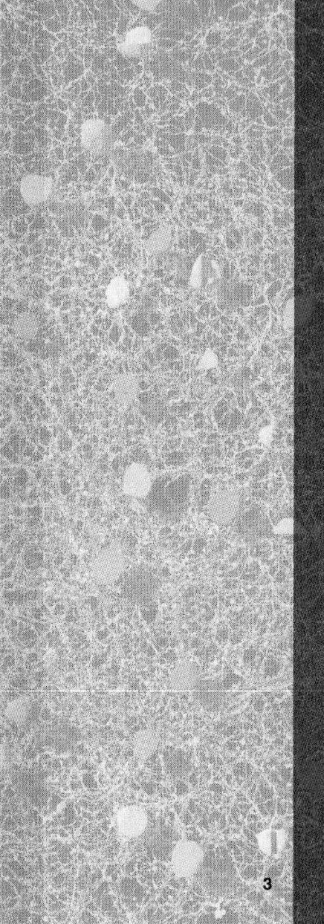

3

BOX 2.2 **BRAIN FOOD**

Expressing One's Mind in the Post-Genomic Era

Sequencing the human genome was a truly monumental achievement, completed in 2003. The Human Genome Project identified all of the approximately 25,000 genes in human DNA. We now live in what has been called the "post-genomic era," in which information about the genes expressed in our tissues can be used to diagnose and treat diseases. Neuroscientists are using this information to tackle long-standing questions about the biological basis of neurological and psychiatric disorders as well as to probe deeper into the origins of individuality. The logic goes as follows. The brain is a product of the genes expressed in it. Differences in gene expression between a normal brain and a diseased brain, or a brain of unusual ability, can be used to identify the molecular basis of the observed symptoms or traits.

The level of gene expression is usually defined as the number of mRNA transcripts synthesized by different cells and tissues to direct the synthesis of specific proteins. Thus, the analysis of gene expression requires comparing the relative abundance of various mRNAs in the brains of two groups of humans or animals. One way to perform such a comparison is to use DNA *microarrays*, which are created by robotic machines that arrange thousands of small spots of synthetic DNA on a microscope slide. Each spot contains a unique DNA sequence that will recognize and stick to a different specific mRNA sequence. To compare the gene expression in two brains, one begins by collecting a sample of mRNAs from each brain. The mRNA of one brain is labeled with a chemical tag that fluoresces green, and the mRNA of the other brain is labeled with a tag that fluoresces red. These samples are then applied to the microarray. Highly expressed genes will produce brightly fluorescent spots, and differences in the relative gene expression between the brains will be revealed by differences in the color of the fluorescence (Figure A).

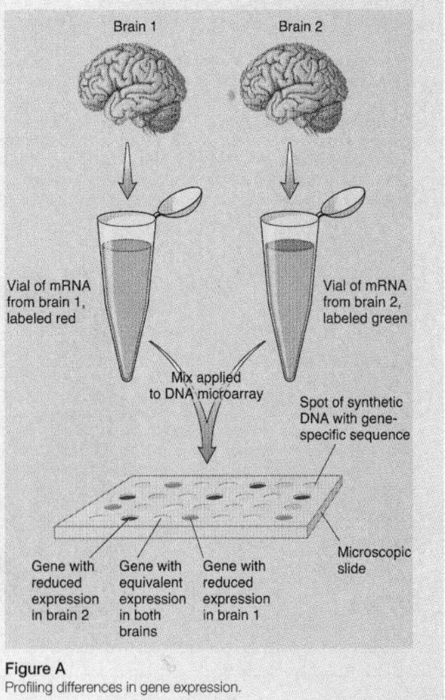

Figure A
Profiling differences in gene expression.

Brain Food Boxes
Want to expand your understanding? These boxes offer optional advanced material so you can expand on what you've learned.

BOX 16.2 OF SPECIAL INTEREST

Marijuana and the Munchies

A well-known consequence of marijuana intoxication is stimulation of appetite, an effect known by users as "the munchies." The active ingredient in marijuana is D^9-tetrahydrocannabinol (THC), which alters neuronal functions by stimulating a receptor called cannabinoid receptor 1 (CB1). CB1 receptors are abundant throughout the brain, so it is overly simplistic to view these receptors as serving only appetite regulation. Nevertheless, "medical marijuana" is often prescribed (where legal) as a means to stimulate appetite in patients with chronic diseases, such as cancer and AIDS. A compound that inhibits CB1 receptors, rimonabant, was also developed as an appetite suppressant. However, human drug trials had to be discontinued because of psychiatric side effects. Although this finding underscores the fact that these receptors do much more than mediate the munchies, it is still of interest to know where in the brain CB1 receptors act to stimulate appetite. Not surprisingly, the CB1 receptors are associated with neurons in many regions of the brain that control feeding, such as the hypothalamus, and some of the orexigenic effects of THC are related to changing the activity of these neurons. However, neuroscientists were surprised to learn in 2014 that much of the appetite stimulation comes from enhancing the sense of smell, at least in mice. Collaborative research conducted by neuroscientists in France and Spain, countries incidentally known for their appreciation of good tastes and smells, revealed that activation of CB1 receptors in the olfactory bulb increases odor detection and is necessary for the increase in food intake stimulated in hungry mice by cannabinoids.

In Chapter 8, we discussed how smells activate neurons in the olfactory bulb which, in turn, relay information to the olfactory cortex. The cortex also sends feedback projections to the bulb that synapse on inhibitory interneurons called *granule cells*. By activating the inhibitory granule cells, this feedback from the cortex dampens ascending olfactory activity. These corticofugal synapses use glutamate as a neurotransmitter. The brain's own endocannabinoids (anandamide and 2-arachidonoylglycerol) are synthesized under fasting conditions, and they inhibit glutamate release by acting on CB1 receptors on the corticofugal axon terminals. Reducing granule cell activation by glutamate in the bulb has the net effect of enhancing the sense of smell (Figure A). It remains to be determined if the munchies arise from enhanced olfaction in marijuana users, but a simple experiment, such as holding your nose while eating, confirms that much of the hedonic value of food derives from the sense of smell.

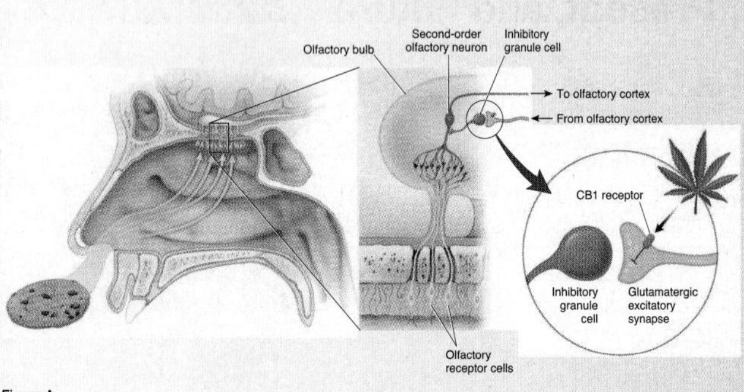

Figure A
Activation of CB1 receptors by THC, the psychoactive ingredient in marijuana, enhances olfaction by suppressing the release of glutamate from corticofugal inputs to inhibitory granule cells in the olfactory bulb. (Source: Adapted from Soria-Gomez et al., 2014.)

Of Special Interest Boxes
Wondering how key concepts appear in the real world? These boxes complement the text by showing some of the more practical applications of concepts. Topics include brain disorders, human case studies, drugs, new technology, and more.

BOX 2.3 PATH OF DISCOVERY

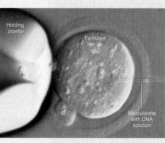

Gene Targeting in Mice

by Mario Capecchi

How did I first get the idea to pursue gene targeting in mice? From a simple observation. Mike Wigler, now at Cold Spring Harbor Laboratory, and Richard Axel, at Columbia University, had published a paper in 1979 showing that exposing mammalian cells to a mixture of DNA and calcium phosphate would cause some cells to take up the DNA in functional form and express the encoded genes. This was exciting because they had clearly demonstrated that exogenous, functional DNA could be introduced into mammalian cells. But I wondered why their efficiency was so low. Was it a problem of delivery, insertion of exogenous DNA into the chromosome, or expression of the genes once inserted into the host chromosome? What would happen if purified DNA was directly injected into the nucleus of mammalian cells in culture?

To find out, I converted a colleague's electrophysiology station into a miniature hypodermic needle to directly inject DNA into the nucleus of a living cell using mechanical micromanipulators and light microscopy (Figure A). The procedure worked with amazing efficiency (Capecchi, 1980). With this method, the frequency of successful integration was now one in three cells rather than one in a million cells as formerly. This high efficiency directly led to the development of transgenic mice through the injection and random integration of exogenous DNA into chromosomes of fertilized mouse eggs, or zygotes. To achieve the high efficiency of expression of the exogenous DNA in the recipient cell, I had to attach small fragments of viral DNA, which we now understand to contain enhancers that are critical in eukaryotic gene expression.

But what fascinated me most was our observation that when many copies of a gene were injected into a cell nucleus, all of these molecules ended up in an ordered head-to-tail arrangement, called a *concatemer* (Figure B). This was astonishing and could not have occurred as a random event. We went on to unequivocally prove that homologous recombination, the process by which chromosomes share genetic information during cell division, was responsible for the incorporation of the foreign DNA (Folger et al., 1982). These experiments demonstrated that all mammalian somatic cells contain a very efficient machinery for swapping segments of DNA that have similar sequences of nucleotides. Injection of a thousand copies of a gene sequence into the nucleus of a cell resulted in chromosomal insertion of a concatemer containing a thousand copies of that sequence, all oriented in the same direction. This simple observation directly led me to

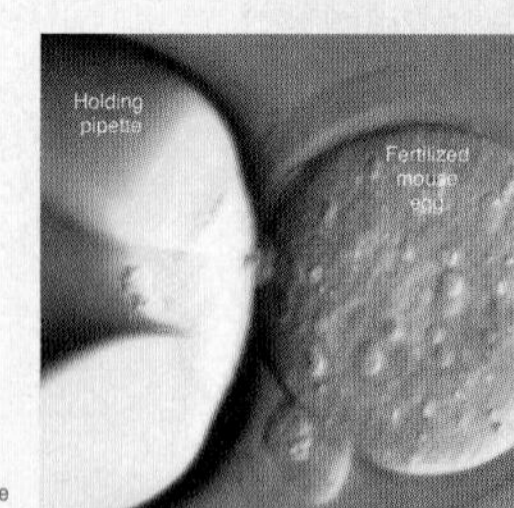

Figure A
Fertilized mouse egg receiving an injection of foreign DNA. (Image courtesy of Dr. Peimin Qi, Division of Comparative Medicine, Massachusetts Institute of Technology.)

Path of Discovery Boxes
Learn about some of the superstars in the field with these boxes. Leading researchers describe their discoveries and achievements and tell the story of how they arrived at them.

the anatomical study of brain cells had to await a method to harden the tissue without disturbing its structure and an instrument that could produce very thin slices. Early in the nineteenth century, scientists discovered how to harden, or "fix," tissues by immersing them in formaldehyde, and they developed a special device called a *microtome* to make very thin slices.

These technical advances spawned the field of **histology**, the microscopic study of the structure of tissues. But scientists studying brain structure faced yet another obstacle. Freshly prepared brain tissue has a uniform, cream-colored appearance under the microscope, with no differences in pigmentation to enable histologists to resolve individual cells. The final breakthrough in neurohistology was the introduction of stains that selectively color some, but not all, parts of the cells in brain tissue.

One stain still used today was introduced by the German neurologist Franz Nissl in the late nineteenth century. Nissl showed that a class of basic dyes would stain the nuclei of all cells as well as clumps of material surrounding the nuclei of neurons (Figure 2.1). These clumps are called *Nissl bodies*, and the stain is known as the **Nissl stain**. The Nissl stain is extremely useful for two reasons: It distinguishes between neurons and ...rrangement, or **cytoarchi-** ... brain. (The prefix *cyto-* is ... cytoarchitecture led to the ...pecialized regions. We now ...nction.

The Nucleus. Its name derived from the Latin word for "nut," the **nucleus** of the cell is spherical, centrally located, and about 5–10 μm across. It is contained within a double membrane called the *nuclear envelope*. The nuclear envelope is perforated by pores about 0.1 μm across.

Within the nucleus are **chromosomes** which contain the genetic material **DNA** (**deoxyribonucleic acid**). Your DNA was passed on to you from your parents and it contains the blueprint for your entire body. The DNA in each of your neurons is the same, and it is the same as the DNA in the cells of your liver and kidney and other organs. What distinguishes a neuron from a liver cell are the specific parts of the DNA that are used to assemble the cell. These segments of DNA are called **genes**.

Each chromosome contains an uninterrupted double-strand braid of DNA, 2 nm wide. If the DNA from the 46 human chromosomes were laid out straight, end to end, it would measure more than 2 m in length. If we were to compare this total length of DNA to the total string of letters that make up this book, the genes would be analogous to the individual words. Genes are from 0.1 to several micrometers in length.

The "reading" of the DNA is known as **gene expression**. The final product of gene expression is the synthesis of molecules called **proteins**, which exist in a wide variety of shapes and sizes, perform many different functions, and bestow upon neurons virtually all of their unique characteristics. **Protein synthesis**, the assembly of protein molecules, occurs in the cytoplasm. Because the DNA never leaves the nucleus, an intermediary must carry the genetic message to the sites of protein synthesis in the cytoplasm. This function is performed by another long molecule called

KEY TERMS

Introduction
neuron (p. 24)
glial cell (p. 24)

The Neuron Doctrine
histology (p. 25)
Nissl stain (p. 25)
cytoarchitecture (p. 25)
Golgi stain (p. 26)
cell body (p. 26)
soma (p. 26)
perikaryon (p. 26)
neurite (p. 26)
axon (p. 26)
dendrite (p. 26)
neuron doctrine (p. 27)

The Prototypical Neuron
cytosol (p. 29)
organelle (p. 29)
cytoplasm (p. 29)
nucleus (p. 29)
chromosome (p. 29)
DNA (deoxyribonucleic acid) (p. 29)
gene (p. 29)
gene expression (p. 29)
protein (p. 29)
protein synthesis (p. 29)
mRNA (messenger ribonucleic acid) (p. 29)
transcription (p. 29)
promoter (p. 31)
transcription factor (p. 31)
RNA splicing (p. 31)
amino acid (p. 32)
translation (p. 32)
genome (p. 32)
genetic engineering (p. 32)
knockout mice (p. 33)
transgenic mice (p. 33)
knock-in mice (p. 33)
ribosome (p. 36)
rough endoplasmic reticulum (rough ER) (p. 36)
polyribosome (p. 36)
smooth endoplasmic reticulum (smooth ER) (p. 36)
Golgi apparatus (p. 36)
mitochondrion (p. 36)
ATP (adenosine triphosphate) (p. 38)
neuronal membrane (p. 38)
cytoskeleton (p. 38)
microtubule (p. 38)
microfilament (p. 39)
neurofilament (p. 39)
axon hillock (p. 39)
axon collateral (p. 39)
axon terminal (p. 41)
terminal bouton (p. 41)
synapse (p. 42)
terminal arbor (p. 42)
innervation (p. 42)
synaptic vesicle (p. 42)
synaptic cleft (p. 43)
synaptic transmission (p. 43)
neurotransmitter (p. 43)
axoplasmic transport (p. 43)
anterograde transport (p. 44)
retrograde transport (p. 44)
dendritic tree (p. 44)
receptor (p. 46)
dendritic spine (p. 46)

Classifying Neurons
unipolar neuron (p. 46)
bipolar neuron (p. 46)
multipolar neuron (p. 46)
stellate cell (p. 46)
pyramidal cell (p. 46)
spiny neuron (p. 46)
aspinous neuron (p. 46)
primary sensory neuron (p. 48)
motor neuron (p. 48)
interneuron (p. 48)
green fluorescent protein (GFP) (p. 48)

Glia
astrocyte (p. 49)
oligodendroglial cell (p. 49)
Schwann cell (p. 49)
myelin (p. 49)
node of Ranvier (p. 49)
ependymal cell (p. 52)
microglial cell (p. 52)

Key Terms
Appearing in bold throughout the text, key terms are also listed at the end of each chapter and defined in the glossary. These can help you study and ensure you've mastered the terminology as you progress through your course.

REVIEW QUESTIONS

1. State the neuron doctrine in a single sentence. To whom is this insight credited?
2. Which parts of a neuron are shown by a Golgi stain that are not shown by a Nissl stain?
3. What are three physical characteristics that distinguish axons from dendrites?
4. Of the following structures, state which ones are unique to neurons and which are not: nucleus, mitochondria, rough ER, synaptic vesicle, Golgi apparatus.
5. What are the steps by which the information in the DNA of the nucleus directs the synthesis of a membrane-associated protein molecule?
6. Colchicine is a drug that causes microtubules to break apart (depolymerize). What effect would this drug have on anterograde transport? What would happen in the axon terminal?
7. Classify the cortical pyramidal cell based on (1) the number of neurites, (2) the presence or absence of dendritic spines, (3) connections, and (4) axon length.
8. Knowledge of genes uniquely expressed in a particular category of neurons can be used to understand how those neurons function. Give one example of how you could use genetic information to study a category of neuron.
9. What is myelin? What does it do? Which cells provide it in the central nervous system?

Review Questions
Test your comprehension of each of the chapter's major concepts with these review questions.

FURTHER READING

De Vos KJ, Grierson AJ, Ackerley S, Miller CCJ. 2008. Role of axoplasmic transport in neurodegenerative diseases. *Annual Review of Neuroscience* 31:151–173.

Eroglu C, Barres BA. 2010. Regulation of synaptic connectivity by glia. *Nature* 468:223–231.

Jones EG. 1999. Golgi, Cajal and the Neuron Doctrine. *Journal of the History of Neuroscience* 8:170–178.

Lent R, Azevedo FAC, Andrade-Moraes CH, Pinto AVO. 2012. How many neurons do you have? Some dogmas of quantitative neuroscience under revision. *European Journal of Neuroscience* 35:1–9.

Nelson SB, Hempel C, Sugino K. 2006. Probing the transcriptome of neuronal cell types. *Current Opinion in Neurobiology* 16:571–576.

Peters A, Palay SL, Webster H deF. 1991. *The Fine Structure of the Nervous System*, 3rd ed. New York: Oxford University Press.

Sadava D, Hills DM, Heller HC, Berenbaum MR. 2011. *Life: The Science of Biology*, 9th ed. Sunderland, MA: Sinauer.

Shepherd GM, Erulkar SD. 1997. Centenary of the synapse: from Sherrington to the molecular biology of the synapse and beyond. *Trends in Neurosciences* 20:385–392.

Wilt BA, Burns LD, Ho ETW, Ghosh KK, Mukamel EA, Schnitzer MJ. 2009. Advances in light microscopy for neuroscience. *Annual Review of Neuroscience* 32:435–506.

Further Reading
Interested in learning more? Recent review articles are identified at the end of each chapter so you can delve further into the content.

(b) Selected Gyri, Sulci, and Fissures. The cerebrum is noteworthy for its convoluted surface. The bumps are called *gyri*, and the grooves are called *sulci* or, if they are especially deep, fissures. The precise pattern of gyri and sulci can vary considerably from individual to individual, but many features are common to all human brains. Some of the important landmarks are labeled here. Notice that the postcentral gyrus lies immediately posterior to the central sulcus, and that the precentral gyrus lies immediately anterior to it. The neurons of the postcentral gyrus are involved in somatic sensation (touch; Chapter 12), and those of the precentral gyrus control voluntary movement (Chapter 14). Neurons in the superior temporal gyrus are involved in audition (hearing; Chapter 11).

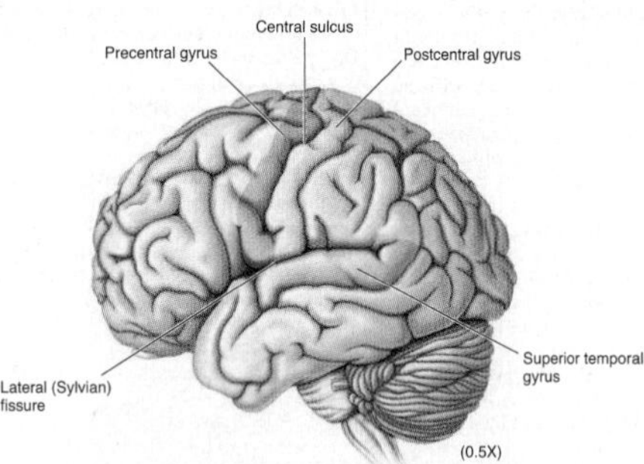

(c) Cerebral Lobes and the Insula. By convention, the cerebrum is subdivided into lobes named after the bones of the skull that lie over them. The central sulcus divides the frontal lobe from the parietal lobe. The temporal lobe lies immediately ventral to the deep lateral (Sylvian) fissure. The occipital lobe lies at the very back of the cerebrum, bordering both parietal and temporal lobes. A buried piece of the cerebral cortex, called the *insula* (Latin for "island"), is revealed if the margins of the lateral fissure are gently pulled apart (inset). The insula borders and separates the temporal and frontal lobes.

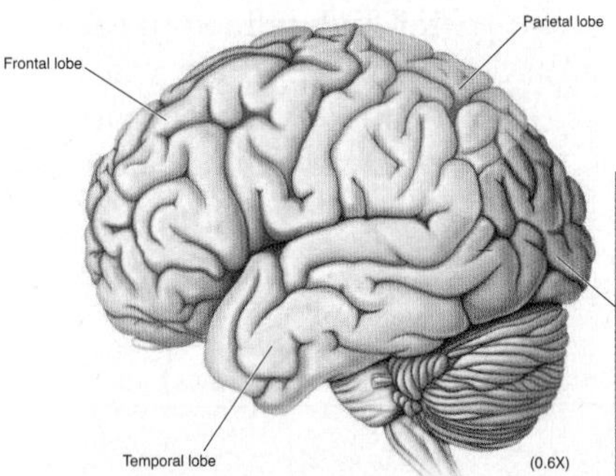

An Illustrated Guide to Human Neuroanatomy
This appendix to Chapter 7 includes an extensive self-quiz with labeling exercises that enable you to assess your knowledge of neuroanatomy.

SELF-QUIZ

This review workbook is designed to help you learn the neuroanatomy that has been presented. Here, we have reproduced the images from the Guide; however, instead of labels, numbered leader lines (arranged in a clockwise fashion) point to the structures of interest. Test your knowledge by filling in the appropriate names in the spaces provided. To review what you have learned, quiz yourself by putting your hand over the names. Experience has shown that this technique greatly facilitates the learning and retention of anatomical terms. Mastery of the vocabulary of neuroanatomy will serve you well as you learn about the functional organization of the brain in the remainder of the book.

The Lateral Surface of the Brain

(a) Gross Features

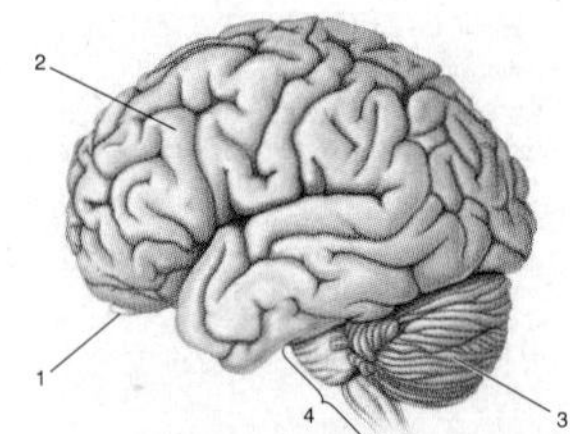

1. ____________
2. ____________
3. ____________
4. ____________

(b) Selected Gyri, Sulci, and Fissures

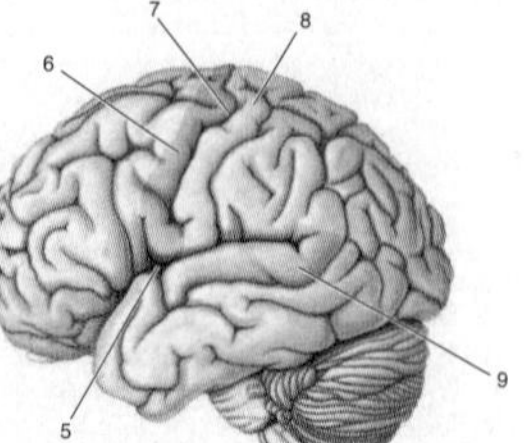

5. ____________
6. ____________
7. ____________
8. ____________
9. ____________

Self-Quiz
Found in Chapter 7, these brief vocabulary reviews can help enhance your understanding of nervous system anatomy.

SELF-QUIZ

Take a few moments right now and be sure you understand the meaning of these terms:

anterior
rostral
posterior
caudal
dorsal
ventral
midline
medial
lateral
ipsilateral
contralateral
midsagittal plane
sagittal plane
horizontal plane
coronal plane

ACKNOWLEDGMENTS

Back in 1993, when we began in earnest to write the first edition of this textbook, we had the good fortune to work closely with a remarkably dedicated and talented group of individuals—Betsy Dilernia, Caitlin and Rob Duckwall, and Suzanne Meagher—who helped us bring the book to fruition. Betsy continued as our developmental editor for the first three editions. We attribute much of our success to her extraordinary efforts to improve the clarity and consistency of the writing and the layout of the book. Betsy's well-deserved retirement caused considerable consternation among the author team, but good fortune struck again with the recruitment of Tom Lochhaas for this new edition. Tom, an accomplished author himself, shares Betsy's attention to detail and challenged us to not rest on our laurels. We are proud of the fourth edition and very grateful to Tom for holding us to a high standard of excellence. We would be remiss for not thanking him also for his good cheer and patience despite a challenging schedule and occasionally distracted authors.

It is noteworthy that despite the passage of time—*21 years!*—we were able to continue working with Caitlin, Rob, and Suzanne in this edition. Caitlin's and Rob's Dragonfly Media Group produced the art, with help and coordination from Jennifer Clements, and the results speak for themselves. The artists took our sometimes fuzzy concepts and made them a beautiful reality. The quality of the art has always been a high priority for the authors, and we are very pleased that they have again delivered an art program that ensures we will continue to enjoy the distinction of having produced the most richly illustrated neuroscience textbook in the world. Finally, we are forever indebted to Suzanne, who assisted us at every step. Without her incredible assistance, loyalty, and dedication to this project, the book would never have been completed. That statement is as true today as it was in 1993. Suzanne, you are—still—the best!

For the current edition, we have the pleasure of acknowledging a new team member, Linda Francis. Linda is an editorial project manager and she worked closely with us from start to finish, helping us to meet a demanding schedule. Her efficiency, flexibility, and good humor are all greatly appreciated. Linda, it has been a pleasure working with you.

In the publishing industry, editors seem to come and go with alarming frequency. Yet one senior editor stayed the course and continued to be an unwavering advocate for our project: Emily Lupash. We thank you Emily and the entire staff under your direction for your patience and determination to get this edition published.

We again would like to acknowledge the architects and current trustees of the undergraduate neuroscience curriculum at Brown University. We thank Mitchell Glickstein, Ford Ebner, James McIlwain, Leon Cooper, James Anderson, Leslie Smith, John Donoghue, Bob Patrick, and John Stein for all they did to make undergraduate neuroscience great at Brown. Similarly, we thank Sebastian Seung and Monica Linden for their innovative contributions to introductory neuroscience at the Massachusetts Institute of Technology. Monica, who is now on the faculty of Brown's Department of Neuroscience, also made numerous suggestions for improvements in the fourth edition of this book for which we are particularly grateful.

We gratefully acknowledge the research support provided to us over the years by the National Institutes of Health, the Whitehall Foundation, the Alfred P. Sloan Foundation, the Klingenstein Foundation, the Charles A. Dana Foundation, the National Science Foundation, the Keck Foundation, the Human Frontiers Science Program, the Office of Naval Research, DARPA, the Simons Foundation, the JPB Foundation, the Picower Institute for Learning and Memory, the Brown Institute for Brain Science, and the Howard Hughes Medical Institute.

We thank our colleagues in the Brown University Department of Neuroscience and in the Department of Brain and Cognitive Sciences at MIT for their ongoing support of this project and helpful advice. We thank the anonymous but very helpful colleagues at other institutions who gave us comments on the earlier editions. We gratefully acknowledge the scientists who provided us with figures illustrating their research results and, in particular, Satrajit Ghosh and John Gabrieli of MIT for providing the striking image that appears on the cover of the new edition (to learn about the image, see p. xxi). In addition, many students and colleagues helped us to improve the new edition by informing us about recent research, pointing out errors in earlier editions, and suggesting better ways to describe or illustrate concepts. Special thanks to Peter Kind of the University of Edinburgh and Weifeng Xu of MIT.

We are very grateful to our many colleagues who contributed "Path of Discovery" stories. You inspire us.

We thank our loved ones, not only for standing by us as countless weekends and evenings were spent preparing this book, but also for their encouragement and helpful suggestions for improving it.

Finally, we wish to thank the thousands of students we have had the privilege to teach neuroscience over the past 35 years.

PATH OF DISCOVERY AUTHORS

Floyd Bloom, M.D.
Scripps Research Institute
La Jolla, California
Exploring the Central Noradrenergic Neurons

Mario Capecchi, Ph.D.
University of Utah
Howard Hughes Medical Institute
Salt Lake City, Utah
Gene Targeting in Mice

Leon N Cooper, Ph.D.
Brown University
Providence, Rhode Island
Memories of Memory

Timothy C. Cope, Ph.D.
Wright State University
Dayton, Ohio
Nerve Regeneration Does Not Ensure Full Recovery

Antonio Damasio, Ph.D.
University of Southern California
Los Angeles, California
Concepts and Names in Everyday Science

Nina Dronkers, Ph.D.
University of California
Davis, California
Uncovering Language Areas of the Brain

Geoffrey Gold, Ph.D.
Monell Chemical Senses Center
Philadelphia, Pennsylvania
Channels of Vision and Smell

Kristen M. Harris, Ph.D.
University of Texas
Austin, Texas
For the Love of Dendritic Spines

Thomas Insel, M.D., Director
United States National Institute of Mental Health
Rockville, Maryland
Bonding with Voles

Stephanie R. Jones, Ph.D.
Brown University
Providence, Rhode Island
The Puzzle of Brain Rhythms

Eric Kandel, M.D.
Columbia University
Howard Hughes Medical Institute
New York, New York
What Attracted Me to the Study of Learning and Memory in Aplysia?

Nancy Kanwisher, Ph.D.
Massachusetts Institute of Technology
Cambridge, Massachusetts
Finding Faces in the Brain

Julie Kauer, Ph.D.
Brown University
Providence, Rhode Island
Learning to Crave

Christof Koch, Ph.D.
Allen Institute for Brain Science
Seattle, Washington
Tracking the Neuronal Footprints of Consciousness

Helen Mayberg, M.D.
Emory University School of Medicine
Atlanta, Georgia
Tuning Depression Circuits

James T. McIlwain, M.D.
Brown University
Providence, Rhode Island
Distributed Coding in the Superior Colliculus

Chris Miller, Ph.D.
Brandeis University
Howard Hughes Medical Institute
Waltham, Massachusetts
Feeling Around Inside Ion Channels in the Dark

Edvard Moser, Ph.D., and May-Britt Moser, Ph.D.
Kavli Institute for Neural Systems
University of Science and Technology
Trondheim, Norway
How the Brain Makes Maps

Georg Nagel, Ph.D.
University of Würzburg
Würzburg, Germany
The Discovery of the Channelrhodopsins

Donata Oertel, Ph.D.
University of Wisconsin School of Medicine and Public Health
Madison, Wisconsin
Capturing the Beat

Pasko Rakic, M.D., Ph.D.
Yale University School of Medicine
New Haven, Connecticut
Making a Map of the Mind

Sebastian Seung, Ph.D.
Princeton University
Princeton, New Jersey
Connecting with the Connectome

Solomon H. Snyder, M.D.
The Johns Hopkins University School of Medicine
Baltimore, Maryland
Finding Opiate Receptors

David Williams, Ph.D.
University of Rochester
Rochester, New York
Seeing Through the Photoreceptor Mosaic

Thomas Woolsey, M.D.
Washington University School of Medicine
St. Louis, Missouri
Cortical Barrels

IMAGES

Cover: An image of a living human brain acquired by magnetic resonance tomography to reveal the diffusion of water molecules. Water diffusion in the brain occurs preferentially along bundles of axons. Axons are the "wires" of the nervous system and conduct electrical impulses generated by brain cells. Thus, this image reveals some of the paths of long-range communication between different parts of the brain. The image, acquired at the Athinoula A. Martinos Center for Biomedical Imaging at the Massachusetts Institute of Technology, was processed by a computer algorithm to display bundles of axons traveling together as pseudo-colored noodles. The colors vary depending on the direction of water diffusion. (Source: Courtesy of Satrajit Ghosh and John Gabrieli, McGovern Institute for Brain Research and Department of Brain and Cognitive Sciences, Massachusetts Institute of Technology.)

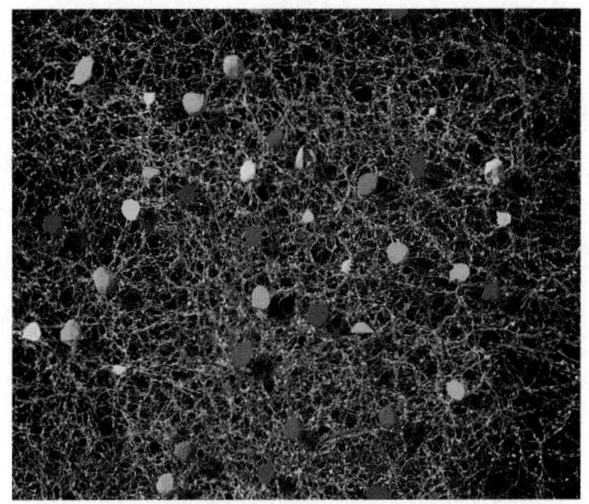

Part One Chapter Opener: Neurons and their neurites. Serial images were taken using an electron microscope of a small piece of the retina as thin slices were shaved off. Then, a computer algorithm, aided by thousands of people worldwide playing an online game called EyeWire, reconstructed each neuron and their synaptic connections—the "connectome" of this volume of tissue. In this image, the neurons are pseudocolored by the computer, and their neurites, the axons and dendrites from each cell, are displayed in their entirety. (Source: Courtesy of Sebastian Seung, Princeton University, and Alex Norton, EyeWire.)

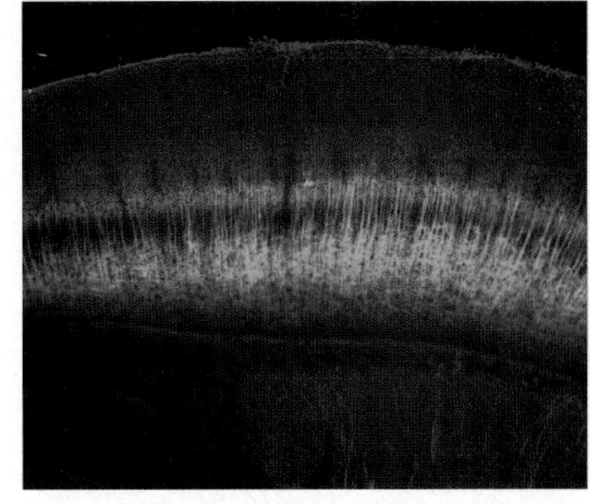

Part Two Chapter Opener: The mouse cerebral cortex. The cerebral cortex lies just under the skull. It is critical for conscious sensory perception and voluntary control of movement. The major subcortical input to the cortex arises from the thalamus, a structure that lies deep inside the brain. Stained red are thalamic axons that bring to the cortex information about the whiskers on the animal's snout. These are clustered into "barrels" that each represent a single whisker. The neurons that project axons back to the thalamus have been genetically engineered to fluoresce green. Blue indicates the nuclei of other cells stained with a DNA marker. (Source: Courtesy of Shane Crandall, Saundra Patrick, and Barry Connors, Department of Neuroscience, Brown University.)

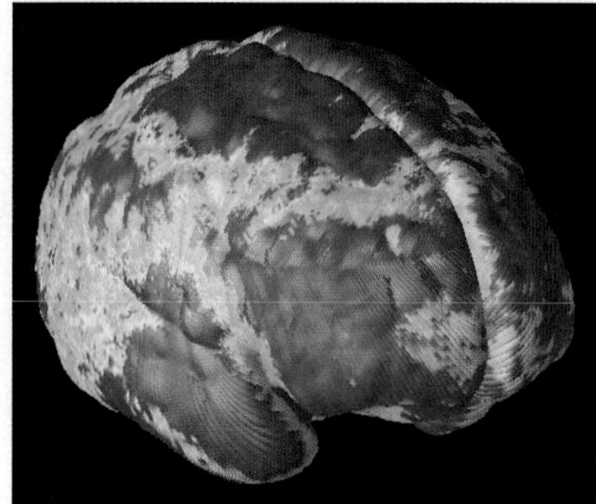

Part Three Chapter Opener: Gray matter loss in the cerebral cortex of adolescents with schizophrenia. Schizophrenia is a severe mental illness characterized by a loss of contact with reality and a disruption of thought, perception, mood, and movement. The disorder typically becomes apparent during adolescence or early adulthood and persists for life. Symptoms may arise in part from shrinkage of specific parts of the brain, including the cerebral cortex. High-resolution magnetic resonance imaging of the brains of adolescents with schizophrenia has been used to track the location and progression of tissue loss. In this image, the regions of gray matter loss are color coded. Severe tissue loss, up to 5% annually, is indicated in red and pink. Regions colored blue are relatively stable over time. (Source: Courtesy of Arthur Toga and Paul Thompson, Keck School of Medicine, University of Southern California.)

Part Four Chapter Opener: Neurons of the hippocampus. The hippocampus is a brain structure that is critical for our ability to form memories. One way that information is stored in the brain is by modification of synapses, the specialized junctions between the axons of one neuron and the dendrites of another. Synaptic plasticity in the hippocampus has been studied to reveal the molecular basis of memory formation. This image shows the neurites of a subset of hippocampal neurons using a time honored method introduced in 1873 by Italian scientist Emilio Golgi. (Source: Courtesy of Miquel Bosch and Mark Bear, The Picower Institute for Learning and Memory and Department of Brain and Cognitive Sciences, Massachusetts Institute of Technology.)

CONTENTS IN BRIEF

EXPANDED CONTENTS

PART THREE **The Brain and Behavior** 519

LIST OF BOXES

PATH OF DISCOVERY

PART ONE

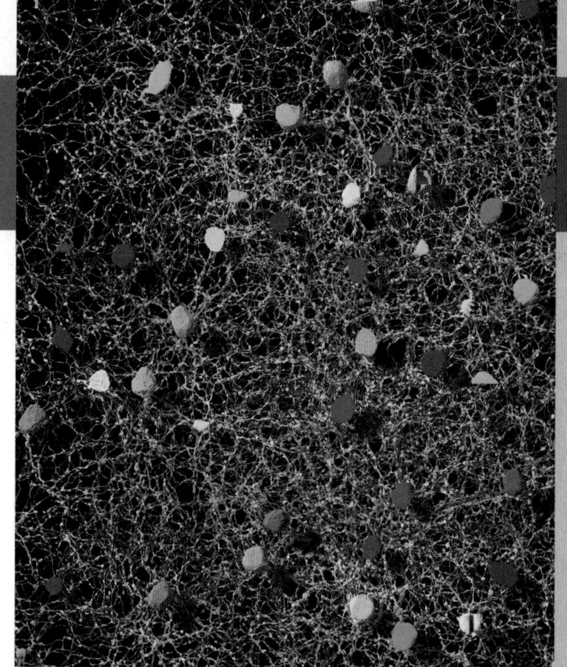

Foundations

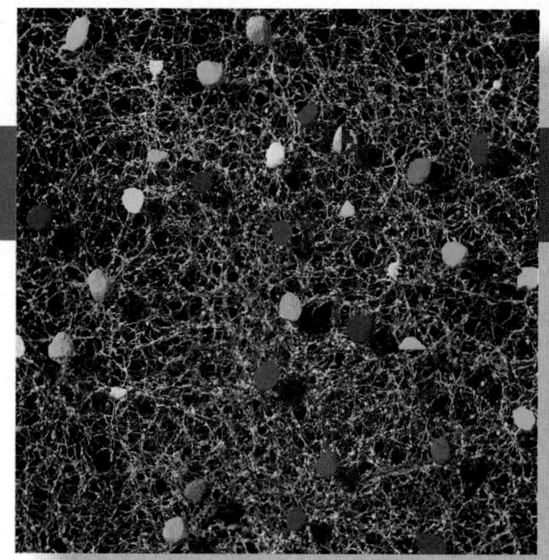

CHAPTER ONE

Neuroscience: Past, Present, and Future

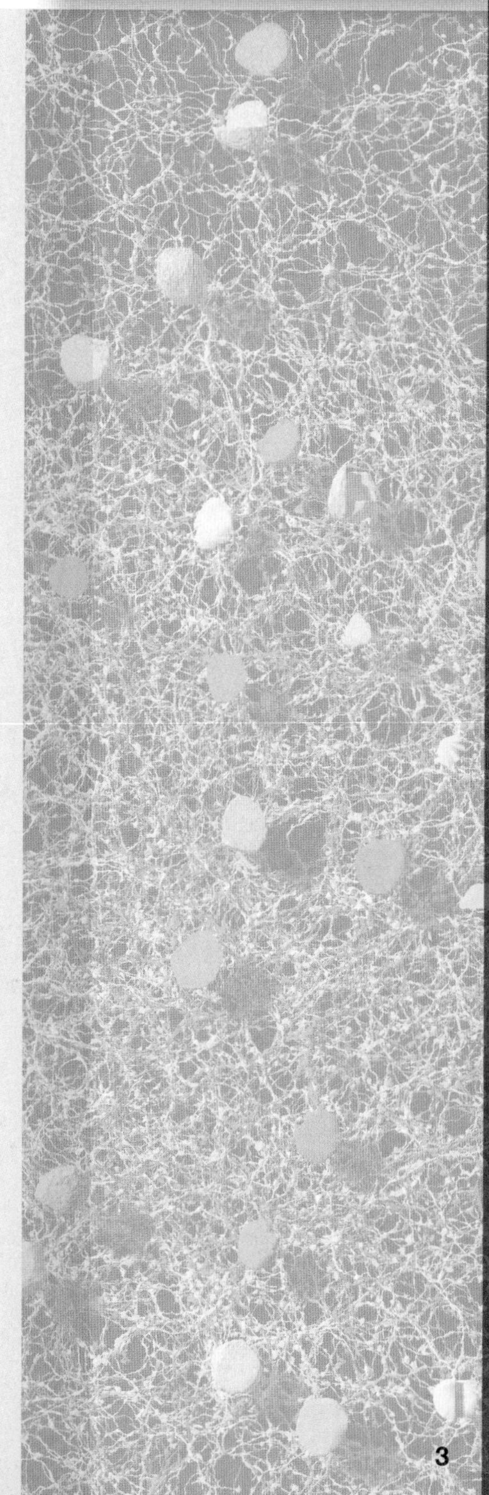

INTRODUCTION

Men ought to know that from nothing else but the brain come joys, delights, laughter and sports, and sorrows, griefs, despondency, and lamentations. And by this, in an especial manner, we acquire wisdom and knowledge, and see and hear and know what are foul and what are fair, what are bad and what are good, what are sweet and what are unsavory. . . . And by the same organ we become mad and delirious, and fears and terrors assail us. . . . All these things we endure from the brain when it is not healthy. . . . In these ways I am of the opinion that the brain exercises the greatest power in the man.

—Hippocrates, *On the Sacred Disease* (Fourth century B.C.E.)

It is human nature to be curious about how we see and hear; why some things feel good and others hurt; how we move; how we reason, learn, remember, and forget; and the nature of anger and madness. Neuroscience research is unraveling these mysteries, and the conclusions of this research are the subject of this textbook.

The word "neuroscience" is young. The Society for Neuroscience, an association of professional neuroscientists, was founded only relatively recently in 1970. The study of the brain, however, is as old as science itself. Historically, the scientists who devoted themselves to an understanding of the nervous system came from different scientific disciplines: medicine, biology, psychology, physics, chemistry, mathematics. The neuroscience revolution occurred when scientists realized that the best hope for understanding the workings of the brain would come from an interdisciplinary approach, a combination of traditional approaches to yield a new synthesis, a new perspective. Most people involved in the scientific investigation of the nervous system today regard themselves as neuroscientists. Indeed, while the course you are now taking may be sponsored by the psychology or biology department at your university or college and may be called biopsychology or neurobiology, you can bet that your instructor is a neuroscientist.

The Society for Neuroscience is one of the largest and fastest growing associations of professional scientists. Far from being overly specialized, the field is as broad as nearly all of natural science, with the nervous system serving as the common point of focus. Understanding how the brain works requires knowledge about many things, from the structure of the water molecule to the electrical and chemical properties of the brain to why Pavlov's dog salivated when a bell rang. This book explores the brain with this broad perspective.

We begin the adventure with a brief tour of neuroscience. What have scientists thought about the brain over the ages? Who are the neuroscientists of today, and how do they approach studying the brain?

THE ORIGINS OF NEUROSCIENCE

You probably already know that the nervous system—the brain, spinal cord, and nerves of the body—is crucial for life and enables you to sense, move, and think. How did this view arise?

Evidence suggests that even our prehistoric ancestors appreciated that the brain was vital to life. The archeological record includes many hominid skulls, dating back a million years and more, that bear signs of fatal cranial damage likely inflicted by other hominids. As early as 7000 years ago, people were boring holes in each other's skulls (a process called *trepanation*), evidently with the aim not to kill but to cure (Figure 1.1). These skulls show signs of healing after the operation, indicating that this procedure had been carried out on live subjects rather than being a ritual conducted after death. Some individuals apparently survived multiple skull surgeries. What those early surgeons hoped to accomplish is not clear, although it has been speculated that this procedure may have been used to treat headaches or mental disorders, perhaps by giving the evil spirits an escape route.

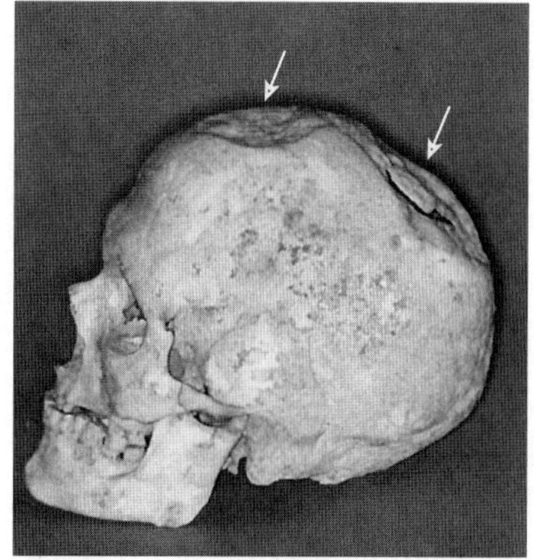

▲ FIGURE 1.1
Evidence of prehistoric brain surgery. This skull of a man over 7000 years old was surgically opened while he was still alive. The arrows indicate two sites of trepanation. (Source: Alt et al., 1997, Fig. 1a.)

Recovered writings from the physicians of ancient Egypt, dating back almost 5000 years, indicate that they were well aware of many symptoms of brain damage. However, it is also very clear that the heart, not the brain, was considered to be the seat of the soul and the repository of memories. Indeed, while the rest of the body was carefully preserved for the afterlife, the brain of the deceased was simply scooped out through the nostrils and discarded! The view that the heart was the seat of consciousness and thought was not seriously challenged until the time of Hippocrates.

Views of the Brain in Ancient Greece

Consider the idea that the different parts of your body look different because they serve different purposes. The structures of the feet and hands are very different, for example, because they perform very different functions: We walk on our feet and manipulate objects with our hands. Thus, there appears to be a very clear *correlation between structure and function*. Differences in appearance predict differences in function.

What can we glean about function from the structure of the head? Quick inspection and a few simple experiments (like closing your eyes) reveal that the head is specialized for sensing the environment with the eyes and ears, nose, and tongue. Even crude dissection can trace the nerves from these organs through the skull into the brain. What would you conclude about the brain from these observations?

If your answer is that the brain is the organ of sensation, then you have reached the same conclusion as several Greek scholars of the fourth century B.C.E. The most influential scholar was Hippocrates (460–379 B.C.E.), the father of Western medicine, who believed that the brain was not only involved in sensation but was also the seat of intelligence.

This view was not universally accepted, however. The famous Greek philosopher Aristotle (384–322 B.C.E.) clung to the belief that the heart was the center of intellect. What function did Aristotle reserve for the brain? He believed it was a radiator for cooling blood that was overheated by the seething heart. The rational temperament of humans was thus explained by the large cooling capacity of our brain.

Views of the Brain During the Roman Empire

The most important figure in Roman medicine was the Greek physician and writer Galen (130–200 C.E.), who embraced the Hippocratic view of brain function. As physician to the gladiators, he must have witnessed the unfortunate consequences of spine and brain injuries. However, Galen's opinions about the brain were probably influenced more by his many careful animal dissections. Figure 1.2 is a drawing of the brain of a sheep, one of Galen's favorite subjects. Two major parts are evident: the *cerebrum* in the front and the *cerebellum* in the back. (The structure of the brain is described in Chapter 7.) Just as we can deduce function

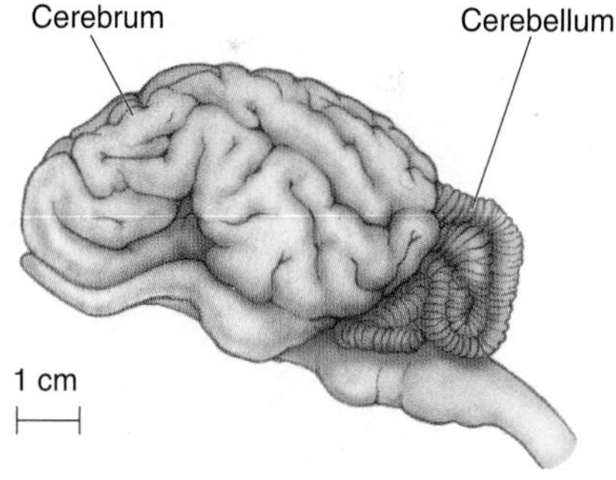

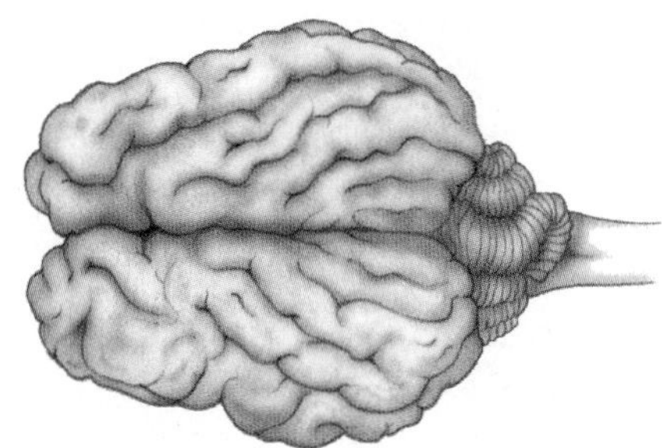

▲ FIGURE 1.2
The brain of a sheep. Notice the location and appearance of the cerebrum and the cerebellum.

from the structure of the hands and feet, Galen tried to deduce function from the structure of the cerebrum and the cerebellum. Poking the freshly dissected brain with a finger reveals the cerebellum is rather hard and the cerebrum rather soft. From this observation, Galen suggested that the cerebrum must receive sensations while the cerebellum must command the muscles. Why such a distinction? He recognized that to form memories, sensations must be imprinted in the brain. Naturally, this must occur in the doughy cerebrum.

As improbable as his reasoning may seem, Galen's deductions were not that far from the truth. The cerebrum, in fact, is largely concerned with sensation and perception, and the cerebellum is primarily a movement control center. Moreover, the cerebrum is a repository of memory. We will see that this is not the only example in the history of neuroscience in which the right general conclusions were reached for the wrong reasons.

How does the brain receive sensations and move the limbs? Galen cut open the brain and found that it is hollow (Figure 1.3). In these hollow spaces, called *ventricles* (like the similar chambers in the heart), there is fluid. To Galen, this discovery fit perfectly with the prevailing theory that the body functioned according to a balance of four vital fluids, or humors. Sensations were registered and movements initiated by the movement of humors to or from the brain ventricles via the nerves, which were believed to be hollow tubes, like the blood vessels.

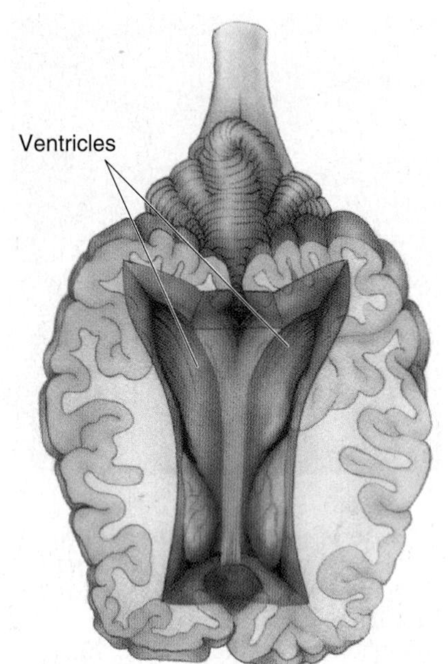

▲ FIGURE 1.3
A dissected sheep brain showing the ventricles.

Views of the Brain from the Renaissance to the Nineteenth Century

Galen's view of the brain prevailed for almost 1500 years. During the Renaissance, the great anatomist Andreas Vesalius (1514–1564) added more detail to the structure of the brain (Figure 1.4). However, the ventricular theory of brain function remained essentially unchallenged. Indeed, the whole concept was strengthened in the early seventeenth century, when French inventors built hydraulically controlled mechanical devices. These devices supported the notion that the brain could be machinelike in its function: Fluid forced out of the ventricles through the nerves might literally "pump you up" and cause the movement of the limbs. After all, don't the muscles bulge when they contract?

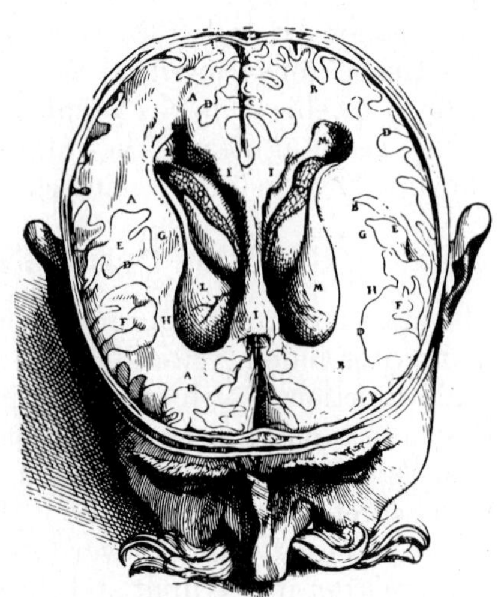

▶ FIGURE 1.4
Human brain ventricles depicted during the Renaissance. This drawing is from *De humani corporis fabrica* by Vesalius (1543). The subject was probably a decapitated criminal. Great care was taken to be anatomically correct in depicting the ventricles. (Source: Finger, 1994, Fig. 2.8.)

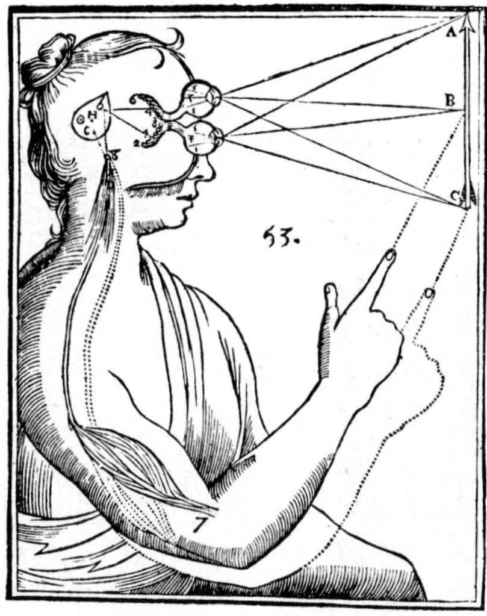

◀ FIGURE 1.5
The brain according to Descartes. This drawing appeared in a 1662 publication by Descartes, who thought that hollow nerves from the eyes projected to the brain ventricles. The mind influenced the motor response by controlling the pineal gland (H), which worked like a valve to control the movement of animal spirits through the nerves that inflated the muscles. (Source: Finger, 1994, Fig. 2.16.)

A chief advocate of this fluid–mechanical theory of brain function was the French mathematician and philosopher René Descartes (1596–1650). Although he thought this theory could explain the brain and behavior of other animals, Descartes believed it could not possibly account for the full range of *human* behavior. He reasoned that unlike other animals, people possess intellect and a God-given soul. Thus, Descartes proposed that brain mechanisms control only human behavior that is like that of the beasts. Uniquely human mental capabilities exist outside the brain in the "mind." Descartes believed that the mind is a spiritual entity that receives sensations and commands movements by communicating with the machinery of the brain via the pineal gland (Figure 1.5). Today, some people still believe that there is a "mind–brain problem," that somehow the human mind is distinct from the brain. However, as we shall see in Part III, modern neuroscience research supports a different conclusion: The mind has a physical basis, which is the brain.

Fortunately, other scientists during the seventeenth and eighteenth centuries broke away from the traditional focus on the ventricles and began examining the brain's substance more closely. They observed, for example, two types of brain tissue: the *gray matter* and the *white matter* (Figure 1.6). What structure–function relationship did they propose? White matter, because it was continuous with the nerves of the body, was correctly believed to contain the fibers that bring information to and from the gray matter.

By the end of the eighteenth century, the nervous system had been completely dissected and its gross anatomy described in detail. Scientists recognized that the nervous system has a central division, consisting of the brain and spinal cord, and a peripheral division, consisting of the network of nerves that course through the body (Figure 1.7). An important breakthrough in neuroanatomy came with the observation that the same general pattern of bumps (called *gyri*) and grooves (called *sulci* and *fissures*) can be identified on the surface of the brain in every individual (Figure 1.8). This pattern, which enables the parceling of the cerebrum into *lobes*, led to speculation that different functions might be localized to the different bumps on the brain. The stage was now set for the era of cerebral localization.

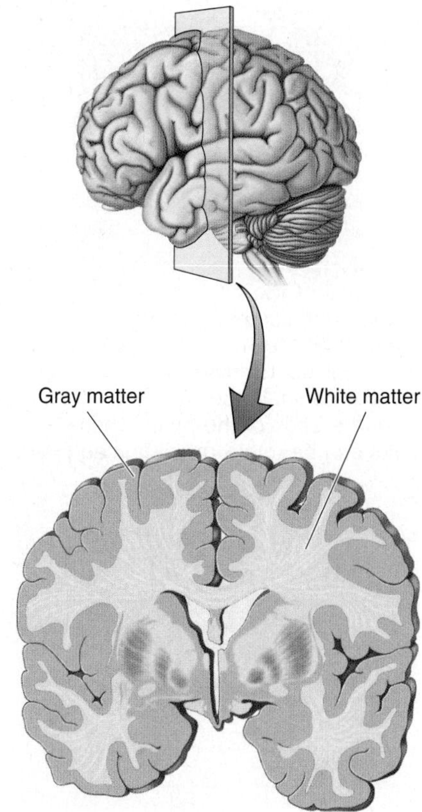

▲ FIGURE 1.6
White matter and gray matter. The human brain has been cut open to reveal these two types of tissue.

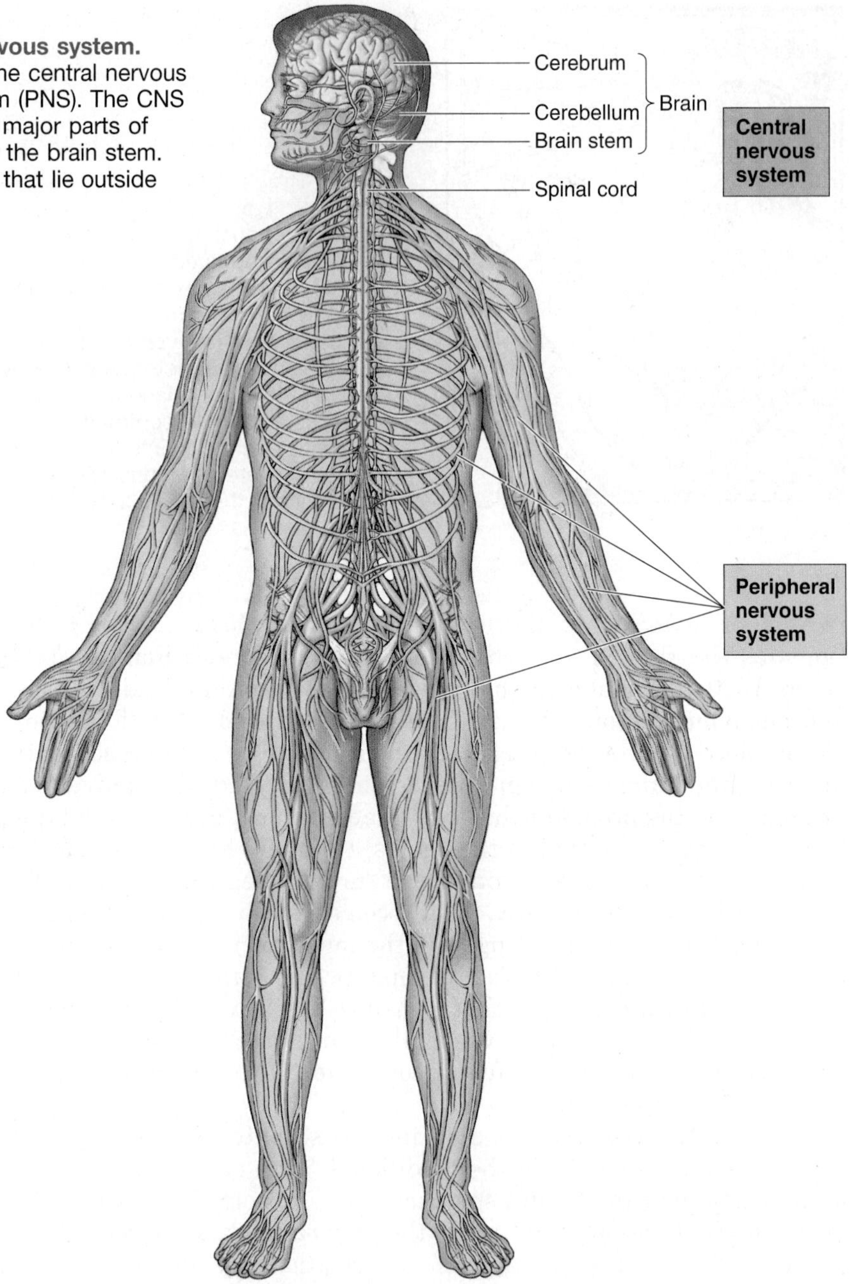

▶ FIGURE 1.7

The basic anatomical subdivisions of the nervous system. The nervous system consists of two divisions, the central nervous system (CNS) and the peripheral nervous system (PNS). The CNS consists of the brain and spinal cord. The three major parts of the brain are the cerebrum, the cerebellum, and the brain stem. The PNS consists of the nerves and nerve cells that lie outside the brain and spinal cord.

Cerebrum
Cerebellum
Brain stem
Brain
Spinal cord
Central nervous system
Peripheral nervous system

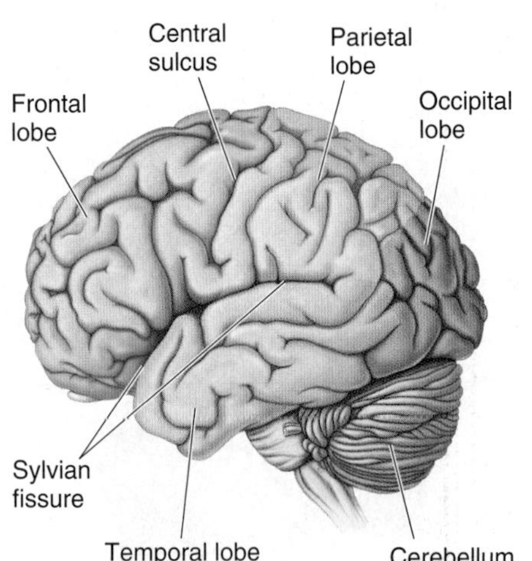

▲ FIGURE 1.8

The lobes of the cerebrum. Notice the deep Sylvian fissure, dividing the frontal lobe from the temporal lobe, and the central sulcus, dividing the frontal lobe from the parietal lobe. The occipital lobe lies at the back of the brain. These landmarks can be found on all human brains.

Nineteenth-Century Views of the Brain

Let's review how the nervous system was understood at the end of the eighteenth century:

- Injury to the brain can disrupt sensations, movement, and thought and can cause death.
- The brain communicates with the body via the nerves.
- The brain has different identifiable parts, which probably perform different functions.
- The brain operates like a machine and follows the laws of nature.

During the next 100 years, more would be learned about the function of the brain than had been learned in all of previous recorded history. This work provided the solid foundation on which modern neuroscience rests. Now we'll look at four key insights gained during the nineteenth century.

Nerves as Wires. In 1751, Benjamin Franklin published a pamphlet titled *Experiments and Observations on Electricity*, which heralded a new understanding of electrical phenomena. By the turn of the century, Italian scientist Luigi Galvani and German biologist Emil du Bois-Reymond had shown that muscles can be caused to twitch when nerves are stimulated electrically and that the brain itself can generate electricity. These discoveries finally displaced the notion that nerves communicate with the brain by the movement of fluid. The new concept was that the nerves are "wires" that conduct electrical signals to and from the brain.

Unresolved was whether the signals to the muscles causing movement use the same wires as those that register sensations from the skin. Bidirectional communication along the same wires was suggested by the observation that when a nerve in the body is cut, there is usually a loss of both sensation and movement in the affected region. However, it was also known that within each nerve of the body there are many thin filaments, or *nerve fibers*, each one of which could serve as an individual wire carrying information in a different direction.

This question was answered around 1810 by Scottish physician Charles Bell and French physiologist François Magendie. A curious anatomical fact is that just before the nerves attach to the spinal cord, the fibers divide into two branches, or roots. The dorsal root enters toward the back of the spinal cord, and the ventral root enters toward the front (Figure 1.9). Bell tested the possibility that these two spinal roots carry information in different directions by cutting each root separately and observing the consequences in experimental animals. He found that cutting only the ventral roots caused muscle paralysis. Later, Magendie was able to show that the dorsal roots carry sensory information into the spinal cord. Bell and Magendie concluded that within each nerve is a mixture of many wires, some of which bring information into the brain and spinal cord and others that send information out to the muscles. In each

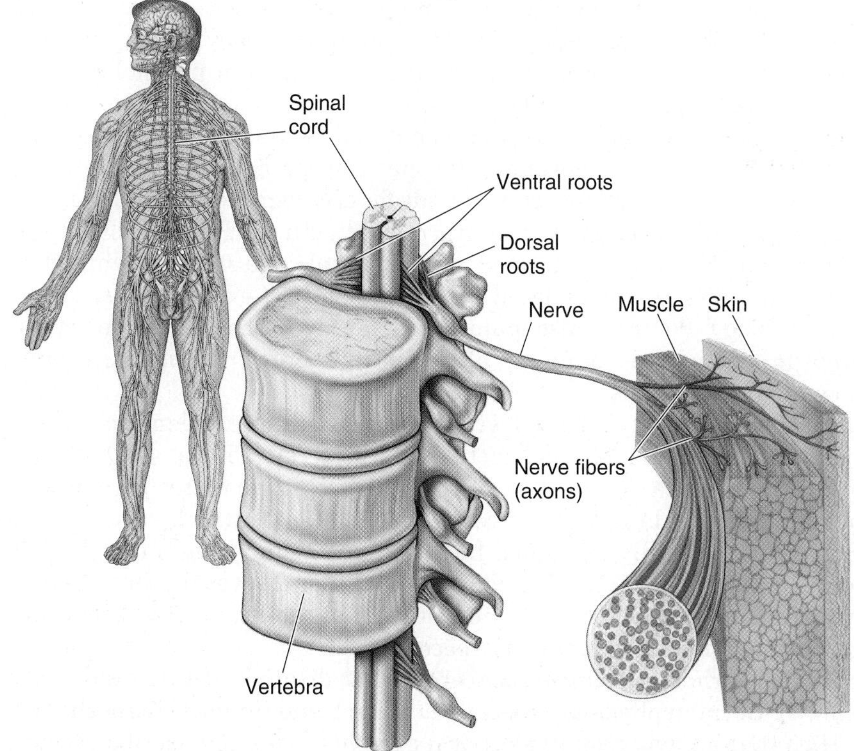

◀ FIGURE 1.9
Spinal nerves and spinal nerve roots. Thirty-one pairs of nerves leave the spinal cord to supply the skin and the muscles. Cutting a spinal nerve leads to a loss of sensation and a loss of movement in the affected region of the body. Incoming sensory fibers (*red*) and outgoing motor fibers (*blue*) divide into spinal roots where the nerves attach to the spinal cord. Bell and Magendie found that the ventral roots contain only motor fibers and the dorsal roots contain only sensory fibers.

sensory and motor nerve fiber, transmission is strictly one-way. The two kinds of fibers are bundled together for most of their length, but they are anatomically segregated where they enter or exit the spinal cord.

Localization of Specific Functions to Different Parts of the Brain. If different functions are localized in different spinal roots, then perhaps different functions are also localized in different parts of the brain. In 1811, Bell proposed that the origin of the motor fibers is the cerebellum and the destination of the sensory fibers is the cerebrum.

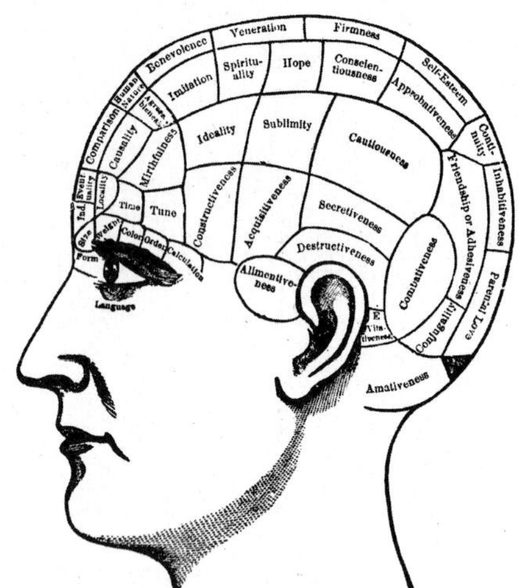

▲ FIGURE 1.10
A phrenological map. According to Gall and his followers, different behavioral traits could be related to the size of different parts of the skull. (Source: Clarke and O'Malley, 1968, Fig. 118.)

How would you test this proposal? One way is to use the same approach that Bell and Magendie employed to identify the functions of the spinal roots: to destroy these parts of the brain and test for sensory and motor deficits. This approach, in which parts of the brain are systematically destroyed to determine their function, is called the *experimental ablation method*. In 1823, the esteemed French physiologist Marie-Jean-Pierre Flourens used this method in a variety of animals (particularly birds) to show that the cerebellum does indeed play a role in the coordination of movement. He also concluded that the cerebrum is involved in sensation and perception, as Bell and Galen before him had suggested. Unlike his predecessors, however, Flourens provided solid experimental support for his conclusions.

What about all those bumps on the brain's surface? Do they perform different functions as well? The idea that they do was irresistible to a young Austrian medical student named Franz Joseph Gall. Believing that bumps on the surface of the skull reflect bumps on the surface of the brain, Gall proposed in 1809 that the propensity for certain personality traits, such as generosity, secretiveness, and destructiveness, could be related to the dimensions of the head (Figure 1.10). To support his claim, Gall and his followers collected and carefully measured the skulls of hundreds of people representing an extensive range of personality types, from the very gifted to the criminally insane. This new "science" of correlating the structure of the head with personality traits was called *phrenology*. Although the claims of the phrenologists were never taken seriously by the mainstream scientific community, they did capture the popular imagination of the time. In fact, a textbook on phrenology published in 1827 sold over 100,000 copies.

One of the most vociferous critics of phrenology was Flourens, the same man who had shown experimentally that the cerebellum and cerebrum perform different functions. His grounds for criticism were sound. For one thing, the shape of the skull is not correlated with the shape of the brain. In addition, Flourens performed experimental ablations showing that particular traits are not isolated to the portions of the cerebrum specified by phrenology. Flourens also maintained, however, that all regions of the cerebrum participate equally in all cerebral functions, a conclusion later shown to be erroneous.

▲ FIGURE 1.11
Paul Broca (1824–1880). By carefully studying the brain of a man who had lost the faculty of speech after a brain lesion (see Figure 1.12), Broca became convinced that different functions could be localized to different parts of the cerebrum. (Source: Clarke and O'Malley, 1968, Fig. 121.)

The person usually credited with tilting the scales of scientific opinion firmly toward localization of functions in the cerebrum was French neurologist Paul Broca (Figure 1.11). Broca was presented with a patient who could understand language but could not speak. After the man's death in 1861, Broca carefully examined his brain and found a lesion in the left frontal lobe (Figure 1.12). Based on this case and several others like it, Broca concluded that this region of the human cerebrum was specifically responsible for the production of speech.

Solid experimental support for cerebral localization in animals quickly followed. German physiologists Gustav Fritsch and Eduard Hitzig showed in 1870 that applying small electrical currents to a circumscribed region of the exposed surface of the brain of a dog could elicit discrete movements. Scottish neurologist David Ferrier repeated these experiments with monkeys. In 1881, he showed that removal of this same region of the cerebrum causes paralysis of the muscles. Similarly, German physiologist Hermann Munk using experimental ablation found evidence that the occipital lobe of the cerebrum was specifically required for vision.

As you will see in Part II of this book, we now know that there is a very clear division of labor in the cerebrum, with different parts performing very different functions. Today's maps of the functional divisions of the cerebrum rival even the most elaborate of those produced by the phrenologists. The big difference is that unlike the phrenologists, scientists today require solid experimental evidence before attributing a specific function to a portion of the brain. All the same, Gall seems to have had in part the right general idea. It is natural to wonder why Flourens, the pioneer of brain localization of function, was misled into believing that the cerebrum acted as a whole and could not be subdivided. This gifted experimentalist may have missed cerebral localization for many different reasons, but it seems clear that one reason was his visceral disdain for Gall and phrenology. He could not bring himself to agree even remotely with Gall, whom he viewed as a lunatic. This reminds us that science, for better or worse, was and still is subject to both the strengths and the weaknesses of human nature.

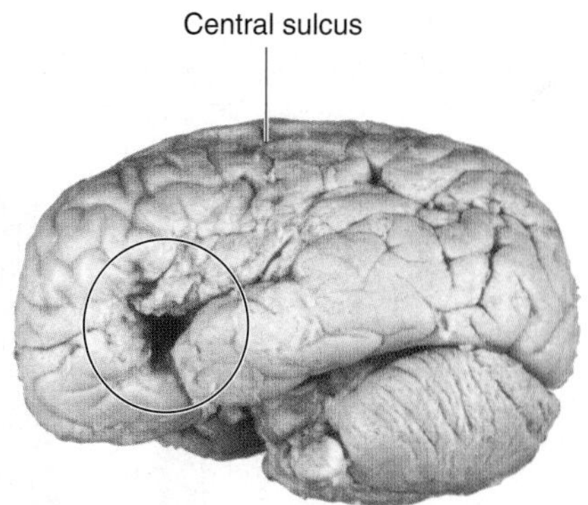

▲ FIGURE 1.12
The brain that convinced Broca of localization of function in the cerebrum. This is the preserved brain of a patient who had lost the ability to speak before he died in 1861. The lesion that produced this deficit is circled. (Source: Corsi, 1991, Fig. III, 4.)

The Evolution of Nervous Systems. In 1859, English biologist Charles Darwin (Figure 1.13) published *On the Origin of Species*. This landmark work articulates a theory of evolution: that species of organisms evolved from a common ancestor. According to his theory, differences among species arise by a process Darwin called *natural selection*. As a result of the mechanisms of reproduction, the physical traits of the offspring are sometimes different from those of the parents. If such traits hold an advantage for survival, the offspring themselves will be more likely to survive to reproduce, thus increasing the likelihood that the advantageous traits are passed on to the next generation. Over the course of many generations, this process led to the development of traits that distinguish species today: flippers on harbor seals, paws on dogs, hands on raccoons, and so on. This single insight revolutionized biology. Today, scientific evidence in many fields ranging from anthropology to molecular genetics overwhelmingly supports the theory of evolution by natural selection.

Darwin included behavior among the heritable traits that could evolve. For example, he observed that many mammalian species show the same reaction when frightened: The pupils of the eyes get bigger, the heart races, hairs stand on end. This is as true for a human as it is for a dog. To Darwin, the similarities of this response pattern indicated that these different species evolved from a common ancestor that possessed the same behavioral trait, which was advantageous presumably because it facilitated escape from predators. Because behavior reflects the activity of the nervous system, we can infer that the brain mechanisms that underlie this fear reaction may be similar, if not identical, across these species.

The idea that the nervous systems of different species evolved from common ancestors and may have common mechanisms is the rationale for relating the results of animal experiments to humans. For example, many of the details of electrical impulse conduction along nerve fibers were discovered first in the squid but are now known to apply equally well to humans. Most neuroscientists today use *animal models* to examine processes they wish to understand in humans. For example, rats show clear signs of addiction if they are repeatedly given the chance to

▲ FIGURE 1.13
Charles Darwin (1809–1882). Darwin proposed his theory of evolution, explaining how species evolve through the process of natural selection. (Source: The Bettman Archive.)

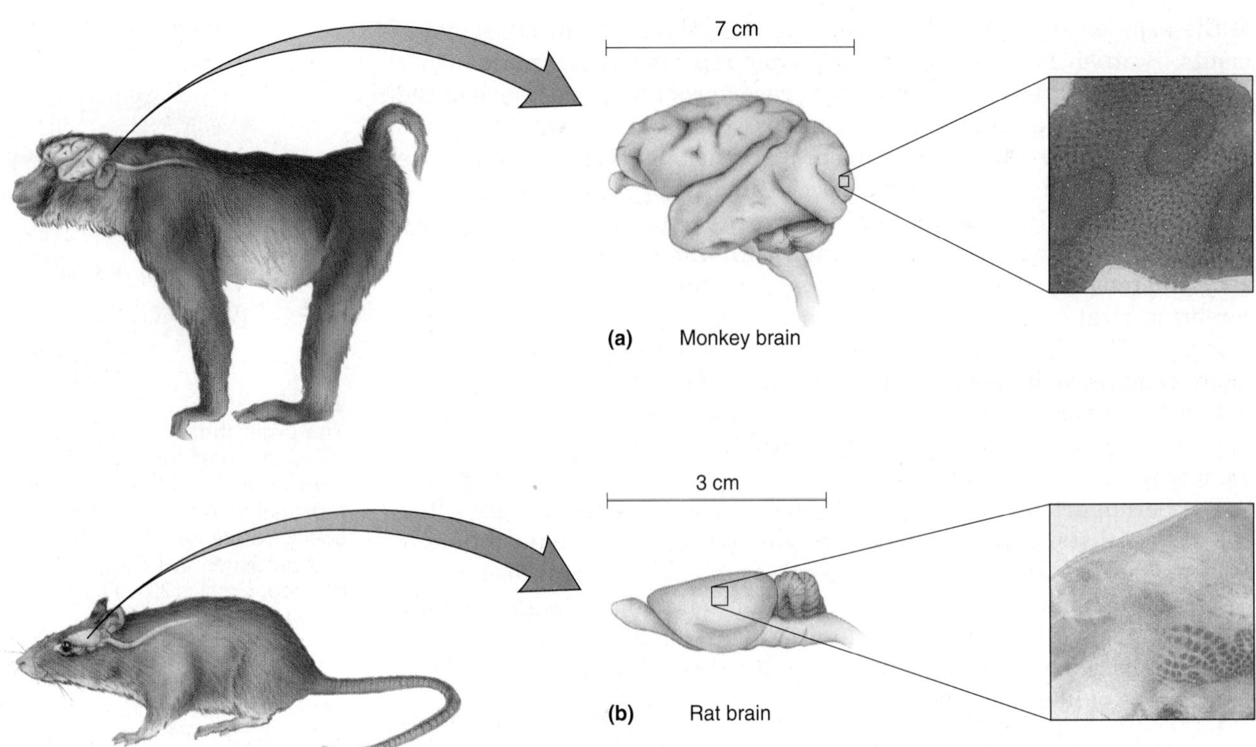

▲ FIGURE 1.14
Different brain specializations in monkeys and rats. (a) The brain of the macaque monkey has a highly evolved sense of sight. The boxed region receives information from the eyes. When this region is sliced open and stained to show metabolically active tissue, a mosaic of "blobs" appears. The neurons within the blobs are specialized to analyze colors in the visual world. (b) The brain of a rat has a highly evolved sense of touch to the face. The boxed region receives information from the whiskers. When this region is sliced open and stained to show the location of the neurons, a mosaic of "barrels" appears. Each barrel is specialized to receive input from a single whisker on the rat's face. (Photomicrographs courtesy of Dr. S.H.C. Hendry.)

self-administer cocaine. Consequently, rats provide a valuable animal model for research focused on understanding how psychoactive drugs exert their effects on the nervous system.

On the other hand, many behavioral traits are highly specialized for the environment (or niche) a species normally occupies. For example, monkeys swinging from branch to branch have a keen sense of sight, while rats slinking through underground tunnels have poor vision but a highly evolved sense of touch using their snout whiskers. Adaptations are reflected in the structure and function of the brain of every species. By comparing the specializations of the brains of different species, neuroscientists have been able to identify which parts of the brain are specialized for different behavioral functions. Examples for monkeys and rats are shown in Figure 1.14.

The Neuron: The Basic Functional Unit of the Brain. Technical advances in microscopy during the early 1800s gave scientists their first opportunity to examine animal tissues at high magnifications. In 1839, German zoologist Theodor Schwann proposed what came to be known as the *cell theory*: All tissues are composed of microscopic units called *cells*.

Although cells in the brain had been identified and described, there was still controversy at that time about whether the individual "nerve cell" was actually the basic unit of brain function. Nerve cells usually have a number of thin projections, or processes, that extend from a central cell body (Figure 1.15). Initially, scientists could not decide whether the processes from different cells fuse together as do blood vessels in the circulatory system. If this were true, then the "nerve net" of connected nerve cells would represent the elementary unit of brain function.

Chapter 2 presents a brief history of how this issue was resolved. Suffice it to say that by 1900, the individual nerve cell, now called the *neuron*, was recognized to be the basic functional unit of the nervous system.

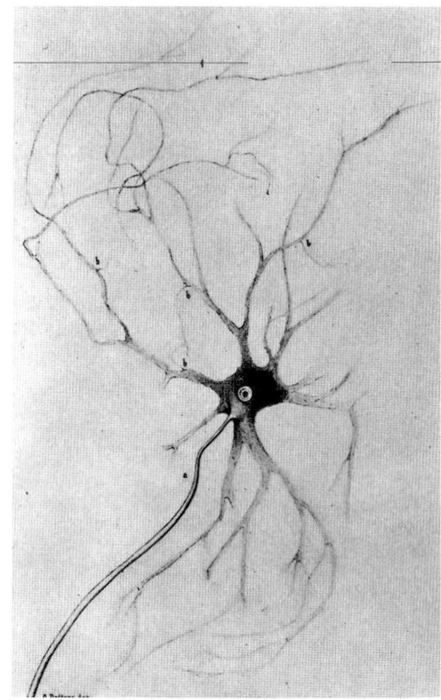

▲ FIGURE 1.15
An early depiction of a nerve cell. Published in 1865, this drawing by German anatomist Otto Deiters shows a nerve cell, or neuron, and its many projections, called *neurites*. For a time it was thought that the neurites from different neurons might fuse together like the blood vessels of the circulatory system. We now know that neurons are distinct entities that communicate using chemical and electrical signals. (Source: Clarke and O'Malley, 1968, Fig. 16.)

NEUROSCIENCE TODAY

The history of modern neuroscience is still being written, and the accomplishments to date form the basis for this textbook. We will discuss the most recent developments in the coming chapters. Before we do, let's take a look at how brain research is conducted today and why it is so important to society.

Levels of Analysis

History has clearly shown that understanding how the brain works is a big challenge. To reduce the complexity of the problem, neuroscientists break it into smaller pieces for systematic experimental analysis. This is called the *reductionist approach*. The size of the unit of study defines what is often called the *level of analysis*. In ascending order of complexity, these levels are molecular, cellular, systems, behavioral, and cognitive.

Molecular Neuroscience. The brain has been called the most complex piece of matter in the universe. Brain matter consists of a fantastic variety of molecules, many of which are unique to the nervous system. These different molecules play many different roles that are crucial for brain function: messengers that allow neurons to communicate with one another, sentries that control what materials can enter or leave neurons, conductors that orchestrate neuron growth, archivists of past experiences. The study of the brain at this most elementary level is called *molecular neuroscience*.

Cellular Neuroscience. The next level of analysis is cellular neuroscience, which focuses on studying how all those molecules work together to give neurons their special properties. Among the questions asked at this level are: How many different types of neurons are there, and how do they differ in function? How do neurons influence other neurons? How do neurons become "wired together" during fetal development? How do neurons perform computations?

Systems Neuroscience. Constellations of neurons form complex circuits that perform a common function, such as vision or voluntary movement. Thus, we can speak of the "visual system" and the "motor system," each of which has its own distinct circuitry within the brain. At this level of analysis, called *systems neuroscience*, neuroscientists study how different neural circuits analyze sensory information, form perceptions of the external world, make decisions, and execute movements.

Behavioral Neuroscience. How do neural systems work together to produce integrated behaviors? For example, are different forms of memory accounted for by different systems? Where in the brain do "mind-altering" drugs act, and what is the normal contribution of these systems to the regulation of mood and behavior? What neural systems account for gender-specific behaviors? Where are dreams created and what do they reveal? These questions are studied in behavioral neuroscience.

Cognitive Neuroscience. Perhaps the greatest challenge of neuroscience is understanding the neural mechanisms responsible for the higher levels of human mental activity, such as self-awareness, imagination, and language. Research at this level, called *cognitive neuroscience*, studies how the activity of the brain creates the mind.

Neuroscientists

"Neuroscientist" sounds impressive, kind of like "rocket scientist." But we were all students once, just like you. For whatever reason—maybe we wanted to know why our eyesight was poor, or why a family member suffered a loss of speech after a stroke—we came to share a thirst for knowledge of how the brain works. Perhaps you will, too.

Being a neuroscientist is rewarding, but it does not come easily. Many years of training are required. One may begin by helping out in a research lab during or after college and then going to graduate school to earn a Ph.D. or an M.D. (or both). Several years of post-doctoral training usually follow, learning new techniques or ways of thinking under the direction of an established neuroscientist. Finally, the "young" neuroscientist is ready to set up shop at a university, institute, or hospital.

Broadly speaking, neuroscience research (and neuroscientists) may be divided into three types: *clinical*, *experimental*, and *theoretical*. Clinical research is mainly conducted by physicians (M.D.s). The main medical specialties associated with the human nervous system are neurology, psychiatry, neurosurgery, and neuropathology (Table 1.1). Many who conduct clinical research continue in the tradition of Broca, attempting to deduce from the behavioral effects of brain damage the functions of various parts of the brain. Others conduct studies to assess the benefits and risks of new types of treatment.

Despite the obvious value of clinical research, the foundation for all medical treatments of the nervous system continues to be laid by experimental neuroscientists, who may hold either an M.D. or a Ph.D. The experimental approaches to studying the brain are so broad that they include almost every conceivable methodology. Neuroscience is highly interdisciplinary; however, expertise in a particular methodology may distinguish one neuroscientist from another. Thus, there are *neuroanatomists*, who use sophisticated microscopes to trace connections in the brain; *neurophysiologists*, who use electrodes to measure the brain's electrical activity; *neuropharmacologists*, who use drugs to study the chemistry of

TABLE 1.1 **Medical Specialists Associated with the Nervous System**

Specialist	Description
Neurologist	An M.D. trained to diagnose and treat diseases of the nervous system
Psychiatrist	An M.D. trained to diagnose and treat disorders of mood and behavior
Neurosurgeon	An M.D. trained to perform surgery on the brain and spinal cord
Neuropathologist	An M.D. or Ph.D. trained to recognize the changes in nervous tissue that result from disease

TABLE 1.2 **Types of Experimental Neuroscientists**

Type	Description
Developmental neurobiologist	Analyzes the development and maturation of the brain
Molecular neurobiologist	Uses the genetic material of neurons to understand the structure and function of brain molecules
Neuroanatomist	Studies the structure of the nervous system
Neurochemist	Studies the chemistry of the nervous system
Neuroethologist	Studies the neural basis of species-specific animal behaviors in natural settings
Neuropharmacologist	Examines the effects of drugs on the nervous system
Neurophysiologist	Measures the electrical activity of the nervous system
Physiological psychologist (biological psychologist, psychobiologist)	Studies the biological basis of behavior
Psychophysicist	Quantitatively measures perceptual abilities

brain function; *molecular neurobiologists*, who probe the genetic material of neurons to find clues about the structure of brain molecules; and so on. Table 1.2 lists some of the types of experimental neuroscientists.

Theoretical neuroscience is a relatively young discipline, in which researchers use mathematical and computational tools to understand the brain at all levels of analysis. In the tradition of physics, theoretical neuroscientists attempt to make sense of the vast amounts of data generated by experimentalists, with the goals of helping focus experiments on questions of greatest importance and establishing the mathematical principles of nervous system organization.

The Scientific Process

Neuroscientists of all stripes endeavor to establish truths about the nervous system. Regardless of the level of analysis they choose, they work according to a scientific process consisting of four essential steps: observation, replication, interpretation, and verification.

Observation. Observations are typically made during experiments designed to test a particular hypothesis. For example, Bell hypothesized that the ventral roots contain the nerve fibers that control the muscles. To test this idea, he performed an experiment in which he cut these fibers and then observed whether or not muscular paralysis resulted. Other types of observation derive from carefully watching the world around us, or from introspection, or from human clinical cases. For example, Broca's careful observations led him to correlate left frontal lobe damage with the loss of the ability to speak.

Replication. Any observation, whether experimental or clinical, must be replicated. Replication simply means repeating the experiment on different subjects or making similar observations in different patients, as many times as necessary to rule out the possibility that the observation occurred by chance.

Interpretation. Once the scientist believes the observation is correct, he or she interprets it. Interpretations depend on the state of knowledge (or ignorance) at the time and on the scientist's preconceived notions (or "mind set"). Interpretations therefore do not always withstand the test of time. For example, at the time he made his observations, Flourens was unaware that the cerebrum of a bird is fundamentally different from that of a mammal. Thus, he wrongly concluded from experimental ablations in birds that there was no localization of certain functions in the cerebrum of mammals. Moreover, as mentioned before, his profound distaste for Gall surely also colored his interpretation. The point is that the correct interpretation often is not made until long after the original observations. Indeed, major breakthroughs sometimes occur when old observations are reinterpreted in a new light.

Verification. The final step of the scientific process is verification. This step is distinct from the replication the original observer performed. Verification means that the observation is sufficiently robust that any competent scientist who precisely follows the protocols of the original observer can reproduce it. Successful verification generally means that the observation is accepted as fact. However, not all observations can be verified, sometimes because of inaccuracies in the original report or insufficient replication. But failure to verify usually stems from the fact that unrecognized variables, such as temperature or time of day, contributed to the original result. Thus, the process of verification, if affirmative, establishes new scientific fact, or, if negative, suggests new interpretations for the original observation.

Occasionally, one reads in the popular press about a case of scientific fraud. Researchers face keen competition for limited research funds and feel considerable pressure to "publish or perish." In the interest of expediency, a few have actually published "observations" they in fact never made. Fortunately, such instances of fraud are rare, thanks to the scientific process. Before long, other scientists find they are unable to verify the fraudulent observations and question how they could have been made in the first place. The fact that we can fill this book with so much knowledge about the nervous system stands as a testament to the value of the scientific process.

The Use of Animals in Neuroscience Research

Most of what we know about the nervous system has come from experiments on animals. In most cases, the animals are killed so their brains can be examined neuroanatomically, neurophysiologically, and/or neurochemically. The fact that animals are sacrificed for the pursuit of human knowledge raises questions about the ethics of animal research.

The Animals. Let's begin by putting the issue in perspective. Throughout history, humans have considered animals and animal products as renewable natural resources that can be used for food, clothing, transportation, recreation, sport, and companionship. The animals used in research, education, and testing have always been a small fraction of those used for other purposes. For example, in the United States, the number of animals used in all types of biomedical research is very small compared to the number killed for food. The number used specifically in neuroscience research is much smaller still.

Neuroscience experiments are conducted using many different species, ranging from snails to monkeys. The choice of species is generally dictated by the question under investigation, the level of analysis, and the extent to which the knowledge gained can be related to humans. As a rule, the more basic the process under investigation, the more distant can be the evolutionary relationship with humans. Thus, experiments aimed at understanding the molecular basis of nerve impulse conduction can be carried out with a distantly related species, such as the squid. On the other hand, understanding the neural basis of movement and perceptual disorders in humans has required experiments with more closely related species, such as the macaque monkey. Today, more than half of the animals used for neuroscience research are rodents—mice and rats—that are bred specifically for this purpose.

Animal Welfare. In the developed world today, most educated adults have a concern for animal welfare. Neuroscientists share this concern and work to ensure that animals are well treated. Society has not always placed such value on animal welfare, however, as reflected in some of the scientific practices of the past. For example, in his experiments early in the nineteenth century, Magendie used unanesthetized puppies (for which he was later criticized by his scientific rival Bell). Fortunately, heightened awareness of animal welfare has more recently led to significant improvements in how animals are treated in biomedical research.

Today, neuroscientists accept certain moral responsibilities toward their animal subjects:

1. Animals are used only in worthwhile experiments that promise to advance our knowledge of the nervous system.
2. All necessary steps are taken to minimize pain and distress experienced by the experimental animals (use of anesthetics, analgesics, etc.).
3. All possible alternatives to the use of animals are considered.

Adherence to this ethical code is monitored in a number of ways. First, research proposals must pass a review by the Institutional Animal Care and Use Committee (IACUC), as mandated by U.S. federal law. Members of this committee include a veterinarian, scientists in other disciplines, and nonscientist community representatives. After passing the IACUC review, proposals are evaluated for scientific merit by a panel of expert neuroscientists. This step ensures that only the most worthwhile projects are carried out. Then, when neuroscientists submit their observations for publication in the professional journals, the papers are carefully reviewed by other neuroscientists for both scientific merit and animal welfare concerns. Reservations about either issue can lead to rejection of the paper, which in turn can lead to a loss of funding for the research. In addition to these monitoring procedures, federal law sets strict standards for the housing and care of laboratory animals.

Animal Rights. Most people accept the necessity for animal experimentation to advance knowledge, as long as it is performed humanely and with the proper respect for animals' welfare. However, a vocal and increasingly violent minority seeks the total abolition of animal use for human purposes, including experimentation. These people subscribe to a philosophical position often called *animal rights*. According to this way of thinking, animals have the same legal and moral rights as humans.

If you are an animal lover, you may be sympathetic to this position. But consider the following questions. Are you willing to deprive yourself and your family of medical procedures that were developed using animals? Is the death of a mouse equivalent to the death of a human being? Is keeping a pet the moral equivalent of slavery? Is eating meat the moral equivalent of murder? Is it unethical to take the life of a pig to save the life of a child? Is controlling the rodent population in the sewers or the roach population in your home morally equivalent to the Holocaust?

If your answer is no to any of these questions, then you do not subscribe to the philosophy of animal rights. *Animal welfare*—a concern that all responsible people share—must not be confused with animal rights.

Animal rights activists have vigorously pursued their agenda against animal research, sometimes with alarming success. They have manipulated public opinion with repeated allegations of cruelty in animal experiments that are grossly distorted or blatantly false. They have vandalized laboratories, destroying years of hard-won scientific data and hundreds of thousands of dollars of equipment (that you, the taxpayer, had paid for). With threats of violence they have driven some researchers out of science altogether.

Fortunately, the tide is turning. Thanks to the efforts of a number of people, scientists and nonscientists alike, the false claims of the extremists have been exposed, and the benefits to humankind of animal research have been extolled (Figure 1.16). Considering the staggering toll in terms of human suffering that results from disorders of the nervous system, neuroscientists take the position that it is our responsibility to wisely use all the resources nature has provided, including animals, to gain an understanding of how the brain functions in health and in disease.

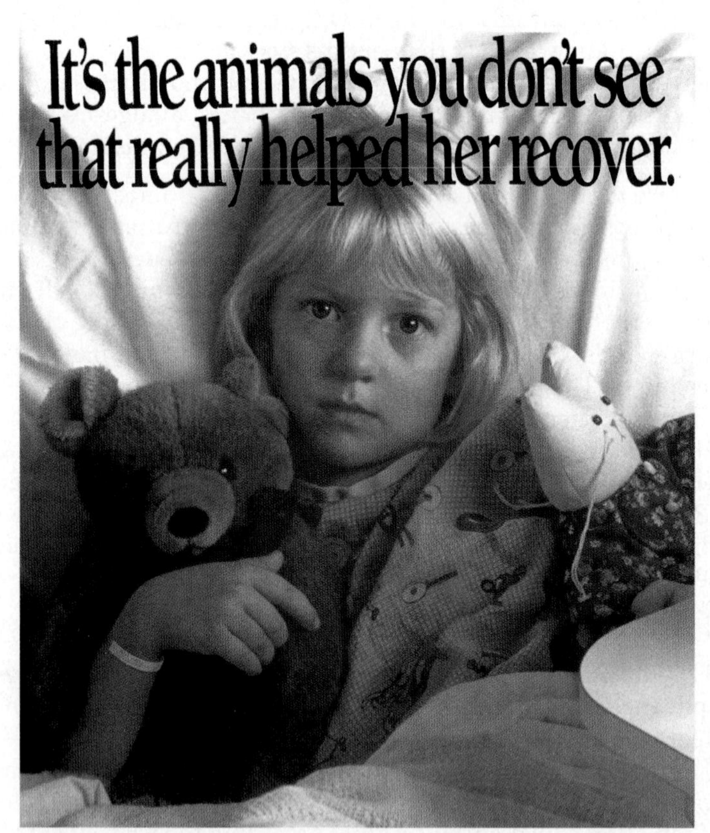

▲ FIGURE 1.16
Our debt to animal research. This poster counters the claims of animal rights activists by raising public awareness of the benefits of animal research. (Source: Foundation for Biomedical Research.)

TABLE 1.3 **Some Major Disorders of the Nervous System**

Disorder	Description
Alzheimer's disease	A progressive degenerative disease of the brain, characterized by dementia and always fatal
Autism	A disorder emerging in early childhood characterized by impairments in communication and social interactions, and restricted and repetitive behaviors
Cerebral palsy	A motor disorder caused by damage to the cerebrum before, during, or soon after birth
Depression	A serious disorder of mood, characterized by insomnia, loss of appetite, and feelings of dejection
Epilepsy	A condition characterized by periodic disturbances of brain electrical activity that can lead to seizures, loss of consciousness, and sensory disturbances
Multiple sclerosis	A progressive disease that affects nerve conduction, characterized by episodes of weakness, lack of coordination, and speech disturbance
Parkinson's disease	A progressive disease of the brain that leads to difficulty in initiating voluntary movement
Schizophrenia	A severe psychotic illness characterized by delusions, hallucinations, and bizarre behavior
Spinal paralysis	A loss of feeling and movement caused by traumatic damage to the spinal cord
Stroke	A loss of brain function caused by disruption of the blood supply, usually leading to permanent sensory, motor, or cognitive deficit

The Cost of Ignorance: Nervous System Disorders

Modern neuroscience research is expensive, but the cost of ignorance about the brain is far greater. Table 1.3 lists some of the disorders that affect the nervous system. It is likely that your family has felt the impact of one or more of these. Let's look at a few brain disorders and examine their effects on society.

Alzheimer's disease and Parkinson's disease are both characterized by progressive degeneration of specific neurons in the brain. Parkinson's disease, which results in a crippling impairment of voluntary movement, currently affects over 500,000 Americans.[1] Alzheimer's disease leads to dementia, a state of confusion characterized by the loss of ability to learn new information and to recall previously acquired knowledge. Dementia affects an estimated 18% of people over age 85.[2] The number of Americans with dementia totals well over 4 million. Indeed, dementia is now recognized as not an inevitable outcome of aging, as was once believed, but as a sign of brain disease. Alzheimer's disease progresses mercilessly, robbing its victims first of their mind, then of control over basic bodily functions, and finally of their life; the disease is always fatal. In the United States, the annual cost of care for people with dementia is greater than $100 billion and rising at an alarming rate.

Depression and schizophrenia are disorders of mood and thought. Depression is characterized by overwhelming feelings of dejection,

[1]National Institute of Neurological Disorders and Stroke. "Parkinson Disease Backgrounder." October 18, 2004.

[2]U.S. Department of Health and Human Services, Agency for Healthcare Research and Quality. "Approximately 5 Percent of Seniors Report One or More Cognitive Disorders." March 2011.

worthlessness, and guilt. Over 30 million Americans will experience a major depressive illness at some time in their lives. Depression is the leading cause of suicide, which claims more than 30,000 lives each year in the United States.[3]

Schizophrenia is a severe psychiatric disorder characterized by delusions, hallucinations, and bizarre behavior. This disease often strikes at the prime of life—adolescence or early adulthood—and can persist for life. Over 2 million Americans suffer from schizophrenia. The National Institute of Mental Health (NIMH) estimates that mental disorders, such as depression and schizophrenia, cost the United States in excess of $150 billion annually.

Stroke is the fourth leading cause of death in the United States. Stroke victims who do not die, over half a million every year, are likely to be permanently disabled. The annual cost of stroke nationwide is $54 billion.[4]

Alcohol or drug addiction affects virtually every family in the United States. The cost in terms of treatment, lost wages, and other consequences exceeds $600 billion per year.[5]

These few examples only scratch the surface. *More Americans are hospitalized with neurological and mental disorders than with any other major disease group, including heart disease and cancer.*

The economic costs of brain dysfunction are enormous, but they pale in comparison with the staggering emotional toll on victims and their families. The prevention and treatment of brain disorders require an understanding of normal brain function, and this basic understanding is the goal of neuroscience. Neuroscience research has already contributed to the development of increasingly effective treatments for Parkinson's disease, depression, and schizophrenia. New strategies are being tested to rescue dying neurons in people with Alzheimer's disease and those who have had a stroke. Major progress has been made in our understanding of how drugs and alcohol affect the brain and how they lead to addictive behavior. The material in this book demonstrates that a lot is known about the function of the brain. But what we know is insignificant compared with what is still left to be learned.

CONCLUDING REMARKS

The historical foundations of neuroscience were established by many people over many generations. Men and women today are working at all levels of analysis, using all types of technology, to shed more light on the functions of the brain. The fruits of this labor form the basis for this textbook.

The goal of neuroscience is to understand how nervous systems function. Many important insights can be gained from a vantage point outside the head. Because the brain's activity is reflected in behavior, careful behavioral measurements inform us of the capabilities and limitations of brain function. Computer models that reproduce the brain's computational properties can help us understand how these properties might arise. From the scalp, we can measure brain waves, which tell us something about the electrical activity of different parts of the brain during

---

[3]National Institute of Mental Health. "Suicide in the U.S.: Statistics and Prevention." September 27, 2010.

[4]American Heart Association/American Stroke Association. "Impact of Stroke (Stroke Statistics)." May 1, 2012.

[5]National Institutes of Health, National Institute of Drug Abuse. "DrugFacts: Understanding Drug Abuse and Addiction." March 2011.

various behavioral states. New computer-assisted imaging techniques enable researchers to examine the structure of the living brain as it sits in the head. And using even more sophisticated imaging methods, we are beginning to see which different parts of the human brain become active under different conditions. But none of these noninvasive methods, old or new, can fully substitute for experimentation with living brain tissue. We cannot make sense of remotely detected signals without being able to see how they are generated and what their significance is. To understand *how* the brain works, we must open the head and examine what's inside—neuroanatomically, neurophysiologically, and neurochemically.

The pace of neuroscience research today is truly breathtaking, raising hopes that soon we will have new treatments for the wide range of nervous system disorders that debilitate and cripple millions of people annually. However, despite the progress in recent decades and the centuries preceding them, we still have a long way to go before we fully understand how the brain performs all of its amazing feats. But this is the fun of being a neuroscientist: Because our ignorance of brain function is so vast, a startling new discovery lurks around virtually every corner.

REVIEW QUESTIONS

1. What are brain ventricles, and what functions have been ascribed to them over the ages?
2. What experiment did Bell perform to show that the nerves of the body contain a mixture of sensory and motor fibers?
3. What did Flourens' experiments suggest were the functions of the cerebrum and the cerebellum?
4. What is the meaning of the term *animal model*?
5. A region of the cerebrum is now called Broca's area. What function do you think this region performs, and why?
6. What are the different levels of analysis in neuroscience research? What questions do researchers ask at each level?
7. What are the steps in the scientific process? Describe each one.

FURTHER READING

Allman JM. 1999. *Evolving Brains*. New York: Scientific American Library.

Clarke E, O'Malley C. 1968. *The Human Brain and Spinal Cord*, 2nd ed. Los Angeles: University of California Press.

Corsi P, ed. 1991. *The Enchanted Loom*. New York: Oxford University Press.

Crick F. 1994. *The Astonishing Hypothesis: The Scientific Search for the Soul*. New York: Macmillan.

Finger S. 1994. *Origins of Neuroscience*. New York: Oxford University Press.

Glickstein M. 2014. *Neuroscience: A Historical Introduction*. Cambridge, MA: MIT Press.

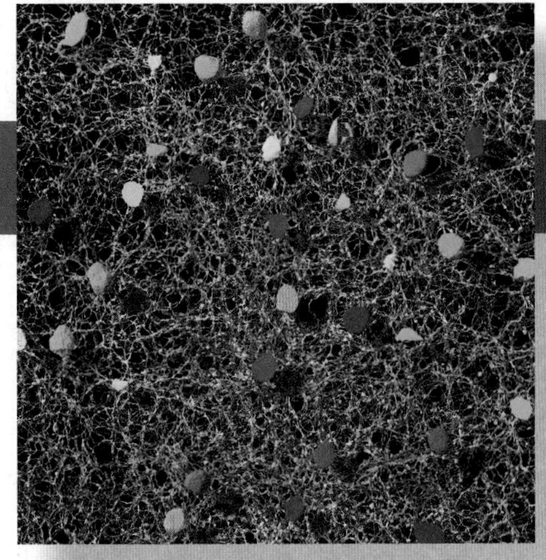

CHAPTER TWO

Neurons and Glia

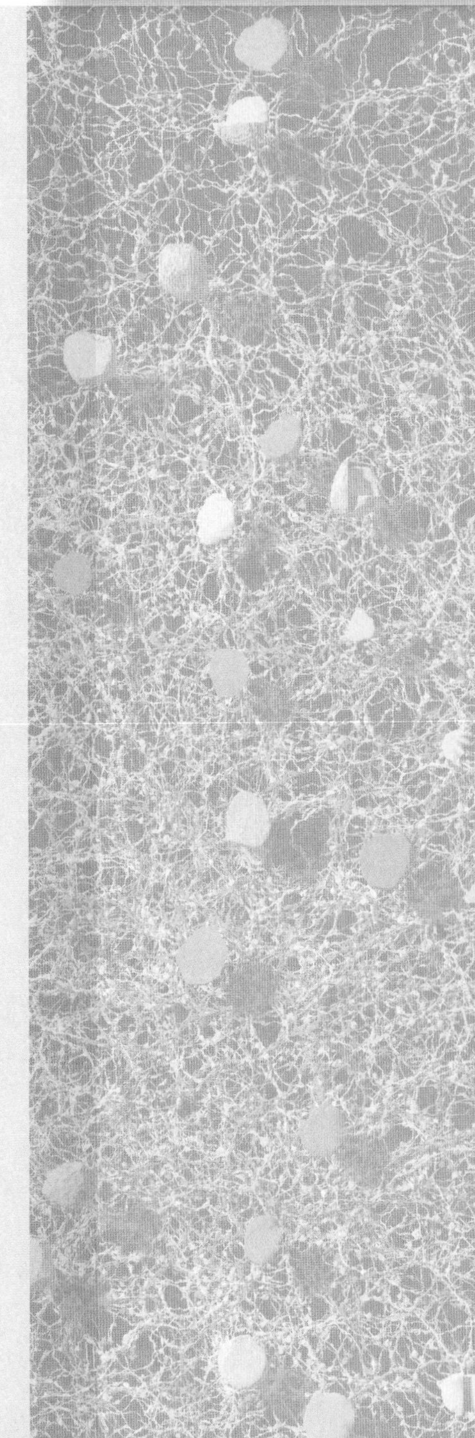

INTRODUCTION

All tissues and organs in the body consist of cells. The specialized functions of cells and how they interact determine the functions of organs. The brain is an organ—to be sure, the most sophisticated and complex organ that nature has devised. But the basic strategy for unraveling its functions is no different from that used to investigate the pancreas or the lung. We must begin by learning how brain cells work individually and then see how they are assembled to work together. In neuroscience, there is no need to separate *mind* from *brain*; once we fully understand the individual and concerted actions of brain cells, we will understand our mental abilities. The organization of this book reflects this "neurophilosophy." We start with the cells of the nervous system—their structure, function, and means of communication. In later chapters, we will explore how these cells are assembled into circuits that mediate sensation, perception, movement, speech, and emotion.

This chapter focuses on the structure of the different types of cells in the nervous system: *neurons* and *glia*. These are broad categories, within which are many types of cells that differ in structure, chemistry, and function. Nonetheless, the distinction between neurons and glia is important. Although there are approximately equal numbers of neurons and glia in the adult human brain (roughly 85 billion of each type), neurons are responsible for most of the unique functions of the brain. It is the **neurons** that sense changes in the environment, communicate these changes to other neurons, and command the body's responses to these sensations. **Glia**, or **glial cells**, contribute to brain function mainly by insulating, supporting, and nourishing neighboring neurons. If the brain were a chocolate chip cookie and the neurons were chocolate chips, the glia would be the cookie dough that fills all the other space and suspends the chips in their appropriate locations. Indeed, the term *glia* is derived from the Greek word for "glue," giving the impression that the main function of these cells is to keep the brain from running out of our ears! Although this simple view belies the importance of glial function, as we shall see later in this chapter, we are confident that neurons perform most information processing in the brain, so neurons receive most of our attention.

Neuroscience, like other fields, has a language all its own. To use this language, you must learn the vocabulary. After you have read this chapter, take a few minutes to review the key terms list and make sure you understand the meaning of each term. Your neuroscience vocabulary will grow as you work your way through the book.

THE NEURON DOCTRINE

To study the structure of brain cells, scientists have had to overcome several obstacles. The first was the small size. Most cells are in the range of 0.01–0.05 mm in diameter. The tip of an unsharpened pencil lead is about 2 mm across; neurons are 40–200 times smaller. (For a review of the metric system, see Table 2.1.) Because neurons cannot be seen by the naked eye, cellular neuroscience could not progress before the development of the compound microscope in the late seventeenth century. Even then, obstacles remained. To observe brain tissue using a microscope, it was necessary to make very thin slices, ideally not much thicker than the diameter of the cells. However, brain tissue has a consistency like a bowl of Jell-O: not firm enough to make thin slices. Thus,

TABLE 2.1 **Units of Size in the Metric System**

Unit	Abbreviation	Meter Equivalent	Real-World Equivalent
Kilometer	km	10^3 m	About two-thirds of a mile
Meter	m	1 m	About 3 feet
Centimeter	cm	10^{-2} m	Thickness of your little finger
Millimeter	mm	10^{-3} m	Thickness of your toenail
Micrometer	μm	10^{-6} m	Near the limit of resolution for the light microscope
Nanometer	nm	10^{-9} m	Near the limit of resolution for the electron microscope

the anatomical study of brain cells had to await a method to harden the tissue without disturbing its structure and an instrument that could produce very thin slices. Early in the nineteenth century, scientists discovered how to harden, or "fix," tissues by immersing them in formaldehyde, and they developed a special device called a *microtome* to make very thin slices.

These technical advances spawned the field of **histology**, the microscopic study of the structure of tissues. But scientists studying brain structure faced yet another obstacle. Freshly prepared brain tissue has a uniform, cream-colored appearance under the microscope, with no differences in pigmentation to enable histologists to resolve individual cells. The final breakthrough in neurohistology was the introduction of stains that selectively color some, but not all, parts of the cells in brain tissue.

One stain still used today was introduced by the German neurologist Franz Nissl in the late nineteenth century. Nissl showed that a class of basic dyes would stain the nuclei of all cells as well as clumps of material surrounding the nuclei of neurons (Figure 2.1). These clumps are called *Nissl bodies*, and the stain is known as the **Nissl stain**. The Nissl stain is extremely useful for two reasons: It distinguishes between neurons and glia, and it enables histologists to study the arrangement, or **cytoarchitecture**, of neurons in different parts of the brain. (The prefix *cyto-* is from the Greek word for "cell.") The study of cytoarchitecture led to the realization that the brain consists of many specialized regions. We now know that each region performs a different function.

The Golgi Stain

The Nissl stain, however, could not tell the whole story. A Nissl-stained neuron looks like little more than a lump of protoplasm containing a

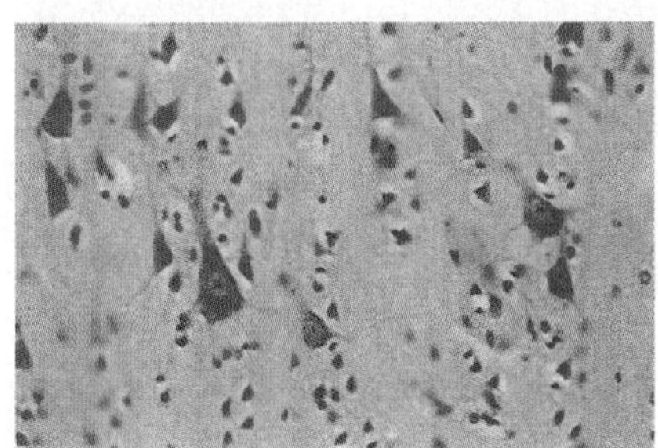

◀ FIGURE 2.1
Nissl-stained neurons. A thin slice of brain tissue has been stained with cresyl violet, a Nissl stain. The clumps of deeply stained material around the cell nuclei are Nissl bodies. (Source: Hammersen, 1980, Fig. 493.)

▲ FIGURE 2.2
Camillo Golgi (1843–1926). (Source: Finger, 1994, Fig. 3.22.)

nucleus. Neurons are much more than that, but how much more was not recognized before Italian histologist Camillo Golgi devised a new method (Figure 2.2). In 1873, Golgi discovered that soaking brain tissue in a silver chromate solution, now called the **Golgi stain**, makes a small percentage of neurons become darkly colored in their entirety (Figure 2.3). This revealed that the neuronal cell body, the region of the neuron around the nucleus that is shown with the Nissl stain, is actually only a small fraction of the total structure of the neuron. Notice in Figures 2.1 and 2.3 how different histological stains can provide strikingly different views of the same tissue. Today, neurohistology remains an active field in neuroscience, along with its credo: "The gain in brain is mainly in the stain."

The Golgi stain shows that neurons have at least two distinguishable parts: a central region that contains the cell nucleus and numerous thin tubes that radiate away from the central region. The swollen region containing the cell nucleus has several names that are used interchangeably: **cell body**, **soma** (plural: somata), and **perikaryon** (plural: perikarya). The thin tubes that radiate from the soma are called **neurites** and are of two types: **axons** and **dendrites** (Figure 2.4).

The cell body usually gives rise to a single axon. The axon is of uniform diameter throughout its length, and any branches from it generally extend at right angles. Because axons can extend over great distances in the body (a meter or more), histologists of the day immediately recognized that axons must act like "wires" that carry the output of the neurons. Dendrites, on the other hand, are rarely longer than 2 mm. Many dendrites extend from the cell body and generally taper to a fine point.

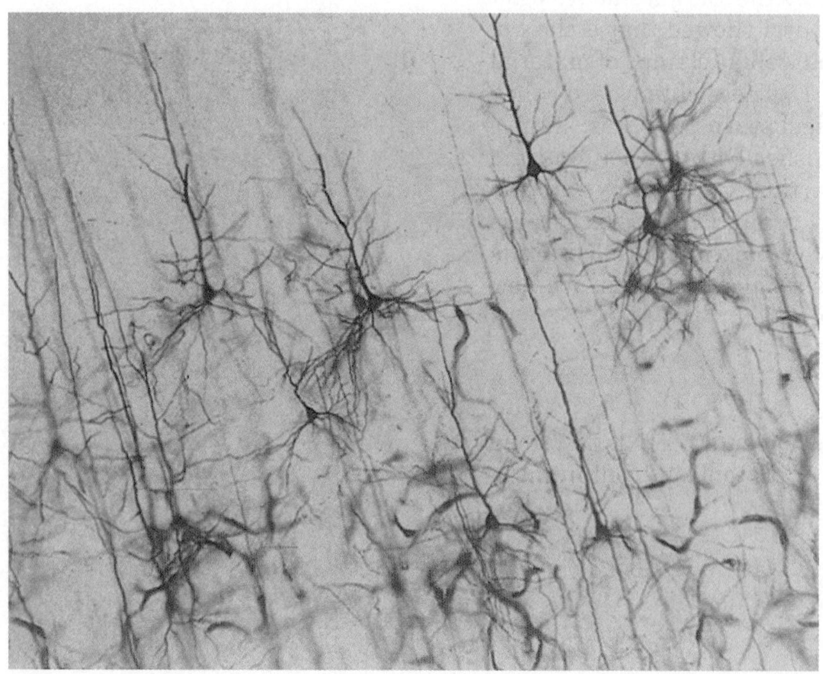

▲ FIGURE 2.3
Golgi-stained neurons. (Source: Hubel, 1988, p. 126.)

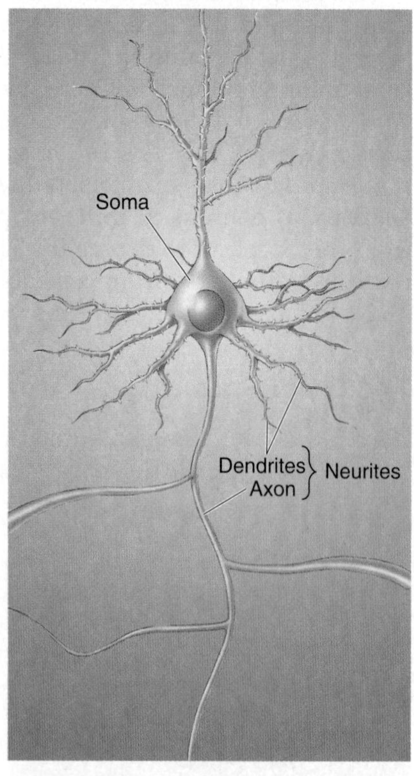

▲ FIGURE 2.4
The basic parts of a neuron.

Early histologists recognized that because dendrites come in contact with many axons, they must act as the antennae of the neuron to receive incoming signals, or input.

▲ FIGURE 2.5
Santiago Ramón y Cajal (1852–1934). (Source: Finger, 1994, Fig. 3.26.)

Cajal's Contribution

Golgi invented the stain, but a Spanish contemporary used it to greatest effect. Santiago Ramón y Cajal was a skilled histologist and artist who learned about Golgi's method in 1888 (Figure 2.5). In a remarkable series of publications over the next 25 years, Cajal used the Golgi stain to work out the circuitry of many regions of the brain (Figure 2.6). Curiously, Golgi and Cajal drew completely opposite conclusions about neurons. Golgi championed the view that the neurites of different cells are fused together to form a continuous reticulum, or network, similar to the arteries and veins of the circulatory system. According to this reticular theory, the brain is an exception to the cell theory, which states that the individual cell is the elementary functional unit of all animal tissues. Cajal, on the other hand, argued forcefully that the neurites of different neurons are not continuous with each other and *communicate by contact, not continuity*. This idea that cell theory also applies to neurons came to be known as the **neuron doctrine**. Although Golgi and Cajal shared the Nobel Prize in 1906, they remained rivals to the end.

The scientific evidence over the next 50 years strongly supported the neuron doctrine, but final proof had to wait for the electron microscope in the 1950s (Box 2.1). With the increased resolving power of the electron microscope, it was finally possible to show that the neurites of different neurons are not continuous with one another (Figure 2.7). Thus, our starting point in the exploration of the brain must be the individual neuron.

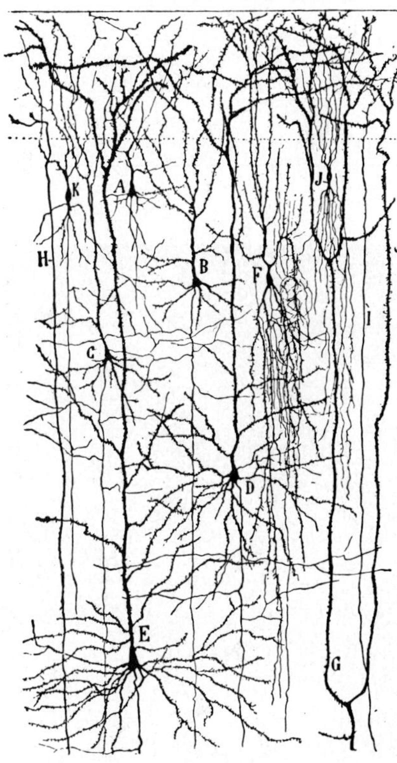

◀ FIGURE 2.6
One of Cajal's many drawings of brain circuitry. The letters label the different elements Cajal identified in an area of the human cerebral cortex that controls voluntary movement. We will learn more about this part of the brain in Chapter 14. (Source: DeFelipe and Jones, 1998, Fig. 90.)

BOX 2.1 OF SPECIAL INTEREST

Advances in Microscopy

The human eye can distinguish two points only if they are separated by more than about one-tenth of a millimeter (100 μm). Thus, we can say that 100 μm is near the limit of resolution for the unaided eye. Neurons have a diameter of about 20 μm, and neurites can be as small as a fraction of a micrometer. The light microscope, therefore, was a necessary development before neuronal structure could be studied. But this type of microscopy has a theoretical limit imposed by the properties of microscope lenses and visible light. With the standard light microscope, the limit of resolution is about 0.1 μm. Because the space between neurons is only 0.02 μm (20 nm), it's no wonder that two esteemed scientists, Golgi and Cajal, disagreed about whether neurites were continuous from one cell to the next. This question could not be answered until about 70 years ago when the electron microscope was developed and applied to biological specimens.

The electron microscope uses an electron beam instead of light to form images, dramatically increasing the resolving power. The limit of resolution for an electron microscope is about 0.1 nm—a million times better than the unaided eye and a thousand times better than a light microscope. Our insights into the fine structure of the inside of neurons—the ultrastructure—have all come from electron microscopic examination of the brain.

Today, microscopes on the leading edge of technology use laser beams to illuminate tissue and computers to create digital images (Figure A). Neuroscientists now routinely introduce into neurons molecules that fluoresce when illuminated by laser light. The fluorescence is recorded by sensitive detectors, and the computer takes these data and reconstructs the image of the neuron. Unlike the traditional methods of light and electron microscopy, which require tissue fixation, these new techniques give neuroscientists the ability to peer into brain tissue that is still alive. Furthermore, they have allowed "super-resolution" imaging, breaking the limits imposed by traditional light microscopy to reveal structures as small as 20 nm across.

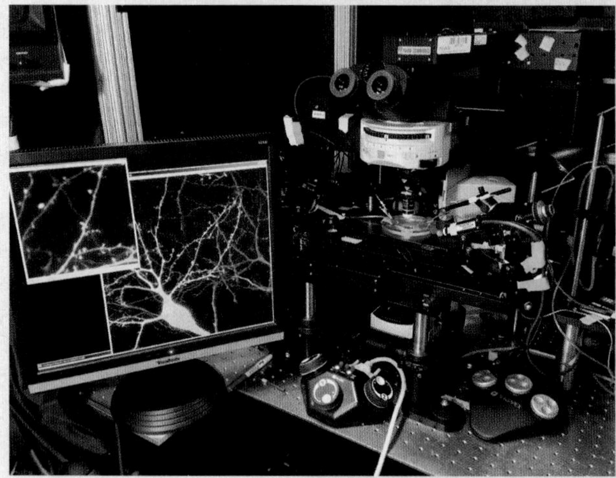

Figure A
A laser microscope and computer display of a fluorescent neuron and dendrites. (Source: Dr. Miquel Bosch, Massachusetts Institute of Technology.)

▶ FIGURE 2.7
Neurites in contact, not continuity. These neurites were reconstructed from a series of images made using an electron microscope. The axon (colored yellow) is in contact with a dendrite (colored blue). (Source: Courtesy of Dr. Sebastian Seung, Princeton University, and Alex Norton, EyeWire.)

THE PROTOTYPICAL NEURON

As we have seen, the neuron (also called a *nerve cell*) consists of several parts: the soma, the dendrites, and the axon. The inside of the neuron is separated from the outside by the *neuronal membrane*, which lies like a circus tent on an intricate internal scaffolding, giving each part of the cell its special three-dimensional appearance. Let's explore the inside of the neuron and learn about the functions of the different parts (Figure 2.8).

The Soma

We begin our tour at the soma, the roughly spherical central part of the neuron. The cell body of the typical neuron is about 20 μm in diameter. The watery fluid inside the cell, called the **cytosol**, is a salty, potassium-rich solution that is separated from the outside by the neuronal membrane. Within the soma are a number of membrane-enclosed structures called **organelles**.

The cell body of the neuron contains the same organelles found in all animal cells. The most important ones are the nucleus, the rough endoplasmic reticulum, the smooth endoplasmic reticulum, the Golgi apparatus, and the mitochondria. Everything contained within the confines of the cell membrane, including the organelles but excluding the nucleus, is referred to collectively as the **cytoplasm**.

The Nucleus. Its name derived from the Latin word for "nut," the **nucleus** of the cell is spherical, centrally located, and about 5–10 μm across. It is contained within a double membrane called the *nuclear envelope*. The nuclear envelope is perforated by pores about 0.1 μm across.

Within the nucleus are **chromosomes** which contain the genetic material **DNA** (**deoxyribonucleic acid**). Your DNA was passed on to you from your parents and it contains the blueprint for your entire body. The DNA in each of your neurons is the same, and it is the same as the DNA in the cells of your liver and kidney and other organs. What distinguishes a neuron from a liver cell are the specific parts of the DNA that are used to assemble the cell. These segments of DNA are called **genes**.

Each chromosome contains an uninterrupted double-strand braid of DNA, 2 nm wide. If the DNA from the 46 human chromosomes were laid out straight, end to end, it would measure more than 2 m in length. If we were to compare this total length of DNA to the total string of letters that make up this book, the genes would be analogous to the individual words. Genes are from 0.1 to several micrometers in length.

The "reading" of the DNA is known as **gene expression**. The final product of gene expression is the synthesis of molecules called **proteins**, which exist in a wide variety of shapes and sizes, perform many different functions, and bestow upon neurons virtually all of their unique characteristics. **Protein synthesis**, the assembly of protein molecules, occurs in the cytoplasm. Because the DNA never leaves the nucleus, an intermediary must carry the genetic message to the sites of protein synthesis in the cytoplasm. This function is performed by another long molecule called **messenger ribonucleic acid**, or **mRNA**. mRNA consists of four different nucleic acids strung together in various sequences to form a chain. The detailed sequence of the nucleic acids in the chain represents the information in the gene, just as the sequence of letters gives meaning to a written word.

The process of assembling a piece of mRNA that contains the information of a gene is called **transcription**, and the resulting mRNA is called

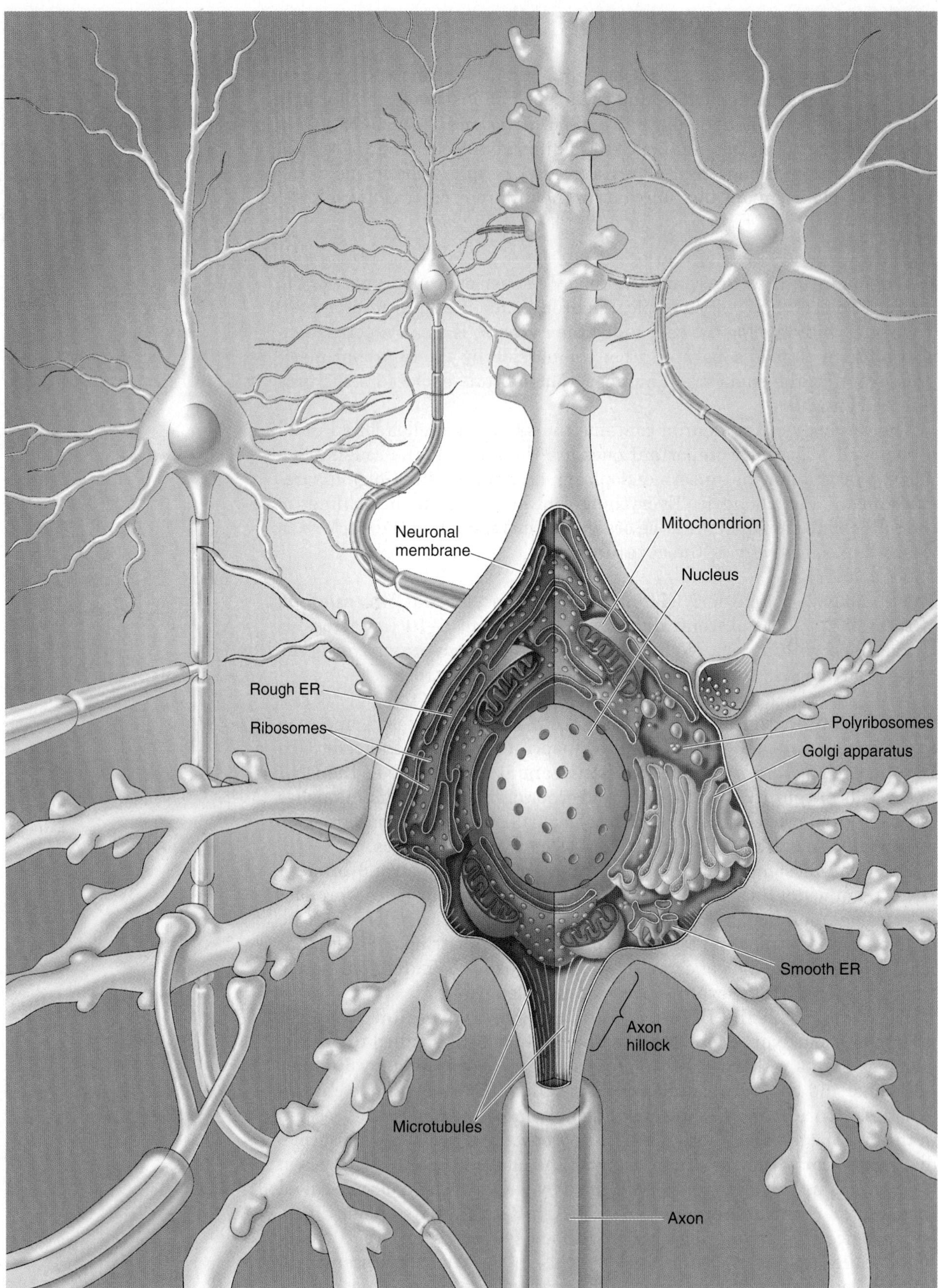

▲ FIGURE 2.8
The internal structure of a typical neuron.

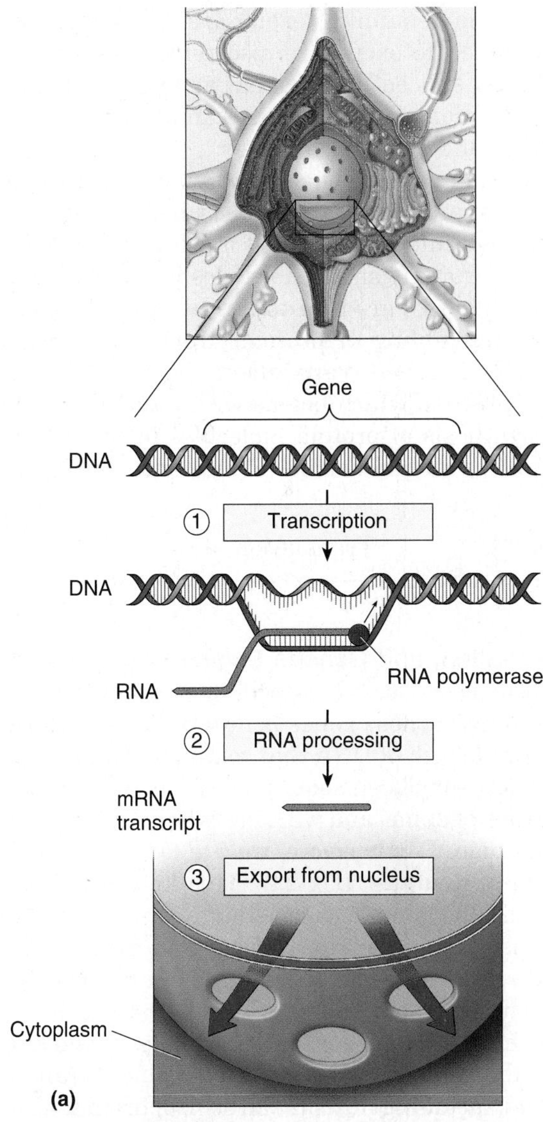

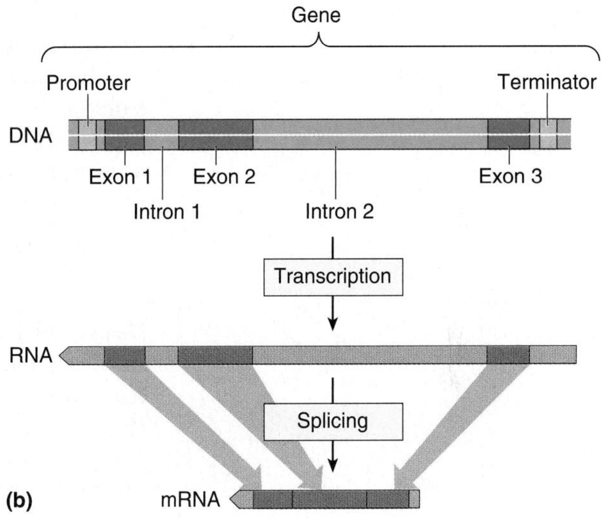

▲ FIGURE 2.9
Gene transcription. (a) RNA molecules are synthesized by RNA polymerase and then processed into mRNA to carry the genetic instructions for protein assembly from the nucleus to the cytoplasm. (b) Transcription is initiated at the promoter region of the gene and stopped at the terminator region. The initial RNA must be spliced to remove the introns that do not code for protein.

the *transcript* (Figure 2.9a). Interspersed between protein-coding genes are long stretches of DNA whose functions remain poorly understood. Some of these regions, however, are known to be important for regulating transcription. At one end of the gene is the **promoter**, the region where the RNA-synthesizing enzyme, *RNA polymerase*, binds to initiate transcription. The binding of the polymerase to the promoter is tightly regulated by other proteins called **transcription factors**. At the other end is a sequence of DNA called the *terminator*, or *stop sequence*, that the RNA polymerase recognizes as the end point for transcription.

In addition to the non-coding regions of DNA that flank the genes, there are often additional stretches of DNA within the gene itself that cannot be used to code for protein. These interspersed regions are called *introns*, and the coding sequences are called *exons*. Initial transcripts contain both introns and exons, but then, by a process called **RNA splicing**, the introns are removed and the remaining exons are fused together (Figure 2.9b). In some cases, specific exons are also removed with the introns, leaving an "alternatively spliced" mRNA that actually encodes a different protein. Thus, transcription of a single gene can ultimately give rise to several different mRNAs and protein products.

mRNA transcripts emerge from the nucleus via pores in the nuclear envelope and travel to the sites of protein synthesis elsewhere in the neuron. At these sites, a protein molecule is assembled much as the mRNA molecule was: by linking together many small molecules into a chain. In the case of protein, the building blocks are **amino acids**, of which there are 20 different kinds. This assembling of proteins from amino acids under the direction of the mRNA is called **translation**.

The scientific study of this process, which begins with the DNA of the nucleus and ends with the synthesis of protein molecules in the cell, is known as *molecular biology*. The "central dogma" of molecular biology is summarized as follows:

$$\text{DNA} \xrightarrow{\text{Transcription}} \text{mRNA} \xrightarrow{\text{Translation}} \text{Protein}$$

Neuronal Genes, Genetic Variation, and Genetic Engineering. Neurons differ from other cells in the body because of the specific genes they express as proteins. A new understanding of these genes is now possible because the human **genome**—the entire length of DNA that comprises the genetic information in our chromosomes—has been sequenced. We now know the 25,000 "words" that comprise our genome, and we know where these genes can be found on each chromosome. Furthermore, we are learning which genes are expressed uniquely in neurons (Box 2.2). This knowledge has paved the way to understanding the genetic basis of many diseases of the nervous system. In some diseases, long stretches of DNA that contain several genes are missing; in others, genes are duplicated, leading to overexpression of specific proteins. These sorts of mishaps, called *gene copy number variations*, often occur at the moment of conception when paternal and maternal DNA mix to create the genome of the offspring. Some instances of serious psychiatric disorders, including autism and schizophrenia, were recently shown to be caused by gene copy number variations in the affected children. (Psychiatric disorders are discussed in Chapter 22.)

Other nervous system disorders are caused by *mutations*—"typographical errors"—in a gene or in the flanking regions of DNA that regulate the gene's expression. In some cases, a single protein may be grossly abnormal or missing entirely, disrupting neuronal function. An example is fragile X syndrome, a disorder that manifests as intellectual disability and autism and is caused by disruption of a single gene (discussed further in Chapter 23). Many of our genes carry small mutations, called *single nucleotide polymorphisms*, which are analogous to a minor misspelling caused by a change in a single letter. These are usually benign, like the difference between "color" and "colour"—different spelling, same meaning. However, sometimes the mutations can affect protein function (consider the difference between "bear" and "bare"—same letters, different meaning). Such single nucleotide polymorphisms, alone or together with others, can affect neuronal function.

Genes make the brain, and understanding how they contribute to neuronal function in both healthy and diseased organisms is a major goal of neuroscience. An important breakthrough was the development of tools for **genetic engineering**—ways to change organisms by design with gene mutations or insertions. This technology has been used most in mice because they are rapidly reproducing mammals with a central nervous

BOX 2.2 **BRAIN FOOD**

Expressing One's Mind in the Post-Genomic Era

Sequencing the human genome was a truly monumental achievement, completed in 2003. The Human Genome Project identified all of the approximately 25,000 genes in human DNA. We now live in what has been called the "post-genomic era," in which information about the genes expressed in our tissues can be used to diagnose and treat diseases. Neuroscientists are using this information to tackle long-standing questions about the biological basis of neurological and psychiatric disorders as well as to probe deeper into the origins of individuality. The logic goes as follows. The brain is a product of the genes expressed in it. Differences in gene expression between a normal brain and a diseased brain, or a brain of unusual ability, can be used to identify the molecular basis of the observed symptoms or traits.

The level of gene expression is usually defined as the number of mRNA transcripts synthesized by different cells and tissues to direct the synthesis of specific proteins. Thus, the analysis of gene expression requires comparing the relative abundance of various mRNAs in the brains of two groups of humans or animals. One way to perform such a comparison is to use DNA *microarrays*, which are created by robotic machines that arrange thousands of small spots of synthetic DNA on a microscope slide. Each spot contains a unique DNA sequence that will recognize and stick to a different specific mRNA sequence. To compare the gene expression in two brains, one begins by collecting a sample of mRNAs from each brain. The mRNA of one brain is labeled with a chemical tag that fluoresces green, and the mRNA of the other brain is labeled with a tag that fluoresces red. These samples are then applied to the microarray. Highly expressed genes will produce brightly fluorescent spots, and differences in the relative gene expression between the brains will be revealed by differences in the color of the fluorescence (Figure A).

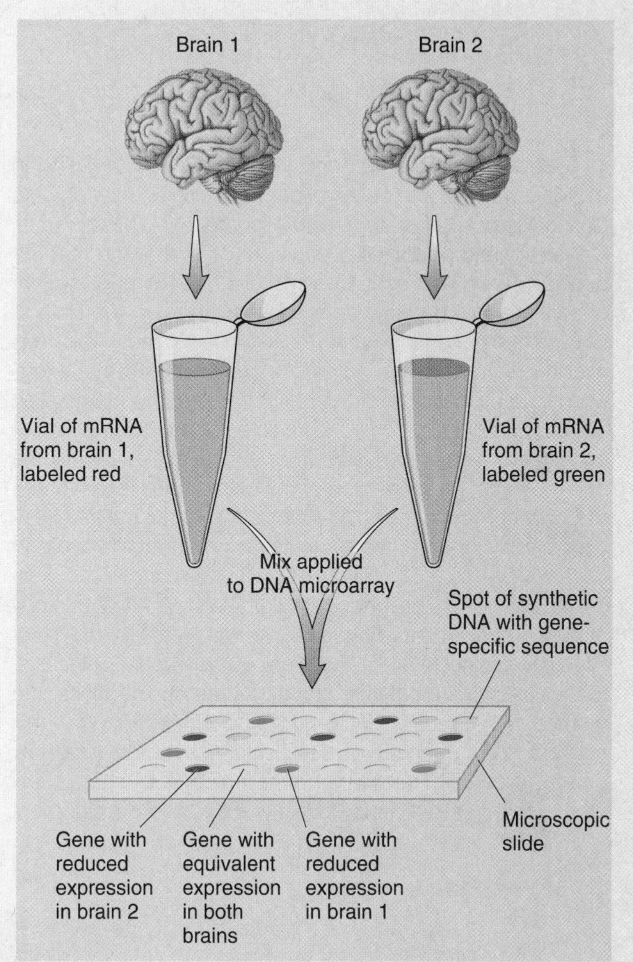

Figure A
Profiling differences in gene expression.

system similar to our own. Today, it is common in neuroscience to hear about **knockout mice**, in which one gene has been deleted (or "knocked out"). Such mice can be used to study the progression of a disease, like fragile X, with the goal of correcting it. Another approach has been to generate **transgenic mice**, in which genes have been introduced and overexpressed; these new genes are called *transgenes*. **Knock-in mice** have also been created in which the native gene is replaced with a modified transgene.

We will see many examples in this book of how genetically engineered animals have been used in neuroscience. The discoveries that allowed genetic modification of mice have revolutionized biology. The researchers who did this work were recognized with the 2007 Nobel Prize in Physiology or Medicine: Martin Evans of Cardiff University, Oliver Smithies of the University of North Carolina at Chapel Hill, and Mario Capecchi of the University of Utah (Box 2.3).

BOX 2.3 PATH OF DISCOVERY

Gene Targeting in Mice

by Mario Capecchi

How did I first get the idea to pursue gene targeting in mice? From a simple observation. Mike Wigler, now at Cold Spring Harbor Laboratory, and Richard Axel, at Columbia University, had published a paper in 1979 showing that exposing mammalian cells to a mixture of DNA and calcium phosphate would cause some cells to take up the DNA in functional form and express the encoded genes. This was exciting because they had clearly demonstrated that exogenous, functional DNA could be introduced into mammalian cells. But I wondered why their efficiency was so low. Was it a problem of delivery, insertion of exogenous DNA into the chromosome, or expression of the genes once inserted into the host chromosome? What would happen if purified DNA was directly injected into the nucleus of mammalian cells in culture?

To find out, I converted a colleague's electrophysiology station into a miniature hypodermic needle to directly inject DNA into the nucleus of a living cell using mechanical micromanipulators and light microscopy (Figure A). The procedure worked with amazing efficiency (Capecchi, 1980). With this method, the frequency of successful integration was now one in three cells rather than one in a million cells as formerly. This high efficiency directly led to the development of transgenic mice through the injection and random integration of exogenous DNA into chromosomes of fertilized mouse eggs, or zygotes. To achieve the high efficiency of expression of the exogenous DNA in the recipient cell, I had to attach small fragments of viral DNA, which we now understand to contain enhancers that are critical in eukaryotic gene expression.

But what fascinated me most was our observation that when many copies of a gene were injected into a cell nucleus, all of these molecules ended up in an ordered head-to-tail arrangement, called a *concatemer* (Figure B). This was astonishing and could not have occurred as a random event. We went on to unequivocally prove that homologous recombination, the process by which chromosomes share genetic information during cell division, was responsible for the incorporation of the foreign DNA (Folger et al., 1982). These experiments demonstrated that all mammalian somatic cells contain a very efficient machinery for swapping segments of DNA that have similar sequences of nucleotides. Injection of a thousand copies of a gene sequence into the nucleus of a cell resulted in chromosomal insertion of a concatemer containing a thousand copies of that sequence, all oriented in the same direction. This simple observation directly led me to

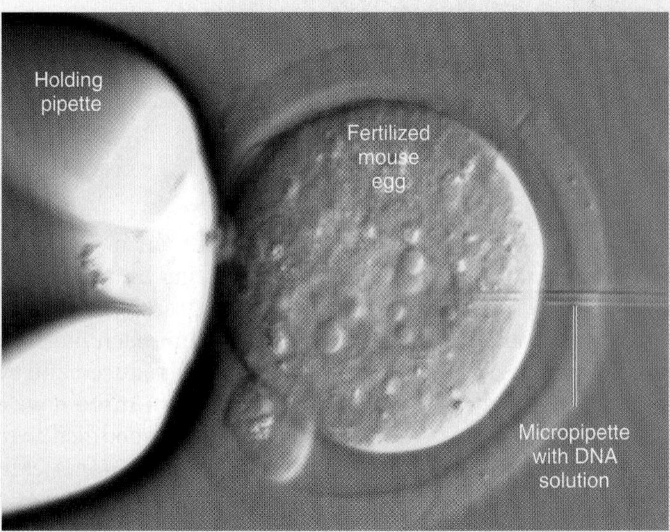

Figure A
Fertilized mouse egg receiving an injection of foreign DNA. (Image courtesy of Dr. Peimin Qi, Division of Comparative Medicine, Massachusetts Institute of Technology.)

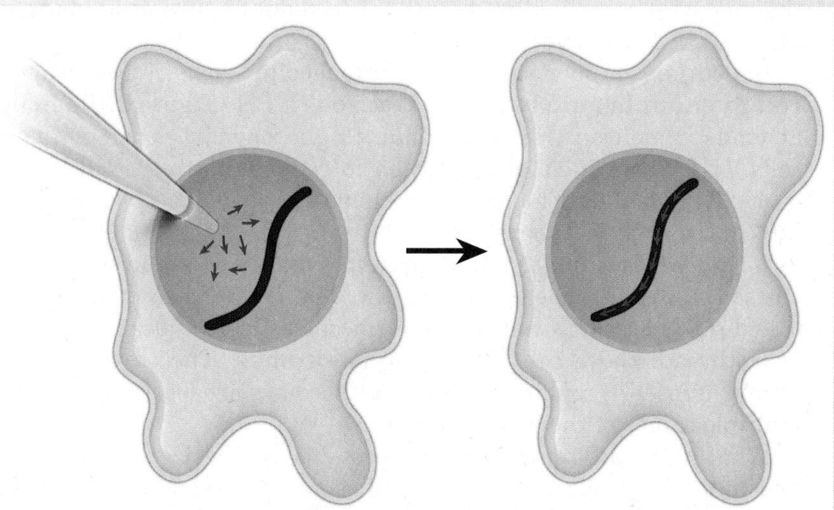

Figure B

envision mutating any chosen gene, in any chosen manner, in living mice by gene targeting.

Excited by this possibility, in 1980, I submitted a grant to the U.S. National Institutes of Health (NIH) proposing to directly alter gene DNA sequences in mammalian cultured cells by homologous recombination. They rejected the proposal, and their arguments were not unreasonable. They argued that the probability of the exogenously added DNA sequence ever finding the DNA sequence similar enough to enable homologous recombination in living mammalian cells (containing 3×10^9 nucleotide base pairs) was vanishingly small. Fortunately, my grant application contained two other proposals that the NIH reviewers liked, and they funded those projects. I used those funds to support the gene targeting project. Four years later, we had results supporting our ability to do gene targeting in cultured mammalian cells. I then resubmitted a new NIH grant application to the same review panel, now proposing to extend gene targeting to generating mutant mice. Their evaluation sheet in response began this way: "We are glad you didn't follow our advice."

It took 10 years to develop gene targeting in mice (Thomas & Capecchi, 1987). Prior to this success, we had to understand the homologous recombination machinery in eukaryotic cells. Also, because the frequency of gene targeting was low, if we were to be successful in transferring our technology to mice, we needed mouse embryonic stem cells capable of contributing to the formation of the germ line—the sperm and eggs—in mature animals. I was getting depressed from our lack of success using cells derived from embryonal carcinoma (EC). Then I heard a rumor that Martin Evans in Cambridge, England was isolating more promising cells, which he called *EK cells*, that resembled EC cells but were derived from a normal mouse embryo rather than from tumors. I called him and asked if the rumor was correct, and he said it was. My next question was whether I could come to his laboratory to learn how to work with those cells, and his answer again was yes. Christmas time, 1985, was beautiful in Cambridge. My wife, who worked with me, and I had a wonderful couple of weeks learning how to maintain these marvelous cells and use them to generate mice capable of germ line transmission.

Investigators often have a preconceived idea about the particular role of their gene of interest in mouse biology, and they are usually very surprised by results when the gene is knocked out. Gene targeting has taken us in many new directions, including most recently pursuing the role of microglia, cells that migrate into the brain after being generated in the bone marrow along with immune and blood cells. Mutating these cells in mice results in a pathology remarkably similar to the human condition called trichotillomania, a type of obsessive-compulsive disorder characterized by strong urges to pull out one's hair. Amazingly, transplanting normal bone marrow into mutant mice permanently cures them of this pathological behavior (Chen et al., 2010). Now, we are deeply immersed in trying to understand the mechanism of how microglia control neural circuit output and, more importantly, exploring the intimate relationship between the immune system (in this case microglia) and neuropsychiatric disorders such as depression, autism, schizophrenia, and Alzheimer's disease.

References

Capecchi MR. 1980. High efficiency transformation by direct microinjection of DNA into cultured mammalian cells. *Cell* 22:479–488.

Chen SC, Tvrdik P, Peden E, Cho S, Wu S, Spangrude G, Capecchi MR. 2010. Hematopoietic origin of pathological grooming in Hoxb8 mutant mice. *Cell* 141(5):775–785.

Folger KR, Wong EA, Wahl G, Capecchi MR. 1982. Patterns of integration of DNA microinjected into cultured mammalian cells: evidence for homologous recombination between injected plasmid DNA molecules. *Molecular and Cellular Biology* 2:1372–1387.

Thomas KR, Capecchi MR. 1987. Site-directed mutagenesis by gene targeting in mouse embryo-derived stem cells. *Cell* 51: 503–512.

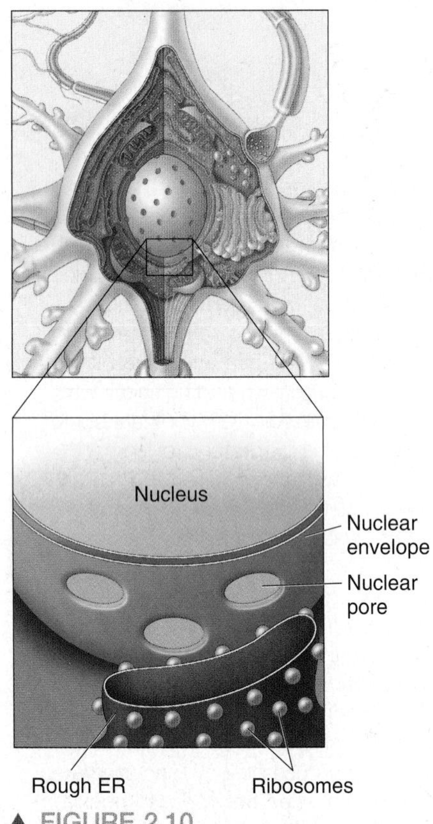

▲ FIGURE 2.10
Rough endoplasmic reticulum, or rough ER.

Rough Endoplasmic Reticulum. Neurons make use of the information in genes by synthesizing proteins. Protein synthesis occurs at dense globular structures in the cytoplasm called **ribosomes**. mRNA transcripts bind to the ribosomes, and the ribosomes translate the instructions contained in the mRNA to assemble a protein molecule. In other words, ribosomes use the blueprint provided by the mRNA to manufacture proteins from raw material in the form of amino acids.

In neurons, many ribosomes are attached to stacks of membrane called **rough endoplasmic reticulum**, or **rough ER** (Figure 2.10). Rough ER abounds in neurons, far more than in glia or most other non-neuronal cells. In fact, we have already been introduced to rough ER by another name: Nissl bodies. This is the organelle stained with the dyes that Nissl introduced over 100 years ago.

Rough ER is a major site of protein synthesis in neurons, but not all ribosomes are attached to rough ER. Many are freely floating and are called *free ribosomes*. Several free ribosomes may appear to be attached by a thread; these are called **polyribosomes**. The thread is a single strand of mRNA, and the associated ribosomes are working on it to make multiple copies of the same protein.

What is the difference between proteins synthesized on the rough ER and those synthesized on the free ribosomes? The answer appears to depend on the intended fate of the protein molecule. If it is destined to reside within the cytosol of the neuron, then the protein's mRNA transcript shuns the ribosomes of the rough ER and gravitates toward the free ribosomes (Figure 2.11a). However, if the protein is destined to be inserted into the membrane of the cell or an organelle, then it is synthesized on the rough ER. As the protein is being assembled, it is threaded back and forth through the membrane of the rough ER, where it is trapped (Figure 2.11b). It is not surprising that neurons have so much rough ER because, as we shall see in later chapters, special membrane proteins are what give these cells their remarkable information-processing abilities.

Smooth Endoplasmic Reticulum and the Golgi Apparatus. The remainder of the cytosol of the soma is crowded with stacks of membranous organelles that look a lot like rough ER without the ribosomes, so much so that one type is called **smooth endoplasmic reticulum**, or **smooth ER**. Smooth ER is heterogeneous and performs different functions in different locations. Some smooth ER is continuous with rough ER and is believed to be a site where the proteins that jut out from the membrane are carefully folded, giving them their three-dimensional structure. Other types of smooth ER play no direct role in the processing of protein molecules but instead regulate the internal concentrations of substances such as calcium. (This organelle is particularly prominent in muscle cells, where it is called *sarcoplasmic reticulum*, as we will see in Chapter 13.)

The stack of membrane-enclosed disks in the soma that lies farthest from the nucleus is the **Golgi apparatus**, first described in 1898 by Camillo Golgi (Figure 2.12). This is a site of extensive "post-translational" chemical processing of proteins. One important function of the Golgi apparatus is believed to be the sorting of certain proteins that are destined for delivery to different parts of the neuron, such as the axon and the dendrites.

The Mitochondrion. Another very abundant organelle in the soma is the **mitochondrion** (plural: mitochondria). In neurons, these sausage-shaped structures are about 1 μm long. Within the enclosure of their outer membrane are multiple folds of inner membrane called *cristae* (singular: crista). Between the cristae is an inner space called *matrix* (Figure 2.13a).

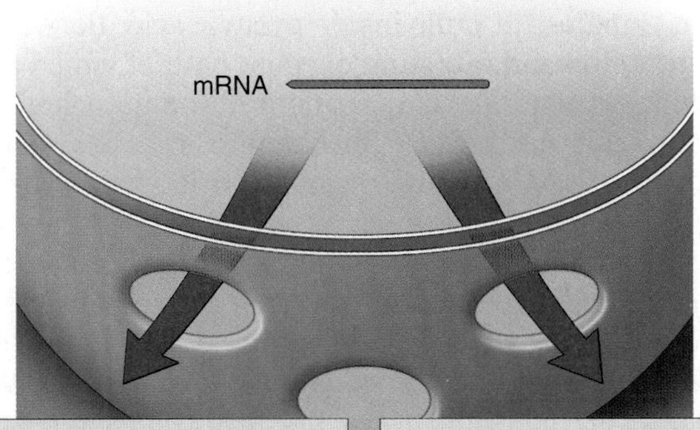

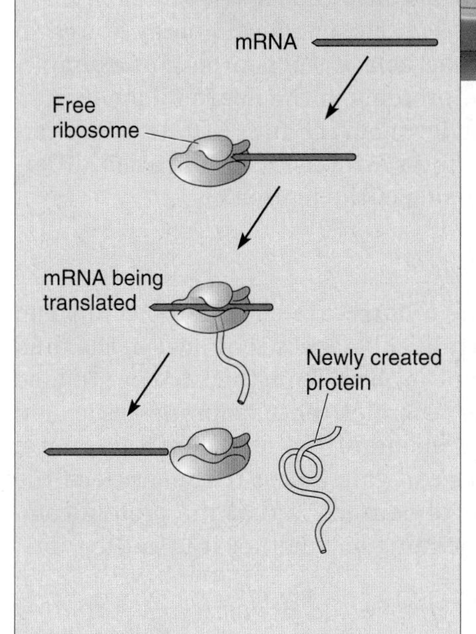

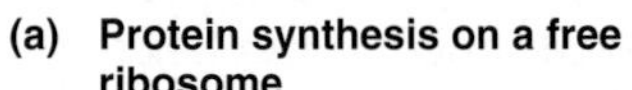

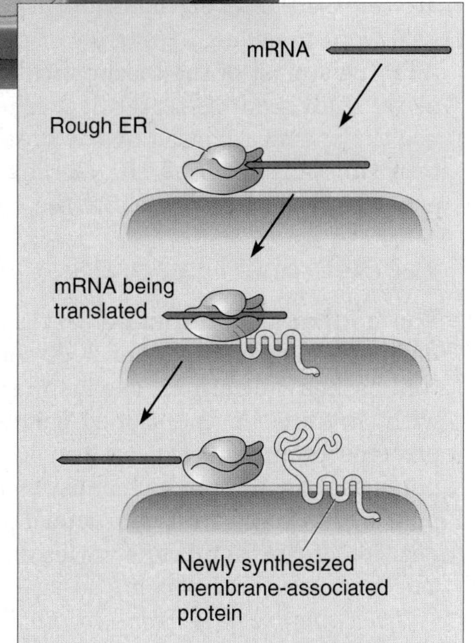

(a) Protein synthesis on a free ribosome

(b) Protein synthesis on rough ER

◀ FIGURE 2.11
Protein synthesis on a free ribosome and on rough ER. mRNA binds to a ribosome, initiating protein synthesis. **(a)** Proteins synthesized on free ribosomes are destined for the cytosol. **(b)** Proteins synthesized on the rough ER are destined to be enclosed by or inserted into the membrane. Membrane-associated proteins are inserted into the membrane as they are assembled.

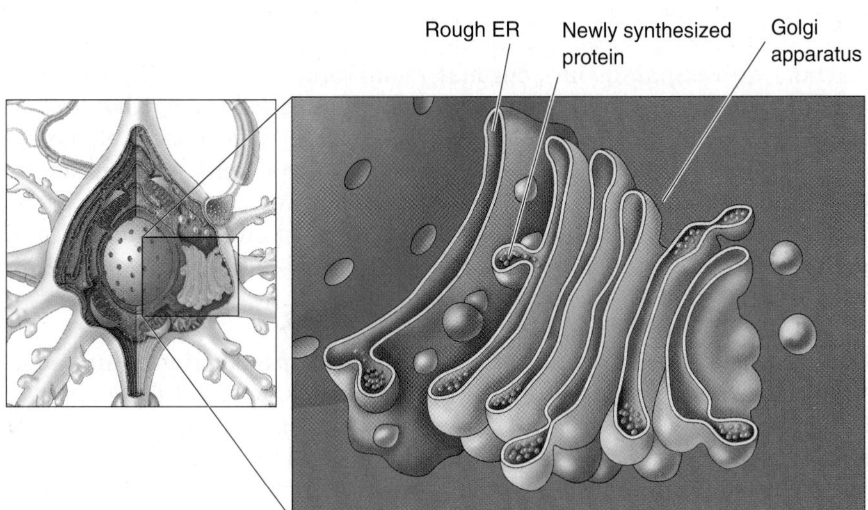

◀ FIGURE 2.12
The Golgi apparatus. This complex organelle sorts newly synthesized proteins for delivery to appropriate locations in the neuron.

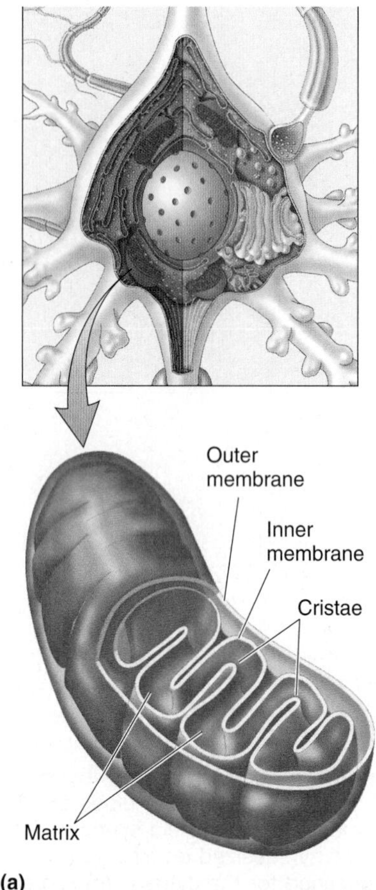

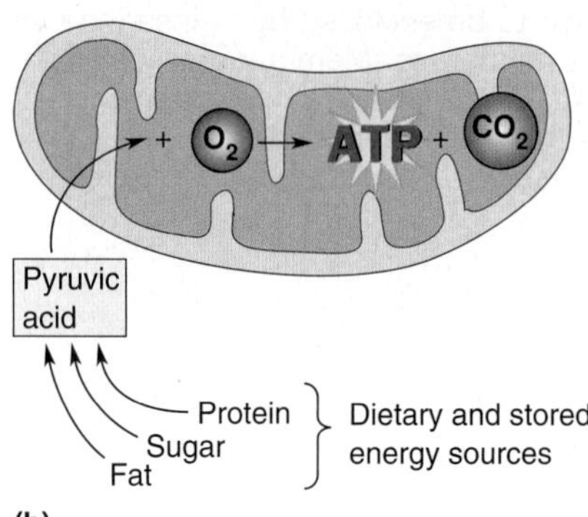

▲ FIGURE 2.13
The role of mitochondria. **(a)** Components of a mitochondrion. **(b)** Cellular respiration. ATP is the energy currency that fuels biochemical reactions in neurons.

Mitochondria are the site of *cellular respiration* (Figure 2.13b). When a mitochondrion "inhales," it pulls inside pyruvic acid (derived from sugars and digested proteins and fats) and oxygen, both of which are floating in the cytosol. Within the inner compartment of the mitochondrion, pyruvic acid enters into a complex series of biochemical reactions called the *Krebs cycle*, named after the German-British scientist Hans Krebs, who first proposed it in 1937. The biochemical products of the Krebs cycle provide energy that, in another series of reactions within the cristae (called the *electron-transport chain*), results in the addition of phosphate to adenosine diphosphate (ADP), yielding **adenosine triphosphate (ATP)**, the cell's energy source. When the mitochondrion "exhales," 17 ATP molecules are released for every molecule of pyruvic acid that had been taken in.

ATP is the energy currency of the cell. The chemical energy stored in ATP fuels most of the biochemical reactions of the neuron. For example, as we shall see in Chapter 3, special proteins in the neuronal membrane use the energy released by the breakdown of ATP into ADP to pump certain substances across the membrane to establish concentration differences between the inside and the outside of the neuron.

The Neuronal Membrane

The **neuronal membrane** serves as a barrier to enclose the cytoplasm inside the neuron and to exclude certain substances that float in the fluid that bathes the neuron. The membrane is about 5 nm thick and is studded with proteins. As mentioned earlier, some of the membrane-associated proteins pump substances from the inside to the outside. Others form pores that regulate which substances can gain access to the inside of the neuron. An important characteristic of neurons is that the protein composition of the membrane varies depending on whether it is in the soma, the dendrites, or the axon.

The function of neurons cannot be understood without understanding the structure and function of the membrane and its associated proteins. In fact, this topic is so important that we'll spend much of the next four chapters looking at how the membrane endows neurons with the remarkable ability to transfer electrical signals throughout the brain and body.

The Cytoskeleton

Earlier, we compared the neuronal membrane to a circus tent draped over an internal scaffolding. This scaffolding is called the **cytoskeleton**, and it gives the neuron its characteristic shape. The "bones" of the cytoskeleton are the microtubules, microfilaments, and neurofilaments (Figure 2.14). Unlike the tent scaffolding, however, the cytoskeleton is not static. Elements of the cytoskeleton are dynamically regulated and are in continual motion. Your neurons are probably squirming around in your head even as you read this sentence.

Microtubules. Measuring 20 nm in diameter, **microtubules** are relatively large and run longitudinally down neurites. A microtubule appears as a straight, thick-walled hollow pipe. The wall of the pipe is composed of smaller strands that are braided like rope around the hollow core. Each of the smaller strands consists of the protein *tubulin*. A single tubulin molecule is small and globular; the strand consists of tubulins stuck together like pearls on a string. The process of joining small proteins to form a long strand is called *polymerization*; the resulting strand is called a *polymer*. Polymerization and depolymerization of microtubules and, therefore, of neuronal shape can be regulated by various signals within the neuron.

One class of proteins that participate in the regulation of microtubule assembly and function are *microtubule-associated proteins*, or *MAPs*. Among other functions (many of which are unknown), MAPs anchor the microtubules to one another and to other parts of the neuron. Pathological changes in an axonal MAP, called *tau*, have been implicated in the dementia that accompanies Alzheimer's disease (Box 2.4).

Microfilaments. Measuring only 5 nm in diameter, **microfilaments** are about the same thickness as the cell membrane. Found throughout the neuron, they are particularly numerous in the neurites. Microfilaments are braids of two thin strands that are polymers of the protein *actin*. Actin is one of the most abundant proteins in cells of all types, including neurons, and is believed to play a role in changing cell shape. Indeed, as we shall see in Chapter 13, actin filaments are critically involved in the mechanism of muscle contraction.

Like microtubules, actin microfilaments are constantly undergoing assembly and disassembly, and this process is regulated by signals in the neuron. In addition to running longitudinally down the core of the neurites like microtubules, microfilaments are also closely associated with the membrane. They are anchored to the membrane by attachments with a meshwork of fibrous proteins that line the inside of the membrane like a spider web.

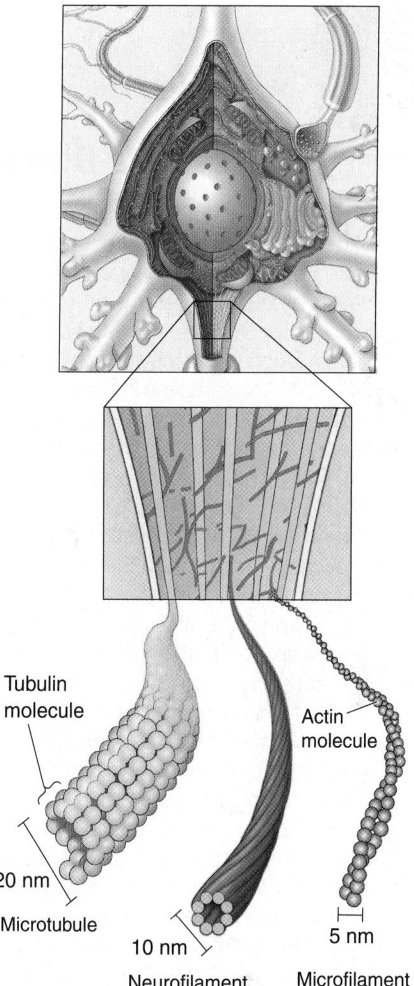

▲ FIGURE 2.14
Components of the cytoskeleton. The arrangement of microtubules, neurofilaments, and microfilaments gives the neuron its characteristic shape.

Neurofilaments. With a diameter of 10 nm, **neurofilaments** are intermediate in size between microtubules and microfilaments. They exist in all cells of the body as *intermediate filaments*; only in neurons are they called *neurofilaments*. The difference in name reflects differences in structure among different tissues. For example, a different intermediate filament, keratin, composes hair when bundled together.

Of the types of fibrous structure we have discussed, neurofilaments most closely resemble the bones and ligaments of the skeleton. A neurofilament consists of multiple subunits (building blocks) that are wound together into a ropelike structure. Each strand of the rope consists of individual long protein molecules, making neurofilaments mechanically very strong.

The Axon

So far, we've explored the soma, organelles, membrane, and cytoskeleton. These structures are not unique to neurons but are found in all the cells in our body. Now we'll look at the axon, a structure found only in neurons and highly specialized for the transfer of information over distances in the nervous system.

The axon begins with a region called the **axon hillock**, which tapers away from the soma to form the initial segment of the axon proper (Figure 2.15). Two noteworthy features distinguish the axon from the soma:

1. No rough ER extends into the axon, and there are few, if any, free ribosomes in mature axons.
2. The protein composition of the axon membrane is fundamentally different from that of the soma membrane.

These structural differences translate into functional distinctions. Because there are no ribosomes, there is no protein synthesis in the axon. This means that all proteins in the axon must originate in the soma. And the different proteins in the axonal membrane enable it to serve as a wire that sends information over great distances.

Axons may extend from less than a millimeter to over a meter long. Axons often branch, and these branches, called **axon collaterals**, can

BOX 2.4 OF SPECIAL INTEREST

Alzheimer's Disease and the Neuronal Cytoskeleton

Neurites are the most remarkable structural feature of a neuron. Their elaborate branching patterns, critical for information processing, reflect the organization of the underlying cytoskeleton. It is therefore no surprise that a devastating loss of brain function can result when the cytoskeleton of neurons is disrupted. An example is *Alzheimer's disease*, which is characterized by the disruption of the cytoskeleton of neurons in the cerebral cortex, a region of the brain crucial for cognitive function. This disorder and its underlying brain pathology were first described in 1907 by the German physician A. Alzheimer in a paper titled "A Characteristic Disease of the Cerebral Cortex." Below are excerpts from the English translation.

> One of the first disease symptoms of a 51-year-old woman was a strong feeling of jealousy toward her husband. Very soon she showed rapidly increasing memory impairments; she could not find her way about her home, she dragged objects to and fro, hid herself, or sometimes thought that people were out to kill her, then she would start to scream loudly.
>
> During institutionalization her gestures showed a complete helplessness. She was disoriented as to time and place. From time to time she would state that she did not understand anything, that she felt confused and totally lost. Sometimes she considered the coming of the doctor as an official visit and apologized for not having finished her work, but other times she would start to yell in the fear that the doctor wanted to operate on her; or there were times that she would send him away in complete indignation, uttering phrases that indicated her fear that the doctor wanted to damage her woman's honor. From time to time she was completely delirious, dragging her blankets and sheets to and fro, calling for her husband and daughter, and seeming to have auditory hallucinations. Often she would scream for hours and hours in a horrible voice.
>
> Mental regression advanced quite steadily. After four and a half years of illness the patient died. She was completely apathetic in the end, and was confined to bed in a fetal position. (Bick et al., 1987, pp. 1–2.)

Following her death, Alzheimer examined the woman's brain under the microscope. He made particular note of changes in the "neurofibrils," elements of the cytoskeleton that can be stained by a silver solution.

> The Bielschowsky silver preparation showed very characteristic changes in the neurofibrils. However, inside an apparently normal-looking cell, one or more single fibers could be observed that became prominent through their striking thickness and specific impregnability. At a more advanced stage, many fibrils arranged parallel showed the same changes. Then they accumulated forming dense bundles and gradually advanced to the surface of the cell. Eventually, the nucleus and cytoplasm disappeared, and only a tangled bundle of fibrils indicated the site where once the neuron had been located.
>
> As these fibrils can be stained with dyes different from the normal neurofibrils, a chemical transformation of the fibril substance must have taken place. This might be the reason why the fibrils survived the destruction of the cell. It seems that the transformation of the fibrils goes hand in hand with the storage of an as yet not closely examined pathological product of the metabolism in the neuron. About one-quarter to one-third of all the neurons of the cerebral cortex showed such alterations. Numerous neurons, especially in the upper cell layers, had totally disappeared. (Bick et al., 1987, pp. 2–3.)

The severity of the dementia in Alzheimer's disease is well correlated with the number and distribution of what are now commonly known as *neurofibrillary tangles*, the "tombstones" of dead and dying neurons (Figure A). Indeed, as Alzheimer speculated, tangle formation in the cerebral cortex very likely causes the symptoms of the disease. Electron microscopy reveals that the major components of the tangles are *paired helical filaments*, long fibrous proteins braided together like strands of a rope (Figure B). It is now understood that these filaments consist of the microtubule-associated protein *tau*.

Tau normally functions as a bridge between the microtubules in axons, ensuring that they run straight and parallel to one another. In Alzheimer's disease, the tau detaches from the microtubules and accumulates in the soma. This disruption of the cytoskeleton causes the axons to wither, thus impeding the normal flow of information in the affected neurons.

travel long distances to communicate with different parts of the nervous system. Occasionally, an axon collateral returns to communicate with the same cell that gave rise to the axon or with the dendrites of neighboring cells. These axon branches are called *recurrent collaterals*.

The diameter of an axon is variable, ranging from less than 1 μm to about 25 μm in humans and to as large as 1 mm in squid. This variation in axon size is important. As will be explained in Chapter 4, the speed

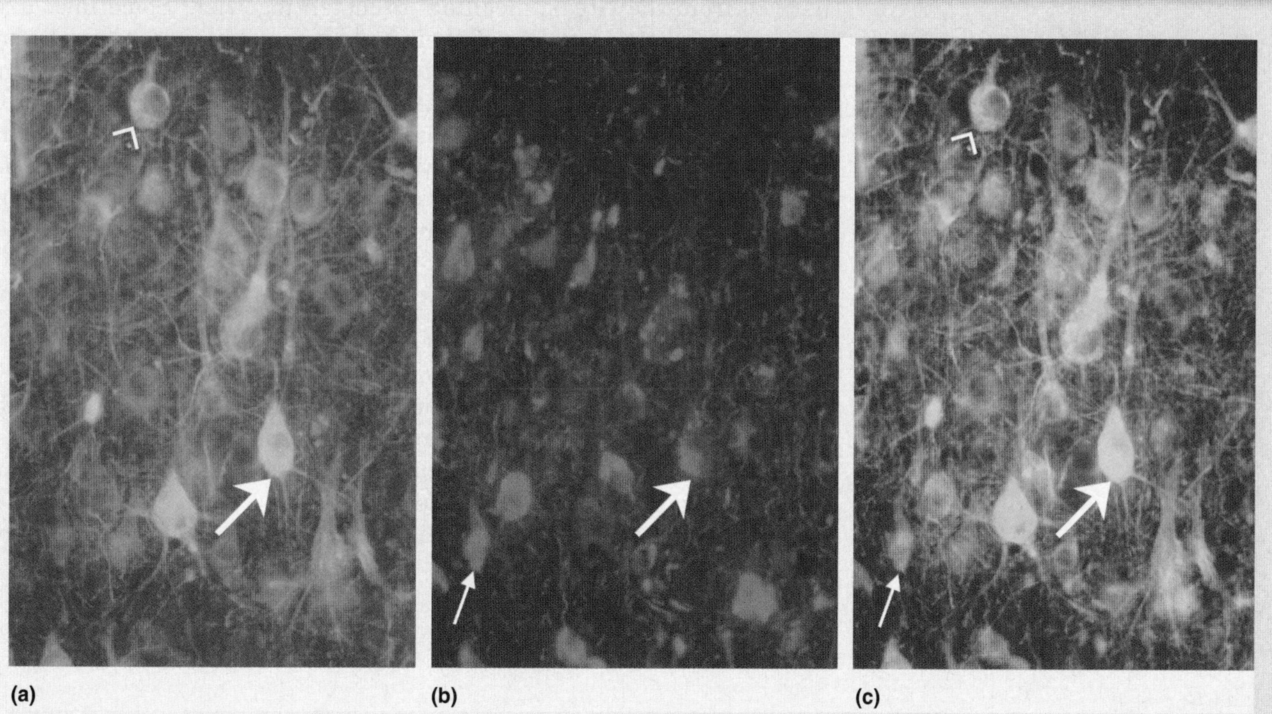

Figure A
Neurons in a human brain with Alzheimer's disease. Normal neurons contain neurofilaments but no neurofibrillary tangles. **(a)** Brain tissue stained by a method that makes neuronal neurofilaments fluoresce green, showing viable neurons. **(b)** The same region of the brain stained to show the presence of tau within neurofibrillary tangles, revealed by red fluorescence. **(c)** Superimposition of images in parts **a** and **b**. The neuron indicated by the arrowhead contains neurofilaments but no tangles and therefore is healthy. The neuron indicated by the large arrow has neurofilaments but also has started to show accumulation of tau and therefore is diseased. The neuron indicated by the small arrow in parts **b** and **c** is dead because it contains no neurofilaments. The remaining tangle is the tombstone of a neuron killed by Alzheimer's disease. (Source: Courtesy of Dr. John Morrison and modified from Vickers et al., 1994.)

What causes such changes in tau? Attention has focused on another protein that accumulates in the brain of Alzheimer's patients, called *amyloid*. Alzheimer's disease research is moving very fast, but the consensus today is that the abnormal secretion of amyloid by neurons is the first step in a process that leads to neurofibrillary tangle formation and dementia. Currently, hope for therapeutic intervention focuses on strategies to reduce the depositions of amyloid in the brain. The need for effective therapy is urgent: In the United States alone, more than 5 million people are afflicted with this tragic disease.

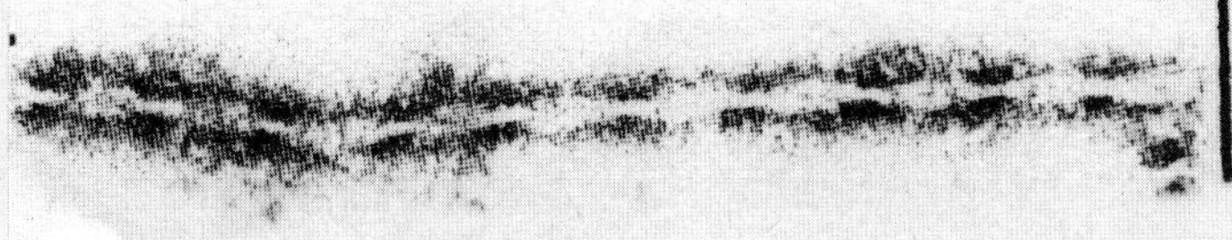

Figure B
Paired helical filaments of a tangle. (Source: Goedert, 1996, Fig. 2b.)

of the electrical signal that sweeps down the axon—the *nerve impulse*—depends on the axonal diameter. The thicker the axon, the faster the impulse travels.

The Axon Terminal. All axons have a beginning (the axon hillock), a middle (the axon proper), and an end. The end is called the **axon terminal** or **terminal bouton** (French for "button"), reflecting the fact that it usually

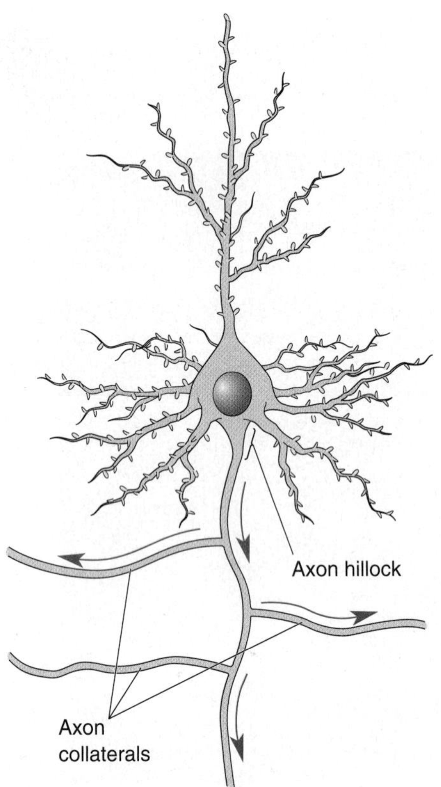

▲ FIGURE 2.15
The axon and axon collaterals. The axon functions like a telegraph wire to send electrical impulses to distant sites in the nervous system. The arrows indicate the direction of information flow.

appears as a swollen disk (Figure 2.16). The terminal is a site where the axon comes in contact with other neurons (or other cells) and passes information on to them. This point of contact is called the **synapse**, a word derived from the Greek, meaning "to fasten together." Sometimes axons have many short branches at their ends, and each branch forms a synapse on dendrites or cell bodies in the same region. These branches are collectively called the **terminal arbor**. Sometimes axons form synapses at swollen regions along their length and then continue on to terminate elsewhere (Figure 2.17). Such swellings are called *boutons en passant* ("buttons in passing"). In either case, when a neuron makes synaptic contact with another cell, it is said to innervate that cell, or to provide **innervation**.

The cytoplasm of the axon terminal differs from that of the axon in several ways:

1. Microtubules do not extend into the terminal.
2. The terminal contains numerous small bubbles of membrane, called **synaptic vesicles**, that measure about 50 nm in diameter.
3. The inside surface of the membrane that faces the synapse has a particularly dense covering of proteins.
4. The axon terminal cytoplasm has numerous mitochondria, indicating a high energy demand.

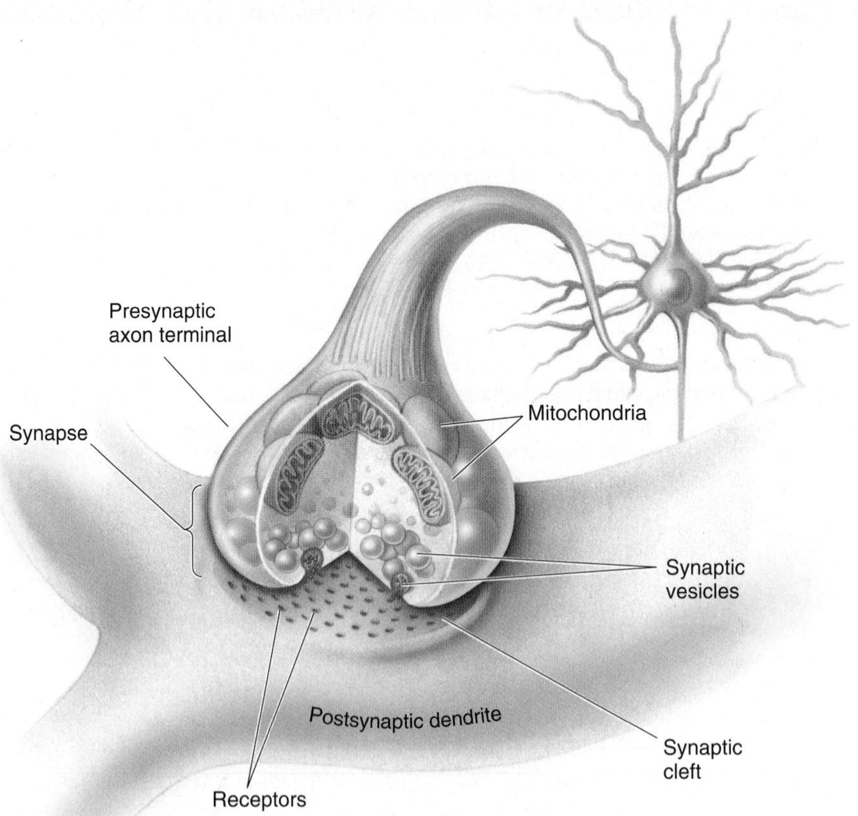

▲ FIGURE 2.16
The axon terminal and the synapse. Axon terminals form synapses with the dendrites or somata of other neurons. When a nerve impulse arrives in the presynaptic axon terminal, neurotransmitter molecules are released from synaptic vesicles into the synaptic cleft. Neurotransmitter then binds to specific receptor proteins, causing the generation of electrical or chemical signals in the postsynaptic cell.

The Synapse. Although Chapters 5 and 6 are devoted entirely to how information is transferred from one neuron to another at the synapse, we'll preview the process here. The synapse has two sides: *presynaptic* and *postsynaptic* (see Figure 2.16). These names indicate the usual direction of information flow from "pre" to "post." The presynaptic side generally consists of an axon terminal, whereas the postsynaptic side may be a dendrite or the soma of another neuron. The space between the presynaptic and postsynaptic membranes is called the **synaptic cleft**. The transfer of information at the synapse from one neuron to another is called **synaptic transmission**.

At most synapses, information in the form of electrical impulses traveling down the axon is converted in the terminal into a chemical signal that crosses the synaptic cleft. On the postsynaptic membrane, this chemical signal is converted again into an electrical one. The chemical signal, called a **neurotransmitter**, is stored in and released from the synaptic vesicles within the terminal. As we will see, different neurotransmitters are used by different types of neurons.

This electrical-to-chemical-to-electrical transformation of information makes possible many of the brain's computational abilities. Modification of this process is involved in memory and learning, and synaptic transmission dysfunction accounts for certain mental disorders. The synapse is also the site of action for many toxins and for most psychoactive drugs.

▲ FIGURE 2.17
A bouton en passant. An axon (colored yellow) makes a synapse on a dendrite (colored blue) as they cross. This synapse was reconstructed from a series of images made using an electron microscope. (Source: Courtesy of Dr. Sebastian Seung, Princeton University, and Alex Norton, EyeWire.)

Axoplasmic Transport. As mentioned, one feature of the cytoplasm of axons, including the terminal, is the absence of ribosomes. Because ribosomes are the protein factories of the cell, their absence means that the proteins of the axon must be synthesized in the soma and then shipped down the axon. Indeed, in the mid-nineteenth century, English physiologist Augustus Waller showed that axons cannot be sustained when separated from their parent cell body. The degeneration of axons that occurs when they are cut is now called *Wallerian degeneration.* Because it can be detected with certain staining methods, Wallerian degeneration is one way to trace axonal connections in the brain.

Wallerian degeneration occurs because the normal flow of materials from the soma to the axon terminal is interrupted. This movement of material down the axon is called **axoplasmic transport**. This was first demonstrated directly by the experiments of American neurobiologist Paul Weiss and his colleagues in the 1940s. They found that if they tied a thread around an axon, material accumulated on the side of the knot closest to the soma. When the knot was untied, the accumulated material continued down the axon at a rate of 1–10 mm per day.

This was a remarkable discovery, but it is not the whole story. If all material moved down the axon by this transport mechanism alone, it would not reach the ends of the longest axons for at least half a year—too long a wait to feed hungry synapses. In the late 1960s, methods were developed to track the movements of protein molecules down the axon into the terminal. These methods entailed injecting the somata of neurons with radioactive amino acids. Recall that amino acids are the building blocks of proteins. The "hot" amino acids were assembled into proteins, and the arrival of radioactive proteins in the axon terminal was timed to calculate the rate of transport. This *fast axoplasmic transport* (so named to distinguish it from *slow axoplasmic transport* described by Weiss) occurred at a rate as high as 1,000 mm per day.

Much is now known about how fast axoplasmic transport works. Material is enclosed within vesicles, which then "walk down" the microtubules of the axon. The "legs" are provided by a protein called *kinesin*, and

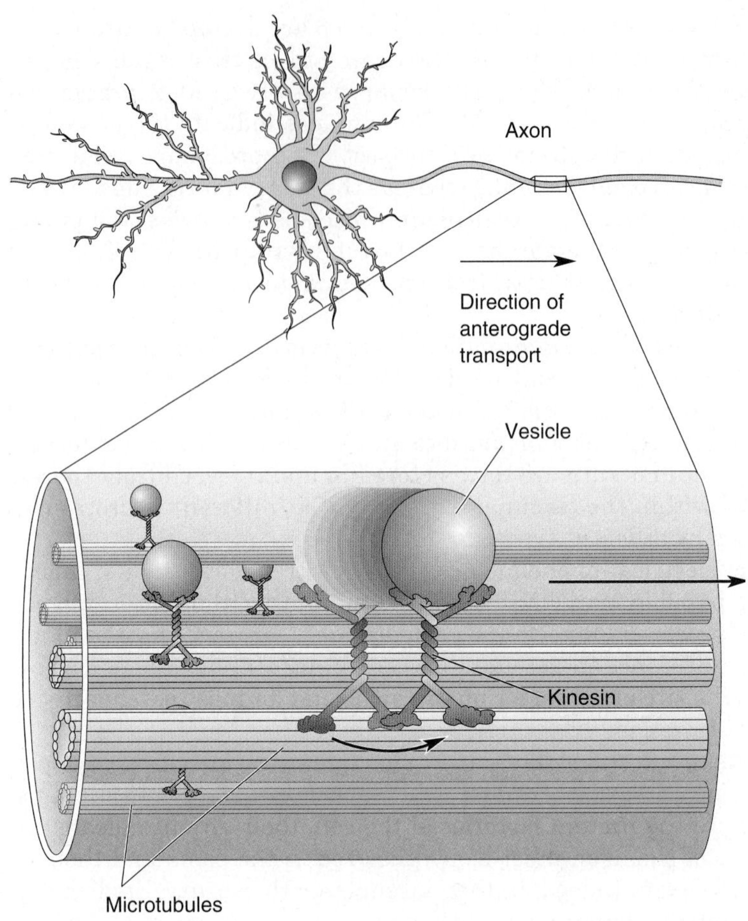

▲ FIGURE 2.18
A mechanism for the movement of material on the microtubules of the axon. Trapped in membrane-enclosed vesicles, material is transported from the soma to the axon terminal by the action of the protein kinesin, which "walks" along microtubules at the expense of ATP.

the process is fueled by ATP (Figure 2.18). Kinesin moves material only from the soma to the terminal. All movement of material in this direction is called **anterograde transport**.

In addition to anterograde transport, there is a mechanism for the movement of material up the axon from the terminal to the soma. This process is believed to provide signals to the soma about changes in the metabolic needs of the axon terminal. Movement in this direction, from terminal to soma, is called **retrograde transport**. The molecular mechanism is similar to anterograde transport, except the "legs" for retrograde transport are provided by a different protein, *dynein*. Both anterograde and retrograde transport mechanisms have been exploited by neuroscientists to trace connections in the brain (Box 2.5).

Dendrites

The term *dendrite* is derived from the Greek for "tree," reflecting the fact that these neurites resemble the branches of a tree as they extend from the soma. The dendrites of a single neuron are collectively called a **dendritic tree**; each branch of the tree is called a *dendritic branch*. The wide variety of shapes and sizes of dendritic trees are used to classify different groups of neurons.

Because dendrites function as the antennae of the neuron, they are covered with thousands of synapses (Figure 2.19). The dendritic membrane

BOX 2.5 OF SPECIAL INTEREST

Hitching a Ride with Retrograde Transport

Fast anterograde transport of proteins in axons was shown by injecting the soma with radioactive amino acids. The success of this method immediately suggested a way to trace connections in the brain. For example, to determine where neurons in the eye send their axons elsewhere in the brain, the eye was injected with radioactive proline, an amino acid. The proline was incorporated into proteins in the somata that were then transported to the axon terminals. By use of a technique called *autoradiography*, the location of radioactive axon terminals could be detected, thereby revealing the extent of the connection between the eye and the brain.

Researchers subsequently discovered that retrograde transport could also be exploited to work out connections in the brain. Strangely enough, the enzyme horseradish peroxidase (HRP) is selectively taken up by axon terminals and then transported retrogradely to the soma. A chemical reaction can then be initiated to visualize the location of the HRP in slices of brain tissue after the experimental animal is euthanized. This method is commonly used to trace connections in the brain (Figure A).

Some viruses also exploit retrograde transport to infect neurons. For example, the oral type of herpesvirus enters axon terminals in the lips and mouth and is then transported back to the parent cell bodies. Here the virus usually remains dormant until physical or emotional stress occurs (as on a first date), at which time it replicates and returns to the nerve ending, causing a painful cold sore. Similarly, the rabies virus enters the nervous system by retrograde transport through axons in the skin. However, once inside the soma, the virus wastes no time in replicating madly, killing its neuronal host. The virus is then taken up by other neurons within the nervous system, and the process repeats itself again and again, usually until the victim dies.

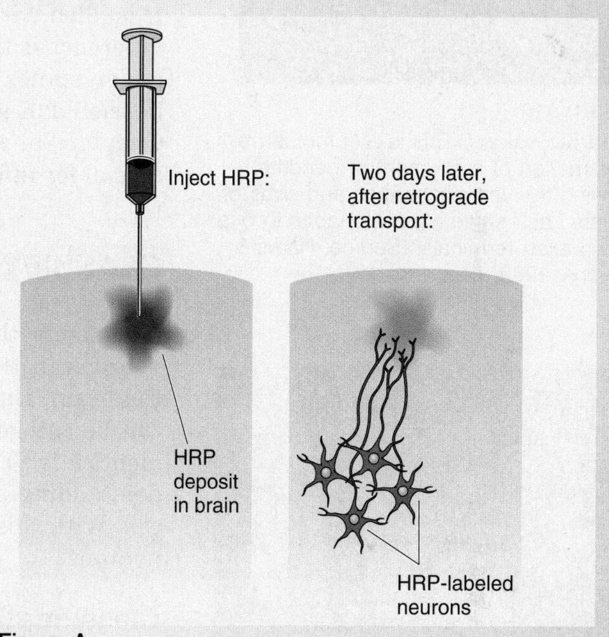

Figure A

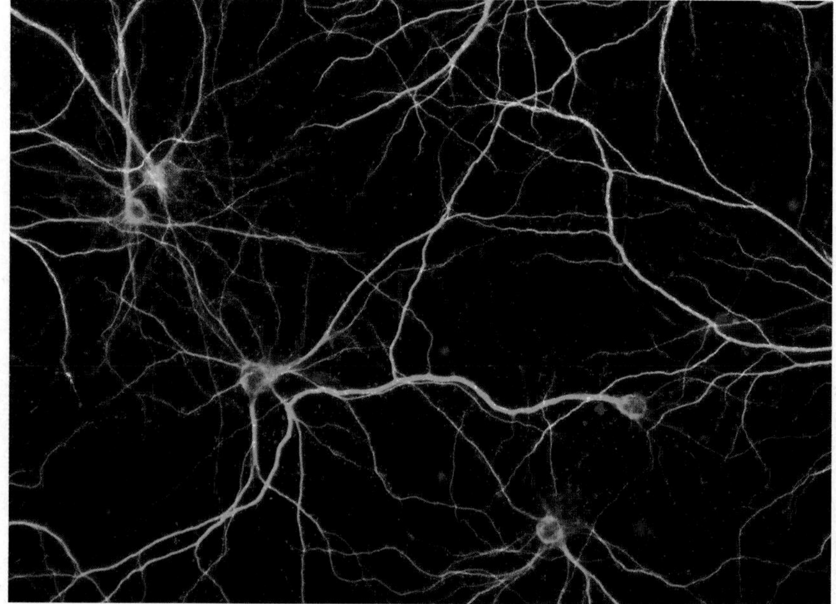

◀ FIGURE 2.19
Dendrites receiving synaptic inputs from axon terminals. Neurons have been made to fluoresce green using a method that reveals the distribution of a microtubule-associated protein. Axon terminals have been made to fluoresce orange-red using a method to reveal the distribution of synaptic vesicles. The nuclei are stained to fluoresce blue. (Source: Dr. Asha Bhakar, Massachusetts Institute of Technology.)

▲ FIGURE 2.20
Dendritic spines. This is a computer reconstruction of a segment of dendrite, showing the variable shapes and sizes of spines. Each spine is postsynaptic to one or two axon terminals. (Source: Harris & Stevens, 1989, cover image.)

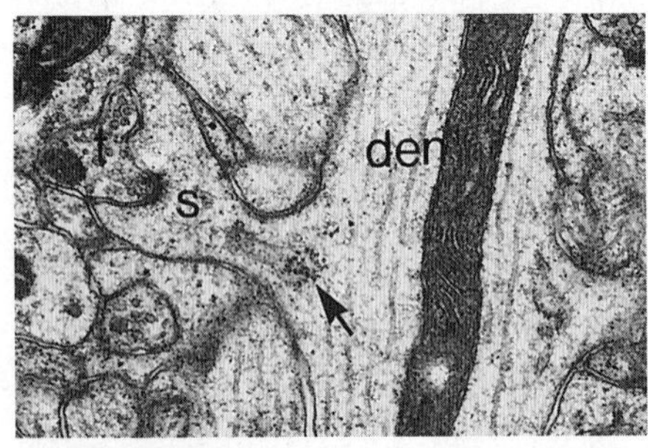

▲ FIGURE 2.21
Postsynaptic polyribosomes. This electron micrograph shows a dendrite (den) with a cluster of polyribosomes (arrow) at the base of a dendritic spine (s) receiving a synapse from an axon terminal (t). (Source: Courtesy of Dr. Oswald Steward, University of California, Irvine.)

under the synapse (the *postsynaptic* membrane) has many specialized protein molecules called **receptors** that detect the neurotransmitters in the synaptic cleft.

The dendrites of some neurons are covered with specialized structures called **dendritic spines** that receive some types of synaptic input. Spines look like little punching bags that hang off the dendrite (Figure 2.20). The unusual morphology of spines has fascinated neuroscientists ever since their discovery by Cajal. They are believed to isolate various chemical reactions that are triggered by some types of synaptic activation. Spine structure is sensitive to the type and amount of synaptic activity. Unusual changes in spines have been shown to occur in the brains of individuals with cognitive impairments (Box 2.6).

For the most part, the cytoplasm of dendrites resembles that of axons. It is filled with cytoskeletal elements and mitochondria. One interesting difference is that polyribosomes can be observed in dendrites, often right under spines (Figure 2.21). Research has shown that synaptic transmission can actually direct local protein synthesis in some neurons. In Chapter 25, we will see that synaptic regulation of protein synthesis is crucial for information storage by the brain.

CLASSIFYING NEURONS

It is likely that we will never understand how each of the 85 billion neurons in the nervous system uniquely contributes to the function of the brain. But what if we could show that all the neurons in the brain can be categorized and that within each category all neurons function identically? The complexity of the problem might then be reduced to understanding the unique contribution of each category rather than each cell. With this hope, neuroscientists have devised schemes for classifying neurons.

Classification Based on Neuronal Structure

Efforts to classify neurons began in earnest with the development of the Golgi stain. These classification schemes, based on the morphology of dendrites, axons, and the structures they innervate, are still in wide use.

Number of Neurites. Neurons can be classified according to the total number of neurites (axons and dendrites) that extend from the soma (Figure 2.22). A neuron with a single neurite is said to be **unipolar**. If there are two neurites, the cell is **bipolar**, and if there are three or more, the cell is multipolar. Most neurons in the brain are **multipolar**.

Dendrites. Dendritic trees can vary widely from one type of neuron to another. Some have inspired names with flourish, like "double bouquet cells" or "chandelier cells." Others have more utilitarian names, such as "alpha cells." Classification is often unique to a particular part of the brain. For example, in the cerebral cortex (the structure that lies just under the surface of the cerebrum), there are two broad classes: **stellate cells** (star shaped) and **pyramidal cells** (pyramid shaped) (Figure 2.23).

Neurons can also be classified according to whether their dendrites have spines. Those that do are called **spiny**, and those that do not are called **aspinous**. These dendritic classification schemes can overlap. For example, in the cerebral cortex, all pyramidal cells are spiny. Stellate cells, on the other hand, can be either spiny or aspinous.

BOX 2.6 OF SPECIAL INTEREST

Intellectual Disability and Dendritic Spines

The elaborate architecture of a neuron's dendritic tree reflects the complexity of its synaptic connections with other neurons. Brain function depends on these highly precise synaptic connections, which are formed during the fetal period and are refined during infancy and early childhood. Not surprisingly, this very complex developmental process is vulnerable to disruption. Intellectual disability is said to have occurred if a disruption of brain development results in subaverage cognitive functioning that impairs adaptive behavior.

According to standardized tests, intelligence in the general population is distributed along a bell-shaped (Gaussian) curve. By convention, the mean intelligence quotient (IQ) is set at 100. About two-thirds of the total population falls within 15 points (one standard deviation) of the mean, and 95% of the population falls within 30 points (two standard deviations). People with intelligence scores below 70 are considered to be intellectually disabled if their cognitive impairment affects their ability to adapt their behavior to the setting in which they live. Some 2–3% of humans fit this description.

Intellectual disability has many causes. The most severe forms are associated with genetic disorders, such as a condition called *phenylketonuria* (*PKU*). The basic abnormality is a deficit in the liver enzyme that metabolizes the dietary amino acid phenylalanine. Infants born with PKU have an abnormally high level of the amino acid in the blood and brain. If the condition goes untreated, brain growth is stunted and severe intellectual disability results. Another example is *Down syndrome*, which occurs when the fetus has an extra copy of chromosome 21, thus disrupting normal gene expression during brain development.

Another cause of intellectual disability is problems during pregnancy that can include a maternal infection, for example with German measles (rubella), and malnutrition. Children born to alcoholic mothers frequently have *fetal alcohol syndrome* comprising a constellation of developmental abnormalities that include intellectual disability. Other causes of intellectual disability are asphyxia of the infant during childbirth and environmental impoverishment—the lack of good nutrition, socialization, and sensory stimulation—during infancy.

Although some forms of intellectual disability have very clear physical correlates (e.g., stunted growth; abnormalities in the structure of the head, hands, and body), most cases have only behavioral manifestations. The brains of these individuals appear grossly normal. How, then, do we account for the profound cognitive impairment? An important clue came in the 1970s from the research of Miguel Marin-Padilla, working at Dartmouth College, and Dominick Purpura, working at the Albert Einstein College of Medicine in New York City. Using the Golgi stain, they studied the brains of intellectually disabled children and discovered remarkable changes in dendritic structure. The dendrites of low-functioning children had many fewer dendritic spines, and the spines that they did have were unusually long and thin (Figure A). The extent of the spine changes was well correlated with the degree of intellectual disability.

Dendritic spines are an important target of synaptic input. Purpura pointed out that the dendritic spines of intellectually disabled children resemble those of the normal human fetus. He suggested that intellectual disability reflects the failure of normal circuits to form in the brain. In the three decades since this seminal work was published, it was established that normal synaptic development, including maturation of the dendritic spines, depends critically on the environment during infancy and early childhood. An impoverished environment during an early critical period of development can lead to profound changes in the circuits of the brain. However, there is some good news. Many of the deprivation-induced changes in the brain can be reversed if intervention occurs early enough. In Chapter 23, we will take a closer look at the role of experience in brain development.

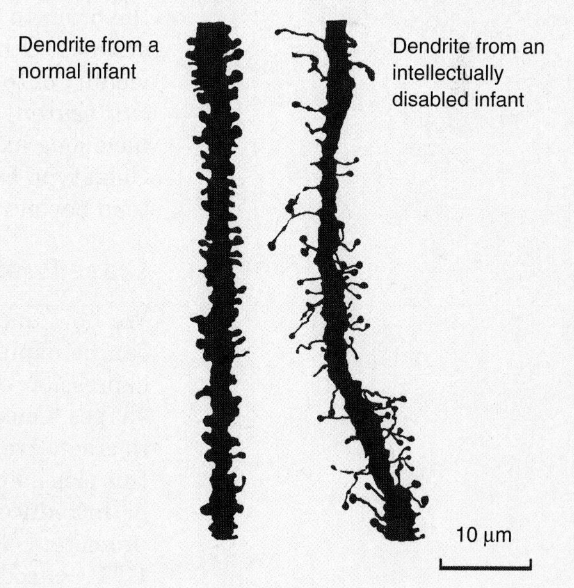

Figure A
Normal and abnormal dendrites. (Source: Purpura, 1974, Fig. 2A.)

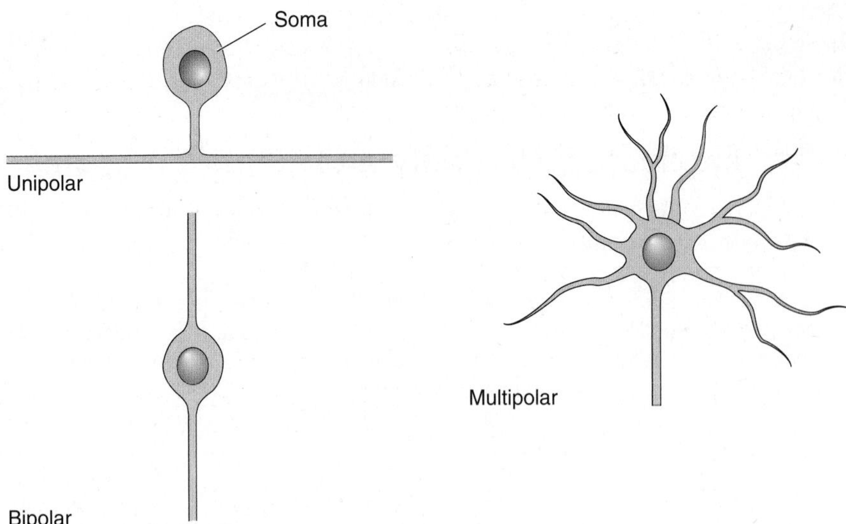

▲ FIGURE 2.22
Classification of neurons based on the number of neurites.

Connections. Information is delivered to the nervous system by neurons that have neurites in the sensory surfaces of the body, such as the skin and the retina of the eye. Cells with these connections are called **primary sensory neurons**. Other neurons have axons that form synapses with the muscles and command movements; these are called **motor neurons**. But most neurons in the nervous system form connections only with other neurons. In this classification scheme, these cells are called **interneurons**.

Axon Length. Some neurons have long axons that extend from one part of the brain to the other; these are called *Golgi type I neurons*, or *projection neurons*. Other neurons have short axons that do not extend beyond the vicinity of the cell body; these are called *Golgi type II neurons*, or *local circuit neurons*. In the cerebral cortex, for example, pyramidal cells usually have long axons that extend to other parts of the brain and are therefore Golgi type I neurons. In contrast, stellate cells have axons that never extend beyond the cerebral cortex and are therefore Golgi type II neurons.

Classification Based on Gene Expression

We now understand that most differences between neurons ultimately can be explained at the genetic level. For example, differences in gene expression cause pyramidal cells and stellate cells to develop different shapes. Once a genetic difference is known, that information can be used to create transgenic mice that allow detailed investigation of neurons in this class. For example, a foreign gene encoding a fluorescent protein can be introduced and placed under the control of a cell type–specific gene promoter. **Green fluorescent protein** (usually simply abbreviated as **GFP**), encoded by a gene discovered in jellyfish, is used commonly in neuroscience research. When illuminated with the appropriate wavelength of light, the GFP fluoresces bright green, allowing visualization of the neuron in which it is expressed. Genetic engineering methods are now commonly used for measuring and manipulating the functions of neurons in different categories (Box 2.7).

We have known for some time that one important way neurons differ is the neurotransmitter they use. Neurotransmitter differences arise because of differences in the expression of proteins involved in transmitter synthesis, storage, and use. Understanding these genetic differences enables a classification of neurons based on their neurotransmitters. For example, the motor neurons that command voluntary movements all release the neurotransmitter *acetylcholine* at their synapses; these motor cells are therefore also classified as *cholinergic*, meaning that they express the genes that enable use of this particular neurotransmitter. Collections of cells that use the same neurotransmitter make up the brain's neurotransmitter systems (see Chapters 6 and 15).

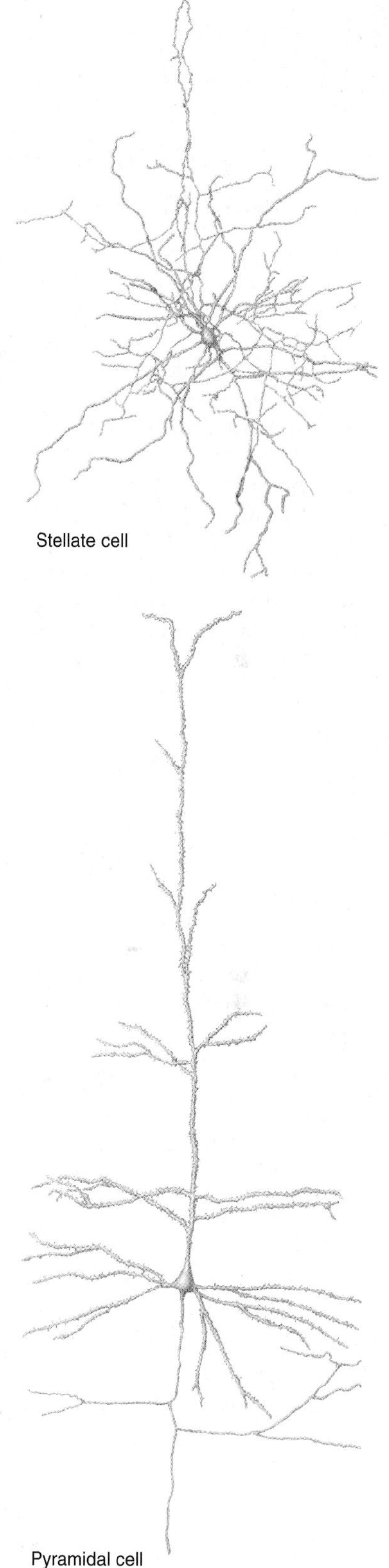

▲ FIGURE 2.23
Classification of neurons based on dendritic tree structure. Stellate cells and pyramidal cells, distinguished by the arrangement of their dendrites, are two types of neurons found in the cerebral cortex.

GLIA

Although most of this chapter is about neurons, as justified by the current state of knowledge, some neuroscientists consider glia the "sleeping giants" of neuroscience. Indeed, we continue to learn that glia contribute much more importantly to information processing in the brain than has been historically appreciated. Nevertheless, the data continue to indicate that glia contribute to brain function mainly by supporting neuronal functions. Although their role may be subordinate, without glia, the brain could not function properly.

Astrocytes

The most numerous glia in the brain are called **astrocytes** (Figure 2.24). These cells fill most of the spaces between neurons. The space that remains between neurons and astrocytes in the brain is only about 20 nm wide. Consequently, astrocytes probably influence whether a neurite can grow or retract.

An essential role of astrocytes is regulating the chemical content of this *extracellular space*. For example, astrocytes envelop synaptic junctions in the brain (Figure 2.25), thereby restricting the spread of neurotransmitter molecules that have been released. Astrocytes also have special proteins in their membranes that actively remove many neurotransmitters from the synaptic cleft. A recent and unexpected discovery is that astrocytic membranes also possess neurotransmitter receptors that, like the receptors on neurons, can trigger electrical and biochemical events inside the glial cell. Besides regulating neurotransmitters, astrocytes also tightly control the extracellular concentration of several substances that could interfere with proper neuronal function. For example, astrocytes regulate the concentration of potassium ions in the extracellular fluid.

Myelinating Glia

Unlike astrocytes, the primary function of **oligodendroglial** and **Schwann cells** is clear. These glia provide layers of membrane that insulate axons. Boston University anatomist Alan Peters, a pioneer in the electron microscopic study of the nervous system, showed that this wrapping, called **myelin**, spirals around axons in the brain (Figure 2.26). Because the axon fits inside the spiral wrapping like a sword in its scabbard, the name *myelin sheath* describes the entire covering. The sheath is interrupted periodically, leaving a short length where the axonal membrane is exposed. This region is called a **node of Ranvier** (Figure 2.27).

We will see in Chapter 4 that myelin serves to speed the propagation of nerve impulses down the axon. Oligodendroglia and Schwann cells differ in their location and some other characteristics. For example, oligodendroglia are found only in the central nervous system (brain and spinal cord), whereas Schwann cells are found only in the peripheral nervous

BOX 2.7 **BRAIN FOOD**

Understanding Neuronal Structure and Function with Incredible Cre

One type of cell in the body can be distinguished from another by the unique pattern of genes it expresses as proteins. Similarly, different classes of neurons in the brain can be identified based on which genes are expressed. With modern methods of genetic engineering, knowledge that a gene is uniquely expressed in one type of neuron can help determine the contributions of this cell type to brain function.

Let's consider as an example the neurons that uniquely express the gene encoding the protein choline acetyltransferase (ChAT). ChAT is an enzyme that synthesizes the neurotransmitter acetylcholine. It is only expressed in "cholinergic neurons" that use acetylcholine because only these neurons have the transcription factors that act on this gene's promoter. If we insert into a mouse's genome a transgene engineered to be under the control of the same promoter, this foreign transgene will also be expressed only in cholinergic neurons. If the transgene expresses the enzyme Cre recombinase, derived from a bacterial virus, we can compel these cholinergic neurons to give up their secrets in myriad ways. Let's examine how.

Cre recombinase recognizes short DNA sequences called *loxP sites* that can be inserted on either side of another gene. The DNA between the loxP sites is said to be *floxed*. The Cre recombinase functions to cut out, or excise, the gene between the loxP sites. By breeding the "Cre mouse" with the "floxed mouse," one can generate mice in which a gene is deleted only in one particular type of neuron.

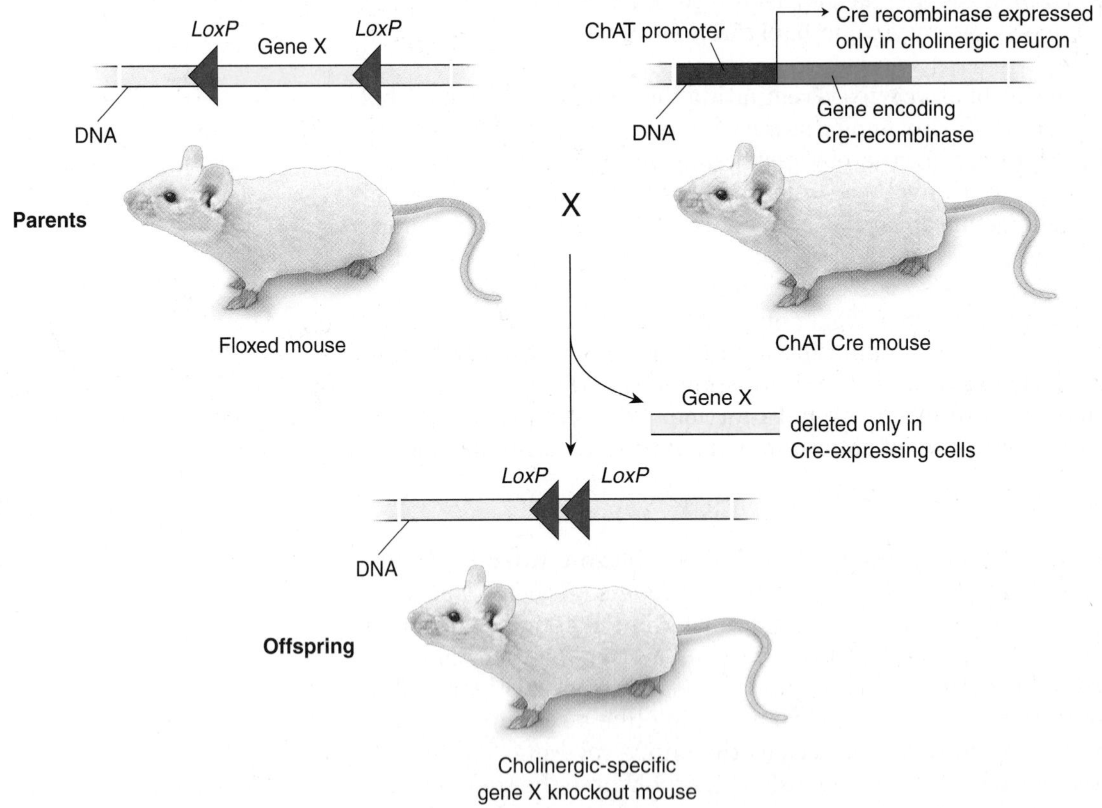

Figure A
Creating a mouse with a gene knockout only in cholinergic neurons is accomplished by crossing a floxed mouse with the gene of interest (gene X) flanked by loxP sites with another mouse in which Cre recombinase is under the control of the ChAT promoter. In the offspring, gene X is cut out only in the cells expressing Cre, namely, the cholinergic neurons.

In a simple example, we can ask how cholinergic neurons react to the deletion of another gene they normally express; let's call this gene X. To answer this question, we cross mice that have had gene X floxed (the "floxed mice") with our mice that express Cre under the control of the ChAT promoter (the "ChAT-Cre mice"). In the offspring, the floxed gene is removed only in those neurons that express Cre; that is, only in the cholinergic neurons (Figure A).

We can also use Cre to cause expression of novel transgenes in cholinergic neurons. Normally, expression of a transgene requires that we include a promoter sequence upstream of the protein-coding region. Transcription of the transgene fails to occur if a stop sequence is inserted between this promoter and the protein-coding sequence. Now consider what happens if we generate a transgenic mouse with this stop sequence flanked by loxP sites. Crossing this mouse with our ChAT-Cre mouse generates offspring in which the transgene is expressed *only* in cholinergic neurons because the stop sequence has been removed only in these neurons (Figure B).

If we design this transgene to encode a fluorescent protein, we can use fluorescence to examine the structure and connections of these cholinergic neurons. If we design this transgene to express a protein that fluoresces only when impulses are generated by the neurons, we can monitor the activity of the cholinergic neurons by measuring light flashes. If we design this transgene to express a protein that kills or silences the neuron, we can see how brain function is altered in the absence of cholinergic neurons. The possible manipulations of cholinergic neurons through this feat of genetic engineering are limited only by the imagination of the scientist.

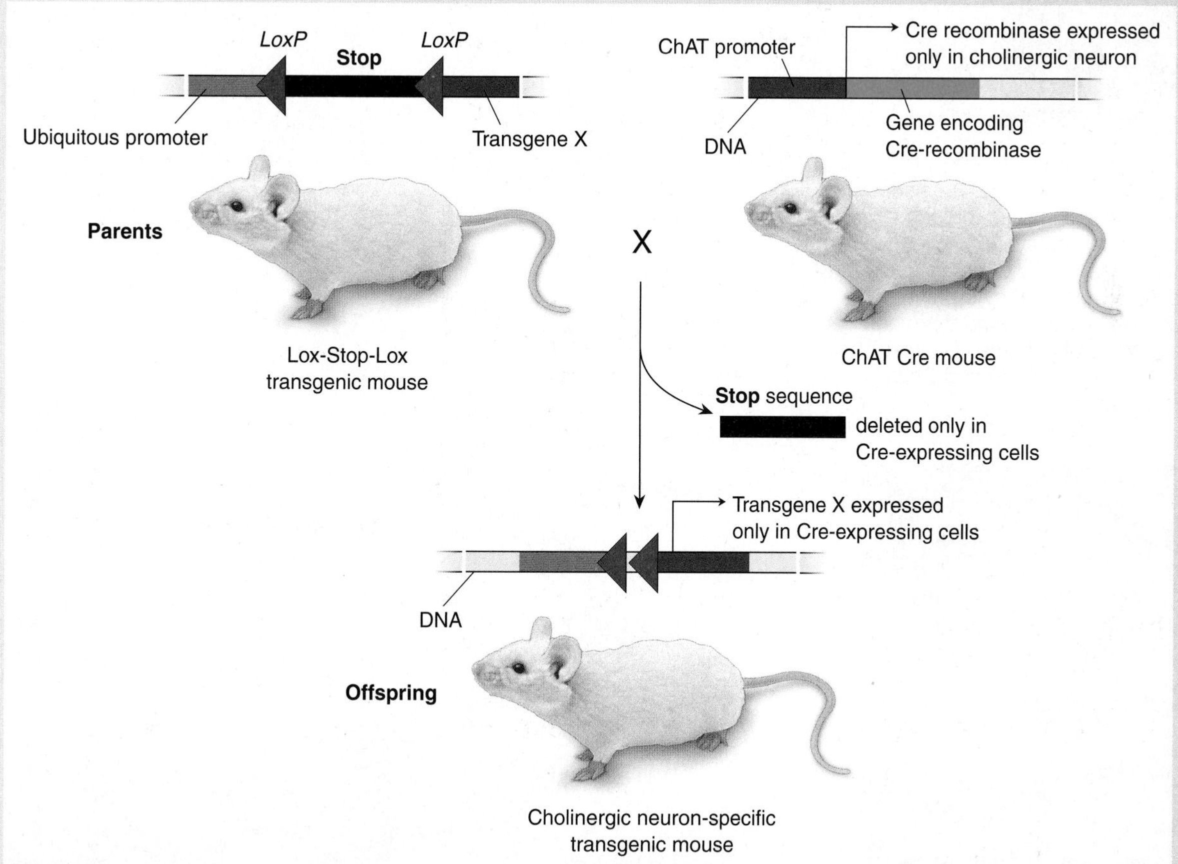

Figure B
A transgene of interest (transgene X) can also be expressed exclusively in cholinergic neurons. First, a mouse is created in which expression of the transgene is prevented by insertion of a floxed stop sequence between a strong, ubiquitous promoter and the coding region of the gene. Crossing this mouse with the ChAT-Cre mouse results in offspring in which the stop sequence has been deleted only in cholinergic neurons, allowing expression of the transgene only in these neurons.

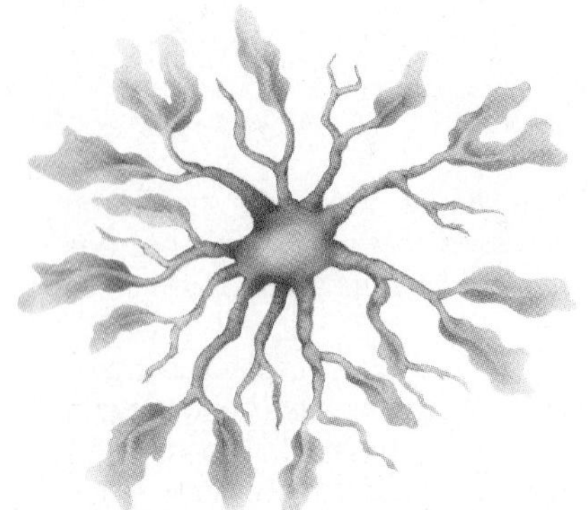

▲ FIGURE 2.24
An astrocyte. Astrocytes fill most of the space in the brain that is not occupied by neurons and blood vessels.

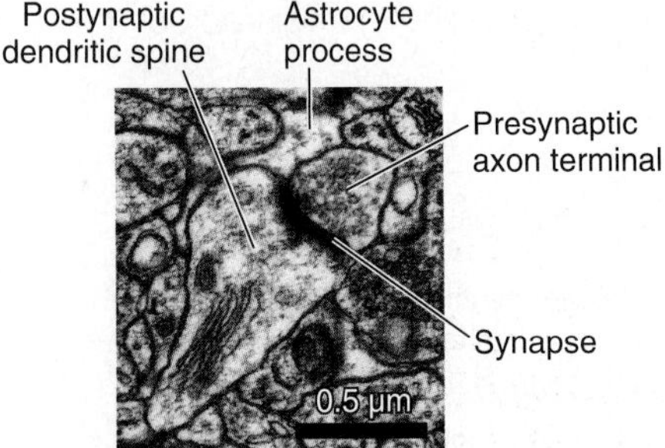

▲ FIGURE 2.25
Astrocytes envelop synapses. An electron micrograph of a thin slice through a synapse showing the presynaptic axon terminal and the postsynaptic dendritic spine (colored green) and an astrocyte process (colored blue) that wraps around them and restricts the extracellular space. (Source: Courtesy of Drs. Cagla Eroglu and Chris Risher, Duke University.)

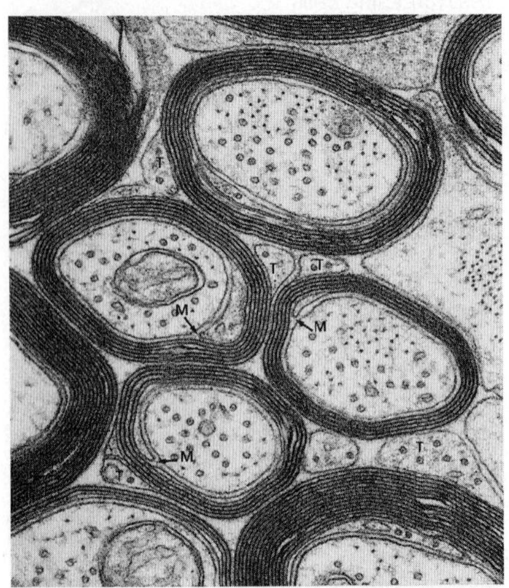

▲ FIGURE 2.26
Myelinated optic nerve fibers cut in cross section. (Source: Courtesy of Dr. Alan Peters.)

system (parts outside the skull and vertebral column). Another difference is that one oligodendroglial cell contributes myelin to several axons, whereas each Schwann cell myelinates only a single axon.

Other Non-Neuronal Cells

Even if we eliminated every neuron, every astrocyte, and every oligodendroglial cell, other cells would still remain in the brain. First, special cells called **ependymal cells** line fluid-filled ventricles within the brain and play a role in directing cell migration during brain development. Second, a class of cells called **microglia** function as phagocytes to remove debris left by dead or degenerating neurons and glia. Microglia have attracted much interest recently, as they appear to be involved in remodeling synaptic connections by gobbling them up. As we saw in Box 2.3, microglia can migrate into the brain from the blood, and disruption of this microglial invasion can interfere with brain functions and behavior. Finally, in addition to glial and ependymal cells, the brain also has vasculature: arteries, veins, and capillaries that deliver via the blood essential nutrients and oxygen to neurons.

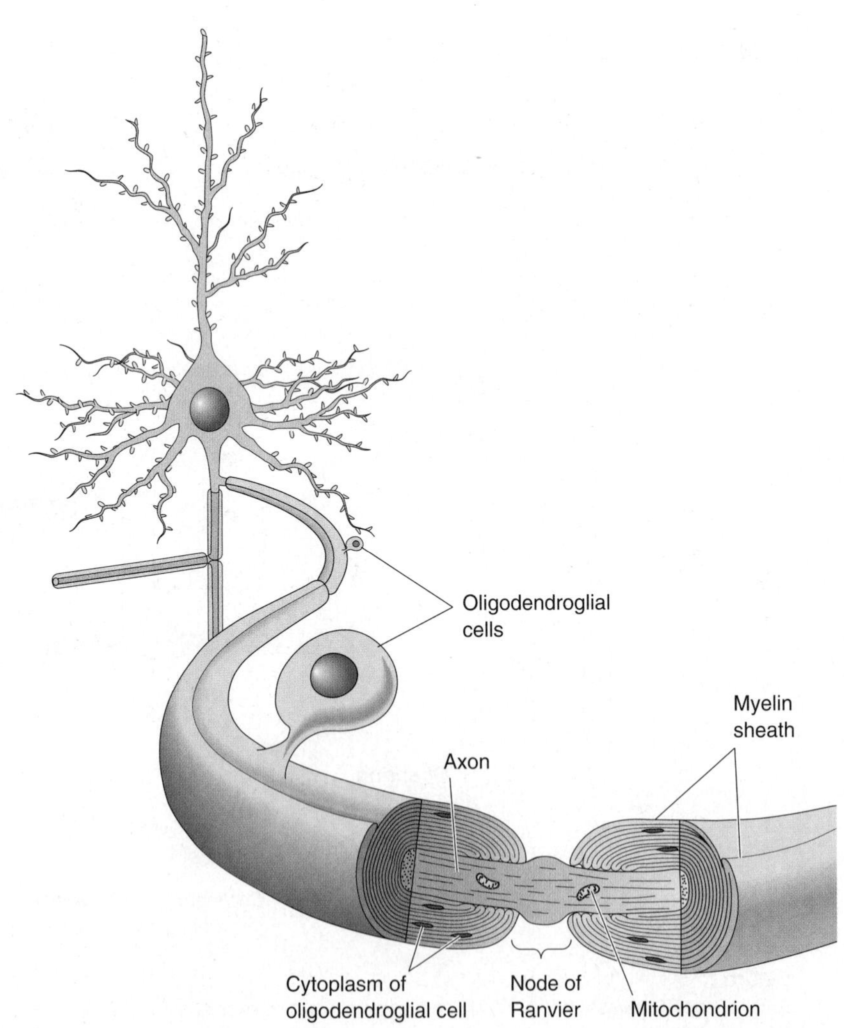

▲ FIGURE 2.27
An oligodendroglial cell. Like the Schwann cells found in the nerves of the body, oligodendroglia provide myelin sheaths around axons in the brain and spinal cord. The myelin sheath of an axon is interrupted periodically at the nodes of Ranvier.

CONCLUDING REMARKS

Learning the structural characteristics of the neuron provides insight into how neurons and their different parts work because structure correlates with function. For example, the absence of ribosomes in the axon correctly predicts that proteins in the axon terminal are provided from the soma via axoplasmic transport. A large number of mitochondria in the axon terminal correctly predicts a high energy demand. The elaborate structure of the dendritic tree appears ideally suited for receiving incoming information, and indeed, this is where most of the synapses are formed with the axons of other neurons.

From the time of Nissl, the rough ER has been recognized as an important feature of neurons. What does this tell us about neurons? Recall that rough ER is a site of the synthesis of proteins destined to be inserted into the membrane. We will next see how the various proteins in the neuronal membrane give rise to the unique capabilities of neurons to transmit, receive, and store information.

KEY TERMS

Introduction

neuron (p. 24)
glial cell (p. 24)

The Neuron Doctrine

histology (p. 25)
Nissl stain (p. 25)
cytoarchitecture (p. 25)
Golgi stain (p. 26)
cell body (p. 26)
soma (p. 26)
perikaryon (p. 26)
neurite (p. 26)
axon (p. 26)
dendrite (p. 26)
neuron doctrine (p. 27)

The Prototypical Neuron

cytosol (p. 29)
organelle (p. 29)
cytoplasm (p. 29)
nucleus (p. 29)
chromosome (p. 29)
DNA (deoxyribonucleic acid) (p. 29)
gene (p. 29)
gene expression (p. 29)
protein (p. 29)
protein synthesis (p. 29)
mRNA (messenger ribonucleic acid) (p. 29)
transcription (p. 29)
promoter (p. 31)
transcription factor (p. 31)
RNA splicing (p. 31)
amino acid (p. 32)
translation (p. 32)
genome (p. 32)
genetic engineering (p. 32)
knockout mice (p. 33)
transgenic mice (p. 33)
knock-in mice (p. 33)
ribosome (p. 36)
rough endoplasmic reticulum (rough ER) (p. 36)
polyribosome (p. 36)
smooth endoplasmic reticulum (smooth ER) (p. 36)
Golgi apparatus (p. 36)
mitochondrion (p. 36)
ATP (adenosine triphosphate) (p. 38)
neuronal membrane (p. 38)
cytoskeleton (p. 38)
microtubule (p. 38)
microfilament (p. 39)
neurofilament (p. 39)
axon hillock (p. 39)
axon collateral (p. 39)
axon terminal (p. 41)
terminal bouton (p. 41)
synapse (p. 42)
terminal arbor (p. 42)
innervation (p. 42)
synaptic vesicle (p. 42)
synaptic cleft (p. 43)
synaptic transmission (p. 43)
neurotransmitter (p. 43)
axoplasmic transport (p. 43)
anterograde transport (p. 44)
retrograde transport (p. 44)
dendritic tree (p. 44)
receptor (p. 46)
dendritic spine (p. 46)

Classifying Neurons

unipolar neuron (p. 46)
bipolar neuron (p. 46)
multipolar neuron (p. 46)
stellate cell (p. 46)
pyramidal cell (p. 46)
spiny neuron (p. 46)
aspinous neuron (p. 46)
primary sensory neuron (p. 48)
motor neuron (p. 48)
interneuron (p. 48)
green fluorescent protein (GFP) (p. 48)

Glia

astrocyte (p. 49)
oligodendroglial cell (p. 49)
Schwann cell (p. 49)
myelin (p. 49)
node of Ranvier (p. 49)
ependymal cell (p. 52)
microglial cell (p. 52)

REVIEW QUESTIONS

1. State the neuron doctrine in a single sentence. To whom is this insight credited?
2. Which parts of a neuron are shown by a Golgi stain that are not shown by a Nissl stain?
3. What are three physical characteristics that distinguish axons from dendrites?
4. Of the following structures, state which ones are unique to neurons and which are not: nucleus, mitochondria, rough ER, synaptic vesicle, Golgi apparatus.
5. What are the steps by which the information in the DNA of the nucleus directs the synthesis of a membrane-associated protein molecule?
6. Colchicine is a drug that causes microtubules to break apart (depolymerize). What effect would this drug have on anterograde transport? What would happen in the axon terminal?
7. Classify the cortical pyramidal cell based on (1) the number of neurites, (2) the presence or absence of dendritic spines, (3) connections, and (4) axon length.
8. Knowledge of genes uniquely expressed in a particular category of neurons can be used to understand how those neurons function. Give one example of how you could use genetic information to study a category of neuron.
9. What is myelin? What does it do? Which cells provide it in the central nervous system?

FURTHER READING

De Vos KJ, Grierson AJ, Ackerley S, Miller CCJ. 2008. Role of axoplasmic transport in neurodegenerative diseases. *Annual Review of Neuroscience* 31:151–173.

Eroglu C, Barres BA. 2010. Regulation of synaptic connectivity by glia. *Nature* 468:223–231.

Jones EG. 1999. Golgi, Cajal and the Neuron Doctrine. *Journal of the History of Neuroscience* 8:170–178.

Lent R, Azevedo FAC, Andrade-Moraes CH, Pinto AVO. 2012. How many neurons do you have? Some dogmas of quantitative neuroscience under revision. *European Journal of Neuroscience* 35:1–9.

Nelson SB, Hempel C, Sugino K. 2006. Probing the transcriptome of neuronal cell types. *Current Opinion in Neurobiology* 16:571–576.

Peters A, Palay SL, Webster H deF. 1991. *The Fine Structure of the Nervous System*, 3rd ed. New York: Oxford University Press.

Sadava D, Hills DM, Heller HC, Berenbaum MR. 2011. *Life: The Science of Biology*, 9th ed. Sunderland, MA: Sinauer.

Shepherd GM, Erulkar SD. 1997. Centenary of the synapse: from Sherrington to the molecular biology of the synapse and beyond. *Trends in Neurosciences* 20:385–392.

Wilt BA, Burns LD, Ho ETW, Ghosh KK, Mukamel EA, Schnitzer MJ. 2009. Advances in light microscopy for neuroscience. *Annual Review of Neuroscience* 32:435–506.

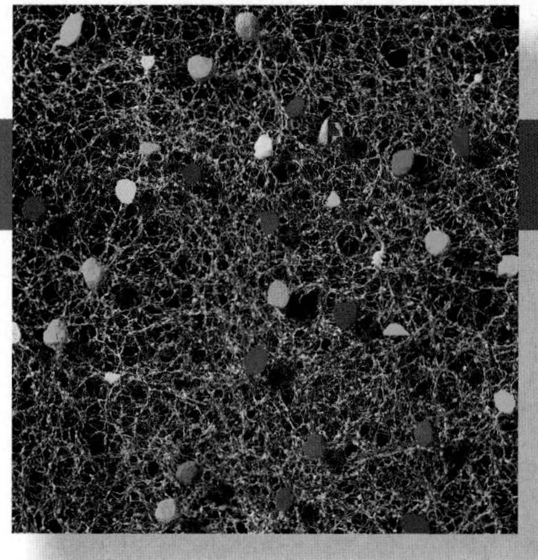

CHAPTER THREE

The Neuronal Membrane at Rest

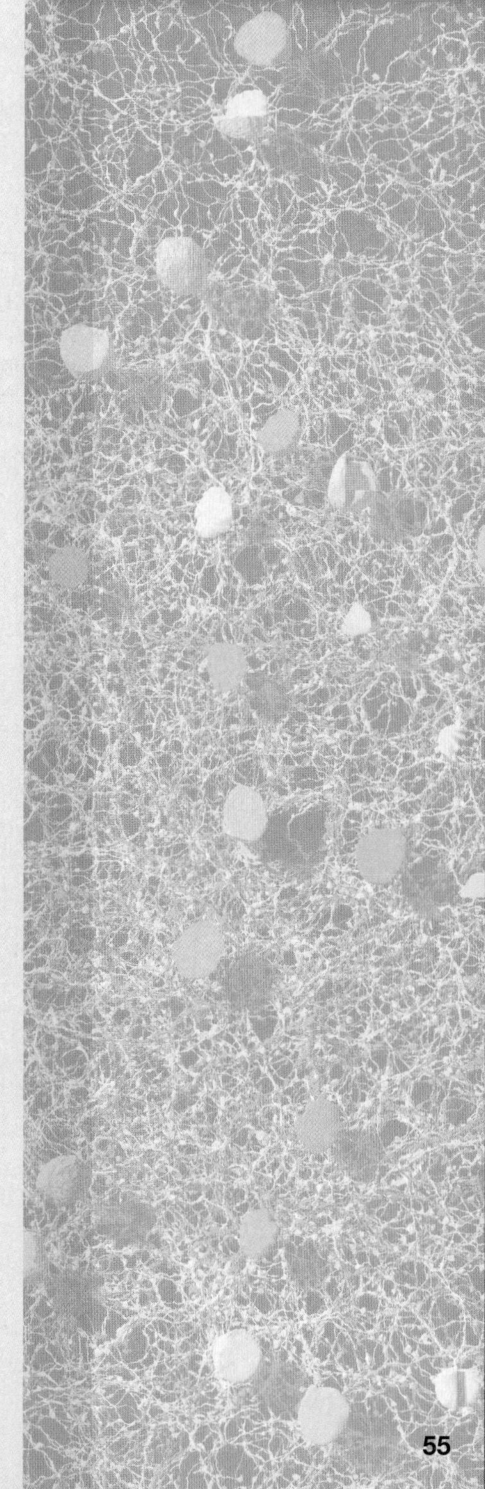

INTRODUCTION

Consider the problem your nervous system confronts when you step on a thumbtack. Your reactions are automatic: You shriek with pain as you jerk up your foot. For this simple response to occur, breaking of the skin by the tack must be translated into neural signals that travel rapidly and reliably up the long sensory nerves of your leg. In the spinal cord, these signals are transferred to interneurons. Some of these neurons connect with the parts of your brain that interpret the signals as being painful. Others connect to the motor neurons that control the leg muscles that withdraw your foot. Thus, even this simple reflex, depicted in Figure 3.1, requires the nervous system to *collect*, *distribute*, and *integrate* information. A goal of cellular neurophysiology is to understand the biological mechanisms that underlie these functions.

The neuron solves the problem of conducting information over a distance by using electrical signals that sweep along the axon. In this sense, axons act like telephone wires. The analogy stops there, however, because the type of signal used by the neuron is constrained by the special environment of the nervous system. In a copper telephone wire, information can be transferred over long distances at a high rate (about half the speed of light) because telephone wire is a superb conductor of electrons, is well insulated, and is suspended in air (air being a poor conductor of

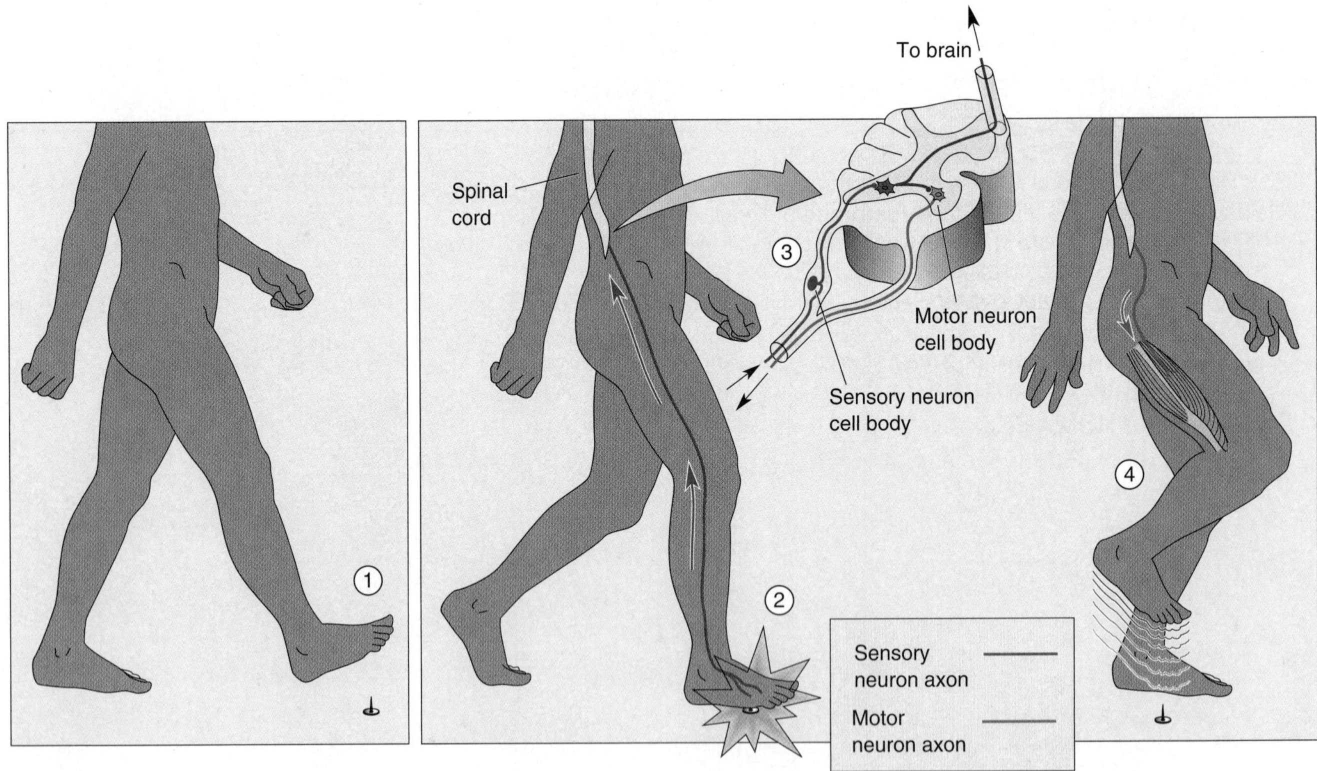

▲ FIGURE 3.1
A simple reflex. ① A person steps on a thumbtack. ② The breaking of the skin is translated into signals that travel up sensory nerve fibers (the direction of information flow, indicated by the arrows). ③ In the spinal cord, the information is distributed to interneurons. Some of these neurons send axons to the brain where the painful sensation is registered. Others synapse on motor neurons, which send descending signals to the muscles. ④ The motor commands lead to muscle contraction and withdrawal of the foot.

electricity). Electrons will, therefore, move within the wire instead of radiating away. In contrast, electrical charge in the cytosol of the axon is carried by electrically charged atoms (ions) instead of free electrons. This makes cytosol far less conductive than copper wire. Also, the axon is not especially well insulated and is bathed in salty extracellular fluid, which conducts electricity. Thus, like water flowing down a leaky garden hose, electrical current flowing down the axon would not go very far before it would leak out.

Fortunately, the axonal membrane has properties that enable it to conduct a special type of signal—the nerve impulse, or **action potential**—that overcomes these biological constraints. As we will see in a moment, the term "potential" refers to the separation of electrical charge across the membrane. In contrast to passively conducted electrical signals, action potentials do not diminish over distance; they are signals of fixed size and duration. Information is encoded in the frequency of action potentials of individual neurons as well as in the distribution and number of neurons firing action potentials in a given nerve. This type of code is somewhat like Morse code sent down an old fashioned telegraph wire; information is encoded in the pattern of electrical impulses. Cells capable of generating and conducting action potentials, which include both nerve and muscle cells, are said to have **excitable membrane**. The "action" in action potentials occurs at the cell membrane.

When a cell with excitable membrane is not generating impulses, it is said to be at rest. In the resting neuron, the cytosol along the inside surface of the membrane has a negative electrical charge compared to the outside. This difference in electrical charge across the membrane is called the **resting membrane potential** (or resting potential). The action potential is simply a brief reversal of this condition, and for an instant—about a thousandth of a second—the inside of the membrane becomes positively charged relative to the outside. Therefore, to understand how neurons signal one another, we must learn how the neuronal membrane at rest separates electrical charge, how electrical charge can be rapidly redistributed across the membrane during the action potential, and how the impulse can propagate reliably along the axon.

In this chapter, we begin our exploration of neuronal signaling by tackling the first question: How does the resting membrane potential arise? Understanding the resting potential is very important because it forms the foundation for understanding the rest of neuronal physiology. And knowledge of neuronal physiology is central to understanding the capabilities and limitations of brain function.

THE CAST OF CHEMICALS

We begin our discussion of the resting membrane potential by introducing the three main players: the salty fluids on either side of the membrane, the membrane itself, and the proteins that span the membrane. Each of these has certain properties that contribute to establishing the resting potential.

Cytosol and Extracellular Fluid

Water is the main ingredient of both the fluid inside the neuron, the intracellular fluid or cytosol, and the outside fluid that bathes the neuron, the extracellular fluid. Electrically charged atoms—ions—that are dissolved in this water are responsible for the resting and action potentials.

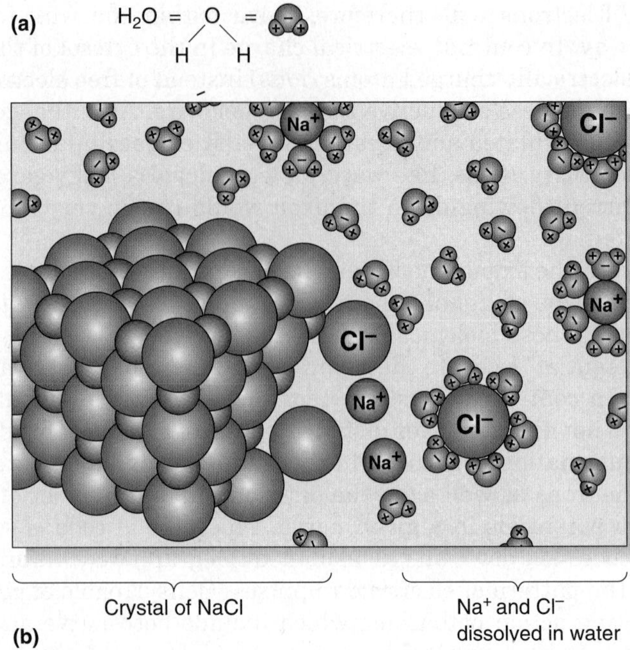

▲ FIGURE 3.2
Water is a polar solvent. **(a)** Different representations of the atomic structure of the water molecule. The oxygen atom has a net negative electrical charge and the hydrogen atoms have a net positive electrical charge, making water a polar molecule. **(b)** A crystal of sodium chloride (NaCl) dissolves in water because the polar water molecules have a stronger attraction for the electrically charged sodium and chloride ions than the ions do for one another.

Water. For our purpose here, the most important property of the water molecule (H_2O) is its uneven distribution of electrical charge (Figure 3.2a). The two hydrogen atoms and the oxygen atom are bonded together covalently, which means they share electrons. The oxygen atom, however, has a greater affinity for electrons than does the hydrogen atom. As a result, the shared electrons spend more time associated with the oxygen atom than with the two hydrogen atoms. Therefore, the oxygen atom acquires a net negative charge (because it has extra electrons), and the hydrogen atoms acquire a net positive charge. Thus, H_2O is said to be a polar molecule, held together by *polar covalent bonds*. This electrical polarity makes water an effective solvent of other charged or polar molecules; that is, other polar molecules tend to dissolve in water.

Ions. Atoms or molecules that have a net electrical charge are known as **ions**. Table salt is a crystal of sodium (Na^+) and chloride (Cl^-) ions held together by the electrical attraction of oppositely charged atoms. This attraction is called an *ionic bond*. Salt dissolves readily in water because the charged portions of the water molecule have a stronger attraction for the ions than the ions have for each other (Figure 3.2b). As each ion breaks away from the crystal, it is surrounded by a sphere of water molecules. Each positively charged ion (Na^+, in this case) is covered by water molecules oriented so that the oxygen atoms (the negative pole) are facing the ion. Likewise, each negatively charged ion (Cl^-) is surrounded by water molecules with the hydrogen atoms (with their net positive charge) facing the chloride ion. These clouds of water that surround each ion are called *spheres of hydration*, and they effectively insulate the ions from one another.

The electrical charge of an atom depends on the difference between its numbers of protons and electrons. When this difference is 1, the ion is said to be *monovalent*; when the difference is 2, the ion is *divalent*; and so on. Ions with a net positive charge are called **cations**; ions with a negative charge are called **anions**. Remember that ions are the major charge carriers in the conduction of electricity in biological systems, including the neuron. The ions of particular importance for cellular neurophysiology are the monovalent cations Na^+ (sodium) and K^+ (potassium), the divalent cation Ca^{2+} (calcium), and the monovalent anion Cl^- (chloride).

The Phospholipid Membrane

As we have seen, substances with a net or uneven electrical charge will dissolve in water because of the polarity of the water molecule. These substances, including ions and polar molecules, are said to be "water-loving," or *hydrophilic*. However, compounds whose atoms are bonded by *nonpolar covalent bonds* have no basis for chemical interactions with water. A nonpolar covalent bond occurs when the shared electrons are distributed evenly in the molecule so that no portion acquires a net electrical charge. Such compounds will not dissolve in water and are said to be "water-fearing," or *hydrophobic*. A familiar example of a hydrophobic substance is olive oil, and, as you know, oil and water don't mix. Another example is *lipid*, a class of water-insoluble biological molecules important to the structure of cell membranes. The lipids of the neuronal membrane contribute to the resting and action potentials by forming a barrier to water-soluble ions and, indeed, to water itself.

The main chemical building blocks of cell membranes are *phospholipids*. Like other lipids, phospholipids contain long nonpolar chains of carbon atoms bonded to hydrogen atoms. In addition, however, a phospholipid has a polar phosphate group (a phosphorus atom bonded to three oxygen atoms) attached to one end of the molecule. Thus, phospholipids are said to have a polar "head" (containing phosphate) that is hydrophilic, and a nonpolar "tail" (containing hydrocarbon) that is hydrophobic.

The neuronal membrane consists of a sheet of phospholipids, two molecules thick. A cross section through the membrane, shown in Figure 3.3, reveals that the hydrophilic heads face the outer and inner watery environments and the hydrophobic tails face each other. This stable arrangement is called a **phospholipid bilayer**, and it effectively isolates the cytosol of the neuron from the extracellular fluid.

Protein

The type and distribution of protein molecules distinguish neurons from other types of cells. The *enzymes* that catalyze chemical reactions in the neuron, the *cytoskeleton* that gives a neuron its special shape, and the *receptors* that are sensitive to neurotransmitters are all made up of protein molecules. The resting potential and action potential depend on special proteins that span the phospholipid bilayer. These proteins provide routes for ions to cross the neuronal membrane.

Protein Structure. In order to perform their many functions in the neuron, different proteins have widely different shapes, sizes, and chemical characteristics. To understand this diversity, let's briefly review protein structure.

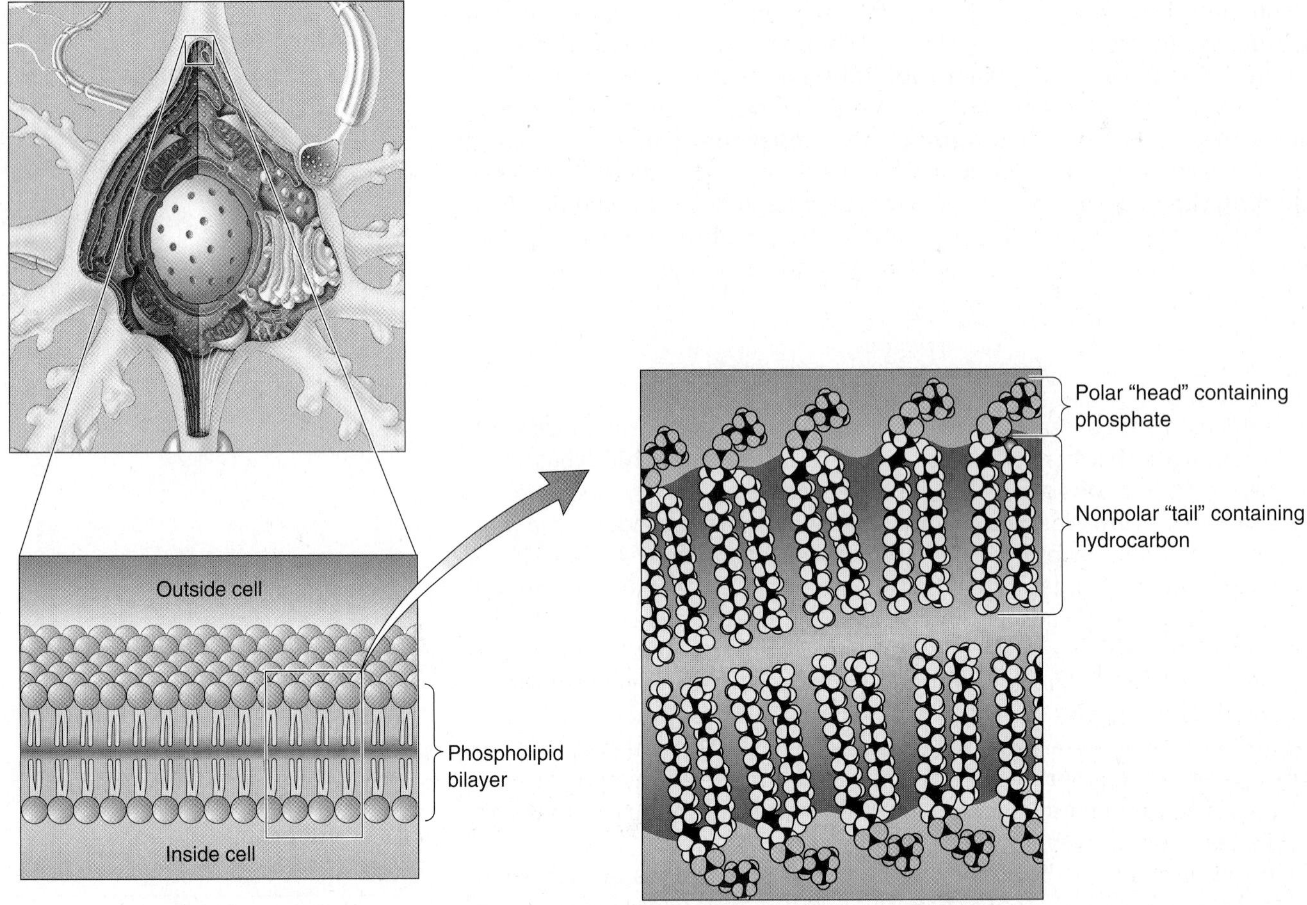

▲ FIGURE 3.3
The phospholipid bilayer. The phospholipid bilayer is the core of the neuronal membrane and forms a barrier to water-soluble ions.

As mentioned in Chapter 2, proteins are molecules assembled from various combinations of 20 different amino acids. The basic structure of an amino acid is shown in Figure 3.4a. All amino acids have a central carbon atom (the alpha carbon), which is covalently bonded to four molecular groups: a hydrogen atom, an amino group (NH_3^+), a carboxyl group (COO^-), and a variable group called the *R group* (R for residue). The differences among amino acids result from differences in the size and nature of these R groups (Figure 3.4b). The properties of the R group determine the chemical relationships in which each amino acid can participate.

Proteins are synthesized by the ribosomes of the neuronal cell body. In this process, amino acids assemble into a chain connected by **peptide bonds**, which join the amino group of one amino acid to the carboxyl group of the next (Figure 3.5a). Proteins made of a single chain of amino acids are also called **polypeptides** (Figure 3.5b).

The four levels of protein structure are shown in Figure 3.6. The *primary structure* is like a chain in which the amino acids are linked together by peptide bonds. As a protein molecule is being synthesized, however, the polypeptide chain can coil into a spiral-like configuration called an *alpha helix*. The alpha helix is an example of what is called the *secondary structure* of a protein molecule. Interactions among the R groups can cause the molecule to change its three-dimensional confor-

$^{+}H_3N-C(H)-COO^-$ with R group

(a)

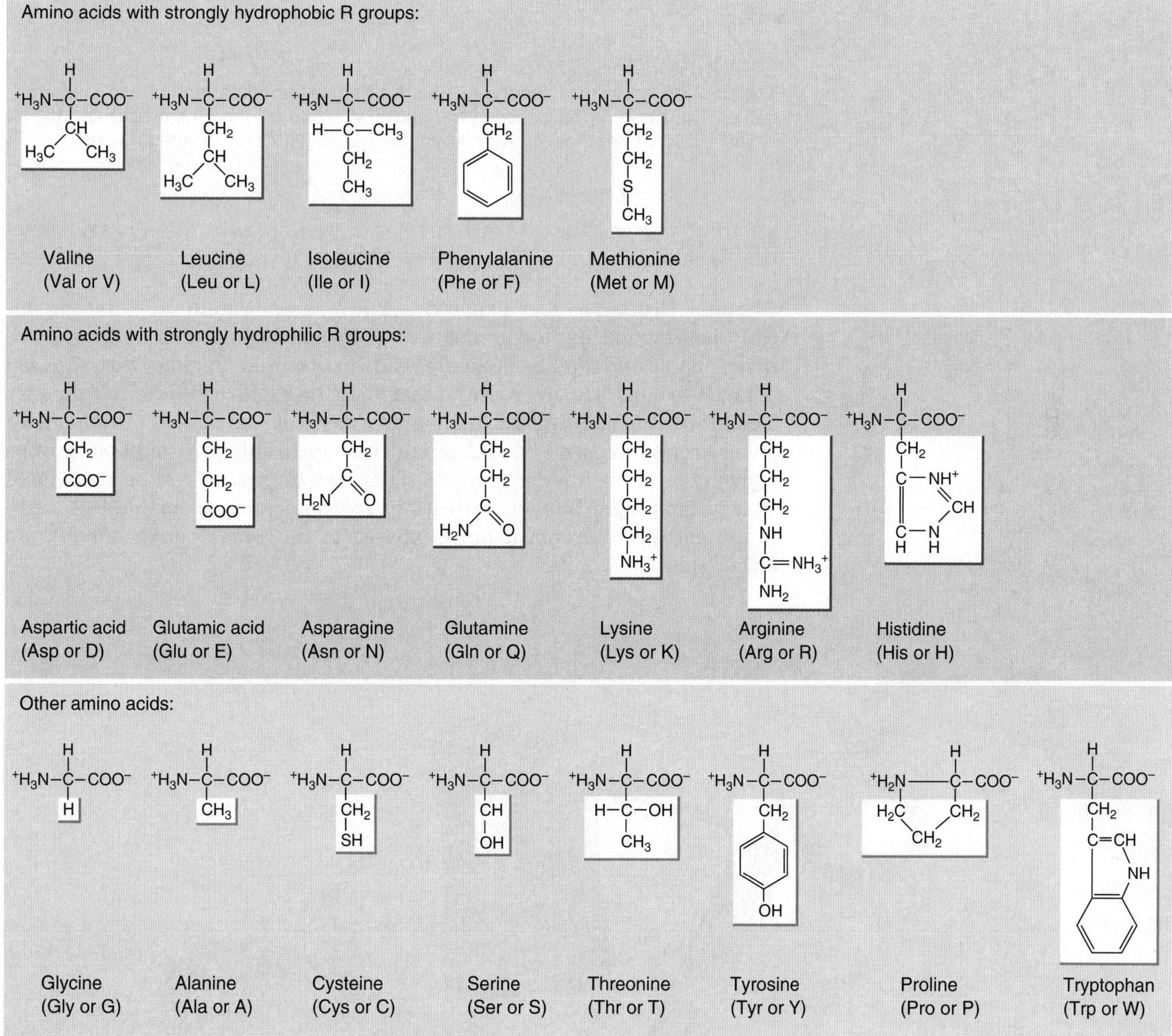

(b)

▲ FIGURE 3.4
Amino acids, the building blocks of protein. **(a)** Every amino acid has in common a central alpha carbon, an amino group (NH_3^+), and a carboxyl group (COO^-). Amino acids differ from one another based on a variable R group. **(b)** The 20 different amino acids that are used by neurons to make proteins. Noted in parentheses are the common abbreviations used for the various amino acids.

mation even further. In this way, proteins can bend, fold, and assume a complex three-dimensional shape. This shape is called *tertiary structure*. Finally, different polypeptide chains can bond together to form a larger molecule; such a protein is said to have *quaternary structure*. Each of the

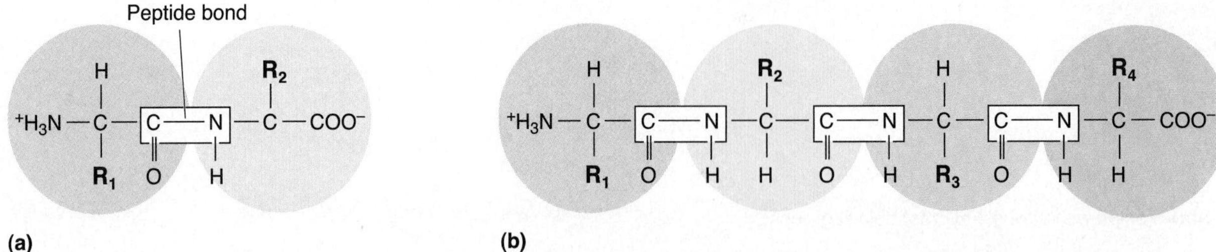

▲ FIGURE 3.5
The peptide bond and a polypeptide. **(a)** Peptide bonds attach amino acids together. The bond forms between the carboxyl group of one amino acid and the amino group of another. **(b)** A polypeptide is a single chain of amino acids.

different polypeptides contributing to a protein with quaternary structure is called a *subunit*.

Channel Proteins. The exposed surface of a protein may be chemically heterogeneous. Regions where nonpolar R groups are exposed are hydrophobic and tend to associate readily with lipid. Regions with exposed polar R groups are hydrophilic and tend to avoid a lipid environment. Therefore, it is not difficult to imagine classes of rod-shaped proteins with polar groups exposed at either end but with only hydrophobic groups showing on their middle surfaces. This type of protein can be suspended in a phospholipid bilayer, with its hydrophobic portion inside the membrane and its hydrophilic ends exposed to the watery environments on either side.

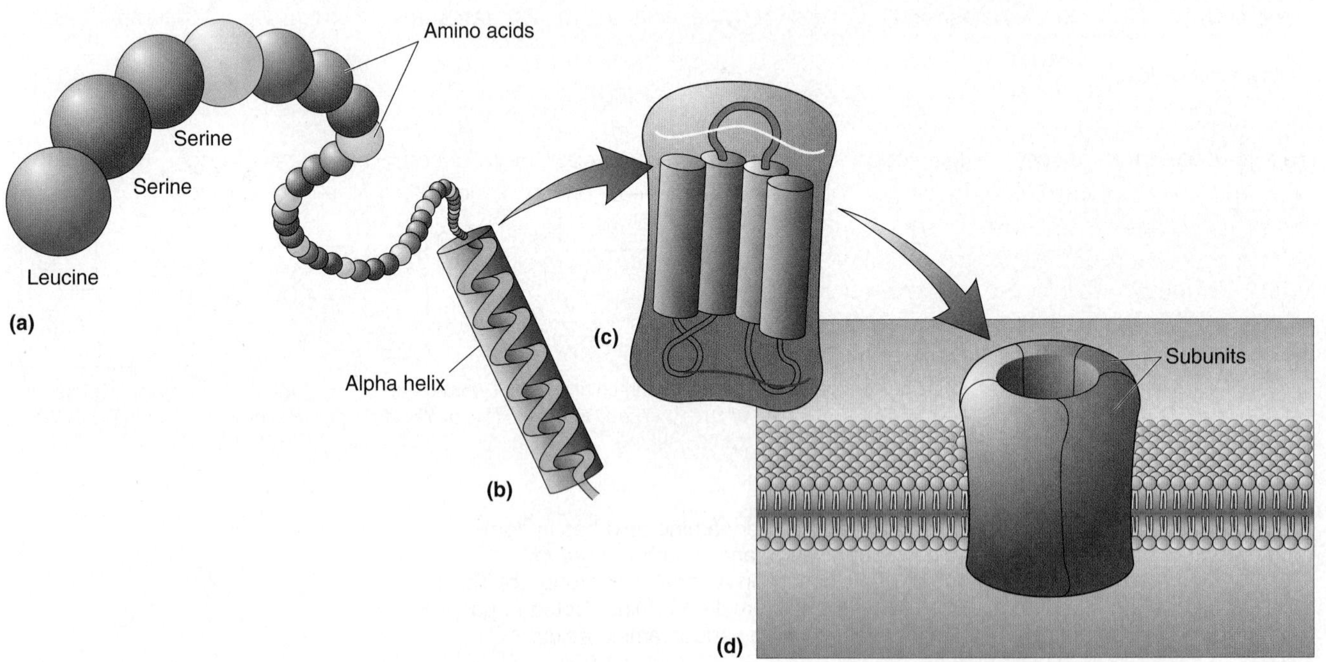

▲ FIGURE 3.6
Protein structure. **(a)** Primary structure: the sequence of amino acids in the polypeptide. **(b)** Secondary structure: coiling of a polypeptide into an alpha helix. **(c)** Tertiary structure: three-dimensional folding of a polypeptide. **(d)** Quaternary structure: different polypeptides bonded together to form a larger protein.

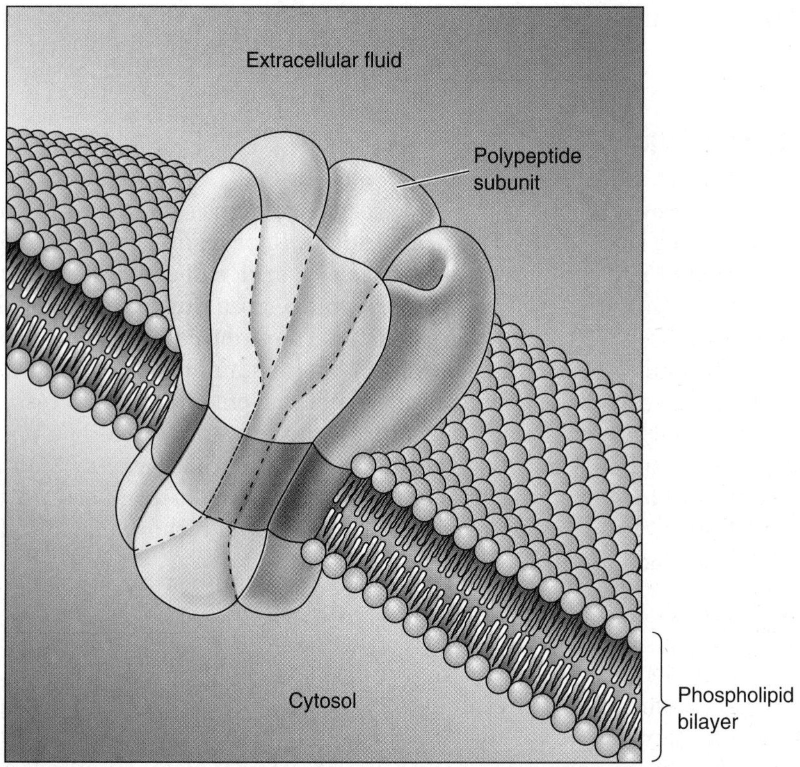

▲ FIGURE 3.7
A membrane ion channel. Ion channels consist of membrane-spanning proteins that assemble to form a pore. In this example, the channel protein has five polypeptide subunits. Each subunit has a hydrophobic surface region (shaded) that readily associates with the phospholipid bilayer.

Ion channels are made from just these sorts of membrane-spanning protein molecules. Typically, a functional channel across the membrane requires that four to six similar protein molecules assemble to form a pore between them (Figure 3.7). The subunit composition varies from one type of channel to the next, and this is what determines their different properties. One important property of most ion channels, specified by the diameter of the pore and the nature of the R groups lining it, is **ion selectivity**. Potassium channels are selectively permeable to K^+. Likewise, sodium channels are permeable almost exclusively to Na^+, calcium channels to Ca^{2+}, and so on. Another important property of many channels is **gating**. Channels with this property can be opened and closed—gated—by changes in the local microenvironment of the membrane.

You will learn much more about channels as you work your way through this book. *Understanding ion channels in the neuronal membrane is key to understanding cellular neurophysiology.*

Ion Pumps. In addition to those that form channels, other membrane-spanning proteins come together to form **ion pumps**. Recall from Chapter 2 that adenosine triphosphate (ATP) is the energy currency of cells. Ion pumps are enzymes that use the energy released by the breakdown of ATP to transport certain ions across the membrane. We will see that these pumps play a critical role in neuronal signaling

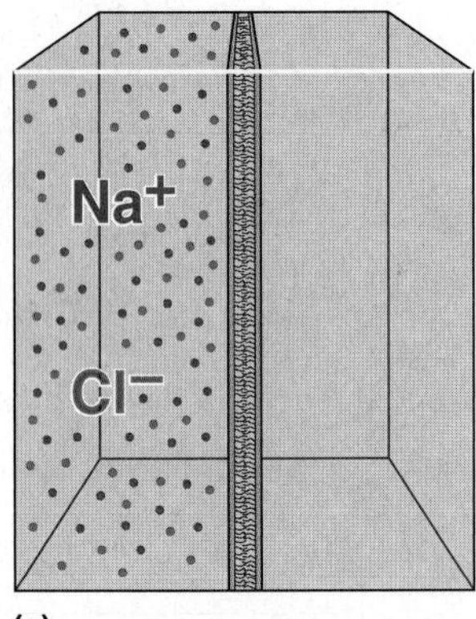

(a)

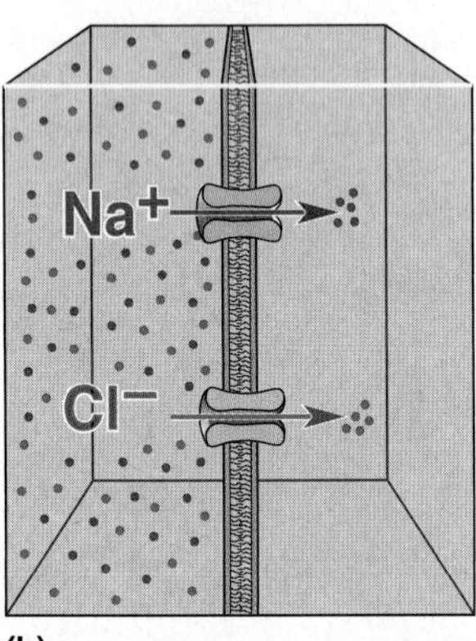

(b)

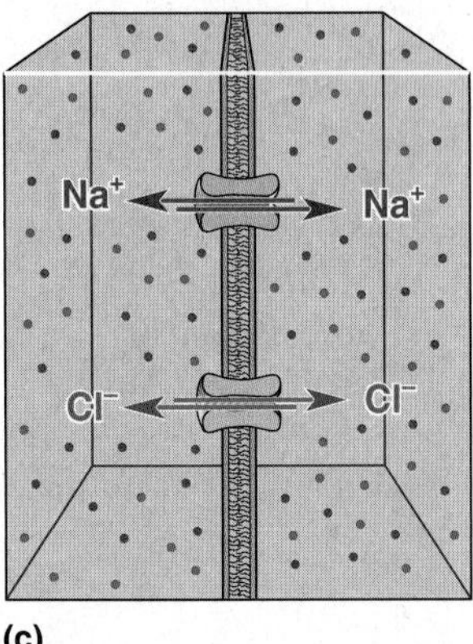

(c)

▲ FIGURE 3.8
Diffusion. **(a)** NaCl has been dissolved on the left side of an impermeable membrane. The sizes of the letters Na^+ and Cl^- indicate the relative concentrations of these ions. **(b)** Channels are inserted in the membrane that allow the passage of Na^+ and Cl^-. Because there is a large concentration gradient across the membrane, there will be a net movement of Na^+ and Cl^- from the region of high concentration to the region of low concentration, from left to right. **(c)** In the absence of any other factors, the net movement of Na^+ and Cl^- across the membrane ceases when they are equally distributed on both sides of the permeable membrane.

by transporting Na^+ and Ca^{2+} from the inside of the neuron to the outside.

THE MOVEMENT OF IONS

A channel across a membrane is like a bridge across a river (or, in the case of a gated channel, like a drawbridge): It provides a path to cross from one side to the other. The existence of a bridge does not necessarily compel us to cross it, however. The bridge we cross during our weekday commute may not be used on the weekend. The same can be said of membrane ion channels. The existence of an open channel in the membrane does not necessarily mean that there is a net movement of ions across the membrane. Such movement also requires that external forces be applied to drive them across. Because the functioning nervous system requires the movement of ions across the neuronal membrane, it is important that we understand these forces. Ionic movements through channels are influenced by two factors: diffusion and electricity.

Diffusion

Ions and molecules dissolved in water are in constant motion. This temperature-dependent, random movement tends to distribute the ions evenly throughout the solution. Therefore, there is a net movement of ions from regions of high concentration to regions of low concentration; this movement is called **diffusion**. For example, when a teaspoon of milk is added to a cup of hot tea, the milk tends to spread evenly through the tea solution. If the thermal energy of the solution is reduced, as with iced tea, the diffusion of milk molecules will take noticeably longer.

Although ions typically do not pass through a phospholipid bilayer directly, diffusion causes ions to be pushed through channels in the membrane. For example, if NaCl is dissolved in the fluid on one side of a permeable membrane (i.e., with channels that allow Na^+ and Cl^- passage), some Na^+ and Cl^- ions will cross until all are evenly distributed in the solutions on both sides (Figure 3.8). Like the milk molecules diffusing in the tea, the net movement is from the region of high concentration to the region of low concentration. (For a review of how concentrations are expressed, see Box 3.1.) Such a difference in concentration is called a **concentration gradient**. Thus, we say that ions will flow down a concentration gradient. The movement of ions across the membrane by diffusion, therefore, happens when (1) the membrane has channels permeable to the ions and (2) there is a concentration gradient across the membrane.

Electricity

In addition to diffusion down a concentration gradient, another way to induce a net movement of ions in a solution is to use an electrical field because ions are electrically charged particles. Consider the situation in Figure 3.9, where wires from the two terminals of a battery are placed in a solution containing dissolved NaCl. Remember, *opposite charges attract and like charges repel*. Consequently, there will be a net movement of Na^+ toward the negative terminal (the cathode) and of Cl^- toward the positive terminal (the anode). The movement of electrical charge is called **electrical current**, represented by the symbol I and measured

BOX 3.1 BRAIN FOOD

A Review of Moles and Molarity

Concentrations of substances are expressed as the number of molecules per liter of solution. The number of molecules is usually expressed in *moles*. One mole is 6.02×10^{23} molecules. A solution is said to be 1 Molar (M) if it has a concentration of 1 mole per liter. A 1 millimolar (mM) solution has 0.001 moles per liter. The abbreviation for concentration is a pair of brackets. Thus, we read [NaCl] = 1 mM as: "The concentration of the sodium chloride solution is 1 millimolar."

in units called *amperes* (amps). According to the convention established by Benjamin Franklin, current is defined as being positive in the direction of positive-charge movement. In this example, therefore, positive current flows in the direction of Na^+ movement, from the anode to the cathode.

Two important factors determine how much current will flow: electrical potential and electrical conductance. **Electrical potential**, also called **voltage**, is the force exerted on a charged particle; it reflects the difference in charge between the anode and the cathode. More current will flow as this difference is increased. Voltage is represented by the symbol V and is measured in units called volts. As an example, the difference in electrical potential between the terminals of a car battery is 12 volts; that is, the electrical potential at one terminal is 12 volts more positive than that at the other.

Electrical conductance is the relative ability of an electrical charge to migrate from one point to another. It is represented by the symbol g and measured in units called *siemens* (S). Conductance depends on the number of ions or electrons available to carry electrical charge, and the ease with which these charged particles can travel through space. A term that expresses the same property in a different way is **electrical resistance**, the relative inability of an electrical charge to migrate. It is represented by the symbol R and measured in units called *ohms* (Ω). Resistance is simply the inverse of conductance (i.e., R = 1/g).

There is a simple relationship between potential (V), conductance (g), and the amount of current (I) that will flow. This relationship, known as **Ohm's law**, may be written I = gV: Current is the product of the conductance and the potential difference. Notice that if the conductance is zero, no current will flow even when the potential difference is very large. Likewise, when the potential difference is zero, no current will flow even when the conductance is very large.

Consider the situation illustrated in Figure 3.10a, in which NaCl has been dissolved in equal concentrations on either side of a phospholipid bilayer. If we drop wires from the two terminals of a battery into the solution on either side, they generate a large potential difference across this membrane. No current will flow, however, because there are no channels to allow migration of Na^+ and Cl^- across the membrane; the conductance of the membrane is zero. Driving an ion across the membrane electrically, therefore, requires that (1) the membrane possesses channels permeable to that ion (to provide conductance)

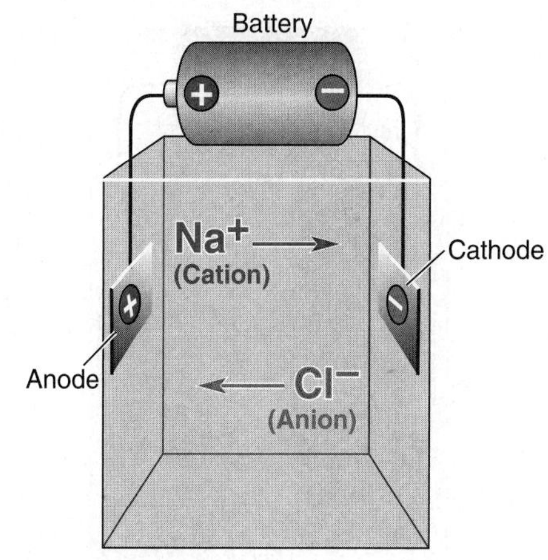

▲ FIGURE 3.9
The movement of ions influenced by an electrical field.

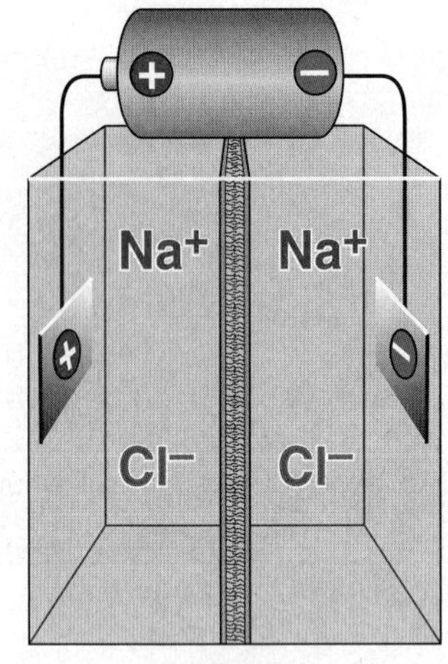

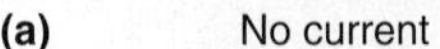

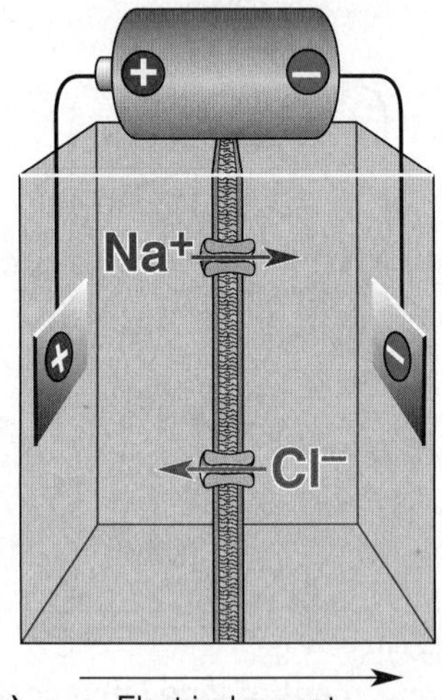

◀ FIGURE 3.10
Electrical current flow across a membrane. **(a)** A voltage applied across a phospholipid bilayer causes no electrical current because there are no channels to allow the passage of electrically charged ions from one side to the other; the conductance of the membrane is zero. **(b)** Inserting channels in the membrane allows ions to cross. Electrical current flows in the direction of cation movement (from left to right, in this example).

and (2) there is an electrical potential difference across the membrane (Figure 3.10b).

The stage is now set. We have electrically charged ions in solution on both sides of the neuronal membrane. Ions can cross the membrane only by way of protein channels. The protein channels can be highly selective for specific ions. The movement of any ion through its channel depends on the concentration gradient and the difference in electrical potential across the membrane. Now let's use this knowledge to explore the resting membrane potential.

THE IONIC BASIS OF THE RESTING MEMBRANE POTENTIAL

The **membrane potential** is the voltage across the neuronal membrane at any moment, represented by the symbol V_m. Sometimes V_m is "at rest"; at other times it is not (such as during an action potential). V_m can be measured by inserting a microelectrode into the cytosol. A typical **microelectrode** is a thin glass tube with an extremely fine tip (diameter 0.5 μm) that can penetrate the membrane of a neuron with minimal damage. It is filled with an electrically conductive salt solution and is connected to a device called a *voltmeter*. The voltmeter measures the electrical potential difference between the tip of this microelectrode and a wire placed outside the cell (Figure 3.11). This method reveals that electrical charge is unevenly distributed across the neuronal membrane. The inside of the neuronal membrane is electrically negative relative to

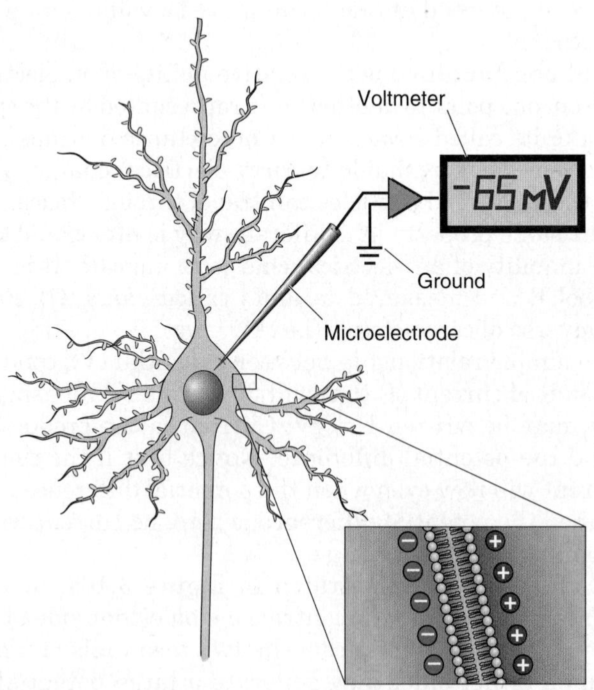

▲ FIGURE 3.11
Measuring the resting membrane potential. A voltmeter measures the difference in electrical potential between the tip of a microelectrode inside the cell and a wire placed in the extracellular fluid, conventionally called "ground" because it is electrically continuous with the earth. Typically, the inside of the neuron is about −65 mV with respect to the outside. This potential is caused by the uneven distribution of electrical charge across the membrane (enlargement).

the outside. This steady difference, the resting potential, is maintained whenever a neuron is not generating impulses.

The resting potential of a typical neuron is about −65 millivolts (1 mV = 0.001 volts). Stated another way, for a neuron at rest, V_m = −65 mV. This negative resting membrane potential of the neuron is an absolute requirement for a functioning nervous system. To understand the negative membrane potential, we look to the ions that are present and how they are distributed inside and outside the neuron.

Equilibrium Potentials

Consider a hypothetical cell in which the inside is separated from the outside by a pure phospholipid membrane with no proteins. Inside this cell, a concentrated potassium salt solution is dissolved, yielding K^+ and A^- anions (any molecules with a negative charge). Outside the cell is a solution with the same salt but diluted twentyfold with water. Although a large concentration gradient exists between the inside of the cell and the outside, there is no net movement of ions because the phospholipid bilayer, having no channel proteins, is impermeable to charged, hydrophilic atoms. Under these conditions, a microelectrode would record no potential difference between the inside and the outside of the cell. In other words, V_m would be equal to 0 mV because the ratio of K^+ to A^- on each side of the membrane equals 1; both solutions are electrically neutral (Figure 3.12a).

Consider how this situation would change if potassium channels were inserted into the phospholipid bilayer. Because of the selective permeability of these channels, K^+ would be free to pass across the membrane, but A^- would not. Initially, diffusion rules: K^+ ions pass through the channels out of the cell, down the steep concentration gradient. Because A^- is left behind, however, the inside of the cell membrane immediately begins to acquire a net negative charge, and an electrical potential difference is established across the membrane (Figure 3.12b). As the inside

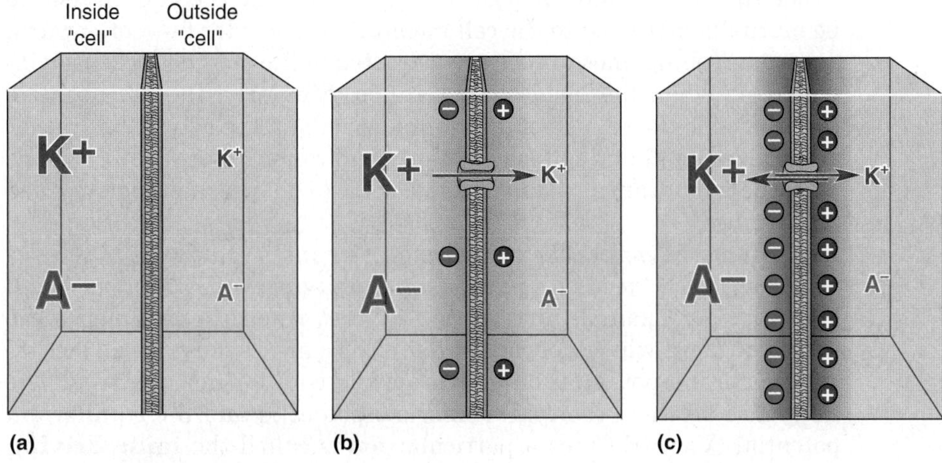

▲ FIGURE 3.12
Establishing equilibrium in a selectively permeable membrane. **(a)** An impermeable membrane separates two regions: one of high salt concentration (inside) and the other of low salt concentration (outside). The relative concentrations of potassium (K^+) and an impermeable anion (A^-) are represented by the sizes of the letters. **(b)** Inserting a channel that is selectively permeable to K^+ into the membrane initially results in a net movement of K^+ down their concentration gradient, from left to right. **(c)** A net accumulation of positive charge on the outside and negative charge on the inside retards the movement of positively charged K^+ from the inside to the outside. Equilibrium is established such that there is no net movement of ions across the membrane, leaving a charge difference between the two sides.

acquires more and more net negative charge, the electrical force starts to pull positively charged K^+ ions back through the channels into the cell. When a certain potential difference is reached, the electrical force pulling K^+ ions inside exactly counterbalances the force of diffusion pushing them out. Thus, an *equilibrium* state is reached in which the diffusional and electrical forces are equal and opposite, and the net movement of K^+ across the membrane ceases (Figure 3.12c). The electrical potential difference that exactly balances an ionic concentration gradient is called an **ionic equilibrium potential**, or simply **equilibrium potential**, and it is represented by the symbol E_{ion}. In this example, the equilibrium potential will be about −80 mV.

The example in Figure 3.12 demonstrates that generating a steady electrical potential difference across a membrane is a relatively simple matter. All that is required is an ionic concentration gradient and selective ionic permeability. Before moving on to the situation in real neurons, however, we can use this example to make four important points.

1. *Large changes in membrane potential are caused by minuscule changes in ionic concentrations.* In Figure 3.12, channels were inserted, and K^+ ions flowed out of the cell until the membrane potential went from 0 mV to the equilibrium potential of −80 mV. How much does this ionic redistribution affect the K^+ concentration on either side of the membrane? Not very much. For a cell with a 50 μm diameter, containing 100 mM K^+, it can be calculated that the concentration change required to take the membrane from 0 to −80 mV is about 0.00001 mM. That is, when the channels were inserted and the K^+ flowed out until equilibrium was reached, the internal K^+ concentration went from 100 to 99.99999 mM—a negligible drop in concentration.
2. *The net difference in electrical charge occurs at the inside and outside surfaces of the membrane.* Because the phospholipid bilayer is so thin (less than 5 nm thick), it is possible for ions on one side to interact electrostatically with ions on the other side. Thus, the negative charges inside the neuron and the positive charges outside the neuron tend to be mutually attracted to the cell membrane. Consider how, on a warm summer evening, mosquitoes are attracted to the outside face of a window pane when the inside lights are on. Similarly, the net negative charge inside the cell is not distributed evenly in the cytosol but rather is localized at the inner face of the membrane (Figure 3.13). In this way, the membrane is said to store electrical charge, a property called *capacitance*.
3. *Ions are driven across the membrane at a rate proportional to the difference between the membrane potential and the equilibrium potential.* Notice in our example in Figure 3.12 that when the channels were inserted, there was a net movement of K^+ only as long as the electrical membrane potential differed from the equilibrium potential. The difference between the real membrane potential and the equilibrium potential ($V_m - E_{ion}$) for a particular ion is called the **ionic driving force**. We'll talk more about this in Chapters 4 and 5 when we discuss the movement of ions across the membrane during the action potential and synaptic transmission.
4. *If the concentration difference across the membrane is known for an ion, the equilibrium potential can be calculated for that ion.* In our example in Figure 3.12, we assumed that K^+ was more concentrated inside the cell. Based on this knowledge, we were able to deduce that the equilibrium potential would be negative if the membrane were selectively permeable to K^+. Let's consider another example, in which Na^+ is more

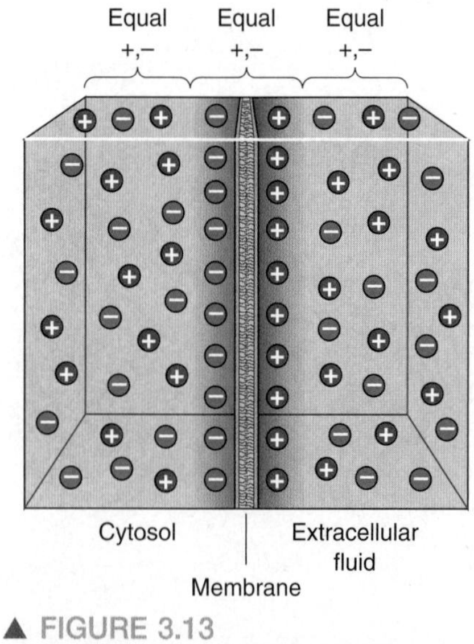

▲ FIGURE 3.13
The distribution of electrical charge across the membrane. The uneven charges inside and outside the neuron line up along the membrane because of electrostatic attraction across this very thin barrier. Notice that the bulk of the cytosol and extracellular fluid is electrically neutral.

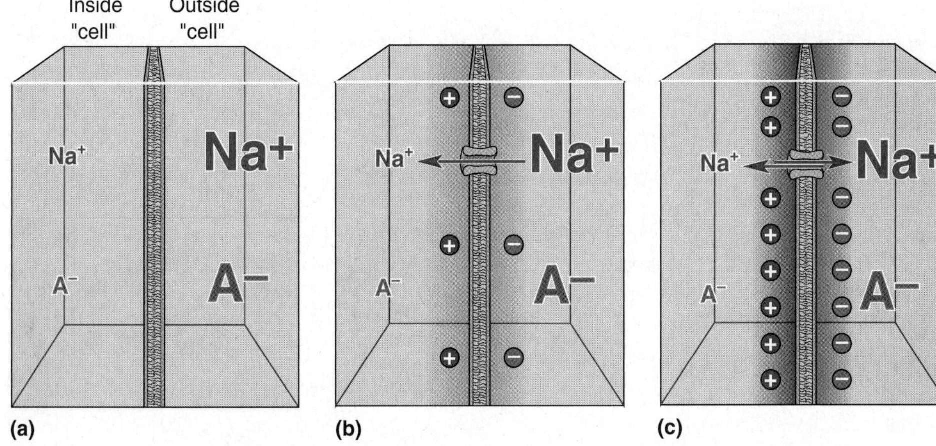

▲ FIGURE 3.14
Another example of establishing equilibrium in a selectively permeable membrane. **(a)** An impermeable membrane separates two regions: one of high salt concentration (outside) and the other of low salt concentration (inside). **(b)** Inserting a channel that is selectively permeable to Na^+ into the membrane initially results in a net movement of Na^+ down its concentration gradient, from right to left. **(c)** A net accumulation of positive charge on the inside and negative charge on the outside retards the movement of positively charged Na^+ from the outside to the inside. Equilibrium is established such that there is no net movement of ions across the membrane, leaving a charge difference between the two sides; in this case, the inside of the cell is positively charged with respect to the outside.

concentrated *outside* the cell (Figure 3.14). If the membrane contains sodium channels, Na^+ would flow down the concentration gradient *into* the cell. The entry of positively charged ions would cause the cytosol on the inner surface of the membrane to acquire a net positive charge. The positively charged interior of the cell membrane would now repel Na^+ ions, tending to push them back out through their channels. At a certain potential difference, the electrical force pushing Na^+ ions out would exactly counterbalance the force of diffusion pushing them in. In this example, the membrane potential at equilibrium would be positive on the inside.

The examples in Figures 3.12 and 3.14 illustrate that if we know the ionic concentration difference across the membrane, we can figure out the equilibrium potential for any ion. Prove it to yourself. Assume that Ca^{2+} is more concentrated on the outside of the cell and that the membrane is selectively permeable to Ca^{2+}. See if you can figure out whether the inside of the cell would be positive or negative at equilibrium. Try it again, assuming that the membrane is selectively permeable to Cl^-, and that Cl^- is more concentrated outside the cell. (Pay attention here; note the charge of the ion.)

The preceding examples show that each ion has its own equilibrium potential—the steady electrical potential that would occur if the membrane were permeable only to that ion. Thus, we can speak of the potassium equilibrium potential, E_K; the sodium equilibrium potential, E_{Na}; the calcium equilibrium potential, E_{Ca}; and so on. And knowing the electrical charge of the ion and the concentration difference across the membrane, we can easily deduce whether the inside of the cell would be positive or negative at equilibrium. In fact, the *exact* value of an equilibrium potential in mV can be calculated using an equation derived from the principles of physical chemistry, the **Nernst equation**, which takes into consider-

BOX 3.2 **BRAIN FOOD**

The Nernst Equation

The equilibrium potential for an ion can be calculated using the Nernst equation:

$$E_{ion} = 2.303 \frac{RT}{zF} \log \frac{[ion]_o}{[ion]_i}$$

where

E_{ion} = ionic equilibrium potential
R = gas constant
T = absolute temperature
z = charge of the ion
F = Faraday's constant
log = base 10 logarithm
$[ion]_o$ = ionic concentration outside the cell
$[ion]_i$ = ionic concentration inside the cell

The Nernst equation can be derived from the basic principles of physical chemistry. Let's see if we can make some sense of it.

Remember that equilibrium is the balance of two influences: diffusion, which pushes an ion down its concentration gradient, and electricity, which causes an ion to be attracted to opposite charges and repelled by like charges. Increasing the thermal energy of each particle increases diffusion and therefore increases the potential difference achieved at equilibrium. Thus, E_{ion} is proportional to T. On the other hand, increasing the electrical charge of each particle decreases the potential difference needed to balance diffusion. Therefore, E_{ion} is inversely proportional to the charge of the ion (z). We need not worry about R and F in the Nernst equation because they are constants.

At body temperature (37°C), the Nernst equation for the important ions—K^+, Na^+, Cl^-, and Ca^{2+} — simplifies to:

$$E_K = 61.54 \text{ mV} \log \frac{[K^+]_o}{[K^+]_i}$$

$$E_{Na} = 61.54 \text{ mV} \log \frac{[Na^+]_o}{[Na^+]_i}$$

$$E_{Cl} = -61.54 \text{ mV} \log \frac{[Cl^-]_o}{[Cl^-]_i}$$

$$E_{Ca} = 30.77 \text{ mV} \log \frac{[Ca^{2+}]_o}{[Ca^{2+}]_i}$$

Therefore, to calculate the equilibrium potential for a certain type of ion at body temperature, all we need to know is the ionic concentrations on either side of the membrane. For instance, in the example we used in Figure 3.12, we stipulated that K^+ was twentyfold more concentrated inside the cell:

If $\frac{[K^+]_o}{[K^+]} = \frac{1}{20}$

and $\log \frac{1}{20} = -1.3$

then $E_K = 61.54 \text{ mV} \times -1.3$

$= -80 \text{ mV}.$

Notice that there is no term in the Nernst equation for permeability or ionic conductance. Thus, calculating the value of E_{ion} does not require knowledge of the selectivity or the permeability of the membrane for the ion. There is an equilibrium potential for each ion in the intracellular and extracellular fluid. E_{ion} is the membrane potential that would just balance the ion's concentration gradient, so that no net ionic current would flow if the membrane were permeable to that ion.

ation the charge of the ion, the temperature, and the ratio of the external and internal ion concentrations. Using the Nernst equation, we can calculate the value of the equilibrium potential for any ion. For example, if K^+ is concentrated twentyfold on the inside of a cell, the Nernst equation tells us that $E_K = -80$ mV (Box 3.2).

The Distribution of Ions Across the Membrane

It should now be clear that the neuronal membrane potential depends on the ionic concentrations on both sides of the membrane. Approximate values for these concentrations appear in Figure 3.15. The important point is that *K^+ is more concentrated on the inside, and Na^+ and Ca^{2+} are more concentrated on the outside.*

How do these concentration gradients arise? Ionic concentration gradients are established by the actions of ion pumps in the neuronal membrane. Two ion pumps are especially important in cellular neurophysiology: the sodium-potassium pump and the calcium pump. The

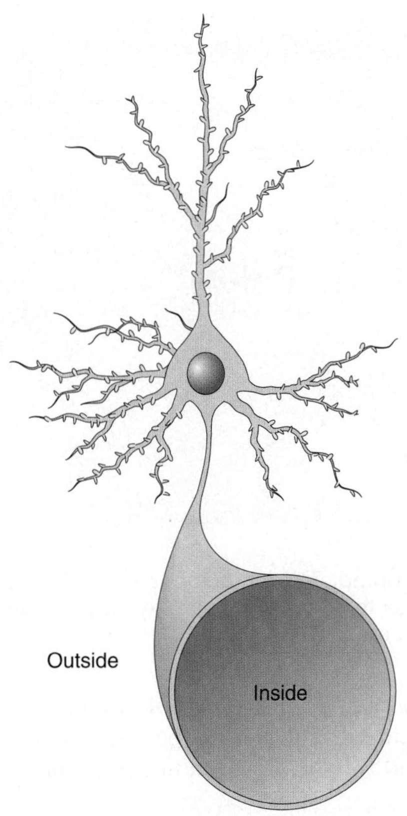

Ion	Concentration outside (in mM)	Concentration inside (in mM)	Ratio Out : In	E_{ion} (at 37∞C)
K^+	5	100	1 : 20	−80 mV
Na^+	150	15	10 : 1	62 mV
Ca^{2+}	2	0.0002	10,000 : 1	123 mV
Cl^-	150	13	11.5 : 1	−65 mV

▲ FIGURE 3.15

Approximate ion concentrations on either side of a neuronal membrane. E_{ion} is the membrane potential that would be achieved (at body temperature) if the membrane were selectively permeable to that ion.

sodium-potassium pump is an enzyme that breaks down ATP in the presence of internal Na^+. The chemical energy released by this reaction drives the pump, which exchanges internal Na^+ for external K^+. The actions of this pump ensure that K^+ is concentrated inside the neuron and that Na^+ is concentrated outside. Notice that the pump pushes these ions across the membrane against their concentration gradients (Figure 3.16). This work requires the expenditure of metabolic energy. Indeed, it has been estimated that the sodium-potassium pump expends as much as 70% of the total amount of ATP utilized by the brain.

The **calcium pump** is also an enzyme that actively transports Ca^{2+} out of the cytosol across the cell membrane. Additional mechanisms decrease intracellular [Ca^{2+}] to a very low level (0.0002 mM); these include intracellular calcium-binding proteins and organelles, such as mitochondria and types of endoplasmic reticulum, which sequester cytosolic calcium ions.

Ion pumps are the unsung heroes of cellular neurophysiology. They work in the background to ensure that the ionic concentration gradients

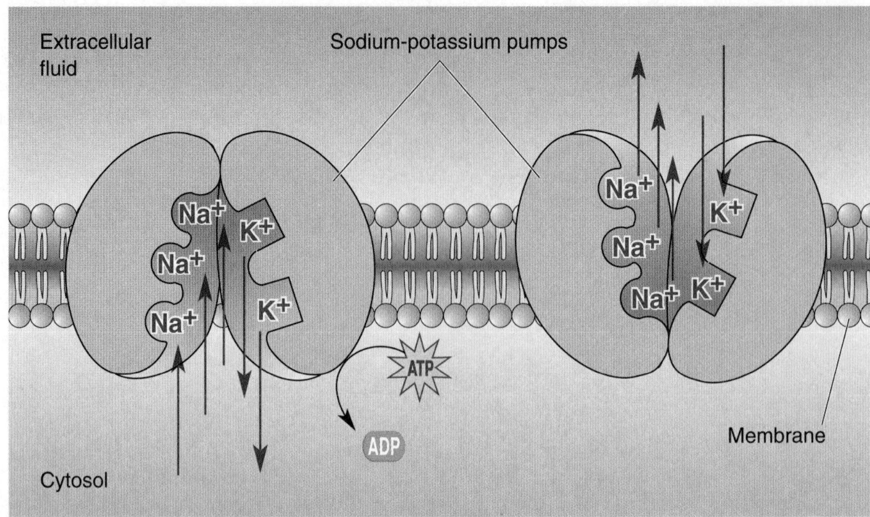

▲ FIGURE 3.16
The sodium-potassium pump. This ion pump is a membrane-associated protein that transports ions across the membrane against their concentration gradients at the expense of metabolic energy.

are established and maintained. These proteins may lack the glamour of a gated ion channel, but without ion pumps, the resting membrane potential would not exist and the brain would not function.

Relative Ion Permeabilities of the Membrane at Rest

The pumps establish ionic concentration gradients across the neuronal membrane. With knowledge of these ionic concentrations, we can use the Nernst equation to calculate equilibrium potentials for the different ions (see Figure 3.15). Remember, though, that an equilibrium potential for an ion is the membrane potential that would result if a membrane were *selectively permeable* to that ion alone. In reality, however, neurons are not permeable to only a single type of ion. How does that affect our understanding?

Let's consider a few scenarios involving K^+ and Na^+. If the membrane of a neuron were permeable only to K^+, the membrane potential would equal E_K, which, according to Figure 3.15, is −80 mV. On the other hand, if the membrane of a neuron were permeable only to Na^+, the membrane potential would equal E_{Na}, 62 mV. If the membrane were equally permeable to K^+ and Na^+, however, the resulting membrane potential would be some average of E_{Na} and E_K. What if the membrane were 40 times more permeable to K^+ than it is to Na^+? The membrane potential again would be between E_{Na} and E_K but much closer to E_K than to E_{Na}. This approximates the situation in real neurons. The actual resting membrane potential of −65 mV approaches, but does not reach, the potassium equilibrium potential of −80 mV. This difference arises because, although the membrane at rest is highly permeable to K^+, there is also a steady leak of Na^+ into the cell.

The resting membrane potential can be calculated using the **Goldman equation**, a mathematical formula that takes into consideration the relative permeability of the membrane to different ions. If we concern ourselves only with K^+ and Na^+, use the ionic concentrations in Figure 3.15, and assume that the resting membrane permeability to K^+ is fortyfold greater than it is to Na^+, then the Goldman equation predicts a resting membrane potential of −65 mV, the observed value (Box 3.3).

BOX 3.3 **BRAIN FOOD**

The Goldman Equation

If the membrane of a real neuron were permeable only to K^+, the resting membrane potential would equal E_K, about −80 mV. But it does not; the measured resting membrane potential of a typical neuron is about −65 mV. This discrepancy is explained because real neurons at rest are not exclusively permeable to K^+; there is also some Na^+ permeability. Stated another way, the *relative permeability* of the resting neuronal membrane is quite high to K^+ and low to Na^+. If the relative permeabilities are known, it is possible to calculate the membrane potential at equilibrium by using the Goldman equation. Thus, for a membrane permeable only to Na^+ and K^+ at 37° C:

$$V_m = 61.54 \text{ mV} \log \frac{P_K[K^+]_o + P_{Na}[Na^+]_o}{P_K[K^+]_i + P_{Na}[Na^+]_i}$$

where V_m is the membrane potential, P_K and P_{Na} are the relative permeabilities to K^+ and Na^+, respectively, and the other terms are the same as for the Nernst equation.

If the resting membrane ion permeability to K^+ is 40 times greater than it is to Na^+, then solving the Goldman equation using the concentrations in Figure 3.15 yields:

$$V_m = 61.54 \text{ mV} \log \frac{40\,(5) + 1\,(150)}{40\,(100) + 1\,(15)}$$

$$= 61.54 \text{ mV} \log \frac{350}{4015}$$

$$= -65 \text{ mV}$$

The Wide World of Potassium Channels. As we have seen, the selective permeability of potassium channels is a key determinant of the resting membrane potential and therefore of neuronal function. What is the molecular basis for this ionic selectivity? Selectivity for K^+ ions derives from the arrangement of amino acid residues that line the pore regions of the channels. It was a major breakthrough in 1987 when researchers succeeded in determining the amino acid sequences of a family of potassium channels in the fruit fly *Drosophila melanogaster*. While these insects may be annoying in the kitchen, they are extremely valuable in the lab because their genes can be studied and manipulated in ways that are not possible in mammals.

Normal flies, like humans, can be put to sleep with ether vapors. While conducting research on anesthetized insects, investigators discovered that flies of one mutant strain responded to the ether by shaking their legs, wings, and abdomen. This strain of fly was designated *Shaker*. Detailed studies soon explained the odd behavior by a defect in a particular type of potassium channel (Figure 3.17a). Using molecular biological techniques, it was possible to map the gene that was mutated in *Shaker*. Knowledge of the DNA sequence of what is now called the *Shake*r potassium channel enabled researchers to find the genes for other potassium channels based on sequence similarity. This analysis has revealed the existence of a very large number of different potassium channels, including those responsible for maintaining the resting membrane potential in neurons.

Most potassium channels have four subunits that are arranged like the staves of a barrel to form a pore (Figure 3.17b). Despite their diversity, the subunits of different potassium channels have common structural features that bestow selectivity for K^+. Of particular interest is a region called the *pore loop*, which contributes to the *selectivity filter* that makes the channel permeable mostly to K^+ (Figure 3.18).

In addition to flies, the deadly scorpion also made an important contribution to the discovery of the pore loop as the selectivity filter. In 1988, Brandeis University biologist Chris Miller and his student Roderick MacKinnon observed that scorpion toxin blocks potassium channels

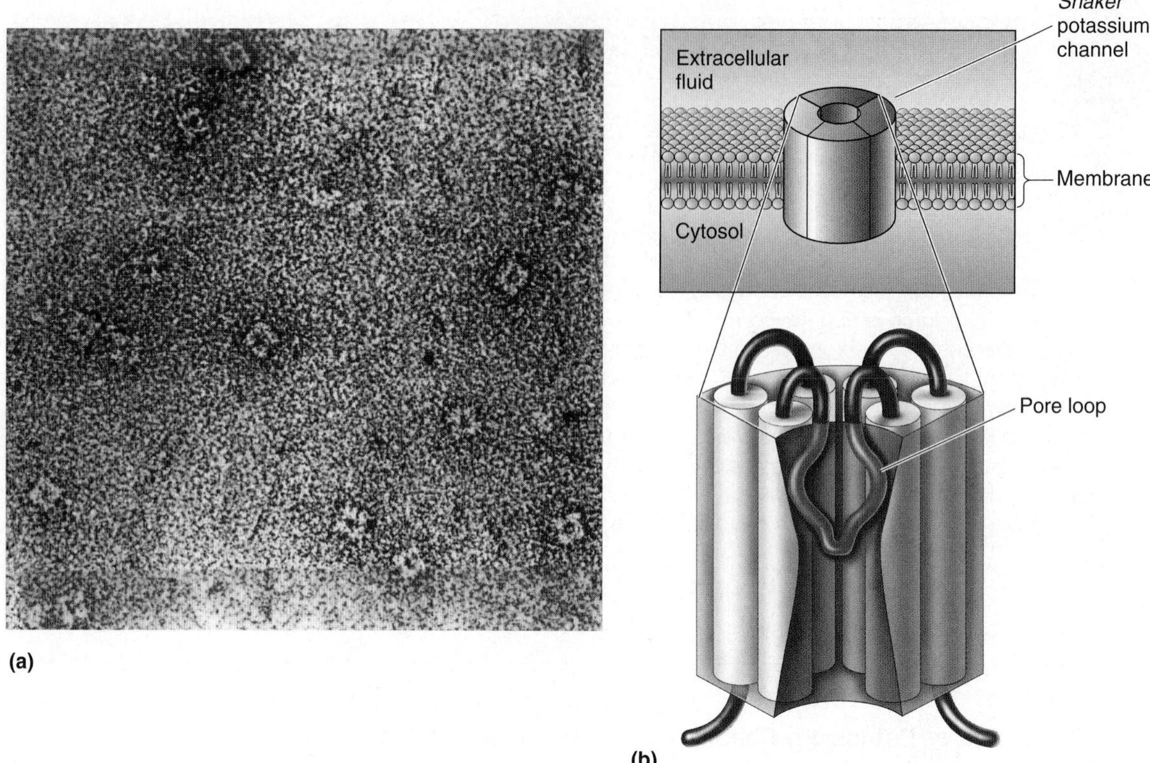

▲ FIGURE 3.17
The structure of a potassium channel. (a) *Shaker* potassium channels in the cell membrane of the fruit fly *Drosophila*, viewed from above with an electron microscope. (Source: Li et al., 1994; Fig. 2.) **(b)** The *Shaker* potassium channel has four subunits arranged like staves of a barrel to form a pore. Enlargement: The tertiary structure of the protein subunit contains a pore loop, a part of the polypeptide chain that makes a hairpin turn within the plane of the membrane. The pore loop is a critical part of the filter that makes the channel selectively permeable to K^+.

(and poisons its victims) by binding tightly to a site within the channel pore. They used the toxin to identify the precise stretch of amino acids that forms the inside walls and selectivity filter of the channel (Box 3.4). MacKinnon went on to solve the three-dimensional atomic structure of a potassium channel. This accomplishment revealed, at long last, the physical basis of ion selectivity and earned MacKinnon the 2003 Nobel Prize in Chemistry. It is now understood that mutations involving only a single amino acid in this region can severely disrupt neuronal function.

An example of this is seen in a strain of mice called *Weaver*. These animals have difficulty maintaining posture and moving normally. The defect has been traced to the mutation of a single amino acid in the pore loop of a potassium channel found in specific neurons of the cerebellum, a region of the brain important for motor coordination. As a consequence of the mutation, Na^+ as well as K^+ can pass through the channel. Increased sodium permeability causes the membrane potential of the neurons to become less negative, thus disrupting neuronal function. (Indeed, the absence of the normal negative membrane potential in these cells is believed to be the cause of their untimely death.) In recent years, it has become increasingly clear that many inherited neurological disorders in humans, such as certain forms of epilepsy, are explained by mutations of specific potassium channels.

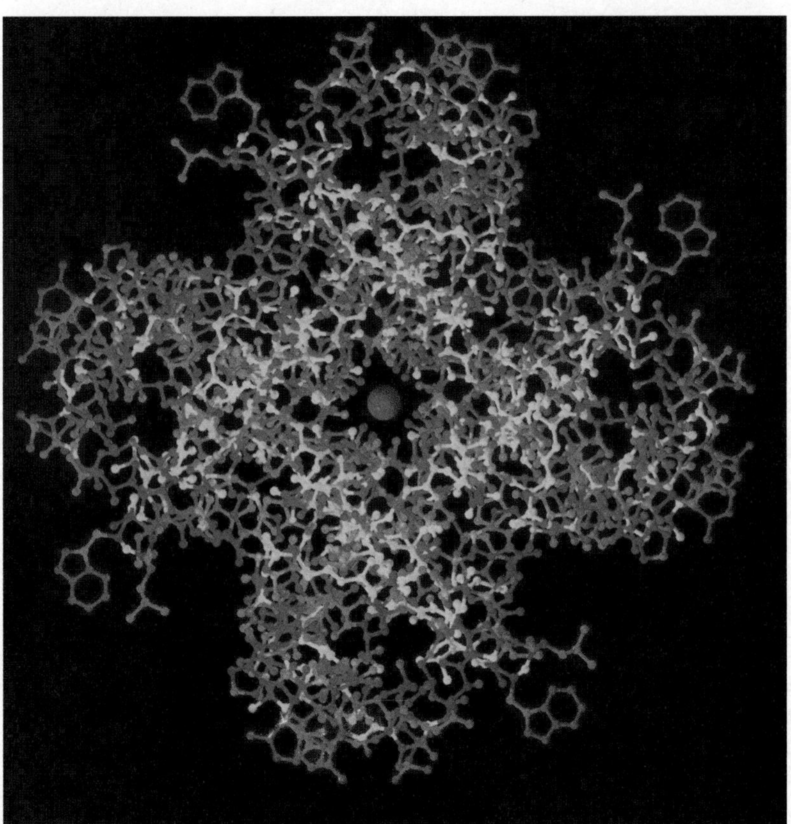

▲ FIGURE 3.18
A view of the potassium channel pore. The atomic structure of potassium-selective ion channels has recently been solved. Here we are looking into the pore from the outside in a three-dimensional model of the atomic structure. The red ball in the middle is a K^+. (Source: Doyle et al., 1998.)

The Importance of Regulating the External Potassium Concentration. Because the neuronal membrane at rest is mostly permeable to K^+, the membrane potential is close to E_K. Another consequence of high K^+ permeability is that the membrane potential is particularly sensitive to changes in the concentration of extracellular potassium. This relationship is shown in Figure 3.19. A tenfold change in the K^+ concentration outside the cell, $[K^+]_o$, from 5 to 50 mM, would change the membrane potential from −65 to −17 mV. A change in membrane potential from the normal resting value (−65 mV) to a less negative value is called a **depolarization** of the membrane. Therefore, *increasing extracellular potassium depolarizes neurons*.

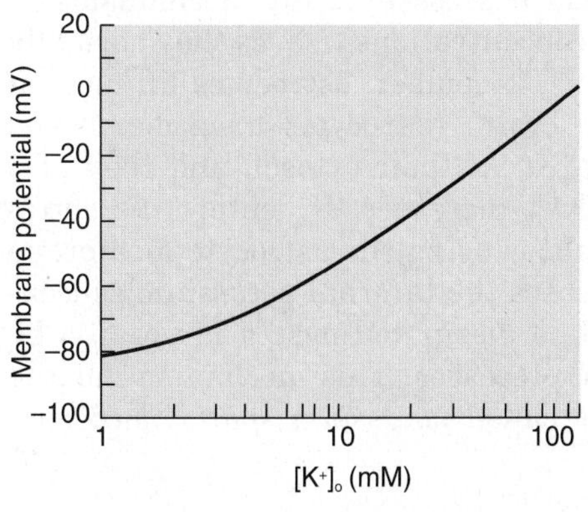

◀ FIGURE 3.19
The dependence of membrane potential on external potassium concentration. Because the neuronal membrane at rest is mostly permeable to potassium, a tenfold change in $[K^+]_o$, from 5 to 50 mM, causes a 48 mV depolarization of the membrane. This function was calculated using the Goldman equation (see Box 3.3).

BOX 3.4 PATH OF DISCOVERY

Feeling Around Inside Ion Channels in the Dark

by Chris Miller

For me, the practice of scientific discovery has always been tightly linked to play. The self-indulgent pleasure of just fiddling around with a problem is what motivated the early stages of every research project I've ever engaged in. Only later comes the intense itch scratching, scholarship, and sweat needed to attack—and sometimes solve—the puzzles presented by nature. The sandbox I've been playing in for the past 40 years contains what are to me the most fascinating of toys: ion channels, the membrane-spanning proteins that literally *make* the electrical signals of neurons, breathing life into the nervous system. To the extent that the brain is a computer—an inaccurate but evocative analogy—the ion channels are the transistors. In response to biological dictates, these tiny proteinaceous pores form diffusion pathways for ions such as Na^+, K^+, Ca^{2+}, H^+, and Cl^-, which carry electrical charge across membranes, thereby generating, propagating, and regulating cell voltage signals. I fell in love with these proteins long ago when I accidentally stumbled upon an unexpected K^+ channel in experiments initially aimed at capturing a completely different sort of beast, a Ca^{2+}-activated enzyme, and over the years that love has only deepened as I've wandered around in a teeming electrophysiological zoo housing many species of ion channel proteins.

An undergraduate background in physics and subsequent experience as a high school math teacher delivered me in the 1970s to graduate school, post-doctoral training, and my own lab at Brandeis with no formal preparation in (and precious little knowledge of) neurobiology or electrophysiology. Picking up bits and pieces of these subjects from reading the literature and osmosing them from my surroundings, I became increasingly fascinated by how ion channels, at that time only just nailed down as proteins, could do their job of producing bioelectricity; in parallel, I grew increasingly horrified by what struck me as the overwhelming complexity of living cells and the ambiguity in molecular interpretation that would inevitably accompany experiments done exclusively on cellular membranes. This combination of fascination and horror provoked my attraction to simplified "artificial membranes" of defined composition, developed by Paul Mueller in the 1960s, with which to follow the electrical activities of ion channels isolated from their complex cellular homes. I worked out a method for inserting single channel molecules from excitable cells into these chemically controllable membranes and used it to record single K^+ channels at a time when card-carrying neurobiologists were beginning to observe single channels in native excitable membranes with the then-new cellular patch-recording methods. I confess that my early technique-building experiments were just play. To watch and control individual protein molecules dancing electrically before my eyes in real time was—and still is—an indescribable thrill, regardless of the particular tasks the channels carry out for the cell.

The sensitivity of the membrane potential to $[K^+]_o$ has led to the evolution of mechanisms that tightly regulate extracellular potassium concentrations in the brain. One of these is the **blood-brain barrier**, a specialization of the walls of brain capillaries that limits the movement of potassium (and other bloodborne substances) into the extracellular fluid of the brain.

Glia, particularly astrocytes, also possess efficient mechanisms to take up extracellular K^+ whenever concentrations rise, as they normally do during periods of neural activity. Remember, astrocytes fill most of the space between neurons in the brain. Astrocytes have membrane potassium pumps that concentrate K^+ in their cytosol, and they also have potassium channels. When $[K^+]_o$ increases, K^+ enters the astrocyte through the potassium channels, causing the astrocyte membrane to depolarize. The entry of K^+ increases the internal potassium concentration, $[K^+]_i$, which is believed to be dissipated over a large area by the extensive network of astrocytic processes. This mechanism for the regulation of $[K^+]_o$ by astrocytes is called *potassium spatial buffering* (Figure 3.20).

Eventually, this play led me to compelling problems that could be advantageously attacked with this reductionist approach. By the mid-1980s, my lab was home to a collection of supremely talented post-docs—Gary Yellen, Rod MacKinnon, and Jacques Neyton among them—going after the remarkable ion selectivity of various K^+ channels: How do they tell the difference between ions as similar as K^+ and Na^+, as they must do if neurons are to fire action potentials, and if we are to think, feel, and act? Having stumbled, while purposelessly fooling around with natural neurotoxins, on a scorpion venom peptide that blocks K^+ channels, we used the power of single-channel analysis to show that this toxin works by plugging up the protein's K^+-selective pore, just like a cork in a bottle (Figure A). In 1988, Rod took our toxin peptide to a Cold Spring Harbor laboratory course he'd signed up for to learn how to express ion channels by recombinant DNA methods. There he made a key discovery: that the toxin also blocks *Shaker*, the first genetically manipulable K^+ channel, cloned the previous year in the lab of Lily and Yuh-Nung Jan. This chance finding led us, by making specific mutations, to a localized region in the channel's amino acid sequence that forms the outer entryway of the K^+ selective pore, a result immediately applicable to the entire family of voltage-dependent K^+, Na^+, and Ca^{2+} channels. A few years later, Rod and Gary, as newly hatched independent investigators, collaboratively homed in on these pore sequences to find the ion-selectivity hot spots, a result that propelled Rod, 7 years later, to bag the first X-ray crystal structure of a K^+ channel and to begin a whole new "structural era" in ion channel studies.

Looking back at my wrestling matches with ion channels, it is clear that the greatest joy I've derived from this endeavor has arisen from seeing—and being surprised by—new and unexpected elements of beauty and coherence in the natural world. This feeling was described by the great theoretical physicist Richard Feynman who, in a riposte to a W.H. Auden poem that dismisses scientific motivation as merely utilitarian, asserted that research scientists, like poets, are driven mainly by aesthetic forces: "We want knowledge so we can love Nature more."

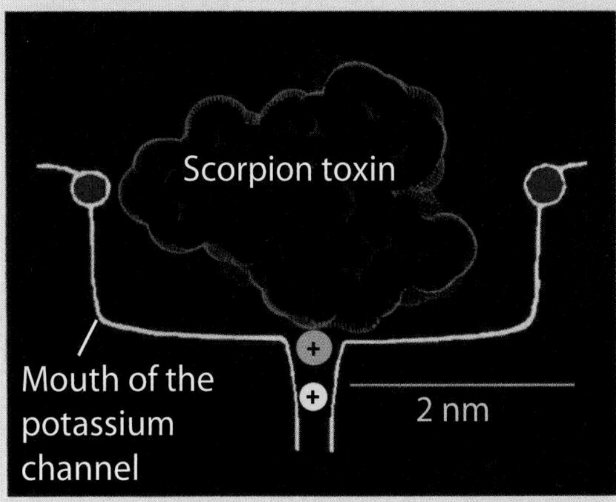

Figure A
The extracellular opening of a K^+ channel with bound scorpion toxin envisioned indirectly in the "pre-structural" days by probing the channel with the toxin of known structure. Points of interaction: site on channel that makes contact with toxin (dark blue circles), key lysine residue on toxin that intrudes into the narrow pore (pale blue circle with +), a K^+ displaced downward into the pore by binding of toxin (yellow circle with +). The yellow scale bar represents 2 nm. (Source: Adapted from Goldstein et al. 1994. *Neuron* 12:1377–1388.)

It is important to recognize that not all excitable cells are protected from increases in potassium. Muscle cells, for example, do not have equivalents to the blood-brain barrier or glial buffering mechanisms. Consequently, although the brain is relatively protected, elevations of $[K^+]$ in the blood can still have serious consequences on body physiology (Box 3.5).

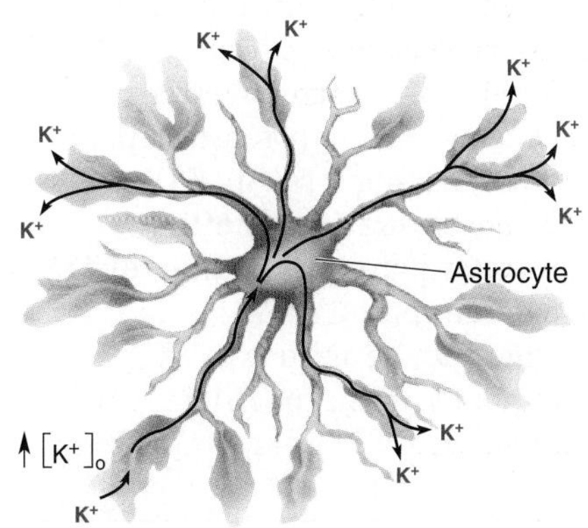

◀ FIGURE 3.20
Potassium spatial buffering by astrocytes. When brain $[K^+]_o$ increases as a result of local neural activity, K^+ enters astrocytes via membrane channels. The extensive network of astrocytic processes helps dissipate the K^+ over a large area.

BOX 3.5 OF SPECIAL INTEREST

Death by Lethal Injection

On June 4, 1990, Dr. Jack Kevorkian shocked the medical profession by assisting in the suicide of Janet Adkins. Adkins, a 54-year-old, happily married mother of three, had been diagnosed with Alzheimer's disease, a progressive brain disorder that always results in senile dementia and death. Mrs. Adkins had been a member of the Hemlock Society, which advocates euthanasia as an alternative to death by terminal illness. Dr. Kevorkian agreed to help Mrs. Adkins take her own life. In the back of a 1968 Volkswagen van at a campsite in Oakland County, Michigan, she was hooked to an intravenous line that infused a harmless saline solution. To choose death, Mrs. Adkins switched the solution to one that contained an anesthetic solution, followed automatically by potassium chloride. The anesthetic caused Mrs. Adkins to become unconscious by suppressing the activity of neurons in part of the brain called the *reticular formation*. Cardiac arrest and death were then caused by the KCl injection. The ionic basis of the resting membrane potential explains why the heart stopped beating.

Recall that the proper functioning of excitable cells (including those of cardiac muscle) requires that their membranes be maintained at the appropriate resting potential whenever they are not generating impulses. The negative resting potential is a result of selective ionic permeability to K^+ and to the metabolic pumps that concentrate potassium inside the cell. However, as Figure 3.19 shows, membrane potential is very sensitive to changes in the extracellular concentration of potassium. A tenfold rise in extracellular K^+ would severely diminish the resting potential. Although neurons in the brain are somewhat protected from large changes in $[K^+]_o$, other excitable cells in the body, such as muscle cells, are not. Without negative resting potentials, cardiac muscle cells can no longer generate the impulses that lead to contraction, and the heart immediately stops beating. Intravenous potassium chloride is, therefore, a lethal injection.

CONCLUDING REMARKS

We have now explored the resting membrane potential. The activity of the sodium-potassium pump produces and maintains a large K^+ concentration gradient across the membrane. The neuronal membrane at rest is highly permeable to K^+, owing to the presence of membrane potassium channels. The movement of K^+ ions across the membrane, down their concentration gradient, leaves the inside of the neuronal membrane negatively charged.

The electrical potential difference across the membrane can be thought of as a battery whose charge is maintained by the work of the ion pumps. In the next chapter, we see how this battery runs our brain.

KEY TERMS

Introduction
action potential (p. 57)
excitable membrane (p. 57)
resting membrane potential (p. 57)

The Cast of Chemicals
ion (p. 58)
cation (p. 59)
anion (p. 59)
phospholipid bilayer (p. 59)
peptide bond (p. 60)
polypeptide (p. 60)
ion channel (p. 63)
ion selectivity (p. 63)
gating (p. 63)
ion pump (p. 63)

The Movement of Ions
diffusion (p. 64)
concentration gradient (p. 64)
electrical current (p. 64)
electrical potential (p. 65)
voltage (p. 65)
electrical conductance (p. 65)
electrical resistance (p. 65)
Ohm's law (p. 65)

The Ionic Basis of the Resting Membrane Potential
membrane potential (p. 66)
microelectrode (p. 66)
ionic equilibrium potential (equilibrium potential) (p. 68)
ionic driving force (p. 68)
Nernst equation (p. 69)
sodium-potassium pump (p. 71)
calcium pump (p. 71)
Goldman equation (p. 72)
depolarization (p. 75)
blood-brain barrier (p. 76)

REVIEW QUESTIONS

1. What two functions do proteins in the neuronal membrane perform to establish and maintain the resting membrane potential?
2. On which side of the neuronal membrane are Na^+ ions more abundant?
3. When the membrane is at the potassium equilibrium potential, in which direction (in or out) is there a net movement of potassium ions?
4. There is a much greater K^+ concentration inside the cell than outside. Why, then, is the resting membrane potential negative?
5. When the brain is deprived of oxygen, the mitochondria within neurons cease producing ATP. What effect would this have on the membrane potential? Why?

FURTHER READING

Hille B. 2001. *Ionic Channels of Excitable Membranes,* 3rd ed. Sunderland, MA: Sinauer.

MacKinnon R. 2003. Potassium channels. *Federation of European Biochemical Societies Letters* 555:62–65.

Nicholls J, Martin AR, Fuchs PA, Brown DA, Diamond ME, Weisblat D. 2011. *From Neuron to Brain*, 5th ed. Sunderland, MA: Sinauer.

Somjen GG. 2004. *Ions in the Brain: Normal Function, Seizures, and Stroke*. New York: Oxford University Press.

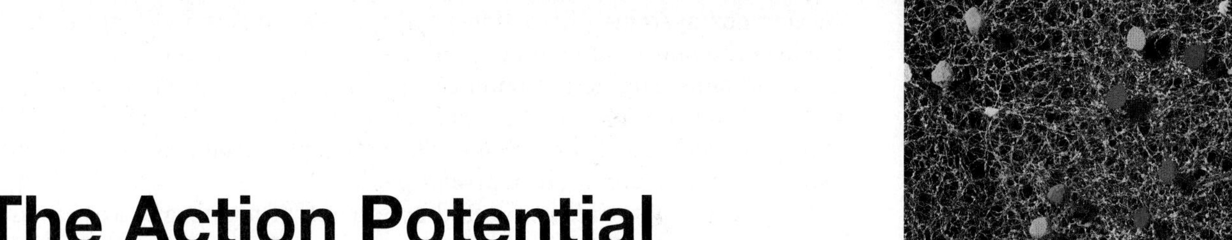

CHAPTER FOUR

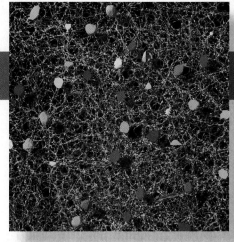

The Action Potential

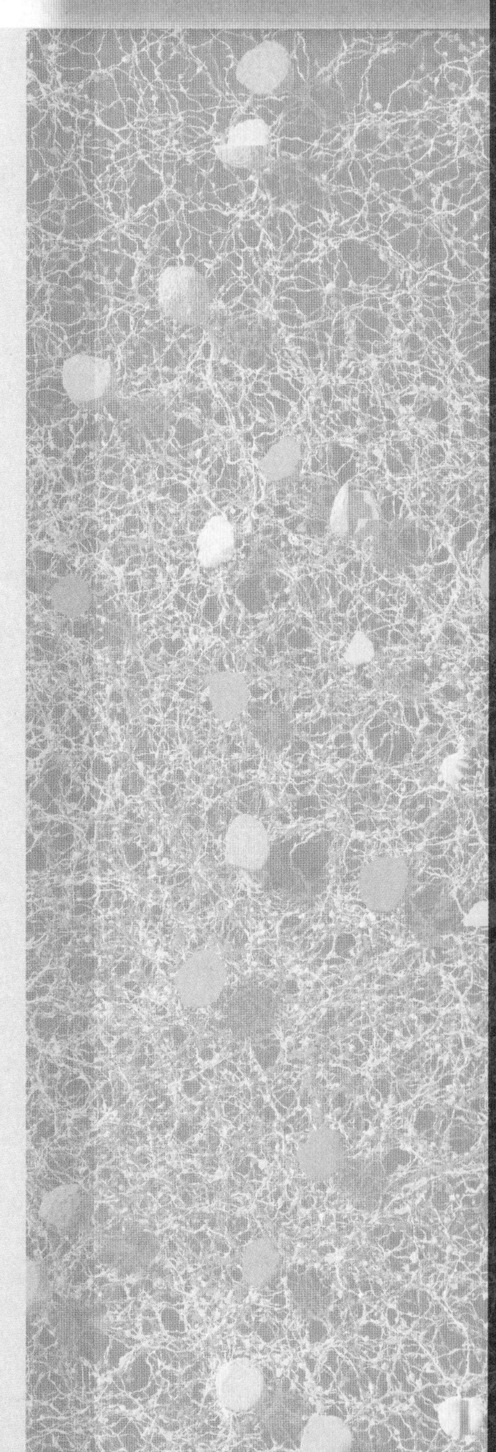

INTRODUCTION

Now we come to the signal that conveys information over distances in the nervous system—the action potential. As we saw in Chapter 3, the inside of the neuronal membrane at rest is negatively charged in relation to the outside. The action potential is a rapid reversal of this situation such that, for an instant, the inside of the membrane becomes positively charged in relation to the outside. The action potential is also often called a *spike*, a *nerve impulse*, or a *discharge*.

The action potentials generated by a patch of membrane are all similar in size and duration, and they do not diminish as they are conducted down the axon. Keep in mind the big picture: The *frequency* and *pattern* of action potentials constitute the code used by neurons to transfer information from one location to another. In this chapter, we discuss the mechanisms that are responsible for the action potential and how it propagates down the axonal membrane.

PROPERTIES OF THE ACTION POTENTIAL

Action potentials have certain universal properties, features that are shared by axons in the nervous systems of every animal, from a squid to a college student. Let's begin by exploring some of these properties. What does the action potential look like? How is it initiated? How rapidly can a neuron generate action potentials?

The Ups and Downs of an Action Potential

In Chapter 3, we saw that the membrane potential, V_m, can be determined by inserting a microelectrode in the cell. A voltmeter is used to measure the electrical potential difference between the tip of this intracellular microelectrode and another placed outside the cell. When the neuronal membrane is at rest, the voltmeter reads a steady potential difference of about −65 mV. During the action potential, however, the membrane potential briefly becomes positive. Because this occurs so rapidly—100 times faster than the blink of an eye—a special type of voltmeter, called an *oscilloscope*, is used to study action potentials. The oscilloscope records the voltage as it changes over time (Box 4.1).

An action potential, as it would appear on the display of an oscilloscope, is shown in Figure 4.1. This graph represents a plot of membrane potential versus time. Notice that the action potential has certain identifiable parts. The first part, called the **rising phase**, is characterized by a rapid depolarization of the membrane. This change in membrane potential continues until V_m reaches a peak value of about 40 mV. The part of the action potential where the inside of the neuron is positively charged with respect to the outside is called the **overshoot**. The **falling phase** of the action potential is a rapid repolarization until the inside of the membrane is actually more negative than the resting potential. This last part of the falling phase is called the **undershoot**, or **after-hyperpolarization**. Finally, there is a gradual restoration of the resting potential. From beginning to end, the action potential lasts about 2 milliseconds (msec).

The Generation of an Action Potential

In Chapter 3, we said that breaking of the skin by a thumbtack was sufficient to generate action potentials in a sensory nerve. Let's continue that example to see how an action potential begins.

BOX 4.1 BRAIN FOOD

Methods of Recording Action Potentials

Methods for studying nerve impulses may be broadly divided into two types: intracellular and extracellular (Figure A). *Intracellular recording* requires impaling the neuron or axon with a microelectrode. The small size of most neurons makes this method challenging, which is why so many early studies of action potentials were performed on the neurons of invertebrates, which can be 50–100 times larger than mammalian neurons. Fortunately, recent technical advances have made even the smallest vertebrate neurons accessible to intracellular recording methods, and these studies have confirmed that much of what was learned in invertebrates is directly applicable to humans.

The goal of intracellular recording is simple: to measure the potential difference between the tip of the intracellular electrode and another electrode placed in the solution bathing the neuron (electrically continuous with the earth, and thus called ground). The intracellular electrode is filled with a concentrated salt solution (often KCl) having a high electrical conductivity. The electrode is connected to an amplifier that compares the potential difference between this electrode and ground. This potential difference can be displayed using an oscilloscope. Early oscilloscopes worked by sweeping a beam of electrons from left to right across a phosphor screen. Vertical deflections of this beam show changes in voltage. Oscilloscopes today take a digital record of voltage across time, but the principle is the same. It is really just a sophisticated voltmeter that can record rapid changes in voltage (such as an action potential).

As we shall see, the action potential is characterized by a sequence of ionic movements across the neuronal membrane. These electrical currents can be detected, without impaling the neuron, by placing an electrode near the membrane. This is the principle behind *extracellular recording*. Again, we measure the potential difference between the tip of the recording electrode and ground. The electrode can be a fine glass capillary filled with a salt solution, but it is often simply a thin insulated metal wire. Normally, in the absence of neural activity, the potential difference between the extracellular recording electrode and ground is zero. However, when the action potential arrives at the recording position, positive charges flow away from the recording electrode into the neuron. Then, as the action potential passes by, positive charges flow out across the membrane toward the recording electrode. Thus, the extracellular action potential is characterized by a brief, alternating voltage difference between the recording electrode and ground. (Notice the different scale of the voltage changes produced by the action potential recorded with intracellular and extracellular recordings.) These changes in voltage can be seen using an oscilloscope, but they can also be heard by connecting the output of the amplifier to a loudspeaker. Each impulse makes a distinctive "pop" sound. Indeed, recording the activity of an active sensory nerve sounds just like popping popcorn.

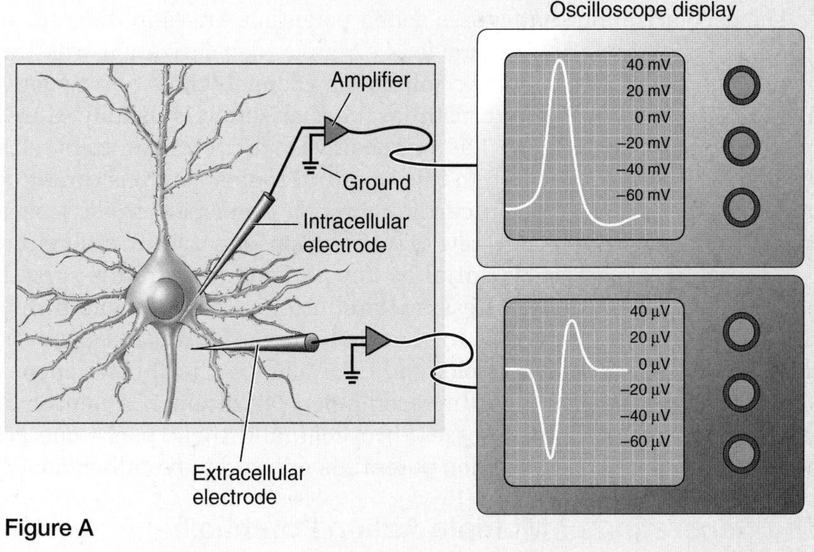

Figure A

The perception of sharp pain when a thumbtack enters your foot is caused by the generation of action potentials in certain nerve fibers in the skin. (We'll learn more about pain in Chapter 12.) The membrane of these fibers has a type of gated sodium channel that opens when the nerve ending is stretched. The chain of events therefore begins this way: (1) The thumbtack enters the skin, (2) the membrane of the nerve fibers in the skin is stretched, (3) and Na^+-permeable channels open. Because of the large concentration gradient and the negative charge of the inside of the membrane, Na^+ crosses the membrane through these channels. The entry of Na^+ depolarizes the membrane; that is, the cytoplasmic (inside) surface of the membrane becomes less negative. If this depolarization, called

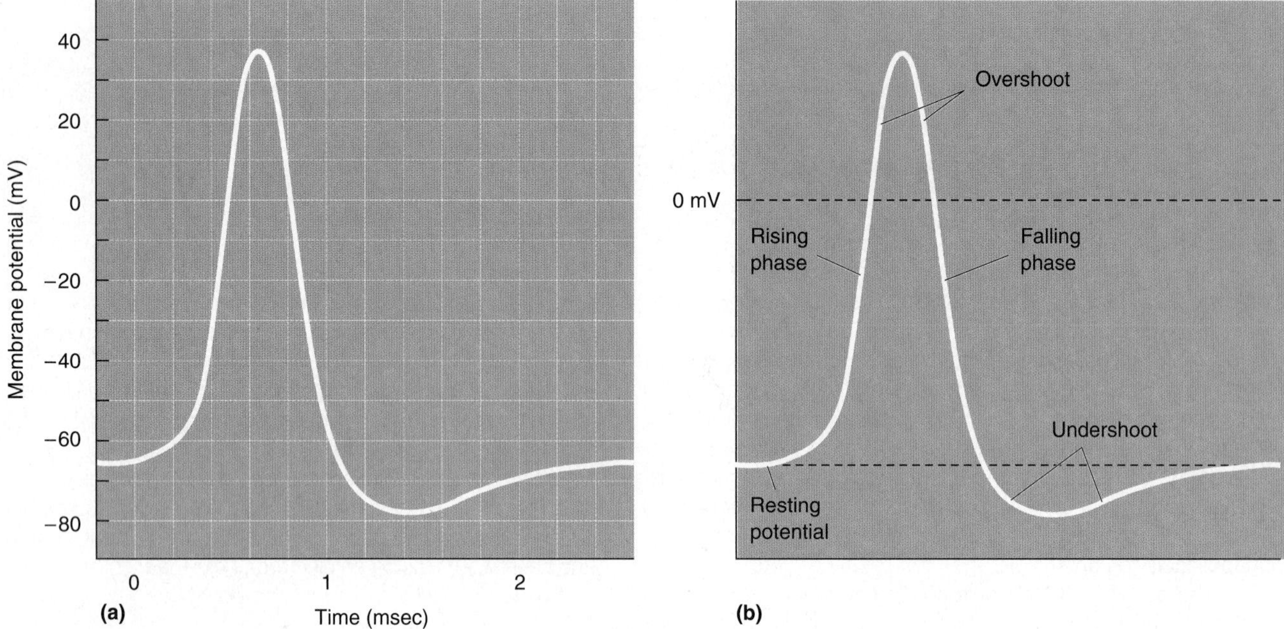

▲ FIGURE 4.1
An action potential. (a) An action potential displayed by an oscilloscope. (b) The parts of an action potential.

a *generator potential,* achieves a critical level, the membrane will generate an action potential. The critical level of depolarization that must be reached in order to trigger an action potential is called **threshold**. *Action potentials are caused by depolarization of the membrane beyond threshold.*

The depolarization that causes action potentials arises in different ways in different neurons. In our previous example, depolarization was caused by the entry of Na^+ through specialized ion channels that were sensitive to membrane stretching. In interneurons, depolarization is usually caused by Na^+ entry through channels that are sensitive to neurotransmitters released by other neurons. In addition to these natural routes, neurons can be depolarized by injecting electrical current through a microelectrode, a method commonly used by neuroscientists to study action potentials in different cells.

Generating an action potential by depolarizing a neuron is something like taking a photograph by pressing the shutter button on an old-fashioned camera. Applying pressure on the button has no effect until it increases to the point of crossing a threshold, and then "click"—the shutter opens and one frame of film is exposed. Increasing depolarization of a neuron similarly has no effect until it crosses threshold, and then "pop"—one action potential. For this reason, action potentials are said to be "all-or-none."

The Generation of Multiple Action Potentials

Earlier we likened the generation of an action potential by depolarization to taking a photograph by pressing the shutter button on a camera. But what if the camera is one of those fancy ones used by fashion and sports photographers where continued pressure on the button causes the camera to shoot frame after frame? The same thing is true for a neuron. If, for example, we pass continuous depolarizing current into a neuron through a microelectrode, we generate not one but many action potentials in succession (Figure 4.2).

The rate of action potential generation depends on the magnitude of the continuous depolarizing current. If we pass enough current through a microelectrode to depolarize just to threshold, but not far beyond, we might find that the cell generates action potentials at a rate of something like one per second, or one hertz (1 Hz). If we crank up the current a little bit more, however, we will find that the rate of action potential generation increases,

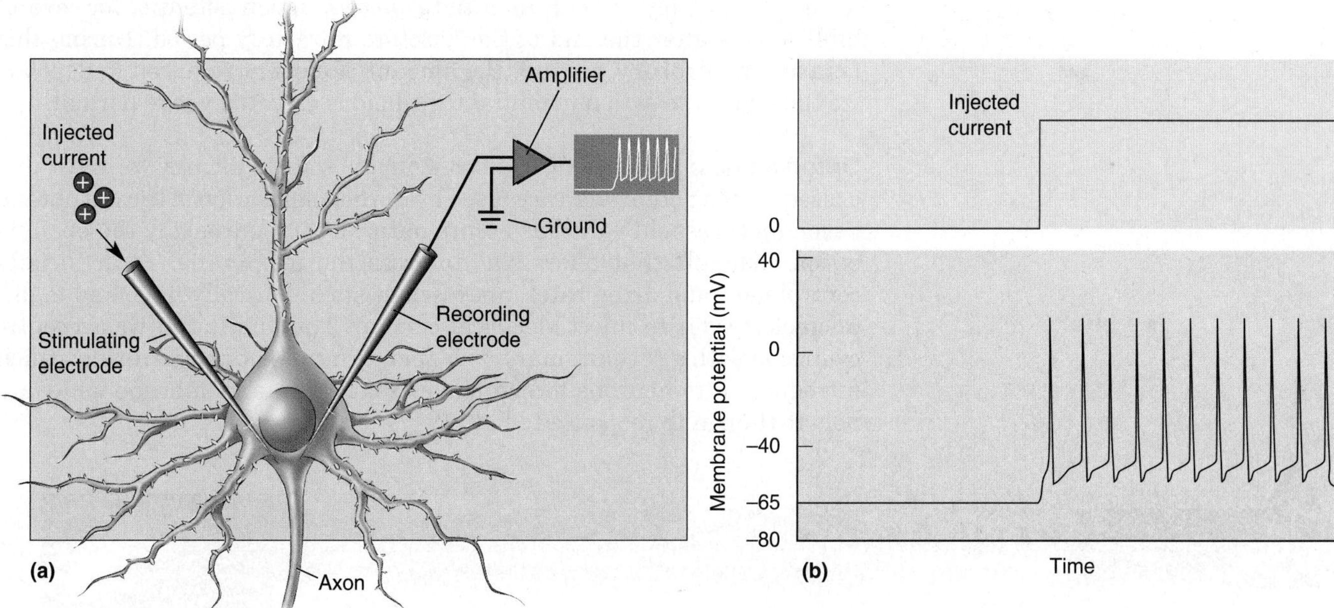

▲ FIGURE 4.2

The effect of injecting positive charge into a neuron. (a) The axon hillock is impaled by two electrodes, one for recording the membrane potential relative to ground and the other for stimulating the neuron with electrical current. **(b)** When electrical current is injected into the neuron (top trace), the membrane is depolarized sufficiently to fire action potentials (bottom trace).

say, to 50 impulses per second (50 Hz). Thus, the *firing frequency* of action potentials reflects the magnitude of the depolarizing current. This is one way that stimulation intensity is encoded in the nervous system (Figure 4.3).

Although firing frequency increases with the amount of depolarizing current, there is a limit to the rate at which a neuron can generate action potentials. The maximum firing frequency is about 1000 Hz; once an action potential is initiated, it is impossible to initiate another for about 1 msec. This period of time is called the **absolute refractory period**. In addition,

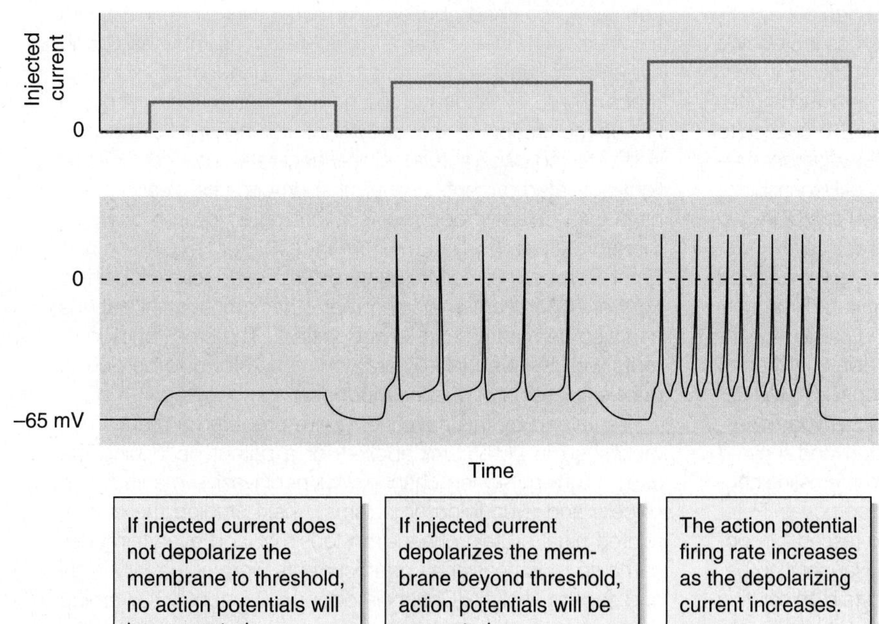

◀ FIGURE 4.3

The dependence of action potential firing frequency on the level of depolarization.

it can be relatively difficult to initiate another action potential for several milliseconds after the end of the absolute refractory period. During this **relative refractory period**, the amount of current required to depolarize the neuron to action potential threshold is elevated above normal.

Optogenetics: Controlling Neural Activity with Light. As we have discussed, action potentials are caused by the depolarization of the membrane beyond a threshold value, as occurs naturally in neurons by the opening of ion channels that allow Na^+ to cross the membrane. To artificially control neuronal firing rates, neuroscientists historically have had to use microelectrodes to inject electrical current. This limitation was recently overcome with a revolutionary new approach called **optogenetics**, which introduces into neurons foreign genes that express membrane ion channels that open in response to light.

BOX 4.2 PATH OF DISCOVERY

The Discovery of the Channelrhodopsins

by Georg Nagel

When I returned in 1992 to the Max Planck Institute of Biophysics in Frankfurt, Germany, from my post-doctoral studies at Yale and Rockefeller University, I was mostly interested in the mechanisms that establish ion gradients across cell membranes. Ernst Bamberg, the director of my department, convinced me to undertake a novel approach in studying microbial rhodopsins—proteins that transport ions across membranes when they absorb light energy. We expressed the gene for bacteriorhodopsin in frog eggs (oocytes), and measured its light-activated electrical current with microelectrodes. In 1995, we demonstrated that illumination of bacteriorhodopsin triggered proton (H^+) pumping across the oocyte membrane. We then went on in 1996 to study the light-activated chloride pump halorhodopsin with this new technique.

We also received DNA for chlamyopsin-1 and -2, proposed to be photoreceptor proteins in the green alga *Chlamydomonas reinhardtii*, from Peter Hegemann at the University of Regensburg. Unfortunately, like all the other labs who received this DNA, we were unable to observe any light-induced electrical signals. Nevertheless, I agreed to test the function of a new presumed rhodopsin from *Chlamydomonas* when Peter called me, announcing that they had found a "real light-gated calcium channel," which he wanted to name chlamyrhodopsin-3. Although this new protein had not been purified, "chlamyopsin-3" was detected in a data bank of DNA sequences from *Chlamydomonas*, produced at the research center in Kazusa, Japan, and showed similarities to bacteriorhodopsin. This made it an interesting candidate for the long-sought rhodopsin in *Chlamydomonas*. Peter requested the DNA from Japan, and I then expressed it in oocytes. Our initial experiments, however, were disappointing because removal or addition of calcium to the oocyte bath solution made no difference to the light-activated electrical current, as would be expected if it actually were a Ca^{2+} permeable channel. The photocurrent itself was rather weak and did not seem to be influenced by any change in the ionic concentrations in the bath solution.

As I still liked the idea of a directly light-gated ion channel, which most other researchers in the field rejected, I continued to test different bath solutions. One evening, I got a stunningly large inward light-activated current with a solution designed to inhibit calcium currents. It turned out, however, that the solution I used was badly buffered; in fact, it was quite acidic with too much H^+! But this was a breakthrough as I now had good evidence for an inward-directed light-dependent H^+ conductance. Then, by acidifying the oocyte (that is, increasing the H^+ concentration of the oocyte interior relative to the outside), I found I was able to reliably generate outward-directed light-activated currents as well. It soon became clear that we have with chlamyrhodopsin–3 a light-gated proton *channel*; therefore, I proposed to my colleagues Peter Hegemann and Ernst Bamberg to call this new protein channelrhodopsin-1. Further experiments revealed that other monovalent cations can also permeate channelrhodopsin-1. The small photocurrents we observed initially are now understood to be due to poor expression of channelrhodopsin-1 in oocytes.

Tantalized by this new finding, we prepared a manuscript (published in 2002) and applied for a patent describing the use of light-gated ion channels for noninvasive manipulation of cells and even living organisms. I next studied the closely related algal protein channelrhodopsin-2, and everything became so much easier as photocurrents were now really large and easy to analyze. Channelrhodopsin-2 (chop2), 737 amino

In Chapter 9, we will discuss how light energy is absorbed by proteins called *photopigments* to generate the neural responses in our retinas that ultimately give us sight. Of course, sensitivity to light is a property of many organisms. In the course of studying light responses in a green alga, researchers working in Frankfurt, Germany, characterized a photopigment they called **channelrhodopsin-2 (ChR2)**. By introducing the *ChR2* gene into mammalian cells, they showed that it encodes a light-sensitive cation channel that is permeable to Na^+ and Ca^{2+} (Box 4.2). The channel opens rapidly in response to blue light, and in neurons the inward flow of cations is sufficient to produce depolarization beyond threshold for action potentials. The enormous potential of optogenetics was subsequently demonstrated by researchers in the United States who showed that the behavior of rats and mice could be dramatically influenced by shining blue light onto neurons in which the *ChR2* gene

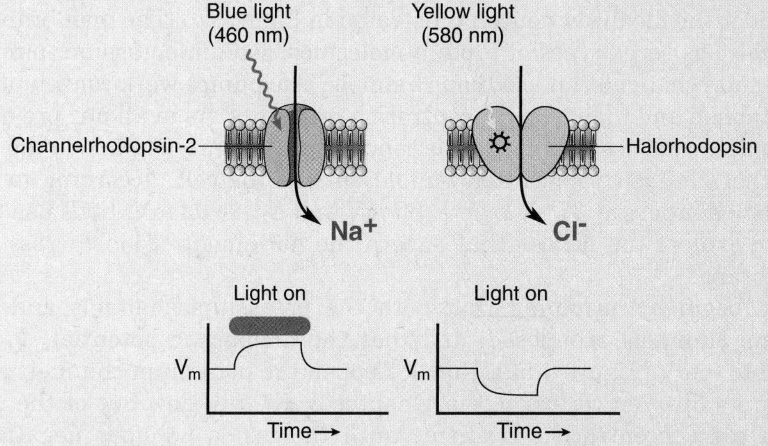

Figure A
Schematic drawings of channelrhodopsin-2 and halorhodopsin in the plasma membrane. Below, the effect of blue and yellow light on membrane potential, mediated by channelrhodopsin-2 and halorhodopsin, respectively.

acids long in its native form, could be shortened to 310 amino acids and attached to yellow fluorescent protein (YFP) to allow visualization of protein expression. After we published the superior features of chop2 in 2003, requests for the DNA started coming in, and we ourselves looked for collaborations with neurobiologists. One of our first "victims" was Alexander Gottschalk at the nearby University of Frankfurt, as he worked with the small translucent nematode worm *Caenorhabditis elegans (C. elegans)*. Unfortunately, I made an error in preparing the DNA such that that the worms, although nicely labeled by YFP, did not react to light. Once I realized my mistake and got chop2–YFP into *C. elegans* muscle cells, we were amazed how easily these little worms could be induced to contract simply by illumination with blue light. At about the same time (April 2004) Karl Deisseroth at Stanford University asked for the DNA and advice on its use in a collaboration, which I happily accepted. Karl quickly demonstrated the power of channelrhodopsin-2 in mammalian neurons. His exciting work with Ed Boyden and Feng Zhang attracted a lot of attention, prompting many requests for the DNA to express this protein in the brain. Many colleagues from Europe only then realized that channelrhodopsins were first characterized in Frankfurt.

The success and ease of application of channelrhodopsin-2 led Karl and Alexander to wonder if there are other rhodopsins that might be used for light-induced *inhibition* of neuronal activity. We told them about bacteriorhodopsin and halorhodopsin, the light-activated proton export and chloride import pumps, respectively. Both pumps render the cell interior more negative (i.e., they are light-activated hyperpolarizers). We recommended halorhodopsin from the microbe *Natronomonas pharaonis* as a light-activated hyperpolarizer. We took advantage of what we had learned back in 1996: that halorhodopsin had a high affinity for chloride and that its expression was stable in animal cells.

As it turned out, light activation of the chloride pump halorhodopsin is sufficient to inhibit action potential firing in mammalian neurons and to inhibit muscle contraction of the nematode *C. elegans*. Ironically, these neurobiological experiments with halorhodopsin (and the same applies for bacteriorhodopsin) could have been done several years earlier, but only the discovery and application of channelrhodopsin-2 encouraged their use and helped create a new field, now called optogenetics. Many neurobiologists are now using these tools, and a few groups, including ourselves, are engaged in further improving and expanding the existing optogenetic tool box.

References:

Nagel G, Szellas T, Huhn W, Kateriya S, Adeishvili N, Berthold P, Ollig D, Hegemann P, Bamberg E. 2003. Channelrhodopsin-2, a directly light-gated cation-selective membrane channel. *Proceedings of the National Academy of Sciences of United States of America* 100:13940–13945.

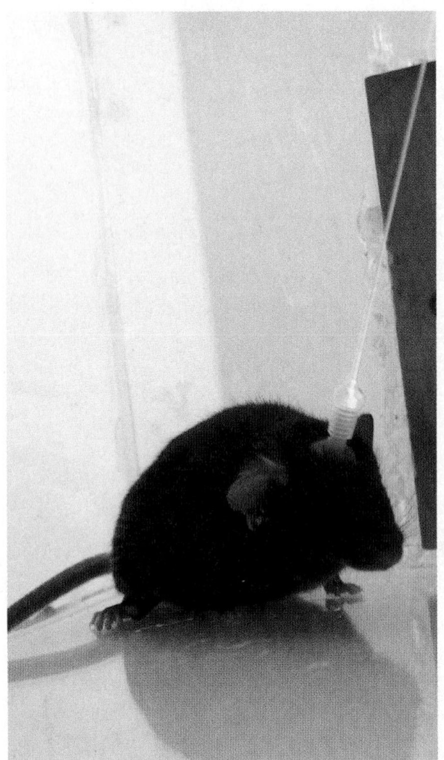

▲ FIGURE 4.4
Optogenetic control of neural activity in a mouse brain. The gene encoding channelrhodopsin-2 was introduced into neurons of this mouse's brain using a virus. The firing of these neurons can now be controlled with blue light delivered via an optic fiber. (Source: Courtesy of Dr. Ed Boyden, Massachusetts Institute of Technology.)

was introduced (Figure 4.4). Newer additions to the "optogenetic toolkit" available to researchers include halorhodopsin, a protein derived from single-cell microbes that will inhibit neurons in response to yellow light.

Understanding how behaviors arise, of course, requires understanding how action potentials arise and propagate through the nervous system. We now take a look at how the movement of ions through the neuron's own specialized protein channels causes a neural signal with these interesting properties.

THE ACTION POTENTIAL, IN THEORY

The action potential is a dramatic redistribution of electrical charge across the membrane. *Depolarization of the cell during the action potential is caused by the influx of sodium ions across the membrane, and repolarization is caused by the efflux of potassium ions.* Let's apply some of the concepts introduced in Chapter 3 to help us understand how ions are driven across the membrane, and how these ionic movements affect the membrane potential.

Membrane Currents and Conductances

Consider the idealized neuron illustrated in Figure 4.5. The membrane of this cell has three types of protein molecules: sodium-potassium pumps, potassium channels, and sodium channels. The pumps work continuously to establish and maintain concentration gradients. As in all our previous examples, we'll assume that K^+ is concentrated twentyfold inside the cell and that Na^+ is concentrated tenfold outside the cell. According to the Nernst equation, at 37°C, $E_K = -80$ mV and $E_{Na} = 62$ mV. Let's use this cell to explore the factors that govern the movement of ions across the membrane.

We begin by assuming that both the potassium channels and the sodium channels are closed, and that the membrane potential, V_m, is equal to 0 mV (Figure 4.5a). Now let's open the potassium channels only (Figure 4.5b). As we learned in Chapter 3, K^+ will flow out of the cell, down the concentration gradient, until the inside becomes negatively charged, and $V_m = E_K$ (Figure 4.5c). Here we want to focus our attention on the movement of K^+ that took the membrane potential from 0 mV to −80 mV. Consider these three points:

1. The net movement of K^+ across the membrane is an electrical current. We can represent this current using the symbol I_K.
2. The number of open potassium channels is proportional to an electrical conductance. We can represent this conductance by the symbol g_K.
3. Membrane potassium current, I_K, will flow only as long as $V_m \neq E_K$. The driving force on K^+ is defined as the difference between the real membrane potential and the equilibrium potential, which can be written as $V_m - E_K$.

There is a simple relationship between the ionic driving force, ionic conductance, and the amount of ionic current that will flow. For K^+, this may be written:

$$I_K = g_K (V_m - E_K).$$

More generally, we write:

$$I_{ion} = g_{ion} (V_m - E_{ion}).$$

If this sounds familiar, that is because it is simply an expression of Ohm's law, $I = gV$, which we learned about in Chapter 3.

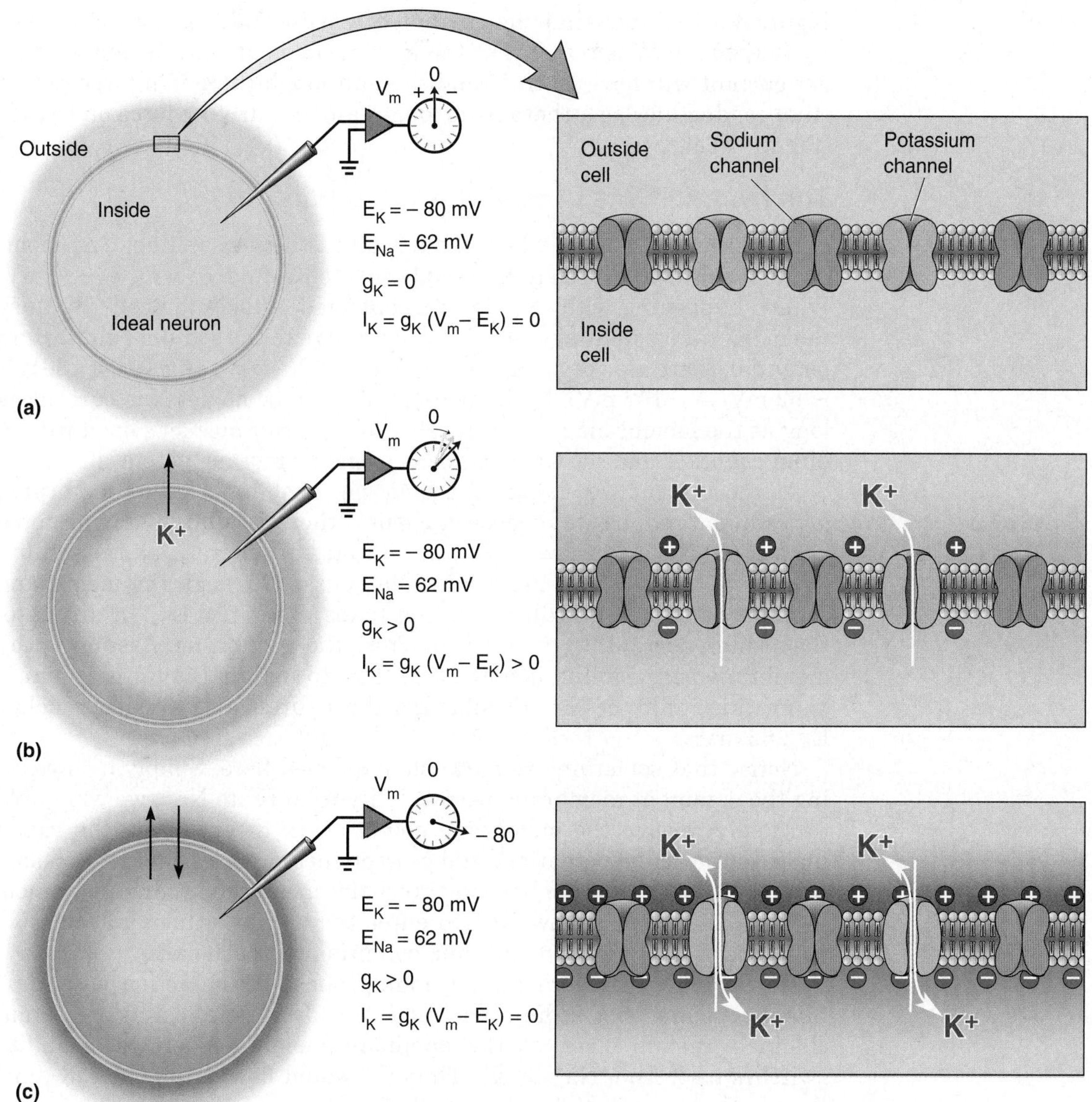

▲ FIGURE 4.5
Membrane currents and conductances. Here is an idealized neuron with sodium-potassium pumps (not shown), potassium channels, and sodium channels. The pumps establish ionic concentration gradients so that K^+ is concentrated inside the cell and Na^+ is concentrated outside the cell. **(a)** Initially, we assume that all channels are closed and the membrane potential equals 0 mV. **(b)** Now we open the potassium channels, and K^+ flows out of the cell. This movement of K^+ is an electrical current, I_K, and it flows as long as the membrane conductance to K^+, g_K, is greater than zero, and the membrane potential is not equal to the potassium equilibrium potential. **(c)** At equilibrium, there is no net potassium current because, although $g_K > 0$, the membrane potential at equilibrium equals E_K. At equilibrium, an equal number of K^+ enters and leaves.

Now let's take another look at our example. Initially we began with $V_m = 0$ mV and no ionic membrane permeability (see Figure 4.5a). There is a large driving force on K^+ because $V_m \neq E_K$; in fact, $(V_m - E_K) =$ 80 mV. However, because the membrane is impermeable to K^+, the potassium conductance, g_K, equals zero. Consequently, $I_K = 0$. Potassium current flows only when the membrane has open potassium channels and therefore $g_K > 0$. Now K^+ flows out of the cell—as long as the membrane potential differs from the potassium equilibrium potential (see

Figure 4.5b). Notice that the current flow is in the direction that takes V_m toward E_K. When $V_m = E_K$, the membrane is at equilibrium, and no net current will flow. In this condition, although there is a large potassium conductance, g_K, there is no longer any net driving force on the K^+ (Figure 4.5c).

The Ins and Outs of an Action Potential

Let's pick up the action where we left off in the last section. The membrane of our ideal neuron is permeable only to K^+, and $V_m = E_K = -80$ mV. What's happening with the Na^+ concentrated outside the cell? Because the membrane potential is so negative relative to the sodium equilibrium potential, there is a very large driving force on Na^+ ($[V_m - E_{Na}] = [-80 \text{ mV} - 62 \text{ mV}] = -142$ mV). Nonetheless, there can be no net Na^+ current as long as the membrane is impermeable to Na^+. But now let's open the sodium channels and see what happens to the membrane potential.

At the instant we change the ionic permeability of the membrane, g_{Na} is high, and, as we discussed earlier, there is a large driving force pushing on Na^+. Thus, we have what it takes to generate a large sodium current, I_{Na}, across the membrane. Na^+ passes through the membrane sodium channels in the direction that takes V_m toward E_{Na}; in this case, the sodium current, I_{Na}, is inward across the membrane. Assuming the membrane permeability is now far greater to sodium than it is to potassium, this influx of Na^+ depolarizes the neuron until V_m approaches E_{Na}, 62 mV.

Notice that something remarkable happened here. Simply by switching the dominant membrane permeability from K^+ to Na^+, we were able to rapidly reverse the membrane potential. In theory, then, the rising phase of the action potential could be explained if, in response to depolarization of the membrane beyond threshold, membrane sodium channels opened. This would allow Na^+ to enter the neuron, causing a massive depolarization until the membrane potential approached E_{Na}.

How could we account for the falling phase of the action potential? Simply assume that sodium channels quickly close and the potassium channels remain open, so the dominant membrane ion permeability switches back from Na^+ to K^+. Then K^+ would flow out of the cell until the membrane potential again equals E_K.

Our model for the ins and outs, ups and downs of the action potential in an idealized neuron is shown in Figure 4.6. The rising phase of the action potential is explained by an inward sodium current, and the falling phase is explained by an outward potassium current. The action potential therefore could be accounted for simply by the movement of ions through channels that are gated by changes in the membrane potential. If you understand this concept, you understand a lot about the ionic basis of the action potential. What's left now is to see how this actually happens—in a real neuron.

THE ACTION POTENTIAL, IN REALITY

Let's quickly review our theory of the action potential. When the membrane is depolarized to threshold, there is a transient increase in g_{Na}. The increase in g_{Na} allows the entry of Na^+, which depolarizes the neuron. And the increase in g_{Na} must be brief in duration to account for the short duration of the action potential. Restoring the negative membrane potential would be further aided by a transient increase in g_K during the falling phase, allowing K^+ to leave the depolarized neuron faster.

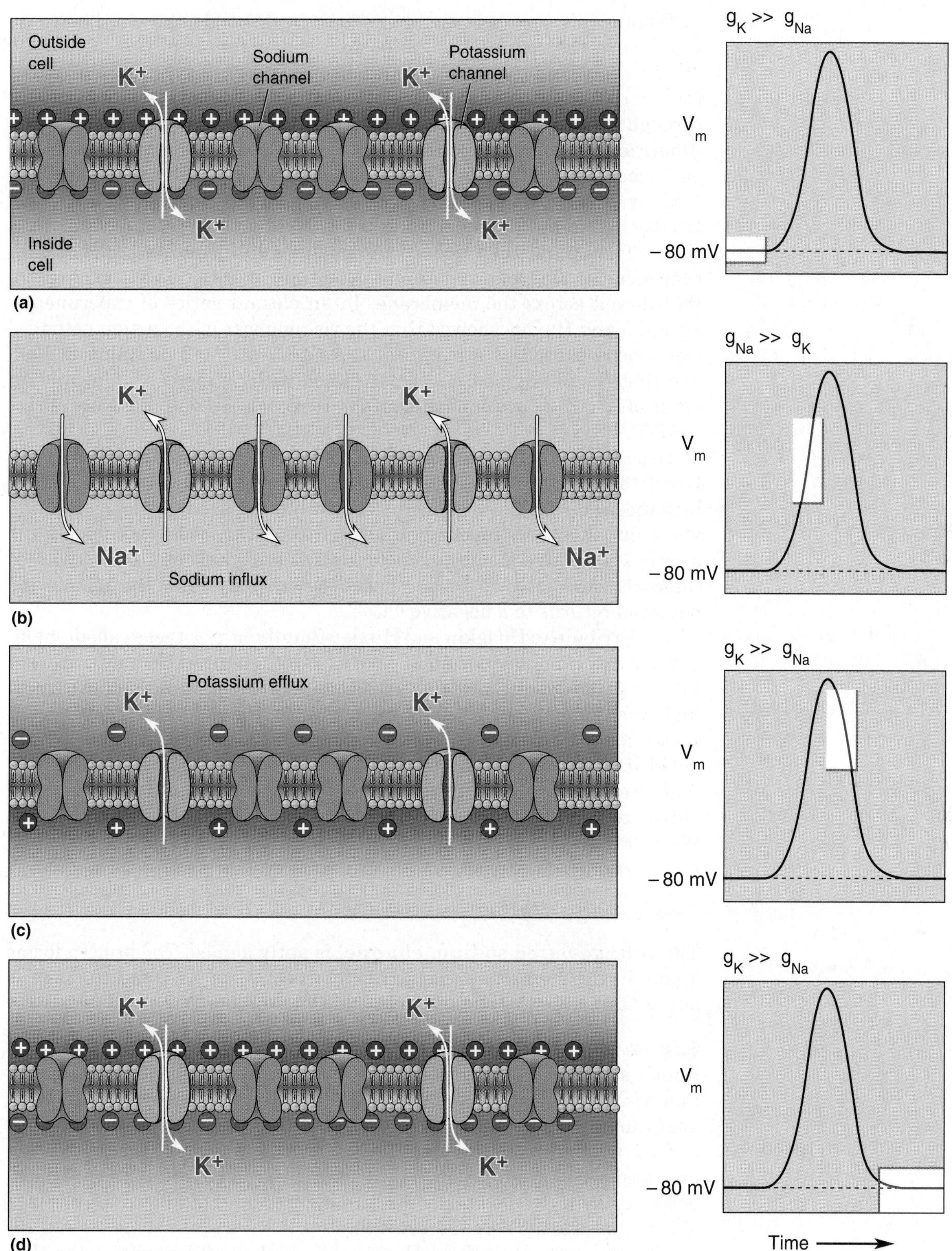

▲ FIGURE 4.6

Flipping the membrane potential by changing the relative ionic permeability of the membrane. **(a)** The membrane of the idealized neuron, introduced in Figure 4.4. We begin by assuming that the membrane is permeable only to K^+ and that $V_m = E_K$. **(b)** We now stipulate that the membrane sodium channels open so that $g_{Na} >> g_K$. There is a large driving force on Na^+, so Na^+ rushes into the cell, taking V_m toward E_{Na}. **(c)** Now we close the sodium channels so that $g_K >> g_{Na}$. Because the membrane potential is positive, there is a large driving force on K^+. The efflux of K^+ takes V_m back toward E_K. **(d)** The resting state is restored where $V_m = E_K$.

Testing this theory is simple enough in principle. All one has to do is measure the sodium and potassium conductances of the membrane during the action potential. In practice, however, such a measurement proved to be quite difficult in real neurons. The key technical breakthrough came with a device called a **voltage clamp**, invented by the American physiologist Kenneth C. Cole and used in decisive experiments performed by Cambridge University physiologists Alan Hodgkin and Andrew Huxley around 1950. The voltage clamp enabled Hodgkin and Huxley to "clamp" the membrane potential of an axon at any value they chose. They could then deduce the changes in membrane conductance that occur at different membrane potentials by measuring the currents that flowed across the membrane. In an elegant series of experiments, Hodgkin and Huxley showed that the rising phase of the action potential was indeed caused by a transient increase in g_{Na} and an influx of Na^+, and that the falling phase was associated with an increase in g_K and an efflux of K^+. Their accomplishments were recognized with the Nobel Prize in 1963.

To account for the transient changes in g_{Na}, Hodgkin and Huxley proposed the existence of sodium "gates" in the axonal membrane. They hypothesized that these gates are "activated" (opened) by depolarization above threshold and "inactivated" (closed and locked) when the membrane acquires a positive membrane potential. These gates are "deinactivated" (unlocked and enabled to be opened again) only after the membrane potential returns to a negative value.

It is a tribute to Hodgkin and Huxley that their hypotheses about membrane gates came more than 20 years before the direct demonstration of voltage-gated channel proteins in the neuronal membrane. We have a new understanding of gated membrane channels, thanks to two more recent scientific breakthroughs. First, new molecular biological techniques have enabled neuroscientists to determine the detailed structure of these proteins. Second, new neurophysiological techniques have enabled neuroscientists to measure the ionic currents that pass through single channels. We will now explore the action potential from the perspective of these membrane ion channels.

The Voltage-Gated Sodium Channel

The **voltage-gated sodium channel** is aptly named. The protein forms a pore in the membrane that is highly selective to Na^+, and the pore is opened and closed by changes in membrane voltage.

Sodium Channel Structure. The voltage-gated sodium channel is created from a single long polypeptide. The molecule has four distinct domains, numbered I–IV; each domain consists of six transmembrane alpha helices, numbered S1–S6 (Figure 4.7). The four domains clump together to form a pore between them. The pore is closed at the negative resting membrane potential. When the membrane is depolarized to threshold, however, the molecule twists into a configuration that allows the passage of Na^+ through the pore (Figure 4.8).

Like the potassium channel, the sodium channel has pore loops that are assembled into a selectivity filter. This filter makes the sodium channel twelve times more permeable to Na^+ than it is to K^+. Apparently, the Na^+ ions are stripped of most, but not all, of their associated water molecules as they pass into the channel. The retained water serves as a sort of molecular chaperone for the ion, and is necessary for the ion to pass the selectivity filter. The ion–water complex can then be used to select Na^+ and exclude K^+ (Figure 4.9).

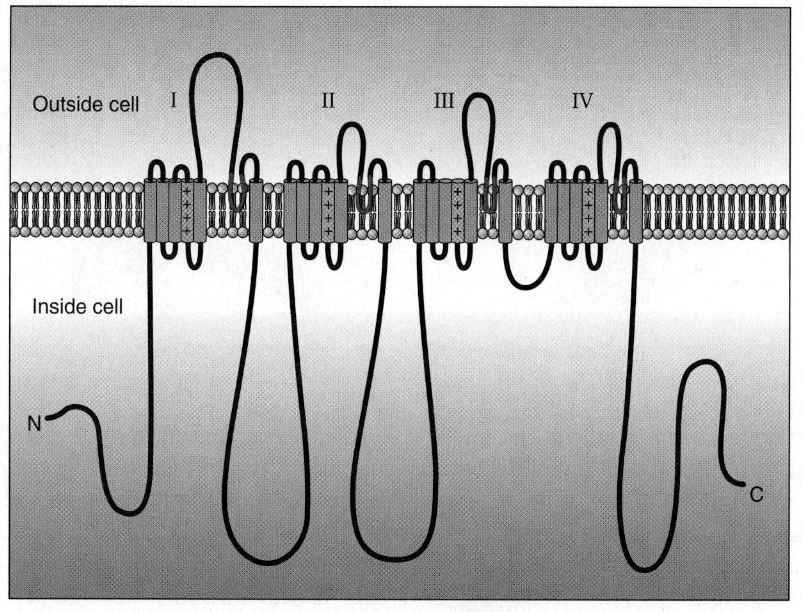

(a)

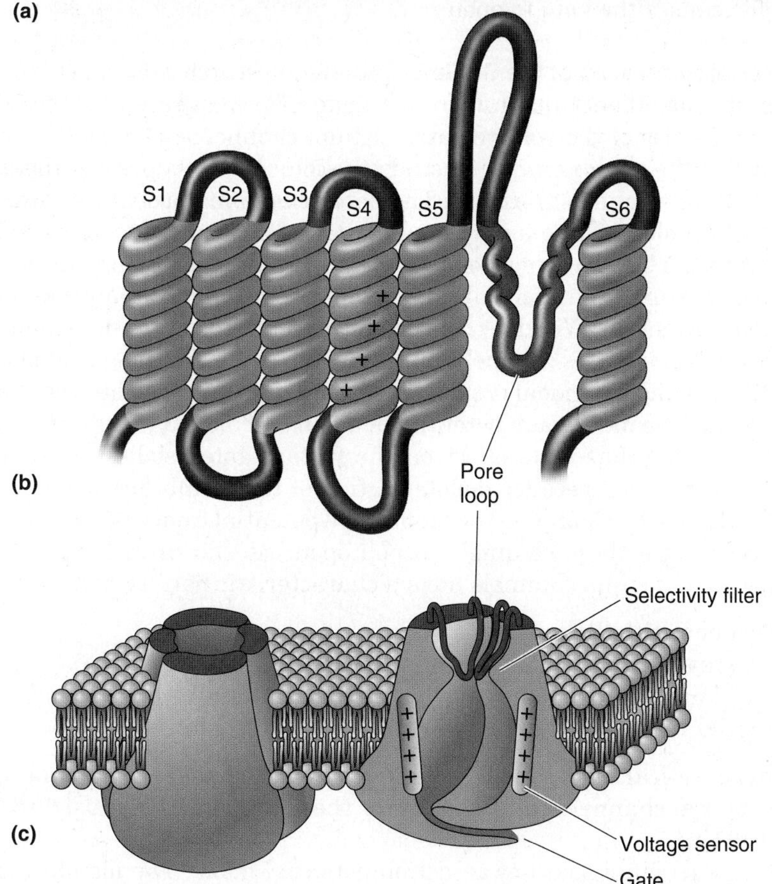

(b)

(c)

▲ FIGURE 4.7

The structure of the voltage-gated sodium channel.

(a) A depiction of how the sodium channel polypeptide chain is believed to be woven into the membrane. The molecule consists of four domains, I–IV. Each domain consists of six alpha helices (represented by the blue and purple cylinders), which pass back and forth across the membrane. **(b)** An expanded view of one domain, showing the voltage sensor of alpha helix S4 and the pore loop (*red*), which contributes to the selectivity filter. **(c)** A view of the molecule showing how the domains may arrange themselves to form a pore between them. (Source: Adapted from Armstrong and Hille, 1998, Fig. 1.)

Closed pore

Open pore

– 65 mV → – 40 mV

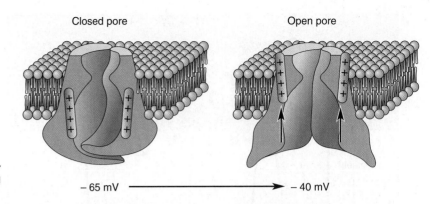

▶ FIGURE 4.8
A hypothetical model for changing the configuration of the sodium channel by depolarizing the membrane.

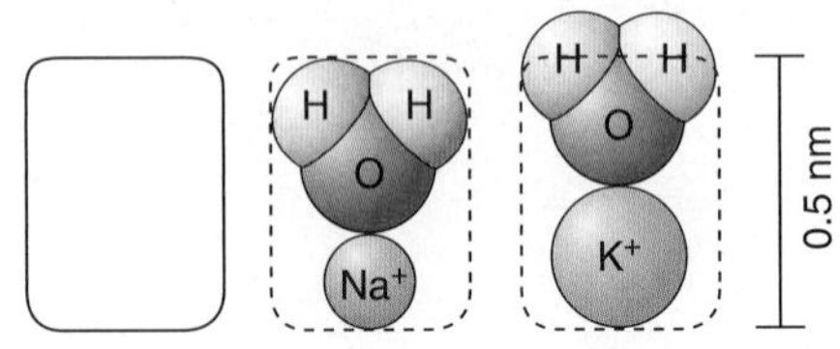

Size of sodium channel selectivity filter

Size of partially hydrated Na^+ ion

Size of partially hydrated K^+ ion

▲ FIGURE 4.9
Dimensions of the sodium channel selectivity filter. Water accompanies the ions as they pass through the channel. Hydrated Na^+ fits; hydrated K^+ does not. (Source: Adapted from Hille, 1992, Figs. 5, 6.)

The sodium channel is gated by a change in voltage across the membrane. It has now been established that the voltage sensor resides in segment S4 of the molecule. In this segment, positively charged amino acid residues are regularly spaced along the coils of the helix. Thus, the entire segment can be forced to move by changing the membrane potential. Depolarization twists S4, and this conformational change in the molecule causes the gate to open.

Functional Properties of the Sodium Channel. Research performed around 1980 at the Max Planck Institute in Goettingen, Germany, revealed the functional properties of the voltage-gated sodium channel. A new method was used, called the **patch clamp**, to study the ionic currents passing through individual ion channels (Box 4.3). The patch-clamp method entails sealing the tip of an electrode to a very small *patch* of neuronal membrane. This patch can then be torn away from the neuron, and the ionic currents across it can be measured as the membrane potential is *clamped* at any value the experimenter selects. With luck, the patch will contain only a single channel, and the behavior of this channel can be studied. Patch clamping enabled investigation of the functional properties of the voltage-gated sodium channel.

Changing the membrane potential of a patch of axonal membrane from −80 to −65 mV has little effect on the voltage-gated sodium channels. They remain closed because depolarization of the membrane has not yet reached threshold. Changing the membrane potential from −65 to −40 mV, however, causes these channels to pop open. As shown in Figure 4.10, voltage-gated sodium channels have a characteristic pattern of behavior:

1. They open with little delay.
2. They stay open for about 1 msec and then close (inactivate).
3. They cannot be opened again by depolarization until the membrane potential returns to a negative value near threshold.

A hypothetical model for how conformational changes in the voltage-gated sodium channel could account for these properties is illustrated in Figure 4.10c.

A single channel does not an action potential make. The membrane of an axon may contain thousands of sodium channels per square micrometer (μm^2), and the concerted action of all these channels is required to generate what we measure as an action potential. Nonetheless, it is interesting to see how many of the properties of the action potential can be explained by the properties of the voltage-gated sodium channel. For example, the fact that single channels do not open until a critical level of membrane depolarization is reached explains the action potential threshold. The rapid opening of the channels in response to depolarization explains why the rising phase of the action potential occurs so quickly. And the short

BOX 4.3 BRAIN FOOD

The Patch-Clamp Method

The very existence of voltage-gated channels in the neuronal membrane was mere conjecture until methods were developed to study individual channel proteins. A revolutionary new method, the patch clamp, was developed by German neuroscientists Bert Sakmann and Erwin Neher in the mid-1970s. In recognition of their contribution, Sakmann and Neher were awarded the 1991 Nobel Prize.

Patch clamping enables one to record ionic currents through single channels (Figure A). The first step is gently lowering the fire-polished tip of a glass recording electrode, 1–5 μm in diameter, onto the membrane of the neuron (part a), and then applying suction within the electrode tip (part b). A tight seal forms between the walls of the electrode and the underlying patch of membrane. This "gigaohm" seal (so named because of its high electrical resistance: $>10^9\ \Omega$) leaves the ions in the electrode only one path to take, through the channels in the underlying patch of membrane. If the electrode is then withdrawn from the cell, the membrane patch can be torn away (part c), and ionic currents can be measured as steady voltages are applied across the membrane (part d).

With a little luck, one can resolve currents flowing through single channels. If the patch contains a voltage-gated sodium channel, for example, then changing the membrane potential from −65 to −40 mV will cause the channel to open, and current (I) will flow through it (part e). The amplitude of the measured current at a constant membrane voltage reflects the channel conductance, and the duration of the current reflects the time the channel is open.

Patch-clamp recordings reveal that most channels flip between two conductance states that can be interpreted as open or closed. The time they remain open can vary, but the single-channel conductance value stays the same and is therefore said to be unitary. Ions can pass through single channels at an astonishing rate—well over a million per second.

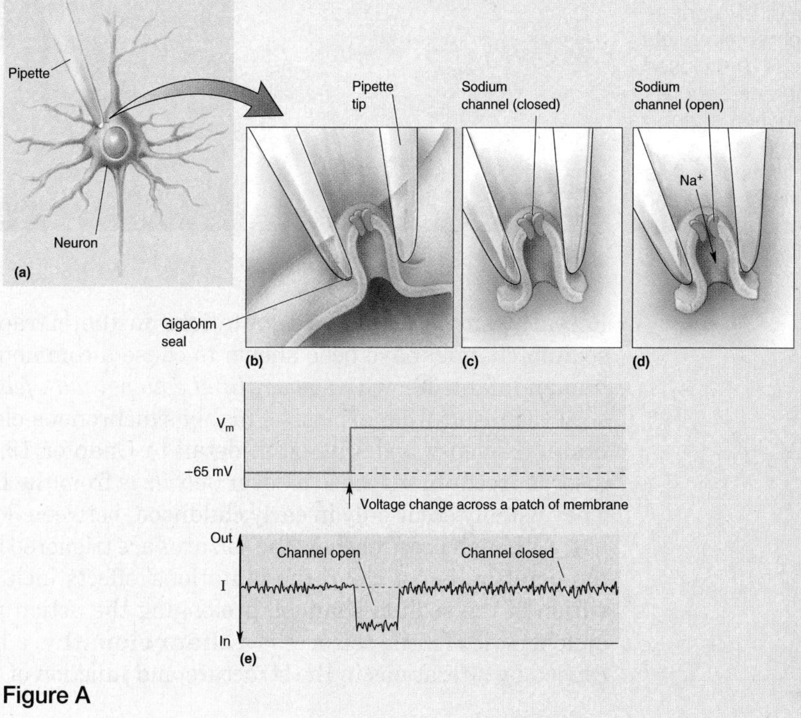

Figure A

time the channels stay open before inactivating (about 1 msec) partly explains why the action potential is so brief. Furthermore, inactivation of the channels can account for the absolute refractory period: Another action potential cannot be generated until the channels are activated.

There are several different sodium channel genes in the human genome. Differences in the expression of these genes among neurons can give rise to subtle but important variations in the properties of the action potential.

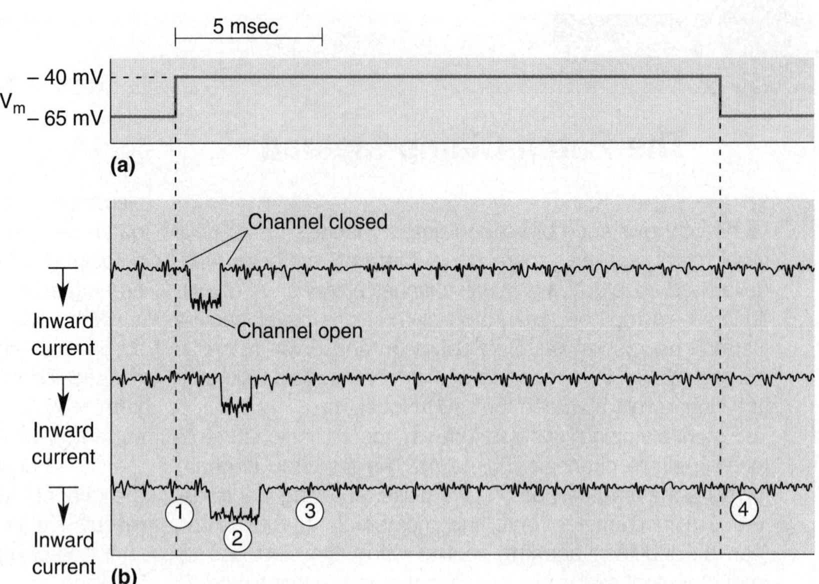

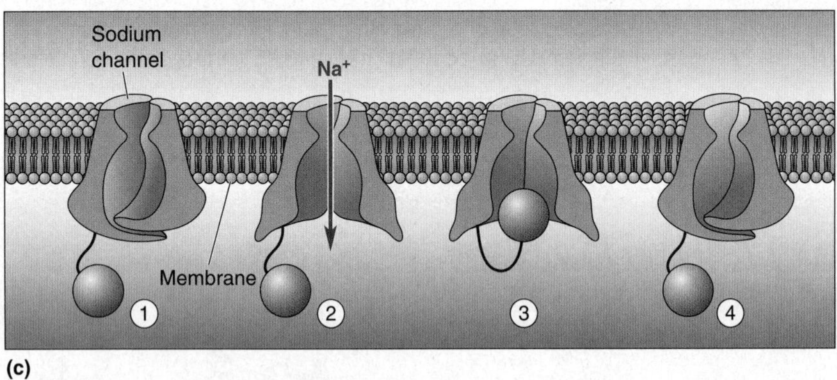

▶ FIGURE 4.10
The opening and closing of sodium channels upon membrane depolarization. (a) This trace shows the electrical potential across a patch of membrane. When the membrane potential is changed from −65 to −40 mV, the sodium channels pop open. (b) These traces show how three different channels respond to the voltage step. Each line is a record of the electrical current that flows through a single channel. ① At −65 mV, the channels are closed, so there is no current. ② When the membrane is depolarized to −40 mV, the channels briefly open and current flows inward, represented by the downward deflection in the current traces. Although there is some variability from channel to channel, all of them open with little delay and stay open for less than 1 msec. Notice that after they have opened once, they close and stay closed as long as the membrane is maintained at a depolarized V_m. ③ The closure of the sodium channel by steady depolarization is called inactivation. ④ To deinactivate the channels, the membrane must be returned to −65 mV again. (c) A model for how changes in the conformation of the sodium channel protein might yield its functional properties. ① The closed channel ② opens upon membrane depolarization. ③ Inactivation occurs when a globular portion of the protein swings up and occludes the pore. ④ Deinactivation occurs when the globular portion swings away and the pore closes by movement of the transmembrane domains.

Recently, single amino acid mutations in the extracellular regions of one sodium channel have been shown to cause a common inherited disorder in human infants known as *generalized epilepsy with febrile seizures*. Epileptic seizures result from explosive, highly synchronous electrical activity in the brain. (Epilepsy is discussed in detail in Chapter 19.) The seizures in this disorder occur in response to fever (*febrile* is from the Latin word for "fever"). They usually occur only in early childhood, between 3 months and 5 years of age. Although precisely how the seizures are triggered by an increase in brain temperature is not clear, the mutations' effects include slowing the inactivation of the sodium channel, prolonging the action potential. Generalized epilepsy with febrile seizures is a **channelopathy**, a human genetic disease caused by alterations in the structure and function of ion channels.

The Effects of Toxins on the Sodium Channel. Researchers at Duke University discovered in the 1960s that a toxin isolated from the ovaries of the puffer fish (Figure 4.11) could selectively block the sodium channel. **Tetrodotoxin (TTX)** clogs the Na^+-permeable pore by binding tightly to a specific site on the outside of the channel. TTX blocks all sodium-dependent action potentials and therefore is usually fatal if ingested. Nonetheless, puffer fish are considered a delicacy in Japan. Sushi chefs licensed by the government train for years to prepare puffer fish in such a way that eating them causes only numbness around the mouth. Talk about adventuresome eating!

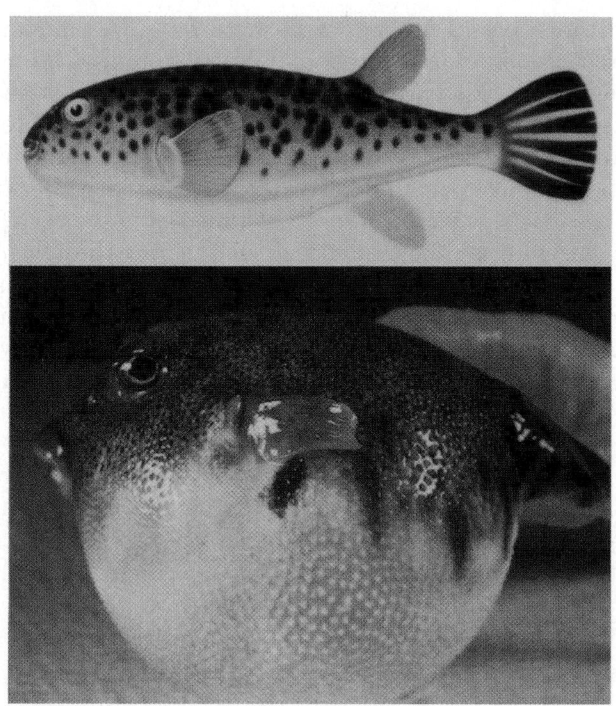

▲ FIGURE 4.11
The puffer fish, source of TTX. (Source: Courtesy of Dr. Toshio Narahashi, Duke University.)

TTX is one of a number of natural toxins that interfere with the function of the voltage-gated sodium channel. Another channel-blocking toxin is *saxitoxin*, produced by dinoflagellates of the genus *Gonyaulax*. Saxitoxin is concentrated in clams, mussels, and other shellfish that feed on these marine protozoa. Occasionally the dinoflagellates bloom, causing what is known as a "red tide." Eating shellfish at these times can be fatal because of the unusually high concentration of the toxin.

In addition to the toxins that block sodium channels, certain compounds interfere with nervous system function by causing the channels to open inappropriately. In this category is *batrachotoxin*, isolated from the skin of a species of Colombian frog. Batrachotoxin causes the channels to open at more negative potentials and to stay open much longer than usual, thus scrambling the information encoded by the action potentials. Toxins produced by lilies (*veratridine*) and buttercups (*aconitine*) have a similar mechanism of action. Sodium channel inactivation is also disrupted by toxins from scorpions and sea anemones.

What can we learn from these toxins? First, the different toxins disrupt channel function by binding to different sites on the protein. Information about toxin binding and its consequences have helped researchers deduce the three-dimensional structure of the sodium channel. Second, the toxins can be used as experimental tools to study the consequences of blocking action potentials. For example, as we shall see in later chapters, TTX is commonly used in experiments that require blocking impulses in a nerve or muscle. The third and most important lesson from studying toxins? Be careful what you put in your mouth!

Voltage-Gated Potassium Channels

Hodgkin's and Huxley's experiments indicated that the falling phase of the action potential was explained only partly by the inactivation of g_{Na}.

They found there was also a transient increase in g_K that functioned to speed the restoration of a negative membrane potential after the spike. They proposed the existence of membrane potassium gates that, like sodium gates, open in response to depolarization of the membrane. Unlike sodium gates, however, potassium gates do not open immediately upon depolarization; it takes about 1 msec for them to open. Because of this delay, and because this potassium conductance serves to rectify, or reset, the membrane potential, they called this conductance the *delayed rectifier*.

We now know that there are many different types of **voltage-gated potassium channels**. Most of them open when the membrane is depolarized and function to diminish any further depolarization by giving K^+ a path to leave the cell across the membrane. The known voltage-gated potassium channels have a similar structure. The channel proteins consist of four separate polypeptide subunits that come together to form a pore between them. Like the sodium channel, these proteins are sensitive to changes in the electrical field across the membrane. When the membrane is depolarized, the subunits are believed to twist into a shape that allows K^+ to pass through the pore.

Putting the Pieces Together

We can now use what we've learned about ions and channels to explain the key properties of the action potential (Figure 4.12):

- *Threshold*. Threshold is the membrane potential at which enough voltage-gated sodium channels open so that the relative ionic permeability of the membrane favors sodium over potassium.
- *Rising phase*. When the inside of the membrane has a negative electrical potential, there is a large driving force on Na^+. Therefore, Na^+ rushes into the cell through the open sodium channels, causing the membrane to rapidly depolarize.
- *Overshoot*. Because the relative permeability of the membrane greatly favors sodium, the membrane potential goes to a value close to E_{Na}, which is greater than 0 mV.
- *Falling phase*. The behavior of two types of channels contributes to the falling phase. First, the voltage-gated sodium channels inactivate. Second, the voltage-gated potassium channels finally open (triggered to do so 1 msec earlier by the depolarization of the membrane). There is a great driving force on K^+ when the membrane is strongly depolarized. Therefore, K^+ rushes out of the cell through the open channels, causing the membrane potential to become negative again.
- *Undershoot*. The open voltage-gated potassium channels add to the resting potassium membrane permeability. Because there is very little sodium permeability, the membrane potential goes toward E_K, causing a hyperpolarization relative to the resting membrane potential until the voltage-gated potassium channels close again.
- *Absolute refractory period*. Sodium channels inactivate when the membrane becomes strongly depolarized. They cannot be activated again, and another action potential cannot be generated, until the membrane potential becomes sufficiently negative to deinactivate the channels.
- *Relative refractory period*. The membrane potential stays hyperpolarized until the voltage-gated potassium channels close. Therefore, more depolarizing current is required to bring the membrane potential to threshold.

We've seen that channels and the movement of ions through them can explain the properties of the action potential. But it is important to remember that the sodium-potassium pump also is working quietly in the background. Imagine that the entry of Na^+ during each action potential is

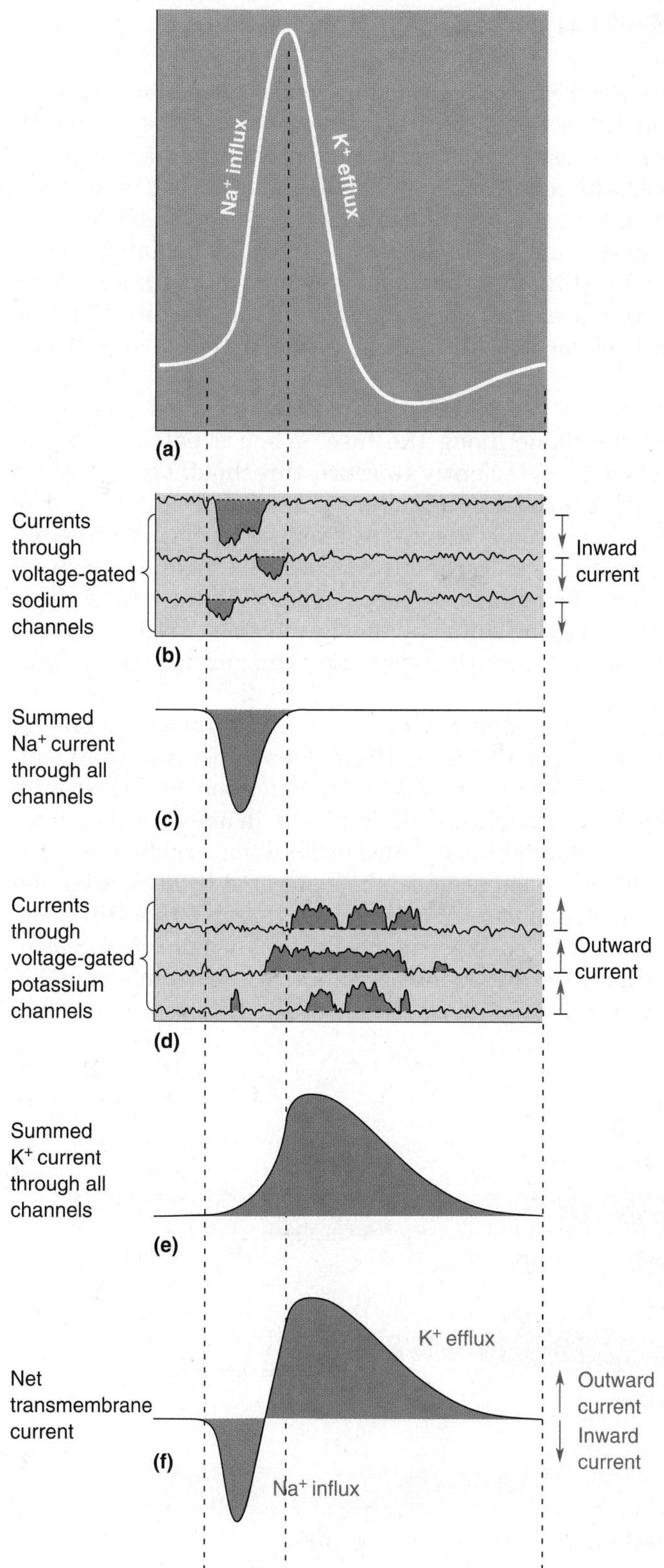

◀ FIGURE 4.12
The molecular basis of the action potential. **(a)** The membrane potential as it changes in time during an action potential. The rising phase of the action potential is caused by the influx of Na^+ through hundreds of voltage-gated sodium channels. The falling phase is caused by sodium channel inactivation and the efflux of K^+ through voltage-gated potassium channels. **(b)** The inward currents through three representative voltage-gated sodium channels. Each channel opens with little delay when the membrane is depolarized to threshold. The channels stay open for no more than 1 msec and then inactivate. **(c)** The summed Na^+ current flowing through all the sodium channels. **(d)** The outward currents through three representative voltage-gated potassium channels. Voltage-gated potassium channels open about 1 msec after the membrane is depolarized to threshold and stay open as long as the membrane is depolarized. The high potassium permeability causes the membrane to hyperpolarize briefly. When the voltage-gated potassium channels close, the membrane potential relaxes back to the resting value, around −65 mV. **(e)** The summed K^+ current flowing through all the potassium channels. **(f)** The net transmembrane current during the action potential (the sum of parts c and e).

like a wave coming over the bow of a boat making way in heavy seas. Like the continuous action of the boat's bilge pump, the sodium-potassium pump works all the time to transport Na^+ back across the membrane. The pump maintains the ionic concentration gradients that drive Na^+ and K^+ through their channels during the action potential.

ACTION POTENTIAL CONDUCTION

To transfer information from one point to another in the nervous system, it is necessary that the action potential, once generated, be conducted down the axon. This process is like the burning of a fuse. Imagine you're holding a firecracker with a burning match held under the end of the fuse. The fuse ignites when it gets hot enough (beyond some threshold). The tip of the burning fuse heats up the segment of fuse immediately ahead of it until it ignites. In this way, the flame steadily works its way down the fuse. Note that the fuse lit at one end only burns in one direction; the flame cannot turn back on itself because the combustible material just behind it is spent.

Propagation of the action potential along the axon is similar to the propagation of the flame along the fuse. When a patch of axonal membrane is depolarized sufficiently to reach threshold, voltage-gated sodium channels pop open, and the action potential is initiated. The influx of positive charge spreads inside the axon to depolarize the adjacent segment of membrane, and when it reaches threshold, the sodium channels in this patch of membrane also pop open (Figure 4.13). In this way, the action potential works its way down the axon until it reaches the axon terminal, thereby initiating synaptic transmission (the subject of Chapter 5).

An action potential initiated at one end of an axon propagates only in one direction; it does not turn back on itself. This is because the membrane just behind it is refractory, due to inactivation of the sodium channels. Normally, action potentials conduct only in one direction, from the soma to the axon terminal; this is called *orthodromic* conduction. But, just like the fuse, an action potential can be generated by depolarization at either end of the axon and can therefore propagate in either direction. Backward propagation, elicited experimentally, is called *antidromic* conduction. Note that because the axonal membrane is excitable (capable of generating action potentials) along its entire length, the impulse will

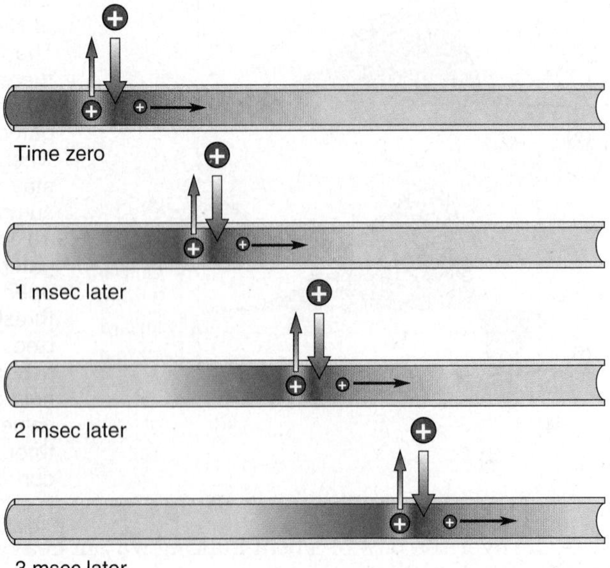

▲FIGURE 4.13
Action potential conduction. The entry of positive charge during the action potential causes the membrane just ahead to depolarize to threshold.

propagate without decrement. The fuse works the same way because it is combustible along its entire length. Unlike the fuse, however, the axon can regenerate its firing ability.

Action potential conduction velocities vary, but 10 m/sec is a typical rate. Remember, from start to finish the action potential lasts about 2 msec. From this, we can calculate the length of membrane that is engaged in the action potential at any instant in time:

$$10 \text{ m/sec} \times 2 \times 10^{-3} \text{ sec} = 2 \times 10^{-2} \text{ m}.$$

Therefore, an action potential traveling at 10 m/sec occurs over a 2 cm length of axon.

Factors Influencing Conduction Velocity

Remember that the inward Na^+ current during the action potential depolarizes the membrane just ahead. When this patch of membrane reaches threshold, the voltage-gated sodium channels will open, and the action potential will "burn" on down the membrane. The speed with which the action potential propagates down the axon depends on how far the depolarization ahead of the action potential spreads, which in turn depends on certain physical characteristics of the axon.

Imagine that the influx of positive charge into an axon during the action potential is like turning on the water to a leaky garden hose. There are two paths the water can take: one, down the inside of the hose; the other, through the perforated wall of the hose. How much water goes along each path depends on their relative resistance; most of the water will take the path of least resistance. If the hose is narrow and the leaks are numerous and large, most of the water will flow out through the leaks. If the hose is wide and the leaks are few and tiny, most of the water will flow down the inside of the hose. The same principles apply to positive current spreading down the axon ahead of the action potential.

There are two paths that positive charge can take: down the inside of the axon, or across the axonal membrane. If the axon is narrow and there are many open membrane pores, most of the current will flow out across the membrane. If the axon is wide and there are few open membrane pores, most of the current will flow down inside the axon. The farther the current goes down the axon, the farther ahead of the action potential the membrane will be depolarized, and the faster the action potential will propagate. As a rule, therefore, action potential conduction velocity increases with increasing axonal diameter.

As a consequence of this relationship between axonal diameter and conduction velocity, neural pathways that are especially important for survival have evolved unusually large axons. An example is the giant axon of the squid, which is part of a pathway that mediates an escape reflex in response to strong sensory stimulation. The squid giant axon can be 1 mm in diameter, so large that originally it was thought to be part of the animal's circulatory system. Neuroscience owes a debt to British zoologist J. Z. Young, who in 1939 called attention to the squid giant axon as an experimental preparation for studying the biophysics of the neuronal membrane. Hodgkin and Huxley used this preparation to elucidate the ionic basis of the action potential, and the giant axon continues to be used today in a wide range of neurobiological studies.

It is interesting to note that axonal size and the number of voltage-gated channels in the membrane also affect axonal excitability. Smaller axons require greater depolarization to reach action potential threshold and are more sensitive to being blocked by local anesthetics (Box 4.4).

BOX 4.4 OF SPECIAL INTEREST

Local Anesthesia

Although you've tried to tough it out, you just can't take it anymore. You finally give in to the pain of the toothache and head for the dentist. Fortunately, the worst part of having a cavity filled is the pinprick in the gum caused by the injection needle. Then your mouth becomes numb and you daydream while the dentist drills and repairs your tooth. What was injected, and how did it work?

Local anesthetics are drugs that temporarily block action potentials in axons. They are called "local" because they are injected directly into the tissue where anesthesia—the absence of sensation—is desired. Small axons, firing a lot of action potentials, are most sensitive to conduction block by local anesthetics.

The first local anesthetic introduced into medical practice was cocaine. The chemical was isolated from the leaves of the coca plant in 1860 by the German physician Albert Niemann. According to the custom of the pharmacologists of his day, Niemann tasted the new compound and discovered that it caused his tongue to go numb. It was soon learned that cocaine also had toxic and addictive properties. (The mind-altering effect of cocaine was studied by another well-known physician of that era, Sigmund Freud. Cocaine alters mood by a mechanism distinct from its local anesthetic action, as we shall see in Chapter 15.)

The search for a suitable synthetic anesthetic as a substitute for cocaine led to the development of lidocaine, which is now the most widely used local anesthetic. Lidocaine can be dissolved into a jelly and smeared onto the mucous membranes of the mouth (and elsewhere) to numb the nerve endings (called *topical anesthesia*); it can be injected directly into a tissue (*infiltration anesthesia*) or a nerve (*nerve block*); it can even be infused into the cerebrospinal fluid bathing the spinal cord (*spinal anesthesia*), where it can numb large parts of the body.

Lidocaine and other local anesthetics prevent action potentials by binding to the voltage-gated sodium channels. The binding site for lidocaine has been identified as the S6 alpha helix of domain IV of the protein (Figure A). Lidocaine cannot gain access to this site from the outside. The anesthetic first must cross the axonal membrane and then pass through the open gate of the channel to find its binding site inside the pore. This explains why active nerves are blocked faster (the sodium channel gates are open more often). The bound lidocaine interferes with the flow of Na^+ that normally results from depolarizing the channel.

Smaller axons are affected by local anesthetics before larger axons because their action potentials have less of a safety margin; more of the voltage-gated sodium channels must function to ensure that the action potential doesn't fizzle out as it conducts down the axon. This increased sensitivity of small axons to local anesthetics is fortuitous in clinical practice. As we will discover in Chapter 12, it is the smaller fibers that convey information about painful stimuli—like a toothache.

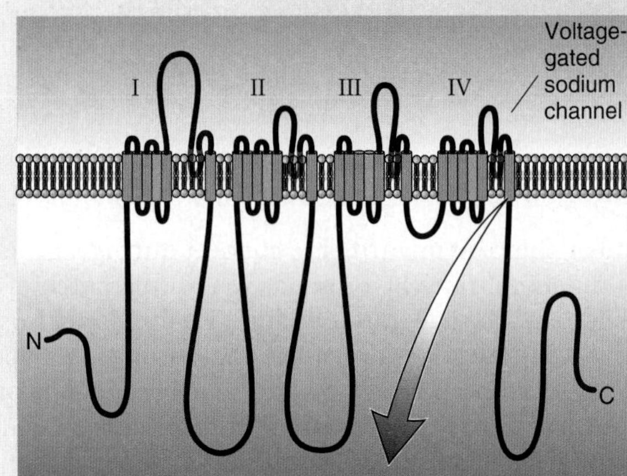

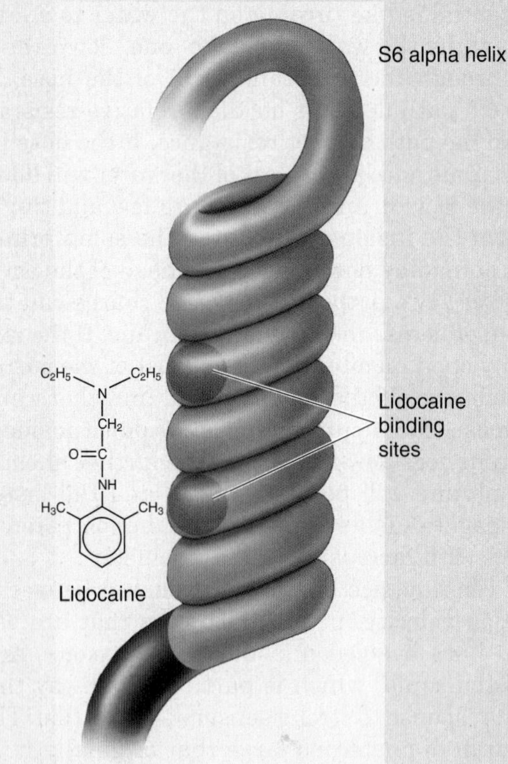

Figure A
Lidocaine's mechanism of action. (Source: Adapted from Hardman, et al., 1996, Fig. 15–3.)

Myelin and Saltatory Conduction

The good thing about fat axons is that they conduct action potentials faster; the bad thing about them is that they take up a lot of space. If all the axons in your brain were the diameter of a squid giant axon, your head would be too big to fit through a barn door. Fortunately, vertebrates evolved another solution for increasing action potential conduction velocity: wrapping the axon with insulation called *myelin* (see Chapter 2). The myelin sheath consists of many layers of membrane provided by glial support cells—Schwann cells in the peripheral nervous system (outside the brain and spinal cord) and oligodendroglia in the central nervous system. Just as wrapping a leaky garden hose with duct tape facilitates water flow down the inside of the hose, myelin facilitates current flow down the inside of the axon, thereby increasing action potential conduction velocity (Box 4.5).

The myelin sheath does not extend continuously along the entire length of the axon. There are breaks in the insulation where ions cross the membrane to generate action potentials. Recall from Chapter 2 that these breaks in the myelin sheath are the nodes of Ranvier (Figure 4.14). Voltage-gated sodium channels are concentrated in the membrane of the nodes. The distance between nodes is usually 0.2–2.0 mm, depending on the size of the axon (fatter axons have larger internodal distances).

Imagine that the action potential traveling along the axon membrane is like you traveling down a sidewalk. Action potential conduction without myelin is like walking down the sidewalk in small steps, heel-to-toe, using every inch of the sidewalk to creep along. Conduction with myelin,

BOX 4.5 OF SPECIAL INTEREST

Multiple Sclerosis, a Demyelinating Disease

The critical importance of myelin for the normal transfer of information in the human nervous system is revealed by the neurological disorder known as *multiple sclerosis* (MS). Victims of MS often complain of weakness, lack of coordination, and impaired vision and speech. The disease is capricious, usually marked by remissions and relapses that occur over a period of many years. Although the precise cause of MS is still poorly understood, the cause of the sensory and motor disturbances is now quite clear. MS attacks the myelin sheaths of bundles of axons in the brain, spinal cord, and optic nerves. The word *sclerosis* is derived from the Greek word for "hardening," which describes the lesions that develop around bundles of axons, and the sclerosis is *multiple* because the disease attacks many sites in the nervous system at the same time.

Lesions in the brain can now be viewed noninvasively using new methods such as magnetic resonance imaging (MRI). However, neurologists have been able to diagnose MS for many years by taking advantage of the fact that myelin serves the nervous system by increasing the velocity of axonal conduction. One simple test involves stimulating the eye with a checkerboard pattern and measuring the elapsed time until an electrical response is noted at the scalp over the part of the brain that is a target of the optic nerve. People who have MS characteristically have a marked slowing of the conduction velocity of their optic nerve.

Another demyelinating disease called *Guillain–Barré syndrome* attacks the myelin of the peripheral nerves that innervate muscle and skin. This disease may follow minor infectious illnesses and inoculations, and it appears to result from an anomalous immunological response against one's own myelin. The symptoms stem directly from the slowing and/or failure of action potential conduction in the axons that innervate the muscles. This conduction deficit can be demonstrated clinically by stimulating peripheral nerves electrically through the skin and measuring the time it takes to evoke a response (a twitch of a muscle, for instance). Both MS and Guillain–Barré syndrome are characterized by a profound slowing of the response time because saltatory conduction is disrupted.

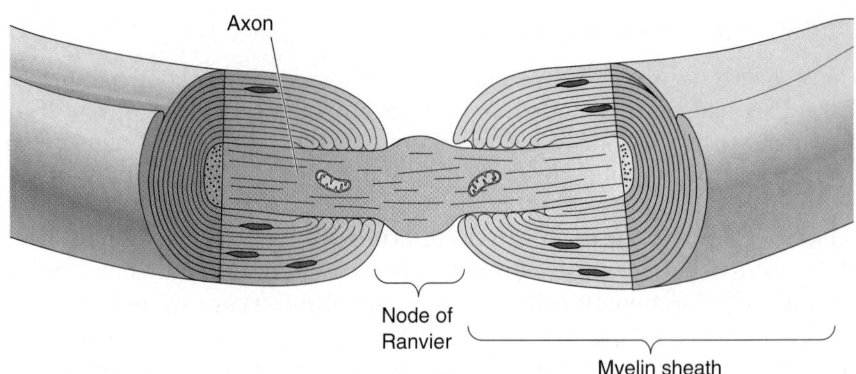

▲ FIGURE 4.14
The myelin sheath and node of Ranvier.
The electrical insulation provided by myelin helps speed action potential conduction from node to node. Voltage-gated sodium channels are concentrated in the axonal membrane at the nodes of Ranvier.

in contrast, is like skipping down the sidewalk. In myelinated axons, action potentials skip from node to node (Figure 4.15). This type of action potential propagation is called **saltatory conduction** (from the Latin meaning "to leap").

ACTION POTENTIALS, AXONS, AND DENDRITES

Action potentials of the type discussed in this chapter are a feature mainly of axons. As a rule, the membranes of dendrites and neuronal cell bodies do not generate sodium-dependent action potentials

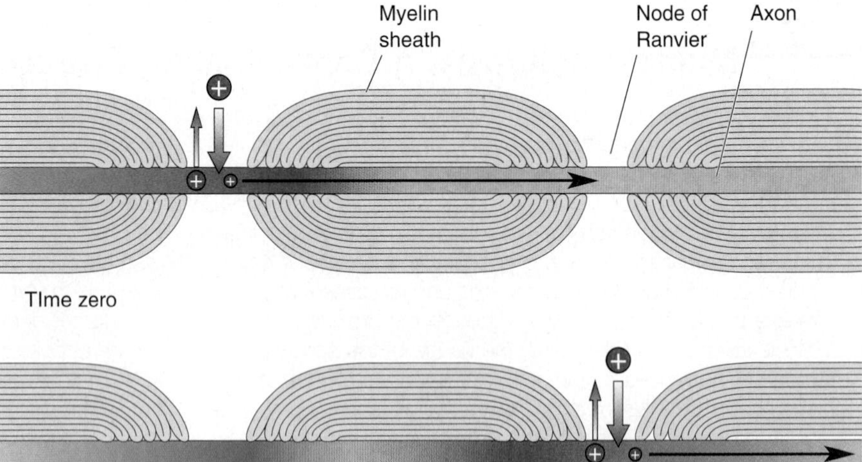

▲ FIGURE 4.15
Saltatory conduction. Myelin allows current to spread farther and faster between nodes, thus speeding action potential conduction. Compare this figure with Figure 4.12.

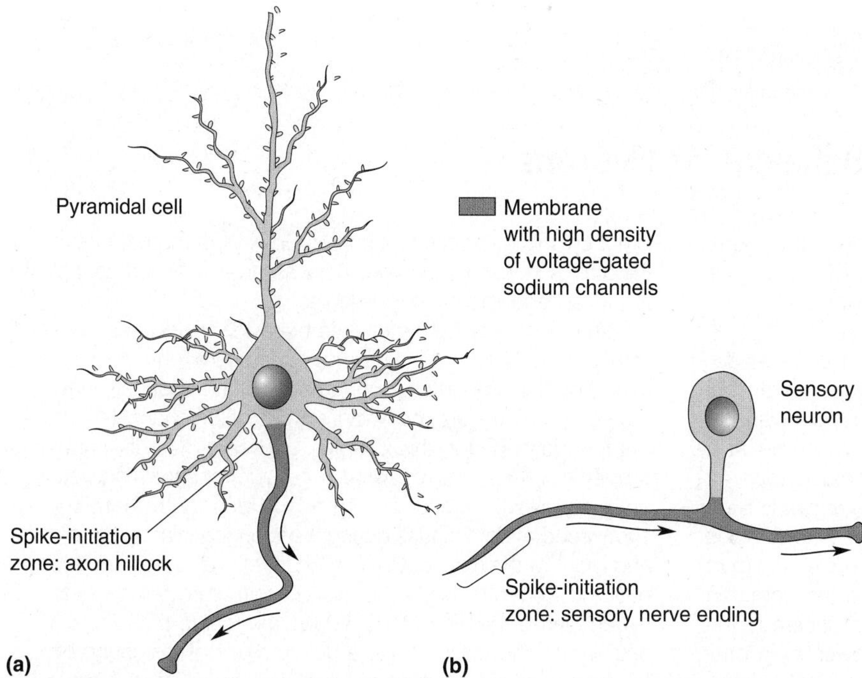

▲ FIGURE 4.16
The spike-initiation zone. Membrane proteins specify the function of different parts of the neuron. Depicted here are **(a)** a cortical pyramidal neuron and **(b)** a primary sensory neuron. Despite the diversity of neuronal structure, the axonal membrane can be identified at the molecular level by its high density of voltage-gated sodium channels. This molecular distinction enables axons to generate and conduct action potentials. The region of membrane where action potentials are normally generated is called the spike-initiation zone. The arrows indicate the normal direction of action potential propagation in these two types of neuron.

because they have very few voltage-gated sodium channels. Only membrane that contains these specialized protein molecules is capable of generating action potentials, and this type of excitable membrane is usually found only in axons. Therefore, the part of the neuron where an axon originates from the soma, the axon hillock, is often also called the **spike-initiation zone**. In a typical neuron in the brain or spinal cord, the depolarization of the dendrites and soma caused by synaptic input from other neurons leads to the generation of action potentials if the membrane of the *axon hillock* is depolarized beyond threshold (Figure 4.16a). In most sensory neurons, however, the spike-initiation zone occurs near the *sensory nerve endings*, where the depolarization caused by sensory stimulation leads to the generation of action potentials that propagate along the sensory nerves (Figure 4.16b).

In Chapter 2, we learned that axons and dendrites differ in their morphology. We now see that they are functionally different, and that this difference in function is specified at the molecular level by the type of protein in the neuronal membrane. Differences in the types and density of membrane ion channels can also account for the characteristic electrical properties of different types of neurons (Box 4.6).

BOX 4.6 OF SPECIAL INTEREST

The Eclectic Electric Behavior of Neurons

Neurons are not all alike; they vary in shape, size, gene expression, and connections. Neurons also differ from one another in their electrical properties. A few examples of the diverse behavior of neurons are shown in Figure A.

The cerebral cortex has two major types of neurons as defined by morphology: aspinous stellate cells and spiny pyramidal cells. A stellate cell typically responds to a steady depolarizing current injected into its soma by firing action potentials at a relatively steady frequency throughout the stimulus (part a). However, most pyramidal cells cannot sustain a steady firing rate. Instead, they fire rapidly at the beginning of the stimulus and then slow down, even if the stimulus remains strong (part b). This slowing over time is called *adaptation*, a very common property among excitable cells. Another firing pattern is the burst, a rapid cluster of action potentials followed by a brief pause. Some cells, including a particular subtype of large pyramidal neuron in the cortex, respond to a steady input with rhythmic, repetitive bursts (part c). Variability of firing patterns is not unique in the cerebral cortex. Surveys of many areas of the brain suggest that neurons have as large an assortment of electrical behaviors as morphologies.

What accounts for the diverse behavior of different types of neurons? Ultimately, each neuron's physiology is determined by the properties and numbers of ion channels in its membrane. There are many more types of ion channels than the few described in this chapter, and each has distinctive properties. For example, some potassium channels activate only very slowly. A neuron with a high density of these will show adaptation because during a prolonged stimulus, more and more of the slow potassium channels will open, and the outward currents they progressively generate will tend to hyperpolarize the membrane. When you realize that a single neuron may have more than a dozen types of ion channels, the source of diverse firing behavior becomes clear. It is the complex interactions among multiple ion channels that create the eclectic electric signature of each class of neuron.

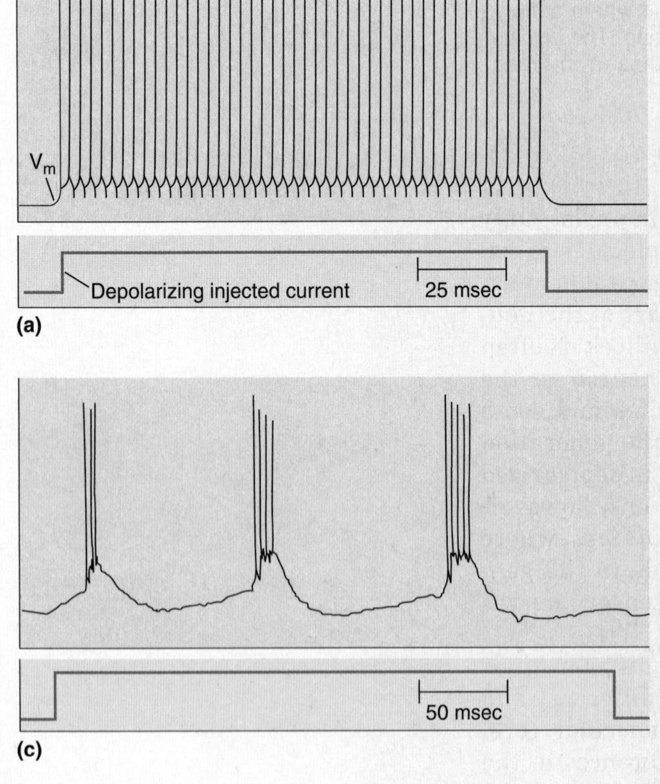

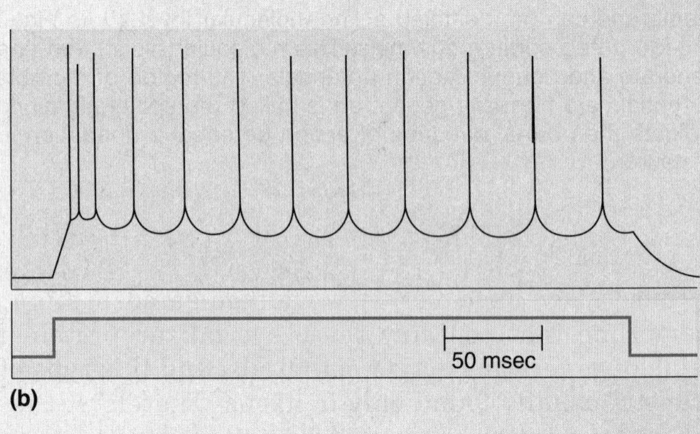

Figure A
The diverse behavior of neurons. (Source: Adapted from Agmon and Connors, 1992.)

CONCLUDING REMARKS

Let's return briefly to the example in Chapter 3 of stepping on a thumbtack. The breaking of the skin caused by the tack stretches the sensory nerve endings of the foot. Special ion channels that are sensitive to the stretching of the membrane open and allow positively charged sodium ions to enter the ends of the axons in the skin. This influx of positive charge depolarizes the membrane to threshold, and the action potential is generated. The positive charge that enters during the rising phase of the action potential spreads along the axon and depolarizes the membrane ahead to threshold. In this way, the action potential is continuously regenerated as it sweeps like a wave along the sensory axon. We now come to the step where this information is distributed to and integrated by other neurons in the central nervous system. This transfer of information from one neuron to another is called *synaptic transmission*, the subject of the next two chapters.

It should come as no surprise that synaptic transmission, like the action potential, depends on specialized proteins in the neuronal membrane. Thus, a picture begins to emerge of the brain as a complicated mesh of interacting neuronal membranes. Consider that a typical neuron with all its neurites has a membrane surface area of about 250,000 μm^2. The surface area of the 85 billion neurons that make up the human brain comes to 21,250 m^2—roughly the size of three soccer fields. This expanse of membrane, with its myriad specialized protein molecules, constitutes the fabric of our minds.

KEY TERMS

Properties of the Action Potential
rising phase (p. 82)
overshoot (p. 82)
falling phase (p. 82)
undershoot (p. 82)
after-hyperpolarization (p. 82)
threshold (p. 84)
absolute refractory period (p. 85)
relative refractory period (p. 86)
optogenetics (p. 86)
channelrhodopsin-2, ChR2 (p. 87)

The Action Potential, in Reality
voltage clamp (p. 92)
voltage-gated sodium channel (p. 92)
patch clamp (p. 94)
channelopathy (p. 96)
tetrodotoxin (TTX) (p. 96)
voltage-gated potassium channel (p. 98)

Action Potential Conduction
saltatory conduction (p. 104)

Action Potentials, Axons, and Dendrites
spike-initiation zone (p. 105)

REVIEW QUESTIONS

1. Define membrane potential (V_m) and sodium equilibrium potential (E_{Na}). Which of these, if either, changes during the course of an action potential?
2. What ions carry the early inward and late outward currents during the action potential?
3. Why is the action potential referred to as "all-or-none"?
4. Some voltage-gated K^+ channels are known as delayed rectifiers because of the timing of their opening during an action potential. What would happen if these channels took much longer than normal to open?
5. Imagine we have labeled tetrodotoxin (TTX) so that it can be seen using a microscope. If we wash this TTX onto a neuron, what parts of the cell would you expect to be labeled? What would be the consequence of applying TTX to this neuron?
6. How does action potential conduction velocity vary with axonal diameter? Why?

FURTHER READING

Boyden ES, Zhang F, Bamberg E, Nagel G, Deisseroth K. 2005. Millisecond-timescale, genetically targeted optical control of neural activity. *Nature Neuroscience* 8:1263–1268.

Hille B. 1992. *Ionic Channels of Excitable Membranes,* 2nd ed. Sunderland, MA: Sinauer.

Hodgkin A. 1976. Chance and design in electrophysiology: an informal account of certain experiments on nerves carried out between 1942 and 1952. *Journal of Physiology* (London) 263:1–21.

Kullmann DM, Waxman SG. 2010. Neurological channelopathies: new insights into disease mechanisms and ion channel function. *Journal of Physiology* (London) 588:1823–1827.

Neher E. 1992. Nobel lecture: ion channels or communication between and within cells. *Neuron* 8:605–612.

Neher E, Sakmann B. 1992. The patch clamp technique. *Scientific American* 266:28–35.

Nicholls J, Martin AR, Fuchs PA, Brown DA, Diamond ME, Weisblat D. 2011. *From Neuron to Brain*, 5th ed. Sunderland, MA: Sinauer.

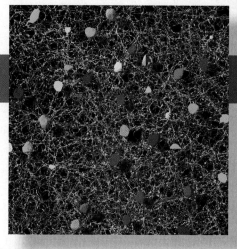

CHAPTER FIVE

Synaptic Transmission

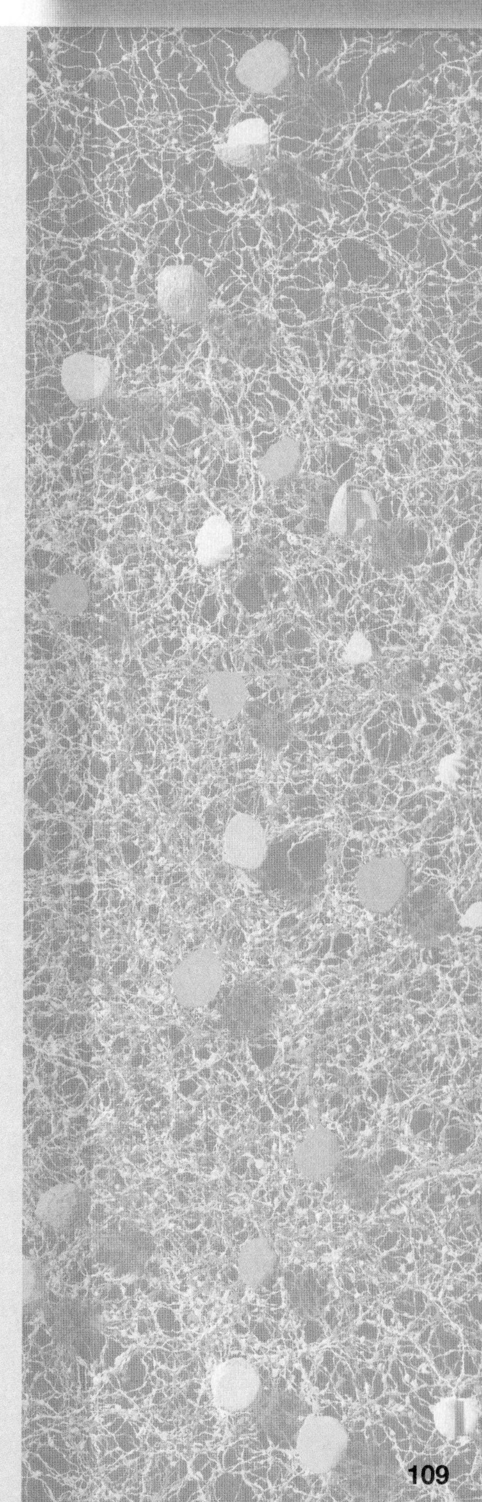

INTRODUCTION

In Chapters 3 and 4, we discussed how mechanical energy, such as a thumbtack entering your foot, can be converted into a neural signal. First, specialized ion channels of the sensory nerve endings allow positive charge to enter the axon. If this depolarization reaches threshold, then action potentials are generated. Because the axonal membrane is excitable and has voltage-gated sodium channels, action potentials can propagate without decrement up the long sensory nerves. For this information to be processed by the rest of the nervous system, however, these neural signals must be passed on to other neurons—for example, the motor neurons that control muscle contraction, as well as neurons in the brain and spinal cord that lead to a coordinated reflex response. By the end of the nineteenth century, it was recognized that this transfer of information from one neuron to another occurs at specialized sites of contact. In 1897, English physiologist Charles Sherrington gave these sites their name: synapses. The process of information transfer at a synapse is called **synaptic transmission**.

The physical nature of synaptic transmission was debated for almost a century. One attractive hypothesis, which nicely explained the speed of synaptic transmission, was that it was simply electrical current flowing from one neuron to the next. The existence of such **electrical synapses** was finally proven in the late 1950s by Edwin Furshpan and David Potter, American physiologists who were studying the nervous system of crayfish at University College London, and Akira Watanabe, who was studying the neurons of lobster at the Tokyo Medical and Dental University. We now know that electrical synapses are common in the brains of invertebrates and vertebrates, including mammals.

An alternative hypothesis about the nature of synaptic transmission, also dating back to the 1800s, was that chemical neurotransmitters transfer information from one neuron to another at the synapse. Solid support for the concept of **chemical synapses** was provided in 1921 by Otto Loewi, then the head of the Pharmacology Department at the University of Graz in Austria. Loewi showed that electrical stimulation of axons innervating the frog's heart caused the release of a chemical that could mimic the effects of neuron stimulation on the heartbeat (Box 5.1). Later, Bernard Katz and his colleagues at University College London conclusively demonstrated that fast transmission at the synapse between a motor neuron axon and skeletal muscle was chemically mediated. By 1951, John Eccles of the Australian National University was studying the physiology of synaptic transmission within the mammalian central nervous system (CNS) using a new tool, the glass microelectrode. These experiments indicated that many CNS synapses also use a chemical transmitter; in fact, chemical synapses comprise the majority of synapses in the brain. During the last decade, new methods of studying the molecules involved in synaptic transmission have revealed that synapses are far more complex than most neuroscientists anticipated.

Synaptic transmission is a large and fascinating topic. The actions of psychoactive drugs, the causes of mental disorders, the neural bases of learning and memory—indeed, all the operations of the nervous system—cannot be understood without knowledge of synaptic transmission. Therefore, we've devoted several chapters to this topic, mainly focusing on chemical synapses. In this chapter, we begin by exploring the basic mechanisms of synaptic transmission. What do different types of synapse look like? How are neurotransmitters synthesized and stored, and how are they released in response to an action potential in the axon terminal? How do neurotransmitters act on the postsynaptic membrane? How do single neurons integrate the inputs provided by the thousands of synapses that impinge upon them?

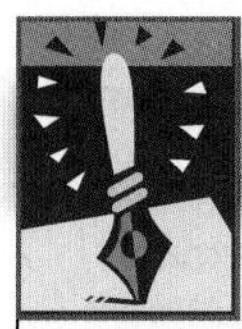

BOX 5.1 OF SPECIAL INTEREST

Otto Loewi's Dream

One of the more colorful stories in the history of neuroscience comes from Otto Loewi, who, working in Austria in the 1920s, showed definitively that synaptic transmission between nerves and the heart is chemically mediated. The heart has two types of innervation; one type speeds the beating of the heart, and the other slows it. The latter type of innervation is supplied by the vagus nerve. Loewi isolated a frog heart with the vagal innervation still intact, stimulated the nerve electrically, and observed the expected effect: the slowing of the heartbeat. The critical demonstration that this effect was chemically mediated came when he applied the solution that had bathed this heart to a second isolated frog heart and found that the beating of this one also slowed.

The idea for this experiment had actually come to Loewi in a dream. Below is his own account:

> In the night of Easter Sunday, 1921, I awoke, turned on the light, and jotted down a few notes on a tiny slip of paper. Then, I fell asleep again. It occurred to me at six o'clock in the morning that during the night I had written down something most important, but I was unable to decipher the scrawl. That Sunday was the most desperate day in my whole scientific life. During the next night, however, I awoke again, at three o'clock, and I remembered what it was. This time I did not take any risk; I got up immediately, went to the laboratory, made the experiment on the frog's heart, described above, and at five o'clock the chemical transmission of the nervous impulse was conclusively proved. . . Careful consideration in daytime would undoubtedly have rejected the kind of experiment I performed, because it would have seemed most unlikely that if a nervous impulse released a transmitting agent, it would do so not just in sufficient quantity to influence the effector organ, in my case the heart, but indeed in such an excess that it could partly escape into the fluid which filled the heart, and could therefore be detected. Yet the whole nocturnal concept of the experiment was based on this eventuality, and the result proved to be positive, contrary to expectation. (Loewi, 1953, pp. 33–34.)

The active compound, which Loewi called *vagusstoff* (literally "vagus substance" in German), turned out to be acetylcholine. As we see in this chapter, acetylcholine is also a transmitter at the synapse between nerve and skeletal muscle. Unlike at the heart, acetylcholine applied to skeletal muscle causes excitation and contraction.

TYPES OF SYNAPSES

We introduced the synapse in Chapter 2. A synapse is the specialized junction where one part of a neuron contacts and communicates with another neuron or cell type (such as a muscle or glandular cell). Information generally flows in one direction, from a neuron to its target cell. The first neuron is said to be *presynaptic*, and the target cell is said to be *postsynaptic*. Let's take a closer look at the different types of synapse.

Electrical Synapses

Electrical synapses are relatively simple in structure and function, and they allow the direct transfer of ionic current from one cell to the next. Electrical synapses occur at specialized sites called **gap junctions**. Gap junctions occur between cells in nearly every part of the body and interconnect many non-neural cells, including epithelial cells, smooth and cardiac muscle cells, liver cells, some glandular cells, and glia.

When gap junctions interconnect neurons, they can function as electrical synapses. At a gap junction, the membranes of two cells are separated by only about 3 nm, and this narrow gap is spanned by clusters of special proteins called *connexins*. There are about 20 different subtypes of connexins, about half of which occur in the brain. Six connexin subunits combine to form a channel called a *connexon*, and two connexons (one from each cell) meet and combine to form a *gap junction channel* (Figure 5.1). The channel allows ions to pass directly from the cytoplasm of one cell

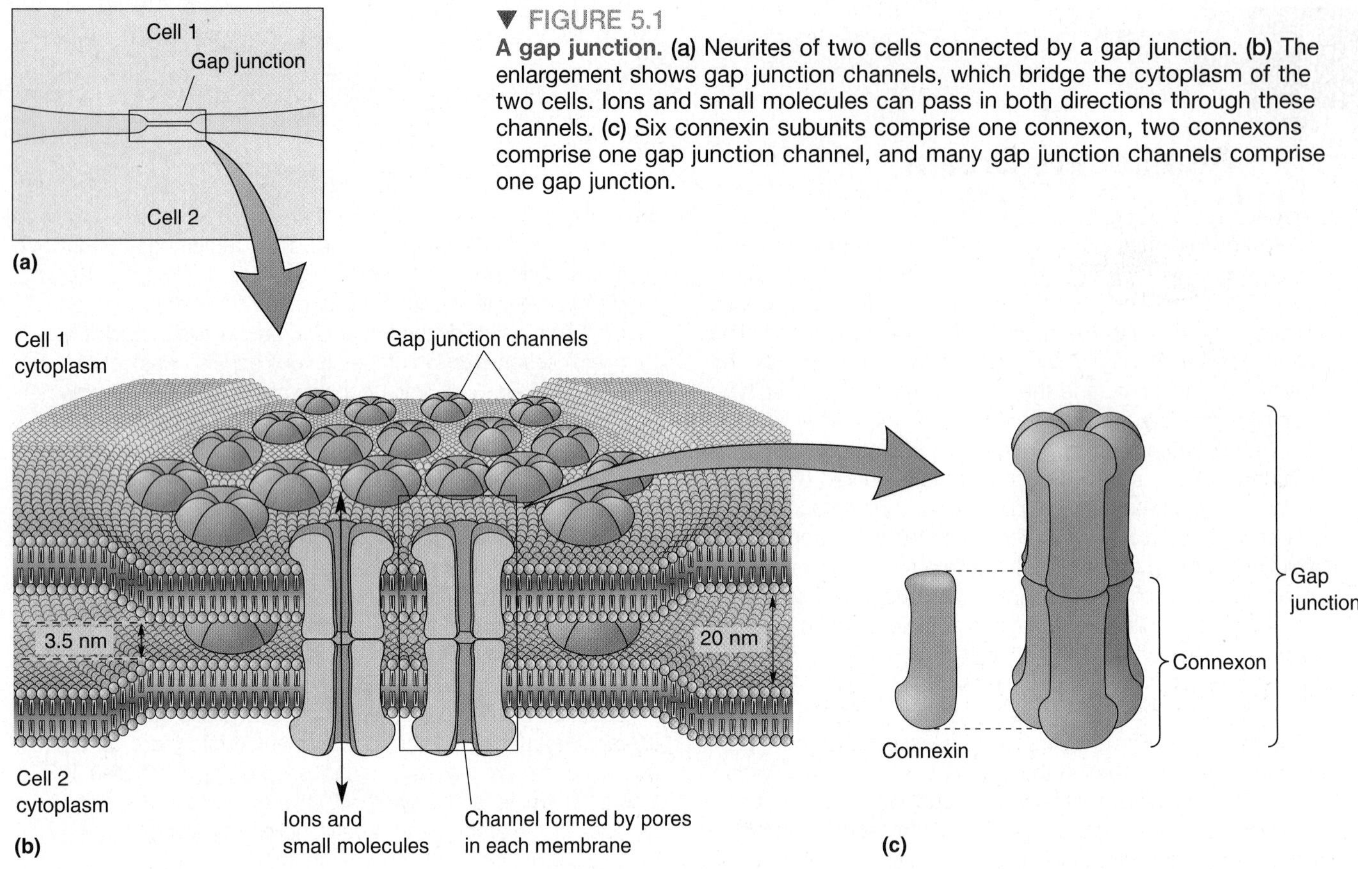

▼ FIGURE 5.1

A gap junction. **(a)** Neurites of two cells connected by a gap junction. **(b)** The enlargement shows gap junction channels, which bridge the cytoplasm of the two cells. Ions and small molecules can pass in both directions through these channels. **(c)** Six connexin subunits comprise one connexon, two connexons comprise one gap junction channel, and many gap junction channels comprise one gap junction.

to the cytoplasm of the other. The pore of most gap junction channels is relatively large. Its diameter is about 1–2 nm, big enough for all the major cellular ions and many small organic molecules to pass through.

Most gap junctions allow ionic current to pass equally well in both directions; therefore, unlike the vast majority of chemical synapses, electrical synapses are bidirectional. Because electrical current (in the form of ions) can pass through these channels, cells connected by gap junctions are said to be *electrically coupled*. Transmission at electrical synapses is very fast and, if the synapse is large, nearly fail-safe. Thus, an action potential in the presynaptic neuron can produce, with very little delay, an action potential in the postsynaptic neuron. In invertebrate species, such as the crayfish, electrical synapses are sometimes found between sensory and motor neurons in neural pathways mediating escape reflexes. This mechanism enables an animal to beat a hasty retreat when faced with a dangerous situation.

Studies in recent years have revealed that electrical synapses are common in every part of the mammalian CNS (Figure 5.2a). When two neurons are electrically coupled, an action potential in the presynaptic neuron causes a small amount of ionic current to flow across the gap junction channels into the other neuron. This current causes an electrically mediated **postsynaptic potential (PSP)** in the second neuron (Figure 5.2b). Note that, because most electrical synapses are bidirectional, when that second neuron generates an action potential, it will in turn induce a PSP in the first neuron. The PSP generated by a single electrical synapse in the mammalian brain is usually small—about 1 mV or less at its peak—and may not, by itself, be large enough to trigger an action

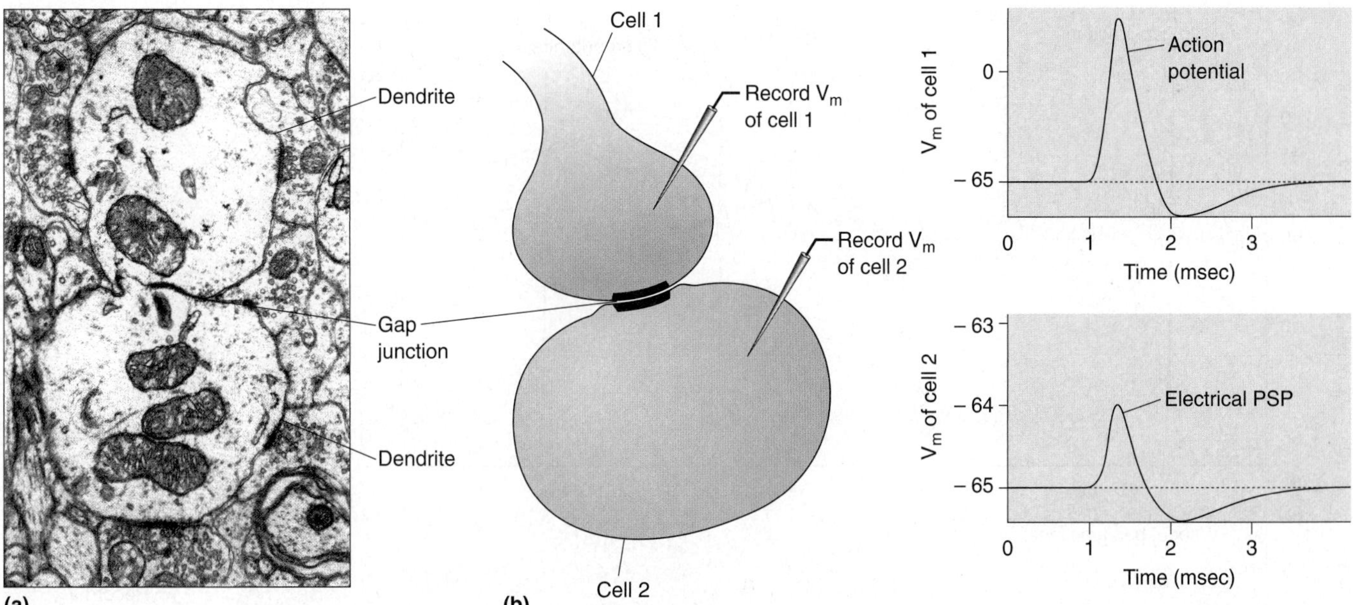

▲ FIGURE 5.2
Electrical synapses. (a) A gap junction interconnecting the dendrites of two neurons constitutes an electrical synapse. (b) An action potential generated in one neuron causes a small amount of ionic current to flow through gap junction channels into a second neuron, inducing an electrical PSP. (Source: Part a from Sloper and Powell, 1978.)

potential in the postsynaptic cell. One neuron usually makes electrical synapses with many other neurons, however, so several PSPs occurring simultaneously may strongly excite a neuron. This is an example of synaptic integration, which is discussed later in the chapter.

The precise roles of electrical synapses vary from one brain region to another. They are often found where normal function requires that the activity of neighboring neurons be highly synchronized. For example, neurons in a brain stem nucleus called the *inferior olive* can generate both small oscillations of membrane voltage and, more occasionally, action potentials. These cells send axons to the cerebellum and are important in motor control. They also make gap junctions with one another. Current that flows through gap junctions during membrane oscillations and action potentials serves to coordinate and synchronize the activity of inferior olivary neurons (Figure 5.3a), and this in turn may help to control the fine timing of motor control. Michael Long and Barry Connors, working at Brown University, found that genetic deletion of a critical gap junction protein called *connexin36* (Cx36) did not alter the neurons' ability to generate oscillations and action potentials but did abolish the synchrony of these events because of the loss of functional gap junctions (Figure 5.3b).

Gap junctions between neurons and other cells are particularly common early in development. Evidence suggests that during prenatal and postnatal brain development, gap junctions allow neighboring cells to share both electrical and chemical signals that may help coordinate their growth and maturation.

Chemical Synapses

Most synaptic transmission in the mature human nervous system is chemical, so the remainder of this chapter and the next will now focus exclusively on chemical synapses. Before we discuss the different types of

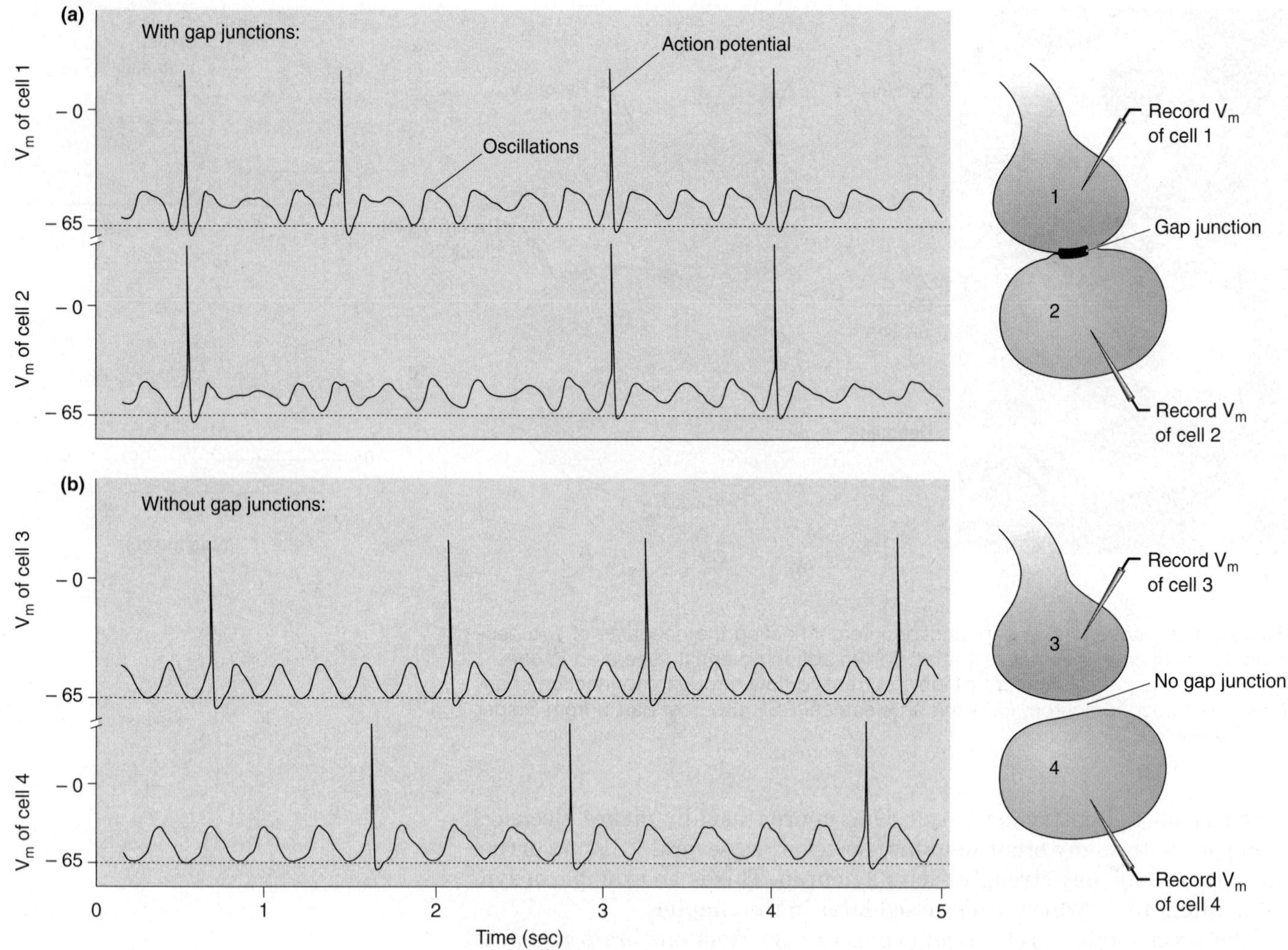

▲ FIGURE 5.3

Electrical synapses can help neurons to synchronize their activity. Certain brain stem neurons generate small, regular oscillations of V_m and occasional action potentials. (a) When two neurons are connected by gap junctions (cells 1 and 2), their oscillations and action potentials are well synchronized. (b) Similar neurons with no gap junctions (cells 3 and 4) generate oscillations and action potentials that are entirely uncoordinated. (Source: Adapted from Long et al., 2002, p. 10903.)

chemical synapses, let's take a look at some of their universal characteristics (Figure 5.4).

The presynaptic and postsynaptic membranes at chemical synapses are separated by a *synaptic cleft* that is 20–50 nm wide, 10 times the width of the separation at gap junctions. The cleft is filled with a matrix of fibrous extracellular protein. One function of this matrix is to serve as a "glue" that binds the pre- and postsynaptic membranes together. The presynaptic side of the synapse, also called the *presynaptic element*, is usually an axon terminal. The terminal typically contains dozens of small membrane-enclosed spheres, each about 50 nm in diameter, called *synaptic vesicles* (Figure 5.5a). These vesicles store neurotransmitter, the chemical used to communicate with the postsynaptic neuron. Many axon terminals also contain larger vesicles, each about 100 nm in diameter, called **secretory granules**. Secretory granules contain soluble protein that appears dark in the electron microscope, so they are sometimes called large, **dense-core vesicles** (Figure 5.5b).

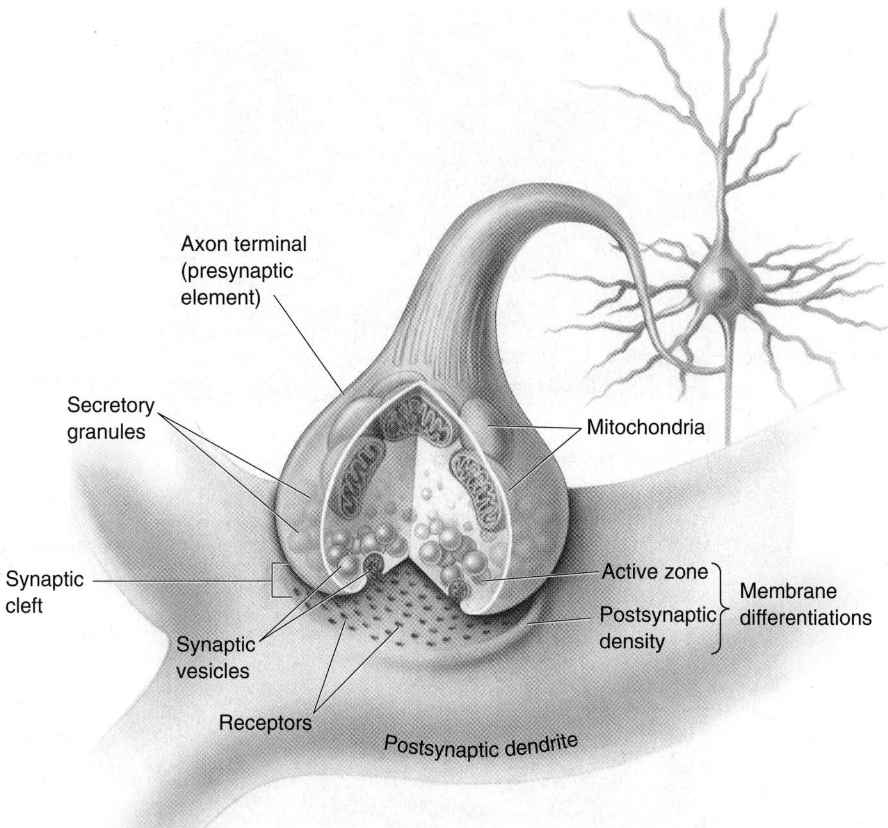

▲ FIGURE 5.4
The components of a chemical synapse.

Dense accumulations of protein adjacent to and within the membranes on either side of the synaptic cleft are collectively called **membrane differentiations**. On the *presynaptic* side, proteins jutting into the cytoplasm of the terminal along the intracellular face of the membrane sometimes look like a field of tiny pyramids. The pyramids, and the membrane associated with them, are the actual sites of neurotransmitter release, called **active zones**. Synaptic vesicles are clustered in the cytoplasm adjacent to the active zones (see Figure 5.4).

The protein thickly accumulated in and just under the *postsynaptic* membrane is called the **postsynaptic density**. The postsynaptic density contains the neurotransmitter receptors, which convert the *intercellular* chemical signal (i.e., neurotransmitter) into an *intracellular* signal (i.e., a change in membrane potential or a chemical change) in the postsynaptic cell. As we shall see, the nature of this postsynaptic response can be quite varied, depending on the type of protein receptor that is activated by the neurotransmitter.

CNS Chemical Synapses. In the CNS, different types of synapse may be distinguished by which part of the neuron is postsynaptic to the axon terminal. If the postsynaptic membrane is on a dendrite, the synapse is said to be *axodendritic*. If the postsynaptic membrane is on the cell body, the synapse is said to be *axosomatic*. In some cases, the postsynaptic membrane is on another axon, and these synapses are called *axoaxonic* (Figure 5.6). When a presynaptic axon contacts a postsynaptic dendritic spine, it is called *axospinous* (Figure 5.7a). In certain specialized neurons, *dendrites* actually form synapses with one another; these are called

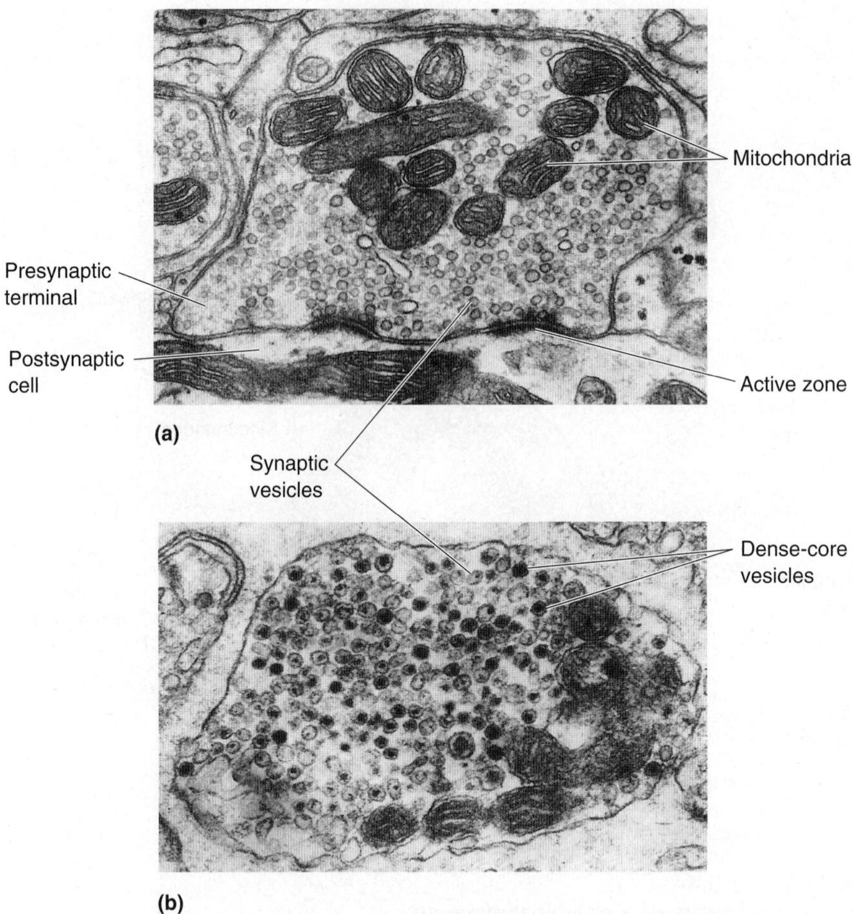

▲ FIGURE 5.5
Chemical synapses, as seen with the electron microscope. **(a)** A fast excitatory synapse in the CNS. **(b)** A synapse in the PNS, with numerous dense-core vesicles. (Source: Part a adapted from Heuser and Reese, 1977, p. 262; part b adapted from Heuser and Reese, 1977, p. 278.)

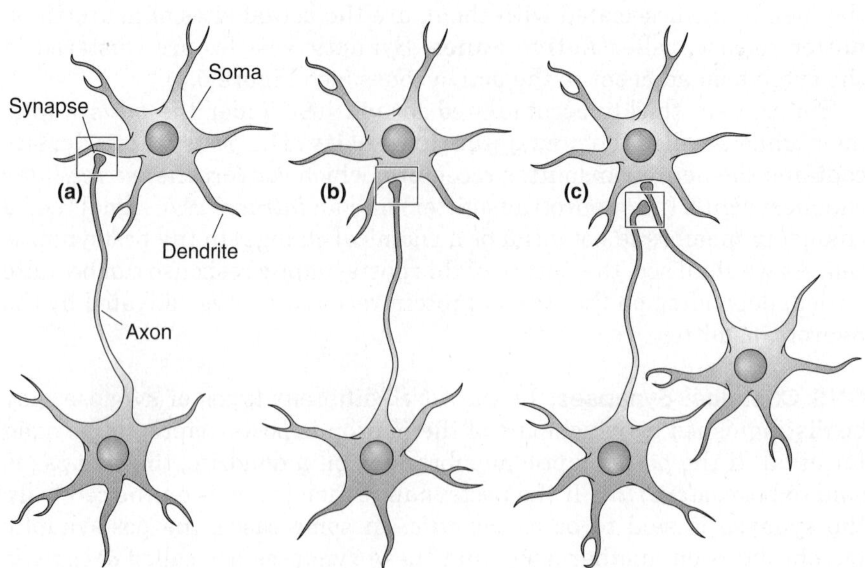

▲ FIGURE 5.6
Synaptic arrangements in the CNS. **(a)** An axodendritic synapse. **(b)** An axosomatic synapse. **(c)** An axoaxonic synapse.

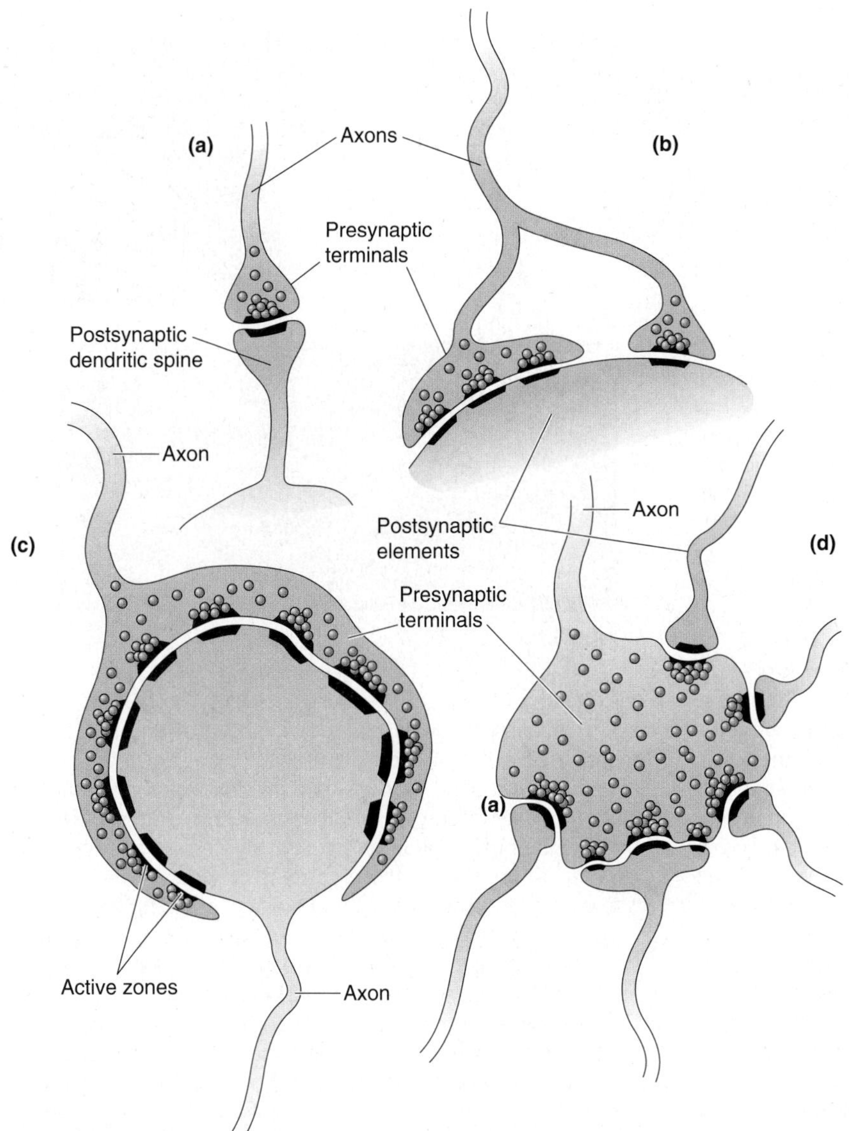

▲ FIGURE 5.7
Various shapes and sizes of CNS synapses. **(a)** Axospinous synapse: A small presynaptic axon terminal contacts a postsynaptic dendritic spine. Notice that presynaptic terminals can be recognized by their many vesicles, and postsynaptic elements have postsynaptic densities. **(b)** An axon branches to form two presynaptic terminals, one larger than the other, and both contact a postsynaptic soma. **(c)** An unusually large axon terminal contacts and surrounds a postsynaptic soma. **(d)** An unusually large presynaptic axon terminal contacts five postsynaptic dendritic spines. Notice that larger synapses have more active zones.

dendrodendritic synapses. The sizes and shapes of CNS synapses also vary widely (Figure 5.7a-d). The finest details of synaptic structure can be studied only under the powerful magnification of the electron microscope (Box 5.2).

CNS synapses may be further classified into two general categories based on the appearance of their presynaptic and postsynaptic membrane differentiations. Synapses in which the membrane differentiation on the postsynaptic side is thicker than that on the presynaptic side are called *asymmetrical synapses*, or *Gray's type I synapses*; those in which the membrane differentiations are of similar thickness are called *symmetrical*

BOX 5.2 PATH OF DISCOVERY

For the Love of Dendritic Spines

by Kristen M. Harris

The first time I looked through the microscope and saw a dendritic spine, it was love at first sight, and the affair has simply never ended. I was a graduate student in the new neurobiology and behavior program at the University of Illinois, and it was indeed an exciting time in neuroscience. The 1979 Society for Neuroscience meeting had only about 5,000 attendees (attendance is now about 25,000), and the member number I obtained during my first year of graduate school was and remains 2500.

I had hoped to discover what a "learned" dendritic spine looks like by training animals and then using the Golgi staining method to quantify changes in spine number and shape. Eagerly, I developed a high-throughput project, preparing the brains from many rats at once, sectioning through the whole brains, checking that the silver staining had worked, and then storing the tissue sections in butanol while engaging undergraduates to help mount them on microscope slides. To our dismay, we found several months later that all the silver had been dissolved out of the cells. There were no cells to see, and the project died an untimely death.

I was fortunate, however, to meet Professor Timothy Teyler at a Gordon Research Conference. He had recently brought the hippocampal slice preparation to the United States from Norway and was moving his lab from Harvard to a new medical school in Rootstown, Ohio. I was completely enamored by the experimental control that brain slices might offer, so I developed a Golgi-slice procedure and moved to complete my PhD with Teyler. This time, I prepared one slice at a time, and as can be seen in Figure A, the spines were exquisitely visible. Unfortunately, accurate counts and shape measurements of the tiny spines were just beyond the resolution of light microscopy.

While I was a graduate student, I talked my way into the esteemed summer course in neurobiology at the Marine Biological Laboratories in Woods Hole, Massachusetts. There I first learned serial-section three-dimensional electron microscopy (3DEM). I was truly hooked. With 3DEM, one could reconstruct dendrites, axons, and glia, and not only measure and count dendritic spines but also see where synapses formed, what was inside them, and how glia associated with synapses (Figure B). The 3DEM platform offers enormous possibilities for discovery. My life continues to be devoted to uncovering the processes of synapse formation and plasticity during learning and memory in the brain.

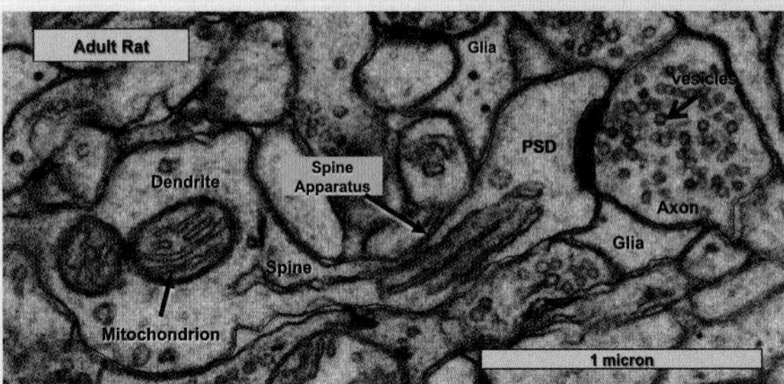

Figure B

Early in my career, while the revolution of molecular biology was sweeping the field, only a rare student or fellow scientist shared my enthusiasm for 3DEM. That bias has shifted dramatically as neuroscientists have come to recognize the importance of understanding how molecules work in consort with intracellular organelles in small spaces like dendrites and spines. Furthermore, all maps of neural circuitry must include synapses. These endeavors have drawn scientists from nearly every field, making 3DEM even more exciting as many of the imaging and reconstruction processes previously done manually are being automated. Figure C shows a recent 3DEM rendering, with color-coding of organelles and synaptic components. It is indeed thrilling to be part of this growth. New findings abound regarding the plasticity of synapse structure during normal changes in brain function and as altered by diseases that tragically affect who we are as human beings.

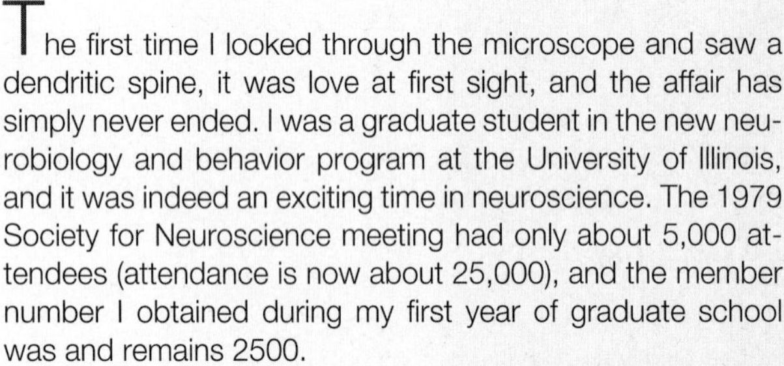

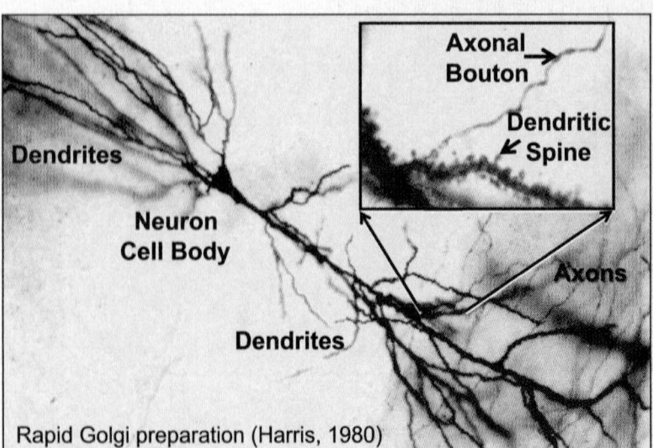

Figure A

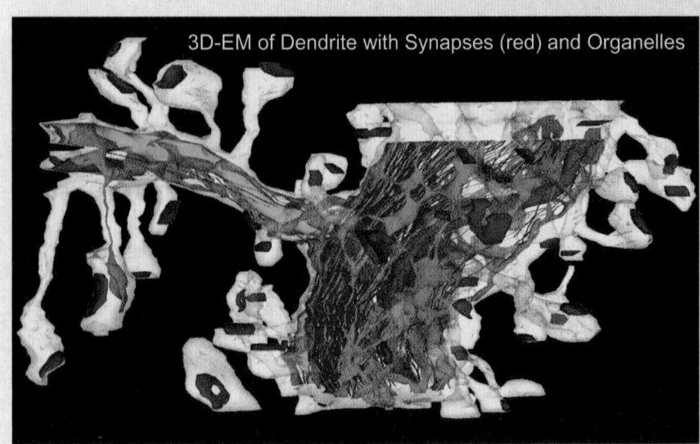

Figure C

synapses, or *Gray's type II synapses* (Figure 5.8). As we shall see later in the chapter, these structural differences reveal functional differences. Gray's type I synapses are usually excitatory, while Gray's type II synapses are usually inhibitory.

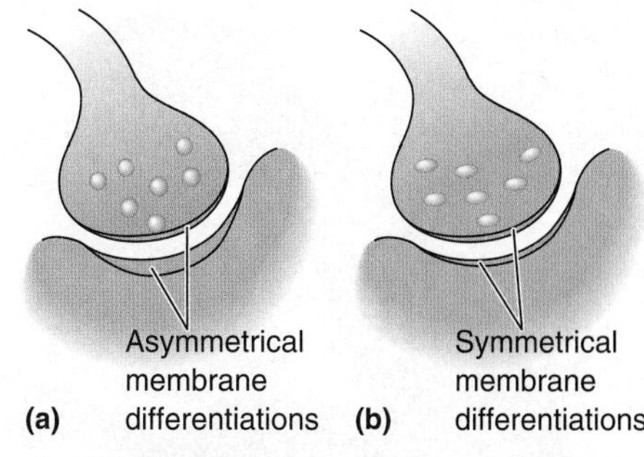

▲ FIGURE 5.8
Two categories of CNS synaptic membrane differentiations. **(a) A Gray's type I synapse is asymmetrical and usually excitatory. (b) A Gray's type II synapse is symmetrical and usually inhibitory.**

The Neuromuscular Junction. Synaptic junctions also exist outside the CNS. For example, axons of the autonomic nervous system innervate glands, smooth muscle, and the heart. Chemical synapses also occur between the axons of motor neurons of the spinal cord and skeletal muscle. Such a synapse is called a **neuromuscular junction**, and it has many of the structural features of chemical synapses in the CNS (Figure 5.9).

Neuromuscular synaptic transmission is fast and reliable. An action potential in the motor axon always causes an action potential in the muscle cell it innervates. This reliability is accounted for, in part, by structural specializations of the neuromuscular junction. Its most important specialization is its size—it is one of the largest synapses in the body. The presynaptic terminal also contains a large number of active zones. In addition, the postsynaptic membrane, also called the **motor endplate**, contains a series of shallow folds. The presynaptic active zones are precisely aligned with these junctional folds, and the postsynaptic membrane of the folds is packed with neurotransmitter receptors. This structure ensures that many neurotransmitter molecules are focally released onto a large surface of chemically sensitive membrane.

Because neuromuscular junctions are more accessible to researchers than CNS synapses, much of what we know about the mechanisms of synaptic transmission was first established here. Neuromuscular junctions are also of considerable clinical significance; diseases, drugs, and poisons that interfere with this chemical synapse have direct effects on vital bodily functions.

PRINCIPLES OF CHEMICAL SYNAPTIC TRANSMISSION

Consider the basic requirements of chemical synaptic transmission. There must be a mechanism for synthesizing neurotransmitter and packing it into the synaptic vesicles, a mechanism for causing vesicles to spill their contents into the synaptic cleft in response to a presynaptic action potential, a mechanism for producing an electrical or biochemical response to neurotransmitter in the postsynaptic neuron, and a mechanism for removing neurotransmitter from the synaptic cleft. And, to be useful for sensation, perception, and the control of movement, all these things must often occur very rapidly, within milliseconds. No wonder physiologists were initially skeptical about the existence of chemical synapses in the brain!

Fortunately, thanks to several decades of research on the topic, we now understand how many of these aspects of synaptic transmission are so efficiently carried out. Here we'll present a general survey of the basic principles. In Chapter 6, we will take a more detailed look at the individual neurotransmitters and their modes of postsynaptic action.

Neurotransmitters

Since the discovery of chemical synaptic transmission, researchers have been identifying neurotransmitters in the brain. Our current understanding is that the major neurotransmitters fall into one of three chemical

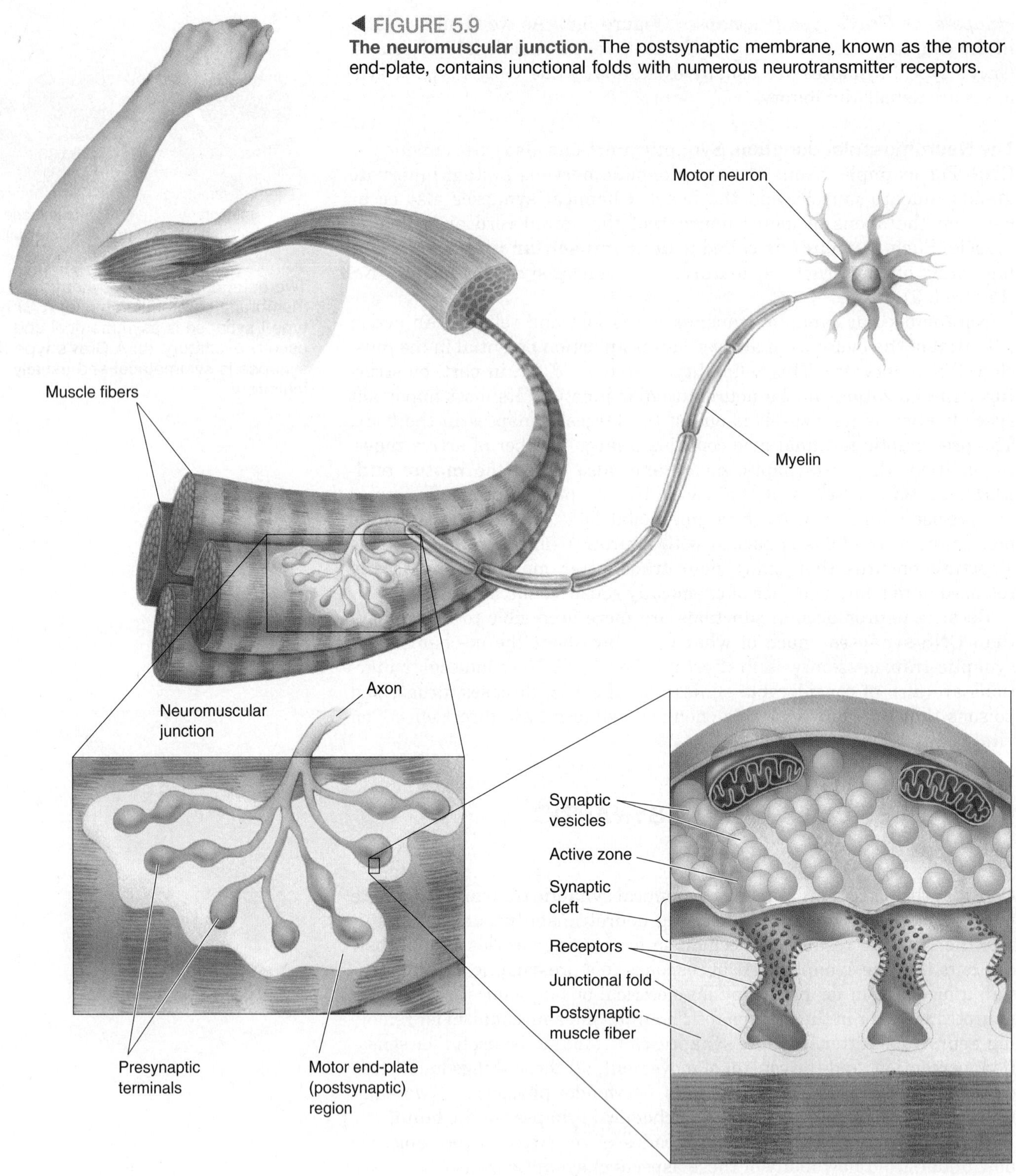

◀ FIGURE 5.9

The neuromuscular junction. The postsynaptic membrane, known as the motor end-plate, contains junctional folds with numerous neurotransmitter receptors.

categories: (1) *amino acids*, (2) *amines*, and (3) *peptides* (Table 5.1). Some representatives of these categories are shown in Figure 5.10. The amino acid and amine neurotransmitters are all small organic molecules containing at least one nitrogen atom, and they are stored in and released from synaptic vesicles. Peptide neurotransmitters are large molecules—chains of amino acids—stored in and released from secretory granules. As discussed earlier, secretory granules and synaptic vesicles are frequently observed in

TABLE 5.1 **The Major Neurotransmitters**

Amino Acids	Amines	Peptides
Gamma-aminobutyric acid (GABA)	Acetylcholine (ACh)	Cholecystokinin (CCK)
Glutamate (Glu)	Dopamine (DA)	Dynorphin
Glycine (Gly)	Epinephrine	Enkephalins (Enk)
	Histamine	*N*-acetylaspartylglutamate (NAAG)
	Norepinephrine (NE)	Neuropeptide Y
	Serotonin (5-HT)	Somatostatin
		Substance P
		Thyrotropin-releasing hormone
		Vasoactive intestinal polypeptide (VIP)

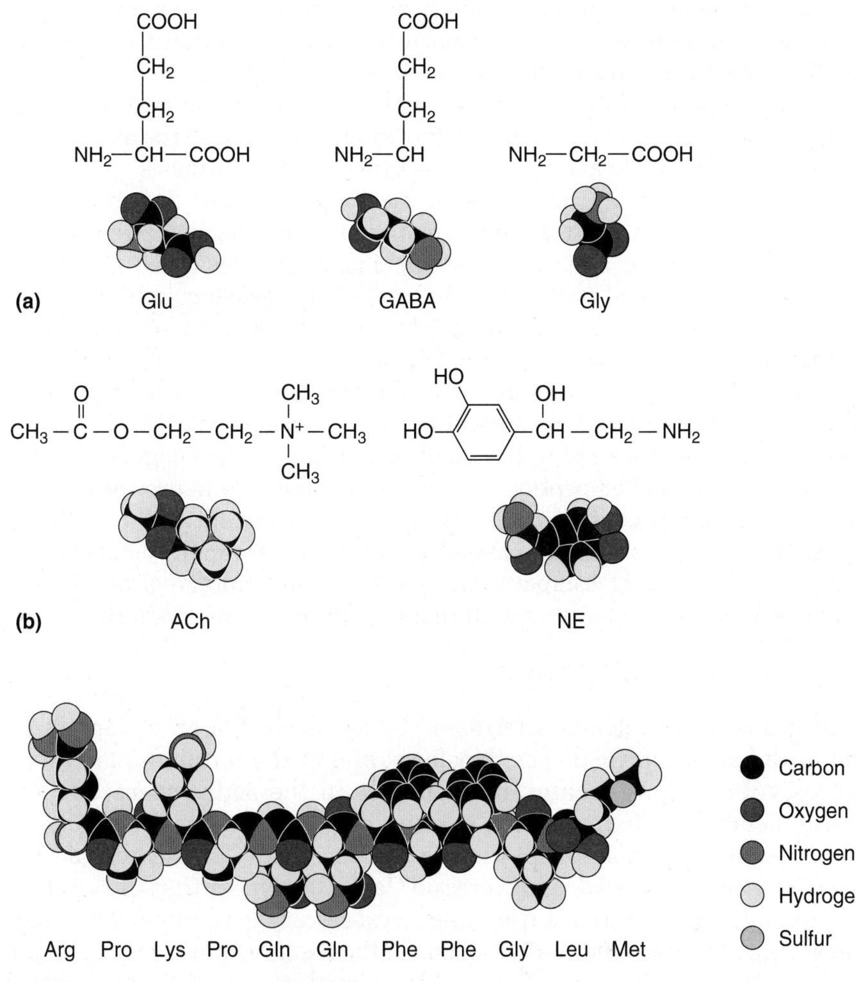

▲ FIGURE 5.10

Representative neurotransmitters. **(a)** The amino acid neurotransmitters glutamate, GABA, and glycine. **(b)** The amine neurotransmitters acetylcholine and norepinephrine. **(c)** The peptide neurotransmitter substance P. (For the abbreviations and chemical structures of amino acids in substance P, see Figure 3.4b.)

the same axon terminals. Consistent with this observation, peptides often exist in the same axon terminals that contain amine or amino acid neurotransmitters. As we'll discuss in a moment, these different neurotransmitters are released under different conditions.

Different neurons in the brain release different neurotransmitters. The speed of synaptic transmission varies widely. Fast forms of synaptic transmission last from about 10–100 msec, and at most CNS synapses are mediated by the amino acids **glutamate (Glu)**, **gamma-aminobutyric acid (GABA)**, or **glycine (Gly)**. The amine **acetylcholine (ACh)** mediates fast synaptic transmission at all neuromuscular junctions. Slower forms of synaptic transmission may last from hundreds of milliseconds to minutes; they can occur in the CNS and in the periphery and are mediated by transmitters from all three chemical categories.

Neurotransmitter Synthesis and Storage

Chemical synaptic transmission requires that neurotransmitters be synthesized and ready for release. Different neurotransmitters are synthesized in different ways. For example, glutamate and glycine are among the 20 amino acids that are the building blocks of protein (see Figure 3.4b); consequently, they are abundant in all cells of the body, including neurons. In contrast, GABA and the amines are made primarily by the neurons that release them. These neurons contain specific enzymes that synthesize the neurotransmitters from various metabolic precursors. The synthesizing enzymes for both amino acid and amine neurotransmitters are transported to the axon terminal, where they locally and rapidly direct transmitter synthesis.

Once synthesized in the cytosol of the axon terminal, the amino acid and amine neurotransmitters must be taken up by the synaptic vesicles. Concentrating these neurotransmitters inside the vesicle is the job of **transporters**, special proteins embedded in the vesicle membrane.

Quite different mechanisms are used to synthesize and store peptides in secretory granules. As we learned in Chapters 2 and 3, peptides are formed when amino acids are strung together by the ribosomes of the cell body. In the case of peptide neurotransmitters, this occurs in the rough ER. Generally, a long peptide synthesized in the rough ER is split in the Golgi apparatus, and one of the smaller peptide fragments is the active neurotransmitter. Secretory granules containing the peptide neurotransmitter bud off from the Golgi apparatus and are carried to the axon terminal by axoplasmic transport. Figure 5.11 compares the synthesis and storage of amine and amino acid neurotransmitters with that of peptide neurotransmitters.

Neurotransmitter Release

Neurotransmitter release is triggered by the arrival of an action potential in the axon terminal. The depolarization of the terminal membrane causes **voltage-gated calcium channels** in the active zones to open. These membrane channels are very similar to the sodium channels we discussed in Chapter 4, except they are permeable to Ca^{2+} instead of Na^{+}. There is a large inward driving force on Ca^{2+}. Remember that the internal calcium ion concentration—$[Ca^{2+}]_i$—at rest is very low, only 0.0002 mM; therefore, Ca^{2+} will flood the cytoplasm of the axon terminal as long as the calcium channels are open. The resulting elevation in $[Ca^{2+}]_i$ is the signal that causes neurotransmitter to be released from synaptic vesicles.

The vesicles release their contents by a process called **exocytosis**. The membrane of the synaptic vesicle fuses to the presynaptic membrane at the active zone, allowing the contents of the vesicle to spill out into the synaptic cleft (Figure 5.12). Studies of a giant synapse in the squid nervous

▲ FIGURE 5.11

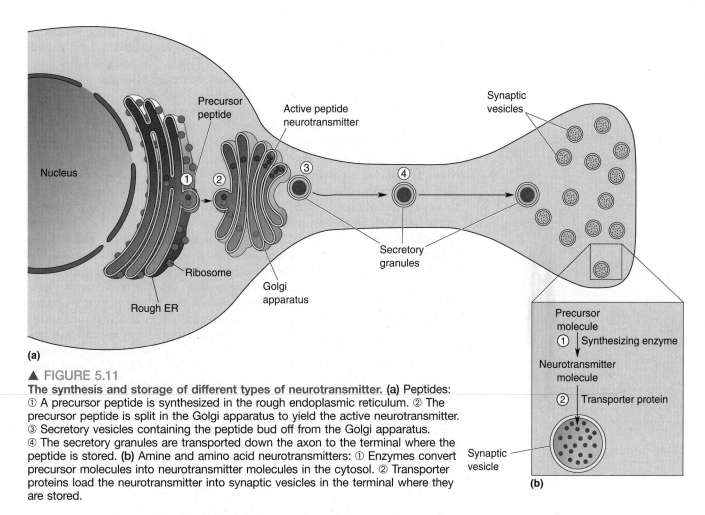

The synthesis and storage of different types of neurotransmitter. (a) Peptides: ① A precursor peptide is synthesized in the rough endoplasmic reticulum. ② The precursor peptide is split in the Golgi apparatus to yield the active neurotransmitter. ③ Secretory vesicles containing the peptide bud off from the Golgi apparatus. ④ The secretory granules are transported down the axon to the terminal where the peptide is stored. **(b)** Amine and amino acid neurotransmitters: ① Enzymes convert precursor molecules into neurotransmitter molecules in the cytosol. ② Transporter proteins load the neurotransmitter into synaptic vesicles in the terminal where they are stored.

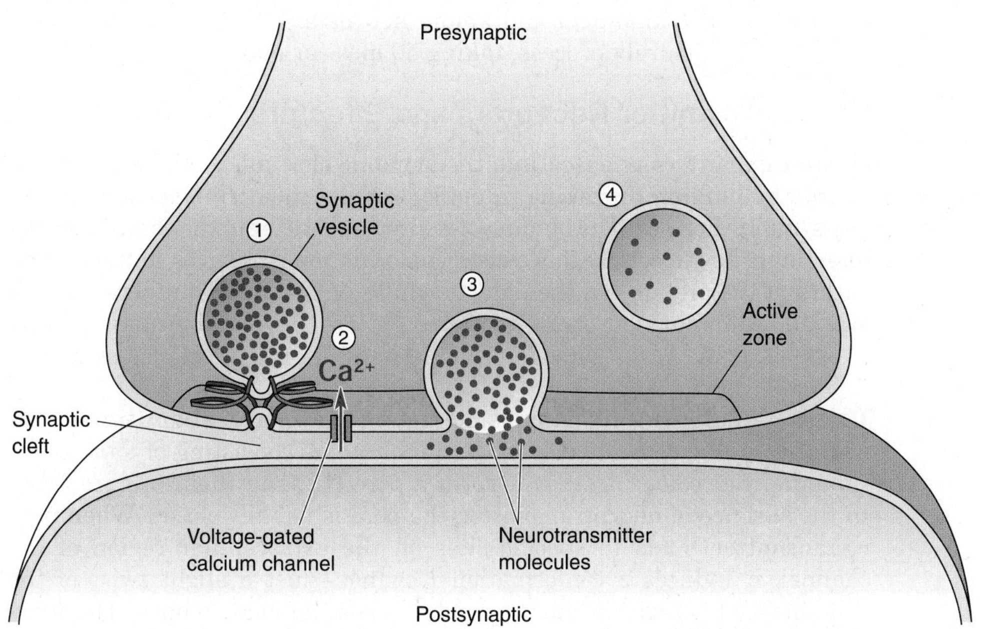

▲ FIGURE 5.12

The release of neurotransmitter by exocytosis. ① A synaptic vesicle loaded with neurotransmitter, in response to ② an influx of Ca^{2+} through voltage-gated calcium channels, ③ releases its contents into the synaptic cleft by the fusion of the vesicle membrane with the presynaptic membrane, and ④ is eventually recycled by the process of endocytosis.

system showed that exocytosis can occur very rapidly, within 0.2 msec of the Ca^{2+} influx into the terminal. Synapses in mammals, which generally occur at higher temperatures, are even faster. Exocytosis is quick because Ca^{2+} enters at the active zone precisely where synaptic vesicles are ready and waiting to release their contents. In this local "microdomain" around the active zone, calcium can achieve relatively high concentrations (greater than about 0.01 mM).

The mechanism by which $[Ca^{2+}]_i$ stimulates exocytosis has been under intensive investigation. The speed of neurotransmitter release suggests that the vesicles involved are those already "docked" at the active zones. Docking is believed to involve interactions between proteins in the synaptic vesicle membrane and the presynaptic cell membrane under the active zone (Box 5.3). In the presence of high $[Ca^{2+}]_i$, these proteins alter their conformation so that the lipid bilayers of the vesicle and presynaptic membranes fuse, forming a pore that allows the neurotransmitter to escape into the cleft. The mouth of this exocytotic fusion pore continues to expand until the membrane of the vesicle is fully incorporated into the presynaptic membrane (Figure 5.13). The vesicle membrane is later recovered by the process of **endocytosis**, and the recycled vesicle is refilled with neurotransmitter (see Figure 5.12). During periods of prolonged stimulation, vesicles are mobilized from a "reserve pool" that is bound to the cytoskeleton of the axon terminal. The release of these vesicles from the cytoskeleton, and their docking to the active zone, is also triggered by elevations of $[Ca^{2+}]_i$.

Secretory granules also release peptide neurotransmitters by exocytosis, in a calcium-dependent fashion, but typically not at the active zones. Because the sites of granule exocytosis occur at a distance from the sites of Ca^{2+} entry, peptide neurotransmitters are usually not released in response to every action potential invading the terminal. Instead, the release of peptides generally requires high-frequency trains of action potentials, so that the $[Ca^{2+}]_i$ throughout the terminal can build to the level required to trigger release away from the active zones. Unlike the fast release of amino acid and amine neurotransmitters, the release of peptides is a leisurely process, taking 50 msec or more.

Neurotransmitter Receptors and Effectors

Neurotransmitters released into the synaptic cleft affect the postsynaptic neuron by binding to specific receptor proteins that are embedded in the postsynaptic density. The binding of neurotransmitter to the receptor is like inserting a key in a lock; this causes conformational changes in the protein such that the protein can then function differently. Although there are well over 100 different neurotransmitter receptors, they can be classified into two types: transmitter-gated ion channels and G-protein-coupled receptors.

Transmitter-Gated Ion Channels. Receptors known as **transmitter-gated ion channels** are membrane-spanning proteins consisting of four or five subunits that come together to form a pore between them (Figure 5.14). In the absence of neurotransmitter, the pore is usually closed. When neurotransmitter binds to specific sites on the extracellular region of the channel, it induces a conformational change—just a slight twist of the subunits—which within microseconds causes the pore to open. The functional consequence of this depends on which ions can pass through the pore.

Transmitter-gated channels generally do not show the same degree of ion selectivity as do voltage-gated channels. For example, the ACh-gated ion channels at the neuromuscular junction are permeable to both Na^+ and K^+. Nonetheless, as a rule, if the open channels are permeable to

BOX 5.3 BRAIN FOOD

How to SNARE a Vesicle

Yeasts are single-cell organisms valued for their ability to make dough rise and grape juice ferment into wine. Remarkably, the humble yeasts have some close similarities to the chemical synapses in our brain. Recent research has shown that the proteins controlling secretion in both yeast cells and synapses have only minor differences. Apparently, these molecules are so generally useful that they have been conserved across more than a billion years of evolution, and they are found in all eukaryotic cells.

The trick to fast synaptic function is to deliver neurotransmitter-filled vesicles to just the right place—the presynaptic membrane—and then cause them to fuse at just the right time, when an action potential delivers a pulse of high Ca^{2+} concentration to the cytosol. This process of exocytosis is a special case of a more general cellular problem, *membrane trafficking*. Cells have many types of membranes, including those enclosing the whole cell, the nucleus, endoplasmic reticulum, Golgi apparatus, and various types of vesicles. To avoid intracellular chaos, each of these membranes needs to be moved and delivered to specific locations within the cell. After delivery, one type of membrane often has to fuse with another type. A common molecular machinery has evolved for the delivery and fusion of all these membranes, and small variations in these molecules determine how and when membrane trafficking takes place.

The specific binding and fusion of membranes seem to depend on the SNARE family of proteins, which were first found in yeast cells. SNARE is an acronym too convoluted to define here, but the name perfectly defines the function of these proteins: SNAREs allow one membrane to snare another. Each SNARE peptide has a lipid-loving end that embeds itself within the membrane and a longer tail that projects into the cytosol. Vesicles have "v-SNAREs," and the outer membrane has "t-SNAREs" (for target membrane). The cytosolic ends of these two complementary types of SNAREs can bind very tightly to one another, allowing a vesicle to "dock" very close to a presynaptic membrane and nowhere else (Figure A).

Although complexes of v-SNAREs and t-SNAREs form the main connection between vesicle membrane and target membrane, a large and bewildering array of other presynaptic proteins stick to this SNARE complex. We still don't understand the functions of all of them, but *synaptotagmin*, a vesicle protein, is the critical Ca^{2+} sensor that rapidly triggers vesicle fusion and thus transmitter release. On the presynaptic membrane side, calcium channels may form part of the docking complex. As the calcium channels are very close to the docked vesicles, inflowing Ca^{2+} can trigger transmitter release with astonishing speed—within about 60 μsec in a mammalian synapse at body temperature. The brain has several varieties of synaptotagmins, including one that is specialized for exceptionally fast synaptic transmission.

We have a way to go before we understand all the molecules involved in synaptic transmission. In the meantime, we can count on yeasts to provide delightful brain food (and drink) for thought.

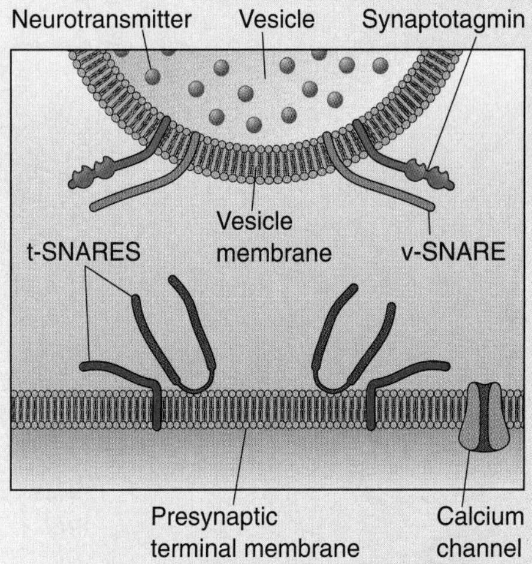

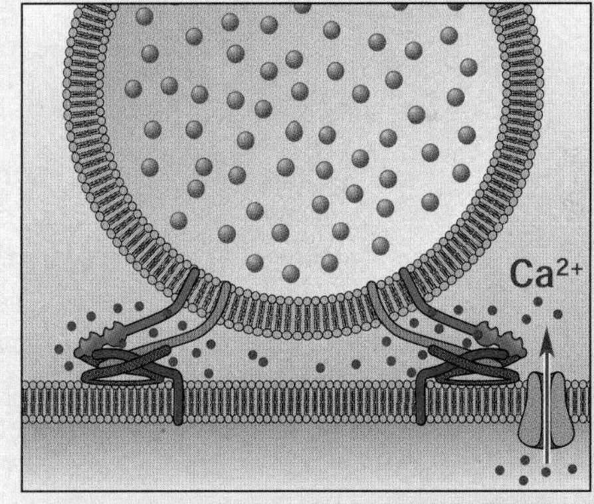

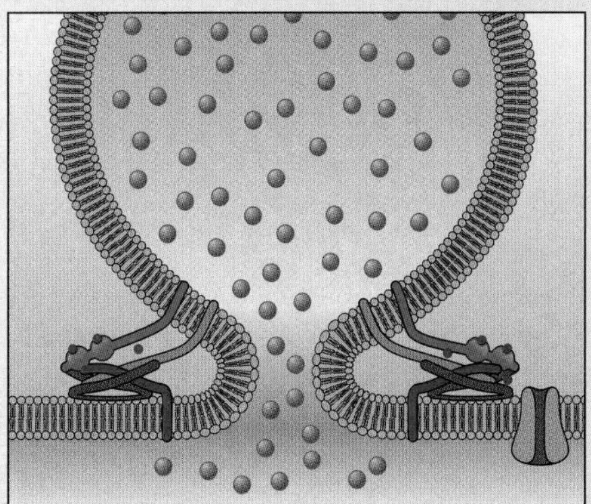

Figure A
SNAREs and vesicle fusion.

▶ FIGURE 5.13
A "receptor's eye" view of neurotransmitter release. (a) This is a view of the extracellular surface of the active zone of a neuromuscular junction in frog. The particles are believed to be calcium channels. (b) In this view, the presynaptic terminal had been stimulated to release neurotransmitter. The exocytotic fusion pores are where synaptic vesicles have fused with the presynaptic membrane and released their contents. (Source: Heuser and Reese, 1973.)

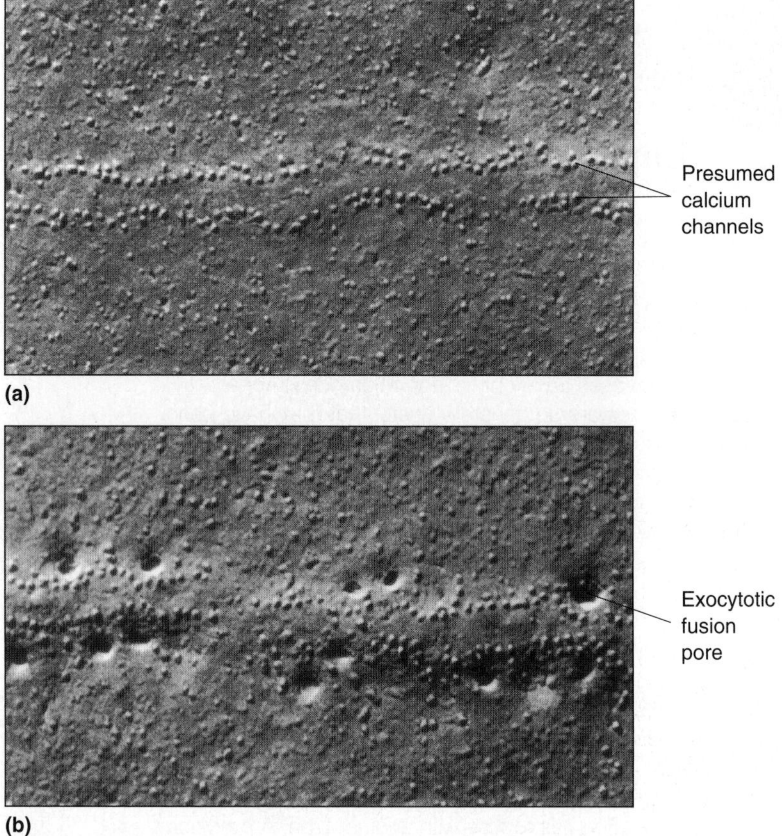

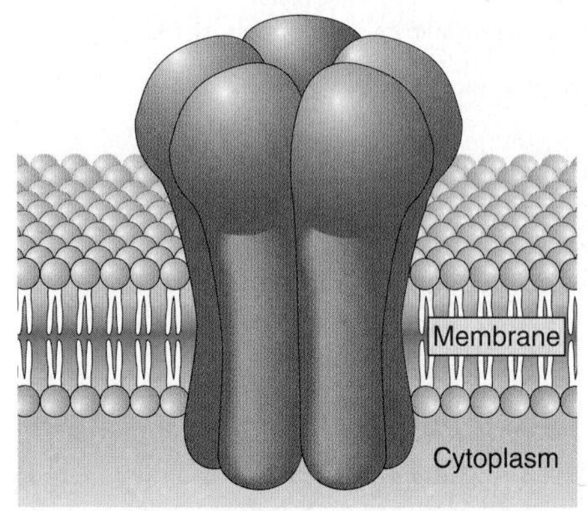

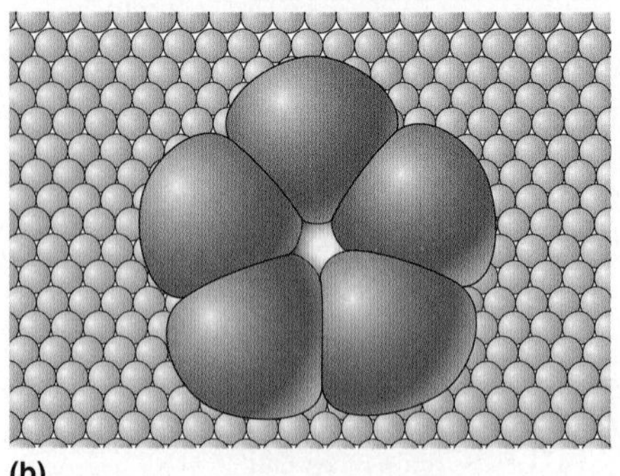

▲ FIGURE 5.14
The structure of a transmitter-gated ion channel. (a) Side view of an ACh-gated ion channel. (b) Top view of the channel, showing the pore at the center of the five subunits.

Na^+, the net effect will be to depolarize the postsynaptic cell from the resting membrane potential (Box 5.4). Because it tends to bring the membrane potential toward threshold for generating action potentials, this effect is said to be *excitatory*. A transient postsynaptic membrane depolarization caused by the presynaptic release of neurotransmitter is called an **excitatory postsynaptic potential (EPSP)** (Figure 5.15). Synaptic activation of ACh-gated and glutamate-gated ion channels causes EPSPs.

If the transmitter-gated channels are permeable to Cl^-, the usual net effect will be to hyperpolarize the postsynaptic cell from the resting membrane potential (because the chloride equilibrium potential is usually negative; see Chapter 3). Because it tends to bring the membrane potential away from threshold for generating action potentials, this effect is said to be *inhibitory*. A transient hyperpolarization of the postsynaptic membrane potential caused by the presynaptic release of neurotransmitter is called an **inhibitory postsynaptic potential (IPSP)** (Figure 5.16). Synaptic activation of glycine-gated or GABA-gated ion channels cause an IPSP. We'll discuss EPSPs and IPSPs in more detail shortly when we explore the principles of synaptic integration.

G-Protein-Coupled Receptors. Fast chemical synaptic transmission is mediated by amino acid and amine neurotransmitters acting on transmitter-gated ion channels. However, all three types of neurotransmitter, acting on **G-protein-coupled receptors**, can also have slower, longer lasting, and much more diverse postsynaptic actions. This type of transmitter action involves three steps:

1. Neurotransmitter molecules bind to receptor proteins embedded in the postsynaptic membrane.

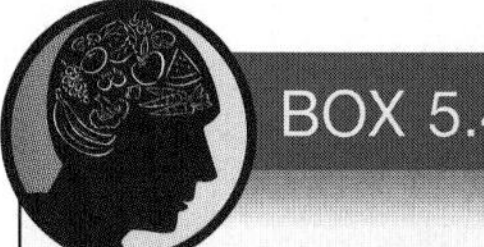

BOX 5.4 BRAIN FOOD

Reversal Potentials

In Chapter 4, we saw that when the membrane voltage-gated sodium channels open during an action potential, Na^+ enters the cell, causing the membrane potential to rapidly depolarize until it approaches the sodium equilibrium potential, E_{Na}, about 40 mV. Unlike the voltage-gated channels, however, many transmitter-gated ion channels are not permeable to a single type of ion. For example, the ACh-gated ion channel at the neuromuscular junction is permeable to both Na^+ and K^+. Let's explore the functional consequence of activating these channels.

In Chapter 3, we learned that the membrane potential, V_m, can be calculated using the Goldman equation, which takes into account the relative permeability of the membrane to different ions (see Box 3.3). If the membrane were equally permeable to Na^+ and K^+, as it would be if many ACh- or glutamate-gated channels were open, then V_m would have a value between E_{Na} and E_K, around 0 mV. Therefore, ionic current would flow through the channels in a direction that brings the membrane potential toward 0 mV. If the membrane potential were $<$0 mV before ACh was applied, as is usually the case, the direction of net current flow through the ACh-gated ion channels would be *inward*, causing a depolarization. However, if the membrane potential were $>$0 mV before ACh was applied, the direction of net current flow through the ACh-gated ion channels would be *outward*, causing the membrane potential to become less positive.

Ionic current flow at different membrane voltages can be graphed, as shown in Figure A. Such a graph is called an *I-V plot* (I: current; V: voltage). The critical value of membrane potential at which the direction of current flow reverses is called the *reversal potential*. In this case, the reversal potential would be 0 mV. The experimental determination of a reversal potential, therefore, helps tell us which types of ions the membrane is permeable to.

If, by changing the relative permeability of the membrane to different ions, a neurotransmitter causes V_m to move toward a value that is more positive than the action potential threshold, the neurotransmitter action would be termed excitatory. As a rule, neurotransmitters that open a channel permeable to Na^+ are excitatory. If a neurotransmitter causes V_m to take on a value that is more negative than the action potential threshold, the neurotransmitter action would be termed inhibitory. Neurotransmitters that open a channel permeable to Cl^- tend to be inhibitory, as are neurotransmitters that open a channel permeable only to K^+.

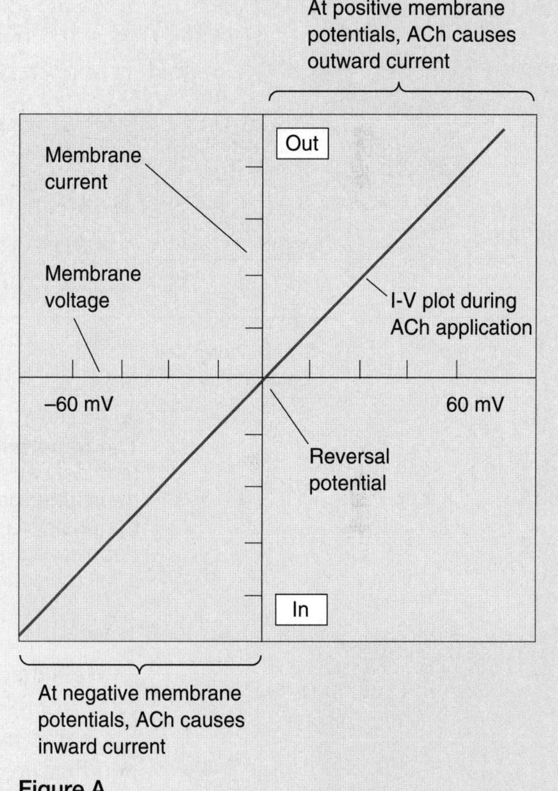

Figure A

2. The receptor proteins activate small proteins, called **G-proteins**, which are free to move along the intracellular face of the postsynaptic membrane.
3. The activated G-proteins activate "effector" proteins.

Effector proteins can be G-protein-gated ion channels in the membrane (Figure 5.17a), or they can be enzymes that synthesize molecules called **second messengers** that diffuse away in the cytosol (Figure 5.17b). Second messengers can activate additional enzymes in the cytosol that can regulate ion channel function and alter cellular metabolism. Because G-protein-coupled receptors can trigger widespread metabolic effects, they are often referred to as **metabotropic receptors**.

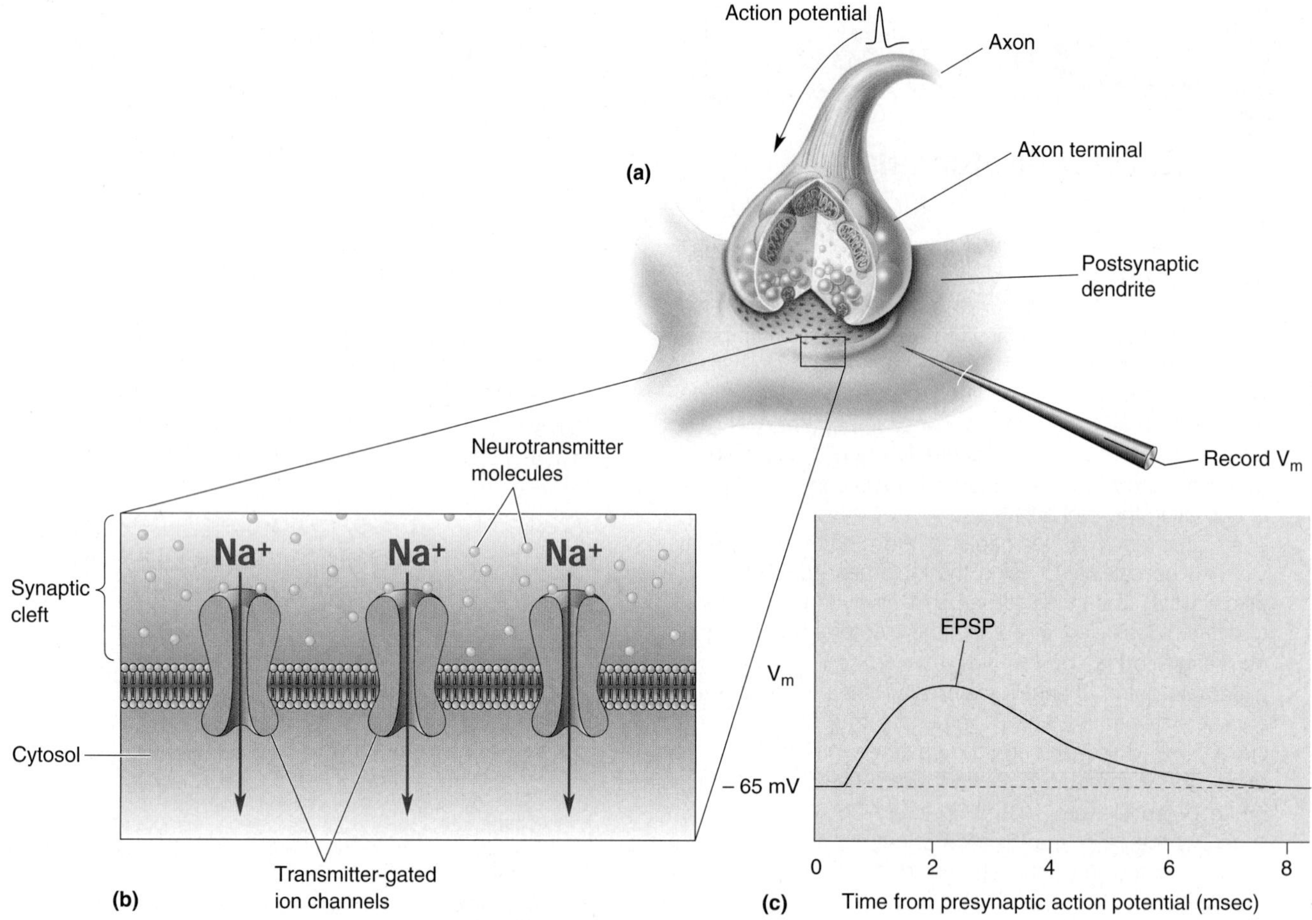

▲ FIGURE 5.15

The generation of an EPSP. (a) An action potential arriving in the presynaptic terminal causes the release of neurotransmitter. **(b)** The molecules bind to transmitter-gated ion channels in the postsynaptic membrane. If Na^+ enters the postsynaptic cell through the open channels, the membrane will become depolarized. **(c)** The resulting change in membrane potential (V_m), as recorded by a microelectrode in the cell, is the EPSP.

We'll discuss the different neurotransmitters, their receptors, and their effectors in more detail in Chapter 6. However, you should be aware that the same neurotransmitter can have different postsynaptic actions, depending on what receptors it binds to. An example is the effect of ACh on the heart and on skeletal muscles. ACh slows the rhythmic contractions of the heart by causing a slow hyperpolarization of the cardiac muscle cells. In contrast, in skeletal muscle, ACh induces contraction by causing a rapid depolarization of the muscle fibers. These different actions are explained by the different receptors involved. In the heart, a metabotropic ACh receptor is coupled by a G-protein to a potassium channel. The opening of the potassium channel hyperpolarizes the cardiac muscle fibers and reduces the rate at which it fires action potentials. In skeletal muscle, the receptor is a transmitter-gated ion channel, specifically an ACh-gated ion channel, permeable to Na^+. The opening of this channel depolarizes the muscle fibers and makes them more excitable.

Autoreceptors. Besides being a part of the postsynaptic density, neurotransmitter receptors are also commonly found in the membrane of the presynaptic axon terminal. Presynaptic receptors that are sensitive to the

▲ FIGURE 5.16

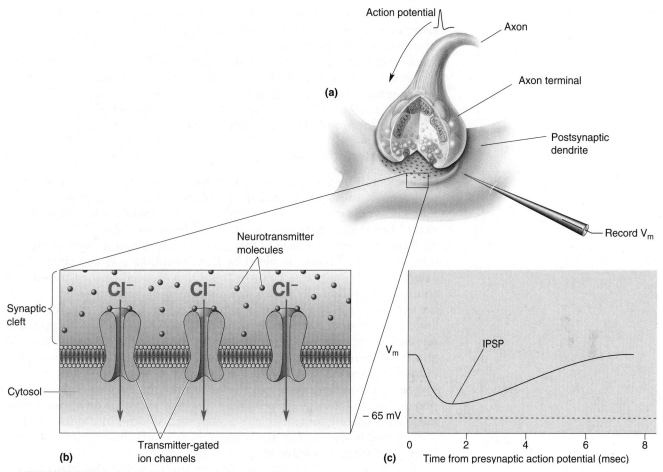

The generation of an IPSP. **(a)** An action potential arriving in the presynaptic terminal causes the release of neurotransmitter. **(b)** The molecules bind to transmitter-gated ion channels in the postsynaptic membrane. If Cl^- enters the postsynaptic cell through the open channels, the membrane will become hyperpolarized. **(c)** The resulting change in membrane potential (V_m), as recorded by a microelectrode in the cell, is the IPSP.

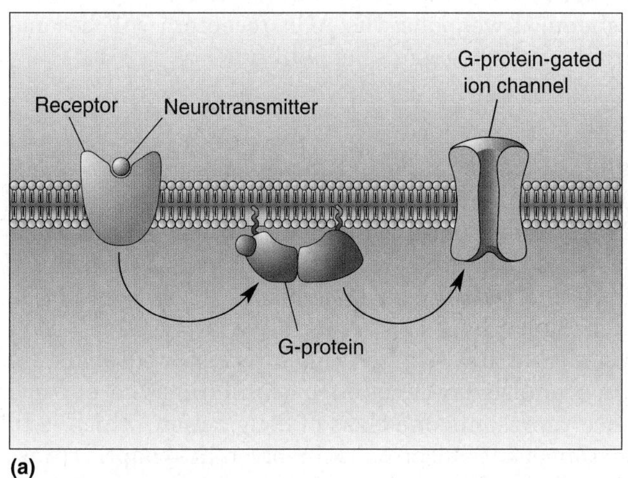

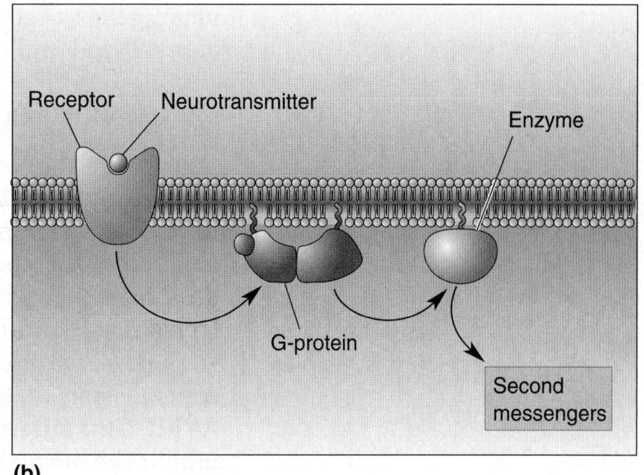

▲ FIGURE 5.17

Transmitter actions at G-protein-coupled receptors. The binding of neurotransmitter to the receptor leads to the activation of G-proteins. Activated G-proteins activate effector proteins, which may be **(a)** ion channels or **(b)** enzymes that generate intracellular second messengers.

neurotransmitter released by the presynaptic terminal are called **autoreceptors**. Typically, autoreceptors are G-protein-coupled receptors that stimulate second messenger formation. The consequences of activating these receptors vary, but a common effect is inhibition of neurotransmitter release and, in some cases, neurotransmitter synthesis. This allows a presynaptic terminal to regulate itself. Autoreceptors appear to function as a sort of safety valve to reduce release when the concentration of neurotransmitter around the presynaptic terminal gets too high.

Neurotransmitter Recovery and Degradation

Once the released neurotransmitter has interacted with postsynaptic receptors, it must be cleared from the synaptic cleft to allow another round of synaptic transmission. One way this happens is by simple diffusion of the transmitter molecules through the extracellular fluid and away from the synapse. For most of the amino acid and amine neurotransmitters, however, diffusion is aided by their reuptake into the presynaptic axon terminal. Reuptake occurs by the action of specific neurotransmitter transporter proteins located in the presynaptic membrane. Once inside the cytosol of the terminal, the transmitters may be reloaded into synaptic vesicles or enzymatically degraded and their breakdown products recycled. Neurotransmitter transporters also exist in the membranes of glia surrounding the synapse, which assist in the removal of neurotransmitter from the cleft.

Neurotransmitter action can also be terminated by enzymatic destruction in the synaptic cleft itself. This is how ACh is removed at the neuromuscular junction, for example. The enzyme acetylcholinesterase (AChE) is deposited in the cleft by the muscle cells. AChE cleaves the ACh molecule, rendering it inactive at the ACh receptors.

The importance of transmitter removal from the cleft should not be underestimated. At the neuromuscular junction, for example, uninterrupted exposure to high concentrations of ACh after several seconds leads to a process called *desensitization*, in which, despite the continued presence of ACh, the transmitter-gated channels close. This desensitized state can persist for many seconds even after the neurotransmitter is removed. The rapid destruction of ACh by AChE normally prevents desensitization from occurring. However, if the AChE is inhibited, as it is by various nerve gases used as chemical weapons, the ACh receptors will become desensitized and neuromuscular transmission will fail.

Neuropharmacology

Each of the steps of synaptic transmission we have discussed so far—neurotransmitter synthesis, loading into synaptic vesicles, exocytosis, binding and activation of receptors, reuptake, and degradation—is chemical, and therefore these steps can be affected by specific drugs and toxins (Box 5.5). The study of the effects of drugs on nervous system tissue is called **neuropharmacology**.

Earlier, we mentioned that nerve gases can interfere with synaptic transmission at the neuromuscular junction by inhibiting the enzyme AChE. This interference represents one class of drug action, which is to inhibit the normal function of specific proteins involved in synaptic transmission; such drugs are called **inhibitors**. Inhibitors of neurotransmitter receptors, called **receptor antagonists**, bind to the receptors and block (antagonize) the normal action of the transmitter. An example of a receptor antagonist is curare, an arrow-tip poison traditionally used by South American natives to paralyze their prey. Curare binds tightly to the ACh

BOX 5.5 OF SPECIAL INTEREST

Bacteria, Spiders, Snakes, and People

What do the bacteria *Clostridium botulinum*, black widow spiders, cobras, and humans have in common? They all produce toxins that attack the chemical synaptic transmission that occurs at the neuromuscular junction. Botulism is caused by several kinds of botulinum neurotoxins that are produced by the growth of *C. botulinum* in improperly canned foods. (The name comes from the Latin word for "sausage" because of the early association of the disease with poorly preserved meat.) Botulinum toxins are very potent blockers of neuromuscular transmission; it has been estimated that as few as 10 molecules of the toxins are enough to inhibit a cholinergic synapse. Botulinum toxins are extraordinarily specific enzymes that destroy certain of the SNARE proteins in the presynaptic terminals, which are critical for transmitter release (see Box 5.3). This specific action of the toxins made them important tools in the early research on SNAREs.

Although its mechanism of action is different, black widow spider venom also exerts deadly effects by affecting transmitter release (Figure A). The venom contains latrotoxin, which first increases, and then eliminates, ACh release at the neuromuscular junction. Electron microscopic examination of synapses poisoned with black widow spider venom reveals that the axon terminals are swollen and the synaptic vesicles are missing. The action of latrotoxin, a protein molecule, is not entirely understood. Venom binds with proteins on the outside of the presynaptic membrane and forms a membrane pore that depolarizes the terminal and allows Ca^{2+} to enter and trigger rapid and total depletion of transmitter. In some cases, the venom can induce transmitter release even without the need for Ca^{2+}, perhaps by interacting directly with neurotransmitter release proteins.

The bite of the Taiwanese cobra also results in the blockade of neuromuscular transmission in its victim, by yet another mechanism. One of the active compounds in the snake's venom, called *α-bungarotoxin*, is a peptide that binds so tightly to the postsynaptic nicotinic ACh receptors that it takes days to be removed. Often, there is not time for its removal, however, because cobra toxin prevents the activation of nicotinic receptors by ACh, thereby paralyzing the respiratory muscles of its victims.

We humans have synthesized a large number of chemicals that poison synaptic transmission at the neuromuscular junction. Originally motivated by the search for chemical warfare agents, this effort led to the development of a new class of compounds called *organophosphates*. These are irreversible inhibitors of AChE. By preventing the degradation of ACh, they cause it to accumulate and probably kill victims by causing a desensitization of ACh receptors. The organophosphates used today as insecticides, like parathion, are toxic to humans only in high doses.

Figure A
Black widow spiders. (Source: Matthews, 1995, p. 174.)

receptors on skeletal muscle cells and blocks the actions of ACh, thereby preventing muscle contraction.

Other drugs bind to receptors, but instead of inhibiting them, they mimic the actions of the naturally occurring neurotransmitter. These drugs are called **receptor agonists**. An example of a receptor agonist is nicotine, derived from the tobacco plant. Nicotine binds to and activates the ACh receptors in skeletal muscle. In fact, the ACh-gated ion channels in muscle are also called **nicotinic ACh receptors**, to distinguish them from other types of ACh receptors, such as those in the heart, that are not activated by nicotine. There are also nicotinic ACh receptors in the CNS, and these are involved in the addictive effects of tobacco use.

The immense chemical complexity of synaptic transmission makes it especially susceptible to the medical corollary of Murphy's law, which states that if a physiological process can go wrong, it will go wrong. When chemical synaptic transmission goes wrong, the nervous system malfunctions. Defective neurotransmission is believed to be the root cause of a large number of neurological and psychiatric disorders. The good news is that, thanks to our growing knowledge of the neuropharmacology of synaptic transmission, clinicians have new and increasingly effective therapeutic drugs for treating these disorders. We'll discuss the synaptic basis of some psychiatric disorders, and their neuropharmacological treatment, in Chapter 22.

PRINCIPLES OF SYNAPTIC INTEGRATION

Most CNS neurons receive thousands of synaptic inputs that activate different combinations of transmitter-gated ion channels and G-protein-coupled receptors. The postsynaptic neuron integrates all these complex ionic and chemical signals to produce a simple form of output: action potentials. The transformation of many synaptic inputs to a single neuronal output constitutes a neural computation. The brain performs billions of neural computations every second we are alive. As a first step toward understanding how neural computations are performed, let's explore some basic principles of synaptic integration. **Synaptic integration** is the process by which multiple synaptic potentials combine within one postsynaptic neuron.

The Integration of EPSPs

The most elementary postsynaptic response is the opening of a single transmitter-gated channel (Figure 5.18). Inward current through these channels depolarizes the postsynaptic membrane, causing the EPSP. The postsynaptic membrane of one synapse may have from a few tens to several thousands of transmitter-gated channels; how many of these are activated during synaptic transmission depends mainly on how much neurotransmitter is released.

Quantal Analysis of EPSPs. The elementary unit of neurotransmitter release is the contents of a single synaptic vesicle. Vesicles each contain about the same number of transmitter molecules (several thousand); the total amount of transmitter released is some multiple of this number.

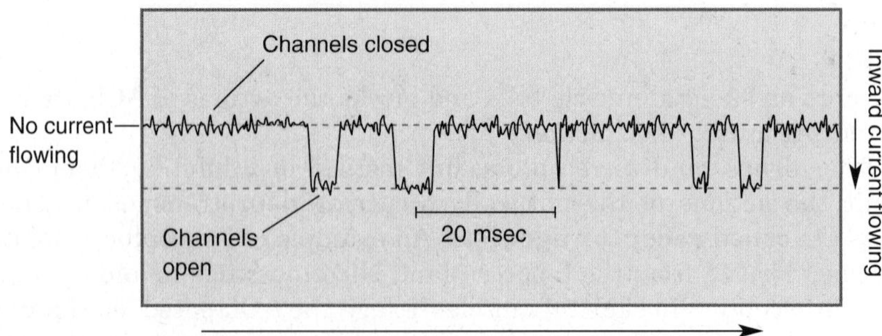

▲ FIGURE 5.18
A patch-clamp recording from a transmitter-gated ion channel. Ionic current passes through the channels when the channels are open. In the presence of neurotransmitter, they rapidly alternate between open and closed states. (Source: Adapted from Neher and Sakmann, 1992.)

Consequently, the amplitude of the postsynaptic EPSP is some multiple of the response to the contents of a single vesicle. Stated another way, postsynaptic EPSPs at a given synapse are *quantized*; they are multiples of an indivisible unit, the *quantum*, which reflects the number of transmitter molecules in a single synaptic vesicle and the number of postsynaptic receptors available at the synapse.

At many synapses, exocytosis of vesicles occurs at some very low rate in the absence of presynaptic stimulation. The size of the postsynaptic response to this spontaneously released neurotransmitter can be measured electrophysiologically. This tiny response is a **miniature postsynaptic potential**, often called simply a *mini*. Each mini is generated by the transmitter contents of one vesicle. The amplitude of the postsynaptic EPSP evoked by a presynaptic action potential, then, is simply an integer multiple (i.e., 1×, 2×, 3×, etc.) of the mini amplitude.

Quantal analysis, a method of comparing the amplitudes of miniature and evoked PSPs, can be used to determine how many vesicles release neurotransmitter during normal synaptic transmission. Quantal analysis of transmission at the neuromuscular junction reveals that a single action potential in the presynaptic terminal triggers the exocytosis of about 200 synaptic vesicles, causing an EPSP of 40 mV or more. At many CNS synapses, in striking contrast, the contents of only a *single vesicle* are released in response to a presynaptic action potential, causing an EPSP of only a few tenths of a millivolt.

EPSP Summation. The difference between excitatory transmission at neuromuscular junctions and CNS synapses is not surprising. The neuromuscular junction has evolved to be fail-safe; it needs to work every time, and the best way to ensure this is to generate an EPSP of a huge size. On the other hand, if every CNS synapse were, by itself, capable of triggering an action potential in its postsynaptic cell (as the neuromuscular junction can), then a neuron would be little more than a simple relay station. Instead, most neurons perform more sophisticated computations, requiring that many EPSPs add together to produce a significant postsynaptic depolarization. This is what is meant by integration of EPSPs.

EPSP summation represents the simplest form of synaptic integration in the CNS. There are two types of summation: spatial and temporal. **Spatial summation** is the adding together of EPSPs generated simultaneously at many different synapses on a dendrite. **Temporal summation** is the adding together of EPSPs generated at the same synapse if they occur in rapid succession, within about 1–15 msec of one another (Figure 5.19).

The Contribution of Dendritic Properties to Synaptic Integration

Even with the summation of several EPSPs on a dendrite, the depolarization still may not be enough to cause the neuron to fire an action potential. The current entering at the sites of synaptic contact must spread down the dendrite and through the soma and cause the membrane of the spike-initiation zone to be depolarized beyond threshold, before an action potential can be generated. The effectiveness of an excitatory synapse in triggering an action potential, therefore, depends on how far the synapse is from the spike-initiation zone and on the properties of the dendritic membrane.

Dendritic Cable Properties. To simplify the analysis of how dendritic properties contribute to synaptic integration, let's assume that dendrites

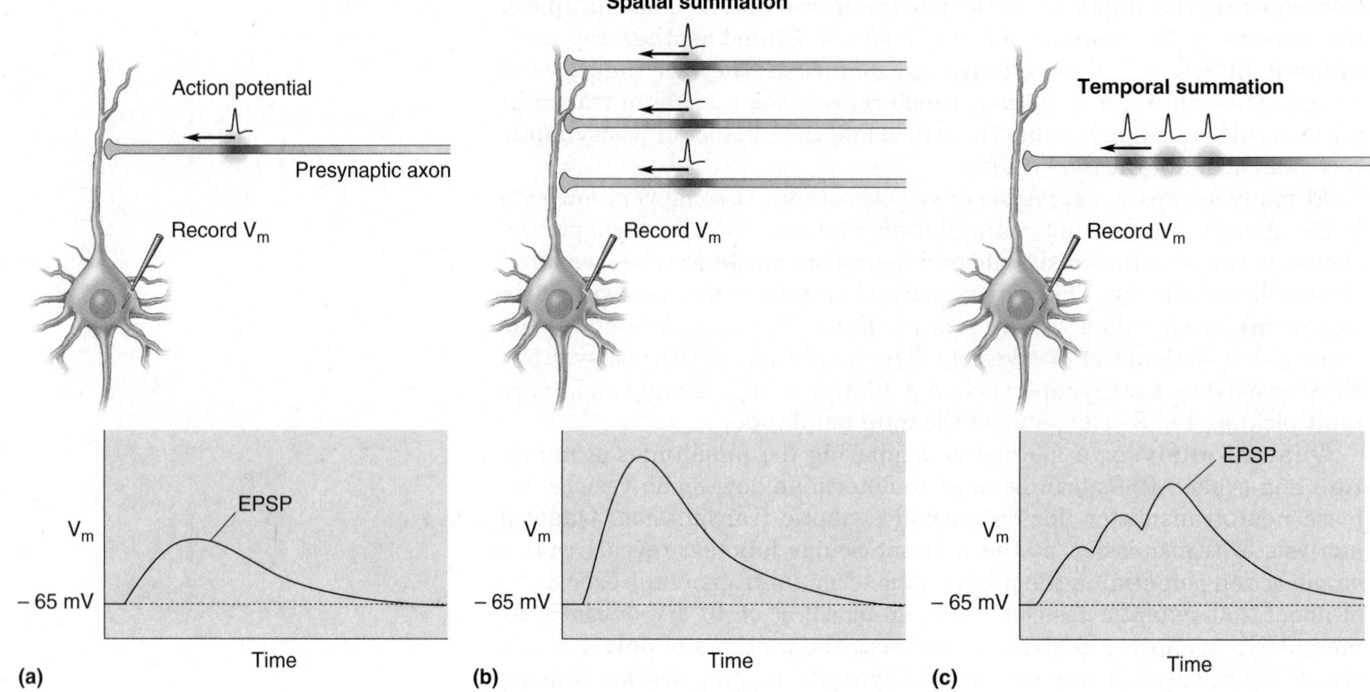

▲ FIGURE 5.19

EPSP summation. **(a)** A presynaptic action potential triggers a small EPSP in a postsynaptic neuron. **(b)** Spatial summation of EPSPs: When two or more presynaptic inputs are active at the same time, their individual EPSPs add together. **(c)** Temporal summation of EPSPs: When the same presynaptic fiber fires action potentials in quick succession, the individual EPSPs add together.

function as cylindrical cables that are electrically passive, that is, lacking voltage-gated ion channels (in contrast, of course, with axons). Using an analogy introduced in Chapter 4, imagine that the influx of positive charge at a synapse is like turning on the water that will flow down a leaky garden hose (the dendrite). There are two paths the water can take: down the inside of the hose or through the leaks. By the same token, there are two paths that synaptic current can take: down the inside of the dendrite or across the dendritic membrane. As the current proceeds down the dendrite and farther from the synapse, the EPSP amplitude will diminish because of the leakage of ionic current through membrane channels. At some distance from the site of current influx, the EPSP amplitude may eventually approach zero.

The decrease in depolarization as a function of distance along a dendritic cable is plotted in Figure 5.20. In order to simplify the mathematics, in this example, we'll assume the dendrite is infinitely long, unbranched, and uniform in diameter. We will also use a microelectrode to inject a long, steady pulse of current to induce a membrane depolarization. Notice that the amount of depolarization falls off exponentially with increasing distance. Depolarization of the membrane at a given distance (V_x) can be described by the equation $V_x = V_o/e^{x/\lambda}$, where V_o is depolarization at the origin (just under the microelectrode), e (= 2.718 . . .) is the base of natural logarithms, x is distance from the synapse, and λ is a constant that depends on the properties of the dendrite. Notice that when $x = \lambda$, then $V_x = V_o/e$. Put another way, $V_\lambda = 0.37\ (V_o)$. This distance λ, where the depolarization is about 37% of that at the origin, is called the dendritic **length constant**. (Remember that this analysis is an oversimplification. Real dendrites have finite

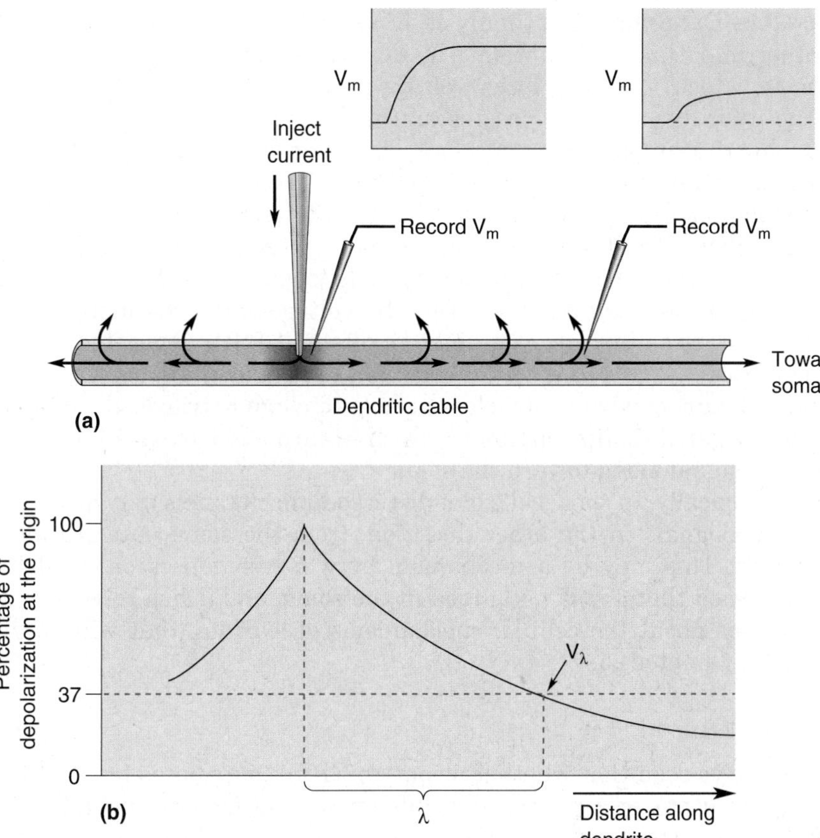

◀ FIGURE 5.20
Decreasing depolarization as a function of distance along a long dendritic cable. **(a)** Current is injected into a dendrite and the depolarization is recorded. As this current spreads down the dendrite, much of it dissipates across the membrane. Therefore, the depolarization measured at a distance from the site of current injection is smaller than that measured right under it. **(b)** A plot of membrane depolarization as a function of distance along the dendrite. At the distance λ, one length constant, the membrane depolarization (V_λ), is 37% of that at the origin.

lengths, have branches, and tend to taper, and EPSPs are transient—all of which affect the spread of current, and thus the effectiveness of synaptic potentials.)

The length constant is an index of how far depolarization can spread down a dendrite or axon. The longer the length constant, the more likely it is that EPSPs generated at distant synapses will depolarize the membrane at the axon hillock. The value of λ in our idealized, electrically passive dendrite depends on two factors: (1) the resistance to current flowing longitudinally down the dendrite, called the **internal resistance** (r_i); and (2) the resistance to current flowing across the membrane, called the **membrane resistance** (r_m). Most current will take the path of least resistance; therefore, the value of λ will increase as membrane resistance increases because more depolarizing current will flow down the inside of the dendrite rather than "leaking" out the membrane. The value of λ will decrease as internal resistance increases because more current will then flow across the membrane. Just as water will flow farther down a wide hose with few leaks, synaptic current will flow farther down a wide dendrite (low r_i) with few open membrane channels (high r_m).

The internal resistance depends only on the diameter of the dendrite and the electrical properties of the cytoplasm; consequently, it is relatively constant in a mature neuron. The membrane resistance, in contrast, depends on the number of open ion channels, which changes from moment to moment depending on what other synapses are active. The dendritic length constant, therefore, is not constant at all! As we will see in a moment, fluctuations in the value of λ are an important factor in synaptic integration.

Excitable Dendrites. Our analysis of dendritic cable properties made another important assumption: The dendrite's membrane is electrically passive, which means it lacks voltage-gated channels. Some dendrites in the brain have nearly passive and inexcitable membranes and thus do follow the simple cable equations. The dendrites of spinal motor neurons, for example, are very close to passive. However, most dendrites are decidedly not passive. A variety of neurons have dendrites with significant numbers of voltage-gated sodium, calcium, and potassium channels. Dendrites rarely have enough ion channels to generate fully propagating action potentials, as axons can. But the voltage-gated channels in dendrites can act as important amplifiers of small PSPs generated far out on dendrites. EPSPs that would diminish to near nothingness in a long, passive dendrite may nevertheless be large enough to trigger the opening of voltage-gated sodium channels, which in turn add current to boost the synaptic signal along toward the soma.

Paradoxically, in some cells, dendritic sodium channels may also carry electrical signals in the other direction, from the soma outward along dendrites. This may be a mechanism by which synapses on dendrites are informed that a spike occurred in the soma, and it has relevance for hypotheses about the cellular mechanisms of learning that will be discussed in Chapter 25.

Inhibition

So far, we've seen that whether or not an EPSP contributes to the action potential output of a neuron depends on several factors, including the number of coactive excitatory synapses, the distance the synapse is from the spike-initiation zone, and the properties of the dendritic membrane. Of course, not all synapses in the brain are excitatory. The action of some synapses is to take the membrane potential away from action potential threshold; these are called *inhibitory synapses*. Inhibitory synapses exert a powerful control over a neuron's output (Box 5.6).

IPSPs and Shunting Inhibition. The postsynaptic receptors under most inhibitory synapses are very similar to those under excitatory synapses; they're transmitter-gated ion channels. The only important differences are that they bind different neurotransmitters (either GABA or glycine) and that they allow different ions to pass through their channels. The transmitter-gated channels of most inhibitory synapses are permeable to only one natural ion, Cl^-. Opening of the chloride channel allows Cl^- to cross the membrane in a direction that brings the membrane potential toward the chloride equilibrium potential, E_{Cl}, about −65 mV. If the membrane potential were less negative than −65 mV when the transmitter was released, activation of these channels would cause a hyperpolarizing IPSP.

Notice that if the resting membrane potential were already −65 mV, no IPSP would be visible after chloride channel activation because the value of the membrane potential would already equal E_{Cl} (i.e., the reversal potential for that synapse; see Box 5.4). If there is no visible IPSP, is the neuron really inhibited? The answer is yes. Consider the situation illustrated in Figure 5.21, with an excitatory synapse on a distal segment of dendrite and an inhibitory synapse on a proximal segment of dendrite, near the soma. Activation of the excitatory synapse leads to the influx of positive charge into the dendrite. This current depolarizes the membrane as it flows toward the soma. At the site of the active inhibitory synapse, however, the membrane potential is approximately equal to

BOX 5.6 OF SPECIAL INTEREST

Startling Mutations and Poisons

A flash of lightning . . . a thunderclap . . . a tap on the shoulder when you think you're alone! If you are not expecting them, any of these stimuli can make you jump, grimace, hunch your shoulders, and breathe faster. We all know the brief but dramatic nature of the startle response.

Luckily, when lightning strikes twice or a friend taps our shoulder again, we tend to be much less startled the second time. We quickly habituate and relax. However, for an unfortunate minority of mice, cows, dogs, horses, and people, life is a succession of exaggerated startle responses. Even normally benign stimuli, such as hands clapping or a touch to the nose, may trigger an uncontrollable stiffening of the body, an involuntary shout, flexion of the arms and legs, and a fall to the ground. Worse yet, these overreactions don't adapt when the stimuli are repeated. The clinical term for startle disease is *hyperekplexia*, and the first recorded cases were members of a community of French–Canadian lumberjacks in 1878. Hyperekplexia is an inherited condition occurring worldwide, and its sufferers are known by colorful local names: the "Jumping Frenchmen of Maine" (Quebec), "myriachit" (Siberia), "latah" (Malaysia), and "Ragin' Cajuns" (Louisiana).

We now know the molecular basis for two general types of startle diseases. Remarkably, both involve defects of inhibitory glycine receptors. The first type, identified in humans and in a mutant mouse called *spasmodic*, is caused by a mutation of a gene for the glycine receptor. The change is the smallest one possible—the abnormal receptors have only one amino acid (out of more than 400) coded wrong—but the result is a chloride channel that opens less frequently when exposed to the neurotransmitter glycine. The second type of startle disease is seen in the mutant mouse *spastic* and in a strain of cattle. In these animals, normal glycine receptors are expressed but in fewer than normal numbers. The two forms of startle disease thus take different routes to the same unfortunate end: The transmitter glycine is less effective at inhibiting neurons in the spinal cord and brain stem.

Most neural circuits depend on a delicate balance of synaptic excitation and inhibition for normal functioning. If excitation is increased or inhibition reduced, then a turbulent and hyperexcitable state may result. An impairment of glycine function yields exaggerated startles; reduced GABA function can lead to the seizures of epilepsy (as discussed in Chapter 19). How can such diseases be treated? There is often a clear and simple logic. Drugs that enhance inhibition can be very helpful.

The genetic mutations of the glycine system resemble strychnine poisoning. Strychnine is a powerful toxin found in the seeds and bark of certain trees and shrubs of the genus *Strychnos*. It was first isolated and identified chemically in the early nineteenth century. Strychnine has traditionally been used by farmers to eradicate pesky rodents and by murderers. It has a simple mechanism of action: It is an antagonist of glycine at its receptor. Mild strychnine poisoning enhances startle and other reflexes and resembles hyperekplexia. High doses nearly eliminate glycine-mediated inhibition in circuits of the spinal cord and brain stem. This leads to uncontrollable seizures and unchecked muscular contractions, spasm and paralysis of the respiratory muscles, and ultimately, death from asphyxiation. It is a painful, agonizing way to die. Since glycine is not a transmitter in the higher centers of the brain, strychnine itself does not impair cognitive or sensory functions.

E_{Cl}, −65 mV. Positive current, therefore, flows outward across the membrane at this site to bring V_m to −65 mV. This synapse acts as an electrical shunt, preventing the current from flowing through the soma to the axon hillock. This type of inhibition is called **shunting inhibition**. The actual physical basis of shunting inhibition is the *inward movement of negatively charged chloride ions,* which is formally equivalent to *outward positive current* flow. Shunting inhibition is like cutting a big hole in the leaky garden hose—more of the water flows down this path of least resistance, out of the hose, before it gets to the nozzle where it can "activate" the flowers in your garden.

Thus, you can see that the action of inhibitory synapses also contributes to synaptic integration. The IPSPs reduce the size of EPSPs, making the postsynaptic neuron less likely to fire action potentials. In addition, shunting inhibition acts to drastically reduce r_m and consequently λ, thus allowing positive current to flow out across the membrane instead of internally down the dendrite toward the spike-initiation zone.

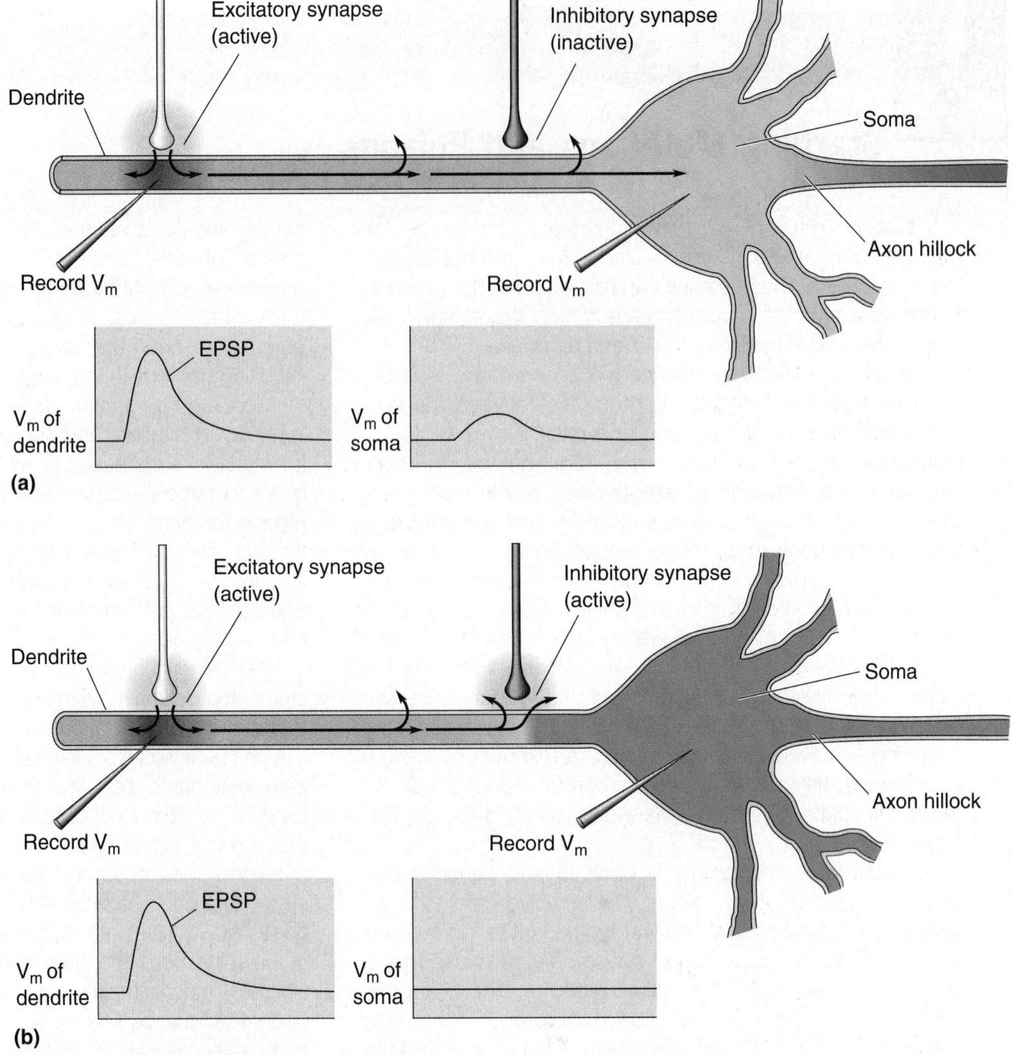

▲ FIGURE 5.21

Shunting inhibition. A neuron receives one excitatory and one inhibitory input. **(a)** Stimulation of the excitatory input causes inward postsynaptic current that spreads to the soma, where it can be recorded as an EPSP. **(b)** When the inhibitory and excitatory inputs are stimulated together, the depolarizing current leaks out before it reaches the soma.

The Geometry of Excitatory and Inhibitory Synapses. Inhibitory synapses in the brain that use GABA or glycine as a neurotransmitter have a morphology characteristic of Gray's type II (see Figure 5.8b). This structure contrasts with excitatory synapses that use glutamate, which have a Gray's type I morphology. This correlation between structure and function has been useful for working out the geometric relationships among excitatory and inhibitory synapses on individual neurons. In addition to being spread over the dendrites, inhibitory synapses on many neurons are found clustered on the soma and near the axon hillock, where they are in an especially powerful position to influence the activity of the postsynaptic neuron.

Modulation

Most of the postsynaptic mechanisms we've discussed so far involve transmitter receptors that are, themselves, ion channels. To be sure, synapses with transmitter-gated channels carry the bulk of the specific information that is processed by the nervous system. However, there are many synapses with G-protein-coupled neurotransmitter receptors that are not directly associated with an ion channel. Synaptic activation of these receptors does not directly evoke EPSPs and IPSPs but instead *modifies* the effectiveness of EPSPs generated by other synapses with transmitter-gated channels. This type of synaptic transmission is called **modulation**. We'll give you a taste for how modulation influences synaptic integration by exploring the effects of activating one type of G-protein-coupled receptor in the brain, the norepinephrine beta (β) receptor.

The binding of the amine neurotransmitter **norepinephrine (NE)** to the β receptor triggers a cascade of biochemical events within the cell. In short, the β receptor activates a G-protein that, in turn, activates an effector protein, the intracellular enzyme adenylyl cyclase. **Adenylyl cyclase** catalyzes the chemical reaction that converts adenosine triphosphate (ATP), the product of oxidative metabolism in the mitochondria, into a compound called **cyclic adenosine monophosphate**, or **cAMP**, that is free to diffuse within the cytosol. Thus, the *first* chemical message of synaptic transmission (the release of NE into the synaptic cleft) is converted by the β receptor into a *second* message (cAMP); cAMP is an example of a second messenger.

The effect of cAMP is to stimulate another enzyme known as a protein kinase. **Protein kinases** catalyze a chemical reaction called **phosphorylation**, the transfer of phosphate groups (PO_3) from ATP to specific sites on cell proteins (Figure 5.22). The significance of phosphorylation is that it can change the conformation of a protein, thereby changing that protein's activity.

In some neurons, one of the proteins that is phosphorylated when cAMP concentration rises is a particular type of potassium channel in the dendritic membrane. Phosphorylation causes this channel to close, thereby reducing the membrane K^+ conductance. By itself, this does not cause any dramatic effects on the neuron. But consider the wider consequence: *Decreasing the K^+ conductance increases the dendritic membrane resistance and therefore increases the length constant.* It is like wrapping the leaky garden hose in duct tape; more water can flow down the inside of the hose and less leaks out the sides. As a consequence of increasing λ, distant or weak excitatory synapses will become more effective in depolarizing the spike-initiation zone beyond threshold; the cell will become *more excitable*. Thus, the binding of NE to β receptors produces little change in membrane potential but greatly increases the response produced by another neurotransmitter at an excitatory synapse. Because this effect

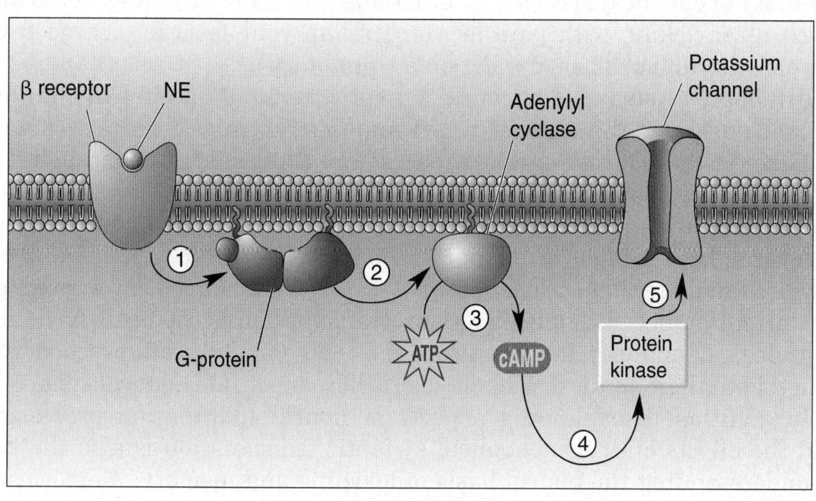

◀ FIGURE 5.22
Modulation by the NE β receptor. ① The binding of NE to the receptor activates a G-protein in the membrane. ② The G-protein activates the enzyme adenylyl cyclase. ③ Adenylyl cyclase converts ATP into the second messenger cAMP. ④ cAMP activates a protein kinase. ⑤ The protein kinase causes a potassium channel to close by attaching a phosphate group to it.

involves several biochemical intermediaries, it can last far longer than the presence of the modulatory transmitter itself.

We have described one particular G-protein-coupled receptor and the consequences of activating it in one type of neuron. But it is important to recognize that other types of receptors can lead to the formation of other types of second messenger molecules. Activation of each of these receptor types will initiate a distinct cascade of biochemical reactions in the postsynaptic neuron that do not always include phosphorylation and decreases in membrane conductance. In fact, cAMP in a different cell type with different enzymes may produce functionally opposite changes in the excitability of cells.

In Chapter 6, we will describe more examples of synaptic modulation and their mechanisms. However, you can already see that modulatory forms of synaptic transmission offer an almost limitless number of ways that information encoded by presynaptic impulse activity can be transformed and used by the postsynaptic neuron.

CONCLUDING REMARKS

This chapter has covered the basic principles of chemical synaptic transmission. The action potential that arose in the sensory nerve when you stepped on that thumbtack in Chapter 3, and that swept along the axon in Chapter 4, has now reached the axon terminal in the spinal cord. The depolarization of the terminal triggered the presynaptic entry of Ca^{2+} through voltage-gated calcium channels, which then stimulated exocytosis of the contents of synaptic vesicles. Liberated neurotransmitter diffused across the synaptic cleft and attached to specific receptors in the postsynaptic membrane. The transmitter (probably glutamate) caused transmitter-gated channels to open, which allowed positive charge to enter the postsynaptic dendrite. Because the sensory nerve was firing action potentials at a high rate, and because many synapses were activated at the same time, the EPSPs summed to bring the spike-initiation zone of the postsynaptic neuron to threshold, and this cell then generated action potentials. If the postsynaptic cell is a motor neuron, this activity will cause the release of ACh at the neuromuscular junction and muscle contraction to jerk your foot away from the tack. If the postsynaptic cell is an interneuron that uses GABA as a neurotransmitter, the activity of the cell will result in inhibition of its synaptic targets. If this cell uses a modulatory transmitter such as NE, the activity could cause lasting changes in the excitability or metabolism of its synaptic targets. It is this rich diversity of chemical synaptic interactions that allows complex behaviors (such as shrieking with pain as you jerk up your foot) to emerge from simple stimuli (such as stepping on a thumbtack).

Although we surveyed chemical synaptic transmission in this chapter, we did not cover the *chemistry* of synaptic transmission in any detail. In Chapter 6, we'll take a closer look at the chemical "nuts and bolts" of different neurotransmitter systems. In Chapter 15, after we've examined the sensory and motor systems in Part II, we'll explore the contributions of several different neurotransmitters to nervous system function and behavior. You'll see that the chemistry of synaptic transmission warrants all this attention because defective neurotransmission is the basis for many neurological and psychiatric disorders. And virtually all psychoactive drugs, both therapeutic and illicit, exert their effects at chemical synapses.

In addition to explaining aspects of neural information processing and the effects of drugs, chemical synaptic transmission is also the key to understanding the neural basis of learning and memory. Memories of past experiences are established by modification of the effectiveness of chemical synapses in the brain. This chapter suggests possible sites of modification, ranging from changes in presynaptic Ca^{2+} entry and neurotransmitter release to alterations in postsynaptic receptors or excitability. As we shall see in Chapter 25, all of these changes are likely to contribute to the storage of information by the nervous system.

KEY TERMS

Introduction
synaptic transmission (p. 110)
electrical synapse (p. 110)
chemical synapse (p. 110)

Types of Synapses
gap junction (p. 111)
postsynaptic potential (PSP) (p. 112)
secretory granule (p. 114)
dense-core vesicle (p. 114)
membrane differentiation (p. 115)
active zone (p. 115)
postsynaptic density (p. 115)
neuromuscular junction (p. 119)
motor end-plate (p. 119)

Principles of Chemical Synaptic Transmission
glutamate (Glu) (p. 122)
gamma-aminobutyric acid (GABA) (p. 122)
glycine (Gly) (p. 122)
acetylcholine (ACh) (p. 122)
transporters (p. 122)
voltage-gated calcium channel (p. 122)
exocytosis (p. 122)
endocytosis (p. 124)
transmitter-gated ion channels (p. 124)
excitatory postsynaptic potential (EPSP) (p. 126)
inhibitory postsynaptic potential (IPSP) (p. 126)
G-protein-coupled receptors (p. 126)
G-proteins (p. 127)
second messengers (p. 127)
metabotropic receptors (p. 127)
autoreceptors (p. 130)
neuropharmacology (p. 130)
inhibitors (p. 130)
receptor antagonists (p. 130)
receptor agonists (p. 131)
nicotinic ACh receptors (p. 131)

Principles of Synaptic Integration
synaptic integration (p. 132)
miniature postsynaptic potential (p. 133)
quantal analysis (p. 133)
EPSP summation (p. 133)
spatial summation (p. 133)
temporal summation (p. 133)
length constant (p. 134)
internal resistance (p. 135)
membrane resistance (p. 135)
shunting inhibition (p. 137)
modulation (p. 139)
norepinephrine (NE) (p. 139)
adenylyl cyclase (p. 139)
cyclic adenosine monophosphate (cAMP) (p. 139)
protein kinases (p. 139)
phosphorylation (p. 139)

REVIEW QUESTIONS

1. What is meant by quantal release of neurotransmitter?
2. You apply ACh and activate nicotinic receptors on a muscle cell. Which way will current flow through the receptor channels when $V_m = -60$ mV? When $V_m = 0$ mV? When $V_m = 60$ mV? Why?
3. This chapter discussed a GABA-gated ion channel that is permeable to Cl^-. GABA also activates a G-protein-coupled receptor, called the *$GABA_B$ receptor*, which causes potassium-selective channels to open. What effect would $GABA_B$ receptor activation have on the membrane potential?
4. You think you have discovered a new neurotransmitter, and you are studying its effect on a neuron. The reversal potential for the response caused by the new chemical is −60 mV. Is this substance excitatory or inhibitory? Why?
5. A drug called *strychnine*, isolated from the seeds of a tree native to India and commonly used as rat poison, blocks the effects of glycine. Is strychnine an agonist or an antagonist of the glycine receptor?
6. How does nerve gas cause respiratory paralysis?
7. Why is an excitatory synapse on the soma more effective in evoking action potentials in the postsynaptic neuron than an excitatory synapse on the tip of a dendrite?
8. What are the steps that lead to increased excitability in a neuron when NE is released presynaptically?

FURTHER READING

Connors BW, Long MA. 2004. Electrical synapses in the mammalian brain. *Annual Review of Neuroscience* 27:393–418.

Cowan WM, Südhof TC, Stevens CF. 2001. *Synapses*. Baltimore: Johns Hopkins University Press.

Kandel ER, Schwartz JH, Jessell TM, Siegelbaum SA, Hudspeth AJ. 2012. *Principles of Neural Science,* 5th ed. New York: McGraw-Hill Professional.

Koch C. 2004. *Biophysics of Computation: Information Processing in Single Neurons*. New York: Oxford University Press.

Nicholls JG, Martin AR, Fuchs PA, Brown DA, Diamond ME, Weisblat D. 2007. *From Neuron to Brain,* 5th ed. Sunderland, MA: Sinauer.

Sheng M, Sabatini BL, Südhof TC. 2012. *The Synapse*. New York: Cold Spring Harbor Laboratory Press.

Stuart G, Spruston N, Hausser M. 2007. *Dendrites,* 2nd ed. New York: Oxford University Press.

Südhof TC. 2013. Neurotransmitter release: the last millisecond in the life of a synaptic vesicle. *Neuron* 80:675–690.

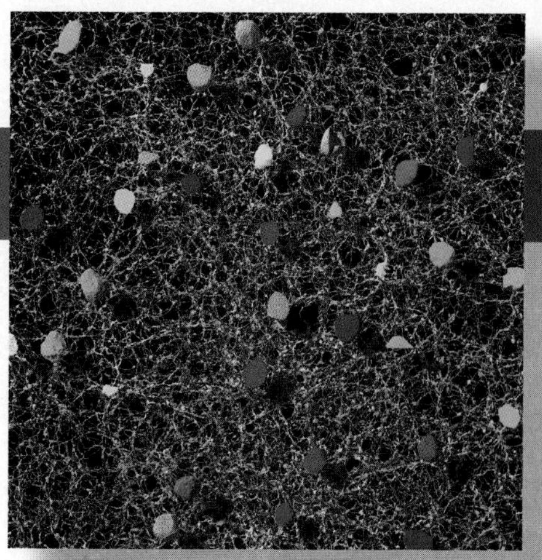

CHAPTER SIX

Neurotransmitter Systems

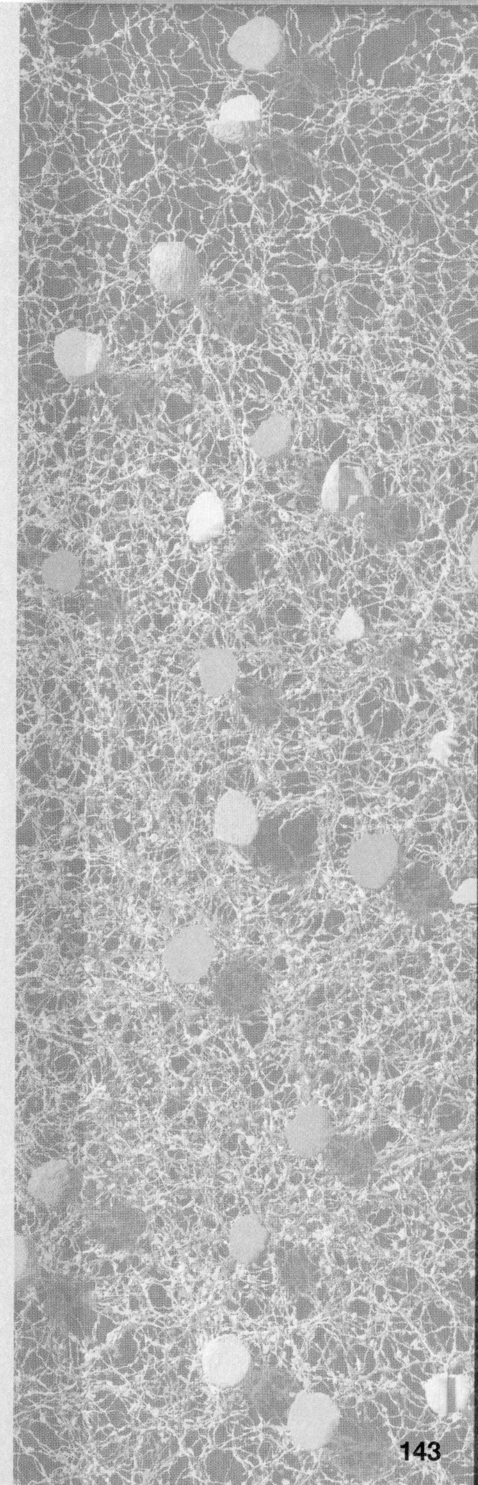

INTRODUCTION

Normal functions of the human brain require an orderly set of chemical reactions. As we have seen, some of the brain's most important chemical reactions are those associated with synaptic transmission. Chapter 5 introduced the general principles of chemical synaptic transmission, using a few specific neurotransmitters as examples. In this chapter, we will explore in more depth the variety and elegance of the major neurotransmitter systems.

Neurotransmitter systems begin with neurotransmitters. In Chapter 5, we discussed the three major classes of neurotransmitters: *amino acids*, *amines*, and *peptides*. Even a partial list of the known transmitters, such as that appearing in Table 5.1, has more than 20 different molecules. Each of these molecules can define a particular transmitter system. In addition to the molecule itself, a neurotransmitter system includes all the molecular machinery responsible for transmitter synthesis, vesicular packaging, reuptake and degradation, and transmitter action (Figure 6.1).

The first molecule positively identified as a neurotransmitter by Otto Loewi in the 1920s was acetylcholine, or ACh (see Box 5.1). To describe the cells that produce and release ACh, British pharmacologist Henry Dale introduced the term **cholinergic**. (Dale shared the 1936 Nobel Prize with Loewi in recognition of his neuropharmacological studies of synaptic transmission.) Dale termed the neurons that use the amine neurotransmitter norepinephrine (NE) **noradrenergic**. (NE is known as noradrenaline in United Kingdom.) The convention of using the suffix *-ergic* continued when additional transmitters were identified.

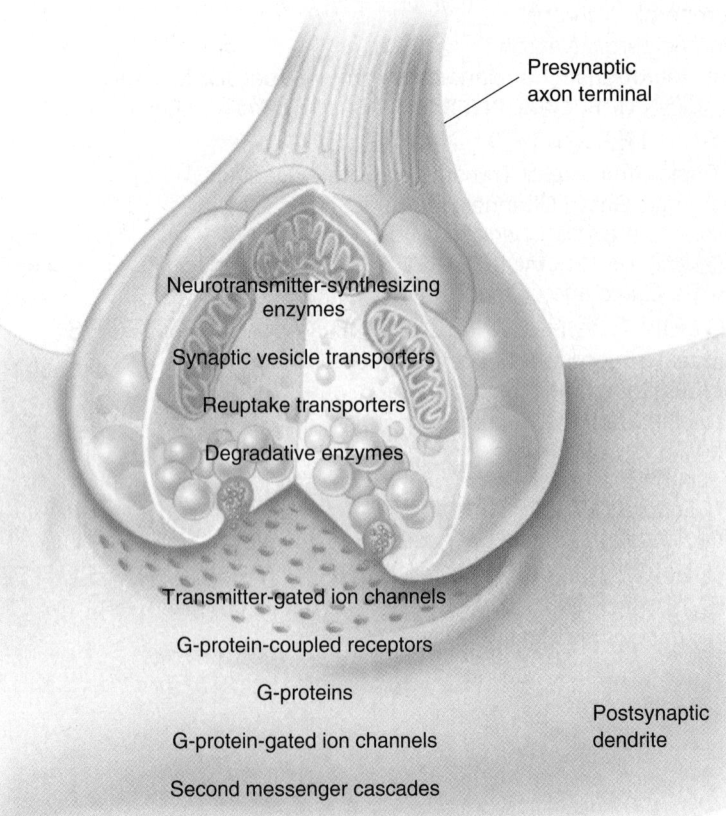

▶ FIGURE 6.1
Elements of neurotransmitter systems.

Therefore, today we speak of **glutamatergic** synapses that use glutamate, **GABAergic** synapses that use GABA, **peptidergic** synapses that use peptides, and so on. These adjectives are also used to identify the various neurotransmitter systems. For example, ACh and all the molecular machinery associated with it are collectively called the *cholinergic system*.

With this terminology in hand, we can begin our exploration of the neurotransmitter systems. We start with a discussion of the experimental strategies that have been used to study transmitter systems. Then we will look at the synthesis and metabolism of specific neurotransmitters and explore how these molecules exert their postsynaptic effects. In Chapter 15, after we have learned more about the structural and functional organization of the nervous system, we'll take another look at specific neurotransmitter systems in the context of their individual contributions to the regulation of brain function and behavior.

STUDYING NEUROTRANSMITTER SYSTEMS

The first step in studying a neurotransmitter system is usually identifying the neurotransmitter. This is no simple task; the brain contains uncountable different chemicals. How can we decide which few chemicals are used as transmitters?

Over the years, neuroscientists have established certain criteria that must be met for a molecule to be considered a neurotransmitter:

1. The molecule must be synthesized and stored in the presynaptic neuron.
2. The molecule must be released by the presynaptic axon terminal upon stimulation.
3. The molecule, when experimentally applied, must produce a response in the postsynaptic cell that mimics the response produced by the release of neurotransmitter from the presynaptic neuron.

Let's start by exploring some of the strategies and methods that are used to satisfy these criteria.

Localization of Transmitters and Transmitter-Synthesizing Enzymes

The scientist often begins with little more than a hunch that a particular molecule may be a neurotransmitter. This idea may be based on observing that the molecule is concentrated in brain tissue or that the application of the molecule to certain neurons alters their action potential firing rate. Whatever the inspiration, the first step in confirming the hypothesis is to show that the molecule is, in fact, localized in, and synthesized by, particular neurons. Many methods have been used to satisfy this criterion for different neurotransmitters. Two of the most important techniques used today are immunocytochemistry and *in situ* hybridization.

Immunocytochemistry. The method of **immunocytochemistry** is used to anatomically localize particular molecules to particular cells. When the same technique is applied to thin sections of tissue, including brain, it is often referred to as *immunohistochemistry*. The principle behind the method is quite simple (Figure 6.2). Once the neurotransmitter candidate has been chemically purified, it is injected under the skin or into the bloodstream of an animal where it stimulates an immune response. (Often, to evoke or enhance the immune response, the molecule is chemically

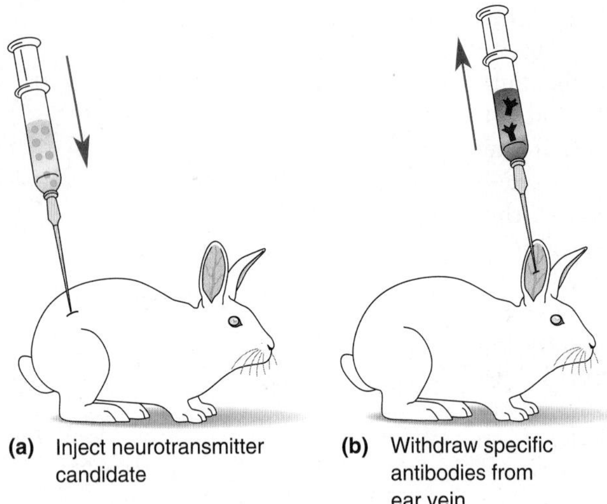

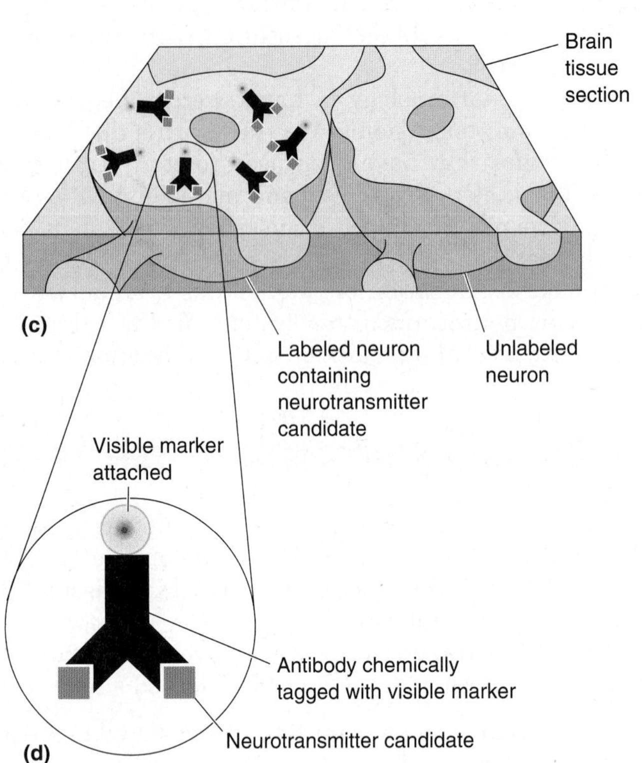

▶ FIGURE 6.2
Immunohistochemistry. This method uses labeled antibodies to identify the location of molecules within cells. **(a)** The molecule of interest (a neurotransmitter candidate) is injected into an animal, causing an immune response and the generation of antibodies. **(b)** Blood is withdrawn from the animal, and the antibodies are isolated from the serum. **(c)** The antibodies are tagged with a visible marker and applied to sections of brain tissue. The antibodies label only those cells that contain the neurotransmitter candidate. **(d)** A close-up of the complex that includes the neurotransmitter candidates, an antibody, and its visible marker.

coupled to a larger molecule.) One feature of the immune response is the generation of large proteins called *antibodies*. Antibodies can bind tightly to specific sites on the foreign molecule, also known as the *antigen*—in this case, the transmitter candidate. The best antibodies for immunocytochemistry bind very tightly to the transmitter of interest and bind very little or not at all to other chemicals in the brain. These specific antibody molecules can be recovered from a blood sample of the immunized animal and chemically tagged with a colorful marker that can be seen with a microscope. When these labeled antibodies are applied to a section of brain tissue, they will color just those cells that contain the transmitter candidate (Figure 6.3a). By using several different antibodies, each labeled with a different marker color, it is possible to distinguish several types of cells in the same region of the brain (Figure 6.3b).

Immunocytochemistry can be used to localize any molecule for which a specific antibody can be generated, including the synthesizing enzymes for transmitter candidates. Demonstration that the transmitter candidate and its synthesizing enzyme are contained in the same neuron—or better yet, in the same axon terminal—can help satisfy the criterion that the molecule be localized in, and synthesized by, a particular neuron.

***In Situ* Hybridization.** The method known as ***in situ* hybridization** is also useful for confirming that a cell synthesizes a particular protein or peptide. Recall from Chapter 2 that proteins are assembled by the ribosomes according to instructions from specific mRNA molecules. There is a unique mRNA molecule for every polypeptide synthesized by a neuron. The mRNA transcript consists of the four different nucleic acids linked together in various sequences to form a long strand. Each nucleic

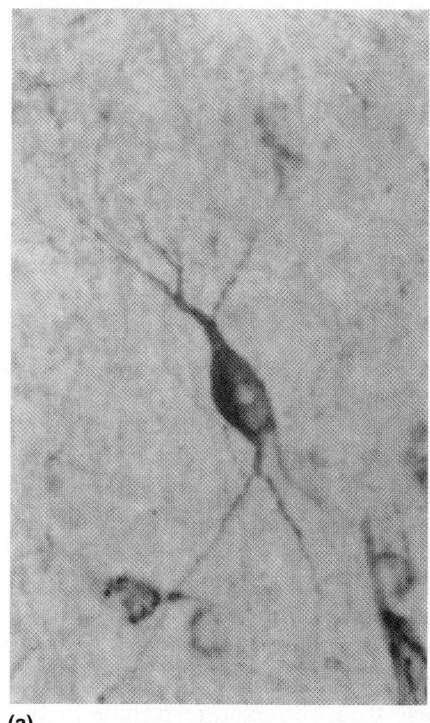
(a)

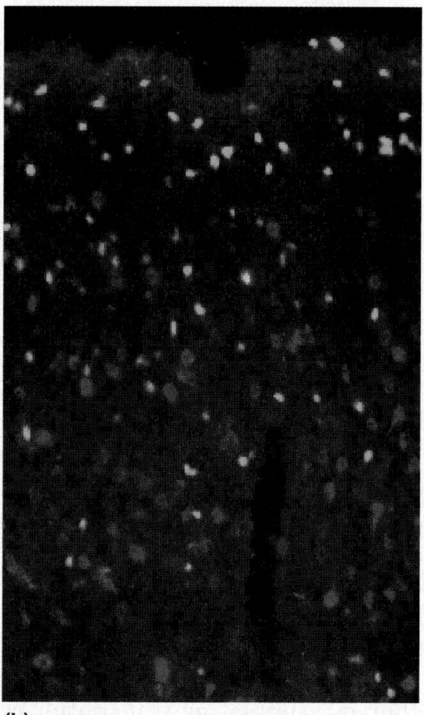
(b)

◀ FIGURE 6.3
Immunohistochemical localization of proteins in neurons. **(a)** A neuron in the cerebral cortex labeled with antibodies that bind to a peptide neurotransmitter. (Source: Courtesy of Dr. Y. Amitai and S. L. Patrick.) **(b)** Three distinct types of neurons in the cerebral cortex, each labeled with a different antibody tagged with a differently colored fluorescent marker (green, red, and blue). (Source: Courtesy of Dr. S.J. Cruikshank and S.L. Patrick.) The image in **a** is shown at a higher magnification than that in **b**.

acid has the unusual property that it will bind most tightly to one other complementary nucleic acid. Thus, if the sequence of nucleic acids in a strand of mRNA is known, it is possible to construct in the lab a complementary strand that will stick, like a strip of Velcro, to the mRNA molecule. The complementary strand is called a *probe*, and the process by which the probe bonds to the mRNA molecule is called *hybridization* (Figure 6.4). In order to see if the mRNA for a particular peptide is localized in a neuron, we chemically label the appropriate probe so it can be detected, apply it to a section of brain tissue, allow time for the probes to stick to any complementary mRNA strands, then wash away all the extra probes that have not stuck. Finally, we search for neurons that contain the label.

To visualize labeled cells after *in situ* hybridization, the probes can be chemically tagged in several ways. A common approach is to make them radioactive. Because we cannot see radioactivity, hybridized probes are detected by laying the brain tissue on a sheet of special film that is sensitive to radioactive emissions. After exposure to the tissue, the film is developed like a photograph, and negative images of the radioactive cells are visible as clusters of small white dots (Figure 6.5). It is also possible to use digital electronic imaging devices to detect the radioactivity. This technique for viewing the distribution of radioactivity is called **autoradiography**. An alternative is to label the probes with brightly colorful fluorescent molecules that can viewed directly with an appropriate microscope. Fluorescence *in situ* hybridization is also known as FISH.

In summary, immunocytochemistry is a method for viewing the location of specific molecules, including proteins, in sections of brain tissue. *In situ* hybridization is a method for localizing specific mRNA transcripts for proteins. Together, these methods enable us to see whether a neuron contains and synthesizes a transmitter candidate and molecules associated with that transmitter.

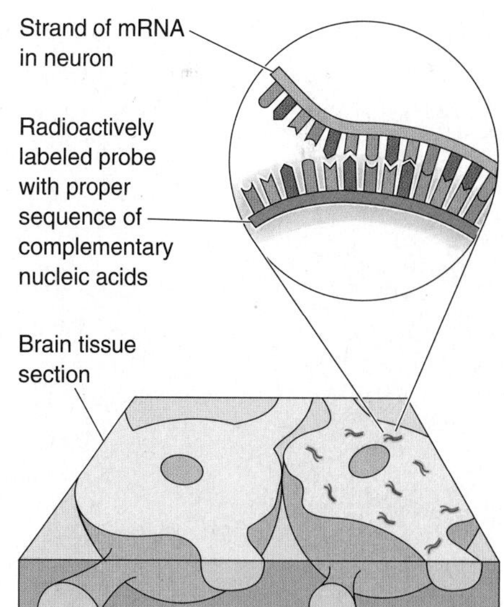

▲ FIGURE 6.4
***In situ* hybridization.** Strands of mRNA consist of nucleotides arranged in a specific sequence. Each nucleotide will stick to one other complementary nucleotide. In the method of *in situ* hybridization, a synthetic probe is constructed containing a sequence of complementary nucleotides that will allow it to stick to the mRNA. If the probe is labeled, the location of cells containing the mRNA will be revealed.

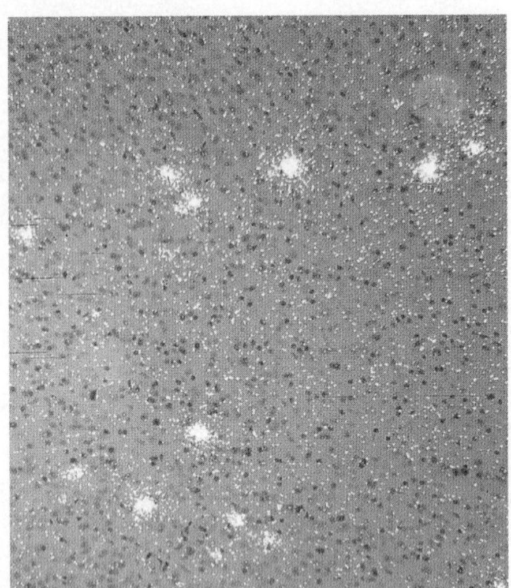

▲ FIGURE 6.5
***In situ* hybridization of the mRNA for a peptide neurotransmitter in neurons, visualized with autoradiography.** Only neurons with the proper mRNA are labeled, visible here as clusters of white dots. (Source: Courtesy of Dr. S. H. C. Hendry.)

Studying Transmitter Release

Once we are satisfied that a transmitter candidate is synthesized by a neuron and localized to the presynaptic terminal, we must show that it is actually released upon stimulation. In some cases, a specific set of cells or axons can be stimulated while taking samples of the fluids bathing their synaptic targets. The biological activity of the sample can then be tested to see if it mimics the effect of the intact synapses, and then the sample can be chemically analyzed to reveal the structure of the active molecule. This general approach helped Loewi and Dale identify ACh as a transmitter at many peripheral synapses.

Unlike the peripheral nervous system (PNS), the nervous system outside the brain and spinal cord studied by Loewi and Dale, most regions of the central nervous system (CNS) contain a diverse mixture of intermingled synapses using different neurotransmitters. Until recently, this often made it impossible to stimulate a single population of synapses containing only a single neurotransmitter. Researchers had to be content with stimulating many synapses in a region of the brain and collecting and measuring all the chemicals that were released. One way to do this is to use brain slices that are kept alive *in vitro*. To stimulate release, the slices are bathed in a solution containing a high K^+ concentration. This treatment causes a large membrane depolarization (see Figure 3.19), thereby stimulating transmitter release from the axon terminals in the tissue. Because transmitter release requires the entry of Ca^{2+} into the axon terminal, it must also be shown that the release of the neurotransmitter candidate from the tissue slice after depolarization occurs only when Ca^{2+} ions are present in the bathing solution. New methods such as optogenetics (see Box 4.2) now make it possible to activate just one specific type of synapse at a time. Genetic methods are used to induce one particular population of neurons to express light-sensitive proteins, and then those neurons can be stimulated with brief flashes of light that have no effect on the surrounding cells. Any transmitters released are likely to have come from the optogenetically selected type of synapse.

Even when it has been shown that a transmitter candidate is released upon depolarization in a calcium-dependent manner, we still cannot be sure that the molecules collected in the fluids were released from the axon terminals; they may have been released as a secondary consequence of synaptic activation. These technical difficulties make the second criterion—that a transmitter candidate must be released by the presynaptic axon terminal upon stimulation—the most difficult to satisfy unequivocally in the CNS.

Studying Synaptic Mimicry

Establishing that a molecule is localized in, synthesized by, and released from a neuron is still not sufficient to qualify it as a neurotransmitter. A third criterion must be met: The molecule must evoke the same response as that produced by the release of the naturally occurring neurotransmitter from the presynaptic neuron.

To assess the postsynaptic actions of a transmitter candidate, a method called **microiontophoresis** is sometimes used. Most neurotransmitter candidates can be dissolved in solutions that will cause them to acquire a net electrical charge. A glass pipette with a very fine tip, just a few micrometers across, is filled with the ionized solution. The tip of the pipette is carefully positioned next to the postsynaptic membrane of the neuron, and the transmitter candidate is ejected in very small amounts by passing electrical current through the pipette. Neurotransmitter candidates can also be ejected

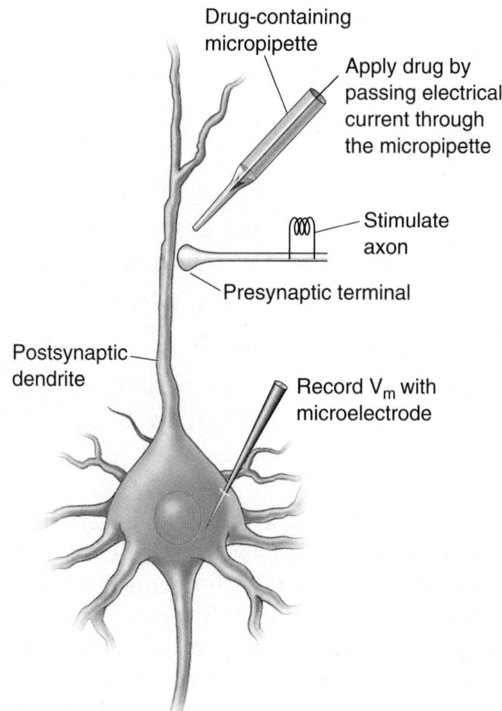

▲ FIGURE 6.6
Microiontophoresis. This method enables a researcher to apply drugs or neurotransmitter candidates in very small amounts to the surface of neurons. The responses generated by the drug can be compared to those generated by synaptic stimulation.

from fine pipettes with pulses of high pressure. A microelectrode in the postsynaptic neuron can be used to measure the effects of the transmitter candidate on the membrane potential (Figure 6.6).

If iontophoretic or pressure application of the molecule causes electrophysiological changes that mimic the effects of transmitter released at the synapse, *and* if the other criteria of localization, synthesis, and release are met, then the molecule and the transmitter are usually considered to be the same chemical.

Studying Receptors

Each neurotransmitter exerts its postsynaptic effects by binding to specific receptors. As a rule, no two neurotransmitters bind to the same receptor; however, one neurotransmitter can bind to many different receptors. Each of the different receptors a neurotransmitter binds to is called a **receptor subtype**. For example, in Chapter 5, we learned that ACh acts on two different cholinergic receptor *subtypes*: One type is present in skeletal muscle, and the other is in heart muscle. Both subtypes are also present in many other organs and within the CNS.

Researchers have tried almost every method of biological and chemical analysis to study the different receptor subtypes of the various neurotransmitter systems. Three approaches have proved to be particularly useful: neuropharmacological analysis of synaptic transmission, ligand-binding methods, and molecular analysis of receptor proteins.

Neuropharmacological Analysis. Much of what we know about receptor subtypes was first learned using neuropharmacological analysis. For instance, skeletal muscle and heart muscle respond differently to various cholinergic drugs. *Nicotine*, derived from the tobacco plant, is a receptor

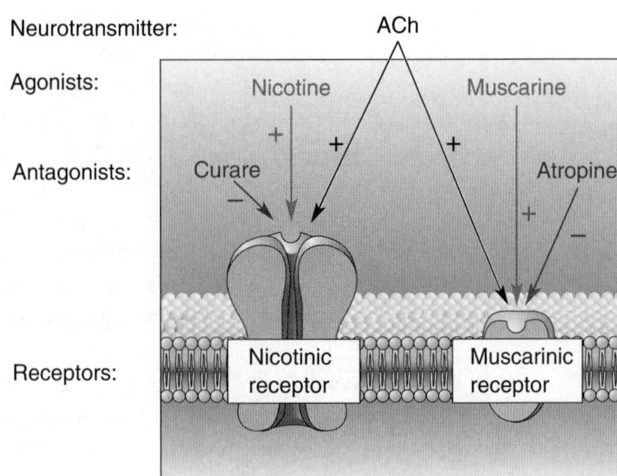

▲ FIGURE 6.7
The neuropharmacology of cholinergic synaptic transmission. Sites on transmitter receptors can bind either the transmitter itself (ACh), an agonist that mimics the transmitter, or an antagonist that blocks the effects of the transmitter and agonists.

agonist in skeletal muscle but has no effect in the heart. On the other hand, *muscarine*, derived from a poisonous species of mushroom, has little or no effect on skeletal muscle but is an agonist at the cholinergic receptor subtype in the heart. (Recall that ACh slows the heart rate; muscarine is poisonous because it causes a precipitous drop in heart rate and blood pressure.) Thus, two ACh receptor subtypes can be distinguished by the actions of different drugs. In fact, the receptors were given the names of their agonists: **nicotinic ACh receptors** in skeletal muscle and **muscarinic ACh receptors** in the heart. Nicotinic and muscarinic receptors also exist in the brain, and some neurons have both types of receptors.

Another way to distinguish receptor subtypes is to use selective antagonists. The South American arrow-tip poison *curare* inhibits the action of ACh at nicotinic receptors (thereby causing paralysis), and *atropine*, derived from belladonna plants (also known as deadly nightshade), antagonizes ACh at muscarinic receptors (Figure 6.7). (The eye drops an ophthalmologist uses to dilate your pupils are related to atropine.)

Different drugs were also used to distinguish several subtypes of glutamate receptors, which mediate much of the synaptic excitation in the CNS. Three subtypes are **AMPA receptors**, **NMDA receptors**, and **kainate receptors**, each named for a different chemical agonist. (AMPA stands for α-amino-3-hydroxy-5-methyl-4-isoxazole propionate, and NMDA stands for *N*-methyl-D-aspartate.) The neurotransmitter glutamate activates all three receptor subtypes, but AMPA acts only at the AMPA receptor, NMDA acts only at the NMDA receptor, and so on (Figure 6.8).

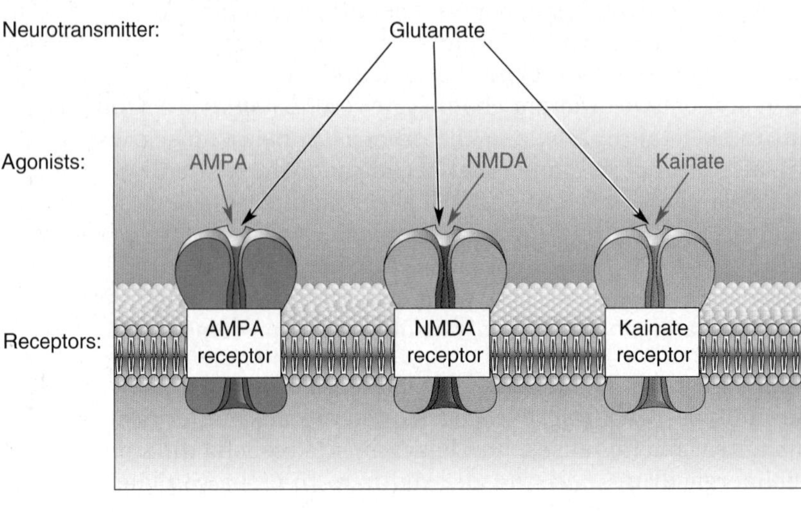

▶ FIGURE 6.8
The neuropharmacology of glutamatergic synaptic transmission. There are three main subtypes of glutamate receptors, each of which binds glutamate and each of which is activated selectively by a different agonist.

TABLE 6.1 **Neurotransmitters, Some Receptors, and Their Pharmacology**

Neurotransmitter	Receptor Subtype	Agonist	Antagonist
Acetylcholine (ACh)	Nicotinic receptor	Nicotine	Curare
	Muscarinic receptor	Muscarine	Atropine
Norepinephrine (NE)	α receptor	Phenylephrine	Phenoxybenzamine
	β receptor	Isoproterenol	Propranolol
Glutamate (Glu)	AMPA	AMPA	CNQX
	NMDA	NMDA	AP5
GABA	$GABA_A$	Muscimol	Bicuculline
	$GABA_B$	Baclofen	Phaclofen
ATP	P2X	ATP	Suramin
Adenosine	A type	Adenosine	Caffeine

Similar pharmacological analyses were used to split the NE receptors into two subtypes, α and β, and to divide GABA receptors into $GABA_A$ and $GABA_B$ subtypes. The same can be said for virtually all the neurotransmitter systems. Thus, selective drugs have been extremely useful for categorizing receptor subclasses (Table 6.1). In addition, neuropharmacological analysis has been invaluable for assessing the contributions of neurotransmitter systems to brain function.

Ligand-Binding Methods. As we said, the first step in studying a neurotransmitter system is usually identifying the neurotransmitter. However, with the discovery in the 1970s that many drugs interact selectively with neurotransmitter receptors, researchers realized that they could use these compounds to begin analyzing receptors even before the neurotransmitter itself had been identified. The pioneers of this approach were Solomon Snyder and his then student Candace Pert at Johns Hopkins University, who were interested in studying compounds called *opiates* (Box 6.1). *Opiates* are a class of drugs, derived from the opium poppy, that are both medically important and commonly abused. *Opioids* are the broader class of opiate-like chemicals, both natural and synthetic. Their effects include pain relief, euphoria, depressed breathing, and constipation.

The question Snyder and Pert originally set out to answer was how heroin, morphine, and other opiates exert their effects on the brain. They and others hypothesized that opiates might be agonists at specific receptors in neuronal membranes. To test this idea, they radioactively labeled opiate compounds and applied them in small quantities to neuronal membranes that had been isolated from different parts of the brain. If appropriate receptors existed in the membrane, the labeled opiates should bind tightly to them. This is just what they found. The radioactive drugs labeled specific sites on the membranes of some, but not all, neurons in the brain (Figure 6.9). Following the discovery of opioid receptors, the search was on to identify endogenous opioids, or *endorphins*, the naturally occurring neurotransmitters that act on these receptors. Two peptides called *enkephalins* were soon isolated from the brain, and they eventually proved to be opioid neurotransmitters.

Any chemical compound that binds to a specific site on a receptor is called a *ligand* for that receptor (from the Latin meaning "to bind"). The technique of studying receptors using radioactively or nonradioactively labeled ligands is called the **ligand-binding method**. Notice that a ligand for a receptor can be an agonist, an antagonist, or the chemical neurotransmitter itself. Specific ligands were invaluable for isolating neurotransmitter receptors and determining their chemical structure.

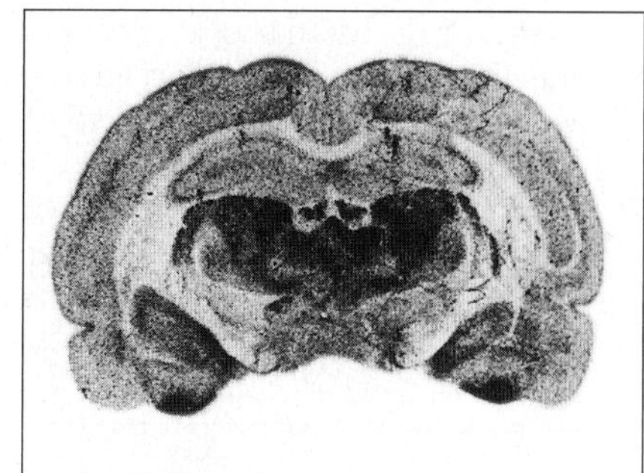

▲ FIGURE 6.9
Opiate receptor binding to a slice of rat brain. Special film was exposed to a brain section that had radioactive opiate receptor ligands bound to it. The dark regions contain more receptors. (Source: Snyder, 1986, p. 44.)

BOX 6.1 PATH OF DISCOVERY

Finding Opiate Receptors

by Solomon H. Snyder

Like so many events in science, identifying the opiate receptors was not simply an intellectual feat accomplished in an ethereal pursuit of pure knowledge. Instead, it all began with President Nixon and his "war on drugs" in 1971, at the height of very well-publicized use of heroin by hundreds of thousands of American soldiers in Vietnam. To combat all this, Nixon appointed as czar of drug abuse research Dr. Jerome Jaffe, a psychiatrist who had pioneered in methadone treatment for heroin addicts. Jaffe was to coordinate the several billions of federal dollars in agencies ranging from the Department of Defense to the National Institutes of Health.

Jerry, a good friend, pestered me to direct our research toward the "poor soldiers" in Vietnam. So I began wondering how opiates act. The notion that drugs act at receptors, specific recognition sites, had been appreciated since the turn of the century. In principle, one could identify such receptors simply by measuring the binding of radioactive drugs to tissue membranes. However, countless researchers had applied this strategy to opiates with no success.

About this time, a new Johns Hopkins faculty member, Pedro Cuatrecasas, located his laboratory adjacent to mine, and we became fast friends. Pedro had recently attained fame for his discovery of receptors for insulin. His success depended upon seemingly simple but important technical advances. Past efforts to identify receptors for hormones had failed because hormones can bind to many nonspecific sites, comprising proteins, carbohydrates, and lipids. The numbers of these nonspecific sites would likely be millions of times greater than the number of specific receptors. To identify the "signal" of insulin receptors binding above the "noise" of nonspecific interactions, Pedro developed a simple filtration assay. Since insulin should adhere more tightly to its receptors than to nonspecific sites, he incubated liver membranes with radioactive insulin and poured the mixture over filters attached to a vacuum that rapidly sucked away the incubation fluid, leaving the membrane with attached insulin stuck to the filters. He then "washed" the filters with large volumes of saline, but did this very rapidly so as to preserve insulin bound to receptors while washing away nonspecific binding.

Despite Pedro's proximity, it did not immediately occur to me that the insulin success could be transferred to the opiate receptor problem. Instead, I had read a paper on nerve growth factor, showing that its amino acid sequence closely resembled that of insulin. Pedro and I soon collaborated in a successful search for the nerve growth factor receptor. Only then did I marshal the courage to extend this approach from proteins such as insulin and nerve growth factor to much smaller molecules such as opiates. Candace Pert, a graduate student in my laboratory, was eager to take on a new research project. We obtained a radioactive drug and monitored its binding to brain membranes using Pedro's magic filter machine. The very first experiment, which took only about two hours, was successful.

Within a few months, we were able to characterize many features of opiate receptors. Knowing the exact sites where receptors are concentrated in the brain explained all the major actions of opiates, such as euphoria, pain relief, depression of breathing, and pupillary constriction. The properties of opiate receptors resembled very much what one would expect for neurotransmitters. Accordingly, we used similar approaches to search for receptors for neurotransmitters in the brain, and within a few years had identified receptors for most of them.

These findings raised an obvious question: Why do opiate receptors exist? Humans were not born with morphine in them. Might the opiate receptor be a receptor for a new transmitter that regulates pain perception and emotional states? We and other groups attempted to isolate the hypothesized, normally occurring, morphine-like neurotransmitters. John Hughes and Hans Kosterlitz in Aberdeen, Scotland, were the first to succeed. They isolated and obtained the chemical structures of the first "endorphins," which are called the *enkephalins*. In our own laboratory, Rabi Simantov and I obtained the structure of the enkephalins soon after the published success of the Scottish group.

From the first experiments identifying opiate receptors until the isolation of the enkephalins, only three years elapsed—an interval of frantic, exhilarating work that changed profoundly how we think about drugs and the brain.

Ligand-binding methods have been enormously important for mapping the anatomical distribution of different neurotransmitter receptors in the brain.

Molecular Analysis. There has been an explosion of information about neurotransmitter receptors in recent decades, thanks to modern methods for studying protein molecules. Information obtained with these methods has enabled us to divide the neurotransmitter receptor proteins into two groups: transmitter-gated ion channels and G-protein-coupled (metabotropic) receptors (see Chapter 5).

Molecular neurobiologists have determined the structure of the polypeptides that make up many proteins, and these studies have led to some startling conclusions. Receptor subtype diversity was expected from the actions of different drugs, but the breadth of the diversity was not appreciated until researchers determined how many different polypeptides could serve as subunits of functional receptors.

Consider as an example the $GABA_A$ receptor, a transmitter-gated chloride channel. Each channel requires five subunits (similar to the ACh-gated ion channel, Figure 5.14), and there are five major classes of subunit proteins, designated α, β, γ, δ, and ρ. At least six different polypeptides (designated α1–6) can substitute for one another as an α subunit. Four different polypeptides (designated β1–4) can substitute as a β subunit, and four different polypeptides (γ1–4) can be used as a γ subunit. Although this is not the complete tally, let's make an interesting calculation. If it takes five subunits to form a $GABA_A$ receptor-gated channel and there are 15 possible subunits to choose from, then there are 151,887 possible combinations and arrangements of subunits. This means there are at least 151,887 potential subtypes of $GABA_A$ receptors!

It is important to recognize that the vast majority of the possible subunit combinations are never manufactured by neurons, and even if they were, they would not work properly. Nonetheless, it is clear that receptor classifications like those appearing in Table 6.1, while still useful, seriously underestimate the diversity of receptor subtypes in the brain.

NEUROTRANSMITTER CHEMISTRY

Research using methods such as those discussed previously has led to the conclusion that the major neurotransmitters are amino acids, amines, and peptides. Evolution is conservative and opportunistic, and it often puts common and familiar things to new uses. This also seems true about the evolution of neurotransmitters. For the most part, they are similar or identical to the basic chemicals of life, the same substances that cells in all species, from bacteria to giraffes, use for metabolism. Amino acids, the building blocks of protein, are essential to life. Most of the known neurotransmitter molecules are either (1) amino acids, (2) amines derived from amino acids, or (3) peptides constructed from amino acids. ACh is one exception, but it is derived from acetyl CoA, a ubiquitous product of cellular respiration in mitochondria, and choline, which is important for fat metabolism throughout the body.

Amino acid and amine transmitters are generally each stored in and released by different sets of neurons. The convention established by Dale classifies neurons into mutually exclusive groups by neurotransmitter (cholinergic, glutamatergic, GABAergic, and so on). The idea that a neuron has only one neurotransmitter is often called **Dale's principle**. Many peptide-containing neurons violate Dale's principle because these cells usually release more than one neurotransmitter: an amino acid or amine *and* a peptide. When two or more transmitters are released from one nerve terminal, they are called **co-transmitters**. Many examples of neurons with co-transmitters have been identified in recent years, including some that release two small transmitters (e.g., GABA and glycine). Nonetheless, most neurons seem to release only a single amino acid or amine neurotransmitter, which can be used to assign them to distinct, nonoverlapping classes. Let's take a look at the biochemical mechanisms that differentiate these neurons.

Cholinergic Neurons

Acetylcholine (ACh) is the neurotransmitter at the neuromuscular junction and is therefore synthesized by all the motor neurons in the spinal cord and brain stem. Other cholinergic cells contribute to the functions of specific circuits in the PNS and CNS, as we will see in Chapter 15.

ACh synthesis requires a specific enzyme, *choline acetyltransferase* (*ChAT*) (Figure 6.10). Like nearly all presynaptic proteins, ChAT is manufactured in the soma and transported to the axon terminal. Only cholinergic neurons contain ChAT, so this enzyme is a good marker for cells that use ACh as a neurotransmitter. Immunohistochemistry with ChAT-specific antibodies, for example, can be used to identify cholinergic neurons. ChAT synthesizes ACh in the cytosol of the axon terminal, and the neurotransmitter is concentrated in synaptic vesicles by the actions of a vesicular ACh **transporter** (Box 6.2).

ChAT transfers an acetyl group from acetyl CoA to choline (Figure 6.11a). The source of choline is the extracellular fluid, where

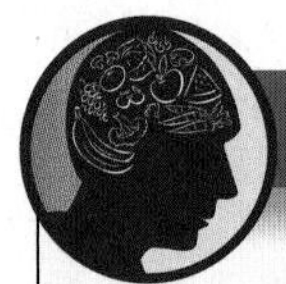

BOX 6.2 BRAIN FOOD

Pumping Ions and Transmitters

Neurotransmitters may lead an exciting life, but the most mundane part would seem to be the steps that recycle them back from the synaptic cleft and eventually into a vesicle. Where synapses are concerned, the exotic proteins of exocytosis and the innumerable transmitter receptors get most of the publicity. Yet, the neurotransmitter transporters are very interesting for at least two reasons: *They succeed at an extraordinarily difficult job, and they are the molecular site at which many important psychoactive drugs act.*

The hard job of transporters is to pump transmitter molecules across membranes so effectively that they become highly concentrated in very small places. There are two general types of neurotransmitter transporters. One type, the *neuronal membrane transporter,* shuttles transmitter from the extracellular fluid, including the synaptic cleft, and concentrates it up to 10,000 times higher within the cytosol of the presynaptic terminal. A second type, the *vesicular transporter*, then crams transmitter into vesicles at concentrations that may be 100,000 times higher than in the cytosol. Inside cholinergic vesicles, for example, ACh may reach the incredible concentration of 1000 mM, or 1 molar—in other words, about twice the concentration of salt in seawater!

How do transporters achieve such dramatic feats of concentration? Concentrating a chemical is like carrying a weight uphill; both are extremely unlikely to occur unless energy is applied to the task. Recall from Chapter 3 that ion pumps in the plasma membrane use ATP as their source of energy to transport Na^+, K^+, and Ca^{2+} against their concentration gradients. These ion gradients are essential for setting the resting potential and for powering the ionic currents that underlie action and synaptic potentials. Similarly, membranes of synaptic vesicles have pumps that use ATP to fuel the transport of H^+ into vesicles. Notice that once ionic gradients are established across a membrane, they can themselves be tapped as sources of energy. Just as the energy spent in pulling up the weights on a cuckoo clock can be reclaimed to turn the gears and hands of the clock (as the weights slowly fall down again), transporters use transmembrane gradients of Na^+ or H^+ as an energy source for moving transmitter molecules up steep concentration gradients. The transporter lets one transmembrane gradient, that of Na^+ or H^+, run down a bit in order to build up another gradient, that of the transmitter.

The transporters themselves are large proteins that span membranes. There can be several transporters for one transmitter (e.g., at least four subtypes are known for GABA). Figure A shows how they work. Plasma membrane transporters use a *cotransport* mechanism, carrying two Na^+ ions along with one transmitter molecule. By contrast, vesicular membrane transporters use a *countertransport* mechanism that trades a transmitter molecule from the cytosol for a H^+ from inside the vesicle. Vesicle membranes have ATP-driven H^+ pumps that keep their contents very acidic, or high in protons (i.e., H^+ ions).

What is the relevance of all this to drugs and disease? Many psychoactive drugs, such as amphetamines and

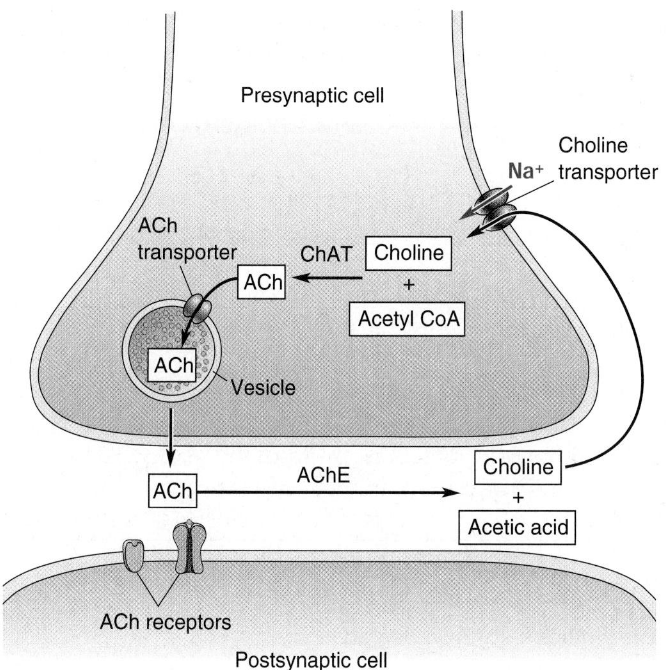

◀ FIGURE 6.10
The life cycle of ACh.

cocaine, potently block certain transporters. By altering the normal recycling process of various transmitters, the drugs lead to chemical imbalances in the brain that can have profound effects on mood and behavior. It is also possible that defects in transporters can lead to psychiatric or neurological disease; certainly, some of the drugs that are therapeutically useful in psychiatry work by blocking transporters. The numerous links between transmitters, drugs, disease, and treatment are tantalizing but complex, and will be discussed further in Chapters 15 and 22.

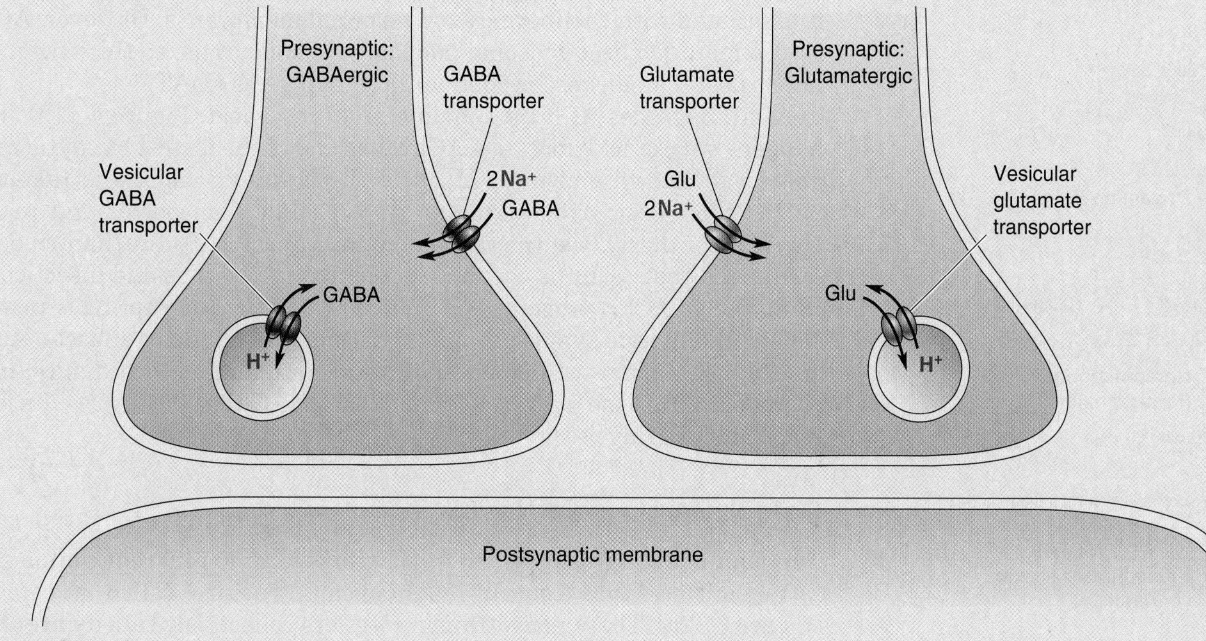

Figure A
Neurotransmitter transporters.

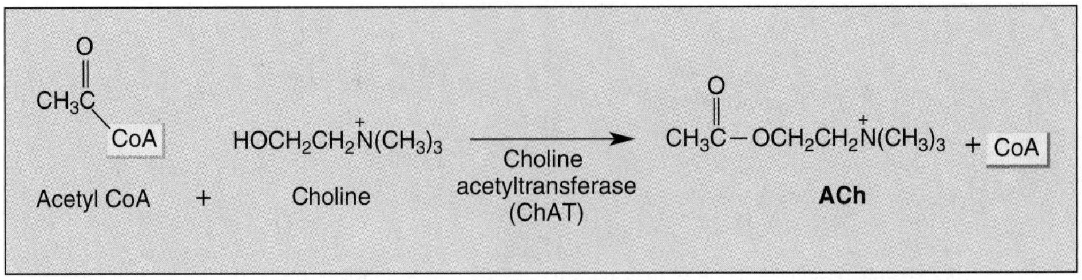

$$\underset{\text{Acetyl CoA}}{CH_3C(=O)\text{–CoA}} + \underset{\text{Choline}}{HOCH_2CH_2\overset{+}{N}(CH_3)_3} \xrightarrow[\text{(ChAT)}]{\text{Choline acetyltransferase}} \underset{\textbf{ACh}}{CH_3C(=O)\text{–}OCH_2CH_2\overset{+}{N}(CH_3)_3} + \text{CoA}$$

(a)

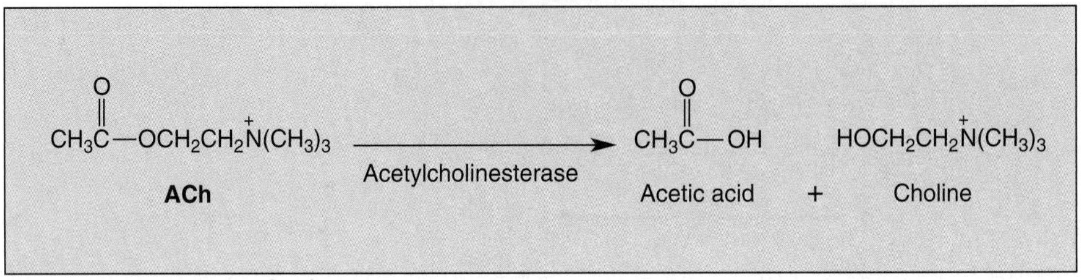

$$\underset{\textbf{ACh}}{CH_3C(=O)\text{—}OCH_2CH_2\overset{+}{N}(CH_3)_3} \xrightarrow{\text{Acetylcholinesterase}} \underset{\text{Acetic acid}}{CH_3C(=O)\text{—}OH} + \underset{\text{Choline}}{HOCH_2CH_2\overset{+}{N}(CH_3)_3}$$

(b)

▲ FIGURE 6.11
Acetylcholine. (a) ACh synthesis. (b) ACh degradation.

it exists in low micromolar concentrations. Choline is taken up by the cholinergic axon terminals via a specific transporter that requires the cotransport of Na^+ to power the movement of choline (see Box 6.2). Because the availability of choline limits how much ACh can be synthesized in the axon terminal, the transport of choline into the neuron is said to be the **rate-limiting step** in ACh synthesis. For certain diseases in which a deficit in cholinergic synaptic transmission has been noted, dietary supplements of choline are sometimes prescribed to boost ACh levels in the brain.

Cholinergic neurons also manufacture the ACh degradative enzyme *acetylcholinesterase (AChE)*. AChE is secreted into the synaptic cleft and is associated with cholinergic axon terminal membranes. However, AChE is also manufactured by some noncholinergic neurons, so this enzyme is not as useful a marker for cholinergic synapses as ChAT.

AChE degrades ACh into choline and acetic acid (Figure 6.11b). This happens very quickly because AChE has one of the fastest catalytic rates among all known enzymes. Much of the resulting choline is taken up by the cholinergic axon terminal via a choline transporter and reused for ACh synthesis (see the red arrow in Figure 6.10). In Chapter 5, we mentioned that AChE is the target of many nerve gases and insecticides. Inhibition of AChE prevents the breakdown of ACh, disrupting transmission at cholinergic synapses on skeletal muscle and heart muscle. Acute effects include marked decreases in heart rate and blood pressure; however, death from the irreversible inhibition of AChE typically results from respiratory paralysis.

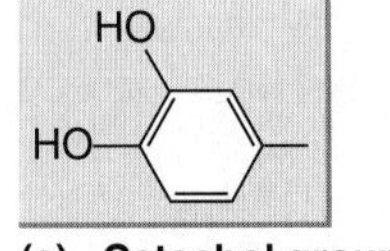

(a) **Catechol group**

HO
HO — $CH_2CH_2NH_2$
Dopamine (DA)

HO
HO — $CHCH_2NH_2$ (OH)
Norepinephrine (NE)
(Noradrenaline)

HO
HO — $CHCH_2NHCH_3$ (OH)
(b) **Epinephrine**
(Adrenaline)

▲ FIGURE 6.12
The catecholamines. (a) A catechol group. (b) The catecholamine neurotransmitters.

Catecholaminergic Neurons

The amino acid tyrosine is the precursor for three different amine neurotransmitters that contain a chemical structure called a *catechol* (Figure 6.12a). These neurotransmitters are collectively called **catecholamines**. The catecholamine neurotransmitters are **dopamine (DA)**, **norepinephrine (NE)**, and **epinephrine**, also called **adrenaline** (Figure 6.12b). Catecholaminergic neurons are found in regions of the

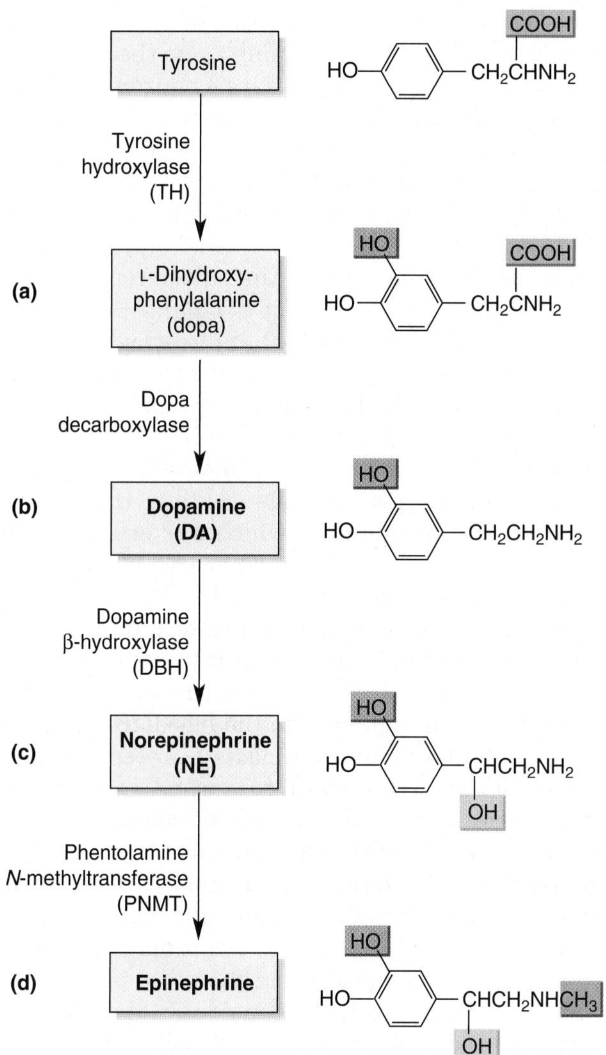

▲ FIGURE 6.13
The synthesis of catecholamines from tyrosine. The catecholamine neurotransmitters are in boldface type.

nervous system involved in the regulation of movement, mood, attention, and visceral function (discussed further in Chapter 15).

All catecholaminergic neurons contain the enzyme *tyrosine hydroxylase (TH)*, which catalyzes the first step in catecholamine synthesis, the conversion of tyrosine to a compound called **dopa** (L-dihydroxyphenylalanine) (Figure 6.13a). The activity of TH is rate limiting for catecholamine synthesis. The enzyme's activity is regulated by various signals in the cytosol of the axon terminal. For example, decreased catecholamine release by the axon terminal causes the catecholamine concentration in the cytosol to rise, thereby inhibiting TH. This type of regulation is called *end-product inhibition.* On the other hand, during periods when catecholamines are released at a high rate, the elevation in $[Ca^{2+}]_i$ that accompanies neurotransmitter release triggers an increase in the activity of TH, so transmitter supply keeps up with demand. In addition, prolonged periods of stimulation actually cause the synthesis of more mRNA that codes for the enzyme.

Dopa is converted into the neurotransmitter DA by the enzyme *dopa decarboxylase* (Figure 6.13b). Dopa decarboxylase is abundant in catecholaminergic neurons, so the amount of DA synthesized depends primarily on the amount of dopa available. In the movement disorder known as Parkinson's disease, dopaminergic neurons in the brain slowly degenerate and eventually die. One strategy for treating Parkinson's disease is the administration of dopa, which causes an increase in DA synthesis in the surviving neurons, increasing the amount of DA available for release. (We will learn more about dopamine and movement in Chapter 14.)

Neurons that use NE as a neurotransmitter contain, in addition to TH and dopa decarboxylase, the enzyme *dopamine β-hydroxylase (DBH)*, which converts DA to NE (Figure 6.13c). Interestingly, DBH is not found in the cytosol but instead is located within the synaptic vesicles. Thus, in noradrenergic axon terminals, DA is transported from the cytosol to the synaptic vesicles, and there it is made into NE.

The last in the line of catecholamine neurotransmitters is epinephrine (adrenaline). Adrenergic neurons contain the enzyme *phentolamine* N-*methyltransferase (PNMT)*, which converts NE to epinephrine (Figure 6.13d). Curiously, PNMT is in the cytosol of adrenergic axon terminals. Thus, NE must first be synthesized in the vesicles and released into the cytosol for conversion into epinephrine, and then the epinephrine must again be transported into vesicles for release. In addition to serving as a neurotransmitter in the brain, epinephrine acts as a hormone when it is released by the adrenal gland into the bloodstream. As we shall see in Chapter 15, circulating epinephrine acts at receptors throughout the body to produce a coordinated visceral response.

The catecholamine systems have no fast extracellular degradative enzyme analogous to AChE. Instead, the actions of catecholamines in the synaptic cleft are terminated by selective uptake of the neurotransmitters back into the axon terminal via Na^+-dependent transporters. This step is sensitive to a number of different drugs. For example, amphetamine and cocaine block catecholamine uptake and therefore prolong the actions of the neurotransmitter in the cleft. Once inside the axon terminal, the catecholamines may be reloaded into synaptic vesicles for reuse, or they may be enzymatically destroyed by the action of *monoamine oxidase (MAO)*, an enzyme found on the outer membrane of mitochondria.

Serotonergic Neurons

The amine neurotransmitter **serotonin**, also called *5-hydroxytryptamine* and abbreviated **5-HT**, is derived from the amino acid tryptophan. **Serotonergic** neurons are relatively few in number, but, as we shall see in Part III, they appear to play an important role in the brain systems that regulate mood, emotional behavior, and sleep.

The synthesis of serotonin occurs in two steps, just like the synthesis of DA (Figure 6.14). Tryptophan is converted first into an intermediary called *5-hydroxytryptophan (5-HTP)* by the enzyme *tryptophan hydroxylase*. The 5-HTP is then converted to 5-HT by the enzyme *5-HTP decarboxylase*. Serotonin synthesis appears to be limited by the availability of tryptophan in the extracellular fluid bathing neurons. The source of brain tryptophan is the blood, and the source of blood tryptophan is the diet (grains, meat, dairy products, and chocolate are particularly rich in tryptophan).

Following release from the axon terminal, 5-HT is removed from the synaptic cleft by the action of a specific transporter. The process of serotonin reuptake, like catecholamine reuptake, is sensitive to a number of different drugs. For example, numerous clinically useful antidepressant

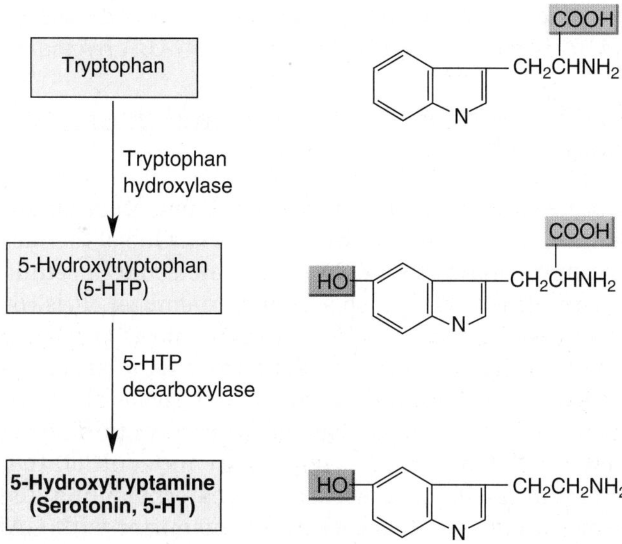

▲ FIGURE 6.14
The synthesis of serotonin from tryptophan.

and antianxiety drugs, including fluoxetine (trade name Prozac), are selective inhibitors of serotonin reuptake. Once it is back in the cytosol of the serotonergic axon terminal, the transmitter is either reloaded into synaptic vesicles or degraded by MAO.

Amino Acidergic Neurons

The amino acids **glutamate (Glu)**, **glycine (Gly)**, and **gamma-aminobutyric acid (GABA)** serve as neurotransmitters at most CNS synapses (Figure 6.15). Of these, only GABA is unique to those neurons that use it as a neurotransmitter; the others are among the 20 amino acids that make up proteins.

Glutamate and glycine are synthesized from glucose and other precursors by the action of enzymes that exist in all cells. Differences among neurons in the synthesis of these amino acids are therefore quantitative rather than qualitative. For example, the average glutamate concentration in the cytosol of glutamatergic axon terminals has been estimated to be about 20 mM, two or three times higher than that in nonglutamatergic cells. The more important distinction between glutamatergic and nonglutamatergic neurons, however, is the transporter that loads the synaptic vesicles. In glutamatergic axon terminals, but not in other types, glutamate transporters concentrate glutamate until it reaches a value of about 50 mM in the synaptic vesicles.

Because GABA is not one of the 20 amino acids used to construct proteins, it is synthesized in large quantities only by the neurons that use it as a neurotransmitter. The precursor for GABA is glutamate, and the key synthesizing enzyme is *glutamic acid decarboxylase (GAD)* (Figure 6.16). GAD, therefore, is a good marker for GABAergic neurons. Immunocytochemical studies have shown that GABAergic neurons are distributed widely in the brain. GABAergic neurons are the major source of synaptic inhibition in the nervous system. Therefore, remarkably, in one chemical step, the major excitatory neurotransmitter in the brain is converted into the major inhibitory neurotransmitter in the brain!

The synaptic actions of the amino acid neurotransmitters are terminated by selective uptake into the presynaptic terminals and glia, once

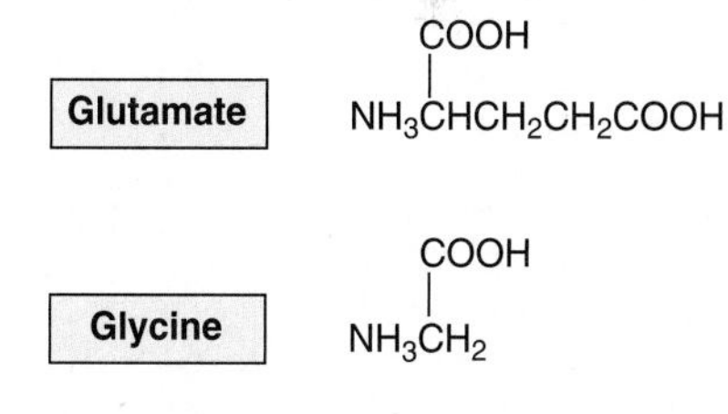

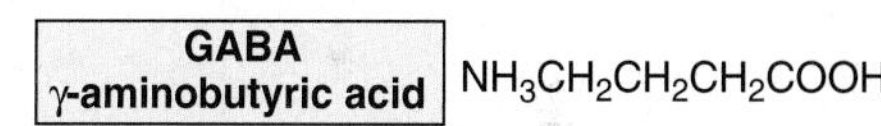

▲ FIGURE 6.15
The amino acid neurotransmitters.

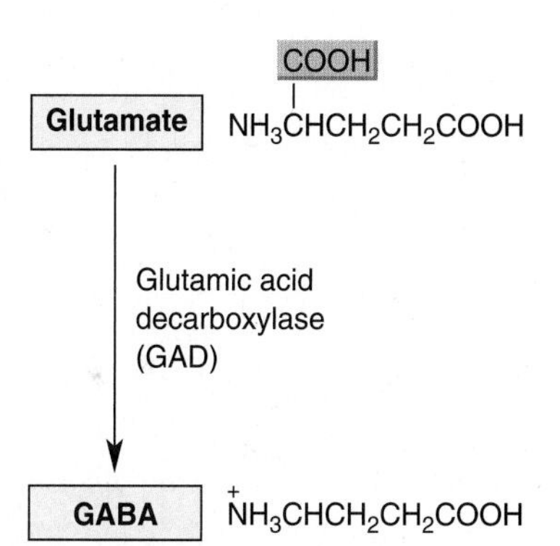

▲ FIGURE 6.16
The synthesis of GABA from glutamate.

again via specific Na^+-dependent transporters. Inside the terminal or glial cell, GABA is metabolized by the enzyme *GABA transaminase*.

Other Neurotransmitter Candidates and Intercellular Messengers

In addition to the amines and amino acids, a few other small molecules serve as chemical messengers between neurons. One of the most common is **adenosine triphosphate (ATP)**, a key molecule in cellular metabolism (see Figure 2.13). ATP is also a neurotransmitter. It is concentrated in all synaptic vesicles in the CNS and PNS, and it is released into the cleft by presynaptic spikes in a Ca^{2+}-dependent manner, just as the classic transmitters are. ATP is often packaged in vesicles along with another classic transmitter. For example, catecholamine-containing vesicles may have 100 mM of ATP, an enormous quantity, in addition to 400 mM of the catecholamine itself. In this case, the catecholamine and ATP are co-transmitters. ATP also occurs as a co-transmitter with GABA, glutamate, ACh, DA, and peptide transmitters in various specialized types of neurons.

ATP directly excites some neurons by gating cation channels. In this sense, some of the neurotransmitter functions of ATP are similar to those of glutamate and ACh. ATP binds to *purinergic receptors*, some of which are transmitter-gated ion channels. There is also a large class of G-protein-coupled purinergic receptors. Following its release from synapses, ATP is degraded by extracellular enzymes, yielding adenosine. Adenosine itself does not meet the standard definition of a neurotransmitter because it is not packaged in vesicles, but it does activate several adenosine-selective receptors.

The most interesting discovery about neurotransmitters in the past few years is that small lipid molecules, called **endocannabinoids** (endogenous cannabinoids), can be released from postsynaptic neurons and act on presynaptic terminals (Box 6.3). Communication in this direction, from "post" to "pre," is called *retrograde signaling*; thus, endocannabinoids are **retrograde messengers**. Retrograde messengers serve as a kind of feedback system to regulate the conventional forms of synaptic transmission, which of course go from "pre" to "post." The details about endocannabinoid signaling are still emerging, but one basic mechanism is now clear (Figure 6.17). Vigorous firing of action potentials in the postsynaptic neuron causes voltage-gated calcium channels to open, Ca^{2+} enters the cell in large quantities, and intracellular $[Ca^{2+}]$ rises. The elevated $[Ca^{2+}]$ then stimulates the synthesis of endocannabinoid molecules from membrane lipids by somehow activating endocannabinoid-synthesizing enzymes. There are several unusual qualities about endocannabinoids:

1. They are not packaged in vesicles like most other neurotransmitters; instead, they are manufactured rapidly and on demand.
2. They are small and membrane permeable; once synthesized, they can diffuse rapidly across the membrane of their cell of origin to contact neighboring cells.
3. They bind selectively to the CB1 type of cannabinoid receptor, which is mainly located on certain presynaptic terminals.

CB1 receptors are G-protein-coupled receptors, and their main effect is often to reduce the opening of presynaptic calcium channels. With its calcium channels inhibited, the ability of the presynaptic terminal to release its neurotransmitter (usually GABA or glutamate) is impaired. Thus, when a postsynaptic neuron is very active, it releases endocannabinoids,

BOX 6.3 OF SPECIAL INTEREST

This Is Your Brain on Endocannabinoids

Most neurotransmitters were discovered long before their receptors, but modern techniques have tended to reverse this tradition. In this story, the receptors were discovered before their transmitters.

Cannabis sativa is the botanical name for hemp, a fibrous plant used through the ages for making rope and cloth. These days, cannabis is much more popular as dope than rope. It is widely, and usually illegally, sold as marijuana or hashish, although medical uses of cannabis-related compounds are slowly being recognized, and medical or recreational use is being legalized in some states and other parts of the world. The Chinese first recognized the potent psychoactive properties of cannabis 4000 years ago, but Western society learned of its intoxicating properties only in the nineteenth century, when Napoleon III's troops returned to France with Egyptian hashish. As a member of Napoleon's Commission of Sciences and Arts reported in 1810, "For the Egyptians, hemp is the plant par excellence, not for the uses they make of it in Europe and many other countries, but for its peculiar effects. The hemp cultivated in Egypt is indeed intoxicating and narcotic" (cited in Piomelli, 2003, p. 873).

At low doses, the effects of cannabis can be euphoria, feelings of calm and relaxation, altered sensations, reduced pain, increased laughter, talkativeness, hunger, and lightheadedness, as well as decreased problem-solving ability, short-term memory, and psychomotor performance (i.e., the skills necessary for driving). High doses of cannabis can cause profound personality changes and even hallucinations. In recent years, forms of cannabis have been approved for limited medicinal use in the United States, primarily to treat nausea and vomiting in cancer patients undergoing chemotherapy and to stimulate appetite in some AIDS patients.

The active ingredient in cannabis is an oily chemical called *Δ^9-tetrahydrocannabinol*, or *THC*. During the late 1980s, it became apparent that THC can bind to specific G-protein-coupled "cannabinoid" receptors in the brain, particularly in motor control areas, the cerebral cortex, and pain pathways. At about the same time, a group at the National Institute of Mental Health cloned the gene for an unknown (or "orphan") G-protein-coupled receptor. Further work showed that the mystery receptor was a cannabinoid (CB) receptor. Two types of cannabinoid receptors are known: CB1 receptors are in the brain, and CB2 receptors are mainly in immune tissues elsewhere in the body.

Remarkably, the brain has more CB1 receptors than *any* other G-protein-coupled receptor. What are they doing there? We are quite certain they did not evolve to bind the THC from hemp. The natural ligand for a receptor is never the synthetic drug, plant toxin, or snake venom that might have helped us identify that receptor in the first place. It is much more likely that the cannabinoid receptors exist to bind some signaling molecule made naturally by the brain: THC-like neurotransmitters called *endocannabinoids*. Research has identified two major endocannabinoids: anandamide (from *ananda*, the Sanskrit word for "internal bliss") and arachidonoylglycerol (2-AG). Anandamide and 2-AG are both small lipid molecules (Figure A), quite different from any other known neurotransmitter.

As the search for new transmitters continues, the hunt is also on for more selective compounds that bind to the CB receptors. Cannabinoids are potentially useful for relieving nausea, suppressing pain, relaxing muscles, treating seizures, and decreasing the intraocular pressure of glaucoma. Antagonists of CB1 receptors have recently been tested as appetite suppressants, but they cause unfortunate side effects. Cannabinoid therapies might be more practical if new drugs can be developed that retain the therapeutic benefits without causing psychoactive and other side effects.

OH
O
^{9}-THC

O
NH
OH
Anandamide

O
OH
O
OH
2-Arachidonoylglycerol (2-AG)

Figure A
Endocannabinoids.

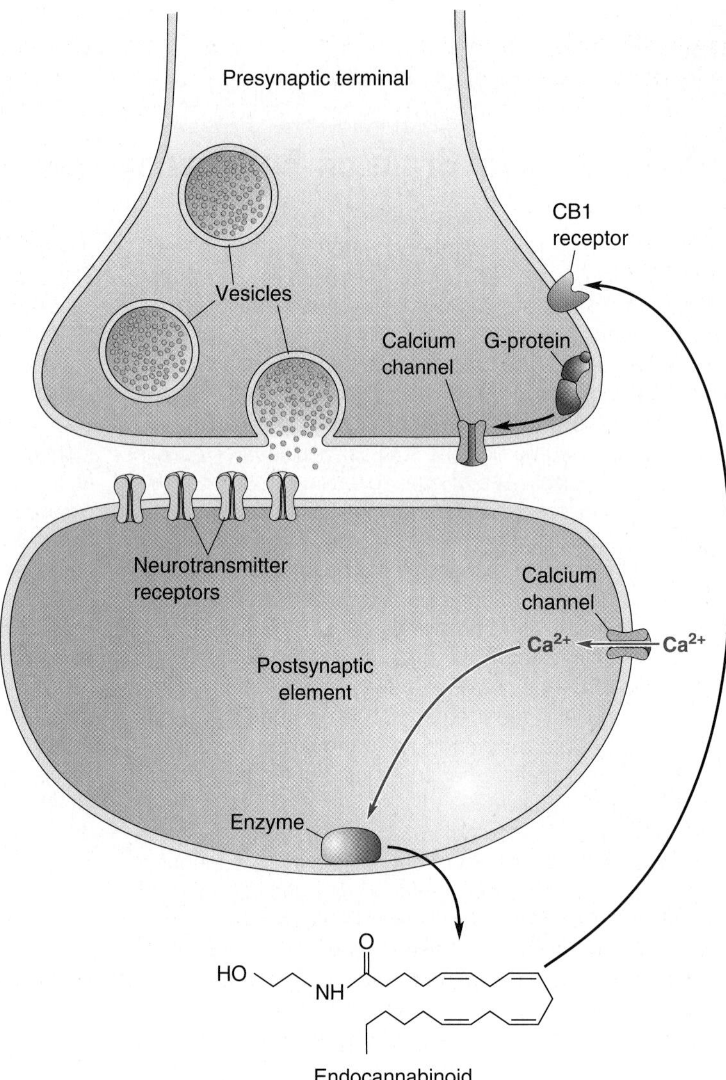

▶ FIGURE 6.17
Retrograde signaling with endocannabinoids.

which suppress either the inhibitory or excitatory drive onto the neuron (depending on which presynaptic terminals have the CB1 receptors). This general endocannabinoid mechanism is used throughout the CNS, for a wide range of functions that we don't completely understand.

One of the more exotic chemical messengers to be proposed for intercellular communication is actually a gaseous molecule, **nitric oxide (NO)**. The gases carbon monoxide (CO) and hydrogen sulfide (H_2S) have also been suggested to be messengers in the brain, although evidence for "gasotransmitter" functions is still sparse. These are the same NO, carbon monoxide, and hydrogen sulfide that are often major air pollutants. NO is synthesized from the amino acid arginine by many cells of the body and has powerful biological effects, particularly in the regulation of blood flow. In the nervous system, NO may be another example of a retrograde messenger. Because NO is small and membrane permeable, similar to endocannabinoids, it can diffuse much more freely than most other transmitter molecules, even penetrating through one cell to affect another beyond it. Its influence may spread throughout a small region of local tissue, rather than being confined to the site of the cells that released it. On the other hand, NO is evanescent and breaks down very rapidly. The functions of gaseous transmitters are being extensively studied and hotly debated.

Before leaving the topic of neurotransmitter chemistry, we point out, once again, that many of the chemicals we call neurotransmitters may also be present in high concentrations in non-neural parts of the body. A chemical may serve dual purposes, mediating communication in the nervous system but doing something entirely different elsewhere. Amino acids, of course, are used to make proteins throughout the body. ATP is the energy source for all cells. NO is released from endothelial cells and causes the smooth muscle of blood vessels to relax. (One consequence in males is penile erection.) The cells with the highest levels of ACh are not in the brain but in the cornea of the eye, where there are no ACh receptors. Likewise, the highest serotonin levels are not in neurons but in blood platelets. These observations underscore the importance of rigorous analysis before a chemical is assigned a neurotransmitter role.

The operation of a neurotransmitter system is like a play with two acts. Act I is presynaptic and culminates in the transient elevation of neurotransmitter concentration in the synaptic cleft. We are now ready to move on to Act II, the generation of electrical and biochemical signals in the postsynaptic neuron. The main players are transmitter-gated channels and G-protein-coupled receptors.

TRANSMITTER-GATED CHANNELS

In Chapter 5, we learned that ACh and the amino acid neurotransmitters mediate fast synaptic transmission by acting on transmitter-gated ion channels. These channels are magnificent minuscule machines. A single channel can be a sensitive detector of chemicals and voltage, it can regulate the flow of surprisingly large currents with great precision, it can sift and select between very similar ions, and it can be regulated by other receptor systems. Yet each channel is only about 11 nm long, just barely visible with the best computer-enhanced electron microscopic methods.

The Basic Structure of Transmitter-Gated Channels

The most thoroughly studied transmitter-gated ion channel is the nicotinic ACh receptor at the neuromuscular junction in skeletal muscle. It is a pentamer, an amalgam of five protein subunits arranged like the staves of a barrel to form a single pore through the membrane (Figure 6.18a). Four different types of polypeptides are used as subunits for the nicotinic receptor, designated α, β, γ, and δ. A complete mature channel is made from two α subunits, and one each of β, γ, and δ (abbreviated $\alpha_2\beta\gamma\delta$). There is one ACh binding site on each of the α subunits; the simultaneous binding of ACh to both sites is required for the channel to open (Figure 6.18b). The nicotinic ACh receptor on neurons is also a pentamer, but, unlike the muscle receptor, most of these receptors are composed of α and β subunits only (in a ratio of $\alpha_3\beta_2$).

Although each type of receptor subunit has a different primary structure, there are stretches where the different polypeptide chains have a similar sequence of amino acids. For example, each subunit polypeptide has four separate segments that will coil into alpha helices (see Figure 6.18a). Because the amino acid residues of these segments are hydrophobic, the four alpha helices are believed to be where the polypeptide is threaded back and forth across the membrane, similar to the pore loops of potassium and sodium channels (see Chapters 3 and 4).

The primary structures of the subunits of other transmitter-gated channels in the brain are also known, and there are obvious similarities

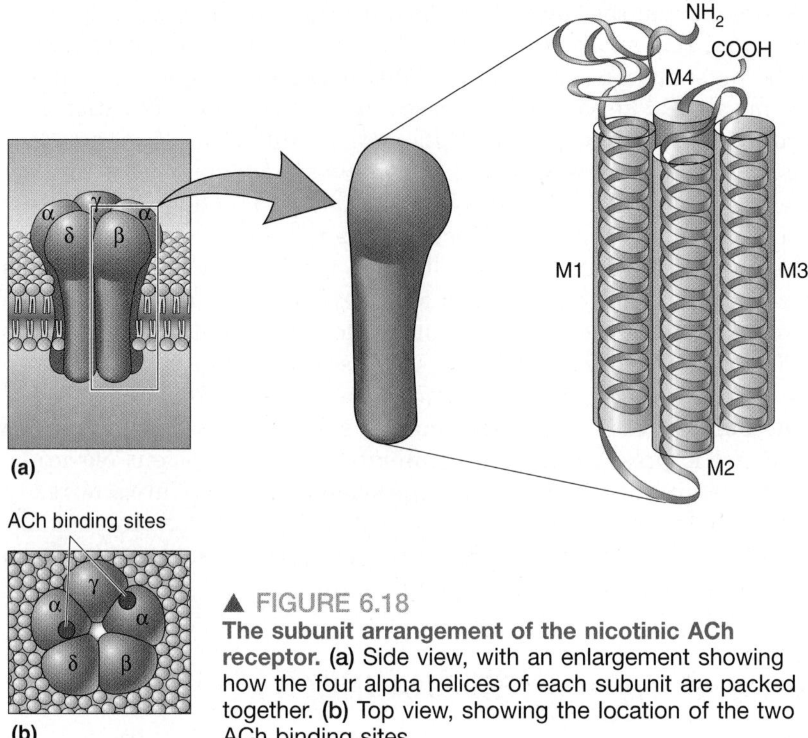

▲ FIGURE 6.18

The subunit arrangement of the nicotinic ACh receptor. **(a)** Side view, with an enlargement showing how the four alpha helices of each subunit are packed together. **(b)** Top view, showing the location of the two ACh binding sites.

(Figure 6.19). Most contain the four hydrophobic segments that span the membrane in subunits of the nicotinic ACh receptor, the GABA$_A$ receptor, and the glycine receptor. These three neurotransmitter receptors are all pentameric complexes of subunits (Figure 6.19b). The glutamate-gated channels are slightly different. Glutamate receptors are tetramers, having four subunits that comprise a functional channel. The M2 region of the glutamate subunits does not span the membrane but instead forms a hairpin that both enters and exits from the inside of the membrane (Figure 6.19c). The structure of the glutamate receptors resembles that of some potassium channels (see Figure 3.17), and this has inspired the surprising hypothesis that glutamate receptors and potassium channels evolved from a common ancestral ion channel. The purinergic (ATP) receptors also have an unusual structure. Each subunit has only two membrane-spanning segments, and three subunits make up a complete receptor.

The most interesting variations among channel structures are the ones that account for their differences. Different transmitter binding sites let one channel respond to Glu while another responds to GABA; certain amino acids around the narrow ion pore allow only Na^+ and K^+ to flow through some channels, Ca^{2+} through others, and only Cl^- through yet others.

Amino Acid-Gated Channels

Amino acid-gated channels mediate most of the fast synaptic transmission in the CNS. Let's take a closer look at their functions because they are central to topics as diverse as sensory systems, memory, and disease. Several properties of these channels distinguish them from one another and define their functions within the brain.

- The *pharmacology* of their binding sites describes which transmitters affect them and how drugs interact with them.
- The *kinetics* of the transmitter binding process and channel gating determine the duration of their effect.

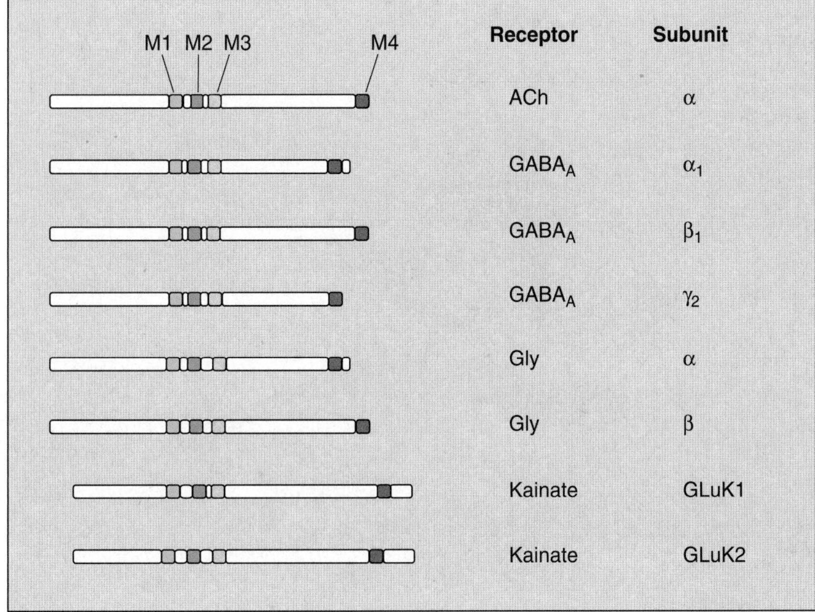

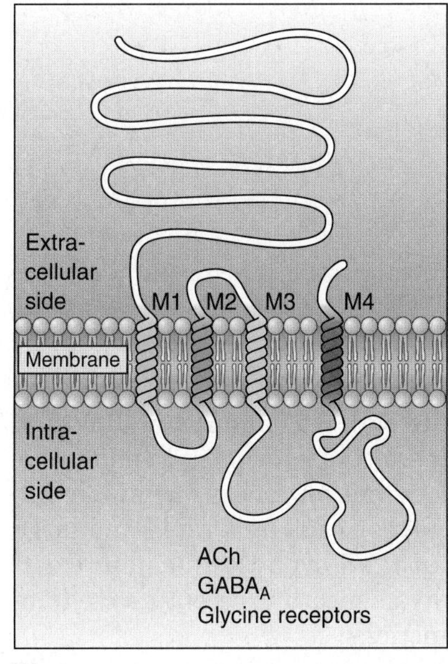

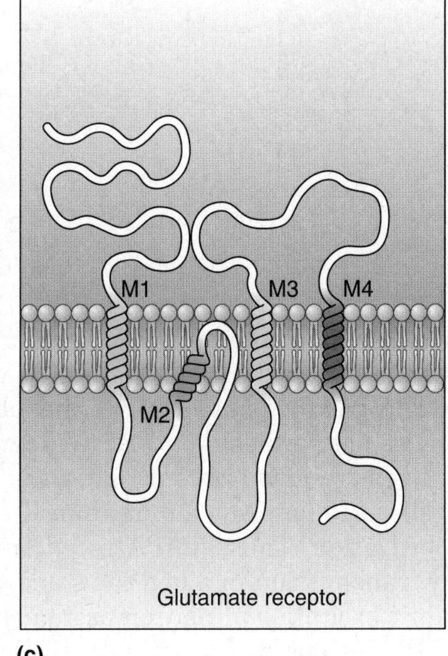

◀ FIGURE 6.19
Similarities in the structure of subunits for different transmitter-gated ion channels. **(a)** If the polypeptides for various channel subunits were stretched out in a line, this is how they would compare. They have in common the four regions called *M1 to M4*, which are segments where the polypeptides will coil into alpha helices to span the membrane. Kainate receptors are subtypes of glutamate receptors. **(b)** M1–M4 regions of the ACh α subunit as they are threaded through the membrane. **(c)** M1–M4 regions of the glutamate receptor subunits; M1, M3, and M4 span the entire thickness of the membrane, whereas M2 penetrates only part way.

- The *selectivity* of the ion channels determines whether they produce excitation or inhibition and whether Ca^{2+} enters the cell in significant amounts.
- The *conductance* of open channels helps determine the magnitude of their effects.

All of these properties are a direct result of the molecular structure of the channels.

Glutamate-Gated Channels. As we discussed previously, three glutamate receptor subtypes bear the names of their selective agonists: AMPA, NMDA, and kainate. Each of these is a glutamate-gated ion channel. The AMPA-gated and NMDA-gated channels mediate the bulk of fast excitatory synaptic transmission in the brain. Kainate receptors also

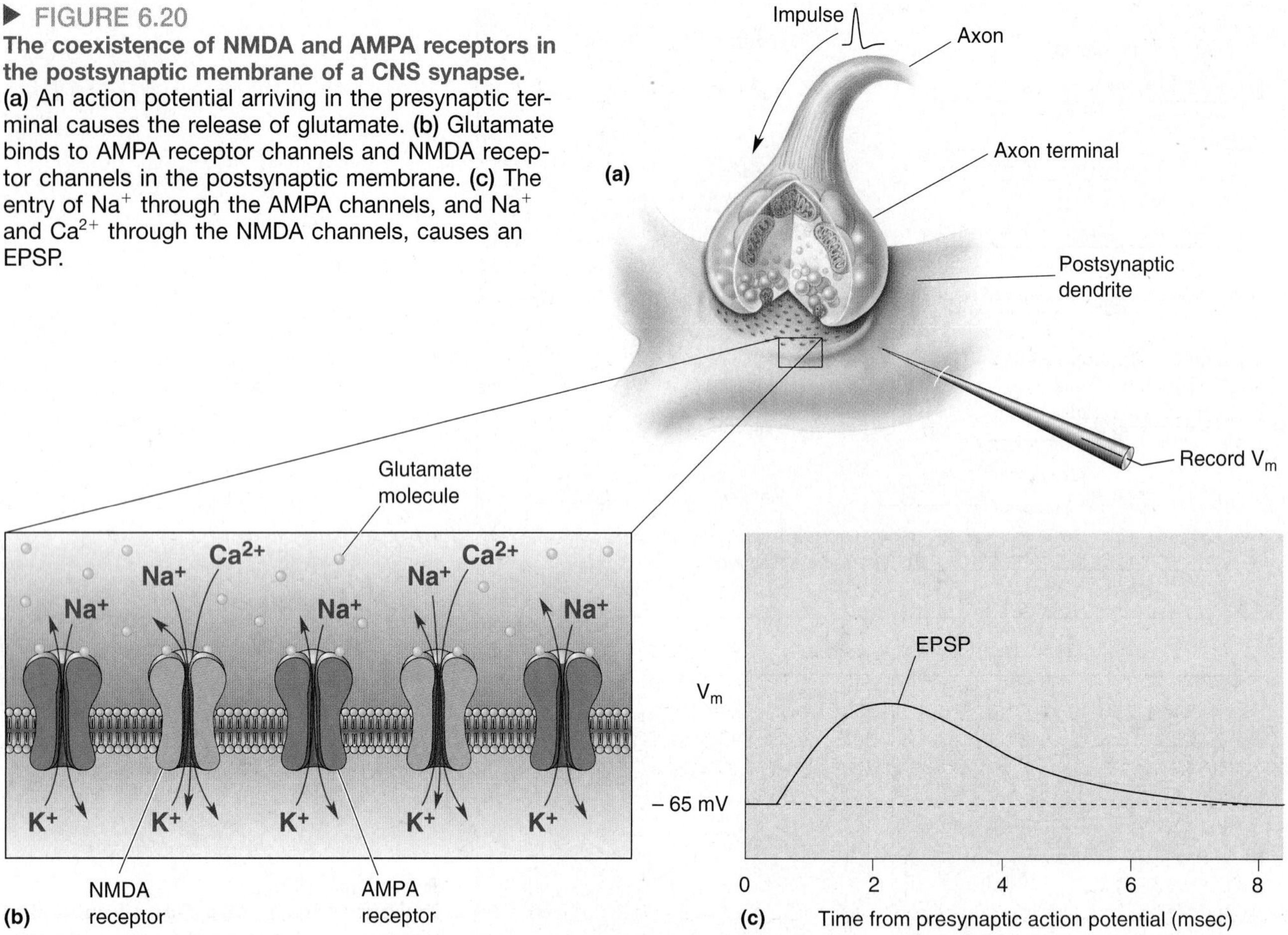

▶ FIGURE 6.20

The coexistence of NMDA and AMPA receptors in the postsynaptic membrane of a CNS synapse. **(a)** An action potential arriving in the presynaptic terminal causes the release of glutamate. **(b)** Glutamate binds to AMPA receptor channels and NMDA receptor channels in the postsynaptic membrane. **(c)** The entry of Na^+ through the AMPA channels, and Na^+ and Ca^{2+} through the NMDA channels, causes an EPSP.

exist throughout the brain, on both presynaptic and postsynaptic membranes, but their functions are not clearly understood.

AMPA-gated channels are permeable to both Na^+ and K^+, and most of them are not permeable to Ca^{2+}. The net effect of activating them at normal, negative membrane potentials is to admit an excess of cations into the cell (i.e., more Na^+ enters than K^+ leaves), causing a rapid and large depolarization. Thus, AMPA receptors at CNS synapses mediate excitatory transmission in much the same way as nicotinic receptors mediate synaptic excitation at neuromuscular junctions.

AMPA receptors coexist with NMDA receptors at many synapses in the brain, so most glutamate-mediated excitatory postsynaptic potentials (EPSPs) have components contributed by both (Figure 6.20). NMDA-gated channels also cause excitation of a cell by admitting an excess of Na^+, but they differ from AMPA receptors in two very important ways: (1) NMDA-gated channels are permeable to Ca^{2+}, and (2) inward ionic current through NMDA-gated channels is voltage dependent. We'll discuss each of these properties in turn.

It is hard to overstate the importance of intracellular Ca^{2+} to cell functions. We have already seen that Ca^{2+} can trigger presynaptic neurotransmitter release. Postsynaptically, Ca^{2+} can also activate many enzymes, regulate the opening of a variety of channels, and affect gene expression; in excessive amounts, Ca^{2+} can even trigger the death of a cell (Box 6.4). Thus, activation of NMDA receptors can, in principle, cause widespread and lasting changes in the postsynaptic neuron. Indeed, as we will see in Chapter 25, Ca^{2+} entry through NMDA-gated channels may cause the changes that lead to long-term memory.

BOX 6.4 OF SPECIAL INTEREST

Exciting Poisons: Too Much of a Good Thing

Neurons of the mammalian brain almost never regenerate, so each dead neuron is one less we have for thinking. One of the fascinating ironies of neuronal life and death is that glutamate, the most essential neurotransmitter in the brain, is also a common killer of neurons. A large percentage of the brain's synapses release glutamate, which is stored in large quantities. Even the cytosol of nonglutamatergic neurons has a very high glutamate concentration, greater than 3 mM. An ominous observation is that when you apply this same amount of glutamate to the outside of isolated neurons, they die within minutes. Mae West once said, "Too much of a good thing can be wonderful," but apparently she wasn't talking about glutamate.

The voracious metabolic rate of the brain demands a continuous supply of oxygen and glucose. If blood flow ceases, as in cardiac arrest, neural activity will stop within seconds, and permanent damage will result within a few minutes. Disease states such as cardiac arrest, stroke, brain trauma, seizures, and oxygen deficiency can initiate a vicious cycle of excess glutamate release. Whenever neurons cannot generate enough ATP to keep their ion pumps working hard, membranes depolarize, and Ca^{2+} leaks into cells. The entry of Ca^{2+} triggers the synaptic release of glutamate. Glutamate further depolarizes neurons, which further raises intracellular Ca^{2+} and causes still more glutamate to be released. At this point, there may even be a *reversal* of the glutamate transporter, further contributing to the cellular leakage of glutamate.

When glutamate reaches high concentrations, it kills neurons by overexciting them, a process called *excitotoxicity*. Glutamate simply activates its several types of receptors, which allow excessive amounts of Na^+, K^+, and Ca^{2+} to flow across the membrane. The NMDA subtype of the glutamate-gated channel is a critical player in excitotoxicity because it is the main route for Ca^{2+} entry. Neuron damage or death occurs because of swelling resulting from water uptake and stimulation by Ca^{2+} of intracellular enzymes that degrade proteins, lipids, and nucleic acids. Neurons literally digest themselves.

Excitotoxicity has been implicated in several progressive neurodegenerative human diseases such as *amyotrophic lateral sclerosis* (ALS, also known as Lou Gehrig disease), in which spinal motor neurons slowly die, and *Alzheimer's disease*, in which brain neurons slowly die. The effects of various environmental toxins mimic aspects of these diseases. Eating large quantities of a certain type of chickpea can cause lathyrism, a degeneration of motor neurons. The pea contains an excitotoxin called *β-oxalylaminoalanine*, which activates glutamate receptors. A toxin called *domoic acid*, found in contaminated shellfish, is also a glutamate receptor agonist. Ingesting small amounts of domoic acid causes seizures and brain damage. And another plant excitotoxin, β-methylaminoalanine, may cause a hideous condition that combines signs of ALS, Alzheimer's disease, and Parkinson's disease in individual patients on the island of Guam.

As researchers sort out the tangled web of excitotoxins, receptors, enzymes, and neurological disease, new strategies for treatment emerge. Glutamate receptor antagonists that can obstruct these excitotoxic cascades and minimize neuronal suicide have shown some clinical promise. Genetic manipulations may eventually thwart neurodegenerative conditions in susceptible people.

When the NMDA-gated channel opens, Ca^{2+} and Na^+ enter the cell (and K^+ leaves), but the magnitude of this inward ionic current depends on the postsynaptic membrane potential in an unusual way, for an unusual reason. When glutamate binds to the NMDA receptor, the pore opens as usual. However, at normal negative resting membrane potentials, the channel becomes clogged by Mg^{2+} ions, and this "magnesium block" prevents other ions from passing freely through the NMDA channel. Mg^{2+} pops out of the pore only when the membrane is depolarized, which usually follows the activation of AMPA channels at the same and neighboring synapses. Thus, inward ionic current through the NMDA channel is *voltage dependent*, in addition to being transmitter gated. Both glutamate and depolarization must coincide before the channel will pass current (Figure 6.21). This property has a significant impact on synaptic integration at many locations in the CNS.

GABA-Gated and Glycine-Gated Channels. GABA mediates most synaptic inhibition in the CNS, and glycine mediates most of the rest. Both the $GABA_A$ receptor and the glycine receptor gate a chloride channel.

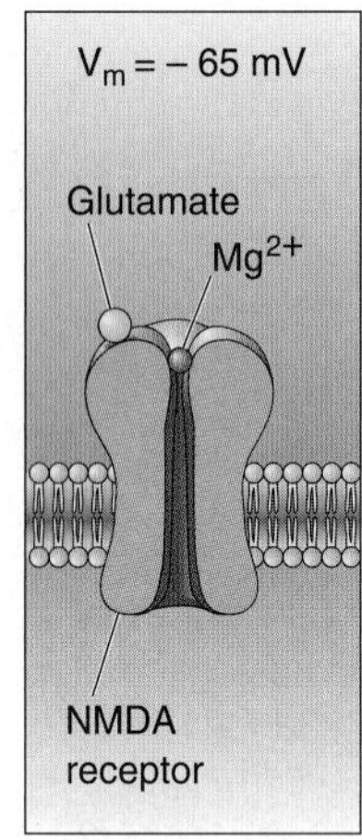

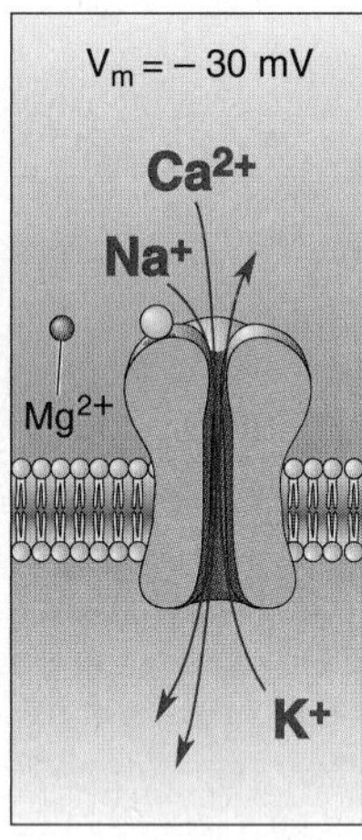

(a) Glutamate **(b)** Glutamate and depolarization

▲ FIGURE 6.21
Inward ionic current through the NMDA-gated channel. (a) Glutamate alone causes the channel to open, but at the resting membrane potential, the pore becomes blocked by Mg^{2+}. (b) Depolarization of the membrane relieves the Mg^{2+} block and allows Na^+ and Ca^{2+} to enter.

Surprisingly, inhibitory $GABA_A$ and glycine receptors have a structure very similar to that of excitatory nicotinic ACh receptors, despite the fact that the first two are selective for anions while the last is selective for cations. Each receptor has α subunits that bind the transmitter and β subunits that do not.

Synaptic inhibition must be tightly regulated in the brain. Too much causes a loss of consciousness and coma; too little leads to a seizure. The need to control inhibition may explain why the $GABA_A$ receptor has, in addition to its GABA binding site, several other sites where chemicals can dramatically modulate its function. For example, two classes of drugs, **benzodiazepines** (such as the tranquilizer diazepam, with the trade name Valium) and **barbiturates** (including phenobarbital and other sedatives and anticonvulsants), each bind to their own distinct site on the outside face of the $GABA_A$ channel (Figure 6.22). By themselves, these drugs do very little to the channel. But when GABA is present, benzodiazepines increase the frequency of channel openings, while barbiturates increase the duration of channel openings. The result in each case is more inhibitory Cl^- current, stronger inhibitory postsynaptic potentials (IPSPs), and the behavioral consequences of enhanced inhibition. The actions of benzodiazepines and barbiturates are selective for the $GABA_A$ receptor, and the drugs have no effect on glycine receptor function. Some of this selectivity can be understood in molecular terms; only receptors with the γ type of $GABA_A$ subunit, in addition to α and β subunits, respond to benzodiazepines.

Another popular drug that strongly enhances the function of the $GABA_A$ receptor is ethanol, the form of alcohol imbibed in beverages. Ethanol has complex actions that include effects on NMDA, glycine, nicotinic ACh, and serotonin receptors. Its effects on $GABA_A$ channels depend on their specific structure. Evidence indicates that particular α, β, and γ subunits are necessary for constructing an ethanol-sensitive $GABA_A$ receptor, similar to the structure that is benzodiazepine sensitive. This explains why ethanol enhances inhibition in some brain areas but not others. By understanding this molecular and anatomical specificity, we can begin to appreciate how drugs like ethanol lead to such powerful, and addictive, effects on behavior.

These myriad drug effects present an interesting paradox. Surely, the $GABA_A$ receptor did not evolve modulatory binding sites just for the benefit of our modern drugs. The paradox has motivated researchers to search for endogenous ligands, natural chemicals that might bind to benzodiazepine and barbiturate sites and serve as regulators of inhibition.

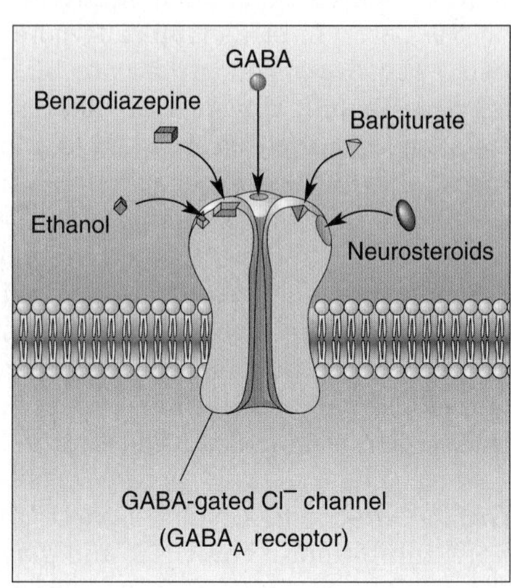

▶ FIGURE 6.22
The binding of drugs to the $GABA_A$ receptor. The drugs by themselves do not open the channel, but they change the effect that GABA has when it binds to the channel at the same time as the drug.

Substantial evidence indicates that natural benzodiazepine-like ligands exist, although identifying them and understanding their functions are proving difficult. Other good candidates as natural modulators of $GABA_A$ receptors are the *neurosteroids*, natural metabolites of steroid hormones that are synthesized from cholesterol primarily in the gonads and adrenal glands, but also in glial cells of the brain. Some neurosteroids enhance inhibitory function while others suppress it, and they seem to do both by binding to their own sites on the $GABA_A$ receptor (see Figure 6.22), distinct from those of the other drugs we've mentioned. The functions of natural neurosteroids are also obscure, but they suggest a means by which brain and body physiology could be regulated in parallel by the same chemicals.

G-PROTEIN-COUPLED RECEPTORS AND EFFECTORS

There are multiple subtypes of G-protein-coupled receptors in every known neurotransmitter system. In Chapter 5, we learned that transmission at these receptors involves three steps: (1) binding of the neurotransmitter to the receptor protein, (2) activation of G-proteins, and (3) activation of effector systems. Let's focus on each of these steps.

The Basic Structure of G-Protein-Coupled Receptors

Most G-protein-coupled receptors are simple variations on a common plan, consisting of a single polypeptide containing seven membrane-spanning alpha helices (Figure 6.23). Two of the extracellular loops of the polypeptide form the transmitter binding sites. Structural variations in this region determine which neurotransmitters, agonists, and antagonists bind to the receptor. Two of the intracellular loops can bind to and activate G-proteins. Structural variations here determine which G-proteins and, consequently, which effector systems are activated in response to transmitter binding.

A very partial list of G-protein-coupled receptors appears in Table 6.2. The human genome has genes coding for about 800 different G-protein-coupled receptors, which are organized into five major families with

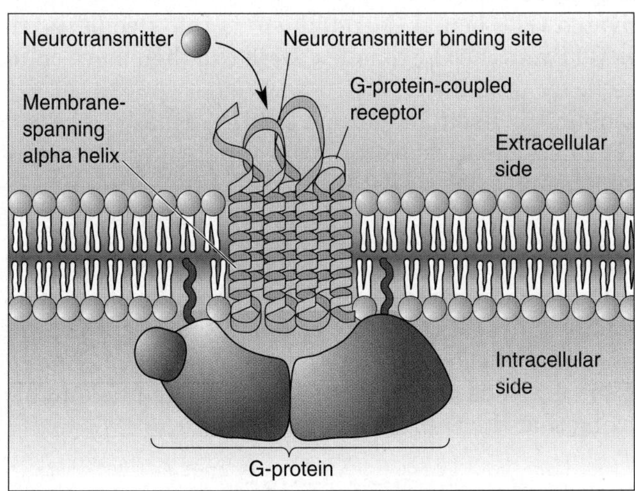

▲ FIGURE 6.23
The basic structure of a G-protein-coupled receptor. Most metabotropic receptors have seven membrane-spanning alpha helices, a transmitter binding site on the extracellular side, and a G-protein binding site on the intracellular side.

TABLE 6.2 **Some G-Protein-Coupled Neurotransmitter Receptors**

Neurotransmitter	Receptors
Acetylcholine (ACh)	Muscarinic receptors (M_1, M_2, M_3, M_4, M_5)
Glutamate (Glu)	Metabotropic glutamate receptors (mGluR1–8)
GABA	$GABA_{B1}$, $GABA_{B2}$
Serotonin (5-HT)	5-HT_{1A}, 5-HT_{1B}, 5-HT_{1D}, 5-HT_{1E}, 5-HT_{2A}, 5-HT_{2B}, 5-HT_4, 5-HT_{5A}
Dopamine (DA)	D1, D2, D3, D4, D5
Norepinephrine (NE)	α_1, α_2, β_1, β_2, β_3
Opioids	μ, δ, κ
Cannabinoid	CB1, CB2
ATP	$P2Y_2$, $P2Y_{11}$, P2T, P2U
Adenosine	A_1, A_{2A}, A_{2B}, A_3

similar structures. Most of these receptors were unknown before the powerful methods of molecular biology were applied to their discovery. It is also important to recall that G-protein-coupled receptors are important in all of the body's cell types, not just neurons.

The Ubiquitous G-Proteins

G-proteins are the common link in most signaling pathways that start with a neurotransmitter receptor and end with effector proteins. G-protein is short for guanosine triphosphate (GTP) binding protein, which is actually a diverse family of about 20 types. There are many more transmitter receptors than G-proteins, so some types of G-proteins can be activated by many receptors.

Most G-proteins have the same basic mode of operation (Figure 6.24):

1. Each G-protein has three subunits, termed α, β, and γ. In the resting state, a guanosine diphosphate (GDP) molecule is bound to the G_α subunit, and the whole complex floats around on the inner surface of the membrane.
2. If this GDP-bound G-protein bumps into the proper type of receptor *and* if that receptor has a transmitter molecule bound to it, then the G-protein releases its GDP and exchanges it for a GTP that it picks up from the cytosol.
3. The activated GTP-bound G-protein splits into two parts: the G_α subunit plus GTP and the $G_{\beta\gamma}$ complex. Both can then move on to influence various effector proteins.
4. The G_α subunit is itself an enzyme that eventually breaks down GTP into GDP. Therefore, G_α eventually terminates its own activity by converting the bound GTP to GDP.
5. The G_α and $G_{\beta\gamma}$ subunits come back together, allowing the cycle to begin again.

The first G-proteins that were discovered had the effect of stimulating effector proteins. Subsequently, it was found that other G-proteins could inhibit these same effectors. Thus, the simplest scheme for subdividing the G-proteins is G_S, designating that the G-protein is stimulatory, and G_i, designating that the G-protein is inhibitory.

G-Protein-Coupled Effector Systems

In Chapter 5, we learned that activated G-proteins exert their effects by binding to either of two types of effector proteins: G-protein-gated ion channels and G-protein-activated enzymes. Because the effects do not

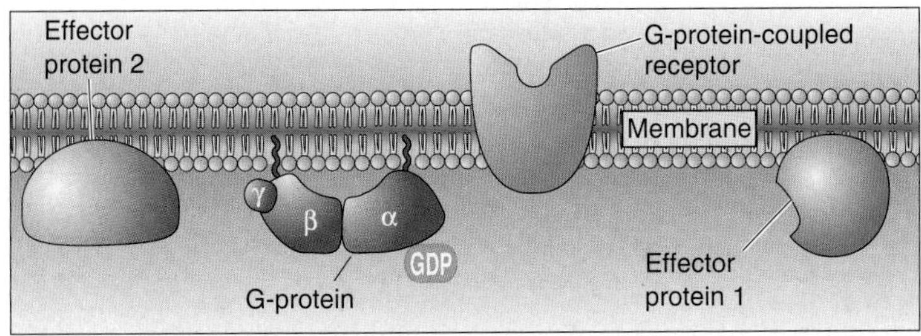

(a)

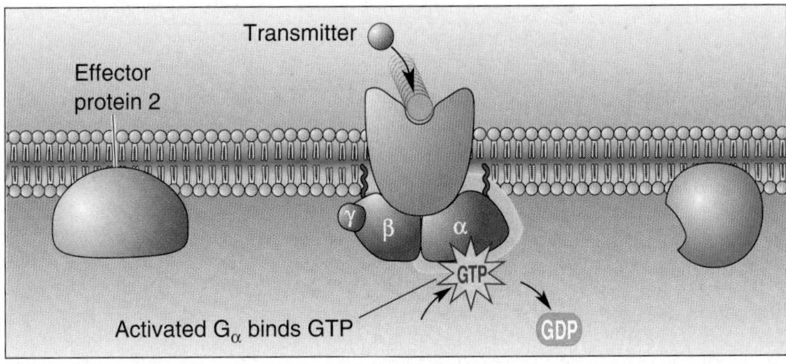

(b)

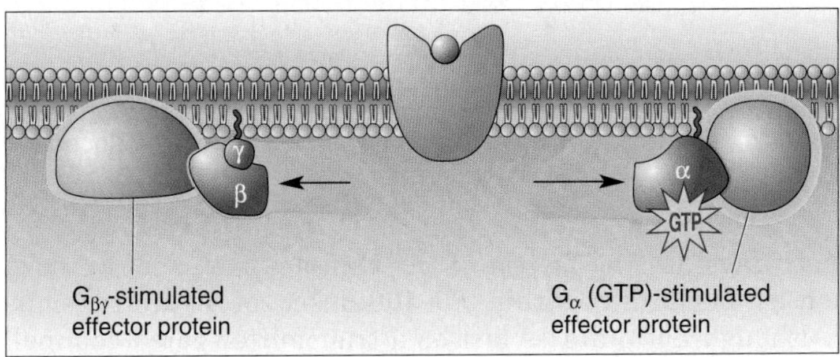

(c)

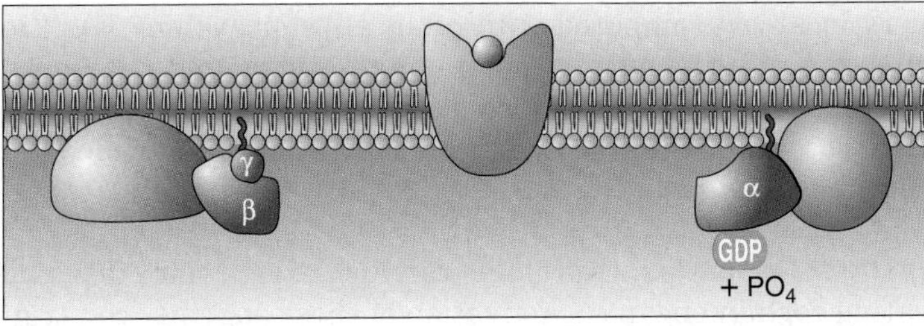

(d)

◀ FIGURE 6.24
The basic mode of operation of G-proteins. (a) In its inactive state, the α subunit of the G-protein binds GDP. (b) When activated by a G-protein-coupled receptor, the GDP is exchanged for GTP. **(c)** The activated G-protein splits, and both the G_α (GTP) subunit and the $G_{\beta\gamma}$ subunit become available to activate effector proteins. **(d)** The G_α subunit slowly removes phosphate (PO_4) from GTP, converting GTP to GDP and terminating its own activity.

involve any other chemical intermediaries, the first route is sometimes called the *shortcut pathway*.

The Shortcut Pathway. A variety of neurotransmitters use the shortcut pathway, from receptor to G-protein to ion channel. One example is the muscarinic receptors in the heart. These ACh receptors are coupled via G-proteins to particular types of potassium channels, explaining why ACh slows the heart rate (Figure 6.25). In this case, the βγ subunits migrate laterally along the membrane until they bind to the right type of potassium channel and induce it to open. Another example is neuronal $GABA_B$ receptors, also coupled by the shortcut pathway to potassium channels.

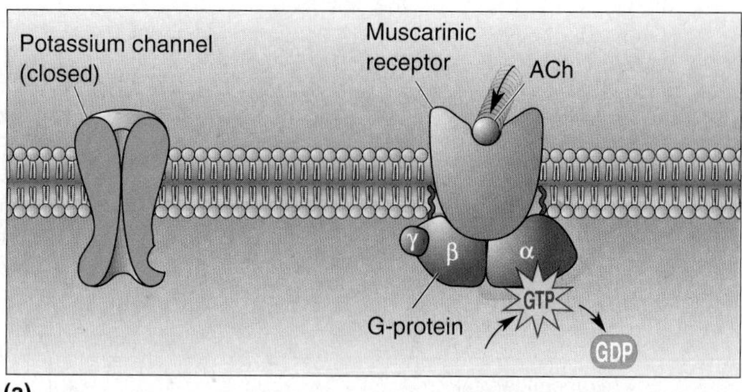

(a)

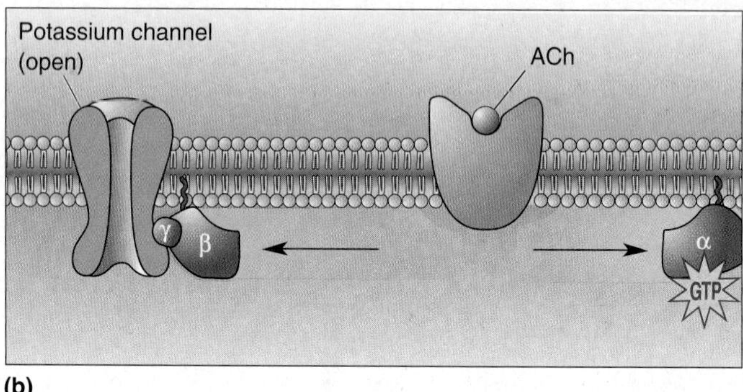

(b)

▲ FIGURE 6.25
The shortcut pathway. **(a)** G-proteins in heart muscle are activated by ACh binding to muscarinic receptors. **(b)** The activated $G_{\beta\gamma}$ subunit directly induces a potassium channel to open.

Shortcut pathways are the fastest of the G-protein-coupled systems, having responses beginning within 30–100 msec of neurotransmitter binding. Although not quite as fast as a transmitter-gated channel, which uses no intermediary between receptor and channel, this is faster than the second messenger cascades we describe next. The shortcut pathway is also very localized compared with other effector systems. As the G-protein diffuses within the membrane, it apparently cannot move very far, so only channels nearby can be affected. Because all the action in the shortcut pathway occurs within the membrane, it is sometimes called the *membrane-delimited pathway*.

Second Messenger Cascades. G-proteins can also exert their effects by directly activating certain enzymes. Activation of these enzymes can trigger an elaborate series of biochemical reactions, a cascade that often ends in the activation of other "downstream" enzymes that alter neuronal function. Between the first enzyme and the last are several *second messengers*. The whole process that couples the neurotransmitter, via multiple steps, to activation of a downstream enzyme is called a **second messenger cascade** (Figure 6.26).

In Chapter 5, we introduced the cAMP second messenger cascade initiated by the activation of the NE β receptor (Figure 6.27a). It begins with the β receptor activating the stimulatory G-protein, G_S, which proceeds to stimulate the membrane-bound enzyme adenylyl cyclase. Adenylyl cyclase converts ATP to cAMP. The subsequent rise of cAMP in the cytosol activates a specific downstream enzyme called **protein kinase A (PKA)**.

Many biochemical processes are regulated with a push–pull method, one to stimulate them and one to inhibit them, and cAMP production is no exception. The activation of a second type of NE receptor, called

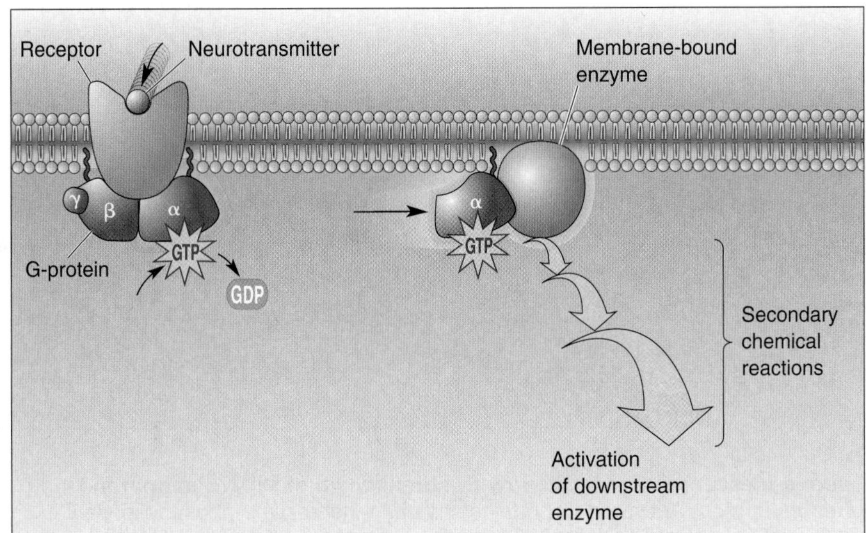

◀ FIGURE 6.26
The components of a second messenger cascade.

the α_2 *receptor*, leads to the activation of G_i (the inhibitory G-protein). G_i suppresses the activity of adenylyl cyclase, and this effect can take precedence over the stimulatory system (Figure 6.27b).

Some messenger cascades can branch. Figure 6.28 shows how the activation of various G-proteins can stimulate **phospholipase C (PLC)**, an enzyme that floats in the membrane-like adenylyl cyclase. PLC acts on a membrane phospholipid (PIP_2, or phosphatidylinositol-4, 5-bisphosphate), splitting it to form two molecules that serve as second messengers: **diacylglycerol (DAG)** and **inositol-1,4,5-triphosphate (IP_3)**. DAG, which is lipid-soluble, stays within the plane of the membrane where it activates a downstream enzyme, **protein kinase C (PKC)**. At the same time, the water-soluble IP_3 diffuses away in the cytosol and binds to specific receptors on the smooth ER and other membrane-enclosed organelles in the cell. These receptors are IP_3-gated calcium channels, so

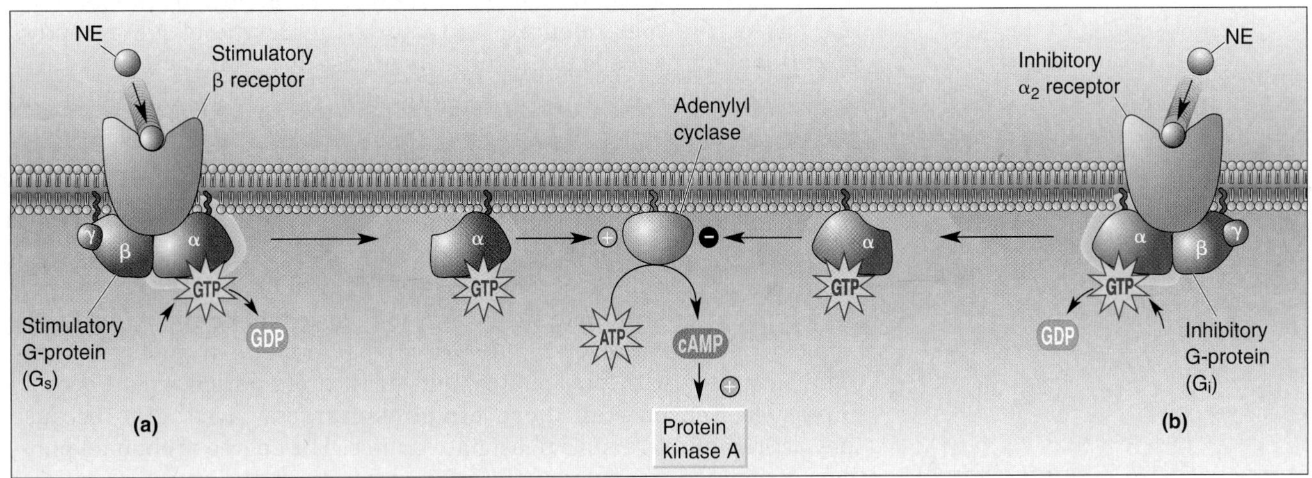

▲ FIGURE 6.27
The stimulation and inhibition of adenylyl cyclase by different G-proteins.
(a) Binding of NE to the β receptor activates G_s, which in turn activates adenylyl cyclase. Adenylyl cyclase generates cAMP, which activates the downstream enzyme protein kinase A. **(b)** Binding of NE to the α_2 receptor activates G_i, which inhibits adenylyl cyclase.

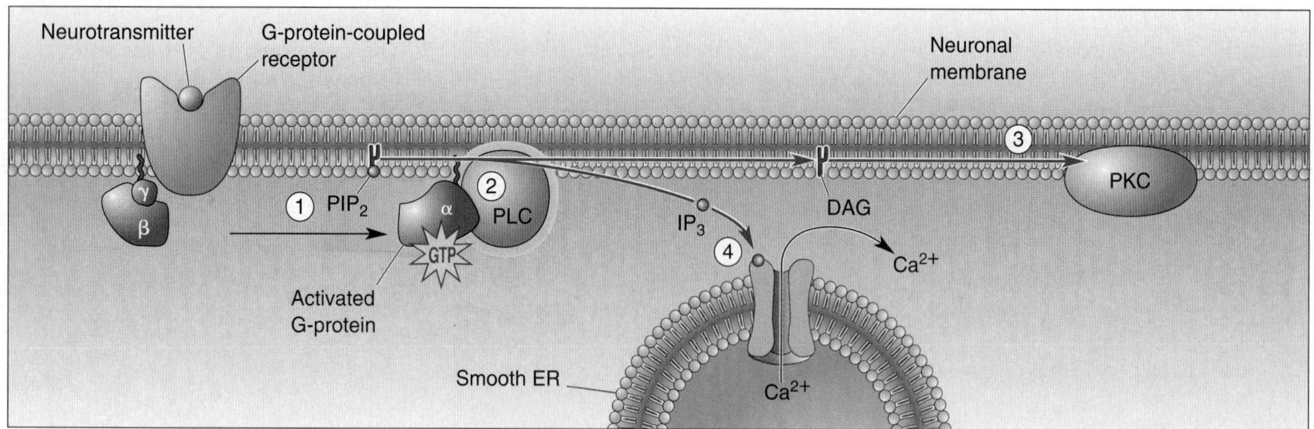

▲ FIGURE 6.28
Second messengers generated by the breakdown of PIP_2, a membrane phospholipid. ① Activated G-proteins stimulate the enzyme phospholipase C (PLC). ② PLC splits PIP_2 into DAG and IP_3. ③ DAG stimulates the downstream enzyme protein kinase C (PKC). ④ IP_3 stimulates the release of Ca^{2+} from intracellular stores. The Ca^{2+} can go on to stimulate various downstream enzymes.

IP_3 causes the organelles to discharge their stores of Ca^{2+}. As we have said, elevations in cytosolic Ca^{2+} can trigger widespread and long-lasting effects. One effect is activation of the enzyme **calcium-calmodulin-dependent protein kinase**, or **CaMK**. CaMK is an enzyme implicated in, among other things, the molecular mechanisms of memory, as we'll see in Chapter 25.

Phosphorylation and Dephosphorylation. The preceding examples show that key downstream enzymes in many of the second messenger cascades are *protein kinases* (PKA, PKC, CaMK). As mentioned in Chapter 5, protein kinases transfer phosphate from ATP floating in the cytosol to proteins, a reaction called *phosphorylation*. The addition of phosphate groups to a protein changes its conformation slightly, thereby changing its biological activity. The phosphorylation of ion channels, for example, can strongly influence the probability that they will open or close.

Consider the consequence of activating the β type of NE receptors on cardiac muscle cells. The subsequent rise in cAMP activates PKA, which phosphorylates the cell's voltage-gated calcium channels, and this *enhances* their activity. More Ca^{2+} flows, and the heart beats more strongly. By contrast, the stimulation of β-adrenergic receptors in many neurons seems to have no effect on calcium channels, but instead causes *inhibition* of certain potassium channels. Reduced K^+ conductance causes a slight depolarization, increases the length constant, and makes the neuron more excitable (see Chapter 5).

If transmitter-stimulated kinases were allowed to phosphorylate without some method of reversing the process, all proteins would quickly become saturated with phosphates, and further regulation would become impossible. Enzymes called **protein phosphatases** save the day because they act rapidly to remove phosphate groups. The degree of channel phosphorylation at any moment therefore depends on the dynamic balance of phosphorylation by kinases and dephosphorylation by phosphatases (Figure 6.29).

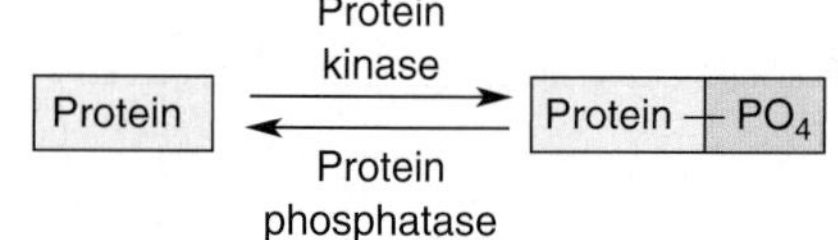

▲ FIGURE 6.29
Protein phosphorylation and dephosphorylation.

The Function of Signal Cascades. Synaptic transmission using transmitter-gated channels is simple and fast. Transmission involving

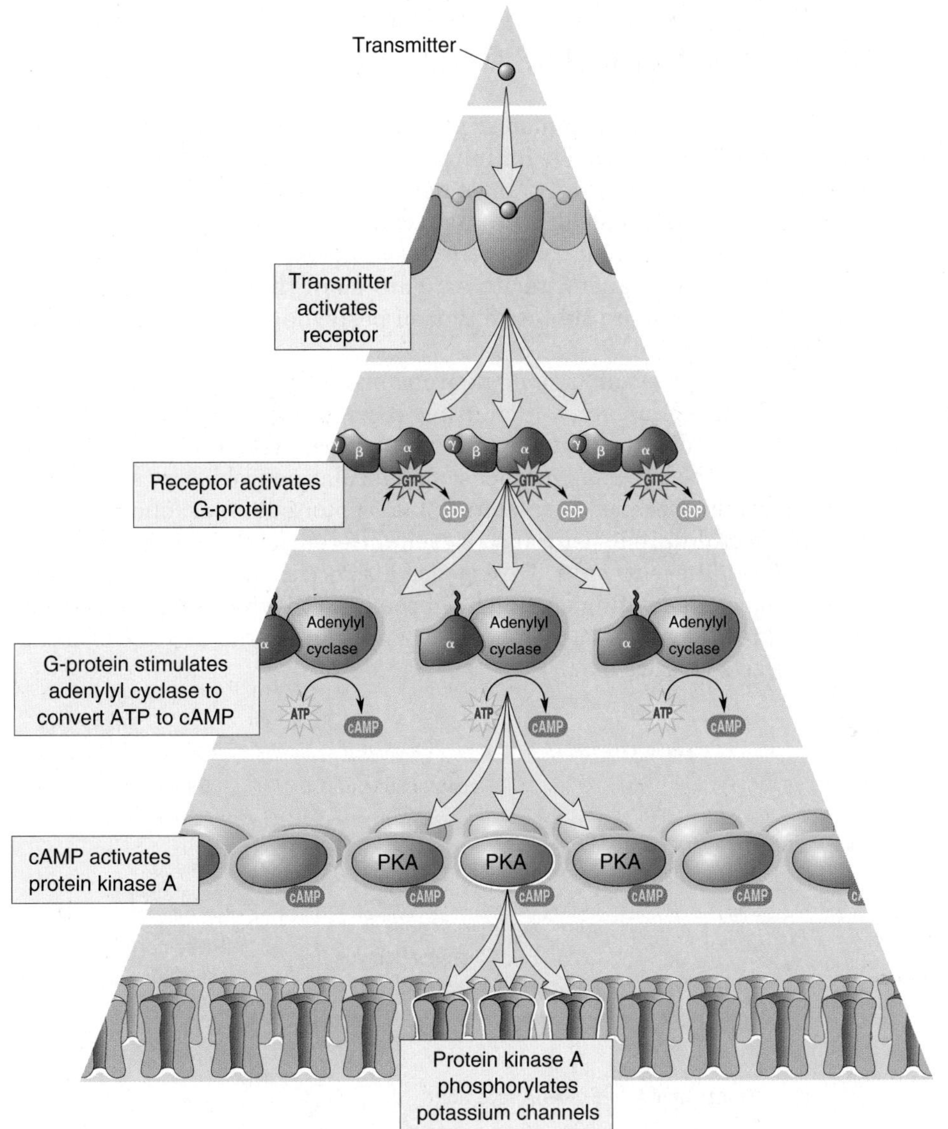
Transmitter
Transmitter activates receptor
Receptor activates G-protein
GTP
GDP
Adenylyl cyclase
G-protein stimulates adenylyl cyclase to convert ATP to cAMP
ATP
cAMP
cAMP activates protein kinase A
PKA
Protein kinase A phosphorylates potassium channels

◀ FIGURE 6.30
Signal amplification by G-protein-coupled second messenger cascades. When a transmitter activates a G-protein-coupled receptor, there can be amplification of the messengers at several stages of the cascade, so that ultimately, many channels are affected.

G-protein-coupled receptors is complex and slow. What are the advantages of having such long chains of command? One important advantage is *signal amplification*: The activation of one G-protein-coupled receptor can lead to the activation of not one, but many, ion channels (Figure 6.30).

Signal amplification can occur at several places in the cascade. A single neurotransmitter molecule, bound to one receptor, can activate perhaps 10–20 G-proteins; each G-protein can activate an adenylyl cyclase, which can make many cAMP molecules that can spread to activate many kinases. Each kinase can then phosphorylate many channels. If all cascade components were tied together in a clump, signaling would be severely limited. The use of small messengers that can diffuse quickly (such as cAMP) also allows signaling at a distance, over a wide stretch of cell membrane. Signal cascades also provide many sites for further regulation, as well as interaction between cascades. Finally, signal cascades can generate very long-lasting chemical changes in cells, which may form the basis for, among other things, a lifetime of memories.

DIVERGENCE AND CONVERGENCE IN NEUROTRANSMITTER SYSTEMS

Glutamate is the most common excitatory neurotransmitter in the brain, while GABA is the pervasive inhibitory neurotransmitter. But this is only part of the story because any single neurotransmitter can have many different effects. A molecule of glutamate can bind to any of several kinds of glutamate receptors, and each of these can mediate a different response. The ability of one transmitter to activate more than one subtype of receptor, and cause more than one type of postsynaptic response, is called *divergence*.

Divergence is the rule among neurotransmitter systems. Every known neurotransmitter can activate multiple receptor subtypes (see Table 6.2), and evidence indicates that the number of known receptors will continue to escalate as the powerful methods of molecular neurobiology are applied to each system. Because of the multiple receptor subtypes, one transmitter can affect different neurons (or even different parts of the same neuron) in very different ways. Divergence also occurs at points beyond the receptor level, depending on which G-proteins and which effector systems are activated. Divergence may occur at any stage in the cascade of transmitter effects (Figure 6.31a).

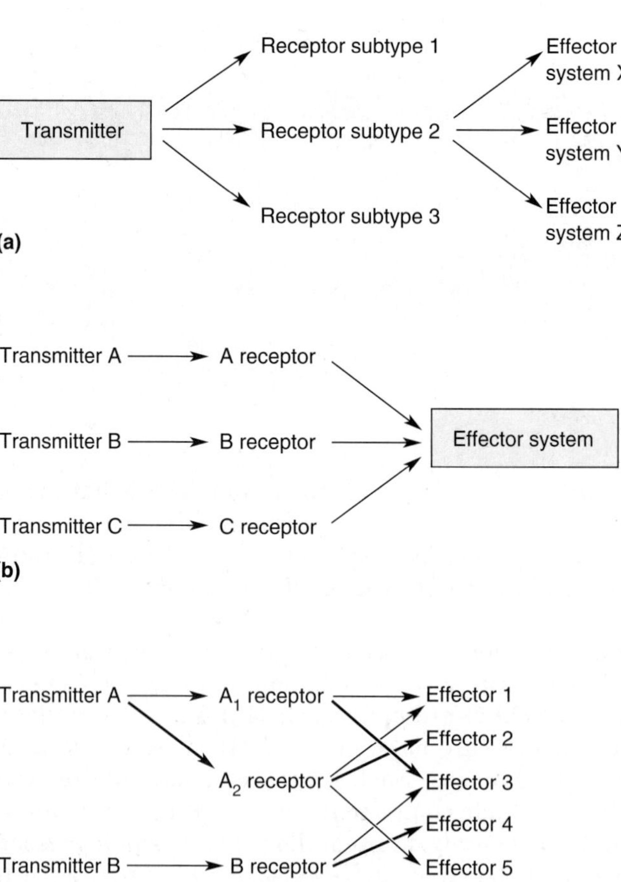

▲ FIGURE 6.31

Divergence and convergence in neurotransmitter signaling systems. **(a)** Divergence. **(b)** Convergence. **(c)** Integrated divergence and convergence.

Neurotransmitters can also exhibit *convergence* of effects. Multiple transmitters, each activating their own receptor type, can converge to influence the same effector system (Figure 6.31b). Convergence in a single cell can occur at the level of the G-protein, the second messenger cascade, or the type of ion channel. Neurons integrate divergent and convergent signaling systems, resulting in a complex map of chemical effects (Figure 6.31c). The wonder is that it ever works; the challenge is to understand how it does.

CONCLUDING REMARKS

Neurotransmitters are the essential links between neurons, and between neurons and other effector cells, such as muscle cells and glandular cells. But it is important to think of transmitters as one link in a chain of events, inciting chemical changes both fast and slow, divergent and convergent. You can envision the many signaling pathways onto and within a single neuron as a kind of information network. This network is in delicate balance, shifting its effects dynamically as the demands on a neuron vary with changes in the organism's behavior.

The signaling network within a single neuron resembles in some ways the neural networks of the brain itself. It receives a variety of inputs, in the form of transmitters bombarding it at different times and places. These inputs cause an increased drive through some signal pathways and a decreased drive through others, and the information is recombined to yield a particular output that is more than a simple summation of the inputs. Signals regulate signals, chemical changes can leave lasting traces of their history, drugs can shift the balance of signaling power—and, in a literal sense, the brain and its chemicals are one.

KEY TERMS

Introduction
cholinergic (p. 144)
noradrenergic (p. 144)
glutamatergic (p. 145)
GABAergic (p. 145)
peptidergic (p. 145)

Studying Neurotransmitter Systems
immunocytochemistry (p. 145)
in situ hybridization (p. 146)
autoradiography (p. 147)
microiontophoresis (p. 148)
receptor subtype (p. 149)
nicotinic ACh receptors (p. 150)
muscarinic ACh receptors (p. 150)
AMPA receptors (p. 150)
NMDA receptors (p. 150)
kainate receptors (p. 150)
ligand-binding method (p. 151)

Neurotransmitter Chemistry
Dale's principle (p. 153)
co-transmitters (p. 153)
acetylcholine (ACh) (p. 154)
transporter (p. 154)
rate-limiting step (p. 156)
catecholamines (p. 156)
dopamine (DA) (p. 156)
norepinephrine (NE) (p. 156)
epinephrine (adrenaline) (p. 156)
dopa (p. 157)
serotonin (5-HT) (p. 158)
serotonergic (p. 158)
glutamate (Glu) (p. 159)
glycine (Gly) (p. 159)
gamma-aminobutyric acid (GABA) (p. 159)
adenosine triphosphate (ATP) (p. 160)
endocannabinoids (p. 160)
retrograde messenger (p. 160)
nitric oxide (NO) (p. 162)

Transmitter-Gated Channels
benzodiazepine (p. 168)
barbiturate (p. 168)

G-Protein-Coupled Receptors and Effectors
second messenger cascade (p. 172)
protein kinase A (PKA) (p. 172)
phospholipase C (PLC) (p. 173)
diacylglycerol (DAG) (p. 173)
inositol-1,4,5-triphosphate (IP_3) (p. 173)
protein kinase C (PKC) (p. 173)
calcium-calmodulin-dependent protein kinase (CaMK) (p. 174)
protein phosphatases (p. 174)

REVIEW QUESTIONS

1. List the criteria that are used to determine whether a chemical serves as a neurotransmitter. What are the various experimental strategies you could use to show that ACh fulfills the criteria of a neurotransmitter at the neuromuscular junction?
2. What are three methods that could be used to show that a neurotransmitter receptor is synthesized or localized in a particular neuron?
3. Compare and contrast the properties of (a) AMPA and NMDA receptors and (b) $GABA_A$ and $GABA_B$ receptors.
4. Synaptic inhibition is an important feature of the circuitry in the cerebral cortex. How would you determine whether GABA or Gly, or both, or neither, is the inhibitory neurotransmitter of the cortex?
5. Glutamate activates a number of different metabotropic receptors. The consequence of activating one subtype is the *inhibition* of cAMP formation. A consequence of activating a second subtype is *activation* of PKC. Propose mechanisms for these different effects.
6. Do convergence and divergence of neurotransmitter effects occur in single neurons?
7. Ca^{2+} are considered to be second messengers. Why?

FURTHER READING

Cooper JR, Bloom FE, Roth RH. 2009. *Introduction to Neuropsychopharmacology*. New York: Oxford University Press.

Cowan WM, Südhof TC, Stevens CF. 2001. *Synapses*. Baltimore: Johns Hopkins University Press.

Katritch V, Cherezov V, Stevens RC. 2012. Diversity and modularity of G protein-coupled receptor structures. *Trends in Pharmacological Sciences* 33:17–27.

Mustafa AK, Gadalla MM, Snyder SH. 2009. Signaling by gasotransmitters. *Science Signaling* 2(68):re2.

Nestler EJ, Hyman SE, Malenka RC. 2008. *Molecular Neuropharmacology: A Foundation for Clinical Neuroscience,* 2nd ed. New York: McGraw-Hill Professional.

Piomelli D. 2003. The molecular logic of endocannabinoid signalling. *Nature Reviews Neuroscience* 4:873–884.

Regehr WG, Carey MR, Best AR. 2009. Activity-dependent regulation of synapses by retrograde messengers. *Neuron* 63:154–170.

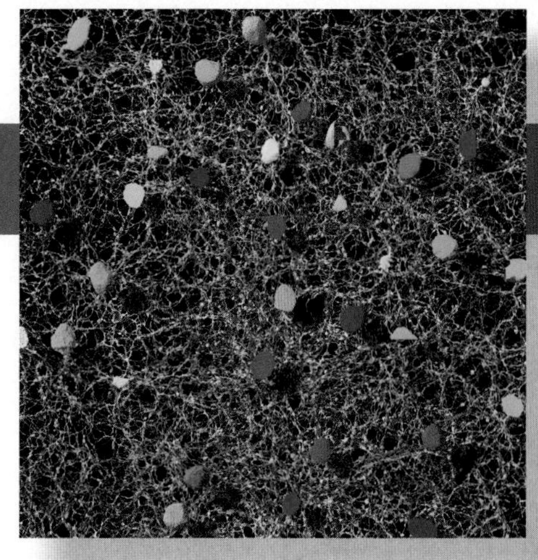

CHAPTER SEVEN

The Structure of the Nervous System

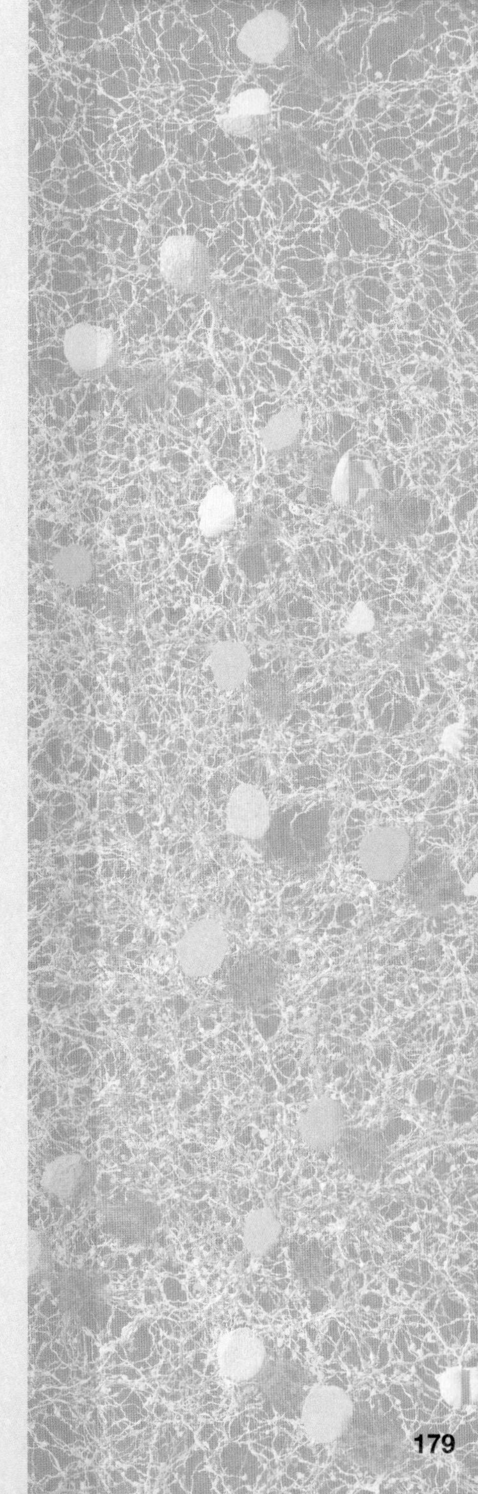

INTRODUCTION

In previous chapters, we saw how individual neurons function and communicate. Now we are ready to assemble them into a nervous system that sees, hears, feels, moves, remembers, and dreams. Just as an understanding of neuronal structure is necessary for understanding neuronal function, we must understand nervous system structure in order to understand brain function.

Neuroanatomy has challenged generations of students—and for good reason: The human brain is extremely complicated. However, our brain is merely a variation on a plan that is common to the brains of all mammals (Figure 7.1). The human brain appears complicated because it is distorted as a result of the selective growth of some parts within the confines of the skull. But once the basic mammalian plan is understood, these specializations of the human brain become clear.

We begin by introducing the general organization of the mammalian brain and the terms used to describe it. Then we take a look at how the three-dimensional structure of the brain arises during embryological and fetal development. Following the course of development makes it easier to understand how the parts of the adult brain fit together. Finally, we explore the cerebral neocortex, a structure that is unique to mammals and proportionately the largest in humans. An Illustrated Guide to Human Neuroanatomy follows the chapter as an appendix.

The neuroanatomy presented in this chapter provides the canvas on which we will paint the sensory and motor systems in Chapters 8–14. Because you will encounter a lot of new terms, self-quizzes within this chapter provide an opportunity for review.

GROSS ORGANIZATION OF THE MAMMALIAN NERVOUS SYSTEM

The nervous system of all mammals has two divisions: the central nervous system (CNS) and the peripheral nervous system (PNS). Here we identify some of the important components of the CNS and the PNS. We also discuss the membranes that surround the brain and the fluid-filled ventricles within the brain. We'll then explore some new methods of examining the structure of the brain. But first, we need to review some anatomical terminology.

Anatomical References

Getting to know your way around the brain is like getting to know your way around a city. To describe your location in the city, you would use points of reference such as north, south, east, and west and up and down. The same is true for the brain, except that the terms—called *anatomical references*—are different.

Consider the nervous system of a rat (Figure 7.2a). We begin with the rat because it is a simplified version that has all the general features of mammalian nervous system organization. In the head lies the brain, and the spinal cord runs down inside the backbone toward the tail. The direction, or anatomical reference, pointing toward the rat's nose is known as **anterior** or **rostral** (from the Latin for "beak"). The direction pointing toward the rat's tail is **posterior** or **caudal** (from the Latin for "tail"). The direction pointing up is known as **dorsal** (from the Latin for "back"), and the direction pointing down is **ventral** (from the Latin for "belly").

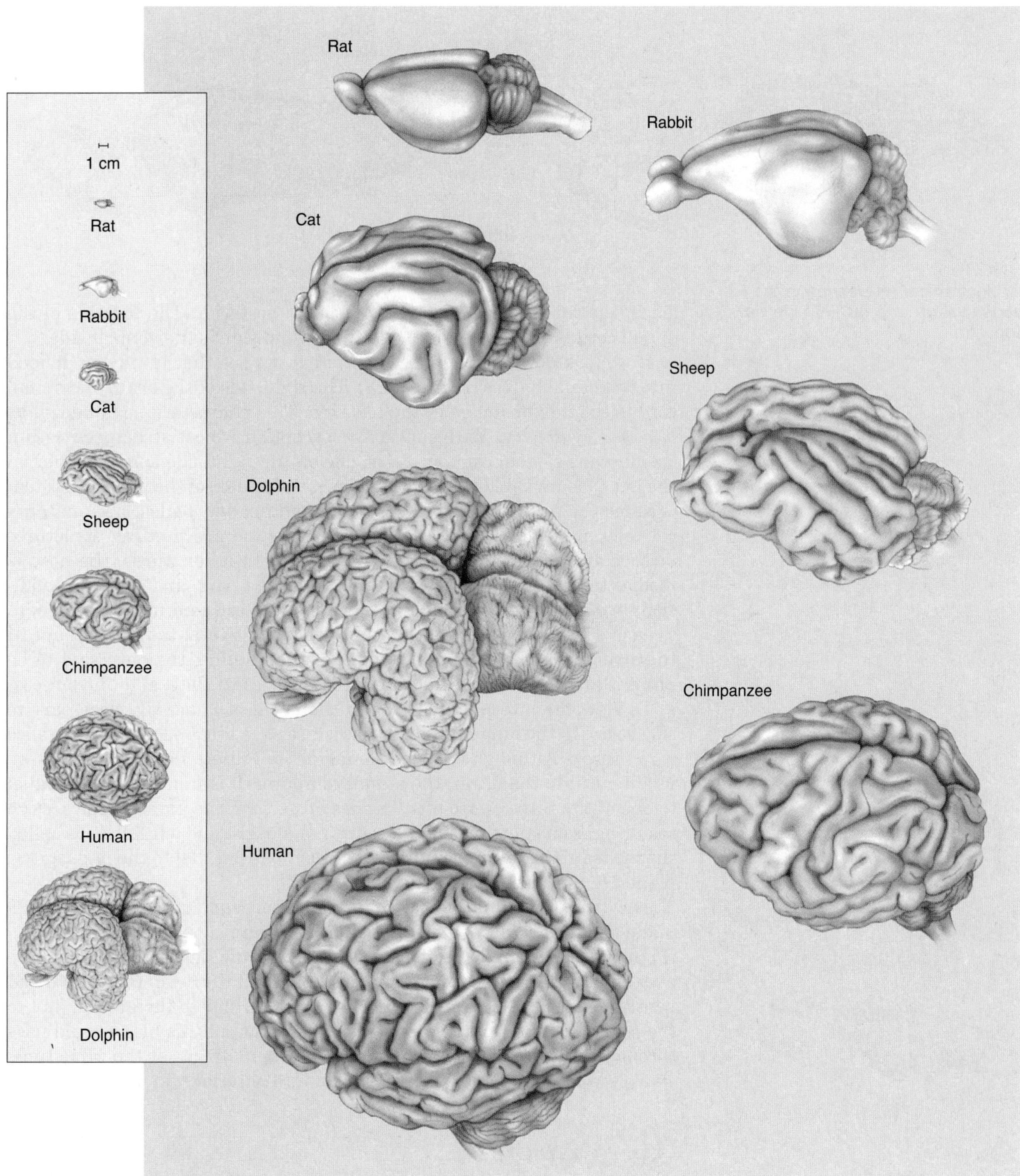

▲ FIGURE 7.1
Mammalian brains. Despite differences in complexity, the brains of all these species have many features in common. The brains have been drawn to appear approximately the same size; their relative sizes are shown in the inset on the left.

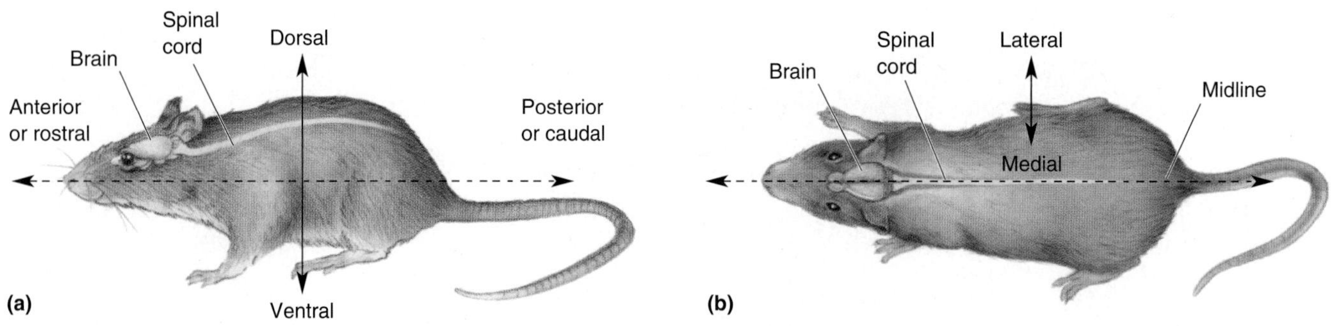

▲ FIGURE 7.2
Basic anatomical references in the nervous system of a rat. (a) Side view. (b) Top view.

Thus, the rat spinal cord runs anterior to posterior. The top side of the spinal cord is the dorsal side, and the bottom side is the ventral side.

If we look down on the nervous system, we see that it may be divided into two equal halves (Figure 7.2b). The right side of the brain and spinal cord is the mirror image of the left side. This characteristic is known as *bilateral symmetry*. With just a few exceptions, most structures within the nervous system come in pairs, one on the right side and the other on the left. The invisible line running down the middle of the nervous system is called the **midline**, and this gives us another way to describe anatomical references. Structures closer to the midline are **medial**; structures farther away from the midline are **lateral**. In other words, the nose is medial to the eyes, the eyes are medial to the ears, and so on. In addition, two structures that are on the same side are said to be **ipsilateral** to each other; for example, the right ear is ipsilateral to the right eye. If the structures are on opposite sides of the midline, they are said to be **contralateral** to each other; the right ear is contralateral to the left ear.

To view the internal structure of the brain, it is usually necessary to slice it up. In the language of anatomists, a slice is called a *section*; to slice is *to section*. Although one could imagine an infinite number of ways we might cut into the brain, the standard approach is to make cuts parallel to one of the three *anatomical planes of section*. The plane of the section resulting from splitting the brain into equal right and left halves is called the **midsagittal plane** (Figure 7.3a). Sections parallel to the midsagittal plane are in the **sagittal plane**.

The two other anatomical planes are perpendicular to the sagittal plane and to one another. The **horizontal plane** is parallel to the ground (Figure 7.3b). A single section in this plane could pass through both the eyes and the ears. Thus, horizontal sections split the brain into dorsal and ventral parts. The **coronal plane** is perpendicular to the ground and to the sagittal plane (Figure 7.3c). A single section in this plane could pass through both eyes or both ears but not through all four at the same time. Thus, the coronal plane splits the brain into anterior and posterior parts.

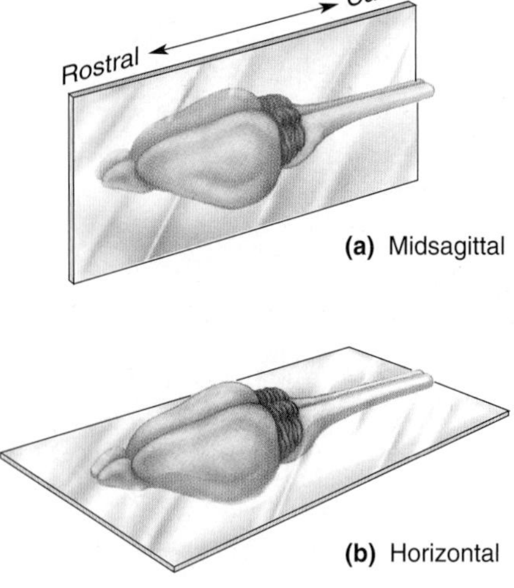

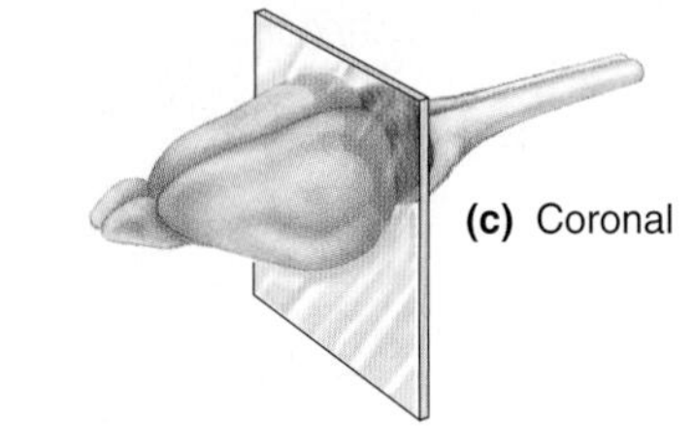

▲ FIGURE 7.3
Anatomical planes of section.

SELF-QUIZ

Take a few moments right now and be sure you understand the meaning of these terms:

anterior	dorsal	lateral	sagittal plane
rostral	ventral	ipsilateral	horizontal plane
posterior	midline	contralateral	coronal plane
caudal	medial	midsagittal plane	

The Central Nervous System

The **central nervous system (CNS)** consists of the parts of the nervous system that are encased in bone: the **brain** and the **spinal cord**. The brain lies entirely within the skull. A side view of the rat brain reveals three parts that are common to all mammals: the cerebrum, the cerebellum, and the brain stem (Figure 7.4a).

The Cerebrum. The rostral-most and largest part of the brain is the **cerebrum**. Figure 7.4b shows the rat cerebrum as it appears when viewed from above. Notice that it is clearly split down the middle into two **cerebral hemispheres**, separated by the deep *sagittal fissure*. In general, the *right* cerebral hemisphere receives sensations from, and controls movements of, the *left* side of the body. Similarly, the *left* cerebral hemisphere is concerned with sensations and movements on the *right* side of the body.

The Cerebellum. Lying behind the cerebrum is the **cerebellum** (the word is derived from the Latin for "little brain"). While the cerebellum is in fact dwarfed by the large cerebrum, it actually contains as many neurons as both cerebral hemispheres combined. The cerebellum is primarily a movement control center that has extensive connections with the cerebrum and the spinal cord. In contrast to the cerebral hemispheres, the left side of the cerebellum is concerned with movements of the left side of the body, and the right side of the cerebellum is concerned with movements of the right side.

The Brain Stem. The remaining part of the brain is the brain stem, best observed in a midsagittal view of the brain (Figure 7.4c). The **brain stem** forms the stalk from which the cerebral hemispheres and the cerebellum sprout. The brain stem is a complex nexus of fibers and cells that in part serves to relay information from the cerebrum to the spinal cord and cerebellum, and vice versa. However, the brain stem is also the site where vital functions are regulated, such as breathing, consciousness, and the control of body temperature. Indeed, while the brain stem is considered the most primitive part of the mammalian brain, it is also the most important to life. One can survive damage to the cerebrum and cerebellum, but damage to the brain stem is usually fatal.

The Spinal Cord. The spinal cord is encased in the bony vertebral column and is attached to the brain stem. The spinal cord is the major conduit of

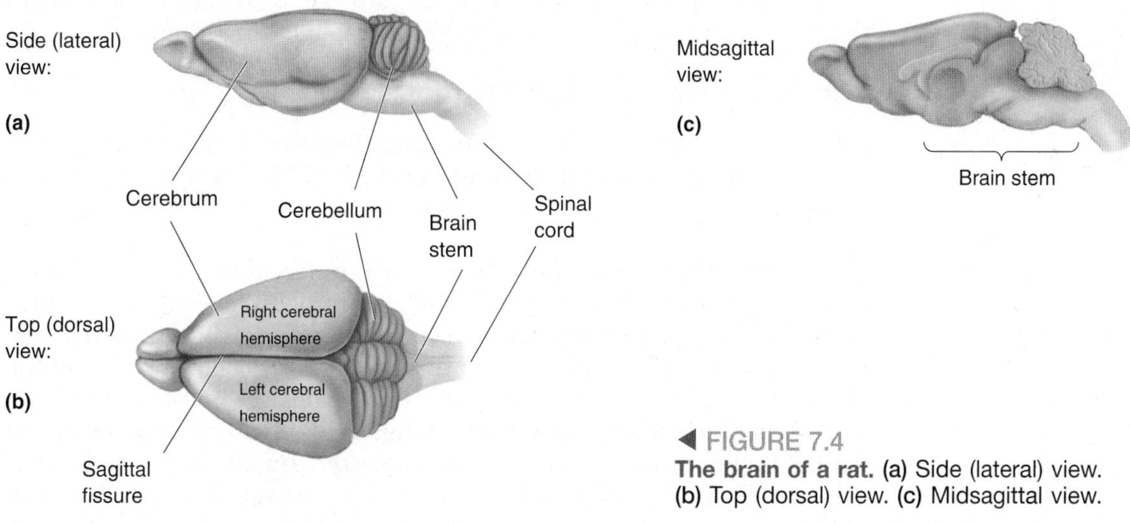

◀ FIGURE 7.4
The brain of a rat. (a) Side (lateral) view. (b) Top (dorsal) view. (c) Midsagittal view.

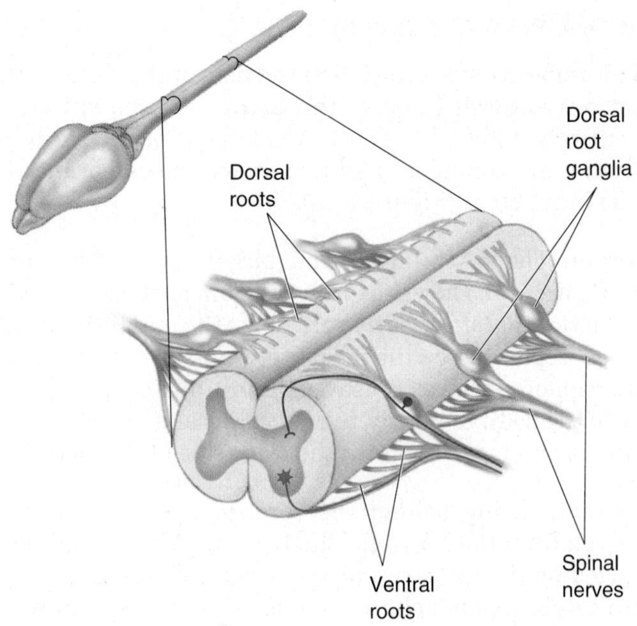

▲ FIGURE 7.5
The spinal cord. The spinal cord runs inside the vertebral column. Axons enter and exit the spinal cord via the dorsal and ventral roots, respectively. These roots come together to form the spinal nerves that course through the body.

information from the skin, joints, and muscles of the body to the brain, and vice versa. A transection of the spinal cord results in anesthesia (lack of feeling) in the skin and paralysis of the muscles in parts of the body caudal to the cut. Paralysis in this case does not mean that the muscles cannot function, but they cannot be controlled by the brain.

The spinal cord communicates with the body via the **spinal nerves**, which are part of the peripheral nervous system (discussed below). Spinal nerves exit the spinal cord through notches between each vertebra of the vertebral column. Each spinal nerve attaches to the spinal cord by means of two branches, the **dorsal root** and the **ventral root** (Figure 7.5). Recall from Chapter 1 that François Magendie showed that the dorsal root contains axons bringing information *into* the spinal cord, such as those that signal the accidental entry of a thumbtack into your foot (see Figure 3.1). Charles Bell showed that the ventral root contains axons carrying information *away from* the spinal cord—for example, to the muscles that jerk your foot away in response to the pain of the thumbtack.

The Peripheral Nervous System

All the parts of the nervous system other than the brain and spinal cord comprise the **peripheral nervous system (PNS)**. The PNS has two parts: the somatic PNS and the visceral PNS.

The Somatic PNS. All the spinal nerves that innervate the skin, the joints, and the muscles that are under voluntary control are part of the **somatic PNS**. The somatic motor axons, which command muscle contraction, derive from motor neurons in the ventral spinal cord. The cell bodies of the motor neurons lie within the CNS, but their axons are mostly in the PNS.

The somatic sensory axons, which innervate and collect information from the skin, muscles, and joints, enter the spinal cord via the dorsal roots. The cell bodies of these neurons lie outside the spinal cord in clusters called **dorsal root ganglia**. There is a dorsal root ganglion for each spinal nerve (see Figure 7.5).

The Visceral PNS. The **visceral PNS**, also called the involuntary, vegetative, or **autonomic nervous system (ANS)**, consists of the neurons that innervate the internal organs, blood vessels, and glands. Visceral sensory axons bring information about visceral function to the CNS, such as the pressure and oxygen content of the blood in the arteries. Visceral motor fibers command the contraction and relaxation of muscles that form the walls of the intestines and the blood vessels (called *smooth muscles*), the rate of cardiac muscle contraction, and the secretory function of various glands. For example, the visceral PNS controls blood pressure by regulating the heart rate and the diameter of the blood vessels.

We will return to the structure and function of the ANS in Chapter 15. For now, remember that when one speaks of an emotional reaction that is beyond voluntary control—like "butterflies in the stomach" or blushing—it usually is mediated by the visceral PNS (the ANS).

Afferent and Efferent Axons. Our discussion of the PNS is a good place to introduce two terms that are used to describe axons in the nervous system. Derived from the Latin, **afferent** ("carry to") and **efferent** ("carry from") indicate whether the axons are transporting information *toward* or *away from* a particular point. Consider the axons in the PNS relative to a point of reference in the CNS. The somatic or visceral sensory axons bringing information *into* the CNS are afferents. The axons that emerge *from* the CNS to innervate the muscles and glands are efferents.

The Cranial Nerves

In addition to the nerves that arise from the spinal cord and innervate the body, there are 12 pairs of **cranial nerves** that arise from the brain stem and innervate (mostly) the head. Each cranial nerve has a name and a number associated with it (originally numbered by Galen, about 1800 years ago, from anterior to posterior). Some of the cranial nerves are part of the CNS, others are part of the somatic PNS, and still others are part of the visceral PNS. Many cranial nerves contain a complex mixture of axons that perform different functions. The cranial nerves and their various functions are summarized in the chapter appendix.

The Meninges

The CNS, that part of the nervous system encased in the skull and vertebral column, does not come in direct contact with the overlying bone. It is protected by three membranes collectively called the **meninges** (singular: meninx), from the Greek for "covering." The three membranes are the dura mater, the arachnoid membrane, and the pia mater (Figure 7.6).

The outermost covering is the **dura mater**, from the Latin words meaning "hard mother," an accurate description of the dura's leatherlike consistency. The dura forms a tough, inelastic bag that surrounds the brain and spinal cord. Just under the dura lies the **arachnoid membrane** (from the Greek for "spider"). This meningeal layer has an appearance and a consistency resembling a spider web. While there normally is no space between the dura and the arachnoid, if the blood vessels passing through the dura are ruptured, blood can collect here and form what is called a *subdural hematoma*. The buildup of fluid in this subdural space can disrupt brain function by compressing parts of the CNS. The disorder is treated by drilling a hole in the skull and draining the blood.

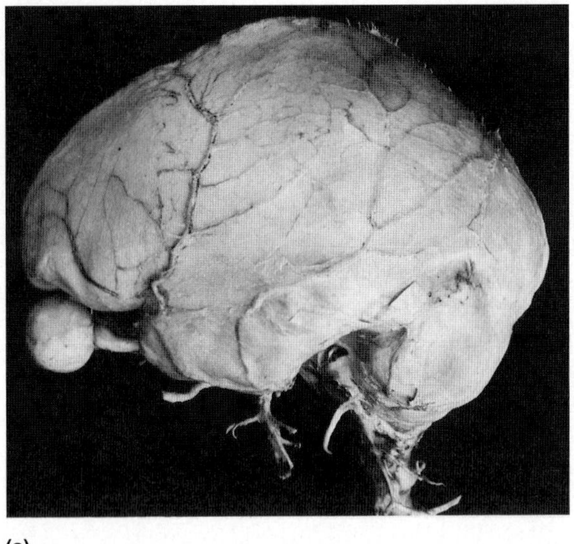

(a)

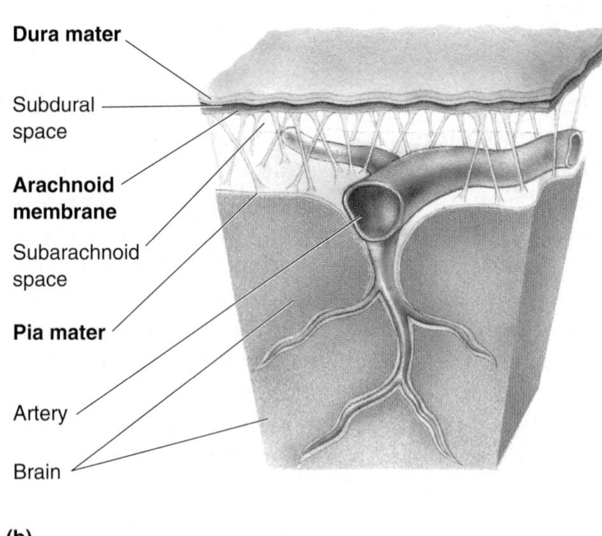

(b)

▲ FIGURE 7.6
The meninges. (a) The skull has been removed to show the tough outer meningeal membrane, the dura mater. (Source: Gluhbegoric and Williams, 1980.) (b) Illustrated in cross section, the three meningeal layers protecting the brain and spinal cord are the dura mater, the arachnoid membrane, and the pia mater.

The **pia mater**, the "gentle mother," is a thin membrane that adheres closely to the surface of the brain. Along the pia run many blood vessels that ultimately dive into the substance of the underlying brain. The pia is separated from the arachnoid by a fluid-filled space. This *subarachnoid space* is filled with salty clear liquid called **cerebrospinal fluid (CSF)**. Thus, in a sense, the brain floats inside the head in this thin layer of CSF.

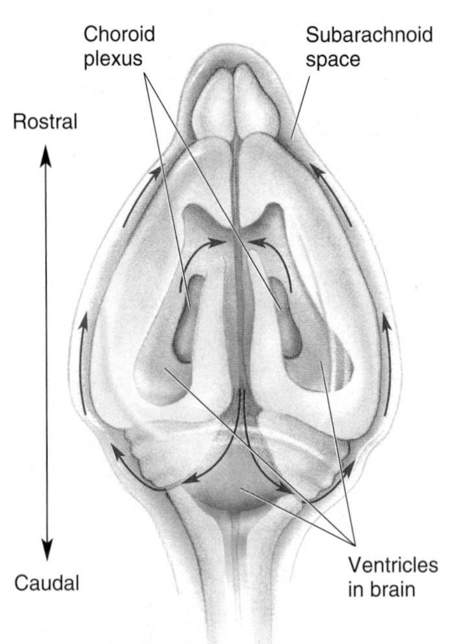

▲ FIGURE 7.7
The ventricular system in a rat brain. CSF is produced in the ventricles of the paired cerebral hemispheres and flows through a series of central ventricles at the core of the brain stem. CSF escapes into the subarachnoid space via small apertures near the base of the cerebellum. In the subarachnoid space, CSF is absorbed into the blood.

The Ventricular System

In Chapter 1, we noted that the brain is hollow. The fluid-filled caverns and canals inside the brain constitute the **ventricular system**. The fluid that runs in this system is CSF, the same as the fluid in the subarachnoid space. CSF is produced by a special tissue, called the *choroid plexus*, in the ventricles of the cerebral hemispheres. CSF flows from the paired ventricles of the cerebrum to a series of connected, central cavities at the core of the brain stem (Figure 7.7). CSF exits the ventricular system and enters the subarachnoid space by way of small openings, or apertures, located near where the cerebellum attaches to the brain stem. In the subarachnoid space, CSF is absorbed by the blood vessels at special structures called *arachnoid villi*. If the normal flow of CSF is disrupted, brain damage can result (Box 7.1).

We will return to fill in some details about the ventricular system in a moment. As we will see, understanding the organization of the ventricular system holds the key to understanding how the mammalian brain is organized.

New Views of the Brain

For centuries, anatomists have investigated the internal structure of the brain by removing it from the skull, sectioning it in various planes, staining the sections, and examining the stained sections. Much has been learned by this approach, but there are some limitations. Among these are the challenges of seeing how parts deep in the brain fit together in

BOX 7.1 OF SPECIAL INTEREST

Water on the Brain

If the flow of CSF from the choroid plexus through the ventricular system to the subarachnoid space is impaired, the fluid will back up and cause a swelling of the ventricles. This condition is called *hydrocephalus*, a term originally meaning "water head."

Occasionally, babies are born with hydrocephalus. However, because the skull is soft and not completely formed, the head will expand in response to the increased intracranial fluid, sparing the brain from damage. Often this condition goes unnoticed until the size of the head reaches enormous proportions.

In adults, hydrocephalus is a much more serious situation because the skull cannot expand, and intracranial pressure increases as a result. The soft brain tissue is then compressed, impairing function and leading to death if left untreated. Typically, this "obstructive" hydrocephalus is also accompanied by severe headache, caused by the distention of nerve endings in the meninges. Treatment consists of inserting a tube into the swollen ventricle and draining off the excess fluid (Figure A).

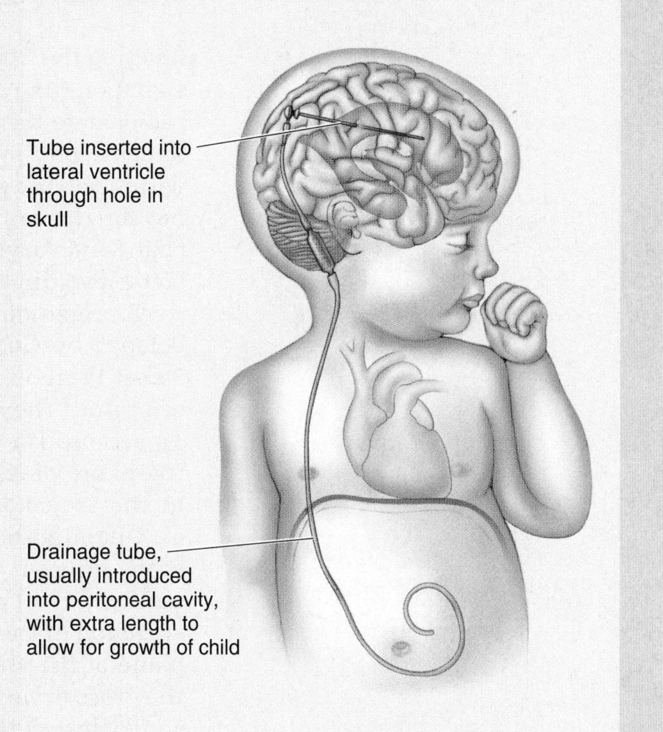

Figure A

three dimensions. A breakthrough occurred in 2013 when researchers at Stanford University introduced a new method, called CLARITY, which allows visualization of deep structures without sectioning the brain. The trick is to soak the brain in a solution that replaces light-absorbing lipids with a water-soluble gel that turns the brain transparent. If such a "clarified" brain contains neurons that are labeled with fluorescent molecules, such as green fluorescent protein (GFP; see Chapter 2), then appropriate illumination will reveal the location of these cells deep inside the brain (Figure 7.8).

(a) (b)

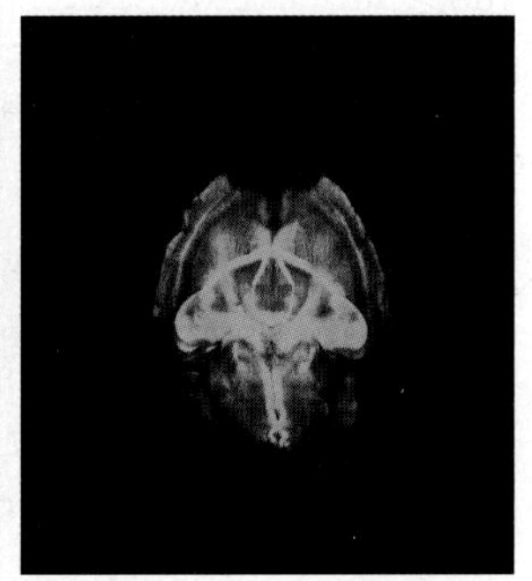

(c)

◀ FIGURE 7.8
A method to turn the brain transparent and visualize fluorescent neurons deep in the brain. **(a)** A mouse brain viewed from above. **(b)** The same brain rendered transparent by replacing lipids with a water-soluble gel. **(c)** The transparent brain illuminated to evoke fluorescence from neurons that express green fluorescent protein. (Source: Courtesy of Dr. Kwanghun Chung, Massachusetts Institute of Technology. Adapted from Chung and Deisseroth. 2013, Figure 2.)

Of course, a clarified brain is still a dead brain. This, to say the least, limits the usefulness of such anatomical methods for diagnosing neurological disorders in living individuals. Thus, it is no exaggeration to say that neuroanatomy was revolutionized by the introduction of several methods that enable one to produce images of the living brain. Here we briefly introduce them.

Imaging the Structure of the Living Brain. Some types of electromagnetic radiation, like X-rays, penetrate the body and are absorbed by various radiopaque tissues. Thus, using X-ray-sensitive film, one can make two-dimensional images of the shadows formed by the radiopaque structures within the body. This technique works well for the bones of the skull, but not for the brain. The brain is a complex three-dimensional volume of slight and varying radiopacity, so little information can be gleaned from a single two-dimensional X-ray image.

An ingenious solution, called *computed tomography (CT),* was developed by Godfrey Hounsfields and Allan Cormack, who shared the Nobel Prize in 1979. The goal of CT is to generate an image of a slice of brain. (The word *tomography* is derived from the Greek for "cut.") To accomplish this, an X-ray source is rotated around the head within the plane of the desired cross section. On the other side of the head, in the trajectory of the X-ray beam, are sensitive electronic sensors of X-irradiation. The information about relative radiopacity obtained with different viewing angles is fed to a computer that executes a mathematical algorithm on the data. The end result is a digital reconstruction of the position and amount of radiopaque material within the plane of the slice. CT scans noninvasively revealed, for the first time, the gross organization of gray and white matter, and the position of the ventricles, in the living brain.

While still used widely, CT is gradually being replaced by a newer imaging method, called *magnetic resonance imaging (MRI)*. The advantages of MRI are that it yields a much more detailed map of the brain than CT, it does not require X-irradiation, and images of brain slices can be made in any plane desired. MRI uses information about how hydrogen atoms in the brain respond to perturbations of a strong magnetic field (Box 7.2). The electromagnetic signals emitted by the atoms are detected by an array of sensors around the head and fed to a powerful computer that constructs a map of the brain. The information from an MRI scan can be used to build a strikingly detailed image of the whole brain.

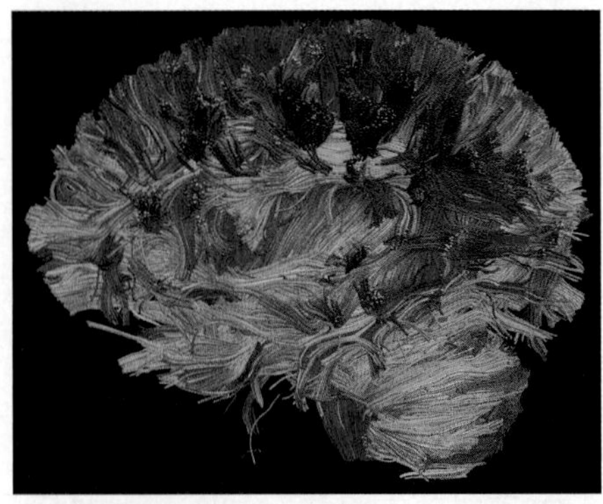

▲ FIGURE 7.9
Diffusion tensor imaging of the human brain. Displayed is a computer reconstruction of axon bundles in a living human brain viewed from the side. Anterior is to the left. The bundles are pseudocolored based on the direction of water diffusion. (Source: Courtesy of Dr. Satrajit Ghosh, Massachusetts Institute of Technology.)

Another application of MRI, called *diffusion tensor imaging* (DTI), enables visualization of large bundles of axons in the brain. By comparing the position of the hydrogen atoms in water molecules at discrete time intervals, the diffusion of water in the brain can be measured. Water diffuses much more readily alongside axon membranes than across them, and this difference can be used to detect axon bundles that connect different regions of the brain (Figure 7.9).

Functional Brain Imaging. CT and MRI are extremely valuable for detecting structural changes in the living brain, such as brain swelling after a head injury and brain tumors. Nonetheless, much of what goes on in the brain—healthy or diseased—is chemical and electrical in nature and not observable by simple inspection of the brain's anatomy. Amazingly, however, even these secrets are beginning to yield to the newest imaging techniques.

BOX 7.2 **BRAIN FOOD**

Magnetic Resonance Imaging

Magnetic resonance imaging (MRI) is a general technique that can be used for determining the amount of certain atoms at different locations in the body. It has become an important tool in neuroscience because it can be used noninvasively to obtain a detailed picture of the nervous system, particularly the brain.

In the most common form of MRI, the hydrogen atoms are quantified—for instance, those located in water or fat in the brain. An important fact of physics is that when a hydrogen atom is put in a magnetic field, its nucleus (which consists of a single proton) can exist in either of two states: a high-energy state or a low-energy state. Because hydrogen atoms are abundant in the brain, there are many protons in each state.

The key to MRI is making the protons jump from one state to the other. Energy is added to the protons by passing an electromagnetic wave (i.e., a radio signal) through the head while it is positioned between the poles of a large magnet. When the radio signal is set at just the right frequency, the protons in the low-energy state will absorb energy from the signal and hop to the high-energy state. The frequency at which the protons absorb energy is called the *resonant frequency* (hence the name magnetic resonance). When the radio signal is turned off, some of the protons fall back down to the low-energy state, thereby emitting a radio signal of their own at a particular frequency. This signal can be picked up by a radio receiver. The stronger the signal, the more hydrogen atoms between the poles of the magnet.

If we used the procedure discussed earlier, we would simply get a measurement of the total amount of hydrogen in the head. However, it is possible to measure hydrogen amounts at a fine spatial scale by taking advantage of the fact that the frequency at which protons emit energy is proportional to the size of the magnetic field. In the MRI machines used in hospitals, the magnetic fields vary from one side of the magnet to the other. This gives a spatial code to the radio waves emitted by the protons: High-frequency signals come from hydrogen atoms near the strong side of the magnet, and low-frequency signals come from the weak side of the magnet.

The last step in the MRI process is to orient the gradient of the magnet at many different angles relative to the head and measure the amount of hydrogen. It takes about 15 minutes to make all the measurements for a typical brain scan. A sophisticated computer program is then used to make a single image from the measurements, resulting in a picture of the distribution of hydrogen atoms in the head.

Figure A is an MRI image of a lateral view of the brain in a living human. In Figure B, another MRI image, a slice has been made in the brain. Notice how clearly you can see the white and gray matter. This differentiation makes it possible to see the effects of demyelinating diseases on white matter in the brain. MRI images also reveal lesions in the brain because tumors and inflammation generally increase the amount of extracellular water.

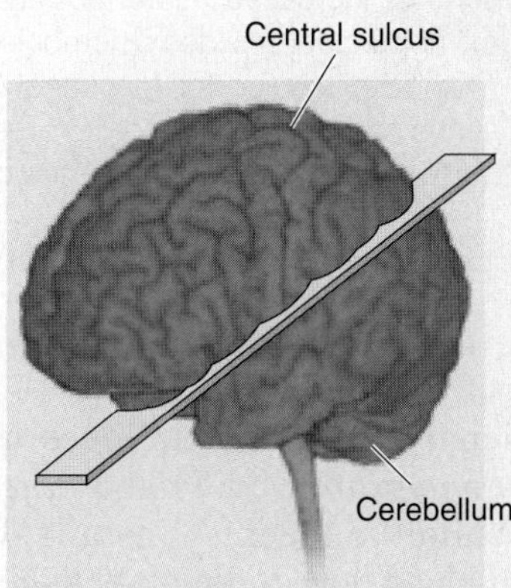

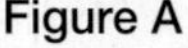
Figure A

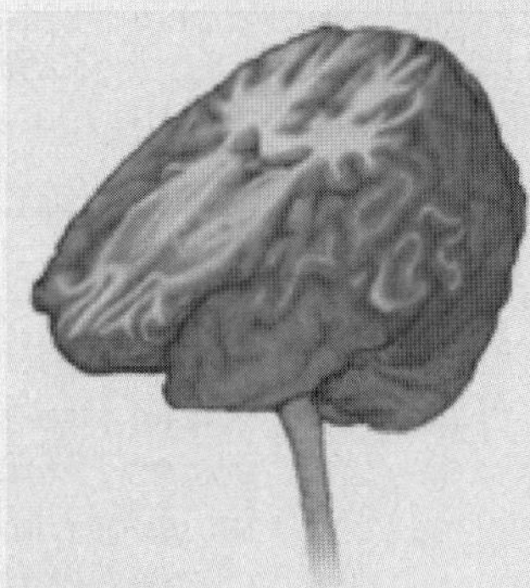

Figure B

BOX 7.3 BRAIN FOOD

PET and fMRI

Until recently, "mind reading" has been beyond the reach of science. However, with the introduction of *positron emission tomography (PET)* and *functional magnetic resonance imaging (fMRI)*, it is now possible to observe and measure changes in brain activity associated with the planning and execution of specific tasks.

PET imaging was developed in the 1970s by two groups of physicists, one at Washington University led by M. M. Ter-Pogossian and M. E. Phelps, and a second at UCLA led by Z. H. Cho. The basic procedure is very simple. A radioactive solution containing atoms that emit positrons (positively charged electrons) is introduced into the bloodstream. Positrons, emitted wherever the blood goes, interact with electrons to produce photons of electromagnetic radiation. The locations of the positron-emitting atoms are found by detectors that pick up the photons.

One powerful application of PET is the measurement of metabolic activity in the brain. In a technique developed by Louis Sokoloff and his colleagues at the National Institute of Mental Health, a positron-emitting isotope of fluorine or oxygen is attached to 2-deoxyglucose (2-DG). This radioactive 2-DG is injected into the bloodstream and travels to the brain. Metabolically active neurons, which normally use glucose, also take up the 2-DG. The 2-DG is phosphorylated by enzymes inside the neuron, and this modification prevents the 2-DG from leaving. Thus, the amount of radioactive 2-DG accumulated in a neuron and the number of positron emissions indicate the level of neuronal metabolic activity.

In a typical PET application, a person's head is placed in an apparatus surrounded by detectors (Figure A). Using computer algorithms, the photons (resulting from positron emissions) reaching each of the detectors are recorded. With this information, levels of activity for populations of neurons at various sites in the brain can be calculated. Compiling these measurements produces an image of the brain activity pattern. The researcher monitors brain activity while the subject performs a task, such as moving a finger or reading aloud. Different tasks "light up" different brain areas. In order to obtain a picture of the activity induced by a particular behavioral or thought task, a subtraction technique is used. Even in the absence of any sensory stimulation, the PET image will contain a great deal of brain activity. To create an image of the brain activity resulting from a specific task, such as a person looking at a picture, this background activity is subtracted out (Figure B).

Although PET imaging has proven to be a valuable technique, it has significant limitations. Because the spatial resolution is only 5–10 mm^3, the images show the activity of many thousands of cells. Also, a single PET brain scan may take one to several minutes to obtain. This, along with concerns about radiation exposure, limits the number of scans that can be obtained from one person in a reasonable time period. Thus, the work of S. Ogawa at Bell Labs, showing that MRI techniques could be used to measure local changes in blood oxygen levels that result from brain activity, was an important advance.

The fMRI method takes advantage of the fact that oxyhemoglobin (the oxygenated from of hemoglobin in the blood) has a magnetic resonance different from that of deoxyhemoglobin (hemoglobin that has donated its oxygen). More active regions of the brain receive more blood, and this blood donates more of its oxygen. Functional MRI detects the locations of increased neural activity by measuring the ratio of oxyhemoglobin to deoxyhemoglobin. It has emerged as the method of choice for functional brain imaging because the scans can be made rapidly (50 msec), they have good spatial resolution (3 mm^3), and they are completely noninvasive.

The two "functional imaging" techniques now in widespread use are *positron emission tomography (PET)* and *functional magnetic resonance imaging (fMRI)*. While the technical details differ, both methods detect changes in regional blood flow and metabolism within the brain (Box 7.3). The basic principle is simple. Neurons that are active demand more glucose and oxygen. The brain vasculature responds to neural activity by directing more blood to the active regions. Thus, by detecting changes in blood flow, PET and fMRI reveal the regions of brain that are most active under different circumstances.

The advent of imaging techniques has offered neuroscientists the extraordinary opportunity of peering into the living, thinking brain. As you can imagine, however, even the most sophisticated brain images are useless unless you know what you are looking at. Next, let's take a closer look at how the brain is organized.

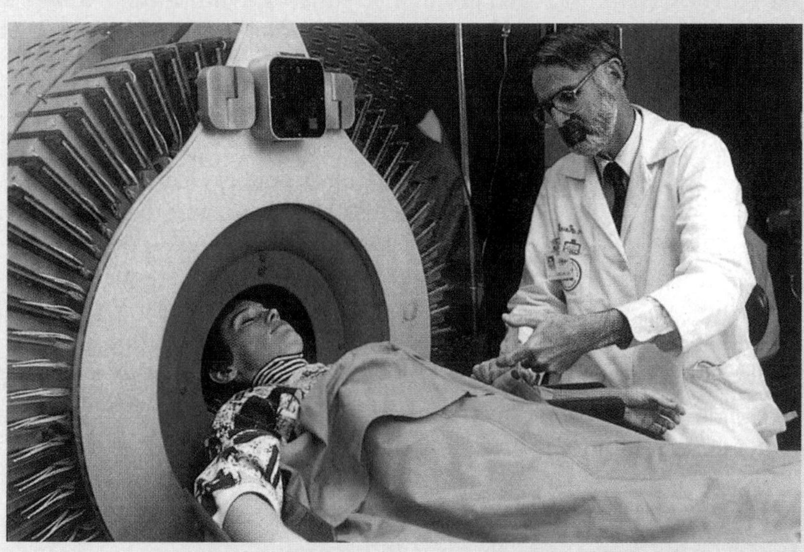

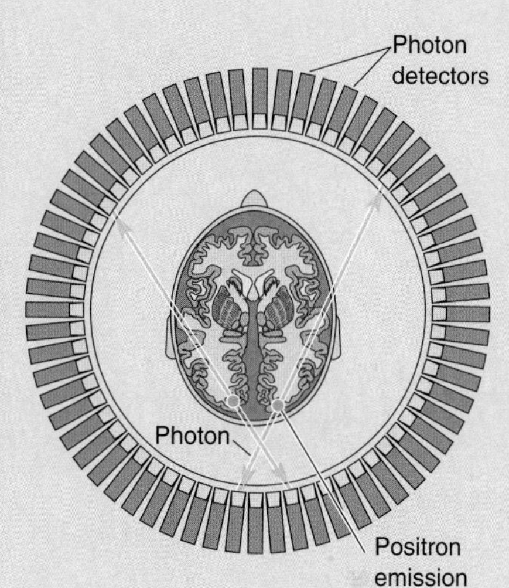

Figure A
The PET procedure. (Source: Posner and Raichle, 1994, p. 61.)

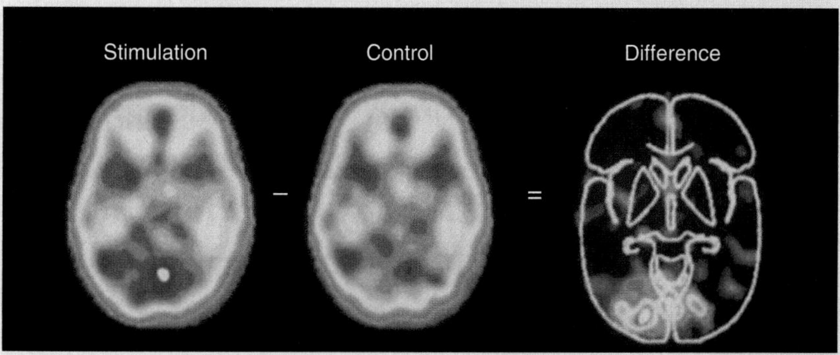

Figure B
A PET image. (Source: Posner and Raichle, 1994, p. 65.)

SELF-QUIZ

Take a few moments right now and be sure you understand the meaning of these terms:

central nervous system (CNS)
brain
spinal cord
cerebrum
cerebral hemispheres
cerebellum
brain stem
spinal nerve
dorsal root
ventral root
peripheral nervous system (PNS)
somatic PNS
dorsal root ganglia
visceral PNS
autonomic nervous system (ANS)
afferent
efferent
cranial nerve
meninges
dura mater
arachnoid membrane
pia mater
cerebrospinal fluid (CSF)
ventricular system

UNDERSTANDING CNS STRUCTURE THROUGH DEVELOPMENT

The entire CNS is derived from the walls of a fluid-filled tube that is formed at an early stage in embryonic development. The inside of the tube becomes the adult ventricular system. Thus, by examining how this tube changes during the course of fetal development, we can understand how the brain is organized and how the different parts fit together. This section focuses on development as a way to understand the structural organization of the brain. Chapter 23 will revisit the topic of development to describe how neurons are born, how they find their way to their final locations in the CNS, and how they make the appropriate synaptic connections with one another.

As you work your way through this section, and through the rest of the book, you will encounter many different names used by anatomists to refer to groups of related neurons and axons. Some common names for describing collections of neurons and axons are given in Tables 7.1 and 7.2. Take a few moments to familiarize yourself with these new terms before continuing.

TABLE 7.1 **Collections of Neurons**

Name	Description and Example
Gray matter	A generic term for a collection of neuronal cell bodies in the CNS. When a freshly dissected brain is cut open, neurons appear gray.
Cortex	Any collection of neurons that form a thin sheet, usually at the brain's surface. *Cortex* is Latin for "bark." Example: *cerebral cortex*, the sheet of neurons found just under the surface of the cerebrum.
Nucleus	A clearly distinguishable mass of neurons, usually deep in the brain (not to be confused with the nucleus of a cell). *Nucleus* is from the Latin word for "nut." Example: *lateral geniculate nucleus*, a cell group in the brain stem that relays information from the eye to the cerebral cortex.
Substantia	A group of related neurons deep within the brain but usually with less distinct borders than those of nuclei. Example: *substantia nigra* (from the Latin for "black substance"), a brain stem cell group involved in the control of voluntary movement.
Locus (plural: loci)	A small, well-defined group of cells. Example: *locus coeruleus* (Latin for "blue spot"), a brain stem cell group involved in the control of wakefulness and behavioral arousal.
Ganglion (plural: ganglia)	A collection of neurons in the PNS. *Ganglion* is from the Greek for "knot." Example: the *dorsal root ganglia*, which contain the cell bodies of sensory axons entering the spinal cord via the dorsal roots. Only one cell group in the CNS goes by this name: the *basal ganglia*, which are structures lying deep within the cerebrum that control movement.

TABLE 7.2 **Collections of Axons**

Name	Description and Example
Nerve	A bundle of axons in the PNS. Only one collection of CNS axons is called a nerve: the *optic nerve*.
White matter	A generic term for a collection of CNS axons. When a freshly dissected brain is cut open, axons appear white.
Tract	A collection of CNS axons having a common site of origin and a common destination. Example: *corticospinal tract,* which originates in the cerebral cortex and ends in the spinal cord.
Bundle	A collection of axons that run together but do not necessarily have the same origin and destination. Example: *medial forebrain bundle*, which connects cells scattered within the cerebrum and brain stem.
Capsule	A collection of axons that connect the cerebrum with the brain stem. Example: *internal capsule*, which connects the brain stem with the cerebral cortex.
Commissure	Any collection of axons that connect one side of the brain with the other side.
Lemniscus	A tract that meanders through the brain like a ribbon. Example: *medial lemniscus*, which brings touch information from the spinal cord through the brain stem.

Anatomy by itself can be pretty dry. It really comes alive only after the functions of different structures are understood. The remainder of this book is devoted to explaining the functional organization of the nervous system. However, we include in this section a preview of some structure-function relationships to provide you with a general sense of how the different parts contribute, individually and collectively, to the function of the CNS.

Formation of the Neural Tube

The embryo begins as a flat disk with three distinct layers of cells called endoderm, mesoderm, and ectoderm. The *endoderm* ultimately gives rise to the lining of many of the internal organs (viscera). From the *mesoderm* arise the bones of the skeleton and the muscles. The nervous system and the skin derive entirely from the *ectoderm*.

Our focus is on changes in the part of the ectoderm that give rise to the nervous system: the *neural plate*. At this early stage (about 17 days from conception in humans), the brain consists only of a flat sheet of cells (Figure 7.10a). The next event of interest is the formation of a

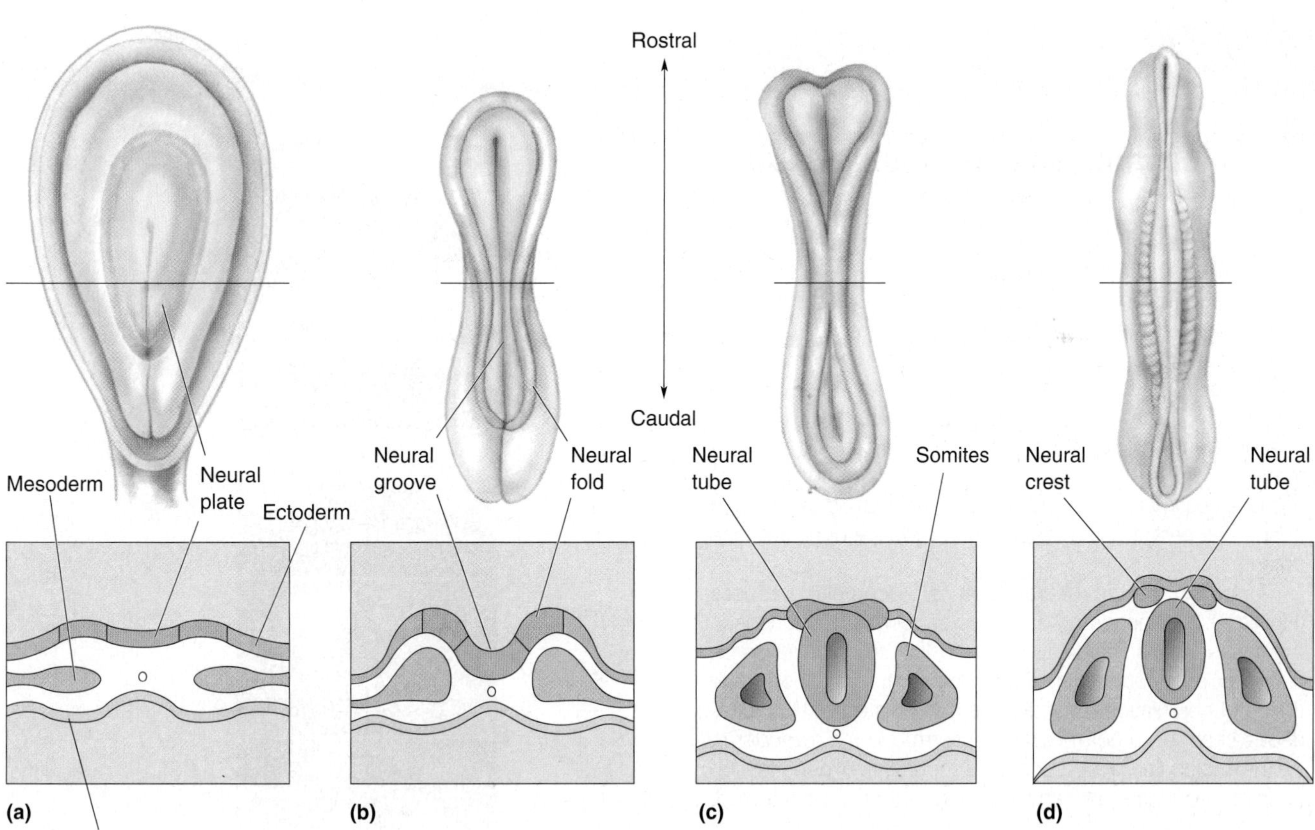

▲ FIGURE 7.10
Formation of the neural tube and neural crest. These schematic illustrations follow the early development of the nervous system in the embryo. The drawings above are dorsal views of the embryo; those below are cross sections. **(a)** The primitive embryonic CNS begins as a thin sheet of ectoderm. **(b)** The first important step in the development of the nervous system is the formation of the neural groove. **(c)** The walls of the groove, called *neural folds*, come together and fuse, forming the neural tube. **(d)** The bits of neural ectoderm that are pinched off when the tube rolls up is called the *neural crest*, from which the PNS will develop. The somites are mesoderm that will give rise to much of the skeletal system and the muscles.

groove in the neural plate that runs rostral to caudal, called the *neural groove* (Figure 7.10b). The walls of the groove are called *neural folds*, which subsequently move together and fuse dorsally, forming the **neural tube** (Figure 7.10c). *The entire central nervous system develops from the walls of the neural tube.* As the neural folds come together, some neural ectoderm is pinched off and comes to lie just lateral to the neural tube. This tissue is called the **neural crest** (Figure 7.10d). *All neurons with cell bodies in the peripheral nervous system derive from the neural crest.*

The neural crest develops in close association with the underlying mesoderm. The mesoderm at this stage in development forms prominent bulges on either side of the neural tube called *somites*. From these somites, the 33 individual vertebrae of the spinal column and the related skeletal muscles will develop. The nerves that innervate these skeletal muscles are therefore called *somatic* motor nerves.

The process by which the neural plate becomes the neural tube is called **neurulation**. Neurulation occurs very early in embryonic

BOX 7.4 OF SPECIAL INTEREST

Nutrition and the Neural Tube

Neural tube formation is a crucial event in the development of the nervous system. It occurs early—only 3 weeks after conception—when the mother may be unaware she is pregnant. Failure of the neural tube to close correctly is a common birth defect, occurring in approximately 1 out of every 500 live births. A recent discovery of enormous public health importance is that many neural tube defects can be traced to a deficiency of the vitamin *folic acid* (or *folate*) in the maternal diet during the weeks immediately after conception. It has been estimated that dietary supplementation of folic acid during this period could reduce the incidence of neural tube defects by 90%.

Formation of the neural tube is a complex process (Figure A). It depends on a precise sequence of changes in the three-dimensional shape of individual cells as well as on changes in the adhesion of each cell to its neighbors. The timing of neurulation must also be coordinated with simultaneous changes in non-neural ectoderm and the mesoderm. At the molecular level, successful neurulation depends on specific sequences of gene expression that are controlled, in part, by the position and local chemical environment of the cell. It is not surprising that this process is highly sensitive to chemicals, or chemical deficiencies, in the maternal circulation.

The fusion of the neural folds to form the neural tube occurs first in the middle, then anteriorly and posteriorly (Figure B). Failure of the anterior neural tube to close results

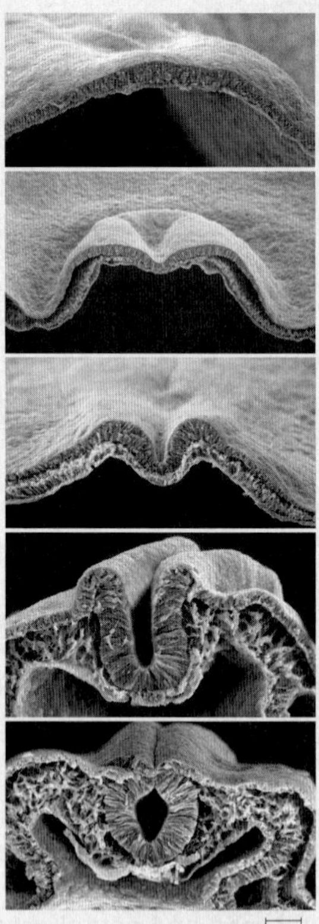

Figure A
Scanning electron micrographs of neurulation.
(Source: Smith and Schoenwolf, 1997.)

development, about 22 days after conception in humans. A common birth defect is the failure of appropriate closure of the neural tube. Fortunately, recent research suggests that most cases of neural tube defects can be avoided by ensuring proper maternal nutrition during this period (Box 7.4).

Three Primary Brain Vesicles

The process by which structures become more complex and functionally specialized during development is called **differentiation**. The first step in the differentiation of the brain is the development, at the rostral end of the neural tube, of three swellings called the primary vesicles (Figure 7.11). *The entire brain derives from the three primary vesicles of the neural tube.*

The rostral-most vesicle is called the *prosencephalon*. *Pro* is Greek for "before"; *encephalon* is derived from the Greek for "brain." Thus, the prosencephalon is also called the **forebrain**. Behind the prosencephalon lies

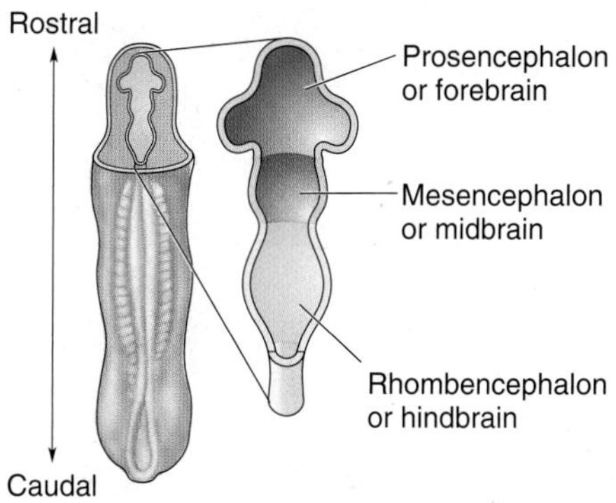

▲ FIGURE 7.11
The three primary brain vesicles. The rostral end of the neural tube differentiates to form the three vesicles that will give rise to the entire brain. This view is from above, and the vesicles have been cut horizontally so that we can see the inside of the neural tube.

in *anencephaly*, a condition characterized by degeneration of the forebrain and skull that is always fatal. Failure of the posterior neural tube to close results in a condition called *spina bifida*. In its most severe form, spina bifida is characterized by the failure of the posterior spinal cord to form from the neural plate (*bifida* is from the Latin word meaning "cleft in two parts"). Less severe forms are characterized by defects in the meninges and vertebrae overlying the posterior spinal cord. Spina bifida, while usually not fatal, does require extensive and costly medical care.

Folic acid plays an essential role in a number of metabolic pathways, including the biosynthesis of DNA, which naturally must occur during development as cells divide. Although we do not precisely understand why folic acid deficiency increases the incidence of neural tube defects, one can easily imagine how it could alter the complex choreography of neurulation. Its name is derived from the Latin word for "leaf," reflecting the fact that folic acid was first isolated from spinach leaves. Besides green leafy vegetables, good dietary sources of folic acid are liver, yeast, eggs, beans, and oranges. Many breakfast cereals are now fortified with folic acid. Nonetheless, the folic acid intake of the average American is only half of what is recommended to prevent birth defects (0.4 mg/day). The U.S. Centers for Disease Control and Prevention recommends that women take multivitamins containing 0.4 mg of folic acid before planning pregnancy.

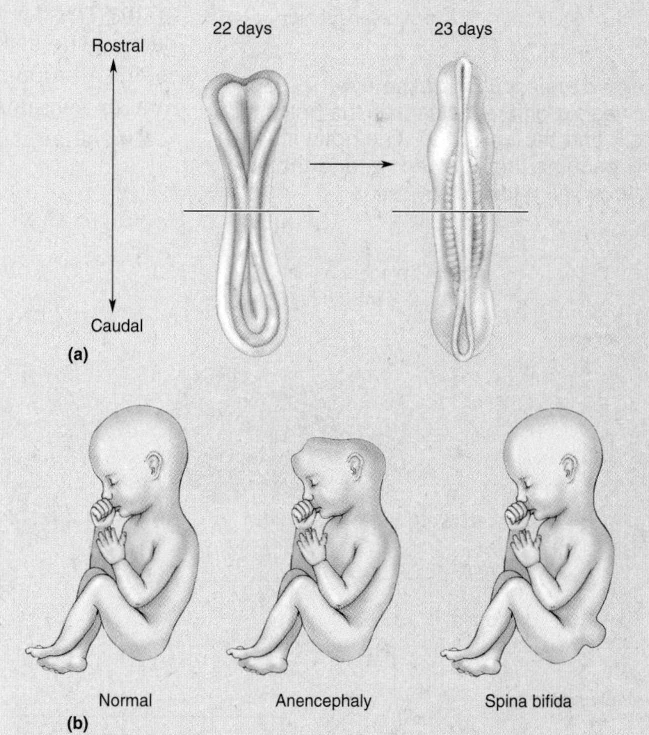

Figure B
(a) Neural tube closure. **(b)** Neural tube defects.

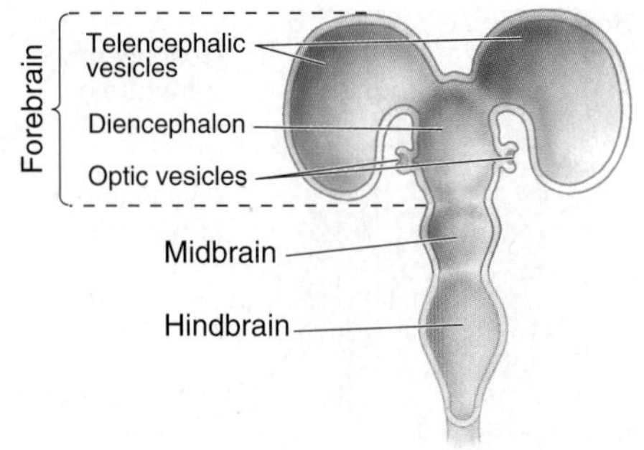

▲ FIGURE 7.12
The secondary brain vesicles of the forebrain. The forebrain differentiates into the paired telencephalic and optic vesicles, and the diencephalon. The optic vesicles develop into the eyes.

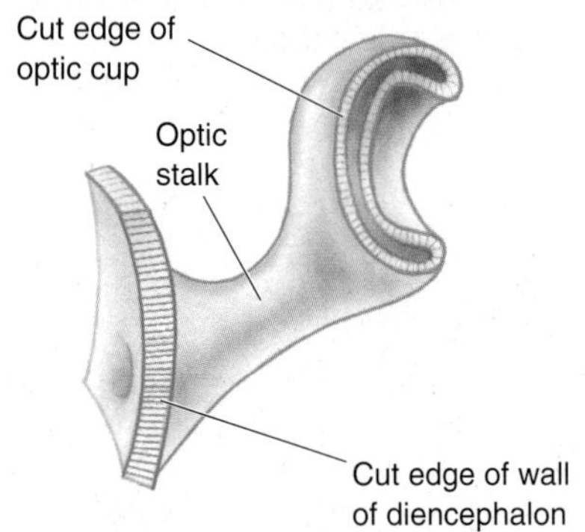

▲ FIGURE 7.13
Early development of the eye. The optic vesicle differentiates into the optic stalk and the optic cup. The optic stalk will become the optic nerve, and the optic cup will become the retina.

another vesicle called the *mesencephalon*, or **midbrain**. Caudal to this is the third primary vesicle, the *rhombencephalon*, or **hindbrain**. The rhombencephalon connects with the caudal neural tube, which gives rise to the spinal cord.

Differentiation of the Forebrain

The next important developments occur in the forebrain, where secondary vesicles sprout off on both sides of the prosencephalon. The secondary vesicles are the *optic vesicles* and the *telencephalic vesicles*. The central structure that remains after the secondary vesicles have sprouted off is called the **diencephalon**, or "between brain" (Figure 7.12). Thus, the forebrain at this stage consists of the two optic vesicles, the two telencephalic vesicles, and the diencephalon.

The optic vesicles grow and invaginate (fold in) to form the optic stalks and the optic cups, which will ultimately become the *optic nerves* and the two *retinas* in the adult (Figure 7.13). The important point is that the retina at the back of the eye, and the optic nerve containing the axons that connect the eye to the diencephalon and midbrain, are part of the brain, not the PNS.

Differentiation of the Telencephalon and Diencephalon. The telencephalic vesicles together form the **telencephalon**, or "endbrain," consisting of the two cerebral hemispheres. The telencephalon continues to develop in four ways. (1) The telencephalic vesicles grow posteriorly so that they lie over and lateral to the diencephalon (Figure 7.14a). (2) Another pair of vesicles sprout off the ventral surfaces of the cerebral hemispheres, giving rise to the **olfactory bulbs** and related structures that participate in the sense of smell (Figure 7.14b). (3) The cells of the walls of the telencephalon divide and differentiate into various structures. (4) White matter systems develop, carrying axons to and from the neurons of the telencephalon.

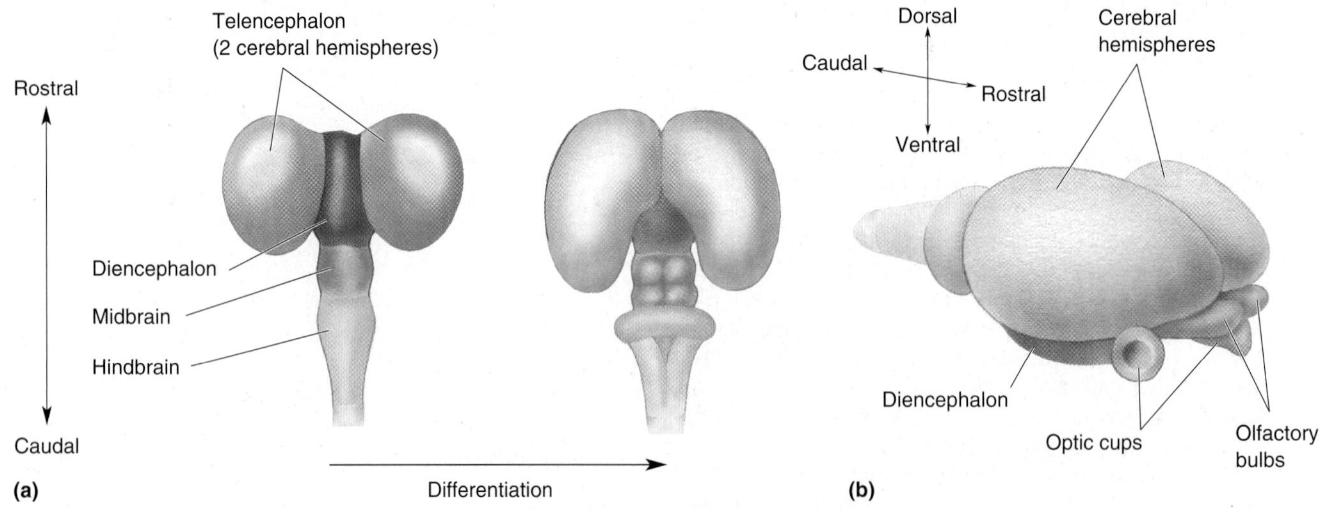

▲ FIGURE 7.14
Differentiation of the telencephalon. **(a)** As development proceeds, the cerebral hemispheres swell and grow posteriorly and laterally to envelop the diencephalon. **(b)** The olfactory bulbs sprout off the ventral surfaces of each telencephalic vesicle.

Figure 7.15 shows a coronal section through the primitive mammalian forebrain, to illustrate how the different parts of the telencephalon and diencephalon differentiate and fit together. Notice that the two cerebral hemispheres lie above and on either side of the diencephalon, and that the ventral–medial surfaces of the hemispheres have fused with the lateral surfaces of the diencephalon (Figure 7.15a).

The fluid-filled spaces within the cerebral hemispheres are called the **lateral ventricles**, and the space at the center of the diencephalon is called the **third ventricle** (Figure 7.15b). The paired lateral ventricles are a key landmark in the adult brain: Whenever you see paired fluid-filled ventricles in a brain section, you know that the tissue surrounding them is in the telencephalon. The elongated, slit-like appearance of the third ventricle in cross section is also a useful feature for identifying the diencephalon.

Notice in Figure 7.15 that the walls of the telencephalic vesicles appear swollen due to the proliferation of neurons. These neurons form two different types of gray matter in the telencephalon: the **cerebral cortex** and the **basal telencephalon**. Likewise, the diencephalon differentiates into two structures: the **thalamus** and the **hypothalamus** (Figure 7.15c). The thalamus, nestled deep inside the forebrain, gets its name from the Greek word for "inner chamber."

The neurons of the developing forebrain extend axons to communicate with other parts of the nervous system. These axons bundle together to form three major white matter systems: the cortical white matter, the corpus callosum, and the internal capsule (Figure 7.15d). The **cortical white matter** contains all the axons that run to and from the neurons in the cerebral cortex. The **corpus callosum** is continuous with the cortical white matter and forms an axonal bridge that links cortical neurons of the two cerebral hemispheres. The cortical white matter is also continu-

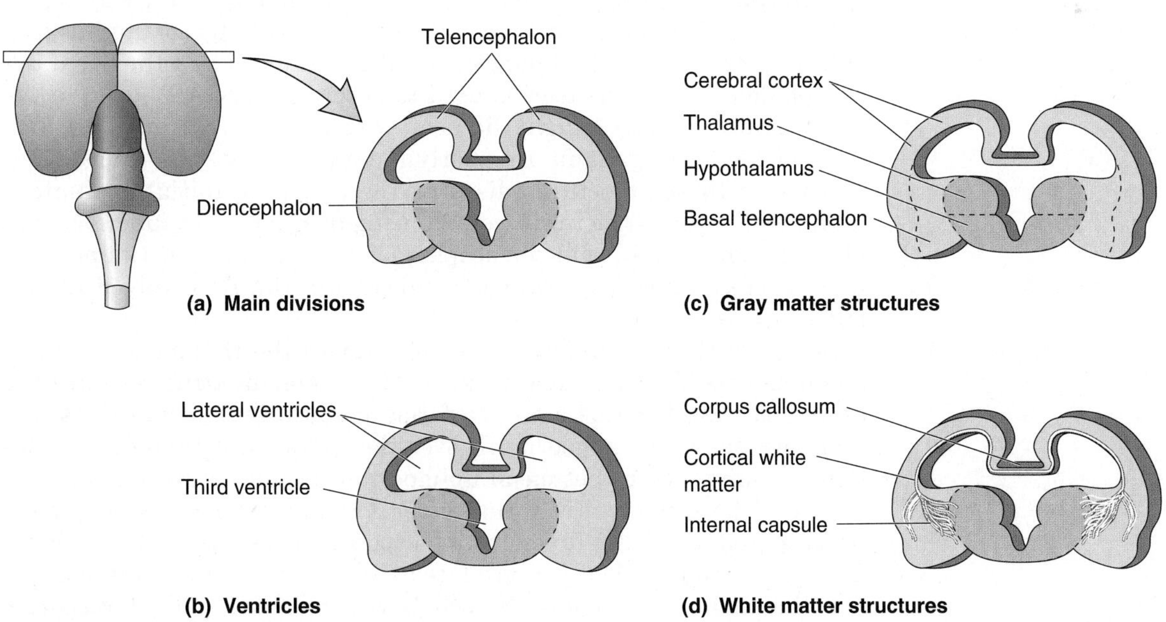

▲ FIGURE 7.15

Structural features of the forebrain. **(a)** A coronal section through the primitive forebrain, showing the two main divisions: the telencephalon and the diencephalon. **(b)** Ventricles of the forebrain. **(c)** Gray matter of the forebrain. **(d)** White matter structures of the forebrain.

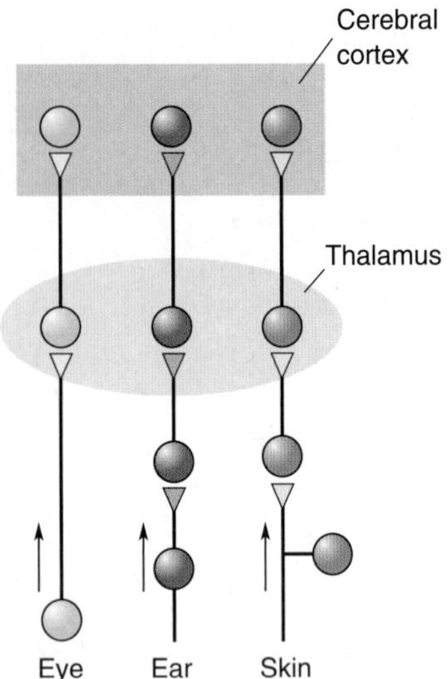

▲ FIGURE 7.16
The thalamus: gateway to the cerebral cortex. The sensory pathways from the eye, ear, and skin all relay in the thalamus before terminating in the cerebral cortex. The arrows indicate the direction of information flow.

ous with the **internal capsule**, which links the cortex with the brain stem, particularly the thalamus.

Forebrain Structure-Function Relationships. The forebrain is the seat of perceptions, conscious awareness, cognition, and voluntary action. All this depends on extensive interconnections with the sensory and motor neurons of the brain stem and spinal cord.

Arguably the most important structure in the forebrain is the cerebral cortex. As we will see later in this chapter, the cortex is the brain structure that has expanded the most over the course of human evolution. Cortical neurons receive sensory information, form perceptions of the outside world, and command voluntary movements.

Neurons in the olfactory bulbs receive information from cells that sense chemicals in the nose (odors), and relay this information caudally to a part of the cerebral cortex for further analysis. Information from the eyes, ears, and skin is also brought to the cerebral cortex for analysis. However, each of the sensory pathways serving vision, audition (hearing), and somatic sensation relays (i.e., synapses upon neurons) in the thalamus en route to the cortex. Thus, the thalamus is often referred to as the gateway to the cerebral cortex (Figure 7.16).

Thalamic neurons send axons to the cortex via the internal capsule. As a general rule, the axons of each internal capsule carry information to the cortex about the contralateral side of the body. Therefore, if a thumbtack entered the *right* foot, it would be relayed to the *left* cortex by the *left* thalamus via axons in the *left* internal capsule. But how does the right foot know what the left foot is doing? One important way is by communication between the hemispheres via the axons in the corpus callosum.

Cortical neurons also send axons through the internal capsule, back to the brain stem. Some cortical axons course all the way to the spinal cord, forming the corticospinal tract. This is one important way cortex can command voluntary movement. Another way is by communicating with neurons in the basal ganglia, a collection of cells in the basal telencephalon. The term *basal* is used to describe structures deep in the brain, and the basal ganglia lie deep within the cerebrum. The functions of the basal ganglia are poorly understood, but it is known that damage to these structures disrupts the ability to initiate voluntary movement. Other structures, contributing to other brain functions, are also present in the basal telencephalon. For example, in Chapter 18, we'll discuss a structure called the *amygdala* that is involved in fear and emotion.

Although the hypothalamus lies just under the thalamus, functionally it is more closely related to certain telencephalic structures like the amygdala. The hypothalamus performs many primitive functions and therefore has not changed much over the course of mammalian evolution. "Primitive" does not mean unimportant or uninteresting, however. The hypothalamus controls the visceral (autonomic) nervous system, which regulates bodily functions in response to the needs of the organism. For example, when you are faced with a threatening situation, the hypothalamus orchestrates the body's visceral fight-or-flight response. Hypothalamic commands to the ANS will lead to (among other things) an increase in the heart rate, increased blood flow to the muscles for escape, and even the standing of your hair on end. Conversely, when you're relaxing after Sunday brunch, the hypothalamus ensures that the brain is well nourished via commands to the ANS, which will increase peristalsis (movement of material through the gastrointestinal tract) and redirect

SELF-QUIZ

Listed below are derivatives of the forebrain that we have discussed. Be sure you know what each of these terms means.

Primary Vesicle	Secondary Vesicle	Some Adult Derivatives
Forebrain (prosencephalon)	Optic vesicle	Retina
		Optic nerve
	Thalamus (diencephalon)	Dorsal thalamus
		Hypothalamus
		Third ventricle
	Telencephalon	Olfactory bulb
		Cerebral cortex
		Basal telencephalon
		Corpus callosum
		Cortical white matter
		Internal capsule

blood to your digestive system. The hypothalamus also plays a key role in motivating animals to find food, drink, and sex in response to their needs. Aside from its connections to the ANS, the hypothalamus also directs bodily responses via connections with the pituitary gland located below the diencephalon. This gland communicates with many parts of the body by releasing hormones into the bloodstream.

Differentiation of the Midbrain

Unlike the forebrain, the midbrain differentiates relatively little during subsequent brain development (Figure 7.17). The dorsal surface of the mesencephalic vesicle becomes a structure called the **tectum** (Latin for "roof").

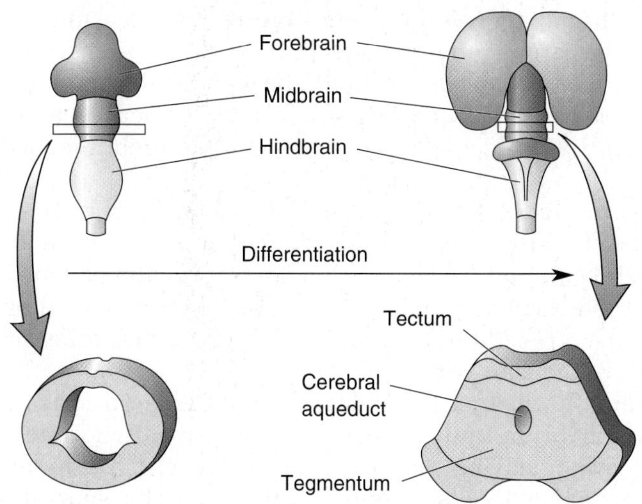

▲ FIGURE 7.17
Differentiation of the midbrain. The midbrain differentiates into the tectum and the tegmentum. The CSF-filled space at the core of the midbrain is the cerebral aqueduct. (Drawings are not to scale.)

The floor of the midbrain becomes the **tegmentum**. The CSF-filled space in between constricts into a narrow channel called the **cerebral aqueduct**. The aqueduct connects rostrally with the third ventricle of the diencephalon. Because it is small and circular in cross section, the cerebral aqueduct is a good landmark for identifying the midbrain.

Midbrain Structure-Function Relationships. For such a seemingly simple structure, the functions of the midbrain are remarkably diverse. Besides serving as a conduit for information passing from the spinal cord to the forebrain and vice versa, the midbrain contains neurons that contribute to sensory systems, the control of movement, and several other functions.

The midbrain contains axons descending from the cerebral cortex to the brain stem and the spinal cord. For example, the corticospinal tract courses through the midbrain en route to the spinal cord. Damage to this tract in the midbrain on one side produces a loss of voluntary control of movement on the opposite side of the body.

The tectum differentiates into two structures: the superior colliculus and the inferior colliculus. The *superior colliculus* receives direct input from the eye, so it is also called the *optic tectum*. One function of the optic tectum is to control eye movements, which it does via synaptic connections with the motor neurons that innervate the eye muscles. Some of the axons that supply the eye muscles originate in the midbrain, bundling together to form cranial nerves III and IV.

The *inferior colliculus* also receives sensory information but from the ear instead of the eye. The inferior colliculus serves as an important relay station for auditory information en route to the thalamus.

The tegmentum is one of the most colorful regions of the brain because it contains both the substantia nigra (the black substance) and the red nucleus. These two cell groups are involved in the control of voluntary movement. Other cell groups scattered in the midbrain have axons that project widely throughout much of the CNS and function to regulate consciousness, mood, pleasure, and pain.

Differentiation of the Hindbrain

The hindbrain differentiates into three important structures: the cerebellum, the **pons**, and the **medulla oblongata**—also simply called the **medulla**. The cerebellum and pons develop from the rostral half of the hindbrain (called the *metencephalon*); the medulla develops from the caudal half (called the *myelencephalon*). The CSF-filled tube becomes the **fourth ventricle**, which is continuous with the cerebral aqueduct of the midbrain.

At the three-vesicle stage, the rostral hindbrain in cross section is a simple tube. In subsequent weeks, the tissue along the dorsal–lateral wall of the tube, called the *rhombic lip*, grows dorsally and medially until it fuses with its twin on the other side. The resulting flap of brain tissue grows into the cerebellum. The ventral wall of the tube differentiates and swells to form the pons (Figure 7.18).

Less dramatic changes occur during the differentiation of the caudal half of the hindbrain into the medulla. The ventral and lateral walls of this region swell, leaving the roof covered only with a thin layer of nonneuronal ependymal cells (Figure 7.19). Along the ventral surface of each side of the medulla runs a major white matter system. Cut in cross section, these bundles of axons appear somewhat triangular in shape, explaining why they are called the *medullary pyramids*.

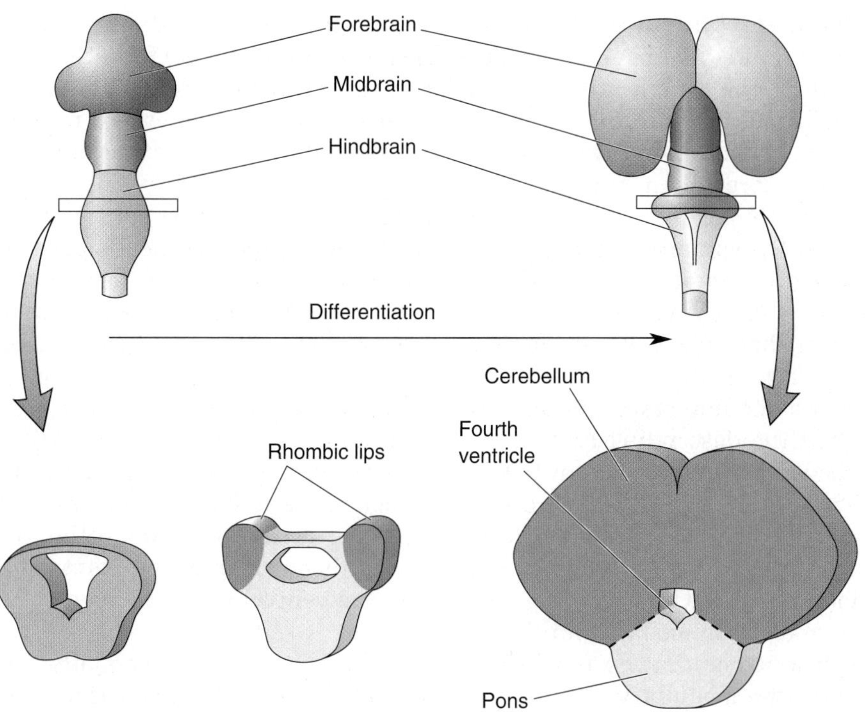

▲ FIGURE 7.18
Differentiation of the rostral hindbrain. The rostral hindbrain differentiates into the cerebellum and pons. The cerebellum is formed by the growth and fusion of the rhombic lips. The CSF-filled space at the core of the hindbrain is the fourth ventricle. (Drawings are not to scale.)

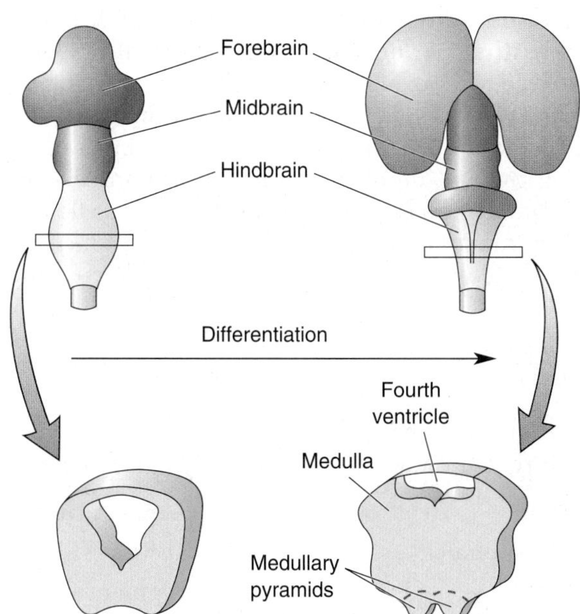

▲ FIGURE 7.19
Differentiation of the caudal hindbrain. The caudal hindbrain differentiates into the medulla. The medullary pyramids are bundles of axons coursing caudally toward the spinal cord. The CSF-filled space at the core of the medulla is the fourth ventricle. (Drawings are not to scale.)

Hindbrain Structure-Function Relationships. Like the midbrain, the hindbrain is an important conduit for information passing from the forebrain to the spinal cord, and vice versa. In addition, neurons of the hindbrain contribute to the processing of sensory information, the control of voluntary movement, and regulation of the autonomic nervous system.

The cerebellum, the "little brain," is an important movement control center. It receives massive axonal inputs from the spinal cord and the pons. The spinal cord inputs provide information about the body's position in space. The inputs from the pons relay information from the cerebral cortex, specifying the goals of intended movements. The cerebellum compares these types of information and calculates the sequences of muscle contractions that are required to achieve the movement goals. Damage to the cerebellum results in uncoordinated and inaccurate movements.

Of the descending axons passing through the midbrain, over 90%—about 20 million axons in the human—synapse on neurons in the pons. The pontine cells relay all this information to the cerebellum on the opposite site. Thus, the pons serves as a massive switchboard connecting the cerebral cortex to the cerebellum. (The word *pons* is from the Latin word for "bridge.") The pons bulges out from the ventral surface of the brain stem to accommodate all this circuitry.

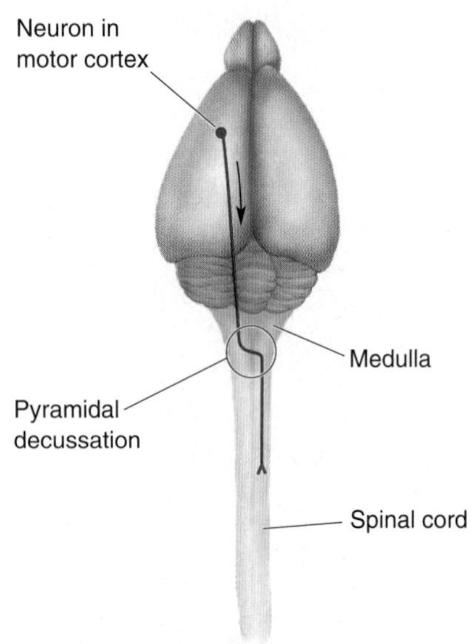

▲ FIGURE 7.20
The pyramidal decussation. The corticospinal tract crosses from one side to the other in the medulla.

The axons that do not terminate in the pons continue caudally and enter the medullary pyramids. Most of these axons originate in the cerebral cortex and are part of the corticospinal tract. Thus, "pyramidal tract" is often used as a synonym for corticospinal tract. Near where the medulla joins with the spinal cord, each pyramidal tract crosses from one side of the midline to the other. A crossing of axons from one side to the other is known as a *decussation*, and this one is called the *pyramidal decussation*. The crossing of axons in the medulla explains why the cortex of one side of the brain controls movements on the opposite side of the body (Figure 7.20).

In addition to the white matter systems passing through, the medulla contains neurons that perform many different sensory and motor functions. For example, the axons of the auditory nerves, bringing auditory information from the ears, synapse on cells in the cochlear nuclei of the medulla. The cochlear nuclei project axons to a number of different structures, including the tectum of the midbrain (the inferior colliculus, discussed above). Damage to the cochlear nuclei leads to deafness.

Other sensory functions of the medulla include touch and taste. The medulla contains neurons that relay somatic sensory information from

SELF-QUIZ

Listed below are derivatives of the midbrain and hindbrain that we have discussed. Be sure you know what each of these terms means.

Primary Vesicle	**Some Adult Derivatives**
Midbrain (mesencephalon)	Tectum
	Tegmentum
	Cerebral aqueduct
Hindbrain (rhombencephalon)	Cerebellum
	Pons
	Fourth ventricle
	Medulla

the spinal cord to the thalamus. Destruction of the cells leads to anesthesia (loss of feeling). Other neurons relay gustatory (taste) information from the tongue to the thalamus. And among the motor neurons in the medulla are cells that control the tongue muscles via cranial nerve XII. (So think of the medulla next time you stick out your tongue!)

Differentiation of the Spinal Cord

As shown in Figure 7.21, the transformation of the caudal neural tube into the spinal cord is straightforward compared to the differentiation of the brain. With the expansion of the tissue in the walls, the cavity of the tube constricts to form the tiny CSF-filled **spinal canal**.

Cut in cross section, the gray matter of the spinal cord (where the neurons are) has the appearance of a butterfly. The upper part of the butterfly's wing is the **dorsal horn**, and the lower part is the **ventral horn**. The gray matter between the dorsal and ventral horns is called the *intermediate zone*. Everything else is white matter, consisting of columns of axons that run up and down the spinal cord. Thus, the bundles of axons running along the dorsal surface of the cord are called the *dorsal columns*, the bundles of axons lateral to the spinal gray matter on each side are called the *lateral columns*, and the bundles on the ventral surface are called the *ventral columns*.

Spinal Cord Structure-Function Relationships. As a general rule, dorsal horn cells receive sensory inputs from the dorsal root fibers, ventral horn cells project axons into the ventral roots that innervate muscles,

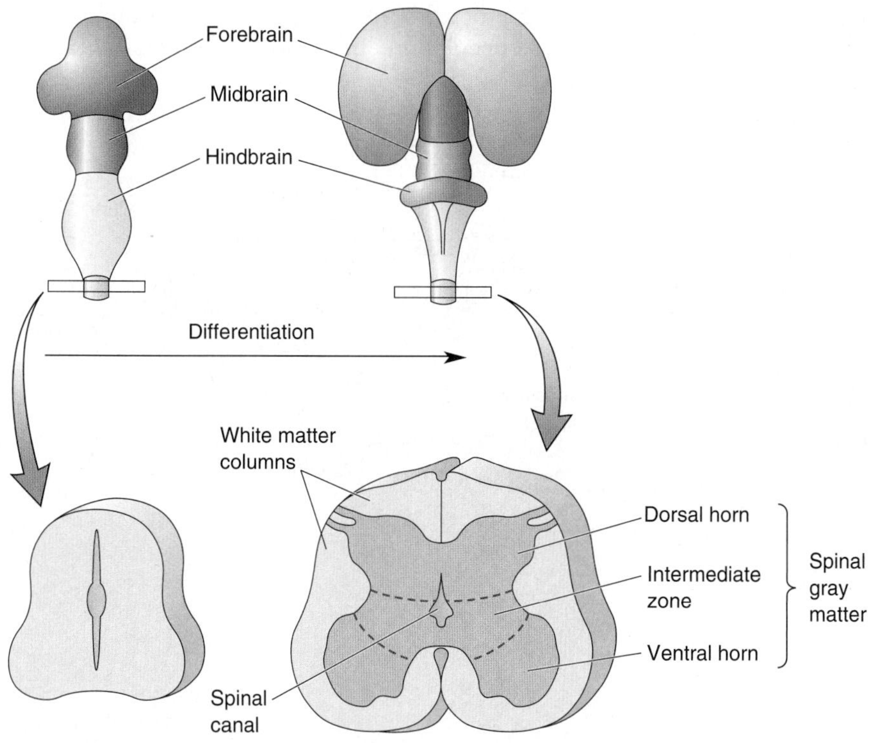

▲ FIGURE 7.21
Differentiation of the spinal cord. The butterfly-shaped core of the spinal cord is gray matter, divisible into dorsal and ventral horns, and an intermediate zone. Surrounding the gray matter are white matter columns running rostrocaudally, up and down the cord. The narrow CSF-filled channel is the spinal canal. (Drawings are not to scale.)

and intermediate zone cells are interneurons that shape motor outputs in response to sensory inputs and descending commands from the brain.

The large dorsal column contains axons that carry somatic sensory (touch) information up the spinal cord toward the brain. It's like a superhighway that speeds information from the ipsilateral side of the body up to nuclei in the medulla. The postsynaptic neurons in the medulla give rise to axons that decussate and ascend to the thalamus on the contralateral side. This crossing of axons in the medulla explains why touching the left side of the body is sensed by the right side of the brain.

The lateral column contains the axons of the descending corticospinal tract, which also cross from one side to the other in the medulla. These axons innervate the neurons of the intermediate zone and ventral horn and communicate the signals that control voluntary movement.

There are at least a half-dozen tracts that run in the columns of each side of the spinal cord. Most are one-way and bring information to or from the brain. Thus, the spinal cord is the major conduit of information from the skin, joints, and muscles to the brain, and vice versa. However, the spinal cord is also much more than that. The neurons of the spinal gray matter begin the analysis of sensory information, play a critical role in coordinating movements, and orchestrate simple reflexes (such as jerking away your foot from a thumbtack).

Putting the Pieces Together

We have discussed the development of different parts of the CNS: the telencephalon, diencephalon, midbrain, hindbrain, and spinal cord. Now let's put all the individual pieces together to make a whole central nervous system.

Figure 7.22 is a highly schematic illustration that captures the basic organizational plan of the CNS of all mammals, including humans. The paired hemispheres of the telencephalon surround the lateral ventricles.

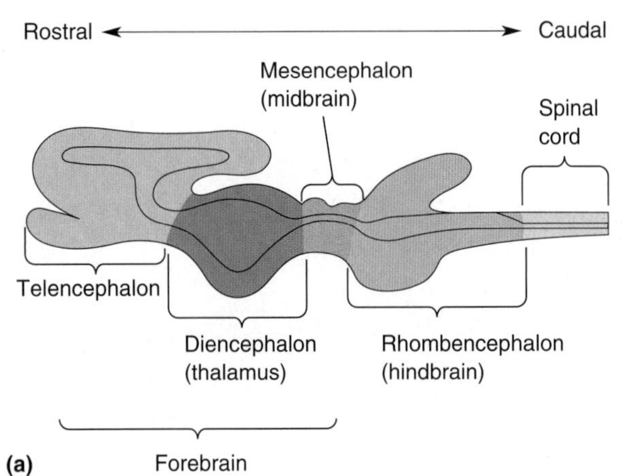

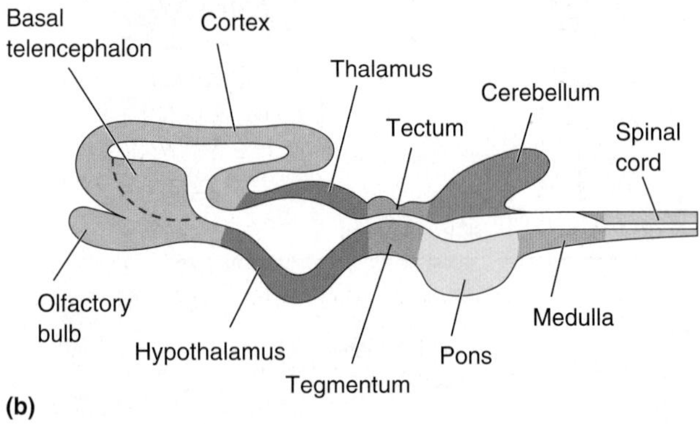

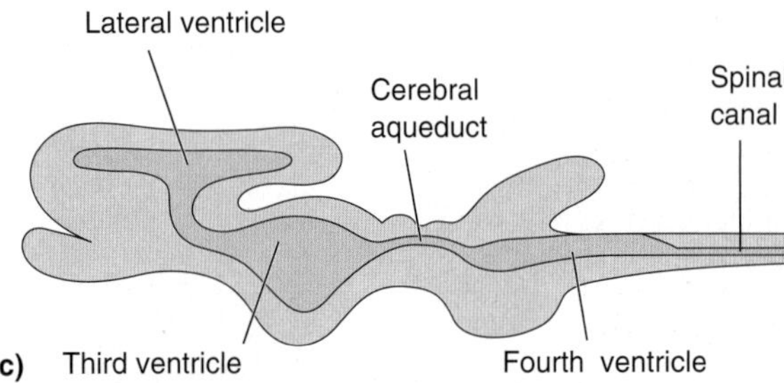

▲ FIGURE 7.22
The "brainship Enterprise." **(a)** The basic plan of the mammalian brain, with the major subdivisions indicated. **(b)** Major structures within each division of the brain. Note that the telencephalon consists of two hemispheres, although only one is illustrated. **(c)** The ventricular system.

TABLE 7.3 **The Ventricular System of the Brain**

Component	Related Brain Structures
Lateral ventricles	Cerebral cortex Basal telencephalon
Third ventricle	Thalamus Hypothalamus
Cerebral aqueduct	Tectum Midbrain tegmentum
Fourth ventricle	Cerebellum Pons Medulla

Dorsal to the lateral ventricles, at the surface of the brain, lies the cortex. Ventral and lateral to the lateral ventricles lies the basal telencephalon. The lateral ventricles are continuous with the third ventricle of the diencephalon. Surrounding this ventricle are the thalamus and the hypothalamus. The third ventricle is continuous with the cerebral aqueduct. Dorsal to the aqueduct is the tectum. Ventral to the aqueduct is the midbrain tegmentum. The aqueduct connects with the fourth ventricle that lies at the core of the hindbrain. Dorsal to the fourth ventricle sprouts the cerebellum. Ventral to the fourth ventricle lie the pons and the medulla.

You should see by now that finding your way around the brain is easy if you can identify which parts of the ventricular system are in the neighborhood (Table 7.3). Even in the complicated human brain, the ventricular system holds the key to understanding brain structure.

Special Features of the Human CNS

So far, we've explored the basic plan of the CNS as it applies to all mammals. Figure 7.23 compares the brains of the rat and the human. You can see immediately that there are indeed many similarities but also some obvious differences.

Let's start by reviewing the similarities. The dorsal view of both brains reveals the paired hemispheres of the telencephalon (Figure 7.23a). A midsagittal view of the two brains shows the telencephalon extending rostrally from the diencephalon (Figure 7.23b). The diencephalon surrounds the third ventricle, the midbrain surrounds the cerebral aqueduct, and the cerebellum, pons, and medulla surround the fourth ventricle. Notice how the pons swells below the cerebellum, and how structurally elaborate the cerebellum is.

Now let's consider some of the structural differences between the rat and human brains. Figure 7.23a reveals a striking one: the many convolutions on the surface of the human cerebrum. The grooves in the surface of the cerebrum are called **sulci** (singular: **sulcus**), and the bumps are called **gyri** (singular: **gyrus**). Remember, the thin sheet of neurons that lies just under the surface of the cerebrum is the cerebral cortex. Sulci and gyri result from the tremendous expansion of the surface area of the cerebral cortex during human fetal development. The adult human cortex, measuring about 1100 cm^2, must fold and wrinkle to fit within the confines of the skull. This increase in cortical surface area is one of the "distortions" of the human brain. Clinical and experimental evidence indicates that the cortex is the seat of uniquely human reasoning and cognition. Without cerebral cortex, a person would be blind, deaf, dumb, and unable to initiate voluntary movement. We will take a closer look at the structure of the cerebral cortex in a moment.

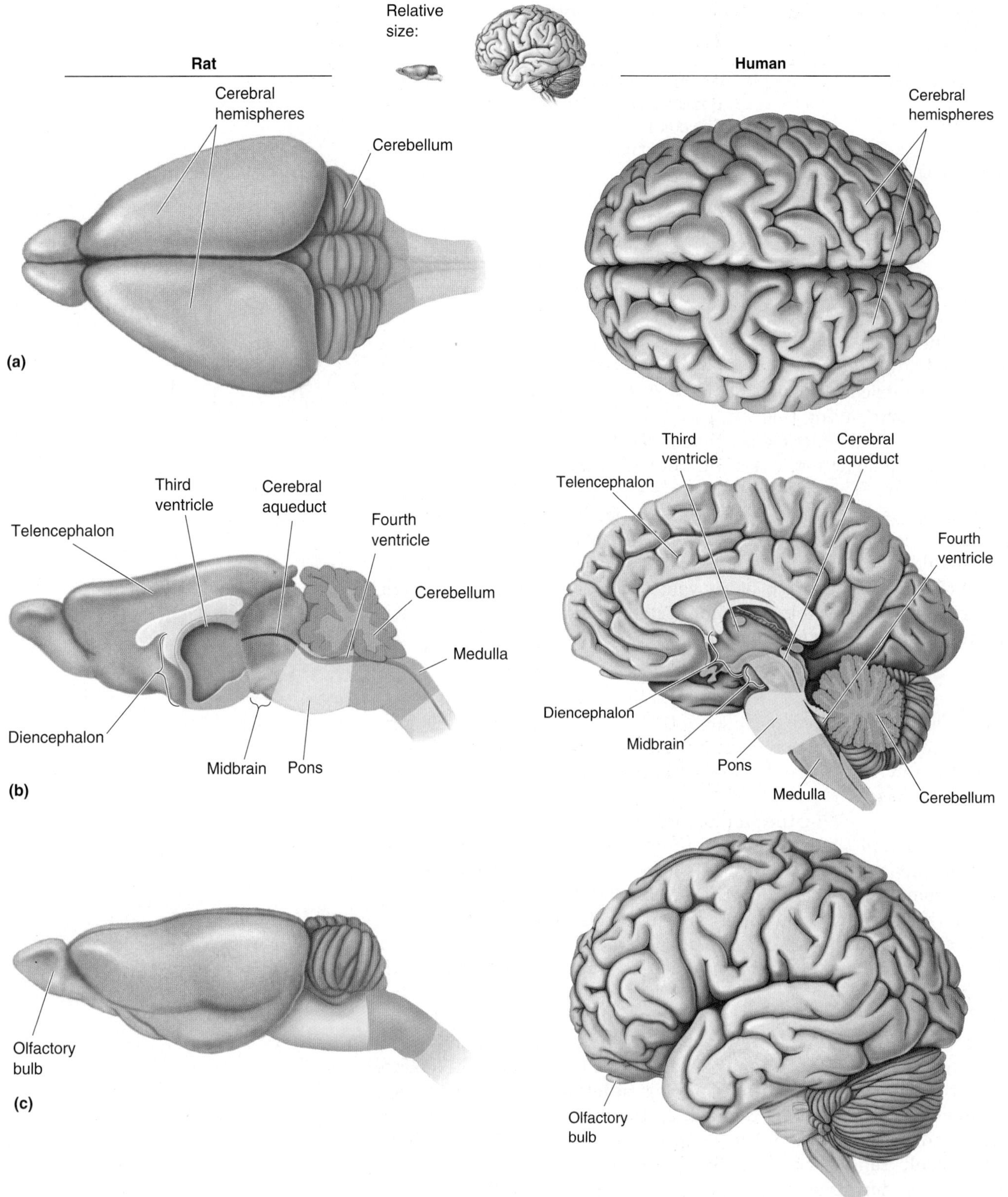

▲ FIGURE 7.23
The rat brain and human brain compared. (a) Dorsal view. (b) Midsagittal view. (c) Lateral view. (Brains are not drawn to the same scale.)

The side views of the rat and human brains in Figure 7.23c reveal further differences in the forebrain. One is the small size of the olfactory bulb in the human relative to the rat. On the other hand, notice again the growth of the cerebral hemisphere in the human. See how the cerebral hemisphere of the human brain arcs posteriorly, ventrolaterally, and then anteriorly to resemble a ram's horn. The tip of the "horn" lies right

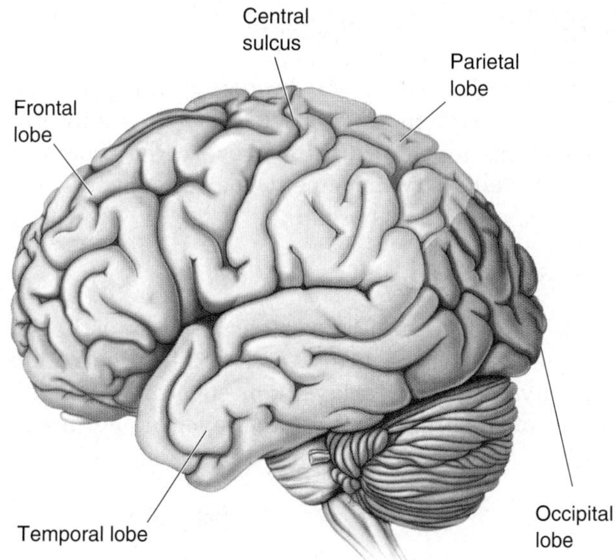

▲ FIGURE 7.24
The lobes of the human cerebrum.

under the temporal bone (temple) of the skull, so this portion of the brain is called the **temporal lobe** (Figure 7.24). Three other lobes (named after skull bones) also describe the parts of human cerebrum. The portion of the cerebrum lying just under the frontal bone of the forehead is called the **frontal lobe**. The deep **central sulcus** marks the posterior border of the frontal lobe, caudal to which lies the **parietal lobe**, under the parietal bone. Caudal to that, at the back of the cerebrum under the occipital bone, lies the **occipital lobe**.

It is important to realize that, despite the disproportionate growth of the cerebrum, the human brain still follows the basic mammalian brain plan laid out during embryonic development. Again, the ventricles are key. Although the ventricular system is distorted, particularly by the growth of the temporal lobes, the relationships that relate the brain to the different ventricles still hold (Figure 7.25).

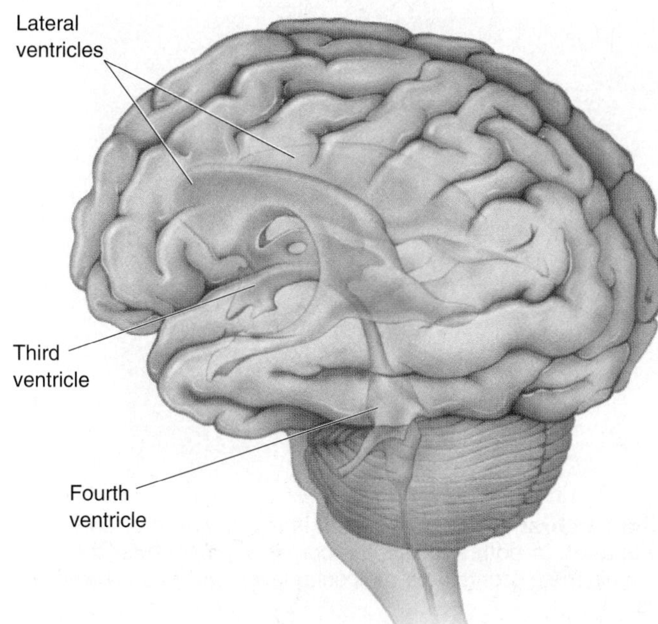

◀ FIGURE 7.25
The human ventricular system. Although the ventricles are distorted by the growth of the brain, the basic relationships of the ventricles to the surrounding brain are the same as those illustrated in Figure 7.22c.

A GUIDE TO THE CEREBRAL CORTEX

Considering its prominence in the human brain, the cerebral cortex deserves further description. As we will see repeatedly in subsequent chapters, the systems in the brain that govern the processing of sensations, perceptions, voluntary movement, learning, speech, and cognition all converge in this remarkable organ.

Types of Cerebral Cortex

Cerebral cortex in the brain of all vertebrate animals has several common features, as shown in Figure 7.26. First, the cell bodies of cortical neurons are always arranged in layers, or sheets, that usually lie parallel to the surface of the brain. Second, the layer of neurons closest to the

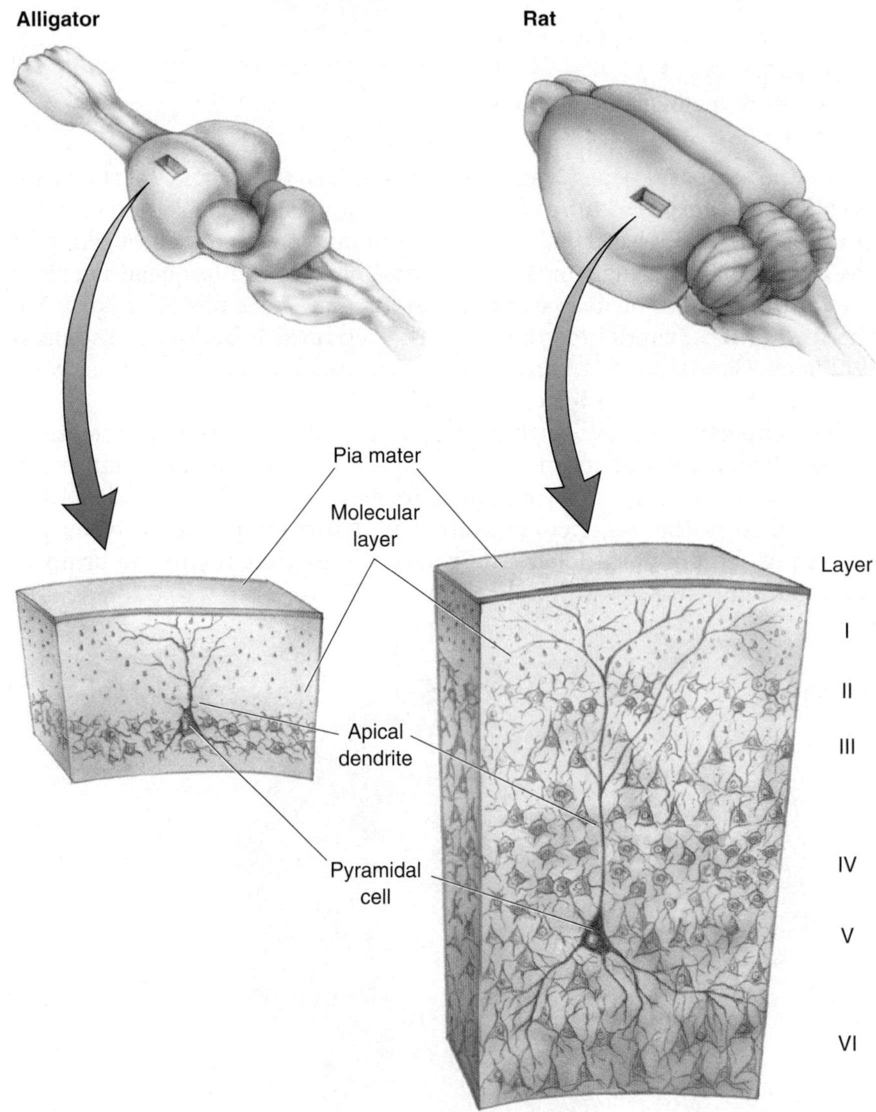

▲ FIGURE 7.26
General features of the cerebral cortex. On the left is the structure of cortex in an alligator; on the right, a rat. In both species, the cortex lies just under the pia mater of the cerebral hemisphere, contains a molecular layer, and has pyramidal cells arranged in layers.

surface (the most superficial cell layer) is separated from the pia mater by a zone that lacks neurons; it is called the molecular layer, or simply *layer I*. Third, at least one cell layer contains pyramidal cells that emit large dendrites, called *apical dendrites*, that extend up to layer I, where they form multiple branches. Thus, we can say that the cerebral cortex has a characteristic cytoarchitecture that distinguishes it, for example, from the nuclei of the basal telencephalon or the thalamus.

Figure 7.27 shows a Nissl-stained coronal section through the caudal telencephalon of a rat brain. You don't need to be Cajal to see that different types of cortex can also be discerned based on cytoarchitecture. Medial to the lateral ventricle is a piece of cortex that is folded onto itself in a peculiar shape. This structure is called the **hippocampus**, which, despite its bends, has only a single cell layer. (The term is from the Greek word for "seahorse.") Connected to the hippocampus ventrally and laterally is another type of cortex that has only two cell layers. It is called the **olfactory cortex** because it is continuous with the olfactory bulb, which sits further anterior. The olfactory cortex is separated by a sulcus, called the *rhinal fissure*, from another more elaborate type of cortex that has many cell layers. This remaining cortex is called **neocortex**. Unlike the hippocampus and olfactory cortex, *neocortex is found only in mammals.*

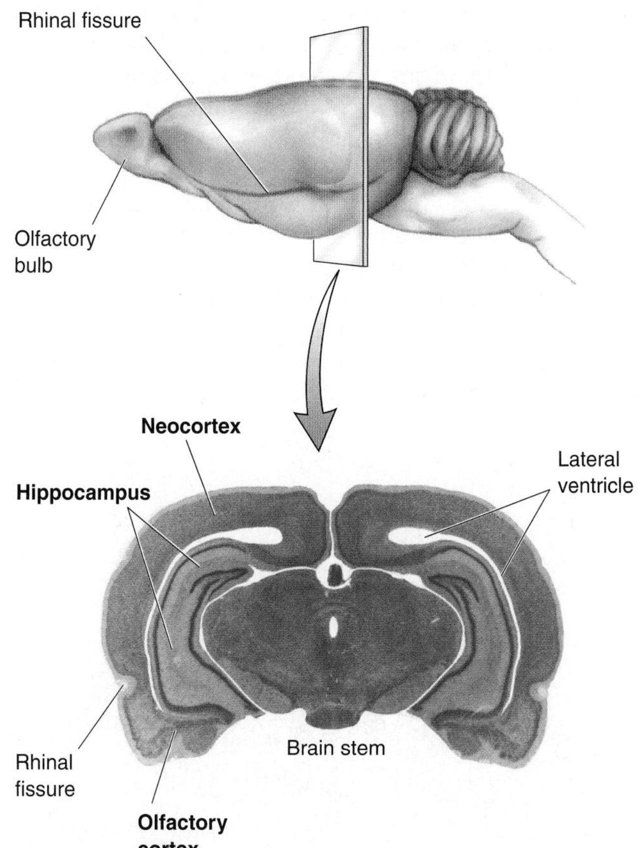

▲ FIGURE 7.27
Three types of cortex in a mammal. In this section of a rat brain, the lateral ventricles lie between the neocortex and the hippocampus on each side. The ventricles are not obvious because they are very long and thin in this region. Below the telencephalon lies the brain stem. What region of brain stem is this, based on the appearance of the fluid-filled space at its core?

Thus, when we said previously that the cerebral cortex has expanded over the course of human evolution, we really meant that the neocortex has expanded. Similarly, when we said that the thalamus is the gateway to the cortex, we meant that it is the gateway to the neocortex. Most neuroscientists are such neocortical chauvinists (ourselves included) that the term *cortex*, if left unqualified, is usually intended to refer to the cerebral neocortex.

In Chapter 8, we will discuss the olfactory cortex in the context of the sense of smell. Further discussion of the hippocampus is reserved until later in this book, when we explore its role in the limbic system (Chapter 18) and in memory and learning (Chapters 24 and 25). The neocortex will figure prominently in our discussions of vision, audition, somatic sensation, and the control of voluntary movement in Part II, so let's examine its structure in more detail.

Areas of Neocortex

Just as cytoarchitecture can be used to distinguish the cerebral cortex from the basal telencephalon, and the neocortex from the olfactory cortex, it can be used to divide the neocortex up into different zones. This is precisely what the famous German neuroanatomist Korbinian Brodmann did at the beginning of the twentieth century. He constructed a **cytoarchitectural map** of the neocortex (Figure 7.28). In this map, each area of cortex having a common cytoarchitecture is given a number. Thus, we have "area 17" at the tip of the occipital lobe, "area 4" just anterior to the central sulcus in the frontal lobe, and so on.

What Brodmann guessed, but could not show, was that cortical areas that look different perform different functions. We now have evidence that this is true. For instance, we can say that area 17 is visual cortex, because it receives signals from a nucleus of the thalamus that is connected to the retina at the back of the eye. Indeed, without area 17, a human is blind. Similarly, we can say that area 4 is motor cortex because neurons in this area project axons directly to the motor neurons of the ventral horn that command muscles to contract. Notice that the different functions of these two areas are specified by their different connections.

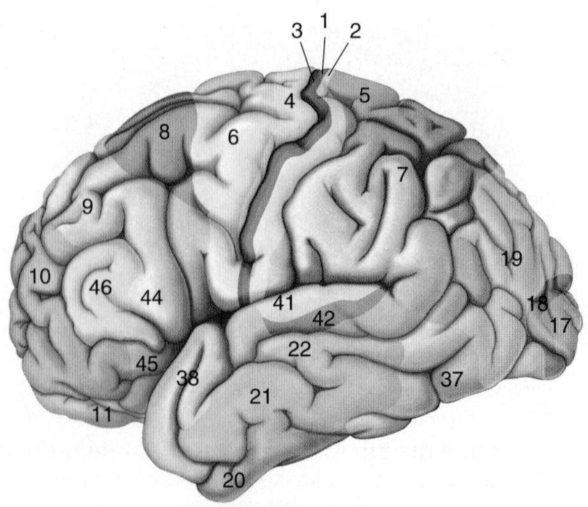

▲ FIGURE 7.28
Brodmann's cytoarchitectural map of the human cerebral cortex.

Neocortical Evolution and Structure-Function Relationships. A problem that has fascinated neuroscientists since the time of Brodmann is how the neocortex has changed over the course of mammalian evolution. The brain is a soft tissue, so there is not a fossil record of the cortex of our early mammalian ancestors. Nonetheless, considerable insight can be gained by comparing the cortex of different living species (see Figure 7.1). The surface area of the cortex varies tremendously among species; for example, a comparison of mouse, monkey, and human cortex reveals differences in size on the order of 1:100:1000. On the other hand, there is little difference in the thickness of the neocortex in different mammals, varying by no more than a factor of two. Thus, we can conclude that the amount of cortex has changed over the course of evolution, but not its basic structure.

The famous Spanish neuroanatomist Santiago Ramon y Cajal, introduced in Chapter 2, wrote in 1899 that "while there are very remarkable differences of organization of certain cortical areas, these points of difference do not go so far as to make impossible the reduction of the cortical structure to a general plan." A challenge that has preoccupied many scientists since then has been to figure out exactly what this plan is. As we will discuss in later chapters, modern thinking is that the smallest functional unit of the neocortex is a cylinder of neurons 2 mm high—the distance from the white matter to the cortical surface—and 0.5 mm in diameter. This cylinder, usually described as a neocortical column, contains on the order of 10,000 neurons and 100 million synapses (approximately 10,000 synapses per neuron). We wish to understand the detailed wiring diagram of how these neurons connect with one another: the **connectome** of the neocortex. This is a tall order because synapses can be identified with confidence only using electron microscopy, which requires very thin (~50 nm) sections of tissue. To give an idea of the magnitude of the challenge, consider the project that South African Nobel laureate Sidney Brenner and his collaborators conducted in the Laboratory of Molecular Biology at the National Institute for Medical Research at Mill Hill, in North London, England. Brenner was convinced that understanding the neural basis of behavior required a circuit diagram, and to tackle this, he chose a simple organism, the 1 mm long flatworm, *Caenorhabditis elegans* (usually abbreviated *C. elegans*)—a far cry from the neocortex, granted, but possibly a tractable problem to solve because the worm has only 302 neurons and about 7000 synapses. Despite this relative simplicity, the "mind of the worm," as they called their project, took over a dozen years to complete. Since the publication of this work in 1986, many of the obstacles to reconstructing a synaptic wiring diagram have begun to yield to advances in technology, including automated sectioning of brain tissue for electron microscopy and computer-aided reconstruction of volumes of tissue from very thin sections (Box 7.5). Although we are not there yet, such advances have spawned optimism that Cajal's dream might soon be realized and not just for the cortex but for the entire brain.

Brodmann proposed that neocortex expanded by the insertion of new areas. Detailed comparisons of cortical structure and function in living species with diverse evolutionary histories suggest that the primordial neocortex of our common mammalian ancestor consisted mainly of three types of cortex. The first type consists of *primary sensory areas*, which are the first to receive signals from the ascending sensory pathways. For example, area 17 is designated as primary visual cortex, or V1, because it receives input from the eyes via a direct path: retina to thalamus to

BOX 7.5 PATH OF DISCOVERY

Connecting with the Connectome

by Sebastian Seung

My career path has been full of zigs and zags. When I was close to completing my Ph.D. in theoretical physics, my advisor sent me to Bell Laboratories in New Jersey for a summer job. As the famous research and development arm of the telecommunication company AT&T, Bell Labs had produced Nobel Prize–winning discoveries and seminal inventions like the transistor. During my summer there, I was supposed to theorize about superconductivity. Instead, I met Haim Sompolinsky, who had just arrived from Israel for a sabbatical year. Haim had previously developed mathematical models of interacting particles in a magnetic field and was now enthusiastically moving on to interacting neurons. I was hooked by this theory of neural networks, so I followed Haim to Jerusalem for post-doctoral training. We applied ideas from statistical physics to understand when artificial neural networks—that is, networks of computational units modeled loosely after neurons—learn not gradually but suddenly, as if with an "aha!" moment. When not engaged in lengthy mathematical calculations, I also learned to speak Hebrew and how to make hummus.

After two years in Jerusalem, I returned to Bell Labs. In the organizational chart, all company departments had a five-digit number. I belonged to Theoretical Physics, Department 11111. That meant we were the smartest of the smart, right? But Bell Labs was under pressure to be useful—to produce not Nobel Prizes but more revenue for AT&T—and some quipped, "The more 1's in your department number, the more useless you are."

Still, Bell Labs was like Disneyland for the mind, jam-packed with researchers working on a dizzying variety of interesting topics. Many left their office doors open, so you could pop in and ask questions any time. Experimental physicists in the Biological Computation Department were pioneering the use of functional MRI and advanced microscopy to observe neural activity. At the other end of the building were computer scientists working in the field of machine learning—a process by which a computer can "learn" from experience rather than being explicitly programmed.

Soon I was inventing algorithms that enabled artificial neural networks to learn, and I developed a mathematical theory of a hindbrain neural circuit called the *oculomotor integrator*. I continued this work after moving to the Massachusetts Institute of Technology as an assistant professor. In 2004, I was tenured and promoted to the rank of full professor. I should have been happy, but instead, I felt depressed. My theory of the oculomotor integrator was interesting and even plausible, judging from experimental tests by my collaborator David Tank at Princeton. But others were continuing to propose alternative theories, and the field showed no sign of converging on a consensus. My theory assumed the existence of recurrent connections between integrator neurons. Yet after a decade of study, I didn't even know for sure whether integrator neurons were connected to each other at all!

When I complained to David, he suggested that I change my research focus. In the 1990s, we had both worked at Bell Labs with Winfried Denk, who had since moved to the Max Planck Institute of Biomedical Research in Heidelberg. There Winfried had built an ingenious automated device that could image the face of a block of brain tissue, and then shave off a thin slice to expose a new face. By repeatedly cutting deeper and deeper into the block, the device could acquire a three-dimensional (3D) image of brain tissue. Because Winfried's device used an electron microscope, the image was sharp enough to reveal all synapses, as well as all neurons in the tissue. (Recall that Cajal could visualize only a small number of neurons with his light microscope and the

cortex. The second type of neocortex consists of *secondary sensory areas*, so designated because of their heavy interconnections with the primary sensory areas. The third type of cortex consists of *motor areas*, which are intimately involved with the control of voluntary movement. These cortical areas receive inputs from thalamic nuclei that relay information from the basal telencephalon and the cerebellum, and they send outputs to motor control neurons in the brain stem and spinal cord. For example, because cortical area 4 sends outputs directly to motor neurons in the ventral horn of the spinal cord, it is designated the primary motor cortex, or M1. It is believed that the common mammalian ancestor had

Golgi stain, and could not see synapses at all.) In principle, from such an image it would be possible to reconstruct the "wiring diagram" of a piece of brain tissue by tracing the paths of neural branches, the "wires" of the brain, and locating the synapses.

The catch was the huge amount of image data that had to be analyzed. Winfried's device had the potential to generate a petabyte of data from a cubic millimeter volume, the equivalent of a billion pictures in your digital photo album. Manual reconstruction of the wiring diagram would be prohibitively time-consuming. I decided to work on the problem of speeding up image analysis by computer automation. In 2006, my laboratory began collaborating with Winfried's laboratory to apply the methods of machine learning to his images. This computational method significantly improved the speed and accuracy of 3D reconstruction of neurons. However, the method still made errors, so it could not completely replace human intelligence. In 2008, we started creating software that would enable humans to work with the machines to reconstruct neural circuits. This eventually turned into the "citizen science" project called *EyeWire*, which has registered over 150,000 players from 100 countries since its 2012 launch (http://blog.eyewire.org/about). "EyeWirers" analyze images by playing a game resembling a 3D coloring book. By coloring, they reconstruct the branches of neurons, which are like the "wires" of the brain (Figure A).

In 2014, *Nature* published the first EyeWire-assisted discovery: a new wiring diagram for a neural circuit in the retina. The discovery suggests a new solution to a problem that has eluded neuroscientists for 50 years: How does the retina detect moving visual stimuli? Researchers are conducting experiments to test our new theory, and only time will tell whether it's correct. But it's already clear that our computational technologies for reconstructing connectivity are accelerating progress towards understanding how neural circuits function. I'm now at the Princeton Neuroscience Institute, where I am continuing to work towards my dream of reconstructing a connectome, a wiring diagram of an entire brain.

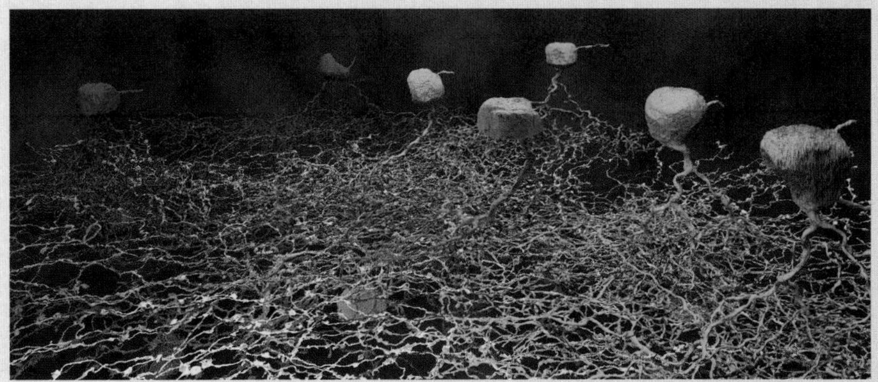

Figure A
Seven neurons in a small volume of retina with their dendrites reconstructed from electron microscopic images. The neurites belonging to each neuron are colored differently. (Source: Courtesy of Dr. Sebastian Seung, Princeton University, and Kris Krug, Pop Tech.)

on the order of about 20 different areas that could be assigned to these three categories.

Figure 7.29 shows views of the brain of a rat, a cat, and a human, with the primary sensory and motor areas identified. It is plain to see that when we speak of the expansion of the cortex in mammalian evolution, what has expanded is the region that lies in between these areas. Much of the "in-between" cortex reflects expansion of the number of secondary sensory areas devoted to the analysis of sensory information. For example, in primates that depend heavily on vision, such as humans, the number of secondary visual areas has been estimated to be between 20

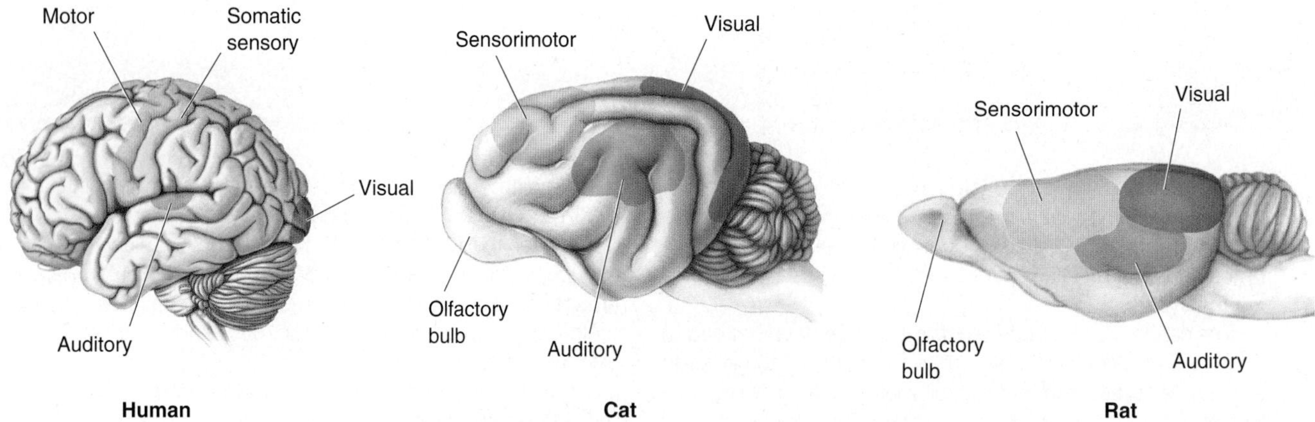

▲ FIGURE 7.29
A lateral view of the cerebral cortex in three species. Notice the expansion of the human cortex that is neither strictly primary sensory nor strictly motor.

and 40. However, even after we have assigned primary sensory, motor, and secondary sensory functions to large regions of cortex, a considerable amount of area remains in the human brain, particularly in the frontal and temporal lobes. These are the *association areas* of cortex. Association cortex is a more recent evolutionary development, a noteworthy characteristic of the primate brain. The emergence of the "mind"—our unique ability to interpret behavior (our own and that of others) in terms of unobservable mental states, such as desires, intentions, and beliefs—correlates best with the expansion of the frontal cortex. Indeed, as we will see in Chapter 18, lesions of the frontal cortex can profoundly alter an individual's personality.

CONCLUDING REMARKS

Although we have covered a lot of new ground in this chapter, we have only scratched the surface of neuroanatomy. Clearly, the brain deserves its status as the most complex piece of matter in the universe. What we have presented here is a shell, or scaffold, of the nervous system and some of its contents.

Understanding neuroanatomy is necessary for understanding how the brain works. This statement is just as true for an undergraduate first-time neuroscience student as it is for a neurologist or a neurosurgeon. In fact, neuroanatomy has taken on a new relevance with the advent of methods of imaging the living brain (Figure 7.30).

An Illustrated Guide to Human Neuroanatomy appears as an appendix to this chapter. Use the guide as an atlas to locate various structures of interest. Labeling exercises are also provided to help you learn the names of the parts of the nervous system you will encounter in this book.

In Part II, Sensory and Motor Systems, the anatomy presented in this chapter and its appendix will come alive, as we explore how the brain goes about the tasks of smelling, seeing, hearing, sensing touch, and moving.

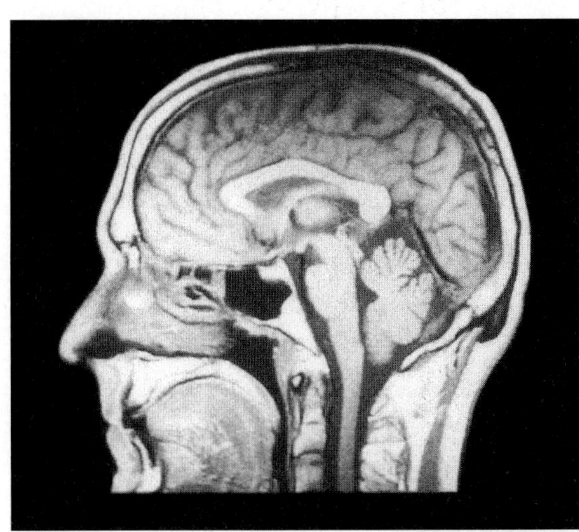

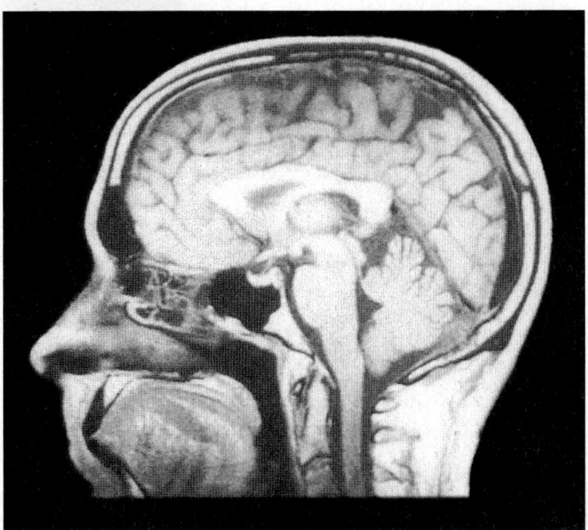

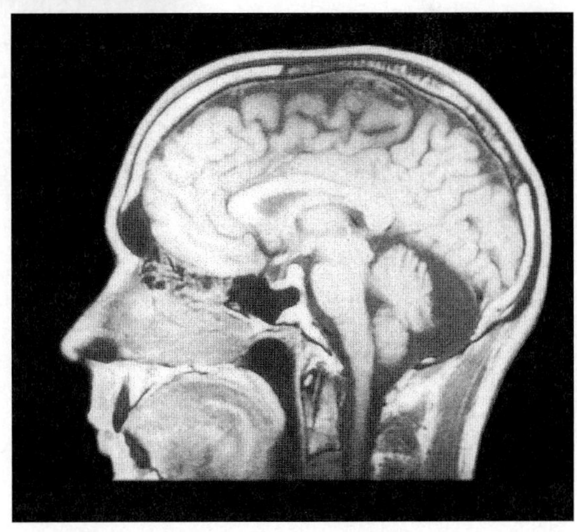

▶ FIGURE 7.30
MRI scans of the authors. How many structures can you label?

KEY TERMS

Gross Organization of the Mammalian Nervous System

anterior (p. 180)
rostral (p. 180)
posterior (p. 180)
caudal (p. 180)
dorsal (p. 180)
ventral (p. 180)
midline (p. 182)
medial (p. 182)
lateral (p. 182)
ipsilateral (p. 182)
contralateral (p. 182)
midsagittal plane (p. 182)
sagittal plane (p. 182)
horizontal plane (p. 182)
coronal plane (p. 182)
central nervous system (CNS) (p. 183)
brain (p. 183)
spinal cord (p. 183)
cerebrum (p. 183)
cerebral hemispheres (p. 183)
cerebellum (p. 183)
brain stem (p. 183)
spinal nerve (p. 184)
dorsal root (p. 184)
ventral root (p. 184)
peripheral nervous system (PNS) (p. 184)
somatic PNS (p. 184)
dorsal root ganglion (p. 185)
visceral PNS (p. 185)
autonomic nervous system (ANS) (p. 185)
afferent (p. 185)
efferent (p. 185)
cranial nerve (p. 185)
meninges (p. 185)
dura mater (p. 185)
arachnoid membrane (p. 185)
pia mater (p. 186)
cerebrospinal fluid (CSF) (p. 186)
ventricular system (p. 186)

Understanding CNS Structure Through Development

gray matter (p. 192)
cortex (p. 192)
nucleus (p. 192)
substantia (p. 192)
locus (p. 192)
ganglion (p. 192)
nerve (p. 192)
white matter (p. 192)
tract (p. 192)
bundle (p. 192)
capsule (p. 192)
commissure (p. 192)
lemniscus (p. 192)
neural tube (p. 194)
neural crest (p. 194)
neurulation (p. 194)
differentiation (p. 195)
forebrain (p. 195)
midbrain (p. 196)
hindbrain (p. 196)
diencephalon (p. 196)
telencephalon (p. 196)
olfactory bulb (p. 196)
lateral ventricle (p. 197)
third ventricle (p. 197)
cerebral cortex (p. 197)
basal telencephalon (p. 197)
thalamus (p. 197)
hypothalamus (p. 197)
cortical white matter (p. 197)
corpus callosum (p. 197)
internal capsule (p. 198)
tectum (p. 199)
tegmentum (p. 200)
cerebral aqueduct (p. 200)
pons (p. 200)
medulla oblongata (medulla) (p. 200)
fourth ventricle (p. 200)
spinal canal (p. 203)
dorsal horn (p. 203)
ventral horn (p. 203)
sulcus (p. 205)
gyrus (p. 205)
temporal lobe (p. 207)
frontal lobe (p. 207)
central sulcus (p. 207)
parietal lobe (p. 207)
occipital lobe (p. 207)

A Guide to the Cerebral Cortex

hippocampus (p. 209)
olfactory cortex (p. 209)
neocortex (p. 209)
cytoarchitectural map (p. 210)
connectome (p. 211)

REVIEW QUESTIONS

1. Are the dorsal root ganglia in the central or peripheral nervous system?
2. Is the myelin sheath of optic nerve axons provided by Schwann cells or oligodendroglia? Why?
3. Imagine that you are a neurosurgeon, about to remove a tumor lodged deep inside the brain. The top of the skull has been removed. What now lies between you and the brain? Which layer(s) must be cut before you reach the CSF?
4. What is the fate of tissue derived from the embryonic neural tube? Neural crest?
5. Name the three main parts of the hindbrain. Which of these is also part of the brain stem?
6. Where is CSF produced? What path does it take before it is absorbed into the bloodstream? Name the parts of the CNS it will pass through in its voyage from brain to blood.
7. What are three features that characterize the structure of cerebral cortex?

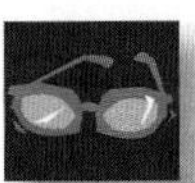

FURTHER READING

Creslin E. 1974. Development of the nervous system: a logical approach to neuroanatomy. *CIBA Clinical Symposium* 26:1–32.

Johnson KA, Becker JA. The whole brain atlas. http://www.med.harvard.edu/AANLIB/home.html.

Krubitzer L. 1995. The organization of neocortex in mammals: are species really so different? *Trends in Neurosciences* 18:408–418.

Nauta W, Feirtag M. 1986. *Fundamental Neuroanatomy*. New York: W.H. Freeman.

Seung S. 2012. *Connectome: How the Brain's Wiring Makes Us Who We Are*. Boston: Houghton Mifflin Harcourt.

Watson C. 1995. *Basic Human Neuroanatomy: an Introductory Atlas*, 5th ed. New York: Little, Brown & Co.

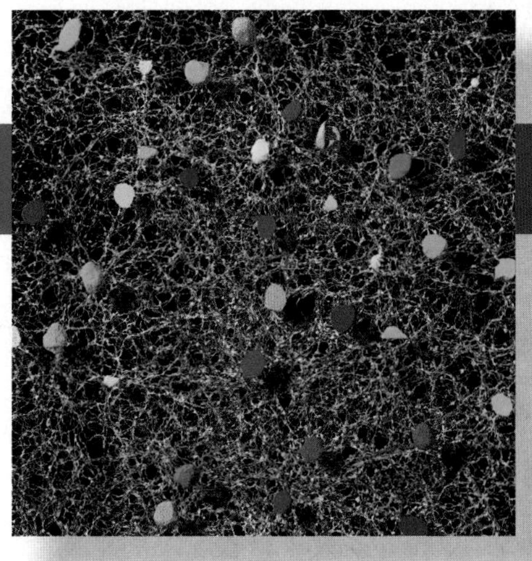

CHAPTER 7 APPENDIX

An Illustrated Guide to Human Neuroanatomy

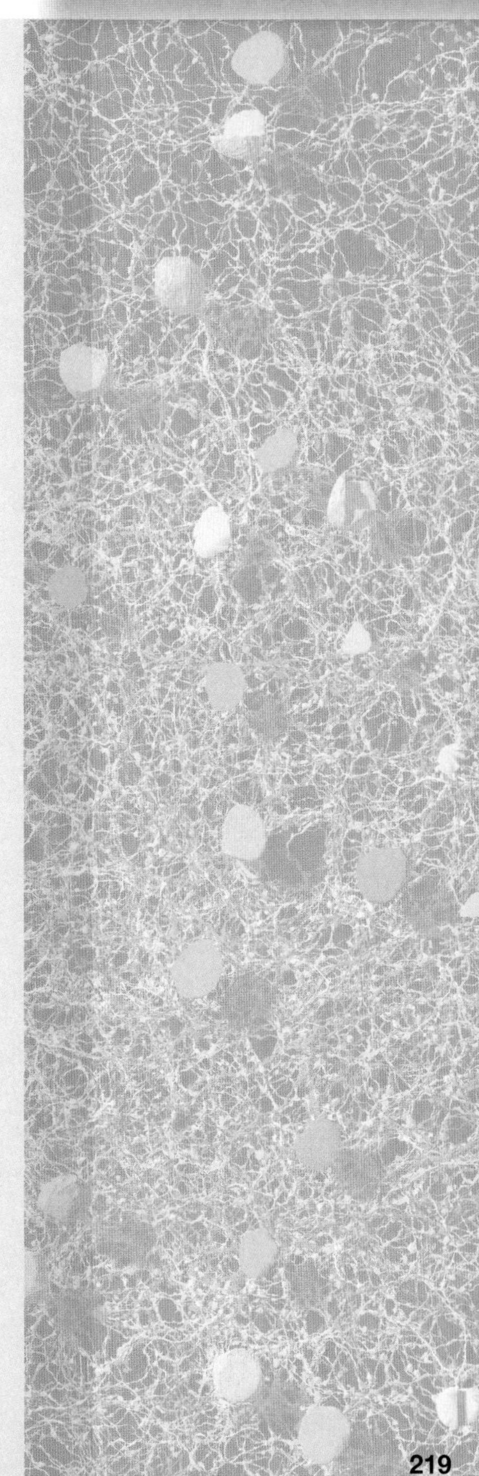

INTRODUCTION

As we will see in the remainder of the book, a fruitful way to explore the nervous system is to divide it up into functional systems. Thus, the *olfactory system* consists of those parts of the brain that are devoted to the sense of smell, the *visual system* includes those parts that are devoted to vision, and so on. While this functional approach to investigating nervous system structure has many merits, it can make the "big picture"—how all these systems fit together inside the box we call the brain—difficult to see. The goal of this Illustrated Guide is to help you learn, in advance, about some of the anatomy that will be discussed in the subsequent chapters. Here, we concentrate on naming the structures and seeing how they are related physically; their functional significance is discussed in the remainder of the book.

The Guide is organized into six main parts. The first part covers the surface anatomy of the brain—the structures that can been seen by inspection of the whole brain, as well as those parts that are visible when the two cerebral hemispheres are separated by a cut in the midsagittal plane. Next, we explore the cross-sectional anatomy of the brain, using a series of slabs that contain structures of interest. The brief third and fourth parts cover the spinal cord and the autonomic nervous system. The fifth part of the Guide illustrates the cranial nerves and summarizes their diverse functions. The last part illustrates the blood supply of the brain.

The nervous system has an astonishing number of bits and pieces. In this Guide, we focus on those structures that will appear later in the book when we discuss the various functional systems. Nonetheless, even this abbreviated atlas of neuroanatomy yields a formidable list of new vocabulary. Therefore, to help you learn the terminology, an extensive self-quiz review is provided at the end, in the form of a workbook with labeling exercises.

SURFACE ANATOMY OF THE BRAIN

Imagine that you hold in your hands a human brain that has been dissected from the skull. It is wet and spongy, and weighs about 1.4 kilograms (3 pounds). Looking down on the brain's dorsal surface reveals the convoluted surface of the cerebrum. Flipping the brain over shows the complex ventral surface that normally rests on the floor of the skull. Holding the brain up and looking at its side—the lateral view—shows the "ram's horn" shape of the cerebrum coming off the stalk of the brain stem. The brain stem is shown more clearly if we slice the brain right down the middle and view its medial surface. In the part of the guide that follows, we will name the important structures that are revealed by such an inspection of the brain. Notice the magnification of the drawings: 1× is life-size, 2× is twice life-size, 0.6× is 60% of life-size, and so on.

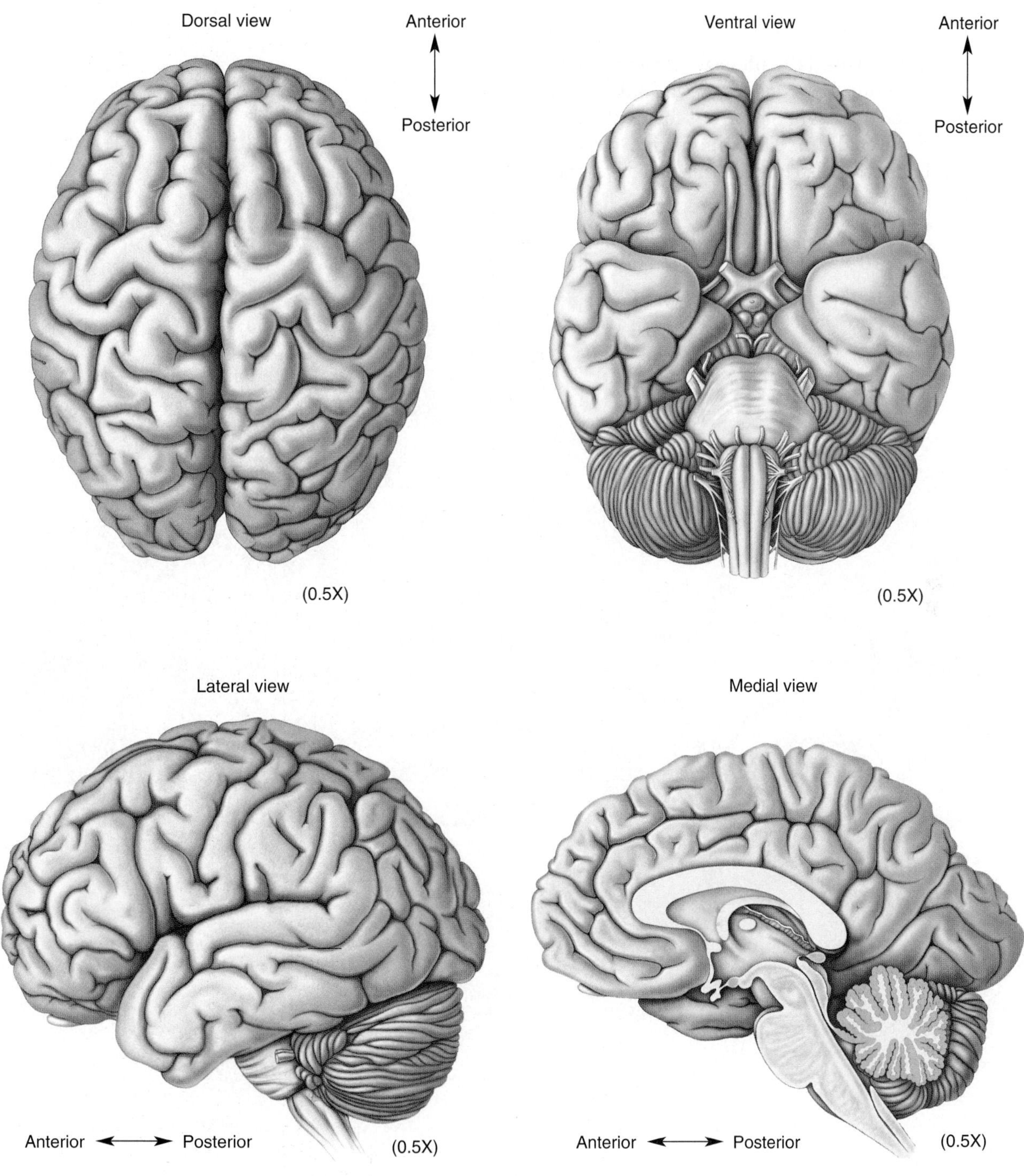
Dorsal view
Anterior
Posterior
(0.5X)
Ventral view
Anterior
Posterior
(0.5X)
Lateral view
Anterior
Posterior
(0.5X)
Medial view
Anterior
Posterior
(0.5X)

The Lateral Surface of the Brain

(a) Gross Features. This is a life-size drawing of the brain. Gross inspection reveals the three major parts: the large cerebrum, the brain stem that forms its stalk, and the rippled cerebellum. The diminutive olfactory bulb of the cerebrum can also be seen in this lateral view.

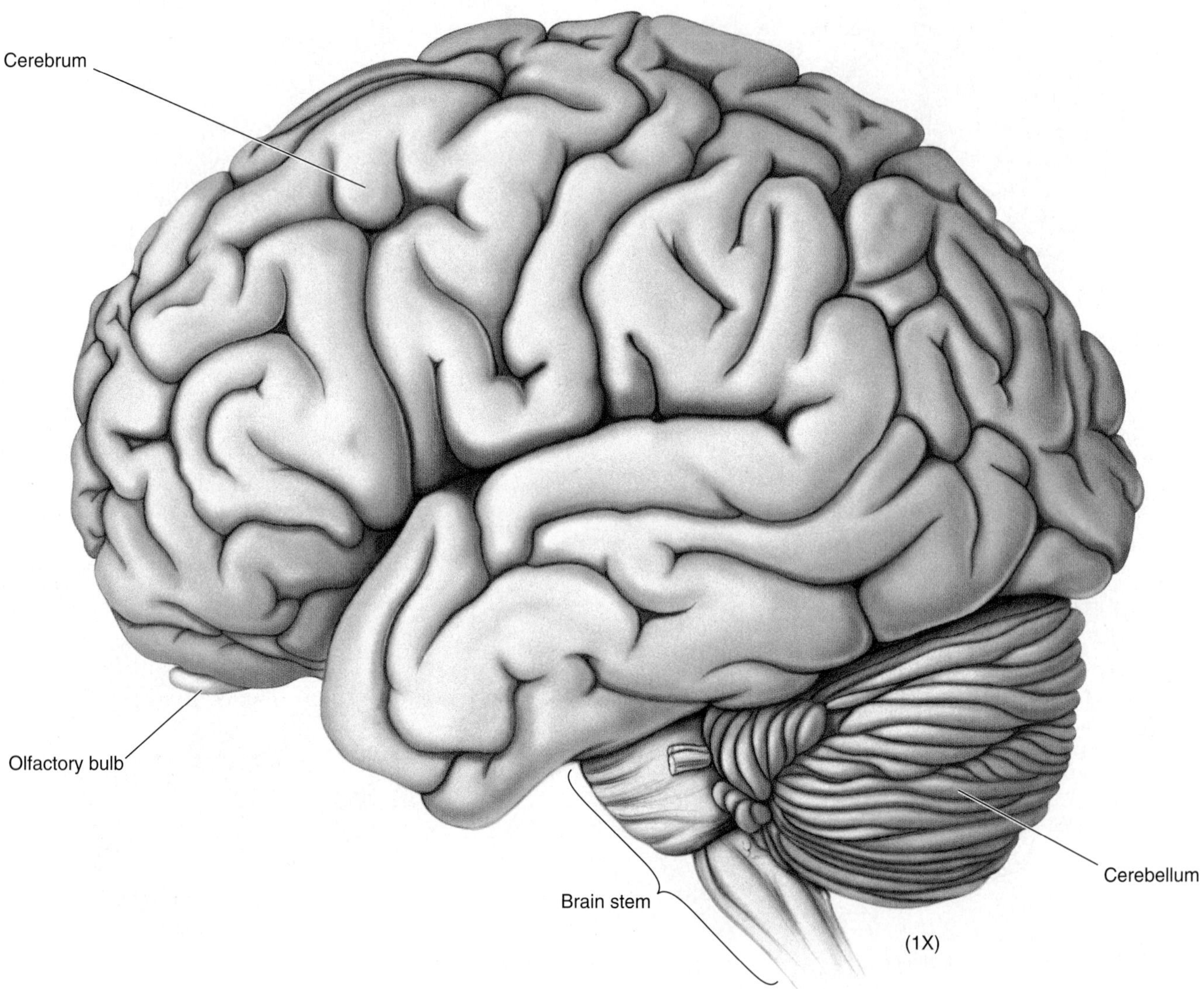

(b) Selected Gyri, Sulci, and Fissures. The cerebrum is noteworthy for its convoluted surface. The bumps are called *gyri*, and the grooves are called *sulci* or, if they are especially deep, fissures. The precise pattern of gyri and sulci can vary considerably from individual to individual, but many features are common to all human brains. Some of the important landmarks are labeled here. Notice that the postcentral gyrus lies immediately posterior to the central sulcus, and that the precentral gyrus lies immediately anterior to it. The neurons of the postcentral gyrus are involved in somatic sensation (touch; Chapter 12), and those of the precentral gyrus control voluntary movement (Chapter 14). Neurons in the superior temporal gyrus are involved in audition (hearing; Chapter 11).

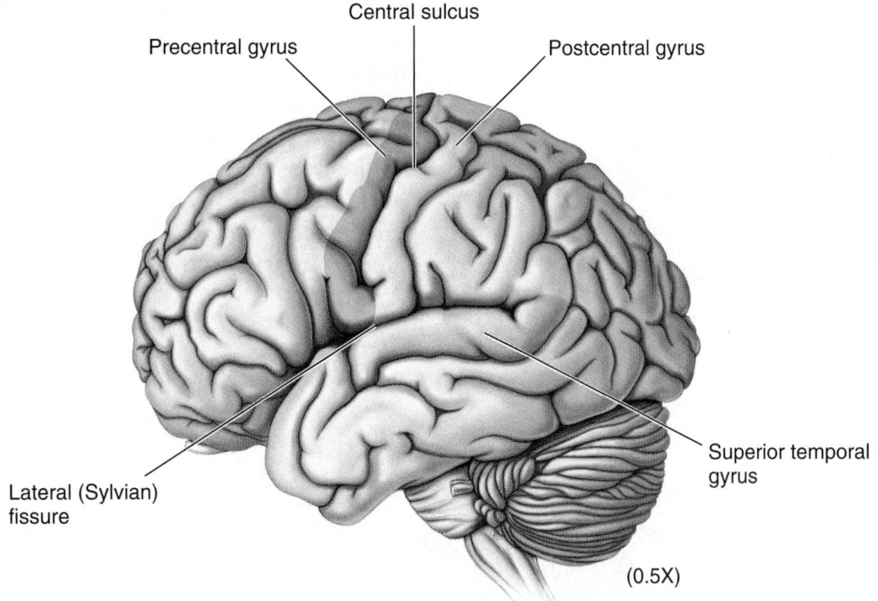

(c) Cerebral Lobes and the Insula. By convention, the cerebrum is subdivided into lobes named after the bones of the skull that lie over them. The central sulcus divides the frontal lobe from the parietal lobe. The temporal lobe lies immediately ventral to the deep lateral (Sylvian) fissure. The occipital lobe lies at the very back of the cerebrum, bordering both parietal and temporal lobes. A buried piece of the cerebral cortex, called the *insula* (Latin for "island"), is revealed if the margins of the lateral fissure are gently pulled apart (inset). The insula borders and separates the temporal and frontal lobes.

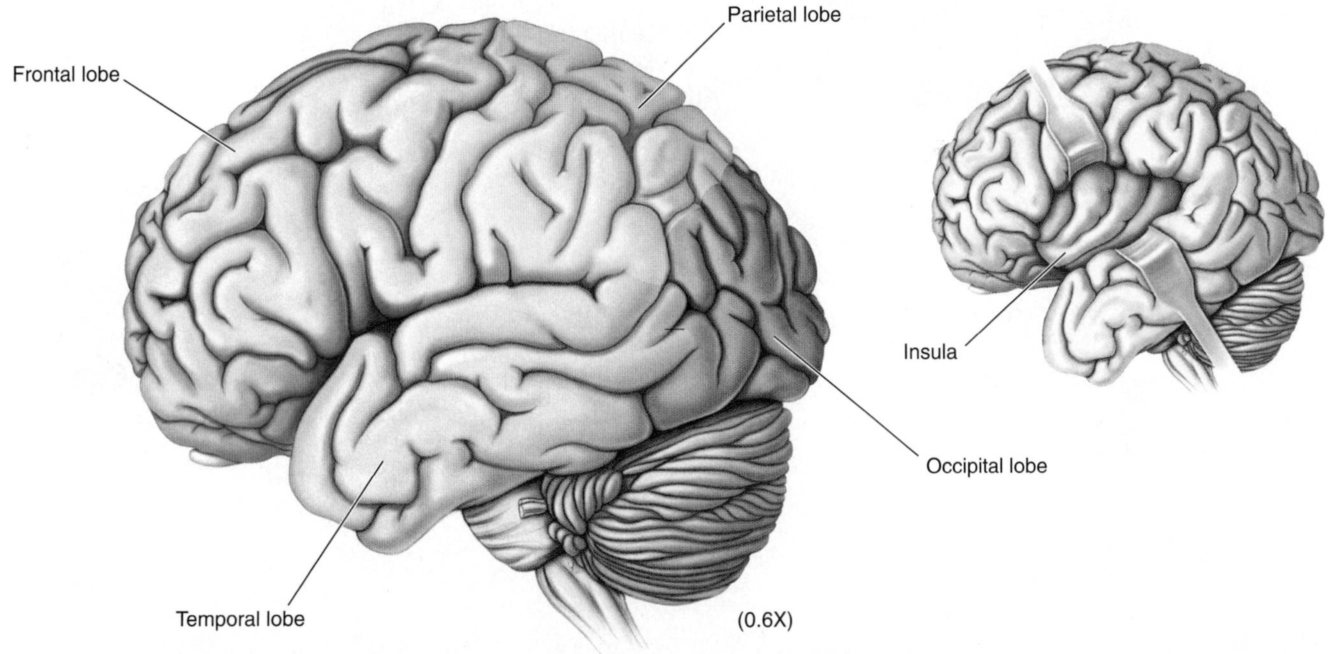

(d) Major Sensory, Motor, and Association Areas of Cortex. The cerebral cortex is organized like a patchwork quilt. The various areas, first identified by Brodmann, differ from one another in terms of microscopic structure and function. Notice that the visual areas (Chapter 10) are found in the occipital lobe, the somatic sensory areas (Chapter 12) are in the parietal lobe, and the auditory areas (Chapter 11) are in the temporal lobe. On the inferior surface of the parietal lobe (the operculum) and buried in the insula is the gustatory cortex, devoted to the sense of taste (Chapter 8).

In addition to the analysis of sensory information, the cerebral cortex plays an important role in the control of voluntary, willful movement. The major motor control areas lie in the frontal lobe, anterior to the central sulcus (Chapter 14). In the human brain, large expanses of cortex cannot be simply assigned to sensory or motor functions. These constitute the association areas of cortex. Some of the more important areas are the prefrontal cortex (Chapters 21 and 24), the posterior parietal cortex (Chapters 12, 21, and 24), and the inferotemporal cortex (Chapters 24 and 25).

Brodmann's map

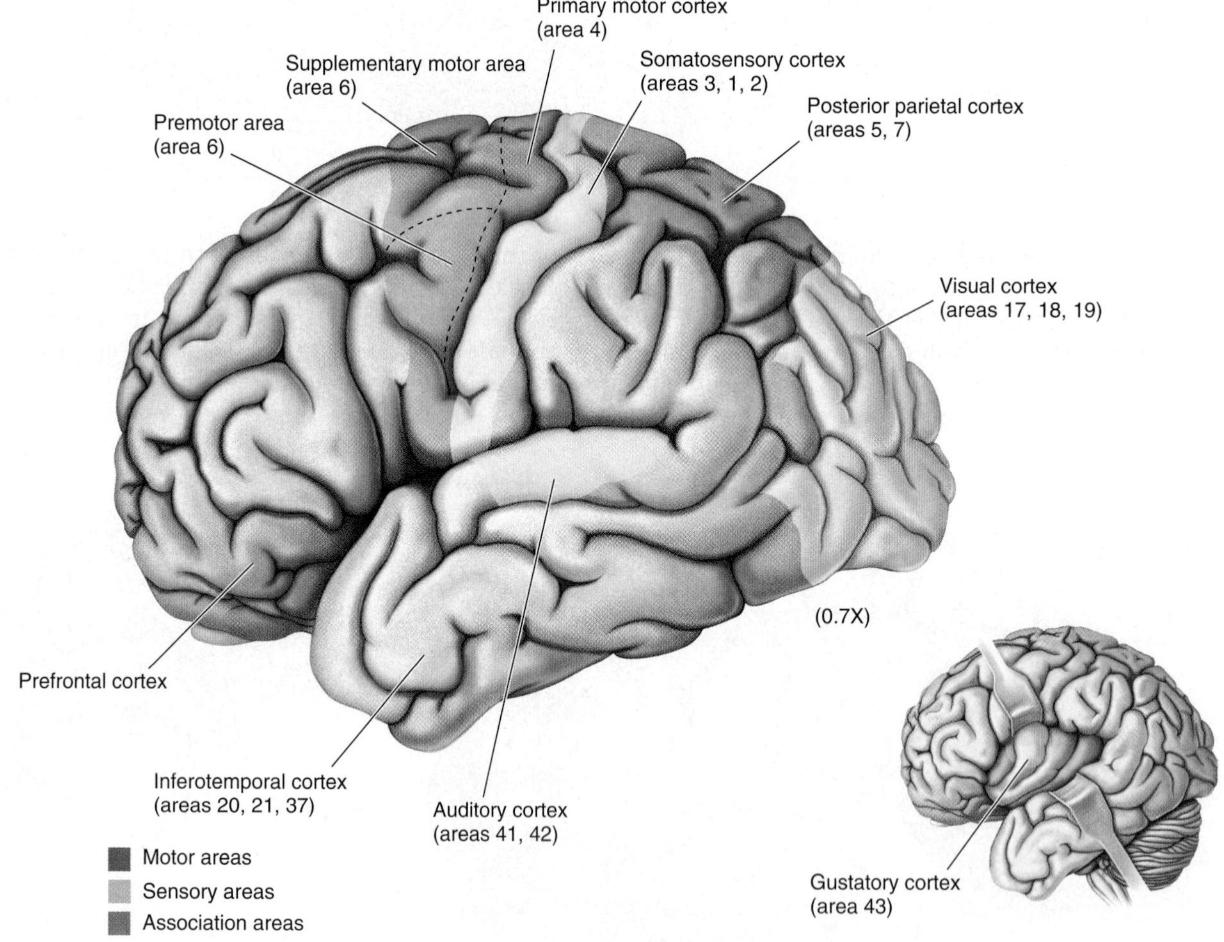

The Medial Surface of the Brain

(a) Brain Stem Structures. Splitting the brain down the middle exposes the medial surface of the cerebrum, shown in this life-size illustration. This view also shows the midsagittal, cut surface of the brain stem, consisting of the diencephalon (thalamus and hypothalamus), the midbrain (tectum and tegmentum), the pons, and the medulla. (It should be noted that some anatomists define the brain stem as consisting only of the midbrain, pons, and medulla.)

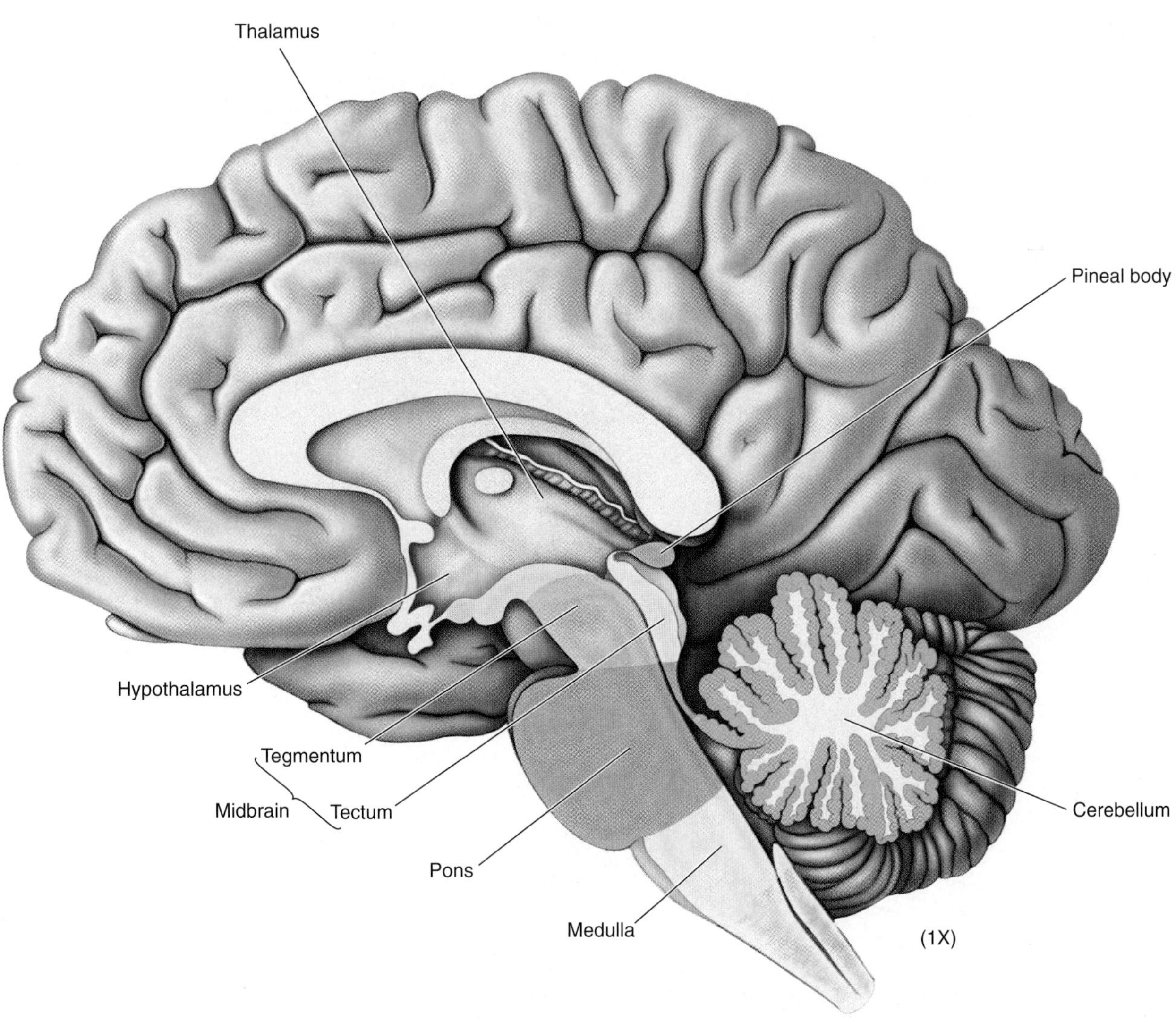

(b) Forebrain Structures. Shown here are the important forebrain structures that can be observed by viewing the medial surface of the brain. Notice the cut surface of the corpus callosum, a huge bundle of axons that connects the two sides of the cerebrum. The unique contributions of the two cerebral hemispheres to human brain function can be studied in patients in which the callosum has been sectioned (Chapter 20). The fornix (Latin for "arch) is another prominent fiber bundle that connects the hippocampus on each side with the hypothalamus. Some of the axons in the fornix regulate memory storage (Chapter 24).

In the lower illustration, the brain has been tilted slightly to show the positions of the amygdala and hippocampus. These are "phantom views" of these structures since they cannot be observed directly from the surface. Both lie deep to the overlying cortex. We will see them again in cross section later in the Guide. The amygdala (from the Latin word for "almond") is an important structure for regulating emotional states (Chapter 18), and the hippocampus is important for memory (Chapters 24 and 25).

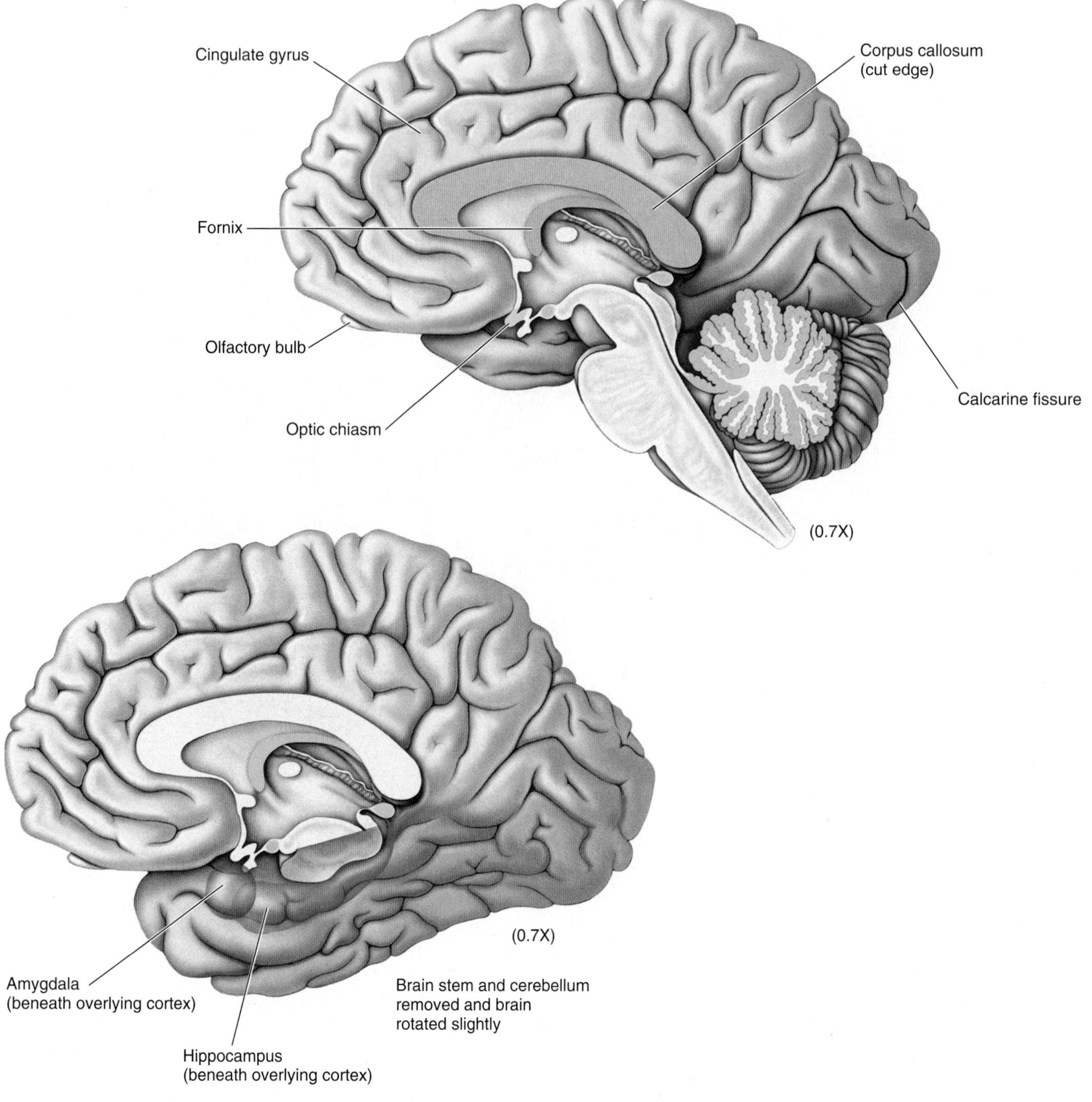

(c) Ventricles. The lateral walls of the unpaired parts of the ventricular system—the third ventricle, the cerebral aqueduct, the fourth ventricle, and the spinal canal—can be observed in the medial view of the brain. These are handy landmarks because the thalamus and hypothalamus lie next to the third ventricle; the midbrain lies next to the aqueduct; the pons, cerebellum, and medulla lie next to the fourth ventricle; and the spinal cord forms the walls of the spinal canal.

The lateral ventricles are paired structures that sprout like antlers from the third ventricle. A phantom view of the right lateral ventricle, which lies underneath the overlying cortex, is shown in the lower illustration. The two cerebral hemispheres surround the two lateral ventricles. Notice how a cross section of the brain at the thalamus–midbrain junction will intersect the "horns" of the lateral ventricle of each hemisphere twice.

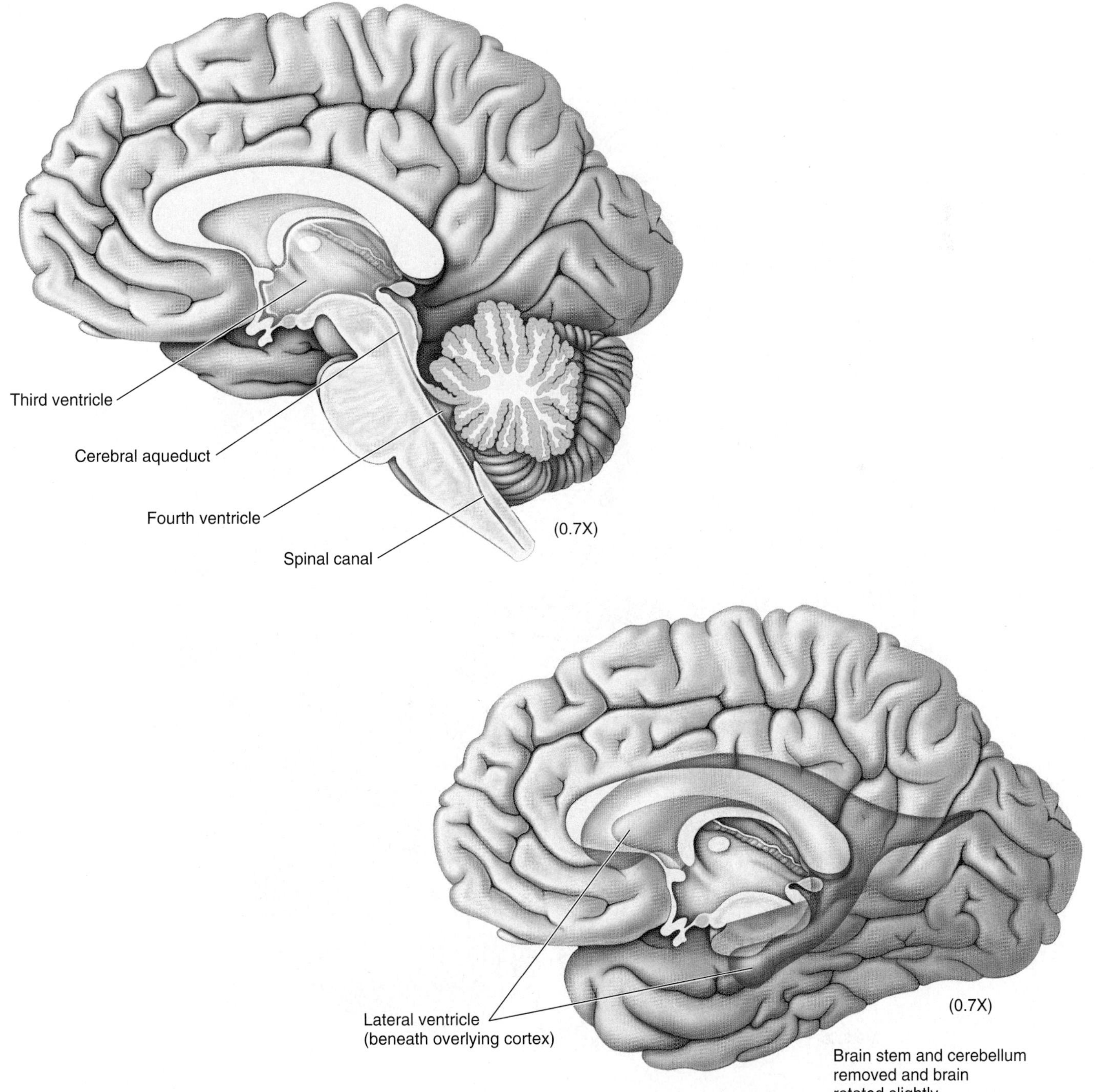

The Ventral Surface of the Brain

The underside of the brain has a lot of distinct anatomical features. Notice the nerves emerging from the brain stem; these are the cranial nerves, which are illustrated in more detail later in the Guide. Also notice the X-shaped optic chiasm, just anterior to the hypothalamus. The chiasm is the place where many axons from the eyes decussate (cross) from one side to another. The bundles of axons anterior to the chiasm, which emerge from the backs of the eyes, are the optic nerves. The bundles lying posterior to the chiasm, which disappear into the thalamus, are called the *optic tracts* (Chapter 10). The paired mammillary bodies (Latin for "nipple") are a prominent feature of the ventral surface of the brain. These nuclei of the hypothalamus are part of the circuitry that stores memory (Chapter 24) and are a major target of the axons of the fornix (seen in the medial view). Notice also the olfactory bulbs (Chapter 8) and the midbrain, pons, and medulla.

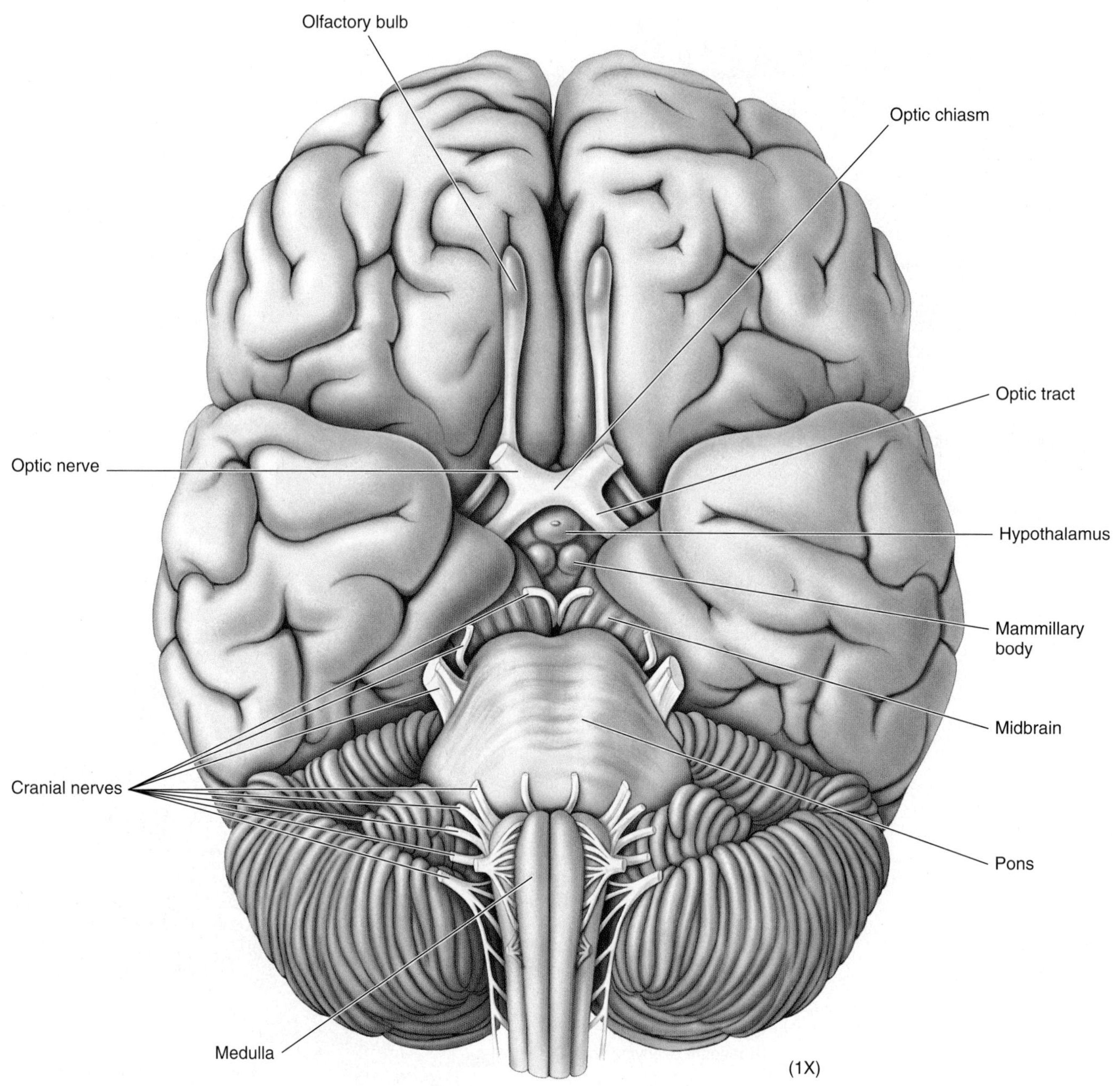

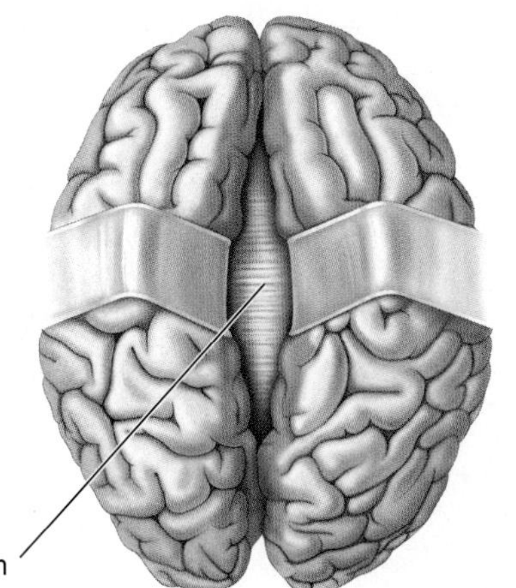

The Dorsal Surface of the Brain

(a) Cerebrum. The dorsal view of the brain is dominated by the large cerebrum. Notice the paired cerebral hemispheres. These are connected by the axons of the corpus callosum (Chapter 20), which can be seen if the hemispheres are retracted slightly. The medial view of the brain, illustrated previously, showed the callosum in cross section.

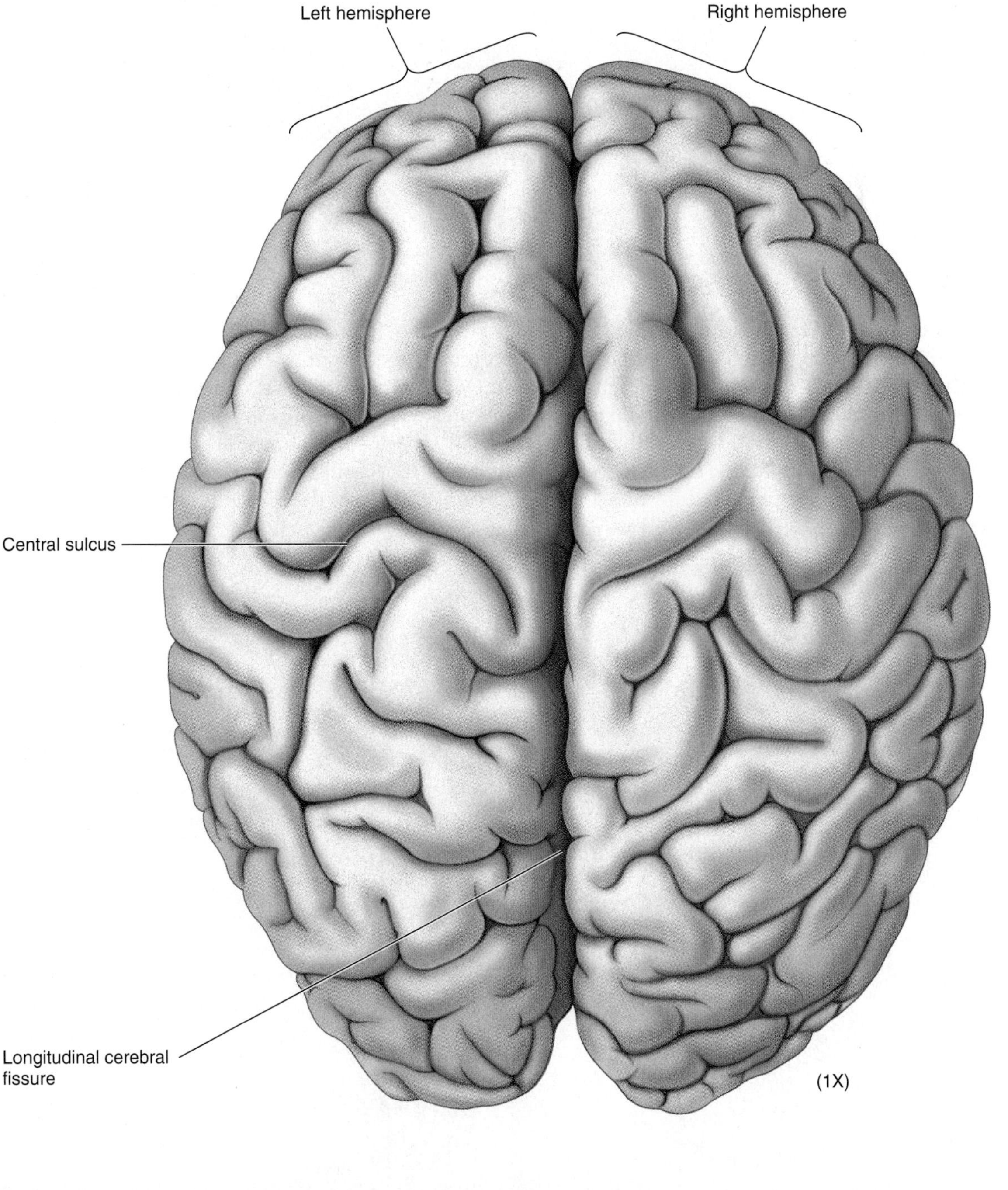

Left hemisphere

Right hemisphere

Central sulcus

Longitudinal cerebral fissure

(1X)

(b) Cerebrum Removed. The cerebellum dominates the dorsal view of the brain if the cerebrum is removed and the brain is tilted slightly forward. The cerebellum is an important motor control structure (Chapter 14), and is divided into two hemispheres and a midline region called the *vermis* (Latin for "worm").

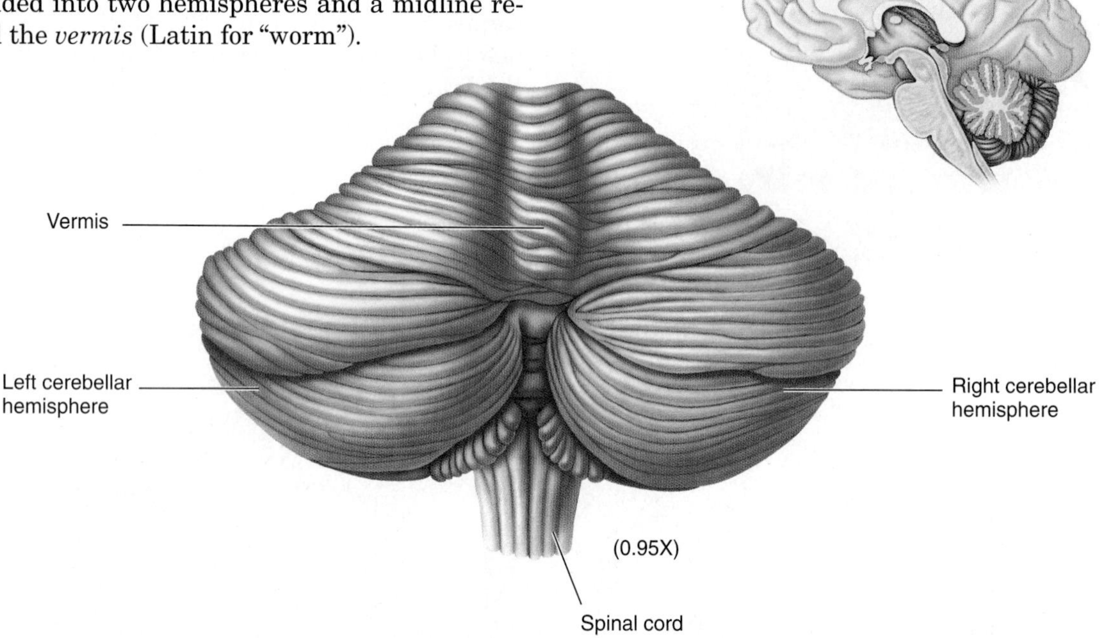

(c) Cerebrum and Cerebellum Removed. The top surface of the brain stem is exposed when both the cerebrum and the cerebellum are removed. The major divisions of the brain stem are labeled on the left side, and some specific structures are labeled on the right side. The pineal body, lying atop the thalamus, secretes melatonin and is involved in the regulation of sleep and sexual behavior (Chapters 17 and 19). The superior colliculus receives direct input from the eyes (Chapter 10) and is involved in the control of eye movements (Chapter 14), while the inferior colliculus is an important component of the auditory system (Chapter 11). (*Colliculus* is Latin for "mound.") The cerebellar peduncles are the large bundles of axons that connect the cerebellum and the brain stem (Chapter 14).

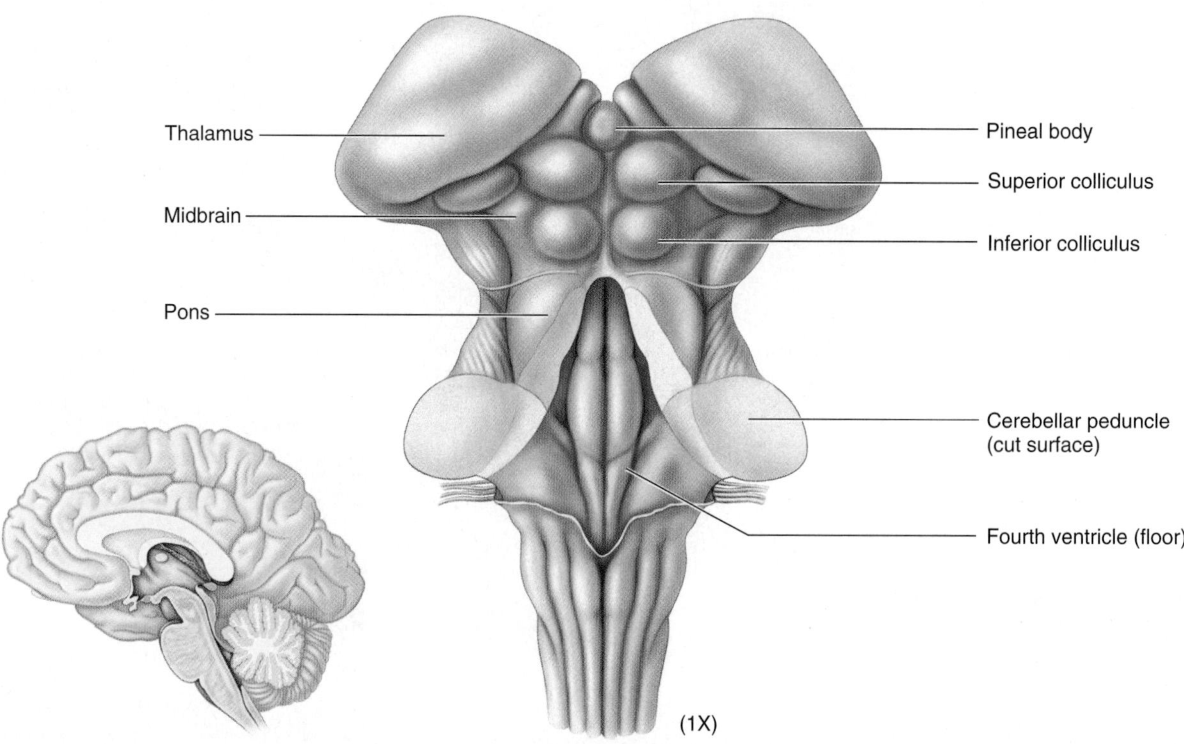

CROSS-SECTIONAL ANATOMY OF THE BRAIN

Understanding the brain requires that we peer inside it, and this is accomplished by making cross sections. Cross sections can be made physically with a knife or, in the case of noninvasive imaging of the living brain, digitally with a magnetic resonance imaging or a computed tomography scan. For learning the internal organization of the brain, the best approach is to make cross sections that are perpendicular to the axis defined by the embryonic neural tube, called the *neuraxis*. The neuraxis bends as the human fetus grows, particularly at the junction of the midbrain and thalamus. Consequently, the best plane of section depends on exactly where we are along the neuraxis.

In this part of the Guide, we take a look at drawings of a series of cross-sectional slabs of the brain, showing the internal structure of the forebrain (cross sections 1–3), the midbrain (cross sections 4 and 5), the pons and cerebellum (cross section 6), and the medulla (cross sections 7–9). The drawings are schematic, meaning that structures within the slab are sometimes projected onto the slab's visible surface.

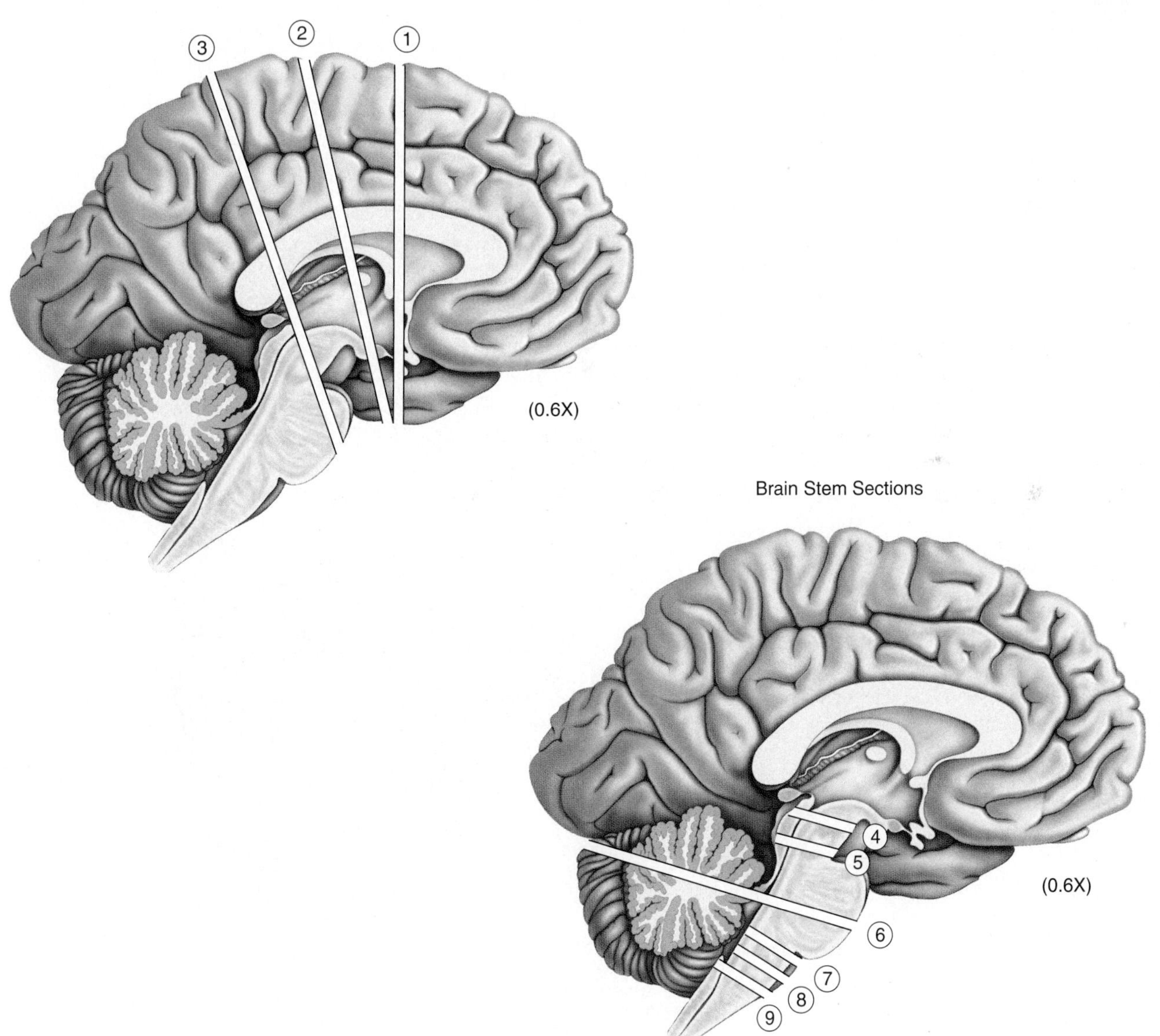

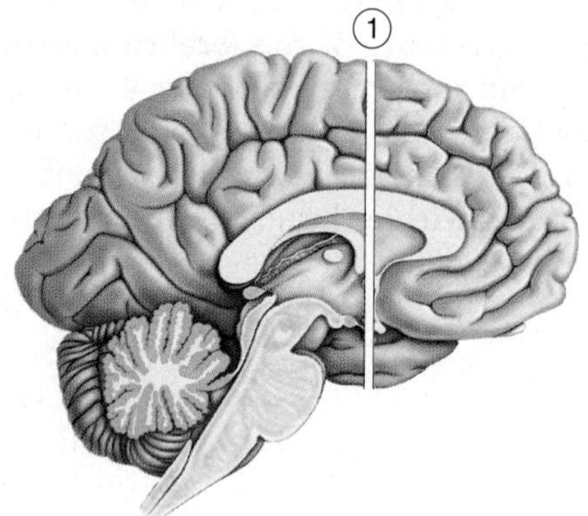

Cross Section 1: Forebrain at Thalamus–Telencephalon Junction

(a) Gross Features. The telencephalon surrounds the lateral ventricles, and the thalamus surrounds the third ventricle. Notice that in this section, the lateral ventricles can be seen sprouting from the slit-like third ventricle. The hypothalamus, forming the floor of the third ventricle, is a vital control center for many basic bodily functions (Chapters 15–17). Notice that the insula (Chapter 8) lies at the base of the lateral (Sylvian) fissure, here separating the frontal lobe from the temporal lobe. The heterogeneous region lying deep within the telencephalon, medial to the insula and lateral to the thalamus, is called the *basal forebrain*.

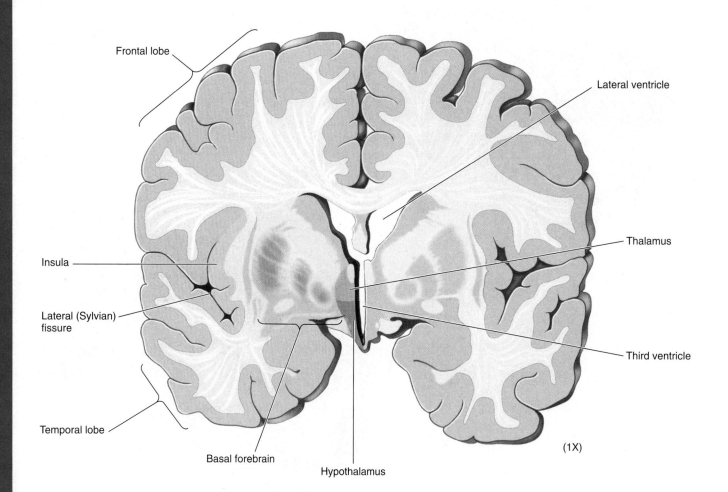

Frontal lobe
Lateral ventricle
Thalamus
Insula
Lateral (Sylvian) fissure
Third ventricle
Temporal lobe
(1X)
Basal forebrain
Hypothalamus

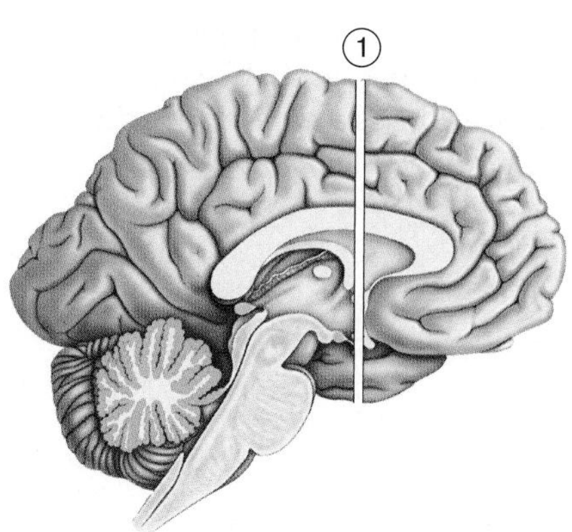

(b) Selected Cell and Fiber Groups. Here, we take a more detailed look at the structures of the forebrain. Notice that the internal capsule is the large collection of axons connecting the cortical white matter with the brain stem, and that the corpus callosum is the enormous sling of axons connecting the cerebral cortex of the two hemispheres. The fornix, shown earlier in the medial view of the brain, is shown here in cross section where it loops around the stalk of the lateral ventricle. The neurons of the closely associated septal area (from *saeptum*, Latin for "partition") contribute axons to the fornix and are involved in memory storage (Chapter 24). Three important collections of neurons in the basal telencephalon are also shown: the caudate nucleus, the putamen, and the globus pallidus. Collectively, these structures are called the *basal ganglia* and are an important part of the brain systems that control movement (Chapter 14).

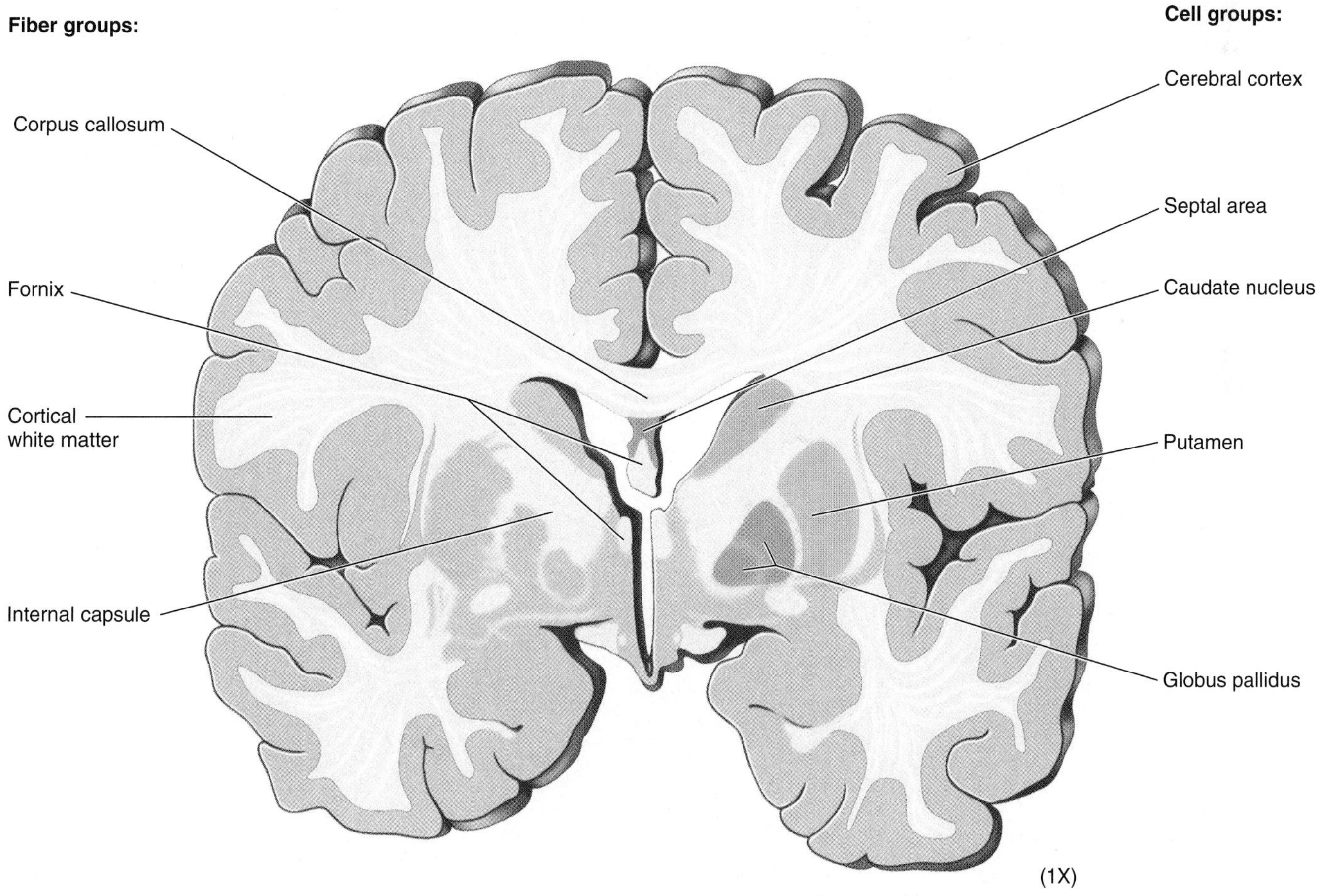

Fiber groups:

Corpus callosum

Fornix

Cortical white matter

Internal capsule

Cell groups:

Cerebral cortex

Septal area

Caudate nucleus

Putamen

Globus pallidus

(1X)

Cross Section 2: Forebrain at Mid-Thalamus

(a) Gross Features. As we move slightly caudal in the neuraxis, we see the heart-shaped thalamus (Greek for "inner chamber") surrounding the small third ventricle at the brain's core. Just ventral to the thalamus lies the hypothalamus. The telencephalon is organized much like what we saw in cross section 1. Because we are slightly posterior, the lateral fissure here separates the parietal lobe from the temporal lobe.

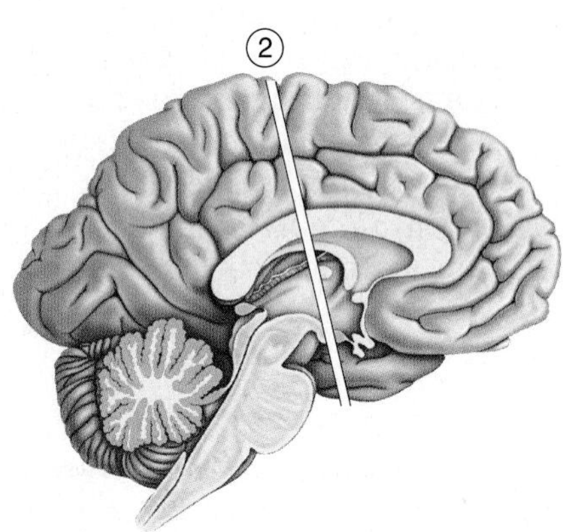

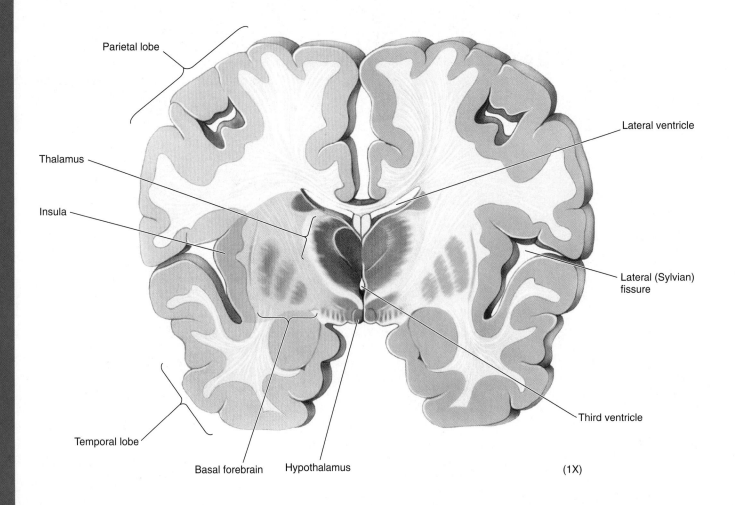

Parietal lobe
Lateral ventricle
Thalamus
Insula
Lateral (Sylvian) fissure
Third ventricle
Temporal lobe
Basal forebrain
Hypothalamus
(1X)

(b) Selected Cell and Fiber Groups. Many important cell and fiber groups appear at this level of the neuraxis. One new structure apparent in the telencephalon is the amygdala, involved in the regulation of emotion (Chapter 18) and memory (Chapter 24). We can also see that the thalamus is divided into separate nuclei, of which two, the ventral posterior nucleus and the ventral lateral nucleus, are labeled. The thalamus provides much of the input to the cerebral cortex, with different thalamic nuclei projecting axons to different areas of cortex. The ventral posterior nucleus is a part of the somatic sensory system (Chapter 12) and projects to the cortex of the postcentral gyrus. The ventral lateral nucleus and closely related ventral anterior nucleus (not shown) are parts of the motor system (Chapter 14) and project to the motor cortex of the precentral gyrus. Visible below the thalamus are the subthalamus and the mammillary bodies of the hypothalamus. The subthalamus is a part of the motor system (Chapter 14), while the mammillary bodies receive information from the fornix and contribute to the regulation of memory (Chapter 24). Because this section also encroaches on the midbrain, a little bit of the substantia nigra ("black substance") can be seen near the base of the brain stem. The substantia nigra is also a part of the motor system (Chapter 14). Parkinson's disease results from the degeneration of this structure.

2

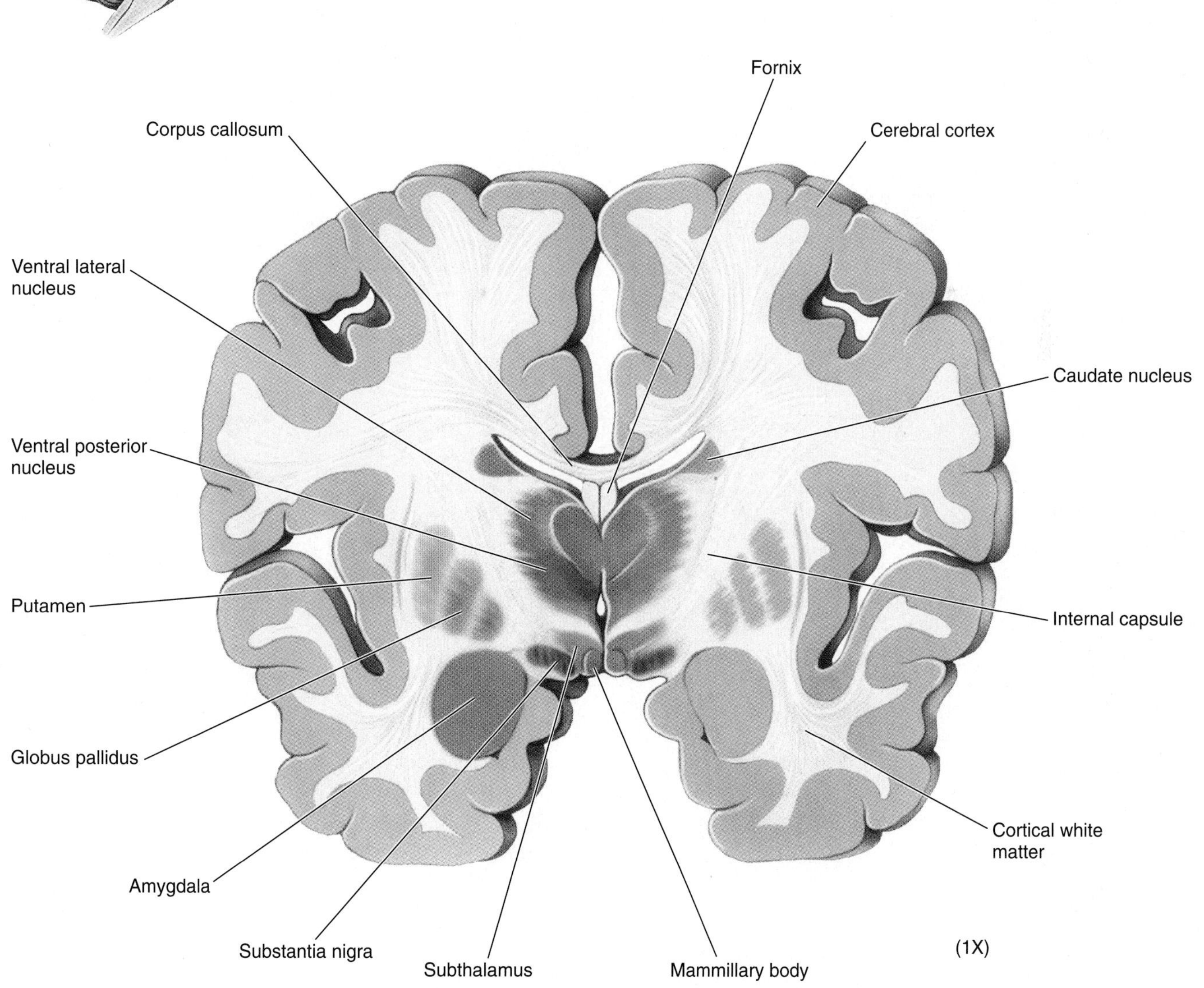

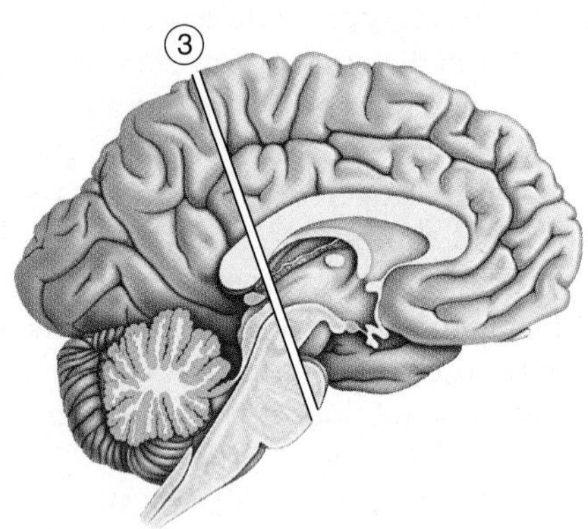

Cross Section 3: Forebrain at Thalamus–Midbrain Junction

(a) Gross Features. The neuraxis bends sharply at the junction of the thalamus and the midbrain. This cross section is taken at a level where the teardrop-shaped third ventricle communicates with the cerebral aqueduct. Notice that the brain surrounding the third ventricle is thalamus, and the brain around the cerebral aqueduct is midbrain. The lateral ventricles of each hemisphere appear twice in this section. You can see why by reviewing the phantom view of the ventricle, shown earlier.

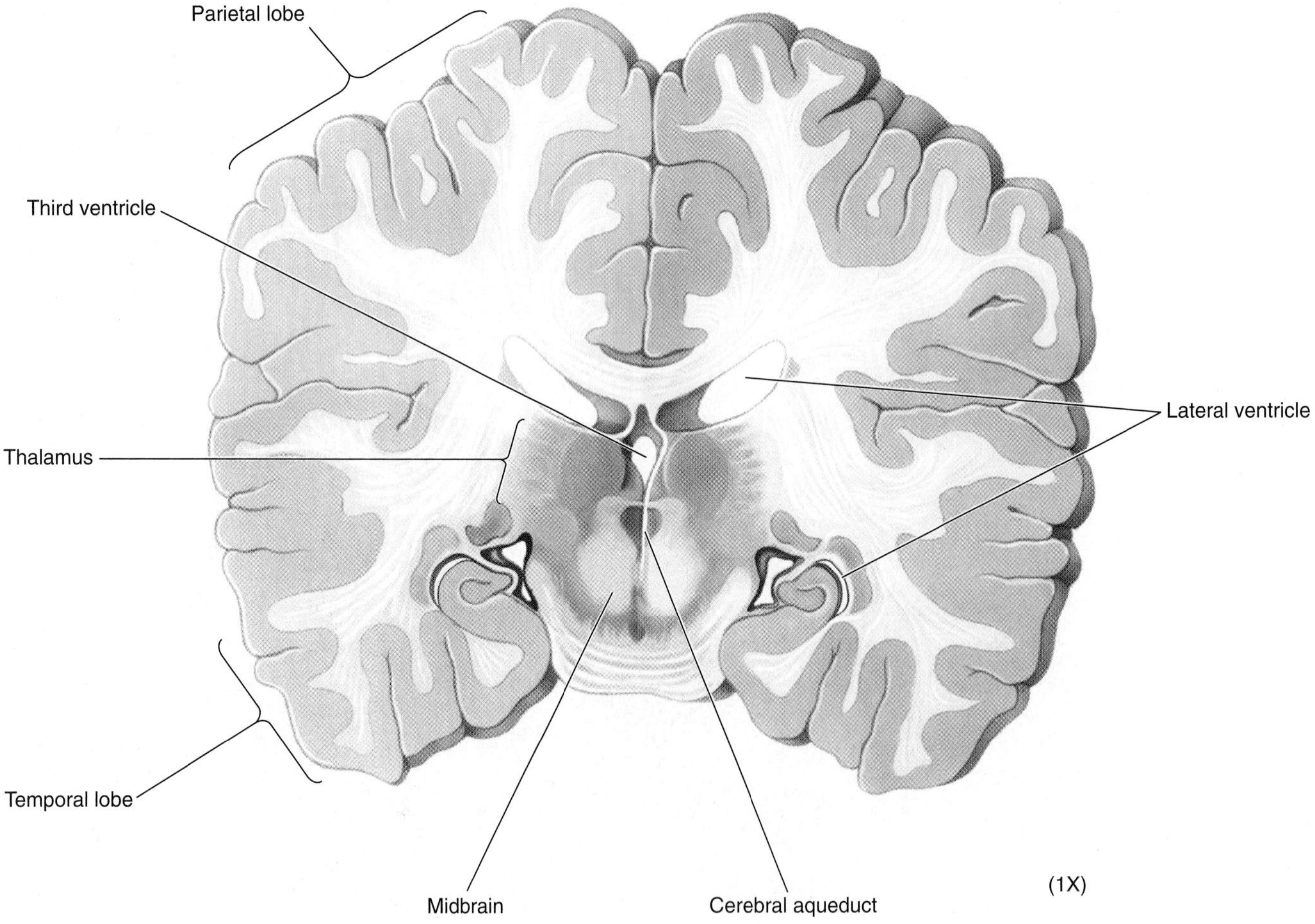

Parietal lobe
Third ventricle
Lateral ventricle
Thalamus
Temporal lobe
Midbrain
Cerebral aqueduct
(1X)

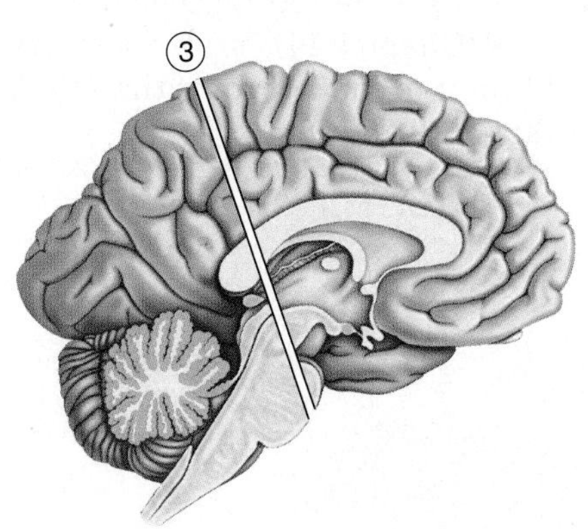

(b) Selected Cell and Fiber Groups. Notice that this section contains three more important nuclei of the thalamus: the pulvinar nucleus and the medial and lateral geniculate nuclei. The pulvinar nucleus is connected to much of the association cortex and plays a role in guiding attention (Chapter 21). The lateral geniculate nucleus relays information to the visual cortex (Chapter 10), and the medial geniculate nucleus relays information to the auditory cortex (Chapter 11). Also notice the location of the hippocampus, a relatively simple form of cerebral cortex bordering the lateral ventricle of the temporal lobe. The hippocampus (Greek for "sea horse") plays an important role in learning and memory (Chapters 24 and 25).

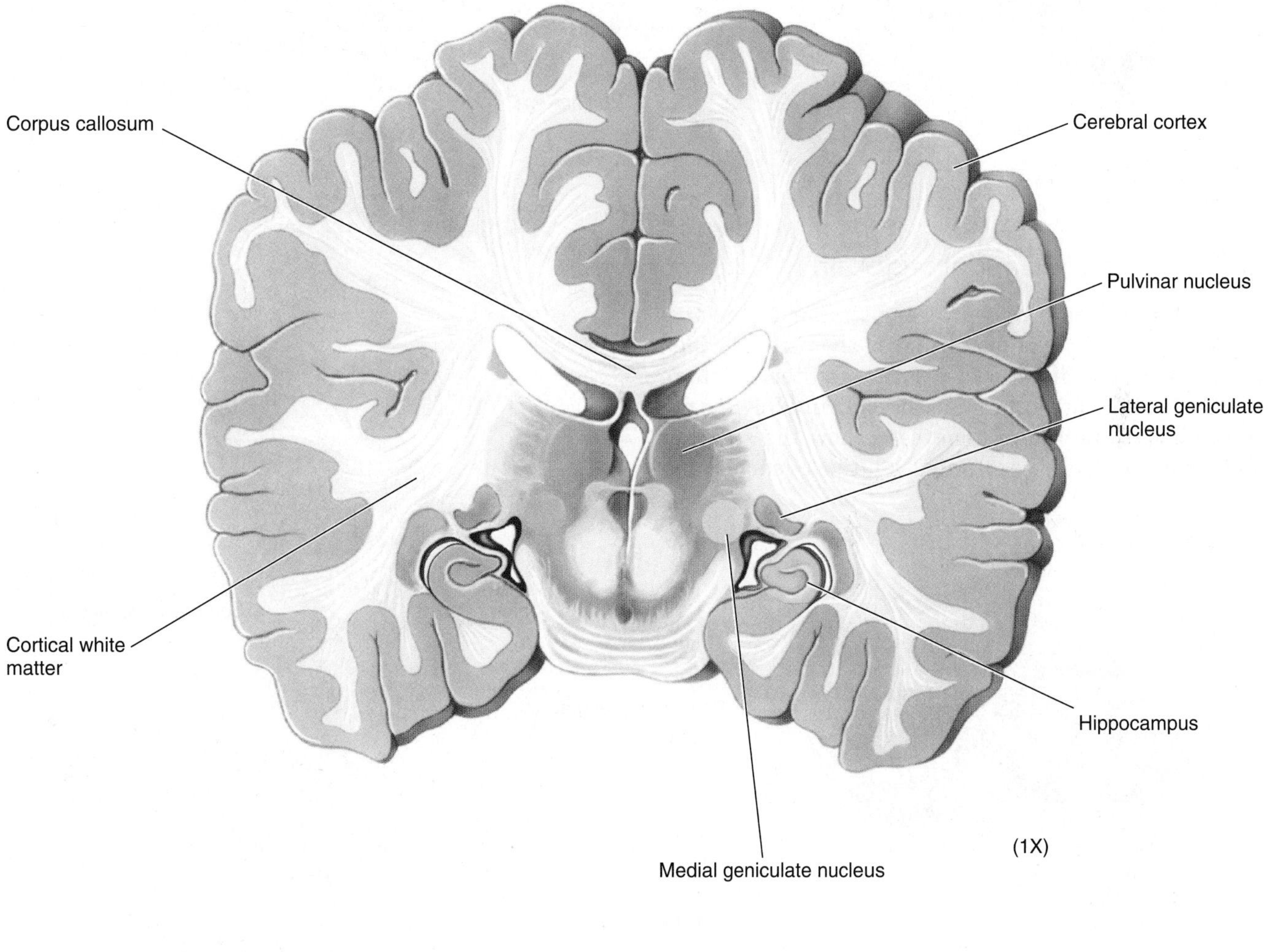

Cross Section 4: Rostral Midbrain

We are now at the midbrain, a part of the brain stem. Notice that the plane of section has been angled relative to the forebrain sections, so that it remains perpendicular to the neuraxis. The core of the midbrain is the small cerebral aqueduct. Here, the roof of the midbrain, also called the *tectum* (Latin for "roof"), consists of the paired superior colliculi. As discussed earlier, the superior colliculus is a part of the visual system (Chapter 10) and the substantia nigra is a part of the motor system (Chapter 14). The red nucleus is also a motor control structure (Chapter 14), while the periaqueductal gray is important in the control of the somatic pain sensations (Chapter 12).

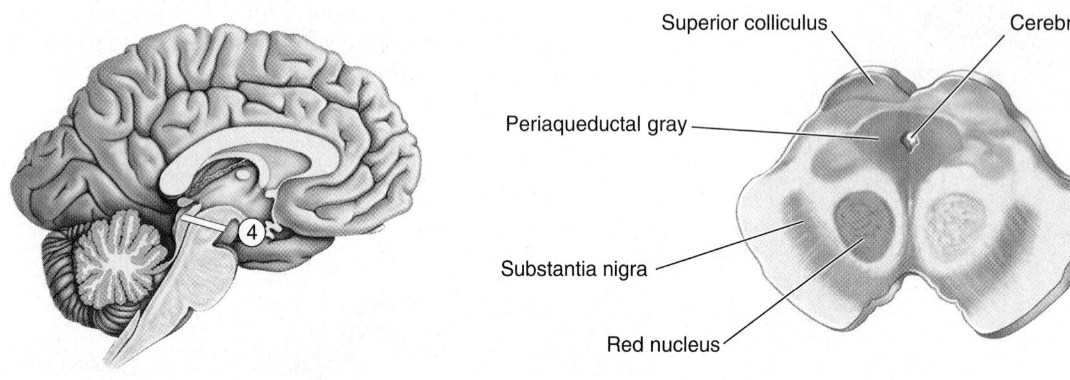

Cross Section 5: Caudal Midbrain

The caudal midbrain appears very similar to the rostral midbrain. However, at this level, the roof is formed by the inferior colliculi (part of the auditory system; Chapter 11) instead of the superior colliculi. Review the dorsal view of the brain stem to see how the superior and inferior colliculi are situated relative to each other.

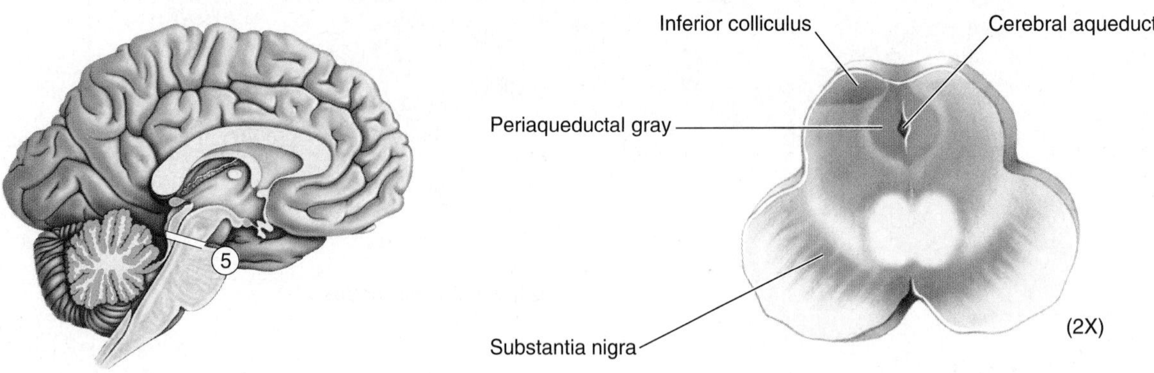

Cross Section 6: Pons and Cerebellum

This section shows the pons and cerebellum, parts of the rostral hindbrain that border the fourth ventricle. As discussed earlier, the cerebellum is important in the control of movement. Much of the input to the cerebellar cortex derives from the pontine nuclei, while the output of the cerebellum is from neurons of the deep cerebellar nuclei (Chapter 14). The reticular formation (*reticulum* is Latin for "net") runs from the midbrain to the medulla at its core, just under the cerebral aqueduct and fourth ventricle. One function of the reticular formation is to regulate sleep and wakefulness (Chapter 19). In addition, a function of the pontine reticular formation is to control body posture (Chapter 14).

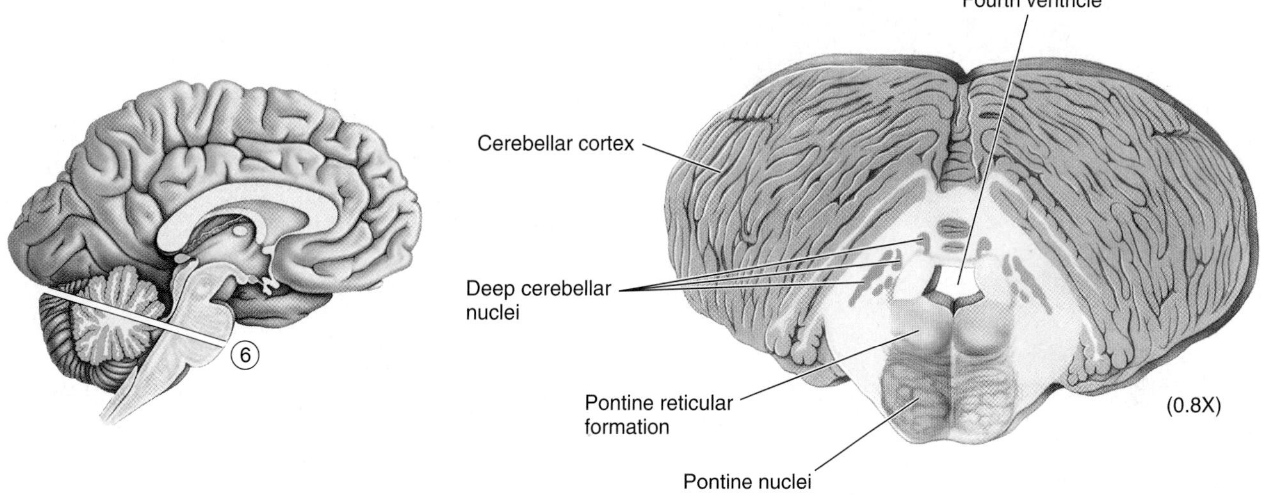

Cross Section 7: Rostral Medulla

As we move further caudally along the neuraxis, the brain surrounding the fourth ventricle becomes the medulla. The medulla is a complex region of the brain. Here, we focus only on those structures whose functions are discussed later in the book. At the very floor of the medulla lie the medullary pyramids, huge bundles of axons descending from the forebrain toward the spinal cord. The pyramids contain the corticospinal tracts, which are involved in the control of voluntary movement (Chapter 14). Several nuclei that are important for hearing are also found in the rostral medulla: the dorsal and ventral cochlear nuclei, and the superior olive (Chapter 11). Also shown are the inferior olive, important for motor control (Chapter 14), and the raphe nucleus, important for the modulation of pain, mood, and wakefulness (Chapters 12, 19, and 22).

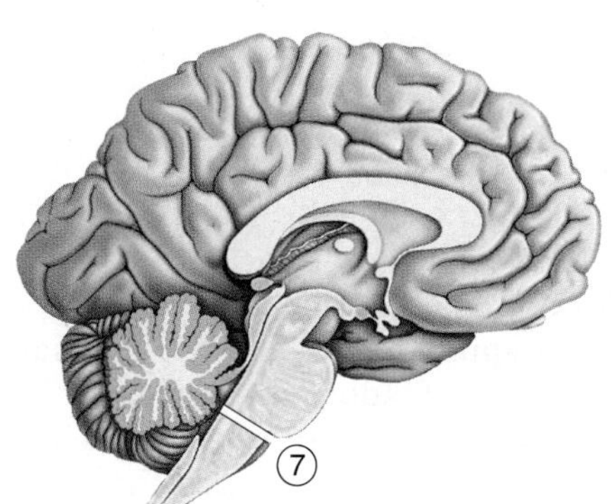

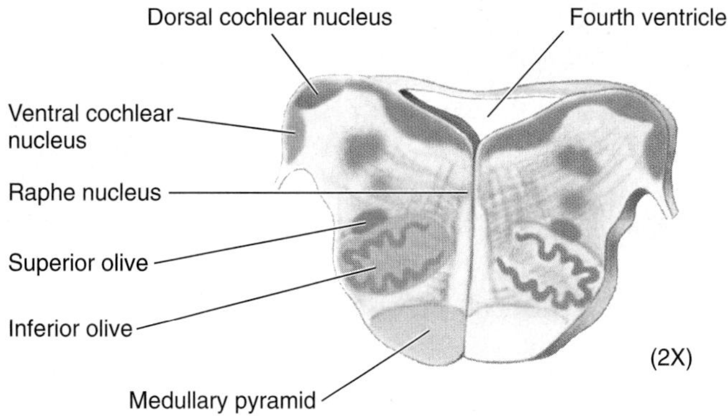

Cross Section 8: Mid-Medulla

The mid-medulla contains some of the same structures labeled in cross section 7. Notice, in addition, the medial lemniscus (Latin for "ribbon"). The medial lemniscus contains axons bringing information about somatic sensation to the thalamus (Chapter 12). The gustatory nucleus, a part of the larger solitary nucleus, serves the sense of taste (Chapter 8). The vestibular nuclei serve the sense of balance (Chapter 11).

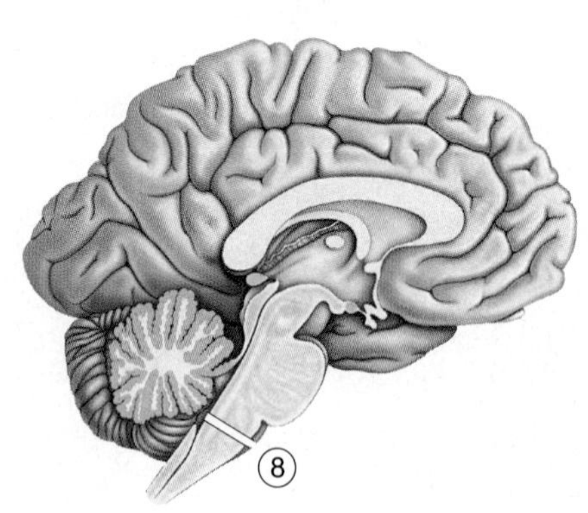

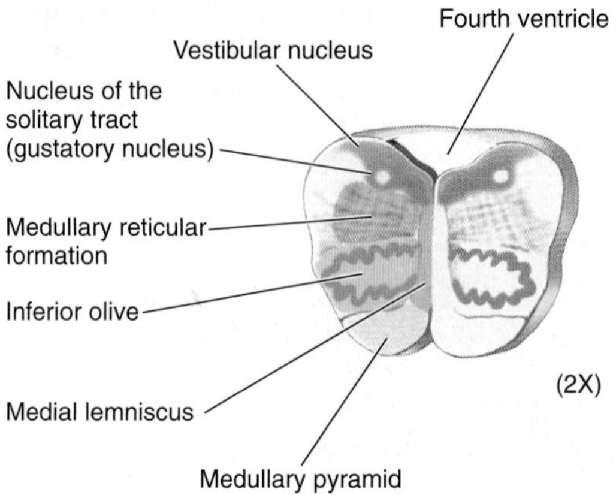

Cross Section 9: Medulla–Spinal Cord Junction

As the medulla disappears, so does the fourth ventricle, now replaced by the beginning of the spinal canal. Notice the dorsal column nuclei, which receive somatic sensory information from the spinal cord (Chapter 12). Axons arising from the neurons in each dorsal column nucleus cross to the other side of the brain (decussate) and ascend to the thalamus via the medial lemniscus.

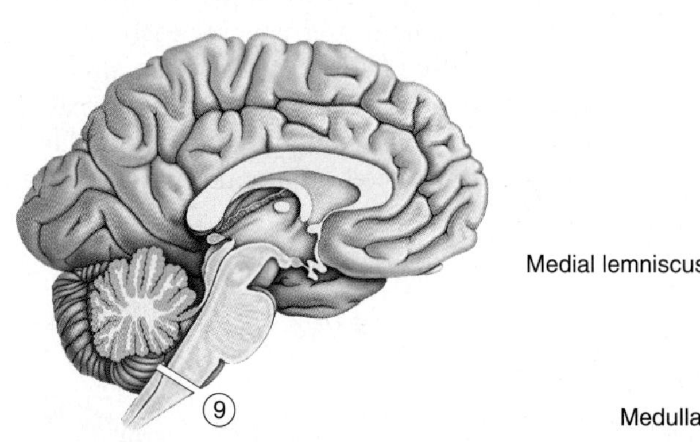

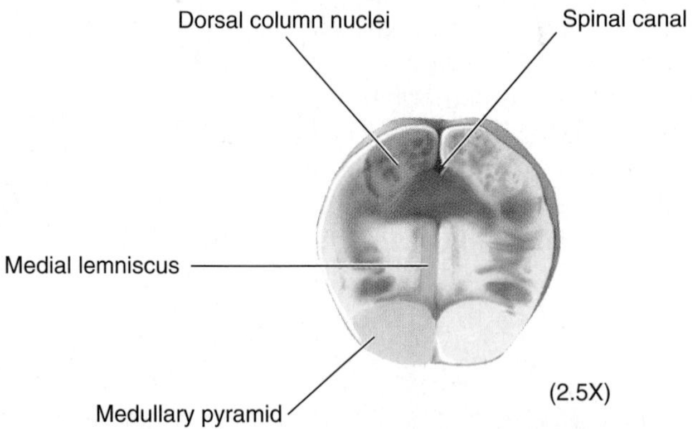

THE SPINAL CORD

The Dorsal Surface of the Spinal Cord and Spinal Nerves

The spinal cord is situated within the vertebral column. The spinal nerves, a part of the somatic peripheral nervous system (PNS), communicate with the cord via notches between the vertebrae. The vertebrae are described based on where they are found. In the neck, they are called *cervical vertebrae* and are numbered from 1 to 7. The vertebrae attached to ribs are called *thoracic vertebrae* and are numbered from 1 to 12. The five vertebrae of the lower back are called *lumbar*, and those within the pelvic area are called *sacral*.

Notice how the spinal nerves and the associated segments of the spinal cord adopt the names of the vertebrae (see how eight cervical nerves are associated with seven cervical vertebrae). Also notice that the spinal cord in the adult human ends at about the level of the third lumbar vertebra. This disparity arises because the spinal cord does not grow after birth, whereas the spinal column does. The bundles of spinal nerves streaming down within the lumbar and sacral vertebral column are called the *cauda equina* (Latin for "horse's tail").

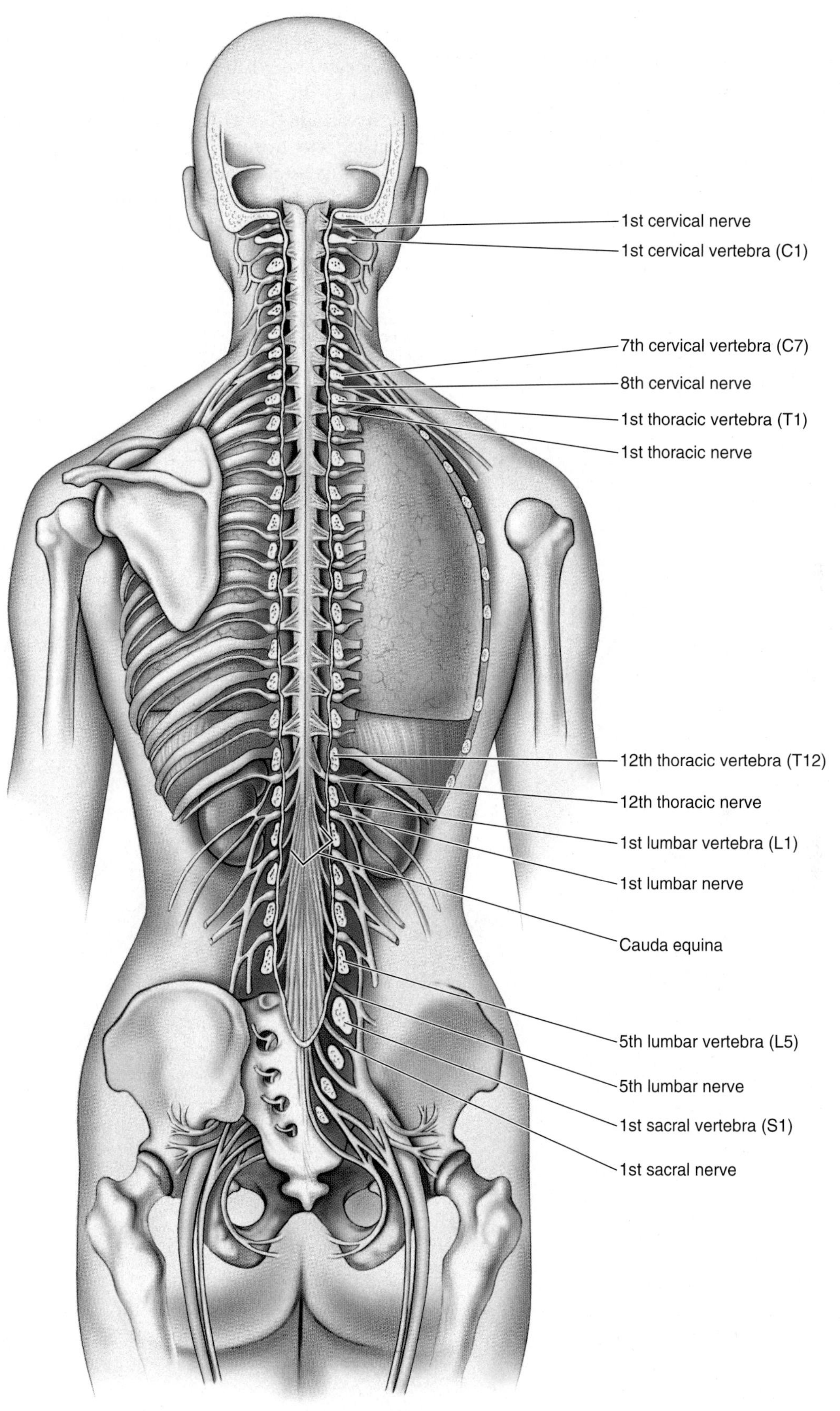
1st cervical nerve
1st cervical vertebra (C1)
7th cervical vertebra (C7)
8th cervical nerve
1st thoracic vertebra (T1)
1st thoracic nerve
12th thoracic vertebra (T12)
12th thoracic nerve
1st lumbar vertebra (L1)
1st lumbar nerve
Cauda equina
5th lumbar vertebra (L5)
5th lumbar nerve
1st sacral vertebra (S1)
1st sacral nerve

The Ventral–Lateral Surface

This view shows how the spinal nerves attach to the spinal cord and how the spinal meninges are organized. As the nerve passes into the vertebral notch, it splits into two roots. The dorsal root carries sensory axons whose cell bodies lie in the dorsal root ganglia. The ventral root carries motor axons arising from the gray matter of the ventral spinal cord. The butterfly-shaped core of the spinal cord is gray matter, consisting of neuronal cell bodies. The gray matter is divided into the dorsal, lateral, and ventral horns. Notice how the organization of gray and white matter in the spinal cord differs from that of the forebrain. In the forebrain, the gray matter surrounds the white matter; in the spinal cord, it is the other way around. The thick shell of white matter, containing the long axons that run up and down the cord, is divided into three columns: the dorsal columns, the lateral columns, and the ventral columns.

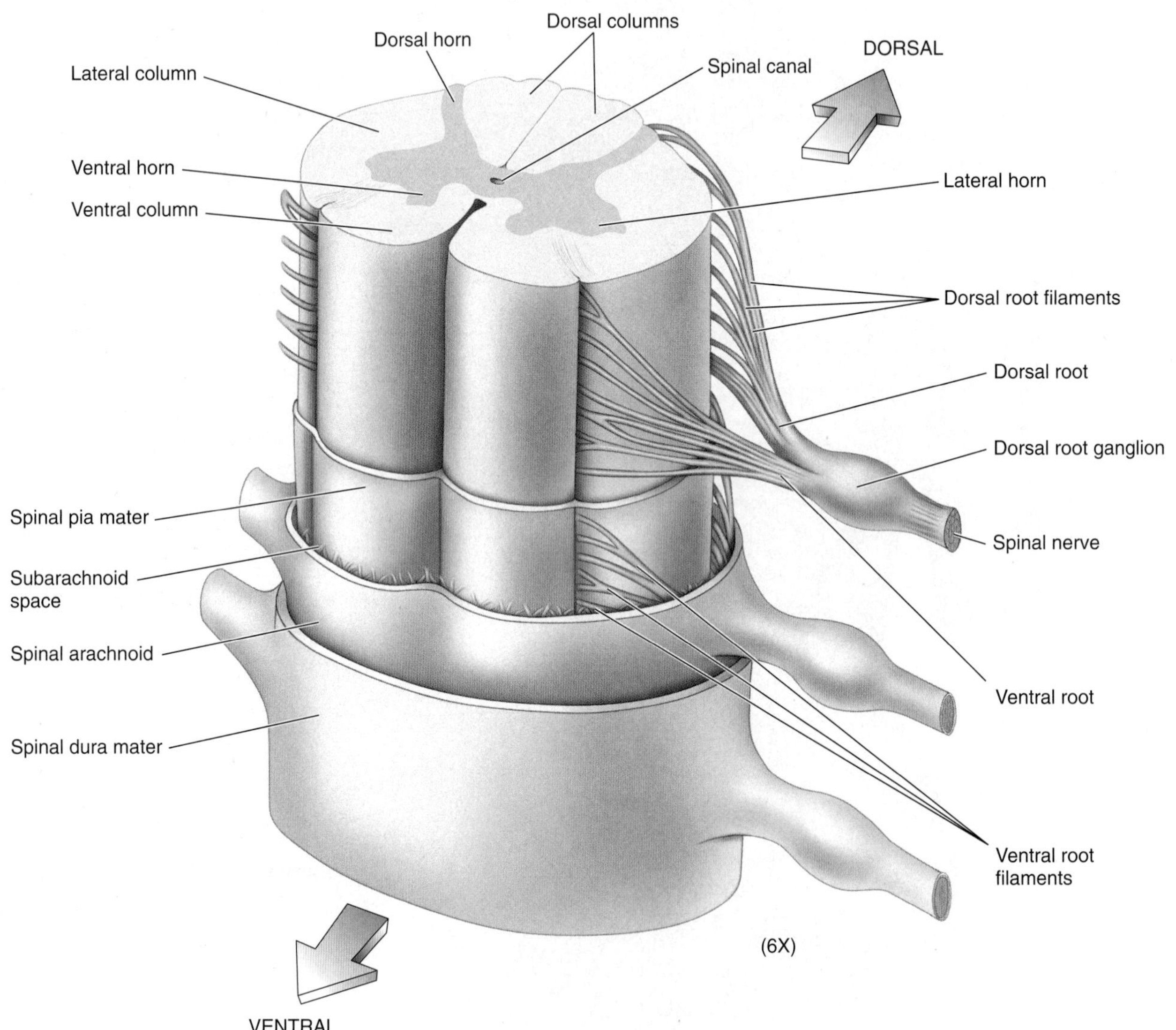

Cross-Sectional Anatomy

Illustrated in this view are some of the important tracts of axons running up and down the spinal cord. On the left side, the major ascending sensory pathways are indicated. Notice that the entire dorsal column consists of sensory axons ascending to the brain. This pathway is important for the conscious appreciation of touch. The spinothalamic tract carries information about painful stimuli and temperature. The somatic sensory system is discussed in Chapter 12. On the right side are some of the descending tracts important for the control of movement (Chapter 14). The names of the tracts accurately describe their origins and terminations (e.g., the vestibulospinal tract originates in the vestibular nuclei of the medulla and terminates in the spinal cord). Notice that the descending tracts contribute to two pathways: the lateral and ventromedial pathways. The lateral pathway carries the commands for voluntary movements, especially of the extremities. The ventromedial pathway participates mainly in the maintenance of posture and certain reflex movements.

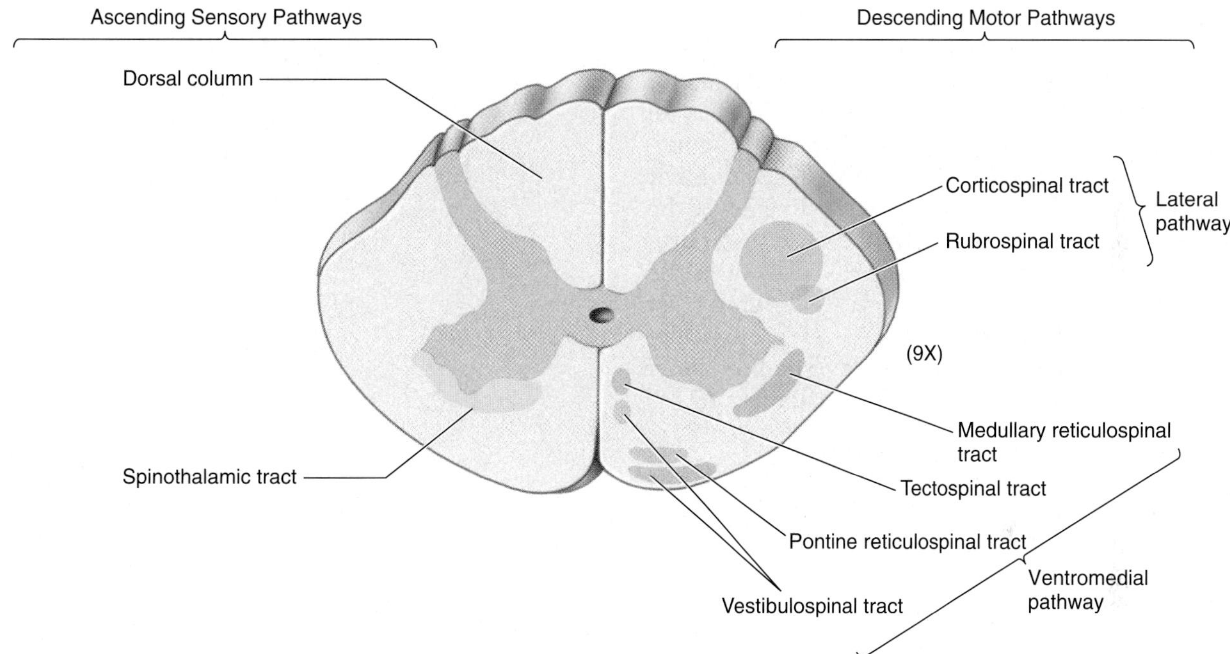

THE AUTONOMIC NERVOUS SYSTEM

In addition to the somatic PNS, which is devoted largely to the voluntary control of movement and conscious skin sensations, there is visceral PNS, devoted to the regulation of the internal organs, glands, and vasculature. Because this regulation occurs automatically and is not under direct conscious control, this system is called the *autonomic nervous system*, or *ANS*. The two most important divisions of the ANS are called the *sympathetic* and *parasympathetic divisions*.

The illustration shows the cavity of the body as it appears when it has been sectioned sagittally at the level of the eye. Notice the vertebral column, which is encased in a thick wall of connective tissue. The spinal nerves can be seen emerging from the column. Notice that the sympathetic division of the ANS consists of a chain of ganglia that runs along the side of the vertebral column. These ganglia communicate with the spinal nerves, with one another, and with a large number of internal organs. The parasympathetic division of the ANS is organized quite differently. Much of the parasympathetic innervation of the viscera arises from the vagus nerve, one of the cranial nerves emerging from the medulla. The other major source of parasympathetic fibers is the sacral spinal nerves. (The functional organization of the ANS is discussed in Chapter 15.)

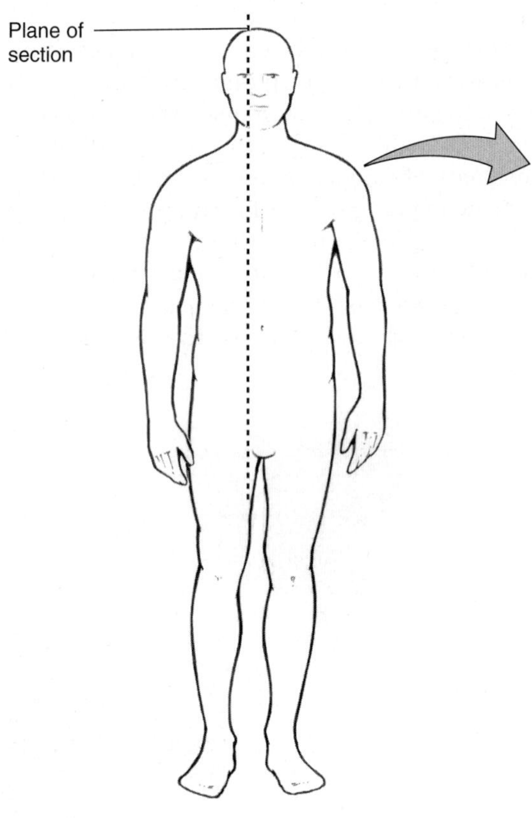

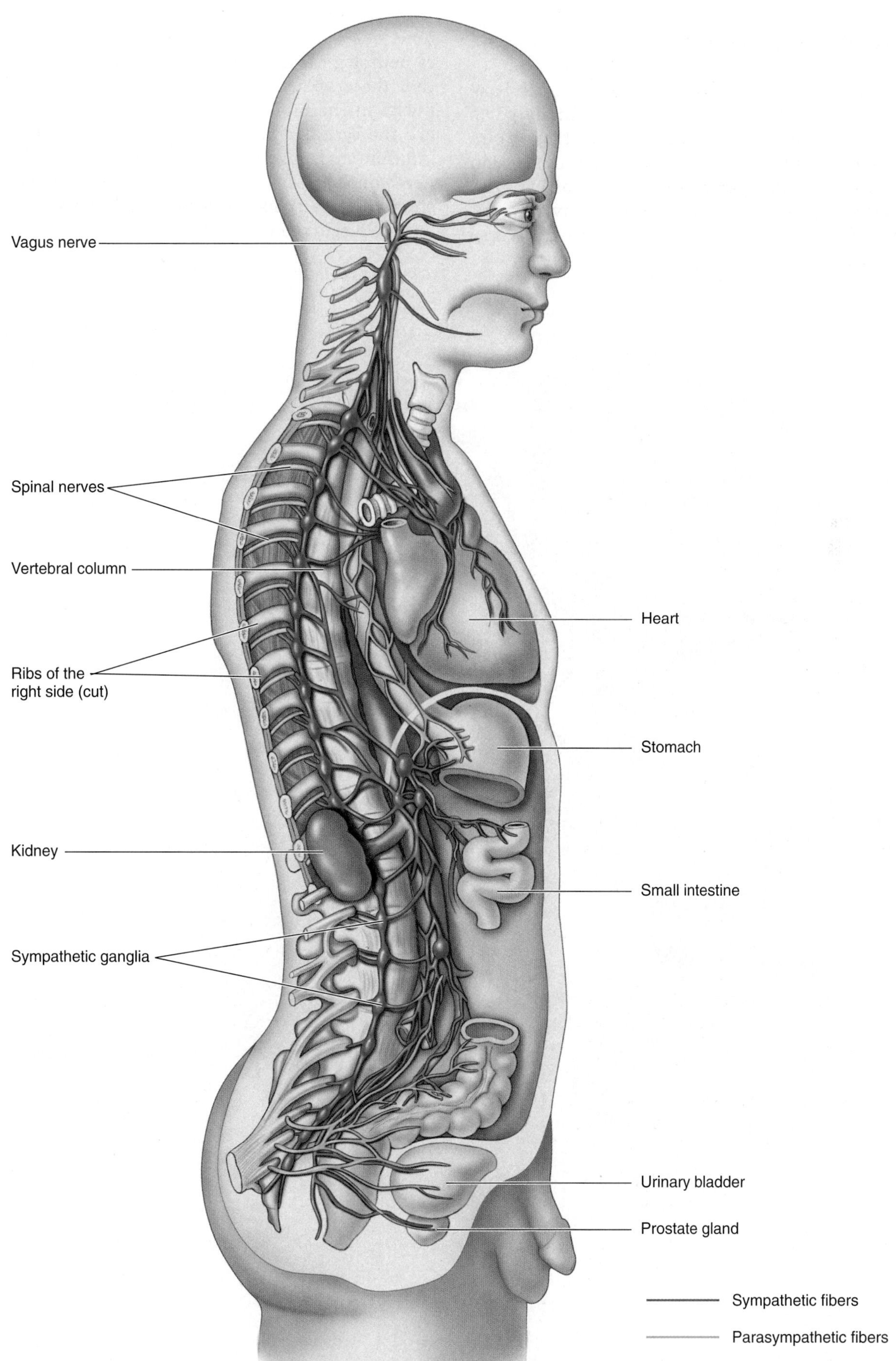
Vagus nerve
Spinal nerves
Vertebral column
Ribs of the right side (cut)
Kidney
Sympathetic ganglia
Heart
Stomach
Small intestine
Urinary bladder
Prostate gland
Sympathetic fibers
Parasympathetic fibers

THE CRANIAL NERVES

Twelve pairs of cranial nerves emerge from the base of the brain. The first two "nerves" are actually parts of the CNS, serving olfaction and vision. The rest are like the spinal nerves, in the sense that they contain axons of the PNS. However, as the illustration shows, a single nerve often has fibers performing many different functions. Knowledge of the nerves and their diverse functions is a valuable aid in the diagnosis of a number of neurological disorders. It is important to recognize that the cranial nerves have associated cranial nerve nuclei in the midbrain, pons, and medulla. Examples are the cochlear and vestibular nuclei, which receive information from cranial nerve VIII. Most of the cranial nerve nuclei were not illustrated or labeled in the brain stem cross sections, however, because their functions are not discussed explicitly in this book.

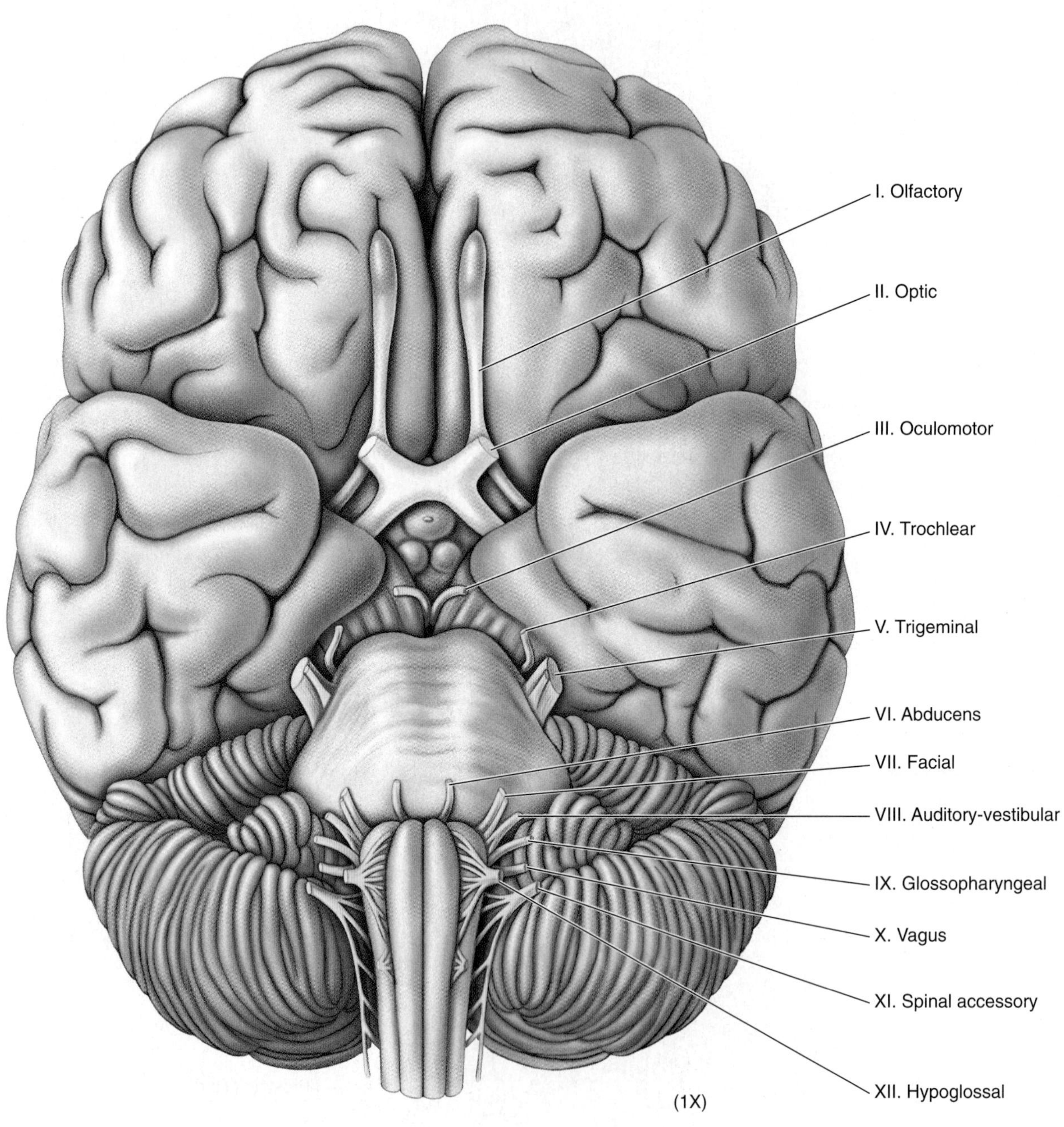

NERVE NUMBER AND NAME	TYPES OF AXONS	IMPORTANT FUNCTIONS
I. Olfactory	Special sensory	Sensation of smell
II. Optic	Special sensory	Sensation of vision
III. Oculomotor	Somatic motor Visceral motor	Movements of the eye and eyelid Parasympathetic control of pupil size
IV. Trochlear	Somatic motor	Movements of the eye
V. Trigeminal	Somatic sensory Somatic motor	Sensation of touch to the face Movement of muscles of mastication (chewing)
VI. Abducens	Somatic motor	Movements of the eye
VII. Facial	Somatic sensory Special sensory	Movement of muscles of facial expression Sensation of taste in anterior two-thirds of the tongue
VIII. Auditory-vestibular	Special sensory	Sensation of hearing and balance
IX. Glossopharyngeal	Somatic motor Visceral motor Special sensory Visceral sensory	Movement of muscles in the throat (oropharynx) Parasympathetic control of the salivary glands Sensation of taste in posterior one-third of the tongue Detection of blood pressure changes in the aorta
X. Vagus	Visceral motor Visceral sensory Somatic motor	Parasympathetic control of the heart, lungs, and abdominal organs Sensation of pain associated with viscera Movement of muscles in the throat (oropharynx)
XI. Spinal accessory	Somatic motor	Movement of muscles in the throat and neck
XII. Hypoglossal	Somatic motor	Movement of the tongue

THE BLOOD SUPPLY OF THE BRAIN

Ventral View

Two pairs of arteries supply blood to the brain: the vertebral arteries and the internal carotid arteries. The vertebral arteries converge near the base of the pons to form the unpaired basilar artery. At the level of the midbrain, the basilar artery splits into the right and left superior cerebellar arteries and the posterior cerebral arteries. Notice that the posterior cerebral arteries send branches, called *posterior communicating arteries*, that connect them to the internal carotids. The internal carotids branch to form the middle cerebral arteries and the anterior cerebral arteries. The anterior cerebral arteries of each side are connected by the anterior communicating artery. Thus, there is a ring of connected arteries at the brain's base, formed by the posterior cerebral and communicating arteries, the internal carotids, and the anterior cerebral and communicating arteries. This ring is called the *circle of Willis*.

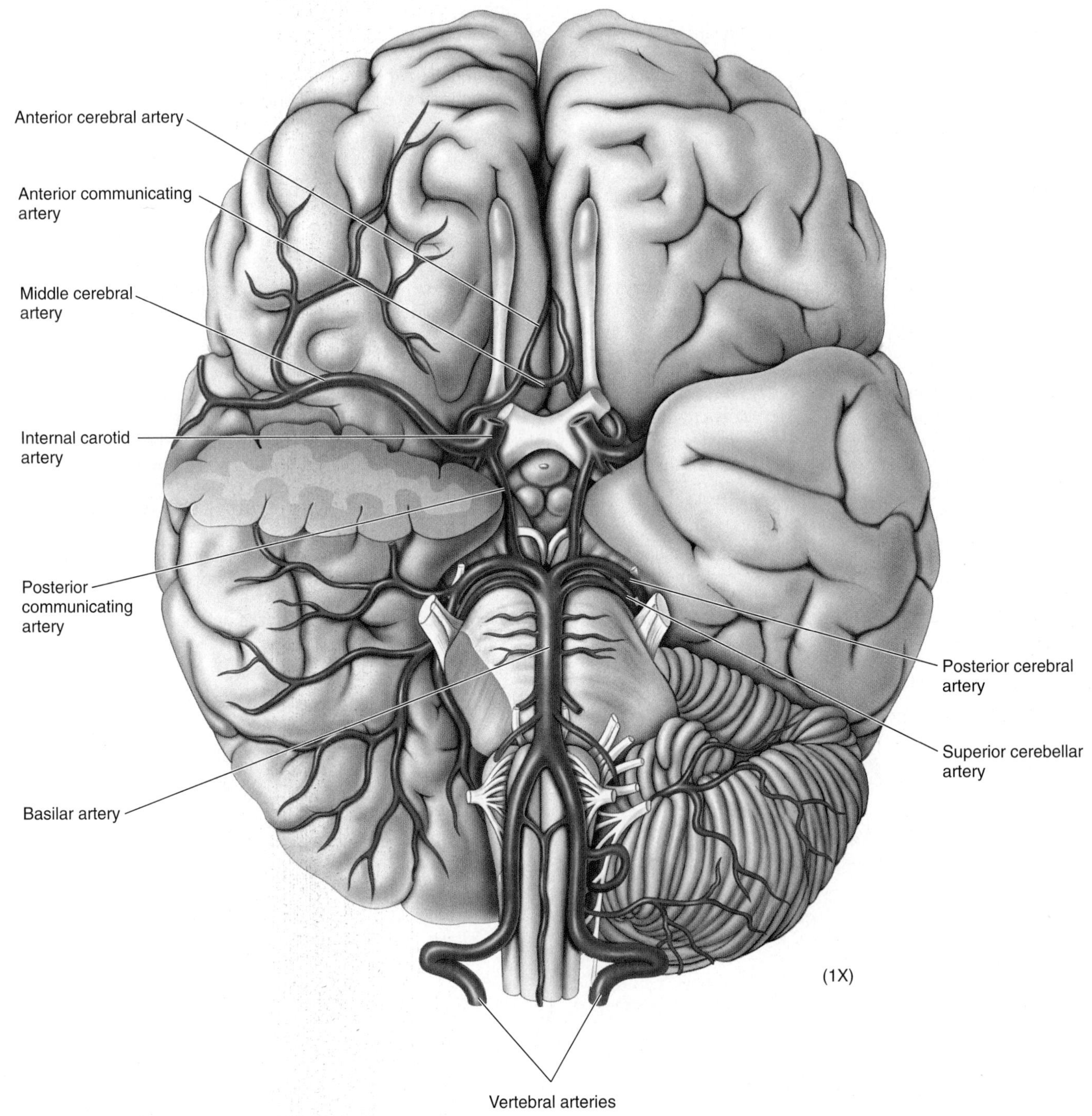

Lateral view

Notice that most of the lateral surface of the cerebrum is supplied by the middle cerebral artery. This artery also feeds the deep structures of the basal forebrain.

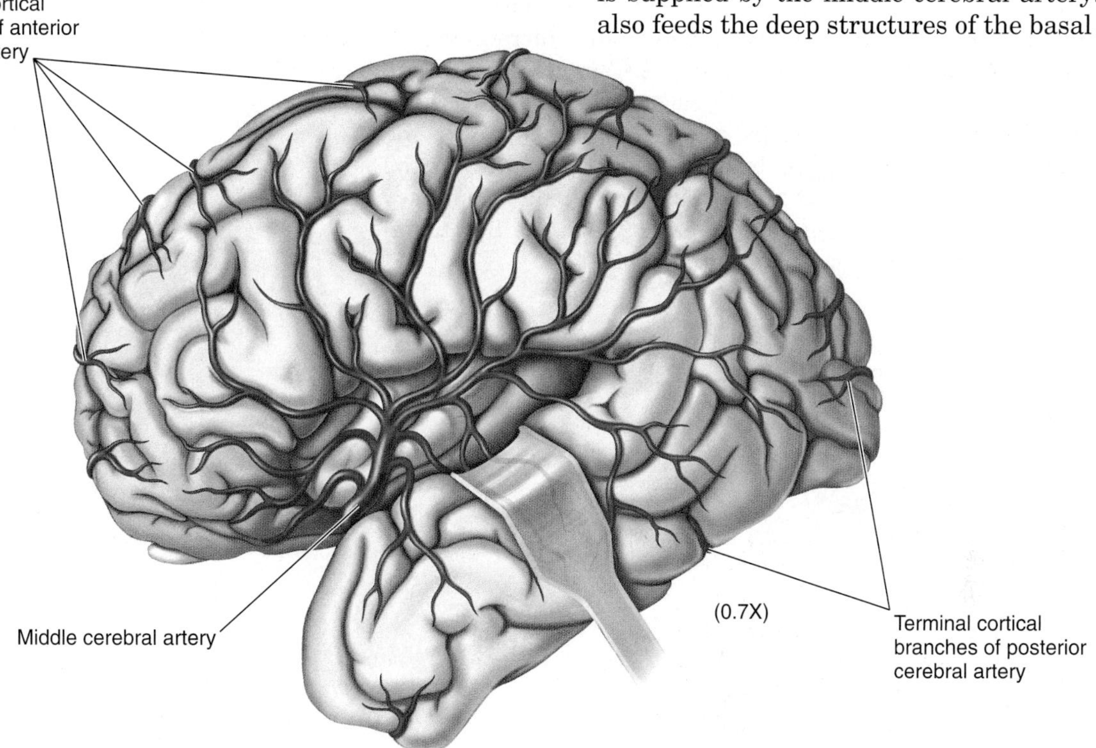

Medial View (Brain Stem Removed)

Most of the medial wall of the cerebral hemisphere is supplied by the anterior cerebral artery. The posterior cerebral artery feeds the medial wall of the occipital lobe and the inferior part of the temporal lobe.

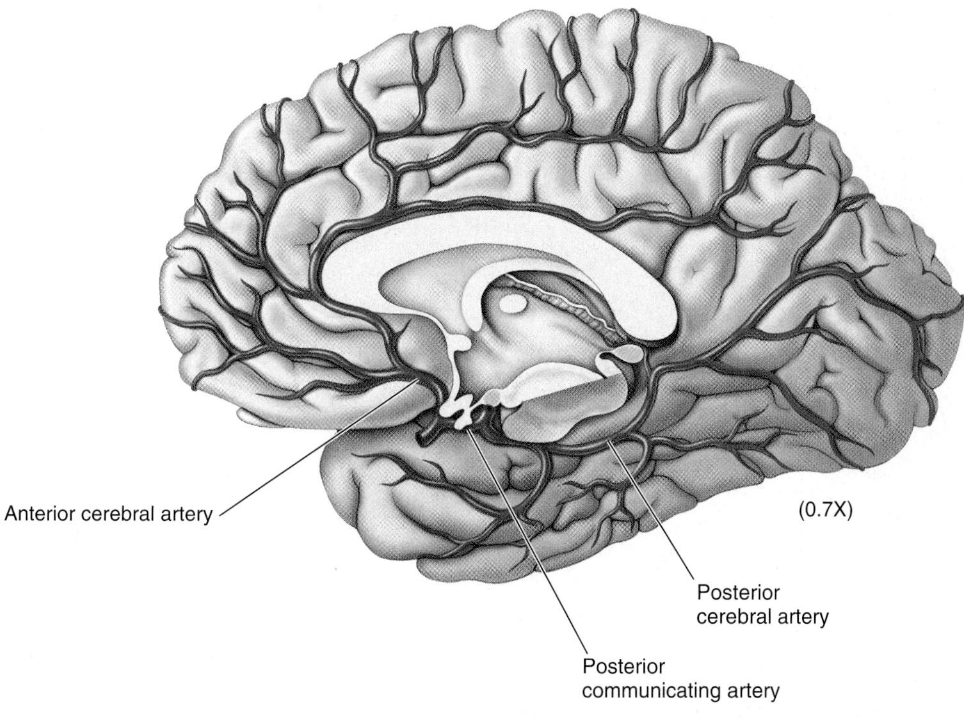

SELF-QUIZ

This review workbook is designed to help you learn the neuroanatomy that has been presented. Here, we have reproduced the images from the Guide; however, instead of labels, numbered leader lines (arranged in a clockwise fashion) point to the structures of interest. Test your knowledge by filling in the appropriate names in the spaces provided. To review what you have learned, quiz yourself by putting your hand over the names. Experience has shown that this technique greatly facilitates the learning and retention of anatomical terms. Mastery of the vocabulary of neuroanatomy will serve you well as you learn about the functional organization of the brain in the remainder of the book.

The Lateral Surface of the Brain

(a) Gross Features

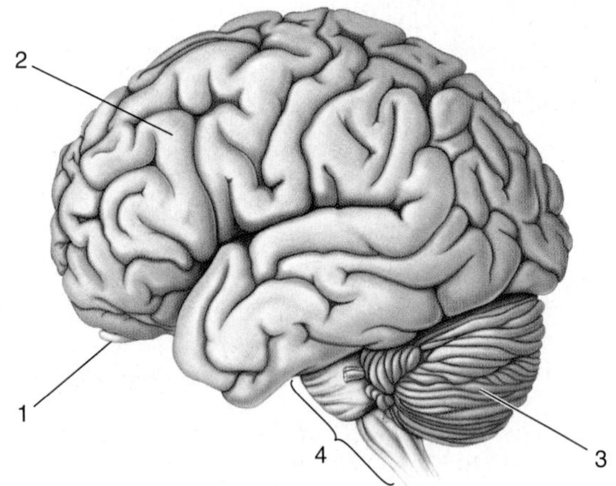

1. ____________________

2. ____________________

3. ____________________

4. ____________________

(b) Selected Gyri, Sulci, and Fissures

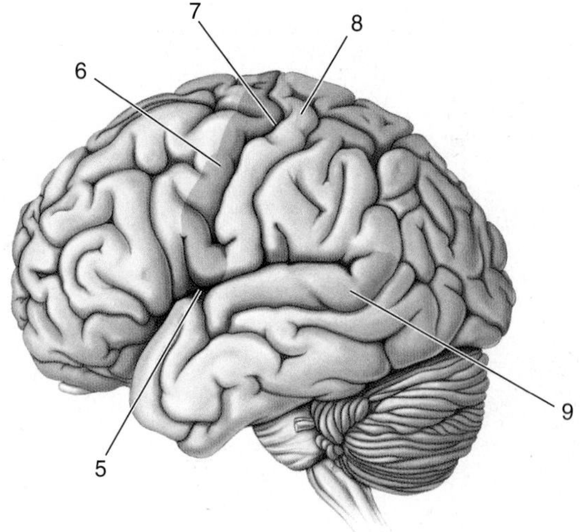

5. ____________________

6. ____________________

7. ____________________

8. ____________________

9. ____________________

The Lateral Surface of the Brain

(c) Cerebral Lobes and the Insula

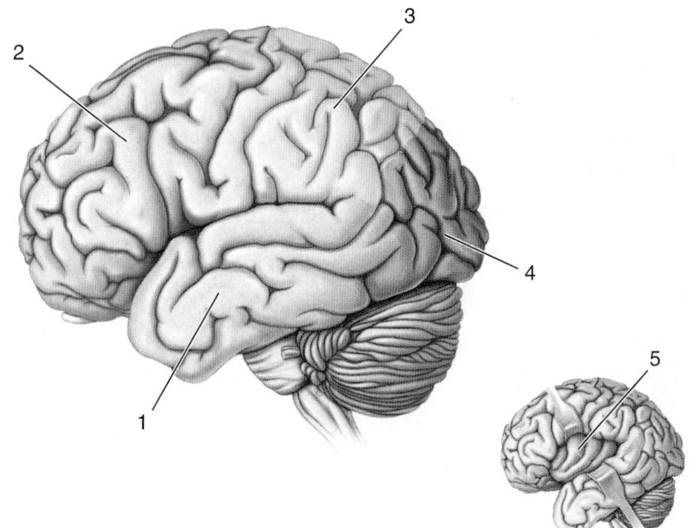

1. ____________________

2. ____________________

3. ____________________

4. ____________________

5. ____________________

(d) Major Sensory, Motor, and Association Areas of Cortex

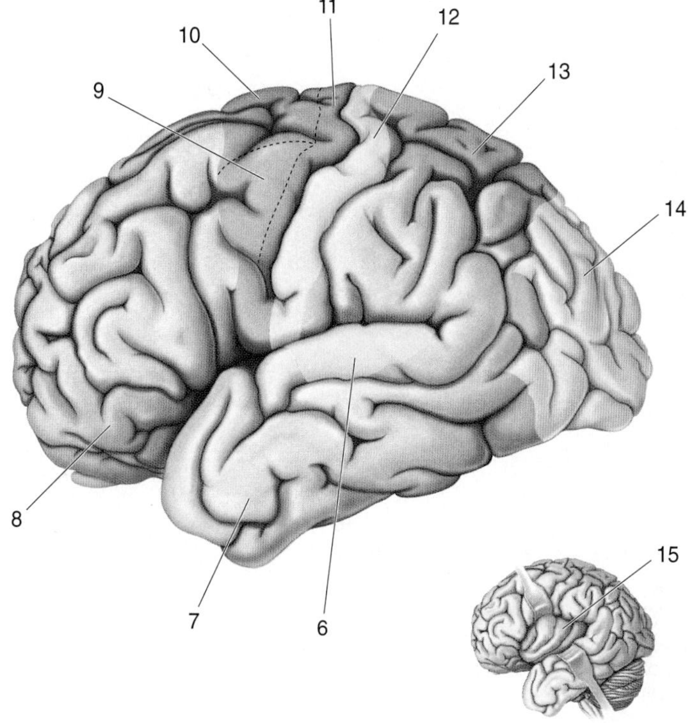

6. ____________________

7. ____________________

8. ____________________

9. ____________________

10. ____________________

11. ____________________

12. ____________________

13. ____________________

14. ____________________

15. ____________________

The Medial Surface of the Brain

(a) Brain Stem Structures

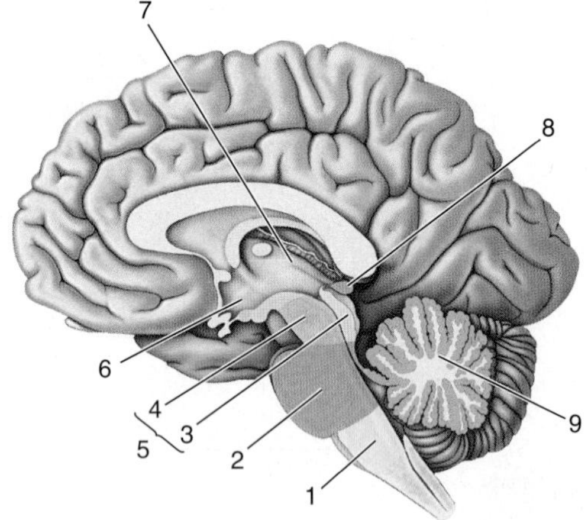

(b) Forebrain Structures

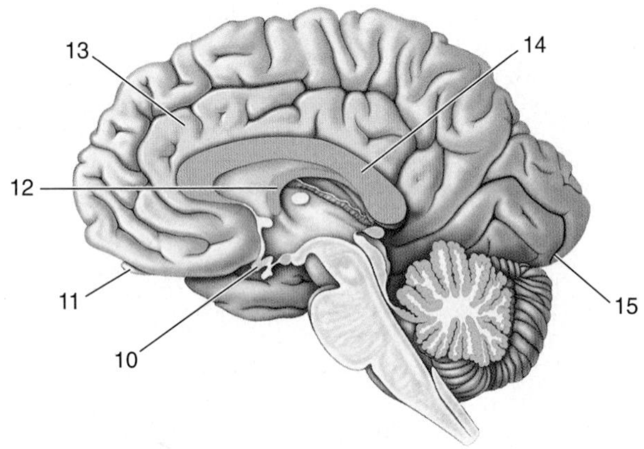

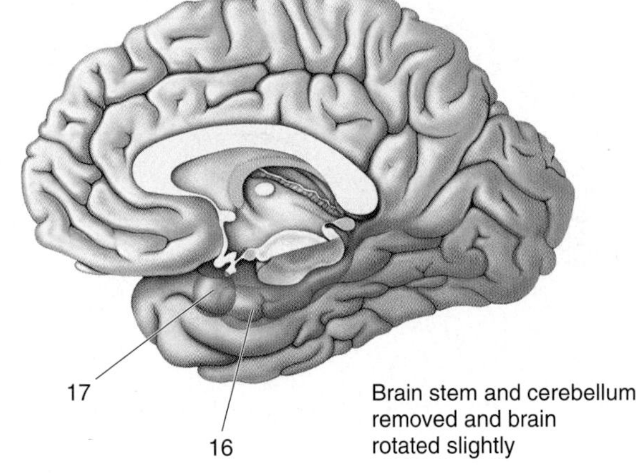

Brain stem and cerebellum removed and brain rotated slightly

1. ______________________

2. ______________________

3. ______________________

4. ______________________

5. ______________________

6. ______________________

7. ______________________

8. ______________________

9. ______________________

10. ______________________

11. ______________________

12. ______________________

13. ______________________

14. ______________________

15. ______________________

16. ______________________

17. ______________________

The Medial Surface of the Brain

(a) Ventricles

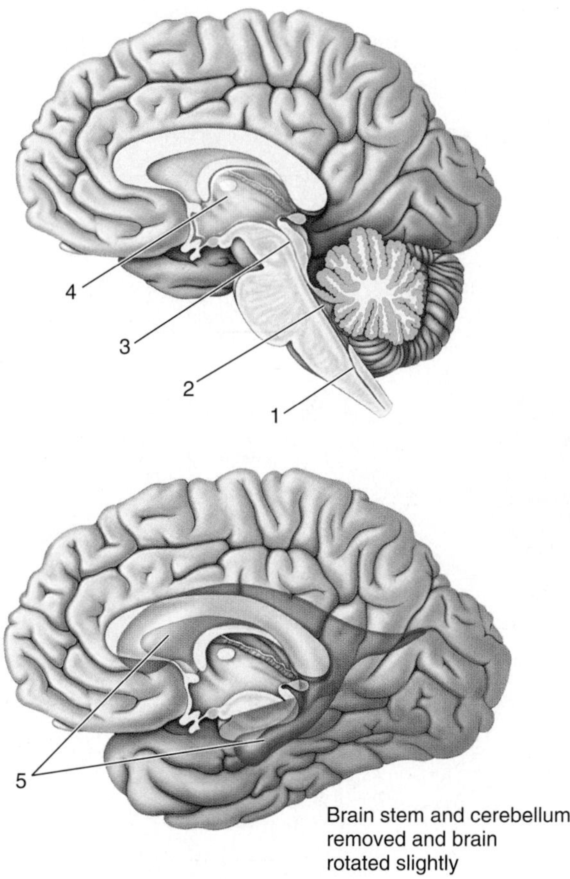

1. ______________________

2. ______________________

3. ______________________

4. ______________________

5. ______________________

The Ventral Surface of the Brain

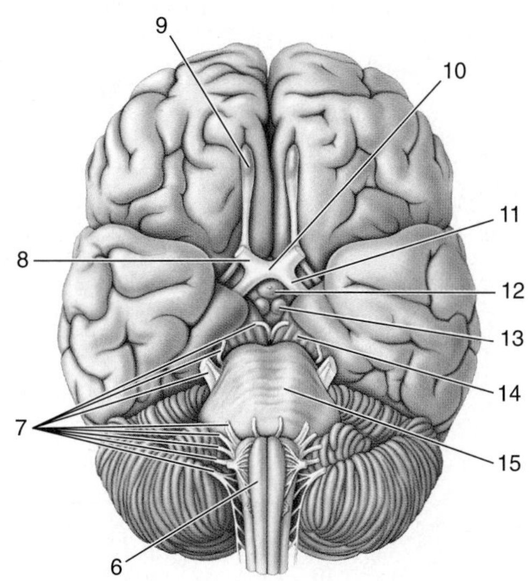

6. ______________________

7. ______________________

8. ______________________

9. ______________________

10. ______________________

11. ______________________

12. ______________________

13. ______________________

14. ______________________

15. ______________________

The Dorsal Surface of the Brain

(a) Cerebrum

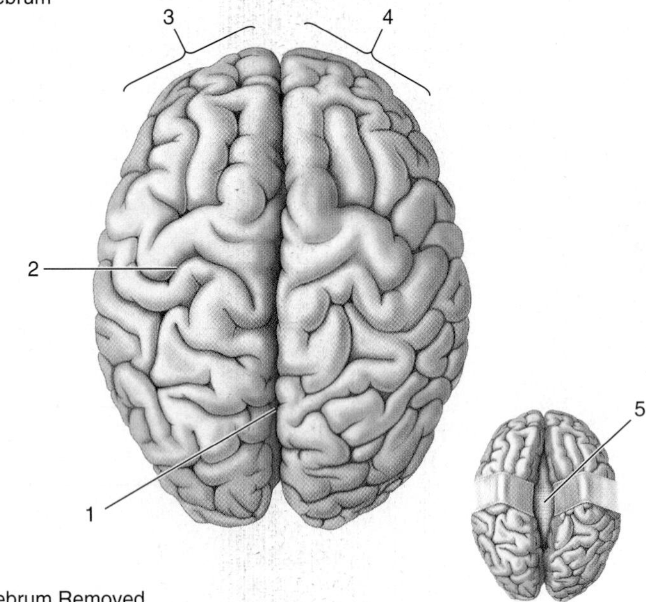

(b) Cerebrum Removed

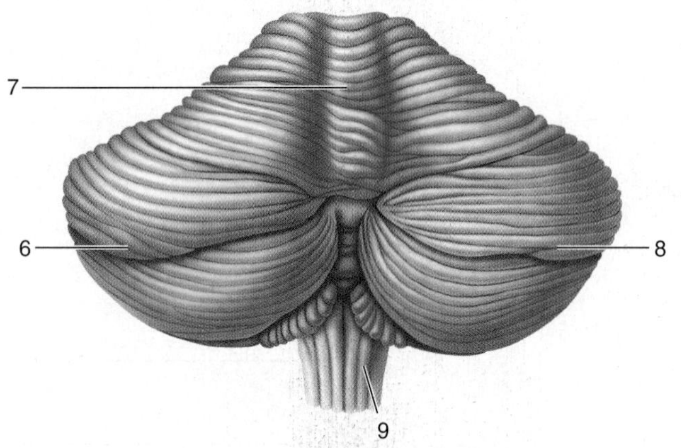

(c) Cerebrum and Cerebellum Removed

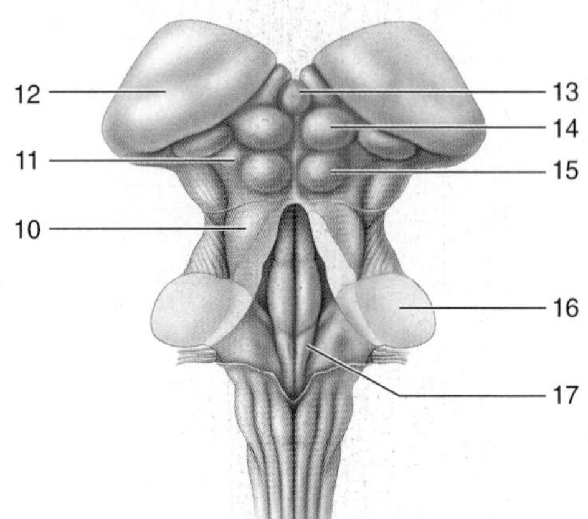

1. ______________________

2. ______________________

3. ______________________

4. ______________________

5. ______________________

6. ______________________

7. ______________________

8. ______________________

9. ______________________

10. ______________________

11. ______________________

12. ______________________

13. ______________________

14. ______________________

15. ______________________

16. ______________________

17. ______________________

Forebrain at Thalamus–Telecephalon Junction

(a) Gross Features

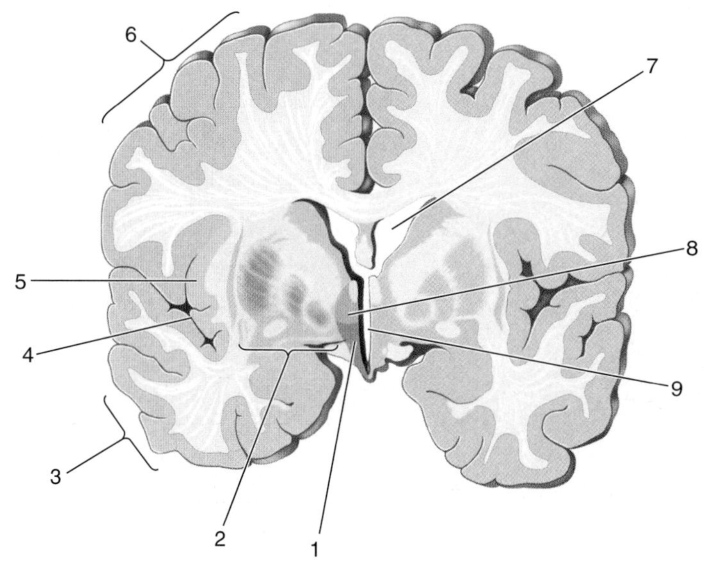

(b) Selected Cell and Fiber Groups

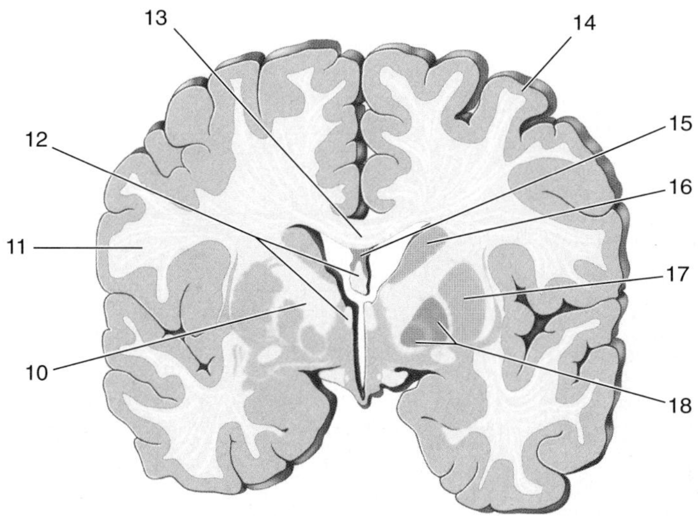

1. ______________________
2. ______________________
3. ______________________
4. ______________________
5. ______________________
6. ______________________
7. ______________________
8. ______________________
9. ______________________
10. ______________________
11. ______________________
12. ______________________
13. ______________________
14. ______________________
15. ______________________
16. ______________________
17. ______________________
18. ______________________

Forebrain at Mid-Thalamus

(a) Gross Features

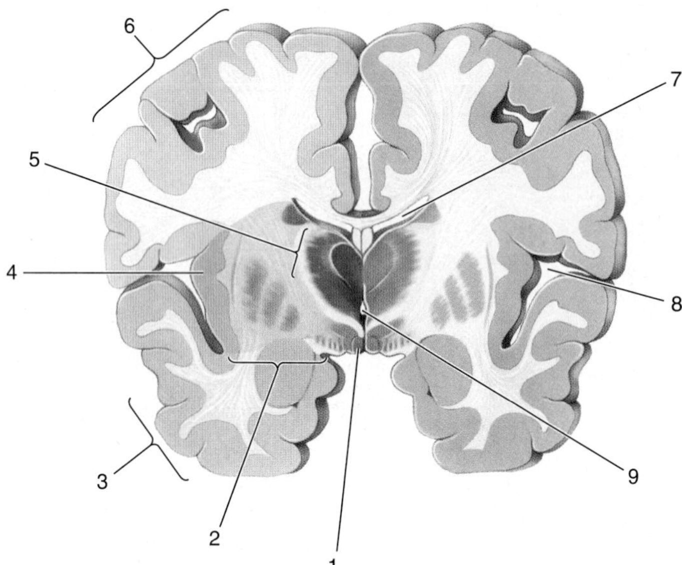

(b) Selected Cell and Fiber Groups

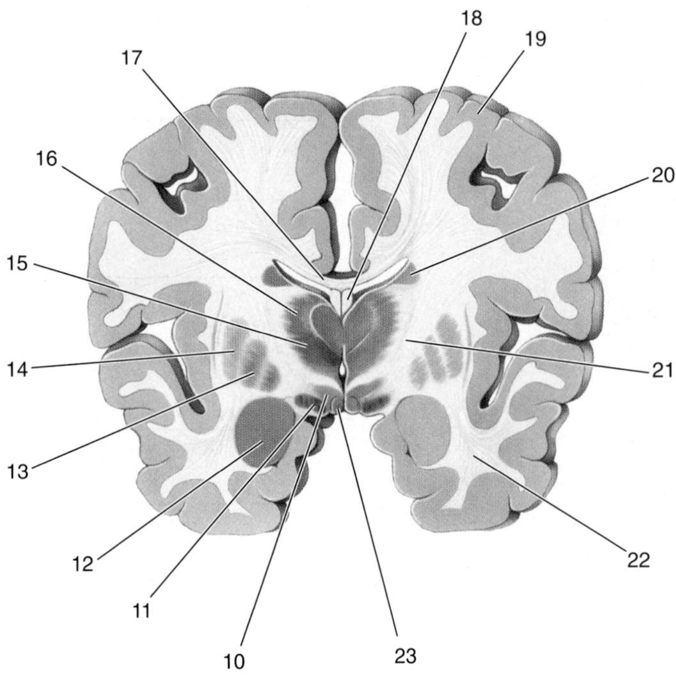

1. ____________________
2. ____________________
3. ____________________
4. ____________________
5. ____________________
6. ____________________
7. ____________________
8. ____________________
9. ____________________
10. ____________________
11. ____________________
12. ____________________
13. ____________________
14. ____________________
15. ____________________
16. ____________________
17. ____________________
18. ____________________
19. ____________________
20. ____________________
21. ____________________
22. ____________________
23. ____________________

Forebrain at Thalamus-Midbrain Junction

(a) Gross Features

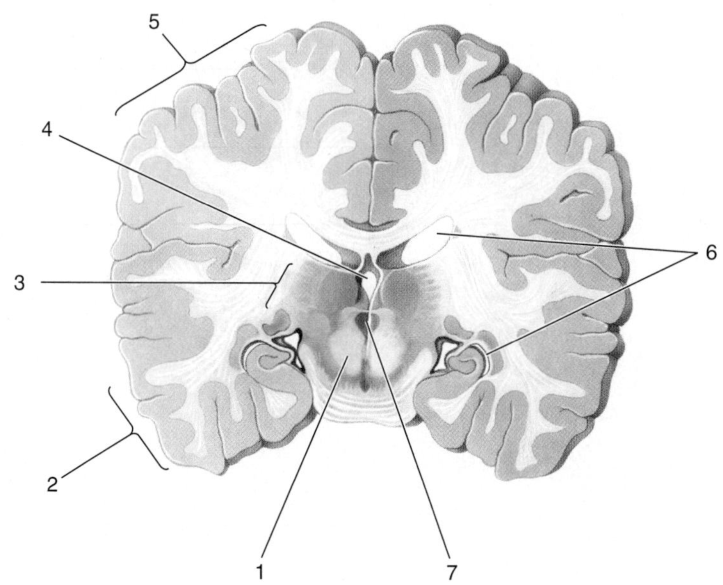

1. ______
2. ______
3. ______
4. ______
5. ______
6. ______
7. ______

(b) Selected Cell and Fiber Groups

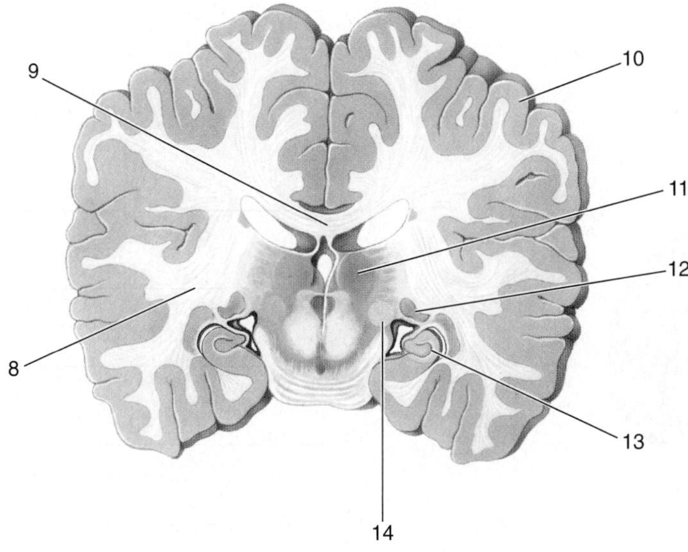

8. ______
9. ______
10. ______
11. ______
12. ______
13. ______
14. ______

Rostral Midbrain

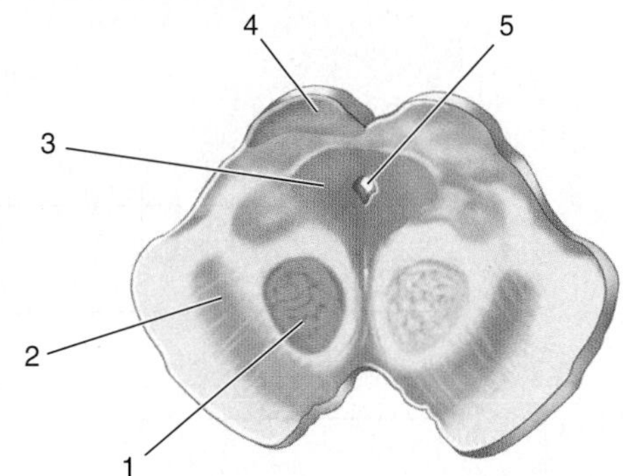

1. ______________________________
2. ______________________________
3. ______________________________
4. ______________________________
5. ______________________________

Caudal Midbrain

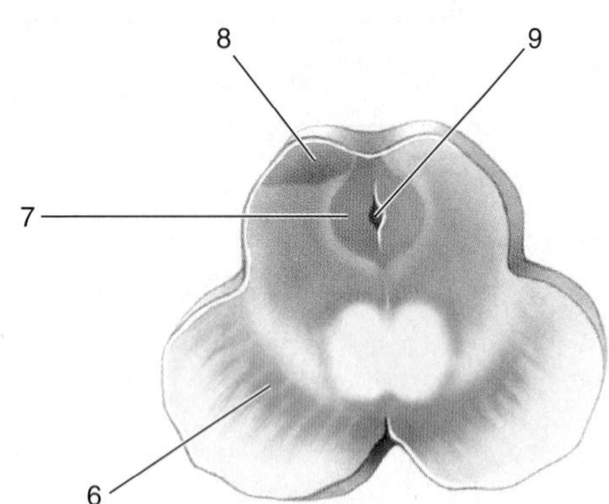

6. ______________________________
7. ______________________________
8. ______________________________
9. ______________________________

Pons and Cerebellum

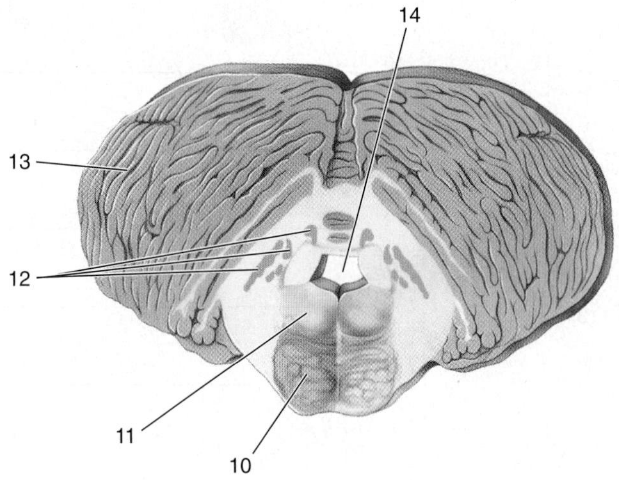

10. ______________________________
11. ______________________________
12. ______________________________
13. ______________________________
14. ______________________________

Rostral Medulla

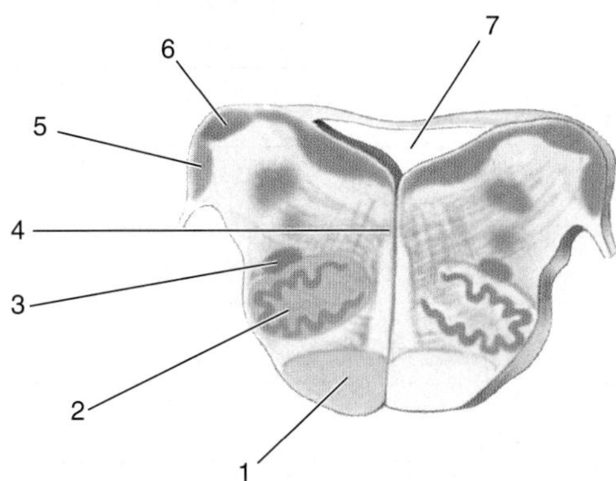

Mid Medulla

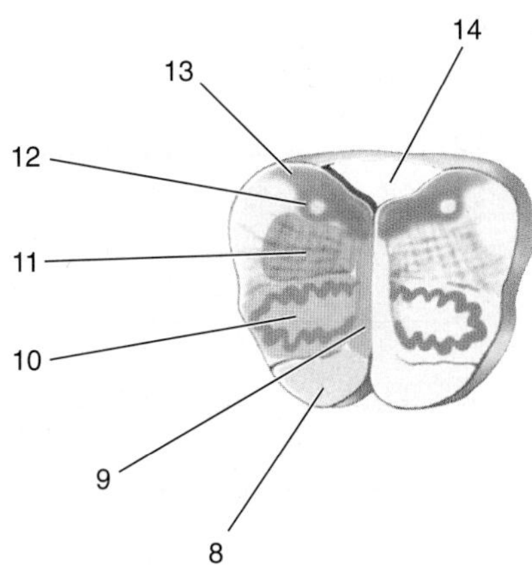

Medulla-Spinal Cord Junction

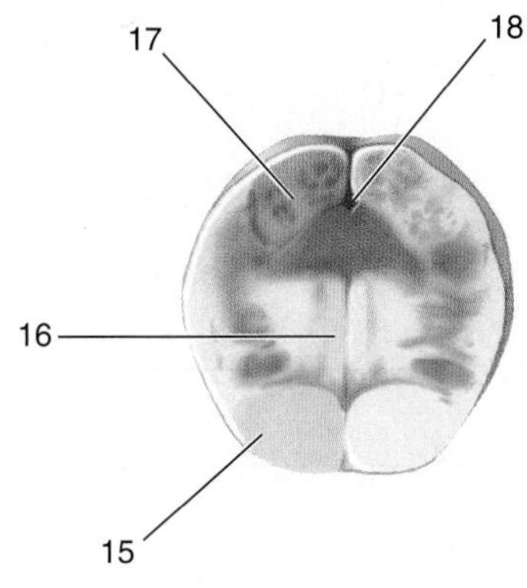

1. __________
2. __________
3. __________
4. __________
5. __________
6. __________
7. __________
8. __________
9. __________
10. __________
11. __________
12. __________
13. __________
14. __________
15. __________
16. __________
17. __________
18. __________

Spinal Cord, Ventral–Lateral Surface

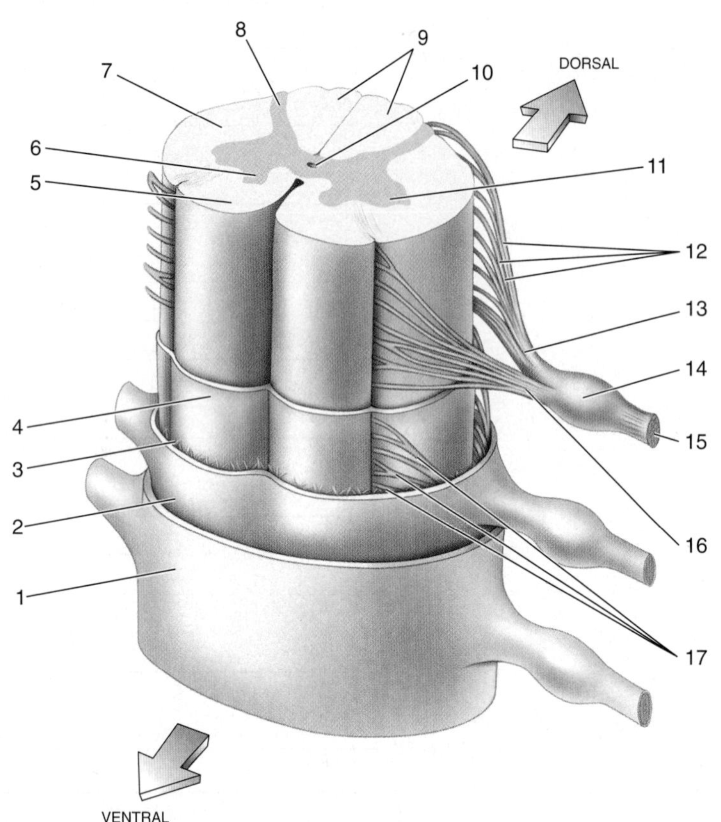

Spinal Cord, Cross-Sectional Anatomy

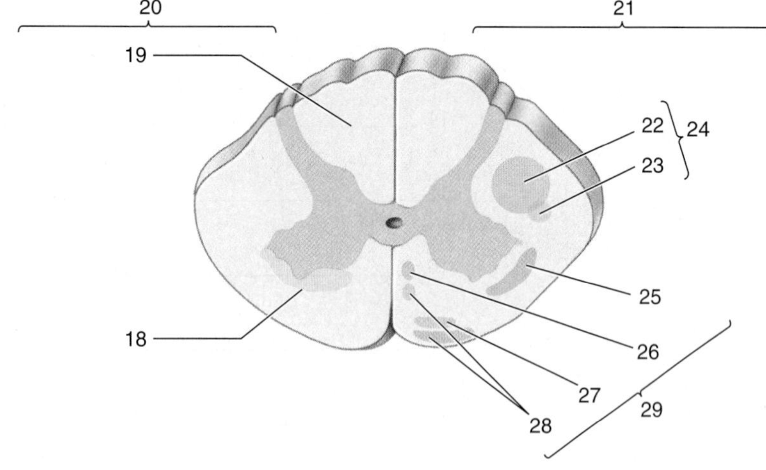

1. ______________________
2. ______________________
3. ______________________
4. ______________________
5. ______________________
6. ______________________
7. ______________________
8. ______________________
9. ______________________
10. ______________________
11. ______________________
12. ______________________
13. ______________________
14. ______________________
15. ______________________
16. ______________________
17. ______________________
18. ______________________
19. ______________________
20. ______________________
21. ______________________
22. ______________________
23. ______________________
24. ______________________
25. ______________________
26. ______________________
27. ______________________
28. ______________________
29. ______________________

The Cranial Nerves

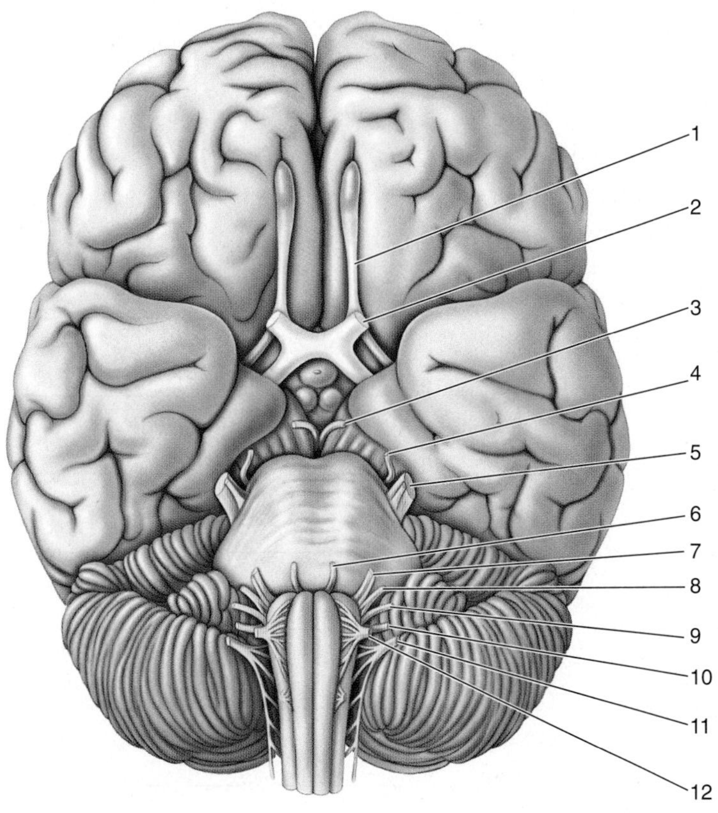

1. ______________________

2. ______________________

3. ______________________

4. ______________________

5. ______________________

6. ______________________

7. ______________________

8. ______________________

9. ______________________

10. ______________________

11. ______________________

12. ______________________

The Blood Supply of the Brain

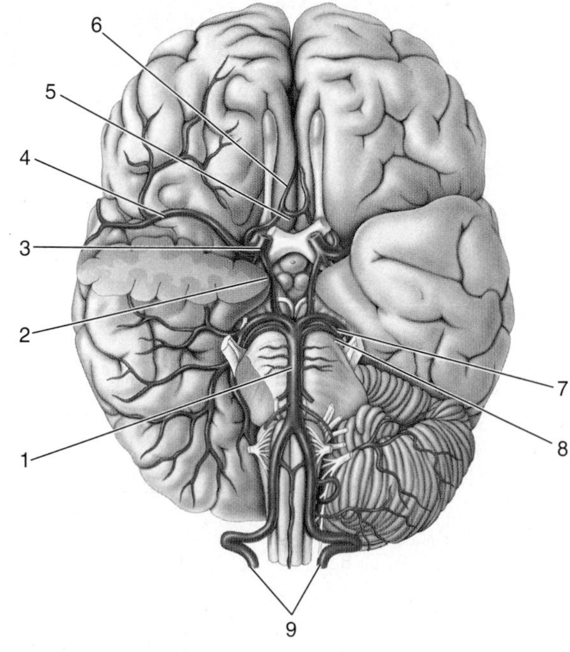

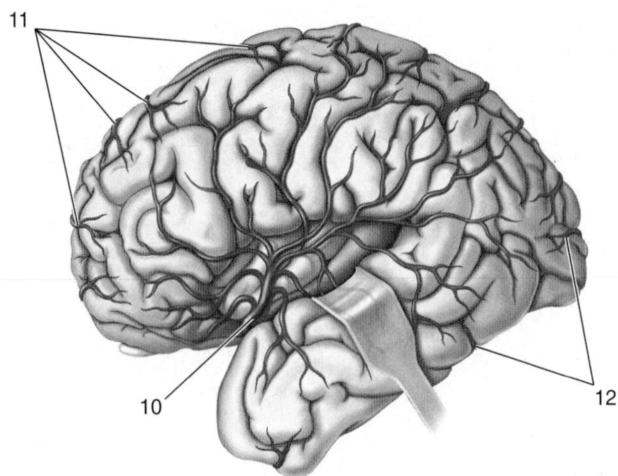

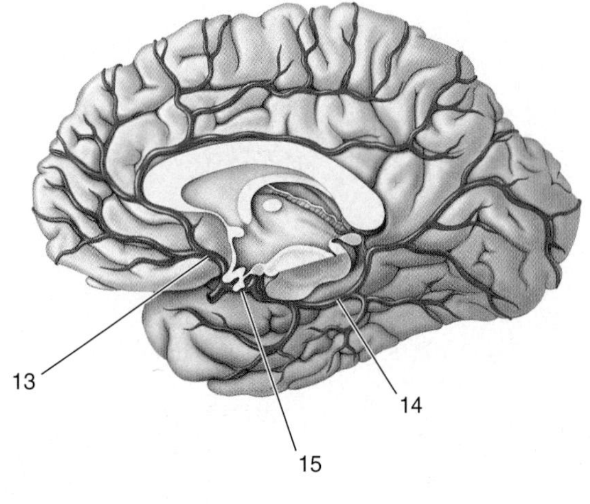

1. ______________________________

2. ______________________________

3. ______________________________

4. ______________________________

5. ______________________________

6. ______________________________

7. ______________________________

8. ______________________________

9. ______________________________

10. ______________________________

11. ______________________________

12. ______________________________

13. ______________________________

14. ______________________________

15. ______________________________

PART TWO

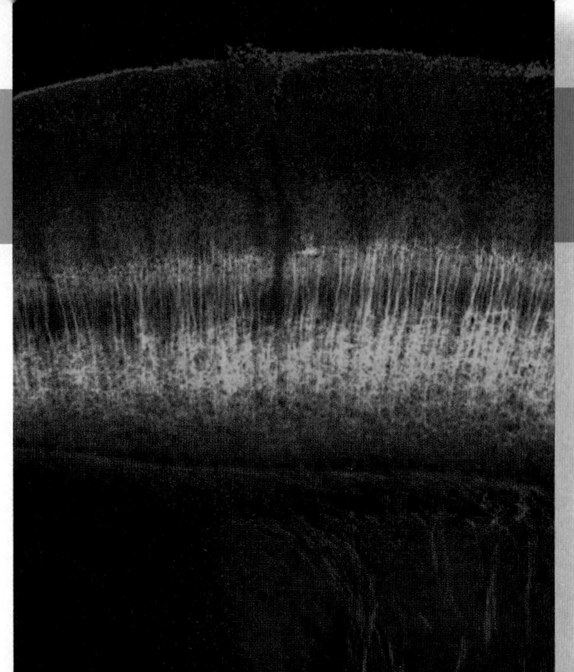

Sensory and Motor Systems

CHAPTER EIGHT

The Chemical Senses

INTRODUCTION

Life evolved in a sea of chemicals. From the beginning, organisms have floated or swum in water containing chemical substances that signal food, poison, or sex. In this respect, things have not changed much in three billion years. Animals, including humans, depend on the chemical senses to help identify nourishment (the sweetness of honey, the aroma of pizza), noxious substances (the bitterness of plant poisons), or the suitability of a potential mate. Of all the sensory systems, chemical sensation is the oldest and most pervasive across species. Even brainless bacteria can detect, and tumble toward, a favorable food source.

Multicellular organisms must detect chemicals in both their internal and their external environments. The variety of chemical detection systems has expanded considerably over the course of evolution. Humans live in a sea of air, full of volatile chemicals; we put chemicals into our mouth for a variety of reasons; and we carry a complex sea within us in the form of blood and the other fluids that bathe our cells. We have specialized detection systems for the chemicals in each environment. The mechanisms of chemical sensation that originally evolved to detect environmental substances now serve as the basis for chemical communication between cells and organs, using hormones and neurotransmitters. Every cell in every organism is responsive to many chemicals.

This chapter considers the most familiar of our chemical senses: taste, or **gustation**, and smell, or **olfaction**. Although taste and smell reach our awareness most often, they are not our only important chemical senses. Many types of chemically sensitive cells, called **chemoreceptors**, are distributed throughout the body. For example, some nerve endings in skin and mucous membranes warn us of irritating chemicals. A wide range of chemoreceptors report subconsciously and consciously about our internal state: Nerve endings in the digestive organs detect many types of ingested substances, receptors in arteries of the neck measure carbon dioxide and oxygen levels in our blood, and sensory endings in muscles respond to acidity, giving us the burning feeling that comes with exertion and oxygen debt.

Gustation and olfaction have a similar task: the detection of environmental chemicals. In fact, only by using both senses together can the nervous system perceive flavor. Gustation and olfaction have unusually strong and direct connections with our most basic internal needs, including thirst, hunger, emotion, sex, and certain forms of memory. However, the systems of gustation and olfaction are separate and different, from the structures and mechanisms of their chemoreceptors, to the gross organization of their central connections, to their effects on behavior. The neural information from each system is processed in parallel and is merged at rather high levels in the cerebral cortex.

TASTE

Humans evolved as omnivores (from the Latin *omnis*, "all," and *vorare*, "to eat"), opportunistically eating the plants and animals they could gather, scavenge, or kill. A sensitive and versatile system of taste was necessary to distinguish between new sources of food and potential toxins. Some of our taste preferences are inborn. We have an innate fondness for sweetness, satisfied by mother's milk. Bitter substances are instinctively rejected, and indeed, many kinds of poisons are bitter. However, experience can strongly modify our instincts, and we can learn to tolerate and even enjoy the bitterness of such substances as coffee and quinine. The body also has the capacity to recognize a deficiency of certain key nutrients and develop an appetite for them. For example, when deprived of essential salt, we may crave salty foods.

The Basic Tastes

Although the number of different chemicals is practically endless and the variety of flavors seems immeasurable, it is likely that we can recognize only a few basic tastes. Most neuroscientists put the number at five. The four obvious taste qualities are saltiness, sourness, sweetness, and bitterness. A fifth taste quality is umami, meaning "delicious" in Japanese; it is defined by the savory taste of the amino acid glutamate; monosodium glutamate, or MSG, is the familiar culinary form. The five major categories of taste qualities seem to be common across human cultures, but there may be additional types of taste qualities (Box 8.1).

The correspondence between chemistry and taste is obvious in some cases. Most acids taste sour, and most salts taste salty. But the chemistry of substances can vary considerably while their basic taste remains the same. Many substances are sweet, from familiar sugars (like fructose, present in fruits and honey, and sucrose, which is white table sugar) to certain proteins (monellin, from the African serendipity berry) to artificial sweeteners (saccharin and aspartame, the second of which is made from two amino acids). Surprisingly, sugars are the least sweet of all of these; gram for gram, the artificial sweeteners and proteins are 10,000–100,000 times sweeter than sucrose. Bitter substances range from simple ions like K^+ (KCl actually evokes both bitter and salty tastes) and Mg^{2+} to complex organic molecules such as quinine and caffeine. Many bitter organic compounds can be tasted even at very low concentrations, down to the nanomolar range. There is an obvious advantage to this, as poisonous substances are often bitter.

With only a handful of basic taste types, how do we perceive the countless flavors of food, such as chocolate, strawberries, and barbecue sauce? First, each food activates a different combination of the basic tastes, helping make it unique. Second, most foods have a distinctive flavor as a result of their combined taste *and* smell occurring simultaneously. For example, without the sense of smell (and sight), a bite of onion can be easily mistaken for the bite of an apple. Third, other sensory modalities contribute to a unique food-tasting experience. Texture and temperature are important, and pain sensations are essential to the hot, spicy flavor of foods laced with capsaicin, the key ingredient in hot peppers. Therefore, to distinguish the unique flavor of a food, our brain actually combines sensory information about its taste, its smell, and its feel.

The Organs of Taste

Experience tells us that we taste with our tongue; but other areas of the mouth, such as the palate, pharynx, and epiglottis, are also involved (Figure 8.1). Odors from the food we are eating can also pass, via the pharynx, into the nasal cavity, where they can be detected by olfactory receptors. The tip of the tongue is most sensitive to sweetness, the back to bitterness, and the sides to saltiness and sourness. This does not mean, however, that we taste sweetness only with the tip of our tongue. Most of the tongue is sensitive to all basic tastes.

Scattered about the surface of the tongue are small projections called **papillae** (Latin for "bumps"). Papillae are shaped like ridges (*foliate papillae*), pimples (*vallate papillae*), or mushrooms (*fungiform papillae*)

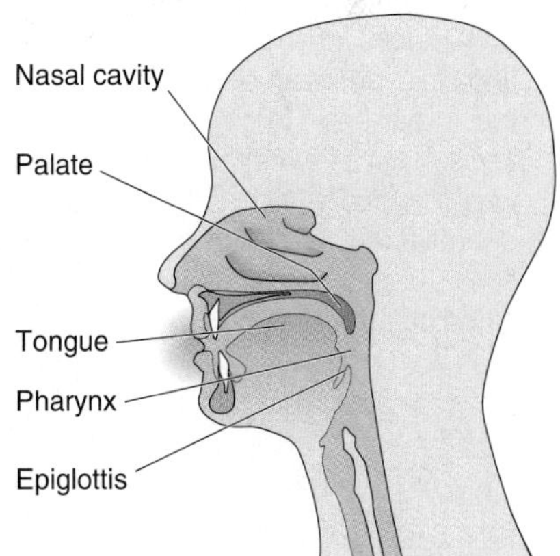

▲ FIGURE 8.1
Anatomy of the mouth, throat, and nasal passages. Taste is primarily a function of the tongue, but regions of the pharynx, palate, and epiglottis also have some sensitivity. Notice how the nasal passages are located so that odors from ingested food can enter through the nose or the pharynx, thereby easily contributing to perceptions of flavor through olfaction.

BOX 8.1 OF SPECIAL INTEREST

Strange Tastes: Fat, Starch, Carbonation, Calcium, Water?

Are there specific taste receptors beyond the classic five: salt, sour, bitter, sweet, umami? The answer is yes, probably. New types of taste receptors have been difficult to identify, but evidence is slowly accumulating.

People love fatty foods and for a good reason. Fat is a concentrated source of calories and essential nutrients. Keen observers as far back as Aristotle have suggested that a taste for fat is basic. But fat stimulates other sensory systems, and this complicates the question of its essential taste. Triglycerides, the fundamental fat molecules, impart a distinct texture to food in the mouth: they feel oily, slippery, creamy. These properties are detected by the somatic sensory system, not taste receptors. Fat also includes many volatile chemicals we can detect with our olfactory system. These odors may be pleasant or foul. Free fatty acids, which are breakdown products of triglycerides, sometimes smell putrid; think of rancid fats. They can also be irritants, sensed again by receptors of the somatic sensory system. But do we also *taste* fats? Yes, probably. Mice prefer water when it is spiked with some types of fatty acids. Mice also have a type of taste cell that is sensitive to fatty acids and expresses a presumed fatty acid receptor protein. A similar receptor is found in some human taste cells, which may be dedicated fat detectors.

People also love starchy foods, such as pasta, bread, and potatoes. Starch is a complex carbohydrate, specifically a polymer of glucose, the essential sugar in our bodies. Perhaps we like starch because we taste the glucose in it? Experiments on rodents suggest this is not the case; a rat's preferences for sugars and glucose polymers seem quite distinct. In a recent study, mice were tested for their ability to detect sugar and starchy molecules after the T1R3 protein—a key subunit of sweet and umami receptors (see Figure 8.6)—was genetically knocked out. The knockout mice seemed to be indifferent to sugar, as expected, but they continued to seek out starchy foods. Perhaps mice, at least, have dedicated starch detectors.

Many people also love carbonated drinks such as soft drinks, club soda, or beer. Water becomes carbonated when substantial amounts of the gas CO_2 are dissolved in it. As with fats, we can often feel carbonation as the fizziness vibrates the skin of the mouth and tongue. Mice, and to a lesser extent people, can also smell CO_2. We may even hear the bubbles bursting. The CO_2 level of blood is a critical measure of respiration, and cells in special arterial detectors sense it. But can we also *taste* carbonation? Yes, probably. Mice have taste cells with an enzyme called *carbonic anhydrase* that catalyzes the combination of CO_2 and H_2O to form protons (H^+) and bicarbonate (HCO_3^-). High levels of protons (i.e., a low pH) taste sour, implying that sour taste cells can detect carbonation. That is at least part of the answer, but how do we discriminate between simple sourness and carbonation? The answer still is not clear. Carbonation sensation may require the proper combination of sour taste and tingling somatic sensation.

People may not love calcium, but they certainly need it for healthy bones, brains, and all other organs. Many animals seem to find calcium salts very tasty when they have been deprived of calcium but reject them when they are sated with calcium. One hypothesis is that Ca^{2+} is sensed as a combination of bitter and sour tastes. Recent experiments suggest a more interesting possibility. Oddly, an aversion to the taste of Ca^{2+} in mice requires the T1R3 protein, and the taste of Ca^{2+} in humans is attenuated by a substance that binds to T1R3. It is possible, although far from proven, that T1R3 is part of a dedicated calcium taste receptor.

Finally, water. Water is crucial for life, and its consumption is regulated by thirst. Wetness, like fattiness and carbonation, can be felt with the somatic sensory system. But can we *taste* water? Distilled water given to humans has been variously described as sweet, salty, or bitter, depending on the test conditions. A specific taste receptor for water would seem to be a useful adaptation, and there is strong evidence for such receptors in insects. But water receptors in mammalian taste cells have not been identified so far.

(Figure 8.2a). Facing a mirror, stick your tongue out and shine a flashlight on it, and you will see your papillae easily—small, rounded ones at the front and sides, and large ones in the back. Each papilla has from one to several hundred **taste buds**, visible only with a microscope (Figure 8.2b). Each taste bud has 50–150 **taste receptor cells**, or taste cells, arranged within the bud like the sections of an orange. Taste cells comprise only about 1% of the tongue epithelium. Taste buds also have basal cells that surround the taste cells, plus a set of gustatory afferent axons (Figure 8.2c). A person typically has 2000–5000 taste buds, although in exceptional cases, there are as few as 500 or as many as 20,000.

Using tiny droplets, it is possible to expose a single papilla on a person's tongue to low concentrations of various basic taste stimuli (something

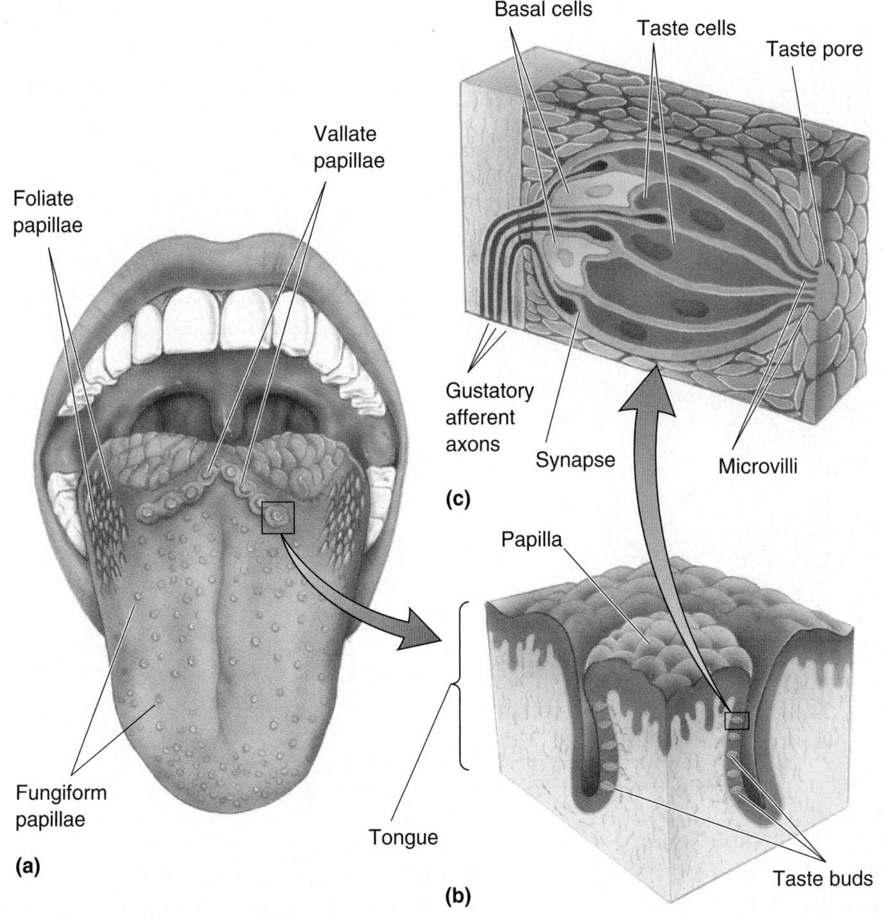

◀ FIGURE 8.2
The tongue, its papillae, and its taste buds. (a) Papillae are the taste-sensitive structures. The largest and most posterior are the vallate papillae. Foliate papillae are elongated. Fungiform papillae are relatively large toward the back of the tongue and much smaller along the sides and tip. **(b)** A cross-sectional view of a vallate papilla, showing the locations of taste buds. **(c)** A taste bud is a cluster of taste cells (the receptor cells), gustatory afferent axons and their synapses with taste cells and basal cells. Microvilli at the apical end of the taste cells extend into the taste pore, the site where chemicals dissolved in saliva can interact directly with taste cells.

almost purely sour, like vinegar, or almost purely sweet, like a sucrose solution). Concentrations too low will not be tasted, but at some critical concentration, the stimulus will evoke a perception of taste; this is the *threshold* concentration. At concentrations just above threshold, most papillae tend to be sensitive to only one basic taste; there are sour-sensitive papillae and sweet-sensitive papillae, for example. When the concentrations of the taste stimuli are increased, however, most papillae become less selective. Whereas a papilla might have responded only to sweet when all stimuli were weak, it may also respond to sour and salt if they are made stronger. We now know that each papilla has multiple types of taste receptor cells and that each receptor type is specialized for a different category of taste.

Taste Receptor Cells

The chemically sensitive part of a taste receptor cell is its small membrane region, called the *apical end*, near the surface of the tongue. The apical ends have thin extensions called *microvilli* that project into the *taste pore*, a small opening on the surface of the tongue where the taste cell is exposed to the contents of the mouth (see Figure 8.2c). Taste receptor cells are not neurons according to standard histological criteria. However, they do form synapses with the endings of the gustatory afferent axons near the bottom of the taste bud. Taste receptor cells also make both electrical and chemical synapses onto some of the basal cells; some basal cells synapse onto the sensory axons, and these may form a simple information-processing circuit within each taste bud. Cells of the

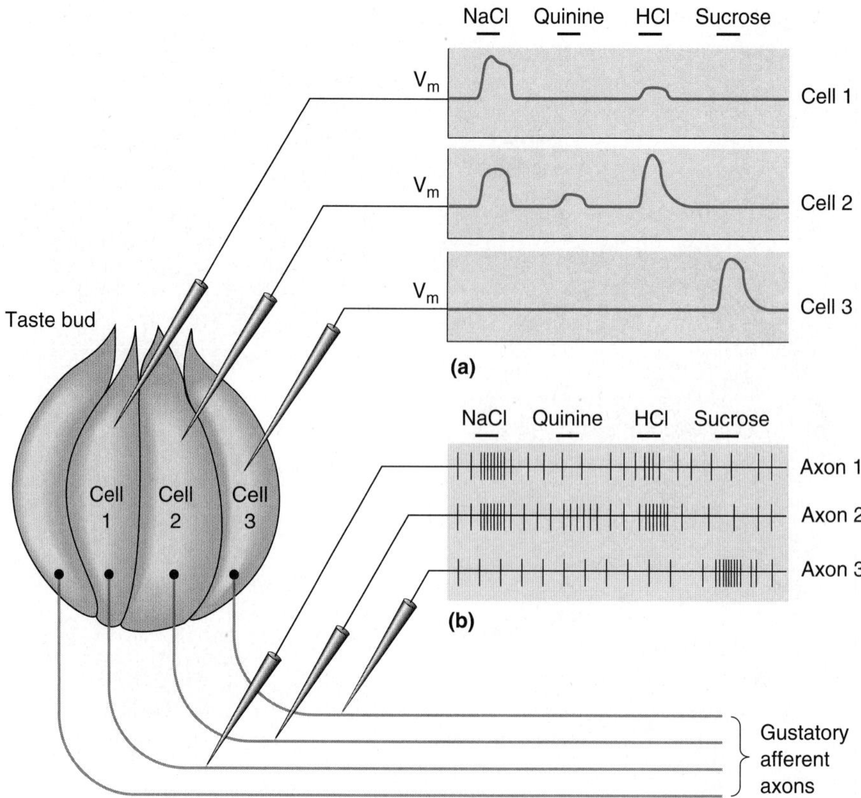

▶ FIGURE 8.3

Taste responsiveness of taste cells and gustatory axons. (a) Three different cells were sequentially exposed to salt (NaCl), bitter (quinine), sour (HCl), and sweet (sucrose) stimuli, and their membrane potential was recorded with electrodes. Notice the different sensitivities of the three cells. **(b)** In this case, the action potential discharge of the sensory axons was recorded. This is an example of extracellular recording of action potentials. Each vertical deflection in the record is a single action potential.

taste bud undergo a constant cycle of growth, death, and regeneration; the lifespan of one taste cell is about 2 weeks. This process depends on an influence of the sensory nerve, because if the nerve is cut, the taste buds will degenerate.

When an appropriate chemical activates a taste receptor cell, its membrane potential changes, usually by depolarizing. This voltage shift is called the **receptor potential** (Figure 8.3a). If the receptor potential is depolarizing and large enough, some taste receptor cells, like neurons, may fire action potentials. In any case, depolarization of the receptor membrane causes voltage-gated calcium channels to open; Ca^{2+} enters the cytoplasm, triggering the release of transmitter molecules. This is basic synaptic transmission, from taste cell to sensory axon. The transmitter released depends on the taste receptor cell type. Sour and salty taste cells release serotonin onto gustatory axons, whereas sweet, bitter, and umami taste cells release adenosine triphosphate (ATP) as their primary transmitter. In both cases, the taste receptor's transmitter excites the postsynaptic sensory axon and causes it to fire action potentials (Figure 8.3b), which communicate the taste signal into the brain stem. Taste cells may also use other transmitters, including acetylcholine, GABA, and glutamate, but their functions are unknown.

Evidence from recent studies of mice suggests that most taste receptor cells respond primarily or even exclusively to just one of the five basic tastes. Examples are cells 1 and 3 in Figure 8.3a, which give strong depolarizing responses to salt (NaCl) and sweet (sucrose) stimuli, respectively. Some taste cells and many gustatory axons show multiple response preferences, however. Each of the gustatory axons in Figure 8.3b is influenced by several of the basic tastes, but each has a clearly different bias.

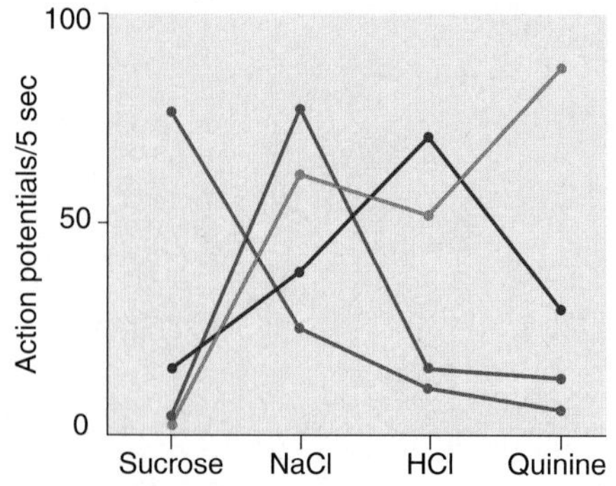

▲ FIGURE 8.4

Action potential firing rates of four different primary gustatory nerve axons in a rat. The taste stimuli were sweet (sucrose), salt (NaCl), sour (HCl), and bitter (quinine). Each colored line represents measurements from a single axon. Notice the differences in selectivity between axons. (Source: Adapted from Sato, 1980, p. 23.)

Figure 8.4 shows the results of similar recordings from four gustatory axons in a rat. One responds strongly only to salt, one only to sweet, and two to all but sweet. Why will one cell respond only to a single chemical type while another responds to three or four categories of chemicals? The answer is that the responses depend on the particular transduction mechanisms present in each cell.

Mechanisms of Taste Transduction

The process by which an environmental stimulus causes an electrical response in a sensory receptor cell is called **transduction** (from the Latin *transducere*, "to lead across"). The nervous system has myriad transduction mechanisms that make it sensitive to chemicals, pressures, sounds, and light. The nature of the transduction mechanism determines the specific sensitivity of a sensory system. We see because our eyes have photoreceptors. If our tongue had photoreceptors, we might see with our mouth.

Some sensory systems have a single basic type of receptor cell that uses one transduction mechanism (e.g., the auditory system). However, taste transduction involves several different processes, and each basic taste uses one or more of these mechanisms. Taste stimuli, or *tastants*, may (1) directly pass through ion channels (salt and sour), (2) bind to and block ion channels (sour), or (3) bind to G-protein-coupled receptors in the membrane that activate second messenger systems that, in turn, open ion channels (bitter, sweet, and umami). These are familiar processes, very similar to the basic signaling mechanisms present in all neurons and synapses, which were described in Chapters 4, 5, and 6.

Saltiness. The prototypical salty chemical is table salt (NaCl), which, apart from water, is the major component of blood, the ocean, and chicken soup. Salt is unusual in that relatively low concentrations (10–150 mM) taste good, whereas higher concentrations tend to be disagreeable and repellent. The taste of salt is mostly the taste of the cation Na^+, but taste receptors use very different mechanisms to detect low and high concentrations of it. To detect low concentrations, salt-sensitive taste cells use a special Na^+-selective channel that is common in other epithelial cells and which is blocked by the drug amiloride (Figure 8.5a). Amiloride is a diuretic (a drug that promotes urine production) used to treat some types of hypertension and heart disease. The amiloride-sensitive sodium channel is quite different from the voltage-gated sodium channel that generates action potentials; the taste channel is insensitive to voltage and generally stays open. When you sip chicken soup, the Na^+ concentration rises outside the receptor cell, and the gradient for Na^+ across the membrane is made steeper. Na^+ then diffuses down its concentration gradient, which means it flows into the cell, and the resulting inward current causes the membrane to depolarize. This depolarization—the receptor potential—in turn causes voltage-gated sodium and calcium channels to open near the synaptic vesicles, triggering the release of neurotransmitter molecules onto the gustatory afferent axon.

Animals avoid very high concentrations of NaCl and other salts, and humans usually report that they taste bad. It appears that high salt levels activate bitter and sour taste cells, which normally trigger avoidance behaviors. How very salty substances stimulate bitter and sour taste cells is still a mystery.

The anions of salts affect the taste of the cations. For example, NaCl tastes saltier than Na acetate, apparently because the larger anion, acetate, *inhibits* the salty taste of the cation. The mechanisms of anion inhibition are poorly understood. Another complication is that as the anions

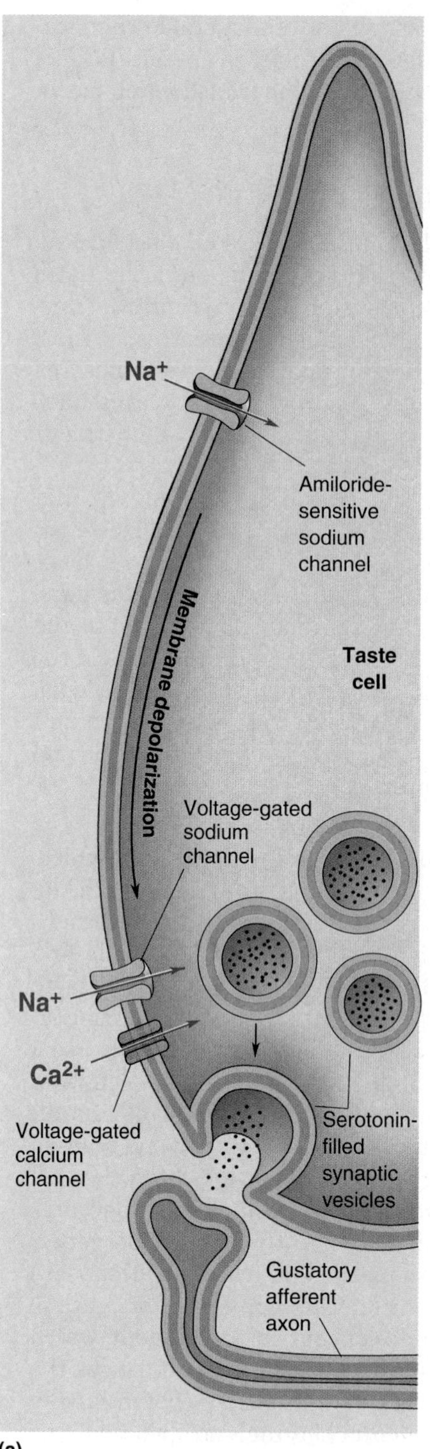

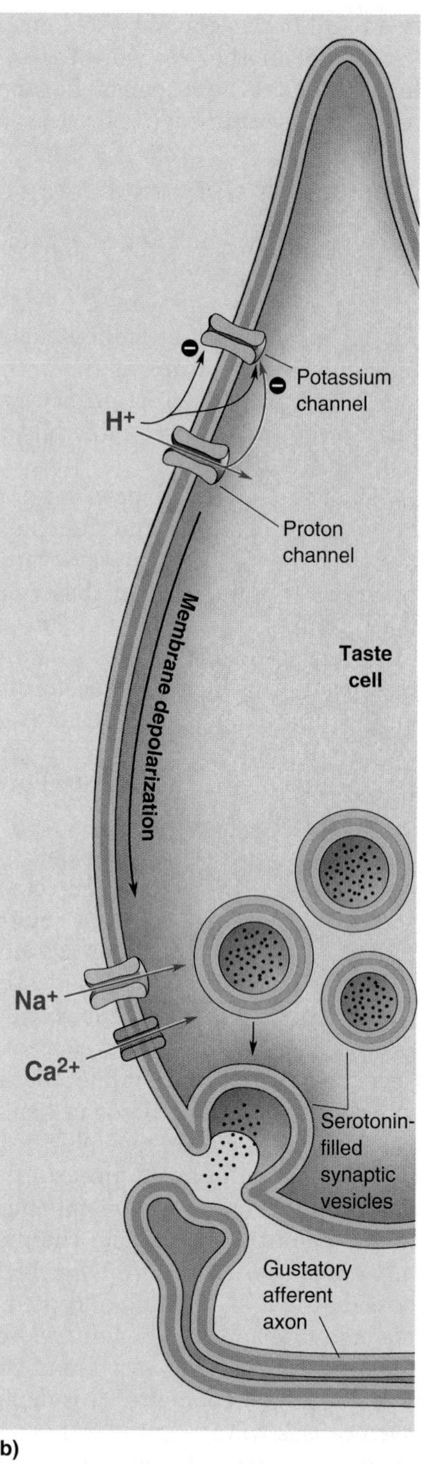

▶ FIGURE 8.5
Transduction mechanisms of (a) salt and (b) sour tastants. Tastants can interact directly with ion channels either by passing through them (Na^+ and H^+) or by blocking them (H^+ blocking the potassium channel). The membrane voltage then influences calcium channels on the basal membrane, which in turn influence the intracellular [Ca^{2+}] and transmitter release.

become larger, they tend to take on tastes of their own. Sodium saccharin tastes sweet because the Na^+ concentrations are far too low for us to taste the saltiness, and the saccharin potently activates sweetness receptors.

Sourness. Foods taste sour because of their high acidity (otherwise known as low pH). Acids, such as HCl, dissolve in water and generate hydrogen ions (protons or H^+). Thus, protons are the causative agents of acidity and sourness. Protons may affect sensitive taste receptors in several ways, from either inside or outside the taste cell membrane, although these processes are poorly understood (Figure 8.5b). It is likely that H^+ can bind to and *block* special K^+-selective channels. When the K^+ permeability of a membrane is decreased, it depolarizes. H^+ may also activate or permeate a special type of ion channel from the superfamily of transient receptor potential (TRP) channels, which are common in many kinds of sensory receptor cells. The cation current through the TRP channels can also depolarize sour receptor cells. pH can affect virtually all cellular processes, and there may be other mechanisms of sour taste transduction. It is possible that a constellation of effects evokes the sour taste.

Bitterness. The transduction processes underlying bitter, sweet, and umami tastes rely on two families of related taste receptor proteins called *T1R* and *T2R*. The various subtypes of T1R and T2R are all G-protein-coupled taste receptors that are very similar to the G-protein-coupled receptors that detect neurotransmitters. There is good evidence that the protein receptors for bitter, sweet, and umami tastes are *dimers;* dimers are two proteins affixed to one another (Figure 8.6). Tightly associated proteins are common in cells (see Figure 3.6); most ion channels (see Figure 3.7) and transmitter-gated channels (see Figure 5.14) consist of several different bound proteins, for example.

Bitter substances are detected by the 25 or so different types of T2R receptors in humans. Bitter receptors are poison detectors, and because we have so many, we can detect a vast array of different poisonous substances. Animals are not very good at telling different bitter tastants apart, however, probably because each bitter taste cell expresses many, and perhaps most, of the 25 bitter receptor proteins. Since each taste cell can send only one type of signal to its afferent nerve, a chemical that can bind to one of its 25 bitter receptors will trigger essentially the same response as a different chemical that binds to another of its bitter receptors. The important message the brain receives from its taste receptors is simply that a bitter chemical is "Bad! Not to be trusted!" And the nervous system apparently does not distinguish one type of bitter substance from another.

Bitter receptors use a second messenger pathway to carry their signal to the gustatory afferent axon. In fact, the bitter, sweet, and umami receptors all seem to use the same second messenger pathway to carry their signals to the afferent axons. The general pathway is illustrated in Figure 8.7. When a tastant binds to a bitter (or sweet or umami) receptor, it activates its G-proteins, which stimulate the enzyme phospholipase C, thereby increasing the production of the intracellular messenger inositol triphosphate (IP_3). IP_3 pathways are ubiquitous signaling systems in cells throughout the body (see Chapter 6). In taste cells, IP_3 activates a special type of ion channel that is unique to taste cells, causing it to open and allow Na^+ to enter, thus depolarizing the taste cell. IP_3 also triggers the release of Ca^{2+} from intracellular storage sites. This Ca^{2+} in turn triggers neurotransmitter release in an unusual way. The taste cells for bitter, sweet, and umami lack conventional transmitter-filled presynaptic vesicles. Instead, increases of intracellular Ca^{2+} activate a special membrane channel that allows ATP to flow out of the cell. The ATP serves as a synaptic transmitter and activates purinergic receptors on postsynaptic gustatory axons.

Sweetness. There are many different sweet tastants, some natural and some artificial. Surprisingly, all of them seem to be detected by the same taste receptor protein. Sweet receptors resemble bitter receptors, in that they are all dimers of G-protein-coupled receptors. A functioning sweet

Bitter receptors: the T2Rs

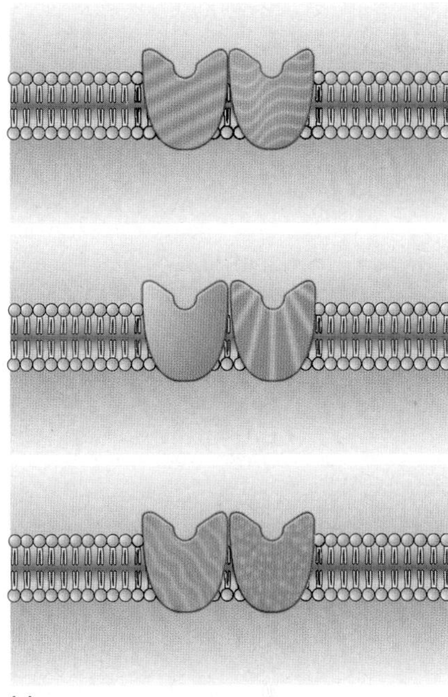

(a)

Sweet receptor: T1R2 + T1R3

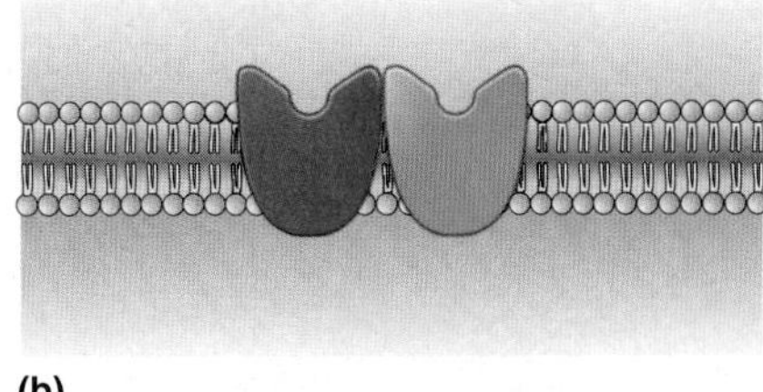

(b)

Umami receptor: T1R1 + T1R3

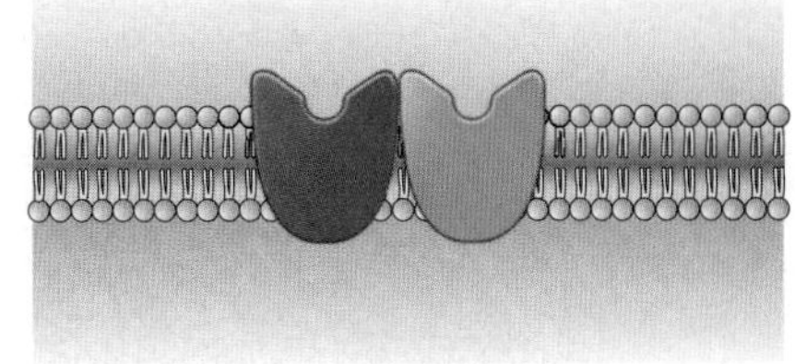

(c)

▲ FIGURE 8.6
Taste receptor proteins. **(a)** There are about 25 types of bitter receptors, comprising a family of T2R proteins. Bitter receptors are probably dimers consisting of two different T2R proteins. **(b)** There is only one type of sweet receptor, formed from the combination of a T1R2 and a T1R3 protein. **(c)** There is only one type of umami receptor, formed from the combination of a T1R1 and a T1R3 protein.

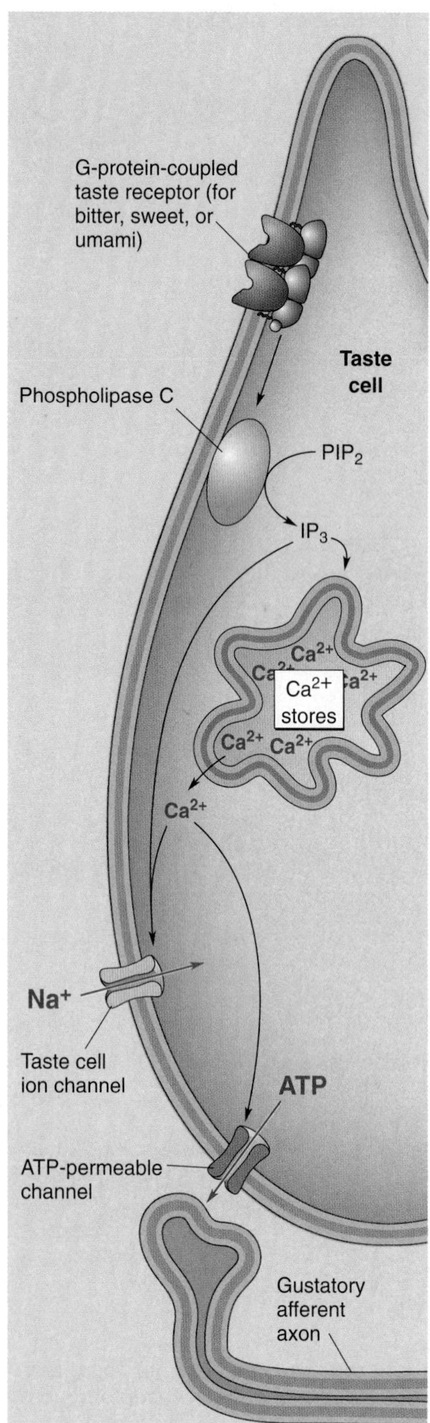

▲ FIGURE 8.7
Transduction mechanisms for bitter, sweet, and umami tastants. Tastants bind directly to G-protein-coupled membrane receptors, which activate phospholipase C, which increases the synthesis of IP_3. IP_3 then triggers the release of Ca^{2+} from internal storage sites, and Ca^{2+} opens a taste-specific ion channel, leading to depolarization and transmitter release. The main transmitter is ATP, which is released from the taste cell by diffusing through ATP-permeable channels.

receptor requires two very particular members of the T1R receptor family: T1R2 and T1R3 (see Figure 8.6). If either one of them is missing or mutated, an animal may not perceive sweetness at all. In fact, all species of cats and some other carnivores lack the genes encoding T1R2 and are indifferent to the taste of most molecules that we consider sweet.

Chemicals binding to the T1R2 + T1R3 receptor (i.e., the sweet receptor) activate exactly the same second messenger system that the bitter receptors activate (Figure 8.7). So, why don't we confuse bitter chemicals with sweet ones? The reason is that bitter receptor proteins and sweet receptor proteins are expressed in different taste cells. Bitter taste cells and sweet taste cells, in turn, connect to different gustatory axons. The activity of different gustatory axons reflects the chemical sensitivities of the taste cells that drive them, so the messages about sweetness and bitterness are delivered to the central nervous system along different transmission lines.

Umami (Amino Acids). "Amino acids" may not be the answer at the tip of your tongue when asked to list your favorite tastes, but recall that proteins are made from amino acids and that they are also excellent energy sources. In short, amino acids are the foods your mother would want you to eat. Most amino acids also taste good, although some taste bitter.

The transduction process for umami is identical to that for sweetness, with one exception. The umami receptor, like the sweet receptor, is composed of two members of the T1R protein family, but in this case, it is T1R1 + T1R3 (see Figure 8.6). The sweet and umami receptors share the T1R3 protein, so it is the other T1R that determines whether the receptor is sensitive to amino acids or sweet tastants. When the gene encoding for the T1R1 protein is removed from mice, they are unable to taste glutamate and other amino acids, although they still demonstrate a sense for sweet chemicals and other tastants.

Similar to other types of taste receptors, the genetics of different mammalian species lead to interesting taste preferences and deficits. Most bats, for example, do not have a functioning T1R1 receptor and presumably cannot taste amino acids. Vampire bats lack functional genes for both umami *and* sweet taste. The ancestors of bats presumably did have umami and sweet receptors; why they were lost is unknown.

Considering how similar the umami receptor is to the sweet and bitter receptors, it will not surprise you that all three use exactly the same second messenger pathways (Figure 8.7). Then why don't we confuse the taste of amino acids with sweet or bitter chemicals? Once again, the taste cells selectively express only one class of taste receptor protein. There are umami-specific taste cells, just as there are sweet-specific taste cells and bitter-specific taste cells. The gustatory axons they stimulate are, in turn, delivering messages of umami, sweetness, or bitterness to the brain.

Central Taste Pathways

The main flow of taste information is from the taste buds to the primary gustatory axons, into the brain stem, up to the thalamus, and to the cerebral cortex (Figure 8.8). Three cranial nerves carry primary gustatory axons and bring taste information to the brain. The anterior two-thirds of the tongue and the palate send axons into a branch of cranial nerve VII, the *facial nerve*. The posterior third of the tongue is innervated by a branch of cranial nerve IX, the *glossopharyngeal nerve*. The regions around the throat, including the glottis, epiglottis, and pharynx, send taste axons to a branch of cranial nerve X, the *vagus nerve*. These nerves

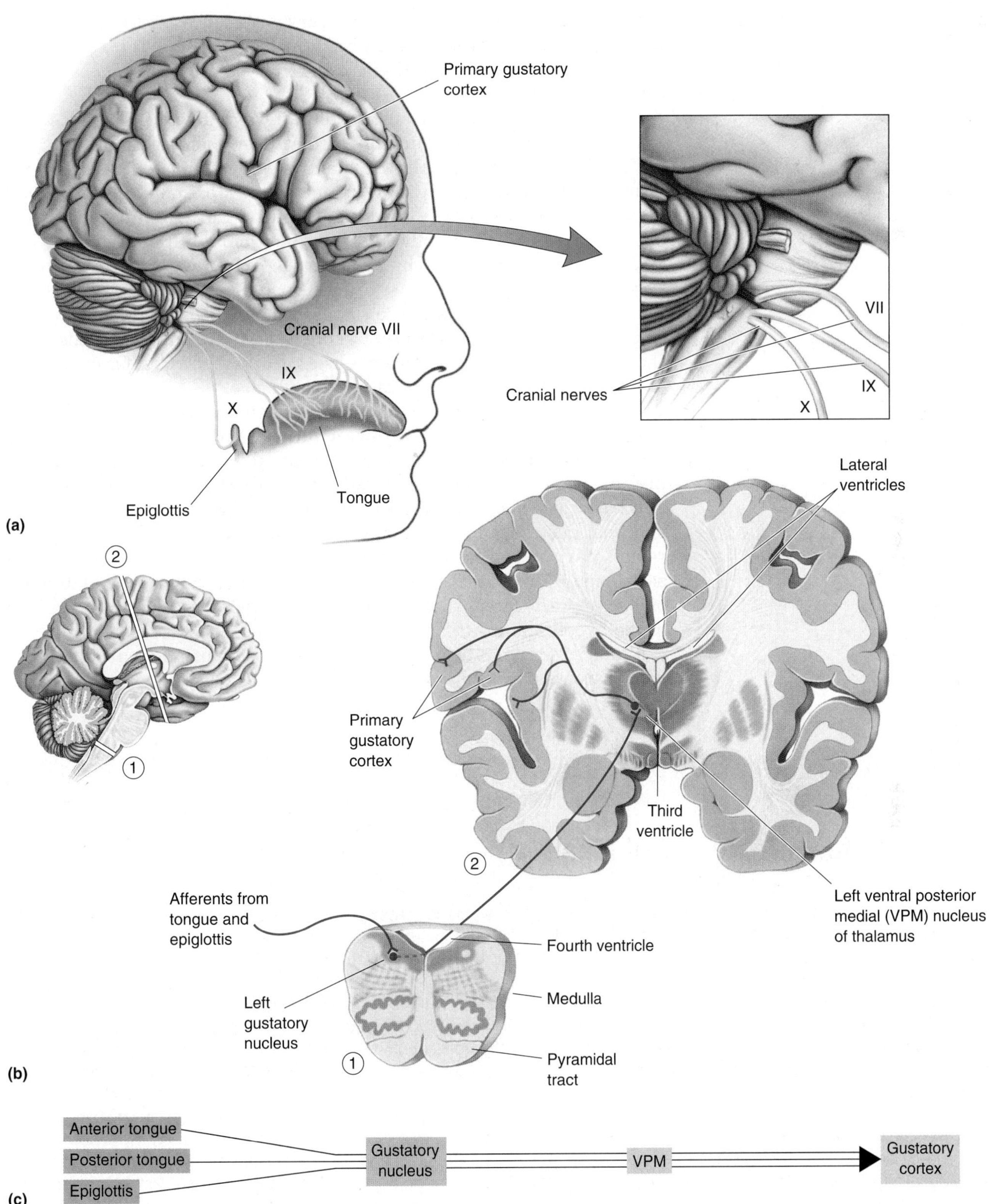

▲ FIGURE 8.8

Central taste pathways. **(a)** Taste information from the tongue and mouth cavity is carried by three cranial nerves (VII, IX, and X) to the medulla. **(b)** Gustatory axons enter the gustatory nucleus within the medulla. Gustatory nucleus axons synapse on neurons of the thalamus, which project to primary gustatory cortex in regions of the postcentral gyrus and insular cortex. The enlargements show planes of section through ① the medulla and ② the forebrain. **(c)** The central taste pathways summarized.

are involved in a variety of other sensory and motor functions, but their taste axons all enter the brain stem, bundle together, and synapse within the slender **gustatory nucleus**, a part of the *solitary nucleus* in the medulla.

From the gustatory nucleus, taste pathways diverge. The conscious experience of taste is presumably mediated by the cerebral cortex. The path to the neocortex via the thalamus is a common one for sensory information. Neurons of the gustatory nucleus synapse on a subset of small neurons in the **ventral posterior medial (VPM) nucleus**, a portion of the thalamus that deals with sensory information from the head. The VPM taste neurons then send axons to the **primary gustatory cortex** (located in Brodmann's area 36 and the insula-operculum regions of cortex). The taste pathways to the thalamus and cortex are primarily ipsilateral to the cranial nerves that supply them. Lesions within the VPM thalamus or the gustatory cortex—as a result of a stroke, for example—can cause *ageusia*, the loss of taste perception.

Gustation is important to basic behaviors such as the control of feeding and digestion, both of which involve additional taste pathways. Gustatory nucleus cells project to a variety of brain stem regions, largely in the medulla, that are involved in swallowing, salivation, gagging, vomiting, and basic physiological functions such as digestion and respiration. In addition, gustatory information is distributed to the hypothalamus and related parts of the basal telencephalon (structures of the limbic system; see Chapter 18). These structures seem to be involved in the palatability of foods and the forces that motivate us to eat (Box 8.2). Localized lesions of the hypothalamus or amygdala, a nucleus of the basal telencephalon, can cause an animal to either chronically overeat or ignore food, or alter its preferences for food types.

BOX 8.2 OF SPECIAL INTEREST

Memories of a Very Bad Meal

When one of us was 14 years old, he ended an entertaining day at an amusement park by snacking on one of his favorite New England foods, fried clams. Within an hour he became nauseated, vomited, and had a most unpleasant bus ride home. Presumably, the clams had been spoiled. Sadly, for years afterward, he could not even imagine eating another fried clam, and the smell of them alone was repulsive. The fried clam aversion was quite specific. It did not affect his enjoyment of other foods, and he felt no prejudice against amusement park rides, buses, or the friends he had been with the day he got sick.

By the time the author reached his thirties, he could happily dine on fried clams again. He also read about research that John Garcia, working at Harvard Medical School, had done just about the same time as the original bad-clam experience. Garcia fed rats a sweet liquid, and, in some cases, he then gave them a drug that made them briefly feel ill. After even one such trial, rats that had received the drug avoided the sweet stimulus forever. The rats' aversion was specific for the taste stimulus; they did not avoid sound or light stimuli under the same conditions.

Extensive research has shown that *flavor aversion learning* results in a particularly robust form of associative memory. It is most effective for food stimuli (taste and smell both contribute); it requires remarkably little experience (as little as one trial); and it can last a very long time—more than 50 years in some people! And learning occurs even when there is a very long delay between the food (the conditioned stimulus) and the nausea (the unconditioned stimulus). This is obviously a useful form of learning in the wild. An animal can't afford to be a slow learner when new foods might be poisonous. For modern humans, this memory mechanism can backfire; many perfectly good fried clams have remained uneaten. Food aversion can be a more serious problem for patients undergoing radiation or chemotherapy for cancer, when the nausea induced by their treatments makes many foods unpalatable. On the other hand, taste aversion learning has also been used to prevent coyotes from stealing domestic sheep and to help people reduce their dependence on alcohol and cigarettes.

The Neural Coding of Taste

If you were going to design a system for coding tastes, you might begin with many specific taste receptors for many basic tastes (sweet, sour, salty, bitter, chocolate, banana, mango, beef, Swiss cheese, etc.). Then you might connect each receptor type, by separate sets of axons, to neurons in the brain that also respond to only one specific taste. All the way up to the cortex, you would expect to find specific neurons responding to "sweet" and "chocolate," and the flavor of chocolate ice cream would involve the rapid firing of these cells and very few of the "salty," "sour," and "banana" cells.

This concept is the *labeled line hypothesis*, and at first it seems simple and rational. At the start of the gustatory system—the taste receptor cells—something like labeled lines are used. As we have seen, individual taste receptor cells are often selectively sensitive to particular classes of stimuli: sweet, bitter, or umami. Some of them, however, are more *broadly tuned* to stimuli; that is, they are less specific in their responses. They may be excited to some extent by both salt and sour, for example (see Figure 8.3). Primary taste axons are even less specific than receptor cells, and most central taste neurons continue to be broadly responsive all the way into the cortex. In other words, the response of a single taste cell is often ambiguous about the food being tasted; the labels on the taste lines are uncertain rather than distinct.

Cells in the taste system are broadly tuned for several reasons. If one taste receptor cell has two different transduction mechanisms, it will respond to two types of tastants (although it may still respond most strongly to one of them). In addition, there is convergence of receptor cell input onto afferent axons. Each receptor cell synapses onto a primary taste axon that also receives input from several other receptor cells in that papilla as well as its neighbors. This means that one axon may combine the taste information from numerous taste cells. If one of those cells is mostly sensitive to sour stimuli and another to salt stimuli, then the axon will respond to salt *and* sour. This pattern continues into the brain: Neurons of the gustatory nucleus receive synapses from many axons of different taste specificities, and they may become less selective for tastes than the primary taste axons.

All of this mixing of taste information might seem like an inefficient way to design a coding system. Why not use many taste cells that are highly specific? In part, the answer might be that we would need an enormous variety of receptor types, and even then we could not respond to new tastes. So, when you taste chocolate ice cream, how does the brain sort through its apparently ambiguous information about the flavor to make clear distinctions between chocolate and thousands of other possibilities? The likely answer is a scheme that includes features of roughly labeled lines and **population coding**, in which the responses of a large number of broadly tuned neurons, rather than a small number of precisely tuned neurons, are used to specify the properties of a particular stimulus, such as a taste.

Population coding schemes seem to be used throughout the sensory and motor systems of the brain, as we shall see in later chapters. In the case of taste, receptor cells are sensitive to a small number of taste types, often only one; gustatory axons and the neurons they activate in the brain tend to respond more broadly—for example, strongly to bitter, moderately to sour and salt, and not at all to sweet (see Figure 8.4). Only with a large population of taste cells, with different response patterns, can the brain distinguish between specific alternative tastes. One food activates a certain subset of neurons, some of them firing very strongly, some moderately, some not at all, others perhaps even inhibited below their spontaneous firing rates (i.e., their nonstimulated rates); a second food excites some of the cells activated by the first food, but also others; and the overall patterns of discharge rates are distinctly different. The relevant population may even include neurons activated by the olfactory, temperature, and textural features of a food; certainly, the coldness and creaminess of chocolate ice cream contribute to our ability to distinguish it from chocolate cake.

SMELL

Olfaction brings both good news and bad news. It combines with taste to help us identify foods, and it increases our enjoyment of many of them. But it can also warn of potentially harmful substances (spoiled meat) or places (smoke-filled rooms). In olfaction, the warnings from bad smells may outweigh the benefits of good smells; by some estimates, we can smell several hundred thousand substances, but only about 20% of them smell pleasant. Practice helps in olfaction, and professional perfumers and whiskey blenders can actually distinguish among thousands of different odors.

Smell is also a mode of communication. Chemicals released by the body, called **pheromones**, are important signals for reproductive behaviors, and they may also be used to mark territories, identify individuals, and indicate aggression or submission. (The term is from the Greek *pherein*, "to carry," and *horman*, "to urge or excite.") Although systems of pheromones are well developed in many other animals, their importance to humans is not clear (Box 8.3).

The Organs of Smell

We do not smell with our nose. Rather, we smell with a small, thin sheet of cells high up in the nasal cavity called the **olfactory epithelium** (Figure 8.9). The olfactory epithelium has three main cell types. *Olfactory*

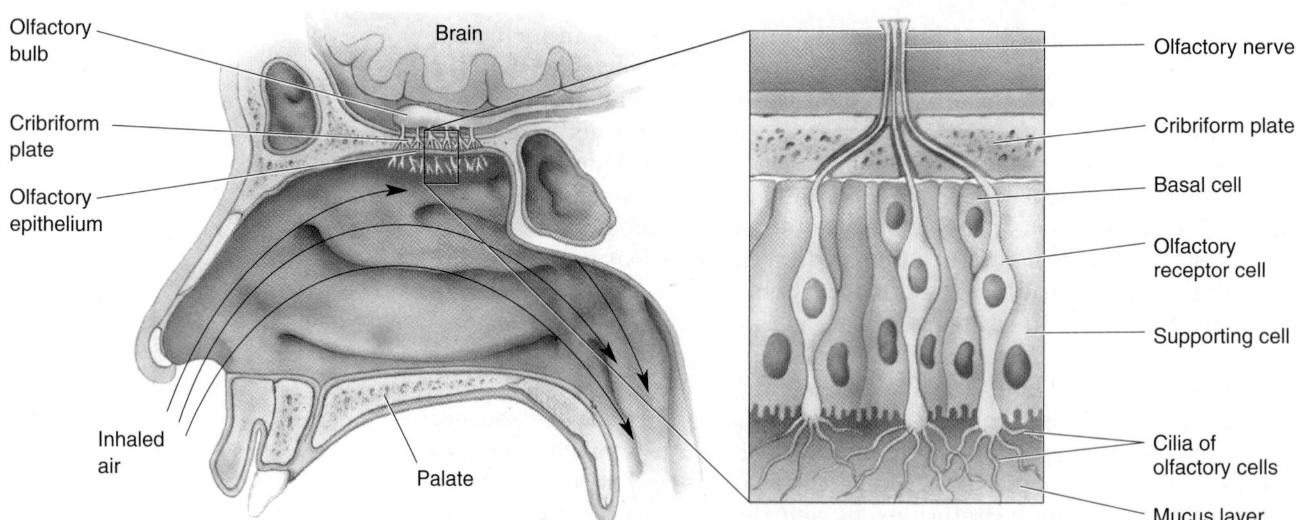

▲ FIGURE 8.9
The location and structure of the olfactory epithelium. The olfactory epithelium consists of a layer of olfactory receptor cells, supporting cells, and basal cells. Odorants dissolve in the mucus layer and contact the cilia of the olfactory cells. Axons of the olfactory cells penetrate the bony cribriform plate on their way to the CNS.

BOX 8.3 OF SPECIAL INTEREST

Human Pheromones?

Smells are surer than sounds and sights to make your heart strings crack.

—*Rudyard Kipling*

Odors can certainly sway emotions and arouse memories, but just how important are they to human behavior? Each of us has a distinctive set of odors that marks our identity as surely as our fingerprints or genes do. In fact, variations in body odor are probably genetically determined. Bloodhounds have great difficulty distinguishing between the smells of identical twins, but not between those of fraternal siblings. For some animals, odor identity is essential: When her lamb is born, the ewe establishes a long-term memory of its specific smell and develops an enduring bond based largely on olfactory cues. In a newly inseminated female mouse, the smell of a strange male (but not the smell of her recent mate, which she remembers) will trigger an abortion of the pregnancy.

Humans have the ability to recognize the scents of other humans. Infants as young as 6 days old show a clear preference for the smell of their own mother's breast over that of other nursing mothers. The mothers, in turn, can usually identify the odor of their own infant from among several choices.

About 30 years ago, researcher Martha McClintock reported that women who spend a lot of time together (e.g., college roommates) often find that their menstrual cycles synchronize. This effect is probably mediated by pheromones. In 1998, McClintock and Kathleen Stern, working at the University of Chicago, found that odorless compounds from one group of women (the "donors") could influence the timing of the menstrual cycles of other women (the "recipients"). Body chemicals were collected by placing cotton pads under the arms of the donors for at least 8 hours. The pads were then wiped under the noses of the recipients, who agreed not to wash their faces for 6 hours. The recipients were not told the source of the chemicals on the pads and did not consciously perceive any odor from them except the alcohol used as a carrier. Nevertheless, depending on the donor's time in her menstrual cycle, the recipient's cycle was either shortened or lengthened. These dramatic results are the best evidence yet that humans can communicate with pheromones.

Many animals use the *accessory olfactory system* to detect pheromones and mediate a variety of social behaviors involving mother, mating, territory, and food. The accessory system runs parallel to the primary olfactory system. It consists of separate chemically sensitive regions in the nasal cavity, in particular the *vomeronasal organ*, which projects to the *accessory olfactory bulb* and from there provides input to the hypothalamus. Alas, it seems likely that the vomeronasal organ is absent or vestigial in most mature people, and even when it is identifiable, it does not appear to have functional receptor proteins or direct connections to the brain. This by itself does not mean humans lack pheromonal signals, since these could pass through the main olfactory organs.

Napoleon Bonaparte once wrote to his love Josephine, asking her not to bathe for the 2 weeks until they would next meet, so he could enjoy her natural aromas. The scent of a woman may indeed be a source of arousal for sexually experienced males, presumably because of learned associations. But there is still no hard evidence for human pheromones that might mediate sexual attraction (for members of either sex) via innate mechanisms. Considering the commercial implications of such a substance, we can be sure the search will continue.

receptor cells are the sites of transduction. Unlike taste receptor cells, olfactory receptors are genuine neurons, with axons of their own that penetrate into the central nervous system. *Supporting cells* are similar to glia; among other things, they help produce mucus. *Basal cells* are the source of new receptor cells. Olfactory receptors (similar to taste receptors) continually grow, die, and regenerate, in a cycle that lasts about 4–8 weeks. In fact, olfactory receptor cells are one of the very few types of neurons in the nervous system that are regularly replaced throughout life.

Sniffing brings air through the convoluted nasal passages, but only a small percentage of that air passes over the olfactory epithelium. The epithelium exudes a thin coating of mucus, which flows constantly and is replaced about every 10 minutes. Chemical stimuli in the air, called *odorants*, dissolve in the mucus layer before they reach receptor cells. Mucus consists of a water base with dissolved mucopolysaccharides (long chains of sugars); a variety of proteins, including antibodies, enzymes, and odorant binding proteins; and salts. The antibodies are critical because olfactory cells can be a direct route by which some viruses (such as the rabies virus) and bacteria enter the brain. Also important are odorant binding proteins, which are small and soluble and which may help concentrate odorants in the mucus.

The size of the olfactory epithelium is one indicator of an animal's olfactory acuity. Humans are relatively weak smellers (although even we can detect some odorants at concentrations as low as a few parts per trillion). The surface area of the human olfactory epithelium is only about 10 cm^2. The olfactory epithelium of certain dogs may be over 170 cm^2, and dogs have over 100 times more receptors in each square centimeter than humans. By sniffing the aromatic air above the ground, dogs can detect the few molecules left by someone walking there hours before. Humans may be able to smell the dog, however, only when it licks our face.

Olfactory Receptor Neurons

Olfactory receptor neurons have a single, thin dendrite that ends with a small knob at the surface of the epithelium (see Figure 8.9). Waving from the knob, within the mucus layer, are several long, thin cilia. Odorants dissolved in the mucus bind to the surface of the cilia and activate the transduction process. On the opposite side of the olfactory receptor cell is a very thin, unmyelinated axon. Collectively, the olfactory axons constitute the *olfactory nerve* (cranial nerve I). The olfactory axons do not all come together as a single nerve bundle, as in other cranial nerves. Instead, after leaving the epithelium, small clusters of the axons penetrate a thin sheet of bone called the *cribriform plate*, then course into the **olfactory bulb** (see Figure 8.9). The olfactory axons are fragile, and in a traumatic injury, such as a blow to the head, the forces between the cribriform plate and surrounding tissue can sever the olfactory axons. After this type of injury, the axons cannot regrow, resulting in *anosmia*, the inability to smell.

Olfactory Transduction. Although taste receptor cells use several different molecular signaling systems for transduction, olfactory receptors probably use only one (Figure 8.10). All of the transduction molecules are located in the thin cilia. The olfactory pathway can be summarized as follows:

Odorants →
- Bind to membrane odorant receptor proteins →
  - Stimulate G-protein (G_{olf}) →
    - Activate adenylyl cyclase →
      - Form cAMP →
        - Bind cAMP to a cyclic nucleotide-gated cation channel →
          - Open cation channels and allow influx of Na^+ and Ca^{2+} →
            - Open Ca^{2+}-activated Cl^- channels →
              - Cause current flow and membrane depolarization (receptor potential).

Once the cation-selective cAMP-gated channels open, current flows inward, and the membrane of the olfactory neuron depolarizes (Figures 8.10 and 8.11). In addition to Na^+, the cAMP-gated channel allows substantial amounts of Ca^{2+} to enter the cilia. In turn, the intracellular Ca^{2+} triggers a Ca^{2+}-activated Cl^- current that may amplify the olfactory receptor potential. (This is a switch from the usual effect of Cl^- currents, which inhibit neurons; in olfactory cells, the internal Cl^- concentration must be unusually high so that a Cl^- current tends to depolarize rather than hyperpolarize the membrane.) If the resulting receptor potential is large

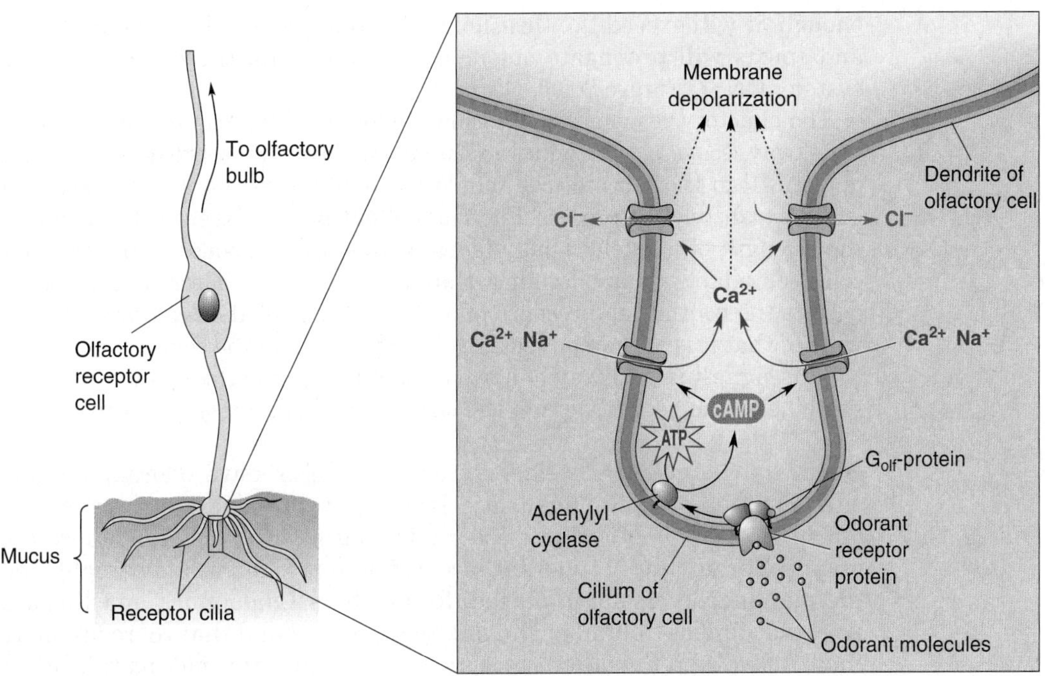

▲ FIGURE 8.10
Transduction mechanisms of vertebrate olfactory receptor cells. The drawing at right shows a single cilium of an olfactory receptor cell and the signaling molecules of olfactory transduction that it contains. G_{olf} is a special form of G-protein found only in olfactory receptor cells.

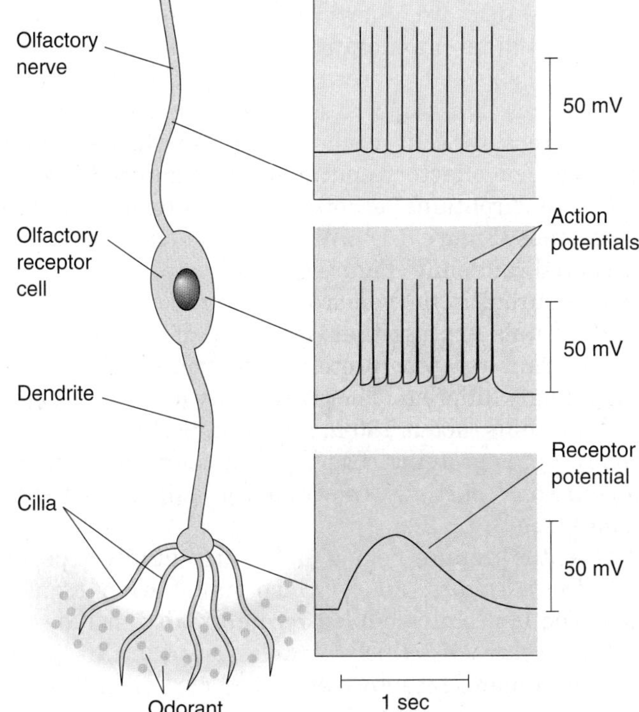

▲ FIGURE 8.11
Voltage recordings from an olfactory receptor cell during stimulation. Odorants generate a slow receptor potential in the cilia; the receptor potential propagates along the dendrite and triggers a series of action potentials within the soma of the olfactory receptor cell. Finally, the action potentials (but not the receptor potential) propagate continuously along the olfactory nerve axon.

enough, it will exceed the threshold for action potentials in the cell body, and spikes will propagate out along the axon into the central nervous system (CNS) (Figure 8.11).

The olfactory response may terminate for several reasons. Odorants diffuse away, scavenger enzymes in the mucus layer often break them down, and cAMP in the receptor cell may activate other signaling pathways that end the transduction process. Even in the continuing presence of an odorant, the strength of a smell usually fades because the response of the receptor cell itself adapts to an odorant within about a minute. Decreased response despite the continuing presence of a stimulus is called *adaptation*, and we will see that it is a common feature of receptors in all the senses.

This signaling pathway has two unusual features: the receptor binding proteins at the beginning and the cAMP-gated channels near the end.

Olfactory Receptor Proteins. Receptor proteins have odorant binding sites on their extracellular surface. Because of your ability to discriminate among thousands of different odorants, you might guess that there are many different types of odorant receptor proteins. You would be right, and the number is very large indeed. Researchers Linda Buck and Richard Axel, working at Columbia University in 1991, found that there are more than 1000 different odorant receptor genes in rodents, making this by far the largest family of mammalian genes yet discovered. This important and surprising discovery earned Buck and Axel the Nobel Prize in 2004.

Humans have fewer odorant receptor genes than rodents—about 350 that code for functional receptor proteins—but this is still an enormous number. Odorant receptor genes comprise about 3–5% of the entire mammalian genome. The receptor genes are scattered about on the genome, and nearly every chromosome has at least a few of them. Each receptor gene has a unique structure, allowing the receptor proteins encoded by these genes to bind different odorants. It is also surprising that each olfactory receptor cell seems to express very few of the many types of receptor genes, in most cases just one. Thus, in mice, there are more than 1000 different types of receptor cells, each identified by the particular receptor gene it expresses. The olfactory epithelium is organized into a few large zones, and each zone contains receptor cells that express a different subset of receptor genes (Figure 8.12). Within each zone, individual receptor types are scattered randomly (Figure 8.13a).

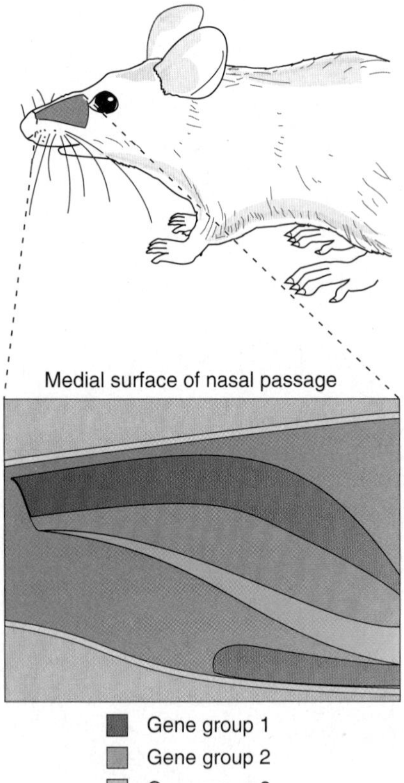

▲ FIGURE 8.12
Maps of the expression of different olfactory receptor proteins on the olfactory epithelium of a mouse. Three different groups of genes were mapped in this case, and each had a different, nonoverlapping zone of distribution. (Source: Adapted from Ressler et al., 1993, p. 602.)

The receptor neurons in the vomeronasal organ of mice, dogs, cats, and many other mammals express their own sets of receptor proteins. The structures of odorant receptor proteins and vomeronasal receptor proteins are surprisingly different. There are many fewer functional vomeronasal receptor proteins (about 180 in mice and perhaps none in humans) than odorant receptor proteins. The types of chemicals vomeronasal receptors detect are only partially known, but it is likely that some of them are pheromones (see Box 8.3).

Olfactory receptor proteins belong to the large family of proteins called *G-protein-coupled receptors*, all of which have seven transmembrane alpha helices. The G-protein-coupled receptors also include a variety of neurotransmitter receptors, which were described in Chapter 6, and the bitter, sweet, and umami receptors described earlier in this chapter. All of these receptors are coupled to G-proteins, which in turn relay a signal to other second messenger systems within the cell (olfactory receptor cells use a particular type of G-protein, called G_{olf}). Increasing evidence indicates that the only second messenger mediating olfactory transduction in vertebrates is cAMP. Some of the most compelling studies have used genetic engineering to produce mice in which critical proteins of the

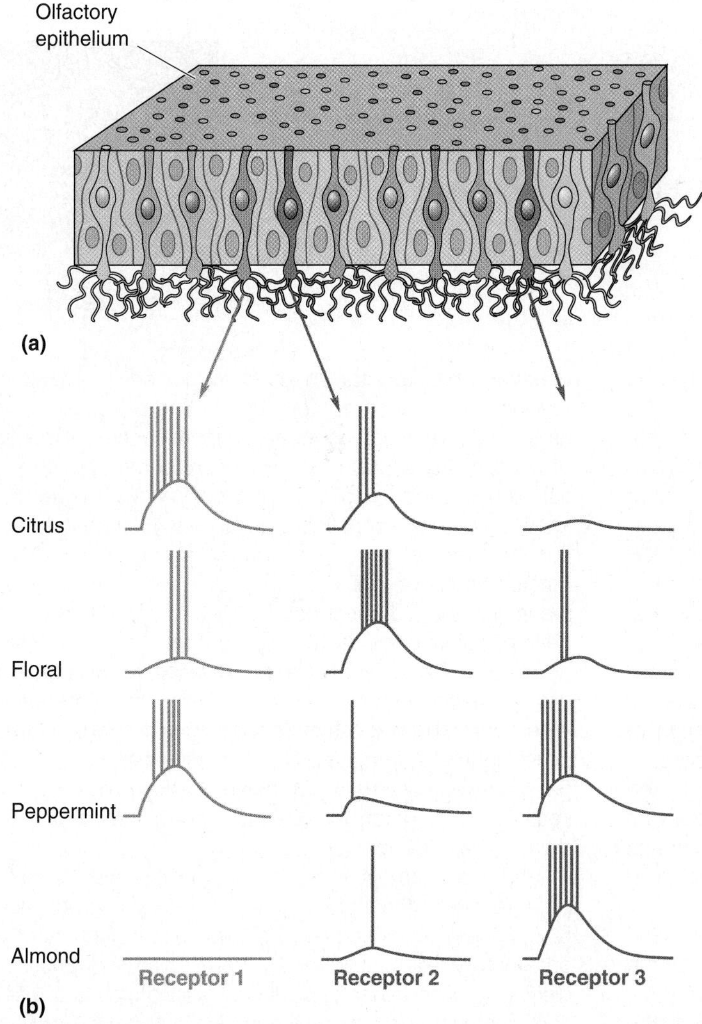

▲ FIGURE 8.13
Broad tuning of single olfactory receptor cells. **(a)** Each receptor cell expresses a single olfactory receptor protein (here coded by cell color), and different cells are randomly scattered within a region of the epithelium. **(b)** Microelectrode recordings from three different cells show that each one responds to many different odors, but with differing preferences. By measuring responses from all three cells, each of the four odors can be clearly distinguished.

olfactory cAMP pathway have been knocked out (e.g., G_{olf}); these mice are inevitably anosmic for a wide variety of odors.

cAMP-Gated Channels. In neurons, cAMP is a common second messenger, but the way it acts in olfactory transduction is quite unusual. Tadashi Nakamura and Geoffrey Gold, working at Yale University in 1987, showed that a population of channels in the cilia of olfactory cells responds directly to cAMP; that is, the channels are cAMP-gated. In Chapter 9, we will see that a similar version of cyclic nucleotide-gated channel is used for visual transduction. This is another demonstration that biology is conservative and evolution recycles its good ideas: Smelling and seeing use some very similar molecular mechanisms (Box 8.4).

How do the 1000 types of receptor cells used by mice discriminate among tens of thousands of odors? Like taste, olfaction involves a population-coding scheme. Each receptor protein binds different odorants more or less readily, so its receptor cell is more or less sensitive to those odorants

BOX 8.4 PATH OF DISCOVERY

Channels of Vision and Smell

by Geoffrey Gold

The discovery of cyclic nucleotide-gated ion channels in olfactory receptor cells provides illuminating examples of how scientific orthodoxy can inhibit progress. Ironically, the story begins with work on vision. Studies of visual transduction began in earnest following the discovery, in 1971, that light caused the breakdown of cyclic guanosine monophosphate (cGMP) in photoreceptors. Yet, it was not until 1985 that the patch-damp method was used to demonstrate a direct effect of cGMP on ion channels from photoreceptors. This long delay was not due to lack of interest because there were at least a dozen labs working on the visual transduction mechanism. Rather, I think the widespread acceptance of protein phosphorylation as the mechanism of action of cyclic nucleotides in most cells effectively suppressed curiosity about other (direct) effects of cyclic nucleotides on ion channels. The cGMP-gated channel of photoreceptors was discovered by a group in the former Soviet Union, perhaps because these scientists were less influenced by the reigning dogma in Western countries.

The discovery of the olfactory cyclic nucleotide-gated channel by Tadashi Nakamura and me also emphasizes the importance of listening to the beat of your own drum. After the odorant-stimulated adenylyl cyclase was discovered in 1985, only a few months after the discovery of the photoreceptor cGMP-gated channel, we (and probably others) thought that olfactory cilia might contain a cyclic nucleotide-gated channel. This was because the biochemical similarities between visual and olfactory transduction suggested an evolutionary relationship between photoreceptors and olfactory receptor cells. Thus, we hypothesized that if the biochemical reactions of sensory transduction were conserved throughout evolution, ion channels might also be conserved. However, we knew that the transduction process was located in the cilia, and structures as small as cilia, which are about 0.2 μm in diameter, had never before been studied with the patch-damp technique. Indeed, most people I talked to thought it would be impossible to excise patches from the cilia. Nevertheless, we reasoned that it should be possible if only we could make patch pipettes with tip openings smaller than the ciliary diameter. This proved easy to accomplish; it only required fire polishing (melting) the tips of the patch pipettes slightly longer than was customary. Once we obtained high-resistance seals on cilia, patch excision and current recording were done in the conventional way.

Perhaps the most ironic thing about this story is that the photoreceptor channel was discovered by a group led by E. E. Fesenko, whose previous (and subsequent) work was on olfactory receptor proteins, whereas our work prior to the discovery of the olfactory channel was on phototransduction. That just goes to show how useful it can be for people to move into new areas. I like to point out that ours was a project that never would have been funded by the conventional grant review process because it was so unlikely to work.

(Figure 8.13b). Some cells are more sensitive to the chemical structure of the odorants they respond to than other cells are, but in general each receptor is quite broadly tuned. A corollary is that each odorant activates many of the 1000 types of receptors. The concentration of odorant is also important, and more odorant tends to generate stronger responses, until response strength saturates. Thus, each olfactory cell yields very ambiguous information about odorant type and strength. It is the job of the central olfactory pathways to respond to the full package of information arriving from the olfactory epithelium—the population code—and use it to classify odors further.

Central Olfactory Pathways

Olfactory receptor neurons send axons into the two olfactory bulbs (Figure 8.14). The bulbs are a neuroscientist's wonderland, full of neural circuits with numerous types of neurons, fascinating dendritic arrangements, unusual reciprocal synapses, and high levels of many different

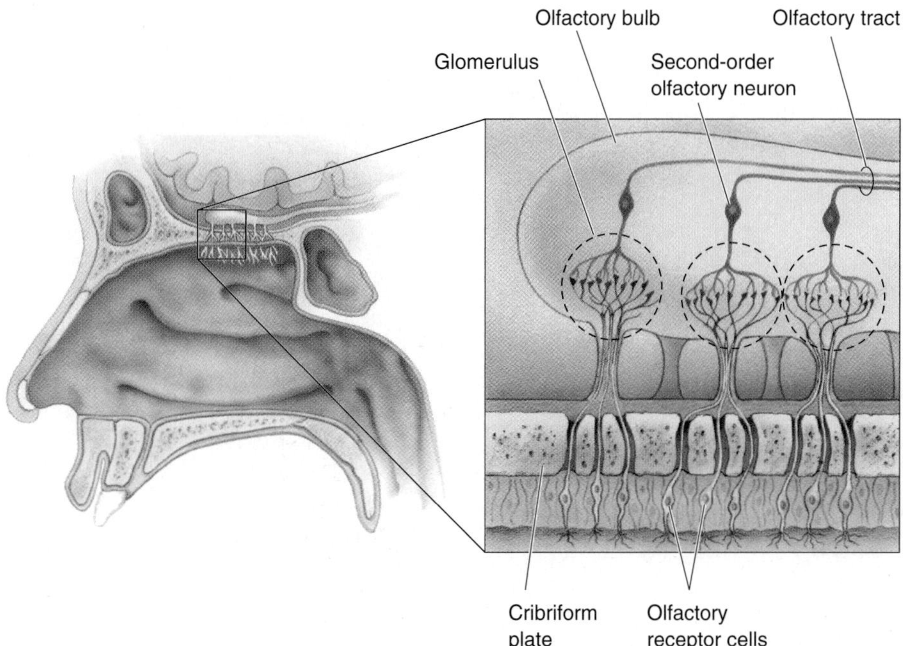

◀ FIGURE 8.14
The location and structure of an olfactory bulb. Axons of olfactory receptor cells penetrate the cribriform plate and enter the olfactory bulb. After multiple branching, each olfactory axon synapses upon second-order neurons within a spherical glomerulus. The second-order neurons send axons through the olfactory tract further into the brain.

neurotransmitters. The input layer of each bulb in mice contains about 2000 spherical structures called **glomeruli**, each about 50–200 μm in diameter. Within each glomerulus, the endings of about 25,000 primary olfactory axons (axons from the receptor cells) converge and terminate on the dendrites of about 100 second-order olfactory neurons.

Recent studies revealed that the mapping of receptor cells onto glomeruli is astonishingly precise. Each glomerulus receives receptor axons from a large region of the olfactory epithelium. When molecular labeling methods are used to tag each receptor neuron expressing one particular receptor gene of the mouse—in this case, a gene called *P2*—we can see that the P2-labeled axons all converge onto only two glomeruli in each bulb, one of which is shown in Figure 8.15a. No axons seem to be out of place, but our knowledge of axonal pathfinding during development cannot yet explain the targeting accuracy of olfactory axons (see Chapter 23).

This precision mapping is also consistent across the two olfactory bulbs; each bulb has only two P2-targeted glomeruli in symmetrical positions (Figure 8.15b). The positions of the P2 glomeruli within each bulb are consistent from one mouse to another. Finally, it seems that each glomerulus receives input from only receptor cells of one particular type. This means that the array of glomeruli within a bulb is a very orderly map of the receptor genes expressed in the olfactory epithelium (Figure 8.16), and, by implication, a map of odor information.

Olfactory information is modified by inhibitory and excitatory interactions within and among the glomeruli and between the two bulbs. Neurons in the bulbs are also subject to modulation from systems of axons that descend from higher areas of the brain. While it is obvious that the elegant circuitry of the olfactory bulbs has important functions, it is not entirely clear what those functions are. It is likely that they begin to segregate odorant signals into broad categories, independent of their strength and possible interference from other odorants. The precise identification of an odor probably requires further processing in the next stages of the olfactory system.

Many brain structures receive olfactory connections. The output axons of the olfactory bulbs course through the olfactory tracts and project

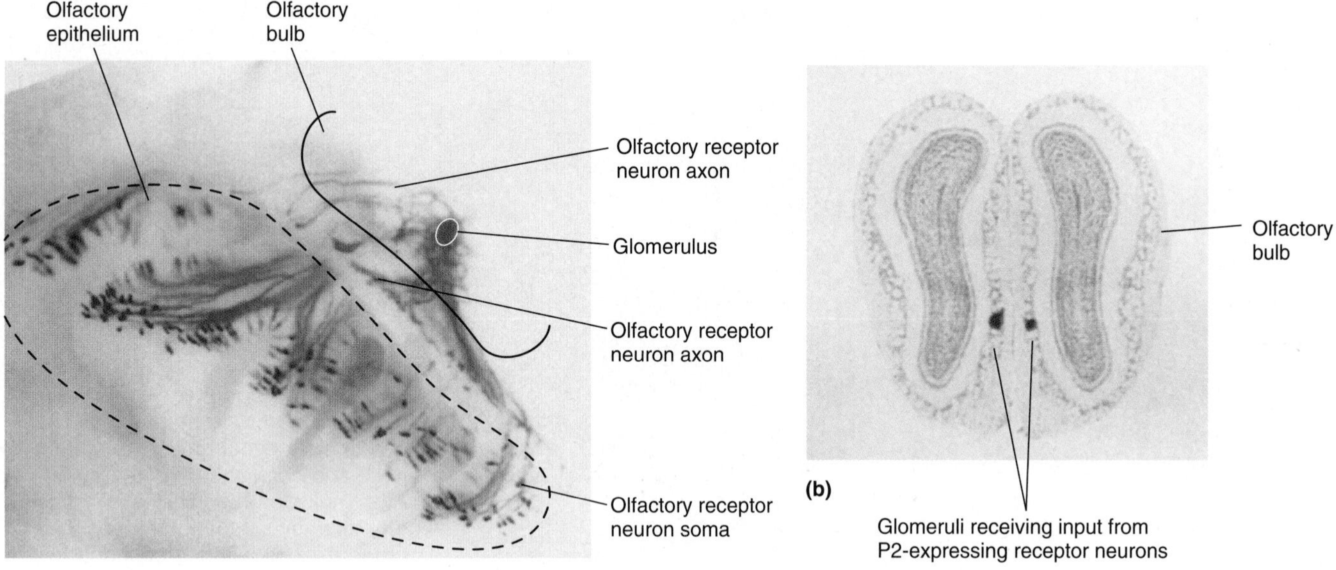

▲ FIGURE 8.15
The convergence of olfactory neuron axons onto the olfactory bulb. Olfactory receptor neurons expressing a particular receptor gene all send their axons to the same glomeruli. **(a)** In a mouse, receptor neurons expressing the P2 receptor gene were labeled blue, and every neuron sent its axon to the same glomeruli in the olfactory bulbs. In this image, only a single glomerulus with P2 axons is visible. **(b)** When the two bulbs were cut in cross section, it was possible to see that the P2-containing receptor axons project to symmetrically placed glomeruli in each bulb. (Source: Adapted from Mombaerts et al., 1996, p. 680.)

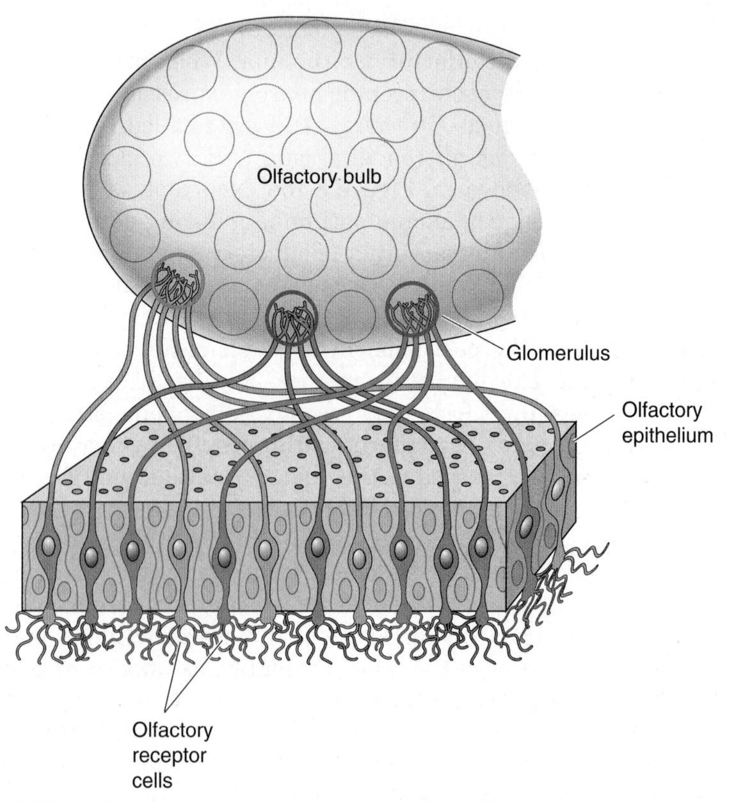

▲ FIGURE 8.16
Specific mapping of olfactory receptor neurons onto glomeruli. Each glomerulus receives input only from receptor cells expressing a particular receptor protein gene. Receptor cells expressing a particular gene are color-coded.

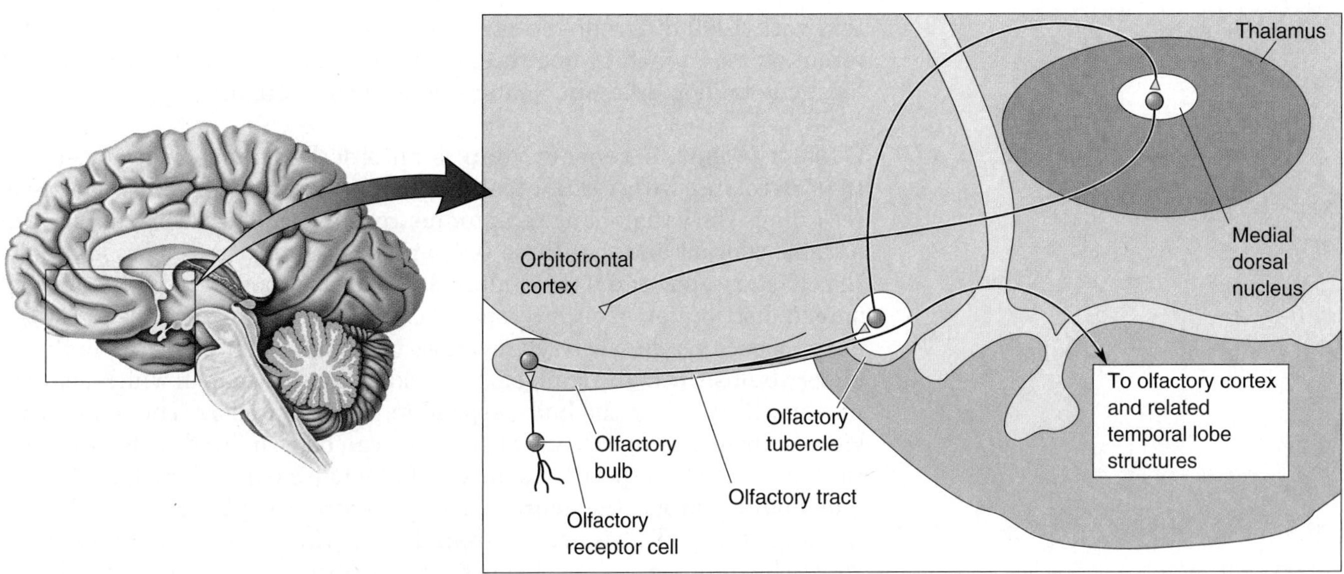

▲ FIGURE 8.17
Central olfactory pathways. Axons of the olfactory tract branch and enter many regions of the forebrain, including the olfactory cortex. The neocortex is reached only by a pathway that synapses in the medial dorsal nucleus of the thalamus.

directly to several targets, some of which are illustrated in Figure 8.17. Among the most important targets are the primitive region of cerebral cortex called the **olfactory cortex** and some of its neighboring structures in the temporal lobes. This anatomy makes olfaction unique. All other sensory systems *first* pass information through the thalamus before projecting it to the cerebral cortex. The olfactory arrangement produces an unusually direct and widespread influence on the parts of the forebrain that have roles in odor discrimination, emotion, motivation, and certain kinds of memory (see Chapters 16, 18, 24 and 25). Conscious perceptions of smell may be mediated by a path from the *olfactory tubercle,* to the *medial dorsal nucleus* of the thalamus, and to the *orbitofrontal cortex* (situated right behind the eyes).

Spatial and Temporal Representations of Olfactory Information

In olfaction, there is an apparent paradox similar to the one in gustation. Individual receptors are broadly tuned to their stimuli; that is, each cell is sensitive to a wide variety of chemicals. However, when we smell those same chemicals, we can easily tell them apart. How is the whole brain doing what single olfactory cells cannot? We will discuss three important ideas: (1) Each odor is represented by the activity of a large population of neurons; (2) the neurons responsive to particular odors may be organized into spatial maps; and (3) the timing of action potentials may be an essential code for particular odors.

Olfactory Population Coding. As in gustation, the olfactory system uses the responses of a large population of receptors to encode a specific stimulus. A simplistic example was shown in Figure 8.13b. When presented with a citrus smell, none of the three different receptor cells can individually distinguish it clearly from the other odors. But by looking at the *combination* of responses from all three cells, the brain could distinguish the citrus smell unambiguously from floral, peppermint, and almond. By using such population coding, you can imagine how an olfactory system with 1000 different receptors might be able to recognize many different odors. In fact, by one recent estimate humans can discriminate at least one trillion different combinations of odor stimuli.

Olfactory Maps. A **sensory map** is an orderly arrangement of neurons that correlates with certain features of the environment. Microelectrode recordings show that many receptor neurons respond to the presentation of a single odorant and that these cells are distributed across a broad area of the olfactory epithelium (see Figure 8.13). This is consistent with the widespread distribution of each receptor gene. However, we have seen that the axons of each receptor cell type synapse upon particular glomeruli in the olfactory bulbs. Such an arrangement yields a sensory map in which neurons in a specific place in the bulb respond to particular odors. The maps of regions activated by one chemical stimulus can be visualized with special recording methods. Experiments reveal that while a particular odor activates many bulb neurons, the neurons' positions form complex but reproducible *spatial* patterns. This is evident from the experiment shown in Figure 8.18, in which a minty-smelling chemical activates one pattern of glomeruli and a fruity scent activates quite a different pattern. Thus, the smell of a particular chemical is converted into a specific map defined by the positions of active neurons within the "neural space" of the bulbs, and the form of the map depends on the nature and concentration of the odorant.

You will see in subsequent chapters that every sensory system uses spatial maps, perhaps for many different purposes. In most cases, the maps correspond obviously to features of the sensory world. For example, in the visual system, there are maps of visual space; in the auditory system, there are maps of sound frequency; and in the somatic sensory system, there are maps of the body surface. The maps of the chemical senses are unusual in that the stimuli themselves have no meaningful

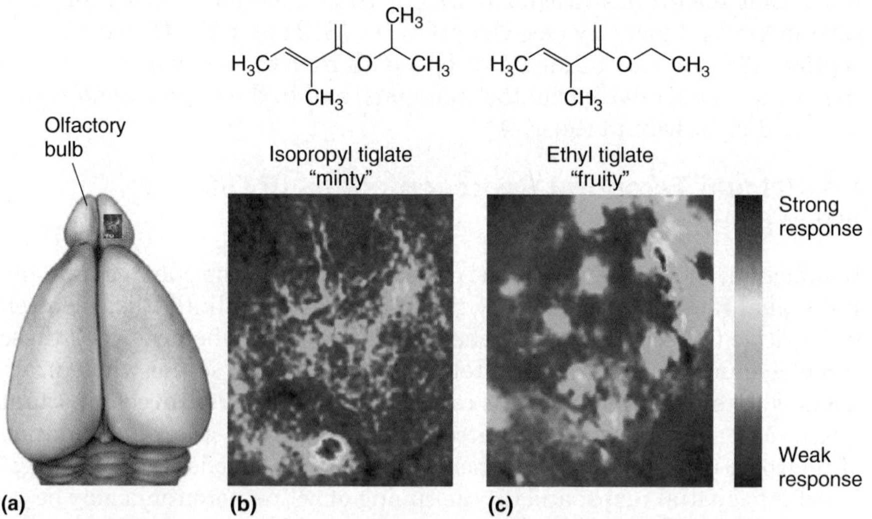

▲ FIGURE 8.18
Maps of neural activation of the olfactory bulb. The activity of neurons in the glomeruli of a mouse olfactory bulb was recorded with a specialized optical method. The cells expressed a fluorescent protein sensitive to intracellular Ca^{2+} levels, and neural activity was then signaled by changes in the amount of light emitted by the protein. The colors on the maps represent differing levels of neural activity; hotter colors (red and orange) imply more activity. Activated glomeruli show up as spots of color. **(a)** The blue box shows the area of one olfactory bulb that was mapped. Different olfactants evoked different spatial patterns of neural activation in the bulb: **(b)** isopropyl tiglate, which smells minty to humans, and **(c)** ethyl tiglate, which smells fruity, activate completely different patterns of glomeruli. (Source: Adapted from Blauvelt et al., 2013, Figure 4.)

spatial properties. Although seeing a skunk walking in front of you may tell you *what* and *where* it is, smell by itself can reveal only the *what*. (By moving your head about, you can localize smells only crudely.) The most critical feature of each odorant is its chemical structure, not its position in space. Because the olfactory system does not have to map the spatial pattern of an odor in the same way that the visual system has to map the spatial patterns of light, neural odor maps may have other functions, such as discrimination among a huge number of different chemicals. Recent studies of the olfactory cortex show that each distinct odor triggers activity in a different subset of neurons. In the experiment shown in Figure 8.19, the orange-smelling octanal excites a group of neurons that are mostly different from the neurons excited by either the piney-scented α-pinene or the grassy-smelling hexanal (Figure 8.19).

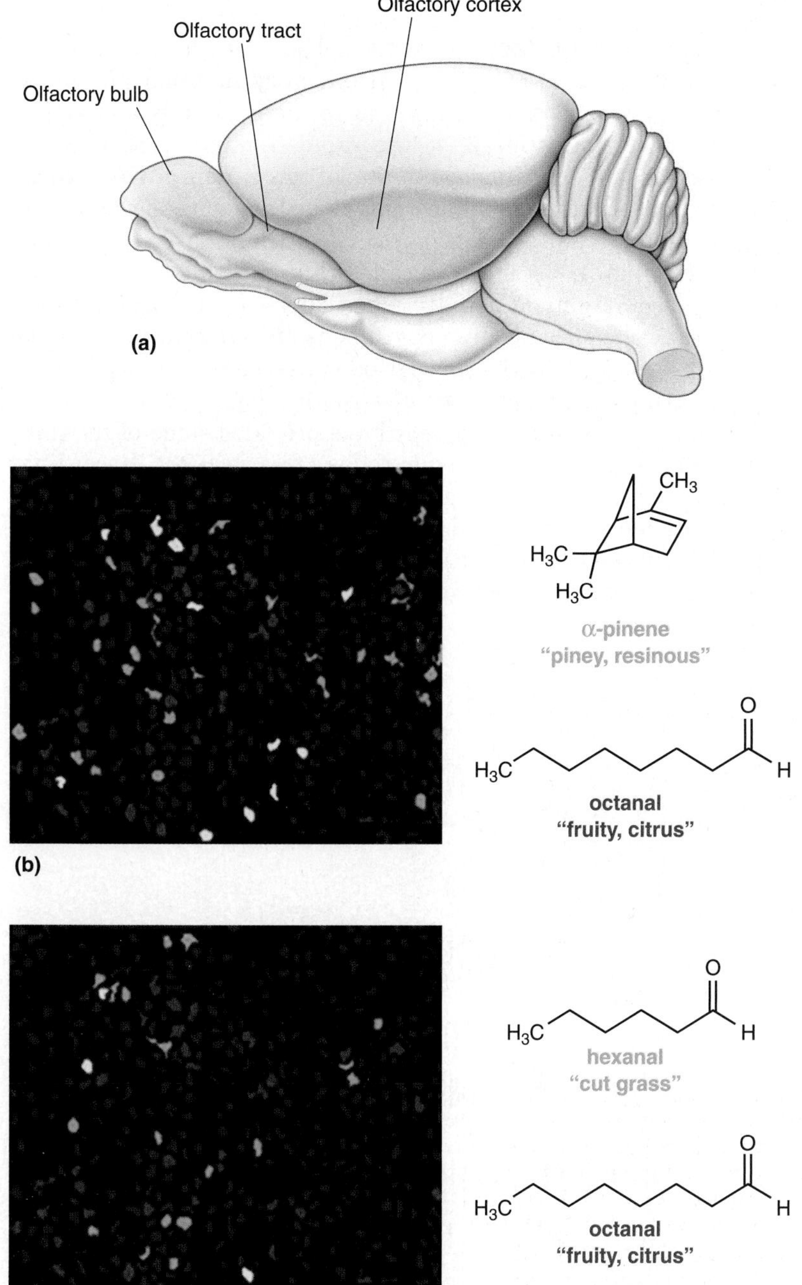

◀ FIGURE 8.19

Maps of neural activation of the olfactory cortex. The activity of many neurons in the mouse olfactory cortex was recorded with a specialized optical method. The cells were loaded with a Ca^{2+}-sensitive fluorescent dye, and neural activity was then signaled by changes in the amount of light emitted. **(a)** The olfactory regions are shaded orange. **(b)** Neurons responsive to the piney scent of α-pinene are color-coded green, neurons responsive to the fruity scent of octanal are red, and cells responsive to both are yellow. **(c)** Neurons responsive to the grassy scent of hexanal are coded green, those responsive to the fruity scent of octanal are red, and cells responsive to both are yellow. Each of the three odorants activated a distinctly different pattern of cortical neurons. (Source: Adapted from Stettler and Axel, 2009, p. 858.)

But are neural odor maps actually used by the brain to distinguish among chemicals? We don't know the answer. For a map to be useful, there must be something that reads and understands it. With practice and very specialized goggles, we might be able to read the "alphabet" of odors mapped on the surface of the olfactory bulb with our eyes. This may roughly approximate what higher regions of the olfactory system do, but so far there is no evidence that the olfactory cortex has this capability. An alternative idea is that spatial maps do not encode odors at all but are simply the most efficient way for the nervous system to form appropriate connections between related sets of neurons (e.g., receptor cells and glomerular cells). With orderly mapping, the lengths of axons and dendrites can be minimized. Neurons with similar functions can interconnect more easily if they are neighbors. The spatial map that results may be simply a side effect of these developmental requirements, rather than a fundamental mechanism of sensory coding itself.

Temporal Coding in the Olfactory System. There is growing evidence that the *temporal* patterns of spiking in olfactory neurons are essential features of olfactory coding. Compared to many sounds and sights, odors are inherently slow stimuli, so the rapid timing of action potentials is not necessary for encoding the timing of odors. **Temporal coding**, which depends on the timing of spikes, might instead encode the quality of odors. Hints about the possible importance of timing are easy to find. Researchers have known for many decades that the olfactory bulb and cortex generate oscillations of activity when odors are presented to the receptors, but the relevance of these rhythms is still unknown. Temporal patterns are also evident in the spatial odor maps, as they sometimes change shape during the presentation of a single odor.

Recent studies of insects and rodents have provided some of the most convincing evidence for temporal odor codes (Figure 8.20). Recordings from the olfactory systems of mice and insects demonstrate that odor information is encoded by the detailed timing of spikes within cells and between groups of cells as well as by the number, temporal pattern, rhythmicity, and cell-to-cell synchrony of spikes.

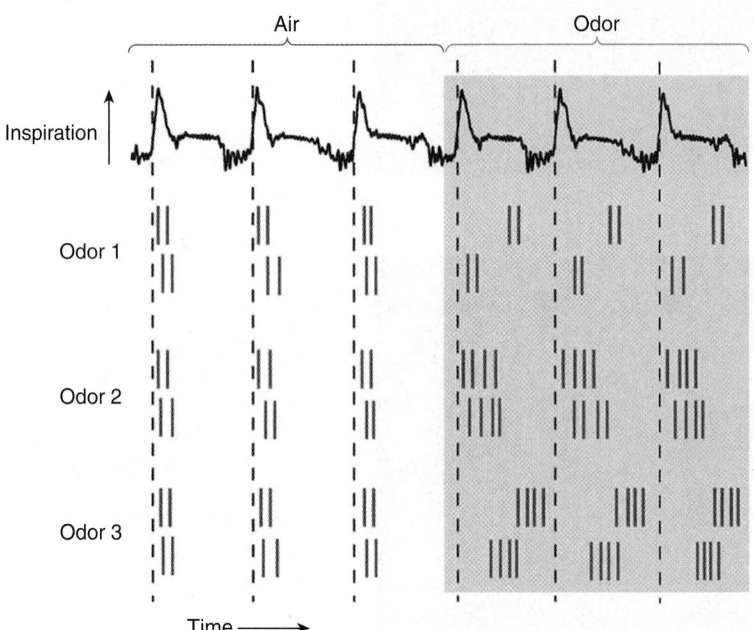

▶ FIGURE 8.20
Spiking patterns may include changes of number, rate, and timing. As a mouse inspired (black trace at top) first air and then air with odorants, the activity of two neurons in the olfactory bulb was recorded. Spikes from the two cells are represented by red and blue lines. Both cells tended to spike twice during inspiration of air alone. Odor 1 did not change the numbers of spikes, but it did cause the red cell to spike much later in the respiratory cycle. Odor 2 increased the numbers of spikes without changing their timing much. Odor 3 increased spike numbers and delayed their onset. (Source: Dhawale et al., 2010, p. 1411.)

As with spatial maps, however, demonstrating that information is carried by spike timing is only a first step; proving that the brain actually *uses* that information is much more difficult. In a fascinating experiment with honeybees, Gilles Laurent and his colleagues at the California Institute of Technology were able to disrupt the rhythmic synchrony of odor responses without otherwise affecting their spiking responses. This loss of synchronous spiking was associated with a loss of the bees' ability to discriminate between similar odors, although not between broad categories of odors. The implication is that the bee analyzes an odor not only by keeping track of *which* olfactory neurons fire but also by *when* they fire. It will be very interesting to learn whether similar processes occur in a mammalian olfactory system.

CONCLUDING REMARKS

The chemical senses are a good place to begin learning about sensory systems because smell and taste are the most basic of sensations. Gustation and olfaction use a variety of transduction mechanisms to recognize the enormous number of chemicals we encounter in the environment. Yet, the molecular mechanisms of transduction are very similar to the signaling systems used in every cell of the body, for functions as diverse as neurotransmission and fertilization. We will see that the transduction mechanisms in other sensory systems are also highly specialized and derive from common cellular processes. Remarkable parallels have been discovered, such as the molecular similarity between the sensory cells of smelling and seeing.

Common sensory principles also extend to the level of neural systems. Most sensory cells are broadly tuned for their stimuli. This means that the nervous system must use population codes to represent and analyze sensory information, resulting in our remarkably precise and detailed perceptions. Populations of neurons are often arranged in sensory maps within the brain. And the timing of action potentials may function to represent sensory information in ways that are not yet understood. In the chapters that follow, we will see similar trends in the anatomy and physiology of systems dealing with light, sound, and pressure.

KEY TERMS

Introduction
gustation (p. 266)
olfaction (p. 266)
chemoreceptors (p. 266)

Taste
papillae (p. 267)
taste buds (p. 268)
taste receptor cells (p. 268)
receptor potential (p. 270)
transduction (p. 271)
gustatory nucleus (p. 276)
ventral posterior medial (VPM) nucleus (p. 276)
primary gustatory cortex (p. 276)
population coding (p. 277)

Smell
pheromones (p. 278)
olfactory epithelium (p. 278)
olfactory bulb (p. 280)
glomeruli (p. 285)
olfactory cortex (p. 287)
sensory map (p. 288)
temporal coding (p. 290)

REVIEW QUESTIONS

1. Most tastes involve some combination of the five basic tastes. What other sensory factors can help define the specific perceptions associated with a particular food?
2. The transduction of saltiness is accomplished, in part, by a Na^+-permeable channel. Why would a sugar-permeable membrane channel be a poor mechanism for the transduction of sweetness?
3. Chemicals that have sweet, bitter, and umami tastes all activate precisely the same intracellular signaling molecules. Given this fact, can you explain how the nervous system can distinguish the tastes of sugars, alkaloids, and amino acids?
4. Why would the size of an animal's olfactory epithelium (and consequently the number of receptor cells) be related to its olfactory acuity?
5. Receptor cells of the gustatory and olfactory systems undergo a constant cycle of growth, death, and maturation. Therefore, the connections they make with the brain must be continually renewed as well. Can you propose a set of mechanisms that would allow the connections to be remade in a specific way, again and again, over the course of an entire lifetime?
6. If the olfactory system does use some kind of spatial mapping to encode specific odors, how might the rest of the brain read the map?

FURTHER READING

Kinnamon SC. 2013. Neurosensory transmission without a synapse: new perspectives on taste signaling. *BMC Biology* 11:42.

Liberles SD. 2014. Mammalian pheromones. *Annual Review of Physiology* 76:151–175.

Liman ER, Zhang YV, Montell C. 2014. Peripheral coding of taste. *Neuron* 81: 984–1000.

Murthy VN. 2011. Olfactory maps in the brain. *Annual Review of Neuroscience* 34:233–258.

Stettler DD, Axel R. 2009. Representations of odor in the piriform cortex. *Neuron* 63: 854–864.

Zhang X, Firestein S. 2002. The olfactory receptor gene superfamily of the mouse. *Nature Neuroscience* 5:124–133.

CHAPTER NINE

The Eye

INTRODUCTION

Vision is remarkable; it lets us detect things as tiny and close as a mosquito on the tip of our nose, or as immense and far away as a galaxy hundreds of thousands of light-years away. Sensitivity to light enables animals, including humans, to detect prey, predators, and mates. Based on the light bounced into our eyes from objects around us, we somehow make sense of a complex world. While this process seems effortless, it is in reality extremely complicated. Indeed, it has proven quite difficult to make computer visual systems with even a small fraction of the capabilities of the human visual system.

Light is electromagnetic energy that is emitted in the form of waves. We live in a turbulent sea of electromagnetic radiation. Like any ocean, this sea has large waves and small waves, short wavelets, and long rollers. The waves crash into objects and are absorbed, scattered, reflected, and bent. Because of the nature of electromagnetic waves and their interactions with the environment, the visual system can extract information about the world. This is a big job, and it requires a lot of neural machinery. However, the mastery of vision over the course of vertebrate evolution has had surprising rewards. It has provided new ways to communicate, given rise to brain mechanisms for predicting the trajectory of objects and events in time and space, allowed for new forms of mental imagery and abstraction, and led to the creation of a world of art. The significance of vision is perhaps best demonstrated by the fact that more than a third of the human cerebral cortex is involved with analyzing the visual world.

The mammalian visual system begins with the eye. At the back of the eye is the **retina**, which contains photoreceptors specialized to convert light energy into neural activity. The rest of the eye acts like a camera and forms crisp, clear images of the world on the retina. Like a camera, the eye automatically adjusts to differences in illumination and focuses on objects of interest. The eye can also track moving objects (by eye movement) and can keep its transparent surfaces clean (by tears and blinking).

While much of the eye functions like a camera, the retina does much more than passively register light levels across space. In fact, as mentioned in Chapter 7, the retina is actually part of the brain. (Think about that the next time you look deeply into someone's eyes.) In a sense, each eye has two overlapping retinas: one specialized for low light levels that we encounter from dusk to dawn, and another specialized for higher light levels and for the detection of color from sunrise to sunset. Regardless of the time of day, however, the output of the retina is not a faithful reproduction of the intensity of the light falling on it. Rather, the retina is specialized to detect *differences* in the intensity of light falling on different parts of it. Image processing is well under way in the retina before any visual information reaches the rest of the brain.

Axons of retinal neurons are bundled into optic nerves, which distribute visual information (in the form of action potentials) to several brain structures that perform different functions. Some targets of the optic nerves are involved in regulating biological rhythms, which are synchronized with the light–dark daily cycle; others are involved in the control of eye position and optics. However, the first synaptic relay in the pathway that serves visual perception occurs in a cell group of the dorsal thalamus called the *lateral geniculate nucleus*, or *LGN*. From the LGN, visual information ascends to the cerebral cortex, where it is interpreted and remembered.

This chapter explores the eye and the retina. We'll see how light carries information to our visual system, how the eye forms images on the retina, and how the retina converts light energy into neural signals that can be used to extract information about luminance and color differences. In Chapter 10, we will pick up the visual pathway at the back of the eye and take it through the thalamus to the cerebral cortex.

PROPERTIES OF LIGHT

The visual system uses light to form images of the world around us. Let's briefly review the physical properties of light and its interactions with the environment.

Light

Electromagnetic radiation is all around us. It comes from innumerable sources, including radio antennas, mobile phones, X-ray machines, and the sun. Light is the electromagnetic radiation that is visible to our eyes. Electromagnetic radiation can be described as a wave of energy. Like any wave, electromagnetic radiation has a *wavelength*, the distance between successive peaks or troughs; a *frequency*, the number of waves per second; and an *amplitude*, the difference between wave trough and peak (Figure 9.1).

Amplitude

Wavelength

▲ FIGURE 9.1
Characteristics of electromagnetic radiation.

The energy content of electromagnetic radiation is proportional to its frequency. Radiation emitted at a high frequency (short wavelengths) has the highest energy content; examples are gamma radiation emitted by some radioactive materials and X-rays used for medical imaging, with wavelengths less than 10^{-9} m (<1 nm). Conversely, radiation emitted at lower frequencies (longer wavelengths) has less energy; examples are radar and radio waves, with wavelengths greater than 1 mm. Only a small part of the electromagnetic spectrum is detectable by our visual system; visible light consists of wavelengths of 400–700 nm (Figure 9.2). As first shown by Isaac Newton early in the eighteenth century, the mix of wavelengths in this range emitted by the sun appears to humans as white, whereas light of a single wavelength appears as one of the colors of the rainbow. It is interesting to note that a "hot" color like red or orange consists of light with a longer wavelength, and hence has *less* energy, than a "cool" color like blue or violet. Clearly, colors are themselves "colored" by the brain, based on our subjective experiences.

Optics

In a vacuum, a wave of electromagnetic radiation travels in a straight line and thus can be described as a *ray*. Light rays in our environment

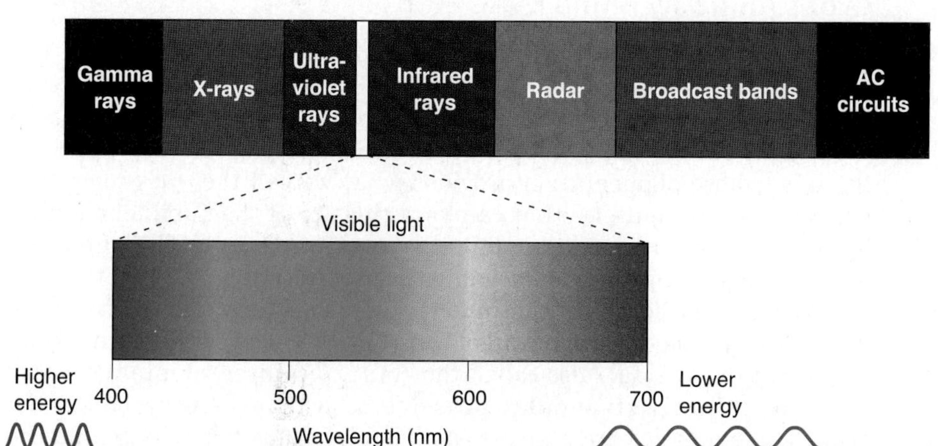

◀ FIGURE 9.2
The electromagnetic spectrum. Only electromagnetic radiation with wavelengths of 400–700 nm is visible to the human eye. Within this visible spectrum, different wavelengths appear as different colors.

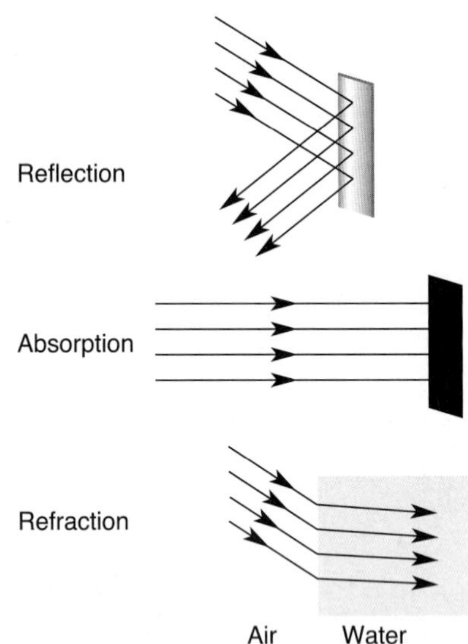

▲ FIGURE 9.3
Interactions between light and the environment. When light strikes an object in our environment, the light may be reflected, absorbed, or some combination of both. Visual perception is based on light coming directly into the eye from a luminous object such as a neon sign or being reflected off objects. Images are formed in the eye by refraction. In this example of light passing through air and then water, the light rays bend toward a line perpendicular to the air–water interface.

also travel in straight lines until they interact with the atoms and molecules of the atmosphere and objects on the ground. These interactions include reflection, absorption, and refraction (Figure 9.3). The study of light rays and their interactions is called *optics*.

Reflection is the bouncing of light rays off a surface. The manner in which a ray of light is reflected depends on the angle at which it strikes the surface. A ray striking a mirror perpendicularly is reflected 180° back upon itself, a ray striking the mirror at a 45° angle is reflected 90°, and so on. Most of what we see is light that has been reflected off objects in our environment.

Absorption is the transfer of light energy to a particle or surface. You can feel this energy transfer on your skin on a sunny day, as visible light is absorbed and warms you up. Surfaces that appear black absorb the energy of all visible wavelengths. Some compounds absorb light energy only in a limited range of wavelengths and reflect the remaining wavelengths. This property is the basis for the colored pigments of paints. For example, a blue pigment absorbs long wavelengths but reflects a range of short wavelengths centered on 430 nm that are perceived as blue. As we will see in a moment, light-sensitive photoreceptor cells in the retina also contain pigments and use the energy absorbed from light to generate changes in membrane potential.

Images are formed in the eye by **refraction**, the bending of light rays that can occur when they travel from one transparent medium to another. When you dangle your leg into a swimming pool, for example, the odd way the leg appears to bend at the surface is a result of refraction. Consider a ray of light passing from the air into a pool of water. If the ray strikes the water surface perpendicularly, it will pass through in a straight line. However, if light strikes the surface at an angle, it will bend toward a line that is perpendicular to the surface. This bending of light occurs because the speed of light differs in the two media; light passes through air more rapidly than through water. The greater the difference between the speed of light in the two media, the greater the angle of refraction. The transparent media in the eye bend light rays to form images on the retina.

THE STRUCTURE OF THE EYE

The eye is an organ specialized for the detection, localization, and analysis of light. Here we introduce the structure of this remarkable organ in terms of its gross anatomy, ophthalmoscopic appearance, and cross-sectional anatomy.

Gross Anatomy of the Eye

When you look into someone's eyes, what are you really looking at? The main structures are shown in Figure 9.4. The **pupil** is the opening that allows light to enter the eye and reach the retina; it appears dark because of the light-absorbing pigments in the retina. The pupil is surrounded by the **iris**, whose pigmentation provides what we call the eye's color. The iris contains two muscles that can vary the size of the pupil; one makes it smaller when it contracts, the other makes it larger. The pupil and iris are covered by the glassy transparent external surface of the eye, the **cornea**. The cornea is continuous with the **sclera**, the "white of the eye," which forms the tough wall of the eyeball. The eyeball sits in a bony eye socket in the skull, also called the *eye's orbit*. Inserted into the sclera are three pairs of **extraocular muscles**, which move the eyeball in the orbit. These muscles are normally not visible because they lie behind the

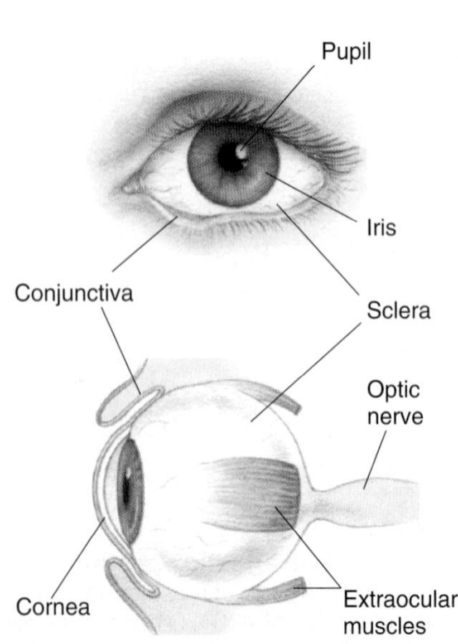

▲ FIGURE 9.4
Gross anatomy of the human eye.

conjunctiva, a membrane that folds back from the inside of the eyelids and attaches to the sclera. The **optic nerve**, carrying axons from the retina, exits the back of the eye, passes through the orbit, and reaches the base of the brain near the pituitary gland.

Ophthalmoscopic Appearance of the Eye

Another view of the eye is afforded by the ophthalmoscope, a device that enables one to peer into the eye through the pupil to the retina (Figure 9.5). The most obvious feature of the retina viewed through an ophthalmoscope is the blood vessels on its surface. These retinal vessels originate from a pale circular region called the **optic disk**, which is also where the optic nerve fibers exit the retina.

It is interesting to note that the sensation of light cannot occur at the optic disk because there are no photoreceptors here, nor can it occur where the large blood vessels exist because the vessels cast shadows on the retina. And yet, our perception of the visual world appears seamless. We are not aware of any holes in our field of vision because the brain fills in our perception of these areas. However, there are tricks by which we can demonstrate the "blind" retinal regions (Box 9.1).

At the middle of each retina is a darker-colored region with a yellowish hue. This is the **macula** (from the Latin word for "spot"), the part of the retina for central vision (as opposed to peripheral vision). Besides its color, the macula is distinguished by the relative absence of large blood vessels. Notice in Figure 9.5 that the vessels arc from the optic disk to the macula; this is also the trajectory of the optic nerve fibers from the macula en route to the optic disk. The relative absence of large blood vessels in this region of the retina is one of the specializations that improves the quality of central vision. Another specialization of the central retina can sometimes be discerned with the ophthalmoscope: the **fovea**, a dark spot about 2 mm in diameter. The term is from the Latin for "pit," and the retina is thinner in the fovea than elsewhere. Because it marks the center of the retina, the fovea is a convenient anatomical reference point. Thus, the part of the retina that lies closer to the nose than the fovea is referred to as *nasal*, the part that lies near the temple is called *temporal*, the part of the retina above the fovea is called *superior*, and the part below is *inferior*.

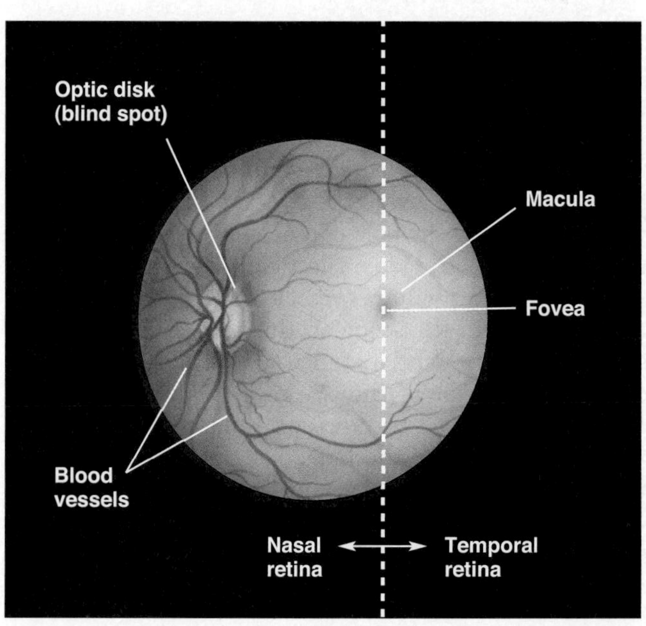

◀ FIGURE 9.5
The retina, viewed through an ophthalmoscope. The dotted line through the fovea represents the demarcation between the side of the eye nearer the nose (nasal retina) and the side of the eye nearer the ear (temporal retina). The imaginary line crosses through the macula, which is in the center of the retina (it appears slightly to one side here because the photograph was taken to include the optic disk off to the nasal side of the retina).

BOX 9.1 OF SPECIAL INTEREST

Demonstrating the Blind Regions of Your Eye

A look through an ophthalmoscope reveals that there is a sizable hole in the retina. The region where the optic nerve axons exit the eye and the retinal blood vessels enter the eye, the optic disk, is completely devoid of photoreceptors. Moreover, the blood vessels coursing across the retina are opaque and block light from falling on photoreceptors beneath them. Although we normally don't notice them, these blind regions can be demonstrated. Look at Figure A. Hold the book about 1.5 feet away, close your right eye, and fixate on the cross with your left eye. Move the book (or your head) around slightly, and eventually you will find a position where the black circle disappears. At this position, the spot is imaged on the optic disk of the left eye. This region of visual space is called the *blind spot* for the left eye.

The blood vessels are a little tricky to demonstrate, but give this a try. Get a standard household flashlight. In a dark or dimly lit room, close your left eye (it helps to hold the eye closed with your finger so you can open your right eye fully). Look straight ahead with the open right eye, and shine the flashlight at an angle into the outer side of the right eye. Jiggle the light back and forth, up and down. If you're lucky, you'll see an image of your own retinal blood vessels. This is possible because the illumination of the eye at this oblique angle causes the retinal blood vessels to cast long shadows on the adjacent regions of retina. For the shadows to be visible, they must be swept back and forth on the retina, hence the jiggling of the light.

If we have all these light-insensitive regions in the retina, why does the visual world appear uninterrupted and seamless? The answer is that mechanisms in the visual cortex appear to "fill in" the missing regions. Perceptual filling-in can be demonstrated with the stimulus shown in Figure B. Fixate on the cross with your left eye and move the book closer and farther from your eye. You'll find a distance at which you will see a continuous uninterrupted line. At this point, the space in the line is imaged on the blind spot, and your brain fills in the gap.

Figure A

Figure B

Cross-Sectional Anatomy of the Eye

A cross-sectional view of the eye shows the path taken by light as it passes through the cornea toward the retina (Figure 9.6). The cornea lacks blood vessels and is nourished by the fluid behind it, the **aqueous humor**. This view reveals the transparent **lens** located behind the iris. The lens is suspended by ligaments (called *zonule fibers*) attached to the **ciliary muscle**, which forms a ring inside the eye. If you imagine using toothpicks to center a strawberry in the hole of a bagel, you have the right mental picture: The strawberry is the lens, the toothpicks represent the zonule fibers, and the bagel is the ciliary muscle that attaches to the sclera. As we shall see, changes in the shape of the lens enable our eyes to adjust their focus to different viewing distances.

The lens also divides the interior of the eye into two compartments containing slightly different fluids. The aqueous humor is the watery fluid that lies between the cornea and the lens. The more viscous, jelly-like **vitreous humor** lies between the lens and the retina; it serves to keep the eyeball spherical.

Although the eyes do a remarkable job of delivering precise visual information to the rest of the brain, a variety of disorders can compromise this ability (Box 9.2).

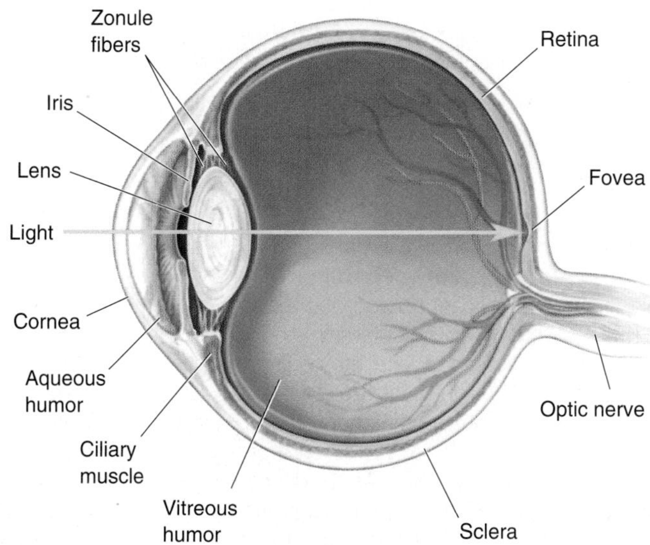

▲ FIGURE 9.6
The eye in cross section. Structures at the front of the eye regulate the amount of light allowed in and refract light onto the retina, which wraps around the inside of the eye.

IMAGE FORMATION BY THE EYE

The eye collects the light rays emitted by or reflected off objects in the environment, and focuses them onto the retina to form images. Bringing objects into focus involves the combined refractive powers of the cornea and lens. You may be surprised to learn that the cornea, rather than the lens, is the site of most of the refractive power of the eyes. This occurs because light reaches the eye from air and the cornea is mostly made of water. A good deal of refraction occurs because light travels significantly slower in water than air. In comparison there is less refraction by the lens because the aqueous humor, lens, and vitreous humor are all composed largely of water.

Refraction by the Cornea

Consider the light emitted from a distant source, perhaps a bright star at night. Light rays are emitted in all directions from the star, but because of the great distance, the rays that reach our eye on earth are virtually parallel. These parallel rays hit our cornea from one side to the other. Nonetheless, we see the star as a point of light, rather than a bright splotch filling our field of view, because the eye uses refraction to focus all the star's light reaching the cornea into a small point on the retina.

Recall that as light passes into a medium where its speed is slowed, it will bend toward a line that is perpendicular to the border, or interface, between the media (see Figure 9.3). This is precisely the situation as light strikes the cornea and passes from the air into the aqueous humor. As shown in Figure 9.7, light that enters the eye perpendicular to the corneal surface passes straight to the retina, but light rays that strike the curved surface of the cornea at angles other than perpendicular are bent such that they converge on the back of the retina. The distance from the refractive surface to the point where parallel light rays converge is called the *focal distance*. Focal distance depends on the curvature of the cornea—the tighter the curve, the shorter the focal distance. The equation in Figure 9.7 shows that the reciprocal of the focal distance in meters is a unit of measurement called the **diopter**. The cornea has a refractive power of about 42 diopters, which

BOX 9.2 OF SPECIAL INTEREST

Eye Disorders

Once you know the basic structure of the eye, you can understand how a partial or complete loss of vision results from abnormalities in various components. For example, if there is an imbalance in the extraocular muscles of the two eyes, the eyes will point in different directions. Such a misalignment or lack of coordination between the two eyes is called *strabismus,* and there are two varieties. In *esotropia*, the directions of gaze of the two eyes cross, and the person is said to be cross-eyed. In *exotropia,* the directions of gaze diverge, and the person is said to be wall-eyed (Figure A). In most cases, both types of strabismus are congenital; it can and should be corrected during early childhood. Treatment usually involves the use of prismatic glasses or surgery to the extraocular muscles to realign the eyes. Without treatment, conflicting images are sent to the brain from the two eyes, degrading depth perception and, more importantly, causing the person to suppress input from one eye. The dominant eye will be normal but the suppressed eye will become *amblyopic*, meaning that it has poor visual acuity. If medical intervention is delayed until adulthood, the condition cannot be corrected.

A common eye disorder among older adults is *cataract*, a clouding of the lens (Figure B). Many people over 65 years of age have some degree of cataract; if it significantly impairs vision, surgery is usually required. In a cataract operation, the lens is removed and replaced with an artificial plastic lens. Although the artificial lens cannot adjust its focus as the normal lens does, it provides a clear image, and glasses can be used for near and far vision (see Box 9.3).

Glaucoma, a progressive loss of vision associated with elevated intraocular pressure, is a leading cause of blindness. Pressure in the aqueous humor plays a crucial role in maintaining the shape of the eye. As this pressure increases, the entire eye is stressed, ultimately damaging the relatively weak point where the optic nerve leaves the eye. The optic nerve axons are compressed, and vision is gradually lost from the periphery inward. Unfortunately, by the time a person notices a loss of more central vision, the damage is advanced and a significant portion of the eye is permanently blind. For this reason, early detection and treatment with medication or surgery to reduce intraocular pressure are essential.

The light-sensitive retina at the back of the eye is the site of numerous disorders that pose a significant risk of blindness. You may have heard of a professional boxer having a *detached retina*. As the name implies, the retina pulls away from the underlying wall of the eye from a blow to the head or by shrinkage of the vitreous humor. Once the retina has started to detach, fluid from the vitreous space flows through small tears in the retina resulting from the trauma, thereby causing more of the retina to separate. Symptoms of retinal detachment include abnormal perception of shadows and flashes of light. Treatment often involves laser surgery to scar the edge of the retinal tear, thereby reattaching the retina to the back of the eye.

Retinitis pigmentosa is characterized by a progressive degeneration of the photoreceptors. The first sign is usually a loss of peripheral vision and night vision. Subsequently, total blindness may result. The cause of this disease is unknown. Some forms clearly have a strong genetic component, and more than 100 genes have been identified that can contain mutations leading to retinitis pigmentosa. There is currently no cure, but taking vitamin A may slow its progression.

In contrast to the tunnel vision typically experienced by patients with retinitis pigmentosa, people with *macular degeneration* lose only central vision. The condition is quite common, affecting more than 25% of all Americans over 65 years of age. While peripheral vision usually remains normal, the ability to read, watch television, and recognize faces is lost as central photoreceptors gradually deteriorate. Laser surgery can sometimes minimize further vision loss, but the disease currently has no known cure.

Figure A
Exotropia. (Source: Newell, 1965, p. 330.)

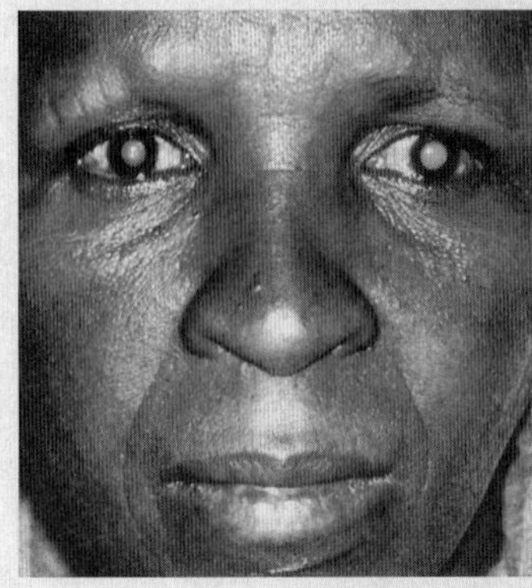

Figure B
Cataract. (Source: Schwab, 1987, p. 22.)

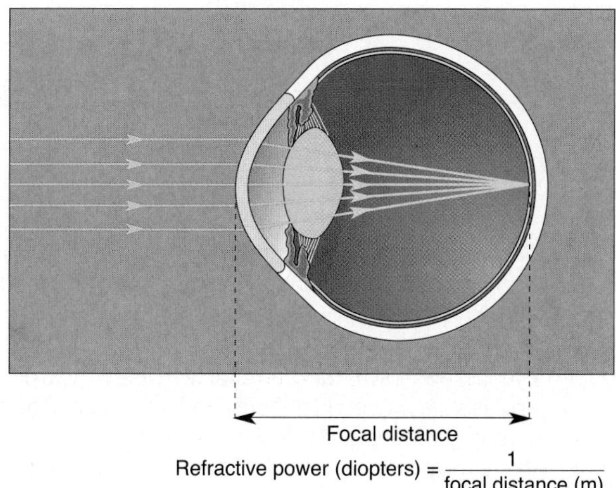

◀ FIGURE 9.7
Refraction by the cornea. The cornea must have sufficient refractive power, measured in diopters, to focus light on the retina at the back of the eye.

means that parallel light rays striking the corneal surface will be focused 0.024 m (2.4 cm) behind it, about the distance from cornea to retina. To get a sense of the large amount of refraction produced by the cornea, note that many prescription eyeglasses have a power of only a few diopters.

Remember that refractive power depends on the slowing of light at the air–cornea interface. If we replace air with a medium that passes light at about the same speed as the internal structures of the eye, the refractive power of the cornea would be eliminated. This is why things look blurry when you open your eyes underwater; the water–cornea interface has very little focusing power. Wearing swimmer's goggles or a scuba mask restores the air–cornea interface and, consequently, the refractive power of the eye.

Accommodation by the Lens

Although the cornea performs most of the eye's refraction, the lens also contributes another dozen or so diopters to the formation of a sharp image of a distant point. However, the lens is involved more importantly in forming crisp images of objects located closer than about 9 m from the eye. With near objects, the light rays originating at a point are no longer parallel. Rather, these rays diverge from a light source or a point on an object, and greater refractive power is required to bring them into focus on the retina. This additional focusing power is provided by changing the shape of the lens, a process known as **accommodation** (Figure 9.8).

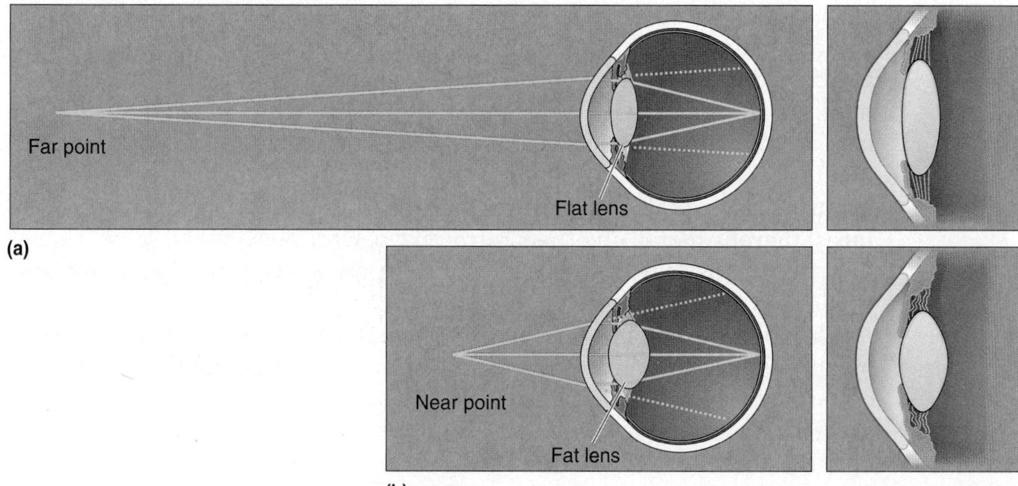

◀ FIGURE 9.8
Accommodation by the lens. **(a)** To focus the eye on a distant point, relatively little refraction is required. The ciliary muscle relaxes, stretching the zonule fibers and flattening the lens. **(b)** Near objects require greater refraction provided by a more spherical lens. This is achieved by contracting the ciliary muscle so there is less tension in the zonule fibers.

BOX 9.3 OF SPECIAL INTEREST

Vision Correction

When the ciliary muscles are relaxed and the lens is flat, the eye is said to be *emmetropic* if parallel light rays from a distant point source are focused sharply on the back of the retina. (The word is from the Greek *emmetros*, meaning "in proper measure," and *ope*, "sight.") Stated another way, the emmetropic eye focuses parallel light rays on the retina without the need for accommodation (Figure A), and accommodation is sufficient to focus the images of objects over a wide range of distances.

Now consider what happens when the eyeball is too short from front to back. Without accommodation, parallel light rays are focused at a point behind the eye. This condition is known as *hyperopia*, or farsightedness. The eye has enough accommodation to achieve good focus for distant objects, but even with the maximum amount of accommodation, near objects are focused at a point behind the retina (Figure B). Farsightedness can be corrected by placing a convex glass or plastic lens in front of the eye (Figure C). The curved front edge of the lens, like the cornea, bends light toward the center of the retina. Also, as the light passes from glass into air as it exits the lens, the back of the lens also increases the refraction (light going from glass to air speeds up and is bent *away* from the perpendicular).

If the eyeball is too long rather than too short, parallel rays will converge before the retina, cross, and again be imaged on the retina as a blurry circle. This condition is known as *myopia,* or nearsightedness. There is more than enough refraction to image close objects, but even with the least amount of accommodation, distant objects are focused in front of the retina (Figure D). For the nearsighted eye to see distant objects clearly, artificial concave lenses must be used to move the image back onto the retina (Figure E).

Some eyes have irregularities such that the curvature and refraction in the horizontal and vertical planes is different. This condition is called *astigmatism*, and it can be corrected with an artificial lens that is curved more along one axis than others.

Even if you are fortunate enough to have perfectly shaped eyeballs and a symmetrical refractive system, you probably will not escape *presbyopia* (from the Greek meaning "old eye"). This condition is a hardening of the lens that accompanies the aging process and is thought to be explained by the fact that while new lens cells are generated throughout life, none are lost. The hardened lens is less elastic, leaving it unable to change shape and accommodate sufficiently to focus on both near and far objects. The correction for presbyopia, first introduced by Benjamin Franklin, is a bifocal lens. These lenses are concave on top to assist far vision and convex on the bottom to assist near vision.

In hyperopia and myopia, the amount of refraction provided by the cornea is either too little or too great for the length of the eyeball. But modern techniques can now change the amount of refraction the cornea provides. In *radial keratotomy*, a procedure to correct myopia, tiny incisions through the peripheral portion of the cornea relax and flatten the central cornea, thus reducing the amount of refraction and minimizing the myopia. The most recent techniques use lasers to reshape the cornea. In *photorefractive keratectomy (PRK)*, a laser is used to reshape the outer surface of the cornea by vaporizing thin layers. *Laser in situ keratomileusis (LASIK)* has become very common; you've probably seen clinics offering this procedure in shopping areas or malls. In LASIK, a microkeratome or laser is used to make a thin flap in the outer cornea. With the cornea flap temporarily lifted, the laser can reshape the cornea from the inside (Figure F). Nonsurgical methods are also being used to reshape the cornea. A person can be fitted with special retainer contact lenses or plastic corneal rings that alter the shape of the cornea and correct refractive errors.

Recall that the ciliary muscle forms a ring around the lens. During accommodation, the ciliary muscle contracts and swells in size, thereby making the area inside the ring smaller (i.e., a smaller hole inside the bagel in our analogy) and decreasing the tension in the suspensory ligaments. Consequently, the lens becomes rounder and thicker because of its natural elasticity. This rounding increases the curvature of the lens surfaces, thereby increasing their refractive power. Conversely, relaxation of the ciliary muscle increases the tension in the suspensory ligaments, and the lens is stretched into a flatter shape.

The ability to accommodate changes with age. An infant's eyes can focus on objects just beyond his or her nose, whereas many middle-aged adults cannot clearly see objects closer than about arm's length. Fortunately, artificial lenses can compensate for this and other defects of the eye's optics (Box 9.3).

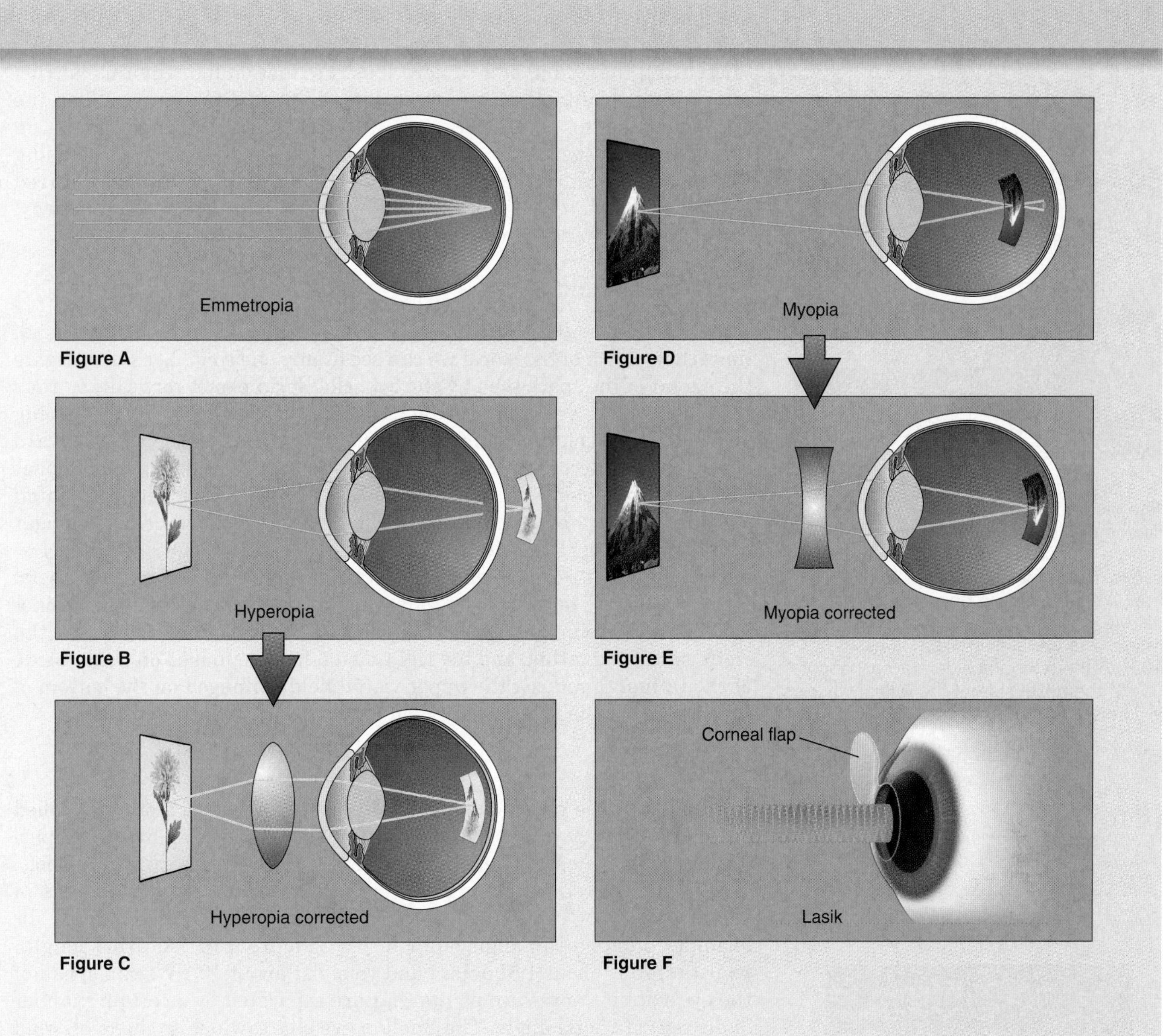

The Pupillary Light Reflex

In addition to the cornea and the lens, the pupil contributes to the optical functioning of the eye by continuously adjusting for different ambient light levels. To check this for yourself, stand in front of a bathroom mirror with the lights out for a few seconds, and then watch your pupils shrink when you turn the lights on. This **pupillary light reflex** involves connections between the retina and neurons in the brain stem that control the muscles that constrict the pupils. An interesting property of this reflex is that it is *consensual*; shining a light into only one eye causes the constriction of the pupils of both eyes. It is unusual, indeed, when the pupils are not the same size; the lack of a consensual pupillary light reflex is often taken as a sign of a serious neurological disorder involving the brain stem.

A beneficial effect of the pupil constriction that results from an increase in light level is an increase in the depth of focus (the range of distances from the eye that appear in focus), just like decreasing the aperture size (increasing the f-stop) of a camera lens. To understand why this is true, consider two points in space, one close and the other far away. When the eye accommodates to the closer point, the image of the farther point on the retina no longer forms a point but rather a blurred circle. Decreasing the aperture—constricting the pupil—reduces the size of this blurred circle so that its image more closely approximates a point. In this way, distant objects appear to be less out of focus.

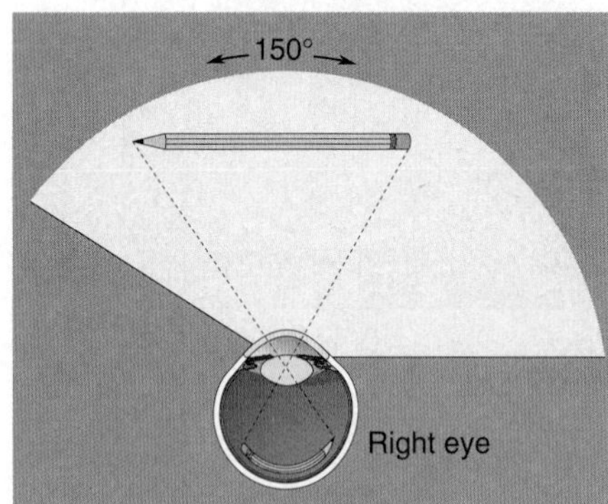

▲ FIGURE 9.9
The visual field for one eye. The visual field is the total amount of space that can be viewed by the retina when the eye is fixated straight ahead. Notice how the image of an object in the visual field (pencil) is left–right reversed on the retina. The visual field extends nearly 100° to the temporal side but only about 60° to the nasal side of the retina, where vision is blocked by the nose.

The Visual Field

The structure of the eyes, and where they are positioned in the head, limits how much of the world we can see at any one time. Let's investigate the extent of the space seen by one eye. Holding a pencil vertically in your right hand, close your left eye and look at a point straight ahead. Keeping your eye fixated on this point, slowly move the pencil to the right (toward your right ear) across your field of view until the pencil disappears. Repeat this exercise, moving the pencil to the left where it will disappear behind your nose, and then up and down. The points where you can no longer see the pencil mark the limits of the **visual field** for your right eye. Now look at the middle of the pencil as you hold it horizontally in front of you. Figure 9.9 shows how the light reflected off this pencil falls on your retina. Notice that the image is inverted; the left visual field is imaged on the right side of the retina, and the right visual field is imaged on the left side of the retina. Similarly, the upper visual field is imaged on the bottom of the retina and the lower visual field is imaged on the upper retina.

Visual Acuity

The ability of the eye to distinguish two points near each other is called **visual acuity**. Acuity depends on several factors but especially on the spacing of photoreceptors in the retina and the precision of the eye's refraction.

Distance across the retina can be described in terms of degrees of **visual angle**. A right angle subtends (spans) 90°, and the moon, for example, subtends an angle of about 0.5° (Figure 9.10). At arm's length, your thumb is about 1.5° across and your fist about 10°. We can speak of the eye's ability to resolve points that are separated by a certain number of degrees of visual angle. The Snellen eye chart, which we have all read at the doctor's office, tests our ability to discriminate letters and numbers at a viewing distance of 20 feet. Your vision is 20/20 when you can recognize a letter that subtends an angle of 0.083° (equivalent to 5 minutes of arc, where 1 minute is 1/60 of a degree).

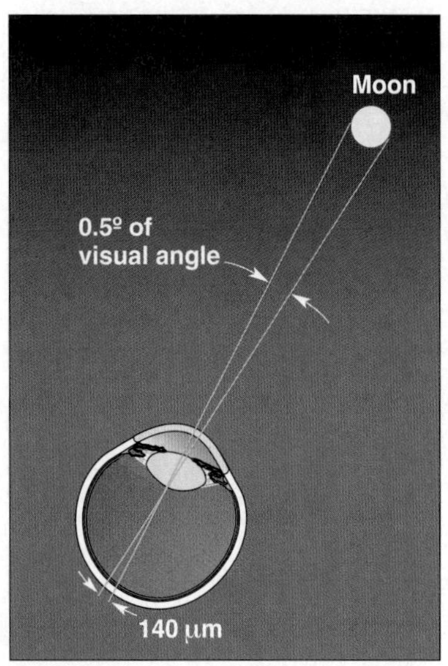

▲ FIGURE 9.10
Visual angle. Distances across the retina can be expressed as degrees of visual angle.

MICROSCOPIC ANATOMY OF THE RETINA

Now that we have an image formed on the retina, we can get to the neuroscience of vision: the conversion of light energy into neural activity. To begin our discussion of image processing in the retina, we must introduce the cellular architecture of this bit of brain.

The basic system of retinal information processing is shown in Figure 9.11. The most direct pathway for visual information to exit the eye is from **photoreceptors** to **bipolar cells** to **ganglion cells**. The photoreceptors respond to light, and they influence the membrane potential of the bipolar cells connected to them. The ganglion cells fire action potentials in response to light, and these impulses propagate along the optic nerve to the rest of the brain. Besides the cells in this direct path from photoreceptor to brain, retinal processing is influenced by two additional cell types. **Horizontal cells** receive input from the photoreceptors and project neurites laterally to influence surrounding bipolar cells and photoreceptors. A wide variety of types of **amacrine cells** generally receive input from bipolar cells and project laterally to influence surrounding ganglion cells, bipolar cells, and other amacrine cells.

There are three important points to remember here:

1. *With one exception, the only light-sensitive cells in the retina are the rod and cone photoreceptors.* All other cells are influenced by light only via direct and indirect synaptic interactions with the photoreceptors. (We will see in a moment that recently discovered types of ganglion cells are also light sensitive, but these uncommon cells do not appear to play a major role in visual perception.)
2. *The ganglion cells are the only source of output from the retina.* No other retinal cell type projects an axon through the optic nerve.
3. *With the exception of certain amacrine cells, ganglion cells are the only retinal neurons that fire action potentials; this is essential for transmitting information outside the eye.* All other retinal cells depolarize or hyperpolarize, with a rate of neurotransmitter release that is proportional to the membrane potential, but they do not fire action potentials.

Now let's take a look at how the different cell types are arranged in the retina.

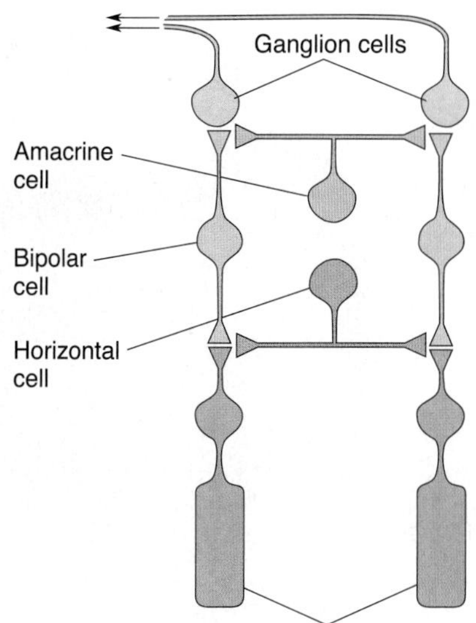

▲ FIGURE 9.11
The basic system of retinal information processing. Information about light flows from the photoreceptors to bipolar cells to ganglion cells, which project axons out of the eye in the optic nerve. Horizontal cells and amacrine cells modify the responses of bipolar cells and ganglion cells via lateral connections.

The Laminar Organization of the Retina

Figure 9.12 shows that the retina has a *laminar organization*: Cells are organized in layers. Notice that the layers are seemingly inside-out; light must pass from the vitreous humor through the ganglion cells and bipolar cells *before* it reaches the photoreceptors. Because the retinal cells above the photoreceptors are relatively transparent, image distortion is minimal as light passes through them. One reason the inside-out arrangement is advantageous is that the *pigmented epithelium* that lies below the photoreceptors plays a critical role in the maintenance of the photoreceptors and photopigments. The pigmented epithelium also absorbs any light that passes entirely through the retina, thus minimizing the scattering of light within the eye that would blur the image. Many nocturnal animals, such as cats and raccoons, have a reflective layer beneath the photoreceptors, called the *tapetum lucidum*, which bounces light back at the photoreceptors if it passes through the retina. The animal is thus more sensitive to low light levels at the expense of reduced acuity. An interesting side effect of the reflective tapetum can be seen when you shine a light at or take a flash photograph of nocturnal animals: There is striking "eyeshine" in which the pupils seem to glow (Figure 9.13).

The cell layers of the retina are named in reference to the middle of the eyeball. Don't get confused by thinking about the head instead of the eye: The photoreceptors are the outermost part of the retina even though they are the farthest from the front of the eye and the deepest inside the head. The innermost retinal layer is the **ganglion cell layer**, which contains the cell bodies of the ganglion cells. Moving outward, there are two other layers that contain the cell bodies of neurons: the **inner nuclear layer**, which contains the cell bodies of the bipolar cells, horizontal cells, and amacrine cells, and the **outer nuclear layer**, which contains the cell bodies of the photoreceptors.

Between the ganglion cell layer and the inner nuclear layer is the **inner plexiform layer** ("plexiform" means a network of connections), which contains the synaptic contacts between bipolar cells, amacrine cells, and

Light

Retina

Optic nerve

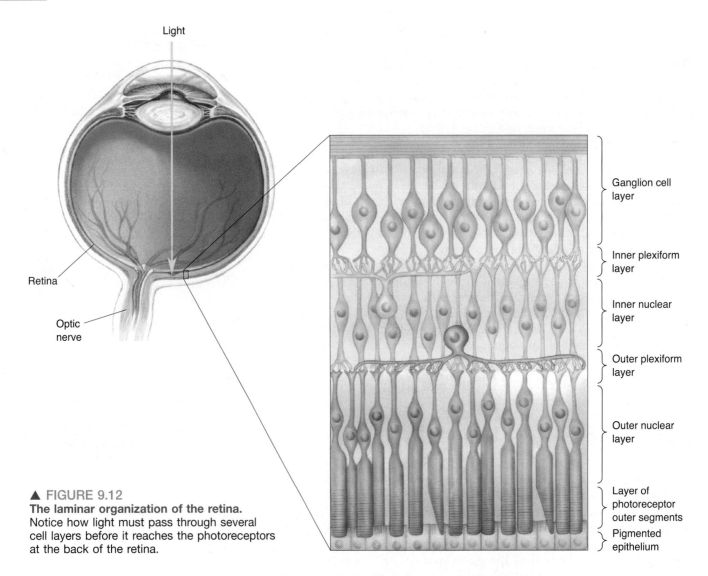

Ganglion cell layer

Inner plexiform layer

Inner nuclear layer

Outer plexiform layer

Outer nuclear layer

Layer of photoreceptor outer segments

Pigmented epithelium

▲ FIGURE 9.12
The laminar organization of the retina. Notice how light must pass through several cell layers before it reaches the photoreceptors at the back of the retina.

ganglion cells. Between the outer and inner nuclear layers is the **outer plexiform layer**, where the photoreceptors make synaptic contact with the bipolar and horizontal cells. Finally, the **layer of photoreceptor outer segments** contains the light-sensitive elements of the retina. The outer segments are embedded in the pigmented epithelium.

Photoreceptor Structure

The conversion of electromagnetic radiation into neural signals occurs in the photoreceptors at the back of the retina. Every photoreceptor has four regions: an outer segment, an inner segment, a cell body, and a synaptic terminal. The outer segment contains a stack of membranous disks. Light-sensitive *photopigments* in the disk membranes absorb light, thereby triggering changes in the photoreceptor membrane potential (discussed later). Figure 9.14 shows the two types of photoreceptors in the retina, easily distinguished by the appearance of their outer segments. **Rod photoreceptors** have a long, cylindrical outer segment, containing many disks. **Cone photoreceptors** have a shorter, tapering outer segment with fewer membranous disks. The greater number of disks and higher photopigment concentration in rods makes them over 1000 times more sensitive to light than cones. It is estimated that there are about 5 million cones and 92 million rods in each human retina.

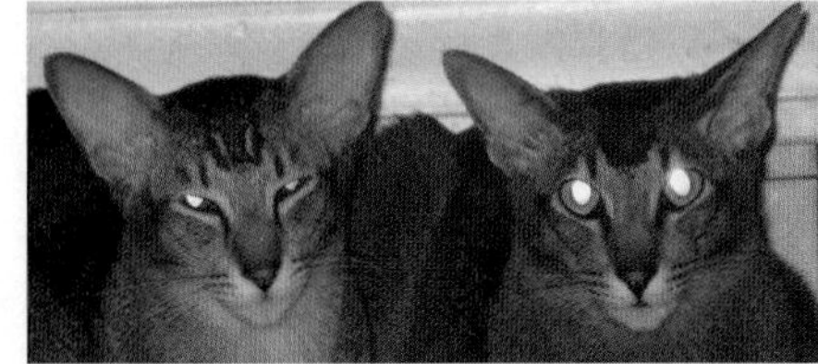

▲ FIGURE 9.13
Eyeshine results from the reflective tapetum in cats.

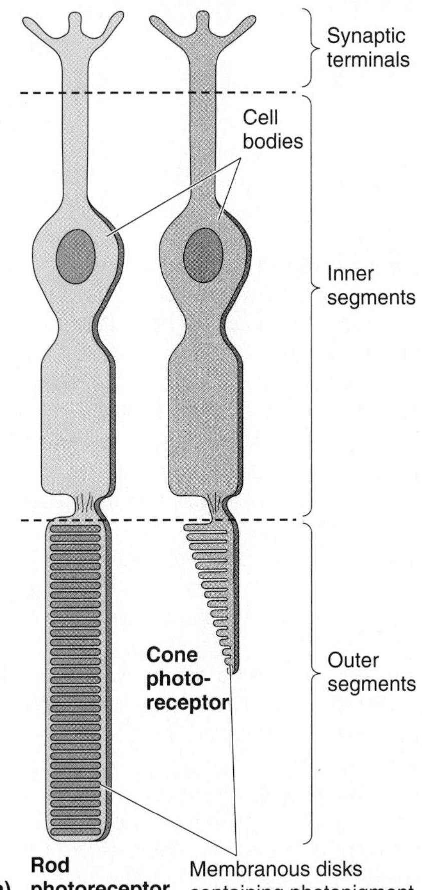

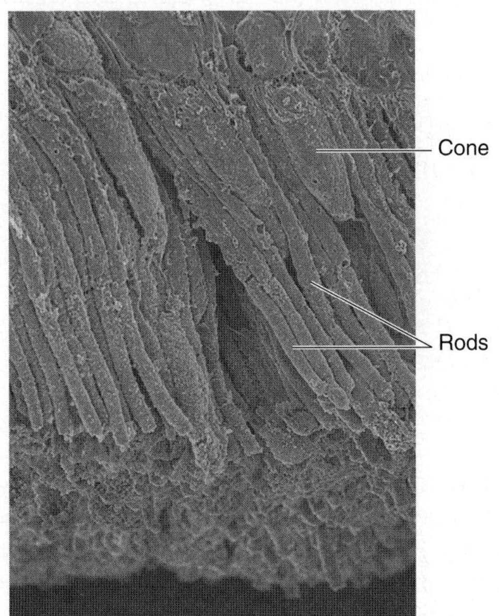

◀ FIGURE 9.14
Rods and cones. **(a)** Rods contain more disks and make vision possible in low light; cones enable us to see in daylight. **(b)** Scanning electron micrograph of rods and cones. (Source: Courtesy of J. Franks and W. Halfter.)

BOX 9.4 PATH OF DISCOVERY

Seeing Through the Photoreceptor Mosaic

by David Williams

When I started graduate school in 1975, we knew almost nothing about the topography of the three classes of cones that form the basis of trichromatic color vision in the human eye. Although Thomas Young had deduced 175 years earlier that color vision depended on three fundamental channels, we did not know the relative numbers of the three kinds of cones or how they were arranged in the retina. With my graduate advisor, Don MacLeod at the University of California, San Diego, I used psychophysical methods to map retinal sensitivity to violet light. We found that the S cones, which are stimulated by this light, are sparsely sprinkled across a mosaic of L and M cones. We also discovered that a person can reliably detect a flash of light that stimulates just one of the 5 million or so cones in the retina.

Later, I kept returning to the topography of the trichromatic mosaic. After several failed attempts over many years, I hit upon a solution from a completely unexpected direction. I'd had a long-standing interest in the limits of visual acuity. While exploring various technologies to prevent the normal optical blurring of the retinal image, I came upon adaptive optics, in which astronomers use a deformable mirror to correct for the blur caused by the turbulent atmosphere when imaging stars with telescopes on earth.

A major obstacle to using adaptive optics in vision science was that such telescope mirrors cost about $1 million. Fortunately, we found an engineer who made us an affordable deformable mirror. We were also fortunate that the military's work with adaptive optics had been recently declassified, so my post-doc, Junzhong Liang, and I were allowed to visit the Starfire Optical Range (SOR), a $16 million satellite-tracking telescope equipped with adaptive optics. I was discouraged by the legions of engineers and expensive optical system needed to run the facility—but then something remarkable happened. Bob Fugate, director of the SOR, was trying to measure atmospheric aberrations by reflecting light from a high-power laser off a mirror left on the moon's surface in the Apollo program. I heard Bob say, "Move that beam over to the right, you missed the whole damn moon!" Suddenly I realized that they were learning from fumbling just the same as in my lab—so there was hope after all.

Liang and I hurried back to the University of Rochester and with another post-doc, Don Miller, built the first adaptive optic system that could correct all of the eye's monochromatic aberrations. This started a minor revolution in optometry and ophthalmology because now many more of the eye's optical defects could be corrected than had been possible before. A person's visual acuity using an adaptive optics system can be better than with the most carefully prescribed spectacles. This led to a better way to correct vision with laser refractive surgery as well as improved designs for contact lenses and intraocular lenses.

We also equipped a camera with adaptive optics to take the sharpest pictures ever of the living human retina, so sharp that the individual cones in the photoreceptor mosaic can be seen. Could we also use adaptive optics to identify which of the three photopigments resides in each cone imaged in the living eye, solving the problem I'd started with in graduate school? Using adaptive optics and another technique called retinal densitometry, two post-docs in my lab, Austin Roorda and later Heidi Hofer, finally answered the question definitively. It turns out the three classes of cones are remarkably disorganized (Figure A), unlike the highly regular mosaics of many insect

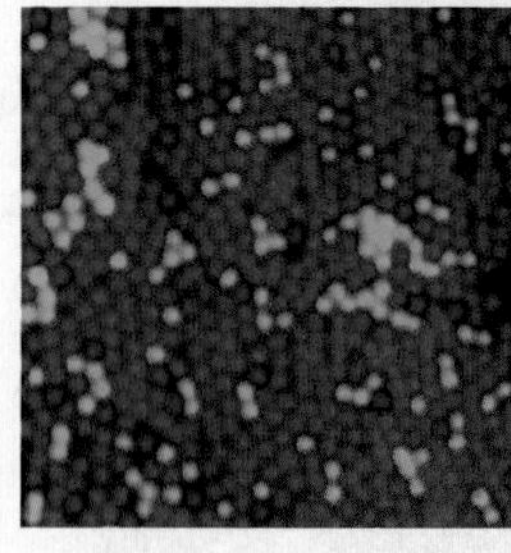

Figure A
The arrangement of the three cone classes in the human eye. (Source: Roorda and Williams, 1999.)

The large differences in the structure and sensitivity of rods and cones led investigators to say humans have a *duplex retina:* essentially two complementary systems in one eye. Some animals have only rods or only cones and thus do not have a duplex retina. The structural differences between rods and cones correlate with important functional differences. For example, in nighttime lighting, or *scotopic* conditions, only rods contribute to vision. Conversely, in daytime lighting, or *photopic* conditions, cones do the bulk of the work. At intermediate light levels (indoor lighting or outdoor traffic lighting at night), or *mesopic* conditions, both rods and cones are responsible for vision.

eyes. Moreover, the relative numbers of the M and L cones vary enormously from one person to another, despite the similarity of color vision in people (Figure B). Joe Carroll, another former postdoc, has gone on to reveal the mosaic organization in colorblind eyes and those with many different genetic mutations.

Adaptive optics is also being used to image many other cells in the retina, including ganglion cells, and is a valuable tool in the diagnosis and treatment of retinal disease. I certainly could never have foreseen that advances in astronomic technology would provide these tools for vision research—or that my graduate school interest in the trichromacy of the cone mosaic would 20 years later spawn these advances in vision correction and single cell imaging.

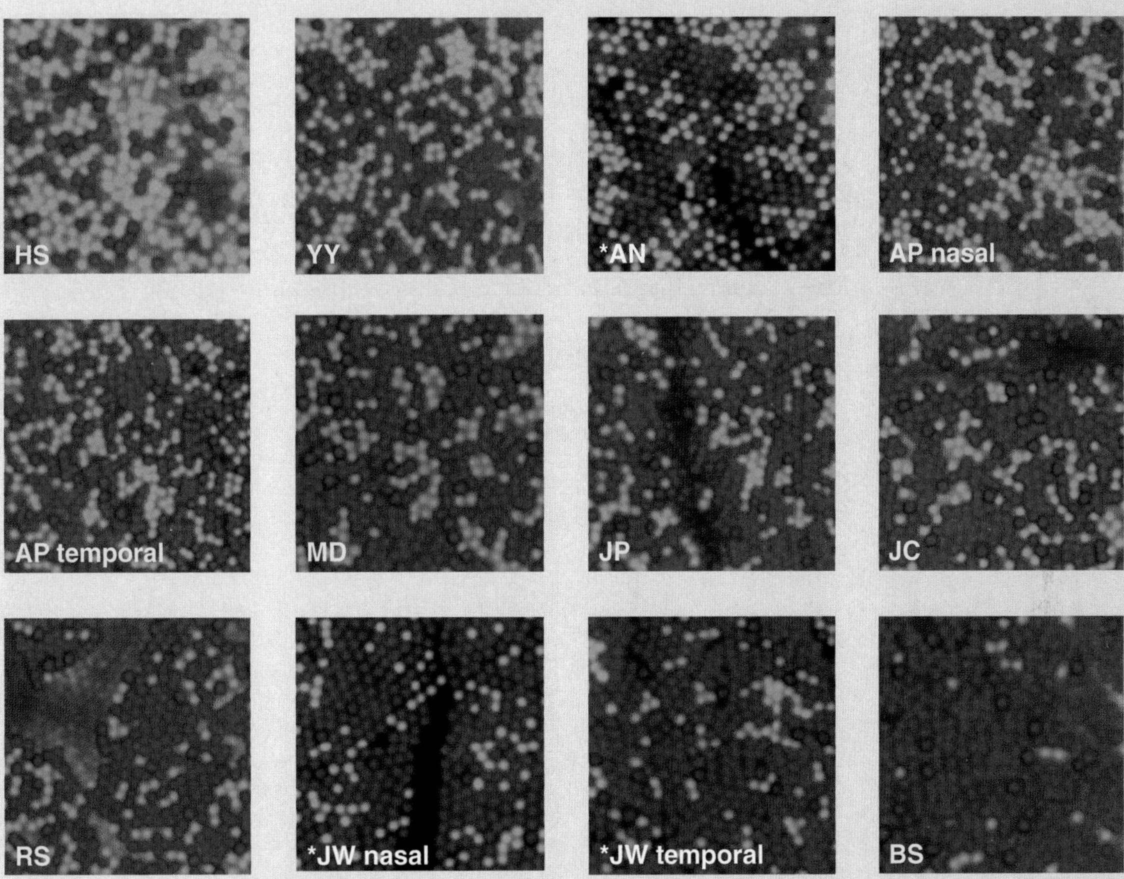

Figure B
Variation in the relative numbers of cones in eyes with normal color vision. (Source: Hofer et al., 2005; Roorda and Williams, 1999.)

Rods and cones differ in other respects as well. All rods contain the same photopigment, but there are three types of cones, each containing a different pigment. The variations among pigments make the different cones sensitive to different wavelengths of light. As we shall see in a moment, only the cones, not the rods, are responsible for our ability to see color. David Williams at the University of Rochester has used clever imaging techniques to reveal the distribution of human cones in exquisite detail. Surprisingly, rather than an arrangement like the neat arrangement of pixels in a computer display, human retinas show striking diversity in the arrangement and distribution of cone photoreceptors (Box 9.4).

Regional Differences in Retinal Structure and Their Visual Consequences

Retinal structure varies from the fovea to the retinal periphery. Most of the 5 million cones are in the fovea, and the proportion diminishes substantially in the retinal periphery. There are no rods in the central fovea, but there are many more rods than cones in the peripheral retina. Rod and cone distributions in the retina are summarized in Figure 9.15.

The differences in rod and cone numbers and distribution across the retina have important visual consequences. At photopic light levels

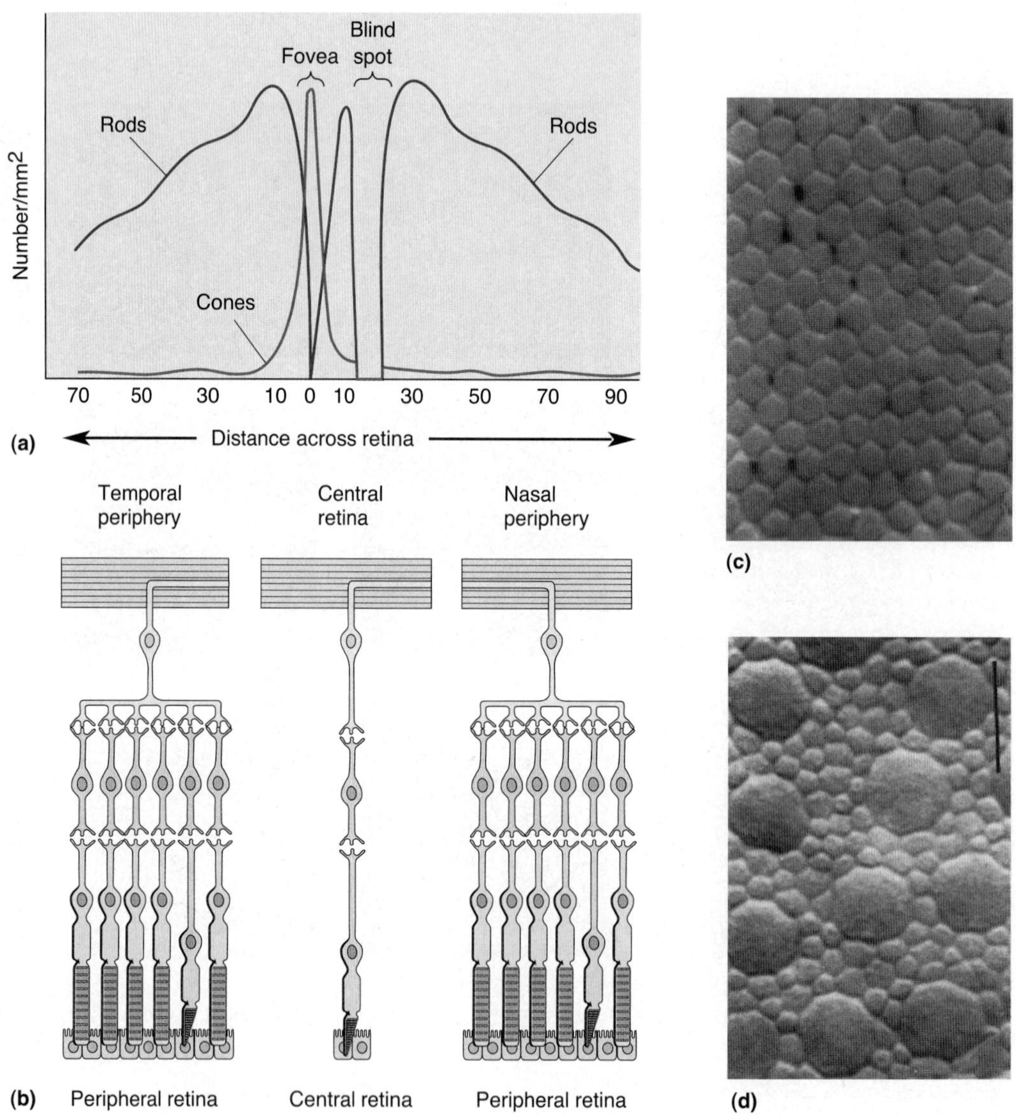

▲ FIGURE 9.15

Regional differences in retinal structure. **(a)** Cones are found primarily in the central retina, within 10° of the fovea. Rods are absent from the central fovea and are found mainly in the peripheral retina. **(b)** In the central retina, relatively few photoreceptors feed information to a ganglion cell; in the peripheral retina, many photoreceptors provide input. This arrangement makes the peripheral retina better at detecting dim light but the central retina better for high-resolution vision. **(c)** This magnified cross section of the human central retina shows the dense packing of cone inner segments. **(d)** At a more peripheral location on the retina, the cone inner segments are larger and appear as islands in a sea of smaller rod inner segments. A 10-micron scale bar for the images in c and d is shown to the right in d. (Source for parts c and d: Curcio et al., 1990, p. 500.)

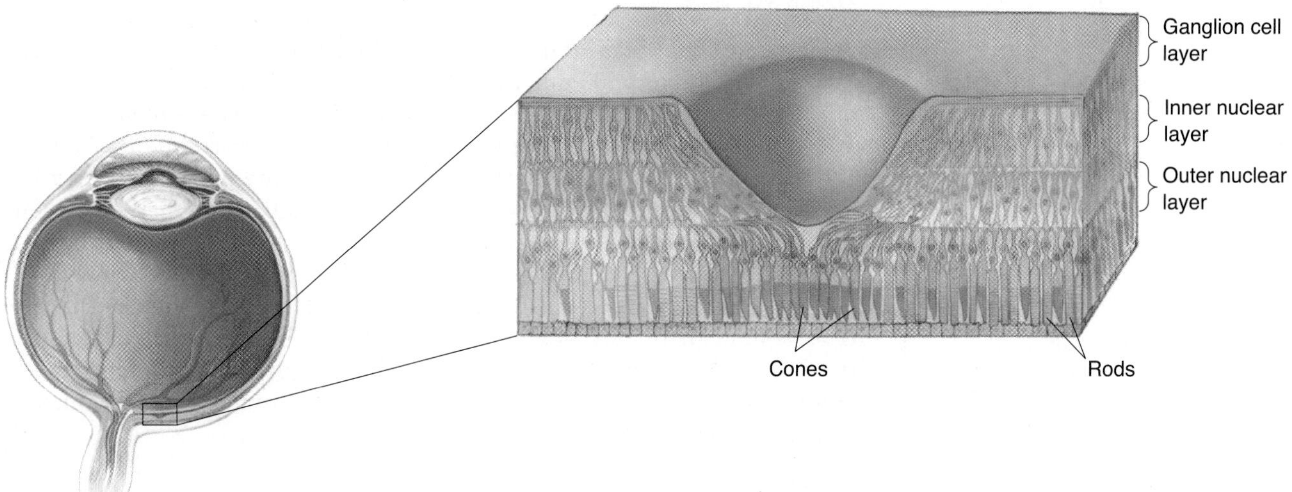

▲ FIGURE 9.16
The fovea in cross section. The ganglion cell layer and the inner nuclear layer are displaced laterally to allow light to strike the foveal photoreceptors directly.

(daylight), the most obvious is that *we have much greater spatial sensitivity on our central retina*. Visual acuity is measured when we look directly at symbols on a test chart, placing the critical features in our cone-rich fovea. Recall that the fovea is a thinning of the retina at the center of the macula. In cross section, the fovea appears as a pit in the retina. This appearance is due to the lateral displacement of the cells above the photoreceptors, allowing light to strike the photoreceptors without passing through the other retinal cell layers (Figure 9.16). This structural specialization maximizes visual acuity at the fovea by pushing aside other cells that might scatter light and blur the image. If you were to take an eye test while looking slightly away from the test chart, or if you try to read the titles of books on a shelf using your peripheral vision, you would need the letters to be much larger for you to read them. Less obvious than the fovea's high spatial acuity is that *we are poorer at discriminating colors on our peripheral retina* because of the smaller number of cones. You might be able to demonstrate this by looking straight ahead and moving a small colored object slowly to the side.

The consequences of rod and cone distribution differences are very different with the light at dim scotopic levels, when we see only with rods. For example, *we are more sensitive to low levels of light on our peripheral retina*. Put another way, *our central vision is blind at scotopic light levels*. This is because rods respond more strongly to low light levels than cones, there are more rods in the peripheral retina (and none in the central fovea), and more rods project to single bipolar and ganglion cells in the peripheral retina (thereby aiding the detection of dim light). You can demonstrate the greater sensitivity of your peripheral retina to yourself on a starry night. (It's fun; try it with a friend.) First, spend about 20 minutes in the dark getting oriented, and then gaze at a bright star. Fixating on this star, search your peripheral vision for a dim star. Then move your eyes to look at this dim star. You will find that the faint star disappears when it is imaged on the central retina (when you look straight at it) but reappears when it is imaged on the peripheral retina (when you look slightly to the side of it).

Because cones alone make the perception of color possible, *we are unable to perceive color differences at night* when rods are active but cones are not. A green tree, a blue car, and a red house all appear to have vaguely the same color (or lack of color). The peak sensitivity of rods is to a wavelength of about 500 nm, and thus at scotopic light levels objects tend to look dark blue-green. The loss of color as the sun goes down is a huge perceptual effect, but one that we hardly notice because of its familiarity.

Nighttime vision for modern humans is not due exclusively to rods, however. In densely populated areas, we actually can perceive some color at night because streetlights and neon signs emit sufficient light to activate the cones. This fact is the basis for differing points of view about the design of automobile dashboard indicator lights. One view is that the lights should be dim blue-green to take advantage of the spectral sensitivity of the rods. An alternate view is that the lights should be bright red because this wavelength affects mainly cones, leaving the rods unsaturated and thus resulting in better night vision.

PHOTOTRANSDUCTION

The photoreceptors convert, or *transduce*, light energy into changes in membrane potential. We begin our discussion of phototransduction with rods, which outnumber cones in the human retina by about 20 to 1. Most of what has been learned about phototransduction by rods has proven to be applicable to cones as well.

Phototransduction in Rods

As we discussed in Part I of this book, one way information is represented in the nervous system is as changes in the membrane potential of neurons. Thus, we look for a mechanism by which the absorption of light energy can be transduced into a change in the photoreceptor membrane potential. In many respects, this process is analogous to the transduction of chemical signals into electrical signals that occurs during synaptic transmission. At a G-protein-coupled neurotransmitter receptor, the binding of transmitter to the receptor activates G-proteins in the membrane, which in turn stimulate various effector enzymes (Figure 9.17a). These enzymes alter the intracellular concentration of cytoplasmic second messenger molecules, which (directly or indirectly) change the conductance of membrane ion channels, thereby altering membrane potential. Similarly, in the photoreceptor, light stimulation of the photopigment activates G-proteins, which in turn activate an effector enzyme that changes the cytoplasmic concentration of a second messenger molecule. This change causes a membrane ion channel to close, and the membrane potential is thereby altered (Figure 9.17b).

Recall from Chapter 3 that a typical neuron at rest has a membrane potential of about −65 mV, close to the equilibrium potential for K^+. In contrast, in complete darkness, the membrane potential of the rod outer segment is about −30 mV. This depolarization is caused by the steady influx of Na^+ through special channels in the outer segment membrane (Figure 9.18a). The movement of positive charge across the membrane, which occurs in the dark, is called the **dark current**. Sodium channels are stimulated to open—are gated—by an intracellular second messenger called **cyclic guanosine monophosphate**, or **cGMP**. cGMP is produced in the photoreceptor by the enzyme guanylyl cyclase, keeping the Na^+ channels open. Light reduces cGMP, causing the Na^+ channels to close, and the membrane potential becomes *more negative* (Figure 9.18b). Thus, *photoreceptors hyperpolarize in response to light* (Figure 9.18c).

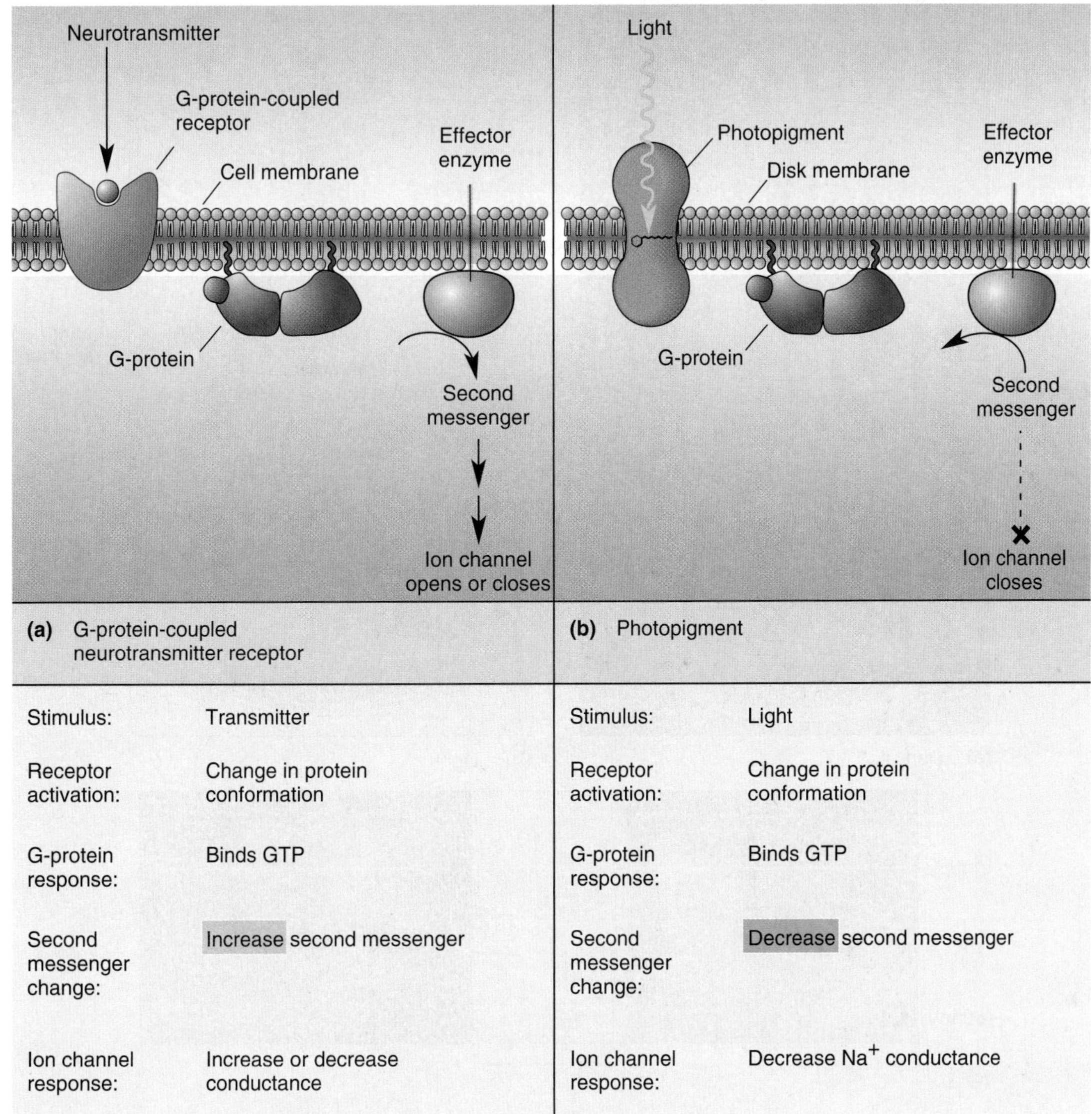

▲ FIGURE 9.17
Light transduction and G-proteins. G-protein-coupled receptors and photoreceptors use similar mechanisms. **(a)** At a G-protein-coupled receptor, the binding of neurotransmitter activates G-proteins and effector enzymes. **(b)** In a photoreceptor, light begins a similar process using the G-protein transducin.

The hyperpolarizing response to light is initiated by the absorption of electromagnetic radiation by the photopigment in the membrane of the stacked disks in the rod outer segments. In the rods, this pigment is called **rhodopsin**. Rhodopsin can be thought of as a receptor protein with a prebound chemical agonist. The receptor protein is called *opsin*, and it has the seven transmembrane alpha helices typical of G-protein-coupled receptors throughout the body. The prebound agonist is called *retinal*, a derivative of vitamin A. The absorption of light causes a change in the conformation of retinal so that it activates the opsin (Figure 9.19). This process is known as bleaching because it changes the wavelengths absorbed by the rhodopsin (the photopigment literally changes color from purple to yellow). The bleaching of rhodopsin stimulates a G-protein called **transducin** in the disk membrane, which in turn activates the effector enzyme **phosphodiesterase (PDE)**, which breaks down the cGMP that is normally present in the cytoplasm of the rod (in the dark).

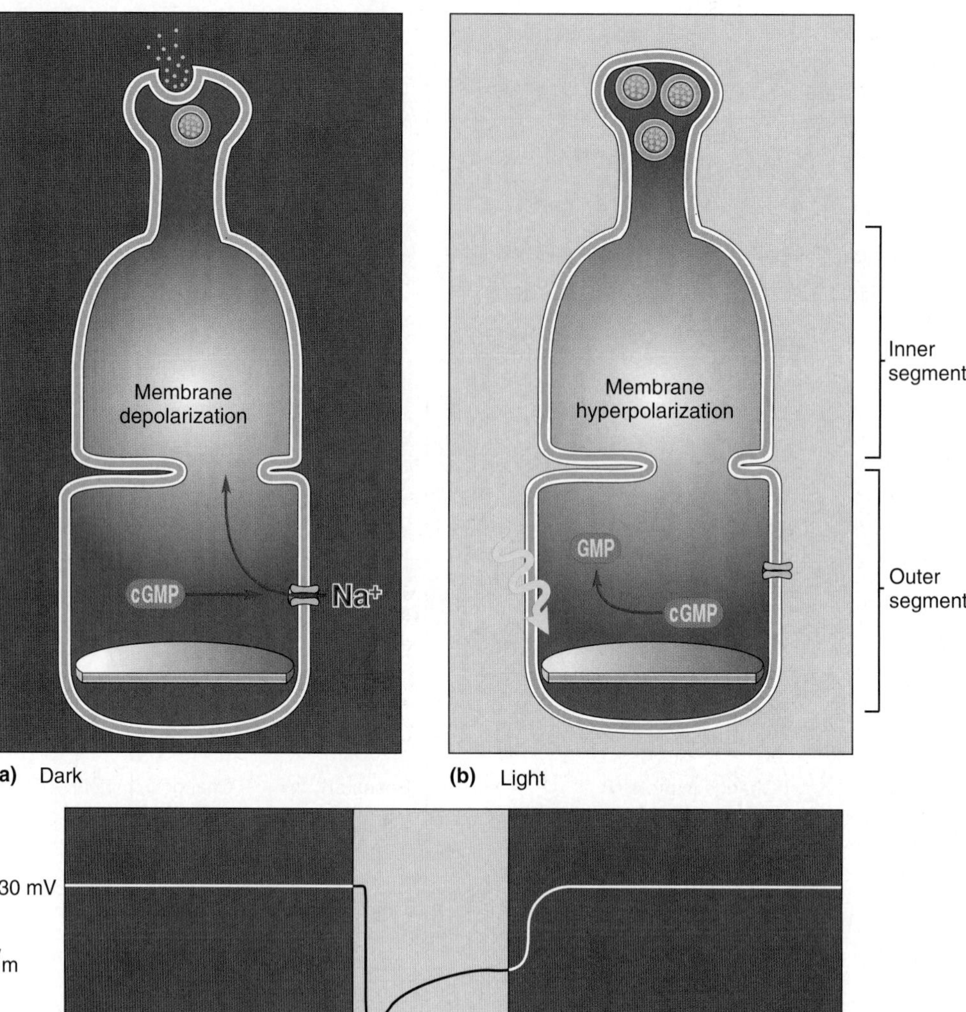

▶ FIGURE 9.18
The hyperpolarization of photoreceptors in response to light. Photoreceptors are continuously depolarized in the dark because of an inward sodium current, the dark current. **(a)** Sodium enters the photoreceptor through a cGMP-gated channel. **(b)** Light leads to the activation of an enzyme that destroys cGMP, thereby shutting off the Na^+ current and hyperpolarizing the cell. **(c)** In a dark auditorium, the membrane potential of our photoreceptors is −30 mV (left). At intermission, we move to a bright lobby and the cells hyperpolarize (middle). The slower depolarization that follows is adaptation. Returning to the dark room sends membrane potential back to −30 mV (right).

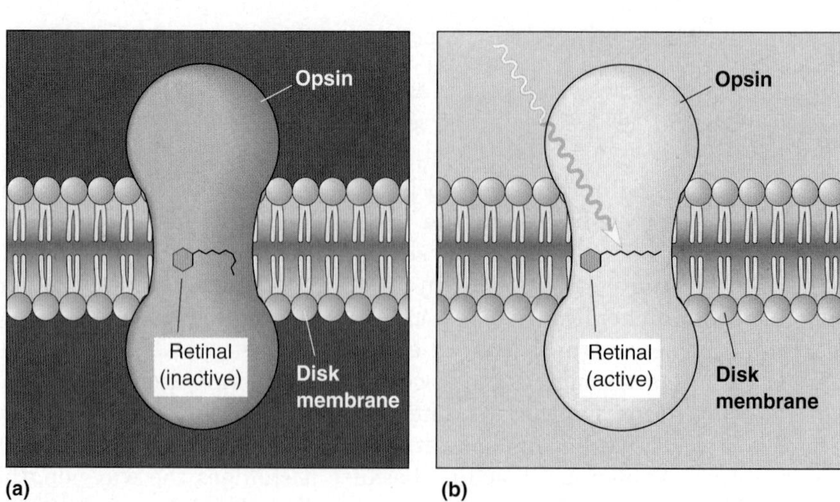

▲ FIGURE 9.19
The activation of rhodopsin by light. Rhodopsin consists of opsin, a protein with seven transmembrane alpha helices, and retinal, a small molecule derived from vitamin A. **(a)** In the dark, retinal is inactive. **(b)** Retinal undergoes a change in conformation when it absorbs light, thereby activating (bleaching) the opsin.

The reduction in cGMP causes the Na^+ channels to close and the membrane to hyperpolarize.

One of the interesting functional consequences of using a biochemical cascade for transduction is signal *amplification*. Many G-proteins are activated by each photopigment molecule, and each PDE enzyme breaks down more than one cGMP molecule. This amplification gives our visual system incredible sensitivity to small amounts of light. Rods are more sensitive to light than cones because they contain more disks in their outer segments and thus more photopigment, and also because they amplify the response to light more than cones do. The combined result is, incredibly, that rods give a measureable response to the capture of a single photon of light, the elementary unit of light energy.

To summarize, here are the steps in the transduction of light by rods:

1. Light activates (bleaches) rhodopsin.
2. Transducin, the G-protein, is stimulated.
3. Phosphodiesterase (PDE), the effector enzyme, is activated.
4. PDE activity reduces the cGMP level.
5. Na^+ channels close, and the cell membrane hyperpolarizes.

The complete sequence of events of phototransduction in rods is illustrated in Figure 9.20.

Phototransduction in Cones

In bright light, cGMP levels in rods fall to the point where the response to light becomes *saturated*; increasing the light level causes no additional hyperpolarization. Thus, vision during the day depends entirely on the cones, whose photopigments require more energy to become bleached.

The process of phototransduction in cones is virtually the same as in rods; the only major difference is in the type of opsins in the membranous disks of the cone outer segments. The cones in our retinas contain one of

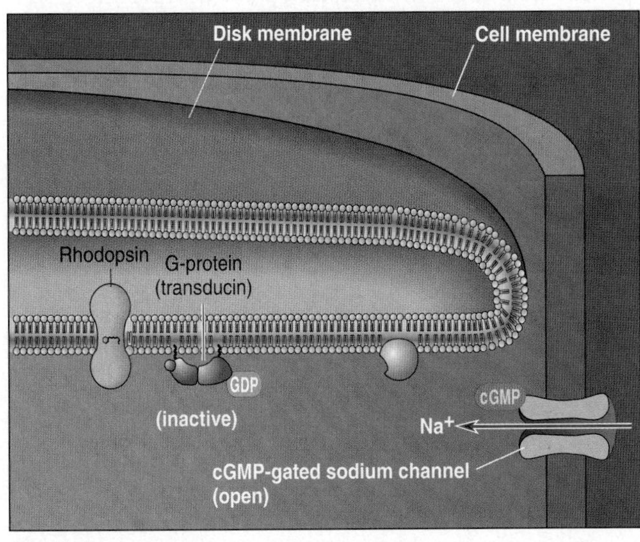

(a) Dark

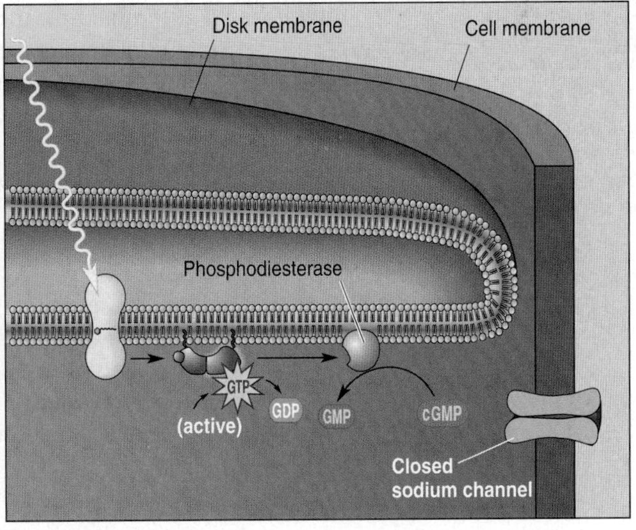

(b) Light

▲ FIGURE 9.20

The light-activated biochemical cascade in a photoreceptor. (a) In the dark, cGMP gates a sodium channel, causing an inward Na^+ current and depolarization of the cell. (b) The activation of rhodopsin by light energy causes the G-protein (transducin) to exchange guanosine diphosphate for guanosine triphosphate (see Chapter 6), which in turn activates the enzyme phosphodiesterase (PDE). PDE breaks down cGMP and shuts off the dark current.

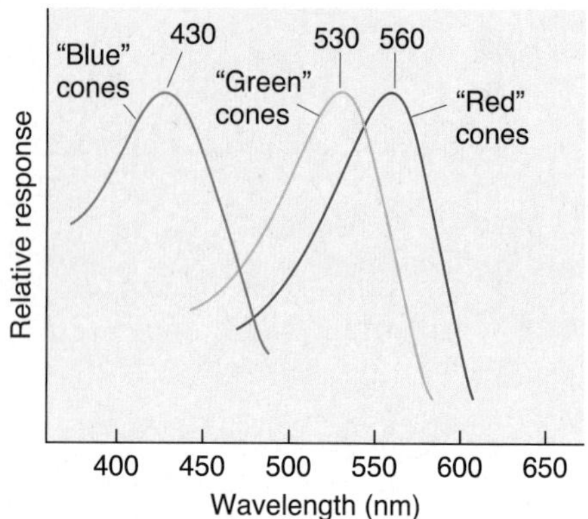

▲ FIGURE 9.21
The spectral sensitivity of the three types of cone pigments. Each photopigment absorbs a wide range of wavelengths from the color spectrum (see Figure 9.2).

three opsins that give the photopigments different spectral sensitivities. Thus, we can speak of short-wavelength or "blue" cones that are maximally activated by light with a wavelength of about 430 nm, medium-wavelength or "green" cones that are maximally activated by light with a wavelength of about 530 nm, and long-wavelength or "red" cones that are maximally activated by light with a wavelength of about 560 nm (Figure 9.21). Note that each type of cone is activated by a broad range of different wavelengths of light, and there is overlap in the wavelengths that affect the three cone types. Commonly the cones are referred to as blue, green, and red, but this can be confusing because different colors are perceived when different wavelengths of light, within the broad sensitive range of a particular cone type, are presented. The short, medium, long terminology is safer.

Color Perception. The color that we perceive is largely determined by the relative contributions of short-, medium-, and long-wavelength cones to the retinal signal. The fact that our visual system perceives colors in this way was actually predicted over 200 years ago by British physicist Thomas Young. Young showed in 1802 that all the colors of the rainbow, including white, could be created by mixing the proper ratio of red, green, and blue light (Figure 9.22). Young proposed, quite correctly, that the retina contains three receptor types, each type being maximally sensitive to a different spectrum of wavelengths. Young's ideas were later championed by Hermann von Helmholtz, an influential nineteenth-century German physiologist. (Among his accomplishments is the invention of the ophthalmoscope in 1851.) This theory of color vision came to be known as the **Young–Helmholtz trichromacy theory**. According to the theory, the brain assigns colors based on a comparative readout of the three cone types. When all types of cones are equally active, as in broad-spectrum light, we perceive "white." Novel colors arise from other mixtures. For example, orange is a mixture of red and yellow, and it looks somewhat red and somewhat yellow (red, orange, and yellow are neighbors in the color spectrum). But note that some color mixtures are perceptually different: No color looks simultaneously red and green or blue and yellow (and these "opponent" colors are not neighbors in the color spectrum). As we will see later, this may be a reflection of further "color opponent" processing by ganglion cells.

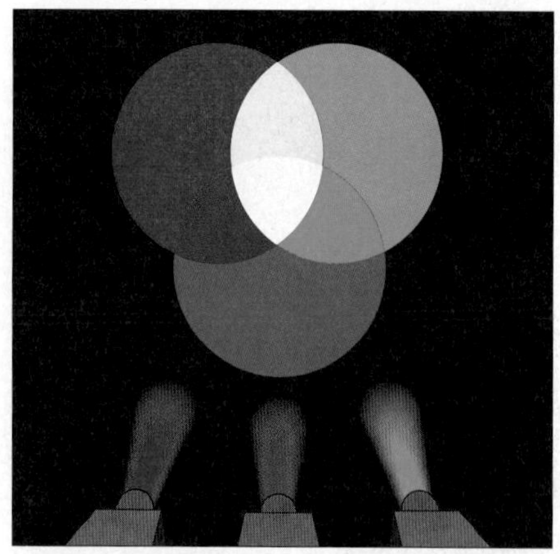

▲ FIGURE 9.22
Mixing colored lights. The mixing of red, green, and blue light causes equal activation of the three types of cones, and the perception of "white" results.

The color vision nomenclature can be puzzling, so be careful to not confuse the color of light with the "color name" of a cone photoreceptor. It is wrong to think that lights perceived as red consist of a single wavelength of light or that this wavelength is absorbed only by long-wavelength cones. The reality is that colored lights generally contain a broad and complex spectrum of wavelengths that may partially activate all three types of cones. The ratios of activations determine color. Various forms of color blindness result when one or more of the cone photopigment types is missing (Box 9.5). And, as discussed earlier, if we had no cones, we would be unable to perceive color differences at all.

Dark and Light Adaptation

The transition from all-cone daytime vision to all-rod nighttime vision is not instantaneous; depending on how high the initial light level is, it can take minutes to nearly an hour to reach the greatest light sensitivity in the dark (hence the time needed to get oriented in the stargazing exercise described earlier). This phenomenon is called **dark adaptation**, or getting used to the dark. Sensitivity to light actually increases a millionfold or more during this period.

BOX 9.5 OF SPECIAL INTEREST

The Genetics of Color Vision

The color we perceive is largely determined by the relative amounts of light absorbed by the red, green, and blue visual pigments in our cones. This means it's possible to perceive any color of the rainbow by mixing different amounts of red, green, and blue light. For example, the perception of yellow light comes from an appropriate mixture of red and green light. Because we use a "three-color" system, humans are referred to as *trichromats*. However, not all normal trichromats perceive colors exactly the same. For example, if a group of people are asked to choose the wavelength of light that most appears green without being yellowish or bluish, there will be small variations in their choices. However, significant abnormalities of color vision extend well beyond this range of normal trichromatic vision.

Most abnormalities in color vision result from small genetic errors that lead to the loss of one visual pigment or a shift in the spectral sensitivity of one type of pigment. The most common abnormalities involve red–green color vision and are much more common in men than in women. The reason for this pattern is that the genes encoding the red and green pigments are on the X chromosome, whereas the gene that encodes the blue pigment is on chromosome 7. Men have abnormal red–green vision if there is a defect on the single X chromosome they inherit from their mother. Women have abnormal red–green vision only if both parents contribute abnormal X chromosomes.

About 6% of men have a red or green pigment that absorbs somewhat different wavelengths of light than do the pigments of the rest of the population. These men are often called "colorblind," but they actually see quite a colorful world. They are more properly referred to as *anomalous trichromats* because they require somewhat different mixtures of red, green, and blue to see intermediate colors (and white) than other people. Most anomalous trichromats have normal genes to encode the blue pigment and either the red or the green pigment, but they also have a hybrid gene that encodes a protein with an abnormal absorption spectrum between that of normal red and green pigments. For example, a person with an anomalous green pigment can match a yellow light with a red–green mixture containing less red than a normal trichromat. Anomalous trichromats perceive the full spectrum of colors that normal trichromats perceive, but in rare instances they disagree about the precise color of an object (e.g., blue versus greenish blue).

About 2% of men actually lack either the red or the green pigment, making them red–green colorblind. Because this leaves them with a "two-color" system, they are referred to as *dichromats*. People lacking the green pigment are less sensitive to green, and they confuse certain red and green colors that appear different to trichromats. A "green dichromat" can match a yellow light with either red or green light; no mixture is needed. In contrast to the roughly 8% of men who either are missing one pigment or have an anomalous pigment, only about 1% of women have such color abnormalities.

People without one color pigment are considered colorblind, but they actually do see colors. Estimates of the number of people lacking all color vision vary, but less than about 0.001% of the population is thought to have this condition. In one type, both red and green cone pigments are missing, in many cases because mutations of the red and green genes make them nonfunctional. These people are called blue cone *monochromats* and live in a world that varies only in lightness, like a trichromat's perception of a black-and-white movie.

Although achromatopsia (lack of color vision) is rare in humans, on the tiny Micronesian island of Pingelap more than 5–10% of the population is colorblind and many more are unaffected carriers. It is known that the underlying basis for the disorder is a genetic mutation associated with incomplete development of cones that leaves them nonfunctional. But why is achromatopsia so common on Pingelap? According to islanders, in the late eighteenth century a typhoon killed all but about 20 of the inhabitants. Those afflicted with achromatopsia appear all to be descendants of one man who was a carrier; in subsequent generations the incidence of achromatopsia grew with inbreeding in the small surviving population.

Recent research has shown that, precisely speaking, there may not be such a thing as normal color vision. In a group of males classified as normal trichromats, it was found that some require slightly more red than others to perceive yellow in a red–green mixture. This difference, which is tiny compared to the deficits discussed above, results from a single alteration of the red pigment gene. The 60% of males who have the amino acid serine at site 180 in the red pigment gene are more sensitive to long-wavelength light than the 40% who have the amino acid alanine at this site. Imagine what would happen if a woman had different red gene varieties on her two X chromosomes. Both red genes should be expressed, leading to different red pigments in two populations of cones. In principle, such women should have a super-normal ability to discriminate colors because of their tetrachromatic color vision, a rarity among all animals.

Dark adaptation is explained by a number of factors. Perhaps the most obvious is dilation of the pupils, which allows more light to enter the eye. However, the diameter of the human pupil only ranges from about 2–8 mm; changes in its size increase sensitivity to light by a factor of only about 10. The larger component of dark adaptation involves the regeneration of unbleached rhodopsin and an adjustment of the functional circuitry of the retina so that information from more rods is available to each ganglion cell. Because of this tremendous increase in sensitivity, when the dark-adapted eye goes back into bright light, it is temporarily saturated. This explains what happens when you first go outside on a bright day. Over the next 5–10 minutes, the eyes undergo **light adaptation**, reversing the changes in the retina that accompanied dark adaptation. This light-dark adaptation in the duplex retina gives our visual system the ability to operate in light intensities ranging from moonless midnight to bright high noon.

Calcium's Role in Light Adaptation. In addition to the factors discussed earlier, the ability of the eye to adapt to changes in light level relies on changes in calcium concentration within the cones. When you step out into bright light from a dark theater, initially the cones are hyperpolarized as much as possible (i.e., to E_K, the equilibrium potential for K^+). If the cones stayed in this state, we would be unable to see further changes in light level. As discussed earlier, the constriction of the pupil helps a bit in reducing the light entering the eye. However, the most important change is a gradual depolarization of the membrane back to about −35 mV (see Figure 9.18c).

The reason this happens stems from the fact that the cGMP-gated sodium channels discussed previously also admit calcium (Figure 9.23). In the dark, Ca^{2+} enters the cones and has an inhibitory effect on the enzyme (guanylyl cyclase) that synthesizes cGMP. When the cGMP-gated channels close, the flow of Ca^{2+} into the photoreceptor is curtailed along with the flow of Na^+; as a result, more cGMP is synthesized (because the synthetic enzyme is less inhibited), thereby allowing the cGMP-gated channels to open again. Stated more simply, when the channels close, a process is initiated that gradually reopens them even if the light level does not change. Calcium also appears to affect photopigments and phosphodiesterase in ways that decrease their response to light. These calcium-based mechanisms ensure that the photoreceptors are always able to register relative changes in light level, though information about the absolute level is lost.

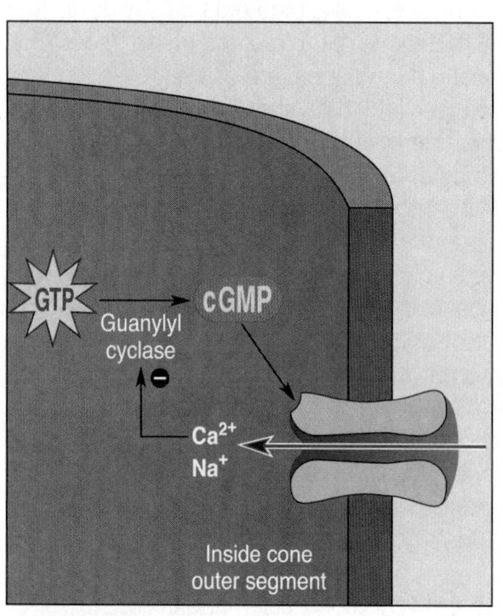

▲ FIGURE 9.23
The role of calcium in light adaptation. Ca^{2+} enters a cone through the same cGMP-gated channels as Na^+; it inhibits the synthesis of cGMP.

Local Adaptation of Dark, Light, and Color. The effect that pupil size has on light and dark adaptation is the same for all photoreceptors. However, photoreceptor bleaching and other adaptational mechanisms, such as calcium's influence on cGMP, can occur on a cone-by-cone basis. You can demonstrate this to yourself with Figure 9.24. First, fixate on the black cross at the center of the gray square in part a for about a minute. Cones imaging the dark spots will become dark-adapted and cones imaging the white spots will be relatively light-adapted. Now look at the cross at the middle of the large light square in part b. Because of the local adaptation of the cones, you should now see white spots where you adapted to black and black spots where you adapted to white. The same idea applies to color adaptation. First look at the yellow or green square in part c or d of the figure, and selectively adapt your cones. Then move your fixation to the light square in part b. You should see blue if you adapted to yellow and red if you adapted to green (the exact colors depend on the inks used in printing). These demonstrations use abnormally long fixations to reveal

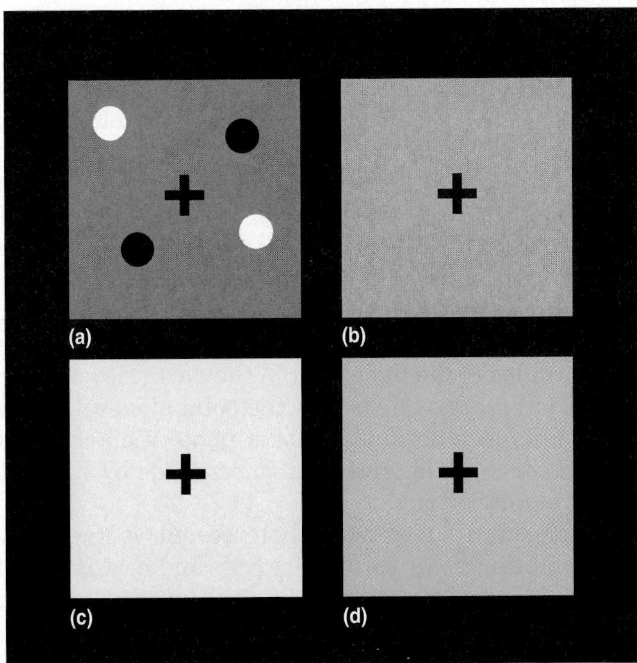

▲ FIGURE 9.24
Light and color adaptation. Fixate on the cross in **(a)** for 1 minute and then look at **(b)** to see the effects of local retinal adaptation to the light and dark spots. After adaptation to the color in **(c)** or **(d)**, the gray in **(b)** will appear to have an opponent color.

the critical adaptational processes that are always at work to keep photoreceptors providing useful information.

RETINAL PROCESSING AND OUTPUT

Now that we have seen how light is first converted into neural activity, we turn to the process by which information about light is passed beyond the eye to the rest of the brain. As the sole source of output from the retina is the action potentials arising from ganglion cells, our goal is to understand what information ganglion cells convey. It is interesting to note that researchers were able to explain some of the ways the retina processes visual images well before the discovery of how photoreceptors work. Since about 1950, neuroscientists have studied the action potential discharges of retinal ganglion cells as the retina is stimulated with light. The pioneers of this approach were neurophysiologists Keffer Hartline and Stephen Kuffler working in the United States and Horace Barlow working in England. Their research uncovered which aspects of a visual image were encoded as ganglion cell output. Early studies of horseshoe crabs and frogs gave way to investigations of cats and monkeys. Researchers learned that similar principles are involved in retinal processing across a wide range of species.

Progress in understanding how ganglion cell properties are generated by synaptic interactions in the retina has been slower. This is so because *only ganglion cells fire action potentials*; all other cells in the retina (except some amacrine cells) respond to stimulation with graded changes in membrane potential. The detection of such graded changes requires technically challenging intracellular recording methods, whereas action potentials can be detected using simple extracellular recording methods (see Box 4.1). It was not until the early 1970s that John Dowling and Frank Werblin at Harvard University were able to show how ganglion cell responses are built from the interactions of horizontal and bipolar cells.

The most direct path for information flow in the retina is from a cone photoreceptor to bipolar cell to ganglion cell. At each synaptic relay, the responses are modified by the lateral connections of horizontal cells and amacrine cells. Photoreceptors, like other neurons, release neurotransmitter when depolarized. The transmitter released by photoreceptors is the amino acid glutamate. As we have seen, photoreceptors are depolarized in the dark and are *hyperpolarized* by light. We thus have the counterintuitive situation in which photoreceptors actually release fewer transmitter molecules in the light than in the dark. However, we can reconcile this apparent paradox if we take the point of view that *dark* rather than *light* is the preferred stimulus for a photoreceptor. Thus, when a shadow passes across a photoreceptor, it responds by depolarizing and releasing more neurotransmitter.

In the outer plexiform layer, each photoreceptor is in synaptic contact with two types of retinal neurons: bipolar cells and horizontal cells. Recall that bipolar cells create the direct pathway from photoreceptors to ganglion cells; horizontal cells feed information laterally in the outer plexiform layer to influence the activity of neighboring bipolar cells and photoreceptors (see Figures 9.11 and 9.12). We now turn to the response properties of bipolar and then ganglion cells by analyzing their receptive fields.

The Receptive Field

Suppose you have a flashlight that can project a very small spot of light onto the retina while you monitor the activity of a visual neuron, such as the output of a retinal ganglion cell. You would find that light applied to only a small portion of the retina would change the firing rate of the neuron (Figure 9.25a). This area of the retina is called the neuron's **receptive field**. Light anywhere else on the retina, outside this receptive field, would have no effect on firing rate. This same procedure can be applied to any neuron in the eye or elsewhere that is involved in vision; its receptive field is specified by the pattern of light on the retina that elicits a neural response. In the visual system, the optics of the eye establish a correspondence between locations on the retina and the visual field. Therefore, it is customary to also describe visual receptive fields as areas of visual space interchangeably with areas of the retina (Figure 9.25b). "Receptive field" is actually a general term useful in describing the stimulus specificity of neurons across sensory systems. For example, we will see in Chapter 12 that receptive fields of neurons in the somatosensory system are small areas of the skin that, when touched, produce a response in a neuron (Figure 9.25c).

As we move farther into the visual pathway, we find that receptive fields change in shape and correspondingly in the sort of stimulus that makes the neurons most active. In the retina, spots of light give optimal responses from ganglion cells, but in different areas of the visual cortex, neurons respond best to bright lines and even complex shapes of biological significance such as hands and faces. These changes may reflect important differences in the kind of information represented at each stage. (We will have a lot more to say about this in Chapter 10.) Receptive fields receive a great deal of attention because of the functional interpretation given to them. An instructive example comes from the early work of Horace Barlow, who made recordings from the frog retina. He found that a frog will jump and snap at a small black spot waved in front of it, and the same stimulus elicits a strong response from retinal ganglion cells. Might this be a "bug detector" used by the frog to hunt? We will see that inferences

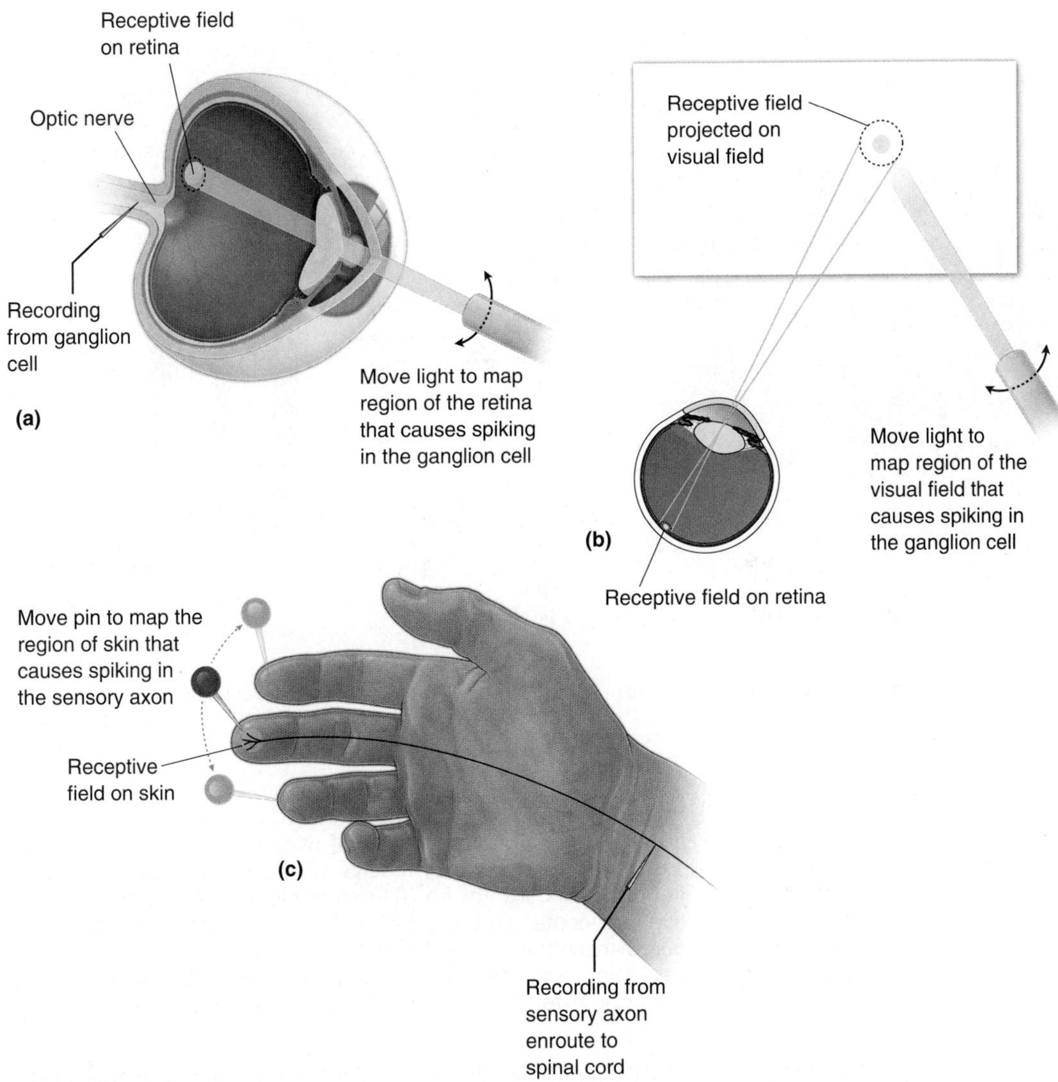

▲ FIGURE 9.25
The receptive field. **(a)** The receptive field of a ganglion cell is located by recording from the neuron's axon in the optic nerve. A small spot of light is projected onto various parts of the retina; the receptive field consists of the locations that increase or decrease the ganglion cell's firing rate. By moving the recording electrode, the same procedure can be used to locate receptive fields for neurons at other locations in the brain and other retinal neurons (seen as changes in membrane potential if they do not fire action potentials). **(b)** A receptive field on the retina corresponds to light coming from a particular location in the visual field. **(c)** The receptive field is a useful concept in other sensory systems; for example, a small patch of skin is a receptive field for touch.

about function based on receptive field properties are also made for monkeys and humans, and they are as exciting as they are speculative.

Bipolar Cell Receptive Fields

Bipolar cells and receptive fields can be categorized into two classes, ON and OFF, based on the response to the glutamate released by photoreceptors. The circuitry that gives rise to bipolar receptive fields consists of direct input from photoreceptors and indirect photoreceptor input relayed by horizontal cells (Figure 9.26a). Let's start by considering only the interactions between cones and bipolar cells in the direct path (no horizontal cells involved), shown in Figure 9.26b. Light shined onto a cone will *hyperpolarize* some bipolar cells. These are called **OFF bipolar cells,**

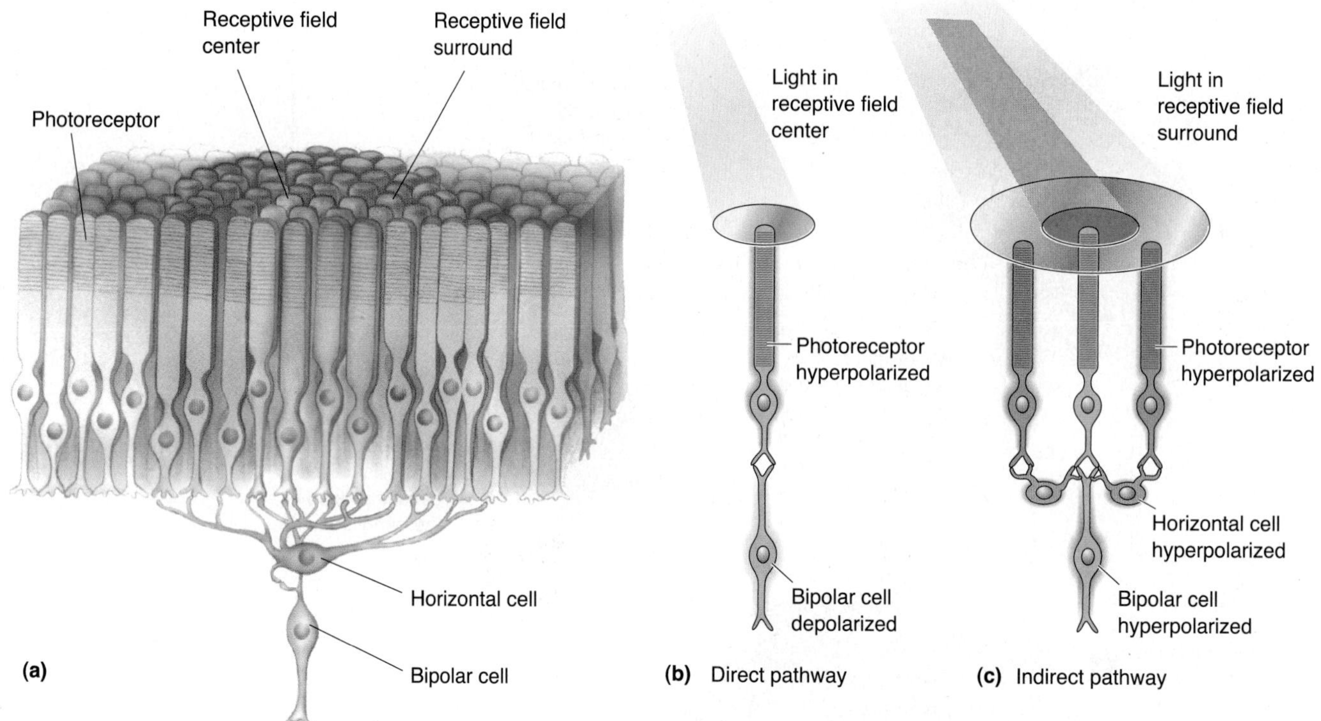

▲ FIGURE 9.26
Direct and indirect pathways from photoreceptors to bipolar cells. (a) Bipolar cells receive direct synaptic input from a cluster of photoreceptors, constituting the receptive field center. In addition, they receive indirect input from surrounding photoreceptors via horizontal cells. (b) An ON-center bipolar cell is depolarized by light in the receptive field center via the direct pathway. (c) Light in the receptive field surround hyperpolarizes the ON-center bipolar cell via the indirect pathway. Because of the intervening horizontal cell, the effect of light on the surround photoreceptors is always opposite the effect of light on the center photoreceptors.

because light effectively turns them off. Conversely, light shined onto a cone will *depolarize* other bipolar cells. These cells "turned on" by light are called **ON bipolar cells**. Evidently the cone-to-bipolar synapse inverts the signal from the cone: The cone hyperpolarizes to light, but the ON bipolar cell depolarizes.

How can different bipolar cells give opposite responses to direct cone input? The answer is that there are two kinds of receptors that receive glutamate released by the photoreceptors. OFF bipolar cells have ionotropic glutamate receptors, and these glutamate-gated channels mediate a classical depolarizing excitatory postsynaptic potential from the influx of Na^+. Hyperpolarization of the cone causes less neurotransmitter to be released, resulting in a more hyperpolarized bipolar cell. On the other hand, ON bipolar cells have G-protein-coupled (metabotropic) receptors and respond to glutamate by hyperpolarizing. Each bipolar cell receives direct synaptic input from a cluster of photoreceptors. The number of photoreceptors in this cluster ranges from one at the center of the fovea to thousands in the peripheral retina.

In addition to direct connections with photoreceptors, bipolar cells also are connected via horizontal cells to a circumscribed ring of photoreceptors that surrounds the central cluster (see Figure 9.26a). The synaptic interactions of photoreceptors, horizontal cells, and bipolar cells are complex, and research is ongoing. For our purposes here, there are two key points. First, when a photoreceptor hyperpolarizes in response to light, output is sent to horizontal cells that also hyperpolarize. Second, the effect of horizontal cell hyperpolarization is to counteract the effect of light on neighboring photoreceptors. In Figure 9.26c, light is shined onto two photoreceptors connected through horizontal cells to a central photoreceptor and bipolar cell. The effect of this indirect path input is to depolarize the central photoreceptor, counteracting the hyperpolarizing effect of light shined directly on it.

Let's synthesize this discussion: The receptive field of a bipolar cell consists of two parts: a circular area of retina providing direct photoreceptor input, the *receptive field center*, and a surrounding area of retina providing input via horizontal cells, the *receptive field surround*. The response of a bipolar cell's membrane potential to light in the receptive field center is opposite to that of light in the surround. Thus, these cells are said to have antagonistic **center-surround receptive fields**. Receptive field dimensions can be measured in millimeters across the retina or, more commonly, in degrees of visual angle. One millimeter on the retina corresponds to a visual angle of about 3.5°. Bipolar cell receptive field diameters range from a fraction of a degree in the central retina to several degrees in the peripheral retina.

The center-surround receptive field organization is passed on from bipolar cells to ganglion cells via synapses in the inner plexiform layer. The lateral connections of the amacrine cells in the inner plexiform layer also contribute to the elaboration of ganglion cell receptive fields and the integration of rod and cone input to ganglion cells. Numerous types of amacrine cells have been identified, which make diverse contributions to ganglion cell responses.

Ganglion Cell Receptive Fields

Most retinal ganglion cells have essentially the same concentric center-surround receptive field organization as bipolar cells. ON-center and OFF-center ganglion cells receive input from the corresponding type of bipolar cell. An important difference is that, unlike bipolar cells, ganglion cells fire action potentials. Ganglion cells actually fire action potentials whether or not they are exposed to light, and light in the receptive field center or surround increases or decreases the firing rate. Thus, an ON-center ganglion cell will be depolarized and respond with a barrage of action potentials when a small spot of light is projected onto the center of its receptive field. An OFF-center cell will fire fewer action potentials when a small spot of light is projected to the center of its receptive field; it will fire more action potentials if a small *dark* spot covers the receptive field center. In both ON and OFF types of cell, the response to stimulation of the center is canceled by the response to stimulation of the surround (Figure 9.27). The surprising implication is that most retinal ganglion cells are not particularly responsive to changes in illumination that include both the receptive field center and the receptive field surround. Rather, it appears that the ganglion cells are mainly responsive to *differences* in illumination that occur within their receptive fields.

To illustrate this point, consider the response generated by an OFF-center cell as a light–dark edge crosses its receptive field (Figure 9.28). Remember that in such a cell, dark in the center of the receptive field causes the cell to depolarize, whereas dark in the surround causes the cell to hyperpolarize. In uniform illumination, the center and surround cancel to yield some low level of response (Figure 9.28a). When the edge enters the surround region of the receptive field without encroaching on the center, the dark area has the effect of hyperpolarizing the neuron, leading to a decrease in

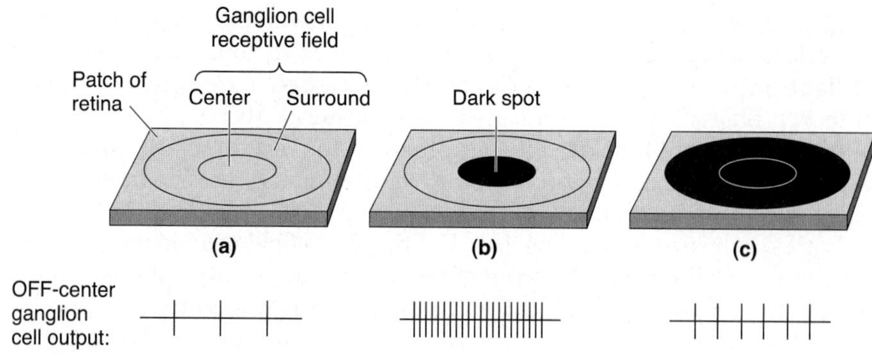

▲ FIGURE 9.27

A center-surround ganglion cell receptive field. **(a, b)** An OFF-center ganglion cell responds with a barrage of action potentials when a dark spot is imaged on its receptive field center. **(c)** If the dark spot is enlarged to include the receptive field surround, the response is greatly reduced.

the cell's firing rate (Figure 9.28b). As the dark area begins to include the center, however, the partial inhibition by the surround is overcome, and the cell response increases (Figure 9.28c). To understand why the response in Figure 28c is increased, consider that there is 100% excitation of the cell because darkness covers the entire receptive field center, but there is only partial inhibition from the portion of the surround in darkness. But when the dark area finally fills the entire surround, the center response is canceled (Figure 9.28d). Notice that the cell response in this example is only slightly different in uniform light and in uniform dark; the response is modulated mainly by the presence of the light–dark edge in its receptive field.

Now let's consider the output of *all* the OFF-center ganglion cells at different positions on the retina that are stimulated by a stationary light–dark edge. The responses will fall into the same four categories illustrated in Figure 9.28. Thus, the cells that will register the presence of the edge are those with receptive field centers and surrounds that are differentially affected by the light and dark areas. The population of cells with receptive field centers "viewing" the light side of the edge will be inhibited (see Figure 9.28b). The population of cells with centers "viewing" the dark side of the edge will be excited (see Figure 9.28c). In this way, the difference in illumination at a light–dark edge is not faithfully represented by the difference in the output of ganglion cells on either side of the edge. Instead, *the center-surround organization of the receptive fields leads to a neural response that emphasizes the contrast at light–dark edges.*

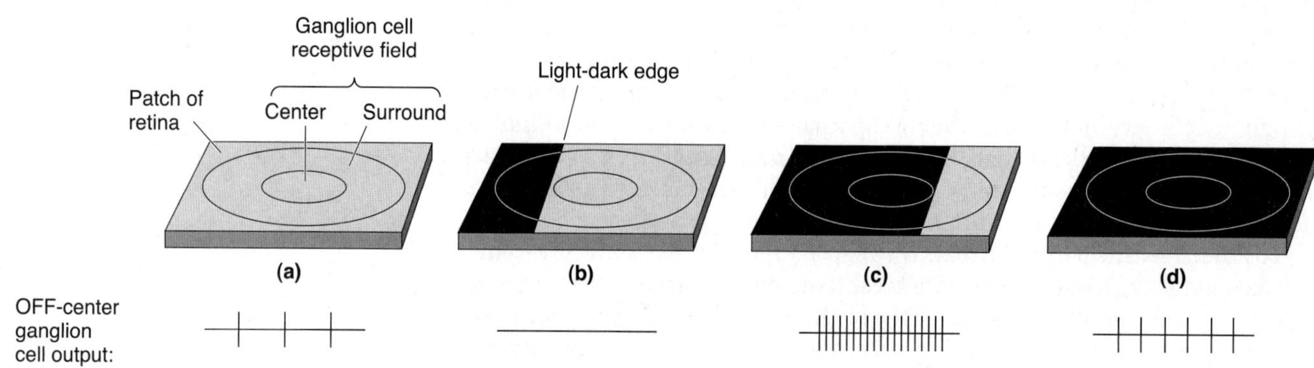

▲ FIGURE 9.28

Responses to a light–dark edge crossing an OFF-center ganglion cell receptive field. The response of the neuron is determined by the fraction of the center and surround that are filled by light and dark. (See text for details.)

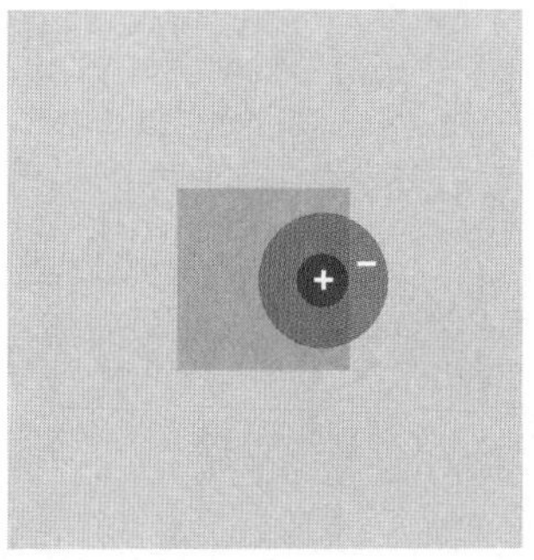

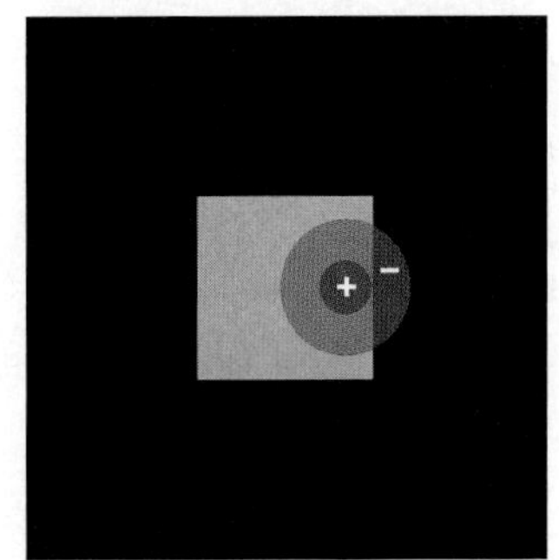

▲ FIGURE 9.29
The influence of contrast on the perception of light and dark. The central boxes are identical shades of gray, but because the surrounding area is lighter on the left, the left central box appears darker. ON-center receptive fields are shown on the left and right of the figure. Which would respond more?

There are many visual illusions involving the perception of light level. The organization of ganglion cell receptive fields suggests an explanation for the illusion shown in Figure 9.29. Even though the two central squares are the same shade of gray, the square on the lighter background appears darker. Consider the two ON-center receptive fields shown on the gray squares. In both cases, the same gray light hits the receptive field center. However, the receptive field on the left has more light in its surround than the receptive field on the right. This will lead to more inhibition and a lower response and may be related to the darker appearance of the left central gray square.

Structure-Function Relationships. Most ganglion cells in the mammalian retina have a center-surround receptive field with either an ON or an OFF center. They can be further categorized based on their appearance, connectivity, and electrophysiological properties. In the macaque monkey retina and human retina, two major types of ganglion cells are distinguished: large **M-type ganglion cells** and smaller **P-type ganglion cells.** (M stands for *magno*, from the Latin for "large"; P stands for *parvo*, from the Latin for "small.") Figure 9.30 shows the relative sizes of M and P ganglion cells at the same location on the retina. P cells constitute about 90% of the ganglion cell population, M cells constitute about 5%, and the remaining 5% are made up of a variety of **nonM–nonP ganglion cell** types that are less well characterized.

The visual response properties of M cells differ from those of P cells in several ways. They have larger receptive fields, they conduct action potentials more rapidly in the optic nerve, and they are more sensitive to low-contrast stimuli. In addition, M cells respond to stimulation of their receptive field centers with a transient burst of action potentials, while P cells respond with a sustained discharge as long as the stimulus is on (Figure 9.31). We will see in Chapter 10 that the different types of ganglion cells appear to play different roles in visual perception.

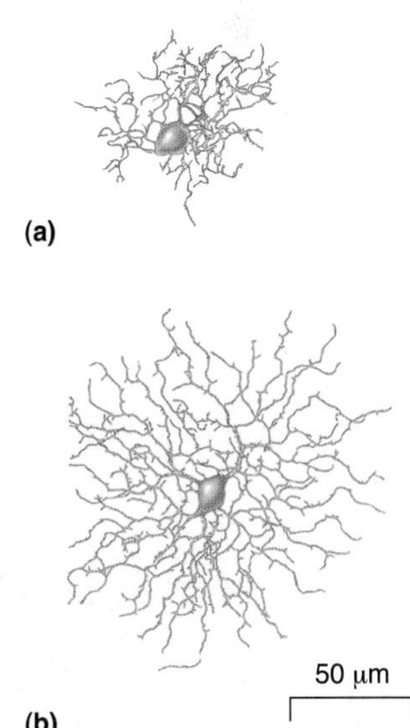

▲ FIGURE 9.30
M-type and P-type ganglion cells in the macaque monkey retina. **(a)** A small P cell from the peripheral retina. **(b)** An M cell from a similar retinal location is significantly larger. (Source: Watanabe and Rodieck, 1989, pp. 437, 439.)

Color-Opponent Ganglion Cells. Another important distinction between ganglion cell types is that some P cells and nonM–nonP cells are sensitive to differences in the wavelength of light. The majority of these color-sensitive neurons are called **color-opponent cells**, reflecting the fact that the response to one color in the receptive field center is canceled by showing another color in the receptive field surround. The two types of opponency are red versus green and blue versus yellow. Consider, for example, a cell with a red ON center and a green OFF surround

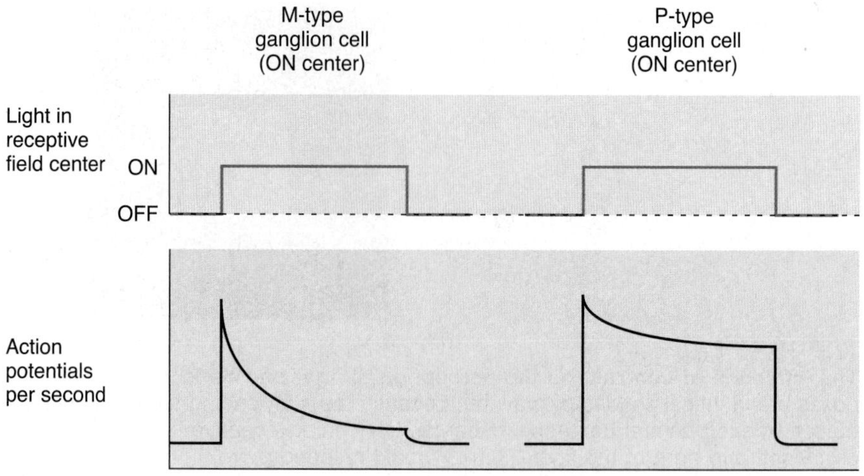

▲ FIGURE 9.31
Different responses to light of M-type and P-type ganglion cells.

(Figure 9.32). The center of the receptive field is fed by red cones, and the surround is fed by green cones via an inhibitory circuit (i.e., the indirect horizontal cell path). To understand how the neuron with this receptive field responds to light, recall Figure 9.21, which shows that red and green cones absorb different, but overlapping, wavelengths of light.

If red light bathes the receptive field center, the neuron responds with a strong burst of action potentials (Fig. 9.32b). If the red light is extended to cover both the center and the surround of the receptive field, the neuron is still excited but much less so (Fig. 9.32c). The reason that red light has an effect in the green OFF surround is that, as shown in Fig. 9.21, red wavelengths of light are partially absorbed by green cones, and their activation inhibits the response of the neuron. To fully activate the inhibitory surround of the receptive, green light is needed. In this case, the red ON center response is canceled by the green OFF surround response (Fig. 9.32d). The shorthand notation for such a cell is R^+G^-, meaning simply that it is optimally excited by red in the receptive field center and inhibited by green in the surround.

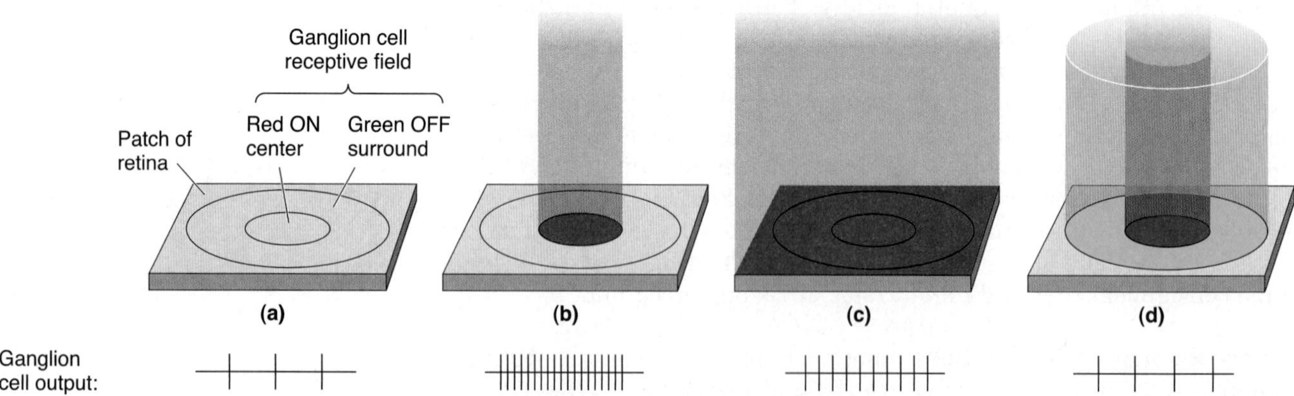

▲ FIGURE 9.32
Color opponency in ganglion cells. **(a)** A color-opponent center-surround receptive field of a P-type ganglion cell. **(b)** A strong response is elicited by red light to the receptive field center, which receives input from long-wavelength-sensitive (red) cones. **(c)** Extending the red light into the surround inhibits the response because the green cones providing input to the surround are also somewhat sensitive to long wavelengths of light. **(d)** Even stronger inhibition is produced by green light in the receptive field surround, which optimally drives the green cones.

What would be the response to white light on the entire receptive field? Because white light contains all visible wavelengths (including both red and green), both center and surround would be equally activated, canceling each other, and the cell would not respond to the light.

Blue–yellow color opponency works the same way. Consider a cell with a blue ON center and a yellow OFF surround (B^+Y^-). The receptive field center receives input from blue cones, and the surround gets input from *both* red and green cones (hence yellow) via an inhibitory circuit. Blue light hitting the receptive field center would give a strong excitatory response that could be canceled by yellow light in the surround. Diffuse blue light covering both the center and the surround of the receptive field would also be an effective stimulus for this cell. Why is diffuse blue light a strong stimulus here, whereas diffuse red light was a much weaker stimulus in R^+G^- cells? The answer lies in the absorption curves (see Figure 9.21); in the R^+G^- cell, red light is absorbed by the "green" photoreceptors, but in the B^+Y^- neuron, very little blue light is absorbed by the "red" and "green" photoreceptors that make up the receptive field surround. Diffuse white light would not be an effective stimulus because it contains red, blue, and green wavelengths, and the center and surround responses would cancel.

Finally, note that M ganglion cells lack color opponency. This doesn't mean they won't respond to colored light; rather, their responses are not color-specific. So, for example, red light in the receptive field center or surround would have the same effect as green light. The lack of color opponency in M- cells is accounted for by the fact that both the center and the surround of the receptive field receive input from more than one type of cone. M-cell receptive fields are therefore denoted as simply either ON center/OFF surround or OFF center/ON surround. The color and light sensitivity of M and P ganglion cells suggest that the overall ganglion cell population sends information to the brain about three different spatial comparisons: light versus dark, red versus green, and blue versus yellow.

Ganglion Cell Photoreceptors

The retinal circuit we have described, in which rods and cones project to bipolar and then ganglion cells, implies that rods and cones are responsible for all phototransduction and that ganglion cells play a different role relaying visual information to the rest of the brain. But even back in the 1980s and 1990s, there were odd findings that were difficult to reconcile with this view. For example, mutant mice lacking rods and cones appeared to synchronize their sleeping and waking with the rise and setting of the sun even though they otherwise behaved as if they were blind. A subset of humans who are totally blind also appear to unconsciously synchronize their behavior to daily changes in sunlight.

The resolution of these mysteries came from the discovery in the 1990s that a small percentage of retinal ganglion cells actually transduce light. These few thousand neurons, known as **intrinsically photosensitive retinal ganglion cells (ipRGCs)**, use melanopsin as a photopigment, an opsin previously studied in frog skin! The ipRGCs function as normal ganglion cells that receive input from rods and cones and send axons out the optic nerve; in addition they are photoreceptors. The photosensitivity of ipRGCs differs from that of rods and cones in important ways, however. Unlike the hyperpolarization that light causes in rods and cones, ipRGCs depolarize to light. The ipRGCs also have very large dendritic fields; because the dendrites are photosensitive, this means the cells sum light input over much larger areas of the retina than rods and cones (Figure 9.33). The small number of ipRGCs and their large receptive fields are not ideal for, and the ipRGCs are not used in, fine pattern vision. As we will see

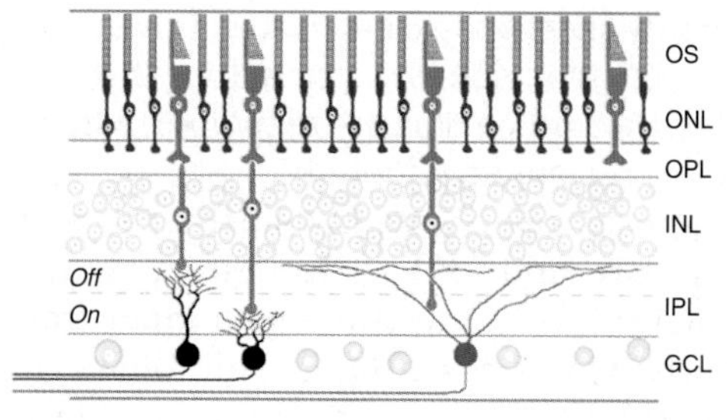

(a)

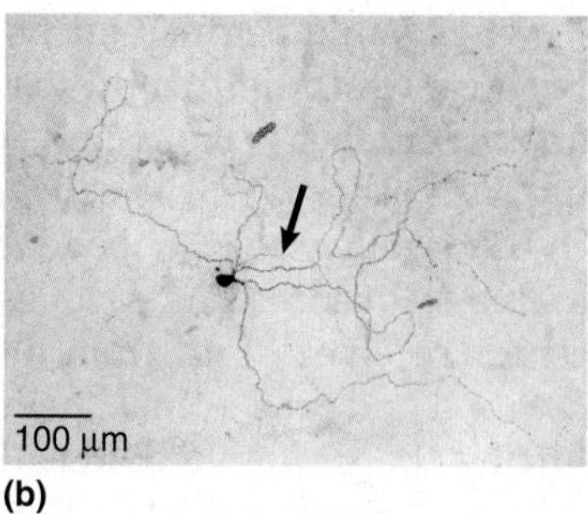

(b)

▲ FIGURE 9.33

Intrinsically photosensitive retinal ganglion cells. **(a)** Rods (blue) and cones (green) project to bipolar cells and then conventional ganglion cells (black) that send axons to the thalamus. In addition to receiving rod and cone inputs, ipRGCs (red) transduce light themselves. Unlike rods and cones, these photosensitive neurons send axons out of the eye without additional neurons and synaptic connections. The dendrites of the ipRGCs spread over a much broader area than conventional ganglion cell dendrites, making these neurons sensitive to light over a larger area. **(b)** A micrograph of an ipRGC shows the long winding dendrites and the axon (arrow). Note how much larger the dendritic field is compared to those of the regular ganglion cells in Figure 9.30. (Source: Berson, 2003, Fig. 1.)

in Chapter 19, an important function of the ipRGCs is providing input to subcortical visual areas that synchronize behavior to daily changes in light level (circadian rhythms). Since their initial discovery, a spectrum of different ipRGCs has been found that vary in their morphology, physiology, and connections to other retinal neurons. Research continues to investigate the multiple roles of these unusual cells in unconscious and conscious vision.

Parallel Processing

One of the important concepts that emerge from our discussion of the retina is the idea of **parallel processing** in the visual system. Parallel processing means that different visual attributes are processed simultaneously using distinct pathways. For example, we view the world with not one but two eyes that provide two parallel streams of information. In the central visual system, these streams are compared to give information about *depth*, the distance of an object from the observer. A second example of parallel processing is the independent streams of information about light and dark that arise from the ON-center and OFF-center ganglion cells in each retina. Finally, ganglion cells of both ON and OFF varieties have different types of receptive fields and response properties. M cells can detect subtle contrasts over their large receptive fields and are likely to contribute to low-resolution vision. P cells have small receptive fields that are well suited for the discrimination of fine detail. P cells and nonM–nonP cells are specialized for the separate processing of red–green and blue–yellow information.

CONCLUDING REMARKS

In this chapter, we have seen how light emitted by or reflected off objects can be imaged by the eye onto the retina. Light energy is first converted into membrane potential changes in the mosaic of photoreceptors. Interestingly, the transduction mechanism in photoreceptors is very similar to that in olfactory receptor cells, both of which involve cyclic nucleotide-gated ion channels. Photoreceptor membrane potential is converted into a chemical signal (the neurotransmitter glutamate), which is again converted into membrane potential changes in the postsynaptic bipolar and horizontal cells. This process of electrical-to-chemical-to-electrical signaling repeats again and again, until the presence of light or dark or color is finally converted to a change in the action potential firing frequency of the ganglion cells.

Information from 97 million photoreceptors is funneled into 1 million ganglion cells. In the central retina, particularly the fovea, relatively few photoreceptors feed each ganglion cell, whereas in the peripheral retina, thousands of receptors do. Thus, the mapping of visual space onto the array of optic nerve fibers is not uniform. Rather, in "neural space," there is an overrepresentation of the central few degrees of visual space, and signals from individual cones are more important. This specialization ensures high acuity in central vision but also requires that the eye move to bring the images of objects of interest onto the fovea.

As we shall see in the next chapter, there is good reason to believe that the different types of information that arise from different types of ganglion cells are, at least in the early stages, processed independently. Parallel streams of information—for example, from the right and left eyes—remain segregated at the first synaptic relay in the lateral geniculate nucleus of the thalamus. The same can be said for the M-cell and P-cell synaptic relays in the LGN. In the visual cortex, it appears that parallel paths may process different visual attributes. For example, the distinction in the retina between neurons that do or do not convey information about color is preserved in the visual cortex. In general, each of the more than two dozen visual cortical areas may be specialized for the analysis of different types of retinal output.

KEY TERMS

Introduction
vision (p. 294)
retina (p. 294)

Properties of Light
refraction (p. 296)

The Structure of the Eye
pupil (p. 296)
iris (p. 296)
cornea (p. 296)
sclera (p. 296)
extraocular muscle (p. 296)
conjunctiva (p. 297)
optic nerve (p. 297)
optic disk (p. 297)
macula (p. 297)
fovea (p. 297)
aqueous humor (p. 298)
lens (p. 298)
ciliary muscle (p. 298)
vitreous humor (p. 298)

Image Formation by the Eye
diopter (p. 299)
accommodation (p. 299)
pupillary light reflex (p. 303)
visual field (p. 304)
visual acuity (p. 304)
visual angle (p. 304)

Microscopic Anatomy of the Retina
photoreceptor (p. 304)
bipolar cell (p. 304)
ganglion cell (p. 304)
horizontal cell (p. 305)
amacrine cell (p. 305)
ganglion cell layer (p. 305)
inner nuclear layer (p. 305)
outer nuclear layer (p. 305)
inner plexiform layer (p. 305)
outer plexiform layer (p. 306)
layer of photoreceptor outer segments (p. 306)
rod photoreceptor (p. 306)
cone photoreceptor (p. 306)

Phototransduction
dark current (p. 3012)
cyclic guanosine monophosphate (cGMP) (p. 312)
rhodopsin (p. 313)
transducin (p. 313)
phosphodiesterase (PDE) (p. 313)
Young–Helmholtz trichromacy theory (p. 316)
dark adaptation (p. 318)
light adaptation (p. 318)

Retinal Processing and Output
receptive field (p. 320)
OFF bipolar cell (p. 321)
ON bipolar cell (p. 322)
center-surround receptive field (p. 323)
M-type ganglion cell (p. 325)
P-type ganglion cell (p. 325)
nonM–nonP ganglion cell (p. 325)
color-opponent cell (p. 325)
intrinsically photosensitive retinal ganglion cell (ipRGC) (p. 327)
parallel processing (p. 328)

REVIEW QUESTIONS

1. What physical property of light is most closely related to the perception of color?
2. Name eight structures in the eye that light passes through before it strikes the photoreceptors.
3. Why is a scuba mask necessary for clear vision underwater?
4. What is myopia, and how is it corrected?
5. Give three reasons explaining why visual acuity is best when images fall on the fovea.
6. How does the membrane potential change in response to a spot of light in the receptive field center of a photoreceptor? Of an ON bipolar cell? Of an OFF-center ganglion cell? Why?
7. What happens in the retina when you "get used to the dark"? Why can't you see color in the dark?
8. In what way is retinal output *not* a faithful reproduction of the visual image falling on the retina?
9. In retinitis pigmentosa, early symptoms include the loss of peripheral vision and night vision. The loss of what type of cells could lead to such symptoms?

FURTHER READING

Arshavsky VY, Lamb TD, Pugh EN. 2002. G proteins and phototransduction. *Annual Review of Physiology* 64:153–187.

Berson DM. 2003. Strange vision: ganglion cells as circadian photoreceptors. *Trends in Neurosciences* 26:314–320.

Field GD, Chichilinsky EJ. 2007. Information processing in the primate retina: circuitry and coding. *Annual Review of Neuroscience* 30:1–30.

Nassi JJ, Callaway EM. 2009. Parallel processing strategies of the primate visual system. *Nature Reviews Neuroscience* 10:360–372.

Solomon SG, Lennie P. 2007. The machinery of colour vision. *Nature Reviews Neuroscience* 8: 276–286.

Wade NJ. 2007. Image, eye, and retina. *Journal of the Optical Society of America* 24:1229–1249.

Wassle H. 2004. Parallel processing in the mammalian retina. *Nature Reviews Neuroscience* 5: 747–757.

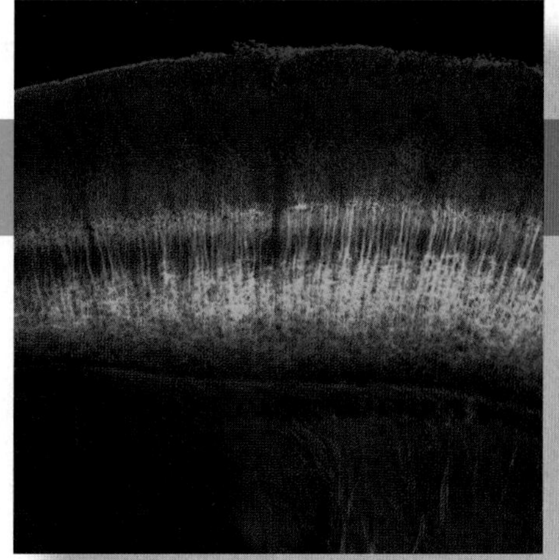

CHAPTER TEN

The Central Visual System

INTRODUCTION

Although our visual system provides us with a unified picture of the world around us, this picture has multiple facets. Objects we see have shape and color. They have a position in space, and sometimes they move. For us to see each of these properties, neurons somewhere in the visual system must be sensitive to them. Moreover, because we have two eyes, we actually have two visual images in our head, and somehow they must be merged.

In Chapter 9, we saw that in many ways the eye acts like a camera. But starting with the retina, the rest of the visual system is far more elaborate, far more interesting, and capable of doing far more than any camera. For example, we saw that the retina does not simply pass along information about the patterns of light and dark that fall on it. Rather, the retina *extracts* information about differences in brightness and color. There are roughly 100 million photoreceptors in the retina, but only 1 million axons leave the eye carrying information to the rest of the brain. What we perceive about the world around us, therefore, depends on what information is extracted by the output cells of the retina and how this information is analyzed and interpreted by the rest of the central nervous system (CNS). An example is color. There is no such thing as color in the physical world; there is simply a spectrum of visible wavelengths of light that are reflected by objects around us. Based on the information extracted by the three types of cone photoreceptors, however, our brain synthesizes a rainbow of colors and fills our world with it.

In this chapter, we explore how the information extracted by the retina is analyzed by the central visual system. The pathway serving conscious visual perception includes the *lateral geniculate nucleus (LGN)* of the

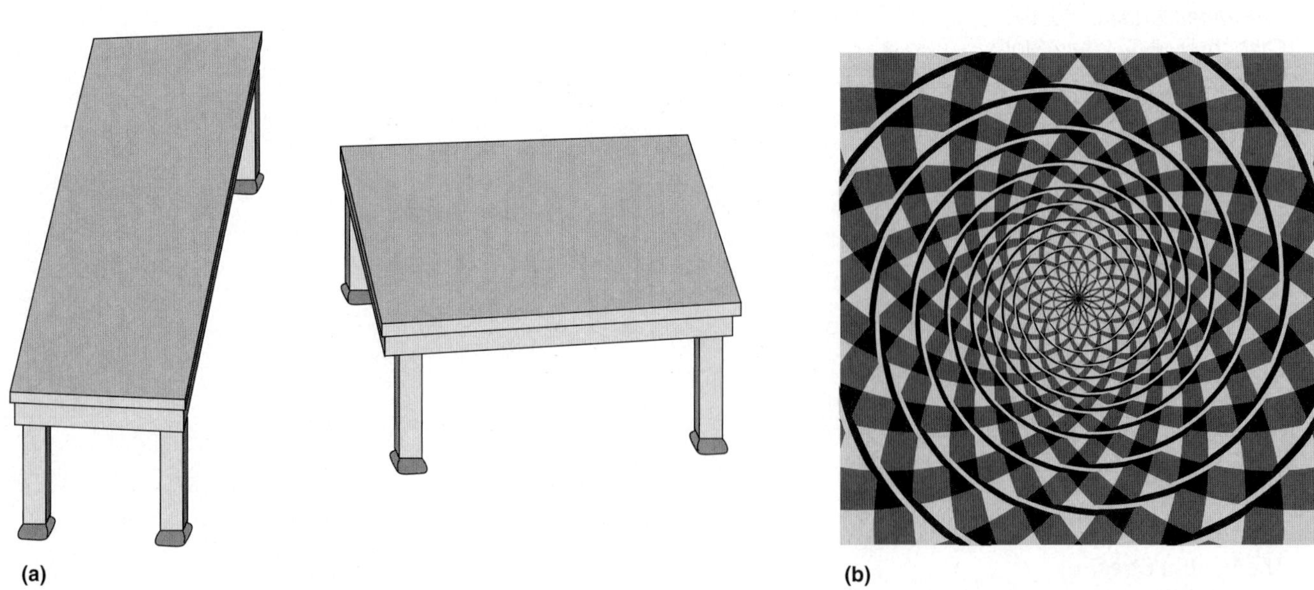

▲ FIGURE 10.1
Perceptual illusions. **(a)** The two tabletops are of identical dimensions and are imaged on similarly sized patches of retina. To prove this to yourself, compare the vertical extent of the left table with the horizontal size of the right table. Because of the brain's 3D interpretation of the 2D image, the perceived sizes are quite different. **(b)** This is an illusory spiral. Try tracing it with your finger. (Source: Part a adapted from R. Shepard, 1990, p. 48; part b adapted from J. Fraser, 1908.)

thalamus and the primary visual cortex, also called *area 17*, *V1*, or the *striate cortex*. We will see that the information funneled through this geniculocortical pathway is processed in parallel by neurons specialized for the analysis of different stimulus attributes. The striate cortex then feeds this information to more than two dozen extrastriate cortical areas in the occipital, temporal, and parietal lobes, and many of these appear to be specialized for different types of analysis.

Much of what we know about the central visual system was first worked out in the domestic cat and then extended to the rhesus monkey, *Macaca mulatta*. The macaque monkey, as it is also called, relies heavily on vision for survival in its habitat, as do we humans. In fact, tests of the performance of this primate's visual system show that in virtually all respects, it rivals that of humans. Thus, although most of this chapter concerns the organization of the macaque visual system most neuroscientists agree that it approximates very closely the situation in our own brain.

Visual neuroscience cannot yet explain every aspect of visual perception (Figure 10.1). However, significant progress has been made in answering a basic question: How do neurons represent the different facets of the visual world? By examining those stimuli that make different neurons in the visual cortex respond, and how these response properties arise, we begin to see how the brain portrays the visual world around us.

THE RETINOFUGAL PROJECTION

The neural pathway that leaves the eye, beginning with the optic nerve, is often referred to as the **retinofugal projection**. The suffix *-fugal* is from the Latin word meaning "to flee" and is commonly used in neuroanatomy to describe a pathway that is directed away from a structure. Thus, a centrifugal projection goes away from the center, a corticofugal projection goes away from the cortex, and the retinofugal projection goes away from the retina.

We begin our tour of the central visual system by looking at how the retinofugal projection courses from each eye to the brain stem on each side, and how the task of analyzing the visual world initially is divided among, and organized within, certain structures of the brain stem. Then, we focus on the major arm of the retinofugal projection that mediates conscious visual perception.

The Optic Nerve, Optic Chiasm, and Optic Tract

The ganglion cell axons "fleeing" the retina pass through three structures before they form synapses in the brain stem. The components of this retinofugal projection are, in order, the optic nerve, the optic chiasm, and the optic tract (Figure 10.2). The **optic nerves** exit the left and right eyes at the optic disks, travel through the fatty tissue behind the eyes in their bony orbits, then pass through holes in the floor of the skull. The optic nerves from both eyes combine to form the **optic chiasm** (named for the X shape of the Greek letter chi), which lies at the base of the brain, just anterior to where the pituitary gland dangles down.

At the optic chiasm, the axons originating in the nasal retinas cross from one side to the other. The crossing of a fiber bundle from one side of the brain to the other is called a **decussation**. Because only the axons

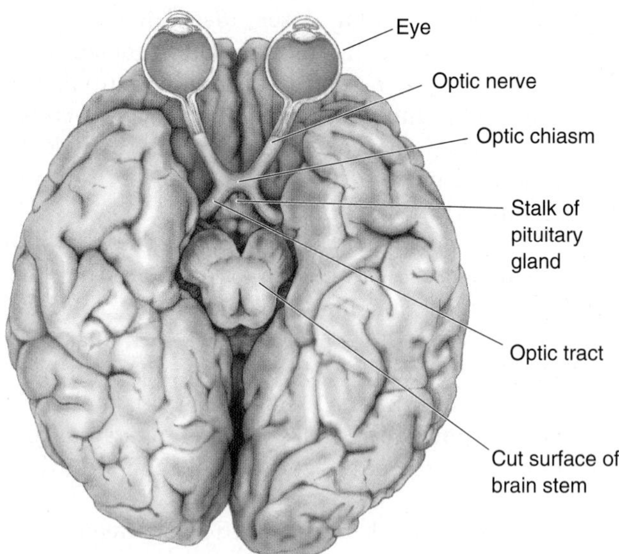

▲ FIGURE 10.2
The retinofugal projection. This view of the base of the brain shows the optic nerves, optic chiasm, and optic tracts.

originating in the nasal retinas cross, we say that a partial decussation of the retinofugal projection occurs at the optic chiasm. Following the partial decussation at the optic chiasm, the axons of the retinofugal projections form the **optic tracts**, which run just under the pia along the lateral surfaces of the diencephalon.

Right and Left Visual Hemifields

To understand the significance of the partial decussation of the retinofugal projection at the optic chiasm, let's review the concept of the visual field introduced in Chapter 9. The full visual field is the entire region of space (measured in degrees of visual angle) that can be seen with both eyes looking straight ahead. Fix your gaze on a point straight ahead. Now imagine a vertical line passing through the fixation point, dividing the visual field into left and right halves. By definition, objects appearing to the left of the midline are in the left **visual hemifield**, and objects appearing to the right of the midline are in the right visual hemifield (Figure 10.3).

By looking straight ahead with both eyes open and then alternately closing one eye and then the other, you will see that the central portion of both visual hemifields is viewed by *both* retinas. This region of space is therefore called the **binocular visual field**. Notice that objects in the binocular region of the left visual hemifield will be imaged on the nasal retina of the left eye and on the temporal retina of the right eye. Because the fibers from the nasal portion of the left retina cross to the right side at the optic chiasm, all the information about the left visual hemifield is directed to the right side of the brain. Remember this rule of thumb: Optic nerve fibers cross in the optic chiasm, such that *the left visual hemifield is "viewed" by the right hemisphere and the right visual hemifield is "viewed" by the left hemisphere.* You may recall from Chapter 7 that there is also a decussation in the descending pyramidal tract such that one side of the brain controls movement of the opposite side of the body. For reasons we do not understand, decussations are common in the sensory and motor systems.

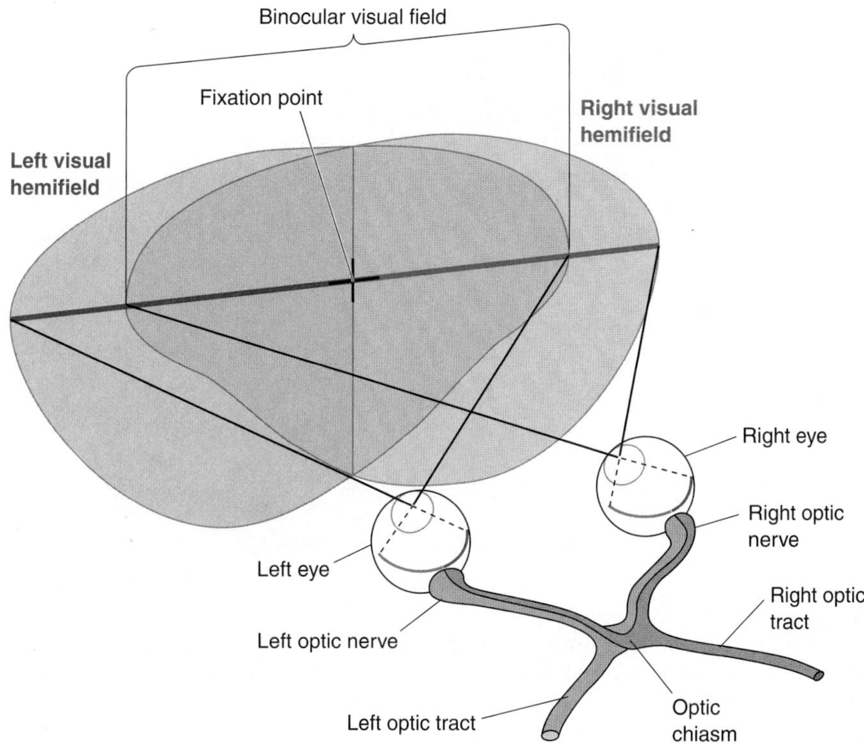

▲ FIGURE 10.3
Right and left visual hemifields. Ganglion cells in both retinas that are responsive to visual stimuli in the right visual hemifield project axons into the left optic tract. Similarly, ganglion cells "viewing" the left visual hemifield project into the right optic tract.

Targets of the Optic Tract

A small number of optic tract axons peel off to form synaptic connections with cells in the hypothalamus, and another 10% or so continue past the thalamus to innervate the midbrain. But most of them innervate the **lateral geniculate nucleus (LGN)** of the dorsal thalamus. The neurons in the LGN give rise to axons that project to the primary visual cortex. This projection from the LGN to the cortex is called the **optic radiation**. Lesions anywhere in the retinofugal projection from the eye to the LGN to the visual cortex in humans cause blindness in part or all of the visual field. Therefore, we know that it is this pathway that mediates conscious visual perception (Figure 10.4).

From our knowledge of how the visual world is represented in the retinofugal projection, we can predict the types of perceptual deficits that would result from its destruction at different levels, as might occur from a traumatic injury to the head, a tumor, or an interruption of the blood supply. As shown in Figure 10.5, while a transection of the left optic *nerve* would render a person blind in the left eye only, a transection of the left optic *tract* would lead to blindness in the right visual field as viewed through either eye. A midline transection of the optic chiasm would affect only the fibers that cross the midline. Because these fibers originate in the nasal portions of both retinas, blindness would result in the regions of the visual field viewed by the nasal retinas—that is, the peripheral visual fields on both sides (Box 10.1). Because unique deficits result from lesions at different sites, neuroophthalmologists and neurologists can locate sites of damage by assessing visual field deficits.

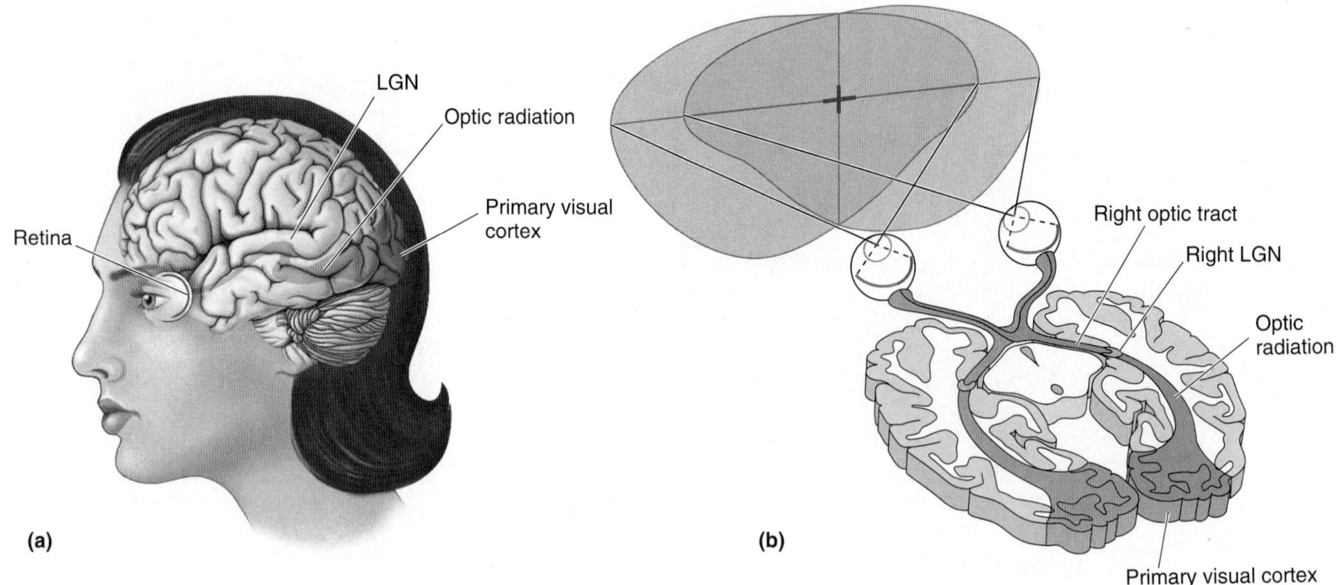

▲ FIGURE 10.4
The visual pathway that mediates conscious visual perception. **(a)** A side view of the brain with the retinogeniculocortical pathway shown inside (blue). **(b)** A horizontal section through the brain exposing the same pathway.

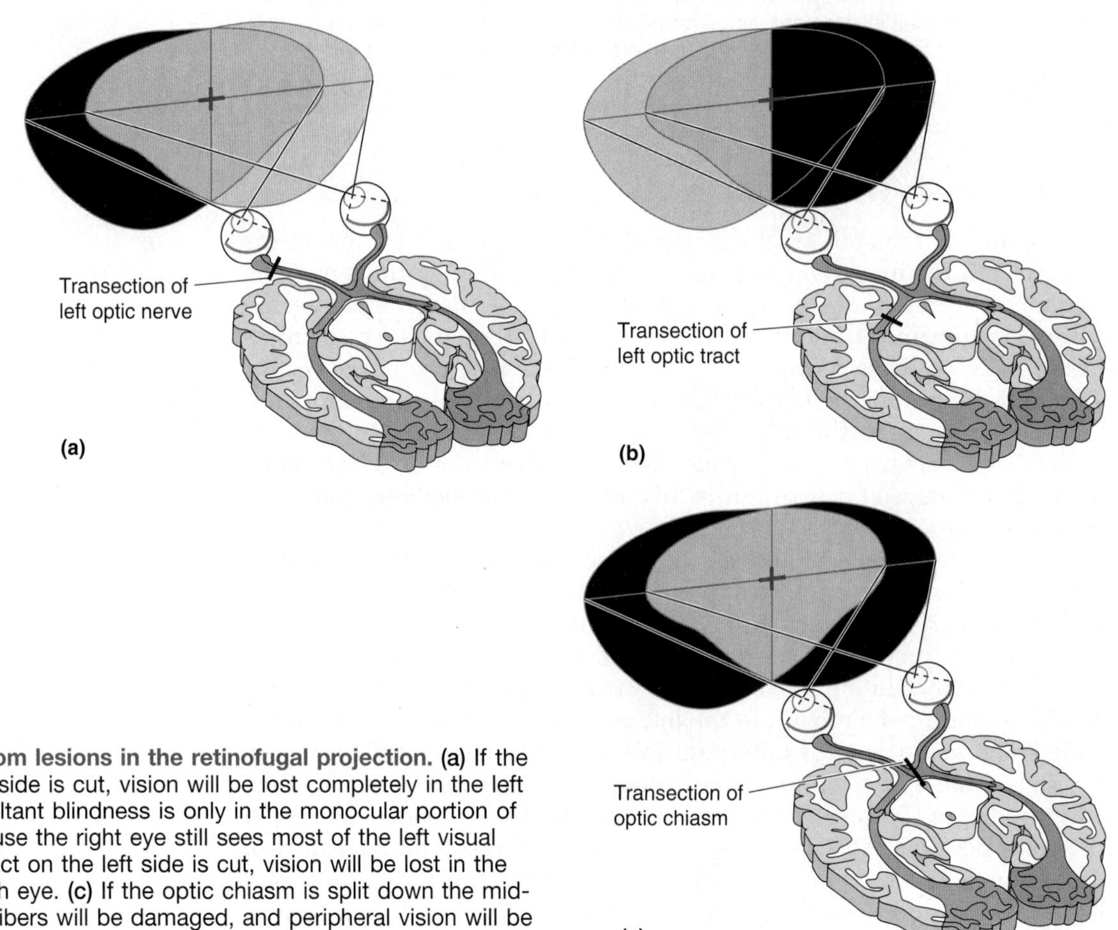

▶ FIGURE 10.5
Visual field deficits from lesions in the retinofugal projection. **(a)** If the optic nerve on the left side is cut, vision will be lost completely in the left eye. Note that the resultant blindness is only in the monocular portion of the left hemifield because the right eye still sees most of the left visual field. **(b)** If the optic tract on the left side is cut, vision will be lost in the right visual field of each eye. **(c)** If the optic chiasm is split down the middle, only the crossing fibers will be damaged, and peripheral vision will be lost in both eyes.

BOX 10.1 OF SPECIAL INTEREST

David and Goliath

Many of you are familiar with the famous story of David and Goliath, which appears in the Hebrew scriptures (Old Testament). The armies of the Philistines and the Israelites were gathered for battle when Goliath, a Philistine, came forth and challenged the Israelites to settle the dispute by sending out their best man to face him in a fight to the death. Goliath, it seems, was a man of great proportions, measuring more than "six cubits" in height. If you consider that a cubit is the distance from the elbow to the tip of the middle finger, about 20 inches, this guy was more than 10 feet tall! Goliath was armed to the teeth with body armor, a javelin, and a sword. To face this giant, the Israelites sent David, a young and diminutive shepherd, armed only with a sling and five smooth stones. Here's how the action is described in the Revised Standard Version of the *Bible* (1 Samuel 17:48):

> When the Philistine arose and came and drew near to meet David, David ran quickly toward the battle line to meet the Philistine. And David put his hand in his bag and took out a stone, and slung it, and struck the Philistine on his forehead; the stone sank into his forehead, and he fell on his face to the ground.

Now why, you might ask, are we giving a theology lesson in a neuroscience textbook? The answer is that our understanding of the visual pathway offers an explanation, in addition to divine intervention, for why Goliath was at a disadvantage in this battle. Body size is regulated by the secretion of growth hormone from the anterior lobe of the pituitary gland. In some cases, the anterior lobe becomes hypertrophied (swollen) and produces excessive amounts of the hormone, resulting in body growth to unusually large proportions. Such individuals are called pituitary giants and can be well over 8 feet tall.

Pituitary hypertrophy also disrupts normal vision. Recall that the optic nerve fibers from the nasal retinas cross in the optic chiasm, which butts up against the stalk of the pituitary. Any enlargement of the pituitary compresses these crossing fibers and results in a loss of peripheral vision called *bitemporal hemianopia*, or *tunnel vision*. (See if you can figure out why this is true from what you know about the visual pathway.) We can speculate that David was able to draw close and smite Goliath because the pituitary giant had completely lost sight of him.

Nonthalamic Targets of the Optic Tract. As we have said, some retinal ganglion cells send axons to innervate structures other than the LGN. Direct projections to part of the hypothalamus play an important role in synchronizing a variety of biological rhythms, including sleep and wakefulness, with the daily dark–light cycle (see Chapter 19). Direct projections to part of the midbrain, called the *pretectum*, control the size of the pupil and certain types of eye movement. And about 10% of the ganglion cells in the retina project to a part of the midbrain tectum called the **superior colliculus** (Latin for "little hill") (Figure 10.6).

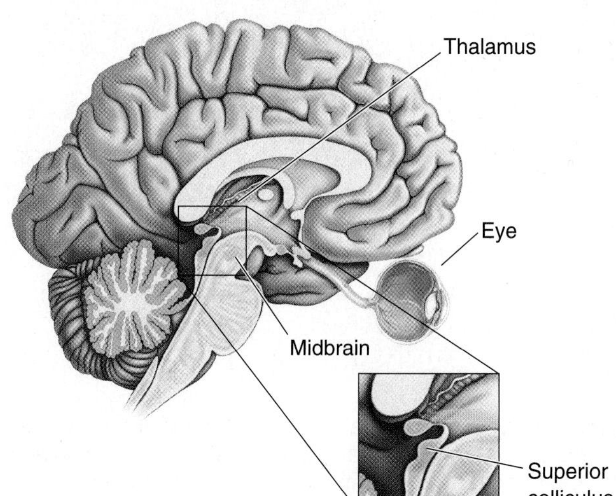

◀ FIGURE 10.6
The superior colliculus. Located in the tectum of the midbrain, the superior colliculus is involved in generating saccadic eye movements, the quick jumps in eye position used to scan across a page while reading.

While 10% may not sound like much of a projection, bear in mind that in primates, this is about 100,000 neurons, which is equivalent to the *total* number of retinal ganglion cells in a cat! In fact, the tectum of the midbrain is the major target of the retinofugal projection in all nonmammalian vertebrates (fish, amphibians, birds, and reptiles). In these vertebrate groups, the superior colliculus is called the **optic tectum**. This is why the projection from the retina to the superior colliculus is often called the **retinotectal projection**, even in mammals.

In the superior colliculus, a patch of neurons activated by a point of light, via indirect connections with motor neurons in the brain stem, commands eye and head movements to bring the image of this point in space onto the fovea. This branch of the retinofugal projection is thereby involved in orienting the eyes in response to new stimuli in the visual periphery. We saw in Chapter 9 that only the fovea has a dense concentration of cones sufficient for high-acuity vision. Therefore, it is critical that eye movements move our fovea to objects in our environment that might be threatening or of interest. We will return to the superior colliculus when we discuss motor systems in Chapter 14.

THE LATERAL GENICULATE NUCLEUS

The right and left lateral geniculate nuclei, located in the dorsal thalamus, are the major targets of the two optic tracts. Viewed in cross section, each LGN appears to be arranged in six distinct layers of cells (Figure 10.7).

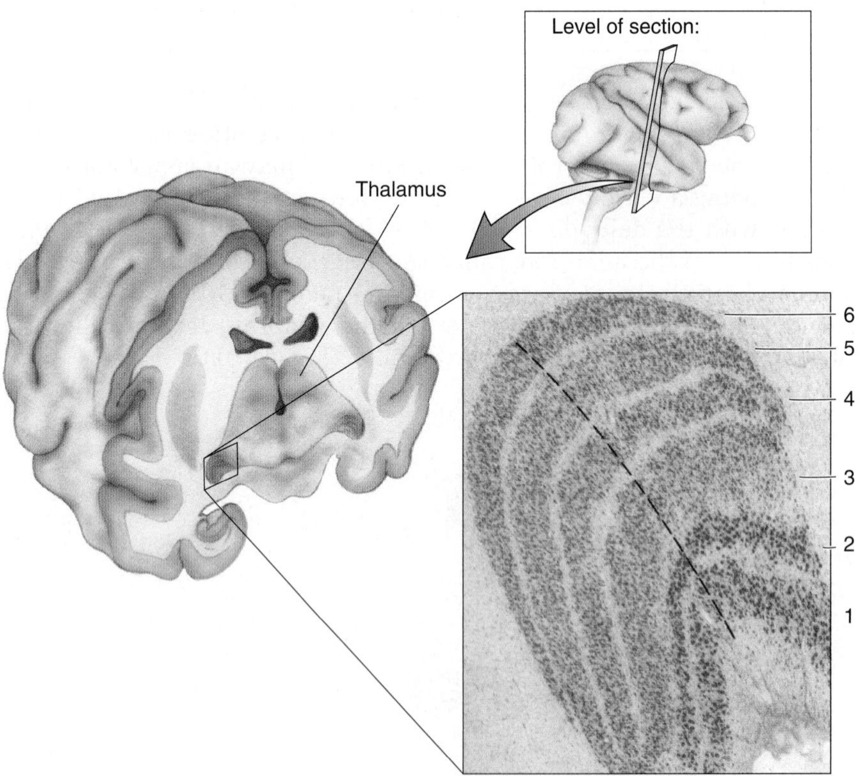

▲ FIGURE 10.7
The LGN of the macaque monkey. The tissue has been stained to show cell bodies, which appear as purple dots. Notice particularly the six principal layers and the larger size of the cells in the two ventral layers (layers 1 and 2). (Source: Adapted from Hubel, 1988, p. 65.)

By convention, the layers are numbered 1 through 6, starting with the most ventral layer, layer 1. In three dimensions, the layers of the LGN are arranged like a stack of six pancakes, one on top of the other. The pancakes do not lie flat, however; they are bent around the optic tract like a knee joint. This shape explains the name geniculate, from the Latin *geniculatus*, meaning "like a little knee."

The LGN is the gateway to the visual cortex and, therefore, to conscious visual perception. Let's explore the structure and function of this thalamic nucleus.

The Segregation of Input by Eye and by Ganglion Cell Type

LGN neurons receive synaptic input from the retinal ganglion cells, and most geniculate neurons project an axon to the primary visual cortex via the optic radiation. The segregation of LGN neurons into layers suggests that different types of retinal information are being kept separate at this synaptic relay, and indeed this is the case: Axons arising from M-type, P-type, and nonM–nonP ganglion cells in the two retinas synapse on cells in different LGN layers.

Recall from our rule of thumb that the *right* LGN receives information about the *left* visual field. The left visual field is viewed by both the nasal left retina and the temporal right retina. At the LGN, input from the two eyes is kept separate. In the right LGN, the right eye (ipsilateral) axons synapse on LGN cells in layers 2, 3, and 5. The left eye (contralateral) axons synapse on cells in layers 1, 4, and 6 (Figure 10.8).

A closer look at the LGN in Figure 10.7 reveals that the two ventral layers, 1 and 2, contain larger neurons, and the four more dorsal

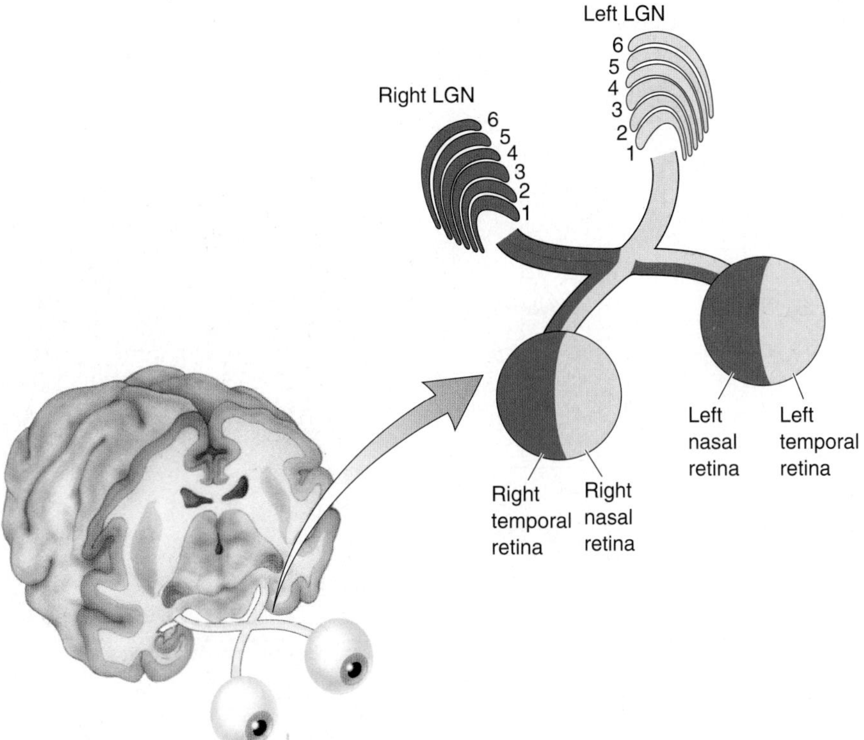

▲ FIGURE 10.8
Retinal inputs to the LGN layers. The retinal axons project such that the LGN is excited by light in the contralateral visual field presented to either eye.

layers, 3 through 6, contain smaller cells. The ventral layers are therefore called **magnocellular LGN layers**, and the dorsal layers are called **parvocellular LGN layers**. Recall from Chapter 9 that ganglion cells in the retina may also be classified into magnocellular and parvocellular groups. As it turns out, P-type ganglion cells in the retina project exclusively to the parvocellular LGN, and M-type ganglion cells in the retina project entirely to the magnocellular LGN.

In addition to the neurons in the six principal layers of the LGN, it was later discovered that there are numerous tiny neurons that lie just ventral to each layer. Cells in these **koniocellular LGN layers** (*konio* is from the Greek for "dust"), sometimes referred to as layers K1–K6, receive input from the nonM–nonP types of retinal ganglion cells and also project to the visual cortex. For the most part, each koniocellular layer gets input from the same eye as the overlying M or P layer. For example, layer K1 receives input from the contralateral eye just as layer 1 neurons do. In Chapter 9, we saw that in the retina, M-type, P-type, and nonM–nonP ganglion cells respond differently to light and color. In the LGN, the different information derived from the three categories of retinal ganglion cells from the two eyes remains largely segregated.

The anatomical organization of the LGN supports the idea that the retina gives rise to streams of information that are processed in parallel. This organization is summarized in Figure 10.9.

Receptive Fields

In Figure 9.25, we saw how the receptive field of a retinal ganglion cell can be mapped out by recording from the neuron while spots of light are shone on the retina. Similarly, by inserting a microelectrode into the LGN, it is possible to study the action potential discharges of a geniculate neuron in response to visual stimuli and map its receptive field. The surprising conclusion of such studies is that the visual receptive fields of

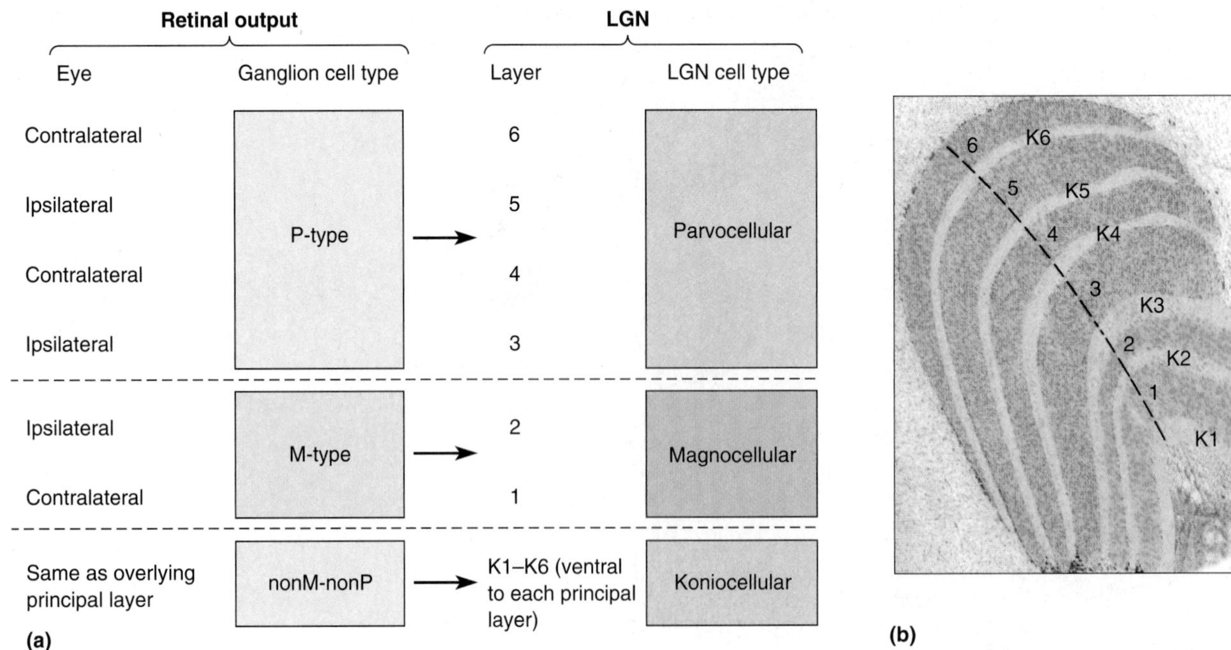

▲ FIGURE 10.9

The organization of the LGN. **(a)** Ganglion cell inputs to the different LGN layers. **(b)** A thin koniocellular layer (shown in pink) is ventral to each of the six principal layers.

LGN neurons are almost identical to those of the ganglion cells that feed them. For example, magnocellular LGN neurons have relatively large center-surround receptive fields, respond to stimulation of their receptive field centers with a transient burst of action potentials, and are insensitive to differences in wavelength. All in all, they are just like M-type ganglion cells. Likewise, parvocellular LGN cells, like P-type retinal ganglion cells, have relatively small center-surround receptive fields and respond to stimulation of their receptive field centers with a sustained increase in the frequency of action potentials; many of them exhibit color opponency. Receptive fields of cells in the koniocellular layers are center-surround and have either light/dark or color opponency. Within all layers of the LGN, the neurons are activated by only one eye (i.e., they are monocular) and ON-center and OFF-center cells are intermixed.

Nonretinal Inputs to the LGN

What makes the similarity of LGN and ganglion cell receptive fields so surprising is that the retina is not the main source of synaptic input to the LGN. In addition to the retina, the LGN receives inputs from other parts of the thalamus and the brain stem. The major input, constituting about 80% of the excitatory synapses, comes from primary visual cortex. Thus, one might reasonably expect that this corticofugal feedback pathway would significantly alter the qualities of the visual responses recorded in the LGN. So far, however, a role for this massive input has not been clearly identified. One hypothesis is that "top–down" modulation from the visual cortex to the LGN gates subsequent "bottom-up" input from the LGN back to the cortex. For example, if we want to selectively pay attention to a portion of our visual field, we might be able to suppress inputs coming from outside the attended area. We'll have more to say about this in our discussion of attention in Chapter 21.

The LGN also receives synaptic inputs from neurons in the brain stem whose activity is related to alertness and attentiveness (see Chapters 15 and 19). Have you ever "seen" a flash of light when you are startled in a dark room? This perceived flash might be a result of the direct activation of LGN neurons by this pathway. Usually, however, this input does not directly evoke action potentials in LGN neurons. But it can powerfully modulate the magnitude of LGN responses to visual stimuli. (Recall modulation from Chapters 5 and 6.) Thus, the LGN is more than a simple relay from the retina to the cortex; it is the first site in the ascending visual pathway where what we see is influenced by how we feel.

ANATOMY OF THE STRIATE CORTEX

The LGN has a single major synaptic target: the primary visual cortex. Recall from Chapter 7 that the cortex may be divided into a number of distinct areas based on their connections and cytoarchitecture. The **primary visual cortex** is Brodmann's **area 17** and is located in the occipital lobe of the primate brain. Much of area 17 lies on the medial surface of the hemisphere, surrounding the calcarine fissure (Figure 10.10). Other terms used interchangeably to describe the primary visual cortex are **V1** and the **striate cortex**. (The term *striate* refers to the fact that area V1 has an unusually dense stripe of myelinated axons running parallel to the surface that appears white in unstained sections.)

We have seen that the axons of different types of retinal ganglion cells synapse on anatomically segregated neurons in the LGN. In the following

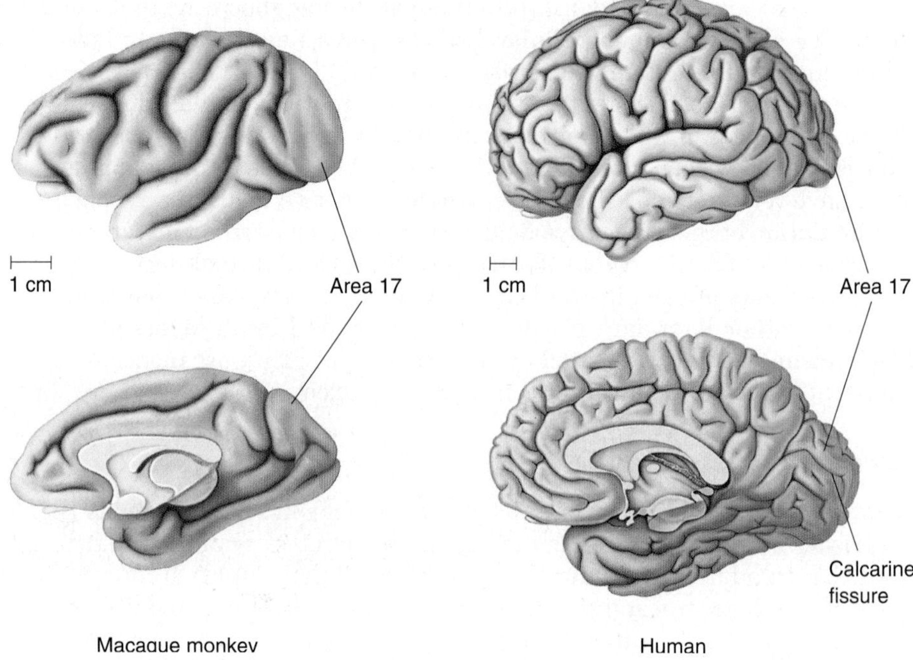

▶ FIGURE 10.10
The primary visual cortex. Top views are lateral; bottom views are medial.

sections, we look at the anatomy of the striate cortex and trace the connections different LGN cells make with cortical neurons. Later, we'll explore how this information is analyzed by cortical neurons. As we did in the LGN, in the striate cortex we'll see a close correlation between structure and function.

Retinotopy

The projection starting in the retina and extending to the LGN and V1 illustrates a general organizational feature of the central visual system called retinotopy. **Retinotopy** is an organization whereby neighboring cells in the retina feed information to neighboring places in their target structures—in this case, the LGN and striate cortex. In this way, the two-dimensional surface of the retina is *mapped* onto the two-dimensional surface of the subsequent structures (Figure 10.11a).

There are three important points to remember about retinotopy. First, the mapping of the visual field onto a retinotopically organized structure is often distorted because visual space is not sampled uniformly by the cells in the retina. Recall from Chapter 9 that there are many more ganglion cells with receptive fields in or near the fovea than in the periphery. Corresponding to this, the representation of the visual field is distorted in the striate cortex: The central few degrees of the visual field are overrepresented, or *magnified*, in the retinotopic map (Figure 10.11b). In other words, there are many more neurons in the striate cortex that receive input from the central retina than from the peripheral retina.

The second point to remember is that a discrete point of light can activate many cells in the retina, and often many more cells in the target structure, due to the overlap of receptive fields. The image of a point of light on the retina actually activates a large population of cortical neurons; every neuron that contains that point in its receptive field is potentially activated. Thus, when the retina is stimulated by a point of light, the activity in the striate cortex is a broad distribution with a peak at the corresponding retinotopic location.

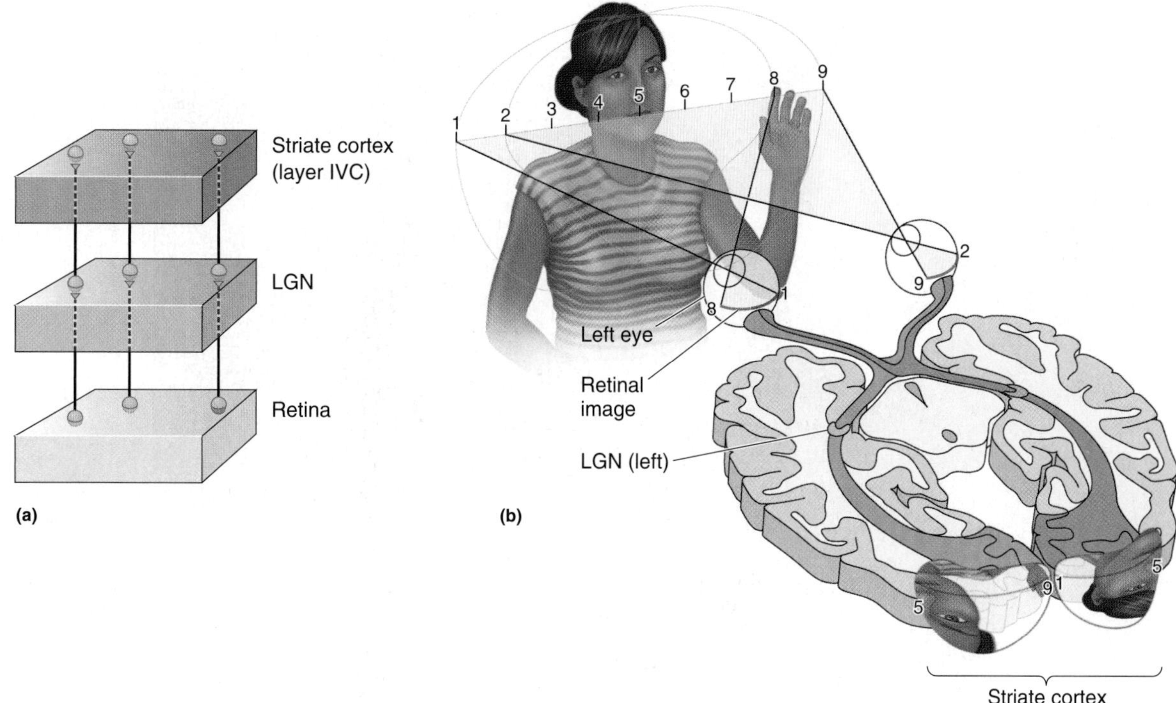

▲ FIGURE 10.11

The retinotopic map in the striate cortex. **(a)** Neighboring locations on the retina project to neighboring locations in the LGN. This retinotopic representation is preserved in the LGN projection to V1. **(b)** The lower portion of V1 represents the top half of visual space, and the upper portion of V1 represents the bottom half of visual space. Notice also that the map is distorted, with more tissue devoted to analysis of the central visual field. Similar maps are found in the superior colliculus, LGN, and other visual cortical areas.

Finally, don't be misled by the word "map." There are no pictures in the primary visual cortex for a little person in our brain to look at. While it's true that the arrangement of connections establishes a mapping between the retina and V1, perception is based on the brain's interpretation of distributed patterns of activity, not literal snapshots of the world. (We discuss visual perception later in this chapter.)

Lamination of the Striate Cortex

The neocortex in general, and the striate cortex in particular, have neuronal cell bodies arranged into about a half-dozen layers. These layers can be seen clearly in a Nissl stain of the cortex, which, as described in Chapter 2, leaves a deposit of dye (usually blue or violet) in the soma of each neuron. Starting at the white matter (containing the cortical input and output fibers), the cell layers are named by Roman numerals VI, V, IV, III, and II. Layer I, just under the pia mater, is largely devoid of neurons and consists almost entirely of axons and dendrites of cells in other layers (Figure 10.12). The full thickness of the striate cortex from white matter to pia is about 2 mm, the height of the lowercase letter m.

As Figure 10.12 shows, describing the lamination of the striate cortex as a six-layer scheme is somewhat misleading. There are actually at least nine distinct layers of neurons. To maintain Brodmann's convention that the neocortex has six layers, however, neuroanatomists combine three sublayers into layer IV, labeled IVA, IVB, and IVC. Layer IVC is further divided into two tiers called IVCα and IVCβ. The anatomical segregation

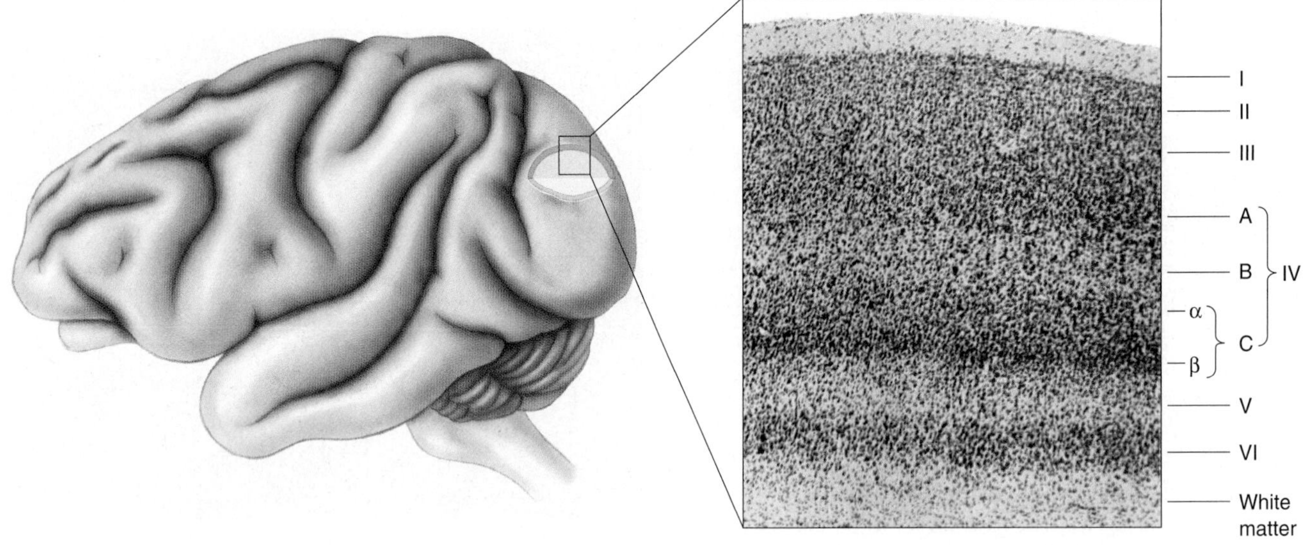

▲ FIGURE 10.12
The cytoarchitecture of the striate cortex. The tissue has been Nissl stained to show cell bodies, which appear as purple dots. (Source: Adapted from Hubel, 1988, p. 97.)

of neurons into layers suggests that there is a division of labor in the cortex, similar to what we saw in the LGN. We can learn a lot about how the cortex handles visual information by examining the structure and connections of its different layers.

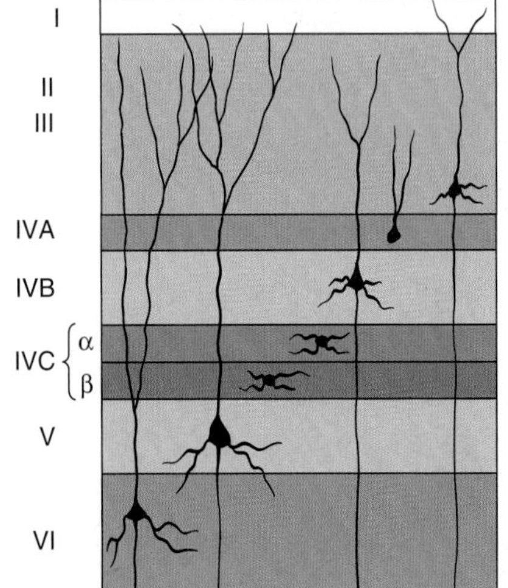

▲ FIGURE 10.13
The dendritic morphology of some cells in the striate cortex. Pyramidal cells are found in layers III, IVB, V, and VI, and spiny stellate cells are found in layer IVC.

The Cells of Different Layers. Many different neuronal shapes have been identified in striate cortex, but here we focus on two principal types, defined by the appearance of their dendritic trees (Figure 10.13). *Spiny stellate cells* are small neurons with spine-covered dendrites that radiate out from the cell body (recall dendritic spines from Chapter 2). They are seen primarily in the two tiers of layer IVC. Outside layer IVC are many *pyramidal cells*. These neurons are also covered with spines and are characterized by a single thick apical dendrite that branches as it ascends toward the pia mater and by multiple basal dendrites that extend horizontally. In Figure 10.13, the axon is the single neurite descending from the soma of each pyramidal cell.

Notice that a pyramidal cell in one layer may have dendrites extending into other layers. For the most part, *only pyramidal cells send axons out of the striate cortex* to form connections with other parts of the brain. The axons of stellate cells, which are indistinguishable from dendrites in Figure 10.13, generally make local connections only within the cortex. An exception to this rule is spiny stellate neurons in layer IVB that project to area V5, which we will discuss shortly.

In addition to the spiny neurons, inhibitory neurons, which lack spines, are sprinkled in all cortical layers as well. These neurons form only local connections.

Inputs and Outputs of the Striate Cortex

The distinct lamination of the striate cortex is reminiscent of the layers we saw in the LGN. In the LGN, every layer receives retinal afferents and sends efferents to the visual cortex. In the visual cortex, the situation is different; only a subset of the layers receives input from the LGN or sends output to a different cortical or subcortical area.

Axons from the LGN terminate in several different cortical layers, with the largest number going to layer IVC. We've seen that the output of the LGN is divided into streams of information, for example, from the magnocellular and parvocellular layers serving the right and left eyes. These streams remain anatomically segregated in layer IVC.

Magnocellular LGN neurons project primarily to layer IVCα, and parvocellular LGN neurons project to layer IVCβ. Imagine that the two tiers of layer IVC are pancakes, stacked one (α) on top of the other (β). Because the input from the LGN to the cortex is arranged topographically, we see that layer IVC contains two overlapping retinotopic maps, one from the magnocellular LGN (IVCα) and the other from the parvocellular LGN (IVCβ). Koniocellular LGN axons follow a different path, making synapses primarily in layers II and III.

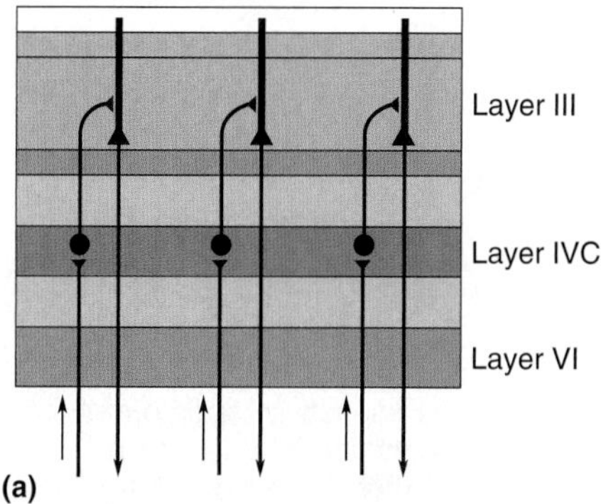

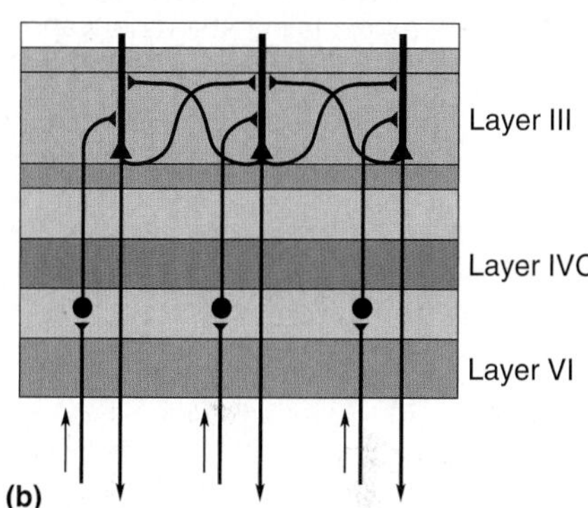

▲ FIGURE 10.14
Patterns of intracortical connections. **(a)** Radial connections. **(b)** Horizontal connections.

Innervation of Other Cortical Layers from Layer IVC. Most intracortical connections extend perpendicular to the cortical surface along radial lines that run across the layers, from white matter to layer I. This pattern of *radial connections* maintains the retinotopic organization established in layer IV. Therefore, a cell in layer VI, for example, receives information from the same part of the retina as does a cell above it in layer IV (Figure 10.14a). However, the axons of some layer III pyramidal cells extend collateral branches that make *horizontal connections* within layer III (Figure 10.14b). Radial and horizontal connections play different roles in the analysis of the visual world, as we'll see later in the chapter.

Leaving layer IV, there continues to be considerable anatomical segregation of the magnocellular and parvocellular processing streams. Layer IVCα, which receives magnocellular LGN input, projects mainly to cells in layer IVB. Layer IVCβ, which receives parvocellular LGN input, projects mainly to layer III. In layers III and IVB, an axon may form synapses with the dendrites of pyramidal cells of all layers.

Ocular Dominance Columns. How are the left eye and right eye LGN inputs arranged when they reach the striate cortex? Do they randomly intermix, or are they kept segregated? The answer was provided by a ground-breaking experiment performed in the early 1970s at Harvard Medical School by neuroscientists David Hubel and Torsten Wiesel. They injected a radioactive amino acid into one eye of a monkey (Figure 10.15). This amino acid was incorporated into proteins by the ganglion cells, and the proteins were transported down the ganglion cell axons into the LGN (recall anterograde transport from Chapter 2). Here, the radioactive proteins spilled out of the ganglion cell axon terminals and were taken up by nearby LGN neurons. But not all LGN cells took up the radioactive material; only those cells that were postsynaptic to the inputs from the injected eye incorporated the labeled protein. These cells then transported the radioactive proteins to their axon terminals in layer IVC of striate cortex. The location of the radioactive axon terminals was visualized by first placing a piece of film over thin sections of striate cortex and later developing the film like a photograph, a process called *autoradiography* (introduced in Chapter 6). The resulting collection of silver grains on the film marked the location of the radioactive LGN inputs.

In sections cut perpendicular to the cortical surface, Hubel and Wiesel observed that the distribution of axon terminals relaying information from the injected eye was not continuous in layer IVC, but rather was split up into a series of equally spaced patches, each about 0.5 mm wide (Figure 10.16a). In later experiments, the cortex was sectioned tangentially, parallel to layer IV. This revealed that the left eye and right eye

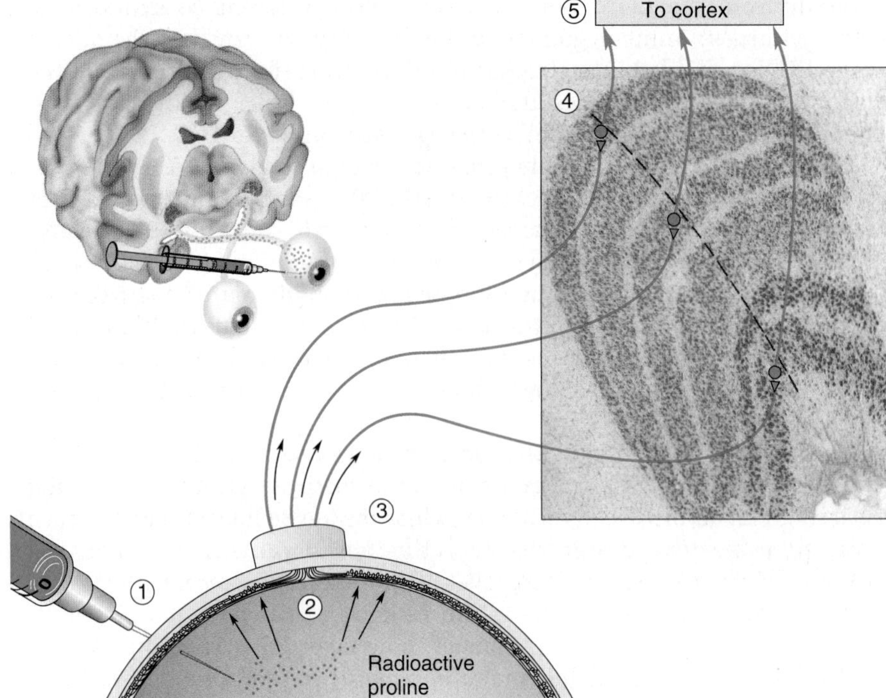

▶ FIGURE 10.15

Transneuronal autoradiography. Radioactive proline is ① injected into one eye, where it is ② taken up by retinal ganglion cells and incorporated into proteins that are ③ transported down the axons to the LGN. Some radioactivity spills out of the retinal terminals and is ④ taken up by LGN neurons that then ⑤ transport it to the striate cortex. The location of radioactivity can be determined using autoradiography.

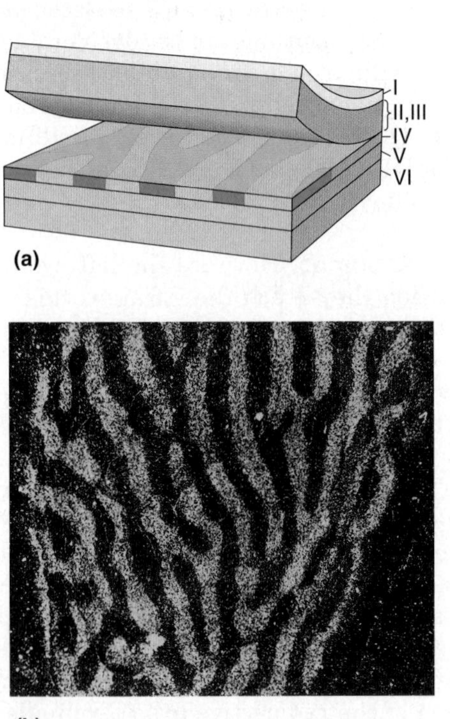

▲ FIGURE 10.16

Ocular dominance columns in layer IV of the striate cortex. **(a)** The organization of ocular dominance columns in layer IV of macaque monkey striate cortex. The distribution of LGN axons serving one eye is shaded blue. In cross section (looking at layer IV from the side), these eye-specific zones appear as patches, each about 0.5 mm wide, in layer IV. Peeled-back layers reveal that the ocular dominance columns in layer IV look like zebra stripes. **(b)** An autoradiograph of a histological section of layer IV viewed from above. Two weeks prior to the experiment, one eye of this monkey was injected with radioactive proline. In the autoradiograph, the radioactive LGN terminals appear bright on a dark background. (Source: LeVay et al., 1980.)

inputs to layer IV are laid out as a series of alternating bands, like the stripes of a zebra (Figure 10.16b). Rather than randomly mixing, neurons connected to the left and right eyes are as distinct in layer IV as they are in the LGN.

Layer IVC stellate cells project axons radially up mainly to layers IVB and III where, for the first time, information from the left eye and right eye begins to mix (Figure 10.17). Whereas all layer IVC neurons receive input from only one eye, most neurons in layers II, III, V, and VI receive some amount of input from each eye. For example, a neuron above a left eye patch of neurons in layer IVC receives input from both left eye and right eye neurons in layer IVC, but more of the projections come from the left eye. It is said that the input to the neuron is "dominated" by the left eye. In Figure 10.17, the red and blue patches of cells in layer III are dominated by the right and left eye, respectively; the purple patches of neurons receive roughly equal input from the two eyes. Because of the alternating patches of left eye and right eye input reaching layer IV and the overall radial projections, neurons outside layer IV are organized into alternating bands dominated by the left and right eye. The bands of cells extending through the thickness of the striate cortex are called **ocular dominance columns**.

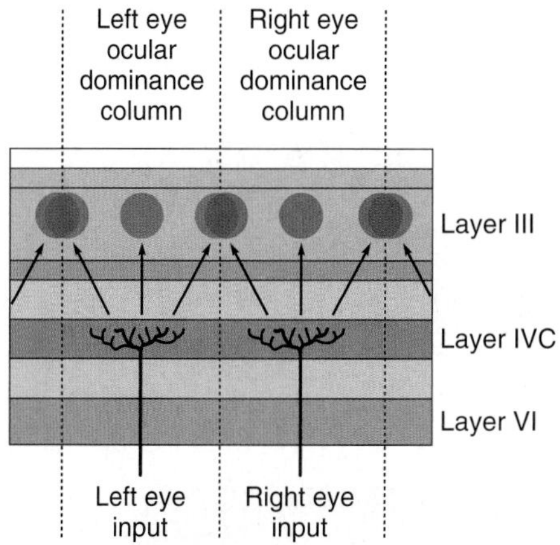

▲ FIGURE 10.17
The mixing of information from the two eyes. Axons project from layer IVC to more superficial layers. Most layer III neurons receive binocular input from both left and right eyes. There are layer III neurons with responses dominated by the right eye (red), left eye (blue), or roughly equally responsive to input from the two eyes (purple). Because of the radial connectivity in striate cortex, neurons in layers above and below layer IV are dominated by the same eye. Ocular dominance columns (between vertical dotted lines) contain neurons with input dominated by one eye, and the columns alternate between left and right eye dominance.

Striate Cortex Outputs. As previously mentioned, the pyramidal cells send axons out of the striate cortex into the white matter. The pyramidal cells in different layers innervate different structures. Layer II, III, and IVB pyramidal cells send their axons to other cortical areas. Layer V pyramidal cells send axons all the way down to the superior colliculus and pons. Layer VI pyramidal cells give rise to the massive axonal projection back to the LGN (Figure 10.18). Pyramidal cell axons in all layers also branch and form local connections in the cortex.

Cytochrome Oxidase Blobs

As we have seen, layers II and III play a key role in visual processing, providing most of the information that leaves V1 for other cortical areas. Anatomical studies suggest that the V1 output comes from two distinct populations of neurons in the superficial layers. When striate cortex tissue is stained to reveal the presence of **cytochrome oxidase**, a mitochondrial enzyme used for cell metabolism, the stain is not uniformly distributed in layers II and III. Rather, the cytochrome oxidase staining in cross sections of striate cortex appears as a colonnade, a series of pillars at regular intervals, running the full thickness of layers II and III and also in layers V and VI (Figure 10.19a). When the cortex is sliced tangentially through layer III, these pillars appear like the spots of a leopard (Figure 10.19b). These pillars of cytochrome oxidase–rich neurons have come to be called **blobs**. The blobs are in rows, each blob centered on an ocular dominance stripe in layer IV. Between the blobs are "interblob" regions. The blobs receive direct LGN input from the koniocellular layers, as well as parvocellular and magnocellular input from layer IVC of striate cortex.

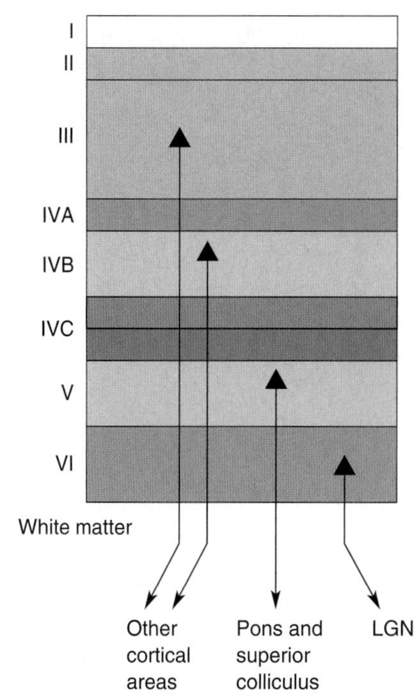

▲ FIGURE 10.18
Outputs from the striate cortex.

PHYSIOLOGY OF THE STRIATE CORTEX

Beginning in the early 1960s, Hubel and Wiesel were the first to systematically explore the physiology of the striate cortex with microelectrodes. They were students of Stephen Kuffler, who was then at Johns

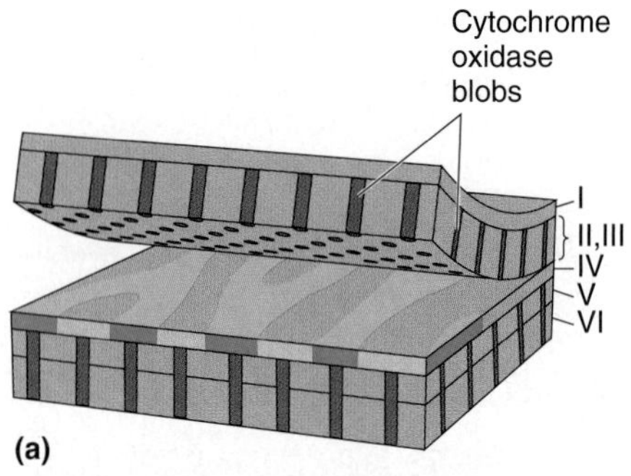

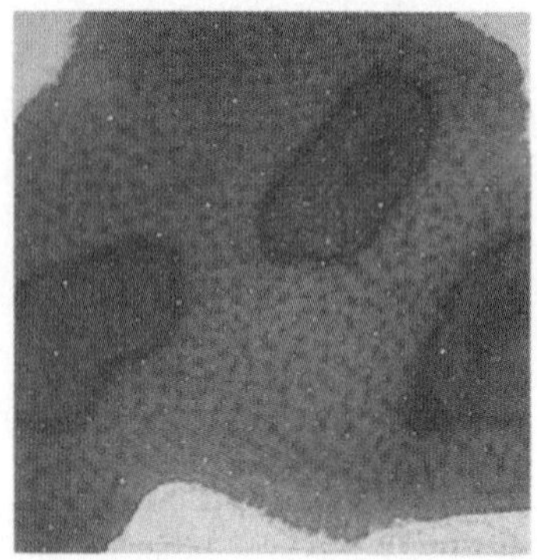

▲ FIGURE 10.19
Cytochrome oxidase blobs. **(a)** The organization of cytochrome oxidase blobs in macaque monkey striate cortex. The tissue that stains darkly for cytochrome oxidase looks like radial pillars in layers II, III, V, and VI. A cross section tangential to the surface shows the discrete patches that gave blobs their name (underside of layer III). **(b)** A photograph of a histological section of layer III, stained for cytochrome oxidase and viewed from above. The dark spots are cytochrome oxidase blobs. (Source: Courtesy of Dr. S.H.C. Hendry.)

Hopkins University and later moved with them to Harvard. They extended Kuffler's innovative methods of receptive field mapping to the central visual pathways. After showing that LGN neurons behave much like retinal ganglion cells, they turned their attention to the striate cortex, initially in cats and later in monkeys. (Here we focus on the monkey cortex.) The work that continues today on the physiology of the striate cortex is built on the solid foundation provided by Hubel and Wiesel's pioneering studies. Their contributions to our understanding of the cerebral cortex were recognized with the Nobel Prize in 1981.

Receptive Fields

By and large, the receptive fields of neurons in layer IVC are similar to the magnocellular and parvocellular LGN neurons providing their input. This means they are generally small monocular center-surround receptive fields. In layer IVCα the neurons are insensitive to the wavelength of light, whereas in layer IVCβ the neurons exhibit center-surround color opponency. Outside layer IVC (and somewhat within), new receptive field characteristics, not observed in the retina or LGN, are present. We will explore these in some depth because they provide clues about the role V1 plays in visual processing and perception.

Binocularity. There is a direct correspondence between the arrangement of connections in V1 and the responses of the neurons to light in the two eyes. Each neuron in layers IVCα and IVCβ receives afferents from a layer of the LGN representing the left or right eye. Physiological recordings confirm that these neurons are monocular, responding to light only in one of the eyes. We have already seen that the axons leaving layer IVC diverge and innervate more superficial cortical layers, mixing the inputs from the two eyes (see Figure 10.17). Microelectrode recordings confirm this anatomical fact; most neurons in layers superficial to IVC are binocular, responding to light in either eye. The ocular dominance columns demonstrated with autoradiography are reflected in the responses of V1 neurons. Above the centers of ocular dominance patches in layer IVC, the layer II and III neurons are more strongly driven by the eye represented in layer IVC (i.e., their response is dominated by one eye even though they are binocular). In areas where there is more equal mixing of left eye and right eye projections from layer IVC, the superficial layer neurons respond about the same to light in either eye.

We say that the neurons have **binocular receptive fields**, meaning that they actually have two receptive fields, one in the ipsilateral eye and one in the contralateral eye. Retinotopy is preserved because the two receptive fields of a binocular neuron are precisely placed on the retinas such that they are "looking" at the same point in in the contralateral visual field. The construction of binocular receptive fields is essential in binocular animals, such as humans. Without them, we would probably be unable to use the inputs from both eyes to form a single image of the world around us and perform fine motor tasks that require stereoscopic vision, such as threading a needle.

Orientation Selectivity. Most of the receptive fields in the retina, LGN, and layer IVC are circular and give their greatest response to a spot of light matched in size to the receptive field center. Outside layer IVC, we encounter cells that no longer follow this pattern. While small spots can elicit a response from many cortical neurons, it is usually possible to produce a much greater response with other stimuli. Rather by accident,

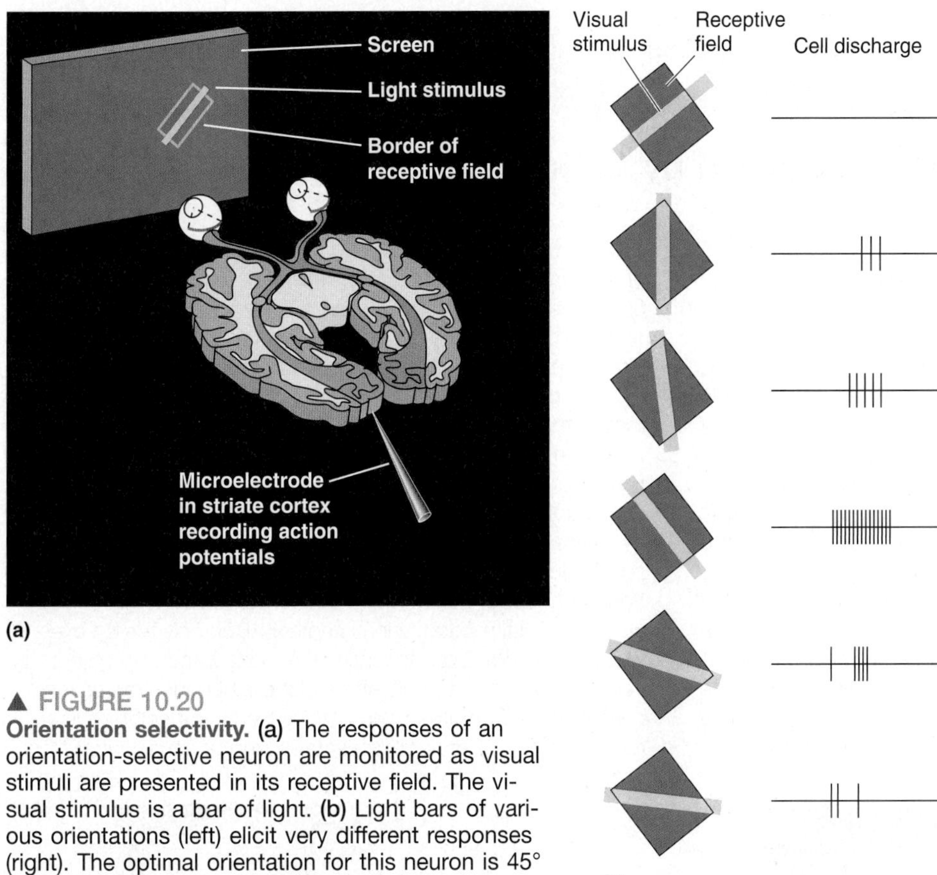

▲ FIGURE 10.20
Orientation selectivity. (a) The responses of an orientation-selective neuron are monitored as visual stimuli are presented in its receptive field. The visual stimulus is a bar of light. (b) Light bars of various orientations (left) elicit very different responses (right). The optimal orientation for this neuron is 45° counterclockwise from vertical.

Hubel and Wiesel found that many neurons in V1 respond best to an elongated bar of light moving across their receptive fields. But the orientation of the bar is critical. The greatest response is given to a bar with a particular orientation; bars perpendicular to the optimal orientation generally elicit much weaker responses (Figure 10.20). Neurons having this type of response are said to exhibit **orientation selectivity**. Most of the V1 neurons outside layer IVC (and some within) are orientation selective. The optimal orientation for a neuron can be any angle around the clock.

If V1 neurons can have any optimal orientation, you might wonder whether the orientation selectivity of nearby neurons is related. From the earliest work of Hubel and Wiesel, the answer to this question was an emphatic yes. As a microelectrode is advanced radially (perpendicular to the surface) from one layer to the next, the preferred orientation remains the same for all the selective neurons encountered from layer II down through layer VI. Hubel and Wiesel called such a radial column of cells an **orientation column**.

As an electrode passes tangentially (parallel to the surface) through the cortex in a single layer, the preferred orientation progressively shifts. We now know, from the use of a technique called *optical imaging*, that there is a mosaic-like pattern of optimal orientations in the striate cortex (Box 10.2). If an electrode is passed at certain angles through this mosaic, the preferred orientation rotates like the sweep of the minute hand of a clock, from the top of the hour to 10 past to 20 past, and so on (Figure 10.21). If the electrode is moved at other angles, more sudden shifts in preferred orientation occur. Hubel and Wiesel found that a complete 180° shift in preferred orientation required a traverse of about 1 mm, on average, within layer III.

BOX 10.2 BRAIN FOOD

Cortical Organization Revealed by Optical and Calcium Imaging

Most of what we know about the response properties of neurons in the visual system, and every other system in the brain, has been learned from intracellular and extracellular recordings with microelectrodes. These recordings give precise information about the activity of one or a few cells. However, unless one inserts thousands of electrodes, it is not possible to observe patterns of activity across large populations of neurons.

A view of neural coding at a scale much larger than individual neurons is provided by optical imaging of brain activity. In one version of optical recording, a voltage-sensitive dye is applied to the surface of the brain. The molecules in the dye bind to cell membranes, and an array of photodetectors or a video camera records changes in the optical properties that are proportional to variations in membrane potential. A second way to optically study cortical activity is to image intrinsic signals. When neurons are active, blood volume and oxygenation change to a degree correlated with neural activity. Blood flow and oxygenation influence the reflection of light from brain tissue, and reflectance changes can be used to indirectly assess neural activity. Light is projected onto the brain, and a video camera records the reflected light. Thus, when intrinsic signals are used to study brain activity, membrane potentials for action potentials are not directly measured.

Figure A is a photograph showing the vasculature in a portion of primary visual cortex. Figure B shows ocular dominance columns in the same patch of striate cortex obtained by optically imaging areas in which blood flow changes occurred during visual stimulation. This figure is actually a subtraction of two images—one made when only the right eye was visually stimulated, minus another when only the left eye was stimulated. Consequently, the dark bands represent cells dominated by the left eye, and the light bands represent cells dominated by the right eye. Figure C is a color-coded representation of preferred orientation in the same patch of

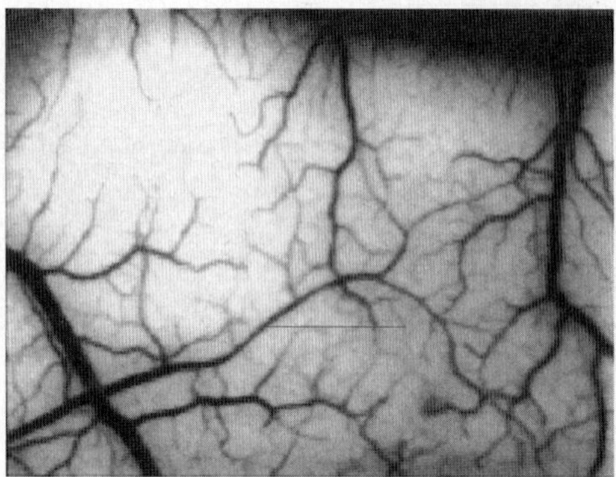

Figure A
Vasculature on the surface of primary visual cortex. (Source: Ts'o et al., 1990, Fig. 1A.)

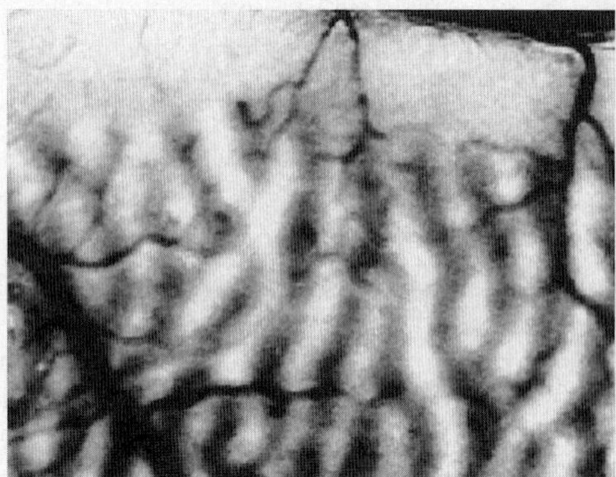

Figure B
Intrinsic signal imaging map of ocular dominance columns. (Source: Ts'o et al., 1990, Fig. 1B.)

The analysis of stimulus orientation appears to be one of the most important functions of the striate cortex. Orientation-selective neurons are thus thought to be specialized for the *analysis of object shape*.

Direction Selectivity. Many V1 receptive fields also exhibit **direction selectivity**; they respond when a bar of light at the optimal orientation moves perpendicular to the orientation in one direction but not in the opposite direction. Direction-selective cells in V1 are a subset of the cells that are orientation selective. Figure 10.22 shows how a direction-selective cell responds to a moving stimulus. Notice in this example that the cell responds to an elongated stimulus swept rightward across the receptive field, but much less with leftward movement. Sensitivity to the direction of stimulus motion is a hallmark of neurons receiving input

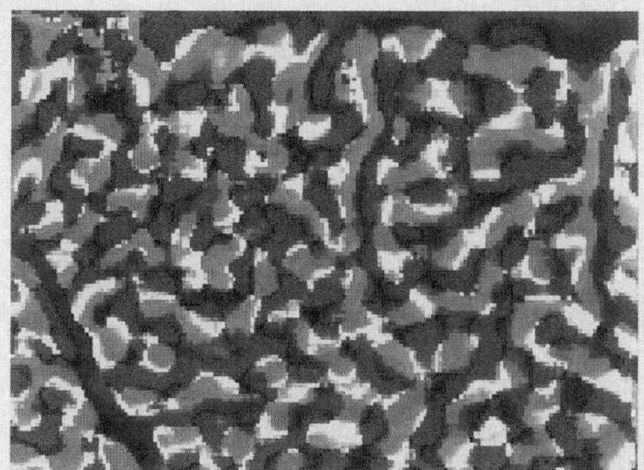

Figure C
Intrinsic signal imaging map of preferred orientations. (Source: Ts'o et al., 1990, Fig. 1C.)

striate cortex. Four different optical images were recorded while bars of light at four different orientations were swept across the visual field. Each location in the figure is colored according to the orientation that produced the greatest response at each location on the brain (blue = horizontal; red = 45°; yellow = vertical; green = 135°). Consistent with earlier results obtained with electrodes (see Figure 10.21), in some regions, the orientation changes progressively along a straight line. However, the optical recording technique reveals that cortical organization based on orientation is much more complex than an idealized pattern of parallel "columns."

Another technique, *in vivo* two-photon calcium imaging, lets us see the activity of thousands of neurons with single-cell resolution. When a neuron fires action potentials, voltage-sensitive calcium channels open, and calcium concentration in the soma increases. These concentration changes can be measured by introducing a calcium-sensitive fluorescent dye into the neurons; the amount of fluorescent light emitted from the neuron is correlated with the amount of calcium in the cell body and, thus, the firing rate of the neuron. To see neural activity at fine spatial and temporal scales, two-photon microscopy is used. The top of Figure D shows a map of orientation preference obtained with intrinsic signal optical imaging from cat visual cortex. The bottom of Figure D shows the orientation preferences of individual neurons based on two-photon calcium imaging. Orientation columns are seen in the clumping of cells with the same color; the results confirm that the optical images result from highly consistent cell-to-cell preferences for orientation. Cells with progressively different orientation preferences are organized in a "pinwheel" fashion, confirmation at the level of single cells of the optical results.

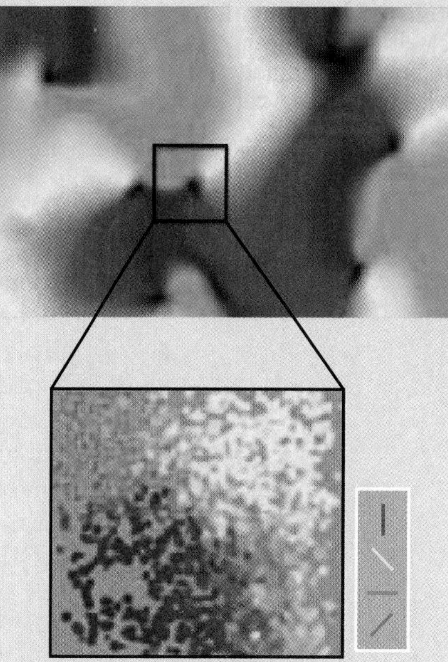

Figure D
A map of preferred orientation based on intrinsic signal optical imaging (top). Two-photo calcium imaging shows the orientation preference of individual neurons (bottom). (Source: Adapted from Ohki and Reid, 2006, Fig. 1.)

from the magnocellular layers of the LGN. Direction-selective neurons are thought to be specialized for the *analysis of object motion*.

Simple and Complex Receptive Fields. Neurons in the LGN have antagonistic center-surround receptive fields, and this organization accounts for the responses of these neurons to visual stimuli. For example, a small spot in the center of the receptive field may yield a much stronger response than a larger spot also covering the antagonistic surround. What do we know about the inputs to V1 neurons that might account for binocularity, orientation selectivity, and direction selectivity in their receptive fields? Binocularity is easy; we have seen that binocular neurons receive afferents from both eyes. The mechanisms underlying orientation and direction selectivity have proven more difficult to elucidate.

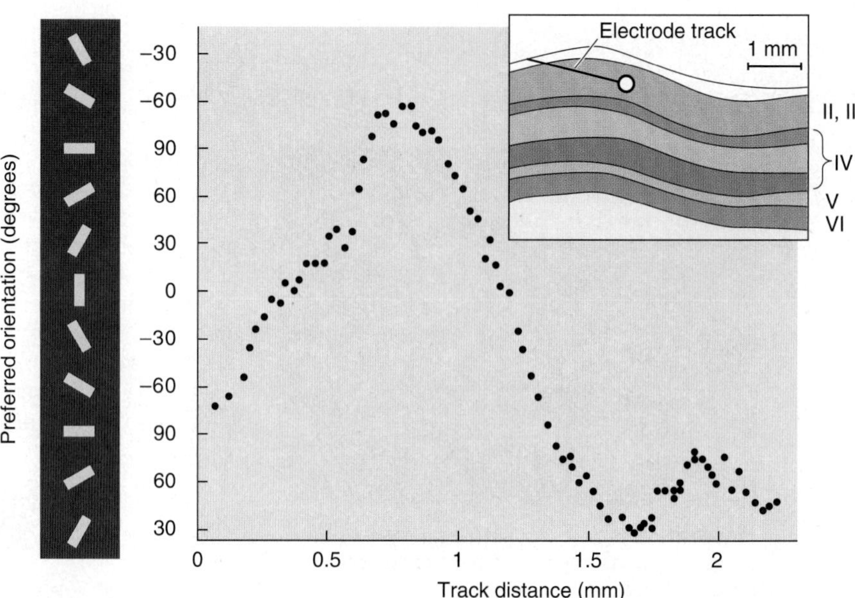

▶ FIGURE 10.21
Systematic variation of orientation preferences across the striate cortex. As an electrode is advanced tangentially across layers II and III of striate cortex, the orientation preference of the neurons encountered is recorded and plotted. In the recording shown, the preferred orientation of the neuron first encountered was near –70°, and as the electrode moved 0.7 mm, the preferred orientations rotated clockwise. As the electrode moved another millimeter, the preferred orientations rotated counterclockwise. (Source: Adapted from Hubel and Wiesel, 1968.)

Many orientation-selective neurons have a receptive field elongated along a particular axis, with an ON-center or OFF-center region flanked on one or both sides by an antagonistic surround (Figure 10.23a). This linear arrangement of ON and OFF areas is analogous to the concentric antagonistic areas seen in retinal and LGN receptive fields. One gets the impression that the cortical neurons receive a converging input from LGN cells with receptive fields that are aligned along one axis (Figure 10.23b). Hubel and Wiesel called neurons of this type **simple cells**. The segregation of ON and OFF regions is a defining property of simple cells, and it is because of this receptive field structure that they are orientation selective.

Other orientation-selective neurons in V1 do not have distinct ON and OFF regions and are therefore not considered simple cells. Hubel and Wiesel called most of these **complex cells**, because their receptive fields appeared to be more complex than those of simple cells. Complex cells give ON and OFF responses to stimuli throughout the receptive field (Figure 10.24). Hubel and Wiesel proposed that complex cells are constructed from the input of several like-oriented simple cells. However, this remains a matter of debate.

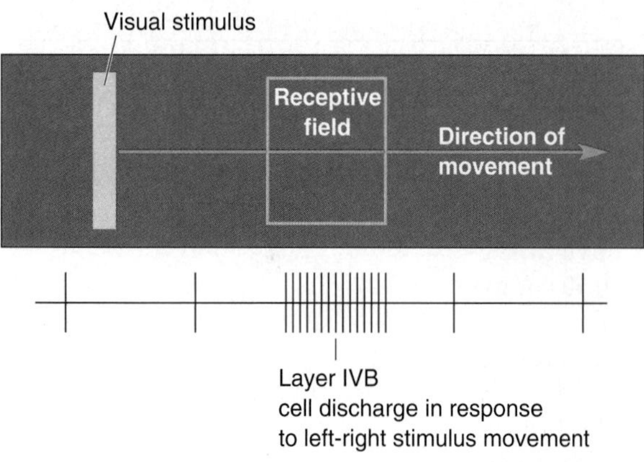

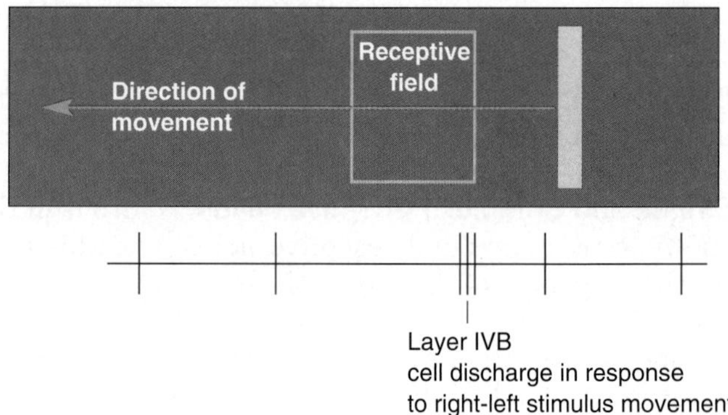

▲ FIGURE 10.22
Direction selectivity. With a bar stimulus at the optimal orientation, the neuron responds strongly when the bar is swept to the right but weakly when it is swept to the left.

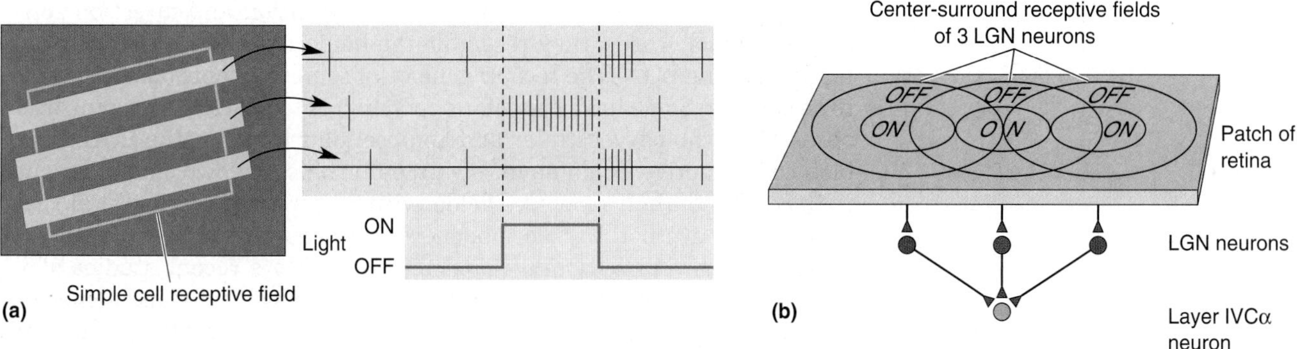

▲ FIGURE 10.23
A simple cell receptive field. **(a)** The response of a simple cell to optimally oriented bars of light at different locations in the receptive field. Notice that the response can be ON or OFF depending on where the bar lies in the receptive field. For this neuron, the middle location gives an ON response and the two flanking locations give OFF responses. **(b)** A simple cell receptive field might be built from convergent inputs of three LGN neurons with aligned center-surround receptive fields.

Simple and complex cells are typically binocular and sensitive to stimulus orientation. Different neurons show a range of sensitivities to color and direction of motion.

Blob Receptive Fields. We have seen repeatedly in the visual system that when two nearby structures label differently with some anatomical technique, there is good reason to suspect the neurons in the structures are functionally different. For example, we have seen how the distinctive layers of the LGN segregate different types of input. Similarly, the lamination of striate cortex correlates with differences in the receptive fields of the neurons. The presence of the distinct cytochrome oxidase blobs outside layer IV of striate cortex immediately raises the question of whether the neurons in the blobs respond differently from interblob neurons. The answer is controversial. The neurons in the interblob areas have some or all of the properties we discussed above: binocularity, orientation selectivity, and direction selectivity. They include both simple cells and complex cells; some are wavelength sensitive and some are not. The blobs receive input directly from the koniocellular layers of the LGN and magnocellular and parvocellular input via layer IVC. Early studies reported that blob cells, unlike interblob cells, are generally wavelength

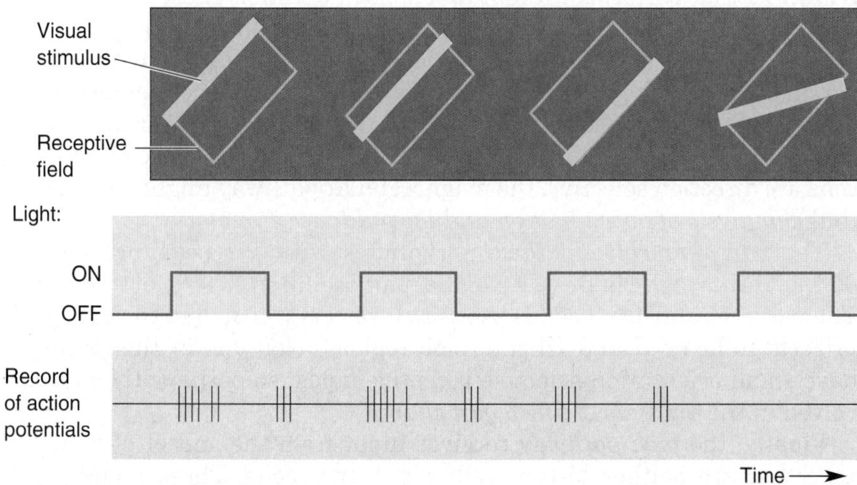

◀ FIGURE 10.24
A complex cell receptive field. Like a simple cell, a complex cell responds best to a bar of light at a particular orientation. However, responses occur to both light ON and light OFF, regardless of position in the receptive field.

sensitive and monocular, and they lack orientation and direction selectivity. In other words, they resemble the koniocellular and parvocellular input from the LGN. The receptive fields of some blob neurons were found to be circular. Some have the color-opponent center-surround organization observed in the parvocellular and koniocellular layers of the LGN. Other blob cell receptive fields have red–green or blue–yellow color opponency in the center of their receptive fields, with no surround regions at all. Still other cells have both a color-opponent center and a color-opponent surround; they are called *double-opponent cells*. More recent studies of V1 have quantified the selectivity of blob and interblob cells and somewhat surprisingly found that, overall, neurons in blobs and interblobs are similar, showing selectivity for both orientation and color.

What should we conclude about the physiological properties of the blob neurons? Despite the distinct cytochrome oxidase labeling, at present there is no simple way to distinguish the receptive field properties of blob cells from neighboring interblob cells. Corresponding to the greater cytochrome oxidase activity in blobs, the firing rates of blob cells are on average higher than those of interblob cells. We can only speculate that future research may discover some receptive field difference that better correlates with the anatomical and firing rate distinctions. It is generally believed that neurons sensitive to wavelength are important for the *analysis of object color*, but we don't know if we would be colorblind without functional cytochrome oxidase blobs.

Parallel Pathways and Cortical Modules

We have seen that neurons in area V1 are clearly not all the same. Anatomical stains show that neurons in different layers, and even within a layer, exhibit a variety of shapes and neurite configurations. Coming into V1 are distinct projections from magnocellular, parvocellular, and koniocellular layers of the LGN. Within V1, cells are selective for different orientations, directions of motion, and colors. Some cells are monocular, while others are binocular. A big question is the extent to which this smorgasbord of neurons is organized into functional pathways that perform unique functions or modules that work cooperatively.

Parallel Pathways. Because there is great interest in how the brain makes sense of our complex visual world, possible systems by which visual analysis might take place have received considerable research attention. An influential model is based on the idea that there are three pathways within V1 that perform different functions in parallel. These can be called the magnocellular pathway, the parvo-interblob pathway, and the blob pathway (Figure 10.25). The *magnocellular pathway* begins with M-type ganglion cells of the retina. These cells send axons to the magnocellular layers of the LGN. These layers project to layer IVCα of striate cortex, which in turn projects to layer IVB. Because many of these cortical neurons are direction selective, the magnocellular pathway might be involved in the *analysis of object motion and the guidance of motor actions*.

The *parvo-interblob pathway* originates with P-type ganglion cells of the retina, which project to the parvocellular layers of the LGN. The parvocellular LGN sends axons to layer IVC$_\beta$ of striate cortex, which projects to layer II and III interblob regions. Neurons in this pathway have small orientation-selective receptive fields, so perhaps they are involved in the *analysis of fine object shape*.

Finally, the *blob pathway* receives input from the subset of ganglion cells that are neither M-type cells nor P-type cells. These nonM–nonP

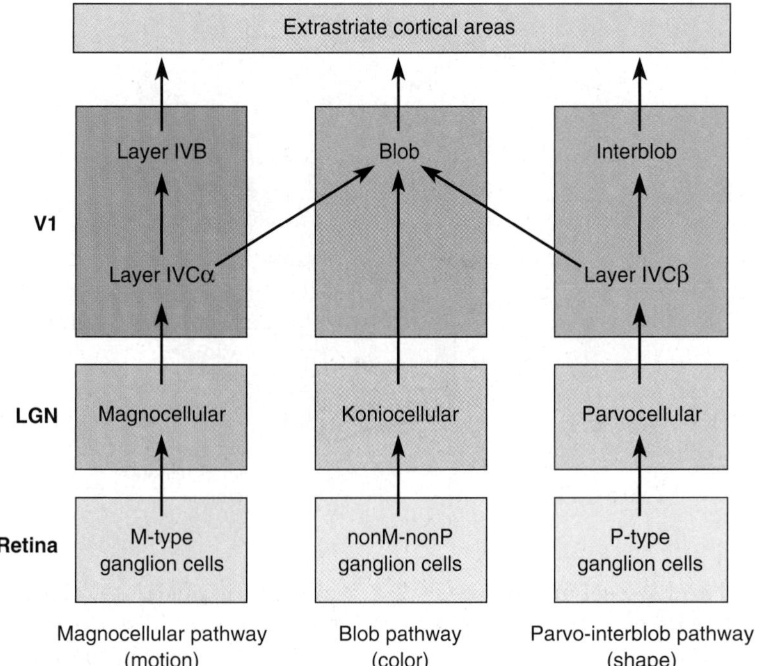

◀ FIGURE 10.25
A hypothetical model of parallel pathways in primary visual cortex. Based on receptive field properties and the pattern of innervation from LGN afferents, it has been suggested that there are three distinct pathways in striate cortex. Below each pathway, a functional role is suggested. Further research demonstrated mixing of magno, parvo, and koniocellular LGN signals and overlap in receptive field properties, raising questions about the distinctions, particularly between the blob and interblob pathways.

cells project to the koniocellular layers of the LGN. The koniocellular LGN projects directly to the cytochrome oxidase blobs in layers II and III. Many neurons in the blobs are color selective, so they might be involved in the *analysis of object color*.

The description earlier is the simple version of the story; the reality is more complicated. Research has shown that the three proposed pathways do not keep magnocellular, parvocellular, and koniocellular signals separate; instead they mix. Also, receptive field properties such as orientation and color tuning are found across the proposed pathways. Thus, it is not the case that magnocellular, parvo-interblob, and blob neurons are strictly segregated and have entirely unique receptive field properties. At present, it appears that striate cortex input reflects the magnocellular, parvocellular, koniocellular segregation also seen in the LGN, but striate cortex output has a different form of parallel processing. For example, layer IVB contains many direction-selective neurons, its output appears to be dominated by magnocellular LGN input, and it projects to cortical areas thought to be involved in motion perception. Collectively, these observations are consistent with the notion that this is an output pathway particularly involved in navigation and the analysis of motion. The case for a distinction between form and color pathways is less compelling. We will see later that beyond the striate cortex, there appear to be two major pathways handling different types of visual information, one extending toward the parietal lobe that deals with motion and the other involved with color and form extending toward the temporal lobe.

Cortical Modules. Receptive fields in primary visual cortex range from a fraction of a degree to several degrees across, and nearby cells have receptive fields that overlap a great deal. For these reasons, even a small spot of light will activate thousands of V1 neurons. Hubel and Wiesel showed that the image of a point in the visual field falls within the receptive fields of neurons in a 2 × 2 mm chunk of macaque striate cortex. Such a block of cortex also contains two complete sets of ocular dominance columns, 16 blobs, and a complete sampling (twice over) of all 180° of possible orientations.

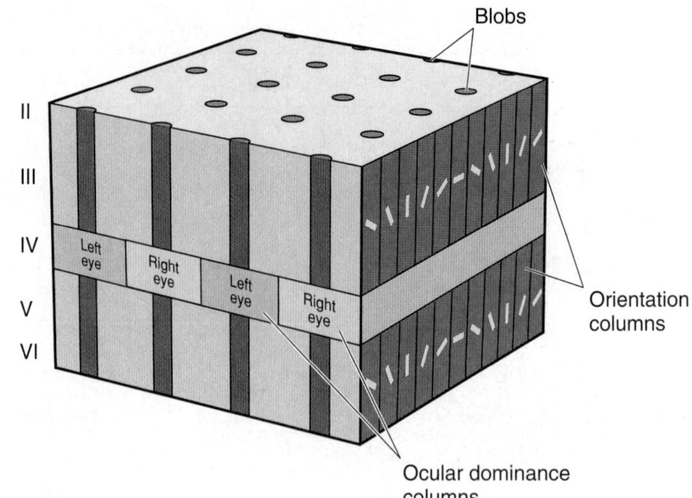

▶ FIGURE 10.26
A cortical module. Each cortical module contains ocular dominance columns, orientation columns, and cytochrome oxidase blobs to fully analyze a portion of the visual field. The idealized cube shown here differs from the actual arrangement, which is not as regular or orderly.

Thus, Hubel and Wiesel argued that a 2 × 2 mm chunk of striate cortex is both necessary and sufficient to analyze the image of a point in space: *necessary* because its removal would leave a blind spot for this point in the visual field, and *sufficient* because it contains all the neural machinery required to analyze the form and color of objects viewed through either eye. Such a unit of brain tissue has come to be called a **cortical module**. Because of the finite size of receptive fields and some scatter in their locations, a cortical module processes information about a small patch of the visual field.

The striate cortex is constructed from perhaps a thousand cortical modules; one is shown in Figure 10.26. We can think of a visual scene being simultaneously processed by these modules, each "looking" at a portion of the scene. Just remember that the modules are an idealization. Optical images of V1 activity reveal that the regions of the striate cortex responding to different eyes and orientations are not nearly as regular as the "ice cube model" in Figure 10.26 suggests.

BEYOND THE STRIATE CORTEX

The striate cortex is called V1, for "visual area one," because it is the first cortical area to receive information from the LGN. Beyond V1 lie another two dozen distinct *extrastriate* areas of cortex that have unique receptive field properties. The contributions to vision of these extrastriate areas are still being vigorously debated. However, it appears that there are two large-scale cortical streams of visual processing, one stretching dorsally from the striate cortex toward the parietal lobe and the other projecting ventrally toward the temporal lobe (Figure 10.27).

The *dorsal stream* appears to serve the analysis of visual motion and the visual control of action. The *ventral stream* is thought to be involved in the perception of the visual world and the recognition of objects. These processing streams have primarily been studied in the macaque monkey brain, where recordings from single neurons can be made. However, functional magnetic resonance imaging (fMRI) research has identified areas in the human brain that have properties analogous to brain areas in the macaque. The locations of some of the human visual areas are shown in Figure 10.28.

The properties of dorsal stream neurons are most similar to those of magnocellular neurons in V1, and ventral stream neurons have properties

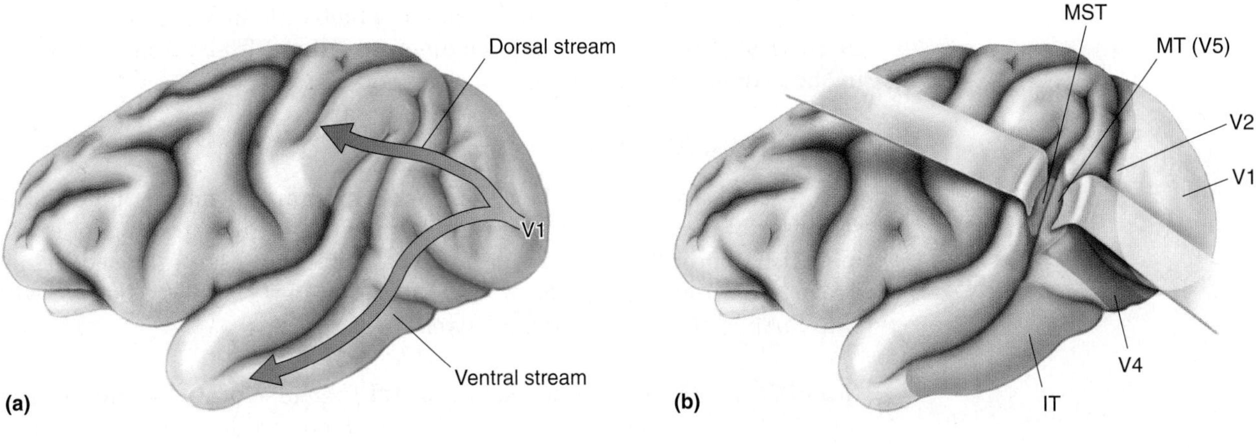

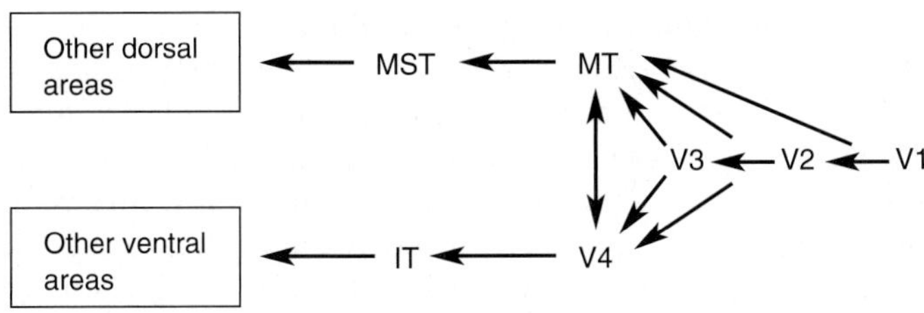

▲ FIGURE 10.27

Beyond the striate cortex in the macaque monkey brain. **(a)** Dorsal and ventral visual processing streams. **(b)** Extrastriate visual areas. **(c)** The flow of information in the dorsal and ventral streams.

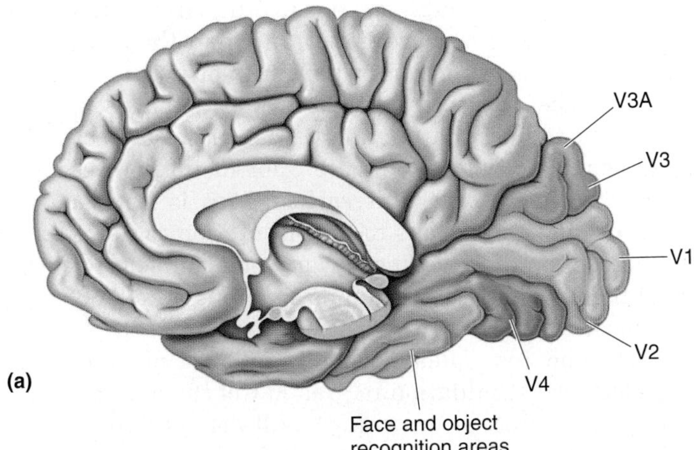

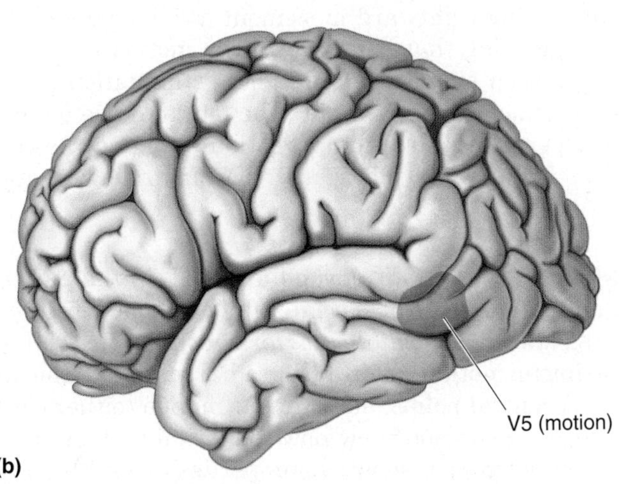

◀ FIGURE 10.28

Visual areas in the human brain. **(a)** Compared to monkeys, human visual areas are shifted more to the medial wall of the occipital lobe, and many are buried in sulci. Early visual areas including V1, V2, V3, V3A, and V4 are retinotopically organized. Higher temporal lobe areas involved in face and object recognition are not retinotopic. **(b)** A variety of areas responsive to visual motion are found on the lateral aspect of the brain. The most studied of these is area V5, also known as MT. (Source: Zeki, 2003, Fig. 2.)

more like features of parvo-interblob and blob cells in V1. However, each extrastriate stream receives some amount of input from all the pathways in the primary visual cortex.

The Dorsal Stream

The cortical areas composing the dorsal stream are not arranged in a strict serial hierarchy, but there does appear to be a progression of areas in which more complex or specialized visual representations develop. Projections from V1 extend to areas designated V2 and V3, but we will skip farther ahead in the dorsal stream.

Area MT. In an area known as V5 or MT (because of its location in the middle temporal lobe in some monkeys), strong evidence indicates that specialized processing of object motion takes place. The location of area MT in the human brain is shown in Figure 10.28b. **Area MT** receives retinotopically organized input from a number of other cortical areas, such as V2 and V3, and it also is directly innervated by cells in layer IVB of the striate cortex. Recall that in layer IVB, the cells have relatively large receptive fields, transient responses to light, and direction selectivity. Neurons in area MT have large receptive fields that respond to stimulus movement in a narrow range of directions. Area MT is most notable for the fact that almost all the cells are direction selective, unlike areas earlier in the dorsal stream or anywhere in the ventral stream.

The neurons in MT also respond to types of motion, such as drifting spots of light, that are not good stimuli for cells in other areas—it appears that the motion of the objects is more important than their structure. Perhaps you have seen illusory motion in paintings or optical illusions; MT has been shown to be activated by some of these images too, suggesting that its neurons tell us what motion we perceive, not necessarily what motion is present. Further specialization for motion processing is evident in the organization of MT. This cortical area is arranged into direction-of-motion columns analogous to the orientation columns in V1. Presumably, the perception of movement at any point in space depends on a comparison of the activity across columns spanning a full 360° range of preferred directions.

William Newsome and his colleagues at Stanford University have shown that weak electrical stimulation in area MT of the macaque monkey appears to alter the direction in which small dots of light are perceived to move. For example, if electrical stimulation is applied to cells in a direction column preferring rightward movement, the monkey makes behavioral decisions suggesting that it has perceived motion in that direction. The artificial motion signal from electrical stimulation in MT appears to combine with visual motion input. The fact that the monkey behaviorally reports a perceived direction of motion based on the combination suggests that MT activity plays an important role in motion perception.

Dorsal Areas and Motion Processing. Beyond area MT, in the parietal lobe, are areas with additional types of specialized movement sensitivity. For example, in an area known as *medial superior temporal (MST)*, there are cells selective for linear motion (as in MT), radial motion (either inward or outward from a central point), and circular motion (either clockwise or counterclockwise). We do not know how the visual system makes use of neurons with complex motion-sensitive properties in MST or of the "simpler" direction-selective cells in V1, MT, and other areas. However, three roles have been proposed:

1. *Navigation:* As we move through our environment, objects stream past our eyes, and the direction and speed of objects in our peripheral vision provide valuable information that can be used for navigation.
2. *Directing eye movements:* Our ability to sense and analyze motion must also be used when we follow objects with our eyes and when we quickly move our eyes to objects in our peripheral vision that catch our attention.
3. *Motion perception:* We live in a world filled with motion, and survival sometimes depends on our interpretation of moving objects.

Striking evidence that cortical areas in the vicinity of MT and MST are critical for motion perception in humans comes from extremely rare cases in which brain lesions selectively disrupt the perception of motion. The clearest case was reported in 1983 by Josef Zihl and his colleagues at the Max Planck Institute for Psychiatry in Munich, Germany. Zihl studied a woman who experienced a stroke at the age of 43 years, bilaterally damaging portions of extrastriate visual cortex known to be particularly responsive to motion (Figure 10.28b). Although some ill effects of the stroke were evident, such as difficulty naming objects, neuropsychological testing showed the patient to be generally normal and to have relatively normal vision, except for one serious deficit: She appeared to be incapable of visually perceiving motion. Before you decide that not seeing motion would be a minor impairment, imagine what it would be like to see the world in snapshots. Zihl's patient complained that when she poured coffee into a cup, it appeared at one moment to be frozen at the bottom of the cup and then suddenly it had overfilled the cup and covered the table. More ominously, she had trouble crossing the street—one moment she would perceive cars to be in the distance, and the next moment they would be right next to her. Clearly, this loss of motion perception had profound ramifications for the woman's lifestyle. The implication of this case is that motion perception may be based on specialized mechanisms located beyond the striate cortex in the dorsal stream.

The Ventral Stream

In parallel with the dorsal stream, a progression of areas from V1, V2, and V3 running ventrally toward the temporal lobes appears specialized for the analysis of visual attributes other than motion.

Area V4. One of the most-studied areas in the ventral stream is **area V4** (see Figures 10.27b and 10.28a for the location of V4 in monkey and human brains). V4 receives input from the blob and interblob regions of the striate cortex via a relay in V2. Neurons in area V4 have larger receptive fields than cells in the striate cortex, and many of the cells are both orientation selective and color selective. Although there is a good deal of ongoing research into the function of V4, this area appears to be important for both shape perception and color perception. If this area is damaged in monkeys, perceptual deficits involving both shape and color result.

A rare clinical syndrome in humans known as *achromatopsia* is characterized by a partial or complete loss of color vision despite the presence of normal functional cones in the retina. People with this condition describe their world as drab, consisting of only shades of gray. Imagine how unappetizing a gray banana would look! Because achromatopsia is associated with cortical damage in the occipital and temporal lobes, without damage to V1, the LGN, or the retina, the syndrome suggests that there is specialized color processing in the ventral stream. Consistent

BOX 10.3 PATH OF DISCOVERY

Finding Faces in the Brain

by Nancy Kanwisher

During my first year in graduate school in 1981, the first functional images of human visual cortex appeared on the cover of *Science*. Captivated by positron emission tomography (PET), a remarkable technology that enabled us for the first time to peer directly into the workings of the normal human brain, I wrote a research proposal using this method to investigate human vision and sent it to all of the PET labs in the world (I think there were five at the time). But it took a decade of pounding on the doors before I gained access to a PET scanner and got to do my first experiment.

Functional MRI (fMRI) was just starting to catch on then, and a few years later, in 1995, I was given the breathtaking privilege of my own weekly slot on the fMRI scanner at Massachusetts General Hospital. Collaborating with an undergraduate, Josh McDermott, and post-doc, Marvin Chun, I spent some of the happiest moments of my life lying inside the scanner bore, biting on a bite bar, and watching Marvin and Josh (upside down) through the mirror over my forehead, as they operated the scanner from out in the console room. What astonishing good fortune to get to use this amazing machine to explore the largely uncharted territory of human visual cortex!

We started out by trying to find brain regions engaged in the perception of object shape. Although we found some intriguing effects, they were weak. As I did not have a grant to pay for scan time, I knew that my scanning privileges would not continue unless I hit a home run—and quickly.

The extensive behavioral literature on normal and brain-damaged individuals strongly suggested that a special part of the brain might exist for face perception. We decided to look for it. Leslie Ungerleider and Jim Haxby and their colleagues at the National Institutes of Health (NIH) had already shown strong activations on the bottom of the temporal lobes when people look at faces. What they had not investigated was whether this response was *specific* to faces, or whether the same region might also be engaged during perception of other complex visual stimuli. This specificity question connected directly to one of the most long-standing and fierce debates in the histories of both cognitive science and neuroscience: To what extent are the mind and brain composed of special-purpose mechanisms, each processing a specific kind of information?

We figured that if a special-purpose part of the brain existed that was selectively involved in face perception, it should produce a stronger response when people look at faces than when they look at objects. To get enough face images, Marvin, Josh, and I went to Harvard's "freshman face capture," where the entire incoming class lined up for ID photos; we asked if we could use their ID photos in our experiments. We then scanned subjects while they looked at these face photos and at photos of common objects.

To our delight, we found that in almost everyone the image showed a nice clear blob on the lateral side of the fusiform gyrus, primarily in the right hemisphere, where the statistics told us that the response was higher when people were looking at faces than when they were looking at objects. However, it was not in exactly the same place in each subject. To deal with this anatomical variability, plus to make our statistical analyses bulletproof, we split the data for each subject in half, using half the data to find the region with our faces-versus-objects contrast, and the other half to quantify the response

with the coexistence of color-sensitive and shape-sensitive cells in the ventral stream, achromatopsia is usually accompanied by deficits in form perception. Some researchers have proposed that V4 is a particularly critical area for color and form perception, but the lesions associated with achromatopsia are generally not limited to V4, and severe visual deficits appear to require damage to other cortical areas in addition to V4.

Area IT. Beyond V4 in the ventral stream are cortical areas that contain neurons with complicated spatial receptive fields. A major output of V4 is an area in the inferior temporal lobe known as **area IT** (see Figure 10.27b and the recognition areas in Figure 10.28a). One reason this area is of particular interest is that it appears to be the farthest extent of visual processing in the ventral stream. A wide variety of colors and abstract shapes have been found to be good stimuli for cells in IT. As we will see in Chapter 24, in that region. This "region of interest" method had already been used successfully by people studying lower level visual areas, and it was not a big leap to extend it to higher-level cortical areas.

Of course, demonstrating that a region of the brain is selectively responsive to faces requires much more than just showing that it responds more strongly to faces than to objects. Over the next few years, we (and others, notably Greg McCarthy and Aina Puce at Yale) tested the face specificity hypothesis against numerous alternative hypotheses. The fusiform face area (FFA) obliged, passing each new test (Figure A).

Other labs have used different methods to make breathtaking discoveries that have greatly extended our understanding of the FFA. After finding face-selective patches in monkeys with fMRI, Doris Tsao and her colleagues at Harvard went on to report that the vast majority of the cells in this region respond nearly exclusively to faces. (Not even I had thought the faces patches were *that* selective!) David Pitcher and his colleagues at University College London briefly disrupted a face-selective region just behind the FFA with transcranial magnetic stimulation, showing that this region is necessary for face perception (but not for perception of objects or bodies). And Yoichi Sugita at the Japan Science and Technology Agency reported that monkeys reared for 2 years without ever seeing a face show adult-like face discrimination abilities in the very first behavioral testing session, suggesting that experience with faces may not be necessary to wire the face-processing system.

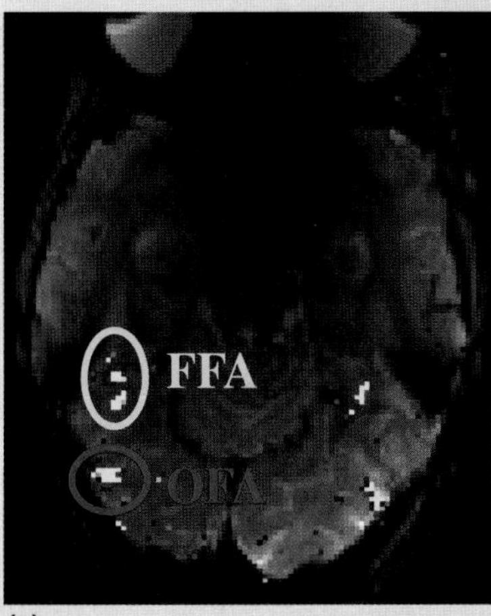

(a)

Figure A
Nancy's FFA, fusiform face area; OFA, occipital face area. (Source: Courtesy of N. Kanwisher.)

Our original decision to work on faces was a pragmatic one (we needed a quick result), and it worked out well for us. But I am just as proud of the completely unanticipated discoveries we made later, such as the scene-selective parahippocampal place area (PPA) working with Russell Epstein, and the body-selective extrastriate body area (EBA) working with Paul Downing. Most astonishing to me was Rebecca Saxe's discovery of a brain region that is selective for thinking about another person's thoughts. (My only role was to tell her this experiment would never work!)

These findings show that the human mind and brain contain at least a few very specialized components, each of which is dedicated to solving a very specific computational problem. These discoveries open up a vast landscape of new questions. What computations go on in each of these regions? How are those computations implemented in neural circuits? What other specialized brain regions exist? How do these specialized regions develop? Why do some mental processes get their own private piece of real estate in the brain, while others do not? Tackling these questions will be a challenge and a thrill.

output from area IT is sent to temporal lobe structures involved in learning and memory; IT itself may be important for both visual perception and visual memory. Recognizing an object clearly involves a combination or comparison of incoming sensory information with stored information.

One of the most intriguing findings concerning IT, as first noted by Charles Gross and his coworkers then at Princeton University, is that a small percentage of IT neurons in monkeys respond strongly to surprisingly complex objects such as pictures of faces. These cells may also respond to stimuli other than faces, but faces produce a particularly vigorous response, and some faces are more effective stimuli than others.

Human brain studies using fMRI appear to be consistent with the findings in monkeys. Nancy Kanwisher and her colleagues at MIT discovered that there is an area in the human brain that is more responsive to faces than to other stimuli (Box 10.3). This area is located on the fusiform gyrus

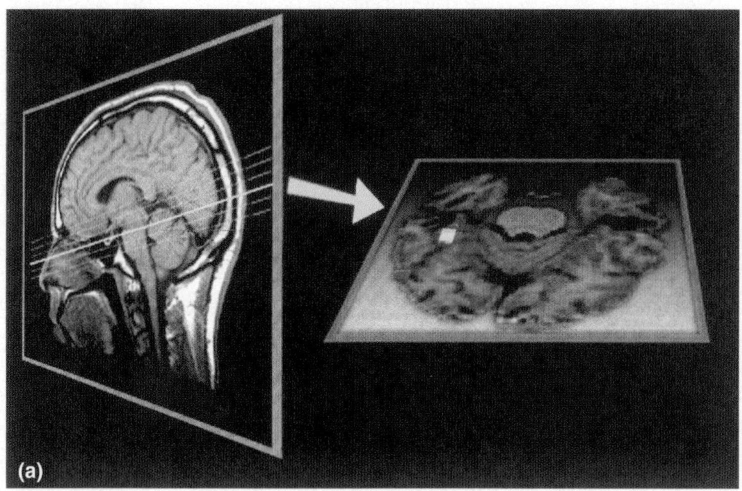

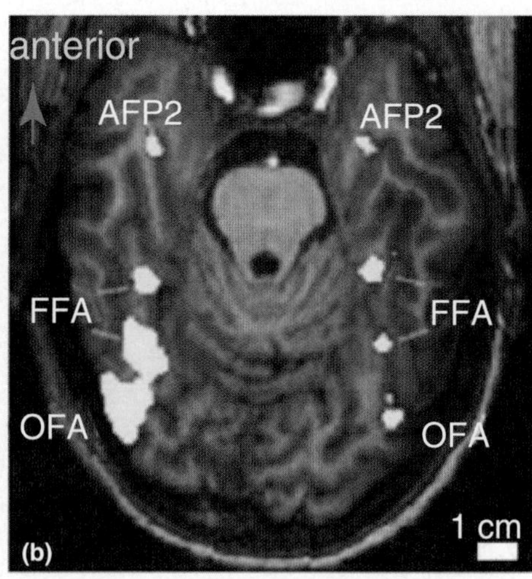

▲ FIGURE 10.29
Human brain activity elicited by pictures of faces. Using fMRI, brain activity was recorded first in response to faces and second in response to non-face stimuli. **(a)** In the horizontal brain section to the right, the red and yellow area on the left side and the symmetrical red area on the right side, known as the fusiform face area, showed significantly greater responses to faces. **(b)** Modified techniques in more recent studies have revealed multiple face-selective areas including occipital face area (OFA), anterior face patch 2 (AFP2), and fusiform face area (FFA). (Source: Part a courtesy of Drs. I. Gauthier, J.C. Gore, and M. Tarr; part b courtesy of Weiner and Grill-Spector, 2012.)

and has come to be called the *fusiform face area* (Figure 10.29a). Might this area play a special role in the ability to recognize faces, which are of great behavioral significance to humans? The finding of face-selective cells and the fusiform face area has sparked much interest, in part because of a syndrome called *prosopagnosia*—difficulty recognizing faces even though vision is otherwise normal. This rare syndrome usually results from a stroke and is associated with damage to extrastriate visual cortex, perhaps including the fusiform face area.

More recent experiments have revealed that there are actually about half a dozen patches of cortex within and near IT that are particularly sensitive to faces, and neurons in each patch have different degrees of sensitivity to face identity (Mary versus Sue) and other attributes such as the side of the head viewed (left, right, front, back) (Figure 10.29b). The implication is that multiple visual areas including portions of area IT may compose a system of areas specialized for facial recognition. In other human brain imaging studies, groups of patchy brain regions have been found that are reportedly involved in representations of color and biological objects.

FROM SINGLE NEURONS TO PERCEPTION

Visual *perception*—the task of identifying and assigning meaning to objects in space—obviously requires the concerted action of many cortical neurons. But which neurons in which cortical areas determine what we perceive? How is the simultaneous activity of widely separated cortical neurons integrated, and where does this integration take place? Neuroscience research is just beginning to tackle these challenging questions. However, sometimes basic observations about receptive fields can give us insight into how we perceive (Box 10.4).

Receptive Field Hierarchy and Perception

Comparing the receptive field properties of neurons at different points in the visual system might provide insight about the basis of perception. The receptive fields of photoreceptors are simply small patches on the retina, whereas those of retinal ganglion cells have a center-surround structure. The ganglion cells are sensitive to variables such as contrast and the wavelength of light. In the striate cortex, we encounter simple and complex receptive fields that have several new properties, including orientation selectivity and binocularity. We have seen that in extrastriate cortical areas, cells are selectively responsive to more complex shapes, object motion, and even faces. It appears that the visual system consists of a hierarchy of areas in which receptive fields become increasingly larger and more complex, moving away from V1 (Figure 10.30). Perhaps our perception of specific objects is based on the excitation of a small number of specialized neurons in some ultimate perceptual area that has not yet been identified. Is a person's recognition of his or her grandmother based on the responses of 5 or 10 cells with receptive field properties so highly refined that the cells respond only to one person? The closest approximation to this is the face-selective neurons in area IT. However, even these fascinating cells do not respond to only one face.

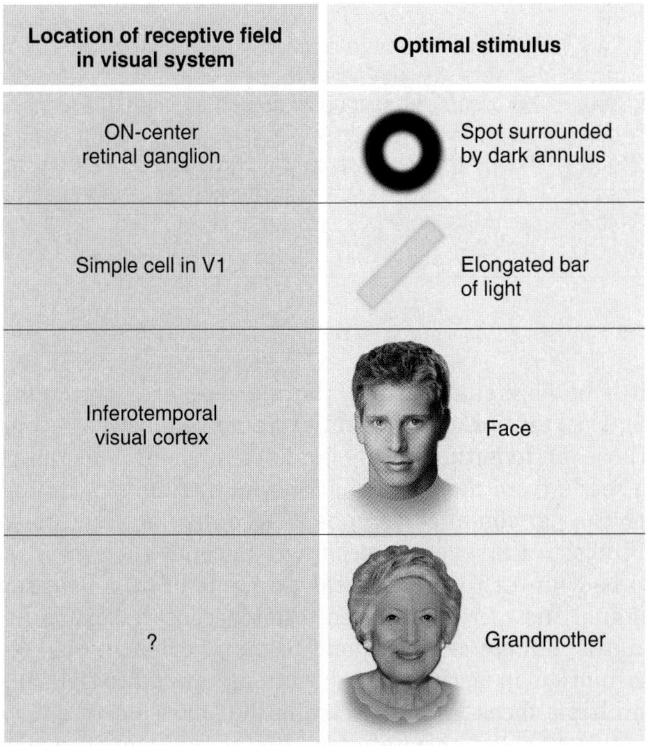

Location of receptive field in visual system	Optimal stimulus
ON-center retinal ganglion	Spot surrounded by dark annulus
Simple cell in V1	Elongated bar of light
Inferotemporal visual cortex	Face
?	Grandmother

▲ FIGURE 10.30
A hierarchy of receptive fields. As one progresses from the retina to extrastriate visual cortex, receptive fields become larger and are selective for more complex shapes. At present, it seems unlikely that perception is based on undiscovered neurons highly selective for every object a person recognizes, such as one's grandmother.

BOX 10.4 OF SPECIAL INTEREST

The Magic of Seeing in 3D

You have probably seen books or posters showing patterns of dots or splotches of color that supposedly contain pictures in 3D, if you contort your eyes just the right way. But how is it possible to see three dimensions on a two-dimensional piece of paper? The answer is based on the fact that our two eyes always see slightly different images of the world because of the distance between them in the head. The closer objects are to the head, the greater the difference in the two images. You can easily demonstrate this to yourself by holding a finger up in front of your eyes and alternately viewing it, at different distances, with the left or right eye closed.

Long before anything was known about binocular neurons in visual cortex, stereograms were a popular form of recreation. Two photographs were taken with lenses separated by a distance roughly the same as that of human eyes. By looking at the left photograph with the left eye and the right photograph with the right eye (by relaxing the eye muscles or with a stereoscope), the brain combines the images and interprets the different views as cues for distance, providing a three-dimensional percept (Figure A).

Figure A
A nineteenth-century stereogram. (Source: Horibuchi, 1994, p. 38.)

In 1960, Bela Julesz, working at the Bell Telephone Laboratories, invented random-dot stereograms (Figure B). These paired images of random dots are, in principle, the same as the nineteenth-century stereograms. The big difference is that the object hidden in the images cannot be seen with normal binocular viewing. To see the 3D object, you must direct your left and right eyes to the left and right images. The principle in constructing the stereo images is to create a background of randomly spaced dots, and wherever an area should be closer or farther away in the fused image, the dots shown to one eye are horizontally shifted relative to those in the other eye. Imagine looking at a white index card covered with random black dots while you hold it in front of a large piece of white paper covered with similar dots. By alternately closing your eyes, the dots on the index card will shift horizontally more than those on the more distant piece of paper. The pair of stereo images captures this difference in viewpoint and erases any other indication that there is a square in front, such as the edge of the index card. Random-dot stereograms shocked many scientists,

While it is by no means settled, there are several arguments against the idea that perception is based on extremely selective receptive fields such as those of hypothetical "grandmother cells." First, recordings have been made from most parts of the monkey brain, but there is no evidence that a portion of cortex has different cells tuned to each of the millions of objects that we all recognize. Second, such great selectivity appears to be counter to the general principle of broad tuning that exists throughout the nervous system. Photoreceptors respond to a range of wavelengths, simple cells respond to many orientations, cells in MT respond to motion in a range of directions, and face cells usually respond to multiple faces. Moreover, cells that are selective for one property—orientation, color, or whatever—are always sensitive also to other properties. For example, we can emphasize the orientation selectivity of V1 neurons and the way in which this might relate to the perception of form, neglecting the fact that the same cells might selectively respond to size, direction of motion, and so on. Finally, it might be too "risky" for the nervous system to rely on extreme selectivity. A blow to

Figure B
A random-dot stereogram and the perception that results from binocularly fusing the images. (Source: Julesz, 1971, p. 21.)

because in 1960, it was commonly thought that depth was perceived only after the objects in each eye were separately recognized.

In the 1970s, Christopher Tyler at the Smith–Kettlewell Eye Research Institute created autostereograms. An autostereogram is a single image that, when properly viewed, gives the perception of objects in 3D (Figure C). The colorful, and sometimes frustrating, autostereograms you see in books are based on an old illusion called the *wallpaper effect*. If you look at wallpaper that contains a repeating pattern, you can cross (or diverge) your eyes and view one piece of the pattern with one eye and the next cycle of the pattern with the other eye. The effect makes the wallpaper appear to be closer (or farther away). In an autostereogram, the wallpaper effect is combined with random-dot stereograms. To see the 3D skull in Figure C, you need to relax your eye muscles so that the left eye looks at the left dot on top and the right eye the right dot. You will know you are getting close when you see three dots at the top of the image. Relax and keep looking, and the picture will become visible.

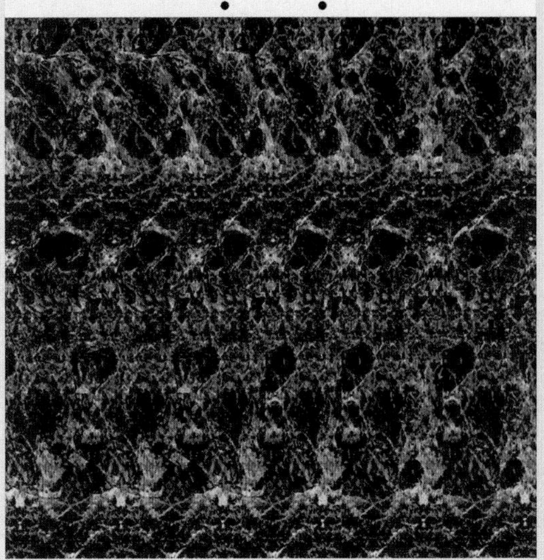

Figure C
An autostereogram. (Source: Horibuchi, 1994, p. 54.)

One of the fascinating things about stereograms is that you often must look at them for tens of seconds or even minutes, while your eyes become "properly" misaligned and your visual cortex "figures out" the correspondence between the left and right eye views. We do not know what is going on in the brain during this period, but presumably it involves the activation of binocular neurons in the visual cortex.

the head might kill all five grandmother cells, and in an instant, one would lose the ability to recognize her. We will have more to say about the robustness of recognition when we explore learning and memory in Chapters 24 and 25.

Parallel Processing and Perception

If we do not rely on "grandmother cells," how does perception work? One alternative hypothesis is formulated around the observation that parallel processing is used throughout the visual system (and other brain systems). We encountered parallel processing in Chapter 9 when we discussed ON and OFF and M and P ganglion cells. In this chapter, we saw three parallel channels coming into V1. Extending away from V1 are dorsal and ventral streams of processing, and different areas in these two streams are biased, or specialized, for various stimulus properties. Perhaps the brain uses a "division of labor" principle for perception. Within a given cortical area, many broadly tuned cells may serve to represent features of objects.

At a bigger scale, a group of cortical areas may contribute to perception, some dealing more with color or form, others more with motion. In other words, perception may be more like the sound produced by an orchestra of visual areas, each with different roles, than by a single musician.

CONCLUDING REMARKS

In this chapter, we have outlined the organization of the sensory pathway from the eye to the thalamus to the cortex. We saw that vision actually involves the perception of numerous different properties of objects—including color, form, and movement—and these properties are processed in parallel by different cells of the visual system. This processing of information evidently requires a strict segregation of inputs at the thalamus, some limited convergence of information in the striate cortex, and finally a massive divergence of information as it is passed on to higher cortical areas. The distributed nature of the cortical processing of visual information is underscored when you consider that the output of a million ganglion cells can recruit the activity of well over a billion cortical neurons throughout the occipital, parietal, and temporal lobes! Somehow, this widespread cortical activity is combined to form a single, seamless perception of the visual world.

Heed the lessons learned from the visual system. As we shall see in later chapters, the basic principles of organization in this system—parallel processing, topographic mappings of sensory surfaces, synaptic relays in the dorsal thalamus, cortical modules, and multiple cortical representations—are also features of the sensory systems devoted to hearing and touch.

KEY TERMS

REVIEW QUESTIONS

1. Following a bicycle accident, you are disturbed to find that you cannot see anything in your left visual field. Where has the retinofugal pathway been damaged?
2. What is the source of most of the input to the *left* LGN?
3. A worm has eaten part of one lateral geniculate nucleus. You can no longer perceive motion in the right visual field of your right eye. What layer(s) of which LGN have most likely been damaged?
4. List the chain of connections that link a cone in the retina to a blob cell in the striate cortex. Is there more than one path by which cones connect to the blob cell?
5. What is meant by the statement that there is a map of the visual world in the striate cortex?
6. What is parallel processing in the visual system? Give two examples.
7. If a child is born cross-eyed and the condition is not corrected before the age of 10 years, binocular depth perception will be lost forever. This is explained by a modification in the circuitry of the visual system. From your knowledge of the central visual system, where do you think the circuitry has been modified?
8. What layers of the striate cortex send efferents to other visual cortex areas?
9. What new receptive field properties are found in the striate cortex and other cortical areas that are not seen in the retina or LGN?
10. What sort of experiment might you perform to investigate the relationship between visual perception and neural activity in the visual cortex?

FURTHER READING

De Haan EHF, Cowey A. 2011. On the usefulness of "what" and "where" pathways in vision. *Trends in Cognitive Science* 15:460–466.

Gegenfurtner KR. 2003. Cortical mechanisms of colour vision. *Nature Reviews Neuroscience* 4: 563–572.

Grill-Spector K, Malach R. 2004. The human visual cortex. *Annual Reviews of Neuroscience* 27: 649–677.

Hendry SHC, Reid RC. 2000. The koniocellular pathway in primate vision. *Annual Reviews of Neuroscience* 23:127–153.

Kreiman G. 2007. Single unit approaches to human vision and memory. *Current Opinion in Neurobiology* 17:471–475.

Milner AD, Goodale MA. 2008. Two visual systems reviewed. *Neuropsychologia* 46: 774–785.

Nasso JJ, Callaway EM. 2009. Parallel processing strategies of the primate visual system. *Nature Reviews Neuroscience* 10:360–372.

Sherman SM. 2012. Thalamocortical interactions. *Current Opinion in Neurobiology* 22: 575–579.

Tsao DY, Moeller S, Freiwald W. 2008. Comparing face patch systems in macaques and humans. *Proceedings of the National Academy of Science* 49:19514–19519.

Zeki S. 2003. Improbable areas in the visual brain. *Trends in Neuroscience* 26:23–26.

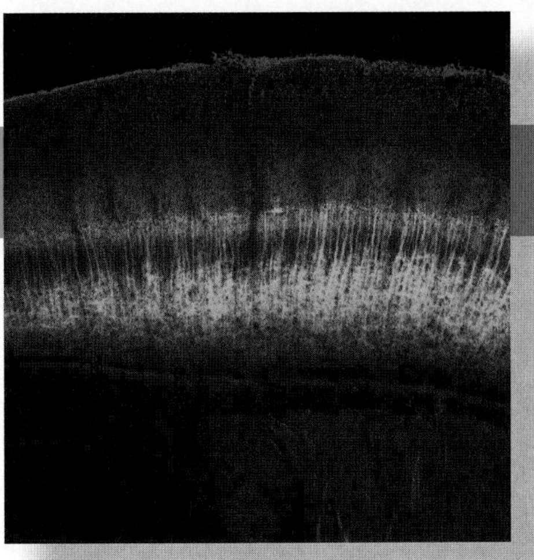

CHAPTER ELEVEN

The Auditory and Vestibular Systems

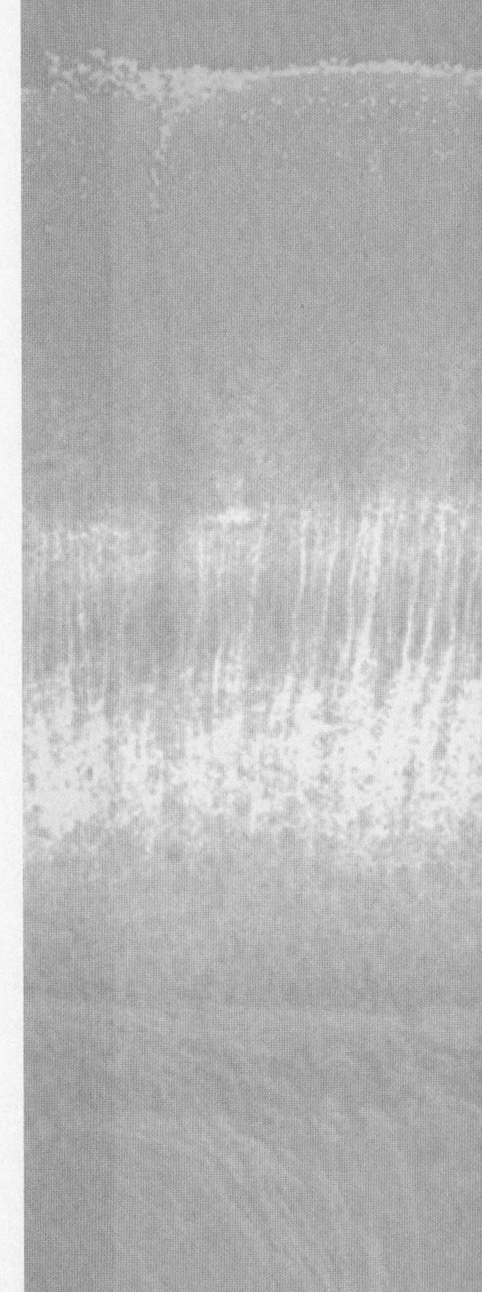

INTRODUCTION

In this chapter, we'll explore two sensory systems that have very different functions but surprising similarities of structure and mechanism: the sense of hearing, also known as **audition**, and the sense of balance, regulated by the **vestibular system**. Hearing is a vivid part of our conscious lives, while balance is something we experience all day but rarely think about.

When we cannot see something or someone, we can often detect its presence, identify its origin, and even receive a message from it just by hearing its sounds. Anyone who has ever hiked through the forest in an area where there are bears or snakes knows that the sound of rustling leaves can be a powerful attention–grabber. Aside from the ability to detect and locate sound, we can perceive and interpret its nuances. We can immediately distinguish the bark of a dog, the voice of a particular friend, the crash of an ocean wave. Because humans are able to produce a wide variety of sounds as well as hear them, spoken language and its reception via the auditory system have become an extremely important means of communication. Audition in humans has even evolved beyond the strictly utilitarian functions of communication and survival; musicians, for example, explore the sensations and emotions evoked by sound.

In contrast to hearing, the sense of balance is strictly a personal, internalized process. The vestibular system informs our nervous system where our head and body are and how they are moving. This information is used, without conscious effort, to control muscular contractions that keep or put our body where we want it to be, to reorient ourselves when something moves us, and to move our eyes so that our visual world stays fixed on our retinas even when our head is bouncing around.

Here, we will explore the mechanisms within the ear and brain that translate the sounds in our environment into meaningful neural signals and the movements of our head into a sense of where we are. We will find that these transformations are carried out in stages rather than all at once. Within the inner ear, neural responses are generated by auditory receptors from the mechanical energy in sound and by vestibular receptors from the tilts and rotations of the head. At subsequent stages in the brain stem and thalamus, signals from the receptors are integrated before they ultimately reach auditory and vestibular cortex. By looking at the response properties of neurons at various points in the system, we begin to understand the relationship between neural activity and our perception of sound and balance.

THE NATURE OF SOUND

Sounds are audible variations in air pressure. Almost anything that can move air molecules can generate a sound, including the vocal cords of the human larynx, the vibration of a string on a guitar, and the explosion of a firecracker. When an object moves toward a patch of air, it compresses the air, increasing the density of the molecules. Conversely, the air is rarefied (made less dense) when an object moves away. This is particularly easy to visualize in the case of a stereo speaker, in which a paper cone attached to a magnet vibrates in and out, alternately rarefying and compressing the air (Figure 11.1). These changes in air pressure are transferred away from the speaker at the speed of sound, which is about 343 m/sec (767 mph) for air at room temperature.

Many sources of sound, such as vibrating strings or a stereo speaker reproducing the sound of a stringed instrument, produce variations in air

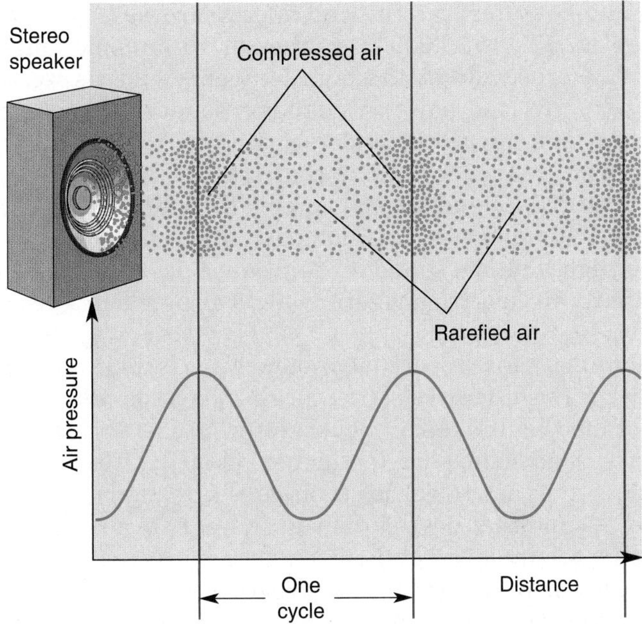

▲ FIGURE 11.1
The production of sound by variations in air pressure. When the paper cone of a stereo speaker pushes out, it compresses the air; when the cone pulls in, it rarefies the air. If the push and pull are rhythmic, there will also be a rhythmic variation in the air pressure, as shown in the graph. The distance between successive compressed (high-pressure) patches of air is one cycle of the sound (indicated by the vertical lines). The sound wave propagates away from the speaker at the speed of sound. The blue line is a graph of air pressure versus distance.

pressure that are rhythmic. The **frequency** of the sound is the number of compressed or rarefied patches of air that pass by our ears each second. One cycle of the sound is the distance between successive compressed patches; the sound frequency, expressed in units called **hertz (Hz)**, is the number of cycles per second. Because sound waves all propagate at the same speed, high-frequency sound waves have more compressed and rarefied regions packed into the same space than low-frequency waves (Figure 11.2a).

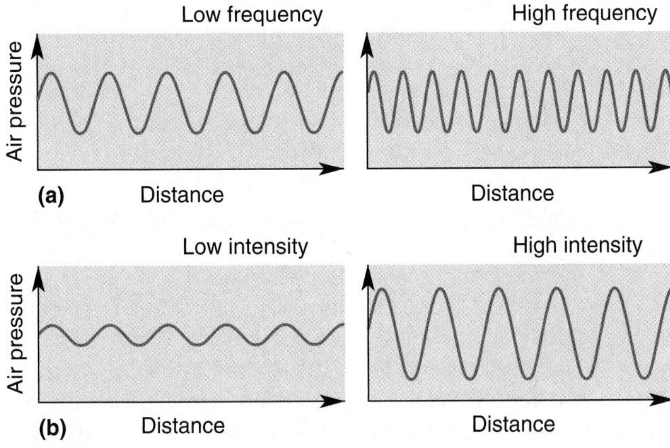

▲ FIGURE 11.2
The frequency and intensity of sound waves. Each graph plots air pressure versus distance for a sound of constant frequency and intensity. Note that the X axis also represents time because the velocity of sound is constant. **(a)** Frequency is the number of sound waves per unit of time or distance. We perceive high-frequency waves as having a higher pitch. **(b)** Intensity is the air pressure difference between peaks and troughs of the sound waves. We perceive high-intensity waves as louder.

Our auditory system can respond to pressure waves over the remarkable range of 20 Hz to 20,000 Hz (although this audible range decreases significantly, especially at the high-frequency end, as we age and expose ourselves to loud noise). Whether a sound is perceived to have a high or low tone, or **pitch**, is determined by the frequency. In order to understand frequency, realize that a room-shaking low note on an organ is about 20 Hz and an ear-piercing high note on a piccolo is about 10,000 Hz. Although humans can hear a great range of frequencies, there are high and low sound wave frequencies our ears cannot hear, just as there are electromagnetic waves of light our eyes cannot see (Box 11.1).

Another important property of a sound wave is its **intensity**, or amplitude, which is the difference in pressure between compressed and rarefied patches of air (Figure 11.2b). Sound intensity determines the *loudness* we perceive, loud sounds having higher intensity. The range of intensities to which the human ear is sensitive is astonishing: The intensity of the loudest sound that doesn't damage our ears is about a trillion times greater than the intensity of the faintest sound that can be heard. If our auditory system were much more sensitive, we would hear a constant roar from the random movement of air molecules.

Real-world sounds rarely consist of simple rhythmic sound waves at one frequency and intensity. It is the simultaneous combination of different frequency waves at different intensities that gives different musical instruments and human voices their unique tonal qualities.

BOX 11.1 OF SPECIAL INTEREST

Ultrasound and Infrasound

Most people are familiar with ***ultrasound*** (sound above the 20 kHz upper limit of our hearing) because it has everyday applications, from ultrasonic cleaners to medical imaging. Many animals can hear these high frequencies. For instance, ultrasonic dog whistles work because dogs can hear up to about 45 kHz. Some bats vocalize at frequencies up to 100 kHz, then listen to the echoes of their calls in order to locate objects (see Box 11.5). Some fish in the shad and herring family can detect sounds as high as 180 kHz, thereby enabling them to hear the echolocating ultrasound generated by dolphins that prey on them. Needless to say, the dolphins can hear their own ultrasonic calls. Similarly, nocturnal moths listen for the ultrasound of hungry bats, so they can evade these predators.

Infrasound is sound at frequencies lower than humans can hear, below about 20 Hz. Some animals can hear infrasonic frequencies; one is the elephant, which can detect 15 Hz tones inaudible to humans. Whales produce low-frequency sounds, which are thought to be a means of communication over distances of many kilometers. The earth also produces low-frequency vibrations, and some animals may sense an impending earthquake by hearing such sounds.

Even though we usually cannot hear very low frequencies with our ears, they are present in the environment and we can sometimes feel them as vibrations with our somatosensory system (see Chapter 12). Infrasound is produced by such devices as air conditioners, boilers, aircraft, and automobiles, and it can have unpleasant subconscious effects. Although even intense infrasound from these machines does not cause hearing loss, it can cause dizziness, nausea, and headache. Many cars produce low-frequency sound when they're moving at highway speeds, making sensitive people carsick. At very high levels, low-frequency sound may also produce resonances in body cavities such as the chest and stomach, which can damage internal organs.

In addition to mechanical equipment, our own bodies generate inaudible low-frequency sound. When muscle changes length, individual fibers vibrate, producing low-intensity sound at about 25 Hz. While we cannot normally hear these sounds, you can demonstrate them to yourself by carefully putting your thumbs in your ears and making a fist with each hand. As you tighten your fist, you can hear a low rumbling sound produced by the contraction of your forearm muscles. Other muscles, including your heart, produce inaudible sound at frequencies near 20 Hz.

THE STRUCTURE OF THE AUDITORY SYSTEM

Before exploring how variations in air pressure are translated into neural activity, let's quickly survey the structure of the auditory system. The components of the ear are shown in Figure 11.3. The visible portion of the ear consists primarily of cartilage covered by skin, forming a sort of funnel called the **pinna** (from the Latin for "wing"), which helps collect sounds from a wide area. The shape of the pinna makes us more sensitive to sounds coming from ahead than from behind. The convolutions in the pinna play a role in localizing sounds, as we'll discuss later on. In humans, the pinna is more or less fixed in position, but animals such as cats and horses have considerable muscular control over the position of their pinna and can orient it toward a source of sound.

The entrance to the internal ear is called the **auditory canal**, which extends about 2.5 cm (1 inch) inside the skull before it ends at the **tympanic membrane**, also known as the *eardrum*. Connected to the medial surface of the tympanic membrane is a series of bones called **ossicles** (from the Latin for "little bones"; the ossicles are indeed the smallest bones in the body). Located in a small air-filled chamber, the ossicles transfer movements of the tympanic membrane into movements of a second membrane covering a hole in the bone of the skull called the **oval window**. Behind the oval window is the fluid-filled **cochlea**, which contains the apparatus for transforming the physical motion of the oval window membrane into a neuronal response. Thus, the first stages of the basic auditory pathway look like this:

Sound wave moves the tympanic membrane. →
Tympanic membrane moves the ossicles. →
Ossicles move the membrane at the oval window. →
Motion at the oval window moves fluid in the cochlea. →
Movement of fluid in the cochlea causes a response in sensory neurons.

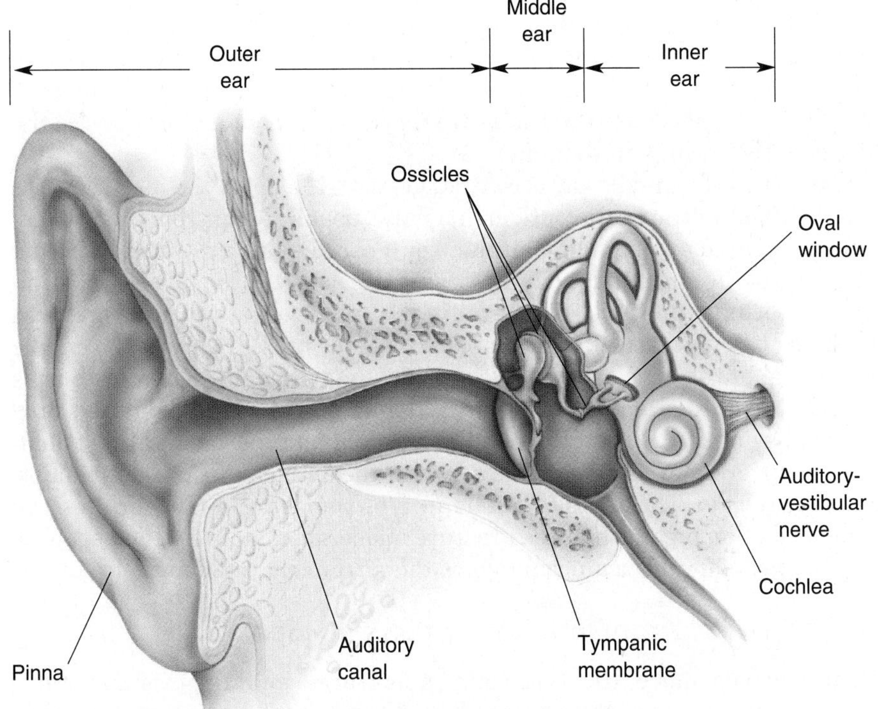

◀ FIGURE 11.3
The outer, middle, and inner ear.

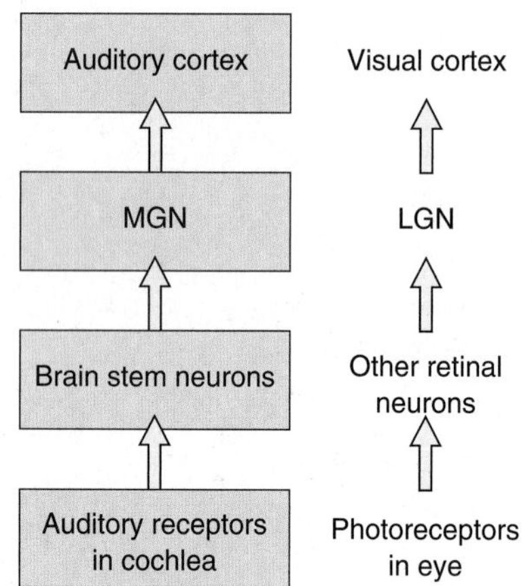

▲ FIGURE 11.4
Auditory and visual pathways compared. Following the sensory receptors, both systems have early integration stages, a thalamic relay, and a projection to sensory cortex.

All the structures from the pinna inward are considered components of the ear, conventionally referenced in three main divisions. The structures from the pinna to the tympanic membrane make up the **outer ear**, the tympanic membrane and the ossicles constitute the **middle ear**, and the apparatus medial to the oval window is the **inner ear**.

Once a neural response to sound is generated in the inner ear, the signal is transferred to and processed by a series of nuclei in the brain stem. Output from these nuclei is sent to a relay in the thalamus, the **medial geniculate nucleus (MGN)**. Finally, the MGN projects to **primary auditory cortex**, or **A1**, located in the temporal lobe. In one sense, the auditory pathway is more complex than the visual pathway because there are more intermediate stages between the sensory receptors and cortex. However, the systems have analogous components. Each starts with sensory receptors, which connect to early integration stages (located in the retina for vision and the brain stem for audition), then to a thalamic relay, and then to sensory cortex (Figure 11.4).

THE MIDDLE EAR

The outer ear funnels sound to the middle ear, an air-filled cavity containing the first elements that move in response to sound. In the middle ear, variations in air pressure are converted into movements of the ossicles. In this section, we'll explore how the middle ear performs an essential transformation of sound energy.

Components of the Middle Ear

The structures within the middle ear are the tympanic membrane, the ossicles, and two tiny muscles that attach to the ossicles. The tympanic membrane is somewhat conical in shape, with the point of the cone extending into the cavity of the middle ear. There are three ossicles, each named (from the Latin) after an object it slightly resembles (Figure 11.5). The ossicle attached to the tympanic membrane is the *malleus* ("hammer"), which forms a rigid connection with the *incus* ("anvil"). The incus forms a flexible connection with the *stapes* ("stirrup"). The flat bottom portion of the stapes, the *footplate*, moves in and out like a piston at the oval window. The movements of the footplate transmit sound vibrations to the fluids of the cochlea in the inner ear.

The air in the middle ear is continuous with the air in the nasal cavities via the **Eustachian tube**, although a valve usually keeps this tube closed. When you're in an ascending airplane or a car heading up a mountain, the pressure of the surrounding air decreases. However, as long as the valve on the Eustachian tube is closed, the air in the middle ear stays at the pressure of the air before you started to climb. Because the pressure inside the middle ear is higher than the air pressure outside, the tympanic membrane bulges out, and you experience unpleasant pressure or pain in the ear. Yawning or swallowing, both of which open the Eustachian tube, can usually equalize the air pressure in the middle ear with the ambient air pressure and so relieve the pain. The opposite can happen as you descend. The air pressure outside is then higher than the pressure inside the middle ear, and opening the Eustachian tube again can relieve the discomfort you may feel.

Sound Force Amplification by the Ossicles

Sound waves move the tympanic membrane, and the ossicles move another membrane at the oval window. Why isn't the ear arranged so that

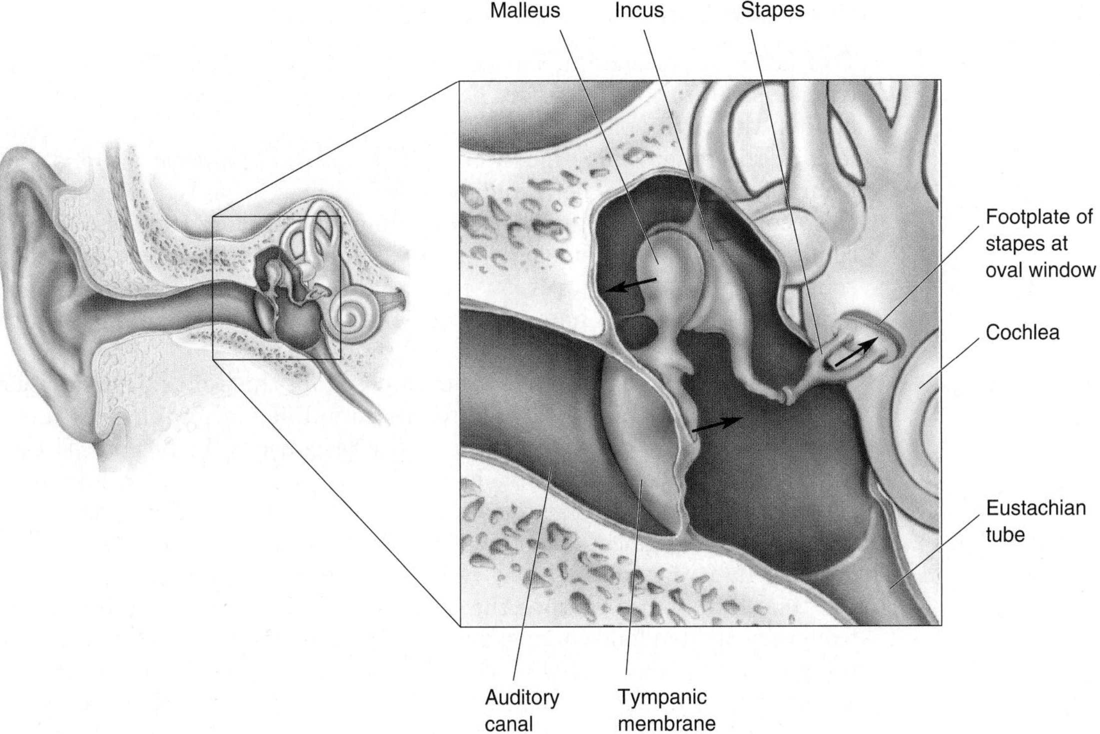

▲ FIGURE 11.5
The middle ear. As the arrows indicate, when air pressure pushes the tympanic membrane, the bottom of the malleus is pushed inward and the lever action of the ossicles makes the footplate of the stapes push inward at the oval window. The pressure pushing at the oval window is greater than that at the tympanic membrane, in part because the surface area of the footplate of the stapes is smaller than the surface area of the tympanic membrane.

sound waves simply directly move the membrane at the oval window? The problem is that the cochlea is filled with fluid, not air. If sound waves impinged directly on the oval window, the membrane would barely move and all but about 0.1% of the sound energy would be reflected away because of the pressure the cochlear fluid exerts at the back of the oval window. If you've ever noticed how quiet it is under water, you know how well water reflects sound coming from above. The fluid in the inner ear resists being moved much more than air does (i.e., fluid has greater inertia), so more pressure is needed to vibrate the fluid than the air can provide. The ossicles provide this necessary amplification in pressure.

To understand the process, consider the definition of pressure. The pressure on a membrane is defined as the force pushing it divided by its surface area. The pressure at the oval window will become greater than the pressure at the tympanic membrane if (1) the force on the oval window membrane is greater than that on the tympanic membrane, or (2) the surface area of the oval window is smaller than the area of the tympanic membrane. The middle ear uses both mechanisms. It increases pressure at the oval window by altering both the force and the surface area. The force at the oval window is greater because the ossicles act like levers. Sound causes large movements of the tympanic membrane, which are transformed into smaller but stronger vibrations of the oval window. And the surface area of the oval window is much smaller than that of the tympanic membrane. These factors combine to make the pressure at the oval window about 20 times greater than at the tympanic membrane, and this increase is sufficient to move the fluid in the inner ear.

The Attenuation Reflex

Two muscles attached to the ossicles have a significant effect on sound transmission to the inner ear. The *tensor tympani muscle* is anchored to bone in the cavity of the middle ear at one end and attaches to the malleus at the other end (Figure 11.6). The *stapedius muscle* also extends from a fixed anchor of bone and attaches to the stapes. When these muscles contract, the chain of ossicles becomes much more rigid, and sound conduction to the inner ear is greatly diminished. The onset of a loud sound triggers a neural response that causes these muscles to contract, a response called the **attenuation reflex**. Sound attenuation is much greater at low frequencies than at high frequencies.

A number of functions have been proposed for this reflex. One function may be to adapt the ear to continuous sound at high intensities. Loud sounds that would otherwise saturate the response of the receptors in the inner ear could be reduced to a level below saturation by the attenuation reflex, thus increasing the dynamic range we can hear. The attenuation reflex also protects the inner ear from loud sounds that would otherwise damage it. Unfortunately, the reflex has a delay of 50–100 msec from the time that sound reaches the ear, so it doesn't offer much protection from very sudden loud sounds; damage might already be done by the time the muscles contract. This is why, despite the best efforts of your attenuation reflex, a loud explosion can still damage your cochlea. Because the attenuation reflex suppresses low frequencies more than high frequencies, it tends to make high-frequency sounds easier to discern in an environment with a lot of low-frequency noise. This capability enables us to understand speech more easily in a noisy environment than we could without the reflex. It is thought that the attenuation reflex is also

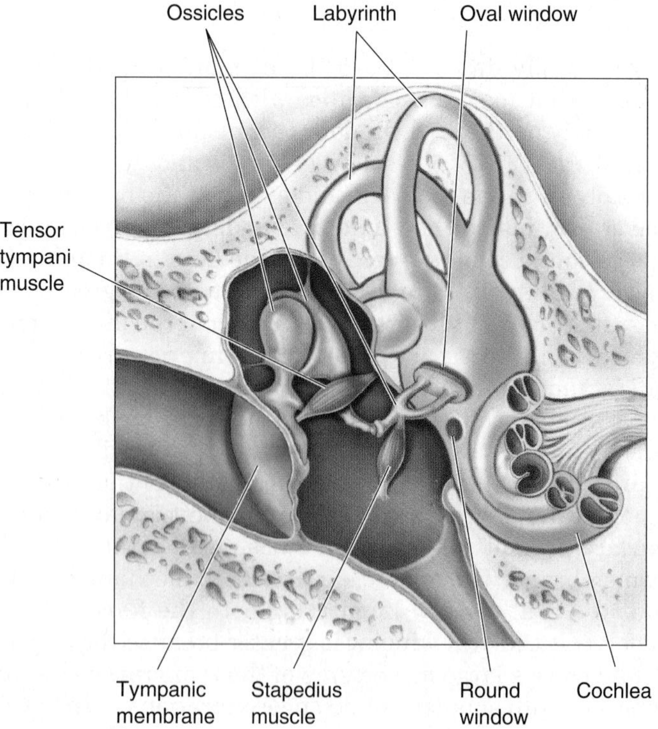

▲ FIGURE 11.6
The middle and inner ear. The stapedius muscle and the tensor tympani muscle are both attached to the wall of the middle ear at one end, and to the ossicles at the other ends.

activated when we speak, so we don't hear our own voices as loudly as we otherwise would.

THE INNER EAR

Although considered part of the ear, not all of the inner ear is concerned with hearing. The inner ear consists of the cochlea, which is part of the auditory system, and the labyrinth, which is not. The labyrinth is an important part of the *vestibular system*, which helps maintain the body's equilibrium. The vestibular system is discussed later in the chapter. Here, we are concerned only with the cochlea and the role it plays in transforming sound into a neural signal.

Anatomy of the Cochlea

The cochlea (from the Latin for "snail") has a spiral shape resembling a snail's shell. Figure 11.6 shows the cochlea cut in half. The structure of the cochlea is similar to a drinking straw wrapped two and a half to three times around the sharpened tip of a pencil. In the cochlea, the hollow tube (represented by the straw) has walls made of bone. The central pillar of the cochlea (represented by the pencil) is a conical bony structure. The actual dimensions are much smaller than the straw-and-pencil model, the cochlea's hollow tube being about 32 mm long and 2 mm in diameter. Rolled up, the human cochlea is about the size of a pea. At the base of the cochlea are two membrane-covered holes: the oval window, which is below the footplate of the stapes, as we have seen, and the **round window**.

If the cochlea is cut in cross section, we can see that the tube is divided into three fluid-filled chambers: the *scala vestibuli*, the *scala media*, and the *scala tympani* (Figure 11.7). The three scalae wrap around inside the cochlea like a spiral staircase (*scala* is from the Latin for "stairway"). *Reissner's membrane* separates the scala vestibuli from the scala media, and the **basilar membrane** separates the scala tympani from the scala media. Sitting upon the basilar membrane is the **organ of Corti**,

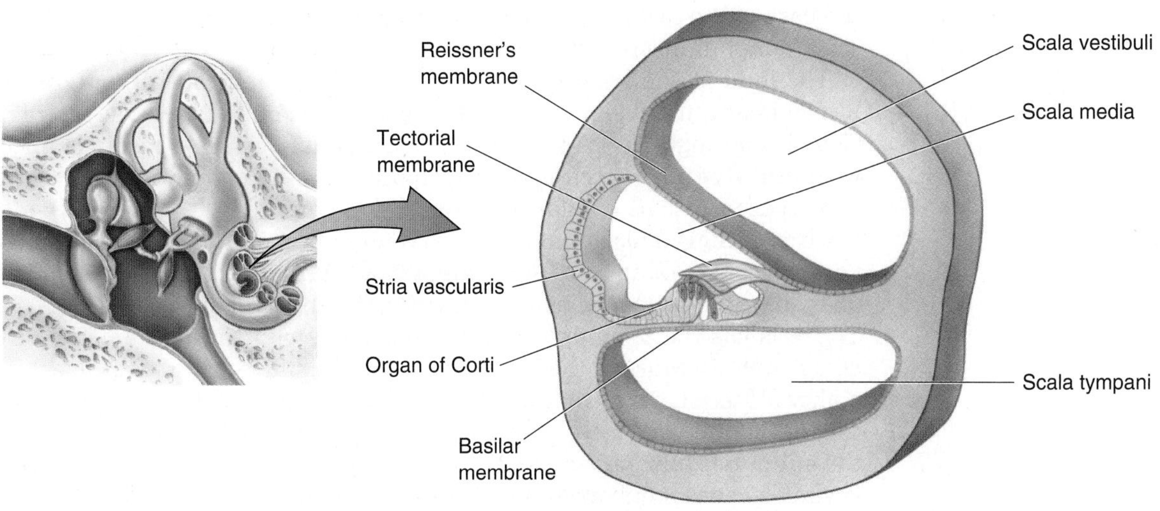

▲ FIGURE 11.7
The three scalae of the cochlea. Viewed in cross section, the cochlea contains three small parallel chambers. These chambers, the scalae, are separated by Reissner's membrane and the basilar membrane. The organ of Corti contains the auditory receptors; it sits upon the basilar membrane and is covered by the tectorial membrane.

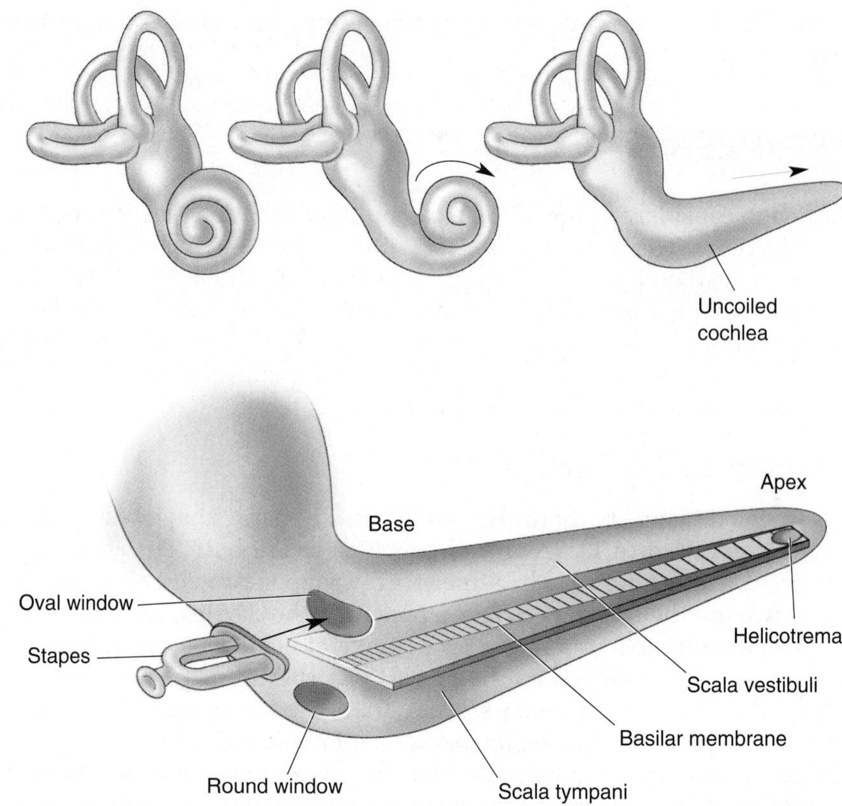

▲ FIGURE 11.8

The basilar membrane in an uncoiled cochlea. Although the cochlea narrows from base to apex, the basilar membrane widens toward the apex. Notice that the basilar membrane is the narrow blue band only. The helicotrema is a hole at the apex of the basilar membrane, which connects the scala vestibuli and scala tympani.

which contains auditory receptor neurons; hanging over this organ is the **tectorial membrane**. At the apex of the cochlea, the scala media is closed off, and the scala tympani becomes continuous with the scala vestibuli at a hole in the membranes called the *helicotrema* (Figure 11.8). At the base of the cochlea, the scala vestibuli meets the oval window and the scala tympani meets the round window.

The fluid in the scala vestibuli and scala tympani, called **perilymph**, has an ionic content similar to that of cerebrospinal fluid: relatively low K^+ (7 mM) and high Na^+ (140 mM) concentrations. The scala media is filled with **endolymph**, which is an unusual extracellular fluid in that it has ionic concentrations similar to intracellular fluid, high K^+ (150 mM) and low Na^+ (1 mM). This difference in ion content is generated by active transport processes taking place at the **stria vascularis**, the endothelium lining one wall of the scala media and contacting the endolymph (see Figure 11.7). The stria vascularis absorbs sodium from, and secretes potassium into, the endolymph. Because of the ionic concentration differences and the permeability of Reissner's membrane, the endolymph has an electrical potential that is about 80 mV more positive than that of the perilymph; this is called the **endocochlear potential**. We shall see that the endocochlear potential is vital because it enhances auditory transduction.

Physiology of the Cochlea

The structure of the cochlea is complex, but its basic operation is fairly simple. Look at Figure 11.8 and imagine what happens when the ossicles move the membrane that covers the oval window. They work like a tiny piston. Inward motion at the oval window pushes perilymph into the scala vestibuli. If the membranes inside the cochlea were completely rigid, then the increase in fluid pressure at the oval window would reach up the scala vestibuli, through the helicotrema, and back down the scala tympani to the round window. Because the fluid pressure has nowhere else to escape, the membrane at the round window would bulge out in response to the inward movement of the membrane at the oval window. Any motion at the oval window must be accompanied by a complementary motion at the round window. Such movement must occur because the cochlea is filled with incompressible fluid held in a solid bony container. The consequence of pushing in at the oval window is a bit like pushing in one end of a tubular water balloon: The other end has to bulge out.

This simple description of the events in the cochlea is complicated by one additional fact: Some structures inside the cochlea are not rigid. Most importantly, the basilar membrane is flexible and bends in response to sound.

The Response of the Basilar Membrane to Sound. The basilar membrane has two structural properties that determine the way it responds to sound. First, the membrane is wider at the apex than at the base by a factor of about 5. Second, the stiffness of the membrane decreases from base to apex, the base being about 100 times stiffer. Think of it as a flipper of the sort used for swimming, with a narrow, stiff base and a wide, floppy apex. When sound pushes the footplate of the stapes at the oval window, perilymph is displaced within the scala vestibuli, and endolymph is displaced within the scala media because Reissner's membrane is very flexible. Sound can also pull the footplate, reversing the pressure gradient. Sound causes a continual push–pull motion of the footplate; again, think of a tiny piston.

We owe much of our understanding of the response of the basilar membrane to the research of Hungarian–American biophysicist Georg von Békésy. Von Békésy determined that the movement of the endolymph makes the basilar membrane bend near its base, starting a wave that propagates toward the apex. The wave that travels up the basilar membrane is similar to the wave that runs along a rope if you hold one end in your hand and give it a snap (Figure 11.9). The distance the wave travels up the basilar membrane depends on the frequency of the sound. If the frequency is high, the stiffer base of the membrane will vibrate a good deal, dissipating most of the energy, and the wave will not propagate very far (Figure 11.10a). However, low-frequency sounds generate

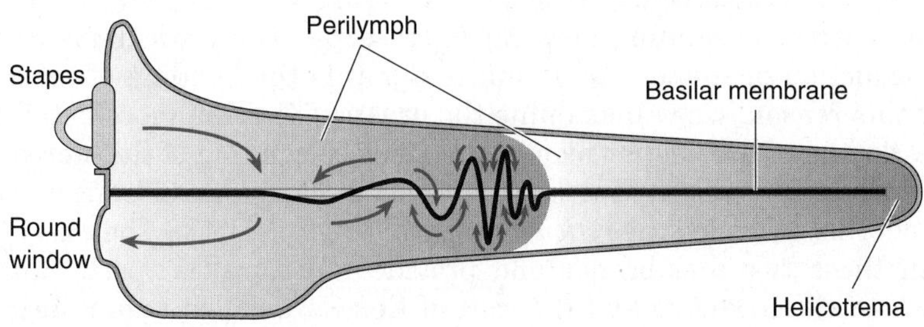

▲ FIGURE 11.9

A traveling wave in the basilar membrane. As the stapes moves in and out, it causes perilymph to flow, as shown by the arrows. This generates a traveling wave in the basilar membrane. (The size of the wave is magnified about 1 million times in this illustration.) At this frequency, 3000 Hz, the fluid and membrane movement end abruptly about halfway between the base and apex. Note that the scala media is not illustrated here. (Source: Adapted from Nobili, Mammano, and Ashmore, 1998, Fig. 1.)

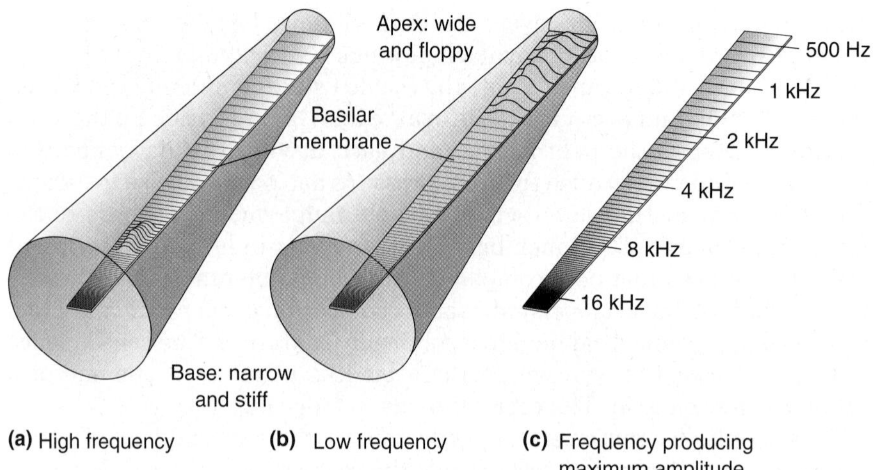

▶ FIGURE 11.10
The response of the basilar membrane to sound. The cochlea is again shown uncoiled. **(a)** High-frequency sound produces a traveling wave, which dissipates near the narrow and stiff base of the basilar membrane. **(b)** Low-frequency sound produces a wave that propagates all the way to the apex of the basilar membrane before dissipating. (The flexing of the basilar membrane is greatly exaggerated for the purpose of illustration.) **(c)** There is a place code on the basilar membrane for the frequency that produces the maximum amplitude deflection.

waves that travel all the way up to the floppy apex of the membrane before most of the energy is dissipated (Figure 10.10b). The response of the basilar membrane establishes a *place code* in which different locations of membrane are maximally deformed at different sound frequencies (Figure 11.10c). Systematic organization of sound frequency within an auditory structure is called **tonotopy**, analogous to retinotopy in the visual system. Tonotopic maps exist on the basilar membrane and within each of the auditory relay nuclei, the MGN, and auditory cortex.

As we shall see, the differences in the traveling waves produced by different sound frequencies are responsible for the neural coding of pitch.

The Organ of Corti and Associated Structures. Everything we have discussed to this point involves the mechanical transformations of sound energy that occur in the middle and inner ear. Now, we come to the point in the system where neurons are first involved. The auditory receptor cells, which convert mechanical energy into a change in membrane polarization, are located in the organ of Corti (named for the Italian anatomist who first identified it). The organ of Corti consists of hair cells, the rods of Corti, and various supporting cells.

The auditory receptors are called **hair cells** because each one has 10–300 hairy-looking **stereocilia** extending from its top. Hair cells are not neurons. They lack axons, and in mammals, they do not generate action potentials. Hair cells are actually specialized epithelial cells. The hair cells and stereocilia are shown in Figure 11.11 as they appear when viewed with a scanning electron microscope. The critical event in the transduction of sound into a neural signal is the bending of these cilia. For this reason, we will examine the organ of Corti in more detail to see how flexing of the basilar membrane leads to bending of the stereocilia.

The hair cells are sandwiched between the basilar membrane and a thin sheet of tissue called the *reticular lamina* (Figure 11.12). The *rods of Corti* span these two membranes and provide structural support. Hair cells between the modiolus and the rods of Corti are called **inner hair cells** (about 4500 form a single row), and cells farther out than the rods of Corti are called **outer hair cells** (in humans, there are about 12,000–20,000 arranged in three rows). The stereocilia at the tops of the hair cells extend above the reticular lamina into the endolymph, and their tips end either in the gelatinous substance of the tectorial membrane (the outer hair cells) or just below the tectorial membrane (the inner hair cells). To keep the membranes within the organ of Corti straight in your mind, remember that the

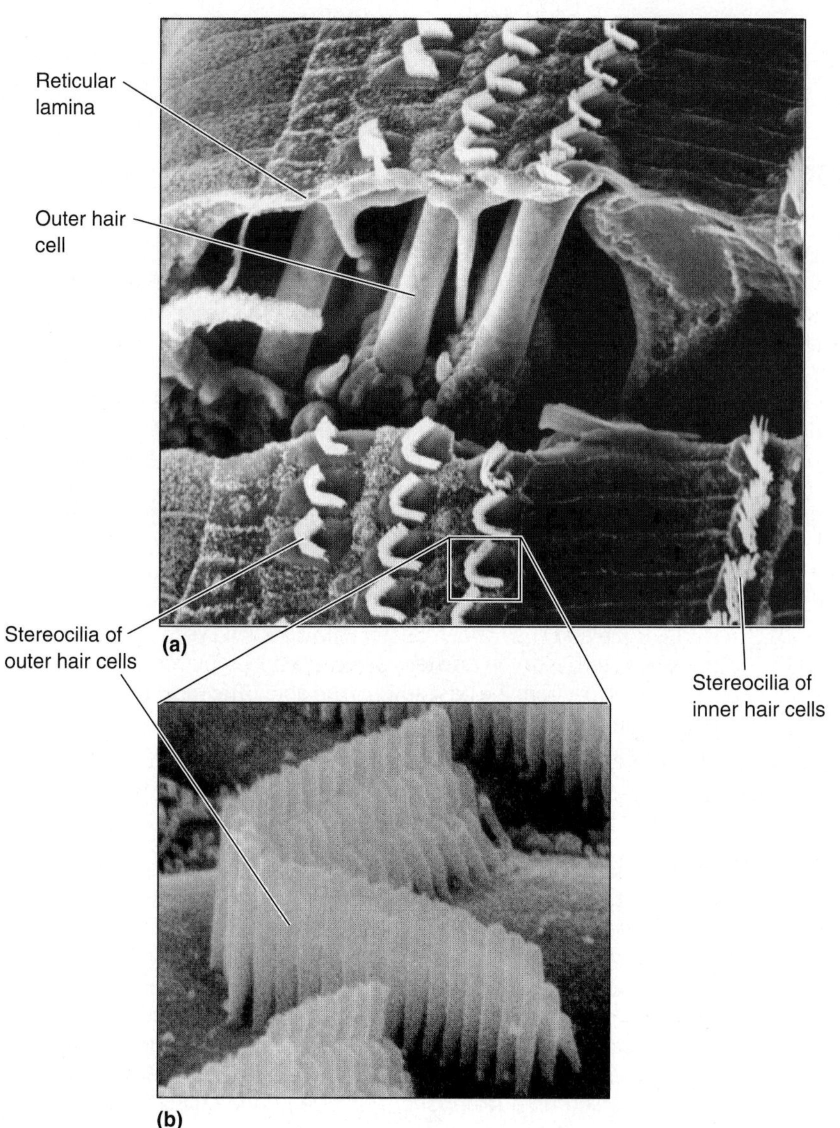

◀ FIGURE 11.11
Hair cells viewed through the scanning electron microscope. **(a)** Hair cells and their stereocilia. **(b)** A higher resolution view of the stereocilia on an outer hair cell. The stereocilia are approximately 5 μm in length. (Source: Courtesy of I. Hunter-Duvar and R. Harrison, The Hospital for Sick Children, Toronto, Ontario, Canada.)

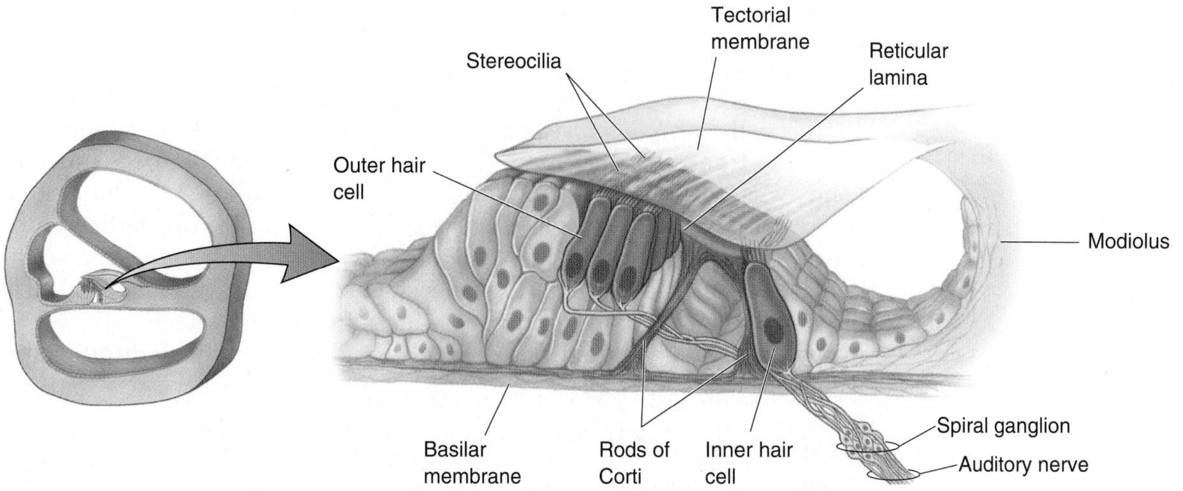

▲ FIGURE 11.12
The organ of Corti. The basilar membrane supports tissue that includes the inner and outer hair cells and the stiff rods of Corti. The tectorial membrane extends from the bony modiolus to cover the stereocilia that protrude from the tops of the hair cells.

BOX 11.2 OF SPECIAL INTEREST

The Deaf Shall Hear: Cochlear Implants

Hair cell damage or death is the most common cause of human deafness (see Box 11.6). In most cases, the auditory nerve remains intact, making it possible to restore some hearing with a *cochlear implant*—essentially an artificial, electronic cochlea. The roots of this technology can be traced back two centuries to the pioneering work of the Italian physicist Alessandro Volta (after whom the electrical unit called the *volt* was named). In 1800, shortly after he invented the battery, Volta gamely (some might say foolishly) inserted the two contacts from a 50 volt battery into his ears. Here's how he described the result:

> At the moment when the circuit was completed, I received a shock in the head, and some moments after I began to hear a sound, or rather noise in the ears, which I cannot well define: it was a kind of crackling with shocks, as if some paste or tenacious matter had been boiling. . . The disagreeable sensation, which I believe might be dangerous because of the shock in the brain, prevented me from repeating this experiment. . . .[1]

We strongly advise you *not* to try this at home.

The art of electrically stimulating the ears has improved considerably since Volta's singular experiment. In fact, in recent years, cochlear implant systems have revolutionized the treatment of inner ear damage for many people. Most of the system is actually external to the body (Figure A). It starts with a headpiece containing a microphone, which receives sound and converts it to an electrical signal. This signal is sent to a battery-powered digital processor. A small radio transmitter placed over the scalp transmits the digital code to a receiver that has been surgically implanted beneath the skin, in the mastoid bone behind the ear. The transmitter and receiver are held close to one another with magnets, and no wires penetrate the skin.

The receiver translates the code into a series of electrical impulses that it sends to the cochlear implant itself—a very thin, flexible bundle of wires that has been threaded through a tiny hole and into the cochlea (Figure B). The cochlear electrode array has about 22 separate stimulation sites that allow it to activate the auditory nerve at various places along the cochlea, from the base toward the apex. The most clever feature of the cochlear implant is that it takes advantage of the tonotopic arrangement of auditory nerve fibers; stimulation near the base of the cochlea evokes a perception of high-frequency sounds, and stimulation toward the apex evokes low-frequency sounds.

By 2012, there were more than 340,000 cochlear implant users in the world, and the popularity of the devices is increasing. About 38,000 children have implants in the U.S. alone. Unfortunately, they are very expensive.

Cochlear implants can provide an extraordinary hearing capacity for many previously deaf people. With several months of training, people can achieve a remarkably good understanding of conversational speech, even when listening on a telephone. Most can understand more than 90% of spoken words when listening in a quiet room.

[1] Quoted in Zeng F-G. 2004. Trends in cochlear implants. *Trends in Amplification* 8:1–34.

basilar is at the *base* of the organ of Corti, the *tectorial* forms a *roof* over the structure, and the *reticular* is in the *middle*, holding onto the hair cells.

Hair cells form synapses on neurons whose cell bodies are located in the **spiral ganglion** within the modiolus. Spiral ganglion cells are bipolar, with neurites extending to the bases and sides of the hair cells, where they receive synaptic input. Axons from the spiral ganglion enter the *auditory nerve*, a branch of the **auditory–vestibular nerve** (cranial nerve VIII), which projects to the cochlear nuclei in the medulla. It is possible to treat certain forms of deafness by using electronic devices to bypass the middle ear and the hair cells, and activate the auditory nerve axons directly (Box 11.2).

Transduction by Hair Cells. When the basilar membrane moves in response to a motion at the stapes, the entire foundation supporting the hair cells moves because the basilar membrane, rods of Corti, reticular lamina, and hair cells are all rigidly connected. These structures move as a unit, pivoting up toward the tectorial membrane or away from it (Figure 11.13). When the basilar membrane moves up, the reticular

The success of the implants varies widely, for reasons that are often unclear. Researchers are working hard to improve the technology of cochlear implants, to reduce their size, and to determine the best ways to train patients in their use.

The best candidates for cochlear implants are young children (optimally as young as one year), and older children or adults whose deafness was acquired after they learned some speech. For adults whose deafness preceded any experience with speech, on the other hand, cochlear implants seem to provide only a crude perception of sounds. It seems that the auditory system, like other sensory systems in the brain, needs to experience normal inputs at a young age in order to develop properly. If it is deprived of exposure to sounds early in life, the auditory system can never develop completely normal function even if hearing is restored later. The concept of critical periods in brain development is described in Chapter 23.

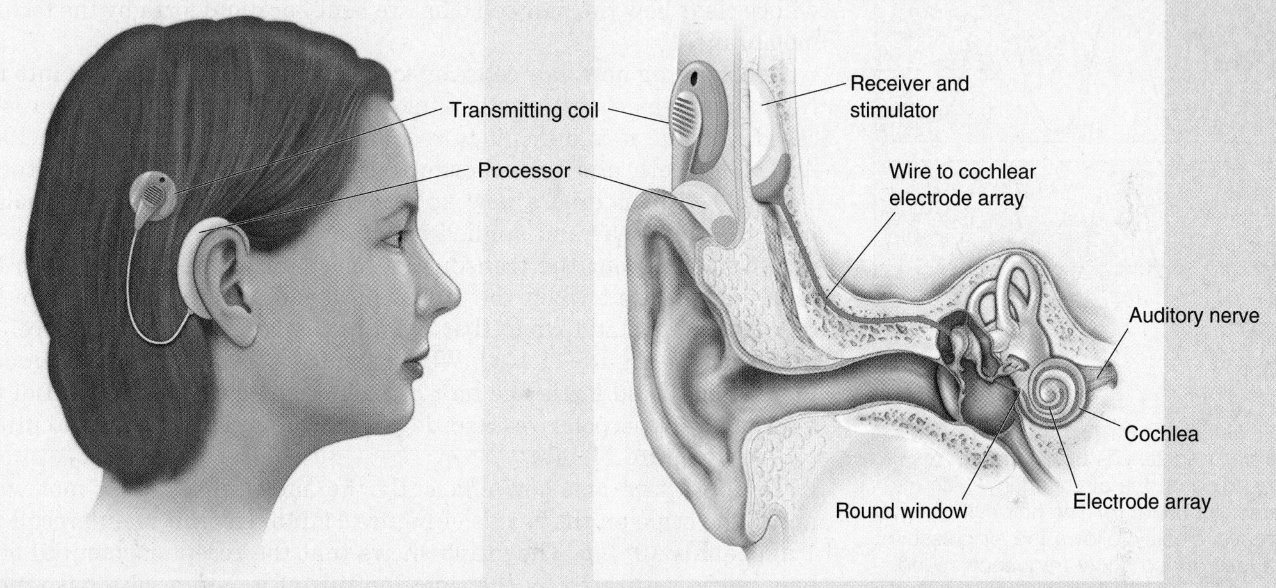

Figure A
The headpiece of a cochlear implant, positioned behind the ear.

Figure B
A cochlear implant in place inside the ear.

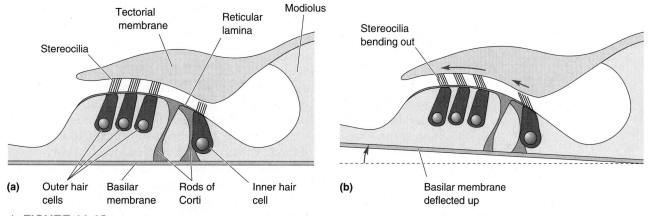

▲ FIGURE 11.13
The bending of stereocilia produced by the upward motion of the basilar membrane. **(a)** At rest, the hair cells are held between the reticular lamina and the basilar membrane, and the tips of the outer hair cell stereocilia are attached to the tectorial membrane. **(b)** When sound causes the basilar membrane to deflect upward, the reticular lamina moves up and inward toward the modiolus, causing the stereocilia to bend outward.

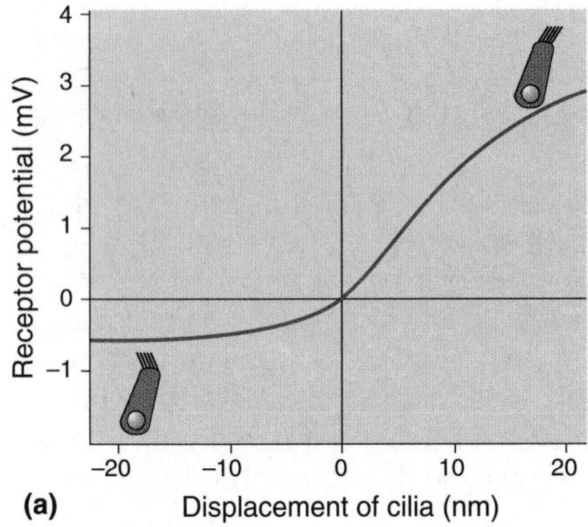

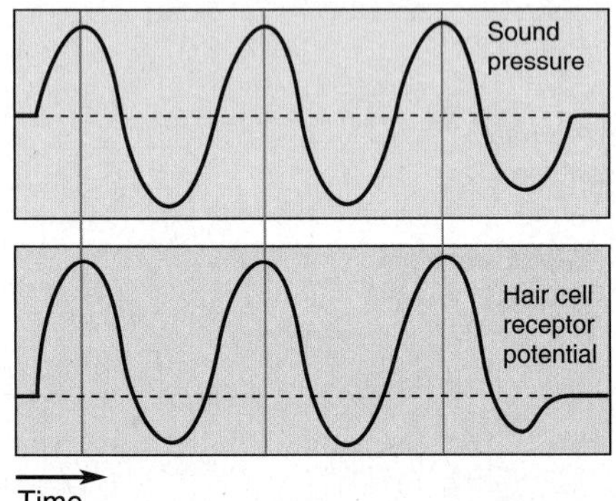

▲ FIGURE 11.14
Hair cell receptor potentials. (a) The hair cell depolarizes or hyperpolarizes, depending on the direction in which the stereocilia bend. **(b)** The hair cell receptor potential closely follows the air pressure changes during a low-frequency sound.

lamina moves up and in toward the modiolus. Conversely, downward motion of the basilar membrane causes the reticular lamina to move down and away from the modiolus. When the reticular lamina moves inward or outward relative to the modiolus, it also moves in or out with respect to the tectorial membrane. Because the tectorial membrane holds the tips of the outer hair cell stereocilia, the lateral motion of the reticular lamina relative to the tectorial membrane bends the stereocilia on the outer hair cells one way or the other. The tips of stereocilia from inner hair cells are also bent, probably because they are pushed by moving endolymph. Aligned actin filaments make stereocilia rigid rods, and they bend only at the base, where they attach to the top of the hair cell. Cross-link filaments make the stereocilia stick to one another, so all the cilia move as a unit. Now imagine a sound wave making the basilar membrane jiggle between the two positions shown in parts a and b of Figure 11.13, and it will be clear how the hair cell cilia are bent back and forth by the tectorial membrane.

Determining how hair cells convert the bending of stereocilia into neural signals was a very challenging problem. Because the cochlea is encased in bone, it is difficult to record from the hair cells. In the 1980s, A. J. Hudspeth and his colleagues, then at the California Institute of Technology, pioneered a new approach in which hair cells are isolated from the inner ear and studied *in vitro*. The *in vitro* technique has revealed much about the transduction mechanism. Recordings from hair cells indicate that when the stereocilia bend in one direction, the hair cell depolarizes, and when they bend in the other direction, the cell hyperpolarizes (Figure 11.14a). When a sound wave causes the stereocilia to bend back and forth, the hair cell generates a receptor potential that alternately hyperpolarizes and depolarizes from the resting potential of −70 mV (Figure 11.14b).

To appreciate just how efficiently the ear works, take a moment to notice the scale on the X axis of Figure 11.14a. Its unit is nm; recall that 1 nm equals 10^{-9} m. The graph shows that the receptor potential of the hair cell is saturated by the time the tips of its stereocilia have moved about 20 nm to the side; this is what an extremely loud sound might do. But the softest sound you can hear moves the stereocilia only 0.3 nm to each side, which is an astoundingly small distance—about the diameter of a large atom! Since each stereocilium is about 500 nm (or 0.5 μm) in diameter, a very soft sound needs to wiggle the stereocilia only about 1/1000 of their diameter in order to produce a perceptible sound. How does the hair cell transduce such infinitesimally small amounts of sound energy?

The tip of each stereocilium has a special type of ion channel that is induced to open and close by the bending of stereocilia. When these mechanosensitive transduction channels are open, an inward ionic current flows and generates the hair cell receptor potential. Despite considerable research effort, the molecular identity of the channels is still uncertain. One reason the channels have been so difficult to identify is that there are so few of them; the tip of each stereocilium has only one or two such channels, and an entire hair cell may have only 100. Some recent experiments suggest that the hair cell transduction channels belong to the *transmembrane protein-like (TMC)* family of proteins, but other studies challenge this conclusion. All we know for sure is that energetic research on this issue will continue.

Figure 11.15 shows how the transduction channels are believed to function. A stiff filament called a *tip link* connects each channel to the upper wall of the adjacent cilium. When the cilia are pointing straight up, the

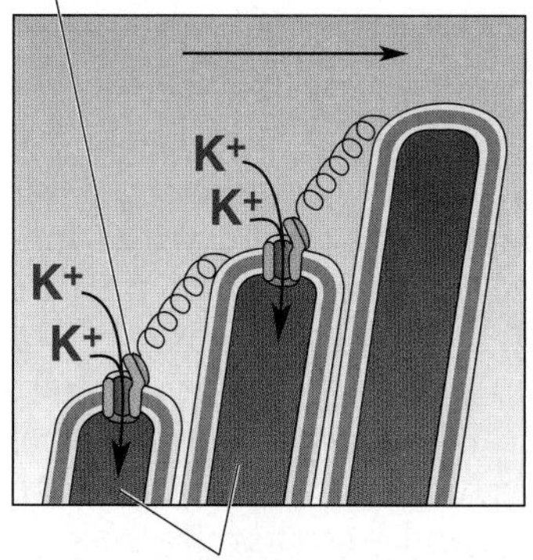

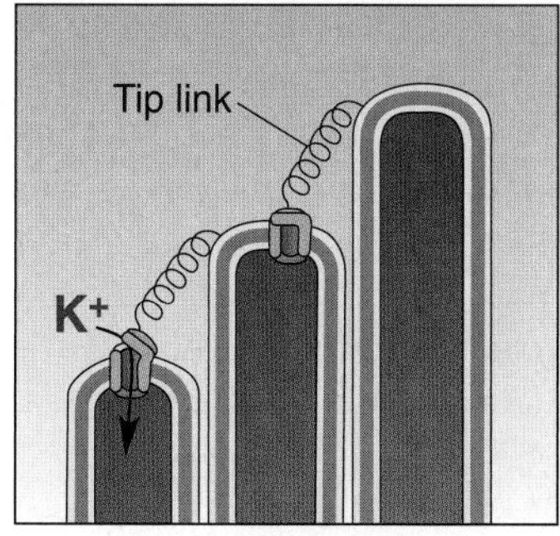

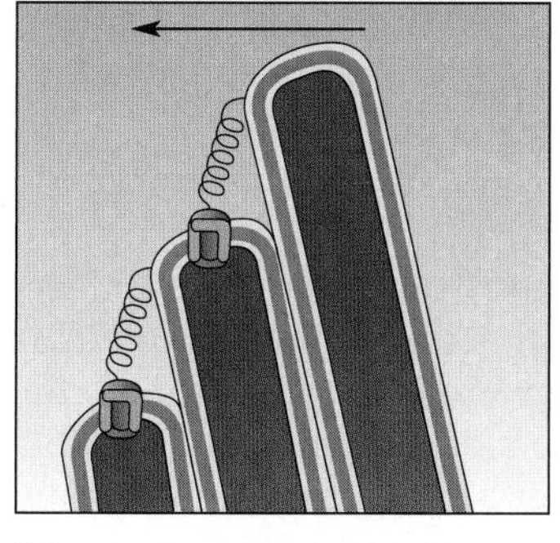

(a)

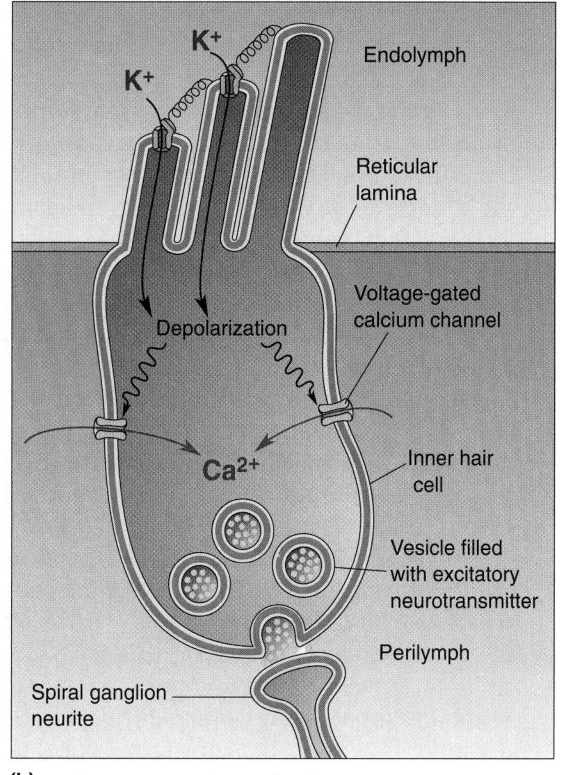

(b)

◀ FIGURE 11.15

Depolarization of a hair cell. **(a)** Ion channels on stereocilia tips are opened when the tip links joining the stereocilia are stretched. **(b)** The entry of K^+ depolarizes the hair cell, which opens voltage-gated calcium channels. Incoming Ca^{2+} leads to the release of neurotransmitter from synaptic vesicles, which then diffuses to the postsynaptic neurite from the spiral ganglion.

tension on the tip link causes the channel to spend part of the time in the opened state, allowing a small amount of K^+ to move from the endolymph into the hair cell. Displacement of the cilia in one direction increases tension on the tip link, increasing the rate of channel openings and the amount of inward K^+ current. Displacement in the opposite direction relieves tension on the tip link, thereby causing the channel to spend more time closed, reducing inward K^+ movement. The entry of K^+ into the hair cell causes a depolarization, which in turn activates voltage-gated calcium channels (Figure 11.15b). The entry of Ca^{2+} triggers the release of the neurotransmitter glutamate, which activates the spiral ganglion fibers lying postsynaptic to the hair cell.

It is interesting that the opening of K^+ channels produces a depolarization of the hair cell, whereas the opening of K^+ channels *hyperpolarizes* most neurons. The reason that hair cells respond differently from neurons is the unusually high K^+ concentration in endolymph, which yields a K^+ equilibrium potential of 0 mV, compared to the equilibrium potential of −80 mV in typical neurons. Another reason that K^+ is driven into hair

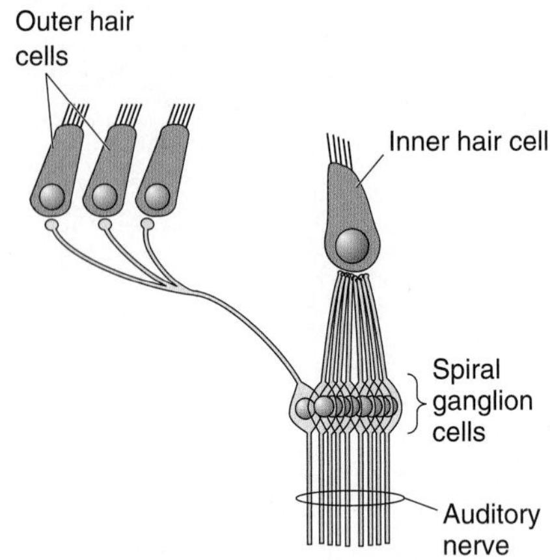

▲ FIGURE 11.16
The innervation of hair cells by neurons from the spiral ganglion.

cells is the 80 mV endocochlear potential, which helps create a 125 mV gradient across the stereocilia membranes.

Hair Cells and the Axons of the Auditory Nerve. The auditory nerve consists of the axons of neurons whose cell bodies are located in the spiral ganglion. Thus, the spiral ganglion neurons, which are the first in the auditory pathway to fire action potentials, provide all the auditory information sent to the brain. Curiously, the numbers of auditory axons contacting the inner and outer hair cells are enormously different. There are about 35,000–50,000 neurons altogether in the spiral ganglion. Despite the fact that inner hair cells are outnumbered by outer hair cells by a factor of 3 to 1, more than 95% of the spiral ganglion neurons communicate with the relatively small number of inner hair cells, and less than 5% receive synaptic input from the more numerous outer hair cells (Figure 11.16). Consequently, one spiral ganglion fiber receives input from only one inner hair cell; moreover, each inner hair cell feeds about 10 spiral ganglion neurites. The situation is the opposite with outer hair cells. Because they outnumber their spiral ganglion cells, one spiral ganglion fiber synapses with numerous outer hair cells.

Simply based on these numbers, we can infer that the vast majority of the information leaving the cochlea comes from inner hair cells. If it's true that the brain is paying very little attention to the outer hair cells, then what are the outer hair cells for?

Amplification by Outer Hair Cells. Given that outer hair cells far outnumber inner hair cells, it seems paradoxical that most of the cochlear output is derived from inner hair cells. However, ongoing research suggests that outer hair cells play a critical role in sound transduction. Ironically, one clue to the nature of this role was the discovery that the ear not only transduces sound; it can create it, too (Box 11.3).

Outer hair cells seem to act like tiny motors that amplify the movement of the basilar membrane during low-intensity sound stimuli. This action of the outer hair cells on the basilar membrane is called the **cochlear amplifier**. There are two molecular mechanisms that probably contribute to this amplifier. The first, and best understood, mechanism involves special *motor proteins* found only in the membranes of outer hair cells (Figure 11.17a). Motor proteins can change the length of outer hair cells, and outer hair cells respond to sound with both a receptor potential and a change in length (Figure 11.17b). The motor proteins do not resemble any other system of cellular movement. The hair cells' motor is driven by the receptor potential, and it does not use adenosine triphosphate (ATP) as an energy source. It is also extremely fast, as it must be able to keep up with the movements induced by high-frequency sounds. The hair cell's primary motor is an unusual protein called *prestin* (from the musical notation *presto*, meaning "fast"). Prestin molecules are tightly packed into the membranes of the outer hair cell bodies, and they are required for outer hair cells to move in response to sound. A second possible molecular mechanism of the cochlear amplifier is located right in the hair bundles. A special type of the contractile protein *myosin* is attached to the upper end of the tip links. Myosin and other tip-link proteins may somehow rapidly enhance the movement of hairs in response to weak sounds; this idea is controversial.

Because outer hair cells are attached to the basilar membrane and reticular lamina, when motor proteins change the length of the hair cell, the basilar membrane is pulled toward or pushed away from the reticular lamina and tectorial membrane. This is why the word "motor" is used;

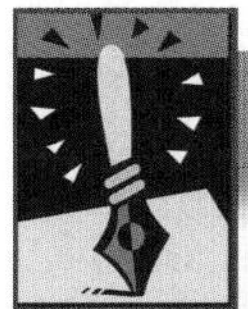

BOX 11.3 OF SPECIAL INTEREST

Hearing with Noisy Ears

Sensory systems are supposed to detect stimulus energy in the environment, not generate it. Can you imagine eyes glowing in the dark, or noses smelling like roses? How about ears buzzing loudly? The truth is that retinas don't radiate light, and olfactory receptors don't emit odors, but some ears can definitely generate sounds loud enough for a bystander to hear! Such sounds are called *otoacoustic emissions*. In one early description, a man sitting next to his dog realized the animal was humming; after some anxious investigation, he discovered that the sound came from one of the dog's ears.

The ears of all vertebrates, including humans, can emit sounds. Presenting a short sound stimulus, such as a click, to a normal human ear causes an "echo" that can be picked up with a sensitive microphone in the auditory canal. We don't usually notice such echoes because they are too faint to be heard over other sounds in the environment.

Ears that emit relatively loud sounds spontaneously, in the absence of any incoming sound, have usually sustained cochlear damage as the result of exposure to extremely loud sounds (from explosions, machines, rock bands), drugs, or disease. If spontaneous otoacoustic emissions are loud enough, they may cause one form of tinnitus—a ringing in the ears (see Box 11.6).

The mechanism that causes the ear to generate its *own* sounds—the cochlear amplifier—is the same one that functions to improve its detection of *environmental* sounds, but operating in reverse. Normal outer hair cells, stimulated with a click, react with a quick movement that drives the cochlear fluids and membranes, which move the ossicles, and ultimately vibrate the tympanic membrane to produce sound in the outside air (the echo). Spontaneous emissions occur because the sensitivity of the cochlear amplifier is very high. Most people with normal hearing can perceive them in an exceptionally quiet environment.

Damaged regions of the cochlea can somehow facilitate the spontaneous movement of some outer hair cells, so that they vibrate all the time. Strangely enough, most people are unaware that their ears are broadcasting sounds. Apparently, their central auditory neurons recognize the spontaneous cochlear activity as noise and suppress the perception of it. The benefit is that they are spared from an otherwise maddening tinnitus, but the cost is a partial hearing loss in the affected frequency range.

Because otoacoustic emissions are a normal attribute of ears, they can be used as a quick and easy test of ear function. A series of sounds is played into the ears, and the echoes they evoke are recorded and analyzed. The characteristics of the echoes can tell us a lot about the function of the middle and inner ears. This is especially useful for testing people who are unable to tell the examiner whether or not they have heard test sounds—newborn babies, for example.

the outer hair cells actively change the physical relationship between the cochlear membranes.

The motor effect of outer hair cells makes a significant contribution to the traveling wave that propagates down the basilar membrane. This was first demonstrated in 1991 by Mario Ruggero and Nola Rich at the University of Minnesota, who administered the chemical furosemide into experimental animals. Furosemide temporarily decreases the transduction that normally results from the bending of stereocilia on hair cells, and it was found to significantly reduce the movement of the basilar membrane in response to sound (Figure 11.17c, d). This effect of furosemide is believed to result from inactivation of the outer hair cell motor proteins and loss of the cochlear amplifier. When the outer hair cells amplify the response of the basilar membrane, the stereocilia on the inner hair cells bend more, and the increased transduction process in the inner hair cells produces a greater response in the auditory nerve. Through this feedback system, therefore, outer hair cells contribute significantly to the output of the cochlea. Without the cochlear amplifier, the peak movement of the basilar membrane would be about 100-fold smaller.

The effect of outer hair cells on the response of inner hair cells can be modified by neurons outside the cochlea. In addition to the spiral ganglion afferents that project from the cochlea to the brain stem, there

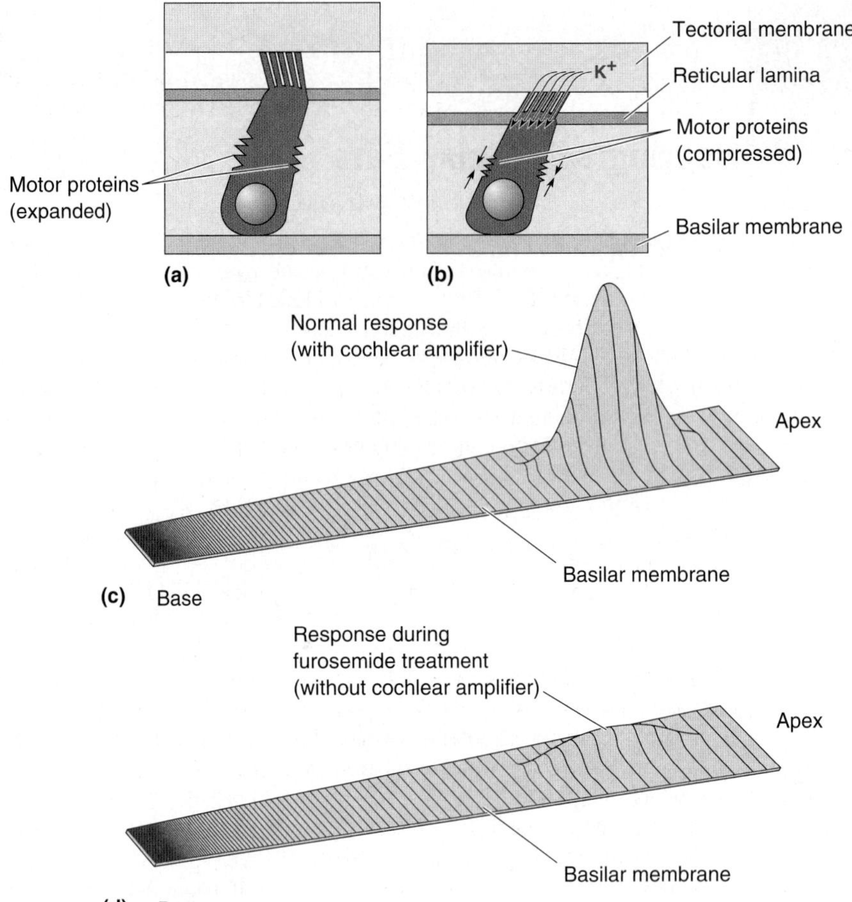

▶ FIGURE 11.17
Amplification by outer hair cells. **(a)** Motor proteins in the membranes of outer hair cells. **(b)** Bending of the stereocilia causes potassium to enter the hair cell, depolarizing it, and triggering motor proteins to activate and shorten the hair cell. **(c)** The shortening and lengthening of the hair cell increase the flexing of the basilar membrane. **(d)** Furosemide decreases hair cell transduction, consequently reducing the flexing of the basilar membrane. (Source: Adapted from Ashmore and Kolston, 1994, Figs. 2, 3.)

are also about 1000 efferent fibers projecting *from* the brain stem *toward* the cochlea. These efferents diverge widely, synapsing onto outer hair cells and releasing acetylcholine. Stimulation of these efferents changes the shape of the outer hair cells, thereby affecting the responses of inner hair cells. In this way, descending input from the brain to the cochlea can regulate auditory sensitivity.

The amplifying effect of outer hair cells explains how certain antibiotics (e.g., kanamycin) that damage hair cells can lead to deafness. After excessive exposure to antibiotics, many inner hair cells are less sensitive to sound. However, the antibiotic almost exclusively damages outer hair cells, not inner hair cells. For this reason, deafness produced by antibiotics is thought to be a consequence of damage to the cochlear amplifier (i.e., outer hair cells), demonstrating just how essential a role the amplifier plays.

Prestin, the protein that is essential for the outer hair cells' motor, is also necessary for the cochlear amplifier to function. When the gene that encodes prestin is eliminated from mice, the animals are nearly deaf; their ears are more than 100-fold less sensitive to sound than normal.

CENTRAL AUDITORY PROCESSES

The auditory pathway appears more complex than the visual pathway because there are more nuclei intermediate between the sensory organ and the cortex. Also, in contrast to the visual system, there are many more alternative pathways by which signals can travel from one nucleus to the next. Nonetheless, the amount of information processing in the two systems is similar when you consider that the cells and synapses of the auditory system in the brain stem are analogous to interactions in the layers of the retina. We will now look at auditory circuitry, focusing on the transformations of auditory information that occur along the way.

The Anatomy of Auditory Pathways

Afferents from the spiral ganglion enter the brain stem in the auditory–vestibular nerve. At the level of the medulla, the axons innervate the **dorsal cochlear nucleus** and **ventral cochlear nucleus** ipsilateral to the cochlea where the axons originated. Each axon branches so that it synapses on neurons in both cochlear nuclei. From this point on, the system gets more complicated, and the connections are less well understood, because there are multiple parallel pathways. Rather than trying to describe all of these connections, we will follow one particularly important pathway from the cochlear nuclei to auditory cortex (Figure 11.18). Cells in the ventral cochlear nucleus send axons that project to the **superior olive** (also called the *superior olivary nucleus*) on both sides of the brain stem. Axons of the olivary neurons ascend in the *lateral lemniscus* (a lemniscus is a collection of axons) and innervate the **inferior colliculus** of the midbrain. Many efferents of the dorsal cochlear nucleus follow a route similar to the pathway from the ventral cochlear nucleus, but the dorsal path bypasses the superior olive. Although there are other routes from the cochlear nuclei to the inferior colliculus, with additional intermediate relays, *all ascending auditory pathways converge onto the inferior colliculus.* The neurons in the inferior colliculus send axons to the medial geniculate nucleus (MGN) of the thalamus, which in turn projects to auditory cortex.

Before moving on to the response properties of auditory neurons, we should make several points:

1. Projections and brain stem nuclei other than the ones described contribute to the auditory pathways. For instance, the inferior colliculus sends axons not only to the MGN but also to the superior colliculus (where the integration of auditory and visual information occurs) and to the cerebellum.
2. There is extensive feedback in the auditory pathways. For instance, brain stem neurons send axons that contact outer hair cells, and auditory cortex sends axons to the MGN and inferior colliculus.
3. Each cochlear nucleus receives input from just the one ear on the ipsilateral side; all other auditory nuclei in the brain stem receive input from both ears. This explains the clinically important fact that the only way by which brain stem damage can produce deafness in one ear is if a cochlear nucleus (or auditory nerve) on one side is destroyed.

Response Properties of Neurons in the Auditory Pathway

To understand the transformations of auditory signals that occur in the brain stem, we must first consider the nature of the input from the neurons in the spiral ganglion of the cochlea. Because most spiral ganglion cells receive input from a single inner hair cell at a particular location on the basilar membrane, they fire action potentials only in response to sound within a limited frequency range. After all, hair cells are excited by deformations of the basilar membrane, and each portion of the membrane is maximally sensitive to a particular range of frequencies.

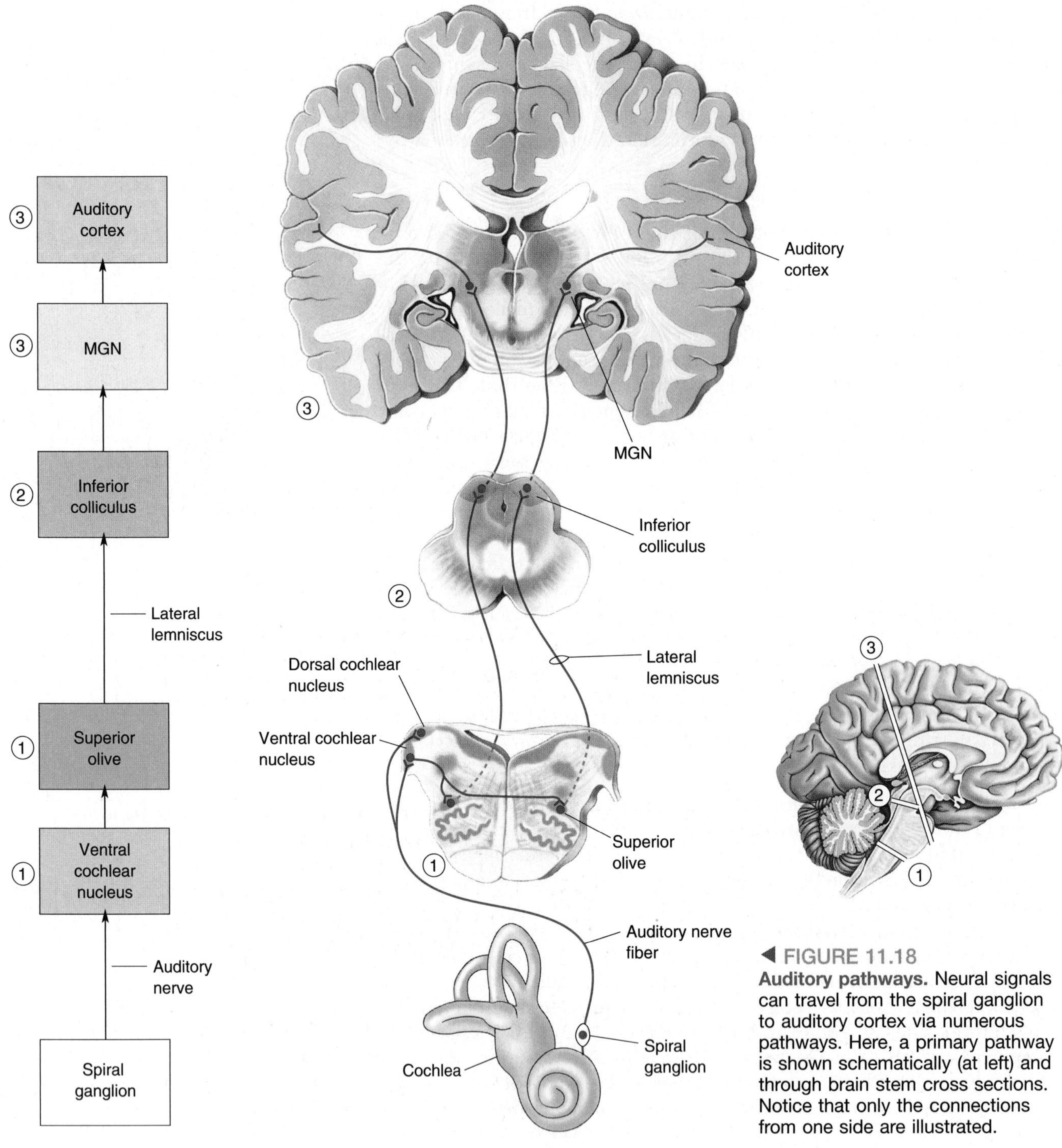

◀ FIGURE 11.18
Auditory pathways. Neural signals can travel from the spiral ganglion to auditory cortex via numerous pathways. Here, a primary pathway is shown schematically (at left) and through brain stem cross sections. Notice that only the connections from one side are illustrated.

Figure 11.19 shows the results of an experiment in which action potentials were recorded from a single auditory nerve fiber (i.e., the axon of a spiral ganglion cell). The graph represents the firing rate in response to sounds at different frequencies. The neuron is most responsive to sound at one frequency, called the neuron's **characteristic frequency**, and it is less responsive at neighboring frequencies. This type of frequency tuning is seen in many neurons in each of the relays from cochlea to cortex.

As one ascends, the auditory pathway in the brain stem, the response properties of the cells become more diverse and complex, just as in the visual pathway. For instance, some cells in the cochlear nuclei are especially sensitive to sounds varying in frequency over time (think of the sound of a trombone as it slides from a low note to a high note). In the MGN, there are cells that respond to fairly complex sounds such as vocalizations, as well as other cells that show simple frequency selectivity, as in the auditory nerve. An important development in the superior olive is that cells receive input from cochlear nuclei on both sides of the brain stem. As discussed later, such binaural neurons are probably important for sound localization.

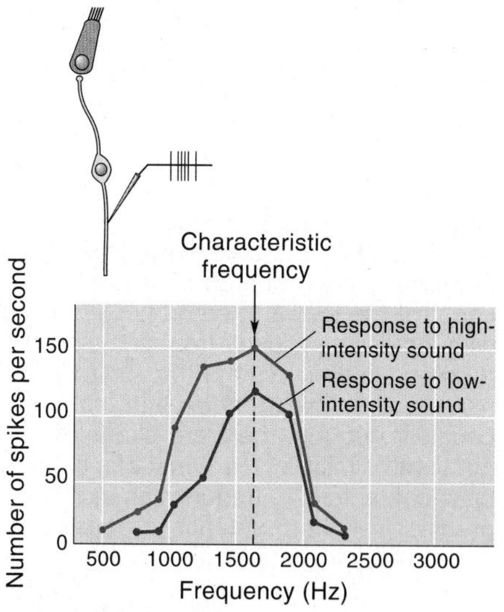

▲ FIGURE 11.19

The response of an auditory nerve fiber to different sound frequencies. This neuron is frequency-tuned and has its greatest response at the characteristic frequency. (Source: Adapted from Rose, Hind, Anderson, and Brugge, 1971, Fig. 2.)

ENCODING SOUND INTENSITY AND FREQUENCY

If you stop reading this book for a moment, you can focus on the many sounds around you. You can probably hear sounds you have been ignoring, and you can selectively pay attention to different sounds occurring at the same time. We are usually bathed in an amazing diversity of sounds—from chattering people to cars to electrical noises to sounds generated within our own bodies—and our brain must be able to analyze just the important sounds while ignoring the noise. We cannot yet account for the perception of each of these sounds by pointing to particular neurons in the brain. However, most sounds have certain features in common, including intensity, frequency, and the location from which they emanate. Each of these features is represented differently in the auditory pathway.

Stimulus Intensity

Information about sound intensity is coded in two interrelated ways: the firing rates of neurons and the number of active neurons. As a stimulus gets more intense, the basilar membrane vibrates with greater amplitude, causing the membrane potential of the activated hair cells to be more depolarized or hyperpolarized. As a result, the nerve fibers with which the hair cells synapse fire action potentials at greater rates. In Figure 11.19, the auditory nerve fiber fires faster to the same sound frequencies when the intensity is increased. In addition, more intense stimuli produce movements of the basilar membrane over a greater distance, which leads to the activation of more hair cells. In a single auditory nerve fiber, this increase in the number of activated hair cells causes a broadening of the frequency range to which the fiber responds. The loudness we perceive is correlated with the number of active neurons in the auditory nerve (and throughout the auditory pathway) and with their firing rates.

Stimulus Frequency, Tonotopy, and Phase Locking

From the hair cells in the cochlea through the various nuclei leading to auditory cortex, most neurons are sensitive to stimulus frequency. They are most sensitive at their characteristic frequency. How is frequency represented in the central nervous system?

Tonotopy. Frequency sensitivity is largely a consequence of the mechanics of the basilar membrane because different portions of the membrane are maximally deformed by sound of different frequencies. Moving from the base to the apex of the cochlea, a progressive decrease occurs in the frequency that produces a maximal deformation of the basilar membrane. This is an example of tonotopy, as we

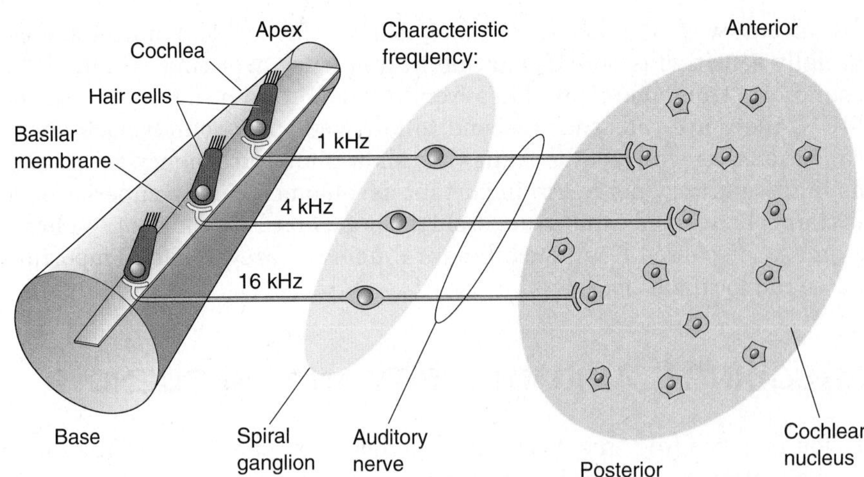

▶ FIGURE 11.20
Tonotopic maps on the basilar membrane and cochlear nucleus. From the base to the apex of the cochlea, the basilar membrane resonates with increasingly lower frequencies. This tonotopy is preserved in the auditory nerve and cochlear nucleus. In the cochlear nucleus, there are bands of cells with similar characteristic frequencies; characteristic frequencies increase progressively from anterior to posterior.

discussed earlier. There is a corresponding tonotopic representation in the auditory nerve; auditory nerve fibers connected to hair cells near the apical basilar membrane have low characteristic frequencies, and those connected to hair cells near the basal basilar membrane have high characteristic frequencies (Figure 11.20). When auditory axons in the auditory–vestibular nerve synapse in the cochlear nuclei, they do so in an organized pattern based on characteristic frequency. Nearby neurons have similar characteristic frequencies, and a systematic relationship exists between position in the cochlear nucleus and characteristic frequency. In other words, there is a map of the basilar membrane within the cochlear nuclei.

Because of the tonotopy present throughout the auditory system, the location of active neurons in auditory nuclei is one indication of the frequency of the sound. However, frequency must be coded in some way other than the site of maximal activation in tonotopic maps, for two reasons. One reason is that these maps do not contain neurons with very low characteristic frequencies, below about 200 Hz. As a result, the site of maximal activation might be the same for a 50 Hz tone as for a 200 Hz tone, so there must be some other way to distinguish them. The second reason that something other than tonotopy is needed is that the region of the basilar membrane maximally displaced by a sound depends on its intensity in addition to its frequency (see Figure 11.19). At a fixed frequency, a more intense sound will produce a maximal deformation at a point further up the basilar membrane than a less intense sound.

Phase Locking. The timing of neural firing provides an additional type of information about sound frequency that complements information derived from tonotopic maps. Recordings made from neurons in the auditory nerve show **phase locking**, the consistent firing of a cell at the same phase of a sound wave (Figure 11.21). If you think of a sound wave as a sinusoidal variation in air pressure, a phase-locked neuron would fire action potentials at the peaks, or the troughs, or some other constant location on the wave. At low frequencies, some neurons fire action potentials every time the sound is in a particular phase (Figure 11.21a). This makes it easy to determine the frequency of the sound; it is the same as the frequency of the neuron's action potentials.

Phase locking can still occur even if an action potential has not fired on every cycle (Figure 11.21b). For instance, a neuron may respond to a

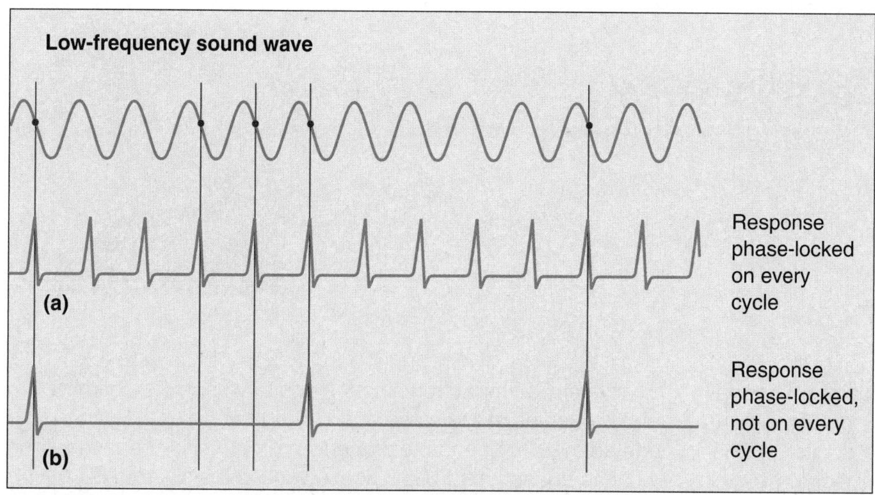

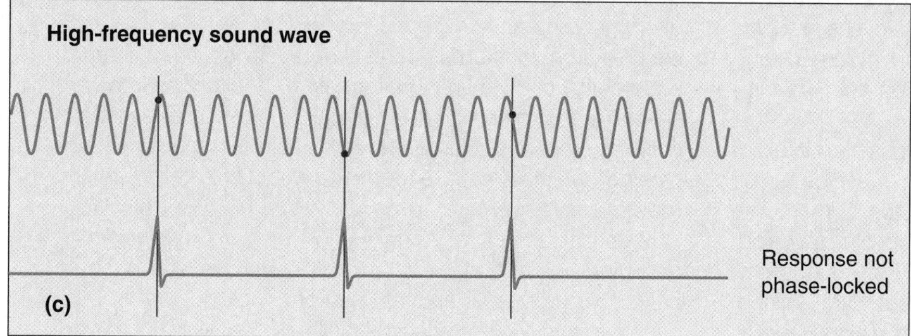

◀ FIGURE 11.21
Phase locking in the response of auditory nerve fibers. Sound at a low frequency can elicit a phase-locked response, either **(a)** on every cycle of the stimulus or **(b)** on some fraction of the cycles. **(c)** At high frequencies, the response does not have a fixed phase relationship to the stimulus.

1000 Hz sound with an action potential on only perhaps 25% of the cycles of the input, but those action potentials still always occur at the same phase of the sound. If you have a group of such neurons, each responding to different cycles of the input signal, it is possible to have a response to every cycle (by some member of the group) and thus a measure of sound frequency. It is likely that intermediate sound frequencies are represented by the pooled activity of a number of neurons, each of which fires in a phase-locked manner; this is called the **volley principle**. Phase locking occurs with sound waves up to about 5 kHz. Above this point, the action potentials fired by a neuron occur at random phases of the sound wave (Figure 11.21c) because the intrinsic variability in the timing of the action potential becomes comparable to the time interval between successive cycles of the sound. In other words, the sound waves cycle too fast for the action potentials of single neurons to accurately represent their timing. Above 5 kHz, frequencies are represented by tonotopy alone.

Many auditory neurons in the brainstem have peculiar membrane properties that make them uniquely sensitive to the precise timing of their synaptic inputs. Adaptations for precise timing are particularly impressive in neurons of the cochlear nuclei, as research by Donata Oertel and her colleagues at the University of Wisconsin has demonstrated (Box 11.4).

To summarize, here is how different sound frequencies are represented by brain stem neurons. At very low frequencies, phase locking is used; at intermediate frequencies, both phase locking and tonotopy are useful; and at high frequencies, tonotopy must be relied on to indicate sound frequency.

BOX 11.4 PATH OF DISCOVERY

Capturing the Beat

by Donata Oertel

While the timing of action potentials carries information everywhere in the brain, the time scales on which neurons work vary dramatically. Neurons in the auditory nuclei of the brain stem can fire with a temporal precision better than 200 μsec. Contrast this with cortical neurons, whose responses to identical stimuli are a hundred-fold less precise. In the auditory system, the timing of firing conveys important sensory information about the pitch of sounds and whether sounds come from the right or left.

In the mid-1960s, computers made possible detailed analyses of the relationship between the waveforms of sounds and the firing of neurons. These studies, some done by my colleagues at the University of Wisconsin, revealed that auditory neurons encode the frequency of sounds not only by their position in the tonotopic map but also by firing in phase with the sounds, by phase locking. This temporal code breaks down at sound frequencies higher than 5 kHz because the firing of neurons is not precise enough to resolve periods shorter than about 200 μsec.

Phase locking at low frequencies is valuable to human beings. First, our impressive ability to distinguish tones as similar as 1000 Hz and 1002 Hz seems to depend on phase locking by neurons in the brain stem. Phase-locked neurons also detect the relative time of arrival of a sound at the two ears with every cycle of the sound, a mechanism important for localizing sounds in the horizontal plane.

How can auditory neurons convey information with a temporal precision of 200 μsec across multiple synapses—hair cells to spiral ganglion cells, to neurons of the cochlear nuclear nucleus, to neurons in the superior olive—using synaptic potentials and action potentials whose durations are in the millisecond range? To achieve this, the firing of postsynaptic neurons must follow the firing of the presynaptic neurons rapidly and with an unvarying delay.

In 1979, Bill Rhode, Phil Smith, and I began to address these issues by making intracellular recordings in anesthetized cats, but those experiments were terribly difficult. One of the problems was that auditory brain stem nuclei are difficult to reach, being surrounded by cerebellum, the inner ear, and the jaw. Another was that blood pulsations and respiratory movements made the microelectrodes unstable. In 1980, I realized I could eliminate these difficulties by making recordings in brain slices, a technique then being used to study the rodent hippocampus and the chicken brain stem. I developed a slice preparation of the mouse cochlear nuclei. There were lots of details to work out. I had to learn how to remove the brain stem without stretching the auditory nerve, to optimize the properties of the saline that bathed the tissue, to generate a flow of saline fast enough to promote efficient gas exchange but not so turbulent that it yanked the electrodes out of cells. One of the first things I discovered is that some auditory neurons have exceptionally low membrane resistance and fast time constants. These properties help them sharpen and convey precise timing information. Auditory neurons are specialized for precision in other ways as well. Their synapses deliver exceptionally large currents using the speediest subtypes of the glutamate receptors to depolarize cells with low resistances rapidly.

A critical function of some auditory neurons is *coincidence detection*—detecting when two inputs arrive at the same time. Two groups of auditory neurons, octopus cells of the cochlear nuclei and principal cells of the superior olivary nucleus, are truly exquisite coincidence detectors. Nace Golding, Ramazan Bal, and Michael Ferragamo, working in my lab, demonstrated that octopus cells have exceptionally large, mutually opposing types of voltage-sensitive ion channels. These give the cells short time constants and allow them to detect coincidence in the submillisecond time range.

Nace and I puzzled over the fact that synaptic potentials sum over a fraction of one millisecond, yet auditory nerve inputs to octopus cells are activated by a traveling wave that sweeps down the cochlea over multiple milliseconds. Matthew McGinley helped to resolve that puzzle. Octopus cells earn their name by extending dendrites in only one direction; cell bodies send dendrites across the tonotopic array of auditory nerve fibers so that the earliest synaptic inputs (tuned to high frequencies) impinge on the tips of dendrites while the latest synaptic inputs (tuned to low frequencies) terminate near the cell body. EPSPs take time to propagate along the dendrites of octopus cells; different dendritic delays thus compensate for the traveling wave delays in the cochlea, allowing octopus cells to signal the occurrence of clicks and onsets of complex sounds with a single, sharply timed action potential. When Nace established his own lab, he studied principal cells of the medial superior olive, which compare the timing of inputs to the two ears. He found that they were in many ways similar to the octopus cells he had worked on as a graduate student. Both cells have specialized ion channels that make them fast, both use their dendrites for detecting coincident EPSPs, and both balance their need to fire action potentials with their need to integrate inputs without the interference of action potentials. And both are still devilishly difficult to record from *in vivo!*

MECHANISMS OF SOUND LOCALIZATION

While the use of frequency information is essential for interpreting sounds in our environment, sound localization can be of critical importance for survival. If a predator is about to eat you, finding the source of a sudden sound and running away are much more important than analyzing the subtleties of the sound. Wild animals do not eat humans very often anymore, but there are other situations in which sound localization can be helpful. If you are carelessly crossing the street, your localization of a car's horn may be all that saves you. Our current understanding of the mechanisms underlying sound localization suggests that we use different techniques for locating sources in the horizontal plane (left–right) and vertical plane (up–down).

If you close your eyes and plug one ear, you can locate a bird singing as it flies overhead almost as well as with both ears open. But if you try to locate the horizontal position of a duck quacking as it swims across a pond, you'll find that you're much less able using only one ear. Thus, good horizontal localization requires a comparison of the sounds reaching the two ears, whereas good vertical localization does not.

Localization of Sound in the Horizontal Plane

An obvious cue to the location of a sound source is the time at which the sound arrives at each ear. If we aren't facing a sound's source directly, it takes the sound longer to reach one ear than the other. For instance, if a sudden noise comes at you from the right, it will reach your right ear first (Figure 11.22a); it will arrive at your left ear later, after what is known as an *interaural time delay*. If the distance between your ears is 20 cm, sound coming from the right, perpendicular to your head, will reach your left ear 0.6 msec after reaching your right ear. If the sound comes from straight ahead, there is no interaural delay; and at angles between straight ahead and perpendicular, the delay will be between 0 and 0.6 msec (Figure 11.22b). Sounds from the left side yield delays opposite to those on the right. Thus, there is a simple relationship between location and interaural delay. Detected by specialized neurons in the brain stem, the delay enables us to locate the source of the sound in the horizontal plane. The interaural delays we can detect are impressively brief. People can discriminate the direction of a sound source in the horizontal plane with a precision of about 2°. This demands that they discriminate the 11 μsec difference between the times it takes a sound to reach their two ears.

If we don't hear the onset of a sound because it is a continuous tone rather than a sudden noise, however, we cannot know the initial arrival times of the sound at the two ears. Thus, continuous tones pose more of a problem for sound localization because they are always present at both ears. However, we can still use arrival time to localize the sound but in a slightly different manner from localizing a sudden sound. The only thing that can be compared with continuous tones is the time at which the same *phase* of the sound wave reaches each ear. Imagine you are exposed to a 200 Hz sound coming from the right. At this frequency, one cycle of the sound covers 172 cm, which is much more than the 20 cm distance between your ears. After a peak in the sound pressure wave passes the right ear, it takes 0.6 msec, the time for sound to travel 20 cm, before you detect the peak at the left ear. Of course, if the sound is straight ahead, peaks in the continuous tone will reach the ears simultaneously. Because the sound wave is much longer than the distance between the ears, we can reliably use the interaural delay of the peak in the wave to determine sound location.

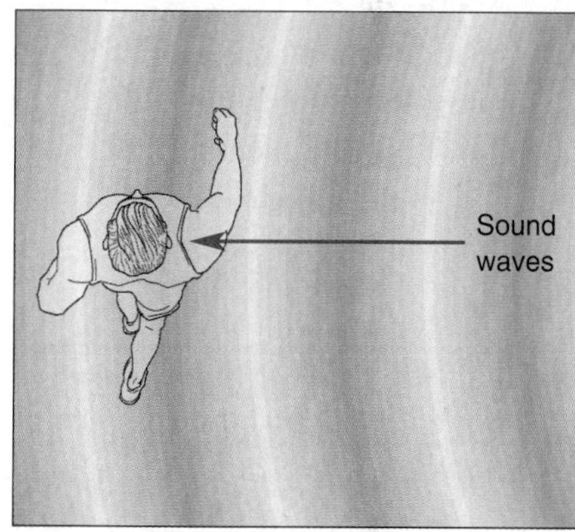

(a)

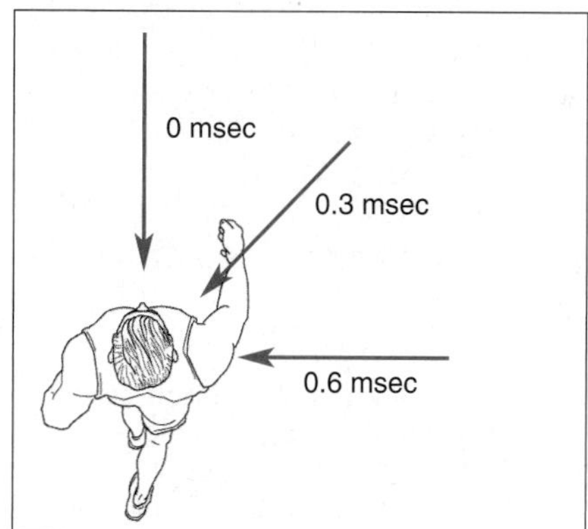

(b)

▲ FIGURE 11.22
Interaural time delay as a cue to the location of sound. **(a)** Sound waves coming from the right side will reach the right ear first, and there will be a large interaural delay before the sound propagates to the left ear. **(b)** If the sound comes from straight ahead, there is no interaural delay. Delays for three different sound directions are shown.

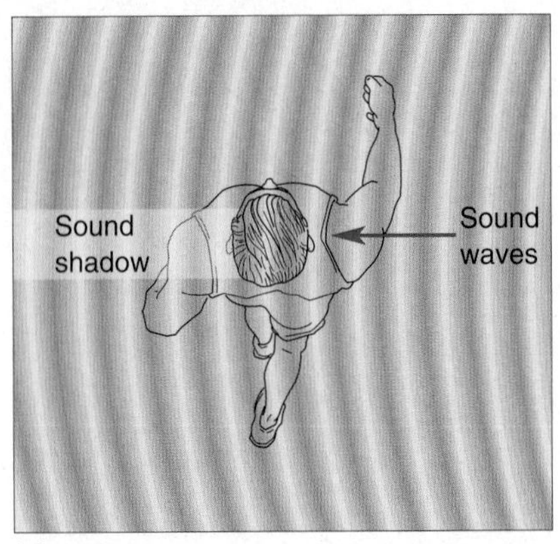

(a)

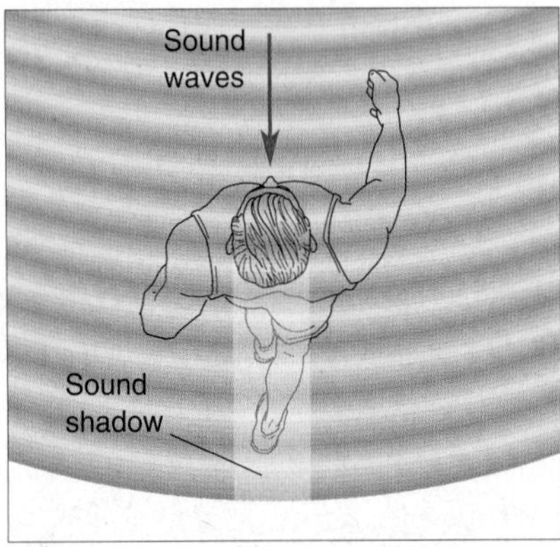

(b)

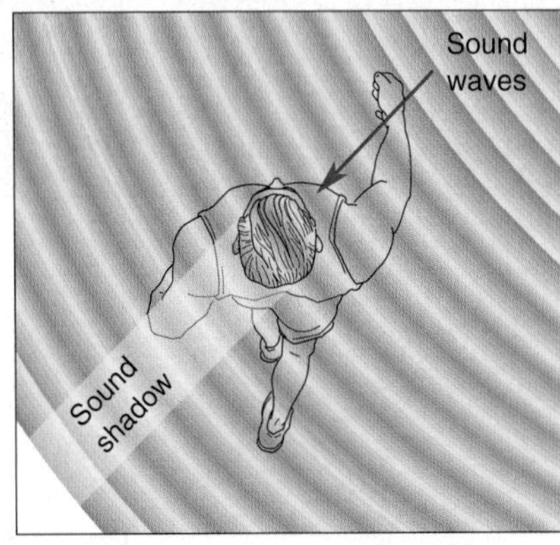

(c)

▲ FIGURE 11.23
Interaural intensity difference as a cue to sound location. **(a)** With high-frequency sound, the head will cast a sound shadow to the left, when sound waves come from the right. Lower intensity sound in the left ear is a cue that the sound came from the right. **(b)** If the sound comes from straight ahead, a sound shadow is cast behind the head, but the sound reaches the two ears with the same intensity. **(c)** Sound coming from an oblique angle will partially shadow the left ear.

Things are more complicated with continuous tones at high frequencies. Suppose that the sound coming from the right now has a frequency of 20,000 Hz, which means that one cycle of the sound covers 1.7 cm. After a peak reaches the right ear, does it still take 0.6 msec before a peak arrives at the left ear? No! It takes a much shorter time because many peaks of such a high-frequency wave will fit between your ears. No longer is there a simple relationship between the direction the sound comes from and the arrival times of the peaks at the two ears. Interaural arrival time is simply not useful for locating continuous sounds with frequencies so high that one cycle of the sound wave is smaller than the distance between your ears (i.e., greater than about 2000 Hz).

Fortunately, the brain has another process for sound localization at high frequencies. An *interaural intensity difference* exists between the two ears because your head effectively casts a sound shadow (Figure 11.23). There is a direct relationship between the direction the sound comes from and the extent to which your head shadows the sound to one ear. If sound comes directly from the right, the left ear will hear a significantly lower intensity (Figure 11.23a). With sound coming from straight ahead, the same intensity reaches the two ears (Figure 11.23b), and with sound coming from intermediate directions, there are intermediate intensity differences (Figure 11.23c). Neurons sensitive to differences in intensity can use this information to locate the sound. Intensity information cannot be used to locate sounds at lower frequencies because sound waves at these frequencies diffract around the head, and the intensities at the two ears are roughly equivalent. There is no sound shadow at low frequencies.

Let's summarize the two processes for localizing sound in the horizontal plane. With sounds in the range of 20–2000 Hz, the process involves *interaural time delay*. From 2000–20,000 Hz, interaural intensity difference is used. Together these two processes constitute the **duplex theory of sound localization**.

The Sensitivity of Binaural Neurons to Sound Location. From our discussion of the auditory pathway, recall that neurons in the cochlear nuclei receive afferents only from the ipsilateral auditory–vestibular nerve. Thus, all of these cells are *monaural neurons*, meaning that they only respond to sound presented to one ear. At all later stages of processing in the auditory system, however, there are *binaural neurons* whose responses are influenced by sound at both ears. The response properties of binaural neurons imply that they play an important role in sound localization in the horizontal plane.

The first structure where binaural neurons are present is the superior olive. Although some controversy exists about the relationship between the activity of such neurons and the behavioral localization of sound, there are several compelling correlations. Neurons in the superior olive receive input from cochlear nuclei on both sides of the brain stem (see Figure 11.18). Cells in the cochlear nuclei that project to the superior olive typically have responses phase locked to lower frequency sound input. Consequently, an olivary neuron receiving spikes from the left and right cochlear nuclei can compute interaural time delay. Recordings made in the superior olive show that each neuron typically gives its greatest response to a particular interaural delay (Figure 11.24). Because interaural delay varies with sound location, each of these neurons may be encoding a particular position in the horizontal plane.

How can a neural circuit produce neurons sensitive to interaural delay? One possibility is to use axons as *delay lines*, and to measure small time differences precisely. A sound hitting the left ear triggers action potentials in the left cochlear nucleus, which propagate along afferent axons into the superior olive (Figure 11.25). Within 0.6 msec of hitting the left ear, that sound reaches the right ear (assuming the sound comes directly from the left) and triggers action potentials in axons from the right cochlear nucleus. However, because of the way the axons and neurons are arranged in the olive, the action potentials from each side take different lengths of time to arrive at the various postsynaptic neurons in the olive. For example, the axon from the left cochlear nucleus has a longer path to travel to neuron 3 in Figure 11.25 than the axon from the right cochlear nucleus; therefore, the arrival of the spike from the left side is delayed just enough that it coincides with the arrival of the spike from the right side. By arriving at precisely the same time, action potentials from the two sides produce excitatory postsynaptic potentials (EPSPs) that summate, yielding a larger EPSP that more strongly excites olivary neuron 3 than an EPSP from each ear could alone. When an interaural delay is more or less than 0.6 msec, the spikes do not arrive together, and thus the EPSPs they trigger do not summate as much.

Other neurons in the superior olive are tuned to other interaural times because of systematic differences in the arrangement of axonal delay

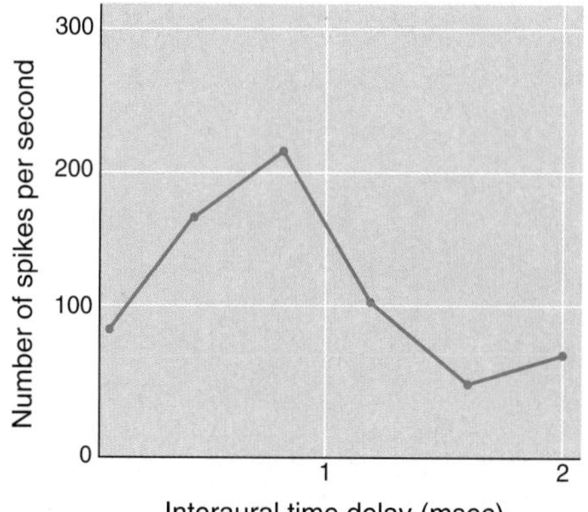

▲ FIGURE 11.24
Responses of a neuron in the superior olive sensitive to interaural time delay. This neuron has an optimal delay of about 1 msec.

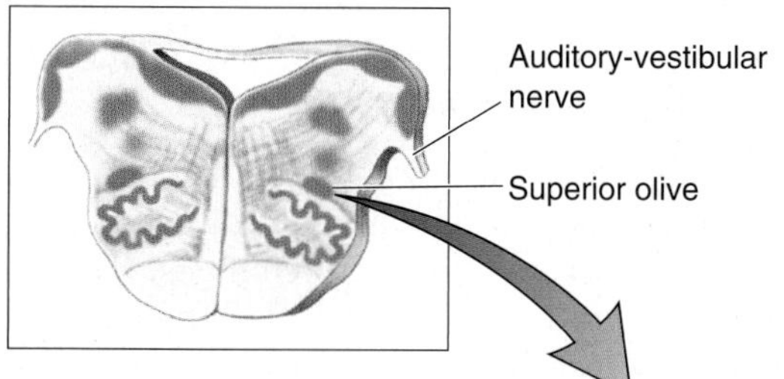

Sound from the left side initiates activity in the left cochlear nucleus; activity is then sent to the superior olive.

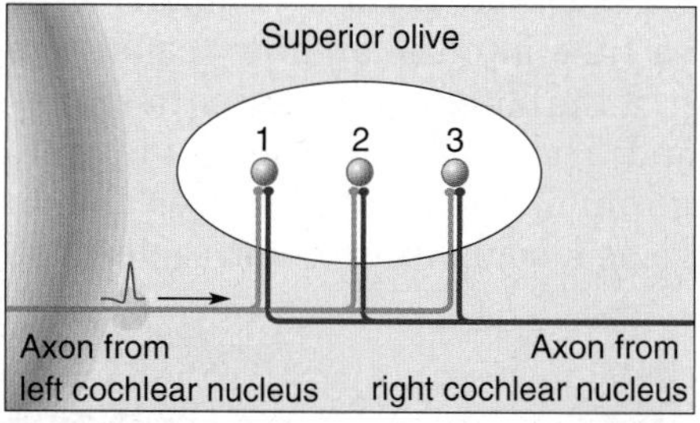

Very soon, the sound reaches the right ear, initiating activity in the right cochlear nucleus. Meanwhile, the first impulse has traveled farther along its axon.

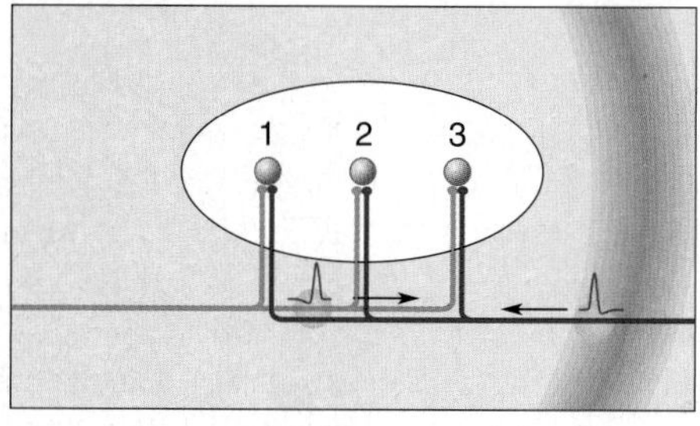

Both impulses reach olivary neuron 3 at the same time, and summation of synaptic potentials generates an action potential.

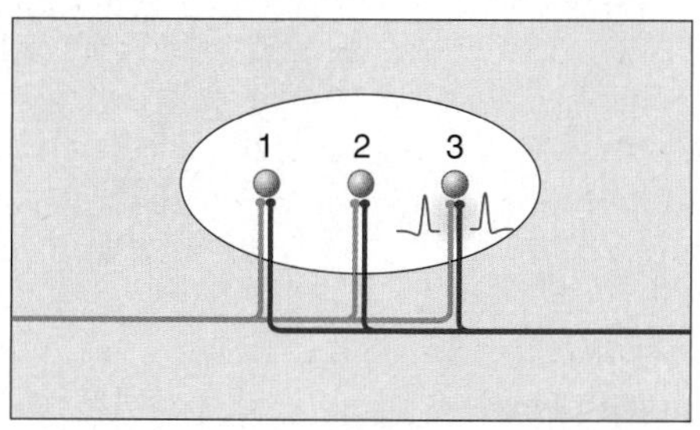

◀ FIGURE 11.25
Delay lines and neuronal sensitivity to interaural delay.

lines. To measure timing differences as accurately as possible, many neurons and synapses of the auditory system are specially adapted for rapid operation; their action potentials and EPSPs are much faster than those of most other neurons in the brain. There are limits to auditory time measurement of this type, however. Phase locking is essential for a precise comparison of the timing of inputs, and because phase locking occurs only at relatively low frequencies, it makes sense that interaural delays are useful only for localizing sounds of relatively low frequency.

The mechanism described in Figure 11.25 is clearly present in bird brains, but it is unlikely that mammals calculate interaural delays exactly this way. Studies on gerbils have suggested that synaptic inhibition, rather than axonal delay lines, generates the sensitivity of superior olivary neurons to interaural delay. It is possible that inhibition and axonal delay lines work together for this purpose.

In addition to their sensitivity to interaural delay, neurons in the superior olive are sensitive to the other sound location cue, interaural intensity. One type of neuron is moderately excited by sound presented to either ear but only gives a maximal response when both ears are stimulated. The other type of neuron is excited by sound in one ear but inhibited by sound in the other ear. Presumably, both types of neurons contribute to horizontal localization of high-frequency sound by encoding differences in interaural intensity.

Localization of Sound in the Vertical Plane

Comparing inputs to both ears is not very useful for localizing sounds in the vertical plane because as a sound source moves up and down, neither the interaural delay nor the interaural intensity changes. This is why, as discussed earlier, localizing sounds in the vertical plane is much less affected by plugging one ear than localizing sounds in the horizontal plane. In order to seriously impair vertical sound localization, one must place a tube into the auditory canal to bypass the pinna. The sweeping curves of the outer ear are essential for assessing the elevation of a source of sound. The bumps and ridges apparently produce reflections of the entering sound. The delays between the direct path and the reflected path change as a sound source moves vertically (Figure 11.26). The combined

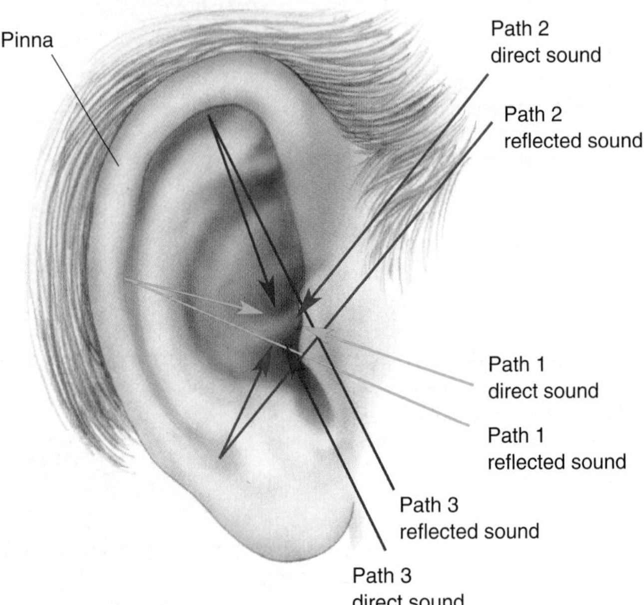

▶ FIGURE 11.26
Vertical sound localization based on reflections from the pinna.

sound, direct and reflected, is subtly different when it comes from above or below. In addition, the outer ear allows higher frequency sounds to enter the auditory canal more effectively when they come from an elevated source. Vertical localization of sound is seriously impaired if the convolutions of the pinna are covered.

Some animals are extremely good at vertical sound localization even though they do not have a pinna. For example, a barn owl can swoop down on a squeaking mouse in the dark, locating accurately by sound, not sight. Although owls do not have a pinna, they can use the same techniques we use for horizontal localization (interaural differences) because their two ears are at different heights on their head. Some animals have a more "active" system for sound localization than humans and owls. Certain bats emit sounds that are reflected off objects, and these echoes are used to locate objects without sight. Many bats detect and capture insects using reflected sound, analogous to the sonar used by ships. In 1989, James Simmons at Brown University made the startling discovery that bats can discriminate time delays that differ by as little as 0.00001 msec. This finding challenges our understanding of how the nervous system, using action potentials lasting almost a millisecond, can perform such fine temporal discriminations.

AUDITORY CORTEX

Axons leaving the MGN project to auditory cortex via the internal capsule in an array called the *acoustic radiation*. Primary auditory cortex (A1) corresponds to Brodmann's area 41 in the temporal lobe (Figure 11.27a). The structure of A1 and the secondary auditory areas is in many ways similar to corresponding areas of the visual cortex. Layer I contains few cell bodies, and layers II and III contain mostly small pyramidal cells. Layer IV, where the medial geniculate axons terminate, is composed of densely packed granule cells. Layers V and VI contain mostly pyramidal

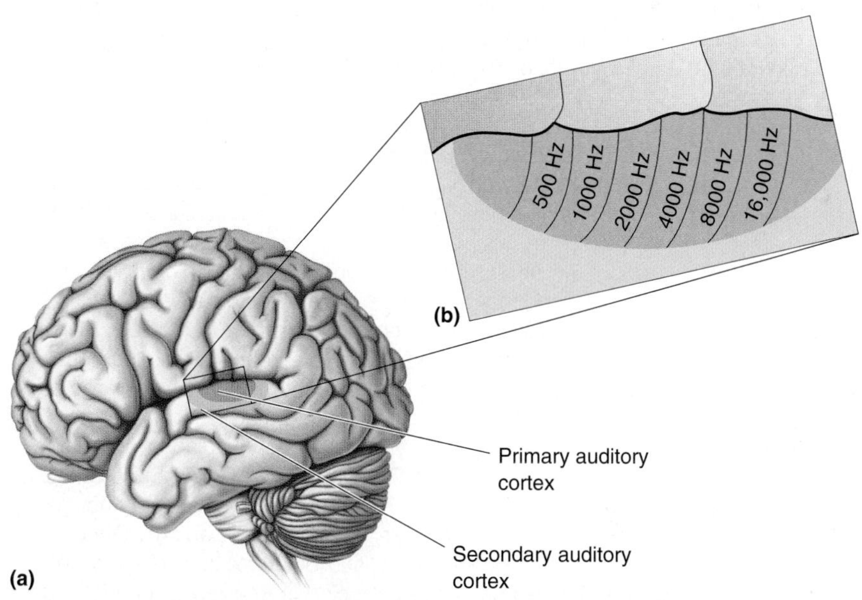

▲ FIGURE 11.27
Primary auditory cortex. (a) Primary auditory cortex (purple) and secondary auditory areas (green) on the superior temporal lobe. (b) Tonotopic organization within primary auditory cortex. The numbers are characteristic frequencies.

cells that tend to be larger than those in the superficial layers. Let's look at how these cortical neurons respond to sound.

Neuronal Response Properties

In general, neurons in monkey (and presumably human) A1 are relatively sharply tuned for sound frequency and possess characteristic frequencies covering the audible spectrum of sound. In electrode penetrations made perpendicular to the cortical surface in monkeys, the cells encountered tend to have similar characteristic frequencies, suggesting a columnar organization on the basis of frequency. In the tonotopic representation in A1, low frequencies are represented rostrally and laterally, whereas high frequencies are represented caudally and medially (Figure 11.27b). Roughly speaking, there are *isofrequency bands* running mediolaterally across A1. In other words, strips of neurons running across A1 contain neurons that have fairly similar characteristic frequencies.

In the visual system, it is possible to describe large numbers of cortical neurons as having some variation on a general receptive field that is either simple or complex. So far, it has not been possible to place the diverse auditory receptive fields into a similarly small number of categories. As they do at earlier stages in the auditory pathway, cortical neurons have

BOX 11.5 OF SPECIAL INTEREST

How Does Auditory Cortex Work? Ask a Specialist

The function of an animal's brain is to help it stay alive and reproduce. Different species have vastly different habits and needs, and some animals have evolved a sensory system specialized for processing its favorite stimuli. The exaggerated systems of the sensory specialists, such as owls and bats, can help us understand how we sensory generalists work.

Barn owls find their prey (a scurrying mouse, for example) in the dark by listening very carefully. They are particularly adept at identifying and localizing faint sounds, and some of the neural mechanisms of sound localization were first understood in owls. Bats have a more unique and active auditory technique. They find their food (a fluttering moth, for example) by echolocating it. A bat emits brief calls and listens for the faint echoes reflected from the target. Bats require their cortex for proper echolocation. Studying bat cortex can certainly provide insight into how auditory cortex works in bats, but it may also enlighten us about the human cortex.

The most interesting stimuli for echolocating bats are their own calls and echoes. A bat's language is very limited. To echolocate, most bats scream loudly at ultrasonic frequencies (20–100 kHz) using essentially a one-word vocabulary. The call of the mustached bat (*Pteronotus parnellii*) is very brief, no more than 20 msec long, and consists of a steady constant frequency (CF) part followed by a sweep of descending frequency—the frequency modulated (FM) part. Figure A is a graph of the bat's call and echo, showing the frequencies of the sounds plotted against time. As it flies, the bat rapidly and continually repeats the call. By listening to its own calls and their echoes, and carefully comparing them in many ways, the bat builds a remarkably detailed auditory image of the nearby world. For example, the *delay* between the call and its echo depends on the distance to a reflecting target (1 msec of delay for each 17 cm of distance). If the target is moving toward or away from the bat, relative to the bat's own movement, the frequency of the echo is *Doppler-shifted* higher or lower (think of the shifting pitch of an ambulance siren as it passes by you; a 1 kHz shift corresponds to a speed of about 3 m/s). A moth's beating wings cause a *rhythm* in the echoes, and that helps the bat know there's a particular kind of insect in front of it rather than something less edible. Many other subtle changes in the echo's frequency, timing, loudness, and pattern tell the bat about other features of targets.

The processing of call-echo information by the auditory cortex of the mustached bat has been studied in great detail by Nobuo Suga at Washington University. Suga found that the bat's auditory cortex is a mixture of distinct auditory areas. Many are specialized for detecting particular features important for echolocating, and others seem more generalized. For example, a large region is devoted to processing Doppler shifts of echoes right around 60 kHz, the CF part of the bat's call; this area processes information about target different temporal response patterns; some have a transient response to a brief sound, and others have a sustained response.

In addition to the frequency tuning that occurs in most cells, some neurons are intensity-tuned, giving a peak response to a particular sound intensity. Even within a vertical column perpendicular to the cortical surface, considerable diversity can exist in the degree of tuning to sound frequency. Some neurons are sharply tuned for frequency, and others are barely tuned at all; the degree of tuning does not seem to correlate well with cortical layers. Other sounds that produce responses in cortical neurons include clicks, bursts of noise, frequency-modulated sounds, and animal vocalizations. Attempting to understand the role of these neurons that respond to seemingly complex stimuli is one of the challenges researchers currently face (Box 11.5).

Given the wide variety of response types that neurophysiologists encounter in studying auditory cortex, you can understand why it is reassuring to see some sort of organization or unifying principle. One organizational principle already discussed is the tonotopic representation in many auditory areas. A second organizational principle is the presence in auditory cortex of columns of cells with similar binaural interaction. As at lower levels in the auditory system, one can distinguish cells that respond more to stimulation of both ears than to either ear separately, as well as cells that are inhibited if both ears are stimulated. As we discussed

velocity and location. Three separate areas detect call-echo delays and yield information about target distance.

The basic features of a bat's calls and a human's spoken words are similar, although human speech is much slower and lower in pitch. Human language syllables consist of particular combinations of CF periods, FM sweeps, brief pauses, and bursts of noise. For example, the syllable "ka" differs from "pa" because their initial FM sweeps bend in different directions (Figure B). The long "a" and the long "i" sound different because each one uses different combinations of CFs. It is very likely that the neuronal circuits that process speech sounds in human auditory cortex use principles very similar to those in the bat's cortex. Interpreting those speech sounds as words, and understanding the concepts they imply, is the realm of language; the brain mechanisms of language will be discussed in Chapter 20.

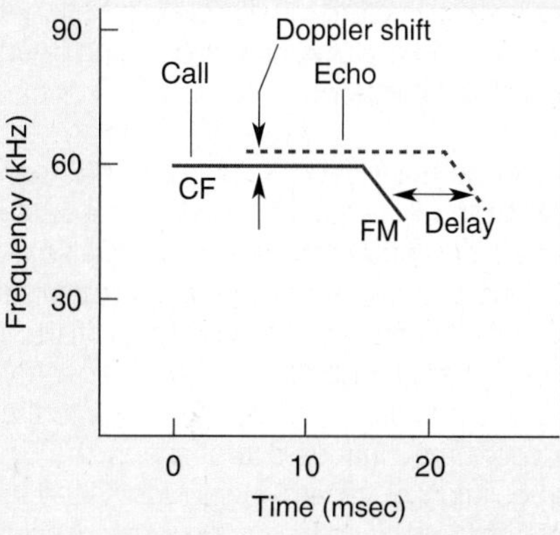

Figure A
A bat's call and echo. (Source: Adapted from Suga, 1995, p. 302.)

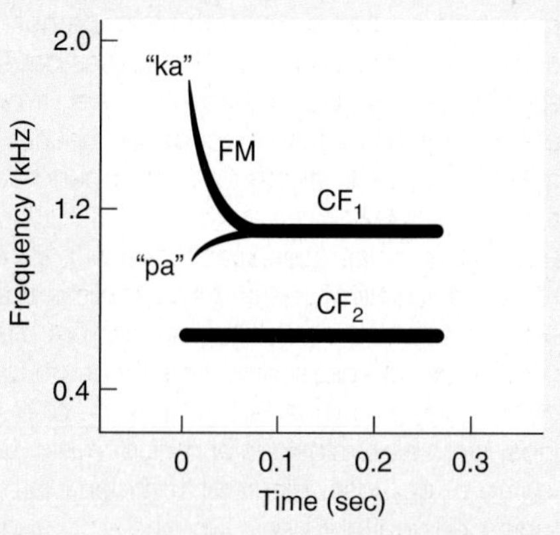

Figure B
A human's spoken words. (Source: Adapted from Suga, 1995, p. 296.)

regarding the superior olive, neurons sensitive to interaural time delays and interaural intensity differences probably play a role in sound localization.

In addition to A1, other cortical areas located on the superior surface of the temporal lobe respond to auditory stimuli. Some of these higher auditory areas are tonotopically organized, and others do not seem to be. As in visual cortex, the stimuli that evoke the strongest responses in higher auditory areas tend to be more complex than those that best excite neurons at lower levels in the system. An example of specialization is Wernicke's area, which we will discuss in Chapter 20. Destruction of this area does not interfere with the sensation of sound, but it seriously impairs the ability to interpret spoken language.

The Effects of Auditory Cortical Lesions and Ablation

Bilateral ablation of auditory cortex leads to deafness, but deafness is more often the consequence of damage to the ears (Box 11.6). A surprising degree of normal auditory function is retained after unilateral lesions in auditory cortex. This outcome is in marked contrast to that of the visual system, in which a unilateral cortical lesion of striate cortex leads to complete blindness in one visual hemifield. The reason for greater

BOX 11.6 OF SPECIAL INTEREST

Auditory Disorders and Their Treatments

Although the effects of cortical lesions provide important information about the role of auditory cortex in perception, the perceptual deficit we all associate with the auditory system, deafness, usually results from problems in or near the cochlea. Deafness is conventionally considered in two categories: conduction deafness and nerve deafness.

Hearing loss caused by a disturbance in the conduction of sound from the outer ear to the cochlea is called *conduction deafness*. Causes of this sensory deficit range from something as simple as excessive wax in the ear to more serious problems such as rupture of the tympanic membrane or pathology of the ossicles. A number of diseases cause binding of the ossicles to the bone of the middle ear, impairing the transfer of sound. Fortunately, most of the mechanical problems in the middle ear that interfere with sound conduction can be treated surgically.

Nerve deafness is deafness associated with the loss of either neurons in the auditory nerve or hair cells in the cochlea. Nerve deafness sometimes results from tumors affecting the inner ear. It also can be caused by drugs that are toxic to hair cells, such as quinine and certain antibiotics, or exposure to loud sounds, such as explosions and loud music. Depending on the degree of cell loss, different treatments are possible. If the cochlea or auditory nerve on one side is completely destroyed, deafness in that ear will be absolute. However, a partial loss of hair cells is more common. In these cases, a hearing aid can be used to amplify the sound for the remaining hair cells. In more serious cases where hearing loss is bilateral and the auditory nerve is intact, cochlear implants are an important option (see Box 11.2).

With deafness, a person hears less sound than normal. With a hearing disorder called *tinnitus*, a person hears noises in the ears even in the absence of any sound stimulus. The subjective sensation can take many forms, including buzzing, humming, and whistling. You may have experienced a mild, temporary form of tinnitus after being at a party with really loud music; your brain may have had fun, but your hair cells were in shock! Tinnitus is a relatively common disorder that can seriously interfere with concentration and work if it persists. You can imagine how distracting it would be if you constantly heard whispering or humming or the crinkling of paper.

Tinnitus can be a symptom of a number of neurological problems. Although it frequently accompanies diseases involving the cochlea or auditory nerve, it may also result from exposure to loud sounds, abnormal vasculature of the neck, or simple aging. It now seems that many of the phantom sounds of tinnitus are caused by changes in central auditory structures, including auditory cortex. Damage to the cochlea or auditory nerve may induce alterations in the brain, such as a downregulation of synaptic inhibition. Although clinical treatment of tinnitus is often only partially successful, using a device that produces a constant sound in the affected ear(s) can often lessen the annoyance of the noise. For unknown reasons, the constant real sound is less annoying than the sound of the tinnitus that gets blocked.

preservation of function after lesions in auditory cortex is that both ears send output to cortex in both hemispheres. In humans, the primary deficit that results from a unilateral loss of A1 is the inability to localize the source of a sound. It may be possible to determine which side of the head a sound comes from, but there is little ability to locate the sound more precisely. Performance on such tasks as frequency or intensity discrimination is near normal.

Studies in experimental animals indicate that smaller lesions can produce rather specific localization deficits. Because of the tonotopic organization of A1, it is possible to make a restricted cortical lesion that destroys neurons with characteristic frequencies within a limited frequency range. Interestingly, there is a localization deficit only for sounds roughly corresponding to the characteristic frequencies of the missing cells. This finding reinforces the idea that information in different frequency bands may be processed in parallel by tonotopically organized structures.

THE VESTIBULAR SYSTEM

Strangely enough, listening to music and balancing on a bicycle both involve sensations that are transduced by hair cells. The vestibular system monitors the position and movement of the head, gives us our sense of balance and equilibrium, and helps coordinate movements of the head and eyes, as well as adjustments to body posture. When the vestibular system operates normally, we are usually unaware of it. When its function is disrupted, however, the results can include the unpleasant, stomach-turning feelings we usually associate with motion sickness—vertigo and nausea, plus a sense of disequilibrium and uncontrollable eye movements.

The Vestibular Labyrinth

The vestibular and auditory systems both use hair cells to transduce movements. Common biological structures often have common origins. In this case, the organs of mammalian balance and hearing both evolved from the *lateral line organs* present in aquatic vertebrates, including fish and some amphibians. Lateral line organs are small pits or tubes along an animal's sides. Each pit contains clusters of hairlike sensory cells whose cilia project into a gelatinous substance that is open to the water in which the animal swims. The purpose of lateral line organs in many animals is to sense vibrations or pressure changes in the water. In some cases, they are also sensitive to temperature or electrical fields. Lateral line organs were lost as reptiles evolved, but the exquisite mechanical sensitivity of hair cells was adopted and adapted for use in the structures of the inner ear that derived from the lateral line.

In mammals, all hair cells are contained within sets of interconnected chambers called the **vestibular labyrinth** (Figure 11.28a). We have already discussed the auditory portion of the labyrinth, the spiraling cochlea (see Figure 11.6). The vestibular labyrinth includes two types of structures with different functions: the **otolith organs**, which detect the force of gravity and tilts of the head, and the **semicircular canals**, which are sensitive to head rotation. The ultimate purpose of each structure is to transmit mechanical energy, derived from head movement, to its hair cells. Each is sensitive to different kinds of movement not because their hair cells differ but because of the specialized structures within which the hair cells reside.

▶ FIGURE 11.28
The vestibular labyrinth. **(a)** Locations of the otolith organs (utricle and saccule) and semicircular canals. **(b)** A vestibular labyrinth resides on each side of the head, with the semicircular canals arranged in parallel planes.

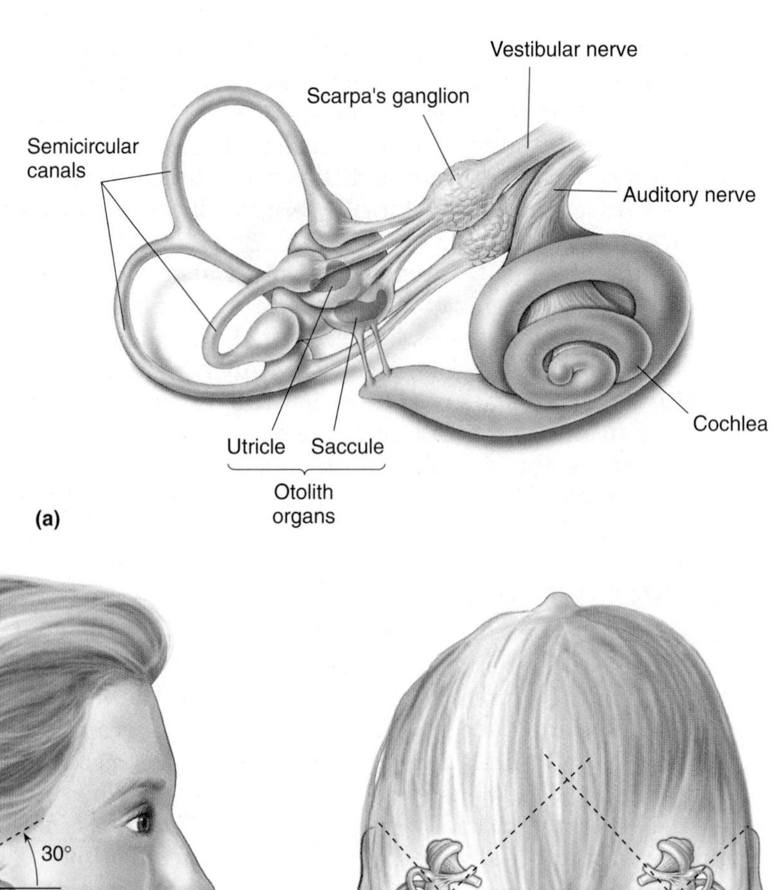

30°
Semicircular canals
(b)

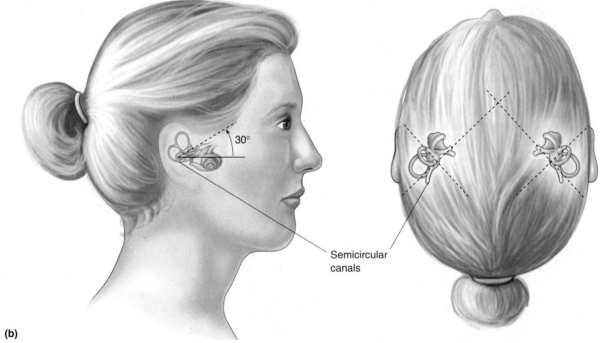

The paired otolith organs, called the *saccule* and the *utricle,* are relatively large chambers near the center of the labyrinth. The semicircular canals are the three arcing structures of the labyrinth. They lie in approximately orthogonal planes, which means there is an angle of about 90° between any pair of them (Figure 11.28b). A set of vestibular organs resides on each side of the head, mirror images of each other.

Each hair cell of the vestibular organs makes an excitatory synapse with the end of a sensory axon from the *vestibular nerve*, a branch of the auditory–vestibular nerve (cranial nerve VIII). There are about 20,000 vestibular nerve axons on each side of the head, and their cell bodies lie in *Scarpa's ganglion*.

The Otolith Organs

The saccule and utricle detect changes of head angle, as well as *linear acceleration* of the head. When you tilt your head, the angle between your otolith organs and the direction of the force of gravity changes. Linear acceleration also generates force, in proportion to the mass of an object. Forces due to linear acceleration are the sort you encounter when you ride in an elevator or a car as it starts or stops. In contrast, when a car or elevator moves smoothly at constant velocity, acceleration is zero, so there is no force (apart from gravitational force). That's why you can fly steadily at 600 mph in a jet yet feel perfectly still; the sudden bouncing you experience during air turbulence, however, is another good example of the forces generated by linear acceleration, and of movements detected by your otolith organs.

Each otolith organ contains a sensory epithelium called a **macula**, which is vertically oriented within the saccule and horizontally oriented within the utricle when the head is upright. (Note that the vestibular macula and the retinal macula are entirely different structures.) The vestibular macula contains hair cells, which lie among a bed of supporting cells with their cilia projecting into a gelatinous cap (Figure 11.29). Movements are transduced by hair cells in the maculae when the hair bundles are deflected. The unique feature of the otolith organs is the tiny crystals of calcium carbonate called *otoconia*, 1–5 μm in diameter. (The word is Greek for "ear stone.") Otoconia encrust the surface of the macula's gelatinous cap, near the tips of the hair bundles, and they are

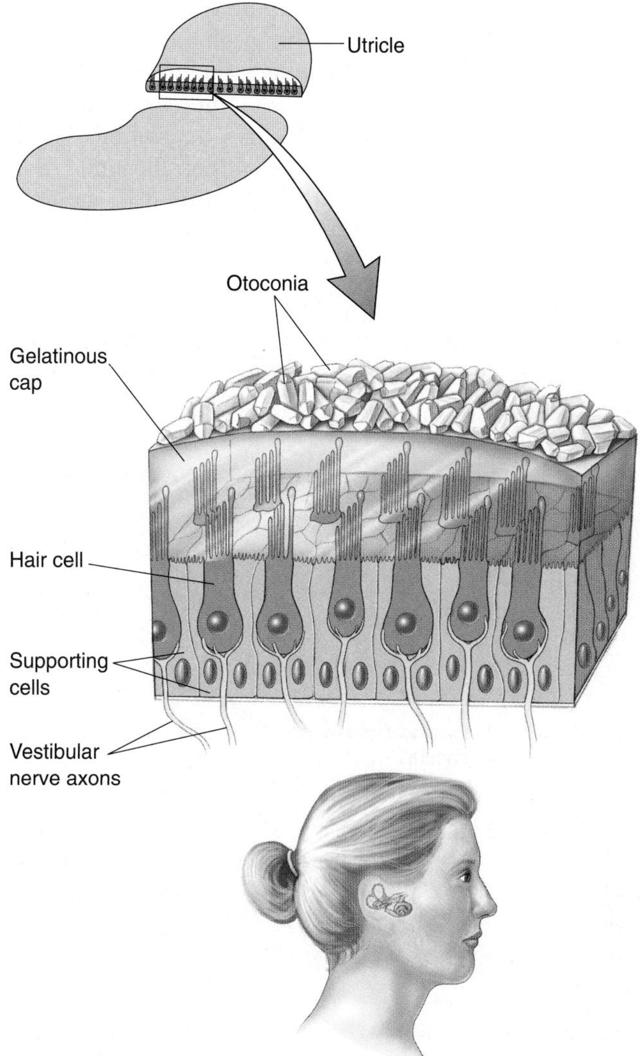

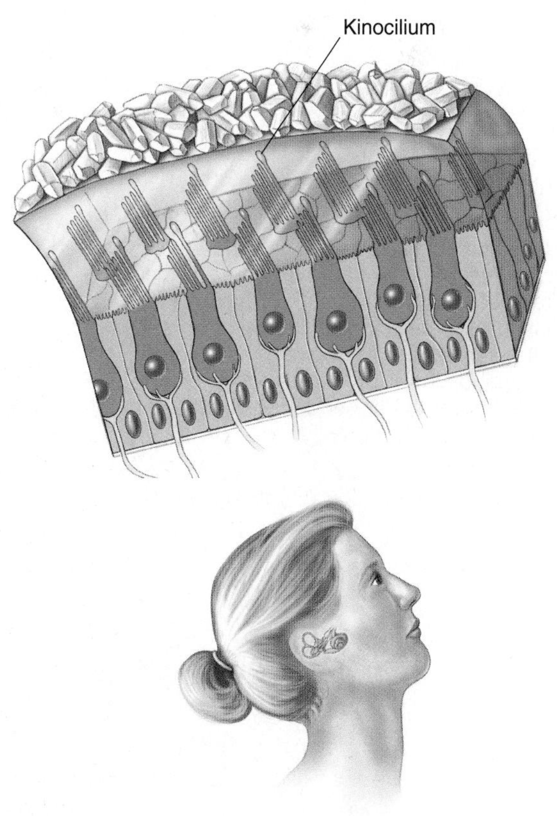

◀ FIGURE 11.29
Macular hair cells responding to tilt. When the utricular macula is level (the head is straight), the cilia from the hair cells also stand straight. When the head and macula are tilted, gravity pulls the otoconia, which deform the gelatinous cap, and the cilia bend.

the key to the tilt sensitivity of the macula. The otoconia have a higher density than the endolymph that surrounds them.

When the angle of the head changes, or when the head accelerates, a force is exerted on the otoconia; this exerts a force in the same direction on the gelatinous cap, which moves slightly, and the cilia of the hair cells bend. Not just any deflection will do, however. Each hair cell has one especially tall cilium, called the *kinocilium*. The bending of hairs toward the kinocilium results in a depolarizing, excitatory receptor potential. Bending the hairs away from the kinocilium hyperpolarizes and inhibits the cell. The cell is exquisitely direction-selective. If the hairs are bent perpendicular to their preferred direction, they barely respond. The transduction mechanism of vestibular hair cells is essentially the same as that in auditory hair cells (see Figure 11.15). As with auditory hair cells, only tiny hair movements are needed. The response saturates when the hairs are bent less than 0.5 μm, about the diameter of one cilium.

The head can tilt and move in any direction, but the hair cells of the utricle and saccule are oriented to transduce all of them effectively. The saccular maculae are oriented more or less vertically, while the utricular maculae are mostly horizontal (Figure 11.30). On each macula, the direction preferences of the hair cells vary in a systematic way. There are enough hair cells in each macula to cover a full range of directions. Because of the mirror-image orientation of the saccule and utricle on each side of the head, when a given head movement excites hair cells on one side, it will tend to inhibit hair cells in the corresponding location on the other. Thus, any tilt or acceleration of the head will excite some hair cells, inhibit others, and have no effect on the rest. The central nervous system, by simultaneously using the information encoded by the full population of otolithic hair cells, can unambiguously interpret all possible linear movements.

The Semicircular Canals

The semicircular canals detect turning movements of the head, such as shaking your head from side to side or nodding up and down. Like the otolith organs, the semicircular canals also sense acceleration, but of a

▶ FIGURE 11.30
Macular orientation. **(a)** The macula in the utricle is horizontal. **(b)** The macula in the saccule is vertical. The arrows on each macula represent how the sheets of hair cells are polarized. Hair cells in the vicinity of an arrow are all polarized the same way; their stereocilia are all oriented so that bending them in the direction of the arrow depolarizes them.

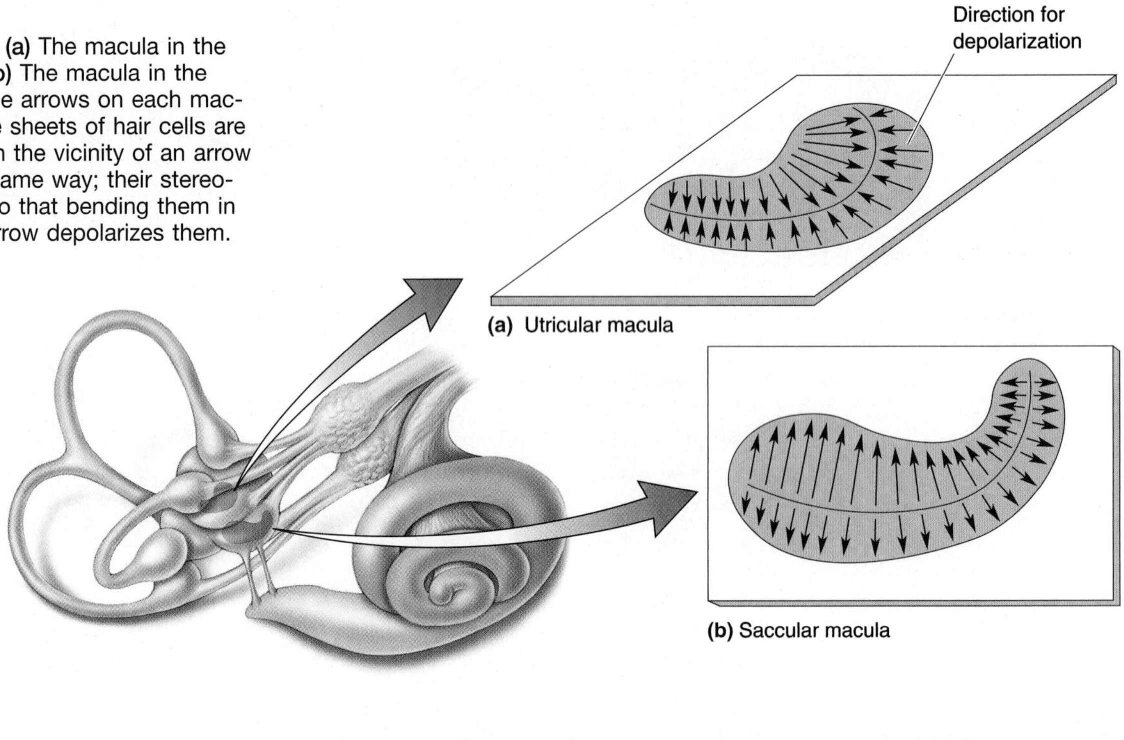

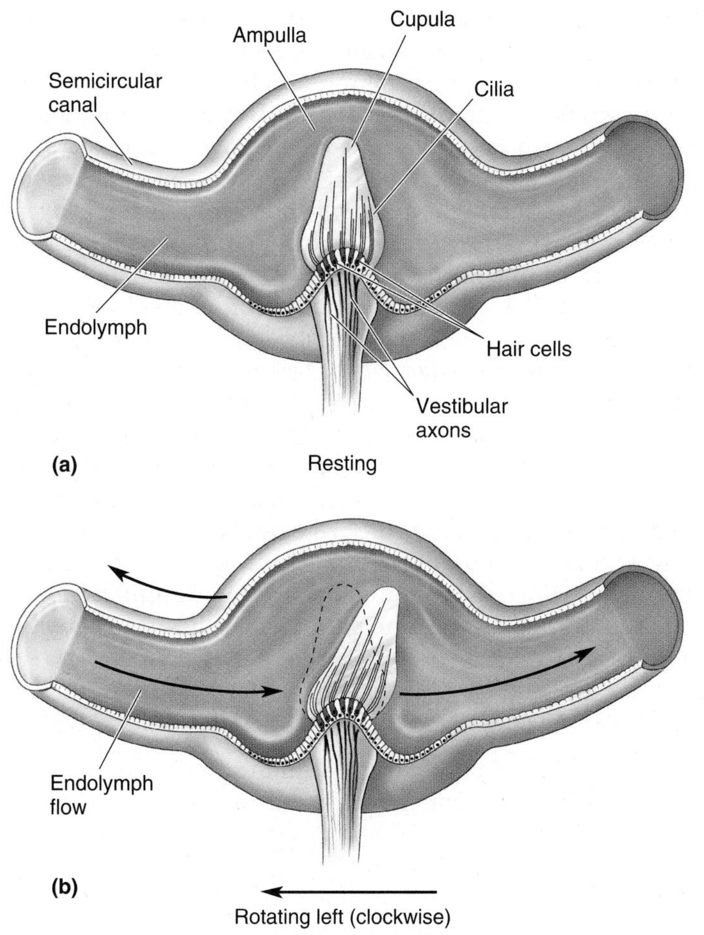

▲ FIGURE 11.31

A cross section through the ampulla of a semicircular canal. **(a)** The cilia of hair cells penetrate into the gelatinous cupula, which is bathed in the endolymph that fills the canals. **(b)** When the canal rotates leftward, the endolymph lags behind and applies force to the cupula, bending the cilia within it.

different kind. *Angular acceleration* is generated by sudden rotational movements, and it is the primary stimulus for the semicircular canals.

The hair cells of the semicircular canals are clustered within a sheet of cells, the *crista*, located within a bulge along the canal called the **ampulla** (Figure 11.31a). The cilia project into the gelatinous *cupula*, which spans the lumen of the canal within the ampulla. All the hair cells in an ampulla have their kinocilia oriented in the same direction, which means that they all get excited or inhibited together. The semicircular canals are filled with endolymph, the same fluid that fills the scala media of the cochlea. Bending of the cilia occurs when the canal is suddenly rotated about its axis like a wheel; as the wall of the canal and the cupula begin to spin, the endolymph tends to stay behind because of inertia. The sluggish endolymph exerts a force upon the cupula, much like wind upon a sail (Figure 11.31b). This force bows the cupula, which bends the cilia, which (depending on the direction of the rotation) either excites or inhibits the release of neurotransmitter from the hair cells onto the vestibular nerve axons.

If head rotation is maintained at a constant velocity, the friction of endolymph with the canal walls eventually makes the two move together, thereby reducing and then eliminating the bending of the cupula after 15–30 seconds. Such adaptation to rotation can be seen clearly in the firing rates of vestibular axons from the canals (Figure 11.32). (This sort

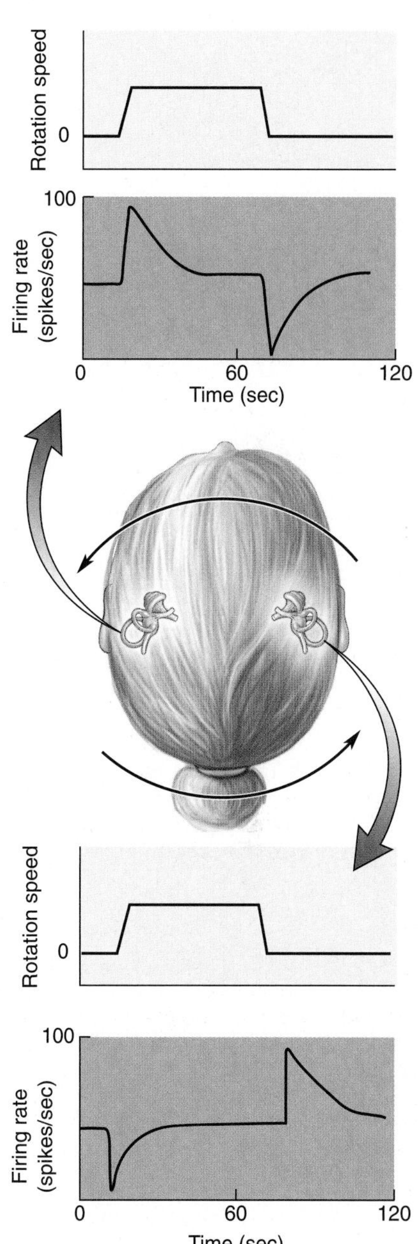

▲ FIGURE 11.32

Push–pull activation of the semicircular canals. Head rotation causes the excitation of hair cells in one horizontal semicircular canal and the inhibition of hair cells in the other. The graphs show that long-lasting head rotation leads to adaptation of the firing in vestibular axons. When rotation is stopped, the vestibular axons from each side begin firing again, but with opposite patterns of excitation and inhibition.

of prolonged head rotation is not something you encounter very often, unless you have a taste for certain amusement park rides.) When rotation of the head (and its canals) is finally stopped, the inertia of the endolymph causes the cupula to bend in the other direction, generating an opposite response from the hair cells and a temporary sensation of counter-rotation. This mechanism explains why you felt dizzy and unbalanced whenever, as a child, you *stopped* spinning your body like a top; your semicircular canals were temporarily sending the message that your body was still spinning, but in the opposite direction.

Together, the three semicircular canals on one side of the head help sense all possible head rotation angles. This is further ensured because each canal is paired with another on the opposite side of the head (see Figure 11.28b). Each member of a pair sits within the same orientation plane as its partner and responds to rotation about the same axis. However, while rotation excites the hair cells of one canal, it inhibits the hair cells of its contralateral partner's canal. Vestibular axons fire at high rates even at rest, so their activity can be driven either up or down depending on the direction of rotation. This "push–pull" arrangement—each rotation causing excitation on one side and inhibition on the other (see Figure 11.32)—optimizes the ability of the brain to detect rotational movements.

Central Vestibular Pathways and Vestibular Reflexes

The central vestibular pathways coordinate and integrate information about head and body movement and use it to control the output of motor neurons that adjust head, eye, and body positions. Primary vestibular axons from cranial nerve VIII make direct connections to the medial and lateral **vestibular nuclei** on the same side of the brain stem, as well as to the cerebellum (Figure 11.33). The vestibular nuclei also receive inputs from other parts of the brain, including the cerebellum, and the visual and somatic sensory systems, thereby combining incoming vestibular information with data about the motor system and other sensory modalities.

The vestibular nuclei in turn project to a variety of targets above them in the brain stem, and below them into the spinal cord (see Figure 11.33). For example, axons from the otolith organs project to the lateral vestibular nucleus, which then projects via the *vestibulospinal tract* to excite

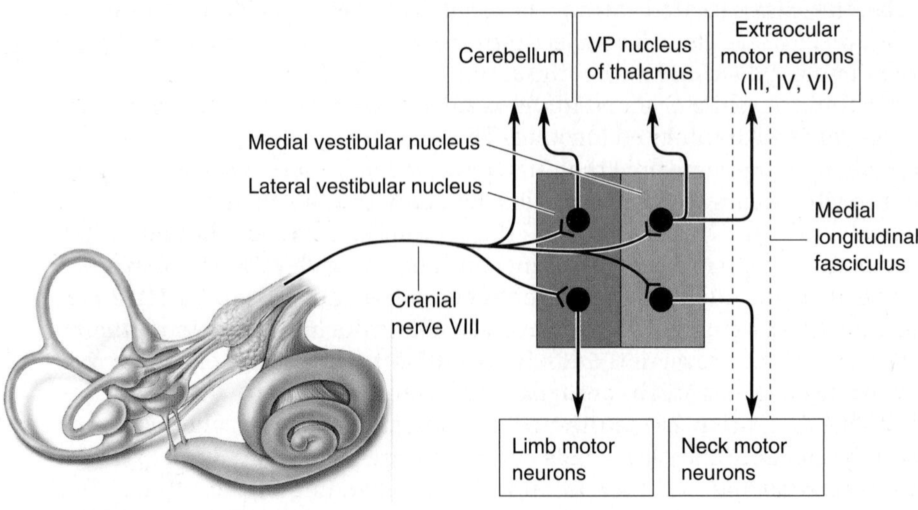

▲ FIGURE 11.33
A summary of the central vestibular connections from one side.

spinal motor neurons controlling muscles in the legs that help maintain posture (see Chapter 14). This pathway helps the body stay upright even on the rolling deck of a boat. Axons from the semicircular canals project to the medial vestibular nucleus, which sends axons via the *medial longitudinal fasciculus* to excite motor neurons of trunk and neck muscles that orient the head. This pathway helps the head stay straight even as the body cavorts around below it.

Similar to the other sensory systems, the vestibular system makes connections to the thalamus and then to the neocortex. The vestibular nuclei send axons into the *ventral posterior (VP) nucleus* of the thalamus, which projects to regions close to the representation of the face in the primary somatosensory and primary motor areas of cortex (see Chapters 12 and 14). At the cortical level, there is considerable integration of information about movements of the body, the eyes, and the visual scene. It is likely that the cortex continually maintains a representation of body position and orientation in space, which is essential for our perception of equilibrium and for planning and executing complex, coordinated movements.

The Vestibulo-Ocular Reflex (VOR). One very important function of the central vestibular system is to keep your eyes pointed in a particular direction, even while you are dancing like a fool. The **vestibulo-ocular reflex (VOR)** performs this function. Recall that accurate vision requires the image to remain stable on the retinas despite movement of the head (see Chapter 9). Each eye can be moved by a set of six extraocular muscles. The VOR works by sensing rotations of the head and immediately commanding a compensatory movement of the eyes in the opposite direction. The movement helps keep your line of sight tightly fixed on a visual target. Because the VOR is a reflex triggered by vestibular input rather than visual input, it works amazingly well even in the dark or when your eyes are closed.

Imagine driving down a very bumpy road. From constant adjustments by the VOR, your view of the world ahead is quite stable because each bump, and its consequent movement of your head, is compensated by an eye movement. To appreciate how effective your VOR is, compare the stability of a passing object during the bumpy drive as you look at it first with your eyes alone, and then with the viewfinder of a simple camera. You'll find that your camera view jumps around hopelessly because your arms are not nearly quick or accurate enough to move the camera with each bump. Many cameras now have an electromechanical equivalent of a VOR that stabilizes an image even when the camera or the photographer holding it is bumping around.

The effectiveness of the VOR depends on complicated connections from the semicircular canals to the vestibular nucleus to the cranial nerve nuclei that excite the extraocular muscles. Figure 11.34, which shows only half of the horizontal component of this circuit, illustrates what happens when the head turns to the left and the VOR induces both eyes to turn right. Axons from the left horizontal canal innervate the left vestibular nucleus, which sends excitatory axons to the contralateral (right) cranial nerve VI nucleus (abducens nucleus). Motor axons from the abducens nucleus in turn excite the lateral rectus muscle of the right eye. Another excitatory projection from the abducens crosses the midline, back to the left side, and ascends (via the medial longitudinal fasciculus) to excite the left cranial nerve III nucleus (oculomotor nucleus), which excites the medial rectus muscle of the left eye.

Mission accomplished, so it would seem: Both eyes are turning right. However, to further ensure speedy operation, the left medial rectus

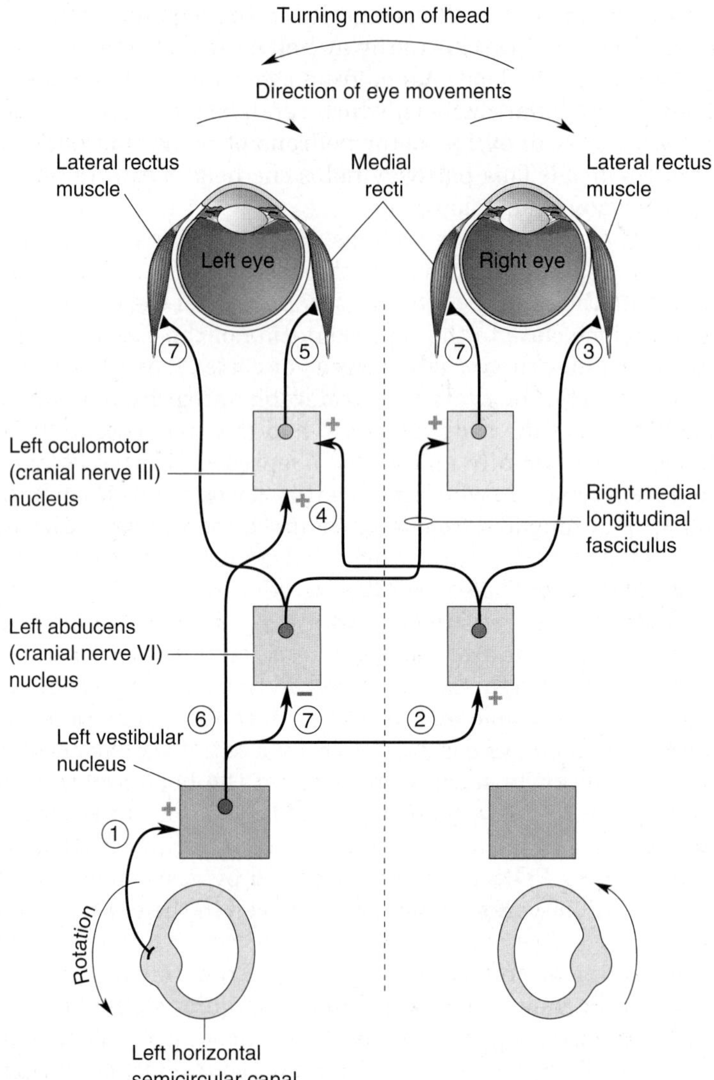

▲ FIGURE 11.34
Vestibular connections mediating horizontal eye movements during the VOR. These pathways are active when the head suddenly turns to the left, causing the eyes to turn to the right. Excitatory connections are denoted by green plus signs; the inhibitory connection is denoted by a red minus sign.

muscle also gets excited via a projection from the vestibular nucleus directly to the left oculomotor nucleus. Speed is also maximized by activating inhibitory connections to the muscles that oppose this movement (the left eye's lateral rectus and right eye's medial rectus, in this case). In order to respond to head rotations in any direction, the complete VOR circuit includes similar connections between the right horizontal canal, the other semicircular canals, and the other extraocular muscles that control eye movements.

Vestibular Pathology

The vestibular system can be damaged in a variety of ways; for example, high doses of antibiotics such as streptomycin can be toxic to hair cells. People with bilateral lesions of the vestibular labyrinths have enormous trouble fixating on visual targets as they move about. Even the minute head pulsations from the blood pressure surges of heartbeats can be disturbing in some cases. When people with vestibular disturbances cannot stabilize an image on their moving retinas, they may also experience the disconcerting feeling that the world is constantly moving around them. The sensation can make walking and standing difficult. Compensatory adjustments come with time, as the brain learns to substitute more visual and proprioceptive cues to help guide smooth and accurate movements.

CONCLUDING REMARKS

Hearing and balance begin with nearly identical sensory receptors, the hair cells, which are exquisitely sensitive to deflections of their stereocilia. These movement detectors are surrounded by three sets of inner ear structures that give them selectivity for three different kinds of mechanical energy: periodic waves of air pressure (sound), rotational forces (head turns), and linear forces (head tilt or acceleration). Except for the similarity in transduction, and the fact that the hair cells of both systems are located in the inner ear, the auditory and vestibular systems are quite different. The sound sensed by audition comes mainly from the external environment, while the vestibular system senses only the movements of itself. Auditory and vestibular pathways are entirely separate except perhaps at the highest levels of the cortex. Auditory information is often at the forefront of our consciousness, while vestibular sensation usually operates unnoticed to coordinate and calibrate our every movement.

We have followed the auditory pathways from the ear to cerebral cortex and seen the ways in which information about sound is transformed. Variations in the density of air are converted to movements of the mechanical components of the middle and inner ear, which are transduced into neural responses. The structures of the ear and cochlea are highly specialized for the transduction of sound. However, this fact should not blind us to the considerable similarities between the organization of the auditory system and that of other sensory systems. Many analogies can be made between the auditory and visual systems. In the sensory receptors of both systems, a spatial code is established. In the visual system, the code in the photoreceptors is retinotopic; the activity of a given photoreceptor indicates light at a particular location. The receptors in the auditory system establish a spatial code that is tonotopic because of the unique properties of the cochlea. In each system, the retinotopy or tonotopy is preserved as signals are processed in secondary neurons, the thalamus, and finally in sensory cortex.

The convergence of inputs from lower levels produces neurons at higher levels that have more complex response properties. Combinations of LGN inputs give rise to simple and complex receptive fields in visual cortex; similarly in the auditory system, the integration of inputs tuned to different sound frequencies yields higher level neurons that respond to complex combinations of frequencies. Another example of increasing visual complexity is the convergence of inputs from the two eyes, which yields binocular neurons that are important for depth perception. Analogously, in the auditory system, input from the two ears is combined to create binaural neurons, which are used for horizontal sound localization. These are just a few of the many similarities in the two systems. Principles governing one system can often help us understand other systems. Keep this in mind while reading about the somatic sensory system in the next chapter, and you'll be able to predict some features of cortical organization based on the types of sensory receptors.

KEY TERMS

Introduction
audition (p. 370)
vestibular system (p. 370)

The Nature of Sound
frequency (p. 371)
hertz (Hz) (p. 371)
pitch (p. 372)
intensity (p. 372)

The Structure of the Auditory System
pinna (p. 373)
auditory canal (p. 373)
tympanic membrane (p. 373)
ossicle (p. 373)
oval window (p. 373)
cochlea (p. 373)
outer ear (p. 374)
middle ear (p. 374)
inner ear (p. 374)
medial geniculate nucleus (MGN) (p. 374)
primary auditory cortex (A1) (p. 374)

The Middle Ear
Eustachian tube (p. 374)
attenuation reflex (p. 376)

The Inner Ear
round window (p. 377)
basilar membrane (p. 377)
organ of Corti (p. 377)
tectorial membrane (p. 378)
perilymph (p. 378)
endolymph (p. 378)
stria vascularis (p. 378)
endocochlear potential (p. 378)
tonotopy (p. 380)
hair cells (p. 380)
stereocilia (p. 380)
inner hair cells (p. 380)
outer hair cells (p. 380)
spiral ganglion (p. 382)
auditory–vestibular nerve (p. 382)
cochlear amplifier (p. 386)

Central Auditory Processes
dorsal cochlear nucleus (p. 389)
ventral cochlear nucleus (p. 389)
superior olive (p. 389)
inferior colliculus (p. 389)
characteristic frequency (p. 390)

Encoding Sound Intensity and Frequency
phase locking (p. 392)
volley principle (p. 393)

Mechanisms of Sound Localization
duplex theory of sound localization (p. 396)

The Vestibular System
vestibular labyrinth (p. 403)
otolith organs (p. 403)
semicircular canals (p. 403)
macula (p. 405)
ampulla (p. 407)
vestibular nuclei (p. 408)
vestibulo-ocular reflex (VOR) (p. 409)

REVIEW QUESTIONS

1. How is the conduction of sound to the cochlea facilitated by the ossicles of the middle ear?
2. Why is the round window crucial for the function of the cochlea? What would happen to hearing if it suddenly didn't exist?
3. Why is it impossible to predict the frequency of a sound simply by looking at which portion of the basilar membrane is the most deformed?
4. Why would the transduction process in hair cells fail if the stereocilia as well as the hair cell bodies were surrounded by perilymph?
5. If inner hair cells are primarily responsible for hearing, what is the function of outer hair cells?
6. Why doesn't unilateral damage to the inferior colliculus or MGN lead to deafness in one ear?
7. What mechanisms function to localize sounds in the horizontal and vertical planes?
8. What symptoms would you expect to see in a person who had recently had a stroke affecting A1 unilaterally? How does the severity of these symptoms compare with the effects of a unilateral stroke involving V1?
9. What is the difference between nerve deafness and conduction deafness?
10. Each macula contains hair cells with kinocilia arranged in all directions. What is the advantage of this arrangement compared to an arrangement of all the cells in the same direction?
11. Imagine a semicircular canal rotating in two different ways: around its axis (like a rolling coin) or end over end (like a flipped coin). How well would its hair cells respond in each case, and why?
12. How would you expect the functions of the otolith organs and the semicircular canals to change in the weightless environment of space?

FURTHER READING

Ashida G, Carr CE. 2011. Sound localization: Jeffress and beyond. *Current Opinion in Neurobiology* 21:745–751.

Cullen KE. 2012. The vestibular system: multimodal integration and encoding of self-motion for motor control. *Trends in Neurosciences* 35:185–196.

Guinan JJ Jr, Salt A, Cheatham MA. 2012. Progress in cochlear physiology after Békésy. *Hearing Research* 293:12–20.

Holt JR, Pan B, Koussa MA, Asai Y. 2014. TMC function in hair cell transduction. *Hearing Research* 311:17–24.

Kazmierczak P, Müller U. 2012. Sensing sound: molecules that orchestrate mechanotransduction by hair cells. *Trends in Neurosciences* 35:220–229.

Oertel D, Doupe AJ. 2013. The auditory central nervous system. In *Principles of Neural Science*, 5th ed., ed. Kandel ER, Schwartz JH, Jessell TM, Siegelbaum SA, Hudspeth AJ. New York: McGraw-Hill Companies, Inc., 682–711.

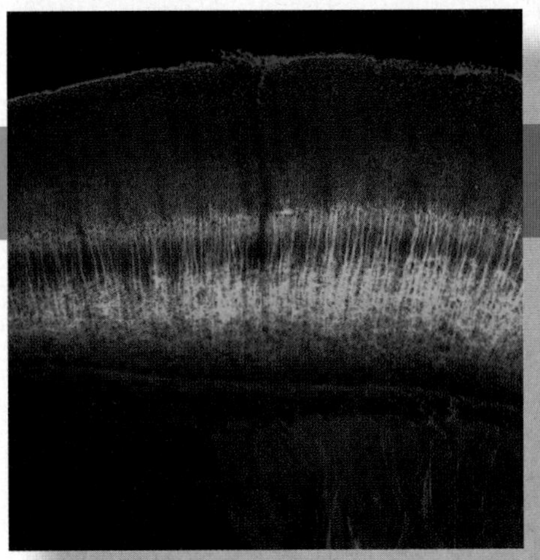

CHAPTER TWELVE

The Somatic Sensory System

INTRODUCTION

The somatic sensory system brings us some of life's most enjoyable experiences, as well as some of the most aggravating. **Somatic sensation** enables our body to feel, to ache, to sense hot or chill, and to know what its parts are doing. It is sensitive to many kinds of stimuli: the pressure of objects against the skin, the position of joints and muscles, distension of the bladder, and the temperature of the limbs and of the brain itself. It is the origin of itch. When stimuli become so strong that they may be damaging, somatic sensation is also responsible for the feeling that is most offensive, but vitally important—pain.

The somatic sensory system is different from other sensory systems in two interesting ways. First, its receptors are distributed throughout the body rather than being concentrated at small, specialized locations. Second, because it responds to many different kinds of stimuli, we can think of it as a group of at least four senses rather than a single one: the senses of touch, temperature, pain, and body position. In fact, those four can in turn be subdivided into many more. The somatic sensory system is really a catch-all name, a collective category for all the sensations that are *not* seeing, hearing, tasting, smelling, and the vestibular sense of balance. The familiar idea that we have only five senses is obviously too simple.

If something touches your finger, you can accurately gauge the place, pressure, sharpness, texture, and duration of the touch. If it is a pinprick, there is no mistaking it for a hammer. If the touch moves from your hand to your wrist, and up your arm to your shoulder, you can track its speed and position. Assuming you are not looking, this information is described entirely by the activity of the sensory nerves in your limb. A single sensory receptor can encode stimulus features such as intensity, duration, position, and sometimes direction. But a single stimulus usually activates many receptors. The CNS interprets the activity of the vast receptor array and uses it to generate coherent perceptions.

In this chapter, we divide our discussion of somatic sensation into two main parts: the sense of touch and the sense of pain. As we shall see, these different categories depend on different receptors, different axonal pathways, and different regions of the brain. We'll also describe sensations of itch, and how we sense changes in temperature. The sense of body position, also called *proprioception*, is discussed in Chapter 13, where we will explore how this type of somatic sensory information is used to control muscle reflexes.

TOUCH

The sensation of touch begins at the skin (Figure 12.1). The two major types of skin are called *hairy* and *glabrous* (hairless), as exemplified by the backs and palms of your hands. Skin has an outer layer, the *epidermis*, and an inner layer, the *dermis*. Skin performs an essential protective function, and it prevents the evaporation of body fluids into the dry environment we live in. But skin also provides our most direct contact with the world; indeed, skin is the largest sensory organ we have. Imagine the beach without the squish of sand between your toes, or consider *watching* a kiss instead of experiencing it yourself. Skin is sensitive enough that a raised dot measuring only 0.006 mm high and 0.04 mm wide can be felt when stroked by a fingertip. In comparison, a Braille dot is 167 times higher.

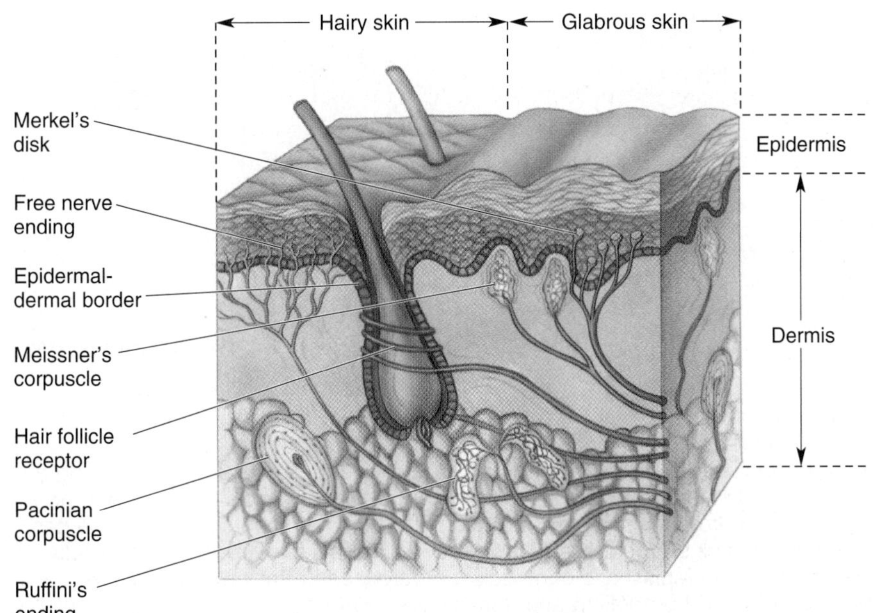

◀ FIGURE 12.1
Somatic sensory receptors in the skin. Hairy skin and glabrous skin have a variety of sensory receptors within the dermal and epidermal layers. Each receptor has an axon and, except for free nerve endings, all of them have associated non-neural tissues.

In this section, we will see how a touch of the skin is transduced into neural signals, how these signals make their way to the brain, and how the brain makes sense of them.

Mechanoreceptors of the Skin

Most of the sensory receptors in the somatic sensory system are **mechanoreceptors**, which are sensitive to physical distortion such as bending or stretching. Present throughout the body, mechanoreceptors monitor skin contact, pressure in the heart and blood vessels, stretching of the digestive organs and urinary bladder, and force against the teeth. At the heart of all mechanoreceptors are unmyelinated axon branches that are sensitive to stretching, bending, pressure, or vibration.

The mechanoreceptors of the skin are shown in Figure 12.1. Most of them are named after the nineteenth century German and Italian histologists who discovered them. The largest and best-studied receptor is the **Pacinian corpuscle**, which lies deep in the dermis and can be as long as 2 mm and almost 1 mm in diameter. Each human hand has about 2500 Pacinian corpuscles, with the highest densities in the fingers. *Ruffini's endings*, found in both hairy and glabrous skin, are slightly smaller than Pacinian corpuscles. *Meissner's corpuscles* are about one-tenth the size of Pacinian corpuscles and are located in the ridges of glabrous skin (the raised parts of your fingerprints, for example). Located within the epidermis, *Merkel's disks* each consist of a nerve terminal and a flattened, non-neural epithelial cell (the Merkel cell). In *Krause end bulbs*, which lie in the border regions of dry skin and mucous membrane (around the lips and genitals, for example), the nerve terminals look like knotted balls of string.

Skin can be vibrated, pressed, pricked, and stroked, and its hairs can be bent or pulled. These are quite different kinds of mechanical energy, yet we can feel them all and easily tell them apart. Accordingly, we have mechanoreceptors that vary in their preferred stimulus frequencies, pressures, and receptive field sizes. Swedish neuroscientist Åke Vallbo and his colleagues developed methods to record from single sensory axons in the human arm, so that they could simultaneously measure the sensitivity of

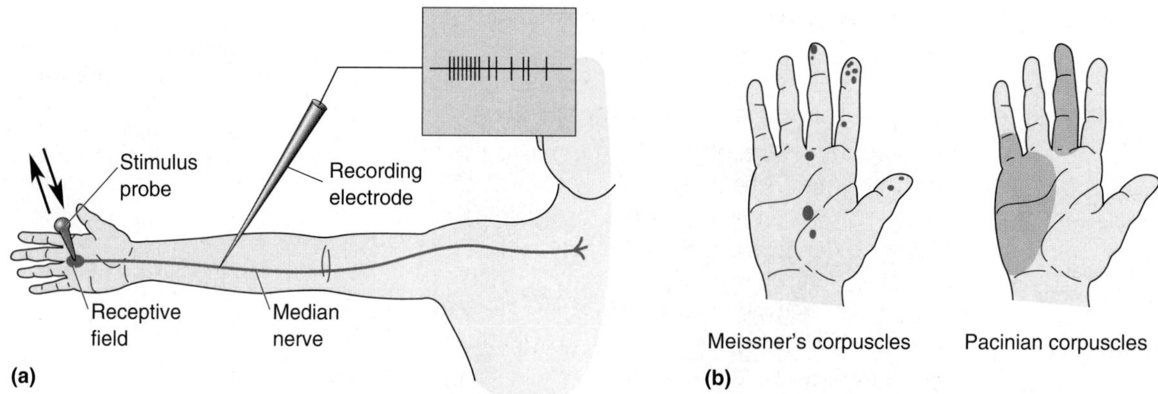

▲ FIGURE 12.2
Testing the receptive fields of human sensory receptors. (a) By introducing a microelectrode into the median nerve of the arm, it is possible to record the action potentials from a single sensory axon and map its receptive field on the hand with a fine stimulus probe. (b) Results show that receptive fields are either relatively small, as in Meissner's corpuscles, or large, as in Pacinian corpuscles. (Source: Adapted from Vallbo and Johansson, 1984.)

mechanoreceptors in the hand *and* evaluate the perceptions produced by various mechanical stimuli (Figure 12.2a). When the stimulus probe was touched to the surface of the skin and moved around, the receptive field of a single mechanoreceptor could be mapped. Meissner's corpuscles and Merkel's disks had small receptive fields, only a few millimeters wide, while Pacinian corpuscles and Ruffini's endings had large receptive fields that could cover an entire finger or half the palm (Figure 12.2b).

Mechanoreceptors also vary in the persistence of their responses to long-lasting stimuli. If a stimulus probe is suddenly pressed against the skin within the receptive field, some mechanoreceptors, such as Meissner's and Pacinian corpuscles, tend to respond quickly at first but then stop firing even though the stimulus continues; these receptors are said to be *rapidly adapting*. Other receptors, such as Merkel's disks and Ruffini's endings, are *slowly adapting*, and generate a more sustained response during a long stimulus. Figure 12.3 summarizes the receptive field size and adaptation rate for four mechanoreceptors of the skin.

Hairs do more than adorn our head and keep a dog warm in winter. Many hairs are part of a sensitive receptor system. To demonstrate this, brush just a single hair on the back of your arm with the tip of a pencil; it feels like an annoying mosquito. For some animals, hair is a major sensory system. Imagine a rat slinking confidently through dark passageways and alleys. The rat navigates in part by waving its facial *vibrissae* (whiskers) to sense the local environment and derive information about the texture, distance, and shape of nearby objects.

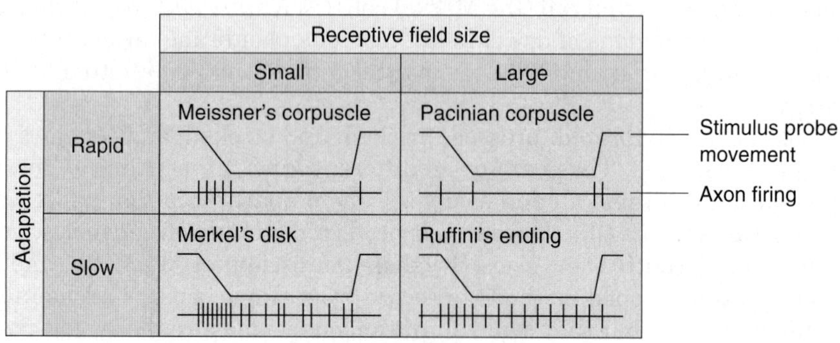

▶ FIGURE 12.3
Variations of receptive field size and adaptation rate for four somatic sensory skin receptors. (Source: Adapted from Vallbo and Johansson, 1984.)

Hairs grow from *follicles* embedded in the skin; each follicle is richly innervated by free nerve endings—the terminations of single axons—that either wrap around the follicle or run parallel to it (see Figure 12.1). There are several types of hair follicles, including some with erectile muscles (essential for mediating the strange sensation we call goose bumps), and the details of their innervation differ. In all cases, the bending of the hair causes a deformation of the follicle and surrounding skin tissues. This, in turn, stretches, bends, or flattens the nearby nerve endings, which then increase or decrease their action potential firing frequency. The mechanoreceptors of hair follicles may be either slowly adapting or rapidly adapting.

The different mechanical sensitivities of mechanoreceptors mediate different sensations. Pacinian corpuscles are most sensitive to vibrations of about 200–300 Hz, while Meissner's corpuscles respond best around 50 Hz (Figure 12.4). Place your hand against a speaker while playing your favorite music loudly; you "feel" the music largely with your Pacinian corpuscles. If you stroke your fingertips across the coarse screen covering the speaker, each point of skin will hit the bumps at frequencies about optimal to activate Meissner's corpuscles. You feel this as a sensation of rough texture. Stimulation at frequencies from about 1 to 10 Hz can also activate Meissner's corpuscles, yielding a "fluttering" feeling.

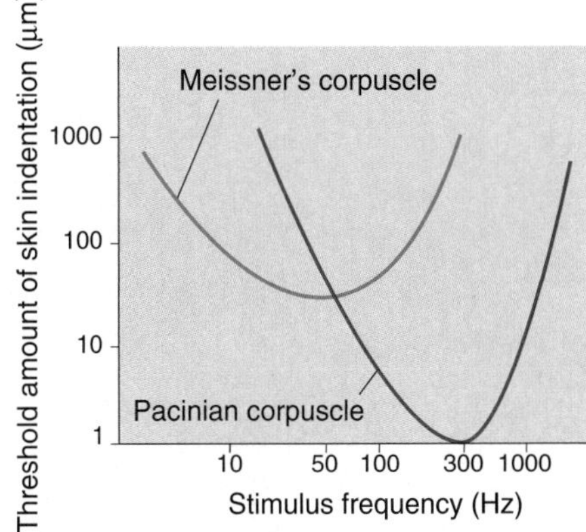

▲ FIGURE 12.4
Frequency sensitivity of two rapidly adapting skin mechanoreceptors. Pacinian corpuscles are most sensitive to high-frequency stimuli, and Meissner's corpuscles are more sensitive to low-frequency stimuli. The skin was indented with a pressure probe, at various frequencies, while recording from the nerve. The amplitude of the stimulus was increased until it generated action potentials; threshold was measured as the amount of skin indentation in micrometers (μm). (Source: Adapted from Schmidt, 1978.)

Vibration and the Pacinian Corpuscle. The selectivity of a mechanoreceptive axon depends primarily on the structure of its special ending. For example, the Pacinian corpuscle has a football-shaped capsule with 20–70 concentric layers of connective tissue, arranged like the layers of an onion, with an axon terminal in the middle (see Figure 12.1). When the capsule is compressed, energy is transferred to the nerve terminal, its membrane is deformed, and mechanosensitive channels open. Current flowing through the channels generates a receptor potential, which is depolarizing (Figure 12.5a). If the depolarization is large enough, the axon

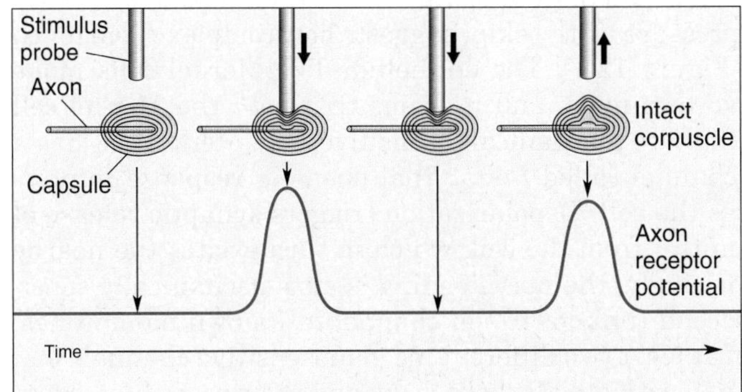

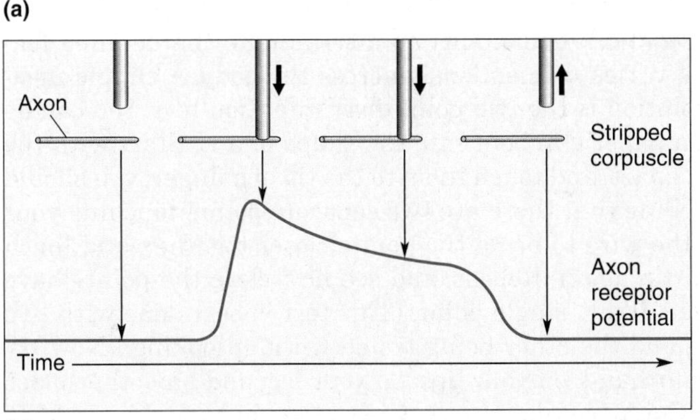

◀ FIGURE 12.5
Adaptation in the Pacinian corpuscle. A single Pacinian corpuscle was isolated and stimulated by a probe that indented it briefly. The receptor potential was recorded from a nearby portion of the axon. **(a)** In the intact corpuscle, a large receptor potential was generated at the onset and offset of the stimulus; during maintained indentation, the receptor potential disappeared. **(b)** The onionlike encapsulation was dissected away, leaving a bare axon ending. When indented by the probe, a receptor potential was again generated, showing the capsule is not necessary for mechanoreception. But while the normal corpuscle responded only to the onset or offset of a long indentation, the stripped version gave a much more prolonged response; its adaptation rate was slowed. Apparently it is the capsule that makes the corpuscle insensitive to low-frequency stimuli.

will fire an action potential. But the capsule layers are slick, with viscous fluid between them. If the stimulus pressure is maintained, the layers slip past one another and transfer the stimulus energy in such a way that the axon terminal is no longer deformed, and the receptor potential dissipates. When pressure is released, the events reverse themselves; the terminal depolarizes again and may fire another action potential.

In the 1960s, Werner Loewenstein and his colleagues, working at Columbia University, stripped away the capsule from single corpuscles and found that the naked nerve terminal became much less sensitive to vibrating stimuli and much more sensitive to steady pressure (Figure 12.5b). Clearly, it is the layered capsule (and not some property of the nerve ending itself) that makes the Pacinian corpuscle exquisitely sensitive to vibrating, high-frequency stimuli and almost unresponsive to steady pressure (see Figure 12.4). In order to communicate information about rapid vibrations to the central nervous system (CNS) in a timely way, the Pacinian corpuscles have some of the largest and fastest axons that originate in the skin.

Mechanosensitive Ion Channels. The mechanoreceptors of the skin all have unmyelinated axon terminals, and the membranes of these axons have *mechanosensitive ion channels* that convert mechanical force into a change of ionic current. Forces applied to these channels alter their gating and either enhance or decrease channel opening. Force can be applied to a channel by the membrane itself when it is stretched or bent, or force may be applied through connections between the channels and extracellular proteins or intracellular cytoskeletal components (e.g., actin, microtubules) (Figure 12.6). Alternatively, mechanical stimuli may somehow trigger the release of second messengers (e.g., DAG, IP_3) that secondarily regulate ion channels.

A variety of ion channel types have been implicated in mechanosensation, but the specific types of channels in most of the somatic sensory receptors are still unidentified. Recent work on Merkel's disks, which are sensitive to gentle pressure on the skin, suggests how complex some touch receptors are (see Figure 12.1). The epithelium-like Merkel cells make synapses onto nerve terminals, and it seems that *both* the Merkel cell and the axon terminal are mechanically sensitive. The Merkel cell has a mechanosensitive channel called *Piezo2* that opens in response to pressure and depolarizes the cell. Depolarization triggers synaptic release of an unknown transmitter from the cell, which in turn excites the nearby nerve ending. Surprisingly, the nerve ending is also mechanically sensitive because of a second (unknown) ion channel in its own membranes. Thus, the actions of at least two different mechanosensitive channels and a synapse cooperate to activate Merkel's disks and their associated axon.

Two-Point Discrimination. Our ability to discriminate the detailed features of a stimulus varies tremendously across the body. A simple measure of spatial resolution is the two-point discrimination test. You can do this yourself with a paper clip bent into the shape of a U. Start with the ends about an inch apart, and touch them to the tip of a finger; you should have no problem telling that there are two separate points touching your finger. Then bend the wire to bring the points closer together, and touch them to your fingertip again. Repeat, and see how close the points have to be before they feel like a single point. (This test is best done with two people, one testing and the other being tested without looking.) Now try it on the back of your hand, on your lips, on your leg, and any other place that interests you. Compare your results with those shown in Figure 12.7.

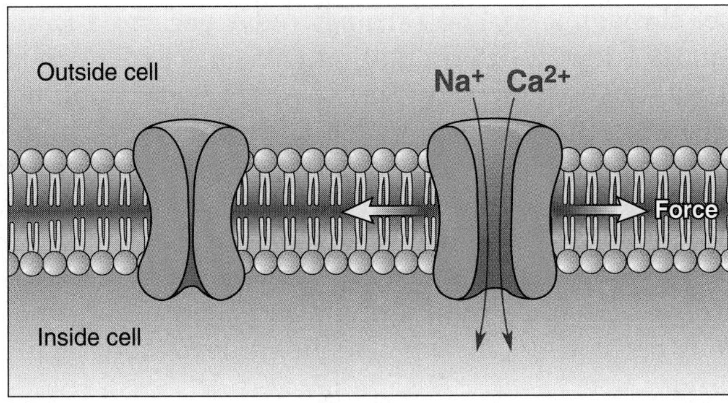

(a)

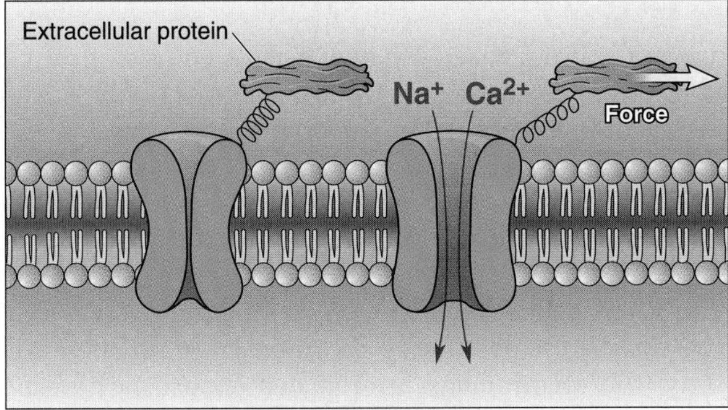

(b)

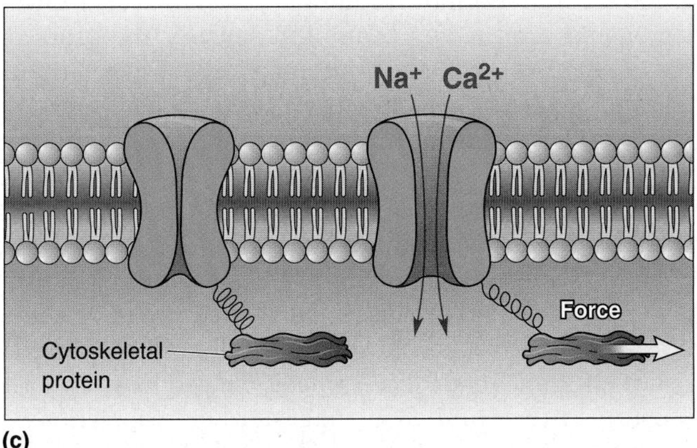

(c)

▲ FIGURE 12.6
Mechanosensitive ion channels. **(a)** Some membrane ion channels are sensitive to stretching of the lipid membrane; tension in the membrane directly induces the channel to open and allow cations to flow. **(b)** Other ion channels open when force is applied to extracellular structures linked to the channels by peptides. **(c)** Mechanically sensitive channels may also be linked to intracellular proteins, especially those of the cytoskeleton; deformation of the cell and stress on its cytoskeleton generate forces that regulate channel gating.

Two-point discrimination varies at least twentyfold across the body. Fingertips have the highest resolution. The dots of Braille are 1 mm high and 2.5 mm apart; up to six dots make a letter. An experienced Braille reader can move an index finger across a page of raised dots and read about 600 letters per minute, which is roughly as fast as someone reading aloud. The Braille reader scans with fingertips because touch is most sensitive when the skin and the stimuli move across each other, compared to simply pressing one against the other. Practice also improves

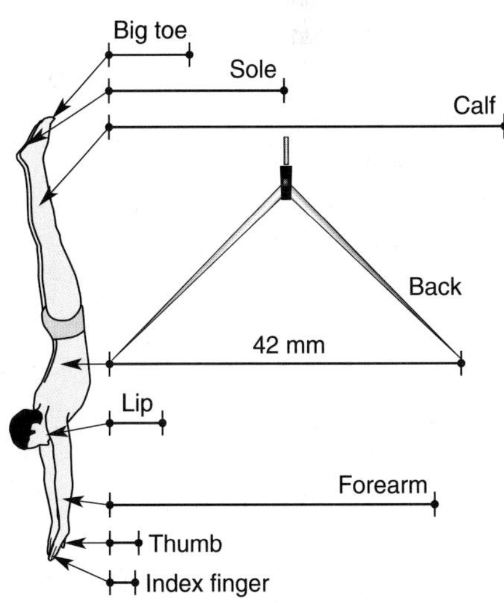

▲ FIGURE 12.7
Two-point discrimination on the body surface. The pairs of dots show the minimum distance necessary to differentiate between two points touching the body simultaneously. Notice the sensitivity of the fingertips compared to the rest of the body. All measurements are shown at their actual scale.

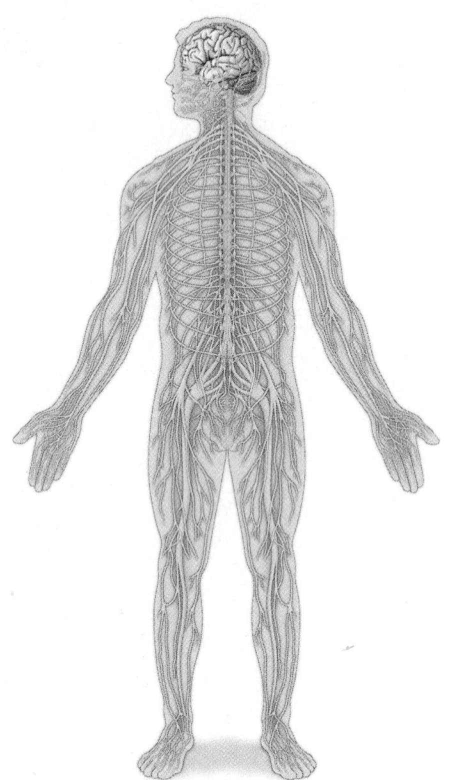

▲ FIGURE 12.8
The peripheral nerves.

performance through a type of learning, and experienced Braille readers are particularly adept at discriminating patterns of small dots.

Several reasons explain why the fingertip is so much better than, say, the elbow for Braille reading: (1) There is a much higher density of mechanoreceptors in the skin of the fingertip than on other parts of the body, (2) the fingertips are enriched in receptor types that have small receptive fields (e.g., Merkel's disks), (3) there is more brain tissue (and thus more raw computing power) devoted to the sensory information of each square millimeter of fingertip than elsewhere, and (4) there may be special neural mechanisms devoted to high-resolution discriminations.

Primary Afferent Axons

The skin is richly innervated by axons that course through the vast network of peripheral nerves on their way to the CNS (Figure 12.8). Axons bringing information from the somatic sensory receptors to the spinal cord or brain stem are the *primary afferent axons* of the somatic sensory system. The primary afferent axons enter the spinal cord through the dorsal roots; their cell bodies lie in the dorsal root ganglia (Figure 12.9).

Primary afferent axons have widely varying diameters, and their size correlates with the type of sensory receptor to which they are attached. Unfortunately, the terminology approaches absurdity here because the different sizes of axons are designated by two sets of names, using Arabic *and* Greek letters *and* Roman numerals. As shown in Figure 12.10, in order of decreasing size, axons from skin sensory receptors are usually designated Aα, Aβ, Aδ, and C; axons of similar size, but innervating the muscles and tendons, are called *groups I, II, III,* and *IV*. Group C (or IV) axons are, by definition, unmyelinated axons, while all the rest are myelinated.

An interesting and simple point is hidden in the many axon names. Recall that the diameter of an axon, together with its myelin, determines its speed of action potential conduction. The smallest axons, the so-called C fibers, have no myelin and are less than about 1 μm in diameter. C fibers mediate pain, temperature sensation, and itch, and they are the slowest of axons, conducting at about 0.5–1 m/sec. To see how slow this is, take a big step, count to two, and then take another step. That's about how fast the action potentials travel along C fibers. On the other hand, touch

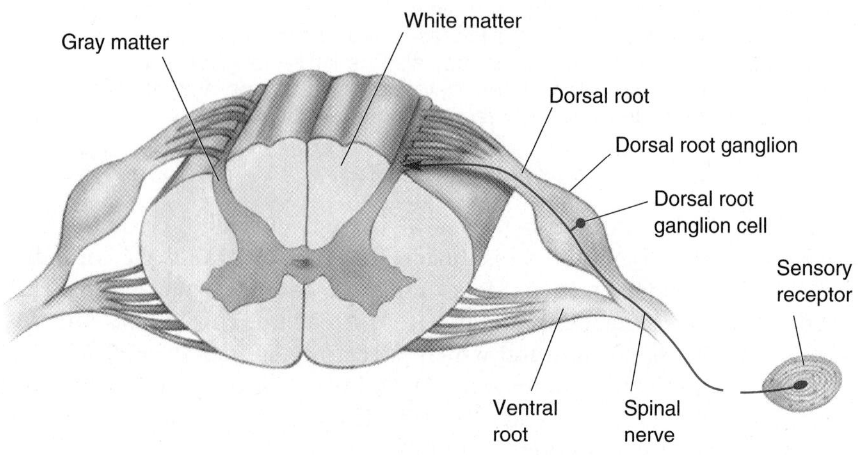

▶ FIGURE 12.9
The structure of a segment of the spinal cord and its roots.

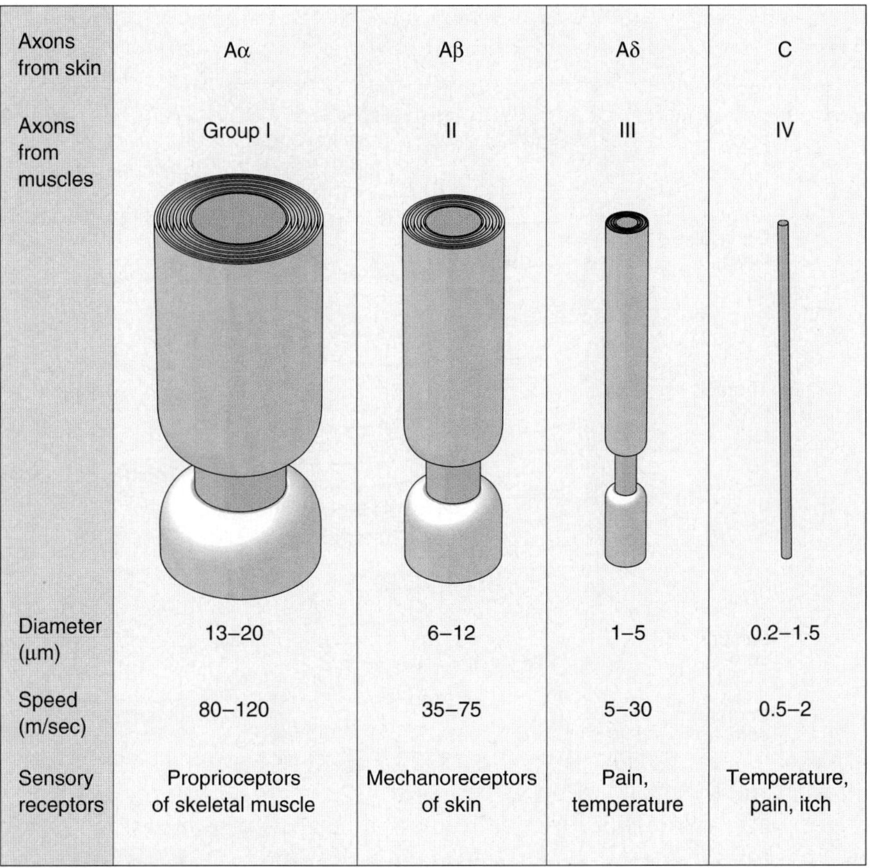

Axons from skin
Aα
Aβ
Aδ
C
Axons from muscles
Group I
II
III
IV
Diameter (μm)
13–20
6–12
1–5
0.2–1.5
Speed (m/sec)
80–120
35–75
5–30
0.5–2
Sensory receptors
Proprioceptors of skeletal muscle
Mechanoreceptors of skin
Pain, temperature
Temperature, pain, itch

▲ FIGURE 12.10
Various sizes of primary afferent axons. The axons are drawn to scale, but they are shown 2000 times larger than life size. The diameter of an axon is correlated with its conduction velocity and with the type of sensory receptor to which it is connected.

sensations, mediated by the cutaneous mechanoreceptors, are conveyed by the relatively large Aβ axons, which can conduct at up to 75 m/sec. For comparison, consider that an exceptional major league baseball pitcher can throw a fastball up to 100 miles per hour, which is about 45 m/sec.

The Spinal Cord

Most peripheral nerves communicate with the CNS via the spinal cord, which is encased in the bony vertebral column.

Segmental Organization of the Spinal Cord. The arrangement of paired dorsal and ventral roots shown in Figure 12.9 is repeated 30 times down the length of the human spinal cord. Each spinal nerve, consisting of dorsal root and ventral root axons, passes through a notch between the vertebrae (the "back bones") of the spinal column. There are as many spinal nerves as there are notches between vertebrae. As shown in Figure 12.11, the 30 **spinal segments** are divided into four groups, and each segment is named after the vertebra adjacent to where the nerves originate: cervical (C) 1–8, thoracic (T) 1–12, lumbar (L) 1–5, and sacral (S) 1–5.

The segmental organization of spinal nerves and the sensory innervation of the skin are related. The area of skin innervated by the right and

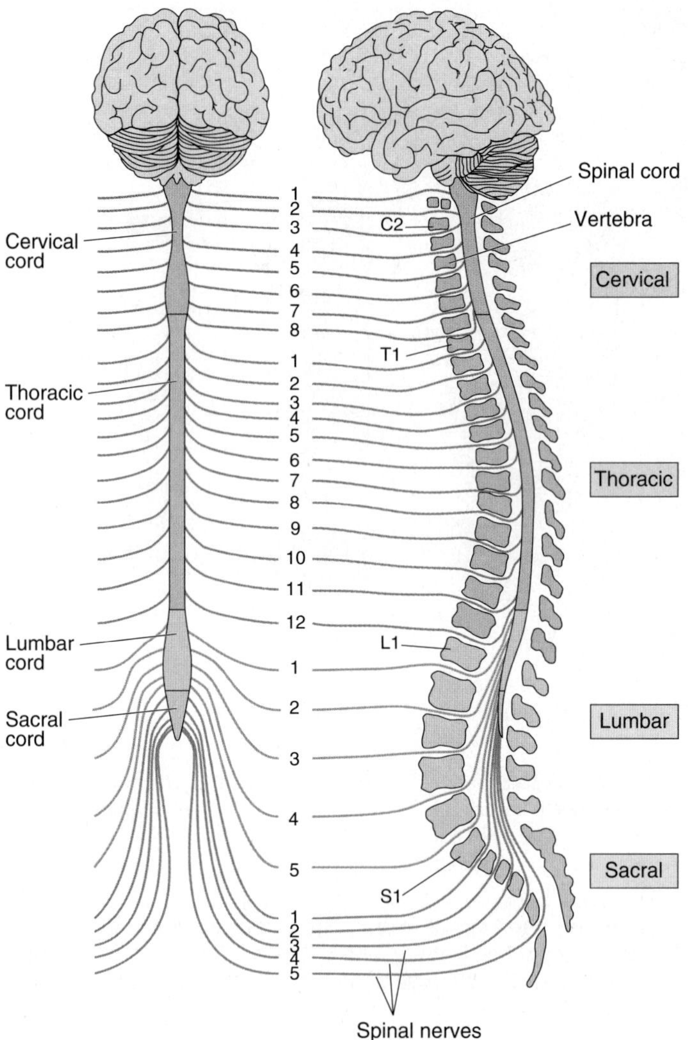

▶ FIGURE 12.11

Segmental organization of the spinal cord. The spinal cord is divided into cervical, thoracic, lumbar, and sacral divisions (left). The cross-sectional view (right) shows the spinal cord within the vertebral column. Spinal nerves are named for the level of the spinal cord from which they exit and are numbered in order from rostral to caudal.

left dorsal roots of a single spinal segment is called a **dermatome**; thus, there is a one-to-one correspondence between dermatomes and spinal segments. When mapped, the dermatomes delineate a set of bands on the body surface, as shown in Figure 12.12. The organization of the dermatomes is best revealed when one bends over to stand on both hands and feet (Figure 12.13). This organization presumably reflects our distant quadrupedal ancestry.

When a dorsal root is cut, the corresponding dermatome on that side of the body does not lose all sensation. The residual somatic sensation is explained by the fact that the adjacent dorsal roots innervate overlapping areas. To lose all sensation in one dermatome, therefore, three adjacent dorsal roots must be cut. However, the skin innervated by the axons of one dorsal root is plainly revealed by a condition called *shingles*, in which all the neurons of a single dorsal root ganglion become infected with a virus (Box 12.1).

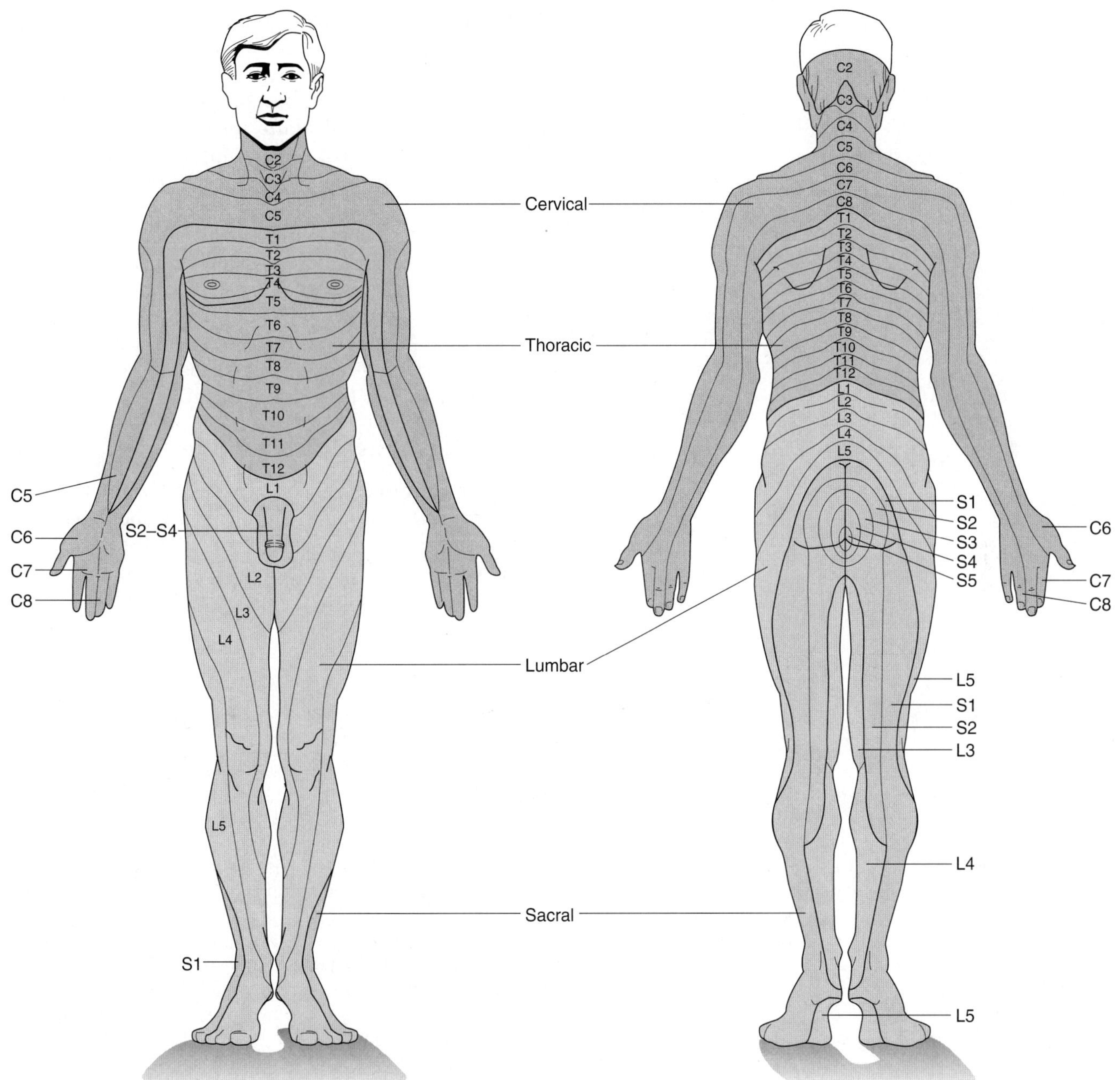

▲ FIGURE 12.12
Dermatomes. These illustrations show the mapping of the approximate boundaries of the dermatomes on the body.

Notice in Figure 12.11 that the spinal cord in the adult ends at about the level of the third lumbar vertebra. The bundles of spinal nerves streaming down within the lumbar and sacral vertebral column are called the *cauda equina* (Latin for "horse's tail"). The cauda equina courses down the spinal column within a sack of dura filled with cerebrospinal fluid (CSF). In a method called *lumbar puncture* (also called a *spinal tap*), used to collect CSF for medical diagnostic tests, a needle is inserted into this CSF-filled cistern at the midline. If the needle is inserted a little off center, however, a nerve can be touched. Not surprisingly, this causes a sensation of sharp pain in the dermatome supplied by that nerve.

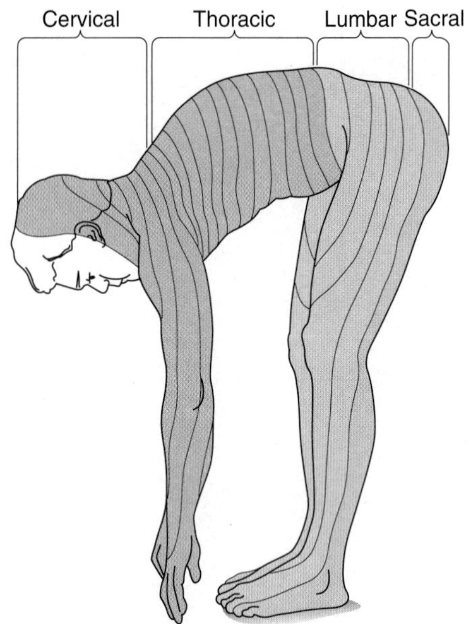

▲ FIGURE 12.13
Dermatomes on all fours.

Sensory Organization of the Spinal Cord. The basic anatomy of the spinal cord was introduced in Chapter 7. The spinal cord is composed of an inner core of gray matter, surrounded by a thick covering of white matter tracts that are often called *columns*. Each half of the spinal gray matter is divided into a *dorsal horn*, an *intermediate zone,* and a *ventral horn* (Figure 12.14). The neurons that receive sensory input from primary afferents are called *second-order sensory neurons*. Most of the second-order sensory neurons of the spinal cord lie within the dorsal horns.

The large, myelinated Aβ axons conveying information about a touch to the skin enter the dorsal horn and branch. One branch synapses in the deep part of the dorsal horn on second-order sensory neurons. These connections can initiate or modify a variety of rapid and unconscious reflexes. The other branch of the Aβ primary afferent axon ascends straight to the brain. This ascending input is responsible for perception, enabling us to form complex judgments about the stimuli touching the skin.

The Dorsal Column–Medial Lemniscal Pathway

Information about touch or vibration of the skin takes a path to the brain that is entirely distinct from that taken by information about pain and temperature. The pathway serving touch is called the **dorsal column–medial lemniscal pathway**, for reasons we will see in a moment. The organization of this pathway is summarized in Figure 12.15.

BOX 12.1 OF SPECIAL INTEREST

Herpes, Shingles, and Dermatomes

Many of us were infected by the varicella zoster virus, a type of herpes virus commonly known as chickenpox, when we were children. After a week or so covered with red, itchy spots on our skin, we usually recovered. Out of sight is not out of body, however. The virus remains in our primary sensory neurons, dormant but viable. Most people never notice it again, but in some cases, the virus revives decades later, wreaking havoc with the somatic sensory system. The result is *shingles*, a condition that can be agonizingly painful for periods of months or even years. The reactivated virus increases the excitability of the sensory neurons, leading to very low thresholds of firing as well as spontaneous activity. The pain is a constant burning, sometimes a stabbing sensation, and the skin is exquisitely sensitive to any stimulus. People with shingles often shun clothes because of their hypersensitivity. The skin itself becomes inflamed and blistered, then scaly—hence, the name (Figure A). Several useful treatments are available that often shorten the outbreak, relieve the pain, and prevent long-term complications.

Fortunately, the varicella zoster virus usually reactivates only in the neurons of one dorsal root ganglion. This means that the symptoms are restricted to the skin innervated by the axons of the affected dorsal root. In effect, the virus performs an anatomical labeling experiment for us by clearly marking the skin territory of one dermatome. Almost any dermatome may be involved, although the thoracic and facial areas are most common. Observations of many shingles patients and their infected areas were actually useful in mapping the dermatomes (see Figure 12.12).

Neuroscientists have learned how to use herpes viruses and other types of viruses to their advantage. Viruses are valuable research tools because they can be used to introduce new genes into neurons.

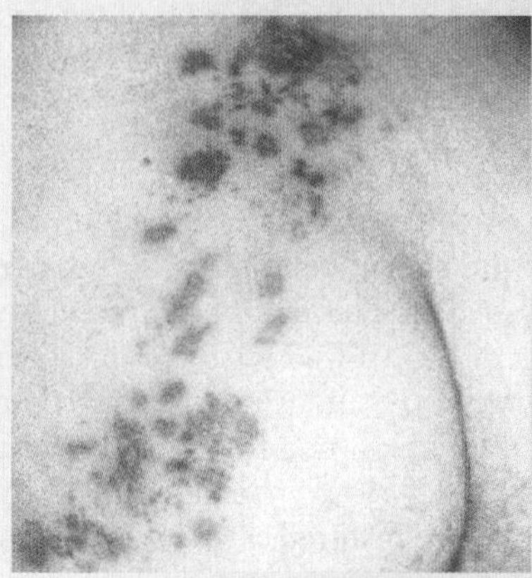

Figure A
Skin lesions caused by shingles, confined to the L4 dermatome on the left side.

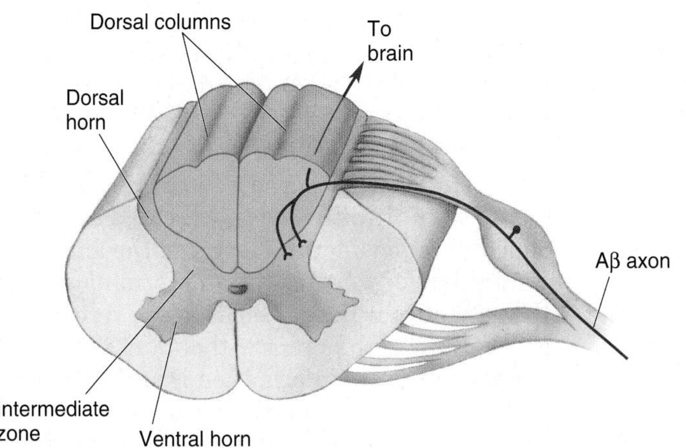

▲ FIGURE 12.14

The trajectory of the touch-sensitive Aβ axons in the spinal cord.

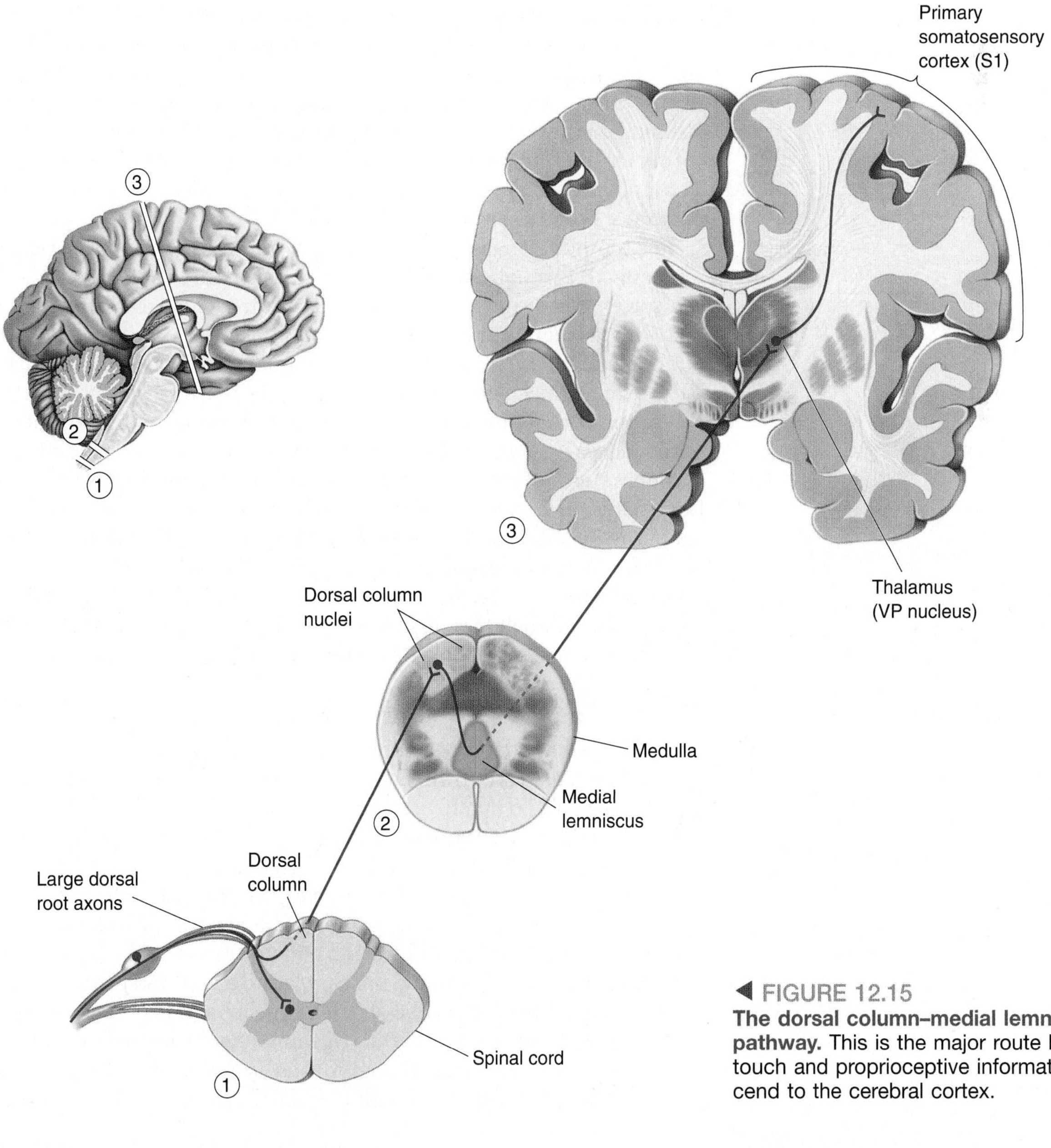

◀ FIGURE 12.15

The dorsal column–medial lemniscal pathway. This is the major route by which touch and proprioceptive information ascend to the cerebral cortex.

The ascending branch of the large sensory axons (Aβ) enters the ipsilateral **dorsal column** of the spinal cord, the white matter tract medial to the dorsal horn (see Figure 12.14). The dorsal columns carry information about tactile sensation (and limb position) toward the brain. They are composed of primary sensory axons, as well as second-order axons from neurons in the spinal gray matter. The axons of the dorsal column terminate in the **dorsal column nuclei**, which lie at the junction of the spinal cord and medulla. Consider that some of the longest axons in your body originate in the skin of your big toe and terminate in the dorsal column nuclei at the base of your head! This is a fast, direct path that brings information from the skin to the brain without an intervening synapse.

At this point in the pathway, information is still represented ipsilaterally; that is, touch information from the *right* side of the body is represented in the activity of cells in the *right* dorsal column nuclei. However, axons from cells of the dorsal column nuclei arch toward the ventral and medial medulla, and decussate. *From this point onward, the somatic sensory system of one side of the brain is concerned with sensations originating from the other side of the body.*

The axons of the dorsal column nuclei ascend within a conspicuous white matter tract called the **medial lemniscus**. The medial lemniscus rises through the medulla, pons, and midbrain, and its axons synapse upon neurons of the **ventral posterior (VP) nucleus** of the thalamus. Remember that almost no sensory information goes directly into the neocortex without first synapsing in the thalamus (olfaction is the exception). Thalamic neurons of the VP nucleus then project to specific regions of **primary somatosensory cortex**, or **S1**.

It is tempting to assume that sensory information is simply transferred, unchanged, through nuclei in the brain stem and thalamus on its way to the cortex, with the actual processing taking place only in the cortex. In fact, this assumption is demonstrated by the term *relay nuclei*, which is often used to describe specific sensory nuclei of the thalamus such as the VP nucleus. Physiological studies prove otherwise, however. In both dorsal column and thalamic nuclei, considerable transformation of information takes place. As a general rule, information is altered every time it passes through a set of synapses in the brain. In particular, inhibitory interactions between adjacent sets of inputs in the dorsal column–medial lemniscal pathway enhance the responses to tactile stimuli (Box 12.2). As we will see later, some synapses in these nuclei can also change their strength, depending on their recent activity. Neurons of both the thalamus and the dorsal column nuclei are also controlled by input from the cerebral cortex. Accordingly, the output of the cortex can influence the input of the cortex!

The Trigeminal Touch Pathway

Thus far, we have described only the part of the somatic sensory system that enters the spinal cord. If this were the whole story, your face and the top of your head would be numb. Somatic sensation of the face is supplied mostly by the large **trigeminal nerves** (cranial nerve V), which enter the brain at the pons (see Chapter 7). (The word is from the Latin *tria*, "three," and *geminus*, "twin.") There are twin trigeminal nerves, one on each side, and each breaks up into three peripheral nerves that innervate the face, mouth areas, the outer two-thirds of the tongue, and the dura mater covering the brain. Additional sensation from the skin around the ears, nasal areas, and pharynx is provided by other cranial nerves: the facial (VII), glossopharyngeal (IX), and vagus (X).

The sensory connections of the trigeminal nerve are analogous to those of the dorsal roots. The large-diameter sensory axons of the trigeminal nerve carry tactile information from skin mechanoreceptors. They synapse onto

BOX 12.2 **BRAIN FOOD**

Lateral Inhibition

Information is usually transformed as it is passed from one neuron to the next in a sensory pathway. One common transformation is the amplification of differences in the activity of neighboring neurons, also known as *contrast enhancement*. We already saw this in retinal ganglion cell receptive fields (see Chapter 9). If all the photoreceptors providing input to a ganglion cell are evenly illuminated, the cell hardly notices. However, if there is a contrast border—a difference in illumination—within the cell's receptive field, the cell's response is strongly modulated. Contrast enhancement is a general feature of information processing in sensory pathways, including the somatic sensory system. One general mechanism underlying contrast enhancement is *lateral inhibition*, whereby neighboring cells inhibit one another. Let's see how this works, using a simple model.

Consider the situation in Figure A. Dorsal root ganglion neurons lettered a through g relay information via excitatory synapses to dorsal column nucleus neurons A through G. All of the neurons fire with baseline rates of 5 spikes/sec, even in the absence of stimulation. Consider what happens when a stimulus is applied to the receptive field of just one sensory neuron, cell d in Figure A. The firing rate of cell d increases to 10 spikes/sec. Let's assume the output of the dorsal column nucleus cells is simply the presynaptic input multiplied by a synaptic gain factor of 1. If the input activity of cell d is 10, the output activity of cell D is also 10. This simple relay does nothing to enhance the difference between the more active neuron, d, and the other neurons. The contrast in activity between neuron D and its neighbors C and E, for example, is 10 versus 5 spikes/sec.

Now consider the situation in Figure B, where inhibitory interneurons have been added that project laterally to inhibit each cell's neighbor. The synaptic gain of the inhibitory synapses (black triangles) is −1, and we've adjusted the gain of the excitatory synapses, as shown in the figure. Calculate the activity of each cell by multiplying the input to each synapse by its synaptic gain, and then summing the effect of all the synapses on the cell. If you perform this calculation when the stimulus is again applied to cell d, you will see that there is significant contrast enhancement: The difference between the activity in cell d and its neighbors has been greatly amplified in the output of cell D. The contrast in activity between neuron D and its neighbors C and E is now 20 versus 0 spikes/sec.

Figure A

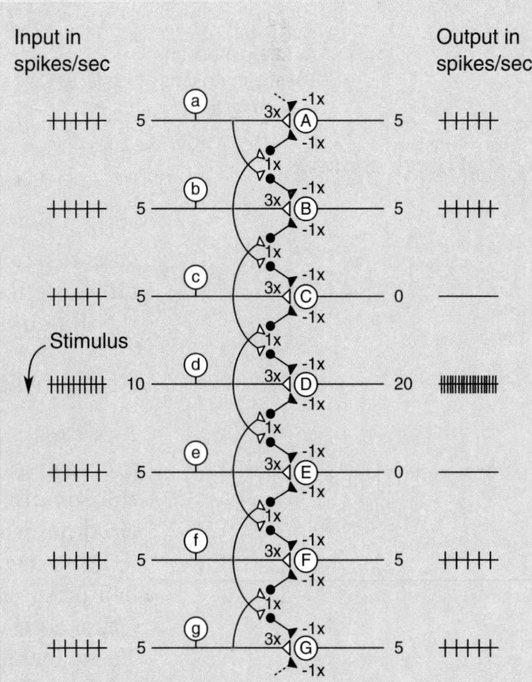

Figure B

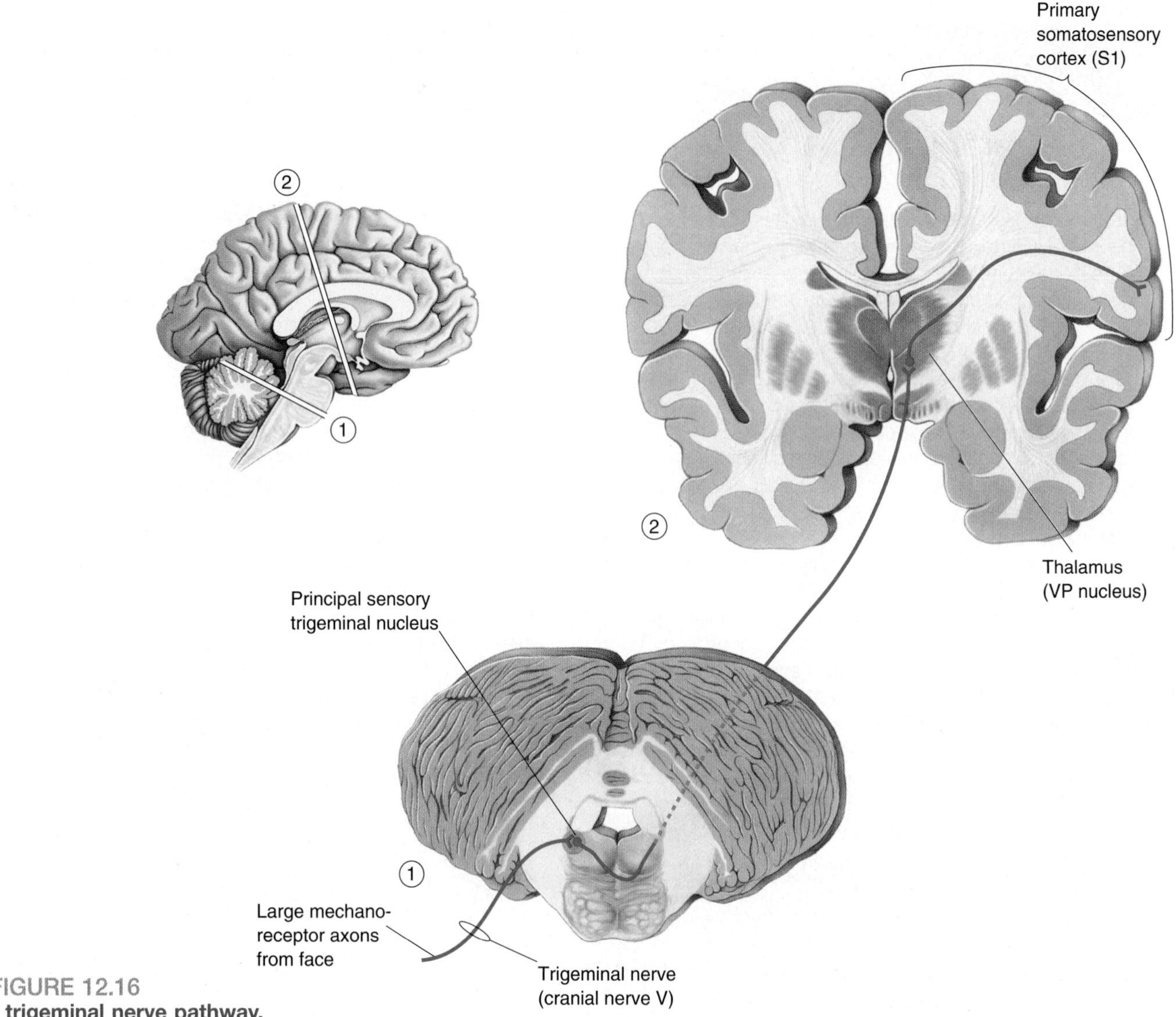

▶ FIGURE 12.16
The trigeminal nerve pathway.

second-order neurons in the ipsilateral trigeminal nucleus, which is analogous to a dorsal column nucleus (Figure 12.16). The axons from the trigeminal nucleus decussate and project into the medial part of the VP nucleus of the thalamus. From here, information is relayed to the somatosensory cortex.

Somatosensory Cortex

As with all other sensory systems, the most complex levels of somatosensory processing occur in the cerebral cortex. Most of the cortex concerned with the somatic sensory system is located in the parietal lobe (Figure 12.17). Brodmann's area 3b, now regarded as the primary somatosensory cortex (S1), is easy to find in humans because it lies on the postcentral gyrus (right behind the central sulcus). (See also Figure 7.28 showing Brodmann's areas.) Other cortical areas that also process somatic sensory information flank S1. These include areas 3a, 1, and 2 on the postcentral gyrus, and areas 5 and 7 on the adjacent **posterior parietal cortex** (see Figure 12.17).

Area 3b is the *primary* somatic sensory cortex because (1) it receives dense inputs from the VP nucleus of the thalamus; (2) its neurons are very responsive to somatosensory stimuli (but not to other sensory stimuli); (3) lesions here impair somatic sensation; and (4) when electrically stimulated, it evokes somatic sensory experiences. Area 3a also receives

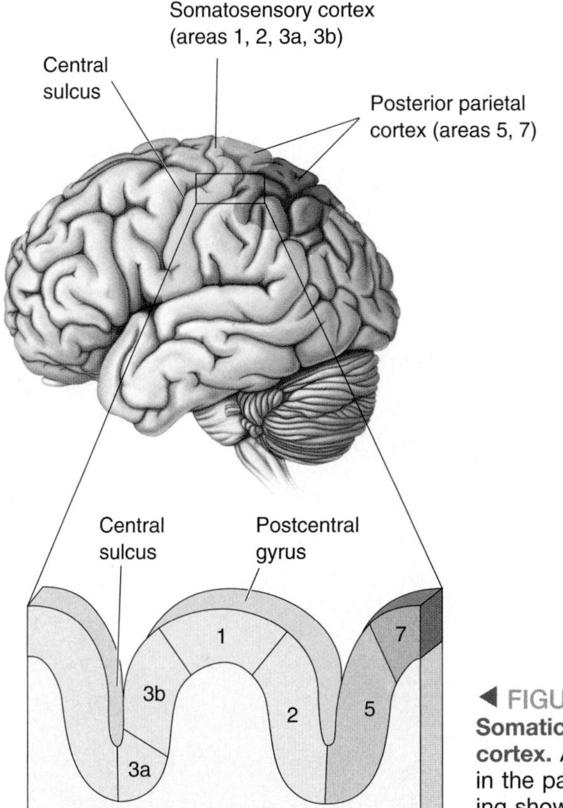

◀ FIGURE 12.17
Somatic sensory areas of the cortex. All of the illustrated areas lie in the parietal lobe. The lower drawing shows that the postcentral gyrus contains S1, area 3b.

a dense input from the thalamus; however, this region is concerned with the sense of body position rather than touch.

Areas 1 and 2 receive dense inputs from area 3b. The projection from 3b to area 1 sends mainly texture information, while the projection to area 2 emphasizes size and shape. Small lesions in area 1 or 2 produce predictable deficiencies in discrimination of texture, size, and shape.

Somatic sensory cortex, like other areas of neocortex, is a layered structure. As is the case for visual and auditory cortex, the thalamic inputs to S1 terminate mainly in layer IV. The neurons of layer IV project, in turn, to cells in the other layers. Another important similarity with other regions of cortex is that S1 neurons with similar inputs and responses are stacked vertically into columns that extend across the cortical layers (Figure 12.18). In fact, the concept of the cortical column, so beautifully elaborated by Hubel and Wiesel in visual cortex, was actually first described in somatic sensory cortex by Johns Hopkins University scientist Vernon Mountcastle.

Cortical Somatotopy. Electrical stimulation of the S1 surface can cause somatic sensations localized to a specific part of the body. Systematically, moving the stimulator around S1 will cause the sensation to move across the body. American-Canadian neurosurgeon Wilder Penfield, working at McGill University from the 1930s through the 1950s, actually used this method to map the cortex of neurosurgical patients. (It is interesting to note that these brain operations can be performed in awake patients, with only local anesthesia of the scalp, because the brain tissue itself lacks the receptors of somatic sensation.) Another way to map the somatosensory cortex is to record the activity of a single neuron and determine the site of its somatosensory receptive field on the body. The receptive fields of many S1 neurons produce an orderly map of the body on the cortex. The mapping of the body's surface sensations onto a structure in the brain is

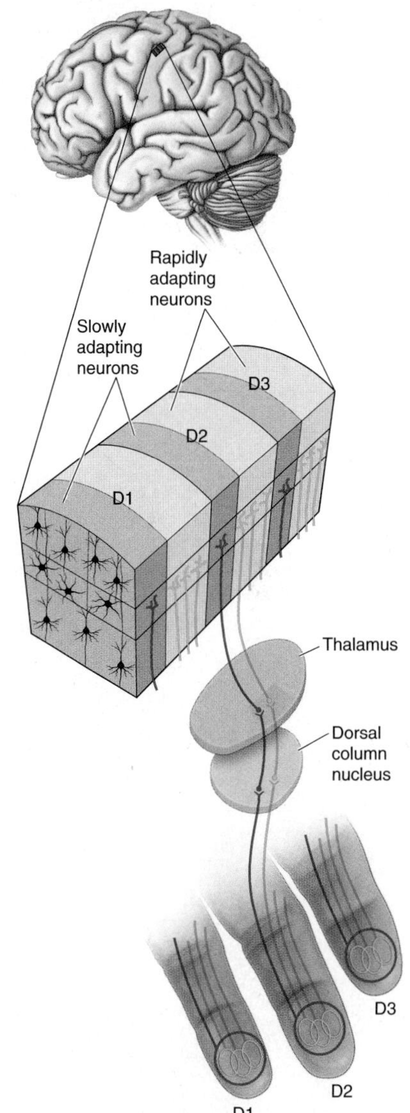

▲ FIGURE 12.18
Columnar organization of S1's area 3b. Each finger (D1–D3) is represented by an adjacent area of cortex. Within the area of each finger representation are alternating columns of cells with rapidly adapting (green) and slowly adapting (red) sensory responses. (Source: Adapted from Kaas et al., 1981, Fig. 8.)

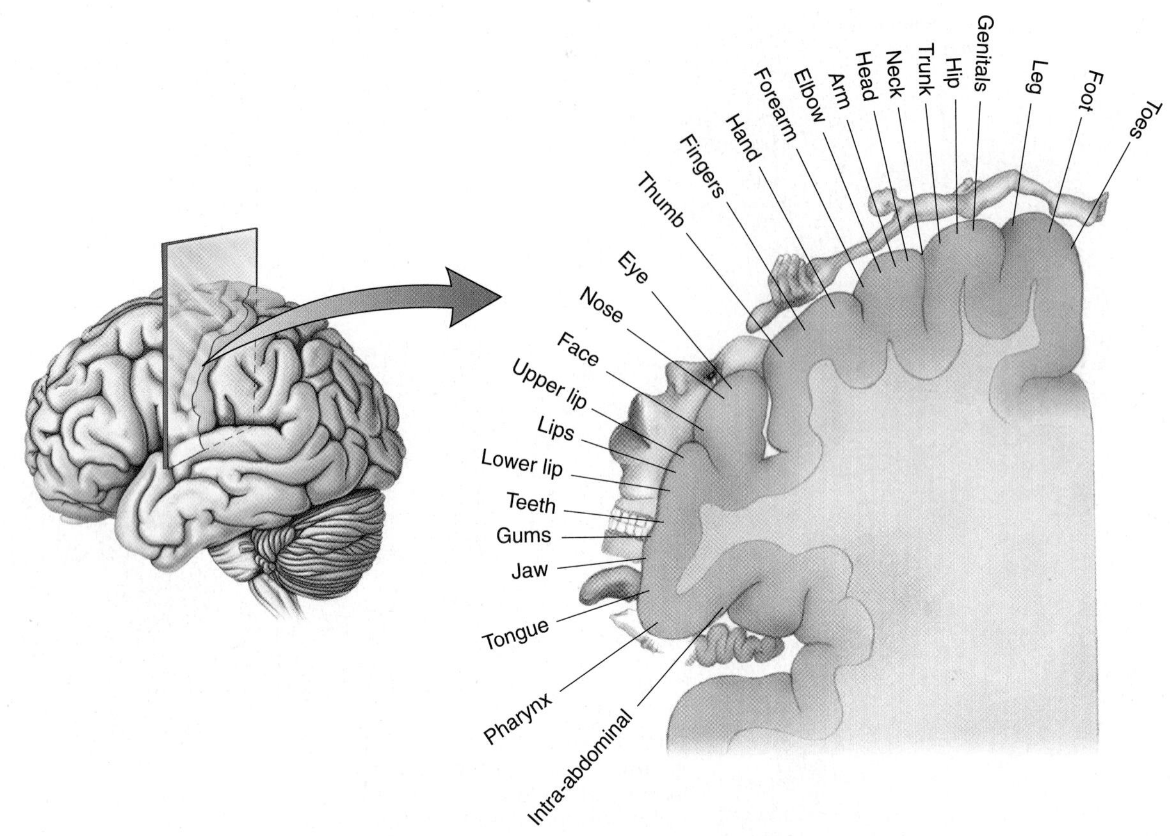

▲ FIGURE 12.19
A somatotopic map of the body surface onto primary somatosensory cortex. This map is a cross section through the postcentral gyrus (shown at top). Neurons in each area are most responsive to the parts of the body illustrated above them. (Source: Adapted from Penfield and Rasmussen, 1952, and from Kell et al., 2005, Fig. 3.)

called **somatotopy**. We have seen previously that the brain has maps of other sensory surfaces, such as the light-sensitive retina in the eye (*retinotopy*) and the sound frequency-sensitive cochlea in the inner ear (*tonotopy*).

Somatotopic maps generated by electrical stimulation and neuronal recording methods are similar. The maps roughly resemble a body with its legs and feet at the top of the postcentral gyrus and its head at the opposite, lower end of the gyrus (Figure 12.19). A somatotopic map is sometimes called a *homunculus* (from the Latin diminutive of "man"; the little man in the brain).

Several things are obvious about the somatotopic map in S1. First, the map is not always continuous but can be broken up. Notice in Figure 12.19 that the representation of the hand separates that of the face and the head. Interestingly, Penfield's original maps suggested that the male genitals were mapped onto the most distant and hidden part of S1, somewhere below the toes. However, a recent study using functional magnetic resonance imaging demonstrated that the penis is actually represented in a less surprising place on the map: in an area between the abdomen and the legs. Unfortunately, neither Penfield nor contemporary researchers have spent much time mapping the somatosensory maps of the female body and its unique features (what some have called the "*her*munculus").

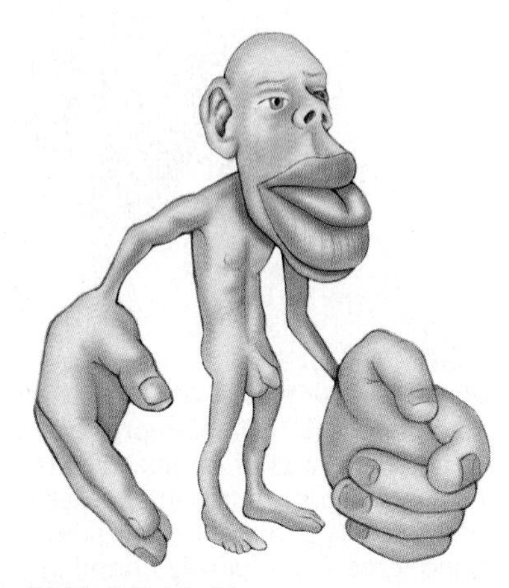

▲ FIGURE 12.20
The homunculus.

Another obvious feature of the somatotopic map is that it is not scaled like the human body. Instead, it looks like a caricature (Figure 12.20): The mouth, tongue, and fingers are incongruously large, while the trunk, arms, and legs are tiny. The relative size of cortex devoted to each body part is correlated with the *density* of sensory input received from that part. Size

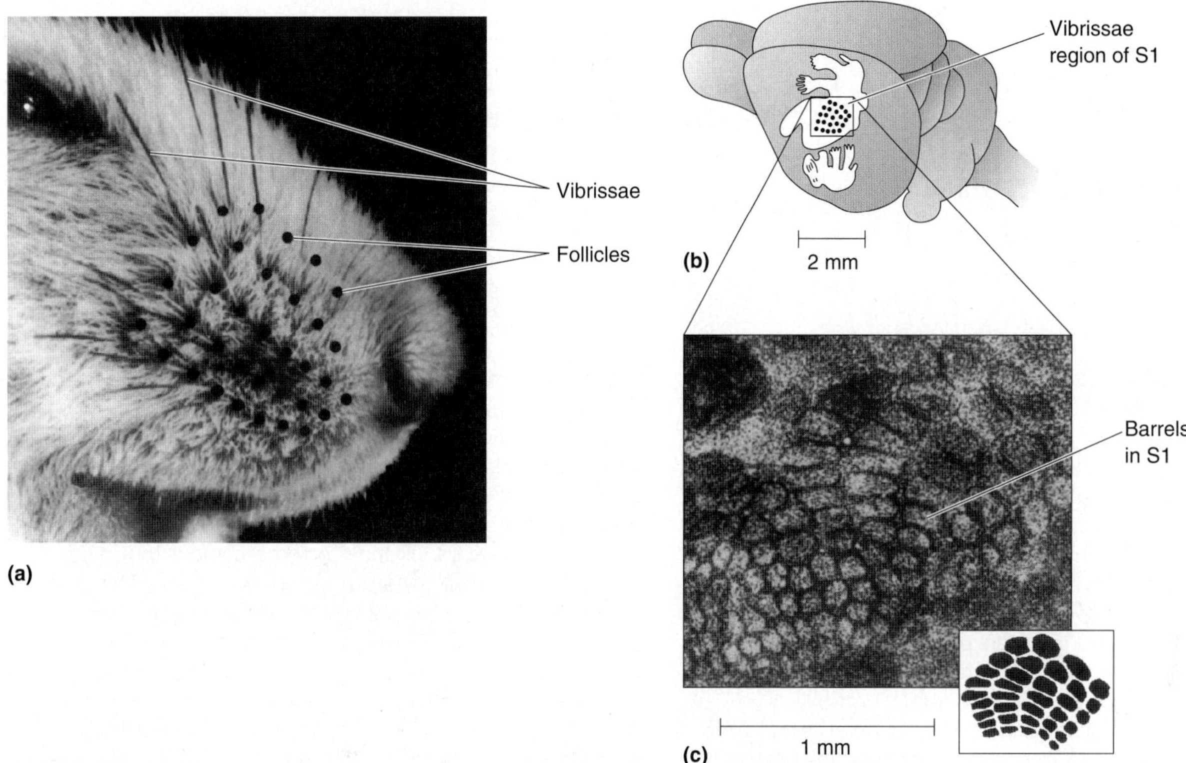

▲ FIGURE 12.21
A somatotopic map of the facial vibrissae on mouse cerebral cortex. **(a)** The positions of the major vibrissae on the face (dots). **(b)** A somatotopic map within S1 of the mouse brain. **(c)** Barrel cortex within S1. The cortex has been thinly sectioned parallel to the surface and Nissl-stained. The inset shows the pattern of barrels, laid out in five rows; compare with the five rows of vibrissae in the photograph in part a. (Source: Adapted from Woolsey and Van der Loos, 1970.)

on the map is also related to the *importance* of the sensory input from that part of the body; information from your index finger is more useful than that from your elbow. The importance of touch information from our hands and fingers is obvious, but why throw so much cortical computing power at the mouth? Two likely reasons are that tactile sensations are important in the production of speech and that your lips and tongue (feeling, as well as tasting) are the last line of defense when deciding if a morsel is delicious, nutritious food, or something that could choke you, break your tooth, or bite back. As we will see in a moment, the importance of an input, and the size of its representation in cortex, are also reflections of how often it is used.

The importance of a body part can vary greatly in different species. For example, the large facial vibrissae (whiskers) of rodents receive a huge share of the territory in S1, while the digits of the paws receive relatively little (Figure 12.21). Remarkably, the sensory signals from each vibrissa follicle go to one clearly defined cluster of S1 neurons; such clusters are called *barrels*. The somatotopic map of rodent vibrissae is easily seen in thin sections of S1; the five rows of cortical barrels precisely match the five rows of facial vibrissae (Box 12.3). Studies of the "barrel cortex" in rats and mice have revealed much about the functions of sensory cortex.

Somatotopy in the cerebral cortex is not limited to a single map. Just as the visual system builds multiple retinotopic maps, the somatic sensory system has several maps of the body. Figure 12.22 shows the detailed somatotopy of S1 in an owl monkey. Carefully compare the maps in areas 3b and 1; they map the same parts of the body, literally in parallel along adjacent strips of cortex. The two somatotopic maps are not identical, but mirror images, as an enlargement of the hand regions makes clear (Figure 12.22b).

BOX 12.3 PATH OF DISCOVERY

Cortical Barrels

by Thomas Woolsey

In the mid-1960s, I completed a physiological study of the organization of touch, hearing, and vision in the mouse brain for an undergraduate research project at the University of Wisconsin. At Wisconsin, histology was done routinely on all such brains. After my first year at medical school, I came back to look at the sections. There was something odd about cortical layer IV just where I had recorded responses to moving the whiskers: The cell bodies were distributed unevenly. This was not new. Several different authors, in nearly forgotten papers written over 50 years earlier, showed this pattern of neurons; but that was before recording was possible, so no one knew the function of the cortex.

Cortex is usually studied in sections cut perpendicular to the brain surface. It occurred to me that cutting the brain parallel to its surface, which had only been done rarely, might give a view of the entire layer IV. The late H. Van der Loos, who taught neuroanatomy, gave me a place to work at Johns Hopkins during an elective period. I prepared specimens in a way so that I could accurately position them for cutting (I knew where I usually got responses to stimulating the face) and cut thicker sections than customary. About 10:00 on a bright late spring morning, after struggling to mount the first sections on slides, I took them down a corridor to the dark student histology lab, where I had a microscope. That first look showed a stunning pattern of cells in layer IV that obviously mimicked the whiskers. There was no doubt about what I had seen; I immediately showed the slides to Van der Loos, who was the second person in the world to know that whiskers are stamped in the mouse brain. We named the cell groups barrels. Later, the hypotheses that each barrel is associated with a single whisker and that each one forms part of a functional cortical column were proven.

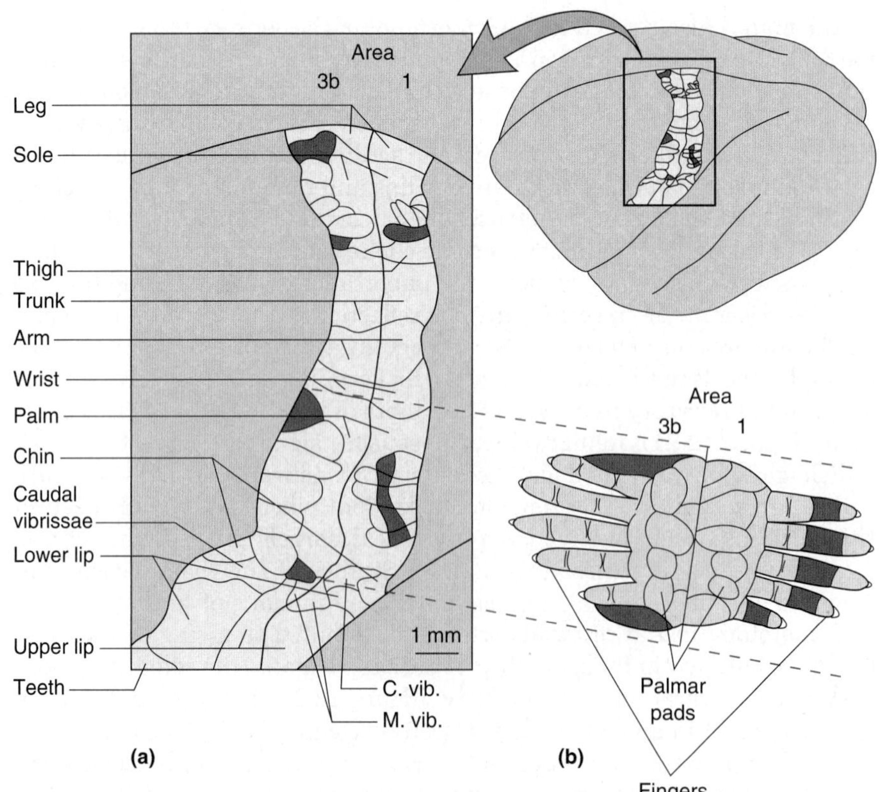

▶ FIGURE 12.22
Multiple somatotopic maps. Recordings were made from areas 3b and 1 of an owl monkey. **(a)** Results show that each area has its own somatotopic map. **(b)** Detailed examination of the hand area shows that the two maps are mirror images. Shaded regions represent the dorsal surfaces of the hands and feet, unshaded regions the ventral surfaces. (Source: Adapted from Kaas et al., 1981.)

Cortical Map Plasticity. What happens to the somatotopic map in cortex when an input, such as the finger, is removed? Does the "finger area" of cortex simply go unused? Does it atrophy? Or is this tissue taken over by inputs from other sources? The answers to these questions could have important implications for the recovery of function after peripheral nerve injury. In the 1980s, neuroscientist Michael Merzenich and his associates at the University of California at San Francisco began a series of experiments to test the possibilities.

Some key experiments are summarized in Figure 12.23. First, the regions of S1 sensitive to stimulation of the hand in an adult owl monkey were carefully mapped with microelectrodes. Then, one finger (digit 3) on the hand was surgically removed. Several months later, the cortex

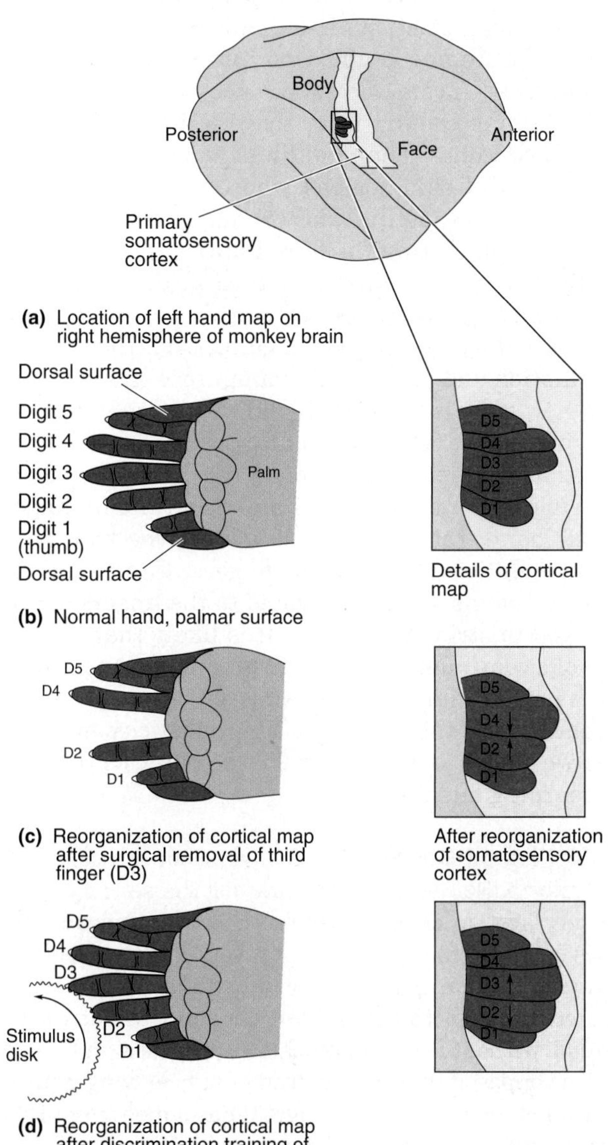

▲ FIGURE 12.23

Somototopic map plasticity. **(a, b)** The fingers of the hand of an owl monkey are mapped onto the surface of S1 cortex. **(c)** If digit 3 is removed, over time, the cortex reorganizes so that the representations of digits 2 and 4 expand. **(d)** If digits 2 and 3 are selectively stimulated, their cortical representations also expand.

was again mapped. The answer? The cortex originally devoted to the amputated digit now responded to stimulation of the adjacent digits (Figure 12.23c). There clearly had been a major rearrangement of the circuitry underlying cortical somatotopy.

In the amputation experiment, the cause of this map rearrangement was the absence of input from the missing digit. What happens when the input activity from a digit is *increased*? To answer this question, monkeys were trained to use selected digits to perform a task for which they received a food reward. After several weeks of this training, microelectrode mapping experiments showed that the representation of the stimulated digits had expanded in comparison with the adjacent, unstimulated ones (Figure 12.23d). These experiments reveal that cortical maps are dynamic and adjust depending on the amount of sensory experience. Subsequent experiments in other areas of cortex (visual, auditory, motor) have shown that this type of map plasticity is widespread in the brain.

The findings of map plasticity in animals have led to a search for similar changes in the human brain. One interesting example comes from studies of amputees. A common experience among amputees is the perception of sensations coming from the missing limb when other body parts are touched. These "phantom limb" sensations are usually evoked by the stimulation of skin regions whose somatotopic representations border those of the missing limb; for example, feeling can be evoked in a phantom arm by stimulating the face. Functional brain imaging reveals that the cortical regions originally devoted to the missing limb are now activated by stimulating the face. While this plasticity may be adaptive in the sense that the cortex does not go unused, the mismatch between sensory stimulation and perception in amputees shows that it can lead to confusion on how signals from S1 should be interpreted.

While having more cortex devoted to a body part may not necessarily be beneficial to amputees, it apparently is to musicians. Violinists and other string instrument players must continually finger the strings with their left hand; the other hand, holding the bow, receives considerably less stimulation of individual fingers. Functional imaging of S1 shows that the amount of cortex devoted to the fingers of the left hand is greatly enlarged in string musicians. It is likely that this is an exaggerated version of a continuous remapping process that goes on in everyone's brain as each person's life experiences vary.

The mechanisms of these types of map plasticity are not understood. However, as we shall see in Chapter 25, they may be related to processes involved in learning and memory.

The Posterior Parietal Cortex. As we have seen, the segregation of different types of information is a general rule for the sensory systems, and the somatic sensory system is no exception. However, information of different sensory types cannot remain separate forever. When we feel for a key in our pocket, we do not ordinarily sense it as a list of traits: a particular size and shape, textured and smooth edges, hard and smooth flat surfaces, a certain weight. Instead, without thinking much about it, we simply confirm with our fingers "key," as opposed to "coin" or "wad of old chewing gum." Separate aspects of a stimulus come effortlessly together as a meaningful object. We have a very poor understanding of how this occurs biologically within any sensory system, much less between sensory systems. After all, many objects have a distinct look, sound, feel, *and* smell, and the melding of these sensations is necessary for the complete mental image of something like your pet cat.

What we do know is that the character of neuronal receptive fields tends to change as information passes through the cortex and receptive fields enlarge. For example, neurons below the cortex and in cortical areas 3a and 3b are not sensitive to the direction of stimulus movement across the skin, but cells in areas 1 and 2 are. The stimuli that neurons prefer become increasingly complex. Certain cortical areas seem to be sites where simple, segregated streams of sensory information converge to generate particularly complex neural representations. When we discussed the visual system, we saw this in the complex receptive fields of area IT. The posterior parietal cortex is also such an area. Its neurons have large receptive fields with stimulus preferences that are a challenge to characterize because they are so elaborate. Moreover, the area is concerned not only with somatic sensation but also with visual stimuli, movement planning, and even a person's state of attentiveness.

Damage to posterior parietal areas can yield some bizarre neurological disorders. Among these is **agnosia**, the inability to recognize objects even though simple sensory skills seem to be normal. People with *astereognosia* cannot recognize common objects by feeling them (e.g., a key), although their sense of touch is otherwise normal and they may have no trouble recognizing the object by sight or sound. Deficits are often limited to the side contralateral to the damage.

Parietal cortical lesions may also cause a **neglect syndrome**, in which a part of the body or a part of the world (the entire visual field left of the center of gaze, for example) is ignored or suppressed, and its very existence is denied (Figure 12.24). Neurologist Oliver Sacks described such a patient in his essay, "The Man Who Fell Out of Bed." After suffering a stroke that presumably damaged his cortex, the man insisted that someone was playing a macabre joke on him by hiding an amputated leg under his blanket. When he tried to remove the leg from his bed, he and the leg ended up on the floor. Of course, the leg in question was his own, still attached, but he was unable to recognize it as part of his body. A neglect syndrome patient may ignore the food on one half of his plate, or attempt to dress only one side of his body. Neglect syndromes are most common following damage to the right hemisphere, and, fortunately, they usually improve or disappear with time.

▲ FIGURE 12.24
Symptoms of a neglect syndrome. A patient who had had a stroke in the right posterior parietal cortex was asked to copy the model drawing but was unable to reproduce many of the features on the left side of the model. (Source: Springer and Deutsch, 1989, p. 193.)

In general, the posterior parietal cortex seems to be essential for the perception and interpretation of spatial relationships, accurate body image, and the learning of tasks involving coordination of the body in space. These functions involve a complex integration of somatosensory information with that from other sensory systems, particularly the visual system.

PAIN

In addition to the mechanosensitive touch receptors we have described so far, somatic sensation depends strongly on **nociceptors**, the free, branching, unmyelinated nerve endings that signal that body tissue is being damaged or is at risk of being damaged. (The word is from the Latin *nocere*, "to hurt.") The information from nociceptors takes a path to the brain that is largely distinct from the path taken by mechanoreceptors; consequently, the subjective experience elicited by activation of these two pathways is different. Selective activation of nociceptors can lead to the conscious experience of pain. Nociception, and pain, are vital to life (Box 12.4).

It is important to realize, however, that nociception and pain are not always the same thing. *Pain* is the feeling, or the perception, of irritating, sore, stinging, aching, throbbing, miserable, or unbearable sensations arising from a part of the body. *Nociception* is the sensory process that provides signals that trigger pain. While nociceptors may fire away wildly

BOX 12.4 OF SPECIAL INTEREST

The Misery of Life Without Pain

Pain teaches us to avoid harmful situations. It elicits withdrawal reflexes from noxious stimuli. It exhorts us to rest an injured part of our body so it can heal. Pain is vital. The most convincing arguments for the functional benefits of pain are the very rare people with a condition called *congenital insensitivity to pain*. They go through life in constant danger of destroying themselves because they do not realize the harm they are doing. They often die young.

A Canadian woman, for example, was born with an indifference to painful stimuli, had no other sensory deficits, and was quite intelligent. Despite early training to avoid damaging situations, she developed progressive degeneration of her joints and spinal vertebrae, leading to skeletal deformation, degeneration, infection, and, finally, death at the age of 28. Apparently, low levels of nociceptive activity are important during everyday tasks to tell us when a particular movement or prolonged posture is putting too much strain on our body. Even during sleep, nociception may be the prod that makes us toss and turn enough to prevent bedsores or skeletal strain.

People with congenital insensitivity to pain reveal that pain is a separate sensation and not simply an excess of the other sensations. Such people usually have a normal ability to perceive other somatic sensory stimuli. The causes of the disorder can include the failure of peripheral nociceptors to develop, altered synaptic transmission in the pain-mediating pathways of the CNS, and genetic mutations. A study of several afflicted families in Pakistan revealed mutations in a gene called *SCN9A*, which codes for a unique type of voltage-gated sodium channel expressed only in nociceptive neurons. The mutation leads to nonfunctional sodium channels, a selective absence of action potentials in nociceptors, and profound insensitivity to pain. Family members with the mutations sustain continual cuts, bruises, bitten lips and tongues, and broken bones.

Clearly, life without pain is *not* a blessing.

and continually, pain may come and go. The opposite may also happen. Pain may be agonizing, even without activity in nociceptors. More than any other sensory system, the cognitive qualities of nociception can be controlled from within, by the brain itself.

Nociceptors and the Transduction of Painful Stimuli

Nociceptors are activated by stimuli that have the potential to cause tissue damage. Tissue damage can result from strong mechanical stimulation, extremes in temperature, oxygen deprivation, and exposure to certain chemicals, among other causes. The membranes of nociceptors contain ion channels that are activated by these types of stimuli.

Consider as an example the events that accompany your stepping on a thumbtack (recall Chapter 3). The simple stretching or bending of the nociceptor membrane activates mechanically gated ion channels that cause the cell to depolarize and generate action potentials. In addition, damaged cells at the site of injury can release a number of substances that cause ion channels on nociceptor membranes to open. Examples of released substances are proteases (enzymes that digest proteins), adenosine triphosphate (ATP), and K^+. Proteases can break down an abundant extracellular peptide called *kininogen* to form another peptide called *bradykinin*. Bradykinin binds to specific receptor molecules that activate ionic conductances in some nociceptors. Similarly, ATP causes nociceptors to depolarize by binding directly to ATP-gated ion channels. And, as we learned in Chapter 3, the elevation of extracellular $[K^+]$ directly depolarizes neuronal membranes.

Now consider leaning against a hot stove. Heat above 43°C causes tissues to burn, and heat-sensitive ion channels in nociceptor membranes open at this temperature. Of course, we also have nonpainful sensations of warmth when the skin is heated from 37 to 43°C. These sensations depend on non-nociceptive thermoreceptors and their CNS connections, which we will discuss in a later section. But for now, note that the sensations of warmth and scalding are mediated by separate neural mechanisms.

Imagine you are a middle-aged runner on the last mile of a marathon. When your tissue oxygen levels do not meet the oxygen demand, your cells use anaerobic metabolism to generate ATP. A consequence of anaerobic metabolism is the release of lactic acid. The buildup of lactic acid leads to an excess of H^+ in the extracellular fluid, and these ions activate H^+-gated ion channels on nociceptors. This mechanism causes the excruciating dull ache associated with very hard exercise.

A bee stings you. Your skin and connective tissue contain *mast cells*, a component of your immune system. Mast cells can be activated by exposure to foreign substances (e.g., bee venom), causing them to release histamine. Histamine binds to specific cell surface receptors on nociceptors and causes membrane depolarization. Histamine also causes blood capillaries to become leaky, which leads to swelling and redness at the site of an injury. Creams containing drugs that block histamine receptors (antihistamines) can be helpful to reduce the pain and the swelling.

Types of Nociceptors. The transduction of painful stimuli occurs in the free nerve endings of unmyelinated C fibers and lightly myelinated Aδ fibers. The majority of nociceptors respond to mechanical, thermal, and chemical stimuli, and are therefore called *polymodal nociceptors*. However, like the mechanoreceptors of touch, many nociceptors show selectivity in their responses to different stimuli. Thus, there are also *mechanical nociceptors*, showing selective responses to strong pressure; *thermal nociceptors*, showing selective responses to burning heat or extreme cold (Box 12.5); and *chemical nociceptors*, showing selective responses to histamine and other chemicals.

Nociceptors are present in most body tissues, including skin, bone, muscle, most internal organs, blood vessels, and the heart. They are notably absent in the brain itself, except for the meninges.

Hyperalgesia and Inflammation. Nociceptors normally respond only when stimuli are strong enough to damage tissue. But we all know that skin, joints, or muscles that have *already* been damaged or inflamed are unusually sensitive. A light, sympathetic mother's touch to a burned area of her child's skin may elicit howls due to unbearable pain. This phenomenon is known as **hyperalgesia**, and it is the most familiar example of our body's ability to control its own pain. Hyperalgesia can be a reduced threshold for pain, an increased intensity of painful stimuli, or even spontaneous pain. *Primary hyperalgesia* occurs within the area of damaged tissue, but tissues surrounding a damaged area may become supersensitive as well, by the process of *secondary hyperalgesia*.

Many different mechanisms appear to be involved in hyperalgesia, some in and around the peripheral receptors and others within the CNS. As noted earlier, when skin is damaged, a variety of substances, sometimes called the *inflammatory soup*, are released. The soup includes certain neurotransmitters (glutamate, serotonin, adenosine, ATP), peptides (substance P, bradykinin), lipids (prostaglandins, endocannabinoids), proteases, neurotrophins, cytokines, and chemokines, ions such as K^+ and H^+, and other substances (Figure 12.25). Together, they can trigger **inflammation**, which is a natural response of the body's tissues as they attempt to eliminate injury and stimulate the healing process. The cardinal signs of inflammation in skin are pain, heat, redness, and swelling. A number of these chemicals can also modulate the excitability of nociceptors, making them more sensitive to thermal or mechanical stimuli (see Box 12.5).

BOX 12.5 OF SPECIAL INTEREST

Hot and Spicy

If you like spicy food, you should know that the active ingredient in a wide variety of hot peppers is *capsaicin* (Figure A). These peppers are "hot" because the capsaicin activates the thermal nociceptors that also signal painful elevations in temperature (above about 43°C). Indeed, it was the peculiar fact that these nociceptive neurons are selectively activated by capsaicin that led to the discovery of the transduction mechanism for the sensation of hot. David Julius at the University of California, San Francisco, found that in some dorsal root ganglion cells, capsaicin activates a particular ion channel, called *TRPV1*, which is also activated by elevations in temperature greater than 43°C. This ion channel causes the neuron to fire by admitting Ca^{2+} and Na^{+} and depolarizing it. TRPV1 is a member of a very large family of related TRP channels, originally identified in photoreceptors of the fruit fly *Drosophila* (TRP stands for transient receptor potential). Different TRP channels contribute to many different types of sensory transduction in organisms from yeast to humans.

Why would a temperature-gated ion channel also be sensitive to hot peppers? Capsaicin appears to mimic the effect of endogenous chemicals released by tissue damage. These chemicals (and capsaicin) cause the TRPV1 channel to open at lower temperatures, explaining the heightened sensitivity of injured skin to increases in temperature. Indeed, inflammation-induced thermal hyperalgesia is absent in mice engineered to lack the TRPV1 channel. While all mammals normally express the TRPV1 channel, birds do not, thus explaining how birds can consume the spiciest of chili peppers. This fact also explains how birdseed laced with capsaicin can be enjoyed by birds without interference from raiding squirrels.

In addition to its protection of birdseed and widespread culinary use, capsaicin has a seemingly paradoxical clinical application. Applied in large quantities, it can cause **analgesia**, the absence of pain. The capsaicin desensitizes pain fibers and depletes the peptide substance P from their nerve terminals. Capsaicin ointments, sprays, and patches are useful treatments for pain associated with arthritis, strains, psoriasis, shingles, and other conditions (see Box 12.1).

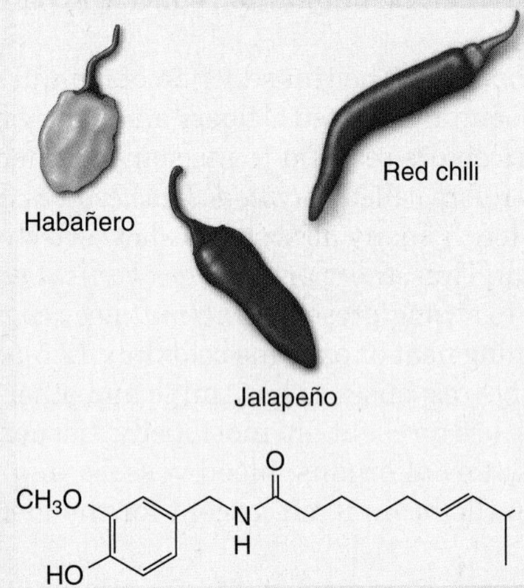

Figure A
Peppers containing capsaicin and the molecule's chemical structure.

Bradykinin was discussed earlier as one of the chemicals that directly depolarizes nociceptors. In addition to this effect, bradykinin stimulates long-lasting intracellular changes that make heat-activated ion channels more sensitive. *Prostaglandins* are chemicals generated by the enzymatic breakdown of lipid membrane. While prostaglandins do not elicit overt pain, they do increase greatly the sensitivity of nociceptors to other stimuli. Aspirin and other nonsteroidal anti-inflammatory drugs are a useful treatment for hyperalgesia because they inhibit the enzymes required for prostaglandin synthesis.

Substance P is a peptide synthesized by the nociceptors themselves. Activation of one branch of a nociceptor axon can lead to the secretion of substance P by the other branches of that axon in the neighboring skin. Substance P causes vasodilation (swelling of the blood capillaries) and the release of histamine from mast cells. Sensitization of other nociceptors around the site of injury by substance P is one cause of secondary hyperalgesia.

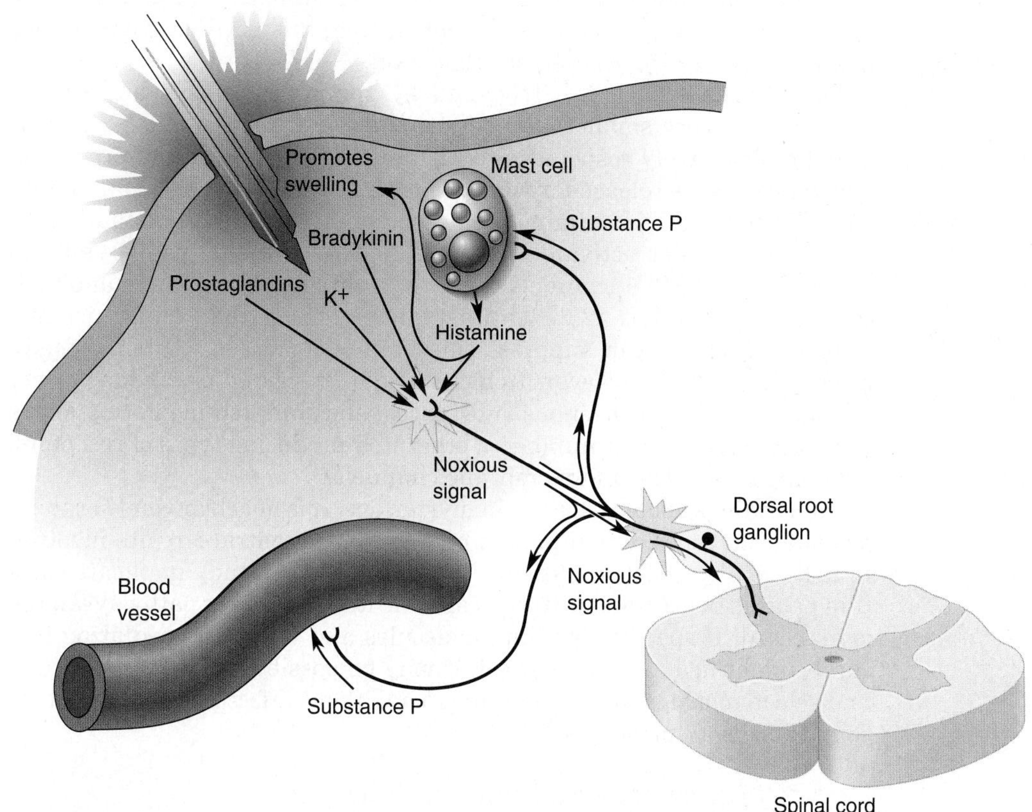

▲ FIGURE 12.25
Peripheral chemical mediators of pain and hyperalgesia.

CNS mechanisms also contribute to secondary hyperalgesia. Following injury, the activation of mechanoreceptive Aβ axons by light touch can evoke pain. Thus, another mechanism of hyperalgesia involves cross-talk between the touch and pain pathways in the spinal cord.

Itch

Itch is defined as a disagreeable sensation that induces a desire or a reflex to scratch. Itch and the scratching it evokes can serve as a natural defense against parasites and plant toxins on the skin and scalp. Itch is usually a brief and minor annoyance. It can also become a chronic and seriously debilitating condition. Chronic itch may be caused by a wide variety of skin conditions such as allergic reactions, infections, infestations, and psoriasis; it can also be triggered by non-skin disorders such as cancers, iron deficiency, hyperthyroidism, liver disease, stress, and psychiatric conditions. Imagine that your worst itch has spread across most of your body and persists through every waking minute. The need to scratch can be unrelenting and irresistible. Chronic itch can be as dreadful as chronic pain, and it is notoriously difficult to treat with current drugs and therapies.

Itch has always been a hard sensation to categorize. Although pain and itch are unmistakably distinct, they also have many similarities. Both are mediated by thin sensory axons, although the axons carrying pain signals seem to be different from those triggering itch. Both can be triggered by various types of stimuli, including chemicals and touch. Some of the drugs and compounds that regulate pain can also trigger itch, and some signaling molecules transduce both sensations. Pain and itch also interact. For example, pain can suppress itch; this is why we sometimes aggressively scratch an itchy patch of skin.

Some types of itch are triggered by specific molecules and neural circuits. The very smallest C fibers (conduction velocity of 0.5 m/sec or less) are selectively responsive to histamine, the natural itch-producing substance that is released by mast cells in the skin during inflammation (see Figure 12.25). Histamine mediates itch by binding to histamine receptors, which then activate TRPV1 channels; surprisingly, these are the same types of TRPV1 channels that are stimulated by capsaicin and high temperature (see Box 12.5). Antihistamines—drugs that antagonize histamine receptors—can suppress this kind of itch. Not all itch is mediated by histamine, however. Itch can also be triggered by a wide variety of endogenous and exogenous substances, and thin itch-mediating axons seem to express a large number of other itch-producing types of receptors, signaling molecules, and membrane channels.

Many things about itch remain mysterious. It is not clear whether there are different types of itch-producing axons. The central circuits involved in itch are also poorly understood. One fascinating study implicated certain peptide neurotransmitters in specific itch-producing pathways in the spinal cord. If specific signaling molecules and receptors mediating itch can be identified and understood, it may be possible to develop effective and selective drugs to treat chronic itch without affecting pain and other somatic sensory processes.

Primary Afferents and Spinal Mechanisms

Aδ and C fibers bring information to the CNS at different rates because of differences in their action potential conduction velocities. Accordingly, the activation of skin nociceptors produces two distinct perceptions of pain: a fast, sharp, *first pain* followed by a duller, longer lasting *second pain*. First pain is caused by the activation of Aδ fibers; second pain is caused by the activation of C fibers (Figure 12.26).

Like the Aβ mechanosensory fibers, the small-diameter fibers have their cell bodies in the segmental dorsal root ganglia, and they enter the dorsal horn of the spinal cord. The fibers branch immediately, travel a short distance up and down the spinal cord in a region called the *zone of Lissauer*, and then synapse on cells in the outer part of the dorsal horn in a region known as the **substantia gelatinosa** (Figure 12.27).

The neurotransmitter of the pain afferents is glutamate; however, as mentioned previously, these neurons also contain the peptide substance P

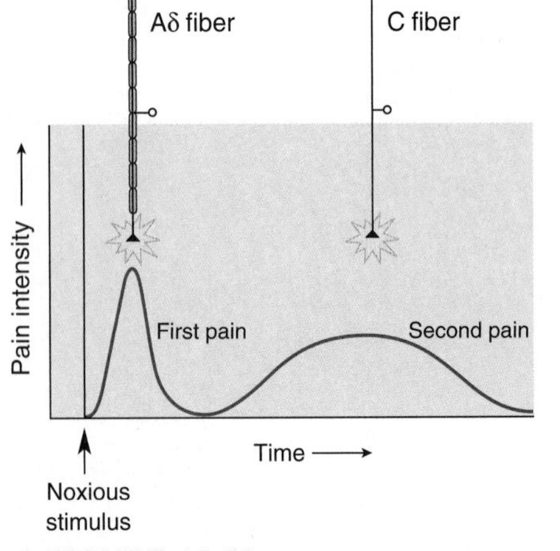

▲ FIGURE 12.26
First and second pain. The first pain sensation registered by noxious stimulation is mediated by fast Aδ axons. The second, longer lasting pain sensation is mediated by slow C fibers.

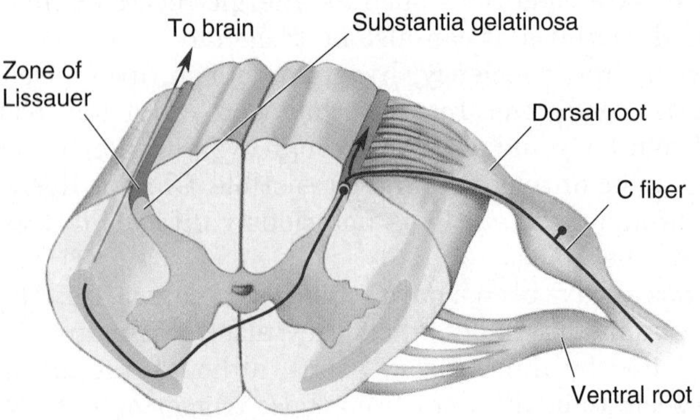

▲ FIGURE 12.27
Spinal connections of nociceptive axons.

(Figure 12.28). Substance P is contained within storage granules in the axon terminals (see Chapter 5) and can be released by high-frequency trains of action potentials. Recent experiments have shown that synaptic transmission mediated by substance P is required to experience moderate to intense pain.

It is interesting to note that nociceptor axons from the viscera enter the spinal cord by the same route as the cutaneous nociceptors. Within the spinal cord, there is substantial mixing of information from these two sources of input (Figure 12.29). This cross-talk gives rise to the phenomenon of **referred pain**, where visceral nociceptor activation is perceived as a cutaneous sensation. The classic example of referred pain is angina, occurring when the heart fails to receive sufficient oxygen. Patients often localize the pain of angina to the upper chest wall and the left arm. Another common example is the pain associated with appendicitis, which in its early stages is referred to the abdominal wall around the navel.

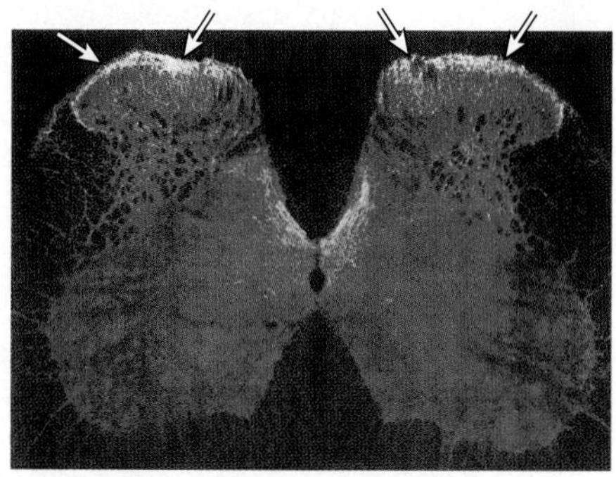

▲ FIGURE 12.28
Immunocytochemical localization of substance P in the spinal cord. Arrows point to strong concentrations of substance P in the substantia gelatinosa. (Source: Mantyh et al., 1997.)

Ascending Pain Pathways

Let's briefly highlight the differences we've encountered between the touch and pain pathways. First, they differ with respect to their nerve endings in the skin. The touch pathway is characterized by specialized structures in the skin; the pain pathway has only free nerve endings. Second, they differ with respect to the diameter of their axons. The touch pathway is swift, using fat, myelinated Aβ fibers; the pain pathway is slow, using thin, lightly myelinated Aδ fibers and unmyelinated C fibers. Third, they differ with respect to their connections in the spinal cord. Branches of the Aβ axons terminate in the deep dorsal horn; the Aδ and C fibers branch, run within the zone of Lissauer, and terminate within the substantia gelatinosa. As we will now see, the two pathways also differ substantially in the way they transmit information to the brain.

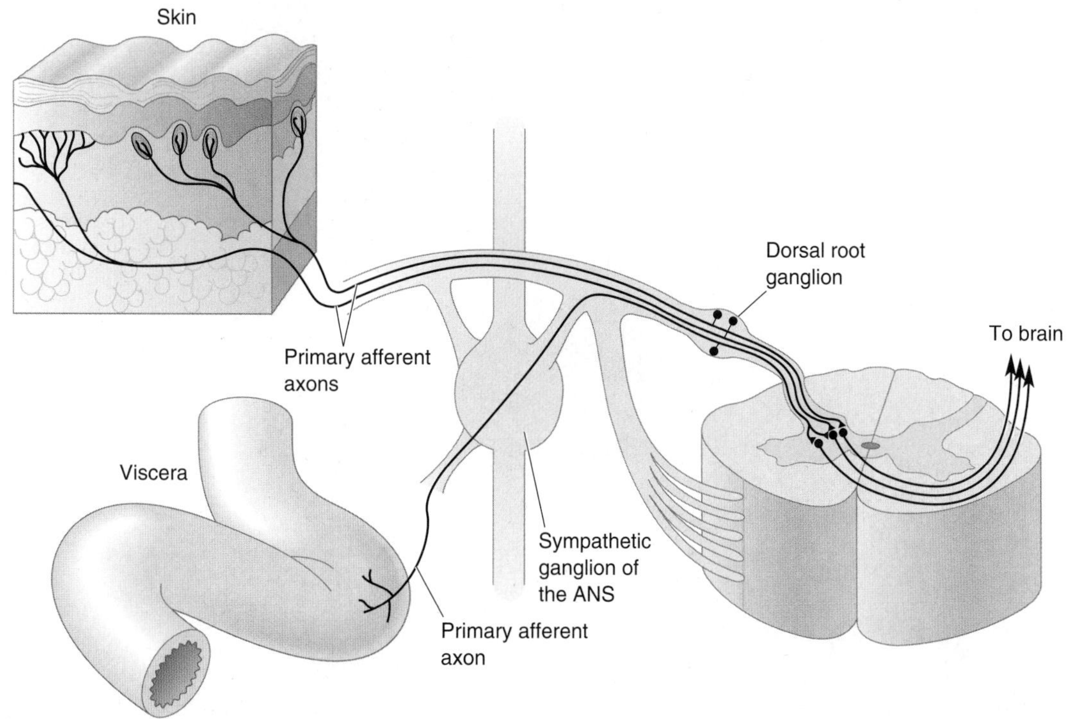

▲ FIGURE 12.29
The convergence of nociceptor input from the viscera and the skin.

The Spinothalamic Pain Pathway. Information about pain (as well as temperature) in the body is conveyed from the spinal cord to the brain via the **spinothalamic pathway**. Unlike the dorsal column–medial lemniscal pathway, the axons of the second-order neurons *immediately decussate* and ascend through the *spinothalamic tract* running along the ventral surface of the spinal cord (compare Figures 12.14 and 12.27). As the name implies, the spinothalamic fibers project up the spinal cord and through the medulla, pons, and midbrain without synapsing, until they reach the thalamus (Figure 12.30). As the spinothalamic axons journey through the brain stem, they eventually come to lie alongside the medial lemniscus, but the two groups of axons remain distinct from each other.

Figure 12.31 summarizes the different ascending pathways for touch and pain information. Notice that information about touch ascends *ipsilaterally*, while information about pain (and temperature) ascends *contralaterally*. This organization can lead to a curious, but predictable, group of deficits when the nervous system is impaired. For example, if half of the spinal cord is damaged, certain deficits of mechanosensitivity occur on the

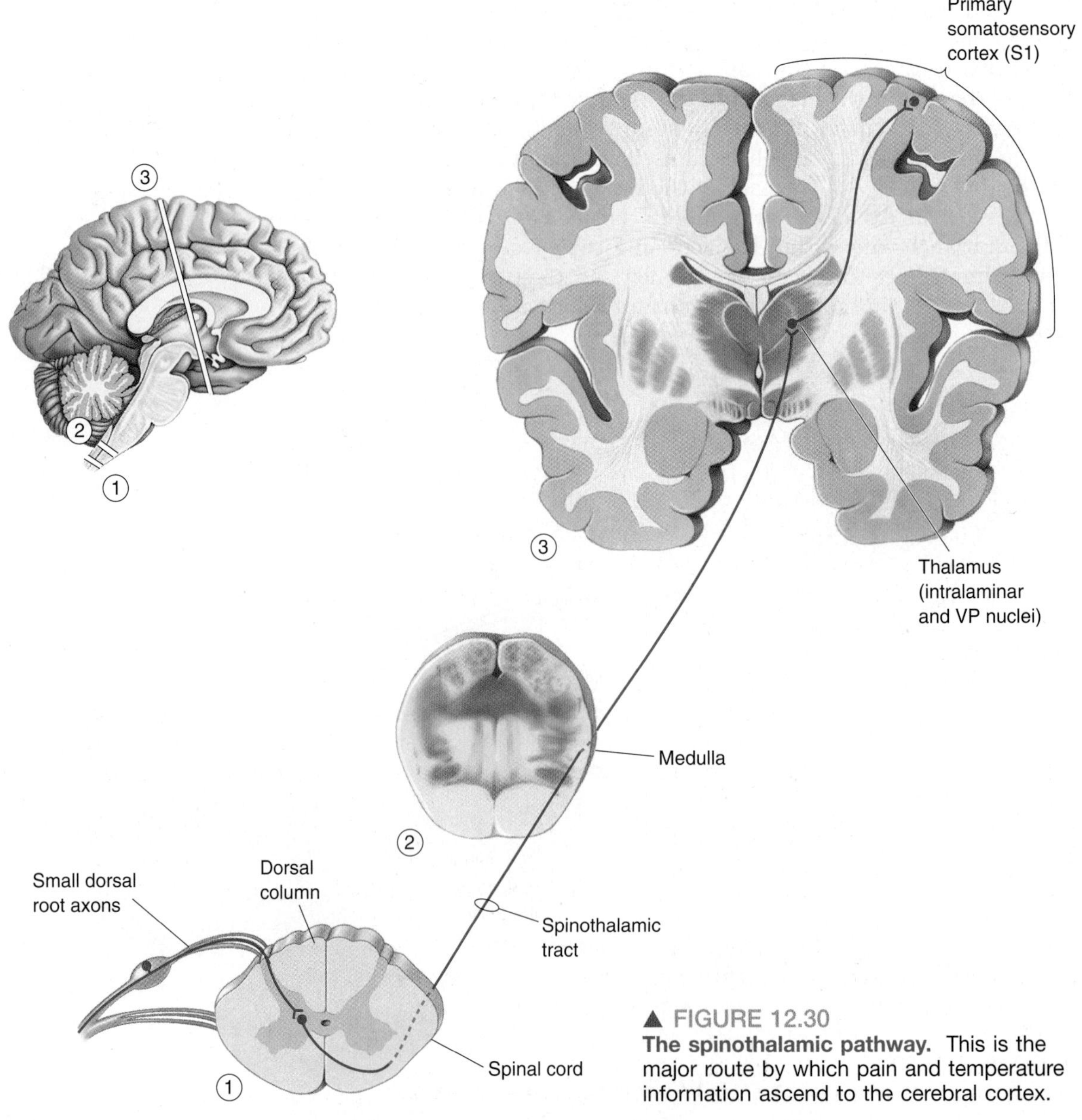

▲ FIGURE 12.30
The spinothalamic pathway. This is the major route by which pain and temperature information ascend to the cerebral cortex.

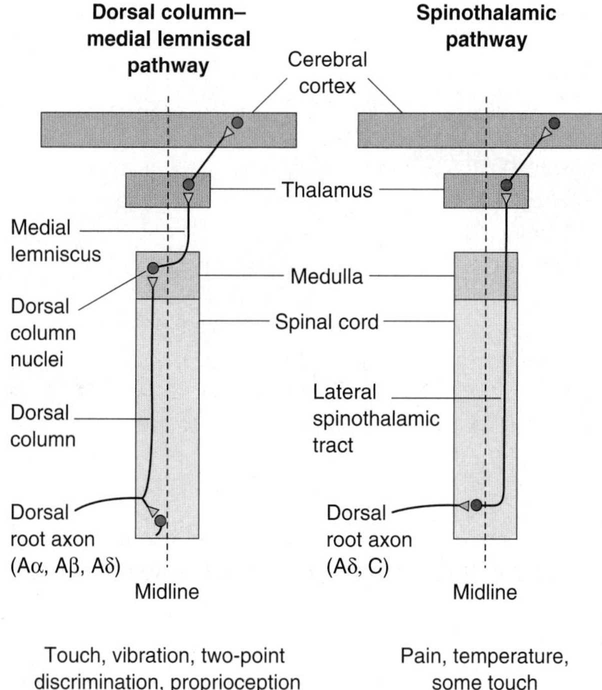

▲ FIGURE 12.31
An overview of the two major ascending pathways of somatic sensation.

same side as the spinal cord damage: insensitivity to light touch, the vibrations of a tuning fork on the skin, the position of a limb. On the other hand, deficits in pain and temperature sensitivity will show up on the side of the body *opposite* the cord damage. Other signs, such as motor deficiency and the exact map of sensory deficits, give additional clues about the site of spinal cord damage. For example, movements will be impaired on the ipsilateral side. The constellation of sensory and motor signs following damage to one side of the spinal cord is called *Brown–Séquard syndrome.*

The Trigeminal Pain Pathway. Pain (and temperature) information from the face and head takes a path to the thalamus that is analogous to the spinal path. The small-diameter fibers in the trigeminal nerve synapse first on second-order sensory neurons in the *spinal trigeminal nucleus* of the brain stem. The axons of these cells cross and ascend to the thalamus in the *trigeminal lemniscus.*

In addition to the spinothalamic and trigeminothalamic pathways, other closely related pain (and temperature) pathways send axons into a variety of structures at all levels of the brain stem, before they reach the thalamus. Some of these pathways are particularly important in generating sensations of slow, burning, agonizing pain, while others trigger a more general state of behavioral arousal and alertness.

The Thalamus and Cortex. The spinothalamic tract and trigeminal lemniscal axons synapse over a wider region of the thalamus than those of the medial lemniscus. Some of the axons terminate in the VP nucleus, just as the medial lemniscal axons do, but the touch and pain systems *still* remain segregated there by occupying separate regions of the nucleus. Other spinothalamic axons end in the small *intralaminar nuclei* of the thalamus (Figure 12.32). From the thalamus, pain and temperature information is projected to various areas of the cerebral cortex. As in the

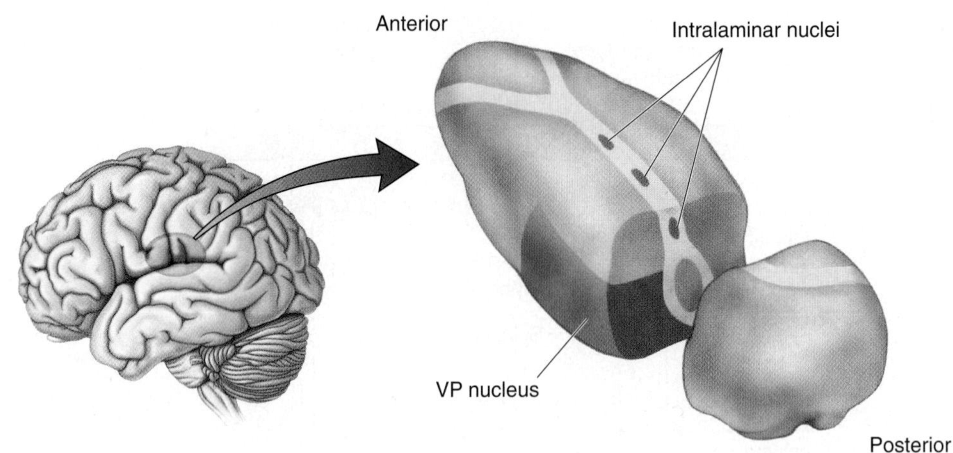

▶ FIGURE 12.32
Somatic sensory nuclei of the thalamus. In addition to the VP nucleus, the intralaminar nuclei relay nociceptive information to a large expanse of the cerebral cortex.

thalamus, this pathway covers a much wider territory than the cortical connections of the dorsal column–medial lemniscal pathway.

The Regulation of Pain

The perception of pain is highly variable. Depending on the concurrent level of nonpainful sensory input and the behavioral context, the same level of nociceptor activity can produce more pain or less pain. Understanding this modulation of pain is of great importance because it may offer new strategies for the treatment of chronic pain, a condition that afflicts up to 20% of the adult population.

Afferent Regulation. We've already seen that light touch can evoke pain via the mechanisms of hyperalgesia. However, pain evoked by activity in nociceptors can also be *reduced* by simultaneous activity in low-threshold mechanoreceptors (Aβ fibers). Presumably, this is why it feels good to rub the skin around your shin when you bruise it. This may also explain an electrical treatment for some kinds of chronic, intractable pain. Wires are taped to the skin surface, and pain is suppressed when the patient simply turns on an electrical stimulator designed to activate large-diameter sensory axons.

In the 1960s, Ronald Melzack and Patrick Wall, then working at MIT, proposed a hypothesis to explain these phenomena. Their *gate theory of pain* suggests that certain neurons of the dorsal horns, which project an axon up the spinothalamic tract, are excited by both large-diameter sensory axons and unmyelinated pain axons. The projection neuron is also inhibited by an interneuron, and the interneuron is both *excited* by the large sensory axon and *inhibited* by the pain axon (Figure 12.33). By this arrangement, activity in the pain axon alone maximally excites the projection neuron, allowing nociceptive signals to rise to the brain. However, if the large mechanoreceptive axon fires concurrently, it activates the interneuron and suppresses nociceptive signals.

Descending Regulation. Stories abound of soldiers, athletes, and torture victims who sustained horrible injuries but apparently felt no pain. Strong emotion, stress, or stoic determination can powerfully suppress feelings of pain. Several brain regions have been implicated in pain suppression (Figure 12.34). One is a zone of neurons in the midbrain called the periventricular and **periaqueductal gray matter (PAG)**. Electrical stimulation of the PAG can cause a profound analgesia that has sometimes been exploited clinically.

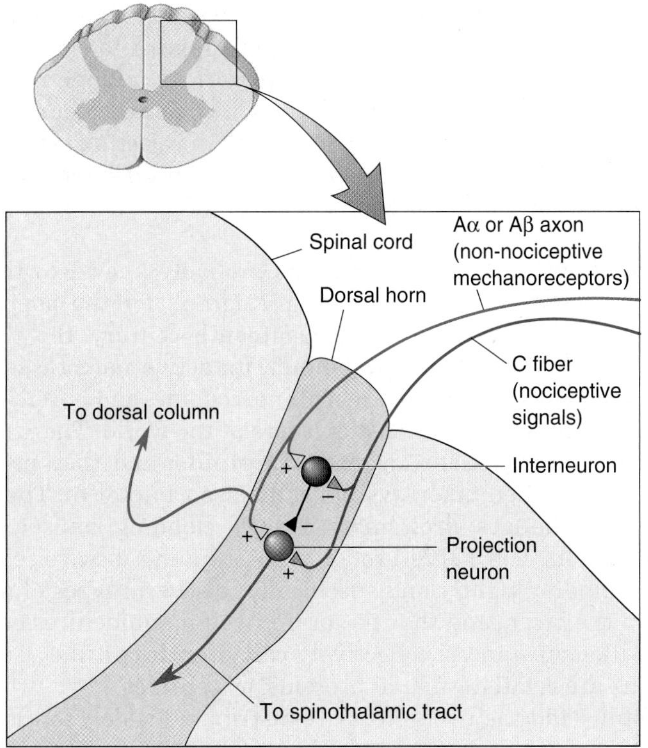

▲ FIGURE 12.33

Melzack's and Wall's gate theory of pain. The relay of nociceptive signals by the projection neuron is gated by the activity of an inhibitory interneuron. Activity in the non-nociceptive mechanoreceptor can suppress, or close the "gate" on, nociceptive signals before they can proceed to the spinothalamic tract. The + signs indicate excitatory synapses and the − signs indicate inhibitory synapses.

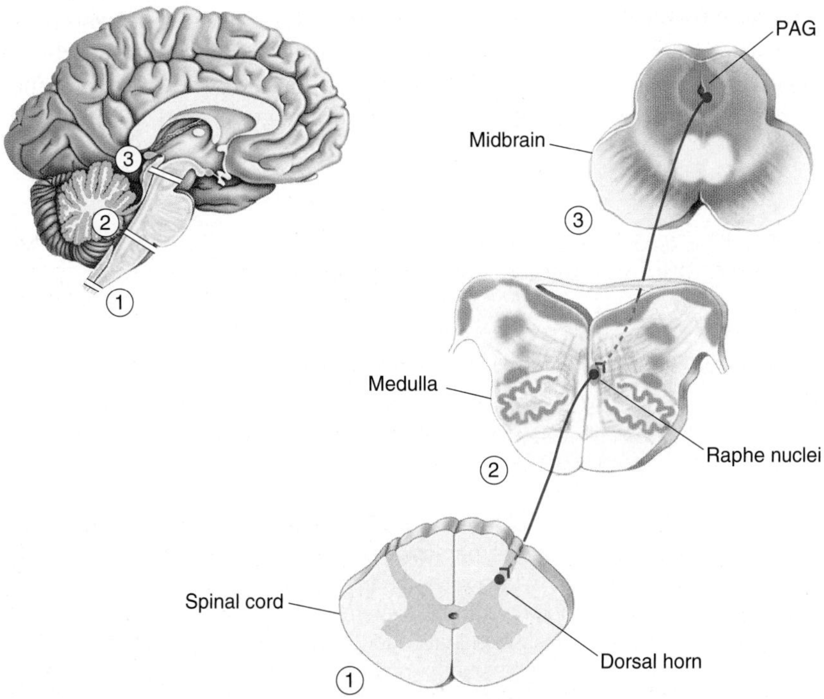

◀ FIGURE 12.34

Descending pain-control pathways. A variety of brain structures, many of which are affected by behavioral state, can influence activity within the periaqueductal gray matter (PAG) of the midbrain. The PAG can influence the raphe nuclei of the medulla, which in turn can modulate the flow of nociceptive information through the dorsal horns of the spinal cord.

The PAG normally receives input from several brain structures, many of them appropriate for transmitting signals related to emotional status. PAG neurons send descending axons into various midline regions of the medulla, particularly to the *raphe nuclei* (which use the neurotransmitter serotonin). These medullary neurons in turn project axons down to the dorsal horns of the spinal cord, where they can effectively depress the activity of nociceptive neurons.

The Endogenous Opioids. Opium was probably known to the ancient Sumerians around 4000 BCE. Their pictograph for the poppy roughly translates to "joy plant." By the seventeenth century, the therapeutic value of opium was undisputed. Opium, its active narcotic ingredients, and their analogues—including morphine, codeine, and heroin—are used and abused widely today, in most cultures of the world. These drugs, and others with similar actions, are called **opioids**, and they produce profound analgesia when taken systemically (see Chapter 6). They can also produce mood changes, drowsiness, mental clouding, nausea, vomiting, and constipation. The 1970s brought the stunning discoveries that opioids act by binding tightly and specifically to several types of **opioid receptors** in the brain, and that the brain itself manufactures endogenous morphine-like substances, collectively called **endorphins** (see Box 6.1). Endorphins are relatively small proteins, or peptides.

Endorphins and their receptors are distributed widely in the CNS, but they are particularly concentrated in areas that process or modulate nociceptive information. Small injections of morphine or endorphins into the PAG, the raphe nuclei, or the dorsal horn can produce analgesia. Because this effect is prevented by administering the specific blocker of opioid receptors, *naloxone*, the injected drugs must have acted by binding to opioid receptors in those areas. Naloxone can also block the analgesic effects induced by electrically stimulating these areas. At the cellular level, endorphins exert multiple effects that include suppressing the release of glutamate from presynaptic terminals and inhibiting neurons by hyperpolarizing their postsynaptic membranes. In general, extensive systems of endorphin-containing neurons in the spinal cord and brain stem prevent the passage of nociceptive signals through the dorsal horn and into higher levels of the brain where the perception of pain is generated (Box 12.6).

BOX 12.6 OF SPECIAL INTEREST

Pain and the Placebo Effect

To test the efficacy of a new drug, clinical trials are often conducted in which one group of subjects receives the drug and the other receives an inert substance. Both groups of subjects may believe they have been given the drug. Surprisingly, the inert substance is often reported to have the effect that patients are told to expect from the drug. The term *placebo* is used to describe such substances (the word is from the Latin meaning, "I shall please"), and the phenomenon is called the *placebo effect*.

Placebos can be highly effective analgesics. A majority of patients suffering from postoperative pain have reported getting relief from an injection of sterile saline! Does this mean that these patients have only imagined their pain? Not at all. The opioid receptor antagonist naloxone can block the analgesic effect of the placebo, just as it antagonizes the effects of morphine, a true analgesic. Apparently, the belief that the treatment will work can be enough to cause activation of the endogenous pain-relief systems of the brain. The placebo effect is a likely explanation for the success of other treatments for pain, such as acupuncture, hypnosis, and, for children, a mother's loving kiss.

TEMPERATURE

As is the case for the sense of touch and pain, nonpainful temperature sensations originate from receptors in the skin (and elsewhere), and they depend on the neocortex for their conscious appreciation. Here, we briefly describe how this system is organized.

Thermoreceptors

Because the rate of chemical reactions depends on temperature, the functioning of all cells is sensitive to temperature. However, **thermoreceptors** are neurons that are exquisitely sensitive to temperature because of specific membrane mechanisms. For example, we can perceive changes in our average skin temperature of as little as 0.01°C. Temperature-sensitive neurons clustered in the hypothalamus and the spinal cord are important in the physiological responses that maintain stable body temperature, but it is the thermoreceptors in the skin that apparently contribute to our perception of temperature.

Temperature sensitivity is not spread uniformly across the skin. With a small cold or warm probe, you can map your skin's sensitivity to temperature changes. Some spots about 1 mm wide are especially sensitive to *either* hot or cold but not both. The fact that the locations of hot and cold sensitivity are different demonstrates that separate receptors encode them. Also, small areas of skin between the hot and cold spots are relatively insensitive to temperature.

The sensitivity of a sensory neuron to a change in temperature depends on the type of ion channels the neuron expresses. Discovery of the channels responsive to painful increases in temperature above 43°C (see Box 12.5) led researchers to wonder if other, closely related channels might be tuned to sense other temperature ranges. Just as the active ingredient in hot peppers was used to identify the "hot" receptor protein, called *TRPV1*, the active ingredient in mint was used to identify a "cold" receptor. Menthol, which produces a sensation of cold, was found to stimulate a receptor, called *TRPM8*, which is also activated by nonpainful decreases in temperature below 25°C.

We now understand that there are six distinct TRP channels in thermoreceptors that confer different temperature sensitivities (Figure 12.35). As a rule, each thermoreceptive neuron appears to express only a single type of channel, thus explaining how different regions of skin can show distinctly different sensitivities to temperature. An exception appears to be some cold receptors that also express TRPV1 and are therefore also sensitive to increases in temperature above 43°C. If such heat is applied to wide areas of skin, it is usually painful, but if the heat is restricted to small regions of skin innervated by a cold receptor, it can produce a paradoxical feeling of cold. This effect emphasizes an important point: The CNS does not know *what* kind of stimulus (in this case, heat) caused the receptor to fire, but it continues to interpret all activity from its cold receptor as a response to cold.

As with mechanoreceptors, the responses of thermoreceptors adapt during long-duration stimuli. Figure 12.36 shows that a sudden drop in skin temperature causes a cold receptor to fire strongly, while it silences a warm receptor. After a few seconds at 32°C, however, the cold receptor slows its firing (but still fires faster than it did at 38°C), while the warm receptor speeds up slightly. Notice that a return to the original warm skin temperature causes opposite responses—transient silence of the cold receptor and a burst of activity in the warm receptor—before both return to their steady, adapted rates. Thus, the differences between the response rates of warm and cold receptors are greatest during, and shortly after,

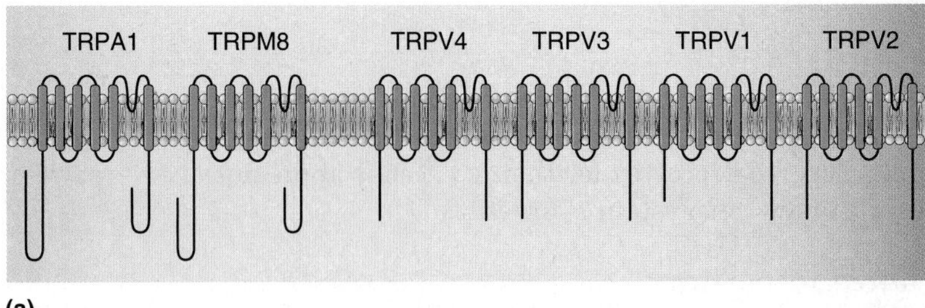

(a)

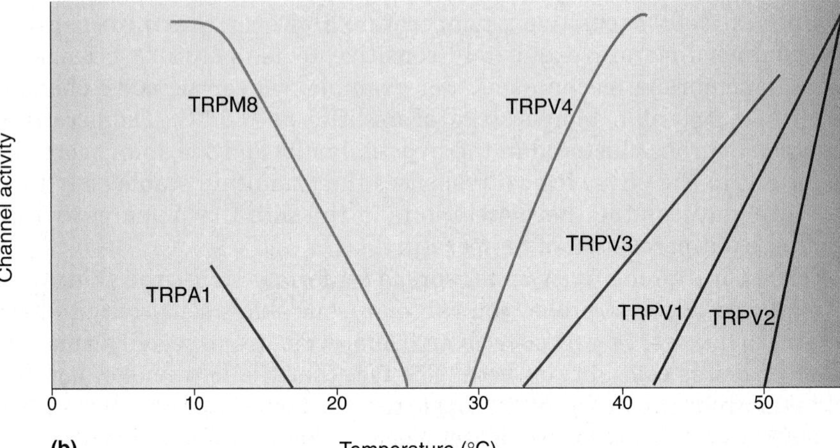

(b)

▶ FIGURE 12.35
Thermoreceptor TRP channels tuned to detect different temperatures. **(a)** The arrangement of the known thermosensitive TRP channel protein molecules in the neuronal membrane. TRPM8 and TRPVI are responsive to menthol and capsaicin, respectively. **(b)** This graph plots the activation of the various TRP channels as a function of temperature. (Source: Adapted from Patapoutian, et al., 2003, Fig. 3.)

temperature changes. Our perceptions of temperature often reflect these skin receptor responses.

Try a simple experiment. Fill two buckets with tap water, one cold and one hot (but not painfully hot). Then plunge your right hand for one minute into each of them in turn. Notice the striking sensations of hot and cold that occur with each change, but also notice how transient the sensations are. With thermoreception, as with most other sensory systems, it is the sudden *change* in the quality of a stimulus that generates the most intense neural and perceptual responses.

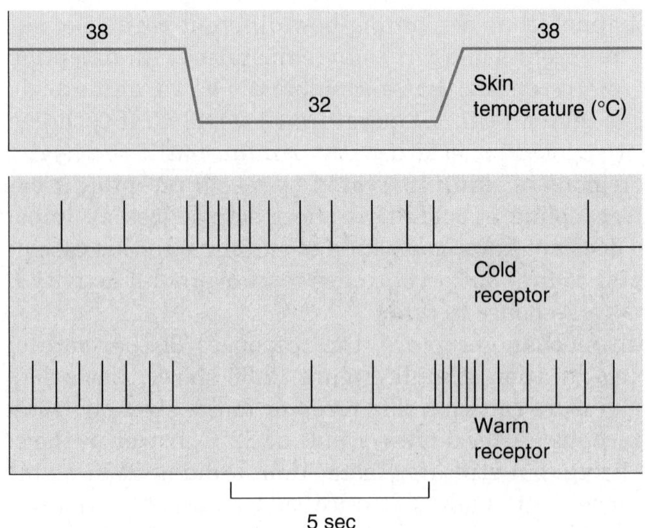

▲ FIGURE 12.36
Adaptations of thermoreceptors. The responses of cold and warm receptors to a step reduction in skin temperature are shown. Both receptors are most responsive to sudden changes in temperature, but they adapt over several seconds.

The Temperature Pathway

At this point, you may be relieved to learn that the organization of the temperature pathway is virtually identical to that of the pain pathway already described. Cold receptors are coupled to Aδ and C fibers, while warm receptors are coupled only to C fibers. As we learned previously, small-diameter axons synapse within the substantia gelatinosa of the dorsal horn. The axons of second-order neurons immediately decussate and ascend in the contralateral spinothalamic tract (see Figure 12.30). Thus, if the spinal cord is transected on one side, there will be a loss of temperature sensitivity (as well as pain) on the opposite side of the body, specifically in those regions of skin innervated by spinal segments below the cut.

CONCLUDING REMARKS

This concludes our discussion of the sensory systems. Although each one has evolved to be the brain's interface with a different form of environmental energy, the systems are strikingly similar in organization and function. Different types of somatic sensory information are necessarily kept separate in the spinal nerves because each axon is connected to only one type of sensory receptor ending. Segregation of sensory types continues within the spinal cord and is largely maintained all the way to the cerebral cortex. In this way, the somatic sensory system repeats a theme common throughout the nervous system: Several flows of related, but distinct, information are passed in parallel through a series of neural structures. The mixing of these streams occurs along the way, but only judiciously, until higher levels of processing are reached in the cerebral cortex. We saw other examples of parallel processing of sensory information in the chemical senses, vision, and audition.

Exactly how the parallel streams of sensory data are melded into perception, images, ideas, and memories remains the Holy Grail of neuroscience. Thus, the perception of any handled object involves the seamless coordination of all facets of somatic sensory information. The bird in hand is rounded, warm, soft, and light in weight; its heartbeat flutters against your fingertips; its claws scratch; and its textured wings brush against your palm. Somehow, your brain knows it's a bird, even without looking or listening, and would never mistake it for a toad. In the chapters that follow, we describe how the brain begins to use sensory information to plan and coordinate movement.

KEY TERMS

Introduction
somatic sensation (p. 416)

Touch
mechanoreceptors (p. 417)
Pacinian corpuscle (p. 417)
spinal segments (p. 423)
dermatome (p. 424)
dorsal column–medial lemniscal pathway (p. 426)
dorsal column (p. 428)
dorsal column nuclei (p. 428)
medial lemniscus (p. 428)
ventral posterior (VP) nucleus (p. 428)
primary somatosensory cortex (S1) (p. 428)
trigeminal nerves (p. 428)
posterior parietal cortex (p. 430)
somatotopy (p. 432)
agnosia (p. 437)
neglect syndrome (p. 437)

Pain
nociceptors (p. 437)
hyperalgesia (p. 439)
inflammation (p. 439)
analgesia (p. 440)
substantia gelatinosa (p. 442)
referred pain (p. 443)
spinothalamic pathway (p. 444)
periaqueductal gray matter (PAG) (p. 446)
opioids (p. 448)
opioid receptors (p. 448)
endorphins (p. 448)

Temperature
thermoreceptors (p. 449)

REVIEW QUESTIONS

1. Imagine rubbing your fingertips across a pane of smooth glass and then across a brick. What kinds of skin receptors help you distinguish the two surfaces? As far as your somatic sensory system is concerned, what is different about the two surfaces?
2. What purpose is served by the encapsulations around some sensory nerve endings in the skin?
3. If someone tossed you a hot potato and you caught it, which information would reach your CNS first: the news that the potato was hot or that it was relatively smooth? Why?
4. At what levels of the nervous system are *all* types of somatic sensory information represented on the contralateral side: the spinal cord, the medulla, the pons, the midbrain, the thalamus, and the cortex?
5. What lobe of the cortex contains the main somatic sensory areas? Where are these areas relative to the main visual and auditory areas?
6. Where within the body can pain be modulated, and what causes its modulation?
7. Where in the CNS does information about touch, shape, temperature, and pain converge?
8. Imagine this experiment: Fill two buckets with water, one relatively cold and one hot. Fill a third bucket with water of an intermediate, lukewarm temperature. Put your left hand into the hot water, your right hand into the cold, and wait one minute. Now quickly plunge both hands into the lukewarm water. Try to predict what sensations of temperature you will feel in each hand. Will they feel the same? Why?

FURTHER READING

Abraira VE, Ginty DD. 2013. The sensory neurons of touch. *Neuron* 79:618–639.

Braz J, Solorzano C, Wang X, Basbaum AI. 2014. Transmitting pain and itch messages: a contemporary view of the spinal cord circuits that generate gate control. *Neuron* 82:522–536.

Di Noto PM, Newman L, Wall S, Einstein G. 2013. The hermunculus: what is known about the representation of the female body in the brain? *Cerebral Cortex* 23:1005–1013.

Eijkelkamp N, Quick K, Wood JN. 2013. Transient receptor potential channels and mechanosensation. *Annual Review of Neuroscience* 36:519–546.

Fain GL. 2003. *Sensory Transduction.* Sunderland, MA: Sinauer.

Hsiao S. 2008. Central mechanisms of tactile shape perception. *Current Opinion in Neurobiology* 18:418–424.

McGlone F, Wessberg J, Olausson H. 2014. Discriminative and affective touch: sensing and feeling. *Neuron* 82:737–755.

Vallbo Å. 1995. Single-afferent neurons and somatic sensation in humans. In *The Cognitive Neurosciences*, ed. Gazzaniga M. Cambridge, MA: MIT Press, pp. 237–251.

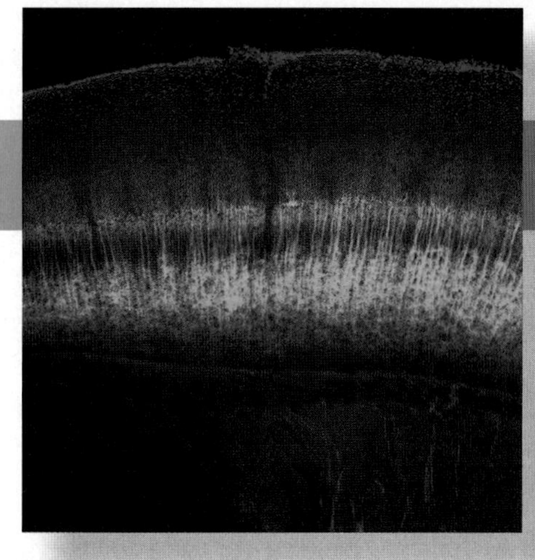

CHAPTER THIRTEEN

Spinal Control of Movement

INTRODUCTION

We are now ready to turn our attention to the system that actually gives rise to behavior. The **motor system** consists of all our muscles and the neurons that control them. The importance of the motor system was summarized by the pioneering English neurophysiologist Charles Sherrington in the Linacre lecture of 1924: "To move things is all that mankind can do . . . for such the sole executant is muscle, whether in whispering a syllable or in felling a forest." A moment's thought will convince you that the motor system is also incredibly complex. Behavior requires the coordinated action of various combinations of almost 700 muscles in a changing and often unpredictable environment.

Have you heard the expression "running around like a chicken with its head cut off"? It comes from the observation that some complex patterns of behavior (running around a barnyard, at least briefly) can be generated without the participation of the brain. There is a considerable amount of circuitry within the spinal cord for the coordinated control of movements, particularly stereotyped (repetitive) ones such as those associated with locomotion. This point was established early in this century by Sherrington and his English contemporary Thomas Graham Brown, who showed that rhythmic movements could be elicited in the hind legs of cats and dogs long after their spinal cords had been severed from the rest of the central nervous system. Today's view is that the spinal cord contains certain *motor programs* for the generation of coordinated movements, and that these programs are accessed, executed, and modified by descending commands from the brain. Thus, motor control can be divided into two parts: (1) the spinal cord's command and control of coordinated muscle contraction, and (2) the brain's command and control of the motor programs in the spinal cord.

In this chapter, we will explore the peripheral somatic motor system: the joints, skeletal muscles, and spinal motor neurons and interneurons, and how they communicate with each other. In Chapter 14, we will take a look at how the brain influences the activity of the spinal cord.

THE SOMATIC MOTOR SYSTEM

Based on their appearance under the microscope, the muscles in the body can be described according to two broad categories: striated and smooth. But they are also distinct in other ways. **Smooth muscle** lines the digestive tract, arteries, and related structures and is innervated by nerve fibers from the autonomic nervous system (see Chapter 15). Smooth muscle plays a role in peristalsis (the movement of material through the intestines) and the control of blood pressure and blood flow. There are two types of **striated muscle**: cardiac and skeletal. **Cardiac muscle** is heart muscle, which contracts rhythmically even in the absence of any innervation. Innervation of the heart from the autonomic nervous system (ANS) functions to accelerate or slow down the heart rate. (Recall Otto Loewi's experiment in Chapter 5.)

Skeletal muscle constitutes the bulk of the muscle mass of the body and functions to move bones around joints, to move the eyes within the head, to inhale and exhale, to control facial expression, and to produce speech. Each skeletal muscle is enclosed in a connective tissue sheath that, at the ends of the muscle, forms the tendons. Within each muscle are hundreds of **muscle fibers**, the cells of skeletal muscle, and each fiber is innervated by a single axon branch from the central nervous system (CNS) (Figure 13.1). Because skeletal muscle is derived embryologically from 33 paired somites (see Chapter 7), these muscles and the parts of

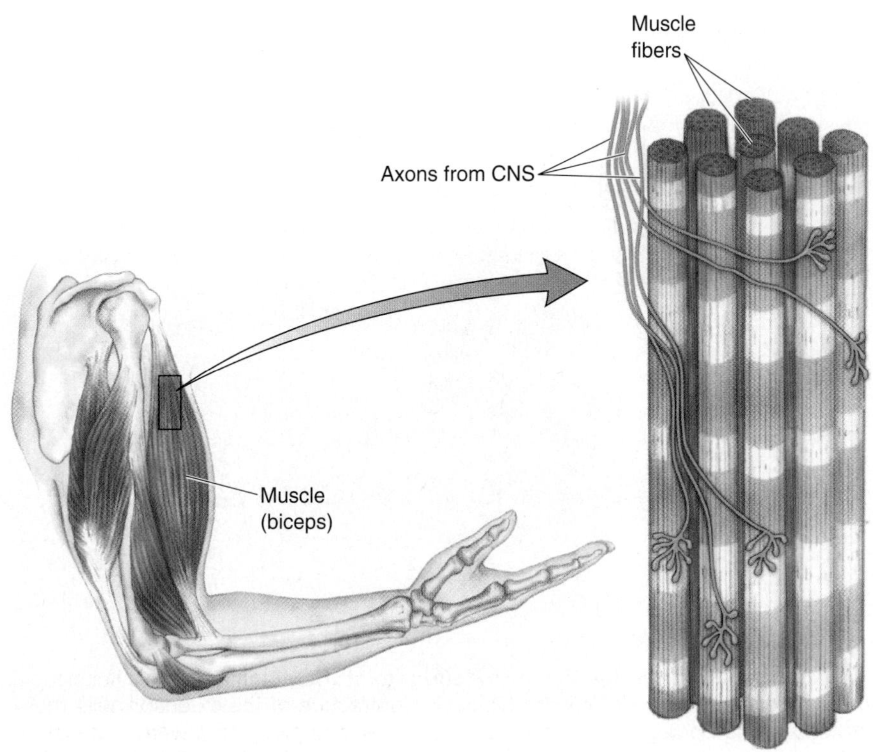

◀ FIGURE 13.1
The structure of skeletal muscle. Each muscle fiber is innervated by a single axon.

the nervous system that control them are collectively called the **somatic motor system**. We focus our attention on this system here because it is under voluntary control and it generates behavior. (The visceral motor system of the ANS will be discussed in Chapter 15.)

Consider the elbow joint (Figure 13.2). This joint is formed where the humerus, the upper arm bone, is bound by fibrous ligaments to the radius and ulna, the bones of the lower arm. The joint functions like a hinge on a pocket knife. Movement in the direction that closes the knife is called **flexion**, and movement in the direction that opens the knife is called **extension**. Note that muscles only pull on a joint; they cannot push. The major muscle that causes flexion is the brachialis, whose tendons insert

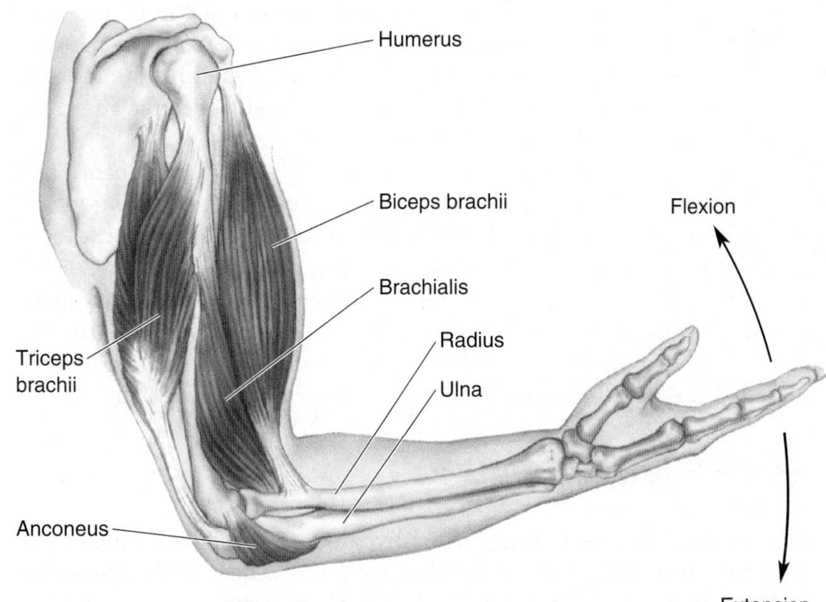

◀ FIGURE 13.2
Major muscles of the elbow joint. The biceps and triceps are antagonistic muscles. Contraction of the biceps causes flexion of the elbow, and contraction of the triceps causes extension.

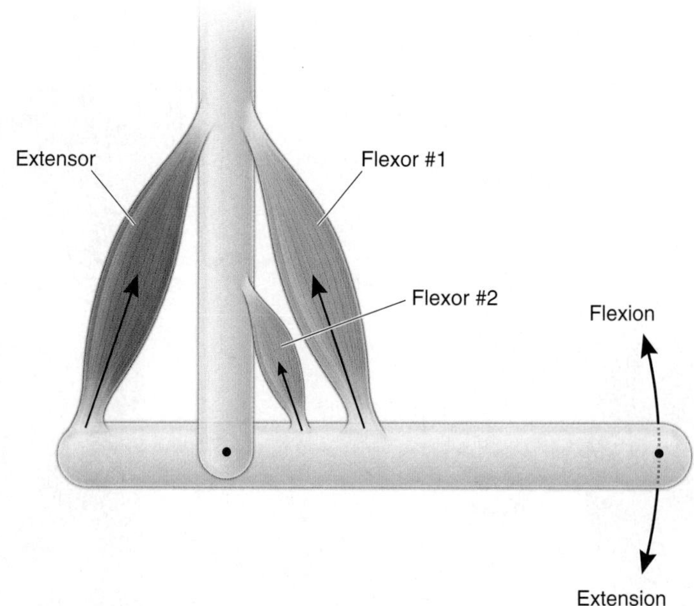

▲ FIGURE 13.3
How contracting muscles flex or extend a joint. Contractions of the flexors pull the right end of the bone upward (flexion). Contraction of the extensor pulls the left end of the bone upward, causing the right end to pivot downward (extension). Flexor #1 and flexor #2 are synergists. Flexors #1 and #2 are antagonist muscles of the extensor.

into the humerus at one end and into the ulna at the other. Two other muscles cause flexion at this joint, the biceps brachii and the coracobrachialis (which lies under the biceps). Together, these muscles are called **flexors** of the elbow joint, and, because the three muscles all work together, they are called **synergists** of one another. The two synergistic muscles that cause extension of the elbow joint are the triceps brachii and the anconeus; these two muscles are called **extensors**. Because the flexors and extensors pull on the joint in opposite directions, they are called **antagonists** to one another. The relationships between these muscles and bones, and the forces and movements they generate, are shown schematically in Figure 13.3. Even the simple flexion of the elbow joint requires the coordinated contraction of the synergistic flexor muscles *and* the relaxation of the antagonistic extensor muscles. Relaxing the antagonists allows movements to be faster and more efficient because the muscles are not working against one another.

Other terms to note about somatic musculature refer to the location of the joints they act on. The muscles that are responsible for movements of the trunk are called **axial muscles**; those that move the shoulder, elbow, pelvis, and knee are called **proximal** (or **girdle**) **muscles**; and those that move the hands, feet, and digits (fingers and toes) are called **distal muscles**. The axial musculature is very important for maintaining posture, the proximal musculature is critical for locomotion, and the distal musculature, particularly of the hands, is specialized for the manipulation of objects.

THE LOWER MOTOR NEURON

The somatic musculature is innervated by the somatic motor neurons in the ventral horn of the spinal cord (Figure 13.4). These cells are sometimes called *lower motor neurons* to distinguish them from the higher

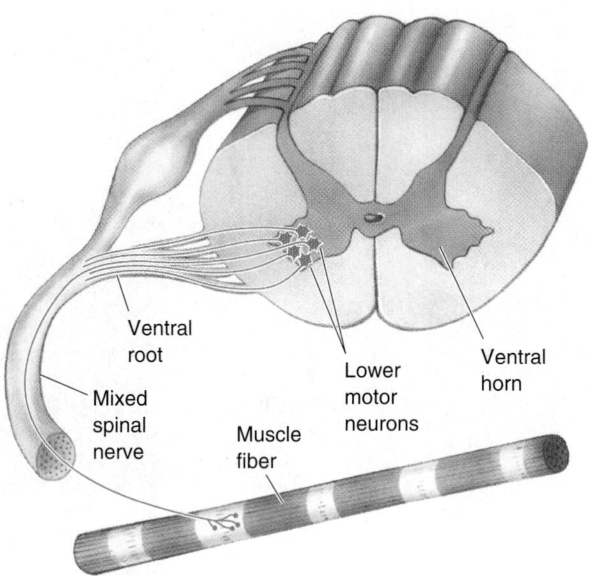

▲ FIGURE 13.4
Muscle innervation by lower motor neurons. The ventral horn of the spinal cord contains motor neurons that innervate skeletal muscle fibers.

order *upper motor neurons* of the brain that supply input to the spinal cord. Remember, only the lower motor neurons directly command muscle contraction. Sherrington called these neurons the *final common pathway* for the control of behavior.

The Segmental Organization of Lower Motor Neurons

The axons of lower motor neurons bundle together to form ventral roots; each ventral root joins with a dorsal root to form a spinal nerve that exits the cord through the notches between vertebrae. Recall from Chapter 12 that there are as many spinal nerves as there are notches between vertebrae; in humans, this adds up to 30 on each side. Because they contain sensory and motor fibers, they are called *mixed spinal nerves.* The motor neurons that provide fibers to one spinal nerve are said to belong to a spinal segment, named for the vertebra where the nerve originates. The segments are cervical (C) 1–8, thoracic (T) 1–12, lumbar (L) 1–5, and sacral (S) 1–5 (see Figure 12.11).

Skeletal muscles are not distributed evenly throughout the body, nor are lower motor neurons distributed evenly within the spinal cord. For example, innervation of the more than 50 muscles of the arm originates entirely from spinal segments C3–Tl. Thus, in this region of the spinal cord, the dorsal and ventral horns appear swollen to accommodate the large number of spinal interneurons and motor neurons that control the arm musculature (Figure 13.5). Similarly, spinal segments Ll–S3 have swollen dorsal and ventral horns because this is where the neurons controlling the leg musculature reside. Thus, we can see that the motor neurons that innervate distal and proximal musculature are found mainly in the cervical and lumbar–sacral segments of the spinal cord, whereas those innervating axial musculature are found at all levels.

The lower motor neurons are also distributed within the ventral horn at each spinal segment in a predictable way, depending on their function. The cells innervating the axial muscles are medial to those innervating the distal muscles, and the cells innervating flexors are dorsal to those innervating extensors (Figure 13.6).

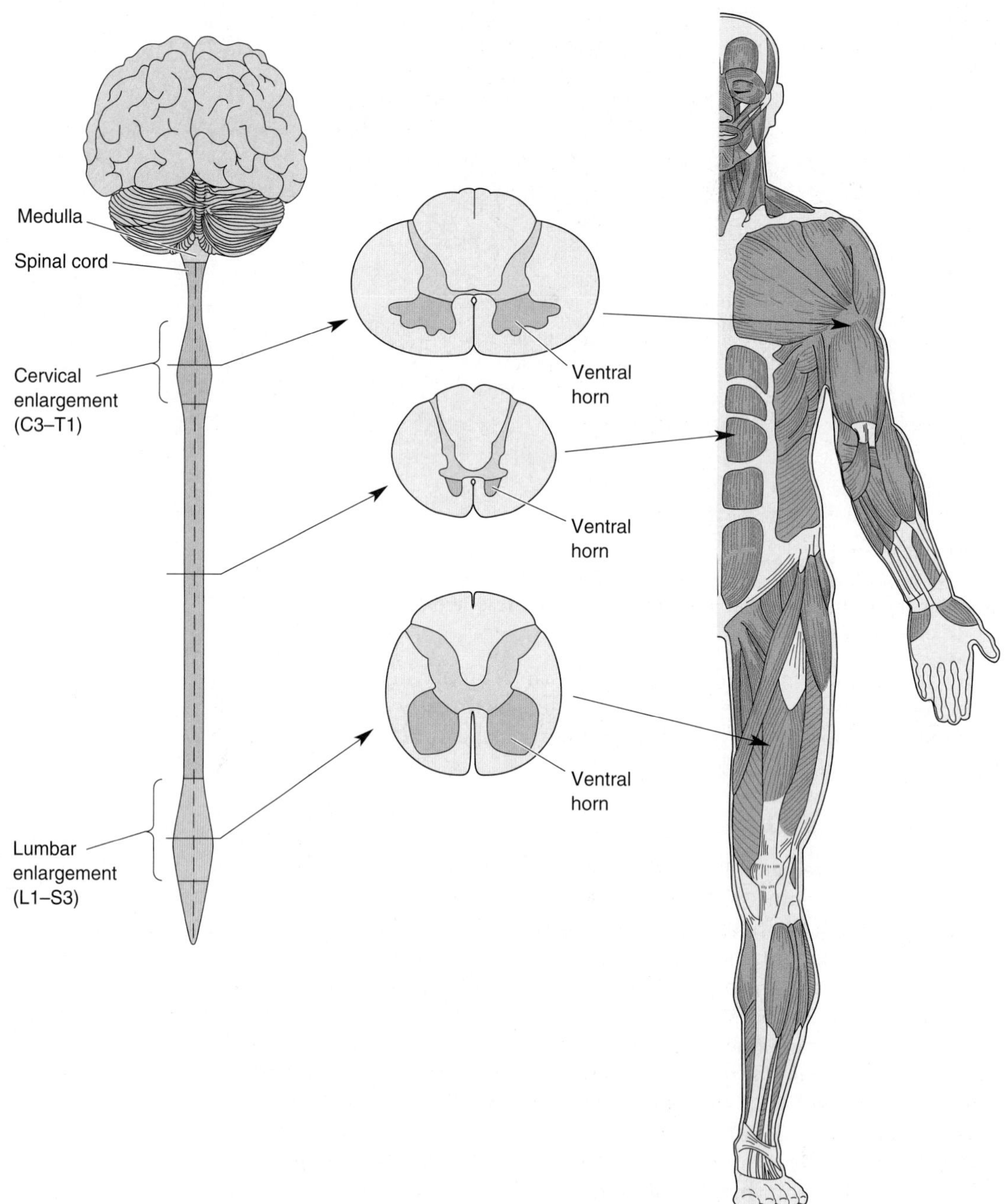

▲ FIGURE 13.5
The distribution of motor neurons in the spinal cord. The cervical enlargement of the spinal cord contains the motor neurons that innervate the arm muscles. The lumbar enlargement contains the neurons that innervate the muscles of the leg.

Alpha Motor Neurons

There are two categories of lower motor neurons of the spinal cord: alpha motor neurons and gamma motor neurons (the latter are discussed later in the chapter). **Alpha motor neurons** directly trigger the generation of force by muscles. One alpha motor neuron and all the muscle fibers it innervates collectively make up the elementary component of motor control; Sherrington called it the **motor unit**. Muscle contraction results from the individual and combined actions of motor units. The collection of alpha motor neurons that innervates a single muscle (e.g., the biceps brachii) is called a **motor neuron pool** (Figure 13.7).

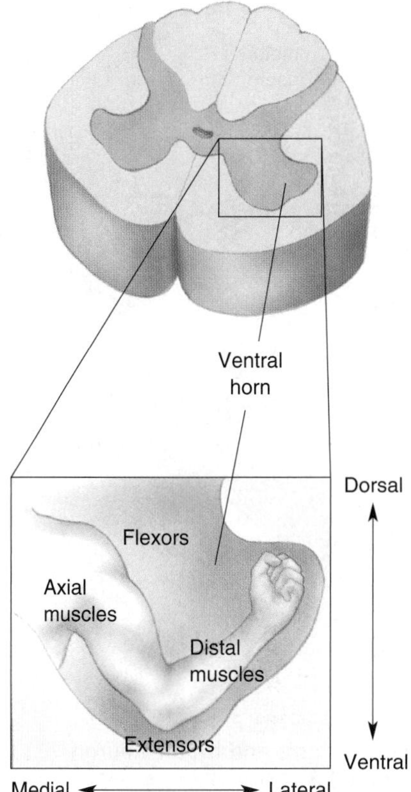

▲ FIGURE 13.6
The distribution of lower motor neurons in the ventral horn. Motor neurons controlling flexors lie dorsal to those controlling extensors. Motor neurons controlling axial muscles lie medial to those controlling distal muscles.

Graded Control of Muscle Contraction by Alpha Motor Neurons. It is important to exert just the right amount of force during movements: Too much, and you'll crush the egg you just picked up, while also wasting metabolic energy. Too little, and you may lose the swim race. Most of the movements we make, such as walking, talking, and writing, require only weak muscle contractions. Now and then we need to jog, hop, or lift a pile of books, and stronger contractions are necessary. We reserve the maximal contraction force of our muscles for rare events, such as competitive sprinting or scrambling up a tree to escape a charging bear. The nervous system uses several mechanisms to control the force of muscle contraction in a finely graded fashion.

The first way the CNS controls muscle contraction is by varying the firing rate of motor neurons. An alpha motor neuron communicates with a muscle fiber by releasing the neurotransmitter acetylcholine (ACh) at the neuromuscular junction, the specialized synapse between a nerve and a skeletal muscle (see Chapter 5). Because of the high reliability of neuromuscular transmission, the ACh released in response to one presynaptic action potential causes an excitatory postsynaptic potential (EPSP) in the muscle fiber (sometimes also called an *end-plate potential*) large enough to trigger one postsynaptic action potential. By mechanisms we will discuss in a moment, a postsynaptic action potential causes a twitch—a rapid sequence of contraction and relaxation—in the muscle fiber. A sustained contraction requires a continual barrage of action potentials. High-frequency presynaptic activity causes temporal summation of the postsynaptic responses, as it does for other types of synaptic transmission. Twitch summation increases the tension in the muscle fibers and smoothes the contraction (Figure 13.8). The rate of firing of motor units is therefore one important way the CNS grades muscle contraction.

A second way the CNS grades muscle contraction is by recruiting additional synergistic motor units. The extra tension provided by the recruitment of an active motor unit depends on how many muscle fibers are in that unit. In the antigravity muscles of the leg (muscles that oppose the force of gravity when standing upright), each motor unit tends to be quite large, with an innervation ratio of over 1000 muscle fibers per single alpha motor neuron. In contrast, the smaller muscles that control the movement of the fingers and the rotation of the eyes are characterized by much smaller innervation ratios, as few as three muscle fibers per alpha motor neuron. In general, muscles with a large number of small motor units can be more finely controlled by the CNS.

Most muscles have a range of motor unit sizes, and these motor units are usually recruited in the order of smallest first, largest last. This orderly recruitment explains why finer control is possible when muscles are under light loads than when they are under greater loads. Small motor units have small alpha motor neurons, and large motor units have large alpha motor neurons. Thus, one way orderly recruitment occurs is that small neurons, as a consequence of the geometry and physiology of their soma and dendrites, are more easily excited by signals descending from the brain. The idea that the orderly recruitment of motor neurons is due to variations in alpha motor neuron size is called the *size principle*, first proposed in the late 1950s by Harvard University neurophysiologist Elwood Henneman.

Inputs to Alpha Motor Neurons. Alpha motor neurons excite skeletal muscles. Therefore, to understand the control of muscles, we must understand what regulates motor neurons. Lower motor neurons are controlled

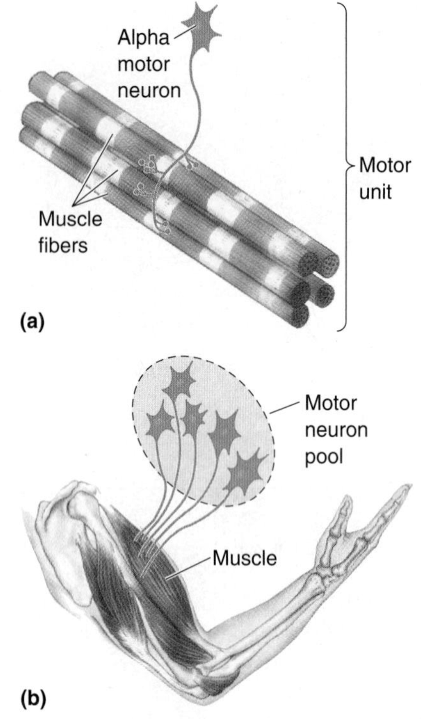

▲ FIGURE 13.7
A motor unit and motor neuron pool. (a) A motor unit is an alpha motor neuron and all the muscle fibers it innervates. (b) A motor neuron pool is all the alpha motor neurons that innervate one muscle.

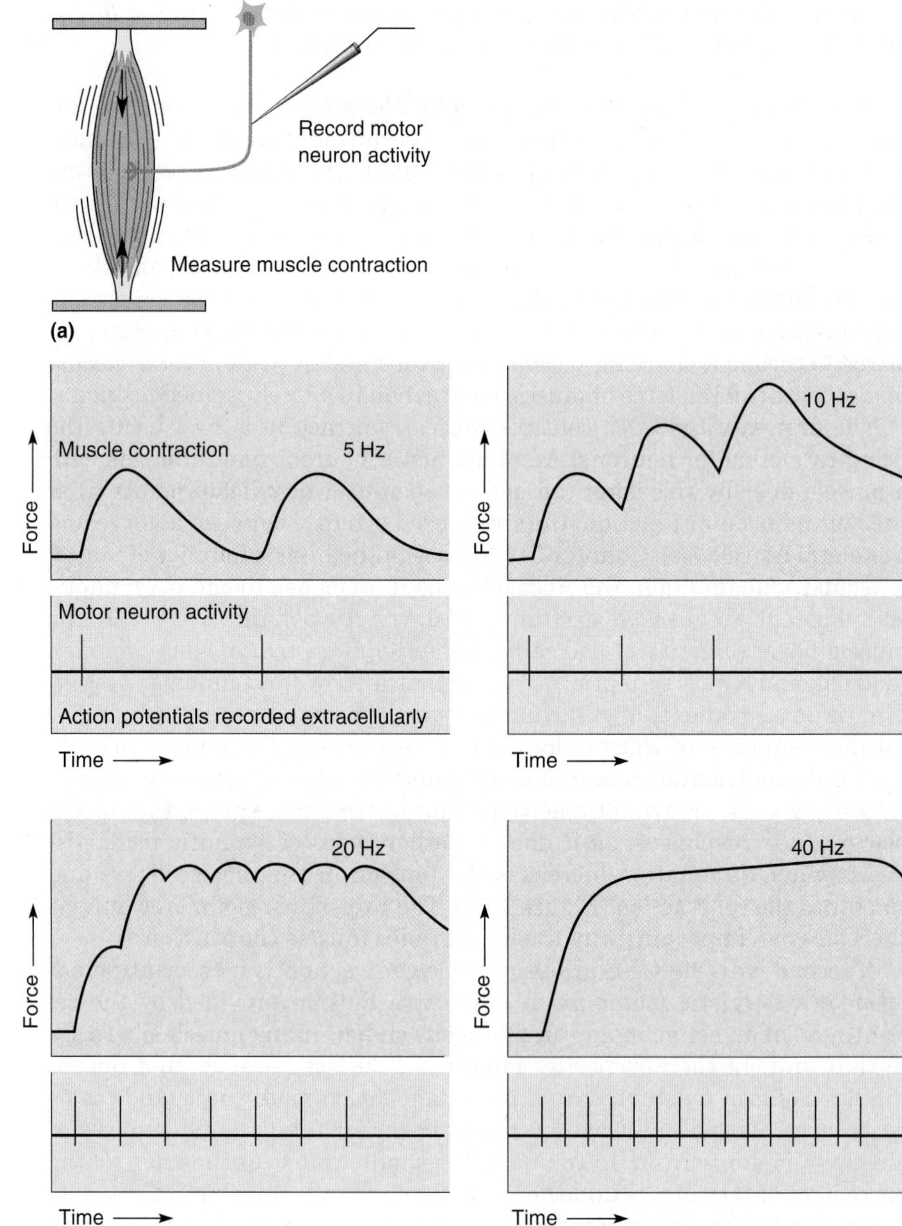

▲ FIGURE 13.8
From muscle twitch to sustained contraction. (a) A single action potential in an alpha motor neuron causes the muscle fiber to twitch. (b) The summation of twitches causes a sustained contraction as the number and frequency of incoming action potentials increase.

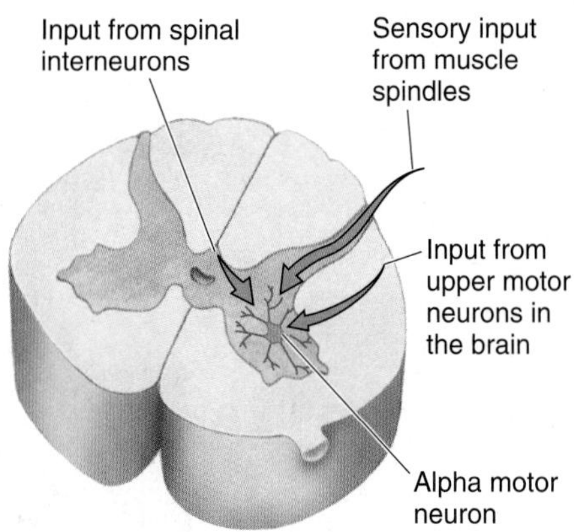

▲ FIGURE 13.9
An alpha motor neuron and its three sources of input.

by synaptic inputs in the ventral horn. *There are only three major sources of input to an alpha motor neuron*, as shown in Figure 13.9. The first source is dorsal root ganglion cells with axons that innervate a specialized sensory apparatus embedded within the muscle known as a *muscle spindle*. As we shall see, this input provides feedback about muscle length. The second source of input to an alpha motor neuron derives from upper motor neurons in the motor cortex and brain stem. This input is important for the initiation and control of voluntary movement and will be discussed in more detail in Chapter 14. The third and largest input to an alpha motor neuron derives from interneurons in the spinal cord.

This input may be excitatory or inhibitory and is part of the circuitry that generates the spinal motor programs.

Types of Motor Units

If you have ever eaten different parts of chicken, you know that not all muscle is the same; the leg has dark meat and the breast and wing have white meat. The different appearance, and taste, of the various muscles are due to the biochemistry of the constituent muscle fibers. The red (dark) muscle fibers are characterized by a large number of mitochondria and enzymes specialized for oxidative energy metabolism. These, sometimes called *slow (S) fibers*, are relatively slow to contract but can sustain contraction for a long time without fatigue. They are typically found in the antigravity muscles of the leg and torso and in the flight muscles of birds that fly (as opposed to domesticated chickens). In contrast, the pale (white) muscle fibers contain fewer mitochondria and rely mainly on anaerobic (without oxygen) metabolism. These fibers, sometimes called *fast (F) fibers*, contract rapidly and powerfully, but they also fatigue more quickly than slow fibers. They are typical of muscles involved in escape reflexes; for example, the jumping muscles of frogs and rabbits. In humans, the arm muscles contain a large number of white fibers. Fast fibers can be further divided into two subtypes: *Fatigue-resistant (FR) fibers* generate moderately strong and fast contractions and are relatively resistant to fatigue; *fast fatigable (FF) fibers* generate the strongest, fastest contractions but are quickly exhausted when stimulated at high frequency for long periods.

Even though all three types of muscle fibers can (and usually do) coexist in a given muscle, each motor unit contains muscle fibers of only a single type. Thus, there is one type of **slow motor unit** that contains only slowly fatiguing red fibers, and there are two types of **fast motor units**, each containing either FR or FF white fibers (Figure 13.10). Just as the muscle fibers of the three types of units differ, so do many of the properties of alpha motor neurons. For example, the motor neurons of the FF units are generally the biggest and have the largest diameter, fastest conducting axons; FR units have motor neurons and axons intermediate in size; and slow units have small-diameter, slowly conducting axons. The firing properties of the three types of motor neuron also differ. Fast (FF) motor neurons tend to generate occasional high-frequency bursts of action potentials (30–60 impulses per second), whereas slow motor neurons are characterized by relatively steady, low-frequency activity (10–20 impulses per second).

Neuromuscular Matchmaking. The precise matching of particular motor neurons to particular muscle fibers raises an interesting question. Since we've been talking about chickens, let's pose the question this way: Which came first, the muscle fiber or the motor neuron? Perhaps during early embryonic development, there is a matching of the appropriate axons with the appropriate muscle fibers. Alternatively, we could imagine that the properties of the muscle are determined solely by the type of innervation it gets. If it receives a synaptic contact from a fast motor neuron, it becomes a fast fiber and vice versa for slow units.

John Eccles and his colleagues, working at the Australian National University, addressed this question with an experiment in which the normal innervation of a fast muscle was removed and replaced with a nerve that normally innervated a slow muscle (Figure 13.11). This procedure resulted in the muscle's acquiring slow properties, including not only the type of contraction (slow, fatigue-resistant) but also a switch in much

▶ FIGURE 13.10
Three types of motor units and their contractile properties. (a) A single action potential triggers contraction strengths of different force and time-course in each of the three types of motor units. **(b)** Repeated trains of action potentials at 40 Hz over many minutes lead to different rates of fatigue in the three types of motor units. (Source: Adapted from Burke et al., 1973.)

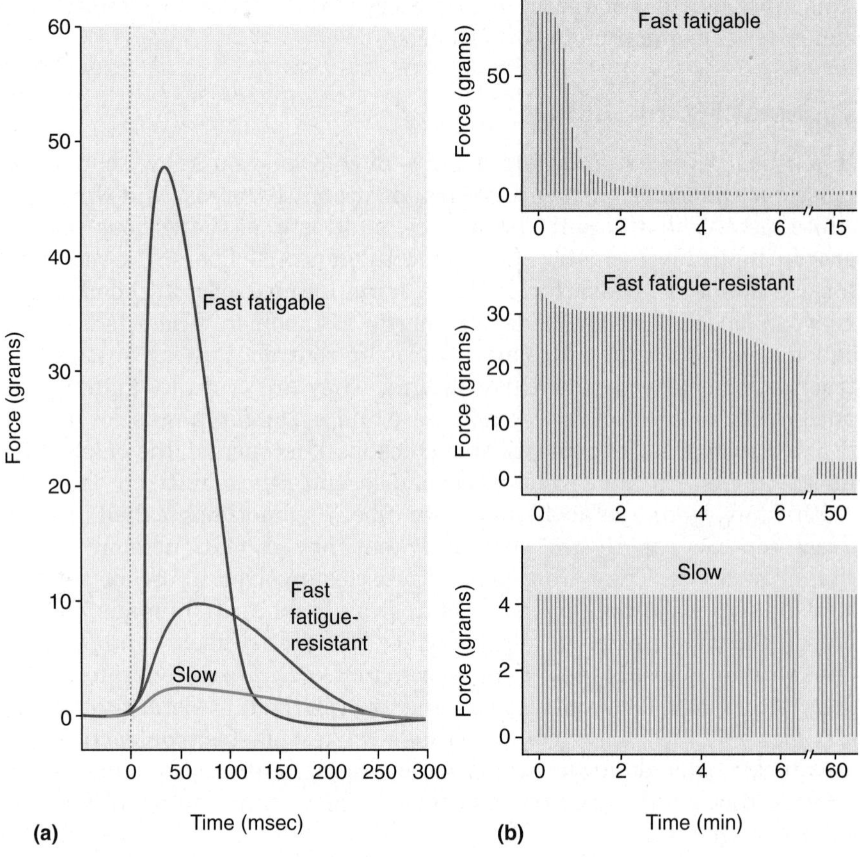

of its underlying biochemistry. This change is referred to as a switch of muscle *phenotype*—its physical characteristics—because the types of proteins expressed by the muscle were altered by the new innervation. Work by Terje Lømo and his colleagues in Norway suggests that this switch in muscle phenotype can be induced simply by changing the activity in the motor neuron from a fast pattern (occasional bursts at 30–60 spikes per second) to a slow pattern (steady activity at 10–20 spikes per second). These findings are particularly interesting because they raise the possibility that *neurons* switch phenotype as a consequence of synaptic activity (experience), and that this may be a basis for learning and memory (see Chapters 24 and 25).

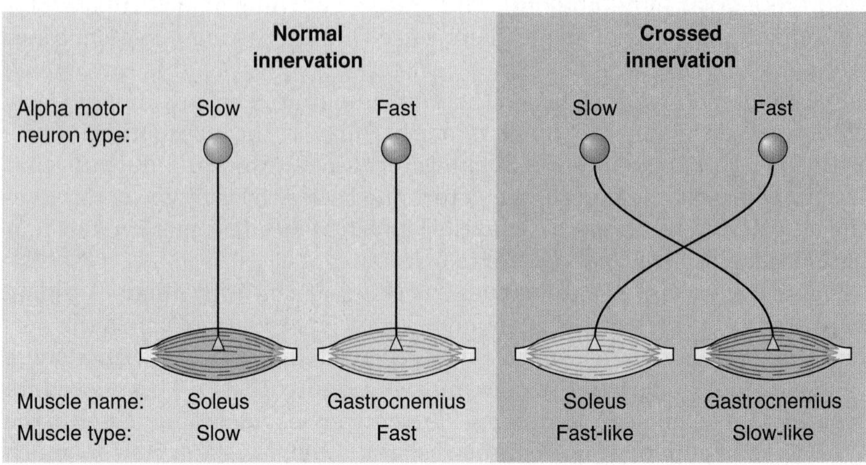

▶ FIGURE 13.11
A crossed-innervation experiment. Forcing slow motor neurons to innervate a fast muscle causes the muscle to switch to assume slow properties.

Besides the alterations imposed by patterns of motor neuron activity, simply varying the absolute amount of activity can also change muscle fibers. A long-term consequence of increased activity (especially due to isometric exercise) is *hypertrophy*, or exaggerated growth, of the muscle fibers as seen in bodybuilders. Conversely, prolonged inactivity leads to *atrophy*, or degeneration, of muscle fibers, which can happen when joints are immobilized in a cast following an injury. Clearly, there is an intimate relationship between the lower motor neuron and the muscle fibers it innervates (Box 13.1).

BOX 13.1 OF SPECIAL INTEREST

ALS: Glutamate, Genes, and Gehrig

Amyotrophic lateral sclerosis (ALS) is a particularly cruel disease that was first described in 1869 by the French neurologist Jean-Martin Charcot. The initial signs of the disease are muscle weakness and atrophy. Usually over the course of 1–5 years, all voluntary movement is lost; walking, speaking, swallowing, and breathing gradually deteriorate. Death is usually caused by failure of the respiratory muscles. Because the disease often has no effect on sensations, intellect, or cognitive function, patients are left to watch their bodies slowly waste away, keenly aware of what is happening. The disease is relatively rare, afflicting one out of approximately every 20,000 individuals. Still, an estimated 30,000 Americans are currently diagnosed with ALS. Its most famous victim was Lou Gehrig, a star baseball player with the New York Yankees, who died of ALS in 1936. In the United States, ALS is often called *Lou Gehrig's disease*.

Muscle weakness and paralysis are characteristics of motor unit damage. Indeed, the main pathology associated with ALS is the degeneration of the large alpha motor neurons. The large neurons of the motor cortex that innervate alpha motor neurons are also affected, but, curiously, other neurons in the CNS are generally spared. The selective damage to motor neurons explains the selective loss of motor functions in ALS patients.

There appear to be many causes of ALS, most of them still unknown. One suspected cause is excitotoxicity. As we learned in Chapter 6, overstimulation by the excitatory neurotransmitter glutamate and closely related amino acids can cause the death of otherwise normal neurons (see Box 6.4). Many ALS patients have elevated levels of glutamate in their cerebrospinal fluid. Excitotoxicity has been implicated in the unusually high incidence of ALS on the island of Guam that occurred before World War II. It has been suggested that one environmental cause in Guam may have been the ingestion of cycad nuts, which contain an excitotoxic amino acid. In addition, research indicates that a glutamate transporter may be defective in ALS, thereby prolonging the exposure of active neurons to extracellular glutamate. Thus, the first drug approved by the U.S. Food and Drug Administration for the treatment of ALS was riluzole, a blocker of glutamate release. The drug treatment can slow the disease by only a few months, however, and unfortunately, the long-term outcome is the same.

Only 10% of ALS cases are obviously inherited, and screens for defective genes have pointed to several mutations that can lead to ALS. The first mutation, discovered in 1993, leads to defects in the enzyme *superoxide dismutase*. A toxic by-product of cellular metabolism is the negatively charged molecule O_2^-, called the *superoxide radical*. Superoxide radicals are extremely reactive and can inflict irreversible cellular damage. Superoxide dismutase is a key enzyme that causes superoxide radicals to lose their extra electrons, converting them back to oxygen. Thus, the loss of superoxide dismutase would lead to a buildup of superoxide radicals and cellular damage, particularly in cells that are metabolically very active. The death of motor neurons seems to depend on the actions of glial cells that surround them.

More recent research has identified mutations of about 15 more genes that can cause inherited forms of ALS. They affect a surprisingly wide variety of basic cellular processes. Some mutations cause defects in proteins that normally bind and regulate RNA during transcription. Others affect proteins involved in the trafficking of vesicles, protein secretion, cell division, ATP production, or the dynamics of the cytoskeleton. Genome-wide association studies, which examine a large number of gene variations to reveal which are associated with a disease, suggest that the coincidence of two mutations of distinctly different genes can also cause ALS. The picture that is emerging is that ALS can have many distinct causes; it is really a group of diseases that happen to share similar clinical characteristics.

There is still much to be learned about selective motor neuron loss in ALS. What we know so far has led to new ideas for possible treatments, including the use of neuronal stem cells to replace lost neurons and glia, and genetics-based strategies to suppress the effects of mutations. Translating these ideas into effective treatments for ALS patients is an exciting but still distant possibility.

EXCITATION–CONTRACTION COUPLING

Muscle contraction is initiated by the release of acetylcholine (ACh) from the axon terminals of alpha motor neurons, as we said. ACh produces a large EPSP in the postsynaptic membrane due to the activation of nicotinic ACh receptors. Because the membrane of the muscle cell contains voltage-gated sodium channels, this EPSP is sufficient to evoke an action potential in the muscle fiber (but see Box 13.2). By the process of **excitation–contraction coupling**, this action potential (the *excitation*) triggers the release of Ca^{2+} from an organelle inside the muscle fiber, which leads to *contraction* of the fiber. Relaxation occurs when the Ca^{2+} levels are lowered by reuptake into the organelle. To understand this process, we must take a closer look at the muscle fiber.

Muscle Fiber Structure

The structure of a muscle fiber is shown in Figure 13.12. Muscle fibers are formed early in fetal development by the fusion of muscle precursor cells, or myoblasts, which are derived from the mesoderm (see Chapter 7). This fusion leaves each cell with more than one cell nucleus, so individual muscle cells are said to be *multinucleated.* The fusion elongates the cells (hence the name fiber), and fibers can range from 1 to 500 mm in length. Muscle fibers are enclosed by an excitable cell membrane called the **sarcolemma**.

BOX 13.2 OF SPECIAL INTEREST

Myasthenia Gravis

The neuromuscular junction is an exceptionally reliable synapse. A presynaptic action potential causes the contents of hundreds of synaptic vesicles to be released into the synaptic cleft. The liberated ACh molecules act at densely packed nicotinic receptors in the postsynaptic membrane, and the resulting EPSP is many times larger than what is necessary to trigger an action potential, and twitch, in the muscle fiber—normally, that is.

In a clinical condition called *myasthenia gravis*, the ACh released is far less effective, and neuromuscular transmission often fails. The name is derived from the Greek for "severe muscle weakness." The disorder is characterized by weakness and fatigability of voluntary muscles, typically including the muscles of facial expression, and it can be fatal if respiration is compromised. The disease strikes roughly one in 10,000 people of all ages and ethnic groups. An unusual feature of myasthenia gravis is that the severity of the muscle weakness fluctuates, even over the course of a single day.

Myasthenia gravis is an *autoimmune disease.* In 1973, Jim Patrick and Jon Lindstrom, working at the Salk Institute in California, discovered that rabbits injected with purified nicotinic ACh receptors generated antibodies to their own ACh receptors and contracted a rabbit version of myasthenia gravis. For reasons we don't understand, the immune systems of most myasthenia-afflicted humans generate antibodies against their own nicotinic ACh receptors. The antibodies bind to the receptors, interfering with the normal actions of ACh at the neuromuscular junctions. In addition, the binding of antibodies to the receptors leads to secondary, degenerative changes in the structure of the neuromuscular junctions that also make transmission much less efficient.

An effective treatment for myasthenia gravis is the administration of drugs that inhibit the enzyme acetylcholinesterase (AChE). Recall from Chapters 5 and 6 that AChE breaks down ACh in the synaptic cleft. In low doses, AChE inhibitors can strengthen neuromuscular transmission by prolonging the life of released ACh. But the drugs are imperfect and the therapeutic window is narrow. As we saw in Box 5.5, too much ACh in the cleft leads to desensitization of the receptors and a block of neuromuscular transmission. Different muscles may respond differently to the same drug dose. The increased levels of ACh can also affect the ANS, leading to side effects such as nausea, vomiting, abdominal cramps, diarrhea, and bronchial secretions. Another common treatment for myasthenia gravis involves suppression of the immune system, either with drugs or by surgical removal of the thymus gland.

With careful and continual medical treatment, the long-term prognosis is good and life expectancy is normal for patients with this disease of the neuromuscular junction.

Within the muscle fiber are a number of cylindrical structures called **myofibrils**, which contract in response to an action potential sweeping down the sarcolemma. Myofibrils are surrounded by the **sarcoplasmic reticulum (SR)**, an extensive intracellular sac that stores Ca^{2+} (similar in appearance to the smooth endoplasmic reticulum of neurons; see Chapter 2). Action potentials sweeping along the sarcolemma gain access to the sarcoplasmic reticulum deep inside the fiber by way of a network of tunnels called **T tubules** (T for transverse). These are like inside-out axons; the interior of each T tubule is continuous with the extracellular fluid.

Where the T tubule comes in close apposition to the SR, there is a specialized coupling of the proteins in the two membranes. A voltage-sensitive cluster of four calcium channels, called a *tetrad*, in the T tubule membrane is linked to a *calcium release channel* in the SR. As illustrated in Figure 13.13, the arrival of an action potential in the T tubule membrane causes a conformational change in the voltage-sensitive tetrad of channels, which opens the calcium release channel in the SR membrane. Some Ca^{2+} flows through the tetrad channels, and even more Ca^{2+} flows through the calcium-release channel, and the resulting increase in free Ca^{2+} within the cytosol causes the myofibril to contract.

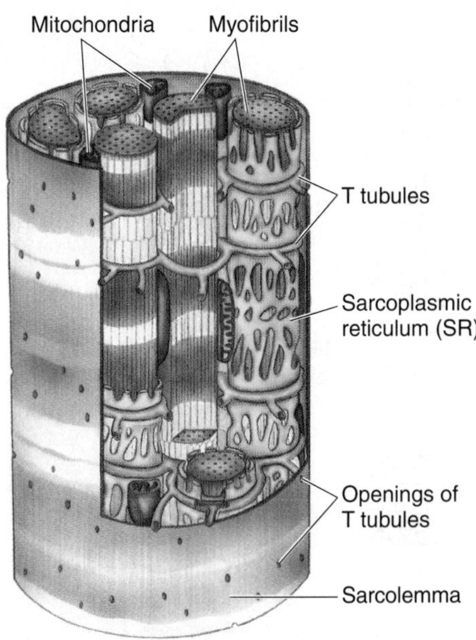

▲ FIGURE 13.12
The structure of a muscle fiber. T tubules conduct electrical activity from the surface membrane into the depths of the muscle fiber.

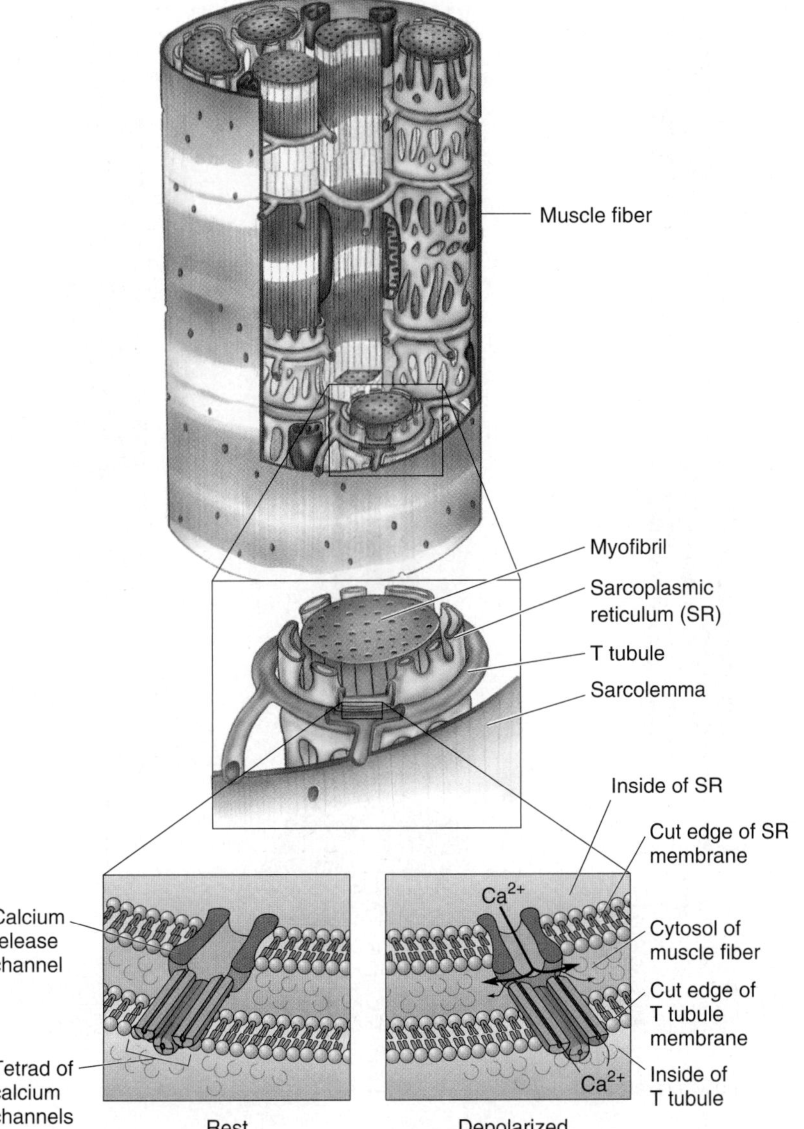

◀ FIGURE 13.13
The release of Ca^{2+} from the sarcoplasmic reticulum. Depolarization of the T tubule membrane causes conformational changes in proteins that are linked to calcium channels in the SR, releasing stored Ca^{2+} into the cytosol of the muscle fiber.

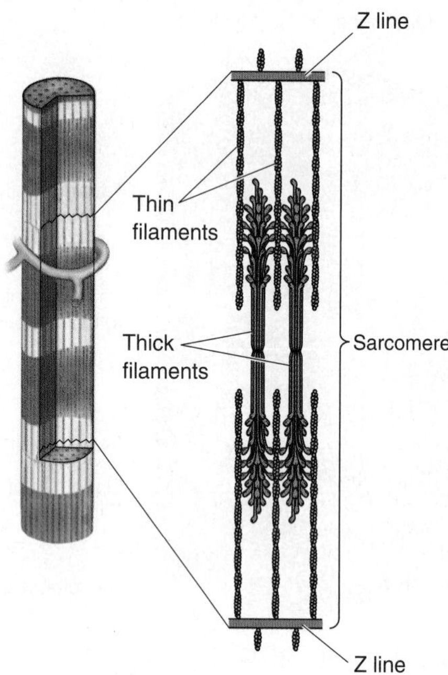

▲ **FIGURE 13.14**
The myofibril: a closer look.

The Molecular Basis of Muscle Contraction

A closer look at the myofibril reveals how Ca^{2+} triggers contraction (Figure 13.14). The myofibril is divided into segments by disks called *Z lines* (named for their appearance when viewed from the side). A segment composed of two Z lines and the myofibril in between is called a **sarcomere**. Anchored to each side of the Z lines is a series of bristles called **thin filaments**. The thin filaments from adjacent Z lines face one another but do not come in contact. Between and among the two sets of thin filaments are a series of fibers called **thick filaments**. Muscle contraction occurs when the thin filaments slide along the thick filaments, bringing adjacent Z lines toward one another. In other words, the sarcomere becomes shorter in length. This *sliding-filament model* of sarcomere shortening is shown in Figure 13.15.

Z line
Thin filaments
Thick filaments
Weight
Fully relaxed → Ca^{2+} → Fully contracted

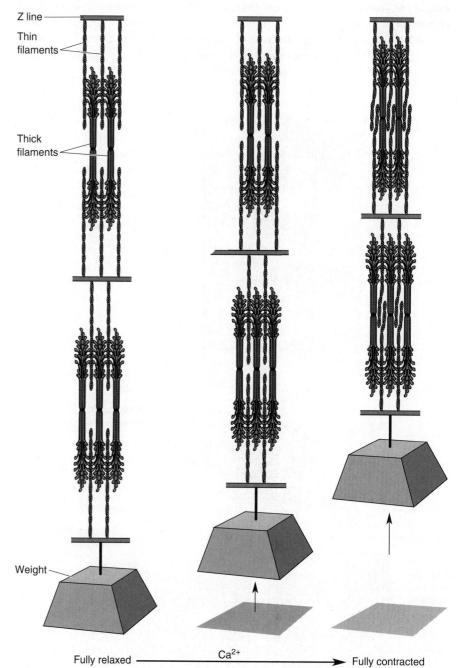

▶ **FIGURE 13.15**
The sliding-filament model of muscle contraction. Myofibrils shorten when the thin filaments slide toward one another on the thick filaments.

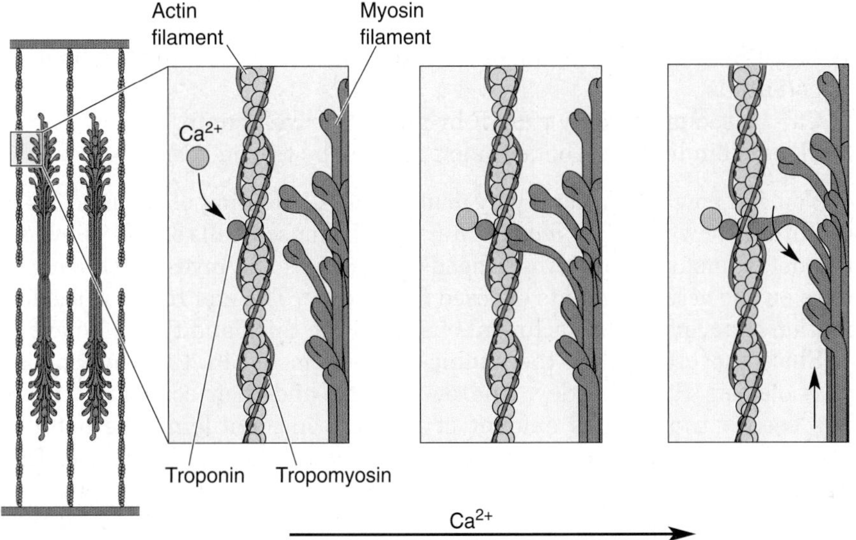

◀ FIGURE 13.16
The molecular basis of muscle contraction. The binding of Ca^{2+} to troponin shifts tropomyosin and allows the myosin heads to bind to the actin filament. Then the myosin heads pivot, causing the filaments to slide with respect to one another.

The sliding of the filaments with respect to one another occurs because of the interaction between the major thick filament protein, **myosin**, and the major thin filament protein, **actin**. The exposed "heads" of the myosin molecules bind actin molecules and then undergo a conformational change that causes them to pivot (Figure 13.16). This pivoting causes the thick filament to move with respect to the thin filament. Adenosine triphosphate (ATP) then binds to the myosin heads and the heads disengage and "uncock" so that the process can repeat itself. Repeating this cycle enables the myosin heads to "walk" along the actin filament.

When the muscle is at rest, myosin cannot interact with actin because the myosin attachment sites on the actin molecule are covered by a complex of two proteins: *tropomyosin* and *troponin*. Ca^{2+} initiates muscle contraction by binding to troponin and causing tropomyosin to shift its position, thereby exposing the sites where myosin binds to actin. Contraction continues as long as Ca^{2+} and ATP are available; relaxation occurs when the Ca^{2+} is sequestered by the SR. The reuptake of Ca^{2+} by the SR depends on the action of a calcium pump and hence also requires ATP.

We can summarize the steps of excitation–contraction coupling as follows:

Excitation

1. An action potential occurs in an alpha motor neuron axon.
2. ACh is released by the axon terminal of the alpha motor neuron at the neuromuscular junction.
3. Nicotinic receptor channels in the sarcolemma open, and the postsynaptic sarcolemma depolarizes (EPSP).
4. Voltage-gated sodium channels in the sarcolemma open and an action potential is generated in the muscle fiber, which sweeps down the sarcolemma and into the T tubules.
5. Depolarization of the T tubules causes Ca^{2+} release from the SR.

Contraction

1. Ca^{2+} binds to troponin.
2. Tropomyosin shifts position and myosin binding sites on actin are exposed.
3. Myosin heads bind actin.
4. Myosin heads pivot.
5. An ATP binds to each myosin head and it disengages from actin.
6. The cycle continues as long as Ca^{2+} and ATP are present.

Relaxation

1. As EPSPs end, the sarcolemma and T tubules return to their resting potentials.
2. Ca^{2+} is sequestered by the SR by an ATP-driven pump.
3. Myosin binding sites on actin are covered by tropomyosin.

You can now understand why death causes stiffening of the muscles, a condition known as *rigor mortis*. Starving the muscle cells of ATP prevents the detachment of the myosin heads and leaves the myosin attachment sites on the actin filaments exposed for binding. The end result is the formation of permanent attachments between the thick and thin filaments.

Since the proposal of the sliding-filament model in 1954 by English physiologists Hugh Huxley, Andrew Huxley, and their colleagues, there has been a tremendous amount of progress in identifying the detailed molecular mechanisms of excitation–contraction coupling in muscle. This progress has resulted from a multidisciplinary approach to the problem, with critical contributions made by the use of electron microscopy as well as biochemical, biophysical, and genetic methods. The application of molecular genetic techniques also has added important new information to our understanding of muscle function, in both health and disease (Box 13.3).

BOX 13.3 OF SPECIAL INTEREST

Duchenne Muscular Dystrophy

Muscular dystrophy is a group of inherited disorders, all of which are characterized by progressive weakness and deterioration of muscle. The most common type, *Duchenne muscular dystrophy*, afflicts about one in 3500 boys before adolescence. The disease is first detected as a weakness of the legs and usually puts its victims in wheelchairs by the time they reach age 12. The disease continues to progress, and afflicted males typically do not survive past the age of 30.

The characteristic hereditary pattern of this disease, which afflicts only males but is passed on from their mothers, led to a search for a defective gene on the X chromosome. Major breakthroughs came in the late 1980s when the defective region of the X chromosome was identified. Researchers discovered that this region contains the gene that codes for a cytoskeletal protein *dystrophin*. The dystrophin gene is enormous—2.6 million base pairs—and its size makes it unusually vulnerable to mutations. Boys with Duchenne muscular dystrophy have an entirely dysfunctional dystrophin gene: They cannot produce the mRNA encoding dystrophin. A milder form of the disease, called *Becker muscular dystrophy*, is associated with an altered mRNA encoding a portion of the dystrophin protein.

Dystrophin is a large protein that helps to link the muscle cytoskeleton, lying just under the sarcolemma, to the extracellular matrix. The protein also seems to be important for helping muscle deal with oxidative stress. Dystrophin must not be strictly required for muscle contraction because movements in afflicted boys appear to be normal during their first few years of life. The absence of dystrophin may lead to secondary changes in the contractile apparatus, eventually resulting in muscle degeneration. It is interesting to note that dystrophin is also concentrated in axon terminals in the brain, where it might contribute to excitation-secretion coupling.

Intensive efforts are underway to find a strategy for treating, or even curing, Duchenne muscular dystrophy with some form of gene therapy. One long-standing idea is to introduce an artificial gene that essentially repairs the patient's defective dystrophin gene or mimics a normal dystrophin gene. A big challenge, as with most attempts at gene therapy, has been to get the artificial gene into dystrophic muscle cells safely and effectively. Specially engineered forms of viruses that carry the gene, infect muscle cells, and induce the cells to express dystrophin are often used. Another approach is to transplant stem cells—immature cells that can grow and differentiate into mature, normal muscle cells that express dystrophin—into dystrophic muscles. Stem cell therapy has been very promising when tested in mouse models of muscular dystrophy. Yet, another strategy is to test small molecules that might minimize muscle degeneration, promote muscle regeneration, mitigate encoding problems by mutant dystrophin genes, or promote the production of other muscle proteins that can substitute for dystrophin.

There is no cure for Duchenne muscular dystrophy as yet, but some of the new treatment strategies have shown promise in clinical trials. It is exciting to think that a devastating genetic disease such as Duchenne muscular dystrophy might soon be treatable.

SPINAL CONTROL OF MOTOR UNITS

We've traced the action potentials sweeping down the axon of the alpha motor neuron and seen how this causes contraction of the muscle fibers in the motor unit. Now let's explore how the activity of the motor neuron is itself controlled. We begin with a discussion of the first source of synaptic input to the alpha motor neuron introduced earlier—sensory feedback from the muscles themselves.

Proprioception from Muscle Spindles

As we mentioned already, deep within most skeletal muscles are specialized structures called **muscle spindles** (Figure 13.17). A muscle spindle, also called a *stretch receptor*, consists of several types of specialized skeletal muscle fibers contained in a fibrous capsule. The middle third of the capsule is swollen, giving the structure the shape for which it is named. In this middle (equatorial) region, group Ia sensory axons wrap around the muscle fibers of the spindle. The spindles and their associated Ia axons, specialized for the detection of changes in muscle length (stretch), are examples of **proprioceptors**. These receptors are a component of the somatic sensory system that is specialized for "body sense," or **proprioception** (from the Latin for "one's own"), which informs us about how our body is positioned and moving in space.

Recall from Chapter 12 that group I axons are the thickest myelinated axons in the body, meaning that they conduct action potentials very rapidly. Within this group, Ia axons are the largest and fastest. Ia axons enter the spinal cord via the dorsal roots, branch repeatedly, and form excitatory synapses upon both interneurons and alpha motor neurons of the ventral horns. The Ia inputs are also very powerful. Neurophysiologist Lorne Mendell, working at Harvard with Henneman, was able to show that a single Ia axon synapses on virtually every alpha motor neuron in the pool innervating the same muscle that contains the spindle.

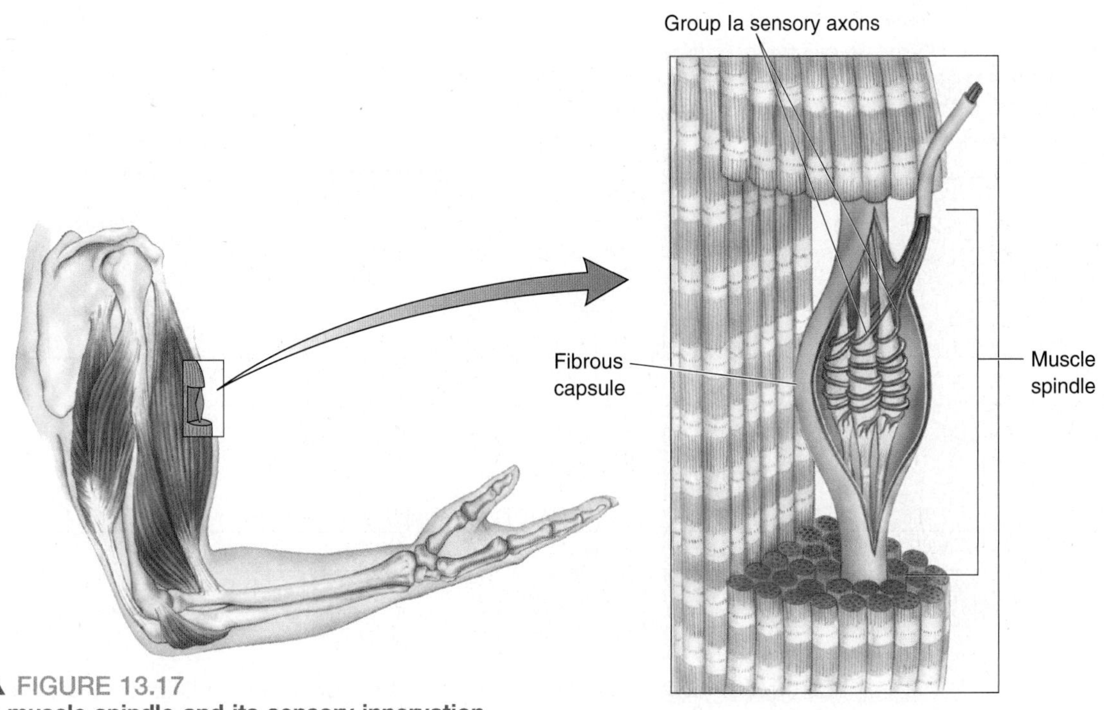

▲ FIGURE 13.17
A muscle spindle and its sensory innervation.

The Stretch Reflex. The function of this sensory input to the spinal cord was first shown by Sherrington, who noted that when a muscle is pulled on, it tends to pull back (contract). The fact that this **stretch reflex**, sometimes called the *myotatic reflex* (*myo* from the Greek for "muscle," *tatic* from the Greek for "stretch"), involves sensory feedback from the muscle was shown by cutting the dorsal roots. Even though the alpha motor neurons were left intact, this procedure eliminated the stretch reflex and caused a loss of muscle tone. Sherrington deduced that the motor neurons must receive a continual synaptic input from the muscles. Later work showed that the discharge of Ia sensory axons is closely related to the length of the muscle. As the muscle is stretched, the discharge rate goes up; as the muscle is shortened and goes slack, the discharge rate goes down.

The Ia axon and the alpha motor neurons on which it synapses constitute the *monosynaptic stretch reflex arc* — "monosynaptic" because only one synapse separates the primary sensory input from the motor neuron output. Figure 13.18 shows how this reflex arc serves as an antigrav-

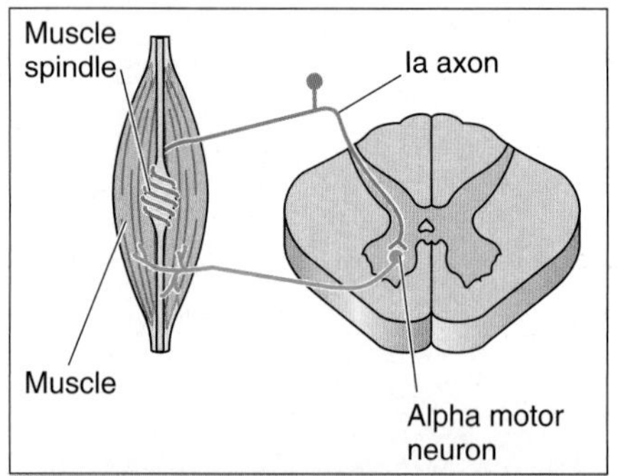

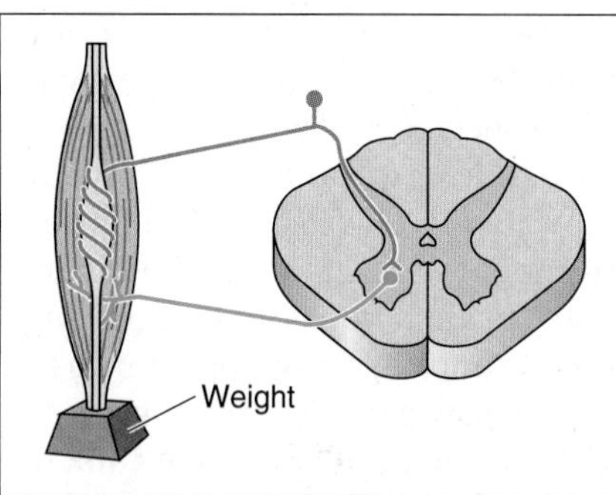

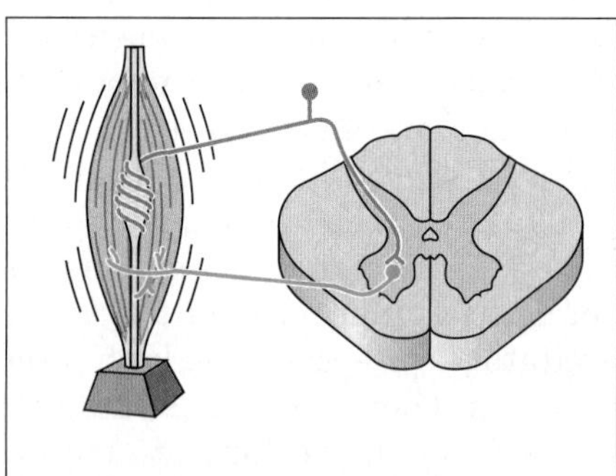

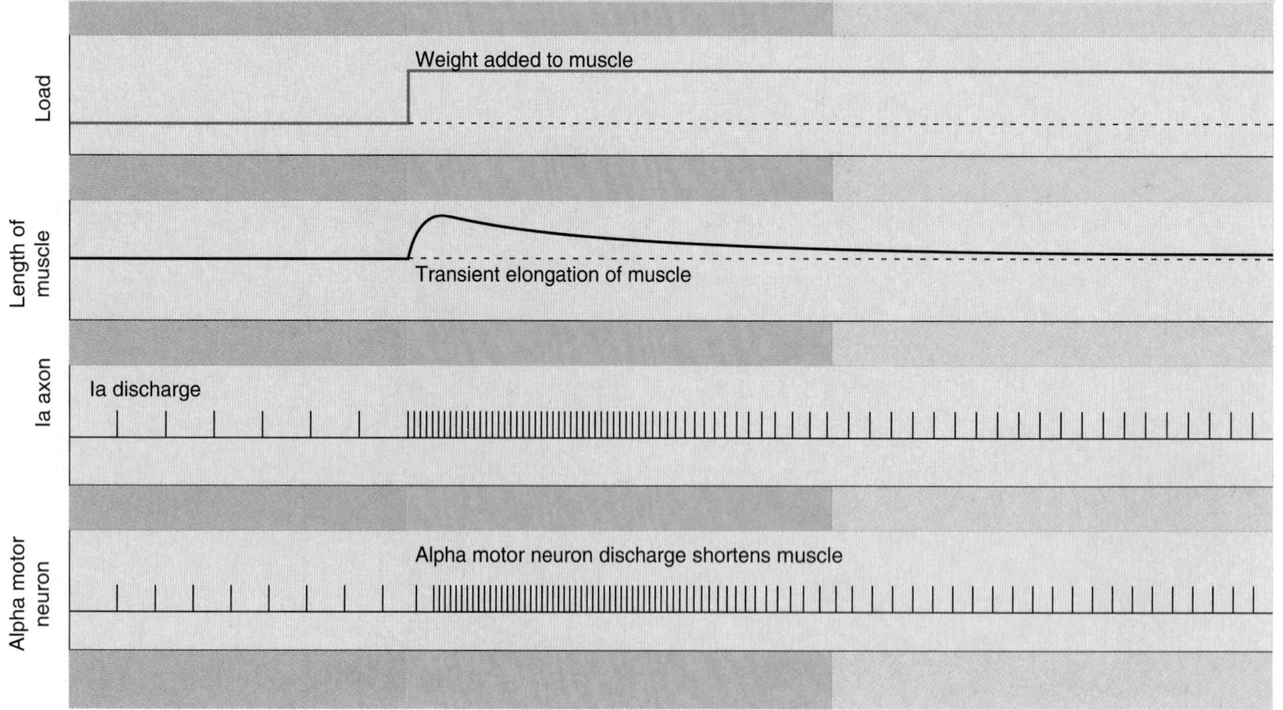

▲ FIGURE 13.18
The stretch reflex. This illustration shows the response of an Ia axon and a motor neuron to the sudden addition of weight that stretches the muscle.

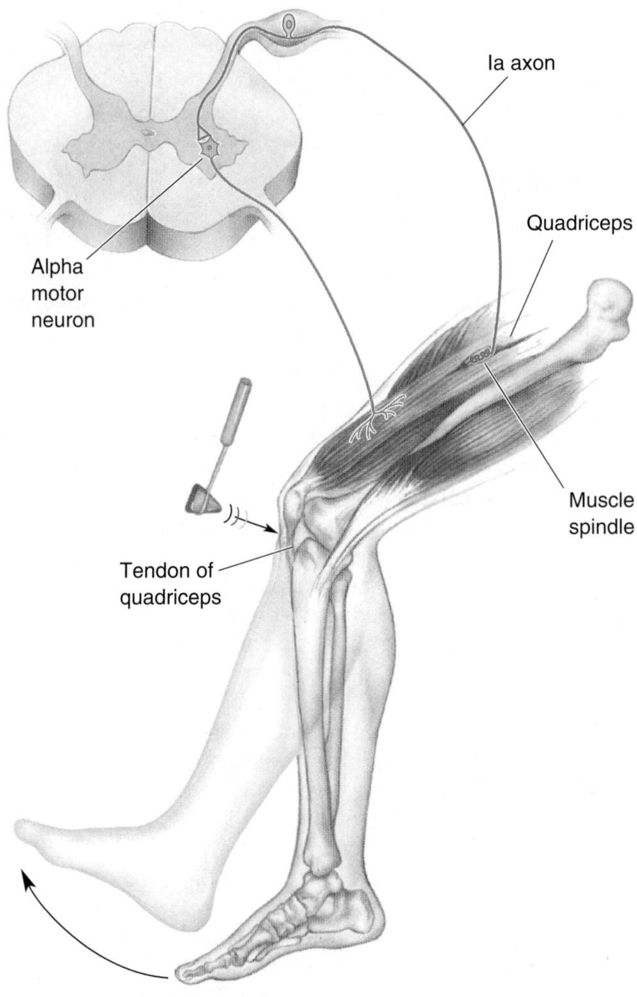

▲ FIGURE 13.19
The knee-jerk reflex.

ity feedback loop. When a weight is placed on a muscle and the muscle starts to lengthen, the muscle spindles are stretched. The stretching of the equatorial region of the spindle leads to depolarization of the Ia axon endings due to the opening of mechanosensitive ion channels (see Chapter 12). The resulting increased action potential discharge of the Ia axons synaptically depolarizes the alpha motor neurons, which respond by increasing their action potential frequency. This causes the muscle to contract, thereby shortening it.

The knee-jerk reflex is one example of the stretch reflex. When your doctor taps the tendon beneath your kneecap, the tendon very briefly stretches the quadriceps muscle of your thigh, which then reflexively contracts and causes your leg to extend (Figure 13.19). The knee-jerk reflex tests the intactness of the nerves and muscles in this reflex arc. Stretch reflexes can also be elicited by stretching muscles of the arm, the ankle, and the jaw.

Peripheral sensory and motor nerves are vulnerable to various type of injury. As we learned earlier, the cut axons of peripheral nerves can often regenerate and reestablish their connections with muscle (see Figure 13.11). Are regenerated axons and synapses as effective as normal axons and synapses? This question has been studied carefully in the neural circuits of the stretch reflex (Box 13.4).

BOX 13.4 PATH OF DISCOVERY

Nerve Regeneration Does Not Ensure Full Recovery

by Timothy C. Cope

Tapping a muscle's tendon can cause the muscle to contract and a leg to jerk. The monosynaptic circuit underlying this stretch reflex is shown in Figure 13.19. You will not be surprised to hear that cutting the sensory and motor nerves interrupts the reflex. Peripheral nerves can often regrow, however. What would you expect after the cut axons grow back into the muscle? The surprising answer is that the stretch reflex does not recover, even though voluntary contractions regain substantial force. It would seem easy to figure out why. Every element of this circuit is accessible to measurements, including the firing patterns of group Ia axons that encode muscle length, firing patterns of the motor neurons, the force of contraction produced by the stretched muscle and its synergists, and even the excitatory postsynaptic potentials (EPSPs) produced by the synapses between Ia axons and motor neurons in the spinal cord. Studies of this problem have fascinated me for more than 20 years. They have given me a rare opportunity to understand how a neural circuit can generate normal behavior, how a circuit responds to injury, and what factors limit the nervous system's capacity to recover from injury.

The recovery problem was most likely on the sensory side of the circuit. It couldn't be a deficit in the motor neurons or muscle or their restored connections because the muscle contracted normally during reflexes triggered by sensory stimuli other than stretch. At first, the most likely alternative hypothesis seemed to be that regenerating sensory axons end up reconnecting with the wrong sensory receptor organs in the periphery. It was well known that sensory axons in severed nerves reconnect somewhat indiscriminately with their targets, meaning that a reduced number of Ia axons would be available after regeneration to detect muscle stretch and to excite motoneurons. Even so, a sizable fraction of Ia axons do manage to reinnervate their normal targets; Lorne Mendell and his coworkers found that nearly 40% of regenerating Ia axons reconnected with muscle spindles. Even if the total excitation from fewer Ia axons were too weak to excite motoneurons to fire during muscle stretch, we would still expect that the action potentials in the reconnected Ia axons would provide a boost to the force of ongoing muscle contraction. In the laboratory, however, Brian Clark and I found no detectable firing modulation of motor units by stretching self-reinnervated muscles. Our colleague Richard Nichols, using different methods, corroborated our findings. The results were clear and puzzling: Muscle stretch utterly fails to recruit motor neurons after recovery from nerve cuts.

So what defect accounts for the nearly complete and persistent absence of stretch reflexes following injury? A key result emerged from our studies of EPSPs recorded from motor neurons during natural stretch of the muscle. The Ia synaptic potentials were weaker, of course, in part because roughly half of the Ia axons were not responding appropriately to muscle stretch. In addition, Edyta Bichler and Katie Bullinger in our lab made the prescient observation that these diminished EPSPs could be found only in about half of the motor neurons studied, while the other half showed no EPSPs at all (Figure A). Normally, Ia axons produce an EPSP in *every* motor neuron that innervates the same muscle. These observations highlighted a key shortcoming of Ia sensory neurons that regenerate their damaged peripheral axons: While some reconnect with muscle spindle receptors in the muscle, they also disconnect from many motor neurons in the spinal cord.

Recently, a structural explanation for the loss of the stretch reflex was provided by Francisco Alvarez and his laboratory in collaboration with ours: A probe that allows microscopic identification of Ia synaptic terminals revealed the loss of more than 70% of Ia synapses on the proximal dendrites of motor neurons. We also found that regenerating Ia axons actually retract their branches from the areas where motor neuron cell bodies and dendrites reside. Synaptic loss and axonal retraction in the spinal cord occur despite successful regeneration of the injured Ia axons' branches in the muscle.

Gamma Motor Neurons

The muscle spindle contains modified skeletal muscle fibers within its fibrous capsule. These muscle fibers are called *intrafusal fibers*, to distinguish them from the more numerous *extrafusal fibers* that lie outside the spindle and form the bulk of the muscle. An important difference between the two types of muscle fibers is that only extrafusal fibers are innervated by alpha motor neurons. Intrafusal fibers receive their motor innervation by another type of lower motor neuron called a **gamma motor neuron** (Figure 13.20).

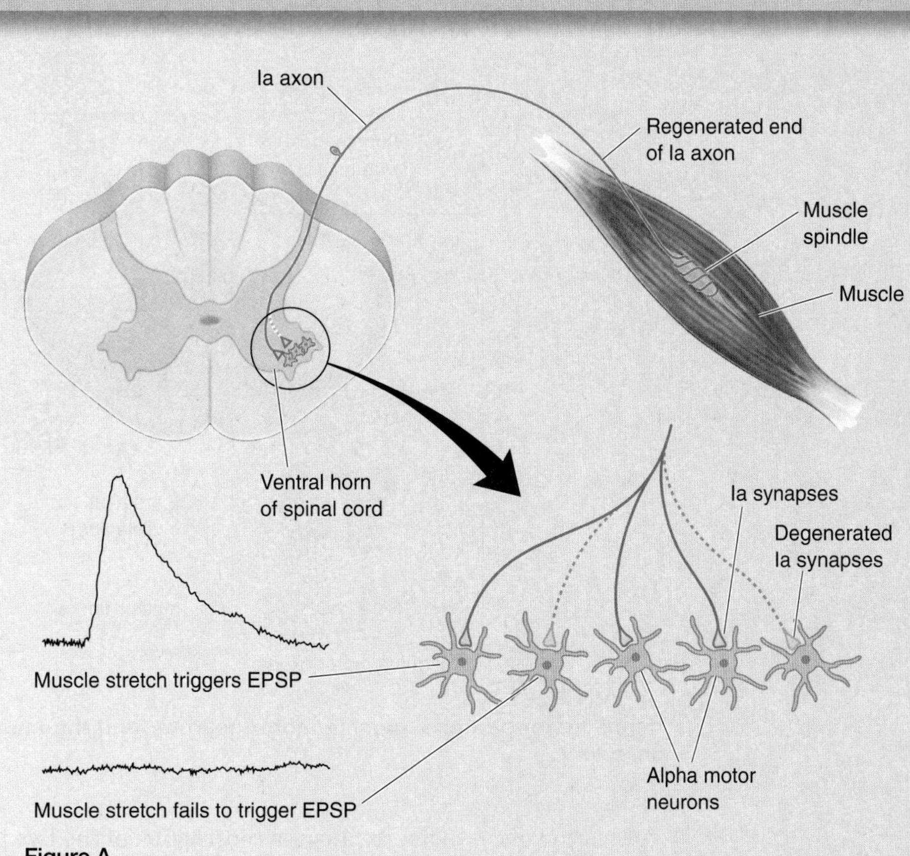

Figure A

What is the value of these findings? The circuits of the stretch reflex play an important role during normal movement by sensing the configuration of the body and limbs and regulating the way they respond to mechanical disturbances. The reorganization of spinal circuits after nerve injury helps us to understand why some movement disorders persist despite the regeneration of axons. Our findings may also apply to circuits beyond the spinal cord. For example, we might consider whether there are similar changes to synapses on corticospinal neurons after disruption of descending motor tracts; this could have implications for therapeutic strategies for dealing with spinal cord injury. Our findings so far strongly motivate us to move forward and learn more about the biological processes that underlie the degeneration of neurons.

Many people participated in these studies, including undergraduate, graduate, and post-doctoral students. Our progress rested on the diverse expertise and hard work of these collaborators. I believe that a team approach is an absolute necessity for tackling the extraordinarily complex functions and malfunctions of the central nervous system. Working collaboratively exposes us to new ideas and promotes our growth as scientists.

REFERENCES

Bullinger KL, Nardelli P, Pinter MJ, Alvarez FJ, Cope TC. 2011. Permanent central synaptic disconnection of proprioceptors after nerve injury and regeneration. II. Loss of functional connectivity with motoneurons. *Journal of Neurophysiology* 106:2471–2485.

Haftel VK, Bichler EK, Wang QB, Prather JF, Pinter MJ, Cope TC. 2005. Central suppression of regenerated proprioceptive afferents. *Journal of Neuroscience* 25:4733–4742.

Imagine a situation in which muscle contraction is commanded by an upper motor neuron. The alpha motor neurons respond, the extrafusal fibers contract, and the muscle shortens. The response of the muscle spindles is shown in Figure 13.21. If they were to become slack, the Ia axons would become silent and the spindle would go “off the air,” no longer providing information about muscle length. This does not happen, however, because the gamma motor neurons are also activated. Gamma motor neurons innervate the intrafusal muscle fiber at the two ends of the muscle spindle.

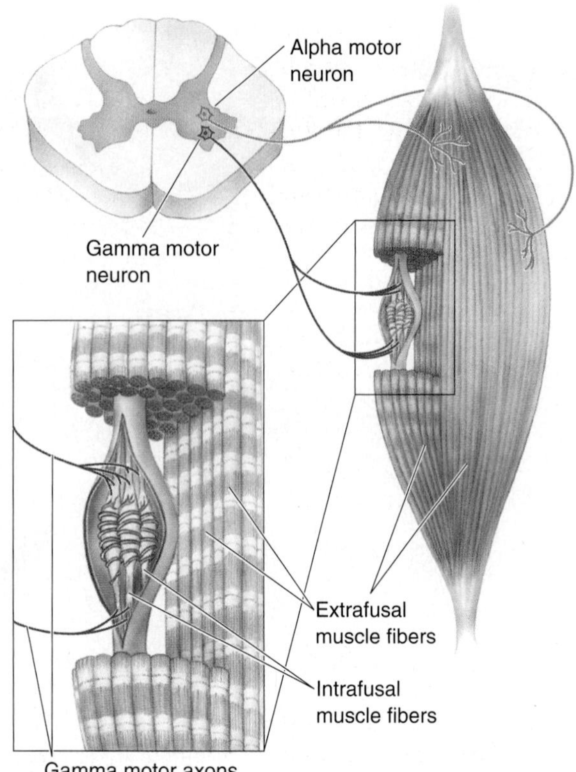

▲ FIGURE 13.20
Alpha motor neurons, gamma motor neurons, and the muscle fibers they innervate.

Activation of these fibers causes a contraction of the two poles of the muscle spindle, thereby pulling on the noncontractile equatorial region and keeping the Ia axons active. Notice that the activation of alpha and gamma motor neurons has opposite effects on Ia output; alpha activation alone decreases Ia activity, while gamma activation alone increases Ia activity.

Recall from our discussion earlier that the monosynaptic stretch reflex arc can be viewed as a feedback loop. The principles of feedback control systems are that a set point is determined (in this case, the desired muscle length), deviations from the set point are detected by a sensor (the Ia axon endings), and deviations are compensated for by an effector system (alpha motor neurons and extrafusal muscle fibers), returning the system to the set point. Changing the activity of the gamma motor neurons

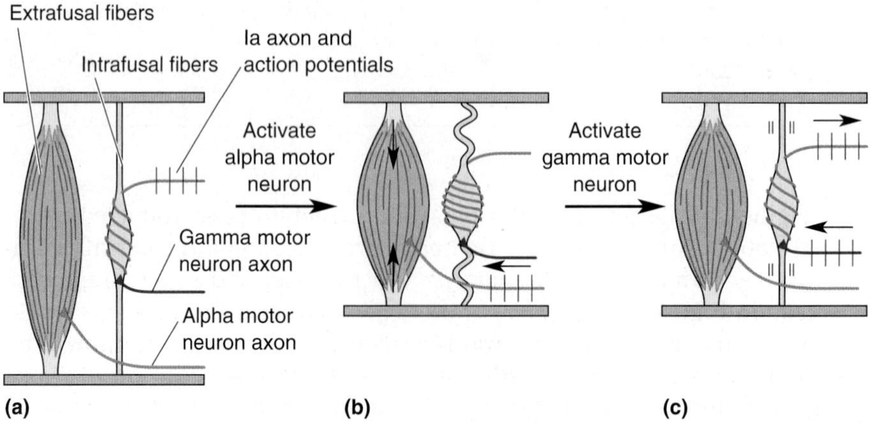

Extrafusal fibers
Intrafusal fibers
Ia axon and action potentials
Activate alpha motor neuron
Gamma motor neuron axon
Alpha motor neuron axon
Activate gamma motor neuron
(a)
(b)
(c)

▶ FIGURE 13.21
The function of gamma motor neurons. (a) Activation of alpha motor neurons causes the extrafusal muscle fibers to shorten. **(b)** If the muscle spindle were to become slack, it would go "off the air" and no longer report the length of the muscle. **(c)** Activation of gamma motor neurons causes the poles of the spindle to contract, keeping it active and "on the air."

changes the set point of the stretch feedback loop. This circuit, gamma motor neuron → intrafusal muscle fiber → Ia afferent axon → alpha motor neuron → extrafusal muscle fibers, is sometimes called the *gamma loop.*

During most normal movements, alpha and gamma motor neurons are simultaneously activated by descending commands from the brain. By regulating the set point of the stretch feedback loop, the gamma loop provides additional control of alpha motor neurons and muscle contraction.

Proprioception from Golgi Tendon Organs

Muscle spindles are not the only source of proprioceptive inputs from the muscles. Another sensor of skeletal muscle is the **Golgi tendon organ**, which acts like a very sensitive strain gauge; that is, it monitors muscle tension, or the force of contraction. Golgi tendon organs are about 1 mm long and 0.1 mm wide. They are located at the junction of the muscle and the tendon and are innervated by group Ib sensory axons, which are slightly smaller than the Ia axons innervating the muscle spindles. Within the Golgi tendon organ, thin branches of the Ib axon entwine among the coils of collagen fibrils (Figure 13.22). When the muscle contracts, the tension on the collagen fibrils increases. As the fibrils straighten and squeeze the Ib axons, their mechanosensitive ion channels are activated and action potentials can be triggered.

It is important to note that while spindles are situated *in parallel* with the muscle fibers, Golgi tendon organs are situated *in series* (Figure 13.23). This different anatomical arrangement helps to determine the types of information these two sensors provide the spinal cord: Ia activity from the spindle encodes *muscle length* information, while Ib activity from the Golgi tendon organ encodes *muscle tension* information.

The Ib axons enter the spinal cord, branch repeatedly, and synapse on special interneurons called *Ib inhibitory interneurons* in the ventral horn. Ib interneurons also receive inputs from other sensory receptors and from descending pathways. Some of the Ib interneurons form inhibitory connections with the alpha motor neurons innervating the same muscle (Figure 13.24). This is the basis for another spinal reflex. In extreme circumstances, this Ib

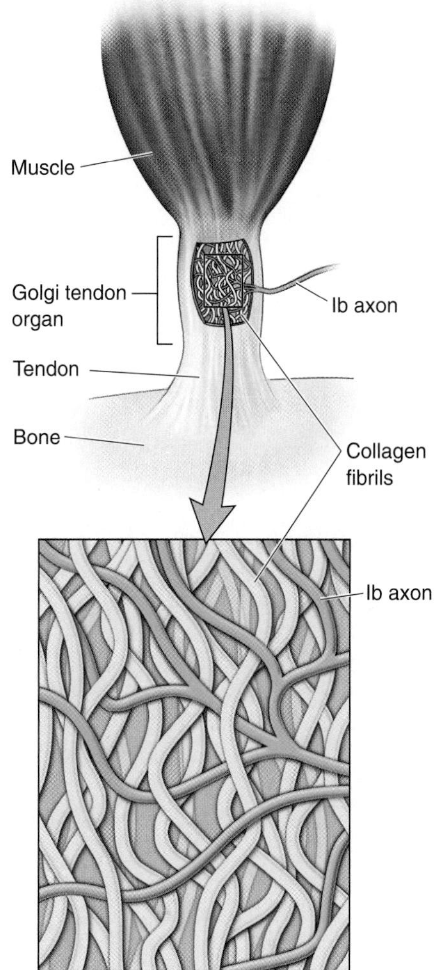

▲ FIGURE 13.22
A Golgi tendon organ.

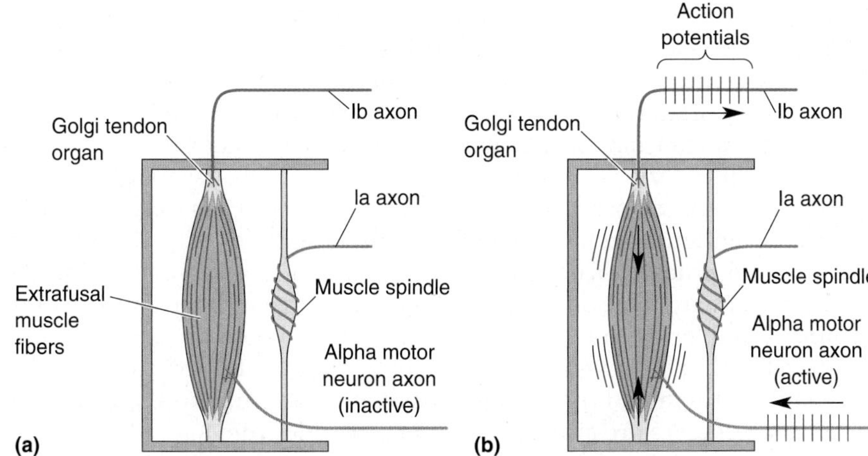

▲ FIGURE 13.23
The organization of muscle proprioceptors. (a) Muscle spindles are arranged parallel to the extrafusal fibers; Golgi tendon organs lie in series, between the muscle fibers and their points of attachment. **(b)** Golgi tendon organs respond to increased tension on the muscle and transmit this information to the spinal cord via group Ib sensory axons. Because the activated muscle does not change length, the Ia axons remain silent in this example.

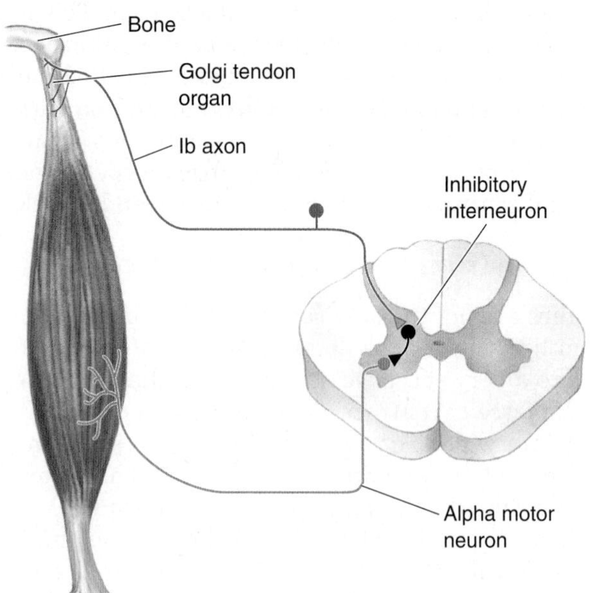

▲ FIGURE 13.24
Golgi tendon organ circuit. The Ib axon of the Golgi tendon organ excites an inhibitory interneuron, which inhibits the alpha motor neurons of the same muscle.

reflex arc may protect the muscle from being overloaded. However, its normal function is to regulate muscle tension within an optimal range. As muscle tension increases, the inhibition of the alpha motor neuron slows muscle contraction; as muscle tension falls, the inhibition of the alpha motor neuron is reduced, and muscle contraction increases. This type of proprioceptive feedback is thought to be particularly important for the proper execution of fine motor acts, such as the manipulation of fragile objects with the hands, which require a steady, but not too powerful, grip.

Proprioception from the Joints. We have focused on the proprioceptors that are involved in reflex control of the spinal motor neurons. However, besides muscle spindles and Golgi tendon organs, a variety of proprioceptive axons are present in the connective tissues of joints, especially within the fibrous tissue surrounding the joints (joint capsules) and ligaments. These mechanosensitive axons respond to changes in the angle, direction, and velocity of movement in a joint. Most are rapidly adapting, meaning that sensory information about a *moving* joint is plentiful, but nerves encoding the *resting* position of a joint are few. We are, nevertheless, quite good at judging the position of a joint, even with our eyes closed. It seems that information from joint receptors, muscle spindles, and Golgi tendon organs, and probably from receptors in the skin, is combined within the CNS to estimate joint angle. Removing one source of information can be compensated for by the use of the other sources. When an arthritic hip is replaced with a steel and plastic one, patients can still tell the angle between their thigh and their pelvis, despite the fact that all their hip joint mechanoreceptors are sitting in a jar of formaldehyde.

Spinal Interneurons

The actions of Ib inputs from Golgi tendon organs on alpha motor neurons are entirely *polysynaptic*; they are all mediated by intervening spinal interneurons. Indeed, most of the input to the alpha motor neurons comes from interneurons of the spinal cord. Spinal interneurons receive synaptic input from primary sensory axons, descending axons from the brain, and collaterals of lower motor neuron axons. The interneurons are themselves networked together in a way that allows coordinated motor programs to be generated in response to their many inputs.

Inhibitory Input. Interneurons play a critical role in the proper execution of even the simplest reflexes. Consider the stretch reflex, for example. Compensation for the lengthening of one set of muscles, such as the flexors of the elbow, involves contraction of the flexors via the stretch reflex but also requires relaxation of the antagonist muscles, the extensors. This process is called **reciprocal inhibition**, the contraction of one set of muscles accompanied by the relaxation of their antagonist muscles. The importance of this is obvious; imagine how hard it would be to lift something by contracting your biceps if its antagonist muscles (e.g., your triceps) were constantly opposing you. In the case of the stretch reflex, reciprocal inhibition occurs because collaterals of the Ia axons synapse on inhibitory spinal interneurons that contact the alpha motor neurons supplying the antagonist muscles (Figure 13.25).

Reciprocal inhibition is also used by descending pathways from the brain to overcome the powerful stretch reflex. Consider a situation in which the flexors of the elbow are voluntarily commanded to contract. You might expect the resulting stretch of the antagonist extensor muscles to activate their stretch reflex arc, which would strongly resist flexion of the joint. However, the descending pathways that activate the alpha motor neurons controlling the flexors also activate interneurons, which inhibit the alpha motor neurons that supply the antagonist muscles.

Excitatory Input. Not all interneurons are inhibitory. An example of a reflex mediated in part by excitatory interneurons is the *flexor reflex*, sometimes called the *flexor withdrawal reflex* (Figure 13.26). This is a complex, polysynaptic reflex arc used to withdraw a limb from an aversive stimulus (such as the withdrawal of your foot from the thumbtack in Chapter 3). The flexor reflex is remarkably specific. The speed of withdrawal depends on how painful the stimulus is. The direction of withdrawal depends on the location of the stimulus; for example, hot stimuli applied to your palm and to the back of your hand trigger withdrawals in opposite directions (as you would hope!).

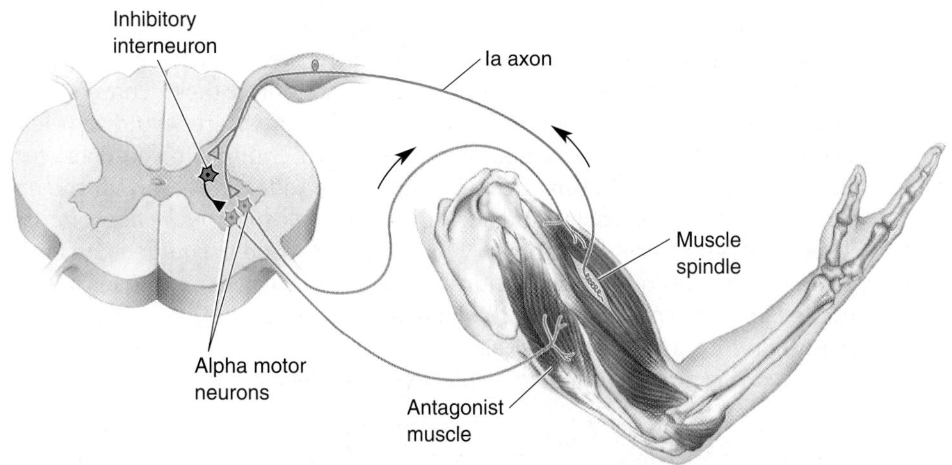

▲ FIGURE 13.25
Reciprocal inhibition of flexors and extensors of the same joint.

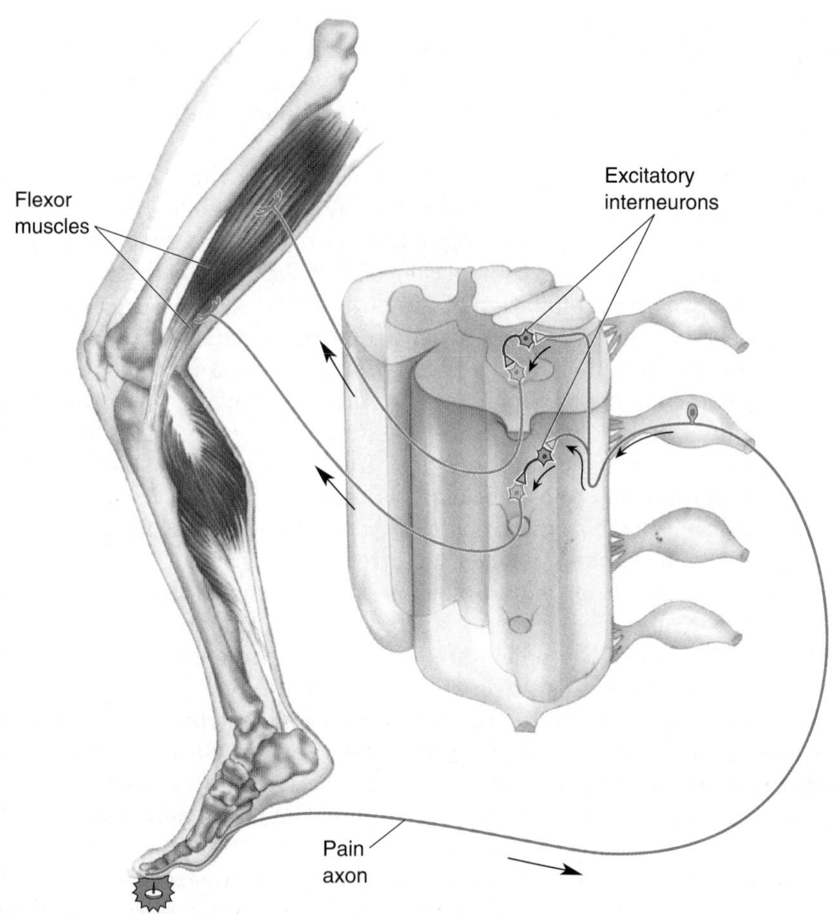

▲ FIGURE 13.26
Circuitry of the flexor withdrawal reflex.

The flexor reflex is far slower than the stretch reflex, indicating that a number of interneurons intervene between the sensory stimulus and the coordinated motor act. The flexor reflex is activated by the small, myelinated Aδ nociceptive axons that trigger pain (see Chapter 12). The nociceptive axons entering the spinal cord branch profusely and activate interneurons in several different spinal segments. These cells eventually excite the alpha motor neurons that control all the flexor muscles of the affected limb (and, needless to say, inhibitory interneurons are also recruited to inhibit the alpha motor neurons that control the extensors).

You're walking around barefoot, and you step on a tack. Thanks to the flexor reflex, you reflexively yank your foot up. But where would that leave the rest of your body if nothing else happened? Falling to the floor, most likely. Luckily, an additional component of the reflex is recruited: the activation of extensor muscles and the inhibition of flexors *on the opposite side.* This is called the *crossed-extensor reflex,* and it is used to compensate for the extra load imposed by limb withdrawal on the antigravity extensor muscles of the opposite leg (Figure 13.27). Notice that this is another example of reciprocal inhibition, but in this case, activation of the flexors on one side of the spinal cord is accompanied by inhibition of the flexors on the opposite side.

The Generation of Spinal Motor Programs for Walking

The crossed-extensor reflex, in which one limb extends as the other limb flexes, seems to provide a building block for locomotion. When you walk,

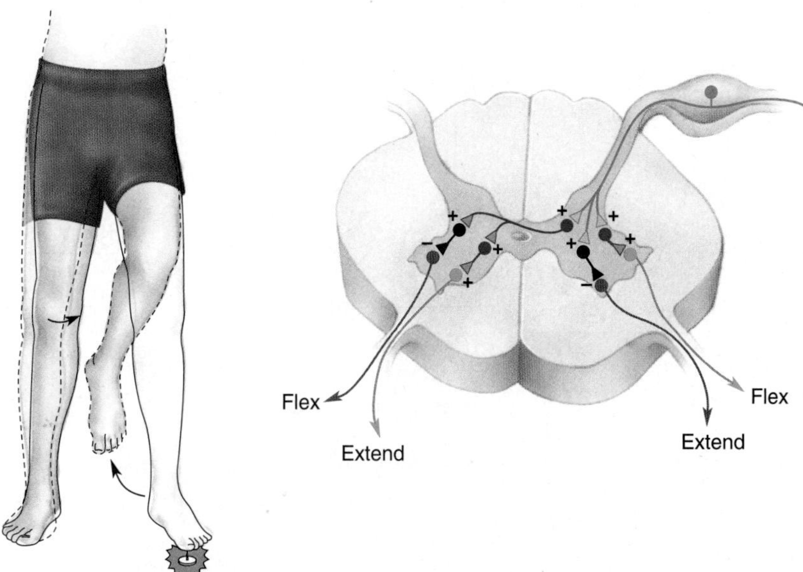

◀ FIGURE 13.27
Circuitry of the crossed-extensor reflex.

you alternately withdraw and extend your two legs. All that's lacking is a mechanism to coordinate the timing. In principle, this could be a series of descending commands from upper motor neurons. However, as we already suspected from our consideration of headless chicken behavior, it seems likely that this control is exerted from within the spinal cord. Indeed, a complete transection of a cat's spinal cord at the mid-thoracic level leaves the hind limbs capable of generating coordinated walking movements. The circuit for the coordinated control of walking must reside, therefore, within the spinal cord. In general, circuits that give rise to rhythmic motor activity are called **central pattern generators**.

How do neural circuits generate rhythmic patterns of activity? Different circuits use different mechanisms. However, the simplest pattern generators are individual neurons whose membrane properties endow them with pacemaker properties. An interesting example comes from the work of Sten Grillner and his colleagues in Stockholm, Sweden. Based on the assumption that the spinal central pattern generators for locomotion in different species are variations on a plan that was established in a common ancestor, Grillner focused on the mechanism for swimming in the lamprey, a jawless fish that has evolved slowly over the course of the past 450 million years. Lampreys swim by undulating their elongated bodies. They lack limbs and even pairs of fins, but the coordinated rhythmic contractions of their body muscles during swimming closely resemble the contraction patterns necessary for terrestrial animals to walk.

The lamprey spinal cord can be dissected and kept alive *in vitro* for several days. Electrical stimulation of the stumps of axons descending from the brain can generate alternating rhythmic activity in the spinal cord, mimicking that which occurs during swimming. In an important series of experiments, Grillner showed that the activation of NMDA receptors on spinal interneurons was sufficient to generate this locomotor activity.

Recall from Chapter 6 that NMDA receptors are glutamate-gated ion channels with two peculiar properties: (1) They allow more current to flow into the cell when the postsynaptic membrane is depolarized, and (2) they admit Ca^{2+} as well as Na^{+} into the cell. In addition to NMDA receptors, spinal interneurons possess calcium-activated potassium channels.

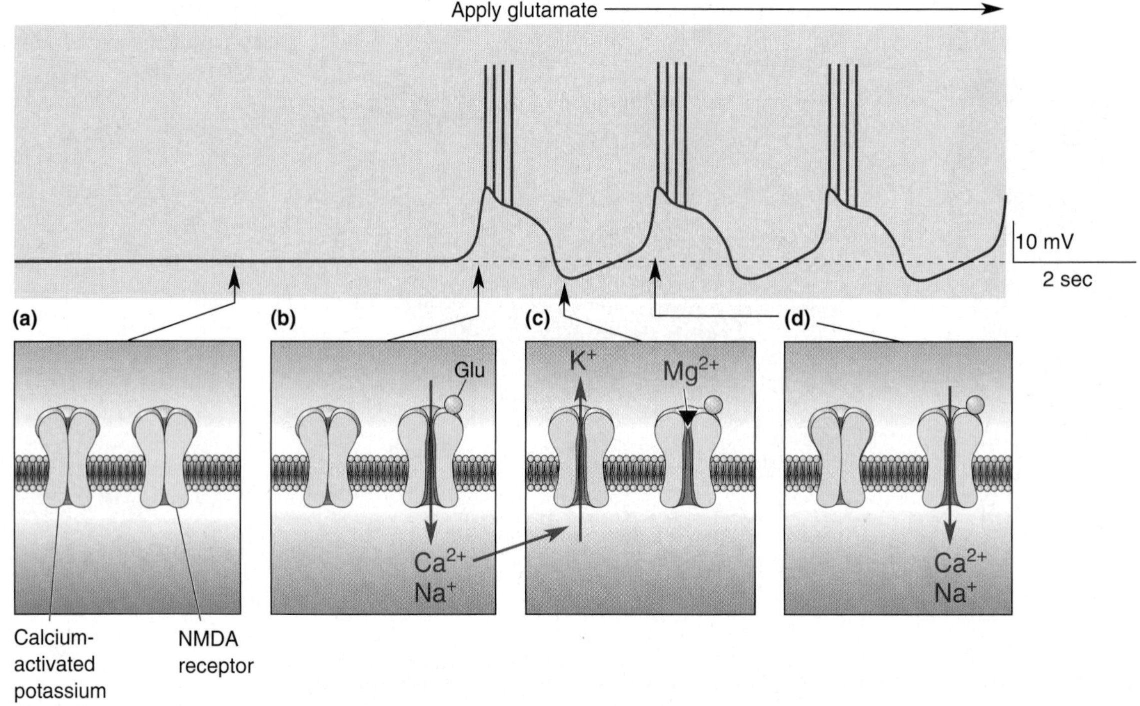

▲ FIGURE 13.28

Rhythmic activity in a spinal interneuron. Some neurons respond to the activation of NMDA receptors with rhythmic depolarization. **(a)** In the resting state, the NMDA receptor channels and the calcium-activated potassium channels are closed. **(b)** Glutamate causes the NMDA receptors to open, the cell membrane to depolarize, and Ca^{2+} to enter the cell. **(c)** The rise in intracellular [Ca^{2+}] causes the Ca^{2+}-activated potassium channels to open. Potassium ions leave the neuron, hyperpolarizing the membrane. The hyperpolarization allows Mg^{2+} to enter and clog the NMDA channel, arresting the flow of Ca^{2+}. **(d)** As [Ca^{2+}] falls, the potassium channels close, resetting the membrane for another oscillation. (Source: Adapted from Wallen and Grillner, 1987.)

Now imagine the cycle that is initiated when NMDA receptors are activated by glutamate (Figure 13.28):

1. The membrane depolarizes.
2. Na^+ and Ca^{2+} flow into the cell through the NMDA receptors.
3. Ca^{2+} activates potassium channels.
4. K^+ flows out of the cell.
5. The membrane hyperpolarizes.
6. Ca^{2+} stops flowing into the cell.
7. The potassium channels close.
8. The membrane depolarizes, and the cycle repeats.

It is easy to imagine how intrinsic pacemaker activity in spinal interneurons might act as the primary rhythmic driving force for sets of motor neurons that in turn command cyclic behaviors like walking. However, pacemaker neurons are not solely responsible for generating rhythms in vertebrates. They are embedded within interconnected circuits, and it is the combination of intrinsic pacemaker properties and synaptic interconnections that produces rhythm.

An example of a possible pattern-generating circuit for walking is shown in Figure 13.29. According to this scheme, walking is initiated when a steady input excites two interneurons that connect to the motor

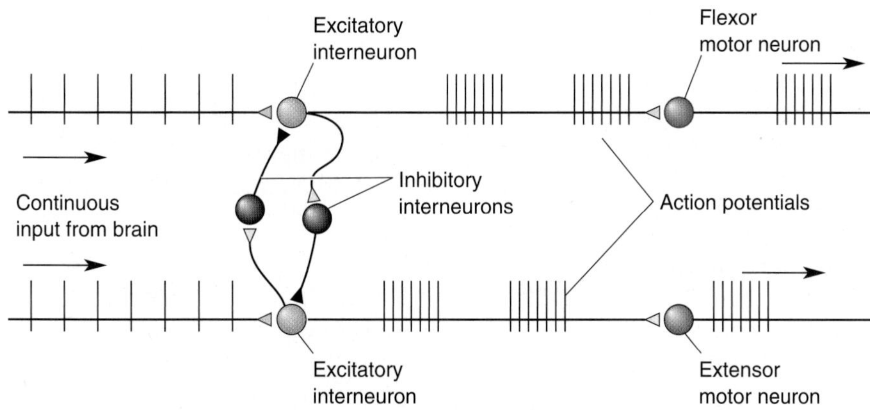

▲ FIGURE 13.29
A possible circuit for rhythmic alternating activity.

neurons controlling the flexors and extensors, respectively. The interneurons respond to a continuous input by generating bursts of outputs (see Figure 13.28). The activity of these two interneurons alternates because they inhibit each other via another set of interneurons, which are inhibitory. Thus, a burst of activity in one interneuron strongly inhibits the other, and vice versa. Then, using the spinal cord circuitry of the crossed-extensor reflex (or a similar circuit), the movements of the opposite limb could be coordinated so that flexion on one side is accompanied by extension on the other. The addition of more interneuronal connections between the lumbar and cervical spinal segments could account for the swinging of the arms that accompanies walking, or the coordination of forelimbs and hind limbs in four-legged animals.

Work on many vertebrate species, from lampreys to humans, has shown that locomotor activity in the spinal cord and its coordination depend on multiple mechanisms. Such complexity is not surprising when we consider the demands on the system—for example, the adjustments necessary when one foot strikes an obstacle while walking, or the changes in output that are necessary to walk forward or backward, or to go from walking, to jogging, to running, to jumping.

CONCLUDING REMARKS

We can draw several conclusions from the preceding discussion of the spinal control of movement. First, a great deal has been learned about movement and its spinal control by working at different levels of analysis, ranging from biochemistry and genetics to biophysics and behavior. Indeed, a complete understanding, whether of excitation–contraction coupling or central pattern generation, requires knowledge derived from every approach. Second, sensation and movement are inextricably linked even at the lowest levels of the neural motor system. The normal function of the alpha motor neuron depends on direct feedback from the muscles themselves and indirect information from the tendons, joints, and skin. Third, the spinal cord contains an intricate network of circuits for the control of movement; it is far more than just a conduit for somatic sensory and motor information.

Evidently, coordinated and complex patterns of activity in these spinal circuits can be driven by relatively crude descending signals. This leaves the question of precisely what the upper motor neurons contribute to motor control—the subject of the next chapter.

KEY TERMS

Introduction
motor system (p. 454)

The Somatic Motor System
smooth muscle (p. 454)
striated muscle (p. 454)
cardiac muscle (p. 454)
skeletal muscle (p. 454)
muscle fibers (p. 454)
somatic motor system (p. 455)
flexion (p. 455)
extension (p. 455)
flexors (p. 456)
synergists (p. 456)
extensors (p. 456)
antagonists (p. 456)
axial muscles (p. 456)
proximal (girdle) muscles (p. 456)
distal muscles (p. 456)

The Lower Motor Neuron
alpha motor neurons (p. 458)
motor unit (p. 458)
motor neuron pool (p. 459)
slow motor unit (p. 461)
fast motor unit (p. 461)

Excitation–Contraction Coupling
excitation–contraction coupling (p. 464)
sarcolemma (p. 464)
myofibrils (p. 465)
sarcoplasmic reticulum (p. 465)
T tubules (p. 465)
sarcomere (p. 466)
thin filaments (p. 466)
thick filaments (p. 466)
myosin (p. 467)
actin (p. 467)

Spinal Control of Motor Units
muscle spindles (p. 469)
proprioceptors (p. 469)
proprioception (p. 469)
stretch reflex (p. 470)
gamma motor neuron (p. 472)
Golgi tendon organ (p. 475)
reciprocal inhibition (p. 477)
central pattern generators (p. 479)

REVIEW QUESTIONS

1. What did Sherrington call the "final common pathway," and why?
2. Define, in one sentence, motor unit. How does it differ from motor neuron pool?
3. Which is recruited first, a fast motor unit or a slow motor unit? Why?
4. When and why does rigor mortis occur?
5. Your doctor taps the tendon beneath your kneecap and your leg extends. What is the neural basis of this reflex? What is it called?
6. What is the function of gamma motor neurons?
7. Lenny, a character in Steinbeck's classic book *Of Mice and Men*, loved rabbits, but when he hugged them, they were crushed to death. Which type of proprioceptive input might Lenny have been lacking?

FURTHER READING

Kernell D. 2006. *The Motoneurone and its Muscle Fibres*. New York: Oxford University Press.

Lieber RL. 2002. *Skeletal Muscle Structure, Function, and Plasticity*, 2nd ed. Baltimore: Lippincott, Williams & Wilkins.

Poppele R, Bosco G. 2003. Sophisticated spinal contributions to motor control. *Trends in Neurosciences* 26:269–276.

Schouenborg J, Kiehn O, eds. 2001. The Segerfalk symposium on principles of spinal cord function, plasticity, and repair. *Brain Research Reviews* 40:1–329.

Stein PSG, Grillner S, Selverston AI, Stuart DG, eds. 1999. *Neurons, Networks, and Motor Behavior*. Cambridge, MA: MIT Press.

Windhorst U. 2007. Muscle proprioceptive feedback and spinal networks. *Brain Research Bulletin* 73:155–202.

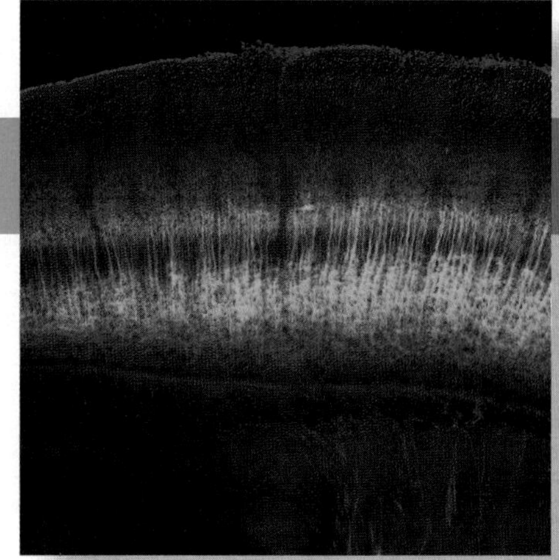

CHAPTER FOURTEEN

Brain Control of Movement

INTRODUCTION

In Chapter 13, we discussed the organization of the peripheral somatic motor system: the joints, skeletal muscles, and their sensory and motor innervation. We saw that the final common pathway for behavior is the alpha motor neuron, that the activity of this cell is under the control of sensory feedback and spinal interneurons, and that reflex movements reveal the complexity of this spinal control system. In this chapter, we'll explore how the brain influences the activity of the spinal cord to command voluntary movements.

The central motor system is arranged as a hierarchy of control levels, with the forebrain at the top and the spinal cord at the bottom. It is useful to think of this motor control hierarchy as having three levels (Table 14.1). The highest level, represented by the association areas of neocortex and basal ganglia of the forebrain, is concerned with *strategy*: the goal of the movement and the movement strategy that best achieves the goal. The middle level, represented by the motor cortex and cerebellum, is concerned with *tactics:* the sequences of muscle contractions, arranged in space and time, required to smoothly and accurately achieve the strategic goal. The lowest level, represented by the brain stem and spinal cord, is concerned with *execution*: activation of the motor neuron and interneuron pools that generate the goal-directed movement and make any necessary adjustments of posture.

To appreciate the different contributions of the three hierarchical levels to movement, consider the actions of a baseball pitcher preparing to pitch to a batter (Figure 14.1). The cerebral neocortex has information—based on vision, audition, somatic sensation, and proprioception—about precisely where the body is in space. Strategies must be devised to move the body from the current state to one in which a pitch is delivered and the desired outcome is attained (a swing and a miss by the batter). Several throwing options are available to the pitcher—a curve ball, a fast ball, a slider, and so on—and these alternatives are filtered through the basal ganglia and back to the cortex until a final decision is made, based in large part on past experience (e.g., "This batter hit a home run last time I threw a fast ball"). The motor areas of cortex and the cerebellum then make the tactical decision (to throw a curve ball) and issue instructions to the brain stem and spinal cord. Activation of neurons in the brain stem and spinal cord then causes the movement to be executed. The properly timed activation of motor neurons in the cervical spinal cord generates a coordinated movement of the shoulder, elbow, wrist, and fingers. Simultaneously, brain stem inputs to the thoracic and lumbar spinal cord command the appropriate leg movements along with postural adjustments that keep the pitcher from falling over during the throw. In addition, brain stem motor neurons are activated to keep the pitcher's eyes fixed on the catcher, his target, as his head and body move about.

According to the laws of physics, the movement of a thrown baseball through space is *ballistic*, referring to a trajectory that cannot be altered.

TABLE 14.1 **The Motor Control Hierarchy**

Level	Function	Structures
High	Strategy	Association areas of neocortex, basal ganglia
Middle	Tactics	Motor cortex, cerebellum
Low	Execution	Brain stem, spinal cord

◀ FIGURE 14.1
The contributions of the motor control hierarchy. As a baseball pitcher plans to pitch a ball to a batter, chooses which pitch to throw, and then throws the ball, he engages the three hierarchical levels of motor control.

The movement of the pitcher's arm that throws the ball is also described as ballistic because it cannot be altered once initiated. This type of rapid voluntary movement is not under the same type of sensory feedback control that regulates antigravity postural reflexes (see Chapter 13). The reason is simple: The movement is too fast to be adjusted by sensory feedback. But the movement does not occur in the absence of sensory information. Sensory information *before* the movement was initiated was crucial in order to decide when to initiate the pitch, to determine the starting positions of the limbs and body, and to anticipate any changes in resistance during the throw. And sensory information *during* the movement is also important, not necessarily for the movement at hand, but for improving subsequent similar movements.

The proper functioning of each level of the motor control hierarchy relies so heavily on sensory information that the motor system of the brain might properly be considered a *sensorimotor system*. At the highest level, sensory information generates a mental image of the body and its relationship to the environment. At the middle level, tactical decisions are based on the memory of sensory information from past movements. At the lowest level, sensory feedback is used to maintain posture, muscle length, and tension before and after each voluntary movement.

In this chapter, we investigate this hierarchy of motor control and how each level contributes to the control of the peripheral somatic motor system. We start by exploring the pathways that bring information to the spinal motor neurons. From there we will ascend to the highest levels of the motor hierarchy, and then we'll fill in the pieces of the puzzle that bring the different levels together. Along the way, we'll describe how pathology in specific parts of the motor system leads to particular movement disorders.

DESCENDING SPINAL TRACTS

How does the brain communicate with the motor neurons of the spinal cord? Axons from the brain descend through the spinal cord along two major groups of pathways, shown in Figure 14.2. One is in the lateral column of the spinal cord, and the other is in the ventromedial column. Remember this rule of thumb: The **lateral pathways** are involved in voluntary movement of the distal musculature and are under direct cortical control, and the **ventromedial pathways** are involved in the control of posture and locomotion and are under brain stem control.

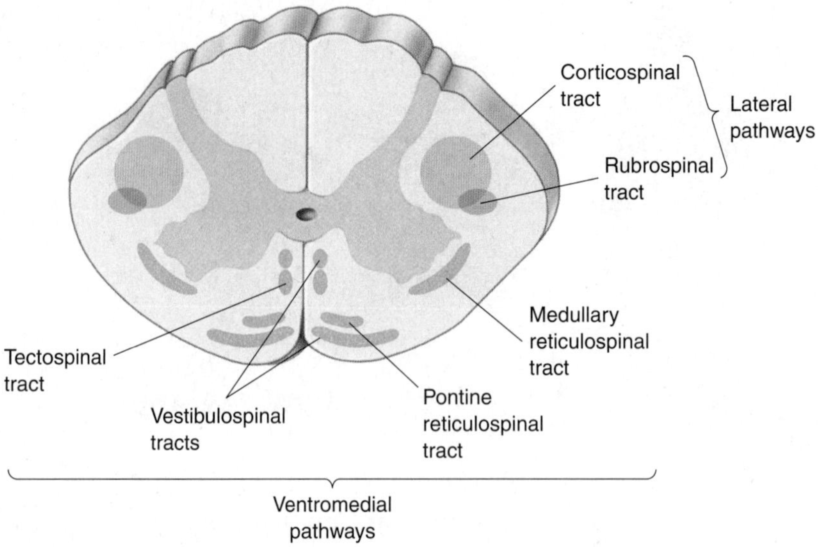

▶ FIGURE 14.2
The descending tracts of the spinal cord. The lateral pathways, consisting of the corticospinal and rubrospinal tracts, control voluntary movements of the distal musculature. The ventromedial pathways, consisting of the medullary reticulospinal, pontine reticulospinal, vestibulospinal, and tectospinal tracts, control postural muscles.

The Lateral Pathways

The most important component of the lateral pathways is the **corticospinal tract** (Figure 14.3a). Originating in the neocortex, it is the longest and one of the largest central nervous system (CNS) tracts (10^6 axons in humans). Two-thirds of the axons in the tract originate in areas 4 and 6 of the frontal lobe, collectively called **motor cortex**. Most of the remaining axons in the corticospinal tract derive from the somatosensory areas of the parietal lobe and serve to regulate the flow of somatosensory information to the brain (see Chapter 12). Axons from the cortex pass through the internal capsule bridging the telencephalon and thalamus, course through the base of the *cerebral peduncle*, a large collection of axons in the midbrain, then pass through the pons, and collect to form a tract at the base of the medulla. The tract forms a bulge, called the *medullary pyramid*, running down the ventral surface of the medulla. When cut, the tract's cross section is roughly triangular, explaining why it is also called the **pyramidal tract**.

At the junction of the medulla and spinal cord, the pyramidal tract crosses, or decussates, at the pyramidal decussation. This means that the *right* motor cortex directly commands the movement of the *left* side of the body, and the *left* motor cortex controls the muscles on the *right* side. As the axons cross, they collect in the lateral column of the spinal cord and form the lateral corticospinal tract. The corticospinal tract axons terminate in the dorsolateral region of the ventral horns and intermediate gray matter, the location of the motor neurons and interneurons that control the distal muscles, particularly the flexors (see Chapter 13).

A much smaller component of the lateral pathways is the **rubrospinal tract**, which originates in the **red nucleus** of the midbrain, named for its distinctive pinkish hue in a freshly dissected brain (*rubro* is from the Latin for "red"). Axons from the red nucleus decussate in the pons, almost immediately, and parallel those in the corticospinal tract in the lateral column of the spinal cord (Figure 14.3b). A major source of input to the red nucleus is the very region of frontal cortex that also contributes to the corticospinal tract. Indeed, it appears that this indirect corticorubrospinal pathway has largely been replaced by the direct corticospinal path over the course of primate evolution Thus, while the rubrospinal tract contributes importantly to motor control in many mammalian species, in humans it appears to be reduced, most of its functions subsumed by the corticospinal tract.

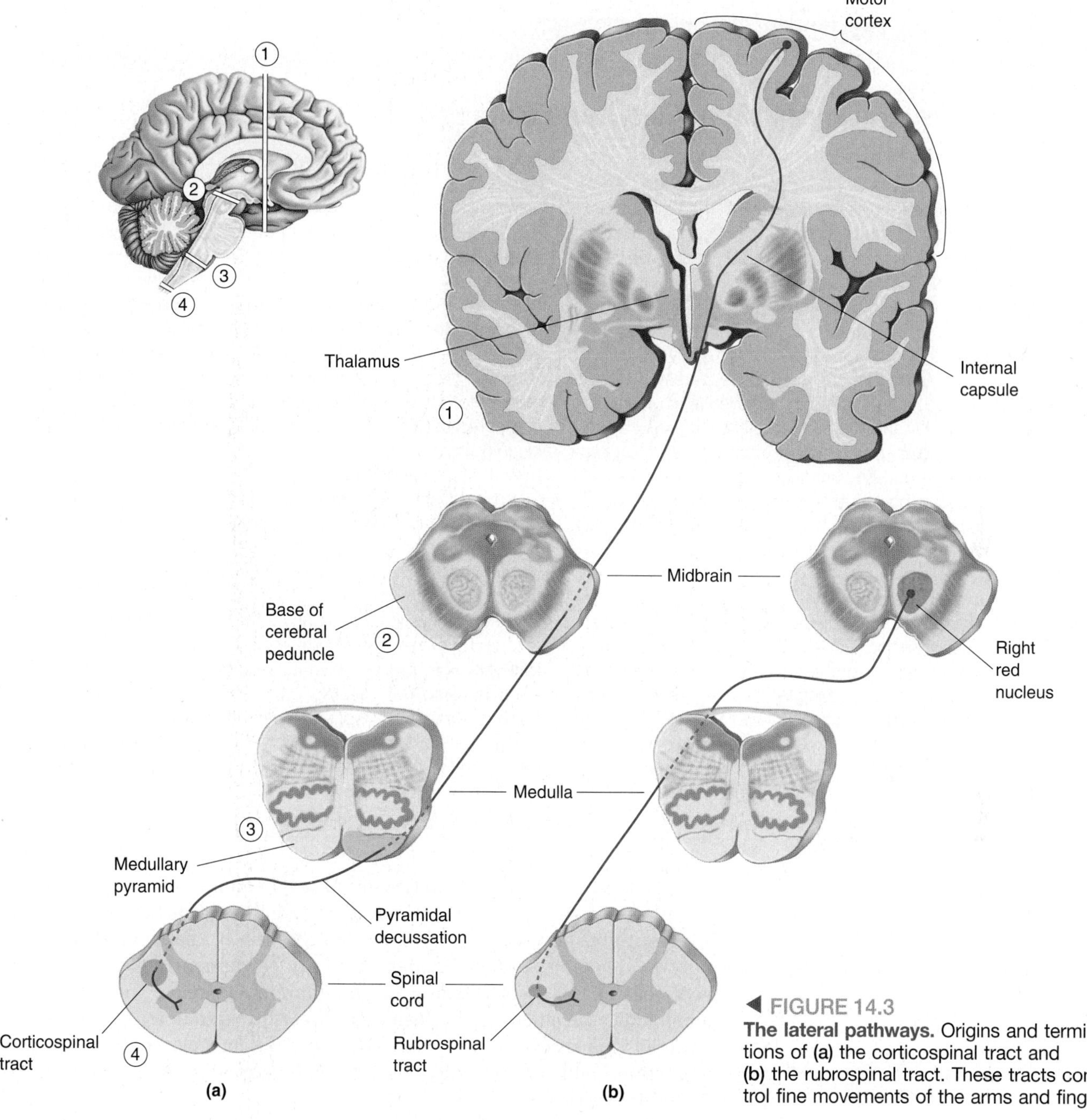

◀ FIGURE 14.3
The lateral pathways. Origins and terminations of **(a)** the corticospinal tract and **(b)** the rubrospinal tract. These tracts control fine movements of the arms and fingers.

The Effects of Lateral Pathway Lesions. Donald Lawrence and Hans Kuypers laid the foundation for the modern view of the functions of the lateral pathways in the late 1960s. Experimental lesions in both corticospinal and rubrospinal tracts in monkeys rendered them unable to make fractionated movements of the arms and hands; that is, they could not move their shoulders, elbows, wrists, and fingers independently. For example, they could grasp small objects with their hands but only by using all the fingers at once. Voluntary movements were also slower and less accurate. Despite this, the animals could sit upright and stand with normal posture. By analogy, a human with a lateral pathway lesion would be able to stand on the pitcher's mound but would be unable to grip the ball properly and throw it accurately.

BOX 14.1 OF SPECIAL INTEREST

Paresis, Paralysis, Spasticity, and Babinski

The neural components of the motor system extend from the highest reaches of the cerebral cortex to the farthest terminals of the motor axons in muscles. Its sheer size makes the motor system uncommonly vulnerable to disease and trauma. The site of motor system damage has a big effect on the types of deficits patients experience.

Damage to the lower parts of the motor system—alpha motor neurons or their motor axons—leads to easily predicted consequences. Partial damage may cause *paresis* (weakness). Complete severing of a motor nerve leads to *paralysis,* a loss of movement of the affected muscles, and *areflexia*, an absence of their spinal reflexes. The muscles also have no *tone* or resting tension; they are flaccid and soft. Damaged motor neurons can no longer exert their trophic influence on muscle fibers (see Chapter 13). The muscles profoundly *atrophy* (decrease in size) with time, losing up to 70–80% of their mass.

Damage to the upper parts of the motor system—the motor cortex or the various motor tracts that descend into the spinal cord—can cause a distinctly different set of motor problems. These are common after a stroke, which damages regions of the cortex or brain stem by depriving them of their blood supply, or traumatic injury, such as a knife or gunshot wound, or even a demyelinating disease that damages axons (see Box 4.5).

Immediately following severe upper motor system damage, there is a period of *spinal shock*: reduced muscle tone (*hypotonia*), areflexia, and paralysis. Paralysis is known as *hemiplegia* if it occurs on one side of the body, *paraplegia* if it involves only the legs, and *quadriplegia* if it involves all four limbs. With the loss of descending brain influences, the functions of the spinal cord appear to shut down. Over the next several days, some of its reflexive functions mysteriously reappear. This is not necessarily a good thing. A condition called *spasticity* sets in, often permanently. Spasticity is characterized by a dramatic and sometimes painful increase of muscle tone (*hypertonia*) and spinal reflexes (*hyperreflexia*), compared to normal. Overactive stretch reflexes often cause *clonus*, rhythmic cycles of contraction and relaxation when limb muscles are stretched.

Another indication of motor tract damage is the *Babinski sign,* described by the French neurologist Joseph Babinski in 1896. Sharply scratching the sole of the foot from the heel toward the toes causes reflexive upward flexion of the big toe and an outward fanning of the other toes. The normal response to this stimulus, for anyone older than about 2 years, is to curl the toes downward. Normal infants also exhibit the Babinski sign, presumably because their descending motor tracts have not yet matured.

By systematically testing a patient's reflexes, muscle tone, and motor ability across his body, a skilled neurologist can often deduce the site and severity of motor system damage with impressive precision.

Lesions in the monkeys' corticospinal tracts alone caused a movement deficit as severe as that observed after lesions in the lateral columns. Interestingly, however, many functions gradually reappeared over the months following surgery. In fact, the only permanent deficit was some weakness of the distal flexors and an inability to move the fingers independently. A subsequent lesion in the rubrospinal tract completely reversed this recovery, however. These results suggest that the corticorubrospinal pathway was able, over time, to partially compensate for the loss of the corticospinal tract input.

Strokes that damage the motor cortex or the corticospinal tract are common in humans. Their immediate consequence can be paralysis on the contralateral side, but considerable recovery of voluntary movements may occur over time (Box 14.1). As in Lawrence and Kuypers' lesioned monkeys, it is the fine, fractionated movements of the fingers that are least likely to recover.

The Ventromedial Pathways

The ventromedial pathways contain four descending tracts that originate in the brain stem and terminate among the spinal interneurons controlling proximal and axial muscles. These tracts are the vestibulospinal tract, the tectospinal tract, the pontine reticulospinal tract, and the medullary reticulospinal tract. The ventromedial pathways use sensory information about balance, body position, and the visual environment to reflexively maintain balance and body posture.

The Vestibulospinal Tracts. The vestibulospinal and tectospinal tracts function to keep the head balanced on the shoulders as the body moves through space and to turn the head in response to new sensory stimuli. The **vestibulospinal tracts** originate in the *vestibular nuclei* of the medulla, which relay sensory information from the vestibular labyrinth in the inner ear (Figure 14.4a). The *vestibular labyrinth* consists of fluid-filled canals and cavities in the temporal bone that are closely associated with the cochlea (see Chapter 11). The motion of the fluid in this labyrinth, which accompanies movement of the head, activates hair cells that signal the vestibular nuclei via cranial nerve VIII.

One component of the vestibulospinal tracts projects bilaterally down the spinal cord and activates the cervical spinal circuits that control neck and back muscles and thus guide head movement. Stability of the head is important because the head contains our eyes, and keeping the eyes stable, even as our body moves, ensures that our image of the world remains stable. Another component of the vestibulospinal tracts projects ipsilaterally as far down as the lumbar spinal cord. It helps us maintain an upright and balanced posture by facilitating extensor motor neurons of the legs.

The Tectospinal Tract. The **tectospinal tract** originates in the superior colliculus of the midbrain, which receives direct input from the retina (Figure 14.4b). (Recall from Chapter 10 that optic tectum is another name

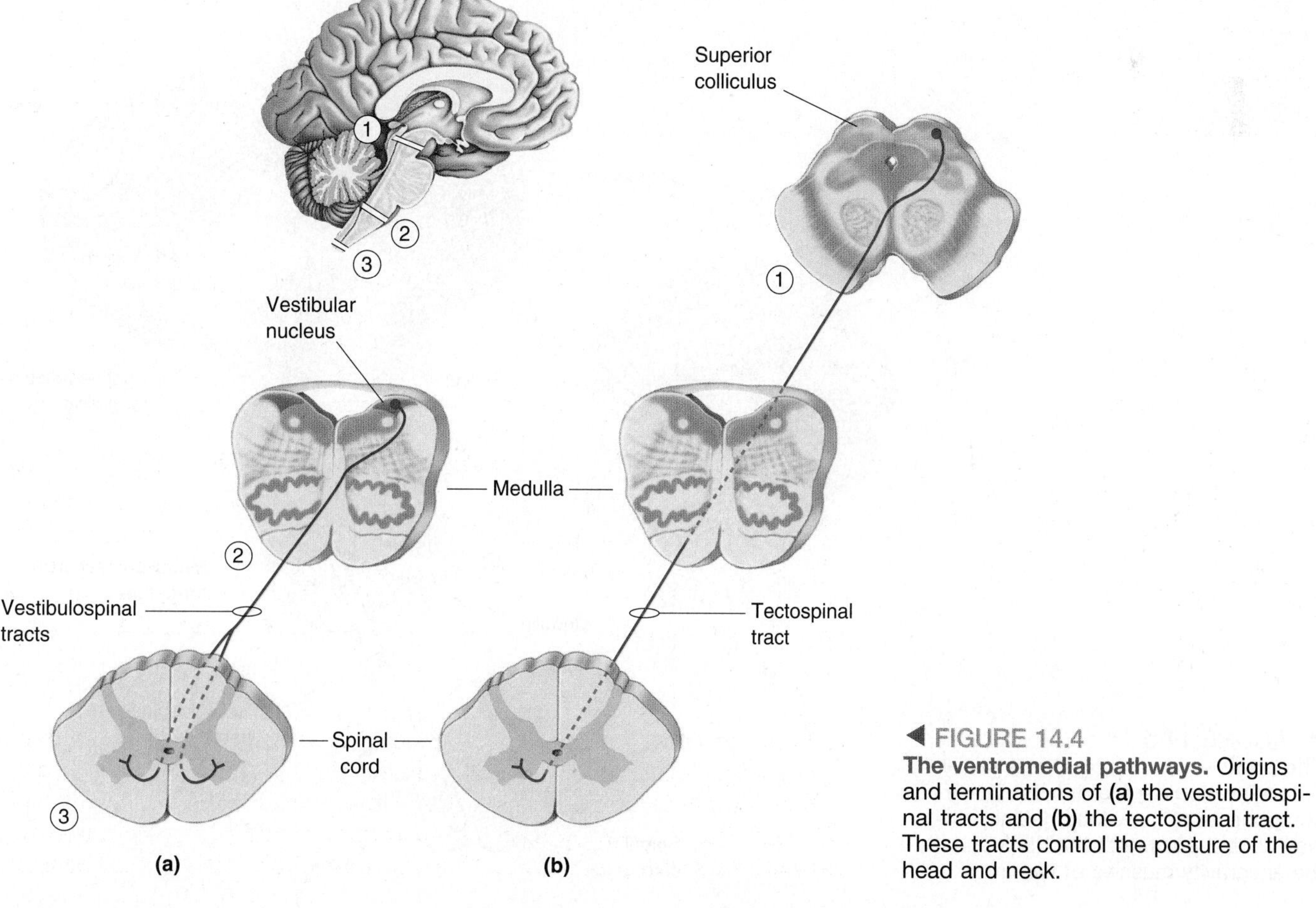

◀ FIGURE 14.4
The ventromedial pathways. Origins and terminations of **(a)** the vestibulospinal tracts and **(b)** the tectospinal tract. These tracts control the posture of the head and neck.

for the superior colliculus.) Besides its retinal input, the superior colliculus receives projections from visual cortex, as well as afferent axons carrying somatosensory and auditory information. From this input, the superior colliculus constructs a map of the world around us; stimulation at one site in this map leads to an orienting response that directs the head and eyes to move so that the appropriate point of space is imaged on the fovea. Activation of the colliculus by an image of a runner sprinting toward second base, for example, would cause the pitcher to orient his head and eyes toward this important new stimulus.

After leaving the colliculus, axons of the tectospinal tract quickly decussate and project close to the midline into cervical regions of the spinal cord, where they help to control muscles of the neck, upper trunk, and shoulders.

The Pontine and Medullary Reticulospinal Tracts. The reticulospinal tracts arise mainly from the **reticular formation** of the brain stem, which runs the length of the brain stem at its core, just under the cerebral aqueduct and fourth ventricle. A complex meshwork of neurons and fibers, the reticular formation receives input from many sources and participates in many different functions. For the purposes of our discussion of motor control, the reticular formation may be divided into two parts that give rise to two different descending tracts: the pontine (medial) reticulospinal tract and the medullary (lateral) reticulospinal tract (Figure 14.5).

The **pontine reticulospinal tract** enhances the antigravity reflexes of the spinal cord. Activity in this pathway, by facilitating the extensors of the lower limbs, helps maintain a standing posture by resisting the effects of gravity. This type of regulation is an important component of motor control: Keep in mind that most of the time, the activity of ventral horn neurons maintains, rather than changes, muscle length and tension.

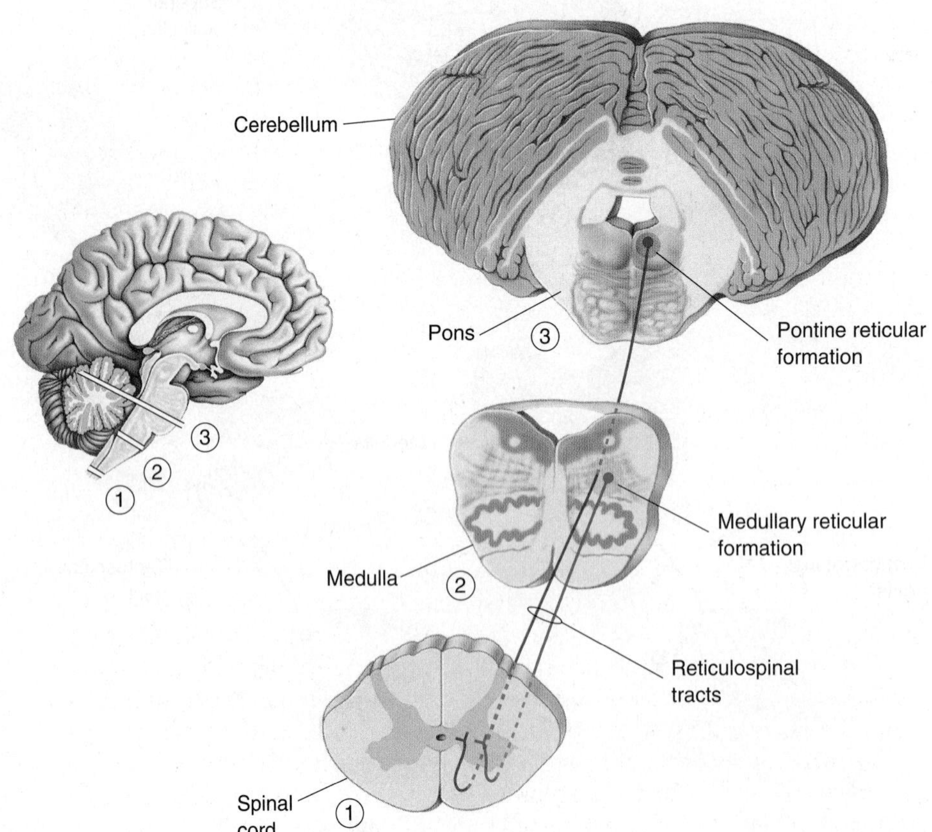

▶ FIGURE 14.5
The pontine (medial) and medullary (lateral) reticulospinal tracts. These components of the ventromedial pathway control posture of the trunk and the antigravity muscles of the limbs.

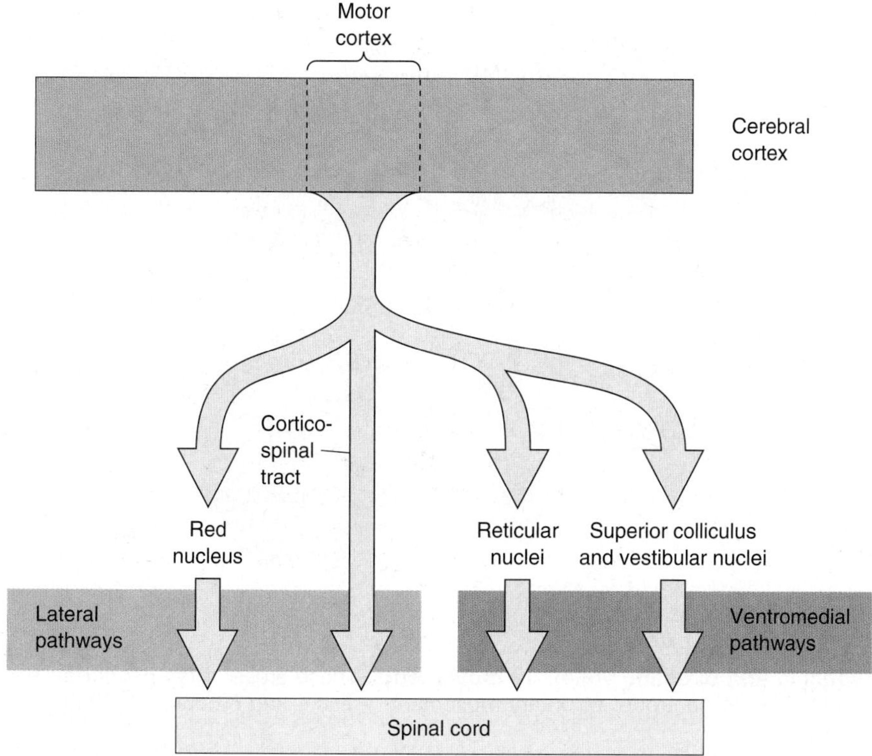

◀ FIGURE 14.6
A summary of the major descending spinal tracts and their origins.

The **medullary reticulospinal tract**, however, has the opposite effect; it liberates the antigravity muscles from reflex control. Activity in both reticulospinal tracts is controlled by descending signals from the cortex. A fine balance between them is required as the pitcher goes from standing on the mound to winding up and throwing the ball.

Figure 14.6 provides a simple summary of the major descending spinal tracts. The ventromedial pathways originate from several regions of the brain stem and participate mainly in the maintenance of posture and certain reflex movements. Initiation of a voluntary, ballistic movement, such as throwing a baseball, requires instructions that descend from the motor cortex along the lateral pathways. The motor cortex directly activates spinal motor neurons and also liberates them from reflex control by communicating with the nuclei of the ventromedial pathways. It is clear that the cortex is critical for voluntary movement and behavior, so we will now focus our attention there.

THE PLANNING OF MOVEMENT BY THE CEREBRAL CORTEX

Although cortical areas 4 and 6 are called *motor cortex*, it is important to recognize that the control of voluntary movement engages almost all of the neocortex. Goal-directed movement depends on knowledge of where the body is in space, where it intends to go, and on the selection of a plan to get it there. Once a plan has been selected, it must be held in memory until the appropriate time. Finally, instructions must be issued to implement the plan. To some extent, these different aspects of motor control are localized to different regions of the cerebral cortex. In this section, we explore some of the cortical areas implicated in motor planning. Later we'll look at how a plan is converted into action.

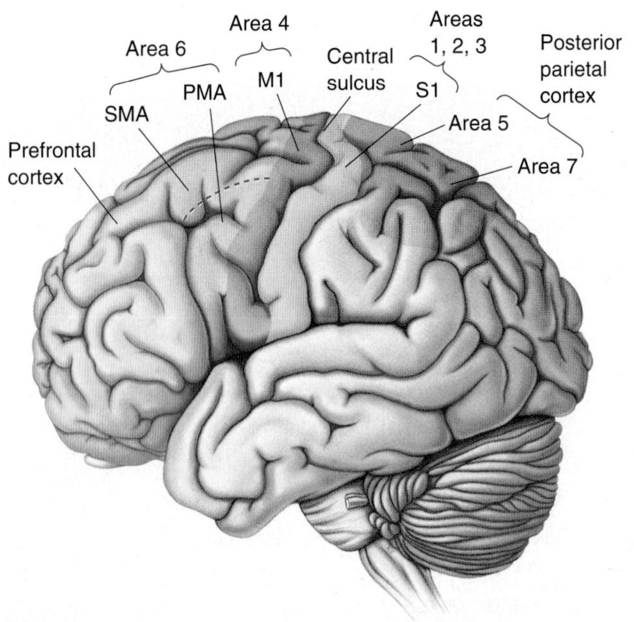

▲ FIGURE 14.7
Planning and directing voluntary movements. These areas of the neocortex are involved in the control of voluntary movement. Areas 4 and 6 constitute the motor cortex.

Motor Cortex

The motor cortex is a circumscribed region of the frontal lobe. Area 4 lies just anterior to the central sulcus on the precentral gyrus, and area 6 lies just anterior to area 4 (Figure 14.7). The definitive demonstration that these areas constitute motor cortex in humans came from the work of neurosurgeon Wilder Penfield. Recall from Chapter 12 that Penfield electrically stimulated the cortex in patients who were undergoing surgery to remove bits of brain thought to be inducing epileptic seizures. The stimulation was used in an attempt to identify which regions of cortex were so critical that they should be spared from the knife. In the course of these operations, Penfield discovered that weak electrical stimulation of area 4 in the precentral gyrus would elicit a twitch of the muscles in a particular region of the body on the contralateral side. Systematic probing of this region established that there is a somatotopic organization in the human precentral gyrus much like that seen in the somatosensory areas of the postcentral gyrus (Figure 14.8). Area 4 is now often referred to as **primary motor cortex** or **M1**.

The foundation for Penfield's discovery had been laid nearly a century before by Gustav Fritsch and Eduard Hitzig, who in 1870 had shown that stimulation of the frontal cortex of anesthetized dogs would elicit movement of the contralateral side of the body (see Chapter 1). Then, around the turn of the century, David Ferrier and Charles Sherrington discovered that the motor area in primates was located in the precentral gyrus. By comparing the histology of this region in Sherrington's apes with that of the human brain, Australian neuroanatomist Alfred Walter Campbell concluded that cortical area 4 is motor cortex.

Campbell speculated that cortical area 6, just rostral to area 4, might be an area specialized for skilled voluntary movement. Penfield's studies 50 years later supported the conjecture that this was a "higher" motor area in humans by showing that electrical stimulation of area 6 could

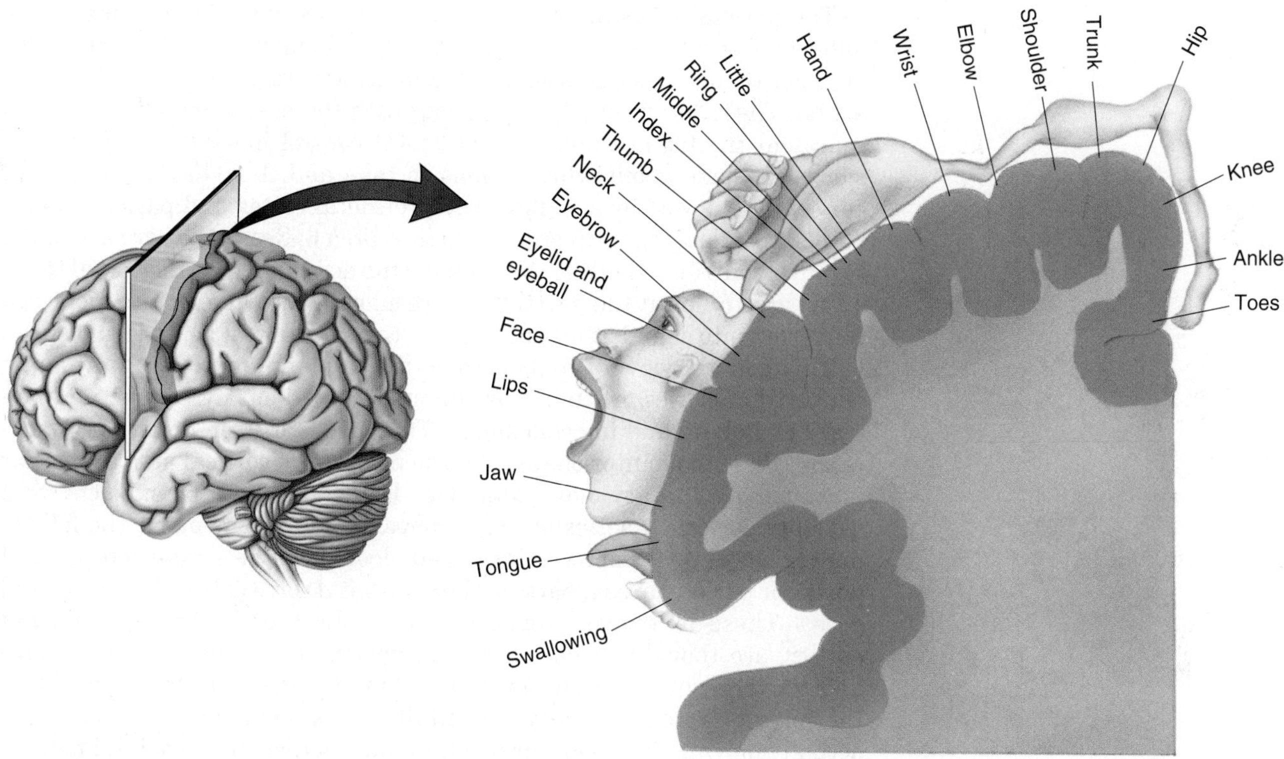

▲ FIGURE 14.8
A somatotopic motor map of the human precentral gyrus. Area 4 of the precentral gyrus is also known as *primary motor cortex* (M1).

evoke complex movements of either side of the body. Penfield found two somatotopically organized motor maps in area 6: one in a lateral region he called the **premotor area (PMA)** and one in a medial region called the **supplementary motor area (SMA)** (see Figure 14.7). These two areas appear to perform similar functions but on different groups of muscles. While the SMA sends axons that innervate distal motor units directly, the PMA connects primarily with reticulospinal neurons that innervate proximal motor units.

The Contributions of Posterior Parietal and Prefrontal Cortex

Recall the baseball player standing on the mound, preparing to pitch. It should be apparent that before the detailed sequence of muscle contractions for the desired pitch can be calculated, the pitcher must have information about the current position of his body in space and how it relates to the positions of the batter and the catcher. This mental body image seems to be generated by somatosensory, proprioceptive, and visual inputs to the posterior parietal cortex.

Two areas are of particular interest in the posterior parietal cortex: area 5, which is a target of inputs from the primary somatosensory cortical areas 3, 1, and 2 (see Chapter 12); and area 7, which is a target of higher order visual cortical areas such as MT (see Chapter 10). Recall that human patients with lesions in these areas of the parietal lobes, as can occur after a stroke, show bizarre abnormalities of body image and the perception of spatial relations. In its most extreme manifestation, the patient will simply neglect the side of the body, and even the rest of the world, opposite the parietal lesion.

The parietal lobes are extensively interconnected with regions in the anterior frontal lobes that in humans are thought to be important for abstract thought, decision making, and anticipating the consequences of action. These "prefrontal" areas, along with the posterior parietal cortex, represent the highest levels of the motor control hierarchy, where decisions are made about what actions to take and their likely outcome (a curve ball followed by a strike). The prefrontal cortex and parietal cortex both send axons that converge on cortical area 6. Recall that areas 6 and 4 together contribute most of the axons to the descending corticospinal tract. Thus, area 6 lies at the junction where signals encoding *what* actions are converted into signals that specify *how* the actions will be carried out.

This general view of higher order motor planning received dramatic support in a series of studies on humans carried out by Danish neurologist Per Roland and his colleagues. They used positron emission tomography (PET) to monitor changes in the patterns of cortical activation that accompany voluntary movements (see Box 7.3). When the subjects were asked to perform a series of finger movements from memory, the following regions of cortex showed increased blood flow: the somatosensory and posterior parietal areas, parts of the prefrontal cortex (area 8), area 6, and area 4. These are the very regions of the cerebral cortex that, as discussed earlier, are thought to play a role in generating the intention to move and converting that intention into a plan of action. Interestingly, when the subjects were asked only to mentally rehearse the movement without actually moving the finger, area 6 remained active but area 4 did not.

Neuronal Correlates of Motor Planning

Experimental work on monkeys further supports the idea that area 6 (SMA and PMA) plays an important role in the planning of movement, particularly complex movement sequences of the distal musculature. Using a method developed in the late 1960s by Edward Evarts at the National Institutes of Health, researchers have recorded the activity of neurons in the motor areas of awake, behaving animals (Box 14.2). Cells in the SMA typically increase their discharge rates about a second before the execution of a hand or wrist movement, consistent with their proposed role in planning movement (recall Roland's findings in humans). An important feature of this activity is that it occurs in advance of the movements of *either* hand, suggesting that the supplementary areas of the two hemispheres are closely linked via the corpus callosum. Indeed, movement deficits observed following an SMA lesion on one side, in both monkeys and humans, are particularly pronounced for tasks requiring the coordinated actions of the two hands, such as buttoning a shirt. In humans, a selective inability to perform complex (but not simple) motor acts is called *apraxia*.

You've heard the expression "ready, set, go." The preceding discussion suggests that readiness ("ready") depends on activity in the parietal and frontal lobes, along with important contributions from the brain centers that control levels of attention and alertness. "Set" may reside in the supplementary and premotor areas, where movement strategies are devised and held until they are executed. A good example is shown in Figure 14.9, based on the work of Michael Weinrich and Steven Wise at the National Institutes of Health. They monitored the discharge of a neuron in the PMA as a monkey performed a task requiring a specific arm movement to a target. The monkey was first given an *instruction stimulus* informing him what the target would be ("Get set, monkey!"), followed after a variable delay by a *trigger stimulus* informing the monkey that it was OK to move ("Go, monkey!"). Successful performance of the task (i.e., waiting for the "go" signal and

BOX 14.2 OF SPECIAL INTEREST

Behavioral Neurophysiology

Showing that a brain lesion impairs movement and that brain stimulation elicits movement does not tell us how the brain *controls* movement. To address this problem, we need to know how the activity of neurons relates to different types of voluntary movement in the intact organism. PET scans and fMRI are extremely valuable for plotting out the distribution of activity in the brain as behaviors are performed, but they lack the resolution to track the millisecond-by-millisecond changes in the activity of individual neurons. The best method for this purpose is extracellular recording with metal microelectrodes (see Box 4.1). But how is this done in awake, behaving animals?

This problem was solved by Edward Evarts and his colleagues at the National Institutes of Health. Monkeys were trained to perform simple tasks; when the tasks were performed successfully, the monkeys were rewarded with a sip of fruit juice. For example, to study the brain's guidance of hand and arm movements, the monkey might be trained to move its hand toward the brightest of several spots on a computer screen. Pointing to the correct spot earned it a juice reward. After training, the animals were anesthetized. In a simple surgical procedure, each monkey was fitted with a small headpiece so that a microelectrode could be introduced into the brain through a small opening in the skull. When the animals recovered from surgery, they showed no signs of discomfort from either the headpiece or the insertion of a microelectrode into the brain (recall from Chapter 12 that there are no nociceptors in the brain). Evarts and his colleagues then recorded the discharges of individual cells in the motor cortex as the animals made voluntary movements. In the example above, one could then see how the neuron's response changes when the animal points to different spots on the screen.

This is an example of what is now called *behavioral neurophysiology*, the recording of cellular activity in the brain of awake, behaving animals. By altering the task that the animal performs, the same method can be applied to the investigation of a wide range of neuroscientific topics, including attention, perception, learning, and movement. Some types of human neurosurgery are also done with the patient awake, at least during part of the procedure. By applying the techniques of behavioral neurophysiology to informed, consenting adults, we have also learned some fascinating information about uniquely human skills.

In recent years, technical developments have made it possible to insert large numbers of microelectrodes into the same or different parts of an animal's brain and to record from dozens or even hundreds of neurons simultaneously. This approach yields a massive amount of information about brain activity and its relationship to behavior. Understanding this relationship is one of the greatest challenges in neuroscience.

then making the movement to the appropriate target) was rewarded with a sip of juice. The neuron in the PMA began firing if the instruction was to move the arm to the left, and it continued to discharge until the trigger stimulus came on and the movement was initiated. If the instruction was to move to the right, this neuron did not fire (presumably another population of PMA cells became active under this condition). Thus, the activity of this PMA neuron reported the direction of the upcoming movement and continued to do so until the movement was made. Although we do not yet understand the details of the coding taking place in the SMA and PMA, the fact that neurons in these areas are selectively active well before movements are initiated is consistent with a role in planning the movement.

Mirror Neurons

We mentioned previously that some neurons in cortical area 6 respond not only when movements are executed but also when the same movement is only imagined—mentally rehearsed. Remarkably, some neurons in motor areas of cortex fire not only when a monkey makes a specific movement himself but also when the monkey simply observes another monkey, or even a human, making the same type of movement (Figure 14.10). These cells were called **mirror neurons** by Giacomo Rizzolatti and his colleagues when they discovered them in the PMA of monkeys at the University of Parma in the early 1990s. Mirror neurons seem to represent

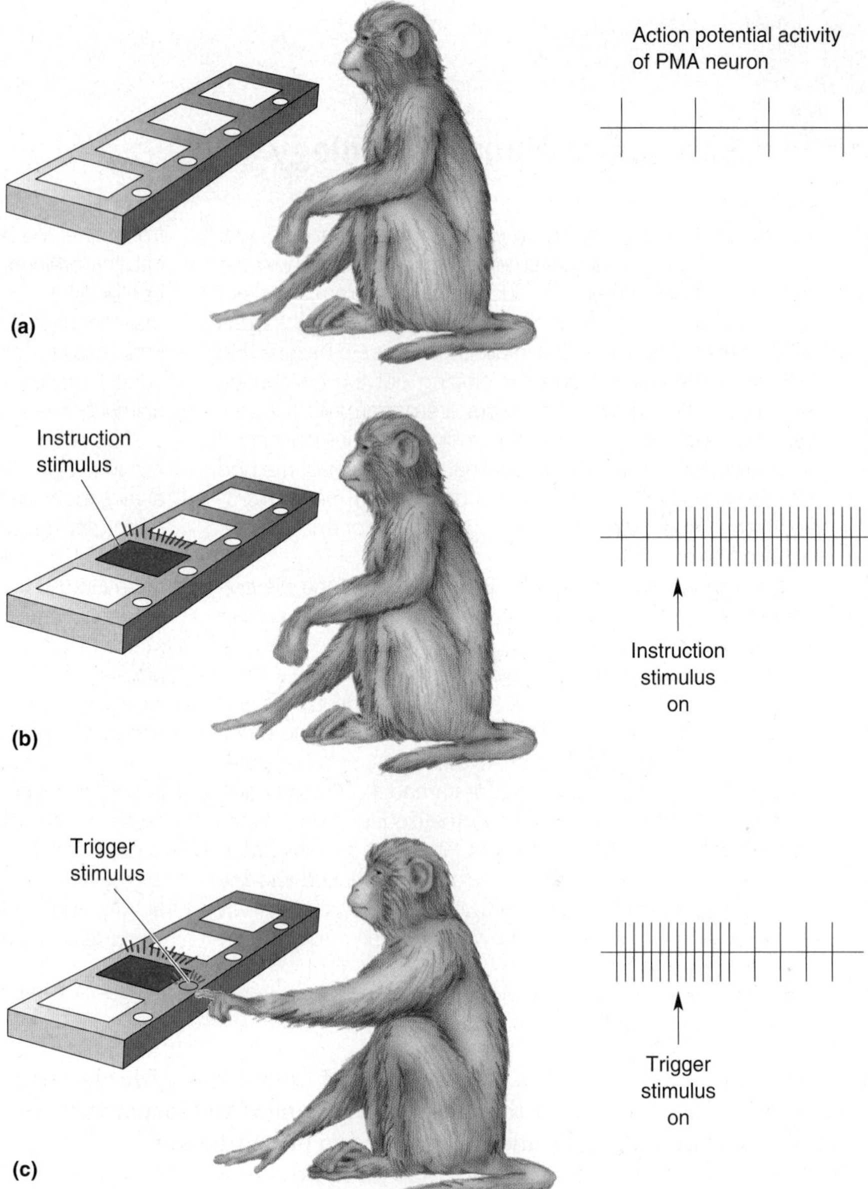

▶ FIGURE 14.9
The discharge of a neuron in the premotor area before a movement. **(a)** *Ready:* A monkey sits before a panel of lights. The task is to wait for an instruction stimulus that will inform him of the movement required to receive a juice reward, then perform the movement when a trigger stimulus goes on. The activity of a neuron in PMA is recorded during the task. **(b)** *Set:* The instruction stimulus (one of the square red lights) occurs at the time indicated by the upward arrow, resulting in the discharge of the neuron in PMA. **(c)** *Go:* A trigger stimulus (a blue light in one of the buttons) tells the monkey when and where to move. Shortly after the movement is initiated, the PMA cell ceases firing. (Source: Adapted from Weinrich and Wise, 1982.)

particular motor acts, such as reaching, grasping, holding, or moving objects, regardless of whether a monkey actually performs the act or merely observes others doing it. Each cell has very specific movement preferences; a mirror neuron that responds when its monkey grasps a food tidbit will also respond to the sight of another monkey making a similar grasp of a tidbit but not when either monkey waves its hand. Many mirror neurons even respond to the unique sounds another monkey produces during a specific movement (e.g., cracking open a peanut), as well as to the sight of that movement. In general, mirror neurons seem to encode the specific goals of motor acts rather than particular sensory stimuli.

It is very likely that humans also have mirror neurons in PMA and other cortical areas, although the evidence for this, mainly from studies using functional magnetic resonance imaging (fMRI) (see Boxes 7.2 and 7.3), is still indirect.

Mirror neurons may be part of an extensive brain system for understanding the actions and even the intentions of others. This is an exciting

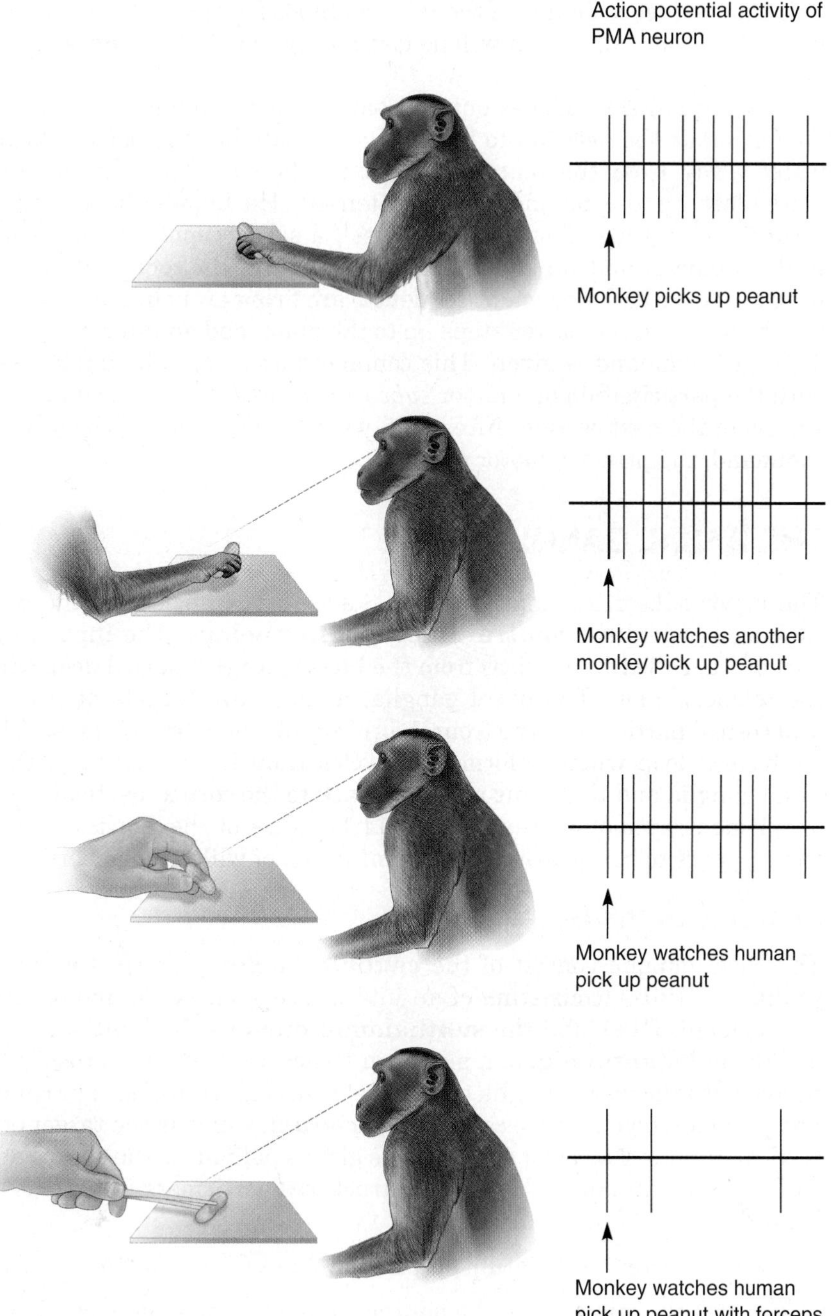
Action potential activity of PMA neuron
Monkey picks up peanut
Monkey watches another monkey pick up peanut
Monkey watches human pick up peanut
Monkey watches human pick up peanut with forceps

◀ FIGURE 14.10
The discharge of a mirror neuron. **(a)** A PMA mirror neuron fires action potentials when a monkey reaches for a peanut. **(b)** The same mirror neuron fires when the monkey sees another monkey reach for a peanut. **(c)** The neuron also fires when the monkey sees a human reach for a peanut. **(d)** When the human reaches for a peanut using a forceps, the mirror neuron is not activated. (Source: Adapted from Rizzolatti et al., 1996.)

and attractive hypothesis. It implies that we use the same motor circuits both for planning our own movements and for understanding the actions and goals of others. When one pitcher watches another pitcher throw a ball, the first pitcher may activate the same motor planning neurons that allow him to throw his own ball. In a sense, he may be experiencing the action of the other pitcher by running his own neural program for the same type of action. More expansive versions of this hypothesis suggest that mirror neurons are also responsible for our ability to read the emotions and sensations of others and to empathize. Some investigators have even suggested that dysfunctional mirror neurons are responsible for certain features of autism, such as the impaired ability to understand the thoughts, intentions, feelings, and ideas of others (see Box 23.4). As intriguing as these hypotheses about mirror neuron functions are, there is still scant evidence for any of them. As methods for recording directly from human neurons improve, it will be fascinating to test these ideas directly.

Now, let's again consider our baseball pitcher standing on the mound. He has made the decision to throw a curve ball, but the batter abruptly walks away from the plate to adjust his helmet. The pitcher stands motionless on the mound, muscles tensed. He knows the batter will return, so he waits. The pitcher is "set"; a select population of neurons in the premotor and supplementary motor cortex (the cells that are planning the curve ball movement sequence) are firing away in anticipation of the throw. Then the batter steps up to the plate, and an internally generated "go" command is given. This command appears to be implemented with the participation of a major *subcortical* input to area 6, which is the subject of the next section. After that, we'll examine the origin of the "go" command, the primary motor cortex.

THE BASAL GANGLIA

The major subcortical input to area 6 arises in a nucleus of the dorsal thalamus, called the **ventral lateral (VL) nucleus**. The input to this part of VL, called *VLo*, arises from the **basal ganglia** buried deep within the telencephalon. The basal ganglia, in turn, are targets of the cerebral cortex, particularly the frontal, prefrontal, and parietal cortex. Thus, we have a loop where information cycles from the cortex through the basal ganglia and thalamus and then back to the cortex, particularly the supplementary motor area (Figure 14.11). One of the functions of this loop appears to be the selection and initiation of willed movements.

Anatomy of the Basal Ganglia

The basal ganglia consist of the **caudate nucleus**, the **putamen**, the **globus pallidus** (consisting of an internal segment, GPi, and an external segment, GPe), and the **subthalamic nucleus**. In addition, we can add the **substantia nigra**, a midbrain structure that is reciprocally connected with the basal ganglia of the forebrain (Figure 14.12). The caudate and putamen together are called the **striatum**, which is the target of the cortical input to the basal ganglia. The globus pallidus is the source of the output to the thalamus. The other structures participate in various side loops that modulate the direct path:

$$\text{Cortex} \rightarrow \text{Striatum} \rightarrow \text{GPi} \rightarrow \text{VLo} \rightarrow \text{Cortex (SMA)}$$

Through the microscope, the neurons of the striatum appear randomly scattered, with no apparent order such as that seen in the layers of the cortex. But this bland appearance hides a degree of complexity in the organization of the basal ganglia that we only partially understand. It appears that the basal ganglia participate in a large number of parallel circuits, only a few of which are strictly motor. Other circuits are involved in certain aspects of memory and cognitive function. We will try to give a concise account of the motor function of the basal ganglia, simplifying this very complex and poorly understood part of the brain.

Direct and Indirect Pathways through the Basal Ganglia

The motor loop through the basal ganglia originates with excitatory connections from the cortex. In the *direct pathway* through the basal ganglia, synapses from cortical cells excite cells in the putamen, which make inhibitory synapses on neurons in the globus pallidus, which in turn make inhibitory

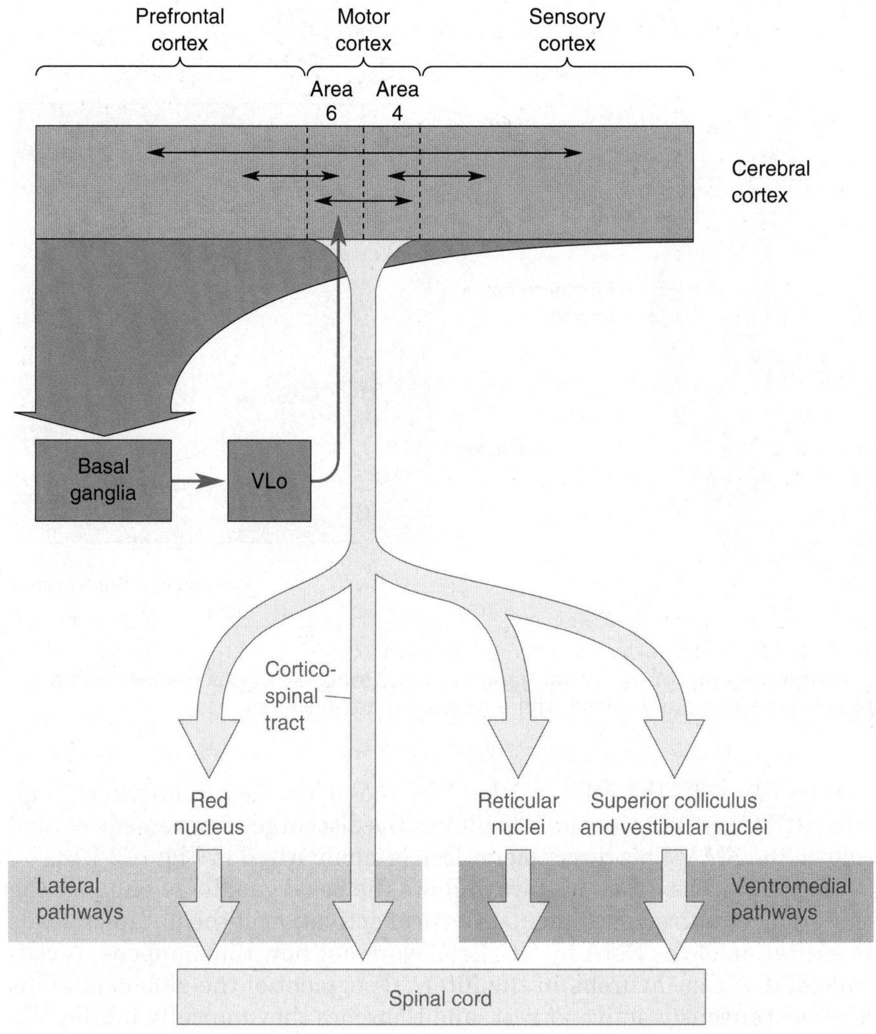

◀ FIGURE 14.11
A summary of the motor loop from the cortex to the basal ganglia to the thalamus and back to area 6.

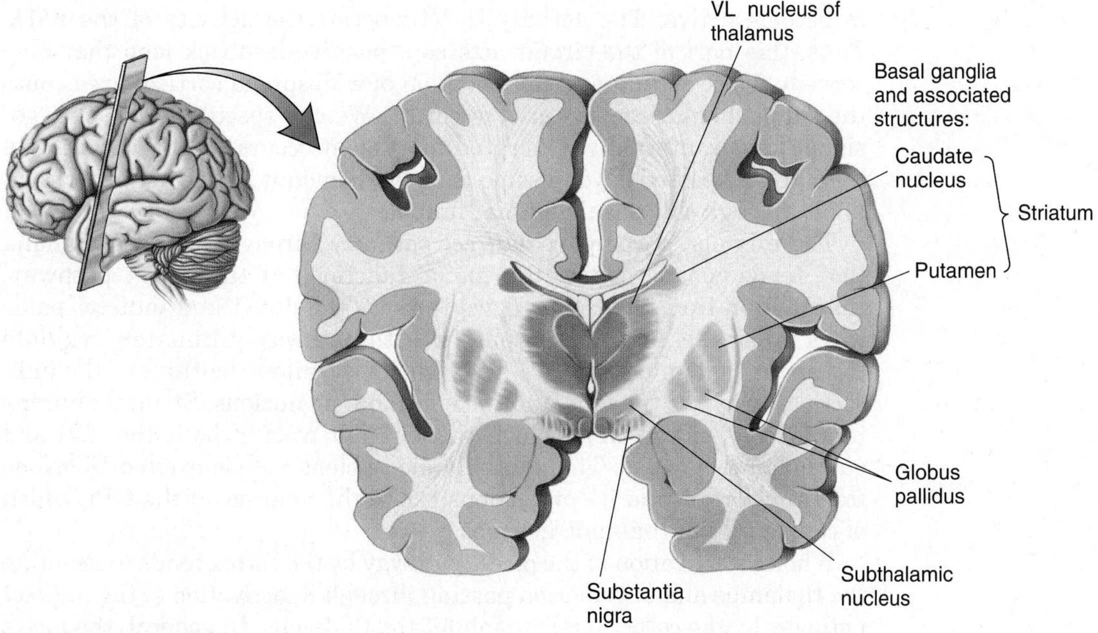

▲ FIGURE 14.12
The basal ganglia and associated structures.

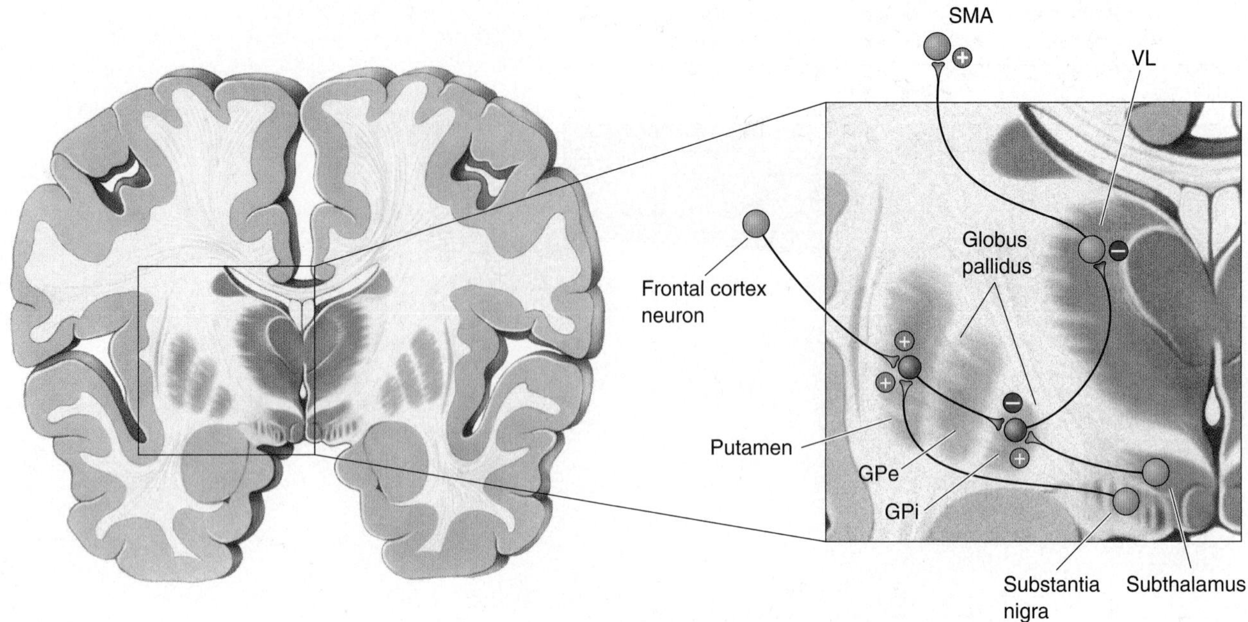

▲ FIGURE 14.13
A wiring diagram of the basal ganglia motor loop. Synapses marked with a plus (+) are excitatory; those with a minus (−) are inhibitory.

connections with the cells in VLo. The thalamocortical connection (from VLo to SMA) is excitatory and facilitates the discharge of movement-related cells in the SMA. This direct motor loop is summarized in Figure 14.13.

In general, the direct pathway allows the basal ganglia to enhance the initiation of desired movements. Cortical activation of the putamen leads to excitation of the SMA by VL. Let's work out how this happens. A critical point is that neurons in the internal segment of the globus pallidus are spontaneously active at rest, and therefore they tonically inhibit VL. Cortical activation (1) excites putamen neurons, which (2) inhibit GPi neurons, which (3) release the cells in VLo from inhibition, allowing them to become active. The activity in VLo boosts the activity of the SMA. Thus, this part of the circuit acts as a positive-feedback loop that may serve to focus, or funnel, the activation of widespread cortical areas onto the supplementary motor area of cortex. We can speculate that the "go" signal for an internally generated movement occurs when activation of the SMA is boosted beyond some threshold amount by the activity reaching it through this basal ganglia "funnel."

There is also a complex *indirect pathway* through the basal ganglia that tends to antagonize the motor functions of the direct pathway. Information from the cortex flows through the direct and indirect pathways in parallel, and the outputs of both pathways ultimately regulate the motor thalamus (Figure 14.14). The most unique features of the indirect pathway are the GPe and the subthalamic nucleus. Striatal neurons inhibit cells of the GPe, which then inhibit cells of both the GPi and subthalamic nucleus. The subthalamic nucleus is also excited by axons from the cortex, and its projections excite the neurons of the GPi, which of course inhibit thalamic neurons.

Whereas activation of the direct pathway by the cortex tends to facilitate the thalamus and information passing through it, activation of the indirect pathway by the cortex tends to inhibit the thalamus. In general, the direct pathway may help to select certain motor actions while the indirect pathway simultaneously suppresses competing, and inappropriate, motor programs.

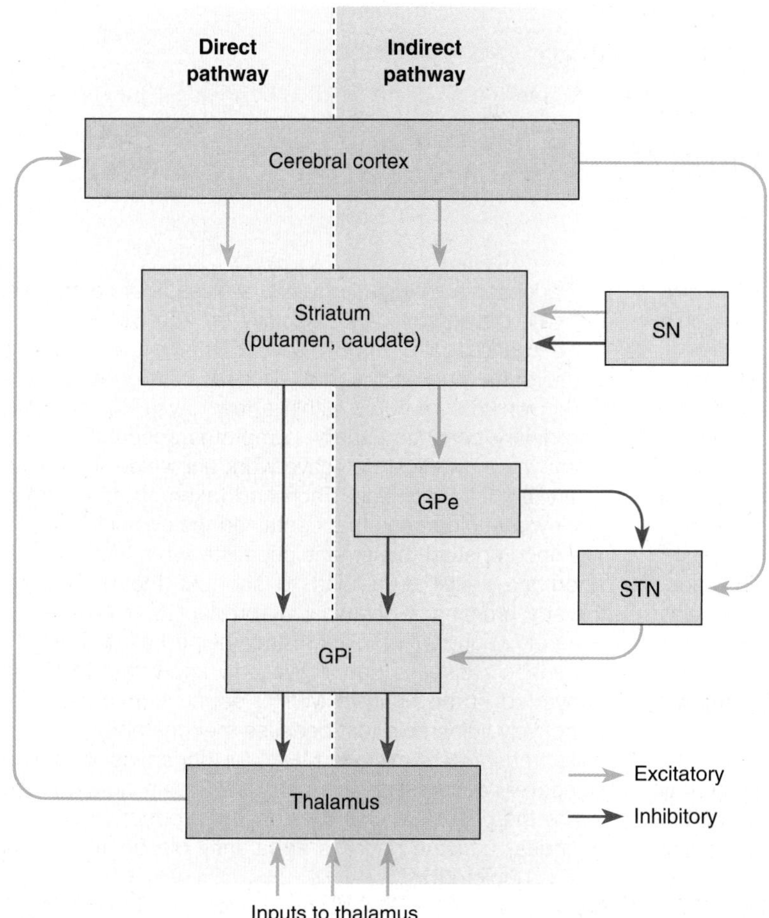

▲ FIGURE 14.14
The direct and indirect pathways through the basal ganglia. Dopaminergic neurons of the substantia nigra (SN) modulate the putamen and caudate nucleus. The GPe and the subthalamic nucleus (STN) are part of the indirect pathway."

Basal Ganglia Disorders. Studies of several human diseases have supported the view that the direct motor loop through the basal ganglia functions to facilitate the initiation of willed movements. According to one model, increased inhibition of the thalamus by the basal ganglia underlies *hypokinesia*, a paucity of movement, whereas decreased basal ganglia output leads to *hyperkinesia*, an excess of movement.

Parkinson's disease exemplifies the first condition. This disorder, which affects about 1% of all people over age 60, is characterized by hypokinesia. Its symptoms include slowness of movement (*bradykinesia*), difficulty in initiating willed movements (*akinesia*), increased muscle tone (*rigidity*), and tremors of the hands and jaw, which are most prominent at rest when the patient is not attempting to move. Many patients also suffer deficits of cognition as the disease progresses. The organic basis of Parkinson's disease is a degeneration of certain substantia nigra neurons and their inputs to the striatum (Box 14.3). These inputs use the neurotransmitter dopamine (DA). The actions of DA are complex because it binds to multiple types of striatal DA receptors that mediate quite different effects (see Figure 14.14). Dopaminergic synapses terminate on striatal neurons closely adjacent to the synaptic inputs from the cortex, and DA can enhance the cortical inputs to the direct pathway. DA facilitates the direct motor loop by activating cells in the putamen (which releases VLo from GPi-induced inhibition). In essence, the depletion of dopamine in Parkinson's disease closes the funnel

BOX 14.3 OF SPECIAL INTEREST

Do Neurons in Diseased Basal Ganglia Commit Suicide?

Several devastating neurological diseases involve the slow, progressive death of neurons. Patients with Parkinson's disease have usually lost more than 80% of the dopamine-utilizing neurons in their substantia nigra (Figure A). Neurons of the striatum and other regions slowly degenerate in sufferers of Huntington's disease (Figure B). Why do these neurons die? Ironically, it may be that natural forms of cell death are involved. A process called *programmed cell death* is essential for normal brain development; certain neurons commit suicide as part of the "program" by which the nervous system forms (see Chapter 23). All cells have several "death genes" that trigger a set of enzymes that destroy cellular proteins and DNA. Some forms of cancer occur when normal programmed cell death is prevented and cells proliferate wildly. Some neurological diseases may result when programmed cell death is unnaturally activated.

Huntington's disease is caused by a dominant gene that codes for a large brain protein called *huntingtin*. The normal molecule has a chain of 10–34 glutamines at one end, but people with a repeat of more than 40 glutamines develop Huntington's disease. The abnormally long huntingtins aggregate; globs of them accumulate and trigger neuronal degeneration. The function of normal huntingtin is unknown, but it may counterbalance the triggers for programmed cell death. Thus, Huntington's disease may arise from normal processes of neuronal degeneration gone awry.

Parkinson's disease is usually a disease of aging, and the vast majority of cases occur after age 60. However, in 1976 and again in 1982, several relatively young drug abusers in Maryland and California developed severe Parkinsonian symptoms within a few days. This was extraordinary because usually, symptoms accumulate over many years. Medical detective work unraveled the cause of the addicts' affliction. Each had taken street versions of a synthetic narcotic that contained the chemical MPTP. The incompetent basement chemists who had synthesized the illegal drug tried to shortcut the procedure, thereby creating a chemical by-product that kills dopaminergic neurons. MPTP has since helped us understand Parkinson's disease better. We now know that MPTP is converted in the brain to MPP^+; dopaminergic cells are selectively vulnerable to it because their membrane dopamine transporters mistake MPP^+ for dopamine, and they selectively accumulate this chemical Trojan horse. Once inside the cell, MPP^+ disrupts energy production in mitochondria, and the neurons apparently die because their ATP is depleted.

The effect of MPTP supports the idea that common forms of Parkinson's disease might be caused by chronic exposure to a slowly acting toxic chemical in the environment. Unfortunately, no one has identified such a toxin. Research has shown that MPTP can induce a form of programmed

that feeds activity to the SMA via the basal ganglia and VLo. At the same time, DA inhibits the neurons in the striatum that send inhibitory outputs, via the indirect pathway, to the GPe.

A central goal of most therapies for Parkinson's disease is to enhance the levels of dopamine delivered to the caudate nucleus and putamen. This is most easily done by administering the compound L-dopa (L-dihydroxyphenylalanine, introduced in Chapter 6), which is a precursor to dopamine. L-dopa crosses the blood–brain barrier and boosts DA synthesis in the cells that remain alive in the substantia nigra, thus alleviating some of the symptoms. DA agonists are also useful drugs in the treatment of Parkinson's disease. However, treatments with L-dopa or DA agonists do not alter the progressive course of the disease, nor do they alter the rate at which substantia nigra neurons degenerate. They also have significant side effects. (We will return to the topic of dopamine neurons in Chapter 15.) The symptoms of some Parkinson's disease patients can also be improved with brain surgery and stimulation (Box 14.4). There are also a variety of experimental treatment strategies. One of them is to graft DA-producing cells into the basal ganglia. A promising possibility is to use human stem cells that have been manipulated developmentally or genetically to produce DA. These may one day provide neuronal death in the substantia nigra. Dopaminergic neurons of Parkinson's patients may degenerate for a similar reason. About 5% of Parkinson's cases are inherited, and mutations in several different genes are now known to cause these rarer types of the syndrome. One hypothesis is that Parkinsonian genes encode mutant proteins that are misfolded, aggregate, accumulate in neurons, and trigger or facilitate the death of dopaminergic neurons.

By understanding how and why neurons self-destruct, we may eventually be able to devise strategies of cellular suicide prevention that halt or avert a variety of terrible neurological diseases.

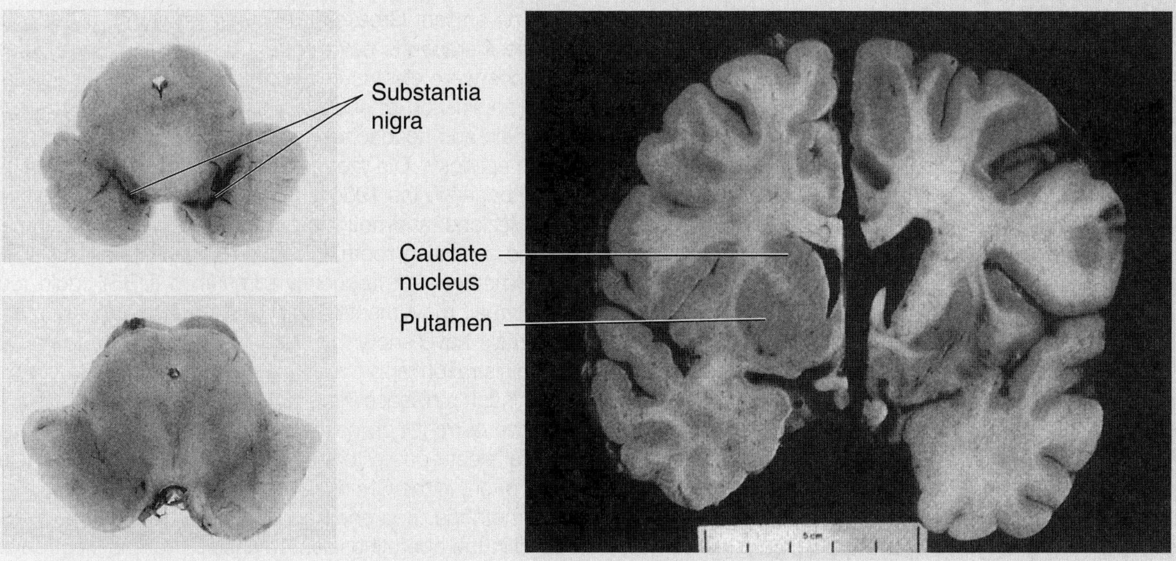

Figure A
Normal (top); Parkinson's disease (bottom). (Source: Strange, 1992, Fig. 10.3.)

Figure B
Normal (left); Huntington's disease (right). (Source: Strange, 1992, Fig. 11.2.)

an effective treatment, perhaps even a cure, for Parkinson's disease, but we are not there yet.

If Parkinson's disease lies at one end of the spectrum of basal ganglia disorders, Huntington's disease lies at the other. **Huntington's disease** is a hereditary, progressive, inevitably fatal syndrome characterized by hyperkinesia and *dyskinesias* (abnormal movements), *dementia* (impaired cognitive abilities), and a disorder of personality. Luckily, it is quite rare, afflicting 5–10 people per 100,000 worldwide. The disease is particularly insidious because its symptoms usually do not appear until well into adulthood. In the past, patients often unwittingly passed the gene on to their children before they knew they had the disease. It is now possible to perform a genetic test that reveals whether a person carries the Huntington gene. People with Huntington's disease exhibit changes in mood, personality, and memory. The most characteristic sign of the disease is *chorea*—spontaneous, uncontrollable, and purposeless movements with rapid, irregular flow and flicking motions of various parts of the body. The most obvious pathology of their brains is a profound loss of neurons in the caudate nucleus, putamen, and globus pallidus, with additional cell loss in the cerebral cortex and elsewhere (see Box 14.3). The damage to structures in the basal ganglia and consequent loss of its

BOX 14.4 OF SPECIAL INTEREST

Destruction and Stimulation: Useful Therapies for Brain Disorders

Brain disorders can be very difficult to treat, and useful therapies are often counterintuitive. Advanced Parkinson's disease, for example, is sometimes treated with small surgical lesions of the brain or by implanting electrodes for *deep brain stimulation (DBS)*. Destruction and stimulation are alternative strategies with the same therapeutic goal—to relieve patients of their severely abnormal movements.

The most common treatment for the early phase of Parkinson's disease, L-dopa, can be tremendously helpful. Unfortunately, with time, the effects of the drug usually diminish, and new types of abnormal and debilitating movements, dyskinesias, may appear. Numerous other drugs can be useful at this stage, but their effectiveness varies and they have side effects of their own.

Surgery for movement disorders began in the 1880s with Victor Horsley, a pioneering British neurosurgeon, who treated a patient's uncontrollable spontaneous movements by removing part of his motor cortex. The abnormal movements ceased, but the patient's limb was paralyzed. Between the 1940s and 1970s, surgeons found that making small lesions in the globus pallidus, thalamus, or subthalamic nucleus could often improve the tremor, rigidity, and akinesia of Parkinson's disease without inducing paralysis. With the introduction of L-dopa in 1968 and a backlash against unjustified types of neurosurgery (see Box 18.4), surgical treatments for Parkinson's disease fell out of favor for a while. Currently, targeted surgical lesions of basal ganglia and thalamus are still used in some Parkinson's patients, but DBS has become an increasingly popular form of treatment.

The ancient Greeks and Egyptians were early advocates of the therapeutic power of electrical shocks. Their medical devices were electric eels and rays, and it was said that direct application of such a stimulating fish could help alleviate pain and headache, hemorrhoids, gout, depression, and even epilepsy. The modern use of DBS for movement disorders began in the 1980s. Taking a cue from their experience with lesions, and noting the promising effects of stimulation in the operating room, surgeons began systematically testing whether high-frequency stimulation (DBS) could reduce abnormal movements over the long term. Several clinical trials have shown that it can. The U.S. Food and Drug Administration approved the use of DBS for treatment of Parkinson's disease in 2002.

The current approach to DBS is to surgically implant bilateral electrodes with their tips in the subthalamic nuclei or, less often, in the GPi nuclei (Figure A). Advanced brain imaging methods, neuronal recordings, and trial stimulation are used in the operating room to ensure that the electrodes are placed precisely. Power and control of the electrodes come from small batteries and computers implanted under the

inhibitory output to the thalamus seem to account for the disorders of movement in Huntington's patients. Cortical degeneration is primarily responsible for their dementia and personality changes.

Hyperkinesia can also result from other types of lesions that affect the basal ganglia. One example is **ballism**, which is characterized by violent, flinging movements of the extremities (somewhat like our baseball pitcher unintentionally throwing the ball while sitting in the dugout). The symptoms usually occur on just one side of the body, and the condition is then called *hemiballismus*. As with Parkinson's disease, the cellular mechanisms associated with ballism are known; it is caused by damage to the subthalamic nucleus (usually resulting from an interruption of its blood supply caused by a stroke). The subthalamic nucleus, part of another side loop within the basal ganglia, excites neurons in the globus pallidus that project to VLo (see Figure 14.14). Remember that excitation of the globus pallidus inhibits VLo (see Figure 14.13). Thus, a loss of excitatory drive to the globus pallidus facilitates VLo, in effect opening the funnel of activity to the SMA.

In summary, the basal ganglia may facilitate movement by focusing activity from widespread regions of cortex onto the SMA. Importantly, however, they also serve as a filter that keeps inappropriate movements from being expressed. We saw in Roland's PET studies that activity in the skin below the collarbone. Postoperatively, therapists work with the patients to tune the properties of the stimulation for optimal effectiveness and minimal side effects.

Considering the complexities of the brain's functions and dysfunctions, DBS is a very crude tool. The most effective stimulation pattern tends to be a continuous stream of brief shocks at very high frequency (130–180 Hz). Since this does not resemble any natural neural pattern in the brain, how does DBS work? Research on this question has been intense, but the answer remains unknown. High-frequency stimulation can block abnormal firing in some cases. Stimulation may also "jam" or suppress abnormal patterns of firing. DBS may activate inhibitory neurons that suppress dysfunctional brain activity. It may trigger release of neurotransmitters that modulate cells and synapses. The mechanism of DBS may also vary with the brain structure being stimulated. It would not be surprising if all of these effects, and more, were important for the efficacy of DBS.

DBS can be quite effective in controlling both hyperkinetic and hypokinetic symptoms and improving patients' quality of life overall. It is not a panacea, however. DBS is not a useful treatment for most of the nonmotor features of the disease, including disturbances of cognition, mood, gait, and speech. There are also side effects and the usual risks of surgery. Batteries need to be surgically replaced every few years, although some DBS systems are now rechargeable.

DBS has therapeutic promise far beyond Parkinson's disease. It can reduce the symptoms of several other movement disorders. It may be helpful for a surprising range of other psychiatric and neurological conditions, including major depression, obsessive-compulsive disorder, Tourette syndrome, schizophrenia, epilepsy, tinnitus, chronic pain, and Alzheimer's disease. The best brain stimulation sites vary for each disorder. DBS is still an experimental treatment for nearly all these conditions, and only further research will determine whether its benefits outweigh its risks and costs.

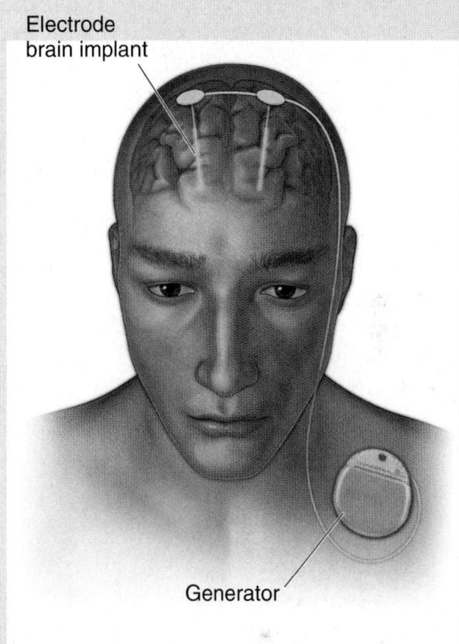

Figure A

SMA does not automatically trigger movement. The initiation of voluntary movement also requires activation of area 4, the subject of the next section.

THE INITIATION OF MOVEMENT BY PRIMARY MOTOR CORTEX

SMA is heavily interconnected with M1, cortical area 4 on the precentral gyrus (see Figure 14.7). The designation of area 4 as primary motor cortex is somewhat arbitrary because this is not the only cortical area that contributes to the corticospinal tract or to movement. Nonetheless, since the time of Sherrington, neuroscientists have recognized that this area has the lowest threshold for the elicitation of movement by electrical stimulation. In other words, stimulation intensities that are unable to evoke movement in other cortical areas are still effective in evoking movement when applied to area 4, meaning that area 4 has dense, strong synaptic connections with the motor neurons and the spinal interneurons that drive them. Focal electrical stimulation of area 4 evokes the contraction of small groups of muscles, and, as we discussed earlier, the somatic musculature is mapped systematically in this area. This ribbon of cortex that stretches the full length of the precentral gyrus is sometimes also called the **motor strip**.

The Input–Output Organization of M1

The pathway by which motor cortex activates lower motor neurons originates in cortical layer V. Layer V has a population of pyramidal neurons, some of which can be quite large (soma diameters approaching 0.1 mm). The largest cells were first described as a separate class by Russian anatomist Vladimir Betz in 1874 and are therefore called *Betz cells*. In humans, many of the large corticospinal cells of layer V project to pools of lower motor neurons and excite them monosynaptically. The same corticospinal axons can also branch and excite local inhibitory interneurons. By controlling selected groups of motor neurons and interneurons, a single corticospinal neuron may generate coordinated effects on antagonist muscles. For example, the motor cortex neurons in Figure 14.15 excite pools of extensor motor neurons and simultaneously inhibit pools of flexor motor neurons. This is similar to the reciprocal inhibition that we saw in the spinal reflex circuitry in Chapter 13 (see Figure 13.25).

The layer V pyramidal cells in M1 receive their inputs primarily from two sources: other cortical areas and the thalamus. The major cortical inputs originate in the areas adjacent to area 4: area 6 immediately anterior; and areas 3, 1, and 2 immediately posterior (see Figure 14.7). The thalamic input to M1 arises mainly from another part of the ventral lateral nucleus, called *VLc*, which relays information from the cerebellum. Besides projecting directly to the spinal cord, layer V pyramidal cells also

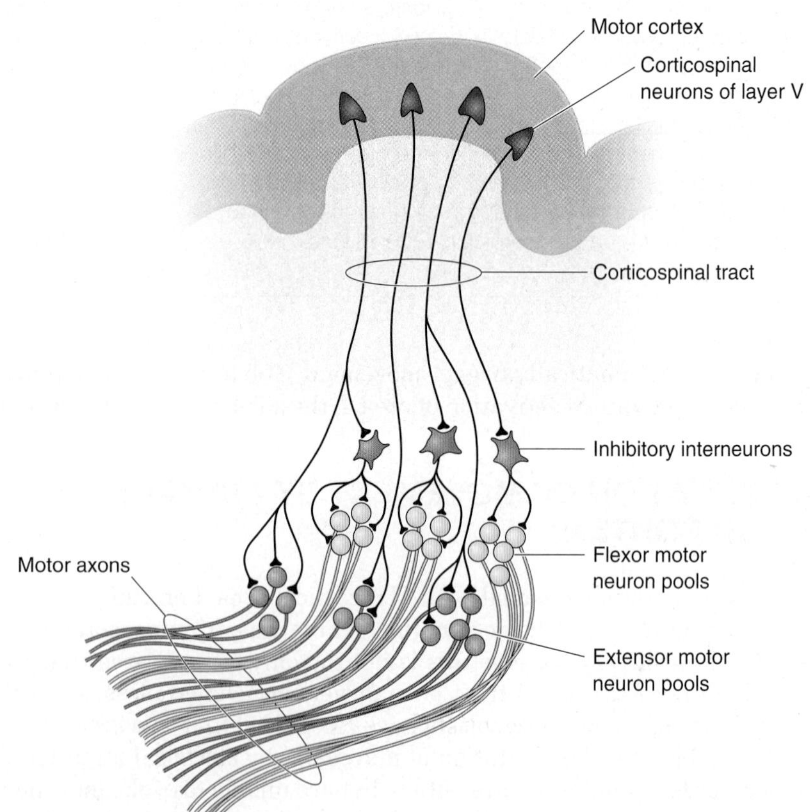

▲ FIGURE 14.15
Corticospinal tract axons control pools of motor neurons. Large pyramidal neurons in layer V of the motor cortex project axons, via the corticospinal tract, to the ventral horns of the spinal cord. In this case, the axons directly excite pools of extensor motor neurons and indirectly (via interneurons) inhibit pools of flexor motor neurons that serve as antagonists of the extensors. (Source: Adapted from Cheney et al., 1985.)

send axon collaterals to many subcortical sites involved in sensorimotor processing, especially the brain stem.

The Coding of Movement in M1

Researchers previously thought the motor cortex consisted of a detailed mapping of the individual muscles, such that the activity of a single pyramidal cell would lead to activity in a single motor neuron pool. However, the view that has emerged from more recent work is that individual pyramidal cells can drive numerous motor neuron pools from a group of different muscles involved in moving a limb toward a desired goal. Recordings from M1 neurons in behaving animals have revealed that a burst of activity occurs immediately before and during a voluntary movement, and that this activity appears to encode two aspects of the movement: force and direction.

Because cortical microstimulation studies had suggested the existence of a fine-grained movement map in M1, the discovery that the movement direction tuning of individual M1 neurons is rather broad came as a surprise. This breadth of tuning is shown clearly in a type of experiment devised by Apostolos Georgopoulos and his colleagues, then working at Johns Hopkins University. Monkeys were trained to move a joystick toward a small light whose position varied randomly around a circle. Some M1 cells fired most vigorously during movement in one direction (180° in the example in Figure 14.16a) but also discharged during movement angles that varied considerably from the preferred direction. The coarseness in the directional tuning of the corticospinal neurons was certainly at

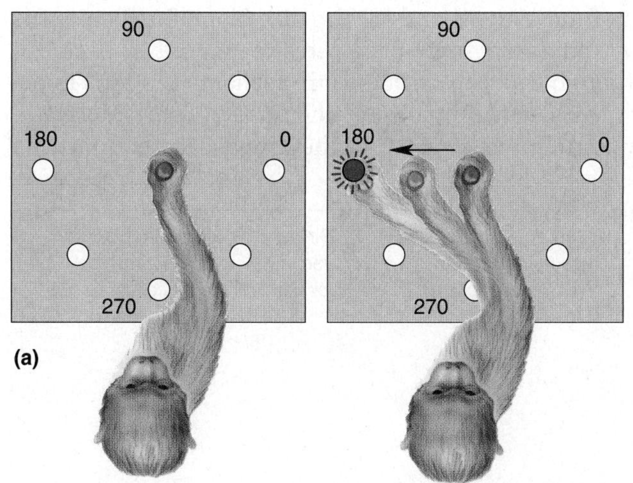

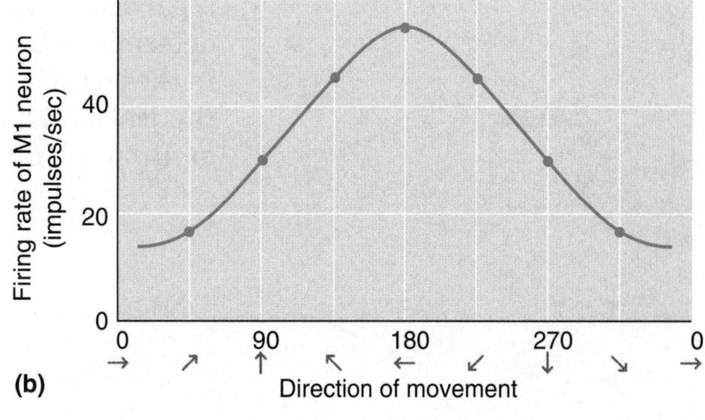

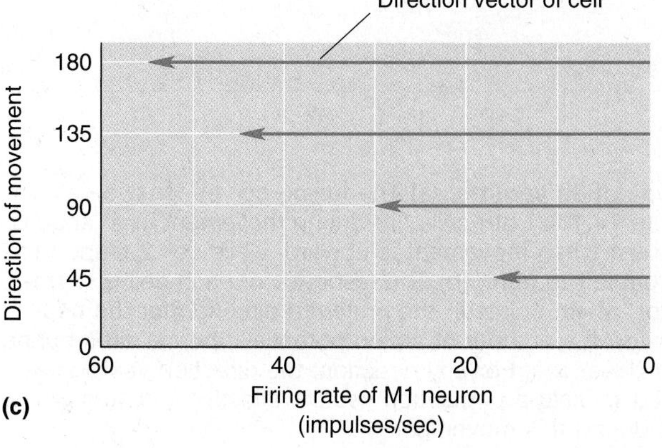

▲ FIGURE 14.16

Responses of an M1 neuron during arm movements in different directions. **(a)** As the monkey moves a handle toward a small light, the responses of an M1 neuron are monitored. When the monkey moves in directions around the clock, the relationship between the cell's discharge rate and movement direction can be determined. **(b)** A tuning curve for an M1 neuron. This cell fires most during movements to the left. **(c)** Because the cell in part b responds best to leftward movement, it is represented by a direction vector pointing in that direction. The length of the vector is proportional to the firing rate of the cell. Notice that as the movement direction changes, the length of the direction vector changes. (Source: Adapted from Georgopoulos et al., 1982.)

odds with the high accuracy of the monkey's movements, suggesting that the direction of movement could not be encoded by the activity of individual cells that command movement in a single direction. Georgopoulos hypothesized that movement direction was encoded instead by the collective activity of a population of neurons. Recall the role of neuronal **population coding** in the sensory systems, where the responses of many broadly tuned neurons are used to specify the properties of a particular stimulus (for example, see Chapter 8). Population coding in the motor system implies that groups of neurons are broadly tuned for the properties of movements.

To test the feasibility of the idea of population coding for movement direction, Georgopoulos and his colleagues recorded from over 200 different neurons in M1; for each cell, they constructed a directional tuning curve such as that shown in Figure 14.16b. From these data, the researchers knew how vigorously each of the cells in the population responded during movement in each direction. The activity of each cell was represented as a *direction vector* pointing in the direction that was best for that cell; the length of the vector represented how active that cell had been during a particular movement (Figure 14.16c). The vectors representing each cell's activity could be plotted together for each direction of movement, then averaged to yield what the researchers called a *population vector* (Figure 14.17). They found a strong correlation between this average vector, representing the activity of the entire population of M1 cells, and the actual direction of movement (Figure 14.18).

These studies suggest three important conclusions about how M1 commands voluntary movement: (1) much of the motor cortex is active for every movement, (2) the activity of each cell represents a single "vote" for a particular direction of movement, and (3) the direction of movement is determined by a tally (and averaging) of the votes registered by each cell in the population. Although this population-coding scheme remains hypothetical in M1, experiments on the superior colliculus by James McIlwain at Brown University and David Sparks at the University of Alabama

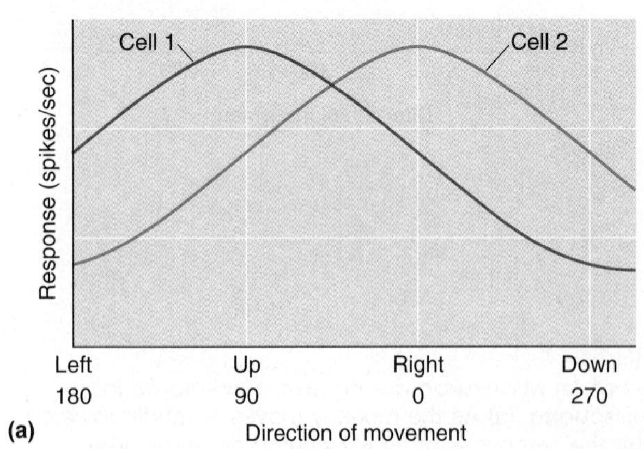

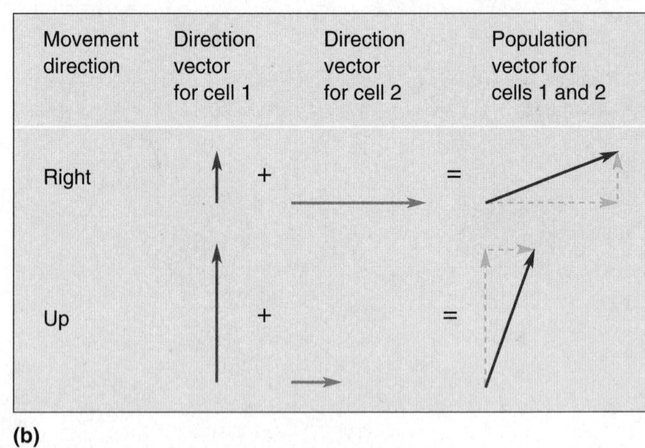

▲ FIGURE 14.17
Direction vectors and population vectors. **(a)** The tuning curves for two cells in the motor cortex (see Figure 14.16). Both cells fire during movement in a range of directions, but cell 1 fires best when movement is upward, while cell 2 responds best when movement is from left to right. **(b)** The response of each cell is represented as a direction vector, which points in the preferred direction for the neuron, but its length depends on the number of action potentials the cell fires during movement over a range of directions. For any direction, the direction vectors of individual cells are averaged to yield a population vector, reflecting the strength of the response of both cells during this movement.

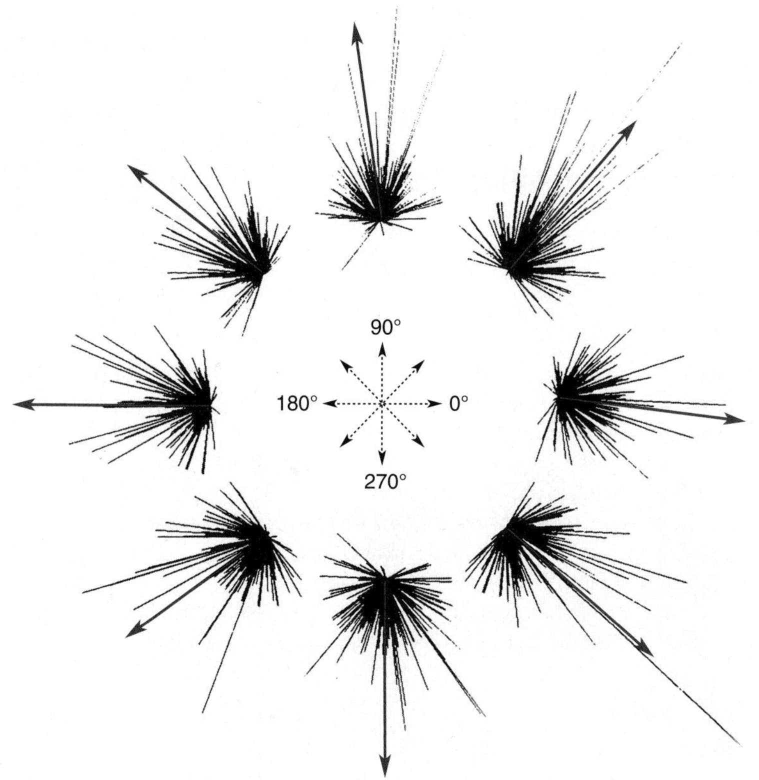

▲ FIGURE 14.18
Predicting the direction of movement by population vectors. Each cluster of lines reflects the direction vectors of many cells in M1. Line length reflects the discharge rate of each cell during a movement in one of eight different directions. Arrows represent the average population vectors, which predict the movement direction of the monkey's arm. (Source: Georgopoulos et al., 1983.)

showed conclusively that a population code is used by this structure to command precisely directed eye movements (Box 14.5).

The Malleable Motor Map. This scheme for motor control leads to an interesting prediction: the larger the population of neurons representing a type of movement, the finer the possible control. From the motor map shown in Figure 14.8, we would predict that finer control should be possible for the hands and the muscles of facial expression, and indeed this is normally the case. Of course, fine movements of other muscles can be learned with experience; consider the finger, wrist, elbow, and shoulder movements of an accomplished cellist. Does this mean that cortical cells in M1 can switch allegiance from participation in one type of movement to another as skills are learned? The answer appears to be yes. John Donoghue, Jerome Sanes, and their students at Brown University collected evidence indicating that such plasticity of the adult motor cortex is possible. For example, in one series of experiments, they used cortical microstimulation in rats and mapped the regions of M1 that normally elicit movements of the forelimb, facial whiskers, or muscles around the eye (Figure 14.19a). Then they cut the motor nerve that supplies the muscles of the snout and its whiskers and found that regions of M1 that had evoked whisker movements now would elicit either forelimb or eye movements (Figure 14.19b). The motor map had been reorganized. These neuroscientists speculated that similar types of cortical reorganization might provide a basis for learning fine motor skills.

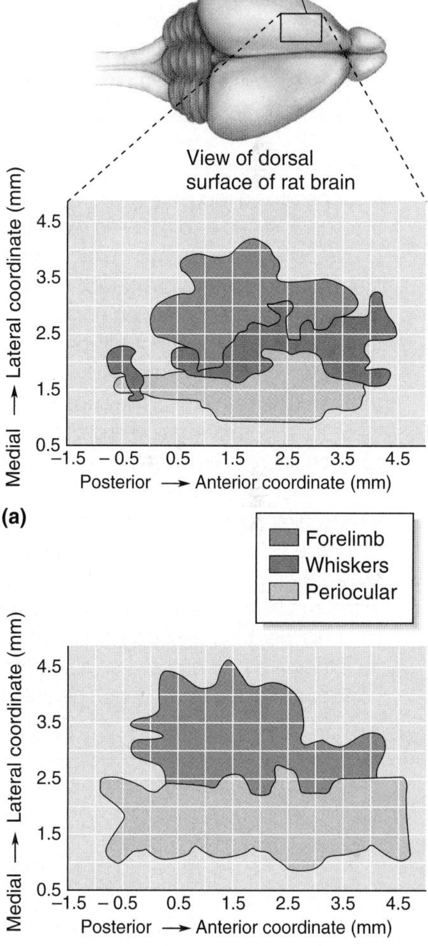

▲ FIGURE 14.19
Malleable motor maps. **(a)** This map represents motor cortex from a normal rat. **(b)** This map represents a rat that had the motor nerve cut that supplies the whiskers (vibrissae). Notice the cortical regions that previously evoked movement of the vibrissae now cause muscle movement in the forelimb or around the eyes (periocular). (Source: Adapted from Sanes and Donoghue, 1997.)

BOX 14.5 PATH OF DISCOVERY

Distributed Coding in the Superior Colliculus

by James T. McIlwain

During the 1960s and 1970s, it became relatively easy to record the electrical activity of single neurons in the brain. The power and promise of this method led to the idea of the "feature-detector neuron," whose discharge was thought to announce the presence of the stimulus feature to which it was most sensitive. This view rarely tempted students of olfaction and gustation because they had found that the discharge of single chemosensory neurons is highly ambiguous with respect to the identity of the stimulus. Not so in vision research, where the cells with the smallest spatial receptive fields and the most refined preferences for specific stimuli seemed most interesting. Any neuron responsive to a range of stimuli was regarded as crudely selective and unsuited to processes requiring high resolution.

Neurophysiologists who examined visual areas of the brain stem soon encountered a paradox. The receptive fields of cells in the superior colliculus turned out to be very large, yet this structure was clearly important for the execution of highly accurate saccadic eye movements, which change the direction of gaze to a stimulus of interest. The superior colliculus receives orderly input from both the retina and the visual cortex, and damage to it impairs an animal's ability to direct its gaze at novel stimuli. Focal electrical stimulation of the colliculus evokes saccades whose directions and amplitudes are correlated with the visual receptive fields of cells at the stimulus site. A small change in the position of the stimulating electrode results in small changes in the saccade's direction and amplitude. Certain collicular neurons discharge in association with saccadic eye movements, as if they are part of the control mechanism that specifies the dimensions of the movement. This activity occurs in association with saccades that terminate across a restricted zone of visual space called the *movement field* of the cell, by analogy to the sensory receptive field. How could such cells specify the target of a saccade with any accuracy if their movement fields and visual receptive fields are very large?

The answer began to emerge from experiments on the cat's superior colliculus in my laboratory at Brown University and from studies of the primate's superior colliculus by David Sparks, then at the University of Alabama. We asked the inverse of the traditional receptive field question or movement field question. From the sensory side, instead of asking where a point of light must be located to activate a collicular neuron, we asked where in the superior colliculus are the cells that have the point in their receptive fields—that "see" the point. Similarly, the key consideration on the motor side is the location of the cells that discharge before a saccade to a given target, rather than the size of their individual movement fields. The analyses from both laboratories revealed that these regions of activity are widespread, occupying considerable fractions of the collicular tissue. As the stimulus or target location moves around in visual space, the corresponding patch of neural activity moves around in the superior colliculus.

A general idea of how a system of such neurons may encode a saccade is shown in Figure A. On the left, each arrow on the retinotopic map of the superior colliculus symbolizes the contribution of its location to the code for the direction of a saccade. The more closely packed the arrows, the stronger the signal from that region to brain stem circuits that shape the motor commands for the saccade. The distribution of the

From the preceding discussion, we can imagine that when the time has come for our baseball pitcher to wind up, his motor cortex generates a torrent of activity in the pyramidal tract. What could appear to be a discordant voice to a neurophysiologist recording from a single M1 neuron is part of a clear chorus of activity to the spinal motor neurons that generate the precise movements necessary to propel the baseball accurately.

THE CEREBELLUM

It is not enough to simply command the muscles to contract. Throwing a ball requires a detailed *sequence* of muscle contractions, each one generating exactly the right amount of force at precisely the right time. These critical motor control functions belong to the **cerebellum** (introduced in Chapter 7). The importance of the cerebellum in this aspect of arrows is consistent with the effects of focal electrical stimulation. Thus, for example, stimulation at successively lower points in the lower half of the map (representing the lower visual field) leads to downwardly directed saccades of increasing amplitude. If the appearance of target 1 in the right side of the figure excited cells in the lower unshaded oval of the map, the collective activity of the cells would specify the downward and horizontal components needed for a saccade to the target. The appearance of target 2 would activate the upper unshaded area, assembling the components of the correct, upwardly directed saccade. In this model, changes in the position of the target modify the composition of the output signal to yield a saccade matching any position of the target.

The large size of the receptive fields and movement fields of collicular neurons means that information about the location of a visual point or a saccade target is distributed among many neurons. The model of Figure A shows how the active population of collicular neurons may indicate the target's position as a motor code. The simple and incomplete model shown here represents only one of several ideas about how the superior colliculus accomplishes its task. It seems certain, though, that target location and saccade measurements are encoded in the distribution of activity across a population of neurons.

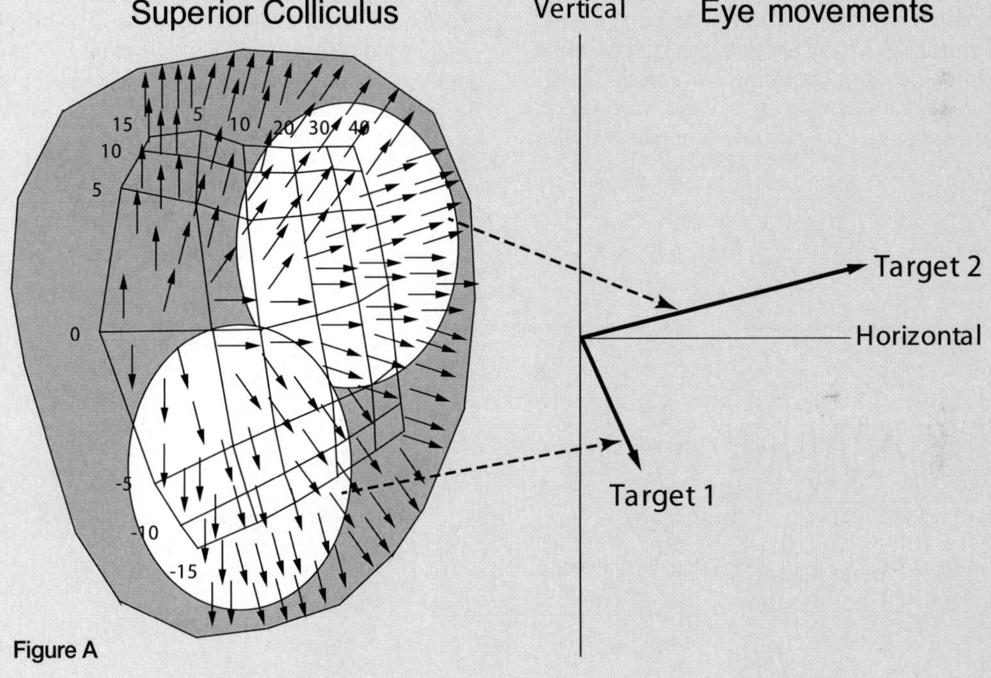

Figure A

motor control is plainly revealed by cerebellar lesions; movements become uncoordinated and inaccurate, a condition known as **ataxia**.

Take this simple test. Lay your arms in your lap for a moment, then touch your nose with one finger. Try it again with your eyes closed. No problem, right? Patients with cerebellar damage are often incapable of performing this simple task. Instead of smoothly and simultaneously moving the shoulder, elbow, and wrist to bring the finger to rest on the nose, they move each joint sequentially—first the shoulder, then the elbow, and finally the wrist. This is called *dyssynergia*, decomposition of synergistic multijoint movement. Another characteristic deficit shown by these patients is that their finger movement will be *dysmetric*; they will either come up short of the nose or shoot past it, poking themselves in the face. You may recognize these symptoms as similar to those that accompany ethanol intoxication. Indeed, much of the clumsiness that accompanies alcohol abuse is a direct consequence of the depression of cerebellar circuits (Box 14.6).

BOX 14.6 OF SPECIAL INTEREST

Involuntary Movements—Normal and Abnormal

Raise your hand in front of your face, and try to hold it as still as possible. You will see a very slight trembling of your fingers. This is called *physiological tremor*, a small, rhythmic oscillation of about 8–12 Hz. It is perfectly normal, and there is nothing you can do to stop it short of resting your hand on the table. A variety of everyday circumstances—stress, anxiety, hunger, fatigue, fever, too much caffeine—can enhance the tremor.

As we have discussed in this chapter, some neurological diseases lead to more dramatic involuntary movements, with distinctive characteristics. Parkinson's disease is often associated with a large *resting tremor* of about 3–5 Hz. Movement is at its worst when the patient is not attempting to move. Strangely enough, the tremor immediately disappears during a voluntary movement. On the other hand, people with cerebellar damage have no abnormal tremor at rest but often show dramatic *intention tremor* when they try to move. Cerebellar tremor is an expression of ataxia, the uncoordinated contractions of the muscles used in the movement. For example, as the patient tries to move her finger from one point in space to another, or track a pathway with her finger, she makes large errors; in trying to correct an error, she makes more errors, and so on, as the finger wobbles inaccurately toward its destination.

Huntington's disease causes *chorea* (from the Greek for "dance")—quick, irregular, involuntary but relatively coordinated movements of the limbs, trunk, head, and face. Other types of basal ganglia disease can lead to *athetosis*—much slower, almost writhing movements of the neck and trunk. The unique properties of each abnormal movement can help in the diagnosis of neurological diseases, and they have taught us volumes about the normal functions of the damaged parts of the brain.

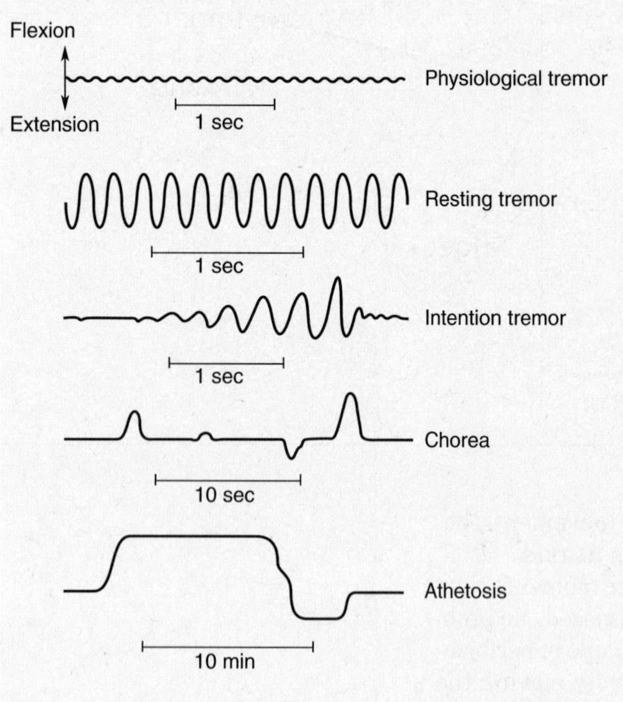

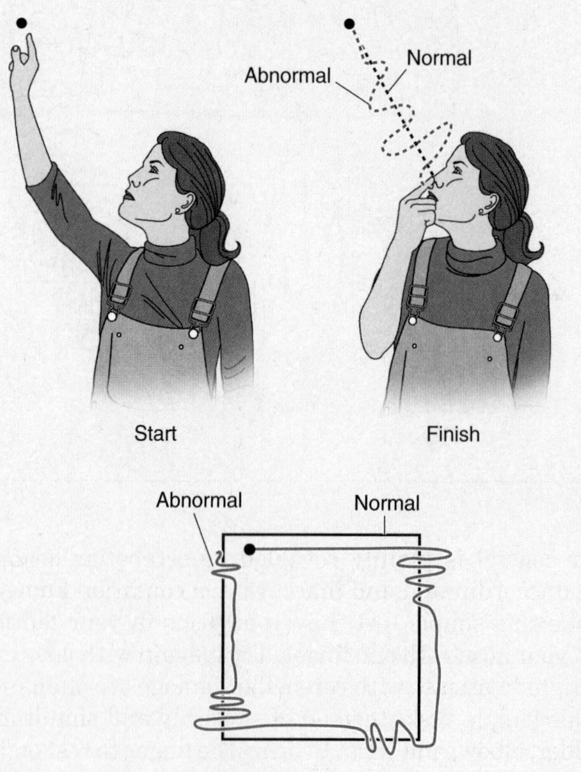

Anatomy of the Cerebellum

Cerebellar anatomy is shown in Figure 14.20. The cerebellum sits on stout stalks of axons called *peduncles* that rise from the pons; the whole structure resembles a piece of cauliflower. The visible part of the cerebellum is actually a thin sheet of cortex, which is repeatedly folded. The dorsal surface is characterized by a series of shallow ridges called *folia* (singular: folium), which run transversely (from side to side). In addition, there are deeper transverse fissures, revealed by making a sagittal slice through the cerebellum; these divide the cerebellum into 10 lobules. Together, folia and lobules serve to greatly increase the surface area of the cerebellar cortex, as the gyri of the cerebrum do for the cerebral cortex. Neurons are also embedded deep within the white matter of the cerebellum, forming the *deep cerebellar nuclei*, which relay most of the cerebellar cortical output to various brain stem structures. The cerebellum constitutes only about 10% of the total volume of the brain, but its cortex has an astonishingly high density of neurons. The vast majority of these are tiny excitatory neurons called *granule cells*, whose somata lie in the granule cell layer (Figure 14.21a,b). The number of granule neurons in the cerebellum is about equal to the number of other neurons in the entire CNS. The largest neuron in the cerebellar cortex is the inhibitory Purkinje cell, which receives excitatory input from granule cells in the molecular layer and sends its inhibitory axons to the deep cerebellar nuclei (Figure 14.21c).

Unlike the cerebrum, the cerebellum is not obviously split down the middle. At the midline, the folia appear to run uninterrupted from one side to the other. The only distinguishing feature of the midline is a bump that runs like a backbone down the length of the cerebellum. This midline region is called the **vermis** (from the Latin for "worm"), and it separates the two lateral **cerebellar hemispheres** from each other. The vermis and the hemispheres represent important functional divisions. The vermis sends output to the brain stem structures that contribute

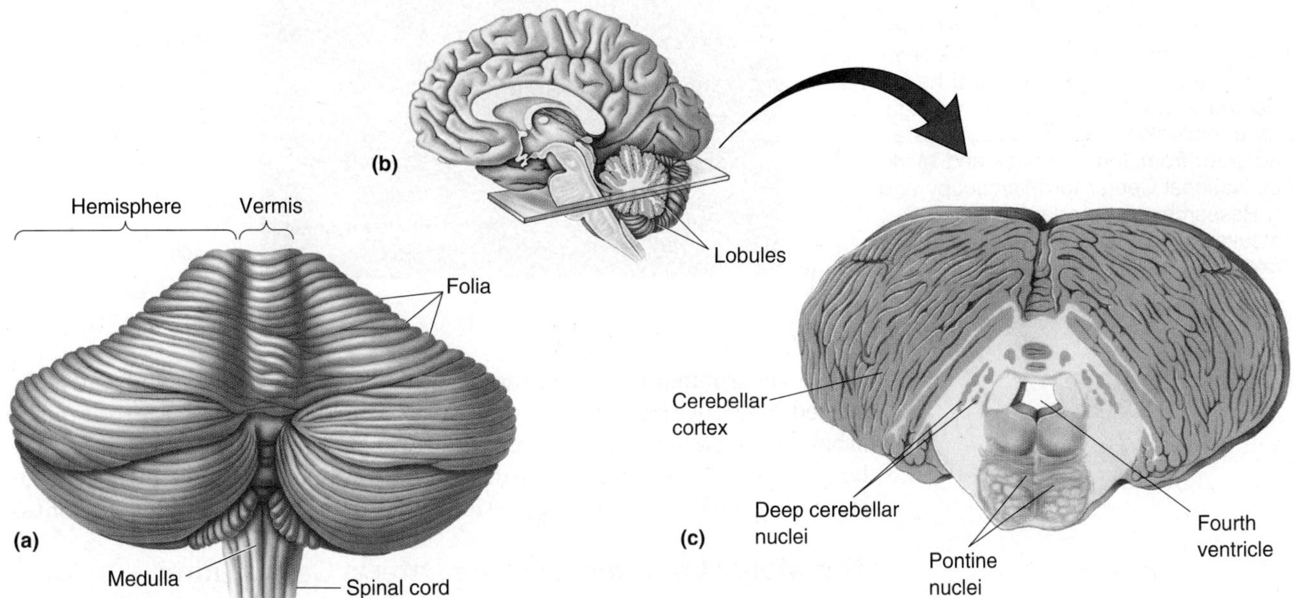

▲ FIGURE 14.20
The cerebellum. **(a)** A dorsal view of the human cerebellum, showing the vermis and hemispheres. **(b)** A midsagittal view of the brain, showing the lobules of the cerebellum. **(c)** A cross section of the cerebellum, showing the cortex and deep nuclei.

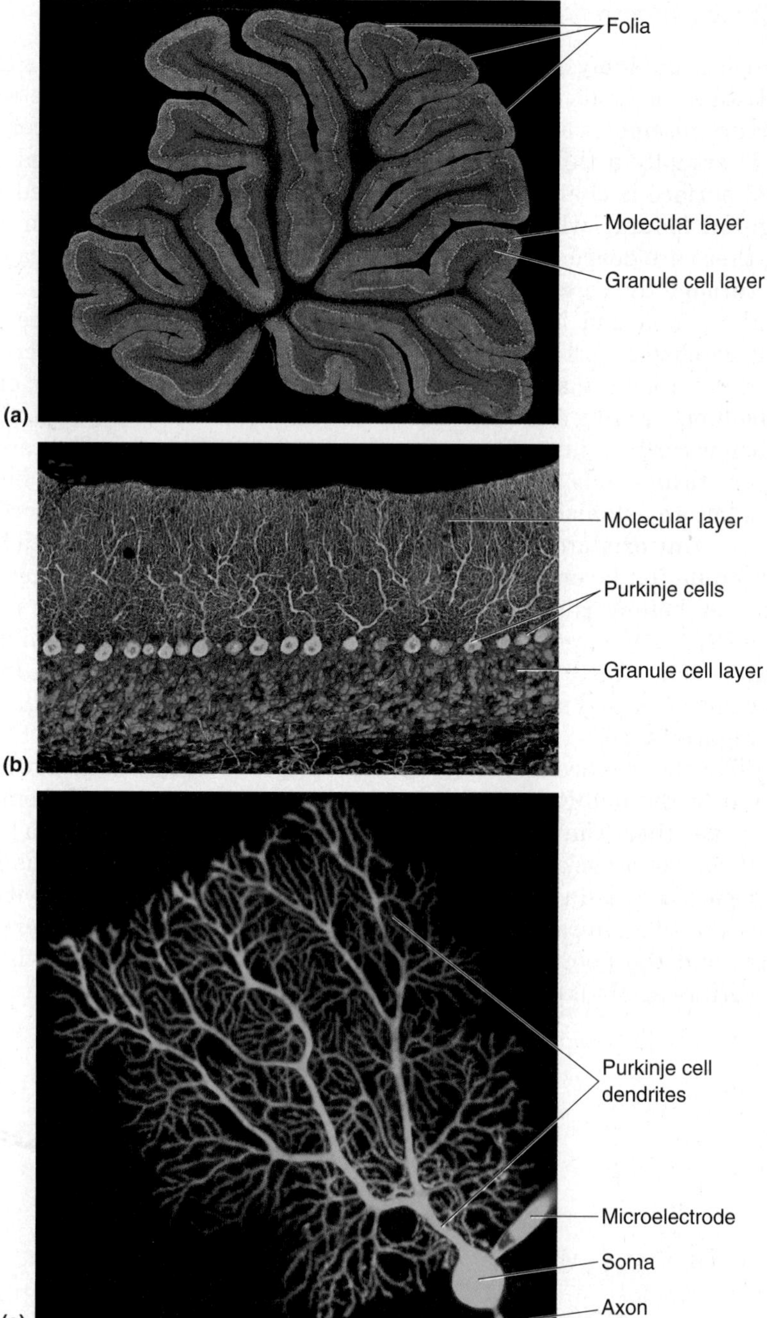

▶ FIGURE 14.21
Neurons of the cerebellar cortex. **(a)** A histological section through the folia of the cerebellar cortex. Fluorescent stains color the molecular layer green and the granular cell layer blue. **(b)** A close-up of the layers of the cerebellar cortex. **(c)** A Purkinje cell after it has been injected with a fluorescent dye through the tip of a microelectrode. (Source: Parts a and b adapted from Tom Deerinck and Mark Ellisman, National Center for Microscopy and Imaging Research; part c adapted from Tetsuya Tatsukawa, RIKEN Brain Science Institute, Wako, Japan)

to the ventromedial descending spinal pathways, which, as already discussed, control the axial musculature. The hemispheres are related to other brain structures that contribute to the lateral pathways, particularly the cerebral cortex. For the purpose of illustration, we'll focus on the lateral cerebellum, which is particularly important for limb movements.

The Motor Loop through the Lateral Cerebellum

The simplest circuit involving the lateral cerebellum constitutes yet another loop, shown schematically in Figure 14.22. Axons arising from layer V pyramidal cells in the sensorimotor cortex—frontal areas 4 and 6, somatosensory areas on the postcentral gyrus, and the posterior pari-

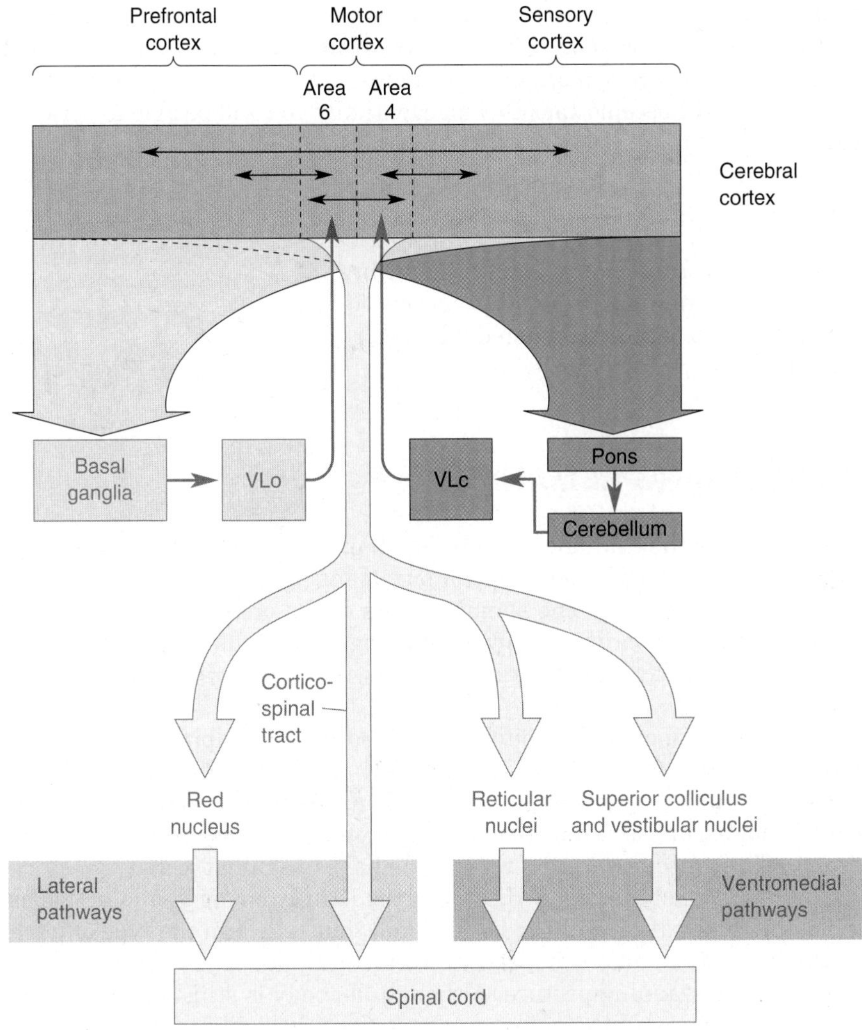

◀ FIGURE 14.22
A summary of the motor loop through the cerebellum.

etal areas—form a massive projection to clusters of cells in the pons, the **pontine nuclei**, which in turn feed the cerebellum. To appreciate the size of this pathway, consider that the corticopontocerebellar projection contains about 20 million axons; that's *20 times* more than in the pyramidal tract! The lateral cerebellum then projects back to the motor cortex via a relay in the ventral lateral nucleus of the thalamus (VLc).

From the effects of lesions in this pathway, we can deduce that it is critical for the proper execution of planned, voluntary, multijoint movements. Indeed, once the cerebellum has received the signal for movement intent, its activity appears to instruct the primary motor cortex with respect to movement direction, timing, and force. For ballistic movements, these instructions are based entirely on predictions about their outcome (because such movements are too fast for sensory feedback to be of much immediate use). Such predictions are based on past experience; that is, they are learned. Therefore, the cerebellum is another important site for motor learning; it is a place where *what is intended* is compared with *what has happened*. When this comparison fails to meet expectations, compensatory modifications are made in certain cerebellar circuits.

Programming the Cerebellum. We will return to cerebellar circuitry and how it is modified by experience in Chapter 25. But for now, think about the process of learning a new motor skill (e.g., skiing, piano playing, juggling, knitting, throwing a curve ball). Early on, you must concentrate on new movements, and you tend to perform them in a disjointed and uncoordinated fashion. Practice makes perfect, however. As you master the skill, the movements will become smooth, and eventually you will be able to perform them almost unconsciously. This process represents the creation of a new motor program that generates the appropriate movement sequences on demand without the need for conscious control.

Recall that the word "cerebellum" derives from the Latin for "little brain." The cerebellum acts as a small brain within the brain, unconsciously determining that the programs for skilled movement are executed properly and are adjusted whenever their execution fails to meet expectations.

CONCLUDING REMARKS

Let's return to the example of the baseball pitcher one last time to put the different pieces of the motor control puzzle together. Imagine the pitcher walking to the mound. The spinal circuits of the crossed extensor reflex are engaged and coordinated by descending commands on the ventromedial pathways. Extensors contract, flexors relax; flexors contract, extensors relax.

Once on the mound, the pitcher is joined by the umpire. Into his outstretched hand the umpire drops a new baseball. The added weight stretches the flexors of the arm. Group Ia axons become more active and cause monosynaptic excitation of the motor neurons innervating the flexors. The muscles contract to hold the ball up against gravity.

He is now ready to pitch. His neocortex is fully engaged and active as he looks at the catcher for the hand signal that tells him the type of pitch to throw. At the same time, the ventromedial pathways are working to maintain his standing posture. Although his body is still, the neurons of the ventral horns of the spinal cord are firing steadily under the influence of the ventromedial pathways, keeping the extensors of the lower leg activated.

The catcher flashes the sign for the curve ball. The sensory information is communicated from the occipital cortex to the parietal and prefrontal cortex. These regions of cortex along with area 6 begin planning the movement strategy.

The batter steps up to the plate and is ready. Activity cycling through the basal ganglia increases, triggering the initiation of the pitch. In response to this input to the cerebral cortex, SMA activity increases, followed immediately by the activation of M1. Now instructions are sweeping down the axons of the lateral pathways. The cerebellum, activated by the corticopontocerebellar inputs, uses these instructions to coordinate the timing and force of the descending activity so the proper sequence of muscle contractions can occur. Cortical input to the reticular formation leads to the release of the antigravity muscles from reflex control. Finally, lateral pathway signals engage the motor neurons and interneurons of the spinal cord, which cause muscles of the arms and legs to contract.

The pitcher winds up and throws. The batter swings. The ball sails over the left-field fence. The crowd jeers; the manager curses; the team owner frowns. Even as the pitcher's cerebellum goes to work making adjustments for the next pitch, his body reacts. His face flushes; he sweats; he's angry and anxious. But these latter reactions are not the stuff of the somatic motor system. These are topics of Part III, The Brain and Behavior.

KEY TERMS

Descending Spinal Tracts
lateral pathway (p. 485)
ventromedial pathway (p. 485)
corticospinal tract (p. 486)
motor cortex (p. 486)
pyramidal tract (p. 486)
rubrospinal tract (p. 486)
red nucleus (p. 486)
vestibulospinal tract (p. 489)
tectospinal tract (p. 489)
reticular formation (p. 490)
pontine reticulospinal tract (p. 490)
medullary reticulospinal tract (p. 491)

The Planning of Movement by the Cerebral Cortex
primary motor cortex (M1) (p. 492)
premotor area (PMA) (p. 493)
supplementary motor area (SMA) (p. 493)
mirror neuron (p. 495)

The Basal Ganglia
ventral lateral (VL) nucleus (p. 498)
basal ganglia (p. 498)
caudate nucleus (p. 498)
putamen (p. 498)
globus pallidus (p. 498)
subthalamic nucleus (p. 498)
substantia nigra (p. 498)
striatum (p. 498)
Parkinson's disease (p. 501)
Huntington's disease (p. 503)
ballism (p. 504)

The Initiation of Movement by the Primary Motor Cortex
motor strip (p. 505)
population coding (p. 508)

The Cerebellum
cerebellum (p. 510)
ataxia (p. 511)
vermis (p. 513)
cerebellar hemispheres (p. 513)
pontine nuclei (p. 515)

REVIEW QUESTIONS

1. List the components of the lateral and ventromedial descending spinal pathways. Which type of movement does each path control?
2. You are a neurologist presented with a patient who has the following symptom: an inability to independently wiggle the toes on the left foot, but with all other movements (walking, independent finger movement) apparently intact. You suspect a lesion in the spinal cord. Where?
3. PET scans can be used to measure blood flow in the cerebral cortex. What parts of the cortex show increased blood flow when a subject is asked to think about moving her right finger?
4. Why is L-dopa used to treat Parkinson's disease? How does it act to alleviate the symptoms?
5. Individual Betz cells fire during a fairly broad range of movement directions. How might they work together to command a precise movement?
6. Sketch the motor loop through the cerebellum. What movement disorders result from damage to the cerebellum?

FURTHER READING

Alstermark B, Isa T. 2012. Circuits for skilled reaching and grasping. *Annual Review of Neuroscience* 35:559–578.

Blumenfeld H. 2011. *Neuroanatomy Through Clinical Cases*, 2nd ed. Sunderland, MA: Sinauer.

Donoghue J, Sanes J. 1994. Motor areas of the cerebral cortex. *Journal of Clinical Neurophysiology* 11:382–396.

Foltynie T, Kahan J. 2013. Parkinson's disease: an update on pathogenesis and treatment. *Journal of Neurology* 260:1433–1440.

Glickstein M, Doron K. 2008. Cerebellum: connections and functions. *Cerebellum* 7:589–594.

Graziano M. 2006. The organization of behavioral repertoire in motor cortex. *Annual Review of Neuroscience* 29:105–134.

Lemon RN. 2008. Descending pathways in motor control. *Annual Review of Neuroscience* 31:195–218.

Rizzolatti G, Sinigaglia C. 2008. *Mirrors in the Brain: How Our Minds Share Actions and Emotions.* New York: Oxford University Press.

PART THREE

The Brain and Behavior

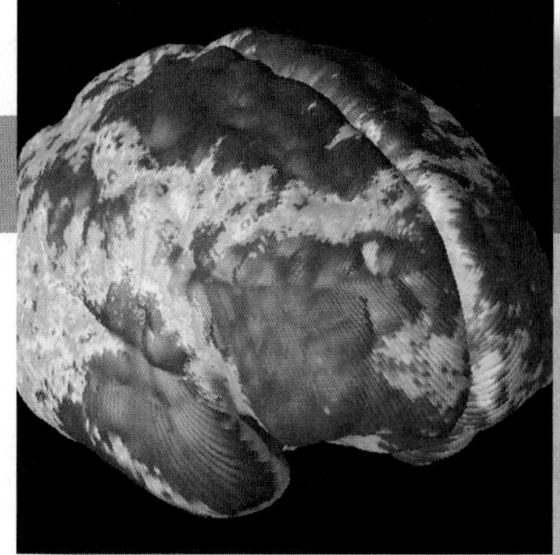

CHAPTER FIFTEEN

Chemical Control of the Brain and Behavior

INTRODUCTION

It should be obvious by now that knowing the organization of synaptic connections is essential to understanding how the brain works. It's not from a love of Greek and Latin that we belabor neuroanatomy! Most of the connections we have described are precise and specific. For example, for you to be able to read these words, there must be a very fine-grained neural mapping of the light falling on your retina—how else could you see the dot in this question mark? The information must be carried centrally and dispersed precisely to many parts of the brain for processing, coordinated with control of the motor neurons that closely regulate the six muscles of each eye as it scans the page.

In addition to anatomical precision, point-to-point communication in the sensory and motor systems requires mechanisms that restrict synaptic communication to the cleft between the axon terminal and its target. It just wouldn't do for glutamate released in the somatosensory cortex to activate neurons in the motor cortex! Furthermore, transmission must be brief enough to allow rapid responses to new sensory inputs. Thus, at these synapses, only minute quantities of neurotransmitter are released with each impulse, and these molecules are then quickly destroyed enzymatically or taken up by neighboring cells. The postsynaptic actions at transmitter-gated ion channels last only as long as the transmitter is in the cleft, a few milliseconds at most. Many axon terminals also possess presynaptic "autoreceptors" that detect the transmitter concentrations in the cleft and inhibit release if they get too high. These mechanisms ensure that this type of synaptic transmission is tightly constrained, in both space and time.

The elaborate mechanisms that constrain point-to-point synaptic transmission are somewhat like those in telecommunications. Telephone systems make possible very specific connections between one place and another so that your mother in Tacoma can talk just to you in Providence, reminding you that her birthday was last week. The telephone lines or cellular transmissions act like precise synaptic connections. The influence of one neuron (your mother) is targeted to a small number of other neurons (in this case, only you). The embarrassing message is limited to your ears only. The influence of a neuron in one of the sensory or motor systems discussed so far usually extends to the few dozen or few hundred cells it synapses on—a conference call, to be sure, but still relatively specific.

Now imagine your mother being interviewed on a television talk show broadcast on a satellite network. The widespread satellite transmission may allow her to tell millions of people that you forgot her birthday, and the loudspeaker in each television set will announce the message to anyone within earshot. Likewise, certain neurons communicate with hundreds of thousands of other cells. These widespread systems tend to act relatively slowly, over seconds to minutes. Because of their broad, protracted actions, such systems in the brain can orchestrate entire behaviors, ranging from falling asleep to falling in love. Indeed, many of the behavioral dysfunctions collectively known as mental disorders are believed to result specifically from imbalances of certain of these chemicals.

In this chapter, we look at three components of the nervous system that operate in expanded space and time (Figure 15.1). One component is the *secretory hypothalamus*. By secreting chemicals directly into the bloodstream, the secretory hypothalamus can influence functions throughout both the brain and the body. A second component, controlled neurally by the hypothalamus, is the *autonomic nervous system (ANS)*, introduced

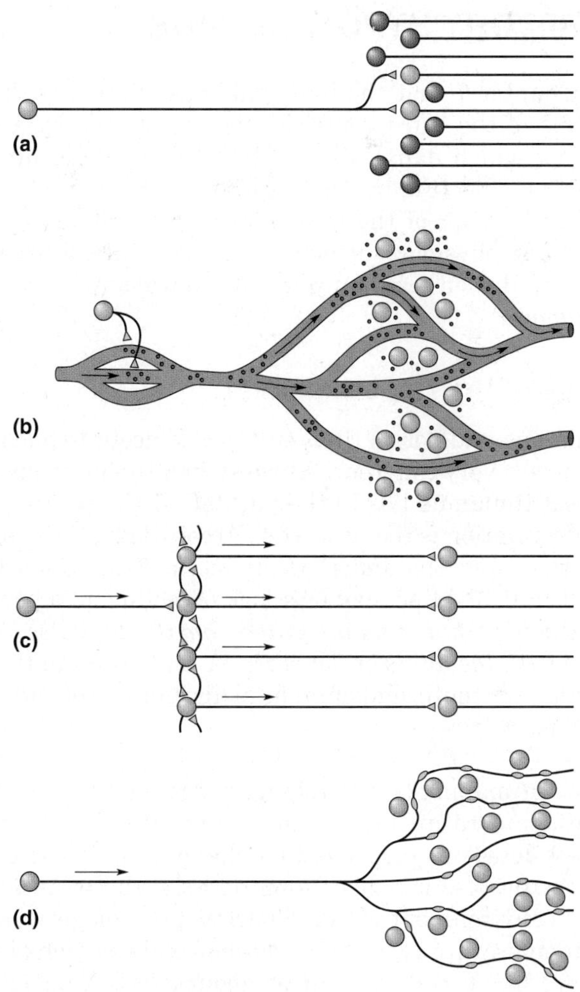

▲ FIGURE 15.1
Patterns of communication in the nervous system. (a) Most of the systems we have discussed in this book may be described as point-to-point. The proper functioning of these systems requires restricted synaptic activation of target cells and signals of brief duration. In contrast, three other components of the nervous system act over great distances and for long periods of time. (b) Neurons of the secretory hypothalamus affect their many targets by releasing hormones directly into the bloodstream. (c) Networks of interconnected neurons of the ANS can work together to activate tissues all over the body. (d) Diffuse modulatory systems extend their reach with widely divergent axonal projections.

in Chapter 7. Through extensive interconnections within the body, the ANS simultaneously controls the responses of many internal organs, blood vessels, and glands. The third component exists entirely within the central nervous system (CNS) and consists of several related cell groups that differ with respect to the neurotransmitter they use. All of these cell groups extend their spatial reach with highly divergent axonal projections and prolong their actions by using metabotropic postsynaptic receptors. Members of this component of the nervous system are called the *diffuse modulatory systems of the brain*. The diffuse systems are believed to regulate, among other things, the level of arousal and mood.

This chapter serves as a general introduction to these systems. Later chapters will explore how they contribute to specific behaviors and brain states: motivation (Chapter 16), sexual behavior (Chapter 17), emotion (Chapter 18), sleep (Chapter 19), and psychiatric disorders (Chapter 22).

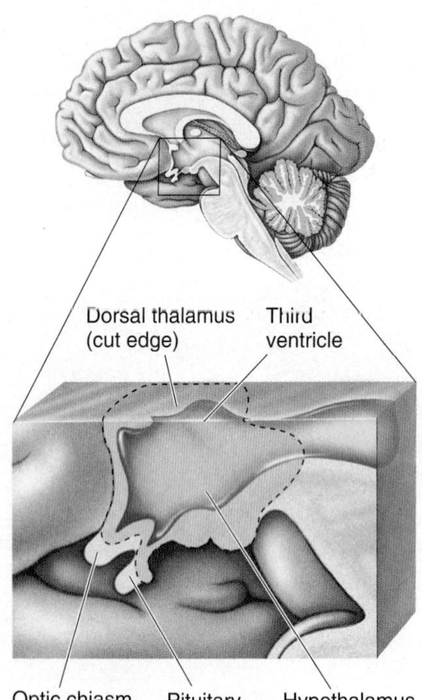

▲ FIGURE 15.2
Locations of the hypothalamus and pituitary. This is a midsagittal section. Notice that the hypothalamus, whose borders are indicated with a dashed line, forms the wall of the third ventricle and sits below the dorsal thalamus.

THE SECRETORY HYPOTHALAMUS

Recall from Chapter 7 that the hypothalamus sits below the thalamus, along the walls of the third ventricle. It is connected by a stalk to the pituitary gland, which dangles below the base of the brain, just above the roof of your mouth (Figure 15.2). Although this tiny cluster of nuclei makes up less than 1% of the brain's mass, the influence of the hypothalamus on body physiology is enormous. Let's take a brief tour of the hypothalamus and then focus on some of the ways in which it exerts its powerful influence.

An Overview of the Hypothalamus

The hypothalamus and dorsal thalamus are adjacent to one another, but their functions are very different. As we saw in the previous seven chapters, the dorsal thalamus lies in the path of all the point-to-point pathways whose destination is the neocortex. Accordingly, the destruction of a small part of the dorsal thalamus can produce a discrete sensory or motor deficit, such as a little blind spot or a lack of feeling on a portion of skin. In contrast, the *hypothalamus integrates somatic and visceral responses in accordance with the needs of the brain*. A tiny lesion in the hypothalamus can produce dramatic and often fatal disruptions of widely dispersed bodily functions.

Homeostasis. In mammals, the requirements for life include a narrow range of body temperatures and blood compositions. The hypothalamus regulates these levels in response to a changing external environment. This regulatory process is called **homeostasis**, the maintenance of the body's internal environment within a narrow physiological range.

Consider temperature regulation. Biochemical reactions in many cells of the body are fine-tuned to occur at about 37°C. A deviation of more than a few degrees in either direction can be catastrophic. Temperature-sensitive cells in the hypothalamus detect changes in brain temperature and orchestrate the appropriate responses. For example, if you stroll naked through the snow, the hypothalamus issues commands that cause you to shiver (generating heat in the muscles), develop goose bumps (a futile attempt to fluff up your nonexistent fur for better insulation—a reflexive remnant from our hairier ancestors), and turn blue (shunting blood *away from* cold surface tissues to keep the sensitive core of the body warmer). In contrast, when you go for a jog in the tropics, the hypothalamus activates heat-loss mechanisms that make you turn red (shunting blood *to* surface tissues where heat can radiate away) and sweat (cooling the skin by evaporation).

Other examples of homeostasis are the tight regulation of blood volume, pressure, salinity, acidity, and blood oxygen and glucose concentrations. The means by which the hypothalamus achieves these different types of regulation are remarkably diverse.

Structure and Connections of the Hypothalamus. Each side of the hypothalamus has three functional zones: lateral, medial, and periventricular (Figure 15.3). The lateral and medial zones have extensive connections with the brain stem and the telencephalon and regulate certain types of behavior, as we will see in Chapter 16. Here we are concerned only with the third zone, which actually receives much of its input from the other two.

The **periventricular zone** is so named because, with the exception of a thin finger of neurons that are displaced laterally by the optic tract

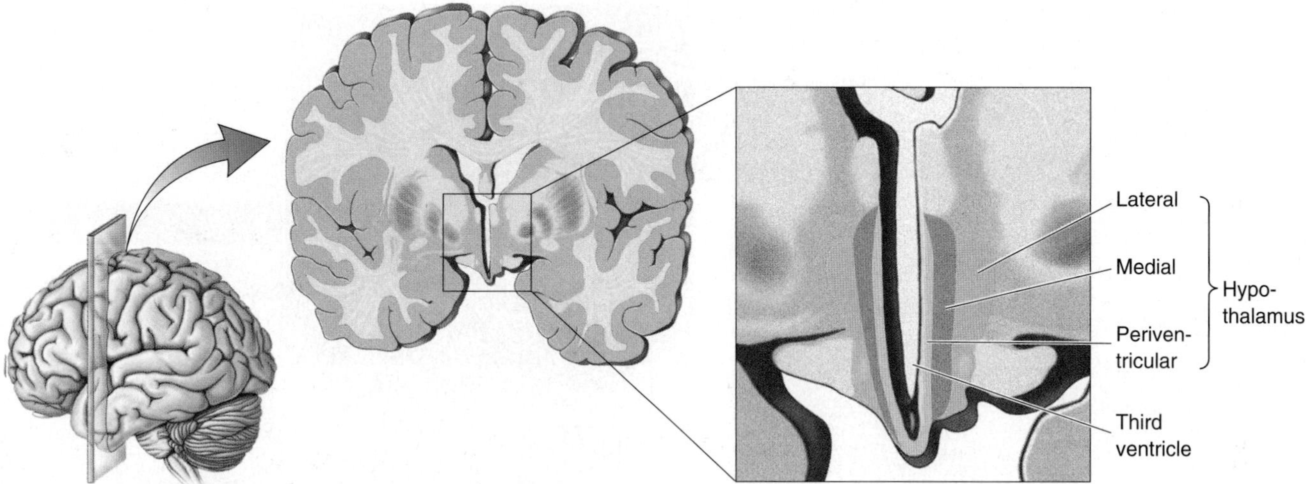

▲ FIGURE 15.3
Zones of the hypothalamus. The hypothalamus has three functional zones: lateral, medial, and periventricular. The periventricular zone receives inputs from the other zones, the brain stem, and the telencephalon. Neurosecretory cells in the periventricular zone secrete hormones into the bloodstream. Other periventricular cells control the autonomic nervous system.

(called the *supraoptic nucleus*), the cells of this region lie right next to the wall of the third ventricle. Within this zone exists a complex mix of neurons with different functions. One group of cells constitutes the *suprachiasmatic nucleus (SCN)*, which lies just above the optic chiasm. These cells receive direct retinal innervation and function to synchronize circadian rhythms with the daily light–dark cycle (see Chapter 19). Other cells in the periventricular zone control the ANS and regulate the outflow of the sympathetic and parasympathetic innervation of the visceral organs. The cells in a third group, called *neurosecretory neurons*, extend axons down toward the stalk of the pituitary gland. These are the cells that now command our attention.

Pathways to the Pituitary

We have said that the pituitary dangles below the base of the brain, which is true when the brain is lifted out of the head. In a living brain, however, the pituitary is gently held in a cradle of bone at the base of the skull. It requires this special protection because it is the "mouthpiece" from which much of the hypothalamus "speaks" to the body. The pituitary has two lobes, posterior and anterior. The hypothalamus controls the two lobes in different ways.

Hypothalamic Control of the Posterior Pituitary. The largest of the hypothalamic neurosecretory cells, **magnocellular neurosecretory cells**, extend axons down the stalk of the pituitary and into the posterior lobe (Figure 15.4). In the late 1930s, Ernst and Berta Scharrer, working at the University of Frankfurt in Germany, proposed that these neurons release chemical substances directly into the capillaries of the posterior lobe. At the time, this was quite a radical idea. It was well established that chemical messengers called *hormones* were released by glands into the bloodstream, but no one had thought that a neuron could act like a gland or that a neurotransmitter could act like a hormone. The Scharrers were correct, however. The substances released into the blood by neurons are now called **neurohormones**.

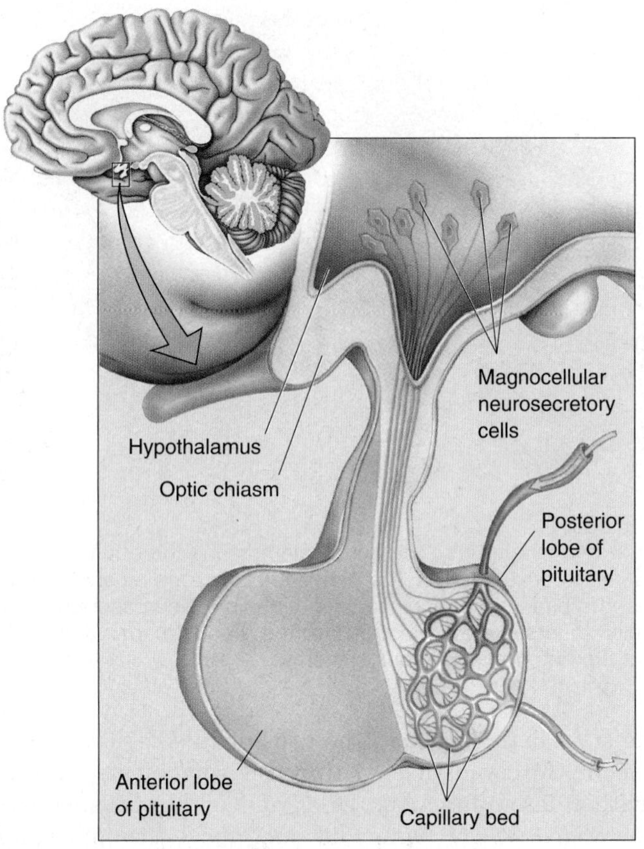

▲ FIGURE 15.4
Magnocellular neurosecretory cells of the hypothalamus. This is a midsagittal view of the hypothalamus and pituitary. Magnocellular neurosecretory cells secrete oxytocin and vasopressin directly into capillaries in the posterior lobe of the pituitary.

The magnocellular neurosecretory cells release two neurohormones into the bloodstream, oxytocin and vasopressin. Both of these chemicals are peptides, each consisting of a chain of nine amino acids. **Oxytocin** has sometimes been called the "love hormone" because levels rise during sexual or intimate behaviors and promote social bonding (discussed further in Chapter 17). In women, it also plays a critical role during the final stages of childbirth by causing the uterus to contract and facilitating the delivery of the newborn. It also stimulates the ejection of milk from the mammary glands. All lactating mothers know about the complex "letdown" reflex that involves the oxytocin neurons of the hypothalamus. Oxytocin release may be stimulated by the somatic sensations generated by a suckling baby. But the sight or sound of a baby (even someone else's) can also trigger the release of milk beyond the mother's conscious control. In each case, information about a sensory stimulus—somatic, visual, or auditory—reaches the cerebral cortex via the usual route, the thalamus, and the cortex ultimately stimulates the hypothalamus to trigger oxytocin release. The cortex can also suppress hypothalamic functions, such as when anxiety inhibits the letdown of milk.

Vasopressin, also called **antidiuretic hormone (ADH)**, regulates blood volume and salt concentration. When the body is deprived of water, the blood volume decreases and blood salt concentration increases. These changes are detected by pressure receptors in the cardiovascular system and salt concentration-sensitive cells in the hypothalamus, respectively.

Vasopressin-containing neurons receive information about these changes and respond by releasing vasopressin, which acts directly on the kidneys and leads to water retention and reduced urine production.

Under conditions of lowered blood volume and pressure, communication between the brain and the kidneys actually occurs in both directions (Figure 15.5). The kidneys secrete an enzyme into the blood called *renin*. Elevated renin sets off a sequence of biochemical reactions in the blood. *Angiotensinogen*, a large protein released from the liver, is converted by renin to *angiotensin I*, which breaks down further to form another small peptide hormone, *angiotensin II*. Angiotensin II has direct effects on the kidney and blood vessels, which help increase blood pressure. But angiotensin II in the blood is also detected by the *subfornical organ*, a part of the telencephalon that lacks a blood-brain barrier. Cells in the subfornical organ project axons into the hypothalamus where they activate, among other things, the vasopressin-containing neurosecretory cells. In addition, the subfornical organ activates cells in the lateral area of the hypothalamus, somehow producing an overwhelming thirst that

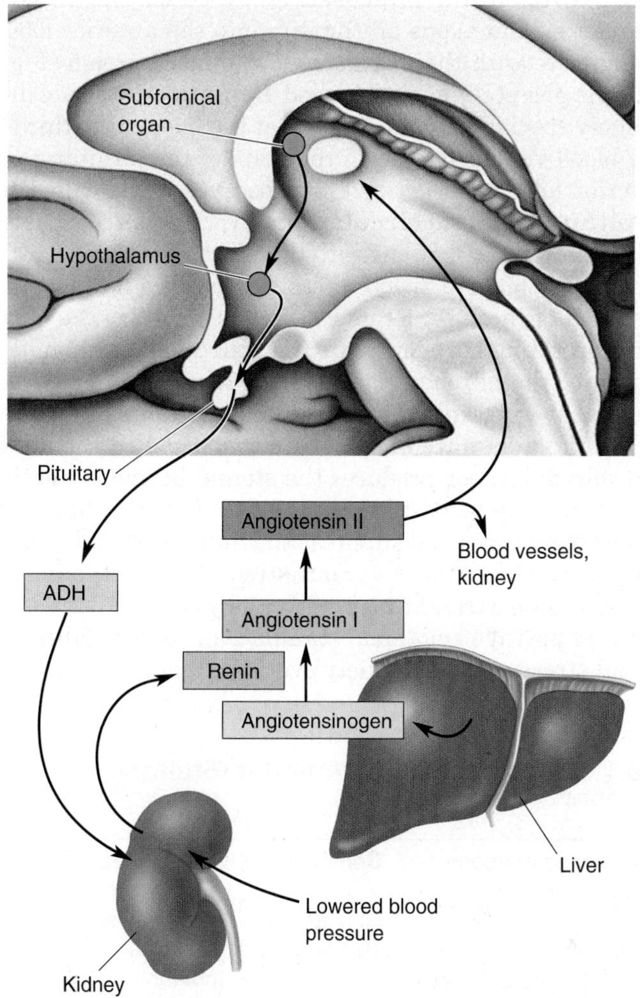

▲ FIGURE 15.5

Communication between the kidneys and the brain. Under conditions of lowered blood volume or pressure, the kidney secretes renin into the bloodstream. Renin in the blood promotes the synthesis of the peptide angiotensin II, which excites the neurons in the subfornical organ. The subfornical neurons stimulate the hypothalamus, causing an increase in vasopressin (ADH) production and a feeling of thirst.

motivates drinking behavior. It may be difficult to accept, but it's true: To a limited extent, our brain is controlled by our kidneys! This example also illustrates that the means by which the hypothalamus maintains homeostasis go beyond control of the visceral organs and can include behavioral responses. In Chapter 16, we will explore in more detail how the hypothalamus incites behavior.

Hypothalamic Control of the Anterior Pituitary. Unlike the posterior lobe, which really is a part of the brain, the anterior lobe of the pituitary is an actual gland. The cells of the anterior lobe synthesize and secrete a wide range of hormones that regulate secretions from other glands throughout the body (together constituting the endocrine system). The pituitary hormones act on the gonads, the thyroid glands, the adrenal glands, and the mammary glands (Table 15.1). For this reason, the anterior pituitary was traditionally described as the body's "master gland." But what controls the anterior pituitary? The secretory hypothalamus. *The hypothalamus itself is the true master gland of the endocrine system.*

The anterior lobe is under the control of neurons in the periventricular area called **parvocellular neurosecretory cells**. These hypothalamic neurons do not extend axons all the way into the anterior lobe; instead, they communicate with their targets via the bloodstream (Figure 15.6). These neurons secrete what are called **hypophysiotropic hormones** into a uniquely specialized capillary bed at the floor of the third ventricle. These tiny blood vessels run down the stalk of the pituitary and branch in the anterior lobe. This network of blood vessels is called the **hypothalamo-pituitary portal circulation**. Hypophysiotropic hormones secreted by hypothalamic neurons into the portal circulation travel downstream until they bind to specific receptors on the surface of pituitary cells. Activation of these receptors causes the pituitary cells to either secrete or stop secreting hormones into the general circulation.

Regulation of the adrenal glands illustrates how this system works. Located just above the kidneys, the adrenal glands consist of two parts, a shell called the **adrenal cortex** and a center called the **adrenal medulla**. The adrenal cortex produces the steroid hormone **cortisol**; when it is released into the bloodstream, cortisol acts throughout the body to mobilize energy reserves and suppress the immune system, preparing us to carry on in the face of life's various stresses. In fact, a good stimulus for cortisol release is stress, ranging from physiological stress, such as a loss of blood; to positive emotional stimulation, such as falling in love; to psychological stress, such as anxiety over an upcoming exam.

TABLE 15.1 **Hormones of the Anterior Pituitary**

Hormone	Target	Action
Follicle-stimulating hormone (FSH)	Gonads	Ovulation, spermatogenesis
Luteinizing hormone (LH)	Gonads	Ovarian and sperm maturation
Thyroid-stimulating hormone (TSH); also called thyrotropin	Thyroid	Thyroxin secretion (increases metabolic rate)
Adrenocorticotropic hormone (ACTH); also called corticotropin	Adrenal cortex	Cortisol secretion (mobilizes energy stores, inhibits immune system, other actions)
Growth hormone (GH)	All cells	Stimulation of protein synthesis
Prolactin	Mammary glands	Growth and milk secretion

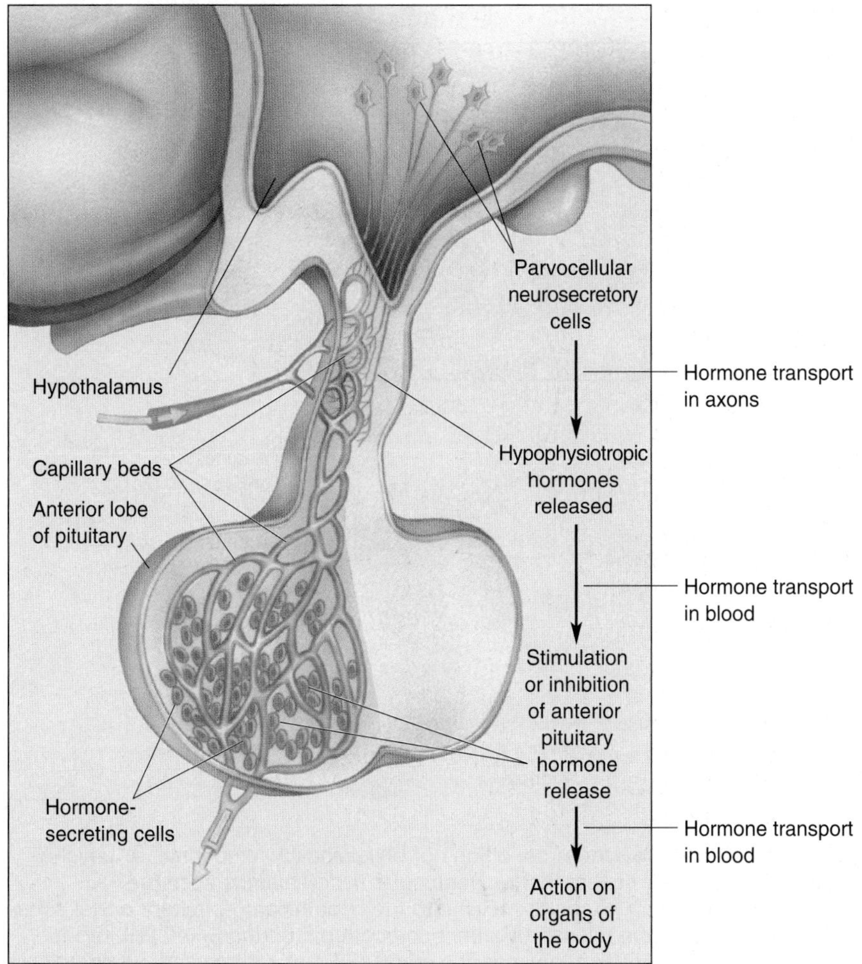

▲ FIGURE 15.6
Parvocellular neurosecretory cells of the hypothalamus. Parvocellular neurosecretory cells secrete hypophysiotropic hormones into specialized capillary beds of the hypothalamo-pituitary portal circulation. These hormones travel to the anterior lobe of the pituitary, where they trigger or inhibit the release of pituitary hormones from secretory cells.

Parvocellular neurosecretory cells that control the adrenal cortex determine whether a stimulus is stressful or not (as defined by the release of cortisol). These neurons lie in the periventricular hypothalamus and release a peptide called *corticotropin-releasing hormone (CRH)* into the blood of the portal circulation. CRH travels the short distance to the anterior pituitary, where, within about 15 seconds, it stimulates the release of *corticotropin*, or *adrenocorticotropic hormone (ACTH)*. ACTH enters the general circulation and travels to the adrenal cortex where, within a few minutes, it stimulates cortisol release (Figure 15.7).

Blood levels of cortisol are, to some extent, self-regulated. Cortisol is a *steroid*, a class of biochemicals related to cholesterol. Thus, cortisol is a lipophilic ("fat-loving") molecule, which dissolves easily in lipid membranes and readily crosses the blood-brain barrier. In the brain, cortisol interacts with specific receptors that lead to inhibition of CRH release, thus ensuring that circulating cortisol levels do not get too high. Physicians need to be mindful of this feedback regulation when they prescribe prednisone, a synthetic form of cortisol. Prednisone is a powerful medicine, frequently used to suppress inflammation. When administered for several days, however, the prednisone circulating in the bloodstream

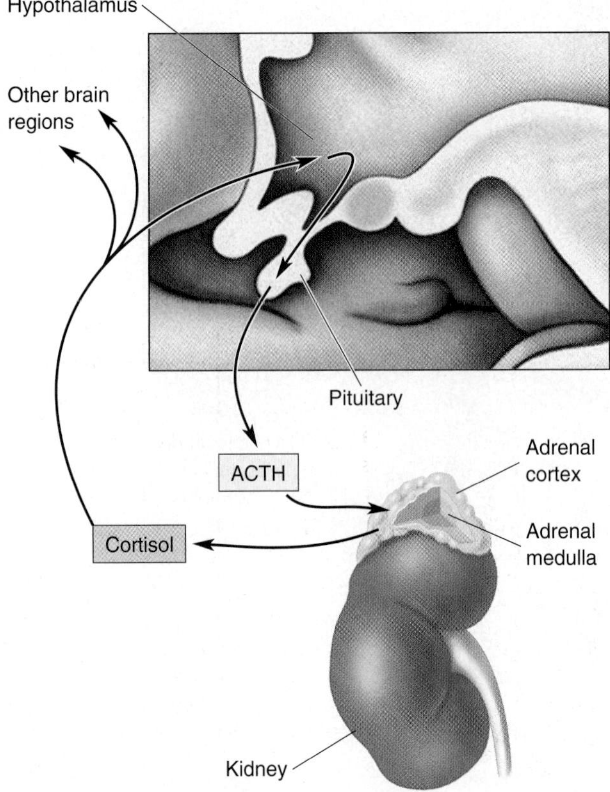

▲ FIGURE 15.7
The stress response. Under conditions of physiological, emotional, or psychological stimulation or stress, the periventricular hypothalamus secretes corticotropin-releasing hormone (CRH) into the hypothalamo-pituitary portal circulation. This triggers the release of adrenocorticotropic hormone (ACTH) into the general circulation. ACTH stimulates the release of cortisol from the adrenal cortex. Cortisol can act directly on hypothalamic neurons, as well as on other neurons elsewhere in the brain.

fools the brain into thinking that naturally released levels of cortisol are too high and shutting down the release of CRH and the adrenal cortex. Abrupt discontinuation of prednisone treatment does not give the adrenal cortex enough time to ramp up cortisol production and can thus result in what is called *adrenal insufficiency*. Among the symptoms of adrenal insufficiency are severe abdominal pain and diarrhea, extremely low blood pressure, and changes in mood and personality. Adrenal insufficiency is also a feature of a rare disorder called *Addison's disease*, named after Thomas Addison, the British physician who first described the condition in 1849. Addison recognized that one cause of this constellation of symptoms is degeneration of the adrenal gland. Perhaps the most famous sufferer of Addison's disease was U.S. President John F. Kennedy. Kennedy required a daily regimen of hormone replacement therapy to compensate for the loss of cortisol, a fact that was concealed during his presidency to protect his youthful and vigorous image.

The flip side of adrenal insufficiency is a condition called *Cushing's disease*, caused by pituitary gland dysfunction that results in elevated levels of ACTH and, consequently, cortisol. The symptoms include rapid weight gain, immune suppression, sleeplessness, memory impairment, and irritability. Not surprisingly, the symptoms of Cushing's disease are a common side effect of prednisone treatment. The myriad behavioral changes caused by too much (or too little) cortisol may be explained by

BOX 15.1 OF SPECIAL INTEREST

Stress and the Brain

Biological stress is created by the brain in response to real or imagined stimuli. The many physiological responses associated with stress help protect the body and the brain from the dangers that triggered the stress in the first place. But stress in chronic doses can have insidious harmful effects as well. Neuroscientists have only begun to understand the relationship between stress, the brain, and brain damage.

Stress leads to the release of the steroid hormone cortisol from the adrenal cortex. Cortisol travels to the brain through the bloodstream and binds to receptors in the cytoplasm of many neurons. The activated receptors travel to the cell nucleus, where they stimulate gene transcription and ultimately protein synthesis. One consequence of cortisol's action is that neurons admit more Ca^{2+} through voltage-gated ion channels. This may be due to a direct change in the channels, or it may be indirectly caused by changes in the cell's energy metabolism. Whatever the mechanism, presumably in the short term cortisol makes the brain better able to cope with the stress—perhaps by helping it figure out a way to avoid it!

But what about the effects of chronic, unavoidable stress? In Chapter 6, we learned that too much calcium can be a bad thing. If neurons become overloaded with calcium, they die (excitotoxicity). The question naturally arises: Can cortisol kill? Bruce McEwen and his colleagues at Rockefeller University, and Robert Sapolsky and his colleagues at Stanford University, have studied this question in the rat brain. They found that daily injections of corticosterone (rat cortisol) for several weeks caused dendrites to wither in many neurons with corticosterone receptors. A few weeks later, these cells started to die. A similar result was found when, instead of daily hormone injections, the rats were stressed every day.

Sapolsky's studies of baboons in Kenya further reveal the scourges of chronic stress. Baboons in the wild maintain a complex social hierarchy, and subordinate males steer clear of dominant males when they can. During one year when the baboon population boomed, local villagers caged many of the animals to prevent them from destroying their crops. Unable to escape the "top baboons" in the cages, many of the subordinate males subsequently died—not from wounds or malnutrition but apparently from severe and sustained stress-induced effects. They had gastric ulcers, colitis, enlarged adrenal glands, and extensive degeneration of neurons in their hippocampus. Subsequent studies suggest that it is the direct effect of cortisol that damages the hippocampus. These effects of cortisol and stress resemble the effects of aging on the brain. Indeed, research has clearly shown that chronic stress causes premature aging of the brain.

In humans, exposure to the horrors of combat, sexual abuse, and other types of extreme violence can lead to post-traumatic stress disorder, with symptoms of heightened anxiety, memory disturbances, and intrusive thoughts. Imaging studies have consistently found degenerative changes in the brains of victims, particularly in the hippocampus. In Chapter 22, we will see that stress, and the brain's response to it, play a central role in several psychiatric disorders.

the fact that neurons with cortisol receptors are found widely distributed in the brain, not just in the hypothalamus. In these other CNS locations, cortisol has been shown to have significant effects on neuronal activity. Thus, we see that the release of hypophysiotropic hormones by cells in the secretory hypothalamus can produce widespread alterations in the physiology of both the body and the brain (Box 15.1).

THE AUTONOMIC NERVOUS SYSTEM

Besides controlling the ingredients of the hormonal soup that flows in our veins, the periventricular zone of the hypothalamus also controls the **autonomic nervous system (ANS)**. The ANS is an extensive network of interconnected neurons that are widely distributed inside the body. From the Greek *autonomia* (roughly meaning "independence"), autonomic functions are usually carried out automatically, without conscious, voluntary control. They are also highly coordinated functions. Imagine a sudden crisis. In a morning class, as you are engrossed in a crossword puzzle, the instructor unexpectedly calls you to the blackboard to solve an impossible-looking equation. You are faced with a classic fight-or-flight situation, and your body reacts accordingly, even as your conscious mind frantically considers whether to blunder through it or beg off in humiliation. Your ANS triggers a host of physiological responses, including increased heart rate and blood pressure, depressed digestive functions, and mobilized glucose reserves. These responses are all produced by the **sympathetic division** of the ANS. Now imagine your relief as the class-ending bell suddenly rings, saving you from acute embarrassment and the instructor's anger. You settle back into your chair, breathe deeply, and read the clue for 24 down. Within a few minutes, your sympathetic responses decrease to low levels, and the functions of your **parasympathetic division** crank up again: Your heart rate slows and blood pressure drops, digestive functions work harder on breakfast, and you stop sweating.

Notice that you may not have moved out of your chair throughout this unpleasant event. Maybe you didn't even move your pencil. But your body's internal workings reacted dramatically. Unlike the *somatic motor system*, whose alpha motor neurons can rapidly excite skeletal muscles with pinpoint accuracy, the actions of the ANS are typically multiple, widespread, and relatively slow. Therefore, the ANS operates in expanded space and time. In addition, unlike the somatic motor system, which can only excite its peripheral targets, the ANS balances synaptic excitation and inhibition to achieve widely coordinated and graded control.

ANS Circuits

Together, the somatic motor system and the ANS constitute the total neural output of the CNS. The somatic motor system has a single function: It innervates and commands skeletal muscle fibers. The ANS has the complex task of commanding *every other* tissue and organ in the body that is innervated. Both systems have upper motor neurons in the brain that send commands to lower motor neurons, which actually innervate the target structures outside the nervous system. However, they have some interesting differences (Figure 15.8). The cell bodies

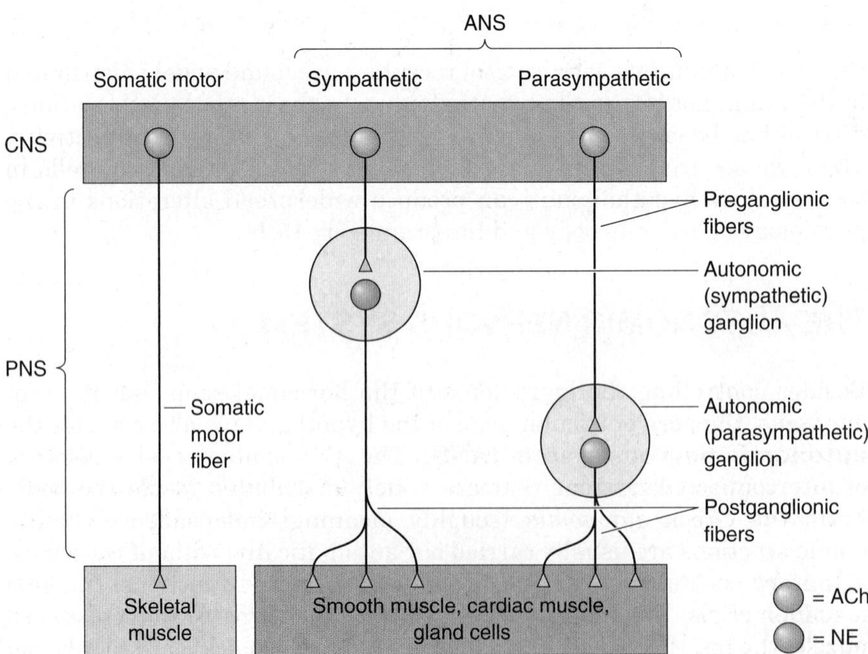

▶ FIGURE 15.8
The organization of the three neural outputs of the CNS. The sole output of the somatic motor system is the lower motor neurons in the ventral horn of the spinal cord and the brain stem, which control skeletal muscle. Visceral functions such as salivating, sweating, and genital stimulation depend on the sympathetic and parasympathetic divisions of the ANS, whose lower motor neurons (i.e., postganglionic neurons) lie outside the CNS in autonomic ganglia.

of all somatic lower motor neurons lie within the CNS in either the ventral horn of the spinal cord or the brain stem. The cell bodies of all autonomic lower motor neurons lie outside the central nervous system, within cell clusters called **autonomic ganglia**. The neurons in these ganglia are called **postganglionic neurons**. Postganglionic neurons are driven by **preganglionic neurons**, whose cell bodies are in the spinal cord and brain stem. Thus, the somatic motor system controls its targets (skeletal muscles) via a *monosynaptic pathway*, while the ANS influences its targets (smooth muscles, cardiac muscle, and glands) using a *disynaptic pathway*.

Sympathetic and Parasympathetic Divisions. The sympathetic and parasympathetic divisions operate in parallel, but they use pathways that are quite distinct in structure and in their neurotransmitter systems. Preganglionic axons of the sympathetic division emerge only from the middle third of the spinal cord (thoracic and lumbar segments). In contrast, preganglionic axons of the parasympathetic division emerge only from the brain stem and the lowest (sacral) segments of the spinal cord, so the two systems complement each other anatomically (Figure 15.9).

The preganglionic neurons of the sympathetic division lie within the *intermediolateral gray matter* of the spinal cord. They send their axons through the ventral roots to synapse on neurons in the ganglia of the **sympathetic chain**, which lies next to the spinal column, or within collateral ganglia found within the abdominal cavity. The preganglionic parasympathetic neurons, on the other hand, sit within a variety of brain stem nuclei and the lower (sacral) spinal cord, and their axons travel within several cranial nerves as well as the nerves of the sacral spinal cord. The parasympathetic preganglionic axons travel much farther than the sympathetic axons because the parasympathetic ganglia are typically located next to, on, or in their target organs (see Figures 15.8 and 15.9).

The ANS innervates three types of tissue: glands, smooth muscle, and cardiac muscle. Thus, almost every part of the body is a target of the ANS, as shown in Figure 15.9. The ANS:

- Innervates the secretory glands (salivary, sweat, tear, and various mucus-producing glands).
- Innervates the heart and blood vessels to control blood pressure and flow.
- Innervates the bronchi of the lungs to meet the oxygen demands of the body.
- Regulates the digestive and metabolic functions of the liver, gastrointestinal tract, and pancreas.
- Regulates the functions of the kidney, urinary bladder, large intestine, and rectum.
- Is essential to the sexual responses of the genitals and reproductive organs.
- Interacts with the body's immune system.

The physiological influences of the sympathetic and parasympathetic divisions generally oppose each other. The sympathetic division tends to be most active during a crisis, real or perceived. The behaviors related to it are summarized in the puerile (but effective) mnemonic used by medical students, called the four Fs: fight, flight, fright, and sex. The parasympathetic division facilitates various non–four-F processes, such as digestion, growth, immune responses, and energy storage. In most cases,

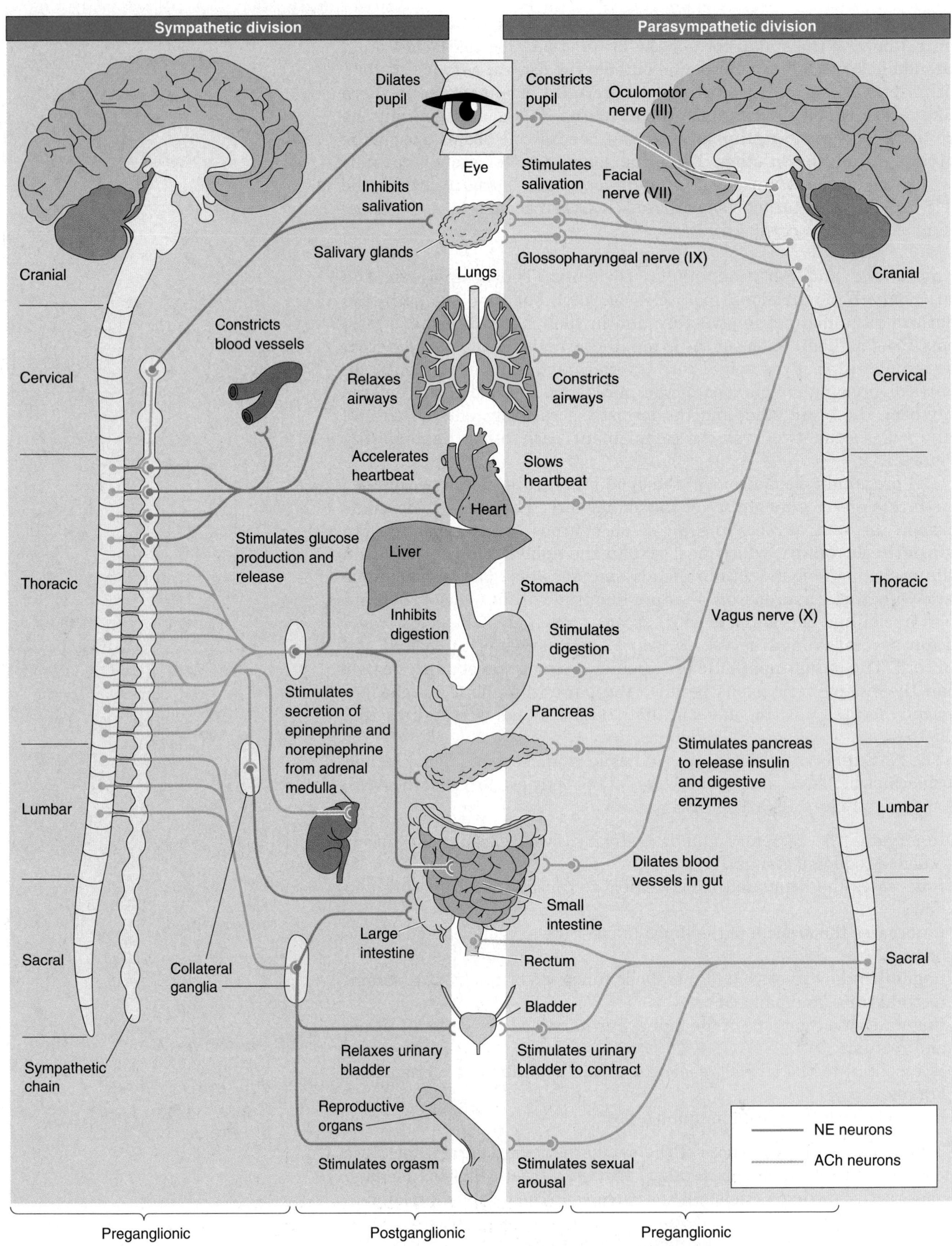

Sympathetic division
Parasympathetic division
Dilates pupil
Eye
Constricts pupil
Oculomotor nerve (III)
Inhibits salivation
Stimulates salivation
Facial nerve (VII)
Salivary glands
Glossopharyngeal nerve (IX)
Cranial
Lungs
Cranial
Constricts blood vessels
Cervical
Relaxes airways
Constricts airways
Cervical
Accelerates heartbeat
Slows heartbeat
Heart
Stimulates glucose production and release
Liver
Thoracic
Stomach
Thoracic
Inhibits digestion
Vagus nerve (X)
Stimulates digestion
Stimulates secretion of epinephrine and norepinephrine from adrenal medulla
Pancreas
Stimulates pancreas to release insulin and digestive enzymes
Lumbar
Lumbar
Dilates blood vessels in gut
Small intestine
Large intestine
Sacral
Rectum
Sacral
Collateral ganglia
Bladder
Relaxes urinary bladder
Stimulates urinary bladder to contract
Sympathetic chain
Reproductive organs
Stimulates orgasm
Stimulates sexual arousal
NE neurons
ACh neurons
Preganglionic neurons
Postganglionic neurons
Preganglionic neurons

the activity levels of the two ANS divisions are reciprocal; when one is high, the other tends to be low, and vice versa. The sympathetic division frenetically mobilizes the body for a short-term emergency at the expense of processes that keep it healthy over the long term. The parasympathetic division works calmly for the long-term good. Both cannot be stimulated strongly at the same time; their general goals are incompatible. Fortunately, neural circuits in the CNS inhibit activity in one division when the other is active.

Some examples illustrate how the balance of activity in the sympathetic and parasympathetic divisions controls organ functions. The pacemaker region of the heart triggers each heartbeat without the help of neurons, but both divisions of the ANS innervate it and modulate it; sympathetic activity results in an increase in the rate of beating, while parasympathetic activity slows it down. The smooth muscles of the gastrointestinal tract are also dually innervated, but the effect of each division is the opposite of its effect on the heart. Intestinal motility, and thus digestion, is stimulated by parasympathetic axons and inhibited by sympathetic axons. Not all tissues receive innervation from both divisions of the ANS, however. For example, blood vessels of the skin, and the sweat glands, are innervated (and excited) only by sympathetic axons. Lacrimal (tear-producing) glands are innervated (and excited) only by parasympathetic input.

Another example of the balance of parasympathetic–sympathetic activity is the curious neural control of the male sexual response. Erection of the human penis is a hydraulic process. It occurs when the penis becomes engorged with blood, which is triggered and sustained by parasympathetic activity. The curious part is that orgasm and ejaculation are triggered by *sympathetic* activity. You can imagine how complicated it must be for the nervous system to orchestrate the entire sexual act; parasympathetic activity gets it going (and keeps it going), but a shift to sympathetic activity is necessary to bring it to a successful conclusion. Anxiety and worry, and their attendant sympathetic activity, tend to inhibit erection and promote ejaculation. Not surprisingly, impotence and premature ejaculation are common complaints of the overstressed male. (We will discuss sexual behavior further in Chapter 17.)

The Enteric Division. The "little brain," as the **enteric division** of the ANS is sometimes called, is a unique neural system embedded in an unlikely place: the lining of the esophagus, stomach, intestines, pancreas, and

◀ FIGURE 15.9

The chemical and anatomical organization of the sympathetic and parasympathetic divisions of the ANS. Notice that the preganglionic inputs of both divisions use ACh as a neurotransmitter. The postganglionic parasympathetic innervation of the visceral organs also uses ACh, but the postganglionic sympathetic innervation uses NE (with the exception of innervation of the sweat glands and vascular smooth muscle within skeletal muscle, which use ACh). The adrenal medulla receives preganglionic sympathetic innervation and secretes epinephrine into the bloodstream when activated. Note the pattern of innervation by the sympathetic division: Target organs in the chest cavity are innervated by postganglionic neurons originating in the sympathetic chain, and target organs in the abdominal cavity are innervated by postganglionic neurons originating in the collateral ganglia.

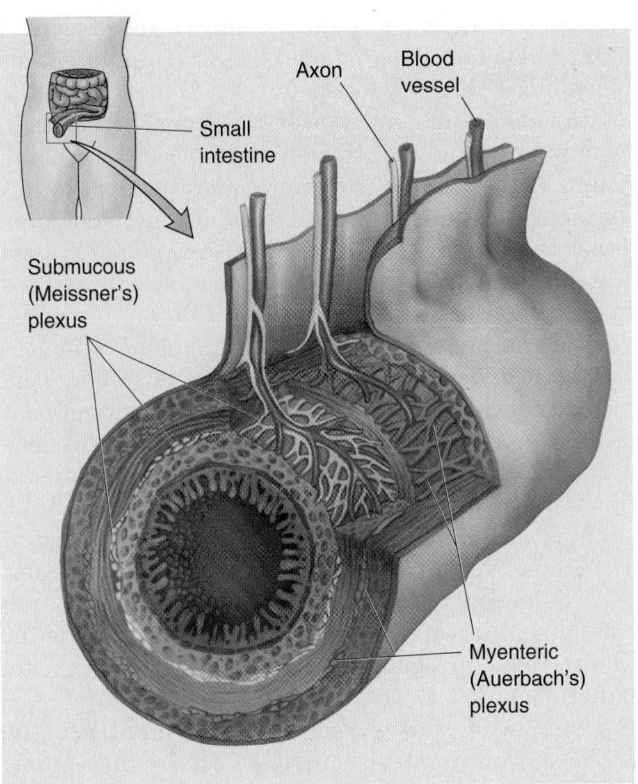

▲ FIGURE 15.10
The enteric division of the ANS. This cross-sectional view of the small intestine shows the two networks of the enteric division: the myenteric plexus and the submucous plexus. They both contain visceral sensory and motor neurons that control the functions of the digestive organs.

gallbladder. It consists of two complicated networks, each with sensory nerves, interneurons, and autonomic motor neurons, called the *myenteric (Auerbach's) plexus* and *submucous (Meissner's) plexus* (Figure 15.10). These networks control many of the physiological processes involved in the transport and digestion of food, from oral to anal openings. The enteric system is not small; it contains about 500 million neurons, the same number of neurons as the entire spinal cord!

If the enteric division of the ANS qualifies as "brain" (which may be overstating the case), it is because it can operate with a great deal of independence. Enteric sensory neurons monitor tension and stretch of the gastrointestinal walls, the chemical status of stomach and intestinal contents, and hormone levels in the blood. This information is used by enteric interneuronal circuits and motor neurons, which also reside in the gut, to govern smooth muscle motility, the production of mucous and digestive secretions, and the diameter of the local blood vessels. For example, consider a partially digested pizza making its way through the small intestine. The enteric nervous system ensures that lubricating mucus and digestive enzymes are delivered, that rhythmic (peristaltic) muscle action works to mix the pizza and enzymes thoroughly, and that intestinal blood flow increases to provide a sufficient fluid source and transport newly acquired nutrients to the rest of the body.

The enteric division is not entirely autonomous. It receives input indirectly from the "real" brain via axons of the sympathetic and parasympathetic divisions. These provide supplementary control and can supersede the functions of the enteric division in some circumstances. For example, the enteric nervous system and digestive functions are inhibited by the strong activation of the sympathetic nervous system that occurs during acute stress.

Central Control of the ANS. As we have said, the hypothalamus is the main regulator of the autonomic preganglionic neurons. Somehow this diminutive structure integrates the diverse information it receives about the body's status, anticipates some of its needs, and provides a coordinated set of both neural and hormonal outputs. Essential to autonomic control are the connections of the periventricular zone to the brain stem and spinal cord nuclei that contain the preganglionic neurons of the sympathetic and parasympathetic divisions. The **nucleus of the solitary tract**, located in the medulla and connected with the hypothalamus, is another important center for autonomic control. In fact, some autonomic functions operate well even when the brain stem is disconnected from all structures above it, including the hypothalamus. The solitary nucleus integrates sensory information from the internal organs and coordinates output to the autonomic brain stem nuclei.

Neurotransmitters and the Pharmacology of Autonomic Function

Even people who have never heard the word *neurotransmitter* know what it means to "get your adrenaline flowing." (In the United Kingdom, this compound is called *adrenaline*, while in the United States, it is called *epinephrine*.) Historically, the autonomic nervous system has probably taught us more than any other part of the body about how neurotransmitters work. Because the ANS is relatively simple compared to the CNS, we understand the ANS much better. In addition, neurons of the peripheral parts of the ANS are outside the blood-brain barrier, so all drugs that enter the bloodstream have direct access to them. The relative simplicity and accessibility of the ANS have led to a deeper understanding of the mechanisms of drugs that influence synaptic transmission.

Preganglionic Neurotransmitters. The primary transmitter of the peripheral autonomic neurons is *acetylcholine (ACh)*, the same transmitter used at skeletal neuromuscular junctions. *The preganglionic neurons of both sympathetic and parasympathetic divisions release ACh.* The immediate effect is that the ACh binds to nicotinic ACh receptors (nAChR), which are ACh-gated channels, and evokes a fast excitatory postsynaptic potential (EPSP) that usually triggers an action potential in the postganglionic cell. This is very similar to the mechanisms of the skeletal neuromuscular junction, and drugs that block nAChRs in muscle, such as curare, also block autonomic output.

Ganglionic ACh does more than neuromuscular ACh, however. It also activates muscarinic ACh receptors (mAChR), which are metabotropic (G-protein-coupled) receptors that can cause both the opening and the closing of ion channels that lead to very slow EPSPs and IPSPs. These slow mAChR events are usually not evident unless the preganglionic nerve is activated repetitively. In addition to ACh, some preganglionic terminals release a variety of small, neuroactive peptides such as *neuropeptide Y (NPY)* and *vasoactive intestinal polypeptide (VIP)*. These also interact with G-protein-coupled receptors and can trigger small EPSPs that last for several minutes. The effects of peptides are modulatory; they do not usually bring the postsynaptic neurons to firing threshold, but they make them more responsive to the fast nicotinic effects when they do come along. Since more than one action potential is required to stimulate the release of these modulatory neurotransmitters, the pattern of firing in preganglionic neurons is an important variable in determining the type of postganglionic activity that is evoked.

Postganglionic Neurotransmitters. Postganglionic cells—the autonomic motor neurons that actually trigger glands to secrete, sphincters to contract or relax, and so on—use different neurotransmitters in the sympathetic and parasympathetic divisions of the ANS. Postganglionic parasympathetic neurons release ACh, but those of most parts of the sympathetic division use *norepinephrine (NE)*. Parasympathetic ACh has a very local effect on its targets and acts entirely through mAChRs. In contrast, sympathetic NE often spreads far, even into the blood where it can circulate widely.

The autonomic effects of a variety of drugs that interact with cholinergic and noradrenergic systems can be confidently predicted once you understand some of the autonomic circuitry and chemistry (see Figure 15.9). In general, drugs that promote the actions of norepinephrine *or* inhibit the muscarinic actions of acetylcholine are *sympathomimetic*; they cause effects that mimic activation of the sympathetic division of the ANS. For example, *atropine*, an antagonist of mAChRs, produces signs of sympathetic activation, such as dilation of the pupils. This response occurs because the balance of ANS activity is shifted toward the sympathetic division when parasympathetic actions are blocked. On the other hand, drugs that promote the muscarinic actions of ACh *or* inhibit the actions of NE are *parasympathomimetic*; they cause effects that mimic activation of the parasympathetic division of the ANS. For example, *propranolol*, an antagonist of the β receptor for NE, slows the heart rate and lowers blood pressure. For this reason, propranolol is sometimes used to prevent the physiological consequences of stage fright.

But what about the familiar flow of adrenaline? Adrenaline (epinephrine) is the compound released into the blood from the adrenal medulla when activated by preganglionic sympathetic innervation. Epinephrine is actually made from norepinephrine (called *noradrenaline* in the United Kingdom), and it has effects on target tissues almost identical to those caused by sympathetic activation. Thus, the adrenal medulla is really nothing more than a modified sympathetic ganglion. You can imagine that as the epinephrine (adrenaline) flows, a coordinated, body-wide set of sympathetic effects kicks in.

THE DIFFUSE MODULATORY SYSTEMS OF THE BRAIN

Consider what happens when you fall asleep. The internal commands "You are becoming drowsy" and "You are falling asleep" are messages that must be received by broad regions of the brain. Dispensing this information requires neurons with a particularly widespread pattern of axons. The brain has several such collections of neurons, each using a particular neurotransmitter and making widely dispersed, diffuse, almost meandering connections. Rather than carrying detailed sensory information, these cells often perform regulatory functions, modulating vast assemblies of postsynaptic neurons (in structures such as the cerebral cortex, the thalamus, and the spinal cord) so that they become more or less excitable, more or less synchronously active, and so on. Collectively, they are a bit like the volume, treble, and bass controls on a radio, which do not change the lyrics or melody of a song but dramatically regulate the impact of both. In addition, different systems appear to be essential for aspects of motor control, memory, mood, motivation, and metabolic state. Many psychoactive drugs affect these modulatory systems, and the systems figure prominently in current theories about the biological basis of certain psychiatric disorders.

Anatomy and Functions of the Diffuse Modulatory Systems

The **diffuse modulatory systems** differ in structure and function, yet they have certain principles in common:

- Typically, the core of each system has a small set of neurons (several thousand).
- Neurons of the diffuse systems arise from the central core of the brain, most of them from the brain stem.
- Each neuron can influence many others because each one has an axon that may contact more than 100,000 postsynaptic neurons spread widely across the brain.
- The synapses made by many of these systems release transmitter molecules into the extracellular fluid, so they can diffuse to many neurons rather than be confined to the vicinity of the synaptic cleft.

We'll focus on the modulatory systems of the brain that use either norepinephrine (NE), serotonin (5-HT), dopamine (DA), or acetylcholine (ACh) as a neurotransmitter. Recall from Chapter 6 that all of these transmitters activate specific metabotropic (G-protein-coupled) receptors, and these receptors mediate most of their effects; for example, the brain has 10–100 times more metabotropic ACh receptors than ionotropic nicotinic ACh receptors.

Because neuroscientists are still working hard to determine the exact functions of these systems in behavior, our explanations here will necessarily be general. It is clear, however, that the functions of the diffuse modulatory systems depend on how electrically active they are, individually and in combination, and on how much neurotransmitter is available for release (Box 15.2).

The Noradrenergic Locus Coeruleus. Besides being a neurotransmitter in the peripheral ANS, NE is also used by neurons of the tiny **locus coeruleus** in the pons (from the Latin for "blue spot" because of the pigment in its cells). Each human locus coeruleus has about 12,000 neurons. We have two of them, one on each side.

A major breakthrough occurred in the mid-1960s, when Nils-Åke Hillarp and Bengt Falck at the Karolinska Institute in Sweden developed a technique that enabled the catecholaminergic (noradrenergic and dopaminergic) neurons to be visualized selectively in histological sections prepared from the brain (Figure 15.11). This analysis revealed that axons leave the locus coeruleus in several tracts but then fan out to innervate just about every part of the brain: all of the cerebral cortex, the thalamus and the hypothalamus, the olfactory bulb, the cerebellum, the midbrain, and the spinal cord (Figure 15.12). The locus coeruleus makes some of the most diffuse connections in the brain, considering that just one of its neurons can make more than 250,000 synapses, and it can have one axon branch in the *cerebral* cortex and another in the *cerebellar* cortex! The organization of this circuitry is so different from what was then known about synaptic connections in the brain that it took many years

BOX 15.2 OF SPECIAL INTEREST

You Eat What You Are

Americans, it seems, are always trying to lose weight. The low-fat, high-carbohydrate diets (think bagels) that were all the rage in the 1990s were replaced by a low-carb craze (think omelets). Changing your diet can alter caloric intake and the body's metabolism, and it can also alter how your brain functions.

The influence of diet on the brain is most clear in the case of the diffuse modulatory systems. Consider serotonin. Serotonin is synthesized in two steps from the dietary amino acid tryptophan (see Figure 6.14). In the first step, a hydroxyl group (OH) is added to tryptophan by the enzyme tryptophan hydroxylase. The low affinity of the enzyme for tryptophan makes this step *rate-limiting* for serotonin synthesis—that is, serotonin can be produced only as fast as this enzyme can hydroxylate tryptophan. And a lot of tryptophan is required to push the synthetic reaction as fast as it can go. However, brain tryptophan levels are well below the level required to saturate the enzyme. Thus, the rate of serotonin synthesis is determined, in part, by the availability of tryptophan in the brain—more tryptophan, more serotonin; less tryptophan, less serotonin.

Brain tryptophan levels are controlled by how much tryptophan is in the blood and by how efficiently it is transported across the blood-brain barrier. Tryptophan in the blood is derived from the proteins we digest in our diet, so a high-protein diet will lead to sharply increased blood levels of tryptophan. Surprisingly, however, there is a decline in brain tryptophan (and serotonin) for several hours after a hearty, high-protein meal. The paradox was resolved by Richard Wurtman and his colleagues at the Massachusetts Institute of Technology who observed that several other amino acids (tyrosine, phenylalanine, leucine, isoleucine, and valine) compete with tryptophan for transport across the blood-brain barrier. These other amino acids are rich in a high-protein diet, and they suppress the entry of the tryptophan into the brain. The situation is reversed with a high-carbohydrate meal that also contains some protein. Insulin, released by the pancreas in response to carbohydrates, decreases the blood levels of the competing amino acids relative to tryptophan. So the tryptophan in the blood is efficiently transported into the brain, and serotonin levels rise.

Increased brain tryptophan correlates with elevated mood, decreased anxiety, and increased sleepiness, likely due to changes in serotonin levels. Inadequate tryptophan may explain the phenomenon of carbohydrate craving that has been reported in humans with seasonal affective disorder—the depression of mood brought on by reduced daylight during winter. It may also explain why clinical trials for treating obesity with extreme carbohydrate deprivation had to be stopped because of complaints of mood disturbances (depression, irritability) and insomnia.

Based on these and other observations, Wurtman and his wife Judith made the intriguing suggestion that our dietary choices may reflect our brain's need for serotonin. Consistent with this notion, drugs that elevate extracellular serotonin can be effective for weight loss (as well as depression), possibly by reducing the body's demand for carbohydrates. We will discuss the involvement of serotonin in appetite regulation further in Chapter 16, and in the regulation of mood in Chapter 22.

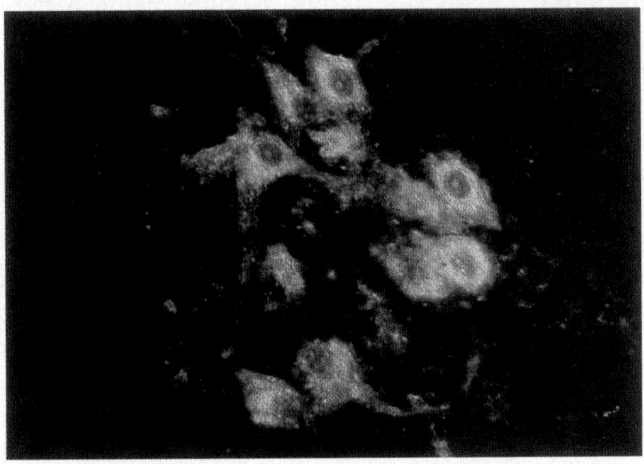

▲ FIGURE 15.11
Norepinephrine-containing neurons of the locus coeruleus. Reaction of noradrenergic neurons with formaldehyde gas causes them to fluoresce green, enabling anatomical investigation of their widespread projections. (Courtesy of Dr. Kjell Fuxe.)

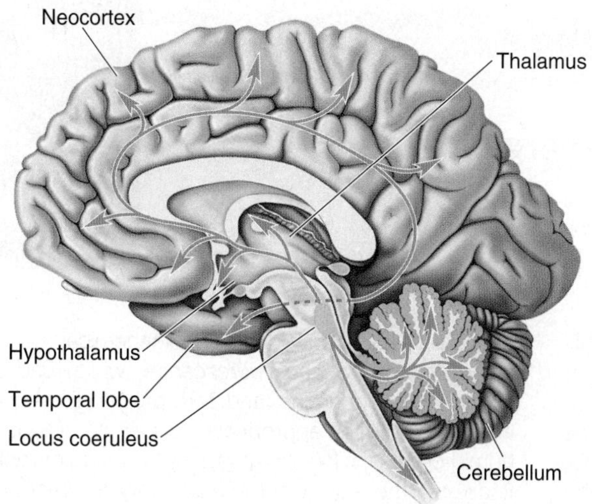

▲ FIGURE 15.12
The noradrenergic diffuse modulatory system arising from the locus coeruleus. The small cluster of locus coeruleus neurons projects axons that innervate vast areas of the CNS, including the spinal cord, cerebellum, thalamus, and cerebral cortex.

of research before the mainstream neuroscience community could accept that NE was a neurotransmitter in the brain (Box 15.3).

Locus coeruleus cells seem to be involved in the regulation of attention, arousal, and sleep–wake cycles as well as learning and memory, anxiety and pain, mood, and brain metabolism. This makes it sound as if the locus coeruleus may run the whole show. But the key word is "involved," which can mean almost anything. For example, our heart, liver, lungs, and kidneys are also involved in every brain function, for without them, the brain cannot survive. Because of its widespread connections, the locus coeruleus can influence virtually all parts of the brain. But to understand its actual functions, we start by determining what activates its neurons. Recordings from awake, behaving rats and monkeys show that locus coeruleus neurons are most strongly activated by new, unexpected, nonpainful sensory stimuli in the animal's environment. They are least active when the animals are not vigilant, just sitting around quietly, digesting a meal. The locus coeruleus may participate in a general arousal of the brain during interesting events in the outside world. Because NE can make neurons of the cerebral cortex more responsive to salient sensory stimuli, the locus coeruleus may function generally to increase brain responsiveness, speeding information processing by the point-to-point sensory and motor systems and making them more efficient.

The Serotonergic Raphe Nuclei. Serotonin-containing neurons are mostly clustered within the nine **raphe nuclei**. *Raphe* means "ridge" or "seam" in Greek, and, indeed, the raphe nuclei lie to either side of the midline of the brain stem. Each nucleus projects to different regions of the CNS (Figure 15.13). Those more caudal, in the medulla, innervate the spinal cord, where they modulate pain-related sensory signals (see Chapter 12). Those more rostral, in the pons and midbrain, innervate most of the brain in much the same diffuse way as do the locus coeruleus neurons.

Similar to neurons of the locus coeruleus, raphe nuclei cells fire most rapidly during wakefulness, when an animal is aroused and active. Raphe

BOX 15.3 PATH OF DISCOVERY

Exploring the Central Noradrenergic Neurons

by Floyd Bloom

Norepinephrine (NE) was accepted as the neurotransmitter for the peripheral autonomic sympathetic nervous system by the 1930s, but identification of this catecholamine's status in the brain remained uncharacterized for another three decades. By the late 1950s, central chemical neurotransmission was conceived as an extension to the brain of the then best studied synapse, the neuromuscular junction. Here acetylcholine had satisfied the four identification criteria of a neurotransmitter: localization, mimicry of nerve action, identical pharmacology, and ionic permeability changes. But what other brain chemicals did the brain use for those synapses not mediated by acetylcholine? NE was detected in the brain and was regionally distributed (rich in hypothalamus, low in cortex), which was incompatible with it being simply the sympathetic innervation of the brain's blood vessels, but what did it do?

When I went to the National Institutes of Health (NIH) in 1962 to avoid the "Doctor's Draft" into the Army, I spent 2 years assessing how neurons in the hypothalamus, olfactory bulb, and striatum respond to NE applied by microionophoresis. The results appeared random: A third fired faster, a third fired slower, and the remaining third were unresponsive. What was missing was knowledge of which neurons, if any, were actually innervated by NE fibers. This critical information was provided when the Swedish scientists Nils-Åke Hillarp and Bengt Falck developed a histochemical method, called *formaldehyde-induced fluorescence*, which caused monoamines (NE, dopamine, and serotonin) to fluoresce when illuminated with the appropriate wavelength of light. But in the humid climate of Washington D.C., I could not replicate their findings. I therefore went to Yale to try different approaches using electron microscopy and autoradiography to see which nerve terminals could concentrate radioactive NE, as Julius Axelrod had done for the sympathetic innervation of the pineal body.

When I returned to NIH in 1968, I had learned enough to suspect that cerebellar Purkinje neurons were targets of synapses utilizing NE as a neurotransmitter. This was also the best understood region of the brain in terms of cellular circuitry. Together with Barry Hoffer, who had studied cerebellar development, and George Siggins, who was expert in the sympathetic innervation of peripheral blood vessels, we set about to test how Purkinje neurons respond to NE. We found that they responded consistently by slowing their spontaneous activity. This effect was blocked by norepinephrine beta receptor antagonists and was prolonged by NE reuptake inhibitors, and both effects were lost when the NE neurons were destroyed with the toxin 6-hydroxydopamine.

While visiting the Karolinska Institute in 1971 (the year Axelrod was awarded the Nobel Prize), I learned from Lars

neurons are the most quiet during sleep. The locus coeruleus and the raphe nuclei are part of a venerable concept called the *ascending reticular activating system*, which implicates the reticular "core" of the brain stem in processes that arouse and awaken the forebrain. This simple idea has been refined and redefined in countless ways since it was introduced in the 1950s, but its basic sense remains. Raphe neurons seem to be intimately involved in the control of sleep–wake cycles, as well as the different stages of sleep. It is important to note that several other transmitter systems are involved in a coordinated way as well. We will discuss the involvement of the diffuse modulatory systems in sleep and wakefulness in Chapter 19.

Serotonergic raphe neurons have also been implicated in the control of mood and certain types of emotional behavior. We will return to serotonin and mood when we discuss clinical depression in Chapter 22.

The Dopaminergic Substantia Nigra and Ventral Tegmental Area. For many years, neuroscientists thought that dopamine existed in the brain only as a metabolic precursor for norepinephrine. However, research

Olson and Kjell Fuxe that the locus coeruleus in the pons provided the NE innervation of the cerebellum as well as the entire forebrain (Figure A). When Siggins, Hoffer, and I electrically stimulated the locus coeruleus, Purkinje cell firing slowed, thus mimicking the effects of NE by microionophoresis. The effect of locus coeruleus stimulation was lost when NE was depleted using inhibitors of tyrosine hydroxylase, or eradicated with 6-hydroxydopamine. At last, we were convinced that NE satisfied the neurotransmitter identification criteria. But clearly, the CNS actions of NE differed dramatically from those of "classical" fast central transmitter systems. Instead of being strictly excitatory or inhibitory, NE seemed to act to enhance the effects of other afferent projections to the same postsynaptic targets. Menahem Segal, working with me at the NIH, reached a similar conclusion for the actions of NE in the hippocampus.

After moving to the Salk Institute, I worked with Steve Foote and Gary Aston-Jones to record the firing patterns of the locus coeruleus neurons in awake behaving rats and squirrel monkeys. These experiments revealed that the locus coeruleus neurons have brief phasic responses to novel sensory signals of all modalities, progressively slowing with loss of attention and becoming silent during rapid eye movement sleep. The phasic and tonic discharge modes correlate with the chemical thresholds for alpha (highly sensitive) and beta (less sensitive) adrenergic receptors.

Subsequently, using immunohistochemistry with antibodies to the enzyme dopamine-beta-hydroxylase, found only in NE-containing neurons, Steve Foote, John Morrison, David Lewis, and I generated detailed NE circuit maps in the nonhuman primate brain. In contrast to the diffuse rodent cortical innervations, their data showed differences in the amount of innervation in architectonically defined areas of the cerebral cortex, particularly in the cingulate and orbitofrontal cortex. This map suggested that the locus coeruleus-NE afferents have a greater influence on spatial and visuomotor detection than on detailed sensory feature detection. My interests in the central catecholamine systems and brain disease continue today and have been intensified by the computational and theoretical concepts now being developed from studies of the roles of these systems in awake behaving primates, including normal cognitive declines with aging.

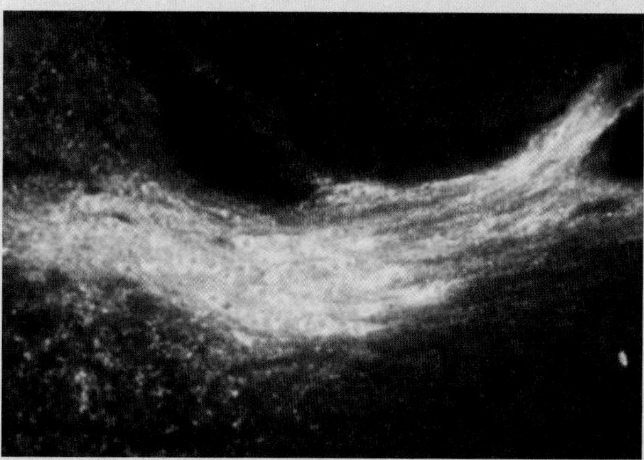

Figure A
Green fluorescent NE neurons in a sagittal section of the rat locus coeruleus. (Source: Courtesy of Dr. Floyd Bloom, The Scripps Research Institute.)

conducted in the 1960s by Arvid Carlsson of the University of Gothenburg in Sweden proved that dopamine was indeed a crucial CNS neurotransmitter. This discovery was honored with the 2000 Nobel Prize in Medicine.

Although there are dopamine-containing neurons scattered throughout the CNS, including some in the retina, the olfactory bulb, and the periventricular hypothalamus, two closely related groups of dopaminergic cells have the characteristics of the diffuse modulatory systems (Figure 15.14). One of these arises in the *substantia nigra* in the midbrain. Recall from Chapter 14 that these cells project axons to the striatum (the caudate nucleus and the putamen), where they somehow facilitate the initiation of voluntary movements. Degeneration of the dopamine-containing cells in the substantia nigra is all that is necessary to produce the progressive, dreadful motor disorders of Parkinson's disease. Although we do not entirely understand the function of DA in motor control, in general it facilitates the initiation of motor responses by environmental stimuli.

The midbrain is also the origin of the other dopaminergic modulatory system, a group of cells that lie very close to the substantia nigra, in the *ventral tegmental area*. Axons from these neurons innervate a

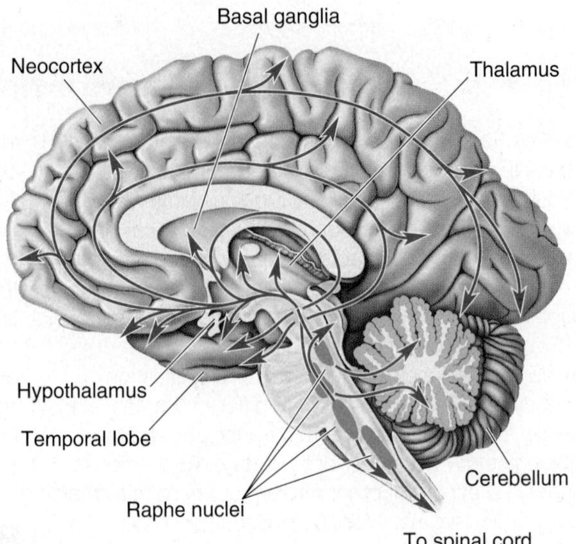

▲ FIGURE 15.13
The serotonergic diffuse modulatory systems arising from the raphe nuclei. The raphe nuclei are clustered along the midline of the brain stem and project extensively to all levels of the CNS.

circumscribed region of the telencephalon that includes the frontal cortex and parts of the limbic system. (The limbic system will be discussed in Chapter 18.) This dopaminergic projection from the midbrain is sometimes called the *mesocorticolimbic dopamine system.* A number of different functions have been ascribed to this complicated projection. For example, evidence indicates that it is involved in a "reward" system that somehow assigns value to, or *reinforces*, certain behaviors that are adaptive. We will see in Chapter 16 that if rats (or humans) are given a chance to do so, they will work to electrically stimulate this pathway. In addition,

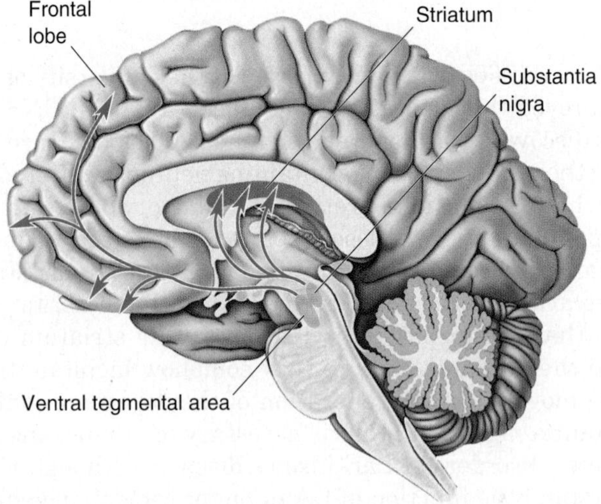

▲ FIGURE 15.14
The dopaminergic diffuse modulatory systems arising from the substantia nigra and the ventral tegmental area. The substantia nigra and ventral tegmental area lie close together in the midbrain. They project to the striatum (caudate nucleus and putamen) and limbic and frontal cortical regions, respectively.

this projection has been implicated in psychiatric disorders, as we will discuss in Chapter 22.

The Cholinergic Basal Forebrain and Brain Stem Complexes. Acetylcholine is the familiar transmitter at the neuromuscular junction, at synapses in autonomic ganglia, and at postganglionic parasympathetic synapses. Cholinergic interneurons also exist within the brain—in the striatum and the cortex, for example. In addition, there are two major diffuse modulatory cholinergic systems in the brain, one of which is called the **basal forebrain complex**. It is a "complex" because the cholinergic neurons lie scattered among several related nuclei at the core of the telencephalon, medial and ventral to the basal ganglia. The best known of these are the *medial septal nuclei*, which provide the cholinergic innervation of the hippocampus, and the *basal nucleus of Meynert*, which provides most of the cholinergic innervation of the neocortex.

The function of the cells in the basal forebrain complex remains mostly unknown. But interest in this region has been fueled by the discovery that these are among the first cells to die during the course of Alzheimer's disease, which is characterized by a progressive and profound loss of cognitive functions. (However, there is widespread neuronal death in Alzheimer's disease, and no specific link between the disease and cholinergic neurons has been established.) Like the noradrenergic and serotonergic systems, the cholinergic system has been implicated in regulating general brain excitability during arousal and sleep–wake cycles. The basal forebrain complex may also play a special role in learning and memory formation.

The second diffuse cholinergic system is called the *pontomesencephalotegmental complex*. These are ACh-utilizing cells in the pons and midbrain tegmentum. This system acts mainly on the dorsal thalamus, where, together with the noradrenergic and serotonergic systems, it regulates the excitability of the sensory relay nuclei. These cells also project up to the telencephalon, providing a cholinergic link between the brain stem and basal forebrain complexes. Figure 15.15 shows the cholinergic systems.

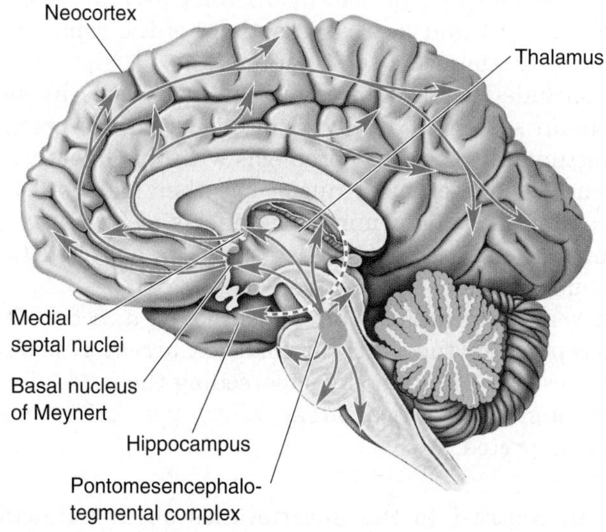

▲ FIGURE 15.15
The cholinergic diffuse modulatory systems arising from the basal forebrain and brain stem. The medial septal nuclei and basal nucleus of Meynert project widely upon the cerebral cortex, including the hippocampus. The pontomesencephalotegmental complex projects to the thalamus and parts of the forebrain.

Drugs and the Diffuse Modulatory Systems

Psychoactive drugs, compounds with "mind-altering" effects, all act on the central nervous system, and most do so by interfering with chemical synaptic transmission. Many abused drugs act directly on the modulatory systems, particularly the noradrenergic, dopaminergic, and serotonergic systems.

Hallucinogens. The use of *hallucinogens*, drugs that produce hallucinations, goes back thousands of years. Hallucinogenic compounds are contained in a number of plants consumed as part of religious ritual, for example, the *Psilocybe* mushroom by the Maya and the peyote cactus by the Aztec. The modern era of hallucinogenic drug use was unwittingly ushered in at the laboratory of Swiss chemist Albert Hofmann. In 1938, Hofmann chemically synthesized a new compound, *lysergic acid diethylamide (LSD)*. For 5 years, the LSD sat on the shelf. Then one day in 1943, Hofmann accidentally ingested some of the powder. His report on the effects attracted the immediate interest of the medical community. Psychiatrists began to use LSD in attempts to unlock the subconscious of mentally disturbed patients. Later the drug was discovered by intellectuals, artists, students, and the U.S. Defense Department, who investigated its "mind-expanding" effects. (A chief advocate of LSD use was former Harvard psychologist Timothy Leary.) In the 1960s, LSD made its way to the street and was widely abused. It is illegal to possess LSD today.

LSD is extremely potent. A dose of only 25 μg is sufficient to produce a full-blown hallucinogenic effect (compared to a normal dose of aspirin at 650 mg, which is 25,000 times larger). Among the reported behavioral effects of LSD are a dreamlike state with heightened awareness of sensory stimuli, often with a mixing of perceptions such that sounds can evoke images, images can evoke smells, and so on.

The chemical structure of LSD (and the active ingredients of *Psilocybe* mushrooms and peyote) is very close to that of serotonin, suggesting that it acts on the serotonergic system. Indeed, LSD is a potent agonist at the serotonin receptors on the presynaptic terminals of neurons in the raphe nuclei. Activation of these receptors markedly inhibits the firing of raphe neurons. Thus, one known CNS effect of LSD is a reduction in the outflow of the brain's serotonergic diffuse modulatory system. It is interesting to note in this regard that decreased activity of the raphe nuclei is also characteristic of dream-sleep (see Chapter 19).

Can we conclude that LSD produces hallucinations by silencing the brain's serotonin systems? If only drug effects on the brain were that simple. Unfortunately, there are problems with this hypothesis. For one, silencing neurons in the raphe nuclei by other means—by destroying them, for example—does not mimic the effects of LSD in experimental animals. Furthermore, animals still respond as expected to LSD after their raphe nuclei have been destroyed.

In recent years, researchers have focused on direct LSD actions at serotonin receptors in the cerebral cortex. Current research suggests that LSD causes hallucinations by superseding the naturally modulated release of serotonin in cortical areas where perceptions normally are formed and interpreted.

Stimulants. In contrast to the uncertainties about hallucinogens and serotonin, it is clear that the powerful CNS stimulants *cocaine* and *amphetamine* both exert their effects at synapses made by dopaminergic and noradrenergic systems. Both drugs give users a feeling of increased alertness and self-confidence, a sense of exhilaration and euphoria, and a decreased appetite. Both are also sympathomimetic—they cause peripheral effects that mimic activation of the sympathetic division of the ANS: increased heart rate and blood pressure, dilation of the pupils, and so on.

Cocaine is extracted from the leaves of the coca plant and has been used by Andean indigenous peoples for hundreds of years. In the mid-nineteenth century, cocaine turned up in Europe and North America as the magic ingredient in a wide range of concoctions touted as having medicinal value. (An example is Coca-Cola, originally marketed in 1886 as a therapeutic agent, which contained both cocaine and caffeine.) Cocaine use fell out of favor early in the twentieth century, only to reemerge in the late 1960s as a popular recreational drug. Ironically, one of the main reasons for the rise in cocaine use during this period was the tightening of regulations against amphetamines. First chemically synthesized in 1887, amphetamines did not come into wide use until World War II, when they were taken by soldiers (particularly aviators) to sustain them in combat. Following the war, amphetamines became available as nonprescription diet aids, as nasal decongestants, and as "pep pills." Regulations were finally tightened after recognition that amphetamines are, like cocaine, highly addictive and dangerous in large doses.

The neurotransmitters dopamine and norepinephrine are *catecholamines*, named for their chemical structure (see Chapter 6). The actions of catecholamines released into the synaptic cleft are normally terminated by specific uptake mechanisms. Cocaine and amphetamine both block this catecholamine uptake (Figure 15.16). However, recent work suggests that cocaine targets DA reuptake more selectively; amphetamine blocks NE and DA reuptake *and* stimulates the release of DA. Thus, these drugs can prolong and intensify the effects of released DA or NE. Is this the means by which cocaine and amphetamine cause their stimulant effects? There is good reason for thinking so. For example, experimental depletion of brain catecholamines by using synthesis inhibitors (such as α-methyltyrosine) will abolish the stimulant effects of both cocaine and amphetamine.

Besides having a similar stimulant effect, cocaine and amphetamine share another, more insidious behavioral action: psychological dependence, or addiction. Users will develop powerful cravings for prolonging and continuing drug-induced pleasurable feelings. These effects are believed to

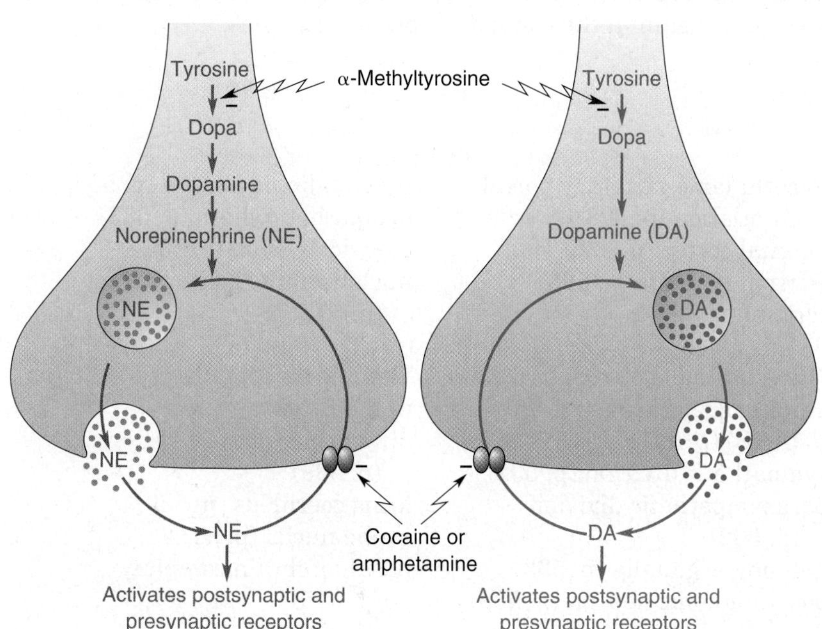

◀ FIGURE 15.16
Stimulant drug action on the catecholamine axon terminal. On the left is a noradrenergic terminal, and on the right is a dopaminergic terminal. Both neurotransmitters are catecholamines synthesized from the dietary amino acid tyrosine. Dopa (3, 4-dihydroxypheynylalanine) is an intermediate in the synthesis of both. The actions of NE and DA are usually terminated by uptake back into the axon terminal. Amphetamine and cocaine block this uptake, thereby allowing NE and DA to remain in the synaptic cleft longer.

result specifically from the enhanced transmission in the mesocorticolimbic dopamine system during drug use. Remember, this system may normally function to reinforce adaptive behaviors. By short-circuiting the system, these drugs instead reinforce drug-seeking behavior. Indeed, just as rats will work to electrically stimulate the mesocorticolimbic projection, they will also work to receive an injection of cocaine. We'll discuss the involvement of dopamine pathways in motivation and addiction further in Chapter 16.

CONCLUDING REMARKS

In this chapter, we have examined three components of the nervous system that are characterized by the great reach of their influences. The secretory hypothalamus and autonomic nervous system communicate with cells all over the body, and the diffuse modulatory systems communicate with neurons in many different parts of the brain. They are also characterized by the duration of their direct effects, which can range from minutes to hours. Finally, they are characterized by their chemical neurotransmitters. In many instances, the transmitter *defines* the system. For example, in the peripheral nervous system, we can use the words "noradrenergic" and "sympathetic" interchangeably. The same thing goes for "raphe" and "serotonin" in the forebrain, and "substantia nigra" and "dopamine" in the basal ganglia. These chemical idiosyncrasies have allowed interpretations of drug effects on behavior that are not possible with most other neural systems. Thus, we have a good idea where in the brain amphetamine and cocaine exert their stimulant effects, and where outside the CNS, they act to raise blood pressure and heart rate.

At a detailed level, each of the systems discussed in this chapter performs different functions. But at a general level, they all *maintain brain homeostasis*: They regulate different processes within a certain physiological range. For example, the ANS regulates blood pressure within a range that is appropriate. Blood pressure variations optimize an animal's performance under different conditions. In a similar way, the noradrenergic locus coeruleus and serotonergic raphe nuclei regulate levels of consciousness and mood. These levels also vary within a range that is adaptive to the organism. In the next several chapters, we will encounter these systems again in the context of specific functions.

KEY TERMS

The Secretory Hypothalamus
homeostasis (p. 524)
periventricular zone (p. 524)
magnocellular neurosecretory cell (p. 525)
neurohormone (p. 525)
oxytocin (p. 526)
vasopressin (p. 526)
antidiuretic hormone (ADH) (p. 526)
parvocellular neurosecretory cell (p. 528)
hypophysiotropic hormone (p. 528)
hypothalamo-pituitary portal circulation (p. 528)
adrenal cortex (p. 528)
adrenal medulla (p. 528)
cortisol (p. 528)

The Autonomic Nervous System
autonomic nervous system (ANS) (p. 531)
sympathetic division (p. 532)
parasympathetic division (p. 532)
autonomic ganglia (p. 533)
postganglionic neuron (p. 533)
preganglionic neuron (p. 533)
sympathetic chain (p. 533)
enteric division (p. 535)
nucleus of the solitary tract (p. 537)

The Diffuse Modulatory Systems of the Brain
diffuse modulatory system (p. 539)
locus coeruleus (p. 539)
raphe nuclei (p. 541)
basal forebrain complex (p. 545)

REVIEW QUESTIONS

1. Battlefield trauma victims who have lost large volumes of blood often express a craving to drink water. Why?
2. You've stayed up all night trying to meet a term paper deadline. You now are typing frantically, keeping one eye on the paper and the other on the clock. How has the periventricular zone of the hypothalamus orchestrated your body's physiological response to this stressful situation? Describe in detail.
3. An "Addisonian crisis" describes a constellation of symptoms that include extreme weakness, mental confusion, drowsiness, low blood pressure, and abdominal pain. What causes these symptoms and what can be done to treat them?
4. Why is the adrenal medulla often referred to as a modified sympathetic ganglion? Why isn't the adrenal cortex included in this description?
5. A number of famous athletes and entertainers have accidentally killed themselves by taking large quantities of cocaine. Usually the cause of death is heart failure. How would you explain the peripheral actions of cocaine?
6. How do the diffuse modulatory and point-to-point synaptic communication systems in the brain differ? List four ways.
7. Under what behavioral conditions are the noradrenergic neurons of the locus coeruleus active? The noradrenergic neurons of the ANS?

FURTHER READING

Bloom FE. 2010. The catecholamine neuron: historical and future perspectives. *Progress in Neurobiology* 90:75–81.

Carlsson A. 2001. A paradigm shift in brain research. *Science* 294:1021–1024.

McEwen BS. 2002. Sex, stress and the hippocampus: allostasis, allostatic load and the aging process. *Neurobiology of Aging* 23(5):921–939.

Meyer JS, Quenzer LF. 2004. *Psychopharmacology: Drugs, the Brain, and Behavior*. Sunderland, MA: Sinauer.

Wurtman RJ, Wurtman JJ. 1989. Carbohydrates and depression. *Scientific American* 260(1): 68–75.

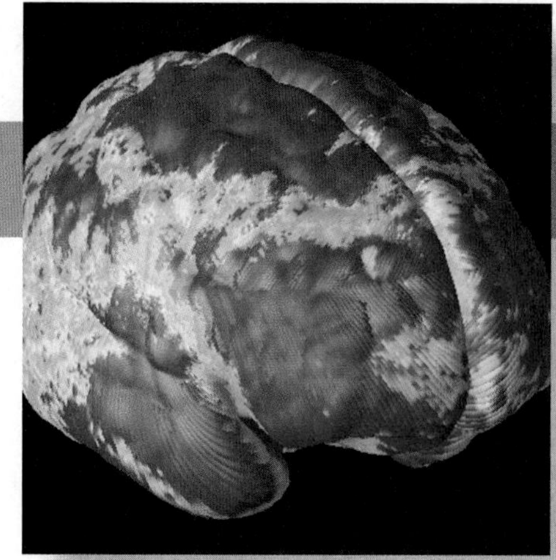

CHAPTER SIXTEEN

Motivation

INTRODUCTION

Behavior happens. But why? In Part II of this book, we discussed various types of motor responses. At the lowest level are unconscious reflexes initiated by sensory stimulation—dilation of the pupils when the lights go out, sudden removal of the foot from a thumbtack, and so on. At the highest level are conscious movements initiated by the neurons of the frontal lobe—for example, the finger movements that tap this text into the computer. Voluntary movements are incited to occur, or *motivated*, in order to satisfy a need. The motivation can be very abstract (the "need" to go sailing on a warm and breezy summer afternoon), but it can also be quite concrete (the need to go to the bathroom when your bladder is full).

Motivation can be thought of as a driving force on behavior. By analogy, consider the driving force on sodium ions to cross the neuronal membrane (an odd analogy, perhaps, but not for a neuroscience text). As we learned back in Chapters 3 and 4, ionic driving force depends on a number of factors, including the concentration of the ion on both sides of the membrane and the electrical membrane potential. Variations in driving force make transmembrane ionic current in a particular direction more or less likely. But the driving force alone does not determine whether the current flows; the transmembrane movement of ions also requires the appropriate gated ion channels to be opened and capable of conducting the current.

Of course, human behavior will never be described by anything as simple as Ohm's law. Still, it is useful to consider that the probability and direction of a behavior will vary with the level of driving force to perform that behavior. And while motivation may be required for a certain behavior, it does not guarantee that behavior. The membrane analogy also allows us to highlight the fact that a crucial part of the control of behavior is to appropriately gate the expression of different motivated actions that have conflicting goals—tapping on the computer keyboard versus spending the afternoon sailing, for example.

Despite tangible progress in recent years, neuroscience cannot yet provide a detailed explanation for why the sailing expedition was abandoned in favor of writing this chapter. Nonetheless, much has been learned about what motivates certain behaviors that are basic to survival.

THE HYPOTHALAMUS, HOMEOSTASIS, AND MOTIVATED BEHAVIOR

The hypothalamus and homeostasis were introduced in Chapter 15. Recall that homeostasis refers to the processes that maintain the internal environment of the body within a narrow physiological range. Although homeostatic reflexes occur at many levels of the nervous system, the hypothalamus plays a key role in the regulation of body temperature, fluid balance, and energy balance.

The hypothalamic regulation of homeostasis starts with sensory transduction. A regulated parameter (e.g., temperature) is measured by specialized sensory neurons, and deviations from the optimal range are detected by neurons concentrated in the periventricular zone of the hypothalamus. These neurons then orchestrate an integrated response to bring the parameter back to its optimal value. The response generally has three components:

1. *Humoral response*: Hypothalamic neurons respond to sensory signals by stimulating or inhibiting the release of pituitary hormones into the bloodstream.

2. *Visceromotor response*: Neurons in the hypothalamus respond to sensory signals by adjusting the balance of sympathetic and parasympathetic outputs of the autonomic nervous system (ANS).
3. *Somatic motor response*: Hypothalamic neurons (particularly within the lateral hypothalamus) respond to sensory signals by inciting an appropriate somatic motor behavioral response.

You are cold, dehydrated, and depleted of energy. The appropriate humoral and visceromotor responses kick in automatically. You shiver, blood is shunted away from the body surface, urine production is inhibited, body fat reserves are mobilized, and so on. But the fastest and most effective way to correct these disturbances of brain homeostasis is to actively seek or generate warmth by moving, to drink water, and to eat. These are examples of **motivated behaviors** generated by the somatic motor system, and they are incited to occur by the activity of the lateral hypothalamus. Our goal in this chapter is to explore the neural basis for this type of motivation. To illustrate, we will concentrate on a subject dear to our hearts: eating.

THE LONG-TERM REGULATION OF FEEDING BEHAVIOR

As you know, even a brief interruption in a person's oxygen supply can lead to serious brain damage and death. You may be surprised to learn that the brain's requirement for food, in the form of glucose, is no less urgent. Only a few minutes of glucose deprivation will lead to a loss of consciousness, eventually followed by death if glucose is not restored. While the external environment normally provides a constant source of oxygen, the availability of food is less assured. Thus, complex internal regulatory mechanisms have evolved to store energy in the body so that it is available when needed. One primary reason we are motivated to eat is to keep these reserves at a level sufficient to ensure that there will not be an energy shortfall.

Energy Balance

The body's energy stores are replenished during and immediately after consuming a meal. This condition, in which the blood is filled with nutrients, is called the *prandial state* (from the Latin word for "breakfast"). During this time, energy is stored in two forms: glycogen and triglycerides (Figure 16.1). Glycogen reserves have a finite capacity, and they are found mainly in the liver and skeletal muscle. Triglyceride reserves are found in adipose (fat) tissue, and they have a virtually unlimited capacity. The assembly of macromolecules such as glycogen and triglycerides from simple precursors is called **anabolism**, or anabolic metabolism.

During the fasting condition between meals, called the *postabsorptive state*, stored glycogen and triglycerides are broken down to provide the body with a continuous supply of the molecules used as fuel for cellular metabolism (glucose for all cells, and fatty acids and ketones for all cells other than neurons). The process of breaking down complex macromolecules is called **catabolism**, or catabolic metabolism; it is the opposite of anabolism. The system is in proper balance when energy reserves are replenished at the same average rate that they are expended. If the intake and storage of energy consistently exceed the usage, the amount of body fat, or *adiposity*, increases, eventually resulting in **obesity**. (The word *obese* is derived from the Latin word for "fat.") If the intake of energy consistently fails to meet the body's demands, loss of fat tissue occurs,

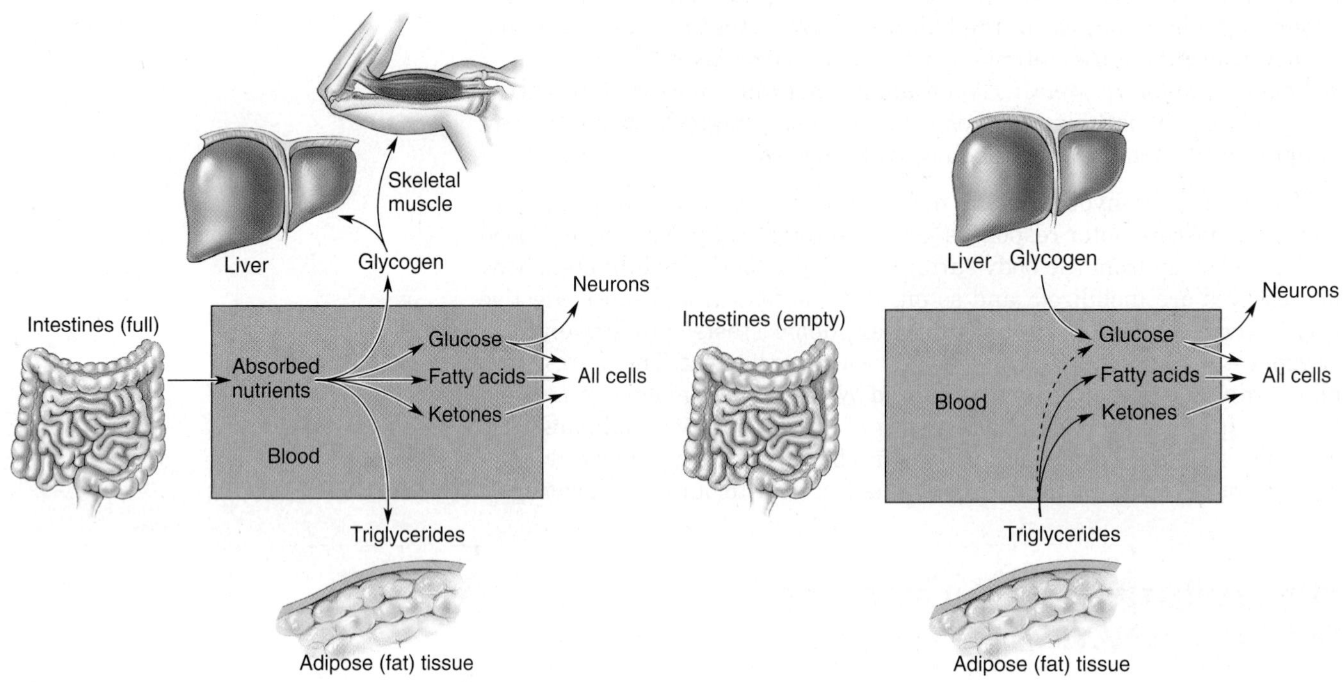

(a) Anabolism during the prandial state

(b) Catabolism during the postabsorptive state

▲ FIGURE 16.1
Loading and emptying the body's energy reserves. **(a)** After a meal, when we are in the prandial state, excess energy is stored as glycogen or as triglycerides. **(b)** During the time between meals, when we are in the postabsorptive state, the glycogen and triglycerides are broken down (catabolized) into smaller molecules that can be used as fuel by the cells of the body.

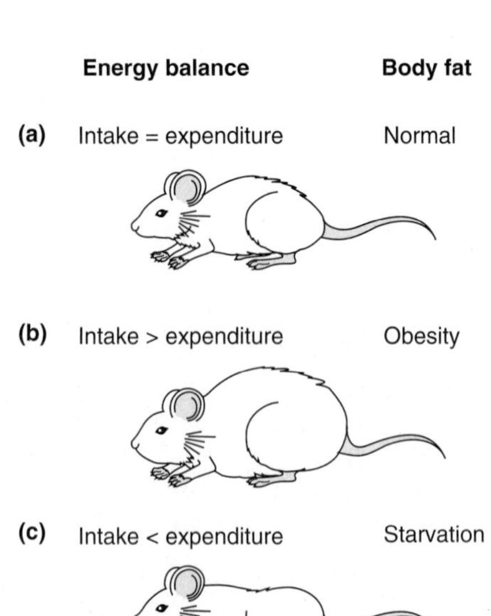

▲ FIGURE 16.2
Energy balance and body fat.
(a) Normal energy balance leads to normal adiposity. **(b)** Prolonged positive energy balance leads to obesity. **(c)** Prolonged negative energy balance leads to starvation.

eventually resulting in **starvation**. Figure 16.2 summarizes the concept of energy balance and body fat.

For the system to stay in balance, there must be some means of regulating feeding behavior, based on the size of the energy reserves and their rate of replenishment. In recent decades, research has made substantial progress in understanding the various means by which this regulation occurs—and none too soon, because eating disorders and obesity are widespread health problems. It is now apparent that there are multiple regulatory mechanisms, some acting over a long period of time to maintain the body's fat reserves, and others acting over a shorter time period to regulate meal size and frequency. We begin our investigation by looking at long-term regulation.

Hormonal and Hypothalamic Regulation of Body Fat and Feeding

The study of the homeostatic regulation of feeding behavior has a long history, but the pieces of the puzzle have only recently fallen into place. As we will see, feeding is stimulated when neurons in the hypothalamus detect a drop in the level of a hormone released by fat cells. These hypothalamic cells are concentrated in the periventricular zone; those neurons that incite feeding behavior are in the lateral hypothalamus.

Body Fat and Food Consumption. If you've ever dieted, you don't need to be told that the body works hard to frustrate any efforts to alter adiposity.

Consider Figure 16.3, showing that a rat can be induced to lose body fat by severely restricting its caloric intake. However, once free access to food is restored, the animal overeats until the original level of body fat has fully returned. It also works the other way around. Animals force fed in order to gain fat mass, once given the chance to regulate their own diet, eat less until their fat levels return to normal. The heavy rat's behavioral response is obviously not a reflection of vanity; it is a mechanism for maintaining energy homeostasis. The idea that the brain monitors the amount of body fat and acts to "defend" this energy store against perturbations, first proposed in 1953 by British scientist Gordon Kennedy, is called the **lipostatic hypothesis**.

The connection between body fat and feeding behavior suggests that there must be communication from adipose tissue to the brain. A bloodborne hormonal signal was immediately suspected, and this suspicion was confirmed in the 1960s by Douglas Coleman and his colleagues at the Jackson Laboratories in Bar Harbor, Maine, working with genetically obese mice. The DNA of one strain of an obese mouse lacked both copies of a gene called *ob* (these mice are therefore called *ob*/*ob* mice). Coleman hypothesized that the protein encoded by the *ob* gene is the hormone telling the brain that fat reserves are normal. Thus, in the *ob*/*ob* mice that lack this hormone, the brain is fooled into thinking that the fat reserves are low, and the animals are abnormally motivated to eat. To test this idea, a parabiosis experiment was performed. *Parabiosis* is the long-term anatomical and physiological union of two animals, as in Siamese twins. Fusion can also be achieved surgically, thereby resulting in parabiosed animals sharing a common blood supply. Coleman and his colleagues found that when *ob*/*ob* animals were parabiosed with normal mice, their feeding behavior and obesity were greatly reduced, as if the missing hormone had been replaced (Figure 16.4).

Then the search was on for the protein encoded by the *ob* gene. In 1994, a group of scientists led by Jeffrey Friedman at Rockefeller University finally isolated the protein, which they called **leptin** (from the Greek for "slender"). Treating *ob*/*ob* mice with leptin completely reverses the obesity and the eating disorder (Figure 16.5). The hormone leptin, released by adipocytes (fat cells), regulates body mass by acting directly on neurons of the hypothalamus that decrease appetite and increase energy expenditure.

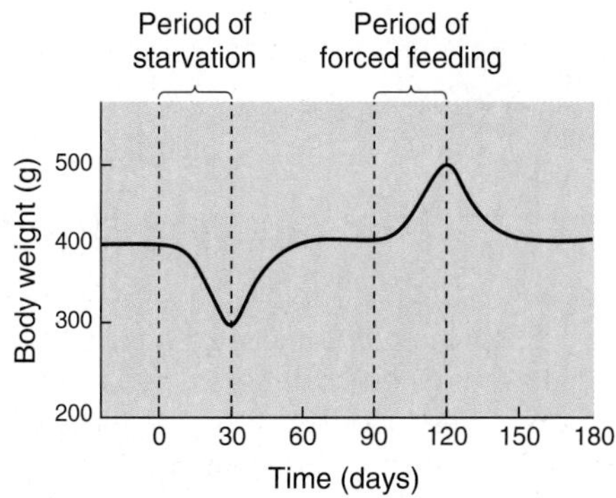

▲ FIGURE 16.3
The maintenance of body weight around a set value. Body weight is normally very stable. Weight lost during a period of starvation is rapidly gained when food is freely available. Similarly, if an animal is force fed, it will gain weight, but the weight is lost as soon as the animal can regulate its own food intake.

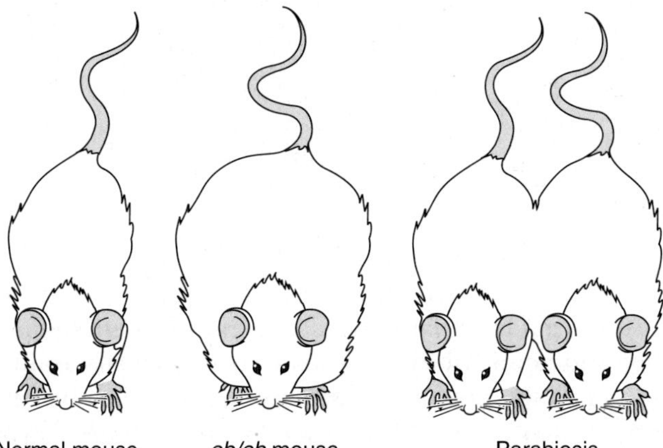

▲ FIGURE 16.4
The regulation of body fat by a circulating hormone. If a genetically obese *ob/ob* mouse is surgically fused with a normal mouse so that bloodborne signals are shared between the animals, the obesity of the *ob/ob* mouse is greatly moderated.

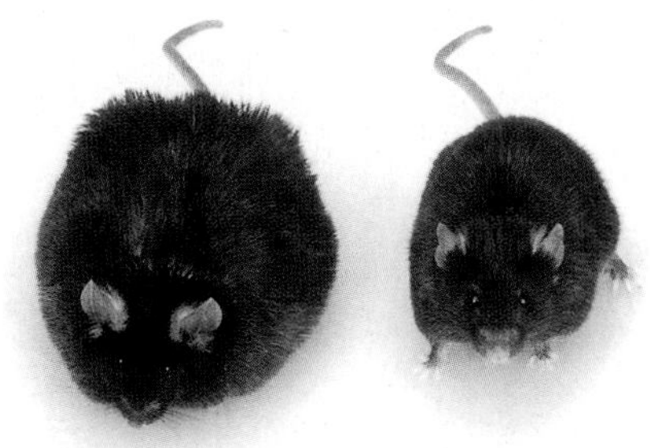

▲ FIGURE 16.5
The reversal of obesity in *ob/ob* mice by leptin. Both of these mice have a defect in the *ob* gene that encodes the fat hormone leptin. The animal on the right received daily hormone replacement treatment, which prevented the obesity that is apparent in the animal on the left. (Courtesy of John Sholtis, Rockefeller University.)

BOX 16.1 OF SPECIAL INTEREST

The Starving Brains of the Obese

Like the *ob/ob* mouse, humans lacking leptin crave food, have slowed metabolism, and become morbidly obese. For these individuals, leptin replacement therapy can be a "miracle cure" (Figure A). While mutations affecting the leptin gene are rare, there is evidence for a genetic basis for many forms of human obesity. The hereditability of obesity is equivalent to that of height and greater than that of many other conditions, including heart disease and breast cancer. Many genes are involved, and the search is on to discover them.

Obesity is a major human health problem. In the United States, two-thirds of the population is overweight, and millions are morbidly obese. Many obese people experience intense cravings for food but at the same time have slowed metabolism. In the case of leptin deficiency, the brain and body respond as if the person is starving, despite massive obesity.

Leptin offered tremendous promise as a treatment for obesity. By supplementing leptin, the logic went, the brain could be fooled into decreasing appetite and increasing metabolism. Unfortunately, other than the rare individuals who congenitally lack the hormone, most obese patients have failed to respond to leptin therapy. Indeed, many have been found to have abnormally elevated blood levels of leptin. It appears that the problem for these patients stems from decreased sensitivity of brain neurons to the leptin circulating in the blood. The problem can arise from decreased penetration of the leptin through the blood-brain barrier, reduced expression of the leptin receptor in the neurons of the periventricular hypothalamus, or altered CNS responses to changes in hypothalamic activity. Intensive efforts are now underway to identify drug targets in the feeding circuits of the brain that are downstream of leptin.

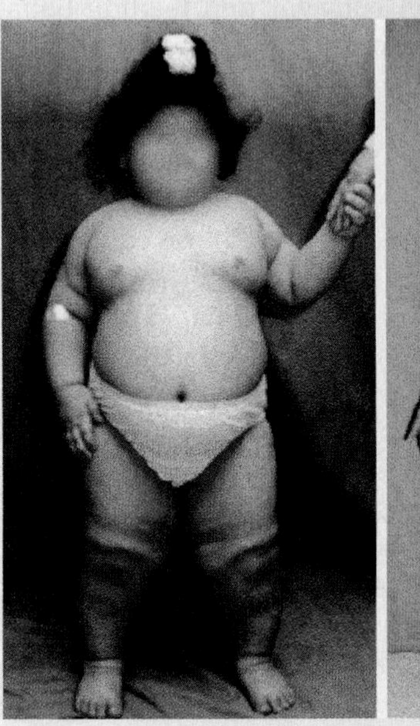

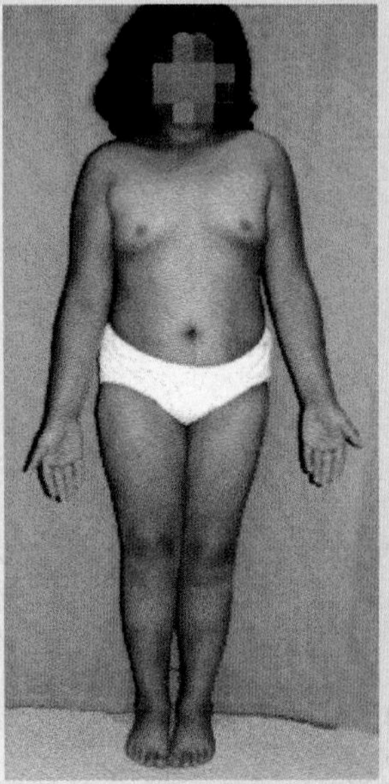

Figure A
The effect of hormone replacement in a leptin-deficient human. Daily leptin treatment begun at age 5 (left) brought this girl's weight down to a near-normal level, shown here at age 9 (right). (Source: Gibson, et al., 2004, p. 4823.)

Well-fed humans tend to focus on how increased leptin can fight obesity (Box 16.1). However, more significant for survival is how leptin depletion fights starvation. Leptin deficiency stimulates hunger and feeding, suppresses energy expenditure, and inhibits reproductive competence—adaptive responses when food is scarce and energy reserves are low.

The Hypothalamus and Feeding. A.W. Hetherington and S.W. Ranson of Northwestern University made the seminal discovery, published in 1940, that small lesions made on both sides of a rat's hypothalamus can have large effects on subsequent feeding behavior and adiposity. Bilateral lesions of the *lateral* hypothalamus caused **anorexia**, a severely diminished appetite for food. In contrast, bilateral lesions of the *ventromedial* hypothalamus caused the animals to overeat and become obese (Figure 16.6). This basic scenario applies to humans as well. Anorexia caused by damage to the lateral hypothalamus is commonly referred to as the **lateral hypothalamic syndrome**; overeating and obesity caused by lesions to the ventromedial hypothalamus is called the **ventromedial hypothalamic syndrome**.

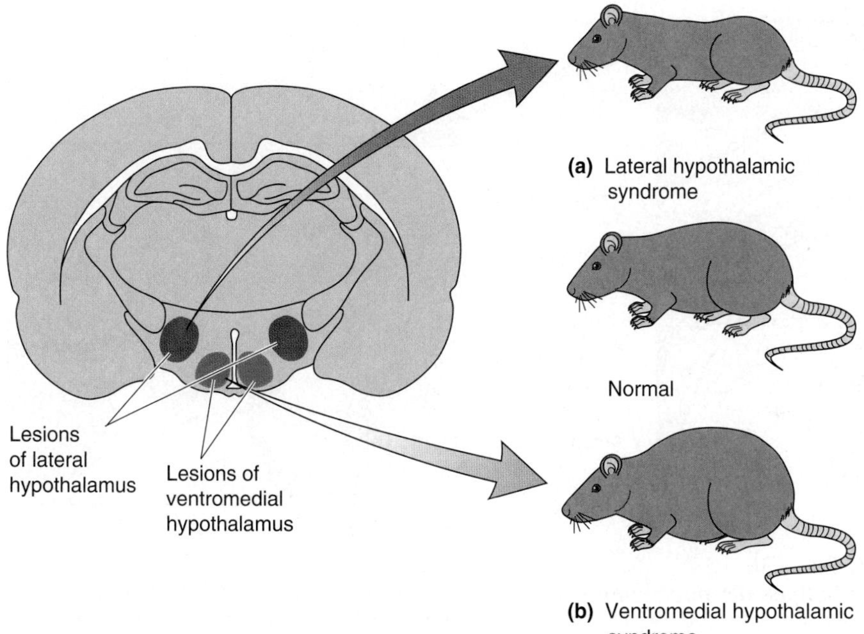

◀ FIGURE 16.6
Altered feeding behavior and body weight resulting from bilateral lesions of the rat hypothalamus. **(a)** The lateral hypothalamic syndrome, characterized by anorexia, is caused by lesions of the lateral hypothalamus. **(b)** The ventromedial hypothalamic syndrome, characterized by obesity, is caused by lesions of the ventromedial hypothalamus.

For a time, it was a popular idea that the lateral hypothalamus is a "hunger center" acting in opposition to the ventromedial hypothalamus "satiety center" and that lesions of the medial or lateral hypothalamus bring the system out of balance. Destruction of the lateral hypothalamus leaves the animals inappropriately satiated, so they do not eat; destruction of the ventromedial hypothalamus leaves the animals insatiable, so they overeat. However, this "dual center" model has proven to be overly simplistic. We now have a better idea why hypothalamic lesions affect body fat and feeding behavior; it has much to do with leptin signaling.

The Effects of Elevated Leptin Levels on the Hypothalamus. Although still sketchy in places, a picture is beginning to emerge of how the hypothalamus participates in body fat homeostasis. First, let's consider the response when leptin levels are high, as they are right after several days of "forced" holiday feasting.

Circulating leptin molecules, released into the bloodstream by fat cells, activate leptin receptors on neurons of the **arcuate nucleus** of the hypothalamus, which lies near the base of the third ventricle (Figure 16.7). The arcuate neurons that are activated by a rise in blood leptin levels contain peptide neurotransmitters called α*MSH* and *CART*, and the levels of these peptides in the brain vary in proportion to the level of leptin in the blood. (To explain the alphabet soup: Peptides are often named by their first discovered function, and these names can lead to confusion when other roles are recognized. Therefore, neuropeptides are usually referred to simply by their abbreviations. For the record, αMSH stands for *alpha-melanocyte-stimulating hormone*, and CART stands for *cocaine- and amphetamine-regulated transcript*. Like other neurotransmitters, the functional role of these molecules depends on the circuits in which they participate.)

Before going any further, let's take a moment to consider the body's integrated response to excessive adiposity, high leptin levels, and activation of the αMSH/CART neurons of the arcuate nucleus. The *humoral response* consists of increased secretion of TSH (thyroid-stimulating hormone) and ACTH (adrenocorticotropic hormone) (see Table 15.1 in Chapter 15). These pituitary hormones act on the thyroid and adrenal glands and have the

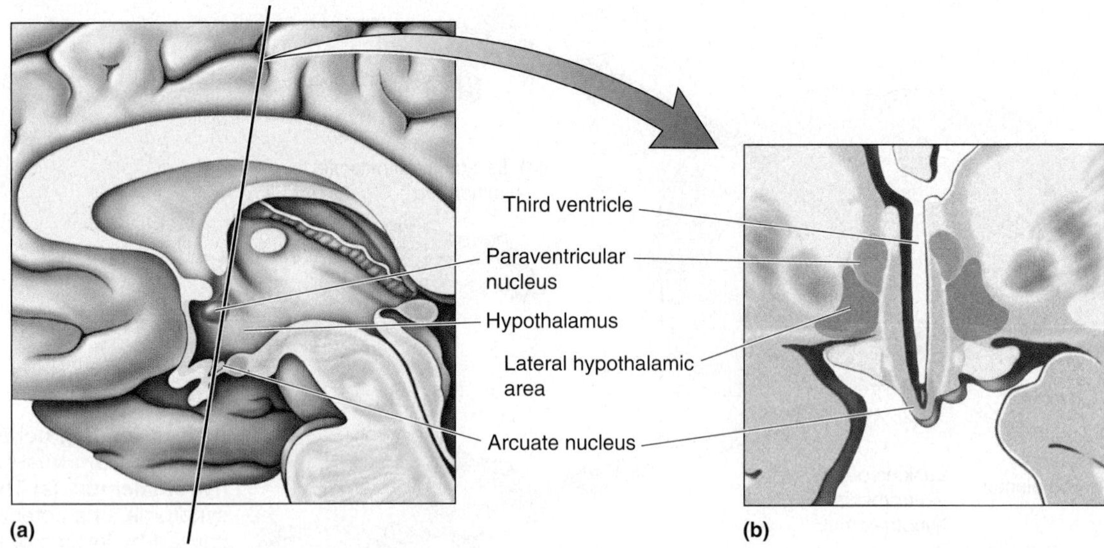

▲ FIGURE 16.7

Hypothalamic nuclei important for the control of feeding. **(a)** A midsagittal view of the human brain, showing the location of the hypothalamus. **(b)** A coronal section, taken in the plane indicated in part a, showing three important nuclei for the control of feeding: the arcuate nucleus, the paraventricular nucleus, and the lateral hypothalamic area.

effect of raising the metabolic rate of cells throughout the body. The *visceromotor response* increases the tone of the sympathetic division of the ANS, which also raises metabolic rate, in part by raising body temperature. The *somatic motor response* decreases feeding behavior. The αMSH/CART neurons of the arcuate nucleus project their axons directly to the regions of the nervous system that orchestrate this coordinated response (Figure 16.8).

αMSH/CART neurons trigger the humoral response by the activating neurons in the **paraventricular nucleus** of the hypothalamus, which in turn causes the release of the hypophysiotropic hormones that regulate the secretion of TSH and ACTH from the anterior pituitary (see Chapter 15). The paraventricular nucleus also controls the activity of the sympathetic division of the ANS with direct axonal projections to neurons in the lower brain stem and to preganglionic neurons in the spinal cord. Additionally, there is also a direct path for arcuate control of the sympathetic response: The αMSH and CART neurons themselves project axons directly down to the intermediolateral gray matter of the spinal cord. Finally, feeding behavior is inhibited via connections of the arcuate nucleus neurons with cells in the lateral hypothalamus. We will take a closer look at the lateral hypothalamus in a moment.

The injection of αMSH or CART into the brain mimics the response to elevated leptin levels. Thus, these are said to be **anorectic peptides**; they diminish appetite. The injection of drugs that block the actions of these peptides increases feeding behavior. These findings suggest that αMSH and CART normally participate in the regulation of energy balance, in part by acting as the brain's own appetite suppressants.

The Effects of Decreased Leptin Levels on the Hypothalamus. In addition to turning off the responses mediated by αMSH/CART neurons, a *fall* in leptin levels actually *stimulates* another type of arcuate nucleus neuron. These neurons contain their own mix of peptides: *NPY* (*neuropeptide Y*) and *AgRP* (*agouti-related peptide*). The NPY/AgRP neurons of the arcuate nucleus also have connections with the paraventricular nucleus and the lateral hypothalamus (Figure 16.9), and the effects of

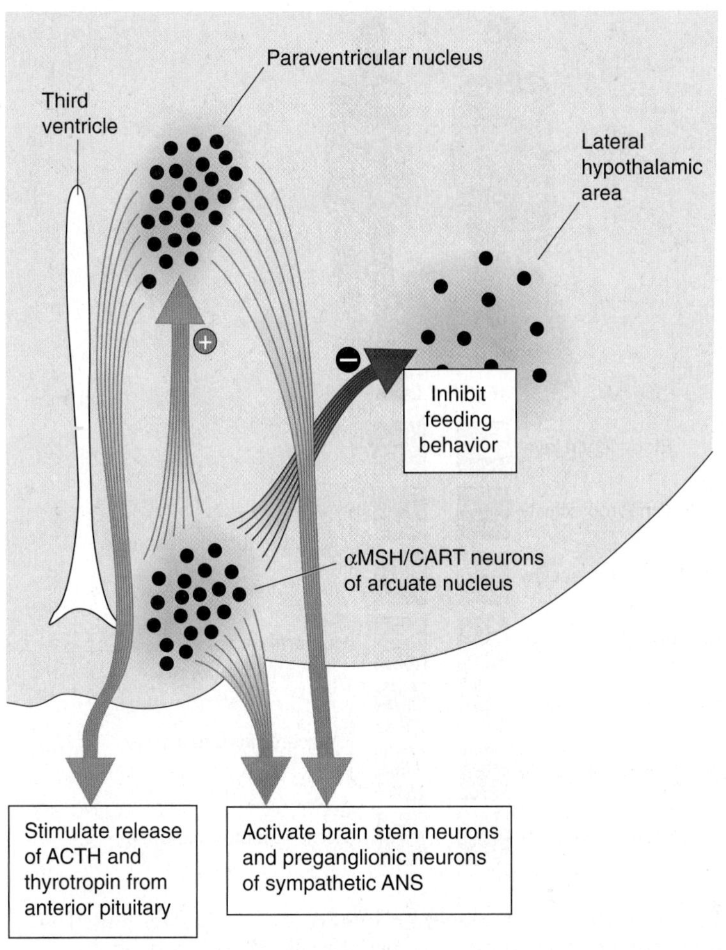

◀ FIGURE 16.8
The response to elevated leptin levels. A rise in leptin levels in the blood is detected by neurons in the arcuate nucleus that contain the peptides αMSH and CART. These neurons project axons to the lower brain stem and spinal cord, the paraventricular nuclei of the hypothalamus, and the lateral hypothalamic area. Each of these connections contributes to the coordinated humoral, visceromotor, and somatic motor responses to increased leptin levels. (Source: Adapted from Sawchenko, 1998, p. 437.)

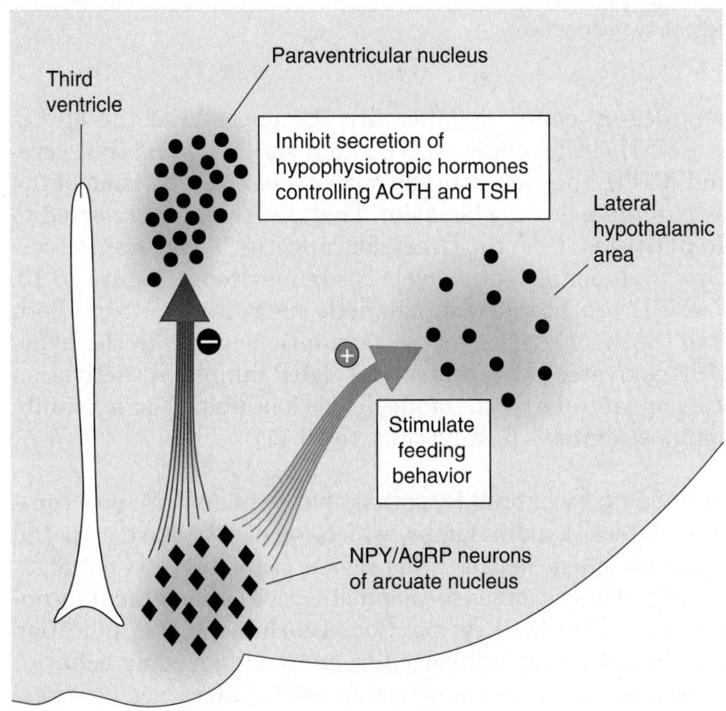

◀ FIGURE 16.9
The response to decreased leptin levels. A reduction in blood levels of leptin is detected by neurons in the arcuate nucleus that contain the peptides NPY and AgRP. These arcuate nucleus neurons inhibit the neurons in the paraventricular nuclei that control the release of TSH and ACTH from the pituitary. In addition, they activate the neurons in the lateral hypothalamus that stimulate feeding behavior. Some of the activated lateral hypothalamic neurons contain the peptide MCH (melanin-concentrating hormone).

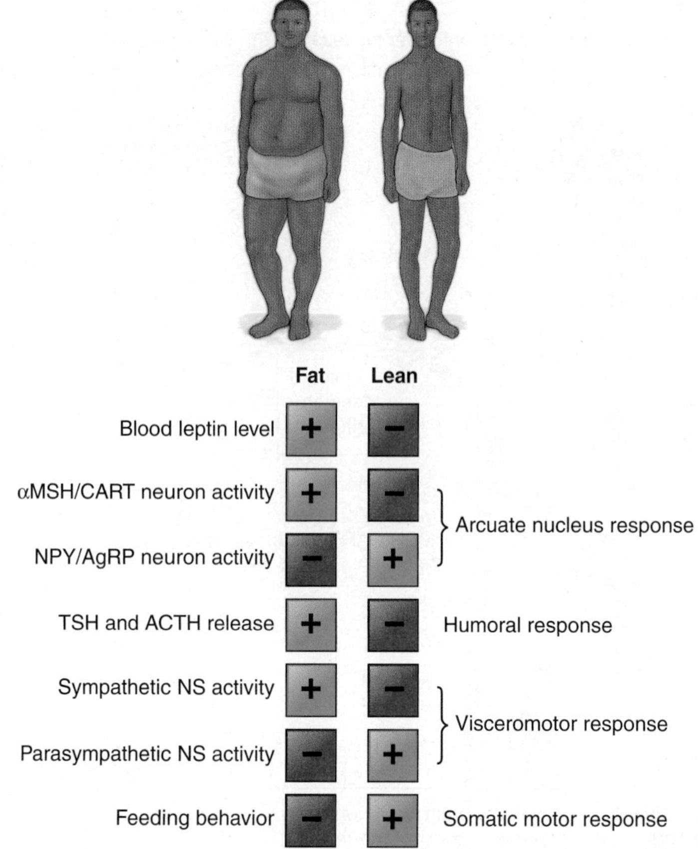

▲ FIGURE 16.10
Summary of the responses to increased and decreased adiposity (fat). The arcuate nucleus senses changes in blood levels of leptin. A rise in leptin increases activity in the aMSH/CART neurons, and a fall in leptin increases activity in the NPY/AgRP neurons. These two populations of arcuate nucleus neurons orchestrate the humoral, visceromotor, and somatic motor responses to increased or decreased adiposity, respectively.

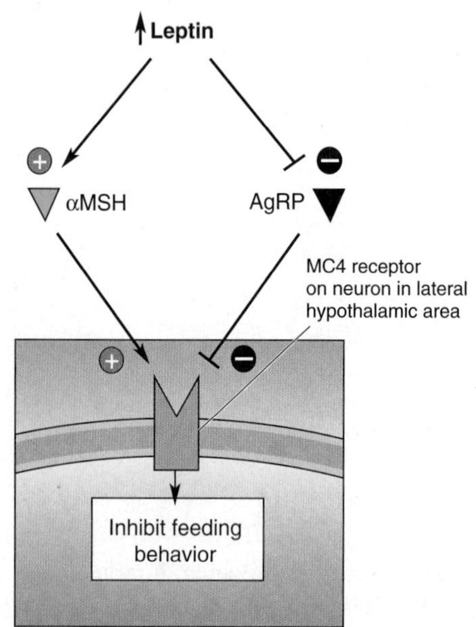

▲ FIGURE 16.11
Competition for activation of the MC4 receptor. One way that αMSH, an anorectic peptide, and AgRP, an orexigenic peptide, exert opposite effects on metabolism and feeding behavior is via an interaction with the MC4 receptor on some hypothalamic neurons. While αMSH stimulates the MC4 receptor, AgRP inhibits it.

these neuropeptides on energy balance are the opposite of the effects caused by the αMSH/CART neurons. NPY and AgRP *inhibit* the secretion of TSH and ACTH, they activate the *parasympathetic* division of the ANS, and they *stimulate* feeding behavior. They are therefore referred to as **orexigenic peptides** (from the Greek for "appetite"). The brain's coordinated response to changing leptin levels is summarized in Figure 16.10.

AgRP and αMSH are literally antagonistic neurotransmitters. Both peptides bind to the *MC4 receptor* on postsynaptic neurons in the hypothalamus. αMSH activates the receptor and AgRP inhibits it. Activation of MC4 receptors on lateral hypothalamic neurons inhibits feeding; inhibiting the receptors stimulates feeding (Figure 16.11).

The Control of Feeding by Lateral Hypothalamic Peptides. We now come to the mysterious lateral hypothalamus, which appears to have a special role in motivating us to eat. Because this region of the brain is not organized into well-defined nuclei, it has the nondistinctive name **lateral hypothalamic area** (see Figure 16.7). As mentioned earlier, the first indication that the lateral hypothalamus is involved in motivating feeding behavior was that a lesion here causes animals to stop eating. Moreover, electrical stimulation of this area triggers feeding behavior, even in satiated animals. These basic findings apply to all mammals that have been examined, including humans. However, crude lesions and electrical stimulation not only affect the neurons with cell bodies in this region but also affect many different axonal pathways passing through the lateral hypothalamus. Modern experiments using optogenetic methods to stimulate and silence specific types of neurons (see Chapter 4) reveal that both neurons intrinsic to the lateral hypothalamus *and* axons passing through the lateral hypothalamus contribute to the motivation of feeding behavior. Let's first concentrate on the role of the neurons within the lateral hypothalamic area.

One group of neurons in the lateral hypothalamus that receives direct input from the leptin-sensitive cells of the arcuate nucleus has yet another peptide neurotransmitter called *MCH* (*melanin-concentrating hormone*). These cells have extremely widespread connections in the brain, including direct monosynaptic innervation of most of the cerebral cortex. The cortex is involved in organizing and initiating goal-directed behaviors, such as raiding the refrigerator. The MCH system is in a strategic position to inform the cortex of leptin levels in the blood and therefore could contribute significantly to motivating the search for food. Supporting this idea, the injection of MCH into the brain stimulates feeding behavior. Moreover, mutant mice that lack this peptide exhibit reduced feeding behavior, have an elevated metabolic rate, and are lean.

A second population of lateral hypothalamic neurons with widespread cortical connections has been identified, containing another peptide called *orexin*. These cells also receive direct inputs from the arcuate nucleus. As is the case for MCH, and as the name suggests, orexin is an orexigenic peptide (i.e., it stimulates feeding behavior). The levels of both MCH and orexin rise in the brain when leptin levels in the blood fall. These two peptides are complementary, not redundant. For example, orexin promotes meal initiation, whereas MCH prolongs consumption. Additionally, orexin, also called *hypocretin*, plays a very important role in the regulation of wakefulness. As we will learn in Chapter 19, gene mutations that disable orexin (hypocretin) signaling not only lead to weight loss but also to excessive daytime sleepiness. Perhaps it is obvious that sleep inhibits feeding behavior; after all, it is difficult to eat when you are asleep. However, you might be surprised to learn that insomnia and obesity also often go together. Orexin (hypocretin) provides an interesting link between these conditions.

To conclude this section, let's briefly summarize hypothalamic responses to blood leptin levels. Remember, leptin levels rise when body fat is increased, and they fall when body fat is decreased:

- A *rise* in leptin levels stimulates the release of αMSH and CART from arcuate nucleus neurons. These anorectic peptides act on the brain, in part by activating the MC4 receptor, to inhibit feeding behavior and increase metabolism.
- A *fall* in leptin levels stimulates the release of NPY and AgRP from arcuate nucleus neurons, and the release of MCH and orexin from neurons in the lateral hypothalamic area. These orexigenic peptides act on the brain to stimulate feeding behavior and decrease metabolism.

THE SHORT-TERM REGULATION OF FEEDING BEHAVIOR

Regulation of the tendency to seek and consume food by the body's levels of leptin is very important, but it is not the whole story. Setting aside social and cultural factors (such as a mother's command: "Eat!"), the motivation to eat depends on how long it has been since the last meal and how much we ate at that time. Moreover, the motivation to continue eating once a meal starts depends on how much food (and what type) has already

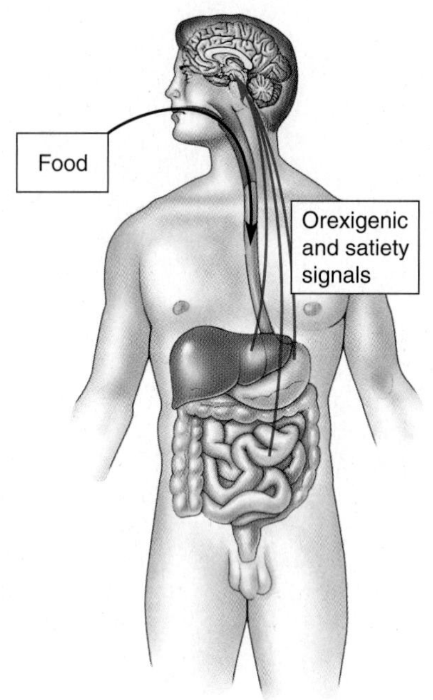

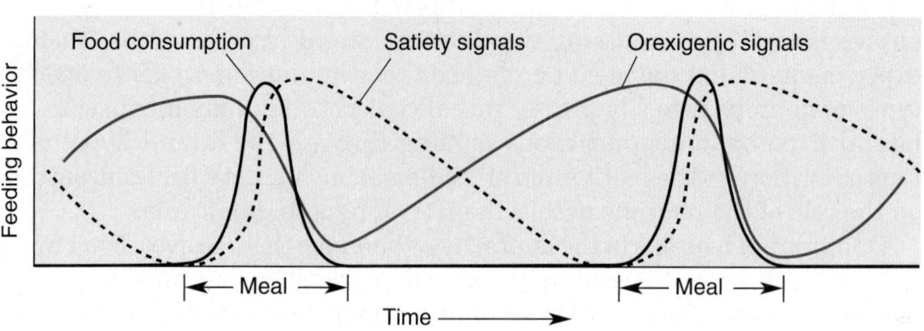

▲ FIGURE 16.12
A hypothetical model for the short-term regulation of feeding behavior. This graph shows a possible means of regulating food consumption by satiety signals. Satiety signals rise in response to feeding. When satiety signals are high, food consumption is inhibited. When the satiety signals fall to zero, the inhibition is eliminated, and food consumption ensues.

been eaten. These are examples of what we are calling the short-term regulation of feeding behavior.

A useful way to think about this regulatory process is to imagine that the drive to eat, which may vary rather slowly with the rise and fall of leptin, is increased by orexigenic signals generated in response to a period of fasting, and inhibited by **satiety signals** that occur when we eat and begin the process of digestion (i.e., the prandial period). These satiety signals both terminate the meal and inhibit feeding for some time afterward. During this postabsorptive (fasting) period, the satiety signals slowly dissipate, and the orexigenic signals build, until the drive to eat again takes over (Figure 16.12). We will use this model to explore the biological basis for the short-term regulation of feeding behavior.

Appetite, Eating, Digestion, and Satiety

You have awakened in the morning after a long night's slumber. You come to the kitchen to find pancakes cooking on the stove; when they are ready, you enthusiastically eat them until you're satiated. Your body's reactions during this process can be divided into three phases: cephalic, gastric, and substrate (also called the *intestinal phase*):

1. *Cephalic phase*. The sight and smell of the pancakes trigger a number of physiological processes that anticipate the arrival of breakfast. The parasympathetic and enteric divisions of the ANS are activated, causing the secretion of saliva into your mouth and digestive juices into your stomach.
2. *Gastric phase*. These responses grow much more intense, when you start chewing, swallowing, and filling your stomach with food.
3. *Substrate phase*. As your stomach fills and the partially digested pancakes move into your intestines, nutrients begin to be absorbed into your bloodstream.

As you pass through these phases, signals that motivate consumption of the pancakes are replaced by those that terminate your meal. Let's look at some of the orexigenic and satiety signals that shape eating behavior during a meal (Box 16.2).

BOX 16.2 OF SPECIAL INTEREST

Marijuana and the Munchies

A well-known consequence of marijuana intoxication is stimulation of appetite, an effect known by users as "the munchies." The active ingredient in marijuana is D[9]-tetrahydrocannabinol (THC), which alters neuronal functions by stimulating a receptor called cannabinoid receptor 1 (CB1). CB1 receptors are abundant throughout the brain, so it is overly simplistic to view these receptors as serving only appetite regulation. Nevertheless, "medical marijuana" is often prescribed (where legal) as a means to stimulate appetite in patients with chronic diseases, such as cancer and AIDS. A compound that inhibits CB1 receptors, rimonabant, was also developed as an appetite suppressant. However, human drug trials had to be discontinued because of psychiatric side effects. Although this finding underscores the fact that these receptors do much more than mediate the munchies, it is still of interest to know where in the brain CB1 receptors act to stimulate appetite. Not surprisingly, the CB1 receptors are associated with neurons in many regions of the brain that control feeding, such as the hypothalamus, and some of the orexigenic effects of THC are related to changing the activity of these neurons. However, neuroscientists were surprised to learn in 2014 that much of the appetite stimulation comes from enhancing the sense of smell, at least in mice. Collaborative research conducted by neuroscientists in France and Spain, countries incidentally known for their appreciation of good tastes and smells, revealed that activation of CB1 receptors in the olfactory bulb increases odor detection and is necessary for the increase in food intake stimulated in hungry mice by cannabinoids.

In Chapter 8, we discussed how smells activate neurons in the olfactory bulb which, in turn, relay information to the olfactory cortex. The cortex also sends feedback projections to the bulb that synapse on inhibitory interneurons called *granule cells*. By activating the inhibitory granule cells, this feedback from the cortex dampens ascending olfactory activity. These corticofugal synapses use glutamate as a neurotransmitter. The brain's own endocannabinoids (anandamide and 2-arachidonoylglycerol) are synthesized under fasting conditions, and they inhibit glutamate release by acting on CB1 receptors on the corticofugal axon terminals. Reducing granule cell activation by glutamate in the bulb has the net effect of enhancing the sense of smell (Figure A). It remains to be determined if the munchies arise from enhanced olfaction in marijuana users, but a simple experiment, such as holding your nose while eating, confirms that much of the hedonic value of food derives from the sense of smell.

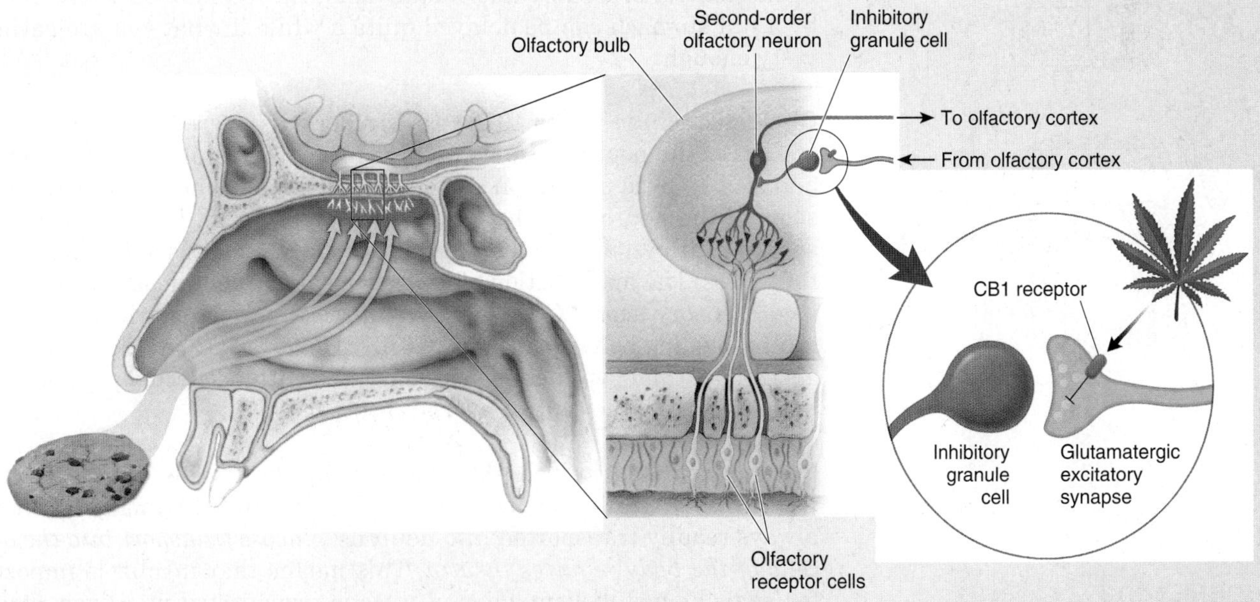

Figure A
Activation of CB1 receptors by THC, the psychoactive ingredient in marijuana, enhances olfaction by suppressing the release of glutamate from corticofugal inputs to inhibitory granule cells in the olfactory bulb. (Source: Adapted from Soria-Gomez et al., 2014.)

Ghrelin. You don't need to be told that the meal begins because you are *hungry*. Until recently, scientists believed that hunger was merely the absence of satiety. This view changed in 1999 with the discovery of a peptide called **ghrelin**. Ghrelin was isolated originally as a factor that stimulates growth hormone release. However, researchers quickly found that the peptide is highly concentrated in the stomach and is released into the bloodstream when the stomach is empty. Your "growling" stomach releases ghrelin ("ghrrrrrrrelin"). Intravenous administration of ghrelin strongly stimulates appetite and food consumption by activating the NPY/AgRP-containing neurons of the arcuate nucleus (the same neurons activated by a drop in leptin in the bloodstream).

Gastric Distension. We all know what it is like to feel "full" after a big meal; as you might expect, the stretching of the stomach wall is a powerful satiety signal. The stomach wall is richly innervated by mechanosensory axons, and most of these ascend to the brain via the **vagus nerve**. Recall from the appendix to Chapter 7 that the vagus nerve (cranial nerve X) contains a mixture of sensory and motor axons, originates in the medulla, and meanders through much of the body cavity (*vagus* is from the Latin for "wandering"). The vagal sensory axons activate neurons in the **nucleus of the solitary tract** in the medulla. These signals inhibit feeding behavior.

You may recall that the nucleus of the solitary tract has been mentioned several times in different contexts. The gustatory nucleus, which receives direct sensory input from the taste buds (see Chapter 8), is actually a subdivision of the nucleus of the solitary tract. The nucleus of the solitary tract is also an important center in the control of the ANS (see Chapter 15). Now we find that the same nucleus receives visceral sensory input from the vagus nerve. It is easy to see how a nucleus with such widespread connections could serve as an important integration center in the control of feeding and metabolism. As you know, satiety induced by a full stomach can be delayed quite a while if what you are eating is tasty enough.

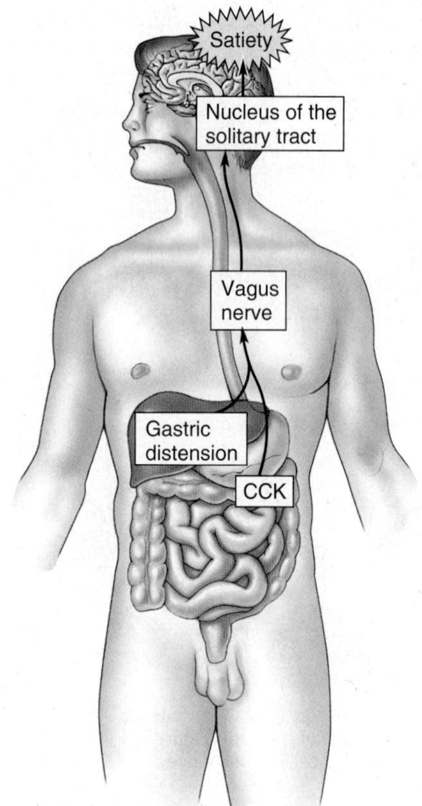

▲ FIGURE 16.13
The synergistic action of gastric distension and CCK on feeding behavior. Both signals converge on axons in the vagus nerve that trigger satiety.

Cholecystokinin. In the 1970s, researchers discovered that the administration of the peptide **cholecystokinin (CCK)** inhibits meal frequency and size. CCK is present in some of the cells that line the intestines and some of the neurons of the enteric nervous system. It is released in response to stimulation of the intestines by certain types of food, especially fatty ones. The major action of CCK as a satiety peptide is exerted on the vagal sensory axons. CCK acts synergistically with gastric distension to inhibit feeding behavior (Figure 16.13). Curiously, CCK, like many other gastrointestinal peptides, is also contained within selected populations of neurons within the central nervous system (CNS).

Insulin. Released into the bloodstream by the β cells of the pancreas, **insulin** is a vitally important hormone (Box 16.3). Although glucose is always readily transported into neurons, *glucose transport into the other cells of the body requires insulin*. This means that insulin is important for anabolic metabolism when glucose is transported into liver, skeletal muscle, and adipose cells for storage as well as for catabolic metabolism when the glucose liberated from storage sites is taken up as fuel by the other cells of the body. Thus, the level of glucose in the blood is tightly regulated by the level of insulin: Blood glucose levels are elevated when

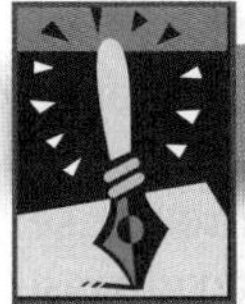

BOX 16.3 OF SPECIAL INTEREST

Diabetes Mellitus and Insulin Shock

Insulin, released by the β cells of the pancreas, plays a pivotal role in maintaining energy balance. After a meal, glucose levels in the blood rise. To be used by the cells of the body, glucose must be shuttled across the plasma membrane by specialized proteins called *glucose transporters*. In all cells other than neurons, the insertion of glucose transporters into the membrane occurs when insulin binds to cell surface insulin receptors. Thus, for the glucose to be utilized or stored by these cells, a rise in blood insulin levels must accompany the rise in blood glucose levels. In the clinical condition known as *diabetes mellitus*, defects in insulin production and release, or in the cellular response to insulin, prevent the normal reaction to elevated glucose. The consequence is elevated blood sugar levels (hyperglycemia) because the glucose absorbed from the intestines cannot be taken up by the cells of the body (other than neurons). The excess glucose passes to the urine, making it sweet. Indeed, the disorder's name is from the Latin for "siphoning honey."

An effective treatment for some types of diabetes mellitus is hypodermic injections of insulin. However, this treatment has risks. An overdose of insulin causes blood glucose levels to plummet (hypoglycemia), starving the neurons of the brain. The resulting condition is called *insulin shock*, characterized by sweating, tremor, anxiety, dizziness, and double vision. If it is not corrected promptly, these early signs are followed by delirium, convulsions, and loss of consciousness. The sudden neurological response to hypoglycemia illustrates how vital energy balance is for the normal functioning of the brain (Figure A).

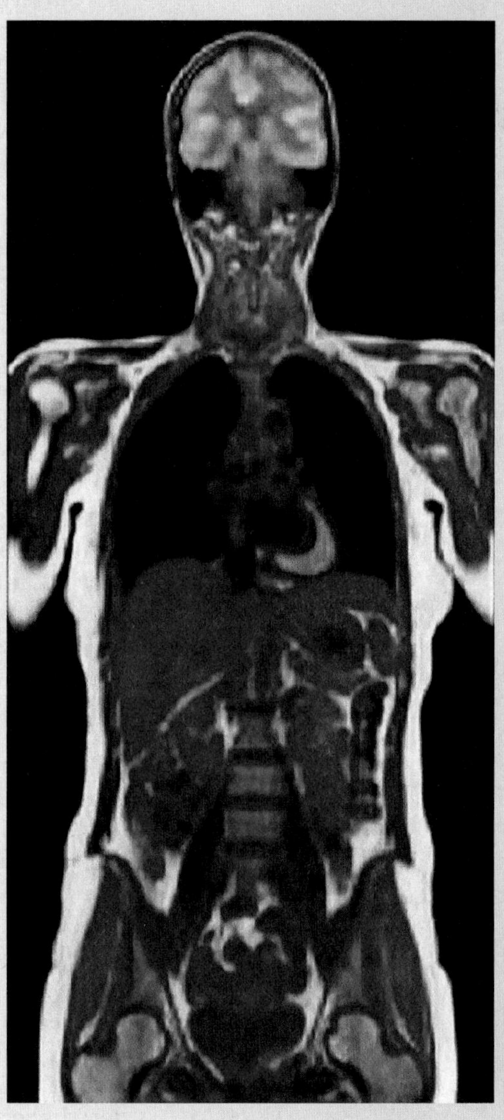

Figure A
A PET image superimposed on an MRI image of the human body. The hot colors (red to yellow) show regions with high glucose utilization. Note that the brain, even at rest, has a very high demand for fuel. When glucose levels in the blood fall, as they do during insulin shock, brain functions are very rapidly lost. (Source: Siemens Healthcare and Professor Marcus Raichle, Washington University, St. Louis.)

insulin levels are reduced; blood glucose levels fall when insulin levels rise.

Insulin release by the pancreas is controlled in a number of ways (Figure 16.14). Consider the example of your pancake breakfast. During the cephalic phase, when you are anticipating food, the parasympathetic innervation of the pancreas (delivered by the vagus nerve) stimulates the β cells to release insulin. In response, blood glucose levels fall slightly, and this change, detected by the neurons of the brain, increases your drive to eat (in part, by activation of the NPY/AgRP neurons of the

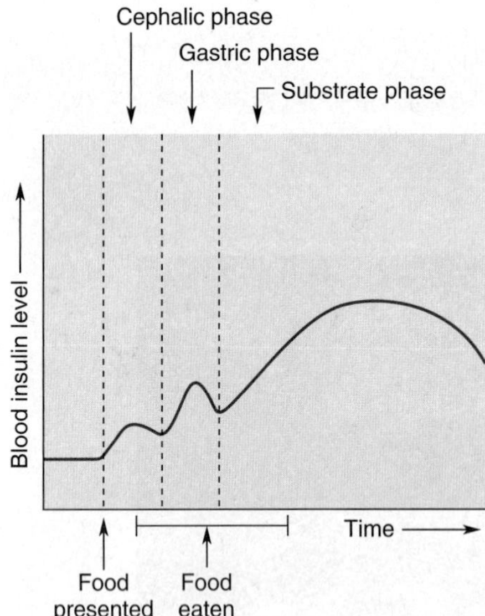

▲ FIGURE 16.14
Changes in blood insulin levels before, during, and after a meal. (Source: Adapted from Woods and Stricker, 1999, p. 1094.)

arcuate nucleus). During the gastric phase, when food enters your stomach, insulin secretion is stimulated further by gastrointestinal hormones, such as CCK. Insulin release is maximal when the food is finally absorbed in the intestines and blood glucose levels rise, during the substrate phase. Indeed, the primary stimulus for insulin release is increased blood glucose levels. This rise in insulin, coupled with the elevated blood glucose levels, is a satiety signal and causes you to stop eating.

In contrast to the other satiety signals we've discussed, which communicate with the brain mainly via the vagus nerve, bloodborne insulin acts to inhibit feeding behavior by acting directly on the arcuate and ventromedial nuclei of the hypothalamus. It appears that insulin acts in much the same way as leptin to regulate feeding behavior.

WHY DO WE EAT?

We have talked about the signals that motivate feeding behavior, but we still have not discussed what that really means in psychological terms. Obviously, we eat because we *like* food. This aspect of motivation is hedonic: It feels good, so we do it. We derive pleasure from the taste, smell, sight, and feel of food and from the act of eating. However, we also eat because we are hungry and we want food. This aspect of motivation can be considered as a *drive reduction*: satisfying a craving. A reasonable assumption is that "liking" and "wanting" are two aspects of a unified process; after all, we typically crave food that we like. However, research on humans and animals suggests that liking and wanting are mediated by separate circuits in the brain.

Reinforcement and Reward

In experiments performed in the early 1950s, James Olds and Peter Milner at McGill University in Montreal, Canada, implanted electrodes in the brains of rats to investigate the effect of electrical brain stimulation on the animals' behavior. The animals were allowed to freely explore a box about 3 ft^2. Every time the rats wandered into one corner of the box, the researchers delivered brain stimulation. They observed that when the electrodes were lodged in certain parts of the brain, the stimulation appeared to cause the animals to spend all their time in the corner that led to stimulation. In a brilliant twist on this experiment, Olds and Milner set up a new box with a lever on one side that, when depressed, caused the brain to be stimulated (Figure 16.15). At first, the rats wandered about the box, stepping on the lever occasionally by accident. But before long, the rats were pressing the lever repeatedly to receive the electrical stimulation. This behavior is called **electrical self-stimulation**. Sometimes the rats would become so involved in pressing the lever that they would shun food and water, stopping only after collapsing from exhaustion (Box 16.4).

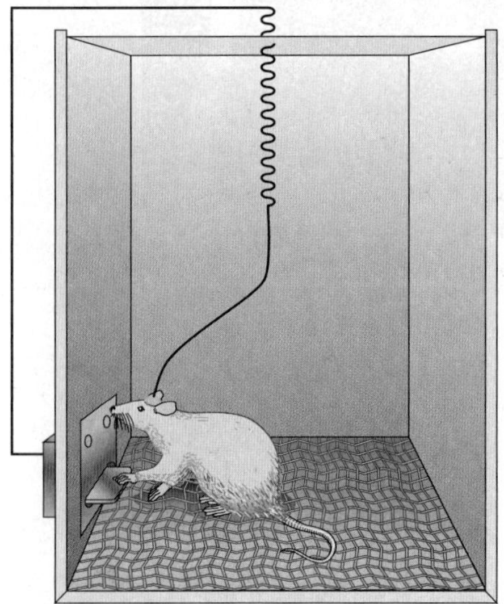

▲ FIGURE 16.15
Electrical self-stimulation by a rat. When the rat presses the lever, it receives a brief electrical current to an electrode in its brain.

Electrical self-stimulation appeared to provide a *reward* that *reinforced* the habit to press the lever. By systematically moving the stimulating electrode to different regions of the brain, researchers were able to identify specific sites that were reinforcing. It became apparent that the most effective sites for self-stimulation fell along the trajectory of dopaminergic axons arising in the ventral tegmental area, projecting through the lateral hypothalamus to several forebrain regions (Figure 16.16). Drugs that block dopamine receptors reduced self-stimulation, suggesting that the animals were working to stimulate the release of dopamine

BOX 16.4 OF SPECIAL INTEREST

Self-Stimulation of the Human Brain

To determine the sensations evoked by brain stimulation, it would be desirable to stimulate a person's brain by inserting electrodes and ask how it feels. Obviously, this is not normally feasible or ethical. However, as treatments of last resort for debilitating medical conditions, humans have occasionally been fitted with intracranial electrodes they can self-stimulate. Let's consider two patients studied by Robert Heath at the Tulane University School of Medicine in the 1960s.

The first patient had severe narcolepsy; he would abruptly go from being awake into a deep sleep. (Narcolepsy and sleep will be discussed in Chapter 19.) The condition significantly interfered with his life and obviously made it difficult to hold a job. He was implanted with 14 electrodes in different areas of the brain in the hope of finding a self-stimulation site that might keep him alert. When he stimulated his hippocampus, he reported feeling mild pleasure. Stimulation of his midbrain tegmentum made him feel alert but unpleasant. The site he chose to frequently self-stimulate was the septal area of the forebrain (Figure A). Stimulating this area made him more alert and gave him a good feeling, which he described as building toward orgasm. He reported that he would sometimes push the button over and over, trying unsuccessfully to achieve orgasm, ultimately ending in frustration.

The second patient's case is a bit more complex. This person had electrodes implanted at 17 brain sites in the hope of learning something about the location of his severe epilepsy. He reported pleasurable feelings with stimulation of the septal area and the midbrain tegmentum. Consistent with the first case above, septal stimulation was associated with sexual feelings. The midbrain stimulation gave him a "happy drunk" feeling. Other mildly positive feelings were produced by stimulation of the amygdala and caudate nucleus. Interestingly, the site he most frequently stimulated was in the medial thalamus, even though stimulation here induced an irritable feeling, one that was less pleasurable than stimulation at other locations. The patient stated that he stimulated this area the most because it gave him the feeling he was about to recall a memory. He repeated the stimulation in a futile attempt to fully bring the memory into his mind, even though, in the end, this process proved to be frustrating.

These two specific cases and many others suggest that self-stimulation is not synonymous with pleasure. Often some reward or anticipated reward is associated with the stimulation, but the experience is not always pleasant.

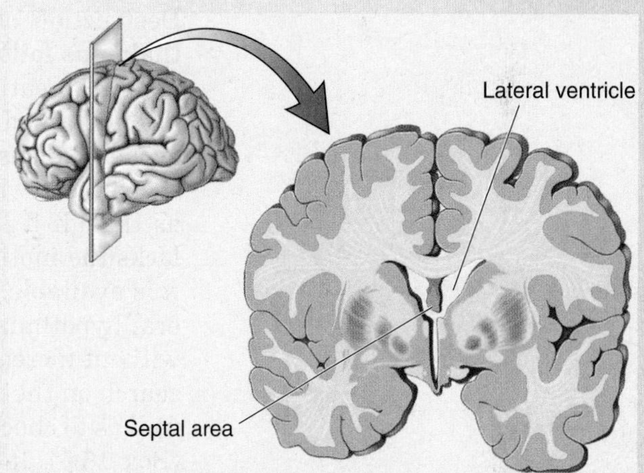

Figure A
The septal area, a site of electrical self-stimulation in humans, is in the rostral forebrain below the lateral ventricle.

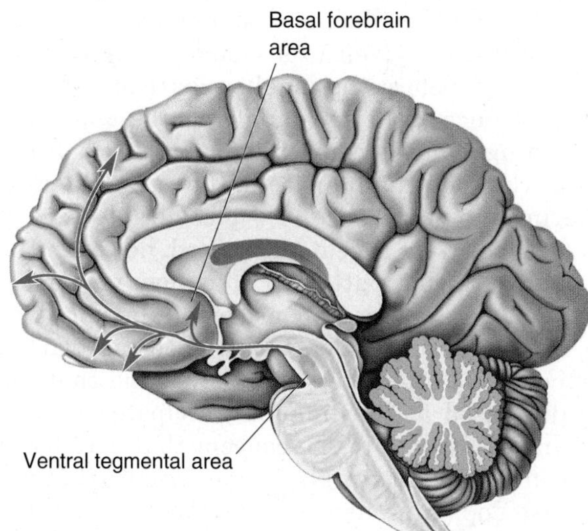

◀ FIGURE 16.16
The mesocorticolimbic dopamine system. Animals are motivated to behave in ways that stimulate the release of dopamine in the basal forebrain area.

in the brain. This idea was further supported when researchers discovered that animals will press a lever to receive an injection of amphetamine, a drug that releases dopamine in the brain. Although there is more to electrical self-stimulation than dopamine, there is little question that dopamine release in the brain will reinforce the behavior that causes it. These experiments suggested a mechanism by which natural rewards (food, water, sex) reinforce particular behaviors. Indeed, a hungry rat will press a lever to receive a morsel of food, and this response is also greatly reduced by dopamine receptor blockers.

The Role of Dopamine in Motivation

For many years, this dopamine projection, from the ventral tegmental area to the forebrain, was believed to serve hedonic reward—in other words, pleasure. In the case of feeding, it was believed that dopamine was released in response to palatable foods, making the sensation pleasurable. Animals were motivated to seek palatable food for the hedonic reward: a squirt of dopamine in the forebrain.

However, this simple idea has been challenged in recent years. Destruction of the dopamine axons passing through the lateral hypothalamus fails to reduce the hedonic responses to food, even though animals stop eating. If a tasty morsel is placed on the tongue of a rat that has sustained such a lesion, the animal will still behave as if the food evokes a pleasurable sensation (the rat equivalent of lip smacking), and the morsel will be consumed. The dopamine-depleted animal behaves as though it *likes* food but does not *want* food. The animal apparently lacks the motivation to seek food, even though it seems to enjoy it when it is available. Conversely, stimulation of the dopamine axons in the lateral hypothalamus of normal rats appears to produce a craving for food without increasing the food's hedonic impact. Not surprisingly, recent research on the cravings associated with addiction (to drugs and alcohol, as well as to chocolate) has focused on the role of this dopaminergic pathway (Box 16.5). It is no coincidence that some of the most highly addictive drugs (cocaine and amphetamine, for example) act directly on dopamine synapses in the brain.

Clues into how dopamine signaling influences behavior have come from animal studies in which the activity of dopamine neurons in the ventral tegmental area of the midbrain is monitored with microelectrodes. In one important study, Wolfram Schultz and colleagues at the University of Cambridge, England, explored what happens to dopamine neurons when a sip of juice is given to a monkey shortly after a light was turned on. Initially, before the monkey learned that the light predicts the delivery of juice, Schultz found that the dopamine neurons had no response to light but became briefly active when the juice was delivered. This is what one might expect if the dopamine neurons were simply registering the occurrence of a pleasurable experience. After the light and the juice were repeatedly paired, however, the dopamine neurons had changed firing patterns. They now responded briefly when the light came on but had no response when the juice was delivered. Furthermore, if Schultz and colleagues tricked the trained monkey and failed to deliver juice after the light, they found that the dopamine neuron firing decreased at the time of anticipated reward (Figure 16.17). These findings have led to the concept that activity of dopamine neurons signals errors in *reward prediction*: Events that are "better than expected" cause dopamine neurons to come to life, those

BOX 16.5 OF SPECIAL INTEREST

Dopamine and Addiction

What do the drugs heroin, nicotine, and cocaine have in common? They act on different neurotransmitter systems in the brain—heroin on the opiate system, nicotine on the cholinergic system, and cocaine on the dopaminergic and noradrenergic systems—and they produce different psychoactive effects. However, all three drugs are highly addictive. This common quality is explained by the fact that they all act on the brain circuitry that motivates behavior—in this case, drug-seeking behavior. We can learn much about the brain mechanisms of motivation by studying drug addiction and vice versa.

Rats, like humans, will self-administer drugs and will develop clear signs of drug dependence. Studies using microinfusions of drugs directly into the brain have mapped out the sites where the drugs cause addiction. In the case of heroin and nicotine, the key site of action is the ventral tegmental area (VTA), home of the dopamine neurons that project axons through the lateral hypothalamus to the forebrain. These dopaminergic neurons have both opiate and nicotinic acetylcholine receptors. In the case of cocaine, a key site of action is the nucleus accumbens, one of the major targets of the ascending dopaminergic axons in the forebrain (Figure A). Recall from Chapter 15 that cocaine prolongs the actions of dopamine at its receptors. Thus, these three drugs either stimulate dopamine release (heroin, nicotine) or enhance dopamine actions (cocaine) in the nucleus accumbens.

The exact role of dopamine in motivating behavior continues to be explored. However, much evidence suggests that animals are motivated to perform behaviors that stimulate dopamine release in the nucleus accumbens and related structures. Behaviors associated with the delivery of drugs that act to stimulate dopamine release are therefore strongly reinforced. However, chronic overstimulation of this pathway causes a homeostatic response: The dopamine "reward" system is downregulated. This adaptation leads to the phenomenon of *drug tolerance*; it takes more and more of the drug to get the desired (or required) effect. Indeed, drug discontinuation in addicted animals is accompanied by a marked decrease in dopamine release and function in the nucleus accumbens. And, of course, one withdrawal symptom is the powerful craving for the discontinued drug.

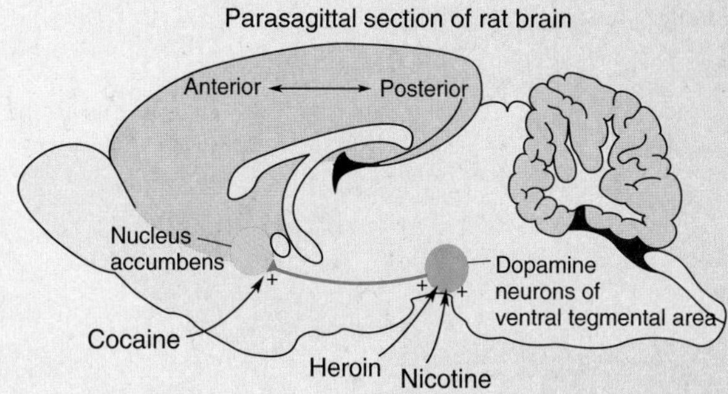

Figure A
Addictive drugs act on the dopaminergic pathway from the ventral tegmental area to the nucleus accumbens. (Source: Adapted from Wise, 1996, p. 248, Fig. 1.)

that are "worse than expected" cause them to be inhibited, and those that occur "as expected" cause no change in firing, even if these events still provide hedonic reward (the juice still tastes good even if you have come to expect it). Behaviors that cause expected or better-than-expected outcomes are repeated; those with outcomes that are worse than expected are not.

Just as the monkey learned that the light predicted delivery of juice, you have learned that the smell or sight of pancakes and coffee predict the delivery of breakfast. This type of learning is integral to the body's "cephalic" preparation for ingestion of a meal. Dopamine is intimately involved in the mechanism behind this learning. Synaptic connections that are active during and shortly before a rise in dopamine are persistently changed to store this memory. While this type of learning is clearly beneficial under normal circumstances, it is hijacked during exposure to addictive drugs, often with devastating consequences. As earlier mentioned, addictive drugs have in common the fact that

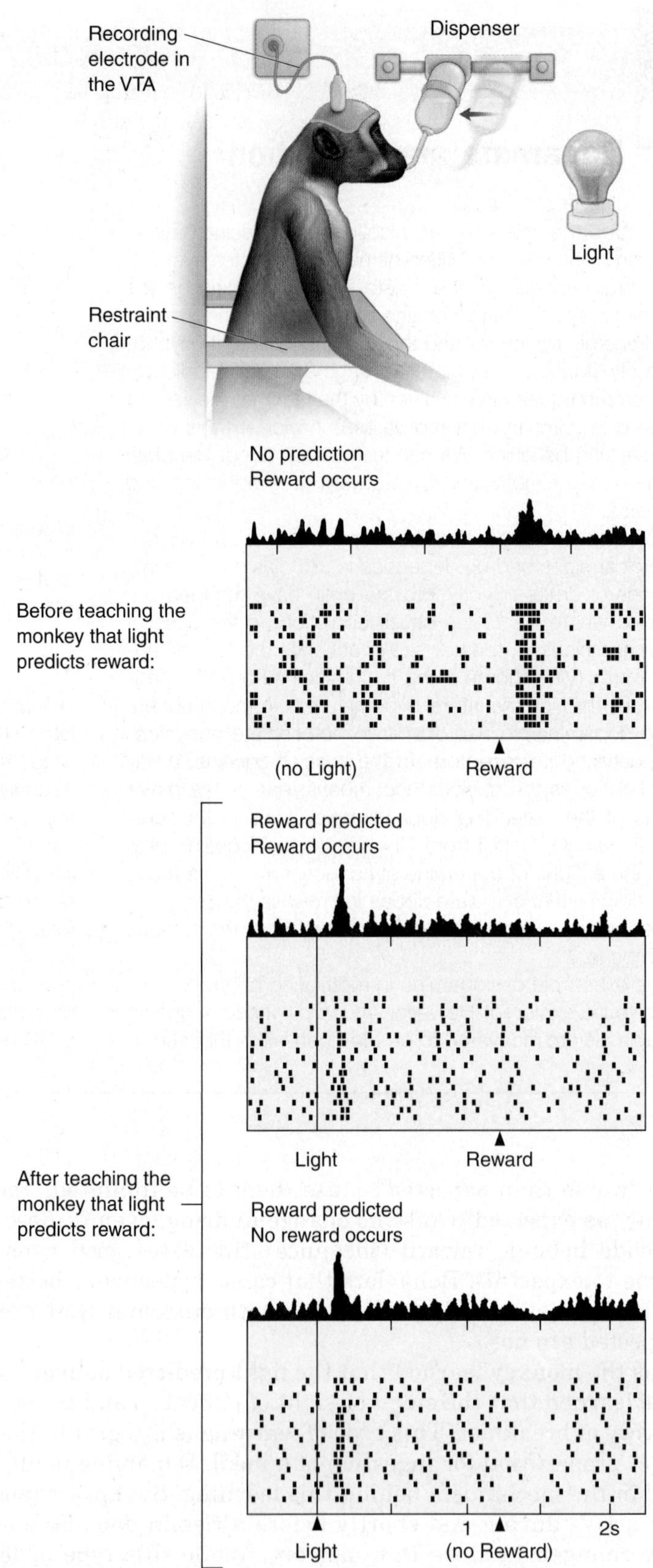

▲ **FIGURE 16.17**
Dopamine neurons in the VTA fire when reward is unexpected.

they act on the central dopaminergic system in the brain. By studying how synapses are modified by drug exposure, researchers have gained insight not only into the neurobiology of addiction and its possible treatments but also into how the brain creates memories (Box 16.6). We will take a closer look at the mechanisms of memory formation in Chapter 25.

Serotonin, Food, and Mood

Mood and food are connected. Consider how grouchy you are when you're on a restricted diet or how good you feel with a whiff and a bite of a freshly baked chocolate chip cookie. As mentioned in Chapter 15, one system in the brain involved in the control of mood uses serotonin as a neurotransmitter. Serotonin provides one of the links between food and mood.

Measurements of serotonin in the hypothalamus reveal that levels are low during the postabsorptive period, rise in anticipation of food, and spike during a meal, especially in response to carbohydrates (Figure 16.18). Serotonin is derived from the dietary amino acid tryptophan, and tryptophan levels in the blood vary with the amount of carbohydrate in the diet (see Box 15.2 in Chapter 15). The rise in blood tryptophan and brain serotonin is one likely explanation for the mood-elevating effects of a chocolate chip cookie. This effect of "carbs" on mood is particularly evident during periods of stress, possibly explaining the food-seeking behavior and subsequent weight gain of many first-year college students.

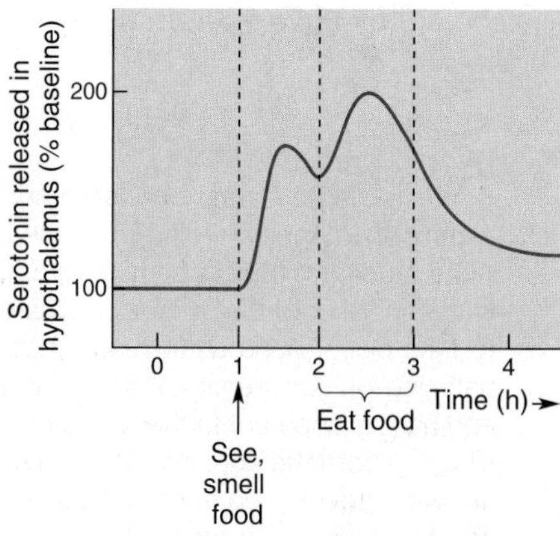

▲ FIGURE 16.18
Changes in hypothalamic serotonin levels before and during a meal. The mood-elevating effects of eating are believed to be related to the release of serotonin in the brain. (Source: Adapted from Schwartz et al., 1990.)

It is interesting to note that drugs that elevate serotonin levels in the brain are powerful appetite suppressants. One of these drugs is dexfenfluramine (trade name Redux), which was used successfully as a treatment for human obesity. Unfortunately, the drug had toxic side effects, leading to its withdrawal from the market in 1997.

Abnormalities in brain serotonin regulation are believed to be one factor that contributes to eating disorders. The defining characteristic of **anorexia nervosa** is a compulsion to maintain body weight at an abnormally low level, while **bulimia nervosa** is characterized by frequent eating binges, often compensated for by forced vomiting. These disorders are also commonly accompanied by *depression*, a severe disturbance of mood that has been linked to lowered brain serotonin levels (we will discuss mood disorders in Chapter 22). The serotonin connection is clearest in the case of bulimia. In addition to depressing mood, low serotonin levels reduce satiety. Indeed, antidepressant drugs that act to elevate brain serotonin levels (e.g., fluoxetine, or Prozac) are also an effective treatment for most bulimia nervosa patients.

OTHER MOTIVATED BEHAVIORS

We have used eating and the regulation of energy balance to give you a fairly detailed picture of the brain mechanisms that incite behavior. The systems involved in motivating several other behaviors that are basic for survival have also been intensively studied. Although we will not cover these other systems in depth, a quick overview will show that the basic principles are the same as those for eating. We will see that the transduction of physiological stimuli in the blood occurs in specialized regions of

BOX 16.6 PATH OF DISCOVERY

Learning to Crave

by Julie Kauer

After college, I was fortunate enough to work as a lab technician in Anne Bekoff's lab at the University of Colorado. Anne studied motor pattern generators, the simple circuits in the spinal cord that allow coordinated muscle movements to take place. Anne and I researched what happens to the hatching pattern generator in chicks after the bird hatches and has no apparent further use for it. When a chick is ready to hatch from the egg, it is tightly curled with its head under the wing pointing up toward the shell. Every 20 seconds or so, it executes two strong leg movements that propel the body slightly within the egg. The beak gradually makes a circular hole, and when this is large enough, the strong leg movements allow the chick to hatch out. To test the fate of the hatching pattern generator, my job was to place recording electrodes in the leg muscles and then carefully fold an already-hatched chick back into the hatching position, this time in a glass egg. Remarkably, the chick became quiet and soon began making leg movements indistinguishable from normal hatching movements. More amazing, we found that chicks even up to 2 months old can be induced to "hatch"; the hatching pattern generator appeared to remain available even weeks after the last time it was needed. While I had a ball putting weeks-old chickens back in glass eggs, I was simultaneously hatching my own scientific approach. I developed a great appreciation for Anne's strategy of asking simple questions that could generate thorough answers, and breaking a complex problem into smaller parts that could be understood clearly. This approach has remained a driving principle in my scientific life ever since.

How does the nervous system store information? This question has been the focus of my work since graduate school, where I first investigated the cellular basis of persistent changes in the nervous system of *Aplysia*, a giant sea slug [will be discussed in Chapter 25]. My fascination with long-lasting changes in neuronal excitability led me to postdoctoral work on long-term potentiation (LTP) of synaptic transmission, a recently discovered phenomenon—and I was hooked forever! Excitatory synapses, when stimulated only for a second or two, increase their strength persistently, for many hours. The opportunity to study how individual synapses are persistently modified was just what I was looking for.

To store information, the brain needs to change in response to environmental stimuli, so it makes sense that many circuits would have the capability of synaptic modification. When I began my own lab in 1991, this idea became more and more interesting to me, and led directly to our ongoing work at Brown University on circuits that underlie motivation. My best friend from graduate school, Marina Wolf, had been studying addiction-related brain alterations and suggested that drugs of abuse might alter synaptic plasticity in the motivational circuit that includes the ventral tegmental area (VTA) and nucleus accumbens. Her hunch launched our lab and others on a quest for the synaptic basis of addictive behaviors.

the hypothalamus, that humoral and visceromotor responses are initiated by activation of the periventricular and medial hypothalamus, and that behavioral action depends on the lateral hypothalamus.

Drinking

Two different physiological signals stimulate drinking behavior. As mentioned in Chapter 15, one of these is a decrease in blood volume, or *hypovolemia*. The other is an increase in the concentration of dissolved substances (solutes) in the blood, or *hypertonicity*. These two stimuli trigger thirst by different mechanisms.

Thirst triggered by hypovolemia is called **volumetric thirst**. In Chapter 15, we used the example of decreased blood volume to illustrate when and how vasopressin is released in the posterior pituitary by the

Animals will self-administer the same drugs that humans abuse, and their drug-seeking behavior closely resembles that of human substance abusers. Rodents will press a lever to receive cocaine, for example, and will do work or even suffer painful shocks in order to press the lever for the drug, much as substance abusers will suffer tremendous personal loss to acquire the drug. A critical idea in the field has been that drugs of abuse hijack the midbrain dopamine neurons, part of the motivational control system, and by doing so produce an overwhelming craving for the drug, analogous perhaps to the craving for water if one is deprived for a long period. Intriguingly, we found that inhibitory, GABAergic synapses on dopamine cells lost their normal ability to exhibit LTP after a single drug exposure. It had been known for some time that all drugs of abuse increase dopamine release from VTA neurons, and the loss of LTP at inhibitory synapses (and net loss of inhibition) on dopamine neurons is likely a contributing factor.

We next made two key findings. First, multiple different drugs of abuse all erased the GABAergic synapse LTP. Secondly, a brief stressor (5-minute exposure to cold water) had exactly the same effect. What could this mean, when the rewarding effects of drugs seem so different from the aversive effects of stress? Previous work had shown that in rats that had "recovered" from cocaine self-administration (they had learned that the lever press no longer caused drug delivery), either a small dose of the drug or a stressful experience restores powerful drug-seeking behavior, a process known as reinstatement. Human patients also report that minimal drug exposure or stress can trigger relapse and drug craving. It has been suggested that by activating the motivational circuitry, either drugs or stress promote drug seeking.

How could our reductionist approach of studying the details of synaptic function tell us anything about a complex disorder like drug addiction? We did many experiments to tease out which molecules and pathways are needed for stress to block LTP at the inhibitory VTA synapses. We found one molecule that was clearly required: the kappa opioid receptor. If we used an inhibitor to block kappa receptors prior to stress, we found that LTP was unaffected by the stressful experience. Thus, we had found a pharmacological tool that prevents this brain alteration triggered by acute stress. Might the kappa receptor blocker affect relapse behavior as well? Our colleagues at the University of Pennsylvania, Chris Pierce and Lisa Briand, taught rats to self-administer cocaine in response to a lever press; then they no longer provided cocaine when the lever was pressed. Over several days, the rats pressed the lever less and less, and as expected, a brief stressful experience at this point restored robust lever pressing, even when no cocaine was forthcoming. If the kappa receptor inhibitor was administered before the stress, however, we saw no such reinstatement! These exciting findings support the idea that kappa opiate receptors are normally activated during a stressful experience and contribute directly to the initiation of drug-seeking behavior in animals, and perhaps to relapse in humans. Kappa receptor inhibitors may therefore have clinical utility in treating stress-induced drug relapse. Despite the brain's complexity, the approach of understanding component parts and processes proved to be powerful in unpredictable and surprising ways.

Working with this team of outstanding scientists for many years has been tremendous fun. Together we have shared ups and downs and dry periods as well as periods of exciting discovery. Our project demonstrates how understanding the building blocks of a complex system not only help us understand how the brain works but can also suggest ways to control brain plasticity. In our case, a reductionist approach gave insight into a possible therapeutic strategy for addicted individuals.

magnocellular neurosecretory cells. Vasopressin (also called *antidiuretic hormone*, or ADH) acts directly on the kidneys to increase water retention and inhibit urine production. The release of vasopressin associated with volumetric thirst is triggered by two types of stimuli (Figure 16.19). First, a rise in blood levels of angiotensin II occurs in response to reduced blood flow to the kidneys (see Figure 15.5 in Chapter 15). The circulating angiotensin II acts on the neurons of the subfornical organ in the telencephalon, which in turn directly stimulate the magnocellular neurosecretory cells of the hypothalamus to release vasopressin. Second, mechanoreceptors in the walls of the major blood vessels and heart signal the loss of blood pressure that accompanies a loss of blood volume. These signals make their way to the hypothalamus via the vagus nerve and the nucleus of the solitary tract.

In addition to this humoral response, reduced blood volume (1) stimulates the sympathetic division of the ANS, which helps correct the drop

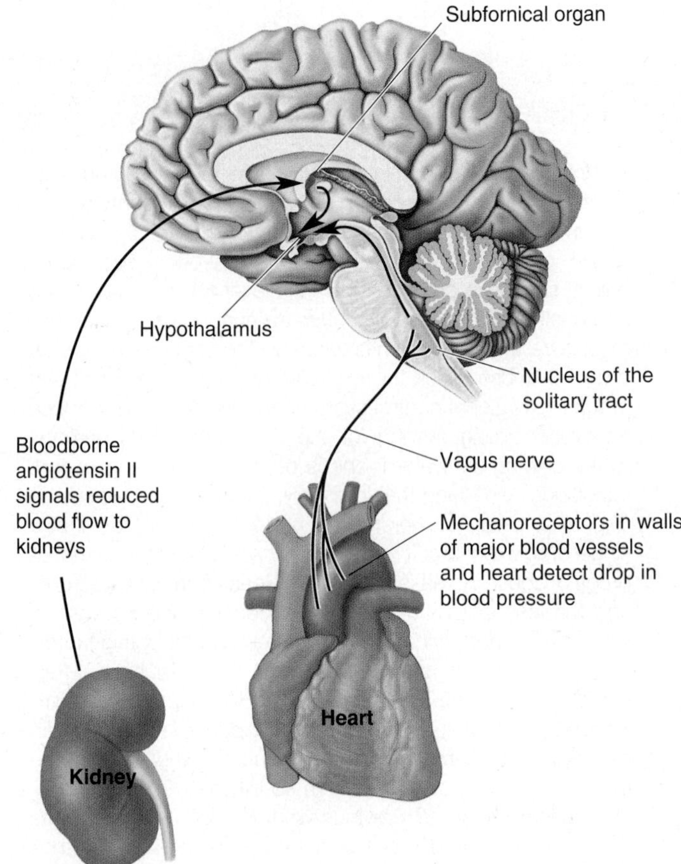

▶ FIGURE 16.19
Pathways triggering volumetric thirst. Hypovolemia is detected in two ways. First, angiotensin II, released into the bloodstream in response to decreased blood flow to the kidneys, activates neurons in the subfornical organ. Second, mechanosensory axons in the vagus nerve, detecting a drop in blood pressure, activate neurons in the nucleus of the solitary tract. The subfornical organ and nucleus of the solitary tract relay this information to the hypothalamus, which orchestrates the coordinated response to reduced blood volume.

in blood pressure by constricting arterioles, and (2) powerfully motivates animals to seek and consume water. Not surprisingly, the lateral hypothalamus has been implicated in inciting the behavioral response, although the details of this process are still poorly understood.

The other stimulus for thirst, hypertonicity of the blood, is sensed by neurons in yet another specialized region of the telencephalon lacking a blood–brain barrier, the **vascular organ of the lamina terminalis (OVLT)**. When the blood becomes hypertonic, water leaves cells by the process of osmosis. This loss of water is transduced by the OVLT neurons into a change in action potential firing frequency. The OVLT neurons (1) directly excite the magnocellular neurosecretory cells that secrete vasopressin, and (2) stimulate **osmometric thirst**, the motivation to drink water when dehydrated (Figure 16.20). Lesions of the OVLT completely prevent the behavioral and humoral responses to dehydration (but not the responses to loss of blood volume).

The motivation to drink and the secretion of vasopressin from the hypothalamus (and the retention of water by the kidneys) normally go hand in hand. However, selective loss of the vasopressin-secreting neurons of the hypothalamus produces a curious condition called *diabetes insipidus*, in which the body works against the brain. As a consequence of the loss of vasopressin, the kidneys pass too much water from the blood to the urine. The resulting dehydration stimulates the strong motivation to drink water; however, the water absorbed from the intestines passes quickly through the kidneys into the urine. Thus, diabetes insipidus is characterized by extreme thirst coupled with frequent excretion of a large amount of pale, watery urine. This condition can be treated by replacing the missing vasopressin.

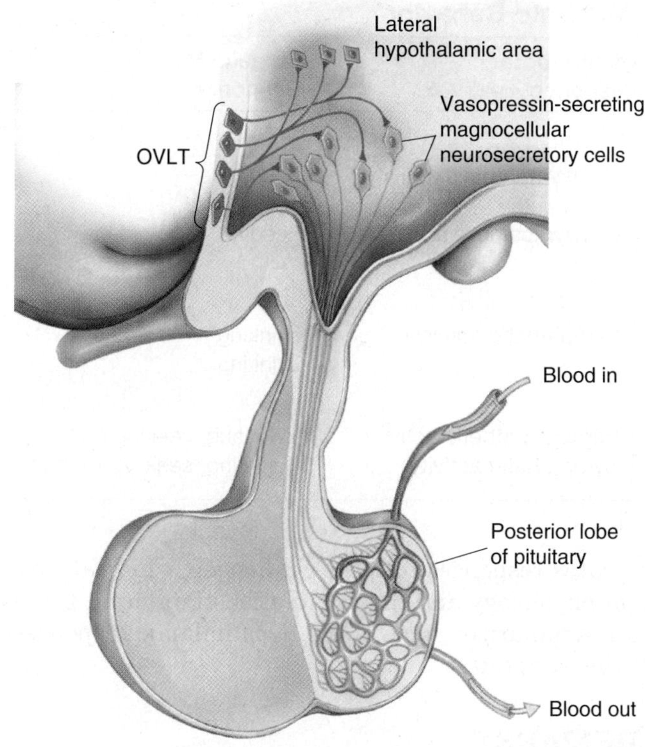

◀ FIGURE 16.20
Osmometric thirst: the hypothalamic response to dehydration. Blood becomes hypertonic when it loses water. Blood hypertonicity is sensed by neurons of the vascular organ of the lamina terminalis (OVLT). The OVLT activates magnocellular neurosecretory cells and cells in the lateral hypothalamus. The neurosecretory cells secrete vasopressin into the blood, and the neurons of the lateral hypothalamus trigger osmometric thirst.

Temperature Regulation

You are hot; you seek a cool place. You are cold; you seek warmth. We are all motivated to interact with our environment to keep our bodies within a narrow range of temperatures. The need for such regulation is clear: The cells of the body are fine-tuned for a constant temperature, 37°C (98.6°F), and deviations from this temperature interfere with cellular functions.

Neurons that change their firing rate in response to small changes in temperature are found throughout the brain and spinal cord. However, the most important neurons for temperature homeostasis are found clustered in the anterior hypothalamus. These cells transduce small changes in blood temperature into changes in their firing rate. Humoral and visceromotor responses are subsequently initiated by neurons in the *medial preoptic area* of the hypothalamus; somatic motor (behavioral) responses are initiated by the neurons of the lateral hypothalamic area. Lesions in these different regions can selectively abolish different components of the integrated response.

A fall in temperature is detected by cold-sensitive neurons of the anterior hypothalamus. In response, TSH is released by the anterior pituitary. TSH stimulates the release of the hormone thyroxin from the thyroid gland, which causes a widespread increase in cellular metabolism. The visceromotor response is constricted blood vessels in the skin and piloerection (goose bumps). An involuntary somatic motor response is shivering (to generate heat in the muscles), and, of course, the other somatic response is to seek warmth.

A rise in temperature is detected by warm-sensitive neurons of the anterior hypothalamus. In response, metabolism is slowed by reducing TSH release, blood is shunted toward the body periphery to dissipate heat, and behavior is initiated to seek shade. In some mammals, an involuntary motor response is panting—in humans, it is sweating—which helps cool the body.

The strong parallels between the hypothalamic control of energy balance, water balance, and temperature should now be clear. In each case, specialized neurons detect variations in the regulated parameter. The

TABLE 16.1 **Hypothalamic Responses to Stimuli That Motivate Behavior**

Bloodborne Stimulus	Site of Transduction	Humoral Response	Visceromotor Response	Somatic Motor Response
Eating signals				
↓ Leptin	Arcuate nucleus	↓ ACTH ↓ TSH	↑ Parasympathetic activity	Feeding
↓ Insulin	Arcuate nucleus	↓ ACTH ↓ TSH	↑ Parasympathetic activity	Feeding
Drinking signals				
↑ Angiotensin II	Subfornical organ	↑ Vasopressin	↑ Sympathetic activity	Drinking
↑ Blood tonicity	OVLT	↑ Vasopressin		Drinking
Thermal signals				
↑ Temperature	Medial preoptic area	↓ TSH	↑ Parasympathetic activity	Sweating, seeking cold
↓ Temperature	Medial preoptic area	↑ TSH	↑ Sympathetic activity	Shivering, seeking warmth

hypothalamus orchestrates responses to these challenges, which always include adjustments in physiology and the stimulation of different types of behavior. Table 16.1 summarizes some of the hypothalamic responses we have discussed in this chapter.

CONCLUDING REMARKS

In the motor system chapters of Part II, we addressed "how" questions related to behavior. How do muscles contract? How is movement initiated? How are the actions of our different muscles coordinated? The discussion of motivation, however, asks a different question: Why? Why do we eat when our energy reserves become depleted? Why do we drink when we are dehydrated? Why do we seek warmth when our blood temperature falls?

Neuroscientists have found concrete answers to both the "how" and the "why" of behavior in the body's periphery. We *move* because of the release of acetylcholine at the neuromuscular junction. We *drink* because we are thirsty, and we are thirsty when angiotensin II levels rise in response to decreased blood flow to the kidneys. However, we remain largely ignorant about the convergence of "how" and "why" in the brain. In this chapter, we chose to focus on feeding behavior, in part because the trail leads farthest into the brain. The discovery of orexigenic peptide neurons in the lateral hypothalamus that respond to changes in leptin levels was a major breakthrough. We are now able to frame the question of how these neurons act elsewhere in the brain to initiate feeding behavior. Advances in this research will have a significant impact on how we interpret our own behavior and the behavior of those around us.

After reading about the bloodborne signals that motivate eating and drinking, you might begin to feel that, indeed, we are ruled by our hormones. However, while bloodborne signals do have a strong effect on the probability of specific types of behavior, we are not their slaves. Clearly, one of the great triumphs of human evolution is the ability to exert cognitive, cortical control over our more primitive instincts. This is not to say that we humans make decisions solely on rational thought, however (Box 16.7). In addition to the powerful forces of self-preservation and heredity, our behaviors are molded by many factors that include our personal fears, ambitions, incentives, and history. In the coming chapters, we will explore additional influences on behavior, including how past experiences leave their mark on the brain.

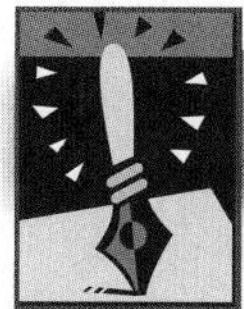

BOX 16.7 OF SPECIAL INTEREST

Neuroeconomics

The field of economics was born in 1776 with the publication of Adam Smith's *The Wealth of Nations*. Among other endeavors, economists attempt to understand how choices are made about the allocation of resources. Economics was called "the dismal science" in the nineteenth century, initially because of dire predictions by economists that humanity was doomed to unending poverty because the food supply could not keep up with population growth. However, that phrase might also apply to how difficult it has been to understand and predict how humans make choices, economic and otherwise (Figure A).

What if we could get under the hood and find out what goes on in the brain during a decision? Advances in the technology of neuroscience, particularly the ability to measure and influence brain activity in awake behaving animals, including humans, make this an attainable goal. In the past decade, economists have increasingly looked to studies of the brain to test the validity of their theoretical assumptions, and neurophysiologists and psychologists have embraced economic theories to interpret their data on the neural basis of choice. The mutual attraction of these disciplines spawned a new field, called *neuroeconomics*. The central goal of neuroeconomics is to combine the tools and insights from economics, neuroscience, and psychology to determine how individuals make economic decisions. The history of science shows that great advances often occur when traditional disciplines come together to solve a common problem. There is perhaps no more urgent scientific challenge than understanding human behavior. More than any other factor, our individual and collective behaviors will determine the destiny of our species and our planet. Although success in this endeavor is by no means assured, it is certain that the understanding of behavior will require the understanding of neuroscience.

Further Reading

Glimcher PW, Fehr E. 2014. *Neuroeconomics: Decision Making and the Brain*, 2nd ed. San Diego, CA: Academic Press.

Figure A
To sail or not to sail?

KEY TERMS

The Hypothalamus, Homeostasis, and Motivated Behavior
motivated behavior (p. 553)

The Long-Term Regulation of Feeding Behavior
anabolism (p. 553)
catabolism (p. 553)
obesity (p. 553)
starvation (p. 554)
lipostatic hypothesis (p. 555)
leptin (p. 555)
anorexia (p. 556)
lateral hypothalamic syndrome (p. 556)
ventromedial hypothalamic syndrome (p. 556)
arcuate nucleus (p. 557)
paraventricular nucleus (p. 558)
anorectic peptide (p. 558)
orexigenic peptide (p. 560)
lateral hypothalamic area (p. 560)

The Short-Term Regulation of Feeding Behavior
satiety signal (p. 562)
ghrelin (p. 564)
vagus nerve (p. 564)
nucleus of the solitary tract (p. 564)
cholecystokinin (CCK) (p. 564)
insulin (p. 564)

Why Do We Eat?
electrical self-stimulation (p. 566)
anorexia nervosa (p. 571)
bulimia nervosa (p. 571)

Other Motivated Behaviors
volumetric thirst (p. 572)
vascular organ of the lamina terminalis (OVLT) (p. 574)
osmometric thirst (p. 574)

REVIEW QUESTIONS

1. A surgical approach to reducing excessive body fat is liposuction—the removal of adipose tissue. Over time, however, body adiposity usually returns to precisely the same value as before surgery. Why does liposuction not work permanently? Contrast this with the effect of gastric surgery to treat obesity.
2. Bilateral lesions of the lateral hypothalamus lead to reduced feeding behavior. Name three types of neurons, distinguished by their neurotransmitter molecules, which contribute to this syndrome.
3. What neurotransmitter agonists and antagonists would you design to treat obesity? Consider drugs that could act on the neurons of the brain as well as drugs that could act on the peripheral nervous system.
4. Name one way the axons of the vagus nerve might stimulate feeding behavior and one way they inhibit it.
5. What does it mean, in neural terms, to be addicted to chocolate? How could chocolate elevate mood?
6. Compare and contrast the functions of these three regions of the hypothalamus: the arcuate nucleus, the subfornical organ, and the vascular organ of the lamina terminalis.

FURTHER READING

Berridge KC. 2009. 'Liking' and 'wanting' food rewards: brain substrates and roles in eating disorders. *Physiology and Behavior* 97: 537–550.

Flier JS. 2004. Obesity wars: molecular progress confronts an expanding epidemic. *Cell* 116:337–350.

Friedman JM. 2004. Modern science versus the stigma of obesity. *Nature Medicine* 10: 563–569.

Gao Q, Hovath TL. 2007. Neurobiology of feeding and energy expenditure. *Annual Review of Neuroscience* 30:367–398.

Kauer JA, Malenka RC. 2007. Synaptic plasticity and addiction. *Nature Reviews Neuroscience* 8:844–858.

Schultz W. 2002. Getting formal with dopamine and reward. *Neuron* 36:241–263.

Wise RA. 2004. Dopamine, learning, and motivation. *Nature Reviews Neuroscience* 5:483–494.

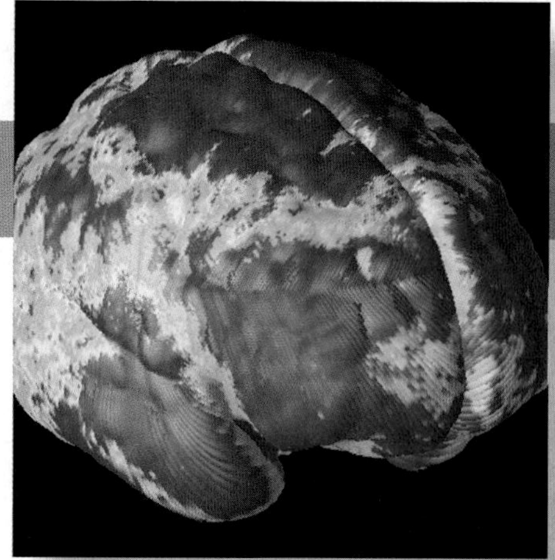

CHAPTER SEVENTEEN

Sex and the Brain

INTRODUCTION

Without sex, there is no human reproduction. And without offspring, no species can survive. Those are the simple facts of life, and over millions of years, the human nervous system evolved for the survival of the species. The drive to reproduce can be compared to the powerful motivation to eat or drink, which we discussed in Chapter 16. For the sake of survival, life-maintaining functions, such as reproduction and eating, are not left entirely to the whims of conscious thought. They are regulated by subcortical structures, and thoughtful conscious control is provided by the cerebral cortex.

In this chapter, we will explore what is known about sex and the brain. Our goal is not a discussion of the birds and the bees; we will assume you have picked up the basics about human sexual behavior from your parents, teachers, friends, or the Internet. Instead, we will look at the neural machinery that makes reproduction possible. For the most part, the neural control of sexual organs uses the same somatosensory and motor pathways we have examined in earlier chapters. Sexual and reproductive behaviors are clearly different in men and women, but just how different are the brains of the two sexes? We will explore this question and see whether brain differences pertain to only reproductive behaviors or more generally to behavior and cognition.

Ultimately, the origin of most distinctions between males and females is the genes inherited from the parents. Under the guidance of certain genes, the human body produces a small number of sex hormones that have powerful effects on the sexual differentiation of the body, as well as adult sexual physiology and behavior. The reproductive organs (the ovaries and testes), which secrete sex hormones, are outside the nervous system, but they are activated by the brain. Recall from Chapter 15 that the hypothalamus controls the release of diverse hormones from the anterior pituitary. In the case of reproductive function, the hormones released by the anterior pituitary regulate secretions from the ovaries and testes. Sex hormones have obvious effects on the human body, but they also influence the brain. There appear to be effects of hormones on gross brain structure and also at the level of neurites on individual neurons. Sex hormones may even influence resistance to certain neurological diseases.

Another point to consider is what it means to be male or female. Is gender determined by genetics, anatomy, or behavior? The answer is not simple; there are cases of gender identities that do not correlate with biological and behavioral factors. And what about sexual orientation? Is an attraction to members of the opposite or same sex determined by experiences in childhood or the structure of the brain? These are challenging questions that address how we perceive ourselves and others. We will examine the extent to which we can answer such questions by looking at the anatomy and physiology of the nervous system.

▲ FIGURE 17.1
Biological and behavioral gender differences. Pheasants are just one of countless animal examples of highly divergent sex traits. The male is dramatically colored and large with a long tail and wattle; it plays little role in raising offspring. The female is small and brownish and an involved mother. (Source: ChrisO at the English Wikipedia.)

SEX AND GENDER

The words *sex* and *gender* both concern distinctions between male and female and are often used synonymously. However, there is disagreement about the meanings of the two terms and the distinction between them. For the sake of clarity, our starting point is the definitions accepted by the World Health Organization. Thus, sex is the biological state of being male or female, and it is determined by chromosomes, hormones, and body anatomy (Figure 17.1). Gender is the set of behaviors and attributes a culture associates with men and women (i.e., masculine and feminine). Of course, it is not always easy or possible to determine whether an expectation or behavior of men and women is a consequence of biology (nature), society (nurture), or both. And, as we will discuss, there are situations in which sex and gender assessments conflict.

The behavioral and cultural implications of a person's sex start at birth. With a newborn we ask the parents, "Is the baby a boy or a girl?" The answer to this question often leads to innumerable assumptions about the life experiences the child will have. We don't typically inquire about an adult's sex because it's usually obvious from appearance. However, identifying someone as female or male still involves many assumptions, as our ideas about sex and gender are associated with numerous biological and behavioral traits. Gender-specific behaviors result from complex interactions among introspection, upbringing, life experiences, societal expectations, genetics, and hormones. These behaviors are related to **gender identity**—our perception of our own gender. In this section, we will discuss some of the genetic and developmental origins of sex.

The Genetics of Sex

Within the nucleus of every human cell, DNA provides a person's genetic blueprint, all the information needed to build an individual. The DNA is organized into 46 chromosomes: 23 from the father and 23 from the mother. Each of us has two versions of the chromosomes 1 through 22, conventionally numbered in order of decreasing size (Figure 17.2). The only exceptions to this pair system are the sex chromosomes, X and Y. Thus, it is usually stated that there are 44 autosomes (22 pairs of matching chromosomes) and two sex chromosomes. Females have two X chromosomes, one from each parent. Males have an X chromosome from the mother and a Y chromosome from the father. Therefore, the female **genotype** is denoted XX and the male genotype XY. These genotypes specify a person's **genetic sex**. Because the mother contributes an X chromosome to every child regardless of sex, the child's genetic sex is determined by the X or Y contribution from the father. In some nonhuman animals, such as birds, it is instead the mother's contribution that determines the genetic sex of the offspring.

The DNA molecules that make up chromosomes are some of the largest molecules known, and they contain genes, the basic units of hereditary information. The piece of DNA comprising a single gene provides the unique information needed to construct a particular protein. There are about 25,000 genes in the human genome, although the number varies depending on the technique used to make the estimate (see Box 2.2).

As you can see in Figure 17.2, the X chromosome is significantly larger than the Y chromosome. Corresponding to this size difference, scientists estimate that the X chromosome contains about 800 genes, whereas the Y chromosome probably contains about 50. You might joke that men are genetically shortchanged, and in a sense, that's right: The XY genotype has serious medical consequences. If a female has a defective gene on an X chromosome, she may experience no negative consequence if her gene on the other X chromosome is normal. However, any defect in the single X chromosome of a male can lead to a developmental defect. Such a defect is called an *X-linked disease*, and there are many. For example, red–green color blindness is relatively common in males (see Box 9.5). Other X-linked diseases that occur more often in men than women are hemophilia and Duchenne muscular dystrophy.

Compared to the X chromosome, the smaller Y chromosome has fewer genes and less diverse functions. Most importantly for sex determination,

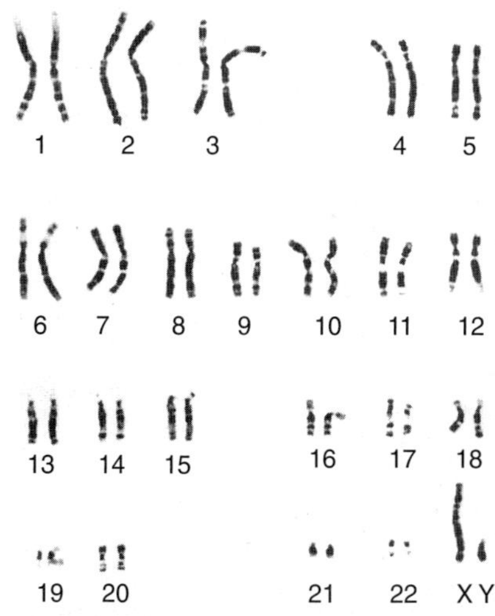

▲ FIGURE 17.2
Human chromosomes. These 23 pairs of chromosomes are from a man. Notice how much smaller the Y chromosome is than the X chromosome. (Source: Yunis and Chandler, 1977.)

it contains a gene called the **sex-determining region of the Y chromosome (SRY)**, which codes for a protein called *testis-determining factor (TDF)*. A human with a Y chromosome and the *SRY* gene develops as a male, and without it, the individual develops as a female. The *SRY* gene was found to be located on the short arm of the Y chromosome in 1990 by Peter Goodfellow, Robin Lovell-Badge, and their colleagues at the Medical Research Council in London (Figure 17.3). If this bit of the Y chromosome is artificially incorporated into the DNA of a fetal XX mouse, the mouse will develop as a male instead of a female. However, this doesn't mean that *SRY* is the only gene involved in sex determination, as *SRY* is known to regulate genes on other chromosomes. Also, male-specific physiology, such as sperm production, relies on other genes on the Y chromosome. Nonetheless, we will see shortly that expression of the *SRY* gene causes the development of the testes, and the hormones from the testes are largely responsible for making a male fetus develop differently from a female fetus.

Sex Chromosome Abnormalities. In rare cases, a person has too few or too many sex chromosomes, with health consequences that range from minimal to lethal. *Turner syndrome* is a partial or complete absence of one X chromosome in a female (XO genotype), affecting about one in 2500 female births. Miscarriage is thought to occur with most XO fetuses. The girls who survive have a variety of characteristics, including short stature, a receding jaw, a webbed neck, and visuospatial and memory difficulties. Their ovaries are abnormal, and estrogen replacement therapy is generally needed for breast development and menstruation. Presumably because the loss of the X chromosome in a male is lethal, there are no known individuals with a YO genotype.

In some cases, people are born with additional sex chromosomes. When this occurs, the sex is always determined by the presence or absence of the Y chromosome. In about one in 1000 male births, there is an extra X chromosome; this defect is known as *Klinefelter syndrome*. These XXY individuals are male because of the presence of the SRY gene on the Y chromosome. In some cases, there are no obvious indications of the

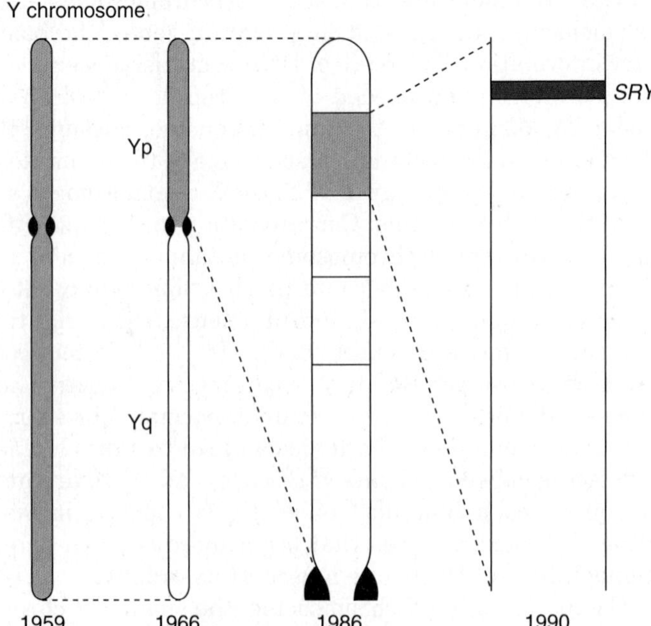

▶ FIGURE 17.3
The location of the *SRY* gene on the Y chromosome. In 1959, researchers found that TDF depended on the Y chromosome, and in 1966, the important location was further restricted to the short (p) arm. Research in the 1980s established that TDF is coded by the *SRY* gene, a small segment near the tip of the short arm of the Y chromosome.
(Source: Adapted from McLaren, 1990, p. 216.)

XXY genotype, but possible symptoms include a less muscular body, less body hair, and increased breast tissue because of lower testosterone production. XYY and XXYY genotypes also occur, and these individuals are male, whereas an XXX person is female.

Sexual Development and Differentiation

Differences between males and females are numerous, from average body size and muscle development to endocrine function. We know it is ultimately the genes of the child that normally determine its sex. But during development, when and how does the fetus differentiate into one sex or the other? How does the genotype of the child lead to the male or female development of the gonads?

The answer involves the unique characteristics of the gonads during development. Unlike organs such as the lung and liver, the rudimentary cells that develop into the gonads are not committed to a single developmental pathway. During the first 6 weeks of pregnancy, the gonads are in an indifferent stage that can develop into either ovaries or testes. The uncommitted gonads possess two key structures, the *Müllerian duct* and the *Wolffian duct* (Figure 17.4). If the fetus has a Y chromosome with an *SRY* gene, testosterone is produced, and the Wolffian duct develops into the

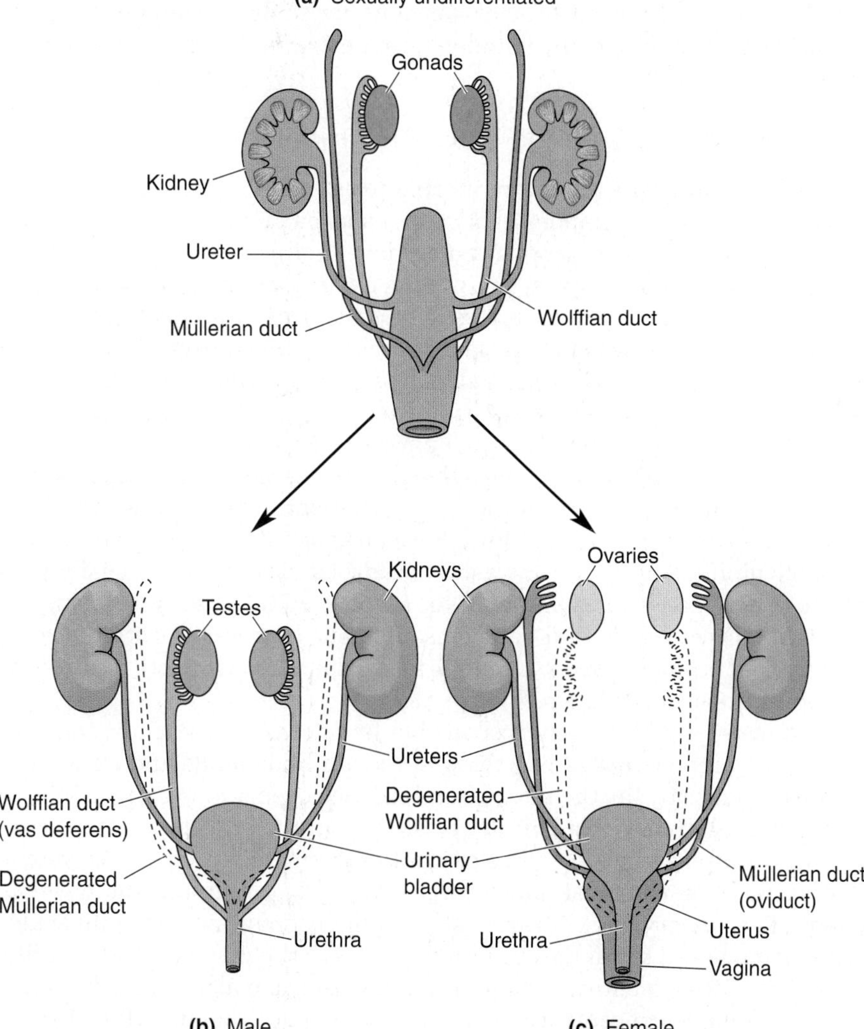

◀ FIGURE 17.4
Development of the reproductive organs. **(a)** The undifferentiated urogenital system has both Müllerian and Wolffian ducts. **(b)** If an *SRY* gene is present, the Wolffian duct develops into the male reproductive organs. **(c)** If there is no *SRY* gene, the Müllerian duct develops into the female reproductive organs. (Source: Adapted from Gilbert, 1994, p. 759.)

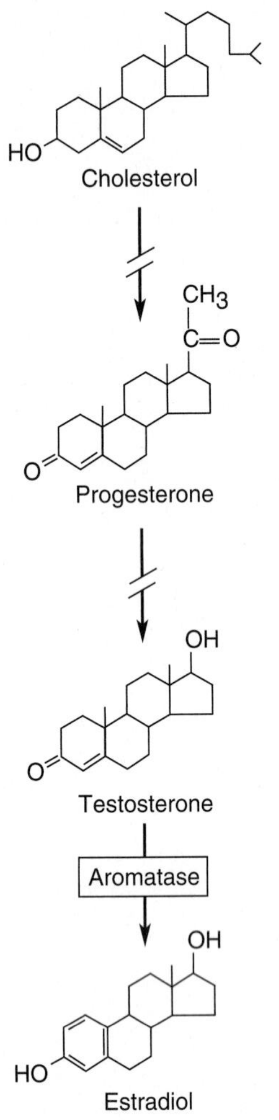

▲ FIGURE 17.5
Cholesterol and the synthesis of the principal steroid sex hormones. Broken arrows indicate where one or more intermediate reactions occur. The enzyme aromatase converts testosterone into estradiol.

male internal reproductive system. At the same time, the Müllerian duct is prevented from developing by a hormone called *Müllerian-inhibiting factor*. Conversely, if there is no Y chromosome and no upsurge of testosterone, the Müllerian duct develops into the female internal reproductive system, and the Wolffian duct degenerates.

The external genitals of both males and females develop from the same undifferentiated urogenital structures. This is why it is possible for a person to be born with genitals intermediate in form between those of typical males and females, a condition known as hermaphroditism.

THE HORMONAL CONTROL OF SEX

Hormones are chemicals, released into the bloodstream, that regulate physiological processes. The endocrine glands we are primarily interested in are the ovaries and testes because they release sex hormones and the pituitary because it regulates this release. The sex hormones are crucial to the development and function of the reproductive system and sexual behavior. The sex hormones are steroids (mentioned in Chapter 15), and some of them are familiar, such as testosterone and estrogen. Steroids are molecules synthesized from cholesterol that have four carbon rings. Small alterations in the basic cholesterol structure have profound consequences for the effects of hormones. For example, testosterone is the most crucial hormone for male development, but it differs from the important female steroid estradiol in only a few places on the molecule.

The Principal Male and Female Hormones

Steroid sex hormones are often referred to as "male" or "female," but men also have "female" hormones and women also have "male" hormones. The designation reflects the fact that men have higher concentrations of **androgens**, or male hormones, and women have more **estrogens**, or female hormones. For example, *testosterone* is an androgen and *estradiol* is an estrogen. In the series of chemical reactions that lead from cholesterol to sex hormones, one of the principal female hormones, estradiol, is actually synthesized from the male hormone testosterone (Figure 17.5). This reaction takes place with the aid of an enzyme called *aromatase*.

Steroids act differently from other hormones because of their structure. Some hormones are proteins and therefore cannot cross the lipid bilayer of a cell membrane. These hormones act at receptors with extracellular binding sites. In contrast, steroids are fatty and can easily pass through cell membranes and bind to receptors within the cytoplasm, giving them direct access to the nucleus and gene expression. Differences in the concentration of various receptors result in steroid effects localized to different areas of the brain (Figure 17.6).

The testes are primarily responsible for the release of androgens, although small amounts are secreted in the adrenal glands and elsewhere. Testosterone is by far the most abundant androgen and is responsible for most masculinizing hormonal effects. Prenatally, elevated testosterone levels are essential for the development of the male reproductive system. Increases in testosterone much later, at puberty, regulate the development of secondary sex characteristics, ranging from increased muscular development and facial hair in human males to the mane of a lion. Oddly, for those with a genetic predisposition, testosterone also causes baldness in men. Female concentrations of testosterone are roughly 10% of those

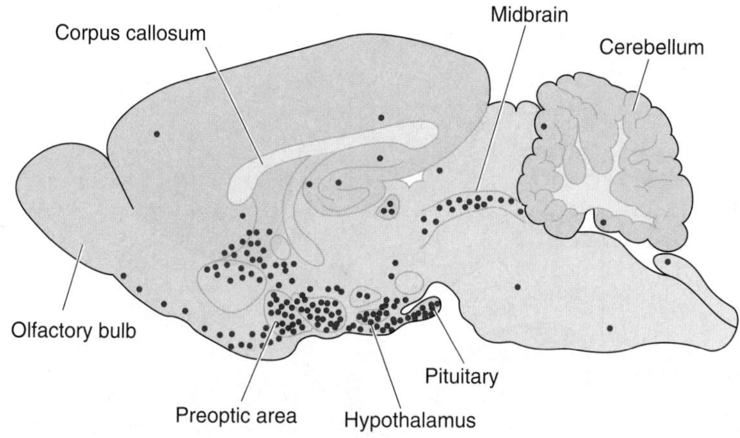

◀ FIGURE 17.6
The distribution of estradiol receptors in a sagittal section of the rat brain. High concentrations of these receptors are found in the pituitary and hypothalamus, including the preoptic area of the anterior hypothalamus. These brain areas are all involved in sexual and reproductive behaviors.

found in males. Male testosterone levels vary during the course of the day because of numerous factors, including stress, exertion, and aggression. It is not clear whether an increase in testosterone is a cause or an effect, but it is correlated with social challenges, anger, and conflict.

The principal female hormones are estradiol and *progesterone*, which are secreted by the ovaries. As already mentioned, estradiol is an estrogen; progesterone is a member of another class of female steroid hormones called *progestins*. Quite low during childhood, estrogen levels increase dramatically at puberty and control the maturation of the female reproductive system and the development of breasts. As in the male, blood concentrations of sex hormones are quite variable in the female. However, whereas in men fluctuations occur rapidly each day, in women, hormonal levels follow a regular cycle of approximately 28 days.

The Control of Sex Hormones by the Pituitary and Hypothalamus

The anterior pituitary gland secretes two hormones that are particularly important for normal sexual development and function in both women and men: **luteinizing hormone (LH)** and **follicle-stimulating hormone (FSH)**. These hormones are also called **gonadotropins**. LH and FSH are secreted by specialized cells scattered throughout the anterior pituitary, comprising about 10% of the total cell population. Recall from Chapter 15 that the secretion of hormones from the anterior pituitary is under the control of hypophysiotropic hormones released by the hypothalamus. **Gonadotropin-releasing hormone (GnRH)** from the hypothalamus does what the name suggests, causing the release of LH and FSH from the pituitary. GnRH is also referred to as LHRH, for luteinizing hormone-releasing hormone because it causes a much greater increase in LH than FSH. Neuronal activity in the hypothalamus is influenced by numerous psychological and environmental factors that indirectly affect the secretion of gonadotropins from the anterior pituitary.

The chain of events from hypothalamic input to the secretion of gonadal hormones is illustrated in Figure 17.7. Neural input from the retina to the hypothalamus causes changes in the release of GnRH based on daily variations in light level. In some nonhuman species, strong seasonal variations in reproductive behavior and gonadotropin secretion occur. Light inhibits the production of the hormone *melatonin* in the pineal gland, increasing gonadotropin secretion because of the inhibitory effect of melatonin on gonadotropin release. By means of this circuit, re-

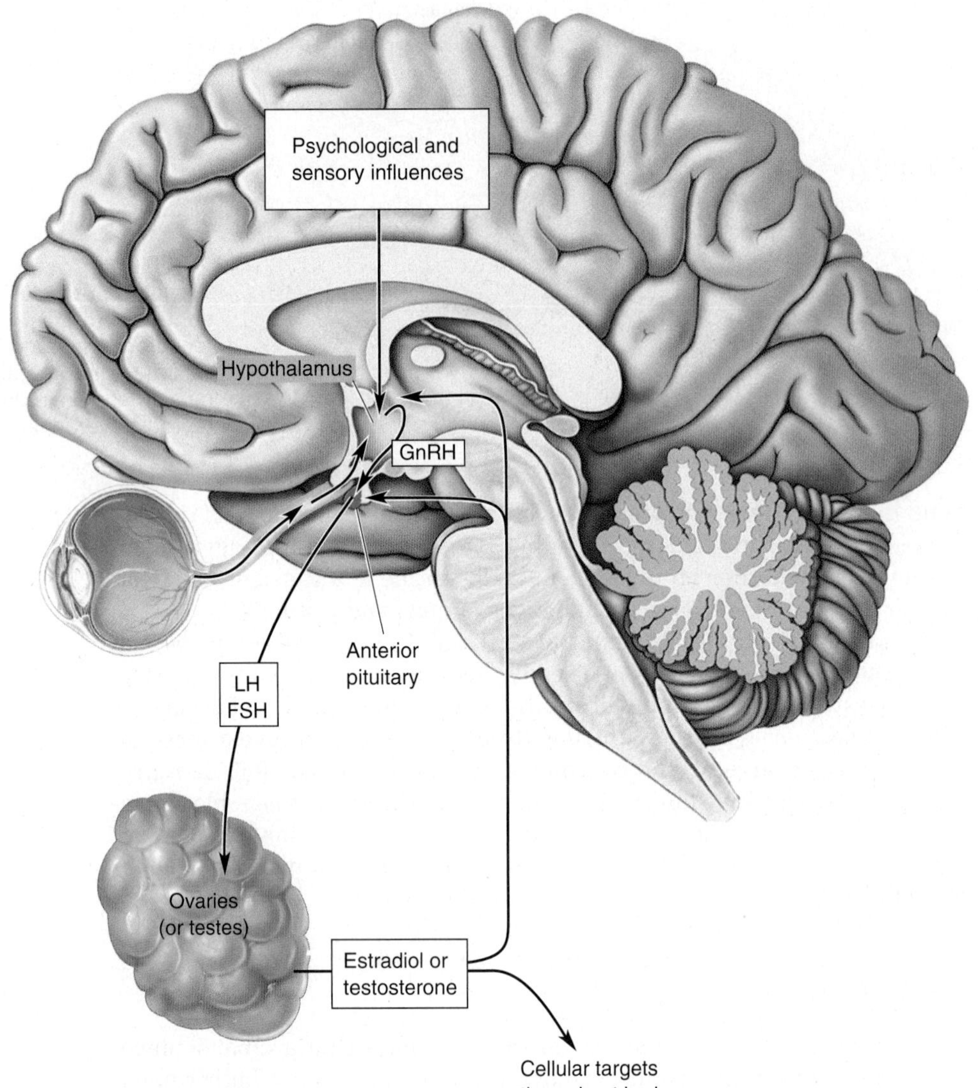

▶ FIGURE 17.7
Bidirectional interactions between the brain and the gonads. The hypothalamus is influenced by both psychological factors and sensory information, such as light hitting the retina. GnRH from the hypothalamus regulates gonadotropin (LH and FSH) release from the anterior pituitary. The testes secrete testosterone and the ovaries secrete estradiol, as directed by the gonadotropins. The sex hormones have diverse effects on the body and also send feedback to the pituitary and hypothalamus.

productive activity can be influenced by the length of daylight during the course of the year, and offspring are born seasonally when they have the best chance of survival. In humans, there is also an inverse relationship between gonadotropin release and melatonin levels, but whether melatonin actually modulates reproductive behavior is not known.

In males, LH stimulates the testes to produce testosterone. FSH is involved in the maturation of sperm cells within the testes. Sperm maturation also requires testosterone, meaning that both LH and FSH play key roles in male fertility. Because there is cortical input to the hypothalamus, it is possible for psychological factors to decrease male fertility by inhibiting gonadotropin secretion and sperm production.

In females, LH and FSH cause the secretion of estrogens from the ovaries. In the absence of gonadotropins, the ovaries are inactive, which is the situation throughout childhood. Cyclic variations in LH and FSH levels in adult females cause periodic changes in the ovaries, and the timing and duration of LH and FSH secretion determine the nature of the reproductive cycle, or **menstrual cycle**. In the follicular phase of the cycle, these hormones (particularly FSH) increase the growth of a small number of follicles, the cavities in the ovaries that enclose and maintain the ova (egg cells). In the luteal phase after egg expulsion, the small cells that surround the egg undergo chemical changes in a process called *luteinization*, which depends on LH release from the pituitary. The duration of the follicular and luteal phases of the reproductive cycle vary significantly for different mammals. The phases are roughly equal in length in the primate menstrual cycle.

In the **estrous cycle** of non-primate mammals, such as rats and mice, the luteal phase is much shorter. In other estrous animals, such as dogs, cats, and farm animals, the phases are more nearly equal in duration. Many estrous animals have only one cycle per year, usually in the spring. Presumably, this timing is for the production of offspring when the weather and food supply are optimal. At the other extreme are animals such as rats, which are said to be *polyestrous* because they have short periods of estrus, or "heat," throughout the year.

THE NEURAL BASIS OF SEXUAL BEHAVIORS

Sexual behavior is a vast, complex, and provocative topic, ranging from the mechanical and biological facts of copulation to the myriad cultural practices of human societies. Here, we will touch on only certain aspects of this subject. We begin with the autonomic and spinal neurons that control the genitals, then discuss various strategies of mating, and conclude with some research about brain mechanisms that are important for monogamy and parenting.

Reproductive Organs and Their Control

Despite the obvious structural differences between female and male reproductive organs, their neural regulation (to the extent it is understood) is surprisingly similar. Sexual arousal of adult men and women can result from erotic psychological and sensory stimuli (including visual, olfactory, and somatosensory), as well as from tactile stimulation of the external sex organs. A full sexual response cycle consists of *arousal* followed by *plateau*, *orgasm*, and *resolution* phases. Although the duration of each phase can vary widely, the physiological changes associated with each are relatively consistent. Neural control of the sexual response comes in part from the cerebral cortex—where erotic thoughts occur—but the spinal cord coordinates this brain activity with sensory information from the genitals and generates the critical outputs that mediate the sexual responses of the genital structures.

The major external and internal sex organs are shown in Figure 17.8. Research on the physiology of the human sexual response has tended to focus unduly on men, but we will try to summarize some of what is known about both sexes. Sexual arousal causes certain parts of the external genitals of both women and men to become engorged with blood, and thus to swell. In women, these structures include the *labia* and the *clitoris*; in men, it is primarily the *penis*. The external genitals are densely innervated by mechanoreceptors, particularly within the clitoris and the glans of the penis. Stimulation of these sensory endings can, by itself, be enough to cause engorgement and erection. The best evidence that engorgement can be generated by a simple spinal reflex is that most men who have suffered a complete transection of the spinal cord at the thoracic or lumbar level can nevertheless experience an erection when their penis is mechanically stimulated. The mechanosensory pathways from the genitals are components of the somatosensory system (see Chapter 12), and their anatomy follows the usual pattern: Axons from mechanoreceptors in the penis and clitoris collect in the dorsal roots of the sacral spinal cord. They then send branches into the dorsal horns of the cord, and into the dorsal columns, through which they project toward the brain.

Engorgement and erection are controlled primarily by axons of the *parasympathetic* division of the ANS (see Figure 15.9). Within the sacral spinal cord, the parasympathetic neurons can be excited by either mechanosensory activity from the genitals (which can directly trigger reflexive erection) or by axons descending from the brain (which account for responses mediated by more cerebral stimuli) (see Figure 17.8). Engorgement of the clitoris and penis depends on dramatic changes in blood flow. Parasympathetic nerve endings are thought to release a potent combination of acetylcholine, vasoactive intestinal polypeptide (VIP), and nitric oxide (NO) directly into the erectile tissues. These neurotransmitters cause the relaxation of smooth muscle cells in the arteries and the spongy substance of the clitoris and penis. The usually flaccid arteries then become filled with blood, thereby distending the organs. (Sildenafil, better known by its trade name Viagra, is a treatment for erectile dysfunction that works by enhancing the effects of NO.) As the penis becomes longer and thicker, the spongy internal tissues swell against two thick, elastic outer coverings of connective tissue that give the erect penis its stiffness. In order to keep the organs sliding easily during copulation throughout the plateau phase, parasympathetic activity also stimulates the secretion of lubricating fluids from the woman's vaginal wall and from the man's bulbourethral gland.

Completing the sexual response cycle requires activity from the *sympathetic* division of the ANS. As sensory axons, particularly from the penis or clitoris, become highly active, they, together with activity descending from the brain, excite sympathetic neurons in the thoracic and lumbar segments of the spinal cord (see Figure 17.8). In men, the sympathetic efferent axons then trigger the process of *emission:* Muscular contractions move sperm from storage sites near the testes through two tubes called the *vas deferens*, combine the sperm with fluids produced by various glands, and propel the resulting mixture (called *semen*) into the *urethra*. During *ejaculation,* a series of coordinated muscular contractions expel the semen from the urethra, usually accompanied by the intense sensations of orgasm. In women, stimulation adequate to trigger orgasm probably also activates the sympathetic system. Sympathetic outflow causes the outer vaginal wall to thicken and, during orgasm itself, triggers a series of strong muscular contractions.

Studies of the neural basis of orgasm are challenging and relatively new. One can only imagine the "technical" challenges that would be encountered fitting two people inside a magnetic resonance imaging (MRI) machine, but a more scientific problem is the investigation of feelings themselves (more on this in Chapters 18 and 21). For example, research has shown that the feelings of orgasm are accompanied by neural activity in widespread cortical and subcortical structures, but we don't know which areas are actually responsible for the feelings and, more generally, it is a complete mystery how patterns of neural activity evoke feelings—why is one activity pattern pleasurable and another painful? Studies in people who experience epileptic seizures give us clues to brain areas particularly relevant for orgasm. In rare cases, the aura that precedes seizures may be sexually arousing, and the loci of such seizures are most commonly in the temporal lobe. In the surgical treatment of epilepsy, electrical stimulation of the medial temporal lobe or the basal forebrain has been reported to cause sexual arousal in some patients. Electrical stimulation of the medial temporal lobe has also been found to produce

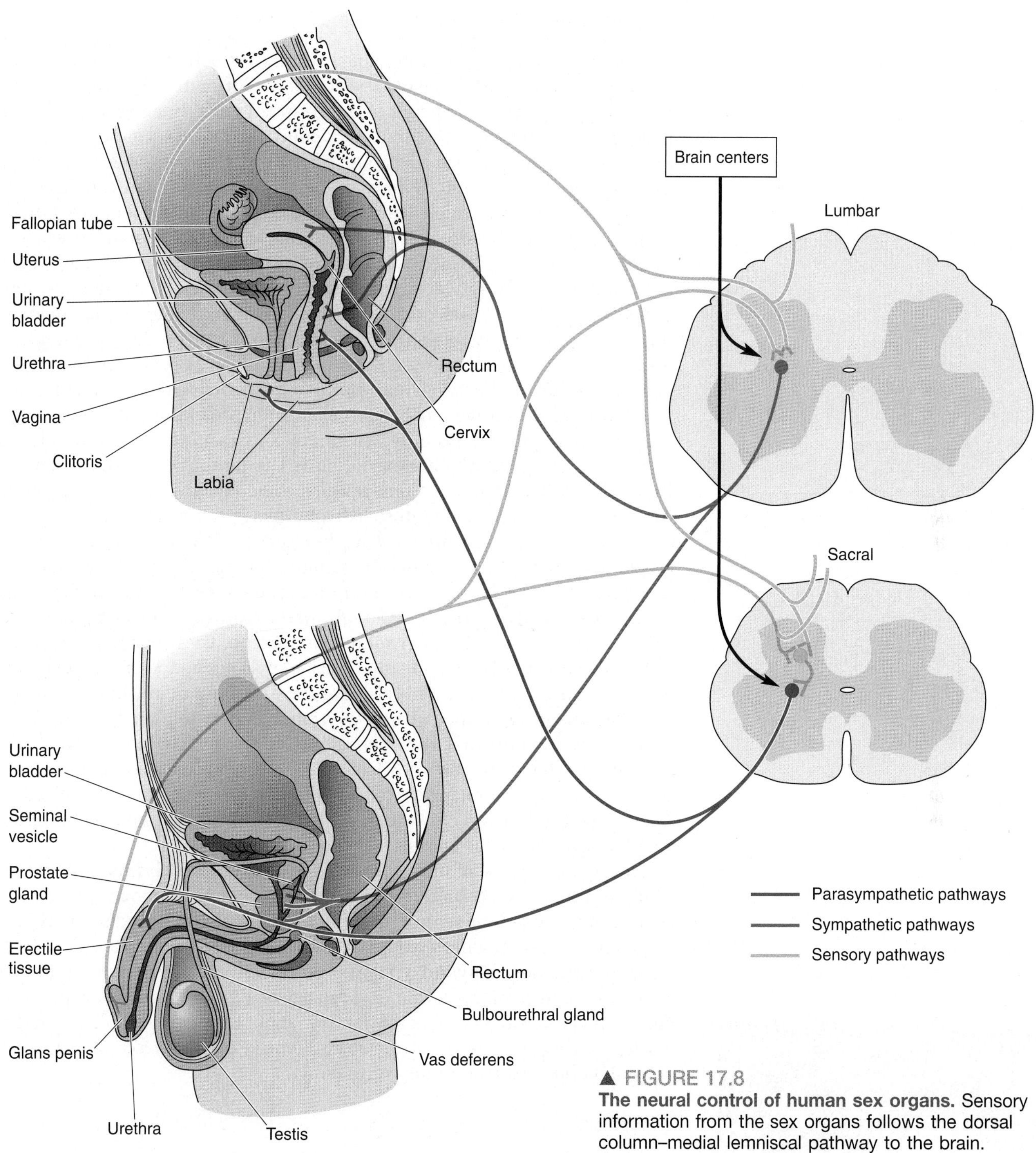

▲ FIGURE 17.8
The neural control of human sex organs. Sensory information from the sex organs follows the dorsal column–medial lemniscal pathway to the brain.

feelings of orgasm in at least a few patients. Studies in additional patients and in non-epileptic brains are needed to confirm the association between orgasm and temporal lobe activation.

Following an orgasm, some time must pass before another orgasm can be triggered in men. The orgasmic experience of women tends to be considerably more variable in frequency and intensity. The resolution phase, which ends the sexual response cycle, includes a draining of blood from the external genitals through veins, and a loss of erection and other signs and sensations of sexual excitement.

Mammalian Mating Strategies

Mammals practice a dazzling range of mating behaviors. Each one is a strategy that ultimately meets the single evolutionary objective: to maximize the survival of offspring and parental genes. Species variations in preferred mating systems seem to depend on the investment that males and females make in raising their offspring, although there are exceptions. Very common among mammals is **polygyny** (from the Greek for "many women"), in which the male mates with many females but the female mates with only one male for one or multiple mating seasons. Polygynous mating (practiced by giraffes, orangutans, and most other mammals) usually has a "one-night stand" quality to it, and the male never looks back to check on the outcome of his many liaisons or the mate. Sometimes, polygyny takes the form of a harem; one male forms a lasting and exclusive association with a group of females, as practiced by gorillas, elephant seals, and a very small number of traditional human cultures.

Polyandry ("many men"), in which one female mates with many males but the males mate with only that female, is rare among mammals and vertebrates in general. One exception is the phalarope, a shorebird that breeds in the cold tundra. Some species practice simultaneous polyandry in which the female bird mates with and lays eggs in the nests of multiple males who raise the young within her territory. Other species practice sequential polyandry in which the female has nothing to do with a mate or offspring once the eggs are laid. Some marmosets and tamarins also appear to be polyandrous. Although historical examples of human polyandry have had a wide geographic distribution, these cases are rare and polyandry appears in only a tiny fraction of societies today. Polygyny and polyandry are both examples of *polygamy*—having more than one mate.

In **monogamy** ("one spouse"), a male and a female form a tightly bound relationship that includes exclusive (or nearly exclusive) mating with each other. Only about 3% of mammalian species are monogamous, although monogamy is practiced by roughly 12% of primate species (and 90% of bird species). The exclusive relationship may last a lifetime or until a new partner is chosen (serial monogamy).

Nearly every type of mating strategy occurs among humans in various cultures and eras. On balance, humans have a strong tendency toward (at least temporary) monogamy, although some cultures condone polygyny. Interestingly, even where polygyny is socially acceptable, most marriages are monogamous. Polyandry for reproductive purposes is rare, and most cultures have penalized women accused of it. Although there has been much speculation about evolutionary explanations for human mating patterns, determining the precise influences of genetics and culture on mating behaviors is virtually impossible.

▲ FIGURE 17.9
Studying reproductive behavior. The prairie vole is a valuable experimental model, representing monogamy and the care of offspring by both parents. (Source: Copyright 2005, Wendy Shattil/Bob Rozinski.)

The Neurochemistry of Reproductive Behavior

Regardless of an animal's choice of reproductive strategy—staying faithful to a mate and devoted to one's children, or wandering promiscuously and abandoning offspring—complex social behaviors are involved. It would be remarkable if the tendency to be monogamous or polygamous were controlled by a few simple brain chemicals. Yet recent work on mouselike rodents called *voles* suggests that certain well-known pituitary hormones do precisely that (at least in voles).

Voles are a wonderful natural experimental model because closely related species of voles have very different reproductive behaviors. The prairie vole (*Microtus ochrogaster*) lives in the American grasslands, practicing solid "family values" (Figure 17.9). It is highly social and as reliably monogamous as any mammal known. After an intense period of initial matings, the male and female form a tight lifelong pair-bond and live together in one nest. The male will fiercely defend his mate, and both parents cooperate in the long-term care of their young. In contrast, the montane vole (*Microtus montanus*) is asocial and promiscuous. Each one lives in an isolated nest, the males take no part in parenting, and the females care for their offspring only briefly before letting them fend for themselves in the world.

Vole pair-bonding has been studied in the laboratory by testing the preference of an animal for spending time with a partner or an unfamiliar animal (Figure 17.10). After mating, a female prairie vole spends more time with its partner than by itself or with a stranger. Female montane voles, on the other hand, spend most of their time in a neutral area alone rather than with their recent mating partner or a stranger.

Because these two vole species are physically and genetically quite similar, relatively few biological factors might account for their different reproductive behaviors. Thomas Insel and his coworkers at Emory University and the National Institute of Mental Health have investigated the subtle brain differences that appear to account for the very different mating strategies in the two species of voles (Box 17.1). Based on clues from previous studies of maternal and territorial behavior, research has focused on the roles of *oxytocin* and *vasopressin* in voles. Recall that these peptide hormones are synthesized in the hypothalamus and can be released into the bloodstream by neurosecretory terminals in the posterior pituitary gland (see Figure 15.4). Circulating vasopressin (also known as ADH, antidiuretic hormone) helps regulate water and salt levels in the

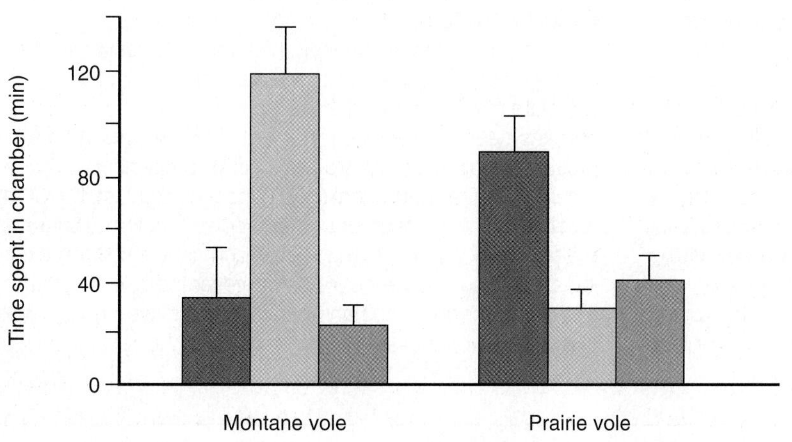

◀ FIGURE 17.10
Pair-bonding in prairie and montane voles. (a) To experimentally measure partner preference, the vole is placed in a neutral chamber and allowed to choose whether to stay alone or visit neighboring chambers, where a partner or a stranger is held. **(b)** After mating, montane voles spend most of their time alone and away from their partner (blue), whereas prairie voles choose to spend most of their time with the partner (purple). (Source: Adapted from Insel and Young, 2001.)

BOX 17.1 PATH OF DISCOVERY

Bonding with Voles

by Thomas Insel

I had never heard of a vole, let alone met one *in vivo*. I trained as a physician and then a psychiatrist. After clinical training, I really had no idea about science. Almost completely by serendipity, I got a job at the National Institutes of Health (NIH) in Bethesda, Maryland. In the early 1980s, NIH seemed to have a Nobel laureate on almost every floor, and the intellectual environment, especially for the fast-emerging field of neuroscience, was infectious. Neuropeptides were the rage, with a new neuropeptide or neuropeptide receptor being discovered nearly every month. And tools for studying slow and fast neurotransmitters were evolving rapidly, so that any young investigator with a new technique could quickly begin to run experiments.

But neuroscience at the NIH in the 1980s also felt a bit crowded. There were lots of talented and technique-savvy scientists working on the neural basis of stress, sadness, and pain. My instinct had always been to go to less crowded places where I could focus on problems without rushing to keep up or ahead of others. And since I had no formal scientific training, I needed time to learn the hard lessons of science. I moved to the Laboratory of Brain Evolution and Behavior in the NIMH, founded by Paul MacLean on a farm in Poolesville, Maryland.

I chose to focus on stress research but decided to study development, focusing on the recently discovered ultrasonic calls rat pups emit when they are separated from their mothers. My behavioral neuroscience career seemed to be progressing until my first post-doctoral fellow arrived, returning with some ambivalence from maternity leave. The idea of listening to rat pups cry after being separated from their mothers was not an ideal project for her. To her great credit and my lasting gratitude, Marianne Wamboldt pointed out that we might study the experience of the mothers and not just the separation distress of the pups.

At that time, very few people were interested in the neurobiology of positive behaviors like parental care, affiliation, or attachment. A robust community of scientists was studying reproductive behavior in rodents and working out the role of gonadal steroids and neuropeptides, but most of this research focused on the motor or sensory aspects of reproduction, not on the emotional or affective experience. With the discovery that neuropeptides, like oxytocin, could modify parental behavior, and with a new post-doc who cared a lot about maternal affect, we were off into a new frontier. Using tools for mapping oxytocin receptors in the brain, we were able to demonstrate the pathways critical for rats becoming maternal, a profound behavioral transition that takes place right at the time of parturition.

These studies helped us to understand the neural mechanisms of maternal care, but what about attachment between adults? Laboratory rats and mice are not ideal for studying attachment. They are highly social but not selective. We needed a species that was monogamous, forming selective and enduring pair bonds. Again, through serendipity, I met a brilliant behavioral endocrinologist, Sue Carter, who was then at the University of Maryland. Sue schooled me in behavioral biology and introduced me to her favorite animal, the prairie vole.

If nature had set out to evolve a species perfect for social neuroscience, prairie voles could well be the result: They are highly affiliative, easy to breed in the lab, and profoundly monogamous. Sue Carter had studied these critters in both the lab and the field, working out simple but rigorous behavioral measures of partner preference and attachment. Bringing the neuroscience expertise from our lab in Poolesville together with the behavioral expertise in Sue's lab, we were able to show the profound effects of oxytocin and vasopressin on affiliative behaviors and attachment.

The story became even more interesting after the lab moved to Emory University in 1994. Joined by Larry Young and Zuoxin Wang, we were able to bring transgenic and viral vector tools to answer questions about the mechanisms by which oxytocin and vasopressin influence social cognition and social behavior. Two insights emerged. First, altering the regional expression of receptors in brain could change social organization, inducing or preventing mating-induced attachment. This was quite a shock because it meant that release of the same peptide had completely different effects in different species. And as we compared monogamous and nonmonogamous species, we noticed a surprising pattern. In monogamous rodents and primates, oxytocin receptors were found in brain areas associated with reward, as if this single receptor linked the social world to the circuitry for motivation. Today, oxytocin is also being studied in autism and schizophrenia.

Of course, the vole work has raised questions about monogamy in humans. I have always been reluctant to extrapolate from voles to mice, so extrapolating from voles to humans seems a fool's errand. But that does not mean the prairie vole is irrelevant. "Nature's gift to social neuroscience" reminds us that neuroanatomy, especially the distribution of receptors, is important for understanding function. Thanks to prairie voles, the neural basis of attachment is now an exciting area of neuroscience. And, whatever the role of oxytocin and vasopressin in human social behavior, we have gained some basic principles for understanding the relationship of form and function in the brain.

body, mainly by affecting the kidneys; oxytocin stimulates smooth muscle, causing uterine contractions during childbirth and milk letdown during lactation. However, oxytocin and vasopressin are also released onto CNS neurons, and, like most signaling molecules, they bind to specific receptors scattered about the brain. Because oxytocin and vasopressin are protein hormones, they bind with extracellular receptors.

As shown in Figure 17.11, maps of these receptors are strikingly different in the brains of prairie voles and montane voles, whereas maps of other types of neurotransmitter and hormone receptors are very similar in the two species. The receptor differences correlate well with reproductive behavior even in other species of voles. Furthermore, the maps are plastic: When the female montane vole gives birth and assumes a maternal role (however briefly), her receptor maps temporarily change to resemble those of the prairie vole.

The distinctive maps of oxytocin and vasopressin receptors tell us that each hormone activates a different network of neurons in the polygynous and monogamous vole brains. This alone does not prove that the hormones have anything to do with sex-related behaviors. But together with the effects of the hormones and drugs that antagonize them, this evidence makes a strong case for cause and effect. When a pair of prairie voles copulates, levels of vasopressin (in males) and oxytocin (in females) rise sharply. Vasopressin antagonists given to a male prairie vole before mating prevent him from forming a pair-bond relationship. This disruption of pair-bonding can be produced with the antagonist selectively infused into the ventral pallidum (the anterior portion of the globus pallidus). Oxytocin antagonists have no such effect. When a male is given vasopressin while he is exposed to a new female, he quickly forms a strong preference for her even without the intense mating that usually precedes pair-bonding. In females, oxytocin appears to be necessary to establish a preference for her mate, while vasopressin has little effect.

A study by Lim et al. provides more direct evidence that vole pair-bonding can be significantly altered by a small change in vasopressin receptors. A virus was used to deliver genes to the ventral pallidum of male montane voles, causing an overexpression of vasopressin receptors. Consequently, the male montane voles had numbers of vasopressin receptors in the ventral pallidum comparable to prairie voles. The manipulated montane voles also pair-bonded like prairie voles. If this cause-and-effect link is supported by further studies, it will dramatically show that a complex social behavior can be altered by the overexpression of a single protein at one location in the brain.

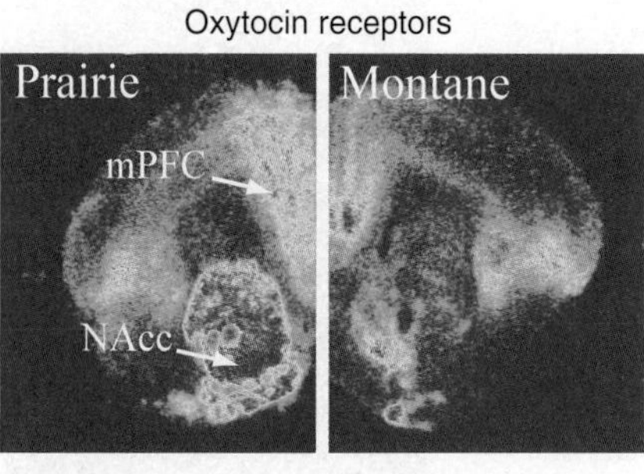

▲ FIGURE 17.11
The role of oxytocin and vasopressin receptors in reproductive behavior. These coronal brain sections show the distribution of oxytocin and vasopressin in the brains of montane voles and prairie voles. The red areas have the highest receptor densities. Compared to montane voles, prairie voles have high vasopressin receptor density in the ventral pallidum (VP) and high oxytocin receptor density in the medial prefrontal cortex (mPFC) and the nucleus accumbens (NAcc). (Source: Young et al., 2011.)

Oxytocin and vasopressin are also involved in vole parenting habits. Vasopressin increases the male prairie vole's paternal proclivities, causing him to spend more time with his pups, and oxytocin similarly stimulates maternal behaviors in the female. The research on voles suggests a very interesting hypothesis about the evolution of complex social behaviors. If genetic mutations change the anatomical distribution of a particular hormone's receptors, then that hormone may evoke an entirely new repertoire of behaviors. Consistent with this idea, administering vasopressin or oxytocin to the naturally promiscuous montane voles does not evoke the effects on pair-bonding and parenting seen in prairie voles, perhaps because they don't have receptors in the necessary places.

Love, Bonding, and the Human Brain

The vole story is a fascinating example of how brain chemicals can regulate critical behaviors. But what does all this have to do with human relationships, faithfulness, and love? It is too early to be sure, but intriguing pieces of evidence suggest that voles may teach us something about the human brain and behavior. For example, there is evidence that human plasma oxytocin levels increase during breastfeeding in mothers and during sexual intercourse in men and women.

In a series of experiments, Andreas Bartels and Semir Zeki at University College London have used functional magnetic resonance imaging (fMRI) to explore human brain activity associated with maternal and romantic love and bonding. In an experiment investigating maternal love, while brain scans were being taken, mothers saw pictures of their child intermixed with pictures of other familiar children. In a second experiment studying romantic love, brain activity was compared when men and women viewed pictures of partners and pictures of friends. The differences in brain activity for one's child versus other children and one's partner versus friends are shown in Figure 17.12. Several brain areas, including the anterior cingulate cortex, the caudate nucleus, and the striatum, are more activated by one's child and partner than by pictures of unrelated people.

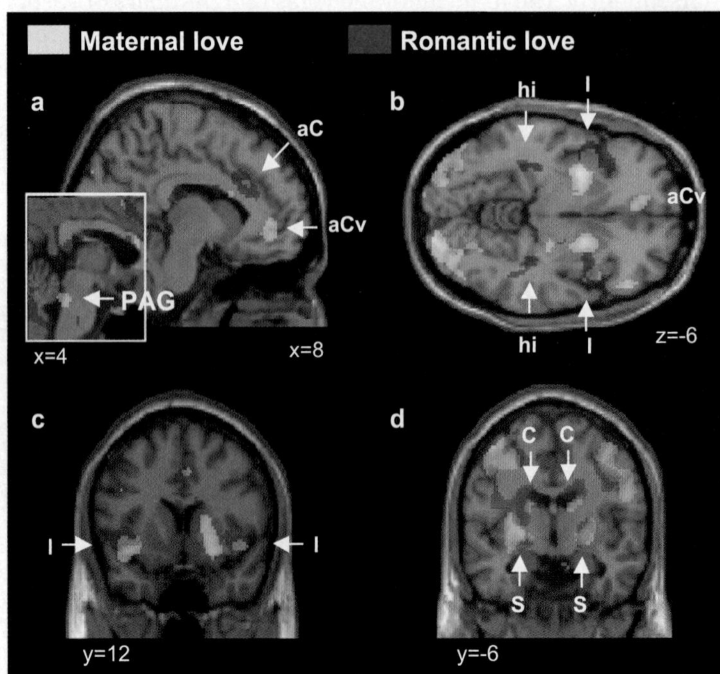

▶ FIGURE 17.12

Imaging maternal and romantic love in the human brain. Brain activation is shown in **(a)** sagittal, **(b)** horizontal, and **(c,d)** two different coronal planes. Yellow areas were more active when mothers saw pictures of their own child than with pictures of other familiar children. Red areas were more activated by pictures of romantic partners than to pictures of friends. Some of the highlighted areas are labeled: *PAG*, periaqueductal gray; *aC*, anterior cingulate cortex; *hi*, hippocampus; *I*, insula; *C*, caudate nucleus; *S*, striatum. (Source: Bartels and Zeki, 2004.)

The heightened responses to child and partner overlap significantly; other brain areas respond differently with the two types of relationship. Many of the areas that are active with both maternal and romantic attachment are part of the brain's reward circuitry (see Chapter 16). We can speculate that the brain activation demonstrates the strong reinforcing nature of partner and parental relationships. Also interesting, and relevant to the vole story, is the finding that many of the brain areas activated by pictures of people's partners and children are rich in oxytocin and vasopressin receptors.

These fMRI studies suggest that oxytocin and vasopressin play roles in human bonding, perhaps similarly to what we see in voles. But surely, the tendency of humans to be monogamous or not isn't as simple as in rodents. Although human behavior undoubtedly involves more complex factors than vole behavior, surprising evidence suggests that vasopressin influences human bonding as well. Hasse Walum and a team of Swedish and American scientists studied 552 pairs of same-sex Swedish twins who were married or had long-term partners. Of particular interest was the gene sequence that codes for vasopressin receptors and the tendency toward monogamy. The DNA sequences that code vasopressin receptors in montane and prairie voles are nearly identical, but the monogamous prairie voles have a DNA sequence adjacent to the gene that encodes the V1aR vasopressin subtype (called a *gene variant*). When this gene variant is transgenically introduced into nonmonogamous mice, their social behavior becomes more like the prairie voles. In the human twin study, the scientists investigated whether vasopressin gene variants might influence pair-bonding in humans too. In women, there was no connection between the vasopressin gene variants and the quality of their marriage as assessed by a variety of questionnaires. In the men, however, an intriguing correlation was found: Men with a particular gene variant scored significantly lower on measures of the quality of their marriage and were twice as likely to report that a marital crisis had occurred in the year before the survey. The wives of the men with the gene variant also reported lower marital quality than the wives of the men without the variant. The function of this gene variant is not known, but these results suggest that, even in humans, vasopressin receptors might play a role in pair-bonding.

WHY AND HOW MALE AND FEMALE BRAINS DIFFER

Sexual reproduction depends on a variety of individual and social behaviors—finding, attracting, and keeping a mate; copulating; giving birth; and nursing and nurturing the offspring—and in each case, the behavior of males and females is usually quite different. Since all behaviors ultimately depend on the structure and function of the nervous system, we can make the strong prediction that male and female brains are also somehow different; that is, they have **sexual dimorphisms** (from the Greek *dimorphos,* "having two forms"). Another good reason to expect that male and female brains differ is simply that male and female bodies differ. The body parts that are unique to each sex require neural systems that have evolved specifically to control them. For example, male rats have a particular muscle at the base of the penis, and their spinal cord has a small cluster of motor neurons that control that muscle; females lack both the muscle and the related motor neurons. Body size and general shape also vary with gender, and thus somatosensory and motor maps must adjust to fit them.

Sexual dimorphisms vary widely across species. In the brain, dimorphisms are sometimes found, but they are significant in some species and

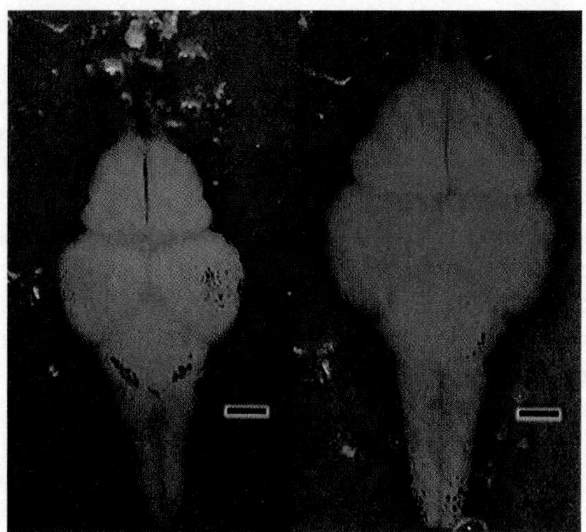

▲ FIGURE 17.13
Dimorphism in brain size. These brains are from adult female (left) and male (right) three-spined stickleback fish that were the same length and weight. The male brain is larger and 23% heavier than the female brain. The scale bar indicates 1 mm. (Source: Kotrschal et al., 2012.)

nonexistent in others. An example of an animal with a large dimorphism is the Icelandic stickleback fish, in which the male brain is much larger than the female brain, perhaps because of the cognitive demands of nest construction, courtship, and childcare (Figure 17.13), which are carried out only by the male. In rodents, the trained eye can tell male from female brains with no ambiguity because of differences in the hypothalamus. The diversity of brain dimorphisms across species is sometimes associated with remarkable variations in sexual behaviors. For example, in some songbird species, only males sing, and, not surprisingly, only males have large singing-related brain nuclei. In human brains, dimorphisms have so far proven to be subtle, few, and of unknown function. Differences between human male and female brains tend to vary along a continuum, with lots of overlap. A particular hypothalamic nucleus might be larger in women than men *on average*, for example, but size variations of the nucleus may be so great that many men have a larger nucleus than many women.

In the rest of this section, we will describe sexual dimorphisms in the nervous systems of humans and other species, focusing on examples that illuminate the relationship between the brain and behavior. We will also discuss some of the neurobiological mechanisms that generate these dimorphisms.

Sexual Dimorphisms of the Central Nervous System

Few dimorphic neural structures are related to their sexual functions in an obvious way. One structure that is related is the collection of spinal motor neurons that innervates the *bulbocavernosus (BC)* muscles surrounding the base of the penis. These muscles have a role in penile erection and help to eject urine. Both women and men have a BC muscle. In women, it surrounds the opening of the vagina and serves to constrict it slightly. The motor neuron pool controlling the BC muscles in humans, called *Onuf's nucleus*, is located in the sacral spinal cord. Onuf's nucleus is moderately dimorphic (there are more motor neurons in men than women) because the male BC muscles are larger than those of females.

The most distinct sexual dimorphisms in the mammalian brain are clustered around the third ventricle, within the *preoptic area of the anterior hypothalamus*. This region seems to have a role in reproductive behaviors. In rats, lesions of the preoptic area disrupt the estrous cycle

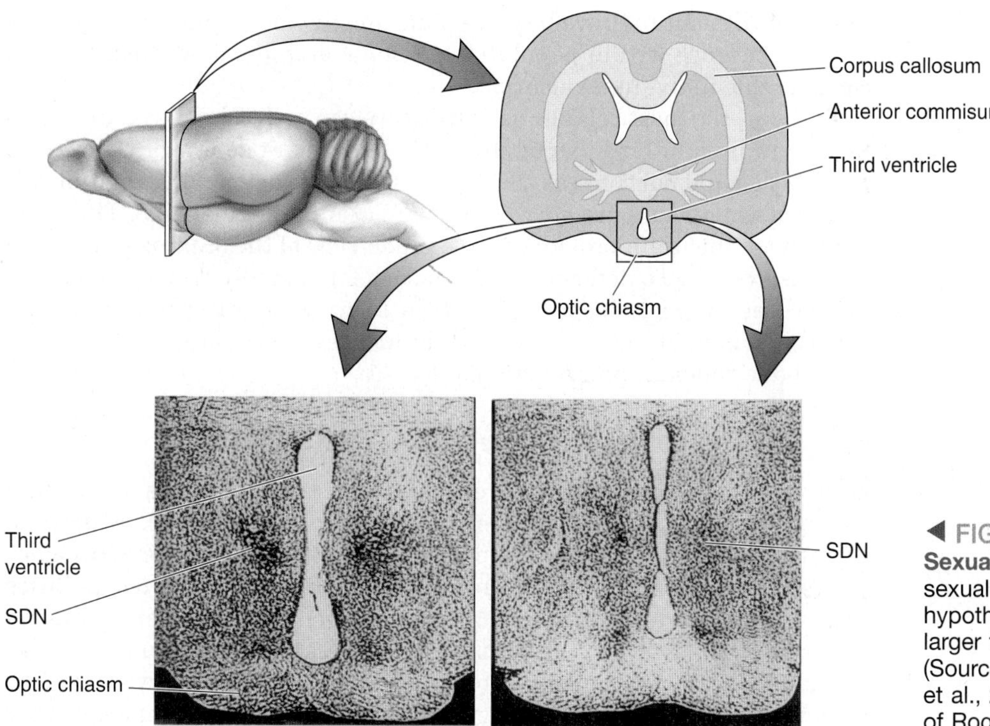

◀ FIGURE 17.14
Sexual dimorphism in rats. The sexually dimorphic nucleus (SDN) in the hypothalamus of male rats (left) is much larger than the SDN in female rats (right). (Source: Adapted from Rosenzweig et al., 2005, Fig 12.21. Photos courtesy of Roger Gorski.)

in females and reduce the frequency of copulation in males. Histological sections of male and female preoptic areas from rats show an obvious difference: The aptly named **sexually dimorphic nucleus (SDN)** is five to eight times larger in males than in females (Figure 17.14).

The preoptic area of humans may also have dimorphisms, but the differences are small and controversial. There are four clusters of neurons called the **interstitial nuclei of the anterior hypothalamus (INAH)**. In different studies INAH-1, INAH-2, and INAH-3 have all been reported to be larger in men than women. INAH-1 may be the human analogue of the rat SDN, but researchers disagree about whether INAH-1 is dimorphic. The clearest dimorphism appears to be in INAH-3, which was first reported to be twice as large in men as women by Laura Allen, Roger Gorski, and their colleagues at UCLA. Evidence of the involvement of the INAH in sexual behavior is thus far inconclusive. Various neurons of the medial preoptic area in male rhesus monkeys fire vigorously during specific phases of sexual behavior, including arousal and copulation. In addition, there may be subtle differences in the size of certain hypothalamic nuclei that correlate with sexual orientation in people.

Human brain dimorphisms outside the hypothalamus have been difficult to prove conclusively, although many have been reported. For example, some studies have found that the corpus callosum is larger in men, but this may be a consequence of men having slightly larger brains (and bodies). In other reports, the posterior end of the corpus callosum, called the *splenium*, is larger in women than men. But even if there is a dimorphism in the size or shape of the corpus callosum, what could it mean? We can only guess. The callosum has no obvious role in mediating specifically sex-related behaviors, but it is important for a variety of cognitive functions that involve coordinated activity between hemispheres. Observations of stroke patients in whom only one hemisphere has been damaged suggest that the functions of female brains may be less lateralized, that is, less dependent on one cerebral hemisphere more than the other. But this conclusion, too, has been challenged. As a rule, sexual dimorphisms of the brain are difficult to prove because male and female brains are very similar, and because within populations of male and female brains, there is considerable individual variation.

Perhaps, the most reliable conclusion we can draw about sexual dimorphisms in human brain structure is that there are so few of them. This probably should not come as a surprise, since the vast majority of women's and men's behaviors are very similar, if not indistinguishable. The gross anatomy of the brain provides only a crude view of the organization of the nervous system. To determine the reasons for sexually dimorphic behavior, we need to look deeper into the patterns of neural connections, the neurochemistry of the brain, and the influence of sex-related hormones on neural development and function.

Sexual Dimorphisms of Cognition

Even if there are no major differences in the brain structures of men and women, there may still be differences in cognitive abilities. Reports of cognitive dimorphisms are sometimes accompanied by an evolutionary explanation: Men evolved as hunters and relied more on their abilities to navigate their environment. Women evolved the behavior of staying closer to home to care for children, so they were more social and verbal.

Numerous studies have reported that women are better at verbal tasks than men. Starting at around age 11 years, girls perform slightly better on tests of comprehension and writing, and this effect is sometimes said to extend through high school and beyond. Perhaps, it reflects a difference in the rates of brain development in the two sexes. Specific tasks at which women excel include naming objects of the same color, listing words beginning with the same letter, and verbal memory (Figure 17.15a).

In other sorts of tasks, men appear to outperform women. Tasks that reportedly favor men include map reading, maze learning, and mathematical reasoning. Researchers speculate that these male advantages evolved from the days when men roamed large areas to hunt wild animals. One of the largest reported differences between the sexes is mental rotation of objects, a task that appears to favor men (Figure 17.15b).

(a) List words beginning with the letter B.

big, bag, bug, boy, banana, bugle, bunny......

(b) Are these two shapes the same?

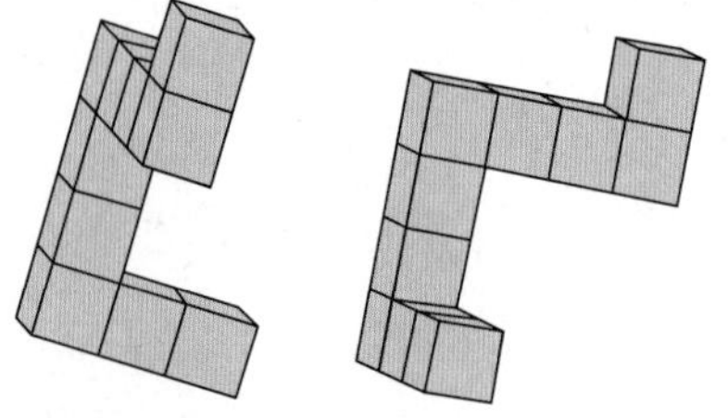

▲ FIGURE 17.15
Cognitive tasks that may favor women or men. **(a)** Women may outperform men in listing words beginning with the same letter. **(b)** Men appear to be somewhat better at spatial rotation tasks, such as deciding whether two three-dimensional objects are the same. (Source: Adapted from Kimura, 1992, p. 120.)

While thinking about dimorphisms of cognition, we need to consider a few things. First, not all studies yield the same results. In some cases, one sex performs better, and in others, there is no difference. Second, across large groups of people of both sexes, there are huge differences in performance. But most of the variation is the result of differences *among individuals*, rather than being sex-specific. Third, it is not clear whether performance differences (or brain dimorphisms) are innate or the result of differences in experience. Typical males and females experience different things and they may, on average, develop slightly different skills. This may in turn influence neural circuitry.

A common interpretation of sex-based differences in performance is that the distinctive hormonal environments of male and female brains make them work somewhat differently. Perhaps, there is a benefit or penalty associated with estrogens or androgens for each task. Consistent with this conjecture are reports that spatial reasoning in women correlates with the menstrual cycle, better performance being observed when estrogen levels are lowest. Evidence also indicates that administering testosterone enhances spatial performance in older men with low testosterone levels. However, cognition cannot be so simply related to hormones, as there is no reliable correlation between performance on verbal or spatial tasks and hormone levels. This doesn't mean that hormones do not affect cognitive function, but we must be cautious about overgeneralizing.

Sex Hormones, the Brain, and Behavior

A variety of factors ranging from genetics to culture and life experiences may make a behavior more common in one sex than the other, but ultimately, all behaviors are controlled by the brain. Even if there are no gross anatomical dimorphisms, male and female brain circuitry must be somewhat different to account for sex-specific behaviors, whether singing by male birds or human sexual behavior. Recall that the kinds of sex hormones circulating in the blood are determined by the gonads, and the dimorphism of the gonads is ordinarily specified by our genes. As described earlier, people with a Y chromosome express a factor (testis-determining factor) that causes the undifferentiated gonads to become testes; people lacking a Y chromosome do not produce TDF, and their gonads differentiate into ovaries. The differentiation of testes or ovaries sets off a cascade of developmental events in the body. Most importantly for the sexual differentiation of the brain, the testes produce androgens, which trigger the masculinization of the nervous system by regulating the expression of a variety of sex-related genes. In the absence of androgens, there is a feminization of the brain through a different pattern of gene expression.

There is nothing fundamentally unique about the brain's sensitivity to hormones. It is just one more body tissue waiting for a hormonal signal to decide its specific pattern of growth and development. Androgens provide a unitary signal for masculinization in the brain, just as in the various other tissues of the body that are sexually dimorphic. Steroids can influence neurons in two general ways (Figure 17.16). First, they can act quickly (within seconds or less) to alter membrane excitability, sensitivity to neurotransmitters, or neurotransmitter release. Steroids do this, in general, by directly binding to, and modulating the functions of, various enzymes, channels, and transmitter receptors. For example, certain metabolites (breakdown products) of progesterone bind to the inhibitory $GABA_A$ receptor and potentiate the amount of chloride current activated by GABA. The effects of these progesterone metabolites are quite similar to the sedative and anticonvulsant effects of the benzodiazepine class of drugs (see Figure 6.22). Second, steroids can diffuse across the outer membrane and bind to specific types of steroid receptors in the cytoplasm and nucleus. Receptors with bound steroid can either promote or inhibit the transcription of specific genes in the nucleus, a process that can take minutes to hours. Specific receptors exist for each type of sex hormone, and the distributions of each receptor type vary widely throughout the brain (see Figure 17.6).

Steroid hormones can exert effects on the brain and body throughout life, but their influence early in development is often fundamentally different from their effects after the animal is mature. For example, the ability of testosterone to alter very young genitals and brain circuitry, leading to distinctly male genitals and masculine behaviors later in life, can be thought of as the **organizational effects** of the hormone. The hormone *organizes* the perinatal tissues in irreversible ways that allow it to generate male functions after sexual maturity has been reached. For a mature animal to express sexual behaviors fully, however, it is often necessary for steroid hormones to circulate *again* during periods of sexual activity, providing **activational effects** on the nervous system. Thus, for example, testosterone levels might surge in the body of a male songbird in the spring, *activating* changes in certain parts of his brain that are essential for normal reproductive behavior (Box 17.2). Activational effects are usually temporary.

Masculinization of the Fetal Brain. Prenatally, elevated testosterone levels are essential for the development of the male reproductive system.

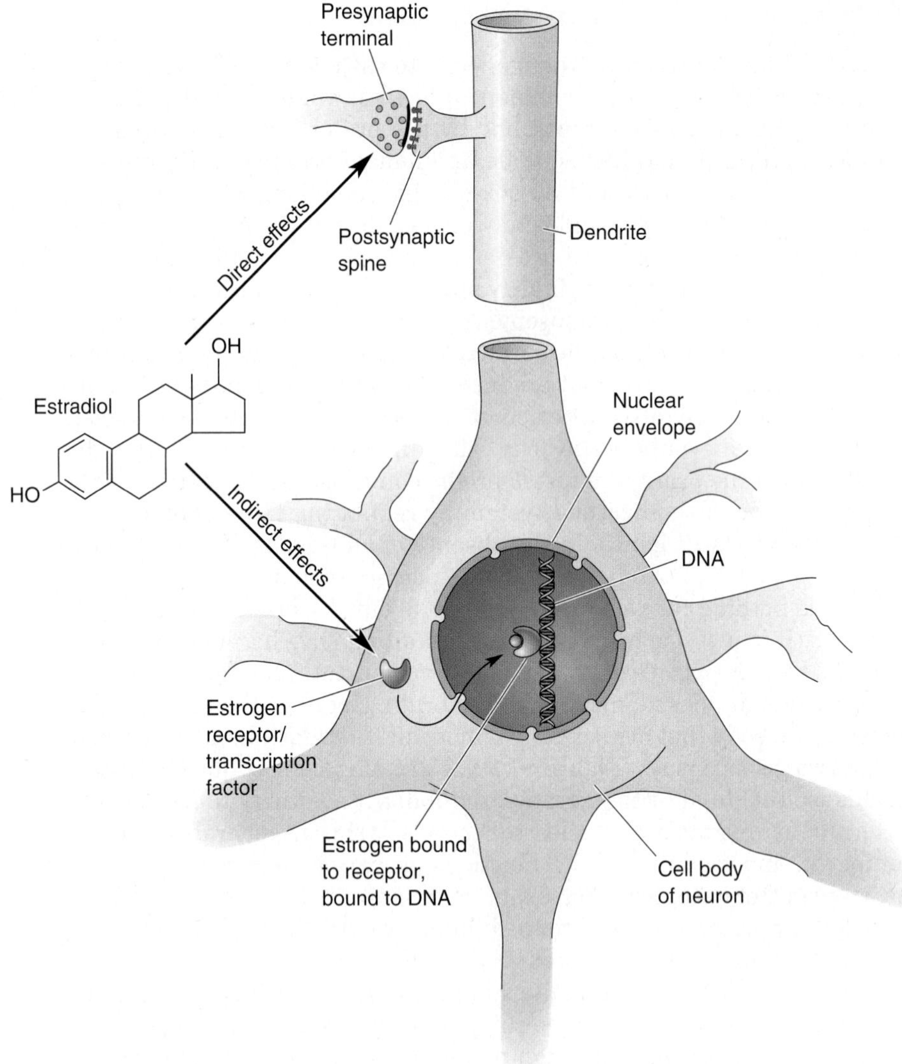

▶ FIGURE 17.16
The direct and indirect effects of steroids on neurons. Steroids can directly affect transmitter synthesis, transmitter release, or postsynaptic transmitter receptors. They can indirectly influence gene transcription.

Ironically, it is a "female" hormone, not testosterone, which causes the changes in gene expression responsible for masculinization of the male brain. Remember that testosterone is converted within neuronal cytoplasm into estradiol in a single chemical step, catalyzed by the enzyme aromatase (see Figure 17.5). The rise in testosterone that occurs prenatally actually leads to an increase in estrogen, which binds to estradiol receptors, thereby triggering masculinization of the developing nervous system. What is not clear is which genes are regulated in various parts of the brain by sex hormones to account for masculinization. Female gonads do not produce an early surge of testosterone or estrogen, so female brains are not affected in the same way as male brains.

An interesting and important complication arises regarding the response of the fetal brain to circulating hormones. In addition to the estrogens and androgens produced by the fetal gonads, hormones coming from the pregnant mother's placenta reach the circulation of the fetus. A reasonable question is why estrogens from the mother do not alter sexual development of the fetal brain. We have said that estrogen, rather than testosterone, is actually responsible for masculinization, so why doesn't the female fetus become masculinized in response to estrogens passed to the fetus from the mother? In rats and mice, the answer to this dilemma is that **α-fetoprotein**,

BOX 17.2 OF SPECIAL INTEREST

Bird Songs and Bird Brains

To our ears, the singing of birds may be simply a pleasant harbinger of spring, but for birds, it is part of the serious business of sex and reproduction. Singing is strictly a male function in many species, performed for the purpose of attracting and keeping a mate and for warning off potential rivals. Studies of two bird species with different habits of reproduction and singing have revealed some fascinating clues about the control and diversity of sexual dimorphisms in the brain.

Zebra finches, which are popular pets, live in the wild in the harsh Australian desert. To breed successfully, birds require dependable sources of food, but in the desert, food comes only with sporadic and unpredictable rains. Zebra finches must therefore be ready and willing to breed whenever food and a mate are available, in any season. Wild canaries, on the other hand, live in the more predictable environment of the Azores and (where else?) the Canary Islands. They breed seasonally during spring and summer and do not reproduce during fall and winter. The males of both species are passionate singers, but they differ greatly in the size of their repertoires. Zebra finches belt out one simple ditty all their lives and cannot learn new ones. Canaries learn many elaborate songs, and they add new ones each spring. The different behaviors of zebra finches and canaries require different mechanisms of neural control.

The birds' sexually dimorphic behavior—singing—is generated by dramatically dimorphic neural structures. Birds sing by forcing air past a special muscularized organ called the *syrinx*, which encircles the air passage. The muscles of the syrinx are activated by motor neurons of the nucleus of cranial nerve XII, which are in turn controlled by a set of higher nuclei collectively called the *vocal control regions*, or *VCRs* (Figure A). In zebra finches and canaries, VCR size is five or more times larger in males than in females.

The development of VCRs and singing behavior is under the control of steroid hormones. However, the very different seasonal requirements of zebra finches and canaries are paralleled by distinctly different modes of steroidal control. Zebra finches apparently require early doses of steroids to *organize* their VCRs, and later androgens to *activate* them. If a hatchling female zebra finch is exposed to testosterone or estradiol, her VCRs will be larger than those of normal females when she reaches adulthood. If the masculinized female is given more testosterone as an adult, her VCRs will grow larger still, and she will then sing like a male. Females that are not exposed to steroids when young are unresponsive to testosterone as adults.

In contrast, the song system in canaries seems to be independent of early steroid exposure, yet it bursts into full service each spring. If female canaries are given androgens for the first time as adults, they will begin singing within a few weeks. The androgens of males surge naturally each spring; their VCRs double in size as neurons grow larger dendrites and more synapses, and singing commences. Remarkably, *neurogenesis*, the birth of neurons, continues throughout adulthood in songbird brains, further contributing to the VCR circuitry during the mating season. By fall, male androgen levels drop, and the canary song system shrinks in size as his singing abates. In a sense, the male canary rebuilds much of his song control system anew each year as courtship begins. This may enable him to learn new songs more easily and, with his enlarged repertoire, gain some advantage in attracting a mate.

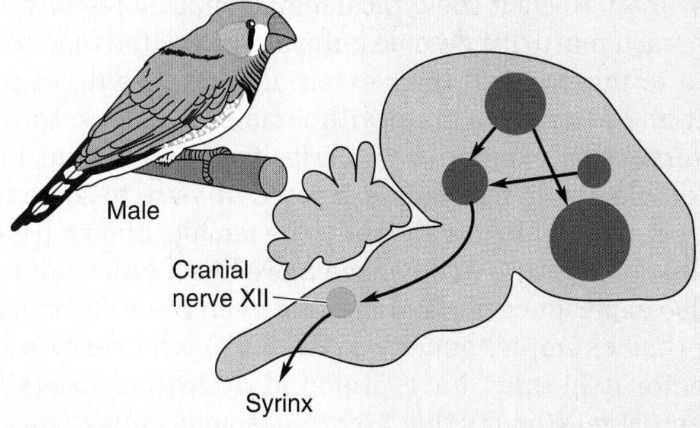

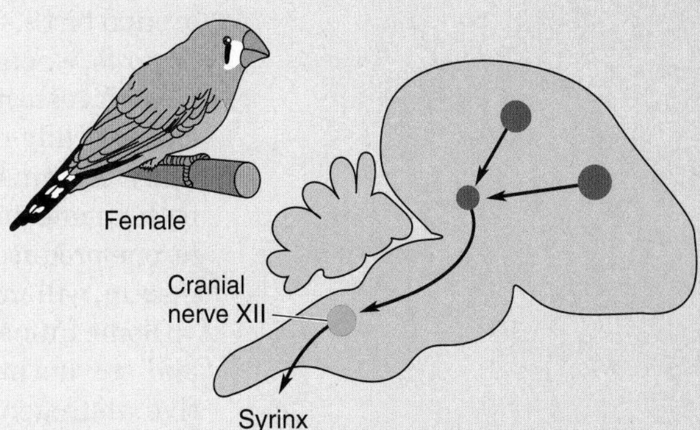

Figure A
Blue circles represent the vocal control regions in male and female zebra finches.

a protein found in high concentrations in fetal blood, binds estrogens and protects the female fetus from masculinization. As odd as it sounds, the female fetus must be protected from "female hormones" to keep its brain from becoming more masculine. Knockout mice that lack α-fetoprotein are sterile and do not exhibit normal sexual behaviors. The role of α-fetoprotein in humans is less clear; there are conflicting reports about whether human α-fetoprotein binds estrogen as it does in rodents. Also noteworthy is the diagnostic measurement of α-fetoprotein levels in maternal blood or amniotic fluid; unusually high levels are a possible indicator of neural tube defects, and unusually low levels are found in *Down syndrome*.

A study by Amateau and McCarthy shows that one factor in brain masculinization "downstream" from testosterone and estradiol is the production of *prostaglandins*. Prostaglandins are compounds derived from arachidonic acid, a fatty acid present in the brain and other organs. One of the enzymes involved in the synthesis of prostaglandins is cyclooxygenase (COX). Prostaglandins have numerous roles; most notably, they are produced after tissue damage and are involved in inducing pain and fever. Amateau and McCarthy found that fetal and neonatal male rats exposed to COX inhibitors showed reduced copulatory behavior as adults. Conversely, female rats treated with COX inhibitors exhibited male-like copulatory behaviors; the female rat behavior, and brain, were partially masculinized despite the lack of the usual sex hormones playing this role, thus moving us one step downstream from estradiol in the chain reaction that causes masculinization. A fascinating side note to this study is that human pain is frequently treated with COX inhibitors such as aspirin. Only time will tell whether a human mother's use of analgesics during pregnancy affects the future sexual behavior of her children.

Mismatches between Genetic Sex and Hormone Action. Under normal circumstances, the genetic sex of an animal or person determines hormonal function and, consequently, the sexual characteristics of the nervous system. However, in situations in which hormonal function is altered, it is possible for genetic males to have female brains and genetic females to have male brains. For example, in all mammalian species studied, treatment with testosterone early in development leads to decreases in at least some features of adult female sexual behavior. Activating fully masculine behavior usually requires extended testosterone treatment before and after birth. If genetically female (XX) rats are exposed to testosterone during the few days around birth, they will fail to elicit the typical female mating posture, called *lordosis*, when they reach maturity. Female guinea pigs treated *in utero* with enough testosterone to masculinize their external genitals will, as adults, energetically mount and attempt to mate with females in estrus. In the absence of human intervention, when a cow carries twin calves that include both a male and a female, the female calf is exposed *in utero* to some testosterone produced by her male twin. As an adult, the female, known as a freemartin, will invariably be infertile and behave more like a bull than a cow.

Some humans also experience mismatches between their chromosomes and sex hormones. For example, genetic males (XY) who carry a defective androgen receptor gene may have profound *androgen insensitivity*. The androgen receptor gene is on the X chromosome; males thus have only one copy of it, and males with the defective gene cannot produce functioning androgen receptors. These individuals develop normal testes but they remain undescended in the abdomen. The testes produce ample testosterone, but outwardly these individuals appear entirely female because their tissues cannot respond to androgen; they have a vagina, a clitoris, and labia, and at puberty, they develop breasts and a female body shape. The testes also produce normal levels of Müllerian-inhibiting factor so the Müllerian duct does not develop into the female reproductive system, the individuals do not menstruate, and they are infertile. Androgen-insensitive genetic males not only look like normal genetic females, but they also behave like them. Even when they understand the circumstances of their biology, they prefer to call themselves women; they dress like women, and they choose men as their sex partners.

Occasionally, genetic females have a condition called *congenital adrenal hyperplasia (CAH)*, which literally means overgrown adrenal glands present at birth. Although they are genetically female, because their adrenal glands secrete unusually large amounts of androgens, CAH females are exposed to abnormally high levels of circulating androgens early in their development. At birth, they have normal ovaries and no testes, but their external genitals are intermediate in size between a normal clitoris and a penis. Surgery and medications are the usual treatments after birth. Nevertheless, CAH girls (and their parents) are more likely to describe their behavior as aggressive and tomboyish. As adults, most CAH women are heterosexual, but, compared to other women, a higher percentage of CAH women are homosexual. Presumably, by analogy to the animal studies, prenatal exposure to high levels of androgens causes a somewhat male-like organization of certain brain circuits in CAH women. We have to be particularly cautious about drawing conclusions about the causes of human behavior, however (Box 17.3). It is very hard to determine whether masculine behavior of a CAH female is due entirely to early androgen exposure and male-like brain dimorphisms, whether her behavior is the product of subtle differences in the way she is treated by others (particularly parents faced with a child who has ambiguous genitals), or both.

Direct Genetic Effects on Behavior and Sexual Differentiation of the Brain

The classic view of sexual differentiation, and the one discussed in this chapter, gives genetics only an indirect role in determining the gender of an individual: Genes direct the development of the gonads, and the hormonal secretions of the gonads control sexual differentiation. While there is no question that hormones are extremely important in sexual development, recent research suggests that genes may sometimes be more directly involved in sexual differentiation, at least in some species. The most compelling evidence comes from studies of birds. In one particularly dramatic study, Agate et al. examined the body and brain of a rare, naturally occurring zebra finch. This bird, technically a *gynandromorph* (meaning that it has both male and female tissues), was genetically female on the left side of its body and brain and genetically male on the right side (Figure 17.17). As both sides of the brain are exposed to the same circulating hormones, they should be equally masculine or feminine if hormones are entirely responsible for sexual differentiation. However, the brain areas associated with singing (see Box 17.2) were masculine on the right side and feminine on the left side, suggesting that different gene expression and not sex hormones in the two brain halves led to the brain dimorphisms. In related studies of gynandromorphic chickens, researchers found that most cells on the male side of the body carry male sex chromosomes and cells on the other side have female chromosomes, the result of a mutation early in development.

Recently, attempts have been made to find direct genetic effects on sexual differentiation in mammals. In a species in which the male and female forms are not as strikingly different as the zebra finch, a gynandromorph

BOX 17.3 OF SPECIAL INTEREST

David Reimer and the Basis of Gender Identity

David Reimer was a normal healthy baby boy when he was born in 1965. But during a circumcision, an accident with an electrocautery device burned his entire penis. David's parents were referred to Johns Hopkins University, where they met Dr. John Money. Because it was not possible to restore David's male genitals, it was recommended that the boy be castrated and undergo cosmetic surgery, followed by estrogen treatment at puberty, to turn him into a girl. Dr. Money's recommendation was based on his hypothesis that, at birth, babies are essentially gender-neutral; their identity as male or female is determined by their subsequent life experiences and identification with their anatomy. Faced with a terrible decision, David's parents were eventually convinced that surgery combined with a female upbringing gave their child the best chance at a normal life.

Dr. Money's accounts of David's life after his transformation sound as if the child adapted well and became a happy normal girl. In publications, Dr. Money refers to "John" having been successfully changed into "Joan." The case even made it into the popular press, as evidenced by a 1973 article in *Time* magazine: "This dramatic case . . . provides strong support . . . that conventional patterns of masculine and feminine behavior can be altered. It also casts doubt on the theory that major sex differences, psychological as well as anatomical, are immutably set by the genes at conception."[1] At that time, dramatic societal changes were taking place in the roles of men and women, and the success of David as a female appeared to confirm that society created gender identity as much as, or more than, biology.

Unfortunately, a follow-up report revealed that David's gender transformation was a disaster from the outset. According to David and his twin brother, David's behavior was always much more like other boys than girls. David rebelled at wearing girls' clothing and playing with traditional girls' toys. Despite cosmetic surgery and female indoctrination, as an adult, he said that he had suspected he was a boy as early as the second grade and imagined growing up to be a muscular man. As a child, David was incessantly teased and ostracized. He knew nothing about the failed circumcision and subsequent surgery, nor the fact that he was genetically a male. However, as he got older, he was more attracted to girls than boys, and he expressed the opinion that he felt like a boy trapped in a girl's body. By the age of 14, after being on estrogen for 2 years, he looked increasingly like a girl, but he stopped living as one (Figure A). David's father finally told him what had happened when he was young. David immediately requested sex-change hormonal therapy and surgery. For years, David dealt with the overwhelming emotional problems resulting from his past. He married, adopted his wife's children, and happily worked a physically demanding job as a janitor in a slaughter house. In the 1990s, David collaborated on a book about his life. Tragically, after numerous traumatic events in his life, including the death of his twin brother and the breakup of his marriage, David committed suicide in 2004 at the age of 38.

David Reimer's experiences demonstrate that, rather than being gender-neutral, he had a "male brain" from the outset. Evidently, his genetically determined sex could not be suppressed even with sex-change surgery, hormonal therapy, and a female upbringing. Clearly, gender identity involves a complex interplay of genetics, hormones, and life experiences.

Figure A
David Reimer (aka John/Joan) and his twin brother Brian shortly before they were told the truth about David's childhood. (Source: Courtesy of Jane Reimer.)

[1]*Time*, Jan. 8, 1973, p. 34.

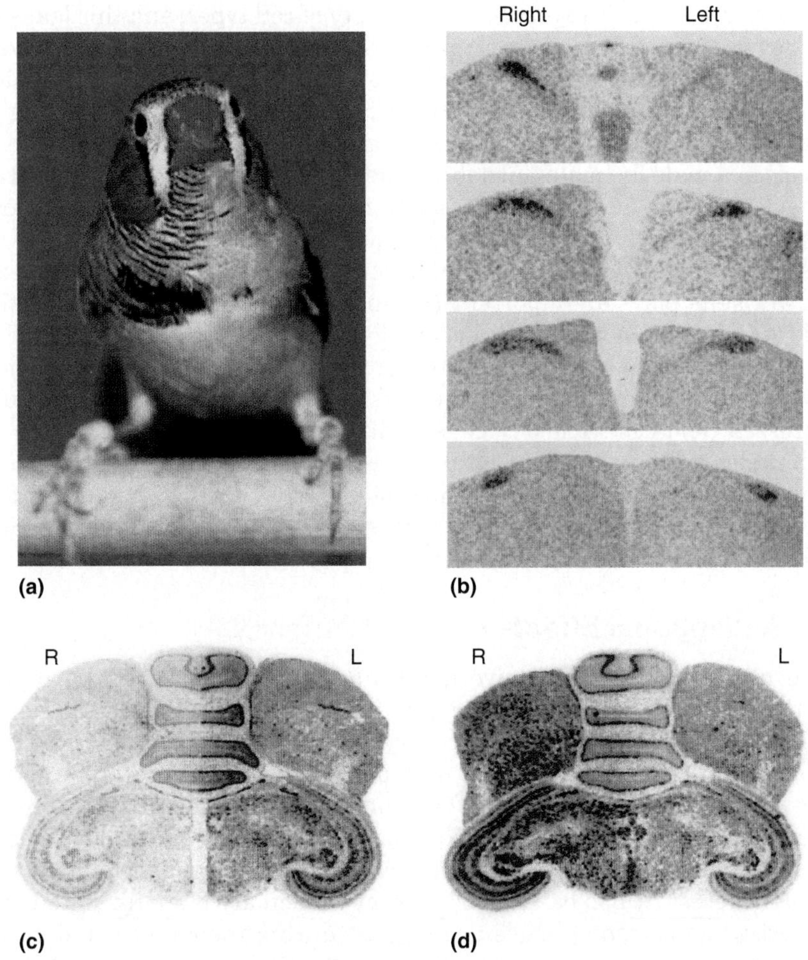

▲ FIGURE 17.17

Brain analysis of a gynandromorphic zebra finch. **(a)** This bird had female plumage on its left side and male plumage on its right. **(b)** The HVC nucleus (hyperstriatum ventrale, pars caudalis) controls singing. It is larger on the genetically male right side (dark patch of labeled neurons). **(c)** An autoradiogram showing the expression of a gene normally expressed only in females labels only the left side of the brain. **(d)** An autoradiogram showing the expression of a gene normally more highly expressed in males labels the right side of the brain more than the left. (Source: Arnold, 2004, Fig. 4.)

might not be as easily recognized. Eric Vilain and his colleagues at UCLA have found 51 genes that are expressed at different levels in male and female mouse brains before the formation of the gonads. The function of these genes is not yet known.

In addition to influencing sexual differentiation, genes can play a role in surprisingly complex sex behaviors. Some of the best evidence comes from studies of the fruit fly *Drosophila melanogaster*. The male courts the female with an enticing set of behaviors, including orienting toward and following her, singing a courtship song, and tapping her with his forelegs before attempting to mate. The female chooses whether to accept or reject the male's advances. Evidently, these behaviors are genetically coded, as males know how to court even if they have never seen courtship by other flies. Many genes may play some role in the courting behaviors, but there appear to be a very small number of critical regulatory genes (genes that regulate the expression of other genes). For example, the *fru* gene (short for fruitless) may be essential for male courtship behaviors. In males, the *fru* gene is expressed in a variety of cell types, and this leads to the development of a male central nervous system. The male CNS develops such that male courtship behaviors automatically occur. In females without *fru* expression, a complete CNS also develops, but its wiring is somewhat different and female behaviors are "built in." If the *fru* gene is absent in a male fly, male courtship behaviors are significantly reduced or nonexistent. Conversely, females made to express *fru* exhibit male courtship behaviors and resist courtship by males.

Another gene involved in sexual differentiation is *dsx* (for double sex). The *dsx* gene plays an important role in the sexual differentiation of the body (the development of male or female genitals) and also interacts with *fru* in controlling sexual differentiation of the CNS and sex-specific behaviors. In the case of *fru*, the gene is either expressed (males) or not expressed (females). The *dsx* gene is different in that it is expressed in both males and females, but alternative splicing leads to the production of male-specific and female-specific proteins. How the *fru*- and *dsx*-influenced structure of the CNS assures sex-specific behaviors is a challenging mystery yet to be explained in detail.

The Activational Effects of Sex Hormones

Long after sex hormones have determined the structure of the reproductive organs, they can have activational effects on the brain. These effects range in scale from temporary modifications in brain organization to changes in the structure of neurites. In men, testosterone has a two-way interaction with sexual behavior. On the one hand, testosterone levels rise in anticipation of a sexual act or even while fantasizing about it. Conversely, reduced testosterone levels are associated with decreased sexual interest. It has been reported that women are more likely to initiate sex when estradiol levels are highest during their menstrual cycle. Through unknown mechanisms, hormone levels in both sexes influence the brain and an individual's interest in sexual behavior.

Brain Changes Associated with Maternal and Paternal Behavior. Patterns of sexual behavior vary over time. In some species, reproduction occurs only during a particular season, and mating may occur only during a specific phase of that season. Obviously, females of all species nurse their offspring only after birth and only temporarily. In most animals, but not in humans, sexual attractiveness and copulation occur only during certain phases of the estrous cycle. Sexually dimorphic changes in the brain are sometimes transient or cyclical, coinciding with the sexual behavior to which they are related.

In Chapter 16, we saw that appetite is controlled in part by blood levels of the hormone leptin that is secreted by fat cells: Higher leptin levels modulate cells in the hypothalamus and suppress eating. During pregnancy, the mother needs more food to supply energy for her growing fetus, and indeed, food intake increases early in pregnancy. Consequently, body fat accumulates and leptin levels rise. Paradoxically, researchers have observed in rats that even though leptin levels rise during pregnancy, appetite and food intake increase rather than decrease. This occurs because hormonal changes associated with pregnancy lead to leptin resistance in the hypothalamus.

Another behavior unique to maternal rats occurs with lactation and nursing. In female rats, the somatosensory cortex contains a sensory representation of the ventral skin surrounding the nipples. Within a few days of the start of nursing, the tactile stimulation leads to a dramatic increase in the representation of the ventral skin (Figure 17.18) and a shrinkage

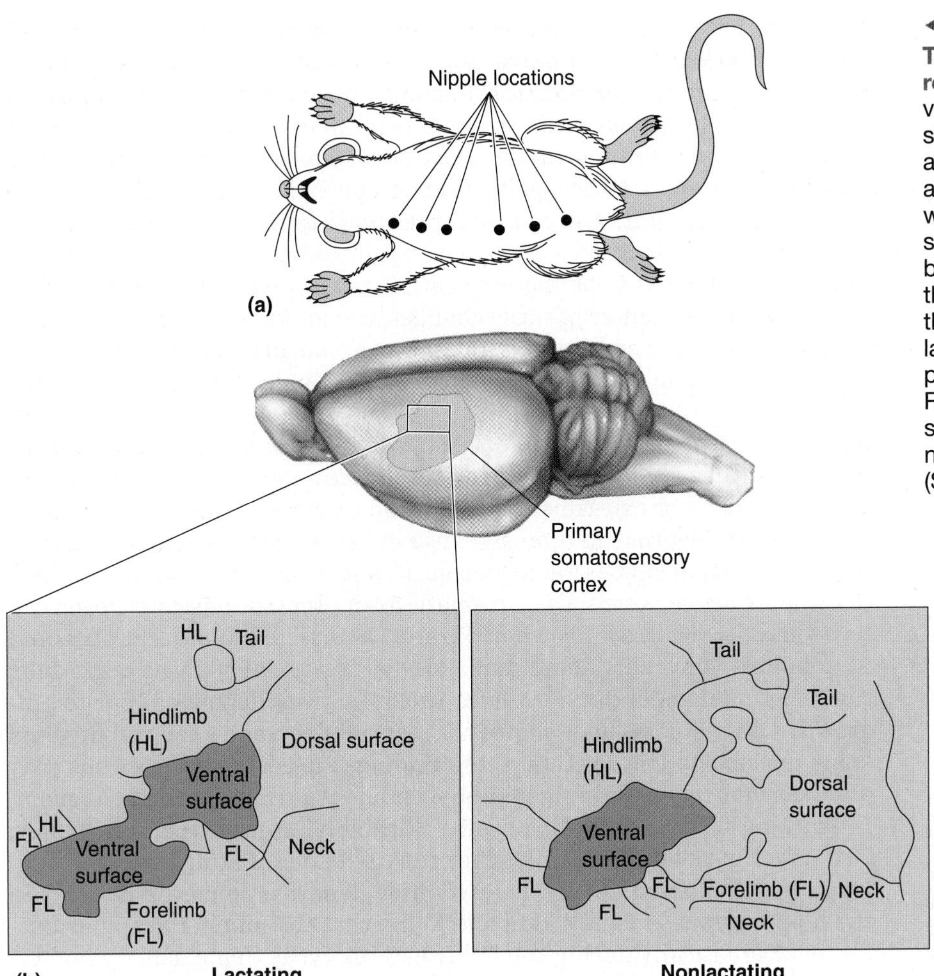

◀ FIGURE 17.18
The effect of lactation on a sensory representation in the cortex. (a) The ventral skin of a nursing rat mother, showing the location of the nipples along the right side. **(b)** The rat's brain and left primary somatosensory cortex, which contains a map of the right ventral skin (top). The boxed area, expanded below, illustrates how the cortical region that responds to the ventral skin around the nipples is enlarged in a postpartum lactating mother (left) compared to a postpartum nonlactating mother (right). Regions of somatosensory cortex subserving other regions of the body were not affected by the lactating state. (Source: Adapted from Xerri et al., 1994.)

of receptive fields to half their normal size. This interesting example of somatosensory map plasticity (see Chapter 12) appears to be temporary, as receptive fields are normal in size several months after weaning.

Lactation also appears to lead to brain changes that reinforce this behavior, which is critical for the survival of mammalian offspring. Despite their very different pharmacological and behavioral effects, all addictive drugs appear to enhance the influence of dopamine released by neurons projecting from the ventral tegmental area (VTA) into the nucleus accumbens (NA) (see Box 16.5). It is becoming increasingly clear that a variety of reinforcing or addictive behaviors also modify this VTA-NA circuit. In one study, fMRI scans of lactating female rats nursing their pups were compared with those of virgin female rats after an injection of cocaine. Brain activation was surprisingly similar in both, with particular activation in the NA. The hypothesis is that in both instances, there is stimulation of the dopamine system associated with reward and addiction. The tactile stimulation of the suckling pups may make nursing a reinforcing behavior to promote mother–infant bonding and, ultimately, the survival of the pups.

Even though fathers do not experience the dramatic body changes associated with pregnancy and lactation, their interactions with children may alter their brains in fundamental ways. A hint of this comes from a study in Elizabeth Gould's lab at Princeton University, where she examined the brains of marmosets. Marmoset fathers are unusually involved in child care; indeed, they carry the babies around for the first few months of life. It is known that across species, the prefrontal cortex is involved in complex, goal-directed behaviors. It is also established that environment can change neurons; for example, dendritic branching and spine density are enhanced when animals are housed in an enriched environment. To see whether fatherhood alters brain structure, Gould's group compared the prefrontal cortex of marmoset fathers with that of non-fathers who were part of a mating pair. Two interesting differences were found: The density of dendritic spines on pyramidal cells was significantly higher in the fathers, and there appeared to be more vasopressin receptors on the spines. The functional consequences of these changes are not known, but they suggest that brains in other species that invest much time in childrearing, whether by the male or female, might be structurally changed by the experience.

Estrogen Effects on Neuron Function, Memory, and Disease. Estrogens have powerful activational effects on the structure and function of neurons. Within minutes of experimental application, estradiol alters the intrinsic excitability of neurons in a broad range of brain areas. By modulating the flow of potassium ions, estradiol depolarizes some neurons and causes them to fire more action potentials. A dramatic example of the effects estrogen can have on cell structure is shown in Figure 17.19. Dominique Toran-Allerand at Columbia University found that estradiol treatment of tissue taken from the hypothalamus of newborn mice causes a great degree of neurite outgrowth. Other studies have shown that estradiol increases cell viability and spine density. Taken together, these findings suggest that estrogens play an important role in forming neuronal circuitry during brain development.

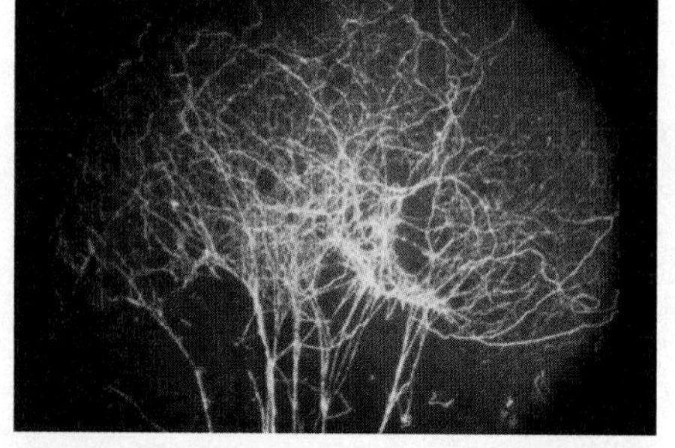

(a)

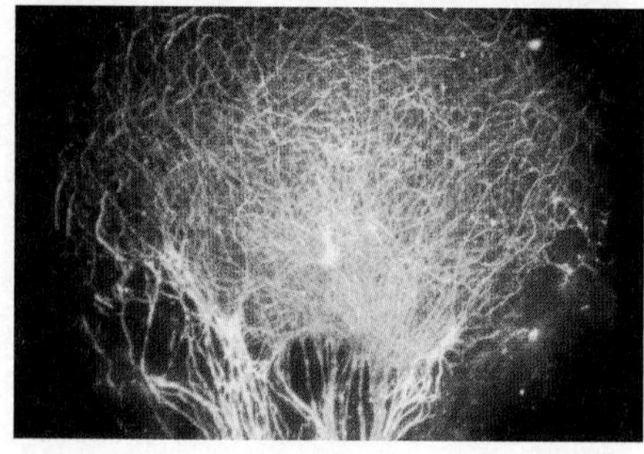

(b)

▲ FIGURE 17.19
The effect of estrogen on neurite growth in the hypothalamus. Below the bottom of each photograph is a piece of hypothalamic tissue from a newborn mouse. **(a)** Without the addition of estrogen, a relatively small number of neurites grow from the tissue. **(b)** With the addition of estrogen, there is an exuberant growth of neurites. (Source: Toran-Allerand, 1980.)

Working at Rockefeller University, Elizabeth Gould, Catherine Woolley, Bruce McEwen, and their colleagues reported a fascinating example of estradiol's activational effects. They counted dendritic spines on neurons in the hippocampus of female rats and found that the number of spines fluctuated dramatically during the 5-day estrous cycle. Spine density and estradiol levels peaked together, and treatment with injected estradiol also increased the numbers of spines in animals whose natural estradiol levels were kept low (Figure 17.20). Since spines are the major site of excitatory synapses on dendrites (see Chapter 2), this provides a possible explanation for the fact that hippocampal excitability also seems to track the estrous cycle. The hippocampus of experimental animals more easily generates seizures, for example, when estrogen levels increase (Figure 17.21). Note that estradiol and progesterone levels peak during the proestrus phase (Figure 17.21a,b) and at this time seizure thresholds are lowest (Figure 17.21c). Woolley and McEwen showed that it is indeed estradiol itself that triggers the increase in spine numbers and that as hippocampal neurons grow more spines, they also grow more excitatory synapses.

How does estradiol increase the numbers of hippocampal spines and excitatory synapses? Although the details of this mechanism are not all clear, it appears that estradiol increases the capacity for synaptic plasticity in the hippocampus in multiple ways. In the presence of estradiol, postsynaptic responses to glutamate are larger than without estradiol. As we will see in Chapter 25, such enhanced responses at excitatory synapses cause the synapses to strengthen. Estradiol may also alter hippocampal function by decreasing synaptic inhibition. Estradiol causes some inhibitory cells to produce less GABA, their neurotransmitter, and therefore synaptic inhibition becomes less effective. Less inhibition increases neural activity, complementing the estradiol effect at excitatory synapses. When the pieces are put together, it seems that estradiol produces a hippocampus with less effective inhibitory synapses and stronger excitatory synapses, thereby triggering an increase in the number of spines on the pyramidal cells.

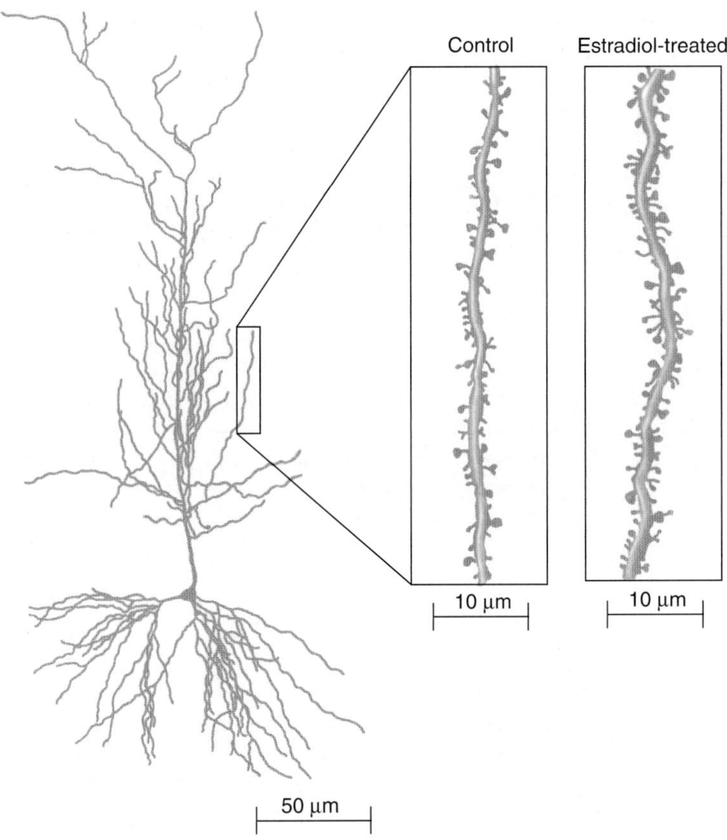

▲ FIGURE 17.20
An activational effect of steroid hormones. Estradiol treatment causes an increase in the number of dendritic spines on hippocampal neurons. (Source: Adapted from Woolley et al., 1997.)

In rats, the hippocampus is particularly important for spatial memory and navigational skills, and several studies demonstrate that estradiol enhances such forms of memory formation. In these experiments, rats are trained to run a maze or remember objects or places. Estradiol administered shortly before or after training enhances performance on these tasks when they are performed again hours later as a test of memory. It is interesting to note that the memory benefit associated with the estradiol disappears if the estrogen is given 2 hours after training. Evidently, estrogens can facilitate memory, but they must be present near in time to the learning experience.

Woolley notes that the peak in hippocampal spine number coincides with the rat's peak in fertility. During that period, the female actively seeks out mates, which may require the heightened spatial ability that might come with a more excitable, NMDA-receptor-filled hippocampus. Thus, the female rat's brain may fine-tune itself on a 5-day cycle in order to meet changing reproductive needs.

Estradiol has also been shown to have a protective effect on neurons that may help combat disease. In neuronal cultures, cells are more likely to survive hypoxia, oxidative stress, and exposure to various neurotoxic agents if they are exposed to estradiol. Clinically, estrogen appears to minimize or delay neural damage in a variety of situations. For example, it may protect against stroke in humans, although the mechanism is unclear. This observation could be related to the finding that tamoxifen, an estrogen receptor antagonist often used in the treatment of breast

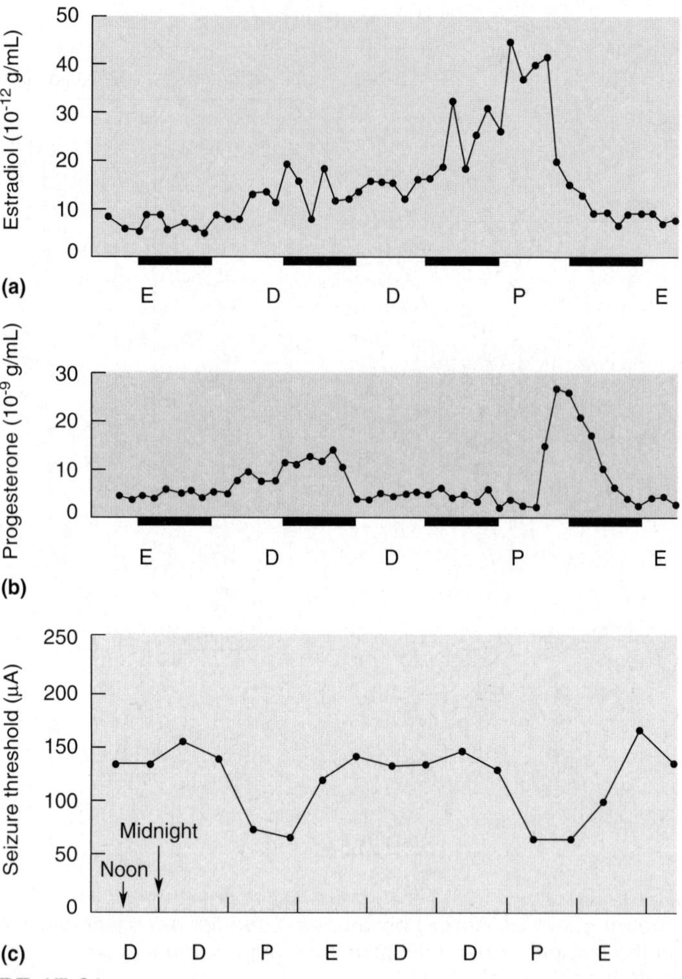

▲ FIGURE 17.21

Fluctuations of hormone levels during the estrous cycle and hippocampal seizure threshold. The circulating levels of **(a)** estradiol and **(b)** progesterone vary during the estrous cycle. Levels of both hormones peak during the proestrus phase. **(c)** The threshold (in terms of stimulating current) for triggering a seizure in the hippocampus of a female rat varies over the estrous cycle and is lowest during the proestrus phase. The phases of the estrous cycle are D, diestrus; P, proestrus; E, estrous. (Sources: Parts a and b, Smith et al., 1975; part c, Terasawa et al., 1968.)

cancer, is associated with an increased risk of stroke in women. Estrogen replacement therapy appears to aid in the treatment of certain neurological disorders. Researchers have observed that increased levels of sex hormones during pregnancy are associated with a reduced severity of multiple sclerosis, and there is some evidence that estrogen may benefit women with multiple sclerosis. It also appears that estrogen replacement therapy might delay the onset of Alzheimer's disease and decrease tremors in Parkinson's disease. The actions of estrogens in these diseases have been difficult to pin down, in part because a variety of cell types express estrogen receptors. Indeed, recent evidence suggests that the benefits of estrogen may come from effects on astrocytes as well as neurons.

Sexual Orientation

Estimates are that about 3–10% of the American population is homosexual. In light of the behavioral differences between, for example, male homosexuals and male heterosexuals, is the anatomy or physiology of the brain different? Is there a biological basis for sexual orientation? In some sense, this must be true if we believe that all behavior is based on brain activity. However, there is no evidence that sexual orientation is related to activational effects of hormones in adults. For example, administering androgens or estrogens to adults, or removing the gonads, has no effect on sexual orientation. As an alternative, perhaps homosexual and heterosexual brains are structurally different due to organizational effects.

We saw earlier that in animals, there are sex differences in the anterior hypothalamus. In rats, the SDN (sexually dimorphic nucleus) in the preoptic area of the anterior hypothalamus is much larger in males than females. After a surgical lesion to this brain area, male rats spend more time with sexually active males than with sexually receptive females, a reversal of their preference before surgery. Another piece of suggestive evidence comes from studies of Rocky Mountain bighorn sheep, in which researchers estimate that about 8% of the male population prefers mounting other males rather than females. The SDN is found to be about half the size in these male-oriented rams compared to female-oriented rams. Thus, it appears that the size of hypothalamic nuclei in some animals may correlate with sexual preference. Unfortunately, the causal relationship between SDN size and sexual orientation is unclear.

In humans, the INAH-3 nucleus (one of the interstitial nuclei of the anterior hypothalamus) is about twice as large in men as in women, a difference that may be related to sexually dimorphic behavior. Some studies of the INAH suggest that there are differences between homosexual and heterosexual brains that might be related to sexual orientation. Simon LeVay, then working at the Salk Institute, found that INAH-3 in gay men is only half the size of the nucleus in straight men (Figure 17.22).

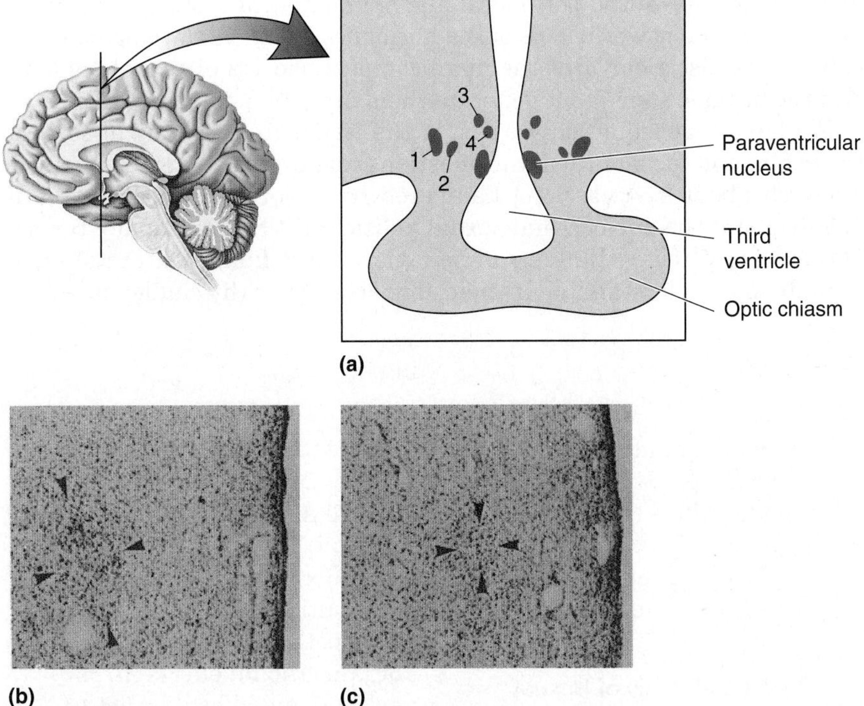

▲ FIGURE 17.22

The location and size of INAH-3. **(a)** The location of the four INAH nuclei in the hypothalamus. In the micrographs, arrows indicate INAH-3 in **(b)** a heterosexual man and **(c)** a homosexual man. In the homosexual, the nucleus is smaller and the cells more scattered. (Source: Micrographs from LeVay, 1991, p. 1035.)

In other words, INAH-3 in homosexual men is similar in size to that of women. While this finding may indicate a biological basis for homosexuality, it is difficult to interpret this in terms of complex human behavior. Moreover, subsequent studies have not always confirmed a correlation between INAH-3 size and sexual orientation.

Other research has found the anterior commissure and suprachiasmatic nucleus to be larger in male homosexuals than in male heterosexuals. One study reported that the bed nucleus of the stria terminalis is larger in men than women and that male-to-female transsexuals have a nucleus comparable in size to females. Collectively, these studies offer the intriguing prospect that complex aspects of human sexual behavior may ultimately be linked to distinct brain organization. However, the difficulties involved in comparing brains, as well as the history of brain dimorphisms, suggest that caution is advised until a research consensus is reached.

CONCLUDING REMARKS

The subject of sex and the brain is complicated by the subtleties of the biological and cultural mechanisms that determine sexual behavior. Particularly in humans, anatomical differences between the female and male nervous systems are not readily apparent, and indeed, most human behavior is not distinctly masculine or feminine. Where small brain differences between the sexes occur, any adaptive purpose they may serve is not clear. And in no case is the neurobiological basis for sex differences in cognition known.

Nevertheless, the essential biological imperative—reproduction—demands sex-specific behaviors, at least for mating and giving birth. For the most concretely sexual structures (such as the muscles and motor neurons controlling the penis or the sensory afferents innervating the clitoris), identifying some of the peripheral and spinal neural systems involved is fairly easy. The powerful role of sex hormones in sexual development and behavior is also clear. But the more complex aspects of sexual behavior and the brain systems that generate them are still quite mysterious.

We have touched on only a few of the issues in the study of sex and the brain, and most of the basic questions remain unanswered. Scientific research about sex was long hampered by society's reluctance to talk openly about the subject, and sexual politics still tend to muddy the scientific waters today. But sexual behavior is a defining feature of being human, and understanding its neural basis is a worthy challenge.

KEY TERMS

Sex and Gender
gender identity (p. 581)
genotype (p. 581)
genetic sex (p. 581)
sex-determining region of the Y chromosome (SRY) (p. 582)

The Hormonal Control of Sex
androgens (p. 584)
estrogens (p. 584)
luteinizing hormone (LH) (p. 585)
follicle-stimulating hormone (FSH) (p. 585)
gonadotropins (p. 585)
gonadotropin-releasing hormone (GnRH) (p. 585)
menstrual cycle (p. 586)
estrous cycle (p. 587)

The Neural Basis of Sexual Behaviors
polygyny (p. 590)
polyandry (p. 590)
monogamy (p. 590)

Why and How Male and Female Brains Differ
sexual dimorphisms (p. 595)
sexually dimorphic nucleus (SDN) (p. 597)
interstitial nuclei of the anterior hypothalamus (INAH) (p. 597)
organizational effects (p. 599)
activational effects (p. 599)
α-fetoprotein (p. 600)

REVIEW QUESTIONS

1. Suppose you have just been captured by aliens who have landed on Earth to learn about humans. The aliens are all one sex, and they are curious about the two human sexes. To earn your freedom, all you must do is tell them how to reliably distinguish between males and females. What biological and/or behavioral tests do you tell them to conduct? Be sure to describe any exceptions that might violate your tests; you don't want the aliens to get angry!
2. Figure 17.18 shows an interesting but unexplained observation: In the brain of a mother rat during periods of lactation, the size of the somatosensory cortex representing the skin around the nipples expands. Speculate about a likely mechanism for this phenomenon. Suggest a reason why such brain plasticity might be advantageous.
3. Estradiol is usually described as a female sex hormone, but it also plays a critical role in the early development of the male brain. Explain how this happens and why the female brain is not similarly affected by estradiol at the same stage of development.
4. Where and how can steroid hormones influence neurons in the brain at the cellular level?
5. What evidence supports the hypothesis that sexual differentiation of the body and brain is not entirely dependent on sex hormones?
6. Suppose that a research team has just claimed that a small, obscure nucleus in the brain stem, nucleus X, is sexually dimorphic and essential for certain "uniquely male" sexual behaviors. Discuss the kinds of evidence you would need to accept these claims about (a) the existence of a dimorphism, (b) the definitions of uniquely male behaviors, and (c) the involvement of nucleus X in these sexual behaviors.

FURTHER READING

Arnold AP. 2004. Sex chromosomes and brain gender. *Nature Reviews Neuroscience* 5:701–708.

Bartels A, Zeki S. 2004. The neural correlates of maternal and romantic love. *Neuroimage* 21:1155–1166.

Colapinto J. 2001. *As Nature Made Him: The Boy Who Was Raised as a Girl*. New York: Harper Collins.

De Boer A, van Buel EM, ter Horst GJ. 2012. Love is more than just a kiss: a neurobiological perspective on love and affection. *Neuroscience* 201:114–124.

Hines M. 2011. Gender development and the human brain. *Annual Review of Neuroscience* 34:69–88.

Pfaus JG. 2009. Pathways of sexual desire. *Journal of Sexual Medicine* 6:1506–1533.

Valente SM, LeVay S. 2003. *Human Sexuality*. Sunderland, MA: Sinauer.

Wooley CS. 2007. Acute effects of estrogen on neuronal physiology. *Annual Review of Pharmacology and Toxicology* 47:657–680.

Wu MV, Shah NM. 2011. Control of masculinization of the brain and behavior. *Current Opinion in Neurobiology* 21:116–123.

Young KA, Gobrogge KL, Liu Y, Wang Z. 2011. The neurobiology of pair bonding: insights from a socially monogamous rodent. *Frontiers in Neuroendocrinology* 32:53–69.

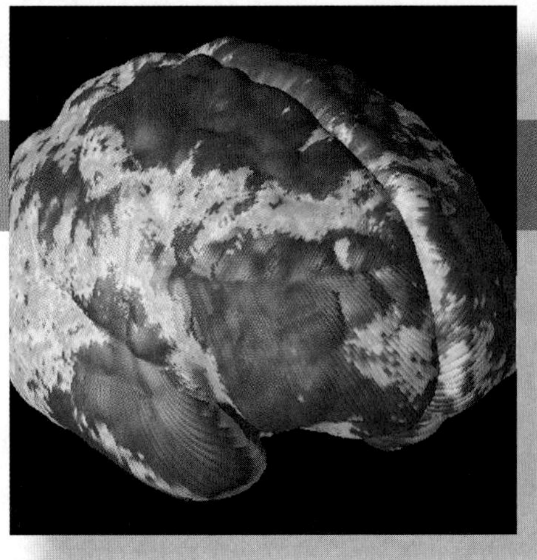

CHAPTER EIGHTEEN

Brain Mechanisms of Emotion

INTRODUCTION

To appreciate the significance of emotions, just imagine life without them. Instead of the daily highs and lows we all experience, life would likely feel like a great empty plain of existence with little significance. Emotional experiences are a large part of being human. In books and movies, aliens and robots may look like people, but they usually seem inhuman simply because they exhibit no emotions.

Affective neuroscience is the investigation of the neural basis of emotion and mood. In this chapter, we explore emotion. Affective disorders, also known as mood disorders, are discussed in Chapter 22. You might be wondering how we can study something as ephemeral as one's feelings. When studying a sensory system, in contrast, you can present a stimulus and seek the neurons that respond to it. You can manipulate the stimulus to determine the stimulus attributes (light intensity, sound frequency, etc.) that are best for evoking a response. It is not as straightforward to study emotions in animals that cannot tell us their subjective feelings, however. What we observe are behaviors presumed to be expressions of internal emotions. We must be careful to distinguish between emotional *experience* (feelings) and emotional *expression*. What we know about the brain mechanisms of emotion has been derived from a synthesis of animal and human studies. In animals, brain activity and the effects of brain lesions on behavior have been noted and interpreted in the context of emotions, even though we cannot determine the animals' feelings. Studies in humans have examined brain activity associated with emotional experience and the recognition of emotion in others.

We are not yet at a stage where we can map out an emotion system the same way that sensory systems have been outlined. Indeed, we will see that earlier ideas about a single emotion system or multiple systems involving brain areas dedicated to particular emotions have been replaced by theories in which emotions are based on distributed networks of brain activity.

EARLY THEORIES OF EMOTION

Emotions—love, hate, happiness, sadness, fear, anxiety, and so on—are feelings we all experience at one time or another. But what precisely defines those feelings? Are they sensory signals from our body, diffuse patterns of activity in our cortex, or something else?

In the nineteenth century, several highly regarded scientists, including Darwin and Freud, considered the role of the brain in the expression of emotion (Figure 18.1). This early research was based on the careful study of emotional expression in animals and humans and emotional

▶ FIGURE 18.1
Expressions of anger in animals and humans. These drawings are from Darwin's book *The Expression of the Emotions in Man and Animals;* they were used to support his claim that there are basic universal emotions. Darwin conducted one of the first extensive studies of emotional expression. Reproduced with permission from John van Wyhe, ed. 2002. *The Complete Work of Charles Darwin Online.* (http://darwin-online.org.uk/)

experience in humans. It may seem like common sense to many now, but Darwin made the important observations that people in different cultures experience the same emotions and that animals appear to express some of the same emotions as humans. Later in the nineteenth and twentieth centuries, scientists developed theories for the physiological basis of emotion and the relationship between emotional expression and experience.

The James–Lange Theory

One of the first well-articulated theories of emotion was proposed in 1884 by the renowned American psychologist and philosopher William James. Similar ideas were proposed by Danish psychologist Carl Lange. This theory, commonly known as the **James–Lange theory** of emotion, proposed that we experience emotion *in response to* physiological changes in our body. To understand why many contemporaries of James and Lange considered this idea counterintuitive, consider an example.

Suppose you wake one morning to find a malicious-looking spider hanging from a web above your bed. If you like many people have arachnophobia, you may experience a fight-or-flight response that involves changes in heart rate, muscle tone, and lung function (see Chapter 15). According to the James–Lange theory, your visual system sends an image of the spider to your brain and in response the brain issues commands to the somatic and autonomic nervous systems that alter muscle and organ function. These responses of the body follow directly from the sensory input without any emotional component. The emotion you experience consists of your feelings that result from the changes in the body. In other words, rather than jumping out of bed in response to being scared, you actually feel scared because you become aware of your racing heart and tensed muscles. This seems like a backward idea to many people today, as it did to many contemporaries of James and Lange. Until this theory was proposed, the common conception was that an emotion is evoked by a situation and the body changes in response to the emotion: You become scared when you see a spider and then your body reacts. The James–Lange theory is the exact opposite.

Consider one of the thought experiments suggested by James. Suppose you are boiling with anger about something that has just happened. Try to strip away all the physiological changes associated with the emotion. Your pounding heart is calmed, your tensed muscles are relaxed, and your flushed face is cooled. As James said, it is hard to imagine maintaining rage in the absence of any of these physiological responses.

Even if it is true that emotion follows from changes in the body's physiological state, this doesn't mean that emotion cannot be felt in the absence of obvious physiological signs (a point even James and Lange would concede). But for strong emotions that are typically associated with physical change, the James–Lange theory says the bodily changes cause the emotion rather than the other way around.

The Cannon–Bard Theory

Although the James–Lange theory became popular in the early twentieth century, it soon came under attack. In 1927, American physiologist Walter Cannon published a paper that offered several compelling criticisms of the James–Lange theory and proposed a new theory. Cannon's theory was modified by Philip Bard and became known as the **Cannon–Bard theory** of emotion. It proposed that emotional experience can occur independently of emotional expression.

One of Cannon's arguments against the James–Lange theory was that emotions can be experienced even if physiological changes cannot be sensed. To support this claim, he offered the cases of animals he and others studied after transection of the spinal cord. Such surgery eliminated body sensations below the level of the cut, but it did not appear to abolish emotion. To the extent possible with muscular control of just the upper body or head, the animals still exhibited signs of experiencing emotions. Similarly, Cannon noted human cases in which a transected spinal cord did not diminish emotion. If emotional experience occurs when the brain senses physiological changes in the body, as the James–Lange theory proposed, then eliminating sensation should also eliminate emotions, and this did not appear to be the case.

A second observation of Cannon's that seems inconsistent with the James–Lange theory is the lack of a reliable correlation between the experience of emotion and the physiological state of the body. For example, fear is accompanied by increased heart rate, inhibited digestion, and increased sweating. However, these same physiological changes accompany other emotions, such as anger, and even nonemotional conditions of illness, such as fever. How can fear be a consequence of the physiological changes when these same changes are associated with states other than fear?

Cannon's new theory focused on the idea that the thalamus plays a special role in emotional sensations. The theory proposes that sensory input is received by the cerebral cortex, which in turn activates certain changes in the body. But according to Cannon, this stimulus–response neural loop is devoid of emotion. Emotions are produced when signals reach the thalamus either directly from the sensory receptors or by descending cortical input. In other words, the character of the emotion is determined by the pattern of activation of the thalamus irrespective of the physiological response to the sensory input. An example may clarify the difference between this and the James–Lange theory. According to James and Lange, you feel sad when you sense that you are crying; if you could prevent the crying, the sadness should go, too. In Cannon's theory, you don't have to cry to feel sad; there simply has to be the appropriate activation of your thalamus in response to the situation. The James–Lange and Cannon–Bard theories of emotion are compared in Figure 18.2.

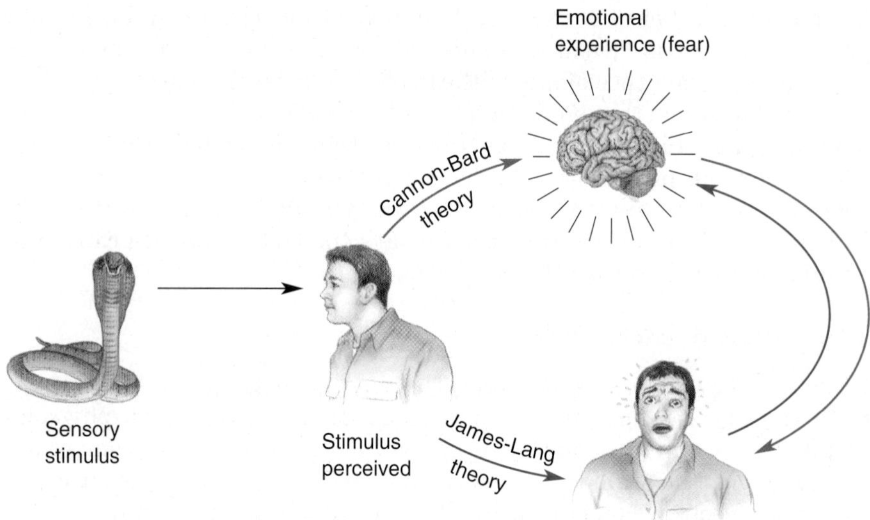

▶ FIGURE 18.2
A comparison of the James–Lange and Cannon–Bard theories of emotion. In the James–Lange theory (red arrows), the man perceives the threatening animal and reacts. When he senses his body's response to the situation, he becomes afraid. In the Cannon–Bard theory (blue arrows), the threatening stimulus first causes the feeling of fear, and the man's reaction follows.

Many theories of emotion have been proposed since the days of the James–Lange and Cannon–Bard theories. Subsequent work has demonstrated that each of the older theories has merits as well as flaws. For instance, contrary to Cannon's statements, fear and rage have been shown to be associated with distinguishable physiological responses, even though they both activate the sympathetic division of the autonomic nervous system (ANS). Although this does not prove that these emotions result from distinct physiological responses, the responses are at least different (Box 18.1). Research has also shown that to some extent we can be aware of our body's autonomic function (called *interoceptive awareness*), a key component of the James–Lange theory. For example, people have been shown to be capable of judging the timing of their heartbeat, and increased activity is found in certain brain areas when this task is performed.

Another interesting challenge to the Cannon–Bard theory, demonstrated by later studies, is that emotion is sometimes affected by damage to the spinal cord. In one study of adult men with spinal injuries, there was a correlation between the extent of sensory loss and reported decreases in emotional experiences, although other studies of people with spinal injuries have not always found a similar correlation. Shortly, we will examine more recent theories of emotion in the context of experimental results that suggest the brain structures involved in the experience and expression of emotion.

Implications of Unconscious Emotion

Although the findings may be counterintuitive, some studies suggest that sensory input can have emotional effects on the brain without our being aware of the stimuli. Several related experiments were conducted by Arne Öhman, Ray Dolan, and their colleagues in Sweden and England. They first showed that if an angry face is briefly flashed and quickly followed by a briefly flashed photo of an expressionless face, subjects report seeing only the expressionless face. The angry face is said to be perceptually "masked," and the expressionless face is the masking stimulus.

In one experiment, subjects were shown a variety of faces without a masking stimulus; and each time an angry face was shown, the subject's finger got a mild electrical shock. After aversive conditioning like this, subjects exhibited altered autonomic activity, such as increased skin conductance (sweaty palms), when the angry faces were shown again. The researchers were interested in what happens when the angry faces are occasionally shown after training but the masking stimulus is reintroduced. Surprisingly, when angry faces were shown, the subjects had an autonomic response (increased skin conductance), even though they were unaware of the angry faces. These findings indicate that the subjects responded to the angry expressions on the aversive face stimuli even though they were not perceptually aware of seeing the faces at all. The concept of an **unconscious emotion** is based on this observation.

In a second experiment, subjects were shown angry faces with or without a loud unpleasant sound (Figure 18.3). As before, the subjects did not perceive the angry faces when a subsequent masking stimulus was presented. Nonetheless, skin conductance showed that the subjects responded to the angry faces that had been paired with the sound. In addition, positron emission tomography (PET) imaging was used to record brain activity while the photos were presented. The brain images revealed that the angry faces conditioned to be unpleasant evoked greater activity in the brain in a particular location, the amygdala. We will have

BOX 18.1 OF SPECIAL INTEREST

Butterflies in the Stomach

Human language has colorful ways of describing emotional experiences. If someone hesitates before bungee jumping off a high bridge, we describe their fear by saying they have "cold feet." At the other end of the temperature spectrum, a person who readily becomes angry is described as a "hot head." Nervous before going out with someone new? You may be experiencing "butterflies in the stomach." These descriptive terms are fun, but do they have the slightest relationship to the physiological experience of emotion?

An intriguing study conducted by scientists at Aalto University in Finland suggests that basic emotions and some other emotions may indeed be associated with unique maps of sensory changes spread across the body. This conclusion was reached based on online testing of over 700 people in Finland, Sweden, and Taiwan. In order to itemize which parts of the body were felt to be affected by an emotion, the experimenters asked participants to color a map of the body using warm colors where they felt an emotion made the body more active and cool colors where the body was made less active. Emotion maps were made in response to a variety of stimuli including emotion words, pictures of emotional facial expressions, emotional experiences in short stories, and emotional scenes in movies. The hope was that by studying participants in different cultures and languages, universal emotional experiences could be mapped rather than cultural stereotypes.

The figure shows maps of presumptive body activity averaged across many observers. Red and yellow indicate elevated activity, and blue reduced activity, relative to a neutral (black) state. Some traits, such as elevated head and chest activity (elevated heart and respiration rates?), were common to multiple emotions. Other characteristics were more unique. Happiness was unusual in the extent to which the entire body showed increased activity, and sadness had a unique lowering of activity in the extremities. The body map for disgust had an odd elevation in activity around the digestive tract and throat (a gag reflex?). What do these colorful maps represent? We can only speculate, but perhaps they are related to patterns of sensation and activation in the autonomic nervous system. Obviously, one has to be cautious interpreting the maps, but it is intriguing that the different emotion maps are distinguishable, and this was even true to some extent for emotions not considered "basic." Also interesting is the finding that the emotion maps are similar across cultures. Even though we can't take a snapshot of butterflies in the stomach, these findings are consistent with Darwin's idea that at least some emotions are unique experiences that are universal across cultures.

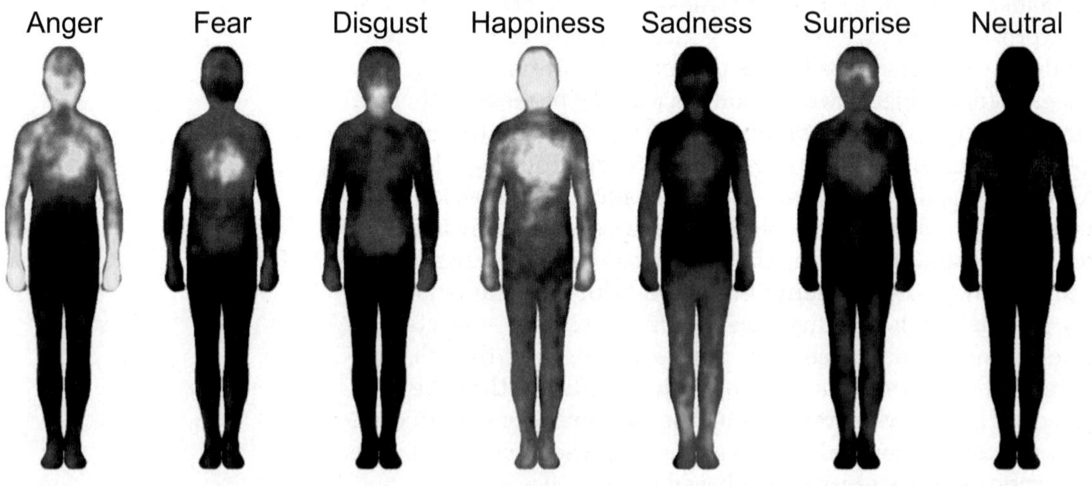

Figure A
Color maps of six basic emotions. Estimates of body activation range from low (blue) to high (yellow). (Source: Adapted from Nummenmaa L, Glerean E, Hari R, Hietanen JK. 2014. Bodily maps of emotions. *Proceedings of the National Academy of Science* 111:646–651, Figure 1.)

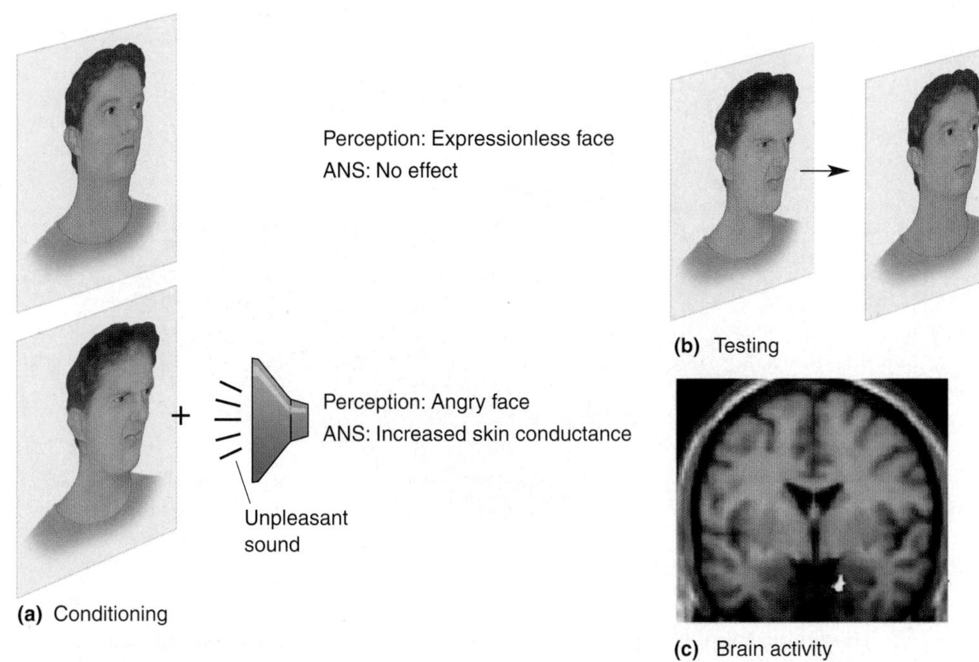

▲ FIGURE 18.3

Unconscious emotional brain activity. (a) Human subjects were conditioned using photos of expressionless and angry faces. Subjects responded to the angry face, paired with a loud unpleasant sound, with increased ANS activity (skin conductance). (b) In the testing phase, an angry face was shown briefly, followed immediately by an expressionless face. Subjects reported seeing only the expressionless face, but increased skin conductance still occurred. (c) Despite the fact that the angry face was not perceived in the testing phase, amygdala activation (red and yellow) occurred only when an angry face preceded the masking stimulus. (Source: Morris, Öhman, and Dolan, 1998.)

more to say about the amygdala later in the chapter. For now, the important point to remember is that measures of both autonomic response and amygdala activity correlate with the presentation of angry faces that are conditioned to be unpleasant despite the fact that the faces are not perceived.

If sensory signals can have emotional impact on the brain without our being aware of it, this seems to rule out theories of emotion in which emotional experience is a prerequisite for emotional expression. But even with this conclusion, there are many possible ways for the brain to process emotional information. We now turn to the pathways in the brain that link sensations (inputs) to the behavioral responses (outputs) that characterize emotional experience. In the remainder of this chapter, we will see that different emotions may depend on different neural circuits, but some parts of the brain are important for multiple emotions.

THE LIMBIC SYSTEM

Previous chapters discussed how sensory information from peripheral receptors is processed along clearly defined, anatomically distinct pathways to the neocortex. The components of a pathway collectively constitute a *system*. For example, neurons located in the retina, lateral geniculate nucleus, and striate cortex work together to serve vision, so we say they are part of the visual system. Is there a system, in this sense, that processes the experiencing of emotions? Beginning around 1930, some scientists argued

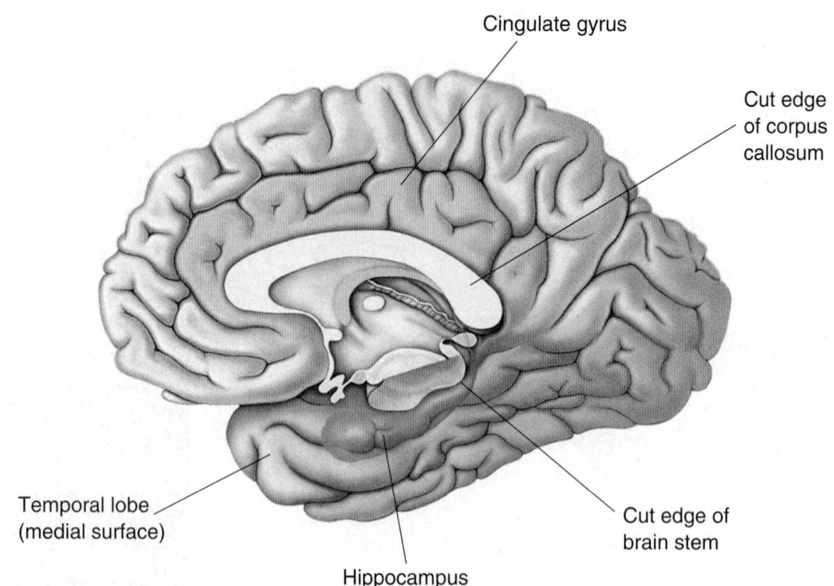

▲ FIGURE 18.4
The limbic lobe. Broca defined the limbic lobe as the structures that form a ring around the brain stem and corpus callosum on the medial walls of the brain. The main structures in the limbic lobe labeled here are the cingulate gyrus, medial temporal cortex, and the hippocampus. The brain stem has been removed in the illustration to make the medial surface of the temporal lobe visible.

that there is, and it came to be known as the limbic system. Shortly, we will discuss the difficulties of trying to define a single system for emotion. But first, let's examine the origin of the limbic system concept.

Broca's Limbic Lobe

In a paper published in 1878, French neurologist Paul Broca noted that, on the medial surface of the cerebrum, all mammals have a group of cortical areas that are distinctly different from the surrounding cortex. Using the Latin word for "border" (*limbus*), Broca named this collection of cortical areas the **limbic lobe** because they form a ring or border around the brain stem (Figure 18.4). According to this definition, the limbic lobe consists of the cortex around the corpus callosum (mainly the cingulate gyrus), the cortex on the medial surface of the temporal lobe, and the hippocampus. Broca did not write about the importance of these structures for emotion, and for some time, they were thought to be primarily involved in olfaction. However, the word *limbic*, and the structures in Broca's limbic lobe, were subsequently closely associated with emotion.

The Papez Circuit

By the 1930s, evidence suggested that a number of limbic structures are involved in emotion. Reflecting on the earlier work of Cannon, Bard, and others, American neurologist James Papez proposed that there is an "emotion system," lying on the medial wall of the brain that links the cortex with the hypothalamus. Figure 18.5 shows the group of structures that have come to be called the **Papez circuit**. Each is connected to another by a major fiber tract.

Papez believed, as do many scientists today, that the cortex is critically involved in the experience of emotion. Following damage to certain cortical areas, there are sometimes profound changes in emotional expression

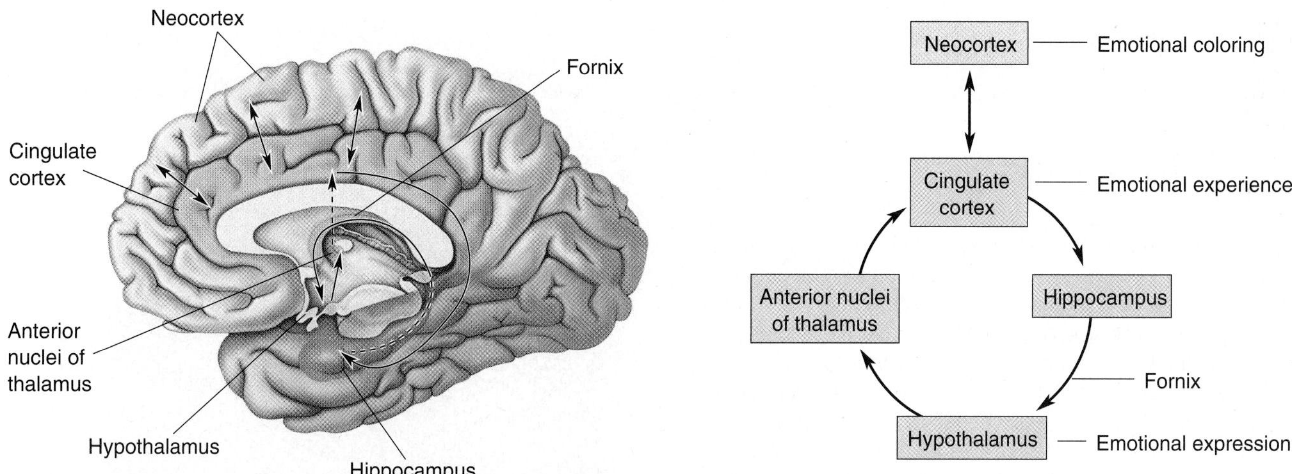

▲ FIGURE 18.5
The Papez circuit. Papez believed that the experience of emotion was determined by activity in the cingulate cortex and, less directly, other cortical areas. Emotional expression was thought to be governed by the hypothalamus. The cingulate cortex projects to the hippocampus, and the hippocampus projects to the hypothalamus by way of the bundle of axons called the fornix. Hypothalamic effects reach the cortex via a relay in the anterior thalamic nuclei.

with little change in perception or intelligence (Box 18.2). Also, tumors located near the cingulate cortex are associated with certain emotional disturbances, including fear, irritability, and depression. Papez proposed that activity evoked in other neocortical areas by projections from the cingulate cortex adds "emotional coloring" to our experiences.

We saw in Chapter 15 that the hypothalamus integrates the actions of the ANS. In the Papez circuit, the hypothalamus governs the behavioral expression of emotion. The hypothalamus and neocortex are arranged so that each can influence the other, thus linking the expression and experience of emotion. In the circuit, the cingulate cortex affects the hypothalamus via the hippocampus and fornix (the large bundle of axons leaving the hippocampus), whereas the hypothalamus affects the cingulate cortex via the anterior thalamus. The fact that communication between the cortex and hypothalamus is bidirectional means that the Papez circuit is compatible with both the James–Lange and the Cannon–Bard theories of emotion.

While anatomical studies demonstrated that the components of the Papez circuit were interconnected as Papez had indicated, there was only suggestive evidence that each was involved in emotion. One reason Papez thought the hippocampus is involved in emotion is that it is affected by the virus responsible for rabies. An indication of rabies infection, and an aid in its diagnosis, is the presence of abnormal cytoplasmic bodies in neurons, especially in the hippocampus. Because rabies is characterized by hyperemotional responses such as exaggerated fear and aggressiveness, Papez reasoned that the hippocampus must be involved in normal emotional experience. Although little evidence related to the role of the anterior thalamus, other clinical reports at the time stated that lesions in this area led to apparent emotional disturbances, such as spontaneous laughing and crying.

You may have noticed the correlation between the elements composing the Papez circuit and Broca's limbic lobe. Because of their similarity, the group of structures in the Papez circuit are often referred to as the **limbic system**, even though Broca's anatomical notion of the limbic lobe originally had nothing to do with emotion. The term *limbic system* was

BOX 18.2 OF SPECIAL INTEREST

Phineas Gage

Brain damage can sometimes have a profound influence on a person's personality. One of the most famous examples is the case of Phineas Gage. On September 13, 1848, while tamping explosive powder into a hole to prepare for blasting at a railroad construction site in Vermont, Phineas made the mistake of looking away from what he was doing. His tamping iron hit a rock and sparked an explosion. The consequences are described by Dr. John Harlow in an 1848 article entitled "Passage of an Iron Rod Through the Head." When the charge went off, it sent Phineas's meter-long, 6-kg iron tamping rod into his head just below his left eye. After passing through his left frontal lobe, the rod exited the top of Gage's head.

Incredibly, after being carried to an ox cart, Gage sat upright on the ride to a nearby hotel and walked up a long flight of stairs to go inside. When Harlow first saw Gage at the hotel, he commented that "the picture presented was, to one unaccustomed to military surgery, truly terrific" (p. 390). As you might imagine, the projectile destroyed a considerable portion of the skull and left frontal lobe, and Gage lost a great deal of blood. The hole through his head was more than 9 cm in diameter. Harlow was able to stick the full length of his index finger into the hole from the top of Gage's head, and also upward from the hole in his cheek. Harlow dressed the wound as best he could. Over the following

Figure A
Phineas Gage and the rod that passed through his brain. (Source: Wikimedia.)

popularized in 1952 by American physiologist Paul MacLean. According to MacLean, the evolution of a limbic system enabled animals to experience and express emotions and freed them from the stereotypical behavior dictated by their brain stem.

Difficulties with the Concept of a Single System for Emotions

We have defined a group of interconnected anatomical structures roughly encircling the brain stem as a limbic system. Experimental work supports the hypothesis that some of the structures in Broca's limbic lobe and the Papez circuit play a role in emotion. On the other hand, some of the components of the Papez circuit are no longer thought to be important for the expression of emotion, such as the hippocampus.

The critical point seems to be conceptual, concerning the definition of an emotion system. Given the diversity of emotions, we experience and weeks, considerable infection developed. No one would have been surprised if the man died. But about a month after the accident, he was out of bed and walking around town.

Harlow corresponded with Gage's family for many years and in 1868 published a second article, "Recovery from the Passage of an Iron Bar Through the Head," describing Gage's life after the accident. After Gage recovered from his wounds, he was apparently normal except for one thing: His personality was drastically and permanently changed. When he tried to return to his old job as construction foreman, the company found he had changed so much for the worse that they wouldn't rehire him. According to Harlow, before the accident Gage was considered "the most efficient and capable foreman. . . . He possessed a well-balanced mind, and was looked upon by those who knew him as a shrewd, smart business man, very persistent in executing all his plans of operation" (pp. 339–340). After the accident, Harlow described him as follows:

> The equilibrium or balance, so to speak, between his intellectual faculties and animal propensities, seems to have been destroyed. He is fitful, irreverent, indulging at times in the grossest profanity (which was not previously his custom), manifesting but little deference for his fellows, impatient of restraint or advice when it conflicts with his desires, at times pertinaciously obstinate, yet capricious and vacillating, devising many plans of future operation, which are no sooner arranged than they are abandoned in turn for others appearing more feasible. . . . His mind was radically changed, so decidedly that his friends and acquaintances said he was "no longer Gage." (pp. 339–340)

Phineas lived for another 12 years; when he died, no autopsy was performed. However, Gage's skull and the tamping iron have been preserved in a museum at Harvard Medical School. In 1994, Hanna and Antonio Damasio and their colleagues at the University of Iowa made new measurements of the skull and used modern imaging techniques to assess the damage to Gage's brain. Their reconstruction of the tamping iron's path is shown in Figure A. The iron rod severely damaged the cerebral cortex in both hemispheres, particularly the frontal lobes. It was presumably this damage that led to Gage's emotional outbursts and the drastic changes in his personality.

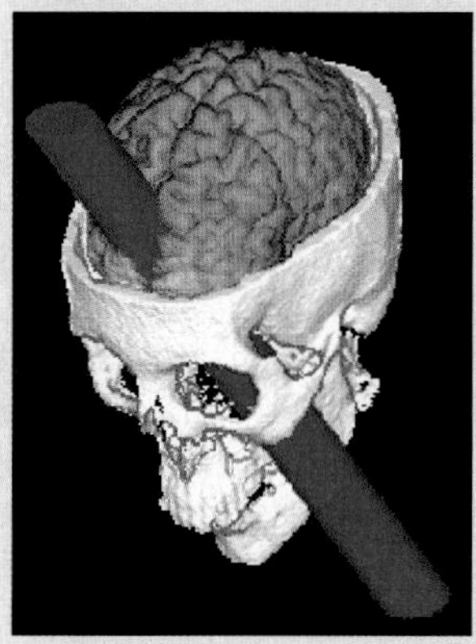

Figure B
The path of the iron rod through Gage's skull. (Source: Damasio et al., 1994, p. 1104.)

the different brain activity associated with each, there is no compelling reason to think that only one system, rather than several, is involved. Conversely, solid evidence indicates that some structures involved in emotion are also involved in other functions; there is not a one-to-one relationship between structure and function in this case. Although the term *limbic system* is still commonly used in discussions of brain mechanisms of emotion, it has become increasingly clear that there is not a single, discrete emotion system.

EMOTION THEORIES AND NEURAL REPRESENTATIONS

Early theories of emotion and subsequent descriptions of the limbic system were built on a combination of introspection and inference based primarily on instances of brain injury and brain disease: If damage to a brain structure alters the experience or expression of emotion, we infer that the structure is important for normal emotional function. Unfortunately, studies of disease and the consequences of lesions are not ideal for revealing *normal* function. Before getting into the nitty-gritty of experiments investigating neural mechanisms of emotion, it may be helpful to consider the representation of emotions from a broad perspective.

Basic Emotion Theories

If the limbic system is not a monolithic system for the experience and expression of all emotions, as now seems the case, another possibility that has been explored is that some emotions are at least associated with distinct patterns of activity in the brain and unique physiological responses in the body (see Box 18.1). In **basic theories of emotion**, certain emotions are thought to be unique, indivisible experiences that are innate and universal across cultures, an idea that seems a logical extension of Darwin's early observations on the universality of a small number of emotions. Commonly, these **basic emotions** are considered anger, disgust, fear, happiness, sadness, and surprise. From a neural perspective, one might hypothesize that basic emotions have distinct representations or circuits in the brain, perhaps analogous to distinct representations for sensory experiences. For example, it has been claimed that sadness correlates best with activity in the medial prefrontal cortex and fear with activity in the amygdala. In a moment, we will take an in-depth look at evidence suggesting a special role for the amygdala in fear. But first, let's look into the general question of brain activity associated with emotion.

One way to get a broad perspective on the representation of emotion is to compare human functional magnetic resonance imaging (fMRI) or PET brain recordings while people experience different emotions. Numerous experiments of this sort have been conducted in which people are induced to experience emotions or shown pictures that evoke different emotions while they lie in a brain imaging machine. Figure 18.6 shows a summary of brain images collected in this way. Several observations can be made from these images. First, there are different "hotspots"—areas of particularly high brain activity associated with each emotion. Second, each emotion is associated with an array of smaller and larger patches of lesser brain activity. Finally, some activated regions are associated with more than one emotion. The bottom portion of the figure compares brain activations for sadness and fear, emotions that could be reliably discriminated from the activation patterns. Consistent with distinct circuits for different emotions, amygdala activity is more associated with fear than with sadness, and medial prefrontal activity is more associated with sadness.

One interpretation of the data in Figure 18.6 is that the most highly activated region uniquely represents an emotion, such as the medial prefrontal cortex for sadness. This might be like the face-selective patches of visual cortex in the temporal lobes (see Chapter 10). Alternatively, the *pattern* of activation could be the basis of the emotion and each active brain region a piece of the puzzle. If either single areas or networks of areas uniquely represent emotions, we might, in principle, be able to scan people's brains and know what they are feeling. This would be consistent with the concept of basic emotions thought to have unique and distinct representations. At this time we do not know which of these interpretations is correct. As we will now see, there are also alternative theories for the nature of the brain's representation of emotion.

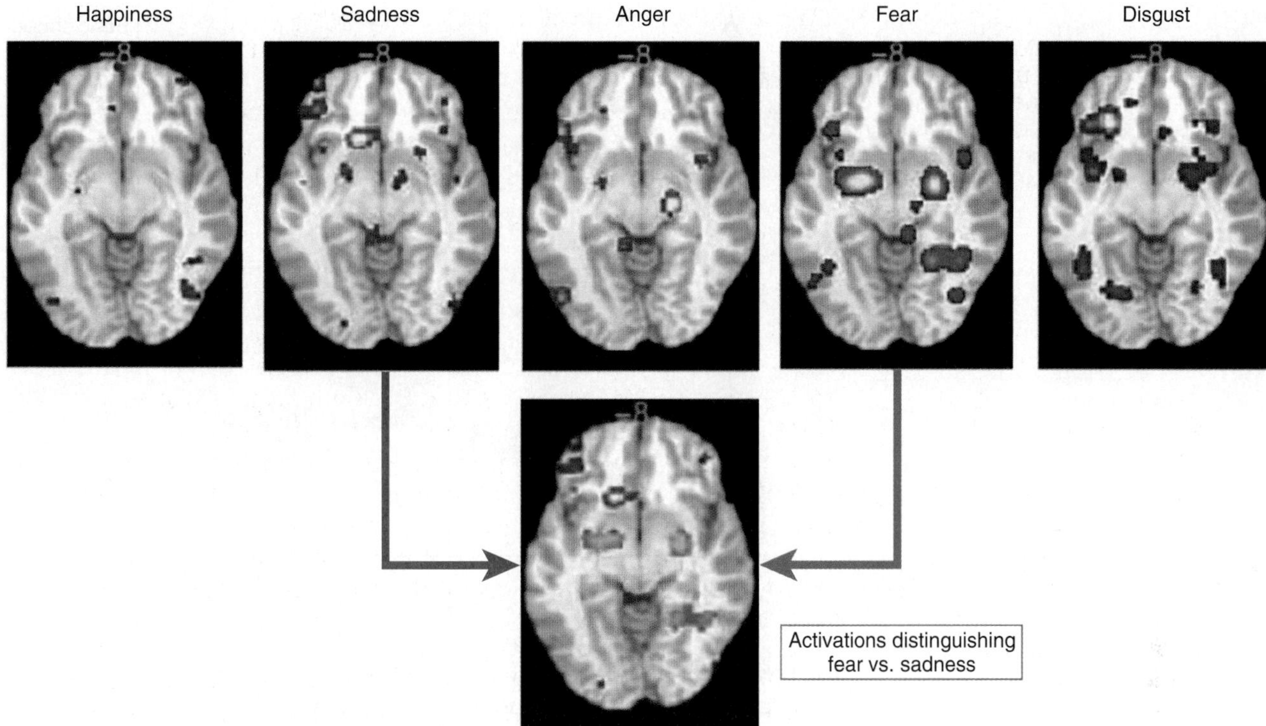

▲ FIGURE 18.6
Brain activation associated with five basic emotions. For each emotion, the strength of brain activation is indicated by color (yellow greater than red). The lower brain image compares activations associated with sadness (red and yellow = greater sadness activity) and fear (blue = greater fear activity). (Source: Hamann, 2012, p. 460.)

Dimensional Emotion Theories

There is an intuitive appeal to the idea that each basic emotion we experience is based on brain activity in a specialized brain area or network of areas; how convenient that would be for us scientists! Unfortunately, we have learned enough about the brain to know it doesn't always do what we find intuitive. An interesting analogy is the coding of body movement. The firing rate of a neuron in the motor cortex might code something rather straightforward such as the contraction properties of a single muscle (e.g., length, force). However, there is evidence that neural activity might represent something more complex, such as the input to a spectrum of muscles that makes up a portion of a complex behavior (e.g., swinging a golf club, dancing a pirouette).

An alternative to basic theories are **dimensional theories of emotion**. These theories are based on the idea that emotions, even basic emotions, can be broken down into smaller fundamental elements combined in different ways and differing amounts, just as all the elements of the periodic table are made of protons, neutrons, and electrons. Examples of proposed affective dimensions are valence ("pleasant–unpleasant") and arousal ("weak emotion–strong emotion"). Imagine a two-dimensional graph with axes labeled in these ways; each emotional experience would be located in a different part of the graph (Figure 18.7). Of course, for any particular emotion, such as happiness, there would be some normal range along a dimension such as emotional strength (arousal). In different theories there are different numbers of dimensions, sometimes with different names. Look again at Figure 18.6, in which we first considered

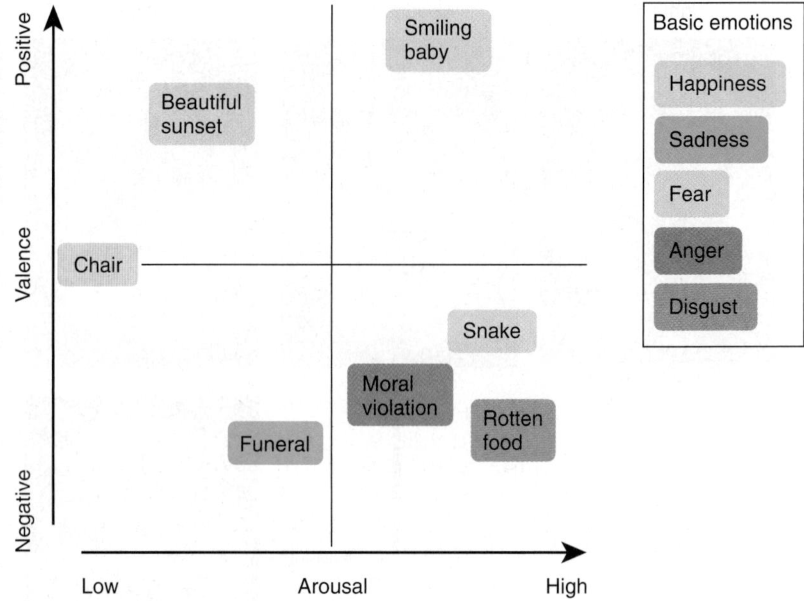

▶ FIGURE 18.7
A dimensional representation of basic emotions. In a dimensional theory, emotions such as happiness and sadness consist of differing amounts of brain activation corresponding to affective dimensions such as valence and arousal. (Source: Hamann, 2012, p. 461.)

the patches of activity in each brain slice to be, as a group, a possible representation of one basic emotion. Might the patches instead be distinct subpatterns, one subpattern associated with the degree of pleasantness, another with emotional strength, and perhaps still others with additional dimensions? The answer to this question is, at present, unclear.

Psychological constructionist theories of emotion are a variation on dimensional theories. These theories are similar to dimensional theories in the sense that emotions are said to consist of smaller building blocks. A key difference is that in the constructionist models, the dimensions do not carry affective weight. Instead of dimensions such as pleasantness, an emotional state is constructed from physiological processes that, on their own, do not concern only emotion. Examples of nonemotional psychological components that construct emotion are things like language, attention, internal sensations from the body, and external sensations from the environment. The emotion is an emergent consequence of the combination of these components, just as a cake results from the combination of ingredients in a recipe.

What is an Emotion?

From before the time of Darwin, there has been speculation about the nature of human emotions. Some researchers contend that a small set of basic emotions has evolved and these emotions are common to humans around the globe and animals as well. Other affective neuroscientists believe that emotions are constructed from building blocks that do or do not have emotional weight themselves. At present, there is a great diversity of perspectives on the nature of emotions, going beyond what we have discussed. One of the leaders in this field is Antonio Damasio at the University of Southern California, who has investigated the nature of emotions, the distinction between emotions and feelings, and the relationship between emotion and other brain functions such as decision making (Box 18.3). Aside from the nature of emotions, a related issue is the neural basis of emotions: Is each emotion represented by activity in a specialized area of the brain, a network of areas, or a more diffuse network of neurons? We do not yet have answers to these questions. Our hope for answering them relies on a convergence of approaches including

BOX 18.3 PATH OF DISCOVERY

Concepts and Names in Everyday Science

by Antonio Damasio

It would appear that the clarity of a concept or of a scientific hypothesis is the trait that counts most in determining the acceptance and impact of an idea. But not so fast: The name given to the concept or hypothesis plays a role in how they succeed, or not. Three examples from my own work illustrate the point.

First: For the past 20 years I have been insisting on a principled distinction between the concepts of emotion and feeling.[1] Emotions are programs of actions that rapidly modify the state of several components of our bodies in response, for instance, to a threat or an opportunity. Quite differently, feelings are the mental experiences of body states, including, of course, those that are caused by emotions. That the two sets of phenomena are distinct is quite clear, and yet the general public, not to mention several scientists, have persisted in lumping them together as if they were one and the same. Worse, when people do make the distinction, they often call the phenomenon by the wrong name (e.g., referring to feeling when they mean emotion or vice versa). Why so much confusion? Surely it cannot be mere carelessness. Well, it so happens that, given the long-standing historical conflation, no distinct words have evolved for the emotion or the feeling of a specific affective state. When I use the word "fear," I could be referring to the actual emotion fear or to the feeling that results from deploying the said emotion. And even worse: One of my intellectual heroes, William James, who is responsible for first sketching a credible physiology of emotion and how it may lead to the feeling experience, is guilty of confusing the two within the very paragraph in which he so well articulated the distinction! One lesson: One should have different and unambiguous terms to designate different phenomena.

Second: Unambiguous naming is only part of what is needed for the success of new ideas. The more transparent one can be about what one means, the more likely it is that people will retain a clear message. About the same time, I began insisting on the emotion–feeling distinction, I also advanced a hypothesis regarding how affect—emotions and feelings, conscious or not—intervene, for better or for worse, in the decision-making process, and, importantly, how they need to be factored in the decision process alongside knowledge and cool reason. I called this the *somatic marker hypothesis.*[2] Why the term "somatic"? Because emotions alter the state of the body, the soma, and feelings originate in that same body, the soma. And why the term "marker"? Because the affective state of the body, by virtue of its natural valence, *marks* a certain option as good, bad, or indifferent. Well, the designation caught. People do refer to it and usually get the gist of the idea from the name. It found a niche.

Third: I had no such luck when I used the terms "convergence" and "divergence" to describe, quite accurately, a connectional neural architecture with two distinct features: (a) neurons project, hierarchically, from a primary sensory cortex to smaller and smaller cortical association fields, thus converging into a narrower brain territory; and (b) other neurons reciprocate the favor in the opposite direction, thus diverging from "convergence–divergence zones" toward the originating points.[3] The reality of this anatomical arrangement in the human brain is unquestionable and is quite evident in the cerebral cortex, for example. The importance of the arrangement to help explain how memory works, in terms of learning and recall, is also high. The correctness of the terms "convergence" and "divergence" is not in question either. And yet, these names did not catch and that did hurt the diffusion of my idea. About the same time, however, the terms "hub" and "spoke" began to be used to designate the same general architecture. Rather than focusing on the actual direction of the neural messages, or the functional role that each direction of the projections played, *hub* and *spoke* succinctly described the resulting flow diagram. Amusingly, after deregulation, U.S. airlines stopped flying everywhere each way and instead operated their flights to and from a few major urban hubs connected by spokes to smaller hubs, in smaller cities, by what else, spokes! Advertisements used "hub and spokes" to refer successfully to the airlines' route system. Guess what: *hub* and *spoke* stuck for the neural architecture as well. The word "hub," in particular, captured in three letters what I described as "convergence–divergence zones and regions."

What's in a name? A lot. A rose by another name is still a rose but it may not smell as well. My prize for the catchiest term to convey a scientific idea goes to mirror neurons. Ironically, mirror neurons depend on a convergence–divergence neuronal architecture and operate in a hub and spoke network![4]

References

1. Damasio AR. 1994. *Descartes' Error*. New York: Penguin Books.
2. Damasio A, Carvalho GB. 2013. The nature of feelings: evolutionary and neurobiological origins. *Nature Reviews Neuroscience* 14: 143–152.
3. Damasio AR. 1996. The somatic marker hypothesis and the possible functions of the prefrontal cortex. *Transactions of the Royal Society* (London) 351:1413–1420.
4. Damasio AR. 1989. Time-locked multiregional retroactivation: a systems level proposal for the neural substrates of recall and recognition. *Cognition* 33:25–62.
5. Meyer K, Damasio A. 2009. Convergence and divergence in a neural architecture for recognition and memory. *Trends in Neurosciences* 32(7):376–382.

behavioral observations, physiological recordings, and studies of the effects of lesions and disease, among others. In the next section, we focus on two emotions, fear and anger/aggression. We could have chosen other emotions just as well, but the research on fear and anger provides good examples that nicely bring together studies in humans and experimental animals.

FEAR AND THE AMYGDALA

As we've seen, there remains considerable uncertainty about the brain's representation of emotions. Human brain imaging gives us pictures of brain activity associated with different emotions, but these pictures cannot tell us how or which brain areas actually contribute to an emotion's experience or expression. That said, there is one brain structure that, more than any other, has a reputation as being critical for emotion: the amygdala. It has been claimed that the amygdala plays a special role in fear. While we explore the evidence connecting the amygdala to fear, keep in mind that other brain structures also appear to be involved in fear and that the amygdala is active also in other emotional states.

The Klüver–Bucy Syndrome

Shortly after Papez's proposal of an emotion circuit in the brain, neuroscientists Heinrich Klüver and Paul Bucy, at the University of Chicago, found that bilateral removal of the temporal lobes, or temporal *lobectomy*, in rhesus monkeys has a dramatic effect on the animals' aggressive tendencies and responses to fearful situations. The surgery produces numerous bizarre behavioral abnormalities collectively referred to as the **Klüver–Bucy syndrome**.

After temporal lobectomy, the monkeys appeared to have good visual perception but poor visual recognition. Placed in a new environment, the monkeys moved about exploring objects they saw. However, unlike normal animals they appeared to rely on placing objects into their mouth to identify them. If a hungry monkey was shown a group of objects it had seen before intermixed with food, the monkey would still go through the process of picking up each object for study before consuming the food bits. A normal hungry monkey in the same situation would make a beeline for the food. The monkeys also showed a markedly increased interest in sex.

The emotional changes in monkeys with Klüver–Bucy syndrome were most dramatically represented by decreases in fear and aggression. For example, a normal wild monkey will avoid humans and other animals. In the presence of an experimenter, it usually crouches in a corner and remains still; if approached, it will dash off to a safer corner or make an aggressive stand. These behaviors were not exhibited by the monkeys with bilateral temporal lobectomies. These otherwise wild monkeys would not only approach and touch the human but would even let the human stroke them and pick them up. They had the same placid demeanor in the presence of other animals that monkeys normally fear. Even after approaching and being attacked by a natural enemy such as a snake, the monkey would go back and try to examine it again. There was also a corresponding decrease in the vocalizations and facial expressions usually associated with fear. It appeared that both the normal experience and the normal expression of fear and aggression were severely decreased by the temporal lobectomy.

Virtually all the symptoms of the Klüver–Bucy syndrome reported in monkeys have also been seen in humans with temporal lobe lesions and, more specifically, amygdala lesions. In addition to visual recognition problems, oral tendencies, and hypersexuality, these people appear to have "flattened" emotions.

Anatomy of the Amygdala

The **amygdala** is situated in the pole of the temporal lobe, just below the cortex on the medial side. Its name is derived from the Greek word for "almond" because of its shape.

The human amygdala is a complex of nuclei that are commonly divided into three groups: the *basolateral nuclei*, the *corticomedial nuclei*, and the *central nucleus* (Figure 18.8). Afferents to the amygdala come from a

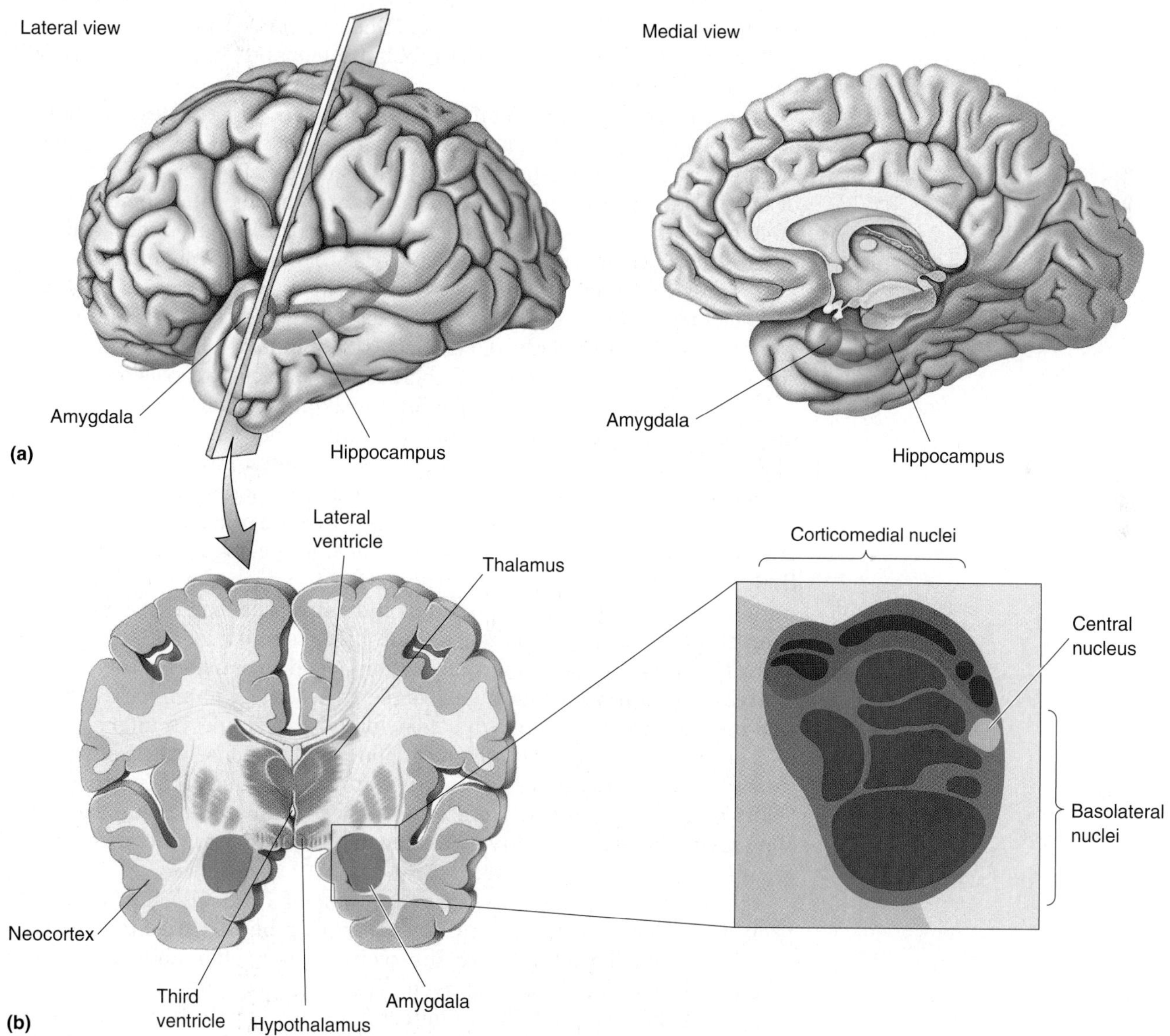

▲ FIGURE 18.8
A cross section of the amygdala. **(a)** Lateral and medial views of the temporal lobe, showing the location of the amygdala in relation to the hippocampus. **(b)** The brain is sectioned coronally to show the amygdala in cross section. The basolateral nuclei (surrounded by red) receive visual, auditory, gustatory, and tactile afferents. The corticomedial nuclei (surrounded by purple) receive olfactory afferents.

large variety of sources, including the neocortex in all lobes of the brain, as well as the hippocampal and cingulate gyri. Of particular interest here is the fact that information from all the sensory systems feeds into the amygdala, particularly the basolateral nuclei. Each sensory system has a different projection pattern to the amygdala nuclei, and interconnections within the amygdala allow the integration of information from different sensory modalities. Two major pathways connect the amygdala with the hypothalamus: the *ventral amygdalofugal pathway* and the *stria terminalis*.

Effects of Amygdala Stimulation and Lesions

Researchers have demonstrated in several species that lesions of the amygdala have the effect of flattening emotion in a manner similar to the Klüver–Bucy syndrome. Bilateral amygdalectomy in animals can profoundly reduce fear and aggression. Reports claim that rats so treated will approach a sedated cat and nibble its ear and that a wild lynx will become as docile as a house cat.

Numerous studies in humans have examined the effect of lesions that include the amygdala on the ability to recognize emotional facial expressions. While there is a consensus that the lesions usually impair the recognition of emotional expression, researchers disagree about which emotions are affected. In different studies, deficits associated with fear, anger, sadness, and disgust have been reported. The variety of deficits probably reflects, in part, differences in the damage; two lesions are rarely alike, and they typically include damage to other brain structures in addition to the amygdala. Nonetheless, the most commonly reported symptom of lesions involving the amygdala is an inability to recognize fear in facial expressions.

Very few cases of humans with bilateral damage isolated to the amygdala have been documented. However, Ralph Adolphs, Antonio Damasio and their colleagues, then at the University of Iowa, studied a 30-year-old woman known as S.M., who had bilateral destruction of the amygdala resulting from Urbach–Wiethe disease, a rare disorder characterized by thickening of the skin, mucus membranes, and certain internal organs. S.M. was somewhat unusual in the extent to which she was indiscriminately friendly and trusting, perhaps indicating she experienced less fear than other people. She had normal intelligence and was perfectly able to identify people from photographs. When asked to categorize the emotion expressed in a person's face, she correctly described happiness, sadness, and disgust. She was somewhat less likely to describe an angry expression as angry, however, and the most abnormal response was that she was much less likely to describe a fearful expression as afraid. Interestingly, S.M. could recognize fear from a person's tone of voice. It appears that the amygdala lesion selectively decreased her ability to recognize fear in faces from visual input alone.

Ten years after the initial examination of S.M., a follow-up study probed her deficit in more detail by comparing her abilities to recognize happiness and fear. In the 10 intervening years, her ability to recognize fear in faces had not improved. The fascinating finding of the later study was that her inability to detect fear and some other emotions resulted from her not looking at the eyes of people in the test photographs. Evidently because she consistently looked at their mouths, she was able to recognize happiness. By comparison, control subjects routinely spent a high percentage of time looking at the eyes as they explored faces. S.M.'s exploratory eye movements were unusual in not fixating on the eyes of the people in the photographs. When she was explicitly instructed to look at a person's eyes, she did so and then was able to correctly recognize fear. Surprisingly, after the more recent tests, she reverted to abnormal eye movements and poor fear recognition. To explain this curious set of results, the scientists hypothesized that fear is normally recognized by a two-way interaction between the amygdala and the visual cortex. Visual information is delivered to the amygdala, which then instructs the visual system to move the eyes and examine the visual input to determine the emotional expression in a face. Without the amygdala, this interaction does not take place, and S.M.'s abnormal eye movements did not allow her to recognize fear.

If removing the amygdala reduces the expression and recognition of fear, what happens if the intact amygdala is electrically stimulated? Depending on the site, amygdala stimulation can lead to different effects, including a state of increased vigilance or attention. Stimulation of the lateral portion of the amygdala in cats can elicit a combination of fear and violent aggression. Electrical stimulation of the amygdala in humans has been reported to lead to anxiety and fear. Not surprisingly, the amygdala figures prominently in current theories about anxiety disorders, as we will see in Chapter 22.

Functional brain imaging demonstrates that neural activity in the amygdala is consistent with its role in fear, as seen in Figure 18.6. For example, in an experiment performed by Breiter et al., subjects were positioned in an fMRI machine and brain activity was monitored as they viewed pictures of neutral, happy, and fearful faces (Figure 18.9a). Brain activity in response to fearful faces showed more amygdala activity than in response to faces with neutral expressions (Figure 18.9b). The amygdala activation was specific to fear, as no difference in activity occurred in response to happy and neutral facial expressions (Figure 18.9c). Other studies have reported amygdala activation in response to other facial expressions including happiness, sadness, and anger. The function the amygdala plays in these various emotions is not yet resolved, but all the evidence together suggests that the amygdala plays a key role in detecting fearful and threatening stimuli.

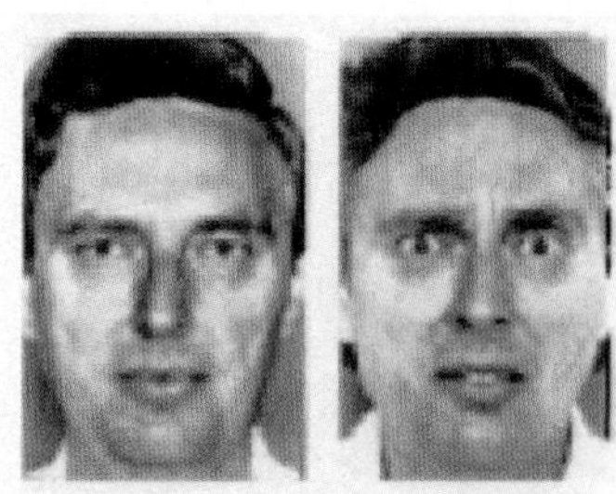

(a)

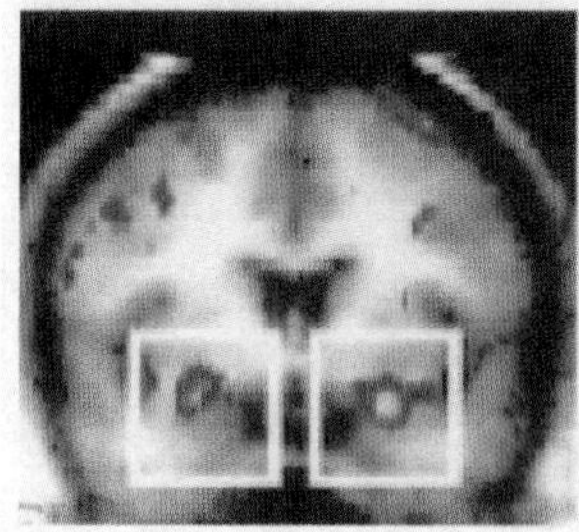

(b)

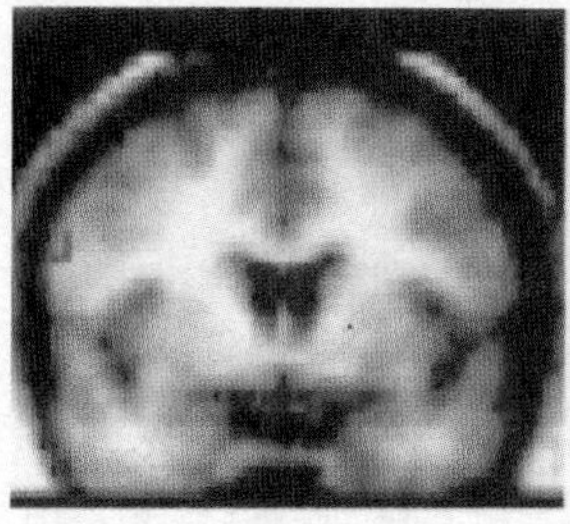

(c)

▲ FIGURE 18.9
Human brain activity in response to emotional stimuli. (a) Neutral and fearful faces were used as visual stimuli. (b) Fearful faces produced greater activity in the amygdala (red and yellow areas within white squares) than neutral faces. (c) No difference in amygdala activity occurred in response to happy and neutral faces. (Source: Breiter et al., 1996.)

A Neural Circuit for Learned Fear

Experiments in animals and humans, as well as introspection, indicate that memories for emotional events are particularly vivid and long-lasting. This is undoubtedly true for **learned fear**. Through socialization or painful experience, we all learn to avoid certain behaviors for fear of being hurt. If you ever received a painful shock as a child by pushing a paper clip into an electrical outlet, you probably never did it again. Memories associated with fear can form quickly and be long-lasting. As we will see in Chapter 22, in post-traumatic stress disorder, intense fear resulting from a traumatic experience can interfere with normal life for many years. Although the amygdala is not thought to be a primary location for memory storage, synaptic changes in the amygdala appear to be involved in forming memories for emotional events.

A number of different experiments suggest that neurons in the amygdala can "learn" to respond to stimuli associated with pain, and after such learning, these stimuli evoke a fearful response. In an experiment performed by Bruce Kapp and his colleagues at the University of Vermont, rabbits were conditioned to associate the sound of a tone with mild pain. A normal sign of fear in rabbits is a change in heart rate. In the experiment an animal was placed in a cage, and at various times it would hear one of two tones. One tone was followed by a mild electrical shock to the feet through the metal floor of the cage; the other tone was benign. After conditioning, Kapp's group found that the rabbit's heart rate developed a fearful response to the tone associated with pain but not to the benign tone. Before conditioning, neurons in the central nucleus of the amygdala had not responded to the tones used in the experiment. After conditioning, however, neurons in the central nucleus of the amygdala responded to the shock-related tone (but not the benign one). Joseph LeDoux of New York University has shown that after this type of fear conditioning, amygdala lesions eliminate the learned visceral responses, such as the changes in heart rate and blood pressure. It appears that the conditioned response in the amygdala arises from synaptic changes in the basolateral nuclei.

Figure 18.10 shows a proposed circuit to account for learned fear. Sensory information, for example, the tone the animal heard and the electrical shock it felt, is sent to the basolateral region of the amygdala, where cells in turn send axons to the central nucleus. The pairing of a benign tone with a painful stimulus leads to changes in synaptic strength that enhance the amygdala's response to the tone after conditioning (Chapters 24 and 25 discuss the neural changes that occur with conditioning). Efferents from the central nucleus project to the hypothalamus, which can alter the state of the ANS, and to the periaqueductal gray matter in the brain stem, which can evoke behavioral reactions via the somatic motor system. The emotional experience is thought to be based on activity in the cerebral cortex.

Recent research suggests that the role of the amygdala in learned fear, first studied in rabbits and rats, extends to humans. In one study, subjects were shown a number of visual stimuli and conditioned to expect a

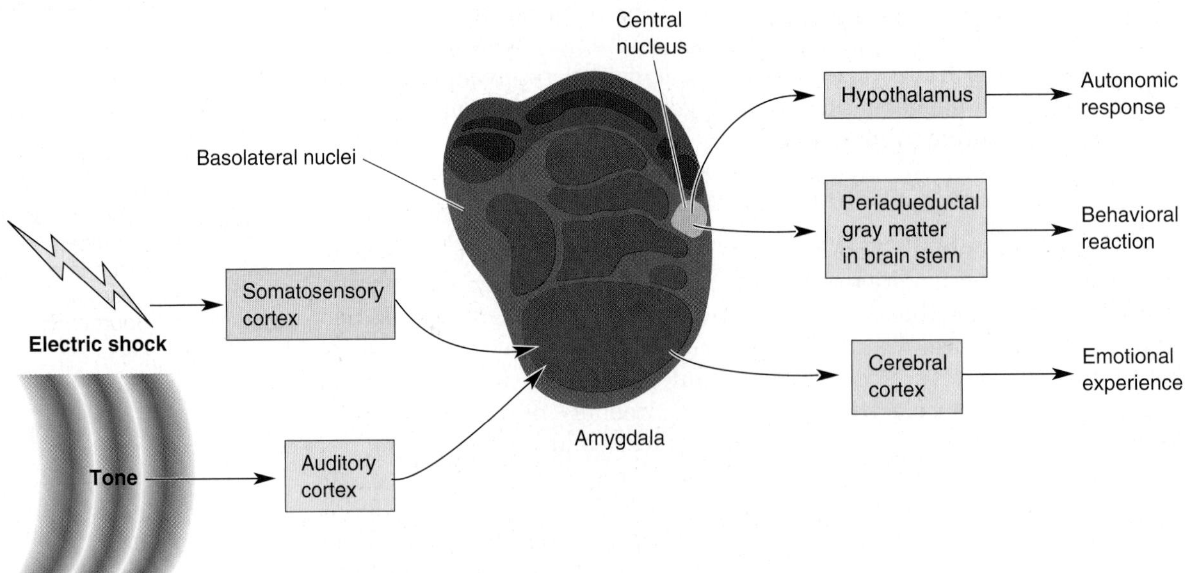

▲ FIGURE 18.10

A neural circuit for learned fear. Through conditioning, a sound tone becomes associated with the pain of an electrical shock. The fear response is mediated by the amygdala. The benign tone and the painful shock reach the basolateral nuclei of the amygdala by way of auditory and somatosensory cortex, and the signal is relayed to the central nucleus. The pairing of these stimuli leads to synaptic change in the amygdala and an enhanced response to the benign tone. Efferents from the amygdala project to the brain stem periaqueductal gray matter, causing the behavioral reaction to the conditioned tone, and to the hypothalamus, resulting in the autonomic response. The experience of an unpleasant emotion presumably involves projections to the cerebral cortex.

mild electrical shock when a particular visual stimulus was presented. An fMRI machine monitored brain activity. The fMRI images show that the "feared" visual stimulus activated the amygdala significantly more than visual stimuli not associated with a shock.

In another study using PET imaging of brain activity conducted by Hamann et al., subjects first viewed a series of pictures. Some of the pictures were pleasant (appealing animals, sexually arousing scenes, appetizing food); some were frightening or aversive (frightening animals, mutilated bodies, violence); and some were neutral (household scenes, plants). Compared to the neutral objects, both the pleasant and unpleasant stimuli affected physiological measures such as heart rate and skin conductance, and they evoked greater activity in the amygdala. These measurements confirm the role of the amygdala in emotional processing, as we have already discussed. In the second phase of the experiment, subjects were put back in the PET imaging machine and shown a variety of pictures. They were asked to use their memory and identify which pictures they had seen in the initial conditioning session. As expected, subjects recalled the emotional pictures better than the neutral ones. The enhanced memories for emotional pictures correlated with recorded amygdala activity (Figure 18.11). There was no such correlation for the neutral pictures.

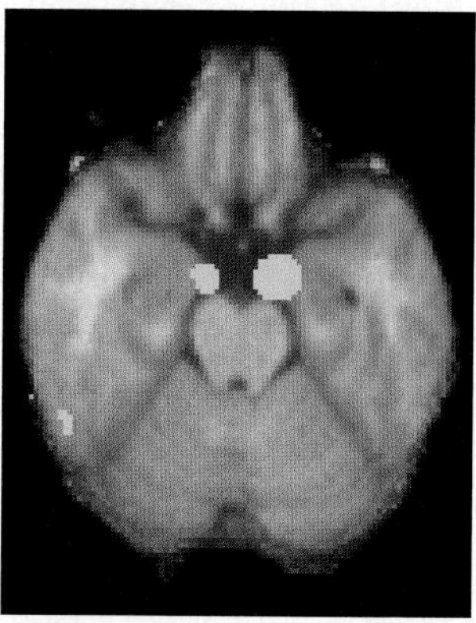

▲ FIGURE 18.11
Amygdala activity associated with enhanced emotional memory. Subjects first viewed pictures of emotional and neutral stimuli while PET imaging recorded brain activity. Later the original and new pictures were viewed. The recall of emotional stimuli was associated with an enhanced response in the amygdala, shown in yellow. (Source: Hamann et al., 1999.)

ANGER AND AGGRESSION

Anger is a basic emotion. Many things can make us angry: frustration, hurt feelings, stress, and so on. Aggression is not an emotion but is one possible behavioral result of anger; an angry drunk might punch someone in the nose. In studies of humans, aggression and the feeling we call anger can be readily distinguished, since people can state that they are angry even if they do not act on that feeling. As we have already seen, emotions are more difficult to study in animals because we cannot ask an animal how it feels but can only measure its physiological or behavioral manifestations. We can infer that an animal is angry only by the aggressive behaviors it exhibits, such as making a loud scary sound, a menacing facial expression, or a threatening posture. Because aggression and anger are often intertwined in animals, we will discuss them together here.

The Amygdala and Aggression

We can distinguish different forms of aggression in humans, ranging from self-defense to murder. Likewise, there are different types of aggression in animals. One animal may act aggressively toward another for many reasons: to kill for food, to defend offspring, to win a mate, or to scare off a potential adversary. There is some evidence that different types of aggression are regulated differently by the nervous system.

Aggression is a multifaceted behavior that is not a product of a single isolated system in the brain. One factor that influences aggression is the level of male sex hormones, or androgens (see Chapter 17). In animals, seasonal androgen levels and aggressive behavior correlate. Consistent with one of the roles of androgens, injections of testosterone can make an immature animal more aggressive, and castration can reduce aggressiveness. In humans, the relationship is less clear, although some have claimed that aggressive behavior in violent criminals is connected to testosterone levels. You may have heard of "roid rage," an uncontrolled outburst of anger and aggression sometimes reported in athletes taking anabolic steroids, which have similar effects on the body as testosterone. In any case, there is strong evidence for a neurobiological component to aggression, which is our focus here.

A useful distinction can be made between predatory aggression and affective aggression. **Predatory aggression** involves attacks against a member of a different species for the purpose of obtaining food, such as a lion hunting a zebra. Attacks of this type are typically accompanied by relatively few vocalizations, and they are aimed at the head and neck of the prey. Predatory aggression is not associated with high levels of activity in the sympathetic division of the ANS. **Affective aggression** is for show rather than to kill for food, and it involves high levels of sympathetic activity. An animal exhibiting affective aggression typically makes vocalizations while adopting a threatening or defensive posture. A cat hissing and arching its back at the approach of a dog is a good example. The behavioral and physiological manifestations of both types of aggression must be mediated by the somatic motor system and the ANS, but the pathways must diverge at some point to account for the dramatic differences in the behavioral responses.

Several lines of evidence indicate that the amygdala is involved in aggressive behavior. American scientist Karl Pribram and his colleagues in 1954 showed that amygdala lesions had a major effect on social interactions in a colony of eight male rhesus monkeys. Having lived together for some time, the animals had established a social hierarchy. The investigators' first intervention was to make bilateral amygdala lesions in the brain of the most dominant monkey. After this animal returned to the colony, it fell to the bottom of the hierarchy, and the monkey previously second in dominance now became dominant. Presumably, the second monkey in the hierarchy discovered that the "top banana" had become more placid and less difficult to challenge. After an amygdalectomy was performed on the new dominant monkey, it likewise fell to the bottom of the hierarchy. This pattern suggested that the amygdala is important for the aggression normally involved in maintaining a position in the social hierarchy. This is consistent with the finding that electrical stimulation of the amygdala can produce a state of agitation or affective aggression.

Surgery to Reduce Human Aggression. In the 1960s, amygdala surgery in violent humans was first performed in the hope that lesions would reduce aggression as they do in animals. It was thought by some that violent behavior frequently resulted from seizures in the temporal lobe. In a human amygdalectomy, electrodes are passed through the brain and down into the temporal lobe. By making neural recordings along the way and imaging the electrodes with X-rays, it is possible to position the tip of the electrode in the amygdala. Electrical current is then passed through the electrode, or a solution is injected, to destroy all or part of the amygdala. The lesions that are produced have a "taming effect" in some patients, reducing the incidence of aggressive outbursts. Brain surgery used as a method of treating psychiatric disorders is called **psychosurgery**. Early in the twentieth century, treating severe disorders involving anxiety, aggression, or neuroses with psychosurgical techniques, including the frontal lobotomy, was a common practice (Box 18.4). By today's standards, psychosurgery is a drastic procedure to be considered only as a treatment of last resort. Although amygdalectomies are still occasionally performed to treat aggressive behavior, medication is the usual treatment.

BOX 18.4 OF SPECIAL INTEREST

The Frontal Lobotomy

Ever since the discoveries by Klüver, Bucy, and others that brain lesions can alter emotional behavior, clinicians have attempted surgery as a means of treating severe behavioral disorders in humans. Today, it is difficult for many people to imagine that destroying a large portion of the brain was once thought to be therapeutic. Indeed, in 1949 the Nobel Prize in Medicine was awarded to Dr. Egas Moniz for his development of the frontal lobotomy technique. Even stranger is the fact that Moniz was shot in the spine and partially paralyzed by a lobotomized patient. Although lobotomies are no longer being performed, tens of thousands were performed following World War II.

Little theory supported the development of the lobotomy. In the 1930s, John Fulton and Carlyle Jacobsen of Yale University reported that frontal lobe lesions had a calming effect in chimpanzees. It has been suggested that frontal lesions have this effect because of the destruction of limbic structures and, in particular, connections with frontal and cingulate cortex. Moniz proposed that ablations of the frontal cortex might be effective in treating psychiatric diseases.

A frightening variety of techniques were used to produce lesions in the frontal lobes. The procedure became more commonplace with the development of a technique known as transorbital lobotomy (Figure A). In this procedure, a leucotome, a 12-cm steel rod that tapers to a point, was driven through the thin bone at the top of the eye's orbit with a hammer. The handle was then swung medially and laterally to destroy cells and interconnecting pathways. Thousands of people were lobotomized with this technique, sometimes called "ice pick psychosurgery"; it was so simple it could be performed in the physician's office. Note that although this surgery left no outward scars, the physician could not see what was being destroyed.

Frontal lobotomy reportedly had beneficial effects on people with a number of disorders, including psychosis, depression, and various neuroses. The effect of the surgery was described as a relief from anxiety and escape from thoughts that were unendurable. Only later did a pattern of less pleasant side effects emerge. While frontal lobotomy can be performed with little decrease in IQ or loss of memory, it did have other profound effects. The changes that appear to be related to the limbic system are a blunting of emotional responses and a loss of the emotional component of thoughts. In addition, lobotomized patients often developed "inappropriate behavior" or an apparent lowering of moral standards. Like Phineas Gage, patients had considerable difficulty planning and working toward goals. Lobotomized patients also had trouble concentrating and were easily distracted.

With our modest understanding of the neural circuitry underlying emotion and other brain functions, it is hard to justify destroying a large portion of the brain. Fortunately, treatment with lobotomy decreased fairly rapidly, and today, instead, drug therapy is primarily used for serious emotional disorders.

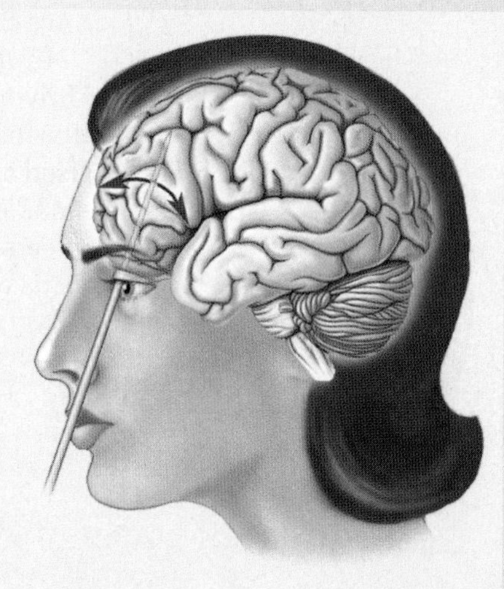

Figure A

Neural Components of Anger and Aggression Beyond the Amygdala

In addition to the amygdala, a variety of brain structures have been reported to be involved in anger and aggression. For example, human brain imaging studies have found that there is greater activity in orbitofrontal cortex and anterior cingulate cortex when subjects recall past experiences that made them angry. Interpreting these patterns of brain activation involves the same challenges we have discussed for other emotions. Historically, studies of anger and aggression have been important for their implications for the involvement of subcortical structures in emotion. We now look at a few of these important research milestones.

Anger, Aggression, and the Hypothalamus. One of the earliest structures linked to anger and aggressive behavior is the hypothalamus. Experiments performed in the 1920s showed that a remarkable behavioral transformation took place in cats or dogs whose cerebral hemispheres had been removed. Animals that were not easy to provoke prior to the surgery would go into a state of violent rage with the least provocation after the surgery. For instance, a violent response might be produced by an act as mild as scratching a dog's back. This state was called **sham rage** because the animal demonstrated all the behavioral manifestations of rage but in a situation that normally would not cause anger. It was also a sham in the sense that the animals would not actually attack as they normally might.

While the extreme behavioral condition called sham rage resulted from removing all of both cerebral hemispheres (the telencephalon), remarkably, the behavioral effect can be reversed by making the lesion just a little bit larger to include portions of the diencephalon, particularly the hypothalamus. Sham rage is observed if the anterior hypothalamus is destroyed along with the cortex but is not seen if the lesion is extended to include the posterior half of the hypothalamus (Figure 18.12). The implication is that the posterior hypothalamus may be particularly important for the expression of anger and aggression and that normally it is inhibited by the telencephalon. But we must bear in mind that these lesions were large, and something other than the posterior hypothalamus may have been destroyed with the larger lesion.

In a series of pioneering studies begun in the 1920s, W.R. Hess at the University of Zurich investigated the behavioral effects of electrically stimulating the diencephalon. Hess made small holes in the skulls of anesthetized cats and implanted electrodes in the brain. After the animal awoke, a small electrical current was passed through the electrodes, and behavioral effects were noted. Various structures were stimulated, but here we will focus on the effects of stimulating different regions of the hypothalamus.

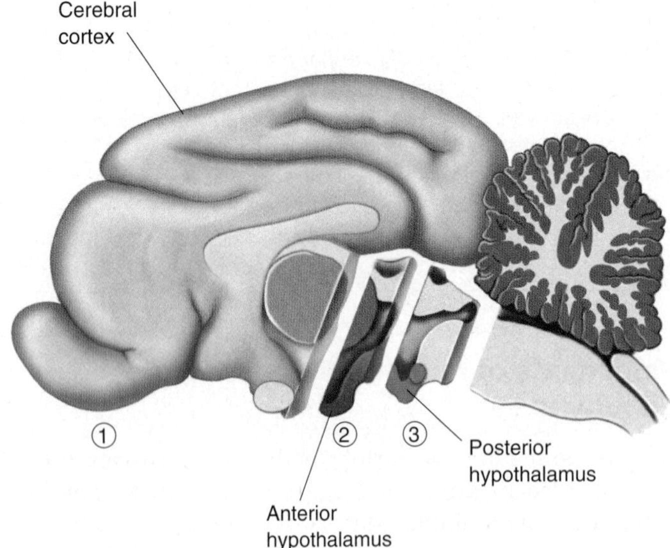

▲ FIGURE 18.12

Brain transections and sham rage. ① If the cerebral hemispheres are removed and the hypothalamus is left intact, sham rage results. ① and ② A similar result is obtained if the anterior hypothalamus is removed in addition to the cerebral cortex. ①, ②, and ③ If the posterior hypothalamus is removed in addition to the anterior hypothalamus, sham rage does not result.

The variety of complex responses to stimulating slightly different portions of the hypothalamus is amazing, considering that it is such a small part of the brain. Depending on where the electrode is placed, stimulation may cause the animal to sniff, pant, eat, or express behaviors characteristic of fear or anger. These reactions illustrate the two primary functions of the hypothalamus discussed in Chapters 15 and 16: homeostasis and the organization of coordinated visceral and somatic motor responses. Responses related to emotional expression can include changes in heart rate, pupillary dilation, and gastrointestinal motility, to name a few. Because stimulation of some parts of the hypothalamus also elicits behavior characteristic of fear and anger, we hypothesize that the hypothalamus is an important component of the system normally involved in expressing these emotions.

The expression of rage Hess evoked by hypothalamic stimulation was similar to the sham rage seen in animals whose cerebral hemispheres had been removed. With a small application of electrical current, a cat would spit, growl, and fold its ears back, and its hair would stand on end. This complex and highly coordinated set of behaviors would normally occur when the cat feels threatened by an enemy. Sometimes the cat would suddenly run as if fleeing an imaginary attacker. If the intensity of the stimulation was increased, the animal might make an actual attack, swatting with a paw or leaping onto an imaginary adversary. When the stimulation was stopped, the rage disappeared as quickly as it started, and the cat might even curl up and go to sleep.

In experiments conducted at the Yale University Medical School in the 1960s, John Flynn found that affective aggression and predatory aggression could be elicited by stimulating different areas of a cat's hypothalamus (Figure 18.13). Affective aggression, also known as a *threat attack*, was observed after stimulating specific sites in the medial hypothalamus. Similar to the rage response reported by Hess, the cat would arch its back, hiss, and spit but would usually not actually attack a victim, such as a nearby rat. Predatory aggression, which Flynn called a *silent-biting attack*, was evoked by stimulating parts of the lateral hypothalamus. While the back might be somewhat arched and the hair slightly on end, predatory aggression was not accompanied by the dramatic threatening gestures of affective aggression. Nonetheless, in this "quiet attack," the cat would move swiftly toward a rat and viciously bite its neck. Despite the crudeness of such experimentation by today's standards, the early research involving lesions and electrical stimulation of the hypothalamus are consistent in suggesting that this structure is important for the expression of anger and aggression in animals.

(a)

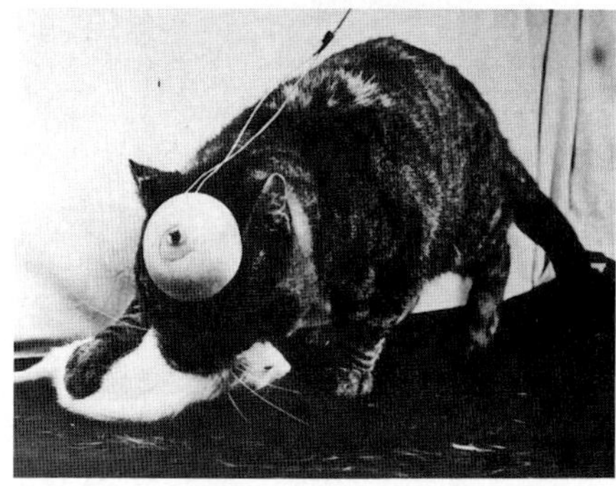

(b)

▲ FIGURE 18.13
Rage reactions in cats with hypothalamic stimulation. **(a)** Stimulation of the medial hypothalamus produces affective aggression (threat attack). **(b)** Stimulation of the lateral hypothalamus evokes predatory aggression (silent-biting attack). (Source: Flynn, 1967, p. 45.)

The Midbrain and Aggression. There are two major pathways by which the hypothalamus sends signals involving autonomic function to the brain stem: the **medial forebrain bundle** and the **dorsal longitudinal fasciculus**. Axons from the lateral hypothalamus make up part of the medial forebrain bundle, and these project to the *ventral tegmental area* in the midbrain. Stimulation of sites within the ventral tegmental area can elicit behaviors characteristic of predatory aggression, just as stimulation of the lateral hypothalamus does. Conversely, lesions in the ventral tegmental area can disrupt offensive aggressive behaviors. One finding suggesting that the hypothalamus influences aggressive behavior via its effect on the ventral tegmental area is that hypothalamic stimulation will not evoke aggression if the medial forebrain bundle is cut. Interestingly, aggressive behavior is not entirely eliminated by this surgery, suggesting that this route is important when the hypothalamus is involved, but that the hypothalamus need not always be involved.

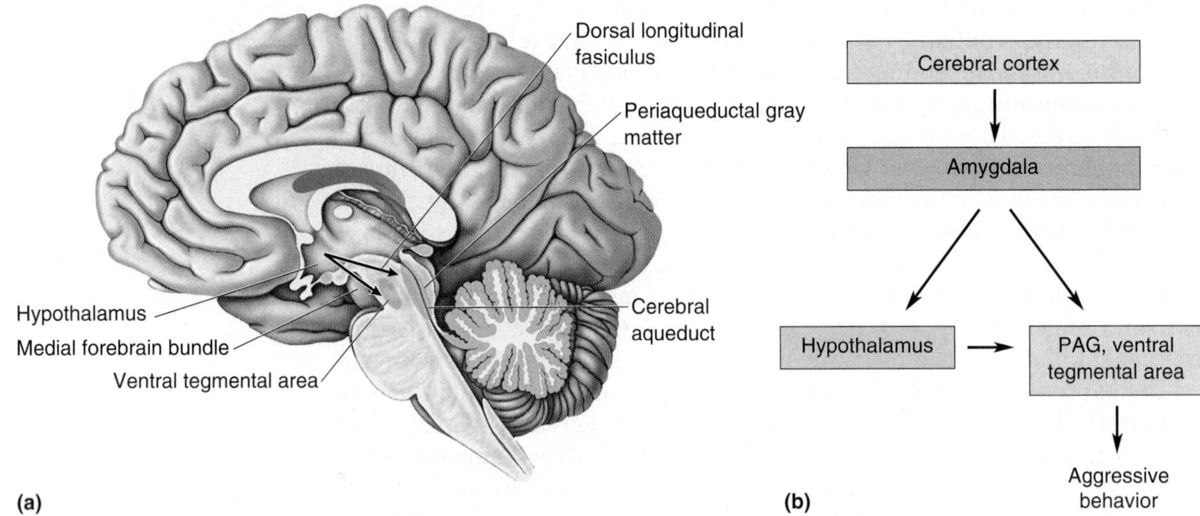

▲ FIGURE 18.14
A neural circuit for anger and aggression. **(a)** The hypothalamus can influence aggressive behavior through projections to the ventral tegmental area and the periaqueductal gray matter. **(b)** In this simplified scheme, the expression of anger and aggression is controlled by a neural pathway from the amygdala through the hypothalamus, periaqueductal gray matter (PAG), and ventral tegmental area.

The medial hypothalamus sends axons to the **periaqueductal gray matter (PAG)** of the midbrain by way of the dorsal longitudinal fasciculus. Electrical stimulation of the PAG can produce affective aggression, and lesions there can disrupt this behavior. Interestingly, the hypothalamus and the midbrain periaqueductal gray matter appear to influence behavior partially based on input from the amygdala. Figure 18.14 shows a simplified circuit for anger and aggression involving the structures we have discussed.

Serotonergic Regulation of Anger and Aggression

A variety of studies suggest that the neurotransmitter serotonin plays an important role regulating anger and aggression. Serotonin-containing neurons are located in the raphe nuclei of the brain stem, and they ascend in the medial forebrain bundle and project to the hypothalamus and various limbic structures involved in emotion (see Figure 15.13). For the most part, experimental evidence supports the **serotonin deficiency hypothesis**, which states that aggression is inversely related to serotonergic activity.

One link between serotonin and aggression comes from studies of induced aggression in rodents. If male mice are isolated in a small cage for several weeks, about half of them will become hyperactive and unusually aggressive when they subsequently encounter other mice. Although the isolation has no effect on the *level* of serotonin in the brain, there is a decrease in the *turnover rate* (the rate of synthesis, release, and resynthesis) of this neurotransmitter. Moreover, this decrease is found only in the mice that later become unusually aggressive and not in those relatively unaffected by the isolation. Also, female mice typically do not become aggressive following isolation, and they show no decrease in serotonin turnover. Evidence indicates that drugs that block the synthesis or release of serotonin increase aggressive behavior. For instance, in one study, when the drug parachlorophenylalanine (PCPA), which blocks serotonin synthesis, was administered, the injected animals increased their attacks on others in their cage.

There are at least 14 serotonin receptor subtypes, and it appears that the 5-HT_{1A} and 5-HT_{1B} subtypes are involved in modulating anger and aggression. For example, a number of experiments have shown in mice that agonists of 5-HT_{1B} receptors decrease aggressiveness while antagonists of these receptors increase aggressiveness. Based on these pharmacological results, one would predict that mice lacking the 5-HT_{1B} receptor would be more aggressive than normal animals. Consistent with this prediction, in some studies, 5-HT_{1B} receptor knockout mice are reported to show more aggressive behavior. Other experiments paint a somewhat different picture, however, suggesting that rather than simply being more aggressive, the knockout mice are more impulsive.

The relationship between serotonin and aggression is similar in primates that have been studied. For example, researchers found that the dominance hierarchy in a colony of vervet monkeys could be manipulated by injecting animals with drugs that either increased or decreased serotonergic activity. The behavior of these animals was consistent: More aggression was associated with less serotonergic activity. However, there was one interesting sociological twist; aggression did not correlate with dominance in the group. If the dominant male was removed, the top position was taken by an animal with artificially *enhanced* serotonergic activity (i.e., one injected with a serotonin precursor or reuptake inhibitor who was *less* aggressive). Conversely, the injection of drugs that reduced serotonin function (serotonin antagonists) was correlated with animals becoming subordinate. The subordinate animals were actually significantly more likely to initiate aggression. Interestingly, the less aggressive dominant male garnered his status by his skills in recruiting females to support his position.

In humans, there are a large number of reports of a negative correlation between serotonin activity and aggression. For example, in a study of men in the military who had been diagnosed with personality disorders, aggression was found to be inversely related to cerebrospinal fluid levels of the serotonin metabolite 5-hydroxyindoleacetic acid (5-HIAA). Questions have been raised, however, about the generality of the correlation between serotonin and aggression when people of different ages and people without personality disorders are examined. As with the animal studies, a correlation is often reported, but the reality is probably more complex.

Many scientists would agree that serotonin is involved in the modulation of anger and aggressive behavior. The evidence presented here suggests a straightforward negative correlation between aggression and activity in the brain's serotonergic system. However, some scientists in the field consider this relationship overly simplistic. Animals exhibit aggressive behaviors for a variety of reasons, and serotonin is not involved equally in all forms. From a mechanistic standpoint, the system is complex. Serotonergic neurons project broadly across the brain. 5-HT_{1A} and 5-HT_{1B} receptors are distributed widely, and they and other serotonin receptors provide interactions with other neurotransmitter systems. As well, there is negative feedback in the system because many of the 5-HT_{1A} and 5-HT_{1B} receptors are autoreceptors (see Chapter 5). Some autoreceptors are presynaptic on the raphe neurons that send serotonin widely to the brain. Activation of them globally inhibits serotonin release. With this negative feedback, serotonin release affects the raphe neurons in such a way as to decrease further release. Because of the diversity of receptor locations and functions, interpretation of pharmacological and knockout experiments is challenging; new approaches are needed to tease out the details of the relationships among serotonin, anger, and aggressive behavior.

CONCLUDING REMARKS

We all know what emotions are—those feelings we have that we call happiness, sadness, and so on. But what exactly are those feelings? As evidenced by the diverse theories we have outlined, there is a great deal of uncertainty. More than a hundred years after the James–Lange theory was proposed, there remains controversy about the extent to which emotions cause changes in the body or bodily changes cause emotions. We do know from brain imaging studies that emotions are associated with widespread brain activation. Some of the structures involved are part of the limbic system, and other structures are not. But even with images of brain activity in various emotional states, understanding the neural basis of emotional experience is challenging. We don't know which of the active areas are responsible for the feelings. Is it the most active area, all of the areas, or something else? What should we make of the observation that some brain structures are activated in multiple emotional states while others are more specific to particular emotions? For that matter, is it even correct to think of brain activity as reflecting feelings, or might feelings be emergent sensations based on combinations of active neurons, none of which independently signals an emotion?

In this chapter, we have focused on a handful of brain structures for which there is particularly strong evidence for involvement in emotion. A way to look at our current state of understanding is that the combined lesion, stimulation, and brain imaging studies have done a good job identifying structures that are candidates for emotional processing. It will take a good deal more work to figure out what various cortical and subcortical areas contribute.

Emotional experiences are the result of complex interactions among sensory stimuli, brain circuitry, past experiences, and the activity of neurotransmitter systems. In light of this complexity, we probably should not be surprised that humans exhibit a broad spectrum of emotional and mood disorders, as we will see in Chapter 22.

When thinking about the neural basis of emotion, keep in mind that the structures apparently involved in emotion also have other functions. For a considerable time after Broca defined the limbic lobe, it was thought to be primarily an olfactory system. And even though our perspective has changed much since Broca's time, parts of the brain involved in olfaction have been included in the definition of the limbic system. We will see in Chapter 24 that some of the limbic structures are also important for learning and memory. Emotions are nebulous experiences that influence our brains and behavior in many ways, so it seems logical that emotional processing should be intertwined with other brain functions.

KEY TERMS

Introduction
affective neuroscience (p. 616)

Early Theories of Emotion
James–Lange theory (p. 617)
Cannon–Bard theory (p. 617)
unconscious emotion (p. 619)

The Limbic System
limbic lobe (p. 622)
Papez circuit (p. 622)
limbic system (p. 623)

Emotion Theories and Neural Representations
basic theories of emotion (p. 626)
basic emotions (p. 626)
dimensional theories of emotion (p. 627)
psychological constructionist theories of emotion (p. 628)

Fear and the Amygdala
Klüver–Bucy syndrome (p. 630)
amygdala (p. 631)
learned fear (p. 633)

Anger and Aggression
predatory aggression (p. 636)
affective aggression (p. 636)
psychosurgery (p. 636)
sham rage (p. 638)
medial forebrain bundle (p. 639)
dorsal longitudinal fasciculus (p. 639)
periaqueductal gray matter (PAG) (p. 640)
serotonin deficiency hypothesis (p. 640)

REVIEW QUESTIONS

1. According to the James–Lange and Cannon–Bard theories of emotion, what is the relationship between the anxiety you would feel after oversleeping through an exam and your physical responses to the situation? Would you experience anxiety before or after the increase in your heart rate?
2. How have the definition of the limbic system and thoughts about its function changed since the time of Broca?
3. What procedures produce an abnormal rage reaction in an experimental animal? Can we know that the animal feels angry?
4. What changes in emotion were observed following temporal lobectomy in monkeys by Klüver and Bucy? Of the numerous anatomical structures they removed, which is thought to be closely related to changes in temperament?
5. Why might performing bilateral amygdalectomy on a dominant monkey in a colony result in that monkey becoming a subordinate?
6. What assumptions about limbic structures underlie the surgical treatment of emotional disorders?
7. The drug fluoxetine (Prozac) is a serotonin-selective reuptake inhibitor. How might this drug affect a person's level of anxiety and aggression?
8. What distinguishes basic emotion, dimensional, and psychological constructionist theories of emotion?
9. How do patterns of brain activation differ for sadness and fear?

FURTHER READING

Barrett LF, Satpute AB. 2013. Large-scale networks in affective and social neuroscience: towards an integrative functional architecture of the brain. *Current Opinion in Neurobiology* 23:361–372.

Dagleish T. 2004. The emotional brain. *Nature Reviews* 5:582–589.

Dolan RJ. 2002. Emotion, cognition, and behavior. *Science* 298:1191–1194.

Duke AA, Bell R, Begue L, Eisenlohr-Moul T. 2013. Revisiting the serotonin-aggression relation in humans: a meta-analysis. *Psychological Bulletin* 139:1148–1172.

Gendron M, Barrett LF. 2009. Reconstructing the past: a century of ideas about emotion in psychology. *Emotion Review* 1:316–339.

Gross CT, Canteras NS. 2012. The many paths to fear. *Nature Reviews Neuroscience* 13:651–658.

Hamann S. 2012. Mapping discrete and dimensional emotions onto the brain: controversies and consensus. *Trends in Cognitive Sciences* 16:458–466.

LeDoux J. 2012. Rethinking the emotional brain. *Neuron* 73:653–676.

Lindquist KA, Wager TD, Kober H, Bliss-Moreau E, Barrett LF. 2012. The brain basis of emotion: a meta-analytic review. *Behavioral and Brain Sciences* 35:121–143.

McGaugh JL. 2004 The amygdala modulates the consolidation of memories of emotionally arousing experiences. *Annual Review of Neuroscience* 27:1–28.

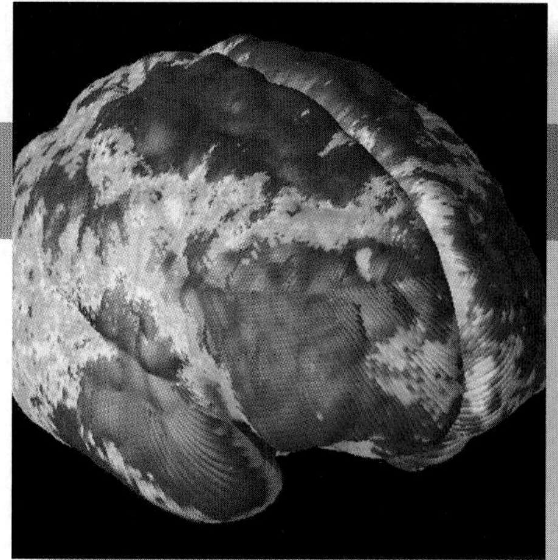

CHAPTER NINETEEN

Brain Rhythms and Sleep

INTRODUCTION

Earth has a rhythmic environment. Temperature, precipitation, and daylight vary with the seasons; light and dark trade places each day; tides ebb and flow. To compete effectively and survive, an animal's behavior must oscillate with the cadences of its environment. Brains have evolved a variety of systems for rhythmic control. Sleeping and waking are the most striking periodic behavior. But some rhythms controlled by the brain have much longer periods, as in hibernating animals, and many have shorter periods, such as the cycles of breathing, the steps of walking, the repetitive stages of one night's sleep, and the electrical rhythms of the cerebral cortex. The functions of some rhythms are obvious, while others are obscure, and some rhythms indicate pathology.

In this chapter, we explore selected brain rhythms, beginning with the fast and proceeding to the slow. The forebrain, especially the cerebral cortex, produces a range of rapid electrical rhythms that are easily measured and that closely correlate with interesting behaviors, including sleep. We discuss the electroencephalogram, or EEG, because it is the classical method of recording brain rhythms and is essential for studying sleep. Sleep is explored in detail because it is complex, ubiquitous, and so dear to our hearts. Finally, we summarize what is known about the timers that regulate the everyday ups and downs of our hormones, body temperature, alertness, and metabolism. Almost all physiological functions change according to daily cycles known as *circadian rhythms*. The clocks that time circadian rhythms are in the brain, calibrated by the sun via the visual system, and they profoundly influence our health and well-being.

THE ELECTROENCEPHALOGRAM

Sometimes the forest is more interesting than the trees. Similarly, we are often less concerned with the activities of single neurons than with understanding the activity of a large population of neurons. The **electroencephalogram (EEG)** is a measurement of electrical activity from the surface of the scalp that enables us to glimpse the generalized activity of the cerebral cortex. The roots of the EEG lie in work done by English physiologist Richard Caton in 1875. Caton made electrical recordings from the surface of dog and rabbit brains using a primitive device sensitive to voltage. The human EEG was first described by Austrian psychiatrist Hans Berger in 1929, who observed that waking and sleeping EEGs are distinctly different. Figure 19.1 shows one of his first published records, taken from the head of his 15-year-old son, Klaus. Today, the EEG is used primarily to help diagnose certain neurological conditions, especially the seizures of epilepsy, and for research purposes, notably to study the stages of sleep and cognitive processes during wakefulness.

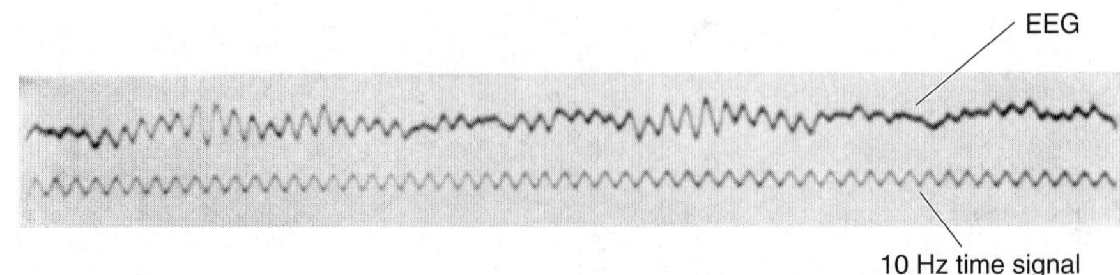

▶ FIGURE 19.1
The first published human EEG rhythm. (Source: Berger, 1929.)

Recording Brain Waves

Recording an EEG is relatively simple. The method is usually noninvasive and is painless. Countless people have slept through entire nights wearing EEG electrodes in the comfort of sleep research laboratories (Figure 19.2). The electrodes are wires taped to the scalp, along with conductive paste to ensure a low-resistance connection. Figure 19.3 shows a common EEG configuration, in which some two dozen electrodes are fixed to standard positions on the head and connected to banks of amplifiers and recording devices. Small voltage fluctuations, usually a few tens of microvolts (μV) in amplitude, are measured between selected pairs of electrodes. Different regions of the brain—anterior and posterior, left and right—can be examined by selecting the appropriate electrode pairs. The typical EEG record is a set of many simultaneous squiggles, indicating voltage changes between pairs of electrodes.

What part of the nervous system generates the fluctuations and oscillations of an EEG? For the most part, an EEG measures voltages generated by the currents that flow during synaptic excitation of the dendrites of many pyramidal neurons in the cerebral cortex, which lies right under the skull and makes up most of the brain's mass. But the electrical contribution of any single cortical neuron is exceedingly small, and the signal must penetrate several layers of non-neural tissue, including the meninges, fluid, bones of the skull, and skin, to reach the electrodes (Figure 19.4). Therefore, it takes many thousands of underlying neurons, activated together, to generate an EEG signal big enough to be measured at all.

This has an interesting consequence: The amplitude of the EEG signal strongly depends, in part, on how *synchronous* is the activity of the underlying neurons. When a group of cells is excited simultaneously, the tiny signals sum to generate one larger surface signal. However, when each cell receives the same amount of excitation but the excitations are spread out in time, the summed signals are meager and irregular (Figure 19.5). Notice that in this case, the *number* of activated cells and the *total amount of excitation* may not have changed, only the timing of the activity. If synchronous excitation of this group of cells is repeated again and again, the resulting EEG will consist of large, rhythmic waves. We often describe

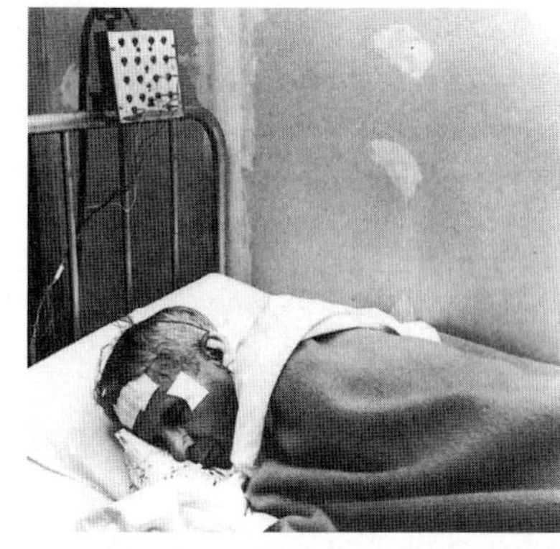

▲ FIGURE 19.2
A subject in a sleep research study. The subject shown here is American sleep researcher Nathaniel Kleitman, codiscoverer of REM sleep. The white patches on his head are pieces of tape holding EEG electrodes, and those next to his eyes hold electrodes that monitor his eye movements. (Source: Carskadon, 1993.)

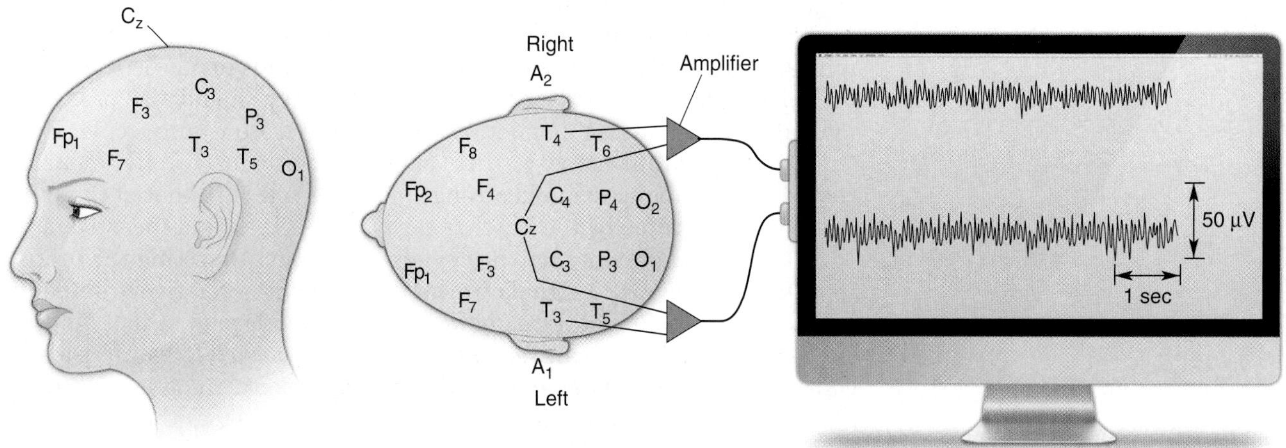

▲ FIGURE 19.3
Standard positions for the placement of EEG electrodes. A, auricle (or ear); C, central; Cz, vertex; F, frontal; Fp, frontal pole; O, occipital; P, parietal; T, temporal. Wires from pairs of electrodes are fed to amplifiers, and each recording measures voltage differences between two points on the scalp. The output of each amplifier is stored in a computer for analysis and display.

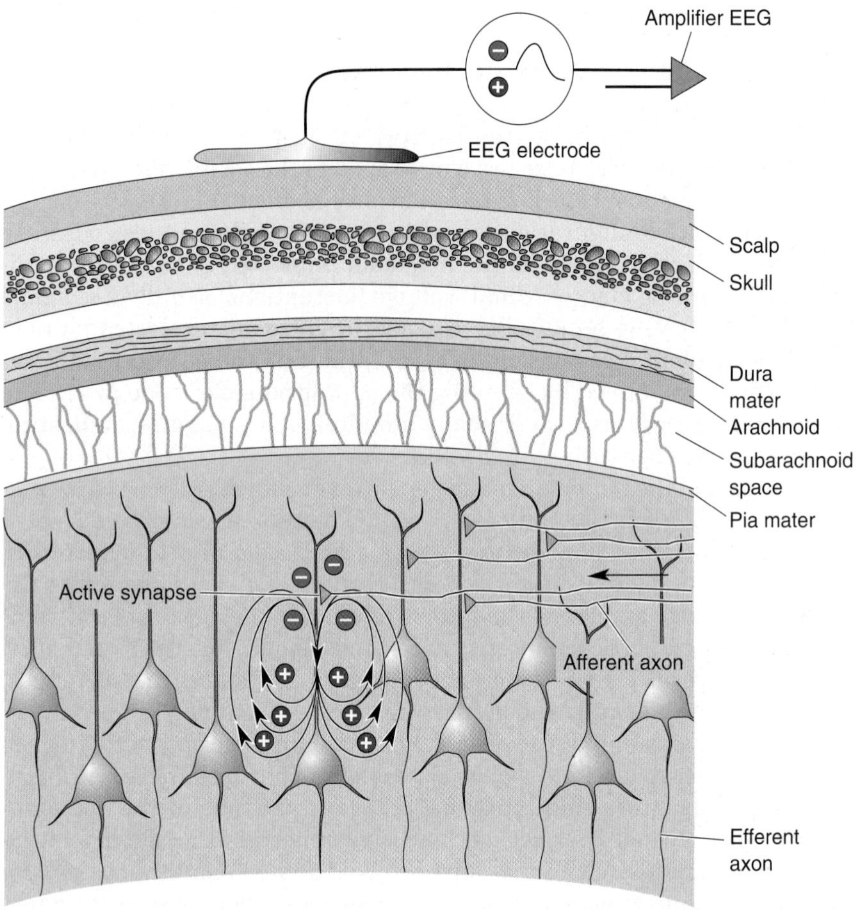

▲ FIGURE 19.4
The generation of very small electrical fields by synaptic currents in pyramidal cells. In this case, the active synapse is on the upper part of the dendrite. When the afferent axon fires, the presynaptic terminal releases glutamate, which opens cation channels. Positive current flows into the dendrite, leaving a slight negativity in the extracellular fluid. Current spreads down the dendrite and escapes from its deeper parts, leaving the fluid slightly positive at those sites. The EEG electrode (referred to a second electrode some distance away) measures this pattern through thick tissue layers. Only if thousands of cells contribute their small voltage is the signal large enough to reach the scalp surface. (Notice the EEG convention of plotting the signals with negativity upward.)

rhythmic EEG signals in terms of their relative amplitude, suggesting how synchronous the underlying activity is (although other factors, especially the number of active neurons, contribute to amplitude as well).

An alternative way to record the rhythms of the cerebral cortex is with **magnetoencephalography (MEG)**. Recall from physics that whenever electrical current flows, a magnetic field is generated according to the "right hand rule" (hold up your right hand loosely; if your thumb points in the direction of electrical current flow, the rest of your curling fingers indicate the direction of the magnetic field). It stands to reason that when neurons generate currents, as in Figure 19.4, they should also produce a magnetic field. But the magnetic field they generate is minuscule. Even the strongest brain activity, with many synchronously active neurons contributing, produces a field strength just one billionth that of the magnetic field generated by the Earth, nearby power lines, and the movement of distant metal objects such as elevators and cars. Detecting the brain's infinitesimal magnetic signals in the midst of those relatively immense sources of environmental magnetic "noise" is analogous to listening for the footsteps of a mouse in the middle of a rock concert! It requires a specially screened room to shield out the magnetic noise and a large, expensive instrument with highly sensitive magnetic detectors that are cooled with liquid helium to –269°C (Figure 19.6).

The capabilities of MEG complement those of other methods that measure brain function. MEG is much better than EEG at localizing the sources of neural activity in the brain, particularly those deep below the surface. Like EEG, MEG can record rapid fluctuations of neural activity that are much too fast to be detected by functional magnetic resonance imaging (fMRI) or

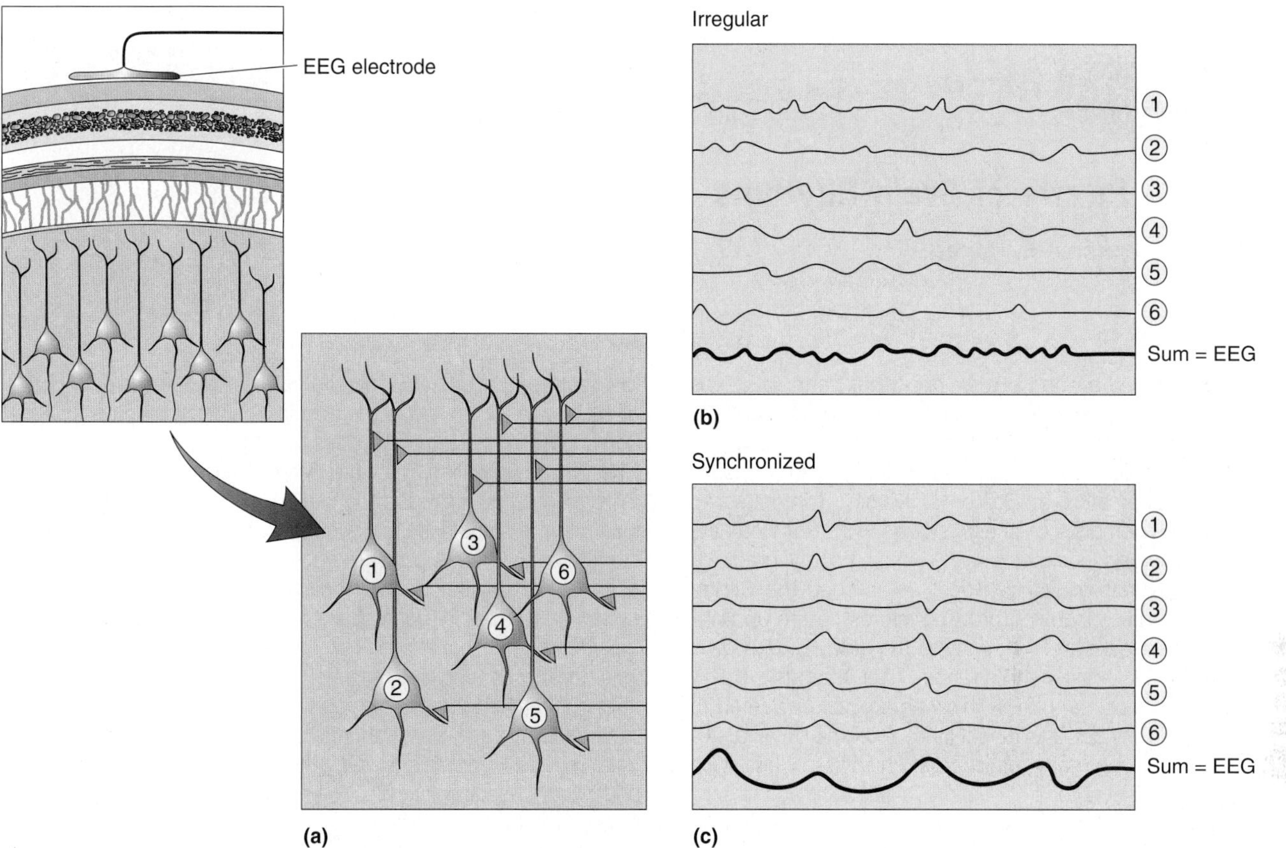

▲ FIGURE 19.5

The generation of large EEG signals by synchronous activity. **(a)** In a population of pyramidal cells located under an EEG electrode, each neuron receives many synaptic inputs. **(b)** If the inputs fire at irregular intervals, the pyramidal cell responses are not synchronized, and the summed activity detected by the electrode has a small amplitude. **(c)** If the same number of inputs fire within a narrow time window so the pyramidal cell responses are synchronized, the resulting EEG is much larger.

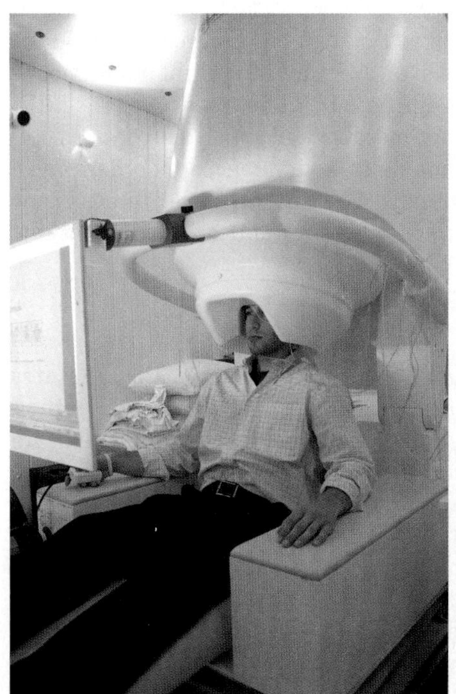

◀ FIGURE 19.6

Magnetoencephalography (MEG). **(a)** A person receiving an MEG scan. **(b)** The tiny magnetic signals generated by neurons on the brain are detected by an array of 150 sensitive magnetic detectors. **(c)** Researchers use the signals to calculate the locations of sources of neural activity (color-coded, in this image). (Sources: part a, http://infocenter.nimh.nih.gov/il/public_il/image_details.cfm?id=80; parts b and c, Los Alamos National Laboratory.)

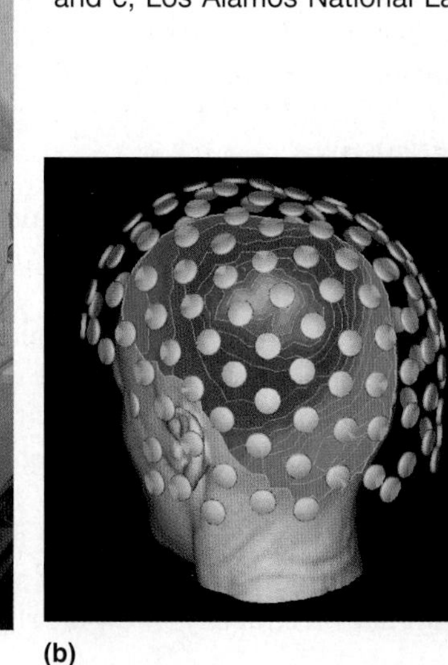

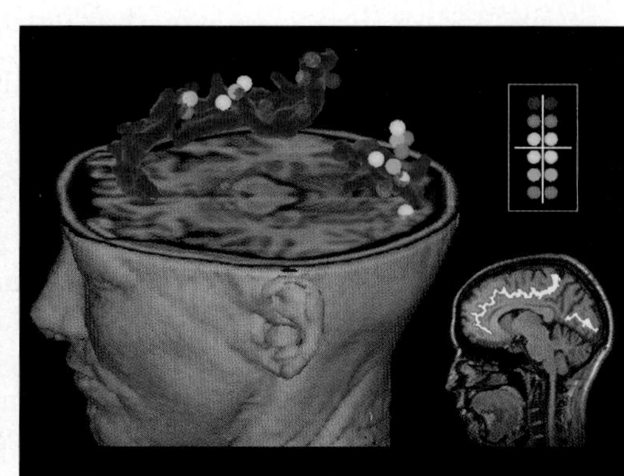

BOX 19.1 PATH OF DISCOVERY

The Puzzle of Brain Rhythms

by Stephanie R. Jones

I've always loved a good puzzle, and what better puzzle is there than understanding how our brains form perceptions and actions? This isn't the puzzle I set out to solve at the start of my career. My analytic nature led me to pursue a Ph.D. in mathematics at Boston University. I intended to study the mathematics of chaos, but as with many careers my path diverged unexpectedly. A year into my graduate study, mathematician Nancy Kopell established the Center for BioDynamics, catalyzing growing interest in the applications of dynamical systems theory to the study of biological phenomena, including neuroscience. After attending a few neuroscience lectures, I knew this was a puzzle I wanted to help solve. To my good fortune, Nancy took me on as a student. I began using mathematics to study rhythmic activity in simplified representations of neural circuits, such as the central pattern-generating network that regulates crayfish swimming. By the time I finished my mathematics Ph.D., I was passionate about neuroscience. I knew I wanted to apply my knowledge to understand human brain dynamics. Little did I know just how many pieces are in that puzzle!

For the next decade, I studied human brain rhythms using magnetoencephalography (MEG) in the brain-imaging center at Massachusetts General Hospital (MGH). At MGH, my path of discovery was again shaped by the gift of fantastic mentors and colleagues. The first is my now close colleague, neurophysiologist Chris Moore, who was himself a postdoctoral fellow at the time. Chris enlightened me to the nuances of neuroscience and to the idea that the somatosensory system was the "ideal system to study" because of its puzzle-like topographical representation of the body, the homunculus (see Figure 12.19). With MEG we began studying tactile perception, such as the detection of light taps to the fingertips, in humans. Chris's notion that this system was "ideal" was fortuitous, as was revealed in a surprising way through the mentorship of physicist Matti Hämäläinen, director of the MEG center. Matti taught me the ins and outs of MEG data collection and, importantly, the electromagnetic physics underlying these brain signals. I learned that the intracellular currents within the long, aligned dendrites of pyramidal neurons are the primary generators of the recorded magnetic field signals. Further, the pyramidal neurons in primary somatosensory cortex (S1) are "ideally" oriented to produce strong MEG signals during finger taps that can be reliably localized to the hand representation in S1. This enabled us to do careful studies of the neural generators of brain rhythms.

positron emission tomography (PET) (see Box 7.3). MEG cannot provide the spatially detailed images of fMRI, however. Another important distinction is that EEG and MEG directly measure the activity of neurons, whereas fMRI and PET detect changes in blood flow or metabolism, which are controlled in part by neuronal activity but which may also be influenced by other physiological factors. MEG is currently being used in experimental studies of the human brain and its cognitive functions and as an aid in the diagnosis of epilepsy and language disorders (Box 19.1).

EEG Rhythms

EEG rhythms vary dramatically and often correlate with particular states of behavior (such as level of attentiveness, sleeping, or waking) and pathology (seizures or coma). Figure 19.7 shows part of a normal EEG. The brain can generate rhythms that are as slow as about 0.05 Hz and as fast as 500 Hz or more. The main EEG rhythms are categorized by their frequency range, and each range is named after a Greek letter. *Delta rhythms* are slow, less than 4 Hz, are often large in amplitude, and are a hallmark of deep sleep. *Theta rhythms* are 4–7 Hz and can occur during both sleeping and waking states. *Alpha rhythms* are about 8–13 Hz, are largest over the occipital cortex, and are associated with quiet, waking states; *Mu rhythms* are similar in frequency to alpha rhythms but are largest over the motor

As with all MEG (and EEG) recordings, the dominant activities from S1 are low-frequency, large-amplitude rhythms, including the beta rhythms of 15–29 Hz. We discovered that when a subject directs her attention to her finger before it is tapped, the beta rhythms in the hand area of S1 decrease compared to when her attention is directed elsewhere. Attention and reduced beta rhythms correlated with an increase in the subject's ability to feel a light tap. Our results were similar to previous findings in the visual cortex, suggesting that beta rhythms may signal inhibitory processes in sensory areas of cortex. But why? What is it about these rhythms, if anything, that links them to decreased perception? And why, in conditions like Parkinson's disease, are beta rhythms over-expressed in motor cortex with a corresponding decrease in motor actions?

To address this piece of the puzzle, I turned to my mathematics roots and began constructing a computational neural model to study the origins of these rhythms. My prior research had given me solid intuitions about how stable rhythms can emerge from neural circuits. However, after much exploration using simplified mathematical representations of neural circuits (e.g., collapsing the activity of an entire neuron to a single point), I realized these models simply could not reproduce signals that resembled the recordings. Next, I drew on the pioneering work of Yoshio Okada, who combined experimental and mathematical modeling to understand MEG signals from pyramidal neurons. Equipped with my new knowledge of the biophysics underlying MEG, I constructed more complex models that included details of the structure and physiology of pyramidal neurons and other cortical neuron types. This endeavor spanned several years that also included the birth of the first of my three children.

To my delight, the detailed model yielded novel and non-intuitive predictions about rhythms. Specifically, it predicted that beta rhythms emerge from the integration of two sets of synaptic inputs that are roughly synchronous and that excite different parts of pyramidal cell dendrites. These inputs drive alternating electrical currents up and down within the dendrites to reproduce rhythms remarkably consistent with recordings. The model not only accounted for many features of the MEG rhythms in S1 but also suggested how these rhythms influence sensory processing. I subsequently tested these secondary predictions with the MEG data, and to my surprise they were confirmed! This discovery was thrilling since the mathematical model was now predicting what the data from new experiments would look like. Finally, pieces of the puzzle were fitting together!

The close agreement between the model's output and the recorded human data gives us confidence in the model's predictions about how neurons generate beta rhythms. More importantly, the model suggests how rhythms influence brain function. Through continued collaboration with Chris Moore and other neurophysiologists and neurosurgeons, we are currently testing model-derived predictions with electrode recordings. We may discover the pieces do not fit together exactly as the model suggests. However, through collaboration and the interplay of interdisciplinary methods, I am convinced we can build interpretive bridges between neural activity and human brain functions. Solving the puzzle of brain rhythms will be an important and exciting step along the way.

and somatosensory areas. *Beta rhythms* are about 15–30 Hz. *Gamma rhythms* are relatively fast, ranging from about 30–90 Hz, and signal an activated or attentive cortex. Additional rhythms include *spindles*, brief 8–14 Hz waves associated with sleep, and *ripples*, brief bouts of 80–200 Hz oscillations. An interesting feature of EEG rhythms is that their characteristics are remarkably similar across mammalian brains from mice to humans, despite 17,000-fold differences in brain mass (Figure 19.8).

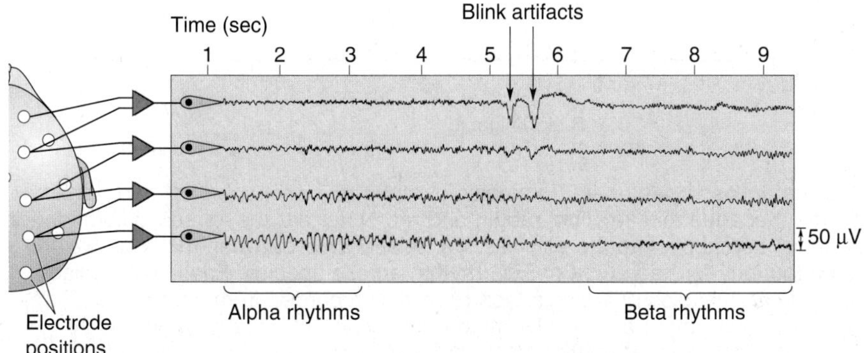

◀ FIGURE 19.7
A normal EEG. The subject is awake and quiet, and recording sites are indicated at the left. The first few seconds show normal alpha activity, which has frequencies of 8–13 Hz and is largest in the occipital regions. About halfway through the recording, the subject opened his eyes, signaled by the large blink artifacts on the top traces (arrows), and alpha rhythms were suppressed.

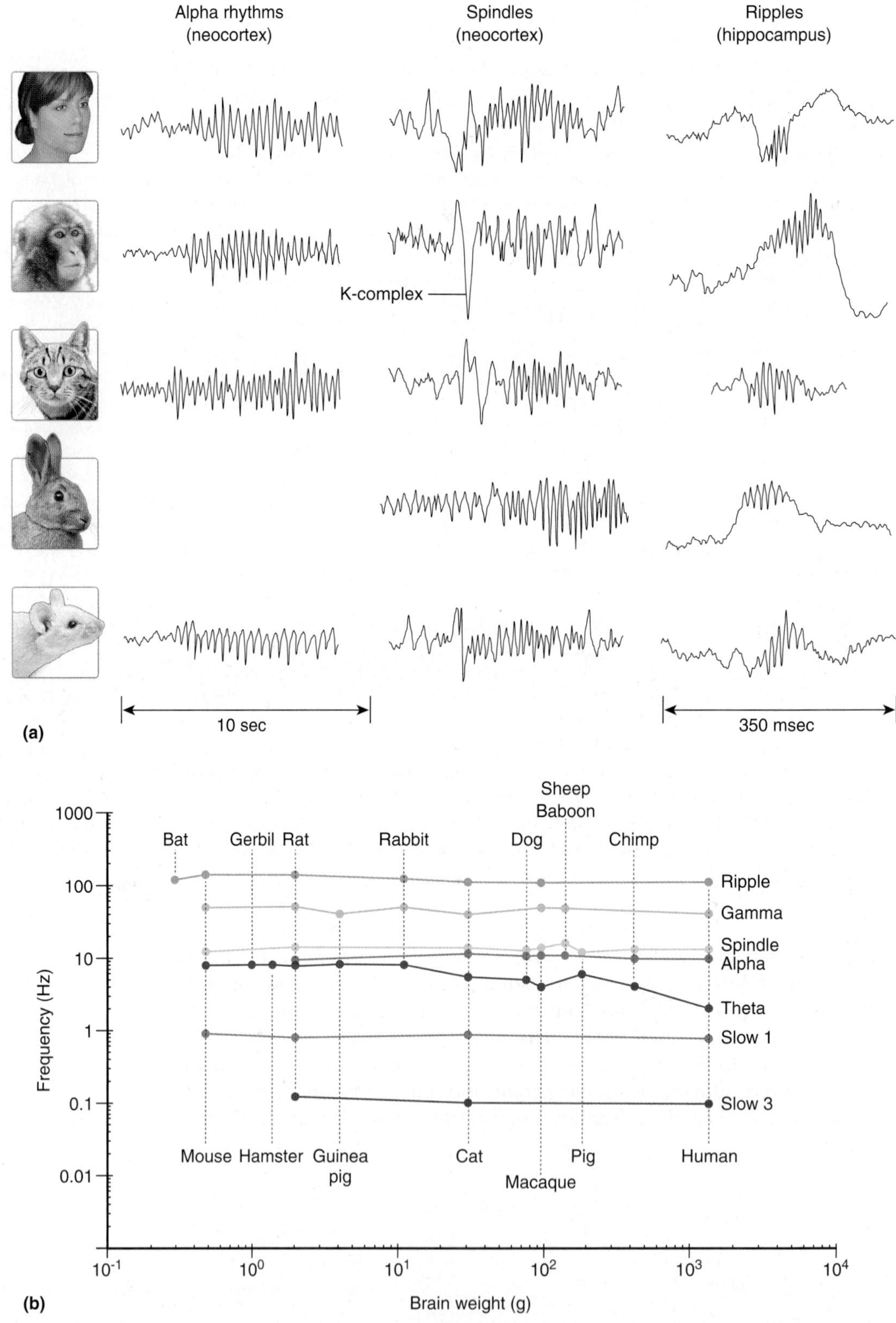

▲ FIGURE 19.8

EEG rhythms across species. (a) Examples of alpha rhythms, spindles, and ripples from human, macaque monkey, cat, rabbit, and rat. Note that the 10 sec calibration for alpha rhythms also applies to the spindles. **(b)** Relationship between the brain weight and main frequency for each type of EEG rhythm across species. Each colored line represents the frequencies of a single type of rhythm recorded from several species (the absence of data about a particular rhythm for a species does not necessarily mean that species lacks that rhythm). Note how little the properties of the EEG rhythms vary despite the vast range of brain sizes. (Source: Buzsáki et al., 2013)

While analysis of an EEG cannot tell us *what* a person is thinking, it can help us know *if* a person is thinking. In general, high-frequency, low-amplitude rhythms are associated with alertness and waking, or the dreaming stages of sleep. Low-frequency, high-amplitude rhythms are associated with nondreaming sleep states, certain drugged states, or the pathological condition of coma. This is logical because when the cortex is most actively engaged in processing information, whether generated by sensory input or by internal processes, the activity level of cortical neurons is relatively high but also relatively unsynchronized. In other words, each neuron, or a very small group of neurons, is vigorously involved in a slightly different aspect of a complex cognitive task; it fires rapidly but not quite simultaneously with most of its neighbors. This leads to low synchrony, so EEG amplitude is low, and gamma and beta rhythms dominate. In contrast, during deep sleep, cortical neurons are not engaged in information processing, and large numbers of them are phasically excited by a common, slow, rhythmic input. In this case, synchrony is high, so EEG amplitude is high.

Mechanisms and Meanings of Brain Rhythms

Electrical rhythms abound in the cerebral cortex. But how are they generated, and what functions, if any, do they perform? Let's take a look at each of these questions.

The Generation of Synchronous Rhythms. The activity of a large set of neurons will produce synchronized oscillations in one of two fundamental ways: (1) They may all take their cues from a central clock, or *pacemaker*, or (2) they may share or distribute the timing function among themselves by mutually exciting or inhibiting one another. The first mechanism is analogous to a band with a leader, with each musician playing in strict time to the beat of the leader's baton (Figure 19.9a). The second mechanism is more subtle because the timing arises from the collective behavior of the cortical neurons themselves. Musically, it is more like a jam session (Figure 19.9b).

The concept of shared synchronous rhythm can be easily demonstrated by a group of people, even nonmusical ones. Simply tell them to begin clapping, but give them no instructions about how fast to clap or whose beat to follow. Almost immediately they will all be clapping in synchrony! How? By listening and watching each other, they will adjust their clapping rates to match. A key factor is person-to-person interaction; in a network of neurons, these interactions occur via synaptic connections. People naturally tend to clap within a narrow range of frequencies, so they don't have to adjust their timing very far in order to clap in synchrony. Likewise, some neurons may fire at certain frequencies much more than others. This kind of collective, organized behavior can generate rhythms of impressive dimensions, which can move in space and time. Have you ever been part of a human wave in the stands of a sold-out football stadium?

Many different circuits of neurons can generate rhythmic activity. A very simple model oscillator, consisting of just one excitatory and one inhibitory neuron, is shown in Figure 19.10. Most real neural oscillators include far more neurons but similar basic features: a source of constant excitatory drive, feedback connections, and synaptic excitation and inhibition.

Within the mammalian brain, rhythmic, synchronous activity is usually coordinated by a combination of the pacemaker and collective methods. For example, the thalamus, with its massive input to all of the

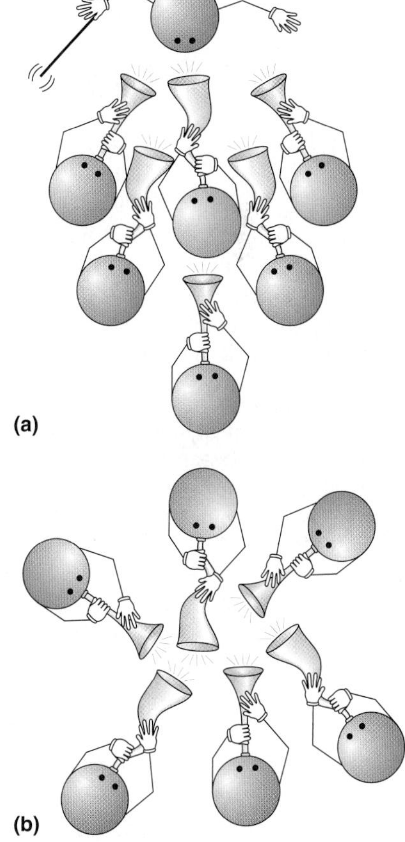

▲ FIGURE 19.9
Two mechanisms of synchronous rhythms. Synchronous rhythms can **(a)** be led by a pacemaker or **(b)** arise from the collective behavior of all participants.

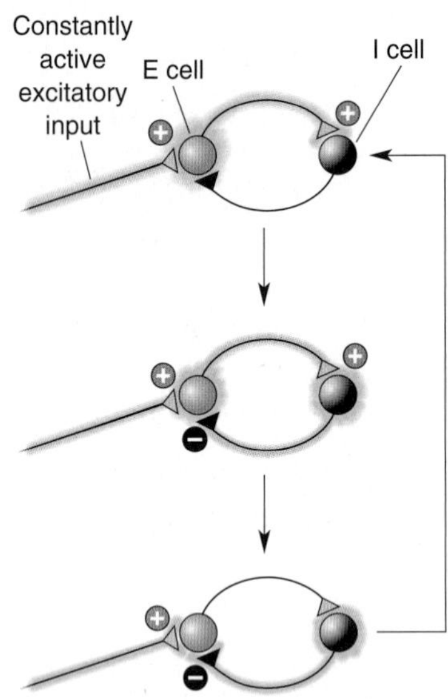

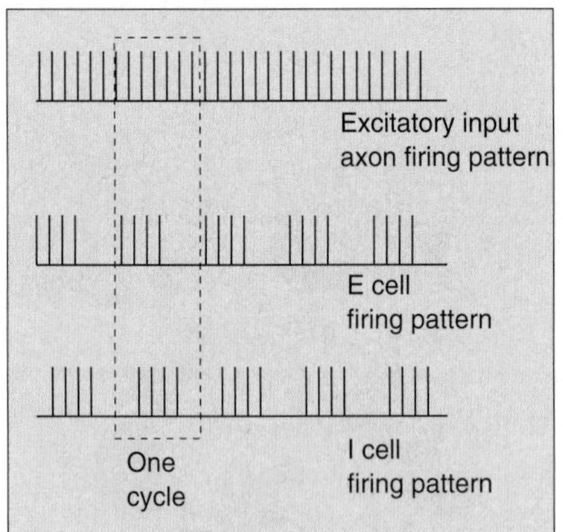

▲ FIGURE 19.10

A two-neuron oscillator. One excitatory cell (E cell) and one inhibitory cell (I cell) synapse upon each another. As long as there is a constant excitatory drive (which does not have to be rhythmic) onto the E cell, activity will tend to trade back and forth between the two neurons. One activity cycle through the network will generate the pattern of firing shown in the dashed box.

cortex, can act as a powerful pacemaker. Under certain conditions, thalamic neurons can generate very rhythmic action potential discharges (Figure 19.11). But how do thalamic neurons oscillate? Some thalamic cells have a particular set of voltage-gated ion channels that allow each cell to generate very rhythmic, self-sustaining discharge patterns even when there is no external input to the cell. The rhythmic activity of each thalamic pacemaker neuron then becomes synchronized with many other thalamic cells via a hand-clapping kind of collective interaction. Synaptic connections between excitatory and inhibitory thalamic neurons force each individual neuron to conform to the rhythm of the group. These coordinated rhythms are then passed to the cortex by the thalamocortical axons, which excite cortical neurons. In this way, a relatively small group of centralized thalamic cells (acting as the band leader) can compel a much larger group of cortical cells (acting as the band) to march to the thalamic beat (Figure 19.12).

Some rhythms of the cerebral cortex do not depend on a thalamic pacemaker but rely instead on the collective, cooperative interactions of cortical neurons themselves. In this case, the excitatory and inhibitory interconnections of the neurons result in a coordinated, synchronous pattern of activity that may remain localized or spread to encompass larger regions of cortex.

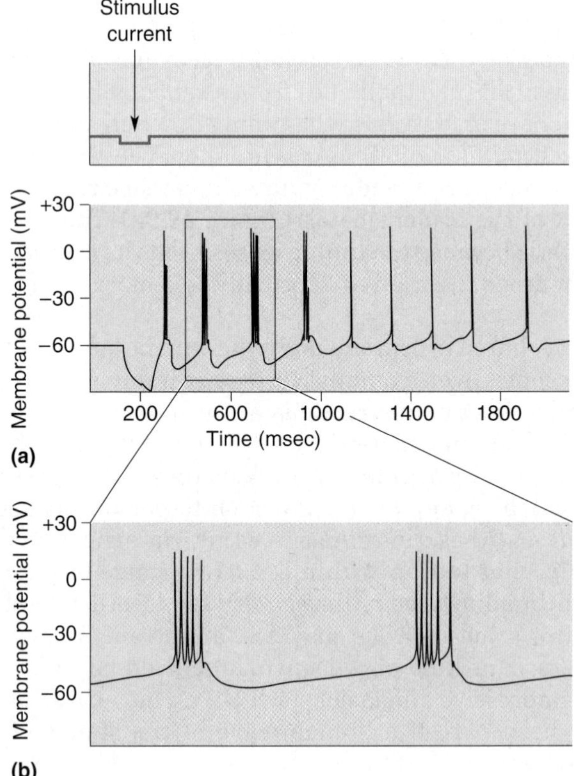

▲ FIGURE 19.11

A one-neuron oscillator. At times during sleep states, thalamic neurons fire in rhythmic patterns that do not reflect their input. Shown here are intracellular recordings of membrane voltage in such a case. **(a)** A short pulse (less than 0.1 second) of stimulus current was applied, and the cell responded with almost 2 seconds of rhythmic firing, first with bursts at about 5 Hz and then with single spikes. **(b)** Two of the bursts expanded in time; each burst is a cluster of five or six action potentials. (Source: Adapted from Bal and McCormick, 1993, Fig. 2.)

Functions of Brain Rhythms. Cortical rhythms are fascinating to watch in an EEG, and they parallel so many interesting human behaviors that we are compelled to ask: Why so many rhythms? More importantly, do they serve a purpose? There are no satisfactory answers yet. Ideas abound, but definitive evidence is scarce. One hypothesis for sleep-related rhythms is that they are the brain's way of disconnecting the cortex from sensory input. When you are awake, the thalamus allows sensory information to pass through it and be relayed up to the cortex. When you are asleep, thalamic neurons enter a self-generated rhythmic state that prevents organized sensory information from being relayed to the cortex. While this idea has intuitive appeal (most people do prefer to sleep in a dark, quiet environment), it does not explain why rhythms are necessary. Why not just steadily inhibit the thalamus and allow the cortex to rest quietly?

A function for fast rhythms in the awake cortex has also been proposed. One scheme for understanding visual perception takes advantage of the fact that cortical neurons responding to the same object are synchronously active. Walter Freeman, a neurobiologist at the University of California, Berkeley, pioneered the idea that neural rhythms are used to coordinate activity between regions of the nervous system. Both sensory and motor systems of the awake brain often generate bursts of synchronous neural activity that give rise to EEG gamma rhythms (30–90 Hz).

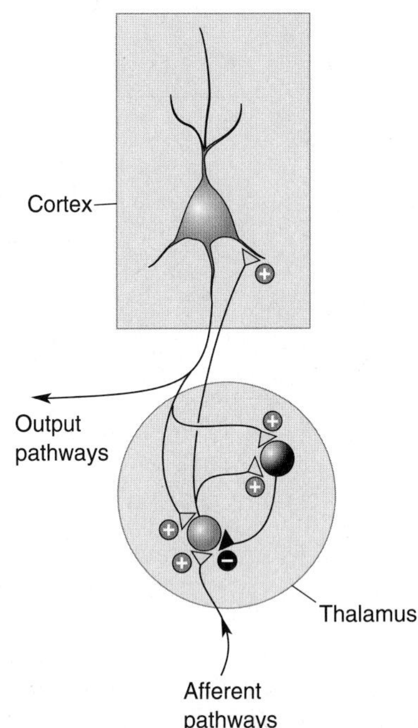

▲ FIGURE 19.12
Rhythms in the thalamus drive rhythms in the cerebral cortex. The thalamus can generate rhythmic activity because of the intrinsic properties of its neurons and because of its synaptic interconnections. In the thalamus, green represents a population of excitatory neurons, and black represents a population of inhibitory neurons.

By momentarily synchronizing the fast oscillations generated by different regions of cortex, perhaps the brain binds together various neural components into a single perceptual construction. For example, when you are trying to catch a basketball, different groups of neurons that simultaneously respond to the specific shape, color, movement, distance, and even the significance of the basketball tend to oscillate synchronously. The fact that the oscillations of these scattered groups of cells (those that together encode "basketballness") are highly synchronous would somehow tag them as a meaningful group, distinct from other nearby neurons, thereby unifying the disjointed neural pieces of the "basketball puzzle." The evidence for this idea is indirect, far from proven, and understandably controversial.

For now, the functions of rhythms in the cerebral cortex are largely a mystery. One plausible hypothesis is that most rhythms have no direct function. Instead, they may be intriguing but unimportant by-products of the tendency for brain circuits to be strongly interconnected, with various forms of excitatory feedback. When something excites itself, whether it is an audio amplifier or the human stadium wave, it often leads to instability or oscillation. Feedback circuits are essential for the cortex to do all the marvelous things it does for us. Oscillations may be the unavoidable consequence of so much feedback circuitry, unwanted but tolerated by necessity. Even if they don't have a function, however, EEG rhythms provide us with a convenient window on the functional states of the brain.

The Seizures of Epilepsy

Seizures, the most extreme form of synchronous brain activity, are always a sign of pathology. A **generalized seizure** involves the entire cerebral cortex of both hemispheres. A **partial seizure** involves only a circumscribed area of the cortex. In both cases, the neurons within the affected areas fire with a synchrony that never occurs during normal behavior. As a consequence, seizures are usually accompanied by very large EEG patterns. The cerebral cortex, probably because of its extensive feedback circuitry, is never far from the runaway excitation we know as a seizure. Isolated seizures are not uncommon during a lifetime, and 7–10% of people in the general population have had at least one. When a person experiences repeated seizures, the condition is known as **epilepsy**. About 0.7% of people worldwide (50 million) have epilepsy. Epilepsy is more common in developing countries, particularly in rural areas, presumably because of higher rates of untreated childhood epilepsy, infections, and poor pre- and postnatal care. The diagnosis of epilepsy occurs most often in young children and among the elderly (Figure 19.13). Childhood epilepsy is usually congenital, caused by genes or a disease or abnormality present at birth, whereas the elderly tend to acquire epilepsy as a consequence of conditions such as stroke, tumors, or Alzheimer's disease.

Epilepsy is more a symptom of disease than a disease itself. Its causes can sometimes be identified, including tumors, trauma, genetics, metabolic dysfunction, infection, and vascular disease, but in many cases, the cause of epilepsy is not known. Different types of seizures have different underlying mechanisms. Some forms of epilepsy show a genetic predisposition, and many of the genes responsible have been identified. These genes code for a diverse array of proteins, including ion channels, transporters, receptors, and signaling molecules. Several mutations of genes that encode for sodium channel proteins, for example, have been linked to rare familial forms of epilepsy. These mutated sodium channels tend to stay open a bit longer than normal, allowing more sodium current to enter the neurons and thus making neurons hyperexcitable. Another group of mutations that lead to epilepsy impair synaptic inhibition mediated by GABA by affecting its receptors, enzymes critical for its synthesis or transport, or proteins involved in its release.

Research suggests that some seizures reflect an upset of the delicate balance of synaptic excitation and inhibition in the brain. Other seizures may be due to excessively strong or dense excitatory interconnections. Drugs that block GABA receptors are very potent *convulsants* (seizure-promoting agents). The withdrawal of chronic depressant drugs, such as alcohol or barbiturates, may also trigger seizures. A variety of drugs are useful for the therapeutic suppression of seizures, and these *anticonvulsants* tend to counter excitability in various ways. For example, some act by prolonging the inhibitory actions of GABA (e.g., barbiturates and benzodiazepines [see Figure 6.22]), while others decrease the tendency for certain neurons to fire action potentials at a high frequency (e.g., phenytoin and carbamazepine).

The behavioral features of a seizure depend on the neurons involved and the patterns of their activity. During most forms of generalized seizures, virtually all cortical neurons participate, so behavior is completely

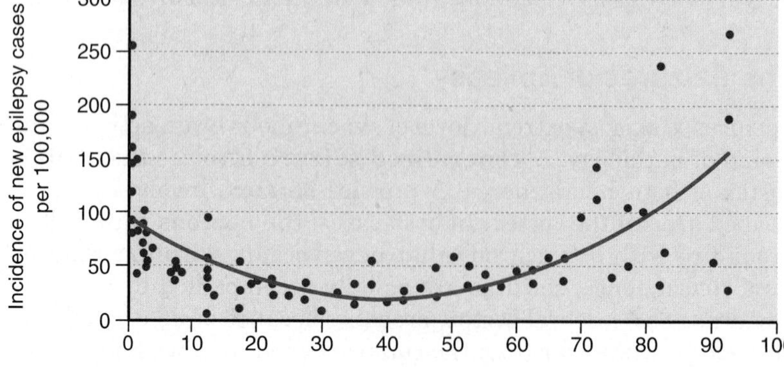

▶ FIGURE 19.13
Incidence of epilepsy by age. The graph plots the number of new cases of epilepsy per 100,000 people, as a function of age at the time of diagnosis. Data were compiled from 12 studies performed in developed countries. (Source: D.J. Thurman, http://iom.edu/~/media/Files/Activity%20 Files/Disease/Epilepsy/Thurman%202.pdf.)

disrupted for many minutes. Consciousness is lost, while all muscle groups may be driven by tonic (ongoing) activity or by clonic (rhythmic) patterns, or by both in sequence, the so-called *tonic–clonic seizure*. *Absence seizures* characterize a childhood form of epilepsy, consisting of less than 30 seconds of generalized, 3 Hz EEG waves accompanied by loss of consciousness. An EEG recorded during an absence seizure illustrates several very striking abnormalities (Figure 19.14). The voltage patterns are extraordinarily large, regular, and rhythmic and are generated synchronously across the entire brain. Despite this dramatic pattern of activity, the motor signs of an absence seizure are strangely subtle, only fluttering eyelids or a twitching mouth.

Partial seizures can be instructive. If they begin in a small area of motor cortex, they can cause clonic movement of part of a limb. In the late 1800s, British neurologist John Hughlings Jackson observed the progression of seizure-related movements across the body, looked at the lesions

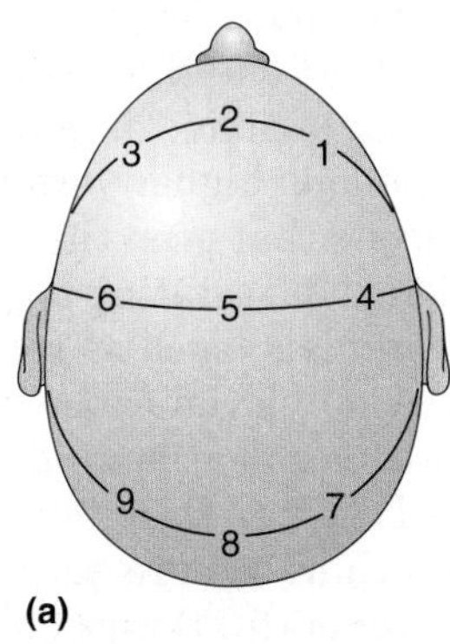

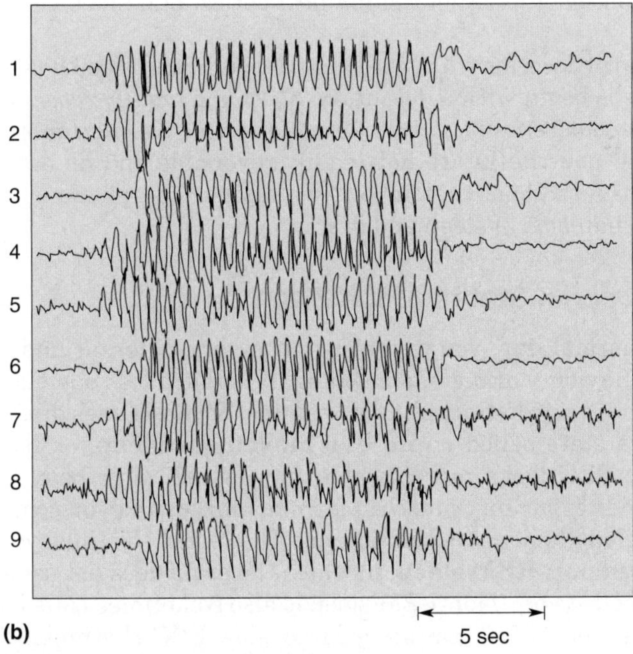

▲ FIGURE 19.14
An EEG of a generalized epileptic seizure. (a) EEG electrodes are placed at various positions on the scalp. (b) They detect a brief absence seizure that begins abruptly, is synchronized across the entire head, generates strong neural activity with rhythms of about 3 Hz, and ends abruptly after about 12 seconds. (Source: J.F. Lambert and N. Chantrier.)

in his patients' brains after they died, and correctly inferred the basic somatotopic map of the motor cortex (see Chapter 14). If seizures begin in a sensory area, they can trigger an abnormal sensation, or *aura*, such as an odd smell or sparkling lights. Most bizarre are the partial seizures that elicit more well-formed auras such as *déjà vu* (the feeling that something has happened before) or hallucinations. Sometimes involving the cortex of the temporal lobes, including the hippocampus and amygdala, they can impair memory, thought, and consciousness. In some cases, partial seizures may spread uncontrollably and become generalized seizures.

SLEEP

Sleep and dreams—they are mysterious, even mystical to some people, and a favorite subject of art and literature, philosophy, and science. Sleep is a powerful master. Each night we abandon our companions, our work, and our play and enter the cloister of sleep. We have only limited control over the decision; we can postpone sleep for a while, but eventually it overwhelms us. We spend about one-third of our lives sleeping, and one-quarter of that time in a state of active dreaming.

Sleep may be universal among higher vertebrates and perhaps among all animals. Research suggests that even the fruit fly, *Drosophila*, sleeps. Prolonged sleep deprivation is devastating to proper functioning, at least temporarily, and in some animals (such as rats and cockroaches, though probably not in humans), it may even cause death. Sleep is essential to our lives, almost as important as eating and breathing. But why do we sleep? What purpose does it serve? Despite many years of research, the joke remains that the only thing we are sure of is that sleep overcomes sleepiness. But one of the wonderful things about science is that the lack of consensus inspires a flourishing of theories, and sleep research is no exception.

We can still describe what we cannot explain, and sleep has been richly studied. Let's begin with a definition: *Sleep is a readily reversible state of reduced responsiveness to, and interaction with, the environment.* (Coma and general anesthesia are not readily reversible and do not qualify as sleep.) In the sections that follow, we discuss the phenomenology and neural mechanisms of sleep and dreaming.

The Functional States of the Brain

During a normal day, you experience two very different and noticeable types of behavior: waking and sleeping. It is much less obvious that your sleep also has very distinct phases or states. Several times during a night, you enter a state called **rapid eye movement sleep**, or **REM sleep**, when your EEG looks more awake than asleep, your body (except for your eye and respiratory muscles) is immobilized, and you conjure up the vivid, detailed illusions we call dreams. The rest of the time, you spend in a state called **non-REM sleep**, in which the brain does not usually generate complex dreams. (Non-REM sleep is also sometimes called *slow-wave sleep* because of its domination by large, slow EEG rhythms.) These fundamental behavioral states—awake, non-REM sleep, and REM sleep—are produced by three distinct states of brain function (Table 19.1). Each behavioral state is also accompanied by large shifts in body function.

Non-REM sleep seems to be a period for rest. Muscle tension throughout the body is reduced, and movement is minimal. It is important to realize that the body is *capable* of movement during non-REM sleep but

TABLE 19.1 **Characteristics of the Three Functional States of the Brain**

Behavior	Awake	Non-REM Sleep	REM Sleep
EEG	Low voltage, fast	High voltage, slow	Low voltage, fast
Sensation	Vivid, externally generated	Dull or absent	Vivid, internally generated
Thought	Logical, progressive	Logical, repetitive	Vivid, illogical, bizarre
Movement	Continuous, voluntary	Occasional, involuntary	Muscle paralysis; movement commanded by the brain but not carried out
Rapid eye movement	Often	Rare	Often

only rarely does the brain command it to move, usually to briefly adjust the body's position. The temperature and energy consumption of the body are lowered. Because of an increase in activity of the parasympathetic division of the ANS, heart rate, respiration, and kidney function all slow down, and digestive processes speed up.

During non-REM sleep, the brain also seems to rest. Its rate of energy use, and the general firing rates of its neurons, is at their lowest point of the day. The slow, large-amplitude EEG rhythms indicate that the neurons of the cortex are oscillating in relatively high synchrony, and experiments suggest that most sensory input cannot even reach the cortex. While there is no way to know for certain what people are thinking when they are asleep, studies indicate that mental processes also hit their daily low during the non-REM state. When awakened, people often recall nothing, or only brief, fragmentary, plausible thoughts with few visual images. Detailed, entertaining, irrational dreams are rare, although not absent, during non-REM sleep. William Dement, a pioneering sleep researcher at Stanford University, characterizes non-REM sleep as *an idling brain in a movable body*.

In contrast, Dement calls REM sleep *an active, hallucinating brain in a paralyzed body*. REM sleep is dreaming sleep. Although the REM period accounts for only a small part of our sleep time, it is the part most researchers get excited about (and this is the state that most excites the brain), perhaps because dreams are so intriguing and enigmatic. If you awaken someone during REM sleep, as Dement, Eugene Aserinsky, and Nathaniel Kleitman first did in the mid-1950s, the person will likely report visually detailed, lifelike episodes, often with bizarre story lines—the kinds of dreams we love to talk about and try to interpret.

The physiology of REM sleep is also bizarre. The EEG looks almost indistinguishable from that of an active, waking brain, with fast, low-voltage fluctuations. This is why REM sleep is sometimes referred to as *paradoxical sleep*. In fact, the oxygen consumption of the brain (a measure of its energy use) is higher in REM sleep than when the brain is awake and concentrating on difficult mathematical problems. The paralysis that occurs during REM sleep is caused by an almost total loss of skeletal muscle tone or **atonia**. Most of the body is actually *incapable* of moving! Respiratory muscles do continue to function but just barely. The muscles controlling eye movement and the tiny muscles of the inner ear are the exceptions; these are strikingly active. With lids closed, the eyes occasionally dart rapidly back and forth. These bursts of rapid eye movement are the best predictors of vivid dreaming, and at least 90% of people awakened during or after them report dreams.

Physiological control systems are dominated by sympathetic activity during REM sleep. Inexplicably, the body's temperature control system simply quits, and core temperature begins to drift downward. Heart and respiration rates increase but become irregular. In healthy people, the clitoris and penis become engorged with blood and erect, although this usually has nothing to do with any sexual content of dreams. Overall, during REM sleep, the brain seems to be doing everything except resting.

The Sleep Cycle

Even a good night's sleep is not a steady, unbroken journey. It normally begins with a period of non-REM sleep. Figure 19.15 shows that a typical full night of sleep includes a regular cycling of eye movements, physiological functions, and penile erections through non-REM and REM periods. It is obvious that sleep takes the brain through a repetitive roller coaster ride of activity, and sometimes the ride is pretty wild (Box 19.2). Roughly 75% of total sleep time is spent in non-REM and 25% in REM, with periodic cycles between these states throughout the night. Non-REM sleep is generally divided into four distinct stages. During a normal night, we slide through the stages of non-REM, then into REM, then back through

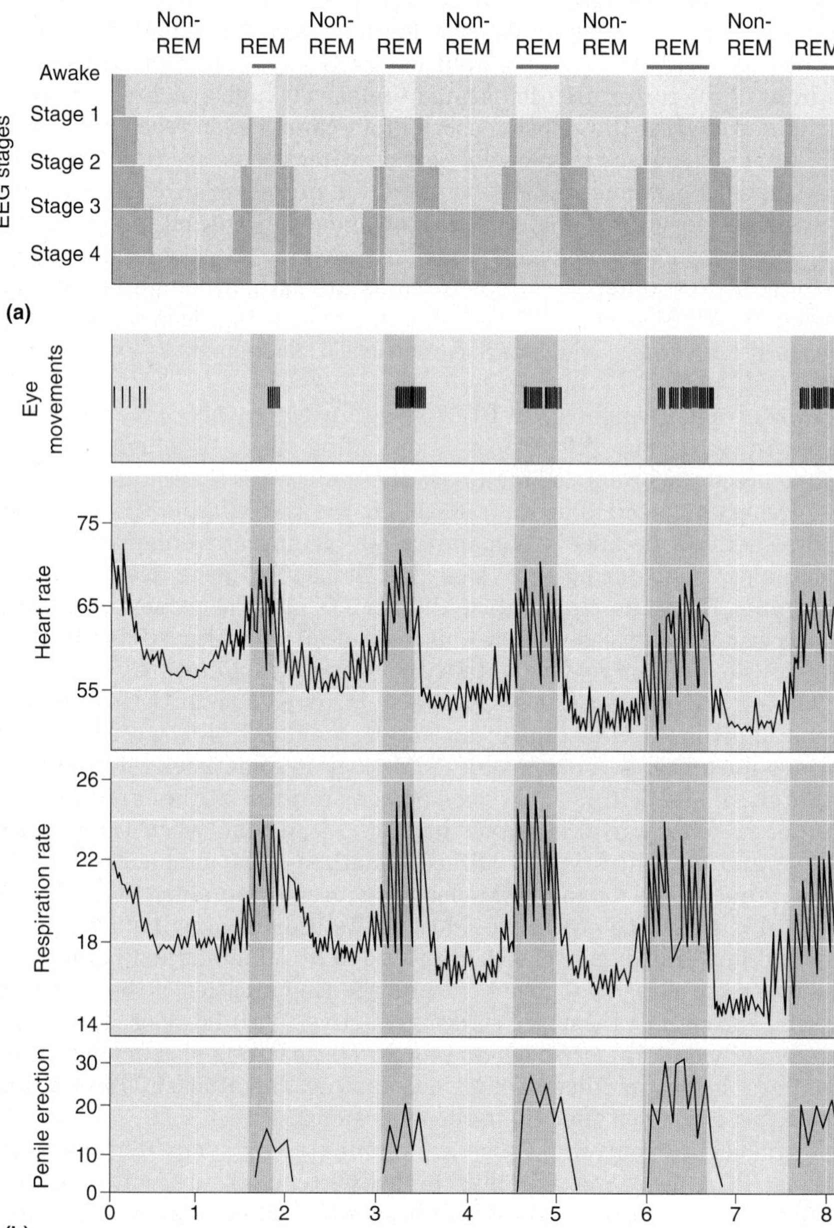

▶ FIGURE 19.15
Physiological changes during non-REM and REM sleep. **(a)** This graph represents one night of sleep, starting with a transition from awake to stage 1 non-REM sleep. The sleep cycle progresses through the deeper stages of non-REM sleep, then into REM sleep. It is repeated several times, but each cycle has shorter and shallower non-REM periods and longer REM periods. **(b)** These graphs show regular increases in heart rate, respiration rate, and penile erection during the REM periods of one night sleep. (Source: Adapted from Purves et al., 2004, Fig. 27.7.)

BOX 19.2 OF SPECIAL INTEREST

Walking, Talking, and Screaming in Your Sleep

Sleep is not always serene and stationary. Talking, walking, and screaming are common, usually occurring during non-REM sleep. If this seems surprising, remember that REM sleep is accompanied by almost total body paralysis. You would therefore be incapable of walking or talking during REM sleep, even if your dream "urged" you to do it.

Sleepwalking, or *somnambulism*, peaks at about age 11. Although 40% of us were sleepwalkers as children, few of us sleepwalk as adults. Sleepwalking usually occurs during the first stage 4 non-REM period of the night. A full-blown sleepwalking incident involves open eyes and movement around the room, the house, or even outside, with enough awareness to avoid objects and climb stairs. Cognitive functioning and judgment are severely impaired. It is often difficult to awaken sleepwalkers because they are in deep, slow-wave sleep. The best treatment is a guiding hand back to bed. Sleepwalkers usually have no memory of the incident the next morning.

Almost everyone practices sleep talking, *somniloquy*, now and then. Unfortunately, sleeping speech is often so garbled or nonsensical that a curious listener is disappointed by its emptiness.

More dramatic are *sleep terrors,* also known as *night terrors,* which are most common in children 5–7 years old. A girl screams in the middle of the night. Her parents rush to her bedside, frantic to know what has alarmed her. The girl cries inconsolably, unable to explain her horrifying experience. After 10 agonizing minutes of shrieking and flailing, she finally sleeps quietly, leaving the parents shaken and baffled. The next morning she is bright and cheerful, with no recollection of the night's misadventure. Sleep terrors are distinctly different from nightmares, which are vivid, complex dreams, outwardly quiet, that occur during REM sleep. By contrast, sleep terrors begin in stage 3 or 4 of non-REM sleep, and the experience is not dreamlike but a feeling of uncontrollable panic, accompanied by greatly increased heart rate and blood pressure. They usually pass with adolescence and are not a symptom of a psychiatric disorder.

the non-REM stages again, repeating the cycle about every 90 minutes. These cycles are examples of **ultradian rhythms**, which have faster periods than circadian rhythms.

EEG rhythms during the stages of sleep are shown in Figure 19.16. An average, healthy adult becomes drowsy and begins to sleep, first entering stage 1 non-REM sleep. Stage 1 is transitional sleep, when the EEG alpha rhythms of relaxed waking become less regular and wane, and the eyes make slow, rolling movements. Stage 1 is fleeting, usually lasting only a few minutes. It is also the lightest stage of sleep, meaning that we are most easily awakened. Stage 2 is slightly deeper and may last 5–15 minutes. Its characteristics include the occasional 8–14 Hz oscillation of the EEG called the *sleep spindle*, which is generated by a thalamic pacemaker (see Figure 19.12). In addition, a high-amplitude sharp wave called the *K complex* is sometimes observed. Eye movements almost cease. Next follows stage 3, and the EEG begins large-amplitude, slow delta rhythms. Eye and body movements are few. Stage 4 is the deepest stage of sleep, with large EEG rhythms of 2 Hz or less. During the first cycle of sleep, stage 4 may persist for 20–40 minutes. Then sleep begins to lighten again, ascends through stage 3 to stage 2 for 10–15 minutes, and suddenly enters a brief period of REM sleep, with its fast EEG beta and gamma rhythms and sharp, frequent eye movements.

As the night progresses, there is a general reduction in the duration of non-REM sleep, particularly in stages 3 and 4, and an increase in the REM periods. Half of the night's REM sleep occurs during its last third, and the longest REM periods may last 30–50 minutes. Still, there seems to be an obligatory refractory period of about 30 minutes between periods of REM; in other words, each REM period is followed by at least 30 minutes of non-REM sleep before the next REM period can begin.

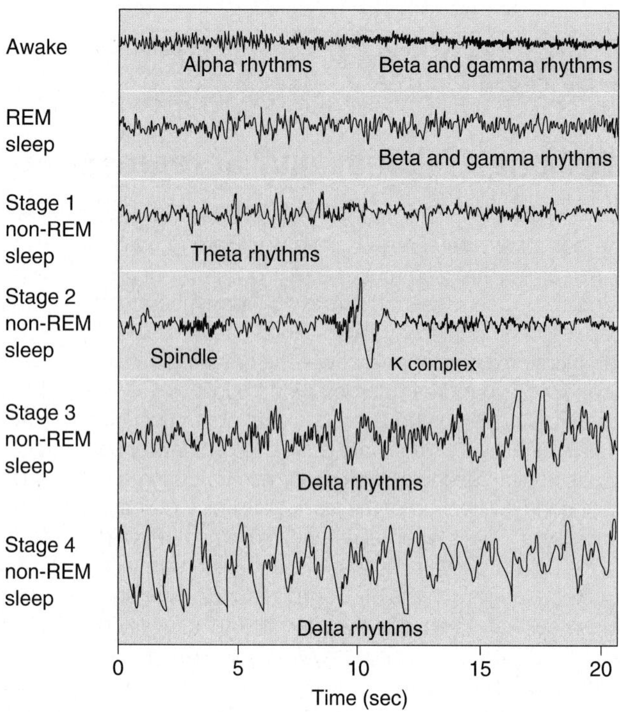

▲ FIGURE 19.16
EEG rhythms during the stages of sleep. (Source: Adapted from Horne, 1988, Fig. 1.1)

What is a normal night's sleep? Your mother may have insisted that you need a "good 8 hours" of sleep each night. Research suggests that normal requirements vary widely among adults, from about 5–10 hours per night. The average length is about 7.5 hours, and the sleep duration of about 68% of young adults is between 6.5 and 8.5 hours. Teenagers may find it especially challenging to get enough sleep. Research by Mary Carskadon at Brown University suggests that sleep requirements do not decrease between preadolescence and early teen years, but changes in circadian timing mechanisms make it progressively harder for teenagers to fall asleep early in the evening. This process often coincides with the move to high school and an earlier start of the school day. As a result, many students are chronically sleep-deprived, which is an unhealthy condition. Too little sleep can reduce cognitive, emotional, and physical well-being.

What is the proper length of sleep time for you? The best measure of successful sleep is the quality of your time awake. You need a certain amount of sleep in order to maintain a reasonable level of alertness. Too much daytime sleepiness can be more than annoying; it can be dangerous if it interferes with driving, for example. Because of the wide variations among individuals, you must decide for yourself how much sleep you need.

Why Do We Sleep?

All mammals, birds, and reptiles appear to sleep, although only mammals and some birds have a REM phase. Sleep time varies widely, from about 18 hours a day in bats and opossums to about 3 hours a day in horses and giraffes. Many people argue that a behavior as pervasive as sleep must have a critical function; otherwise, some species would have

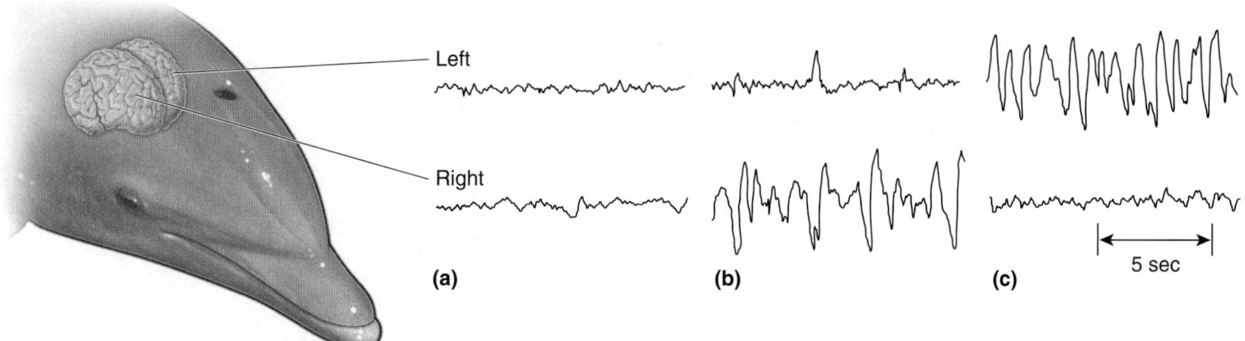

▲ FIGURE 19.17
Sleep in the bottlenose dolphin. These EEG patterns were recorded from the right and left hemispheres of swimming dolphins. **(a)** High-frequency activity on both sides during alert wakefulness. **(b)** Large delta rhythms of deep sleep only on the right side, with fast activation on the left. **(c)** The patterns shift to opposite hemispheres some time later. (Source: Lyamin et al., 2008, Fig. 1.)

lost the need to sleep through evolution. Whatever the function, there is good reason to believe sleep is mainly for the brain. Cognitive impairment is the most immediate and obvious consequence of sleep deprivation. A restful 8 hours in bed without sleep might allow your body to recover from physical exertion, but you would not be at your best mentally the next day.

Some animals apparently have more reason *not* to sleep than others. Imagine living your entire life in deep or turbulent water, yet needing to breathe air every minute or so. Even a quick nap would be awkward, at best. This is precisely the situation with dolphins and whales, yet they sleep about as much as humans do. Remarkably, bottlenose dolphins sleep with only one cerebral hemisphere at a time: about 2 hours of non-REM sleep on just one side, then 1 hour awake on both sides, 2 hours of non-REM sleep on the other side, and so on, for a total of about 12 hours per night (Figure 19.17). (This gives new meaning to the phrase "being half asleep.") There is no evidence that dolphins or whales have REM sleep. Another unusual sleep strategy is used by the blind Indus River dolphin of Pakistan. This dolphin uses sonar to navigate through muddy, turbid, sweeping currents, and during monsoon season it must never stop swimming or it will come to grief on the rocks and debris of the flooded estuary it calls home. Still, the Indus River dolphin seems to sleep, snatching "microsleeps" 4–6 seconds long while continuing to swim slowly. Its many microsleeps add up to about 7 hours in a 24-hour day.

Dolphins have evolved extraordinary sleep mechanisms that adapt them to a demanding environment. But the fact that dolphins are not sleepless reinforces our question: What is so important about sleeping?

No single theory of the function of sleep is widely accepted, but the most reasonable ideas fall into two categories: theories of *restoration* and theories of *adaptation*. The first category is a commonsense explanation: We sleep in order to rest and recover and to prepare to be awake again. The second category is less obvious: We sleep to keep us out of trouble, to hide from predators when we are most vulnerable or from other harmful features of the environment, or to conserve energy.

If sleep is restorative, what is it restoring? Quiet rest is certainly not a substitute for sleep. Sleeping does something more than simple resting. Prolonged sleep deprivation can lead to serious physical and behavioral problems (Box 19.3). Unfortunately, no one has yet identified a particular

BOX 19.3 OF SPECIAL INTEREST

The Longest All-Nighter

In 1963, Randy Gardner was a 17-year-old high school student with an ambitious idea for a San Diego Science Fair project. On December 28 he awoke at 6 a.m. to begin. When he finished 11 days (264 hours) later, he had broken the world's record for nonstop wakefulness, under the continuous scrutiny of two friends and, during the last 5 days, fascinated sleep researchers. He had used no drugs, not even caffeine.

The experience was not pleasant. After 2 days without sleep, Randy became irritable and nauseated, had trouble remembering, and could not even watch television. By the fourth day, he had mild delusions and overwhelming fatigue, and by the seventh day, he had tremors, his speech was slurred, and his EEG no longer showed alpha rhythms. At times he was paranoid or hallucinating. Fortunately, he did not become psychotic, despite the predictions of some experts. On the contrary, on his last awake night, he beat one of his better rested observers at an arcade baseball game, and he gave a coherent account of himself at a national press conference.

When he finally went to bed, he slept for almost 15 hours straight, then stayed awake 23 hours to wait for nightfall, and slept for another 10.5 hours. After the first sleep, his symptoms had mostly disappeared, and within a week, he was sleeping and behaving normally.

One of the most interesting things about Randy's ordeal is that there were no lasting harmful effects. The same is not true for some animals deprived of sleep. If rats are kept awake for long periods, they progressively lose weight while consuming much more food, become weak, accumulate stomach ulcers and internal hemorrhages, and even die. They seem to suffer from an impairment of their ability to regulate body temperature and metabolic needs. Total sleep deprivation is not necessary. Prolonged REM sleep loss alone is detrimental. These results may imply that sleep provides something physiologically essential.

physiological process that is clearly restored by sleep or an essential substance that is made or a toxin that is destroyed while sleeping. Sleep does prepare us to be effectively awake again. But does sleep renew us in the same way that eating and drinking do, by replacing essential substances, or the way the healing of a wound repairs damaged tissues? For the most part, evidence indicates that sleep is not a time of increased tissue repair for the body. However, it is possible that brain regions such as the cerebral cortex can achieve some form of essential "rest" only during non-REM sleep.

Adaptation theories of sleep take many forms. Some large animals eat small animals; a stroll in the moonlight is far too risky for a squirrel living in owl and fox territory. The squirrel's best strategy may be to stay safely tucked away in an underground burrow during the night, and sleep is a good way to enforce such isolation. At the same time, sleep may be an adaptation for conserving energy. While sleeping, the body does only just enough work to stay alive, core temperature drops, temperature regulation is depressed, and the rate of calories burned is kept low.

Functions of Dreaming and REM Sleep

In many ancient cultures, people believed that dreams were a window on some higher world and a source of information, guidance, power, or enlightenment. Perhaps they were right, but the collective wisdom of the past does not agree on exactly how to interpret the meaning of dreams. Today we must take a step backward and first ask whether dreams even *have* meaning. Dreams are difficult to study. Obviously, we can't directly observe the dreams of someone else, and even the dreamer has access to them only after he or she has awakened and perhaps forgotten or distorted the experience. Modern explanations of dreaming lean heavily on studies of REM sleep rather than dreaming because the phenomena of REM can be objectively measured. But it is important to remember that the two are not synonymous. Some dreams can occur outside of REM sleep, and REM sleep has many peculiar features that have nothing to do with dreaming.

Do we need to dream? No one knows, but the body does seem to crave REM sleep. It is possible to deprive sleepers of REM sleep specifically by waking them every time they enter the REM state; when they fall asleep a minute or two later, it is inevitably into a non-REM state, and they can accumulate an entire night of relatively pure non-REM sleep. As Dement first observed, after several days of this annoying treatment, sleepers attempt to enter the REM state much more frequently than normal. When they are finally allowed to sleep undisturbed, they experience *REM rebound* and spend more time in REM proportional to the duration of their deprivation. Most studies have found that REM deprivation does not cause any major psychological harm during the daytime. Again, it is important not to interpret REM deprivation as dream deprivation, since during REM deprivation, dreams may continue to occur during sleep onset and during non-REM periods.

Sigmund Freud suggested many functions for dreams. For Freud, dreams were disguised wish fulfillment, an unconscious way for us to express our sexual and aggressive fantasies, which are forbidden while we are awake. Bad dreams might help us conquer the anxiety-provoking events of life. Recent theories of dreaming are more biologically based. Allan Hobson and Robert McCarley of Harvard University propose an "activation–synthesis hypothesis," which explicitly rejects freudian, psychological interpretations. Instead, dreams, or at least some of their bizarre features, are seen as the associations and memories of the cerebral cortex that are elicited by the random discharges of the pons during REM sleep. Thus, the pontine neurons, via the thalamus, *activate* various areas of the cerebral cortex, elicit well-known images or emotions, and the cortex then tries to *synthesize* the disparate images into a sensible whole. Not surprisingly, the "synthesized" dream product may be quite bizarre and even nonsensical because it is triggered by the semirandom activity of the pons. Evidence for the activation–synthesis hypothesis is mixed. It does predict the weirdness of dreams and their correlation with REM sleep. But it does not explain how random activity can trigger the complex and fluid stories that many dreams contain nor how it can evoke dreams that recur night after night.

Many researchers have suggested that REM sleep, and perhaps dreams themselves, have an important role in memory. None of the evidence is definitive, but intriguing hints indicate that REM sleep somehow aids the integration or consolidation of memories. Depriving humans or rats of REM sleep can impair their ability to learn a variety of tasks. Some studies show an increase in the duration of REM sleep after an intense learning experience. In one study, Israeli neuroscientist Avi Karni and his colleagues trained people to identify the orientation of a small line in their peripheral visual field. The task was made difficult by presenting the visual stimulus for a very short period of time. With repeated practice over days, people got much better at this task; surprisingly, their performance also improved between evening and morning, after a night's sleep. Karni found that if people were deprived of REM sleep, their learning of the task did not improve overnight. Depriving them of non-REM sleep, on the other hand, actually enhanced their performance. Karni hypothesizes that this kind of memory requires a period of time to strengthen and that REM sleep is particularly effective for this purpose.

You may have heard about sleep learning, the notion that you can study for an exam by simply listening to a tape of the material while you blissfully snooze away. Sounds like a student's fantasy, right? Unfortunately, it is exactly that and no more. There is no scientific evidence for sleep learning, and careful studies have shown that the very few things recalled the next morning were heard when the subjects briefly woke up. In fact, sleep is a profoundly amnesic state. Most of our dreams, for example, seem to be lost forever. Although we dream profusely during each of the four or five REM periods every night, we usually remember only the last dream before waking. Also, when we briefly wake up to do something in the middle of the night, we have often forgotten the incident by morning.

At this point you are probably confused about the functions of dreaming and REM sleep. So are we. Unfortunately, there is not enough evidence to support or dismiss any of the theories we have discussed. There are also many other creative and plausible ideas that we do not have the space to present here.

Neural Mechanisms of Sleep

Until the 1940s, it was generally believed that sleep was a passive process: Deprive the brain of sensory input, and it will fall asleep. However, when the sensory afferents to an animal's brain are blocked, the animal continues to have cycles of waking and sleeping. We now know that sleep is an active process that requires the participation of a variety of brain regions. As we saw in Chapter 15, wide expanses of the cortex are actually controlled by very small collections of neurons much deeper in the brain. These cells act like the switches or tuners of the forebrain, altering cortical excitability and gating the flow of sensory information into it. The full details of these control systems are complex and not fully understood. But we can summarize a few basic principles:

1. The neurons most critical to the control of sleeping and waking are part of the diffuse modulatory neurotransmitter systems (see Chapter 15, Figures 15.12 to 15.15).
2. The brain stem modulatory neurons using norepinephrine and serotonin fire during waking and enhance the awake state; some neurons using acetylcholine enhance critical REM events, and other cholinergic neurons are active during waking.
3. The diffuse modulatory systems control the rhythmic behaviors of the thalamus, which in turn controls many EEG rhythms of the cerebral cortex; slow, sleep-related rhythms of the thalamus apparently block the flow of sensory information up to the cortex.
4. Sleep also involves activity in descending branches of the diffuse modulatory systems, such as the inhibition of motor neurons during dreaming.

There are three basic kinds of evidence for the localization of sleep mechanisms in the brain. Lesion data reveal changes in function after a part of the brain is removed, results of stimulation experiments identify changes following the activation of a brain region, and recordings of neural activity determine the relationship between that activity and different brain states.

Wakefulness and the Ascending Reticular Activating System. Lesions in the brain stem of humans can cause sleep and coma, suggesting that the brain stem has neurons whose activity is essential to keeping us awake. Italian neurophysiologist Giuseppe Moruzzi and his colleagues, working

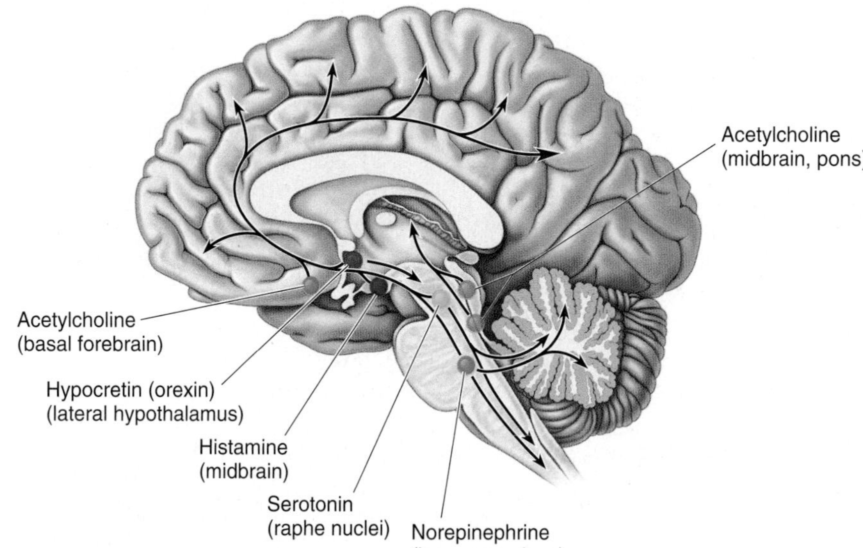

◀ FIGURE 19.18
Key components of the modulatory systems that regulate waking and sleeping.

in the 1940s and 1950s, began to sort out the neurobiology of the brain stem's control of waking and arousal. They found that lesions in the midline structures of the brain stem caused a state similar to non-REM sleep, but lesions in the lateral tegmentum, which interrupted ascending sensory inputs, did not. Conversely, electrical stimulation of the midline tegmentum of the midbrain, within the reticular formation, transformed the cortex from the slow, rhythmic EEGs of non-REM sleep to a more alert and aroused state with an EEG similar to that of waking. Moruzzi called this ill-defined region of stimulation the *ascending reticular activating system* (mentioned in Chapter 15). This area is now much better defined, anatomically and physiologically, and it is clear that Moruzzi's stimulation was affecting many different sets of ascending modulatory systems.

Several sets of neurons increase their firing rates in anticipation of awakening and during various forms of arousal. They include cells of the locus coeruleus, which contain norepinephrine, serotonin-containing cells of the raphe nuclei, acetylcholine-containing cells of the brain stem and basal forebrain, midbrain neurons that use histamine as a neurotransmitter, and neurons of the hypothalamus that use *hypocretin (orexin)* as a transmitter (Figure 19.18). Collectively, these neurons synapse directly on the entire thalamus, cerebral cortex, and many other brain regions. The general effects of their transmitters are a depolarization of neurons, an increase in their excitability, and a suppression of rhythmic forms of firing. These effects are most clearly seen in the relay neurons of the thalamus (Figure 19.19).

Hypocretin (also known as *orexin*; see Chapter 16) is a small peptide neurotransmitter expressed mainly by neurons whose cell bodies are in the lateral hypothalamus. The axons of hypocretin (orexin)-secreting neurons project widely in the brain, and they strongly excite cells of the cholinergic, noradrenergic, serotonergic, dopaminergic, and histaminergic modulatory systems. When the peptide was first discovered, researchers thought hypocretin (orexin) was involved specifically in feeding behavior (see Chapter 16), but it clearly has a more general role. The peptide also promotes wakefulness, inhibits REM sleep, facilitates neurons that enhance certain kinds of motor behavior, and is involved in the regulation of neuroendocrine and autonomic systems. The loss of hypocretin (orexin) neurons leads to a sleep disorder called *narcolepsy* (Box 19.4).

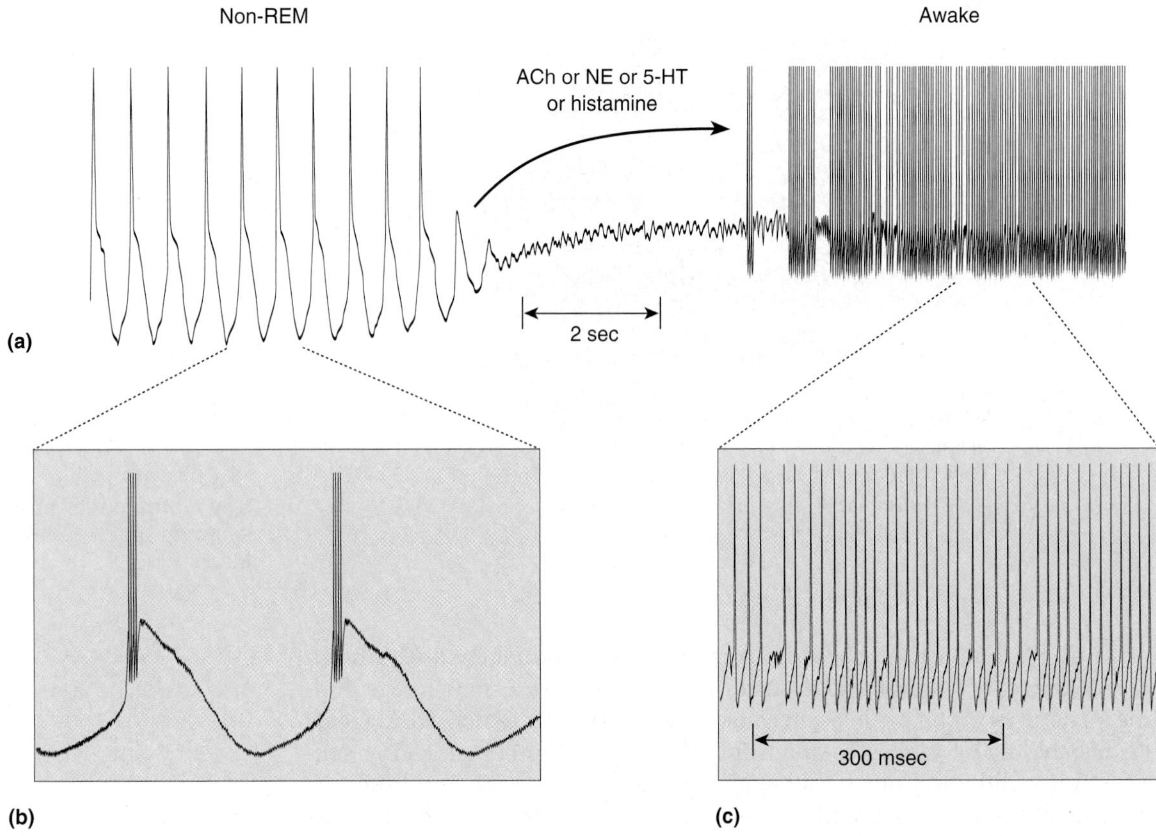

▲ FIGURE 19.19
Modulating thalamic rhythmicity during waking and sleeping. **(a)** Thalamic neurons at rest have a tendency to generate slow, delta frequency rhythms of intrinsic burst-firing (left). Under the influence of several neuromodulators such as ACh, NE, and histamine, neurons depolarize and switch to a more excitable single-spiking mode (right). This may resemble what happens during transitions from non-REM sleep to the waking state. Expanded views of rhythmic bursting **(b)** and single-spiking **(c)** are also shown. (Source: Adapted from McCormick and Pape, 1990, Fig. 14.)

Falling Asleep and the Non-REM State. Falling asleep involves a progression of changes over several minutes, culminating in the non-REM state. It is not entirely clear what initiates non-REM sleep, although certain sleep-promoting factors contribute (as we will describe later), and there is a general decrease in the firing rates of most brain stem modulatory neurons (those using NE, 5-HT, and ACh). Although most regions of the basal forebrain seem to promote alertness and arousal, a subset of its cholinergic neurons increases their firing rate with the onset of non-REM sleep and are silent during wakefulness.

Early stages of non-REM sleep include the EEG sleep spindles, described earlier, which are generated in part by the inherent rhythmicity of thalamic neurons (see Figure 19.11). As non-REM sleep progresses, spindles disappear and are replaced by slow delta rhythms (less than 4 Hz). Delta rhythms may also be a product of thalamic cells, occurring when their membrane potentials become even more negative than during spindle rhythms (and much more negative than during waking). Synchronization of activity during spindle or delta rhythms is due to neural interconnections within the thalamus and between the thalamus and cortex. Because of the strong, two-way excitatory connections between the thalamus and cortex, rhythmic activity in one is often strongly and widely projected upon the other.

BOX 19.4 OF SPECIAL INTEREST

Narcolepsy

Narcolepsy is a bizarre and disabling disturbance of sleeping and waking. Despite the sound of the name, it is not a form of epilepsy. It can include some or all of the following manifestations.

Excessive daytime sleepiness can be severe and often leads to unwanted "sleep attacks." *Cataplexy* is a sudden muscular paralysis while consciousness is maintained. In the middle of a normal day, sufferers suddenly collapse into a state similar to REM sleep. Cataplexy is often brought on by strong emotional expression, such as laughter or tears, or by surprise or sexual arousal, and it usually lasts less than a minute. *Sleep paralysis*, a similar loss of muscle control, occurs during the transition between sleeping and waking. Sometimes occurring in the absence of narcolepsy, it can be very disconcerting; even though conscious, a person may be unable to move or speak for several minutes. *Hypnagogic hallucinations* are graphic dreams, often frightening, that can accompany sleep onset and may occur following sleep paralysis. Sometimes such dreams flow smoothly with real events that occurred just prior to falling asleep.

EEG monitoring reveals a distinct difference between narcoleptic and normal sleep. A narcoleptic person goes directly from waking into a REM phase, whereas normal adult sleepers always enter a long period of non-REM sleep first. Most narcolepsy symptoms might be interpreted as an abnormal intrusion of the characteristics of REM sleep into waking.

The prevalence of narcolepsy varies widely, affecting about 1 in 1000–2000 people in the U.S. population but only 1 in 500,000 in Israel, for example. The typical age of onset is 12–16 years. The disorder has a genetic component, and a high percentage of narcoleptics have a particular form of the human leukocyte antigen (HLA) gene. However, about 25% of the general population has the narcoleptic form of the HLA gene, yet the large majority do not develop narcolepsy. Environmental factors may also play an important role. A recent study in China found that the onset of narcolepsy in children varies with the seasons and tends to be highest following winter-related respiratory infections. There was a particularly sharp rise in cases of narcolepsy just after the H1N1 influenza pandemic in 2009–2010, followed by a decrease in the two years following. Narcolepsy rates increased both in Europe, where many people were vaccinated against H1N1, and in China, where vaccines were not available.

Narcolepsy occurs in goats, donkeys, ponies, and more than a dozen breeds of dogs. In 1999, Emmanuel Mignot, Seiji Nishino, and their research team at Stanford University found that canine narcolepsy is caused by a mutation of the gene for a hypocretin receptor. Also in 1999, Masashi Yanagisawa and his group at the University of Texas Southwestern Medical Center deleted the genes responsible for the peptide neurotransmitter hypocretin in mice and found that the animals were narcoleptic. Basic animal research of this sort quickly inspired important studies of human narcolepsy.

In 2000, two research teams discovered that the brains of human narcoleptics have about 10% or less of the normal complement of hypocretin-containing neurons (Figure A). Their CSF has immeasurably low levels of hypocretin, whereas hypocretin is found at normal levels in nearly every other neurological disease. Human narcolepsy almost certainly results from the selective death of hypocretin-containing neurons in most cases. Unlike in some animal versions of the disease, hypocretin deficiency is rarely caused by mutations of the hypocretin or hypocretin receptor genes. The reason hypocretin neurons die in narcoleptic patients is unknown, although there is strong evidence that some kind of autoimmune process is involved. Fragments of viral proteins may mimic hypocretin, somehow priming immune cells to attack hypocretin-releasing cells.

There is no cure for narcolepsy yet, and current treatments aim only to relieve the symptoms. Frequent naps, amphetamines, and a drug called *modafinil* may help daytime sleepiness, while tricyclic antidepressant drugs (which have REM-suppressant effects) may reduce cataplexy and sleep paralysis. The discovery that hypocretin deficiency underlies narcolepsy suggests an obvious potential treatment; administer hypocretin or its agonists. Results from human trials have so far been disappointing. One problem is that hypocretin does not penetrate the blood–brain barrier very well. Transplantation of hypocretin neurons has shown some promise in animal studies, but no human trials have been attempted.

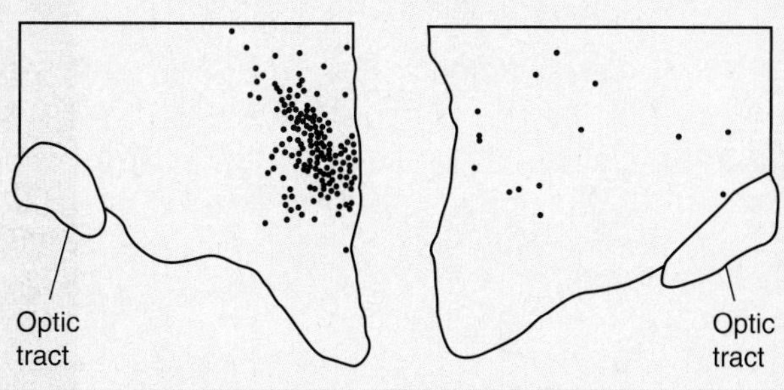

Figure A
Hypocretin (orexin)-containing neurons in the hypothalamus of a normal brain (left) and a narcoleptic brain (right). (Source: Adapted from Thannickal et al., 2000, Fig. 1)

Mechanisms of REM Sleep. REM is such a different state from non-REM that we would expect some clear neural distinctions. Many cortical areas are at least as active during REM sleep as they are during waking. For example, neurons of the motor cortex fire rapidly and generate organized motor patterns that attempt to command the entire body but succeed only with a few muscles of the eye and inner ear and those essential for respiration. The elaborate dreams of REM sleep certainly require the cerebral cortex. However, the cortex is not necessary for the *production* of REM sleep.

The use of PET and fMRI imaging in the waking and sleeping human brain has provided fascinating glimpses of the activity patterns that distinguish waking from REM and non-REM sleep. Figure 19.20a shows the difference in brain activity between REM sleep and waking. Some areas, including primary visual cortex, were about equally active in the two states. However, extrastriate cortical areas and portions of the limbic system were significantly more active during REM sleep than waking. Conversely, regions of the frontal lobes were noticeably less active during REM. Figure 19.20b contrasts the brain activity in REM and non-REM sleep. The primary visual cortex and a number of other areas are significantly less active during REM sleep, but extrastriate cortex is more active than it is in non-REM. These results paint an intriguing picture of what happens when we sleep. During REM, there is an explosion of extrastriate activity, presumably during the times when we dream. However, there is no corresponding activity increase in the primary visual cortex, suggesting that the extrastriate excitation is internally generated. The emotional component of dreams might derive from the heightened limbic activation. The low activity in the frontal lobe suggests that high-level integration or interpretation of the extrastriate visual information might not take place, leaving us with a buzz of uninterpreted visual imagery.

The control of REM sleep, as with the other functional brain states, derives from diffuse modulatory systems in the core of the brain stem, particularly the pons. The firing rates of the two major systems of the

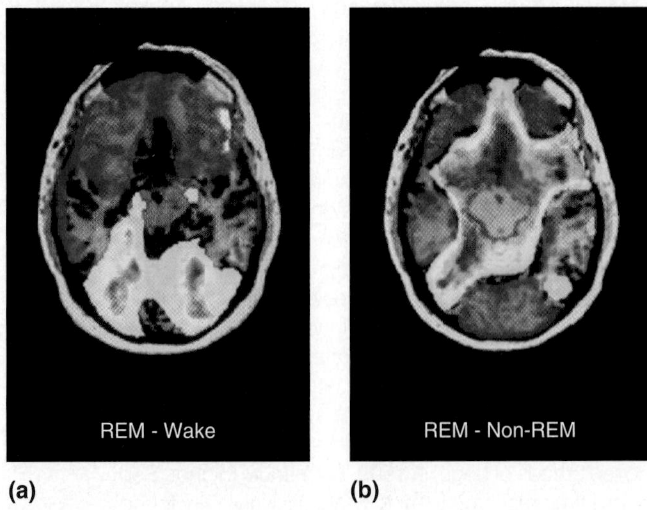

▲ FIGURE 19.20

PET images of the waking and sleeping human brain. These images show brain activity in horizontal sections. **(a)** Colors represent differences in activity between REM sleep and waking; green, yellow, and red indicate higher activation during REM, and purples indicate lower activation during REM. Note the dark notch at the bottom (posterior) edge of the section, indicating that striate cortex is equally active in the two states. **(b)** REM sleep compared with non-REM sleep. In REM, striate cortex is less active. (Source: Braun et al., 1998, Fig. 1.)

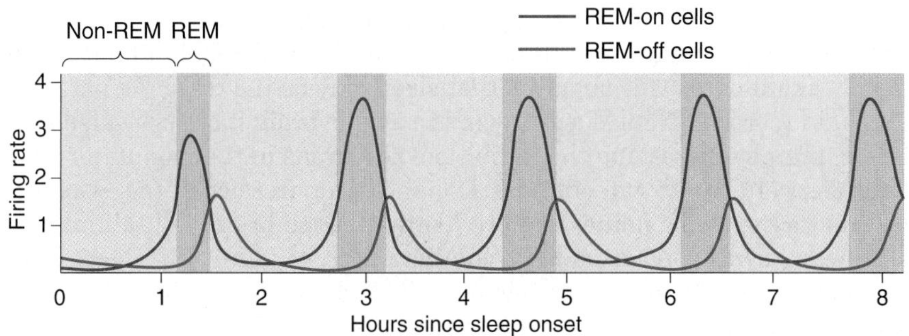

◀ FIGURE 19.21
Control of the onset and offset of REM periods by brain stem neurons. This graph shows the relative firing rates of REM-associated neurons during a single night. Periods of REM sleep are green. REM-on cells are cholinergic neurons of the pons, and they increase their firing rates just before the onset of REM sleep (red line). REM-off cells are noradrenergic and serotonergic neurons of the locus coeruleus and raphe nuclei, respectively, and their firing rates increase just before the end of REM sleep (blue line). (Source: McCarley and Massaquoi, 1986, Fig. 4B.)

upper brain stem, the locus coeruleus and the raphe nuclei, decrease to almost nothing before the onset of REM (Figure 19.21). However, there is a concurrent sharp increase in the firing rates of ACh-containing neurons in the pons, and some evidence suggests that cholinergic neurons induce REM sleep. It is probably the action of ACh during REM sleep that causes the thalamus and cortex to behave so much like they do in the waking state.

Why don't we act out our dreams? The same core brain stem systems that control the sleep processes of the forebrain also actively inhibit our spinal motor neurons, preventing the descending motor activity from expressing itself as actual movement. This is clearly an adaptive mechanism, protecting us from ourselves. In rare cases, usually elderly men, dreamers seem to act out their dreams; they have a hazardous condition known as *REM sleep behavior disorder*. These people often sustain repeated injuries, and even their spouses have fallen victim to their nocturnal flailings. One man dreamed he was in a football game and tackled his bedroom bureau. Another imagined he was defending his wife from attack, when in fact he was beating her in her bed. The basis for this REM disorder seems to be disruption of the brain stem systems that normally mediate REM atonia. Experimental lesions in certain parts of the pons can cause a similar condition in cats. During REM periods, they may seem to chase imaginary mice or investigate invisible intruders. Disorders of REM control mechanisms, caused by a deficiency of hypocretin (orexin), also contribute to the problems of people with narcolepsy (see Box 19.4).

Sleep-Promoting Factors. Sleep researchers have searched intensively for chemicals in the blood or cerebrospinal fluid (CSF) that promote or even cause sleep. Many sleep-promoting substances have been identified in sleep-deprived animals. We will describe some of the major ones here. One key sleep substance is adenosine. *Adenosine* is used by all cells to build some of the most basic molecules of life, including DNA, RNA, and adenosine triphosphate (ATP). Adenosine is also released by some neurons and glia and acts as a neuromodulator at synapses throughout the brain. It's a substance that might appeal to the millions who drink coffee, tea, and cola. From ancient times, antagonists of adenosine receptors, such as caffeine and theophylline, have been used to keep people awake. Conversely, the administration of adenosine or its agonists increases sleep. Extracellular levels of natural brain adenosine are higher during waking than while sleeping. Levels progressively increase during prolonged waking periods and sleep deprivation, and they gradually decrease during sleep. Waking-related changes in adenosine levels occur not in the entire brain, but only in certain sleep-related regions. These two properties of adenosine—its sleep-promoting effect and the levels that track the need for sleep—strongly suggest it is an important sleep-promoting factor.

How might adenosine promote sleep? Adenosine has an inhibitory effect on the diffuse modulatory systems for ACh, NE, and 5-HT that tend to promote wakefulness. This suggests that sleep may be the result of a molecular chain reaction. Neural activity in the awake brain increases adenosine levels, thereby increasing the inhibition of neurons in the modulatory systems associated with wakefulness. Enhanced suppression of the "wakeful" modulatory systems makes it more likely that the brain will fall into the slow-wave synchronous activity characteristic of non-REM sleep. After sleep begins, adenosine levels slowly fall, and activity in the modulatory systems gradually increases until we wake up to start the cycle anew.

Another important sleep-promoting factor is *nitric oxide (NO)*. Recall that NO is a small, mobile, gaseous molecule that can diffuse easily across membranes and serves as a retrograde (postsynaptic to presynaptic) messenger between certain neurons (see Chapter 6). The wake-promoting cholinergic neurons of the brain stem express particularly high levels of the synthesizing enzyme for NO. Brain NO levels are highest during waking, and they rise rapidly with sleep deprivation. How does NO promote sleep? Studies have shown that NO triggers the release of adenosine. As we have seen, adenosine promotes non-REM sleep by suppressing the activity of neurons that help to sustain waking.

Sleepiness is one of the most familiar consequences of infectious diseases, such as the common cold and the flu. There are direct links between the immune response to infection and the regulation of sleep. In the 1970s, physiologist John Pappenheimer of Harvard University identified a muramyl dipeptide in the spinal fluid of sleep-deprived goats that facilitated non-REM sleep. Muramyl peptides are usually produced only by the cell walls of bacteria, not brain cells, and they also cause fever and stimulate immune cells of the blood. It is not clear how they appear in CSF, but they may be synthesized by bacteria in the intestines. More recent research has implicated several *cytokines*, small signaling peptides involved in the immune system, in the regulation of sleep. One of them is *interleukin-1*, which is synthesized in the brain by glia, and by macrophages, cells throughout the body that scavenge foreign material. Like adenosine and NO, interleukin-1 levels increase during waking, and in humans its levels peak just before the onset of sleep. Interleukin-1 promotes non-REM sleep even when the immune system has not been challenged. When given to humans, it induces fatigue and sleepiness. Interleukin-1 also stimulates the immune system.

Another endogenous sleep-promoting substance is *melatonin*, a hormone secreted by the pea-sized pineal body (see the appendix to Chapter 7). Melatonin is a derivative of the amino acid tryptophan. It has been called the "Dracula of hormones" because it is released only when the environment darkens—normally at night—and its release is inhibited by light. In humans, melatonin levels tend to rise around the time we become sleepy in the evening, peak in the early morning hours, and then fall to baseline levels by the time we awaken. Evidence suggests that melatonin helps initiate and maintain sleep, but its precise role in natural sleep–wake cycles is not clear. In recent years, melatonin has become popular as an over-the-counter sleep-promoting drug. Although it has shown some promise in treating the symptoms of jet lag and the insomnia that affects some older adults, the general effect of melatonin on improving sleep remains debatable.

Gene Expression during Sleeping and Waking. Research into the neural function of sleep has benefited from studies at various levels of analysis, including sleep behavior, brain physiology, and the action of diffuse modulatory systems. Methods from molecular neurobiology have also contributed interesting facts. While the pieces do not all fit together quite yet, it is clear that the behavioral states of sleeping and waking are different even at the molecular level. For example, in the macaque monkey, most areas of cerebral cortex show higher rates of protein synthesis in deep sleep than in light sleep. In rats, levels of cAMP in several brain areas have been found to be lower during sleeping than waking.

Research has demonstrated that sleeping and waking are associated with differences in the expression of certain genes. Chiara Cirelli and Giulio Tononi, working at the Neurosciences Institute in San Diego and at the University of Wisconsin, have studied the expression of thousands of genes in rats that were awake or asleep. The vast majority of genes were expressed at the same level in the two states. However, the 0.5% of genes that showed different levels of expression may provide insight into what happens in the brain during sleep. Most of the genes that were more highly expressed in the awake brain could be placed into one of three groups. One group includes what are called *immediate early genes*, genes that code for transcription factors that affect the expression of other genes. Some of these genes appear to be related to changes in synaptic strength. The low expression of these genes during sleep may be associated with the fact that learning and memory formation are largely absent in this state. The second group of genes comes from mitochondria. Increased expression of these genes may play a role in satisfying the higher metabolic demands of the awake brain. The third group includes genes related to responses to cellular stress.

A different group of genes was most highly expressed during sleep, and some of them might contribute to increased protein synthesis and synaptic plasticity mechanisms that complement those most prevalent during waking. An important point is that the sleep-related changes in gene expression were specific to the brain, and they did not change in other tissues, such as the liver and skeletal muscle. This is consistent with the widely held hypothesis that sleep is a process generated by the brain for the benefit of the brain.

CIRCADIAN RHYTHMS

Almost all land animals coordinate their behavior according to **circadian rhythms**, the daily cycles of daylight and darkness that result from the spin of the Earth. (The term is from the Latin *circa*, "approximately," and *dies*, "day.") The precise schedules of circadian rhythms vary among species. Some animals are active during daylight hours, others only at night, and others mainly at the transitional periods of dawn and dusk. Most physiological and biochemical processes in the body also rise and fall with daily rhythms; body temperature, blood flow, urine production, hormone levels, hair growth, and metabolic rate all fluctuate (Figure 19.22). In humans, there is an approximately inverse relationship between the propensity to sleep and body temperature.

When the cycles of daylight and darkness are removed from an animal's environment, circadian rhythms continue on more or less the same schedule because the primary clocks for circadian rhythms are not astronomical (the sun and Earth) but biological in the brain. Brain clocks, like all clocks, are imperfect and require occasional resetting. Now and then you readjust your watch to keep it in sync with the rest of the world (or at least the time on your computer). Similarly, external stimuli, such as light and dark, or daily temperature changes, help adjust the brain's clocks to keep them synchronized with the coming and going of the sunlight.

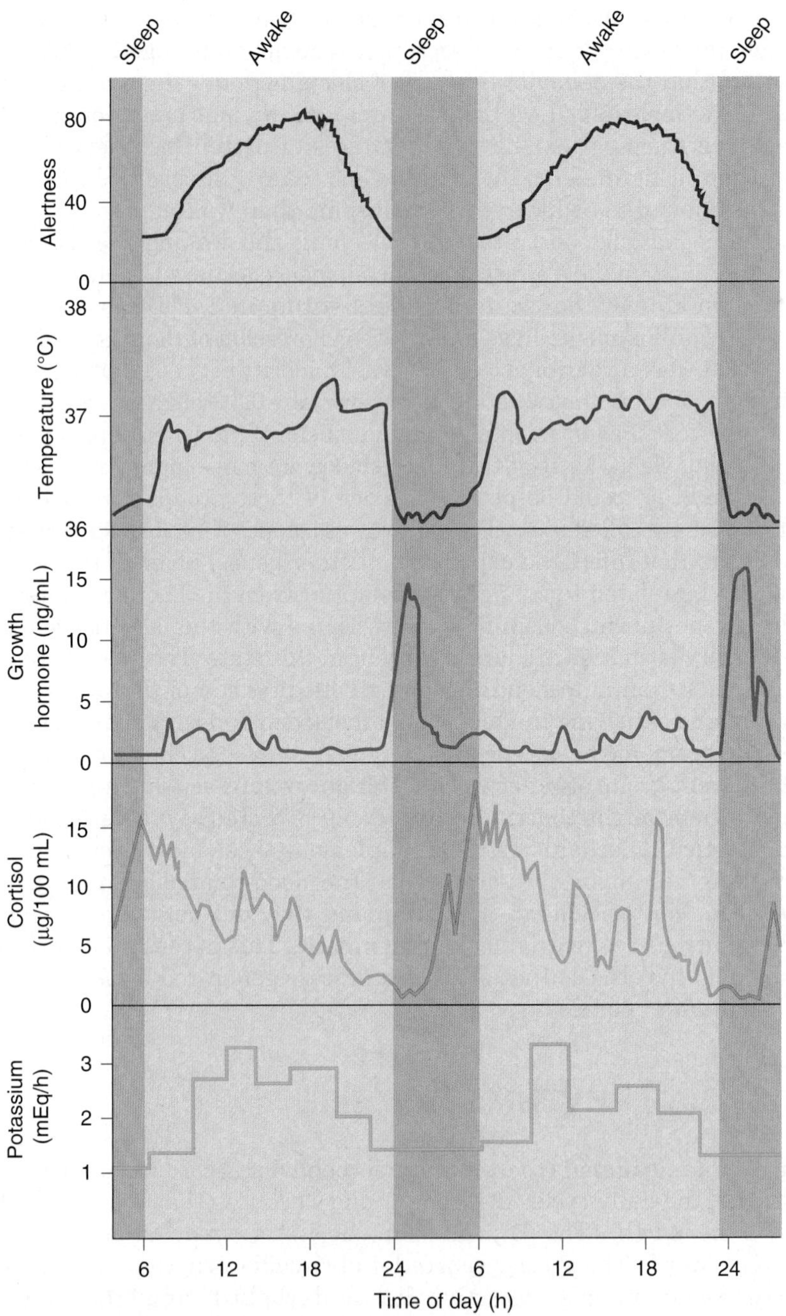

▲ FIGURE 19.22

Circadian rhythms of physiological functions. Fluctuations over two consecutive days are shown here. Alertness and core body temperature vary similarly, but growth hormone and cortisol levels in the blood are highest during sleep, although at different times. The bottom graph shows the excretion of potassium by the kidneys, which is highest during the day. (Source: Adapted from Coleman, 1986, Fig. 2.1.)

Circadian rhythms have been well studied at behavioral, cellular, and molecular levels. Brain clocks are an interesting example of the link between the activity of specific neurons and behavior.

Biological Clocks

The first evidence for a biological clock came from a brainless organism, the mimosa plant. The mimosa raises its leaves during the day and lowers them at night. It seemed obvious to many people that the plant simply reacted to sunlight with some kind of reflex movement. In 1729, French physicist Jean Jacques d'Ortous de Mairan tested the obvious; he put some mimosa plants in a dark closet and found that they continued to raise and lower their leaves. But a surprising new observation can still lead to a wrong conclusion. It was de Mairan's opinion that the plant was still somehow sensing the sun's movement, even in the darkness. More than a century later, Swiss botanist Augustin de Candolle showed that a similar plant in the dark moved its leaves up and down every 22 hours, rather than every 24 hours, according to the sun's movement. This implied that the plant was not responding to the sun and very likely had an internal biological clock.

Environmental time cues (light/dark, temperature and humidity variations) are collectively termed **zeitgebers** (German for "time givers"). In the presence of zeitgebers, animals become *entrained* to the day–night rhythm and maintain an activity cycle of exactly 24 hours. Obviously, even small, consistent errors of timing could not be tolerated for long. A 24.5-hour cycle would, within 3 weeks, completely shift an animal from daytime to nighttime activity. When mammals are completely deprived of zeitgebers, they settle into a rhythm of activity and rest that often has a period more or less than 24 hours, in which case their rhythms are said to *free-run*. In mice, the natural free-running period is about 23 hours, in hamsters it is close to 24 hours, and in humans it tends to be 24.5–25.5 hours (Figure 19.23).

It is quite difficult to separate a human from all possible zeitgebers. Even inside a laboratory, society provides many subtle time cues, such as the sounds of machinery, the comings and goings of people, and the on–off cycling of heating and air conditioning. Some of the most secluded environments are deep caves, which have been the sites for several isolation

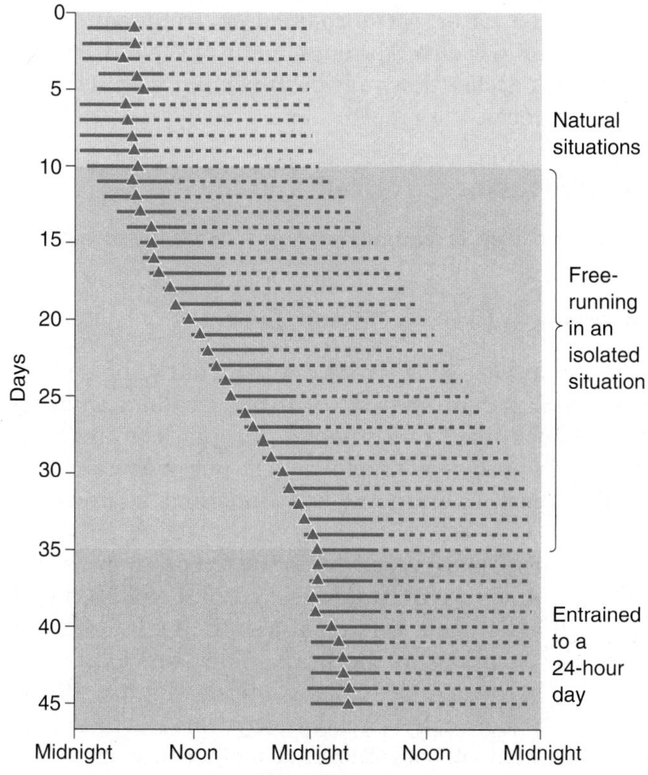

◀ FIGURE 19.23
Circadian rhythms of sleep and wakefulness. This is a plot of one person's daily sleep-wake cycles over a period of 45 days. Each horizontal line is a day; solid lines indicate sleep, and dashed lines indicate waking. A triangle indicates the point of the day's lowest body temperature. The subject was first exposed to 9 days of natural 24-hour cycles of light and dark, noise and quiet, and air temperature. During the middle 25 days, all time cues were removed, and the subject was free to set his own schedule. Notice that the sleep–wake cycles remained stable, but each lengthened to about 25 hours. The subject was now free-running. Notice also that the low point of body temperature shifted from the end of the sleep period to the beginning. During the last 11 days, a 24-hour cycle of light and meals was reintroduced, the subject again entrained to a day-long rhythm, and body temperature gradually shifted back to its normal point in the sleep cycle. (Source: Adapted from Dement, 1976, Fig. 2.)

studies. When people in caves are allowed to set their own schedules of activity for months on end—waking and sleeping, turning lights on and off, and eating when they choose—they initially settle into roughly a 25-hour rhythm. But after days to weeks, their activity may begin to free-run with a surprisingly long period of 30–36 hours. They may stay awake for about 20 hours straight, then sleep for about 12 hours, and this pattern seems perfectly normal to them at the time.

In isolation experiments, behavior and physiology do not always continue to cycle together. Recent studies have found that body temperature and other physiological measures may continue to change reliably over a 24-hour cycle, even if people are entrained on a 20-hour or 28-hour "day" with artificial lighting. This means that the rhythms of temperature and sleeping–waking, which are normally synchronized to a 24-hour period, become desynchronized. In the cave experiments described earlier, there can be even larger differences in the periods of behavioral and physiological cycles, when people are allowed to set their own schedules. Normally, our lowest body temperature occurs shortly before we awaken in the morning, but when desynchronized, this temperature nadir can drift, first moving earlier into the sleep period, and then into waking time. Sleep quality and waking comfort are impaired when cycles are desynchronized. One implication of this desynchronization is that the body has more than one biological clock because sleeping–waking and temperature can cycle at their own pace, uncoupled from one another.

Desynchronization may occur temporarily when we travel and force our bodies suddenly into a new sleep–wake cycle. This is the familiar experience known as *jet lag*, and the best cure is bright light, which helps resynchronize our biological clocks.

The primary zeitgeber for mature mammals is the light–dark cycle. A mother's hormone levels may be the first zeitgeber for some mammals, however, already entraining their activity levels in the womb. In studies of various adult animals, effective zeitgebers have also included the periodic availability of food or water, social contact, environmental temperature cycles, and noise–quiet cycles. Although many of these are much less effective than light–dark cycles, they may be important for particular species, in certain circumstances.

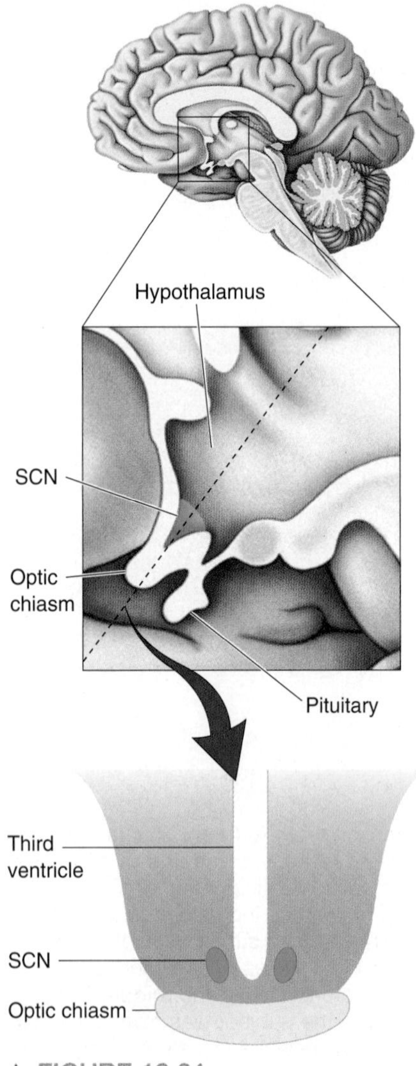

▲ FIGURE 19.24
The human suprachiasmatic nuclei. Two SCN reside within the hypothalamus, just above the optic chiasm and next to the third ventricle. The sagittal view is followed by a frontal view, sectioned at the dashed line.

The Suprachiasmatic Nucleus: A Brain Clock

A biological clock that produces circadian rhythms consists of several components:

$$\text{Light sensor} \rightarrow \text{Clock} \rightarrow \text{Output pathway}$$

One or more input pathways are sensitive to light and dark and entrain the clock and keep its rhythm coordinated with the circadian rhythms of the environment. The clock itself continues to run and keep its basic rhythm even when the input pathway is removed. Output pathways from the clock allow it to control certain brain and body functions according to the timing of the clock.

Mammals have a tiny pair of neuron clusters in the hypothalamus that serves as a biological clock: the **suprachiasmatic nuclei (SCN)**, introduced in Chapter 15. Each SCN has a volume of less than 0.3 mm^3, and its neurons are among the smallest in the brain. They are located on either side of the midline, bordering the third ventricle (Figure 19.24). When the SCN is stimulated electrically, circadian rhythms can be shifted in a predictable way. Removal of both nuclei abolishes the circadian rhythmicity of physical activity, sleeping and waking, and feeding and

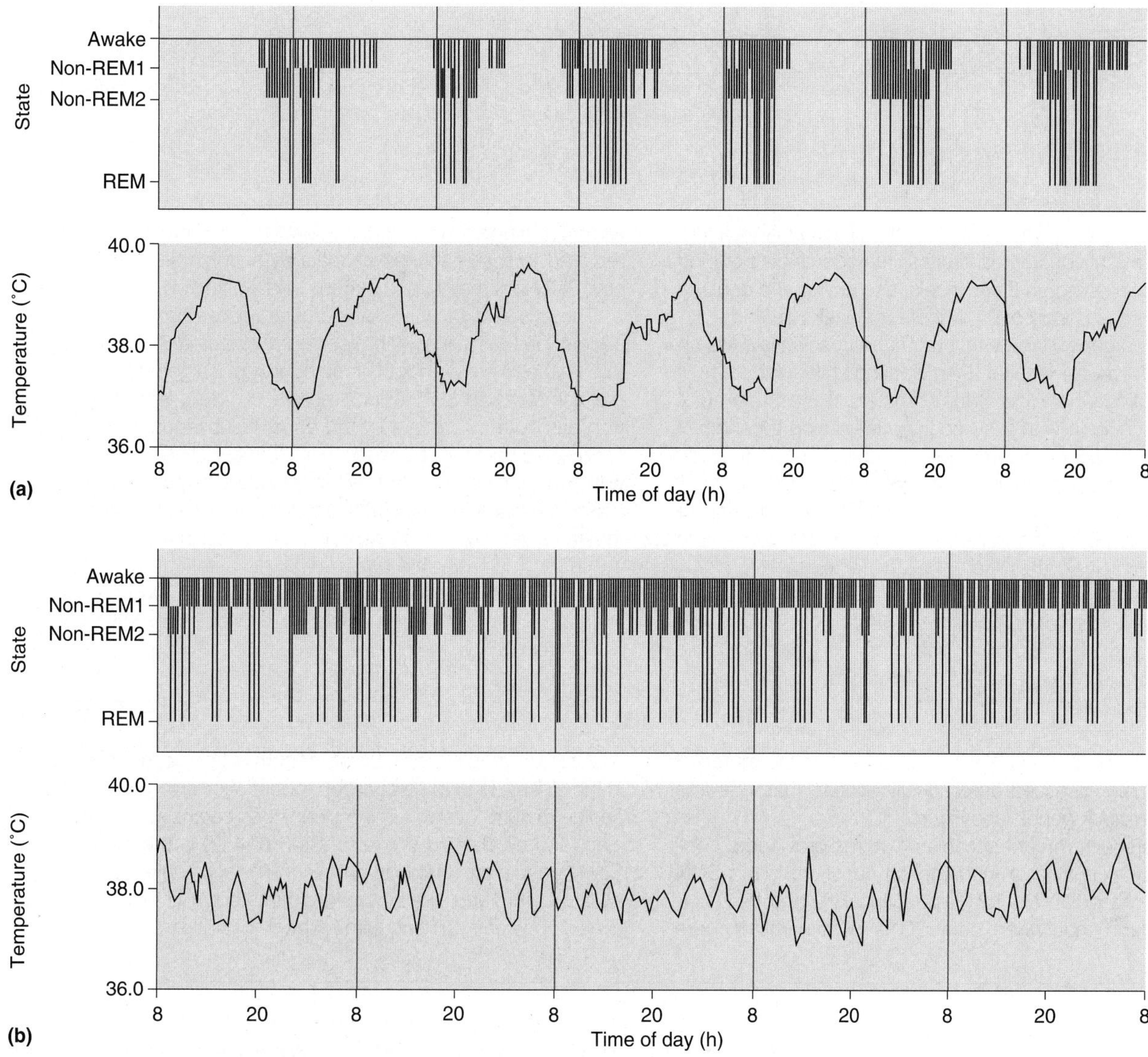

▲ FIGURE 19.25

The SCN and circadian rhythms. (a) Normal squirrel monkeys kept in a constantly lit environment display circadian rhythms of about 25.5 hours. The graph shows the stages of waking–sleeping and concurrent variations in body temperature. The animals' activity states were defined as awake, two levels of non-REM sleep (non-REM1 or non-REM2), or REM sleep. **(b)** Circadian rhythms are abolished in monkeys with lesions in both SCN, kept in the same constant light environment. Notice that persistent high-frequency rhythms of both activity and temperature result from SCN lesions. (Source: Adapted from Edgar et al., 1993, Figs. 1, 3.)

drinking (Figure 19.25). In hamsters, the transplantation of a new SCN can restore rhythms within 2–4 weeks (Box 19.5). The brain's internal rhythms never return without an SCN. Lesions in the SCN do not abolish sleeping, however, and animals will continue to coordinate their sleeping and waking with light–dark cycles if they are present. Sleep appears to be regulated by a mechanism other than the circadian clock, which depends primarily on the amount and timing of prior sleep.

Because behavior is normally synchronized with light–dark cycles, there must also be a photosensitive mechanism for resetting the brain clock. The SCN accomplishes this via the retinohypothalamic tract: Axons from ganglion cells in the retina synapse directly on the dendrites of SCN neurons. This input from the retina is necessary and sufficient to entrain

BOX 19.5 OF SPECIAL INTEREST

Mutant Hamster Clocks

Golden hamsters are perfectionists of circadian timing. When placed in constant darkness, they continue sleeping and waking, running on their wheels, and eating and drinking over an average period of 24.1 hours for weeks on end.

It was this dependability that made neuroscientists Martin Ralph and Michael Menaker, then working at the University of Oregon, notice when one of the hamsters in their laboratory began punching in with 22.0-hour cycles during a period of 3 weeks in the dark. This maverick male was bred with three females of unimpeachable circadian character (their free-running periods were 24.01, 24.03, and 24.04 hours—quite normal). When 20 pups from the three resulting litters were tested in the dark, their free-running periods were evenly split into two narrow groups. Half had periods of 24.0 hours and half had periods of 22.3 hours. Further cross-breeding showed that the hamsters with the shorter circadian periods had one mutant copy of a gene (*tau*) that was dominant over their normal gene. After further breeding, Ralph and Menaker found that animals with two copies of the mutant *tau* gene had free-running periods of only 20 hours! The *tau* mutation was eventually identified as a specific kinase that interacts with certain clock genes (see Figure 19.27).

Hamsters with mutant circadian rhythms provided a convincing way to answer a fundamental question: Is the SCN the brain's circadian clock? Ralph, Menaker, and their colleagues found that when both SCNs of a hamster were ablated, rhythms were entirely lost. But rhythms could be restored to these ablated animals by simply transplanting a new SCN into their hypothalamus and waiting about a week.

The key finding was that hamsters receiving transplants adopted the circadian rhythm of the *transplanted* SCN, not the rhythm they were born with. In other words, if a genetically normal hamster with SCN lesions received an SCN from a donor with one copy of the mutant *tau* gene, it subsequently cycled at about 22 hours. If its transplanted SCN came from an animal with two mutant *taus*, it cycled at 20 hours. This is very compelling evidence that the SCN is the master circadian clock in the hamster brain and probably in our own brain as well.

Short circadian periods were often devastating to a mutant hamster's lifestyle when it was placed into normal 24-hour light and dark cycles. A hamster's normal preference is to be active at night, but most *tau* animals could not completely entrain to the 24-hour rhythm. Instead, they found their activity periods continually shifting through various parts of the light–dark cycle.

A similar problem sometimes occurs in people, most often in the elderly. Due to an age-dependent shortening of the circadian rhythm, overwhelming sleepiness begins in early evening, and awakening comes at 3:00 or 4:00 in the morning. Some people are unable to entrain their sleep–wake cycles to a daily rhythm and, like the mutant hamsters, find their activity cycles constantly shifting with respect to daylight.

sleeping and waking cycles to night and day. When recordings are made from neurons of the SCN, many are indeed sensitive to light. Unlike the more familiar neurons of the visual pathways (see Chapter 10), SCN neurons have very large, nonselective receptive fields and respond to the luminance of light stimuli rather than their orientation or motion.

Research in the last decade suggests, surprisingly, that the retinal cells synchronizing the SCN are neither rods nor cones. It had long been known that eyeless mice cannot use light to reset their clocks, but mutant mice with intact retinas that lack rods and cones can! Since rods and cones were the only known photoreceptors in mammals, it remained a mystery how light could affect the circadian clock without them.

This mystery was solved by David Berson of Brown University and his colleagues. They discovered a new photoreceptor in the retina that was not at all like rods and cones but was, remarkably, a very specialized type of ganglion cell. Recall from Chapter 9 that ganglion cells are retinal neurons whose axons send visual information to the rest of the brain; ganglion cells, like nearly all other neurons in the brain, were not supposed to be directly sensitive to light. The light-sensitive ganglion cells, however, express a unique type of photopigment called *melanopsin*, which is not present in rods and cones. These neurons are very slowly excited by light, and their axons send a signal directly to the SCN that can reset the circadian clock that resides there.

Output axons of the SCN mainly innervate nearby parts of the hypothalamus, but some also go to the midbrain and other parts of the diencephalon. Because almost all SCN neurons use GABA as their primary neurotransmitter, presumably they inhibit the neurons they innervate. It is not yet clear how the SCN sets the timing of so many important behaviors. Extensive lesions in the efferent SCN pathways disrupt circadian rhythms. In addition to the axonal output pathways, SCN neurons may rhythmically secrete the peptide neuromodulator vasopressin (see Chapter 15).

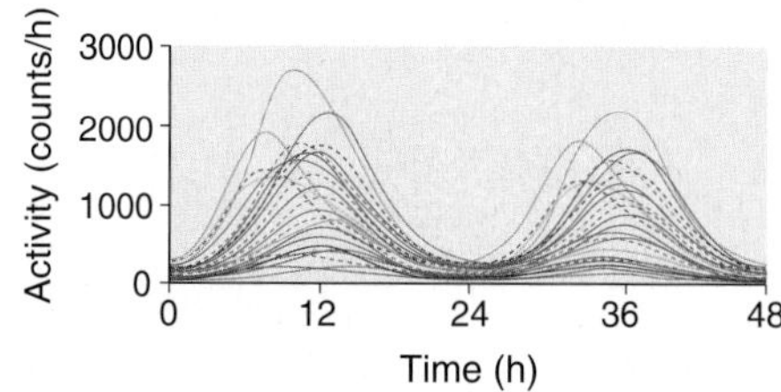

▲ FIGURE 19.26
Circadian rhythms of the SCN isolated from the rest of the brain. The activity of a clock gene was monitored in 100 individual SCN neurons maintained in tissue culture. Each neuron generates a strong circadian rhythm that is well coordinated with the other neurons. (Source: Adapted from Yamaguchi et al., 2003, Fig. 1)

SCN Mechanisms

How do neurons of the SCN keep time? It's clear that each SCN cell is a minuscule clock that keeps time with regular ticks and tocks of its molecular machinery. The ultimate isolation experiment has been simply to remove neurons from the SCN of an animal and grow each cell alone in a tissue culture dish, segregating them from the rest of the brain and from each other. Nevertheless, their rates of action potential firing, glucose utilization, vasopressin production, and protein synthesis continue to vary with rhythms of about 24 hours, just as they do in the intact brain (Figure 19.26). SCN cells in culture can no longer be entrained to light–dark cycles (input from the eyes is necessary for this), but their basic rhythmicity remains intact and expresses itself just as it does when an animal is deprived of zeitgebers.

SCN cells communicate their rhythmic message to the rest of the brain through efferent axons, using action potentials in the usual way, and rates of SCN cell firing vary with a circadian rhythm. However, action potentials are not necessary for SCN neurons to maintain their rhythm. When tetrodotoxin (TTX), a blocker of sodium channels, is applied to SCN cells, it blocks their action potentials but has no effect on the rhythmicity of their metabolism and biochemical functions. When the TTX is removed, action potentials resume firing with the same phase and frequency they had originally, before the TTX, implying that the SCN clock keeps running even without action potentials. SCN action potentials are like the hands of a clock; removing the clock's hands does not stop the innards of the clock from working, but it does make it impossible to read the time in the usual way.

What is the nature of this clock that functions without action potentials? Research in a wide range of species indicates that it is a molecular cycle based on gene expression. The molecular clock used in humans is similar to those found in mice, fruit flies (*Drosophila*), and even bread mold. In *Drosophila* and mice, the system involves a variety of **clock genes.** Some of the more important genes in mammals are known as *period (per)*, *cryptochrome*, and *clock*. Although the details vary across species, the basic scheme is a negative feedback loop. Many of the details were first worked out in experiments performed by Joseph Takahashi and his colleagues at Northwestern University, who named the *clock* gene (an acronym for *circadian locomotor output cycles kaput*). A clock gene is transcribed to produce mRNA that is then translated into proteins. After a delay, the newly manufactured proteins send feedback and interact with the transcription mechanism, causing a decrease in gene expression. As a consequence of decreased transcription, less protein is produced, and gene expression again increases to start the cycle anew. This entire cycle takes about 24 hours, and thus it is a circadian rhythm (Figure 19.27).

If each SCN neuron is a clock, there must be a mechanism to coordinate the thousands of cellular clocks so that the SCN as a whole transmits a

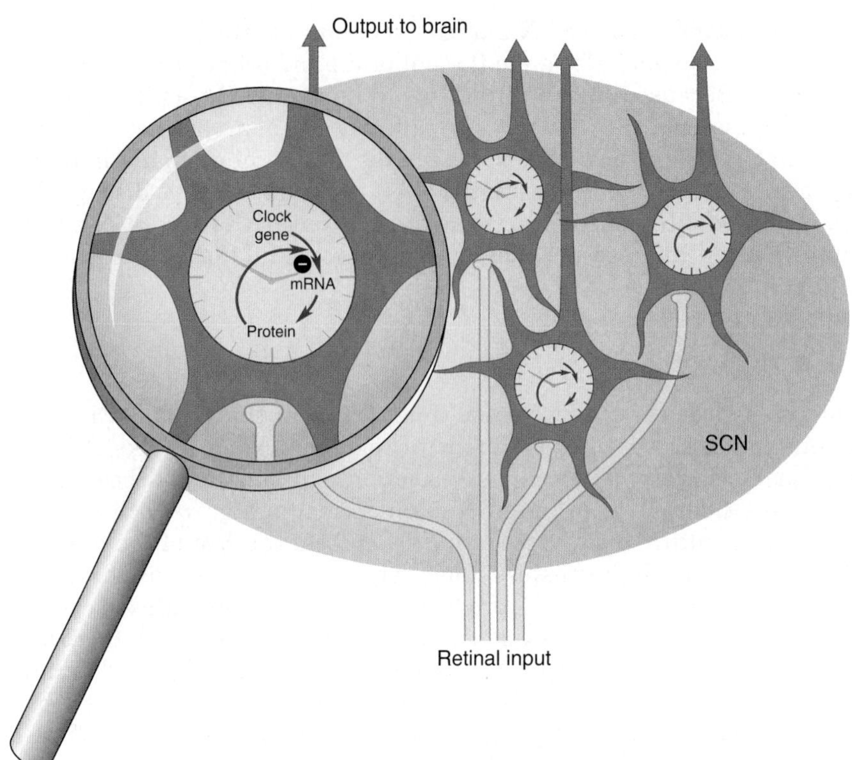

▶ FIGURE 19.27
Clock genes. In the SCN, clock genes produce proteins that inhibit further transcription. Gene transcription and the firing rate of individual SCN neurons cycle up and down over 24 hours. The cycles of many cells are synchronized by light exposure (input from the retina) and by interactions of the SCN neurons.

single, clear message about time to the rest of the brain. Light information from the retina serves to reset the clocks in the SCN neurons each day, but the SCN neurons also communicate directly with each other. Surprisingly, the coordination of rhythms *between* SCN cells seems to be independent of action potentials and normal synaptic transmission because TTX does not block them. Also, the SCN of the very young rat brain coordinates circadian rhythms perfectly well, even before it has developed any chemical synapses. The nature of neuron-to-neuron communication within the SCN is poorly understood, but in addition to classical chemical synapses, it includes other chemical signals, electrical synapses (gap junctions), and the participation of glia.

Research has shown that nearly every cell of the body, including those in the liver, kidney, and lung, has a circadian clock. The same types of gene transcription feedback loops that drive the SCN clock also drive the clocks in these peripheral tissues. When cells from liver, kidney, or lung are grown in isolation, each exhibits a circadian rhythm of its own. Under normal conditions in an intact body, however, all cells' clocks are under the master control of the SCN. How does the SCN govern the innumerable clocks scattered throughout the body's organs? Several signaling pathways seem to be important. The SCN has a strong circadian influence on the autonomic nervous system, core body temperature, adrenal gland hormones such as cortisol, and neural circuits that control feeding, movement, and metabolism (Figure 19.28). Each of these processes in turn regulates many of the body's circadian clocks. Body temperature, for example, has a powerful effect on the clocks of peripheral tissues. It drops sharply by about 1°C each night under the influence of the SCN (see Figure 19.22). This pulse of cooling helps to ensure that the clocks of the internal organs all remain set to the daily rhythms of the SCN and thus to the cycles of environmental dark and light. Interestingly, the SCN's circadian clock is very resistant

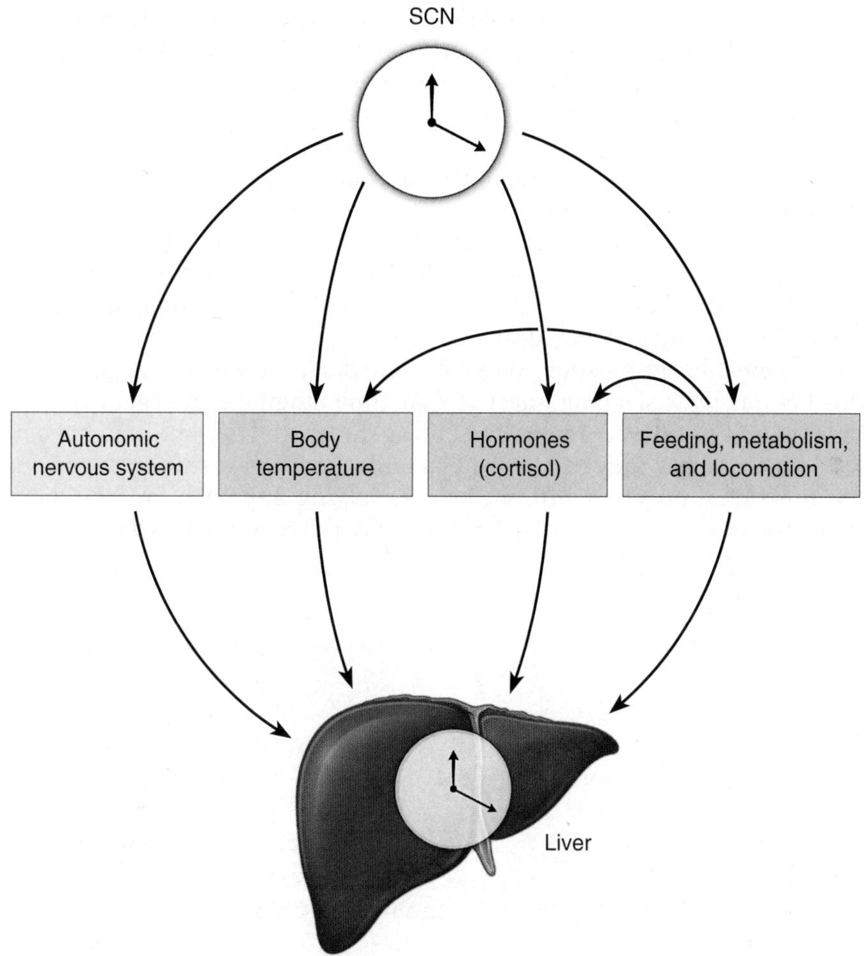

▲ FIGURE 19.28
Control pathways from the SCN to peripheral circadian clocks. The SCN regulates circadian clocks throughout the body (including the liver shown here) via its control over the ANS; core body temperature; cortisol and other hormones; and feeding, locomotion, and metabolism. (Source: Adapted from Mohawk et al., 2012, Fig. 3.)

to temperature changes; this makes sense because it ensures that the SCN, which controls core temperature changes, is not destabilized by its own control signals.

The complex systems that coordinate the body's clocks are not perfect. Odd feeding schedules, chronic doses of methamphetamine and, as we mentioned previously, extreme living conditions (long-term cave dwelling) can all desynchronize the body's circadian clocks.

CONCLUDING REMARKS

Rhythms are ubiquitous in the vertebrate central nervous system. They also span a broad range of frequencies, from more than 500 Hz in the cortical EEG to once per year (0.00000003 Hz) for many seasonal behaviors, such as the autumn mating of deer, the winter hibernation of chipmunks, and the instinct that drives migrating swallows to return to Capistrano, California every March 19. According to local legend, in 200 years, these swallows have missed the date only twice. In some cases, these rhythms are based on intrinsic brain mechanisms; in some, they result from environmental factors; and in others, such as the SCN clock, they represent an interaction of a neural process and zeitgebers.

While the purpose of some rhythms is obvious, the functions of many neural rhythms are unknown. Indeed, some rhythms may have no function at all but arise as a secondary consequence of neural interconnections that are essential for other, nonrhythmic, purposes.

Among the most conspicuous yet inexplicable of brain rhythms is sleep. Sleep provides a fascinating set of problems for neuroscience. Unlike most studies of single ion channels, single neurons, or the systems mediating perception and movement, sleep research begins with profound ignorance about a most basic question: Why? We still don't know why we spend one-third of our lives sleeping, most of that time languid and vegetative and the rest of it paralyzed and hallucinating. Sleep and dreams may have no vital function, but they can be studied and enjoyed nevertheless. Ignoring the functional question will not be a satisfying approach for long, however. For most neuroscientists, asking "Why?" remains the deepest and most challenging problem of all.

KEY TERMS

The Electroencephalogram
electroencephalogram (EEG) (p. 646)
magnetoencephalography (MEG) (p. 648)
generalized seizure (p. 655)
partial seizure (p. 655)
epilepsy (p. 656)

Sleep
rapid eye movement sleep (REM sleep) (p. 658)
non-REM sleep (p. 658)
atonia (p. 659)
ultradian rhythm (p. 661)

Circadian Rhythms
circadian rhythm (p. 673)
zeitgeber (p. 675)
suprachiasmatic nucleus (SCN) (p. 676)
clock gene (p. 679)

REVIEW QUESTIONS

1. Why do EEGs with relatively fast frequencies tend to have smaller amplitudes than EEGs with slower frequencies?
2. The human cerebral cortex is very large and must be folded extensively to fit within the skull. What do the foldings of the cortical surface do to the brain signals that are recorded by an EEG electrode at the scalp?
3. Sleep seems to be a behavior of every species of mammal, bird, and reptile. Does this mean that sleep performs a function essential for the life of these higher vertebrates? If you do not think so, what might be an explanation for the abundance of sleep?
4. An EEG during REM sleep is very similar to an EEG when awake. How do the brain and body in REM sleep *differ* from the brain and body when awake?
5. What is a likely explanation for the brain's relative insensitivity to sensory input during REM sleep compared to the waking state?
6. The SCN receives direct input from the retina, via the retinohypothalamic tract, and this is how light–dark cycles can entrain circadian rhythms. If the retinal axons were somehow disrupted, what would be the likely effect on a person's circadian rhythms of sleeping and waking?
7. What differences would there be in the behavioral consequences of a free-running circadian clock versus no clock at all?

FURTHER READING

Brown RE, Basheer R, McKenna JT, Strecker RE, McCarley RW. 2012. Control of sleep and wakefulness. *Physiological Reviews* 92: 1087–1187.

Buzsáki G. 2006. *Rhythms of the Brain*. New York: Oxford University Press.

Carskadon MA, ed. 1993. *Encyclopedia of Sleep and Dreaming*. New York: Macmillan.

Fries P. 2009. Neuronal gamma-band synchronization as a fundamental process in cortical computation. *Annual Review of Neuroscience* 32:209–224.

Goldberg EM, Coulter DA. 2013. Mechanisms of epileptogenesis: a convergence on neural circuit dysfunction. *Nature Reviews Neuroscience* 14:337–349.

Mohawk JA, Green CB, Takahashi JS. 2012. Central and peripheral circadian clocks in mammals. *Annual Review of Neuroscience* 35:445–462.

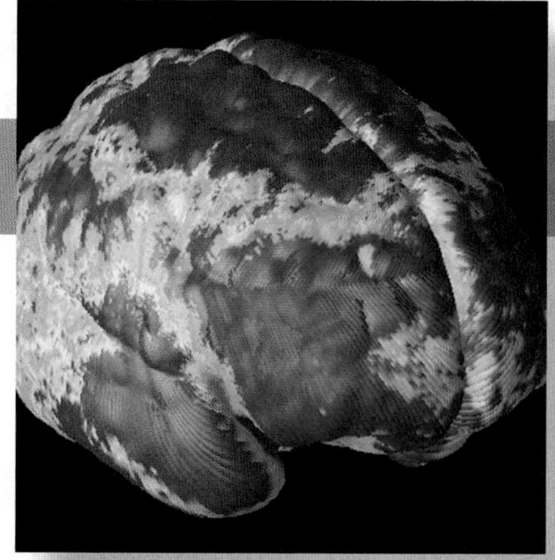

CHAPTER TWENTY

Language

INTRODUCTION

Language is a remarkable system for communication that has an enormous impact on our lives. You can walk into a café and order a large skinny mocha cappuccino with a dash of vanilla and be reasonably certain you will not be handed a bucket of mud. You can speak on the telephone with someone thousands of miles away and explain to them both the complexities of quantum physics and the emotional toll the physics course is having on your social life. There are endless debates about whether animals also have language, but there is no question that the complex and flexible system of language we use is unique to humans. Without language, we could not learn most of what we study in school, and this would greatly limit what we could accomplish.

More than just sounds, language is a system by which sounds, symbols, and gestures are used for communication. Language comes into our brains through the visual and auditory systems, and we produce speech and writing with our motor system. But the brain processing between the sensory and motor systems is the essence of language. Because animals are of limited use in studying human language, for many years, language was studied primarily by linguists and psychologists rather than neuroscientists. Much of what we know about the brain mechanisms responsible for language is derived from studies of language deficits resulting from brain damage. Numerous different aspects of language can be selectively disrupted, including speech, comprehension, and naming, suggesting that language is processed in multiple, anatomically distinct, stages. More recently, the imaging of activity in the human brain with fMRI and PET has given us intriguing insights into the complex circuits underlying language.

Language is universal in human society, perhaps because of specialized brain organization. It is estimated that there are over 5000 languages and dialects throughout the world, and languages differ in many ways, such as the order in which nouns and verbs are arranged. But despite differences in syntax, from Patagonia to Katmandu, all languages convey the subtleties of human experience and emotion. Consider the fact that no mute tribe of people has ever been found, not even in the remotest corner of the world. Many scientists believe the universality of language is a consequence of the fact that the human brain has evolved special language-processing systems.

WHAT IS LANGUAGE?

Language is a system for representing and communicating information that uses words combined according to grammatical rules. Language can be expressed in a variety of ways including gestures, writing, and speech. **Speech** is an audible form of communication built on the sounds humans produce. Speech comes naturally to humans: Even with no formal training, children raised in a normal language environment will invariably learn to understand spoken language and to speak. Reading and writing, on the other hand, require years of formal training, and more than 10% of the world's population is illiterate.

Human Sound and Speech Production

Across the animal kingdom, a variety of systems are used to produce sounds, but we'll focus on the basics of sound production in humans

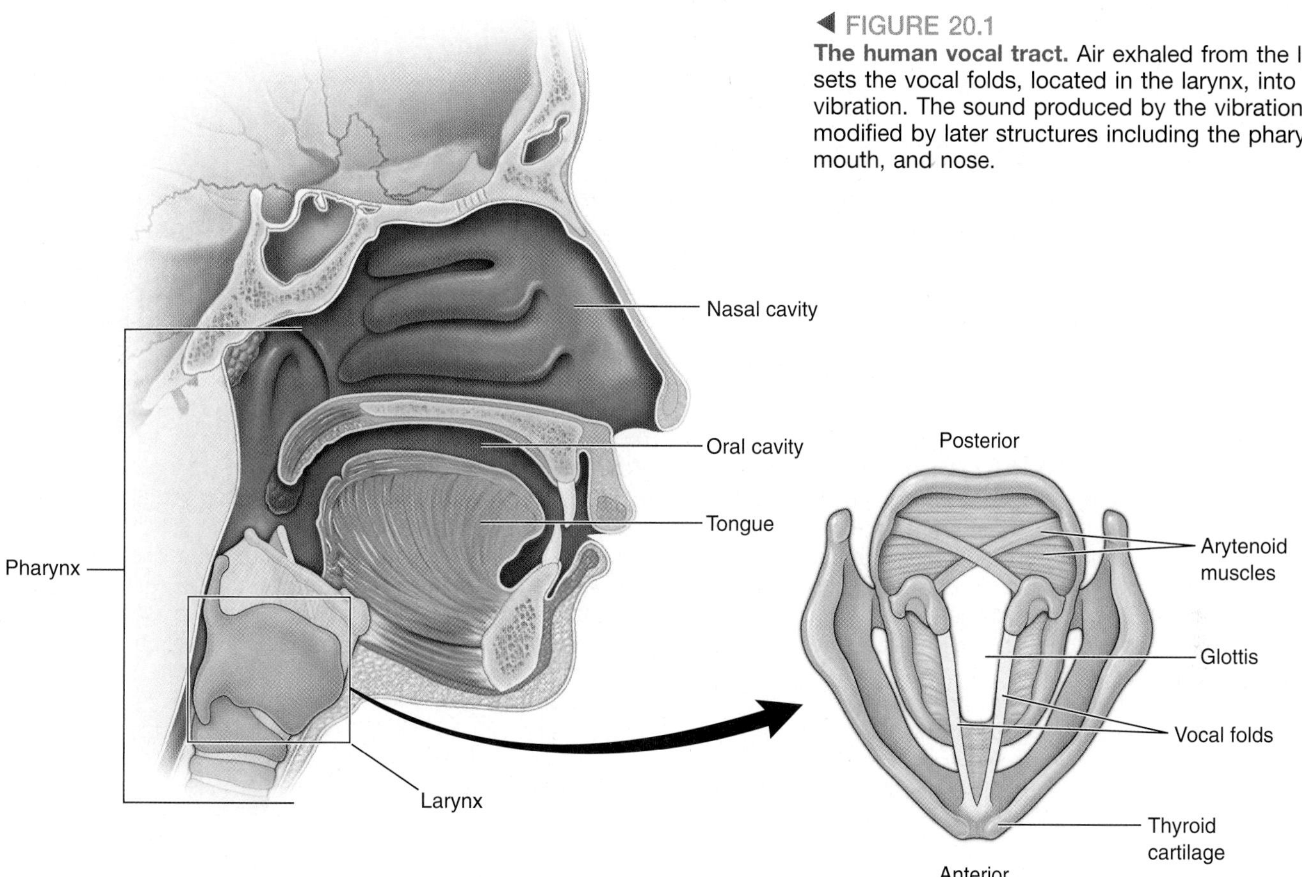

◀ FIGURE 20.1
The human vocal tract. Air exhaled from the lungs sets the vocal folds, located in the larynx, into vibration. The sound produced by the vibration is modified by later structures including the pharynx, mouth, and nose.

(Figure 20.1). Human speech involves a remarkable coordination of over 100 muscles ranging from those controlling the lungs to those of the larynx and the mouth. Ultimately all of these muscles are controlled by motor cortex, something we will have a good deal more to say about. Human sound begins when a person exhales air out of the lungs. The air passes through the *larynx*, also known as the voice box. What we call the Adam's apple in the neck is the larynx's anterior wall of cartilage. Within the larynx are the **vocal folds**, two bands of muscle also called *vocal chords*, which form a V. The space between the vocal folds is the *glottis*. Sounds are produced by vibrations in tightened vocal folds, somewhat as you can make a buzzing sound by blowing on a tightly held blade of grass. If the vocal fold muscles are relaxed, no sound will be produced, just as blowing on a loose blade of grass will make no sound; this is what happens when we exhale without speaking. The pitch of the sound results from the frequency of the vocal fold vibrations: Greater tension in the vocal folds produces higher-frequency vibrations and higher pitch sound. The sound is modified at further stages of the vocal tract, including the pharynx (especially the throat between larynx and mouth), mouth, and nose. Finally, rapid changes in the positions of the tongue, lips, and soft palate modulate sound for speech. The fundamental sounds that a language uses to communicate are called **phonemes**. Different spoken languages have different phonemes that build words unique to the language. Interestingly, research suggests that the words used in a language may have subtle effects on how people using that language think (Box 20.1).

BOX 20.1 OF SPECIAL INTEREST

Thinking in Different Languages

There are hundreds of different human cultures around the globe, each with its own customs, beliefs, and ways of life. We are used to the idea that people's attitudes about what is beautiful, delicious, and appropriate vary a good deal from place to place. Thinking may be affected by culture, but it shouldn't be influenced by language, which is simply the way people express themselves, right? Is it possible that the language a person uses alters how the person thinks? In the 1950s, Benjamin Lee Whorf proposed that the language people speak constrains their thoughts, perceptions, and actions, an extreme stance that has been largely abandoned. However, language does appear to impact thought in a few subtle and intriguing situations. Consider the use of gender in language. Most European languages, unlike English, assign a gender to inanimate objects. English speakers may occasionally assign gender to an object—"I like that car over there, she's a beauty!"—but this is rare. Contrast that with Italian, in which objects are masculine or feminine: teeth, flowers, and the sea—all masculine, who knows why. Or consider French, in which the same three things are feminine. The German dinner table is a funny arrangement with a masculine spoon, a feminine fork, and a neutral knife. It's enough to make a college language student's head spin.

Research suggests that the seemingly random gender assignments used in many languages may influence the way people think about the objects. In one study, French and Spanish speakers were asked to help make a movie in which objects would come to life. The subjects were shown pictures of the objects and asked to assign a man's or woman's voice. The objects were a mixture of ones that have the same gender in French and Spanish (e.g., "ballerina") and others that have the opposite gender (e.g., "broom"). The idea of using pictures was to avoid the bias that might come from saying the masculine or feminine words of each language. The experimental results showed that both French and Spanish speakers were significantly more likely to say an object should use a female voice when the object has a female gender in their language. This was true when the languages agreed about gender, but when these languages had different genders for the same object, the French and Spanish speakers disagreed about the appropriate gender for the voice.

In another experiment, Spanish and German speakers were asked to memorize object–name pairs in which each inanimate object that had male or female gender was paired with a human male or female name. Even though the experimental subjects were not asked any questions about the qualities of the inanimate objects, speakers of both languages found it more difficult to remember object–name pairs when the paired human name conflicted with the gender of the object in their language. The researchers speculated that the gender associated with the objects affected the subjects' memory of them as well as their thinking about them. Perception of masculine and feminine properties of inanimate objects is just one example of the subtle effects of language. Other effects may include influences on people's descriptions of color, time, and spatial location.

Further Reading

Deutscher G. 2010. *Through the Looking Glass: Why the World Looks Different in Other Languages*. New York: Picador.

Language in Animals

Animals communicate with each other in a variety of ways, from the dance of a honey bee to the watery bellow of a whale. Animals and humans also communicate with each other in many ways, such as the verbal command we give our dogs, if we're lucky, to jump down from the mailman. But do animals use language as we humans do? Human language is a remarkably complex, flexible, and powerful *system* for communication that involves the creative use of words according to the rules of a systematic grammar. Do other animals have anything similar? Actually, there are two questions we want to examine: Do animals naturally use language? Can animals be taught human language? These questions are difficult to investigate, but the answers have major implications for the evolution of human language.

Let's first consider the possible use of vocalizations for language in nonhuman primates. In the wild, chimpanzees have been reported to make tens of different vocalizations. There are alarm calls that express fear and warn others of a predator and pant-hoots that announce a chimp's presence and excitement. However, compared to humans, nonhuman primates appear to have a very limited range of vocalizations, and there is little evidence that they are based on rules (phonological rules) as in humans. Most of the vocalizations chimpanzees make may be stereotyped responses to behavioral situations. In comparison, human language is highly creative; limited only by the rules of grammar, it is effectively infinite. New word combinations and sentences are constantly being made by humans, and the combinations have clear meaning according to the meaning of the individual words plus the rules used in arranging them.

But maybe we haven't been fair in our comparison; perhaps animal speech requires that they have a language tutor, just as human children must be exposed to language. A variety of animals, including bottlenose dolphins and chimpanzees, have undergone training in an attempt to get them to speak human language. The harbor seal known as Hoover raised by fishermen in Maine learned to utter phrases that sounded something like a drunken man with a New England accent saying "hey hey hello there" or "get outta there." In the 1940s, several psychologists tried raising baby chimpanzees just like human children, including teaching them to speak. Despite extensive training, the chimps and other animals never learned to utter anything like the range of sounds and words used by humans. In the 1960s, the physician and inventor John Lily, famous for his later development of the sensory deprivation tank and studies with psychedelic drugs, flooded a small house with a few feet of water so that a dolphin could live around the clock with humans. As the trainer moved between a wet bed and a floating desk, she tried to teach the dolphin to speak, such as counting with numbers. Despite positive reports, later experiments did not corroborate the results.

The lack of a rich chimpanzee spoken language and their inability to speak human language are not surprising, given that their vocal tracts are not structured to make the sounds humans make. For example, in chimpanzees and other nonhuman primates, the larynx is much higher, closer to the mouth, making it impossible to make the range of sounds used in human speech. An alternative or adjunct to vocalizations for communication in chimpanzees is the use of gestures and facial expressions. There is evidence indicating that chimpanzees make gestures with the intention of influencing the behavior of other animals. In a recent study, Catherine Hobaiter and Richard Byrne from the University of St. Andrews reported the results of analyzing thousands of gestures categorized into 66 types. Behavioral reactions in other chimpanzees viewing the gestures were used to infer the meaning and intended outcome of each gesture. Examples are "groom me," "follow me," and "stop that." Some gestures appeared to have a particular intention, and others were used more flexibly. This is clearly a much more elaborate system of communication than the dance of a honey bee.

To test and quantify the abilities of animals, a range of studies have attempted to teach them nonverbal communication using words represented by American Sign Language gestures, plastic objects with various patterns and shapes, or keys with different colors and patterns on a keyboard. Well-known examples include the chimpanzee named Washoe, trained by Allen and Beatrix Gardner; the gorilla named Koko, trained by Francine Patterson; and the bonobo named Kanzi, raised by Sue Savage-Rumbaugh. Without question, these animals learned the meanings of gestures or symbols. They showed an ability to understand the meaning of phrases in human language, and they were able to use the improvised communication systems to request objects and actions from the scientists.

▲ FIGURE 20.2
Language in animals.

Whether we can conclude from such studies that animals use or can use language is controversial. Animals certainly communicate, and to some scientists, their systems of communication are sophisticated enough to be considered rudimentary language. These simpler systems may hint at the origins of human language. Other scientists think the distance between human and animal language is too great, that animals do not use language, defined as communication that is flexible, able to describe new things, and systematic according to rules of grammar. Regardless of your conclusion, it is important to distinguish between language, thought, and intelligence. Language is not a requirement for intelligence or thought. Nonhuman primates, dolphins, and humans raised without any language can do many things requiring abstract reasoning. Many creative people say they do some of their best thinking without words. Albert Einstein claimed that many of his ideas about relativity came from visually thinking of himself riding on a beam of light while looking at clocks and other objects. In any event, Fido does think even if he doesn't use language as we do (Figure 20.2).

Language Acquisition

Language processing in the adult human brain relies on carefully orchestrated interactions among a number of cortical areas and subcortical structures that we will look at in a moment. But how does the brain learn to use language? Learning a language, **language acquisition**, is a remarkable and fascinating process that proceeds in a similar manner in all cultures. The gurgles of newborns turn into babbling around 6 months of age. By 18 months of age, children understand about 150 words and can speak about 50. Interestingly, even at this early age, infants start to lose the ability to distinguish sounds they could discriminate earlier; an

(a) Spoken

"There are no silences between words"

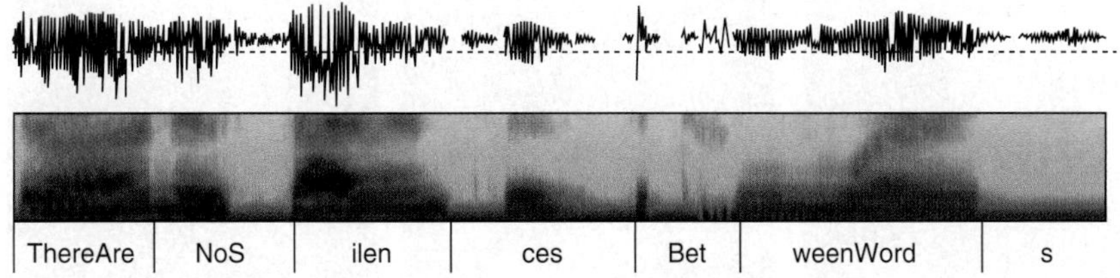

(b) Printed text

THEREDONATEAKETTLEOFTENCHIPS

THE RED ON A TEA KETTLE OFTEN CHIPS or THERE, DON ATE A KETTLE OF TEN CHIPS

◀ FIGURE 20.3
Word boundaries in spoken and written English. **(a)** Acoustic analysis of a spoken sentence demonstrates that word boundaries cannot be determined simply from sounds. **(b)** An analogous situation might be reading text without spaces between words. Indeed, some patterns of letters could form more than one sentence. (Source: Kuhl, 2004.)

example is a Japanese child's difficulty discriminating the English "r" and "l" sounds because these sounds are not used in Japanese. By 1–2 years of age, children's speech has the tones, rhythm, and accent of the language they are exposed to. A 3-year-old can produce full sentences and knows roughly 1000 words. By adulthood, a person knows tens of thousands of words. On the other hand, after puberty, learning a second language becomes more difficult. A critical period for language acquisition is suggested by the difficulty older children have learning a second language compared to the first and the difficulty acquiring a first language if they were not exposed to any spoken language before puberty.

The speed at which infants learn language belies the challenges involved. When we first hear a foreign language, it sounds rapidly spoken, and it is difficult to determine where one word stops and the next begins. This is one of the problems faced by infants learning their native tongue. By 1 year of age, however, infants can already recognize the sounds of their language and words even though they don't understand the words. Spoken language does not reliably indicate the divisions between words; it's rather like reading text without the spaces (Figure 20.3). Yet, infants must learn to understand thousands of words that are all constructed from the same small pool of language-specific sounds. Jenny Saffran and her colleagues at the University of Wisconsin found that this is achieved by statistical learning in the infant. In other words, the child learns that some combinations of sounds are far more likely than others. When a low probability combination occurs, it suggests the possible location of a word boundary. For example, in the phrase "pretty baby," the probability that "ty" follows "pret" in a single word is higher than "ba" following "ty" in a single word. Another cue that infants learn to use is the syllable emphasis most common in the language. For example in English, the stress is usually on the first syllable, and this helps determine where words start and stop. Adults of both genders when talking to infants often use "motherese," in which the speech is slower and exaggerated and vowel sounds are more clearly articulated. Motherese may assist the child in learning speech sounds.

We do not yet know the brain mechanisms by which infants learn to distinguish and speak words. However, Ghislaine Dehaene-Lambertz et al. have found using fMRI that even at the age of 3 months, the brain response to spoken words is distributed in a manner similar to that in adults (Figure 20.4). Listening to speech activates extensive areas in the temporal lobe, with the activation strongly biased toward the left hemisphere. These findings do not show that the infant brain processes language the same as the adult brain, but they indicate a similar early organization of auditory areas and lateralization for language.

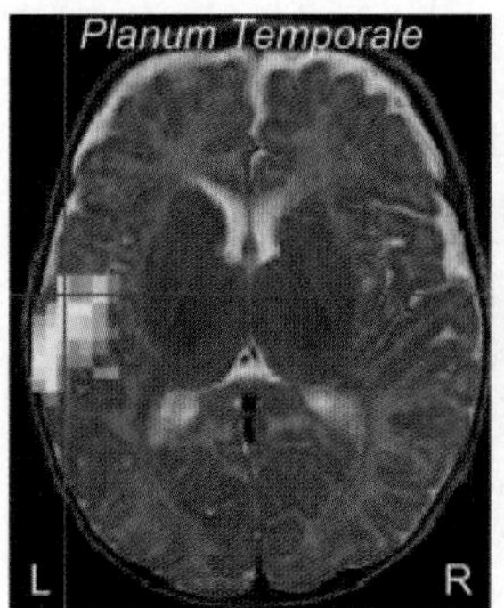

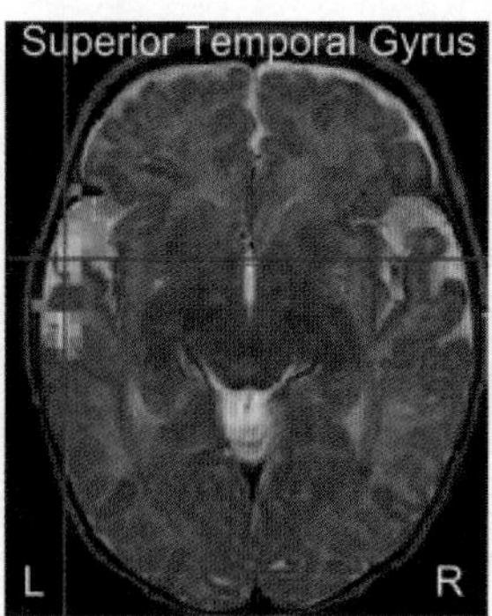

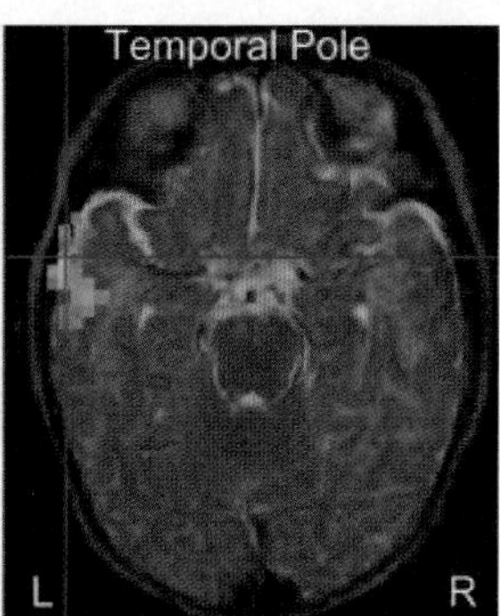

▶ FIGURE 20.4
Brain activity in a 3-month-old infant listening to speech. Horizontal sections show the planum temporale, superior temporal gyrus, and the pole of the temporal lobe that were all significantly activated when the infant listened to speech. In the fMRI images, red, orange, and yellow indicate increasingly greater brain activity. (Source: Dehaene-Lambertz et al., 2002.)

Genes Involved in Language

Speech and language disorders run in families and are more likely to co-occur in identical twins than in fraternal twins. These observations suggest that genetic factors play an important role in the susceptibility to language disorders. However, for many years the complex patterns of inheritance in language disorders made it difficult to implicate particular genes.

***FOXP2* and Verbal Dyspraxia.** The view of how genetics may affect language changed dramatically in 1990 with the first publications describing a British family known only as KE. In three generations of the KE family, about half had **verbal dyspraxia**, an inability to produce the coordinated muscular movements needed for speech (Figure 20.5a). Their speech was

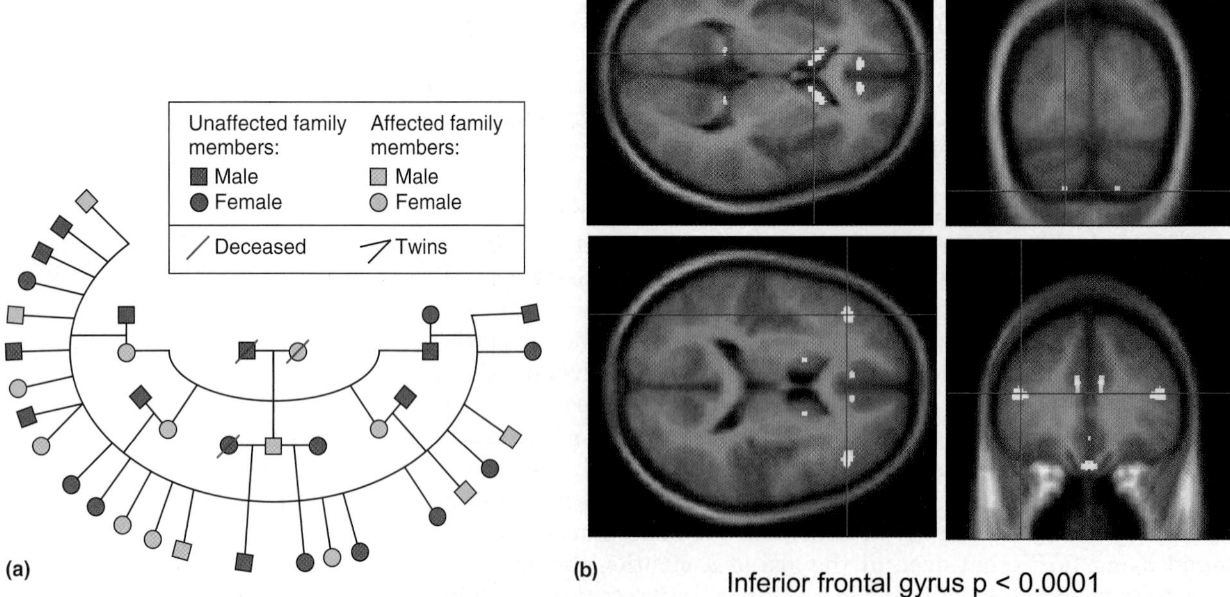

▲ FIGURE 20.5
***FOXP2* mutations in the KE family.** **(a)** Inheritance of language deficits in three generations of the KE family. **(b)** In affected KE family members, reduced gray matter was found in the caudate nucleus (upper left), cerebellum (upper right), and Broca's area in the frontal lobe (bottom). (Sources: a adapted from Watkins et al., 2002; b adapted from Vargha-Khadem et al., 2005.)

largely unintelligible to both the general public and family members, and they developed hand signs to supplement spoken language. In addition to dyspraxia, the affected members of the KE family had broader difficulties involving grammar and language and a lower IQ than unaffected members. The deficit was thought to be language-specific rather than a more general cognitive impairment because the language problems were observed even in affected family members with normal IQs. Brain scans revealed that affected members of the KE family had structural abnormalities in a variety of motor structures, including the motor cortex, cerebellum, and striatum (caudate and putamen) compared to unaffected members (Figure 20.5b).

What is known about the genetics underlying this striking familial language disorder? The first thing to note is that unlike previously observed inherited language disorders that appeared to involve multiple genes, the inheritance pattern seen in the KE family was consistent with the mutation of a single gene. This gene appears to affect the development of the motor cortex, cerebellum, and striatum; there are particular deficits in the muscular control of the lower face. The hunt for the culprit gene was aided by the discovery of an unrelated boy, known as CS, who had a language disorder resembling that in the KE family. Combining what was known about CS and the KE family, the mutated gene was eventually identified as *FOXP2*, which codes a transcription factor responsible for turning other genes on and off. It would be incorrect to call *FOXP2* *the* language gene, but it does appear to be one critical gene involved in language. From our two parents, we all have two copies of the *FOXP2* gene, but a mutation in either one is sufficient to produce severe language deficits. It is a striking finding that the change in a single gene can affect a complex behavior such as speech. It should be recognized, however, that through its action as a transcription factor, *FOXP2* is capable of influencing hundreds of other genes that might be involved in language.

Versions of *FOXP2* are found in many animals. It is interesting to note that in highly vocal songbirds, *FOXP2* is strongly expressed in brain areas involved in song learning. An important question is what might be special in humans that conceivably underlies our extensive language abilities compared to those of nonhuman primates. Only two amino acids distinguish the human form of the *FOXP2* protein from the chimpanzee, gorilla, and rhesus monkey form. The evolutionary paths leading to humans and chimpanzees diverged about six million years ago, but it has been estimated that the mutations differentiating human and nonhuman primate *FOXP2* genes occurred about 200,000 years ago. An exciting speculation is that a small and relatively recent mutation in the *FOXP2* gene set humans on a path toward developing language that was needed for higher cognitive function and the development of human culture.

Genetic Factors in Specific Language Impairment and Dyslexia. After the KE family was studied and the *FOXP2* gene implicated in their verbal dyspraxia, other individuals unrelated to the KE family were identified who had various mutations of *FOXP2*. Their speech impairments were consistent with the idea that a mutation in the *FOXP2* gene alone can disrupt the normal development of speech. Affected individuals also exhibited other grammatical and cognitive deficits, but it is not clear if these are distinct deficits or somehow related to the verbal dyspraxia.

Spurred on by the *FOXP2* findings, growing numbers of genes have been identified that are potentially involved in common language disorders. For example, **specific language impairment (SLI)** is found in about 7% of all 6-year-olds in the United States. This condition consists of a developmental delay in the mastery of language that may persist into adulthood and that is not associated with hearing difficulty or more general developmental delays. These children have difficulty learning and using words, especially verbs. Because more than 50% of children with SLI have a parent or sibling with the condition, there appears to be a strong genetic component.

Genetic studies of children with SLI have identified a handful of genes that may be involved. Often mentioned, in addition to *FOXP2*, are the genes *CNTNAP2* and *KIAA0319*. More interesting than the long acronyms denoting these genes are their functions. *CNTNAP2* codes a neurexin protein; these are proteins on the presynaptic side of synapses that serve to hold the presynaptic and postsynaptic elements together. The *CNTNAP2* neurexin plays an important role in brain development; it appears to be involved in properly locating potassium channels in developing neurons. *KIAA0319* is thought to be critical for neuronal migration during neocortical development as well as normal function of adult neurons. We have not reached the point of knowing the specific neural abnormalities underlying SLI, but the candidate genes have focused attention on key aspects of neuronal migration and development.

Another common disorder associated with language is **dyslexia,** which involves a difficulty learning to read despite normal intelligence and training. Estimates are that dyslexia occurs in 5–10% of people and is somewhat more common in males than in females. The disorder appears to have a strong genetic link, as children of a dyslexic parent have a roughly 30% chance of being dyslexic, and 30–50% of the siblings of a dyslexic person also have dyslexia. A gene often associated with dyslexia is *KIAA0319*, one of the genes thought to be involved in specific language impairment. Interestingly, dyslexia is quite often found in individuals with SLI. Comorbidity of the disorders is about 40–50%, suggesting that they may have similar causes or be different manifestations of the same deficit. Like SLI, dyslexia appears to involve deviations from the normal pattern of neocortical development.

THE DISCOVERY OF SPECIALIZED LANGUAGE AREAS IN THE BRAIN

As in many other areas of neuroscience, only in the last century have we come to understand a clear relationship between language and the brain. Much of what we know about the importance of certain brain areas is derived from studies of aphasia. **Aphasia** is the partial or complete loss of language abilities following brain damage, often without the loss of cognitive faculties or the ability to move the muscles used in speech.

During the Greek and Roman Empires, it was commonly thought that the tongue controls speech and that speech disorders originate there rather than in the brain. If a head injury resulted in a loss of speech, the treatment involved special gargles or massage of the tongue. By the sixteenth century, it had been noted that a person could suffer speech impairment without paralysis of the tongue. However, in spite of this step forward, treatment still included such procedures as cutting the tongue, drawing blood, and applying leeches.

Around 1770, Johann Gesner published a relatively modern theory of aphasia, describing it as the inability to associate images or abstract ideas with their expressive verbal symbols. He attributed this loss to brain damage resulting from disease. Gesner's definition makes the important observation that in aphasia, cognitive ability may remain intact but some function specific to verbal expression is lost. Despite the incorrect association Franz Joseph Gall and later phrenologists made between skull shape and brain function (see Chapter 1), they too made an important observation about aphasia. They reasoned that cases of brain lesions in which speech is lost but other mental faculties are retained suggest that there is a specific region of the brain used for speech.

In 1825, based on many case studies, the French physician Jean-Baptiste Bouillaud proposed that speech is specifically controlled by the frontal lobes. But it took another four decades before this idea was generally accepted. In 1861, Simon Alexandre Ernest Aubertin, Bouillaud's son-in-law, described the case of a man who had shot away his frontal skull bone in a failed suicide attempt. While treating this man, Aubertin discovered that if a spatula was pressed against the exposed frontal lobe while the man was speaking, his speech immediately halted and did not resume until the pressure was released. He inferred that the pressure on the brain interfered with the normal function of a cortical area in the frontal lobe.

Broca's Area and Wernicke's Area

Also in 1861, French neurologist Paul Broca had a patient who was almost entirely unable to speak (the man was called Tan because "tan" was the only sound he could produce). The patient died shortly after Broca met him, and at the postmortem Broca found a lesion in the frontal lobes. Perhaps because of a change in the scientific climate, Broca's case study appears to have swung popular opinion around to the idea that there is a language center in the brain. In 1863, Broca published a paper describing eight cases in which language was disturbed by damage to the frontal lobe in the left hemisphere. Additional similar cases, along with reports that speech was not disturbed by right hemispheric lesions, led Broca in 1864 to propose that language expression is controlled by only one hemisphere, almost always the left. This view is supported by results from the **Wada procedure**, in which a single hemisphere of the brain is anesthetized. In most cases, anesthesia of the left hemisphere, but not the right, disrupts speech. In the 1990s, functional brain imaging began to replace the Wada procedure for assessing the dominant hemisphere for language, and the findings are the same (Box 20.2).

If one hemisphere is thought to be more heavily involved in a particular task, it is said to be *dominant*. The region of the dominant left frontal lobe that Broca identified as critical for articulate speech has come to be called **Broca's area** (Figure 20.6). Broca's work has considerable significance as the first clear demonstration that brain functions can be anatomically localized.

In 1874, German neurologist Karl Wernicke reported that lesions in the left hemisphere, in a region distinct from Broca's area, also disrupt normal speech. Located on the superior surface of the temporal lobe between the auditory cortex and the angular gyrus, this region is now commonly called **Wernicke's area** (see Figure 20.6). The nature of the aphasia Wernicke observed is different from that associated with damage to Broca's area. Having established that there are two language areas in the left hemisphere, Wernicke and others proceeded to construct maps of language processing in the brain. Interconnections between the auditory cortex, Wernicke's area, Broca's area, and the muscles required for speech were hypothesized, and different types of language disabilities were attributed to damage in different parts of this system.

Although the terms *Broca's area* and *Wernicke's area* are still commonly used, the boundaries of these areas are not clearly defined, and they appear to be quite variable from one person to the next. We will also see that each area may be involved in more than one language function, although that more recent finding will make sense only after we look at the aphasias produced by damage to Broca's and Wernicke's areas.

BOX 20.2 OF SPECIAL INTEREST

Assessing Hemispheric Language Dominance

The first reports that one of the cerebral hemispheres is dominant for language came from studies of brain-damaged patients. A simple procedure used for studying the function of a single cerebral hemisphere in people without brain damage is the *Wada procedure,* developed by Japanese-Canadian neurologist Juhn Wada. A fast-acting barbiturate, such as sodium amytal, is injected into the carotid artery on one side of the neck (Figure A). The drug is preferentially carried in the bloodstream to the hemisphere ipsilateral to the injection, where it acts as an anesthetic for about 10 minutes. The effects are sudden and dramatic. Within a matter of seconds, the limbs on the side of the body contralateral to the injection become paralyzed along with loss of somatic sensation.

By asking the patient to answer questions, one can assess his or her ability to speak. If the injected hemisphere is dominant for speech, the patient will be completely unable to talk until the anesthesia wears off. If the injected hemisphere is not dominant, the person can continue to speak throughout the procedure.

Table A shows that in 96% of right-handed people and 70% of left-handed people, the left hemisphere is dominant for speech. Because 90% of all people are right-handed, this means that the left hemisphere is dominant for language in roughly 93% of people. While small but significant numbers of people with either handedness have a dominant right hemisphere, only in left-handers are bilateral representations of speech seen. In the Wada procedure, this is indicated when an injection into either hemisphere has some disruptive effect on speech, although specifics of the disruption may be different for the two hemispheres.

More recently, fMRI has been used to assess hemispheric dominance for language. As fMRI is not invasive and it is not limited by the short duration of the anesthetic effect of sodium amytal, it has advantages over the Wada procedure. The brain images in Figure B were collected while a subject was given a word and asked to select a synonym from four word options. The brain scans show that frontal, temporal, and parietal areas are activated exclusively in the left hemisphere, which is thus dominant for language in this person. (Note that the convention with MRI images is to show the left hemisphere on the right side of the image.)

TABLE A **Hemispheric Control of Speech in Relation to Handedness**

Handedness	Number of Cases	Speech Representation (%)		
		Left	Bilateral	Right
Right	140	96	0	4
Left	122	70	15	15

(Source: Rasmussen and Milner, 1977, Table 1.)

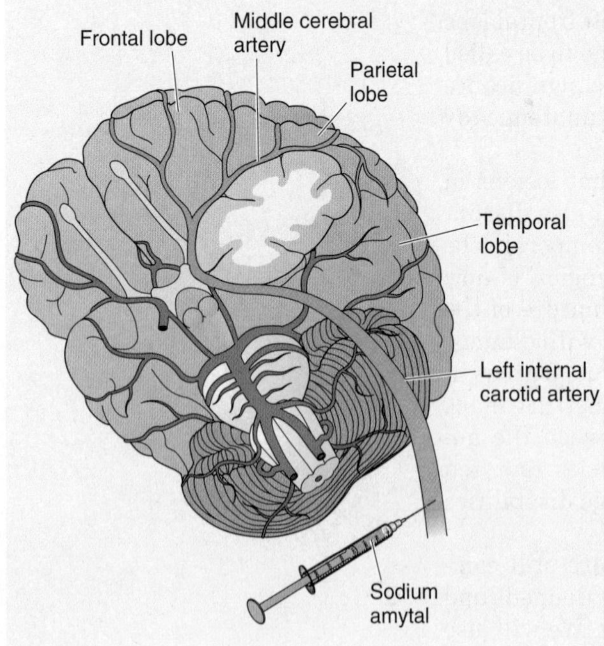

Figure A

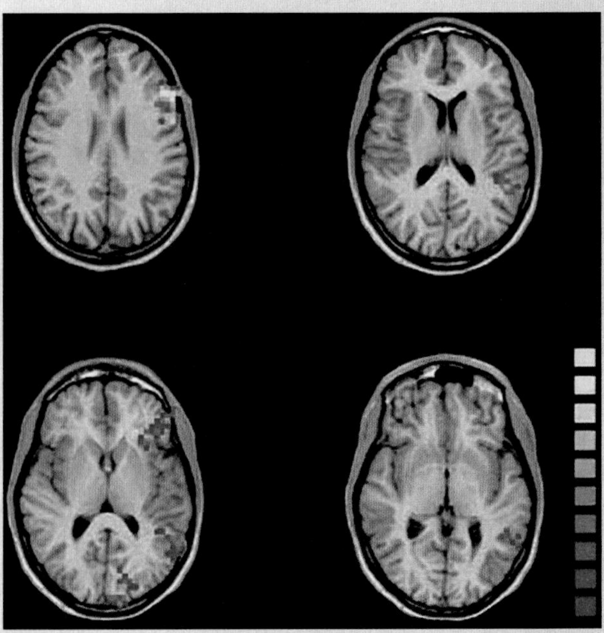

Figure B
(Source: Spreer et al., 2002, Fig. 4.)

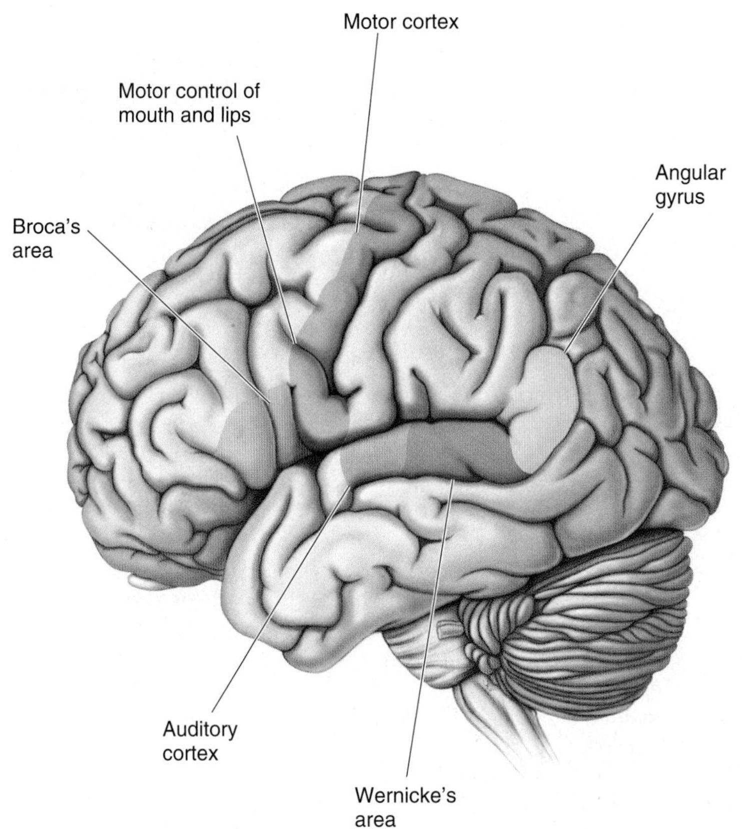

◀ FIGURE 20.6
Key components of the language system in the left hemisphere. In the frontal lobe, Broca's area lies next to the area that controls the mouth and lips in the motor cortex. Wernicke's area, on the superior surface of the temporal lobe, is situated between the auditory cortex and the angular gyrus.

LANGUAGE INSIGHTS FROM THE STUDY OF APHASIA

As in the studies of Broca and Wernicke, the oldest technique for studying relationships between language and the brain involves correlating functional deficits with lesions in particular brain areas. The occurrence of distinct types of aphasia, as shown in Table 20.1, suggests that language is processed in several stages at different locations in the brain. By examining language deficits that result from damage to different areas of the brain, Nina Dronkers at the University of California at Davis, has clarified much of the neural machinery of language (Box 20.3).

TABLE 20.1 **Characteristics of Types of Aphasia**

Type of Aphasia	Site of Brain Damage	Comprehension	Speech	Impaired Repetition	Paraphasic Errors
Broca's	Motor association cortex of frontal lobe	Good	Nonfluent, agrammatical	Yes	Yes
Wernicke's	Posterior temporal lobe	Poor	Fluent, grammatical, meaningless	Yes	Yes
Conduction	Arcuate fasciculus	Good	Fluent, grammatical	Yes	Yes
Global	Portions of temporal and frontal lobes	Poor	Very little	Yes	—
Transcortical motor area	Frontal lobe anterior to Broca's	Good	Nonfluent, agrammatical	No	Yes
Transcortical sensory	Cortex near the junction of temporal, parietal, and occipital lobes	Poor	Fluent, grammatical, meaningless	No	Yes
Anomic	Inferior temporal lobe	Good	Fluent, grammatical	No	

BOX 20.3 PATH OF DISCOVERY

Uncovering Language Areas of the Brain

by Nina Dronkers

My passion for clinical neuroscience began one day in a class at the University of California at Berkeley where I was pondering my "senioritis" and what to do with my life after college. My professor showed a video of a man who couldn't read a handwritten message. Paradoxically, the man had just written the message himself! Such cases of language disturbance after brain injury were what sparked my interest in how the brain processes language and continued to fascinate me for the next 30 years.

In working with individuals who have sustained brain injuries, I have had the unique opportunity to evaluate the relationships between affected areas of the brain (as imaged with brain scans) and the speech and language deficits (aphasia) that result from the injury. The first thing that struck me in working with individuals who have aphasia was that the classic relationship between aphasia syndromes and injury to certain language areas was not always as I had learned. I saw patients with Broca's aphasia who did not necessarily have lesions in Broca's area, and those with lesions in Broca's area did not always have Broca's aphasia. The same discrepancy held for other aphasia syndromes as well. Soon, my colleagues and I realized that some deficits could still be "localized," but that these needed to be narrowed down into smaller components of the speech and language system, rather than by entire syndromes. Deficits such as coordinating complex articulatory movements could be related to lesions in a small part of the insula, problems with the verbatim repetition of low-frequency sentences were seen after injury to the posterior superior temporal gyrus, and difficulty recognizing the syntactic structure of a sentence could be related to the lesions in the anterior superior temporal gyrus. We found that fiber pathways in the brain also play an important role in language production and comprehension. Destruction of the arcuate fasciculus, for example, can lead to a severe speech production disorder. It became clear that while certain individual brain structures can play a specific role in speech or language functions, aphasia syndromes are caused by injury to large swaths of brain tissue as well as the fiber pathways that connect them. In the normal brain, all of these structures work together in a complex network that helps to support the extraordinary language functions we all take for granted.

A particularly exciting time of my life occurred in Paris, France, where I had the extraordinary opportunity to study the brains of Paul Broca's first two patients. These were the cases of aphasia he had examined as a surgeon in 1861 and whose deficits led Broca to believe that the inferior part of the frontal lobe was important for spoken language. Since so much had been written about Broca's area, particularly with the growth of functional neuroimaging with PET and fMRI in cognitive neuroscience, it was clearly necessary to revisit these cases and see which anatomical areas were actually affected in these historic brains. Luckily, the brains had never been dissected or discarded, and my colleague, Odile Plaisant, and I were tremendously fortunate in being able to examine these brains more closely. We could see right away that the area we now think of as Broca's area was only partially affected in these two cases. We were curious to know how deeply the injuries extended, and with the expertise of Marie-Thérèse Iba-Zizen and Emanuel Cabanis, both neuroradiologists, we were actually able to scan these two important brains in an MRI scanner and acquired high-resolution images with exquisite detail.

What astonished us was the degree of involvement of other regions of the brain, particularly in the insula and in the fiber tracts that travel throughout the brain. The case of Mr. Leborgne, or "Tan," Broca's first and most famous case, had extensive involvement of the insula but only a portion of what we now call Broca's area. In addition, the major fiber bundles, including the arcuate and superior longitudinal fasciculi that travel between the frontal and posterior parts of the brain, were completely destroyed. Broca's second case, Mr. Lelong, had atrophy in the insula, but when the scanner advanced into the deeper parts of the brain, we saw several small lesions again in the arcuate and superior longitudinal fasciculi. This had never been seen before, and we were quite thrilled to see it unfold as we watched. Thus, we saw that the damage that produced the aphasia in these two cases was more extensive than previously thought but, in fact, consistent with what we see in our current cases of severe Broca's aphasia.

As a neuroscientist, I feel extraordinarily lucky to be able to work with individuals with aphasia who have taught us so much about this remarkable part of the body. The brain is, in many ways, still an open frontier, with much to be learned about its functions, its mechanisms, and its potential for recovery. The next generations of neuroscientists will have much to contribute to our knowledge of the brain and will surely experience the same excitement of discovery as those who came before them.

Broca's Aphasia

The syndrome called **Broca's aphasia** is also known as motor or nonfluent aphasia, because the person has difficulty speaking even though he or she can understand language heard or read. The case of David Ford is typical. Ford was a radio operator in the Coast Guard when, at age 39 years, he suffered a stroke. He remained an intelligent man, but he had little control over his right arm and leg (demonstrating that his lesion was in the left hemisphere). His speech was also abnormal, as the following discussion with psychologist Howard Gardner illustrates:

> "I asked Mr. Ford about his work before he entered the hospital.
>
> I'm a sig. . . no. . . man. . . uh, well,. . . again." These words were emitted slowly, and with great effort. The sounds were not clearly articulated; each syllable was uttered harshly, explosively, in a throaty voice. With practice, it was possible to understand him, but at first I encountered considerable difficulty in this.
>
> "Let me help you," I interjected. "You were a signal. . . ."
>
> "A signal man. . . right," Ford completed my phrase triumphantly.
>
> "Were you in the Coast Guard?"
>
> "No, er, yes, yes. . . ship. . . Massachu. . . chusetts. . . Coastguard. . .years." He raised his hands twice, indicating the number "nineteen."
>
> "Could you tell me, Mr. Ford, what you've been doing in the hospital?"
>
> "Yes, sure. Me go, er, uh, P.T. nine o'cot, speech. . . two times. . . read. . . wr. . . ripe, er, rike, er, write. . . practice. . . get-ting better."
>
> "And have you been going home on weekends?"
>
> "Why, yes. . . Thursday, er, er, er, no, er, Friday. . . Bar-ba-ra. . . wife. . . and, oh, car. . . drive. . . purnpike. . . you know. . . rest and. . . tee-vee."
>
> "Are you able to understand everything on television?"
>
> "Oh, yes, yes. . . well. . . almost." Ford grinned a bit. (Gardner, 1974, pp. 60–61)

People with Broca's aphasia have difficulty saying anything, often pausing to search for the right word. The inability to find words is called **anomia** (literally meaning "no name"). Interestingly, there are certain "overlearned" things Broca aphasics can say without much hesitation, such as the days of the week and the American Pledge of Allegiance. The hallmark of Broca's aphasia is a telegraphic style of speech, in which mainly *content words* (nouns, verbs, and adjectives carrying content specific to the sentence) are used. For instance, when Mr. Ford was asked about being in the Coast Guard, his answer contained the words "ship," "Massachusetts," "Coast Guard," and "years," but little else. Many *function words* (articles, pronouns, and conjunctions connecting the parts of the sentence grammatically) are left out (there are no ifs, ands, or buts). As well, verbs are frequently not conjugated. In the jargon of aphasia deficits, the inability to construct grammatically correct sentences is called *agrammatism.* There are some peculiar nuances to the agrammatical tendencies in Broca's aphasia. Ford, for example, could read and use the words "bee" and "oar" but had difficulty with the more common words "be" and "or." This problem is related not to the word's sound but to whether or not it is a noun. In a similar vein, Broca aphasics have difficulty repeating things spoken to them, although they tend to be better with familiar nouns such as "book" and "nose." Sometimes they substitute incorrect sounds or words (Ford said "purnpike" for "turnpike"); these are called *paraphasic errors.*

In contrast to the speech difficulties in Broca's aphasia, comprehension is generally quite good. In the dialogue above, Ford seemed to understand the questions asked of him, and for the most part, he said he understood what he saw on television. In Gardner's study, Ford was able to answer simple questions, such as "Does a stone float on water?" However, more difficult questions demonstrated that he did not have completely normal comprehension abilities. If he was told "The lion was killed by the tiger; which animal is dead?" or "Put the cup on top of the fork and place the knife inside the cup," he had difficulty understanding. This was probably related to the fact that he generally had trouble with the function words "by" in the first example and "on top of" in the second example.

Because the most obvious difficulty is in producing speech, Broca's aphasia is considered a language disturbance toward the motor end of the language system. Language is understood but not easily produced. While it is true that Broca aphasics are worse at speech than other types of aphasics, several characteristics suggest that there is more to the syndrome. As pointed out earlier, comprehension is generally good, but comprehension deficits can be demonstrated by tricky questions. Also, a simple motor deficit would not explain the ability to say "bee" but not "be." Finally, patients sometimes have considerable anomia, suggesting that they have problems "finding" words as well as making the appropriate sounds.

Wernicke suggested that the area damaged in Broca's aphasia contains memories for the fine series of motor commands required for articulating word sounds. Because Broca's area is near the part of the motor cortex that controls the mouth and lips, there is an appealing logic to this idea. Wernicke's theory is still held by some, but there are other ways of looking at the problem. For instance, the difference in the aphasic's ability to use content words and function words suggests that Broca's area and nearby cortex may be specifically involved in making grammatical sentences out of words. This might explain why Mr. Ford could produce sounds such as "bee" and "oar" when they represent content words but not when the sounds represent the function words "be" and "or."

Wernicke's Aphasia

When Wernicke noted that superior temporal lesions could lead to aphasia, the syndrome he observed was quite distinct from Broca's aphasia. Indeed, Wernicke suggested there were two general types of aphasia. In Broca's aphasia, speech is disturbed but comprehension is relatively intact. In **Wernicke's aphasia**, speech is fluent but comprehension is poor. (Although these descriptions are oversimplified, they are useful for remembering these syndromes.)

Let's consider the case of Philip Gorgan, another patient studied by Gardner.

> "What brings you to the hospital?" I asked the 72-year-old retired butcher 4 weeks after his admission to the hospital.
>
> "Boy, I'm sweating, I'm awful nervous, you know, once in a while I get caught up, I can't mention the tarripoi, a month ago, quite a little, I've done a lot well, I impose a lot, while, on the other hand, you know what I mean, I have to run around, look it over, trebbin and all that sort of stuff."
>
> I attempted several times to break in, but was unable to do so against this relentlessly steady and rapid outflow. Finally, I put up my hand, rested it on Gorgan's shoulder, and was able to gain a moment's reprieve.
>
> "Thank you, Mr. Gorgan. I want to ask you a few—"
>
> "Oh sure, go ahead, any old think you want. If I could I would. Oh, I'm taking the word the wrong way to say, all of the barbers here whenever they stop you it's going around and around, if you know what I mean, that is tying and tying for repucer, repuceration, well, we were trying the best that we could while another time it was with the beds over there the same thing. . . ." (Gardner, 1974, pp. 67–68)

Clearly, Mr. Gorgan's speech is altogether different from that of Mr. Ford. Gorgan's speech is fluent, and he has no trouble using function words as well as content words. If you didn't understand English, his speech would probably sound normal because of its fluency. However, the content does not make much sense. It is a strange mixture of clarity and gibberish. Along with their far greater output of speech compared to Broca aphasics, Wernicke aphasics also make far more paraphasic errors. Gorgan would sometimes use the correct sounds but in an incorrect sequence, such as "plick" instead of "clip." Occasionally, he would stumble around the correct sound or word, as when, in another conversation, he called a piece of paper "piece of handkerchief, pauper, hand pepper, piece of hand paper." Interestingly, he would sometimes use an incorrect word but one categorically similar to the correct word, such as "knee" instead of "elbow."

Because of the stream of unintelligible speech, it is difficult to assess with speech alone whether Wernicke aphasics comprehend what they hear or read. Indeed, one of the intriguing things about Wernicke aphasics is that they frequently appear undisturbed by the sound of their own speech and the speech of others, even though they probably don't understand either. Comprehension is usually assessed by asking the patient to respond in a nonverbal manner. For instance, the patient could be asked to put object A on top of object B. Questions and commands of this sort quickly lead to the conclusion that Wernicke aphasics do not understand most instructions. They are completely unable to comprehend questions of the sort understood by Broca aphasics. When Gorgan was presented with commands written on cards ("Wave goodbye," "Pretend to brush your teeth"), he was often able to read the words but never acted as if he understood what they meant.

Gorgan's strange speech was mirrored in his writing and his ability to play music. When Gardner gave him a pencil, he spontaneously took it and wrote "Philip Gorgan. This is a very good beautifyl day is a good day, when the wether has been for a very long time in this part of the campaning. Then we want on a ride and over to for it culd be first time. . . ." (p. 71). Likewise, when he sang or played the piano, pieces of the appropriate song were intermixed with musical gibberish, and he had a difficult time ending, just as in his speech.

Insight about the possible function of Wernicke's area is provided by its location on the superior temporal gyrus near the primary auditory cortex. Wernicke's area may play a critical role in relating incoming sounds to their meaning. In other words, it is an area specialized for storing memories of the sounds that make up words. Some have suggested that Wernicke's area is a high-order area for sound recognition in the same sense that the inferior temporal cortex is thought to be a high-order area for visual recognition. A sound recognition deficit would explain why Wernicke aphasics don't comprehend speech well. However, there must be more to Wernicke's area to account for the odd speech patterns and comprehension deficits. Speech in Wernicke's aphasia suggests that Broca's area and the system responsible for speech production are running without control over content. The speech zooms along, swerving in every direction like a car with a sleepy driver at the wheel.

The Wernicke–Geschwind Model of Language and Aphasia

Shortly after making his observations about what came to be called Wernicke's aphasia, Wernicke proposed a model for language processing in the brain. Later extended by Norman Geschwind at Boston University,

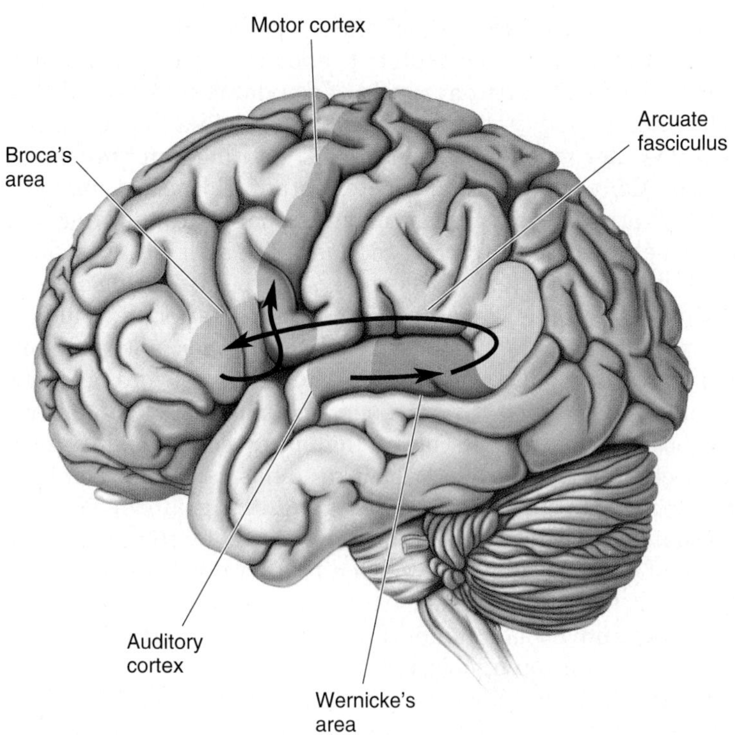

▶ FIGURE 20.7
The pathway involved in repeating a spoken word, according to the Wernicke–Geschwind model.

this model is known as the **Wernicke–Geschwind model**. The key anatomical elements in the system are Broca's area, Wernicke's area, the *arcuate fasciculus* (a bundle of axons connecting the two cortical areas), and the *angular gyrus*. The model also includes sensory and motor areas involved in receiving and producing language. To understand what the model entails, we'll consider the performance of two tasks.

The first task is the repetition of spoken words (Figure 20.7). When incoming speech sounds reach the ear, the auditory system processes the sounds, and neural signals eventually reach the auditory cortex. According to the model, the sounds are not understood as meaningful words until they are processed in Wernicke's area. For the person to be able to repeat the words, word-based signals are passed to Broca's area from Wernicke's area via the arcuate fasciculus. In Broca's area, the words are converted to a code for the muscular movements required for speech. Output from Broca's area is sent to the nearby motor cortical areas responsible for moving the lips, tongue, larynx, and so on.

The second task we'll consider is reading written text aloud (Figure 20.8). In this case, the incoming information is processed by the visual system through the striate cortex and higher-order visual cortical areas. The visual signals are then passed to the angular gyrus at the junction of the occipital, parietal, and temporal lobes. In the cortex of the angular gyrus, it is assumed that a transformation occurs so that the output evokes the same pattern of activity in Wernicke's area as if the words were spoken rather than written. From this point, the processing follows the same progression as in the first example: Wernicke's area to Broca's area to motor cortex.

This model offers simple explanations for key elements of both Broca's and Wernicke's aphasia. A lesion in Broca's area seriously interferes with speech production because the proper signals can no longer be sent to the motor cortex. On the other hand, comprehension is relatively intact because Wernicke's area is undisturbed. A lesion in Wernicke's area

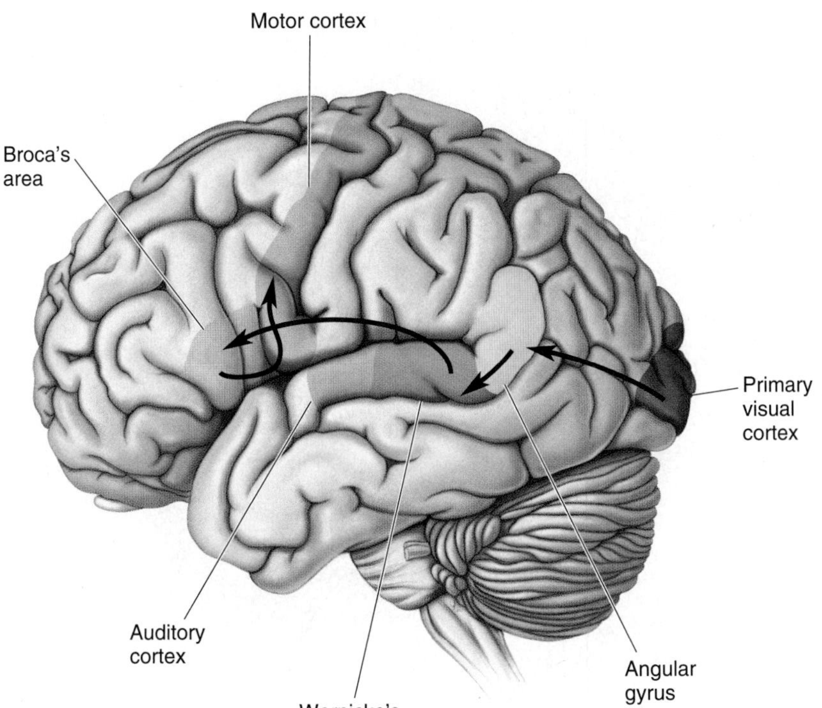

◀ FIGURE 20.8
The pathway involved in repeating aloud a written word, according to the Wernicke–Geschwind model.

produces great comprehension problems because this is the site of the transformation of sounds into words. The ability to speak is unaffected because Broca's area is still able to drive the muscles required for speech.

The Wernicke–Geschwind model has several errors and oversimplifications, however. For example, words read do not have to be transformed into a pseudo-auditory response, as suggested in the reading task described earlier. In fact, visual information can reach Broca's area from the visual cortex without making a stop at the angular gyrus. One of the dangers inherent in any model is overstating the significance of a given cortical area for a particular function. It has been found that the severity of Broca's and Wernicke's aphasias depends on how much cortex is damaged beyond the limits of Broca's and Wernicke's areas. Also, aphasia is influenced by damage to subcortical structures such as the thalamus and caudate nucleus, which are not in the model. In surgical cases where parts of cortex are removed, the resulting language deficits are usually milder than the deficits resulting from stroke, which affects both cortical and subcortical structures.

Another important factor is that there is often significant recovery of language function after a stroke, and it appears that other cortical areas can sometimes compensate for what is lost. As with many neurological syndromes, young children generally recover very well, but even adults, especially left-handers, can show good recovery of function.

A final problem with the Wernicke–Geschwind model is that most aphasias involve both comprehension and speech deficits. Mr. Ford, with Broca's aphasia, had good comprehension but was confused by complex questions. Conversely, Mr. Gorgan, with Wernicke's aphasia, had several speech abnormalities in addition to a severe lack of comprehension. Therefore, in cortical processing, the sharp functional distinctions between regions as implied by the model do not exist. Despite its problems, the Wernicke–Geschwind model continues to be of clinical use because of its simplicity and approximate validity. In the latter half of the twentieth

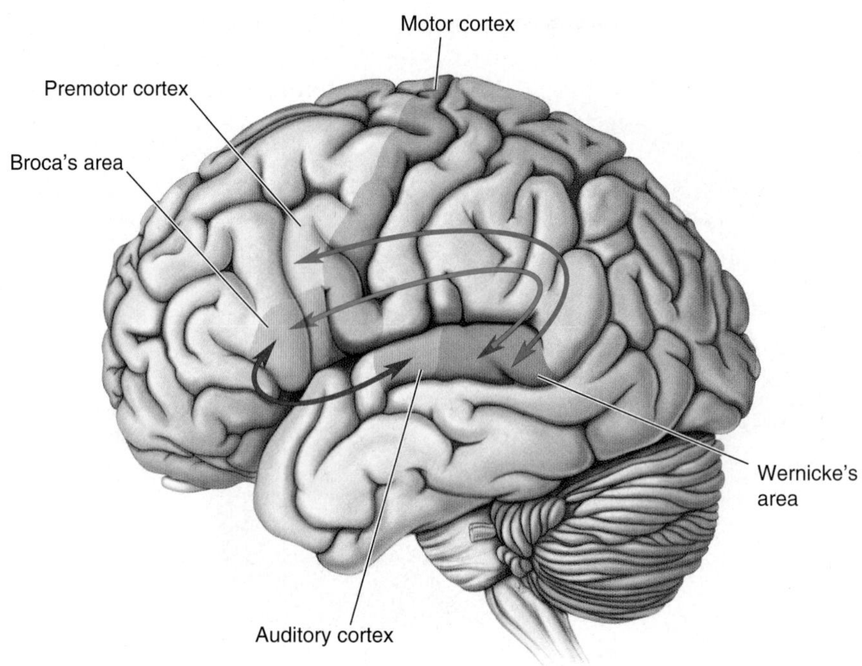

▲ FIGURE 20.9
Parallel language pathways. Current models of language processing emphasize multiple streams of processing, much like the dorsal and ventral streams described for vision. The model here includes two dorsal pathways and a ventral pathway. Note that unlike in the Wernicke–Geschwind model, language here is not based on a single stream connecting Wernicke's area with Broca's area via the arcuate fasciculus. One dorsal pathway (blue) connects the superior temporal gyrus (Wernicke's and auditory areas) with premotor cortex and is involved in speech production and repeating words. The other dorsal pathway (green) connects the superior temporal gyrus with Broca's area and is thought to be involved in processing complex syntactic structure—that is, the analysis of words arranged according to a grammar system. The ventral stream (red) takes the sounds of speech and extracts their meaning. (Adapted from Berwick et al., 2013, Fig. 2.)

century, numerous more elaborate language models were developed to account for the complexities of language and brain processing as well as the shortcomings of the Wernicke–Geschwind model. Similar to parallel pathways that have been outlined in the visual system, these models involve a variable number of parallel pathways with distinct but interacting functions (Figure 20.9).

Conduction Aphasia

The value of any model involves not only its ability to account for previous observations but also its ability to predict. Based on his observation that distinct forms of aphasia result from lesions in areas of frontal and superior temporal cortex, Wernicke predicted that a unique form of aphasia should result from a lesion that disconnects these two brain areas but leaves the areas themselves intact. In the Wernicke–Geschwind model, a lesion in the fibers composing the arcuate fasciculus could accomplish this. In reality, such disconnection lesions usually involve damage to parietal cortex in addition to the arcuate fasciculus, but Broca's and Wernicke's areas are spared.

Wernicke's prediction turned out to be correct; aphasia from a disconnection lesion was demonstrated and is now known as **conduction aphasia.** As the model predicts, based on the preservation of Broca's and

Wernicke's areas, comprehension is good and speech is fluent. Patients are typically able to express themselves through speech without difficulty. The deficit that chiefly characterizes conduction aphasia is difficulty in repeating words. In response to hearing a few words, the patient may attempt to repeat what was said, but the repetition will substitute words, omit words, and include paraphasic errors. Repetition is usually best with nouns and short common expressions, but it may fail entirely if the spoken words are function words, polysyllabic words, or nonsense sounds. Interestingly, a person with conduction aphasia comprehends sentences he or she reads aloud, even though what is said aloud contains many paraphasic errors. This is consistent with the idea that comprehension is good and the deficit occurs between the regions involved in comprehension and speech.

One of the sad yet fascinating things about aphasia is the diversity of syndromes that occur following strokes. While these syndromes challenge language models, each one offers a clue to our understanding of language processing. Characteristics of a few other aphasias are listed in Table 20.1.

Aphasia in Bilinguals and Deaf People

Cases of aphasia in bilingual people and deaf people provide fascinating insight about language processing in the brain. Suppose a person knows two languages before having a stroke. Does the stroke produce aphasia in one language and not the other, or do both languages suffer equally? The answer depends on several factors, including the order in which the languages were learned, the fluency achieved in each language, and how recently the language was used. The consequences of a stroke are not always predictable, but the language learned more fluently and earlier in life tends to be relatively more preserved. If the person learned two languages at the same time to equivalent levels of fluency, a lesion will probably produce similar deficits in both languages. If the languages were learned at different times in life, it is likely that one language will be affected more than the other. The implication is that the second language may make use of different, although overlapping, populations of neurons from those used by the first.

The study of language deficits in those who are deaf and/or know sign language suggests that there is some universality to language processing in the brain. American Sign Language uses hand gestures to express the ideas and emotions most of us convey with spoken language (Figure 20.10). Left hemispheric lesions in people who use sign language appear to cause language deficits similar to those occurring in verbal aphasics. There are cases analogous to Broca's aphasia in which comprehension is good but the ability to "speak" through sign language is severely impaired. Importantly, the ability to move the hands is not impaired (i.e., the problem is not with motor control). Rather, the deficit is specific to the use of hand movements for the expression of language.

There are also sign language versions of Wernicke's aphasia in which the patient signs fluently but with many mistakes while also having difficulty comprehending the signing of others. In one unusual case, a hearing man who was the child of deaf parents learned both sign language and verbal language. A left hemispheric stroke initially gave him global aphasia, but his condition significantly improved with time. An important observation was that his verbal and sign languages recovered together, as if overlapping brain areas were used. While there do appear to be aphasias in those using sign language analogous to speech aphasias, there is also evidence that signing aphasia and speaking aphasia can be produced by left hemispheric lesions in somewhat different locations.

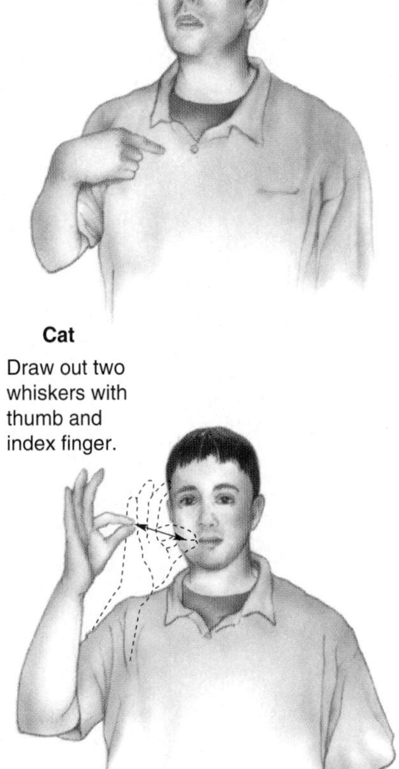
Me
Index finger points to and touches chest.
Cat
Draw out two whiskers with thumb and index finger.

▲ FIGURE 20.10
"Speaking" in American Sign Language.

ASYMMETRICAL LANGUAGE PROCESSING IN THE TWO CEREBRAL HEMISPHERES

We have seen that damage to certain parts of the brain leads to a variety of aphasias. As the early work of Broca indicated, language is usually not handled equally by the two cerebral hemispheres. Some of the most valuable and fascinating findings on the language differences of the two hemispheres come from **split-brain studies** in which the hemispheres are surgically disconnected. Communication between the cerebral hemispheres is normally served by several bundles of axons known as commissures. Recall from Chapter 7 that the largest of these is the great cerebral commissure, also called the **corpus callosum** (Figure 20.11). The corpus callosum consists of about 200 million axons crossing between the hemispheres. Surely such a huge bundle of fibers must be of great importance. Surprisingly, until about 1950, researchers had been unable to demonstrate any important role of the corpus callosum.

In split-brain procedures, the skull is opened and the axons making up the corpus callosum are severed (Figure 20.12). The hemispheres may retain some communication via the brain stem or smaller commissures (if they aren't also severed), but most of the intercerebral communication is lost. In the 1950s, Roger Sperry and his colleagues at the University of Chicago and later at the California Institute of Technology performed a series of experiments, using split-brain animals, to explore the function of the corpus callosum and the separated cerebral hemispheres. Sperry's group confirmed earlier reports that cutting the corpus callosum in a cat or monkey has no noticeable effect on the animal's behavior. Temperament is unchanged, and the animals appeared to be normal in coordination, reaction to stimuli, and ability to learn. However, in cleverly devised experiments, Sperry's group showed that the animals sometimes acted as if they had two separate brains. For example, in one experiment, circle and cross visual stimuli were both shown to a monkey's left eye and the animal was trained to select the circle. In alternate trials, it was trained to choose the cross when the same visual stimuli were presented to the right eye. With both eyes open, it is not possible for the monkey (or humans) to know which

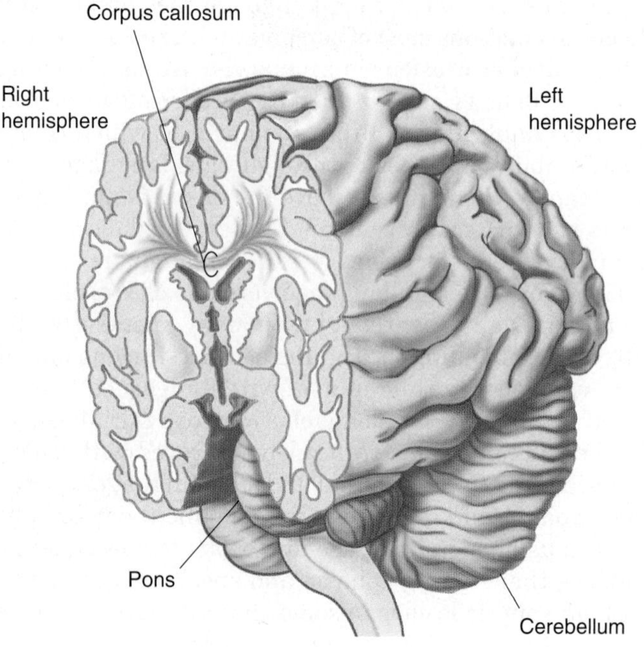

▶ FIGURE 20.11
The corpus callosum. The corpus callosum is the largest bundle of axons providing communication between the cerebral hemispheres.

eye sees a stimulus, so it appeared that the two cerebral hemispheres were learning opposite discriminations at the same time—thinking with two brains. You might wonder what happens if both eyes simultaneously see the stimuli. The answer is that the monkey hesitates and then chooses either the cross or the circle and sticks with this choice for a series of trials before switching to the opposite stimulus. The scientists speculated that the two cerebral hemispheres competed and, on any given trial, one won.

Language Processing in Split-Brain Humans

Because split-brain monkeys did not appear to have any major deficits, surgeons felt they were justified in cutting the corpus callosum as a last resort in treating certain types of severe epilepsy in humans. They hoped to prevent the spread of epileptic activity from one hemisphere to the other. It may seem questionable to cut 200 million axons on the assumption that they are not very important, but the surgery is often beneficial in restoring a seizure-free life. Michael Gazzaniga, then at New York University, studied a number of these people. Gazzaniga had initially worked with Sperry, and his techniques were modifications of those used with experimental animals.

One key methodological feature of studying split-brain humans involves careful control to present visual stimuli to only one cerebral hemisphere. Gazzaniga did this by taking advantage of the fact that only the right hemisphere sees objects to the left of the point of fixation, and only the left hemisphere sees objects to the right (see Figure 10.3), as long as the eyes can't move to bring the image onto the fovea (Figure 20.13). Pictures or words were flashed on for a fraction of a second using a device with a camera-like shutter. Note that when a shutter opens it does not present a stimulus to one eye and not the other; instead, it presents a stimulus to both eyes in such a way that only one cerebral hemisphere "sees" the stimulus. Because the images were presented for a shorter time than that required to move the eyes, the images were seen by only one hemisphere.

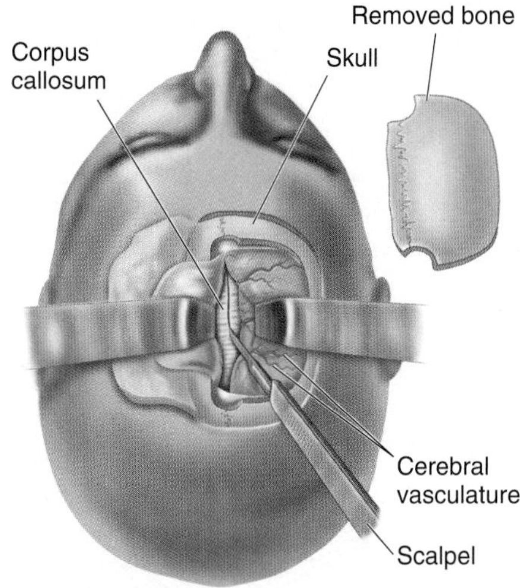

▲ FIGURE 20.12
Split-brain surgery in a human.
To reach and cut the corpus callosum, a portion of the skull is removed and the cerebral hemispheres are retracted.

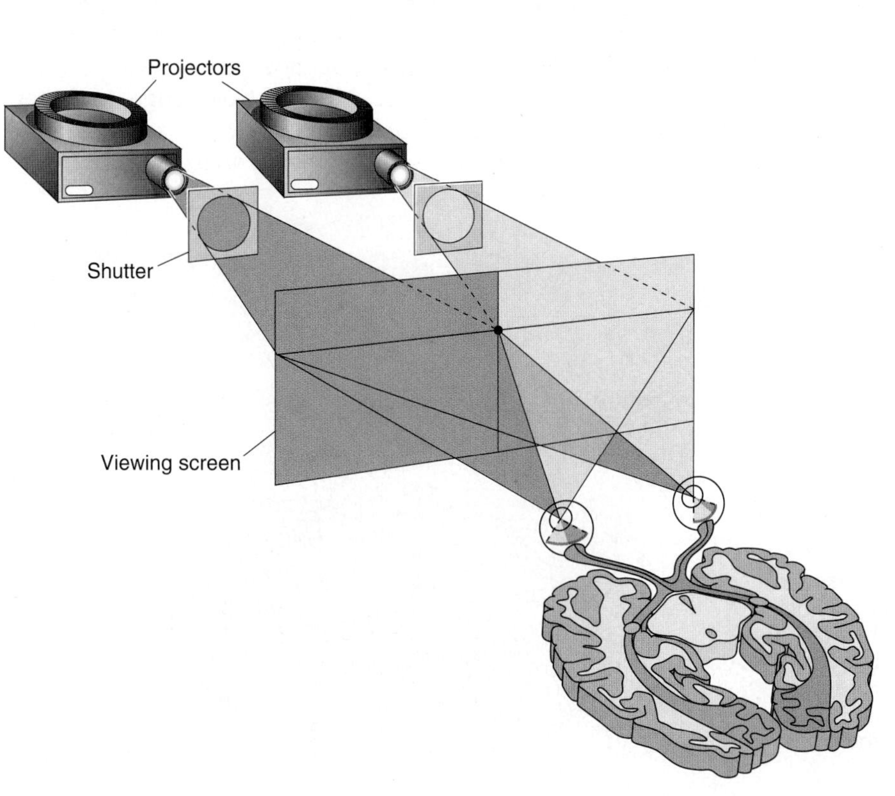

◄ FIGURE 20.13
Visual stimulation of one hemisphere in humans. A visual stimulus is briefly flashed to the left or right visual field by means of a shutter. Each projector shows an image to both eyes in such a way that only one cerebral hemisphere "sees" the image. The display time is shorter than the time needed to generate a saccadic eye movement, thus assuring that only one hemisphere sees the stimulus.

Left Hemisphere Language Dominance. Although split-brain humans are normal in most every way, there is a striking asymmetry in their ability to verbalize answers to questions posed separately to the two hemispheres. For instance, they can repeat or describe numbers, words, and pictures visually presented only in the right visual field because the left hemisphere is usually dominant for language. Likewise, they can describe objects being manipulated only by the right hand (out of view of both eyes). These findings would be entirely unremarkable except for the fact that such simple verbal descriptions of sensory input are impossible for the right hemisphere.

If an image is shown only in the left visual field or an object is felt only by the left hand, a split-brain person cannot describe it and usually says that nothing is there (Figure 20.14). An object could be covertly placed in a patient's left hand, and there would be no verbal indication of even noticing. This absence of response is a consequence (and demonstration) of the fact that the left hemisphere controls speech in most people. If you think about the implications for split-brain people, you'll realize that they have an unusual existence. Following their surgery, they are unable to describe anything to the left of their visual fixation point: the left side of a painting, the left side of a room, and so on. What is startling is that this doesn't seem to disturb the patients.

Language Functions of the Right Hemisphere. While there is a dramatic inability of the right hemisphere to speak, this does not mean it knows nothing of language. It can be demonstrated that the right hemisphere can read and understand numbers, letters, and short words as long as the required response is nonverbal. In one experiment, the right hemisphere is shown a word that is a noun. As already mentioned, the person will say nothing is seen. Of course, that's the talkative left hemisphere speaking, and *it* didn't see anything. But if the person is urged to use the left hand to select a card containing a picture corresponding to the word shown, or pick out an object by touch, the person can do it (see Figure 20.14). The right hemisphere cannot do this with more complex words or sentences, but the results clearly imply that the right hemisphere does have language comprehension.

A study conducted by Kathleen Baynes, Michael Gazzaniga, and their colleagues then at the University of California at Davis suggests that the right hemisphere may sometimes also be able to write even if it can't

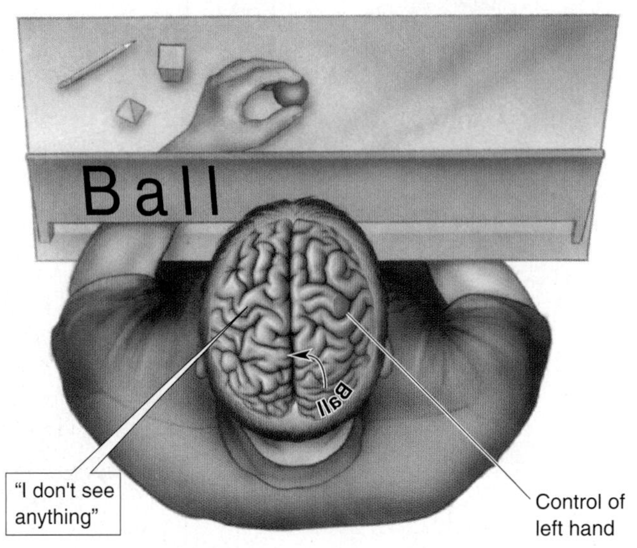

▶ FIGURE 20.14
Demonstrating language comprehension in the right hemisphere. A split-brain person shown a word in the left visual field will say nothing is seen. This is because the left hemisphere, which usually controls speech, did not see the word, and the right hemisphere, which saw the word, cannot speak. Information about the word cannot be transferred to the verbal left hemisphere without the corpus callosum. However, the left hand, which is controlled by the right hemisphere, can pick out the object corresponding to the word by touch alone.

speak. In most people, reading, speaking, and writing are all controlled by the left hemisphere. However, in a split-brain woman known as V.J., this was found not to be true. Words were flashed to either her left or her right hemisphere. Words seen by her left hemisphere could be spoken but not written. Conversely, she could write but not speak words shown to her right hemisphere. While this separation of function may be abnormal, the case of V.J. indicates that there isn't necessarily a single brain system for all aspects of language located in one hemisphere.

Evidence also suggests that the right hemisphere understands complex pictures despite its inability to say so. In one experiment, a subject was shown a series of pictures in her left visual field, and at one point a nude photo appeared in the series. When the researcher asked what she saw, she said "Nothing," but then she began to laugh. She told the researcher that she didn't know what was funny, but that perhaps it was the machine used in the experiment.

The right hemisphere appears to be more skilled at certain tasks than the left hemisphere. For instance, even though the split-brain patients were right-handed and thus their left hemispheres were much more practiced at drawing, the left hand controlled by the right hemisphere was better at drawing or copying figures containing three-dimensional perspective. These patients were also better at solving complex puzzles with their left hand. It has also been reported that the right hemisphere is somewhat better at perceiving nuances in sound.

In some split-brain studies, the two hemispheres initiated conflicting behaviors, apparently because they were thinking differently. In one task, a patient was asked to arrange a group of blocks to match a pattern on a small card. He was told to do this using only his right hand (left hemisphere), which is not generally good at this type of task. As the right hand struggled to arrange the blocks, the left hand (right hemisphere), which knew how to do it, reached in to take over. Only the restraint of the experimenter kept the left hand from pushing the right one out of the way to solve the puzzle. Another patient studied by Gazzaniga would sometimes find himself pulling his pants down with one hand while pulling them up with the other. These bizarre behaviors make a strong case that there are two independent brains controlling the two sides of the body.

The results of these split-brain studies demonstrate that the two hemispheres can function as independent brains and that they have different language abilities. Although the left hemisphere is usually dominant for language, the right hemisphere has significant skills in comprehending language. It is important to keep in mind that the split-brain studies test the ability of each hemisphere to perform on its own. Presumably in the intact brain, the callosum allows for synergistic interactions between the hemispheres for language and other functions.

Anatomical Asymmetry and Language

In the nineteenth century, there were reports of anatomical differences between the two hemispheres. For example, it was noted that the left Sylvian fissure is longer and less steep than the right (Figure 20.15). However, as recently as the 1960s, there was considerable doubt about the existence of significant cortical asymmetries. Because of the strikingly asymmetrical control of speech demonstrated by the Wada procedure, it would be interesting to know if the two cerebral hemispheres are anatomically different. Some of the first good quantitative data demonstrating hemispheric differences came from the work of Geschwind and his colleague Walter Levitsky. Initial observations were made of postmortem

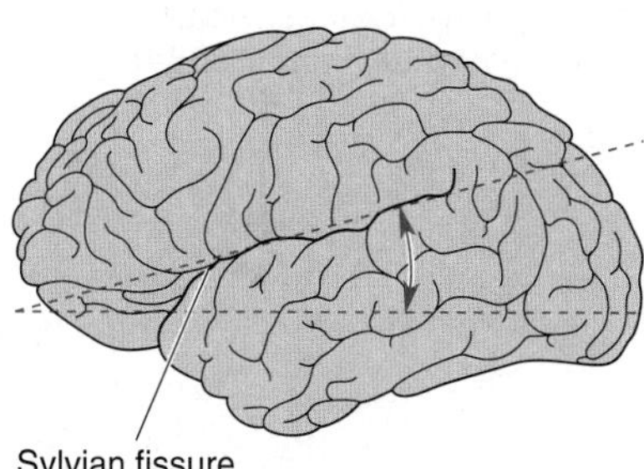

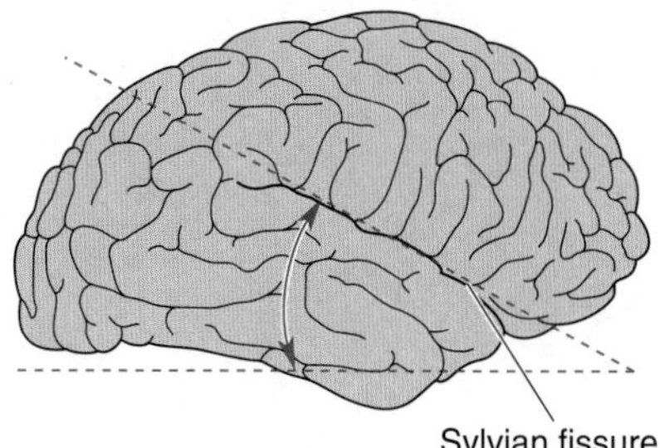

▲ FIGURE 20.15
Asymmetry of the Sylvian fissure. In most right-handed people, the Sylvian fissure in the left hemisphere is longer and runs at a shallower angle than the fissure in the right hemisphere. (Source: Adapted from Geschwind, 1979, p. 192.)

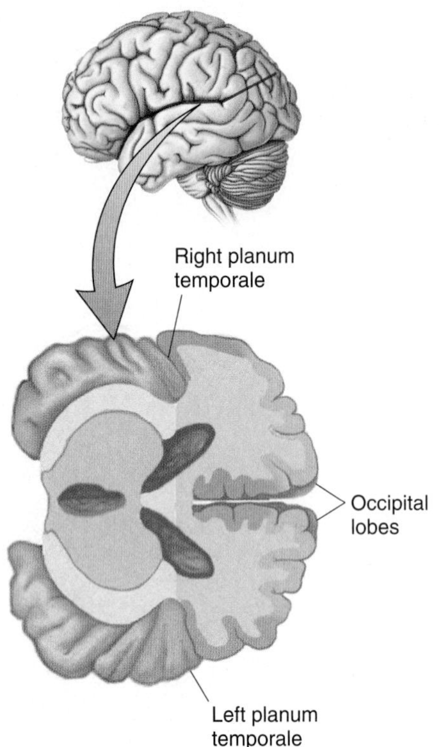

▲ FIGURE 20.16
Asymmetry of the planum temporale. This region on the superior temporal lobe is usually significantly larger in the left hemisphere. (Source: Adapted from Geschwind and Levitsky, 1968, Fig. 1.)

brains, but more recently, the results were confirmed with magnetic resonance imaging (MRI) (see Box 7.2).

The most significant difference seen was in a region called the **planum temporale**, which is a part of Wernicke's area on the superior surface of the temporal lobe (Figure 20.16). Based on measurements of 100 brains, Geschwind and Levitsky found that in about 65% of the brains, the left planum temporale was larger than the right, whereas in only about 10%, the right was larger. In some instances, the left area was more than five times larger than the right. Interestingly, the asymmetry in this area is seen even in the human fetus, suggesting that it is not a developmental consequence of the use of the left hemisphere for speech. Indeed, apes also have a larger left planum temporale. This suggests the possibility that speech became dominant in the left hemisphere because of a preexisting size difference. Other studies found that a portion of Broca's area also tends to be larger in the left hemisphere. Are these larger areas in the left hemisphere related to the common dominance of the left hemisphere for speech?

More recent investigations of gray matter volume have been made using MRI in living subjects, making it possible to test correlations among brain anatomy, asymmetries, and language dominance. One challenge in these studies is finding enough people who are right hemisphere dominant for language. Several language areas, including the planum temporale, Broca's area, and the insula, are generally larger in the left hemisphere than the right, and this is true in people with a left or right hemisphere dominant for language. The big question is whether the hemisphere dominant for language can be predicted from the degree to which the structure on the left is larger than on the right. Perhaps some structure on the left side is much bigger in a person with a dominant left hemisphere but only a little bigger or smaller in a person with a dominant right hemisphere.

There have been mixed reports about a correlation between the size of the left and right planum temporale and the hemisphere dominant for language. There have also been reports about a correlation between the language-dominant hemisphere and the relative size of Broca's area on each side of the brain. At present, it appears there may be some correlation between the asymmetric sizes of Broca's area and the planum temporale with the hemisphere dominant for language, but the correlation is not strong enough to allow one to predict the language-dominant hemisphere from anatomical measurements alone. The brain region that presently appears to best predict which hemisphere is dominant for language is the **insula**, which is the cerebral cortex within the lateral sulcus that is between the temporal and parietal lobes (Figure 20.17). Even though the insula has been thought to be involved in human language for some time, the relationship between its size and language lateralization is somewhat surprising, as the language functions of the insula have been studied less and are less well understood than other language areas in the brain. Also, the insula appears to be involved in numerous brain functions ranging from taste to emotion. Further research is needed to clarify its role in language and its relationship with hemispheric dominance.

It has probably occurred to you that a functional human asymmetry more obvious than language is handedness. More than 90% of humans are right-handed and usually relatively uncoordinated with their left hand, implying that in some way the left hemisphere is specialized for fine motor control. Is this related to the left hemispheric dominance for language? The answer is not known, but it is interesting that humans are different from nonhuman primates in handedness as well as language. While animals of many species show a consistent preference for using one hand, there are typically equal numbers of left-handers and right-handers.

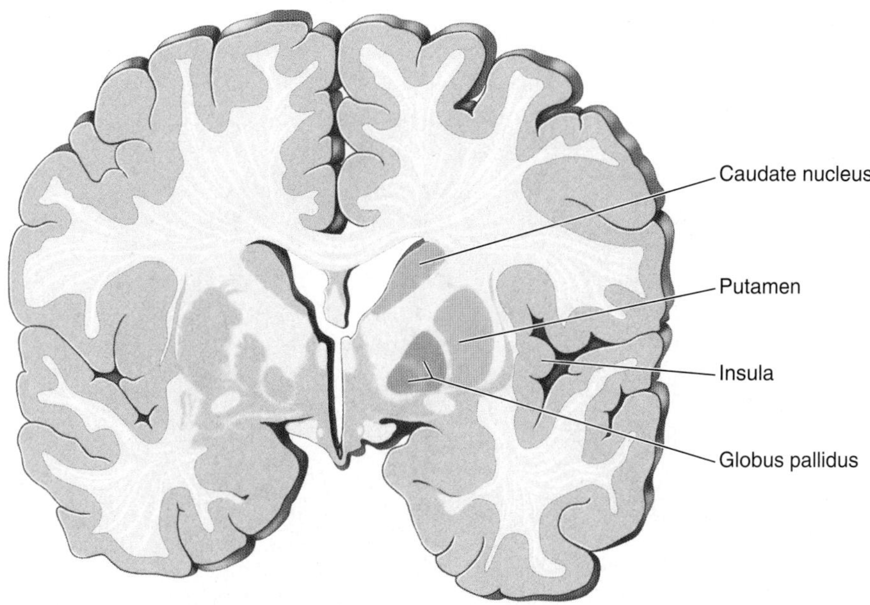

◀ FIGURE 20.17
The insula. The insula, also called insular cortex, lies within the lateral sulcus between the temporal and parietal lobes.

LANGUAGE STUDIES USING BRAIN STIMULATION AND HUMAN BRAIN IMAGING

Until late in the twentieth century, language processing in the brain was examined mainly by correlating language deficits with postmortem analysis of brain damage. But now, aspects of language processing have been revealed by electrical brain stimulation and brain imaging with fMRI and PET in living humans.

The Effects of Brain Stimulation on Language

At several points in this book, we have discussed the electrical brain stimulation studies of Wilder Penfield. Without general anesthesia, patients could report the effects of stimulation at different cortical sites. In these experiments, Penfield noted that stimulation at certain locations affected speech. These effects occurred in three main categories: vocalizations, speech arrest, and speech difficulties similar to aphasia.

Stimulation of motor cortex in the area that controls the mouth and lips caused immediate speech arrest (Figure 20.18). Such a response is logical because the activated muscles sometimes pulled the mouth to one side or clenched the jaw shut. Stimulation of motor cortex occasionally evoked cries or rhythmic vocalizations; importantly, these effects occurred with electrical stimulation of motor cortex on either side of the brain. Penfield found three other areas where electrical stimulation interfered with speech, but these were only in the dominant left hemisphere. One of these areas appeared to correspond to Broca's area. If this area was stimulated while a person was speaking, speech either stopped entirely (with strong stimulation) or became hesitant (with weaker stimulation). Some patients were unable to name objects they could name before and after the brain stimulation. Occasionally, they substituted an incorrect word. They apparently experienced a mild transient form of anomia. In some subjects, word confusion and speech arrest also resulted from stimulation at two other sites, one in the posterior parietal lobe near the Sylvian fissure and the other in the temporal lobe. These two areas were in the vicinity of the arcuate fasciculus and Wernicke's area, although not perfectly aligned with estimated locations of those areas.

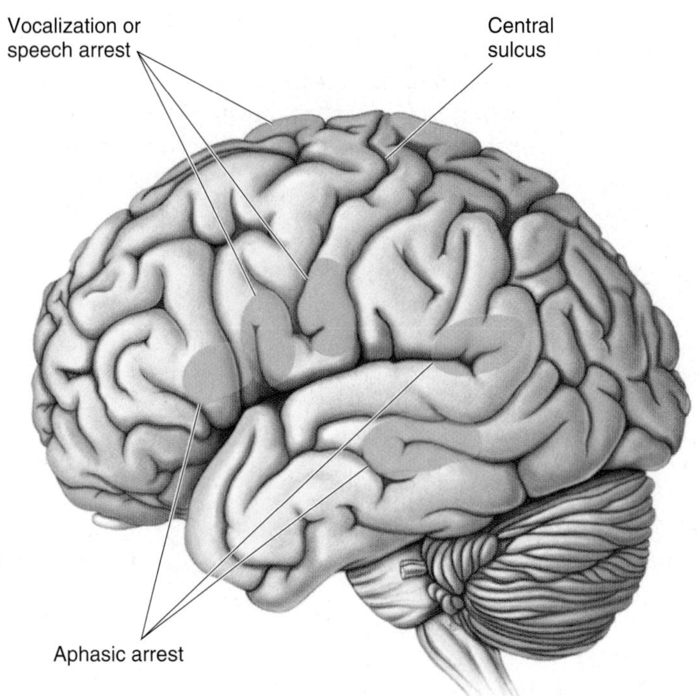

▲ FIGURE 20.18
Sites where electrical brain stimulation affects language. Stimulation of motor cortex causes vocalizations or speech arrest by activating facial muscles. At other sites, stimulation causes an aphasic arrest in which language is agrammatical or anomia is observed. (Source: Adapted from Penfield and Rasmussen, 1950, Fig. 56.)

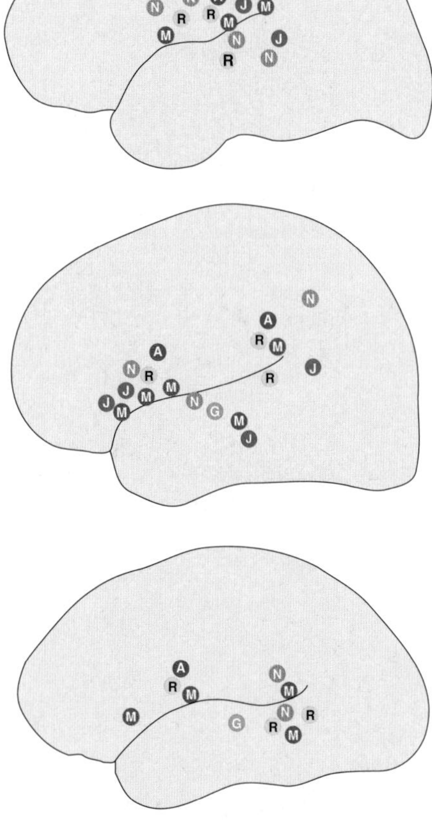

▲ FIGURE 20.19
The effects of brain stimulation in three patients being treated for epilepsy. Patients were awake, and difficulties in speaking or reading were noted. N, naming difficulty with intact speech (anomia); A, arrest of speech; G, grammatical errors; J, jargon (fluent speech with frequent errors); R, failure to read; M, facial movement errors. (Source: Adapted from Ojemann and Mateer, 1979, Fig. 1.)

It is reassuring to find that electrical stimulation selectively affects speech in the brain areas roughly corresponding to those responsible for aphasia. However, the results of stimulation vary surprisingly between nearby cortical sites and between subjects. In studies similar to those of Penfield, neurosurgeon George Ojemann, at the University of Washington, found that the effects of stimulation are sometimes quite specific. For example, stimulation of small parts of cortex at different locations can interfere with naming, reading, or repetition of facial movements (Figure 20.19). The data support several important conclusions. First, across different people there is clearly considerable variability in the brain areas at which electrical stimulation affects language. Second, between patches of cortex involved in different aspects of language are other areas not affected by stimulation. We don't know if further testing would have revealed that these in-between patches serve some untested role in language or whether these areas are actually not involved in language. Third, electrical stimulation at nearby sites can evoke quite different effects, and conversely, stimulation at distant sites can have the same effect. These findings suggest that language areas in the brain are much more complex than is implied by the Wernicke–Geschwind model. Areas involved in language are also more extensive than simply Broca's and Wernicke's areas, as they have been found to include other cortical areas as well as parts of the thalamus and striatum. Within Broca's area and Wernicke's area there may be specialized regions, possibly on the scale of functional columns in somatosensory cortex or ocular dominance columns in visual cortex. It appears that the large language areas identified on the basis of aphasic syndromes may well encompass a good deal of finer structure.

Imaging of Language Processing in the Human Brain

With the advent of modern imaging techniques, it has become possible to observe language processing in the human brain. With positron emission tomography (PET) and functional magnetic resonance imaging (fMRI), the level of neural activity in different parts of the brain is inferred from regional blood flow (see Box 7.3). In many ways, brain imaging confirms what was already known about language areas in the brain. For example, various language tasks make different portions of cerebral cortex active, and the activated areas are generally consistent with areas implicated by studies of aphasia.

However, brain imaging suggests that language processing is more complex. In an experiment conducted by Lehericy and associates, brain activity was recorded while subjects performed three different language tasks (Figure 20.20). In the first task, subjects were instructed to generate as many words as possible in a particular category such as fruits or animals (see Figure 20.20a). In the second task, subjects simply listened to a story being read aloud (see Figure 20.20b). The third task required that they silently repeat to themselves a sentence they previously heard read aloud. (see Figure 20.20c). Note that the location of activated brain areas is roughly consistent with the temporal and parietal

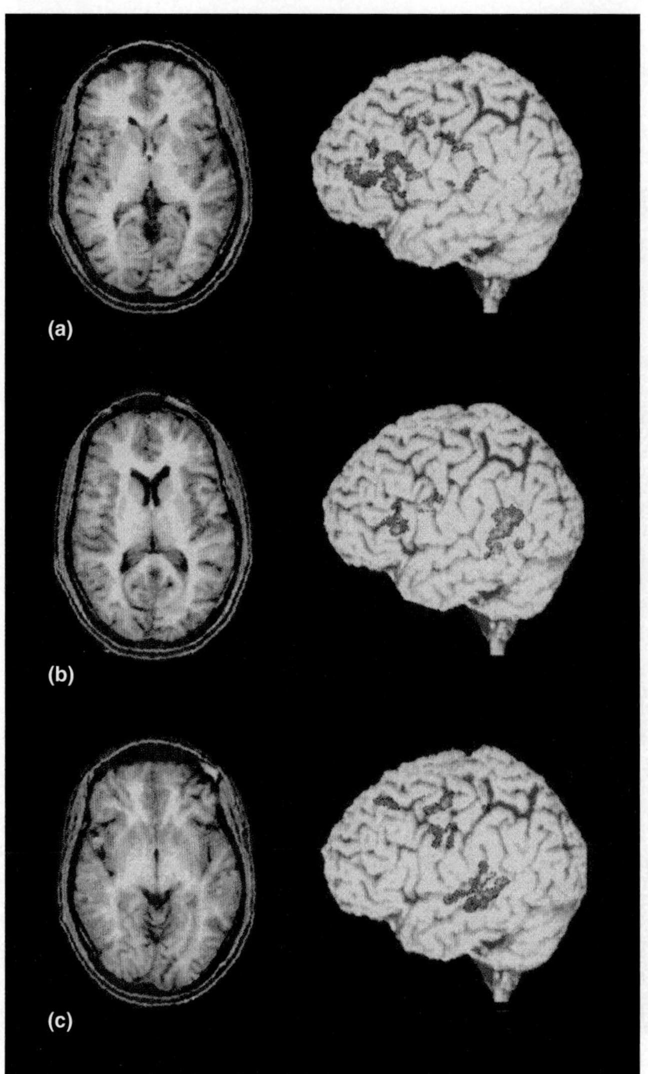

◀ FIGURE 20.20
Bilateral brain activation shown by fMRI. Based on a Wada procedure, the subject illustrated here had a strongly dominant left hemisphere for language. fMRI shows significant bilateral activation of language areas on a **(a)** word generation task, **(b)** passive story listening, and **(c)** silent sentence repetition. (Source: Adapted from Lehericy et al., 2000, Fig 1.)

language areas discovered by examining aphasia resulting from brain lesions. More surprising is the extent to which the brain activation is bilateral. Based on a Wada procedure for language lateralization, the subject shown in Figure 20.20 had a strongly dominant left hemisphere. The fMRI results suggest that there is more going on in the nondominant hemisphere than the Wada procedure suggests. Significant bilateral activation is commonly observed in fMRI studies, with ongoing debate about its significance. Recent PET and fMRI studies also suggest fascinating similarities and differences in language processing of spoken language, sign language, and Braille (Box 20.4).

In another study, researchers used PET imaging to observe differences in brain activity between the sensory responses to words and the production of speech. They began by measuring cerebral blood flow with the subject at rest. They then had the person either look at words presented on a computer display ("seeing words") or listen to words being read aloud

BOX 20.4 OF SPECIAL INTEREST

Hearing Sight and Seeing Touch

The human brain is a remarkably adaptable organ, and some of the most dramatic examples of brain reorganization come from studies of human language processing. The top panel of Figure A, from an fMRI study, shows brain areas activated when normally hearing English speakers read sentences in English. Areas in red are those most highly

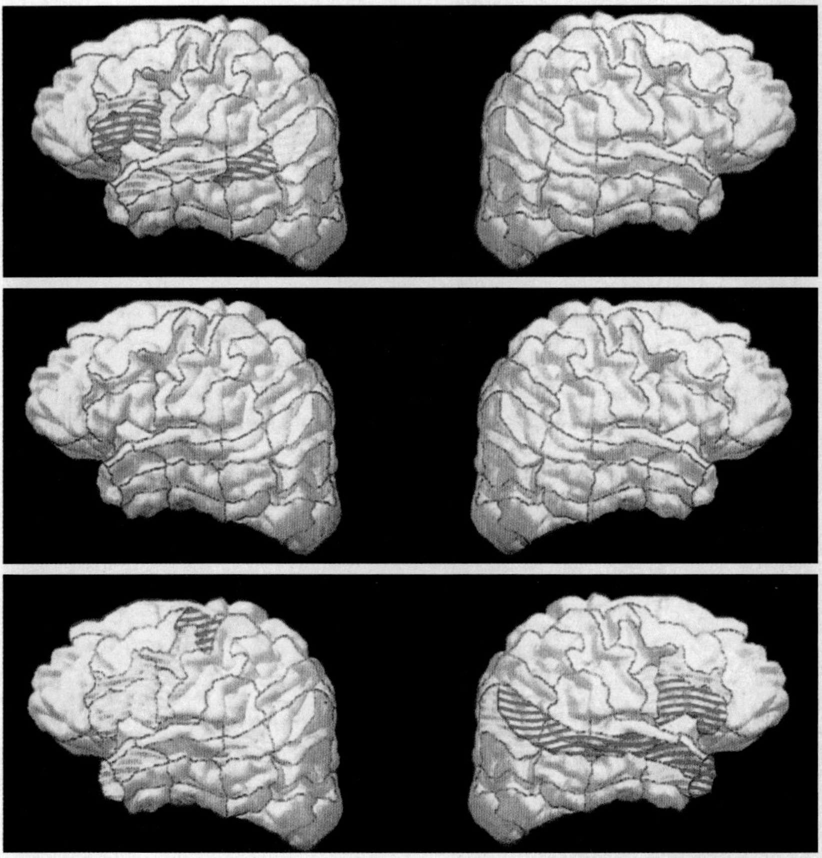

("hearing words"). By subtracting the levels of blood flow at rest from the levels during hearing or seeing, blood flow levels were obtained specifically corresponding to the activity evoked by the sensory input. The results are shown in the top half of Figure 20.21. Not surprisingly, the visual stimuli evoked increased brain activity in striate cortex and extrastriate cortex, and the auditory stimuli elicited activity in primary and secondary auditory cortex. However, the areas activated in extrastriate cortex and secondary auditory cortex did not respond to visual and auditory stimuli that were not words. These areas may be specialized for the coding of words either seen or heard. The visual stimuli did not evoke noticeably increased activity in the area of the angular gyrus and Wernicke's area, as one would expect based on the Wernicke–Geschwind model.

Another task studied with PET imaging was repeating words. To know what words to repeat, subjects must perceive and process the words by either the visual or the auditory system. Thus, the brain activity seen

specifically activated by language, and areas in yellow are somewhat less activated. (Visual activity not associated with language has been subtracted out.) There is significant activity in the standard language areas of the left hemisphere, including Broca's area and Wernicke's area, but little activity in the right hemisphere.

The center panel of Figure A is the activity seen when hearing English speakers saw sentences in American Sign Language (ASL). These people who did not understand ASL showed no specific activity after their response to meaningless hand signs was subtracted. On the other hand, the bottom panel of Figure A shows the response to ASL in deaf subjects who were raised with it as their sole language. The activation includes Broca's area and Wernicke's area in the left hemisphere, indicating that ASL activates the same key language areas activated by spoken English in hearing subjects. Even more surprising is the great deal of activity in the right hemisphere. Remarkably, the superior temporal gyrus is active in response to ASL in deaf subjects, an area that would normally respond to spoken language in hearing subjects. As these areas are also active in hearing subjects who are bilingual in ASL, it appears that something about the ASL language recruits right hemisphere auditory areas in addition to the usual left hemisphere language areas.

A different form of brain reorganization is seen in blind subjects who read Braille. A system of writing that uses patterns of small bumps on paper to represent letters, Braille is read by scanning the fingertips across the bumps. As you would expect, Braille reading activates the somatosensory cortex, but other brain areas also activated are a real surprise. Figure B shows a PET image obtained from a person reading Braille. There is significant activity at the occipital pole of the brain (yellow)—unmistakably portions of visual cortex. Through a process of brain reorganization, these blind subjects have a brain that uses traditional visual areas to process Braille, similar in spirit to the auditory areas used for ASL in deaf subjects. (The chapters in Part IV explore the mechanisms by which sensory experience can affect brain organization and lead to learning and memory.)

◀ **Figure A**
Top: Written English, read by hearing subjects. Center: ASL, seen by English-speaking hearing subjects. Bottom: ASL, seen by deaf subjects. (Source: Neville et al., 1998.)

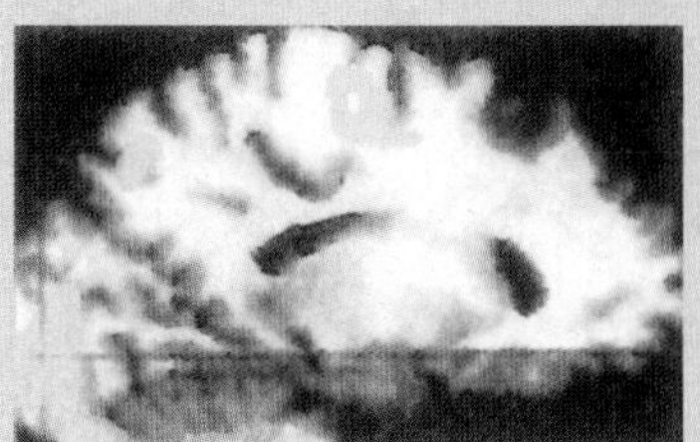

Figure B
Reading Braille. (Source: Sadato et al., 1996).

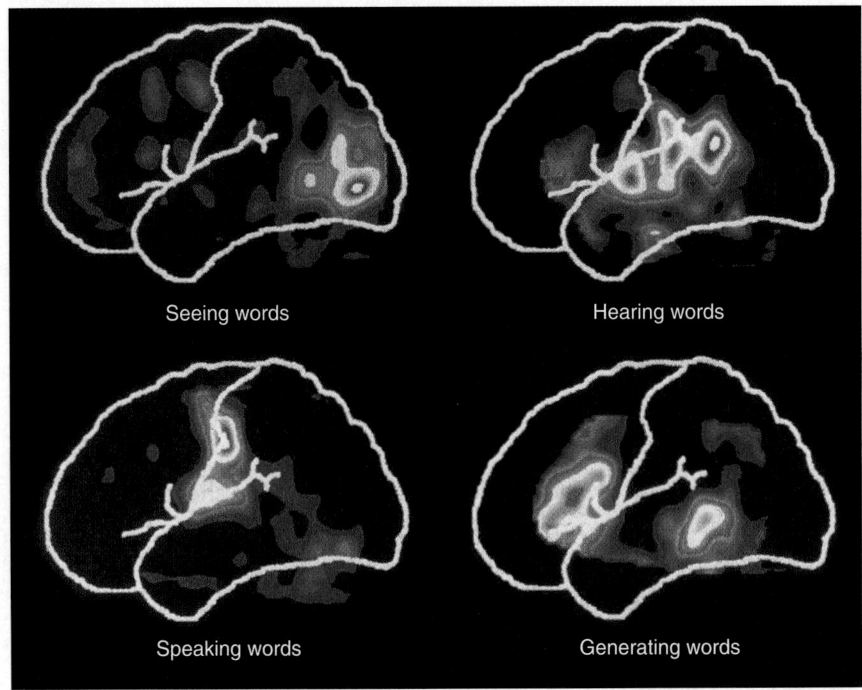

▲ FIGURE 20.21
PET imaging of sensation and speech. Relative levels of cerebral blood flow are color coded. Red indicates the highest levels, and progressively lower levels are represented by orange, yellow, green, and blue. (Source: Posner and Raichle, 1994, p. 115.)

in the repetition task should include a component associated with the basic perceptual process as well as a component associated with speech. To isolate the speech component, the response pattern previously obtained in the simple sensory task was subtracted out. In other words, the image shown for "speaking words" equals an image corresponding to "repeating spoken words" minus an image corresponding to "listening to words." After the subtraction, the blood flow pattern indicated high levels of activity in primary motor cortex and the supplementary motor area (Figure 20.21, bottom left). There was also increased blood flow around the Sylvian fissure near Broca's area. However, the PET images showed such activity *bilaterally,* and it was even observed when the subjects were instructed to move their mouth and tongue without speaking. Since there is good evidence that Broca's area is unilateral, it may not be showing up in these images, for reasons unknown.

The final task required the subjects to think a little. For each word presented, the subject had to state a use for the word (e.g., if "cake" was presented, the subject said "eat"). To isolate activity specific to this noun–verb association task ("generating words"), the blood flow pattern obtained previously for speaking words was subtracted. Areas that were activated in the association task are located in the left inferior frontal area, the anterior cingulate gyrus, and the posterior temporal lobe (see Figure 20.21, bottom right). The activity in frontal and temporal cortex is believed to be related to the performance of the word-association task, whereas the activity in cingulate cortex may be related to attention.

There is also evidence from numerous PET and fMRI studies that distributed and distinct brain areas store information about different categories of objects. These reports appear consistent with observations showing that brain damage sometimes leads to category-specific losses in the ability to name objects. For example, with a brain lesion a person may be able to name tools and some living things such as fruits and vegetables but show a large deficit in the ability to name animals. One such patient called a giraffe a kangaroo and a goat a chicken. In a PET imaging experiment, different regions of the temporal lobe were more active when a person was asked to name people, animals, and tools. In other studies, overlapping but different brain activity patterns were found for concrete nouns (e.g., "door"), abstract nouns (e.g., "despair"), concrete verbs (e.g., "speak"), and abstract verbs (e.g., "suffer"). The findings in these experiments raise many questions for further research to answer. How might the brain process different categories of words differently and integrate the results in a unified understanding? What is the distinction between brain areas involved in recognition of sensory input and areas that assign names or meaning to objects perceived?

CONCLUDING REMARKS

Language was one of the most important steps in human evolution. Communication between people is such a fundamental aspect of being human that it is difficult to imagine life without language. Current estimates are that the human capacity for language evolved relatively recently, around 100,000 years ago. While animals use a great diversity of sounds and behaviors to communicate, none of these come close to the elaborate and flexible system of language and speech used by humans. Aspects of language acquisition and use have been fruitfully studied in songbirds and nonhuman primates, but unlike other brain systems, the study of human language requires experiments and observations in humans. For this reason, experimental approaches are largely limited to behavioral studies of language acquisition and function, the consequences of lesions, effects of brain stimulation, and brain imaging with fMRI and PET. Even with these few techniques, however, a great deal has been learned. Consistent with the locations of sensory and motor areas in the brain, the basics of language organization can be understood. Early studies focused attention on Broca's area, which is near the motor cortex and associated with speech production, and Wernicke's area, which is near the auditory cortex and associated with speech comprehension. These observations are still of clinical use today.

More recent research has shown that language processing is far more complex and engages far more of the brain, than implied by the Wernicke–Geschwind model. Brain imaging and stimulation studies have revealed widespread brain areas in both hemispheres that are involved in language and that vary from one person to the next. From our perspective today, the complexity of language and its extensive representation in the brain are not so surprising because language involves many different components, such as understanding the basis of words in sound, the meanings of words, the grammar used to combine words into meaningful statements, naming objects, producing speech, and on and on. As in research on other brain systems for sensation, motor output, emotion, and so on, we are interested in the extent to which language processing involves a collection of interacting subsystems for different language skills. There is clearly still much to be learned. Further brain imaging studies will hopefully clarify the organization of language systems in the brain at a finer scale than was possible from studying the consequences of brain lesions and perhaps identify distinct circuits that serve different functions.

KEY TERMS

What Is Language?
language (p. 686)
speech (p. 686)
vocal folds (p. 687)
phonemes (p. 687)
language acquisition (p. 690)
verbal dyspraxia (p. 692)
specific language impairment (SLI) (p. 693)
dyslexia (p. 694)

The Discovery of Specialized Language Areas in the Brain
aphasia (p. 694)
Wada procedure (p. 695)
Broca's area (p. 695)
Wernicke's area (p. 695)

Language Insights from the Study of Aphasia
Broca's aphasia (p. 699)
anomia (p. 699)
Wernicke's aphasia (p. 700)
Wernicke–Geschwind model (p. 702)
conduction aphasia (p. 704)

Asymmetrical Language Processing in the Two Cerebral Hemispheres
split-brain study (p. 706)
corpus callosum (p. 706)
planum temporale (p. 710)
insula (p. 710)

REVIEW QUESTIONS

1. How is it possible for a split-brain human to speak intelligibly if the left hemisphere controls speech? Isn't this inconsistent with the fact that the left hemisphere must direct motor cortex in both hemispheres to coordinate movements of the mouth?
2. What can you conclude about the normal function of Broca's area from the observation that there are usually some comprehension deficits in Broca's aphasia? Must Broca's area itself be directly involved in comprehension?
3. Pigeons can be trained to press one button when they want food and to press other buttons when they see particular visual stimuli. This means the bird can "name" things it sees. How would you determine whether or not the pigeon is using a new language—"button-ese"?
4. What does the Wernicke–Geschwind language processing model explain? What data are inconsistent with this model?
5. In what ways is the left hemisphere usually language dominant? What does the right hemisphere contribute?
6. What evidence is there that Broca's area is not simply a premotor area for speech?

FURTHER READING

Berwick RC, Friederici AD, Chomsky N, Bolhuis JJ. 2013. Evolution, brain, and the nature of language. *Trends in Cognitive Sciences* 17:89–98.

Bookeheimer S. 2002. Functional MRI of language: new approaches to understanding the cortical organization of semantic processing. *Annual Review of Neuroscience* 25:51–188.

Friederici, AD. 2012. The cortical language circuit: from auditory perception to sentence comprehension. *Trends in Cognitive Sciences* 16:262–268.

Graham SA, Fisher SE. 2013. Decoding the genetics of speech and language. *Current Opinion in Neurobiology* 23:43–51.

Kuhl PK. 2010. Brain mechanisms in early language acquisition. *Neuron* 67:713–727.

Saffran EM. 2000. Aphasia and the relationship of language and brain. *Seminars in Neurology* 20:409–418.

Scott SK, Johnsrude IS. 2002. The neuroanatomical and functional organization of speech perception. *Trends in Neurosciences* 26:100–107.

Vargha-Khadem F, Gadian DG, Copp A, Mishkin M. 2005. *FOXP2* and the neuroanatomy of speech and language. *Nature Reviews Neuroscience* 6:131–138.

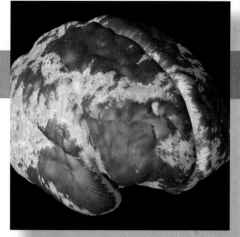

CHAPTER TWENTY-ONE

The Resting Brain, Attention, and Consciousness

INTRODUCTION

Imagine yourself at the beach, lying in the sand with your feet in the waves. As you take a sip of a cold drink, you stare out at the sky daydreaming. This peaceful moment is suddenly interrupted when your attention is grabbed by the dorsal fin of a shark sticking out of the water and moving toward you. You jump up and are about to run when you become aware of the fact that the "shark" is actually a child wearing a fake fin.

This imaginary scene includes three mental functions we will examine in this chapter. The first is the brain at rest. You might logically think the brain activity of a person daydreaming at the beach would be about as interesting as looking at a blank piece of paper. On the contrary, recent research indicates that in the brain "at rest" a network of areas is busy doing things such as diffusely monitoring our surroundings and processing daydreams.

When we become more active, the brain must deal with the enormous volume of information coming in through our senses. Rather than trying to process all of these signals simultaneously, we selectively focus on things that catch our attention, such as a shark fin in the water or an object important to us, like a cold drink about to fly out of our hand. Selective attention, or simply **attention**, is the ability to focus on one aspect of sensory input. In the visual system, attention enables us to concentrate on one object over many others in our visual field. Interactions between modalities also occur. For example, if you are performing an attention-demanding visual task, such as reading a book at a coffee shop, you will be less sensitive to the sounds of people talking around you. Amidst all the sights, sounds, and tastes coming into our brain, we are able to preferentially process some information and ignore the rest. We will see that attention has significant effects on perception and that there are corresponding changes in the sensitivity of neurons at many locations in the brain.

A brain function related to attention is consciousness. In general use, **consciousness** means awareness of something (the fake shark fin in our example). For centuries, philosophers have wrestled with the meaning of consciousness, and more recently, neuroscientists have devised experiments to reveal the neural basis of the conscious brain. The link between attention and consciousness appears to be tight, as we are generally aware of what we pay attention to. However, we will see they are actually distinctly different processes.

RESTING STATE BRAIN ACTIVITY

If you go into a quiet room, lie down, and close your eyes (but stay awake), what do you suppose your brain is doing? If your answer is "not much," you are probably in good company. In our discussions of various brain systems, we have described how neurons become active in response to incoming sensory information or the generation of movement. Modern brain imaging techniques are consistent with this view that, in response to behavioral demands, neurons become more active in cortical areas that process ongoing perceptual or motor information. It is reasonable to infer that the brain is quiet in the absence of active processing. However, when the entire brain is imaged with positron emission tomography (PET) or functional magnetic resonance imaging (fMRI), it is found that its **resting state activity** includes some regions that really are fairly quiet and others that are surprisingly active. An important question is what, if anything, does the resting activity signify?

The Brain's Default Mode Network

Human imaging studies suggest that the difference between the brain's resting state and the activity recorded while a person performs a task may teach us important lessons about the nature of the resting brain and the functions that it performs. The existence of resting state activity does not in itself allow us to conclude much. Conceivably, the resting activity might vary randomly from moment to moment and person to person, and activations associated with behavioral tasks would be superimposed on this random background. However, this does not seem to be the case. When a person engages in a perceptual or behavioral task, there are decreases in the activity of some brain areas at the same time that task-relevant brain areas become more active. One possibility is that both the decreases and increases in activity are related to the task. For example, if a person is required to perform a difficult visual task and ignore irrelevant sounds, we might expect the visual cortex to become more active and the auditory cortex less active.

Two further observations suggest that there is something fundamental and significant about the resting brain activity. First, the areas that show decreased activity compared to the resting state are consistent when the nature of the task is changed. It appears that the areas showing decreased activity during behavioral tasks are always active at rest and become less active during *any* task. Figure 21.1 summarizes data from experiments using nine different tasks involving vision, language, and memory. The blue and green patches in the figure show brain areas in which activity decreased from the resting state when humans engaged in any of the nine tasks. The particular task does not seem to account for the activity changes. Second, the patterns in brain activity changes are consistent across human subjects. These observations suggest that the brain might be "busy" even in the state we call rest, that the resting activities are consistent, and that these activities are decreased when a task is performed.

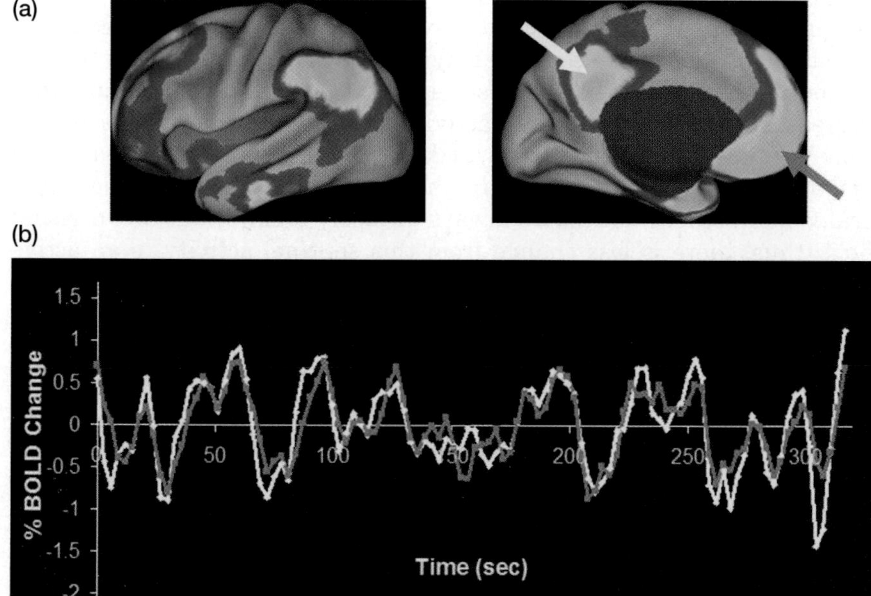

◀ FIGURE 21.1
The default mode network. **(a)** Data from nine PET imaging studies involving different behavioral tasks were averaged to produce these lateral and medial views of brain activity. The brains have been computer "inflated" so activity in the sulci can be seen. Brain areas colored blue and green were more active during quiet rest periods than during the behavioral tasks. **(b)** Slow fluctuations in brain activity are correlated between the medial prefrontal cortex and the cingulate cortex (arrows in a). These fMRI recordings were made while a person quietly fixated on a small cross on a visual display. (Source: Raichle et al., 2007, Fig. 1.)

Brain areas that show more activity in the resting state than during behavioral tasks include the medial prefrontal cortex, posterior cingulate cortex, posterior parietal cortex, hippocampus, and lateral temporal cortex. Together these areas are called the **default mode network**, or default network, to indicate that the brain defaults to activity in this group of interconnected areas when it is not engaged in an overt task. Some scientists believe this network of areas defines a system or a group of interacting systems in the same sense as we define sensory and motor systems. A finding consistent with this idea is the striking degree of correlation in brain activity between components of the default mode network. Figure 21.1b shows a 5-minute recording made from the two brain areas (medial prefrontal, cingulate cortex) indicated by arrows in Figure 21.1a. The subject was lying in an fMRI machine doing nothing more than fixating on a small crosshair on a visual display. For reasons unknown, there were continuous variations in the fMRI signal and there was a remarkable degree of correlation between activity in the distant brain areas. Whether these fluctuations are related to thought is not known, but they suggest coordination or interaction between brain areas.

Establishing the function of the default mode network is challenging because the brain areas involved participate in diverse activities. It is tempting to view the resting state activity as an indication of the inner life of the brain. When we relax, it is common to daydream, remember, and imagine, things referred to as *spontaneous cognition*. As the default mode network is deactivated during most tasks, it is difficult to conduct experiments examining its function. However, we can infer possible functions of the default network by considering its components and the handful of tasks that do activate it. For starters, the absence of primary sensory and motor areas is consistent with the idea that the default network is not chiefly concerned with taking in sensory information or controlling movement.

Functions of the Default Network. A variety of hypotheses have been considered for the function of the default network. The two we will consider here are the *sentinel hypothesis* and the *internal mentation hypothesis*. The idea behind the sentinel hypothesis is that even when we rest, we must broadly monitor (pay attention to) our environment; in comparison, when we are active, we focus our attention on the matter at hand. If you imagine our ancient ancestors living in a world harboring constant threats, it makes sense that we may have evolved to be always "on the lookout." One experiment with results consistent with this idea found that decreases from default network activity are less when a person switches from rest to a peripheral vision task than from rest to a foveal vision task. Perhaps the decrease is less when the active task involves peripheral vision because at rest we always diffusely monitor our broad visual field (thus, there is less change from this sentinel activity to an active task involving peripheral vision). Another study reported that the default network became activated in an experiment requiring people to broadly monitor their peripheral visual field for stimuli at random locations but not when they were instructed to focus on one location where a stimulus might appear. Also relevant to the sentinel hypothesis is a rare disorder known as *simultagnosia* (a component of Bálint's syndrome), in which people have normal visual fields and are able to perceive individual objects, but are unable to integrate simultaneous information to understand a complex scene. For example, shown a picture of an animal, a patient said, "There is a round head joining what looks like a powerful body; there are four shortish legs; it doesn't say anything to me; ah, but there is a

PAST AND FUTURE EVENT ELABORATION

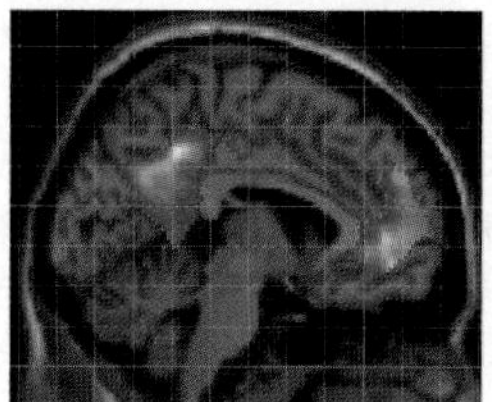

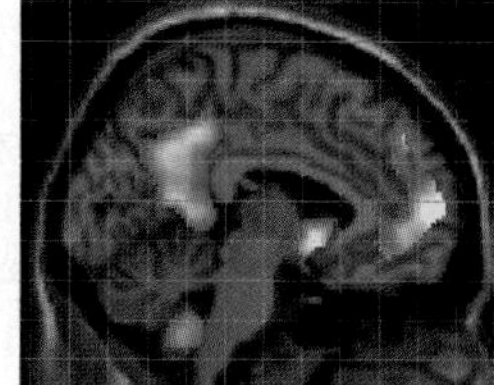

◀ FIGURE 21.2
Activation of the default mode network. In the main experimental conditions, subjects were asked to recall a past event or imagine an event in the future after seeing a cue word (e.g., dress). In control tasks, subjects either generated a sentence or named objects in response to a cue. Recordings with fMRI show that similar posterior cingulate and medial prefrontal components of the default network were activated in the autobiographical memory tasks more than in the control tasks. (Source: Addis et al., 2007, Fig. 2.)

small and curly tail so I think it must be a pig." In other words, this person was not able to put the pieces mentally together and recognized the pig only because of its distinctive tail. Evidence suggests that the posterior cingulate cortex, a part of the default network, may play a role in diffusely monitoring the visual field for stimuli. The speculation is that this brain area, which is part of the default network and is damaged in simultagnosia, is involved in sentinel activities.

The internal mentation hypothesis says that the default mode network supports thinking and remembering, the sort of daydreaming we do while sitting quietly. In an experiment that suggests this hypothesis, brain activity is imaged while subjects are asked to silently recall past events in their lives or imagine an event that might happen to them in the future. For example, "recall a past event from last week" or "envisage a future event in the next 5–20 years." Brain activity in these autobiographical memory tasks is contrasted with activity recorded in control tasks that involve the simple use of facts rather than recall and thinking about autobiographical information. In one control task, subjects construct a sentence using a cue word, and in another, they imagine objects bigger and smaller than an object named by a cue word. In the memory tasks, the hippocampus and neocortical areas in the default network become more active; in the control tasks, these brain areas were not activated. The hypothesis is that the memory tasks activate the brain in a manner similar to daydreaming about one's life, unlike the structured use of facts in the control tasks. Figure 21.2 shows that recalling past events and imagining new ones activate similar regions of medial prefrontal and posterior cingulate cortex.

Although not all scientists agree with the definition of a default mode network, considerable evidence suggests that certain components of the brain are busier while we rest and that they do different things from brain areas that become engaged during active tasks. The overall hypothesis is that, when the situation requires us to become actively involved in a perceptual or motor task, we switch modes from sentinel and internal mentation activities (high default network activity) to focused processing of sensory input (low default network activity and increased sensory–motor activity). Studying functions of the resting brain is challenging because typical experimental tasks ("do this," "look at that") shut down the areas being studied. One thing that seems clear is that the transition from brain activity in the default network at rest to sensory processing in active tasks involves a change in the focus of attention—our next topic to explore.

ATTENTION

Picture yourself at a crowded party surrounded by loud music and the chatter of hundreds of people. Even though you are bombarded by sound from all directions, you are somehow able to concentrate on the conversation you're having and ignore most of the other noise and talking. You are *paying attention* to the one conversation. Behind you, you hear someone mention your name, and you decide to eavesdrop. Without turning around, you start focusing your attention on this other conversation to find out what's being said about you. This common experience of filtering auditory input, called the *cocktail party effect*, is an example of attention that we use within and across sensory modalities. Because its behavioral and neural functions have been studied in detail, we will focus on visual attention.

Studies of attention often describe it as a limited resource or bottleneck in brain processing. It is common to use the term "selective attention" to emphasize that it is directed to select objects, unlike overall arousal, which is unselective. For brevity, we will use "attention," but this term should be taken to mean selective attention. The limit that attention puts on brain processing is probably a good thing; one can only imagine how overwhelming sensory input might be if we were to simultaneously focus on every portion of our visual environment, every sound, and every smell. The limited capacity of attention probably explains why traffic accidents are much more common while people are texting or talking on cell phones. As we will see, attention has significant effects on behavioral speed and precision. While it is not simply a problem of attention, attention-deficit hyperactivity disorder demonstrates how critical attentional mechanisms are (Box 21.1).

BOX 21.1 OF SPECIAL INTEREST

Attention-Deficit Hyperactivity Disorder

It is the last lecture of the school year, and you are failing miserably at focusing on the instructor as you stare longingly at the green grass and trees outside the window. At times, we all have difficulty concentrating on our work, sitting still, and resisting the urge to play. But for millions of people, the syndrome that has come to be called **attention-deficit hyperactivity disorder (ADHD)** routinely and seriously interferes with their ability to get things done.

The three traits commonly associated with ADHD are inattention, hyperactivity, and impulsiveness. Children normally exhibit these traits more than adults, but if their behaviors are pronounced, ADHD may be diagnosed. Estimates are that 5–10% of all school-age children worldwide have ADHD, and the disorder interferes with their schoolwork and interactions with classmates. Follow-up studies show that many people diagnosed with ADHD continue to exhibit some symptoms as adults.

We do not know what causes ADHD, but there are several clues. For example, it has been reported that in MRI scans, several brain structures, including prefrontal cortex and the basal ganglia, are smaller in children with ADHD. It is not known whether these differences are behaviorally significant, and they are not reliable enough to be a basis for diagnosing the disorder. However, the possible involvement of these structures is intriguing because they have long been implicated in the regulation and planning of behavior. You might recall Phineas Gage from Box 18.2 in Chapter 18, who experienced great difficulty making and carrying out plans after a severe lesion to his prefrontal cortex.

Several pieces of evidence suggest that heredity plays a significant role in ADHD. Children of parents with ADHD are more likely to develop it, and a child is much more likely to have it if an identical twin does. Nongenetic factors, such as brain injury and premature birth, may also be involved. Several genes related to the function of dopaminergic neurons have been reported to be abnormal in people with ADHD. These include the D4 dopamine receptor gene, the D2 dopamine receptor gene, and the dopamine transporter gene. We have seen in several previous chapters how important dopaminergic transmission is for a variety of behaviors, so it will be a challenge to clarify dopamine's involvement in ADHD.

At present the most common treatment for ADHD, aside from behavioral therapy, is psychostimulant drugs such as Ritalin. Ritalin is a mild central nervous system stimulant similar to amphetamines. It also inhibits the dopamine transporter, increasing the postsynaptic effect of dopamine. In many children, Ritalin successfully decreases impulsiveness and inattention, although questions about long-term use remain.

◀ FIGURE 21.3
Visual "pop out." A salient difference in color automatically draws your attention. It may take a moment to even notice the faces of the men wearing red. (Source: Courtesy of Steve McCurry photographer, Magnum Photos.)

Daily life suggests that attention is directed in two different ways. Suppose you are walking through a field of grass where amidst the sea of green grows a single bright-yellow dandelion. Your attention may be automatically drawn to the dandelion because the color "pops out" from the background. We say the dandelion "caught our attention." Certain visual features, such as a distinctive color, movement, or a flashing light, draw our attention automatically (Figure 21.3). This is called **exogenous attention** or **bottom–up attention** because the stimulus attracts our attention without any cognitive input. Presumably a process like this is used by many animals to rapidly detect and evade predators. Quite different is **top–down attention**, also called **endogenous attention**, in which attention is deliberately directed by the brain to some object or place to serve a behavioral goal. You might be flipping through this book looking for a passage that you know is in the upper-right corner of a page. The search is made easier by the allocation of attention specifically to the page corner.

Behavioral Consequences of Attention

In most situations, if we want to visually scrutinize something, we move our eyes so that the object of interest is imaged on the fovea in each eye. Implicit in this behavior is the fact that most of the time we pay attention to the object we are looking at. However, it is also possible to shift attention to objects imaged on parts of the retina outside the fovea; this phenomenon of "looking out of the corner of one's eyes" is called *covert attention* because our gaze doesn't reveal what we are attending to. Whether on the fovea or a more peripheral portion of the retina, focusing attention enhances visual processing at that location in several ways. Two ways we'll look at are enhanced visual sensitivity and faster reaction times.

Attention Enhances Visual Sensitivity. Figure 21.4 shows an experiment for studying the effects of directing visual attention to different locations. The observer fixated on a central point, and her task was to say whether a target stimulus was flashed on at a location to the left of the fixation point, to the right, or not at all. The task was difficult because the target was small and flashed on very briefly. The experiment included several special procedures for identifying the effects of attention. Each trial began with the presentation of a cue stimulus at the fixation point. The cue was either a plus sign, an arrow pointing left, or an arrow pointing right. After the cue was extinguished, a variable delay period followed, during which

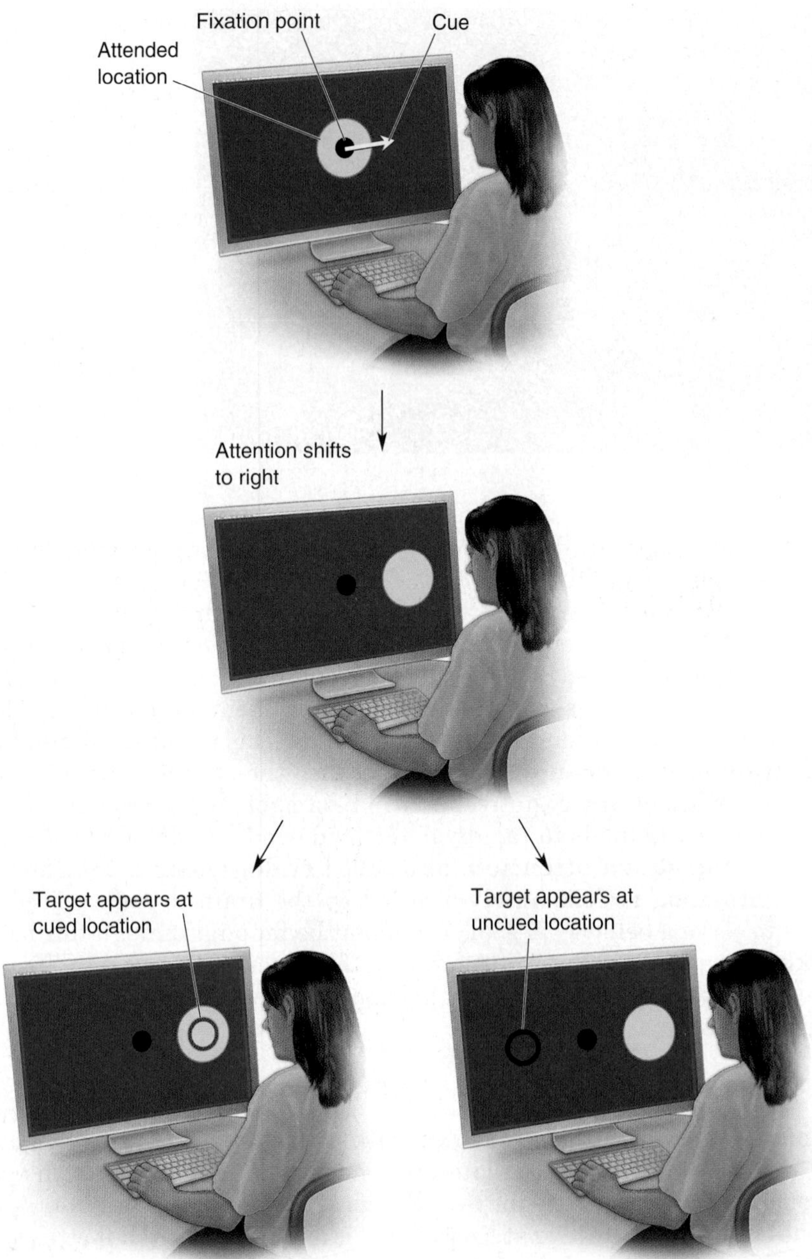

▲ FIGURE 21.4
An experiment to measure the effect of attention on visual detection. While an observer maintains steady fixation, a cue directs her to shift her attention to one side of the computer screen. In each trial, the observer indicates whether a circular target is seen on either side of the screen.

only the fixation point was seen. In half the trials, there was no further stimulus, and in the other half, a small target circle was flashed on for 15 msec at either the left or right position.

A key element of the experiment is that the cue was used to direct attention. If the central cue was a plus sign, it was equally likely that a little circle would appear to either the left or the right. The plus sign was thus a "neutral cue." If the cue was a left arrow, it was four times more likely that the target would appear on the left than on the right. If the cue was a right arrow, it was four times more likely that the target would appear on the right than on the left. If the target appeared to the cued side, the cue was "valid"; the cue was "invalid" if it pointed away from where the target would appear. The observer was required to keep her eyes pointing straight ahead, but in order to make the most correct responses on the difficult task of detecting the flashed circle targets, it would be advantageous to make use of the cue. For instance, if the cue was a right arrow, it would be beneficial to try to covertly attend to the right target location more than the left.

For each of the subjects used in this experiment, the data collected consisted of the percentage of the time a circle was correctly detected. Because there was no target circle in half the trials, the observers could not get a high percentage correct by "cheating" (i.e., by saying there was always a target at the side where the arrow pointed). In trials where the central cue was a plus sign, the observers detected the target stimulus in about 60% of the trials in which one was presented. When the cue was a right arrow, the observers detected the target stimulus to the right on about 80% of the trials in which one was presented there. However, when the cue pointed right, the observers detected the target stimulus to the left in only about 50% of the trials in which one was presented there. With the appropriate left–right reversal, the results were about the same with left arrows. The results are summarized in Figure 21.5.

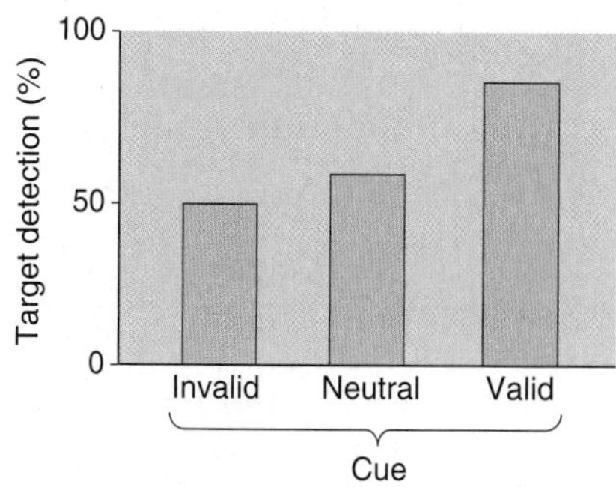

▲ FIGURE 21.5
The effect of cueing on target detection. Detection of a visual target is improved if the location of the upcoming target is validly cued. Performance with an invalid cue is worse than with a neutral cue that gives no indication of where the target will appear.

What do these data mean? To answer, we must imagine what the observer was doing. Evidently, the expectation of the observer based on the cues influenced her ability to detect the subsequent targets. It appears that the arrow cues caused the observer to shift her attention to the side where the arrow pointed, even though her eyes didn't move. Presumably, this covert shift of attention made it easier to detect the flashed targets compared to the trials when the central cue was a plus sign. Conversely, the observer was less sensitive to the targets on the side of fixation opposite to where the arrow cue pointed. Based on these results and those from many other similar experiments, our first conclusion about the behavioral effects of attention is that it increases our visual sensitivity, making things easier to detect. This is probably one of the reasons we can listen in on one conversation among many when we give it our attention.

Attention Speeds Reaction Times. Using an experimental technique similar to the one discussed earlier, it has been demonstrated that attention increases the speed of our reactions to sensory events. In a typical experiment, an observer fixated on a central point on a computer screen, and target stimuli were presented to either the left or the right of the fixation point. However, in this experiment, the observer was told to wait until he perceived a stimulus at either location and to press a button when he did. A measurement was made of how long it took the observer to react to the presentation of a stimulus and press a button. Preceding the target was a cue stimulus, either a plus sign or an arrow pointing left or right. The arrows indicated the side to which a stimulus was more likely to appear, whereas the plus sign meant that either side was equally likely.

Results from this experiment demonstrated that an observer's reaction times were influenced by where the central cue told him to direct his attention. When the central cue was a plus sign, it took about 250–300 msec to press the button. When an arrow cue correctly indicated where a target would appear (e.g., right arrow and right target), reaction times were 20–30 msec faster. Conversely, when the arrow cue pointed in one direction and the target appeared at the opposite location, it took 20–30 msec longer to react to the target and press the button. The reaction time included time for transduction in the visual system, time for visual process-

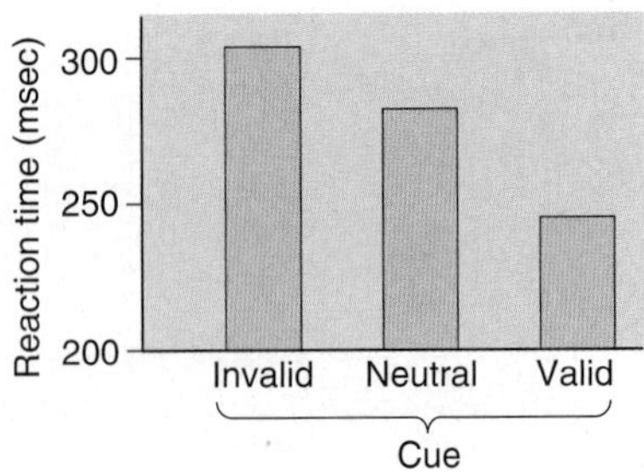

▲ FIGURE 21.6
The effect of cueing on reaction time. In neutral cue trials, the cue was a plus sign, which gave no indication of the likely location of the following target. In valid cue trials, the arrow cue pointed to the location where the target later appeared, speeding reactions to the targets. When the cue was invalid, pointing in a direction opposite to where the target later appeared, reaction times were slower. (Source: Adapted from Posner, Snyder, and Davidson, 1980, Fig. 1.)

ing, time to make a decision, time to code for the finger movement, and time to press the button. Nonetheless, there was a small but reliable effect based on which direction the arrows directed the observer's attention (Figure 21.6). If we assume that attention to visual objects does not have a direct effect on visual transduction or motor coding, we are left with the hypothesis that attention can alter the speed of visual processing, or the time to make a decision about pressing the button. An example from daily experience underscores the behavioral implications of a reaction time delay due to attention: If you are driving a car at 60 mph and attending to something other than the road, a 30 msec delay translates into applying the car brakes about 25 feet further down the road, perhaps too late to avoid hitting a car or person.

Physiological Effects of Attention

What is happening in the brain when we shift our attention to something? For example, in the behavioral studies just discussed, is a subject's performance enhanced because neural activity in a particular brain area is somehow "better"? While it's conceivable that attention is strictly a high-level cognitive process, experiments demonstrate that the effects of attention can be observed in numerous sensory areas stretching from the lateral geniculate nucleus (LGN) to visual cortical areas in the parietal and temporal lobes. We will look at human brain imaging studies that show activity changes associated with the allocation of attention and then turn to animal studies that reveal the effects of attention on individual neurons. These experiments show the consequences of allocating attention to a location or feature.

Functional MRI Imaging of Human Attention to Location. A key observation made in behavioral studies of visual attention is that enhancements in detection and reaction time are selective for spatial location. When we know where an important stimulus is more likely to appear, we move our attention to it and process the sensory information with greater sensitivity and speed. A common analogy is that there is a **spotlight of attention** that moves to illuminate objects of particular interest or significance. Experiments using fMRI imaging of the human brain suggest that there may be selective changes in brain activity associated with spatial shifts in attention.

In one experiment, subjects in the fMRI machine viewed a stimulus consisting of patches of colored lines arranged in 24 sectors as shown at the bottom of Figure 21.7a. The upper panels in Figure 21.7a show a sequence of four sectors, moving out from the fixation point, that a subject was cued to attend to. The location of the cued sector changed every 10 seconds. During the 10-second period, the color and orientation of the line segments in all of the sectors changed every 2 seconds. Each time the line segments changed, the subject's task was to press one button if the lines were blue and horizontal or orange and vertical, and a second button if the lines were blue and vertical or orange and horizontal. The reason for having the subjects perform this task was to force them to attend to a particular sector of the stimulus and ignore the rest. Remember that subjects always kept their gaze fixed at the center of the bull's-eye stimulus.

The fascinating aspect of this experiment involves what happens when the location of the sector being attended to changes. Figure 21.7b shows brain activity recorded with the attended sector at four locations at increasing distance from the fixation point. Notice how the areas of enhanced brain activity (red and yellow) move away from the occipital pole

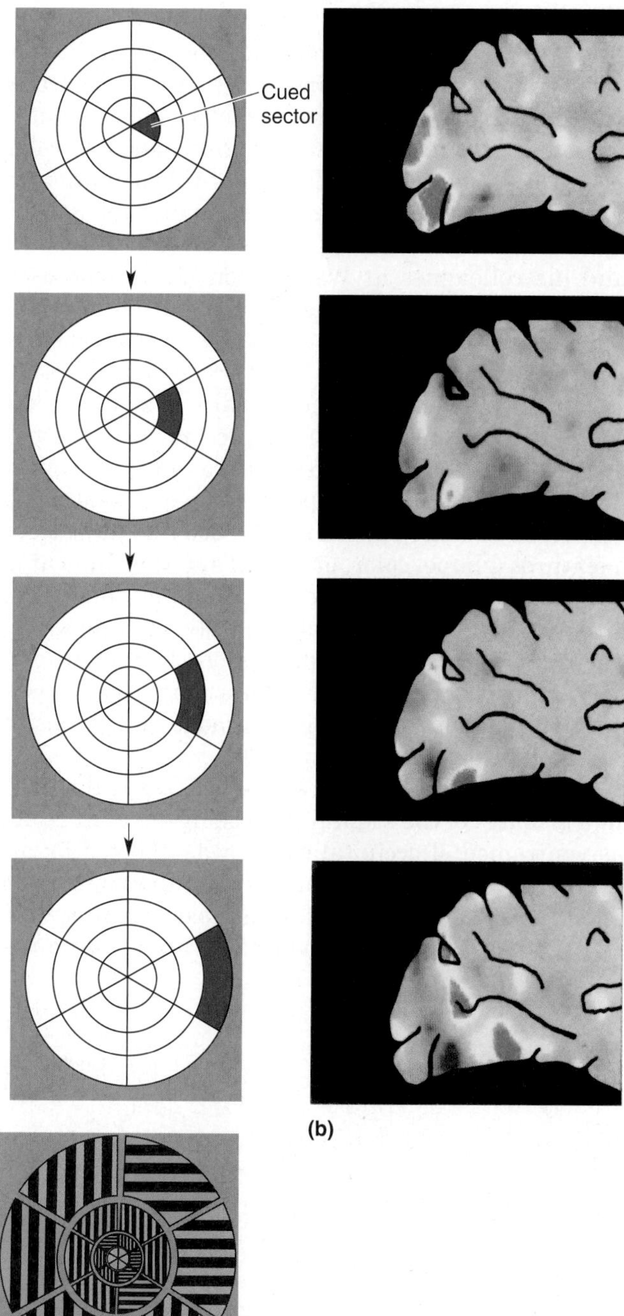

▲ FIGURE 21.7
The spotlight of attention. **(a)** The stimulus (bottom) consisted of patches of vertical and horizontal blue and orange lines arranged into 24 sectors radiating out from the central fixation point. The orientation and color of each sector changed randomly every 2 seconds. From top to bottom, the upper four bull's-eye patterns indicate in red a sequence of four sectors that a subject was cued to attend to. **(b)** Superimposed on a schematic image of the brain, red and yellow areas show locations of enhanced brain activity associated with attention to the sector to the left in (a). As attention was directed from central to more peripheral vision, cortical enhancement moved away from the pole of the occipital lobe (Source: Courtesy of J. A. Brefczynski and E. A. DeYoe.)

as the attended sector moves out from the fovea. The pattern of brain activity shifts retinotopically, even though the visual stimuli are the same regardless of which sector is attended. The hypothesis is that these images show the neural effect of the spotlight of attention moving to different locations.

PET Imaging of Human Attention to Features. The fMRI results discussed above appear consistent with the behavioral observation that visual attention can be moved independently of eye position. But attention

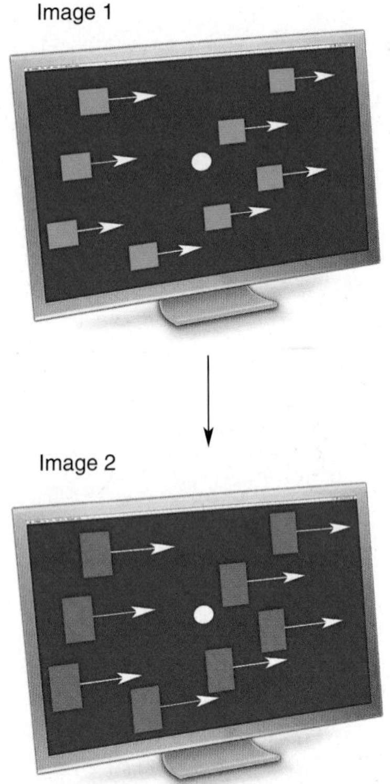

▲ FIGURE 21.8
Same–different stimuli used for PET imaging. The observer sees image 1 followed by image 2. The moving elements in the stimuli can change in shape, color, and speed of motion from image 1 to image 2. The observer responds by indicating whether the stimuli in the two images are the same or different.

involves more than just location. Imagine walking down a crowded city sidewalk in the winter looking for someone. Everyone is bundled in heavy coats, but you know your friend will be wearing a red hat. By mentally "focusing" on red, it will be much easier to pick out your friend. Evidently, we are able to pay particular attention to visual features such as color to enhance our performance. Is there any reflection of this attention to features in brain activity? The answer has come from studies using PET imaging in humans.

Steven Petersen and his colleagues at Washington University used PET imaging while humans performed a same–different discrimination task (Figure 21.8). An image was flashed on a computer screen for about half a second; after a delay period, another image was flashed. Each image was composed of small elements that could vary in shape, color, and speed of motion. The task of the observer was to indicate whether the two successive images were the same or different. To isolate the effect of attention, two different versions of the experiment were conducted. In *selective-attention* experiments, subjects were instructed to pay attention to just one of the features (shape, color, or speed) and they indicated whether that feature was the same or different in the two images. In *divided-attention* experiments, subjects simultaneously monitored all features and based their same–different judgments on changes in any feature. The researchers then subtracted the divided-attention responses from the selective-attention responses to obtain an image of changes in brain activity associated with attention to one feature.

Figure 21.9 illustrates the results. Different areas of cortex had higher activity when different attributes of the stimuli were being discriminated. For instance, whereas ventromedial occipital cortex was affected by attention in color (blue spots) and shape (orange spots) discrimination tasks, it was not affected in the speed discrimination task (green spots). Conversely, areas in parietal cortex were influenced by attention to the

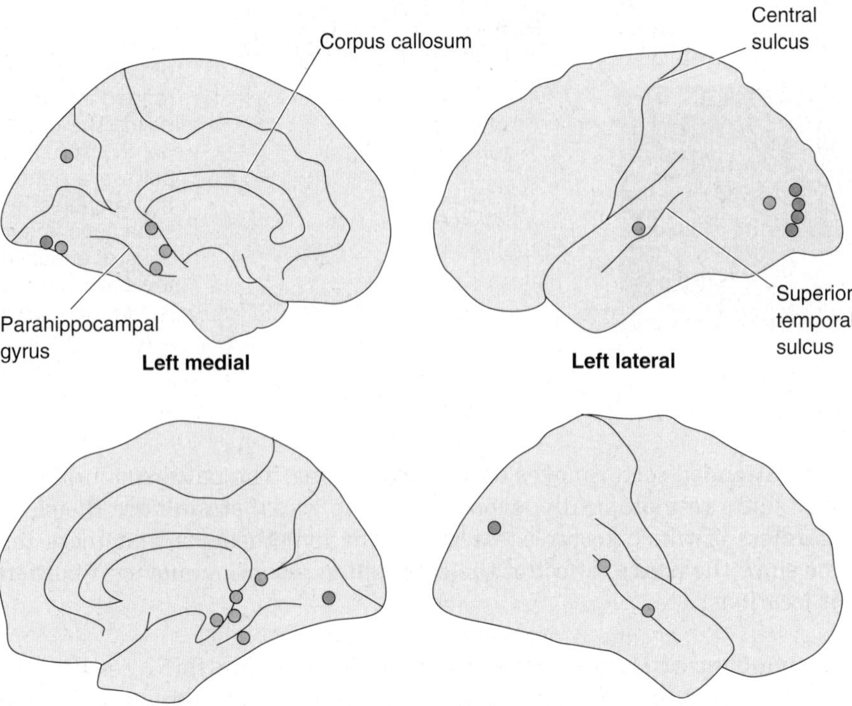

▶ FIGURE 21.9
Feature-specific effects of visual attention. Symbols indicate where activity in PET images was higher in selective-attention experiments relative to divided-attention experiments. Selective-attention was associated with enhanced activity in different brain areas when attention was directed to speed (green), color (blue), or shape (orange). (Source: Adapted from Corbetta et al., 1990, Fig. 2.)

motion task but not the other tasks. While it is not possible to know with certainty which cortical areas were highlighted in these experiments, areas of heightened activity in the color and shape tasks may have corresponded to areas V4 and IT and other visual cortical areas in the temporal lobe. The area most affected by performing the motion task was near area MT. These effects of attention to different features are roughly consistent with the tuning properties of neurons in extrastriate visual areas discussed in Chapter 10.

The important points to take from these and other brain imaging studies are that numerous brain areas appear to be affected by attention and that the particular areas depend on the nature of the behavioral task performed. We'll now examine two of these areas in detail and see how studies in behaving monkeys have clarified the neural effects of attention.

Attention Enhances Responses of Neurons in Parietal Cortex. The perceptual studies discussed earlier show that attention can be moved independently of eye position. But what happens normally when you move your eyes to explore your environment? Let's say you are scrutinizing an object imaged on your fovea when a bright light is flashed in your peripheral visual field. What happens to your attention before, during, and after you make a saccadic eye movement to the flashed light? Behavioral studies show that shifts in attention can occur in about 50 msec, whereas saccades take about 200 msec. It appears that attention, initially focused at the fovea, shifts to the peripheral retina and is followed by the saccade.

The assumption that attention changes location prior to an eye movement underlies an experiment performed by neurophysiologists Robert Wurtz, Michael Goldberg, and David Robinson at the National Institutes of Health. They recorded from several brain areas in monkeys to determine whether shifts of attention are associated with changes in neural activity. Because of the close relationship between attention and eye movements, the investigation started with parts of the brain involved in generating saccades.

The researchers recorded from neurons in the posterior parietal cortex of monkeys while the animals performed a simple behavioral task (Figure 21.10). This cortical area is thought to be involved in directing eye movements, in part because electrical stimulation here will evoke saccades. In the experiment, a monkey fixated on a spot on a computer display and a stimulus was flashed on the peripheral retina at the location of the receptive field under study. The monkey was cued to either hold fixation on the initial spot or saccade to the flashed stimulus. In either case, the parietal neuron was excited by the stimulus flashing in its receptive field (Figure 21.11a). The observation that makes this experiment important is that the response to the flashed stimulus was significantly enhanced when the animal made a saccade to foveate it compared to trials in which the monkey held fixation at the initial location (Figure 21.11b). Remember, the stimulus is the same in both cases. The enhancement effect was seen only when a saccade was made to the receptive field location but not to other locations and even though the saccade was made after the neuron responded to the target stimulus. This suggests that attention moves to the end of the planned saccade before the eyes move and only neurons with receptive fields at that location have responses enhanced by the attention shift that precedes the saccade (Figure 21.11c). A second interpretation that has to be considered is that the enhanced response was a premotor signal related to coding for the

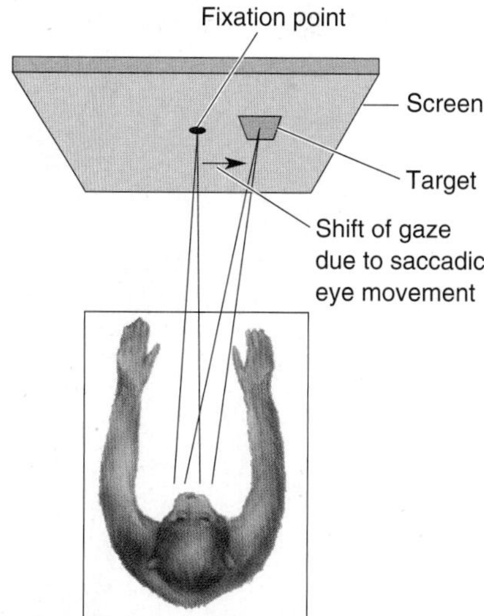

▲ FIGURE 21.10
A behavioral task for directing a monkey's attention. While recordings are made from the posterior parietal cortex, the monkey fixates on a point on a computer screen. When a peripheral target appears (usually in a neuron's receptive field), the animal makes a saccade to the target. (Source: Adapted from Wurtz, Goldberg, and Robinson, 1982, p. 128.)

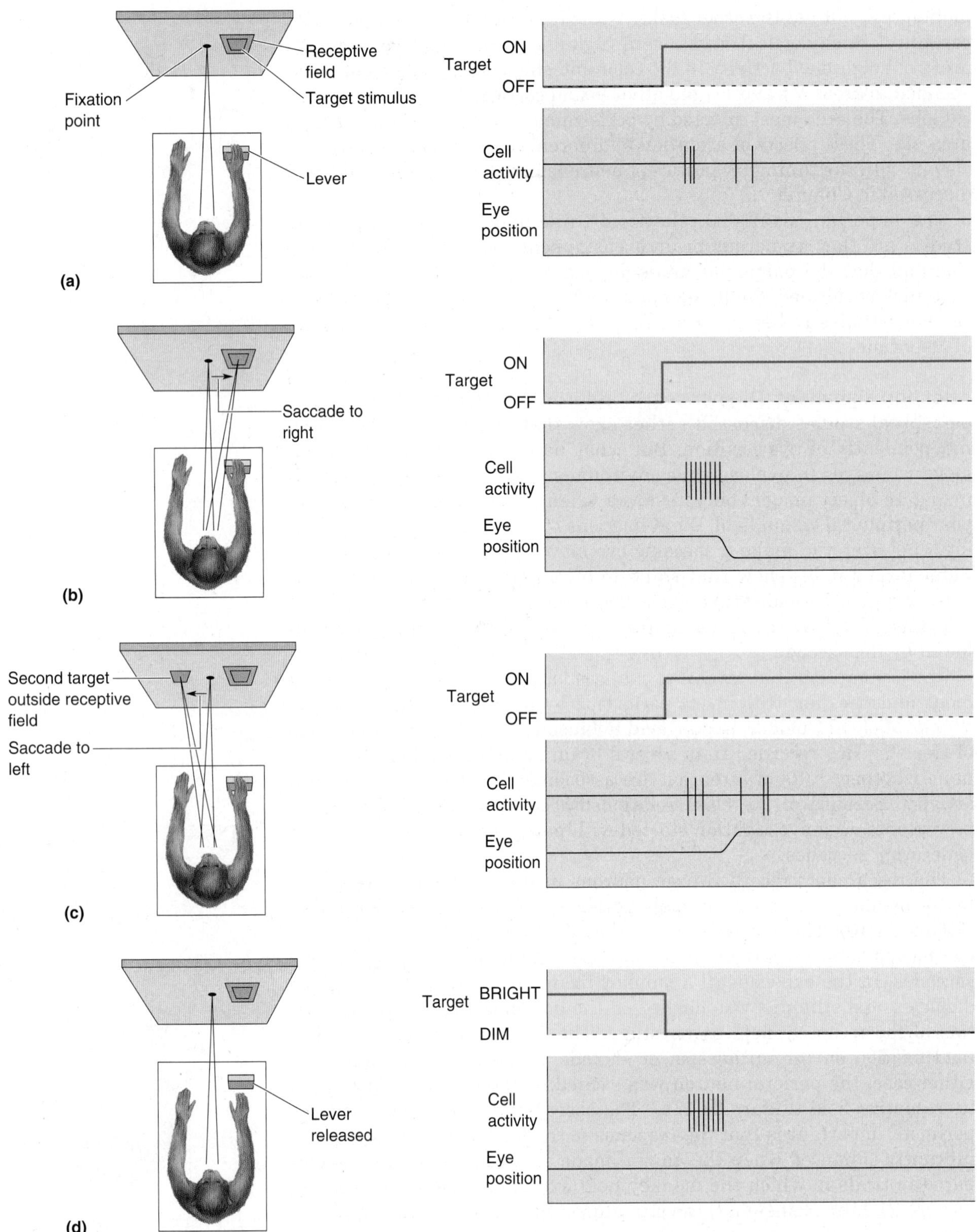

▲ FIGURE 21.11

The effect of attention on the response of a neuron in posterior parietal cortex. **(a)** A neuron in posterior parietal cortex responds to a target stimulus in its receptive field. **(b)** The response is enhanced if the target presentation is followed by a saccade to the target. **(c)** The enhancement effect is spatially selective, as it is not seen if a saccade is made to a stimulus not in the receptive field. **(d)** Enhancement is also seen when the task requires the animal to release a hand lever when the peripheral spot dims. (Source: Adapted from Wurtz, Goldberg, and Robinson, 1982, p. 128.)

subsequent eye movement, just as neurons in motor cortex fire before hand movements. To address this possibility, the researchers performed a variation of the experiment in which the animal moved its hand rather than its eyes to indicate the location of the peripherally flashed stimulus (Figure 21.11d). Even without a saccade there was an enhanced response to the target in the receptive field suggesting that, rather than a premotor signal, the enhanced response was a result of an attention shift needed to accurately perform the task.

It's easy to see how response enhancement of the kind observed in posterior parietal cortex might be involved in the behavioral benefits of attention discussed earlier. If attention drawn to one location in the visual field by a cueing stimulus increases the response to other stimuli near that location, this could account for the spatially selective improvement in the ability to detect a target. Likewise, the augmented response might lead to more rapid visual processing and ultimately faster reaction times, as seen in the perceptual experiments.

Attention Focuses Receptive Fields in Area V4. In a fascinating series of experiments, Robert Desimone and his colleagues, then at the National Institute of Mental Health, revealed surprisingly specific effects of attention on the receptive fields of neurons in visual cortical area V4. In one experiment, monkeys performed a same–different task with pairs of stimuli within the receptive fields of V4 neurons. For example, suppose that a particular V4 cell responded strongly to vertical and horizontal red bars of light in its receptive field but did not respond to vertical or horizontal green bars. The red bars were "effective" stimuli and the green bars were "ineffective" stimuli. While the monkey fixated, two stimuli (each either effective or ineffective) were briefly presented at different locations in the receptive field, and after a delay period, two more stimuli were presented at the same locations. In an experimental session, the animal was cued to base its same–different judgments on the successive stimuli at one of the two locations within the receptive field. In other words, the animal had to pay attention to one location in the receptive field but ignore the other in order to perform the task. The animal pushed a lever one way with its hand if the successive stimuli at the attended location were the same, and the opposite way if the stimuli were different.

Consider what happened in a trial when effective stimuli appeared at the attended location and ineffective stimuli appeared at the other location (Figure 21.12a). Not surprisingly, the V4 neuron responded strongly in this situation because there were perfectly good "effective" stimuli in the receptive field. The monkey was then cued to base its same–different judgments on the stimuli at the other location in the receptive field (Figure 21.12b). At this location, only green ineffective stimuli were presented. The response of the neuron should have been the same as before because exactly the same stimuli were in the receptive field, right? Surprisingly, that is not what they observed. Even though the stimuli were identical, on average the responses of V4 neurons were less than half as great when the animal attended to the area in the neuron's receptive field containing the ineffective stimuli. It's as if the receptive field contracted around the attended area, decreasing the response to the effective stimuli at the unattended location. The location-specific effect that attention has on neural activity in this experiment may be directly related to the specificity discussed earlier in the human detection experiment.

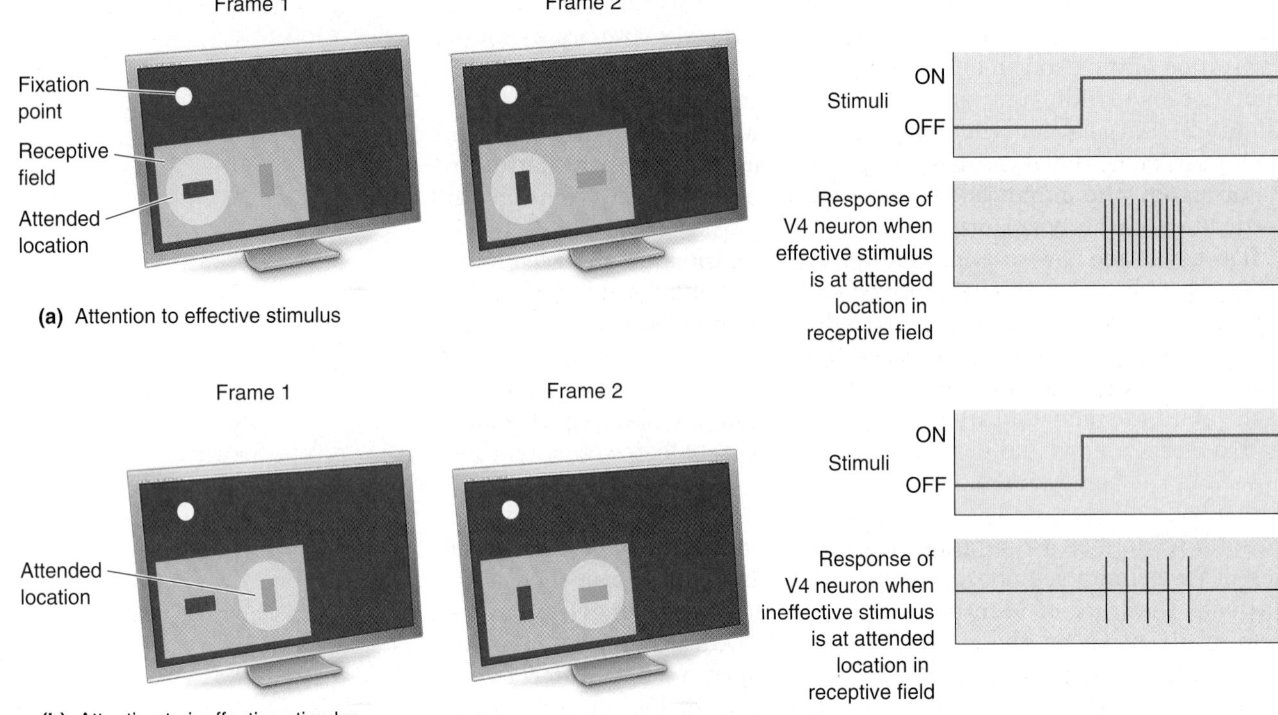

▲ FIGURE 21.12
The effect of attention in visual cortical area V4. The yellow circle indicates whether the monkey is attending to **(a)** the left or **(b)** the right location in the receptive field. For this neuron, red bars of light are effective in producing a response and green bars are ineffective. Even though the stimuli are always the same, the neuron's response is greater when attention is directed to the effective stimuli. (Source: Adapted from Moran and Desimone, 1985, p. 782.)

Brain Circuits for the Control of Attention

We have seen that attention has beneficial effects on visual processing and that it alters the sensitivity of visual neurons. These are the *consequences* of attention. We now turn to the brain mechanisms that *guide* attention, a topic that is more difficult to study because the networks of cortical and subcortical structures involved are distributed across the brain. Numerous experiments suggest that the brain circuitry responsible for saccadic eye movements plays a critical role in guiding attention. This link is consistent with human behavior because we saccade to objects that are either salient or of behavioral interest. We won't examine every structure thought to be involved in attention but will highlight a few and suggest how control circuits might be organized.

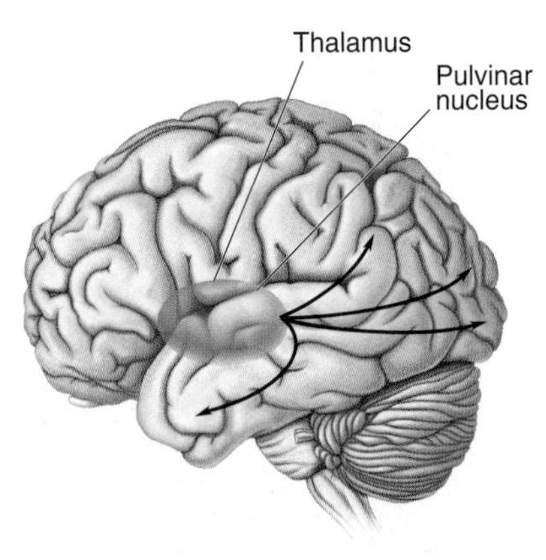

▲ FIGURE 21.13
Pulvinar projections to the cortex. The pulvinar nucleus is in the posterior thalamus. It sends widespread efferents to areas of cerebral cortex, including areas V1, V2, MT, parietal cortex, and inferior temporal cortex.

The Pulvinar, a Subcortical Component. One structure that has been studied for its possible role in guiding attention is the **pulvinar nucleus** of the thalamus. Several properties of the pulvinar make it interesting. As in other neocortical areas we have discussed, pulvinar neurons respond more robustly when a monkey attends to a stimulus in the receptive field than they do to the same stimulus when attention is directed elsewhere. Also, the pulvinar has reciprocal connections with most visual cortical areas of the occipital, parietal, and temporal lobes, giving it the potential to modulate widespread cortical activity (Figure 21.13). Consistent with this anatomical observation, it has been found in monkeys that when attention is drawn to a pulvinar receptive field, there is increased synchronization between neural activity in the pulvinar, area V4, and area IT. As the pulvinar provides input to V4 and IT, the hypothesis is that the pulvinar regulates information flow in areas of visual cortex.

Humans with pulvinar lesions respond abnormally slowly to stimuli on the contralateral side, particularly when there are competing stimuli on the ipsilateral side. It has been proposed that such a deficit reflects a reduced ability to focus attention on objects in the contralateral visual field. A similar phenomenon has been observed in monkeys. When muscimol, an agonist of the inhibitory neurotransmitter GABA, is unilaterally injected into the pulvinar, the activity of the neurons is suppressed. Behaviorally, the injection produces difficulty in shifting attention to contralateral stimuli, which seems similar to the effect of pulvinar lesions in humans. Interestingly, injection of the GABA antagonist bicuculline appears to facilitate shifting attention to the contralateral side.

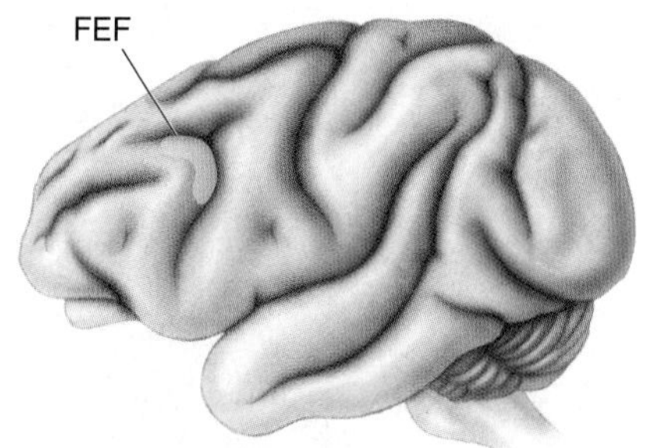

▲ FIGURE 21.14
The frontal eye fields (FEF) in a macaque brain. The FEFs are involved in the production of saccadic eye movements and may play a role in the guidance of attention.

The Frontal Eye Fields, Eye Movements, and Attention. Tirin Moore and his colleagues, then at Princeton University, examined a cortical area in the frontal lobe known as the **frontal eye fields** or **FEF** (Figure 21.14). There are direct connections between the FEF and numerous areas known to be influenced by attention, including areas V2, V3, V4, MT, and the parietal cortex. Neurons in the FEF have *motor fields*, which are small areas in the visual field. If a sufficient electrical current is passed into the FEF, the eyes rapidly make a saccade to the motor field of the stimulated neurons.

In one experiment, Moore et al. trained monkeys to look at a computer display that contained numerous small spots of light. They placed an electrode into the FEF and determined the motor field of the neurons at the electrode's tip. The animal's task was to fixate on the center of the visual display but pay attention to one of the spots, the "target" spot, specified by the experimenter. On each experimental trial, if the target spot dimmed, the monkey moved a lever with its hand. If the spot did not dim, the monkey did not move the lever. By varying the light level, the experimenters measured the minimum light difference or threshold needed by the monkey to detect the dimming. The task was made difficult for the monkey by "distracter" spots that blinked on and off at random times (Figure 21.15a).

Unbeknownst to the monkey, on some trials a very small amount of electrical current was passed into the FEF electrode. Importantly, the current was insufficient to drive the eyes to the motor field, and the animal continued to look at the central fixation point. The aim of the experiment was to determine whether the small electrical stimulation could enhance the animal's ability to detect the dimming of the target spot, a sort of artificial attentional "boost." The results are summarized in Figure 21.15b. The histogram shows that when the target stimulus was located inside the motor field, the threshold light difference needed to detect dimming of the target was about 10% less with electrical stimulation than without. The right side of this histogram shows that performance was not enhanced and may actually have been impaired by electrical stimulation if the target was outside the motor field (as if attention was drawn to the motor field and away from the target). As predicted, electrical stimulation in the FEF improved performance in a manner similar to added attention. Moreover, the effect of electrical stimulation was location specific just as attentional modulation normally is.

If Moore et al.'s results mean that the FEF is part of a system for directing attention and enhancing visual performance in a location-specific

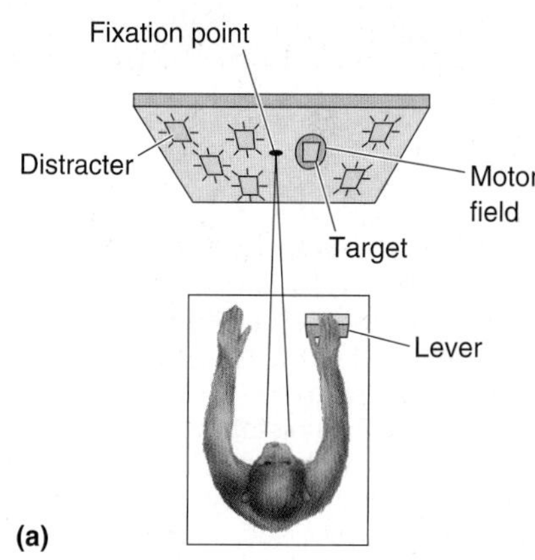

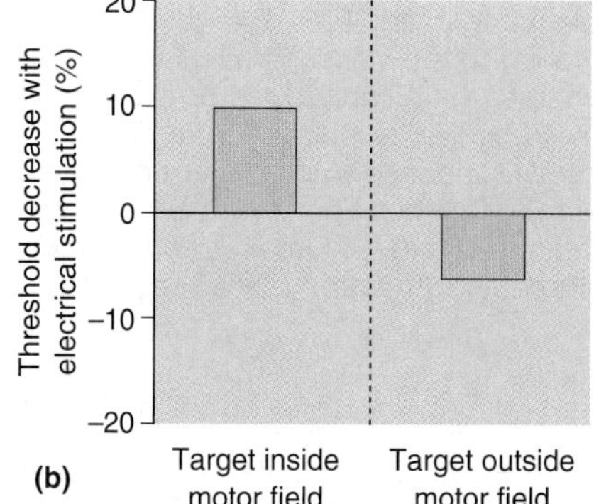

▲ FIGURE 21.15
FEF stimulation alters perceptual thresholds. **(a)** A monkey views spots on a visual display; all of the spots blink on and off except for the target spot. The monkey releases a lever if the target spot dims. **(b)** If the target spot is in the motor field of neurons under study, electrical stimulation in the FEF reduces the threshold light difference needed to detect that the target spot dimmed. If the target is outside the motor field, electrical stimulation slightly increases the threshold. (Source: Adapted from Moore and Fallah, 2001, Fig. 1.)

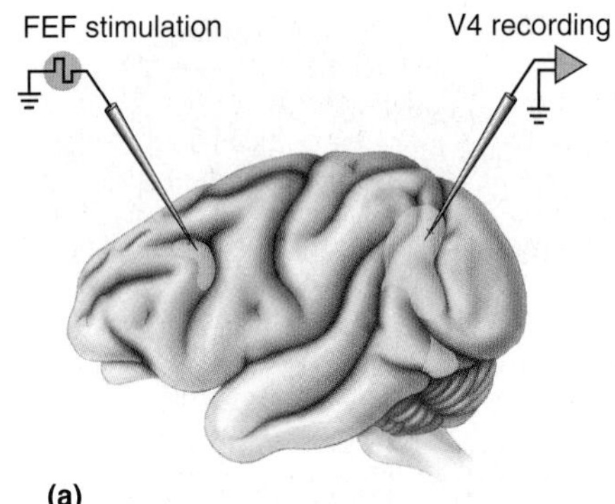

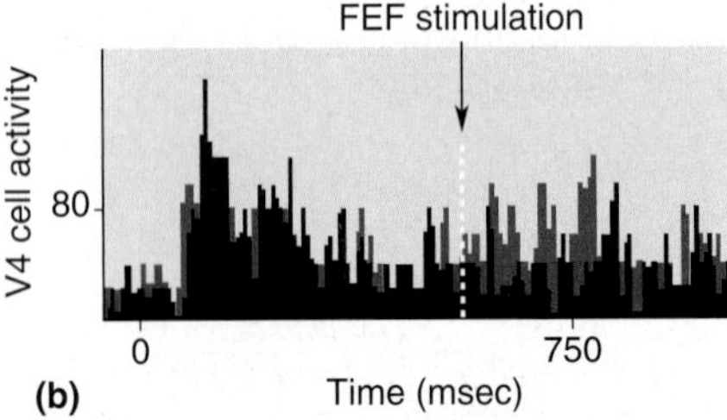

▲ FIGURE 21.16
The effect of FEF stimulation on neuron activity in area V4 in the monkey's brain. **(a)** A small electrical current is passed into the FEF while the activity of a neuron in V4 is recorded. **(b)** A stimulus is presented in the V4 receptive field at time zero. The histogram shows that the response to the visual stimulus peaks after a short delay and then declines. After 500 msec, the FEF is electrically stimulated (downward arrow) on some trials (red) but not on other trials (black). Prior to 500 msec, the V4 response to the visual stimulus was similar whether the stimulus was or was not followed by FEF stimulation. After 500 msec, the V4 response was greater on trials with FEF stimulation (red) than on trials without (black). (Source: Adapted from Moore and Armstrong, 2003, p. 371.)

manner, how might this work? One possibility is that FEF activity indicating the location of a potential future saccadic eye movement feeds back to the cortical areas it is connected to, enhancing activity there. Moore's research group tested this hypothesis by recording in area V4 during electrical stimulation of the FEF. They located electrodes in the two areas so that the motor field of the neurons in the FEF overlapped the visual receptive fields of the V4 neurons. A visual stimulus was used to excite the V4 neuron, and after a 500 msec delay the FEF was electrically stimulated on some trials. Figure 21.16 shows that when the FEF was stimulated (with a current insufficient to evoke a saccade), the visual response of the V4 neuron was increased (red) compared to trials without FEF stimulation (black). Without a visual stimulus to excite the V4 neuron, there was no effect of FEF stimulation on the V4 response, suggesting that the increased V4 activity was an enhancement of a visual response rather than a direct consequence of electrical stimulation.

Taken together, Moore's experiments suggest that FEF stimulation mimics both the physiological and behavioral effects of attention. Other scientists have found similar results with electrical stimulation of the superior colliculus, another structure involved in generating saccadic eye movements. These findings make a compelling case that the guidance of attention is integrated with a system involving the FEF and superior colliculus that is used to move the eyes.

Directing Attention with Salience and Priority Maps. In the search for brain processes involved in directing attention, we must consider both bottom–up attention drawn by a stimulus and top–down attention that moves to objects of behavioral importance. A now-common hypothesis to explain how certain visual features grab your attention (e.g., the yellow dandelion surrounded by green grass) is a **salience map**, an idea introduced by Laurent Itti and Christof Koch at Caltech. Rather than a brain map showing the locations of objects, a salience map shows the locations of conspicuous features. This concept is illustrated in Figure 21.17. In Chapter 10, we saw that the visual system has neurons selective for a variety of stimulus attributes, such as orientation, color, and motion, and that visual cortex is organized on the basis of these features (e.g., orientation columns). The first stage in the salience map model consists of maps of individual features that locate areas of high feature contrast (e.g., changes from rightward to leftward movement or red to green). Through neural interactions within a map, a form of competition might suppress responses associated with lower feature contrast. The locations of high contrast in each feature map feed into a salience map that locates areas of high contrast irrespective of the specific features. Competition between the locations with high contrast leads to a winning location to which attention is moved. To prevent attention from getting stuck at the single most salient location, however, "inhibition of return" prevents successive attentional loci from being the same.

As described, this model accounts only for bottom–up guidance of attention. We can add top–down attentional modulation to the model in Figure 21.17 by inserting top–down cognitive input to either the feature maps ("I'm looking for a friend wearing a red hat") or the salience map ("I remember the key figure was on the right side of the page in the textbook"). With this addition, our model no longer simply indicates salience (a bottom–up property of stimuli) but instead indicates attentional priority. A **priority map** is a map showing locations where attention should be directed based on stimulus salience and cognitive input. In other words, a priority map is a salience map plus top–down effects.

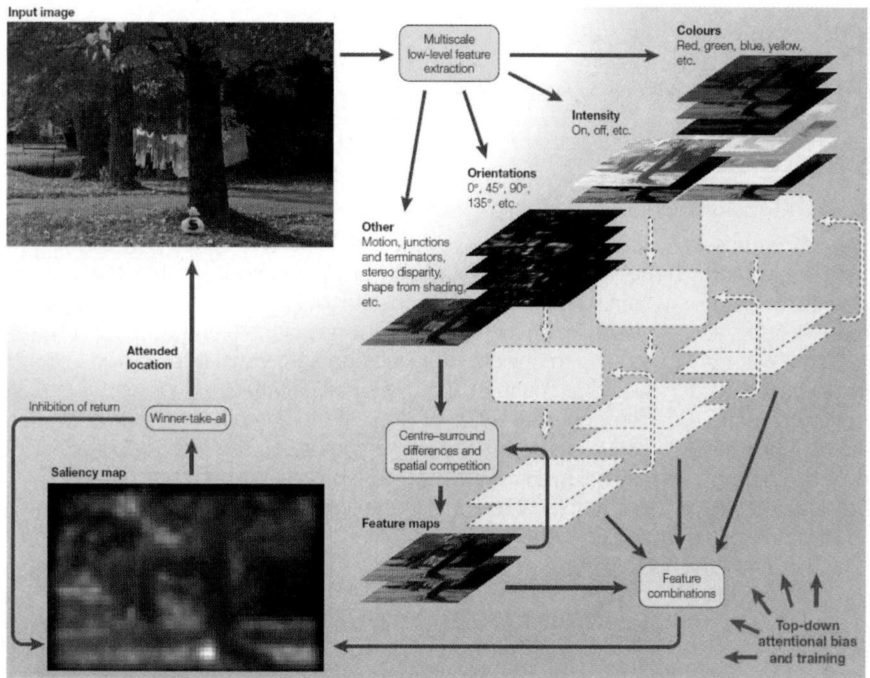

◀ FIGURE 21.17
Guiding attention with a salience map. An influential hypothesis for the guidance of attention is that the brain uses a salience map to determine where and what we should attend to. The visual input is analyzed by neurons that are sensitive to the spatial scale of stimulus elements and other stimulus features such as color, intensity, orientation, motion, and so on. The maps for these individual features determine where there are significant changes in each parameter, such as a transition from one color to another, light to dark, or contours with different orientation. The feature maps are combined into a salience map that identifies a "winner"—the most conspicuous object across feature maps that will become the next attended location, in this case a bag of money. To ensure the system does not become locked on a single salient object, the "inhibition of return" feedback ensures that a current target of attention is inhibited from being the next. As indicated at the lower right, attention is also influenced by top–down factors. (Source: Itti and Koch 2001, Fig. 1.)

A Priority Map in the Parietal Lobe. Salience and priority maps have been studied in visual cortical areas (e.g., V1, V4) as well as cortical areas in the parietal and frontal lobes. In a series of studies, Michael Goldberg at Columbia University, James Bisley at the University of California, Los Angeles, and their colleagues have shown that the **lateral intraparietal cortex (area LIP)** appears to construct a priority map based on both bottom–up and top–down inputs (Figure 21.18). Area LIP plays an important role in directing eye movements, a function clearly related to the guidance of attention. Lesions in the parietal cortex are also associated with the **neglect syndrome**, in which there is an inability to attend to half of the environment (Box 21.2).

An experiment demonstrates a salience effect in LIP that is akin to the experience you might have walking into a familiar room: You might not pay any attention to an old sofa or light fixture on the wall of your living room at home, but your attention would be captured by a new puppy jumping on the floor. In the experiment, a monkey first sees a computer display, and a neuron in area LIP is shown to respond when an object, say a star, is flashed in its receptive field (Figure 21.19a). In the second experiment, eight objects are on the computer display and one of them is the star. The animal initially fixates a point on the bottom of the display such that the LIP neuron's receptive field does not encompass any of the objects. When the fixation point is moved to the center of the display, the animal makes a saccade that brings the star into the receptive field. Figure 21.19b shows that there is little response in this situation. In a third experiment, the animal fixates the same location at the bottom of the computer display, and all of the stimuli are shown except for the star (none of the stimuli are in the receptive field). Then, about 500 msec before the animal makes a saccade to the central fixation point, the star is turned on. When the central fixation point is displayed, the animal makes a saccade to it and the neuron responds vigorously to the star (Figure 21.19c). Note that in this last version of the experiment, the star is not in the receptive field when it is turned on. By the time the eyes move and the star enters the receptive field, the eight objects on the display are

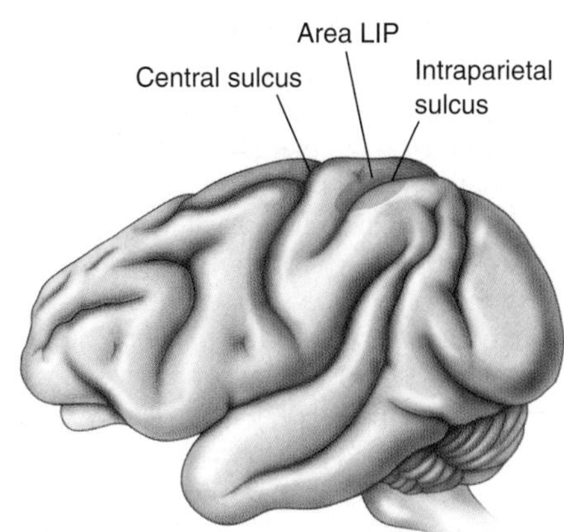

▲ FIGURE 21.18
Area LIP, buried in the intraparietal sulcus of the macaque brain. LIP neurons are involved in guiding eye movements and attention.

BOX 21.2 OF SPECIAL INTEREST

Hemispatial Neglect Syndrome

In Chapter 12, we briefly discussed the *neglect syndrome*, in which a person appears to ignore objects, people, and sometimes their own body to one side of their center of gaze. Some have argued that this syndrome is a unilateral deficit in attention. The manifestations of a neglect syndrome can be so bizarre that they're hard to believe if not directly observed. In mild cases, the behavior may not be apparent with casual observation. But in severe cases, patients act as if half the universe no longer exists. They may shave only one side of their face, brush the teeth on only one side of their mouth, dress only one side of their body, and eat food from only one side of the plate.

Because neglect syndrome is less common following left hemispheric damage, it has primarily been studied in regard to neglect of the left half of space as a result of damage to right cerebral cortex. In addition to neglecting objects to the left side, some patients exhibit denial. For instance, they may say that their left hand isn't really paralyzed or, in extreme cases, may refuse to believe that a limb on their left side is part of their body. Refer to Figure 12.24 in Chapter 12 as a typical example of the distorted sense of space these patients have. If asked to make a drawing, they may crowd everything into the right half, leaving the left half blank. A particularly dramatic example is the paintings shown in Figure A, which were painted by an artist as he recovered from a stroke.

If patients with a neglect syndrome are asked to close their eyes and point toward the midline of their body, they typically point too far to the right, as if there has been shrinkage of the left half. If blindfolded and asked to explore objects placed on a table before them, patients behave normally in exploring objects to the right but are haphazard about probing to the left. All of these examples point toward a problem in relating to the space around them.

Neglect syndrome is most commonly associated with lesions in posterior parietal cortex in the right hemisphere, but it has also been reported to occur following damage to right hemisphere prefrontal cortex, cingulate cortex, and other areas. It has been proposed that the posterior parietal cortex is involved in attending to objects at different positions in extrapersonal space. If this is true, then neglect syndrome might be a disruption of the ability to shift attention. One piece of evidence supporting this hypothesis is that objects in the right visual field of patients with a neglect syndrome are sometimes abnormally effective in capturing attention, and patients may experience difficulty disengaging their attention from an object on this side.

It is not clear why the syndrome more often accompanies right hemispheric damage than left hemispheric damage. The right hemisphere appears to be dominant for understanding spatial relationships, and in split-brain studies, it has been shown to be superior at solving complex puzzles. This finding seems consistent with the greater loss of spatial sense after right hemispheric lesions. One hypothesis is that the left hemisphere is involved in attending to objects in the right visual field, whereas the right hemisphere is involved in attending to objects in the left and right visual fields. While this would account for the asymmetrical effects of left and right hemispheric lesions, at present there is only suggestive evidence in support of such a hypothesis. One final riddle about neglect syndrome is that there is partial or complete recovery in a matter of months (in the figure, note the recovery in the self-portraits).

Figure A
Self-portraits during recovery from a stroke that caused a neglect syndrome. Two months after suffering a stroke affecting parietal cortex on the right side, the artist made the upper-left portrait. There is virtually no left side to the face in the painting. About 3.5 months after the stroke, there is some detail on the left side but not nearly as much as on the right side (upper right). At 6 months (lower left) and 9 months (lower right) after the stroke, there is increasingly more treatment of the left side of the painting. (Source: Posner and Raichle, 1994, p. 152.)

Star flashes in RF

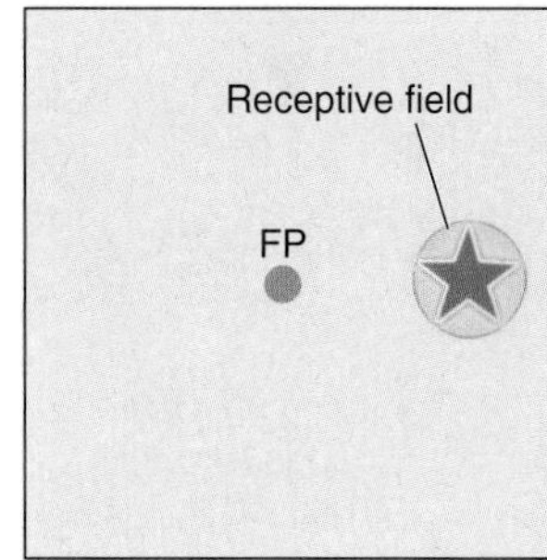

Saccade brings stable stimulus into RF

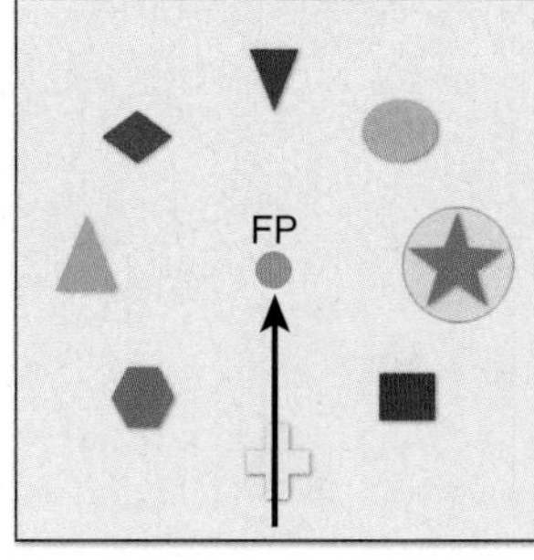

Saccade brings stimulus into RF after flash on

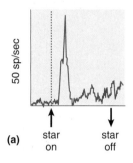

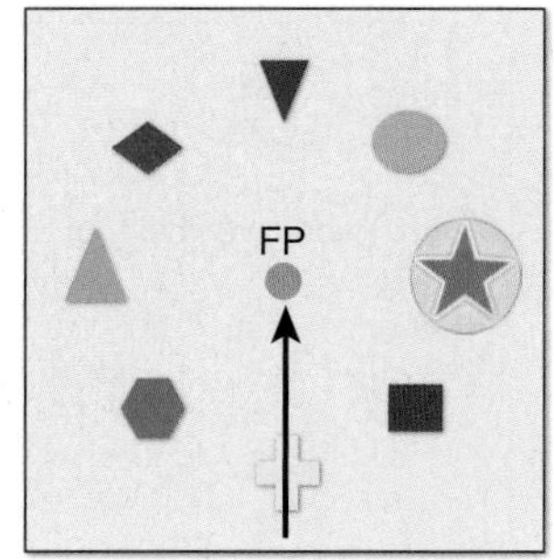

50 sp/sec

(a) star on star off

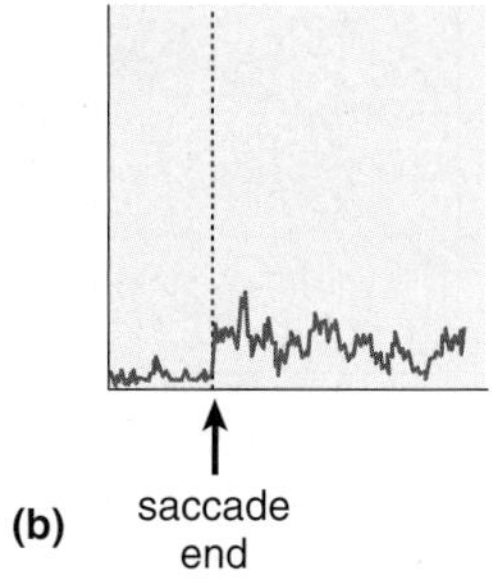

(b)

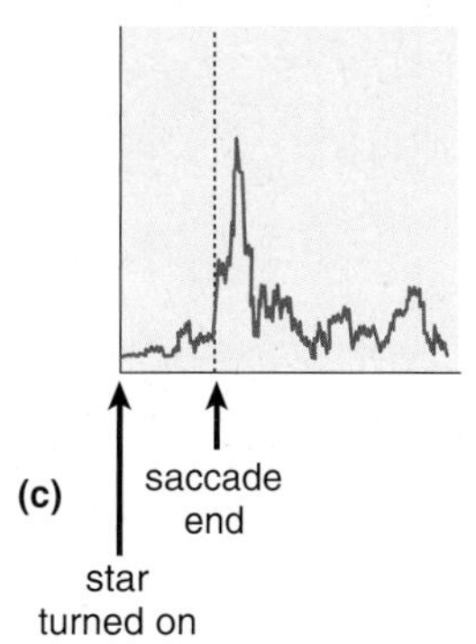

(c)

◀ FIGURE 21.19
Evidence for a bottom–up priority map in area LIP. (a) An LIP neuron responds when an effective stimulus, a star, is flashed in its receptive field. **(b)** The LIP neuron responds little if all eight stimuli are present before a saccade brings the star into the receptive field. **(c)** If the star is turned on 500 msec before the saccade, the LIP neuron strongly responds after the saccade. FP, fixation point. (Source: Adapted from Bisley and Goldberg, 2010, Fig. 2.)

identical to the second experiment. It appears that the large response in the third experiment is a consequence of turning on the star just before it entered the receptive field. The hypothesis is that the stimulus onset catches the animal's attention and this enhances the response of the LIP neuron. This effect is consistent with a salience map in area LIP in that the neuron's response is strongly modulated by a conspicuous bottom–up stimulus.

A variation of this experiment brings out a top–down attention effect. This study uses the same eight stimuli, but now they are all always shown (no flashing stimuli). As before, the animal fixates on a point on the computer display such that none of the stimuli are in the receptive field. A small cue stimulus is flashed on and off, and this indicates to the animal which of the eight stimuli is of behavioral significance in this experimental trial. In the case illustrated, the LIP neuron does not respond to the star cue because it is outside the receptive field (Figure 21.20a). The fixation point then moves to the center of the display and the monkey makes a saccade to this location, bringing the star into the receptive field and the neuron responds to the star (Figure 21.20b). Finally, the animal makes a saccade to the star stimulus and the LIP response ends (Figure 21.20c). Compare this response pattern with that observed with identical stimuli when the cue does not match the stimulus that enters the receptive field. As before there is no response to the cue—now a triangle rather than a star (Figure 21.20d). When the animal makes the first saccade, the star enters the receptive field, but the response is much less than before (Figure 21.20e). Finally, the animal makes a second saccade to the triangle (Figure 21.20f).

Note that in both the first and second experiments there was a condition in which the animal made a saccade that brought a stable (not previously flashed) star stimulus into the receptive field. From Figure 21.19, we conclude that without something like a flash to increase its salience, the LIP neuron does not respond very much to the star in its receptive field.

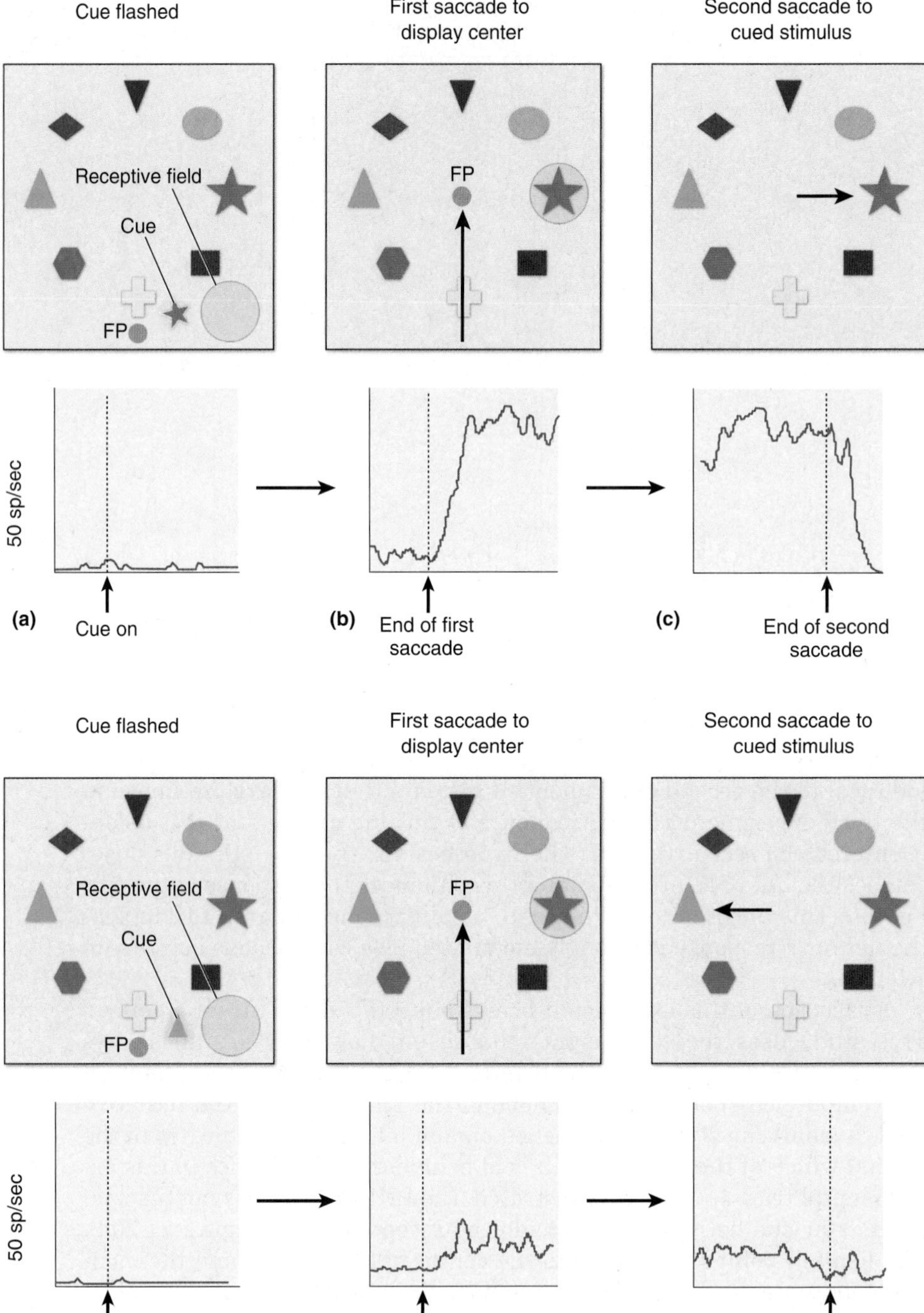

▶ FIGURE 21.20
Evidence for a top–down priority map in area LIP. **(a)** A small star-shaped cue stimulus flashes on to inform the monkey which stimulus is important. The cue is outside the receptive field and there is no response in the LIP neuron. **(b)** A first saccade is made to the center of the computer display, bringing the star stimulus into the receptive field. The neuron responds to the star. **(c)** The monkey makes a second saccade to the cued star stimulus. **(d)** In this second experiment with the same LIP neuron, the cue is a triangle. **(e)** There is a significantly reduced response to the star when the cue is the triangle. **(f)** The monkey saccades to the cued triangle stimulus. (Source: Adapted from Bisley and Goldberg, 2010, Fig. 4.)

In Figure 21.20b, there is a much greater response than in Figure 21.20e presumably because of a top–down signal informing the LIP neuron that in the former case the star is important (for planning the last saccade) even though it isn't flashing. A variety of experiments along these lines suggest that LIP neurons carry information appropriate for a priority map of visual attention.

The Frontoparietal Attention Network. As more is learned about brain areas that are affected by attention and areas that appear to hold saliency or priority maps, an outline of circuits involved in attention is emerging.

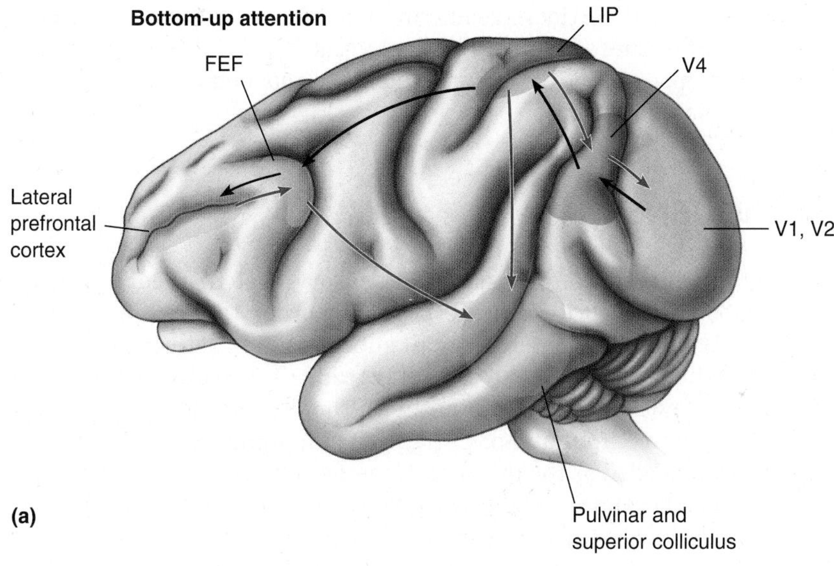

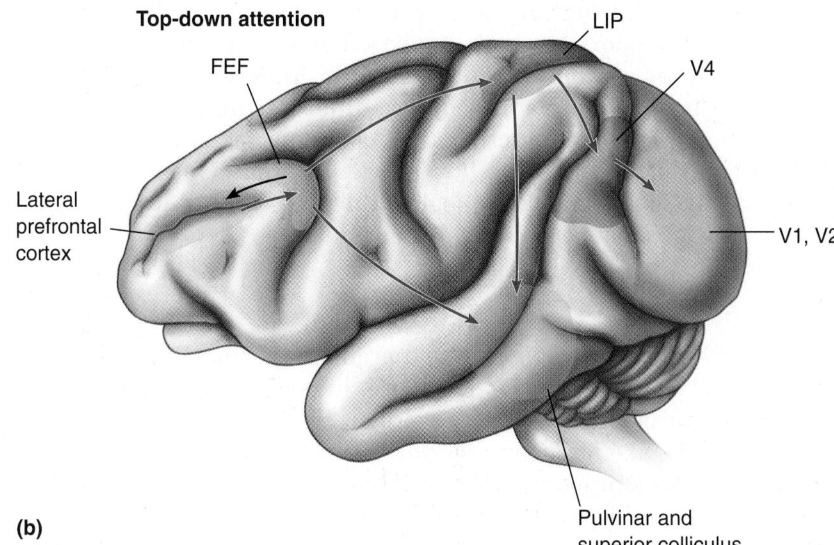

◀ FIGURE 21.21
The frontoparietal attention network in the macaque brain. **(a)** In bottom–up attention, information about a conspicuous object is passed from visual areas in the occipital lobe to area LIP where a salience map is constructed. Early attention signals are also seen in prefrontal cortex and the frontal eye fields that interact with LIP. Signals sent from LIP and FEF may direct the eyes and enhance visual processing in occipital visual cortex. **(b)** In top–down attention, frontal lobe areas show attentional modulation at the earliest times and signals sent to other structures influence eye movements and perception. Black arrows: bottom–up signals; Red arrows: top–down signals.

The involved brain areas compose the **frontoparietal attention network** (Figure 21.21).

In bottom–up attention, input from visual areas in the occipital lobe reaches area LIP where the first critical step may be the construction of a salience map based on conspicuous objects in the field of view (see Figure 21.21a). The frontal eye fields also contain a salience map, but salience is signaled there after LIP. Through feedback to visual areas and connections with eye movement structures, visual processing of a salient object is enhanced and the eyes may move to foveate the object.

Top–down control of attention is directed by behavioral goals, and it is fitting that the cortical areas in the frontal lobe appear to be critical. Recordings across a range of brain areas show that attention effects follow a temporal sequence, occurring first in the frontal lobe (prefrontal cortex and the frontal eye fields) and followed progressively by areas LIP, V4 and MT, V2, and then V1 (see Figure 21.21b). The causal relationships in these brain areas are still being investigated, but we can speculate that behavioral goals are established in the frontal and parietal areas, information is processed to create a priority map in LIP and FEF, and modulation of visual cortical areas enhances the perception of selected objects. Several of these brain areas, including LIP, FEF, and the superior colliculus, also serve to guide saccadic eye movements to scrutinize the objects allocated attention.

CONSCIOUSNESS

In previous chapters, we discussed the sensory systems that bring information about our world into the brain. In order to serve behavioral goals, we focus our attention on a small subset of the vast sensory input. Presumably other animals make a similar tradeoff: Broadly monitor sensory input at a low resolution, perhaps with the default mode network, or filter out much of the input and pay attention at a higher resolution to only the information needed at the moment. Consider the next step in the chain of brain events—becoming consciously aware of the world around us.

It is probably fair to say neuroscientists tend to be *materialists* in their attitudes about consciousness, meaning that consciousness arises from physical processes: Like every other product of the brain, consciousness can ultimately be understood as being based on the structure and function of the nervous system. An alternative to materialism is *dualism*, which states that the mind and body are different things and one cannot be fully explained by the other (i.e., consciousness cannot be fully explained by physical processes). If it is true that consciousness is based on standard physical principles, a logical inference is that it should one day be possible to construct a conscious machine.

What Is Consciousness?

The nature of human consciousness is a problem that has vexed philosophers and scientists for centuries. There are challenges right at the outset; even defining consciousness is controversial. Suffice it to say numerous definitions have been offered over the years, and numerous models of consciousness have been proposed. Our intent is not to jump into this controversy. However, considering some of the background issues will lead us to a discussion of the type of neuroscientific investigations of consciousness that might be fruitful. Think of the ways we use the word itself. We say that a person who is given a general anesthetic or who is asleep is unconscious and that they become conscious when they wake up. If our hair looks weird one day, we might be self-conscious. A person under the influence of a hallucinogenic drug is said to be in an altered state of consciousness. When long wavelength light hits our retina, we have the conscious experience of the color red. But does "conscious" mean the same thing in all of these examples? It seems we use the word in different ways, and understanding these facets of consciousness may involve distinct lines of neuroscientific inquiry.

In 1995, the philosopher David Chalmers at the University of California, Santa Cruz, proposed a distinction that is a helpful starting point. He outlined what he called the easy problems of consciousness and the hard problem. What Chalmers meant by the **easy problems of consciousness** are phenomena that seem answerable by standard scientific methodology. For example, what is the difference between being awake and asleep? We don't know the full answer to this question, but as we saw in Chapter 19, research is revealing much and may someday fully outline the nature of conscious wakefulness. Another example comes from studies of attention. It is sometimes said that we are conscious of things we pay attention to. Thus experiments on attention may tell us something about consciousness. Other brain functions that may give us insight into consciousness include our ability to integrate information from sensory systems, make decisions about sensory input, and so on.

The **hard problem of consciousness** is the experience itself. We experience the emotion called happiness, the sound of a saxophone, the color blue. Why and how do these subjective experiences arise from physical processes? When a baby cries, a mother's soothing touch evokes some pattern of activity in the child's brain, but why is the internal experience a pleasant one rather than a feeling of pain, the smell of burnt toast, or the sound of a car's horn? We can look for neural activity associated with these experiences (the easy part of the problem), but understanding why the experience is the way it is seems much harder. In reality, none of these problems we've mentioned is easy; it may have been more appropriate to refer to the hard problems of consciousness and the seemingly impossible problem! At any rate, our discussion here will be limited to the "easy" problems.

Neural Correlates of Consciousness

For centuries, the study of consciousness was in the hands of philosophers; it was widely considered beyond the reach of experimental science. In recent years, this attitude has changed, and a spectrum of scientists are cutting trails through the challenging landscape of consciousness. To make headway, we should ask questions that offer some hope of solution rather than going immediately after the mysteries of internal experience. Christof Koch and Francis Crick, who was awarded a Nobel Prize for his work on the structure of DNA, were two pioneers who collaborated to bring a neuroscientific approach to the study of consciousness (Box 21.3). Koch defined **neural correlates of consciousness (NCC)** as the minimal neuronal events sufficient for a specific conscious percept. In other words, what must happen in which neurons for you to experience the taste of a strawberry or the feeling of joy?

A general experimental approach that has been taken starts with visual images presented to the eyes that can be seen in two different ways, called bistable images. Well-known examples are shown in Figure 21.22. The question we are interested in is what happens to brain activity when a person or animal switches from one percept to another? For example, in Figure 21.22c, at one moment, you see a bunny and later you see a duck (but not both at the same time). Since the image is always the same, the hypothesis is that changes in neural activity that correlate with changes in perception may be related to our conscious awareness of one object or the other. Experiments along these lines have been conducted both in animals with single neuron recordings and in humans with PET and fMRI.

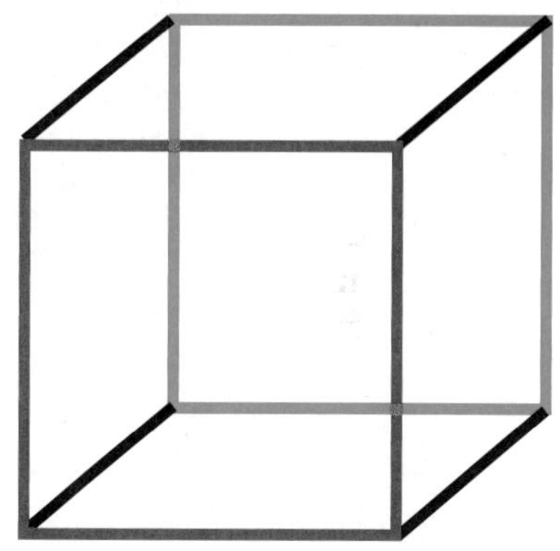

▲ FIGURE 21.22
Bistable images can be perceived two different ways. **(a)** The face-vase illusion. **(b)** The Necker cube can appear to have either the surface bordered by green or the surface bordered by red closest to the observer. **(c)** A duck or a bunny?

Neuronal Correlates of Alternating Perception in Binocular Rivalry. Binocular rivalry is a visual effect that has been put to good use exploring the neural basis of conscious awareness. In **binocular rivalry**, different images are seen by the two eyes, and perceptual awareness alternates between the two images. For example, if one eye sees vertical lines and the other sees horizontal lines, a person will perceive alternations at random times between only vertical lines, only horizontal lines, and sometimes a patchwork of the two orientations. The two images are always the same and both eyes are always open, so what and where is something switching inside the brain?

An experiment along these lines was conducted by David Sheinberg and Nikos Logothetis, then at the Baylor College of Medicine. Recordings were

BOX 21.3 PATH OF DISCOVERY

Tracking the Neuronal Footprints of Consciousness

by Christof Koch

It was an everyday occurrence in the summer of 1988 that set my life on a new path while I was teaching summer school in Woods Hole on Cape Cod. I had taken an aspirin, but my toothache persisted. Lying in bed, I couldn't sleep because of the pounding in my lower molar. Trying to distract myself from the pain, I wondered how it came to hurt so much. I knew that an inflammation of the tooth pulp sends electrical activity up one of the branches of the trigeminal nerve. After the signal passes through several relay stations, neocortical neurons become active and discharge their electrical impulses. Such bioelectrical activity in this part of the brain goes hand in hand with the consciousness of pain, including its awful, aching feeling. But how could this physical process trigger elusive, nonphysical feelings? It's just a bunch of ions—Na^+, K^+, Cl^-, Ca^{2+} and so on—moving across membranes, in principle no different from similar ions sloshing around in my liver or electrons moving onto and off transistor gates in my laptop. As a physicist, I knew that neither quantum mechanics nor general relativity, the two most powerful scientific theories that all matter has to obey, make any mention of consciousness. By dint of what natural law could a highly excitable chunk of organized matter exude something nonphysical and give rise to subjective states, to ephemeral feelings? And so my toothache on that distant summer day set me on a course to explore the seas of consciousness, with the brain as my lodestar.

I started what became a 16-year collaboration with Francis Crick at the Salk Institute in La Jolla, California. Coauthoring over 20 papers and book chapters, we advocated for an empirical research program focused on isolating those neurons and brain regions that are involved in generating a specific content of consciousness, such as seeing a horizontal grating rather than a vertical one or seeing red versus green. We felt that no matter what philosophical position people adopt in regard to the mind–body problem, finding such neuronal correlates would be a critical step on the way to an ultimate theory of consciousness.

Back in 1990, when Francis and I published our first paper on the topic, we were energized by the rediscovery, by Wolf Singer and Charlie Gray in Frankfurt, Germany, of 40 Hz synchronized oscillations in the firing pattern of neurons in the visual cortex of cats. We argued that this so-called "gamma band" activity was one of the hallmarks of consciousness. Reality has turned out more complex than that. Today, it is known that such oscillatory activity, widespread in the cortex of all species investigated so far, is probably more closely tied to selective attention than to consciousness, even though these two processes are often intimately relayed.

We debated endlessly between ourselves and with a small number of colleagues willing to publicly think about this hitherto banned subject—Nikos Logothetis, Wolf Singer, David Chalmers, Patricia Churchland, Giulio Tononi, and V.S. Ramachandran, to mention a few—about nerve cells and their circuits, both those that must be involved in consciousness and those that give rise to the myriad types of unconscious behaviors (which we dubbed "zombie systems") such as speed typing, moving the eyes, adjusting limbs in a dynamic environment, and so on. Today, the study of the

made from neurons in inferotemporal cortex (area IT), which we introduced as a high-level vision area in Chapter 10. Instead of vertical and horizontal lines (or ducks and bunnies), stimuli were used that excited IT neurons. Before the experiment, a monkey was trained to pull a left lever if it saw an object from the "left object group" and a right lever if it saw an object from the "right object group." For the experiment shown in Figure 21.23, the left object group consisted of pictures of starbursts and the right object group was composed of pictures of animal and human faces.

Once the animal reliably indicated whether an object from the left or right group was seen without rivalry, neural recordings in IT were made in a rivalry situation. From baseline recordings, the scientists knew that the particular cell under study gave a strong response to the presentation of a monkey face (to either eye) and little or no response to the starburst pattern. In the rivalry experiment shown in the figure, a starburst neuronal mechanisms underlying both states and content of consciousness is carried out by many in both the clinic and the laboratory, but in 1990 it helped to have a famous Nobel Prize–winning biologist as a coauthor (in particular before attaining the holy state of tenure).

We continued to work together until Francis's death on July 28, 2004. Two days before, he phoned me on the way to the hospital, calmly informing me that his comments on our last manuscript—on a rather obscure brain region called the *claustrum*, an elongated sheet of neurons underneath the neocortex, and its putative role in consciousness—would be delayed. His wife, Odile Crick, recounted how in the hours before his death Francis hallucinated about rapidly firing claustrum neurons—a scientist to the end. Today, as I write these lines, a new clinical case study of an epileptic patient has just been published. The neurologists stimulated electrodes implanted into the patient's brain to discover the location where her seizures originated. One electrode, located close to the left claustrum, turned the patient's consciousness immediately off, repeatedly, and reversibly, for as long as the electrical stimulation lasted. The patient stared, became unresponsive to commands, and later had no recollection of these episodes. Ever data-hungry, how Francis would have loved this!

References:

Koubeissi MZ, et al. 2014. Electrical stimulation of a small brain area reversibly disrupts consciousness. *Epilepsy and Behavior* 37:32–35.

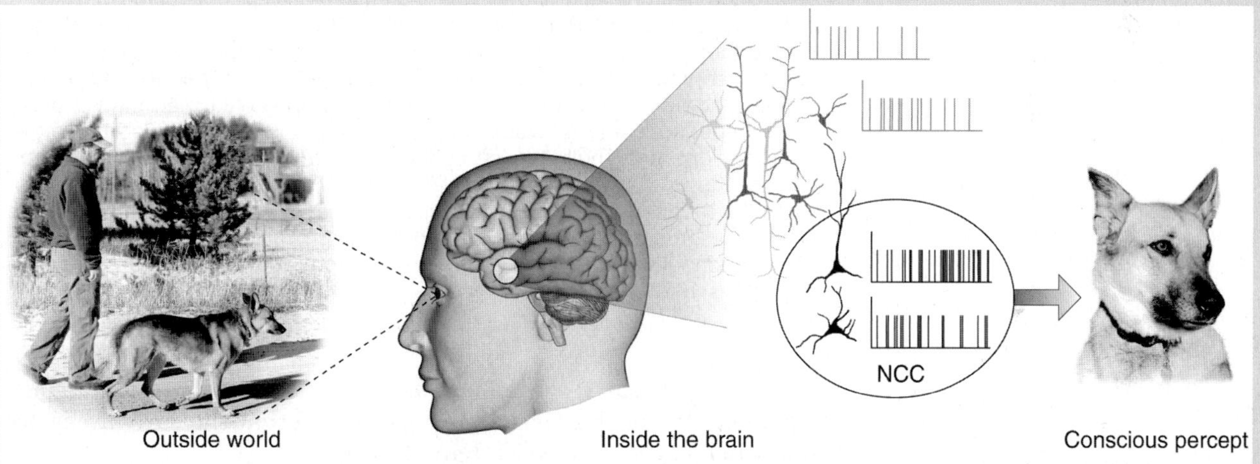

Figure A
The neuronal correlates of consciousness (NCC) are defined as the minimal neuronal events jointly sufficient for any one specific conscious percept (here, seeing a German Shepherd). (Source: Courtesy of Christof Koch).

stimulus was presented to the left eye and a monkey face to the right eye. Based on its training, the monkey alternately pulled the left or right lever suggesting that it alternately perceived the starburst or monkey face. The striking result seen in the neural recording is that the response of the IT neuron fluctuated between low and high activity roughly in sync with the animal pulling the left or right lever, even though the stimulus was fixed.

The results from this experiment and others indicate that there is a correspondence between changes in the activity of IT neurons and perception. The implication is that binocular rivalry produces an alternation in the conscious awareness the monkey has of the rivalrous images and that neural activity in IT may be a neural correlate of this awareness. Similar experiments in other brain areas of the monkey showed that with the rivalry paradigm the neural–perceptual correspondence was relatively

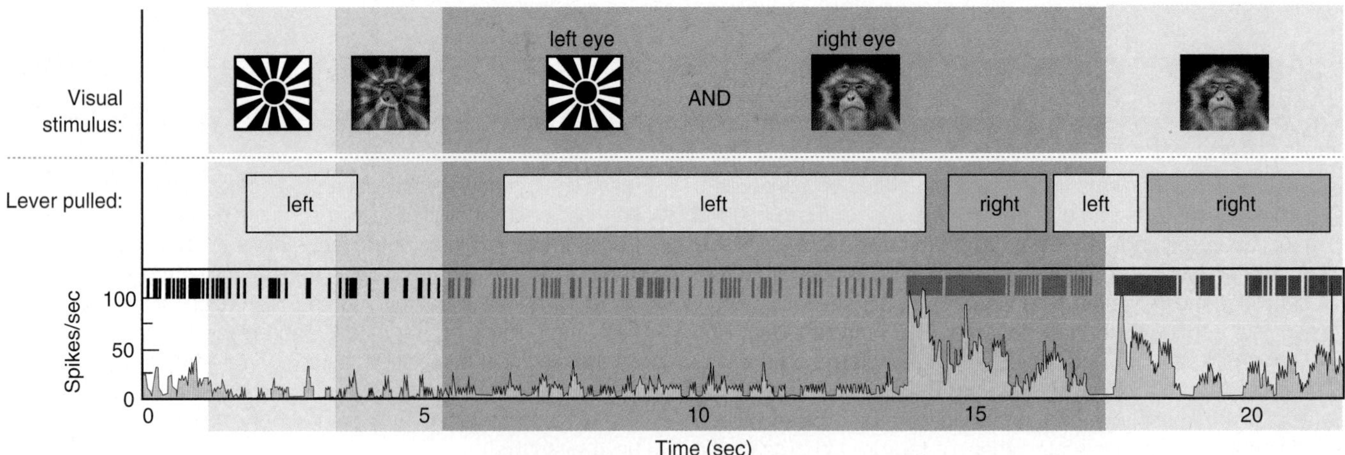

▲ FIGURE 21.23

Responses of a monkey IT neuron during binocular rivalry. Preliminary testing showed that this neuron in inferotemporal cortex was excited by a photograph of a monkey face but not by a starburst pattern. The top row shows the visual stimulus the monkey saw. First, the starburst was shown (blue shading) followed by an ambiguous stimulus (pink shading); both of these stimuli were ineffective in driving the neuron. Next shown was the critical rivalrous condition with the starburst to the left eye and the monkey face to the right eye (orange shading). Finally, the monkey face was shown alone (blue shading). The second row, below the dashed line, shows which lever the monkey pulled. When the starburst or monkey face was shown alone, the animal appropriately pulled the left or right lever, respectively. In the rivalrous condition, the animal first pulled the left lever, then the right, and finally the left again. The bottom row, below the solid horizontal line, shows that with fixed visual input the IT neuron was much less active when the monkey pulled the left lever than when it pulled the right lever. The vertical line segments show individual extracellularly recorded action potentials above a smoothed response histogram. (Source: Adapted from Sheinberg and Logothetis 1997, Fig. 3.)

uncommon in early areas such as V1 and V2 and became almost universal in IT. For this reason it is speculated that the early areas are less likely to be part of an NCC than area IT.

Visual Awareness and Human Brain Activity. Rivalry experiments have also been conducted with human subjects while brain activity is recorded using fMRI. An example is shown in Figure 21.24. Rather than using separate images presented to the left and right eye, a composite image was shown to both eyes; glasses with red and green filters assured that one eye saw only the image of the face and the other eye only the house. Subjects indicated which percept they had at different times. Recordings were made from two areas in the temporal lobe. We discussed the fusiform face area (FFA) in Chapter 10 because it appears to respond preferentially to pictures of faces. The parahippocampal place area (PPA) responds to pictures of houses and other places but not to other classes of stimuli. As subjects reported alternations between the face and house percepts, FFA and PPA activity was averaged at each transition. In the rivalry condition, house-to-face perceptual transitions were accompanied by decreases in PPA activity (red lines) and increases in FFA activity (blue lines) (see Figure 21.24a). At face-to-house transitions, FFA activity went down and PPA activity up. Similar alternations in FFA and PPA activity were seen in non-rivalry conditions in which only the face or house was shown at a time to one eye (see Figure 21.24b). The alternating patterns of brain activity in FFA and PPA seen in Figure 21.24a occur during

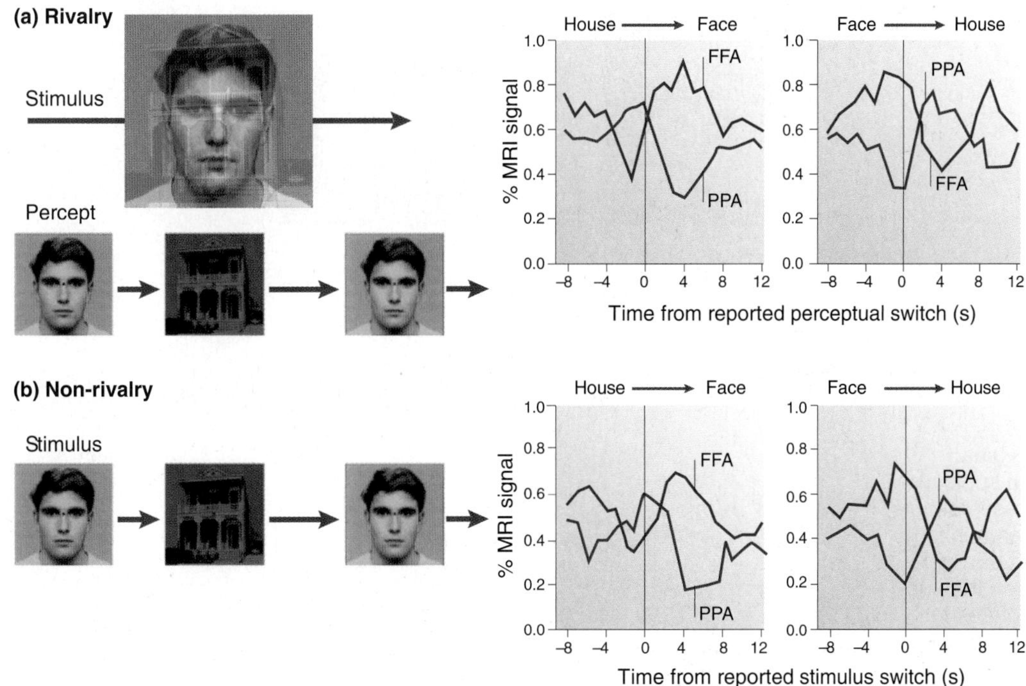

▲ FIGURE 21.24

Human brain activity during binocular rivalry recorded with fMRI. **(a)** In the rivalry condition a subject viewed the top visual stimulus through glasses with red and green lenses so that one eye saw a face and the other eye saw a house. The subject alternately perceived the face and house ("percept"). fMRI was used to measure brain activity in two temporal lobe areas: the fusiform face area (FFA) that responds more to faces than other stimuli, and the parahippocampal place area (PPA) that responds to houses and places but not to faces. The data are averaged over many perceptual transitions from house to face and face to house. Even though the stimulus is fixed, the FFA is more active when the face is perceived (blue line) and the PPA is more active when the house is seen (red line). **(b)** In the non-rivalry condition, the face and house stimuli were shown alternately to one eye. Responses from the FFA and PPA are consistent with the presentation of the face or house stimulus. (Source: Rees et al., 2002, Fig. 4.)

constant visual input, suggesting that activity in these areas may be an NCC for faces and houses.

A variety of approaches other than binocular rivalry have also been used to explore neural correlates of consciousness. An interesting example is visual imagery in which a person is instructed to imagine a visual image in their "mind's eye." Give it a try: create an image of the house you live in and imagine walking around the house counting the windows as you go. There is evidence suggesting that imagery activates some of the same visual processes driven by external visual stimuli, so imagery seems a valid method for exploring visual consciousness. Gabriel Kreiman and Christof Koch at Caltech with Itzhak Fried at the University of California, Los Angeles, conducted an imagery experiment that included recordings from human neurons. For the clinical assessment of epileptic foci, electrodes were placed in a variety of brain structures. The neuron shown in Figure 21.25 was located in entorhinal cortex, an area in the medial temporal lobe that provides input to the hippocampus. After trying a variety of visual stimuli, it was determined that the neuron responded vigorously to a photograph of dolphins, but little response was evoked by a picture of a girl's face (see Figure 21.25a). The subject was then asked to close his eyes. His task was to imagine the photo of the dolphins when a high tone was presented and imagine the girl's face with a low tone. Figure 21.25b

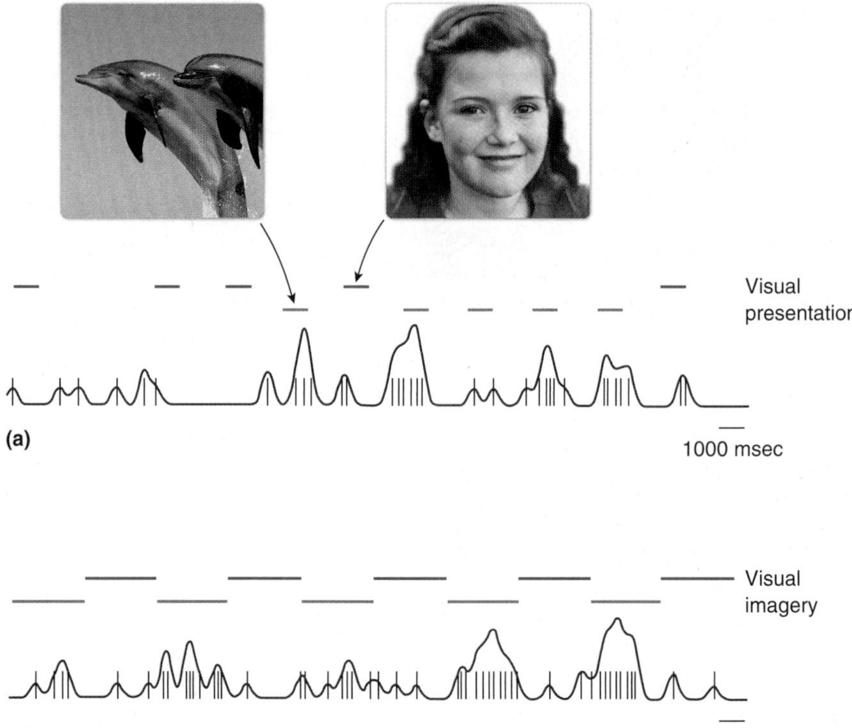

▶ FIGURE 21.25
A human neuron active during visual imagery. **(a)** A neuron in a human subject's entorhinal cortex is excited by presentations of a photograph of dolphins (green horizontal lines) but not by a photograph of a girl's face (red horizontal lines). **(b)** When the subject was asked to imagine the pictures in his "mind's eye" in response to auditory cues, the neuron was much more active while the dolphins were imagined than the face. (Source: Adapted from Crick et al., 2004, Fig. 5.)

shows the response of the entorhinal neuron during imagery. The responses aren't identical to the visually evoked responses, but there is clearly more activity when the person imagines the dolphin photo than the face photo. In this case, might entorhinal cortex be part of an NCC?

Challenges in the Study of Consciousness. Consciousness is a subject of great interest and discussion, but it is slippery to pin down. In an effort to be concrete, our discussion has focused on neural correlates of consciousness (the "easy" problems). We have seen several brain areas in which activity changes in a manner correlated with conscious awareness. These candidate NCC areas are joined by more that we have not discussed. In addition to the fMRI studies, electrode recordings made in a range of cortical areas in animals and humans show responses correlated with awareness even in single cells. Taken together, the experimental findings are encouraging that baby steps may take us closer to an understanding of consciousness, at least in a limited sense.

That said, challenges abound in the interpretation of consciousness studies. The goal in the search for neural correlates of consciousness is to find the minimal brain activity sufficient for some conscious experience. The word "minimal" poses some problems. When NCC areas are proposed, one must always consider the possibility of "contamination." Might the neural activity under study be a prerequisite for the conscious experience or a consequence of the experience but not the neural substrate of the experience? Might the NCC actually be coordinated or correlated activity spread across multiple brain areas rather than the activity of any individual area? As attention is usually linked to conscious awareness, might activity in a candidate NCC confound attention with awareness?

The connection between attention and consciousness is important and controversial. For many years, these words were taken to be nearly synonymous. However, recent perceptual experiments have demonstrated that it is possible to have your attention drawn to an object but still not perceive it. The implication is that attention is or can be independent of awareness. There are also studies reporting the opposite, that awareness can occur without attention, although that conclusion remains contentious.

Finally, there is uncertainty about where in the brain NCCs are observed. We discussed several experiments that used binocular rivalry to locate NCCs. Unfortunately, the various studies do not fit neatly together. For example, it has been found using fMRI that brain activity stretching from early visual areas such as the LGN and V1 to later visual areas such as IT changes systematically when people and animals report switches in rivalrous percepts. Contrast this with animal experiments using electrodes that show single cell response modulation in extrastriate cortex but much less in earlier structures. The difference may be a technical factor associated with the recording techniques or perhaps the spatial scale and magnitude of conscious modulation. Another conclusion reached in a recent fMRI study is that some investigations of NCCs do not adequately control attention. In other words, awareness may be reflected only in later areas; what looks like awareness correlations in earlier areas may actually reflect changes in attention rather than awareness.

Despite the concerns we've discussed, the neuroscientific investigation of consciousness has made huge strides in the past 20 years. Examining situations in which a single sensory input, visual or otherwise, gives rise to more than one percept offers fascinating opportunities for identifying neural correlates of consciousness. This leaves open the "hard problem" of consciousness, but it is a start.

CONCLUDING REMARKS

In this chapter, we have explored the dynamics of brain activity at a broad scale. Consistent changes between the brain at rest and the behaviorally active brain have been used to define a default mode network. Moving from a restful state to an active one appears to involve a global switch in brain function from the default pattern to processing tailored to behavioral needs. We cannot say with certainty what the resting state activities are, but they likely include monitoring the environment and daydreaming.

When we studied sensory and motor systems, we did so in isolation. The reality of behavior is obviously different: Sensory information comes in, we attend to a small portion of it that is momentarily important, and we generate motor outputs. Attention is a crucial link in this process. Some animals can surely function without attention, having nervous systems that hardwire behavioral responses to specific types of sensory input (e.g., threats). However, attention confers flexibility. In some situations, attention is "grabbed," but in many others, we use attention as a tool to focus our mental resources. We have seen that this involves a network of brain areas that, based on thoughts and goals, constructs priority maps for the allocation of attention followed by selective enhanced processing in sensory cortex.

How we become consciously aware of the information we attend to remains a mystery. We have sidestepped the "hard problem of consciousness," the reasons why experiences feel the way they do. On the other hand, progress is being made in the hunt for neural correlates of consciousness. As consciousness involves holding information in mind, it surely involves interactions with the memory systems we will explore in Chapter 24.

KEY TERMS

Introduction
attention (p. 720)
consciousness (p. 720)

Resting State Brain Activity
resting state activity (p. 720)
default mode network (p. 722)

Attention
exogenous attention (p. 725)
bottom–up attention (p. 725)
top–down attention (p. 725)
endogenous attention (p. 725)
spotlight of attention (p. 728)
pulvinar nucleus (p. 734)
frontal eye fields (FEF) (p. 735)
salience map (p. 736)
priority map (p. 736)
lateral intraparietal cortex (area LIP) (p. 737)
neglect syndrome (p. 737)
frontoparietal attention network (p. 741)

Consciousness
easy problems of consciousness (p. 742)
hard problem of consciousness (p. 743)
neural correlates of consciousness (p. 743)
binocular rivalry (p. 743)

Box 21.1
attention-deficit hyperactivity disorder (ADHD) (p. 724)

REVIEW QUESTIONS

1. What areas of the resting brain are active, and what might they be doing?
2. What behavioral advantages are produced by attention?
3. What neurophysiological data are consistent with the concept of a spotlight of attention?
4. How are shifts in attention and eye movements related?
5. How might a salience map guide bottom–up attention?
6. In what ways is hemispatial neglect different from blindness in half of the visual field?
7. Why can't the identification of neural correlates of attention answer the "hard problem of consciousness"?
8. How is binocular rivalry used to explore conscious awareness?

FURTHER READING

Bisley JW, Goldberg ME. 2010. Attention, intention, and priority in the parietal lobe. *Annual Review of Neuroscience* 33:1–21.

Buckner RL, Andrews-Hanna JR, Schacter DL. 2008. The brain's default network: anatomy, function, and relevance to disease. *Annals of the New York Academy of Sciences* 1124:1–38.

Cohen MA, Dennett DC. 2011. Consciousness cannot be separated from function. *Trends in Cognitive Science* 15:358–364.

Koch C, Greenfield S. 2007. How does consciousness happen? *Scientific American* 297:76–83.

Miller EK, Buschman TJ. 2013. Cortical circuits for the control of attention. *Current Opinion in Neurobiology* 23:216–222.

Noudoost B, Chang MH, Steimetz NA, Moore T. 2010. Top-down control of visual attention. *Current Opinion in Neurobiology* 20:183–190.

Raichle ME, Snyder AZ. 2007. A default mode of brain function: a brief history of an evolving idea. *Neuroimage* 37:1083–1090.

Shipp S. 2004. The brain circuitry of attention. *Trends in Cognitive Science* 8:223–230.

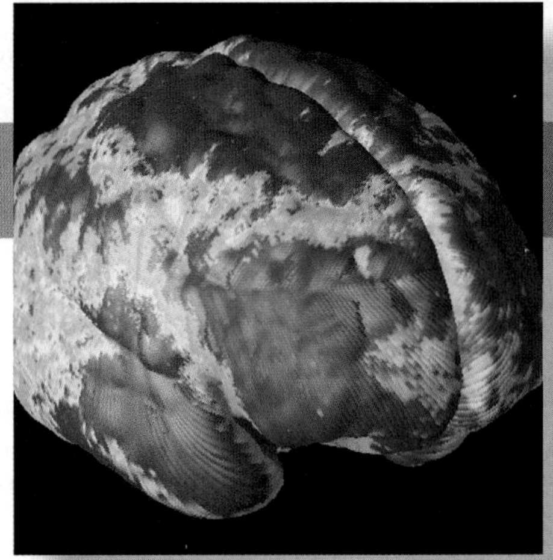

CHAPTER TWENTY-TWO

Mental Illness

INTRODUCTION

Neurology is a branch of medicine concerned with the diagnosis and treatment of nervous system disorders. We have discussed many neurological disorders in this book, ranging from multiple sclerosis to aphasia. While they are significant and fascinating in their own right, neurological disorders also help illustrate the role of physiological processes in normal brain function, such as the importance of myelin for action potential conduction and the role of the frontal lobe in language.

Psychiatry, on the other hand, has a different focus. This branch of medicine is concerned with the diagnosis and treatment of disorders that affect the *mind*, or *psyche.* (In Greek mythology, the beautiful young woman Psyche was the personification of the human soul.) Aspects of brain function that are disturbed by mental illness—our fears, moods, and thoughts—were once considered beyond the reach of neuroscience. But, as we saw in the earlier chapters of Part III, many higher brain functions have begun to yield their secrets. Today, there is hope that neuroscience will also solve the riddle of mental illness.

In this chapter, we will discuss some of the most severe and prevalent psychiatric disorders: anxiety disorders, affective disorders, and schizophrenia. Once again, we will see that a great deal can be learned about the nervous system by studying what happens when things go wrong.

MENTAL ILLNESS AND THE BRAIN

Human behavior is the product of brain activity, and the brain is the product of two interacting factors: heredity and environment. Obviously, one important determinant of your individualism is your complement of DNA, which, unless you have an identical twin, is unique. This means that physically your brain, like your fingerprints, is different from all others. A second factor that makes your brain unique is your history of personal experience. Experiences can include trauma and disease, but, as we saw in the case of somatosensory map plasticity (see Chapter 12), the sensory environment itself can leave a permanent mark on the brain. (We'll return to this theme in Part IV of this book when we discuss development, learning, and memory.) Thus, despite the gross physical similarities you might share with a genetic twin, at a fine scale, neither your brains nor your behaviors are identical. To complicate matters further, variations in genetic makeup and past experience make the brain differentially susceptible to modification by subsequent experiences. These genetic and experiential variations, all ultimately expressed as physical changes in the brain, give rise to the full range of behaviors exhibited by the human population.

Health and illness are two points along a continuum of bodily function, and the same can be said for mental health and mental illness. While we all have our odd characteristics, an individual is said to be "mentally ill" at the point when the person has a diagnosable disorder of thought, mood, or behavior that causes distress or impaired functioning. An unfortunate legacy of our past ignorance about brain function is the common distinction drawn between "physical" and "mental" health. The philosophical roots of this distinction can be traced to Descartes's proposed separation of body and mind (see Chapter 1). Disorders of the body (which, for Descartes, included the brain) had an organic basis and were the concern of physicians and medicine. Disorders of the mind, on the other hand, were considered spiritual or moral and were the concern of clergymen and religion. The fact that most disorders of mood, thought, and behavior have, until very recently, remained resistant to biological explanations or treatments has reinforced this dichotomy.

▲ FIGURE 22.1
Sigmund Freud. Freud proposed psychoanalytic theories of mental illness.

Psychosocial Approaches to Mental Illness

An important advance in the secularization of mental illness was the emergence of the medical discipline of psychiatry, devoted to treating disorders of human behavior. The Austrian neurologist and psychiatrist Sigmund Freud (1856–1939) had an enormous impact on the new field, especially in the United States (Figure 22.1). Freud's theory of *psychoanalysis* is based on two major assumptions that (1) much of mental life is unconscious (beyond awareness), and (2) past experiences, particularly in childhood, shape how a person will feel and respond throughout life. According to Freud, mental illness results when the unconscious and conscious elements of the psyche come into conflict. The way to resolve the conflict, and to treat the illness, is to help the patient unearth the hidden secrets of the unconscious. Often, these dark secrets are related to incidents (e.g., physical, mental, or sexual abuse) that occurred during childhood and were suppressed from consciousness.

A different theory of personality, championed by Harvard University psychologist B. F. Skinner (1904–1990), is based on the assumption that many behaviors are learned responses to the environment. *Behaviorism* rejects the notions of underlying conflicts and the unconscious and focuses instead on observable behaviors and their control by the environment. In Chapter 16, we learned about some of the forces that motivate behavior. The probability of a type of behavior increases when it satisfies a craving or produces a pleasurable sensation (positive reinforcement), and it decreases when the consequences are deemed unpleasant or unsatisfactory (negative reinforcement). According to this theory, mental disorders may represent maladaptive behaviors that are learned. Treatment consists of active attempts to "unlearn" through behavior modification, either by introducing new types of behavioral reinforcement or by providing an opportunity to observe and recognize behavioral responses that are appropriate.

Such "psychosocial" approaches to treating mental illness have a sound neurobiological basis. The brain is structurally modified through learning and early experience, and these modifications will alter behavioral responses. Treatment relies on *psychotherapy*, the use of verbal communication to help the patient. Of course, "talk therapy" is not appropriate for all mental disorders, any more than a particular antibiotic is appropriate for all infections. However, until the revolution in biological psychiatry, variations in psychotherapy were the only tools available to psychiatrists. Moreover, despite the shift in "blame" away from one's moral character and toward early childhood experience, psychotherapy contributed to the stigmatizing notion that mental illness (in contrast to physical illness) could be overcome by willpower alone. Freud himself recognized the shortcomings of psychotherapy, stating that the "deficiencies in our [the psychoanalytic] description would probably vanish if we were already in a position to replace the psychological terms by physiological or chemical ones" (1920, p. 54). Now, nearly a century later, neuroscience has advanced to a point where this goal seems attainable.

Biological Approaches to Mental Illness

A spectacular success in the early biological diagnosis and treatment of mental illness actually occurred during Freud's time. A major psychiatric disorder at the turn of the twentieth century was called *general paresis of the insane*, afflicting 10–15% of all institutionalized psychiatric patients. The disorder had a progressive course, starting with symptoms of mania—excitement, euphoria, and grandiose delusions—and evolved to cognitive deterioration and, ultimately, paralysis and death. Initially blamed on psychological factors, the cause was eventually traced to infection of the brain with *Treponema pallidum*, the microorganism that causes syphilis. Once the cause was known, increasingly effective treatments quickly followed. By 1910, German microbiologist Paul Ehrlich had established that the drug arsphenamine could act as a "magic bullet," killing the *T. pallidum* in the blood without damaging its human host. Eventually, the antibiotic penicillin (discovered in 1928 by British microbiologist Alexander Fleming) was found to be so effective in killing the microorganism that established brain infections could be completely eradicated. Thus, when penicillin became widely available by the end of World War II, a major psychiatric disorder was virtually eliminated.

A number of other mental illnesses can be traced directly to biological causes. For example, a dietary deficiency in niacin (a B vitamin) can cause agitation, impaired reasoning, and depression. The penetration of HIV (the AIDS virus) into the brain causes progressive cognitive and behavioral impairments. A form of obsessive-compulsive disorder (discussed later) has been linked to an autoimmune response triggered by streptococcal pharyngitis (strep throat) in children. Understanding the causes of these diseases will lead to treatments and, ultimately, to cures of the associated mental disorders.

The Promise and Challenge of Molecular Medicine in Psychiatry. Of course, serious mental disorders also occur in well-nourished and infection-free individuals. Although the causes in most cases remain to be determined, it is safe to say that the roots of these disorders lie in altered brain anatomy, chemistry, and function. An exciting new way to understand brain malfunction has been unleashed by knowledge of the human genome. As in other complex diseases like cancer, gene mutations can cause or confer risk for psychiatric disease, and major efforts are now well underway to identify these genes. The approach of using genetic information to develop a treatment is sometimes referred to as **molecular medicine**.

A path from gene to treatment is illustrated in Figure 22.2. Searching the DNA of individuals with a psychiatric disease may reveal causative gene mutations that can be reproduced in genetically engineered mice. By comparing the neurobiology of these animals with normal, "wild type" mice, researchers can determine how the brain functions differently in association with these mutations. Discovery of an abnormal physiological condition, or **pathophysiology**, may suggest biological processes that can be targeted with drug therapy—for example, too much or too little of a neurotransmitter. If drug candidates succeed in human clinical trials, then new therapeutics can be introduced to treat the disease.

Despite the enormous promise of molecular medicine, brain diseases present some unique challenges. First, mental disorders are diagnosed by clinicians based on how they appear or are described by the patient (signs and symptoms), not by knowledge of their underlying cause (etiology). It is now understood that the same diagnosis may arise from many causes, so no single treatment approach is likely to succeed in all patients, thus complicating clinical trials. Second, not all mental illnesses have a clear genetic basis, and for those that do, a large number of genes have been implicated. In some cases, it appears that pathophysiology may be caused by inheritance of numerous small mutations in many different genes. In these cases, although no single mutation has much effect, together they

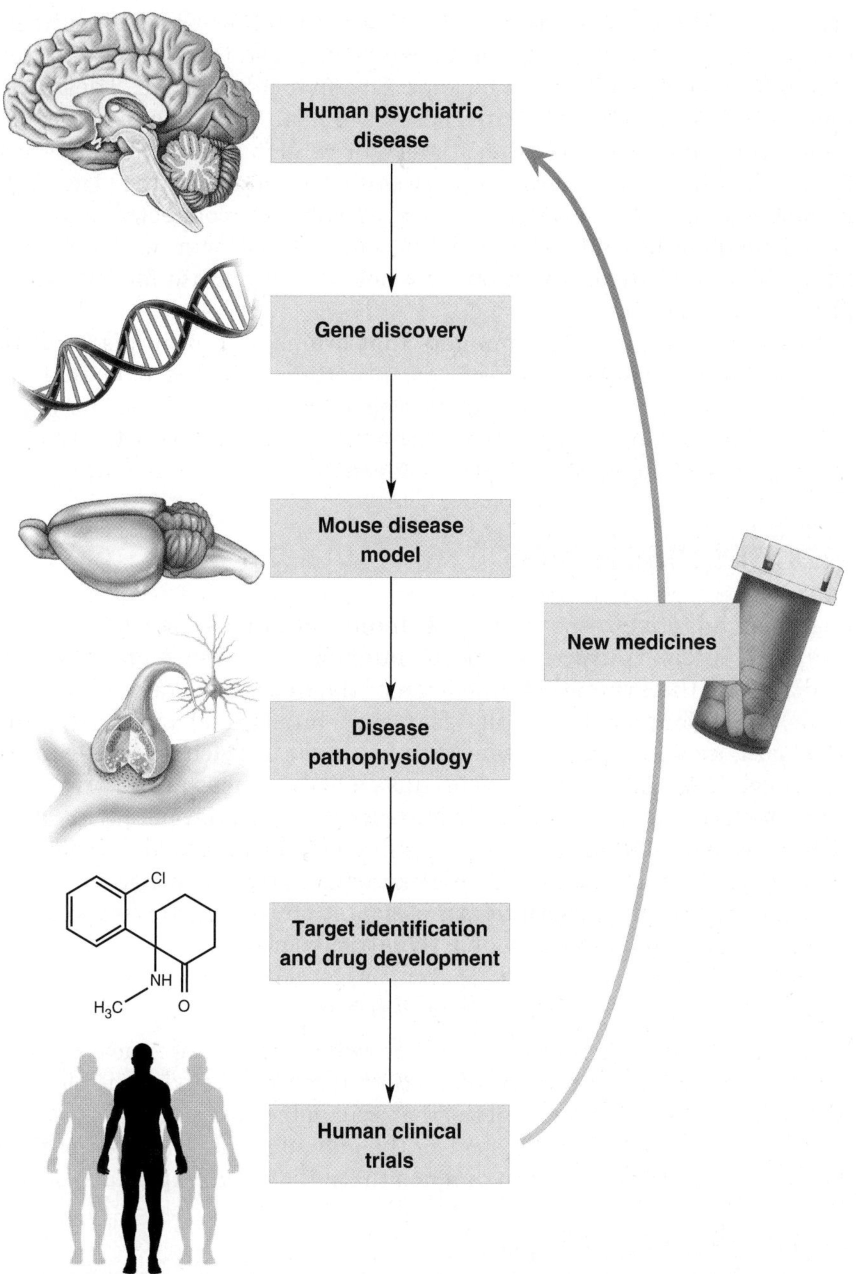

◀ FIGURE 22.2
Molecular medicine. A path from genes to treatments for psychiatric disorders.

greatly increase the risk for a mental illness (metaphorically, death by a thousand small knife cuts). In other cases, duplication or deletion of a gene or segment of genes, called *gene copy number variants*, might be the single cause of the diagnosis. Although each specific variant occurs rarely in the human population, variation in many different segments of DNA can result in the same diagnosis (by analogy, death by a gunshot wound; although the end result is the same, each fatal wound can uniquely affect a different part of the body). This genetic complexity interferes with development of broadly useful animal models.

A radical new approach to overcome these challenges is to study the pathophysiology of neurons from individual patients. Don't worry; this does not entail a brain biopsy! Rather, the approach takes advantage of the recent discovery that if skin cells scraped from a patient are treated with the correct mixture of chemicals, they can be transformed into what are called **induced pluripotent stem cells**, or simply iPSCs. Treatment with another mixture of chemicals can then cause these cells to differentiate into neurons that can be kept alive in a culture dish. These neurons can then be compared with those from healthy people to determine their pathophysiology. However, the major challenge with this approach is that the brain is far more complicated than a single neuron. The brain is composed of myriad cell types that are richly interconnected, and gene mutations manifest differently in different types of neurons. Treatment for pathophysiology of a neuron may not be appropriate for the pathophysiology of a brain.

Despite these sobering reminders that brains and brain diseases are extraordinarily complex, there is much optimism in the field that the challenges can and soon will be overcome. Now let's explore the major psychiatric disorders and see how neuroscience has already both provided insight about their possible causes and contributed to their treatment.

ANXIETY DISORDERS

Fear is an adaptive response to threatening situations. As we learned in Chapter 18, fear is expressed by the autonomic fight-or-flight response, mediated by the sympathetic division of the autonomic nervous system (ANS) (see Chapter 15). Many fears are innate and species-specific. A mouse does not need to be taught to fear a cat. But fear is also learned. One touch is usually all it takes to cause a horse to fear an electric fence. The adaptive value of fear is obvious. As the old aviation saying goes, "There are old pilots, and there are bold pilots, but there are no old bold pilots." But fear is not an appropriate or adaptive response in all circumstances. An inappropriate expression of fear characterizes **anxiety disorders**, the most common of psychiatric disorders.

A Description of Anxiety Disorders

It has been estimated that in any given year, over 15% of Americans will suffer from one of the recognized anxiety disorders listed in Table 22.1. Although they differ in terms of the real or imagined stimuli that evoke the anxiety, and the behavioral responses the individual uses to attempt to reduce it, these disorders have in common the pathological expression of fear.

TABLE 22.1 **Anxiety Disorders**

Name	Description
Panic disorder	Frequent panic attacks consisting of discrete periods with the sudden onset of intense apprehension, fearfulness, or terror, often associated with feelings of impending doom
Agoraphobia	Anxiety about, or the avoidance of, places or situations from which escape might be difficult or embarrassing, or in which help may not be available in the event of a panic attack
Generalized anxiety disorder	At least 6 months of persistent and excessive anxiety and worry
Specific phobias	Clinically significant anxiety provoked by exposure to a specific feared object or situation, often leading to avoidance behavior
Social phobia	Clinically significant anxiety provoked by exposure to certain types of social or performance situations, often leading to avoidance behavior

Source: Adapted from American Psychiatric Association, 2013.

Panic Disorder. *Panic attacks* are sudden feelings of intense terror that occur without warning. The symptoms include palpitations, sweating, trembling, shortness of breath, chest pain, nausea, dizziness, tingling sensations, and chills or blushing. Most people report an overwhelming fear that they are dying or "going crazy," and flee from the place where the attack begins, often seeking emergency medical assistance. The attacks are short-lived, however, usually lasting less than 30 minutes. Panic attacks can occur in response to specific stimuli. They may be a feature of a number of anxiety disorders, but they can also occur spontaneously.

The condition that psychiatrists call **panic disorder** is characterized by recurring, seemingly unprovoked panic attacks and a persistent worry about having further attacks. About 2% of the population suffers from panic disorder, which is twice as common in women as in men. The onset of the disorder is most common after adolescence but before the age of 50 years. Half the individuals who have panic disorder will also suffer from major depression (see below), and 25% of them will become alcoholic or develop substance-abuse problems.

Agoraphobia. Severe anxiety about being in situations where escape might be difficult or embarrassing is characteristic of *agoraphobia* (from the Greek, for "fear of an open marketplace"). The anxiety leads to avoidance of situations irrationally perceived as threatening, such as being alone outside the home, in a crowd of people, in a car or airplane, or on a bridge or elevator. Agoraphobia is often an adverse outcome of panic disorder, as the situation in Box 22.1 describes. About 5% of the population is agoraphobic, with the incidence among women being twice that of men.

Other Disorders Characterized by Increased Anxiety

Several disorders that are no longer classified by the American Psychiatric Association as "anxiety disorders" are nevertheless characterized by increased anxiety. Two of the most prevalent are post-traumatic stress disorder and obsessive-compulsive disorder.

Post-Traumatic Stress Disorder. To a pathologist, trauma refers to a wound caused by sudden violence. In the realm of psychiatry, the term refers to the psychological wounds of experiencing or witnessing a shocking event or events. A long-lasting consequence can be post-traumatic stress disorder, or PTSD. The symptoms of PTSD can include increased anxiety, intrusive memories, dreams or flashbacks of the traumatic experiences, irritability, and emotional numbness. PTSD affects 3.5% of the adult population in the United States.

Obsessive-Compulsive Disorder. People with **obsessive-compulsive disorder (OCD)** have *obsessions*, which are recurrent, intrusive thoughts, images, ideas, or impulses that the person perceives as being inappropriate, grotesque, or forbidden. Common themes are thoughts of contamination with germs or body fluids, thoughts that the sufferer has unknowingly caused harm to someone, and violent or sexual impulses. These thoughts are recognized by the affected individual as being foreign, and they evoke considerable anxiety. People with OCD also have *compulsions*, which are repetitive behaviors or mental acts that are performed to reduce the anxiety associated with obsessions. Examples are repeated hand-washing, counting, and checking to make sure that something is not out of place. OCD affects over 2% of the population, with an equal incidence among men and women. The disorder usually appears in young adult life, and the symptoms fluctuate in response to stress levels.

BOX 22.1 OF SPECIAL INTEREST

Agoraphobia with Panic Attacks

To appreciate the distress and disruption caused by anxiety disorders, consider the following case history from Nancy C. Andreasen's book, *The Broken Brain*.

> Greg Miller is a 27-year-old unmarried computer programmer. When asked about his main problem, he replied, "I am afraid to leave my house or drive my car."
>
> The patient's problems began approximately one year ago. At that time he was driving across the bridge that he must traverse every day in order to go to work. While driving in the midst of the whizzing six-lane traffic, he began to think (as he often did) about how awful it would be to have an accident on that bridge. His small, vulnerable VW convertible could be crumpled like an aluminum beer can, and he could die a bloody, painful death or be crippled for life. His car could even hurtle over the side of the bridge and plunge into the river.
>
> As he thought about these possibilities, he began to feel increasingly tense and anxious. He glanced back and forth at the cars on either side of him and became frightened that he might run into one of them. Then he experienced an overwhelming rush of fear and panic. His heart started pounding and he felt as if he were going to suffocate. He began to take deeper and deeper breaths, but this only increased his sense of suffocation. His chest felt tight and he wondered if he might be about to die of a heart attack. He certainly felt that something dreadful was going to happen to him quite soon. He stopped his car in the far right lane in order to try to regain control of his body and his feelings. Traffic piled up behind with many honking horns, and drivers pulled around him yelling obscenities. On top of his terror, he experienced mortification. After about three minutes, the feeling of panic slowly subsided, and he was able to proceed across the bridge and go to work. During the remainder of the day, however, he worried constantly about whether or not he would be able to make the return trip home across the bridge without a recurrence of the same crippling fear.
>
> He managed to do so that day, but during the next several weeks he would begin to experience anxiety as he approached the bridge, and on three or four occasions he had a recurrence of the crippling attack of panic. The panic attacks began to occur more frequently so that he had them daily. By this time he was overwhelmed with fear and began to stay home from work, calling in sick each day. He knew that his main symptom was an irrational fear of driving across the bridge, but he suspected that he might also have some type of heart problem. He saw his family doctor, who found no evidence of any serious medical illness, and who told him that his main problem was excessive anxiety. The physician prescribed a tranquilizer for him and told him to try to return to work.
>
> For the next six months, Greg struggled with his fear of driving across the bridge. He was usually unsuccessful and continued to miss a great deal of work. Finally, he was put on disability for a few months and told by the company doctor to seek psychiatric treatment. Greg was reluctant and embarrassed to do this, and instead he stayed home most of the time, reading books, listening to records, playing chess on his Apple computer, and doing various "handy-man" chores around the house. As long as he stayed home, he had few problems with anxiety or the dreadful attacks of panic. But when he tried to drive his car, even to the nearby shopping center, he would sometimes have panic attacks. Consequently, he found himself staying home nearly all the time and soon became essentially housebound. (Andreasen, 1984, pp. 65–66)

Biological Bases of Anxiety Disorders

A genetic predisposition has been established for many anxiety disorders. Other anxiety disorders appear to be rooted more in the occurrence of stressful life events.

Fear is normally evoked by a threatening stimulus, called a *stressor*, and it is manifest by a response known as the stress response. As mentioned previously, the stimulus–response relationship can be strengthened by experience (recall the horse and the electric fence), but it can also be weakened. Consider, for example, an expert skier who no longer views a precipitous drop as fearful. A healthy person regulates the stress response through learning. The hallmark of anxiety disorders is the occurrence of an inappropriate stress response either when a stressor is not present or when it is not immediately threatening. Thus, a key to understanding anxiety is to understand how the stress response is regulated by the brain.

The Stress Response. The *stress response* is the coordinated reaction to threatening stimuli. It is characterized by the following:

- Avoidance behavior
- Increased vigilance and arousal
- Activation of the sympathetic division of the ANS
- Release of cortisol from the adrenal glands

It should come as no surprise that the hypothalamus is centrally involved in orchestrating appropriate humoral, visceromotor, and somatic motor responses (see Chapters 15 and 16). To get an idea of how this response is regulated, let's focus on the humoral response, which is mediated by the **hypothalamic-pituitary-adrenal (HPA) axis** (Figure 22.3).

As we learned in Chapter 15, the hormone cortisol (a glucocorticoid) is released from the adrenal cortex in response to an elevation in the blood level of **adrenocorticotropic hormone (ACTH)**. ACTH is released by the anterior pituitary gland in response to **corticotropin-releasing hormone (CRH)**. CRH is released into the blood of the portal circulation by parvocellular neurosecretory neurons in the paraventricular nucleus of the hypothalamus. Thus, this arm of the stress response can be traced back to activation of the CRH-containing neurons of the hypothalamus. Much can be learned about anxiety disorders by understanding how the activity of these neurons is regulated. For example, when CRH is overexpressed in genetically engineered mice, the animals display increased anxiety-like behaviors. When the receptors for CRH are genetically eliminated from mice, they have less anxiety-like behavior than normal mice.

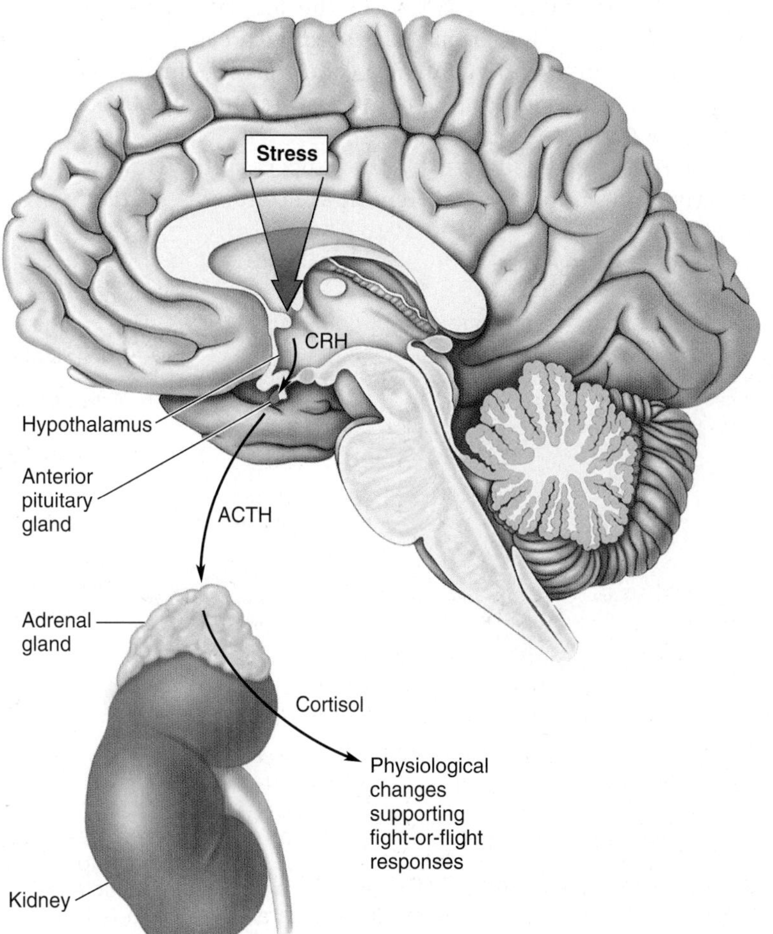

◀ FIGURE 22.3
The hypothalamic-pituitary-adrenal axis. The HPA axis regulates the secretion of cortisol from the adrenal gland in response to stress. CRH is the chemical messenger between the paraventricular nucleus of the hypothalamus and the anterior pituitary gland. ACTH released by the pituitary gland travels in the bloodstream to the adrenal gland lying atop the kidney, where it stimulates cortisol release. Cortisol contributes to the body's physiological response to stress.

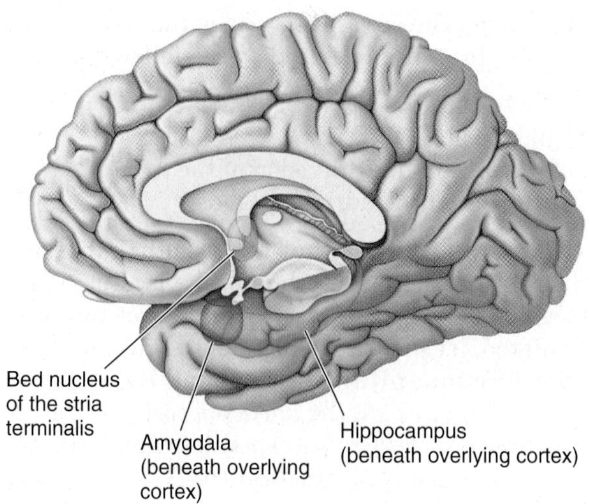

▲ FIGURE 22.4
The location of the amygdala and hippocampus.

Regulation of the HPA Axis by the Amygdala and Hippocampus. The CRH neurons of the hypothalamus are regulated by two structures that were introduced in earlier chapters: the *amygdala* and the *hippocampus* (Figure 22.4). As we learned in Chapter 18, the amygdala is critical to fear responses. Sensory information enters the basolateral amygdala, where it is processed and relayed to neurons in the central nucleus. When the central nucleus of the amygdala becomes active, the stress response ensues (Figure 22.5). Inappropriate activation of the amygdala, as measured using fMRI (see Box 7.3), has been associated with some anxiety disorders. Downstream from the amygdala is a collection of neurons called the *bed nucleus of the stria terminalis*. The bed nucleus neurons activate the HPA axis and the stress response.

The HPA axis is also regulated by the hippocampus. However, hippocampal activation suppresses, rather than stimulates, CRH release.

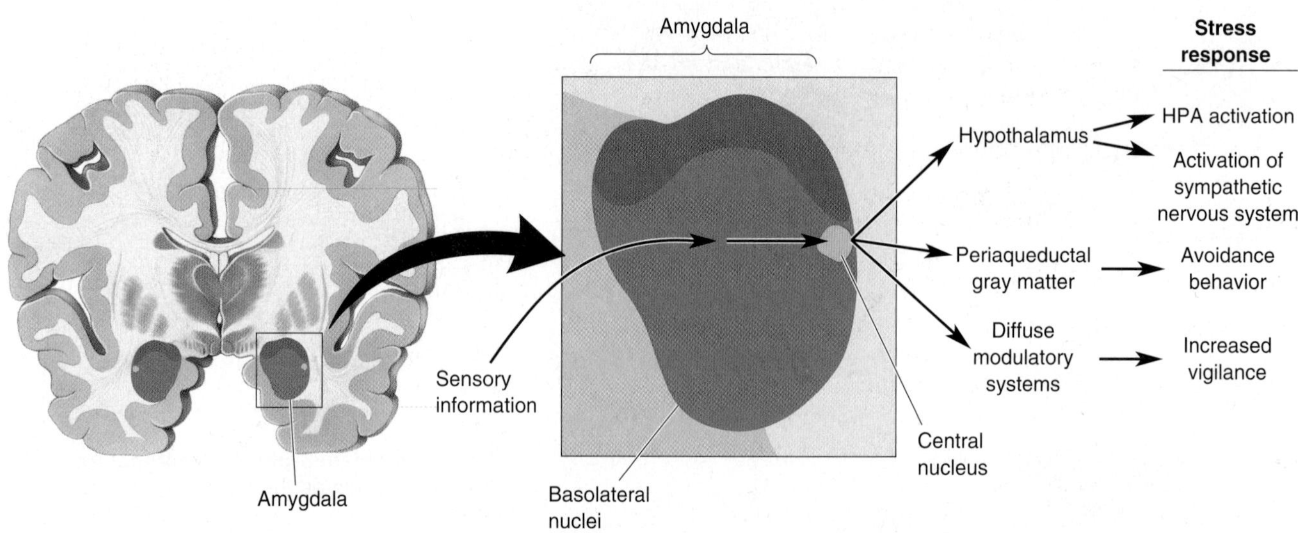

▲ FIGURE 22.5
Control of the stress response by the amygdala. The amygdala receives ascending sensory information from the thalamus as well as descending inputs from the neocortex. This information is integrated by the basolateral nuclei and is relayed to the central nucleus. Activation of the central nucleus leads to the stress response.

The hippocampus contains numerous **glucocorticoid receptors** that respond to the cortisol released from the adrenal gland in response to HPA system activation. Thus, the hippocampus normally participates in the feedback regulation of the HPA axis by inhibiting CRH release (and the subsequent release of ACTH and cortisol) when circulating cortisol levels get too high. Continuous exposure to cortisol, such as during periods of chronic stress, can cause hippocampal neurons to wither and die in experimental animals (see Box 15.1). This degeneration of the hippocampus may set off a vicious cycle, in which the stress response becomes more pronounced, leading to even greater cortisol release and more hippocampal damage. Human brain imaging studies have shown a decrease in the volume of the hippocampus in some people suffering from PTSD.

To summarize, the amygdala and the hippocampus regulate the HPA axis and the stress response in a push–pull fashion (Figure 22.6). Anxiety disorders have been related to both hyperactivity of the amygdala and diminished activity of the hippocampus. It is important to keep in mind, however, that the amygdala and hippocampus both receive highly processed information from the neocortex. Indeed, another consistent finding in humans with anxiety disorders has been elevated activity of the prefrontal cortex.

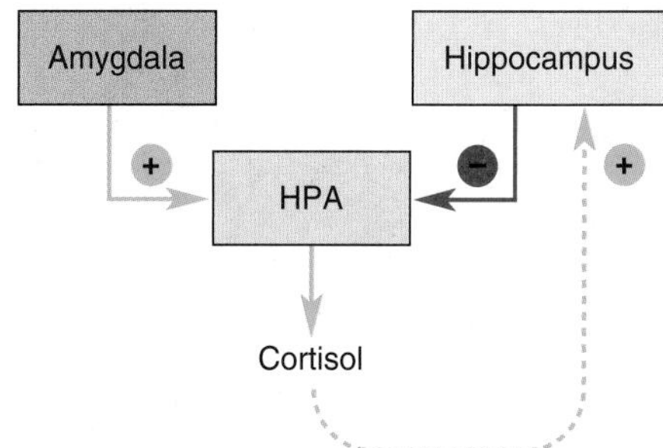

▲ FIGURE 22.6
Push–pull regulation of the HPA axis by the amygdala and hippocampus. Amygdala activation stimulates the HPA system and the stress response (green lines). Hippocampal activation, on the other hand, suppresses the HPA system (red line). Because the hippocampus has glucocorticoid receptors that are sensitive to circulating cortisol, it is important in the feedback regulation of the HPA axis in preventing excessive cortisol release.

Treatments for Anxiety Disorders

Several treatments are available for anxiety disorders. In many cases, patients respond well to psychotherapy and counseling; in other cases, specific medications are more effective.

Psychotherapy. We have seen that there is a strong learning component to fear, so it should not be surprising that psychotherapy can be an effective treatment for many of the anxiety disorders. The therapist gradually increases the exposure of the patient to the stimuli that produce anxiety, reinforcing the notion that the stimuli are not dangerous. At the neurobiological level, the aim of the psychotherapy is to alter connections in the brain such that the real or imagined stimuli no longer evoke the stress response.

Anxiolytic Medications. Medications that reduce anxiety, described as **anxiolytic drugs**, act by altering chemical synaptic transmission in the brain. The major classes of drugs currently used in the treatment of anxiety disorders are benzodiazepines and serotonin-selective reuptake inhibitors.

Recall that GABA is an important inhibitory neurotransmitter in the brain. $GABA_A$ receptors are GABA-gated chloride channels that mediate fast inhibitory postsynaptic potentials (see Chapter 6). The proper action of GABA is critical to the proper functioning of the brain: Too much inhibition results in coma, and too little results in seizures. In addition to its GABA-binding site, the $GABA_A$ receptor contains sites where chemicals can act to powerfully modulate channel function. **Benzodiazepines** bind to one of these sites and act to make GABA much more effective in opening the channel and producing inhibition (Figure 22.7). The site on the receptor that binds benzodiazepines is believed to be used normally by a naturally occurring brain chemical, although the identity of the endogenous molecule has not been established.

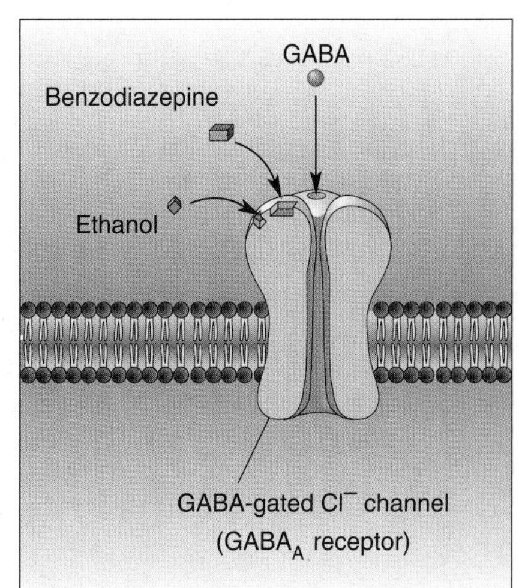

▲ FIGURE 22.7
The action of benzodiazepine. Benzodiazepines bind to a site on the $GABA_A$ receptor that makes it much more responsive to GABA, the major inhibitory neurotransmitter in the forebrain. A different site can bind ethanol and also make the receptor more responsive to GABA.

Benzodiazepines, of which Valium (diazepam) is perhaps the most well known, are highly effective treatments for acute anxiety. Indeed, virtually all drugs that stimulate GABA actions are anxiolytic, including the active ingredient in alcoholic beverages, ethanol. A reduction in anxiety is likely to explain, at least in part, the widespread social use of alcohol. The anxiolytic effects of alcohol are also an obvious reason that anxiety disorders and alcohol abuse often go hand-in-hand.

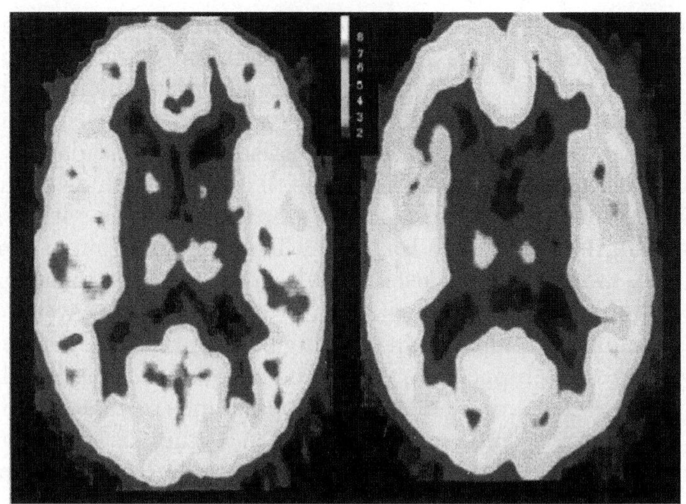

▲ FIGURE 22.8
Diminished binding of radioactive benzodiazepine in a patient with panic disorder. PET scans in the horizontal plane of the brain of a healthy person (left) and the brain of a person suffering from panic disorder (right). The color-coding indicates the number of benzodiazepine binding sites in the brain (hot colors indicate more; cool colors indicate fewer). The frontal cortex, at the top of the scan, shows many fewer binding sites in the individual with panic disorder. (Source: Nutt DJ, Malizia AL. 2001. New insights into the role of the GABAA-benzodiazepine receptor in psychiatric disorder. *The British Journal of Psychiatry* 179:390–396.)

We may infer that the calming actions of benzodiazepines are due to the suppression of activity in the brain circuits used in the stress response. Benzodiazepine treatment might be required to restore normal function to these circuits. Indeed, a study of patients with panic disorder using positron emission tomography (PET) imaging (see Box 7.3) demonstrated that the number of benzodiazepine binding sites was reduced in regions of the frontal cortex that show hyperactive responsiveness during anxiety (Figure 22.8). These results are promising not only because they might reveal the sites of benzodiazepine action in the brain but also because they suggest that an alteration in the endogenous regulation of GABA receptors is a cause of the anxiety disorder.

Serotonin-selective reuptake inhibitors (SSRIs) are widely used in the treatment of mood disorders, as we will discuss in a moment. However, SSRIs are also highly effective for treating other psychiatric disorders, notably including OCD. Recall that serotonin is released throughout the brain by a diffuse modulatory system originating in the raphe nuclei of the brain stem (see Figure 15.13). The actions of serotonin are mediated by G-protein-coupled receptors and are terminated by reuptake, via serotonin transporter proteins, into the axon terminal. Thus, as the name implies, SSRIs act to prolong the actions of released serotonin at their receptors by inhibiting reuptake. In a recent study, the presence in some families of a rare mutation in the serotonin transporter gene was associated with a high incidence of OCD, further implicating serotonin in the origins of this disease.

Unlike the benzodiazepines, however, the anxiolytic actions of the SSRIs are not immediate. Therapeutic effects develop slowly, over a period of weeks, in response to regular daily dosing. This finding indicates that the immediate rise in extracellular serotonin caused by the SSRI is *not* responsible for the anxiolytic effect. Rather, the effect appears to be due to an adaptation of the nervous system to chronically elevated brain serotonin, via some structural or functional change that is not understood.

We will return to a discussion of the actions of SSRIs when we discuss depression. However, it is very interesting in the context of anxiety disorders that one adaptive response to SSRIs is an increase in the glucocorticoid receptors in the hippocampus. SSRIs might act to dampen anxiety by enhancing the feedback regulation of the CRH neurons in the hypothalamus (see Figure 22.6).

Although benzodiazepines and SSRIs have proven to be effective in treating a wide variety of anxiety disorders, novel drugs are now being developed based on our new understanding of the stress response. One promising drug target is the receptors for CRH. Not only is CRH used by hypothalamic neurons to control ACTH release from the pituitary but it is also used as a neurotransmitter in some of the central circuits involved in the stress response. For example, some neurons of the central nucleus of the amygdala contain CRH, and injections of CRH into the brain can produce the full-blown stress response and signs of anxiety. Thus, there is hope that antagonists of CRH receptors will be useful for the treatment of some anxiety disorders.

AFFECTIVE DISORDERS

Affect is the medical term for emotional state or mood; **affective disorders** are disorders of mood. In a given year, over 9% of the population will suffer from one of the mood disorders.

A Description of Affective Disorders

An occasional, brief feeling of low mood—getting "the blues"—is a common response to life's events, such as suffering a loss or disappointment, and we would not call this a disorder. However, the affective disorder that psychiatrists and psychologists call *depression* is something more prolonged and much more severe, characterized by a feeling that one's emotional state is no longer under one's control. It can occur suddenly, often without obvious external cause, and if left untreated, it usually lasts 4–12 months.

Depression is a serious disease. It is a main precipitating cause of suicide, which claims more than 38,000 lives each year in the United States. Depression is also widespread. Perhaps as many as 20% of the population will suffer a major, incapacitating episode of depression during their lifetime. In a subset of patients with bipolar disorder, bouts of depression are punctuated with emotional highs that can also be highly disruptive.

Major Depression. The mental illness known as **major depression** is the most common mood disorder, affecting 6% of the population every year. The cardinal symptoms are lowered mood and decreased interest or pleasure in all activities. For a diagnosis of major depression, these symptoms must be present every day for a period of at least 2 weeks and not be obviously related to bereavement. Other symptoms also occur, including:

- Loss of appetite (or increased appetite)
- Insomnia (or hypersomnia)
- Fatigue
- Feelings of worthlessness and guilt
- A diminished ability to concentrate
- Recurrent thoughts of death

Episodes of major depression usually don't last longer than 2 years, although the disease has a chronic, unremitting course in about 17% of patients. Without treatment, however, depression will recur in 50% of cases, and after three or more episodes, the odds of recurrence increase to over 90%. Another expression of depression, afflicting 2% of the adult population, is called *dysthymia*. Although milder than major depression, dysthymia has a chronic, "smoldering" course, and it seldom disappears spontaneously. Major depression and dysthymia are twice as common in women as men.

Bipolar Disorder. Like major depression, **bipolar disorder** is a recurrent mood disorder. It consists of repeated episodes of mania, or mixed episodes of mania and depression, and therefore is also called *manic-depressive disorder*. **Mania** (derived from a French word meaning "crazed" or "frenzied") is a distinct period of abnormally and persistently elevated, expansive, or irritable mood. During the manic phase, other common symptoms include:

- Inflated self-esteem or grandiosity
- A decreased need for sleep
- Increased talkativeness or feelings of pressure to keep talking
- Flight of ideas, or a subjective experience that thoughts are racing
- Distractibility
- Increased goal-directed activity

Another symptom is impaired judgment. Spending sprees, offensive or disinhibited behavior, promiscuity, or other reckless behaviors are common.

According to current diagnostic criteria, there are two types of bipolar disorder. Type I bipolar disorder is characterized by the manic episodes just described (with or without incidents of major depression), and occurs in about 1% of the population, equally among men and women. Type II bipolar disorder, affecting about 0.6% of the population, is characterized by *hypomania*, a milder form of mania that is not associated with marked impairments in judgment or performance. Indeed, hypomania in some may take the form of a marked increase in efficiency, accomplishment, or creativity (Box 22.2). However, type II bipolar disorder is also always associated with episodes of major depression. When hypomania alternates with periods of depression that are not severe enough to warrant the description "major" (i.e., fewer symptoms and shorter duration), the disorder is called *cyclothymia*.

Biological Bases of Affective Disorders

Like most other mental illnesses, affective disorders reflect the altered functioning of many parts of the brain at the same time. How else can we explain the coexistence of symptoms ranging from eating and sleeping disorders to a loss of the ability to concentrate? For this reason, research has focused on the role of the diffuse modulatory systems, with their wide reach and diverse effects. However, in the last few years, disruption of the HPA system and related cortical areas has also been implicated as playing an important role in depression. Let's take a closer look at the neurobiology of mood disorders.

The Monoamine Hypothesis. The first real indication that depression might result from a problem with the central diffuse modulatory systems came in the 1960s. A drug called *reserpine*, introduced to control high blood pressure, caused severe depression in about 20% of cases. Reserpine depletes central catecholamines and serotonin by interfering with their loading into synaptic vesicles. Another class of drugs that were introduced to treat tuberculosis caused a marked mood elevation. These drugs inhibit

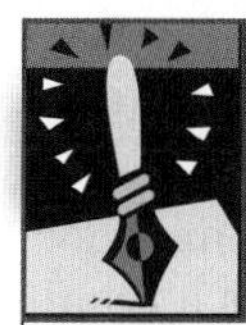

BOX 22.2 OF SPECIAL INTEREST

A Magical Orange Grove in a Nightmare

Winston Churchill called it his "black dog."[1] The writer F. Scott Fitzgerald often found himself "hating the night when I couldn't sleep and hating the day because it went toward night."[2] It was the most "terrible of all the evils of existence" for the composer Hector Berlioz.[3] They were speaking of their lifelong bouts with depression. From the Scottish poet Robert Burns to the American grunge rocker Kurt Cobain, extraordinarily creative people have suffered inordinately from affective disorders. Biographical studies of accomplished artists have been consistent and alarming; their estimated rates of major depression are about 10 times higher than in the general population, and their rates of bipolar disorder may be up to 30 times higher.

Many artists have eloquently described their misfortunes. But can mood disorders actually reinforce great talent and creative productivity? Certainly, most people with mood disorders are not artistic or unusually imaginative, and most artists are not manic-depressive. However, artists with bipolar disorders can sometimes draw vigor and inspiration from their condition. Edgar Allan Poe wrote of his cycles of depression and mania, "I am excessively slothful, and wonderfully industrious—by fits."[4] The poet Michael Drayton mused about "that fine madness . . . which rightly should possess a poet's brain."[5] Studies have suggested that hypomania can heighten certain cognitive processes, increase original and idiosyncratic thought, and even enhance linguistic skills. Manic states can also reduce the need for sleep, foster intense and obsessive concentration, create unmitigated self-confidence, and eliminate concern for social norms—just what you need, perhaps, to push the envelope of artistic creativity.

The poet's madness is much more often a scourge than an inspiration. For Robert Lowell, manic experiences were "a magical orange grove in a nightmare."[6] Virginia Woolf's husband described how "she talked almost without stopping for two or three days, paying no attention to anyone in the room or anything said to her."[7] It is hard to overstate the depths of melancholy that can accompany major depression. The suicide rate among accomplished poets is said to be 5 to 18 times higher than in the general population. Poet John Keats once wrote desperately, "I am in that temper that if I were under water I would scarcely kick to come to the top."[8] But when Keats's mood pitched the other way, he wrote most of his best poetry during a 9-month period in 1819, before dying of tuberculosis at age 25. Figure A shows how Robert Schumann's wildly fluctuating output of musical compositions coincided with the oscillations of his manic-depressive episodes.

The psychiatrist Kay Redfield Jamison has suggested that "depression is a view of the world through a dark glass, and mania is that seen through a kaleidoscope—often brilliant but fractured."[9] Today, we are lucky to have effective treatments for both conditions, for the dark glass and the kaleidoscope carry a heavy price.

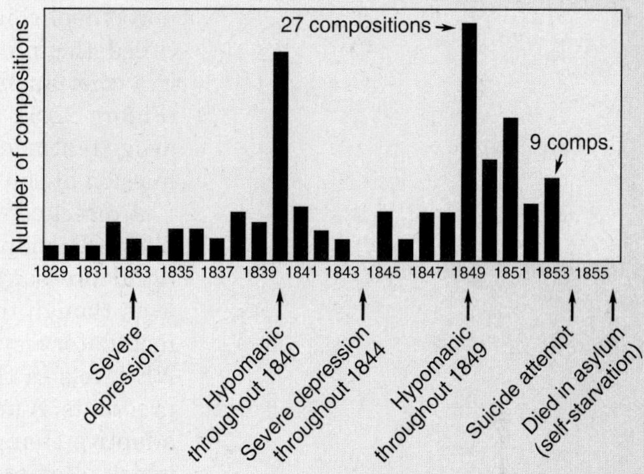

Figure A
Schumann's output of musical composition. (Source: Adapted from Slater and Meyer, 1959).

[1] Quoted in Ludwig AM. 1995. *The Price of Greatness: Resolving the Creativity and Madness Controversy*. New York: Guilford Press, p. 174.

[2] F. Scott Fitzgerald. 1956. The Crack-Up. In *The Crack-Up and Other Stories*. New York: New Directions, pp. 69–75.

[3] Hector Berlioz. 1970. *The Memoirs of Hector Berlioz,* trans. David Cairns. St. Albans, England: Granada, p. 142.

[4] Edgar Allan Poe. 1948. Letter to James Russell Lowell, June 2, 1844. In *The Letters of Edgar Allan Poe,* Vol. 1, ed. John Wand Ostrom. Cambridge, MA: Harvard University Press, p. 256.

[5] Michael Drayton. 1753. "To my dearly beloved Friend, Henry Reynolds, Esq.; of Poets and Poesy," lines 109–110, *The Works of Michael Drayton, Esq.*, vol. 4, London: W. Reeve.

[6] Ian Hamilton. 1982. *Robert Lowell: A Biography*. New York: Random House, p. 228.

[7] Leonard Woolf. 1964. *Beginning Again: An Autobiography of the Years 1911 to 1918*. New York: Harcourt Brace, pp. 172–173.

[8] Quoted by Kay Jamison in a presentation at the Depression and Related Affective Disorders Association/Johns Hopkins Symposium, Baltimore, Maryland, April 1997.

[9] Jamison KR. Manic-depressive illness and creativity. *Scientific American* 272: 62–67.

monoamine oxidase (MAO), the enzyme that destroys catecholamines and serotonin. Another piece of the puzzle fell into place when neuroscientists recognized that the drug imipramine, introduced some years earlier as an antidepressant, inhibits the reuptake of released serotonin and norepinephrine, thus promoting their action in the synaptic cleft. As a result of these observations, researchers developed the hypothesis that mood is closely tied to the levels of released "monoamine" neurotransmitters—norepinephrine and/or serotonin—in the brain. According to this idea, called the **monoamine hypothesis of mood disorders**, depression is a consequence of a deficit in one of these diffuse modulatory systems (Figure 22.9). Indeed, as we will see in a moment, many of the modern drug treatments for depression have in common enhanced neurotransmission at central serotonergic and/or noradrenergic synapses.

A direct correlation between mood and modulator, however, is too simplistic. Perhaps the most striking problem is the clinical finding that the antidepressant action of all of these drugs takes several weeks to develop, even though they have almost immediate effects on transmission at the modulatory synapses. Another concern is that other drugs that raise NE levels in the synaptic cleft, like cocaine, are not effective as antidepressants. A new hypothesis is that the effective drugs promote long-term adaptive changes in the brain, involving alterations in gene expression, which alleviate the depression. One adaptation occurs in the HPA axis, which, as we'll discuss next, has also been implicated in mood disorders.

The Diathesis–Stress Hypothesis. Evidence clearly indicates that mood disorders run in families and that our genes predispose us to this type of mental illness. The medical term for a predisposition for a certain disease is *diathesis*. However, researchers have also established that early childhood abuse or neglect and other stresses of life are important risk factors in the development of mood disorders in adults. According to the **diathesis–stress hypothesis of affective disorders**, the HPA axis is the main site where genetic and environmental influences converge to cause mood disorders.

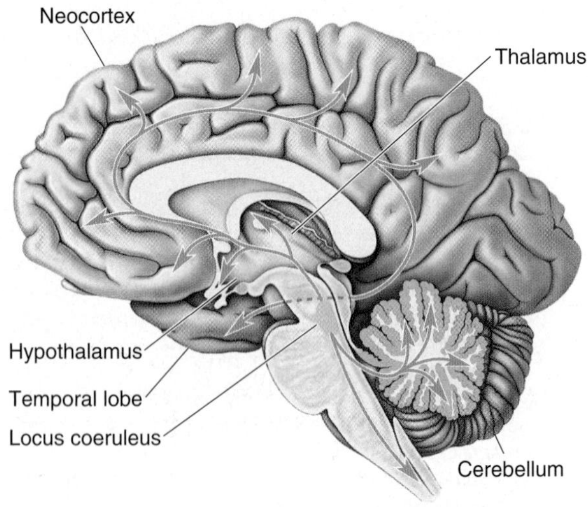

▲ FIGURE 22.9
The diffuse modulatory systems implicated in affective disorders. The norepinephrine and serotonin systems, introduced in Chapter 15, are characterized by the broad reach of their axonal projections.

As we have seen, exaggerated activity of the HPA system is associated with anxiety disorders. However, anxiety and depression often coexist (in fact, this "comorbidity" is the rule rather than the exception). Indeed, one of the most robust findings in all of biological psychiatry is hyperactivity of the HPA axis in severely depressed patients: Blood cortisol levels are elevated, as is the concentration of CRH in the cerebrospinal fluid. Could this hyperactive HPA system, and the resulting deleterious effects on brain function, be the *cause* of depression? Animal studies are highly suggestive. Injected CRH into the brains of animals produces behavioral effects that are similar to those of major depression: insomnia, decreased appetite, decreased interest in sex, and, of course, an increased behavioral expression of anxiety.

Recall that the activation of the hippocampal glucocorticoid receptors by cortisol normally leads to feedback inhibition of the HPA axis (see Figure 22.6). In depressed patients, this feedback is disrupted, explaining why HPA function is hyperactive. A molecular basis for the diminished hippocampal response to cortisol is a decreased number of glucocorticoid receptors. What regulates glucocorticoid receptor number? In a fascinating parallel with the factors implicated in mood disorders, the answer is genes, monoamines, and early childhood experience.

Glucocorticoid receptors, like all proteins, are a product of gene expression. In rats, it has been shown that the amount of glucocorticoid receptor gene expression is regulated by early sensory experience. Rats that received a lot of maternal care as pups express more glucocorticoid receptors in their hippocampus, less CRH in their hypothalamus, and reduced anxiety as adults. The maternal influence can be replaced by increasing the tactile stimulation of the pups. Tactile stimulation activates the ascending serotonergic inputs to the hippocampus, and the serotonin triggers a long-lasting increase in the expression of the glucocorticoid receptor gene. More glucocorticoid receptors equip the animal to respond to stressors as adults. However, the beneficial effect of experience is restricted to a critical period of early postnatal life; stimulation of the rats as adults does not have the same effect. Childhood abuse and neglect, in addition to genetic factors, are known to put people at risk for developing mood and anxiety disorders, and these animal findings suggest one cause. Elevations in brain CRH, and decreased feedback inhibition of the HPA system, may make the brain especially vulnerable to depression.

Anterior Cingulate Cortex Dysfunction. Functional brain imaging studies have consistently found increased resting-state metabolic activity in the **anterior cingulate cortex** of depressed patients (Figure 22.10). This region of the brain is considered to be a "node" in an extensive network of interconnected structures that include other regions of the frontal cortex, hippocampus, amygdala, hypothalamus, and brain stem. The hypothesis that anterior cingulate cortex dysfunction contributes to the symptoms of major depression is supported by a number of findings, including studies that have shown that activity here is increased by autobiographical recall of a sad event and is decreased following successful medical treatment for depression. Based on these findings, the anterior cingulate cortex is considered to be an important link between an internally generated emotional state and the HPA.

Treatments for Affective Disorders

Mood disorders are very common, and the burden they impose on human health, happiness, and productivity is enormous. Fortunately, a number of helpful treatments are available.

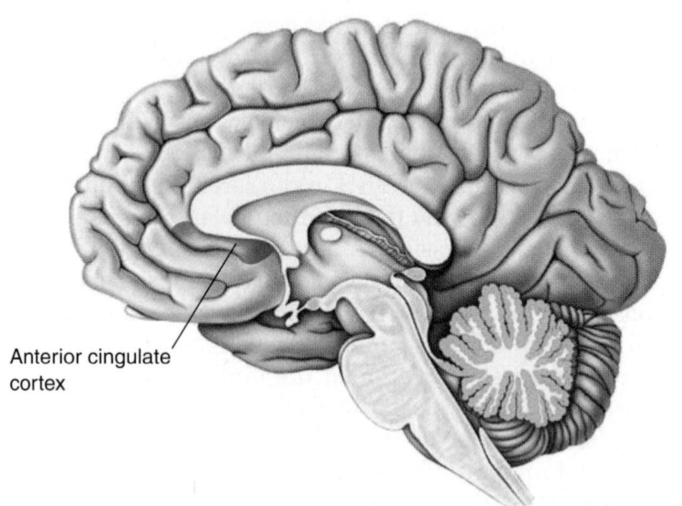

▶FIGURE 22.10
The anterior cingulate cortex. Activity in this region assessed by PET or fMRI imaging is increased in patients suffering from major depression and reduced by successful treatments.

Electroconvulsive Therapy. It might surprise you to learn that one of the most effective treatments for depression and mania involves inducing seizure activity in the temporal lobes. In **electroconvulsive therapy (ECT)**, electrical currents are passed between two electrodes placed on the scalp. Localized electrical stimulation triggers seizure discharges in the brain, but the patient is given anesthesia and muscle relaxants to prevent violent movements during treatment. An advantage of ECT is that relief can occur quickly, sometimes after the first treatment session. This attribute of ECT is especially important in cases where suicide risk is high. An adverse effect of ECT, however, is memory loss. As we will see in Chapter 24, temporal lobe structures (including the hippocampus) play a vital role in memory. ECT usually disrupts memories for events that occurred before treatment, and this can extend back as far as 6 months. In addition, ECT can temporarily impair the storage of new information.

The mechanism by which ECT relieves depression is unknown. However, as mentioned earlier, one temporal lobe structure affected by ECT is the hippocampus, which we have seen is involved in regulating CRH and the HPA axis.

Psychotherapy. Psychotherapy can be effective in treating mild to moderate cases of depression. The main goal of psychotherapy is to help depressed patients overcome negative views of themselves and their future. The neurobiological basis of the treatment has not been established, although we can infer that it relates to establishing cognitive, neocortical control over the activity patterns in disturbed circuits.

Antidepressants. A number of highly effective pharmacological treatments are available for mood disorders. **Antidepressant drugs** include (1) tricyclic compounds (named for their chemical structure), such as imipramine, which among other actions, block the reuptake of both norepinephrine and serotonin by transporters; (2) SSRIs, such as fluoxetine, which act only on serotonin terminals; (3) NE- and 5-HT-selective reuptake inhibitors, such as venlafaxine; and (4) MAO inhibitors, such as phenelzine, which reduce the enzymatic degradation of serotonin and norepinephrine (Figure 22.11). All of these drugs elevate the levels of monoamine neurotransmitters in the brain; however, as mentioned, their therapeutic actions take weeks to develop.

The adaptive response in the brain for the clinical effectiveness of these drugs has not been established with certainty. Nonetheless, an intriguing

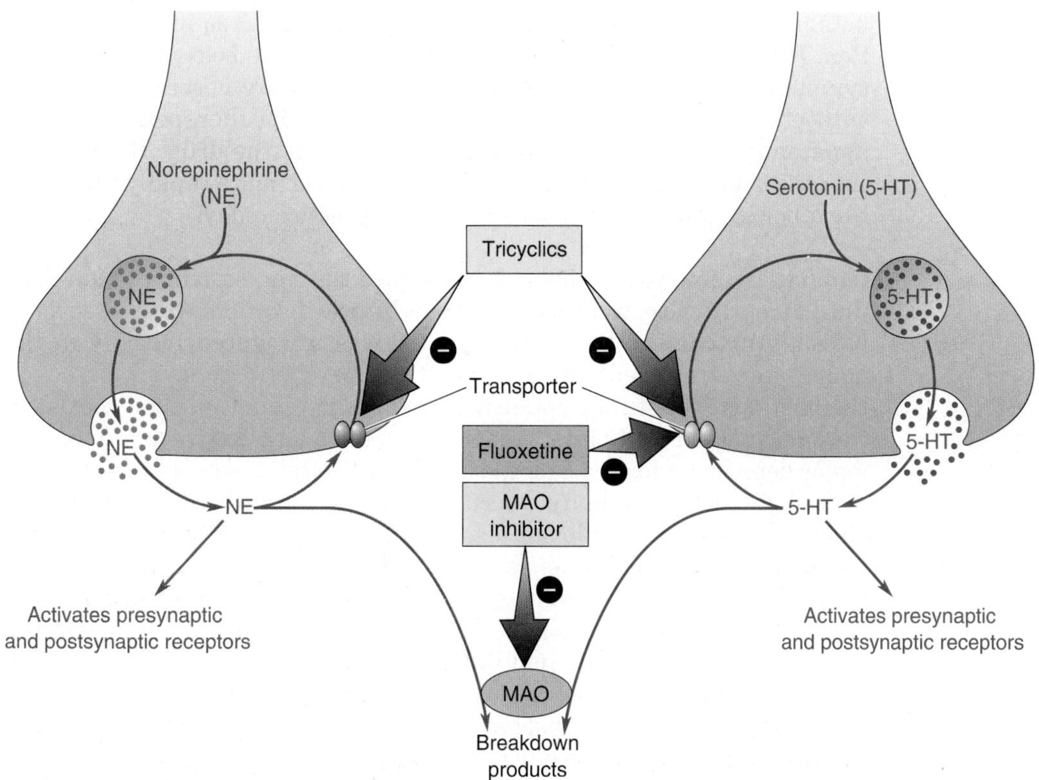

▲ FIGURE 22.11
Antidepressant drugs and the biochemical life cycles of norepinephrine and serotonin. MAO inhibitors, tricyclics, and SSRIs are used as antidepressants. MAO inhibitors enhance the actions of NE and 5-HT by preventing their enzymatic destruction. Tricyclics enhance NE and 5-HT action by blocking uptake. SSRIs act the same way but are selective for serotonin.

finding is that clinically effective treatment with antidepressants dampens the hyperactivity of the HPA system and the anterior cingulate cortex in humans. Animal studies suggest that this effect may be due, in part, to increased glucocorticoid receptor expression in the hippocampus, which occurs in response to a long-term elevation in serotonin. Recall that CRH plays a crucial role in the stress response of the HPA axis. New drugs that act as CRH receptor antagonists are currently under development and testing. Recent research has also shown that prolonged treatment with SSRIs increases *neurogenesis*, the proliferation of new neurons, in the hippocampus. (Neurogenesis is discussed further in Chapter 23.) Remarkably, this proliferation may be important for the beneficial behavioral effects of SSRIs, presumably in part by boosting the control of the HPA by the hippocampus.

The long delay between treatment onset and antidepressant effect presents not only a scientific mystery but also a clinical challenge. Patients can feel discouraged when their expectations for improvement are not met, and this can temporarily exacerbate the depression. This is a serious limitation, particularly in cases where there is a high risk of suicide. Thus, the search is on for rapidly acting antidepressant medications that do not require weeks of treatment to become effective. The hope for attaining this goal has been fueled by recent findings that a single intravenous dose of the anesthetic drug *ketamine* can rapidly alleviate symptoms of depression for several days. Although these findings support the concept of a rapidly acting antidepressant, ketamine itself is not a practical treatment for depression. As we will discuss later in the context of schizophrenia, ketamine can cause severe psychotic episodes that require hospitalization. It is only after the drug is eliminated from the body, and psychotic symptoms diminish, that antidepressant effects are observed. Thus, as with other medical treatments for depression, the therapeutic effect is apparently caused by some adaptive response to the drug. However, in the case of ketamine, this adaptation occurs much more rapidly than with the other antidepressants used in clinical practice today.

Lithium. By now, you probably have formed the (correct) impression that, until recently, most treatments for psychiatric disorders were discovered virtually by chance. For example, ECT was introduced initially in the 1930s as a treatment of last resort for psychotic behavior based on the mistaken belief that epilepsy and schizophrenia could not coexist in the same person. Only later was it shown to be an effective treatment for major depression for reasons that still remain unknown.

"Enlightened serendipity" was again at work in the discovery of a highly effective treatment for bipolar disorder. Working in the 1940s, Australian psychiatrist John Cade was searching for psychoactive substances in the urine of manic patients. He injected guinea pigs with urine or urinary constituents and observed their behavioral effects. Cade wanted to test the effect of uric acid, but he had difficulty getting it into solution. Instead, he used lithium urate because it dissolved easily and was readily available in the pharmacy. He observed, quite unexpectedly, that this treatment calmed the guinea pigs (he had predicted the opposite effect). Because other lithium salts also produced this behavioral effect, he concluded that it was the lithium, not a constituent of urine, that was responsible. He went on to test lithium treatment on patients with mania, and, amazingly, it worked. Subsequent studies showed that lithium is highly effective in stabilizing the mood of patients with bipolar disorder, by preventing not only the recurrence of mania but also the episodes of depression (Figure 22.12).

Lithium affects neurons in many ways. In solution, it is a monovalent cation that passes freely through neuronal sodium channels. Inside the neuron, lithium prevents the normal turnover of phosphatidyl inositol (PIP_2), a precursor for important second messenger molecules that are generated in response to activation of some G-protein-coupled neurotransmitter receptors (see Chapter 6). Lithium also interferes with the actions of adenylyl cyclase, essential for the generation of the second messenger cyclic-AMP, and glycogen synthase kinase, a critical enzyme in cellular energy metabolism. Why lithium is such an effective treatment for

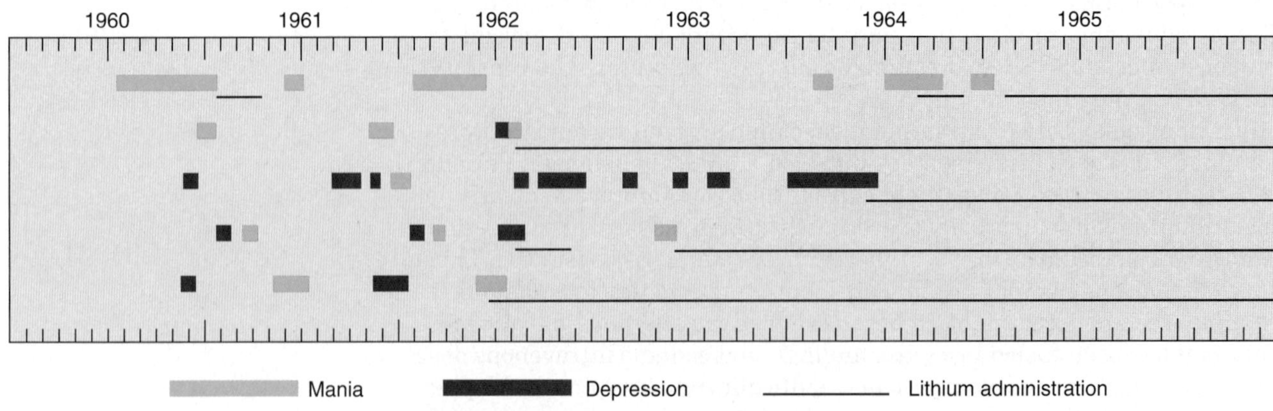

▲ FIGURE 22.12
The mood-stabilizing effect of lithium treatment in five patients. (Source: Adapted from Barondes, 1993, p. 139.)

bipolar disorder, however, remains completely unknown. Like other antidepressants, the therapeutic effects of lithium require long-term use. The answer, again, appears to lie in an adaptive change in the central nervous system (CNS), but the nature of this change remains to be determined.

Deep Brain Stimulation. In a substantial fraction of patients, severe depression fails to respond to ECT, medicine, or talk therapy. In such cases, more drastic measures are called for, and one entails undergoing a surgical procedure in which an electrode is implanted deep in the brain. This approach to treat depression was pioneered by Helen Mayberg, a neurologist at Emory University (Box 22.3). Recall that activity in the anterior cingulate cortex is increased by sadness and decreased by successful treatment with standard antidepressant medications. The observation that activity in this region fails to decrease in patients with unrelenting, treatment-resistant depression inspired Mayberg to contemplate using direct brain stimulation to modulate activity here. Although it seems counterintuitive, electrical stimulation can actually *decrease* activity in brain circuits that are chronically overactive (the reasons remain unclear but likely include recruitment of inhibitory neurons). Indeed, Mayberg and a team of neurosurgeons at the University of Toronto found that electrical stimulation of a circumscribed region of the anterior cingulate cortex, comprising Brodmann's area 25, could produce immediate relief from depression.

Recall that during most neurosurgical procedures, the patient remains awake, which is possible because there are no pain receptors in the brain. Thus, the patients in Mayberg's study could report the effect of stimulation during the operation. They described a "sudden calmness" or "lightness" and "disappearance of the void" when the stimulator was turned on. These patients were discharged from the hospital with the implanted electrodes connected to a battery-operated stimulator that continuously applied electric pulses. The majority of patients experienced continued relief from their depression.

These findings have generated considerable excitement in the field but are still considered to be preliminary. Additional studies are underway to confirm these initial results. Obviously, brain surgery is always considered a treatment of last resort.

SCHIZOPHRENIA

Although their severity might be hard to fully comprehend, we all have some idea of what mood and anxiety disorders are like because they are extremes in the spectrum of brain states that are part of normal experience. The same cannot be said for schizophrenia. This severe mental disorder distorts thoughts and perceptions in ways that healthy people find difficult to understand. Schizophrenia is a major public health problem, affecting 1% of the adult population. Over 2 million people suffer from the disorder in the United States alone.

A Description of Schizophrenia

Schizophrenia is characterized by a loss of contact with reality and a disruption of thought, perception, mood, and movement. The disorder typically becomes apparent during adolescence or early adulthood and usually persists for life. The name, introduced in 1911 by Swiss psychiatrist Eugen Bleuler, roughly means "divided mind," because of his observation that many patients seemed to oscillate between normal and abnormal states.

BOX 22.3 PATH OF DISCOVERY

Tuning Depression Circuits

by Helen Mayberg

It was never my plan to study depression. I am a neurologist, and depression was generally considered to be beyond the purview of my medical discipline. While many patients with neurological disorders develop depression, it was often seen as a nonspecific response to a distressing diagnosis (stroke, Parkinson's disease, Alzheimer's disease, and the like). Furthermore, the notion that a global change like depression could be localized to specific brain regions, the way a language deficit might be traced to disruption of specific parts of the frontal or temporal lobes, was not intuitive. For the most part, strategies to study and treat depression in neurological patients mirrored those in patients with depression without identified neurological disease—focusing on brain chemistry—that is, until the early 1990s when advances in neuroimaging changed the playing field.

By 2001, we had learned a lot about the functional neuroanatomy of depression. Using positron emission tomography and functional magnetic resonance imaging, we had identified activity patterns that subdivided depressed patients by their symptom clusters. We also studied changes that distinguished the response to antidepressant drugs from that of psychotherapy and identified baseline patterns that might guide treatment selection for each treatment. A brain wiring diagram of depression was emerging.

It was around this time that we had the opportunity to directly examine the role of the subcallosal cingulate region (Brodmann's area 25) in our evolving depression circuit (Figure A). We had converging evidence of common changes in this region across a wide variety of effective antidepressant treatments. We also knew that failure to effect changes in this region were associated with treatment nonresponse. We hypothesized that relief from major depression could be achieved with focal brain stimulation using a well-established neurosurgical technique used to treat Parkinson's disease—deep brain stimulation (DBS). Inserting electrodes into our intended target, the subcallosal cingulate white matter, was not technically more difficult or of higher risk than the basal ganglia implantations used for Parkinson's disease, or so said the surgeon. We became convinced it should be attempted, but what patient would be appropriate for such a procedure?

Treatment-resistant depression is a dire condition defined by failure to respond to multiple available antidepressant treatments including electroconvulsive therapy. What I had not appreciated in my years of studying depression was the banality of our definitions and rating scales, as they failed to capture the degree of suffering experienced by patients in what can only be described as a malignant condition, a pervasive state of sustained mental pain and physical immobility with no "off switch."

I can still remember that first case on the morning of May 23, 2003. We were prepared technically: where to implant, what side effects to watch for. But otherwise, we had no expectations. How could there be when what you are doing is something that has never been done before? Our patient was awake (DBS electrodes are implanted using local anesthesia), and it was easy enough to monitor the obvious—discomfort, pain, general distress. The primary goal was to get the electrodes implanted and then turn them on and make sure nothing bad happened. Our mindset going in was that the real work would come later as we tested various stimulation parameters to achieve clinical effects—a process we thought would take weeks, like other antidepressant treatments.

The plan was to observe, keep the patient safe, and if something didn't seem right, turn it off. So we weren't expecting it when the patient's mood abruptly lifted during testing of the second contact on the left electrode. As the current

However, there are many variations in the manifestations of schizophrenia, including those that show a steadily deteriorating course. Indeed, it is still not clear whether what is called schizophrenia is a single disease or several.

The symptoms of schizophrenia fall into two categories: positive and negative. *Positive symptoms* reflect the presence of abnormal thoughts and behaviors, such as:

- Delusions
- Hallucinations
- Disorganized speech
- Grossly disorganized or catatonic behavior

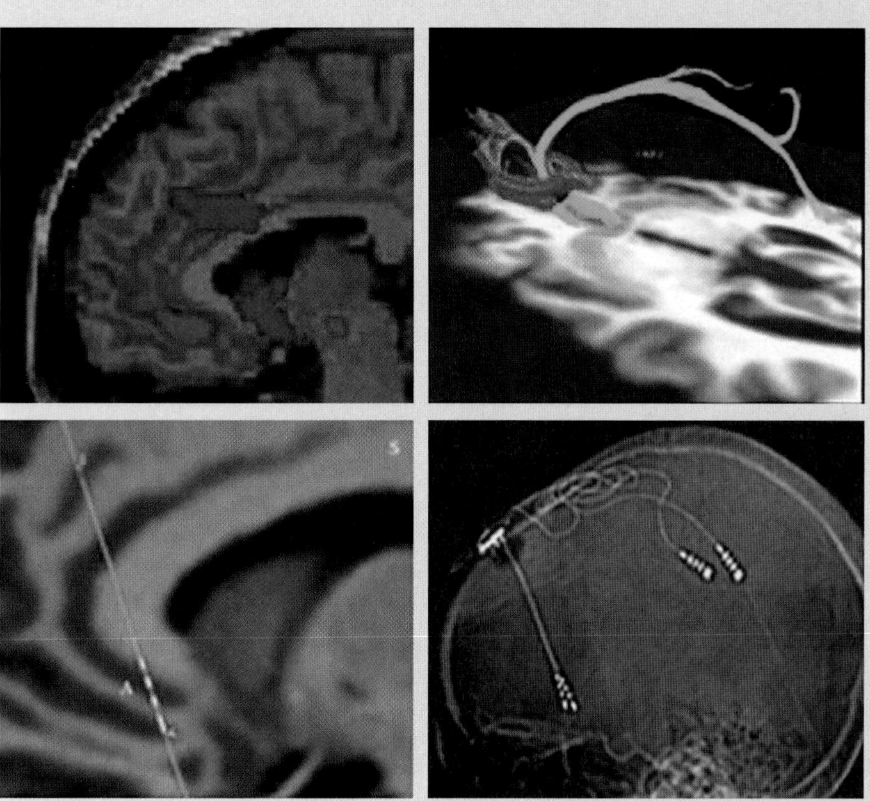

Figure A
Abnormal activity in the anterior cingulate cortex and the use of DBS to correct it. Upper left: PET scan of a depressed patient demonstrating increased blood flow—indicating hyperactivity—in the subcallosal cingulate cortex (red). DBS quiets this region. Upper right: Diffusion-weighted MRI scan used prior to surgery to identify the intersection of three white matter bundles passing through the subcallosal cingulate region thus defining the optimal location to implant the DBS electrode. Lower left: Structural MRI scan used in the operating room to plan and verify the targeted location of the implanted DBS electrode. Lower right: Postsurgical skull X-ray showing the actual implanted DBS electrodes. (Source: Courtesy of Dr. Helen Mayberg.)

was turned up, the patient suddenly asked if we had done something different. She felt calm, with a lightness and serenity she hadn't felt in a long time. I was looking right at her on the nonsterile side of the surgical table. Her eyes were wider, looking around; her speech was noticeably louder and less halting; and she was more engaged with the room and with me. It was as if we had hit a spot and literally turned her "negative" feeling off, releasing the rest of her brain to go about doing whatever it wanted to do. And then we turned down the current back to zero; the relief faded and the void returned. That moment changed everything I knew about depression and how to study it.

Negative symptoms reflect the absence of responses that are normally present. These symptoms include:

- Reduced expression of emotion
- Poverty of speech
- Difficulty in initiating goal-directed behavior
- Memory impairment

Individuals affected by schizophrenia often have delusions organized around a theme; for example, they may believe that powerful adversaries are out to get them. These are often accompanied by auditory hallucinations (such as hearing imaginary voices) related to the same delusional theme. There can also be a lack of emotional expression (called a "flat affect"), coupled with disorganized behavior and incoherent speech. Speech may be accompanied by silliness and laughter that appear to have no relation to what is being said. In some cases, schizophrenia is accompanied by peculiarities of voluntary movement, such as immobility and stupor (catatonia), bizarre posturing and grimacing, and senseless, parrot-like repetition of words or phrases.

Biological Bases of Schizophrenia

Understanding the neurobiological basis for schizophrenia represents one of the greatest challenges of neuroscience because the disorder affects many of the characteristics that make us human: thought, perception, self-awareness. Although considerable progress has been made, we still have much more to learn.

Genes and the Environment. Schizophrenia runs in families. As shown in Figure 22.13, the likelihood of having the disorder varies in relation to the number of genes that are shared with an affected family member. If your identical twin has schizophrenia, the probability is about 50% that you will also have it. The chances you will have the disease decline as the number of genes you share with an affected family member decreases. These findings argue that schizophrenia is primarily a genetic disorder. Recently, researchers have identified several specific genes that seem to increase susceptibility to schizophrenia. Nearly all of these genes have important roles in synaptic transmission, its plasticity, or the growth of synapses.

Remember, however, that identical twins have exactly the same genes. So why, in 50% of cases, is one sibling spared when the other has

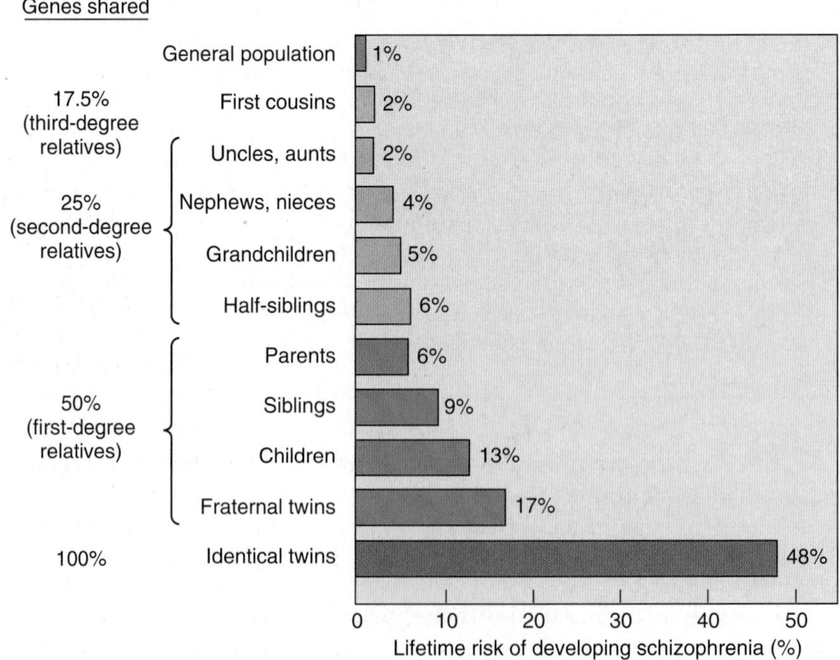

▲ FIGURE 22.13

The familial nature of schizophrenia. The risk of developing schizophrenia increases with the number of shared genes, suggesting a genetic basis for the disease. (Source: Adapted from Gottesman, 1991, p. 96.)

schizophrenia? The answer must lie in the environment. In other words, faulty genes seem to make some people vulnerable to environmental factors that cause schizophrenia. Although the symptoms may not appear until a person reaches his or her twenties, considerable evidence indicates that the biological changes causing the condition begin early in development, perhaps prenatally. Viral infections during fetal and infant development have been implicated as contributing causes, as has poor maternal nutrition. In addition, environmental stresses throughout life are known to exacerbate the course of the disorder. A number of studies have suggested that use of marijuana increases the risk for developing schizophrenia in genetically vulnerable adolescents.

Schizophrenia is associated with physical changes in the brain. An interesting example appears in Figure 22.14. The figure shows brain scans of identical twins, one with schizophrenia and one without. Normally, the structures of the brains of identical twins are nearly identical. However, in this case, the brain of the schizophrenic sibling shows enlarged lateral ventricles, which reflects the shrinkage of brain tissue around them. This difference is consistent when large numbers of people are sampled; the brains of schizophrenics have, on average, a significantly larger ventricle-to-brain-size ratio than people who do not have the disorder.

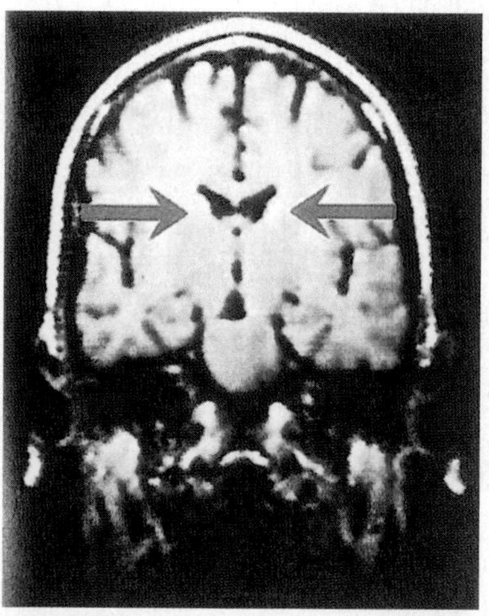

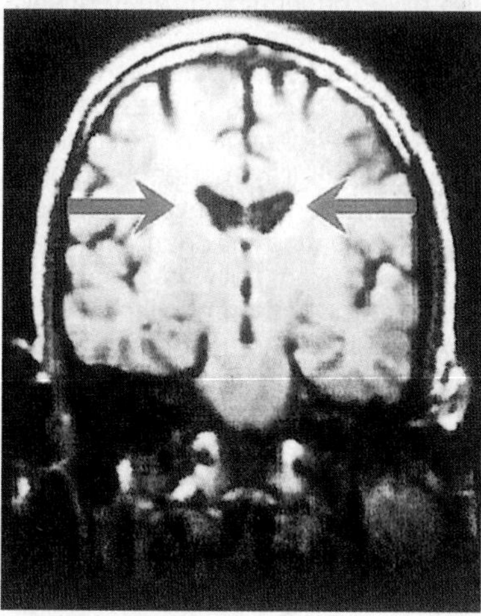

▲ FIGURE 22.14
Enlarged lateral ventricles in schizophrenia. These MRI scans are from the brains of identical twins. The sibling on the top was normal; the one on the bottom was diagnosed with schizophrenia. Notice the enlarged lateral ventricles in the schizophrenic sibling, indicating a loss of brain tissue. (Source: Barondes, 1993, p. 153.)

Such pronounced structural changes are not always apparent in the brains of schizophrenics, however. Important physical changes in their brains also occur in the microscopic structure and function of cortical connections. For example, schizophrenics often have defects in the myelin sheaths surrounding axons in their cerebral cortex, although it is not clear whether this is a cause or consequence of the disease. Another common finding in schizophrenia is reduced cortical thickness and abnormal neuronal lamination (Figure 22.15). Changes in synapses and several neurotransmitter systems have also been implicated. As we'll see next, particular attention has focused on alterations in chemical synaptic transmission mediated by dopamine and glutamate.

The Dopamine Hypothesis. Recall that dopamine is the neurotransmitter used by another of the diffuse modulatory systems (Figure 22.16). A link between the *mesocorticolimbic dopamine system* and schizophrenia has been made on the basis of two main observations. The first relates to the effects of amphetamine in otherwise healthy people. Remember from our discussion in Chapter 15 that amphetamine enhances neurotransmission at catecholamine-utilizing synapses and causes the release of dopamine. Amphetamine's normal stimulant action bears little resemblance to schizophrenia. However, because of its addictive properties, users of amphetamines often risk taking more and more to satisfy their cravings. The resulting overdose can lead to a psychotic episode with positive symptoms that are virtually indistinguishable from those of schizophrenia. This suggests that psychosis is somehow related to too much catecholamine in the brain.

A second reason to associate dopamine with schizophrenia relates to the CNS effects of drugs that are effective in reducing the positive symptoms of the disorder. In the 1950s, researchers discovered that the drug *chlorpromazine*, initially developed as an antihistamine, could prevent the positive symptoms in schizophrenia. Chlorpromazine and other related antipsychotic drugs, collectively called **neuroleptic drugs**, were later found to be potent blockers of dopamine receptors, specifically the D_2 receptor. When a large number of neuroleptics are examined, the correlation between the dosage effective for controlling schizophrenia and their

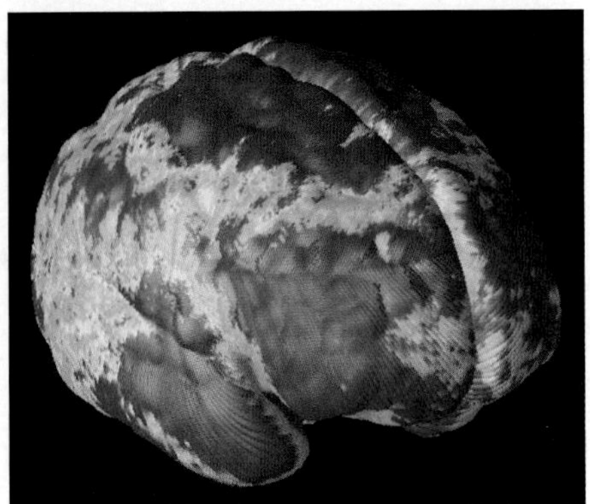

▲ FIGURE 22.15
Loss of cortical gray matter in schizophrenics during adolescence.. The brains of 12 patients with early-onset schizophrenia were imaged repeatedly over the course of 5 years, between the ages of 13 to 18. This image shows the average annual change in the thickness of their cortical gray matter, with red colors indicating regions of greatest loss and blue indicating no change. Severe loss (up to 5% annually) is observed in parietal, motor and anterior temporal cortex. (Source: Thompson et al., 2001, Figure 1, with permission.)

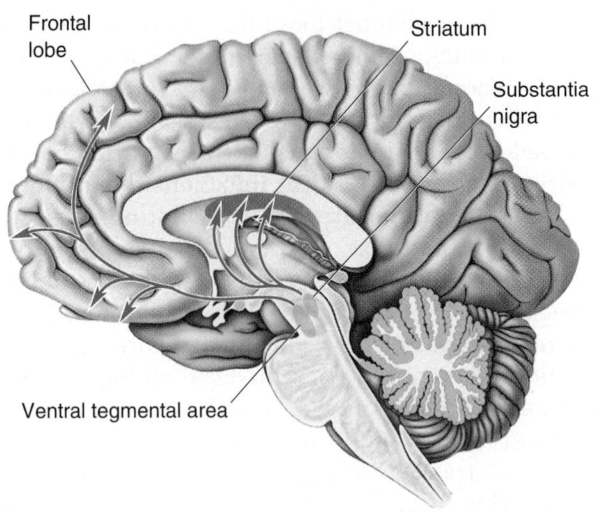

▲ FIGURE 22.16
The dopaminergic diffuse modulatory systems of the brain. The mesocorticolimbic dopamine system, which arises in the ventral tegmental area, has been implicated in the cause of schizophrenia. A second dopaminergic system arising from the substantia nigra is involved in the control of voluntary movement by the striatum.

ability to bind to D_2 receptors is impressive (Figure 22.17). Indeed, these same drugs are effective in the treatment of amphetamine and cocaine psychoses. According to the **dopamine hypothesis of schizophrenia**, psychotic episodes in schizophrenia are triggered specifically by the activation of dopamine receptors.

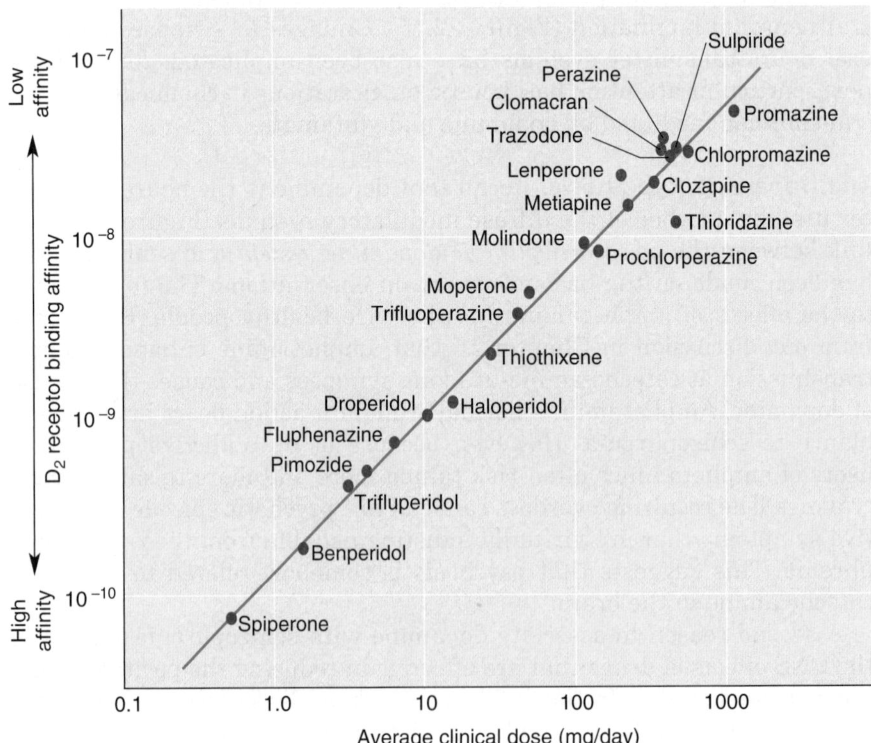

▲ FIGURE 22.17
Neuroleptics and D_2 receptors. The neuroleptic dosages effective in controlling schizophrenia correlate well with the binding affinities of the drugs for D_2 receptors. The units on the Y axis are the molar concentrations of drug that inhibit half of the D_2 receptors in the brain. Higher affinity drugs block the receptors at lower concentrations. (Source: Adapted from Seeman, 1980.)

Despite the tantalizing link between the positive symptoms of schizophrenia and dopamine, there seems to be more to the disorder than an overactive dopamine system. One indication is that newly developed antipsychotic drugs, like *clozapine*, have little effect on D_2 receptors. These drugs are called *atypical neuroleptics*, indicating that they act in a novel way. The mechanism by which these compounds exert their neuroleptic effect has not been established with certainty, but an interaction with serotonin receptors is suspected.

The Glutamate Hypothesis. Another indication that there is more to schizophrenia than dopamine comes from the behavioral effects of *phencyclidine (PCP)* and *ketamine*. These drugs were introduced in the 1950s as anesthetics. However, many patients experienced adverse side effects, sometimes lasting for days, which included hallucinations and paranoia. Although it is no longer used clinically, PCP is now a common illegal drug of abuse, known by users as "angel dust" or "hog." Ketamine, which is still used in veterinary medicine, has also made it to the street where it is referred to as "special K" or "vitamin K." PCP and ketamine intoxication cause many of the symptoms of schizophrenia, both positive and negative. However, neither drug has an effect on dopaminergic transmission; they affect synapses that use glutamate as a neurotransmitter.

Recall from Chapter 6 that glutamate is the main fast excitatory neurotransmitter in the brain, and that NMDA receptors are one subtype of glutamate receptor. PCP and ketamine act by inhibiting NMDA receptors (Figure 22.18). Thus, according to the **glutamate hypothesis of schizophrenia**, the disorder reflects diminished activation of NMDA receptors in the brain.

In order to study the neurobiology of schizophrenia, neuroscientists have attempted to establish animal models of the disorder. Low doses of PCP administered chronically to rats produce changes in brain biochemistry and behavior that resemble those in schizophrenic patients. Mice that have been genetically engineered to express fewer NMDA receptors also display some schizophrenia-like behaviors, including repetitive

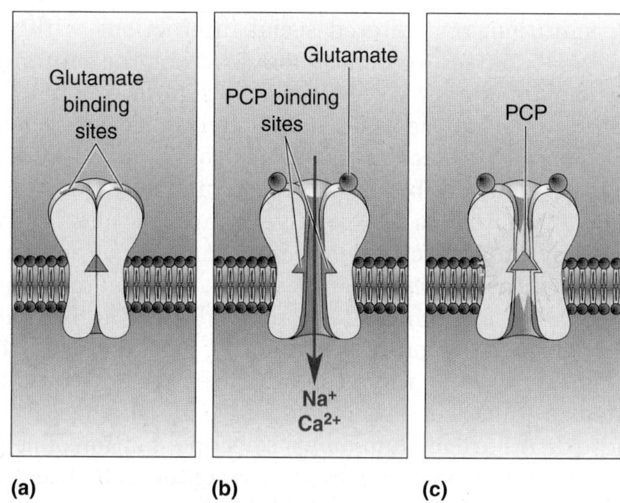

▲ FIGURE 22.18
Blocking of the NMDA receptor by PCP. NMDA receptors are glutamate-gated ion channels. **(a)** In the absence of glutamate, the channel is closed. **(b)** In the presence of glutamate, the channel is open, exposing PCP binding sites. **(c)** The channel is blocked when PCP enters and binds. The blockade of brain NMDA receptors by PCP produces effects on behavior that resemble the symptoms of schizophrenia.

▲ FIGURE 22.19
Social withdrawal in mutant mice with reduced numbers of NMDA receptors. The mice on the left have a normal number of NMDA receptors. The photographs were taken 30 minutes apart over 2 hours to monitor social behavior. These mice tended to nest together. The mice on the right have been genetically altered to express fewer NMDA receptors. Notice that these mice tended to avoid social contact with one another. (Source: Mohn et al., 1999, p. 432.)

movements, agitation, and altered social interactions with other mice (Figure 22.19). Of course, we don't know whether the mutant mice feel paranoid or hear imaginary voices. But it is significant that the observable behavioral abnormalities can be lessened by treating the mice with either conventional or atypical neuroleptic drugs.

While all drugs that inhibit NMDA receptors will impair memory and cognition, not all reproduce the positive symptoms of schizophrenia in humans. The key difference lies in the mechanism of action. PCP and ketamine do not interfere with the binding of glutamate to the receptor like other NMDA receptor inhibitors. Instead, they act by entering the channel and clogging the pore. Consequently, blockade by PCP and ketamine is possible only when the receptors are active and the channels are open. This feature has led researchers to wonder if the psychomimetic effects of these drugs are mediated by a select population of neurons with high ongoing activity and tonic NMDA receptor activation. One such population is comprised of cortical GABAergic neurons in the cerebral cortex. Inhibition of the NMDA receptors on these neurons might distort thinking and the processing of sensory information. Notably, postmortem examination of the brains of individuals with schizophrenia has found the cortex to be deficient in many interneurons.

Treatments for Schizophrenia

The treatment of schizophrenia consists of drug therapy combined with psychosocial support. As mentioned earlier, the conventional neuroleptics, such as chlorpromazine and haloperidol, act at D_2 receptors. These drugs reduce the positive symptoms of schizophrenia in the majority of patients. Unfortunately, the drugs also have numerous side effects related to their actions on the dopaminergic input to the striatum that arises from the substantia nigra (see Chapter 14). Not surprisingly, the effects of blocking dopamine receptors in the striatum resemble the symptoms of Parkinson's disease, including rigidity, tremor, and difficulty initiating movements. Chronic treatment with conventional neuroleptics also can result in the emergence of *tardive dyskinesia*, which is characterized by involuntary movements of the lips and jaw. Many of these side effects are avoided by using atypical neuroleptics, such as clozapine and risperidone, because they do not act directly on the dopamine receptors in the striatum. These medications are also more effective against the negative symptoms of schizophrenia.

The newest focus of drug research is the NMDA receptor. Investigators hope that increasing NMDA receptor responsiveness in the brain, perhaps in combination with decreasing D_2 receptor activation, will further alleviate the symptoms of schizophrenia.

CONCLUDING REMARKS

Neuroscience has had a huge impact on psychiatry. Mental illness is now recognized as the consequence of pathologic modifications of the brain, and psychiatric treatments today are focused on correcting these changes. Just as importantly, neuroscience has changed how society views people who suffer from mental illness. Suspicion about mentally ill people is slowly giving way to compassion. Mental illnesses today are recognized as diseases of the body, just like hypertension or diabetes.

Despite remarkable progress in treating psychiatric disorders, we have a very incomplete understanding of how current treatments work their magic on the brain. In the case of drug therapy, we know with great precision about how chemical synaptic transmission is affected. But we do not know why, in many cases, the therapeutic effect of a drug takes weeks to emerge. Even less is known about how psychosocial treatments act on the brain. In general, the answer seems to lie in adaptive changes that occur in the brain in response to treatment.

We also do not know the causes of most mental disorders. It is clear that our genes either put us at risk or protect us. However, the environment also plays an important role. Environmental stresses before birth may contribute to schizophrenia, and those after birth may precipitate depression. Not all environmental effects are bad, however. Appropriate sensory stimulation, especially in early childhood, can apparently produce adaptive changes that help protect us from developing mental illnesses later in life.

Psychiatric disorders and their treatment illustrate that our brains and behaviors are influenced by past experience, whether it is exposure to inescapable stress or to pharmacologically elevated levels of serotonin. Of course, much more subtle sensory experiences also leave their mark on the brain. In Part IV, we will explore how sensory experiences modify the brain during development and during learning.

KEY TERMS

Mental Illness and the Brain
molecular medicine (p. 754)
pathophysiology (p. 754)
induced pluripotent stem cell (iPSC) (p. 756)

Anxiety Disorders
anxiety disorder (p. 756)
panic disorder (p.757)
obsessive-compulsive disorder (OCD) (p. 757)
hypothalamic-pituitary-adrenal (HPA) axis (p. 759)
adrenocorticotropic hormone (ACTH) (p. 759)
corticotropin-releasing hormone (CRH) (p. 759)
glucocorticoid receptor (p. 761)
anxiolytic drug (p. 761)
benzodiazepine (p. 761)
serotonin-selective reuptake inhibitor (SSRI) (p. 762)

Affective Disorders
affective disorder (p. 763)
major depression (p. 763)
bipolar disorder (p. 764)
mania (p. 764)
monoamine hypothesis of affective disorders (p. 766)
diathesis–stress hypothesis of affective disorders (p. 766)
anterior cingulate cortex (p. 767)
electroconvulsive therapy (ECT) (p. 768)
antidepressant drug (p. 768)
lithium (p. 770)

Schizophrenia
schizophrenia (p. 771)
neuroleptic drug (p. 775)
dopamine hypothesis of schizophrenia (p. 776)
glutamate hypothesis of schizophrenia (p. 777)

REVIEW QUESTIONS

1. How and where in the brain do benzodiazepines act to reduce anxiety?
2. Depression is often accompanied by bulimia nervosa, which is characterized by frequent eating binges followed by purging. Where does the regulation of mood and appetite converge in the brain?
3. Snuggling with your mom as a baby might help you cope with stress better as an adult. Why?
4. What three types of drugs are used to treat depression? What do they have in common?
5. Psychiatrists often refer to the dopamine theory of schizophrenia. Why do they believe dopamine is linked to schizophrenia? Why must we be cautious about accepting a simple correlation between schizophrenia and too much dopamine?

FURTHER READING

American Psychiatric Association. 2013. *Diagnostic and Statistical Manual of Mental Disorders*, 5th ed. Arlington, VA: American Psychiatric Association.

Andreasen NC. 2004. *Brave New Brain: Conquering Mental Illness in the Era of the Genome*. New York: Oxford University Press.

Charney DS, Nestler EJ, eds. 2004. *Neurobiology of Mental Illness*, 2nd ed. New York: Oxford University Press.

Harrison PJ, Weinberger DR. 2005. Schizophrenia genes, gene expression, and neuropathology: on the matter of their convergence. *Molecular Psychiatry* 10:40–68.

Holtzheimer PE, Mayberg HS. 2011. Deep brain stimulation for psychiatric disorders. *Annual Review of Neuroscience* 34:289–307.

Insel TR. 2012. Next generation treatments for psychiatric disorders. *Science Translational Medicine* 4:1–9.

PART FOUR

The Changing Brain

CHAPTER TWENTY-THREE

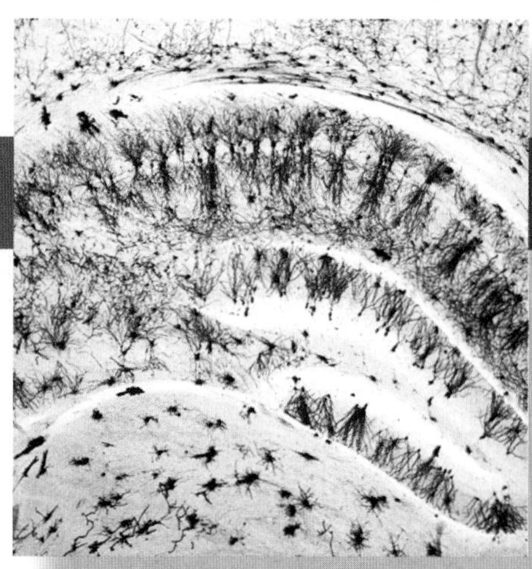

Wiring the Brain

INTRODUCTION

We have seen that most of the operations of the brain depend on remarkably precise interconnections among its 85 billion neurons. As an example, consider the precision in the wiring of the visual system, from retina to lateral geniculate nucleus (LGN) to cortex, shown in Figure 23.1. All retinal ganglion cells extend axons into the optic nerve, but only ganglion cell axons from the nasal retinas cross at the optic chiasm. Axons from the two eyes are mixed in the optic tract, but in the LGN, they are sorted out again (1) by ganglion cell type, (2) by eye of origin (ipsilateral or contralateral), and (3) by retinotopic position. LGN neurons project axons into the optic radiations that travel via the internal capsule to the primary visual (striate) cortex. Here, they terminate (1) only in cortical area 17, (2) only in specific cortical layers (mainly layer IV), and (3) again according to cell type and retinotopic position. Finally, the neurons in layer IV make very specific connections with cells in other cortical layers that are appropriate for binocular vision and are specialized to enable the detection of contrast borders. How did such precise wiring arise?

Back in Chapter 7, we looked at the embryological and fetal development of the nervous system to understand how it changed from a simple tube in the early embryo into the structures we recognize in the adult

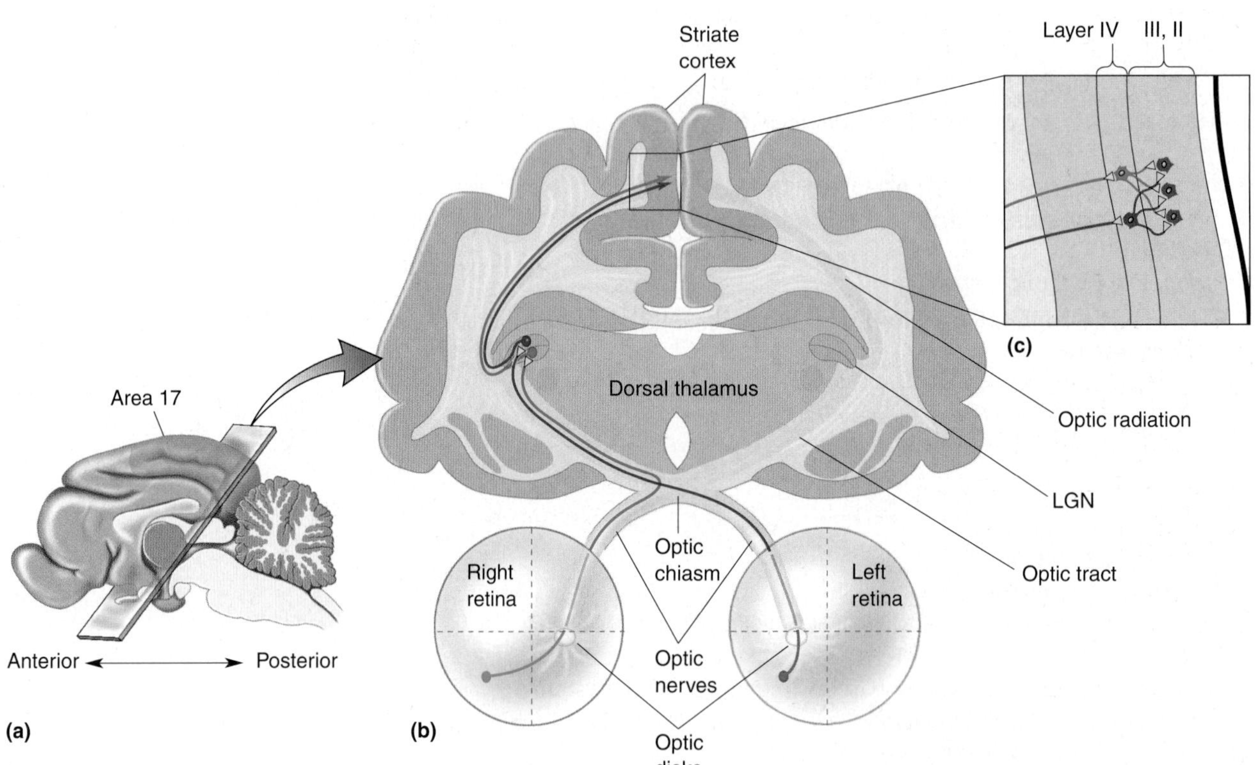

▲ FIGURE 23.1
Components of the mature mammalian retinogeniculocortical pathway. **(a)** A midsagittal view of a cat brain, showing the location of primary visual cortex (striate cortex, area 17). **(b)** Components of the ascending visual pathway. Notice that the right eye's temporal retina and the left eye's nasal retina project axons via the optic nerve and optic tract to the LGN of the right dorsal thalamus. Inputs from the eyes remain segregated in separate layers at the level of this synaptic relay. LGN neurons project to striate cortex via the optic radiations. These axons terminate mainly in layer IV, where inputs serving the eyes continue to be segregated. **(c)** The first site of major convergence of inputs from both eyes is in the projection of layer IV cells onto cells in layer III.

as the brain and spinal cord. Here, we'll take another look at brain development, this time to see how connections are formed and modified as the brain matures. We will discover that most of the wiring in the brain is specified by genetic programs that allow axons to detect the correct pathways and the correct targets. However, a small but important component of the final wiring depends on sensory information about the world around us during early childhood. In this way, "nurture and nature" both contribute to the final structure and function of the nervous system. We will be using the central visual system as an example whenever possible, so you may want to quickly review Chapter 10 before continuing.

THE GENESIS OF NEURONS

The first step in wiring the nervous system together is the generation of neurons. Consider as an example the striate cortex. In the adult, there are six cortical layers, and the neurons in each of these layers have characteristic appearances and connections that distinguish striate cortex from other areas. Neuronal structure develops in three major stages: cell proliferation, cell migration, and cell differentiation.

Cell Proliferation

Recall from Chapter 7 that the brain develops from the walls of the five fluid-filled vesicles. These fluid-filled spaces remain in the adult and constitute the ventricular system. Very early in development, the walls of the vesicles consist of only two layers: the ventricular zone and the marginal zone. The *ventricular zone* lines the inside of each vesicle, and the *marginal zone* faces the overlying pia. Within these layers of the telencephalic vesicle, a cellular ballet is performed that gives rise to all the neurons and glia of the visual cortex. The choreography of cell proliferation is described later, and the five "positions" correspond to the circled numbers in Figure 23.2a:

1. *First position:* A cell in the ventricular zone extends a process that reaches upward toward the pia.
2. *Second position:* The nucleus of the cell migrates upward from the ventricular surface toward the pial surface; the cell's DNA is copied.
3. *Third position:* The nucleus, containing two complete copies of the genetic instructions, settles back to the ventricular surface.
4. *Fourth position:* The cell retracts its arm from the pial surface.
5. *Fifth position:* The cell divides in two.

These dividing cells—the *neural progenitors* that give rise to all the neurons and astrocytes of the cerebral cortex—are called **radial glial cells**. For many years it was believed these cells served only as a temporary scaffold to guide newly formed neurons to their final destinations. We now understand that the radial glial cells also give rise to most of the neurons of the central nervous system.

Early in embryonic development, the radial glial cells number in the hundreds. To give rise to the billions of neurons in the adult brain, these multipotent stem cells—meaning they can assume several different destinies—divide to expand the population of neural progenitors via a process called *symmetrical cell division* (Figure 23.2b). Later in development, *asymmetrical cell division* is the rule. In this case, one "daughter" cell migrates away to take up its position in the cortex, where it will never divide again. The other daughter remains in the ventricular zone to undergo more divisions (Figure 23.2c). Radial glial cells repeat this pattern until all the neurons and glia of the cortex have been generated.

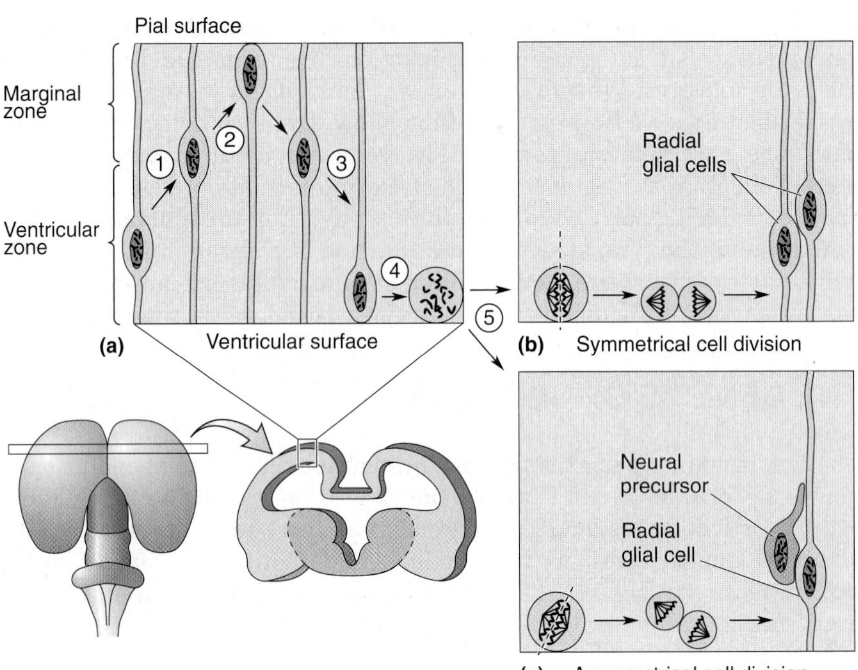

▶ FIGURE 23.2
The choreography of cell proliferation. **(a)** The wall of the brain vesicles initially consists of only two layers, the marginal zone and the ventricular zone. Each cell performs a characteristic "dance" as it divides, shown here from left to right. The circled numbers correspond to the five "positions" described in the text. The fate of the daughter cells depends on the plane of cleavage during division. **(b)** After symmetrical cell division, both daughters remain in the ventricular zone to divide again. **(c)** After asymmetrical cell division, the daughter farthest away from the ventricle ceases further division and migrates away.

In humans, the vast majority of neocortical neurons are born between the fifth week and the fifth month of gestation (pregnancy), peaking at the astonishing rate of 250,000 new neurons per minute. Although most of the action is over well before birth, some restricted regions of the adult brain retain some capacity to generate new neurons (Box 23.1). However, it is important to realize that once a daughter cell commits to a neuronal fate, it will never divide again. Furthermore, in most parts of the brain, the neurons you are born with are all you will have in your lifetime.

How is a cell's fate determined? Remember that all of our cells contain the same complement of DNA we inherited from our parents, so every daughter cell has the same genes. The factor that makes one cell different from another is the specific genes that generate mRNA and, ultimately, protein. Thus, cell fate is regulated by differences in gene expression during development. Recall from Chapter 2 that gene expression is regulated by cellular proteins called *transcription factors*. If transcription factors, or the "upstream" molecules that regulate them, are unevenly distributed within a cell, then the cleavage plane during asymmetrical cell division can determine which factors are passed on to the daughter cells and this can determine their fate (Figure 23.3).

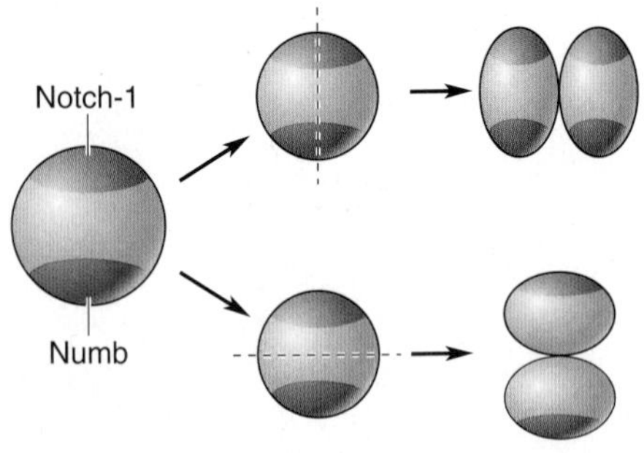

▲ FIGURE 23.3
The distribution of cell constituents in precursor cells. The proteins notch-1 and numb are differentially distributed in the precursor cells of the developing neocortex. Symmetrical cleavage partitions these proteins equally in the daughters, but asymmetrical cleavage does not. Differences in the distribution of proteins in the daughters causes them to have different fates.

Mature cortical cells can be classified as glia or neurons, and the neurons can be further classified according to the layer in which they reside, their dendritic morphology and axonal connections, and the neurotransmitter they use. Conceivably, this diversity could arise from different types of precursor cell in the ventricular zone. In other words, there could be one class of precursor cell that gives rise only to layer VI pyramidal cells, another which gives rise to layer V cells, and so on. However, this is not the case. Multiple cell types, including neurons and glia, can arise from the same precursor cell depending on what genes are transcribed during early development.

The ultimate fate of the migrating daughter cell is determined by a combination of factors, including the age of the precursor cell, its position within the ventricular zone, and its environment at the time of division. Cortical pyramidal neurons and astrocytes derive from the dorsal

BOX 23.1 OF SPECIAL INTEREST

Neurogenesis in Adult Humans (or How Neuroscientists Learned to Love the Bomb)

For many years, neuroscientists believed that neurogenesis—the generation of new neurons—was restricted to early brain development. But new findings have challenged this view. It now appears that new neurons are continuously generated by neural progenitors in the adult brain.

Cell division requires the synthesis of DNA, which can be detected by feeding the cells chemically labeled DNA precursor molecules. Cells undergoing division at the time the precursor is available incorporate the chemical label into their DNA. In the mid-1980s, Fernando Nottebohm of Rockefeller University used this approach to prove that new neurons are generated in the brains of adult canaries, particularly in regions associated with song learning. This finding resurrected interest in adult neurogenesis in mammals, which had actually first been described in 1965 by Joseph Altman and Gopal Das of the Massachusetts Institute of Technology. Research in the past few years by Fred Gage at the Salk Institute has established definitively that new neurons are generated in the adult rat hippocampus, a structure that is important for learning and memory (as we will see in Chapter 24). Interestingly, the number of new neurons goes up in this region if the animal is exposed to an enriched environment, filled with toys and playmates. In addition, rats given the chance to have a daily run on an exercise wheel show enhanced neurogenesis. In both cases, the increased number of neurons correlates with enhanced performance on memory tasks that require the hippocampus.

Until very recently, however, it has been unclear if neurogenesis also continues in the adult human brain. A definitive answer was finally obtained by the analysis of an experiment that several governments, most prominently those of the United States and the Soviet Union, unwittingly performed on the world population during the Cold War. In the years between 1955 and 1963, hundreds of nuclear bombs were detonated in atmospheric tests (Figure A), causing the widespread dissemination of radioactive fallout. There was a spike in the environmental levels of the radioactive isotope of carbon, ^{14}C, which was incorporated into the biological molecules of all living things, including the replicating DNA of human neurons. This radioactivity put a time stamp on every cell born during the "bomb pulse." Inspired by Gage's findings in rodents, Kirsty Spalding, Jonas Frisén, and their colleagues working at the Karolinska Institutet in Stockholm, Sweden developed methods to detect this carbon dating in the neurons of human postmortem brains. They discovered that the neurons of the neocortex were as old as the individual, meaning no new cells had been generated as adults, consistent with dogma. However, the data showed that hippocampal neurons were continuously generated across the lifespan. According to their calculations, in the adult human brain, 700 new neurons are added to the hippocampus every day. About as many are also lost, keeping the total number of hippocampal cells roughly constant. The annual turnover rate is almost 2%. Your hippocampus is not the same hippocampus you had a year ago.

Neurogenesis in the adult brain appears to be a peculiarity of the hippocampus and is still far too limited to repair CNS damage. However, understanding how adult neurogenesis is regulated—for example, by the quality of the environment—might suggest ways it can be harnessed to promote regeneration of the hippocampus after brain injury or disease.

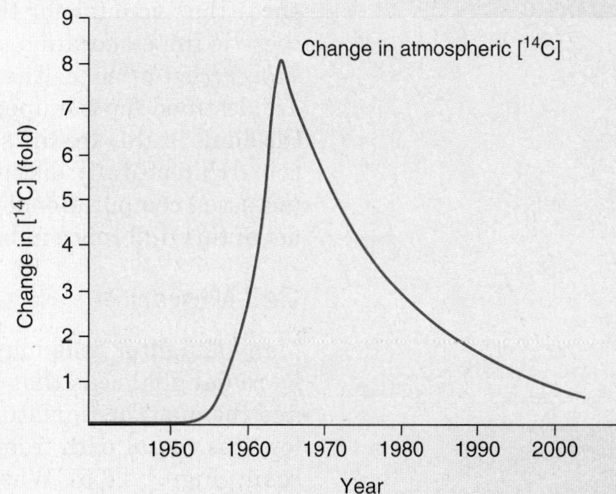

Figure A

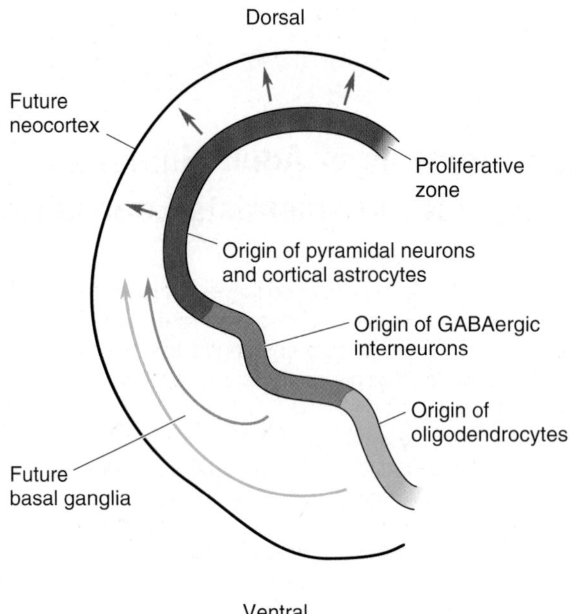

▲ FIGURE 23.4
The sources of cortical cells. Proliferation of cortical pyramidal neurons and astrocytes occurs in the ventricular zone of the dorsal telencephalon. However, inhibitory interneurons and oligodendroglia are generated in the ventricular zone of the ventral telencephalon; consequently, these cells must migrate laterally over some distance to arrive at their final destination in the cortex. (Source: Adapted from Ross, et al., 2003.)

ventricular zone, whereas inhibitory interneurons and oligodendroglia derive from the ventral telencephalon (Figure 23.4). The first cells to migrate away from the dorsal ventricular zone are destined to reside in a layer called the **subplate**, which eventually disappears as development proceeds. The next cells to divide become layer VI neurons, followed by the neurons of layers V, IV, III, and II.

It is worth noting that most of what we understand about cortical development has come from studies on rodents. The general principles appear to apply to primates such as ourselves, but there are some differences that account for the complexity of the primate neocortex. One of these is the elaboration of a second proliferative layer of cells, called the *subventricular zone*. The neurons deriving from the subventricular zone are destined for the upper layers of the cortex (layers II–III), which, in the adult brain, are the source of corticocortical connections that connect cytoarchitecturally distinct areas. It is reasonable to speculate that the increased computational powers of the primate brain are, in part, a product of this difference in brain development.

Cell Migration

Many daughter cells migrate by slithering along the thin fibers emitted by radial glial cells that span the distance between the ventricular zone and the pia. The immature neurons, called **neural precursor cells**, follow this radial path from the ventricular zone toward the surface of the brain (Figure 23.5). When cortical assembly is complete, the radial glia withdraw their radial processes. Not all migrating cells follow the path provided by the radial glial cells, however. About one-third of the neural precursor cells wander horizontally on their way to the cortex.

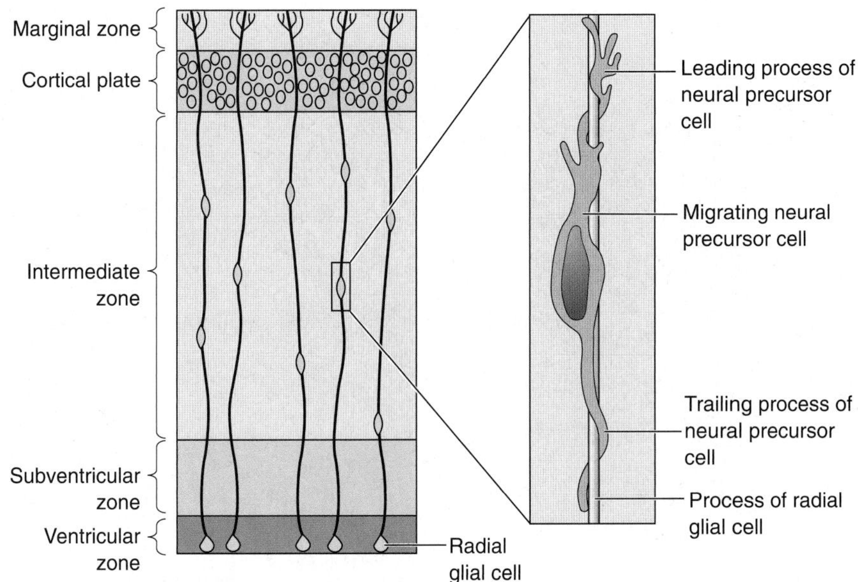

◀ FIGURE 23.5
The migration of neural precursor cells to the cortical plate. This is a schematic section through the dorsal telencephalon early in development. The expanded view shows a neural precursor cell crawling along the thin processes of the radial glia en route to the cortical plate, which forms just under the marginal zone.

The neural precursor cells destined to become subplate cells are among the first to migrate away from the ventricular zone. Neural precursor cells destined to become the adult cortex migrate next. They cross the subplate and form another cell layer called the **cortical plate**. The first cells to arrive in the cortical plate are those that will become layer VI neurons. Next come the layer V cells, followed by layer IV cells, and so on. Notice that each new wave of neural precursor cells migrates right past those in the existing cortical plate. In this way, the cortex is said to be assembled *inside out* (Figure 23.6). This orderly process can be disrupted by a number of gene mutations. For example, in a mutant mouse called *reeler* (describing the mouse's wobbly appearance), the neurons of the cortical plate are unable to pass through the subplate and pile up below it. Subsequent discovery of the affected gene revealed one of the factors, a protein called *reelin* that regulates the assembly of the cortex.

Cell Differentiation

The process by which a cell takes on the appearance and characteristics of a neuron is called *cell differentiation*. Differentiation is the consequence of a specific spatiotemporal pattern of gene expression. As we have seen, neural precursor cell differentiation begins as soon as the precursor cells divide with the uneven distribution of cell constituents. Further neuronal differentiation occurs when the neural precursor cell arrives in the cortical plate. Thus, layer V and VI neurons have differentiated into recognizable pyramidal cells even before layer II cells have migrated into the cortical plate. Neuronal differentiation occurs first, followed by astrocyte differentiation that peaks at about the time of birth. Oligodendrocytes are the last cells to differentiate.

Differentiation of the neural precursor cell into a neuron begins with the appearance of neurites sprouting off the cell body. At first, these neurites all appear about the same, but soon one becomes recognizable as the axon and the others as dendrites. Differentiation will occur even if the neural precursor cell is removed from the brain and placed in a tissue culture. For example, cells destined to become neocortical pyramidal cells will often assume the same characteristic dendritic architecture in the tissue culture. This means that differentiation is programmed well before

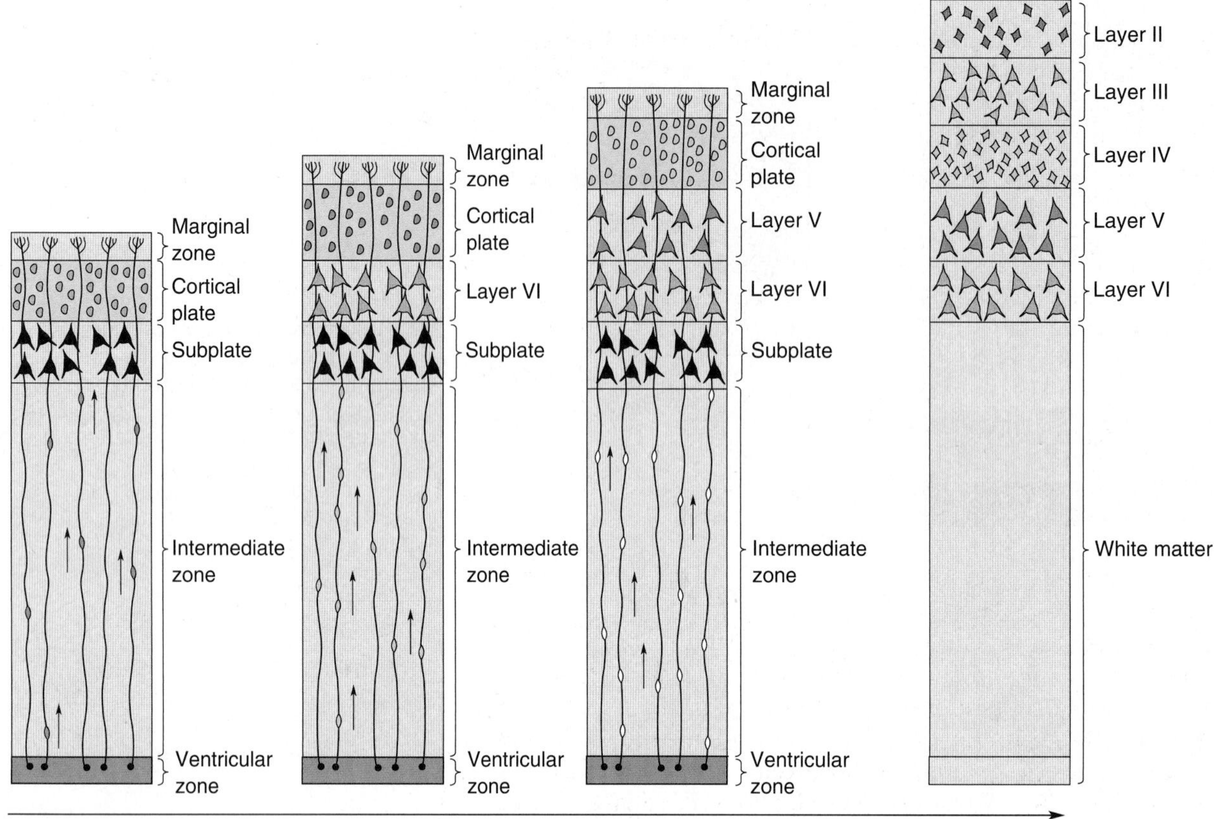

▲ FIGURE 23.6
Inside out development of the cortex. The first cells to migrate to the cortical plate are those that form the subplate. As these differentiate into neurons, the neural precursor cells destined to become layer VI cells migrate past and collect in the cortical plate. This process repeats again and again until all layers of the cortex have differentiated. The subplate neurons then disappear.

the neural precursor cell arrives at its final resting place. However, the stereotypical architecture of cortical dendrites and axons also depends on intercellular signals. As we have learned, pyramidal neurons are characterized by a large apical dendrite that extends radially, toward the pia, and an axon that projects in the opposite direction. Research has shown that a protein called *semaphorin 3A* is secreted by cells in the marginal zone. The protein acts first to repel growing pyramidal cell axons, causing them to stream away from the pial surface, and second to attract the growing apical dendrites, causing them to stream toward the brain surface (Figure 23.7). We will see that the oriented growth of neurites in response to diffusible molecules is a recurring theme in neural development.

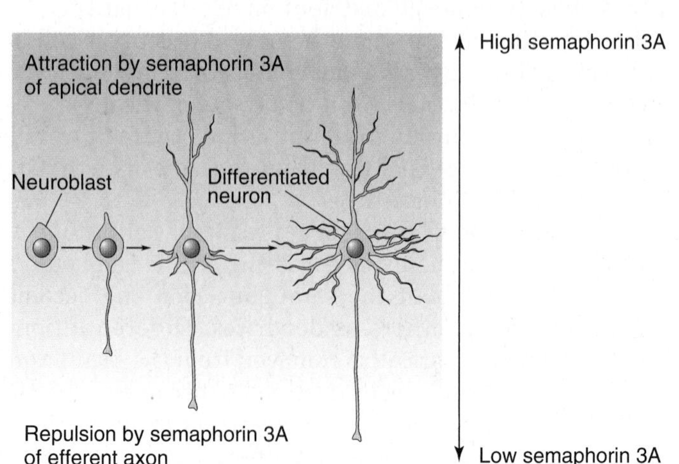

▶ FIGURE 23.7
The differentiation of a neural precursor cell into a pyramidal neuron. Semaphorin 3A, a protein secreted by cells in the marginal zone, repels the growing axon and attracts the growing apical dendrite, giving the pyramidal neuron its characteristic polarity.

Differentiation of Cortical Areas

The neocortex is often described as a sheet of tissue. In reality, however, cortex is much more like a patchwork quilt, with many structurally distinct areas stitched together. One of the consequences of human evolution was the creation of new neocortical areas that are specialized for increasingly sophisticated analysis. It is natural to wonder exactly how all these areas arise during development.

As we have seen, most cortical neurons are born in the ventricular zone and then migrate along radial glia to take up their final position in one of the cortical layers. Thus, it seems reasonable to conclude that cortical areas in the adult brain simply reflect an organization that is already present in the ventricular zone of the fetal telencephalon. According to this idea, the ventricular zone contains something like a film record of the future cortex, which is projected onto the wall of the telencephalon as development proceeds.

The idea of such a cortical "protomap," proposed by Yale University neuroscientist Pasko Rakic (Box 23.2), is based on the assumption that migrating neural precursor cells are precisely guided to the cortical plate by the network of radial glial fibers. If migration is strictly radial, we might expect that all the offspring of a single neural progenitor cell would migrate to exactly the same neighborhood of the cortex. Indeed, this has proven to be the case for the majority of cortical neurons. The concept that an entire radial column of cortical neurons originates from the same birthplace in the ventricular zone, called *the radial unit hypothesis*, suggests a basis for the dramatic expansion of the human neocortex over the course of evolution. The surface area of the human cerebral cortex is 1000 × greater than that of the mouse and 10 × greater than a macaque monkey, but differs in thickness by less than a factor of two. These differences in surface area arise from the size of the proliferative ventricular zone, which in turn can arise from differences in the duration of the period of symmetrical cell division early in gestation. An appealing hypothesis is that one happy accident of human evolution was the chance mutation of genes that regulate the kinetics of cell proliferation, allowing for an increase in the number of proliferative radial glial cells and consequently an enlarged surface of the neocortex.

As mentioned earlier, however, one-third of all neural precursor cells stray considerable distances as they migrate toward the cortical plate. How do they find their final resting place? One solution to this puzzle is suggested by the finding that neurons in different regions of the cortex have distinct molecular identities. For example, two complementary gradients of transcription factors, called *Emx2* and *Pax6*, have been discovered along the anterior–posterior axis of the ventricular zone of the developing neocortex (Figure 23.8). Neurons destined for the anterior region of neocortex express higher levels of Pax6, and neurons destined for posterior cortex express higher levels of Emx2. Recall that differences in transcription factors lead to differences in gene expression and protein production; these can be used as signals to attract neural precursor cells to the appropriate destinations. Indeed, if mice are genetically engineered to produce less Emx2, there is an expansion of the anterior cortical areas, such as the motor cortex, and a shrinkage of posterior cortical areas, such as the visual cortex. Conversely, if Pax6 is knocked out, there is an expansion of visual cortex and a shrinkage of frontal cortex.

BOX 23.2 PATH OF DISCOVERY

Making a Map of the Mind

by Pasko Rakic

My interest in the development of a cerebral cortical map began in the mid-1960s while I was a resident in neurosurgery at the Belgrade University. My professors repeatedly warned me to be extremely conservative when cutting the cerebral cortex "because unlike other organs, it is a map of the different areas that are precisely wired for specific functions, and once removed, cannot be replaced or regenerated." When I inquired how the map was formed, I was referred to nineteenth century literature since little had been learned since then. This is when I decided to abandon neurosurgery until I found an answer to my question. I was fortunate to receive a U.S. Fogarty International Fellowship, which took me to Harvard where I met Paul Yakovlev, a giant figure in developmental neuropathology. I learned from him about the old Wilhelm His hypothesis that cortical neurons in humans originate near the cerebral cavity. Experimental proof, however, was lacking.

Upon returning to Belgrade from the fellowship, I made slices of fresh human embryonic forebrain tissue at various prenatal ages and placed them into a dish of culture medium containing radioactive thymidine, one of the building blocks of DNA. This specific DNA replication marker was impossible to get in Eastern Europe, but I succeeded in bringing it in from the United States unnoticed. To my knowledge, this experiment was the first use of a slice preparation to study cortical development. Since cells continue to divide and synthesize DNA supravitally (after death), I was able to localize them close to the ventricular cavity and in the layer just above, which I named the ventricular (VZ) and subventricular (SVZ) zones, terms later adopted by the Boulder Nomenclature Committee for the neurogenic zones in all vertebrates. Most importantly, I did not find incorporation of radioactivity into cells in the cortical plate, providing the first experimental evidence that, indeed, newly generated neurons are programed to migrate outward to the developing cortex situated below the cerebral surface. This finding was part of my doctoral thesis on the development of the human brain that not only opened a new field of inquiry but also prompted an offer from Professor Raymond Adams to join the faculty at Harvard Medical School in 1969.

After establishing my laboratory at Harvard, I initiated a comprehensive analysis of when neurons are born, migrate, and differentiated in the cerebral cortex of the macaque monkey, chosen because of its slowly developing brain similar to the human brain. I learned that, even in this large, convoluted cerebrum, neurons migrate and settle in columns in which each new generation of neurons bypasses the previous one. Furthermore, since at mid-gestation in this species postmitotic neurons require more than 2 weeks to migrate to their final destination, I was able to explore the mechanism of how they find their final position in the increasingly distant and convoluted cortex. For example, reconstruction of electron microscopic images of serially sectioned tissue revealed selective attachment of migrating neurons to radial glial cells. In primates, these transient cells are distinct and more differentiated than in other mammals, and their elongated shaft spans the entire thickness of the fetal cerebral wall (see the animated Figure A at http://rakiclab.med.yale.edu/research/CorticalNeuronMigration.aspx). Since this is a huge distance for a small migrating neuron, we performed a full reconstruction of monkey fetal cerebral wall at various ages, each requiring thousands of serial electron micrographs. To create an automated 3D reconstruction in the era before microcomputers, we were given free access to the NASA computers used for the Apollo Project.

These discoveries inspired a new research field and led me to postulate the *radial unit* and *protomap hypotheses* of how the complex three-dimensional organization of the cortex is built from a two-dimensional layer of dividing neural stem cells in the proliferative ventricular and subventricular zones (see the animated Figure B at http://rakiclab.med.yale.edu/research/RadialMigration.aspx). These hypotheses suggested a mechanism for the evolutionary expansion of the surface, rather than the thickness, of the cerebral cortex. The protomap hypothesis also explained how genetic modifications could induce arrays of different radial units, giving rise to different cortical areas. Experiments with transgenic mice have provided further supporting evidence for both models.

The realization that the largest structure of our brain receives all its neurons by orderly, long-distance migration fascinated me so much that after moving to Yale University in 1979, I decided to focus on the molecular mechanisms underlying the coordination of these complex processes. My strategy has been to perform comparative studies of cortical development in rodents, nonhuman primates, and humans using a variety of *in vitro* and *in vivo* assays, including genetic manipulations in animals together with mRNA profiling in embryonic human brain slices following laser microdissection. I began with the idea of differential cell adhesion and searched for molecules that would enable a migrating neuron to recognize the surface of the radial glial cell shaft, similar to an antigen–antibody interaction. We have identified a number of genes and signaling molecules involved in the regulation of the proliferation and migration of cortical neurons to their proper areal laminar and columnar positions. By manipulating neuronal migration using genetic and environmental factors, we discovered hidden abnormalities of neuronal positioning that cannot be discerned by routine postmortem examination, opening new insights into the pathogenesis of brain disorders (see Box 23.4).

Over the years, I have come to recognize that the development of the cortex is a complex, multipronged process involving many genes, regulatory elements, and signaling molecules. Thus, even after five decades of effort, I am still as committed as ever to my quest to find out how the cortical map is formed, not only because it is the organ that holds the secret to what distinguishes us from all other species but also because it is also the site of devastating mental disorders yet to be fully understood.

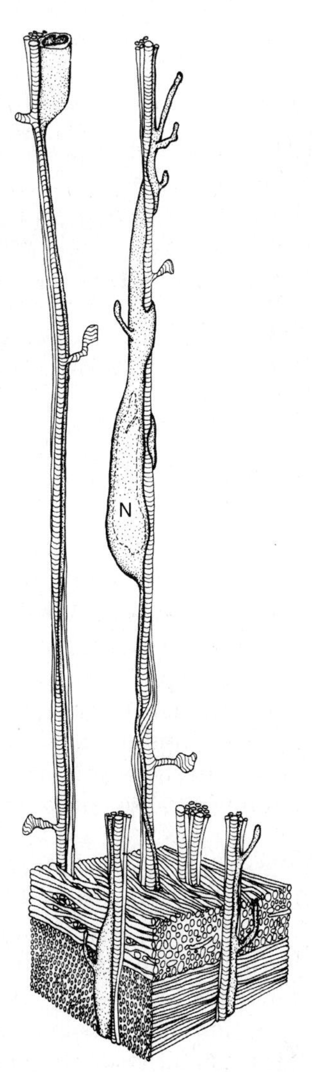

Figure A

This drawing is based on a 3D reconstruction of thousands of electron microscopic images, showing a neural precursor cell (labeled N) migrating along a radial glial fiber. (Source: Courtesy of Dr. Pasko Rakic.)

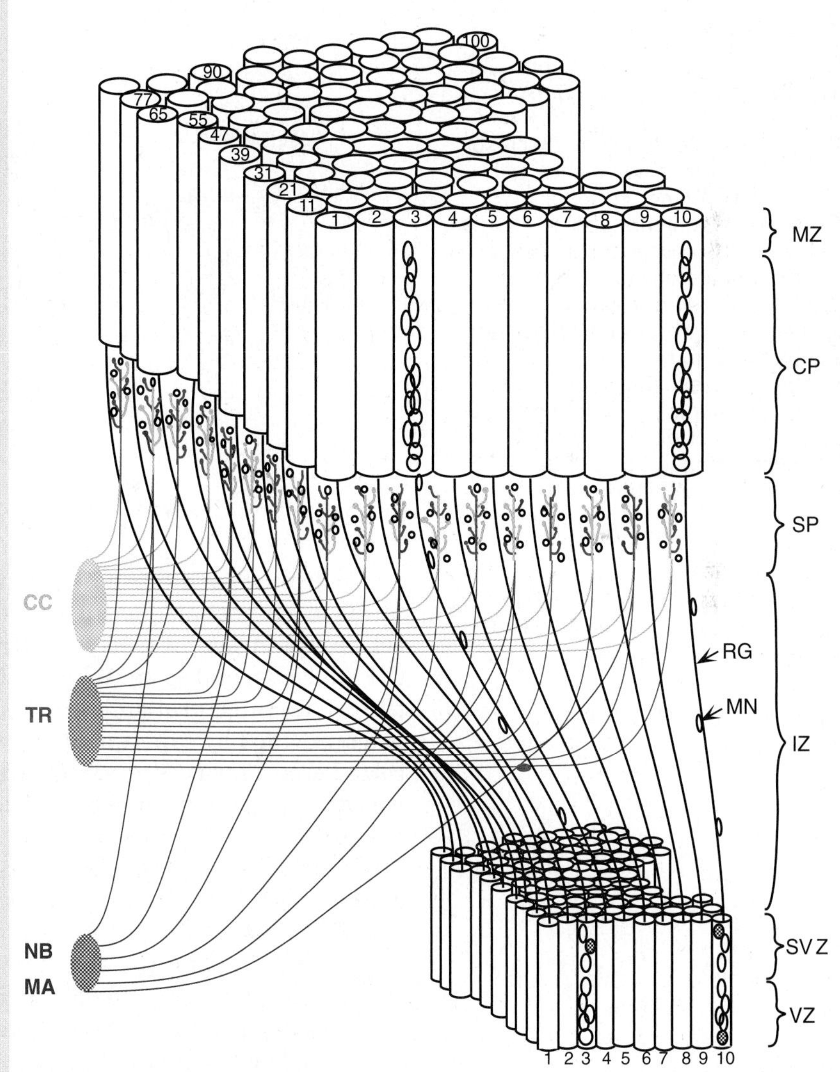

Figure B

This drawing shows how the protomap in the ventricular (VZ) and subventricular zones (SVZ) is related to the mature cerebral cortex. (Abbreviations: IZ, intermediate zone; SP, subplate; CP, cortical plate; MZ, marginal zone; CC, corpus callosum; TR, thalamic radiations; MA, monoamine input; NB, nucleus basalis input; RG, radial glia; MN, migrating neural precursor cell. (Source: Courtesy of Dr. Pasko Rakic.)

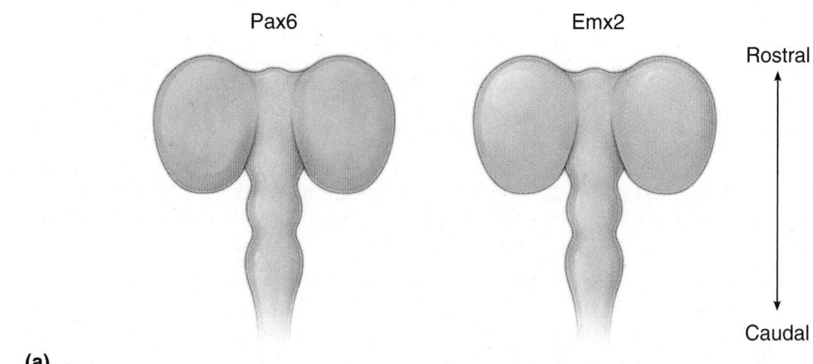

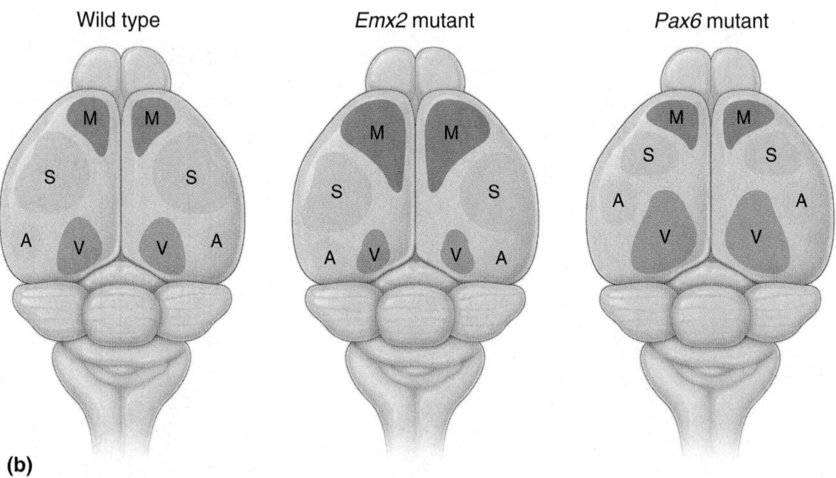

▶ FIGURE 23.8
Gradients of transcription factors control the size of cortical areas. **(a)** In the fetal telencephalon, Pax6 and Emx2 are expressed by neural precursor cells in complementary gradients, with Pax6 highest in anterior cortex and Emx2 highest in posterior cortex. **(b)** The sizes of different cortical areas change if these gradients are changed. In mice genetically engineered to produce less Emx2, there is an expansion in the size of anterior areas. In mice with reduced Pax6, there is an expansion of posterior areas. Abbreviations: M, motor cortex; S, somatosensory cortex; A, auditory cortex; V, visual cortex. (Source: Adapted from Hamasaki et al., 2004.)

Cortical areas differ not only in terms of cytoarchitecture but also in terms of connections, particularly with the dorsal thalamus. Area 17 receives input from the LGN, area 3 receives input from the ventral posterior (VP) nucleus, and so on. What is the contribution of the thalamic input to the cytoarchitectural differentiation of cortical areas? A clear answer has been provided by experiments in which the LGN input to monkey striate cortex area 17 was eliminated early in fetal life. In these animals, area 17 was greatly reduced in size, with a concomitant increase in the size of the extrastriate cortex (Figure 23.9).

Thalamic input is clearly necessary, but is it sufficient to induce cytoarchitectural differentiation in a cortical area? Researchers Brad Schlaggar and Dennis O'Leary of the Salk Institute addressed this question in a clever way. In rats, the thalamic fibers wait in the cortical white matter and do not enter the cortex until a few days after birth. Schlaggar and O'Leary peeled off the parietal cortex in newborn rats and replaced it with a piece of occipital cortex. This created a situation in which the thalamic fibers from the VP nucleus were waiting under what would have been visual cortex. Remarkably, the fibers invaded the new piece of cortex and it assumed the cytoarchitecture that is characteristic of the rodent somatosensory cortex (the "barrels"; see Figure 12.21). Together, these results suggest that the thalamus is important for specifying the pattern of cortical areas.

But how did the appropriate thalamic axons come to lie in wait under the parietal cortex in the first place? The answer, apparently, lies in the subplate. Subplate neurons, which have a more strictly radial migra-

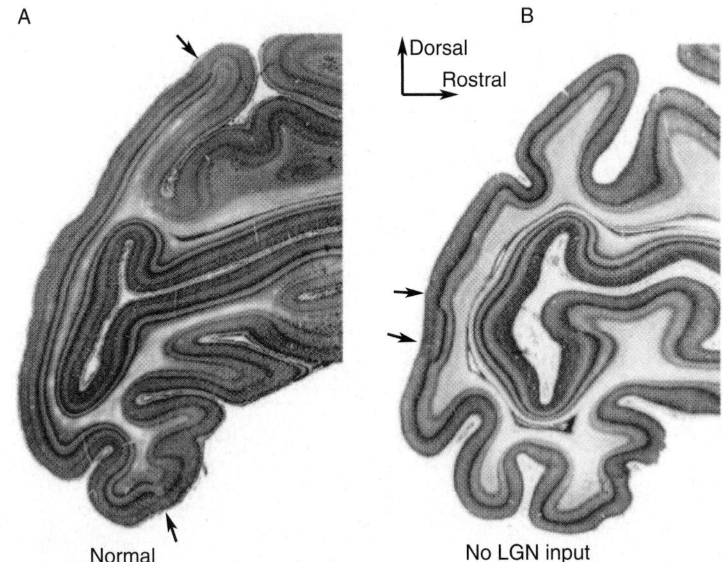

▲ FIGURE 23.9
Differentiation of monkey striate cortex requires LGN input during fetal development. The arrowheads indicate the borders between areas 17 and 18 in **(a)** a normal monkey, and **(b)** a monkey in which the LGN input degenerated early in fetal development. (Source: Dehay and Kennedy, 2007.)

tion pattern, attract the appropriate thalamic axons to different parts of the developing cortex: LGN axons to occipital cortex, VP nucleus axons to parietal cortex, and so on. The area-specific thalamic axons initially innervate distinct populations of subplate cells. When the overlying cortical plate grows to a sufficient size, the axons invade the cortex. The arrival of the thalamic axons causes the cytoarchitectural differentiation we recognize in the adult brain. Thus, the subplate layer of earliest born neurons seems to contain the instructions for the assembly of the cortical quilt.

THE GENESIS OF CONNECTIONS

As neurons differentiate, they extend axons that must find their appropriate targets. Think of this development of long-range connections, or pathway formation, in the central nervous system (CNS) as occurring in three phases: pathway selection, target selection, and address selection. Let's understand the meaning of these terms in the context of the development of the visual pathway from the retina to the LGN, as shown in Figure 23.10.

Imagine for a moment that you must lead a growing retinal ganglion cell axon to the correct location in the LGN. First you travel down the optic stalk toward the brain. But soon you reach the optic chiasm at the base of the brain and must decide which fork in the road to take. You have three choices: You can enter the optic tract on the same side, you can enter the optic tract on the opposite side, or you can dive into the other optic nerve. The correct path depends on the location in the retina of your ganglion cell and on the cell type. If you came from the nasal retina, you would cross over at the chiasm into the contralateral optic tract, but if you came from the temporal retina, you would stay in the tract on the same side. And in no case would you enter the other optic nerve. These

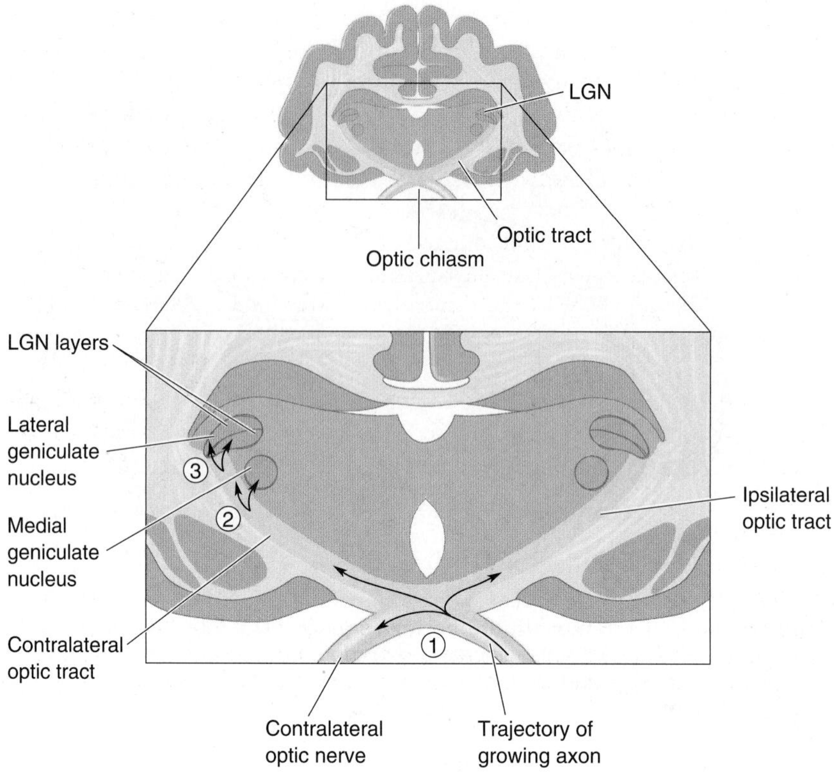

▶ FIGURE 23.10
The three phases of pathway formation. The growing retinal axon must make several "decisions" to find its correct target in the LGN. ① During pathway selection, the axon must choose the correct path. ② During target selection, the axon must choose the correct structure to innervate. ③ During address selection, the axon must choose the correct cells to synapse with in the target structure.

are examples of the "decisions" that must be made by the growing axon during *pathway selection*.

Having forged your way into the dorsal thalamus, you are now confronted with the choice of which thalamic nucleus to innervate. The correct choice, of course, is the lateral geniculate nucleus. This decision is called *target selection*.

But finding the correct target still isn't enough. You must now find the correct layer of the LGN. You also must make sure that you sort yourself out with respect to other invading retinal axons so that retinotopy in the LGN is established. These are examples of the decisions that must be made by the growing axon during *address selection*.

We will see that each of the three phases of pathway formation depends critically on communication between cells. This communication occurs in several ways: direct cell–cell contact, contact between cells and the extracellular secretions of other cells, and communication between cells over a distance via diffusible chemicals. As the pathways develop, the neurons also begin to communicate via action potentials and synaptic transmission.

▲ FIGURE 23.11
The growth cone. The filopodia probe the environment and direct the growth of the neurite towards attractive cues.

The Growing Axon

Once the neural precursor cell has migrated to take up its appropriate position in the nervous system, the neuron differentiates and extends the processes that will ultimately become the axon and dendrites. At this early stage, however, the axonal and dendritic processes appear quite similar and collectively are still called *neurites*. The growing tip of a neurite is called a **growth cone** (Figure 23.11).

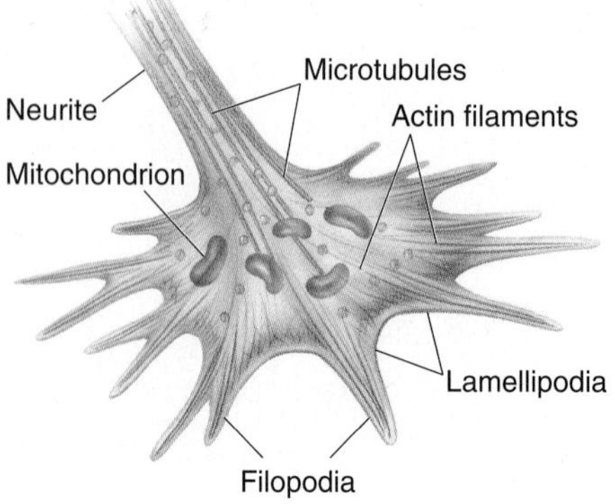

The growth cone is specialized to identify an appropriate path for neurite elongation. The leading edge of the growth cone consists of flat sheets of membrane called *lamellipodia* that undulate in rhythmic waves like the wings of a stingray swimming along the ocean bottom. Extending from the lamellipodia are thin spikes called *filopodia*, which constantly probe the environment, moving in and out of the lamellipodia. Growth of the neurite occurs when a filopodium, instead of retracting, takes hold of the substrate (the surface on which it is growing) and pulls the advancing growth cone forward.

Obviously, axonal growth cannot occur unless the growth cone is able to advance along the substrate. An important substrate consists of fibrous proteins that are deposited in the spaces between cells, the **extracellular matrix**. Growth occurs only if the extracellular matrix contains the appropriate proteins. An example of a permissive substrate is the glycoprotein *laminin*. The growing axons express special surface molecules called *integrins* that bind laminin, and this interaction promotes axonal elongation. Permissive substrates, bordered by repulsive ones, can provide corridors that channel axon growth along specific pathways.

Travel down such molecular highways is also aided by **fasciculation**, a mechanism that causes axons growing together to stick together (Figure 23.12). Fasciculation is due to the expression of specific surface molecules called **cell-adhesion molecules (CAMs)**. The CAMs in the membrane of neighboring axons bind tightly to one another, causing the axons to grow in unison.

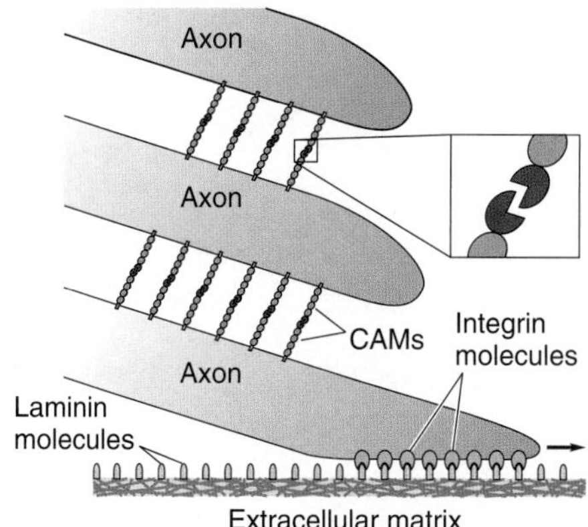

▲ FIGURE 23.12
Fasciculation. The bottom axon grows along the molecular "highway" of the extracellular matrix. The other axons ride piggyback, sticking to one another by the interaction of cell-adhesion molecules (CAMs) on their surfaces.

Axon Guidance

Wiring the brain appears to be a formidable challenge, particularly in view of the long distances that many axons traverse in the mature nervous system. Remember, though, that distances are not nearly as great early in development, when the entire nervous system is no more than a few centimeters long. A common mode of pathway formation is the initial establishment of connections by *pioneer axons*. These axons "stretch" as the nervous system expands and guide their later developing neighbor axons to the same targets. Still, the question remains of how the pioneer axons grow in the correct direction, along the correct paths, to the correct targets. The answer appears to be that the trajectory of the axon is broken into short segments that may only be a few hundred microns long. The axon concludes a segment when it arrives at an intermediate target. The interaction of the axon and the intermediate target throws a molecular switch that sends the axon onward to another intermediate target. Thus, by "connecting the dots," the axon eventually arrives at its final destination.

Guidance Cues. Growth cones differ in terms of the molecules they express on their membranes. Interactions of these cell surface molecules with *guidance cues* in the environment determine the direction and amount of growth. Guidance cues can be attractive or repulsive, depending on the receptors expressed by the axons.

A **chemoattractant** is a diffusible molecule that acts over a distance to attract growing axons toward their targets, like the aroma of freshly brewed java might attract a coffee lover. Although the existence of such chemoattractants was proposed over a century ago by Cajal and was inferred by many experimental studies since then, only very recently have attractant molecules been identified in mammals. The first to be discovered is a protein called **netrin**. Netrin is secreted by neurons in the

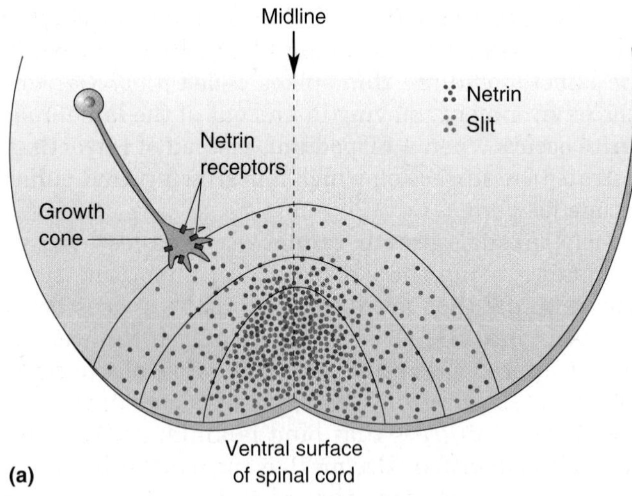

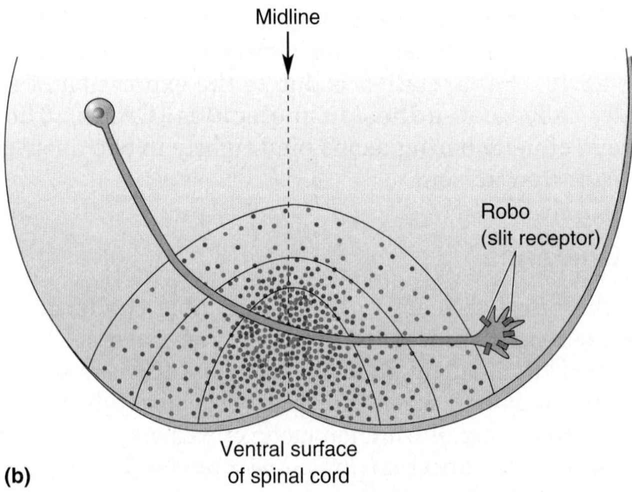

▲ FIGURE 23.13

Chemoattraction and chemorepulsion. Axons grow across the midline in two stages. First they are attracted to the midline, and then they are repelled from the midline. **(a)** The protein called netrin is secreted by cells in the ventral midline of the spinal cord. Axons with the appropriate netrin receptors are attracted to the region of highest netrin concentration. **(b)** The protein called slit is also secreted by midline cells. Axons that express the protein called robo, the slit receptor, grow away from the region of highest slit concentration. Up-regulation of robo by axons that cross the midline ensures that they keep growing away from the midline.

ventral midline of the spinal cord (Figure 23.13). The gradient of netrin attracts the axons of dorsal horn neurons that will cross the midline to form the spinothalamic tract. These axons possess netrin receptors, and the binding of netrin to the receptor spurs growth toward the source of netrin.

But that's only half the story. Once the decussating axons cross the midline, they need to escape the powerful siren song of netrin. This escape is enabled by the action of *slit*, another protein secreted by midline cells. Slit is an example of a **chemorepellent**, a diffusible molecule that chases axons away. For slit to exert this action, however, the axon must express on its surface the slit receptor, a protein called *robo*. The growth cones that are attracted to the midline by netrin express little robo and are therefore insensitive to repulsion by slit. However, once they cross the midline, they encounter a signal that causes robo to be up-regulated. Now slit repels the axons so they grow away from the midline.

This example shows how axons can be "pulled" and "pushed" by the coordinated actions of chemoattractants and chemorepellents. The trajectory of the axons to and from the midline is also constrained by the permissive substrates that are available for growth. In this example, the cells of the midline are an intermediate target—one of the "dots"—along the molecular highway that spans the midline. These cells serve to alternately attract and then repel the growing axon as it crosses from one side of the CNS to the other.

Establishing Topographic Maps. Let's return to the example of the growing retinogeniculate axon (see Figure 23.10). These axons grow along the substrate provided by the extracellular matrix of the ventral wall of the optic stalk. An important "choice point" occurs at the optic chiasm. Axons from the nasal retina cross and ascend in the contralateral optic tract, while axons from the temporal retina remain in the ipsilateral optic tract. From our discussion so far, we can infer that nasal and temporal retinal axons express different receptors to cues secreted at the midline.

Once the axons from the retinas are sorted out at the midline, they continue on to innervate targets such as the LGN and the superior colliculus. Sorting of the axons occurs again, this time to establish a retinotopic map in the target structure. If we accept the notion that axons differ depending on their position in the retina (as they must, to account for the partial decussation at the optic chiasm), then we have a potential molecular basis for the establishment of retinotopy. This idea, that chemical markers on growing axons are matched with complementary chemical markers on their targets to establish precise connections, is called the **chemoaffinity hypothesis**.

This hypothesis was first tested in the 1940s by Roger Sperry, at the California Institute of Technology, in an important series of experiments using the retinotectal projection in frogs. The tectum is the amphibian homologue of the mammalian superior colliculus. The tectum receives retinotopically ordered input from the contralateral eye and uses this information to organize movements in response to visual stimulation, such as lunging after a fly passing overhead. Thus, this system can be used to investigate the mechanisms that generate orderly maps in the CNS.

Another advantage of amphibians is that their CNS axons will regenerate after being cut, which is not true for mammals (Box 23.3). Sperry took advantage of this property to investigate how the retinotopic map was established in the tectum. In one experiment, Sperry cut the optic nerve, rotated the eye 180° in the orbit, and then allowed the upside-down nerve to regenerate. Despite the fact that the axons in the optic nerve were now scrambled from where they would occur naturally, the axons grew into the tectum to exactly the same sites that they occupied originally. Now, when a fly passed overhead, these frogs lunged down instead of up because their eyes were providing the brain a mirror image of the world.

What factors control the guidance of retinal axons to the correct part of the tectum? When the axons arrive at the tectum, they must grow along the membranes of tectal cells. The axons from the nasal retina cross the anterior part of the tectum and innervate the neurons in the posterior part. The axons of the temporal retina, in contrast, grow into the anterior tectum and stop there (Figure 23.14a). Why? Experiments have shown that the cell membranes of anterior and posterior tectal neurons differentially express factors that allow the growth of nasal and temporal retinal axons. Nasal axons grow well on the substrate provided by both anterior and posterior tectal membranes (Figure 23.14b). However, temporal axons grow only on anterior tectal membranes; the posterior membranes are repulsive (Figure 23.14c). Research has led to the discovery that proteins called

BOX 23.3 OF SPECIAL INTEREST

Why Our CNS Axons Don't Regenerate

Compared with other vertebrates, mammals are fortunate in many ways. We have computing power and behavioral flexibility that our distant aquatic cousins, fish and amphibians, utterly lack. However, in one interesting respect, fish and frogs have a distinct advantage—the growth of axons in the adult CNS after injury. Cut the optic nerve in a frog, and it grows back. Do the same thing in a human, and the person is blind forever. Of course, our CNS axons do grow over long distances early in development. But something happens shortly after birth that makes the CNS, especially the white matter, a hostile environment for axon growth.

When an axon is cut, the distal segment degenerates because it is isolated from the soma. However, the severed tip of the proximal segment initially responds by emitting growth cones. In the adult mammalian CNS, sadly, this growth is aborted. Not in the mammalian PNS, though; if you've ever had a deep cut that severed a peripheral nerve, you know that eventually sensations can come back in the denervated skin. This happens because PNS axons are capable of regeneration over long distances.

Surprisingly, the critical difference between the mammalian PNS and CNS is not the neurons. A PNS dorsal root ganglion cell axon regenerates in the peripheral nerve, but when it hits the environment of the CNS, in the dorsal horn of the spinal cord, growth ceases. Conversely, if a CNS alpha motor neuron axon is cut in the periphery, it grows back to its target. If it is cut in the CNS, no regeneration occurs. Thus, the critical difference seems to be the different environments of the CNS and PNS. This idea was tested in a very important series of experiments, beginning in the early 1980s, performed on adult rodents by Albert Aguayo and his colleagues at Montreal General Hospital. They showed that crushed optic nerve axons can grow long distances if they are given a peripheral nerve graft to grow along (Figure A). However, as soon as the axon hit the CNS target of the nerve graft, growth ceased.

What is different about peripheral nerves? The type of myelinating glial cell varies: oligodendroglia in the CNS and Schwann cells in the PNS (see Chapter 2). Experiments performed by Martin Schwab of the University of Zurich showed that CNS neurons grown in tissue culture will extend axons along substrates prepared from Schwann cells but not along CNS oligodendroglia and myelin. This finding led to the search for glial factors that inhibit axon growth, and a molecule called *nogo* was finally identified early in 2000. Nogo is apparently released when oligodendroglia are damaged.

Antibodies raised against nogo neutralize the molecule's growth-suppressing activity. Schwab and his colleagues have injected the anti-nogo antibody (called *IN-1*) into adult rats after spinal cord injury. This treatment enabled about 5% of the severed axons to regenerate—a modest effect, perhaps, but sufficient for the animals to show a remarkable functional recovery. The same antibodies have also been used to localize nogo in the nervous system. The protein is made by oligodendroglia in mammals, but not in fish, and it is not found in Schwann cells.

One of the last steps in wiring the mammalian brain is wrapping the young axons in myelin. This has the beneficial effect of speeding action potential conduction, but it comes with a heavy cost—the inhibition of axon growth after injury. The lack of axon regeneration in the adult CNS was accepted by neurologists in the last century as a dismal fact of life. However, our recent understanding of molecules with the power to stimulate or inhibit CNS axon growth offers hope for the twenty-first century that treatments can be devised to promote axon regeneration in the damaged human brain and spinal cord.

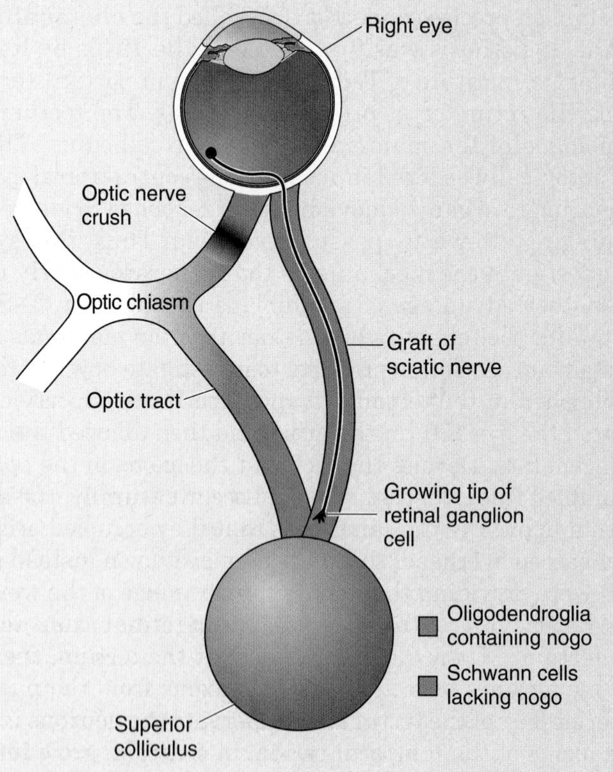

Figure A

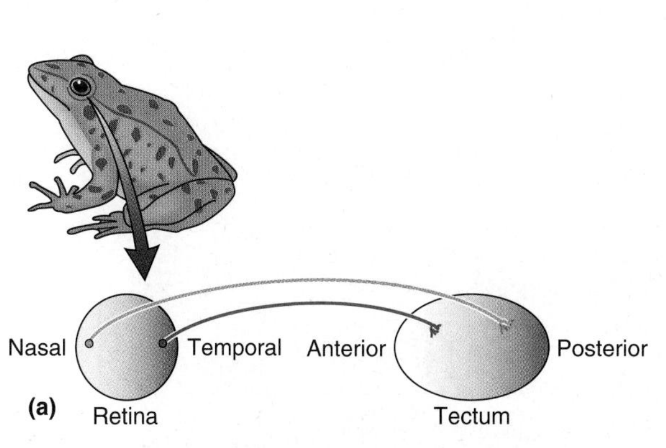

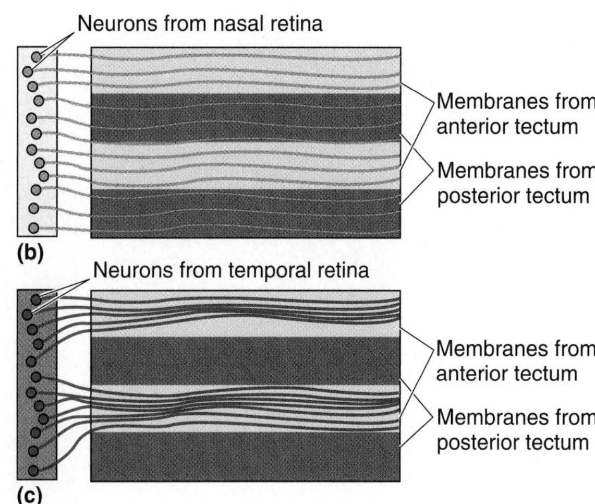

▲ FIGURE 23.14
Establishing retinotopy in the frog retinotectal projection. **(a)** Retinotopy is established when the nasal retina projects to the posterior tectum and the temporal retina projects to the anterior tectum. **(b)** To reveal how this retinotopy is established, membranes of cells from anterior and posterior tectum are removed from the frog and deposited in a striped pattern at the bottom of a Petri dish. Experiments show that nasal retinal axons *in vitro* grow equally well on anterior and posterior membranes. **(c)** Temporal axons, in contrast, are repelled by membranes from posterior tectum and grow only on anterior tectal membranes.

ephrins are one repulsive signal for temporal retinal axons. Specific ephrin molecules are secreted in a gradient across the surface of the tectum, with the highest levels found on posterior tectal cells. An ephrin interacts with a receptor, called *eph*, on the growing axon. The interaction of ephrin with its receptor inhibits further axonal growth, similar to the slit–robo interaction discussed earlier.

Such gradients in the expression of guidance cues and their axonal receptors can impose considerable topographic order on the wiring of the retina to its targets in the brain. However, as we will see in a moment, the final refinement of connections often requires neural activity.

Synapse Formation

When the growth cone comes in contact with its target, a synapse is formed. Most of what is known about this process comes from studies of the neuromuscular junction. The first step appears to be the induction of a cluster of postsynaptic receptors under the site of nerve–muscle contact. This clustering is triggered by an interaction between proteins secreted by the growth cone and the target membrane. At the neuromuscular junction, one of these proteins, called *agrin*, is deposited in the extracellular space at the site of contact (Figure 23.15). The layer of proteins in this space is called the *basal lamina*. Agrin in the basal lamina binds to a receptor in the muscle cell membrane called *muscle-specific kinase* or *MuSK*. MuSK communicates with another molecule, called *rapsyn*, which appears to act like a shepherd to gather the postsynaptic acetylcholine receptors (AChRs) at the synapse. The size of the "flock" of receptors is regulated by another molecule released by the axon, called *neuregulin*, which stimulates the receptor gene expression in the muscle cell.

The interaction between axon and target occurs in both directions, and the induction of a presynaptic terminal also appears to involve proteins in the basal lamina. Basal lamina factors provided by the target cell evidently can stimulate Ca^{2+} entry into the growth cone, which triggers

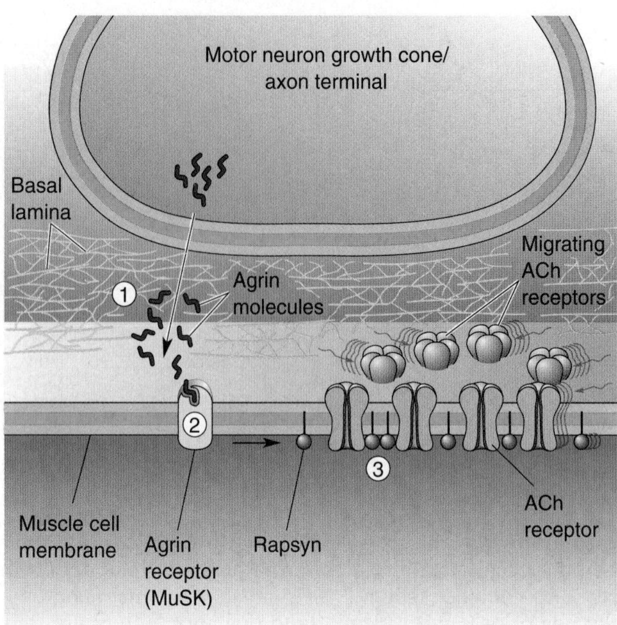

▲ FIGURE 23.15
Steps in the formation of a neuromuscular synapse. ① The growing motor neuron secretes the protein agrin into the basal lamina. ② Agrin interacts with MuSK in the muscle cell membrane. This interaction leads to ③ the clustering of ACh receptors in the postsynaptic membrane via the actions of rapsyn.

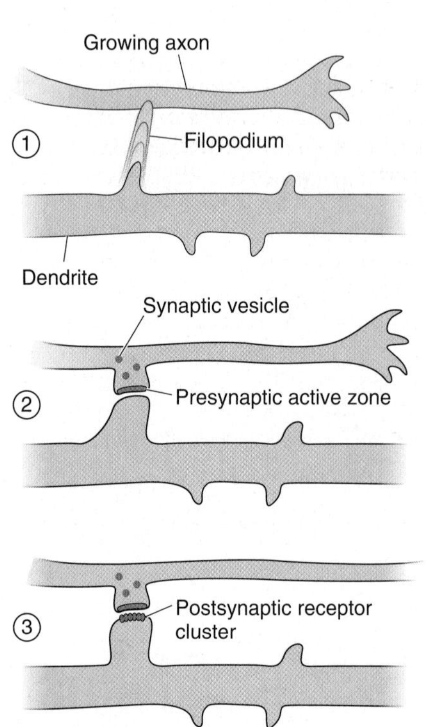

▲ FIGURE 23.16
Steps in the formation of a CNS synapse. ① A dendritic filopodium contacts an axon. ② Contact leads to the recruitment of synaptic vesicles and active zone proteins to the presynaptic membrane. ③ Neurotransmitter receptors accumulate postsynaptically.

neurotransmitter release. Thus, although the final maturation of synaptic structure may take a matter of weeks, rudimentary synaptic transmission appears very rapidly after contact is made. Besides mobilizing transmitter, Ca^{2+} entry into the axon also triggers changes in the cytoskeleton that cause it to assume the appearance of a presynaptic terminal and to adhere tightly to its postsynaptic partner.

Similar steps are involved in synapse formation in the CNS, but these may occur in a different order and they definitely use distinct molecules (Figure 23.16). Microscopic imaging of neurons in tissue culture reveals that filopodia are continually being formed and retracted from neuronal dendrites seeking innervation. Synapse formation begins when such a dendritic protrusion reaches out and touches an axon that might be passing by. This interaction appears to cause a preassembled presynaptic active zone to be deposited at the site of contact followed by the recruitment of neurotransmitter receptors to the postsynaptic membrane. In addition, specific adhesion molecules are expressed by both presynaptic and postsynaptic membranes that serve to glue the partners together.

THE ELIMINATION OF CELLS AND SYNAPSES

The mechanisms of pathway formation we've discussed are sufficient to establish considerable order in the connections of the fetal brain. For example, in the visual system, these mechanisms ensure that (1) retinal axons reach the LGN, (2) geniculate axons reach layer IV of striate cortex, and (3) both of these sets of axons form synapses in their target structures in proper retinotopic order. But the job of wiring together the nervous system isn't finished yet. A prolonged period of development follows, from before birth all the way through adolescence, in which these connections are refined. It may come as a surprise that one of the most

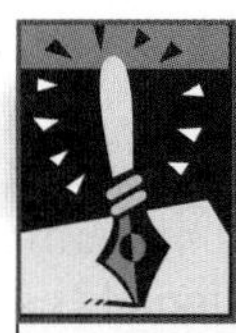

BOX 23.4 OF SPECIAL INTEREST

The Mystery of Autism

Autism is a developmental disorder in humans characterized by repetitive or stereotyped patterns of behavior and impairments in communication and social interactions. Although affected children appear to be normal at birth, symptoms gradually appear over the course of the first 3 years. Among the signs first noticed by the parents of autistic children are a failure to speak by 16 months of age, poor eye contact, an inability to play with toys, an obsessive attachment to a toy or object, and a failure to smile. Although all individuals with an autism diagnosis will show these traits, the severity varies considerably from one person to the next, as does the association or "comorbidity" with other diagnosable disorders such as intellectual disability and seizures. In recognition of this diversity, clinicians typically use the term "autism spectrum disorder" or ASD to describe this condition. Individuals at one end of the spectrum may never develop language and exhibit severe cognitive impairment. At the other end, individuals may grow up to be socially awkward but intellectually gifted.

ASD is a highly heritable disorder, but the genetics are complex. In some cases, the gene mutations conferring risk for autism occur *de novo,* meaning that they occur sporadically either in the sperm or egg cells of the parents. One risk factor for such sporadic mutations is advanced parental age, especially of fathers. In other cases, the cause seems to be many small mutations passed on from the parents that only manifest as ASD in offspring that get a "double hit." Advances in DNA sequencing technology have enabled the discovery of many of the inherited and sporadic mutations in ASD. The affected genes number in the hundreds, suggesting that disruption of many different cellular processes during brain development can manifest as ASD. Thus, as for the other psychiatric disorders discussed in Chapter 22, a diagnosis of ASD alone does not identify the cause, or etiology, of the underlying disease. The diversity of genetic etiologies partly explains why the symptoms vary so much from one person to the next.

Although abnormal behaviors emerge gradually after birth, there is evidence that in some cases, the stage may be set for ASD during fetal development. For example, researchers recently discovered in postmortem brain samples from autistic children that small patches of frontal cortex had disorganized cortical layers which, as we have learned in this chapter, are formed early in development. Furthermore, many genes implicated in ASD are also known to be important for mid-gestational cortical development.

Imaging studies have shown that autistic children also tend to have accelerated growth of the brain, both gray and white matter, after birth. This finding suggests the brains of autistic infants have too many neurons and too many axons, although changes in glia are also possible. Brain growth is controlled by balancing the genesis and destruction of cells, axons, and synapses and the proteins that comprise them. Mutations that bring this process out of balance, by excessive genesis or reduced destruction, could lead to the abnormal brain growth that is ultimately expressed as the impairments in behavior, communication, and social interactions that characterize autism.

Neuroscientists hope that understanding how the brain normally becomes wired together will suggest therapies to correct the altered trajectory of brain growth in children at risk for autism. Studies of a disease called *fragile X syndrome* (FXS) provide a case in point. FXS, characterized by intellectual disability and ASD, is caused by disruption of the *FMR1* gene that encodes a protein called *FMRP* (introduced in Chapter 2). By knocking this gene out in mice and fruit flies, researchers have been able to identify how brains function differently with this mutation. These studies have shown that FMRP normally serves as a brake on protein synthesis in neurons. In the absence of FMRP, too many proteins are produced. Remarkably, treatments designed to dampen down this excessive protein synthesis have been shown to correct many of the deficits caused by deletion of FMRP in the animal models. These studies have raised the tantalizing possibility that the veil of autism and intellectual disability might be lifted in some cases with appropriate drug therapy.

significant refinements is a large-scale *reduction* in the numbers of all those newly formed neurons and synapses. The development of proper brain function requires a careful balance between the genesis and elimination of cells and synapses (Box 23.4).

Cell Death

Entire populations of neurons are eliminated during pathway formation by a process known as *programmed cell death*. After axons have reached their targets and synapse formation has begun, there is a progressive decline in the number of presynaptic axons and neurons. Cell death reflects competition for **trophic factors**, life-sustaining substances that are provided in limited quantities by the target cells. This process is believed

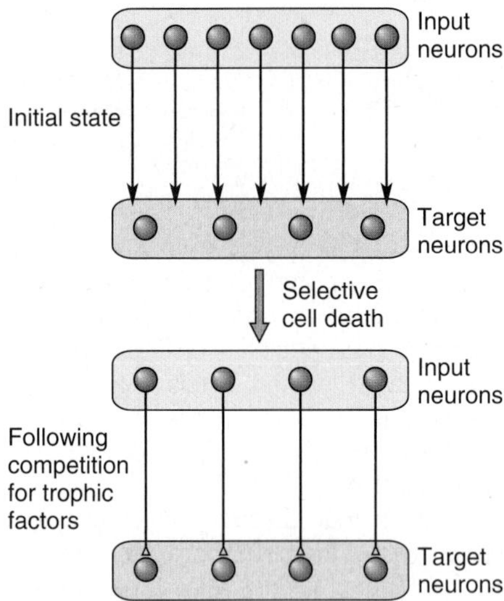

▲ FIGURE 23.17
Matching inputs with targets by selective cell death. The input neurons are believed to compete with one another for limited quantities of trophic factors produced by the target neurons.

to produce the proper match in the number of presynaptic and postsynaptic neurons (Figure 23.17).

A peptide called **nerve growth factor (NGF)** was the first trophic factor to be identified in the 1940s by Italian biologist Rita Levi-Montalcini. NGF is produced by the targets of axons in the sympathetic division of the ANS. Levi-Montalcini and Stanley Cohen found that the injection of antibodies to NGF into newborn mice resulted in total degeneration of the sympathetic ganglia. NGF, produced and released by the target tissue, is taken up by the sympathetic axons and transported retrogradely, where it acts to promote neuronal survival. Indeed, if axoplasmic transport is disrupted, the neurons will die despite the release of NGF by the target tissue. Their pioneering work earned Levi-Montalcini and Cohen the 1986 Nobel Prize.

NGF is one of a family of related trophic proteins collectively called the **neurotrophins**. Family members include the proteins *NT-3*, *NT-4*, and *brain-derived neurotrophic factor (BDNF)*, which is important for the survival of visual cortical neurons. Neurotrophins act at specific cell surface receptors. Most of the receptors are neurotrophin-activated protein kinases, called *trk receptors*, that phosphorylate tyrosine residues on their substrate proteins (recall phosphorylation from Chapter 6). This phosphorylation reaction stimulates a second messenger cascade that ultimately alters gene expression in the cell's nucleus.

The description of cell death during development as "programmed" reflects the fact that it is actually a consequence of genetic instructions to self-destruct. The important discovery of cell death genes by Robert Horvitz at the Massachusetts Institute of Technology was recognized with the 2004 Nobel Prize. It is now understood that neurotrophins save neurons by switching off this genetic program. The expression of cell death genes causes neurons to die by a process called **apoptosis**, the systematic disassembly of the neuron. Apoptosis differs from *necrosis*, which is the accidental cell death resulting from injury to cells. Research on neuronal cell death is proceeding at a rapid pace, fueled by the hope that it might be possible to rescue dying neurons in neurodegenerative disorders, such as Alzheimer's disease (see Box 2.4) and amyotrophic lateral sclerosis (see Box 13.1).

Changes in Synaptic Capacity

Each neuron can receive on its dendrites and soma a finite number of synapses. This number is the *synaptic capacity* of the neuron. Throughout the nervous system, synaptic capacity peaks early in development and then declines as the neurons mature. For example, in the striate cortex of all species examined so far, the synaptic capacity of immature neurons exceeds that of adult cells by about 50%. In other words, visual cortical neurons in the infant brain receive one-and-a-half times as many synapses as do the neurons in adults.

When do cortical neurons lose all those synapses? Yale University scientists Jean-Pierre Bourgeois and Pasko Rakic conducted a detailed study to address this question in the striate cortex of the macaque monkey. They discovered that synaptic capacity was remarkably constant in the striate cortex from infancy until the time of puberty. However, during the subsequent adolescent period, synaptic capacity declined sharply—by almost 50% in just over 2 years. A quick calculation revealed the following startling fact: The loss of synapses in the primary visual cortex during adolescence occurs at an average rate of *5000 per second*. (No wonder adolescence is such a trying time!)

Once again, the neuromuscular junction has provided a useful model for studying synaptic elimination. Initially, a muscle fiber may receive

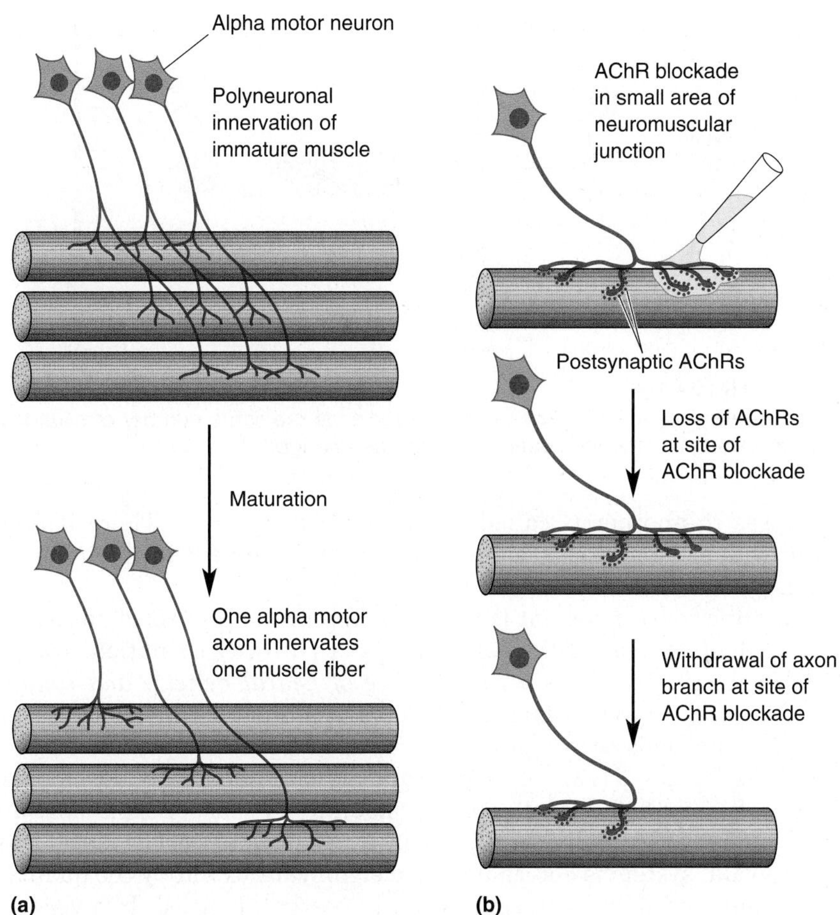

◀ FIGURE 23.18
Synapse elimination. (a) Initially, each muscle fiber receives inputs from several alpha motor neurons. Over the course of development, all inputs but one are lost. (b) Normally, postsynaptic AChR loss precedes the withdrawal of the axon branch. Simply blocking a subset of receptors with α-bungarotoxin can also stimulate synapse elimination.

input from several different motor neurons. Eventually, however, this polyneuronal innervation is lost, and each muscle fiber receives synaptic input from a single alpha motor neuron (Figure 23.18a). This process is regulated by electrical activity in the muscle. Silencing the activity of the muscle fiber leads to a retention of polyneuronal innervation, while stimulation of the muscle accelerates the elimination of all but one input.

Careful observations have revealed that the first change during synapse elimination is the loss of postsynaptic AChRs, followed by the disassembly of the presynaptic terminal and retraction of the axon branch. What causes the receptors to disappear? The answer appears to be insufficient receptor activation in an otherwise active muscle. If receptors are partially blocked with α-bungarotoxin (see Box 5.5), they are internalized and the overlying axon terminal withdraws (Figure 23.18b). However, if *all* the AChRs are blocked, the synapses remain because the muscle is also silent. As we will see in a moment, a similar process appears to occur during the refinement of connections in the CNS.

ACTIVITY-DEPENDENT SYNAPTIC REARRANGEMENT

Imagine a neuron that has a synaptic capacity of six synapses and receives inputs from two presynaptic neurons, A and B (Figure 23.19). One arrangement is that each of the presynaptic neurons provides three synapses. Another arrangement is that neuron A provides one synapse and neuron B provides five. A change from one such pattern of

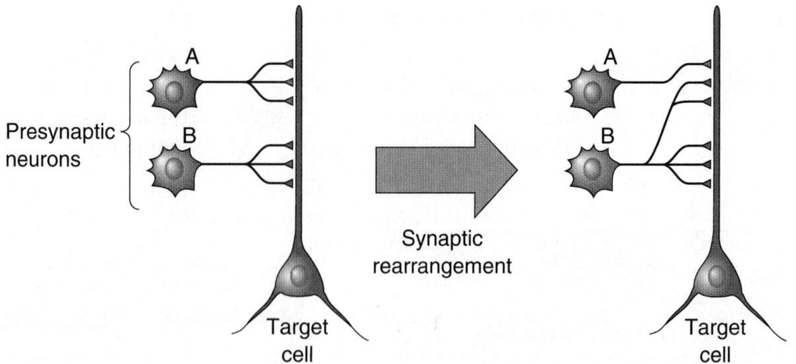

▲ FIGURE 23.19
Synaptic rearrangement. The target cell receives the same number of synapses in both cases, but the innervation pattern has changed.

synapses to another is called *synaptic rearrangement*. There is abundant evidence for widespread synaptic rearrangement in the immature brain.

Synaptic rearrangement is the final step in the process of address selection. Unlike most of the earlier steps of pathway formation, *synaptic rearrangement occurs as a consequence of neural activity and synaptic transmission*. In the visual system, some of this activity-dependent shaping of connections occurs prior to birth in response to spontaneous neuronal discharges. However, significant activity-dependent development occurs after birth and is influenced profoundly by sensory experience during childhood. Thus, we will find that the ultimate performance of the adult visual system is determined to a significant extent by the quality of the visual environment during the early postnatal period. In a very real sense, *we learn to see during a critical period of postnatal development*.

The neuroscientists who pioneered this field were none other than David Hubel and Torsten Wiesel who, you'll recall from Chapter 10, also laid the foundation for our current understanding of the central visual system in the adult brain. In 1981, they shared the Nobel Prize with Roger Sperry. Macaque monkeys and cats were used by Hubel and Wiesel as models for studies of activity-dependent visual system development because, like humans, both of these species have good binocular vision. Recent studies have used rodents because they are better suited for investigation of the underlying molecular mechanisms.

Synaptic Segregation

The precision of wiring achieved by chemical attractants and repellents can be impressive. In some circuits, however, the final refinement of synaptic connections appears to require neural activity. A classic example is the segregation of eye-specific inputs in the cat LGN.

Segregation of Retinal Inputs to the LGN. The first axons to reach the LGN are usually those from the contralateral retina, and they spread out to occupy the entire nucleus. Somewhat later, the ipsilateral projection arrives and intermingles with the axons of the contralateral eye. Then the axons from the two eyes segregate into the eye-specific domains that are characteristic of the adult nucleus. Silencing retinal activity with TTX (tetrodotoxin) prevents this process of segregation (recall that TTX blocks action potentials). What is the source of the activity, and how does it orchestrate segregation?

Since segregation occurs in the womb, prior to the development of photoreceptors, the activity cannot be driven by light stimulation. Rather, it appears that ganglion cells are spontaneously active during this period of fetal development. This activity is not random, however. Studies by Carla Shatz and her colleagues at Stanford University indicate that ganglion cells fire in quasisynchronous "waves" that spread across the retina. The origin of the wave and its direction of propagation may be random, but during each wave, the activity in a ganglion cell is highly correlated with the activity of its nearest neighbors. And because these waves are generated independently in the two retinas, the activity patterns arising in the two eyes are not correlated with respect to each other.

Segregation is thought to depend on a process of synaptic stabilization whereby only retinal terminals that are active at the same time as their postsynaptic LGN target neuron are retained. This hypothetical mechanism of synaptic plasticity was first articulated by Canadian psychologist Donald Hebb in the 1940s. Consequently, synapses that can be modified in this way are called **Hebb synapses**, and synaptic rearrangements of this sort are called **Hebbian modifications**. According to this hypothesis, whenever a wave of retinal activity drives a postsynaptic LGN neuron to fire action potentials, the synapses between them are stabilized (Figure 23.20). Because the activity from the two eyes does not occur at the same time, the inputs will compete on a "winner-takes-all" basis until one input is retained and the other is eliminated. Stray retinal inputs in the inappropriate LGN layer are the losers because their activity does not consistently correlate with the strongest postsynaptic response (which is evoked by the activity of the other eye). In a moment, we'll explore some potential mechanisms for such correlation-based synaptic modification.

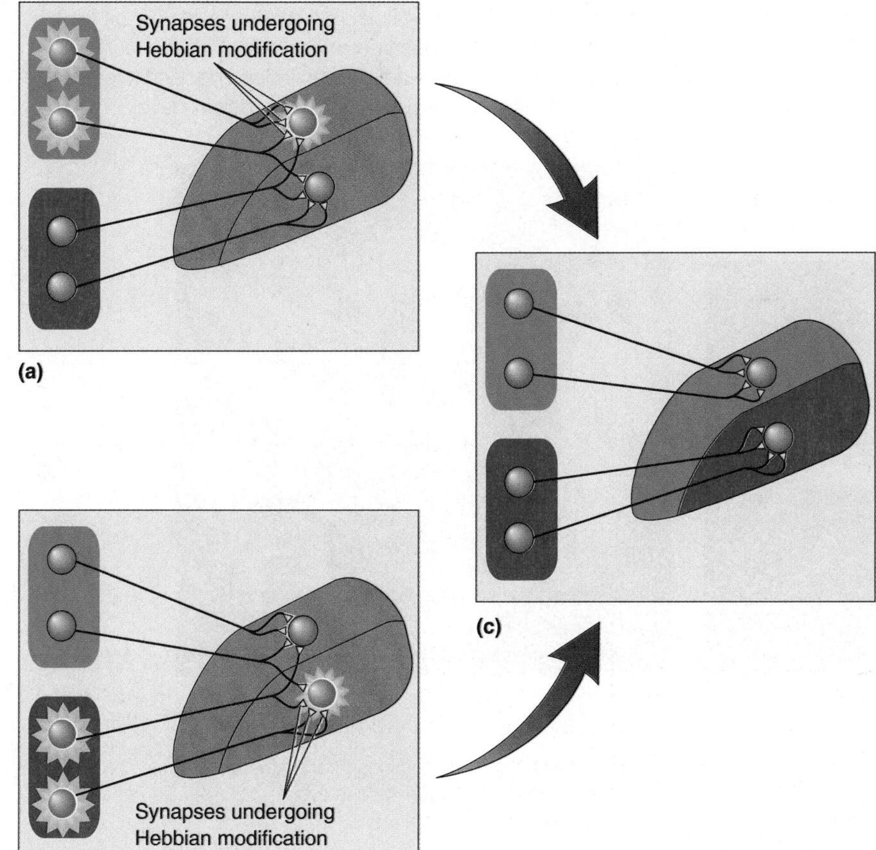

◀ FIGURE 23.20
Plasticity at Hebb synapses. Two target neurons in the LGN have inputs from different eyes. Inputs from the two eyes initially overlap and then segregate under the influence of activity. **(a)** The two input neurons in one eye (top) fire at the same time. This is sufficient to cause the top LGN target neuron to fire but not the bottom one. The active inputs onto the active target undergo Hebbian modification and become more effective. **(b)** This is the same situation as in part **a**, except that now the two input neurons in the other eye (bottom) are active simultaneously, causing the bottom target neuron to fire. **(c)** Over time, neurons that fire together wire together. Notice also that input cells that fire out of sync with the target lose their link.

Segregation of LGN Inputs in the Striate Cortex. In visual cortex of monkeys and cats (but not in most other species), the inputs from LGN neurons serving the two eyes are segregated into ocular dominance columns. This segregation occurs before birth and appears to be due to a combination of molecular guidance cues and retinal activity differences (Box 23.5).

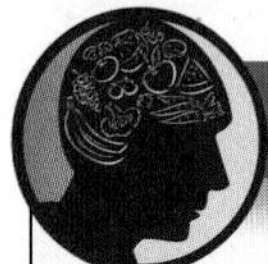

BOX 23.5 BRAIN FOOD

Three-Eyed Frogs, Ocular Dominance Columns, and Other Oddities

Ocular dominance columns (stripes, or bands, depending on how they are viewed) are a peculiar feature of some primates, most notably humans and macaque monkeys, and some carnivores, notably cats and ferrets. For many years, researchers believed that inputs from the two eyes initially overlapped in layer IV of visual cortex of these species and that segregation into alternating columns was based on the comparison of activity generated in the retinas. However, this notion was challenged by the observation that eye-specific inputs to the visual cortex of ferrets can be detected even when development proceeds without any retinal activity at all. This finding suggests that molecular guidance mechanisms, rather than activity patterns, cause segregation into ocular dominance stripes.

It is important to recognize, however, that some problems in development can have more than one solution. The branches of the mammalian family tree leading to modern carnivores and primates diverged very early in evolutionary history, about 95 million years ago. Since most other mammals lack ocular dominance columns, evolutionary biologists believe that the columns in carnivores and primates evolved independently, so we must be cautious when generalizing about the mechanisms of ocular dominance formation.

This point is nicely illustrated by studies of three-eyed frogs conducted in the 1980s by Martha Constantine-Paton and her students, then at Princeton University. Frogs don't have three eyes, of course. Normally they have two eyes, and each retina projects axons exclusively to the contralateral optic tectum. However, by transplanting the eye bud from one embryo into the forebrain area of another, the researchers were able to create a situation where two retinal projections were forced to grow into the same tectum (Figure A, part a). Amazingly, the input segregated into stripes that look very much like the ocular dominance patterns in monkey striate cortex (Figure A, part b). If activity in the retinas was blocked, however, the axons from the two eyes rapidly became intermingled. This experiment proves that differences in activity can indeed be used to segregate inputs, as suggested by Hebbian models of development.

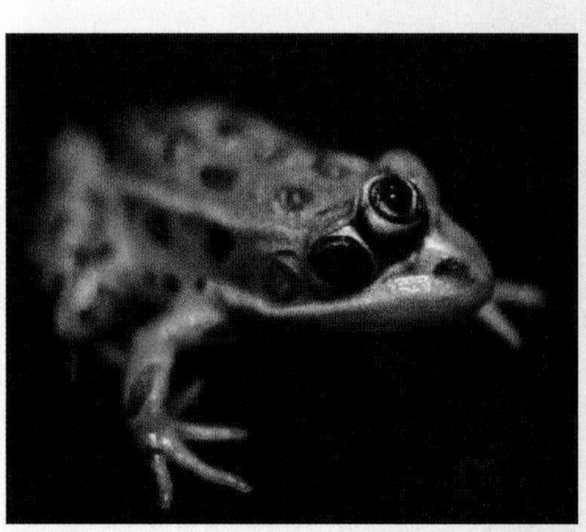

(a)

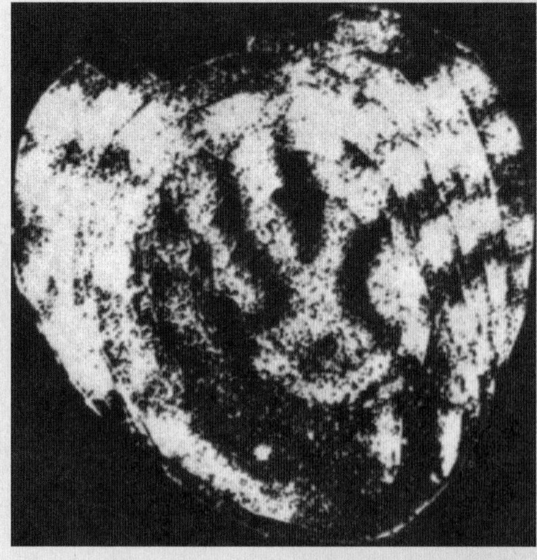

(b)

Figure A

(a) This frog's third eye formed from a transplanted embryonic eye bud. **(b)** Tangential sections through the tectum of a three-eyed frog, illuminated to show the distribution of radioactive axon terminals serving one eye.

(Source: Courtesy of Dr. Martha Constantine-Paton.)

Regardless of how ocular dominance columns form, however, the appearance of segregation does not mean that the axons lose their ability to grow and retract. "Plasticity" of ocular dominance columns in macaques after birth can be dramatically demonstrated by an experimental manipulation used by Wiesel and Hubel, called **monocular deprivation**, in which one eyelid of a young monkey is sealed closed. If monocular deprivation is begun shortly after birth, the striking result is that the "open-eye" columns expand in width while the "closed-eye" columns shrink (Figure 23.21). Moreover, these effects of monocular deprivation can be reversed simply by closing the previously open eye and opening the previously closed eye. The result of this "reversed-occlusion" manipulation is that the shrunken ocular dominance columns of the formerly closed eye expand, and the expanded columns of the formerly open eye shrink. Thus, LGN axons and their synapses in layer IV are highly dynamic, even after birth. Notice that this type of synaptic rearrangement is not just activity dependent but is also *experience dependent* because it relies on the quality of the sensory environment.

The plasticity of ocular dominance columns does not occur throughout life, however. Hubel and Wiesel found that if the deprivation is begun later in life, these anatomical effects are not observed in layer IV. Thus, a **critical period** exists for this type of structural modification. In the macaque monkey, the critical period for anatomical plasticity in layer IV lasts until about 6 weeks of age. After this critical period, the LGN afferents apparently lose their capacity for growth and retraction and, in a sense, are cemented in place.

It is important to appreciate that there are many "critical periods" during development—specific times when developmental fate is influenced by the environment (Box 23.6). In the visual cortex, the end of the critical period for anatomical plasticity in layer IV does not spell the end of the influence of visual experience on cortical development. As we will now see, the synapses in striate cortex remain modifiable by experience until adolescence and beyond.

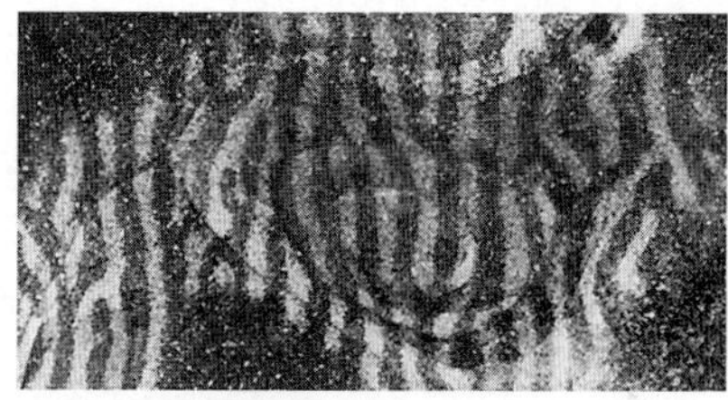
(a)

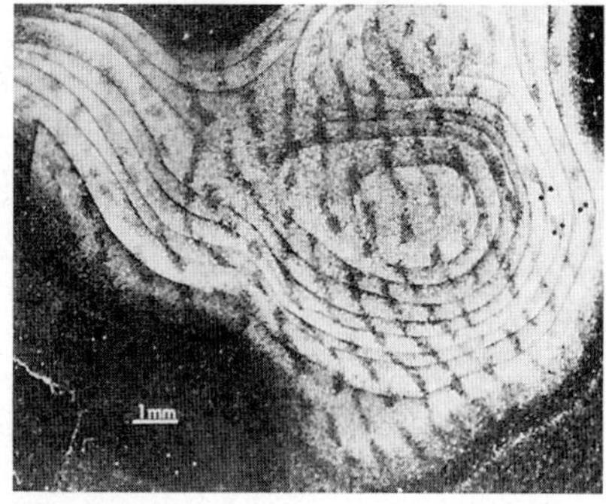

(b)

▲ FIGURE 23.21
Modification of ocular dominance stripes after monocular deprivation. Tangential sections through layer IV of macaque monkey striate cortex illuminated to show the distribution of radioactive LGN terminals serving one eye. **(a)** A normal monkey. **(b)** A monkey that had been monocularly deprived for 22 months, starting at 2 weeks of age. The nondeprived eye had been injected, revealing expanded ocular dominance columns in layer IV. (Source: Wiesel, 1982, p. 585.)

Synaptic Convergence

Although the streams of information from the two eyes are initially segregated, eventually they must be combined to make binocular vision possible. The anatomical basis for binocular vision in species with ocular dominance columns is the convergence of inputs from layer IV cells serving the right and left eyes onto cells in layer III. These are among the last connections to be specified during the development of the retinogeniculocortical pathway. Again, experience-dependent synaptic rearrangement plays a major role in this process.

Binocular connections are formed and modified under the influence of the visual environment during infancy and early childhood. Unlike segregation of eye-specific domains, which evidently depends on *asynchronous* patterns of activity spontaneously generated in the two eyes, the *establishment of binocular receptive fields depends on correlated patterns of activity that arise from the two eyes as a consequence of vision*. This has been demonstrated clearly by experiments that bring the patterns of activity from the two eyes out of register. For example, monocular deprivation, which replaces patterned activity in one eye with random activity, profoundly disrupts the binocular connections in striate cortex. Neurons that normally have binocular receptive fields instead respond only to stimulation of the nondeprived eye. This change in the

BOX 23.6 BRAIN FOOD

The Critical Period Concept

A critical period of development may be defined as a period of time in which intercellular communication alters a cell's fate. The concept is usually credited to the experimental embryologist Hans Spemann. Working around the turn of the twentieth century, Spemann showed that transplantation of a piece of early embryo from one location to another often caused the "donor" tissue to take on the characteristics of the "host," but only if transplantation had taken place during a well-defined time period. Once the transplanted tissue had been induced to change its developmental fate, the outcome could not be reversed. The intercellular communication that altered the physical characteristics (the phenotype) of the transplanted cells was shown to be mediated by both contact and chemical signals.

The term took on new significance with respect to brain development as a result of the work of Konrad Lorenz in the mid-1930s. Lorenz was interested in the process by which young graylag geese become socially attached to their mother. He discovered that in the absence of the mother, the young geese formed social attachment to a wide variety of moving objects, including Lorenz himself (Figure A). Once imprinted on an object, the goslings would follow it and behave toward it as if it were their mother. Lorenz used the word "imprinting" to suggest that this first visual image was somehow permanently etched in the young bird's nervous system. Imprinting was also found to be limited to a finite window of time (the first 2 days after hatching), which Lorenz called the "critical period" for social attachment. Lorenz himself drew the analogy between this process of imprinting the external environment on the nervous system and the induction of tissue to change its developmental fate during critical periods of embryonic development.

Figure A
Konrad Lorenz with graylag geese. (Source: Nina Leen/Time Pix.)

This work had a tremendous impact in the field of developmental psychology. The very terms *imprinting* and *critical period* suggested that changes in the behavioral phenotype caused by early sensory experience were permanent and irreversible later in life, much like the determination of tissue phenotype during embryonic development. Numerous studies extended the critical period concept to aspects of mammalian psychosocial development. The fascinating implication was that the fate of neurons and neural circuits in the brain depended on the experience of the animal during early postnatal life. It is not difficult to appreciate why research in this area took on social as well as scientific significance.

By necessity, the effects of experience on neuronal fate must be exercised by neural activity generated at the sensory epithelia and communicated by chemical synaptic transmission. The idea that synaptic activity can alter the fate of neuronal connectivity during CNS development first received solid neurobiological support from the study of mammalian visual system development, beginning with the experiments of Hubel and Wiesel. Using anatomical and neurophysiological methods, they found that visual experience (or lack thereof) was an important determinant of the state of connectivity in the central visual pathways, and that this environmental influence was restricted to a finite period of early postnatal life. A great deal of work has been devoted to the analysis of experience-dependent plasticity of connections in the visual system. Thus, the visual system is an excellent model for illustrating the principles of critical periods in nervous system development.

binocular organization of the cortex is called an **ocular dominance shift** (Figure 23.22).

These effects of monocular deprivation are not merely a passive reflection of the anatomical changes in layer IV discussed earlier. First, ocular dominance shifts can occur very rapidly, within hours of monocular deprivation, before any gross changes in axonal arbors can be detected (Figure 23.23). Such rapid changes reflect changes in the structure and molecular composition of synapses without a substantial remodeling of axons. Second, ocular dominance shifts can occur at ages well beyond the critical period for changes in LGN axonal arbors.

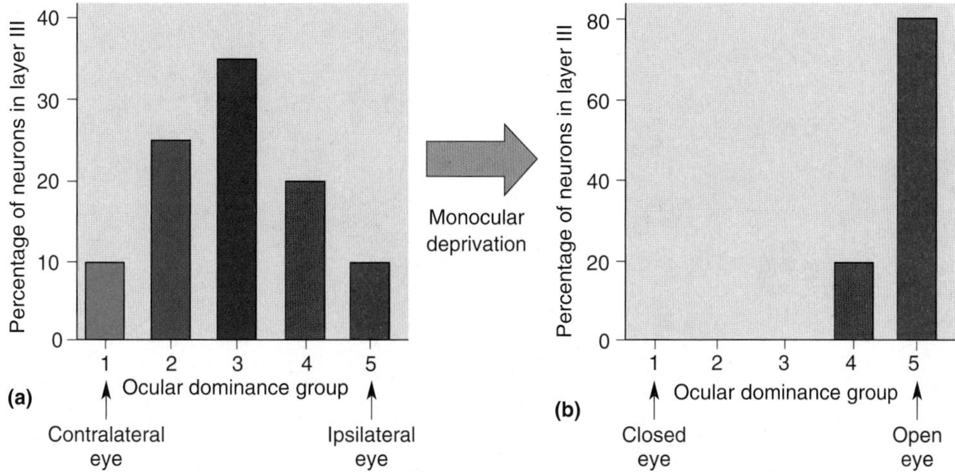

▲ FIGURE 23.22

The ocular dominance shift. These ocular dominance histograms were constructed after electrophysiological recording from neurons in the striate cortex of **(a)** normal cats and **(b)** cats after a period of monocular deprivation early in life. The bars show the percentage of neurons in each of five ocular dominance categories. Cells in groups 1 and 5 are activated by stimulation of either the contralateral or the ipsilateral eye, respectively, but not both. Cells in group 3 are activated equally well by either eye. Cells in groups 2 and 4 are binocularly activated but show a preference for either the contralateral or the ipsilateral eye, respectively. The histogram in part a reveals that the majority of neurons in the visual cortex of a normal animal are driven binocularly. The histogram in part b shows that a period of monocular deprivation leaves few neurons responsive to the deprived eye.

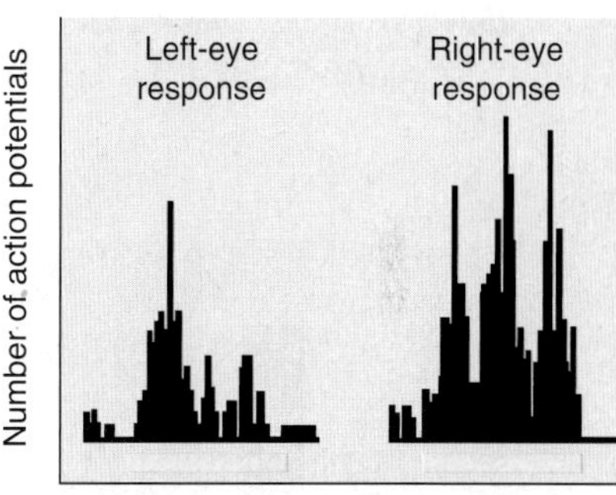

(a) Initial state

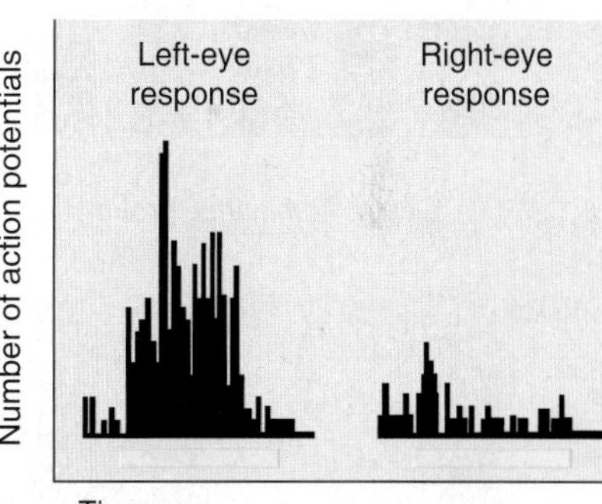

(b) After 17-hour right-eye deprivation

▲ FIGURE 23.23

Rapid shifts in ocular dominance. These histograms show the number of action potentials generated by a single neuron in the visual cortex of a young kitten over time. A visual stimulus was presented at the times indicated by the yellow bars, first to the left eye and then to the right eye. **(a)** Initial responses, before a period of monocular deprivation. Notice that although there is a slight ocular dominance favoring the right eye, each eye evokes a strong response. **(b)** The same neuron, recorded after reopening the right eye following 17 hours of monocular deprivation. The eye that had been deprived is no longer able to evoke a response. (Source: Adapted from Mioche and Singer, 1989.)

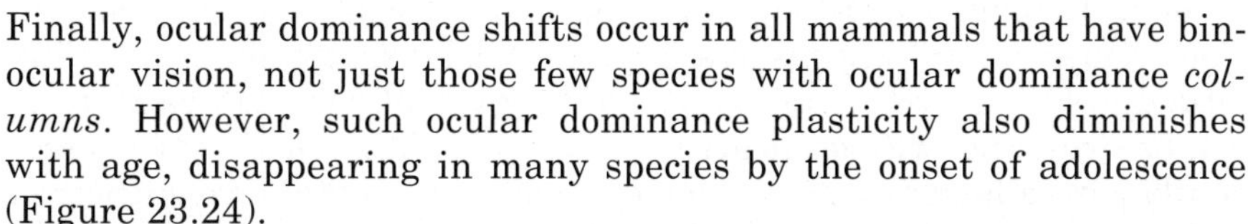

Finally, ocular dominance shifts occur in all mammals that have binocular vision, not just those few species with ocular dominance *columns*. However, such ocular dominance plasticity also diminishes with age, disappearing in many species by the onset of adolescence (Figure 23.24).

The critical period of maximal ocular dominance plasticity coincides with the time of greatest growth of the head and eyes. Thus, it is believed that the plasticity of binocular connections is normally required for maintaining good binocular vision throughout this period of rapid growth. The hazard associated with such activity-dependent fine-tuning is that these connections are also highly susceptible to deprivation.

Synaptic Competition

As you well know, a muscle that is not used regularly will atrophy and lose strength, thus the saying "Use it or lose it." Is the disconnection of activity-deprived synapses simply a consequence of disuse? This does not appear to be the case in striate cortex, because the disconnection of a deprived eye input requires that the open-eye inputs be active. Rather, a process of **binocular competition** evidently occurs, whereby the inputs from the two eyes actively compete for synaptic control of the postsynaptic neuron. If the activity of the two eyes is correlated and equal in strength, the two inputs will be retained on the same cortical cell. However, if this balance is disrupted by depriving one eye, the more active input will somehow displace the deprived synapses or cause them to be less effective.

Competition in visual cortex is demonstrated by the effects of **strabismus**, a condition in which the eyes are not perfectly aligned

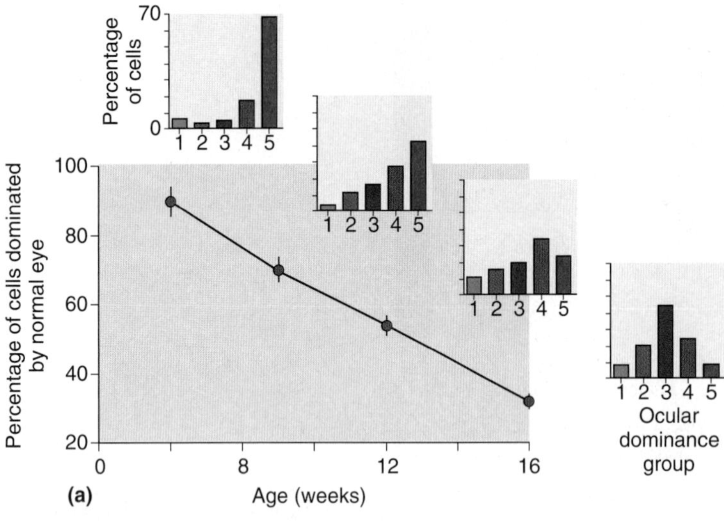

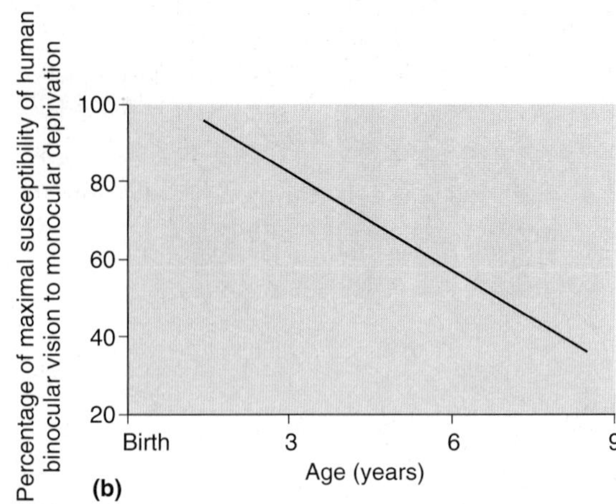

▲ FIGURE 23.24
The critical period for plasticity of binocular connection. These graphs show the sensitivity of binocular connections to monocular deprivation (of the contralateral eye) initiated at different postnatal ages in a kitten. **(a)** The ocular dominance shifts recorded in kittens in response to 2 days of monocular deprivation. The line graph plots the decline in plasticity as a function of age, and the histograms show the corresponding ocular dominance shifts. **(b)** An estimate of the developmental decline of plasticity of binocular connections in humans. (Source: Adapted from Mower, 1991.)

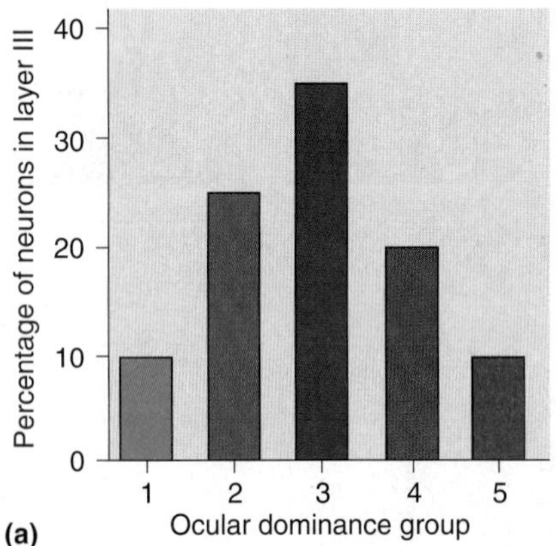

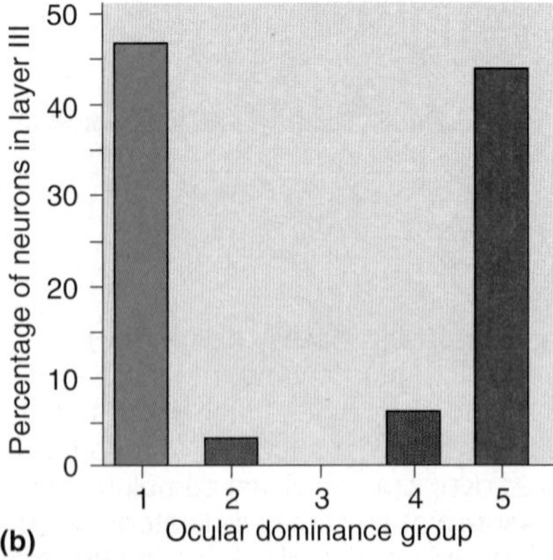

▲ FIGURE 23.25
The effects of strabismus on cortical binocularity. **(a)** An ocular dominance histogram from a normal animal like that in Figure 23.22a. **(b)** In this case, the eyes have been brought out of alignment by cutting one of the eye muscles. After a brief period of strabismus, binocular cells are almost completely absent. The cells in visual cortex are driven by either the right or the left eye but not by both.

(i.e., they are "cross-eyed" or "wall-eyed"). This common visual disorder in humans can result in the permanent loss of stereoscopic vision. Experimental strabismus is produced by surgically or optically misaligning the two eyes, thereby causing visually evoked patterns of activity from the two eyes to arrive out of sync in the cortex. If you press gently with a finger alongside one eye, you can see the consequences of misalignment of the two eyes. A total loss of binocular receptive fields occurs following a period of strabismus, even though the two eyes retain equal representation in the cortex (Figure 23.25). This is a clear demonstration that the disconnection of inputs from one eye results from competition rather than disuse (the two eyes are equally active, but for each cell, one "winner takes all"). If produced early enough, strabismus can also sharpen the segregation of ocular dominance columns in layer IV.

The changes in ocular dominance and binocularity after deprivation have clear behavioral consequences. An ocular dominance shift after monocular deprivation leaves the animal visually impaired in the deprived eye, and the loss of binocularity associated with strabismus completely eliminates stereoscopic depth perception. However, neither of these effects is irreversible if corrected early enough in the critical period. The clinical lesson is clear: Congenital cataracts or ocular misalignment must be corrected in early childhood, as soon as surgically feasible, to avoid permanent visual disability.

Modulatory Influences

With increasing age, there appear to be additional constraints on the forms of activity that will cause modifications of cortical circuits. Before birth, spontaneously occurring bursts of retinal activity are sufficient to orchestrate aspects of address selection in the LGN and cortex. After birth, an interaction with the visual environment is of critical importance. However, even visually driven retinal activity may be insufficient for modifications of binocularity during this critical period. Increasing experimental evidence indicates that such modifications also require the animal to pay attention to visual stimuli and use vision to guide behavior. For example, modifications of binocularity following monocular stimulation do not occur when an animal is kept anesthetized, even though it is known that cortical neurons respond briskly to visual stimulation under this condition. These and related observations have led to the proposal that synaptic plasticity in the cortex requires the release of "enabling factors" that are linked to behavioral state (level of alertness, for example).

Some progress has been made in identifying the physical basis of these enabling factors. Recall that a number of diffuse modulatory systems innervate the cortex (see Chapter 15). These include the noradrenergic inputs from the locus coeruleus and the cholinergic inputs from the basal forebrain. The effects of monocular deprivation have been studied in animals in which these modulatory inputs to striate cortex were eliminated. This was found to cause a substantial impairment of ocular dominance plasticity, even though transmission in the retinogeniculocortical pathway was unaffected by the lesion (Figure 23.26).

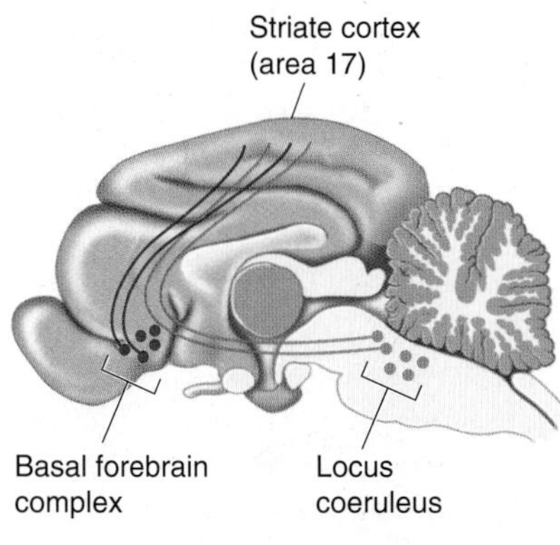

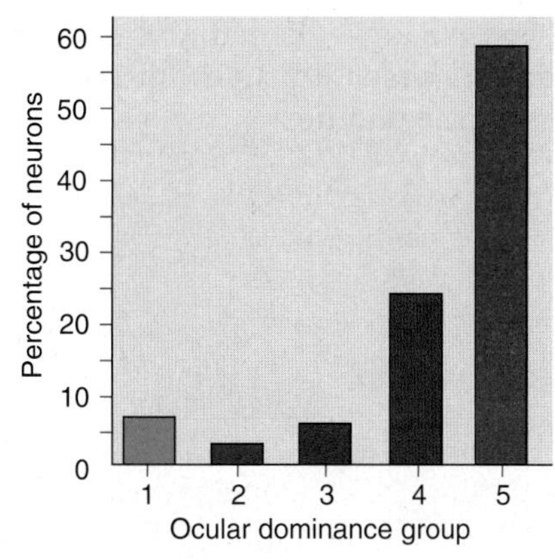

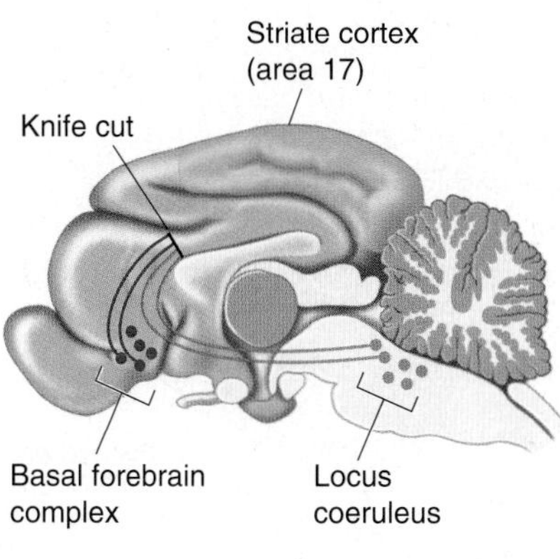

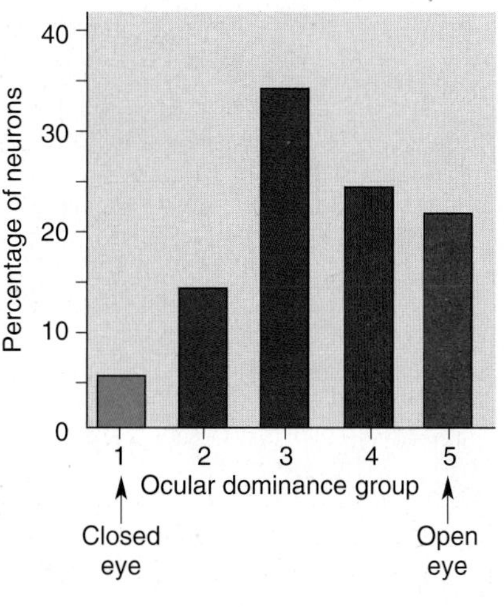

◄ FIGURE 23.26
The dependence of plasticity of binocular connections on modulatory inputs. **(a)** A midsagittal view of a cat brain, showing the trajectory of two types of modulatory input to striate cortex. One, colored green, arises in the locus coeruleus and uses NE as a transmitter, and the other, colored red, arises in the basal forebrain complex and uses ACh as a transmitter. The activity of both of these inputs is related to levels of attention and alertness. If these systems are intact, monocular deprivation will produce the expected ocular dominance shift, shown in the histogram at the right. **(b)** The result of depleting the cortex of these modulatory inputs. Monocular deprivation has little effect on the binocular connections in striate cortex. (Source: Adapted from Bear and Singer, 1986.)

ELEMENTARY MECHANISMS OF CORTICAL SYNAPTIC PLASTICITY

Synapses form in the absence of any electrical activity. However, as we have seen, the awakening of synaptic transmission during development plays a vital role in the final refinement of connections. Based on the analysis of experience-dependent plasticity in the visual cortex and elsewhere, we can formulate two simple "rules" for synaptic modification:

1. When the presynaptic axon is active and, at the same time, the postsynaptic neuron is *strongly activated* under the influence of other inputs, then the synapse formed by the presynaptic axon is strengthened. This is another way of stating Hebb's hypothesis, mentioned previously. In other words, *neurons that fire together wire together*.
2. When the presynaptic axon is active and, at the same time, the postsynaptic neuron is *weakly activated* by other inputs, then the synapse formed by the presynaptic axon is weakened. In other words, *neurons that fire out of sync lose their link*.

The key appears to be *correlation*. Remember that in most locations in the CNS, including the visual cortex, a single synapse has little influence on the firing rate of the postsynaptic neuron. To be "heard," the activity of the synapse must be correlated with the activity of many other inputs converging on the same postsynaptic neuron. When the activity of the synapse consistently correlates with a strong postsynaptic response (and, therefore, with the activity of many other inputs), the synapse is retained and strengthened. When synaptic activity consistently fails to correlate with a strong postsynaptic response, the synapse is weakened and eliminated. In this way, synapses are "validated" based on their ability to participate in the firing of their postsynaptic partner.

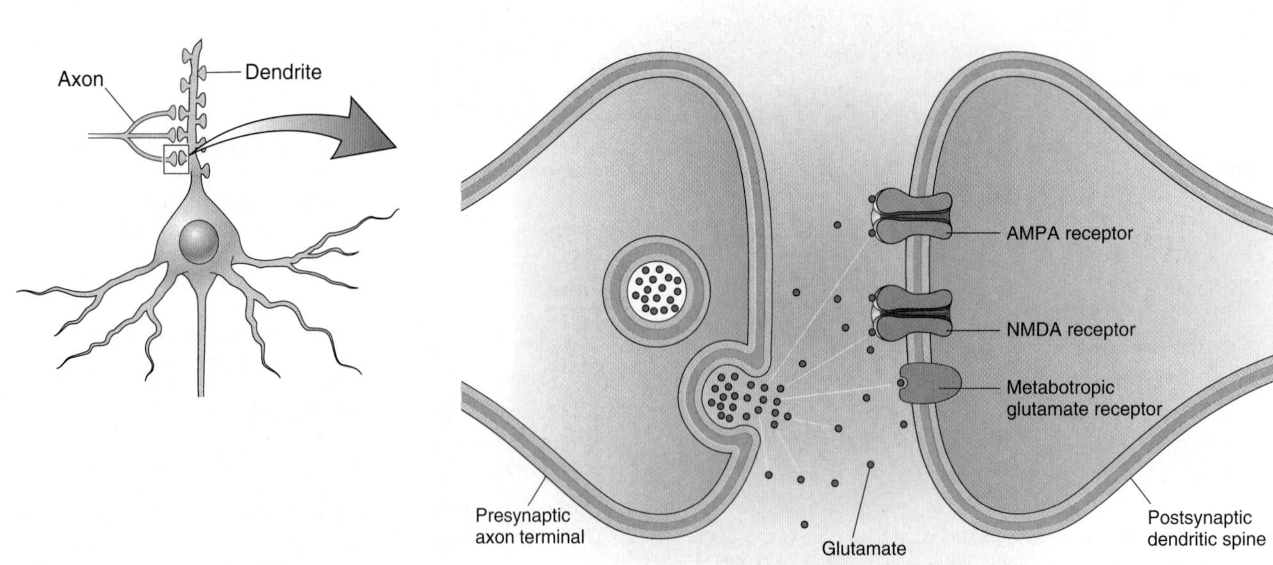

▲ FIGURE 23.27
Glutamate receptors at excitatory synapses.

What mechanisms underlie such correlation-based synaptic modifications? The answer requires knowledge of the mechanisms of excitatory synaptic transmission in the brain.

Excitatory Synaptic Transmission in the Immature Visual System

Glutamate is the transmitter at all of the modifiable synapses we have discussed (retinogeniculate, geniculocortical, and corticocortical), and it activates several subtypes of postsynaptic receptors. Recall from Chapter 6 that neurotransmitter receptors may be classified into two broad categories: G-protein-coupled, or metabotropic, receptors; and transmitter-gated ion channels (Figure 23.27). Postsynaptic glutamate-gated ion channels allow the passage of positively charged ions into the postsynaptic cell and may be further classified as either AMPA receptors or NMDA receptors. AMPA and NMDA receptors are colocalized at many synapses.

An NMDA receptor has two unusual features that distinguish it from an AMPA receptor (Figure 23.28). First, NMDA receptor conductance is voltage-gated, owing to the action of Mg^{2+} at the channel. At the resting membrane potential, the inward current through the NMDA receptor is interrupted by the movement of Mg^{2+} into the channel, where they become lodged. As the membrane is depolarized, however, the Mg^{2+} block is displaced from the channel, and current is free to pass into the cell. Thus, substantial current through the NMDA receptor channel requires the concurrent release of glutamate by the presynaptic terminal *and* depolarization of the postsynaptic membrane. The other distinguishing feature of an NMDA receptor is that its channel conducts Ca^{2+}. Therefore, *the magnitude of the Ca^{2+} flux passing through the NMDA receptor channel specifically signals the level of pre- and postsynaptic coactivation.*

Curiously, when a glutamatergic synapse first forms, only NMDA receptors appear in the postsynaptic membrane. As a consequence, released glutamate at a single synapse evokes little response when the postsynaptic membrane is at the resting potential. Such "silent" synapses announce their presence only when enough of them are active at the same time to cause sufficient depolarization to relieve the Mg^{2+} block of the NMDA receptor channels. In other words, "silent" synapses "speak" only when there is highly correlated activity, the necessary condition for synaptic enhancement during development.

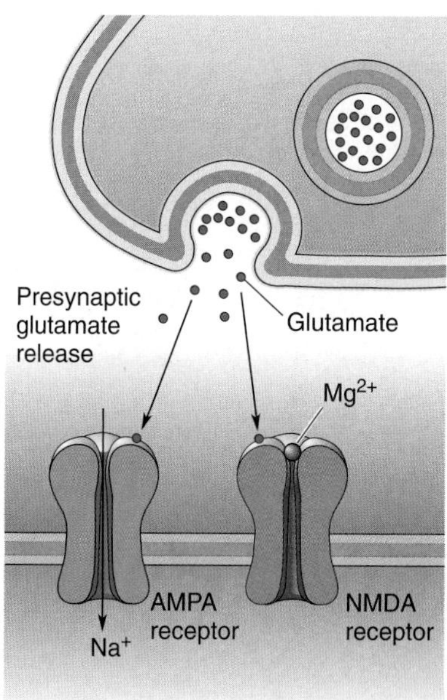

(a) Postsynaptic membrane at **resting** potential

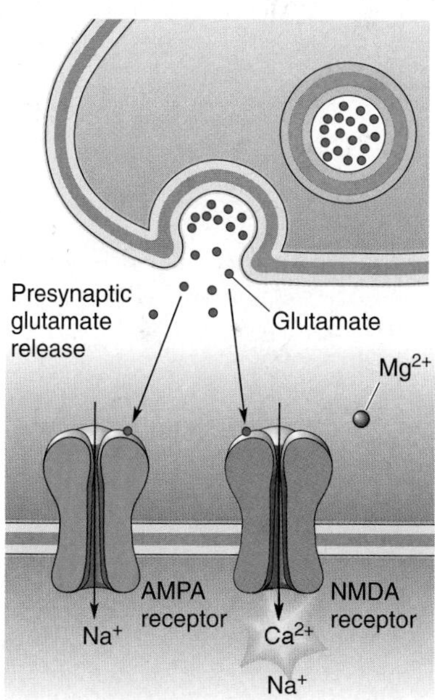

(b) Postsynaptic membrane at **depolarized** potential

▲ FIGURE 23.28

NMDA receptors activated by simultaneous presynaptic and postsynaptic activity. (a) Presynaptic activation causes the release of glutamate, which acts on postsynaptic AMPA receptors and NMDA receptors. At the negative resting membrane potential, the NMDA receptors pass little ionic current because they are blocked with Mg^{2+}. **(b)** If glutamate release coincides with depolarization sufficient to displace the Mg^{2+}, then Ca^{2+} will enter the postsynaptic neuron via the NMDA receptor. Hebbian modification could be explained if the Ca^{2+} admitted by the NMDA receptor were to trigger enhanced synaptic effectiveness.

Long-Term Synaptic Potentiation

Perhaps NMDA receptors serve as Hebbian detectors of simultaneous presynaptic and postsynaptic activity, and Ca^{2+} entry through the NMDA receptor channel triggers the biochemical mechanisms that modify synaptic effectiveness. Tests of this hypothesis have been performed by electrically stimulating axons to monitor the strength of synaptic transmission before and after an episode of strong NMDA receptor activation (Figure 23.29a, b). Results consistently indicate that a consequence of strong NMDA receptor activation is a strengthening of synaptic transmission called **long-term potentiation (LTP)**.

What accounts for LTP of the synapse? One consequence of the strong NMDA receptor activation, and the resulting flood of Ca^{2+} into the postsynaptic dendrite, is the insertion of new AMPA receptors into the synaptic membrane (Figure 23.29c). Such "AMPAfication" of the synapse

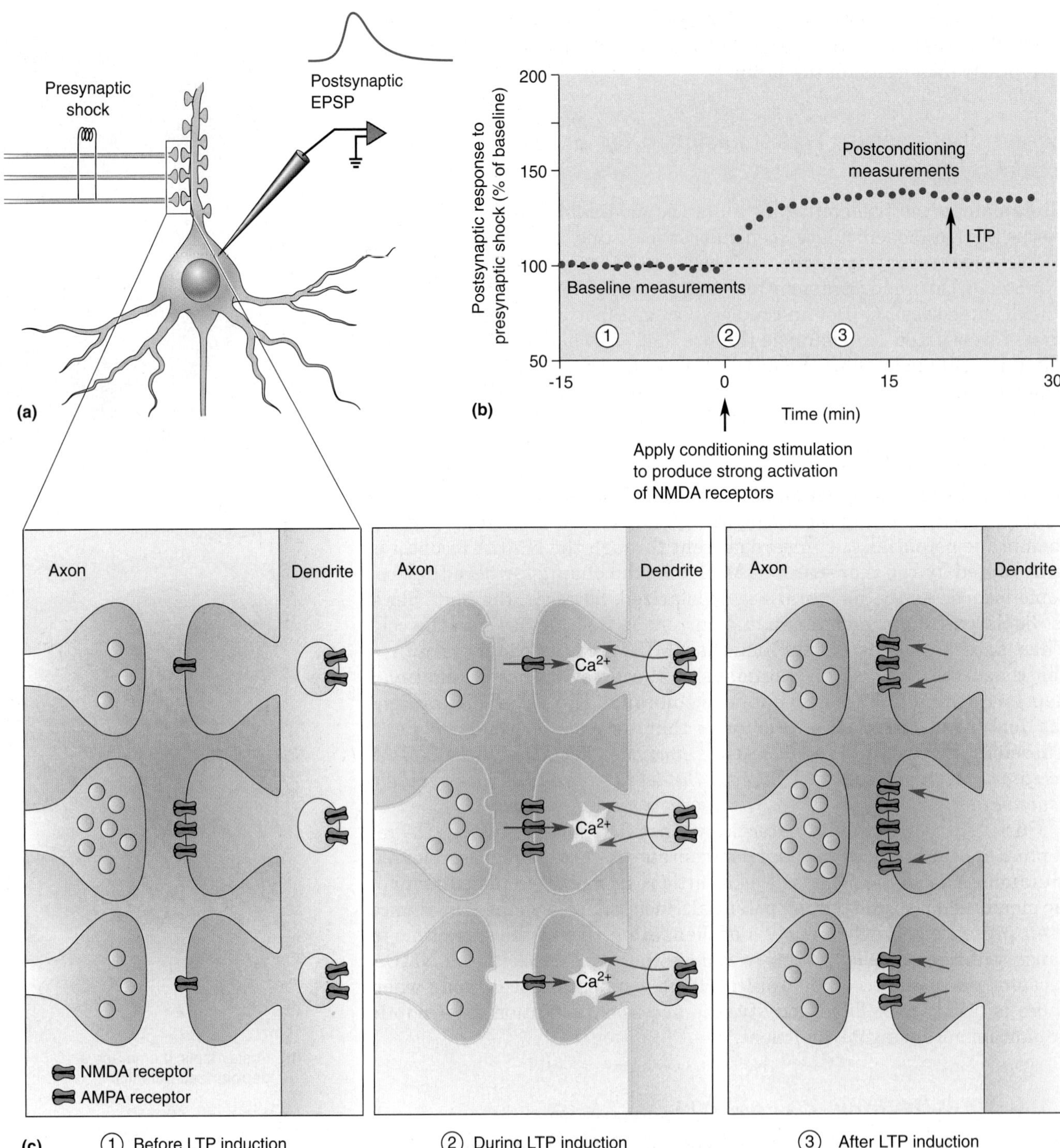

▲ FIGURE 23.29

The lasting synaptic effects of strong NMDA receptor activation. (a) An experiment in which presynaptic axons are stimulated electrically to evoke an action potential and microelectrode recordings of the resulting EPSPs are made from the postsynaptic neuron. **(b)** This graph shows how the strength of synaptic transmission is changed by strong NMDA receptor activation. The conditioning stimulation consists of depolarizing the postsynaptic neuron via current injection through the microelectrode, at the same time that the synapses are repeatedly stimulated. LTP is the resulting enhancement of synaptic transmission. **(c)** LTP at many synapses is associated with the insertion of AMPA receptors into synapses that previously had none. The circled numbers correspond to the times before and after LTP in part b.

makes transmission stronger. In addition to this change in the complement of glutamate receptors, recent evidence suggests that the synapses can actually split in half following LTP induction, forming two different sites of synaptic contact.

Cortical neurons grown in a tissue culture form synapses with one another and become electrically active. The immature synapses contain clusters of NMDA receptors but few AMPA receptors. Consistent with the idea that LTP is a mechanism for synaptic maturation, electrically active synapses gain AMPA receptors over the course of development in cell culture. This change fails to occur, however, if NMDA receptors are blocked with an antagonist. Thus, the strong activation of NMDA receptors that occurs when pre- and postsynaptic neurons fire together appears to account, at least in part, for why they wire together during visual system development. (We will discuss LTP and its molecular basis further in Chapter 25.)

Long-Term Synaptic Depression

Neurons that fire out of sync lose their link. In the case of strabismus, for example, synapses whose activity fails to correlate with that of the postsynaptic cell are weakened and then eliminated. Similarly, during monocular deprivation, the residual activity in the deprived retina fails to correlate with the responses evoked in cortical neurons by the seeing eye, and the deprived-eye synapses are weakened. What mechanism accounts for this form of synaptic plasticity?

In principle, weak coincidences could be signaled by lower levels of NMDA receptor activation and less Ca^{2+} influx. Indeed, experiments suggest that the lower level of Ca^{2+} admitted under these conditions triggers an opposite form of synaptic plasticity, **long-term depression (LTD)**, whereby the active synapses are decreased in effectiveness. One consequence of LTD induction is a loss of AMPA receptors from the synapse, and one long-term consequence of LTD is synapse elimination. Recall that at the neuromuscular junction, the loss of postsynaptic receptors also stimulates the physical retraction of the presynaptic axon.

Studies in the visual cortex of rats and mice have confirmed that AMPA receptors are lost from the surface of visual cortical neurons during monocular deprivation. This change, like the loss of visual responsiveness, requires residual activity in the deprived retina and activation of cortical NMDA receptors. Furthermore, selective inhibition of NMDA receptor-dependent AMPA receptor internalization prevents the ocular dominance shift after monocular deprivation. Thus, it is possible to reconstruct, at least in rough outline, what happens when an animal is monocularly deprived by closing one eyelid (Figure 23.30). Eyelid closure prevents proper image formation on the retina; therefore, it replaces well-correlated retinal ganglion cell activity with less-correlated activity that can be considered as static or noise. This activity, presynaptic to neurons in visual cortex, rarely correlates with a strong postsynaptic response and therefore only weakly activates NMDA receptors. The modest entry of Ca^{2+} through the NMDA receptors initiates a cascade of molecular events that results in the removal of AMPA receptors from the visually deprived synapse. With fewer AMPA receptors, these synapses lose influence over responses of cortical neurons.

• • •

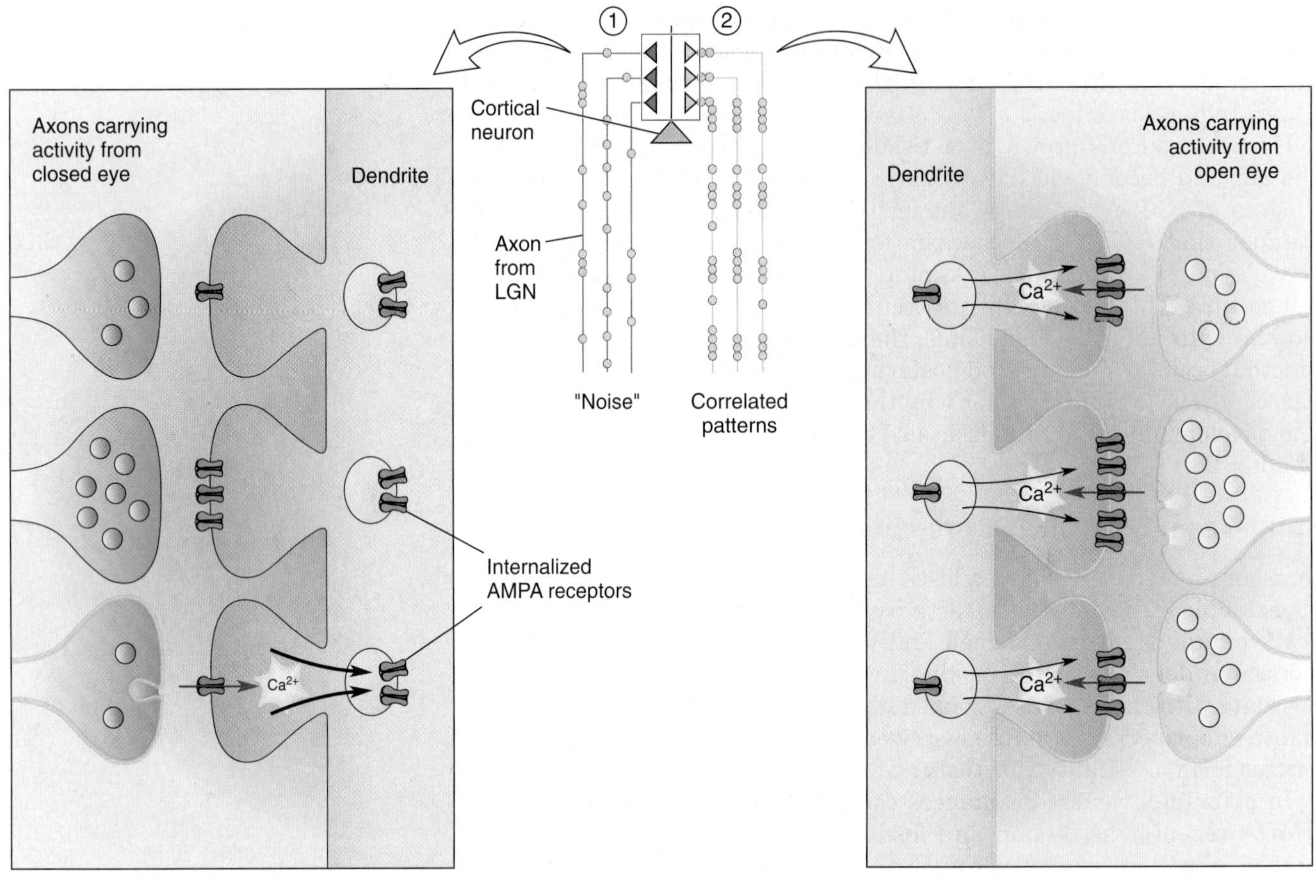

▲ FIGURE 23.30

How brief monocular deprivation leads to reduced visual responsiveness. Closing one eye replaces well-correlated presynaptic action potentials (denoted as yellow dots) with less correlated "noise." In the visual cortex, illustrated here, the noise weakly activates NMDA receptors, and the resulting modest increases in Ca^{2+} cause internalization of AMPA receptors (left enlargement). On the other hand, the well-correlated activity strongly depolarizes the postsynaptic neurons and stimulates large increases in Ca^{2+}, which stimulate AMPA receptor delivery to the synapse (right enlargement).

How are presynaptic and postsynaptic correlations used to refine synaptic connections in the visual system? The data accumulated thus far suggest that the maintenance of some connections formed during development depends on their success in evoking an NMDA receptor-mediated response beyond some threshold level. Failure to achieve this threshold leads to disconnection. Both processes depend on activity originating in the retina, NMDA receptor activation, and postsynaptic Ca^{2+} entry.

WHY CRITICAL PERIODS END

Although plasticity of visual connections persists in the adult brain, the range over which this plasticity occurs constricts with increasing age. Early in development, gross rearrangements of axonal arbors are possible, while in the adult, plasticity appears to be restricted to local changes in synaptic efficacy. In addition, the adequate stimulus for evoking a change also appears to be increasingly constrained as the brain matures. An obvious example is the fact that simply patching one eye will cause a profound alteration in the binocular connections of the superficial layers during infancy, but by adolescence, this type of experience typically fails to cause a lasting alteration in cortical circuitry.

Why do critical periods end? Here are three current hypotheses:

1. *Plasticity diminishes when axon growth ceases.* We've seen that there is a period of several weeks when the geniculate arbors can contract and expand within layer IV under the influence of visual experience. Thus, a factor that limits the critical period in layer IV could be a loss of the capability for changes in axon length, which in turn may be due to changes in the extracellular matrix or the myelination of the axons by oligodendroglia.
2. *Plasticity diminishes when synaptic transmission matures.* The end of a critical period may reflect changes in the elementary mechanisms of synaptic plasticity. Evidence indicates that glutamate receptors change during postnatal development. For example, it has been shown that the activation of metabotropic glutamate receptors stimulates very different postsynaptic responses in striate cortex during the critical period when binocular connections are most susceptible to monocular deprivation. In addition, the molecular composition and properties of NMDA receptors change during the course of the critical period. Accordingly, the properties of LTP and LTD vary with age, and at some synapses, they seem to disappear altogether.
3. *Plasticity diminishes when cortical activation is constrained.* As development proceeds, certain types of activity may be filtered by successive synaptic relays to the point where they no longer activate NMDA receptors, or other elementary mechanisms, sufficiently to trigger plasticity. As mentioned previously, ACh and NE facilitate synaptic plasticity in the superficial cortical layers, perhaps simply by enhancing polysynaptic intracortical transmission. A decline in the effectiveness of these neurotransmitters, or a change in the conditions under which they are released, may contribute to the decline in plasticity. Indeed, evidence suggests that supplementing the adult cortex with NE can restore some degree of modifiability.

Evidence also indicates that intrinsic inhibitory circuitry is late to mature in the striate cortex. Consequently, patterns of activity that may have gained access to modifiable synapses in superficial layers early in postnatal development might be tempered by inhibition in the adult. Consistent with the idea that inhibition regulates the duration of the critical period, recent research in mice has shown that genetic manipulations that accelerate the maturation of GABAergic inhibition in visual cortex also shorten the duration of the critical period for ocular dominance plasticity. Conversely, manipulations that slow the development of inhibition can prolong the critical period.

The question of why critical periods end is important. Synaptic modification and rewiring of circuitry provide the capacity for some recovery of function after CNS damage. However, such recovery is disappointingly limited in the adult brain. On the other hand, the recovery of function after brain injury can be nearly 100% in the immature nervous system when synaptic rearrangements are widespread. Thus, an understanding of how plasticity is regulated during normal development may suggest ways to promote recovery from damage later in life.

CONCLUDING REMARKS

We have seen that the generation of circuitry during brain development occurs mostly prior to birth and is guided by cell-to-cell communication through physical contact and by diffusible chemical signals. Nonetheless, while most of the "wires" find their proper place before birth, the final refinement of synaptic connections, particularly in the cortex, occurs during infancy and is influenced by the sensory environment. Although we have focused on the visual system, other sensory and motor systems are also readily modified by the environment during critical periods of early childhood. In this way, our brain is a product not only of our genes but also of the world in which we grow up.

The end of developmental critical periods does not signify an end to experience-dependent synaptic plasticity in the brain. Indeed, the environment must modify the brain throughout life, or there would be no basis for memory formation. In the next two chapters, we'll explore the neurobiology of learning and memory. We will see that the mechanisms of synaptic plasticity proposed to account for learning bear a close resemblance to those believed to play a role in synaptic rearrangement during development.

KEY TERMS

The Genesis of Neurons
radial glial cell (p. 785)
subplate (p. 788)
neural precursor cell (p. 788)
cortical plate (p. 789)

The Genesis of Connections
growth cone (p. 796)
extracellular matrix (p. 797)
fasciculation (p. 797)
cell-adhesion molecule (CAM) (p. 797)
chemoattractant (p. 797)
netrin (p. 797)
chemorepellent (p. 798)
chemoaffinity hypothesis (p. 799)
ephrin (p. 801)

The Elimination of Cells and Synapses
trophic factor (p. 803)
nerve growth factor (NGF) (p. 804)
neurotrophin (p. 804)
apoptosis (p. 804)

Activity-Dependent Synaptic Rearrangement
Hebb synapse (p. 807)
Hebbian modification (p. 807)
monocular deprivation (p. 809)
critical period (p. 809)
ocular dominance shift (p. 810)
binocular competition (p. 811)
strabismus (p. 811)

Elementary Mechanisms of Cortical Synaptic Plasticity
long-term potentiation (LTP) (p. 815)
long-term depression (LTD) (p. 817)

REVIEW QUESTIONS

1. What do we mean by saying that the cortex develops "inside out"?
2. Describe the three phases of pathway formation. In which phase (or phases) does neural activity play a role?
3. What are three ways that Ca^{2+} is thought to contribute to the processes of synapse formation and rearrangement?
4. How are the elimination of polyneuronal innervation of a muscle fiber and the segregation of retinal terminals in the LGN similar? How do these processes differ?
5. Not long ago, when a child was born with strabismus, the defect was usually not corrected until after adolescence. Today, surgical correction is always attempted during early childhood. Why? How does strabismus affect the connections in the brain, and how does it affect vision?
6. Children are often able to learn several languages apparently without effort, while most adults must struggle to master a second language. From what you know about brain development, why would this be true?
7. Neurons that fire out of sync lose their link. How does this occur?

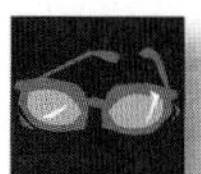

FURTHER READING

Cooke SF, Bear MF. 2014. How the mechanisms of long-term synaptic potentiation and depression serve experience-dependent plasticity in primary visual cortex. *Philosophical Transactions of the Royal Society of London*. Series B, Biological sciences 369, 20130284.

Dehay C, Kennedy H. 2007. Cell-cycle control and cortical development. *Nature Reviews Neuroscience* 8(6):438–450.

Goda Y, Davis GW. 2003. Mechanisms of synapse assembly and disassembly. *Neuron* 40:243–264.

Katz LC, Crowley JC. 2002. Development of cortical circuits: lessons from ocular dominance columns. *Nature Reviews Neuroscience* 3(1):34–42.

McLaughlin T, O'Leary DDM. 2005. Molecular gradients and development of retinotopic maps. *Annual Reviews of Neuroscience* 28:327–355.

Price DJ, Jarman AP, Mason JO, Kind PC 2011. *Building Brains: An Introduction to Neural Development*. Boston: Wiley-Blackwell.

Wiesel T. 1982. Postnatal development of the visual cortex and the influence of the environment. *Nature* 299:583–592.

CHAPTER TWENTY-FOUR

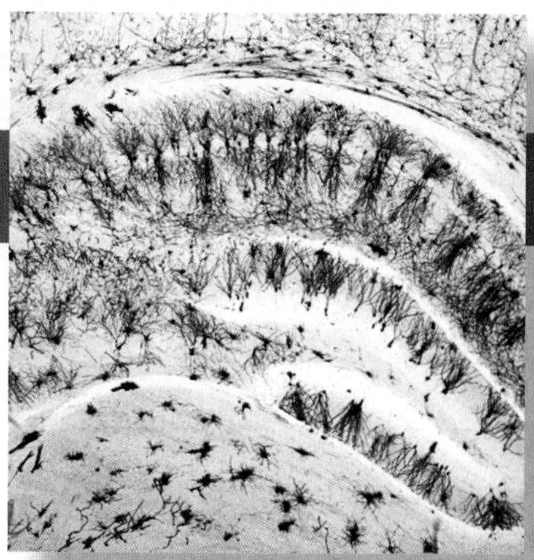

Memory Systems

INTRODUCTION

The brain has numerous systems for performing functions related to sensation, action, and emotion, and each system contains billions of neurons with enormous numbers of interconnections. In Chapter 23, we explored the mechanisms that guide the construction of these systems during brain development. But as impressive and orderly as prenatal development is, no one would confuse a newborn baby and a Nobel Prize winner. Much of the difference between the two comes down to what has been learned and remembered. From the moment we take our first breath, and possibly before, the sensory stimuli we experience modify our brain and influence our behavior. We learn an enormous number of things, some straightforward (e.g., snow is cold), and others more abstract (e.g., an isosceles triangle has two sides of equal length). Some of the things we learn are easily stated facts, while others, such as driving or playing soccer, involve ingrained motor patterns. We will see that brain lesions differentially affect different types of remembered information, suggesting that there is more than one memory system.

There is a close relationship between what we called experience-dependent brain development in Chapter 23 and what we call learning in this chapter. Visual experience during infancy is essential for the normal development of the visual cortex, but it also allows us to recognize an image of our mother's face. Visual development and learning probably involve similar mechanisms, but at different times and in different cortical areas. Understood in this way, learning and memory are lifelong adaptations of brain circuitry to the environment. They enable us to respond appropriately to situations we have experienced before.

In this chapter, we discuss the anatomy of memory—the different parts of the brain involved in storing particular types of information. Chapter 25 will then focus on the elementary molecular mechanisms that can store information in the brain.

TYPES OF MEMORY AND AMNESIA

Learning is the acquisition of new knowledge or skills. **Memory** is the retention of learned information. We learn and remember lots of different things, and it is important to appreciate that these various things might not be processed and stored by the same neural hardware. No single brain structure or cellular mechanism accounts for all learning. Moreover, the way in which information of a particular type is stored may change over time.

Declarative and Nondeclarative Memory

Psychologists have studied learning and memory extensively and have distinguished what appear to be different types. Useful for our purposes is the distinction between declarative memory and nondeclarative memory.

During the course of our lives, we learn many facts (e.g., the capital of Thailand is Bangkok, Darth Vader is Luke Skywalker's father). We also store memories of life's events (e.g., "Yesterday's neuroscience exam was fun!" or "I went swimming with my pet dog named Axon when I was five years old.") Memory of facts and events is called **declarative memory** (Figure 24.1). A declarative memory distinction we will examine later on is between *episodic memory* for autobiographical life experiences and *semantic memory* for facts. Declarative memory is what people usually

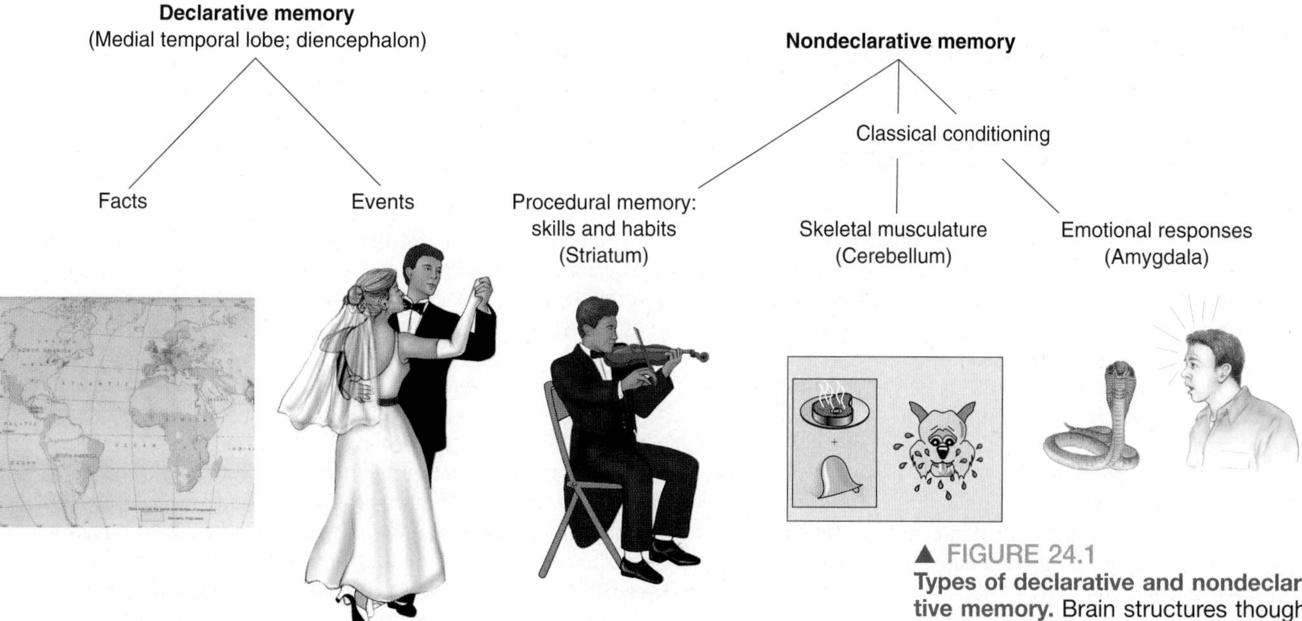

▲ FIGURE 24.1
Types of declarative and nondeclarative memory. Brain structures thought to be involved in each type of memory are indicated. (Note that this is not a complete representation of all types of memory.)

mean in everyday uses of the word "memory," but we actually remember many other things too. These **nondeclarative memories** fall into several categories. The type we are most concerned with here is **procedural memory**, or memory for skills, habits, and behaviors. We learn to play the piano, throw a Frisbee, or tie our shoes, and somewhere that is stored in our brain.

Generally, declarative memories are accessed for conscious recollection, whereas the tasks we learn, as well as the reflexes and emotional associations we have formed, operate smoothly without conscious recollection. As the old saying goes, you never forget how to ride a bicycle. You may not explicitly remember the day you first rode a two-wheeler on your own (the declarative part of the memory), but your brain remembers what to do when you're on one (the procedural part of the memory). Nondeclarative memory is also frequently called *implicit memory,* because it results from direct experience, and declarative memory is often called *explicit memory*, because it results from more conscious effort.

Another distinction is that declarative memories are often easy to form and are easily forgotten. In contrast, forming nondeclarative memories usually require repetition and practice over a longer period of time, but these memories are less likely to be forgotten. Think of the difference between memorizing the names of people you meet at a party (declarative) and learning to ski (nondeclarative). While there is no clear limit to the number of declarative memories the brain can store, there is great diversity in the ease and speed with which such new information is acquired. Studies of humans with abnormally good memories suggest that the limit on the storage of declarative information is remarkably high (Box 24.1).

Types of Procedural Memory

The type of nondeclarative memory we will focus on is procedural memory, which involves learning a motor response (procedure) in reaction to a sensory input. The formation of procedural memories occurs through two categories of learning: nonassociative learning and associative learning.

BOX 24.1 OF SPECIAL INTEREST

Extraordinary Memory

Some people have astonishing memory abilities, and these cases suggest that human memory capacity may, in general, be incredibly large. For example, the British artist Stephen Wiltshire draws enormous cityscapes from memory; his most elaborate is a 10-meter-wide accurate drawing of Tokyo that he drew in 7 days after only a 30-minute helicopter ride over the city. In rare cases of hyperthymesia, also known as superior autobiographical memory, people have explicit memories for almost every day of their lives. The American actress Marylu Henner has this ability.

One of the oldest and best documented cases of an extraordinary memory was documented by the Russian psychologist Alexander Luria. In the 1920s, a man named Solomon Shereshevsky came to see Luria, beginning a 30-year study of the uncommon memory of this man Luria referred to simply as S. Luria published his fascinating description of this study in his book, *The Mind of a Mnemonist*. Luria initially studied S. by giving him conventional tests, such as memorizing lists of words, numbers, or nonsense syllables. He'd read the list once and then ask S. to repeat it. Much to Luria's surprise, he couldn't come up with a test that S. could not pass. Even when 70 words were read in a row, S. could repeat them forward, backward, and in any other order. During the many years they worked together, Luria never found a limit to S.'s memory. In tests of his retention, S. demonstrated that he remembered lists he had previously seen even 15 years earlier!

How did he do it? S. described several factors that may have contributed to his great memory. One was his unusual sensory response to stimuli; he retained vivid images of things he saw. When shown a table of 50 numbers, he claimed it was easy to later read off numbers in one row or along the diagonal because he simply called up a visual image of the entire table. Interestingly, when he occasionally made errors in recalling tables of numbers written on a chalkboard, they appeared to be "reading" errors rather than memory errors. For instance, if the handwriting was sloppy, he would mistake a 3 for an 8 or a 4 for a 9. It was as if he were seeing the chalkboard and numbers all over again when he was recalling the information.

Another interesting aspect of S.'s sensory response to stimuli was a powerful form of synesthesia. *Synesthesia* is a phenomenon in which sensory stimuli evoke sensations usually associated with stimuli of a different sense or different stimuli in the same sense. For example, when S. heard a sound, in addition to hearing, he would see splashes of colored light and perhaps have a certain taste in his mouth. The multimodal response to sensory input may have caused the brain to form particularly strong memory traces.

After learning that his memory was unusual, S. left his job as a reporter and became a professional stage performer—a mnemonist. In order to remember huge lists of numbers or tables of words given by members of the audience trying to stump him, he complemented his lasting sensory responses to stimuli and his synesthesia with memory "tricks." To remember a long list of items, he made use of the fact that each item evoked some sort of visual image. As the list was read or written, S. imagined himself walking through his home town; as each item was given, he placed its evoked image along his walk—the image evoked by item 1 by the mailbox, the image for item 2 by a bush, and so on. To later recall the items, he walked the same route and picked up the imaginary items he had put down. Though we may not have the complex synesthetic sensations of S., this ancient technique of making associations with familiar objects is one we all can use.

But not everything about S.'s memory was to his advantage. While the complex sensations evoked by stimuli helped him remember lists of numbers and words, they interfered with his ability to integrate and remember more complex things. He had trouble recognizing faces because each time a person's expression changed, he would also "see" changing patterns of light and shade, which would confuse him. He also wasn't very good at following a story read to him. Rather than ignoring the exact words and focusing on the important ideas, S. was overwhelmed by an explosion of sensory experiences. Imagine how bewildering it would be to be bombarded by constant visual images evoked by each word, plus sounds and images evoked by the tone of voice of the person reading the story.

S. also experienced the inability to forget. This became a particular problem when performing as a professional mnemonist and he was asked to remember things written on a chalkboard. He would see things that had been written there on many different occasions. Although he tried various tricks to try to forget old information, such as mentally erasing the board, nothing worked. Only by the strength of his attention, and by actively telling himself to let information slip away, was he able to forget. It was as if the effort most of us use for remembering and the ease with which we forget were reversed for S.

We don't know the neural basis for S.'s remarkable memory. Perhaps he lacked the segregation most of us have between sensations in different sensory systems. This may have contributed to an uncommonly strong multimodal coding of memories. Maybe his synapses were more malleable than normal. Unfortunately, we'll never know.

Nonassociative Learning. **Nonassociative learning** describes a change in behavioral response that occurs over time in response to a single type of stimulus. There are two types: habituation and sensitization.

Suppose you live in a house with a single telephone. When the phone rings, you run to answer it, but every time the call is for someone else. Over time, you stop reacting to the ringing of the phone and eventually no longer even notice it. This type of learning, **habituation**, is learning to ignore a stimulus that lacks meaning (Figure 24.2a). You are habituated to a lot of stimuli. Perhaps as you read this sentence, cars and trucks are passing by outside, a dog is barking, your roommate is playing the same tune for the hundredth time—and all this goes on without your really noticing. You have habituated to these stimuli.

Now suppose you're walking down the sidewalk on a well-lit city street at night, and suddenly there is a blackout. You hear footsteps behind you, and though this wouldn't normally disturb you, now you nearly jump out of your skin. Car headlights appear, and you react by side-stepping away from the street. The strong sensory stimulus (the blackout) caused **sensitization**, a form of learning that intensifies your response to all stimuli, even ones that previously evoked little or no reaction (Figure 24.2b).

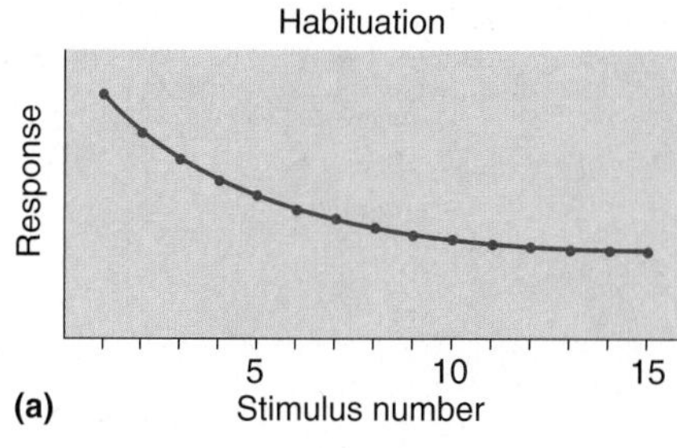

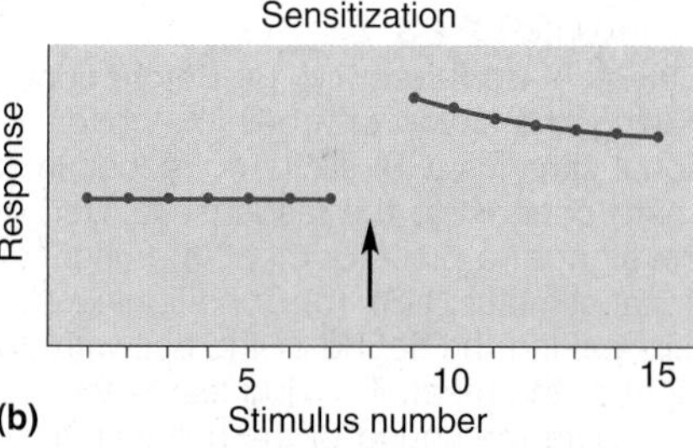

▲ FIGURE 24.2
Types of nonassociative learning. **(a)** In habituation, repeated presentation of the same stimulus produces a progressively smaller response. **(b)** In sensitization, a strong stimulus (arrow) results in an exaggerated response to all subsequent stimuli.

Associative Learning. In **associative learning** behavior is altered by the formation of associations between events; this is in contrast to a changed response to a single stimulus in nonassociative learning. Two types of associative learning are usually distinguished: classical conditioning and instrumental conditioning.

Classical conditioning was discovered and characterized in dogs by the famous Russian physiologist Ivan Pavlov around the turn of the nineteenth century. Classical conditioning involves associating a stimulus that evokes a measurable response with a second stimulus that normally does not evoke this response. The first type of stimulus, the one that normally evokes the response, is called the *unconditional stimulus (US)* because no training (conditioning) is required for it to yield a response. In Pavlov's experiments, the US was the sight of a piece of meat, and the dog's response was salivation. The second type of stimulus, the one that normally does not evoke this same response, is called the *conditional stimulus (CS)* because this one requires training (conditioning) before it will yield this response. In Pavlov's experiments, the CS was an auditory stimulus, such as the sound of a bell. Training consisted of repeatedly *pairing* the presentation of the meat with the sound of the bell (Figure 24.3a). After many of these pairings the meat was withheld, and the animal salivated to the sound alone. The dog had learned an association between the sound (CS) and the presentation of meat (US) (Figure 24.3b). The learned response to the conditioned stimulus is called the *conditioned response (CR).*

Instrumental conditioning was discovered and studied by Columbia University psychologist Edward Thorndike early in the last century. In instrumental conditioning, an individual learns to associate a response, a motor act, with a meaningful stimulus, typically a reward such as food. For example, consider what happens when a hungry rat is placed in a box with a lever that dispenses food. In the course of exploring the box, the rat bumps the lever and out pops a piece of food. After this happy accident occurs a few more times, the rat learns that pressing the lever leads to a food reward. The rat will then work the lever (and eat the food) until it is no longer hungry. As in classical conditioning, a predictive relationship is learned during instrumental conditioning. In classical conditioning, the subject learns that one stimulus (CS) predicts another stimulus (US). In instrumental conditioning, the subject learns that a particular behavior is

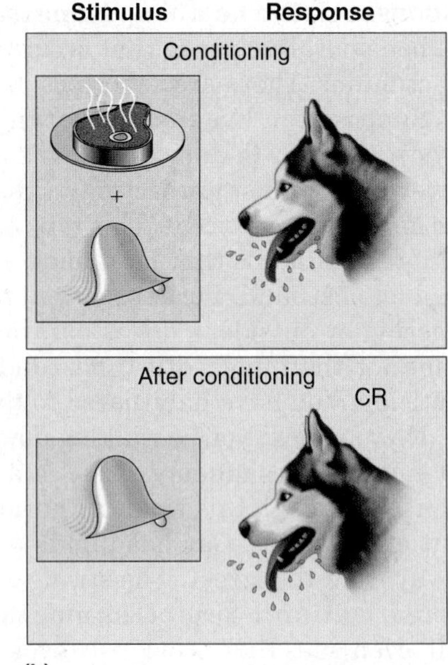

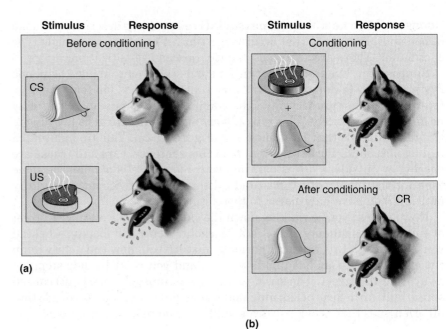

▶ FIGURE 24.3
Classical conditioning. **(a)** Before conditioning, the sound of a bell (the conditional stimulus, CS) elicits no response, in sharp contrast to the response elicited by the sight of a piece of meat (the unconditional stimulus, US). **(b)** Conditioning entails pairing the sound of the bell with the sight of the meat. The dog learns to associate the sound of the bell with the meat and after conditioning will salivate when the bell rings without the meat.

associated with a particular consequence. Because motivation plays such a large part in instrumental conditioning (after all, only a hungry rat will lever-press for a food reward), the underlying neural circuits are considerably more complex than those involved in simple classical conditioning.

Types of Declarative Memory

From daily experience we know that some memories last longer than others. **Long-term memories** are those that you can recall days, months, or years after they were originally stored. The information that makes it into long-term memory, of course, represents only a fraction of what we experience every day. Most information is held by the brain only temporarily, on the order of hours. These **short-term memories** have in common the property that they are vulnerable to disruption. For example, short-term memory can be erased by head trauma or electroconvulsive therapy (ECT) used to treat psychiatric illness. But the same trauma and ECT do not affect long-term memories, which were stored long ago (e.g., childhood memories). These observations have led to the idea that facts and events are stored in short-term memory and a subset are converted into long-term memories via a process called **memory consolidation** (Figure 24.4).

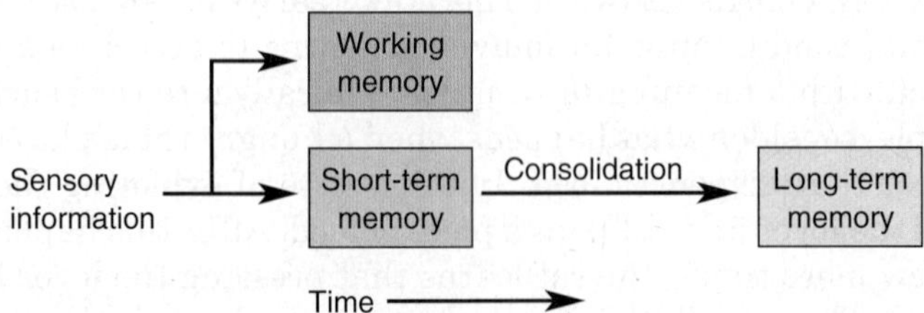

▲ FIGURE 24.4
Memory Consolidation. Sensory information can be temporarily stored in short-term memory that is susceptible to disruption. Stable long-term memories are formed by consolidation. Another type of memory, working memory, is used to hold information "in mind."

A second, entirely distinct form of temporary storage, lasting on the order of seconds, is **working memory**. Unlike the short-term memory discussed above, working memories are sharply limited in capacity and require rehearsal. It is often said that working memory is information held "in mind." When someone tells you his or her phone number, you can retain it for a limited period of time by repeating the number to yourself. Keeping a memory alive through repetition is a hallmark of working memory. If the number is too long (e.g., a phone number with extra numbers for a foreign country), you may have trouble remembering the number at all. Eventually the number may be consolidated into a long-term memory. Working memory is commonly studied by measuring a person's *digit span*, the maximum number of randomly chosen numbers a person can repeat back after hearing a list read. The normal digit span is seven plus or minus two. Working memory is distinguished from short-term memory by the very limited capacity, the need for repetition, and the very short duration.

Interestingly, there are reports of humans with cortical lesions who have normal memory for information coming from one sensory system (e.g., they can remember the same number of visually seen numbers as other people) but a profound deficit when information comes from another sensory system (e.g., they cannot remember more than one number spoken to them). These different digit spans in different modalities are consistent with the notion of multiple temporary storage areas in the brain.

Amnesia

As we all know, in daily life, forgetting happens nearly as often as learning. Less commonly, certain diseases and injuries to the brain cause a serious loss of memory and/or the ability to learn called **amnesia**. Concussion, chronic alcoholism, encephalitis, brain tumor, and stroke can all disrupt memory. You've probably seen a movie or television show in which a person experiences some trauma and wakes up the next day not knowing who he or she is and not remembering the past. That kind of absolute amnesia for past events and information is actually extremely rare. It is more common for trauma to cause limited amnesia along with other nonmemory deficits. If amnesia is not accompanied by any other cognitive deficit, it is known as *dissociated amnesia* (i.e., the memory problems are dissociated from any other problems). We will focus on cases of dissociated amnesia because a clear relationship can be drawn between memory deficits and brain injury.

Following trauma to the brain, two different types of memory loss may occur: retrograde amnesia and anterograde amnesia (Figure 24.5).

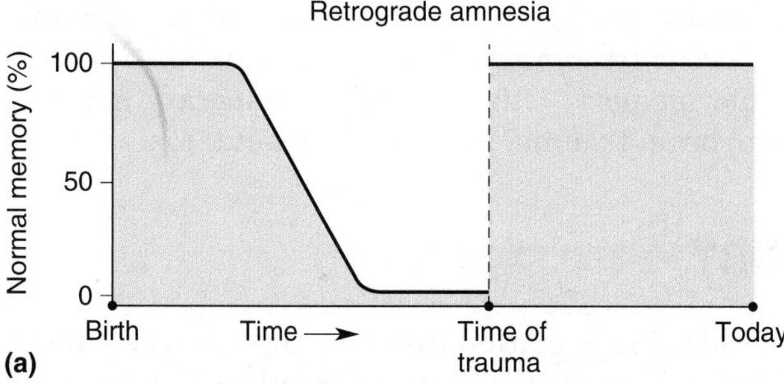

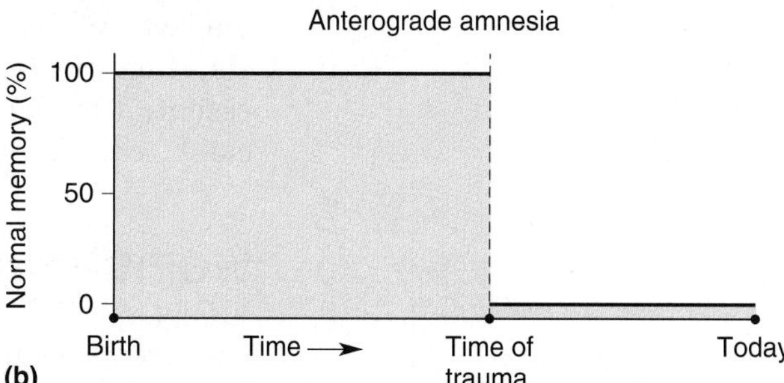

▲ FIGURE 24.5

Amnesia produced by trauma to the brain. **(a)** In retrograde amnesia, events for a period of time prior to the trauma are forgotten, but memories from the distant past and the period following the trauma are intact. **(b)** In anterograde amnesia, events prior to the trauma can be remembered, but there are no memories for the period following the trauma.

Retrograde amnesia is characterized by memory loss for events prior to the trauma; you forget things you already knew. In severe cases, there might be complete amnesia for all declarative information learned before the trauma. More often, retrograde amnesia follows a pattern in which events of the months or years preceding the trauma are forgotten, but memory is increasingly strong for older memories. The graded loss of old memories across time apparently reflects the changing nature of memory storage, a topic we explore in Chapter 25. **Anterograde amnesia** is an inability to form new memories following brain trauma. If the anterograde amnesia is severe, a person might be completely incapable of learning and remembering anything new. In milder cases, learning may be slower and require more repetition than normal. In clinical cases, there is often a mixture of retrograde and anterograde amnesias of different degrees of severity.

An example will help clarify. Suppose that on the last day of freshman year in college you are walking past a friend's dormitory. In a fit of excitement about the end of the semester, your friend throws her books out the window, crashing on your head. If this trauma causes you to have retrograde amnesia, you may be unable to remember the final exam you took the day before or, in a more serious case, any of the courses you took freshman year. If you experience anterograde amnesia, you may recall the exams you took before the accident, but when you graduate from college you might be unable to recall the ambulance ride to the hospital after the accident, your friend's endless apologies, or even the summer you spent recovering after freshman year.

A form of amnesia that involves a much shorter period of time is called *transient global amnesia*. This is a sudden onset of anterograde amnesia that lasts for only a period of minutes to days, often accompanied by retrograde amnesia for recent events preceding the attack. During the spell, the person may appear disoriented and ask the same questions repeatedly, but he or she is conscious, and measures of working memory, such as digit span, are normal. In a matter of hours the attack usually subsides, and the person is left with a permanent memory gap.

Transient global amnesia can be frightening to both the person experiencing it and those witnessing it. Although the cause has not been clearly established, brief cerebral ischemia, in which the blood supply to the brain is temporarily reduced, or concussion to the head from trauma, such as a car accident or a hard blow while playing football, might be implicated. There have been reports of transient global amnesia brought on by seizures, physical stress, drugs, cold showers, and even sex, presumably because all of these affect cerebral blood flow. Many cases were linked to use of the antidiarrheal drug clioquinol (which was taken off the market). While we don't know exactly what causes transient global amnesia, it may be a consequence of temporary blood deprivation to structures essential for learning and memory. Other forms of temporary amnesia can be caused by disease, brain trauma, and environmental toxins.

WORKING MEMORY

Our brains acquire all kinds of information through our sensory systems, but as discussed in Chapter 21, we pay attention to only a fraction of it. To serve immediate behavioral needs, some of this sensory information is "held in mind" by working memory, such as a phone number that we must remember in order to call. Unlike long-term memory, working memory has a very small capacity, as shown by the digit span described earlier.

However, there are subtleties to the quantification of working memory capacity. For example, more words can be held in memory if they are short common words. Also, more words and numbers can be held in working memory if they can be chunked into meaningful groups (e.g., a 12-digit number is easily held when chunked into three years, such as 1945 1969 2001). Working memory can be thought of as a limited resource that can be used in a variety of ways; there are tradeoffs in the amount and precision of stored information that are influenced by the behavioral significance of the information.

Information held in working memory might be converted into long-term memories, but most of it is discarded when no longer needed. How is information retained in the brain by working memory long enough to be useful? Research in both animals and humans suggests that, rather than a single system, working memory is a capability of neocortex found in numerous locations in the brain. To illustrate, we look at a couple of examples of working memory in the frontal and parietal cortex.

The Prefrontal Cortex and Working Memory

One of the most obvious anatomical differences between primates (especially humans) and other mammals is that primates have a large frontal lobe. The rostral end of the frontal lobe, the **prefrontal cortex**, is particularly highly developed (Figure 24.6). Compared to the sensory and motor cortical areas, the function of the prefrontal cortex is relatively poorly understood. But because it is so well developed in humans, the prefrontal cortex is often assumed to be involved in those characteristics that distinguish us from other animals, such as self-awareness and the capacity for complex planning and problem solving.

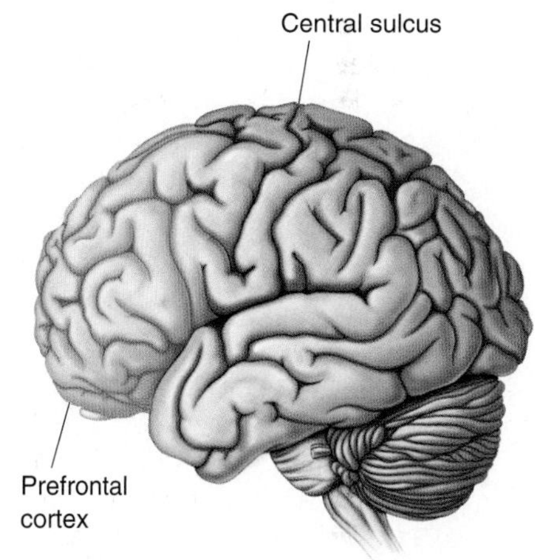

▲ FIGURE 24.6
Prefrontal cortex. The brain rostral to the central sulcus is the frontal lobe. The prefrontal cortex is the anterior portion of the frontal lobe that receives afferents from the medial dorsal nucleus of the thalamus.

Some of the first evidence suggesting that the frontal lobe is important for learning and memory came from experiments performed in the 1930s using a *delayed-response task*. A monkey was first shown food being placed in a well below one of two identical covers in a table. A delay period followed, during which the animal could not see the table. Finally, the animal was allowed to see the table again and received the food as a reward if it chose the correct well. Large prefrontal lesions seriously degraded performance in this delayed-response task, as well as other tasks including a delay period. Moreover, the monkeys performed increasingly poorly as the delay period was lengthened. These results imply that the prefrontal cortex may normally be involved in retaining information in working memory.

Experiments conducted more recently suggest that the prefrontal cortex is involved with working memory for problem solving and the planning of behavior. One piece of evidence comes from the behavior of humans with lesions in the prefrontal cortex. Recall the case of Phineas Gage, discussed in Chapter 18. Having sustained severe frontal lobe damage by an iron bar passing through the head, Gage had a difficult time maintaining a course of behavior. Although he could carry out behaviors appropriate for different situations, he had difficulty planning and organizing these behaviors, perhaps because of the damage to his frontal lobe.

The Wisconsin card-sorting test can demonstrate problems associated with prefrontal cortical damage. A person is asked to sort a deck of cards with a variable number of colored geometric shapes (Figure 24.7). The cards can be sorted by color, shape, or number of symbols, but at the beginning of the test, the subject isn't told which category to use. The subject begins putting cards into stacks and is informed when errors occur, by which the subject learns what sorting category is to be used. Then, after

▲ FIGURE 24.7

The Wisconsin card-sorting test. Cards containing various numbers of colored symbols must first be sorted by color. After a string of correct responses is made, the sorting category is changed to shape.

ten correct card placements have been made, the sorting category is changed, and the subject starts over again. To perform well on this test, a person must use memory of previous cards and errors in order to plan the next card placement. People with prefrontal lesions have great difficulty on this task when the sorting category is changed; they continue to sort according to a rule that no longer applies. It appears that they have a working memory deficit that limits their ability to make use of recent information to change their behavior.

The same kind of deficit is seen in other tasks. For example, a person with a prefrontal lesion might be asked to trace a path through a maze drawn on a piece of paper. Although the patient understands the task, he or she will repeatedly make the same mistakes, returning to blind alleys. In other words, these patients do not learn from their recent experience in the same way as a normal person, suggesting a working memory deficit.

The neurons in prefrontal cortex have a variety of response types, some of which may reflect a role in working memory. Figure 24.8 shows two response patterns obtained while a monkey performed a delayed-response task. The neuron in the top trace responded while the animal first saw the food wells, was unresponsive during the delay interval, and responded again when the animal saw the food wells again (Figure 24.8a). The response of the neuron simply correlates with visual stimulation. More interesting is the response pattern of the other neuron, which fired only during the delay interval (Figure 24.8b). This cell was not directly activated by the stimuli in the first or second interval in which the monkey saw the food wells. The increased activity during the delay period may be related to the retention of information needed to make the correct choice after the delay (i.e., working memory).

Imaging Working Memory in the Human Brain. Human brain imaging experiments suggest that numerous brain areas in the prefrontal cortex are involved in working memory. In one study by Courtney et al., brain activity was recorded by positron emission tomography (PET) while subjects performed two working memory tasks. In the identity task, three face photographs were briefly shown in succession; each image was at

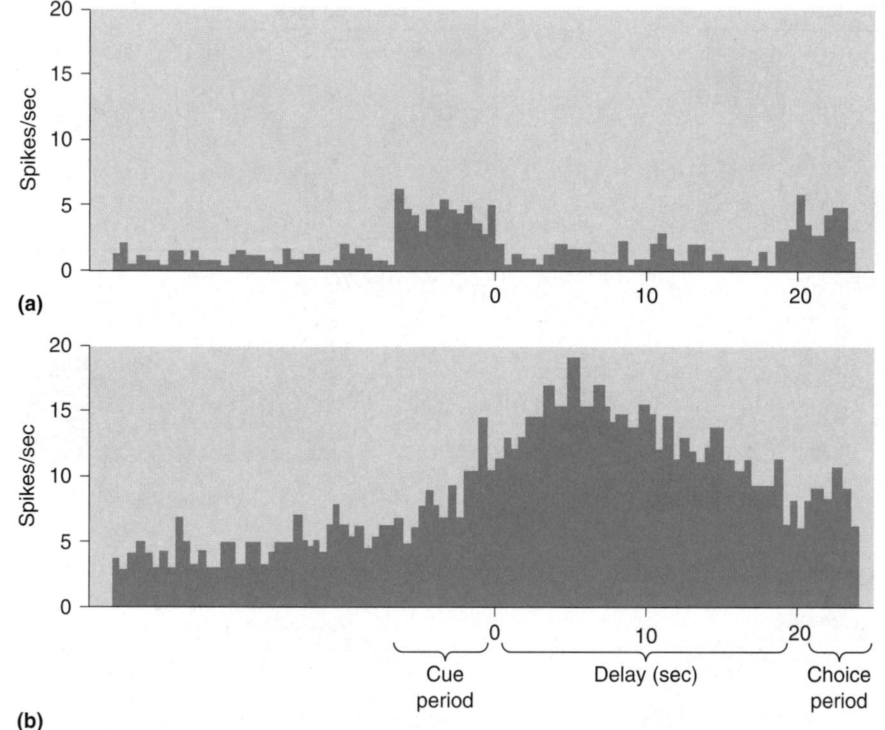

◀ FIGURE 24.8
Working memory activity in monkey prefrontal cortex. The two histograms show the activity of cells in prefrontal cortex recorded while the animal performed a delayed-response task. During a cue period of 7 seconds, food is placed into one of two wells within the monkey's view. During the delay period, the animal cannot see the food wells; after the delay, it is allowed to choose a well to receive a food reward (the choice period). **(a)** This cell responds when the animal first sees the food wells and when it sees them again after the delay period. **(b)** This cell responds strongest during the delay period, when there is no visual stimulus. (Source: Adapted from Fuster, 1973, Fig. 2.)

a different location and the subject looked at each face to memorize it. In the test phase, a face picture was shown at a new location and the subject indicated whether the face was the same as one of the memorized faces (Figure 24.9a). In the location task a similar paradigm was used, but the subject's task was to memorize the locations of the three faces presented before the delay, the identities of the faces were irrelevant. In the test phase a fourth face was shown and the subject answered whether that face was at the same location as one of the memorized locations (Figure 24.9b). Both experiments looked for brain activity during the delay interval between the memorization and test phases during which the subject had to hold information in mind. In the first experiment, this was information about faces; in the second experiment, it was information about spatial locations.

The brain areas that demonstrated significant working memory activity in these experiments are shown in Figure 24.9c and Figure 24.9d. Six areas in the frontal lobe showed significant sustained activity during the delay period, suggesting a role in working memory. Three areas exhibited stronger sustained activity for facial identity than spatial location, one area was more responsive to spatial memory, and two areas were active equally in facial and spatial memory tasks. An interesting unanswered question is whether working memory for other types of information is held in the same or different brain areas.

Area LIP and Working Memory

Cortical areas outside the frontal lobe have also been found to contain neurons that appear to retain working memory information. In Chapter 14, we saw an example in area 6 (see Figure 14.9). Another example is provided by the **lateral intraparietal cortex (area LIP)**, buried in the intraparietal sulcus (see Figure 21.18). Area LIP is thought to be involved in guiding eye movements because electrical stimulation here elicits

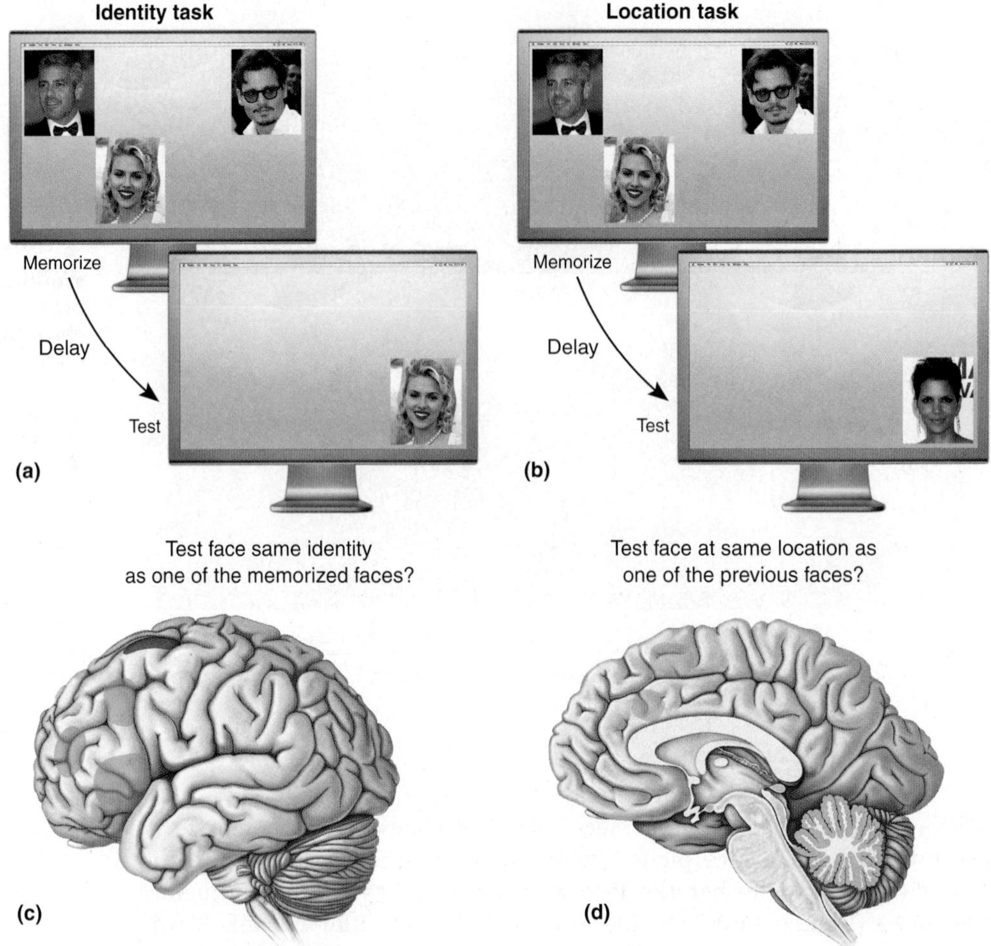

▲ FIGURE 24.9
Human brain activity in two working memory tasks. **(a)** In an identity task, subjects saw three photographs of faces in succession (shown together in the figure). The faces were memorized, and after a delay, a test face was presented at a new location. Subjects indicated whether the test face was the same as one of the memorized faces. **(b)** In a location task, three faces were shown in succession and the subjects memorized their locations on the display. In the test phase, a fourth face was shown and subjects indicated whether it appeared at the same location as one of the faces in the memorization phase. **(c)** Lateral and **(d)** medial views of brain activation in the two tasks. Six areas in the frontal lobe showed sustained activity correlated with working memory. The three blue areas showed greater activity in the facial identity task, the two green areas were equally active in facial identity and spatial location tasks, and the red area was more active in the location task. (Source: Adapted from Haxby et al., 2000, Fig. 5.)

saccades in monkeys. The responses of many neurons in area LIP of monkeys suggest that they are also involved in a type of working memory. This pattern is evident in a *delayed-saccade task*, in which the animal fixates on a point on a computer screen and a target is briefly flashed at a peripheral location (Figure 24.10a). After the target goes off, there is a variable length delay. At the end of the delay period, the fixation point disappears, and the animal's eyes make a saccadic movement to the remembered location of the target. The response of an LIP neuron while a monkey performs this task is shown in Figure 24.10b. The neuron begins firing shortly after the peripheral target is presented; this seems like a normal stimulus-evoked response. But the cell keeps firing throughout the delay period in which there is no stimulus. The neuron stops firing

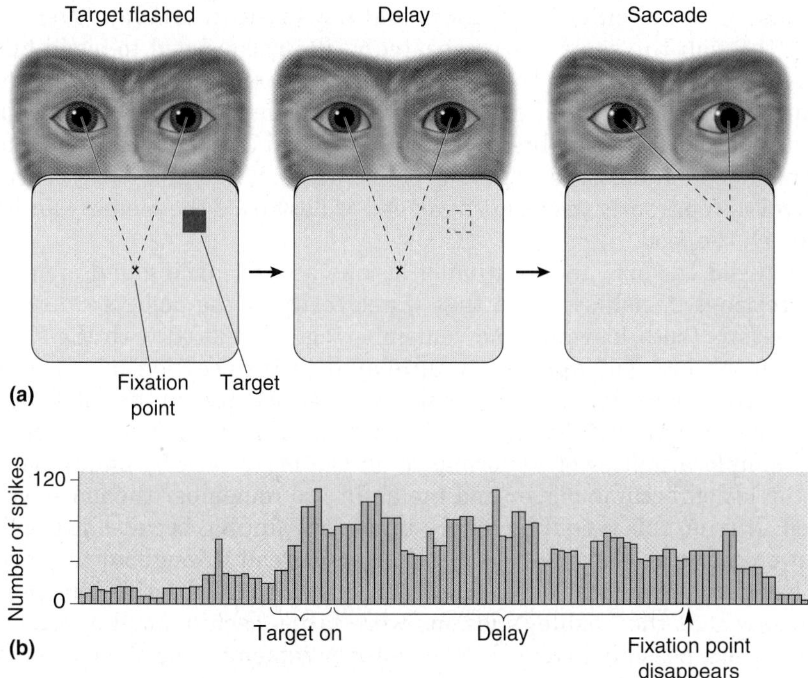

◀ FIGURE 24.10
The delayed-saccade task. (a) To obtain a juice reward, the monkey is trained to perform the following actions. First, the animal fixates a central point while a peripheral target flashes on and off. During a delay period after the target goes off, the monkey continues to fixate the central point (the dashed square shows the remembered location where the target had been). At the end of the delay, the fixation point disappears, and the animal saccades to the remembered location of the target. **(b)** The histogram shows the response of an LIP neuron. The neuron begins firing when the target is presented and continues firing through the delay period until after the fixation point is gone and the saccadic eye movement begins. (Source: Adapted from Goldman-Rakic, 1992, Fig. p. 113, and Gnadt and Andersen, 1988, Fig. 2.)

only after the saccadic eye movement begins. Further experiments using this delayed-saccade task suggest that the response of the LIP neuron is temporarily holding information that will be used to produce the saccade.

Other areas in the parietal and temporal cortex have been shown to have analogous working memory responses. These areas seem to be modality-specific, just as the responses in area LIP are specific to vision. This is consistent with the clinical observation that there are distinct auditory and visual working memory deficits in humans produced by cortical lesions.

DECLARATIVE MEMORY

We've seen that sensory information can be temporarily held in mind by working memory, but how does the brain retain information for a longer time? Even before humans evolved to the point that we could cram for neuroscience exams by drawing cartoons of the brain, we needed to remember many things—the location of the river to drink from, where to find food, which cave to call home. To understand the neural basis of declarative memory storage, we first need to examine *where* in the brain it is stored. In other words, we must explore the location of a memory, known as an **engram** or **memory trace**. For example, when you learn the meaning of a word in a foreign language, where in your brain is this information stored; where is the engram?

The Neocortex and Declarative Memory

In the 1920s, American psychologist Karl Lashley conducted experiments to study the effects of brain lesions on learning in rats. Well aware of the cytoarchitecture of the neocortex, Lashley set out to determine whether engrams resided in particular association areas of cortex (see Chapter 7), as was widely believed at the time.

In a typical experiment, he trained a rat to run through a maze to get a food reward. On the first trial, the rat was slow getting to the food because it would enter blind alleys and have to turn around. After running through the same maze repeatedly, the rat learned to avoid blind alleys and go straight to the food. Lashley was investigating how performance on this task was affected by different lesions in the rat's cortex. He found that a rat given a brain lesion after it had learned to run the maze then made mistakes and went down blind alleys it had previously learned to avoid. Apparently the lesion damaged or destroyed the memory for how to reach the food.

How did the size and location of lesions affect learning and memory? Interestingly, Lashley found that the severity of the deficits caused by the lesions (both learning and remembering) correlated with the *size* of the lesions but was apparently unrelated to the *location* of the lesion within the cortex. Based on these findings, he speculated that all cortical areas contribute equally (are *equipotential)* to learning and memory; it was simply a matter of performance on the maze task becoming poorer as the lesion became bigger and the ability to remember the maze worsened. If true, this would be a very important finding because it implied that engrams are based on neural changes spread throughout the cortex rather than being localized to one area. The problem with this interpretation was that the Lashley's lesions were large, each damaging multiple brain areas possibly involved in learning or remembering the maze task. Another problem was that the rats may have solved the maze in several different ways—by sight, feel, and smell—and the loss of one memory might have been compensated for by another.

Subsequent research has proven Lashley's conclusions to be incorrect. All cortical areas do not contribute equally to all memories. Nonetheless, his conclusions that all of the cortex participates in memory storage, and that an engram can be widely distributed in the brain, are correct and important. Lashley had a major impact on the study of learning and memory because he led other scientists to consider ways in which memories might be distributed among the vast number of neurons of the cerebral cortex.

Hebb and the Cell Assembly. Lashley's most famous student was Donald Hebb, introduced in Chapter 23. Hebb reasoned that it was crucial to understand how external events are represented in the activity of the brain before one can hope to understand how and where these representations are stored. In his remarkable book *The Organization of Behavior*, published in 1949, Hebb proposed that the internal representation of an object consists of all the cortical cells that are activated by the external stimulus (e.g., the circle in Figure 24.11). Hebb called this group of simultaneously active neurons a **cell assembly** (Figure 24.11a). Hebb imagined that all these cells were reciprocally interconnected. The internal representation of the object was held in working memory as long as activity reverberated through the connections of the cell assembly. Hebb further hypothesized that if activation of the cell assembly persisted long enough, consolidation would occur by a "growth process" that made these reciprocal connections more effective; neurons that fired together would wire together (Figure 24.11b). Subsequently, if only a fraction of the cells of the assembly were activated by a later stimulus (e.g., segments of a circle), the now-powerful reciprocal connections would cause the whole assembly to become active again, thus recalling the entire internal representation of the external stimulus—in this case, the circle (Figure 24.11c).

Hebb's important message about the engram was twofold: (1) It could be widely distributed among the connections that link the cells of the assembly, and (2) it could involve the same neurons that are involved in sensation and perception. Destruction of only a fraction of the cells of

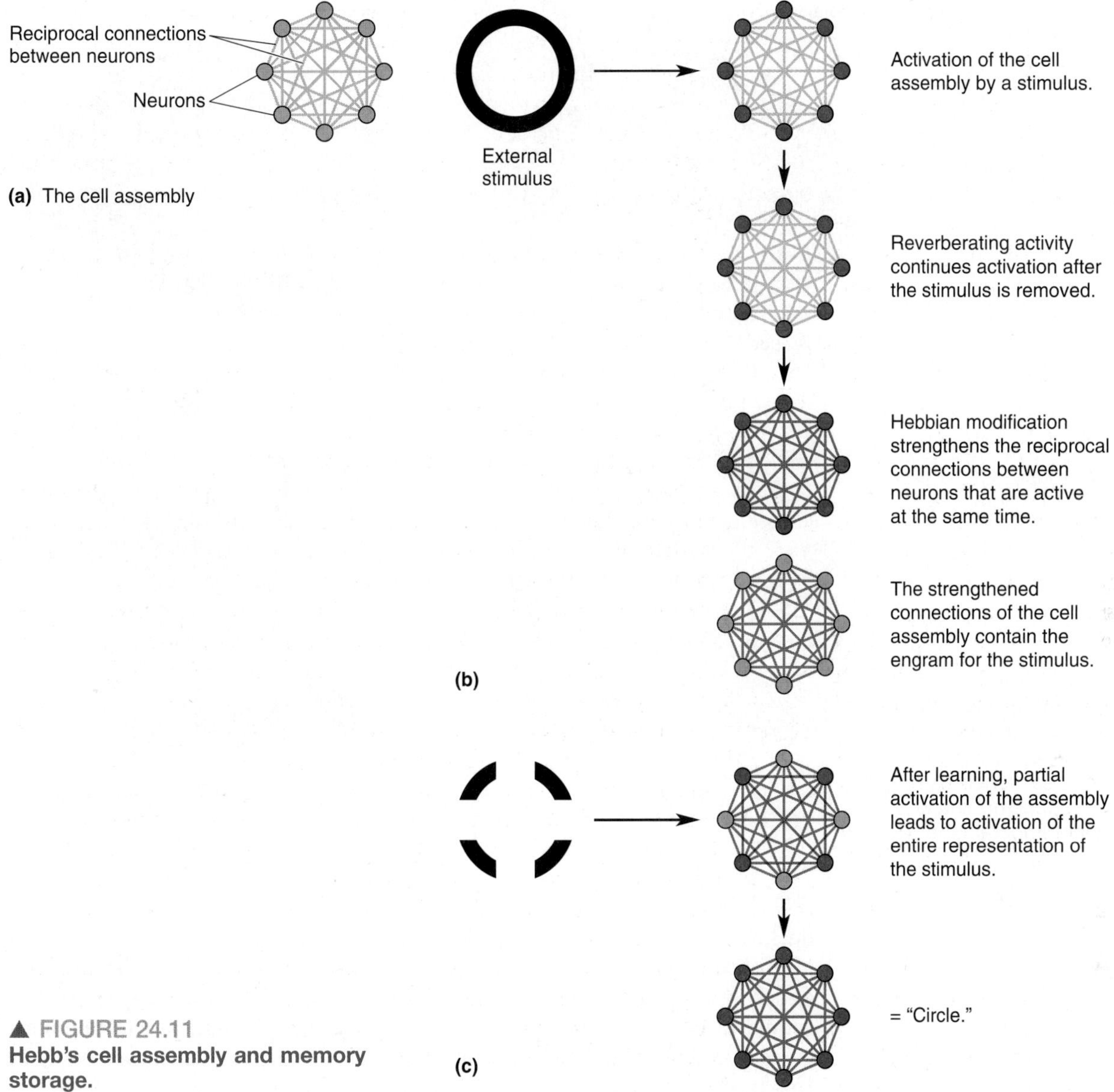

Reciprocal connections between neurons
Neurons
(a) The cell assembly
External stimulus
Activation of the cell assembly by a stimulus.
Reverberating activity continues activation after the stimulus is removed.
Hebbian modification strengthens the reciprocal connections between neurons that are active at the same time.
The strengthened connections of the cell assembly contain the engram for the stimulus.
(b)
After learning, partial activation of the assembly leads to activation of the entire representation of the stimulus.
= "Circle."
(c)

▲ FIGURE 24.11
Hebb's cell assembly and memory storage.

the assembly would not be expected to eliminate the memory, possibly explaining Lashley's results. Hebb's ideas stimulated the development of neural network computer models. Although his original assumptions had to be modified slightly, we will see in Chapter 25 that these models have successfully reproduced many features of human memory.

Where is the engram for a foreign language? Look to the regions of the brain in the temporal and parietal lobes that normally process language. A lesion here can disrupt your memory of a foreign word but leave intact the memory of your foreign-born grandmother's face. However, although declarative memories may finally reside in many areas of the neocortex, decades of research indicate that to get there, they must pass through structures in the medial temporal lobes. Let's explore the evidence.

Studies Implicating the Medial Temporal Lobes

A variety of experiments indicate that structures in the medial temporal lobe are particularly important for the consolidation and storage of

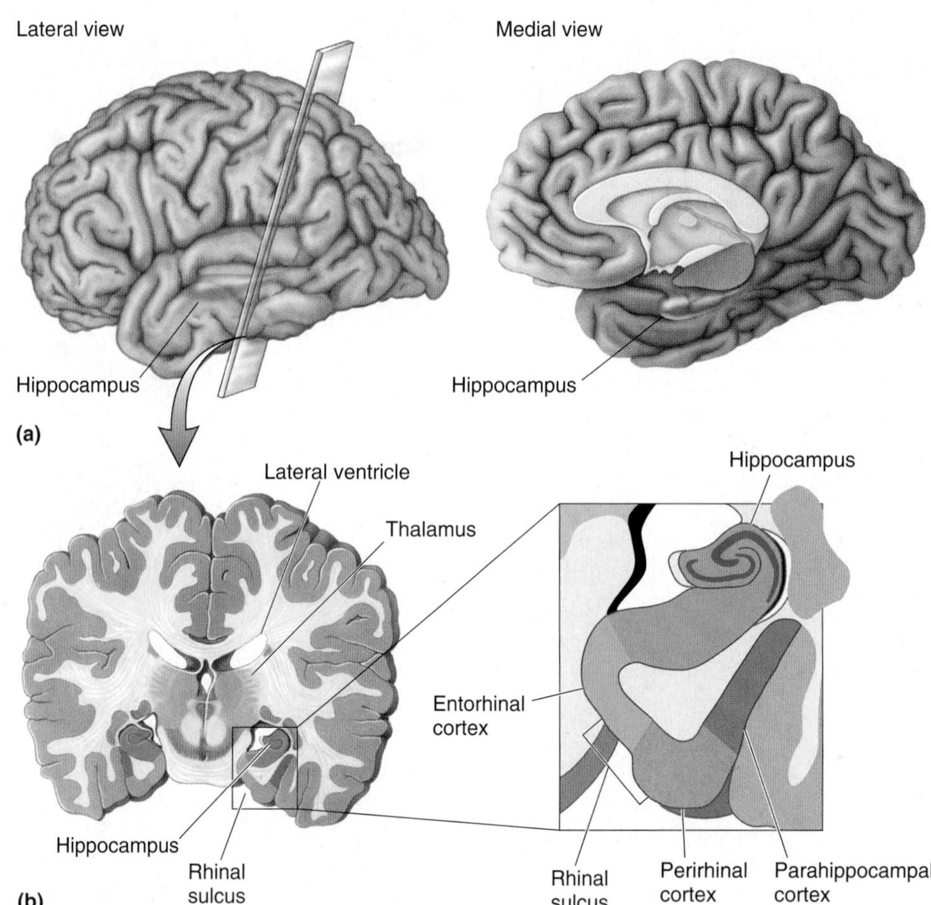

▶ FIGURE 24.12
Structures in the medial temporal lobe involved in declarative memory formation. **(a)** Lateral and medial views show the location of the hippocampus in the temporal lobe. **(b)** The brain is sectioned coronally to show the hippocampus and cortex of the medial temporal lobe.

declarative memories. Examples are experiments that used electrical stimulation or neural recordings from the temporal lobe. Further evidence came from cases of amnesia that resulted from damage to the temporal lobes. Before considering the studies that suggest that memory storage involves the medial temporal lobes, let's look at the anatomy of this brain region.

Anatomy of the Medial Temporal Lobe. The temporal lobe is located under the temporal bone, so named because the hair of the temples is often the first to go gray with the passage of time (*tempus* is Latin for "time"). The association of the temporal lobe with time was fortuitous as this region of the brain is important for recording past events. The medial portion of the temporal lobe contains the temporal neocortex, which may be a site of long-term memory storage, and a group of structures interconnected with neocortex that are critical for the formation of declarative memories.

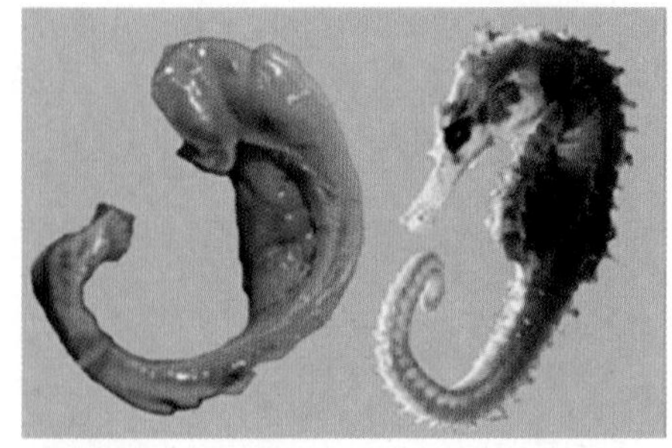

▲ FIGURE 24.13
The hippocampus. In Greek mythology, the hippocampus was a creature combining the front of a horse with the hind end of a dolphin or fish. Here a dissected hippocampus is shown next to a seahorse. (Source: Laszlo Seress/Wikimedia Commons.)

The key structures are the hippocampus, the nearby cortical areas, and the pathways that connect these structures with other parts of the brain (Figure 24.12). As we saw in Chapter 7, the **hippocampus** is a folded structure situated medial to the lateral ventricle. The name means "seahorse," a resemblance you can see in Figure 24.13. Ventral to the hippocampus are three important cortical regions that surround the rhinal sulcus: the **entorhinal cortex**, which occupies the medial bank of the rhinal sulcus; the **perirhinal cortex**, which occupies the lateral bank; and the **parahippocampal cortex**, which lies lateral to the rhinal

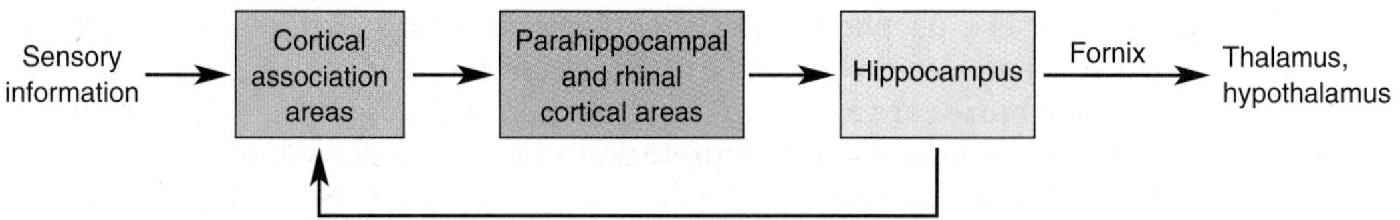

▲ FIGURE 24.14
Information flow through the medial temporal lobe.

sulcus. (We'll refer to entorhinal cortex and perirhinal cortex collectively as rhinal cortex.)

Inputs to the medial temporal lobe come from the association areas of the cerebral cortex, containing highly processed information from all sensory modalities (Figure 24.14). For instance, inferotemporal visual cortex (area IT) projects to the medial temporal lobe, but low-order visual areas such as striate cortex do not. This means that the input contains complex representations, perhaps of behaviorally important sensory information, rather than responses to simple features such as light–dark borders. Input first reaches the rhinal and parahippocampal cortex before being passed to the hippocampus. A major output pathway from the hippocampus is the **fornix**, which loops around the thalamus before terminating in the hypothalamus.

Electrical Stimulation of the Human Temporal Lobes. One of the most intriguing and controversial studies implicating the neocortex of the temporal lobe in the storage of declarative memory traces involved electrical stimulation of the human brain. In Chapters 12 and 14 we discussed the work of Wilder Penfield, in which patients' brains, as part of surgical treatment for severe epilepsy, were electrically stimulated at numerous locations prior to ablation of the seizure-prone region. Stimulation of somatic sensory cortex caused the patient to experience tingling sensations in regions of skin, whereas stimulation of motor cortex caused a certain muscle to twitch.

Electrical stimulation of the temporal lobe occasionally produced more complex sensations than stimulation in other brain areas. In a number of cases, Penfield's patients described sensations that sounded like hallucinations or recollections of past experiences. This is consistent with reports that epileptic seizures of the temporal lobes can evoke complex sensations, behaviors, and memories. Here is part of Penfield's account of one operation:

> At the time of operation, stimulation of a point on the anterior part of the first temporal convolution on the right caused him [the patient] to say, "I feel as though I were in the bathroom at school." Five minutes later, after negative stimulations elsewhere, the electrode was reapplied near the same point. The patient then said something about "street corner." The surgeon asked him, "where" and he replied, "South Bend, Indiana, corner of Jacob and Washington." When asked to explain, he said he seemed to be looking at himself—at a younger age. (Penfield, 1958, p. 25.)

Another patient reported a similar sense of experiencing flashbacks. When her temporal cortex was stimulated, she said, "I think I heard a mother calling her little boy somewhere. It seemed to be something that happened years ago." With stimulation at another location, she said, "Yes, I hear voices. It is late at night around the carnival somewhere—some sort of traveling circus. . . I just saw lots of big wagons that they use to haul animals in."

Are these people reexperiencing events from earlier in their life because memories are evoked by the electrical stimulation? Does this mean that memories are stored in the neocortex of the temporal lobe? Those are tough questions! One interpretation is that the sensations are recollections of past experiences. The fact that such elaborate sensations resulted only when the temporal lobe was stimulated suggests that the temporal lobe may play a special role in memory storage. However, other aspects of the findings do not clearly support the hypothesis that engrams are being electrically activated. For instance, some brain-stimulated patients said they saw themselves, something that we cannot normally experience. Also, it is important to appreciate that complex sensations were reported only by a minority of the patients, and all of these patients had an abnormal cortex associated with their epilepsy.

There is no way to prove whether the complex sensations evoked by temporal lobe stimulation are recalled memories. However, it is clear that the consequences of temporal lobe stimulation and temporal lobe seizures can be qualitatively different from stimulation of other areas of the neocortex.

Neural Recordings from the Human Medial Temporal Lobe. Electrical stimulation of the temporal lobe sometimes produces memory-like experiences, and as we will see, lesions here disrupt memory. But what are the medial temporal neurons normally doing? A glimpse at the normal function of these neurons comes from studies using microelectrodes to record from the brains of living humans who suffered from epileptic seizures that were not alleviated by drugs. As in Penfield's studies, plans were made to surgically remove the abnormally active part of the brain after localizing the area with electrode recordings. Because temporal lobe seizures are common, the electrodes were often placed in the hippocampus and surrounding structures. In some ways such recordings are reminiscent of studies of visual neurons in monkey inferotemporal cortex. As seen in monkeys, neurons were found that preferentially responded to categories of objects, including faces, household objects, and outdoor scenes (recall face-selective neurons in Chapter 10). These neurons are said to be invariant because they responded to quite a variety of visual images that are structurally or conceptually related.

In further studies, even more selectivity was discovered in a small percentage of the neurons examined. For example, individual hippocampal neurons were found in one patient that responded selectively to pictures of the actress Jennifer Aniston or the basketball player Michael Jordan. Figure 24.15 shows a neuron in the hippocampus that responded to various stimuli associated with the actress Halle Berry. The diversity of the effective stimuli was striking, including photographs of Ms. Berry, a drawing of her face, and even her printed name. The neuron was activated by pictures of Ms. Berry dressed for her role in the movie *Catwoman*, yet it did not

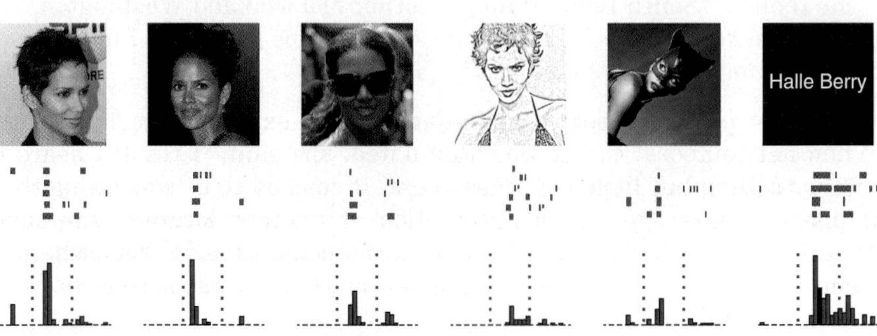

▶ FIGURE 24.15
A patient's hippocampal neuron selectively responds to actress Halle Berry. This neuron responded to photos and drawings of Halle Berry as well as her written name. The cell responded less or not at all to photos, drawings, and written names of other people. (Adapted from: Quiroga et al., 2005, Fig. 2.)

respond to other women similarly dressed. Other neurons were selective for landmarks such as the Eiffel Tower and the Leaning Tower of Pisa.

What are we to make of these neurons? One way to think about them is that they are somewhere on a continuum between purely visual coding in the lateral temporal lobe and memory coding in the medial temporal lobe. We can't be sure, but such neurons are probably not essential for recognition because common objects and famous faces are recognized even after lesions to the hippocampus; even H.M. recognized people and things that existed before his surgery. Recognition may rely on portions of the temporal lobe that are more lateral and posterior. The highly selective neurons in the hippocampus may serve a role in the formation of new memories of people and things we already recognize, such as the patient's memory of Halle Berry. Many questions remain. Might less specific responses have been found in these experiments if more stimuli were used (a cell that responds to Justin Timberlake, canned peas, and doorknobs)? Are there neurons selectively activated by every object that we recognize, or are the examples we've discussed rare cases related to repeated exposure of very famous people or things? Do these findings even apply to normal brains, since it is conceivable that the seizure-prone brains were abnormally organized and responsive?

Temporal Lobe Amnesia

If the temporal lobe is particularly important for learning and memory, one would expect that removing both temporal lobes would have a profound effect on these functions. Studies in both humans and animals show that this is indeed the case.

The Case of H.M.: Temporal Lobectomy and Amnesia. A renowned case of amnesia resulting from temporal lobe damage provides further evidence for the importance of this region in memory. This case concerns the memory of Henry Molaison, whose name was made public only after his death in 2008 (Figure 24.16). For the previous half century, studies referred to him as H.M., probably the most famous initials in the history of neuroscience. H.M. experienced minor epileptic seizures beginning around age 10, and as he aged, they became more serious generalized seizures involving convulsions, tongue biting, and loss of consciousness. Although the cause of the seizures is not known, they may have resulted from damage sustained in a bicycle accident at the age of 9 that left him unconscious for 5 minutes. After graduating from high school he got a job, but despite heavy medication with anticonvulsants, his seizures increased in frequency and severity to the point that he was unable to work. In 1953, at the age of 27, H.M. had an operation in which an 8 cm length of medial temporal lobe was bilaterally excised, including cortex, the underlying amygdala, and the anterior two-thirds of the hippocampus, in a last-ditch attempt to assuage the seizures. The surgery was successful in alleviating the seizures.

The removal of much of the temporal lobes had little effect on H.M.'s perception, intelligence, or personality, but the surgery left him with profound and debilitating anterograde amnesia. Drs. Brenda Milner and Suzanne Corkin, initially at the Montreal Neurological Institute, worked with H.M. for 50 years, but, incredibly, they had to introduce themselves to him every time they met. They found that H.M. would forget events almost as quickly as they occurred. With repetition he could remember a number for a short time, but if he was distracted he would not only forget the number, he would also forget that he had even been asked to remember one.

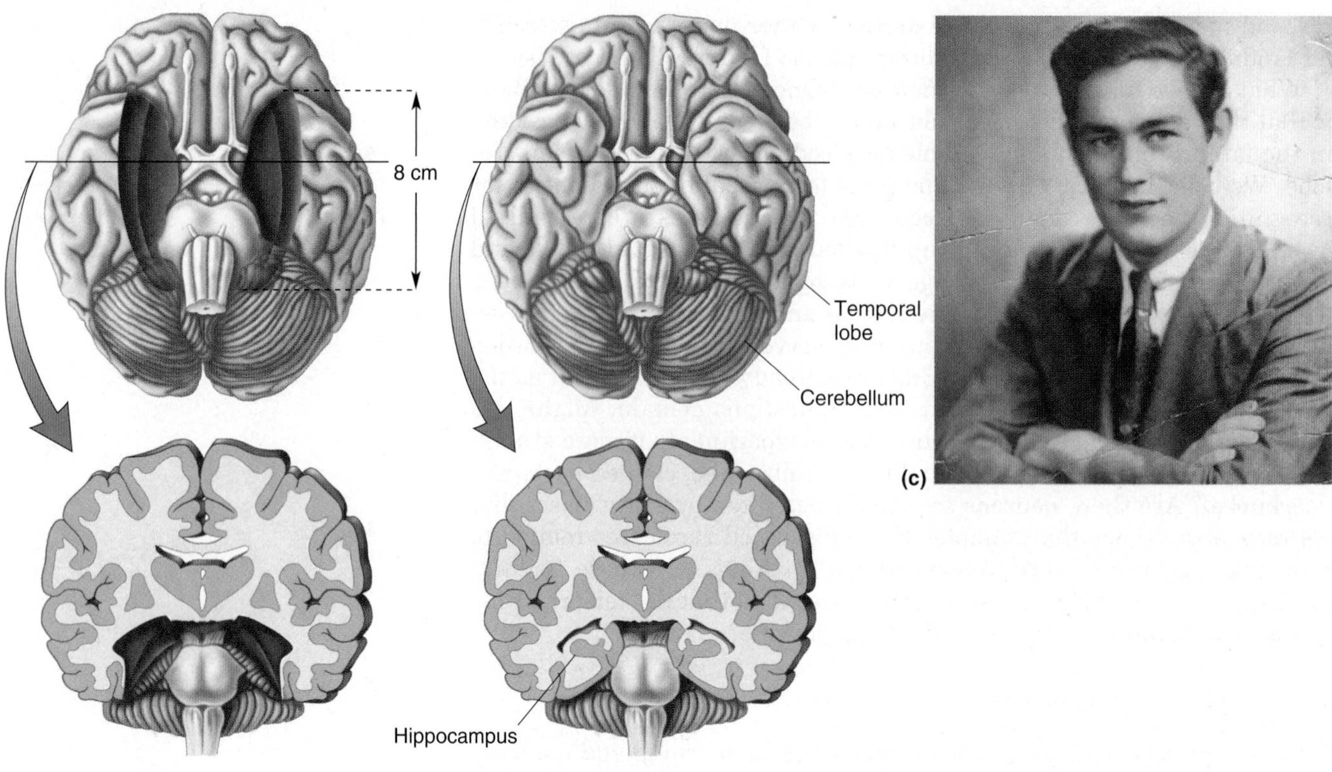

▲ FIGURE 24.16
The brain lesion in patient H.M. that produced profound anterograde amnesia. **(a)** The medial temporal lobe was removed from both hemispheres in H.M.'s brain to alleviate epileptic seizures. **(b)** A normal brain, showing the location of the hippocampus and cortex that were removed from H.M.'s brain. (Source: Adapted from Scoville and Milner, 1957, Fig. 2.) **(c)** Henry Molaison as a high school student before the surgery. (Source: Photo courtesy of Suzanne Corkin. Copyright © by Suzanne Corkin used by permission of The Wylie Agency LLC.)

To be clear about the nature of H.M.'s amnesia, we must contrast what was lost with what was retained. In addition to anterograde amnesia, he had some degree of retrograde amnesia. He retained some memories of his childhood but little or no memory for events just before his surgery. Testing shortly after the surgery suggested that H.M. had retrograde amnesia for events extending back several years before his surgery. Later studies suggested that his retrograde amnesia might extend back decades. H.M.'s working memory was largely normal. For instance, with constant rehearsal he could remember a list of six numbers, although any interruption would cause him to forget. H.M. was actually able to learn a very small number of declarative facts after his surgery. For example, he could recognize and name a few people who became famous after his surgery, such as U.S. President John Kennedy. He also learned the floor plan of a home he moved to after the surgery. These rare remembered facts probably resulted from extensive daily repetition. H.M. was also able to learn new tasks (i.e., form new *procedural* memories). For example, he was taught to draw by looking at his hand in a mirror, a task that takes a good deal of practice for anyone. The odd thing is that he learned to perform new tasks despite the fact that he had no recollection of the specific experiences in which he was taught to do them (the declarative component of the learning).

To appreciate the significance of H.M., realize that prior to his surgery, little was known about the function of the hippocampus and surrounding structures. Considering H.M.'s amnesia in the context of earlier research, we can conclude that the medial temporal lobe is critical for memory consolidation but not for the retrieval of memories. Although there is some controversy about the temporal extent of H.M.'s retrograde amnesia, he clearly retained declarative memories for many things from before the surgery, such as famous faces and the meanings of words. This implies that medial temporal structures do not store all memories even though engrams for some things may be located there. The fact that his working memory was largely intact means this does not rely on the medial temporal lobe. Finally, H.M's amnesia indicates that the formation and retention of procedural memories use brain structures distinct from those involved in declarative memory consolidation and perhaps storage.

An Animal Model of Human Amnesia. The amnesia of H.M. makes a strong case that one or more structures in the medial temporal lobe are essential for the formation of declarative memories. If these structures are damaged, severe anterograde amnesia results. Experiments have mostly used the experimental ablation technique to assess whether the removal of various parts of the temporal lobe affects memory.

Because the macaque monkey brain is similar in many ways to the human brain, macaques are frequently studied to further our understanding of human amnesia. The monkeys are often trained to perform tasks called delayed match-to-sample and **delayed non-match to sample (DNMS)** (Figure 24.17). In this type of experiment, a monkey faces a table that has several small wells in its surface. It first sees the table with one object on it covering a well. The object might be a wooden block or a chalkboard eraser (the sample stimulus). The monkey is trained to displace the object so that it can grab a food reward in the well under the object. After getting the food, a screen prevents the monkey from seeing the table for some period of time (the delay interval). Finally, the animal gets to see the table again, but now there are two objects on it: One is the same as before, and another is new. If a match-to-sample experiment is being conducted, the animal must displace the object it recognizes to get a food reward. In DNMS, the monkey's task is to displace the new object (the non-matching object) in order to get a food reward in the well below it. Normal monkeys are relatively easy to train on the non-matching task and get very good at it, perhaps because it exploits their natural curiosity for novel objects. With delays between the two stimulus presentations of anywhere from a few seconds to 10 minutes, the monkey correctly displaces the non-matching stimulus on about 90% of the trials. Memory required in the DNMS task has been called **recognition memory** because it involves the ability to judge whether a stimulus has been seen before.

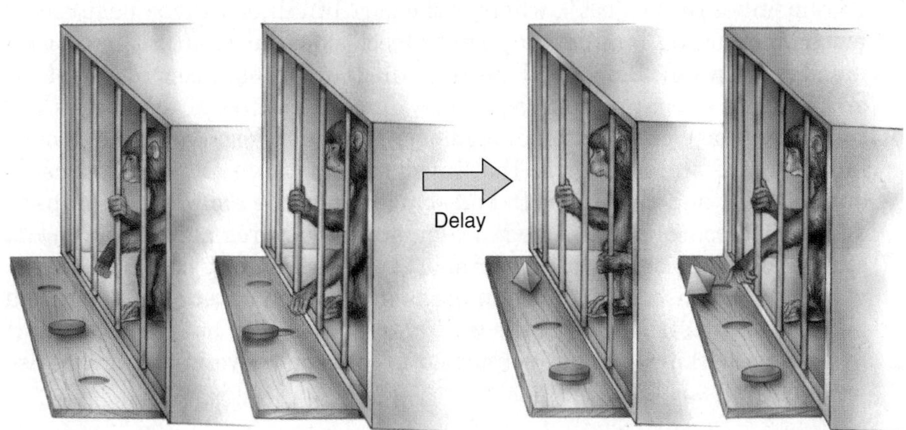

◀ FIGURE 24.17
The delayed non-match to sample (DNMS) task. A monkey first displaces a sample object to obtain a food reward. After a delay, two objects are shown, and recognition memory is tested by having the animal choose the object that does not match the sample. (Source: Adapted from Mishkin and Appenzeller, 1987, p. 6.)

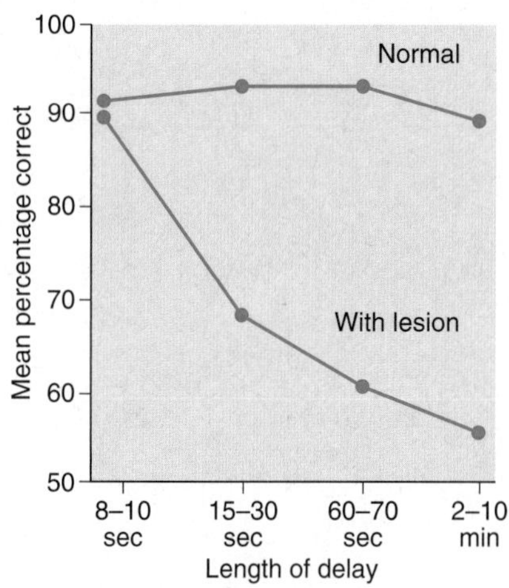

▲ FIGURE 24.18
The effect of medial temporal lesions on DNMS performance. The Y axis shows the percentage of correct choices made by monkeys as a function of the length of the delay interval. The performance of normal monkeys is compared to that of monkeys with large bilateral medial temporal lesions. (Source: Adapted from Squire, 1987, Fig. 49.)

In the early 1980s, experiments performed by Mortimer Mishkin and his colleagues at the National Institute of Mental Health, and Larry Squire and his coworkers at the University of California, San Diego, demonstrated that severe deficits on the DNMS task result from bilateral medial temporal lesions in macaque monkeys. Performance was close to normal if the delay between the sample stimulus and the two test stimuli was short (a few seconds). This is important because it indicates that the monkey's perception was still intact after the ablation and the animal remembered the DNMS procedure. But when the delay was increased from a few seconds to a few minutes, the monkey made increasingly more errors choosing the non-matching stimulus (Figure 24.18). With the lesion, the animal was no longer as good at remembering what the sample stimulus was in order to choose the other object. This behavior suggests that it forgot the sample stimulus if the delay was too long. The deficit in recognition memory produced by the lesion was not specific to the visual modality, since this deficit was also observed if the monkey was allowed to touch but not see the objects.

The monkeys with medial temporal lesions appeared to provide a good model of human amnesia. As with H.M., the amnesia was anterograde, it involved declarative rather than procedural memory, working memory was intact, and consolidation was severely impaired. Note that the surgical lesions that produced recognition memory deficits in these monkeys were quite large. They included the hippocampus, amygdala, and rhinal cortex. At one time it was thought that the key structures damaged in such lesions were the hippocampus and amygdala. Recall from Chapter 18 that the amygdala plays a special role in memory for emotional experiences. However, research has now shown that selective amygdala lesions have no effect on recognition memory, and lesions of the hippocampus alone produce only relatively mild amnesia. For example, Squire studied a man known as R.B. who had bilateral hippocampal damage as a result of oxygen deprivation during surgery. Although R.B. clearly had difficulty forming new memories, his anterograde amnesia was not nearly as severe as that seen in H.M. The most severe memory deficits result from damage to the perirhinal cortex. The anterograde amnesia resulting from perirhinal lesions is not specific to information from a particular sensory modality, reflecting the convergence of input from association cortex of multiple sensory systems.

Together with the hippocampus, the cortex in and around the rhinal sulcus evidently performs a critical transformation of the information coming from association cortex. Some studies suggest that the hippocampus and rhinal cortex are involved in different facets of memory; the hippocampus may signal that a particular object had been seen before ("I remember that object"), whereas the perirhinal cortex may be involved more in signaling familiarity ("That object looks familiar but I don't remember it specifically"). Still, such distinctions remain controversial. At any rate, it appears that, collectively, medial temporal structures are critical for the consolidation of memory. They may also have an essential intermediate processing role that involves something more than consolidation. H.M. and possibly R.B. had some retrograde amnesia. Perhaps in addition to consolidation, medial temporal structures play a role in the storage of memories for a long or short time (depending on which expert you ask). Our discussion of brain areas involved in anterograde amnesia has focused on structures in the medial temporal lobe, but it is important to note that lesions to interconnected areas elsewhere in the brain also produce amnesia (Box 24.2).

BOX 24.2 OF SPECIAL INTEREST

Korsakoff's Syndrome and the Case of N.A.

In Chapter 18 we learned about the Papez circuit, a series of strongly interconnected structures that ring the diencephalon. A major component of this circuit is a massive bundle of axons called the fornix that connects the hippocampus with the mammillary bodies in the hypothalamus (Figure A). The mammillary bodies, in turn, send a strong projection to the anterior nuclei of the thalamus. The dorsomedial nucleus of the thalamus also receives input from temporal lobe structures, including the amygdala and inferotemporal neocortex, and it projects to virtually all of the frontal cortex.

Considering the central role of the temporal lobes in memory processing, it is perhaps not surprising that damage to these connected diencephalic structures can also cause amnesia.

A particularly dramatic example of the amnesic effects of damage to the diencephalon in humans is the case of a man known as N.A. In 1959 at the age of 21, N.A. was a radar technician in the U.S. Air Force. One day he was sitting down assembling a model in his barracks, while behind him a roommate played with a miniature fencing foil. N.A. turned at the wrong moment and was stabbed. The foil went through his right nostril and took a leftward course into his brain. Many years later when a CT scan was performed, the only obvious damage was a lesion in his left dorsomedial thalamus, though there may have been other damage.

After his recovery, N.A.'s cognitive ability was normal, but his memory was impaired. He had relatively severe anterograde amnesia as well as retrograde amnesia for a period of about 2 years preceding the accident. While he could remember some faces and events from the years following his accident, even these memories were sketchy. He had difficulty watching television because during commercials, he'd forget what had previously happened. In a sense, he lived in the past and preferred to wear old familiar clothes and keep his hair in an older style.

Although N.A.'s amnesia was less severe than H.M.'s, its quality was strikingly similar. He had preservation of short-term memory, recollection of old memories, and general intelligence. Along with difficulty forming new declarative memories, he had retrograde amnesia for 2 years preceding the accident that produced the amnesia. The similarities in the effects of medial temporal and diencephalic lesions suggests that these interconnected areas are part of a system serving the common function of memory consolidation.

Further support for a role of the diencephalon in memory comes from studies of Korsakoff's syndrome. Usually resulting from chronic alcoholism, **Korsakoff's syndrome** is characterized by confusion, confabulations, severe memory impairment, and apathy. As a result of poor nutrition, alcoholics may develop a thiamin deficiency, which can lead to such symptoms as abnormal eye movements, loss of coordination, and tremors. This condition can be treated with supplemental thiamin. If left untreated, however, thiamin deficiency can lead to structural brain damage that produces Korsakoff's syndrome. Although all cases of Korsakoff's syndrome are not associated with damage to the same parts of the brain, there are usually lesions in the dorsomedial thalamus and the mammillary bodies.

In addition to anterograde amnesia, Korsakoff's syndrome can involve more severe retrograde amnesia than observed in N.A. and H.M. No strong correlation exists between the severities of anterograde amnesia and retrograde amnesia in Korsakoff's syndrome. This is consistent with the other studies of amnesia we've discussed, suggesting that the mechanisms involved in consolidation (disrupted in anterograde amnesia) are largely distinct from the processes used to recall memories (disrupted in retrograde amnesia). Based on a small number of cases such as that of N.A., researchers suspect that anterograde amnesia associated with diencephalic lesions results from damage to the thalamus and mammillary bodies. Although it is not clear which damage causes the retrograde amnesia, in addition to diencephalic lesions, Korsakoff's patients sometimes have damage to the cerebellum, brain stem, and neocortex.

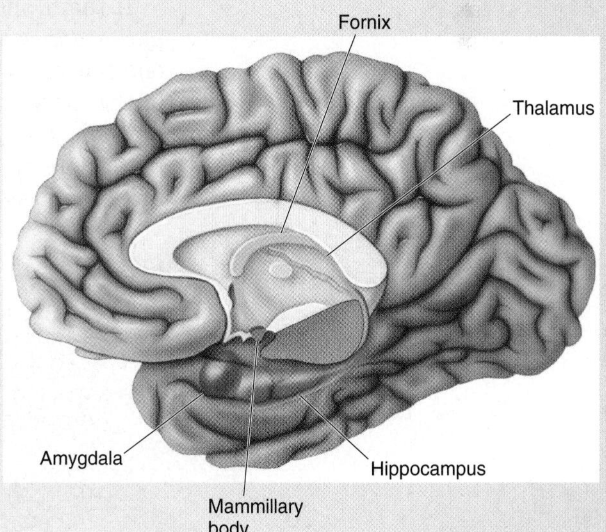

Figure A
Midline brain structures involved in memory. Temporal lobe structures including the hippocampus, amygdala, and inferotemporal cortex project to the thalamus and hypothalamus, including the mammillary bodies, in the diencephalon.

Memory Functions of the Hippocampal System

Memory formation, retention, and retrieval involve a system of interconnected brain areas. Considerable evidence points to the importance of the medial temporal lobe for declarative memory, and within this region of the brain the hippocampus has received the greatest attention. It's not simple to pin down just what the hippocampus does, however, as it is involved in various memory functions at multiple time scales. To avoid getting lost, keep a few basic points in mind. First, the hippocampus appears to play a critical role in binding sensory information for the purpose of memory consolidation. Second, a lot of research, particularly in rodents, has shown that the hippocampus supports spatial memory of the location of objects of behavioral importance. This may be one of the specialized functions of the hippocampus, or it might be an example of the binding of sensory information. Finally, the hippocampus is involved in the storage of memories for some length of time, though the time duration is controversial.

The Effects of Hippocampal Lesions in Rats. Rodents have played an important role elucidating the memory functions of the hippocampus. In one type of experiment, rats are trained to get food in a *radial arm maze*, a device devised by David Olton and his colleagues at Johns Hopkins University. This apparatus consists of arms, or passageways, radiating from a central platform (Figure 24.19a). If a normal rat is put in such a maze, it will explore until it finds the food at the end of each arm. With practice, the rat becomes efficient at finding all the food, going down each arm of the maze just once (Figure 24.19b). To run through the maze without going twice into any of the arms, the rat uses visual or other cues around the maze to remember where it has already been. Working memory is presumably used to retain information about which arms have been visited.

If the hippocampus is destroyed before the rat is put in the maze, its performance will differ from normal behavior in an interesting way. In one sense, rats with lesions seem normal; they learn to go down the arms of the maze and eat the food placed at the end of each arm. But unlike normal rats, they never learn to do this efficiently. Rats with hippocampal lesions go down the same arms more than once, only to find no food after the first trip, and they leave other arms containing food unexplored

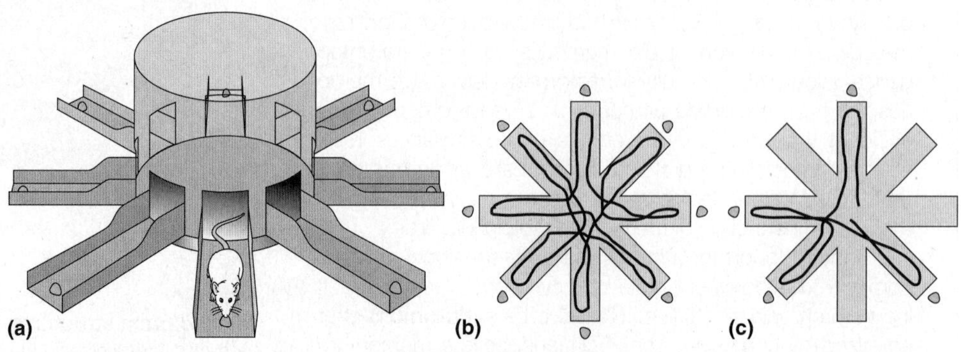

▲ FIGURE 24.19
Following a rat through a radial arm maze. (a) An eight-arm radial maze. (b) The path of a rat through a maze in which all the arms contain food. (c) If a rat learns that four of the eight arms never contain food, it will ignore these and follow a path to only the baited arms. (Source: Parts b and c, adapted from Cohen and Eichenbaum, 1993, Fig. 7.4.)

for an abnormally long time. It appears that the rats can learn the task in the sense that they go down the arms in search of food (the procedural memory). But they cannot seem to remember which arms they've already visited.

A variation on the radial arm experiment illustrates an important subtlety in the deficit produced by destroying the hippocampus. Instead of placing food at the end of all the arms of the maze, food is placed only at the ends of certain arms and never in the other ones. After a bit of practice, a normal rat learns to avoid going down the arms that never contain food (Figure 24.19c). At the same time, the rat learns to efficiently get the food in the other arms, entering each arm just once. How do rats with hippocampal lesions perform on this task? Interestingly, just like normal rats, they are able to learn to avoid the arms that never contain food. But they still are not able to get the food from the other arms without wasting time going down the same arms more than once. So how can we argue that the lesion disrupts the ability to learn the locations of arms that have already been entered, even though the rat can learn to avoid arms that never contain food? Evidently, the key to making sense of these findings is that the information about the no-food arms is always the same each time the rat goes in the maze (i.e., no-food arms are memorized as part of the "procedure"), whereas the information about which arms the rat has already entered requires working memory and varies from one trial to the next.

Spatial Memory, Place Cells, and Grid Cells. Several lines of evidence suggest that the hippocampus is particularly important for spatial memory. The **Morris water maze**, a commonly used test of spatial memory in rats, was devised by Richard Morris of the University of Edinburgh. In this test, a rat is placed in a pool filled with cloudy water (Figure 24.20). Submerged just below the surface in one location is a small platform that allows the rat to escape the water. A rat placed in the water the first time will swim around until it bumps into the hidden platform, and then it will climb onto it. Normal rats quickly learn the spatial location of the platform and on subsequent trials waste no time swimming straight to it. Moreover, once they have figured out what to search for, rats put in a maze with the platform at a different location learn the task much faster. But rats with bilateral hippocampal damage never seem to figure out the game or remember the location of the platform.

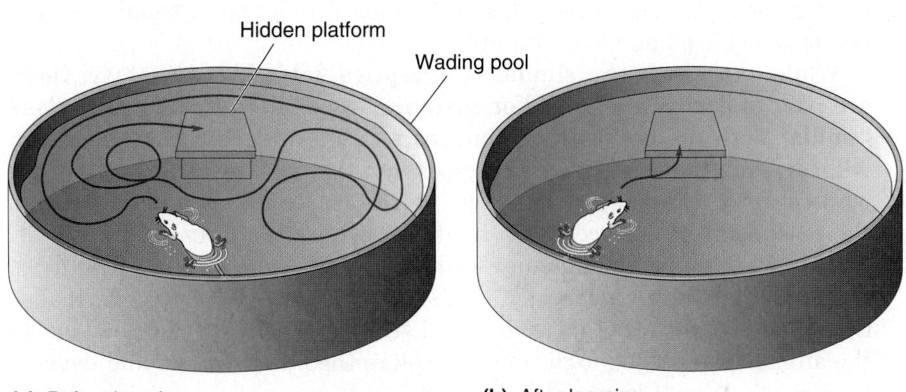

▲ FIGURE 24.20
The Morris water maze. **(a)** The trajectory a rat might take to find a hidden platform the first time the rat is placed in the pool. **(b)** After repeated trials, the rat knows where the platform is and swims straight to it.

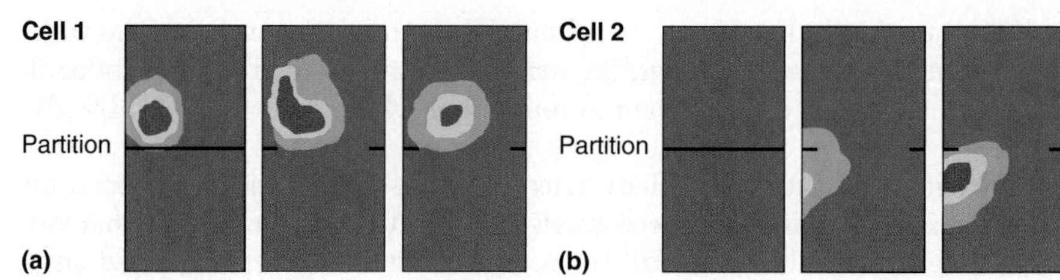

▶ FIGURE 24.21
Place cells in the hippocampus. A rat explores a small box for 10 minutes (left panels). Then the partition is removed, so the rat can explore a larger area (center and right panels). **(a)** Color coding indicates the area in the box where one place cell in the hippocampus responds: red, large response; yellow, moderate response; blue, no response. This cell has a place field in the smaller upper box; when the partition is removed, it stays in the same location. **(b)** In this case, an electrode is next to a cell in the hippocampus that does not respond when the rat is in the smaller upper box (left). In the first 10 minutes after the partition is removed, the cell also does not respond (center). But after another 10 minutes, a place field develops in the new larger box (right). (Source: Adapted from Wilson and McNaughton, 1993, Fig. 2.)

What properties of hippocampal neurons aid them in their spatial navigation and memory? In a fascinating series of experiments begun in the early 1970s, John O'Keefe and his colleagues at University College London showed that many neurons in the hippocampus selectively respond when a rat is in a particular location in its environment. Suppose we have a microelectrode implanted in the hippocampus of a rat while it scurries about inside a large box. At first the cell is quiet, but when the rat moves into the northwest corner of the box, the cell starts firing. When it moves out of the corner, the firing stops; when it returns, the cell starts firing again. The cell responds only when the rat is in that one portion of the box (Figure 24.21a). This location, which evokes the greatest response, is called the neuron's *place field*. We try recording from another hippocampal cell and it too has a place field, but this one fires only when the rat goes to the center of the box. For obvious reasons, these neurons are called **place cells**.

In some ways, place fields are similar to the receptive fields of neurons in the sensory systems. For instance, the location of the place field is related to sensory input such as visual stimuli in the environment. In our experiment with the rat in the box, we could paint images above the four corners, such as a star above the northwest corner, a triangle above the southeast corner, and so on. Consider a cell that responds only when the rat is in the northwest corner of the box, near the painted star. Suppose we take the rat out of the box and blindfold it. We then secretly go back and rotate the box 180° so that now the northwest corner has the triangle and the southeast corner has the star. Will the cell we were previously studying respond when the animal is in the northwest corner, or will it respond in the corner where the star is now located (the southeast corner)? We put the rat back in the box and take off the blindfold. It starts exploring, and the neuron becomes active when the rat goes to the corner near the star. This demonstrates that, at least under some conditions, the response is based on visual input.

While place cells are similar to receptive fields in some ways, there are also major differences. For instance, once the animal has become familiar with the box with the images painted in each corner, a neuron will continue to fire when the rat goes into the northwest corner, even if we turn off the lights so it cannot see the location markers. Evidently, the responses of place cells are related to where the animal *thinks* it is. If there are obvious visual cues (such as the star and triangle), the place fields will be based on these cues. But if there are no cues (e.g., because the lights are out), the place cells will still be location-specific as long as the animal has had enough time to explore the environment and develop a sense of where it is.

Performance in the radial arm maze, discussed above, may utilize these place cells that code for location. Of particular importance in this regard is the finding that place fields are dynamic. For instance, if the box the animal is in is stretched along one axis, place fields will stretch in the same direction. In another manipulation, we first let a rat explore a small box and determine the place fields of several cells. Then we cut a hole in a side of the box so the animal can explore a larger area. Initially, there are no place fields outside the smaller box, but after the rat has explored its new expanded environment, some cells will develop place fields outside the smaller box (Figure 24.21b). These cells seem to *learn* in the sense that they alter their receptive fields to suit behavioral needs in the new larger environment. It's easy to imagine how these sorts of cells could be involved in remembering arms already visited in the radial arm maze, just as you might return from a long hike by following markers you left when you first walked through the woods. If hippocampal place cells are involved in running the maze, it makes sense that performance is degraded by destroying the hippocampus.

Whether or not there are place cells in the human brain is not known. However, PET imaging studies show that the human hippocampus is activated in situations involving virtual or imagined navigation through the environment. In one experiment, subjects were positioned in a PET machine while playing a video game. They could navigate a virtual town on the game monitor by using buttons for forward movement, backward movement, and turning (Figure 24.22a). After learning their way around the virtual town, their brain activity was recorded while they navigated from an arbitrary starting point to a chosen destination. In a control condition, subjects moved through the virtual environment from the same start to finish locations, but arrows in the town always pointed them in the correct direction. In this condition, they did not have to think about how to navigate.

(a)

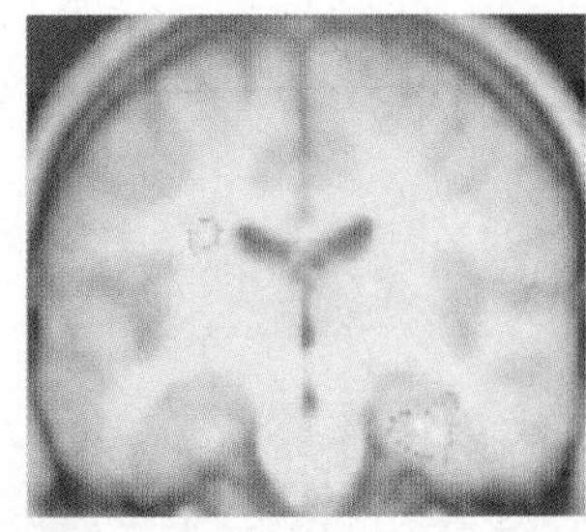

(b)

▲ FIGURE 24.22
Activity in the human brain related to spatial navigation. **(a)** A virtual town was shown on a computer monitor, and subjects in a PET imaging machine used buttons to navigate the virtual environment. **(b)** Increased brain activity associated with spatial navigation was observed in the right hippocampus and left tail of the caudate (yellow).
(Source: Maguire et al., 1998, Fig. 1.)

Figure 24.22b shows the difference in brain activity between the navigation condition and the control condition with directional arrows. When the person had to navigate the environment, there was increased activation of the right hippocampus and the left tail of the caudate. The asymmetry in the activation of the left and right hemispheres is an interesting frequent observation, but our primary point here is that the hippocampus is particularly active in this spatial navigation task with humans, just as it is in rats. The caudate activation is thought to reflect movement planning.

The hippocampus has also been studied in London taxi drivers who must learn the locations of innumerable city sites and roughly 25,000 streets to pass rigorous exams and acquire a license. One study showed that, compared to a control group, taxi drivers have a larger posterior and smaller anterior hippocampus. The size of the posterior hippocampus also seems to correlate with the length of experience as a taxi driver.

If the human hippocampus is used for spatial navigation, do hippocampal lesions impair navigation? An interesting case is a man known as T.T. who suffered bilateral hippocampal damage from encephalitis after a career of nearly 40 years as a London taxi driver. After his hippocampal damage, T.T. was very good at recognizing city landmarks and their topographical layout. A virtual reality simulation of driving through London was used to test his navigation skills. The researchers found that sometimes T.T. could "drive" efficiently from one point to another in the city while at other times he deviated from the ideal route. They found that T.T. did a good job when he could stick to main roads but got lost when smaller roads were required, as if he had lost the fine-grained knowledge of city topography he once had. The studies of humans navigating space in video games and the streets of London suggest that the human hippocampus is important for spatial memory, a finding reminiscent of the rat lesion experiments.

In addition to place cells in the hippocampus, recordings in rodents have identified neurons in the entorhinal cortex that are called **grid cells**. These cells, discovered by Edvard and May-Britt Moser and their colleagues at the Norwegian University of Science and Technology (Box 24.3), are also spatially selective. However, unlike place cells, grid cells respond when the animal is at multiple locations that form a hexagonal grid (Figure 24.23). Cells in different portions of entorhinal cortex differ in the spacing between "hotspots" in the grid, but the sensitivity grid for each cell tiles the environment the rodent is in.

A recent experiment suggests that there may also be grid cells in human entorhinal cortex. If you imagine drawing lines through the centers of grid cell hotspots in Figure 24.23, you will notice that along some axes you can connect many of the hotspots, such as with diagonal lines

BOX 24.3 PATH OF DISCOVERY

How the Brain Makes Maps

by Edvard and May-Britt Moser

We both grew up on remote islands off the west coast of Norway, a couple of hundred miles north of Bergen. This was not exactly a center of academic ferment or intellectual competition. Still, our scientific interests were nurtured by parents who'd had no opportunity to get an education themselves. We went to the same high school but didn't really get to know each other until we met again at the University of Oslo in the 1980s.

With no clear career plan, and with different scientific backgrounds, we met in an undergraduate course in psychology. Psychology kindled and reinforced our fascination with the brain, and we jointly decided to learn more about the neural basis of behavior. The university had no neuroscience curriculum at the time, but Carl-Erik Grenness, who taught an undergraduate course on behavior analysis, alerted us to the pioneering work on brain–behavior relationships taking place at this time. He also gave us a copy of a special issue about the brain published by *Scientific American* in 1979. During our wandering in wilderness, this was like manna from heaven. The issue conveyed the enthusiasm of the field and strongly attracted us to this evolving scientific discipline. Among the advances reviewed there were Kandel's demonstration of synaptic mechanisms of memory in *Aplysia californica* and Hubel and Wiesel's characterization of the mechanism for feature analysis in the visual cortex.

Grenness also sent us to Terje Sagvolden, the only psychologist at the university with research projects in neuroscience at that time. We worked on neurochemical mechanisms of attention deficit disorder for two years, in parallel with studies in psychology, and learned the basics of animal behavior and experimental design. This triggered our interest in animal learning. That brought us to visit Per Andersen, the grand neurophysiologist of Norway. We sat there for hours, trying to persuade him to take us on as graduate students. He really couldn't get us out of his office, and we wouldn't take no for an answer. In the end, he yielded to our combination of furious curiosity and unwavering determination and took us on.

Per Andersen became our Ph.D. supervisor and introduced us to the mysteries of the brain. We learned to focus on basic questions with broad implications. Through Per, we came in touch with Richard Morris at the University of Edinburgh and John O'Keefe at University College London. Richard and John were the best mentors we could have had. They guided us into the mysteries at the intersection of behavior and neuroscience. During our studies we visited Richard several times to participate in work on the functions of the hippocampus and the role of hippocampal long-term potentiation in memory formation. After our Ph.D. defense late in 1995, we spent a few very rewarding months with John to learn place cell recording in the hippocampus. This was probably the most intense learning experience in our lives. But then, in 1996, we were almost ambushed with a job offer in Trondheim. We could not move there if only one of us got a job, so we negotiated two jobs as well as equipment to start up a new lab. We literally started our lab in a bomb shelter in a basement at the university. Our few months of postdoctoral experience had been brief, but with a decent start-up package we now had the opportunity to combine what we had learned about animal behavior and neurophysiology, fulfilling our dream from the early 1980s. We began from the lower left to the upper right. If the drawn line is then rotated around the clock, there will be a periodic variation in the number of hotspots connected. This suggests that if a rat or human were to walk in various directions, grid cells would be activated more often and there would be more overall activity in the entorhinal cortex in some directions than others. This idea was tested by Christian Doeller, Caswell Barry, and Neil Burgess at University College London by having human subjects navigate a virtual reality game while fMRI images of their brain activity were recorded. The recordings made in this clever experiment showed a sinusoidal variation in the size of the fMRI signal from the entorhinal cortex when subjects navigated in different directions around the clock. This implies the presence of human grid cells that have their hotspots aligned similarly in space.

recording the activity of individual cells in the hippocampus, with electrodes implanted in the brains of rats as they roamed a square black box.

Our start-up in Trondheim was tough but enjoyable. There were no animal housing facilities, no workshops, and no technicians. We did all the work on our own; we cleaned rat cages, changed bedding, sliced brains, and repaired cables. Starting from scratch gave us the opportunity to shape the lab exactly as we wanted it.

As we began, we got a grant from the European Commission to coordinate a consortium of seven groups that collectively aimed to perform one of the first integrated neural network studies of hippocampal memory. This was virgin territory in the late 1990s. One of the aims was to determine how the position code of the hippocampus is computed. It had been known since John O'Keefe's studies in 1971 that the hippocampus has place cells that fire if and only if an animal is in a certain place. But it was unclear whether those place signals originated in the hippocampus itself or came from the outside. To address this question, we made intrahippocampal lesions that disconnected the output stage of the circuit—CA1—from the earlier stages. To our surprise, this did not abolish place coding in CA1. Then we had to grapple with the idea that the spatial signal might originate from somewhere else, most likely the surrounding cortex, through connections that went around the intrahippocampal circuit. The most likely candidate was the entorhinal cortex, a cortical region with major direct connections to the CA1 area of the hippocampus.

We started recording in this region, with invaluable help from Menno Witter, a neuroanatomist who was then located at the Free University of Amsterdam but later moved to become a part of the Kavli Institute in Trondheim. Witter had by that time worked out much of the connectivity between the entorhinal cortex and hippocampus and helped us in the delicate task of guiding electrodes to the right spot. By 2002, our research group had grown, and we now had an outstanding team of students working side by side with us in the lab and on the computer.

Sometimes scientific discoveries are portrayed as "Eureka" moments, where the researcher suddenly understands the significance of what he or she has found. In our case, it didn't quite work that way: We didn't immediately realize that the cells we recorded from were grid cells. At first we noticed that many entorhinal cells spiked every time a rat went to a particular spot, like the place cells in the hippocampus. However, each cell had multiple firing locations. After seeing the firing locations in sufficiently large environments, we became convinced that those locations formed a peculiarly regular pattern—a hexagonal grid—much like the arrangement of marbles in a Chinese checker board. Every cell did it this way, with actual firing locations differing between cells. The cells were organized topographically in the sense that the size of and distance between grid fields increased from dorsal to ventral. Moreover, cells maintained firing relationships from one environment to the next, suggesting that we were on track of a universal type of spatial map, a map whose activity pattern in many ways disregarded the fine details of the environment. With their strict regularity, the cells had the metrics of the spatial map that had not been found in the hippocampus.

These discoveries were published in a series of papers that began in 2004, only two years after we published the hippocampal disconnection study. The grid pattern itself was published in 2005. Since then we have continued to explore how grid cells operate, how they are generated, and how they interact with other spatial cell types. There is still a lot to find out. Grid cells have helped us better understand the neural representation of space, but they also provide a window into some of the innermost workings of the brain. Perhaps the most fascinating thing is that the hexagonal pattern is generated by the cortex itself. There is no grid pattern in the outside world; this is made by the brain alone. Because the pattern is so reliable and so regular, it may put us on the track of understanding the fundamental computations of the cortex.

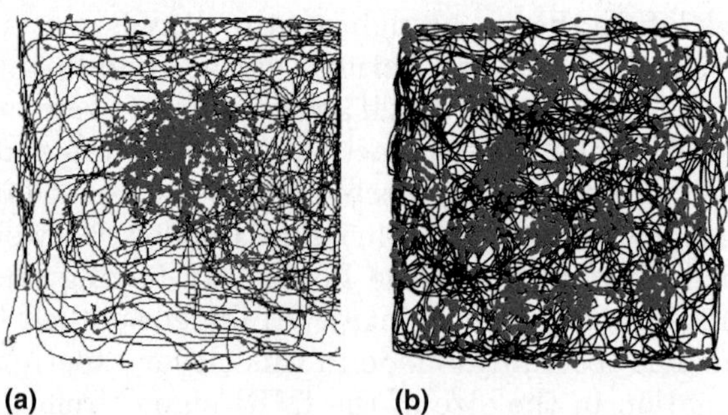

▲ FIGURE 24.23
A rat place cell and a grid cell. The black lines show the path a rat took through a square enclosure. Red spots indicate locations of the rat associated with neural activity. **(a)** A place cell in the hippocampus responds when the rat is in one region of the enclosure. This is the cell's place field. **(b)** A grid cell in entorhinal cortex is active when the rat is at multiple locations that form a grid pattern. (Source: Moser et al., 2008, Fig. 1.)

Recall that the entorhinal cortex provides input to the hippocampus. Models suggest that place fields in the hippocampus may result from summation of inputs from grid cells. The single place field of a hippocampal neuron would be the location at which the grid locations of multiple input grid cells align. Like place cells, grid cells continue to fire when the animal is at the same grid locations even when the lights are turned off. This suggests that rather than being simply a sensory receptive field, the cell's response is based on where the animal thinks it is. Taken together, place cells, grid cells, and other hippocampal system neurons showing sensitivity for head direction make a compelling case that this brain region is highly specialized for spatial navigation.

Hippocampal Functions Beyond Spatial Memory. From our discussion of the hippocampus to this point, it may seem that its role is easily defined. First, we saw that performance in a radial arm maze, which requires memory for the locations of arms already explored, is disrupted by hippocampal lesions. Second, the responses of place cells in the hippocampus, considered along with grid cells in the entorhinal cortex, suggests that these neurons are specialized for location memory. This is consistent with the **cognitive map theory** proposed by O'Keefe and his colleague Lynn Nadel, which states that the hippocampus is specialized for creating a spatial map of the environment. In one sense it is undeniable that the hippocampus, at least in rats, seems to play an important role in spatial memory.

Others argue, however, that this is not the only or best description of what the hippocampus does. In his original studies using the radial arm maze, Olton described the result of hippocampal lesions as a deficit in working memory. The rats were not able to retain recently acquired information concerning arms already explored. Thus, working memory might be one aspect of hippocampal function. This would explain why the rats with lesions could avoid going down arms that never contained food but still not remember which arms they had recently visited. Presumably after training, the information about no-food arms was saved in long-term memory, but working memory was still required to avoid the arms where food had already been retrieved.

Other theories are built on observations that the hippocampus integrates or associates sensory input of behavioral importance. For example, as you read this book, you may form memories relating multiple things: specific facts, illustrations that catch your eye, interesting passages, the arrangement of material on the page, and information about the sounds or events going on around you as you read. Maybe you've tried finding a passage in a book by searching for a page that looks a particular way. Another common example is how remembering one thing, such as the theme song to an old television show, can bring back a flood of related associations—the characters in the show, your living room at home, the friends you watched with, and so on. Interconnectedness is a key feature of declarative memory storage.

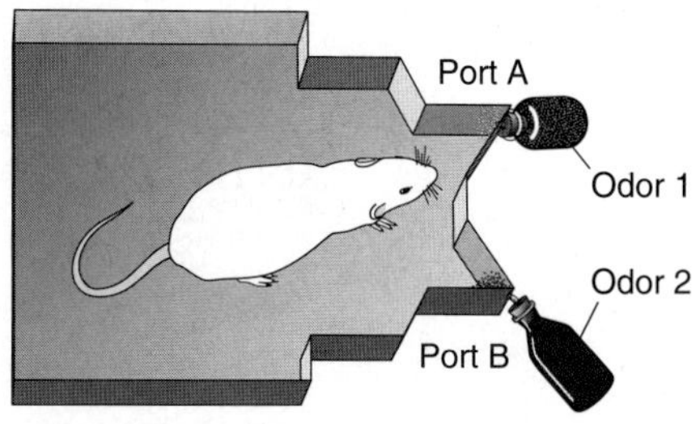

▲ FIGURE 24.24
An odor discrimination experiment used to study relational memory. For various combinations of odors, rats were trained to move toward a port emitting one odor and avoid the other port. (Source: Adapted from Eichenbaum et al., 1988, Fig. 1.)

Odor discrimination provides an example of the involvement of the hippocampus in tasks that are not entirely based on spatial memory. In one such task, a rat's cage had two ports at which the animal smelled two different odors (Figure 24.24). For each pair of odors, the animal was trained to go toward the port releasing one odor and to avoid the other port. The researchers found that some neurons in the hippocampus became selectively responsive for certain pairs of odors. Moreover, the neurons were particular about which odor was at which port; they would respond strongly with odor 1 at port A and odor 2 at port B, but not with the odors switched to the opposite ports. This indicates that the response of the hippocampal neurons related the specific odors, their spatial locations, and whether they were presented separately or together. It was also shown that hippocampal lesions produced deficits on this discrimination task.

Let's summarize the diverse studies of the hippocampus we've discussed. First, research since the time of H.M. indicates that the hippocampus is critical for memory consolidation of facts and events. Strong evidence suggests that in rodents, and in people, the hippocampus is especially important for spatial memory. In recordings from human hippocampal neurons there is sometimes surprising selectivity for people or objects we are familiar with. Finally, hippocampal cells appear to form associations between sensory stimuli even when the information is not about space. A thread that runs through these various studies is that the hippocampus links different experiences together. It receives a huge spectrum of sensory inputs and may construct new memories by integrating the varied sensory experiences associated with an event (e.g., the theme music of a television show is integrated with memories of people and place). The hippocampus may also be essential for building or enhancing memories by connecting new sensory input with existing knowledge. It has been suggested that the input from the grid cells in the entorhinal cortex provides "where" information to the hippocampus, while other input contains information about "what." The neural associations constructed and then consolidated in the hippocampus might effectively establish memories for "what happened where."

Consolidating Memories and Retaining Engrams

There is compelling evidence that declarative memory formation involves a system of interconnected brain structures that take in sensory information, make associations between related information, consolidate learned information, and store engrams for later recall. Components of this system include the hippocampus, cortical areas around the hippocampus, the diencephalon, neocortex, and more. The questions we want to look at now concern the timing of these things: When and where are memories stored in a permanent form? How long does it take memories to become "permanent"? Do engrams change location over time? Can subsequent experiences alter, enhance, or degrade memories?

Standard and Multiple Trace Models of Consolidation. Since the time of H.M., a view of memory consolidation and storage has developed that has come to be called the **standard model of memory consolidation**. In this model, information comes through neocortex areas associated with sensory systems and is then sent to the medial temporal lobe for processing (especially the hippocampal system). As we will discuss in more detail in Chapter 25, changes in synapses create a memory trace via a process sometimes called **synaptic consolidation** (Figure 24.25a). After synaptic consolidation, or perhaps overlapping with it in time, **systems consolidation** occurs in which engrams are moved gradually over time into distributed areas of the neocortex (Figure 24.25b). It is in a variety of neocortical areas that permanent engrams are stored. Before systems consolidation, memory retrieval requires the hippocampus, but after systems consolidation is complete, the hippocampus is no longer needed.

Many observations about memory formation are consistent with the standard model, but questions have been raised about whether it provides the most accurate account of consolidation. A key point is the duration of retrograde amnesia. For example, early accounts of H.M.'s amnesia reported that his retrograde amnesia extended back a few years. An interpretation of this observation is that synaptic consolidation is complete quickly but systems consolidation takes years to complete and the retrograde memories H.M. lost were ones not "fully baked" (i.e., engrams still dependent on the hippocampus). Later studies examined H.M.'s retrograde amnesia in more detail and found that it extended back decades. Perhaps systems consolidation is a very slow process that takes decades. However, some scientists have asked whether this makes sense in a species whose individuals, not so long ago, lived only for a few decades. As if that question isn't confusing enough, it should be noted that later studies of H.M. suggested that he had retrograde amnesia for episodic memories that extended back virtually his entire life. This implies that the hippocampus, perhaps in concert with other medial temporal structures, may be involved with memories for a lifetime.

Alternatives to the standard model have been proposed, most notably the **multiple trace model of consolidation** proposed by Lynn Nadel of the University of Arizona and Morris Moscovitch of the University of Toronto. The multiple trace model was proposed as a way to avoid the necessity of the decades-long systems consolidation process the standard model needs to account for extended retrograde amnesia. If hippocampal damage disrupts episodic memories going back decades or a lifetime, perhaps the hippocampus is always involved in memory storage. In other words, systems consolidation does not ever relinquish engrams entirely to neocortex.

According to this theory, engrams involve neocortex, but even old memories also involve the hippocampus (Figure 24.25c). The term "multiple trace" refers to the way the model allows for retrograde amnesia resulting from hippocampal damage to sometimes be graded in time. The hypothesis is that each time an episodic memory is retrieved, it occurs in a context different from the initial experience and the recalled information combines with new sensory input to form a new memory trace involving both the hippocampus and neocortex. This creation of multiple memory traces presumably gives the memory a more solid foundation and makes it easier to recall. Because retrieval requires the hippocampus, complete loss of the hippocampus should cause retrograde amnesia for all episodic memories no matter how old. If there is partial damage, then the memories that are intact would be the ones with multiple traces. To the extent that older memories would have been recalled more times than recent

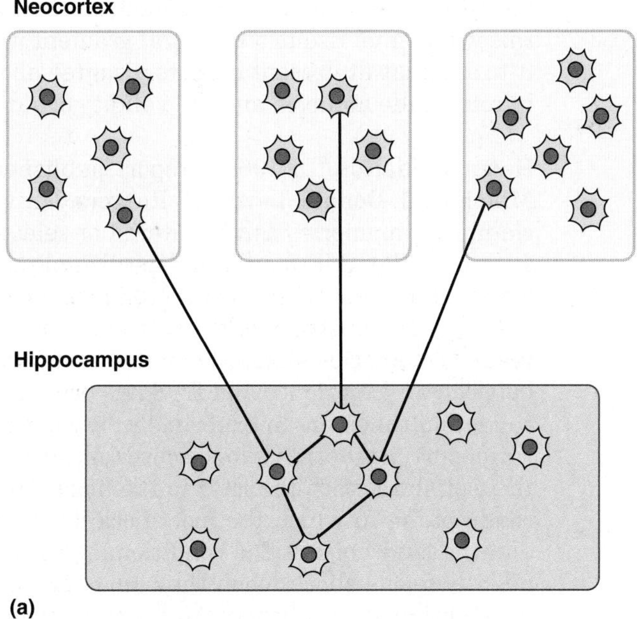

Neocortex

Hippocampus

(b)

Neocortex

Hippocampus

(c)

▲ FIGURE 24.25

Two models of memory consolidation. **(a)** In both standard and multiple trace models, memory formation is initiated by synaptic changes in the hippocampus. In this schematic, the hippocampal neurons interact with neurons in three distributed areas of neocortex. **(b)** In the standard model, a temporary memory trace is formed in the hippocampus through synaptic consolidation and engrams later develop in neocortex through systems consolidation. Over time, memory depends more on connections in neocortex (solid lines) and less on the hippocampus (dashed lines). **(c)** In the multiple trace model, engrams for episodic memories always involve both the hippocampus and neocortex (all lines are solid). The red and green lines indicate two traces for the same memory that were made in somewhat different sensory contexts.

memories, they would be more likely to survive hippocampal damage, and this would give rise to a temporal gradient in retrograde amnesia. Suffice it to say that at present experts disagree about assessments of gradients in retrograde amnesia and the validity of various models of consolidation.

Reconsolidation. In 1968 a report published by James Misanin, Ralph Miller, and Donald Lewis of Rutgers University made the surprising claim that memories can be altered or selectively erased even after they are consolidated. From the perspective of the standard model, this is unexpected because memories should be sensitive to disruption, such as by electrical convulsive shock (ECS), only before consolidation. After consolidation, memories should be stable and fixed. Indeed, varying the time between an experience and ECS has been used to measure how long consolidation takes. In an experiment by Misanin and colleagues, rats were trained by presenting a loud noise followed by a foot shock (don't confuse the foot shock with the ECS to the head). This is an example of classical conditioning in which the foot shock is the unconditional stimulus (US) and the loud noise is the conditional stimulus (CS). Trained rats subsequently fear a shock when they hear the noise. After the loud noise, the unconditioned rats licked rapidly from a drinking tube, but the trained animals drank much slower, an indication that they feared the foot shock. If an animal was conditioned but immediately given electroconvulsive shock, it would lick quickly the next day, just as if it had never been conditioned. This is evidence of retrograde amnesia for the conditioning. If there was a delay of 24 hours between the conditioning and the electroconvulsive shock, there was no fear attenuation. Apparently by 24 hours the fear memory had been consolidated and amnesia for the conditioning could no longer be caused by ECS.

On the second day of the experiment, presumably after consolidation of the fear memory, some of the conditioned animals were presented the loud noise immediately followed by the ECS. On the third day of the experiment the rats were tested once more. Surprisingly, the rats that received the noise–foot shock combination on the second day licked as much as rats that had received ECS on the first day immediately after training (i.e., the animals with amnesia that didn't fear the loud noise). They also licked much more than rats given ECS on the second day but without the loud noise. What these results suggest is that the sound on the second day reactivated the fear memory, and, once reactivated, this memory was wiped out by ECS. If the memory was not reactivated (the animals that got ECS on the second day but without the loud sound), then ECS had no effect. The implication of the study is very important because it suggests that reactivating a memory makes it sensitive just as it had been immediately after the memory was first formed (before consolidation). For this reason the reactivation effect is called **reconsolidation**.

In recent years there has been increased interest in reconsolidation, which appears to also occur in human memory. The original rat experiment on reconsolidation involved procedural learning (classical conditioning), but it now appears that reconsolidation is also seen with human episodic memory. In one experiment, college students were instructed to memorize 20 objects (e.g., balloon, crayon, spoon) that they were shown one at a time and then placed in a basket. Each student practiced naming all the items until they remembered at least 17 of them from this day 1 list. One day later, presumably after memory consolidation, some of the students were given a reminder by showing them the empty basket and asking them to describe what happened the day before without naming the objects. The aim was to subtly reactivate the memories, possibly opening them up to reconsolidation. A control group of students did not receive this reminder, with the hope of not reactivating their day 1 memories. Both the reminded and the not-reminded groups were then asked to memorize a second group of 20 objects on day 2. Finally, on day 3 of the experiment, all the subjects were asked to recall the objects from the first group on day 1 of the study.

On average, subjects recalled about eight of the objects from day 1 regardless of whether they had received a reminder on day 2. The interesting finding was that the subjects given the reminder on day 2 "accidentally" included numerous objects from day 2 in their recall (five objects on average), whereas these intrusions rarely occurred in subjects not given the reminder on day 2. Based on that experiment and several other variations, it appeared that the reminder may have served to reactivate the consolidated day 1 memories, making them labile again. These memories were then reconsolidated on day 2, erroneously mixing in new sensory information associated with the day 2 objects. A variety of experiments looking at reconsolidation in humans have been performed and have even shown hippocampal activity in the reactivation condition. Evidently, when we recall a memory, it becomes susceptible to change and reconsolidation. This fascinating finding has profound implications for the treatment of post-traumatic stress associated with unpleasant memories and even for the reliability of our normal recollections (Box 24.4).

PROCEDURAL MEMORY

Thus far, we have focused on the brain systems involved in the formation and retention of declarative memories, partly because declarative information is what we usually mean when we say we remember something. In addition, however, the neural basis of nondeclarative memory is complex because different types seem to involve different brain structures. As indicated in Figure 24.1, various kinds of nondeclarative memory are thought to involve different parts of the brain. As an example of nondeclarative memory, we will take a look at evidence supporting the involvement of the striatum in habit learning and procedural memory.

Recall from Chapter 14 that the basal ganglia are important for the control of voluntary movements. Two elements of the basal ganglia are the caudate nucleus and the putamen, together called the **striatum**. The striatum sits at a key location in the motor loop, receiving input from frontal and parietal cortex and sending output to thalamic nuclei and cortical areas involved in movement. Several lines of evidence in studies of rodents and humans suggest that the striatum is critical for the procedural memory involved in forming behavioral habits.

The Striatum and Procedural Memory in Rodents

The amnesia experienced by H.M. was surprising, in part, because he was able to learn new habits despite his complete inability to form new declarative memories. Indeed, this is one of the most compelling reasons for hypothesizing that procedural memory uses distinct circuitry. In the monkey model of amnesia, we saw that the formation of new declarative memories could be disrupted by making small lesions in the rhinal cortex of the medial temporal lobe. Such a lesion had relatively little effect on procedural memory, which raises an obvious question: Are there comparable lesions that disrupt procedural memory without affecting declarative memory? In rodents, lesions to the striatum have this effect.

BOX 24.4 OF SPECIAL INTEREST

Introducing False Memories and Erasing Bad Memories

If reconsolidation can alter existing memories, how can we ever be sure what we remember is correct? It may sound like science fiction, but it has been reported that memories already consolidated can be altered and memories for things that never happened can be introduced into the brain.

Experiments testing the ability to tinker with memories have largely been conducted with mice. In a frightening situation a mouse will "freeze" in place, a reaction that is presumably a form of "playing dead" to avoid detection or consumption by a predator. In a laboratory experiment a mouse is placed in a box that the animal can recognize based on its visual appearance and smell. When the mouse enters the box it receives an electrical foot shock. If the mouse is removed from the box and returned to it the next day, it will freeze when it recognizes the box, in anticipation of a foot shock even when none is given. The animal does not freeze when placed in a distinctly different box, indicating that the animal has formed a memory associating the shock with only the box where it received the shock.

A team of scientists at the Massachusetts Institute of Technology recently used transgenic mice in a clever variation on this fear-conditioning experiment to study the malleability of memories. The mice explored a box, and a small percentage of hippocampal neurons became activated by the visual and olfactory experiences associated with the box. Critically, in these transgenic animals the scientists had the ability to essentially turn a switch on and off chemically so that active neurons would or would not express channelrhodopsin-2 (ChR2), as described in Chapter 4. Neurons with ChR2 could then be activated at a later time by exposing them to blue light. With these abilities in place, the experimental procedure went like this (Figure A):

1. On day 1, expose animals to box A with the label-active-cells switch on so that neurons activated by the box A sensory stimulation express ChR2. No foot shock is given, and animals do not freeze in box A.
2. On day 2, expose animals to the different sensory context of box B. On this day the cell label switch is off, so ChR2 is not expressed in the hippocampal neurons activated by the box B context. While the animals are in box B, blue light was delivered to the hippocampus through a fine fiber optic cable, which reactivated the neurons from the previous day that encoded sensory information associated with box A. At the same time, a foot shock was delivered. Remember, this all occurs when the animal is in box B. The hypothesis was that the reactivated memory of box A would be reconsolidated on the second day in a manner that links in the painful foot shock.

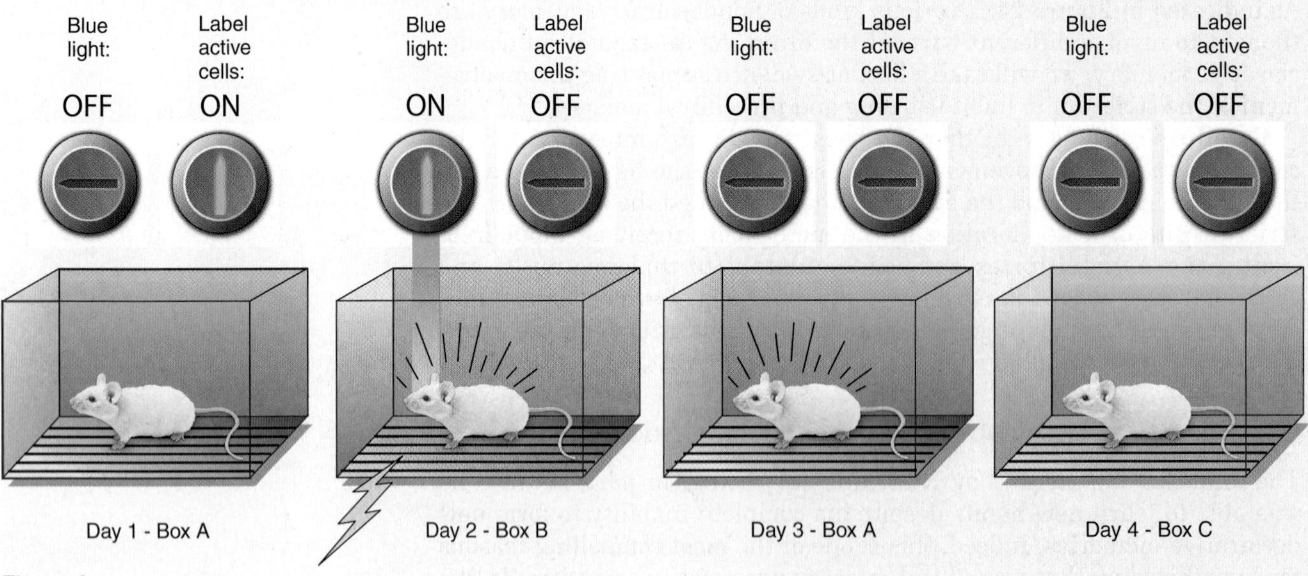

Figure A

3. The moment of truth! On day 3 the mouse is put back in box A, and as predicted, it freezes even though it has never actually experienced a foot shock in box A. When the animal is placed in an unfamiliar context, box C, it does not freeze.

It appears that a false memory has been created such that the animal fears box A even though the electrical shock occurred only in box B. The lack of freezing behavior in box C indicates that the false memory is specific to box A, presumably because neurons encoding box A information were reactivated by the blue light while the animal was in box B. You have probably heard of people convicted of crimes based on eyewitness testimony and the convicted person is later released from prison based on DNA evidence proving their innocence. Evidently the memory of the eyewitness had been wrong. Might this happen in some cases because witnesses are coached and the coached information interacts and reconsolidates with the events of the recalled crime? Ongoing research is examining the conditions under which reconsolidation can occur; the results have important implications for the judicial system and our ability to trust our own memories.

If we are able to alter memories after they have been consolidated, perhaps a process can be devised to treat people with memories that haunt them. We all have embarrassing moments we would like to forget, but some people have memories that are so disturbing that they interfere with daily life. In post-traumatic stress disorder (PTSD), an earlier traumatic event has serious deleterious effects on later behavior, mood, and social interactions, even in nonthreatening situations. An example is a war veteran who experiences stress and fear in daily life long after the war is over. What if there was a way to erase or at least weaken such unpleasant memories? Studies exploring this question suggest that it might be possible.

One approach takes advantage of the observation that administration of the beta-adrenergic antagonist propranolol shortly after a traumatic event will reduce physiological responses (e.g., heart rate) to later recollection of the event. It is thought that propranolol may counter the effect of stress hormones usually activated by the fearful experience. Unfortunately, it is generally not possible to mediate with someone immediately after a traumatic experience. The important question for a potential PTSD treatment is whether it is possible to make use of reconsolidation to weaken the traumatic memory well after the event. In one study looking into this question, people suffering from chronic PTSD were instructed to describe the trauma they experienced. They were simultaneously given either propranolol or a placebo. One week later when they were asked to recall their traumatic event, physiological responses were lower in the people given propranolol than in the placebo group. Perhaps the drug administration at the time of memory reactivation led to a reconsolidated memory with weakened emotional impact. Note that in this case, the propranolol treatment affected the emotional weight of the memory but not the declarative memory itself.

We don't know if mice experience PTSD, but a recent study by Tsai and her colleagues at MIT attempted to reduce an unpleasant memory in mice by targeting neural plasticity rather than affecting whole-body physiology as with propranolol. As in experiments we have already examined, mice are made to fear a loud sound by pairing the sound with an electrical foot shock. When mice later hear the sound they freeze in fear even if the foot shock is not given. The usual way to decrease the strength of the fear reaction is to repeatedly expose the mice to the sound without giving an electrical shock (similar to the treatment humans with PTSD are given in which they recall a traumatic memory in a safe environment). This extinction therapy in the mice is found to reduce or eliminate fear associated with the chamber if the therapy is started one day after the traumatic experience but not 30 days later. With an eye toward the delayed treatment of PTSD in humans, Tsai et al aimed to reduce the memory in mice a month after the electrical shock, a time at which extinction therapy alone is ineffective. This was accomplished by combining the fear-inducing sound with administration of a drug that inhibits the HDAC2 (histone deacetylase 2) enzyme. This enzyme, which switches neuroplasticity genes off in the nucleus of neurons (discussed in Chapter 25), was found to be inactive the day after the electrical shock but active a month later. By inhibiting HDAC2, the plasticity genes were activated at the later time. With the genes active and the traumatic memories revived by the loud sound, it was possible to reconsolidate the memory in a less fearful form. After just one dose, the mice no longer froze when the sound was heard. We don't know if this or a related approach will work to treat human PTSD, but there is hope that memory reconsolidation might be of use in alleviating this disorder.

Further Reading:

Brunet A, Orr, SP, Tremblay J, Robertson K, Nader K, Pitman RK. 2008. Effect of post-retrieval propranolol on psychophysiologic responding during subsequent script-driven traumatic imagery in post-traumatic stress disorder. *Journal of Psychiatric Research* 42:503–506.

Graff J, Joseph NF, Horn, ME, Samiei A, Meng J, Seo J, et al. 2014. Epigenetic priming of memory updating during reconsolidation to attenuate remote fear memories. *Cell* 156:261–276.

Ramirez S, Liu X, Lin P, Suh J, Pignatelli M, Redondo RL, et al. 2013. Creating a false memory in the hippocampus. *Science* 341: 387–391.

In one study, rats had to learn two versions of the radial arm maze task. The first was the standard version described earlier, in which the rat learned to move as efficiently as possible when retrieving the food from each of the baited arms of the maze. In the second version, small lights were illuminated above one or more arms containing food, and the unlit arms had no food. The lights could be turned on or off at any time. In this case, optimal performance meant that the animal kept returning to retrieve food from lit arms as long as they were lit, and avoided arms that were never lit. The standard maze task was designed to require the use of declarative memory. The "light" version of the task was intended to draw on procedural memory because of the consistent association between the presence of food and the illuminated lights. The rat did not have to remember which arms it had already explored; it simply had to form a habit based on the association of the light with food. The rat's performance on the light task was analogous to the habits H.M. was able to form, such as mirror drawing.

Performance on the two versions of the radial arm maze task were affected in markedly different ways by two types of brain lesions. If the hippocampal system was damaged (in this case, with a lesion to the fornix that sends hippocampal output), performance was degraded on the standard maze task but was relatively unaffected on the light version. Conversely, a lesion in the striatum impaired performance of the light task but had little effect on the standard task. This "double dissociation" of the lesion site and the behavioral deficit suggests that the striatum is part of a procedural memory system but is not crucial for the formation of declarative memories.

Recordings made from the rat striatum in other experiments showed that neural responses changed as the animals learned a procedure associated with a food reward. In a simple T-maze task, rats were placed at the end of one arm of the T and, as they moved away from the arm's end, a tone sounded (Figure 24.26a). A low tone was used to train the

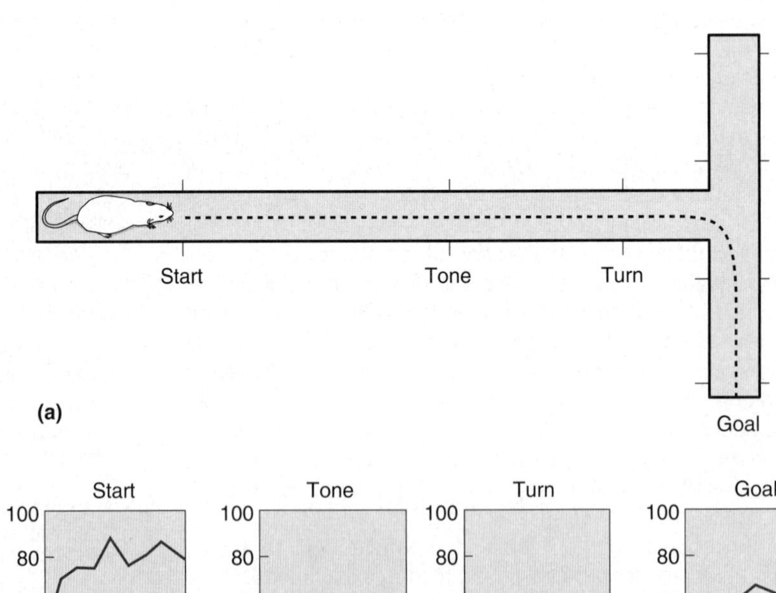

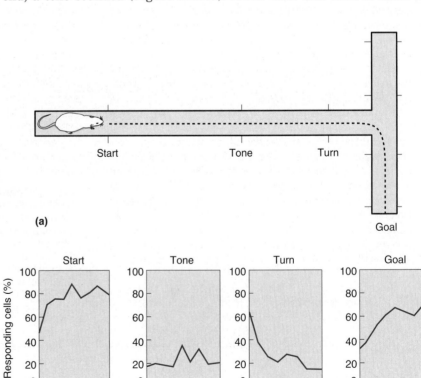

▶ FIGURE 24.26
Changing responses in rat striatum during the learning of a habit. (a) The rat began at the end of the long arm of a T-maze and turned left or right depending on the pitch of a tone. **(b)** The percentage of neurons that responded during several phases of the task: start position, tone sounding, turning into reward arm, and reaching the reward (goal). Each stage of training and testing consisted of 40 runs through the maze. Over the stages of learning and mastering the maze, more cells responded to the start and the goal and fewer to the turn. (Source: Adapted from Jog et al., 1999, Figs. 1 and 2.)

animal to turn left into the next arm of the T to get a chocolate snack, and a high tone was used to train the animal to turn right for the reward. Figure 24.26b shows the percentage of neurons that responded when the animal was in several phases of the task: start position, tone sounding, turning into reward arm, and reaching the reward.

When the rats first performed the task (stage 1), the highest percentage of neurons responded when the animals turned into the reward arm. However, in later stages of the experiment, as training and testing progressed, this percentage decreased significantly. As the rats mastered the procedure, increasingly more neurons became responsive at the start and completion of the task. Also, increasing numbers of neurons responded during more than one stage of the task. One possible interpretation of these changes in response patterns is that they reflect the formation of a habit for which the striatum codes a sequence of behaviors initiated in the T-maze situation. At present this is only a hypothesis, but it is intriguing because of the connectivity of the striatum—taking in highly processed sensory information and sending out signals involved in motor responses.

Habit Learning in Humans and Nonhuman Primates

Studies with monkeys indicate that the effects of selective brain lesions on memory are comparable in rodents and primates. In primates, there is a similar dissociation between the effects of lesions to the hippocampal system and the striatum. As we have already seen, lesions to the medial temporal lobe significantly impair performance on the delayed non-match to sample task that uses declarative memory. However, consider another task in which the animal repeatedly sees two visual stimuli, such as a square and a cross, and must learn to associate a food reward with only the cross (i.e., instrumental conditioning). This kind of habit learning is relatively unaffected by medial temporal lesions. Such preservation of habit learning in monkeys is analogous to the rat's ability to retrieve food consistently associated with a light above an arm of a maze, even after a fornix lesion.

In monkeys, lesions that involve the striatum or connections to it have quite different effects from medial temporal lesions. When the striatum is damaged, there is no effect on the performance of the DNMS task, demonstrating both that declarative memory formation is still possible and that the animal can discriminate visual stimuli. But when the striatum is damaged, the animal is unable to form the habit of always retrieving food associated with one visual stimulus rather than another. Repeated exposure to this fixed stimulus–reward situation just doesn't seem to sink in. Thus, there appear to be somewhat distinct anatomical systems for declarative memory and procedural memory, and behaviors such as learned habits utilize the striatum.

Several diseases in humans attack the basal ganglia, and certain of their effects on memory appear consistent with the striatum's role in procedural memory. For example, Huntington's disease kills neurons throughout the brain, but the striatum is a focus of the attack. Huntington's patients have been shown to have difficulty learning tasks in which a motor response is associated with a stimulus. Although these people generally have motor dysfunction, difficulty in learning the stimulus–response habit does not correlate with the severity of the motor deficits, suggesting that it is an independent consequence of the disease.

Further evidence for the involvement of the striatum in habit learning comes from comparisons of patients with Parkinson's disease with amnesic patients. As we saw in Chapter 14, Parkinson's disease is characterized by

(a)

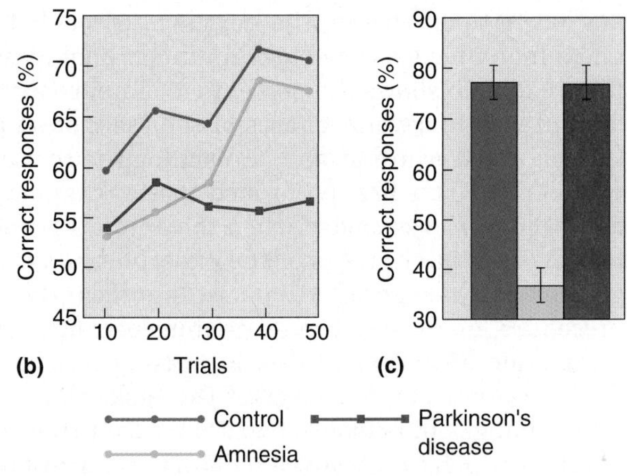

▲ FIGURE 24.27

The performance of patients with amnesia and Parkinson's disease on two memory tasks. **(a)** Four cue cards were presented in various combinations associated with the icons indicating sun or rain. Based on repeated exposure to the combinations, patients had to learn to predict sun or rain by inferring the associations. **(b)** With successive trials, control subjects and amnesic patients improved on the association task. Parkinson's patients showed little improvement. **(c)** On a test of declarative memory formation (a questionnaire), Parkinson's patients performed similar to control subjects, and amnesia patients were greatly impaired. (Source: Adapted from Knowlton et al., 1996, Figs. 1 and 2.)

degeneration of the substantia nigra inputs to the striatum. In one study, people were tested on two tasks. In the first task, patients saw one, two, or three out of four possible cues in one of 14 possible combinations. They then had to guess whether this combination was arbitrarily associated with a prediction of sunny or rainy weather (Figure 24.27a). For each patient, the experimenter assigned different probabilities that various cues were associated with sun or rain. By being told when they guessed correctly or incorrectly about the predicted weather, the patients slowly built up an association between the cues and the weather. The idea behind this task was that it draws on the formation of a stimulus–response habit. In the second task, declarative memory was tested by having patients answer multiple-choice questions about the appearance of the cues and the computer screen.

The Parkinson's patients had great difficulty learning the weather forecasting task (Figure 24.27b) but performed at normal levels on the declarative memory questionnaire (Figure 24.27c). Conversely, amnesic patients had no trouble learning the weather classification, but they performed significantly worse than either Parkinson's patients or normal controls on the questionnaire. These results suggest that the striatum in humans may play a role in procedural memory as part of a system distinct from the medial temporal system used for declarative memory.

CONCLUDING REMARKS

Far from being a computer with fixed connections, the human brain is constantly changing as a result of experience. We use working memory to temporarily hold onto information, and the patterns of sensory input from some of our experiences are assembled into permanent engrams. As a child you learned to do a summersault, and the sequence of movements was unconsciously stored for use time after dizzying time. You learned the structure of the brain and are able to impress Aunt Tilly by making a sketch showing the location of the medulla oblongata. We cannot identify the precise neurons and synapses involved in storing nondeclarative or declarative memories, but research is moving us closer to such an understanding. We know that learning and memory involve changes widely across the brain. Structures in the medial temporal lobe and diencephalon are critical for memory consolidation, and engrams are stored in the neocortex through interactions with the hippocampus and other structures. Specifying precisely what each brain structure contributes to learning and memory continues to challenge researchers.

We have seen that memories can be classified based on duration, the kind of information stored, and the brain structures involved. Early brain research relied on interpreting the effects of brain lesions on amnesia. From the case of H.M. alone, a tremendous amount was learned about memory in the human brain. The distinct types of memory, and the fact that one type can be disrupted without affecting others in amnesia, indicate that multiple brain systems are used for memory storage. More recent research uses human brain imaging and molecular genetic techniques to examine memory formation and sort out the temporal processes and multiple systems. There is even hope that one day there will be a treatment to significantly reduce the deleterious consequences of traumatic memories.

In this chapter we have focused on questions about where memories are stored and how different brain structures interact. But what is the physiological basis for the memory storage? When we try to remember a phone number, an interruption can make us forget, suggesting that memories are initially held in a particularly fragile form. Long-term memory is much more robust; it can survive interruption, anesthesia, and the normal bumps and traumas of life. Because of this robustness, it is thought that memories are ultimately stored in structural brain changes. The nature of these structural changes in the brain is the subject of Chapter 25.

KEY TERMS

Types of Memory and Amnesia
learning (p. 824)
memory (p. 824)
declarative memory (p. 824)
nondeclarative memory (p. 825)
procedural memory (p. 825)
nonassociative learning (p. 827)
habituation (p. 827)
sensitization (p. 827)
associative learning (p. 827)
classical conditioning (p. 827)
instrumental conditioning (p. 827)
long-term memory (p. 828)
short-term memory (p. 828)
memory consolidation (p. 828)
working memory (p. 829)
amnesia (p. 829)
retrograde amnesia (p. 830)
anterograde amnesia (p. 830)

Working Memory
prefrontal cortex (p. 831)
lateral intraparietal cortex (area LIP) (p. 833)

Declarative Memory
engram (p. 835)
memory trace (p. 835)
cell assembly (p. 836)
hippocampus (p. 838)
entorhinal cortex (p. 838)
perirhinal cortex (p. 838)
parahippocampal cortex (p. 838)
fornix (p. 839)
delayed non-match to sample (DNMS) (p. 843)
recognition memory (p. 843)
Korsakoff's syndrome (p. 845)
Morris water maze (p. 847)
place cell (p. 848)
grid cell (p. 850)
cognitive map theory (p. 852)
standard model of memory consolidation (p. 854)
synaptic consolidation (p. 854)
systems consolidation (p. 854)
multiple trace model of consolidation (p. 854)
reconsolidation (p. 856)

Procedural Memory
striatum (p. 857)

REVIEW QUESTIONS

1. If you try to recall how many windows there are in your house by mentally walking from room to room, are you using declarative memory, procedural memory, or both?
2. What kind of experiment might you conduct to find the place in the brain that people use to hold a phone number in mind?
3. In what brain areas have neural correlates of working memory been observed?
4. What structures in the medial temporal lobe are thought to be involved in memory?
5. Why did Lashley conclude that all cortical areas contribute equally to learning and memory? Why was this conclusion later called into question?
6. What arguments can you think of for and against the idea that Wilder Penfield's electrical brain stimulation evoked memories?
7. What evidence is there that declarative and nondeclarative memory use distinctly different circuits?
8. In the famous amnesic known as H.M., what types of memory were lost after temporal lobe surgery? What kinds were preserved?
9. What are place cells and grid cells? Where have they been observed?
10. What evidence indicates that long-term memories are stored in the neocortex?
11. The multiple trace model of memory consolidation was proposed to deal with what concern(s) about the standard model of memory consolidation?
12. Where is it thought that procedural memories are stored?

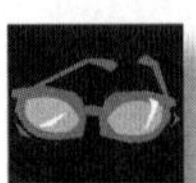

FURTHER READING

Corkin S. 2013. *Permanent Present Tense: The Unforgettable Life of the Amnesic Patient H.M.* New York: Basic Books.

Kandel ER, Dudai Y, Mayford MR. 2014. The molecular and systems biology of memory. *Cell* 157:163–186.

Ma WJ, Husain M, Bays PM. 2014. Changing concepts of working memory. *Nature Neuroscience* 17:347–356.

McKenzie S, Eichenbaum H. 2011. Consolidation and reconsolidation: two lives of memories? *Neuron* 71:224–233.

Moser EI, Kropff E, Moser M. 2008. Place cells, grid cells, and the brain's spatial representation system. *Annual Review of Neuroscience* 31:69–89.

Nadel L, Hardt O. 2011. Update on memory systems and processes. *Neuropsychopharmacology* 36:251–273.

Quiroga RQ, Kreiman G, Koch C, Fried I. 2008. Sparse but not "grandmother-cell" coding in the medial temporal lobe. *Trends in Cognitive Sciences* 12:87–91.

Squire LR, Wixted JT. 2011. The cognitive neuroscience of human memory since H.M. *Annual Review of Neuroscience* 34:259–288.

Wang S, Morris RGM. 2010. Hippocampal-neocortical interactions in memory formation, consolidation, and reconsolidation. *Annual Review of Psychology* 61:49–79.

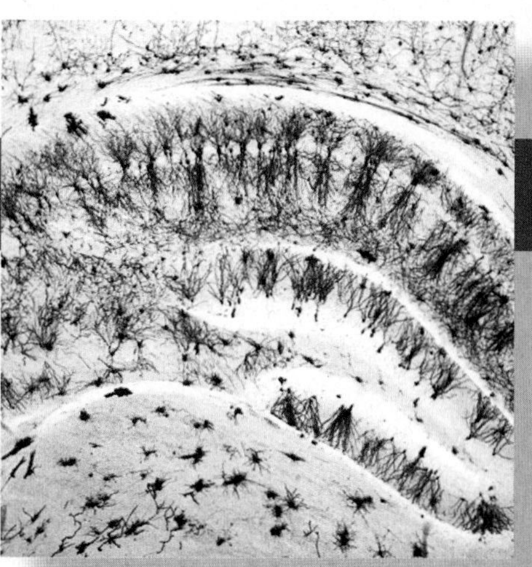

CHAPTER TWENTY-FIVE

Molecular Mechanisms of Learning and Memory

INTRODUCTION

An important first step in understanding the neurobiology of memory is identifying *where* different types of information are stored. As we saw in Chapter 24, basic neuroscientific research is beginning to answer this question. However, an equally important question concerns *how* information is stored. As Hebb pointed out, memories can result from subtle alterations in synapses, and these alterations can be widely distributed in the brain. This insight helps narrow the search for a physical basis of memory, synaptic modifications, but it also raises a dilemma. The synaptic modifications that underlie memory may be too small and too widely distributed to be observed and studied experimentally.

These considerations inspired some researchers, led by Eric Kandel of Columbia University, to study the nervous systems of simple invertebrate animals for insights into the molecular mechanisms of memory. Through the history of neuroscience, researchers have used a large menagerie of invertebrate creatures for neurobiological experiments. You are already familiar with the squid and the contribution of its giant axon and giant synapse to our understanding of cellular neurophysiology (see Chapters 4 and 5). Other experimental invertebrates are lobsters, crayfish, cockroaches, flies, bees, leeches, and nematode worms. The reason for using them is that invertebrates have some important experimental advantages, including small nervous systems with large neurons, known and reproducible connections between neurons, and simple genetics.

Invertebrates can be particularly useful for analyzing the neural basis of behavior. Although the behavioral repertoire of the average invertebrate is rather limited, many invertebrate species exhibit some of the simple forms of learning that were introduced in the last chapter. One species in particular has been used to study the neurobiology of learning, the sea snail *Aplysia californica*. Kandel shared the 2000 Nobel Prize in Physiology or Medicine for his seminal contributions to the understanding of memory mechanisms in this creature. Invertebrate research has clearly shown that Hebb was right: Memories *can* reside in synaptic alterations. Moreover, it has been possible to identify some of the molecular mechanisms that lead to this synaptic plasticity. Although nonsynaptic changes have also been found that may account for some types of memory, invertebrate research leaves little doubt that the synapse is an important site of information storage.

The past several decades have seen rapid advances in understanding how our own brains form memories. This progress has come from the study of neural activity in regions of the mammalian brain associated with different types of memory. Insight from theoretical analysis of neural networks helped focus attention on modifications most likely to store information, and new technologies have made feasible the detection of candidate mechanisms. One fruitful approach has been the use of electrical brain stimulation to produce measurable synaptic alterations whose mechanisms can be studied. Researchers could then ask if these same mechanisms contribute to natural memory formation. One of the interesting conclusions of this research is that the mechanisms of activity-dependent synaptic plasticity and memory formation in the adult brain have much in common with those operating during development for wiring the brain.

There is a growing sense of optimism among neuroscientists that we may soon know the physical basis of learning and memory. This investigation has benefited from the combined approaches of researchers in disciplines ranging from psychology to molecular biology. In this chapter, we'll take a look at some of their discoveries.

MEMORY ACQUISITION

It is useful to consider learning and memory as occurring in two stages: (1) the acquisition of a short-term memory and (2) the consolidation of a long-term memory (Figure 25.1). In this context, memory acquisition (learning) occurs by a *physical modification of the brain* caused by incoming sensory information. This is different from working memory introduced in Chapter 24, which is vulnerable to erasure by distraction and has a very limited capacity (think about holding a phone number in mind). Working memory can be achieved by keeping neural activity going with continuous rehearsal, and does not require any lasting physical change in the brain. In contrast, short-term memory survives distraction, has a large capacity, and can last minutes to hours with no conscious effort. Do you remember what you had for breakfast this morning or for dinner last night? These memories persist for some time without rehearsal but are considered to be "short term" because they will be forgotten unless they are consolidated into long-term memory. Thus, you probably do not readily remember what you had for dinner two weeks ago on Tuesday because the brain changes that encoded this information have since faded away.

Memory consolidation, introduced in Chapter 24, is the process by which some experiences, held temporarily by transient modifications of neurons, are selected for permanent storage in long-term memory. Perhaps last Tuesday's dinner coincided with an emotionally charged event, like a first date with the love of your life. In that case, it would not be surprising if every detail of the evening was etched into your long-term memory. This example makes the point that not all memories are created equal. The brain has mechanisms that ensure that some experiences are retained while others are lost.

We will divide our discussion of memory mechanisms into those responsible for the initial acquisition of short-term memory and those acting to convert a temporary change into a permanent one. We'll see that acquisition occurs by modifying synaptic transmission between neurons and that synaptic consolidation requires, in addition, new gene expression and protein synthesis.

Cellular Reports of Memory Formation

"While it might appear that I am doing nothing, at the cellular level I am really quite busy." We don't know to whom this statement should be attributed, but it certainly applies to memory. In the last chapter we discussed different types of memories and where they are stored. For example, we learned that declarative memories (facts, events, places, faces) ultimately reside in the cerebral cortex. However, when it comes to information storage, no neuron is spared. Virtually every neuron in the nervous system can form a memory of recent patterns of activity. By the same token, countless molecular mechanisms participate in information

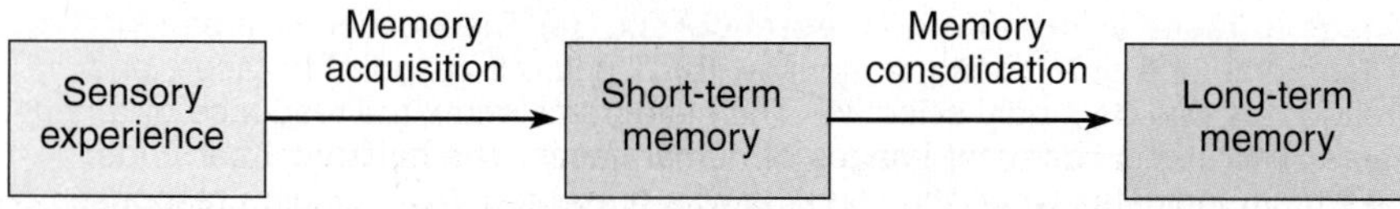

▲ FIGURE 25.1
The flow of sensory information into long-term memory. The first step is memory acquisition, by which experiences are encoded by synaptic modifications. The second step is memory consolidation, by which temporary synaptic changes are made permanent.

▶ FIGURE 25.2
Famous people who may be familiar to you. What happened in your brain when you first saw photos or video of these people and formed a memory?

storage of various kinds, so our coverage here will necessarily be selective. As a general example, let's consider what happens in the cerebral cortex when a novel experience becomes familiar (Figure 25.2).

We and other primates are experts at using vision to recognize and discriminate differences between familiar objects and individuals. Where is this information stored? According to Hebb, if an engram is based on information from only one sensory modality, it should be possible to localize it within the regions of cortex that serve this modality (see Chapter 24). For example, if the engram relies only on visual information, then we would expect it to reside within the visual cortex. Studies of visual discrimination in monkeys are consistent with this proposal.

Macaque monkeys can be trained to discriminate images of objects and associate them with a food reward. However, they lose this ability when lesions are made in the inferotemporal cortex. This region contains area

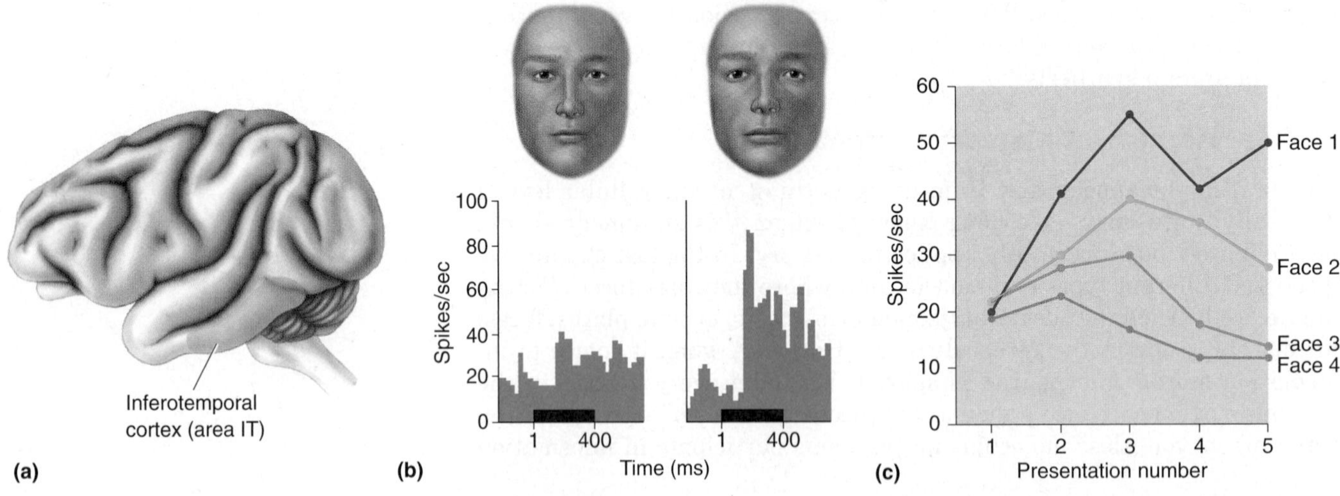

▲ FIGURE 25.3
Responses to faces in the inferotemporal cortex. (a) The location of area IT in the inferior temporal lobe of a macaque monkey. (b) IT neurons respond to faces, and these responses can be highly selective. The histograms show the response of a neuron in area IT to slightly different images of human faces. The horizontal bar under each histogram indicates when the stimulus was presented. (c) Changing responses of a cell as novel faces become familiar. When the four faces are presented for the first time, there is a moderate response to each. With subsequent presentations, the cell becomes more responsive to faces 1 and 2 and less responsive to faces 3 and 4. Acquisition of face selectivity correlates with the ability of the animal to recognize and distinguish among these faces. (Sources: Part b, adapted from Leopold et al., 2006, Fig. 6; part c, adapted from Rolls et al., 1989, Fig. 1.)

IT (Figure 25.3a) which we learned in Chapter 10 is part of the "ventral stream," a series of higher order visual areas concerned with visual perception. After lesions of inferotemporal cortex, monkeys appear to be unable to recognize familiar objects, even though basic visual capacities remain intact. IT therefore appears to be both a visual area and an area involved in memory storage. This conclusion is further supported by a fascinating clinical condition called *prospagnosia*, a selective amnesia for familiar faces (including one's own) that can result from damage to the inferotemporal cortex in humans.

Like most cortical neurons, IT neurons typically show the property of *stimulus selectivity*; that is, they respond with a barrage of action potentials to the presentation of some but not all stimuli. As we learned back in Chapter 10, IT neurons have the distinction of responding to complex images and shapes that can include familiar faces. In a typical experiment, an electrode is used to record from an IT neuron in an alert monkey. When shown a series of images of familiar faces (other monkeys in the colony or the experimenters), the neuron responds vigorously to some but not all images: The neuron shows stimulus (face) selectivity (Figure 25.3b).

Now, what happens to an IT neuron as a visual recognition memory is formed, when a novel set of faces becomes familiar? The first time new faces are seen, the cell responds at about the same moderate level to all of them: There are responses but no selectivity (Figure 25.3c, presentation 1). However, with repeated presentations, the responses change and selectivity emerges. The response of the neuron grows to some faces and diminishes to others. With continued presentation of the same group of faces, the response of the neuron becomes more stable and more selective (Figure 25.3c, presentations 4 and 5). Other nearby neurons in IT show similar changes, but their responses grow and diminish to different faces. Are we observing the birth of a memory trace? There are good reasons to think so. Shifts in the selectivity of cortical neurons are a very common cellular correlate of memories formed in other modalities (audition, somatic sensation, etc.) as well.

Distributed Memory Storage. Analysis of a simple "neural network" model helps illustrate what is behind an experience-dependent shift in neuronal selectivity. Consider the network of connected neurons depicted in Figure 25.4. Three stimuli (say, the faces of Mark, Barry, and Mike) are conveyed by separate inputs to three postsynaptic cortical neurons (call them A, B, and C). Initially, in our first experience of these gentlemen, we find that neurons A, B, and C respond moderately to Mark, Barry, and Mike. There is no selectivity, and there is no neural response that could be used to distinguish one face from another. However, after repeated exposure to Mark, Barry, and Mike, the neurons of the network acquire selectivity; although all neurons respond to all faces, neuron A responds best to Mark, neuron B responds best to Barry, and neuron C responds best to Mike. This transformation of the responses to (now) familiar faces occurred by adjustments in the strength or "weights" of the three synaptic inputs converging on the cortical neurons.

Where is the "memory" in this network? Put another way, how does output from our three cortical neurons uniquely represent Mark, Barry, or Mike? The answer is that after learning, there is a unique *pattern* or ratio of activity in the three neurons for each face. Mark is represented by high activity in neuron A, moderate activity in neuron B, and weak activity in neuron C. We call this a **distributed memory**. By analogy, consider how a color is represented in the visual system, not solely by the output of any one type of cone photoreceptor, but by a comparison of the activity in all the three types of cone photoreceptor (Chapter 9).

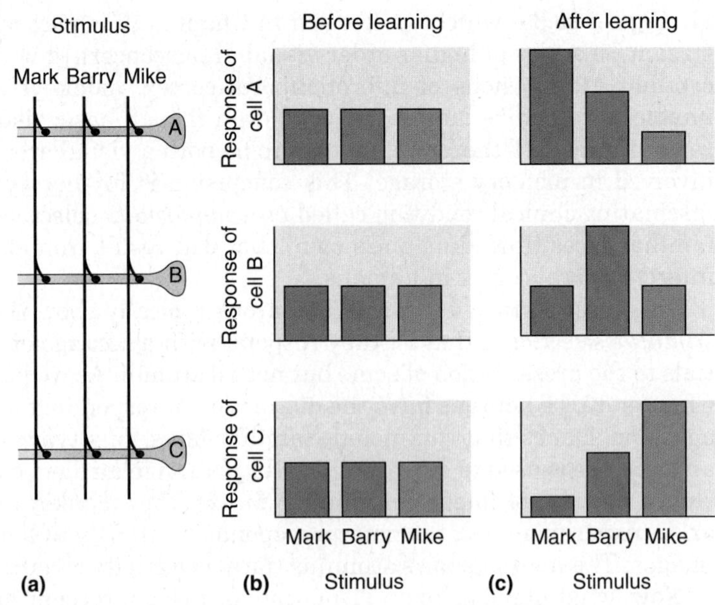

▶ FIGURE 25.4
A model of a distributed memory. (a) In this simple neural network, three inputs conveying information about the appearance of three faces (Mark, Barry, and Mike) synapse on three cortical neurons: cells A, B, vand C. **(b)** Before learning to recognize these faces, every neuron in the network responds moderately to every face. There is no selectivity for one face over another. **(c)** After learning, the neurons display a face preference. Cell A prefers Mark, cell B prefers Barry, and cell C prefers Mike. Notice that comparing the relative responses of all three neurons can determine which face is being viewed. For example, Mark evokes a strong response in A, a moderate response in B, and a weak response in C. Even if neuron A were to die, Mark would continue to be represented by a specific pattern of activity in neurons B and C.

To understand the advantages of this type of memory storage, let's consider an alternative in which the memory were encoded solely by neuron A: When it is active, Mark would be recalled. After learning, neuron A would become a "Mark detector" that certainly could store the memory of Mark, but what would happen if neuron A died from a bump on the head or some other mishap of normal life? *Poof*, there goes Mark. The distributed memory avoids this problem because no single neuron represents Mark; he is represented by the pattern of activity across all neurons in the cortical network. If neuron A met an early demise, there is still a unique pattern or ratio of activity in neurons B and C that can represent Mark. The more neurons there are in the network, the more unique memories can be stored and the more resistant the memories are to damage to individual neurons. This is a good thing, because although they are numerous, neurons in the brain die every day.

Using artificial neural network models created in the laboratory with a computer, researchers can ask what happens when neurons in the network are gradually removed. The answer is that memories show what is called *graceful degradation*. Instead of a catastrophic loss of any one memory, representations tend to blend together as neurons are lost, such that one memory gets confused with another. This type of memory loss is similar to what often happens in old age or following the death of a large number of neurons due to a brain disease.

Neural network models can reproduce the experimental observations of experience-dependent shifts in neuronal selectivity, thereby yielding insights into how memory is stored. As we have said, one such insight is that memories are distributed and show graceful degradation in response to a loss of neurons. Another key insight is that *the physical change that leads to memory can be the modification of synaptic weight* that changes the input–output relationships of neurons. Synapses store memories.

The notion of a synaptic basis for memory received strong experimental support from Eric Kandel's studies of the marine snail *Aplysia*. Kandel and colleagues were able to show that simple forms of learning, such as habituation and sensitization, were accompanied by changes in the strength of synaptic transmission between sensory neurons and motor neurons. Moreover, they were able to dissect many of the molecular mechanisms that underlie these changes. These studies provided a strong foundation for subsequent analysis of synaptic modification in the mammalian brain (Box 25.1).

BOX 25.1 **PATH OF DISCOVERY**

What Attracted Me to the Study of Learning and Memory in *Aplysia*?

by Eric Kandel

There was little in my early life to indicate that the biology of mind would become the passion of my academic career. In fact, there was little to suggest that I would have an academic career. Rather, my early life was shaped in large part by the traumatic events that occurred in the place of my birth: Vienna, Austria.

I was born in November 1929. In March 1938, when I was eight years old, Hitler entered Austria and was received by the Viennese with enormous enthusiasm. Within hours, that enthusiasm turned into an almost indescribable outburst of anti-Semitic violence. After a humiliating and frightening year, my older brother, Ludwig, and I were able to leave Vienna in April 1939. The two of us crossed the Atlantic by ourselves to live with our grandparents in New York. Our parents joined us six months later.

The spectacle of Vienna under the Nazis presented me for the first time with the dark side of human behavior. How is one to understand the sudden viciousness of so many people? How could a highly cultivated society listen to Haydn, Mozart, and Beethoven one day and the next day embrace the brutality of Kristallnacht? This question still haunted and fascinated me while in college at Harvard, where I majored in twentieth-century history and literature. I wrote my honors dissertation on the attitude of three German writers toward National Socialism, and I intended to do graduate work in modern European intellectual history. But at the end of my junior year, I decided that to obtain insights into the human mind and its capability for good and evil, it would be better to become a psychoanalyst rather than an intellectual historian.

I entered medical school in the fall of 1952, dedicated to becoming a psychoanalyst. While in medical school I loved the clinical work but had no particular interest in basic science. In my senior year, however, I decided that perhaps even a New York psychoanalyst should know something about the brain, so I took an elective at Columbia University with the neurophysiologist Harry Grundfest.

In Grundfest's laboratory I was astonished to discover that science in the laboratory is dramatically different from taking courses and reading books.

Knowing of my interest in behavior, Grundfest suggested that I set up an electrophysical system to record from the large axon of the crayfish, which controls the animal's tail and thus its escape from predators. I learned how to manufacture glass microelectrodes for insertion into individual nerve cells of crayfish and how to obtain and interpret electrical recordings from them. It was in the course of those experiments, which were almost laboratory exercises, since I was not exploring new ground scientifically or conceptually, that I first began to feel the excitement of working on my own. Whenever I penetrated a cell, I, too, could hear the crack of an action potential. I am not fond of the sound of gunshots, but I found the "bang! bang! bang!" of action potentials intoxicating. The idea that I had successfully impaled a cell and was actually listening in on the brain of the crayfish as it conveyed messages seemed marvelously intimate. I was becoming a true psychoanalyst: I was listening to the deep, hidden thoughts of my crayfish!

Had I not been exposed to the excitement of actually doing research, of carrying out experiments to discover something new, I would have ended up with a very different career and, I presume, a very different life.

I began to realize that what makes science so distinctive is not just the experiments themselves but also the social context, the sense of equality between student and teacher, and the open, ongoing, and brutally frank exchange of ideas and criticism.

Based on my six-month stay in his laboratory, Grundfest nominated me for a research position at the National Institutes of Health. I arrived at the NIH in July 1957, just after Brenda Milner published her classic research showing that complex memories—for people, places, and objects—are localized in the hippocampus. I realized that the problems of memory storage, once the exclusive province of psychologists and psychoanalysts, were now approachable with the methods of cell biology. What are the cellular mechanisms for that storage? I wondered. No one knew anything about the nerve cells of the hippocampus at that time. I thought perhaps the nerve cells that participate in memory storage would have novel properties that would speak to me of memory!

Together with Alden Spencer, a young colleague from NIH, I set out to study the properties of hippocampal nerve cells. We were the first scientists in the world to record signals from those cells. Our work showed, surprisingly, that these cells from the region of the brain that encodes our dearest memories function pretty much the same way as other nerve cells in the brain. I now realized that these cells did not speak to us about memory. We had climbed Mt. Everest, but we had no view.

I further realized that to explore memory I would need to study not nerve cells per se, but nerve cells during a learning experience that leads to the formation of a memory. This was too difficult to do in a complex structure like the hippocampus: In the late 1950s, we did not even know what sensory

(continues on page 8)

BOX 25.1 PATH OF DISCOVERY (continued)

input affected hippocampal cells. Alden and I tried visual, tactile, and auditory inputs, all without effect. I became convinced that to succeed in bringing the power of cell biology to bear on the study of learning and memory, I would first have to take a very different approach, a reductionist approach. My first step had to be to study not the most complex case but the *simplest* case of memory storage—and to study it in the simplest, most tractable experimental animal available.

While a reductionist strategy was within the realm of traditional biology, most investigators were reluctant to apply it to mental processes such as learning and memory. From the outset it seemed to me that the mechanisms of memory storage are so important for survival that they must have been conserved through evolution. Moreover, a molecular analysis of learning, no matter how simple the animal or the task, was likely to reveal those mechanisms.

I needed to develop an experimental system in which a simple reflex behavior, controlled by a small number of large, accessible nerve cells, could be modified by a simple form of learning like classical conditioning. Only then could I relate the animal's overt learned behavior to cellular and molecular events occurring in the neurons that control the behavior.

After looking at crayfish, lobsters, worms, and flies, I settled on the marine snail *Aplysia*, which has extremely large nerve cells that are conducive to recording. One of the two people in the world working on *Aplysia* at that time was Ladislav Tauc, so I spent 1962–1963 in Paris with him, and I have worked on *Aplysia* ever since.

In the early 1960s we had no frame of reference for studying the biological basis of memory formation and storage. Two conflicting theories prevailed. One was the aggregate field approach, which assumed that information is stored in the bioelectric field generated by the aggregate activity of many neurons. The other was the cellular connectionist approach, which derived from Santiago Ramón y Cajal's idea that memory is stored as an anatomical change in the strength of synaptic connections between nerve cells (Cajal, 1894). In 1948 Jerzy Konorski renamed Cajal's concept "synaptic plasticity" (Konorski, 1948).

In my studies of *Aplysia* my work focused on the cellular substrates of the gill-withdrawal reflex that occurs when the siphon of the animal is touched (Figure A). This reflex undergoes sensitization (a simple form of learning) when a noxious stimulus is applied to the tail of the animal. I found that short-term memory results from a transient strengthening of preexisting synaptic connections, due to the modification of preexisting proteins, whereas long-term memory results from a persistent strengthening of synaptic connections brought about by alterations in gene expression, the synthesis of new proteins, and the growth of new synaptic connections. I discovered that the transient strengthening results in an increase

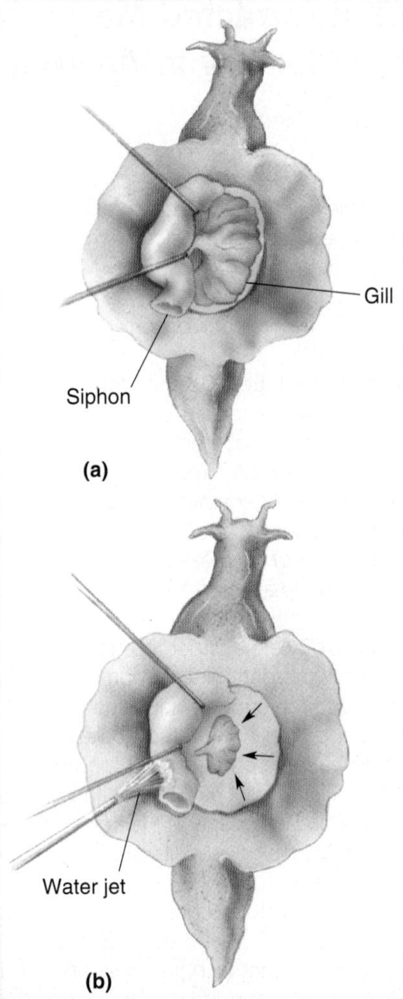

Figure A
The gill-withdrawal reflex in *Aplysia*. **(a)** The mantle is held aside to show the gill in its normal position. **(b)** The gill retracts when water is sprayed on the siphon.

in the amount of transmitter released by the sensory neuron onto the motor neuron that controls the gill musculature. This increase is produced by activation by the tail stimulus of serotonergic modulatory neurons (Figure B part a). Serotonin increases the strength of the synapse between sensory and motor neurons by increasing the concentration of cAMP, an intracellular signaling molecule in sensory neurons that activates protein kinase A (PKA). When we similarly simply injected cAMP directly into the sensory neuron, it resulted in an increase in the release of the transmitter (glutamate) into

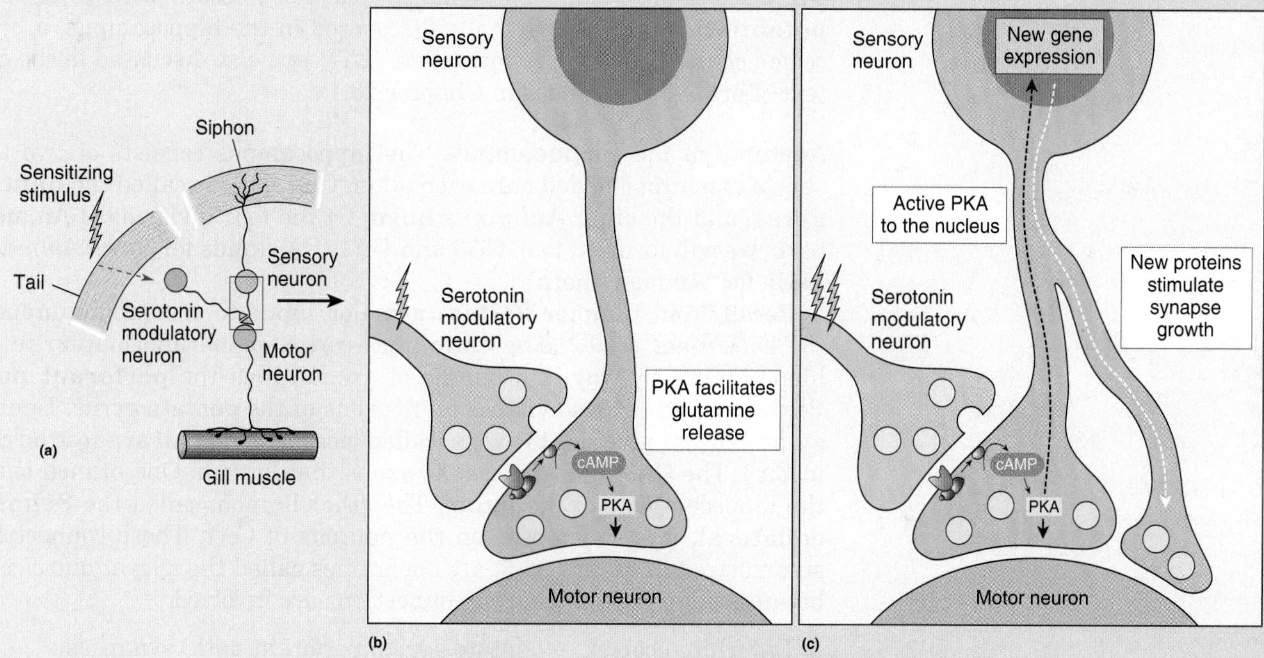

Figure B
A mechanism for sensitization of the gill-withdrawal reflex. **(a)** A simple wiring diagram showing the minimal circuitry for sensitization of the gill withdrawal reflex. A noxious stimulus to the tail activates serotonergic modulatory neurons that influence synaptic transmission at the sensory–motor synapse. **(b)** Serotonin stimulates a rise in cAMP and activation of PKA in the sensory nerve terminal, causing an increase in the amount of glutamate released when the siphon is touched. **(c)** Repeated activation of the serotonergic modulatory neurons causes long-term sensitization, requiring new nuclear gene expression and protein synthesis.

the synaptic cleft, thus temporarily strengthening the connection with the motor neuron (Figure B part b).

Beginning in 1980, the insights and methods of molecular biology enabled us to identify common mechanisms of short-term memory in different animals and to explore how short-term memory is converted to long-term memory. We found that, following long-term sensitization, PKA moves into the nucleus and activates gene expression, leading to the synthesis of new proteins and a twofold increase in the number of synaptic connections made by *Aplysia*'s sensory neurons (Figure B part c). Moreover, the dendrites of the motor neurons, which receive the signals from the sensory neurons, grow and remodel to accommodate the additional sensory input.

Together, these early cellular studies of simple behaviors provided direct evidence supporting Cajal's suggestion that synaptic connections between neurons are not immutable; they can be modified in learning, and those anatomical modifications are likely to subserve memory storage. In the gill-withdrawal reflex of *Aplysia*, changes in synaptic strength occur not only in the connections between sensory and motor neurons but also in the connections between sensory neurons and interneurons. Thus, even in a simple reflex, memory appears to be distributed among multiple sites. Studies showed further that a single synaptic connection is capable of being modified in opposite ways by different forms of learning and for different periods of time, paralleling the different stages of memory.

By 1980 my progress on *Aplysia* had been so heartening that I summoned up the courage to return to the hippocampus. There I found, much as Charles Darwin might have predicted, that once nature finds a solution that works, it tends to hold on to it. In other words, the same general principles that govern short- and long-term memory storage in simple animals also apply to complex ones.

References

Cajal SR. 1894. The Croonian Lecture: la fine structure des centres nerveux. *Proceedings of the Royal Society, London* 55:344–468.
Konorski J. 1948. *Conditioned Reflexes and Neuron Organization*. Cambridge, MA: University Press.

Strengthening Synapses

Consideration of neural network models, such as the one shown in Figure 25.4, indicates that both increases and decreases in synaptic weights can shift neuronal selectivity and store information. We begin our discussion of how this synaptic plasticity occurs with **long-term potentiation (LTP)**, originally discovered in the hippocampus, a brain region critical for memory formation. (LTP was also discussed in the context of brain development in Chapter 23.)

Anatomy of the Hippocampus. The hippocampus consists of two thin sheets of neurons folded onto each other. One sheet is called the **dentate gyrus**, and the other **Ammon's horn**. Of the four divisions of Ammon's horn, we will focus on two: **CA3** and **CA1** (CA stands for *cornu Ammonis*, Latin for "Ammon's horn").

Recall from Chapter 24 that a major input to the hippocampus is the *entorhinal cortex*. The entorhinal cortex sends information to the hippocampus by way of a bundle of axons called the **perforant path**. Perforant path axons synapse on neurons of the dentate gyrus. Dentate gyrus neurons give rise to axons (called mossy fibers) that synapse on cells in CA3. The CA3 cells give rise to axons that branch. One branch leaves the hippocampus via the fornix. The other branch, called the **Schaffer collateral**, forms synapses on the neurons of CA1. These connections, summarized in Figure 25.5, are sometimes called the *trisynaptic circuit*, because three sets of synaptic connections are involved:

1. Entorhinal cortex → dentate gyrus (perforant path) synapses
2. Dentate gyrus → CA3 (mossy fiber) synapses
3. CA3 → CA1 (Schaffer collateral) synapses

Because of its very simple architecture and organization, the hippocampus is an ideal place to study synaptic transmission in the mammalian brain. In the late 1960s, researchers discovered that the hippocampus

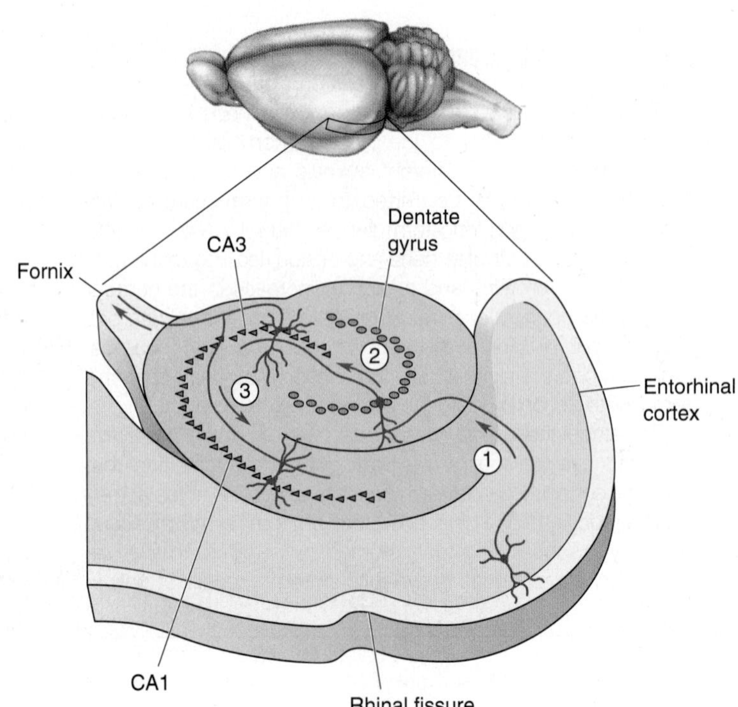

▶ FIGURE 25.5
Some microcircuits of the hippocampus. ① Information flows from the entorhinal cortex via the perforant path to the dentate gyrus. ② The dentate gyrus granule cells emit axons called mossy fibers that synapse upon pyramidal neurons in area CA3. ③ Axons from the CA3 neurons, called the Schaffer collaterals, synapse upon pyramidal neurons in area CA1.

could actually be removed from the brain (usually in experimental animals) and cut up like a loaf of bread, and that the resulting slices could be kept alive *in vitro* for many hours. In such a *brain slice preparation*, fiber tracts can be stimulated electrically and synaptic responses recorded. Because cells in the slice can be observed, stimulating and recording electrodes can be positioned with the precision previously available only in invertebrate preparations. This brain slice preparation has greatly facilitated the study of LTP.

Properties of LTP in CA1. In 1973, an important discovery was made in the hippocampus by Timothy Bliss and Terje Lømo, working together in Norway. They found that brief, high-frequency electrical stimulation of the perforant path synapses on the neurons of the dentate gyrus produced LTP. It was subsequently shown that most excitatory (and many inhibitory) synapses support LTP, and that the mechanisms can vary from one synapse type to another. However, the most sophisticated understanding of LTP has come from studying the Schaffer collateral synapses on the CA1 pyramidal neurons in brain slice preparations. This will be our focus.

In a typical experiment, the effectiveness of the Schaffer collateral synapse is monitored by giving a bundle of presynaptic axons a brief electrical stimulus, then measuring the size of the resulting EPSP in a postsynaptic CA1 neuron (Figure 25.6). Usually such a test stimulation is given every minute or so for 15–30 minutes to ensure that the baseline response is stable. Next, to induce LTP, the same axons are given a **tetanus**, a brief burst of high-frequency stimulation (typically 50–100 stimuli at a rate of

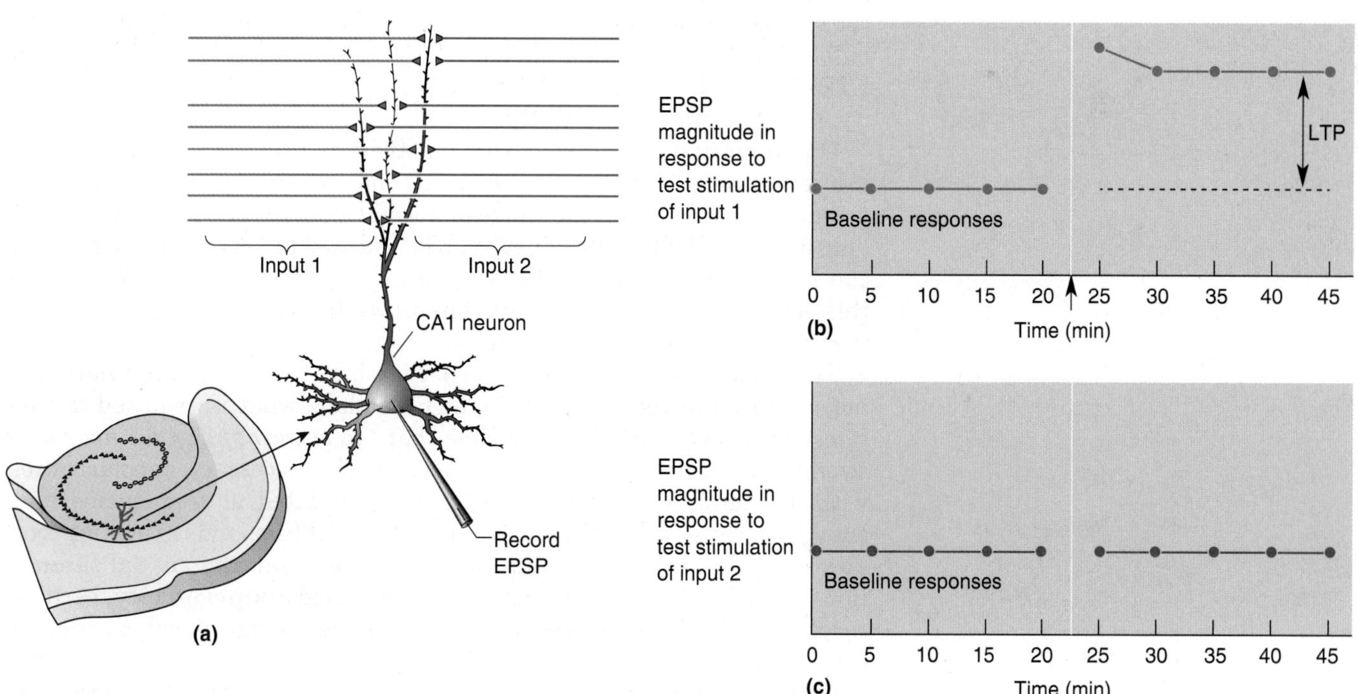

▲ FIGURE 25.6
Long-term potentiation in CA1. **(a)** The response of a CA1 neuron is monitored as two inputs are alternately stimulated. LTP is induced in input 1 by giving this input a tetanus. **(b)** The graph shows a record of the experiment. The tetanus to input 1 (arrow) yields a potentiated response to stimulation of this input. **(c)** LTP is input-specific, so there is no change in the response to input 2 after a tetanus to input 1.

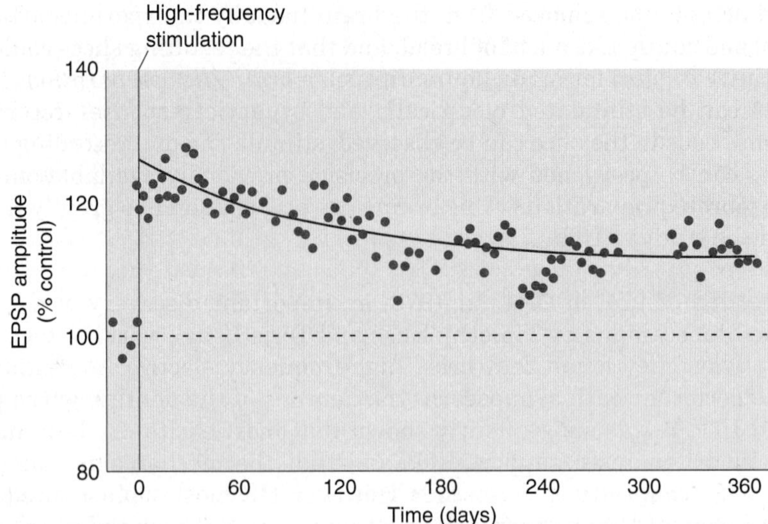

▲ FIGURE 25.7

LTP can last a long, long time. In this experiment, LTP was induced with tetanic stimulation using electrodes implanted into the hippocampus of an awake rat. Each data point is the amplitude of the EPSP evoked with electrical stimulation of the synapses that had been tetanized. The LTP was still evident a year later. (Source: Adapted from Abraham et al., 2002.)

100/sec). Usually this tetanus induces LTP, and subsequent test stimulation evokes an EPSP that is much greater than it was during the initial baseline period. In other words, the tetanus has caused a modification of the stimulated synapses so they are more effective. Other synaptic inputs onto the same neuron that did not receive tetanic stimulation do not show LTP. This property, that only the active inputs show the synaptic plasticity, is called **input specificity**.

One remarkable feature of this plasticity is that it can be induced by a brief tetanus, lasting less than a second, consisting of stimulation at frequencies well within the range of normal axon firing. A second remarkable feature of LTP is its longevity. LTP induced in CA1 of awake animals can last many weeks, possibly even a lifetime (Figure 25.7). No wonder this form of synaptic plasticity has attracted interest as a candidate mechanism for declarative memory.

Subsequent research has shown that high-frequency stimulation is not an absolute requirement for LTP. Rather, what is required is that *synapses be active at the same time that the postsynaptic CA1 neuron is strongly depolarized*. In order to achieve the necessary depolarization with a tetanus, (1) synapses must be stimulated at frequencies high enough to cause temporal summation of the EPSPs, and (2) enough synapses must be active simultaneously to cause significant spatial summation of EPSPs. This second requirement is called **cooperativity,** because coactive synapses must cooperate to produce enough depolarization to cause LTP.

Consider for a moment how the cooperativity property of hippocampal LTP could be used to form associations. Imagine a hippocampal neuron receiving synaptic inputs from three sources: I, II, and III. Initially, no single input is strong enough to evoke an action potential in the postsynaptic neuron. Now imagine that inputs I and II repeatedly fire at the same time. Because of spatial summation, inputs I and II are now capable of firing the postsynaptic neuron and of causing LTP. Only the active synapses will be potentiated, and these, of course, are those belonging to inputs I and II. Now, because of potentiation of their synapses, *either* input I *or* input II can fire the postsynaptic neuron (but not input III). Thus, LTP has caused an association of inputs I and II. In this way, the sight of a duck could be associated with the quack of a duck (they often occur at the same time), but never with the bark of a dog.

Speaking of associations, remember the idea of a Hebb synapse, introduced in Chapter 23, to account for aspects of visual development? LTP in CA1 is Hebbian: Inputs that fire together wire together.

Mechanisms of LTP in CA1. Excitatory synaptic transmission in the hippocampus is mediated by glutamate receptors. Na^+ ions passing through the AMPA subclass of glutamate receptor cause the EPSP at the Schaffer collateral–CA1 pyramidal cell synapse. However, CA1 neurons also have postsynaptic *NMDA receptors*. Recall that these glutamate receptors have the unusual property that they conduct Ca^{2+} ions, but only when glutamate binds *and* the postsynaptic membrane is depolarized enough to displace Mg^{2+} ions that clog the channel (Figure 25.8). Thus, Ca^{2+} entry through the NMDA receptor specifically signals when presynaptic and postsynaptic elements are active at the same time (Box 25.2).

Considerable evidence now links this rise in postsynaptic $[Ca^{2+}]_i$ to the induction of LTP. For example, LTP induction is prevented if NMDA receptors are pharmacologically inhibited, or if rises in postsynaptic $[Ca^{2+}]_i$

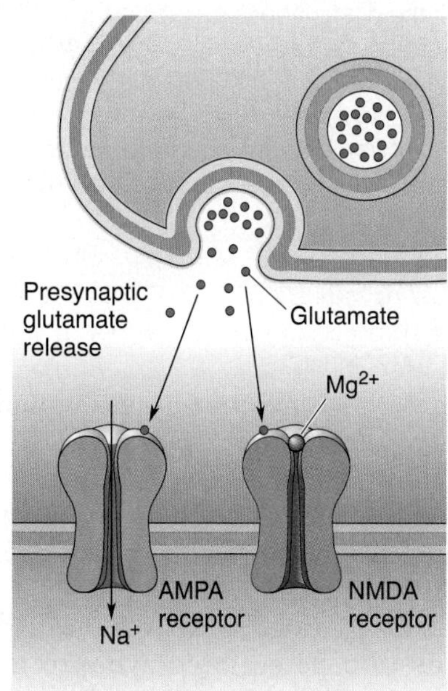

(a) Postsynaptic membrane at **resting** potential

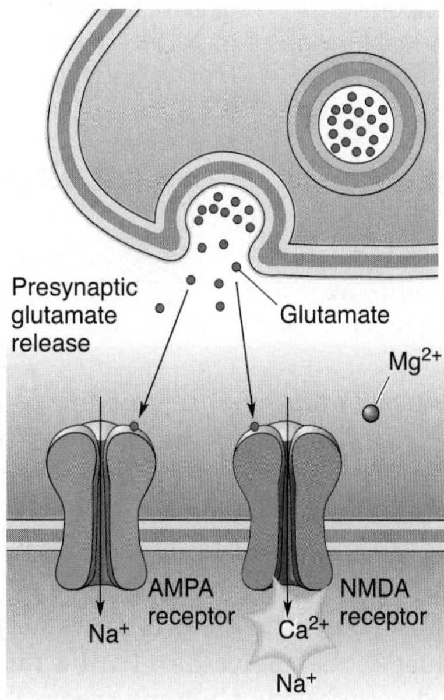

(b) Postsynaptic membrane at **depolarized** potential

▲ FIGURE 25.8

NMDA receptors activated by simultaneous presynaptic and postsynaptic activity. (a) Presynaptic activation causes the release of glutamate, which acts on postsynaptic AMPA receptors and NMDA receptors. At the negative resting membrane potential, the NMDA receptors pass little ionic current because they are blocked with Mg^{2+}. (b) When glutamate release coincides with depolarization sufficient to displace the Mg^{2+}, then Ca^{2+} enters the postsynaptic neuron via the NMDA receptor.

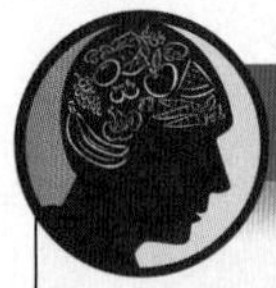

BOX 25.2 BRAIN FOOD

Synaptic Plasticity: Timing Is Everything

When enough synapses are active at the same time, the postsynaptic neuron will be depolarized sufficiently to fire an action potential. Donald Hebb proposed that each individual synapse grows a little stronger when it successfully participates in the firing of the postsynaptic neuron. The phenomenon of LTP comes close to satisfying Hebb's ideal. The synapse gets stronger when the glutamate released by the presynaptic terminal binds to postsynaptic NMDA receptors *and* the postsynaptic membrane is depolarized strongly enough to displace Mg^{2+} from the NMDA receptor channel.

Is there a role for postsynaptic action potentials in this "strong" depolarization? The first evidence that appropriate timing of a postsynaptic action potential might be important for LTP was obtained in the early 1980s by William Levy and Oswald Steward at the University of Virginia. They found that LTP occurred if a postsynaptic action potential occurred simultaneously with, or slightly after, presynaptic release of glutamate. However, action potentials are generated in the soma in response to depolarization of the membrane beyond threshold. Because this happens far away from the synapses located out on the dendritic tree, it was assumed for a time that the actual occurrence of the spike was not important for the mechanism of synaptic potentiation. The important thing was the strong depolarization in the dendrite, due to summed synaptic currents, which, coincidentally, was also usually sufficient to evoke a postsynaptic action potential.

While it remains true that the key is strong postsynaptic depolarization, researchers took another look at the role of the postsynaptic spike in LTP. This new attention resulted from the discovery that action potentials generated in the soma can actually "back-propagate" into the dendrites of some cells. Thus, Henry Markram, Bert Sakmann, and their colleagues at the Max Planck Institute investigated what happens when a postsynaptic spike is generated (via a microelectrode) at various time intervals before or after an EPSP. Remarkably, they found that if a postsynaptic action potential follows the EPSP within about 50 msec, the synapse potentiates. Nothing happens in response to the spike or the EPSP alone; LTP results specifically from the precise timing of EPSP and spike, just as Hebb suggested! In addition, the timing requirements for LTP in these studies agreed very well with those originally reported by Levy and Steward. This is an example of what is now referred to as *spike timing–dependent plasticity*.

What accounts for the LTP-promoting effect of a back-propagating action potential? The answer, of course, is strong depolarization. NMDA receptors have a very high affinity for glutamate, so the transmitter remains bound to the receptor for many tens of milliseconds. However, this bound glutamate does nothing if the postsynaptic membrane is not depolarized strongly, because the channel is clogged with Mg^{2+}. The timely occurrence of the action potential is sufficient to awaken these dormant channels by ejecting the Mg^{2+}. Then, as long as glutamate is still bound to the receptor, Ca^{2+} will enter the cell and trigger the mechanism of LTP.

are prevented by the injection of a Ca^{2+} chelator into the postsynaptic neuron. The rise in $[Ca^{2+}]_i$ activates two protein kinases: *protein kinase C* and *calcium-calmodulin-dependent protein kinase II*, also known as CaMKII (pronounced "cam-K-two"). Recall from Chapters 5 and 6 that protein kinases regulate other proteins by phosphorylating (attaching phosphate groups to) them.

Following the rise in postsynaptic $[Ca^{2+}]_i$ and the activation of the kinases, the molecular trail that leads to a potentiated synapse gets harder to follow. Current research suggests that this trail may actually branch (Figure 25.9). One path appears to lead toward an increased effectiveness of existing postsynaptic AMPA receptors by way of phosphorylation. Phosphorylation of the AMPA receptor, by either protein kinase C or CaMKII, leads to a change in the protein that increases the ionic conductance of the channel. The other path leads to the insertion of entirely new AMPA receptors into the postsynaptic membrane. According to a current model, vesicular organelles studded with AMPA receptors lie in wait near the postsynaptic membrane. In response to CaMKII activation, the vesicle membrane fuses with the postsynaptic membrane, and the new AMPA receptors are thereby delivered to the

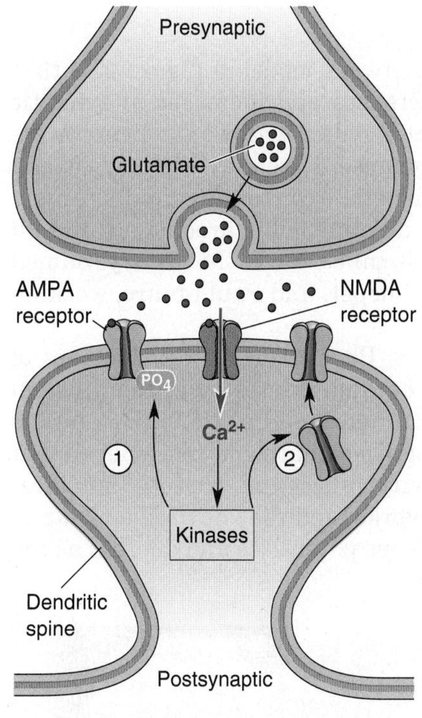

◀ FIGURE 25.9
Routes for the expression of LTP in CA1. Ca^{2+} entering through the NMDA receptor activates protein kinases. This can cause LTP by ① changing the effectiveness of existing postsynaptic AMPA receptors or ② stimulating the insertion of new AMPA receptors.

synapse. This addition of new membrane causes the spines to swell (Figure 25.10).

Evidence also indicates that synaptic structure changes following LTP. In particular, postsynaptic dendritic spines appear to bud and form new synaptic contacts with axons. Thus, following LTP, a single axon can make multiple synapses on the same postsynaptic neuron, which is not the normal pattern in CA1. This sprouting of synapses not only increases the responsive postsynaptic surface but also increases the probability that an action potential in the axon will trigger presynaptic glutamate release.

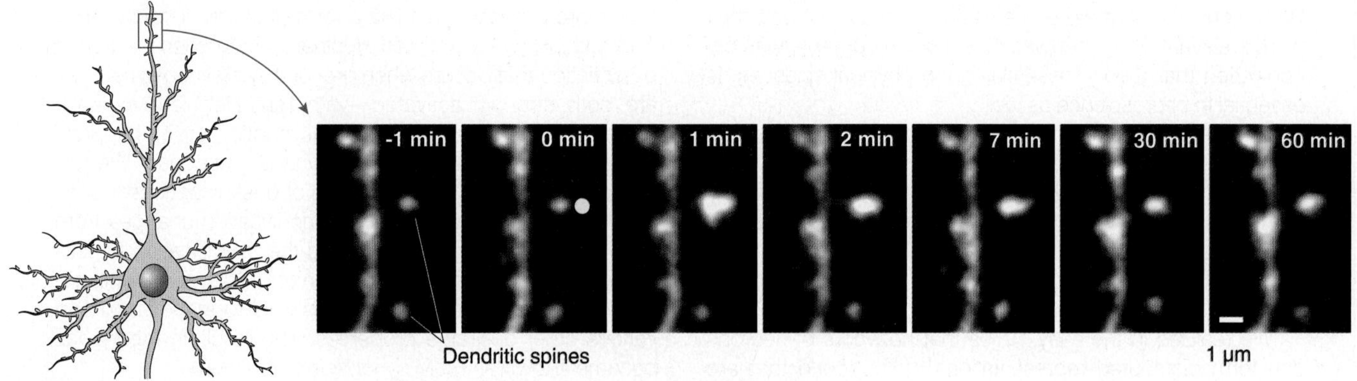

▲ FIGURE 25.10
The growth of spines following LTP. A segment of dendrite was filled with a fluorescent dye and imaged in living tissue using a special microscope. After LTP, the spines grew and sometimes sprouted to accommodate new synapses. Each frame is a snapshot of the dendrite at a different time, indicated in the upper-right corner (in minutes). At the time marked 0 min, the yellow dot indicates that this spine was repetitively activated with glutamate to induce LTP. After LTP, the spine grew to accommodate more AMPA receptors. (Photos courtesy of Dr. Miquel Bosch, Massachusetts Institute of Technology.)

Weakening Synapses

We have seen from our simple neural network model in Figure 25.4 that information can be stored as both decreases and increases in synaptic effectiveness. Recall Hebb's theory that a synapse grows stronger, or potentiates, when the activity of that synapse correlates with the strong activation of the postsynaptic neuron by other converging inputs. An extension of Hebb's theory, designed to account for bidirectional (up and down) regulation of synaptic strength, is called the **BCM theory**, named for its authors, Elie Bienenstock, Leon Cooper, and Paul Munro, working at Brown University.

After sharing the 1972 Nobel Prize in Physics for his development of a theory of superconductivity, Cooper became interested in the problem of memory storage by large networks of neurons (Box 25.3). He and his students Bienenstock and Munro recognized that experience-dependent changes in neuronal stimulus selectivity reflect the synaptic modifications that store memory in neural networks, and they devised a synaptic "learning rule" to account for how synapses potentiate and depress as the

BOX 25.3 PATH OF DISCOVERY

Memories of Memory

by Leon Cooper

I have been asked many times: "What led you from physics to neuroscience?" The best I can come up with is to repeat Humphrey Bogart's response to Claude Rains in *Casablanca*: "I was misinformed." After the publication of our theory on superconductivity, I worked on other "many-electron" problems. I came to believe that the mathematical techniques used there could be applied to "many-neuron" problems. While some of them did prove useful, for the most part they were irrelevant. But what was most useful, perhaps, was my conviction that theory, essential in the physical sciences, is essential in neuroscience as well.

So when I strayed from the lofty realm of theoretical physics to the earthy problem of the brain, my first effort was to attempt to construct networks of neurons that would display some of the qualitative features associated with what I called animal memory. Initial success was the seduction that lured me further into this exotic domain. It is this journey that I will describe.

We realized in the early 1970s that networks of neurons can form distributed representations of the world that are "associative" (recollection of one memory can lead to the recollection of another linked to it by experience) and "content addressable" (memories are accessed by content rather than by a physical address in the network). Such representations are resistant to the loss of individual neurons and synapses and thus provide a candidate substrate for memory storage in the animal brain. But how can these representations be constructed in networks of neurons? That is, how can the strengths of the vast numbers of synapses that make up neuronal networks be adjusted to obtain a mapping that corresponds to an appropriate memory?

This might result if synaptic modification (or learning) follows the famous Hebbian rule. But Hebbian synaptic modification needs stabilization. Elie Bienenstock, Paul Munro, and I therefore proposed in 1982 a form of bidirectional synaptic modification that combines Hebbian modification—synaptic potentiation that occurs when pre- and postsynaptic neurons are both strongly activated—with "anti-Hebbian" synaptic weakening that occurs when presynaptic activity happens in the absence of a strong postsynaptic response. We further proposed that the critical level of postsynaptic response at which the polarity of synaptic modification changes from weakening to strengthening, called the "modification threshold," varies according to the history of postsynaptic cell activity. Together, these assumptions lead to stabilization and various other desirable properties. The resulting theory has become known as BCM synaptic modification.

In the late 1980s I began a long and very fruitful collaboration with Mark Bear who was then also at Brown. Mark and his students performed experiments to test the validity of BCM assumptions at excitatory glutamatergic synapses in the cerebral cortex. The shape of the BCM synaptic modification function was first confirmed by Dudek and Bear (1992) in hippocampus, and Kirkwood and Bear (1994) showed that sensory environment is changed. A key assumption of the BCM theory, published in 1982, is that synapses will undergo synaptic weakening instead of LTP when they are active at the same time the postsynaptic cell is only weakly depolarized by other inputs. This idea inspired a search for long-term synaptic depression in hippocampal area CA1, using stimuli that were designed to evoke a modest postsynaptic response. In 1992 Serena Dudek and Mark Bear, working together at Brown University, showed that tetanic stimulation of the Schaffer collaterals at low frequencies (1–5 Hz) indeed produces synaptic weakening (Figure 25.11). Because it occurs only at the stimulated synapses, it is often referred to as *homosynaptic* **long-term depression (LTD)**.

It has now been established experimentally that bidirectional plasticity of many cortical synapses is indeed governed by two simple rules:

1. Synaptic transmission occurring at the same time as strong depolarization of the postsynaptic neuron causes LTP of the active synapses.
2. Synaptic transmission occurring at the same time as weak or modest depolarization of the postsynaptic neuron causes LTD of the active synapses.

the result was the same in visual cortex. Since then, similar findings have been confirmed in many different regions of neocortex in many species of both young and old animals. Of particular interest are data showing that the same principles of synaptic plasticity apply in the human inferotemporal cortex, a region believed to be a repository of visual memories. Together, the data support the idea that very similar principles guide synaptic plasticity in many species in widely different regions of the brain.

According to BCM, the modification threshold, θ_m, must vary depending on the history of postsynaptic cortical activity. An experimental test of this hypothesis was first reported by Kirkwood, Marc Rioult, and Bear (1996). They compared the synaptic modification function in the visual cortex of normal animals with that in the visual cortex of animals reared in complete darkness and found a shift of this function in accordance with the theoretical postulate. Elizabeth Quinlan, Ben Philpot, and Bear, in collaboration with Richard Huganir at Johns Hopkins School of Medicine, went on to show in 1999 that the ratio of two distinct subunits of the cortical NMDA receptor is set according to the activation history of the cortex, providing a potential mechanism for the sliding modification threshold.

Consequences of BCM synaptic modification in networks modeled after visual cortex have been shown by my students Nathan Intrator, Harel Shouval, Brian Blais, and many others, using analysis and simulations, to be in agreement with experimental observations on the shifts in neuronal selectivity that have been observed by rearing animals in various visual environments. Thus, the BCM theory provides a bridge between the molecular mechanisms of synaptic modification and the systems-level properties of distributed information storage.

Given the level of skepticism displayed when synaptic modification ideas were discussed 40 years ago, I think it is reasonable to say that we have made considerable progress. Our initial aim to build a theoretical structure relevant to a fundamental brain process that was sufficiently concrete so that it could be tested by experiment has been accomplished. It is particularly gratifying that theory has inspired experiments that, in addition to confirming the various postulates and predictions of our theory, have led to the discovery of new phenomena such as homosynaptic long-term depression and metaplasticity. Possibly most important, we have an excellent example of the fruitful interaction of theory with experiment in neuroscience.

References

Bienenstock EL, Cooper LN, Munro PW. 1982. Theory for the development of neuron selectivity: orientation specificity and binocular interaction in visual cortex. *Journal of Neuroscience* 2:32–48.

Blais B, Cooper LN, Shouval H. 2000. Formation of direction selectivity in natural scene environments. *Neural Computation* 12:1057–1066.

Blais BS, Intrator N, Shouval HZ, Cooper LN. 1998. Receptive field formation in natural scene environments: comparison of single-cell learning rules. *Neural Computation* 10:1797–1813.

Dudek SM, Bear MF. 1992. Homosynaptic long-term depression in area CA1 of hippocampus and effects of N-methyl-D-aspartate receptor blockade. *Proceedings of the National Academy of Sciences USA* 89:4363–4367.

Kirkwood A, Bear MF. 1994. Homosynaptic long-term depression in the visual cortex. *Journal of Neuroscience* 14:3404–3412.

Kirkwood A, Rioult MC, Bear MF. 1996. Experience-dependent modification of synaptic plasticity in visual cortex. *Nature* 381:526–528.

Quinlan EM, Philpot BD, Huganir RL, Bear MF. 1999. Rapid, experience-dependent expression of synaptic NMDA receptors in visual cortex in vivo. *Nature Neuroscience* 2:352–357.

Shouval H, Intrator N, Cooper LN. 1997. BCM network develops orientation selectivity and ocular dominance in natural scene environment. *Vision Research* 37:3339–3342.

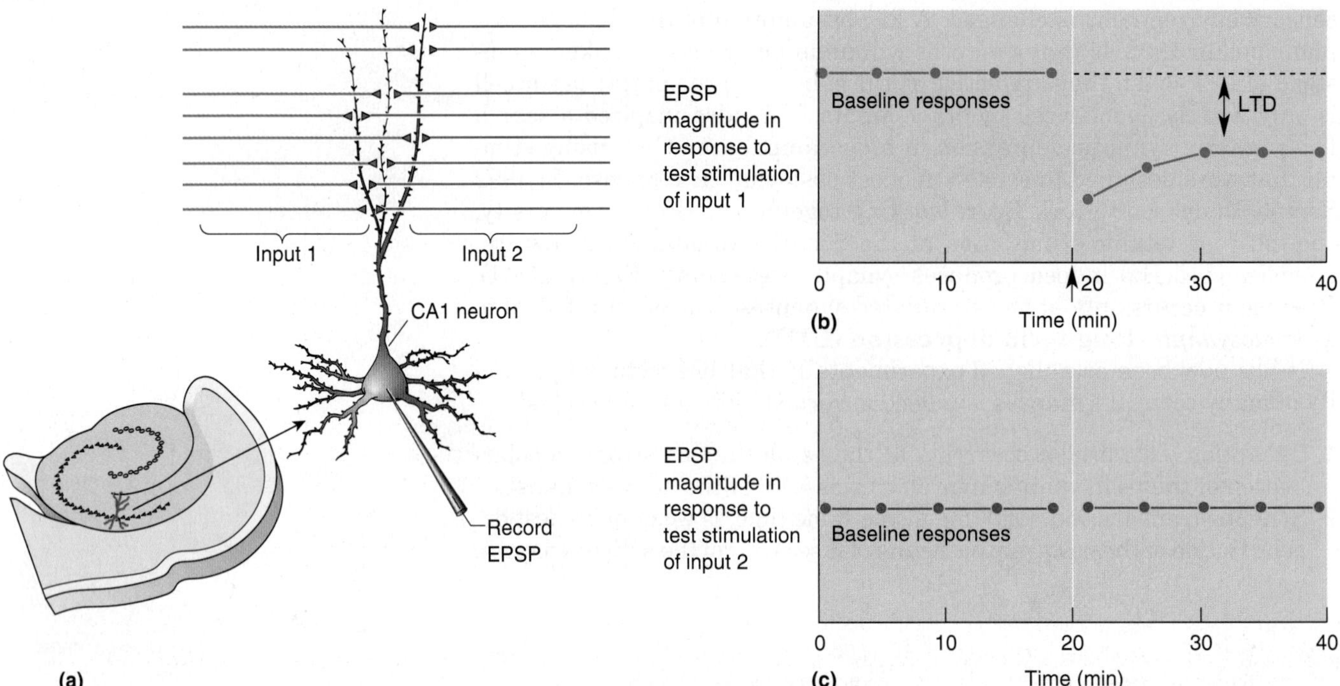

▲ FIGURE 25.11
Homosynaptic LTD in the hippocampus. **(a)** The response of a CA1 neuron is monitored as two inputs are alternately stimulated. LTD is induced in input 1 by giving this input a 1 Hz tetanus. **(b)** The graph shows a record of the experiment. The low-frequency tetanus to input 1 (arrow) yields a depressed response to stimulation of this input. **(c)** LTD is input-specific, so there is no change in the response to input 2 after a tetanus to input 1.

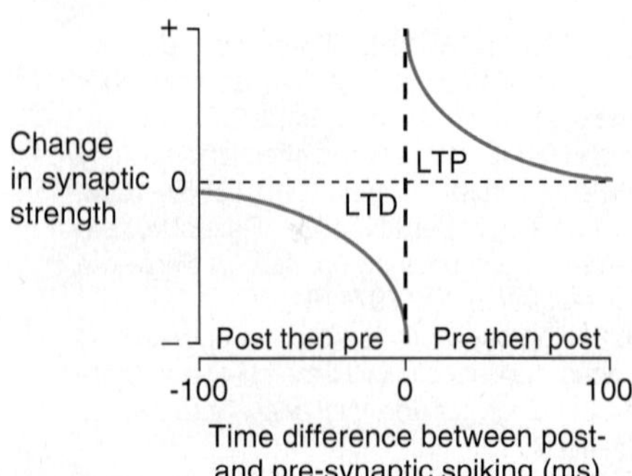

▲ FIGURE 25.12
Spike timing–dependent plasticity. When postsynaptic spikes consistently follow the EPSPs induced by presynaptic spiking, the synapse grows stronger. However, when the postsynaptic spikes consistently precede the EPSPs, the synapse grows weaker. This graph relates the change in synaptic strength to the relative timing difference.

Although these rules apply to many cortical synapses, it is important to appreciate that LTD is a widespread form of synaptic plasticity. The properties and mechanisms of LTD vary from one synapse type to the next (Box 25.4). At some synapses, the timing of pre- and postsynaptic actions potentials is a key variable. As discussed in Box 25.2, LTP can result when the EPSP caused by synaptic glutamate release *precedes* an action potential in the postsynaptic neuron; this is an example of **spike timing–dependent plasticity**. At many of these same synapses, LTD can result instead when the EPSP caused by glutamate release *follows* a postsynaptic action potential (Figure 25.12).

As is the case for LTP, we know most about the mechanism of homosynaptic LTD in area CA1 of the hippocampus, so this will be our focus.

Mechanisms of LTD in CA1. At the Schaffer collateral–CA1 synapse, two distinct forms of homosynaptic LTD have been described. The first one to be discovered depends on activation of the NMDA receptor. The second form, discovered a few years later, requires activation of G-protein coupled *metabotropic glutamate receptors (mGluRs)*. Here we will focus on NMDA receptor-dependent LTD.

Because NMDA receptors admit Ca^{2+} to the postsynaptic neuron, it came as no surprise that a rise in postsynaptic $[Ca^{2+}]$ is necessary to trigger LTD. But how can the same signal, Ca^{2+} entry through the NMDA receptor, trigger *both* LTP *and* LTD? The key difference lies in the *level* of

BOX 25.4 **BRAIN FOOD**

The Wide World of Long-Term Synaptic Depression

We saw in Chapter 14 that the cerebellum is important for learning and remembering motor skills. The unusual circuitry of the cerebellar cortex suggested to David Marr at the University of Cambridge how this learning might occur. The output of the cerebellar cortex arises from large neurons called *Purkinje cells*, and these cells receive two converging inputs. Each Purkinje cell receives input from a single *climbing fiber* that arises from a nucleus in the medulla called the inferior olive. The climbing fiber synapses are very powerful and always cause the Purkinje cell to fire action potentials. *Parallel fibers* arising from cerebellar granule cells provide the second input, and the organization is very different. Each Purkinje cell receives weak parallel fiber synapses from as many as 100,000 different granule cells. Marr proposed that this unusual convergence of parallel and climbing fiber inputs onto Purkinje cell dendrites serves motor learning. He proposed that (1) the climbing fiber input carries error signals indicating a movement has failed to meet expectations, and (2) corrections are made by adjusting the effectiveness of the parallel fiber inputs to the Purkinje cell. The theory was modified by James Albus at the Goddard Space Flight Center in Greenbelt, Maryland, to explicitly predict LTD of the parallel fiber synapse *if it is active at the same time as the climbing fiber input to the postsynaptic Purkinje cell*.

Masao Ito and his colleagues at the University of Tokyo tested this idea by pairing electrical stimulation of the climbing fibers with stimulation of the parallel fibers. Remarkably, they found that after this pairing procedure, activation of the parallel fibers alone resulted in a smaller postsynaptic response in the Purkinje cell (Figure A). It is now understood that the requirements for induction of this form of LTD are a large surge in postsynaptic [Ca^{2+}] caused by activation of the climbing fiber, coincident with activation of metabotropic glutamate receptor 1 (mGluR1) by the parallel fibers. This conjunction triggers the internalization of AMPA receptors and the depression of synaptic transmission at the parallel fiber synapses. A mechanistically similar form of mGluR-dependent LTD was later discovered in the hippocampus, although it does not require a Ca^{2+} surge.

At other synapses in the brain, activation of mGluRs triggers LTD by a different mechanism altogether. For example, in the nucleus accumbens, activation of postsynaptic mGluR5 stimulates the synthesis of endocannabinoids, which travel retrogradely to the presynaptic terminal and cause a persistent depression of glutamate release. (Endocannabinoids were introduced in Chapter 6; see Box 6.2.)

Yet another LTD variant has been observed in the neocortex. Endocannabinoids in some neocortical pyramidal neurons are released in response to dendritic action potentials. If these endocannabinoids impinge on glutamatergic axon terminals at the same time they are releasing glutamate, then these synapses will be depressed. This mechanism yields a timing requirement on LTD, such that it is induced when a postsynaptic spike (causing the release of endocannabinoids) precedes the presynaptic spike by a few tens of milliseconds.

Each mechanism for LTD imposes different rules on the patterns of activity that yield synaptic plasticity. We can speculate that these have evolved to optimize the contribution of synaptic plasticity to the functions of different brain circuits.

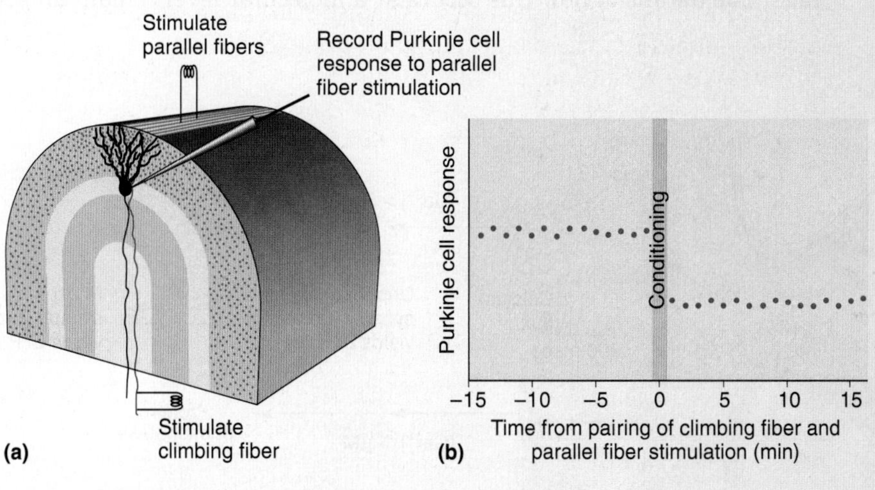

Figure A Cerebellar LTD. **(a)** The experimental arrangement for demonstrating LTD. The magnitude of the Purkinje cell response to stimulation of a "beam" of parallel fibers is monitored. Conditioning involves pairing parallel fiber stimulation with climbing fiber stimulation. **(b)** A graph of an experiment performed in this way. After the pairing, LTD of the response to parallel fiber stimulation results.

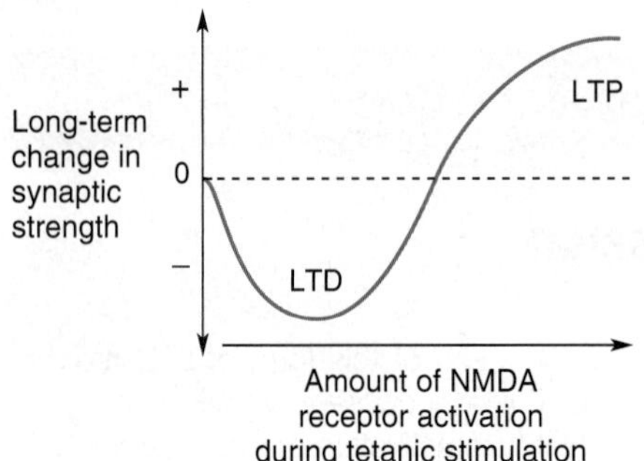

▲ FIGURE 25.13
NMDA receptor activation and bidirectional synaptic plasticity. The long-term change in synaptic transmission is graphed as a function of the level of NMDA receptor activation during conditioning stimulation. The level of NMDA receptor activation at which the polarity of synaptic modification switches from LTD to LTP is called the modification threshold.

NMDA receptor activation (Figure 25.13). When the postsynaptic neuron is only weakly depolarized, the partial blocking of the NMDA receptor channels by Mg^{2+} prevents all but a trickle of Ca^{2+} into the postsynaptic neuron. On the other hand, when the postsynaptic neuron is strongly depolarized, the Mg^{2+} block is displaced entirely, and Ca^{2+} floods into the postsynaptic neuron. These different types of Ca^{2+} response selectively activate different types of enzymes. Instead of the kinases that are activated by high $[Ca^{2+}]_i$, modest and prolonged elevations in $[Ca^{2+}]_i$ activate *protein phosphatases*, enzymes that pluck phosphate groups off proteins. Therefore, if LTP is putting phosphate groups on, LTD apparently is taking them off. Indeed, biochemical evidence now indicates that AMPA receptors are dephosphorylated in response to stimulation that induces LTD (Figure 25.14). Moreover, the induction of hippocampal LTD can also be associated with the internalization of AMPA receptors at the synapse. Thus, LTP and LTD appear to reflect the bidirectional regulation of both the phosphorylation and the number of postsynaptic AMPA receptors.

Glutamate Receptor Trafficking. Besides revealing much about the likely synaptic basis for learning and memory, studies of LTP and LTD have led to a much deeper understanding of how synaptic transmission is maintained in the brain. Current research suggests that AMPA receptors in the postsynaptic membrane are continually being added and removed even in the absence of synaptic activity. Researchers estimate that half the synaptic AMPA receptors are replaced every 15 minutes! However, despite this remarkable turnover, synaptic transmission will remain stable as long as one receptor is added whenever one receptor is removed. LTP and LTD disrupt this equilibrium, leading to a net increase or decrease in the capacity of the synaptic membrane for AMPA receptors.

The capacity of the postsynaptic membrane is determined by the size of a scaffold of what has been termed *slot proteins*. Imagine the scaffold is like an egg carton, and the slot proteins form each of the egg cups. AMPA receptors are the eggs that fill the carton. As long as the size of the carton doesn't change, synaptic transmission is stable even if the individual eggs are continually replaced (Figure 25.15).

Stable LTP requires increasing the size of the carton and providing new eggs. The details of how this occurs at a molecular level remain an active

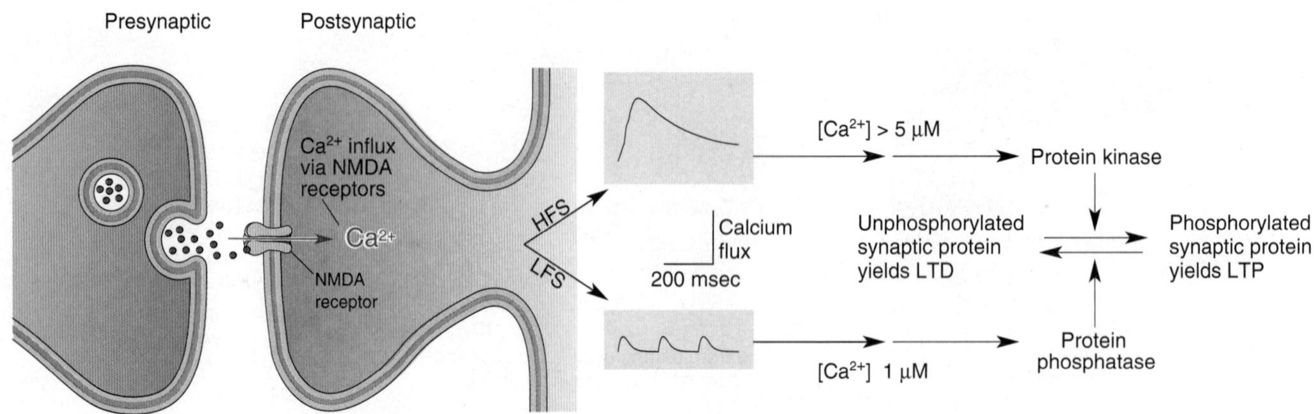

▲ FIGURE 25.14
A model for how Ca^{2+} can trigger both LTP and LTD in the hippocampus. High-frequency stimulation (HFS) yields LTP by causing a large elevation of $[Ca^{2+}]$. Low-frequency stimulation (LFS) yields LTD by causing a smaller elevation of $[Ca^{2+}]$. (Source: Adapted from Bear and Malenka, 1994, Fig. 1.)

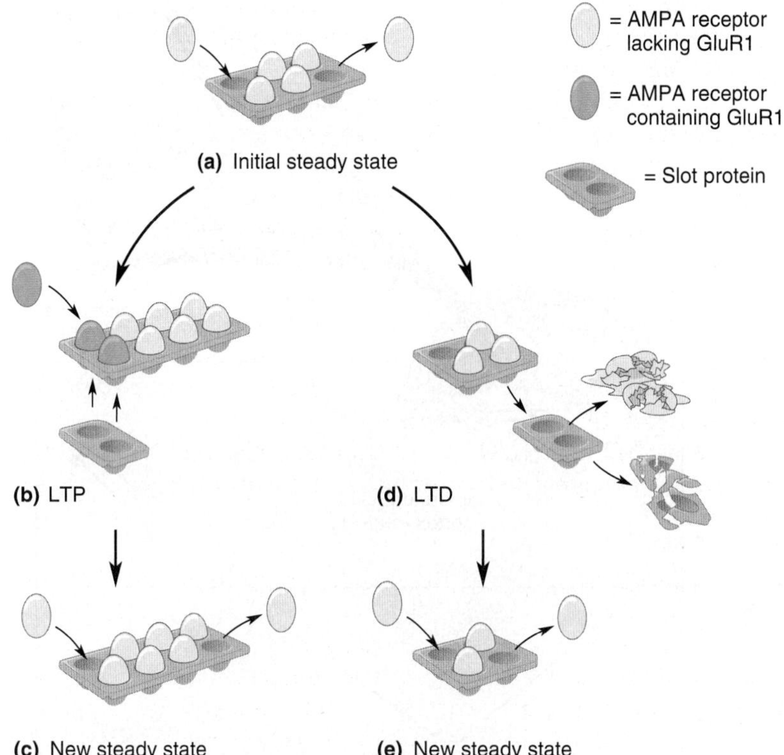

▲ FIGURE 25.15

An egg carton model of AMPA receptor trafficking at the synapse. Each egg represents an AMPA receptor, and the carton represents the capacity of the synapse for receptors, which might be determined by the amount of PSD-95. **(a)** The initial steady state. Each AMPA receptor that is removed is replaced with a new receptor. **(b)** LTP. More PSD-95 is added, increasing the synaptic capacity for AMPA receptors. The new receptors (blue) contain the GluR1 subunit. **(c)** The new steady state. Over time, ongoing turnover of receptors replaces those with GluR1. **(d)** LTD. Some PSD-95 is destroyed, decreasing the synaptic capacity for AMPA receptors. **(e)** The new steady state following LTD.

area of research, and today's conclusions may be overturned by tomorrow's experiments. However, there is evidence that a protein called *PSD-95* (a postsynaptic density protein with a molecular weight of 95 kilodaltons) may comprise the egg carton. Increasing the expression of PSD-95 in neurons increases the synaptic capacity for AMPA receptors. In addition, there is evidence that the new eggs may be AMPA receptors that contain a distinctive subunit called *GluR1*. LTP can selectively increase in the number of GluR1-containing AMPA receptors in the membrane. Over time, these receptors are replaced by those that lack GluR1. By analogy, imagine the neuron contains a stash of blue Easter eggs that can be delivered to the carton in response to LTP-inducing stimulation. Over time, the blue eggs are replaced by uncolored eggs. But because the size of the carton is increased, there continues to be a net increase in the number of eggs.

Conversely, stable LTD requires reducing the size of the egg carton, which reduces the capacity for eggs. Indeed, research has shown that LTD-inducing stimulation leads to both destruction of PSD-95 and a net loss of AMPA receptors from the postsynaptic membrane.

LTP, LTD, and Memory

LTP and LTD have attracted a lot of interest because theoretical work shows that these mechanisms of synaptic plasticity can contribute to

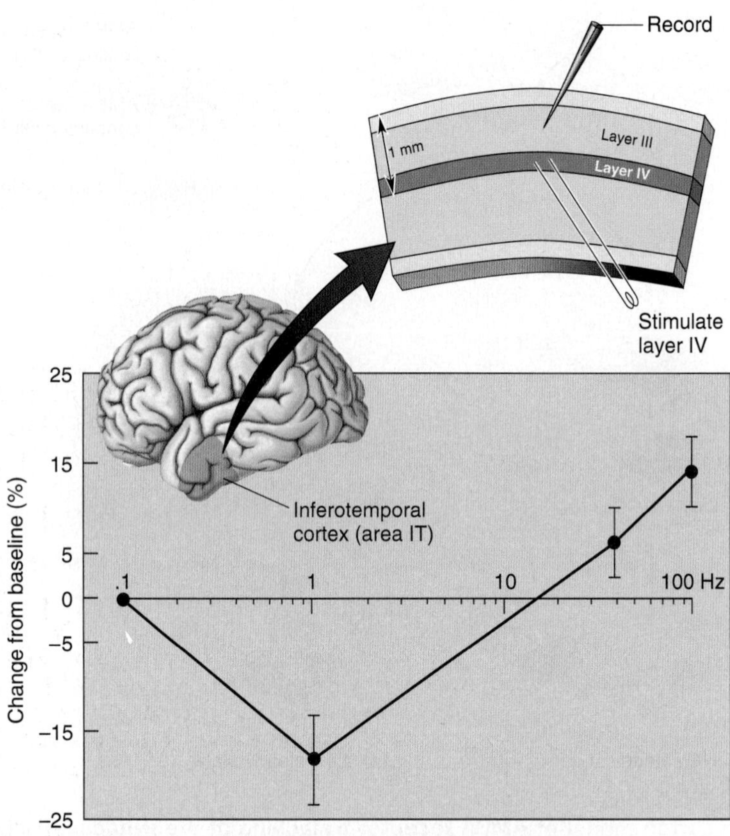

▲ FIGURE 25.16

Bidirectional synaptic modifications in human area IT. Slices of human temporal cortex, removed during the course of surgery to gain access to deeper structures, were maintained *in vitro*. Synaptic responses were monitored following various types of tetanic stimulation. As in rat CA1, stimulation of 1 Hz produced LTD, while 100 Hz stimulation produced LTP. (Source: Adapted from Chen et al., 1996.)

the formation of declarative memories. Recent research indicates that the types of NMDA receptor-dependent synaptic plasticity that have been characterized in the hippocampus also occur throughout the neocortex, including area IT where memories of familiar faces are created (Figure 25.16). It appears that plasticity at many synapses in the cerebral cortex may be governed by similar rules and might use similar mechanisms. (But remember, there are many exceptions to these "rules," and they do not apply to all synapses, even within a single structure.)

LTP and LTD are clearly appealing models, but what evidence links them to memory? So far, all we've described is a possible neural basis for a memory of having one's brain electrically stimulated! One approach has been to insert stimulating and recording electrodes in the hippocampus and use these to monitor the state of synaptic transmission during learning. Because of the distributed nature of memory, success with this approach required the use of a particularly robust type of learning called *inhibitory avoidance*. In this experiment, a rat learns to associate a place (the dark side of a box) with an aversive experience (a foot shock) (Figure 25.17a). Animals of all types (from flies to humans) will learn to avoid the place they received the shock after only one trial (depending, of course, on the strength of the shock). This type of learning is not subtle, and neither are the patterns of hippocampal activation it produces. The widespread activation

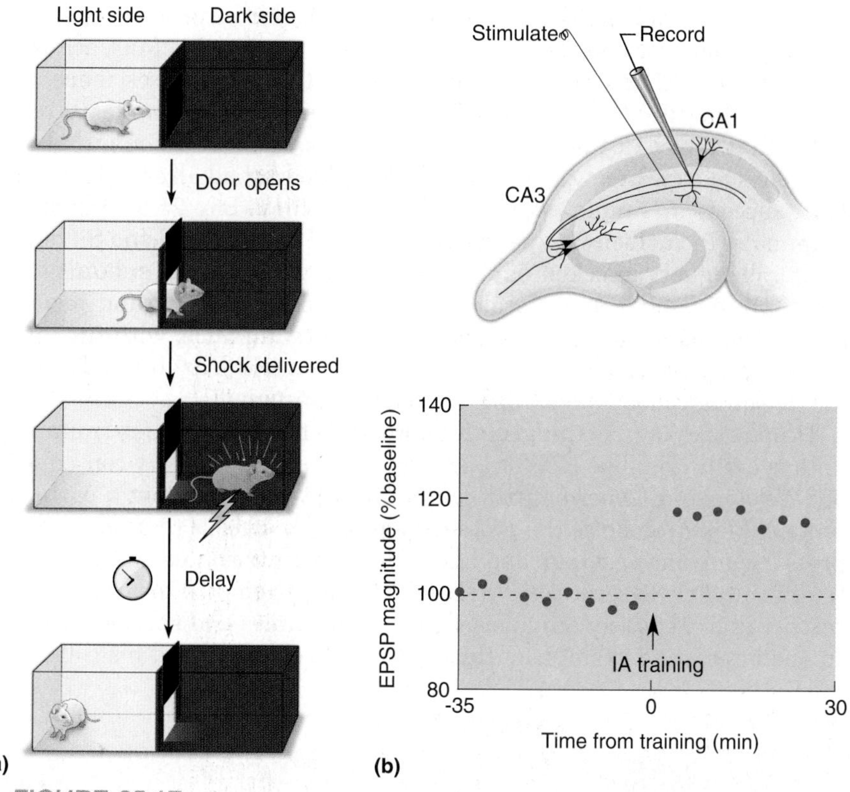

▲ FIGURE 25.17
LTP in CA1 induced by learning. In this experiment, electrodes were implanted in the rat's hippocampus to monitor the strength of synaptic transmission before and after inhibitory avoidance training. **(a)** The rat is placed in a box divided by a closed door that separates the light side from the dark side. When the door is opened, the rat scurries into the dark side to avoid the light. On the dark side, a foot shock is delivered to the rat. To test for the creation of a memory trace, one can measure the time it takes for the rat to re-enter the dark side at various time points after the initial experience. **(b)** Recordings of synaptic transmission in area CA1 showed evidence of LTP when this type of memory was formed. (Source: Adapted from Whitlock et al., 2006.)

of the hippocampus after inhibitory avoidance training gave researchers the opportunity to detect changes in synaptic transmission at Schaffer collateral–CA1 synapses, and *voilà!*—LTP was observed (Figure 25.17b). In other experiments, exposing animals to a novel environment without a foot shock caused LTD instead. These experiments tell us that learning does indeed induce LTP and LTD at hippocampal synapses.

Another approach has been to see whether the molecules involved in LTP and LTD are also involved in learning and memory. For example, both forms of synaptic plasticity may require activation of the NMDA receptors. To assess the possible role of hippocampal NMDA receptors in learning, researchers injected an NMDA receptor blocker into the hippocampus of rats undergoing inhibitory avoidance training. This treatment prevented formation of a memory of the aversive experience. These experiments built on pioneering studies performed by Richard Morris in the late 1980s at the University of Edinburgh, in which NMDA receptor blockers were infused into the hippocampus of rats while they were being trained in a water maze (see Figure 24.20). Unlike normal animals, these rats failed to learn the rules of the game or the location of the escape platform. This finding provided the first evidence that NMDA-receptor-dependent processes play a role in memory.

A revolutionary new approach to the molecular basis of learning and memory was introduced by Susumu Tonegawa at the Massachusetts Institute of Technology. Tonegawa, who switched to neuroscience after winning the 1987 Nobel Prize for his research in immunology, recognized that molecules and behavior could be connected by manipulating the genes of experimental animals. This approach had already been tried with success in simple organisms like fruit flies (Box 25.5), but not in mammals. In their first experiment in mice, Tonegawa, Alcino Silva, and their colleagues "knocked out" (deleted) the gene for one subunit (α) of CaMKII, and found parallel deficits in hippocampal LTP and memory. Since then, many genes have been manipulated in mice, with the aim of assessing the role of LTP and LTD mechanisms in learning. LTP, LTD, and learning clearly have many common requirements.

Despite the power of this genetic approach, it has some serious limitations. Loss of a function, like LTP or learning, might be a secondary consequence of developmental abnormalities caused by growing up without a particular protein. Moreover, since the protein is missing in all cells that normally express it, pinpointing where and how a molecule contributes to learning can be difficult. For these reasons, researchers have attempted to devise ways to restrict their genetic manipulations to specific times and specific locations. In one interesting example of this approach, Tonegawa and his colleagues

BOX 25.5 OF SPECIAL INTEREST

Memory Mutants

Of the several hundred thousand proteins manufactured by a neuron, some may be more important than others when it comes to learning. It is even possible that some proteins are *uniquely* involved in learning and memory. Needless to say, we could gain considerable insight about the molecular basis of learning and learning disorders if such hypothetical "memory molecules" could be identified.

Recall that each protein molecule is the readout of a gene. One way to identify a "memory protein" is to delete genes one at a time and see if specific learning deficits result. This is precisely the strategy that Seymour Benzer, Yadin Dudai, and their colleagues at the California Institute of Technology tried using the fruit fly *Drosophila melanogaster*. *Drosophila* has long been a favorite species of geneticists, but one might reasonably question to what extent a fruit fly learns. Fortunately, *Drosophila* can perform the same tricks that other invertebrate species like *Aplysia* have mastered. For example, fruit flies can learn that a particular odor predicts a shock. They demonstrate this memory after training by flying away when the odor is presented. The strategy is to produce mutant flies by exposing them to chemicals or X-rays. They are then bred and their offspring are screened for behavioral deficits. The first mutant displaying a fairly specific learning deficit was described in 1976 and called Dunce. Other memory-deficient mutants were later described and given vegetable names, such as Rutabaga and Cabbage. The next challenge was to identify exactly which proteins had been deleted. It turned out that all three of these memory mutants lacked particular enzymes in intracellular signaling pathways.

In these early *Drosophila* studies, the mutations that were induced occurred at random, followed by extensive screening, first to find a learning deficit and then to determine exactly which gene was missing. More recently, however, genetic engineering techniques have made it possible to make very specific deletions of known genes, not only in *Drosophila* but also in mammals. Thus, for example, in 1992 Susumu Tonegawa, Alcino Silva, and their colleagues at the Massachusetts Institute of Technology were able to isolate and delete one subunit (α) of the calcium-calmodulin-dependent protein kinase II in mice. Experiments had already suggested that this enzyme is critical for the induction of long-term potentiation. Sure enough, these mice had a clear deficit in LTP in the hippocampus and the neocortex. And, when tested in the Morris water maze, they were found to have a severe memory deficit. Thus, these mice were memory mutants, just like their distant cousins Dunce, Rutabaga, and Cabbage.

Are we to conclude that the missing proteins in these mutants are the elusive "memory molecules"? No. All the mutants displayed other behavioral deficits in addition to memory. We can only conclude, at present, that animals growing up without these proteins are unusually poor learners. However, the studies do underscore the critical importance of specific second messenger pathways in translating a fleeting experience into a lasting memory.

found a way to restrict the genetic deletion of NMDA receptors to the CA1 region in mice, starting at about 3 weeks of age. These animals showed a striking deficit in LTP, LTD, and water maze performance, thus revealing an essential role for CA1 NMDA receptors in this type of learning.

If too little hippocampal NMDA receptor activation is bad for learning and memory, what would happen if we boosted the number of NMDA receptors? Amazingly, animals engineered to produce more than the normal number of NMDA receptors show *enhanced* learning ability in some tasks. Taken together, the pharmacological and genetic studies show that hippocampal NMDA receptors play a key role not only in synaptic modification, such as LTP and LTD, but also in learning and memory.

Synaptic Homeostasis

Synaptic plasticity is widespread in the brain, and analysis by theoretical neuroscientists has revealed that this can present a problem. To illustrate, let's consider Hebbian synaptic strengthening. Synapses potentiate when they are active at the same time as their postsynaptic target neuron. As they undergo LTP, these synapses exert more influence on the postsynaptic cell, making it more likely to respond and thereby causing further potentiation of all synapses that are active at the same time. Computer simulations show that eventually all synapses on the neuron will potentiate, and stimulus selectivity (and memory) will be lost. A similar problem can arise with synaptic weakening: By reducing postsynaptic activity, LTD makes synapses more likely to be weakened until they eventually disappear altogether. Unchecked synaptic plasticity can therefore lead to unstable neuronal responses. As we learned in Chapter 15, *homeostasis* is the term used to describe regulatory processes that maintain the internal environment of the body within a narrow physiological range. There must be homeostatic mechanisms that provide stability and keep synaptic weights within a useful dynamic range. We will discuss two such mechanisms here.

Metaplasticity. Consider again the graph in Figure 25.13. It shows that weak NMDA receptor activation causes LTD and strong NMDA receptor activation causes LTP. At some level of moderate NMDA receptor activation, between that required for LTD and for LTP, there is no net change. This value is called the *synaptic modification threshold*. The BCM theory proposed that the value of the modification threshold adjusts depending on the history of integrated postsynaptic activity. Therefore, when activity rises, due perhaps to too much LTP, the modification threshold slides up, making LTP more difficult to produce. Conversely, if activity levels fall, due perhaps to too much LTD, the modification threshold slides down, making LTD less likely and LTP easier to produce. This general concept, that the rules of synaptic plasticity change depending on the history of synaptic or cellular activity, is called **metaplasticity**. Computer simulations show that ongoing adjustments of the value of the modification threshold ensure that synaptic modifications are constrained to maintain neuronal stimulus selectivity and memory.

Research inspired by the BCM theory has confirmed the existence of metaplasticity. Although many different mechanisms contribute to the sliding modification threshold, one appears to be adjustments in the molecular composition of NMDA receptors themselves. NMDA receptors are composed of four subunits: two NR1 subunits and two NR2 subunits. At many synapses in the cerebral cortex, two types of NR2 subunits are used to construct the receptor: NR2A and NR2B. The ratio of NR2A to NR2B subunits determines the properties of the receptor, including how much Ca^{2+} can pass and what intracellular enzymes are activated. LTP is favored when more NR2B-containing receptors are expressed at the synapse, whereas LTD is favored when more NR2A-containing receptors are expressed. The ratio of NR2A- to NR2B-containing receptors depends in part on the relative abundance of these proteins in the neuron. Research has shown that after a period of high cortical activity, NR2A levels increase and NR2B levels decrease, promoting LTD over LTP. On the other hand, NR2B levels increase and NR2A levels decrease after a period of low cortical activity, promoting LTP over LTD (Figure 25.18).

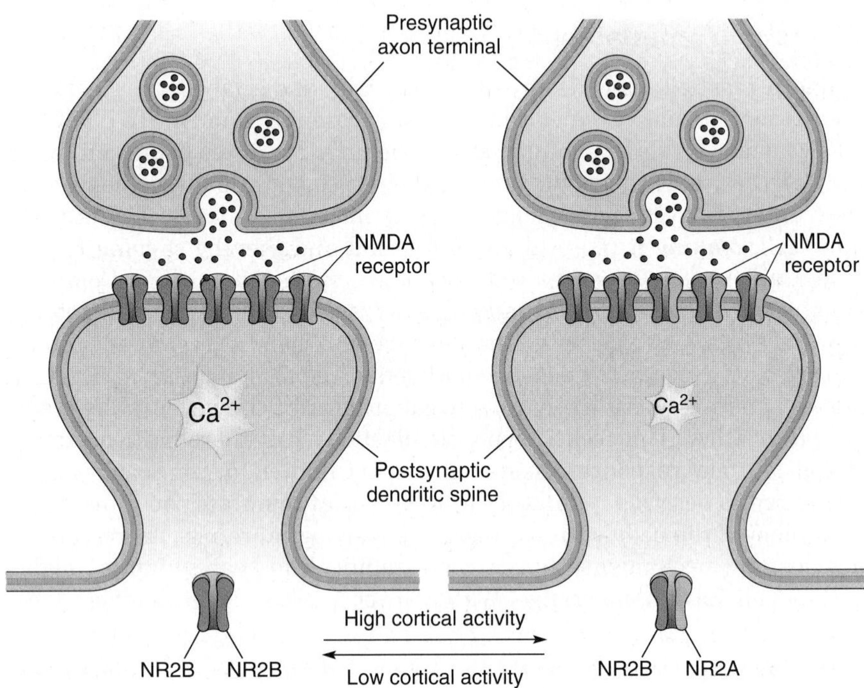

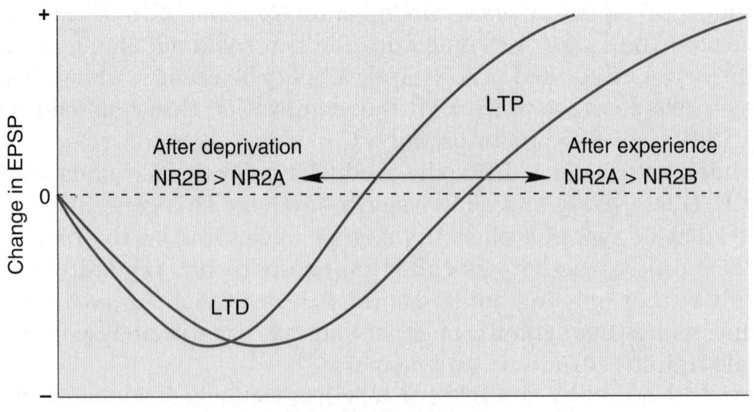

▲ FIGURE 25.18

The sliding modification threshold. Experiments in which cortical activity is reduced for several days reveal a shift in the curve that relates stimulation frequency to LTD and LTP. Lowering activity favors LTP over LTD, whereas raising activity favors LTD over LTP. This shift is accounted for in part by changes in the subunit composition of NMDA receptors. NMDA receptors with more NR2B admit more Ca^{2+}. (Source: Adapted from Bear, 2003.)

These changes in NMDA receptor subunit composition occur relatively slowly, over the course of hours, presumably because they depend on the synthesis of new protein subunits.

Synaptic Scaling. In a classic series of experiments dating back to the 1930s, the eminent physiologist Walter Cannon (introduced in Chapter 18) showed that cutting the nerve to a muscle leads to an increase in the electrical excitability and sensitivity of the muscle to ACh, the neurotransmitter of the neuromuscular junction. This phenomenon, called *denervation supersensitivity*, was later shown to be a widespread response of neurons to the loss of synaptic input. Denervation is not necessary to induce supersensitivity, however. A similar response occurs if the neurotransmitter receptors are blocked pharmacologically or if muscles or neurons are electrically silenced with tetrodotoxin (TTX). Cannon suggested this was likely to represent a homeostatic response of excitable cells to the loss of input.

An analogous phenomenon occurs in cortical neurons after manipulations of overall synaptic input. When cortical neurons are silenced with TTX, their electrical excitability increases, as does the strength of excitatory synapses that impinge on them. But what does this gross adjustment of overall synaptic strength do to the carefully tuned patterns of synaptic weights that have stored memories? Gina Turrigiano and her colleagues at Brandeis University discovered that *relative differences* in the strengths of synapses on a neuron are unchanged, even as the absolute levels go up or down; that is, the neuron adjusts by multiplying (or dividing) the values of all synaptic weights by the same number. This adjustment of absolute synaptic effectiveness that preserves the relative distribution of synaptic weights is called **synaptic scaling**.

As with metaplasticity, multiple mechanisms contribute to synaptic scaling. One appears to use Ca^{2+} entering the soma through voltage-gated Ca^{2+} channels and activation of calcium-calmodulin-dependent kinase IV (CaMKIV, a close relative of CaMKII) to regulate gene expression. A period of elevated activity increases CaMKIV-dependent gene expression, whereas a period of inactivity decreases it. The ultimate consequence of these changes in gene expression is the cellwide removal or insertion of glutamate receptors at synapses (both NMDA receptors and AMPA receptors). Like metaplasticity, scaling occurs over a longer time period (hours to days) than does the induction of LTP or LTD (seconds to minutes). This time is necessary for the synthesis (or degradation) of proteins required to adjust the strengths of the thousands of synapses impinging on the neuron.

With metaplasticity and scaling, the neuron keeps a lid on the roiling boil of ongoing synaptic plasticity. When activity is too high for too long, these mechanisms kick in to promote LTD and scale down synaptic weights. When activity is too low, they promote LTP and scale up synaptic weights. Proper neuronal function, experience-dependent shifts in selectivity, and learning and memory all require the appropriate balance of synaptic change and stability.

MEMORY CONSOLIDATION

We have seen that memory can result from experience-dependent alterations in synaptic transmission. In most examples of synaptic plasticity, transmission is initially modified as a result of changing the number of phosphate groups that are attached to proteins in the synaptic membrane.

In the case of LTD and LTP, this occurs at the postsynaptic AMPA receptor and at proteins regulating AMPA receptor number at the synapse.

Adding phosphate groups to a protein can change synaptic effectiveness and form a memory, but only as long as the phosphate groups remain attached to that protein. Phosphorylation as a long-term memory consolidation mechanism is problematic for two reasons:

1. Phosphorylation of a protein is not permanent. Over time, the phosphate groups are removed, thereby erasing the memory.
2. Protein molecules themselves are not permanent. Most proteins in the brain have a lifespan of less than 2 weeks and are undergoing a continual process of replacement. Memories tied to changes in individual protein molecules would not be expected to survive this rate of molecular turnover.

Thus, we need to consider mechanisms that might convert what initially is a change in synaptic protein phosphorylation to a form that can last a lifetime.

Persistently Active Protein Kinases

Phosphorylation of synaptic proteins, and memory, could be maintained if the kinases, the enzymes that attach phosphate groups to proteins, were made to stay "on" all the time. Normally the kinases are tightly regulated and are "on" only in the presence of a second messenger. But what if learning changed these kinases so they no longer required the second messenger? The relevant synaptic proteins would remain phosphorylated all the time.

Recent evidence suggests that some kinases can become independent of their second messengers. Let's consider as examples the changes that occur in two protein kinases during LTP in the hippocampus.

CaMKII. Recall that the entry of Ca^{2+} into the postsynaptic cell and the activation of CaMKII are required for the induction of LTP in CA1. Research has shown that CaMKII stays "on" long after $[Ca^{2+}]_i$ has fallen back to a low level.

CaMKII consists of ten subunits arranged in a rosette pattern. Each subunit catalyzes the phosphorylation of substrate proteins in response to a rise in Ca^{2+}-calmodulin. How might CaMKII be switched permanently on? The answer requires some knowledge of how this enzyme is normally regulated (Figure 25.19). Each subunit is built like a pocket knife, with two parts connected by a hinge. One part, the *catalytic region*, performs the phosphorylation reaction. The other part is called the *regulatory region*. Normally, in the absence of the appropriate second messenger, the knife is closed and the catalytic region is covered by the regulatory region. This keeps the enzyme "off." The normal action of the second messenger (Ca^{2+}-calmodulin) is to pry the knife open, but only as long as the messenger is present. When the messenger is removed, the molecule usually snaps shut, and the kinase turns off again. After LTP, however, it appears that the knife fails to close completely in the α subunits of CaMKII. The exposed catalytic region continues to phosphorylate CaMKII substrates.

How is the hinge of the protein kinase molecule kept open? The answer lies in the fact that CaMKII is an *autophosphorylating protein kinase*; each subunit within the CaMKII molecule can be phosphorylated by a neighboring subunit. The consequence of subunit phosphorylation is that the hinge stays open. If the initial activation of CaMKII by Ca^{2+}-calmodulin is sufficiently strong, autophosphorylation will occur at a faster rate than

(a)

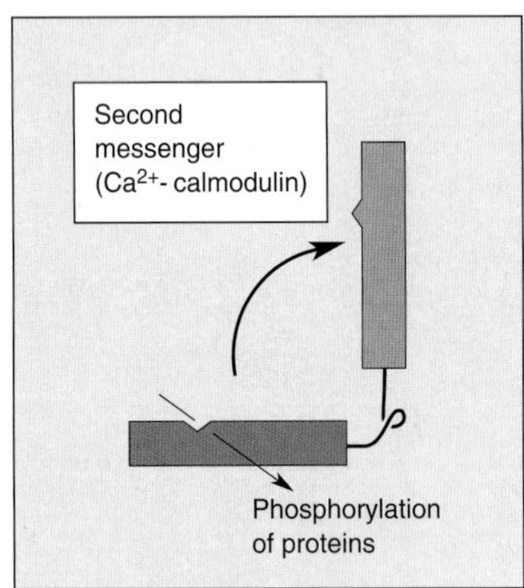

(b)

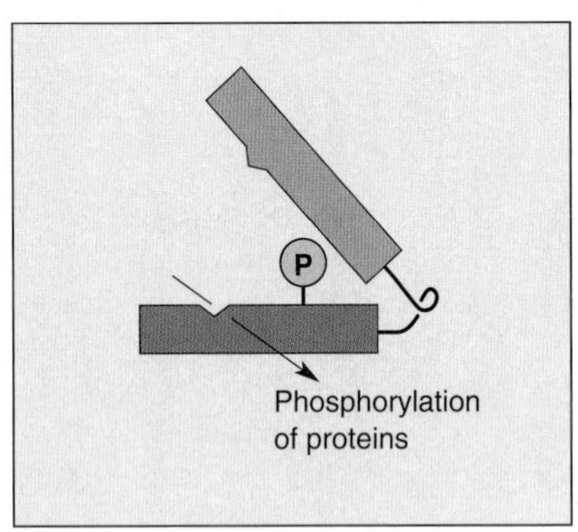

(c)

▲ FIGURE 25.19
The regulation of CaMKII. **(a)** The hingelike subunit of CaMKII is normally "off" when the catalytic region is covered by the regulatory region. **(b)** The hinge opens upon activation of the molecule by Ca^{2+}-bound calmodulin, freeing the catalytic region to add phosphate groups to other proteins. **(c)** A large elevation of Ca^{2+} can cause phosphorylation (P) of one subunit by another (autophosphorylation), which enables the catalytic region to stay "on" permanently.

dephosphorylation, and the molecule will be switched on. Persistent activity of CaMKII could contribute to the maintenance of synaptic potentiation, for example, by keeping the postsynaptic AMPA receptors phosphorylated. The general idea that an autophosphorylating kinase could store information at the synapse, initially proposed by John Lisman at Brandeis University, is called the **molecular switch hypothesis**.

Protein Kinase M Zeta. Recent research has implicated a new player in the maintenance of LTP and certain forms of memory, *protein kinase M zeta (PKMζ)*. Excitement about the role of this protein kinase originates with the work of Todd Sacktor at the State University of New York Downstate Medical Center. Sacktor and his colleagues showed that an intracerebral injection of a small peptide called ZIP, designed to specifically inhibit PKMζ, can erase LTP and memories established many days before the injection. Simply put, ZIP zaps memories. This intriguing finding suggests that persistent PKMζ activity maintains changes in synaptic strength by continuing to phosphorylate its substrates. ZIP, by temporarily inhibiting the kinase, allows these substrates to become dephosphorylated, thus erasing the memory trace.

How does PKMζ become persistently active in response to synaptic activity? A current model suggests that the mRNA for this kinase exists at synapses but is normally not translated into protein. Strong synaptic activation, and the corresponding rise in [Ca^{2+}], trigger a burst of synaptic protein synthesis and the birth of new PKMζ molecules. PKMζ phosphorylates synaptic proteins involved in regulating AMPA receptor number and, in addition, proteins involved in the regulation of mRNA translation at the synapse. By switching translation on in the absence of elevated [Ca^{2+}], PKMζ levels can be replenished in the face of ongoing degradation of the kinase molecules.

The erasure of memory by ZIP has been reproduced by several laboratories, but at this point it is unclear if it acts specifically by inhibiting PKMζ. It is certain, however, that unlocking the mystery of how ZIP works will reap great rewards in our understanding of memory mechanisms.

Protein Synthesis and Memory Consolidation

The idea that protein synthesis is important for memory consolidation is not new. It has been investigated extensively since the introduction of drugs in the 1960s that selectively inhibit the assembly of protein from messenger RNA. Protein synthesis inhibitors can be injected into the brains of experimental animals as they are trained to perform a task, and deficits in learning and memory can be assessed. These studies reveal that if brain protein synthesis is inhibited at the time of training, the animals learn normally but fail to remember when tested days later. A deficit in long-term memory is also often observed if the inhibitors are injected shortly after training. However, the memories become increasingly resistant to the inhibition of protein synthesis as the interval between the training and the injection of inhibitor is increased. These findings indicate a requirement for new protein synthesis during the period of memory consolidation, when short-term memories are converted into long-term ones.

Consider the example of inhibitory avoidance memory (see Figure 25.17a). As we discussed, the memory is created in a single trial and can be measured by how much a rat avoids the location where it received a foot shock (usually the dark side of a box with two rooms separated by a door). Normally, this memory is very stable and lasts a long time (days to weeks depending on the strength of the shock). If the animal is injected with a protein synthesis inhibitor shortly before the training, learning occurs normally, as measured by the animal's immediate avoidance of the dark side. However, this memory fades within a day without new protein synthesis. Similarly, inhibiting protein synthesis at the time of a tetanus has no effect on the induction of LTP in the hippocampus, but instead of lasting days to months, the synaptic potentiation gradually disappears over a few hours.

Synaptic Tagging and Capture. From what we have learned so far, memory formation appears initially to involve the rapid modification of existing synaptic proteins. These modifications, perhaps with the help of persistently active kinases, work against the factors that would erase a memory (such as molecular turnover). It's a losing battle, unless a new protein arrives at the modified synapse and converts the temporary change in the synapse to a more permanent one. But how do the proteins required for consolidating synaptic changes and memories find the modified synapses? An answer is suggested by a clever series of experiments performed collaboratively in the late 1990s by Julietta Frey in Magdeburg, Germany, and Richard Morris in Edinburgh, Scotland.

Frey had shown previously that LTP caused by "weak" tetanic stimulation, which activates briefly only a small number of synapses, decays back to baseline over the course of an hour or two because it fails to trigger protein synthesis. On the other hand, repeated episodes of "strong" stimulation, recruiting a larger number of synapses, induces long-lasting LTP because it stimulates new protein synthesis (Figure 25.20a,b). Frey and Morris asked if the newly synthesized proteins acted only on the synapses whose activity triggered their synthesis. They found that the wave of protein synthesis triggered by strong stimulation of one synaptic input to a hippocampal neuron would also consolidate the LTP caused by the weak stimulation of a second input (Figure 25.20c). It appeared that the weak stimulation endows synapses with a *tag* that enables them to *capture* the newly synthesized proteins that consolidate LTP. By varying the interval between the weak and strong stimulation of the two inputs, Frey and Morris were able to determine that the tag lasts about 2 hours. In this way, a trivial event that would otherwise be forgotten, like last Tuesday's dinner, might be seared into long-term memory if it occurs within 2 hours of a momentous event that triggers a wave of new protein synthesis, like the first kiss of the love of your life. The molecular mechanism for synaptic tagging has not been fully elucidated, but it should come as no surprise that it is believed to involve phosphorylation of synaptic proteins by various kinases, including CaMKII and PKMζ.

CREB and Memory. What regulates the protein synthesis that is required for memory consolidation? Recall that the very first step in protein synthesis is the generation of an mRNA transcript of a gene (see Figure 2.9). This process of gene expression is regulated by *transcription factors* in the nucleus. One transcription factor is called the **cyclic AMP response element binding protein (CREB)**. CREB is a protein that binds to specific segments of DNA, called *cyclic AMP response elements (CREs)*, and functions to regulate the expression of neighboring genes (Figure 25.21). There are two forms of CREB: CREB-2 represses gene expression when it binds to the CRE; CREB-1 activates transcription, but only when it is phosphorylated by protein kinase A. In a seminal study published in 1994, Tim Tully and Jerry Yin at Cold Spring Harbor Laboratory showed that CREB regulates the gene expression required for memory consolidation in the fruit fly *Drosophila melanogaster* (see Box 25.5).

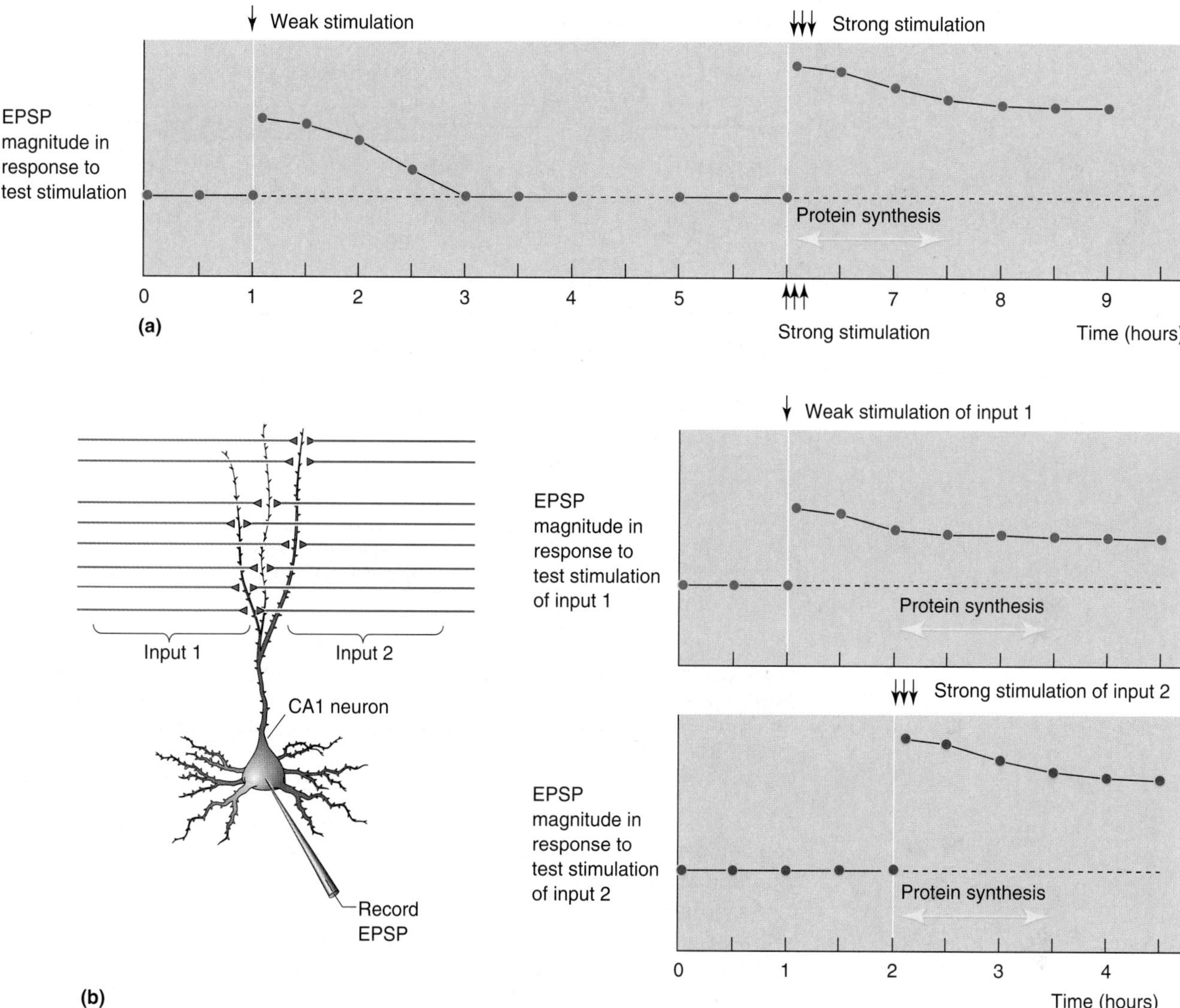

▲ FIGURE 25.20

Synaptic tagging and capture. **(a)** The persistence of LTP depends on whether synaptic stimulation is strong enough to trigger protein synthesis in the postsynaptic neuron. Weak stimulation induces LTP of the EPSP that rapidly decays. Strong stimulation induces LTP *and* stimulates the protein synthesis that converts a temporary synaptic change into a long-lasting one. **(b)** Two inputs to the same neuron are stimulated in alternation. Weak stimulation of input 1 induces LTP that would normally fade, but if it is followed an hour later by strong stimulation of input 2, the wave of new proteins can be captured by synapses tagged by the weak stimulation of input 1. The timely arrival of new protein converts the short-lasting LTP into long-lasting LTP.

In their first series of experiments, Tully and Yin bred *Drosophila* that would make extra copies of the fly's version of CREB-2 (called dCREBb) when the animal was warmed up (a miracle of fly genetic engineering that is not possible in mammals). This manipulation repressed all gene expression that is regulated by the CREs, and also blocked memory consolidation in a simple memory task. Thus, CREB-regulated gene expression is critical for memory consolidation in fruit flies. More interesting, however, is what they found when they generated flies that could make extra copies of fly CREB-1 (called dCREBa). Now, tasks that would take normal flies many trials to learn could be remembered after a single training trial. These mutant flies had a "photographic" memory! And these results are not peculiar to flies; CREB has been implicated in regulating the

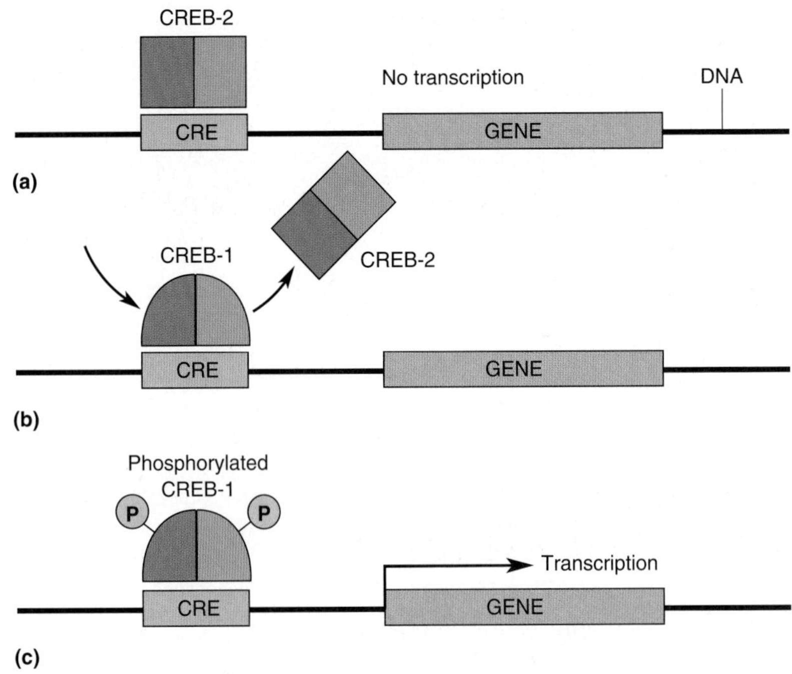

▲ FIGURE 25.21
The regulation of gene expression by CREB. Shown here is a piece of DNA containing a gene whose expression is regulated by the interaction of a CREB protein with a CRE on the DNA. **(a)** CREB-2 functions as a repressor of gene expression. **(b)** CREB-1, an activator of gene expression, can displace CREB-2. **(c)** When CREB-1 is phosphorylated by protein kinase A (and other kinases), transcription can ensue.

consolidation of sensitization in *Aplysia*, as well as long-term potentiation and spatial memory in mice.

As we have discussed, not all experiences are remembered equally. Some are seared permanently into our memories. Others remain with us only briefly and then fade away. The modulation of gene expression by CREB provides a molecular mechanism that can control the strength of a memory.

A failure of memories to consolidate is a feature of numerous brain disorders as well as the aging process. The recent understanding of how consolidation is regulated has spawned a new industry focused on developing memory-enhancing drugs. Such drugs could greatly improve the quality of life for individuals with neurological diseases, such as Alzheimer's. However, there is also the allure of performance enhancement in healthy people. By analogy, consider the widespread use of such drugs as Viagra introduced to treat male erectile dysfunction. "Viagra for the brain," as such memory enhancers have been whimsically called, continues to be pursued. Although compounds have been discovered that boost memory consolidation, their side effects have thus far prevented their development into drugs. As was the case for drugs that enhance athletic performance, the ethics of using memory boosters in the absence of a clear medical justification are certain to be hotly debated.

Structural Plasticity and Memory. How does the synapse make use of the timely occurrence of gene expression and the arrival of a new protein? One possibility is that newly synthesized proteins (such as PKMζ) switch local synaptic protein synthesis on in order to maintain a synaptic change. However, to account for the fact that blocking protein synthesis fails to disrupt memories that have already been consolidated, we would have to further posit that these newly synthesized proteins have a lifespan that is long enough to survive the temporary inhibition of protein synthesis.

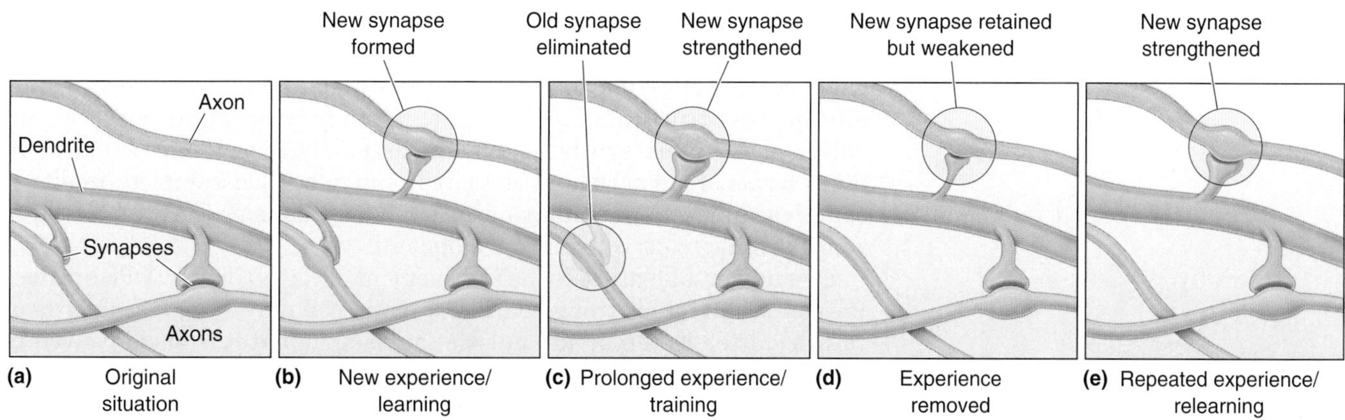

▲ FIGURE 25.22
Synaptic remodeling in the cerebral cortex during learning and memory. This illustration summarizes some of the structural changes that have been observed in the neocortex when mice are exposed to new sensory environments that are encoded as memory. (Source: Adapted from Hofer and Bonhoeffer, 2010, Fig. 1.)

Another possibility is that the lasting imprint of new protein synthesis is the construction (or demolition) of synapses. Work on the invertebrate *Aplysia* has shown that some types of long-term (but not short-term) memory can cause a *doubling* of the number of synapses made by some neurons!

Do similar structural changes occur in the mammalian nervous system after learning? This is a difficult problem to solve experimentally because of the complexity of the mammalian brain and the distributed nature of memory. One approach has been to compare brain structure in animals that have had ample opportunity to learn with that of animals that have had little chance to learn. For example, putting a laboratory rat in a "complex" environment filled with toys and playmates (other rats) has been shown to increase the number of synapses per neuron in the occipital cortex by about 25%. Very recently, advances in microscopy and methods for cell labeling (see Box 2.1) have made it possible for researchers to image the same neuron in living mice over the course of many days. Altering the visual or tactile environment stimulates the formation of new dendritic spines, important sites of excitatory synaptic transmission, in the visual and somatosensory cortex, respectively (Figure 25.22). As exposure to this new environment is increased, these new synapses grow larger and other synapses on the same dendrites are eliminated, as would be expected if the mechanisms of LTP and LTD are at work to encode a memory. The new spines may shrink if the animal is returned to the original environment but do not disappear, consistent with the interpretation that these persistent spines contribute to a long-term memory of the altered environment.

It is important to recognize that there are limits to structural plasticity in the adult brain. As we discussed in Chapter 23, large changes in brain circuitry are generally confined to critical periods of early life. The growth and retraction of most axons in the adult central nervous system (CNS) are restricted to no more than a few tens of micrometers. But it is now very clear that the end of a critical period does not necessarily signify an end to changes in the structure of axon terminals or the effectiveness of their synapses.

CONCLUDING REMARKS

Learning and memory can occur at synapses. Regardless of the species, brain location, and memory type, many of the underlying mechanisms appear to be universal. Events are represented first as changes in the electrical activity of the brain, then as intracellular second messengers, and next as modifications of existing synaptic proteins. These temporary changes are converted to permanent ones, and long-term memories, by altering the structure of the synapse. In many forms of memory, this involves new protein synthesis and the assembly of new microcircuits. In other forms of memory, existing circuits may be disassembled. In either case, learning requires many of the same mechanisms that were used to refine brain circuitry during development.

One universal feature is the involvement of Ca^{2+}. Clearly, calcium does far more than build strong bones and teeth. Not only is it critical for neurotransmitter secretion and muscle contraction, but it is also involved in nearly every form of synaptic plasticity. Because it is a charge-carrying ion on the one hand and a potent second messenger substance on the other, Ca^{2+} has a unique ability to directly couple electrical activity with long-term changes in the brain.

Can basic neuroscience research take us from ions to intelligence? From calcium to cognition? If you remember what you have learned in this textbook, and if synaptic plasticity truly is the basis of declarative memory, it would appear that the answer is yes.

KEY TERMS

Memory Acquisition

memory consolidation (p. 867)
distributed memory (p. 869)
long-term potentiation (LTP) (p. 874)
dentate gyrus (p. 874)
Ammon's horn (p. 874)
CA3 (p. 874)
CA1 (p. 874)
perforant path (p. 874)
Schaffer collateral (p. 874)
tetanus (p. 875)
input specificity (p. 876)
cooperativity (p. 876)
BCM theory (p. 880)
long-term depression (LTD) (p. 881)
spike timing–dependent plasticity (p. 882)
metaplasticity (p. 889)
synaptic scaling (p. 891)

Memory Consolidation

molecular switch hypothesis (p. 893)
cyclic AMP response element binding protein (CREB) (p. 894)

REVIEW QUESTIONS

1. What is the most common cellular correlate of memory formation in the cerebral cortex? What does this say about how memories are stored?
2. How does one account for the graceful degradation of memory as neurons die over the lifespan?
3. How can LTD contribute to memory?
4. Sketch the trisynaptic circuit of the hippocampus.
5. How can the mechanisms of LTP serve associative memory?
6. What property of the NMDA receptor makes it well suited to detect coincident presynaptic and postsynaptic activity? How could Ca^{2+} entering through the NMDA receptor possibly trigger both LTP and LTD in CA1 and the neocortex?
7. Compare and contrast metaplasticity and synaptic scaling.
8. In H.M and R.B. (see Chapter 24), destruction of the hippocampus appears to have impaired the mechanism that "fixes" new memories in the neocortex. Propose a mechanism involving CREB explaining why this might be true.

FURTHER READING

Abraham WC, Robins A. 2005. Memory retention: the synaptic stability versus plasticity dilemma. *Trends in Neuroscience* 28:73–78.

Bear MF. 1996. A synaptic basis for memory storage in the neocortex. *Proceedings of the National Academy of Sciences USA* 93:13453–13459.

Cooper LN, Bear MF. 2012. The BCM theory of synapse modification at 30: interaction of theory and experiment. *Nature Reviews Neuroscience* 13:798–810.

Kandel ER. 2006. *In Search of Memory: The Emergence of a New Science of Mind.* New York: Norton.

Kessels HW, Malinow R. 2009. Synaptic AMPA receptor plasticity and behavior. *Neuron* 61:340–350.

Malenka RC, Bear MF. 2004. LTP and LTD: an embarrassment of riches. *Neuron* 44:5–21.

GLOSSARY

5-HT *See* serotonin.

A1 *See* primary auditory cortex.

absolute refractory period The period of time, measured from the onset of an action potential, during which another action potential cannot be triggered.

accommodation The focusing of light by changing the shape of the eye's lens.

acetylcholine (ACh) An amine that serves as a neurotransmitter at many synapses in the peripheral and central nervous systems, including the neuromuscular junction.

ACh *See* acetylcholine.

ACTH *See* adrenocorticotropic hormone.

actin A cytoskeletal protein in all cells and the major thin filament protein in a skeletal muscle fiber; causes muscle contraction by specific chemical interactions with myosin.

action potential A brief fluctuation in membrane potential caused by the rapid opening and closing of voltage-gated ion channels; also known as spike, nerve impulse, or discharge. Action potentials sweep like a wave along axons to transfer information from one place to another in the nervous system.

activational effect The ability of a hormone to activate reproductive processes or behaviors in the mature organism.

active zone A presynaptic membrane differentiation that is the site of neurotransmitter release.

adenosine triphosphate (ATP) The molecule that is the cell's energy source. The hydrolysis of ATP to produce adenosine diphosphate (ADP) releases energy that fuels most of the biochemical reactions of the neuron. ADP is converted back to ATP in the mitochondria.

adenylyl cyclase An enzyme that catalyzes the conversion of adenosine triphosphate (ATP) to cyclic adenosine monophosphate (cAMP), a second messenger.

adrenal cortex The outer segment of the adrenal gland; releases cortisol when stimulated by the pituitary adrenocorticotropic hormone.

adrenal medulla The inner segment of the adrenal gland, innervated by preganglionic sympathetic fibers; releases epinephrine.

adrenaline A catecholamine neurotransmitter synthesized from norepinephrine; also called epinephrine.

adrenocorticotropic hormone (ACTH) A hormone released by the anterior pituitary in response to corticotropin-releasing hormone; stimulates the release of cortisol from the adrenal gland.

affective aggression A threatening or defensive form of aggression accompanied by vocalizations and a high level of ANS activity.

affective disorder A psychiatric condition characterized by disordered emotions; also called mood disorder. Examples are major depression and bipolar disorder.

affective neuroscience The study of the neural basis of mood and emotion.

afferent An axon coursing toward and innervating a given structure. *See also* efferent.

after-hyperpolarization The hyperpolarization that follows strong depolarization of the membrane; the last part of an action potential, also called undershoot.

agnosia The inability to recognize objects, even though simple sensory skills appear to be normal; most commonly caused by damage to posterior parietal areas of the brain.

agoraphobia A mental disorder characterized by severe anxiety about being in situations in which escape might be difficult or embarrassing.

alpha motor neuron A neuron that innervates the extrafusal fibers of skeletal muscle.

amacrine cell A neuron in the retina of the eye that projects neurites laterally in the inner plexiform layer.

amino acid A chemical building block of protein molecules, containing a central carbon atom, an amino group, a carboxyl group, and a variable R group.

Ammon's horn A layer of neurons in the hippocampus that sends axons into the fornix.

amnesia A severe loss of memory or the ability to learn. *See also* anterograde amnesia, retrograde amnesia.

AMPA receptor A subtype of glutamate receptor; a glutamate-gated ion channel that is permeable to Na^+ and K^+.

ampulla The bulge along a semicircular canal, which contains the hair cells that transduce rotation.

amygdala An almond-shaped nucleus in the anterior temporal lobe thought to be involved in emotion and certain types of learning and memory.

anabolism The biosynthesis of organic molecules from nutritive precursors; also called anabolic metabolism. *See also* catabolism.
analgesia The absence of normal sensations of pain.
androgens Male sex steroidal hormones, the most important of which is testosterone.
anion A negatively charged ion. *See also* cation.
anomia The inability to find words.
anorectic peptide A neuroactive peptide that acts to inhibit feeding behavior; examples are cholecystokinin (CCK), alpha-melanocyte-stimulating hormone (αMSH), and cocaine- and amphetamine-regulated transcript peptide (CART).
anorexia A state of negative energy balance in which energy expenditure exceeds energy intake.
anorexia nervosa A psychiatric disorder characterized by an obsession with food, an intense fear of gaining weight, and voluntary maintenance of weight at below-normal levels.
ANS *See* autonomic nervous system.
antagonist muscle A muscle that acts against another at the same joint.
anterior An anatomical reference meaning toward the nose or rostral.
anterior cingulate cortex Region of the cerebral cortex, just anterior to the descending limb of the corpus callosum, which has been implicated in the pathophysiology of mood disorders.
anterograde amnesia The inability to form new memories.
anterograde transport Axoplasmic transport from a neuron's soma to the axon terminal.
antidepressant drug A drug that treats the symptoms of depression by elevating brain levels of monoamine neurotransmitters; examples are tricyclics, monoamine oxidase (MAO) inhibitors, and SSRIs.
antidiuretic hormone (ADH) *See* vasopressin.
anxiety disorder One of a group of mental disorders characterized by irrational or inappropriate expressions of fear, including panic disorder, agoraphobia, and OCD.
anxiolytic drug A medication that reduces anxiety; examples are benzodiazepines and SSRIs.
aphasia A partial or complete loss of language abilities following brain damage. *See also* Broca's aphasia, conduction aphasia, Wernicke's aphasia.
apoptosis A mechanism of orderly, genetically programmed cell death.
aqueous humor The fluid between the cornea and the lens of the eye.
arachnoid membrane The middle of the three meninges, the three membranes that cover the surface of the central nervous system.
arcuate nucleus A nucleus in the periventricular area of the hypothalamus containing a large number of neurons sensitive to changes in leptin levels, contributing to the regulation of energy balance.
area 17 Primary visual cortex.
area IT An area of neocortex, on the inferior surface of the temporal lobe, that is part of the ventral visual processing stream; contains neurons with responses to complex objects, including faces, and appears to be involved in visual memory.
area LIP *See* lateral intraparietal cortex.
area MT An area of neocortex, at the junction of the parietal and temporal lobes, that receives input from primary visual cortex and appears to be specialized for the processing of object motion; also called V5.
area V4 An area of neocortex, anterior to striate cortex, that is in the ventral visual processing stream and appears to be important for both shape perception and color perception.
aspinous neuron A neuron lacking dendritic spines.
associative learning The learning of associations between events; two types are usually distinguished: classical conditioning and instrumental conditioning.
astrocyte A glial cell in the brain that supports neurons and regulates the extracellular ionic and chemical environment.
ataxia Abnormally uncoordinated and inaccurate movements, often associated with cerebellar dysfunction.
atonia The absence of muscle tone.
ATP *See* adenosine triphosphate.
attention The state of selectively allocating mental energy to a sensory stimulus.
attention-deficit hyperactivity disorder (ADHD) A behavioral condition characterized by inattention, hyperactivity, and impulsiveness.
attenuation reflex The contraction of muscles in the middle ear, resulting in a reduction in auditory sensitivity.
audition The sense of hearing.
auditory canal A channel leading from the pinna to the tympanic membrane; the entrance to the internal ear.
auditory–vestibular nerve Cranial nerve VIII, consisting of axons projecting from the spiral ganglion to the cochlear nuclei.
autonomic ganglia Peripheral ganglia of the sympathetic and parasympathetic divisions of the autonomic nervous system.
autonomic nervous system (ANS) A system of central and peripheral nerves that innervates the internal organs, cardiovascular system, and glands; also called visceral PNS. The ANS consists of sympathetic, parasympathetic, and enteric divisions.
autoradiography A method for visualizing sites of radioactive emissions in tissue sections.
autoreceptor A receptor in the membrane of a presynaptic axon terminal that is sensitive to the neurotransmitter released by that terminal.

axial muscle A muscle that controls movements of the trunk of the body.
axon A neurite specialized to conduct nerve impulses, or action potentials, normally away from the soma.
axon collateral A branch of an axon.
axon hillock A swelling of the axon where it joins the soma.
axon terminal The end region of an axon, usually a site of synaptic contact with another cell; also called terminal bouton or presynaptic terminal.
axoplasmic transport The process of transporting materials down an axon.

ballism A movement disorder caused by damage to the subthalamus, characterized by violent, flinging movements of the extremities.
barbiturate A class of drugs with sedative, general anesthetic, and anticonvulsant effects; barbiturates act in part by binding to $GABA_A$ receptors and prolonging their inhibitory actions.
basal forebrain complex Several cholinergic nuclei of the telencephalon, including the medial septal nuclei and basal nucleus of Meynert.
basal ganglia A collection of associated cell groups in the basal forebrain, including the caudate nucleus, putamen, globus pallidus, and subthalamus.
basal telencephalon The region of the telencephalon lying deep in the cerebral hemispheres.
basic emotions A small set of emotions thought to be unique indivisible experiences that are innate and universal across cultures.
basic theories of emotion Explanations of emotions based on the principle that there are a small set of hard-wired emotions that are universal across cultures.
basilar membrane A membrane separating the scala tympani and scala media in the cochlea in the inner ear.
BCM theory A theory proposing that synapses are bidirectionally modifiable. Synaptic potentiation results when presynaptic activity correlates with a strong postsynaptic response, and synaptic depression results when presynaptic activity correlates with a weak postsynaptic response. An extension of the Hebb synapse concept, proposed by Bienenstock, Cooper, and Munro at Brown University. *See also* Hebb synapse, Hebbian modification.
benzodiazepine A class of drugs with antianxiety, sedative, muscle-relaxing, and anticonvulsant effects; acts by binding to GABAA receptors and prolonging their inhibitory actions.
binocular competition A process believed to occur during the development of the visual system whereby the inputs from the two eyes actively compete to innervate the same cells.
binocular receptive field The receptive field of a neuron that responds to stimulation of either eye.
binocular rivalry Perception that alternates in time between the image seen by one eye and a dissimilar image seen by the other eye.
binocular visual field The portion of the visual field viewed by both eyes.
bipolar cell In the retina, a cell that connects photoreceptors to ganglion cells.
bipolar disorder A psychiatric affective disorder characterized by episodes of mania, sometimes interspersed with episodes of depression; also called manic-depressive disorder.
bipolar neuron A neuron with two neurites.
blob A collection of cells, mainly in primary visual cortical layers II and III, characterized by a high level of the enzyme cytochrome oxidase.
blob pathway A visual information-processing pathway that passes through the parvocellular and koniocellular layers of the LGN and converges on the blobs of striate cortical layer III; believed to process information about object color.
blood–brain barrier A specialization of the walls of brain capillaries that limits the movement of bloodborne substances into the extracellular fluid of the brain.
bottom-up attention Attention reflexively directed to a salient external stimulus; also called exogenous attention.
brain The part of the central nervous system contained in the skull, consisting of the cerebrum, cerebellum, brain stem, and retinas.
brain stem The diencephalon, midbrain, pons, and medulla. (Some anatomists exclude the diencephalon.)
Broca's aphasia A language disturbance in which a person has difficulty speaking or repeating words but can understand language; also known as motor or nonfluent aphasia.
Broca's area A region of the frontal lobe associated with Broca's (motor) aphasia when damaged.
bulimia nervosa A psychiatric disorder characterized by large, uncontrolled eating binges followed by compensatory behavior, such as forced vomiting.
bundle A collection of axons that run together but do not necessarily have the same origin and destination.

CA1 A region of Ammon's horn in the hippocampus that receives input from the neurons of CA3.
CA3 A region of Ammon's horn in the hippocampus that receives input from the neurons of the dentate gyrus.
calcium-calmodulin-dependent protein kinase (CaMK) A protein kinase activated by elevations of internal Ca^{2+} concentration.
calcium pump An ion pump that removes cytosolic Ca^{2+}.
CAM *See* cell-adhesion molecule.
cAMP *See* cyclic adenosine monophosphate.

Cannon–Bard theory A theory of emotion proposing that emotional experience is independent of emotional expression and is determined by the pattern of thalamic activation.

capsule A collection of axons that connect the cerebrum with the brain stem.

cardiac muscle A type of striated muscle found only in the heart; it contracts rhythmically even in the absence of innervation.

catabolism The breaking down of complex nutrient molecules into simpler molecules; also called catabolic metabolism. *See also* anabolism.

catecholamines The neurotransmitters dopamine, norepinephrine, and epinephrine.

cation A positively charged ion. *See also* anion.

caudal An anatomical reference meaning toward the tail, or posterior.

caudate nucleus A part of the basal ganglia in the basal forebrain, involved in motor control.

CCK *See* cholecystokinin.

cell-adhesion molecule (CAM) A molecule on the cell surface that causes cells to adhere to one another.

cell assembly A group of simultaneously active neurons that represents an object held in memory.

cell body The central region of the neuron containing the nucleus; also called soma or perikaryon.

center-surround receptive field A visual receptive field with a circular center region and a surround region forming a ring around the center; stimulation of the center produces a response opposite that generated by stimulation of the surround.

central nervous system (CNS) The brain (including the retinas) and spinal cord. *See also* peripheral nervous system.

central pattern generator A neural circuit that gives rise to rhythmic motor activity.

central sulcus The sulcus in the cerebrum that divides the frontal lobe from the parietal lobe.

cerebellar cortex A sheet of gray matter lying just under the pial surface of the cerebellum.

cerebellar granule cell A neuron in the cerebellar cortex that receives input from mossy fibers and gives rise to parallel fibers that innervate Purkinje cells. Plasticity of the synapses between granule cells and Purkinje cells is believed to be important for motor learning.

cerebellar hemispheres The lateral regions of the cerebellum.

cerebellum A structure derived from the rhombencephalon, attached to the brain stem at the pons; an important movement control center.

cerebral aqueduct A canal filled with cerebrospinal fluid within the midbrain.

cerebral cortex The layer of gray matter that lies just under the surface of the cerebrum.

cerebral hemispheres The two sides of the cerebrum, derived from the paired telencephalic vesicles.

cerebrospinal fluid (CSF) In the central nervous system, the fluid produced by the choroid plexus that flows through the ventricular system to the subarachnoid space.

cerebrum The largest part of the forebrain; also called telencephalon.

cGMP *See* cyclic guanosine monophosphate.

channelopathy A human genetic disease caused by alterations in the structure and function of ion channels.

channelrhodopsin-2 (ChR2) A light-sensitive cation channel, originally isolated from green algae, which can be expressed in neurons and used to control their activity using light.

characteristic frequency The sound frequency to which a neuron in the auditory system gives its greatest response.

chemical synapse A synapse in which presynaptic activity stimulates the release of neurotransmitter, which activates receptors in the postsynaptic membrane.

chemoaffinity hypothesis The hypothesis that chemical markers on growing axons are matched with complementary chemical markers on their targets.

chemoattractant A diffusible molecule that acts over a distance to attract growing axons.

chemoreceptor Any sensory receptor selective for chemicals.

chemorepellent A diffusible molecule that acts over a distance to repel growing axons.

cholecystokinin (CCK) A peptide found within some neurons of the central and peripheral nervous systems and in some endothelial cells lining the upper gastrointestinal tract; a satiety signal that inhibits feeding behavior, in part, by acting on axons of the vagus nerve that respond to gastric distension.

cholinergic Describing neurons or synapses that produce and release acetylcholine.

chromosome A structure in the cell nucleus containing a single linear thread of DNA.

ciliary muscle A muscle that controls the shape of the eye's lens.

circadian rhythm Any rhythm with a period of about 1 day.

classical conditioning A learning procedure in which a stimulus that evokes a measurable response is associated with another stimulus that normally does not evoke this response.

climbing fiber An axon of an inferior olive neuron that innervates a Purkinje cell of the cerebellum. Climbing fiber activity is an important trigger for LTD, a form of synaptic plasticity believed to be important for motor learning.

clock gene A gene that is critically involved in the molecular mechanism of the circadian rhythm; clock genes are translated into proteins that regulate their own transcription, and their expression rises and falls over a cycle of about 24 hours.

CNS *See* central nervous system.

cochlea A spiral bony structure in the inner ear that contains the hair cells that transduce sound.

cochlear amplifier Outer hair cells, including the motor proteins in the outer hair cell membrane, that amplify displacements of the basilar membrane in the cochlea.

cochlear nucleus *See* dorsal cochlear nucleus, ventral cochlear nucleus.

cognitive map theory The idea that the hippocampus is specialized to form a spatial map of the environment.

color-opponent cell A cell in the visual system with an excitatory response to wavelengths of light of one color and an inhibitory response to wavelengths of another color; the color pairs that cancel each other are red–green and blue–yellow.

commissure Any collection of axons that connect one side of the brain with the other side.

complex cell A type of visual cortical neuron that has an orientation-selective receptive field without distinct ON and OFF subregions.

concentration gradient A difference in concentration from one region to another. Ionic concentration gradients across the neuronal membrane help determine the membrane potential.

conductance *See* electrical conductance.

conduction aphasia A type of aphasia associated with damage to the arcuate fasciculus, characterized by good comprehension and speech but difficulty repeating words.

cone photoreceptor A photoreceptor in the retina containing one of three photopigments that are maximally sensitive to different wavelengths of light. Cones are concentrated in the fovea, specialized for daytime vision, and responsible for all color vision. *See also* rod photoreceptor.

conjunctiva The membrane that folds back from the eyelids and attaches to the sclera of the eye.

connectome A detailed wiring diagram of how neurons connect with one another via synapses.

consciousness Awareness of external stimuli and internal thoughts and feelings.

contralateral An anatomical reference meaning on the opposite side of the midline.

cooperativity The property of long-term potentiation, reflecting the requirement that many inputs be active at the same time during a tetanus to induce LTP. *See also* long-term potentiation.

cornea The transparent external surface of the eye.

coronal plane An anatomical plane of section that divides the nervous system into anterior and posterior parts.

corpus callosum The great cerebral commissure, consisting of axons connecting the cortex of the two cerebral hemispheres.

cortex Any collection of neurons that forms a thin sheet, usually at the brain's surface.

cortical module The unit of cerebral cortex that is necessary and sufficient to analyze one discrete point in a sensory surface.

cortical plate A cell layer of the immature cerebral cortex containing undifferentiated neurons.

cortical white matter A collection of axons lying just below the cerebral cortex.

corticospinal tract The tract that originates in the neocortex and terminates in the spinal cord; involved in the control of voluntary movement.

corticotropin-releasing hormone (CRH) A hormone released by neurons in the paraventricular nucleus of the hypothalamus; stimulates the release of ACTH from the anterior pituitary.

cortisol A steroid hormone released by the adrenal cortex; mobilizes energy reserves, suppresses the immune system, and has direct actions on some CNS neurons.

co-transmitter One of two or more different neurotransmitters that are released from a single presynaptic nerve terminal.

cranial nerves Twelve pairs of nerves that arise from each side of the brain stem, numbered from anterior to posterior. Cranial nerve I is actually the olfactory tract, and cranial nerve II is the optic nerve; both are parts of the central nervous system. Cranial nerves III–XII, which are in the peripheral nervous system, perform many diverse functions.

CRH *See* corticotropin-releasing hormone.

critical period A limited period of time when a particular aspect of brain development is sensitive to a change in the external environment.

CSF *See* cerebrospinal fluid.

cyclic adenosine monophosphate (cAMP) A second messenger formed from adenosine triphosphate by the action of the enzyme adenylyl cyclase.

cyclic AMP response element binding protein (CREB) A protein that binds to specific regions of DNA (cyclic AMP response elements) and functions to regulate gene transcription; a key regulator of protein synthesis–dependent memory consolidation.

cyclic guanosine monophosphate (cGMP) A second messenger formed from guanosine triphosphate by the action of the enzyme guanylyl cyclase.

cytoarchitectural map A map, usually of the cerebral cortex, based on cytoarchitectural differences.

cytoarchitecture The arrangement of neuronal cell bodies in various parts of the brain.

cytochrome oxidase A mitochondrial enzyme concentrated in cells that form the blobs in primary visual cortex.
cytoplasm Cellular material contained by the cell membrane, including the organelles but excluding the nucleus.
cytoskeleton The internal scaffolding that gives a cell its characteristic shape; consists of microtubules, neurofilaments, and microfilaments.
cytosol The watery fluid inside a cell.

DA *See* dopamine.
DAG *See* diacylglycerol.
Dale's principle The idea that a neuron has a unique identity with respect to neurotransmitter.
dark adaptation The process by which the retina becomes more sensitive to light in dim light.
dark current The inward sodium current that occurs in photoreceptors in the dark.
declarative memory Memory for facts and events.
default mode network An interconnected group of brain areas that are consistently more active when the brain is at rest than during active behavioral tasks.
delayed non-match to sample (DNMS) A behavioral task in which animals are trained to displace one of two alternative objects that does not match a previously seen sample object.
dendrite A neurite specialized to receive synaptic inputs from other neurons.
dendritic spine A small sac of membrane that protrudes from the dendrites of some cells and receives synaptic input.
dendritic tree All the dendrites of a single neuron.
dense-core vesicle *See* secretory granule.
dentate gyrus A layer of neurons in the hippocampus that receives input from the entorhinal cortex.
deoxyribonucleic acid *See* DNA.
depolarization A change in membrane potential, taking it from the value at rest (e.g., –65 mV) to a less negative value (e.g., 0 mV).
dermatome A region of skin innervated by the pair of dorsal roots from one spinal segment.
diacylglycerol (DAG) A second messenger molecule formed by the action of phospholipase C on the membrane phospholipid phosphatidylinositol-4, 5-bisphosphate. DAG activates the enzyme protein kinase C.
diathesis–stress hypothesis of mood disorders A hypothesis suggesting that depression is caused by a combination of genetic predisposition and environmental stress.
diencephalon A region of the brain stem derived from the prosencephalon (forebrain). Diencephalic structures include the thalamus and hypothalamus.
differentiation During embryonic development, the process by which structures become more complex and functionally specialized.
diffuse modulatory system One of several systems of CNS neurons that project widely and diffusely onto large areas of the brain and use modulatory neurotransmitters, including dopamine, norepinephrine, serotonin, and acetylcholine.
diffusion The temperature-dependent movement of molecules from regions of high concentration to regions of low concentration, resulting in a more even distribution.
dimensional theories of emotion Explanations of emotions in which each emotion is built from emotional components such as level of arousal and emotional strength.
diopter A unit of measurement for the refractive power of the eye; the reciprocal of the focal distance.
direction selectivity The property of cells in the visual system that respond only when stimuli move within a limited range of directions.
distal muscle A muscle that controls the hands, feet, or digits.
distributed memory The concept that memories are encoded by widespread synaptic modifications of many neurons, not by a single synapse or cell.
DNA (deoxyribonucleic acid) A double-stranded molecule constructed from four nucleic acids that contains the genetic instructions for a cell.
DNMS *See* delayed non-match to sample.
dopa A chemical precursor of dopamine and the other catecholamines.
dopamine (DA) A catecholamine neurotransmitter synthesized from dopa.
dopamine hypothesis of schizophrenia A hypothesis suggesting that schizophrenia is caused by excessive activation of D2 receptors in the mesocorticolimbic dopamine system in the brain.
dorsal An anatomical reference meaning toward the back.
dorsal cochlear nucleus A nucleus in the medulla that receives afferents from the spiral ganglion in the cochlea of the inner ear.
dorsal column A white matter tract on the dorsal side of the spinal cord, carrying touch and proprioceptive axons to the brain stem.
dorsal column–medial lemniscal pathway An ascending somatic sensory pathway that mediates information about touch, pressure, vibration, and limb proprioception.
dorsal column nucleus One of a pair of nuclei located in the posterior medulla; target of dorsal column axons, mediating touch and proprioceptive input from the limbs and trunk.
dorsal horn The dorsal region of the spinal cord containing neuronal cell bodies.
dorsal longitudinal fasciculus A bundle of axons reciprocally connecting the hypothalamus and midbrain periaqueductal gray matter.

dorsal root A bundle of sensory neuron axons that emerges from a spinal nerve and attaches to the dorsal side of the spinal cord. Dorsal root axons bring information into the spinal cord. *See also* ventral root.
dorsal root ganglion A collection of cell bodies of the sensory neurons that are part of the somatic PNS. There is one dorsal root ganglion for each spinal nerve.
duplex theory of sound localization The principle that two schemes function in sound localization: interaural time delay at low frequencies and interaural intensity difference at high frequencies.
dura mater The outermost of the three meninges, the membranes that cover the surface of the central nervous system.
dyslexia Difficulty learning to read despite normal intelligence and training.

easy problems of consciousness Phenomena related to consciousness that can be studied by scientific methods; not the hard problem of the neural basis of conscious experience.
EEG *See* electroencephalogram.
efferent An axon originating in and coursing away from a given structure. *See also* afferent.
electrical conductance The relative ability of an electrical charge to migrate from one point to another, represented by the symbol g and measured in siemens (S). Conductance is the inverse of resistance and is related to electrical current and voltage by Ohm's law.
electrical current The rate of movement of electrical charge, represented by the symbol I and measured in amperes (amp).
electrical potential The force exerted on an electrically charged particle, represented by the symbol V and measured in volts; also called voltage or potential difference.
electrical resistance The relative inability of an electrical charge to migrate from one point to another, represented by the symbol R and measured in ohms (Ω). Resistance is the inverse of conductance and is related to electrical current and voltage by Ohm's law.
electrical self-stimulation Electrical stimulation that an animal can voluntarily deliver to a portion of its brain.
electrical synapse A synapse in which electrical current flows directly from one cell to another via a gap junction.
electroconvulsive therapy (ECT) A treatment for major depression that involves eliciting electrical seizure activity in the brain.
electroencephalogram (EEG) A measurement of electrical activity generated by the brain and recorded from the scalp.
endocannabinoid A natural (endogenous) chemical that binds to, and activates, cannabinoid (CB) receptors.
endocochlear potential The voltage difference between the endolymph and the perilymph, about 80 mV.
endocytosis The process by which a bit of the cell membrane is pinched off, internalized, and converted to an intracellular vesicle. *See also* exocytosis.
endogenous attention Attention voluntarily directed by the brain to serve a behavioral goal; also called top–down attention.
endolymph The fluid that fills the scala media in the cochlea of the inner ear, containing high K^+ and low Na^+ concentrations.
endorphin One of many endogenous opioid peptides with actions similar to those of morphine; present in many brain structures, particularly those related to pain.
engram The physical representation or location of a memory; also called memory trace.
enteric division A division of the autonomic nervous system that innervates the digestive organs; consists of the myenteric and submucous plexuses.
entorhinal cortex A cortical region in the medial temporal lobe that occupies the medial bank of the rhinal sulcus; provides input to the hippocampus.
ependymal cell A type of glial cell that provides the lining of the brain's ventricular system.
ephrin A protein secreted by neurons in many parts of the developing nervous system that helps establish topographic axonal connections.
epilepsy A chronic brain disorder characterized by recurrent seizures.
epinephrine A catecholamine neurotransmitter synthesized from norepinephrine; also called adrenaline.
EPSP *See* excitatory postsynaptic potential.
EPSP summation A simple form of synaptic integration whereby excitatory postsynaptic potentials combine to produce a larger postsynaptic depolarization.
equilibrium potential *See* ionic equilibrium potential.
estrogens Female steroidal hormones, the most important of which are estradiol and progesterone.
estrous cycle The female reproductive cycle in most nonprimate mammals in which there are periodic episodes of estrus or "heat."
eustachian tube An air-filled tube connecting the middle ear to the nasal cavities.
excitable membrane Any membrane capable of generating action potentials. The membrane of axons and muscle cells is excitable.
excitation–contraction coupling The physiological process by which the excitation of a muscle cell leads to its contraction.
excitatory postsynaptic potential (EPSP) Depolarization of the postsynaptic membrane potential by the action of a synaptically released neurotransmitter.

exocytosis The process whereby material is released from an intracellular vesicle into the extracellular space by fusion of the vesicle membrane with the cell membrane. *See also* endocytosis.

exogenous attention Attention reflexively directed to an external stimulus because of its salience; also called bottom-up attention.

extension The direction of movement that opens a joint.

extensor A muscle that causes extension when it contracts.

extracellular matrix The network of fibrous proteins deposited in the space between cells.

extrafusal fiber A muscle fiber in skeletal muscle that lies outside muscle spindles and receives innervation from alpha motor neurons.

extraocular muscle A muscle that moves the eye in the orbit.

falling phase The part of an action potential characterized by a rapid fall of membrane potential from positive to negative.

fasciculation A process in which axons growing together stick to one another.

fast motor unit A motor unit with a large alpha motor neuron innervating rapidly contracting and rapidly fatiguing white muscle fibers.

flexion The direction of movement that closes a joint.

flexor A muscle that causes flexion when it contracts.

follicle-stimulating hormone (FSH) A hormone secreted by the anterior pituitary; its diverse roles include the growth of follicles in the ovaries and the maturation of sperm in the testes.

forebrain The region of the brain derived from the rostral primary embryonic brain vesicle; also called prosencephalon. Forebrain structures include the telencephalon and the diencephalon.

fornix A bundle of axons that originates in the hippocampal formation, loops around the thalamus, and terminates in the diencephalon.

fourth ventricle The CSF-filled space within the hindbrain.

fovea The pit or depression in the retina at the center of the macula; in humans, the fovea contains only cone photoreceptors and is specialized for high-acuity vision.

frequency The number of sound waves or other discrete events per second, expressed in hertz (Hz).

frontal eye field (FEF) A cortical area in the frontal lobe involved in the generation of saccadic eye movements.

frontal lobe The region of the cerebrum lying anterior to the central sulcus under the frontal bone.

frontoparietal attention network A group of interconnected brain areas involved in guiding visual attention.

FSH *See* follicle-stimulating hormone.

GABA *See* gamma-aminobutyric acid.

GABAergic Describing neurons or synapses that produce and release gamma-aminobutyric acid.

gamma-aminobutyric acid (GABA) An amino acid synthesized from glutamate; the major inhibitory neurotransmitter in the central nervous system.

gamma motor neuron A motor neuron that innervates intrafusal muscle fibers.

ganglion A collection of neurons in the peripheral nervous system. Plural: ganglia.

ganglion cell A cell in the retina that receives input from bipolar cells and sends an axon into the optic nerve.

ganglion cell layer A layer of the retina closest to the center of the eye, containing ganglion cells.

gap junction A specialized junction where a narrow gap between two cells is spanned by protein channels (connexons) that allow ions to pass directly from one cell to another.

gating A property of many ion channels, making them open or closed in response to specific signals, such as membrane voltage or the presence of neurotransmitters.

gender identity A person's perception of his or her maleness or femaleness.

gene A unit of heredity; a sequence of DNA that encodes a single polypeptide or protein.

gene expression The process of transcribing the information from a gene into messenger RNA; a gene is a segment of DNA carrying the instructions for a single protein.

generalized seizure Pathologically large and synchronous neural activity that spreads to encompass the entire cerebral hemispheres. *See also* partial seizure.

genetic engineering The manipulation of an organism's genome by inserting or deleting DNA.

genetic sex The sex of an animal or person based solely on genotype.

genome The total content of an organism's genetic material.

genotype The genetic makeup of an animal or person.

ghrelin A peptide secreted by cells in the stomach that stimulates appetite by activating orexigenic neurons in the hypothalamus.

glial cell A support cell in the nervous system. Glia are classified into four categories: astrocytes, oligodendroglia, Schwann cells, and microglia. Astrocytes regulate the extracellular environment of the brain, oligodendroglia and Schwann cells provide myelin, and microglia scavenge debris.

globus pallidus A part of the basal ganglia in the basal forebrain; consists of external (GPe) and internal (GPi) segments. It is involved in motor control.

glomerulus A cluster of neurons in the olfactory bulb that receives input from olfactory receptor neurons.

glucocorticoid receptor A receptor activated by cortisol released from the adrenal gland.

glutamate (Glu) An amino acid; the major excitatory neurotransmitter in the central nervous system.

glutamate hypothesis of schizophrenia A hypothesis suggesting that schizophrenia is caused by the reduced activation of NMDA receptors in the brain.

glutamatergic Describing neurons or synapses that produce and release glutamate.

glycine (Gly) An amino acid; an inhibitory neurotransmitter at some locations in the central nervous system.

GnRH See gonadotropin-releasing hormone.

Goldman equation A mathematical relationship used to predict membrane potential from the concentrations and membrane permeabilities of ions.

Golgi apparatus An organelle that sorts and chemically modifies proteins that are destined for delivery to different parts of the cell.

Golgi stain A method of staining brain tissue that shows neurons and all of their neurites; named for its discoverer, Italian histologist Camillo Golgi (1843–1926).

Golgi tendon organ A specialized structure within the tendons of skeletal muscle that senses muscle tension.

gonadotropin-releasing hormone (GnRH) A hypophysiotropic hormone secreted by the hypothalamus; regulates the release of luteinizing hormone and follicle-stimulating hormone from the anterior pituitary.

gonadotropins Hormones secreted by the anterior pituitary that regulate the release of androgens and estrogens from the testes and ovaries.

G-protein A membrane-enclosed protein that binds guanosine triphosphate (GTP) when activated by a membrane receptor. Active G-proteins can stimulate or inhibit other membrane-enclosed proteins.

G-protein-coupled receptor A membrane protein that activates G-proteins when it binds neurotransmitter.

gray matter A generic term for a collection of neuronal cell bodies in the central nervous system. When a freshly dissected brain is cut open, neurons appear gray. *See also* white matter.

green fluorescent protein (GFP) A jellyfish protein that can be expressed in mammalian neurons by genetic engineering, causing these neurons to fluoresce bright green when illuminated by the appropriate wavelength of light.

grid cell Neurons in the entorhinal cortex that have multiple place fields arranged in a hexagonal grid.

growth cone The specialized tip of a growing neurite.

gustation The sense of taste.

gustatory nucleus A nucleus in the brain stem that receives primary taste input.

gyrus A bump or bulge lying between the sulci of the cerebrum. Plural: gyri.

habituation A type of nonassociative learning leading to decreased behavioral responses to repeated stimulation.

hair cell An auditory cell that transduces sound into a change in membrane potential or a vestibular cell that transduces head movements into a change in membrane potential.

hard problem of consciousness Why and how subjective conscious experiences arise from physical processes.

Hebb synapse A synapse that exhibits Hebbian modifications.

Hebbian modification An increase in the effectiveness of a synapse caused by the simultaneous activation of presynaptic and postsynaptic neurons.

helicotrema A hole at the apex of the cochlea in the inner ear that connects the scala tympani to the scala vestibuli.

hertz (Hz) The unit of sound frequency equivalent to cycles per second.

hindbrain The region of the brain derived from the caudal primary embryonic brain vesicle; also called rhombencephalon. Hindbrain structures include the cerebellum, pons, and medulla.

hippocampus A region of the cerebral cortex lying adjacent and medial to the olfactory cortex. In humans, the hippocampus is in the temporal lobe and plays important roles in learning and memory and the regulation of the hypothalamic-pituitary axis.

histology The microscopic study of the structure of tissues.

homeostasis The balanced functioning of physiological processes and maintenance of an organism's internal environment within a narrow range.

horizontal cell A cell in the retina of the eye that projects neurites laterally in the outer plexiform layer.

horizontal plane An anatomical plane of section that divides the nervous system into dorsal and ventral parts.

HPA axis *See* hypothalamic-pituitary-adrenal axis.

Huntington's disease A hereditary, progressive, inevitably fatal condition characterized by dyskinesias, dementia, and personality disorders; associated with profound degeneration of neurons in the basal ganglia and cerebral cortex.

hyperalgesia A reduced threshold for pain, an increased response to painful stimuli, or a spontaneous pain that follows localized injury.

hypophysiotropic hormone A peptide hormone, such as corticotropin-releasing hormone or gonadotropin-releasing hormone, released into the blood by the parvocellular neurosecretory cells of the hypothalamus; stimulates or inhibits the secretion of hormones from the anterior pituitary.

hypothalamic-pituitary-adrenal (HPA) axis A system of CNS neurons and endocrine cells that regulates the release of cortisol from the adrenal gland. Dysfunction of the HPA system has been implicated in anxiety disorders and affective disorders.

hypothalamo-pituitary portal circulation A system of blood vessels that carries hypophysiotropic hormones from the hypothalamus to the anterior pituitary.

hypothalamus The ventral part of the diencephalon, involved in the control of the autonomic nervous system and the pituitary gland.

immunocytochemistry An anatomical method that uses antibodies to study the location of molecules within cells.

incus An ossicle in the middle ear whose shape somewhat resembles an anvil.

induced pluripotent stem cells Stem cells with the potential to differentiate into any cell type, including neurons, that are chemically transformed from mature cells derived from a person.

inferior colliculus A nucleus in the midbrain from which all ascending auditory signals project to the medial geniculate nucleus.

inferior olive A nucleus of the medulla that gives rise to climbing fiber input to the cerebellar cortex. Climbing fiber activity is an important trigger for LTD, a form of synaptic plasticity believed to be important for motor learning.

inflammation A natural protective response of tissues to harmful stimuli. The cardinal signs of inflammation in skin include heat, redness, swelling, and pain.

inhibitor A drug or toxin that blocks the normal action of a protein or a biochemical process.

inhibitory postsynaptic potential (IPSP) A change in the postsynaptic membrane potential by the action of a synaptically released neurotransmitter, making the postsynaptic neuron less likely to fire action potentials.

inner ear The cochlea, which is part of the auditory system, plus the labyrinth, which is part of the vestibular system.

inner hair cell An auditory cell located between the modiolus and the rods of Corti; the primary transducer of sound into an electrochemical signal.

inner nuclear layer A layer of the retina of the eye containing the cell bodies of bipolar, horizontal, and amacrine cells.

inner plexiform layer A layer of the retina of the eye, located between the ganglion cell layer and the inner nuclear layer; contains the neurites and synapses between bipolar cells, amacrine cells, and ganglion cells.

innervation The provision of synaptic input to a cell or collection of cells.

inositol-1, 4, 5-triphosphate (IP_3) A second messenger molecule formed by the action of phospholipase C on the membrane phospholipid phosphatidylinositol-4, 5-bisphosphate. IP_3 causes the release of Ca^{2+} from intracellular stores.

input specificity A property of synapse plasticity that ensures that modifications induced by stimulation of one input onto a neuron do not spread to other unstimulated inputs on the same neuron.

***in situ* hybridization** A method for localizing strands of messenger RNA within cells.

instrumental conditioning A learning procedure in which a response, such as a motor act, is associated with a stimulus reward, such as food.

insula Cerebral cortex, also known as insular cortex, that lies within the lateral sulcus between the temporal and parietal lobes.

insulin A hormone released by the β cells of the pancreas; regulates blood glucose levels by controlling the expression of glucose transporters in the plasma membrane of non-neuronal cells.

intensity The amplitude of a sound wave. Sound intensity is the amplitude of the pressure differences in a sound wave that perceptually determines loudness.

internal capsule A large collection of axons that connects the telencephalon with the diencephalon.

internal resistance The resistance to electrical current flows longitudinally down a cable or neurite, represented by the symbol r_i.

interneuron Any neuron that is not a sensory or motor neuron; also describes a CNS neuron whose axon does not leave the structure in which it resides.

interstitial nuclei of the anterior hypothalamus (INAH) Four neuron clusters in the preoptic area of the anterior hypothalamus in humans, some of which may be sexually dimorphic.

intrafusal fiber A specialized muscle fiber within a muscle spindle that receives motor innervation from gamma motor neurons.

intrinsically photosensitive retinal ganglion cells Light-sensitive neurons in the ganglion cell layer of the retina that transduce light with the photopigment melanopsin.

ion An atom or molecule that has a net electrical charge because of a difference in the number of electrons and protons.

ion channel A membrane-spanning protein that forms a pore that allows the passage of ions from one side of the membrane to the other.

ion pump A protein that transports ions across a membrane at the expense of metabolic energy.

ion selectivity A property of ion channels that are selectively permeable to some ions and not to others.

ionic driving force The difference between the real membrane potential, V_m, and the ionic equilibrium potential, E_{ion}.

ionic equilibrium potential The electrical potential difference that exactly balances an ionic concentration gradient, represented by the symbol E_{ion}; also known as equilibrium potential.

IP_3 *See* inositol-1, 4, 5-triphosphate.

ipsilateral An anatomical reference meaning on the same side of the midline.

IPSP *See* inhibitory postsynaptic potential.

iris The circular, pigmented muscle that controls the size of the pupil in the eye.

James–Lange theory A theory proposing that the subjective experience of emotion is a consequence of physiological changes in the body.

kainate receptor A subtype of glutamate receptor; a glutamate-gated ion channel that is permeable to Na^+ and K^+.

Klüver–Bucy syndrome A constellation of symptoms resulting from bilateral temporal lobectomy in humans and monkeys that includes decreased fear and aggression (flattened emotions), the tendency to identify objects by oral examination rather than visual inspection, and altered sexual behavior.

knock-in mice Mice in which a gene has been replaced by another gene engineered to function differently.

knockout mice Mice in which a gene of interest has been silenced or deleted by genetic engineering.

koniocellular LGN layer A layer of the lateral geniculate nucleus containing very small cells, lying just ventral to each magnocellular and parvocellular layer.

Korsakoff's syndrome A neurological syndrome resulting from chronic alcoholism, characterized by confusion, confabulations, apathy, and amnesia.

language A system for communicating information that uses words or signs combined according to grammatical rules.

language acquisition The process by which humans learn to understand language and speak.

lateral An anatomical reference meaning away from the midline.

lateral geniculate nucleus (LGN) A thalamic nucleus that relays information from the retina to the primary visual cortex.

lateral hypothalamic area A poorly defined region of the hypothalamus that has been implicated in the motivation of behavior.

lateral hypothalamic syndrome Anorexia associated with lesions of the lateral hypothalamic area.

lateral intraparietal cortex (area LIP) A cortical area buried in the intraparietal sulcus that is involved in guiding eye movements; the responses of LIP neurons suggest that they are involved in working memory.

lateral pathway Axons in the lateral column of the spinal cord that are involved in the control of voluntary movements of the distal musculature and are under direct cortical control.

lateral ventricle The CSF-filled space within each cerebral hemisphere.

layer of photoreceptor outer segments A layer of the retina farthest from the center of the eye containing the light-sensitive elements of the photoreceptors.

learning The acquisition of new knowledge or skills.

lemniscus A tract that meanders through the brain like a ribbon.

length constant A parameter used to describe how far changes in membrane potential can passively spread down a cable such as an axon or a dendrite, represented by the symbol λ. The length constant λ is the distance at which the depolarization falls to 37% of its original value; it depends on the ratio of membrane resistance (r_m) to internal resistance (r_i).

lens The transparent structure lying between the aqueous humor and the vitreous humor that enables the eye to adjust its focus to different viewing distances.

leptin A protein hormone released by adipocytes (fat cells) that communicates with neurons of the arcuate nucleus of the hypothalamus.

LGN *See* lateral geniculate nucleus.

ligand-binding method A method that uses radioactive receptor ligands (agonists or antagonists) to locate neurotransmitter receptors.

light adaptation The process by which the retina becomes less sensitive to light in bright light conditions.

limbic lobe The hippocampus and cortical areas bordering the brain stem in mammals, which Broca proposed as a distinct lobe of the brain.

limbic system A group of structures, including those in the limbic lobe and Papez circuit, that are anatomically interconnected and are probably involved in emotion, learning, and memory.

lipostatic hypothesis A hypothesis proposing that body fat is maintained homeostatically at a specific level.

lithium An element, existing in solution as a monovalent cation, that is effective in the treatment of bipolar disorder.

locus A small, well-defined group of cells. Plural: loci.

locus coeruleus A small nucleus located bilaterally in the pons; using NE as their neurotransmitter, its neurons project widely to all levels of the CNS.

long-term depression (LTD) A long-lasting decrease in the effectiveness of synaptic transmission that follows certain types of conditioning stimulation.

long-term memory Information storage that is relatively permanent and does not require continual rehearsal.

long-term potentiation (LTP) A long-lasting enhancement of the effectiveness of synaptic transmission that follows certain types of conditioning stimulation.

LTD *See* long-term depression.

LTP *See* long-term potentiation.

luteinizing hormone (LH) A hormone secreted by the anterior pituitary; its diverse roles include the stimulation of testosterone production in males and the facilitation of follicle development and ovulation in females.

M1 Primary motor cortex, area 4.

macula (1) In the eye, a yellowish spot in the middle of the retina with relatively few large blood vessels; contains the fovea. (2) In the ear, a sensory epithelium in the otolith organs whose hair cells transduce head tilt and acceleration.

magnetoencephalography (MEG) A measurement of electrical activity generated by the brain and recorded by detecting associated magnetic field fluctuations with sensors surrounding the scalp.

magnocellular LGN layer A layer of the lateral geniculate nucleus receiving synaptic input from M-type retinal ganglion cells.

magnocellular neurosecretory cell A large neuron of the periventricular and supraoptic nuclei of the hypothalamus that projects to the posterior pituitary and secretes oxytocin or vasopressin into the blood.

magnocellular pathway A visual information-processing pathway that begins with M-type retinal ganglion cells and leads to layer IVB of striate cortex; believed to process information about object motion and motor actions.

major depression An affective disorder characterized by prolonged, severe impairment of mood; may include anxiety, sleep disturbances, and other physiological disturbances.

malleus An ossicle in the middle ear attached to the tympanic membrane; shaped somewhat like a hammer.

mania An elevated, expansive, or irritable mood that is characteristic of bipolar disorder.

Marr–Albus theory of motor learning The theory that parallel fiber synapses on Purkinje cells are modified when their activity coincides with climbing fiber activity.

mechanoreceptor Any sensory receptor selective for mechanical stimuli, such as hair cells of the inner ear, various receptors of the skin, and stretch receptors of skeletal muscle.

medial An anatomical reference meaning toward the midline.

medial forebrain bundle A large bundle of axons coursing through the hypothalamus carrying efferents from the dopaminergic, noradrenergic, and serotonergic neurons in the brain stem and fibers interconnecting the hypothalamus, limbic structures, and midbrain tegmental area.

medial geniculate nucleus (MGN) A relay nucleus in the thalamus through which all auditory information passes on its way from the inferior colliculus to the auditory cortex.

medial lemniscus A white matter tract of the somatic sensory system carrying axons from dorsal column nuclei to the thalamus.

medulla oblongata The part of the hindbrain caudal to the pons and cerebellum; also called medulla.

medullary reticulospinal tract A tract originating in the medullary reticular formation and terminating in the spinal cord; involved in the control of movement.

membrane differentiation A dense accumulation of protein adjacent to and within the membrane on either side of a synaptic cleft.

membrane potential The voltage across a cell membrane; represented by the symbol Vm.

membrane resistance The resistance to electrical current flow across a membrane; represented by the symbol rm.

memory The retention of learned information.

memory consolidation The process by which short-term memories lasting hours to days are converted into long-term memories lasting weeks to years.

memory trace The physical representation or location of a memory; also known as an engram.

meninges Three membranes that cover the surface of the central nervous system: the dura mater, arachnoid membrane, and pia mater. Singular: meninx.

menstrual cycle The female reproductive cycle in primates.

messenger RNA (mRNA) A molecule constructed from four nucleic acids that carries the genetic instructions for the assembly of a protein from the nucleus to the cytoplasm.

metabotropic receptor A G-protein-coupled receptor whose primary action is to stimulate an intracellular biochemical response.

metaplasticity Activity-dependent modification of the rules of synaptic plasticity.

MGN *See* medial geniculate nucleus.

microelectrode A probe used to measure the electrical activity of cells. Microelectrodes have a very fine tip and can be fashioned from etched metal or glass pipettes filled with electrically conductive solutions.

microfilament A polymer of the protein actin, forming a braided strand 5 nm in diameter; a component of the cytoskeleton.

microglial cell A type of cell that functions as a phagocyte in the nervous system to remove debris left by dead or dying neurons and glia.

microionophoresis A method of applying drugs and neurotransmitters in very small quantities to cells.

microtubule A polymer of the protein tubulin, forming a straight, hollow tube 20 nm in diameter. Microtubules, a component of the cytoskeleton, play an important role in axoplasmic transport.

midbrain The region of the brain derived from the middle primary embryonic brain vesicle; also called mesencephalon. Midbrain structures include the tectum and the tegmentum.

middle ear The tympanic membrane plus the ossicles.

midline An invisible line that bisects the nervous system into right and left halves.

midsagittal plane An anatomical plane of section through the midline that is perpendicular to the ground. A section in the midsagittal plane divides the nervous system into right and left halves.

miniature postsynaptic potential A change in postsynaptic membrane potential caused by the action of neurotransmitter released from a single synaptic vesicle.

mirror neuron A neuron of the cerebral cortex that fires when an animal performs a motor act and when the animal merely observes the same act performed by another.

mitochondrion An organelle responsible for cellular respiration. Mitochondria generate adenosine triphosphate using the energy produced by the oxidation of food.

modulation A term used to describe the actions of neurotransmitters that do not directly evoke postsynaptic potentials but modify the cellular response to excitatory postsynaptic potentials and inhibitory postsynaptic potentials generated by other synapses.

molecular medicine The approach of using genetic information to develop medical treatments for disease.

molecular switch hypothesis The idea that protein kinases can be switched "on" by autophosphorylation to a state in which they no longer require the presence of a specific second messenger to be active. Such persistently active kinases may hold the memory of an episode of strong synaptic activation. Initially proposed by John Lisman at Brandeis University.

monoamine hypothesis of mood disorders A hypothesis suggesting that depression is a consequence of a reduction in the levels of monoamine neurotransmitters, particularly serotonin and norepinephrine, in the brain.

monocular deprivation An experimental manipulation that deprives one eye of normal vision.

monogamy Mating behavior in which two individuals form a tightly bound relationship that includes exclusive or nearly exclusive mating with each other.

Morris water maze A task used to assess spatial memory in which a rodent must swim to a hidden platform below the surface of a pool of water.

mossy fiber An axon of a pontine neuron that innervates cerebellar granule cells. This term is also used to describe the axons from dentate gyrus granule cells that innervate area CA3 of the hippocampus.

motivated behavior Behavior that is incited to achieve a goal.

motor cortex Cortical areas 4 and 6, which are directly involved in the control of voluntary movement.

motor end-plate The postsynaptic membrane at the neuromuscular junction.

motor neuron A neuron that synapses on a muscle cell and causes muscle contraction.

motor neuron pool All the alpha motor neurons innervating the fibers of a single skeletal muscle.

motor strip A name for area 4 on the precentral gyrus; also called primary motor cortex.

motor system All skeletal muscles and the parts of the central nervous system that control them.

motor unit One alpha motor neuron and all the skeletal muscle fibers it innervates.

mRNA *See* messenger RNA.

M-type ganglion cell A type of ganglion cell in the retina characterized by a large cell body and dendritic arbor, a transient response to light, and no sensitivity to different wavelengths of light; also called M cell.

multiple trace model of consolidation An alternative to the standard model of memory consolidation in which the hippocampus participates indefinitely in memory storage along with the neocortex; in this model, each time an episodic memory is recalled in a new context, an additional memory trace is formed.

multipolar neuron A neuron with three or more neurites.

muscarinic ACh receptor A subtype of acetylcholine receptor that is G-protein-coupled.

muscle fiber A multinucleated skeletal muscle cell.

muscle spindle A specialized structure within skeletal muscles that senses muscle length; provides sensory information to neurons in the spinal cord via group Ia axons; also called stretch receptor.

myelin A membranous wrapping, or sheath, around axons provided by oligodendroglia in the central nervous system and Schwann cells in the peripheral nervous system.
myofibril A cylindrical structure within a skeletal muscle fiber that contracts in response to an action potential.
myosin A cytoskeletal protein in all cells and the major thick filament protein in a skeletal muscle fiber; causes muscle contraction by chemical interaction with actin.
myotatic reflex A reflex that leads to muscle contraction in response to muscle stretch, mediated by the monosynaptic connection between group Ia axons from a muscle spindle and an alpha motor neuron innervating the same muscle; also called stretch reflex.

NE *See* norepinephrine.
neglect syndrome A neurological disorder in which a part of the body or a part of the visual field is ignored or suppressed; most commonly associated with damage to posterior parietal areas of the brain.
neocortex The cerebral cortex, with six or more layers of neurons, found only in mammals.
Nernst equation A mathematical relationship used to calculate an ionic equilibrium potential.
nerve A bundle of axons in the peripheral nervous system.
nerve growth factor (NGF) A neurotrophin required for survival of the cells of the sympathetic division of the ANS; also important for aspects of CNS development.
netrin An axonal guidance module; a protein secreted by cells in specific locations in the developing CNS that can act as either an axonal attractant or a repellent, depending on the type of netrin receptor expressed on the growing axon.
neural correlates of consciousness The minimal neuronal events sufficient for a specific conscious percept.
neural crest The primitive embryonic peripheral nervous system, consisting of neural ectoderm that pinches off laterally as the neural tube forms.
neural precursor cell An immature neuron before cell differentiation.
neural tube The primitive embryonic central nervous system, consisting of a tube of neural ectoderm.
neurite A thin tube extending from a neuronal cell body; the two types are axons and dendrites.
neurofilament A type of intermediate filament found in neurons, 10 nm in diameter; an important component of the neuronal cytoskeleton.
neurohormone A hormone released by neurons into the bloodstream.
neuroleptic drug An antipsychotic drug used to treat schizophrenia by blocking dopamine receptors; examples are chlorpromazine and clozapine.
neuromuscular junction A chemical synapse between a spinal motor neuron axon and a skeletal muscle fiber.
neuron The information-processing cell of the nervous system; also called nerve cell. Most neurons use action potentials to send signals over a distance, and all neurons communicate with one another using synaptic transmission.
neuron doctrine The concept that the neuron is the elementary functional unit of the brain and that neurons communicate with each other by contact, not continuity.
neuronal membrane The barrier, about 5 nm thick, that separates the inside of a nerve cell from the outside; consists of a phospholipid bilayer with proteins embedded in it; encloses the intracellular organelles and vesicles.
neuropharmacology The study of the effects of drugs on nervous system tissue.
neurotransmitter A chemical released by a presynaptic element upon stimulation that activates postsynaptic receptors.
neurotrophin A member of a family of related neuronal trophic factors, including nerve growth factor and brain-derived neurotrophic factor.
neurulation The formation of the neural tube from the neural ectoderm during embryonic development.
NGF *See* nerve growth factor.
nicotinic ACh receptor A class of acetylcholine-gated ion channel found in various locations, notably at the neuromuscular junction.
Nissl stain A class of basic dyes that stain the somata of neurons; named for its discoverer, German histologist Franz Nissl (1860–1919).
nitric oxide (NO) A gas produced from the amino acid arginine that serves as an intercellular messenger.
NMDA receptor A subtype of glutamate receptor; a glutamate-gated ion channel that is permeable to Na^+, K^+, and Ca^{2+}. Inward ionic current through the *N*-methyl-D-aspartate receptor is voltage dependent because of a magnesium block at negative membrane potentials.
nociceptor Any receptor selective for potentially harmful stimuli; may induce sensations of pain.
node of Ranvier A space between two consecutive myelin sheaths where an axon comes in contact with the extracellular fluid.
nonassociative learning A change in the behavioral response that occurs over time in response to a single type of stimulus; the two types are habituation and sensitization.
nondeclarative memory Memory for skills, habits, emotional responses, and some reflexes.
nonM–nonP ganglion cell A ganglion cell in the retina that is not of the M type or P type, based on cell morphology and response properties. Of the variety of cell types in this category, some are known to be sensitive to the wavelength of light.

non-REM sleep A stage of sleep characterized by large, slow EEG waves, a paucity of dreams, and some muscle tone. *See also* rapid eye movement sleep.

noradrenergic Describing neurons or synapses that produce and release norepinephrine.

norepinephrine (NE) A catecholamine neurotransmitter synthesized from dopamine; also called noradrenaline.

nucleus (1) The roughly spherical organelle in a cell body containing the chromosomes. (2) A clearly distinguishable mass of neurons, usually deep in the brain.

nucleus of the solitary tract A brain stem nucleus that receives sensory input and uses it to coordinate autonomic function via its outputs to other brain stem and forebrain nuclei and to the hypothalamus.

obesity A state of positive energy balance in which energy intake and storage exceed energy expenditure, resulting in an increase in body fat. *See also* starvation.

obsessive-compulsive disorder (OCD) A mental disorder that includes obsessions (recurrent, intrusive thoughts, images, ideas, or impulses that are perceived as being inappropriate, grotesque, or forbidden) and compulsions (repetitive behaviors or mental acts that are performed to reduce the anxiety associated with obsessions).

occipital lobe The region of the cerebrum lying under the occipital bone.

OCD *See* obsessive-compulsive disorder.

ocular dominance column A region of striate cortex receiving information predominantly from one eye.

ocular dominance shift A change in visual cortex interconnections that makes neurons more responsive to one eye or the other.

OFF bipolar cell A bipolar cell of the retina that depolarizes in response to dark (light OFF) in the center of its receptive field.

Ohm's law The relationship between electrical current (I), voltage (V), and conductance (g): $I = gV$. Because electrical conductance is the inverse of resistance (R), Ohm's law may also be written: $V = IR$.

olfaction The sense of smell.

olfactory bulb A bulb-shaped brain structure derived from the telencephalon that receives input from olfactory receptor neurons.

olfactory cortex The region of the cerebral cortex connected to the olfactory bulb and separated from the neocortex by the rhinal fissure.

olfactory epithelium A sheet of cells lining part of the nasal passages that contains olfactory receptor neurons.

oligodendroglial cell A glial cell that provides myelin in the central nervous system.

ON bipolar cell A bipolar cell of the retina that depolarizes in response to light (light ON) in the center of its receptive field.

opioid receptor A membrane protein that selectively binds natural (e.g., endorphin) and synthetic (e.g., morphine) opioid substances.

opioids A class of drugs, including morphine, codeine, and heroin, that can produce analgesia as well as mood changes, drowsiness, mental clouding, nausea, vomiting, and constipation.

optic chiasm The structure in which the right and left optic nerves converge and partially decussate (cross) to form the optic tracts.

optic disk The location on the retina where optic nerve axons leave the eye.

optic nerve The bundle of ganglion cell axons that passes from the eye to the optic chiasm.

optic radiation A collection of axons coursing from the lateral geniculate nucleus to the visual cortex.

optic tectum A term used to describe the superior colliculus, particularly in nonmammalian vertebrates.

optic tract A collection of retinal ganglion cell axons stretching from the optic chiasm to the brain stem. Important targets of the optic tract are the lateral geniculate nucleus and superior colliculus.

optogenetics A method that allows the control of neuronal activity, comprising the introduction of foreign genes into neurons that express membrane ion channels that open in response to light.

orexigenic peptide A neuroactive peptide that stimulates feeding behavior; examples are neuropeptide Y (NPY), agouti-related peptide (AgRP), melanin-concentrating hormone (MCH), and orexin.

organ of Corti An auditory receptor organ that contains hair cells, rods of Corti, and supporting cells.

organelle A membrane-enclosed structure inside a cell; examples are the nucleus, mitochondrion, endoplasmic reticulum, and Golgi apparatus.

organizational effect The ability of a hormone to influence the prenatal development of sex organs and the brain.

orientation column A column of visual cortical neurons stretching from layer II to layer VI that responds best to the same stimulus orientation.

orientation selectivity The property of a cell in the visual system that responds to a limited range of stimulus orientations.

osmometric thirst The motivation to drink water as a result of an increase in blood tonicity.

ossicle One of three small bones in the middle ear.

otolith organ The utricle or the saccule, organs of the vestibular labyrinth in the inner ear that transduce head tilt and acceleration.

outer ear The pinna plus the auditory canal.

outer hair cell An auditory receptor cell located farther from the modiolus than the rods of Corti in the inner ear.

outer nuclear layer A layer of the retina of the eye containing the cell bodies of photoreceptors.

outer plexiform layer A layer of the retina of the eye between the inner nuclear layer and the outer nuclear layer; contains the neurites and synapses between photoreceptors, horizontal cells, and bipolar cells.

oval window A hole in the bony cochlea of the inner ear, where movement of the ossicles is transferred to movement of the fluids in the cochlea.

overshoot The part of an action potential when the membrane potential is more positive than 0 mV.

oxytocin A small peptide hormone released from the posterior pituitary by magnocellular neurosecretory cells; stimulates uterine contractions and milk letdown from the mammary glands.

Pacinian corpuscle A mechanoreceptor of the deep skin, selective for high-frequency vibrations.

PAG *See* periaqueductal gray matter.

panic disorder A mental disorder characterized by recurring, seemingly unprovoked panic attacks and a persistent worry about having further attacks.

Papez circuit A circuit of structures interconnecting the hypothalamus and cortex, proposed by Papez to be an emotion system.

papilla A small protuberance on the surface of the tongue that contains taste buds.

parahippocampal cortex A cortical region in the medial temporal lobe that lies lateral to the rhinal sulcus.

parallel fiber An axon of a cerebellar granule cell that innervates Purkinje cells. Plasticity of the synapse between a parallel fiber and a Purkinje cell is believed to be important for motor learning.

parallel processing The idea that different stimulus attributes are processed by the brain in parallel using distinct pathways.

parasympathetic division A division of the autonomic nervous system that maintains heart rate and respiratory, metabolic, and digestive functions under normal conditions; its peripheral axons emerge from the brain stem and sacral spinal cord. *See also* sympathetic division.

paraventricular nucleus A region of the hypothalamus involved in the regulation of the autonomic nervous system and in controlling the secretion of thyroid-stimulating hormone and adrenocorticotropic hormone from the anterior pituitary.

parietal lobe The region of the cerebrum lying under the parietal bone.

Parkinson's disease A movement disorder caused by damage to the substantia nigra, characterized by paucity of movement, difficulty in initiating willed movement, and resting tremor.

partial seizure Pathologically large and synchronous neural activity that remains localized to a relatively small region of the brain. *See also* generalized seizure.

parvocellular LGN layer A layer of the lateral geniculate nucleus receiving synaptic input from P-type retinal ganglion cells.

parvocellular neurosecretory cell A small neuron of the medial and periventricular hypothalamus that secretes hypophysiotropic peptide hormones into the hypothalamo-pituitary portal circulation to stimulate or inhibit the release of hormones from the anterior pituitary.

parvo-interblob pathway A visual information-processing pathway that begins with P-type retinal ganglion cells and leads to the interblob regions of striate cortical layer III; believed to process information about fine object shape.

patch clamp A method that enables an investigator to hold constant the membrane potential of a patch of membrane while current through membrane channels is measured.

pathophysiology Abnormal physiology that causes the symptoms of disease.

PDE *See* phosphodiesterase.

peptide bond The covalent bond between the amino group of one amino acid and the carboxyl group of another.

peptidergic Describing neurons or synapses that produce and release peptide neurotransmitters.

perforant path The axonal pathway from the entorhinal cortex to the dentate gyrus of the hippocampus. Perforant path synapses exhibit LTP and LTD, forms of synaptic plasticity believed to be important for memory formation.

periaqueductal gray matter (PAG) A region surrounding the cerebral aqueduct in the core of the midbrain, with descending pathways that can inhibit the transmission of pain-causing signals.

perikaryon The central region of the neuron containing the nucleus; also called soma or cell body.

perilymph The fluid that fills the scala vestibuli and scala tympani in the cochlea in the inner ear, containing low K^+ and high Na^+ concentrations.

peripheral nervous system (PNS) The parts of the nervous system other than the brain and spinal cord. The PNS includes all the spinal ganglia and nerves, cranial nerves III–XII, and the autonomic nervous system. *See also* central nervous system.

perirhinal cortex A cortical region in the medial temporal lobe that occupies the lateral bank of the rhinal sulcus. Lesions to this area in humans produce profound anterograde amnesia.

periventricular zone A hypothalamic region that lies most medially, bordering the third ventricle.

phase locking The consistent firing of an auditory neuron at the same phase of a sound wave.

pheromone An olfactory stimulus used for chemical communication between individuals.

phonemes The set of distinct sounds used in a language.

phosphodiesterase (PDE) An enzyme that breaks down the cyclic nucleotide second messengers cyclic adenosine monophosphate (cAMP) and cyclic guanosine monophosphate (cGMP).

phospholipase C (PLC) An enzyme that cleaves the membrane phospholipid phosphatidylinositol-4, 5-bisphosphate to form the second messengers diacylglycerol (DAG) and inositol triphosphate (IP_3).

phospholipid bilayer The arrangement of phospholipid molecules that forms the basic structure of the cell membrane. The core of the bilayer is lipid, creating a barrier to water and to water-soluble ions and molecules.

phosphorylation A biochemical reaction in which a phosphate group (PO_4^{2-}) is transferred from adenosine triphosphate (ATP) to another molecule. Phosphorylation of proteins by protein kinases changes their biological activity.

photoreceptor A specialized cell in the retina that transduces light energy into changes in membrane potential.

pia mater The innermost of the three meninges, the membranes that cover the surface of the central nervous system.

pinna The funnel-shaped outer ear, consisting of cartilage covered by skin.

pitch The perceptual quality of a sound determined by the sound's frequency.

PKA *See* protein kinase A.

PKC *See* protein kinase C.

place cell A neuron in the rat hippocampus that responds only when the animal is in a certain region of space.

planum temporale An area on the superior surface of the temporal lobe that is frequently larger in the left than in the right hemisphere.

PLC *See* phospholipase C.

PNS *See* peripheral nervous system.

polyandry Mating behavior in which one female mates with more than one male.

polygyny Mating behavior in which one male mates with more than one female.

polypeptide A string of amino acids held together by peptide bonds.

polyribosome A collection of several ribosomes floating freely in the cytoplasm.

pons The part of the rostral hindbrain that lies ventral to the cerebellum and the fourth ventricle.

pontine nuclei The clusters of neurons that relay information from the cerebral cortex to the cerebellar cortex.

pontine reticulospinal tract A tract originating in the pontine reticular formation and terminating in the spinal cord, involved in the control of movement.

population coding The representation of sensory, motor, or cognitive information by activity distributed over a large number of neurons. An example is color, which is encoded by the relative activity of the types of retinal cones.

posterior An anatomical reference meaning toward the tail or caudal.

posterior parietal cortex The posterior region of the parietal lobe, mainly Brodmann's areas 5 and 7, involved in visual and somatosensory integration and attention.

postganglionic neuron A peripheral neuron of the sympathetic and parasympathetic divisions of the autonomic nervous system; its cell body lies in autonomic ganglia, and its axons terminate on peripheral organs and tissues.

postsynaptic density A postsynaptic membrane differentiation that is the site of neurotransmitter receptors.

postsynaptic potential (PSP) A change in the postsynaptic membrane potential by the presynaptic action of an electrical synapse or a synaptically released neurotransmitter.

predatory aggression Attack behavior, often with the goal of obtaining food, accompanied by few vocalizations and low ANS activity.

prefrontal cortex A cortical area at the rostral end of the frontal lobe that receives input from the dorsomedial nucleus of the thalamus.

preganglionic neuron A neuron of the sympathetic and parasympathetic divisions of the autonomic nervous system; its cell body lies in the CNS (spinal cord or brain stem), and its axons extend peripherally to synapse on postganglionic neurons in the autonomic ganglia.

premotor area (PMA) The lateral part of cortical area 6, involved in the control of voluntary movement.

primary auditory cortex Brodmann's area 41 on the superior surface of the temporal lobe; also called A1.

primary gustatory cortex The area of neocortex that receives taste information from the ventroposterior medial nucleus.

primary motor cortex Brodmann's area 4, located on the precentral gyrus; the region of cortex that, when weakly stimulated, elicits localized muscle contractions; also called M1.

primary sensory neuron A neuron specialized to detect environmental signals at the body's sensory surfaces.

primary somatosensory cortex Brodmann's area 3b located in the postcentral gyrus; also called S1.

primary visual cortex Brodmann's area 17, located at the pole of the occipital lobe; also called striate cortex and V1.

priority map A map of visual space that shows locations where attention should be directed based on stimulus salience or cognitive input.
procedural memory Memory for skills and behaviors.
promoter A region of DNA that binds RNA polymerase to initiate gene transcription.
proprioception The sensation of body position and movement using sensory signals from muscles, joints, and skin.
proprioceptor A sensory receptor from the muscles, joints, and skin that contributes to proprioception.
protein A polymer of amino acids strung together by peptide bonds.
protein kinase A class of enzyme that phosphorylates proteins, a reaction that changes the conformation of the protein and its biological activity.
protein kinase A (PKA) A protein kinase activated by the second messenger cAMP.
protein kinase C (PKC) A protein kinase activated by the second messenger DAG.
protein phosphatase An enzyme that removes phosphate groups from proteins.
protein synthesis The assembly of protein molecules in the cell's cytoplasm according to genetic instructions.
proximal (girdle) muscle A muscle that controls the shoulder or pelvis.
psychological constructionist theories of emotion Explanations of emotions in which each emotion is an emergent consequence of combining non-emotional psychological components such as body sensations and attention.
psychosurgery Brain surgery used to treat mental or behavioral disorders.
P-type ganglion cell A type of ganglion cell in the retina characterized by a small cell body and dendritic arbor, a sustained response to light, and sensitivity to different wavelengths of light; also called P cell.
pulvinar nucleus A mass of neurons in the posterior thalamus that have widespread reciprocal connections with areas across the cerebral cortex.
pupil The opening that allows light to enter the eye and strike the retina.
pupillary light reflex An adjustment by the pupil to different levels of ambient light; the pupil's diameter becomes larger in dim light and smaller in bright light, in response to retinal inputs to brain stem neurons that control the iris.
Purkinje cell A cell in the cerebellar cortex that projects an axon to the deep cerebellar nuclei.
putamen A part of the basal ganglia in the basal forebrain; involved in motor control.
pyramidal cell A neuron characterized by a pyramid-shaped cell body and elongated dendritic tree; found in the cerebral cortex.
pyramidal tract A tract running along the ventral medulla that carries corticospinal axons.

quantal analysis A method of determining how many vesicles release neurotransmitter during normal synaptic transmission.

radial glial cell A glial cell in the embryonic brain extending a process from the ventricular zone to the surface of the brain; immature neurons and glia migrate along this process.
raphe nuclei Clusters of serotonergic neurons that lie along the midline of the brain stem from the midbrain to the medulla and project diffusely upon all levels of the CNS.
rapid eye movement sleep (REM sleep) A stage of sleep characterized by low-amplitude, high-frequency EEG waves, vivid dreams, rapid eye movements, and atonia. *See also* non-REM sleep.
rate-limiting step In the series of biochemical reactions that leads to the production of a chemical, the one step that limits the rate of synthesis.
receptive field The region of a sensory surface (retina, skin) that, when stimulated, changes the membrane potential of a neuron.
receptor (1) A specialized protein that detects chemical signals, such as neurotransmitters, and initiates a cellular response. (2) A specialized cell that detects environmental stimuli and generates neural responses.
receptor agonist A drug that binds to a receptor and activates it.
receptor antagonist A drug that binds to a receptor and inhibits its function.
receptor potential A stimulus-induced change in the membrane potential of a sensory receptor.
receptor subtype One of several receptors to which a neurotransmitter binds.
reciprocal inhibition The process whereby the contraction of one set of muscles is accompanied by the relaxation of antagonist muscles.
recognition memory Memory required to perform a delayed non-match to sample task.
reconsolidation The process of retrieving, modifying, and storing a memory that was previously consolidated.
red nucleus A cell group in the midbrain involved in the control of movement.
referred pain Pain that is perceived as coming from a site other than its true origin. Nociceptor activation within visceral organs is typically perceived as pain originating in skin or skeletal muscle.
refraction The bending of light rays that can occur when they travel from one transparent medium to another.

Reissner's membrane The cochlear membrane in the inner ear that separates the scala vestibuli from the scala media.

relational memory A type of memory in which all of the events occurring at a given time are stored in a manner linking them.

relative refractory period The period of time following an action potential during which more depolarizing current than usual is required to achieve threshold.

REM sleep *See* rapid eye movement sleep.

resistance *See* electrical resistance.

resting membrane potential The membrane potential, or membrane voltage, maintained by a cell when it is not generating action potentials; also called resting potential. Neurons have a resting membrane potential of about –65 mV.

resting state activity Activity in the brain during quiet restful wakefulness.

reticular formation A region of the brain stem ventral to the cerebral aqueduct and fourth ventricle; involved in many functions, including the control of posture and locomotion.

reticular lamina A thin sheet of tissue in the inner ear that holds the tops of hair cells in the organ of Corti.

retina A thin layer of cells at the back of the eye that transduces light energy into neural activity.

retinofugal projection A neural pathway that carries information away from the eye.

retinotectal projection A neural pathway that carries information from the retina to the superior colliculus.

retinotopy The topographic organization of visual pathways in which neighboring cells on the retina send information to neighboring cells in a target structure.

retrograde amnesia Memory loss for events before an illness or brain trauma.

retrograde messenger Any chemical messenger that communicates information from the postsynaptic side of a synapse to the presynaptic side.

retrograde transport Axoplasmic transport from an axon terminal to the soma.

rhodopsin The photopigment in rod photoreceptors.

ribosome A cellular organelle that assembles new proteins from amino acids according to the instructions carried by messenger RNA.

rising phase The first part of an action potential, characterized by a rapid depolarization of the membrane.

RNA splicing The process by which introns, the regions of a primary RNA transcript that are not used to code protein, are removed.

rod photoreceptor A photoreceptor in the retina containing rhodopsin and specialized for low light levels. *See also* cone photoreceptor.

rostral An anatomical reference meaning toward the nose or anterior.

rough endoplasmic reticulum (rough ER) A membrane-enclosed cellular organelle with ribosomes attached to its outer surface; a site of synthesis for proteins destined to be inserted into membrane or to be enclosed by membrane.

round window A membrane-covered hole in the bony cochlea of the inner ear that is continuous with the scala tympani in the cochlea.

rubrospinal tract A tract originating in the red nucleus and terminating in the spinal cord; involved in the control of movement.

S1 *See* primary somatosensory cortex.

sagittal plane An anatomical plane of section that is parallel to the midsagittal plane.

salience map A map of visual space that highlights the locations of conspicuous objects.

saltatory conduction The propagation of an action potential down a myelinated axon.

sarcolemma The outer cell membrane of a skeletal muscle fiber.

sarcomere The contractile element between Z lines in a myofibril; contains the thick and thin filaments that slide along one another to cause muscle contraction.

sarcoplasmic reticulum (SR) An organelle within a skeletal muscle fiber that stores Ca^{2+} and releases it when stimulated by an action potential in T tubules.

satiety signal A factor that reduces the drive to eat without causing sickness; examples are gastric distension and cholecystokinin released by the intestinal cells in response to food.

scala media A chamber in the cochlea that lies between the scala vestibuli and the scala tympani.

scala tympani A chamber in the cochlea that runs from the helicotrema to the round window.

scala vestibuli A chamber in the cochlea that runs from the oval window to the helicotrema.

Schaffer collateral An axon of a CA3 neuron that innervates neurons in CA1 of the hippocampus. Schaffer collateral synapses exhibit LTP and LTD, forms of synaptic plasticity believed to be important for memory formation.

schizophrenia A mental disorder characterized by a loss of contact with reality; fragmentation and disruption of thought, perception, mood, and movement; delusions; hallucinations; and disordered memory.

Schwann cell A glial cell that provides myelin in the peripheral nervous system.

sclera The tough outer wall of the eyeball; the white of the eye.

SCN *See* suprachiasmatic nucleus.

second messenger A short-lived chemical signal in the cytosol that can trigger a biochemical response. Second messenger formation is usually stimulated by a first messenger (a neurotransmitter or hormone) acting at a G-protein-coupled cell surface receptor. Examples of second messengers are cyclic adenosine monophosphate (cAMP), cyclic guanosine monophosphate (cGMP), and inositol-1,4,5-triphosphate (IP_3).

second messenger cascade A multistep process that couples activation of a neurotransmitter receptor to activation of intracellular enzymes.

secretory granule A spherical membrane-enclosed vesicle about 100 nm in diameter containing peptides intended for secretion by exocytosis; also called dense-core vesicle.

semicircular canal A component of the vestibular labyrinth in the inner ear that transduces head rotation.

sensitization A type of nonassociative learning leading to an intensified response to all stimuli.

sensory map A representation of sensory information within a neural structure that preserves the spatial organization of that information established on the sensory organ. Examples are retinotopic maps in the superior colliculus, lateral geniculate nucleus, and visual cortex, where neurons in specific places respond selectively to stimulation of specific parts of the retina.

serotonergic Describing neurons or synapses that produce and release serotonin.

serotonin (5-HT) An amine neurotransmitter, 5-hydroxytryptamine.

serotonin deficiency hypothesis The idea that aggression is inversely related to serotonergic activity.

serotonin-selective reuptake inhibitor (SSRI) A drug, such as fluoxetine (Prozac), that prolongs the actions of synaptically released serotonin by preventing reuptake; used to treat depression and obsessive-compulsive disorder.

sex-determining region of the Y chromosome (SRY) A gene on the Y chromosome that codes for testis-determining factor; essential for normal male development.

sexual dimorphism A sex-related difference in structure or behavior.

sexually dimorphic nucleus (SDN) A neuron cluster in the preoptic area of the anterior hypothalamus that in rats is significantly larger in males than in females.

sham rage A display of great anger in a situation that would not normally cause anger; behavior produced by brain lesions.

short-term memory Retention of information about recent events or facts that is not yet consolidated into long-term memory.

shunting inhibition A form of synaptic inhibition in which the main effect is to reduce membrane resistance, thereby shunting depolarizing current generated at excitatory synapses.

simple cell A neuron found in primary visual cortex that has an elongated orientation-selective receptive field with distinct ON and OFF subregions.

skeletal muscle A type of striated muscle that is under voluntary control and that functions to move bones around joints; derived from mesodermal somites.

slow motor unit A motor unit with a small alpha motor neuron innervating slowly contracting and slowly fatiguing red muscle fibers.

SMA *See* supplementary motor area.

smooth endoplasmic reticulum (smooth ER) A membrane-enclosed cellular organelle that is heterogeneous and performs different functions in different locations.

smooth muscle A type of muscle in the digestive tract, arteries, and related structures; innervated by the autonomic nervous system and not under voluntary control.

sodium-potassium pump An ion pump that removes intracellular Na^+ and concentrates intracellular K^+, using adenosine triphosphate as its energy source.

soma The central region of the neuron containing the nucleus; also called cell body or perikaryon.

somatic motor system The skeletal muscles and the parts of the nervous system that control them; the system that generates behavior.

somatic PNS The part of the peripheral nervous system that innervates the skin, joints, and skeletal muscles.

somatic sensation The senses of touch, temperature, body position, and pain.

somatotopy The topographic organization of somatic sensory pathways in which neighboring receptors in the skin feed information to neighboring cells in a target structure.

spatial summation The combining of excitatory postsynaptic potentials generated at more than one synapse on the same cell. *See also* temporal summation.

specific language impairment A delay in the mastery of language in the absence of hearing loss or more general developmental delays.

speech Spoken language.

spike-initiation zone A region of the neuronal membrane where action potentials are normally initiated, characterized by a high density of voltage-gated sodium channels.

spike timing–dependent plasticity Bidirectional modification of synaptic strength induced by varying the relative timing of the presynaptic and postsynaptic spikes.

spinal canal The CSF-filled space within the spinal cord.

spinal cord The part of the central nervous system in the vertebral column.

spinal nerve A nerve attached to the spinal cord that innervates the body.

spinal segment One set of dorsal and ventral roots plus the portion of spinal cord related to them.

spinothalamic pathway An ascending somatic sensory pathway traveling from the spinal cord to the thalamus via the lateral spinothalamic columns; mediates information about pain, temperature, and some forms of touch.

spiny neuron A neuron with dendritic spines.

spiral ganglion A collection of neurons in the modiolus of the cochlea that receives input from hair cells and sends output to the cochlear nuclei in the medulla via the auditory nerve.

split-brain study An examination of behavior in animals or humans whose cerebral hemispheres have been disconnected by cutting the corpus callosum.

spotlight of attention The ability of visual attention to shift to different objects like the way a spotlight moves to explore a dark room.

SRY *See* sex-determining region of the Y chromosome.

SSRI *See* serotonin-selective reuptake inhibitor.

standard model of memory consolidation An explanation of memory formation in which sensory information is processed by the hippocampus and later transferred to neocortex for permanent storage.

stapes An ossicle in the middle ear attached to the oval window that somewhat resembles a stirrup.

starvation A state of negative energy balance in which energy intake fails to meet the body's demands, resulting in a loss of fat tissue. *See also* obesity.

stellate cell A neuron characterized by a radial, star-like distribution of dendrites.

stereocilium A hairlike cilium attached to the top of a hair cell in the inner ear.

strabismus A condition in which the eyes are not perfectly aligned.

striate cortex Primary visual cortex, Brodmann's area 17; also called V1.

striated muscle A type of muscle with a striated, or striped, appearance; the two types are skeletal and cardiac.

striatum A collective term for the caudate nucleus and putamen; involved in the initiation of willed movements of the body; plays a role in procedural memory.

stria vascularis A specialized endothelium lining one wall of the scala media and responsible for secreting endolymph.

subplate A layer of cortical neurons lying below the cortical plate early in development; when the cortical plate has differentiated into the six layers of the neocortex, the subplate disappears.

substantia A group of related neurons deep within the brain, usually with less distinct borders than those of nuclei.

substantia gelatinosa A thin dorsal part of the dorsal horn of the spinal cord that receives input from unmyelinated C fibers; important in the transmission of nociceptive signals.

substantia nigra A cell group in the midbrain that uses dopamine as a neurotransmitter and innervates the striatum.

subthalamic nucleus A part of the basal ganglia in the basal forebrain; involved in motor control.

sulcus A groove in the surface of the cerebrum running between neighboring gyri. Plural: sulci.

superior colliculus A structure in the tectum of the midbrain that receives direct retinal input and controls saccadic eye movements.

superior olive A nucleus in the caudate pons that receives afferents from the cochlear nuclei and sends efferents to the inferior colliculus; also called superior olivary nucleus.

supplementary motor area (SMA) The medial part of cortical area 6; involved in the control of voluntary movement.

suprachiasmatic nucleus (SCN) A small nucleus of the hypothalamus just above the optic chiasm that receives retinal innervation and synchronizes circadian rhythms with the daily light–dark cycle.

sympathetic chain A series of interconnected sympathetic ganglia of the autonomic nervous system, adjacent to the vertebral column, that receive input from preganglionic sympathetic fibers and project postganglionic fibers to target organs and tissues.

sympathetic division A division of the autonomic nervous system that in fight-or-flight situations activates physiological responses, including increased heart rate, respiration, blood pressure, and energy mobilization and decreased digestive and reproductive functions; its peripheral axons emerge from the thoracic and lumbar spinal cord. *See also* parasympathetic division.

synapse The region of contact where a neuron transfers information to another cell.

synaptic cleft The region separating the presynaptic and postsynaptic membranes of neurons.

synaptic consolidation The transformation of sensory information into a temporary memory trace in the hippocampus.

synaptic integration The process by which multiple EPSPs and/or IPSPs combine within one postsynaptic neuron, in some cases triggering one or more action potentials.

synaptic scaling A cell-wide adjustment of synaptic strengths in response to a change in the average firing rate of the postsynaptic neuron.

synaptic transmission The process of transferring information from one cell to another at a synapse.
synaptic vesicle A membrane-enclosed structure, about 50 nm in diameter, containing neurotransmitter and found at a site of synaptic contact.
synergist muscle A muscle that contracts with other muscles to produce movement in one direction.
systems consolidation The transformation of a temporary hippocampal memory trace into a permanent engram in neocortex.

T tubule A membrane-enclosed tunnel running within a skeletal muscle fiber that links excitation of the sarcolemma with the release of Ca^{2+} from the sarcoplasmic reticulum.
taste bud A cluster of cells, including taste receptor cells, in papillae of the tongue.
taste receptor cell A modified epithelial cell that transduces taste stimuli.
tectorial membrane A sheet of tissue that hangs over the organ of Corti in the cochlea in the inner ear.
tectospinal tract A tract originating in the superior colliculus and terminating in the spinal cord; involved in the control of head and neck movement.
tectum The part of the midbrain lying dorsal to the cerebral aqueduct.
tegmentum The part of the midbrain lying ventral to the cerebral aqueduct.
telencephalon A region of the brain derived from the prosencephalon (forebrain). Telencephalic structures include the paired cerebral hemispheres that contain cerebral cortex and the basal telencephalon.
temporal coding The representation of information by the timing of action potentials rather than by their average rate.
temporal lobe The region of the cerebrum lying under the temporal bone.
temporal summation The combining of excitatory postsynaptic potentials generated in rapid succession at the same synapse. *See also* spatial summation.
terminal arbor Branches at the end of an axon terminating in the same region of the nervous system.
terminal bouton The end region of an axon, usually a site of synaptic contact with another cell; also called axon terminal.
tetanus A type of repetitive stimulation.
tetrodotoxin (TTX) A toxin that blocks Na^+ permeation through voltage-gated sodium channels, thereby blocking action potentials.
thalamus The dorsal part of the diencephalon, highly interconnected with the cerebral neocortex.
thermoreceptor A sensory receptor selective for temperature changes.
thick filament A part of the cytoskeleton of a muscle cell containing myosin, lying between and among thin filaments and sliding along them to cause muscle contraction.
thin filament A part of the cytoskeleton of a muscle cell containing actin, anchored to Z lines and sliding along thick filaments to cause muscle contraction.
third ventricle The CSF-filled space within the diencephalon.
threshold A level of depolarization sufficient to trigger an action potential.
tonotopy The systematic organization within an auditory structure on the basis of characteristic frequency.
top–down attention Attention voluntarily directed by the brain to serve a behavioral goal; also called endogenous attention.
tract A collection of central nervous system axons with a common site of origin and a common destination.
transcription The process of synthesizing a messenger RNA molecule according to genetic instructions encoded in DNA.
transcription factor A protein that regulates the binding of RNA polymerase to a gene promoter.
transducin The G-protein that couples rhodopsin to the enzyme phosphodiesterase in rod photoreceptors.
transduction The transformation of sensory stimulus energy into a cellular signal, such as a receptor potential.
transgenic mice Mice in which extra genes have been introduced by genetic engineering.
translation The process of synthesizing a protein molecule according to genetic instructions carried by a messenger RNA molecule.
transmitter-gated ion channel A membrane protein forming a pore that is permeable to ions and gated by neurotransmitter.
transporter A membrane protein that transports neurotransmitters, or their precursors, across membranes to concentrate them in either presynaptic cytosol or synaptic vesicles.
trigeminal nerve Cranial nerve V; attaches to the pons and carries primarily sensory axons from the head, mouth, and dura mater and motor axons of mastication.
trophic factor Any molecule that promotes cell survival.
troponin A protein that binds Ca^{2+} in a skeletal muscle cell, thereby regulating the interaction of myosin and actin.
TTX *See* tetrodotoxin.
tympanic membrane A membrane at the internal end of the auditory canal that moves in response to variations in air pressure; also called eardrum.

ultradian rhythm Any rhythm with a period significantly less than 1 day. *See also* circadian rhythm.
unconscious emotion The experience or expression of emotion in the absence of conscious awareness of the stimulus that evoked the emotion.
undershoot The part of an action potential when the membrane potential is more negative than at rest; also called after-hyperpolarization.
unipolar neuron A neuron with a single neurite.

V1 Primary visual cortex or striate cortex.
vagus nerve Cranial nerve X, arising from the medulla and innervating the viscera of the thoracic and abdominal cavities; a major source of preganglionic parasympathetic visceromotor axons.
vascular organ of the lamina terminalis (OVLT) A specialized region of the hypothalamus containing neurons that are sensitive to blood tonicity; they activate magnocellular neurosecretory cells to release vasopressin into the blood, triggering osmometric thirst.
vasopressin A small peptide hormone released from the posterior pituitary by magnocellular neurosecretory cells; promotes water retention and decreased urine production by the kidney; also called antidiuretic hormone (ADH).
ventral An anatomical reference meaning toward the belly.
ventral cochlear nucleus A nucleus in the medulla that receives afferents from the spiral ganglion in the cochlea of the inner ear.
ventral horn The ventral region of the spinal cord containing neuronal cell bodies.
ventral lateral (VL) nucleus A nucleus of the thalamus that relays information from the basal ganglia and cerebellum to the motor cortex.
ventral posterior medial (VPM) nucleus The part of the ventral posterior nucleus of the thalamus that receives somatosensory input from the face, including afferents from the tongue.
ventral posterior (VP) nucleus The main thalamic relay nucleus of the somatic sensory system.
ventral root A bundle of motor neuron axons that emerges from the ventral spinal cord and joins sensory fibers to form a spinal nerve. Ventral root axons carry information away from the spinal cord. *See also* dorsal root.
ventricular system The cerebrospinal fluid–filled spaces inside the brain, consisting of the lateral ventricles, third ventricle, cerebral aqueduct, and fourth ventricle.
ventromedial hypothalamic syndrome Obesity associated with lesions of the lateral hypothalamic area.
ventromedial pathway Axons in the ventromedial column of the spinal cord that are involved in the control of posture and locomotion and are under brain stem control.
verbal dyspraxia An inability to produce the coordinated muscular movements needed for speech in the absence of damage to nerves or muscles.
vermis The midline region of the cerebellum.
vestibular labyrinth A part of the inner ear specialized for the detection of head motion; consists of the otolith organs and semicircular canals.
vestibular nucleus A nucleus in the medulla that receives input from the vestibular labyrinth of the inner ear.
vestibular system The neural system that monitors and regulates the sense of balance and equilibrium.
vestibulo-ocular reflex (VOR) A reflexive movement of the eyes stimulated by rotational movements of the head; stabilizes the visual image on the retinas.
vestibulospinal tract A tract originating in the vestibular nuclei of the medulla and terminating in the spinal cord; involved in the control of movement and posture.
visceral PNS The part of the peripheral nervous system that innervates the internal organs, blood vessels, and glands; also called autonomic nervous system.
vision The sense of sight.
visual acuity The ability of the visual system to distinguish between two nearby points.
visual angle A way to describe distance across the retina; an object that subtends an angle of 3.5° will form an image on the retina that is 1 mm across.
visual field The total region of space viewed by both eyes when the eyes are fixated on a point.
visual hemifield The half of the visual field to one side of the fixation point.
vitreous humor The jellylike substance filling the eye between the lens and the retina.
VL nucleus *See* ventral lateral nucleus.
vocal folds Two bands of muscle within the larynx, also known as vocal chords, that vibrate to produce the human voice.
volley principle The idea that high sound frequencies are represented in the pooled activity of a number of neurons, each of which fires in a phase-locked manner.
voltage The force exerted on an electrically charged particle, represented by the symbol V and measured in volts; also called electrical potential or potential difference.
voltage clamp A device that enables an investigator to hold the membrane potential constant while transmembrane currents are measured.
voltage-gated calcium channel A membrane protein forming a pore that is permeable to Ca^{2+} and gated by depolarization of the membrane.
voltage-gated potassium channel A membrane protein forming a pore that is permeable to K^+ and gated by depolarization of the membrane.

voltage-gated sodium channel A membrane protein forming a pore that is permeable to Na^+ and gated by depolarization of the membrane.

volumetric thirst The motivation to drink water as a result of a decrease in blood volume.

VOR *See* vestibulo-ocular reflex.

VPM nucleus *See* ventral posterior medial nucleus.

VP nucleus *See* ventral posterior nucleus.

Wada procedure A procedure in which one cerebral hemisphere is anesthetized to enable testing of the function of the other hemisphere.

Wernicke–Geschwind model A model for language processing involving interactions of Broca's area and Wernicke's area with sensory and motor areas.

Wernicke's aphasia A language disturbance in which speech is fluent but comprehension is poor.

Wernicke's area An area on the superior surface of the temporal lobe between auditory cortex and the angular gyrus; associated with Wernicke's aphasia when damaged.

white matter A generic term for a collection of central nervous system axons. When a freshly dissected brain is cut open, axons appear white. *See also* gray matter.

working memory Information storage that is temporary, limited in capacity, and requires continual rehearsal.

Young–Helmholtz trichromacy theory The theory that the brain assigns colors based on a comparison of the readout of the three types of cone photoreceptors.

zeitgeber Any environmental cue, such as the light–dark cycle, which signals the passage of time.

Z line A band delineating sarcomeres in a myofibril of a muscle fiber.

REFERENCES

Chapter 1

Allman JM. 1999. *Evolving Brains*. New York: Scientific American Library.

Alt KW, Jeunesse C, Buitrago-Téllez CH, Wächter R, Boës E, Pichler SL. 1997. Evidence for stone age cranial surgery. *Nature* 387:360.

Alzheimer's Association. http://www.alz.org

American Stroke Association. http://www.strokeassociation.org/

Centers for Disease Control and Prevention, National Center for Injury Prevention and Control. http://www.cdc.gov

Clarke E, O'Malley C. 1968. *The Human Brain and Spinal Cord*, 2nd ed. Los Angeles: University of California Press.

Corsi P, ed. 1991. *The Enchanted Loom*. New York: Oxford University Press.

Crick F. 1994. *The Astonishing Hypothesis: The Scientific Search for the Soul*. New York: Macmillan.

Finger S. 1994. *Origins of Neuroscience*. New York: Oxford University Press.

Glickstein M. 2014. *Neuroscience: A Historical Introduction*. Cambridge: MIT Press.

Hall MJ, DeFrances CJ. 2003. *National Hospital Discharge Survey. Advance Data from Vital and Health Statistics No. 332*. Hyattsville, MD: National Center for Health Statistics.

Kessler RC, Berglund P, Demler O, Jin R, Koretz D, Merikangas KR, Rush AJ, Walters EE, Wang PS. 2003. The epidemiology of major depressive disorder: results from the National Comorbidity Survey Replication (NCS-R). *Journal of the American Medical Association* 289(23):3095–3105.

Mościcki EK. 1997. Identification of suicide risk factors using epidemiologic studies. *Psychiatric Clinics of North America* 20(3):499–517.

National Academy of Sciences Institute of Medicine. 1991. *Science, Medicine, and Animals.* Washington, DC: National Academy Press.

National Institute of Mental Health. http://www.nimh.nih.gov/index.shtml

National Institute on Drug Abuse. http://www.drugabuse.gov

National Parkinson Foundation. http://www.parkinson.org.

National Stroke Association. http://www.stroke.org.

Shepherd GM, Erulkar SD. 1997. Centenary of the synapse: from Sherrington to the molecular biology of the synapse and beyond. *Trends in Neurosciences* 20:385–392.

U.S. Department of Health and Human Services. 2004. *Mental Health: A Report of the Surgeon General.* Rockville, MD: U.S. Department of Health and Human Services, Substance Abuse and Mental Health Services Administration, Center for Mental Health Services, National Institutes of Health, National Institute of Mental Health.

U.S. Office of Science and Technology Policy. 1991. Decade of the Brain 1990–2000: *Maximizing Human Potential.* Washington, DC: Subcommittee on Brain and Behavioral Sciences.

Chapter 2

Alberts B, Johnson A, Lewis J, Raff M, Roberts K, Walter P. 2008. *Molecular Biology of the Cell*, 5th ed. New York: Garland.

Bick K, Amaducci L, Pepeu G. 1987. *The Early Story of Alzheimer's Disease.* New York: Raven Press.

Capecchi MR. 1980. High efficiency transformation by direct microinjection of DNA into cultured mammalian cells. *Cell* 22:479–488.

Chen SC, Tvrdik P, Peden E, Cho S, Wu S, Spangrude G, Capecchi MR. 2010. Hematopoietic origin of pathological grooming in Hoxb8 mutant mice. *Cell* 141(5):775–785.

DeFelipe J, Jones EG. 1998. *Cajal on the Cerebral Cortex*. New York: Oxford University Press.

De Vos KJ, Grierson AJ, Ackerley S, Miller CCJ. 2008. Role of axoplasmic transport in neurodegenerative diseases. *Annual Review of Neuroscience* 31:151–173.

Eroglu C, Barres BA. 2010. Regulation of synaptic connectivity by glia. *Nature* 468: 223–231.

Finger S. 1994. *Origins of Neuroscience*. New York: Oxford University Press.

Folger KR, Wong EA, Wahl G, Capecchi MR. 1982. Patterns of integration of DNA microinjected into cultured mammalian cells: evidence for homologous recombination between injected plasmid DNA molecules. *Molecular and Cellular Biology* 2:1372–1387.

Goedert M, Spillantini MG, Hasegawa M, Jakes R, Crowther RA, Krug A. 1996. Molecular dissection of the neurofibrillary lesions of Alzheimer's disease. *Cold Spring Harbor Symposia on Quantitative Biology,* Vol. LXI. Cold Spring Harbor, NY: Cold Spring Harbor Laboratory Press.

Grafstein B, Forman DS. 1980. Intracellular transport in neurons. *Physiological Reviews* 60:1167–1283.

Hammersen F. 1980. *Histology*. Baltimore: Urban & Schwarzenberg.

Harris KM, Stevens JK. 1989. Dendritic spines of CA1 pyramidal cells in the rat hippocampus: serial electron microscopy with reference to their biophysical characteristics. *Journal of Neuroscience* 9:2982–2997.

Hubel DH. 1988. *Eye, Brain and Vision.* New York: Scientific American Library.

Jones EG. 1999. Golgi, Cajal and the neuron doctrine. *Journal of the History of Neuroscience* 8:170–178.

Lent R, Azevedo FAC, Andrade-Moraes CH, Pinto AVO. 2012. How many neurons do you have? Some dogmas of quantitative neuroscience under revision. *European Journal of Neuroscience* 35: 1–9.

Levitan I, Kaczmarek L. 2002. *The Neuron: Cell and Molecular Biology*, 3rd ed. New York: Oxford University Press.

Nelson SB, Hempel C, Sugino K. 2006. Probing the transcriptome of neuronal cell types. *Current Opinion in Neurobiology* 16:571–576.

Peters A, Palay SL, Webster H deF. 1991. *The Fine Structure of the Nervous System*, 3rd ed. New York: Oxford University Press.

Purpura D. 1974. Dendritic spine "dysgenesis" and mental retardation. *Science* 20:1126–1128.

Sadava D, Hills DM, Heller HC, Berenbaum MR. 2011. *Life: The Science of Biology*, 9th ed. Sunderland, MA: Sinauer.

Shepherd GM, Erulkar SD. 1997. Centenary of the synapse: from Sherrington to the molecular biology of the synapse and beyond. *Trends in Neurosciences* 20:385–392.

Steward O, Schuman EM. 2001. Protein synthesis at synaptic sites on dendrites. *Annual Review of Neuroscience* 24:299–325.

Thomas KR, Capecchi MR. 1987. Site-directed mutagenesis by gene targeting in mouse embryo-derived stem cells. *Cell* 51:503–512.

Vickers JC, Riederer BM, Marugg RA, Buee-Scherrer V, Buee L, Delacourte A, Morrison JH. 1994. Alterations in neurofilament protein immunoreactivity in human hippocampal neurons related to normal aging and Alzheimer's disease. *Neuroscience* 62:1–13.

Wilt BA, Burns LD, Ho ETW, Ghosh KK, Mukamel EA, Schnitzer MJ. 2009. Advances in light microscopy for neuroscience. *Annual Review of Neuroscience* 32: 435–506.

Chapter 3

Doyle DA, Cabral JM, Pfuetzner RA, Kuo A, Gulbis JM, Cohen SL, Chait BT, MacKinnon R. 1998. The structure of the potassium channel: molecular basis of K^+ conduction and selectivity. *Science* 280:69–77.

Goldstein SA, Pheasant DJ, Miller C. 1994. The charybdotoxin receptor of a Shaker K^+ channel: peptide and channel residues mediating molecular recognition. *Neuron* 12:1377–1388.

Hille B. 2001. *Ionic Channels of Excitable Membranes,* 3rd ed. Sunderland, MA: Sinauer.

Jan L, Jan YN. 1997. Cloned potassium channels from eukaryotes and prokaryotes. *Annual Review of Neuroscience* 20:91–123.

Levitan I, Kaczmarek L. 2002. *The Neuron: Cell and Molecular Biology*, 3rd ed. New York: Oxford University Press.

Li M, Unwin N, Staufer KA, Jan YN, Jan L. 1994. Images of purified *Shaker* potassium channels. *Current Biology* 4:110–115.

MacKinnon R. 1995. Pore loops: an emerging theme in ion channel structure. *Neuron* 14:889–892.

MacKinnon R. 2003. Potassium channels. *Federation of European Biochemical Societies Letters* 555:62–65.

Miller C. 1988. *Shaker* shakes out potassium channels. *Trends in Neurosciences* 11:185–186.

Nicholls J, Martin AR, Fuchs PA, Brown DA, Diamond ME, Weisblat D. 2011. *From Neuron to Brain*, 5th ed. Sunderland, MA: Sinauer.

Ransom BR, Goldring S. 1973. Slow depolarization in cells presumed to be glia in cerebral cortex of cat. *Journal of Neurophysiology* 36:869–878.

Sanguinetti MC, Spector PS. 1997. Potassium channelopathies. *Neuropharmacology* 36:755–762.

Shepherd G. 1994. *Neurobiology*, 3rd ed. New York: Oxford University Press.

Somjen GG. 2004. *Ions in the Brain: Normal Function, Seizures, and Stroke.* New York: Oxford University Press.

Stoffel M, Jan LY. 1998. Epilepsy genes: excitement traced to potassium channels. *Nature Genetics* 18:6–8.

Chapter 4

Agmon A, Connors BW. 1992. Correlation between intrinsic firing patterns and thalamocortical synaptic responses of neurons in mouse barrel cortex. *Journal of Neuroscience* 12:19–329.

Armstrong CM, Hille B. 1998. Voltage-gated ion channels and electrical excitability. *Neuron* 20:371–380.

Boyden ES, Zhang F, Bamberg E, Nagel G, Deisseroth K. 2005. Millisecond-timescale, genetically targeted optical control of neural activity. *Nature Neuroscience* 8:1263–1268.

Brunton L, Chabner B, Knollman B. 2011. *Goodman and Gilman's the Pharmacological Basis of Therapeutics*, 12th ed. New York: McGraw-Hill.

Cole KS. 1949. Dynamic electrical characteristics of the squid axon membrane. *Archives of Scientific Physiology* 3:253–258.

Connors B, Gutnick M. 1990. Intrinsic firing patterns of diverse neocortical neurons. *Trends in Neurosciences* 13:99–104.

Hille B. 2001. *Ionic Channels of Excitable Membranes*, 3rd ed. Sunderland, MA: Sinauer.

Hodgkin A. 1976. Chance and design in electrophysiology: an informal account of certain experiments on nerves carried out between 1942 and 1952. *Journal of Physiology (London)* 263:1–21.

Hodgkin AL, Huxley AF, Katz B. 1952. Measurement of current voltage relations in the membrane of the giant axon of Loligo. *Journal of Physiology (London)* 116:424–448.

Huguenard J, McCormick D. 1994. *Electrophysiology of the Neuron*. New York: Oxford University Press.

Kullmann DM, Waxman SG. 2010. Neurological channelopathies: new insights into disease mechanisms and ion channel function. *Journal of Physiology (London)* 588:1823–1827.

Levitan I, Kaczmarek L. 2002. *The Neuron: Cell and Molecular Biology*, 3rd ed. New York: Oxford University Press.

Llinás R. 1988. The intrinsic electrophysiological properties of mammalian neurons: insights into central nervous system function. *Science* 242:1654–1664.

Nagel G, Szellas T, Huhn W, Kateriya S, Adeishvili N, Berthold P, Ollig D, Hegemann P, Bamberg E. 2003. Channelrhodopsin-2, a directly light-gated cation-selective membrane channel. *Proceedings of the National Academy of Sciences of the United States of America* 100:13940–13945.

Narahashi T. 1974. Chemicals as tools in the study of excitable membranes. *Physiology Reviews* 54:813–889.

Narahashi T, Deguchi T, Urakawa N, Ohkubo Y. 1960. Stabilization and rectification of muscle fiber membrane by tetrodotoxin. *American Journal of Physiology* 198:934–938.

Narahashi T, Moore JW, Scott WR. 1964. Tetrodotoxin blockage of sodium conductance increase in lobster giant axons. *Journal of General Physiology* 47:965–974.

Neher E. 1992. Nobel lecture: ion channels or communication between and within cells. *Neuron* 8:605–612.

Neher E, Sakmann B. 1992. The patch clamp technique. *Scientific American* 266:28–35.

Nicholls J, Martin AR, Fuchs PA, Brown DA, Diamond ME, Weisblat D. 2011. *From Neuron to Brain*, 5th ed. Sunderland, MA: Sinauer.

Noda M, Shimizu S, Tanabe T, Takai T, Kayano T, Ikeda T, Takahashi H, Nakayama H, Kanaoka Y, Minamino N, et al. 1984. Primary structure of *Electrophorus electricus* sodium channel deduced from cDNA sequence. *Nature* 312:121–127.

Shepherd G. 1994. *Neurobiology*, 3rd ed. New York: Oxford University Press.

Sigworth FJ, Neher E. 1980. Single Na^+ channel currents observed in cultured rat muscle cells. *Nature* 287:447–449.

Unwin N. 1989. The structure of ion channels in membranes of excitable cells. *Neuron* 3:665–676.

Watanabe A. 1958. The interaction of electrical activity among neurons of lobster cardiac ganglion. *Japanese Journal of Physiology* 8:305–318.

Chapter 5

Bloedel JR, Gage PW, Llinás R, Quastel DM. 1966. Transmitter release at the squid giant synapse in the presence of tetrodotoxin. *Nature* 212:49–50.

Chouquet D, Triller A. 2013. The dynamic synapse. *Neuron* 80:691–703.

Colquhoun D, Sakmann B. 1998. From muscle endplate to brain synapses: a short history of synapses and agonist-activated ion channels. *Neuron* 20:381–387.

Connors BW, Long MA. 2004. Electrical synapses in the mammalian brain. *Annual Review of Neuroscience* 27:393–418.

Cowan WM, Südhof TC, Stevens CF. 2001. *Synapses*. Baltimore: Johns Hopkins University Press.

Fatt P, Katz B. 1951. An analysis of the end-plate potential recorded with an intracellular electrode. *Journal of Physiology (London)* 115:320–370.

Furshpan E, Potter D. 1959. Transmission at the giant motor synapses of the crayfish. *Journal of Physiology (London)* 145:289–325.

Harris KM, Weinberg RJ. 2012. Ultrastructure of synapses in the mammalian brain. *Cold Spring Harbor Perspectives in Biology* 4:a005587.

Heuser J, Reese T. 1973. Evidence for recycling of synaptic vesicle membrane during transmitter release at the frog neuromuscular junction. *Journal of Cell Biology* 57:315–344.

Heuser J, Reese T. 1977. Structure of the synapse. In *Handbook of Physiology—Section 1. The Nervous System, Vol. I. Cellular Biology of Neurons,* eds. Brookhart JM, Mountcastle VB. Bethesda, MD: American Physiological Society, pp. 261–294.

Johnston D, Wu SM-S. 1994. *Foundations of Cellular Neurophysiology*. Cambridge, MA: MIT Press.

Kandel ER, Schwartz JH, Jessell TM, Siegelbaum SA, Hudspeth AJ. 2012. *Principles of Neural Science,* 5th ed. New York: McGraw-Hill Professional.

Koch C. 2004. *Biophysics of Computation: Information Processing in Single Neurons*. New York: Oxford University Press.

Llinás R, Sugimori M, Silver RB. 1992. Microdomains of high calcium concentration in a presynaptic terminal. *Science* 256:677–679.

Loewi O. 1953. *From the Workshop of Discoveries*. Lawrence: University of Kansas Press.

Long MA, Deans MR, Paul DL, Connors BW. 2002. Rhythmicity without synchrony in the electrically uncoupled inferior olive. *Journal of Neuroscience* 22:10898-10905.

Matthews R. 1995. *Nightmares of Nature*. London: Harper Collins.

Neher E. 1998. Vesicle pools and Ca^{2+} microdomains: new tools for understanding their roles in neurotransmitter release. *Neuron* 20:389–399.

Neher E, Sakmann B. 1992. The patch clamp technique. *Scientific American* 266:44–51.

Nicholls JG, Martin AR, Fuchs PA, Brown DA, Diamond ME, Weisblat D. 2011. *From Neuron to Brain,* 5th ed. Sunderland, MA: Sinauer.

Rajendra S, Schofield PR. 1995. Molecular mechanisms of inherited startle syndromes. *Trends in Neurosciences* 18:80–82.

Rothman JE. 2002. Lasker Basic Medical Research Award. The machinery and principles of vesicle transport in the cell. *Nature Medicine* 8:1059–1062.

Sheng M, Sabatini BL, Südhof TC. 2012. *The Synapse.* New York: Cold Spring Harbor Laboratory Press.

Shepherd GM. 2003. *The Synaptic Organization of the Brain*. New York: Oxford University Press.

Sherrington C. 1906. *Integrative Action of the Nervous System*. New Haven: Yale University Press.

Siksou L, Triller A, Marty S. 2011. Ultrastructural organization of presynaptic terminals. *Current Opinion in Neurobiology* 21:261–268.

Sloper JJ, Powell TP. 1978. Gap junctions between dendrites and somata of neurons in the primate sensori-motor cortex. *Proceedings of the Royal Society, Series B* 203:39–47.

Stuart G, Spruston N, Hausser M. 2007. *Dendrites,* 2nd ed. New York: Oxford University Press.

Südhof TC. 2013. Neurotransmitter release: the last millisecond in the life of a synaptic vesicle. *Neuron* 80:675–690.

Unwin N. 1993. Neurotransmitter action: opening of ligand-gated ion channels. *Cell* 72:31–41.

Watanabe A. 1958. The interaction of electrical activity among neurons of lobster cardiac ganglion. *Japanese Journal of Physiology* 8:305-318.

Chapter 6

Attwell D, Mobbs P. 1994. Neurotransmitter transporters. *Current Opinion in Neurobiology* 4:353–359.

Brezina V, Weiss KR. 1997. Analyzing the functional consequences of transmitter complexity. *Trends in Neurosciences* 20:538–543.

Burnstock G, Krügel U, Abbracchio MP, Illes P. 2011. Purinergic signalling: from normal behaviour to pathological brain function. *Progress in Neurobiology* 95:229–274.

Castillo PE, Younts TJ, Chávez AE, Hashimotodani Y. 2012. Endocannabinoid signaling and synaptic function. *Neuron* 76:70–81.

Changeux J-P. 1993. Chemical signaling in the brain. *Scientific American* 269:58–62.

Colquhoun D, Sakmann B. 1998. From muscle endplate to brain synapses: a short history of synapse and agonist-activated ion channels. *Neuron* 20:381–387.

Cowan WM, Südhof TC, Stevens CF. 2001. *Synapses*. Baltimore: Johns Hopkins University Press.

Gilman AG. 1995. Nobel lecture: G proteins and regulation of adenylyl cyclase. *Bioscience Report* 15:65–97.

Gudermann T, Schöneberg T, Schultz G. 1997. Functional and structural complexity of signal transduction via G-protein-couple receptors. *Annual Review of Neuroscience* 20:399–427.

Hille B. 2001. *Ionic Channels of Excitable Membranes*, 3rd ed. Sunderland, MA: Sinauer.

Iversen LL, Iversen SD, Bloom FE, Roth RH. 2008. *Introduction to Neuropsychopharmacology*. New York: Oxford University Press.

Jiang J, Amara SG. 2011. New views of glutamate transporter structure and function: advances and challenges. *Neuropharmacology* 60:172–181.

Katritch V, Cherezov V, Stevens RC. 2012. Diversity and modularity of G protein-coupled receptor structures. *Trends in Pharmacological Sciences* 33:17–27.

Krnjević K. 2010. When and why amino acids? *Journal of Physiology (London)* 588:33–44.

Kumar J, Mayer ML. 2013. Functional insights from glutamate receptor ion channel structures. *Annual Review of Physiology* 75:313–337.

Matsuda LA. 1997. Molecular aspects of cannabinoid receptors. *Critical Reviews in Neurobiology* 11:143–166.

Mayer ML, Armstrong N. 2004. Structure and function of glutamate receptor ion channels. *Annual Review of Physiology* 66:161–181.

Meyer JS, Quenzer LF. 2013. *Psychopharmacology: Drugs, the Brain, and Behavior*, 2nd ed. Sunderland, MA: Sinauer.

Mustafa AK, Gadalla MM, Snyder SH. 2009. Signaling by gasotransmitters. *Science Signaling* 2(68):re2.

Nestler EJ, Hyman SE, Malenka RC. 2008. *Molecular Neuropharmacology: A Foundation for Clinical Neuroscience,* 2nd ed. McGraw-Hill Professional.

Nicholls JG, Martin AR, Fuchs PA, Brown DA, Diamond ME, Weisblat D. 2011. *From Neuron to Brain*, 5th ed. Sunderland, MA: Sinauer.

Nicoll R, Malenka R, Kauer J. 1990. Functional comparison of neurotransmitter receptor subtypes in the mammalian nervous system. *Physiological Reviews* 70:513–565.

Palczewski K, Orban T. 2013. From atomic structures to neuronal functions of G protein–coupled receptors. *Annual Review of Neuroscience* 36:139–164.

Pierce KL, Premont RT, Lefkowitz RJ. 2002. Seven-transmembrane receptors. *Nature Reviews Molecular and Cell Biology* 3:639–650.

Piomelli D. 2003. The molecular logic of endocannabinoid signalling. *Nature Reviews Neuroscience* 4:873–884.

Regehr WG, Carey MR, Best AR. 2009. Activity-dependent regulation of synapses by retrograde messengers. *Neuron* 63:154–170.

Siegel GJ, Agranoff BW, Albers RW, Fisher SK, Uhler MD, eds. 1998. *Basic Neurochemistry: Molecular, Cellular and Medical Aspects*, 6th ed. Baltimore: Lippincott Williams & Wilkins.

Snyder S. 1986. *Drugs and the Brain*. New York: W.H. Freeman.

Squire LR, Berg D, Bloom FE, du Lac S, Ghosh A, Spitzer NC. 2012. *Fundamental Neuroscience*, 4th ed. San Diego: Academic Press.

Wilson RI, Nicoll RA. 2002. Endocannabinoid signaling in the brain. *Science* 296:678–682.

Wollmuth LP, Sobolevsky AI. 2004. Structure and gating of the glutamate receptor ion channel. *Trends in Neurosciences* 27:321–328.

Chapter 7

Butterworth CE, Bendich A. 1996. Folic acid and the prevention of birth defects. *Annual Review of Nutrition* 16:73–97.

Cajal SR. 1899. *Clark University, 1889–1899: Decennial Celebration,* ed. Story WE. Worcester: Clark University Press, pp. 311–382.

Chung K, Deisseroth K. 2013. CLARITY for mapping the nervous system. *Nature Methods* 10:508–513.

Creslin E. 1974. Development of the nervous system: a logical approach to neuroanatomy. *CIBA Clinical Symposium* 26:1–32.

Frackowick RSJ. 1998. The functional architecture of the brain. *Daedalus* 127:105–130.

Gilbert SF. 2003. *Developmental Biology,* 7th ed. Sunderland, MA: Sinauer.

Gluhbegoric N, Williams TH. 1980. *The Human Brain: A Photographic Guide*. Philadelphia: Lippincott.

Johnson KA, Becker JA. The whole brain atlas. http://www.med.harvard.edu/AANLIB/home/html

Kaas JH. 1995. The evolution of neocortex. *Brain, Behavior and Evolution* 46:187–196.

Kaas JH. 2013. The evolution of brains from early mammals to humans. *Wiley Interdisciplinary Reviews. Cognitive Science* 4:33–45.

Posner MI, Raichle M. 1994. *Images of Mind*. New York: Scientific American Library.

Povinelli DJ, Preuss TM. 1995. Theory of mind: evolutionary history of a cognitive specialization. *Trends in Neurosciences* 18:414–424.

Seung S. 2012. *Connectome: How the Brain's Wiring Makes Us Who We Are*. Boston: Joughton Mifflin Harcourt.

Smith JL, Schoenwolf GC. 1997. Neurulation: coming to closure. *Trends in Neurosciences* 20:510–517.

Watson C. 1995. *Basic Human Neuroanatomy: An Introductory Atlas*, 5th ed. Baltimore: Lippincott Williams & Wilkins.

Chapter 8

Belluscio L, Gold GH, Nemes A, Axel R. 1998. Mice deficient in G(olf) are anosmic. *Neuron* 20:69–81.

Blauvelt DG, Sato TF, Wienisch M, Knöpfel T, Murthy VN. 2013. Distinct spatiotemporal activity in principal neurons of the mouse olfactory bulb in anesthetized and awake states. *Frontiers in Neural Circuits* 7:46.

Brennan PA, Keverne EB. 2004. Something in the air? New insights into mammalian pheromones. *Current Biology* 14:R81–R89.

Buck LB. 1996. Information coding in the vertebrate olfactory system. *Annual Review of Neurosciences* 19:517–554.

Buck LB, Axel R. 1991. A novel multigene family may encode odorant receptors: a molecular basis for odor recognition. *Cell* 65:175–187.

Bushdid C, Magnasco MO, Vosshall LB, Keller A. 2014. Humans can discriminate more than 1 trillion olfactory stimuli. *Science* 343:1370–1372.

Dhawale AK, Hagiwara A, Bhalla US, Murthy VN, Albeanu DF. 2010. Non-redundant odor coding by sister mitral cells revealed by light addressable glomeruli in the mouse. *Nature Neuroscience* 13:1404–1412.

Dorries KM. 1998. Olfactory coding: time in a model. *Neuron* 20:7–10.

Engen T. 1991. *Odor Sensation and Memory*. New York: Praeger.

Fain GL. 2003. *Sensory Transduction*. Sunderland, MA: Sinauer.

Getchell TV, Doty RL, Bartoshuk LM, Snow JB. 1991. *Smell and Taste in Health and Disease*. New York: Raven Press.

Jones G, Teeling EC, Rossiter SJ. 2013. From the ultrasonic to the infrared: molecular evolution and the sensory biology of bats. *Frontiers in Physiology* 4:117.

Kauer JS. 1991. Contributions of topography and parallel processing to odor coding in the vertebrate olfactory pathway. *Trends in Neurosciences* 14:79–85.

Kinnamon SC. 2013. Neurosensory transmission without a synapse: new perspectives on taste signaling. *BMC Biology* 11:42.

Laurent G. 2002. Olfactory network dynamics and the coding of multidimensional signals. *Nature Reviews Neuroscience* 3:884–895.

Laurent G, Wehr M, Davidowitz H. 1996. Temporal representations of odors in an olfactory network. *Journal of Neuroscience* 16:3837–3847.

Liberles SD. 2014. Mammalian pheromones. *Annual Review of Physiology* 76:151–175.

Liman ER, Zhang YV, Montell C. 2014. Peripheral coding of taste. *Neuron* 81:984–1000.

Luo M, Katz LC. 2004. Encoding pheromonal signals in the mammalian vomeronasal system. *Current Opinion in Neurobiology* 14:429–434.

Mattes RD. 2009. Is there a fatty acid taste? *Annual Review of Nutrition* 29:305–27.

McClintock MK. 1971. Menstrual synchrony and suppression. *Nature* 229:244–245.

Meredith M. 2001. Human vomeronasal organ function: a critical review of best and worst cases. *Chemical Senses*. 26:433–445.

Mombaerts P. 2004. Genes and ligands for odorant, vomeronasal and taste receptors. *Nature Reviews Neuroscience* 5:263–278.

Murthy VN. 2011. Olfactory maps in the brain. *Annual Review of Neuroscience* 34:233–258.

Nakamura T, Gold GH. 1987. A cyclic nucleotide-gated conductance in olfactory receptor cilia. *Nature* 325:442–444.

Nelson G, Hoon MA, Chandrashekar J, Zhang Y, Ryba NJ, Zuker CS. 2001. Mammalian sweet taste receptors. *Cell* 106:381–390.

Ressler J, Sullivan SL, Buck LB. 1993. A zonal organization of odorant receptor gene expression in the olfactory epithelium. *Cell* 73:597–609.

Sato T. 1980. Recent advances in the physiology of taste cells. *Progress in Neurobiology* 14:25–67.

Stern K, McClintock MK. 1998. Regulation of ovulation by human pheromones. *Nature* 392:177–179.

Stettler DD, Axel R. 2009. Representations of odor in the piriform cortex. *Neuron* 63:854–864.

Stewart RE, DeSimone JA, Hill DL. 1997. New perspectives in gustatory physiology: transduction, development, and plasticity. *American Journal of Physiology* 272:C1–C26.

Stopfer M, Bhagavan S, Smith BH, Laurent G. 1997. Impaired odour discrimination on desynchronization of odour-encoding neural assemblies. *Nature* 390:70–74.

Strausfeld NJ, Hildebrand JG. 1999. Olfactory systems: common design, uncommon origins? *Current Opinion in Neurobiology* 9:634–639.

Wysocki CJ, Preti G. 2004. Facts, fallacies, fears, and frustrations with human pheromones. *Anatomical Record* 281A:1201–1211.

Zhang X, Firestein S. 2002. The olfactory receptor gene superfamily of the mouse. *Nature Neuroscience* 5:124–133.

Zhao GQ, Zhang Y, Hoon MA, Chandrashekar J, Erlenbach I, Ryba NJ, Zuker CS. 2003. The receptors for mammalian sweet and umami taste. *Cell* 115:255–266.

Chapter 9

Arshavsky VY, Lamb TD, Pugh EN. 2002. G proteins and phototransduction. *Annual Review of Physiology* 64:153–187.

Barlow H. 1953. Summation and inhibition in the frog's retina. *Journal of Physiology (London)* 119:69–78.

Baylor DA. 1987. Photoreceptor signals and vision. *Investigative Ophthalmology and Visual Science* 28:34–49.

Berson DM. 2003. Strange vision: ganglion cells as circadian photoreceptors. *Trends in Neurosciences* 26:314–320.

Burns ME, Baylor DA. 2001. Activation, deactivation, and adaptation in vertebrate photoreceptor cells. *Annual Review of Neuroscience* 24:779–805.

Curcio CA, Sloan KR, Kalina RE, Hendrickson AE. 1990. Human photoreceptor topography. *Journal of Comparative Neurology* 292:497–523.

Dacey DM, Packer OS. 2003. Colour coding in the primate retina: diverse cell types and cone-specific circuitry. *Current Opinion in Neurobiology* 13:421–427.

Dowling JE. 2012. *The Retina: An Approachable Part of the Brain*. Revised ed. Cambridge, MA: Harvard University Press.

Dowling JE, Werblin FS. 1971. Synaptic organization of the vertebrate retina. *Vision Research Suppl* 3:1–15.

Fesenko EE, Kolesnikov SS, Lyubarsky AL. 1985. Induction by cyclic GMP of cationic conductance in plasma membrane of retinal rod outer segment. *Nature* 313:310–313.

Field GD, Chichilinsky EJ. 2007. Information processing in the primate retina: circuitry and coding. *Annual Review of Neuroscience* 30:1–30.

Gegenfurtner KR, Kiper DC. 2003. Color vision. *Annual Review of Neuroscience* 26:181–206.

Hofer H, Carroll J, Neitz J, Neitz M, Williams DR. 2005. Organization of the human trichromatic cone mosaic. *Journal of Neuroscience* 25:9669–9679.

Kuffler S. 1953. Discharge patterns and functional organization of the mammalian retina. *Journal of Neurophysiology* 16:37–68.

Masland RH. 2001. The fundamental plan of the retina. *Nature Neuroscience* 4:877–886.

Nassi JJ, Callaway EM. 2009. Parallel processing strategies of the primate visual system. *Nature Reviews Neuroscience* 10:360–372.

Nathans J. 1999. The evolution and physiology of human color vision: insights from molecular genetic studies of visual pigments. *Neuron* 24:299–312.

Neitz J, Jacobs GH. 1986. Polymorphism of the long-wavelength cone in normal human colour vision. *Nature* 323:623–625.

Newell FW. 1996. *Ophthalmology*, 8th ed. St. Louis: Mosby.

Rodieck RW. 1998. *The First Steps in Seeing*. Sunderland, MA: Sinauer.

Roorda A, Williams DR. 1999. The arrangement of the three cone classes in the living human eye. *Nature* 397:520–522.

Schmidt TM, Chen S, Hattar S. 2011. Intrinsically photosensitive retinal ganglion cells: many subtypes, diverse functions. *Trends in Neurosciences* 34:572–580.

Schnapf JL, Baylor DA. 1987. How photoreceptor cells respond to light. *Scientific American* 256:40–47.

Schwab L. 1987. *Primary Eye Care in Developing Nations*. New York: Oxford University Press.

Smith SO. 2010. Structure and activation of the visual pigment rhodopsin. *Annual Review of Neuroscience* 39:309–328.

Solomon SG, Lennie P. 2007. The machinery of colour vision. *Nature Reviews Neuroscience* 8:276–286.

Wade NJ. 2007. Image, eye, and retina. *Journal of the Optical Society of America* 24:1229–1249.

Wässle H. 2004. Parallel processing in the mammalian retina. *Nature Reviews Neuroscience* 5:747–757.

Wässle H, Boycott B. 1991. Functional architecture of the mammalian retina. *Physiological Reviews* 71:447–480.

Watanabe M, Rodieck RW. 1989. Parasol and midget ganglion cells of the primate retina. *Journal of Comparative Neurology* 289:434–454.

Weiland JD, Liu W, Humayun MS. 2005. Retinal prosthesis. *Annual Review of Biomedical Engineering* 7:361–401.

Chapter 10

Alonso JM. 2002. Neural connections and receptive field properties in the primary visual cortex. *Neuroscientist* 8:443–456.

Barlow H. 1972. Single units and sensation: a neuron doctrine for perceptual psychology? *Perception* 1:371–394.

Callaway EM. 1998. Local circuits in primary visual cortex of the macaque monkey. *Annual Review of Neuroscience* 21:47–74

Casagrande VA, Xu X. 2004. Parallel visual pathways: a comparative perspective. In *The Visual Neurosciences*, eds. Chalupa L, Werner JS. Cambridge, MA: MIT Press, pp. 494–506.

Courtney SM, Ungerleider LG. 1997. What fMRI has taught us about human vision. *Current Opinion in Neurobiology* 7:554–561.

De Haan EHF, Cowey A. 2011. On the usefulness of "what" and "where" pathways in vision. *Trends in Cognitive Sciences* 15:460–466.

Desimone R, Albright TD, Gross CG, Bruce C. 1984. Stimulus-selective properties of inferior temporal neurons in the macaque. *Journal of Neuroscience* 4:2051–2062.

Fraser J. 1908. A new visual illusion of direction. *British Journal of Psychology* 2:307–320.

Gauthier I, Tarr MJ, Anderson AW, Skudlarski P, Gore JC. 1999. Activation of the middle fusiform "face area" increases with expertise in recognizing novel objects. *Nature Neuroscience* 2:568–573.

Gegenfurtner KR. 2003. Cortical mechanisms of colour vision. *Nature Reviews Neuroscience* 4:563–572.

Goodale MA, Westwood DA. 2004. An evolving view of duplex vision: separate but interacting cortical pathways for perception and action. *Current Opinion in Neurobiology* 14:203–211.

Grill-Spector K, Malach R. 2004. The human visual cortex. *Annual Reviews of Neuroscience* 27:649–677.

Gross CG, Rocha-Miranda CE, Bender DB. 1972. Visual properties of neurons in inferotemporal cortex of the macaque. *Journal of Neurophysiology* 35:96–111.

Hendry SHC, Reid RC. 2000. The koniocellular pathway in primate vision. *Annual Reviews of Neuroscience* 23:127–153.

Horibuchi S, ed. 1994. *Stereogram*. Tokyo: Shogakukan.

Hubel D. 1982. Explorations of the primary visual cortex, 1955–78 (Nobel lecture). *Nature* 299:515–524.

Hubel D. 1988. *Eye, Brain, and Vision*. New York: W.H. Freeman.

Hubel D, Wiesel T. 1962. Receptive fields, binocular interaction and functional architecture in the cat's visual cortex. *Journal of Physiology (London)* 160:106–154.

Hubel D, Wiesel T. 1968. Receptive fields and functional architecture of monkey striate cortex. *Journal of Physiology (London)* 195:215–243.

Hubel D, Wiesel T. 1977. Functional architecture of the macaque monkey visual cortex (Ferrier lecture). *Proceedings of the Royal Society of London Series B* 198:1–59.

Julesz B. 1971. *Foundations of Cyclopean Perception*. Chicago: University of Chicago Press.

Kenichi O, Reid RC. 2007. Specificity and randomness in the visual cortex. *Current Opinion in Neurobiology* 17:401–407.

Kourtzi Z, Connor CE. 2011. Neural representations for object perception: structure, category, and adaptive coding. *Annual Review of Neuroscience* 34:45–67.

Kravitz DJ, Kadharbatcha SS, Baker CI, Ungerleider LG, Mishkin M. 2013. The ventral visual pathway: an expanded neural framework for the processing of object quality. *Trends in Cognitive Sciences* 17:26–49.

Kreiman G. 2007. Single unit approaches to human vision and memory. *Current Opinion in Neurobiology* 17:471–475.

Kuffler SW. 1953. Discharge patterns and functional organization of mammalian retina. *Journal of Neurophysiology* 16:37–68.

Leopold DA. 2012. Primary visual cortex: awareness and blindsight. *Annual Review of Neuroscience* 35:91–109.

LeVay S, Wiesel TN, Hubel DH. 1980. The development of ocular dominance columns in normal and visually deprived monkeys. *Journal of Comparative Neurology* 191:1–51.

Livingstone M, Hubel D. 1984. Anatomy and physiology of a color system in the primate visual cortex. *Journal of Neuroscience* 4:309–356.

Martin K. 1994. A brief history of the "feature detector." *Cerebral Cortex* 4:1–7.

Milner AD, Goodale MA. 2008. Two visual systems re-viewed. *Neuropsychologia* 46:774–785.

Nassi JJ, Callaway EM. 2009. Parallel processing strategies of the primate visual system. *Nature Reviews Neuroscience* 10:360–372.

Ohki K, Chung S, Ch'ng YH, Kara P, Reid RC. 2005. Functional imaging with cellular resolution reveals precise microarchitecture in visual cortex. *Nature* 433:597–603.

Ohki K, Chung S, Kara P, Hubener M, Bonhoeffer T, Reid C. 2006. Highly ordered arrangement of single neurons in orientation pinwheels. *Nature* 442:925–928.

Ohki K, Reid RC. 2007. Specificity and randomness in the visual cortex. *Current Opinion in Neurobiology* 17:401–407.

Orban B. 2011. The extraction of 3D shape in the visual system of human and nonhuman primates. *Annual Review of Neuroscience* 34:361–388.

Palmer SE. 1999. *Vision Science: Photons to Phenomenology*. Cambridge, MA: MIT Press.

Paradiso MA. 2002. Perceptual and neuronal correspondence in primary visual cortex. *Current Opinion in Neurobiology* 12:155–161.

Salzman C, Britten K, Newsome W. 1990. Cortical microstimulation influences perceptual judgments of motion detection. *Nature* 346:174–177.

Sereno MI, Dale AM, Reppas JB, Kwong KK, Belliveau JW, Brady TJ, Rosen BR, Tootell RB. 1995. Borders of multiple visual areas in humans revealed by functional magnetic resonance imaging. *Science* 268:889–893.

Sharpee TO. 2013. Computational identification of receptive fields. *Annual Review of Neuroscience* 36:103–120.

Shepard RN. 1990. *Mind Sights: Original Visual Illusions, Ambiguities, and other Anomalies*. New York: W.H. Freeman.

Sherman SM. 2012. Thalamocortical interactions. *Current Opinion in Neurobiology* 22:575–579.

Sherman SM, Guillery RW. 2002. The role of the thalamus in the flow of information to the cortex. *Philosophical Transactions of the Royal Society of London B* 357:1695–1708.

Sincich LC, Horton JC. 2005. The circuitry of V1 and V2: integration of color, form, and motion. *Annual Review of Neuroscience* 28:303–326.

Singer W, Gray CM. 1995. Visual feature integration and the temporal correlation hypothesis. *Annual Review of Neuroscience* 18:555–586.

Tsao DY, Moeller S, Freiwald W. 2008. Comparing face patch systems in macaques and humans. *Proceedings of the National Academy of Science* 49:19514–19519.

Ts'o DY, Frostig RD, Lieke EE, Grinivald A. 1990. Functional organization of primate visual cortex revealed by high resolution optical imaging. *Science* 249:417–420.

Tyler C, Clarke MB. 1990. The autostereogram. *Proceedings of the International Society for Optical Engineering* 1256:182–197.

Weiner KS, Grill-Spector K. 2012. The improbable simplicity of the fusiform face area. *Trends in Cognitive Sciences* 16:251–254.

Zeki S. 1993. *A Vision of the Brain*. London: Blackwell Scientific.

Zeki S. 2003. Improbable areas in the visual brain. *Trends in Neuroscience* 26:23–26.

Zihl J, von Cramon D, Mai N. 1983. Selective disturbance of movement vision after bilateral brain damage. *Brain* 106:313–340.

Chapter 11

Ashida G, Carr CE. 2011. Sound localization: Jeffress and beyond. *Current Opinion in Neurobiology* 21:745–751.

Baloh RW, Honrubia V. 2001. *Clinical Neurophysiology of the Vestibular System*, 3rd ed. New York: Oxford University Press.

Brandt T. 1991. Man in motion: historical and clinical aspects of vestibular function. A review. *Brain* 114:2159–2174.

Copeland BJ, Pillsbury HC 3rd. 2004. Cochlear implantation for the treatment of deafness. *Annual Review Medicine* 55:157–167.

Cullen KE. 2012. The vestibular system: multimodal integration and encoding of self-motion for motor control. *Trends in Neurosciences* 35:185–196.

Eatock RA, Songer JE. 2011. Vestibular hair cells and afferents: two channels for head motion signals. *Annual Review of Neuroscience* 34:501–534.

Goldberg JM. 1991. The vestibular end organs: morphological and physiological diversity of afferents. *Current Opinion in Neurobiology* 1:229–235.

Guinan JJ Jr, Salt A, Cheatham MA. 2012. Progress in cochlear physiology after Békésy. *Hearing Research* 293:12–20.

Holt JR, Pan B, Koussa MA, Asai Y. 2014. TMC function in hair cell transduction. *Hearing Research* 311:17–24..

Hudspeth AJ. 1997. How hearing happens. *Neuron* 19:947–950.

Joris PX, Schreiner CE, Rees A. 2004. Neural processing of amplitude-modulated sounds. *Physiological Reviews* 84:541–577.

Kazmierczak P, Müller U. 2012. Sensing sound: molecules that orchestrate mechanotransduction by hair cells. *Trends in Neurosciences* 35:220–229.

Kemp DT. 2002. Otoacoustic emissions, their origin in cochlear function, and use. *British Medical Bulletin* 63:223–241.

Knipper M, Van Dijk P, Nunes I, Rüttiger L, Zimmermann U. 2013. Advances in the neurobiology of hearing disorders: recent developments regarding the basis of tinnitus and hyperacusis. *Progress in Neurobiology* 111:17–33.

Konishi M. 2003. Coding of auditory space. *Annual Review of Neuroscience* 26:31–55.

Liberman MC, Gao J, He DZ, Wu X, Jia S, Zuo J. 2002. Prestin is required for electromotility of the outer hair cell and for the cochlear amplifier. *Nature* 419:300–304.

McAlpine D, Grothe B. 2003. Sound localization and delay lines—do mammals fit the model? *Trends in Neurosciences* 26:347–350.

Middlebrooks JC, Green DM. 1991. Sound localization by human listeners. *Annual Review of Psychology* 42:135–159.

Oertel D. 1997. Encoding of timing in the brain stem auditory nuclei of vertebrates. *Neuron* 19:959–962.

Oertel D, Doupe AJ. 2013. The auditory central nervous system. In *Principles of Neural Science,* 5th ed., eds. Kandel ER, Schwartz JH, Jessell TM, Siegelbaum SA, Hudspeth AJ. New York: McGraw-Hill Companies, Inc., pp. 682–711.

Palmer AR. 2004. Reassessing mechanisms of low-frequency sound localisation. *Current Opinion in Neurobiology* 14:457–460.

Rose JE, Hind JE, Anderson DJ, Brugge JF. 1971. Some effects of stimulus intensity on response of auditory nerve fibers in the squirrel monkey. *Journal of Neurophysiology* 24:685–699.

Ruggero MA, Rich NC. 1996. Furosemide alters organ of Corti mechanics: evidence for feedback of outer hair cells upon the basilar membrane. *Journal of Neuroscience* 11:1057–1067.

Santos-Sacchi J. 2003. New tunes from Corti's organ: the outer hair cell boogie rules. *Current Opinion in Neurobiology* 13:459–468.

Shamma SA, Micheyl C. 2010. Behind the scenes of auditory perception. *Current Opinion in Neurobiology* 20:361–366.

Simmons JA. 1989. A view of the world through the bat's ear: the formation of acoustic images in echolocation. *Cognition* 33:155–199.

Suga N. 1995. Processing of auditory information carried by species-specific complex sounds. In *The Cognitive Neurosciences*, ed. Gazzaniga MS. Cambridge, MA: MIT Press, pp. 295–314.

Trussell LO. 1999. Synaptic mechanisms for coding timing in auditory neurons. *Annual Review of Physiology* 61:477-496.

Volta A. 1800. On the electricity excited by mere contact of conducting substances of different kinds. *Philosophical Transactions of the Royal Society of London* 90:403–431.

von Békésy G. 1960. *Experiments in Hearing*, ed. and trans. Wever EG. New York: McGraw-Hill.

Zeng F-G. 2004. Trends in cochlear implants. *Trends in Amplification* 8:1–34.

Zenner H-P, Gummer AW. 1996. The vestibular system. In *Comprehensive Mammalian Physiology. From Cellular Mechanisms to Integration, Vol. 1*, eds. Greger R, Windhorst U. Berlin: Springer-Verlag, pp. 697–710.

Chapter 12

Abraira VE, Ginty DD. 2013. The sensory neurons of touch. *Neuron* 79:618–639.

Bautista DM, Wilson SR, Hoon MA. 2014. Why we scratch an itch: the molecules, cells and circuits of itch. *Nature Neuroscience* 17:175–182.

Braz J, Solorzano C, Wang X, Basbaum AI. 2014. Transmitting pain and itch messages: a contemporary view of the spinal cord circuits that generate gate control. *Neuron* 82:522–536.

Chen R, Corwell B, Yaseen Z, Hallett M, Cohen L. 1998. Mechanisms of cortical reorganization in lower-limb amputees. *Journal of Neuroscience* 18(9):3443–3450.

Coste B, Mathur J, Schmidt M, Earley TJ, Ranade S, Petrus MJ, Dubin AE, Patapoutian A. 2010. Piezo1 and Piezo2 are essential components of distinct mechanically activated cation channels. *Science* 330:55–60.

Costigan M, Scholz J, Woolf CJ. 2009. Neuropathic pain: a maladaptive response of the nervous system to damage. *Annual Review of Neuroscience* 32:1–32.

Cox JJ, Reimann F, Nicholas AK, Thornton G, Roberts E, Springell K, Karbani G, Jafri H, Mannan J, Raashid Y, Al-Gazali L, Hamamy H, Valente EM, Gorman S, Williams R, McHale DP, Wood JN, Gribble FM, Woods CG. 2006. An SCN9A channelopathy causes congenital inability to experience pain. *Nature* 444:894–898.

DeFelipe C, Huerrero J, O'Brien J, Palmer J, Doyle C. 1998. Altered nociception, analgesia and aggression in mice lacking the receptor for substance P. *Nature* 392:394–397.

Diamond ME. 2010. Texture sensation through the fingertips and the whiskers. *Current Opinion in Neurobiology* 20:319–327.

Di Noto PM, Newman L, Wall S, Einstein G. 2013. The hermunculus: what is known about the representation of the female body in the brain? *Cerebral Cortex* 23:1005–1013.

Eijkelkamp N, Linley JE, Torres JM, Bee L, Dickenson AH, Gringhuis M, Minett MS, Hong GS, Lee E, Oh U, Ishikawa Y, Zwartkuis FJ, Cox JJ, Wood JN. 2013. A role for Piezo2 in EPAC1-dependent mechanical allodynia. *Nature Communications* 4:1682.

Eijkelkamp N, Quick K, Wood JN. 2013. Transient receptor potential channels and mechanosensation. *Annual Review of Neuroscience* 36:519–546.

Elbert T, Pantev C, Wienbruch C, Rockstroh B, Taub E. 1995. Increased cortical representation of the fingers of the left hand in string players. *Science* 270:305–306.

Fain GL. 2003. *Sensory Transduction.* Sunderland, MA: Sinauer.

Fields H. 2004. State-dependent opioid control of pain. *Nature Reviews Neuroscience* 5:565–575.

Gawande A. 2008, June 30. The itch. *The New Yorker*. 58–65.

Hsiao S. 2008. Central mechanisms of tactile shape perception. *Current Opinion in Neurobiology* 18:418–424.

Jenkins WM, Merzenich MM, Ochs MT, Allard T, Guic-Robles E. 1990. Functional reorganization of primary somatosensory cortex in adult owl monkeys after behaviorally controlled tactile stimulation. *Journal of Neurophysiology* 63:82–104.

Johnson KO, Hsiao SS. 1992. Neural mechanisms of tactile form and texture perception. *Annual Review of Neuroscience* 15:227–250.

Julius D, Basbaum AL. 2001. Molecular mechanisms of nociception. *Nature* 413:203–210.

Kaas SH, Nelson RH, Sur M, Merzenich MM. 1981. Organization of somatosensory cortex in primates. In *The Organization of the Cerebral Cortex*, eds. Schmitt FO, Worden FG, Adelman G, Dennis SG. Cambridge, MA: MIT Press, pp. 237–262.

Kass J. 1998. Phantoms of the brain. *Nature* 391:331–333.

Kell CA, von Kriegstein K, Rösler A, Kleinschmidt A, Laufs H. 2005. The sensory cortical representation of the human penis: revisiting somatotopy in the male homunculus. *Journal of Neuroscience* 25:5984–5987.

Loewenstein WR, Mendelson M. 1965. Components of receptor adaptation in a Pacinian corpusule. *Journal of Physiology* 177:377–397.

Maksimovic S, Nakatani M, Baba Y, Nelson AM, Marshall KL, Wellnitz SA, Firozi P, Woo SH, Ranade S, Patapoutian A, Lumpkin EA. 2014. Epidermal Merkel cells are mechanosensory cells that tune mammalian touch receptors. *Nature* 509:617–621.

Mantyh PW, Rogers SD, Honore P, Allen BJ, Ghilardi JR, Li J, et al. 1997. Inhibition of hyperalgesia by ablation of lamina I spinal neurons expressing the substance P receptor. *Science* 278:275–279.

McGlone F, Wessberg J, Olausson H. 2014. Discriminative and affective touch: sensing and feeling. *Neuron* 82:737–755.

Melzack R, Wall P. 1983. *The Challenge of Pain*. New York: Basic Books.

Mendelson M, Loewenstein WR. 1964. Mechanisms of receptor adaptation. *Science* 144:554–555.

Merzenich MM, Nelson RJ, Stryker MP, Cynader MS, Schoppman A. 1984. Somatosensory cortical map changes following digit amputation in adult monkeys. *Journal of Comparative Neurology* 224:591–605.

Mountcastle VB. 1997. The columnar organization of the neocortex. *Brain* 120:701–722.

Patapoutian A, Peier AM, Story GM, Viswanath V. 2003. ThermoTrp channels and beyond: mechanisms of temperature sensation. *Nature Reviews Neuroscience* 4:529–539.

Penfield W, Rasmussen T. 1952. *The Cerebral Cortex of Man*. New York: Macmillan.

Ramachandran VS. 1998. Consciousness and body image: lessons from phantom limbs, Capgras syndrome and pain asymbolia. *Philosophical Transactions of the Royal Society of London. Series B, Biological Sciences* 353(1377):1851–1859.

Sacks O. 1985. *The Man Who Mistook His Wife for a Hat and Other Clinical Tales*. New York: Summit.

Sadato N, Pascual-Leone A, Grafman J, Ibanez V, Delber M-P. 1996. Activation of the primary visual cortex by Braille reading in blind subjects. *Nature* 380:526–527.

Schmidt RF. 1978. *Fundamentals of Sensory Physiology*. New York: Springer-Verlag.

Springer SP, Deutsch G. 1989. *Left Brain, Right Brain*. New York: W.H. Freeman.

Taddese A, Nah S-Y, McCleskey E. 1995. Selective opioid inhibition of small nociceptive neurons. *Science* 270:1366–1369.

Tsunozaki M, Bautista DM. 2009. Mammalian somatosensory mechanotransduction. *Current Opinion in Neurobiology* 19:1–8.

Vallbo Å. 1995. Single-afferent neurons and somatic sensation in humans. In *The Cognitive Neurosciences*, ed. Gazzaniga M. Cambridge, MA: MIT Press, pp. 237–251.

Vallbo Å, Johansson R. 1984. Properties of cutaneous mechanoreceptors in the human hand related to touch sensation. *Human Neurobiology* 3:3–14.

Wall P. 1994. The placebo and the placebo response. In *Textbook of Pain*, eds. Wall P, Melzack R. Edinburgh: Churchill Livingstone, pp. 1297–1308.

Woo SH, Ranade S, Weyer AD, Dubin AE, Baba Y, Qiu Z, Petrus M, Miyamoto T, Reddy K, Lumpkin EA, Stucky CL, Patapoutian A. 2014. Piezo2 is required for Merkel-cell mechanotransduction. *Nature* 509:622–626.

Woolsey TA, Van Der Loos H. 1970. The structural organization of layer IV in the somatosensory region (S1) of mouse cerebral cortex: the description of a cortical field composed of discrete cytoarchitectonic units. *Brain Research* 17:205–242.

Chapter 13

Brown T. 1911. The intrinsic factors in the act of progression in the mammal. *Proceedings of the Royal Society of London Series B* 84:308–319.

Buller A, Eccles J, Eccles R. 1960. Interactions between motoneurons and muscles in respect to the characteristic speeds of their responses. *Journal of Physiology (London)* 150:417–439.

Bullinger KL, Nardelli P, Pinter MJ, Alvarez FJ, Cope TC. 2011. Permanent central synaptic disconnection of proprioceptors after nerve injury and regeneration. II. Loss of functional connectivity with motoneurons. *Journal of Neurophysiology* 106:2471–2485.

Burke RE, Levine DN, Tsairis P, Zajac FE 3rd. 1973. Physiological types and histochemical profiles in motor units of the cat gastrocnemius. *Journal of Physiology (London)* 234:723–748.

Dalkilic I, Kunkel LM. 2003. Muscular dystrophies: genes to pathogenesis. *Current Opinion in Genetics and Development* 13:231–238.

Eccles JC. 1974. Trophic interactions in the mammalian central nervous system. *Annals of the New York Academy of Sciences* 228:406–423.

Enoka RM, Pearson KG. 2013. The motor unit and muscle action. In *Principles of Neural Science*, 5th ed., eds. Kandel ER, Schwartz JH, Jessell TM, Siegelbaum SA, Hudspeth AJ. New York: McGraw-Hill.

Grillner S, Ekeberg Ö, El Manira A, Lansner A, Parker D, Tegnér J, Wallén P. 1998. Intrinsic function of a neuronal network: a vertebrate central pattern generator. *Brain Research Reviews* 26:184–197.

Haftel VK, Bichler EK, Wang QB, Prather JF, Pinter MJ, Cope TC. 2005. Central suppression of regenerated proprioceptive afferents. *Journal of Neuroscience* 25:4733–4742.

Henneman E, Somjen G, Carpenter D. 1965. Functional significance of cell size in spinal motoneurons. *Journal of Neurophysiology* 28:560–580.

Huxley A, Niedergerke R. 1954. Structural changes in muscle during contraction. Interference microscopy of living muscle fibres. *Nature* 173:971–973.

Huxley H, Hanson J. 1954. Changes in cross-striations of muscle during contraction and stretch and their structural interpretation. *Nature* 173:973–976.

Kernell D. 2006. *The Motoneurone and its Muscle Fibres*. New York: Oxford University Press.

Leung DG, Wagner KR. 2013. Therapeutic advances in muscular dystrophy. *Annals of Neurology* 74:404–411.

Lieber RL. 2002. *Skeletal Muscle Structure, Function, and Plasticity*, 2nd ed. Baltimore: Lippincott Williams & Wilkins.

Lømo T, Westgaard R, Dahl H. 1974. Contractile properties of muscle: control by pattern of muscle activity in the rat. *Proceedings of the Royal Society of London Series B* 187:99–103.

Mendell L, Henneman E. 1968. Terminals of single Ia fibers: distribution within a pool of 300 homonymous motor neurons. *Science* 160:96–98.

Nicolle MW. 2002. Myasthenia gravis. *The Neurologist* 8:2–21.

Patrick J, Lindstrom J. 1973. Autoimmune response to acetylcholine receptor. *Science* 180:871–872

Pette D. 2001. Historical perspectives: plasticity of mammalian skeletal muscle. *Journal of Applied Physiology* 90:1119–1124.

Poppele R, Bosco G. 2003. Sophisticated spinal contributions to motor control. *Trends in Neurosciences* 26:269–276.

Renton AE, Chiò A, Traynor BJ. 2014. State of play in amyotrophic lateral sclerosis genetics. *Nature Neuroscience* 17:17–23.

Rotterman TM, Nardelli P, Cope TC, Alvarez FJ. 2014. Normal distribution of VGLUT1 synapses on spinal motoneuron dendrites and their reorganization after nerve injury. *Journal of Neuroscience* 34:3475–3492.

Schouenborg J, Kiehn O, eds. 2001. The Segerfalk symposium on principles of spinal cord function, plasticity, and repair. *Brain Research Reviews* 40:1–329.

Sherrington C. 1947. *The Integrative Action of the Nervous System,* 2nd ed. New Haven: Yale University Press.

Sherrington C. 1979. 1924 Linacre lecture. In *Sherrington: His Life and Thought*, eds. Eccles JC, Gibson WC. New York: Springer-Verlag, p. 59.

Silvestri NJ, Wolfe GI. 2012. Myasthenia gravis. *Seminars in Neurology* 32:215–226.

Stein PSG, Grillner S, Selverston AI, Stuart DG, eds. 1999. *Neurons, Networks, and Motor Behavior*. Cambridge, MA: MIT Press.

Vucic S, Rothstein JD, Kiernan MC. 2014. Advances in treating amyotrophic lateral sclerosis: insights from pathophysiological studies. *Trends in Neuroscience* 37:433–442.

Wallen P, Grillner S. 1987. *N*-methyl-D-aspartate receptor-induced, inherent oscillatory activity in neurons active during fictive locomotion in the lamprey. *Journal of Neuroscience* 7:2745–2755.

Windhorst U. 2007. Muscle proprioceptive feedback and spinal networks. *Brain Research Bulletin* 73:155–202.

Chapter 14

Alstermark B, Isa T. 2012. Circuits for skilled reaching and grasping. *Annual Review of Neuroscience* 35:559–578.

Andersen RA, Musallam S, Pesaran B. 2004. Selecting the signals for a brain-machine interface. *Current Opinion in Neurobiology* 14:720–726.

Betz W. 1874. Anatomischer Nachweis zweier Gehirncentra. *Centralblatt für die medizinischen Wissenschaften* 12:578–580, 595–599.

Blumenfeld H. 2011. *Neuroanatomy through Clinical Cases,* 2nd ed. Sunderland, MA: Sinauer.

Campbell A. 1905. *Histological Studies on the Localization of Cerebral Function*. Cambridge, England: Cambridge University Press.

The Cerebellum: Development, Physiology, and Plasticity. 1998. *Trends in Neurosciences* 21:367–419 (special issue).

Cheney PD, Fetz EE, Palmer SS. 1985. Patterns of facilitation and suppression of antagonist forelimb muscles from motor cortex sites in the awake monkey. *Journal of Neurophysiology* 53:805–820.

Dauer W, Przedborski S. 2003. Parkinson's disease: mechanisms and models. *Neuron* 39:889–909.

Donoghue JP. 2002. Connecting cortex to machines: recent advances in brain interfaces. *Nature Neuroscience* 5(Suppl):1085–1088.

Evarts EV. 1973. Brain mechanisms in movement. *Scientific American* 229:96–103.

Feigin A. 1998. Advances in Huntington's disease: implications for experimental therapeutics. *Current Opinion in Neurology* 11:357–362.

Ferrier D. 1890. The Croonian lectures on cerebral localisation. Delivered before the Royal College of Physicians, June 1890. London: Smith Elder, 61:152.

Foltynie T, Kahan J. 2013. Parkinson's disease: an update on pathogenesis and treatment. *Journal of Neurology* 260:1433–1440.

Fritsch G, Hitzig E. 1870/1960. On the electrical excitability of the cerebrum, trans. von Bonin G. In *Some Papers on the Cerebral Cortex*, Springfield: Thomas, 1960:73–96. Originally published in 1870.

Georgopoulos A, Caminiti R, Kalaska J, Massey J. 1983. Spatial coding of movement: a hypothesis concerning the coding of movement direction by motor control populations. *Experimental Brain Research* Suppl 7:327–336.

Georgopoulos A, Kalaska J, Caminiti R, Massey J. 1982. On the relations between the direction of two-dimensional arm movements and cell discharge in primate motor cortex. *Journal of Neuroscience* 2:1527–1537.

Glickstein M, Doron K. 2008. Cerebellum: connections and functions. *Cerebellum* 7:589–594.

Graziano M. 2006. The organization of behavioral repertoire in motor cortex. *Annual Review of Neuroscience* 29:105–134.

Kilner JM, Lemon RN. 2013. What we know currently about mirror neurons. *Current Biology* 23:R1057–R1062.

Langston JW, Palfreman J. 1995. *The Case of the Frozen Addicts*. New York: Pantheon.

Lawrence D, Kuypers H. 1968. The functional organization of the motor system in the monkey: I. The effects of bilateral pyramidal lesions. *Brain* 91:1–14.

Lawrence D, Kuypers H. 1968. The functional organization of the motor system in the monkey: II. The effects of lesions of the descending brain-stem pathways. *Brain* 91:15–36.

Lemon RN. 2008. Descending pathways in motor control. *Annual Review of Neuroscience* 31:195–218.

Lozano AM, Lipsman N. 2013. Probing and regulating dysfunctional circuits using deep brain stimulation. *Neuron* 77:406–424.

Porter R, Lemon R. 1993. *Corticospinal Function and Voluntary Movement.* Oxford, England: Clarendon Press.

Rizzolatti G, Fadiga L, Gallese V, Fogassi L. 1996. Premotor cortex and the recognition of motor actions. *Brain Research: Cognitive Brain Research* 3:131–141.

Rizzolatti G, Sinigaglia C. 2008. *Mirrors in the Brain: How Our Minds Share Actions and Emotions.* New York: Oxford University Press.

Roland PE, Zilles K. 1996. Functions and structures of the motor cortices in humans. *Current Opinion in Neurobiology* 6:773–781.

Roland P, Larsen B, Lassen N, Skinhøf E. 1980. Supplementary motor area and other cortical areas in organization of voluntary movements in man. *Journal of Neurophysiology* 43:118–136.

Sanes JN, Donoghue JP. 1997. Static and dynamic organization of motor cortex. *Advances in Neurology* 73:277–296.

Sanes JN, Donoghue JP. 2000. Plasticity and primary motor cortex. *Annual Review of Neuroscience* 23:393–415.

Schwalb JM, Hamani C. The history and future of deep brain stimulation. *Neurotherapeutics* 5:3–13.

Shadmehr R, Smith MA, Krakauer JW. 2010. Error correction, sensory prediction, and adaptation in motor control. *Annual Review of Neuroscience* 33:89–108.

Strange PG. 1992. *Brain Biochemistry and Brain Disorders*. New York: Oxford University Press.

Weinrich M, Wise S. 1982. The premotor cortex of the monkey. *Journal of Neuroscience* 2:1329–1345.

Wichmann T, DeLong MR. 2003. Functional neuroanatomy of the basal ganglia in Parkinson's disease. *Advances in Neurology* 91:9–18.

Chapter 15

Aghananian GK, Marek GJ. 1999. Serotonin and hallucinogens. *Neuropsychopharmacology* 2(Suppl):16S–23S.

Appenzeller O. 1990. *The Autonomic Nervous System: An Introduction to Basic and Clinical Concepts*, 4th ed. New York: Elsevier.

Aston-Jones G, Bloom FE. 1981. Norepinephrine-containing locus coeruleus neurons in behaving rats exhibit pronounced responses to non-noxious environmental stimuli. *Journal of Neuroscience* 1:887–900.

Bloom FE. 2010. The catecholamine neuron: historical and future perspectives. *Progress in Neurobiology* 90:75–81.

Bloom FE, Hoffer BJ, Siggins GR. 1972. Norepinephrine mediated cerebellar synapses: a model system for neuropsychopharmacology. *Biological Psychiatry* 4:157–177.

Carlsson A. 2001. A paradigm shift in brain research. *Science* 294:1021–1024.

Cooper JR, Bloom FE, Roth RH. 2002. *The Biochemical Basis of Neuropharmacology*. New York: Oxford University Press.

Dahlstroem A, Fuxe K. 1964. A method for the demonstration of monoamine-containing nerve fibers in the central nervous system. *Acta Physiologica Scandinavica* 60:293–294.

Falck B, Hillarp NA. 1959. On the cellular localization of catechol amines in the brain. *Acta Anatomica* 38:277–279.

Foote SL, Bloom FE, Aston-Jones G. 1983. Nucleus locus ceruleus: new evidence of anatomical and physiological specificity. *Physiological Reviews* 63:844–914.

Furness, JB. 2012. The enteric nervous system and neurogastroenterology. *National Review of Gastroenterological Hepatology* 9:286–294.

Hofmann A. 1979. How LSD originated. *Journal of Psychedelic Drugs* 11:1–2.

Jänig W, McLachlan EM. 1992. Characteristics of function-specific pathways in the sympathetic nervous system. *Trends in Neurosciences* 15:475–481.

Kerr DS, Campbell LW, Applegate MD, Brodish A, Landsfield PW. 1991. Chronic stress-induced acceleration of electrophysiologic and morphometric biomarkers of hippocampal aging. *Journal of Neuroscience* 11:1316–1324.

Kerr DS, Campbell LW, Hao S-Y, Landsfield PW. 1989. Corticosteroid modulation of hippocampal potentials: increased effect with aging. *Science* 245:1505–1509.

Koob GF. 1992. Drugs of abuse: anatomy, pharmacology and function of reward pathways. *Trends in Pharmacological Sciences* 13:177–184.

McEwen BS. 2002. Sex, stress and the hippocampus: allostasis, allostatic load and the aging process. *Neurobiological Aging* 23(5):921–939.

McEwen BS, Schmeck HM. 1994. *The Hostage Brain*. New York: Rockefeller University Press.

Meyer JS, Quenzer LF. 2013. *Psychopharmacology: Drugs, the Brain, and Behavior*, 2nd ed. Sunderland, MA: Sinauer.

Moore RY, Bloom FE. 1979. Central catecholamine neuron systems: anatomy and physiology of the norepinephrine and epinephrine systems. *Annual Review of Neuroscience* 2:113–168.

Morrison JH, Foote SL, O'Connor D, Bloom FE. 1982. Laminar, tangential and regional organization of the noradrenergic innervation of monkey cortex: dopamine-beta-hydroxylase immunohistochemistry. *Brain Research Bulletin* 9:309–319.

Sapolsky RM. 1994. *Why Zebras Don't Get Ulcers: A Guide to Stress, Stress-Related Diseases, and Coping*. New York: W.H. Freeman.

Sapolsky RM, Krey LC, McEwen BS. 1986. The neuroendocrinology of stress and aging: the glucocorticoid cascade hypothesis. *Endocrine Reviews* 7:284–301.

Scharrer E, Scharrer B. 1939. Secretory cells within the hypothalamus. *Research Publications—Association for Research in Nervous and Mental Disease* 20:179–197.

Snyder SH. 1986. *Drugs and the Brain*. New York: Scientific American Books.

Watanabe Y, Gould E, McEwen BS. 1992. Stress induces atrophy of apical dendrites of hippocampal CA3 pyramidal neurons. *Brain Research* 588:341–345.

Wurtman RJ, Wurtman JJ. 1989. Carbohydrates and depression. *Scientific American* 260:68–75.

Chapter 16

Berridge KC. 2004. Motivation concepts in behavioral neuroscience. *Physiology & Behavior* 81:179–209.

Berridge KC. 2009. "Liking" and "wanting" food rewards: brain substrates and roles in eating disorders. *Physiology & Behavior* 97:537–550.

Berridge KC, Robinson TE. 1998. What is the role of dopamine in reward: hedonic impact, reward learning, or incentive salience? *Brain Research Review* 28:308–367.

Di Marzo V, Ligresti A, Cristino L. 2009. The endocannabinoid system as a link between homoeostatic and hedonic pathways involved in energy balance regulation. *International Journal of Obesity (London)* 33(Suppl 2):S18–S24.

Flier JS. 2004. Obesity wars: molecular progress confronts an expanding epidemic. *Cell* 116:337–350.

Friedman JM. 2004. Modern science versus the stigma of obesity. *Nature Medicine* 10:563–569.

Friedman JM. 2009. Leptin at 14 y of age: an ongoing story. *The American Journal of Clinical Nutrition* 89:973S–979S.

Gao Q, Hovath TL. 2007. Neurobiology of feeding and energy expenditure. *Annual Review of Neuroscience* 30:367–398.

Gibson WT, Farooqui IS, Moreau M, DePaoli AM, Lawrence E, O'Rahilly S, Trussell RA. 2004. Congenital leptin deficiency due to homozygosity for the delta 133 mutation: report of another case and evaluation response to four years of leptin therapy. *Journal of Clinical Endocrinology and Metabolism* 89:4821–4826.

Glimcher PW, Fehr E. 2014. *Neuroeconomics: Decision Making and the Brain*, 2nd ed. San Diego, CA: Academic Press.

Heath RG. 1963. Electrical self-stimulation of the brain in man. *American Journal of Psychiatry* 120:571–577.

Hoebel BG. 1997. Neuroscience and appetitive behavior research: 25 years. *Appetite* 29:119–133.

Kauer JA, Malenka RC. 2007. Synaptic plasticity and addiction. *Nature Reviews Neuroscience* 8:844–858.

Koob GF, Sanna PP, Bloom FE. 1998. Neuroscience of addiction. *Neuron* 21:467–476.

Navakkode S, Korte M. 2014. Pharmacological activation of cb1 receptor modulates long term potentiation by interfering with protein synthesis. *Neuropharmacology* 79:525–533.

Olds J, Milner P. 1954. Positive reinforcement produced by electrical stimulation of the septal area and other regions of the rat brain. *Journal of Comparative Physiological Psychology* 47:419–427.

Saper CB, Chou TC, Elmquist JK. 2002. The need to feed: homeostatic and hedonic control of eating. *Neuron* 36:199–211.

Sawchenko PE. 1998. Toward a new neurobiology of energy balance, appetite, and obesity: the anatomists weigh in. *Journal of Comparative Neurology* 402:435–441.

Schultz W. 2002. Getting formal with dopamine and reward. *Neuron* 36:241–263.

Schultz W. 2013. Updating dopamine reward signals. *Current Opinion in Neurobiology* 23:229–238.

Schwartz DH, Hernandez L, Hoebel BG. 1990. Serotonin release in lateral and medial hypothalamus during feeding and its anticipation. *Brain Research Bulletin* 25:797–802.

Soria-Gomez E, Bellocchio L, Reguero L, Lepousez G, Martin C, Bendahmane M,Ruehle S, Remmers F, Desprez T, Matias I, Wiesner T, Cannich A, Nissant A, Wadleigh A, Pape HC, Chiarlone AP, Quarta C, Verrier D, Vincent P, Massa F, Lutz B, Guzmán M, Gurden H, Ferreira G, Lledo PM, Grandes P, Marsicano G. 2014. The endocannabinoid system controls food intake via olfactory processes. *Nature Neuroscience* 17:407–415.

Squire LR, Berg D, Bloom FE, du Lac S, Ghosh A, Spitzer NC. 2012. *Fundamental Neuroscience*, 4th ed. San Diego: Academic Press.

Wise RA. 2004. Dopamine, learning, and motivation. *Nature Reviews Neuroscience* 5:483–494.

Woods SC, Seeley RJ, Porte D, Schwartz MW. 1998. Signals that regulate food intake and energy homeostasis. *Science* 280:1378–1382.

Woods SC, Stricker EM. 1999. Food intake and metabolism. In *Fundamental Neuroscience*, eds. Zigmond MJ, Bloom FE, Landis SC, Roberts JL, Squire LR. New York: Academic Press, pp. 1091–1109.

Chapter 17

Agate RJ, Grisham W, Wade J, Mann S, Wingfield J, Schanen C, Palotie A, Arnold AP. 2003. Neural, not gonadal, origin of brain sex differences in a gynandromorphic finch. *Proceedings of the National Academy of Sciences USA* 100:4873–4878.

Allen LS, Richey MF, Chai YM, Gorski RA. 1991. Sex differences in the corpus callosum of the living human being. *Journal of Neuroscience* 11:933–942.

Alvarez-Buylla A, Kirn JR. 1997. Birth, migration, incorporation, and death of vocal control neurons in adult songbirds. *Journal of Neurobiology* 33:585–601.

Amateau SK, McCarthy MM. 2004. Induction of PGE2 by estradiol mediates developmental masculinization of sex behavior. *Nature Neuroscience* 7:643–650.

Arnold AP. 2004. Sex chromosomes and brain gender. *Nature Reviews Neuroscience* 50:701–708.

Bakker J, Baum MJ. 2008. Role for estradiol in female-typical brain and behavioral sexual differentiation. *Frontiers in Neuroendocrinology* 29:1–16.

Bakker J, de Mees C, Douhard Q, Balthazart J, Gabant P, Szpirer J, Szpirer C. 2006. Alpha-fetoprotein protects the developing female mouse brain from masculinization and defeminization by estrogens. *Nature Neuroscience* 9:220–226.

Bartels A, Zeki S. 2004. The neural correlates of maternal and romantic love. *Neuroimage* 21:1155–1166.

Berne RM, Levy MN. 2009. *Physiology* 6th ed. St. Louis: Mosby.

Blum D. 1997. *Sex on the Brain: The Biological Differences Between Men and Women*. New York: Viking.

Breedlove SM. 1994. Sexual differentiation in the human nervous system. *Annual Review of Psychology* 45:389–418.

Colapinto J. 2001. *As Nature Made Him: The Boy Who Was Raised as a Girl*. New York: Harper Collins.

De Boer A, van Buel EM, ter Horst GJ. 2012. Love is more than just a kiss: a neurobiological perspective on love and affection. *Neuroscience* 201:114–124.

Dewing P, Shi T, Horvath S, Vilain E. 2003. Sexually dimorphic gene expression in mouse brain precedes gonadal differentiation. *Molecular Brain Research* 118:82–90.

Diamond J. 1997. *Why Is Sex Fun? The Evolution of Human Sexuality*. New York: Basic Books.

Fausto-Sterling A. 1992. *Myths of Gender: Biological Theories About Women and Men*. New York: Basic Books.

Fausto-Sterling A. 2000. *Sexing the Body*. New York: Basic Books.

Ferris CF, Kulkarni P, Sullivan JM Jr, Harder JA, Messenger TL, Febo M. 2005. Pup suckling is more rewarding than cocaine: evidence from functional magnetic resonance imaging and three-dimensional computational analysis. *Journal of Neuroscience* 25:149–156.

Garcia-Segura LM, Azcoitia I, DonCarlos LL. 2001. Neuroprotection by estradiol. *Progress in Neurobiology* 63:29–60.

Gilbert SF. 2013. *Developmental Biology*, 10th ed. Sunderland, MA: Sinauer.

Gould E, Woolley CS, Frankfurt M, McEwen BS. 1990. Gonadal steroids regulate spine density on hippocampal pyramidal cells in adulthood. *Journal of Neuroscience* 10:1286–1291.

Hamer DH, Hu S, Magnuson VL, Hu N, Pattatucci AM. 1993. A linkage between DNA markers on the X chromosome and male sexual orientation. *Science* 261:321–327.

Hines M. 2011. Gender development and the human brain. *Annual Review of Neuroscience* 34:69–88.

Insel TR, Young LJ. 2001. The neurobiology of attachment. *Nature Reviews Neuroscience* 2:129–136.

Kimura D. 1992. Sex differences in the brain. *Scientific American* 267:119–125.

Kimura D. 1996. Sex, sexual orientation and sex hormones influence human cognitive function. *Current Opinion in Neurobiology* 6:259–263.

Koopman P, Gubbay J, Vivian N, Goodfellow P, Lovell-Badge R. 1991. Male development of chromosomally female mice transgenic for Sry. *Nature* 351:117–121.

Kotrschal A, Rasanen K, Kristjansson BK, Senn M, Kolm N. 2012. Extreme sexual brain size dimorphism in sticklebacks: a consequence of the cognitive challenges of sex and parenting? *PLoS ONE* 7:1–4.

Kozorovitsky Y, Hughes M, Lee K, Gould E. 2006. Fatherhood affects dendritic spines and vasopressin V1a receptors in the primate prefrontal cortex. *Nature Neuroscience* 9:1094–1095.

LeVay S. 1991. A difference in hypothalamic structure between heterosexual and homosexual men. *Science* 253:1034–1037.

LeVay S. 1993. *The Sexual Brain.* Cambridge, MA: MIT Press.

LeVay S, Baldwin J. 2011. *Human Sexuality*, 4th ed. Sunderland, MA: Sinauer.

Lim MM, Wang Z, Olazabal DE, Ren X, Terwilliger EF, Young LJ. 2004. Enhanced partner preference in a promiscuous species by manipulating the expression of a single gene. *Nature* 429:754–757.

Maggi A, Ciana P, Belcredito S, Vegeto E. 2004. Estrogens in the nervous system: mechanisms and nonreproductive functions. *Annual Review of Physiology* 66:291–313.

McEwen BS. 1976. Interactions between hormones and nerve tissue. *Scientific American* 235:48–58.

McEwen BS. 1999. Permanence of brain sex differences and structural plasticity of the adult brain. *Proceedings of the National Academy of Sciences USA* 96:7128–7130.

McEwen BS, Akama KT, Spencer-Segal JL, Milner TA, Waters EM. 2012. Estrogen effects on the brain: actions beyond the hypothalamus via novel mechanisms. *Behavioral Neuroscience* 126:4–16.

McEwen BS, Davis PG, Parsons BS, Pfaff DW. 1979. The brain as a target for steroid hormone action. *Annual Review of Neuroscience* 2:65–112.

McLaren A. 1990. What makes a man a man? *Nature* 346:216–217.

Morris JA, Jordan CL, Breedlove SM. 2004. Sexual differentiation of the vertebrate nervous system. *Nature Neuroscience* 7:1034–1039.

Murphy DD, Cole NB, Greenberger V, Segal M. 1998. Estradiol increases dendritic spine density by reducing GABA neurotransmission in hippocampal neurons. *Journal of Neuroscience* 18:2550–2559.

Nottebohm F, Arnold AP. 1976. Sexual dimorphism in vocal control areas of the songbird brain. *Science* 194:211–213.

Pfaus JG. 1999. Neurobiology of sexual behavior. *Current Opinion in Neurobiology* 9:751–758.

Pfaus JG. 2009. Pathways of sexual desire. *Journal of Sexual Medicine* 6:1506–1533.

Rinn JL, Snyder M. 2005. Sexual dimorphism in mammalian gene expression. *Trends in Genetics* 21:298–305.

Roselli CE, Reddy RC, Kaufman KR. 2011. The development of male-oriented behavior in rams. *Frontiers in Neuroendocrinology* 32:164–169.

Rosenzweig MR, Breedlove SM, Watson NV. 2013. *Biological Psychology,* 7th ed. Sunderland, MA: Sinauer.

Sinclair AH, Berta P, Palmer MS, Hawkins JR, Griffiths BL, Smith MJ,Foster JW, Frischauf AM, Lovell-Badge R, Goodfellow PN. 1990. A gene from the human sex-determining region encodes a protein with homology to a conserved DNA-binding motif. *Nature* 346:240–242.

Smith MS, Freeman ME, Neill JD. 1975. The control of progesterone secretion during the estrous cycle and early pseudopregnancy in the rat: prolactin, gonadotropin and steroid levels associated with rescue of the corpus luteum of pseudopregnancy. *Endocrinology* 96:219–226.

Terasawa E, Timiras PS. 1968. Electrical activity during the estrous cycle of the rat: cyclical changes in limbic structures. *Endocrinology* 83:207–216.

Toran-Allerand CD. 1980. Sex steroids and the development of the newborn mouse hypothalamus and preoptic area in vitro. II. Morphological correlates and hormonal specificity. *Brain Research* 189:413–427.

Walum H, Westberg L, Henningsson S, Neiderhiser JM, Reiss D, Igl W, Ganiban JM, Spotts EL, Pedersen NL, Eriksson E, Lichtenstein P. 2008. Genetic variation in the vasopressin receptor 1a gene (AVPR1A) associates with pair-bonding behavior in humans. *Proceedings of the National Academy of Sciences* 105:14153–14156.

Woolley CS. 1999. Effects of estrogen in the CNS. *Current Opinion in Neurobiology* 9:349–354.

Woolley CS. 2007. Acute effects of estrogen on neuronal physiology. *Annual Review of Pharmacology and Toxicology* 47:657–680.

Woolley CS, Schwartzkroin PA. 1998. Hormonal effects on the brain. *Epilepsia* 39(Suppl):S2–S8.

Woolley CS, Weiland N, McEwen BS, Schwartzkroin PA. 1997. Estradiol increases the sensitivity of hippocampal CA1 pyramidal cells to NMDA receptor-mediated synaptic input: correlation with dendritic spine density. *Journal of Neuroscience* 17:1848–1859.

Wu MV, Shah NM. 2011. Control of masculinization of the brain and behavior. *Current Opinion in Neurobiology* 21:116–123.

Xerri C, Stern JM, Merzenich MM. 1994. Alterations of the cortical representation of the rat ventrum induced by nursing behavior. *Journal of Neuroscience* 14:1710–1721.

Young KA, Gobrogge KL, Liu Y, Wang Z. 2011. The neurobiology of pair bonding: insights from a socially monogamous rodent. *Frontiers in Neuroendocrinology* 32:53–69.

Young LJ, Wang Z, Insel TR. 1998. Neuroendocrine bases of monogamy. *Trends in Neurosciences* 21:71–75.

Yunis JJ, Chandler ME. 1977. The chromosomes of man—clinical and biologic significance: a review. *American Journal of Pathology* 88:466–495.

Zhao D, McBride D, Nandi S, McQueen HA, McGrew MJ, Hocking PM, Lewis PD, Sang HM, Clinton M. 2010. Somatic sex identity is cell autonomous in the chicken. *Nature* 464:237–243.

Chapter 18

Adolphs R. 2002. Neural systems for recognizing emotion. *Current Opinion in Neurobiology* 12:169–177.

Adolphs R, Tranel D, Damasio H, Damasio A. 1994. Impaired recognition of emotion in facial expressions following bilateral damage to the human amygdala. *Nature* 372:669–672.

Aggleton JP. 1993. The contribution of the amygdala to normal and abnormal emotional states. *Trends in Neurosciences* 16:328–333.

Bard P. 1934. On emotional expression after decortication with some remarks on certain theoretical views. *Psychological Reviews* 41:309–329.

Barrett LF, Satpute AB. 2013. Large-scale networks in affective and social neuroscience: towards an integrative functional architecture of the brain. *Current Opinion in Neurobiology* 23:361–372.

Breiter HC, Etcoff NL, Whalen PJ, Kennedy WA, Rauch SL, Buckner RL, Strauss MM, Hyman SE, Rosen BR. 1996. Response and habituation of the human amygdala during visual processing of facial expression. *Neuron* 17:875–887.

Broca P. 1878. Anatomie compare de circonvolutions cérébrales. Le grand lobe limbique et la scissure limbique dans la série des mammiféres. *Revue d'Anthropologie* 1:385–498.

Büchel C, Morris J, Dolan RJ, Friston KJ. 1998. Brain systems mediating aversive conditioning: an event-related fMRI study. *Neuron* 20:947–957.

Cannon WB. 1927. The James-Lange theory of emotion. *American Journal of Psychology* 39:106–124.

Dagleish T. 2004. The emotional brain. *Nature Reviews* 5:582–589.

Damasio A, Carvalho GB. 2013. The nature of feelings: evolutionary and neurobiological origins. *Nature Reviews Neuroscience* 14:143–152.

Damasio AR. 1989. Time-locked multiregional retroactivation: a systems level proposal for the neural substrates of recall and recognition. *Cognition* 33:25–62.

Damasio AR. 1994. *Descartes' Error*, 1st ed. New York: Penguin Books.

Damasio AR. 1996. The somatic marker hypothesis and the possible functions of the prefrontal cortex. *Transactions of the Royal Society (London)* 351:1413–1420.

Damasio H, Grabowski T, Frank R, Galaburda AM, Damasio AR. 1994. The return of Phineas Gage: clues about the brain from the skull of a famous patient. *Science* 264:1102–1105.

Darwin C. 1872/1955. *The Expression of the Emotions in Man and Animals*. New York: The Philosophical Library, 1955. Originally published in 1872.

Davis M. 1992. The role of the amygdala in fear and anxiety. *Annual Review of Neuroscience* 15:353–375.

Dolan RJ. 2002. Emotion, cognition, and behavior. *Science* 298:1191–1194.

Duke AA, Bell R, Begue L, Eisenlohr-Moul T. 2013. Revisiting the serotonin-aggression relation in humans: a meta-analysis. *Psychological Bulletin* 139:1148–1172.

Edwards DH, Kravitz EA. 1997. Serotonin, social status and aggression. *Current Opinion in Neurobiology* 7:812–819.

Flynn JP. 1967. The neural basis of aggression in cats. In *Neurophysiology and Emotion*, ed. Glass DC. New York: Rockefeller University Press.

Fulton JF. 1951. *Frontal Lobotomy and Affective Behavior. A Neurophysiological Analysis*. New York: Norton.

Gallagher M, Chiba AA. 1996. The amygdala and emotion. *Current Opinion in Neurobiology* 6:221–227.

Gendron M, Barrett LF. 2009. Reconstructing the past: a century of ideas about emotion in psychology. *Emotion Review* 1:316–339.

Gross CT, Canteras NS. 2012. The many paths to fear. *Nature Reviews Neuroscience* 13:651–658.

Hamann S. 2012. Mapping discrete and dimensional emotions onto the brain: controversies and consensus. *Trends in Cognitive Sciences* 16:458–466.

Hamann SB, Ely TD, Grafton ST, Kilts CD. 1999. Amygdala activity related to enhanced memory for pleasant and aversive stimuli. *Nature Neuroscience* 2:289–293.

Harlow JM. 1848. Passage of an iron rod through the head. *Boston Medical and Surgical Journal* 39:389–393.

Harlow JM. 1868. Recovery from the passage of an iron bar through the head. *Publication of the Massachusetts Medical Society* 2:329–347.

Heisler LK, Chu HM, Brennan TJ, Danao JA, Bajwa P, Parsons LH, Tecott LH. 1998. Elevated anxiety and antidepressant-like responses in serotonin 5-HT1A receptor mutant mice. *Proceedings of the National Academy of Sciences USA* 95:15049–15054.

Hess WR. 1954. *Diencephalon: Autonomic and Extrapyramidal Functions*. New York: Grune & Stratton.

Jacobsen CF, Wolf JB, Jackson TA. 1935. An experimental analysis of the functions of the frontal association areas in primates. *Journal of Nervous and Mental Disease* 82:1–14.

James W. 1884. What is an emotion? *Mind* 9:188–205.

Julius D. 1998. Serotonin receptor knockouts: a moody subject. *Proceedings of the National Academy of Sciences USA* 95:15153–15154.

Kalin NH. 1993. The neurobiology of fear. *Scientific American* 268:94–101.

Kapp BS, Pascoe JP, Bixler MA. 1984. The amygdala: a neuroanatomical systems approach to its contributions to aversive conditioning. In *Neuropsychology of Memory*, eds. Butler N, Squire LR. New York: Guilford.

Klüver H, Bucy PC. 1939. Preliminary analysis of functions of the temporal lobes in monkeys. *Archives of Neurology and Psychiatry* 42:979–1000.

LaBar KS, Gatenby JC, Gore JC, LeDoux JE, Phelps EA. 1998. Human amygdala activation during conditioned fear acquisition and extinction: a mixed-trial fMRI study. *Neuron* 20:937–945.

Lange CG. 1887. *Uber Gemuthsbewegungen*. Liepzig: T. Thomas.

LeDoux JE. 1994. Emotion, memory and the brain. *Scientific American* 270:50–57.

LeDoux JE. 2012. Rethinking the emotional brain. *Neuron* 73:653–676.

Lindquist KA, Wager TD, Kober H, Bliss-Moreau E, Barrett LF. 2012. The brain basis of emotion: a meta-analytic review. *Behavioral and Brain Sciences* 35:121–143.

MacLean PD. 1955. The limbic system ("visceral brain") and emotional behavior. *Archives of Neurology and Psychiatry* 73:130–134.

McGaugh JL. 2004. The amygdala modulates the consolidation of memories of emotionally arousing experiences. *Annual Review of Neuroscience* 27:1–28.

Meyer K, Damasio A. 2009. Convergence and divergence in a neural architecture for recognition and memory. *Trends in Neurosciences* 32(7):376–382.

Morris JS, Öhman A, Dolan RJ. 1998. Conscious and unconscious emotional learning in the human amygdala. *Nature* 393:467–470.

Nummenmaa L, Glerean E, Hari R, Hietanen JK. 2014. Bodily maps of emotions. *Proceedings of the National Academy of Sciences USA* 111:646–651.

Olivier B, van Oorschot R. 2005. 5-HT_{1B} receptors and aggression: a review. *European Journal of Pharmacology* 526:207–217.

Papez JW. 1937. A proposed mechanism of emotion. *Archives of Neurology and Psychiatry* 38:725–743.

Pare D, Quirk GJ, LeDoux JE. 2004. New vistas on amygdala networks in conditioned fear. *Journal of Neurophysiology* 92:1–9.

Pessoa L, Ungerleider LG. 2005. Neuroimaging studies of attention and the processing of emotion-laden stimuli. *Progress in Brain Research* 144:171–182.

Pribram KH. 1954. Towards a science of neuropsychology (method and data). In *Current Trends in Psychology and the Behavioral Sciences*, ed. Patton RA. Pittsburgh: University of Pittsburgh Press.

Raleigh MJ, McGuire MT, Brammer GL, Pollack DB, Yuwiler A. 1991. Serotonergic mechanisms promote dominance acquisition in adult vervet monkeys. *Brain Research* 559:181–190.

Saudou F, Amara DA, Dierich A, LeMeur M, Ramboz S, Segu L, Buhot MC, Hen R. 1994. Enhanced aggressive behavior in mice lacking 5-HT1B receptor. *Science* 265:1875–1878.

van Wyhe J, ed. 2002. *The Complete Work of Charles Darwin Online.* http://darwin-online.org.uk/

Chapter 19

Albrecht U. 2012. Timing to perfection: the biology of central and peripheral circadian clocks. *Neuron* 74:246–260.

Bal T, McCormick DA. 1993. Mechanisms of oscillatory activity in guinea-pig nucleus reticularis thalami *in vitro*: a mammalian pacemaker. *Journal of Physiology (London)* 468:669–691.

Basheer R, Strecker RE, Thakkar MM, McCarley RW. 2004. Adenosine and sleep-wake regulation. *Progress in Neurobiology* 73:379–396.

Berger H. 1929. Über das elektroenkephalogramm des menschen. *Archiv für Psychiatrie und Nervenkrankheiten* 87:527–570. Translated and republished as: Berger H. 1969. On the electroencephalogram of man. *Electroencephalography Clinical Neurophysiology* Suppl 28:37–73.

Berson DM. 2003. Strange vision: ganglion cells as circadian photoreceptors. *Trends in Neurosciences* 26:314–320.

Braun AR, Balkin TJ, Wesensten NJ, Gwadry F, Carson RE, Varga M, Baldwin P, Belenky G, Herscovitch P. 1998. Dissociated pattern of activity in visual cortices and their projections during human rapid eye movement sleep. *Science* 279:91–95.

Brown RE, Basheer R, McKenna JT, Strecker RE, McCarley RW. 2012. Control of sleep and wakefulness. *Physiological Reviews* 92:1087–1187.

Brzezinski A, Vangel MG, Wurtman RJ, Norrie G, Zhdanova I, Ben-Shushan A, Ford I. 2005. Effects of exogenous melatonin on sleep: a meta-analysis. *Sleep Medicine Reviews* 9:41–50.

Buzsáki G. 2006. *Rhythms of the Brain.* New York: Oxford University Press.

Buzsáki G, Anastassiou CA, Koch C. 2012. The origin of extracellular fields and currents—EEG, ECoG, LFP and spikes. *Nature Reviews Neuroscience* 13:407–420.

Buzsáki G, Logothetis N, Singer W. 2013. Scaling brain size, keeping timing: evolutionary preservation of brain rhythms. *Neuron* 80:751–764.

Carskadon MA, ed. 1993. *Encyclopedia of Sleep and Dreaming.* New York: Macmillan.

Carskadon MA, Acebo C, Jenni OG. 2004. Regulation of adolescent sleep: implications for behavior. *Annals of the New York Academy of Sciences* 1021:276–291.

Caton R. 1875. The electric currents of the brain. *British Medical Journal* 2:278.

Chemelli RM, Willie JT, Sinton CM, Elmquist JK, Scammell T, Lee C, Richardson JA, Williams SC, Xiong Y, Kisanuki Y, Fitch TE, Nakazato M, Hammer RE, Saper CB, Yanagisawa M. 1999. Narcolepsy in orexin knockout mice: molecular genetics of sleep regulation. *Cell* 98:437–451.

Cirelli C, Gutierrez CM, Tononi G. 2004. Extensive and divergent effects of sleep and wakefulness on brain gene expression. *Neuron* 41:35–43.

Cirelli C, Tononi G. 2008. Is sleep essential? *PLoS Biology* 6:e216.

Coleman RM. 1986. *Wide Awake at 3:00 A.M. by Choice or by Chance?* New York: W.H. Freeman.

Czeisler CA, Duffy JF, Shanahan TL, Brown EN, Mitchell JF, Rimmer DW, Ronda JM, Silva EJ, Allan JS, Emens JS, Dijk DJ, Kronauer RE. 1999. Stability, precision, and near-24-hour period of the human circadian pacemaker. *Science* 284:2177–2181.

Dement WC. 1976. *Some Must Watch While Some Must Sleep.* San Francisco: San Francisco Book Company.

Do MT, Yau KW. 2010. Intrinsically photosensitive retinal ganglion cells. *Physiological Reviews* 90:1547–1581.

Edgar DM, Dement WC, Fuller CA. 1993. Effect of SCN lesions on sleep in squirrel monkeys: evidence for opponent processes in sleep-wake regulation. *Journal of Neuroscience* 13:1065–1079.

Engel AK, Fries P. 2010. Beta-band oscillations—signaling the status quo? *Current Opinion in Neurobiology* 20:156–165.

Engel AK, Fries P, Singer W. 2001. Dynamic predictions: oscillations and synchrony in top-down processing. *Nature Reviews Neuroscience* 2:704–716.

Freeman W. 1991. The physiology of perception. *Scientific American* 264:78–85.

Fries P. 2009. Neuronal gamma-band synchronization as a fundamental process in cortical computation. *Annual Review of Neuroscience* 32:209–224.

Gekakis N, Staknis D, Nguyen HB, Davis FC, Wilsbacher LD, King DP, Takahashi JS, Weitz CJ. 1998. Role of the CLOCK protein in the mammalian circadian mechanism. *Science* 280:1564–1568.

Goldberg EM, Coulter DA. 2013. Mechanisms of epileptogenesis: a convergence on neural circuit dysfunction. *Nature Reviews Neuroscience* 14:337–349.

Greene R, Siegel J. 2004. Sleep: a functional enigma. *NeuroMolecular Medicine* 5:59–68.

Hobson JA. 1993. Sleep and dreaming. *Current Opinion in Neurobiology* 10:371–382.

Horne JA. 1988. *Why We Sleep: The Functions of Sleep in Humans and Other Mammals.* New York: Oxford University Press.

Jackson JH. 1932. *Selected Writings of John Hughlings Jackson,* ed. Taylor J. London: Hodder and Stoughton.

Jacobs MP, Leblanc GG, Brooks-Kayal A, Jensen FE, Lowenstein DH, Noebels JL, Spencer DD, Swann JW. 2009. Curing epilepsy: progress and future directions. *Epilepsy and Behavior* 14:438–445.

Karni A, Tanne D, Rubenstein BS, Akenasy JJM, Sagi D. 1994. Dependence on REM sleep of overnight performance of a perceptual skill. *Science* 265:679–682.

Kisanuki Y, Fitch TE, Nakazato M, Hammer RE, Saper CB, Yanagisawa M. 1999. Narcolepsy in orexin knockout mice: molecular genetics of sleep regulation. *Cell* 98:437–451.

Lamberg L. 1994. *Bodyrhythms: Chronobiology and Peak Performance.* New York: Morrow.

Lin L, Faraco J, Li R, Kadotani H, Rogers W, Lin X, Qiu X, de Jong PJ, Nishino S, Mignot E. 1999. The sleep disorder canine narcolepsy is caused by a mutation in the hypocretin (orexin) receptor 2 gene. *Cell* 98:365–376.

Lowrey PL, Takahashi JS. 2004. Mammalian circadian biology: elucidating genome-wide levels of temporal organization. *Annual Review of Genomics and Human Genetics* 5:407–441.

Lyamin OI, Manger PR, Ridgway SH, Mukhametov LM, Siegel JM. 2008. Cetacean sleep: an unusual form of mammalian sleep. *Neuroscience Biobehavioral Reviews* 32:1451–1484.

McCarley RW, Massaquoi SG. 1986. A limit cycle reciprocal interaction model of the REM sleep oscillator system. *American Journal of Physiology* 251:R1011.

McCormick DA, Bal T. 1997. Sleep and arousal: thalamocortical mechanisms. *Annual Review of Neuroscience* 20:185–216.

McCormick DA, Pape H-C. 1990. Properties of a hyperpolarization-activated cation current and its role in rhythmic oscillation in thalamic relay neurones. *Journal of Physiology (London)* 431:291–318.

Mohawk JA, Green CB, Takahashi JS. 2012. Central and peripheral circadian clocks in mammals. *Annual Review of Neuroscience* 35:445–462.

Moruzzi G. 1964. Reticular influences on the EEG. *Electroencephalography and Clinical Neurophysiology* 16:2–17.

Mukhametov LM. 1984. Sleep in marine mammals. In *Sleep Mechanisms*, eds. Borbély AA, Valatx JL. Munich: Springer-Verlag, pp. 227–238.

Noebels JL. 2003. The biology of epilepsy genes. *Annual Review of Neuroscience* 26:599–625.

Novarino G, Baek ST, Gleeson JG. 2013. The sacred disease: the puzzling genetics of epileptic disorders. *Neuron* 80:9–11.

Obal F Jr, Krueger JM. 2003. Biochemical regulation of non-rapid eye-movement sleep. *Frontiers in Bioscience* 8:520–550.

Pappenheimer JR, Koski G, Fencl V, Karnovsky ML, Krueger J. 1975. Extraction of sleep-promoting factor S from cerebrospinal fluid and from brains of sleep-deprived animals. *Journal of Neurophysiology* 38:1299–1311.

Partinen M, Kornum BR, Plazzi G, Jennum P, Julkunen I, Vaarala O. 2014. Narcolepsy as an autoimmune disease: the role of H1N1 infection and vaccination. *Lancet Neurology* 13:600–613.

Porkka-Heiskanen T, Strecker RE, Thakkar M, Bjorkum AA, Greene RW, McCarley RW. 1997. Adenosine: a mediator of the sleep inducing effects of prolonged wakefulness. *Science* 276:1265–1268.

Purves D, Augustine GJ, Fitzpatrick D, Hall WC, LaMantia AS, McNamara JO, Williams SM. 2004. *Neuroscience*, 3rd ed. Sunderland MA: Sinauer.

Ralph MR, Foster RG, Davis FC, Menaker M. 1990. Transplanted suprachiasmatic nucleus determines circadian period. *Science* 247:975–978.

Ralph MR, Menaker M. 1988. A mutation of the circadian system in golden hamsters. *Science* 241:1225–1227.

Savage N. 2014. The complexities of epilepsy. *Nature* 511:S2–S3.

Shaw PJ, Cirelli C, Greenspan RJ, Tononi G. 2000. Correlates of sleep and waking in *Drosophila melanogaster. Science* 287:1834–1837.

Siegel JM. 2004. Hypocretin (orexin): role in normal behavior and neuropathology. *Annual Review of Psychology* 55:125–148.

Steriade M, McCormick DA, Sejnowski TJ. 1993. Thalamocortical oscillations in the sleeping and aroused brain. *Science* 262:679–685.

Thannickal TC, Moore RY, Nienhuis R, Ramanathan L, Gulyani S, Aldrich M, Cornford M, Siegel JM. 2000. Reduced number of hypocretin neurons in human narcolepsy. *Neuron* 27:469–474.

Wheless JW, Castillo E, Maggio V, Kim HL, Breier JI, Simos PG, Papanicolaou AC. 2004. Magnetoencephalography (MEG) and magnetic source imaging (MSI). *The Neurologist* 10:138–153.

Winson J. 1993. The biology and function of rapid eye movement sleep. *Current Opinion in Neurobiology* 3:243–248.

Yamaguchi S, Isejima H, Matsuo T, Okura R, Yagita K, Kobayashi M, Okamura H. 2003. Synchronization of cellular clocks in the suprachiasmatic nucleus. *Science* 302:1408–1412.

Chapter 20

Baynes K, Eliassen JC, Lutsep HL, Gazzaniga MS. 1988. Modular organization of cognitive systems masked by interhemispheric integration. *Science* 280:902–905.

Berwick RC, Friederici AD, Chomsky N, Bolhuis JJ. 2013. Evolution, brain, and the nature of language. *Trends in Cognitive Sciences* 17:89–98.

Binder JR, Frost JA, Hammeke TA, Cox RW, Rao SM, Prieto T. 1997. Human brain language areas identified by functional magnetic resonance imaging. *Journal of Neuroscience* 17:353–362.

Bookeheimer S. 2002. Functional MRI of language: new approaches to understanding the cortical organization of semantic processing. *Annual Review of Neuroscience* 25:51–188.

Broca P. 1861. Perte de la parole, ramollissement chronique et destruction partielle du lobe anterieur gauche du cerveau. *Bulletins de la Societe d'Anthropologie* 2:235–238.

Damasio AR, Damasio H. 1992. Brain and language. *Scientific American* 267:88–95.

Dehaene-Lambertz G, Dehaene S, Hertz-Pannier L. 2002. Functional neuroimaging of speech perception in infants. *Science* 298:2013–2015.

Deutscher G. 2010. *Through the Looking Glass: Why the World Looks Different in Other Languages*. New York: Picador.

Dronkers NF, Plaisant O, Iba-Zizen MT, Cabanis EA. 2007. Paul Broca's historic cases high resolution MR imaging of the brains of Leborgne and Lelong. *Brain* 130:1432–1441.

Friederici, AD. 2012. The cortical language circuit: from auditory perception to sentence comprehension. *Trends in Cognitive Sciences* 16:262–268.

Fromkin V, Rodman R. 2013. *An Introduction to Language*, 10th ed. New York: Wadsworth Publishing Company.

Gardner H. 1974. *The Shattered Mind*. New York: Vintage Books.

Gardner RA, Gardner B. 1969. Teaching sign language to a chimpanzee. *Science* 165:664–672.

Gazzaniga MS. 1970. *The Bisected Brain*. New York: Appleton-Century-Crofts.

Geschwind N. 1979. Specializations of the human brain. *Scientific American* 241:180–199.

Geschwind N, Levitsky W. 1968. Human-brain: left-right asymmetries in temporal speech region. *Science* 161:186–187.

Graham SA, Fisher SE. 2013. Decoding the genetics of speech and language. *Current Opinion in Neurobiology* 23:43–51.

Hobaiter C, Byrne RW. 2014. The meanings of chimpanzee gestures. *Current Biology* 24:1596–1600.

Kuhl PK. 2004. Early language acquisition: cracking the speech code. *Nature Reviews Neuroscience* 5:831–843.

Kuhl PK. 2010. Brain mechanisms in early language acquisition. *Neuron* 67:713–727.

Lehericy S, Cohen L, Bazin B, Samson S, Giacomini E, Rougetet R, Hertz-Pannier L, Le Bihan D, Marsault C, Baulac M. 2000. Functional MR evaluation of temporal and frontal language dominance compared with the Wada test. *Neurology* 54:1625–1633.

Neville HJ, Bavelier D, Corina D, Rauschecker J, Karni A, Lalwani A, Braun A, Clark V, Jezzard P, Turner R. 1998. Cerebral organization for language in deaf and hearing subjects: biological constraints and effects of experience. *Proceedings of the National Academy of Sciences USA* 95:922–929.

Ojemann G, Mateer C. 1979. Human language cortex: localization of memory, syntax, and sequential motor-phoneme identification systems. *Science* 205:1401–1403.

Patterson FG. 1978. The gestures of a gorilla: language acquisition in another pongid. *Brain and Language* 5:56–71.

Penfield W, Rasmussen T. 1950. *The Cerebral Cortex of Man*. New York: Macmillan.

Petersen SE, Fox PT, Posner MI, Mintum M, Raichle ME. 1988. Positron emission tomographic studies of the cortical anatomy of single-word processing. *Nature* 331:585–589.

Pinker S. 1994. *The Language Instinct*. New York: Morrow.

Posner MI, Raichle M. 1994. *Images of Mind*. New York: Scientific American Library.

Rasmussen T, Milner B. 1977. The role of early left-brain injury in determining lateralization of cerebral speech functions. *Annals of New York Academy of Sciences* 299:355–369.

Sadato N, Pascual-Leone A, Grafman J, Ibanez V, Deiber M, Dold G, Hallett M. 1996. Activation of the primary visual cortex by Braille reading in blind subjects. *Nature* 380:526–528.

Saffran EM. 2000. Aphasia and the relationship of language and brain. *Seminars in Neurology* 20:409–418.

Saffran JR, Aslin RN, Newport EL. 1996. Statistical learning by 8-month-old infants. *Science* 274:1926–1928.

Saygin AP, Dick F, Wilson SW, Dronkers NF. Bates E. 2003. Neural resources for processing language and environmental sounds: evidence from aphasia. *Brain* 126:928–945.

Scott SK, Johnsrude IS. 2002. The neuroanatomical and functional organization of speech perception. *Trends in Neurosciences* 26:100–107.

Sperry RW. 1964. The great cerebral commissure. *Scientific American* 210:42–52.

Spreer J, Arnold S, Quiske A, Wohlfarth R, Ziyeh S, Altenmuller D, Herpers M, Kassubek J, Klisch J, Steinhoff BJ, Honegger J, Schulze-Bonhage A, Schumacher M. 2002. Determination of hemisphere dominance for language: comparison of frontal and temporal fMRI activation with intracarotid amytal testing. *Neuroradiology* 44:467–474.

Vargha-Khadem F, Gadian DG, Copp A, Mishkin M. 2005. *FOXP2* and the neuroanatomy of speech and language. *Nature Reviews Neuroscience* 6:131–138.

Wada JA, Clarke R, Hamm A. 1975. Cerebral hemispheric asymmetry in humans. Cortical speech zones in 100 adults and 100 infant brains. *Archaeological Neurology* 32:239–246.

Watkins KE, Dronkers NF, Vargha-Khadem F. 2002. Behavioural analysis of an inherited speech and language disorder: comparison with acquired aphasia. *Brain* 125:452–464.

Wernicke C. 1874/1977. Der aphasische symptomenkomplex: eine psychologische studie auf anatomischer basis, trans. Eggert GH. In *Wernicke's Works on Aphasia: A Sourcebook and Review*. The Hague: Mouton, 1977. Originally published in 1874.

Chapter 21

Addis DR, Wong AT, Schacter DL. 2007. Remembering the past and imagining the future: common and distinct neural substrates during event construction and elaboration. *Neuropsychologia* 45:1363–1377.

Behrmann M, Geng JJ, Shomstein S. 2004. Parietal cortex and attention. *Current Opinion in Neurobiology* 14:212–217.

Bisley JW, Goldberg ME. 2010. Attention, intention, and priority in the parietal lobe. *Annual Review of Neuroscience* 33:1–21.

Borji A, Itti L. 2013. State-of-the-art in visual attention modeling. *IEEE Transactions on Pattern Analysis and Machine Intelligence* 35:185–207.

Brefczynski JA, DeYoe EA. 1999. A physiological correlate of the "spotlight" of visual attention. *Nature Neuroscience* 2:370–374.

Buckner RL, Andrews-Hanna JR, Schacter DL. 2008. The brain's default network: anatomy, function, and relevance to disease. *Annals of the New York Academy of Sciences* 1124:1–38.

Chalmers DJ. 1995. Facing up to the problem of consciousness. *Journal of Consciousness Studies* 2:200–219.

Cohen MA, Dennett DC. 2011. Consciousness cannot be separated from function. *Trends in Cognitive Science* 15:358–364.

Corbetta M, Miezin FM, Dobmeyer S, Shulman GL, Petersen SE. 1990. Attentional modulation of neural processing of shape, color, and velocity in humans. *Science* 248:1556–1559.

Courtney SM, Ungerleider LG, Keil K, Haxby JV. 1997. Transient and sustained activity in a distributed neural system for human working memory. *Nature* 386:608–611.

Crick R. 1994. *The Astonishing Hypothesis: The Scientific Search for the Soul*. New York: Scribner's.

Crick F, Koch C, Kreiman G, Fried I. 2004. Consciousness and neurosurgery. *Neurosurgery* 55:273–282.

Fried F, MacDonald KA, Wilson CL. 1997. Single neuron activity in human hippocampus and amygdala during recognition of faces and objects. *Neuron* 18:753–765.

Itti L, Koch C. 2001. Computational modeling of visual attention. *Nature Reviews Neuroscience* 2:194–203.

Koch C, Greenfield S. 2007. How does consciousness happen? *Scientific American* 297:76–83.

Koubeissi MZ, Bartolomei F, Beltagy A, Picard F. 2014. Electrical stimulation of a small brain area reversibly disrupts consciousness. *Epilepsy & Behavior* 37:32–35.

Kreiman G, Koch C, Fried I. 2000. Imagery neurons in the human brain. *Nature* 408:357–361.

Mesulam MM. 1999. Spatial attention and neglect: parietal, frontal and cingulate contributions to the mental representation and attentional targeting of salient extrapersonal events. *Philosophical Transactions of the Royal Society of London. Series B, Biological Sciences* 1387:1325–1346.

Miller EK, Buschman TJ. 2013. Cortical circuits for the control of attention. *Current Opinion in Neurobiology* 23:216–222.

Moore T, Armstrong KM. 2003. Selective gating of visual signals by microstimulation of frontal cortex. *Nature* 421:370–373.

Moore T, Fallah M. 2001. Control of eye movements and spatial attention. *Proceedings of the National Academy of Sciences USA* 98:1273–1276.

Moran J, Desimone R. 1985. Selective attention gates visual processing in the extrastriate cortex. *Science* 229:782–784.

Noudoost B, Chang MH, Steimetz NA, Moore T. 2010. Top-down control of visual attention. *Current Opinion in Neurobiology* 20:183–190.

Olfson M. 2004. New options in the pharmacological management of attention-deficit/hyperactivity disorder. *American Journal of Managed Care* 10(4)(Suppl):S117–S124.

Pessoa L, Ungerleider LG. 2004. Neuroimaging studies of attention and the processing of emotion-laden stimuli. *Progress in Brain Research* 144:171–182.

Petersen SE, Fox PT, Posner MI, Mintum M, Raichle ME. 1988. Positron emission tomographic studies of the cortical anatomy of single-word processing. *Nature* 331:585–589.

Posner MI, Petersen SE. 1990. The attention system of the human brain. *Annual Review of Neuroscience* 13:25–42.

Posner MI, Raichle M. 1994. *Images of Mind*. New York: Scientific American Library.

Posner MI, Snyder CRR, Davidson BJ. 1980. Attention and the detection of signals. *Journal of Experimental Psychology General* 109:160–174.

Ptak R. 2012. The frontoparietal attention network of the brain: action, saliency, and a priority map of the environment. *Neuroscientist* 18:502–515.

Raichle ME, Snyder AZ. 2007. A default mode of brain function: a brief history of an evolving idea. *Neuroimage* 37:1083–1090.

Rees G, Kreiman G, Koch C. 2002. Neural correlates of consciousness in humans. 2002. *Nature Reviews Neuroscience* 3:261–270.

Sheinberg DL, Logothetis NK. 1997. The role of temporal cortical areas in perceptual organization. *Proceedings of the National Academy of Sciences* 94:3408–3413.

Shipp S. 2004. The brain circuitry of attention. *Trends in Cognitive Science* 8:223–230.

Tong F, Nakayama K, Vaughn JT, Kanwisher N. 1998. Binocular rivalry and visual awareness in human extrastriate cortex. *Neuron* 21:753–759.

Treue S. 2003. Visual attention: the where, what, how and why of saliency. *Current Opinion in Neurobiology* 13:428–432.

Wurtz RH, Goldberg ME, Robinson DL. 1982. Brain mechanisms of visual attention. *Scientific American* 246:124–135.

Chapter 22

American Psychiatric Association. 2013. *Diagnostic and Statistical Manual of Mental Disorders*, 5th ed. Arlington, VA: American Psychiatric Association.

Andreasen NC. 1984. *The Broken Brain*. New York: Harper Collins.

Andreasen NC. 2004. *Brave New Brain: Conquering Mental Illness in the Era of the Genome*. New York: Oxford University Press.

Barondes SH. 1993. *Molecules and Mental Illness*. New York: W.H. Freeman.

Cade JFJ. 1949. Lithium salts in the treatment of psychotic excitement. *Medical Journal of Australia* 36:349–352.

Callicott JH. 2003. An expanded role for functional neuroimaging in schizophrenia. *Current Opinion in Neurobiology* 13:256–260.

Charney DS, Nestler EJ, eds. 2009. *Neurobiology of Mental Illness*, 3rd ed. New York: Oxford University Press.

Corfas G, Roy K, Buxbaum JD. 2004. Neuregulin 1-erbB signaling and the molecular/cellular basis of schizophrenia. *Nature Neuroscience* 7:575–580.

Davidson RJ, Abercrombie H, Nitschke JB, Putnam K. 1999. Regional brain function, emotion and disorders of emotion. *Current Opinion in Neurobiology* 9:228–234.

Fogel BS, Schiffer RB, Rao SM, Fogel BS. 2003. *Neuropsychiatry: A Comprehensive Textbook* 2nd ed. Baltimore: Lippincott Williams & Wilkins.

Freud S. 1920/1990. *Beyond the Pleasure Principle*. New York: Norton, 1990. Originally published in 1920.

Gordon JA, Hen R. 2004. Genetic approaches to the study of anxiety. *Annual Review of Neuroscience* 27:193–222.

Gottesman II. 1991. *Schizophrenia Genesis*. New York: W.H. Freeman.

Harrison PJ, Weinberger DR. 2005. Schizophrenia genes, gene expression, and neuropathology: on the matter of their convergence. *Molecular Psychiatry* 10:40–68.

Heuser I. 1998. The hypothalamic-pituitary-adrenal system in depression. Anna-Monika-Prize paper. *Pharmacopsychiatry* 31:10–13.

Holtzheimer PE, Mayberg HS. 2011. Deep brain stimulation for psychiatric disorders. *Annual Review of Neuroscience* 34:289–307.

Insel TR. 2012. Next generation treatments for psychiatric disorders. *Science Translational Medicine* 4:1–9.

Lewis DA, Levitt P. 2002. Schizophrenia as a disorder of neurodevelopment. *Annual Review of Neuroscience* 25:409–432.

Liu D, Diorio J, Tannenbaum B, Caldji C, Francis D, Freedman A, Sharma S, Pearson D, Plotsky PM, Meaney MJ. 1997. Maternal care, hippocampal glucocorticoid receptors, and hypothalamic-pituitary-adrenal responses to stress. *Science* 277:1659–1662.

Malberg JE, Eisch AJ, Nestler EJ, Duman RS. 2000. Chronic antidepressant treatment increases neurogenesis in adult rat hippocampus. *Journal of Neuroscience* 20:9104–9110.

Malizia AL, Cunningham VJ, Bell CJ, Liddle PF, Jones T, Nutt DJ. 1998. Decreased brain GABA(A)-benzodiazepine receptor binding in panic disorder: preliminary results from a quantitative PET study. *Archives General Psychiatry* 55:715–720.

Mayberg HS. 2009. Targeted electrode-based modulation of neural circuits for depression. *The Journal of Clinical Investigation.* 119:717–725.

Mayberg HS, Lozano AM, Voon V, McNeely HE, Seminowicz D, Hamani C, Schwalb JM, Kennedy SH. 2005. Deep brain stimulation for treatment-resistant depression. *Neuron.* 45:651–660.

McCarthy SE, McCombie WR, Corvin A. 2014. Unlocking the treasure trove: from genes to schizophrenia biology. *Schizophrenia Bulletin* 40:492–496.

Moghaddam B, Wolf ME, eds. 2003. Glutamate and disorders of cognition and motivation. *Annals of the New York Academy of Sciences* 1003:1–484.

Mohn AR, Gainetdinov RR, Caron MG, Koller BH. 1999. Mice with reduced NMDA receptor expression display behaviors related to schizophrenia. *Cell* 98:427–436.

Morris BJ, Cochran SM, Pratt JA. 2005. PCP: from pharmacology to modelling schizophrenia. *Current Opinion in Pharmacology* 5:101–106.

Nemeroff CB. 1998. The neurobiology of depression. *Scientific American* 278(6):42–49.

Santarelli L, Saxe M, Gross C, Surget A, Battaglia F, Dulawa S, Weisstaub N, Lee J, Duman R, Arancio O, Belzung C, Hen R. 2003. Requirement of hippocampal neurogenesis for the behavioral effects of antidepressants. *Science* 301:805–809.

Satcher D. 1999. *Mental Health: A Report of the Surgeon General*. Washington, DC: US Government Printing Office.

Seeman P. 1980. Brain dopamine receptors. *Pharmacological Reviews* 32:229–313.

Slater E, Meyer A. 1959. Contributions to a pathology of the musicians. *Confinia Psychiatrica* 2:65–94.

Thompson PM, Vidal C, Giedd JN, Gochman P, Blumenthal J, Nicolson R, Toga AW, Rapoport JL. 2001. Mapping adolescent brain change reveals dynamic wave of accelerated gray matter loss in very early-onset schizophrenia. *Proceedings of the National Academy of Sciences USA* 98:11650–11655.

Winterer G, Weinberger DR. 2004. Genes, dopamine and cortical signal-to-noise ratio in schizophrenia. *Trends in Neurosciences* 27:683–690.

Wong ML, Licinio J. 2001. Research and treatment approaches to depression. *Nature Reviews Neuroscience* 2:343–351.

Zhang X, Gainetdinov RR, Beaulieu JM, Sotnikova TD, Burch LH, Williams RB, Schwartz DA, Krishnan KR, Caron MG. 2005. Loss-of-function mutation in tryptophan hydroxylase-2 identified in unipolar major depression. *Neuron* 45:11–16.

Chapter 23

Altman J, Das GD. 1965. Autoradiographic and histological evidence of postnatal hippocampal neurogenesis in rats. *Journal of Comparative Neurology* 124:319–335.

Balice-Gordon RJ, Lichtman JW. 1994. Long-term synapse loss induced by focal blockade of postsynaptic receptors. *Nature* 372:519–524.

Bear MF. 2003. Bidirectional synaptic plasticity: from theory to reality. *Philosophical Transactions of the Royal Society of London. Series B, Biological Sciences* 358:649–655.

Bear MF, Huber KM, Warren ST. 2004. The mGluR theory of fragile X mental retardation. *Trends in Neurosciences* 27:370–377.

Bear MF, Kleinschmidt A, Gu Q, Singer W. 1990. Disruption of experience-dependent synaptic modifications in striate cortex by infusion of an NMDA receptor antagonist. *Journal of Neuroscience* 10:909–925.

Bear MF, Singer W. 1986. Modulation of visual cortical plasticity by acetylcholine and noradrenaline. *Nature* 320:172–176.

Bourgeois J, Rakic P. 1993. Changes of synaptic density in the primary visual cortex of the macaque monkey from fetal to adult stage. *Journal of Neuroscience* 13:2801–2820.

Cooke SF, Bear MF. 2014. How the mechanisms of long-term synaptic potentiation and depression serve experience-dependent plasticity in primary visual cortex. *Philosophical Transactions of the Royal Society of London. Series B, Biological Sciences* 369:20130284.

Dehay C, Kennedy H. 2007. Cell-cycle control and cortical development. *National Review of Neuroscience* 8(6):438–450.

Dudek SM, Bear MF. 1989. A biochemical correlate of the critical period for synaptic modification in the visual cortex. *Science* 246:673–675.

Espinosa JS, Styker MP. 2012. Development and plasticity of the primary visual cortex. *Neuron* 75:230-249.

Fish JL, Dehay C, Kennedy H, Huttner WB. 2008. Making bigger brains-the evolution of neural-progenitor-cell division. *Journal of Cell Science* 121(Pt 17):2783–2793.

Fixsen W, Sternberg P, Ellis H, Horvitz R. 1985. Genes that affect cell fates during the development of *Caenorhabditis elegans*. *Cold Spring Harbor Symposium on Quantitative Biology* 50:99–104.

Ghosh A, Carnahan J, Greenberg M. 1994. Requirement for BDNF in activity-dependent survival of cortical neurons. *Science* 263:1618–1623.

Goda Y, Davis GW. 2003. Mechanisms of synapse assembly and disassembly. *Neuron* 40:243–264.

Goldberg JL, Barres BA. 2000. Nogo in nerve regeneration. *Nature* 403:369–370.

Hamasaki T, Leingartner A, Ringstedt T, O'Leary DD. 2004. EMX2 regulates sizes and positioning of the primary sensory and motor areas in neocortex by direct specification of cortical progenitors. *Neuron* 43(3):359–372.

Harris WC, Holt CE. 1999. Slit, the midline repellent. *Nature* 398:462–463.

Hebb DO. 1949. *Organization of Behavior*. New York: Wiley.

Heynen AJ, Yoon BJ, Liu CH, Chung HJ, Huganir RL, Bear MF. 2003. Molecular mechanism for loss of visual responsiveness following brief monocular deprivation. *Nature Neuroscience* 6:854–862.

Honda T, Tabata H, Nakajima K. 2003. Cellular and molecular mechanisms of neuronal migration in neocortical development. *Seminars in Cellular and Developmental Biology* 14:169–174.

Huang ZJ, Kirkwood A, Pizzorusso T, Porciatti V, Morales B, Bear MF, Maffei L, Tonegawa S. 1999. BDNF regulates the maturation of inhibition and the critical period of plasticity in mouse visual cortex. *Cell* 98:39–55.

Katz LC, Crowley JC. 2002. Development of cortical circuits: lessons from ocular dominance columns. *Nature Reviews Neuroscience* 3(1):34–42.

Kempermann G, Wiskott L, Gage FH. 2004. Functional significance of adult neurogenesis. *Current Opinion in Neurobiology* 14:186–191.

Kennedy T, Serafini T, Torre JDL, Tessier-Lavigne M. 1994. Netrins are diffusible chemotropic factors for commissural axons in the embryonic spinal cord. *Cell* 78:425–435.

Law MI, Constantine-Paton M. 1981. Anatomy and physiology of experimentally produced striped tecta. *Journal of Neuroscience* 1:741–759.

LeVay S, Stryker MP, Shatz CJ. 1978. Ocular dominance columns and their development in layer IV of the cat's visual cortex: a quantitative study. *Journal of Comparative Neurology* 179:223–244.

Levi-Montalcini R, Cohen S. 1960. Effects of the extract of the mouse submaxillary salivary glands on the sympathetic system of mammals. *Annals of the New York Academy of Sciences* 85:324–341.

Liao D, Zhang X, O'Brien R, Ehlers MD, Huganir RL. 1999. Regulation of morphological postsynaptic silent synapses in developing hippocampal neurons. *Nature Neuroscience* 2:37–43.

McConnel SK. 1995. Constructing the cerebral cortex: neurogenesis and fate determination. *Neuron* 15:761–768.

McLaughlin T, O'Leary DDM. 2005. Molecular gradients and development of retinotopic maps. *Annual Review of Neuroscience* 28:327–355.

Meister M, Wong R, Baylor D, Shatz C. 1991. Synchronous bursts of action potentials in ganglion cells of the developing mammalian retina. *Science* 252:939–943.

Mioche L, Singer W. 1989. Chronic recordings from single sites of kitten striate cortex during experience-dependent modifications of receptive-field properties. *Journal of Neurophysiology* 62:185–197.

Mower GD. 1991. The effect of dark rearing on the time course of the critical period in cat visual cortex. *Developmental Brain Research* 58:151–158.

Paton JA, Nottebohm FN. 1984. Neurons generated in the adult brain are recruited into functional circuits. *Science* 225:1046–1048.

Price DJ, Jarman AP, Mason JO, Kind PC. 2011. *Building Brains: An Introduction to Neural Development*. Boston: Wiley-Blackwell.

Richardson PM, McGuinness UM, Aguayo AJ. 1980. Axons from CNS neurons regenerate into PNS grafts. *Nature* 284:264–265.

Ross SE, Greenberg ME, Stiles CD. 2003. Basic helix-loop-helix factors in cortical development. *Neuron* 39:13–25.

Schlagger BL, O'Leary DD. 1991. Potential of visual cortex to develop an array of functional units unique to somatosensory cortex. *Science* 252:1556–1560.

Spalding KL, Bergmann O, Alkass K, Bernard S, Salehpour M, Huttner HB, Boström E, Westerlund I, Vial C, Buchholz BA, Possnert G, Mash DC, Druid H, Frisén J. 2013. Dynamics of hippocampal neurogenesis in adult humans. *Cell* 153:1219–1227.

Sperry R. 1963. Chemoaffinity in the orderly growth of nerve fiber patterns and connections. *Proceedings of the National Academy of Sciences USA* 4:703–710.

Stoner R, Chow ML, Boyle MP, Sunkin SM, Mouton PR, Roy S, Wynshaw-Boris A, Colamarino SA, Lein ES, Courchesne E. 2014. Patches of disorganization in the neocortex of children with autism. *The New England Journal of Medicine* 370:1209–1219.

Tessier-Lavigne M, Goodman CS. 1996. The molecular biology of axon guidance. *Science* 274:1123–1133.

Walsh C, Cepko C. 1992. Widespread dispersion of neuronal clones across functional regions of the cerebral cortex. *Science* 255:434.

Whitford KL, Pijkhuizen P, Polleux F, Ghosh A. 2002. Molecular control of cortical dendrite development. *Annual Review of Neuroscience* 25:127–149.

Wiesel T. 1982. Postnatal development of the visual cortex and the influence of the environment. *Nature* 299:583–592.

Wiesel TN, Hubel DH. 1963. Single cell responses in striate cortex of kittens deprived of vision in one eye. *Journal of Neurophysiology* 26:1003–1017.

Chapter 24

Baddeley A. 2003. Working memory: looking back and looking forward. *Nature Reviews Neuroscience* 4:829–839.

Brunet A, Orr, SP, Tremblay J, Robertson K, Nader K, Pitman RK. 2008. Effect of post-retrieval propranolol on psychophysiologic responding during subsequent script-driven traumatic imagery in post-traumatic stress disorder. *Journal of Psychiatric Research* 42:503–506.

Cohen NJ, Eichenbaum H. 1993. *Memory, Amnesia, and the Hippocampal System*. Cambridge, MA: MIT Press.

Corkin S. 2002. What's new with the amnesic patient H.M.? *Nature Reviews Neuroscience* 3:153–160.

Corkin S. 2013. *Permanent Present Tense: The Unforgettable Life of the Amnesic Patient H.M.* New York: Basic Books.

Courtney SM, Ungerleider LG, Keil K, Haxby JV. 1996. Object and spatial visual working memory activate separate neural systems in human cortex. *Cerebral Cortex* 6:39–49.

Courtney SM, Ungerleider LG, Keil K, Haxby JV. 1997. Transient and sustained activity in a distributed neural system for human working memory. Nature 386:608–611.

Desimone R, Albright TD, Gross CG, Bruce C. 1984. Stimulus selective properties of inferior temporal neurons in the macaque. *Journal of Neuroscience* 4:2051–2062.

Doeller CF, Barry C, Burgess N. 2010. Evidence for grid cells in a human memory network. *Nature* 463:657–661.

Eichenbaum H. 2000. A cortical hippocampal system for declarative memory. *Nature Reviews Neuroscience* 1:41–50.

Eichenbaum H. 2011. *The Cognitive Neuroscience of Memory*. New York: Oxford University Press.

Eichenbaum H, Dudchenko P, Wood E, Shapiro M, Tanila H. 1999. The hippocampus, memory, and place cells: is it spatial memory or a memory space? *Neuron* 23:209–226.

Eichenbaum H, Fagan H, Mathews P, Cohen NJ. 1988. Hippocampal system dysfunction and odor discrimination learning in rats: impairment or facilitation depending on representational demands. *Behavioral Neuroscience* 102:331–339.

Fried F, MacDonald KA, Wilson CL. 1997. Single neuron activity in human hippocampus and amygdala during recognition of faces and objects. *Neuron* 18:753–765.

Fuster JM. 1973. Unit activity in prefrontal cortex during delayed response performance: neuronal correlates of transient memory. *Journal of Neurophysiology* 36:61–78.

Fuster JM. 1995. *Memory in the Cerebral Cortex*. Cambridge, MA: MIT Press.

Gauthier I, Skularski P, Gore JC, Anderson AW. 2000. Expertise for cars and birds recruits brain areas involved in face recognition. *Nature Neuroscience* 3:191–197.

Gnadt JW, Andersen RA. 1988. Memory related motor planning activity in posterior parietal cortex of macaque. *Experimental Brain Research* 70:216–220.

Goldman-Rakic P. 1992. Working memory and the mind. *Scientific American* 267:111–117.

Graff J, Joseph NF, Horn ME, Samiei A, Meng J, Seo J, Rei D, Bero AW, Phan TX, Wagner F, Holson E, Xu J, Sun J, Neve RL, Mach RH, Haggarty SJ, Tsai LH. 2014. Epigenetic priming of memory updating during reconsolidation to attenuate remote fear memories. *Cell* 156:261–276.

Haxby JV, Petit L, Ungerleider LG, Courtney SM. 2000. Distinguishing the functional roles of multiple regions in distributed neural systems for visual working memory. *Neuroimage* 11:380–391.

Hebb DO. 1949. *The Organization of Behavior: A Neuropsychological Theory*. New York: Wiley.

Jog MS, Kubota Y, Connolly CI, Hillegaart V, Graybiel AM. 1999. Building neural representations of habits. *Science* 286:1745–1749.

Kandel ER, Dudai Y, Mayford MR. 2014. The molecular and systems biology of memory. *Cell* 157:163–186.

Knowlton BJ, Mangels JA, Squire LR. 1996. A neostriatal habit learning system in humans. *Science* 273:1399–1402.

Lashley KS. 1929. *Brain Mechanisms and Intelligence.* Chicago: University of Chicago Press.

Luria A. 1968. *The Mind of a Mnemonist*. Cambridge, MA: Harvard University Press.

Ma WJ, Husain M, Bays PM. 2014. Changing concepts of working memory. *Nature Neuroscience* 17:347–356.

Maguire EA, Burgess N, Donnett JG, Frackowiak RS, Frith CD, O'Keefe J. 1998. Knowing where and getting there: a human navigation network. *Science* 280:921–924.

McKenzie S, Eichenbaum H. 2011. Consolidation and reconsolidation: two lives of memories? *Neuron* 71:224–233.

Mishkin M, Appenzeller T. 1987. The anatomy of memory. *Scientific American* 256:80–89.

Morris RGM. 1984. Developments of a water-maze procedure for studying spatial learning in the rat. *Journal of Neuroscience Methods* 11:47–60.

Moser EI, Kropff E, Moser M. 2008. Place cells, grid cells, and the brain's spatial representation system. *Annual Review of Neuroscience* 31:69–89.

Nadel L, Hardt O. 2011. Update on memory systems and processes. *Neuropsychopharmacology* 36:251–273.

O'Keefe JA. 1979. Place units in the hippocampus of the freely moving rat. *Experimental Neurology* 51:78–109.

O'Keefe JA, Nadel L. 1978. *The Hippocampus as a Cognitive Map*. London: Oxford University Press.

Olton DS, Samuelson RJ. 1976. Remembrance of places passed: spatial memory in rats. *Journal of Experimental Psychology* 2:97–116.

Passingham D, Sakai K. 2004. The prefrontal cortex and working memory: physiology and brain imaging. *Current Opinion in Neurobiology* 14:163–168.

Penfield W. 1958. *The Excitable Cortex in Conscious Man.* Liverpool: Liverpool University Press.

Quiroga RQ, Kreiman G, Koch C, Fried I. 2008. Sparse but not "grandmother-cell" coding in the medial temporal lobe. *Trends in Cognitive Sciences* 12:87–91.

Quiroga RQ, Reddy L, Kreiman G, Koch C, Fried I. 2005. Invariant visual representation by single neurons in the human brain. *Nature* 435:1102–1107.

Ramirez S, Liu X, Lin P, Suh J, Pignatelli M, Redondo RL, Ryan TJ, Tonegawa S. 2013. Creating a false memory in the hippocampus. *Science* 341:387–391.

Scoville WB, Milner B. 1957. Loss of recent memory after bilateral hippocampal lesions. *Journal of Neurology, Neurosurgery, and Psychiatry* 20:11–21.

Squire LR, Stark CEL, Clark RE. 2004. The medial temporal lobe. *Annual Review of Neuroscience* 27:279–306.

Squire LR, Wixted JT. 2011. The cognitive neuroscience of human memory since H.M. *Annual Review of Neuroscience* 34:259–288.

Wang S, Morris RGM. 2010. Hippocampal-neocortical interactions in memory formation, consolidation, and reconsolidation. *Annual Review of Psychology* 61:49–79.

Wilson MA, McNaughton BL. 1993. Dynamics of the hippocampal ensemble code for space. *Science* 261:1055–1058.

Zola-Morgan S, Squire LR, Amaral DG, Suzuki WA. 1989. Lesions of perirhinal and parahippocampal cortex that spare the amygdala and hippocampal formation produce severe memory impairment. *Journal of Neuroscience* 9:4355–4370.

Chapter 25

Abraham WC, Logan B, Greenwood JM, Dragunow M. 2002. Induction and experience-dependent consolidation of stable long-term potentiation lasting months in the hippocampus. *Journal of Neuroscience* 22:9626–9634.

Abraham WC, Robins A. 2005. Memory retention: the synaptic stability versus plasticity dilemma. *Trends in Neuroscience* 28:73–78.

Bailey CH, Kandel ER. 1993. Structural changes accompanying memory storage. *Annual Review of Neuroscience* 55:397–426.

Bear MF. 1996. A synaptic basis for memory storage in the cerebral cortex. *Proceedings of the National Academy of Sciences USA* 93:13453–13459.

Bear MF. 2003. Bidirectional synaptic plasticity: from theory to reality. *Philosophical Transactions of the Royal Society of London. Series B, Biological Sciences* 358:649–655.

Bienenstock EL, Cooper LN, Munro PW. 1982. Theory for the development of neuron selectivity: orientation specificity and binocular interaction in visual cortex. *Journal of Neuroscience* 2:32–48.

Blais BA, Cooper LN, Shouval H. 2000. Formation of direction selectivity in natural scene environments. *Neural Computation* 12:1057–1066.

Blais BS, Intrator N, Shouval HZ, Cooper LN. 1998. Receptive field formation in natural scene environments: comparison of single-cell learning rules. *Neural Computation* 10:1797–1813.

Bliss TVP, Collingridge GL. 1993. A synaptic model of memory: long-term potentiation in the hippocampus. *Nature* 361:31–39.

Bredt DS, Nicoll RA. 2003. AMPA receptor trafficking at excitatory synapses. *Neuron* 40:361–379.

Cajal SR. 1894. The Croonian Lecture: la fine structure des centres nerveux. *Philosophical Transactions of the Royal Society of London. Series B, Biological Sciences* 55:344–468.

Carew TJ, Sahley CL. 1986. Invertebrate learning and memory: from behavior to molecules. *Annual Review of Neuroscience* 9:435–487.

Castellucci VF, Kandel ER. 1974. A quantal analysis of the synaptic depression underlying habituation of the gill-withdrawal reflex in *Aplysia*. *Proceedings of the National Academy of Sciences USA* 77:7492–7496.

Chen WR, Lee S, Kato K, Spencer DD, Shepherd GM, Williamson A. 1996. Long-term modifications of synaptic efficacy in the human inferior and middle temporal cortex. *Proceedings of the National Academy of Sciences USA* 93:8011–8015.

Colledge M, Snyder EM, Crozier RA, Soderling JA, Jin Y, Langeberg LK, Lu H, Bear MF, Scott JD. 2003. Ubiquitination regulates PSD-95 degradation and AMPA receptor surface expression. *Neuron* 40:595–607.

Cooper LN, Bear MF. 2012. The BCM theory of synapse modification at 30: interaction of theory and experiment. *Nature Reviews Neuroscience* 13:798–810.

Davis HP, Squire LR. Protein synthesis and memory. 1984. *Psychological Bulletin* 96:518–559.

Dudai Y, Jan YN, Byers D, Quinn WG, Benzer S. 1976. Dunce, a mutant of *Drosophila* deficient in learning. *Proceedings of the National Academy of Sciences USA* 73:1684–1688.

Dudek SM, Bear MF. 1992. Homosynaptic long-term depression in area CA1 of hippocampus and effects of N-methyl-D-aspartate receptor blockade. *Proceedings of the National Academy of Sciences USA* 89:4363–4367.

Hofer SB, Bonhoeffer T. 2010. Dendritic spines: the stuff that memories are made of? *Current Biology* 20:R157–R159.

Ito M. 1982. Experimental verification of Marr-Albus' plasticity assumption for the cerebellum. *Acta Biology* 33:189–199.

Kandel ER. 1970. Nerve cells and behavior. *Scientific American* 223:57–67.

Kandel ER. 2001. The molecular biology of memory storage: a dialogue between genes and synapses. *Science* 294:1030–1038.

Kandel ER. 2006. *In Search of Memory: The Emergence of a New Science of Mind.* New York: Norton.

Kessels HW, Malinow R. 2009. Synaptic AMPA receptor plasticity and behavior. *Neuron* 61:340–350.

Kirkwood A, Bear MF. 1994. Homosynaptic long-term depression in the visual cortex. *Journal of Neuroscience* 14:3404–3412.

Kirkwood A, Rioult MC, Bear MF. Experience-dependent modification of synaptic plasticity in visual cortex. *Nature* 1996;381:526–528.

Konorski J. 1948. *Conditioned Reflexes and Neuron Organization*. Cambridge, MA: University Press.

Leopold DA, Bondar IV, Giese MA. 2006. Norm-based face encoding by single neurons in the monkey inferotemporal cortex. *Nature* 442:572–575.

Levy WB, Steward O. 1983. Temporal contiguity requirements for long-term associative potentiation/depression in the hippocampus. *Neuroscience* 8:791–797.

Linden DJ, Connor JA. 1993. Cellular mechanisms of long-term depression in the cerebellum. *Current Opinion in Neurobiology* 3:401–406.

Lisman JE, Fallon JR. 1999. What maintains memories? *Science* 283:339–340.

Lisman J, Schulman H, Cline H. 2002. The molecular basis of CaMKII function in synaptic and behavioural memory. *Nature Reviews Neuroscience* 3:175–190.

Lynch G, Baudry M. 1984. The biochemistry of memory: a new and specific hypothesis. *Science* 224(4653):1057–1063.

Malenka RC, Bear MF. 2004. LTP and LTD: an embarrassment of riches. *Neuron* 44:5–21.

Malinow R. 2003. AMPA receptor trafficking and long-term potentiation. *Philosophical Transactions of the Royal Society of London. Series B, Biological Sciences* 358:707–714.

Markram H, Lubke J, Frotscher M, Sakmann B. 1997. Regulation of synaptic efficacy by coincidence of postsynaptic APs and EPSPs. *Science* 275:213–215.

Marr D. 1969. A theory of cerebellar cortex. *Journal of Physiology* 202:437–470.

Morris RGM, Anderson E, Lynch GS, Baudry M. 1986. Selective impairment of learning and blockade of long-term potentiation by an *N*-methyl-D-aspartate receptor antagonist, AP5. *Nature* 319:774–776.

Quinlan EM, Olstein DH, Bear MF. 1999. Bidirectional, experience-dependent regulation of N-methyl-D-aspartate receptor subunit composition in the rat visual cortex during postnatal development. *Proceedings of the National Academy of Sciences USA* 96:12876–12880.

Quinlan EM, Philpot BD, Huganir RL, Bear MF. 1999. Rapid, experience-dependent expression of synaptic NMDA receptors in visual cortex in vivo. *Nature Neuroscience* 2:352–357.

Roberts AC, Glanzman DL. 2003. Learning in *Aplysia*: looking at synaptic plasticity from both sides. *Trends in Neurosciences* 26:662–670.

Rolls ET, Baylis GC, Hasselmo ME, Nalwa V. 1989. The effect of learning on the face selective responses of neurons in the cortex in the superior temporal sulcus of the monkey. *Experimental Brain Research* 76:153–164.

Schwartz JH. 1993. Cognitive kinases. *Proceedings of the National Academy of Sciences USA* 90:8310–8313.

Shouval H, Intrator N, Cooper LN. 1997. BCM network develops orientation selectivity and ocular dominance in natural scene environment. *Vision Research* 37:3339–3342.

Silva AJ, Paylor R, Wehner JM, Tonegawa S. 1992. Impaired spatial learning in alpha-calcium-calmodulin kinase II mutant mice. *Science* 257:206–211.

Silva AJ, Stevens CF, Tonegawa S, Wang Y. 1992. Deficient hippocampal long-term potentiation in alpha-calcium-calmodulin kinase II mutant mice. *Science* 257:201–206.

Tang YP, Shimizu E, Dube GR, Rampon C, Kerchner GA, Zhuo M, Liu G, Tsien JZ. 1999. Genetic enhancement of learning and memory in mice. *Nature* 401:63–69.

Thorndike EL. 1911. *Animal Intelligence: Experimental Studies*. New York: Macmillan.

Tsien JZ, Huerta PT, Tonegawa S. 1996. The essential role of hippocampal CA1 NMDA receptor dependent synaptic plasticity in spatial memory. *Cell* 87:1327–1338.

Whitlock JR, Heynen AJ, Shuler MG, Bear MF. 2006. Learning induces long-term potentiation in the hippocampus. *Science* 313:1093–1097.

Yin JC, Tully T. 1996. CREB and the formation of long-term memory. *Current Opinion in Neurobiology* 6:264–268.

INDEX

Note: Page numbers in *italics* refer to figures; those followed by *b* and *t* refer to boxes and tables, respectively.

C